Developmental Mathematics for College Students

In memory of
Nancy B. Kinsell
and in honor of
Robert C. Kinsell,
whose dedication to their family
and their country has been
an inspiration to me.

AST

Books in the Tussy and Gustafson Series

In paperback:

Basic Mathematics for College Students, Second Edition
Student edition: ISBN 0-534-37643-6
Instructor's edition: ISBN 0-534-38585-0

Prealgebra, Second Edition
Student edition: ISBN 0-534-37642-8
Instructor's edition: ISBN 0-534-38309-2

Developmental Mathematics for College Students
Student edition: ISBN 0-534-38031-X
Instructor's edition: ISBN 0-534-38584-2

Introductory Algebra, Second Edition
Student edition: ISBN 0-534-37641-X
Instructor's edition: ISBN 0-534-38571-0

Intermediate Algebra, Second Edition
Student edition: ISBN 0-534-37640-1
Instructor's edition: ISBN 0-534-38586-9

In hardcover:

Elementary Algebra, Second Edition
Student edition: ISBN 0-534-38629-6
Instructor's edition: ISBN 0-534-39119-2

Intermediate Algebra, Second Edition
Student edition: ISBN 0-534-38628-8
Instructor's edition: ISBN 0-534-39120-6

Elementary and Intermediate Algebra, Second Edition
Student edition: ISBN 0-534-38627-X
Instructor's edition: ISBN 0-534-39118-4

Developmental Mathematics for College Students

Alan S. Tussy
Citrus College

R. David Gustafson
Rock Valley College

BROOKS/COLE

THOMSON LEARNING ™

Australia • Canada • Mexico • Singapore • Spain • United Kingdom • United States

BROOKS/COLE

THOMSON LEARNING

Sponsoring Editor: *Jennifer Huber*
Assistant Editor: *Rachael Sturgeon*
Editorial Assistant: *Lisa Jones*
Marketing: *Leah Thomson*
Marketing Communications: *Samantha Cabaluna*
Marketing Assistant: *Maria Salinas*
Production Editor: *Ellen Brownstein*
Production Service: *Hoyt Publishing Services*
Manuscript Editors: *Penelope Suess* and
   David Hoyt

Permissions Editor: *Sue Ewing*
Interior Design: *Vernon T. Boes* and *John Edeen*
Cover Design: *Roy R. Neuhaus*
Cover Illustration: *Goerge Abe*
Interior Illustration: *Lori Heckelman*
Print Buyer: *Vena Dyer*
Typesetting: *The Clarinda Company*
Cover Printing: *Phoenix Color Corp.*
Printing and Binding: *Quebecor World Book*
   Services–Dubuque

For more information about this or any other Brooks/Cole product, contact:
BROOKS/COLE
511 Forest Lodge Road
Pacific Grove, CA 93950 USA
www.brookscole.com
1-800-423-0563 (Thomson Learning Academic Resource Center)

For permission to use material from this work, contact us at
www.thomsonrights.com
fax: 1-800-730-2215
phone: 1-800-730-2214

Printed in the United States of America

10   9   8   7   6   5   4   3   2   1

Images provided by PhotoDisc © 2000

Library of Congress Cataloging-in-Publication Data
Tussy, Alan S., [date]
   Developmental mathematics for college students / Alan S. Tussy, R. David Gustafson.
     p. cm.
   Includes index.
   ISBN 0-534-38031-X (pbk.: alk. paper)
   1. Mathematics.   I. Gustafson, R. David (Roy David), [date] II. Title.
QA39.3 .T88 2002
510—dc21
                                    2001035840

CONTENTS

1 Whole Numbers 1

1.1 An Introduction to the Whole Numbers 2
1.2 Adding and Subtracting Whole Numbers 11
1.3 Multiplying and Dividing Whole Numbers 21
　　Estimation 34
1.4 Prime Factors and Exponents 36
1.5 Order of Operations 43
1.6 Solving Equations by Addition and Subtraction 50
1.7 Solving Equations by Division and Multiplication 59
　　Key Concept: Variables 67
　　Accent on Teamwork 68
　　Chapter Review 69
　　Taking a Math Test 74
　　Chapter Test 75

2 The Integers 77

2.1 An Introduction to the Integers 78
2.2 Addition of Integers 87
2.3 Subtraction of Integers 95
2.4 Multiplication of Integers 102
2.5 Division of Integers 111
2.6 Order of Operations and Estimation 116
2.7 Solving Equations Involving Integers 122
　　Key Concept: Signed Numbers 132
　　Accent on Teamwork 133
　　Chapter Review 134
　　Chapter Test 139
　　Cumulative Review Exercises 141

3 *Fractions and Mixed Numbers*   *143*

3.1 The Fundamental Property of Fractions   144
3.2 Multiplying Fractions   152
3.3 Dividing Fractions   161
3.4 Adding and Subtracting Fractions   168
The LCM and the GCF   177
3.5 Multiplying and Dividing Mixed Numbers   179
3.6 Adding and Subtracting Mixed Numbers   187
3.7 Order of Operations and Complex Fractions   195
3.8 Solving Equations Containing Fractions   202
Key Concept: The Fundamental Property of Fractions   210
Accent on Teamwork   211
Chapter Review   212
Chapter Test   217
Cumulative Review Exercises   219

4 *Decimals*   *221*

4.1 An Introduction to Decimals   222
4.2 Addition and Subtraction with Decimals   231
4.3 Multiplication with Decimals   238
4.4 Division with Decimals   246
Estimation   253
4.5 Fractions and Decimals   255
4.6 Solving Equations Containing Decimals   263
4.7 Square Roots   268
Key Concept: The Real Numbers   274
Accent on Teamwork   275
Chapter Review   276
Chapter Test   281
Cumulative Review Exercises   283

5 *Percent*   *285*

5.1 Percents, Decimals, and Fractions   286
5.2 Solving Percent Problems   295
5.3 Applications of Percent   306
Estimation   316
5.4 Interest   318
Key Concept: Percent   326
Accent on Teamwork   327
Chapter Review   328

Chapter Test   333
Cumulative Review Exercises   335

6 *Descriptive Statistics*   *337*

6.1 Reading Graphs and Tables   338
6.2 Mean, Median, and Mode   349
         Key Concept: Mean, Median, and Mode   356
         Accent on Teamwork   357
         Chapter Review   358
         Chapter Test   361
         Cumulative Review Exercises   363

7 *Introduction to Geometry*   *365*

7.1 Some Basic Definitions   366
7.2 Parallel and Perpendicular Lines   375
7.3 Polygons   381
7.4 Properties of Triangles   388
7.5 Perimeters and Areas of Polygons   395
7.6 Circles   407
7.7 Surface Area and Volume   415
         Key Concept: Formulas   425
         Accent on Teamwork   426
         Chapter Review   427
         Chapter Test   435
         Cumulative Review Exercises   437

8 *Algebraic Expressions, Equations, and Inequalities*   *440*

8.1 Algebraic Expressions   441
8.2 Simplifying Algebraic Expressions   452
8.3 Solving Equations   462
8.4 Formulas   470
8.5 Problem Solving   483
8.6 Inequalities   496
         Key Concept: Simplify and Solve   509
         Accent on Teamwork   510
         Chapter Review   511
         Chapter Test   517
         Cumulative Review Exercises   519

9 *Graphs, Linear Equations, and Functions*   *521*

9.1 Graphing Using the Rectangular Coordinate System   522
9.2 Equations Containing Two Variables   532
9.3 Graphing Linear Equations   544
9.4 Rate of Change and the Slope of a Line   556
9.5 Describing Linear Relationships   568
9.6 Writing Linear Equations   578
9.7 Functions   587
Key Concept: Describing Linear Relationships   597
Accent on Teamwork   598
Chapter Review   599
Chapter Test   605
Cumulative Review Exercises   607

10 *Exponents and Polynomials*   *609*

10.1 Natural-Number Exponents   610
10.2 Zero and Negative Integer Exponents   618
10.3 Scientific Notation   626
10.4 Polynomials   632
10.5 Adding and Subtracting Polynomials   638
10.6 Multiplying Polynomials   645
10.7 Dividing Polynomials by Monomials   654
10.8 Dividing Polynomials by Polynomials   660
Key Concept: Polynomials   666
Accent on Teamwork   667
Chapter Review   668
Chapter Test   673
Cumulative Review Exercises   675

11 *Factoring and Quadratic Equations*   *678*

11.1 Factoring Out the Greatest Common Factor and Factoring by Grouping   679
11.2 Factoring Trinomials of the Form $x^2 + bx + c$ 688
11.3 Factoring Trinomials of the Form $ax^2 + bx + c$ 695
11.4 Special Factorizations and a Factoring Strategy   703
11.5 Quadratic Equations   712
Key Concept: Factoring   723
Accent on Teamwork   724
Chapter Review   725

Chapter Test   729
Cumulative Review Exercises   731

12 *Rational Expressions and Equations*   733

12.1 Simplifying Rational Expressions   734
12.2 Multiplying and Dividing Rational Expressions   742
12.3 Adding and Subtracting Rational Expressions   750
12.4 Complex Fractions   760
12.5 Rational Equations and Problem Solving   766
12.6 Proportions and Similar Triangles   777
12.7 Variation   787
Key Concept: Expressions and Equations   795
Accent on Teamwork   796
Chapter Review   797
Chapter Test   801
Cumulative Review Exercises   803

13 *Solving Systems of Equations and Inequalities*   806

13.1 Solving Systems of Equations by Graphing   807
13.2 Solving Systems of Equations by Substitution   817
13.3 Solving Systems of Equations by Addition   824
13.4 Applications of Systems of Equations   831
13.5 Graphing Linear Inequalities   841
13.6 Solving Systems of Linear Inequalities   850
Key Concept: Systems of Equations and Inequalities   859
Accent on Teamwork   860
Chapter Review   861
Chapter Test   867
Cumulative Review Exercises   869

14 *Roots and Radicals*   871

14.1 Square Roots   872
14.2 Higher-Order Roots; Radicands That Contain Variables   882
14.3 Simplifying Radical Expressions   889
14.4 Adding and Subtracting Radical Expressions   898
14.5 Multiplying and Dividing Radical Expressions   904
14.6 Solving Radical Equations; the Distance Formula   913
14.7 Rational Exponents   922
Key Concept: Inverse Operations   929

Accent on Teamwork   930
Chapter Review   931
Chapter Test   935
Cumulative Review Exercises   937

---

15 *Quadratic Equations*   *941*

15.1 Completing the Square   942
15.2 The Quadratic Formula   951
15.3 Graphing Quadratic Functions   962
Key Concept: Quadratic Equations   973
Accent on Teamwork   974
Chapter Review   975
Chapter Test   979
Cumulative Review Exercises   981

Appendix I
Inductive and Deductive Reasoning   *A-1*

Appendix II
Measurement   *A-8*
II.1 American Units of Measurement   A-8
II.2 Metric Units of Measurement   A-18
II.3 Converting between American and Metric Units   A-28

Appendix III
Solving Absolute Value Equations and Inequalities   *A-35*

Appendix IV
Probability   *A-45*

Appendix V
Roots and Powers   *A-49*

Appendix VI
Answers to Selected Exercises   *A-50*

Index   *I-1*

For the Instructor

An increasing number of schools are offering an arithmetic course and an introductory algebra course in combination. There are several advantages in doing this:

- It eliminates much of the redundancy involved in teaching the arithmetic/algebra sequence as two separate courses. As a result, students have more time to master the material.
- A combined approach allows for true integration of the arithmetic and the algebra topics.
- For students, purchasing a single textbook saves money.

However, offering a combination course raises several concerns:

- The textbook used in such a course must include enough arithmetic so that students master the prerequisite skills necessary for success in algebra.
- The introductory algebra material should not get too difficult to fast.
- The course must cover additional topics, such as statistics, geometry, probability, inductive and deductive reasoning, and measurement, in order to satisfy many state-mandated educational requirements.

Developmental Mathematics for College Students has been written to address these concerns. The first five chapters provide a complete course in arithmetic, in which the standard arithmetic topics are presented at a reasonable pace. The fundamental algebra concepts of variables and equations, as well as signed numbers, are introduced early and are revisited frequently. This ensures a smooth transition into the complete course in introductory algebra found in Chapters 8–15. Chapter 6, Descriptive Statistics, and Chapter 7, Introduction to Geometry, add breadth to the text. Material on probability, inductive and deductive reasoning, and measurement is included in a strong complement of appendices.

Our goal has been to write a book that is interesting and enjoyable to read—one that will attract and keep the attention of college students of all ages. A variety of instructional approaches are used, reflecting the recommendations of NCTM and AMATYC. In combination with the student and instructor supplements that are available, *Developmental Mathematics for College Students* can be used in lecture, laboratory, or self-study formats.

Features of the text

A Blend of the Traditional and Reform Approaches
We have used a combination of instructional methods from the traditional and reform approaches, endeavoring to write a book that contains the best of both. You will find the

vocabulary, practice, and well-defined pedagogy of a traditional basic mathematics book. The text also features problem solving, reasoning, communicating, and technology, as emphasized by the reform movement.

Thorough Coverage of Arithmetic

This book provides thorough coverage of the arithmetic of whole numbers, fractions, and decimals. Other topics traditionally taught in an arithmetic course are also included, such as percent, ratio and proportion, and measurement.

Arithmetic and Algebra Are Integrated Throughout

To prepare students for introductory algebra, this book provides a review of arithmetic while introducing basic algebraic concepts. For example, Chapter 1 covers whole-number arithmetic, but it also introduces the concept of a variable, develops the geometric formulas for perimeter and area, and shows how to simplify numerical expressions. In Chapter 1, we also lay the groundwork for rectangular coordinate graphing and solve some simple equations. Additionally, we establish a five-step problem-solving strategy and use it to solve real-world problems.

Competency with Signed Numbers

The rules for adding, subtracting, multiplying, and dividing integers are introduced in Chapter 2. Students apply these rules again in Chapter 3 with signed fractions, and in Chapter 4 with signed decimals.

We feel this spiral approach is superior to that of introducing signed numbers in a single chapter at, or near, the end of the arithmetic review. Revisiting these rules in several different contexts builds a thorough understanding of signed numbers, and this pays great dividends when the student makes the transition to the introductory algebra portion of the text.

Instructional Support for the Visual Learner

A colorful design has been used to attract and keep the attention of students. For those who are visual learners, we have included many illustrations and a number of diagrams to help clarify important arithmetic and algebraic concepts.

Solving equations

Since the solution of an equation is usually not given, we must develop a process to find it. This process is called *solving the equation*. To develop an understanding of the properties and procedures used to solve an equation, we will examine $x + 2 = 5$ and make some observations as we solve it in a practical way.

We can think of the scales shown in Figure 8-1(a) as representing the equation $x + 2 = 5$. The weight (in grams) on the left-hand side of the scales is $x + 2$, and the weight (in grams) on the right-hand side is 5. Because these weights are equal, the scales are in balance. To find x, we need to isolate it. That can be accomplished by removing 2 grams from the left-hand side of the scales. Common sense tells us that we must also remove 2 grams from the right-hand side if the scales are to remain in balance. In Figure 8-1(b), we can see that x grams will be balanced by 3 grams. We say that we have *solved* the equation and that the *solution* is 3.

FIGURE 8-1

Writing improper fractions as mixed numbers

To write an improper fraction as a mixed number, we must find two things: the *whole-number part* and the *fractional part* of the mixed number. To develop a procedure to do this, let's consider the improper fraction $\frac{7}{3}$. To find the number of groups of 3 in 7, we can divide 7 by 3. This will find the whole-number part of the mixed number. The remainder is the numerator of the fractional part of the mixed number.

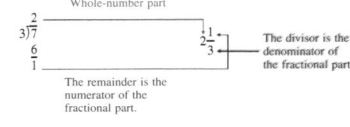

 Diagrams and color are used to help the visual learner.

Interactivity

Author's notes are used to explain the steps in the solutions of examples. The notes are extensive so as to increase the students' ability to read and write mathematics. Most worked examples in the text are accompanied by Self Checks. This feature allows students to practice skills discussed in the example by working a similar problem. Because the Self-Check problems are adjacent to the worked examples, students can easily refer to the solution and author's notes of the example as they solve the Self Check.

Example titles highlight the ▶
concept being discussed.

Author's notes explain the steps ▶
in the solution process.

The Self Check answers ▶
are provided.

EXAMPLE 1 *Order of operations.* Evaluate: $-4(-3)^2 - (-2)$.

Solution
This expression contains the operations of multiplication, raising to a power, and subtraction. The rules for the order of operations tell us to find the power first.

$$-4(-3)^2 - (-2) = -4(9) - (-2) \quad \text{Evaluate the exponential expression: } (-3)^2 = 9.$$
$$= -36 - (-2) \quad \text{Do the multiplication: } -4(9) = -36.$$
$$= -36 + 2 \quad \text{To do the subtraction, add the opposite of } -2.$$
$$= -34 \quad \text{Do the addition.}$$

Self Check
Evaluate: $-5(-2)^2 - (-6)$.

Answer: -14 ∎

Study Sets—More Than Just Exercises

The problems at the end of each section are called Study Sets. Each Study Set includes Vocabulary, Notation, and Writing problems designed to help students improve their ability to read, write, and communicate mathematical ideas. The problems in the Concepts section of the Study Sets encourage students to engage in independent thinking and reinforce major ideas through exploration. In the Practice section, students get the drill necessary to master the material. In the Applications section, students deal with real-life situations that involve the topics being studied. Each Study Set concludes with a Review section consisting of problems based on material from previous sections.

◀ Each Study Set contains Vocabulary, Concepts, Notation, Practice, Applications, Writing, and Review sections.

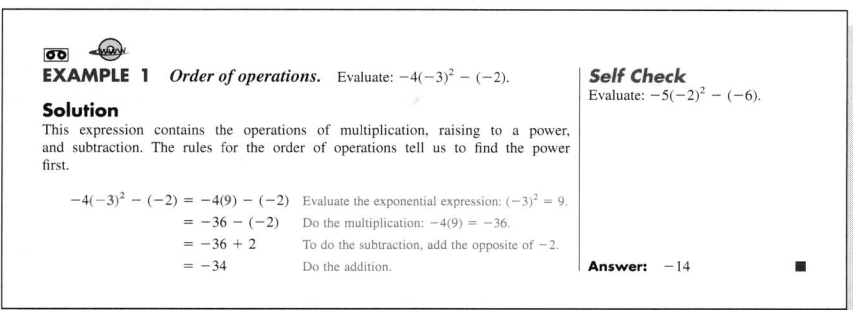

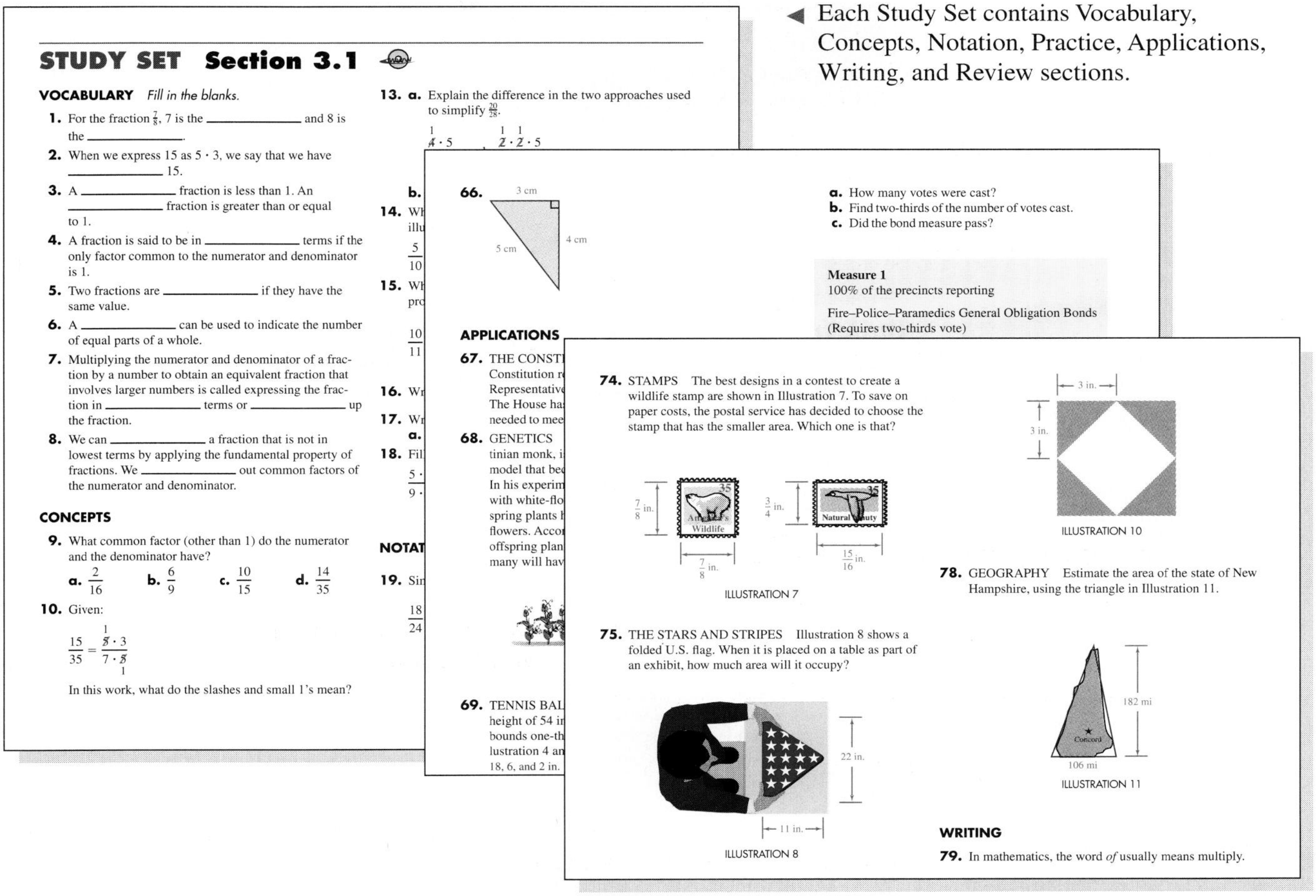

STUDY SET   Section 3.1

VOCABULARY *Fill in the blanks.*

1. For the fraction $\frac{7}{8}$, 7 is the _____ and 8 is the _____.

2. When we express 15 as $5 \cdot 3$, we say that we have _____ 15.

3. A _____ fraction is less than 1. An _____ fraction is greater than or equal to 1.

4. A fraction is said to be in _____ terms if the only factor common to the numerator and denominator is 1.

5. Two fractions are _____ if they have the same value.

6. A _____ can be used to indicate the number of equal parts of a whole.

7. Multiplying the numerator and denominator of a fraction by a number to obtain an equivalent fraction that involves larger numbers is called expressing the fraction in _____ terms or _____ up the fraction.

8. We can _____ a fraction that is not in lowest terms by applying the fundamental property of fractions. We _____ out common factors of the numerator and denominator.

CONCEPTS

9. What common factor (other than 1) do the numerator and the denominator have?
a. $\frac{2}{16}$ **b.** $\frac{6}{9}$ **c.** $\frac{10}{15}$ **d.** $\frac{14}{35}$

10. Given:

$$\frac{15}{35} = \frac{\overset{1}{\cancel{5}} \cdot 3}{7 \cdot \underset{1}{\cancel{5}}}$$

In this work, what do the slashes and small 1's mean?

13. a. Explain the difference in the two approaches used to simplify $\frac{20}{28}$.

$$\frac{1}{4 \cdot 5} \qquad \frac{1 \quad 1}{2 \cdot 2 \cdot 5}$$

b.

14. Wh illu

$\frac{5}{10}$

15. Wh pro

$\frac{10}{11}$

NOTAT

16. Wr

17. Wr
a.

18. Fil
$\frac{5}{9} \cdot$

19. Sir

$\frac{18}{24}$

APPLICATIONS

66. 3 cm
5 cm   4 cm

67. THE CONSTI
Constitution r
Representative
The House ha
needed to mee

68. GENETICS
tinian monk, i
model that be
In his experim
with white-flo
spring plants h
flowers. Accor
offspring plan
many will hav

69. TENNIS BAL
height of 54 in
bounds one-th
lustration 4 an
18, 6, and 2 in.

a. How many votes were cast?
b. Find two-thirds of the number of votes cast.
c. Did the bond measure pass?

Measure 1
100% of the precincts reporting
Fire–Police–Paramedics General Obligation Bonds
(Requires two-thirds vote)

74. STAMPS   The best designs in a contest to create a wildlife stamp are shown in Illustration 7. To save on paper costs, the postal service has decided to choose the stamp that has the smaller area. Which one is that?

$\frac{7}{8}$ in.   American Wildlife $\frac{3}{4}$ in.   Natural Beauty
$\frac{7}{8}$ in. $\frac{15}{16}$ in.

ILLUSTRATION 7

75. THE STARS AND STRIPES   Illustration 8 shows a folded U.S. flag. When it is placed on a table as part of an exhibit, how much area will it occupy?

22 in.
◀ 11 in. ▶

ILLUSTRATION 8

3 in.
3 in.

ILLUSTRATION 10

78. GEOGRAPHY   Estimate the area of the state of New Hampshire, using the triangle in Illustration 11.

182 mi
Concord
106 mi

ILLUSTRATION 11

WRITING

79. In mathematics, the word *of* usually means multiply.

Applications and Connections to Other Disciplines

A distinguishing feature of this book is its wealth of application problems. We have included numerous applications from disciplines such as science, economics, business, manufacturing, history, and entertainment, as well as mathematics.

Every application problem has a title. ▶

85. SCRABBLE Illustration 3(a) shows part of the game board before and Illustration 3(b) shows it after the words *brick* and *aphid* were played. Determine the scoring for each word. (The number on each tile gives the point value of the letter.)

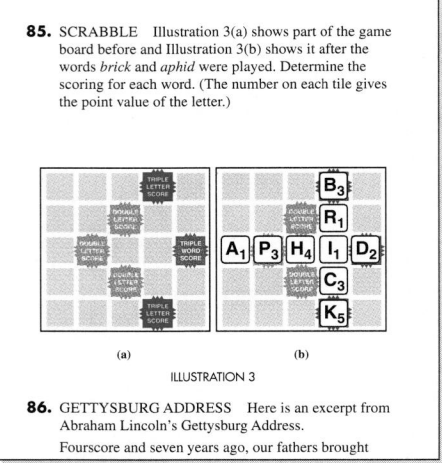

(a)　(b)

ILLUSTRATION 3

86. GETTYSBURG ADDRESS Here is an excerpt from Abraham Lincoln's Gettysburg Address.

Fourscore and seven years ago, our fathers brought

71. WEATHER MAP Illustration 7 shows the predicted Fahrenheit temperatures for a day in mid-January.

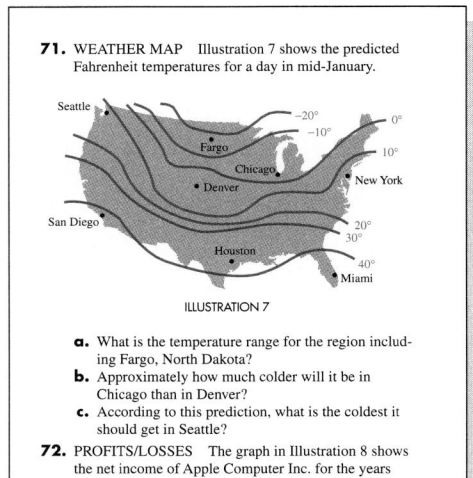

ILLUSTRATION 7

a. What is the temperature range for the region including Fargo, North Dakota?
b. Approximately how much colder will it be in Chicago than in Denver?
c. According to this prediction, what is the coldest it should get in Seattle?

72. PROFITS/LOSSES The graph in Illustration 8 shows the net income of Apple Computer Inc. for the years

Application problems are drawn from a wide variety of disciplines. ▶

63. BLUEPRINT The scale for the drawing in Illustration 6 tells the reader that a $\frac{1}{4}$-inch length $\left(\frac{1}{4}''\right)$ on the drawing corresponds to an actual size of 1 foot $(1'0'')$. Suppose the length of the kitchen is $2\frac{1}{2}$ inches on the blueprint. How long is the actual kitchen?

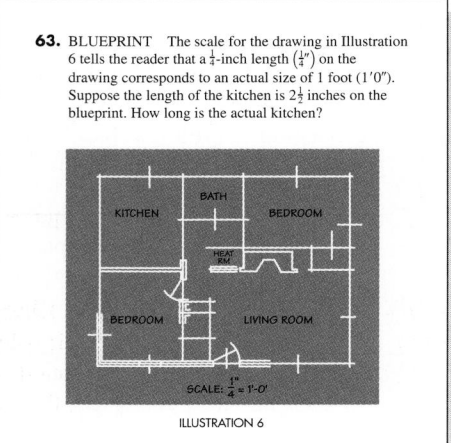

ILLUSTRATION 6

83. BIOLOGY DNA is found in cells. It is referred to as the genetic "blueprint." In humans, it determines such traits as eye color, hair color, and height. A model of DNA appears in Illustration 8. If Å = 0.000000004 inch, determine the three dimensions shown in the illustration.

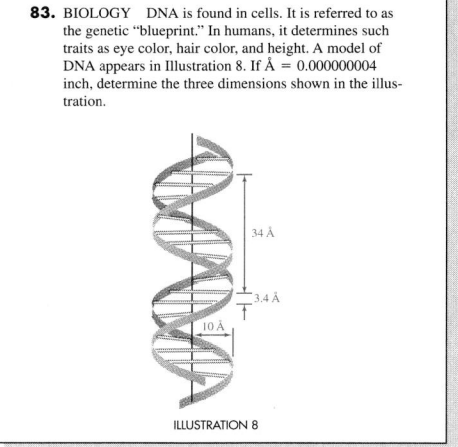

ILLUSTRATION 8

Building a Foundation for Graphing

Graphing on the number line is introduced in Chapter 2 when working with integers. The topic is revisited in Chapters 3 and 4 when the students work with signed fractions and decimals.

Graphing on the number line prepares students for rectangular coordinate graphing. ▶

Graphing fractions and mixed numbers

Earlier, we graphed whole numbers and integers on a number line. Fractions and mixed numbers can also be graphed on a number line.

EXAMPLE 3 *Graphing fractions and mixed numbers.* Graph $-2\frac{3}{4}$, $-1\frac{1}{2}$, $-\frac{1}{8}$, and $\frac{13}{5}$ on a number line.

Solution
To help locate the graph of each number, we make some observations:

• Since $-2\frac{3}{4} < -2$, the graph of $-2\frac{3}{4}$ is to the left of -2.
• The number $-1\frac{1}{2}$ is between -1 and -2.
• The number $-\frac{1}{8}$ is less than 0.
• Expressed as a mixed number, $\frac{13}{5} = 2\frac{3}{5}$.

Self Check

Graph $-1\frac{7}{8}$, $-\frac{2}{3}$, and $\frac{9}{4}$ on a number line.

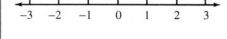

Answer:

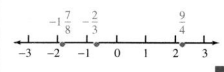

Problem-Solving Strategy

One of the major objectives of this textbook is to make students better problem solvers. To this end, we use a five-step problem-solving strategy throughout the book. The five steps are: *Analyze the problem, Form an equation, Solve the equation, State the conclusion,* and *Check the result.*

A five-step problem-solving ▶ strategy is used.

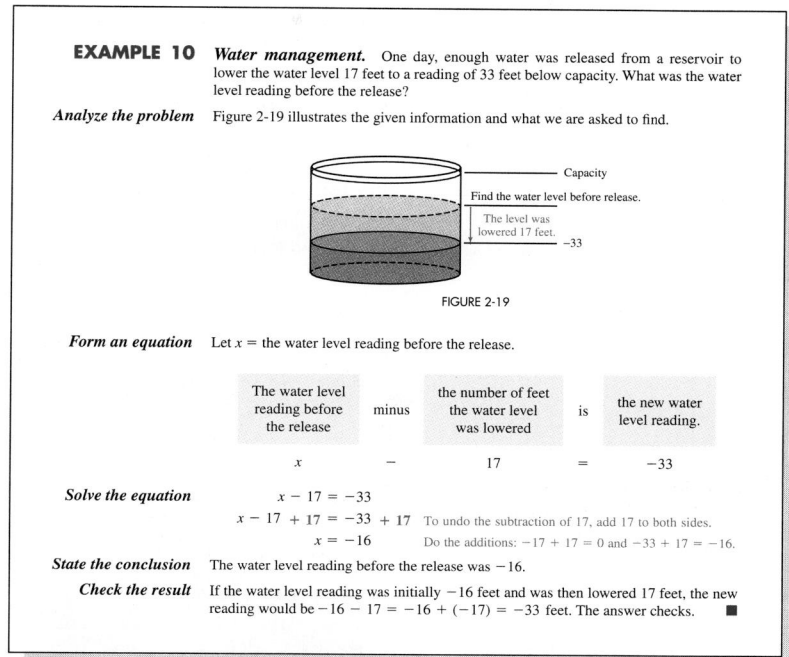

EXAMPLE 10 *Water management.* One day, enough water was released from a reservoir to lower the water level 17 feet to a reading of 33 feet below capacity. What was the water level reading before the release?

Analyze the problem Figure 2-19 illustrates the given information and what we are asked to find.

FIGURE 2-19

Form an equation Let x = the water level reading before the release.

The water level reading before the release	minus	the number of feet the water level was lowered	is	the new water level reading.
x	$-$	17	$=$	-33

Solve the equation
$$x - 17 = -33$$
$$x - 17 + 17 = -33 + 17 \quad \text{To undo the subtraction of 17, add 17 to both sides.}$$
$$x = -16 \quad \text{Do the additions: } -17 + 17 = 0 \text{ and } -33 + 17 = -16.$$

State the conclusion The water level reading before the release was -16.

Check the result If the water level reading was initially -16 feet and was then lowered 17 feet, the new reading would be $-16 - 17 = -16 + (-17) = -33$ feet. The answer checks. ∎

Constructing Charts, Tables, and Graphs; Statistics

Many problems require students to present their solutions in the form of a table, graph, or chart. Often, students must examine such data displays to obtain necessary information to solve a problem. Chapter 6, on descriptive statistics, covers the mean, median, and mode.

Real data ▶ are integrated throughout the text.

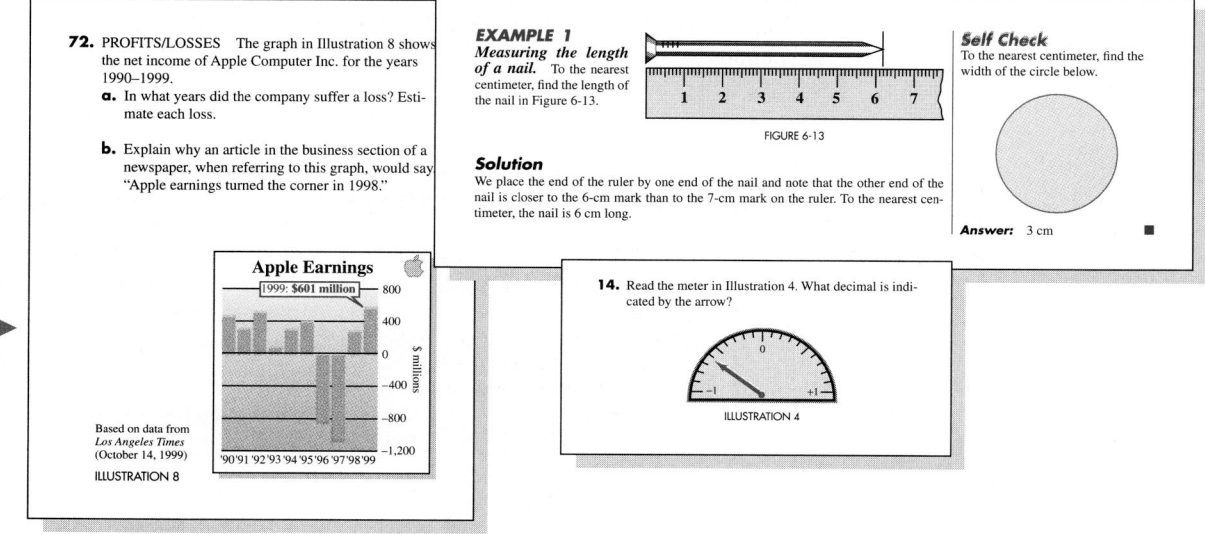

Calculators Are Optional

For those instructors who wish to use calculators as part of the instruction in the arithmetic portion of the course, the text includes an Accent on Technology feature that introduces keystrokes and shows how scientific calculators can be used to solve application problems. Some Study Sets include problems that are to be solved using a calculator; these problems are indicated by the calculator logo ▦. Instructors who do not wish to introduce calculators can skip that material without interrupting the flow of ideas.

In the introductory algebra portion of the text, the Accent on Technology features give keystrokes for both scientific and graphing calculators. Also included in Section 9.2 is a special feature that explains how a graphing calculator can be used to graph equations.

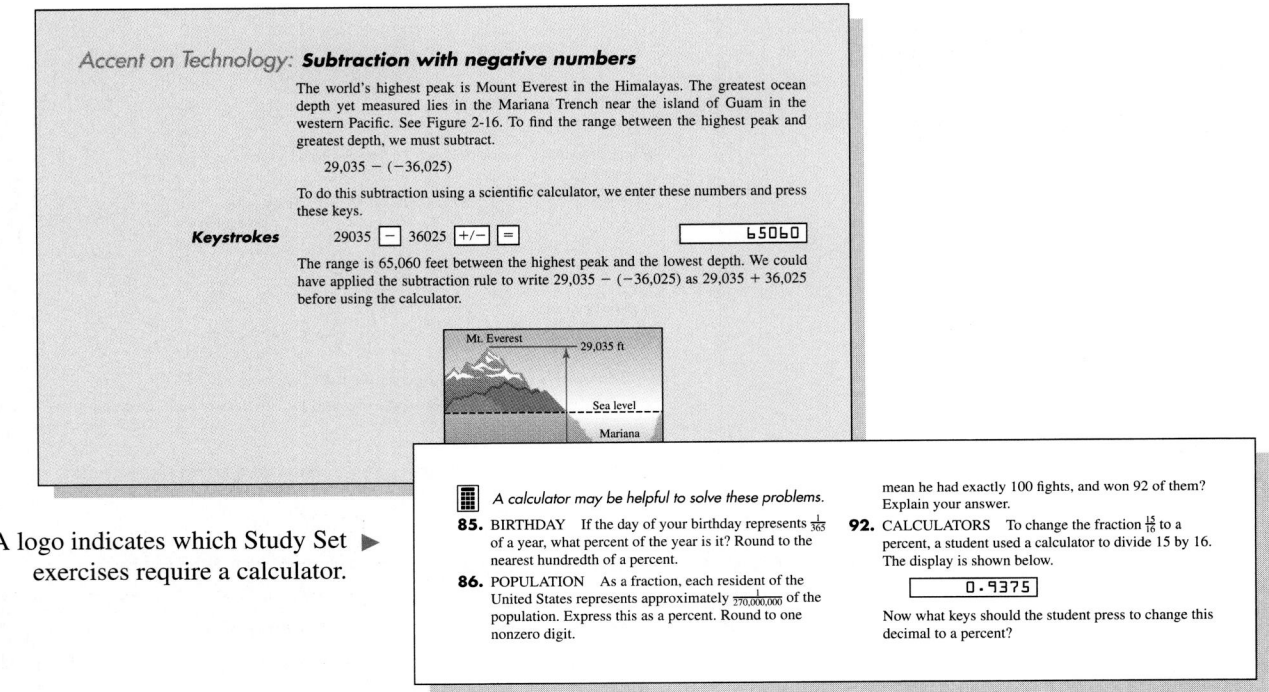

A logo indicates which Study Set ▶ exercises require a calculator.

Estimation

Special two-page Estimation features appear in the chapters on Whole Numbers, Decimals, and Percent. In these features, estimation procedures are introduced and put to use in real-life situations that require only approximate answers.

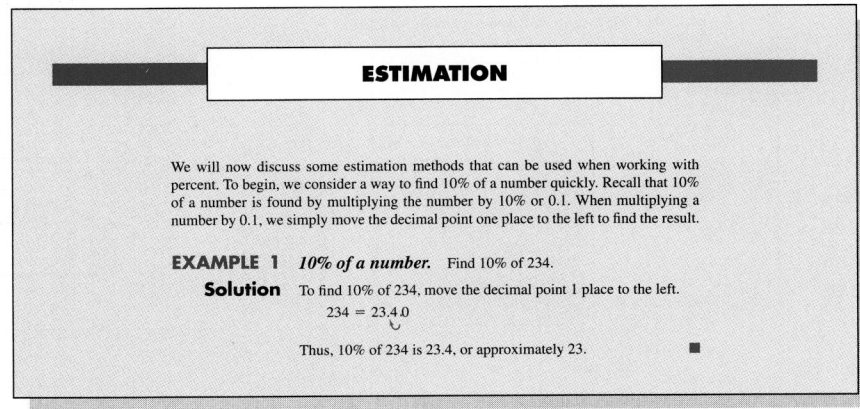

Key Concepts

Fifteen important mathematical concepts are highlighted in one-page Key Concept features, appearing near the end of each chapter. Each Key Concept page summarizes a concept and gives students an opportunity to review the role it plays in the overall picture.

Group Work

A one-page feature called Accent on Teamwork appears near the end of each chapter. It gives a set of problems that the instructor can assign as group work or to individual students as outside-of-class projects.

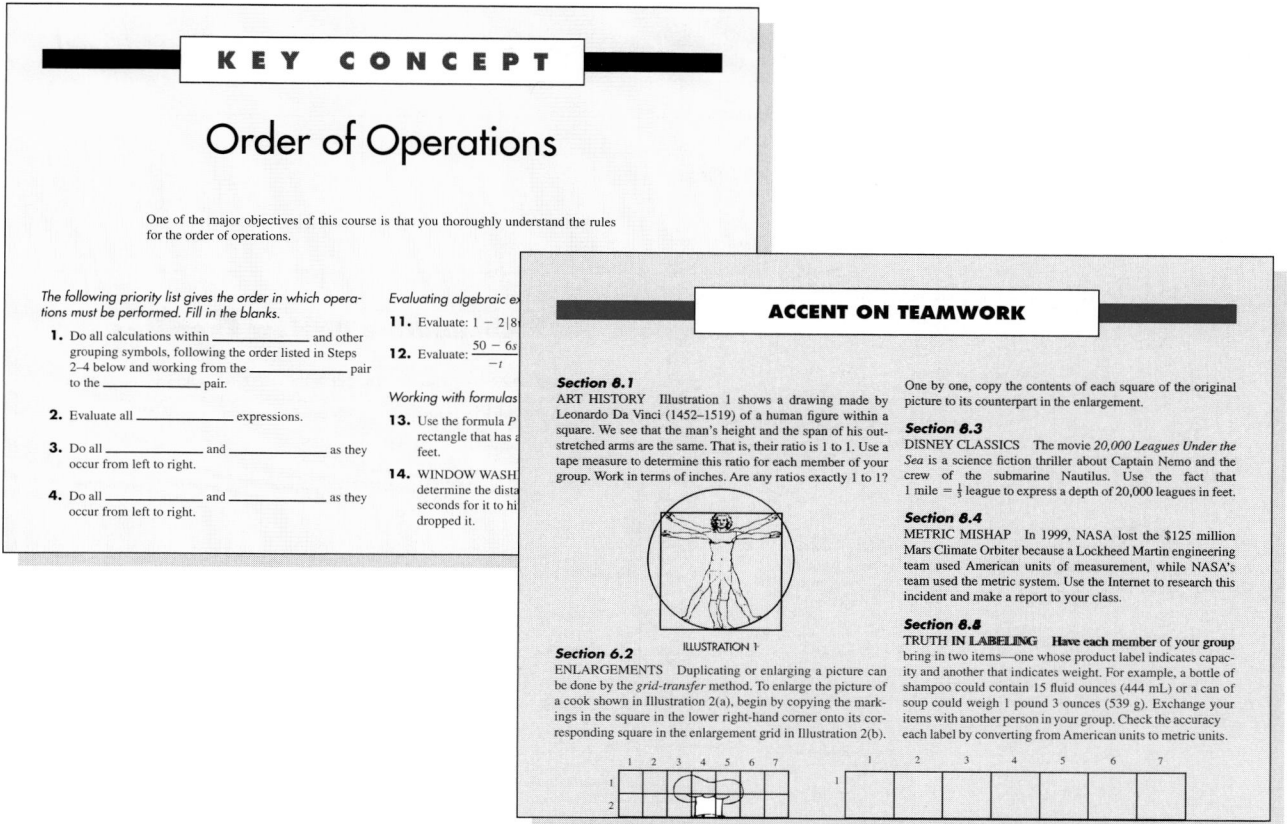

KEY CONCEPT

Order of Operations

One of the major objectives of this course is that you thoroughly understand the rules for the order of operations.

The following priority list gives the order in which operations must be performed. Fill in the blanks.

1. Do all calculations within _____ and other grouping symbols, following the order listed in Steps 2–4 below and working from the _____ pair to the _____ pair.

2. Evaluate all _____ expressions.

3. Do all _____ and _____ as they occur from left to right.

4. Do all _____ and _____ as they occur from left to right.

Evaluating algebraic e...

11. Evaluate: $1 - 2|8...$

12. Evaluate: $\dfrac{50 - 6s}{-1}$

Working with formulas

13. Use the formula P... rectangle that has ... feet.

14. WINDOW WASH... determine the dista... seconds for it to hi... dropped it.

ACCENT ON TEAMWORK

Section 8.1
ART HISTORY Illustration 1 shows a drawing made by Leonardo Da Vinci (1452–1519) of a human figure within a square. We see that the man's height and the span of his outstretched arms are the same. That is, their ratio is 1 to 1. Use a tape measure to determine this ratio for each member of your group. Work in terms of inches. Are any ratios exactly 1 to 1?

ILLUSTRATION 1

Section 6.2
ENLARGEMENTS Duplicating or enlarging a picture can be done by the *grid-transfer* method. To enlarge the picture of a cook shown in Illustration 2(a), begin by copying the markings in the square in the lower right-hand corner onto its corresponding square in the enlargement grid in Illustration 2(b).

One by one, copy the contents of each square of the original picture to its counterpart in the enlargement.

Section 6.3
DISNEY CLASSICS The movie *20,000 Leagues Under the Sea* is a science fiction thriller about Captain Nemo and the crew of the submarine Nautilus. Use the fact that 1 mile = $\frac{1}{3}$ league to express a depth of 20,000 leagues in feet.

Section 8.4
METRIC MISHAP In 1999, NASA lost the $125 million Mars Climate Orbiter because a Lockheed Martin engineering team used American units of measurement, while NASA's team used the metric system. Use the Internet to research this incident and make a report to your class.

Section 8.8
TRUTH IN LABELING Have each member of your group bring in two items—one whose product label indicates capacity and another that indicates weight. For example, a bottle of shampoo could contain 15 fluid ounces (444 mL) or a can of soup could weigh 1 pound 3 ounces (539 g). Exchange your items with another person in your group. Check the accuracy of each label by converting from American units to metric units.

In-Depth Coverage of Geometry

The concepts of perimeter and area are introduced in Chapter 1 and revisited throughout the book. We have also included a wide variety of plane and solid figures in the Study Sets. Since many of the students taking this course did not take a geometry class in high school, Chapter 7 offers an overview of some of the most important geometry topics. The material is presented in a way that reinforces algebraic concepts such as formula, evaluation, and problem solving.

Geometry topics are presented ▶ in a practical setting.

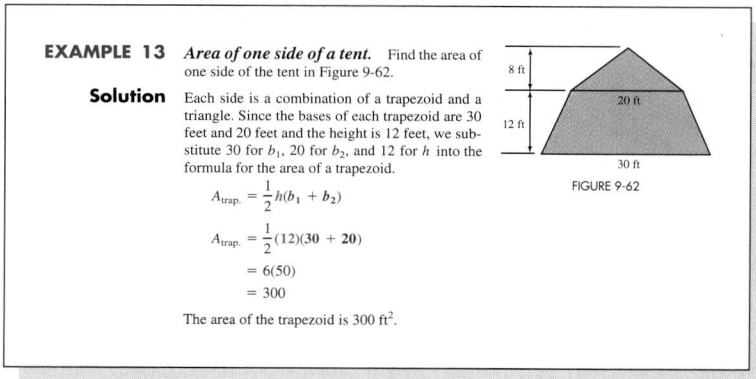

EXAMPLE 13   *Area of one side of a tent.*   Find the area of one side of the tent in Figure 9-62.

Solution   Each side is a combination of a trapezoid and a triangle. Since the bases of each trapezoid are 30 feet and 20 feet and the height is 12 feet, we substitute 30 for b_1, 20 for b_2, and 12 for h into the formula for the area of a trapezoid.

$$A_{\text{trap.}} = \frac{1}{2}h(b_1 + b_2)$$

$$A_{\text{trap.}} = \frac{1}{2}(12)(30 + 20)$$

$$= 6(50)$$

$$= 300$$

The area of the trapezoid is 300 ft^2.

FIGURE 9-62

Systematic Review

Each Study Set ends with a Review section that contains problems similar to those in previous sections. Each chapter ends with a Chapter Review and a Chapter Test. The chapter reviews have been designed to be user friendly. In a unique format, the reviews lists the important concepts of each section of the chapter in one column, with appropriate review problems running parallel in a second column. In addition, Cumulative Review Exercises appear at the end of each chapter (except Chapter 1).

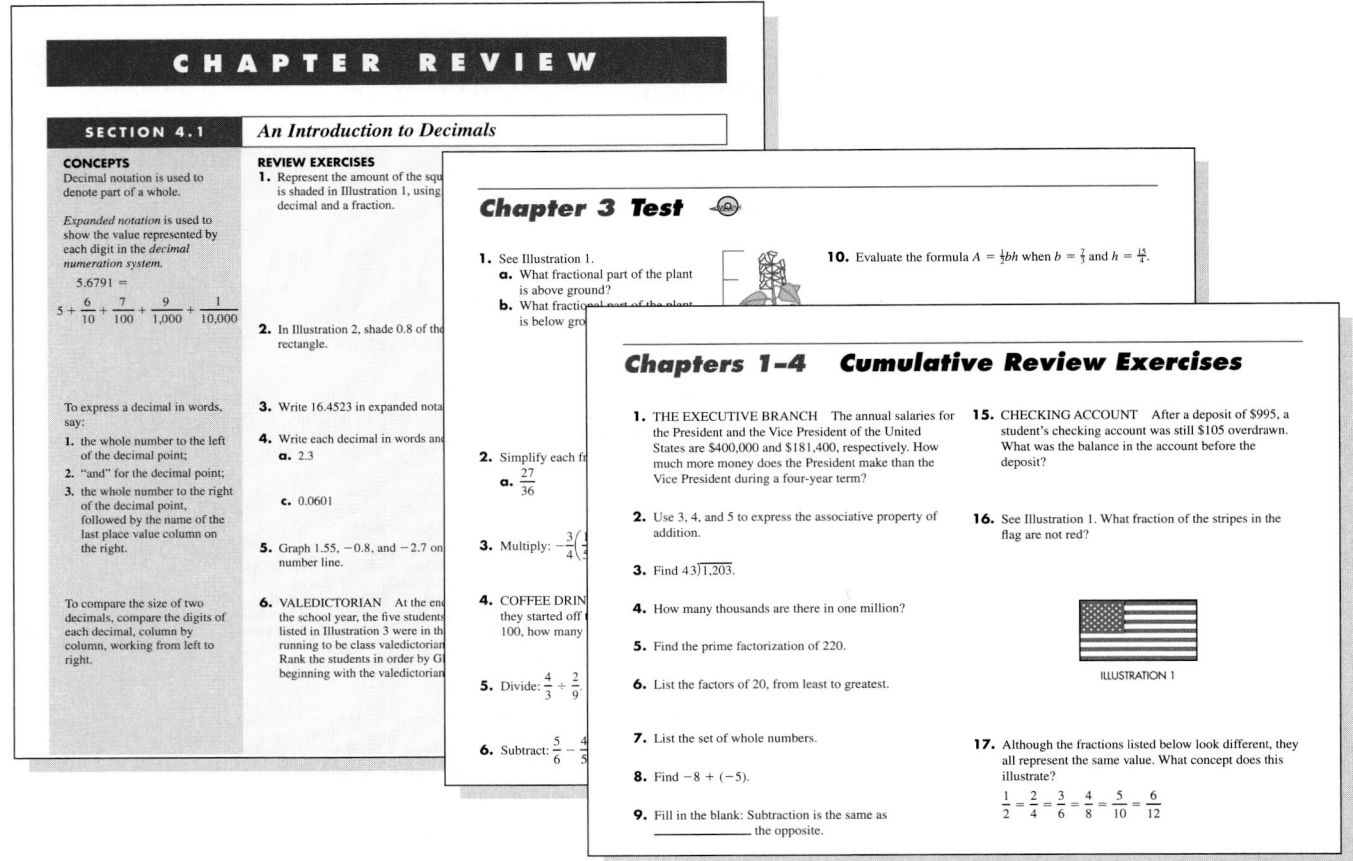

Student support

We have included many features that make *Developmental Mathematics for College Students* very accessible to students.

Worked Examples

The text contains more than 600 worked examples, many with several parts. Explanatory notes make the examples easy to follow.

Videotapes

The videotape series that accompanies this book shows students the steps in solving many examples in the text. A video logo 📼 placed next to an example indicates that the example is taught on tape. In addition, the tapes present the solutions of many Study Set problems.

Author's Notes

Author's notes, printed in red, are used to explain the steps in the solutions of examples. The notes are extensive; complete sentences are used so as to increase the students' ability to read and write mathematics.

A special logo shows which examples are included in the ▶ videotape series.

Another logo shows ▼ which examples are included in the *Interactive Video Skillbuilder* CD.

Each step is explained using ▶ detailed author's notes.

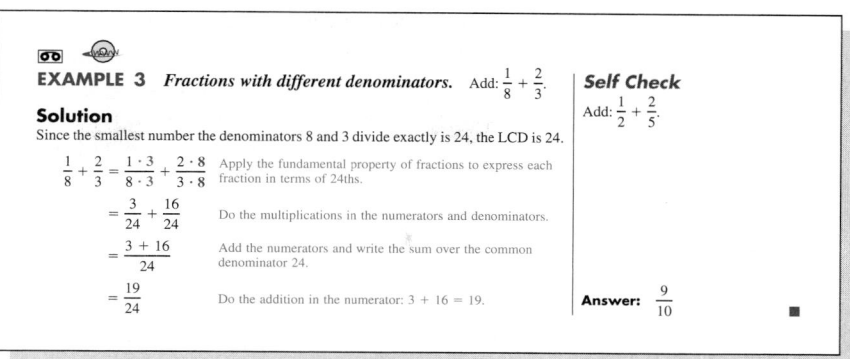

Self Checks

There are more than 525 Self Check problems, which allow students to practice the skills demonstrated in the worked examples.

Comments

Throughout the text, Comments call attention to common mistakes and explain how to avoid them.

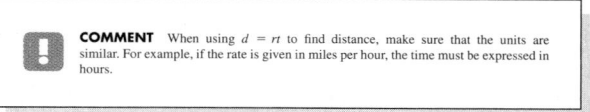

Functional Use of Color

For easy reference, definitions, strategies, rules, and properties are printed in blue boxes. In addition, the book uses color to highlight terms and expressions that you would point to in a classroom discussion.

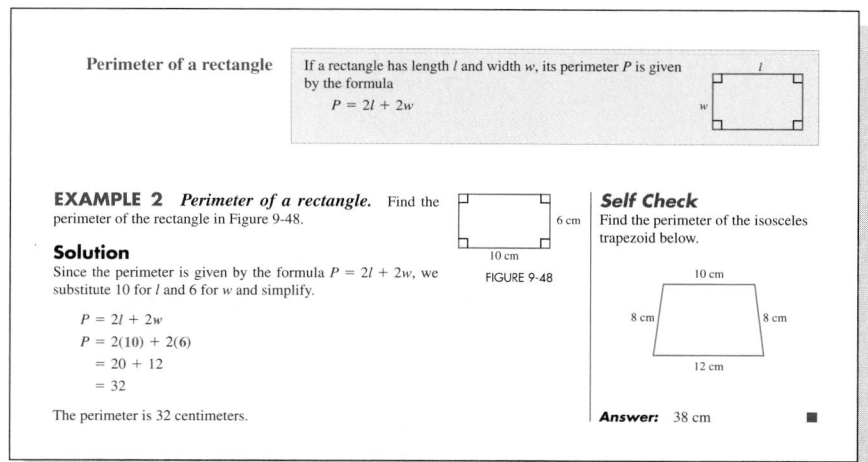

Problems and Answers

The book includes thousands of carefully graded exercises. Appendix VI provides the answers to the odd-numbered exercises in the Study Sets as well as all the answers to the Chapter Review, Chapter Test, and Cumulative Review problems.

Reading and Writing Mathematics

Also included (on pages xxiii-xxiv) are two features to help students improve their ability to read and write mathematics. "Reading Mathematics" helps students get the most out of the examples in this book by showing them how to read the solutions properly. "Writing Mathematics" highlights the characteristics of a well-written solution.

Study Skills and Math Anxiety

These two topics are discussed in detail in the section entitled "For the Student" at the end of this preface. Additionally, in "Success in Mathematics," students are asked to design a strategy for studying and learning the material. "Taking a Math Test," on page 74, helps students prepare for a test and then gives them suggestions for improving their performance.

Ancillaries for the instructor

Annotated Instructor's Edition

This is a special version of the complete student text, with all answers printed in blue next to the respective exercises.

Complete Solutions Manual

The *Complete Solutions Manual* provides worked-out solutions to all the exercises.

BCA Testing

Brooks/Cole Assessment is a text-specific, Internet-ready testing suite that allows instructors to customize exams and track student progress in a browser-based format. BCA offers full algorithmic generation of problems and free-response questions. The testing and course-management components simplify routine tasks. Test results flow automatically into the gradebook, and the instructor can easily communicate with individuals, sections, or entire courses.

Text-Specific Videotapes

A set of videotapes is available free to a school upon adoption of the text. Each tape covers one chapter of the text, broken into problem-solving sessions of 10 to 20 minutes. Examples from each section of the chapter are covered, as well as exercises from each Study Set. Where an example is taught on tape, a special logo 📼 is printed next to the example in the text.

Ancillaries for the student

Student Solutions Manual

The *Student Solutions Manual* provides worked-out solutions to the odd-numbered exercises in the text.

BCA Tutorial

This text-specific interactive software is delivered via the Web (at http://bca. brookscole.com). It is offered in both student's and instructor's versions. Because it is browser-based, it can serve as an intuitive guide even for students who have little technological proficiency. BCA Tutorial allows students to work with real math notation in real time, providing instant analysis and feedback. In the instructor's version, a built-in tracking program enables instructors to monitor student progress.

Interactive Video Skillbuilder CD

Packaged with each book, this single CD-ROM contains more than eight hours of video instruction. The problems worked during each video lesson are listed next to the viewing screen, so that students can work them ahead of time if they choose. To help students evaluate their progress, each section contains a 10-question Web quiz, and each chapter contains a chapter test, with answers provided.

Acknowledgments

We are grateful to the instructors who have reviewed the text at various stages of its development. Their comments and suggestions have proven invaluable in making this a better book. We sincerely thank them for lending their time and talent to this project.

Linda Beattie
Western New Mexico University

Joan Bookbinder
University of San Diego

Gerald Busald
San Antonio College

Karen Clark
Tacoma Community College

Linda Clay
Albuquerque TVI

Gudryn Doherty
Community College of Denver

David L. Fama
Germanna Community College

Barbara Gentry
Parkland College

Maria Gushanas
Seton Hall University

Laurie Hoecherl
Kishwaukee College

Judy Holcomb
Florida Community College

Judith M. Jones
Valencia Community College

Joanne Juedes
University of Wisconsin-Marathon County

Dennis Kimzey
Rogue Community College

Jan Alicia Nettler
Holyoke Community College

Scott Perkins
Lake-Sumter Community College

Angela Peterson
Portland Community College

Leah C. Pierce
Crafton Hills College

J. Doug Richey
Northeast Texas Community College

Martha Scarbrough
Motlow State Community College

Allison Sloan
Valencia West Campus

Renee Starr
Beaver College

Gwen Terwilliger
University of Toledo

Betty Truitt
Black Hawk College

Thomas Vanden Eynden
Thomas More College

Jane Weber
University of Alaska, Fairbanks

Annette Wiesner
University of Wisconsin-Parkside

Mary Jane Wolfe
University of Rio Grande

Ronald W. Yates
Community College of Southern Nevada

We want to express our gratitude to Elizabeth Morrison, Judy Jones, Sally Lesik, Linda Rottman, and Mike Murphy for their input and suggestions. We offer special thanks to Barbara Rugeley, Eagle Zhuang, Tom Gerfen, Arnold Kondo, Rob Everest, Sheila White, Doug Keebaugh, Ron Livingston, Eric Rabitoy, Robin Carter, Karl Hunsicker, Dave Ryba, Terry Damron, Marion Hammond, Linda Humphrey, Cathy Gong, Bob Billups, Liz Tussy, Tanja Rinkel, Dennis Korn, and Laureen Breegle for their assistance with some of the application problems.

Without the talents and dedication of the editorial, marketing, and production staff of Brooks/Cole, this first edition of *Developmental Mathematics for College Students* could not have been so well accomplished. We express our sincere appreciation for the hard work of Bob Pirtle, Jennifer Huber, Rachael Sturgeon, Leah Thomson, Samantha Cabaluna, Ellen Brownstein, Vernon Boes, Micky Lawler, and Vena Dyer, as well as the freelance talents of David Hoyt, Lori Heckelman, Penny Suess, Dana Gurnee, and Roy Neuhaus and the superb typesetting of the Clarinda Company.

Alan S. Tussy
R. David Gustafson

For the Student

Success in mathematics

To be successful in mathematics, you need to know how to study it. The following checklist will help you develop your own personal strategy to study and learn the material. The following suggestions require some time and self-discipline on your part, but it will be worth the effort. This will help you get the most out of this course.

As you read each of the following statements, place a check mark in the box if you can truthfully answer Yes. If you can't answer Yes, think of what you might do to make the suggestion part of your personal study plan. You should go over this checklist several times during the semester to be sure you are following it.

To follow each of these suggestions will take time. It takes a lot of practice to learn mathematics, just as with any other skill.

Preparing for the Class

- ☐ I have made a commitment to myself to give this course my best effort.
- ☐ I have the proper materials: a pencil with an eraser, paper, a notebook, a ruler, a calculator, and a calendar or day planner.
- ☐ I am willing to spend a minimum of two hours doing homework for every hour of class.
- ☐ I will try to work on this subject every day.
- ☐ I have a copy of the class syllabus. I understand the requirements of the course and how I will be graded.
- ☐ I have scheduled a free hour after the class to give me time to review my notes and begin the homework assignment.

Class Participation

- ☐ I know my instructor's name.
- ☐ I will regularly attend the class sessions and be on time.
- ☐ When I am absent, I will find out what the class studied, get a copy of any notes or handouts, and make up the work that was assigned when I was gone.
- ☐ I will sit where I can hear the instructor and see the chalkboard.
- ☐ I will pay attention in class and take careful notes.
- ☐ I will ask the instructor questions when I don't understand the material.
- ☐ When tests, quizzes, or homework papers are passed back and discussed in class, I will write down the correct solutions for the problems I missed so that I can learn from my mistakes.

Study Sessions

- ☐ I will find a comfortable and quiet place to study.
- ☐ I realize that reading a math book is different from reading a newspaper or a novel. Quite often, it will take more than one reading to understand the material.
- ☐ After studying an example in the textbook, I will work the accompanying Self Check.
- ☐ I will begin the homework assignment only after reading the assigned section.
- ☐ I will try to use the mathematical vocabulary mentioned in the book and used by my instructor when I am writing or talking about the topics studied in this course.
- ☐ I will look for opportunities to explain the material to others.
- ☐ I will check all my answers to the problems with those provided in the back of the book (or with the *Student Solutions Manual*) and reconcile any differences.
- ☐ My homework will be organized and neat. My solutions show all the necessary steps.
- ☐ I will work some review problems every day.
- ☐ After completing the homework assignment, I will read the next section to prepare for the coming class session.
- ☐ I will keep a notebook containing my class notes, homework papers, quizzes, tests, and any handouts—all in order by date.

Special Help

- ☐ I know my instructor's office hours and am willing to go in to ask for help.
- ☐ I have formed a study group with classmates that meets regularly to discuss the material and work on problems.
- ☐ When I need additional explanation of a topic, I view the tutorial CD and videos and check the website.
- ☐ I make use of extra tutorial assistance that my school offers for mathematics courses.
- ☐ I have purchased the *Student Solutions Manual* that accompanies this text, and I use it.

To follow each of these suggestions will take time. It takes a lot of practice to learn mathematics, just as with any other skill.

No doubt, you will sometimes become frustrated along the way. This is natural. When it occurs, take a break and come back to the material after you have had time to clear your thoughts. Keep in mind that the skills and discipline you learn in this course will help make for a brighter future. Good luck!

Alan S. Tussy
R. David Gustafson

Reading mathematics

To get the most out of this book, you need to learn how to read it correctly. A mathematics textbook must be read differently than a novel or a newspaper. For one thing, you need to read it slowly and carefully. At times, you will have to reread a section to understand its content. You should also have pencil and paper, so that you can work along with the text to understand the concepts presented.

Perhaps the most informative parts of a mathematics book are its examples. Each example in this textbook consists of a problem and its corresponding solution. One form of solution that is used many times in this book is shown in the diagram below. It is important that you follow the "flow" of its steps if you are to understand the mathematics involved. For this solution form, the basic idea is this:

- A property, rule, or procedure is applied to the original expression to obtain an equivalent expression. We show that the two expressions are equivalent by writing an equals sign between them. The property, rule, or procedure that was used is then listed next to the equivalent expression in the form of an author's note, printed in red.

- The process of writing equivalent expressions and explaining the reasons behind them continues, step by step, until the final result is obtained.

The solution in the following diagram consists of three steps, but solutions have varying lengths.

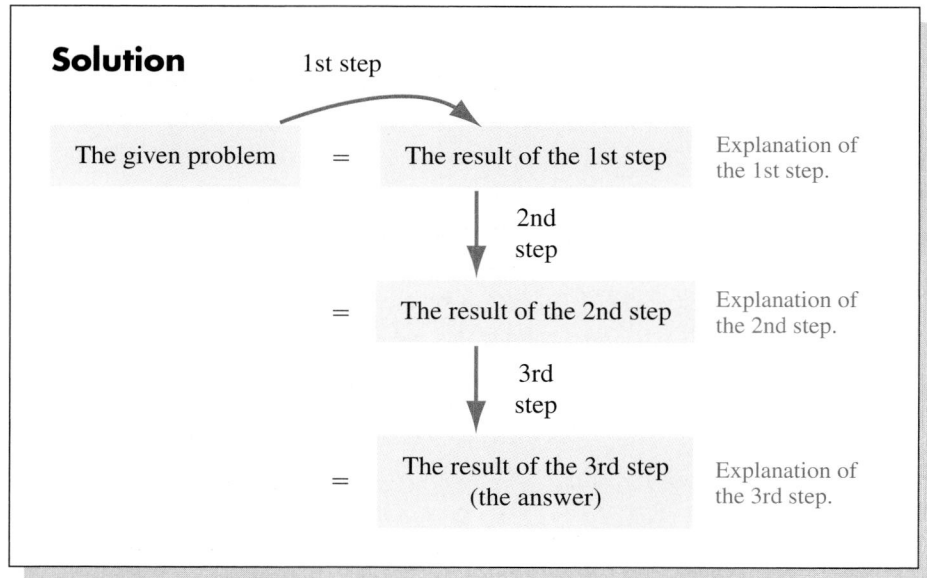

Writing mathematics

One of the major objectives of this course is for you to learn how to write solutions to problems properly. A written solution to a problem should explain your thinking in a series of neat and organized mathematical steps. Think of a solution as a mathematical essay—one that your instructor and other students should be able to read and understand. Some solutions will be longer than others, but they must all be in the proper format and use the correct notation. To learn how to do this will take time and practice.

To give you an idea of what will be expected, let's look at two samples of student work. In the first, we have highlighted some important characteristics of a well-written solution. The second sample is poorly done and would not be acceptable.

Evaluate $35 - 2^2 \cdot 3$.

A well-written solution:

The problem has been ▶ copied from the textbook.

$$35 - 2^2 \cdot 3 = 35 - 4 \cdot 3$$
$$= 35 - 12$$
$$= 23$$

◀ The first step of the solution is written here.

◀ The steps are written under each other in a neat, organized manner.

▲ The equals signs are lined up vertically.

A poorly written solution:

The problem has not ▶ been copied from the text.

$$2^2 = 4 = 35 - 4 \cdot 3$$
$$\underbrace{}_{12}$$

SUB: 35
     −12
      23 ⟶ = ⃝23

◀ An equals sign is improperly used.

◀ The work is disorganized and difficult to follow.

Whole Numbers

1

1.1 An Introduction to the Whole Numbers

1.2 Adding and Subtracting Whole Numbers

1.3 Multiplying and Dividing Whole Numbers

Estimation

1.4 Prime Factors and Exponents

1.5 Order of Operations

1.6 Solving Equations by Addition and Subtraction

1.7 Solving Equations by Division and Multiplication

Key Concept: Variables

Accent on Teamwork

Chapter Review

Taking a Math Test

Chapter Test

IN THIS CHAPTER, WE BEGIN OUR MATHEMATICS STUDY BY EXAMINING THE PROCEDURES USED TO SOLVE PROBLEMS THAT INVOLVE WHOLE NUMBERS.

1.1   An Introduction to the Whole Numbers

In this section, you will learn about

- Sets of numbers • Place value • Expanded notation
- Graphing on the number line • Ordering of the whole numbers
- Rounding whole numbers • Tables and graphs

INTRODUCTION. In this section, we will discuss the natural numbers and the whole numbers. These numbers are used to answer questions such as "How many?" "How fast?" "How heavy?" "How far?"

- The movie *Saving Private Ryan* won 5 Academy Awards.
- The speed limit on interstate highways in Wisconsin is 65 mph.
- The Statue of Liberty weighs 225 tons.
- The driving distance between Chicago and Houston is 1,067 miles.

Sets of numbers

A **set** is a collection of objects. Two important sets in mathematics are the natural numbers (the numbers that we count with) and the whole numbers. When writing a set, we use **braces** {   } to enclose the **members** (or **elements**) of the set.

The set of natural numbers   {1, 2, 3, 4, 5, 6, 7, 8, 9, 10, 11, 12, . . .}

The set of whole numbers   {0, 1, 2, 3, 4, 5, 6, 7, 8, 9, 10, 11, 12, . . .}

The three dots in the previous lists indicate that these sets continue on forever. There is no largest natural number or whole number.

Since every natural number is also a whole number, we say that the set of natural numbers is a **subset** of the set of whole numbers. However, not all whole numbers are natural numbers, because 0 is a whole number but not a natural number.

Place value

When we express a whole number with a *numeral* containing the *digits* 0, 1, 2, 3, 4, 5, 6, 7, 8, 9, we say that we have written the number in **standard notation.** The position of a digit in a numeral determines its value. In the numeral 325, the 5 is in the *ones column,* the 2 is in the *tens column,* and the 3 is in the *hundreds column.*

$$3\ 2\ 5$$

Hundreds column ⌐┘ ↑ └⌐ Ones column

Tens column

To make a numeral easy to read, we use commas to separate its digits into groups of three, called **periods.** Each period has a name, such as *ones, thousands, millions,* and so on. The following table shows the place value of each digit in the numeral 345,576,402,897,415, which is read as

three hundred forty-five trillion, five hundred seventy-six billion, four hundred two million, eight hundred ninety-seven thousand, four hundred fifteen

345 trillion			576 billion			402 million			897 thousand			4 hundred fifteen		
3	4	5	5	7	6	4	0	2	8	9	7	4	1	5
Trillions			Billions			Millions			Thousands			Ones		
Hundreds	Tens	Ones	Hundreds	Tens	Ones	Hundreds	Tens	Ones	Hundreds	Tens	Ones	Hundreds	Tens	Ones

As we move to the left in this table, the place value of each column is 10 times greater than the column to its right. This is why we call our number system a *base-10 number system.*

EXAMPLE 1   *TV news.*   The cable network CNN is carried by 11,528 cable systems. Which digit tells the number of hundreds?

Solution
In 11,528, the hundreds column is the third column from the right. The digit 5 tells the number of hundreds.

Self Check

The Fox Family Channel is carried by 13,820 cable systems. Which digit tells the number of ten thousands?

Answer: 1   ■

Expanded notation

In the numeral 6,352, the digit 6 is in the thousands column, 3 is in the hundreds column, 5 is in the tens column, and 2 is in the ones (or units) column. The meaning of 6,352 becomes clear when we write it in **expanded notation.**

6 thousands + 3 hundreds + 5 tens + 2 ones

We read the numeral 6,352 as "six thousand, three hundred fifty-two."

EXAMPLE 2   *Expanded notation.*   Write each number in expanded notation: **a.** 63,427   and   **b.** 1,251,609.

Solution
a. 6 ten thousands + 3 thousands + 4 hundreds + 2 tens + 7 ones

We read this number as "sixty-three thousand, four hundred twenty-seven."

b. 1 million + 2 hundred thousands + 5 ten thousands + 1 thousand + 6 hundreds + 0 tens + 9 ones

Since 0 tens is zero, the expanded notation can also be written as

1 million + 2 hundred thousands + 5 ten thousands + 1 thousand + 6 hundreds + 9 ones

We read this number as "one million, two hundred fifty-one thousand, six hundred nine."

Self Check

Write 808,413 in expanded notation.

Answer:
8 hundred thousands + 8 thousands + 4 hundreds + 1 ten + 3 ones. Read as "eight hundred eight thousand, four hundred thirteen."   ■

EXAMPLE 3   *Translating to standard notation.*   Write twenty-three thousand forty in standard notation.

Solution

In expanded notation, the number is written as

> 2 ten thousands + 3 thousands + 4 tens   There are 0 hundreds and 0 ones.

In standard notation, this is written as 23,040.

Self Check

Write seventy-six thousand three in standard notation.

Answer: 76,003   ■

Graphing on the number line

Whole numbers can be illustrated by drawing points on a **number line.** A number line is a horizontal or vertical line that is used to represent numbers graphically. Like a ruler, a number line is straight and has uniform markings. (See Figure 1-1.) To construct a number line, we begin on the left with a point on the line representing the number 0. This point is called the **origin.** We then proceed to the right, drawing equally spaced marks and labeling them with whole numbers that increase progressively in size. The arrowhead at the right indicates that the number line continues forever.

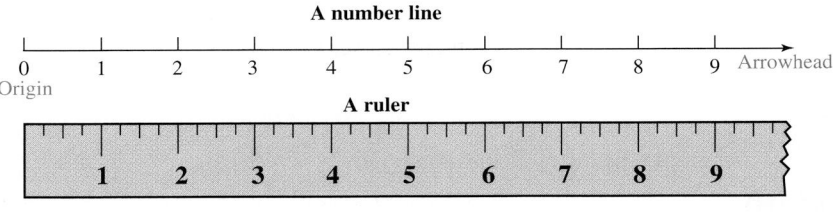

FIGURE 1-1

Using a process known as **graphing,** a single number or a set of numbers can be represented on a number line. *The graph of a number* is the point on the number line that corresponds to that number. *To graph a number* means to locate its position on the number line and then to highlight it using a heavy dot. Figure 1-2 shows the graphs of the whole numbers 5 and 8.

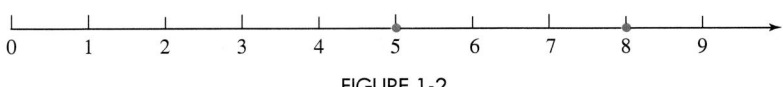

FIGURE 1-2

Ordering of the whole numbers

As we move to the right on the number line, the numbers get larger. Because 8 lies to the right of 5, we say that 8 is greater than 5. The **inequality symbol** > ("is greater than") can be used to write this fact.

> 8 > 5   Read as "8 is greater than 5."

Since 8 > 5, it is also true that 5 < 8. (Read as "5 is less than 8.")

   COMMENT   To distinguish between these two inequality symbols, remember that they always point to the smaller of the two numbers involved.

> 8 > 5   5 < 8
>
>   Points to the
>   smaller number.

EXAMPLE 4 *Inequality symbols.* Place an $<$ or an $>$ symbol in the box to make a true statement: **a.** 3 ▨ 7 and **b.** 18 ▨ 16.

Solution

a. Since 3 is to the left of 7 on the number line, $3 < 7$.

b. Since 18 is to the right of 16 on the number line, $18 > 16$.

Self Check

Place an $<$ or an $>$ symbol in the box to make a true statement.

a. 12 ▨ 4 **b.** 7 ▨ 10

Answers: a. $>$, **b.** $<$ ■

Rounding whole numbers

When we don't need exact results, we often round numbers. For example, when a teacher with 36 students in his class orders 40 textbooks, he has rounded the actual number to the *nearest ten,* because 36 is closer to 40 than it is to 30.

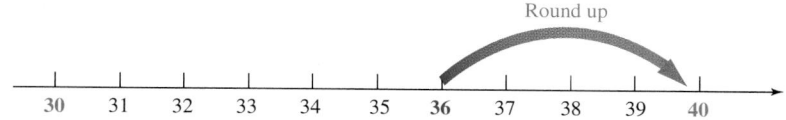

When a geologist says that the height of Alaska's Mount McKinley is "about 20,300 feet," she has rounded to the *nearest hundred,* because its actual height of 20,320 feet is closer to 20,300 than it is to 20,400.

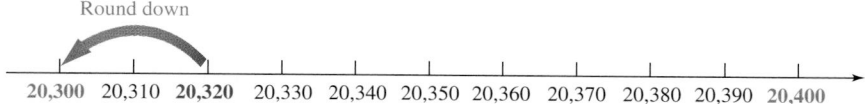

To round a whole number, we follow an established set of rules. To round a number to the nearest ten, for example, we begin by locating the **rounding digit** in the tens column. If the **test digit** to the right of that column (the digit in the ones column) is 5 or greater, we *round up* by increasing the tens digit by 1 and placing a 0 in the ones column. If the test digit is less than 5, we *round down* by leaving the tens digit unchanged and placing a 0 in the ones column.

EXAMPLE 5 *Rounding to the nearest ten.* Round each number to the nearest ten: **a.** 3,764 and **b.** 12,087.

Solution

a. We find the rounding digit in the tens column, which is 6.

   ┌— Rounding digit
  3,764
    └— Test digit

We then look at the test digit to the right of 6, the 4 in the ones column. Since $4 < 5$, we round down by leaving the 6 unchanged and replacing the test digit with 0. The rounded answer is 3,760.

b. We find the rounding digit in the tens column, which is 8.

    ┌— Rounding digit
  12,087
    └— Test digit

We then look at the test digit to the right of 8, the 7 in the ones column. Because $7 > 5$, we round up by adding 1 to 8 and replacing the test digit with 0. The rounded answer is 12,090.

Self Check

Round each number to the nearest ten:

a. 35,642

b. 3,756

Answers: a. 35,640,
b. 3,760 ■

A similar procedure is used to round numbers to the nearest hundred, the nearest thousand, the nearest ten thousand, and so on.

Rounding a whole number

1. To round a number to a certain place, locate the rounding digit in that place.
2. Look at the test digit to the right of the rounding digit.
3. If the test digit is 5 or greater, round up by adding 1 to the rounding digit and changing all of the digits to the right of the rounding digit to 0.

   If the test digit is less than 5, round down by keeping the rounding digit and changing all of the digits to the right of the rounding digit to 0.

EXAMPLE 6   *Rounding to the nearest hundred.*   Round 7,960 to the nearest hundred.

Solution

First, we find the rounding digit in the hundreds column. It is 9.

```
    ┌── Rounding digit
7,960
    └── Test digit
```

We then look at the 6 to the right of 9. Because $6 > 5$, we round up and increase 9 in the hundreds column by 1. Since the 9 in the hundreds column represents 900, increasing 9 by 1 represents increasing 900 to 1,000. Thus, we replace the 9 with a 0 and add 1 to the 7 in the thousands column. Finally, we replace the two rightmost digits with 0's. The rounded answer is 8,000.

Self Check

Round 365,283 to the nearest hundred.

Answer: 365,300 ▪

EXAMPLE 7   *U.S. cities.*   In 1999, Denver was the nation's 27th largest city. Round the population of Denver given in Figure 1-3 **a.** to the nearest thousand   and   **b.** to the nearest ten thousand.

Denver
CITY LIMIT
Pop. 506,250   Elev. 5,280

FIGURE 1-3

Solution

a. The rounding digit in the thousands column is 6. The test digit, 2, is less than 5, so we round down. To the nearest thousand, Denver's population was 506,000 in 1999.

b. The rounding digit in the ten thousands column is 0. The test digit, 6, is more than 5, so we round up. To the nearest ten thousand, Denver's population was 510,000 in 1999. ▪

Tables and graphs

The table in Figure 1-4(a) is an example of the use of whole numbers. It shows the number of women elected to the United States House of Representatives in the Congressional elections held every two years from 1990 to 1998.

Table

Year	Number of women elected
1990	29
1992	48
1994	53
1996	51
1998	55

(a)

Bar graph

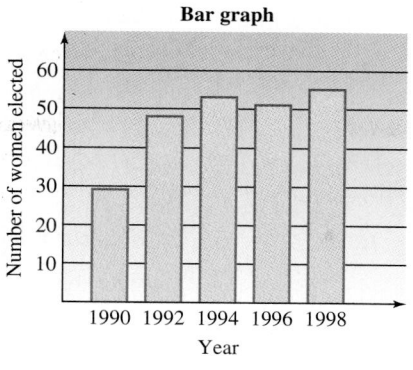

(b)

Line graph

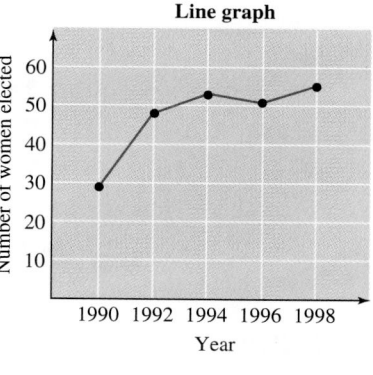

(c)

FIGURE 1-4

In Figure 1-4(b), the election results are presented in a **bar graph.** The horizontal scale is labeled "Year" and scaled in units of 2 years. The vertical scale is labeled "Number of women elected" and scaled in units of 10. The bar directly over each year extends to a height indicating the number of women elected to Congress that year.

Another way to present the information in the table is with a **line graph.** Instead of using a bar to denote the number of women elected, we use a heavy dot drawn at the correct height. After drawing data points for 1990, 1992, 1994, 1996, and 1998, we connect the points with line segments to create the line graph in Figure 1-4(c).

STUDY SET Section 1.1

VOCABULARY *Fill in the blanks.*

1. A _____ is a collection of objects.

2. The set of _____ numbers is {1, 2, 3, 4, 5, . . .}, and the set of _____ numbers is {0, 1, 2, 3, 4, 5, . . .}.

3. When 297 is written as 2 hundreds + 9 tens + 7 ones, it is written in _____ notation.

4. If we _____ 627 to the nearest ten, we get 630.

5. Using a process known as graphing, whole numbers can be represented as points on a _____ line.

6. The symbols > and < are _____ symbols.

CONCEPTS *In Exercises 7–10, consider the numeral 57,634.*

7. What digit is in the tens column?

8. What digit is in the thousands column?

9. What digit is in the hundreds column?

10. What digit is in the ten thousands column?

11. What set of numbers is obtained when 0 is combined with the natural numbers?

12. Place the numbers 25, 17, 37, 15, 45 in order, from smallest to largest.

13. Graph 1, 3, 5, and 7.

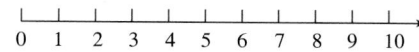

14. Graph 0, 2, 4, 6, and 8.

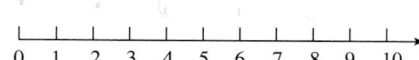

15. Graph the whole numbers less than 6.

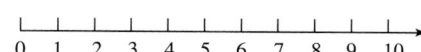

16. Graph the whole numbers between 2 and 8.

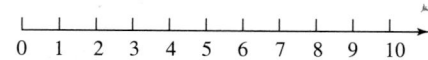

Place an > or an < symbol in the box to make a true statement.

17. 47 ☐ 41

18. 53 ☐ 67

19. 309 ☐ 300

20. 841 ☐ 814

21. 2,052 ☐ 2,502

22. 999 ☐ 998

23. Since 4 < 7, it is also true that 7 ☐ 4.

24. Since 9 > 0, it is also true that 0 ☐ 9.

NOTATION   *Fill in the blanks.*

25. The symbols { }, called _____, are used when writing a set.

26. The symbol > means _____, and the symbol < means _____.

PRACTICE   *Write each number in expanded notation and then write it in words.*

27. 245

28. 508

29. 3,609

30. 3,960

31. 32,500

32. 73,009

33. 104,401

34. 570,003

Write each number in standard notation.

35. 4 hundreds + 2 tens + 5 ones

36. 7 hundreds + 7 tens + 7 ones

37. 2 thousands + 7 hundreds + 3 tens + 6 ones

38. 7 billions + 3 hundreds + 5 tens

39. Four hundred fifty-six

40. Three thousand seven hundred thirty-seven

41. Twenty-seven thousand five hundred ninety-eight

42. Seven million, four hundred fifty-two thousand, eight hundred sixty

43. Nine thousand one hundred thirteen

44. Nine hundred thirty

45. Ten million, seven hundred thousand, five hundred six

46. Eighty-six thousand four hundred twelve

Round 79,593 to the nearest . . .

47. ten

48. hundred

49. thousand

50. ten thousand

Round 5,925,830 to the nearest . . .

51. thousand

52. ten thousand

53. hundred thousand

54. million

Round $419,161 to the nearest . . .

55. $10

56. $100

57. $1,000

58. $10,000

APPLICATIONS

59. EATING HABITS   The following list shows the ten countries with the largest per-person annual consumption of meat. Construct a two-column table that presents the data in order, beginning with the largest per-person consumption. (The abbreviation "lb" means "pounds.")

Australia: 239 lb	New Zealand: 259 lb
Austria: 229 lb	Saint Lucia: 222 lb
Canada: 211 lb	Spain: 211 lb
Cyprus: 236 lb	Uruguay: 230 lb
Denmark: 219 lb	United States: 261 lb

60. PRESIDENTS   The following list shows the ten youngest U.S. presidents and their ages (in years/days) when they took office. Construct a two-column table that presents the data in order, beginning with the youngest president.

C. Arthur 50 yr/350 days	U. Grant 46 yr/236 days
G. Cleveland 47 yr/351 days	J. Kennedy 43 yr/236 days
W. Clinton 46 yr/154 days	F. Pierce 48 yr/101 days
M. Filmore 50 yr/184 days	J. Polk 49 yr/122 days
J. Garfield 49 yr/105 days	T. Roosevelt 42 yr/322 days

61. MISSIONS TO MARS   The United States, Russia, and Japan have launched Mars space probes. The graph in Illustration 1 shows the success rate of the missions, by decade.

   a. What decade had the greatest number of successful missions?

b. What decade had the greatest number of unsuccessful missions?

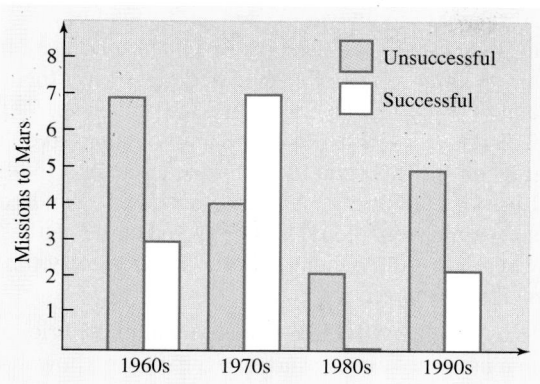

Source: *USA TODAY* (December 8, 1999)

ILLUSTRATION 1

62. BANKING CRISES   Illustration 2 shows the number of banks that were closed or taken over by federal agencies during the years 1934–1995.
  a. During what two time spans was there an upsurge in bank failures?
  b. In what year were there the most bank failures? Estimate the number of banks that failed that year.

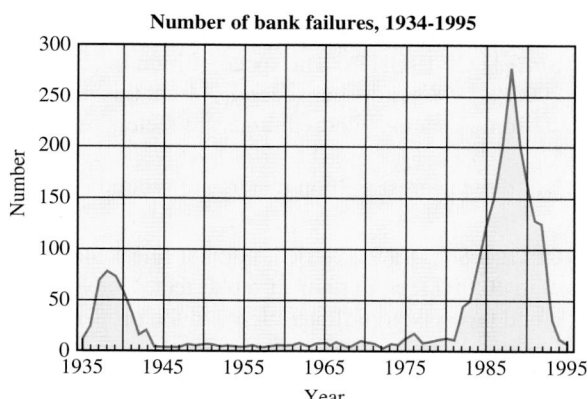

Source: FDIC Division of Research and Statistics

ILLUSTRATION 2

63. ENERGY RESERVES Construct a bar graph using the data shown in Illustration 3.

Natural gas reserves, 1998 (in trillion cubic feet)	
United States	167
Venezuela	143
Canada	65
Mexico	64
Argentina	24

Source: *Oil and Gas Journal*

ILLUSTRATION 3

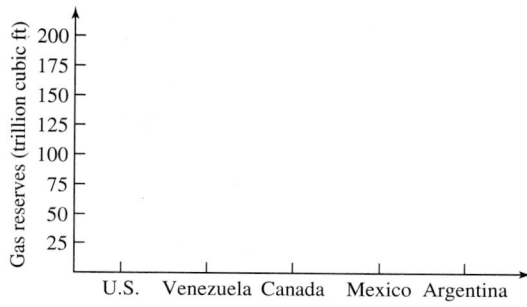

64. ENERGY RESERVES   Construct a line graph using the data in Illustration 3.

65. COFFEE   Construct a line graph using the data shown in Illustration 4.

Starbucks locations	
Year	**Number**
1992	165
1993	272
1994	425
1995	676
1996	1,015
1997	1,412
1998	1,886
1999	2,200

Source: Starbucks Company

ILLUSTRATION 4

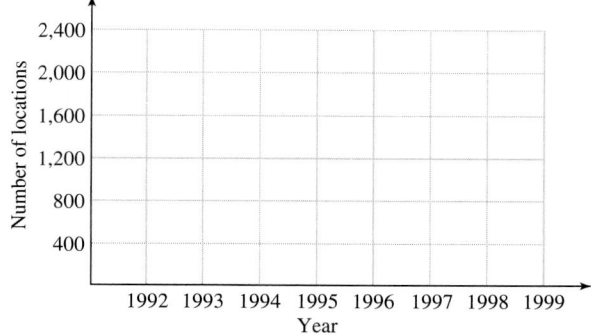

66. COFFEE   Construct a bar graph using the data shown in Illustration 4 on the preceding page.

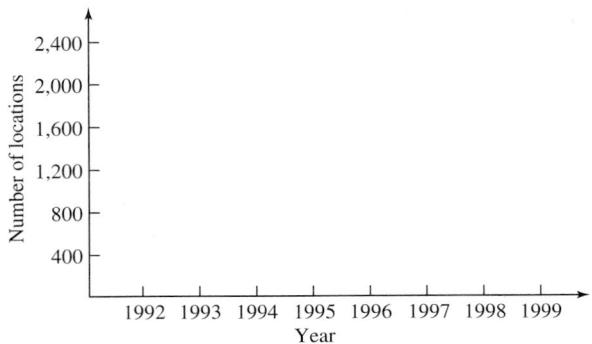

67. Complete each check by writing the amount in words on the proper line.

a.

No. 201	March 9 , 20 *00*
Payable to _____ Davis Chevrolet _____	$ 15,601 *00/100*
_____ DOLLARS	
45-365-02	Don Smith

b.

No. 7890	Aug. 12 , 20 *00*
Payable to _____ Dr. Anderson _____	$ 3,433 *46/100*
_____ DOLLARS	
45-828-02	Juan Decito

68. FORMAL ANNOUNCEMENTS   One style used when printing formal invitations and announcements is to write all numbers in words. Use this style to write each of the following phrases.

a. This diploma awarded this 27th day of June, 1996.

b. The suggested contribution for the fundraiser is $850 a plate, or an entire table may be purchased for $5,250.

69. EDITING   Edit this excerpt from a history text by circling all numbers written in words and rewriting them using digits.

Abraham Lincoln was elected with a total of one million, eight hundred sixty-five thousand, five hundred ninety-three votes—four hundred eighty-two thousand, eight hundred eighty more than the runner-up, Stephen Douglas. He was assassinated after having served a total of one thousand five hundred three days in office. Lincoln's Gettysburg Address, a mere two hundred sixty-nine words long, was delivered at the battle site where forty-three thousand four hundred forty-nine casualties occurred.

70. READING A METER   The amount of electricity used in a household is measured by a meter in kilowatt-hours (kwh). Determine the reading on the meter shown in Illustration 5. (When the pointer is between two numbers, read the lower number.)

| Thousands | Hundreds | Tens | Units |
| of kwh | of kwh | of kwh | of kwh |

ILLUSTRATION 5

71. SPEED OF LIGHT   The speed of light in a vacuum is 299,792,458 meters per second. Round this number
a. to the nearest hundred thousand meters per second.

b. to the nearest million meters per second.

72. CLOUDS   Draw a vertical number line scaled from 0 to 40,000 feet, in units of 5,000 feet. Graph each cloud type given in Illustration 6 at the proper altitude.

Cloud type	Altitude (ft)
Altocumulus	21,000
Cirrocumulus	37,000
Cirrus	38,000
Cumulonimbus	15,000
Cumulus	8,000
Stratocumulus	9,000
Stratus	4,000

ILLUSTRATION 6

WRITING

73. Explain why the natural numbers are called the counting numbers.

74. Explain how you would round 687 to the nearest ten.

75. The houses in a new subdivision are priced "in the low 130s." What does this mean?

76. A million is a thousand thousands. Explain why this is so.

1.2 Adding and Subtracting Whole Numbers

In this section, you will learn about

- Properties of addition • Adding whole numbers
- Perimeter of a rectangle and a square • Subtracting whole numbers
- Combinations of operations

INTRODUCTION. Mastering the operations of addition and subtraction with whole numbers enables us to solve problems from geometry, business, and science. For example, to find the distance around a rectangle, we need to add the lengths of its four sides. To prepare an annual budget, we need to add separate line items. To find the difference between two temperatures, we need to subtract one from the other.

Properties of addition

Addition is the process of finding the total of two (or more) numbers. It can be illustrated using a number line (see Figure 1-5). For example, to compute $4 + 5$, we begin at zero and draw an arrow 4 units long, extending to the right. This represents 4. From the tip of that arrow, we draw another arrow 5 units long, also extending to the right. The second arrow points to 9. This result corresponds to the addition fact $4 + 5 = 9$, where 4 and 5 are called **addends** and 9 is called the **sum.**

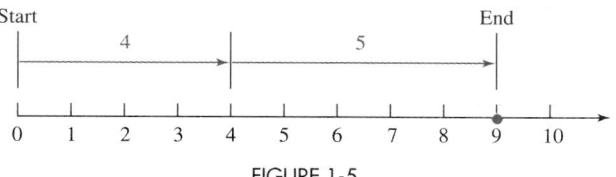

FIGURE 1-5

We have used a number line to find that $4 + 5 = 9$. If we add 4 and 5 in the opposite order, Figure 1-6 shows that we get the same result: $5 + 4 = 9$.

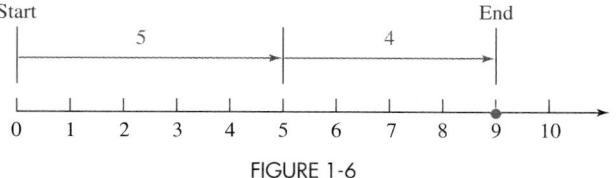

FIGURE 1-6

These examples illustrate that two whole numbers can be added in either order to get the same sum. The order in which we add two numbers does not affect the result. This property is called the **commutative property of addition.** To state the commutative property of addition concisely, we can use variables.

Variables

> A **variable** is a letter that is used to stand for a number.

We now use the variables *a* and *b* to state the commutative property of addition.

Commutative property of addition

> If *a* and *b* represent numbers, then
> $$a + b = b + a$$

To find the sum of three whole numbers, we add two of them and then add the third to that result. In the following examples, we add $3 + 4 + 2$ in two ways. We will use the **grouping symbols (   )**, called **parentheses,** to show this. We must do the operation within parentheses first.

Method 1: Group 3 + 4

$(3 + 4) + 2 = 7 + 2$ Because of the parentheses, add 3 and 4 first to get 7.

$\qquad\qquad\quad = 9$ Then add 7 and 2 to get 9.

Method 2: Group 4 + 2

$3 + (4 + 2) = 3 + 6$ Because of the parentheses, add 4 and 2 to get 6.

$\qquad\qquad\quad = 9$ Then add 3 and 6 to get 9.

Either way, the sum is 9. It does not matter how we group (or associate) numbers in addition. This property is called the **associative property of addition.**

Associative property of addition

> If *a*, *b*, and *c* represent numbers, then
> $$(a + b) + c = a + (b + c)$$

Whenever we add 0 to a number, the number remains the same. For example,

$$3 + 0 = 3, \qquad 5 + 0 = 5, \qquad \text{and} \qquad 9 + 0 = 9$$

These examples suggest the **addition property of 0.**

Addition property of 0

> If *a* represents any number, then
> $$a + 0 = a \qquad \text{and} \qquad 0 + a = a$$

EXAMPLE 1   *Properties of addition.*   Find each sum: **a.** $8 + 9$ and $9 + 8$,   **b.** $5 + (1 + 8)$ and $(5 + 1) + 8$,   and   **c.** $(3 + 0) + 4$.

Solution

a. $8 + 9 = 17$ and $9 + 8 = 17$.   The results are the same.

b. In each case, we do the addition within parentheses first.

$5 + (1 + 8) = 5 + 9 \qquad (5 + 1) + 8 = 6 + 8$

$\qquad\qquad\quad = 14 \qquad\qquad\qquad\quad = 14$

The results are the same.

c. $(3 + 0) + 4 = 3 + 4$ Do the addition within parentheses first: $3 + 0 = 3$.

$\qquad\qquad\quad = 7$

Self Check

Find each sum.

a. $6 + 7$ and $7 + 6$
b. $2 + (6 + 3)$ and
$\qquad (2 + 6) + 3$
c. $3 + (0 + 4)$

Answers: a. 13, 13,
b. 11, 11,   **c.** 7

Adding whole numbers

We can add whole numbers greater than 10 by using a vertical format that adds digits with the same place value. Because the additions within each column often exceed 9, it is sometimes necessary to *carry* the excess to the next column to the left. For example, to add 27 and 15, we write the numerals with the digits of the same place value aligned vertically.

$$\begin{array}{r} 2\,7 \\ +\,1\,5 \\ \hline \end{array}$$

We begin by adding the digits in the ones column: $7 + 5 = 12$. Because $12 =$ 1 ten and 2 ones, we place a 2 in the ones column of the answer and carry 1 to the tens column.

$$\begin{array}{r} 1 \\ 2\,7 \\ +\,1\,5 \\ \hline 2 \end{array}$$ Add the digits in the ones column: $7 + 5 = 12$. Carry 1 (shown in blue) to the tens column.

Then we add the digits in the tens column.

$$\begin{array}{r} 1 \\ 2\,7 \\ +\,1\,5 \\ \hline 4\,2 \end{array}$$ Add 1, 2, and 1. Place the result, 4, in the tens column of the answer.

Thus, $27 + 15 = 42$.

EXAMPLE 2   *Carrying.*   Add: $9,834 + 692$.

Solution
We write the numerals with their corresponding digits aligned vertically. Then we add the numbers, one column at a time, working from right to left.

$$\begin{array}{r} 9,8\,3\,4 \\ +\quad 6\,9\,2 \\ \hline 6 \end{array}$$ Add the digits in the ones column and place the result in the ones column of the answer.

$$\begin{array}{r} 1 \\ 9,8\,3\,4 \\ +\quad 6\,9\,2 \\ \hline 2\,6 \end{array}$$ Add the digits in the tens column. The result, 12, exceeds 9. Place the 2 in the tens column of the answer and carry 1 (shown in blue) to the hundreds column.

$$\begin{array}{r} 1\,1 \\ 9,8\,3\,4 \\ +\quad 6\,9\,2 \\ \hline 5\,2\,6 \end{array}$$ Add the digits in the hundreds column. Since the result, 15, exceeds 9, place the 5 in the hundreds column of the answer and carry 1 (shown in green) to the thousands column.

$$\begin{array}{r} 1\,1 \\ 9,8\,3\,4 \\ +\quad 6\,9\,2 \\ \hline 1\,0,5\,2\,6 \end{array}$$ Since the sum of the digits in the thousands column is 10, write 0 in the thousands column and 1 in the ten thousands column of the answer.

Thus, $9,834 + 692 = 10,526$.

Self Check
Add: $675 + 1,497$.

Answer: 2,172   ■

To see if the result in Example 2 is reasonable, we can **estimate** the answer. 9,834 is a little less than 10,000, and 692 is a little less than 700. We estimate that the answer will be a little less than $10,000 + 700$, or 10,700. An answer of 10,526 is reasonable. Estimation is discussed in more detail later in this chapter.

Words such as *increase, gain, credit, up, forward, rise, in the future,* and *to the right* are used to indicate addition.

EXAMPLE 3   *Calculating temperatures.*   At noon, the temperature in Helena, Montana, was 31°. By 1:00 P.M., the temperature had increased 5°, and by 2:00 P.M., it had risen another 7°. Find the temperature at 2:00 P.M.

Solution   To the temperature at noon, we add the two increases.

$$31 + 5 + 7$$

The two additions are done working from left to right.

$$31 + 5 + 7 = 36 + 7$$
$$= 43$$

The temperature at 2:00 P.M. was 43°.   ■

EXAMPLE 4   *U.S. history.*   The populations of four American colonies in 1630 are shown in Figure 1-7. Find the total population.

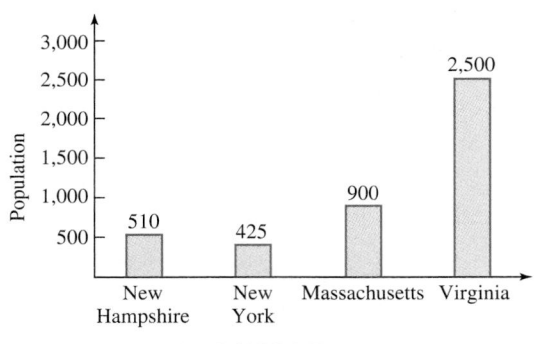

FIGURE 1-7

Self Check

By 1700, the populations of the four colonies were New Hampshire 5,000, New York 19,100, Massachusetts 55,900, and Virginia 58,600. Find the total population.

Solution

The word *total* indicates that we must add the populations of the individual colonies.

$$
\begin{array}{r}
\overset{2}{}510 \\
425 \\
900 \\
+\,2,500 \\
\hline
4,335
\end{array}
$$

Align the numerals vertically. Add the digits, one column at a time, working from right to left.

The total population was 4,335.

Answer: 138,600   ■

Perimeter of a rectangle and a square

A **rectangle** is a four-sided figure (like a dollar bill) whose opposite sides are of equal length. Either of the longer sides is called its **length,** and either of the shorter sides is called its **width.** A rectangle with all four sides of equal length is called a **square.** (See Figure 1-8.)

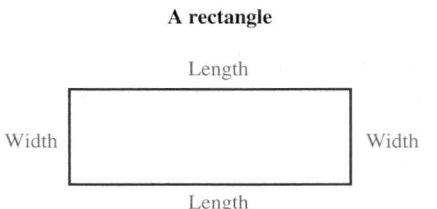

FIGURE 1-8

The distance around a rectangle or a square is called its **perimeter.** To find the perimeter of a rectangle, we add the lengths of its four sides.

The perimeter of a rectangle	=	length	+	length	+	width	+	width

To find the perimeter of a square, we add the lengths of its four sides.

The perimeter of a square	=	side	+	side	+	side	+	side

EXAMPLE 5 *Perimeter.* Find the perimeter of the dollar bill shown in Figure 1-9 (mm stands for millimeters).

Self Check

A Monopoly game board is square-shaped, with sides 19 inches long. Find the perimeter of the board.

Solution

To find the perimeter of the rectangular bill, we add the lengths of its four sides.

$$\begin{array}{r} 22 \\ 156 \\ 156 \\ 65 \\ + \ 65 \\ \hline 442 \end{array}$$

Width = 65 mm

Length = 156 mm

FIGURE 1-9

The perimeter is 442 mm.

To see whether this result is reasonable, we estimate the answer. Because the rectangle is about 150 mm by 70 mm, its perimeter is approximately 150 + 150 + 70 + 70, or 440 mm. An answer of 442 mm is reasonable.

Answer: 76 in. ■

Subtracting whole numbers

Subtraction is the process of finding the difference between two numbers. It can be illustrated using a number line (see Figure 1-10). For example, to compute $9 - 4$, we begin at zero and draw an arrow 9 units long, extending to the right. From the tip of that arrow, we draw another arrow 4 units long, but extending to the left. (This represents taking away 4.) The second arrow points to 5, indicating that $9 - 4 = 5$. In this subtraction fact, 9 is called the **minuend,** 4 is called the **subtrahend,** and 5 is called the **difference.**

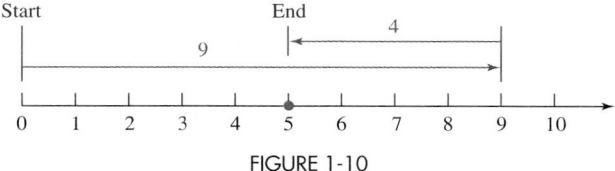

FIGURE 1-10

With whole numbers, we cannot subtract in the opposite order and find the difference $4 - 9$, because we cannot take away 9 objects from 4 objects. Since subtraction of whole numbers cannot be done in either order, subtraction is not commutative.

Subtraction is not associative either, because if we group in different ways, we get different answers.

$$\begin{array}{lll} (9 - 5) - 1 = 4 - 1 & \text{but} & 9 - (5 - 1) = 9 - 4 \\ \qquad\qquad\quad = 3 & & \qquad\qquad\quad = 5 \end{array}$$

EXAMPLE 6 *Subtracting whole numbers.* Do the subtractions: $9 - (6 - 3)$ and $(9 - 6) - 3$.

Solution
In each case, we do the subtraction in parentheses first.

$$9 - (6 - 3) = 9 - 3 \qquad (9 - 6) - 3 = 3 - 3$$
$$= 6 \qquad\qquad\qquad = 0$$

We note that the results are different.

Whole numbers can be subtracted using a vertical format. Because subtractions often require subtracting a larger digit from a smaller digit, we may need to *borrow*. For example, to subtract 15 from 32, we write the minuend, 32, and the subtrahend, 15, in a vertical format, aligning the digits with the same place value.

$$\begin{array}{r} 3\,2 \\ -1\,5 \\ \hline \end{array}$$ Write the numerals in a column, with corresponding digits aligned vertically.

Since 5 can't be subtracted from 2, we borrow from the tens column of 32.

$$\begin{array}{r} {\scriptstyle 2\ 12} \\ \not{3}\ 2 \\ -1\ 5 \\ \hline 7 \end{array}$$ To subtract in the ones column, borrow 1 ten from the tens column. We show this by drawing a slash through the 3 and writing a 2 above it. Add 10 to the 2 in the ones column, which gives 12. Then subtract: $12 - 5 = 7$.

$$\begin{array}{r} {\scriptstyle 2\,12} \\ \not{3}\ 2 \\ -1\ 5 \\ \hline 1\ 7 \end{array}$$ Subtract in the tens column: $2 - 1 = 1$.

Thus, $32 - 15 = 17$. To check the result, we add the difference, 17, and the subtrahend, 15. We should obtain the minuend, 32.

$$\textit{Check:}\quad \begin{array}{r} {\scriptstyle 1} \\ 17 \\ +15 \\ \hline 32 \end{array}$$

EXAMPLE 7 *Borrowing.*   Subtract 576 from 2,021.

Solution

$$\begin{array}{r} 2{,}0\,2\,1 \\ -\ \ 5\,6\,7 \\ \hline \end{array}$$ Write the numerals in a column, with the digits of the same place value aligned vertically.

$$\begin{array}{r} {\scriptstyle 1\ 11} \\ 2{,}0\,2\,\not{1} \\ -\ \ 5\,7\,6 \\ \hline 5 \end{array}$$ To subtract in the ones column, borrow 1 ten from the tens column and add it to the ones column. Then subtract: $11 - 6 = 5$.

Since we can't subtract 7 from 1 in the tens column, we borrow. Because there is a 0 in the hundreds column of 2,021, we must borrow from the thousands column. We can take 1 thousand from the thousands column (leaving 1 thousand behind) and write it as 10 hundreds, placing a 10 in the hundreds column. From these 10 hundreds, we take 1 hundred (leaving 9 hundreds behind) and think of it as 10 tens. We add these 10 tens to the 1 ten that is already in the tens column to get 11 tens. From these 11 tens, we subtract 7 tens: $11 - 7 = 4$.

$$\begin{array}{r} {\scriptstyle 9} \\ {\scriptstyle 1\ \not{10}1111} \\ 2{,}\not{0}\,2\,\not{1} \\ -\ \ 5\,7\,6 \\ \hline 4\ 5 \end{array}$$ To subtract in the tens column, borrow 10 hundreds from the thousands digit and add it to the hundreds digit. Borrow 10 tens from the hundreds digit and add it to the tens digit. Then subtract: $11 - 7 = 4$.

$$
\begin{array}{r}
9 \\
1\ \cancel{1}01111 \\
2,\cancel{0}\ 2\ \cancel{1} \\
-\quad 5\ 7\ 6 \\
\hline
4\ 4\ 5
\end{array}
$$
Subtract in the hundreds column: $9 - 5 = 4$.

$$
\begin{array}{r}
9 \\
1\ \cancel{1}01111 \\
2,\cancel{0}\ 2\ \cancel{1} \\
-\quad 5\ 7\ 6 \\
\hline
1,4\ 4\ 5
\end{array}
$$
Subtract in the thousands column: $1 - 0 = 1$.

Thus, $2,021 - 576 = 1,445$. Check the result using addition.

Answer: 576; $576 + 1,445 = 2,021$ ■

Words such as *minus, decrease, loss, debit, down, backward, fall, reduce, in the past,* and *to the left* indicate subtraction.

EXAMPLE 8   *Drinking and driving.*   In 1996, there were 17,126 alcohol-related traffic deaths in the United States. That number declined in 1997, dropping by 937. In 1998, it fell by an additional 253. How many alcohol-related traffic fatalities were there in 1998?

Solution

The words *dropping* and *fell* indicate subtraction. We can show the calculations necessary to solve this example in a single expression:

$$17,126 - 937 - 253$$

The two subtractions are done working from left to right.

$$
\begin{aligned}
17,126 - 937 - 253 &= 16,189 - 253 \\
&= 15,936
\end{aligned}
$$

There were 15,936 alcohol-related traffic deaths in 1998. ■

Combinations of operations

Additions and subtractions often appear in the same problem. It is important to read the problem carefully, extract the useful information, and organize it correctly.

EXAMPLE 9   *Bus passengers.*   Twenty-seven people were riding a bus on Route 47. At the Seventh Street stop, 16 riders got off the bus and 5 got on. How many riders were left on the bus?

Solution

The route and street number are not important. The phrase *got off the bus* indicates subtraction, and the phrase *got on* indicates addition. The number of riders on the bus can be found by calculating $27 - 16 + 5$. Working from left to right, we have

$$
\begin{aligned}
27 - 16 + 5 &= 11 + 5 \\
&= 16
\end{aligned}
$$

There were 16 riders left on the bus.

Self Check

One share of ABC Corporation stock cost $75. The price fell $7 per share. However, it recovered and rose $13 per share. What is its current price?

Answer: $81 ■

 COMMENT   When doing the calculation in Example 9, we must perform the subtraction first. If the addition is done first, we obtain an incorrect answer of 6. For expressions containing addition and subtraction, perform them as they occur from left to right.

$$
\begin{aligned}
27 - 16 + 5 &= 27 - 21 \\
&= 6
\end{aligned}
$$

Accent on Technology: **Calculators**

A calculator can be helpful when checking an answer or when performing a tedious computation. Before making regular use of one, make sure that you have mastered the fundamentals of arithmetic.

Several brands of calculators are available. For specific details about the operation of your calculator, please consult the owner's manual.

To check the addition done in Example 4 (U.S. history) using a scientific calculator, we enter these numbers and press these keys.

Keystrokes   510 $+$ 425 $+$ 900 $+$ 2500 $=$ | 4335 |

The display shows that in 1630, the total population of the four colonies was 4,335.

We can use a scientific calculator to check the subtraction performed in Example 8 (drinking and driving) by entering these numbers and pressing these keys.

Keystrokes   17126 $-$ 937 $-$ 253 $=$ | 15936 |

The display shows that there were 15,936 alcohol-related traffic deaths in 1998.

STUDY SET Section 1.2

VOCABULARY   *Fill in the blanks.*

1. When two numbers are added, the result is called a _____. The numbers that are to be added are called _____.

2. A _____ is a letter that stands for a number.

3. A _____ is a four-sided figure (like a dollar bill) whose opposite sides are of equal length.

4. A _____ is a rectangle with all sides of equal length.

5. When two numbers are subtracted, the result is called a _____. In a subtraction problem, the _____ is subtracted from the _____.

6. The property that guarantees that we can add two numbers in either order and get the same sum is called the _____ property of addition.

7. The property that allows us to group numbers in an addition in any way we want is called the _____ property of addition.

8. The distance around a rectangle (or a square) is called its _____.

CONCEPTS   *In Exercises 9–12, tell which property of addition guarantees that the quantities are equal.*

9. $3 + 4 = 4 + 3$

10. $(3 + 4) + 5 = 3 + (4 + 5)$

11. $7 + (8 + 2) = (7 + 8) + 2$

12. $(8 + 5) + 1 = 1 + (8 + 5)$

13. a. Use the variables x and y to write the commutative property of addition.

 b. Use the variables x, y, and z to write the associative property of addition.

14. Show how to check the result:

$$\begin{array}{r} 74 \\ - 29 \\ \hline 45 \end{array}$$

15. Any number added to ▨ stays the same.

16. a. In calculating $12 + (8 + 5)$, which numbers should be added first?

 b. In calculating $60 - 15 + 4$, which operation should be performed first?

17. What addition fact is illustrated below?

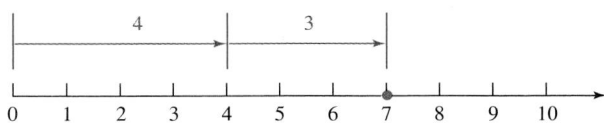

18. What subtraction fact is illustrated below?

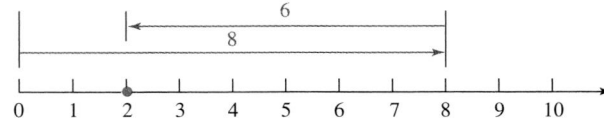

NOTATION *Fill in the blanks.*

19. The grouping symbols ( ) are called _____.

20. The minus sign − means _____.

Complete each solution.

21. (36 + 11) + 5 = ▮▮ + 5
          = 52

22. 12 + (15 + 2) = 12 + ▮▮
          = 29

PRACTICE *Do each addition.*

23. 25 + 13

24. 47 + 12

25. 156 + 305

26. 647 + 38

27. 19 + 39 + 53

28. 27 + 16 + 48

29. (95 + 16) + 39

30. 832 + (97 + 27)

31. 25 + (321 + 17)

32. (4,231 + 213) + 5,234

33.   632
        +347

34.   423
        +570

35.   1,372
        + 613

36.   2,477
        + 693

37.   6,427
        +3,573

38.   3,567
        +8,778

39.   8,539
        +7,368

40.   5,799
        +6,879

41.   1,246
           578
        +   37

42.   4,689
          3,422
        +   26

43.   3,156
          1,578
        + 578

44.   2,379
          4,779
        +2,339

Find the perimeter of each rectangle or square.

45.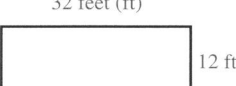
32 feet (ft)
12 ft

46.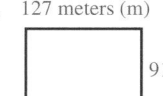
127 meters (m)
91 m

47.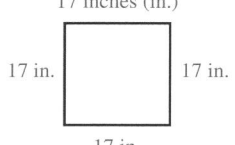
17 inches (in.)
17 in.   17 in.
17 in.

48.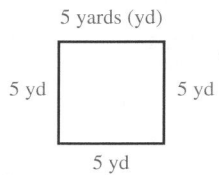
5 yards (yd)
5 yd   5 yd
5 yd

Do each subtraction.

49. 17 − 14

50. 42 − 31

51. 39 − 14

52. 45 − 32

53. 174 − 71

54. 257 − 155

55. 633 − (598 − 30)

56. 600 − (497 − 60)

57. 160 − 15 − 4

58. 498 − 17 − 162

59. 29 − 17 − 12

60. 53 − 26 − 27

61.   367
        −343

62.   224
        −122

63.   423
        −305

64.   330
        −270

65.   1,537
        − 579

66.   2,470
        − 863

67.   4,267
        −2,578

68.   7,356
        −3,578

69.   17,246
        − 6,789

70.   34,510
        −27,593

71.   15,700
        −15,397

72.   35,021
        −23,999

Do the computations.

73. 43 − 12 + 9

74. 59 − 16 + 2

75. 120 + 30 − 40

76. 600 + 99 − 54

APPLICATIONS

77. TAXIS For a 17-mile trip, Wanda paid the taxi driver $23. If $5 was a tip, how much was the fare?

78. SPACE FLIGHT Astronaut Walter Schirra's first space flight orbited the Earth 6 times and lasted 9 hours. His second flight orbited the Earth 16 times and lasted 26 hours. How long was Schirra in space?

79. MAGAZINE CIRCULATION In 1997, the monthly circulation of *Ebony* magazine grew by 15,865. In 1998, the monthly circulation decreased by 69,404. If the monthly circulation in 1996 was 1,803,566, what was it in 1998?

80. JEWELRY MAKING Gold melts at about 1,947°F. The melting point of silver is 183°F lower. What is the melting point of silver?

81. BANKING A savings account contained $370. After a deposit of $40 and a withdrawal of $197, how much is in the account?

82. TRAVEL   A student wants to make the 2,221-mile trip from Detroit to Seattle in three days. If she drives 751 miles on the first day and 875 miles on the second day, how far must she travel on the third day?

83. TAX DEDUCTION   For tax purposes, a woman kept the mileage records shown in Illustration 1. Find the total number of miles that she drove in the first 6 months of the year.

84. COMPANY BUDGET   A department head prepared an annual budget with the line items shown in Illustration 2. Find the projected number of dollars to be spent.

Month	Miles driven
January	2,345
February	1,712
March	1,778
April	445
May	1,003
June	2,774

ILLUSTRATION 1

Line item	Amount
Equipment	17,242
Contractual	5,443
Travel	2,775
Supplies	10,553
Development	3,225
Maintenance	1,075

ILLUSTRATION 2

In Exercises 85–86, refer to Illustration 3. To use this salary schedule, note that the annual salary of a third-year teacher with 15 units of course work beyond a Bachelor's degree is $30,887 per year (Step 3/Column 2).

Teachers' Salary Schedule			
Years teaching	Column 1: B.D.	Column 2: B.D. + 15	Column 3: B.D. + 30
Step 1	$26,785	$28,243	$29,701
Step 2	$28,107	$29,565	$31,023
Step 3	$29,429	$30,887	$32,345
Step 4	$30,751	$32,209	$33,667
Step 5	$32,073	$33,531	$34,989

ILLUSTRATION 3

85. INCOME   How much money will a new teacher make in his first five years of teaching if he begins at
a. Step 1/Column 1?
b. Step 1/Column 3?

86. PAY INCREASE   If a teacher is now on Step 2/Column 2, how much more money will she make next year when she
a. gains one year of experience?
b. completes 15 units of course work?

87. BLUEPRINT   Find the length of the house shown in Illustration 4.

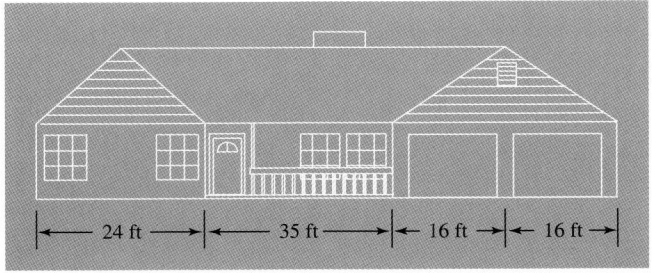

ILLUSTRATION 4

88. MACHINERY   Find the length of the motor on the machine shown in Illustration 5.

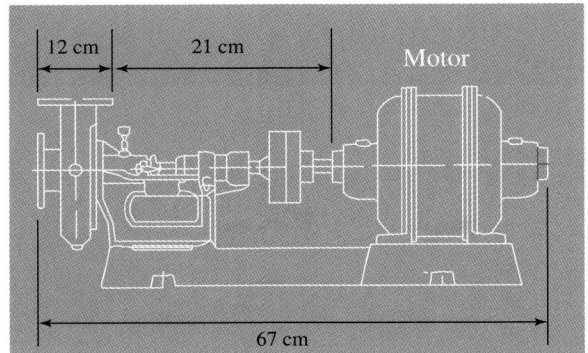

ILLUSTRATION 5

89. CAR EMISSIONS   Illustration 6 shows the number of tons of hydrocarbons and nitrogen oxides that have been removed from the air because of antismog legislation in California. How many tons have been removed daily because of this legislation?

Step taken	Tons removed daily
State gasoline reformulation	215
Federal gasoline reformulation	85
Auto nitrogen oxide standard	117
Auto hydrocarbon standard	35
Diesel fuel reformulation	70
Smog check	150
Gas pump nozzles	120

ILLUSTRATION 6

90. DALMATIANS   See Illustration 7. How many fewer Dalmatians were registered in 1998 than in the year when they were at their height of popularity?

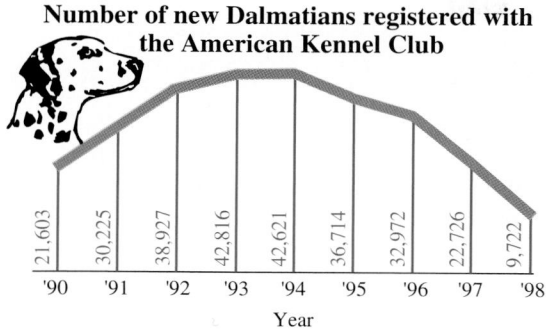

Number of new Dalmatians registered with the American Kennel Club

21,603	30,225	38,927	42,816	42,621	36,714	32,972	22,726	9,722
'90	'91	'92	'93	'94	'95	'96	'97	'98

Year

ILLUSTRATION 7

91. CITY FLAG   To decorate a city flag, yellow fringe is to be sewn around its outside edges, as shown in Illustration 8. The fringe comes on long spools and is sold by the inch. How many inches of fringe must be purchased to complete the project?

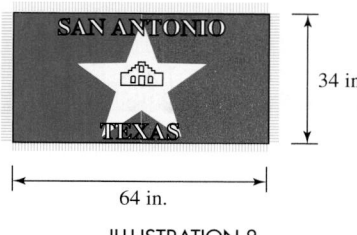

34 in.

64 in.

ILLUSTRATION 8

92. BOXING   How much padded rope is needed to create the square boxing ring in Illustration 9 if each side is 24 feet long?

ILLUSTRATION 9

WRITING

93. Explain why the operation of addition is commutative.

94. Explain how addition can be used to check subtraction.

REVIEW   *Write each numeral in expanded notation.*

95. 3,125

96. 60,037

Round 6,354,784 to the specified place.

97. nearest ten

98. nearest hundred

99. nearest ten thousand

100. nearest hundred thousand

1.3   *Multiplying and Dividing Whole Numbers*

In this section, you will learn about

- Properties of multiplication • Multiplying whole numbers
- Finding the area of a rectangle • Division • Properties of division
- Dividing whole numbers

INTRODUCTION. Mastering the operations of multiplication and division with whole numbers enables us to find areas of geometric figures and to solve business and transportation problems. For example, to find the area of a rectangle, we need to multiply its length by its width. To figure a paycheck, we need to multiply the number of hours worked by the hourly rate of pay. To calculate the fuel economy of a bus, we need to divide the miles it travels by the number of gallons of gas that are used.

Properties of multiplication

There are several symbols used to indicate multiplication.

Symbols that are used for multiplication

Symbol		Examples		
$\times$	times sign	4×5	or	$\begin{array}{r} 4 \\ \times\ \underline{5} \end{array}$
$\cdot$	raised dot	$4 \cdot 5$		
$(\ )$	parentheses	$(4)(5)$ or $4(5)$ or $(4)5$		

Recall that a variable is a letter that stands for a number. We will often multiply a variable by another number or multiply a variable by another variable. When we do this, we don't need to use a symbol for multiplication.

$5a$ means $5 \cdot a$, ab means $a \cdot b$,   and xyz means $x \cdot y \cdot z$

 COMMENT   In this book, we will seldom use the $\times$ sign, because it can be confused with the letter x.

Multiplication is repeated addition. For example, $4 \cdot 5$ means the sum of four 5's:

$$\overbrace{4 \cdot 5 = 5 + 5 + 5 + 5}^{\text{The sum of four 5's.}}$$
$$= 20$$

In the above multiplication, the result of 20 is called a **product.** The numbers that were multiplied (4 and 5) are called **factors.**

$$\begin{array}{ccc} \text{Factor} & \text{Factor} & \text{Product} \\ \downarrow & \downarrow & \downarrow \\ 4 & \cdot\quad 5\quad = & 20 \end{array}$$

The multiplication $5 \cdot 4$ means the sum of five 4's:

$$\overbrace{5 \cdot 4 = 4 + 4 + 4 + 4 + 4}^{\text{The sum of five 4's.}}$$
$$= 20$$

We see that $4 \cdot 5 = 20$ and $5 \cdot 4 = 20$. The results are the same. These examples illustrate that the order in which we multiply two numbers does not affect the result. This property is called the **commutative property of multiplication.**

Commutative property of multiplication

If a and b represent numbers, then

$a \cdot b = b \cdot a$ or, more simply, $ab = ba$

Table 1-1 summarizes the basic multiplication facts.

- To find the product of 6 and 8 using the table, we find the intersection of the 6th row and the 8th column. The product is 48.
- To find the product of 8 and 6, we find the intersection of the 8th row and the 6th column. Once again, the product is 48.

In the table, we see that the set of answers above and the set of answers below the diagonal line in bold print are identical. This further illustrates that multiplication is commutative.

·	0	1	2	3	4	5	6	7	8	9
0	**0**	0	0	0	0	0	0	0	0	0
1	0	**1**	2	3	4	5	6	7	8	9
2	0	2	**4**	6	8	10	12	14	16	18
3	0	3	6	**9**	12	15	18	21	24	27
4	0	4	8	12	**16**	20	24	28	32	36
5	0	5	10	15	20	**25**	30	35	40	45
6	0	6	12	18	24	30	**36**	42	48	54
7	0	7	14	21	28	35	42	**49**	56	63
8	0	8	16	24	32	40	48	56	**64**	72
9	0	9	18	27	36	45	54	63	72	**81**

TABLE 1-1

From the table, we see that whenever we multiply a number by 0, the product is 0. For example,

$$0 \cdot 5 = 0, \qquad 0 \cdot 8 = 0, \qquad \text{and} \qquad 9 \cdot 0 = 0$$

We also see that whenever we multiply a number by 1, the number remains the same. For example,

$$3 \cdot 1 = 3, \qquad 7 \cdot 1 = 7, \qquad \text{and} \qquad 1 \cdot 9 = 9$$

These examples suggest the multiplication properties of 0 and 1.

Multiplication properties of 0 and 1

If a represents any number, then
$$a \cdot 0 = 0 \qquad \text{and} \qquad 0 \cdot a = 0$$
$$a \cdot 1 = a \qquad \text{and} \qquad 1 \cdot a = a$$

EXAMPLE 1 *Computing daily wages.* Raul worked an 8-hour day at an hourly rate of $9. How much money did he earn?

Solution

For each of the 8 hours, Raul earned $9. His total pay for the day is the sum of eight 9's: $9 + 9 + 9 + 9 + 9 + 9 + 9 + 9$. This can be calculated by multiplication.

Total wages = $8 \cdot 9$

= 72   See the multiplication table.

Raul earned $72.

Self Check

At a rate of $8 per hour, how much will a school bus driver earn if she works from 8:00 A.M. until noon?

Answer: $32   ∎

To multiply three numbers, we first multiply two of them and then multiply that result by the third number. In the following examples, we multiply $3 \cdot 2 \cdot 4$ in two ways. The parentheses show us which multiplication to do first.

Method 1: Group 3 · 2

$(3 \cdot 2) \cdot 4 = 6 \cdot 4$ Multiply 3 and 2 to get 6.

= 24   Then multiply 6 and 4 to get 24.

Method 2: Group 2 · 4

$3 \cdot (2 \cdot 4) = 3 \cdot 8$ Multiply 2 and 4 to get 8.

$\qquad\qquad\quad = 24$ Then multiply 3 and 8 to get 24.

The answers are the same. This illustrates that changing the grouping when multiplying numbers does not affect the result. This property is called the **associative property of multiplication.**

Associative property of multiplication

> If a, b, and c represent numbers, then
>
> $\qquad (a \cdot b) \cdot c = a \cdot (b \cdot c)$ or, more simply, $(ab)c = a(bc)$

Multiplying whole numbers

To find the product $8 \cdot 47$, it is inconvenient to add up eight 47's. Instead, we find the product by a multiplication process.

$$\begin{array}{r} 4\,7 \\ \times\quad 8 \\ \hline \end{array}$$ Write the factors in a column, with the corresponding digits aligned vertically.

$$\begin{array}{r} 5 \\ 4\,7 \\ \times\quad 8 \\ \hline 6 \end{array}$$ Multiply 7 by 8. The product is 56. Place 6 in the ones column of the answer and carry 5 (in blue) to the tens column.

$$\begin{array}{r} 5 \\ 4\,7 \\ \times\quad 8 \\ \hline 3\,7\,6 \end{array}$$ Multiply 4 by 8. The product is 32. To the 32, add the carried 5 to get 37. Place the 7 in the tens column and the 3 in the hundreds column of the answer.

The product is 376.

To find the product $23 \cdot 435$, we use the multiplication process. Because $23 = 20 + 3$, we multiply 435 by 20 and by 3 and then add the products. To do this, we write the factors in a column, with the corresponding digits aligned vertically. We then begin the process by multiplying 435 by 3:

$$\begin{array}{r} 1 \\ 4\,3\,5 \\ \times\quad 2\,3 \\ \hline 5 \end{array}$$ Multiply 5 by 3. The product is 15. Place 5 in the ones column and carry 1 (in blue) to the tens column.

$$\begin{array}{r} 1\,1 \\ 4\,3\,5 \\ \times\quad 2\,3 \\ \hline 0\,5 \end{array}$$ Multiply 3 by 3. The product is 9. To the 9, add the carried 1 to get 10. Place the 0 in the tens column and carry the 1 (in green) to the hundreds column.

$$\begin{array}{r} 1\,1 \\ 4\,3\,5 \\ \times\quad 2\,3 \\ \hline 1\,3\,0\,5 \end{array}$$ Multiply 4 by 3. The product is 12. Add the 12 to the carried 1 to get 13. Write 13.

We continue by multiplying 435 by 2 tens, or 20:

$$\begin{array}{r} 1 \\ 4\,3\,5 \\ \times\quad 2\,3 \\ \hline 1\,3\,0\,5 \\ 0 \end{array}$$ Multiply 5 by 2. The product is 10. Write 0 in the tens column and carry 1 (in purple).

$$\begin{array}{r} 1 \\ 4\,3\,5 \\ \times\quad 2\,3 \\ \hline 1\,3\,0\,5 \\ 7\,0 \end{array}$$ Multiply 3 by 2. The product is 6. Add 6 to the carried 1 to get 7. Write the 7. There is no carry.

$$\begin{array}{r} \overset{1}{4}\,3\,5 \\ \times\ 2\,3 \\ \hline 1\,3\,0\,5 \\ 8\,7\,0 \end{array}$$

Multiply 4 by 2. The product is 8. There is no carry to add. Write the 8.

$$\begin{array}{r} 4\,3\,5 \\ \times\ 2\,3 \\ \hline 1\,3\,0\,5 \\ 8\,7\,0 \\ \hline 1\,0\,0\,0\,5 \end{array}$$

Draw another line beneath the two completed rows. Add the two rows. This sum gives the product of 435 and 23.

Thus, $23 \cdot 435 = 10{,}005$.

EXAMPLE 2 *Mileage.* Specifications for a 1999 Ford Explorer 4 × 4 are shown in the table below. For city driving, how far can it travel on a tank of gas? (The abbreviation "mpg" means "miles per gallon.")

Engine	4.0 L V6
Fuel capacity	21 gal
Fuel economy (mpg)	15 city/19 hwy

Solution

For city driving, each of the 21 gallons of gas that the tank holds enables the Explorer to go 15 miles. The total distance it can travel is the sum of twenty-one 15's. This can be calculated by multiplication: $21 \cdot 15$.

$$\begin{array}{r} \overset{1}{1}\,5 \\ \times\,2\,1 \\ \hline 1\,5 \\ 3\,0 \\ \hline 3\,1\,5 \end{array}$$

For city driving, the Explorer can go 315 miles on a tank of gas.

Self Check

For highway driving, how far can the Explorer travel on a tank of gas?

Answer: 399 mi ■

EXAMPLE 3 *Calculating production.* The labor force of an electronics firm works two 8-hour shifts each day and manufactures 53 television sets each hour. Find how many sets will be manufactured in 5 days.

Solution

The number of TV sets manufactured in 5 days is given by the product:

2 shifts per day	8 hr per shift	53 sets per hr	5 days
↓	↓	↓	↓
2	· 8	· 53	· 5

This could also be written 2(8)(53)(5).

We perform the multiplications working from left to right.

$$2 \cdot 8 \cdot 53 \cdot 5 = \mathbf{16} \cdot 53 \cdot 5 \qquad \text{Multiply 2 and 8.}$$
$$= 848 \cdot 5 \qquad \text{Multiply 16 and 53.}$$
$$= 4{,}240$$

4,240 television sets will be manufactured in 5 days. ■

Accent on Technology: **Checking an answer**

We can use a scientific calculator to check the multiplication performed in Example 3. To find the product $2 \cdot 8 \cdot 53 \cdot 5$, we enter these numbers and press these keys.

Keystrokes   2 $\times$ 8 $\times$ 53 $\times$ 5 $=$

$$\boxed{4240}$$

The display verifies that the multiplication was done correctly in Example 3.

We can use multiplication to count objects arranged in rectangular patterns. For example, the display on the left below shows a rectangular array consisting of 5 rows of 7 stars. The product $5 \cdot 7$, or 35, indicates the total number of stars.

Because multiplication is commutative, the array on the right below, consisting of 7 rows of 5 stars, contains the same number of stars.

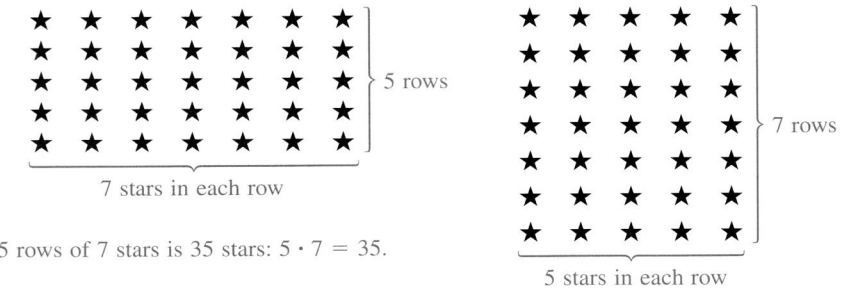

5 rows of 7 stars is 35 stars: $5 \cdot 7 = 35$.

7 rows of 5 stars is 35 stars: $7 \cdot 5 = 35$.

EXAMPLE 4 *Computer science.* To draw graphics on a computer screen, a computer controls each *pixel* (one dot on the screen). See Figure 1-11. A standard computer graphics image is 800 pixels wide and 600 pixels high. How many pixels does the computer control?

Solution

The graphics image is a rectangular array of pixels. Each of its 600 rows consists of 800 pixels. The total number of pixels is the product of 600 and 800:

$$600 \cdot 800 = 480,000$$

The computer controls 480,000 pixels.

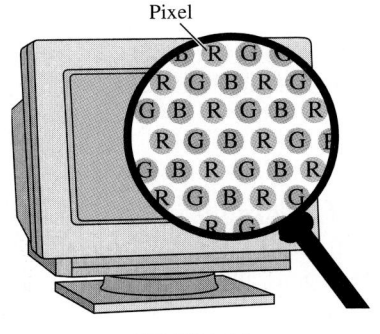

Pixel

FIGURE 1-11

Self Check

On a color monitor, each of the pixels can be red, green, or blue. How many colored pixels does the computer control?

Answer: 1,440,000 ■

Finding the area of a rectangle

One important application of multiplication is finding the area of a rectangle. The **area of a rectangle** is the measure of the amount of surface it encloses. Area is measured in square units, such as square inches (denoted as in.^{2}) or square centimeters (denoted as cm^2). (See Figure 1-12.)

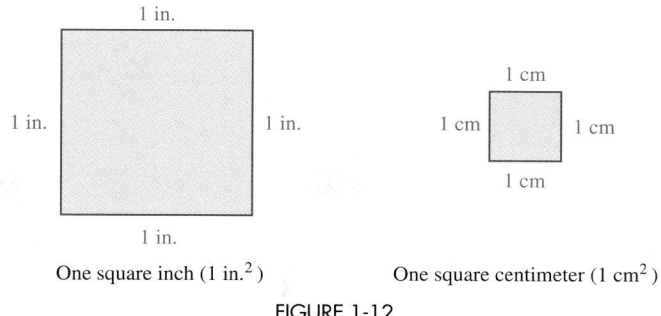

One square inch (1 in.^{2})   One square centimeter (1 cm^2)

FIGURE 1-12

The rectangle in Figure 1-13 has a length of 5 centimeters and a width of 3 centimeters. Each small square covers an area of one square centimeter (1 cm^2). The small squares form a rectangular pattern, with 3 rows of 5 squares.

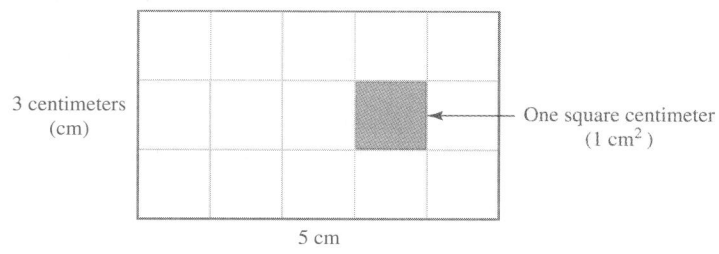

FIGURE 1-13

Because there are 5 · 3, or 15, small squares, the area of the rectangle is 15 cm^2. This suggests that the area of any rectangle is the product of its length and its width.

Area of a rectangle	=	length	·	width

Using the variables *l* and *w* to represent the length and width, we can write this formula in simpler form.

Area of a rectangle

> The area *A* of a rectangle is the product of the rectangle's length *l* and its width *w*:
> $$A = l \cdot w \quad \text{or, more simply,} \quad A = lw$$

EXAMPLE 5 *Wrapping paper.* When completely unrolled, a long sheet of gift wrapping paper has the dimensions shown in Figure 1-14. How many square feet of gift wrap are on the roll?

Solution

To find the number of square feet of paper, we need to find the area of the rectangle shown in the figure.

$A = lw$ The formula for the area of a rectangle.

$A = 12 \cdot 3$ Replace *l* with 12 and *w* with 3.

$ = 36$ Do the multiplication.

There are 36 square feet (ft^2) of wrapping paper on the roll.

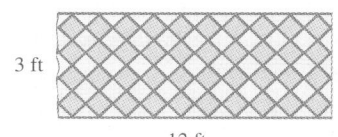

3 ft

12 ft

FIGURE 1-14

Self Check

A package of 9-inch-by-12-inch paper contains 500 sheets. How many square inches of paper does one package contain?

Answer: 54,000 in.^{2} ■

COMMENT Remember that the perimeter of a rectangle is the distance around it. The area of a rectangle is a measure of the surface it encloses.

Division

If $12 is distributed equally among 4 people, we must divide to see that each person would receive $3.

$$4\overline{)12}^{\,3}$$

Several symbols can be used to indicate division.

Symbols that are used for division

Symbol		Example
$\div$	division sign	$12 \div 4$
$\overline{)}$	long division	$4\overline{)12}$
$\underline{}$	fraction bar	$\dfrac{12}{4}$

In a division, the number that is being divided is called the **dividend.** The number that we are dividing by is called the **divisor,** and the answer is called the **quotient.**

$$\text{Dividend} \div \text{divisor} = \text{quotient} \qquad \text{Divisor}\overline{)\text{dividend}}^{\,\text{quotient}} \qquad \dfrac{\text{Dividend}}{\text{Divisor}} = \text{quotient}$$

Division can be thought of as repeated subtraction. To divide 12 by 4 is to ask, "How many 4's can be subtracted from 12?" Exactly three 4's can be subtracted from 12 to get 0:

$$12 - \overbrace{4 - 4 - 4}^{\text{Three 4's}} = 0$$

Thus, $12 \div 4 = 3$.

Division is also related to multiplication.

$$\dfrac{12}{4} = 3 \quad \text{because} \quad 4 \cdot 3 = 12 \qquad \text{and} \qquad \dfrac{20}{5} = 4 \quad \text{because} \quad 5 \cdot 4 = 20$$

Properties of division

We will now consider three types of division that involve zero. In the first case, we will examine a division of zero; in the second, a division by zero; in the third, a division of zero by zero.

Division statement	Corresponding multiplication statement	Result
$\dfrac{0}{2} = ?$	$2(?) = 0$ $\uparrow$ This must be 0 if the product is to be 0.	$\dfrac{0}{2} = 0$
$\dfrac{2}{0} = ?$	$0(?) = 2$ $\uparrow$ There is no number that gives 2 when multiplied by 0.	There is no quotient.
$\dfrac{0}{0} = ?$	$0(?) = 0$ $\uparrow$ Any number times 0 is 0.	Any number can be the quotient.

We see that $\frac{0}{2} = 0$. This suggests that the quotient of 0 divided by any nonzero number is 0. Since $\frac{2}{0}$ does not have a quotient, we say that division of 2 by 0 is *undefined*. In general, division of any nonzero number by 0 is undefined. Since $\frac{0}{0}$ can be any number, we say that $\frac{0}{0}$ is undetermined.

Division with 0

1. If a represents any nonzero number, $\dfrac{0}{a} = 0$.

2. If a represents any nonzero number, $\dfrac{a}{0}$ is undefined.

3. $\dfrac{0}{0}$ is undetermined.

The example $\frac{12}{1} = 12$ illustrates that *any number divided by 1 is the number itself.* The example $\frac{12}{12} = 1$ illustrates that *any number (except 0) divided by itself is 1.*

Division properties

If a represents any number,

$$\frac{a}{1} = a \quad \text{and} \quad \frac{a}{a} = 1 \text{ (provided that } a \neq 0) \quad \text{Read} \neq \text{as "is not equal to."}$$

Dividing whole numbers

We can use a process called **long division** to divide whole numbers. To divide 832 by 23, for example, we proceed as follows:

Quotient $\longrightarrow$

Divisor $\longrightarrow$ $2\,3\overline{)8\,3\,2}$ Place the divisor and the dividend as indicated. The quotient will appear above the long division symbol.

$\uparrow$

Dividend

We will find the quotient using the following division process.

$\dfrac{4}{2\,3\overline{)8\,3\,2}}$ Ask: "How many times will 23 divide 83?" Because an estimate is 4, place 4 in the tens column of the quotient.

$\begin{array}{r} 4 \\ 2\,3\overline{)8\,3\,2} \\ 9\,2 \end{array}$ Multiply $23 \cdot 4$ and place the answer, 92, under the 83. Because 92 is larger than 83, our estimate of 4 for the tens column of the quotient was too large.

$\begin{array}{r} 3 \\ 2\,3\overline{)8\,3\,2} \\ 6\,9\downarrow \\ \hline 1\,4\,2 \end{array}$ Revise the estimate of the quotient to be 3. Multiply $23 \cdot 3$ to get 69, place 69 under the 83, draw a line, and subtract.

Bring down the 2 in the ones column.

$\begin{array}{r} 3\,7 \\ 2\,3\overline{)8\,3\,2} \\ 6\,9 \\ \hline 1\,4\,2 \\ 1\,6\,1 \end{array}$ Ask, "How many times will 23 divide 142?" The answer is approximately 7. Place 7 in the ones column of the quotient. Multiply $23 \cdot 7$ to get 161. Place 161 under 142. Because 161 is larger than 142, the estimate of 7 is too large.

$\begin{array}{r} 3\,6 \\ 2\,3\overline{)8\,3\,2} \\ 6\,9 \\ \hline 1\,4\,2 \\ 1\,3\,8 \\ \hline 4 \end{array}$ Revise the estimate of the quotient to be 6. Multiply: $23 \cdot 6 = 138$.

Place 138 under 142 and subtract.

The quotient is 36, and the leftover 4 is the **remainder.** We can write this result as 36 R 4.

To check the result of a division, we multiply the divisor by the quotient and then add the remainder. The result should be the dividend.

$$\textit{Check:}\quad \underset{36}{\text{Quotient}} \;\cdot\; \underset{23}{\text{divisor}} \;+\; \underset{4}{\text{remainder}} \;=\; \underset{832}{\text{dividend}}$$

$$828 + 4 = 832$$

$$832 = 832$$

EXAMPLE 6   *Managing a soup kitchen.*   A soup kitchen plans to feed 1,990 people. Because of space limitations, only 165 people can be served at one time. How many seatings will be necessary to feed everyone? How many will be served at the last seating?

Solution
The 1,990 people can be fed 165 at a time. To find the number of seatings, we divide.

```
        12
   165)1,990
       1 65↓
        340
        330
         10
```

The quotient is 12, and the remainder is 10. Thirteen seatings will be needed: 12 full-capacity seatings and one partial seating to serve the remaining 10 people.

Self Check

Each gram of fat in a meal provides 9 calories. A fast-food meal contains 243 calories from fat. How many grams of fat does the meal contain?

Answer: 27

Accent on Technology: **Yard sale**

At a yard sale, a family sold old shirts, pants, blouses, and dresses for 75¢ each. They made $80.25 from the sale of the clothes. To find the number of items they sold, we divide the total sales (8,025¢) by the cost of each item (75¢). We can use a scientific calculator to find 8,025 ÷ 75 by entering these numbers and pressing these keys.

Keystrokes   8025 ÷ 75 =   107

They sold 107 items of clothing at the yard sale.

STUDY SET   Section 1.3

VOCABULARY   *Fill in the blanks.*

1. _____ is repeated addition.

2. Numbers that are to be multiplied are called _____. The result of a multiplication is called a _____.

3. The statement $ab = ba$ expresses the _____ property of multiplication.

4. The statement $(ab)c = a(bc)$ expresses the _____ property of multiplication.

5. If a square measures one inch on each side, its area is 1 _____.

6. In a division, the dividend is divided by the _____. The result of a division is called a _____.

CONCEPTS

7. Write $8 + 8 + 8 + 8$ as a multiplication.

8. a. Use the variables x and y to write the commutative property of multiplication.
   b. Use the variables x, y, and z to write the associative property of multiplication.

9. How do we find the amount of surface enclosed by a rectangle?

10. Tell whether *perimeter* or *area* is the concept that should be applied to find each of the following.
   a. The amount of floor space to be carpeted
   b. The amount of clear glass to be tinted
   c. The amount of lace needed to trim the sides of a handkerchief

11. Do each multiplication.
   a. $1 \cdot 25$ **b.** $62(1)$
   c. $10 \cdot 0$ **d.** $0(4)$

12. Do each division.
   a. $25 \div 1$ **b.** $\dfrac{7}{1}$
   c. $\dfrac{0}{1}$ **d.** $\dfrac{5}{0}$
   e. $\dfrac{0}{0}$ **f.** $\dfrac{0}{2,757}$

13. Write a multiplication statement that finds the number of red squares.

14. Consider

$$\begin{array}{r} 12 \\ 15\overline{)182} \\ \underline{15} \\ 32 \\ \underline{30} \\ 2 \end{array}$$

Fill in the blanks:

$12 \cdot \boxed{} + \boxed{} = \boxed{}$

NOTATION

15. a. Write three symbols that are used for multiplication.
   b. Write three symbols that are used for division.

16. Write each multiplication in simpler form.
   a. $8 \cdot x$ **b.** $l \cdot w$

17. What does ft^2 mean?

18. Draw a figure having an area of 1 square inch.

PRACTICE   *Do each multiplication.*

19. $12 \cdot 7$ **20.** $15 \cdot 8$
21. $27(12)$ **22.** $35(17)$
23. $9 \cdot (4 \cdot 5)$ **24.** $(3 \cdot 5) \cdot 12$
25. $5 \cdot 7 \cdot 3$ **26.** $7 \cdot 6 \cdot 8$

27. $\begin{array}{r} 99 \\ \times 77 \\ \hline \end{array}$ **28.** $\begin{array}{r} 73 \\ \times 59 \\ \hline \end{array}$

29. $\begin{array}{r} 20 \\ \times 53 \\ \hline \end{array}$ **30.** $\begin{array}{r} 78 \\ \times 20 \\ \hline \end{array}$

31. $\begin{array}{r} 112 \\ \times\ 23 \\ \hline \end{array}$ **32.** $\begin{array}{r} 232 \\ \times\ 53 \\ \hline \end{array}$

33. $\begin{array}{r} 207 \\ \times\ 97 \\ \hline \end{array}$ **34.** $\begin{array}{r} 768 \\ \times\ 70 \\ \hline \end{array}$

35. $13,456 \cdot 217$ **36.** $17,456 \cdot 257$

37. $3,302 \cdot 358$ **38.** $123,112 \cdot 46$

Find the area of each rectangle or square.

39.
6 in.
14 in.

40.
50 m
22 m

41.
12 in.
12 in.

42.
20 cm
20 cm

Do each division.

43. $40 \div 5$ **44.** $40 \div 8$
45. $42 \div 14$ **46.** $65 \div 13$
47. $132 \div 11$ **48.** $132 \div 12$
49. $\dfrac{221}{17}$ **50.** $\dfrac{221}{13}$
51. $13\overline{)949}$ **52.** $73\overline{)949}$
53. $33\overline{)1,353}$ **54.** $41\overline{)1,353}$
55. $39\overline{)7,995}$ **56.** $71\overline{)7,313}$
57. $29\overline{)6,090}$ **58.** $13\overline{)7,410}$

Do each division and give the quotient and the remainder.

59. $31\overline{)273}$ **60.** $25\overline{)290}$
61. $37\overline{)743}$ **62.** $79\overline{)931}$

63. $42)\overline{1,273}$ **64.** $83)\overline{3,280}$

65. $57)\overline{1,795}$ **66.** $99)\overline{9,876}$

APPLICATIONS

67. FIGURING WAGES   A cook worked 12 hours at $11 per hour. How much did she earn?

68. SQUARES ON A CHESSBOARD   A chessboard consists of 8 rows, with 8 squares in each row. How many squares are there on a chessboard?

69. FINDING DISTANCE   A car with a tank that holds 14 gallons of gasoline goes 29 miles on one gallon. How far can the car go on a full tank?

70. RENTING APARTMENTS   Mia owns an apartment building with 18 units. Each unit generates a monthly income of $450. Find her total monthly income.

71. CONCERT ATTENDANCE   A jazz quartet gave two concerts in each of 37 cities. Approximately 1,700 fans attended each concert. How many persons heard the group?

72. CEREAL   A cereal maker advertises "Two cups of raisins in every box." Find the number of cups of raisins in a case of 36 boxes of cereal.

73. ORANGE JUICE   It takes 13 oranges to make one can of orange juice. Find the number of oranges used to make a case of 24 cans.

74. ROOM CAPACITY   A college lecture hall has 17 rows of 33 seats. A sign on the wall reads, "Occupancy by more than 570 persons is prohibited." If the seats are filled and there is one instructor, is the college breaking the rule?

75. CAPACITY OF AN ELEVATOR   There are 14 people in an elevator with a capacity of 2,000 pounds. If the average weight of a person on the elevator is 150 pounds, is the elevator overloaded?

76. CHANGING UNITS   There are 12 inches in 1 foot. How many inches are in 80 feet?

77. WORD PROCESSING   A student used the option shown in Illustration 1 when typing a report. How many entries will the table hold?

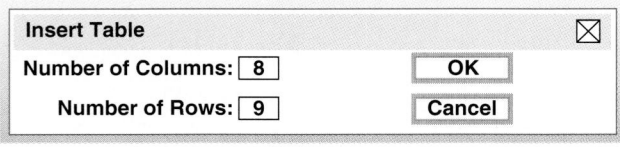

ILLUSTRATION 1

78. FILLING PRESCRIPTIONS   How many tablets should a pharmacist put in the container shown in Illustration 2?

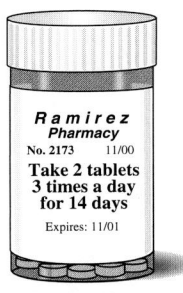

ILLUSTRATION 2

79. DISTRIBUTING MILK   A first grade class received 73 half-pint cartons of milk to distribute evenly to the 23 students. How many cartons were left over?

80. LIFT SYSTEM   If the bus shown in Illustration 3 weighs 58,000 pounds, how much weight is on each jack?

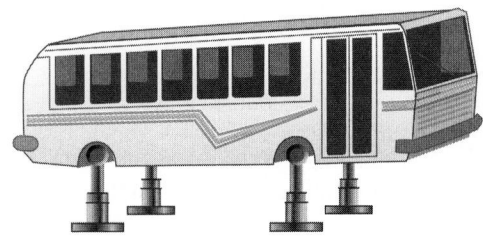

ILLUSTRATION 3

81. MILEAGE   A touring rock group travels in a bus that has a range of 700 miles on one tank (140 gallons) of gasoline. How far can the bus travel on one gallon of gas?

82. RUNNING   Brian runs 7 miles each day. In how many days will Brian run 371 miles?

83. How many feet more than two miles is 11,000 feet? (*Hint:* 5,280 feet = 1 mile.)

84. ORDERING DOUGHNUTS   How many dozen doughnuts must be ordered for a meeting if 156 people are expected to attend, and each person will be served one doughnut?

85. PRICE OF A TEXTBOOK   An author knows that her publisher received $954,193 on the sale of 23,273 textbooks. What is the price of each book?

86. WATER DISCHARGE   The Susquehanna River discharges 38,200 cubic feet of water per second into the Chesapeake Bay. How long will it take for the river to discharge 1,719,000 cubic feet?

87. VOLLEYBALL LEAGUE   A total of 216 girls tried out for a city volleyball program. How many girls should be put on each team roster if the following requirements must be met?

- All the teams are to have the same number of players.
- A reasonable number of players on a team is 7 to 10.
- For scheduling purposes, there must be an even number of teams.

88. AREA OF WYOMING   The state of Wyoming is a rectangle 360 miles long and 270 miles wide. Find its perimeter and its area.

89. COMPARING ROOMS   Which has the greater area, a rectangular room that is 14 feet by 17 feet or a square room that is 16 feet on each side? Which has the greater perimeter?

90. MATTRESSES   A queen-size mattress measures 60 inches by 80 inches, and a full-size mattress measures 54 inches by 75 inches. How much more sleeping surface is there on a queen-size mattress?

91. GARDENING   A rectangular garden is 27 feet long and 19 feet wide. A path in the garden uses 125 square feet of space. How many square feet are left for planting?

92. TENNIS   See Illustration 4.
  a. Find the number of square feet of court area a singles tennis player must defend.
  b. Do the same for a doubles player.
  c. What is the difference between the two results?

WRITING

93. Explain why the division of two numbers is not commutative.

94. Explain the difference between what perimeter measures and what area measures.

95. Explain the difference between 1 foot and 1 square foot.

96. When two numbers are multiplied, the result is 0. What conclusion can be drawn about the numbers?

REVIEW

97. Consider 372,856. What digit is in the hundreds column?

98. Round 45,995 to the nearest thousand.

99. Add 357, 39, and 476.

100. DISCOUNT   A car, originally priced at $17,550, is being sold for $13,970. By how many dollars has the price been decreased?

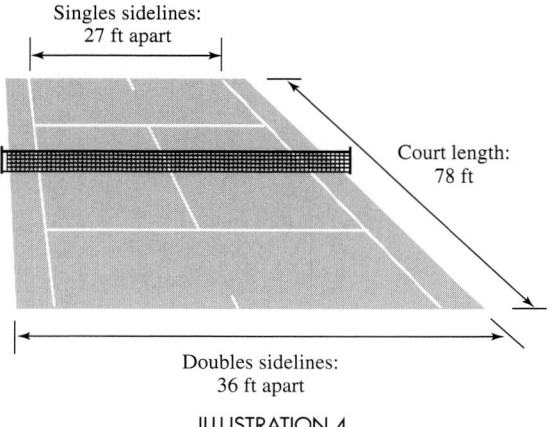

Singles sidelines: 27 ft apart

Court length: 78 ft

Doubles sidelines: 36 ft apart

ILLUSTRATION 4

ESTIMATION

In the previous two sections, we have used **estimation** as a means of checking the reasonableness of an answer. We now take a more in-depth look at the process of estimating.

Estimation is used to find an *approximate* answer to a problem. Estimates can be helpful in two ways. First, they serve as an accuracy check that can detect major computational errors. If an answer does not seem reasonable when compared to the estimate, the original problem should be reworked. Second, some situations call for only an approximate answer rather than the exact answer.

There are several ways to estimate, but there is one overriding theme of all the methods: The numbers in the problem are simplified so that the computation can be made easily and quickly. The first method we will study uses what is called **front-end rounding.** Each number is rounded to its largest place value, so that all but the first digit of each number is zero.

EXAMPLE 1 *Estimating sums, differences, and products.*

a. Estimate the sum: $3,714 + 2,489 + 781 + 5,500 + 303$.

Solution: Use front-end rounding.

3,714 ⟶	4,000
2,489 ⟶	2,000
781 ⟶	800
5,500 ⟶	6,000
+ 303 ⟶	+ 300
	13,100

Each number is rounded to its largest place value. All but the first digit is zero.

The estimate is 13,100.

If we compute $3,714 + 2,489 + 781 + 5,500 + 303$, the sum is 12,787. We can see that our estimate is close; it's just 313 more than 12,787. This example illustrates the tradeoff when using estimation: The calculations are easier to perform and they take less time, but the answers are not exact.

b. Estimate the difference: $46,721 - 13,208$.

Solution: Use front-end rounding.

46,721 ⟶	50,000
−13,208 ⟶	−10,000
	40,000

Only the first digit is nonzero.

The estimate is 40,000.

c. Estimate the product: $334 \cdot 59$.

Solution: Use front-end rounding.

334 ⟶	300
× 59 ⟶	× 60
	18,000

334 rounds to 300, and 59 rounds to 60.

The estimate is 18,000.

Self Check

a. Estimate the sum.

$$6,780$$
$$3,278$$
$$566$$
$$4,230$$
$$+1,923$$

b. Estimate the difference.

$$89,070$$
$$-15,331$$

c. Estimate the product.

$$707$$
$$\times251$$

Answers: **a.** 16,600, **b.** 70,000, **c.** 210,000

34

To estimate quotients, we will use a method that approximates both the dividend and the divisor so that they will divide easily. With this method, some insight and intuition are needed. There is one rule of thumb for this method: If possible, round both numbers up or both numbers down.

EXAMPLE 2   *Estimating quotients.*   Estimate the quotient: 170,715 ÷ 57.

Solution

Both numbers are rounded up. The division can then be done in your head.

```
        ┌─ The dividend is ─┐
        │   approximately   │
170,715 ÷ 57      180,000 ÷ 60 = 3,000
        └─ The divisor is ──┘
           approximately
```

The estimate is 3,000.

STUDY SET   *Use front-end rounding to find an estimate to check the reasonableness of each answer. Write* yes *if it appears reasonable and* no *if it does not.*

1.
```
  25,405
  11,222
   8,909
   1,076
  14,595
 +33,999
  73,206
```

2.
```
  568,334
 − 31,225
  497,109
```

3.
```
   451
 ×  73
 39,923
```

4.
```
   616
 ×  98
 60,368
```

Use estimation to check the reasonableness of each answer.

5. 57,238 ÷ 28 = 200

6. 322)13,202 (quotient 41)

Use an estimation procedure to answer each problem.

7. CAMPAIGNING   The number of miles flown each day by a politician on a campaign swing are shown here. Estimate the number of miles she flew during this time.

Day 1	3,546 miles
Day 2	567
Day 3	1,203
Day 4	342
Day 5	2,699

8. SHOPPING MALL   The total sales income for a downtown mall in its first three years in operation are shown here.

1998	$5,234,301
1999	$2,898,655
2000	$6,343,433

Estimate the difference in income for 1999 and 2000 as compared to the first year, 1998.

9. GOLF COURSE   Estimate the number of bags of grass seed needed to plant a fairway whose area is 86,625 square feet if the seed in each bag covers 2,850 square feet.

10. CENSUS   Estimate the total population of the ten largest counties in the United States as of 1998.

Largest counties, by population	
1. Los Angeles, CA	9,213,533
2. Cook, IL	5,189,689
3. Harris, TX	3,206,063
4. Maricopa, AZ	2,784,075
5. San Diego, CA	2,780,592
6. Orange, CA	2,721,701
7. Kings, NY	2,267,942
8. Miami-Dade, FL	2,152,437
9. Wayne, MI	2,118,129
10. Dallas, TX	2,050,865

11. CURRENCY   Estimate the number of $5 bills in circulation as of March 1, 1999, if the total value of the currency was $7,733,317,335.

12. REVENUES   In 1998, the revenues of IBM were $81,667,000,000. Approximately how many times larger was this than the revenues of Sun Microsystems, which took in $9,791,000,000?

1.4 Prime Factors and Exponents

In this section, you will learn about

- Factoring whole numbers • Even and odd whole numbers • Prime numbers
- Composite numbers • Finding prime factorizations with the tree method
- Exponents • Finding prime factorizations with the division method

INTRODUCTION. In this section, we will learn how to represent whole numbers in alternative forms. The procedures used to find these forms involve multiplication and division. We will then discuss exponents, a shortcut way to represent repeated multiplication.

Factoring whole numbers

The statement $3 \cdot 2 = 6$ has two parts: the numbers that are being multiplied, and the answer. The numbers that are being multiplied are *factors,* and the answer is the *product.* We say that 3 and 2 are factors of 6.

Factors

Numbers that are multiplied together are called **factors.**

EXAMPLE 1   *Finding factors of a whole number.*   Find the factors of 12.

Solution
We need to find the possible ways that we can multiply two whole numbers to get a product of 12.

$$1 \cdot 12 = 12, \qquad 2 \cdot 6 = 12, \qquad \text{and} \qquad 3 \cdot 4 = 12$$

In order, from least to greatest, the factors of 12 are 1, 2, 3, 4, 6, and 12.

Self Check
Find the factors of 20.

Answer: 1, 2, 4, 5, 10, and 20

Example 1 shows that 1, 2, 3, 4, 6, and 12 are the factors of 12. This observation was established by using multiplication facts. Each of these factors is related to 12 by division as well. Each of them divides 12, leaving a remainder of 0. Because of this fact, we say that 12 is **divisible** by each of its factors. When a division ends with a remainder of 0, we say that the division comes out even or that one of the numbers divides the other *exactly.*

Divisibility

One number is **divisible** by another if, when dividing them, the remainder is 0.

When we say that 3 is a factor of 6, we are using the word *factor* as a noun. The word *factor* is also used as a verb.

Factoring a whole number

To **factor** a whole number means to express it as the product of other whole numbers.

EXAMPLE 2 *Factoring a whole number.* Factor 40 using **a.** two factors and **b.** three factors.

Solution

a. There are several possibilities:

$$40 = 1 \cdot 40, \qquad 40 = 2 \cdot 20, \qquad 40 = 4 \cdot 10, \qquad \text{or} \qquad 40 = 5 \cdot 8$$

b. Again, there are several possibilities. Two of them are

$$40 = 5 \cdot 4 \cdot 2 \qquad \text{and} \qquad 40 = 2 \cdot 2 \cdot 10$$

Even and odd whole numbers

Even and odd whole numbers

> If a whole number is divisible by 2, it is called an **even** number.
> If a whole number is not divisible by 2, it is called an **odd** number.

The even whole numbers are the numbers

0, 2, 4, 6, 8, 10, 12, 14, 16, 18, . . .

The odd whole numbers are the numbers

1, 3, 5, 7, 9, 11, 13, 15, 17, 19, . . .

There are infinitely many even and infinitely many odd whole numbers.

Prime numbers

EXAMPLE 3 *Finding the factors of a whole number.* Find the factors of 17.

Solution

$$1 \cdot 17 = 17$$

The only factors of 17 are 1 and 17.

In Example 3 and its Self Check, we saw that the only factors of 17 are 1 and 17, and the only factors of 23 are 1 and 23. Numbers that have only two factors, 1 and the number itself, are called **prime numbers.**

Prime numbers

> A **prime number** is a whole number, greater than 1, that has only 1 and itself as factors.

The prime numbers are the numbers

2, 3, 5, 7, 11, 13, 17, 19, 23, 29, 31, . . .

The dots at the end of the list indicate that there are infinitely many prime numbers.

Note that the only even prime number is 2. Any other even whole number is divisible by 2, and thus has 2 as a factor, in addition to 1 and itself. Also note that not all odd whole numbers are prime numbers. For example, since 15 has factors of 1, 3, 5, and 15, it is not a prime number.

Composite numbers

The set of whole numbers contains many prime numbers. It also contains many numbers that are not prime.

Composite numbers

> The **composite numbers** are whole numbers, greater than 1, that are not prime.

The composite numbers are the numbers

4, 6, 8, 9, 10, 12, 14, 15, 16, 18, . . .

The three dots at the end of the list indicate that there are infinitely many composite numbers.

EXAMPLE 4   *Prime and composite numbers.*
a. Is 37 a prime number?   **b.** Is 45 a prime number?

Solution
a. Since 37 is a whole number greater than 1 and its only factors are 1 and 37, it is prime.
b. The factors of 45 are 1, 3, 5, 9, 15, and 45. Since there are factors other than 1 and 45, 45 is not prime. It is a composite number.

Self Check

a. Is 57 a prime number?

b. Is 39 a prime number?

Answers: **a.** no,   **b.** no   ■

 COMMENT   The numbers 0 and 1 are neither prime nor composite, because neither is a whole number greater than 1.

Finding prime factorizations with the tree method

Every composite number can be formed by multiplying a specific combination of prime numbers. The process of finding that combination is called **prime factorization.**

Prime factorization

> To find the **prime factorization** of a whole number means to write it as the product of only prime numbers.

Two methods can be used to find the prime factorization of a number. The first is called the **tree method.** We will use the tree method to find the prime factorization of 90 in two ways.

1. Factor 90 as 9 · 10.
2. Factor 9 and 10.
3. The process is complete when only prime numbers appear.

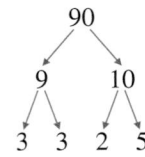

1. Factor 90 as 6 · 15.
2. Factor 6 and 15.
3. The process is complete when only prime numbers appear.

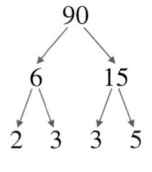

In either case, the prime factors are 2 · 3 · 3 · 5. Thus, the prime-factored form of 90 is 2 · 3 · 3 · 5. As we have seen, it does not matter how we factor 90. We will always get the same set of prime factors. No other combination of prime factors will multiply together and produce 90. This example illustrates an important fact about composite numbers.

Fundamental theorem of arithmetic

> Any composite number has exactly one set of prime factors.

EXAMPLE 5 *Factoring whole numbers with factor trees.* Use a factor tree to find the prime factorization of 210.

Solution

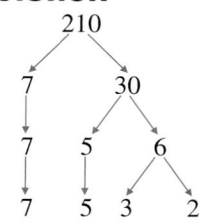

Factor 210 as $7 \cdot 30$.

Bring down the 7. Factor 30 as $5 \cdot 6$.

Bring down the 7 and the 5. Factor 6 as $3 \cdot 2$.

The prime factorization of 210 is $7 \cdot 5 \cdot 3 \cdot 2$. Writing the prime factors in order, from least to greatest, we have $210 = 2 \cdot 3 \cdot 5 \cdot 7$.

Self Check
Use a factor tree to find the prime factorization of 120.

Answer: $2 \cdot 2 \cdot 2 \cdot 3 \cdot 5$ ■

Exponents

In the Self Check of Example 5, we saw that the prime factorization of 120 is $2 \cdot 2 \cdot 2 \cdot 3 \cdot 5$. Because this factorization has three factors of 2, we call 2 a *repeated factor.* To express a repeated factor, we can use an **exponent.**

Exponent and base

An **exponent** is used to indicate repeated multiplication. It tells how many times the **base** is used as a factor.

The exponent is 3.

$$\underbrace{2 \cdot 2 \cdot 2} = 2^3$$

Repeated factors. The base is 2.

Read 2^3 as "2 to the third power" or "2 cubed."

The prime factorization of 120 can be written in a more compact form using exponents: $2 \cdot 2 \cdot 2 \cdot 3 \cdot 5 = 2^3 \cdot 3 \cdot 5$.

In the **exponential expression** a^n, a is the base, and n is the exponent. The expression is called a **power of a.**

EXAMPLE 6 *Exponents.* Use exponents to write each prime factorization: **a.** $5 \cdot 5 \cdot 5$, **b.** $7 \cdot 7 \cdot 11$, and **c.** $2(2)(2)(2)(3)(3)(3)$

Solution
a. $5 \cdot 5 \cdot 5 = 5^3$ 5 is used as a factor 3 times.
b. $7 \cdot 7 \cdot 11 = 7^2 \cdot 11$ 7 is used as a factor 2 times.
c. $2(2)(2)(2)(3)(3)(3) = 2^4(3^3)$ 2 is used as a factor 4 times, and 3 is used as a factor 3 times.

Self Check
Use exponents to write each prime factorization.

a. $3 \cdot 3 \cdot 7$
b. $5(5)(7)(7)$
c. $2 \cdot 2 \cdot 2 \cdot 3 \cdot 3 \cdot 5$

Answers: **a.** $3^2 \cdot 7$, **b.** $5^2(7^2)$, **c.** $2^3 \cdot 3^2 \cdot 5$ ■

EXAMPLE 7 *Exponential expressions.* Find the value of each expression.

a. $7^2 = 7 \cdot 7 = 49$ Read 7^2 as "7 to the second power" or "7 squared."
b. $2^5 = 2 \cdot 2 \cdot 2 \cdot 2 \cdot 2 = 32$ Read 2^5 as "2 to the fifth power."
c. $10^4 = 10 \cdot 10 \cdot 10 \cdot 10 = 10,000$ Read 10^4 as "10 to the fourth power."
d. $6^1 = 6$ Read 6^1 as "6 to the first power."

Self Check
Which of the numbers 3^5, 4^4, and 5^3 is the largest?

Answer: $4^4 = 256$ ■

EXAMPLE 8 *Evaluating exponential expressions.* The prime factorization of a number is $2^3 \cdot 3^4 \cdot 5$. What is the number?

Solution

To find the number, we find the value of each power and then do the multiplication.

$$2^3 \cdot 3^4 \cdot 5 = \mathbf{8 \cdot 81 \cdot 5} \quad 2^3 = 8 \text{ and } 3^4 = 81.$$
$$= 648 \cdot 5 \quad \text{Do the multiplications, working from left to right.}$$
$$= 3{,}240$$

The number is 3,240.

Self Check

The prime factorization of a number is $3^3 \cdot 4^2 \cdot 5^2$. What is the number?

Answer: 10,800 ■

 COMMENT Note that 5^3 means $5 \cdot 5 \cdot 5$. It does not mean $5 \cdot 3$. That is, $5^3 = 125$ and $5 \cdot 3 = 15$.

Accent on Technology: **Bacterial growth**

At the end of one hour, a culture contains two bacteria. Suppose the number of bacteria doubles every hour thereafter. Use exponents to determine how many bacteria the culture will contain after 24 hours.

We can use Table 1-2 to help model the situation. From the table, we see a pattern developing: The number of bacteria in the culture after 24 hours will be 2^{24}. We can evaluate this exponential expression using the exponential key $\boxed{y^x}$ on a scientific calculator ($\boxed{x^y}$ on some models).

To find the value of 2^{24}, we enter these numbers and press these keys.

Time	Number of bacteria
1 hr	$2 = 2^1$
2 hr	$4 = 2^2$
3 hr	$8 = 2^3$
4 hr	$16 = 2^4$
24 hr	$? = 2^{24}$

TABLE 1-2

Keystrokes   2 $\boxed{y^x}$ 24 $\boxed{=}$ $\boxed{16777216}$

Since $2^{24} = 16{,}777{,}216$, there will be 16,777,216 bacteria after 24 hours.

Finding prime factorizations with the division method

We can also find the prime factorization of a whole number by division. For example, to find the prime factorization of 363, we begin the division method by choosing the *smallest* prime number that will divide the given number exactly. We continue this "inverted division" process until the result of the division is a prime number.

Step 1: The prime number 2 doesn't divide 363 exactly, but 3 does. The result is 121, which is not prime. We continue the division process.

$$3\overline{)363} \\ \quad 121$$

Step 2: Next, we choose the smallest prime number that will divide 121. The primes 2, 3, 5, and 7 don't divide 121 exactly, but 11 does. The result is 11, which is prime. We are done.

$$3\overline{)363} \\ 11\overline{)121} \\ \quad 11$$

$$363 = 3 \cdot 11 \cdot 11$$

Using exponents, we can write the prime factorization of 363 as $3 \cdot 11^2$.

EXAMPLE 9   *Factoring with the division method.*   Use the division method to find the prime factorization of 100. Use exponents to express the result.

Solution

2 divides 100 exactly. The result is 50, which is not prime. ────────► $2\lfloor\underline{100}$

2 divides 50 exactly. The result is 25, which is not prime. ────────► $2\lfloor\underline{50}$

5 divides 25 exactly. The result is 5, which is prime. We are done. ────────► $5\lfloor\underline{25}$

$\qquad\qquad\qquad\qquad\qquad\qquad\qquad\qquad\qquad\qquad\qquad\qquad\qquad 5$

The prime factorization of 100 is $2^2 \cdot 5^2$.

Self Check

Use the division method to find the prime factorization of 108. Use exponents to express the result.

Answer: $2^2 \cdot 3^3$ ∎

STUDY SET   Section 1.4

VOCABULARY   *Fill in the blanks.*

1. Numbers that are multiplied together are called _____.

2. One number is _____ by another if the remainder is 0 when they are divided. When a division ends with a remainder of 0, we say that one of the numbers divides the other _____.

3. To _____ a whole number means to express it as the product of other whole numbers.

4. A _____ number is a whole number, greater than 1, that has only 1 and itself as factors.

5. Whole numbers, greater than 1, that are not prime numbers are called _____ numbers.

6. An _____ whole number is exactly divisible by 2. An _____ whole number is not exactly divisible by 2.

7. To prime factor a number means to write it as a product of only _____ numbers.

8. An _____ is used to represent repeated multiplication.

9. In the exponential expression 6^4, 6 is called the _____, and 4 is called the _____.

10. Another way to say "5 to the second power" is 5 _____. Another way to say "7 to the third power" is 7 _____.

CONCEPTS

11. Write 27 as the product of two factors.

12. Write 30 as the product of three factors.

13. The complete list of the factors of a whole number is given. What is the number?
 a. 2, 4, 22, 44, 11, 1
 b. 20, 1, 25, 100, 2, 4, 5, 50, 10

14. a. Find the factors of 24.
  b. Find the prime factorization of 24.

15. Find the factors of each number.
 a. 11
 b. 23
 c. 37
 d. From the results obtained in parts a–c, what can be said about 11, 23, and 37?

16. Suppose a number is divisible by 10. Is 10 a factor of the number?

17. If 4 is a factor of a whole number, will 4 divide the number exactly?

18. Give examples of whole numbers that have 11 as a factor.

In Exercises 19–22, the prime factorization of a whole number is given. Find the number.

19. $2 \cdot 3 \cdot 3 \cdot 5$ **20.** $3^3 \cdot 2$

21. $11^2 \cdot 5$ **22.** $2 \cdot 2 \cdot 2 \cdot 7$

23. Can we change the order of the base and the exponent in an exponential expression and obtain the same result? In other words, does $3^2 = 2^3$?

24. Find the prime factors of 30 and 165. What prime factors do they have in common?

25. Find the prime factors of 30 and 242. What prime factor do they have in common?

26. Find the prime factors of 20 and 35. What prime factor do they have in common?

27. Find the prime factors of 20 and 50. What prime factors do they have in common?

28. Find 1^2, 1^3, and 1^4. From the results, what can be said about any power of 1?

29. Finish the process of prime factoring 150. Compare the results.

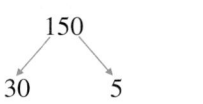
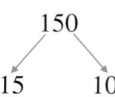

30. Find three whole numbers, less than 10, that would fit at the top of this tree diagram.

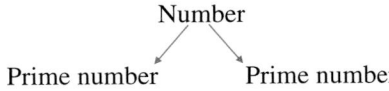

31. Complete the table

Product of factors of 12	Sum of factors of 12
1 · 12	
2 · 6	
3 · 4	

32. Consider 1, 4, 9, 16, 25, 36, 49, 64, 81, 100. Of the numbers listed, which is the *largest* factor of
a. 18 **b.** 24 **c.** 50

33. When using the division method to find the prime factorization of an even number, what is an obvious choice with which to start the division process?

34. When using the division method to find the prime factorization of a number ending in 5, what is an obvious choice with which to start the division process?

NOTATION *Write each expression without using exponents.*

35. 7^3 **36.** 8^4

37. 3^5 **38.** 4^6

39. $5^2(11)$ **40.** $2^3 \cdot 3^2$

41. 10^1 **42.** 2^1

Use exponents to write each expression in simpler form.

43. $2 \cdot 2 \cdot 2 \cdot 2 \cdot 2$ **44.** $3 \cdot 3 \cdot 3 \cdot 3 \cdot 3 \cdot 3$

45. $5 \cdot 5 \cdot 5 \cdot 5$ **46.** $9(9)(9)$

47. $4(4)(5)(5)$ **48.** $12 \cdot 12 \cdot 12 \cdot 16$

PRACTICE *Find the factors of each whole number.*

49. 10   **50.** 6

51. 40   **52.** 75

53. 18   **54.** 32

55. 44   **56.** 65

57. 77   **58.** 81

59. 100   **60.** 441

Write each number in prime-factored form.

61. 39   **62.** 20

63. 99   **64.** 105

65. 162   **66.** 400

67. 220   **68.** 126

69. 64   **70.** 243

71. 147   **72.** 98

Evaluate each exponential expression.

73. 3^4 **74.** 5^3

75. 2^5 **76.** 10^5

77. 12^2 **78.** 7^3

79. 8^4 **80.** 9^5

81. $3^2(2^3)$ **82.** $3^3(4^2)$

83. $2^3 \cdot 3^3 \cdot 4^2$ **84.** $3^2 \cdot 4^3 \cdot 5^2$

85. 🖩 234^3 **86.** 🖩 51^4

87. 🖩 $23^2 \cdot 13^3$ **88.** 🖩 $12^3 \cdot 15^2$

APPLICATIONS

89. PERFECT NUMBERS A whole number is called a **perfect number** when the sum of its factors that are less than the number equals the number. For example, 6 is a perfect number, because $1 + 2 + 3 = 6$. Find the factors of 28. Then use addition to show that 28 is also a perfect number.

90. CRYPTOGRAPHY Information is often transmitted in code. Many codes involve writing products of large primes, because they are difficult to factor. To see how difficult, try finding two prime factors of 7,663. (*Hint:* Both primes are greater than 70.)

91. LIGHT Illustration 1 shows that the light energy that passes through the first unit of area, 1 yard away from the bulb, spreads out as it travels away from the source. How much area does that energy cover 2 yards, 3 yards, and 4 yards from the bulb? Express each answer using exponents.

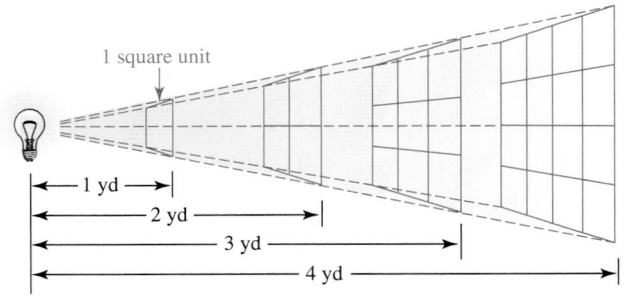

ILLUSTRATION 1

92. CELL DIVISION   After one hour, a cell has divided to form another cell. In another hour, these two cells have divided so that four cells exist. In another hour, these four cells divide so that eight exist.

a. How many cells exist at the end of the fourth hour?

b. The number of cells that exist after each division can be found using an exponential expression. What is the base?

c. Use a calculator to find the number of cells after 12 hours.

WRITING

93. Explain how to test a number to see whether it is prime.

94. Explain how to test a number to see whether it is even.

95. Explain the difference between the *factors* of a number and the *prime factorization* of the number.

96. Explain why it would be incorrect to say that the area of the square shown in Illustration 2 is 25^2 ft. How should we express its area?

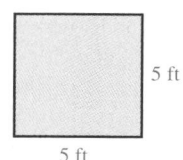

5 ft

5 ft

ILLUSTRATION 2

REVIEW

97. Round 230,999 to the nearest thousand.

98. Write the set of whole numbers.

99. What is $0 \div 15$?

100. Find $15 \cdot (6 \cdot 9)$.

101. What is the formula for the area of a rectangle?

102. MARCHING BANDS   When a university band lines up in eight rows of 15 musicians, there are five musicians left over. How many band members are there?

1.5 *Order of Operations*

In this section, you will learn about

- Order of operations • Evaluating expressions with no grouping symbols
- Evaluating expressions containing grouping symbols
- The arithmetic mean (average)

INTRODUCTION. Punctuation marks, such as commas, quotations, and periods, serve an important purpose when writing compositions. They determine the way in which sentences are to be read and interpreted. To read and interpret mathematical expressions correctly, we must use an agreed-upon set of priority rules for the *order of operations*.

Order of operations

Suppose you are asked to contact a friend if you see a certain type of watch for sale while you are traveling in Europe. While in Switzerland, you spot the watch and send the following E-mail message.

E-Mail
WATCH $4,500.   SHOULD I BUY IT FOR YOU?

The next day, you get this response from your friend.

E-Mail
NO PRICE TOO HIGH! REPEAT... NO! PRICE TOO HIGH.

Something is wrong. One statement says to buy the watch at any price. The other says not to buy it, because it's too expensive. The placement of the exclamation point makes us read these statements differently, resulting in different interpretations.

When reading a mathematical statement, the same kind of confusion is possible. For example, we consider

$$3 + 2 \cdot 5$$

This expression contains two operations: addition and multiplication. We can evaluate it (find its value) in two different ways. We can do the addition first and then do the multiplication. Or we can do the multiplication first and then do the addition. However, we get different results.

Method 1		**Method 2**	
$3 + 2 \cdot 5 = 5 \cdot 5$	Add first: $3 + 2 = 5$.	$3 + 2 \cdot 5 = 3 + 10$	Multiply first: $2 \cdot 5 = 10$.
$= 25$	Multiply 5 and 5.	$= 13$	Add 3 and 10.

———— Different results ————

If we don't establish an order of operations, the expression $3 + 2 \cdot 5$ will have two different answers. To avoid this possibility, we will evaluate expressions in the following order.

Order of operations

> 1. Do all calculations within parentheses and other grouping symbols following the order listed in Steps 2–4, working from the innermost pair to the outermost pair.
> 2. Evaluate all exponential expressions.
> 3. Do all multiplications and divisions as they occur from left to right.
> 4. Do all additions and subtractions as they occur from left to right.
>
> When all grouping symbols have been removed, repeat Steps 2–4 to complete the calculation.
>
> If a fraction bar is present, evaluate the expression above the bar (called the **numerator**) and the expression below the bar (called the **denominator**) separately. Then do the division indicated by the fraction bar, if possible.

To evaluate $3 + 2 \cdot 5$ correctly, we must apply the rules for the order of operations. Since there are no grouping symbols and there are no powers, we perform the multiplication first and then do the addition.

Ignore the addition for now   and do the multiplication first: $2 \cdot 5 = 10$.

$$3 + 2 \cdot 5 = 3 + 10$$
$$= 13 \qquad \text{Now do the addition.}$$

Using the rules for the order of operations, we see that the correct answer is 13.

Evaluating expressions with no grouping symbols

EXAMPLE 1   *Order of operations.*   Evaluate $2 \cdot 4^2 - 8$.

Solution

Since the expression does not contain any grouping symbols, we begin with Step 2 of the rules for the order of operations.

$$2 \cdot 4^2 - 8 = 2 \cdot 16 - 8 \qquad \text{Evaluate the exponential expression: } 4^2 = 16.$$
$$= 32 - 8 \qquad \text{Do the multiplication: } 2 \cdot 16 = 32.$$
$$= 24 \qquad \text{Do the subtraction.}$$

Self Check
Evaluate $4 \cdot 3^3 - 6$.

Answer: 102

EXAMPLE 2   *Order of operations.*   Evaluate $8 - 3 \cdot 2 + 16$.

Solution

Since the expression does not contain grouping symbols and since there are no powers to find, we look for multiplications or divisions to perform.

$$8 - 3 \cdot 2 + 16 = 8 - 6 + 16 \qquad \text{Do the multiplication: } 3 \cdot 2 = 6.$$
$$= 2 + 16 \qquad \text{Working from left to right, do the subtraction:}$$
$$8 - 6 = 2.$$
$$= 18 \qquad \text{Do the addition.}$$

Self Check

Evaluate $10 - 2 \cdot 3 + 24$.

Answer: 28

   COMMENT   Some students incorrectly think that additions are always done before subtractions. As Example 2 shows, this is not true. Working from left to right, we do the additions or subtractions *in the order in which they occur*. The same is true for multiplications and divisions.

EXAMPLE 3   *Order of operations.*   Evaluate $192 \div 6 - 5(3)2$.

Solution

Although this expression contains parentheses, there are no calculations to perform within them. Since there are no powers, we do multiplications and divisions as they are encountered from left to right.

$$192 \div 6 - 5(3)2 = 32 - 5(3)2 \qquad \text{Working from left to right, do the division:}$$
$$192 \div 6 = 32.$$
$$= 32 - 15(2) \qquad \text{Working from left to right, do the multiplication:}$$
$$5(3) = 15.$$
$$= 32 - 30 \qquad \text{Do the multiplication: } 15(2) = 30.$$
$$= 2 \qquad \text{Do the subtraction.}$$

Self Check

Evaluate $36 \div 9 + 4(2)3$.

Answer: 28

EXAMPLE 4   *Phone bill.*   Figure 1–15 shows the rates for international telephone calls charged by a 10-10 long-distance company. A businesswoman calls Germany for 20 minutes, South Korea for 5 minutes, and Mexico City for 35 minutes. What is the total cost of the calls?

All rates are per minute.	
Canada	10¢
Germany	23¢
Jamaica	68¢
Mexico City	42¢
South Korea	29¢

FIGURE 1–15

Solution

We can find the cost of a call (in cents) by multiplying the rate charged per minute by the length of the call (in minutes). To find the total cost, we add the costs of the three calls.

The cost of the call to Germany.   The cost of the call to South Korea.   The cost of the call to Mexico City.

$$23(20) \qquad + \qquad 29(5) \qquad + \qquad 42(35)$$

To evaluate this expression, we apply the rules for the order of operations.

$$23(20) + 29(5) + 42(35) = 460 + 145 + 1{,}470 \qquad \text{Do the multiplications.}$$
$$= 2{,}075 \qquad \text{Do the additions.}$$

The total cost of the calls is 2,075 cents, or \$20.75.

Evaluating expressions containing grouping symbols

Grouping symbols serve as mathematical punctuation marks. They help determine the order in which an expression is to be evaluated. Examples of grouping symbols are parentheses (   ), brackets [   ], and the fraction bar ——.

In the next example, we have two similar-looking expressions. However, because of the parentheses, we evaluate them in a different order.

EXAMPLE 5   *Working with grouping symbols.*   Evaluate each expression: a. $12 - 3 + 5$ **and   b.** $12 - (3 + 5)$.

Solution
a. We perform the additions and subtractions as they occur, from left to right.

$$12 - 3 + 5 = \mathbf{9} + 5 \qquad \text{Do the subtraction: } 12 - 3 = 9.$$
$$= 14 \qquad \text{Do the addition.}$$

b. This expression contains parentheses. We must do the calculation within the parentheses first.

$$12 - (3 + 5) = 12 - \mathbf{8} \qquad \text{Do the addition: } 3 + 5 = 8.$$
$$= 4 \qquad \text{Do the subtraction.}$$

Self Check
Evaluate each expression.

a. $20 - 7 + 6$

b. $20 - (7 + 6)$

Answers: a. 19,   **b.** 7   ■

EXAMPLE 6   *Working with grouping symbols.*   Evaluate $(2 + 6)^3$.
Solution
We begin by doing the calculation within the parentheses.

$$(2 + 6)^3 = \mathbf{8}^3 \qquad \text{Do the addition.}$$
$$= 512 \qquad \text{Evaluate the exponential expression: } 8^3 = 8 \cdot 8 \cdot 8 = 512.$$

Self Check
Evaluate $(1 + 2)^4$.

Answer: 81   ■

EXAMPLE 7   *Order of operations within grouping symbols.*
Evaluate $5 + 2(13 - 5 \cdot 2)$.
Solution
This expression contains grouping symbols. We will apply the rules for the order of operations within the parentheses first, to evaluate $13 - 5 \cdot 2$.

$$5 + 2(13 - 5 \cdot 2) = 5 + 2(13 - \mathbf{10}) \qquad \text{Do the multiplication within the parentheses.}$$
$$= 5 + 2(\mathbf{3}) \qquad \text{Do the subtraction within the parentheses.}$$
$$= 5 + \mathbf{6} \qquad \text{Do the multiplication: } 2(3) = 6.$$
$$= 11 \qquad \text{Do the addition.}$$

Self Check
Evaluate $25 - 2(12 - 5 \cdot 2)$.

Answer: 21   ■

Sometimes an expression contains two or more sets of grouping symbols. Since it can be confusing to read an expression such as $16 + 2(14 - 3(5 - 2))$, we often use brackets in place of the second pair of parentheses.

$$16 + 2[14 - 3(5 - 2)]$$

If an expression contains more than one pair of grouping symbols, we always begin by working within the innermost pair and then work to the outermost pair.

Innermost parentheses
↓   ↓
$$16 + 2[14 - 3(5 - 2)]$$
↑   ↑
Outermost brackets

EXAMPLE 8 *Grouping symbols within grouping symbols.* Evaluate $16 + 2[14 - 3(5 - 2)]$.

Solution

$$16 + 2[14 - 3(5 - 2)] = 16 + 2[14 - 3(3)]$$ Do the subtraction within the parentheses.

$$= 16 + 2(14 - 9)$$ Do the multiplication within the brackets. Since only 1 set of grouping symbols is now needed, write $14 - 9$ within parentheses.

$$= 16 + 2(5)$$ Do the subtraction within the parentheses.

$$= 16 + 10$$ Do the multiplication: $2(5) = 10$.

$$= 26$$ Do the addition.

Self Check

Evaluate $46 - 2[4 + 3(6 - 2)]$.

Answer: 14 ■

EXAMPLE 9 *Working with a fraction bar.* Evaluate $\dfrac{2(13) - 2}{3(2^3)}$.

Solution

A fraction bar is a grouping symbol. We evaluate the numerator and denominator separately and then do the indicated division.

$$\frac{2(13) - 2}{3(2^3)} = \frac{26 - 2}{3(8)}$$ In the numerator, do the multiplication.
In the denominator, do the calculation within the parentheses.

$$= \frac{24}{24}$$ In the numerator, do the subtraction.
In the denominator, do the multiplication.

$$= 1$$ Do the division.

Self Check

Evaluate $\dfrac{3(14) - 6}{2(3^2)}$.

Answer: 2 ■

The arithmetic mean (average)

The **arithmetic mean,** or **average,** of several numbers is a value around which the numbers are grouped. It gives you an indication of the "center" of the set of numbers. When finding the mean of a set of numbers, we usually need to apply the rules for the order of operations.

Finding an arithmetic mean | To find the mean of a set of scores, divide the sum of the scores by the number of scores.

EXAMPLE 10 *NCAA basketball.* In 1998, the Lady Vols of the University of Tennessee won the women's basketball championship, capping a perfect 39-0 season. Find their average margin of victory in their last four tournament games shown below.

Regional	**Regional final**	**Semifinal**	**Championship**
beat Rutgers by 32 points	beat North Carolina by 6 points	beat Arkansas by 28 points	beat Louisiana Tech by 18 points

Solution

To find the average margin of victory, add the margins of victory and divide by 4.

$$\text{Average} = \frac{32 + 6 + 28 + 18}{4}$$

$$= \frac{84}{4}$$

$$= 21$$

Their average margin of victory was 21 points.

Self Check

The University of Kentucky won the 1998 NCAA men's basketball championship. Find their average margin of victory in their last five tournament games, which they won by 27, 26, 2, 1, and 9 points.

Answer: 13 points ■

Accent on Technology: **Order of operations and parentheses**

Scientific calculators have the rules for order of operations built in. Even so, some evaluations require the use of a left parenthesis key $($ and a right parenthesis key $)$. For example, to evaluate $\frac{240}{20-15}$, we enter these numbers and press these keys.

Keystrokes   240 $\div$ $($ 20 $-$ 15 $)$ $=$ | 48 |

STUDY SET   Section 1.5

VOCABULARY   *Fill in the blanks.*

1. The grouping symbols ( ) are called
_____, and the symbols [ ] are called
_____.

2. The expression above a fraction bar is called the
_____. The expression below a fraction bar
is called the _____.

3. To _____ $2 + 5 \cdot 4$ means to find its value.

4. To find the _____ of several values, we add
the values and divide by the number of values.

CONCEPTS

5. Consider $5(2)^2 - 1$. How many operations need to be
performed to evaluate the expression? List them in the
order in which they should be performed.

6. Consider $15 - 3 + (5 \cdot 2)^3$. How many operations
need to be performed to evaluate this expression? List
them in the order in which they should be
performed.

7. Consider $\frac{5 + 5(7)}{2 + (8 - 4)}$. In the numerator, what operation
should be done first? In the denominator, what opera-
tion should be done first?

8. In the expression $\frac{3 - 5(2)}{5(2) + 4}$, the bar is a grouping symbol.
What does it separate?

9. Explain the difference between $2 \cdot 3^2$ and $(2 \cdot 3)^2$.

10. Use brackets to write $2(12 - (5 + 4))$ in better
form.

NOTATION   *Complete each solution.*

11. $28 - 5(2)^2 = 28 - 5(\ \ )$
$\qquad\qquad = 28 - \rule{1cm}{0.4pt}$
$\qquad\qquad = 8$

12. $2 + (5 + 6 \cdot 2) = 2 + (5 + \rule{0.8cm}{0.4pt}\ )$
$\qquad\qquad\qquad = 2 + \rule{1cm}{0.4pt}$
$\qquad\qquad\qquad = 19$

13. $[4(2 + 7)] - 6 = [4(\ \ )] - 6$
$\qquad\qquad\qquad = \rule{1cm}{0.4pt} - 6$
$\qquad\qquad\qquad = 30$

14. $\dfrac{5(3) + 12}{9 - 6} = \dfrac{\rule{0.8cm}{0.4pt} + 12}{\rule{1cm}{0.4pt}}$

$\qquad\qquad = \dfrac{\rule{1cm}{0.4pt}}{\rule{1cm}{0.4pt}}$

$\qquad\qquad = 9$

PRACTICE   *Evaluate each expression.*

15. $7 + 4 \cdot 5$ **16.** $10 - 2 \cdot 2$

17. $2 + 3(0)$ **18.** $5(0) + 8$

19. $20 - 10 + 5$ **20.** $80 - 5 + 4$

21. $25 \div 5 \cdot 5$ **22.** $6 \div 2 \cdot 3$

23. $7(5) - 5(6)$ **24.** $4 \cdot 2 + 2 \cdot 4$

25. $4^2 + 3^2$ **26.** $12^2 - 5^2$

27. $2 \cdot 3^2$ **28.** $3^3 \cdot 5$

29. $3 + 2 \cdot 3^4 \cdot 5$ **30.** $3 \cdot 2^3 \cdot 4 - 12$

31. $5 \cdot 10^3 + 2 \cdot 10^2 + 3 \cdot 10^1 + 9$

32. $8 \cdot 10^3 + 0 \cdot 10^2 + 7 \cdot 10^1 + 4$

33. $3(2)^2 - 4(2) + 12$

34. $5(1)^3 + (1)^2 + 2(1) - 6$

35. $(8 - 6)^2 + (4 - 3)^2$

36. $(2 + 1)^2 + (3 + 2)^2$

37. $60 - (6 + \dfrac{40}{8})$ **38.** $7 + (5^3 - \dfrac{200}{2})$

39. $6 + 2(5 + 4)$ **40.** $3(5 + 1) + 7$

41. $3 + 5(6 - 4)$ **42.** $7(9 - 2) - 1$

43. $(7 - 4)^2 + 1$ **44.** $(9 - 5)^3 + 8$

45. $6^3 - (10 + 8)$ **46.** $5^2 - (9 + 3)$

47. $50 - 2(4)^2$ **48.** $30 + 2(3)^3$

49. $16^2 - 4(2)(5)$ **50.** $8^2 - 4(3)(1)$

51. $39 - 5(6) + 9 - 1$ **52.** $15 - 3(2) - 4 + 3$

53. $(18 - 12)^3 - 5^2$ **54.** $(9 - 2)^2 - 3^3$

55. $2(10 - 3^2) + 1$ **56.** $1 + 3(18 - 4^2)$

57. $6 + \dfrac{25}{5} + 6(3)$ **58.** $15 - \dfrac{24}{6} + 8 \cdot 2$

59. $3\left(\dfrac{18}{3}\right) - 2(2)$ **60.** $2\left(\dfrac{12}{3}\right) + 3(5)$

61. $(2 \cdot 6 - 4)^2$ **62.** $2(6 - 4)^2$

63. $4[50 - (3^3 - 5^2)]$ **64.** $6[15 + (5 \cdot 2^2)]$

65. $80 - 2[12 - (5 + 4)]$ **66.** $15 + 5[12 - (2^2 + 4)]$

67. $2[100 - (5 + 4)] - 45$ **68.** $8[6(6) - 6^2] + 4(5)$

69. $\dfrac{10 + 5}{6 - 1}$ **70.** $\dfrac{18 + 12}{2(3)}$

71. $\dfrac{5^2 + 17}{6 - 2^2}$ **72.** $\dfrac{3^2 - 2^2}{(3 - 2)^2}$

73. $\dfrac{(3 + 5)^2 + 2}{2(8 - 5)}$ **74.** $\dfrac{25 - (2 \cdot 3 - 1)}{2 \cdot 9 - 8}$

75. $\dfrac{(5 - 3)^2 + 2}{4^2 - (8 + 2)}$ **76.** $\dfrac{(4^3 - 2) + 7}{5(2 + 4) - 7}$

77. $12{,}985 - (1{,}800 + 689)$ **78.** $\dfrac{897 - 655}{88 - 77}$

79. $3{,}245 - 25(16 - 12)^2$ **80.** $\dfrac{24^2 - 4^2}{22 + 58}$

APPLICATIONS   *In Exercises 81–86, write an expression to solve each problem. Then evaluate the expression.*

81. BUYING GROCERIES   At the supermarket, Carlos has 2 cases of soda, 4 bags of potato chips, and 2 cans of dip in his cart. Each case of soda costs $6, each bag of chips costs $2, and each can of dip costs $1. Find the total cost of the groceries.

82. JUDGING   The scores received by a junior diver are as follows:

5	2	4	6	3	4

The formula for computing the overall score for the dive is as follows:

1. Throw out the lowest score.

2. Throw out the highest score.

3. Divide the sum of the remaining scores by 4.

Find the diver's score.

83. BANKING   When a customer deposits cash, a teller must complete a "currency count" on the back of the deposit slip. In Illustration 1, what is the total amount of cash being deposited?

24			
—			
6			
10			
12			
2			
1			

ILLUSTRATION 1

84. WRAPPING GIFTS   How much ribbon is needed to wrap the package shown in Illustration 2 if 15 inches of ribbon are needed to make the bow?

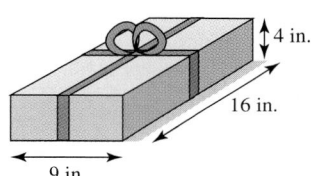

ILLUSTRATION 2

85. SCRABBLE   Illustration 3(a) shows part of the game board before and Illustration 3(b) shows it after the words *brick* and *aphid* were played. Determine the scoring for each word. (The number on each tile gives the point value of the letter.)

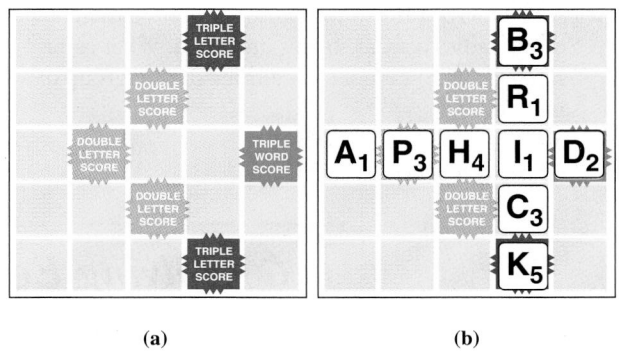

(a)   (b)

ILLUSTRATION 3

86. GETTYSBURG ADDRESS   Here is an excerpt from Abraham Lincoln's Gettysburg Address.

Fourscore and seven years ago, our fathers brought forth on this continent a new nation, conceived in liberty, and dedicated to the proposition that all men are created equal.

Lincoln's comments refer to the year 1776, when the United States declared its independence. If a score is 20 years, in what year did Lincoln deliver the Gettysburg Address?

87. CLIMATE   One December week, the temperatures in Honolulu, Hawaii were 75°, 80°, 83°, 80°, 77°, 72°, and 86°. Find the week's average (mean) temperature.

88. GRADES   In a psychology class, a student had test scores of 94, 85, 81, 77, and 89. He also overslept, missed the final exam, and received a 0 on it. What was his test average in the class?

89. NATURAL NUMBERS   What is the average (mean) of the first nine natural numbers?

90. ENERGY USAGE   See Illustration 4. Find the average number of therms of natural gas used per month. Then draw a dashed line across the graph showing the average.

| Acct 45-009 | 2001 Energy Audit | Tri-City Gas Co. |
| Janice C. Milton | 23 N. State St. Apt. B | Salem, OR |

ILLUSTRATION 4

91. FAST FOOD   Illustration 5 shows the sandwiches Subway advertises as its low-fat menu. What is the average (mean) number of calories for the group of sandwiches?

6-inch subs	Calories	Fat (g)
Veggie Delite	237	3
Turkey Breast	289	4
Turkey Breast & Ham	295	5
Ham	302	5
Roast Beef	303	5
Subway Club	312	5
Roasted Chicken Breast	348	6

ILLUSTRATION 5

92. TV RATINGS   The list below shows the number of people watching *Who Wants to Be a Millionaire?* on five weeknights in November of 1999. How large was the average audience?

Monday	26,800,000
Tuesday	24,900,000
Wednesday	22,900,000
Thursday	25,900,000
Friday	21,900,000

WRITING

93. Explain why rules for the order of operations are necessary.

94. Explain the difference between the steps used to evaluate $5 \cdot 2^3$ and $(5 \cdot 2)^3$.

95. Explain the process of finding the mean of a large group of numbers. What does an average tell you?

96. What does it mean when we say to do all additions and subtractions *as they occur from left to right*?

REVIEW   *Do the operations.*

97.
$$\begin{array}{r} 4,029 \\ +3,271 \\ \hline \end{array}$$

98.
$$\begin{array}{r} 4,263 \\ -3,764 \\ \hline \end{array}$$

99.
$$\begin{array}{r} 417 \\ \times\ 23 \\ \hline \end{array}$$

100. $82\overline{)50,430}$

1.6 *Solving Equations by Addition and Subtraction*

In this section, you will learn about

• Equations • Checking solutions • Solving equations
• Problem solving with equations

INTRODUCTION. The language of mathematics is *algebra*. The word *algebra* comes from the title of a book written by the Arabian mathematician Al-Khowarazmi around

A.D. 800. Its title, *Ihm aljabr wa'l muqabalah,* means restoration and reduction, a process then used to solve equations. In this section, we will begin discussing equations, one of the most powerful ideas in algebra.

Equations

An **equation** is a statement indicating that two expressions are equal. Some examples of equations are

$$x + 5 = 21, \qquad 16 + 5 = 21, \qquad \text{and} \qquad 10 + 5 = 21$$

Equations | **Equations** are mathematical sentences that contain an $=$ sign.

In the equation $x + 5 = 21$, the expression $x + 5$ is called the **left-hand side,** and 21 is called the **right-hand side.** The letter x is the **variable** (or the **unknown**).

An equation can be true or false. For example, $16 + 5 = 21$ is a true equation, whereas $10 + 5 = 21$ is a false equation. An equation containing a variable can be true or false, depending upon the value of the variable. If $x = 16$, the equation $x + 5 = 21$ is true, because

$$16 + 5 = 21 \qquad \text{Substitute 16 for } x.$$

However, this equation is false for all other values of x.

Any number that makes an equation true when substituted for its variable is said to *satisfy* the equation. Such numbers are called **solutions** or **roots.** Because 16 is the only number that satisfies $x + 5 = 21$, it is the only solution of the equation.

Checking solutions

EXAMPLE 1   *Checking a solution.* Verify that 18 is a solution of the equation $x - 3 = 15$.

Solution

We substitute 18 for x in the equation and verify that both sides of the equation are equal.

$$x - 3 = 15 \qquad \text{The given equation.}$$
$$18 - 3 \stackrel{?}{=} 15 \qquad \text{Substitute 18 for } x. \text{ Read } \stackrel{?}{=} \text{ as "is possibly equal to."}$$
$$15 = 15 \qquad \text{Do the subtraction.}$$

Since $15 = 15$ is a true equation, 18 is a solution of $x - 3 = 15$.

Self Check

Is 8 a solution of $x + 17 = 25$?

Answer: yes   ■

EXAMPLE 2   *Checking a solution.* Is 23 a solution of $32 = y + 10$?

Solution

We substitute 23 for y and simplify.

$$32 = y + 10 \qquad \text{The given equation.}$$
$$32 \stackrel{?}{=} 23 + 10 \qquad \text{Substitute 23 for } y.$$
$$32 \neq 33 \qquad \text{Do the addition.}$$

Since the left-hand and right-hand sides are not equal, 23 is not a solution.

Self Check

Is 5 a solution of $20 = y + 17$?

Answer: no   ■

Solving equations

Since the solution of an equation is usually not given, we must develop a process to find it. This process is called *solving the equation.* To develop an understanding of the properties and procedures used to solve an equation, we will examine $x + 2 = 5$ and make some observations as we solve it in a practical way.

We can think of the scale shown in Figure 1-16(a) as representing the equation $x + 2 = 5$. The weight (in grams) on the left-hand side of the scale is $x + 2$, and the weight (in grams) on the right-hand side is 5. Because these weights are equal, the scale is in balance. To find x, we need to isolate it. That can be accomplished by removing 2 grams from the left-hand side of the scale. Common sense tells us that we must also remove 2 grams from the right-hand side if the scale is to remain in balance. In Figure 1-16(b), we can see that x grams will be balanced by 3 grams. We say that we have *solved* the equation and that the *solution* is 3.

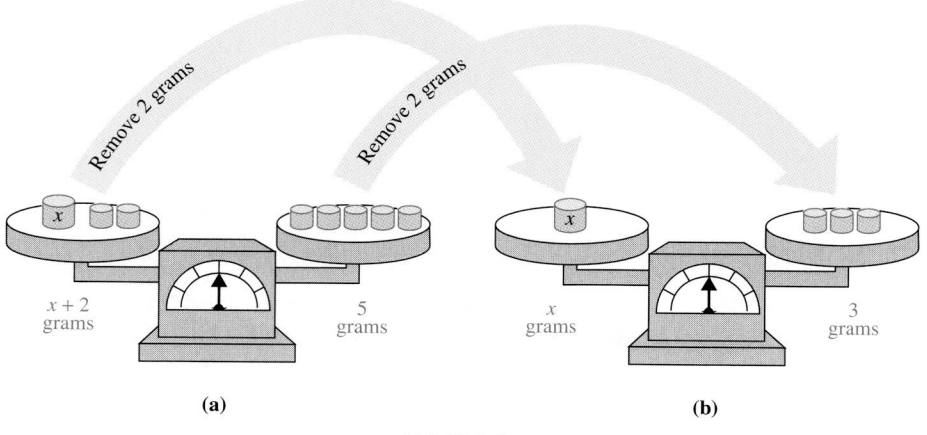

(a)   **(b)**

FIGURE 1-16

From this example, we can make some observations about solving an equation.

- To find the value of x, we needed to isolate it on the left-hand side of the scale.
- To isolate x, we had to undo the addition of 2 grams. This was accomplished by subtracting 2 grams from the left-hand side.
- We wanted the scale to remain in balance. When we subtracted 2 grams from the left-hand side, we subtracted the same amount from the right-hand side.

The observations suggest a property of equality: *If the same quantity is subtracted from equal quantities, the results will be equal quantities.* We can express this property in symbols.

Subtraction property of equality

> Let a, b, and c represent numbers.
>   If $a = b$, then $a - c = b - c$.

When we use this property, the resulting equation will be equivalent to the original equation.

Equivalent equations

> Two equations are **equivalent equations** when they have the same solutions.

In the previous example, we found that $x + 2 = 5$ is equivalent to $x = 3$. This is true because these equations have the same solution, $x = 3$.

We now show how to solve $x + 2 = 5$ using an algebraic approach.

EXAMPLE 3 *Solving an equation.* Solve $x + 2 = 5$.

Solution

To isolate x on the left-hand side of the equation, we undo the addition of 2 by subtracting 2 from both sides of the equation.

$$x + 2 = 5 \qquad \text{The equation to solve.}$$
$$x + 2 - 2 = 5 - 2 \qquad \text{Subtract 2 from both sides.}$$
$$x = 3 \qquad \text{On the left-hand side, subtracting 2 undoes the addition of}$$
$$\text{2 and leaves } x. \text{ On the right-hand side, } 5 - 2 = 3.$$

We check by substituting 3 for x in the original equation and simplifying. If 3 is the solution, we will obtain a true statement.

$$x + 2 = 5$$
$$3 + 2 \overset{?}{=} 5 \qquad \text{Substitute 3 for } x.$$
$$5 = 5 \qquad \text{Do the addition.}$$

Since the resulting equation is true, 3 is a solution.

Self Check

Solve $x + 7 = 14$ and check the result.

Answer: 7 ∎

A second property that we will use to solve equations involves addition. It is based on the following idea: *If the same quantity is added to equal quantities, the results will be equal quantities.* In symbols, we have the following property.

Addition property of equality

Let a, b, and c represent numbers.

If $a = b$, then $a + c = b + c$.

We can think of the scale shown in Figure 1-17(a) as representing the equation $x - 2 = 3$. To find x, we need to use the addition property of equality and add 2 grams of weight to each side. The scale will remain in balance. From the scale in Figure 1-17(b), we can see that x grams will be balanced by 5 grams. The solution of $x - 2 = 3$ is therefore $x = 5$.

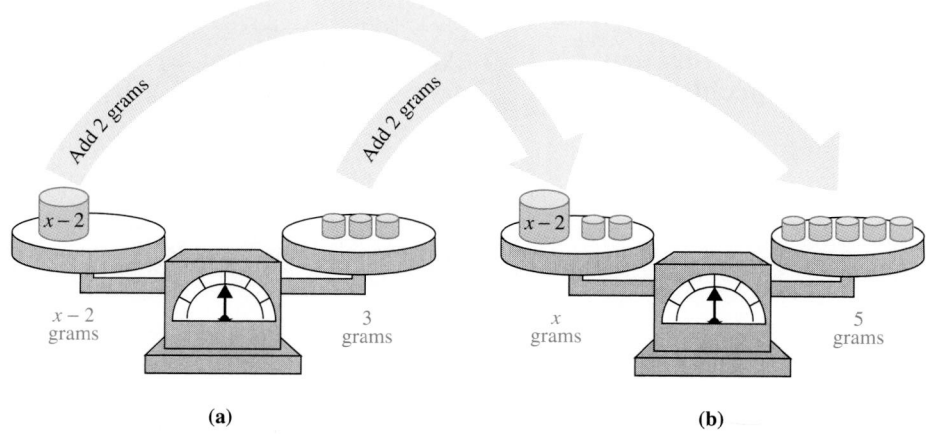

FIGURE 1-17

To solve $x - 2 = 3$ algebraically, we apply the addition property of equality. We can isolate x on the left-hand side of the equation by adding 2 to both sides.

$$x - 2 = 3$$
$$x - 2 + 2 = 3 + 2 \qquad \text{To undo the subtraction of 2, add 2 to both sides.}$$
$$x = 5 \qquad \text{On the left-hand side, adding 2 undoes the subtraction of 2 and}$$
$$\text{leaves } x. \text{ On the right-hand side, } 3 + 2 = 5.$$

To check this result, we substitute 5 for *x* in the original equation and simplify.

$$x - 2 = 3$$
$$5 - 2 \stackrel{?}{=} 3 \quad \text{Substitute 5 for } x.$$
$$3 = 3 \quad \text{Do the subtraction.}$$

Since this is a true statement, $x = 5$ is a solution.

EXAMPLE 4   *Isolating the variable on the right-hand side.*   Solve $19 = y - 7$ and check the result.

Solution

To isolate the variable *y* on the right-hand side, we use the addition property of equality. We can undo the subtraction of 7 by adding 7 to both sides.

$$19 = y - 7$$
$$19 + 7 = y - 7 + 7 \quad \text{Add 7 to both sides.}$$
$$26 = y \qquad \text{On the left-hand side, } 19 + 7 = 26. \text{ On the right-hand side,}$$
$$\text{adding 7 undoes the subtraction of 7 and leaves } y.$$
$$y = 26 \qquad \text{When stating a solution, it is common practice to write the}$$
$$\text{variable first. If } 26 = y, \text{ then } y = 26.$$

We check by substituting 26 for *y* in the original equation and simplifying.

$$19 = y - 7 \quad \text{The original equation.}$$
$$19 \stackrel{?}{=} 26 - 7 \quad \text{Substitute 26 for } y.$$
$$19 = 19 \quad \text{Do the subtraction.}$$

Since this is a true statement, 26 is a solution.

Self Check

Solve $75 = b - 38$ and check the result.

Answer: 113

Problem solving with equations

The key to problem solving is to understand the problem and then to devise a plan for solving it. The following list of steps provides a good strategy to follow.

Strategy for problem solving

> 1. **Analyze the problem** by reading it carefully to understand the given facts. What information is given? What vocabulary is given? What are you asked to find? Often a diagram will help you visualize the facts of a problem.
> 2. **Form an equation** by picking a variable to represent the quantity to be found. Then express all other unknown quantities as expressions involving that variable. Finally, write an equation expressing a quantity in two different ways.
> 3. **Solve the equation.**
> 4. **State the conclusion.**
> 5. **Check the result** in the words of the problem.

We will now use this five-step strategy to solve problems. The purpose of the following examples is to help you learn the strategy, even though you can probably solve these examples without it.

EXAMPLE 5   *Financial data.*   Figure 1-18 shows the 1999 quarterly net income for Nike, the athletic shoe company. What was the company's total net income for 1999?

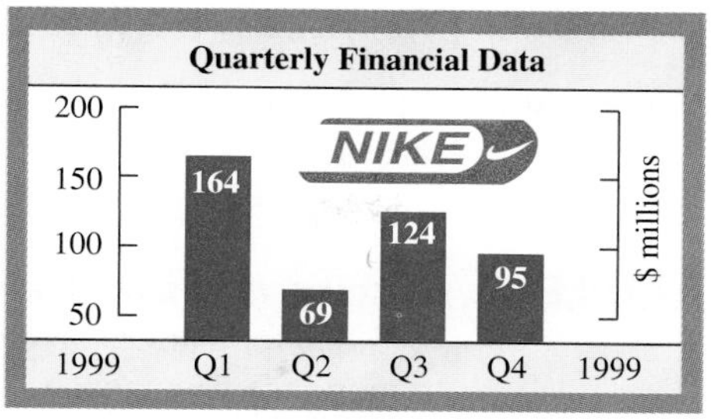

FIGURE 1-18

Analyze the problem
- We are given the net income for each quarter.
- We are asked to find the total net income.

Form an equation   Let n = the total net income for 1999. To form an equation involving n, we look for a key word or phrase in the problem.

Key word: *total*   **Translation:** *addition*

We can express the total income in two ways.

Total net income	is	1st qtr. net income	plus	2nd qtr. net income	plus	3rd qtr. net income	plus	4th qtr. net income.
n	$=$	164	$+$	69	$+$	124	$+$	95

Solve the equation
$$n = 164 + 69 + 124 + 95 \quad \text{We are working in millions of dollars.}$$
$$n = 452 \quad \text{Do the additions.}$$

State the conclusion   Nike's total net income was $452 million.

Check the result   We can check the result by estimation. To estimate, we round the net income from each quarter and add.

$$160 + 70 + 120 + 100 = 450$$

The answer, 452, is reasonable.   ■

EXAMPLE 6   *Small business.*   Last year a hairdresser lost 17 customers who moved away. If he now has 73 customers, how many did he have originally?

Analyze the problem
- We know that he started with some unknown number of customers, and after 17 moved away, 73 were left.
- We are asked to find the number of customers he had before any moved away.

Form an equation   We can let c = the original number of customers.
To form an equation involving c, we look for a key word or phrase in the problem.

Key phrase: *moved away*   **Translation:** *subtract*

We can express the remaining number of customers in two ways.

The original number of customers	minus	17	is	the remaining number of customers.
c	$-$	17	$=$	73

Solve the equation
$$c - 17 = 73$$
$$c - 17 + \mathbf{17} = 73 + \mathbf{17} \quad \text{To undo the subtraction of 17, add 17 to both sides.}$$
$$c = 90 \quad \text{Simplify each side of the equation.}$$

State the conclusion	He originally had 90 customers.
Check the result	The hairdresser had 90 customers. After losing 17, he has $90 - 17$, or 73 left. The answer, 90, checks. ■

  EXAMPLE 7   ***Buying a house.***   Sue wants to buy a house that costs \$87,000. Since she has only \$15,000 for a down payment, she will have to borrow some additional money by taking a mortgage. How much will she have to borrow?

Analyze the problem
- The house costs \$87,000.
- Sue has \$15,000 for a down payment.
- We must find how much money she needs to borrow.

Form an equation   We can let $x = $ the money that she needs to borrow.
To form an equation involving x, we look for a key word or phrase in the problem.

Key phrase: *borrow some additional money*   **Translation:** *addition*

We can express the total cost of the house in two ways.

The amount Sue has	plus	the amount she borrows	is	the total cost of the house.
15,000	+	x	=	87,000

Solve the equation

$$15,000 + x = 87,000$$
$$15,000 + x - \mathbf{15,000} = 87,000 - \mathbf{15,000}$$ To undo the addition of 15,000, subtract 15,000 from both sides.
$$x = 72,000$$ Do the subtractions.

State the conclusion   She must borrow \$72,000.

Check the result   With a \$72,000 mortgage, she will have \$15,000 + \$72,000, which is the \$87,000 that is necessary to buy the house. The answer, 72,000, checks. ■

STUDY SET Section 1.6

VOCABULARY *Fill in the blanks.*

1. An equation is a statement that two expressions are _____. An equation contains an _____ sign.

2. A _____ of an equation is a number that satisfies the equation.

3. The answer to an equation is called a _____ or a _____.

4. A letter that is used to represent a number is called a _____.

5. _____ equations have exactly the same solutions.

6. To solve an equation, we _____ the variable on one side of the equals sign.

CONCEPTS *In Exercises 7–8, complete the properties of equality.*

7. If $x = y$ and c is any number, then $x + c = $ _____.

8. If $x = y$ and c is any number, then $x - c = $ _____.

9. In $x + 6 = 10$, what operation is performed on the variable? How do we undo that operation to isolate the variable?

10. In $9 = y - 5$, what operation is performed on the variable? How do we undo that operation to isolate the variable?

NOTATION *Complete each solution to solve the given equation.*

11.
$$x + 8 = 24$$
$$x + 8 - \boxed{} = 24 - \boxed{}$$
$$x = 16$$

Check: $x + 8 = 24$
$$\boxed{} + 8 \stackrel{?}{=} 24$$
$$\boxed{} = 24$$

So $\boxed{}$ is a solution.

12.
$$x - 8 = 24$$
$$x - 8 + \boxed{} = 24 + \boxed{}$$
$$x = 32$$

Check: $x - 8 = 24$
$$\boxed{} - 8 \stackrel{?}{=} 24$$
$$\boxed{} = 24$$

So $\boxed{}$ is a solution.

PRACTICE *Tell whether each statement is an equation.*

13. $x = 2$
14. $y - 3$
15. $7x < 8$
16. $7 + x = 2$
17. $x + y = 0$
18. $3 - 3y > 2$
19. $1 + 1 = 3$
20. $5 = a + 2$

For each equation, is the given number a solution?

21. $x + 2 = 3; 1$
22. $x - 2 = 4; 6$
23. $a - 7 = 0; 7$
24. $x + 4 = 4; 0$
25. $8 - y = y; 5$
26. $10 - c = c; 5$
27. $x + 32 = 0; 16$
28. $x - 1 = 0; 4$
29. $z + 7 = z; 7$
30. $n - 9 = n; 9$
31. $x = x; 0$
32. $x = 2; 0$

Use the addition or subtraction property of equality to solve each equation. Check each answer.

33. $x - 7 = 3$
34. $y - 11 = 7$
35. $a - 2 = 5$
36. $z - 3 = 9$
37. $1 = b - 2$
38. $0 = t - 1$
39. $x - 4 = 0$
40. $c - 3 = 0$
41. $y - 7 = 6$
42. $a - 2 = 4$
43. $70 = x - 5$
44. $66 = b - 6$
45. $312 = x - 428$
46. $x - 307 = 113$
47. $x - 117 = 222$
48. $y - 27 = 317$
49. $x + 9 = 12$
50. $x + 3 = 9$
51. $y + 7 = 12$
52. $c + 11 = 22$
53. $t + 19 = 28$
54. $s + 45 = 84$
55. $23 + x = 33$
56. $34 + y = 34$
57. $5 = 4 + c$
58. $41 = 23 + x$

59. $99 = r + 43$
60. $92 = r + 37$
61. $512 = x + 428$
62. $x + 307 = 513$
63. $x + 117 = 222$
64. $y + 38 = 321$
65. $3 + x = 7$
66. $b - 4 = 8$
67. $y - 5 = 7$
68. $z + 9 = 23$
69. $4 + a = 12$
70. $5 + x = 13$
71. $x - 13 = 34$
72. $x - 23 = 19$

APPLICATIONS *Complete each solution.*

73. ARCHAEOLOGY   A 1,700-year-old manuscript is 425 years older than the clay jar in which it was found. How old is the jar?

Analyze the problem
  • The manuscript is _____ old.
  • The manuscript is _____ older than the jar.
  • We are asked to find _____.

Form an equation   Since we want to find the age of the jar, we can let $x =$ _____. Now we look for a key word or phrase in the problem.
  Key phrase: _____
  Translation: _____
We can express the age of the manuscript in two ways.

	is	425	plus	the age of the jar.
$\boxed{}$	$=$	425	$+$	$\boxed{}$

Solve the equation
$$\boxed{} = 425 + x$$
$$1{,}700 - \boxed{} = 425 + x - \boxed{}$$
$$\boxed{} = x$$

State the conclusion _____.

Check the result   If the jar is $\boxed{}$ years old, then the manuscript is $1{,}275 + 425 = \boxed{}$ years old. The answer checks.

74. BANKING   After a student wrote a $1,500 check to pay for a car, he had a balance of $750 in his account. How much did he have in the account before he wrote the check?

Analyze the problem
  • A _____ check was written.
  • The balance became _____.
  • We are asked to find
    _____.

Form an equation   Since we want to find his balance before he wrote the check, we let

$x = $ _____ Now we look for a key word or phrase in the problem.

Key phrase: _____
Translation: _____

We can express the balance now in the account in two ways.

The original balance in the account	minus	1,500	is	750.
x	$-$		$=$	750

Solve the equation

$$-1,500 = 750$$
$$x - 1,500 + = 750 + $$
$$x = $$

State the conclusion   _____

Check the result   The original balance was _____. After writing a check, his balance was $2,250 - \$1,500$, or _____. The answer checks.

In Exercises 75–86, let a variable represent the unknown quantity. Then write and solve an equation to answer the question.

75. ELECTIONS   Illustration 1 shows the votes received by the three major candidates running for President of the United States in 1996. Find the total number of votes cast for them.

Bill Clinton (D)	47,401,185
Bob Dole (R)	39,197,469
H. Ross Perot (RF)	8,085,294

ILLUSTRATION 1

76. HIT RECORDS   The oldest artist to have a number 1 single was Louis Armstrong at age 67, with *Hello Dolly*. The youngest artist to have the number 1 single was 12-year-old Jimmy Boyd, with *I Saw Mommy Kissing Santa Claus*. What is the difference in their ages?

77. PARTY INVITATIONS   Three of Mia's party invitations were lost in the mail, but 59 were delivered. How many invitations did she send?

78. HEARING PROTECTION   The sound intensity of a jet engine is 110 decibels. What noise level will an airplane mechanic experience if the ear plugs she is wearing reduce the sound intensity by 29 decibels?

79. FAST FOOD   The franchise fee and startup costs for a Taco Bell restaurant are $287,000. If an entrepreneur has $68,500 to invest, how much money will she need to borrow to open her own Taco Bell restaurant?

80. BUYING GOLF CLUBS   A man needs $345 for a new set of golf clubs. How much more money does he need if he now has $317?

81. CELEBRITY EARNINGS   *Forbes* magazine estimates that in 1998, Celine Dion earned $56 million. If this was $69 million less than Oprah Winfrey's earnings, how much did Oprah earn in 1998?

82. HELP WANTED   From the following advertisement from the classified section of a newspaper, determine the value of the benefit package. ($45K means $45,000.)

★**ACCOUNTS PAYABLE**★
2-3 yrs exp as supervisor. Degree a +. High vol company. Good pay, $45K & xlnt benefits; total compensation worth $52K. Fax resume.

83. POWER OUTAGE   The electrical system in a building automatically shuts down when the meter shown in Illustration 2 reads 85. By how much must the current reading increase to cause the system to shut down?

ILLUSTRATION 2

84. VIDEO GAMES   After a week of playing Sega's *Sonic Adventure*, a boy scored 11,053 points in one game—an improvement of 9,485 points over the very first time he played. What was his score for his first game?

85. AUTO REPAIR   A woman paid $29 less to have her car fixed at a muffler shop than she would have paid at a gas station. At the gas station, she would have paid $219. How much did she pay to have her car fixed?

86. BUS RIDER   A man had to wait 20 minutes for a bus today. Three days ago, he had to wait 15 minutes longer than he did today, because four buses passed by without stopping. How long did he wait three days ago?

WRITING

87. Explain what it means for a number to satisfy an equation.

88. Explain how to tell whether a number is a solution of an equation.

89. Explain what Figure 1-16 (page 52) is trying to show.

90. Explain what Figure 1-17 (page 53) is trying to show.

91. When solving equations, we *isolate* the variable. Write a sentence in which the word *isolate* is used in a different setting.

92. Think of a number. Add 8 to it. Now subtract 8 from that result. Explain why we will always obtain the original number.

REVIEW

93. Round 325,784 to the nearest ten

94. Find 1^5.

95. Evaluate: $2 \cdot 3^2 \cdot 5$.

96. Represent $4 + 4 + 4$ as a multiplication.

97. Evaluate: $8 - 2(3) + 1^3$.

98. Write 1,055 in words.

1.7 *Solving Equations by Division and Multiplication*

In this section, you will learn about

- The division property of equality
- The multiplication property of equality
- Problem solving with equations

INTRODUCTION. In the previous section, we solved equations of the forms

$$x - 4 = 10 \quad \text{and} \quad x + 5 = 16$$

by using the addition and subtraction properties of equality. In this section, we will learn how to solve equations of the forms

$$2x = 8 \quad \text{and} \quad \frac{x}{3} = 25$$

by using the division and multiplication properties of equality.

The division property of equality

To solve many equations, we must divide both sides of the equation by the same nonzero number. The resulting equation will be equivalent to the original one. This idea is summed up in the division property of equality: *If equal quantities are divided by the same nonzero quantity, the results will be equal quantities.*

Division property of equality

Let a, b, and c represent numbers.

If $a = b$, then $\dfrac{a}{c} = \dfrac{b}{c}$. $(c \neq 0)$

We will now consider how to solve the equation $2x = 8$. You will recall that $2x$ means $2 \cdot x$. Therefore, the given equation can be rewritten as $2 \cdot x = 8$. We can think of the scale in Figure 1-19(a) as representing the equation $2 \cdot x = 8$. The weight (in grams) on the left-hand side of the scale is $2 \cdot x$, and the weight (in grams) on the right-hand side is 8. Because these weights are equal, the scale is in balance. To find x, we need to isolate it. That can be accomplished by using the division property of equality to remove half of the weight from each side. The scale will remain in balance. From the scale shown in Figure 1-19(b), we see that x grams will be balanced by 4 grams.

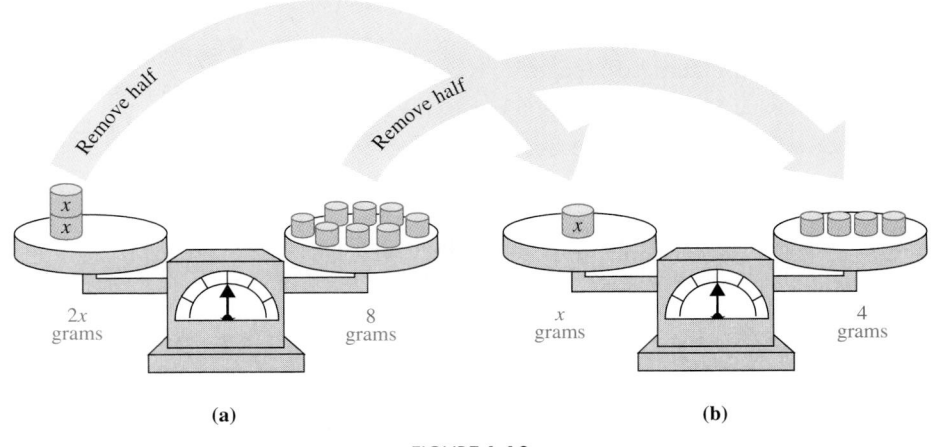

$2x$ grams	8 grams
x grams	4 grams
(a)	**(b)**

FIGURE 1-19

We now show how to solve $2x = 8$ using an algebraic approach.

EXAMPLE 1   *Solving equations.*   Solve $2x = 8$ and check the result.

Solution

Recall that $2x = 8$ means $2 \cdot x = 8$. To isolate x on the left-hand side of the equation, we undo the multiplication by 2 by dividing both sides of the equation by 2.

$2x = 8$ The equation to solve.

$\dfrac{2x}{2} = \dfrac{8}{2}$ To undo the multiplication by 2, divide both sides by 2.

$x = 4$ When x is multiplied by 2 and that product is then divided by 2, the result is x. Do the division: $8 \div 2 = 4$.

To check this result, we substitute 4 for x in $2x = 8$.

$2x = 8$

$2 \cdot 4 \overset{?}{=} 8$ Substitute 4 for x.

$8 = 8$ Do the multiplication.

Since $8 = 8$ is a true statement, 4 is a solution.

Self Check

Solve $17x = 153$ and check the result.

Answer: 9

The multiplication property of equality

We can also multiply both sides of an equation by the same nonzero number to get an equivalent equation. This idea is summed up in the multiplication property of equality: *If equal quantities are multiplied by the same nonzero quantity, the results will be equal quantities.*

Multiplication property of equality

Let a, b, and c represent numbers.

If $a = b$, then $c \cdot a = c \cdot b$ or, more simply, $ca = cb$ $(c \neq 0)$.

We can think of the scale shown in Figure 1-20(a) as representing the equation $\frac{x}{3} = 25$. The weight on the left-hand side of the scale is $\frac{x}{3}$ grams, and the weight on the right-hand side is 25 grams. Because these weights are equal, the scale is in balance. To find *x*, we can use the multiplication property of equality to triple (or multiply by 3) the weight on each side. The scale will remain in balance. From the scale shown in Figure 1-20(b), we can see that *x* grams will be balanced by 75 grams.

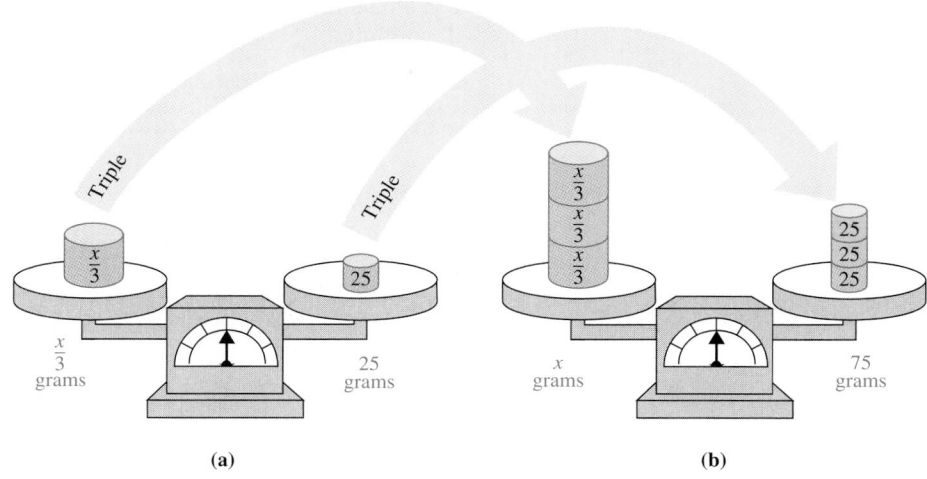

(a)   **(b)**

FIGURE 1-20

We now show how to solve $\frac{x}{3} = 25$ using an algebraic approach.

EXAMPLE 2   *Solving equations.* Solve $\frac{x}{3} = 25$ and check the result.

Self Check

Solve $\frac{x}{12} = 24$ and check the result.

Solution

To isolate *x* on the left-hand side of the equation, we undo the division of the variable by 3 by multiplying both sides by 3.

$$\frac{x}{3} = 25 \quad \text{The equation to solve.}$$

$$3 \cdot \frac{x}{3} = 3 \cdot 25 \quad \text{To undo the division by 3, multiply both sides by 3.}$$

$$x = 75 \quad \text{When } x \text{ is divided by 3 and that quotient is then multiplied by 3, the result is } x. \text{ Do the multiplication: } 3 \cdot 25 = 75.$$

Check:

$$\frac{x}{3} = 25$$

$$\frac{75}{3} \stackrel{?}{=} 25 \quad \text{Substitute 75 for } x.$$

$$25 = 25 \quad \text{Do the division: } 75 \div 3 = 25. \text{ The answer checks.}$$

Answer: 288

Problem solving with equations

As before, we can use equations to solve problems. Remember that the purpose of these early examples is to help you learn the strategy, even though you can probably solve the examples without it.

EXAMPLE 3   *Buying electronics.* The owner of an apartment complex bought six television sets that were on sale for $499 each. What was the total cost?

Analyze the problem
- 6 television sets were bought.
- They cost $499 each.
- We are asked to find the total cost.

Form an equation   We let c = the total cost of the TVs. To form an equation, we look for a key word or phrase in the problem. We can add 499 six times or multiply 499 by 6. Since it is easier, we will multiply.

Key phrase: *six TVs, each costing $499*   **Translation:** *multiplication*

We can express the total cost in two ways.

The total number of TVs	multiplied by	the cost of each TV	is	the total cost.
6	·	499	=	c

Solve the equation
$$6 \cdot 499 = c$$
$$2,994 = c \quad \text{Do the multiplication.}$$

State the conclusion   The total cost will be $2,994.

Check the result   We can check by estimation. Since each TV costs a little less than $500, we would expect the total cost to be a little less than 6 · $500, or $3,000. An answer of $2,994 is reasonable. ■

  **EXAMPLE 4**   *Splitting an inheritance.*   If seven brothers inherit $343,000 and split the money evenly, how much will each brother get?

Analyze the problem
- There are 7 brothers.
- They split $343,000 evenly.
- We are asked to find how much each brother will get.

Form an equation   We can let g = the number of dollars each brother will get. To form an equation, we look for a key word or phrase in the problem.

Key phrase: *split the money evenly*   **Translation:** *division*

We can express the share each brother will get in two ways.

The total amount of the inheritance	divided by	the number of brothers	is	the share each brother will get.
343,000	÷	7	=	g

Solve the equation
$$\frac{343,000}{7} = g \quad 343,000 \div 7 \text{ can be written as } \tfrac{343,000}{7}.$$
$$49,000 = g \quad \text{Do the division.}$$

State the conclusion   Each brother will get $49,000.

Check the result   If we multiply $49,000 by 7, we get $343,000. ■

EXAMPLE 5   *Traffic violations.*   For a speeding ticket, a motorist had to pay a fine of $592. The violation occurred on a stretch of highway posted with special signs like that shown in Figure 1-21. What would the fine have been if such signs were not posted?

> **TRAFFIC FINES DOUBLED IN CONSTRUCTION ZONE**

FIGURE 1-21

Analyze the problem
- The motorist was fined $592.
- The fine was double what it would normally have been.
- We are asked to find what the fine would have been, had the area not been posted.

Form an equation
We can let f = the amount that the fine would normally have been. To form an equation, we look for a key word or phrase in the problem or analysis.

Key word: *double*   **Translation:** *multiply by 2*

We can express the amount of the new fine in two ways.

Two	times	the normal speeding fine	is	the new fine.
2	$\cdot$	f	=	592

Solve the equation

$2f = 592$ Write $2 \cdot f$ as $2f$.

$\dfrac{2f}{2} = \dfrac{592}{2}$ To undo the multiplication by 2, divide both sides by 2.

$f = 296$ Do the division: $592 \div 2 = 296$.

State the conclusion   The fine would normally have been $296.

Check the result   If we double $296, we get 2 ($296) = $592. The answer checks.   ■

EXAMPLE 6   ***Entertainment costs.*** A five-piece band worked on New Year's Eve. If each player earned $120, what fee did the band charge?

Analyze the problem
- There were 5 players in the band.
- Each player made $120.
- We are asked to find the band's fee. We know that the fee divided by the number of players will give each person's share.

Form an equation
We can let f = the band's fee. To form an equation, we look for a key word or phrase. In this case, we find it in the analysis of the problem.

Key phrase: *divided by*   **Translation:** *division*

We can express each person's share in two ways.

The band's fee	divided by	the number in the band	is	each person's share.
f	$\div$	5	=	120

Solve the equation

$\dfrac{f}{5} = 120$ Write $f \div 5$ as $\frac{f}{5}$.

$5 \cdot \dfrac{f}{5} = 5 \cdot 120$ To undo the division by 5, multiply both sides by 5.

$f = 600$ Do the multiplication: $5 \cdot 120 = 600$.

State the conclusion   The band's fee was $600.

Check the result   If we divide $600 by 5, we get each person's share: $120.   ■

STUDY SET Section 1.7

VOCABULARY *Fill in the blanks.*

1. According to the _____ property of equality, "If equal quantities are divided by the same nonzero quantity, the results will be equal quantities."

2. According to the _____ property of equality, "If equal quantities are multiplied by the same nonzero quantity, the results will be equal quantities."

CONCEPTS *In Exercises 3–6, fill in the blanks.*

3. If we multiply x by 6 and then divide that product by 6, what is the result?

4. If we divide x by 8 and then multiply that quotient by 8, what is the result?

5. If $x = y$, then $\frac{x}{z} = \quad$ ($z \neq 0$).

6. If $x = y$, then $zx = \quad$ ($z \neq 0$).

7. In the equation $4t = 40$, what operation is being performed on the variable? How do we undo it?

8. In the equation $\frac{t}{15} = 1$, what operation is being performed on the variable? How do we undo it?

9. Name the first step in solving each of the following equations.
a. $x + 5 = 10$ **b.** $x - 5 = 10$

c. $5x = 10$ **d.** $\frac{x}{5} = 10$

10. For each of the following equations, check the given answer.
a. $16 = t - 8; t = 33$
b. $16 = t + 8; t = 8$
c. $16 = 8t; t = 128$
d. $16 = \frac{t}{8}; t = 2$

NOTATION *Complete each solution.*

11. $3x = 12$

$\dfrac{3x}{\quad} = \dfrac{12}{\quad}$

$x = 4$

Check: $3x = 12$

$3 \cdot \quad \overset{?}{=} 12$

$\quad = 12$

So   is a solution.

12. $\dfrac{x}{5} = 9$

$\quad \cdot \dfrac{x}{5} = \quad \cdot 9$

$x = 45$

Check: $\dfrac{x}{5} = 9$

$\dfrac{\quad}{5} \overset{?}{=} 9$

$\quad = 9$

So   is a solution.

PRACTICE *Use the division or the multiplication property of equality to solve each equation. Check each answer.*

13. $3x = 3$ **14.** $5x = 5$

15. $2x = 192$ **16.** $4x = 120$

17. $17y = 51$ **18.** $19y = 76$

19. $34y = 204$ **20.** $18y = 90$

21. $100 = 100x$ **22.** $35 = 35y$

23. $16 = 8r$ **24.** $44 = 11m$

25. $\dfrac{x}{7} = 2$ **26.** $\dfrac{x}{12} = 4$

27. $\dfrac{y}{14} = 3$ **28.** $\dfrac{y}{13} = 5$

29. $\dfrac{a}{15} = 5$ **30.** $\dfrac{b}{25} = 5$

31. $\dfrac{c}{13} = 3$ **32.** $\dfrac{d}{100} = 11$

33. $1 = \dfrac{x}{50}$ **34.** $1 = \dfrac{x}{25}$

35. $7 = \dfrac{t}{7}$ **36.** $4 = \dfrac{m}{4}$

37. $9z = 90$ **38.** $3z = 6$

39. $7x = 21$ **40.** $13x = 52$

41. $86 = 43t$ **42.** $288 = 96t$

43. $21s = 21$ **44.** $31x = 155$

45. $\dfrac{d}{20} = 2$ **46.** $\dfrac{x}{16} = 4$

47. $400 = \dfrac{t}{3}$ **48.** $250 = \dfrac{y}{2}$

APPLICATIONS *Complete each solution.*

49. NOBEL PRIZE   In 1998, three Americans, Louis Ignarro, Robert Furchgott, and Dr. Fred Murad, were awarded the Nobel Prize for Medicine. They shared the prize money. If each person received $318,500, what was the Nobel Prize cash award?

Analyze the problem
- ____ people shared the cash award.
- Each person received ____.
- We are asked to find the _____.

Form an equation
Since we want to find what the Nobel Prize cash award was, we let $c =$ _____. To form an equation, we look for a key word or phrase in the problem.

Key phrase: _____

Translation: _____

We can now form the equation.

The Nobel Prize cash award	divided by	the number of recipients	was	$318,500.
	÷	3	=	$318,500

Solve the equation
$$\frac{x}{3} = 318{,}500$$
$$\cdot \frac{x}{3} = \cdot 318{,}500$$
$$x =$$

State the conclusion
_____

Check the result
If we divide the Nobel Prize cash award by 3, we have
$$\frac{}{3} = .$$ This was the amount each person received. The answer checks.

50. INVESTING An investor has watched the value of his portfolio double in the last 12 months. If the current value of his portfolio is $274,552, what was its value one year ago?

Analyze the problem
- The value of the portfolio ____ in 12 months.
- The current value is ____.
- We must find _____.

Form an equation
We can let $x =$ _____. We now look for a key word or phrase in the problem.

Key phrase: _____

Translation: _____

We can now form the equation.

2	times	the value of the portfolio one year ago	is	the current value of the portfolio.
2	·		=	$274,552

Solve the equation
$$2x =$$
$$\frac{2x}{} = \frac{274{,}552}{}$$
$$x =$$

State the conclusion
_____

Check the result
If the value of the portfolio one year ago was ____ and it doubled, its current value would be ____. The answer checks.

In Exercises 51–60, let a variable represent the unknown quantity. Then write and solve an equation to answer the question.

51. SPEED READING An advertisement for a speed reading program claimed that successful completion of the course could triple a person's reading rate. If Alicia can currently read 130 words a minute, at what rate can she expect to read after taking the classes?

52. COST OVERRUN Lengthy delays and skyrocketing costs caused a rapid-transit construction project to go over budget by a factor of 10. The final audit showed the project costing $540 million. What was the initial cost estimate?

53. STAMPS Large sheets of commemorative stamps honoring Marilyn Monroe are to be printed. See Illustration 1. On each sheet, there are 112 stamps, with 8 stamps per row. How many rows of stamps are on a sheet?

ILLUSTRATION 1

54. SPREADSHEET The grid shown in Illustration 2 is a Microsoft Excel spreadsheet. The rows are labeled with numbers, and the columns are labeled with letters. Each empty box of the grid is called a *cell*. Suppose a certain project calls for a spreadsheet with 294 cells, using columns A through F. How many rows will need to be used?

	Microsoft Excel-Book 1					
	File	Edit	View	Insert	Format	Tools
	A	**B**	**C**	**D**	**E**	**F**
1						
2						
3						
4						
5						
6						
7						
8						

Sheet 1 / Sheet 2 / Sheet 3 / Sheet 4 / Sheet 5

ILLUSTRATION 2

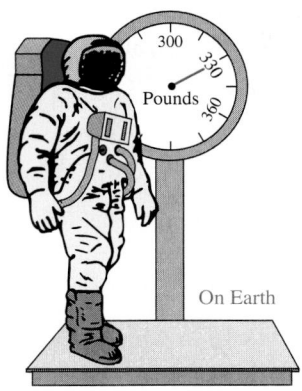

On Earth

ILLUSTRATION 3

55. PHYSICAL EDUCATION   A high school PE teacher had the students in her class form three-person teams for a basketball tournament. Thirty-two teams participated in the tournament. How many students were in the PE class?

56. LOTTO WINNERS   The grocery store employees listed below pooled their money to buy $120 worth of lottery tickets each week, with the understanding they would split the prize equally if they happened to win. One week they did have the winning ticket and won $480,000. What was each employee's share of the winnings?

Sam M. Adler	Ronda Pellman	Manny Fernando
Lorrie Jenkins	Tom Sato	Sam Lin
Kiem Nguyen	H. R. Kinsella	Tejal Neeraj
Virginia Ortiz	Libby Sellez	Alicia Wen

57. ANIMAL SHELTER   The number of phone calls to an animal shelter quadrupled after the evening news aired a segment explaining the services the shelter offered. Before the publicity, the shelter received 8 calls a day. How many calls did the shelter receive each day after being featured on the news?

58. OPEN HOUSE   The attendance at an elementary school open house was only half of what the principal had expected. If 120 people visited the school that evening, how many had she expected to attend?

59. GRAVITY   The weight of an object on Earth is 6 times greater than what it is on the moon. The situation shown in Illustration 3 took place on the Earth. If it took place on the moon, what weight would the scale register?

60. INFOMERCIAL   The number of orders received each week by a company selling skin care products increased fivefold after a Hollywood celebrity was added to the company's infomercial. After adding the celebrity, the company received about 175 orders each week. How many orders were received each week before the celebrity took part?

WRITING

61. Explain what Figure 1-19 (page 60) is trying to show.

62. Explain what Figure 1-20 (page 61) is trying to show.

63. What does it mean to solve an equation?

64. Think of a number. Double it. Now divide it by 2. Explain why you always obtain the original number.

REVIEW

65. Find the perimeter of a rectangle with sides measuring 8 cm and 16 cm.

66. Find the area of a rectangle with sides measuring 23 inches and 37 inches.

67. Find the prime factorization of 120.

68. Find the prime factorization of 150.

69. Evaluate $3^2 \cdot 2^3$.

70. Evaluate $5 + 6 \cdot 3$.

71. FUEL ECONOMY   Five basic models of automobiles made by Saturn have city mileage ratings of 24, 22, 28, 29, and 27 miles per gallon. What is the average (mean) city mileage for the five models?

72. Solve the equation $x - 4 = 20$.

Variables

One of the major objectives of this course is for you to become comfortable working with **variables.** You will recall that a variable is a letter that stands for a number.

The application problems of Sections 1.6 and 1.7 were solved with the help of a variable. In these problems, we let the variable represent an unknown quantity such as the number of customers a hairdresser used to have, the age of a jar, and the cash award given a Nobel Prize winner. We then wrote an equation to describe the situation mathematically and solved the equation to find the value represented by the variable.

In Exercises 1–6, suppose that you are going to solve the following problems. What quantity should be represented by a variable? State your response in the form "Let $x = \ldots$."

1. The monthly cost to lease a van is $120 less than to buy it. To buy it, the monthly payments are $290. How much does it cost to lease the van each month?

2. One piece of pipe is 10 feet longer than another. Together, their lengths total 24 feet. How long is the shorter piece of pipe?

3. The length of a rectangular field is 50 feet. What is its width if it has a perimeter of 200 feet?

4. If one hose can fill a vat in 2 hours and another can fill it in 3 hours, how long will it take to fill the vat if both hoses are used?

5. Find the distance traveled by a motorist in three hours if her average speed was 55 miles per hour.

6. In what year was a couple married if their 50th anniversary was in 1988?

Variables can also be used to state properties of mathematics in a concise, "shorthand" notation. In Exercises 7–14, state each property using mathematical symbols and the given variable(s).

7. Use the variables a and b to state that two numbers can be added in either order to get the same sum.

8. Use the variable x to state that when 0 is subtracted from a number, the result is the same number.

9. Use the variable b to state that the result when dividing a number by 1 is the same number.

10. Use the variable x to show that the sum of a number and 1 is greater than the number.

11. Using the variable n, state the fact that when 1 is subtracted from any number, the difference is less than the number.

12. State the fact that the product of any number and 0 is 0, using the variable a.

13. Use the variables r, s, and t to state that the way we group three numbers when adding them does not affect the answer.

14. Using the variable n, state the fact that when a number is multiplied by 1, the result is the number.

ACCENT ON TEAMWORK

Section 1.1

PLACE VALUE Have each student in your group bring a calculator to class so that you can examine several different models. For each model, determine the largest number (if there is one) that can be entered on the display of the calculator. Then press the appropriate calculator keys to add 1 to that number. What does the display show?

LARGE NUMBERS Bill Gates, founder of Microsoft Corporation, is said to be a billionaire. How many millions make one billion?

Section 1.2

READING THE PROBLEM CAREFULLY In reading Example 9 of Section 1.2, you will notice that it contains several facts that are not used in the solution of the problem. Have each person in your group write a similar problem that requires careful reading to extract the useful information. Then have each person share his or her problem with the other students in the group.

Section 1.3

DIVISIBILITY TESTS Certain tests can help us decide whether one whole number is divisible by another.

- A number is divisible by 2 if the last digit of the number is 0, 2, 4, 6, or 8.
- A number is divisible by 3 if the sum of the digits is divisible by 3.
- A number is divisible by 4 if the number formed by the last two digits is divisible by 4.
- A number is divisible by 5 if the last digit of the number is 0 or 5.
- A number is divisible by 6 if the last digit of the number is 0, 2, 4, 6, or 8 and the sum of the digits is divisible by 3.
- A number is divisible by 8 if the number formed by the last three digits is divisible by 8.
- A number is divisible by 9 if the sum of the digits is divisible by 9.
- A number is divisible by 10 if the last digit of the number is 0.
- Determine whether each number is divisible by 2, 3, 4, 5, 6, 8, 9, and/or 10.

    a. 660   **b.** 2,526

    c. 11,523   **d.** 79,503

    e. 135,405   **f.** 4,444,440

Section 1.4

COMMON FACTORS The prime factorizations of 36 and 126 are shown below. The prime factors that are common to 36 and 126 (highlighted in color) are 2, 3, and 3.

$$36 = 2 \cdot 2 \cdot 3 \cdot 3$$
$$126 = 2 \cdot 3 \cdot 3 \cdot 7$$

Find the common prime factors for each of the following pairs of numbers.

a. 25, 45   **b.** 24, 60

c. 18, 45   **d.** 40, 112

e. 180, 210   **f.** 242, 198

Section 1.5

ORDER OF OPERATIONS Consider the expression

$$5 + 8 \cdot 2^3 - 3 \cdot 2$$

Insert a set of parentheses somewhere in the expression so that, when it is evaluated, you obtain

a. 63   **b.** 132

c. 21   **d.** 127

Section 1.6

SOLVING EQUATIONS Borrow a scale and some weights from the chemistry department. Use them as part of a class presentation to explain how the subtraction property of equality is used to solve the equation $x + 2 = 5$. See the discussion and Figure 1-16 on page 52 for some suggestions on how to do this.

Section 1.7

FORMING AN EQUATION Reread Example 4 in Section 1.7. This problem could have been solved by forming an equation involving the operation of multiplication instead of the operation of division.

The number of brothers	times	the share each brother will get	is	the total amount of the inheritance.
7	$\cdot$	g	$=$	343,000

For Examples 5 and 6 in Section 1.7, write another equation that could be used to solve the problem. Then solve the equation and state the result.

SECTION 1.1 — *An Introduction to the Whole Numbers*

CONCEPTS

A *set* is a collection of objects.

The set of *natural numbers* is
{1, 2, 3, 4, 5, . . .}

The set of *whole numbers* is
{0, 1, 2, 3, 4, 5, . . .}

Whole numbers are often used in tables, bar graphs, and line graphs.

REVIEW EXERCISES

1. Graph each set.

a. The natural numbers less than 5

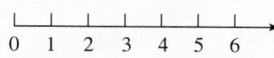

b. The whole numbers between 0 and 3

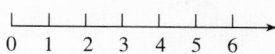

2. FARMING   The table below shows the size of the average U.S. farm (in acres) for the period 1940–2000, in 20-year increments.

Year	1940	1960	1980	2000
Average size (acres)	174	297	426	432

a. Construct a bar graph of the data.

b. Construct a line graph of the data.

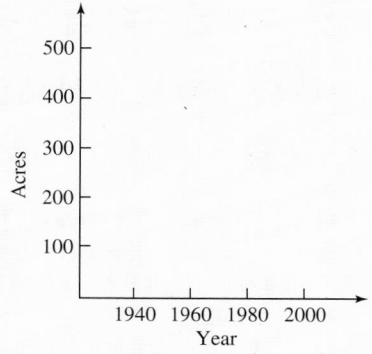

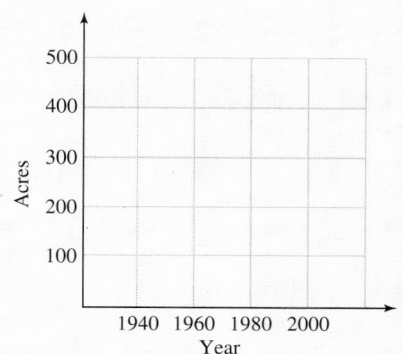

The digits in a whole number have *place value*.

3. Consider the number 2,365,720. Which digit is in the

a. ten thousands column?

b. hundreds column?

A whole number is written in *expanded notation* when its digits are written with their place values.

4. Write each number in expanded notation.

a. 570,302

b. 37,309,054

We use the digits 0, 1, 2, 3, 4, 5, 6, 7, 8, and 9 to write a number in *standard notation*.

5. Write each number in standard notation.

a. 3 thousands + 2 hundreds + 7 ones

b. twenty-three million, two hundred fifty-three thousand, four hundred twelve

c. sixteen billion

The symbol < means "is less than." The symbol > means "is greater than."

6. Place an < or > symbol in the box to make a true statement.

a. 9 ☐ 7

b. 3 ☐ 5

To give approximate answers, we often use *rounded numbers*.

7. Round 2,507,348 to the specified place.
   a. nearest hundred
   b. nearest ten thousand
   c. nearest ten
   d. nearest hundred thousand

SECTION 1.2

Adding and Subtracting Whole Numbers

Addition is the process of finding the total of two (or more) numbers. Do additions within parentheses first.

8. Find each sum.
   a. 56 + 22
   b. 137 + 0
   c. 15 + (27 + 13)
   d. 82 + 17 + 50
   e. (111 + 222) + 444
   f. 0 + 2,332

The *commutative* and *associative properties of addition:*
$$a + b = b + a$$
$$(a + b) + c = a + (b + c)$$

9. Do each addition.
   a.   236
      +282
   b.   5,345
      + 655
   c. 135 + 213 + 615 + 47
   d. 4,447 + 7,478 + 13,061

10. What property of addition is shown?
   a. 12 + 8 = 8 + 12
   b. 12 + (8 + 2) = (12 + 8) + 2

The *perimeter* of a rectangle or a square is the distance around it.

11. Find the perimeter of a square with sides 24 inches long.

Subtraction is the process of finding the difference between two numbers.

12. Do each subtraction.
   a. 18 − 5
   b. 9 − (7 − 2)
   c. 22 − 5 − 6
   d. 5,231 − 5,177
   e.   343
      −269
   f.   17,800
      −15,725

13. Give the addition or subtraction fact that is illustrated by each figure.
   a.
   b.

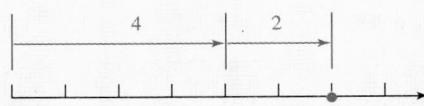
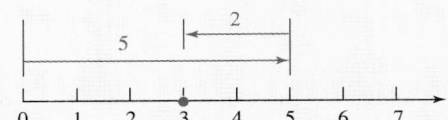

14. TRAVEL   A direct flight from Omaha to San Francisco costs $237. Another flight with one stop in Reno costs $192. How much can be saved by taking the less expensive flight?

15. SAVINGS ACCOUNTS   A savings account contains $931. If the owner deposits $271 and makes withdrawals of $37 and $380, find the final balance.

16. REBATE   The price of a new Chevrolet Camaro was advertised in a newspaper as $21,991*. A note at the bottom of the ad read, "*Reflects $1,550 factory rebate." What was the car's original sticker price?

SECTION 1.3 | *Multiplying and Dividing Whole Numbers*

Multiplication is repeated addition. For example,

The sum of four 6's
$$4 \cdot 6 = 6 + 6 + 6 + 6$$
$$= 24$$

The result, 24, is called the *product,* and the 4 and 6 are called *factors.*

The *commutative* and *associative properties of multiplication.*
$$a \cdot b = b \cdot a$$
$$(a \cdot b) \cdot c = a \cdot (b \cdot c)$$

The *area A of a rectangle* is the product of its length *l* and its width *w.*
$$A = l \cdot w$$

17. Do each multiplication.
   a. $8 \cdot 7$ **b.** $7(8)$
   c. $8 \cdot 0$ **d.** $7 \cdot 1$
   e. $10 \cdot 8 \cdot 7$ **f.** $5 \cdot (7 \cdot 6)$

18. Do each multiplication.
   a. $157 \cdot 21$ **b.** $3{,}723 \cdot 48$
   c. $\begin{array}{r} 356 \\ \times\ \ 89 \\ \hline \end{array}$ **d.** $\begin{array}{r} 5{,}624 \\ \times\ \ \ 81 \\ \hline \end{array}$

19. What property of multiplication is shown?
   a. $12 \cdot (8 \cdot 2) = (12 \cdot 8) \cdot 2$
   b. $12 \cdot 8 = 8 \cdot 12$

20. WAGES   If a math tutor worked for 38 hours and was paid $9 per hour, how much did she earn?

21. HORSESHOES   Find the perimeter and the area of the rectangular horseshoe court shown in Illustration 1.

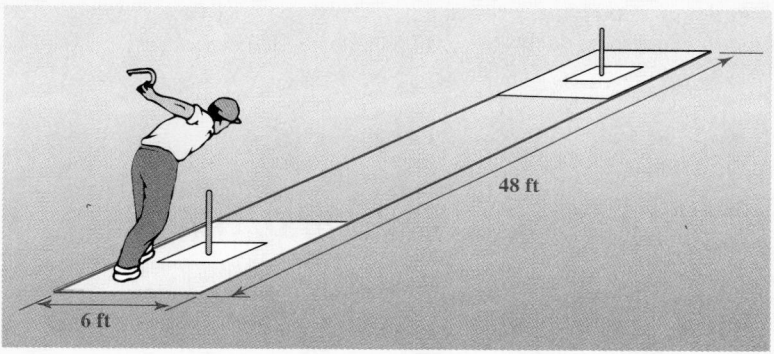

48 ft

6 ft

ILLUSTRATION 1

22. PACKAGING   There are 12 eggs in one dozen, and 12 dozen in one gross. How many eggs are in a shipment of 5 gross?

Division is an operation that determines how many times a number (the *divisor*) is contained in another number (the *dividend*). *Remember that you can never divide by 0.*

23. Do each division, if possible.
   a. $\dfrac{6}{3}$ **b.** $\dfrac{15}{1}$
   c. $73 \div 0$ **d.** $\dfrac{0}{8}$
   e. $357 \div 17$ **f.** $1{,}443 \div 39$
   g. $21\overline{)405}$ **h.** $54\overline{)1{,}269}$

24. TREATS   If 745 candies are divided equally among 45 children, how many will each child receive? How many candies will be left over?

25. COPIES   An elementary school teacher had copies of a 3-page social studies test made at Quick Copy Center. She was charged for 84 sheets of paper. How many copies of the test were made?

   Prime Factors and Exponents

Numbers that are multiplied together are called *factors*.

A *prime number* is a whole number greater than 1 that has only 1 and itself as factors. Whole numbers greater than 1 that are not prime are called *composite numbers*.

Whole numbers divisible by 2 are *even* numbers. Whole numbers not divisible by 2 are *odd* numbers.

The *prime factorization* of a whole number is the product of its prime factors.

An *exponent* is used to indicate repeated multiplication. In the exponential expression a^n, a is the *base*, and n is the exponent.

26. Find all of the factors of each number.
   a. 18   **b.** 25

27. Identify each number as a prime, composite, or neither.
   a. 31   **b.** 100
   c. 1   **d.** 0
   e. 125   **f.** 47

28. Identify each number as an even or odd number.
   a. 171   **b.** 214
   c. 0   **d.** 1

29. Find the prime factorization of each number.
   a. 42   **b.** 375

30. Write each expression using exponents.
   a. $6 \cdot 6 \cdot 6 \cdot 6$ **b.** $5 \cdot 5 \cdot 5 \cdot 13 \cdot 13$

31. Evaluate each expression.
   a. 5^3 **b.** 11^2
   c. $2^3 \cdot 5^2$ **d.** $2^2 \cdot 3^3 \cdot 5^2$

   Order of Operations

Do mathematical operations in the following order:

1. Do all calculations within parentheses and other grouping symbols.
2. Evaluate all exponential expressions.
3. Do all multiplications and divisions in order from left to right.
4. Do all additions and subtractions in order from left to right.

To evaluate an expression containing grouping symbols, do all calculations within each pair of grouping symbols, working from the innermost pair to the outermost pair.

The *arithmetic mean* (average) is a value around which numbers are grouped.

32. Evaluate each expression.
   a. $13 + 12 \cdot 3$ **b.** $35 - 15 \div 5$
   c. $(13 + 12)3$ **d.** $(8 - 2)^2$
   e. $8 \cdot 5 - 4 \div 2$ **f.** $8 \cdot (5 - 4 \div 2)$
   g. $2 + 3(10 - 4 \cdot 2)$ **h.** $4(20 - 5 \cdot 3 + 2) - 4$
   i. $3^3\left(\dfrac{12}{6}\right) - 1^4$ **j.** $\dfrac{12 + 3 \cdot 7}{5^2 - 14}$
   k. $7 + 3[10 - 3(4 - 2)]$ **l.** $5 + 2[(15 - 3 \cdot 4) - 2]$

33. DICE GAME   Write an expression which finds the total of all the dice shown in Illustration 2. Then evaluate the expression.

34. YAHTZEE   See Illustration 3. Find the player's average (mean) score for the 6 games.

ILLUSTRATION 2

	Yahtzee SCORE CARD				
Game #1	Game #2	Game #3	Game #4	Game #5	Game #6
159	244	184	240	166	213

ILLUSTRATION 3

SECTION 1.6 | *Solving Equations by Addition and Subtraction*

An *equation* is a statement that two expressions are equal. A *variable* is a letter that stands for a number.

Two equations with exactly the same solutions are called *equivalent equations*.

To solve an equation, isolate the variable on one side of the equation by undoing the operation performed on it.

If the same number is added to (or subtracted from) both sides of an equation, an equivalent equation results:
If $a = b$, then $a + c = b + c$.
If $a = b$, then $a - c = b - c$.

Problem-solving strategy:

1. Analyze the problem.
2. Form an equation.
3. Solve the equation.
4. State the conclusion.
5. Check the result.

35. Tell whether the given number is a solution of the equation. Explain why or why not.
   a. $x + 2 = 13; x = 5$ **b.** $x - 3 = 1; x = 4$

36. Identify the variable in each equation.
   a. $y - 12 = 50$ **b.** $114 = 4 - t$

37. Solve the equation and check the result.
   a. $x - 7 = 2$ **b.** $x - 11 = 20$
   c. $225 = y - 115$ **d.** $101 = p - 32$
   e. $x + 9 = 18$ **f.** $b + 12 = 26$
   g. $175 = p + 55$ **h.** $212 = m + 207$
   i. $x - 7 = 0$ **j.** $x + 15 = 1,000$

In Exercises 38–39, let a variable represent the unknown quantity. Then write and solve an equation to answer the question.

38. FINANCING   A newly married couple made a $25,500 down payment on a $122,750 house. How much did they need to borrow?

39. DOCTOR'S PATIENTS   After moving his office, a doctor lost 13 patients. If he had 172 patients left, how many did he have originally?

SECTION 1.7 | *Solving Equations by Division and Multiplication*

If both sides of an equation are divided by (or multiplied by) the same nonzero number, an equivalent equation results:
If $a = b$, then $\frac{a}{c} = \frac{b}{c}$ ($c \neq 0$).

If $a = b$, then $a \cdot c = b \cdot c$ ($c \neq 0$).

40. Solve the equation and check the result.
   a. $3x = 12$ **b.** $15y = 45$
   c. $105 = 5r$ **d.** $224 = 16q$
   e. $\frac{x}{7} = 3$ **f.** $\frac{a}{3} = 12$
   g. $15 = \frac{s}{21}$ **h.** $25 = \frac{d}{17}$
   i. $12x = 12$ **j.** $\frac{x}{12} = 12$

In Exercises 41–42, let a variable represent the unknown quantity. Then write and solve an equation to answer the question.

41. CARPENTRY   If you cut a 72-inch board into three equal pieces, how long will each piece be? Disregard any loss due to cutting.

42. JEWELRY   Four sisters split the cost of a gold chain evenly. How much did the chain cost if each sister's share was $32?

TAKING A MATH TEST

The best way to relieve anxiety about taking a mathematics test is to know that you are well-prepared for it and that you have a plan. Before any test, ask yourself three questions. When? What? How?

When will I study?

1. When is the test?

2. When will I begin to review for the test?

3. What are the dates and times that I will reserve for studying for the test?

What will I study?

1. What sections will the test cover?

2. Has the instructor indicated any types of problems that are guaranteed to be on the test?

How will I prepare for the test?

Put a check mark by each method you will use to prepare for the test.

☐ Review the class notes.

☐ Outline the chapter(s) to see how the topics relate to one another.

☐ Recite the important formulas, definitions, vocabulary, and rules into a tape recorder.

☐ Make flash cards for the important formulas, definitions, vocabulary, and rules.

☐ Rework problems from the homework assignments.

☐ Rework each of the Self Check problems in the text.

☐ Form a study group to discuss and practice the topics to be tested.

☐ Complete the appropriate Chapter Review(s) and the Chapter Test(s).

☐ Review the Comments given in the text.

☐ Work on improving my speed in answering questions.

☐ Review the methods that can be used to check my answers.

☐ Write a sample test, trying to think of the questions the instructor will ask.

☐ Complete the appropriate Cumulative Review Exercises.

☐ Get organized the night before the test. Have materials ready to go so that the trip to school will not be hurried.

☐ Take some time to relax immediately before the test. Don't study right up to the last minute.

Taking the test

Here are some tips that can help improve your performance on a mathematics test.

- Write down any formulas or rules as soon as you receive the test.

- When you receive the test, scan it, looking for the types of problems you had expected to see. Do them first.

- Read the instructions carefully.

- Don't spend too much time on any one problem until you have attempted all the problems.

- If your instructor gives partial credit, at least try to begin a solution.

- Don't be afraid to skip a problem and come back to it later.

- Save the most difficult problems for last.

- If you finish early, go back over your work and look for mistakes.

Chapter 1 Test

1. Graph the whole numbers less than 5.

$$\underset{0\quad 1\quad 2\quad 3\quad 4\quad 5\quad 6\quad 7}{\xrightarrow{\hspace{3cm}}}$$

2. Write "five thousand two hundred sixty-six" in expanded notation.

3. Write "7 thousands + 5 hundreds + 7 ones" in standard notation.

4. Round 34,752,341 to the nearest million.

In Problems 5–6, refer to the data in the table.

Lot number	1	2	3	4
Defective bolts	7	10	5	15

5. Use the data to make a bar graph.

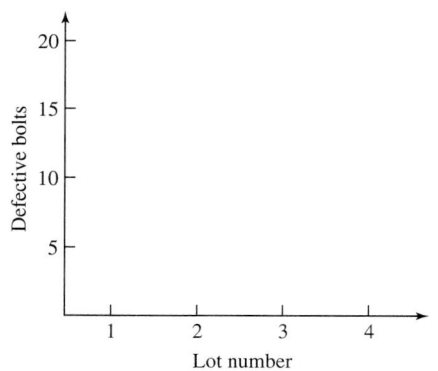

6. Use the data to make a line graph.

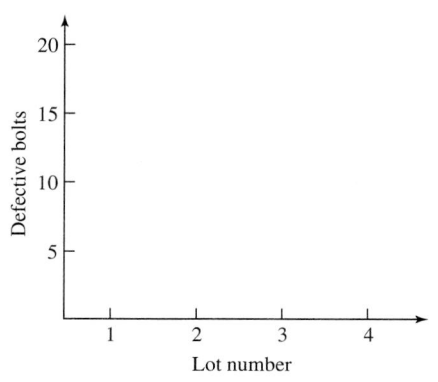

Place an < or > symbol in the box to make a true statement.

7. 15 ☐ 10

8. 12 ☐ 17

9. Add 327 + 435 + 123 + 606.

10. Subtract 287 from 535.

11. Add: 44,526
+13,579

12. Subtract: 4,521
−3,579

13. STOCKS   On Tuesday, a share of KBJ Company was selling at $73. The price rose $12 on Wednesday and fell $9 on Thursday. Find its price on Thursday.

14. List the factors of 20 in order, from least to greatest.

Do each operation.

15. Multiply: 53
× 8

16. Multiply: 367(73).

17. Divide: $63\overline{)4,536}$

18. Divide: $73\overline{)8,379}$

19. FURNITURE SALE   See the advertisement in Illustration 1. Find the perimeter of the rectangular space under the tent. Then fill in the blank in the advertisement.

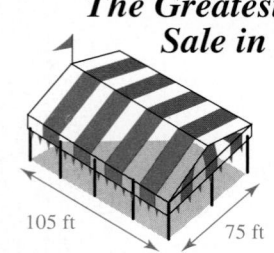

The Greatest Parking Lot Tent Sale in Our History!

| ? | square feet in our "outdoor showroom" 2 DAYS ONLY!

Thomastown Home Furnishings

105 ft   75 ft

ILLUSTRATION 1

20. If 3,451 students are placed in groups of 74, how many will be left over?

21. COLLECTIBLES   There are 12 baseball cards in every pack. There are 24 packs in every box. There are 12 boxes in every case. How many cards are in a case?

22. Find the prime factorization of 252.

23. Evaluate $9 + 4 \cdot 5$.

24. Evaluate $\dfrac{3 \cdot 4^2 - 2^2}{(2 - 1)^3}$.

25. Evaluate $10 + 2[12 - 2(6 - 4)]$.

26. GRADES   A student scored 73, 52, and 70 on three exams and received 0 on two missed exams. Find his average (mean) score.

27. Is 3 a solution of the equation $x + 13 = 16$? Explain why or why not.

Solve each equation. Check the result.

28. $100 = x + 1$

29. $y - 12 = 18$

30. $5t = 55$

31. $\dfrac{q}{3} = 27$

In Problems 32–33, let a variable represent the unknown quantity. Then write and solve an equation to answer the question.

32. PARKING   After many student complaints, a college decided to commit funds to double the number of parking spaces on campus. This increase would bring the total number of spaces up to 6,200. How many parking spaces does the college have at this time?

33. LIBRARY   A library building is 6 years shy of its 200th birthday. How old is the building at this time?

34. Explain what it means to *solve* an equation.

The Integers

2

2.1 An Introduction to the Integers

2.2 Addition of Integers

2.3 Subtraction of Integers

2.4 Multiplication of Integers

2.5 Division of Integers

2.6 Order of Operations and Estimation

2.7 Solving Equations Involving Integers

Key Concept: Signed Numbers

Accent on Teamwork

Chapter Review

Chapter Test

Cumulative Review Exercises

IN THIS CHAPTER, THE CONCEPT OF A NEGATIVE NUMBER IS INTRODUCED, AS WE EXPLORE AN EXTENSION OF THE SET OF WHOLE NUMBERS CALLED THE INTEGERS.

2.1 An Introduction to the Integers

In this section, you will learn about

- The integers • Extending the number line • Inequality symbols
- Absolute value • The opposite of a number • The − symbol

INTRODUCTION. Whole numbers are not adequate to describe many situations that arise in everyday life.

The record cold temperature in the state of Florida was 2 degrees *below* zero on February 13, 1899, in Tallahassee.

NUMBER	DATE	DESCRIPTION OF TRANSACTION	PAYMENT/DEBIT (−)		FEE (IF ANY) (−)	DEPOSIT/CREDIT (+)		BALANCE
		RECORD ALL CHARGES OR CREDITS THAT AFFECT YOUR ACCOUNT					$	450 00
1207	5/2	Wood's Auto Repair Transmission	$ 500 00		$	$		

A check for $500 was written when there was only $450 in the account. The checking account is *overdrawn*.

The American lobster is found off the East Coast of North America at depths as much as 600 feet *below* sea level.

In this section, we will see how negative numbers called *integers* (read as "int-i-jers") can be used to describe these three situations, as well as many others.

The integers

To describe a temperature of 2 degrees below zero, or $50 overdrawn, or 600 feet below sea level, we need to use negative numbers. **Negative numbers** are numbers less than 0, and they are written using a **negative sign, −**.

In words	In symbols	Read as
2 degrees below zero	−2	"negative two"
$50 overdrawn	−50	"negative fifty"
600 feet below sea level	−600	"negative six hundred"

Positive and negative numbers

> **Positive numbers** are greater than 0. **Negative numbers** are less than 0.

COMMENT   Zero is neither positive nor negative.

Positive and negative numbers are often referred to as **signed numbers.** Negative numbers must always be written with a negative sign. However, positive numbers are not always written with a **positive sign,** +. For example, the elevation of Mexico City, which is 7,110 feet above sea level, can be written as +7,110 feet or as 7,110 feet.

The collection of all positive whole numbers, negative whole numbers, and 0 is called the set of **integers.**

The set of integers | $\{ \ldots, -5, -4, -3, -2, -1, 0, 1, 2, 3, 4, 5, \ldots \}$

Since every natural number is also an integer, the natural numbers form a subset of the integers. Likewise, the whole numbers form a subset of the integers.

 COMMENT Note that not all integers are natural numbers, and that not all integers are whole numbers. For example, -2 is an integer, but it is neither a natural number nor a whole number.

Extending the number line

An excellent way to learn about negative numbers is with the help of a number line. Negative numbers can be represented on a number line by extending the line to the left. Beginning at the origin (the 0 point), we move left, marking equally spaced points and then labeling them with progressively smaller negative whole numbers. (See Figure 2-1.) As you move to the right on a number line, the values of the numbers increase. As you move to the left, the values decrease.

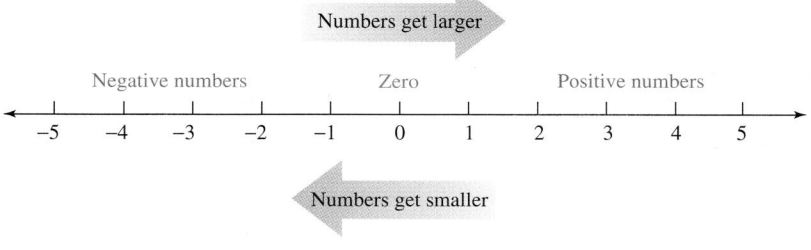

FIGURE 2-1

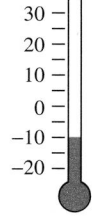

A thermometer is an example of a number line. The thermometer on the left is scaled in degrees, and it shows a temperature of $-10°$. In the study set, you will see examples of number lines illustrating historical and scientific situations that involve negative numbers.

EXAMPLE 1   *Graphing.*   Graph the integers $2, -3, 4,$ and -1.

Solution

To graph each integer, we locate its position on the number line and draw a heavy dot.

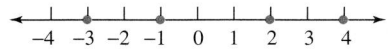

Self Check

Graph the integers $3, -4, 1,$ and -2.

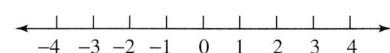

Answer:

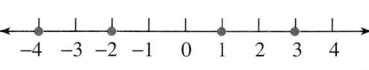

By extending the number line to include negative numbers, we can represent more situations using bar graphs and line graphs. For example, the bar graph shown in Figure 2-2 illustrates the annual profits *and losses* of Toys R Us over a five-year period. Note that the profit in 1998 was $490 million, and that the loss in 1999 was $132 million.

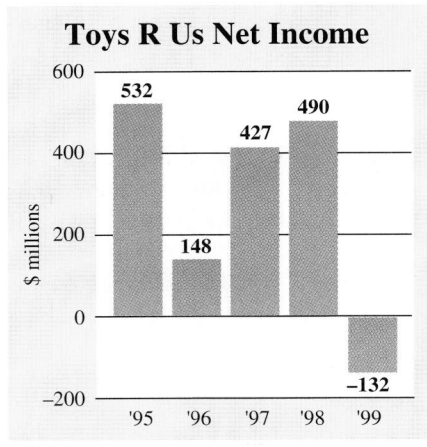

FIGURE 2-2

Inequality symbols

Figure 2-3 shows the graph of the integers -2 and 1. Because -2 is to the left of 1 on the number line, -2 is less than 1. We can use an inequality symbol to state this fact: $-2 < 1$. Since $-2 < 1$, it is also true that $1 > -2$.

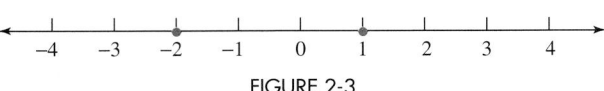

FIGURE 2-3

EXAMPLE 2   *Inequality symbols.* Use one of the symbols $>$ or $<$ to make each statement true: **a.** 4 ⬜ -5 and   **b.** -4 ⬜ -2.

Solution
a. Since 4 is to the right of -5 on the number line, $4 > -5$.
b. Since -4 is to the left of -2 on the number line, $-4 < -2$.

Self Check

Use one of the symbols $>$ or $<$ to make each statement true.

a. 6 ⬜ -6

b. -6 ⬜ -5

Answers: a. $>$,   **b.** $<$ ■

Absolute value

Using a number line, we can see that the numbers 3 and -3 are both a distance of 3 units away from 0, as shown in Figure 2-4.

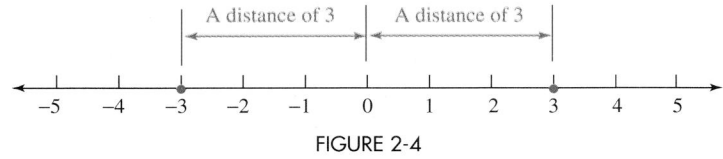

FIGURE 2-4

The **absolute value** of a number gives the distance between the number and 0 on a number line. To indicate absolute value, the number is inserted between two vertical bars. For the previous example, we would write $|-3| = 3$. This is read as "The absolute value of negative 3 is 3," and it tells us that the distance between -3 and 0 is 3 units. In the example, we also see that $|3| = 3$.

Absolute value

> The **absolute value** of a number is the distance on a number line between the number and 0.

 COMMENT Absolute value expresses distance. The absolute value of a number is always positive or zero, but never negative!

EXAMPLE 3 *Absolute value.* Evaluate each absolute value: **a.** $|8|$, **b.** $|-5|$, and **c.** $|0|$.

Solution

a. On a number line, the distance between 8 and 0 is 8. Therefore,

$$|8| = 8$$

b. On a number line, the distance between -5 and 0 is 5. Therefore,

$$|-5| = 5$$

c. On a number line, the distance between 0 and 0 is 0. Therefore,

$$|0| = 0$$

Self Check

Evaluate each absolute value:

a. $|-9|$

b. $|4|$

Answers: a. 9, **b.** 4 ∎

The opposite of a number

Opposites or negatives

> Two numbers represented by points on a number line that are the same distance from 0, but on opposite sides of it, are called **opposites** or **negatives.**

The numbers 4 and -4 are opposites because they are the same distance from zero. (See Figure 2-5.)

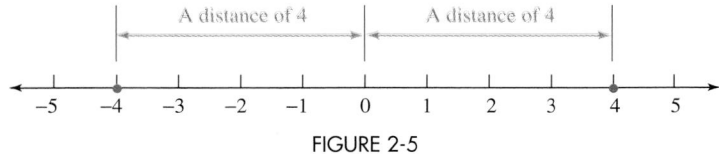

FIGURE 2-5

To write the opposite of a number, a $-$ symbol is used. For example, the opposite of 5 is -5 (read as "negative 5"). Parentheses are needed to express the opposite of a negative number. The opposite of -5 is written as $-(-5)$. Since 5 and -5 are the same distance from zero, the opposite of -5 is 5. Therefore, $-(-5) = 5$. This leads to the following conclusion.

Opposite of an opposite

> If a represents a number, then $-(-a) = a$.

In words, this rule says that *the opposite of the opposite of a number is that number.*

Number	Opposite	
57	−57	Read as "negative fifty-seven."
−8	−(−8) = 8	Read as "the opposite of negative eight." Apply the double negative rule.
0	−0 = 0	The opposite of 0 is 0.

The concept of opposite can also be applied to an absolute value. For example, the opposite of the absolute value of −8 can be written as $-|-8|$. Think of this as a two-step process. Find the absolute value first, and then attach the − to that result.

First, find the absolute value.

$$-\,|-8| \quad = \quad -8$$

Then attach a − sign.

EXAMPLE 4   *The opposite of a number.*   Simplify each expression:
a. −(−44)   and   **b.** $-|-225|$.

Solution

a. −(−44) means the opposite of −44. Since the opposite of −44 is 44, we write

$$-(-44) = 44$$

b. $-|-225|$ means the opposite of $|-225|$. Since $|-225| = 225$, and the opposite of 225 is −225, we write

$$-|-225| = -225$$

Self Check

Simplify each expression:
a. −(−1)   and   **b.** $-|-99|$.

Answers: **a.** 1,   **b.** −99   ■

The − symbol

The − symbol is used to indicate a negative number, the opposite of a number, and the operation of subtraction. The key to interpreting the − symbol correctly is to examine the context in which it is used.

Interpreting the − symbol

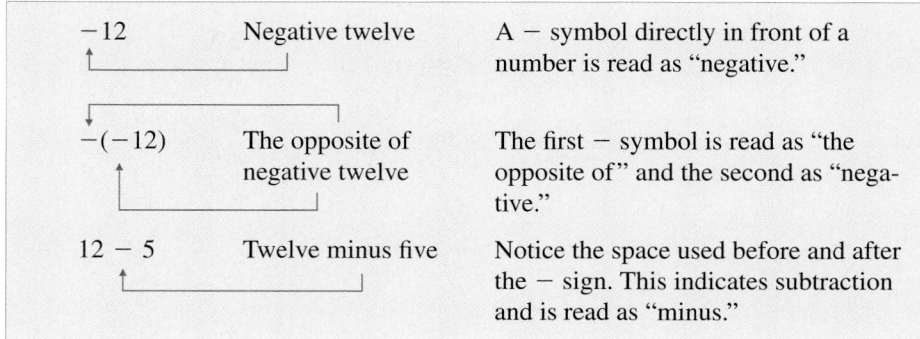

−12	Negative twelve	A − symbol directly in front of a number is read as "negative."
−(−12)	The opposite of negative twelve	The first − symbol is read as "the opposite of" and the second as "negative."
12 − 5	Twelve minus five	Notice the space used before and after the − sign. This indicates subtraction and is read as "minus."

STUDY SET   Section 2.1

VOCABULARY   *Fill in the blanks.*

1. _____ numbers are less than 0.

2. The collection of all positive whole numbers, negative whole numbers, and 0 is called the set of _____.

3. Numbers can be represented by points equally spaced on a _____ line.

4. To _____ a number means to locate it on a number line and highlight it with a dot.

5. The symbols $>$ and $<$ are called _____ symbols.

6. The _____ value of a number is the distance between it and zero on a number line.

7. Two numbers on a number line that are the same distance from zero, but on opposite sides of the origin, are called _____.

8. The opposite of the _____ of a number is that number.

CONCEPTS

9. As we move to the left on a number line, how do the values of the numbers change?

10. Tell what is wrong with each number line.

a.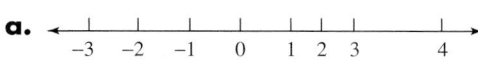
$$-3 \quad -2 \quad -1 \quad 0 \quad 1 \ 2 \ 3 \quad\quad 4$$

b.
$$-3 \quad -2 \quad -1 \quad 0 \quad 2 \quad 4 \quad 6 \quad 8$$

c.
$$-3 \quad -2 \quad -1 \quad 1 \quad 2 \quad 3 \quad 4 \quad 5$$

d.
$$-3 \quad -2 \quad -1 \quad 0 \quad 1 \quad 2 \quad 3 \quad 4$$

11. Does every integer have an opposite?

12. Is the absolute value of a number ever negative?

13. Which of the following contains a minus sign: $15 - 8$, $-(-15)$, or -15?

14. Is there a number that is both greater than 10 and less than 10 at the same time?

15. Express the fact $12 < 15$ using the $>$ symbol.

16. Express the fact $5 > 4$ using the $<$ symbol.

17. Represent each of these situations using a signed number.
  a. $225 overdrawn
  b. 10 seconds before liftoff
  c. 3 degrees below normal
  d. A trade deficit of $12,000

18. Represent each of these situations using a signed number, and then describe its opposite in words.
  a. A bacteria count 70 more than the standard

  b. A profit of $67
  c. A business $1 million in the "black"

  d. 20 units over their quota

19. On a number line, what number is 3 units to the right of -7?

20. On a number line, what number is 4 units to the left of 2?

21. What two numbers on a number line are a distance of 5 away from -3?

22. What two numbers on a number line are a distance of 4 away from 3?

23. Which number is closer to -3 on the number line, 2 or -7?

24. Which number is farther from 1 on the number line, -5 or 8?

25. Give examples of the $-$ symbol used in three different ways.

26. What is the opposite of 0?

NOTATION

27. Translate each phrase to mathematical symbols.
  a. The opposite of negative eight
  b. The absolute value of negative eight
  c. Eight minus eight
  d. The opposite of the absolute value of negative eight

28. Write the set of integers.

PRACTICE   *Simplify each expression.*

29. $|9|$

30. $|12|$

31. $|-8|$

32. $|-1|$

33. $|-14|$

34. $|-85|$

35. $-|20|$

36. $-|110|$

37. $-|-6|$

38. $|0|$

39. $|203|$

40. $-|-11|$

41. -0

42. $-|0|$

43. $-(-11)$

44. $-(-1)$

45. $-(-4)$

46. $-(-9)$

47. $-(-1,201)$

48. $-(-255)$

Graph each set of numbers on a number line labeled from -5 to 5.

49. $\{-3, 0, 3, 4, -1\}$

50. $\{-4, -1, 2, 5, 1\}$

51. The opposite of -3, the opposite of 5, and the absolute value of -2

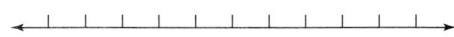

52. The absolute value of 3, the opposite of 3, and the number that is 1 less than −3

Insert one of the symbols > or < in the blank to make a true statement.

53. −5 ____ 5

54. 0 ____ −1

55. −12 ____ −6

56. −6 ____ −7

57. −10 ____ −11

58. −11 ____ −20

59. |−2| ____ 0

60. |−30| ____ −40

61. −1,255 ____ −(−1,254)

62. 0 ____ −3

63. −|−3| ____ 4

64. −|−163| ____ −150

APPLICATIONS

65. FLIGHT OF A BALL   A boy throws a ball from the top of a building, as shown in Illustration 1. At the instant he does this, his friend starts a stopwatch and keeps track of the time as the ball rises to a peak and then falls to the ground. Use the vertical number line to complete the table at the top of the page by finding the position of the ball at the specified times.

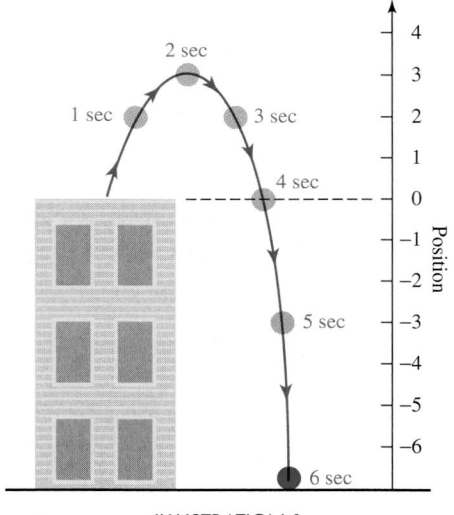

ILLUSTRATION 1

Time	Position of ball
1 sec	
2 sec	
3 sec	
4 sec	
5 sec	
6 sec	

66. SHOOTING GALLERY   At an amusement park, a shooting gallery contains moving ducks. The path of one duck is shown in Illustration 2, along with the time it takes the duck to reach certain positions on the gallery wall. Complete the table using the horizontal number line in the illustration.

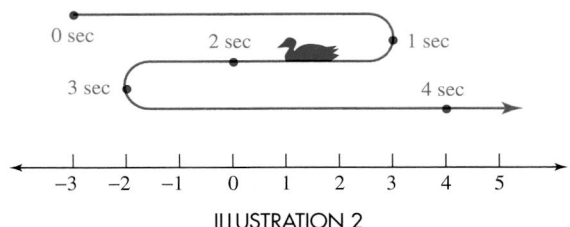

ILLUSTRATION 2

Time	Position of duck
0 sec	
1 sec	
2 sec	
3 sec	
4 sec	

67. TECHNOLOGY   The readout from a testing device is shown in Illustration 3. It is important to know the height of each of the three "peaks" and the depth of each of the three "valleys." Use the vertical number line to find these numbers.

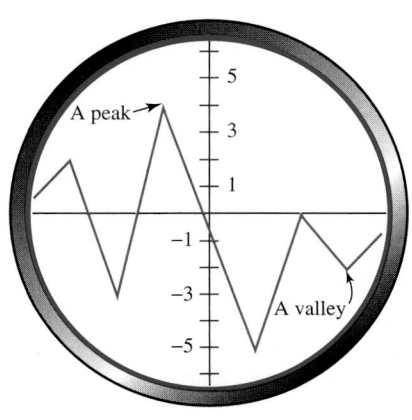

ILLUSTRATION 3

68. FLOODING   A week of daily reports listing the height of a river in comparison to flood stage is given in the table. Complete the bar graph in Illustration 4.

Flood stage report	
Sun.	2 ft below
Mon.	3 ft over
Tue.	4 ft over
Wed.	2 ft over
Thu.	1 ft below
Fri.	3 ft below
Sat.	4 ft below

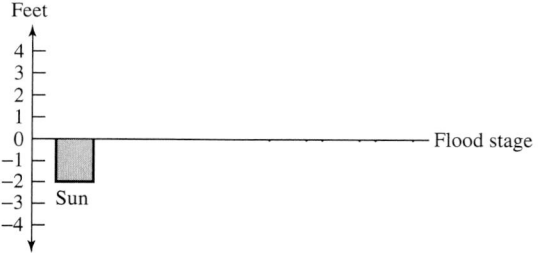

ILLUSTRATION 4

69. GOLF   In golf, *par* is the standard number of strokes considered necessary on a given hole. A score of −2 indicates that a golfer used 2 strokes less than par. A score of +2 means 2 more strokes than par were used. In Illustration 5, each golf ball represents the score of a professional golfer on the 16th hole of a certain course.

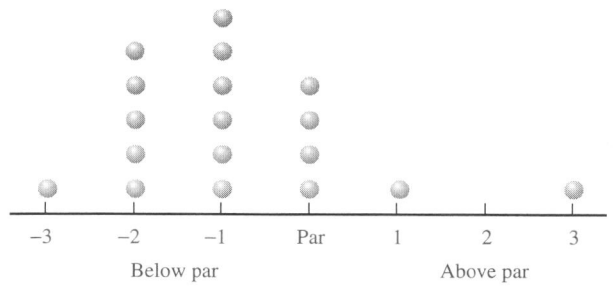

ILLUSTRATION 5

a. What score was shot most often on this hole?

b. What was the best score on this hole?

c. Explain why this hole appears to be too easy for a professional golfer.

70. PAYCHECK   Examine the items listed on the paycheck stub in Illustration 6. Then write two columns on your paper—one headed "positive" and the other "negative." List each item under the appropriate heading.

Tom Dryden Dec. 01		Christmas bonus	$100
Gross pay	$2,000	**Reductions**	
Overtime	$300	Retirement	$200
Deductions		**Taxes**	
Union dues	$30	Federal withholding	$160
U.S. Bonds	$100	State withholding	$35

ILLUSTRATION 6

71. WEATHER MAP   Illustration 7 shows the predicted Fahrenheit temperatures for a day in mid-January.

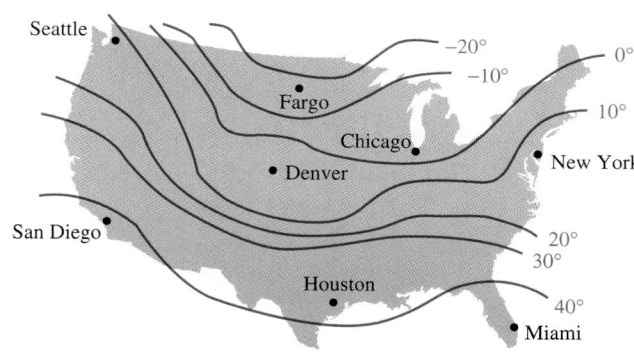

ILLUSTRATION 7

a. What is the temperature range for the region including Fargo, North Dakota?

b. Approximately how much colder will it be in Chicago than in Denver?

c. According to this prediction, what is the coldest it should get in Seattle?

72. PROFITS/LOSSES   The graph in Illustration 8 shows the net income of Apple Computer Inc. for the years 1990–1999.

a. In what years did the company suffer a loss? Estimate each loss.

b. Explain why an article in the business section of a newspaper, when referring to this graph, would say, "Apple earnings turned the corner in 1998."

Based on data from
Los Angeles Times
(October 14, 1999)

ILLUSTRATION 8

73. HISTORICAL TIME LINE   Number lines can be used to display historical data. Some important world events are shown on the time line in Illustration 9.

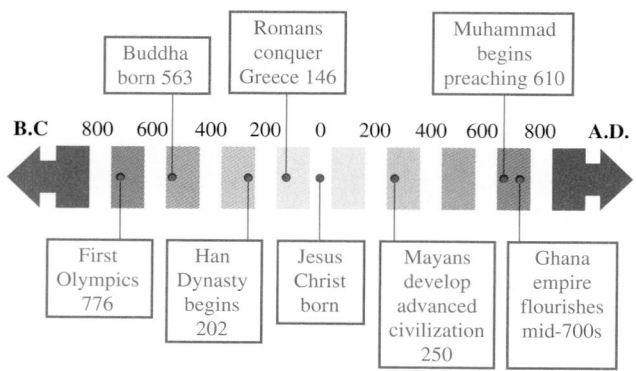

ILLUSTRATION 9

a. What basic unit was used to scale this time line?

b. What could be thought of as positive numbers?

c. What could be thought of as negative numbers?

d. What important event distinguishes the positive from the negative numbers?

74. ASTRONOMY   Astronomers use a type of number line called the *apparent magnitude scale* to denote the brightness of objects in the sky. The brighter an object appears to an observer on Earth, the more negative is its apparent magnitude. Graph each of the following on the scale in Illustration 10.

- Full moon -12
- Pluto $+15$
- Sirius (brightest star) -2
- Sun -26
- Venus -4
- Visual limit of binoculars $+10$
- Visual limit of large telescope $+20$
- Visual limit of naked eye $+6$

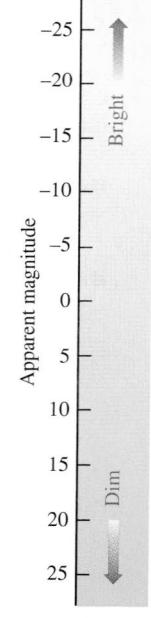

ILLUSTRATION 10

75. LINE GRAPH   Each thermometer in Illustration 11 gives the daily high temperature in degrees Fahrenheit. Plot each daily high temperature on the grid and then construct a line graph.

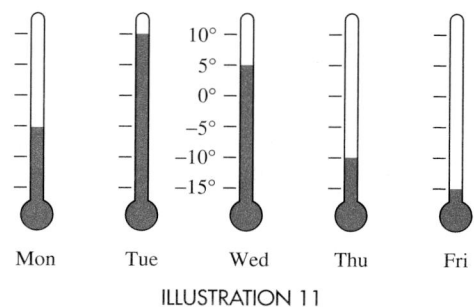

ILLUSTRATION 11

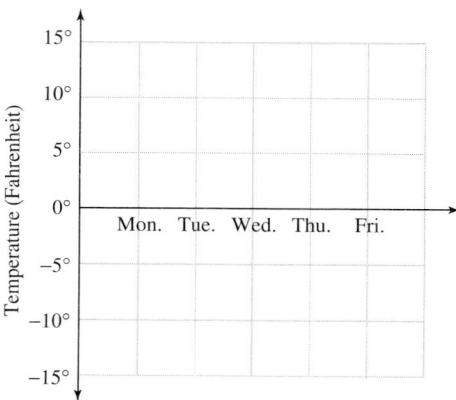

76. GARDENING   Illustration 12 shows the depths at which the bottoms of various types of flower bulbs should be planted. (The symbol ″ represents inches.)

a. At what depth should a tulip bulb be planted?

b. How much deeper are hyacinth bulbs planted than gladiolus bulbs?

c. Which bulb must be planted the deepest? How deep?

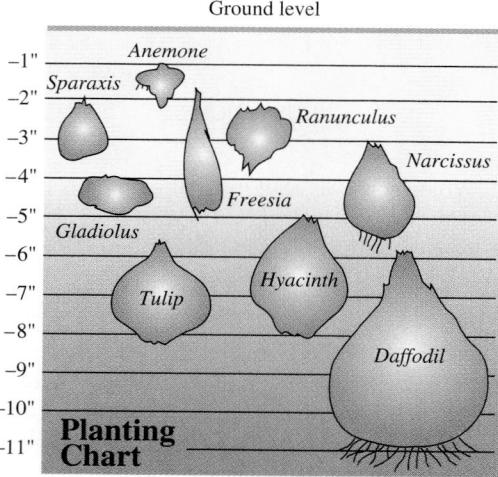

ILLUSTRATION 12

WRITING

77. Explain the concept of the opposite of a number.

78. What real-life situation do you think gave rise to the concept of a negative number?

79. Explain why the absolute value of a number is never negative.

80. Give an example of the use of a number line that you have seen in another course.

81. DIVING   Divers use the terms *positive buoyancy, neutral buoyancy,* and *negative buoyancy* as shown in Illustration 13. What do you think each of these terms means?

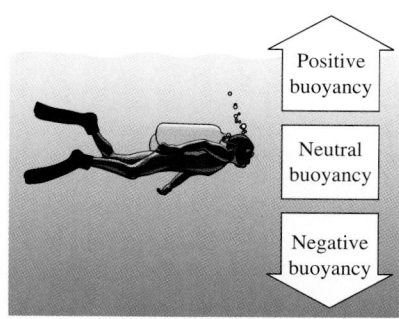

ILLUSTRATION 13

82. NEW ORLEANS   The city of New Orleans, Louisiana, lies largely below sea level. Find out why the city is not under water.

REVIEW

83. Round 23,456 to the nearest hundred.

84. Evaluate: $19 - 2 \cdot 3$.

85. Solve $2x = 34$.

86. Is this statement true or false? $345 < 354$

87. Give the name of the property illustrated here: $(13 \cdot 2) \cdot 5 = 13 \cdot (2 \cdot 5)$

88. Write four times five using three different notations.

2.2 *Addition of Integers*

In this section, you will learn about

- Adding two integers with the same sign
- Adding two integers with different signs • The addition property of zero
- The additive inverse of a number

INTRODUCTION. A dramatic change in temperature occurred in 1943 in Spearfish, South Dakota. On January 22, at 7:30 A.M., the temperature was $-4°$F. In just two minutes, the temperature rose 49 degrees! To calculate the temperature at 7:32 A.M., we need to add 49 to -4.

$$-4 + 49$$

To perform this addition, we must know how to add positive and negative integers. In this section, we will develop rules to help us make such calculations.

? — | 7:32 A.M.

49° increase

$-4°$ F — | 7:30 A.M.

Adding two integers with the same sign

$4 + 3$
both positive

To explain addition of signed numbers, we can use a number line. (See Figure 2-6 on the next page.) To compute $4 + 3$, we begin at the **origin** (the point labeled 0) and draw an arrow 4 units long, pointing to the right. This represents positive 4. From that point, we draw an arrow 3 units long, pointing to the right, to represent positive 3. The second arrow points to the answer. Therefore, $4 + 3 = 7$.

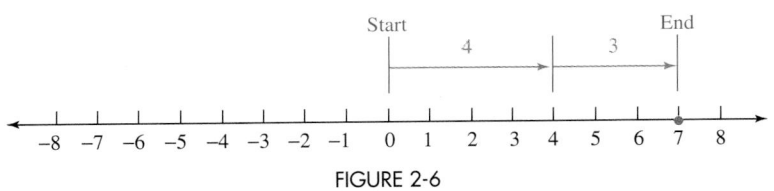

FIGURE 2-6

To check our work, let's think of the problem in terms of money. If you had $4 and earned $3 more, you would have a total of $7.

$$-4 + (-3)$$
both negative

To compute $-4 + (-3)$, we begin at the origin and draw an arrow 4 units long, pointing to the left. (See Figure 2-7.) This represents -4. From there, we draw an arrow 3 units long, pointing to the left, to represent -3. The second arrow points to the answer: $-4 + (-3) = -7$.

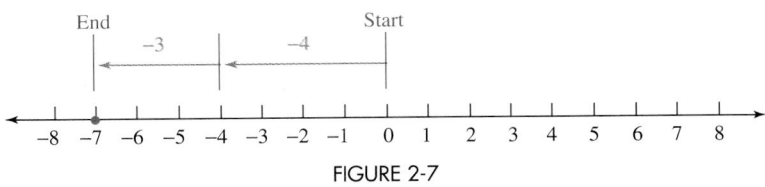

FIGURE 2-7

Let's think of this problem in terms of money. If you had a debt of $4 (negative 4) and then incurred $3 more debt (negative 3), you would be in debt $7 (negative 7).

Here are some observations about the process of adding two numbers that have the same sign, using a number line.

- Both arrows point in the same direction and "build" on each other.

- The answer has the same sign as the two numbers being added.

$$\begin{array}{ccccccc} 4 & + & 3 & = & 7 \\ \text{positive} & + & \text{positive} & = & \text{positive} \\ & & & & \text{answer} \end{array} \qquad \begin{array}{ccccccc} -4 & + & (-3) & = & -7 \\ \text{negative} & + & \text{negative} & = & \text{negative} \\ & & & & \text{answer} \end{array}$$

These observations suggest the following rule.

Adding two integers with the same sign

To add two integers with the **same sign,** add their absolute values and attach their common sign to the sum. If both integers are positive, their sum is positive. If both integers are negative, their sum is negative.

COMMENT   When writing additions that involve signed numbers, write negative numbers within parentheses to separate the negative sign $-$ from the plus sign $+$.

$$9 + (-4) \qquad \cancel{9 + -4} \qquad \text{and} \qquad -9 + (-4) \qquad \cancel{-9 + -4}$$

EXAMPLE 1   *Adding two negative integers.*   Find the sum: $-9 + (-4)$.

Solution   *Step 1:* To add two integers with the same sign, we first add the absolute values of the integers. Since $|-9| = 9$ and $|-4| = 4$, we begin by adding 9 and 4.

$$9 + 4 = 13$$

Step 2: We then attach the common sign (which is negative) to this result. Therefore,

$$-9 + (-4) = -13$$

└── Make the answer negative.

After some practice, you will be able to do this kind of problem in your head. It will not be necessary to show all the steps as we have done here. ∎

EXAMPLE 2   *Adding two negative integers.*   Find the sum: $-80 + (-60)$.

Solution

Since both integers are negative, the answer will be negative.

$-80 + (-60) = -140$ Because the numbers have the same sign, add their absolute values, 80 and 60, to get 140. Then attach the common negative sign.

Self Check
Find the sum: $-300 + (-100)$.

Answer: -400 ∎

Adding two integers with different signs

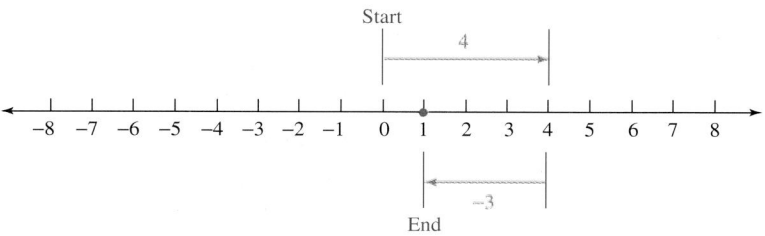

$4 + (-3)$
one positive, one negative

To compute $4 + (-3)$, we start at the origin and draw an arrow 4 units long, pointing to the right. (See Figure 2-8.) This represents positive 4. From there, we draw an arrow 3 units long, pointing to the left, to represent -3. The second arrow points to the answer: $4 + (-3) = 1$.

Start

4

−8 −7 −6 −5 −4 −3 −2 −1 0 1 2 3 4 5 6 7 8

−3

End

FIGURE 2-8

In terms of money, if you had \$4 (positive 4) and then incurred a debt of \$3 (negative 3), you would have \$1 (positive 1) left.

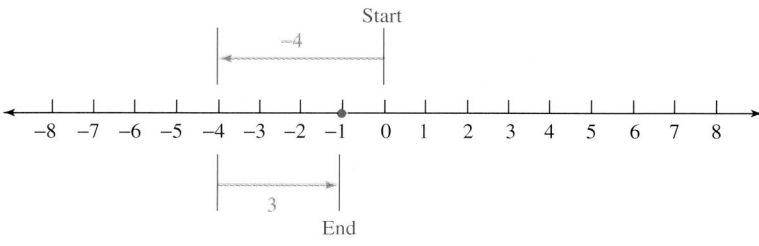

$-4 + 3$
one positive, one negative

The problem $-4 + 3$ can be illustrated by drawing an arrow 4 units long from the origin, pointing to the left. (See Figure 2-9.) This represents -4. From there, we draw an arrow 3 units long, pointing to the right, to represent positive 3. The second arrow points to the answer: $-4 + 3 = -1$.

Start

−4

−8 −7 −6 −5 −4 −3 −2 −1 0 1 2 3 4 5 6 7 8

3

End

FIGURE 2-9

This problem can be thought of as owing \$4 (negative 4) and then paying back \$3 (positive 3). You will still owe \$1 (negative 1).

The last two examples lead us to some observations about adding two integers with different signs, using a number line.

• The arrows representing the integers point in opposite directions.
• The longer of the two arrows determines the sign of the answer.

These observations suggest the following rule.

Adding two integers with different signs

> To add two integers with **different signs,** subtract their absolute values, the smaller from the larger. Then attach to that result the sign of the integer with the larger absolute value.

EXAMPLE 3 *Adding a positive and a negative integer.* Find the sum: $5 + (-7)$.

Solution *Step 1:* To add two integers with different signs, we first subtract the smaller absolute value from the larger absolute value. Since $|5|$, which is 5, is smaller than $|-7|$, which is 7, we begin by subtracting 5 from 7.

$$7 - 5 = 2$$

Step 2: -7 has the larger absolute value, so we attach a negative sign to the result from Step 1. Therefore,

$$5 + (-7) = -2$$

— Make the answer negative.

EXAMPLE 4 *Adding a positive and a negative integer.* Find the sum:
a. $-8 + 5$ and **b.** $11 + (-5)$.

Solution

a. Since -8 has the larger absolute value, the answer will be negative.

$-8 + 5 = -3$ Because the signs of the numbers are different, subtract their absolute values, 5 from 8, to get 3. Attach a negative sign to that result.

b. Since 11 has the larger absolute value, the answer will be positive.

$11 + (-5) = 6$ Subtract the absolute values, 5 from 11, to get 6. The answer is positive.

Self Check
Find the sum:

a. $-2 + 7$

b. $6 + (-9)$

Answers: a. 5, **b.** -3

EXAMPLE 5 *Temperature change.* In the introduction to this section, we learned that at 7:30 A.M. on January 22, 1943, in Spearfish, South Dakota, the temperature was $-4°$. The temperature then rose 49 degrees in just two minutes. What was the temperature at 7:32 A.M.?

Solution The phrase *temperature rose 49 degrees* indicates addition. We need to add 49 to -4.

$-4 + 49 = 45$ Subtract the smaller absolute value, 4, from the larger absolute value, 49. The sum is positive.

At 7:32 A.M., the temperature was $45°$F.

EXAMPLE 6 *Adding several integers.* Add: $-3 + 5 + (-12) + 2$.

Solution
This expression contains four integers. We add them, working from left to right.

$$\begin{aligned}
-3 + 5 + (-12) + 2 &= 2 + (-12) + 2 \quad \text{Add: } -3 + 5 = 2. \\
&= -10 + 2 \quad\quad\quad \text{Add: } 2 + (-12) = -10. \\
&= -8
\end{aligned}$$

Self Check
Add: $-12 + 8 + (-6) + 1$.

Answer: -9

An alternate approach to problems like Example 6 is to add all the positive numbers, add all the negative numbers, and then add those results.

EXAMPLE 7 *Adding several integers.* Find the sum: $-3 + 5 + (-12) + 2$.

Solution

We can use the commutative property of addition to reorder the numbers and use the associative property of addition to group the positives together and the negatives together.

$$-3 + 5 + (-12) + 2 = 5 + 2 + (-3) + (-12) \qquad \text{Reorder the numbers.}$$
$$= (5 + 2) + [(-3) + (-12)] \quad \text{Group the positives.}$$
$$\text{Group the negatives.}$$

We do the operations inside the grouping symbols first.

$$(5 + 2) + [(-3) + (-12)] = 7 + (-15) \qquad \text{Add the positives. Add the negatives.}$$
$$= -8 \qquad \text{Add the numbers with different signs.}$$

Self Check

Find the sum:

$$-12 + 8 + (-6) + 1$$

Answer: -9 ∎

Accent on Technology: **Entering negative numbers**

The United States' largest trading partner in Africa is Nigeria. To calculate the 1998 U.S. trade balance with Nigeria, we add the $816,800,000 worth of exports to Nigeria (considered positive) to the $4,194,000,000 worth of imports from Nigeria (considered negative). We can use a scientific calculator to do the addition: $816,800,000 + (-4,194,000,000)$.

- We do not have to do anything special to enter a positive number. When we key in 816,800,000, a positive number is entered.
- To enter $-4,194,000,000$, we press the change-of-sign key $\boxed{+/-}$ *after* entering 4,194,000,000. Note that the change-of-sign key is different from the subtraction key $\boxed{-}$.

Keystrokes   816800000 $\boxed{+}$ 4194000000 $\boxed{+/-}$ $\boxed{=}$ $\boxed{-3377200000}$

In 1998, the United States had a trade balance of $-\$3,377,200,000$ with Nigeria. Because the result is negative, it is called a *trade deficit.*

The addition property of zero

When 0 is added to a number, the number remains the same. For example, $5 + 0 = 5$, and $0 + (-4) = -4$. Because of this, we call 0 the **additive identity.**

Addition property of zero

> For any number a,
>
> $$a + 0 = a \qquad \text{and} \qquad 0 + a = a$$

The additive inverse of a number

A second fact concerning 0 and the operation of addition can be demonstrated by considering the sum of a number and its opposite. To illustrate this, we use the number line in Figure 2-10 on page 92 to add 6 and its opposite, -6. We see that $6 + (-6) = 0$.

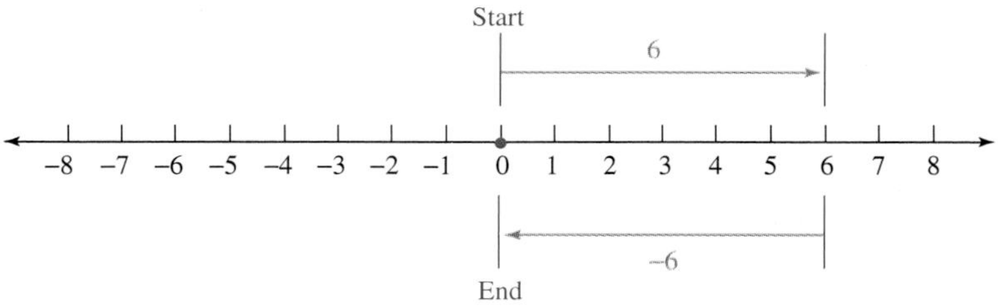

FIGURE 2-10

If the sum of two numbers is 0, the numbers are said to be **additive inverses** of each other. Since $6 + (-6) = 0$, we say that 6 and -6 are additive inverses.

We can now classify a pair of numbers such as 6 and -6 in three ways: as opposites, negatives, or additive inverses.

The additive inverse of a number

For any numbers a and b, if $a + b = 0$, then a and b are called **additive inverses.**

EXAMPLE 8 *The additive inverse.* What is the additive inverse of -3? Justify your result.

Solution

The additive inverse of -3 is its opposite, 3. To justify the result, we add and show that the sum is 0.

$$-3 + 3 = 0$$

Self Check

What is the additive inverse of 12? Justify your result.

Answer: -12;
$12 + (-12) = 0$ ∎

STUDY SET Section 2.2

VOCABULARY *Fill in the blanks.*

1. When 0 is added to a number, the number remains the same. We call 0 the additive _____.

2. Since $-5 + 5 = 0$, we say that 5 is the additive _____ of -5. We can also say that 5 and -5 are _____.

CONCEPTS *In Exercises 3–6, find each answer using a number line.*

3. $-3 + 6$

4. $-3 + (-2)$

5. $-5 + 3$

6. $-1 + (-3)$

7. a. Is the sum of two positive integers always positive?
   b. Is the sum of two negative integers always negative?

8. a. What is the sum of a number and its additive inverse?
   b. What is the sum of a number and its opposite?

9. Find each absolute value.
   a. $|-7|$ **b.** $|10|$

10. If the sum of two numbers is 0, what can be said about the numbers?

Fill in the blanks.

11. To add two integers with unlike signs, _____ their absolute values, the smaller from the larger. Then attach to that result the sign of the number with the _____ absolute value.

12. To add two integers with like signs, add their _____ values and attach their common _____ to the sum.

NOTATION *Complete each solution.*

13. Evaluate $-16 + (-2) + (-1)$.
$$-16 + (-2) + (-1) = \boxed{} + (-1)$$
$$= -19$$

14. Evaluate $-8 + (-2) + 6$.
$$-8 + (-2) + 6 = \boxed{} + 6$$
$$= -4$$

15. Evaluate $(-3 + 8) + (-3)$.
$$(-3 + 8) + (-3) = \boxed{} + (-3)$$
$$= 2$$

16. Evaluate $-5 + [2 + (-9)]$.
$$-5 + [2 + (-9)] = -5 + (\boxed{})$$
$$= -12$$

17. Explain why the expression $-6 + -5$ is not written correctly. How should it be written?

18. What mathematical symbol is implied when the word *sum* is used?

PRACTICE *Find the additive inverse of each number.*

19. -11 **20.** 9
21. -23 **22.** -43
23. 0 **24.** 1
25. 99 **26.** 250

Find each sum.

27. $-6 + (-3)$ **28.** $-2 + (-3)$
29. $-5 + (-5)$ **30.** $-8 + (-8)$
31. $-6 + 7$ **32.** $-2 + 4$
33. $-15 + 8$ **34.** $-18 + 10$
35. $20 + (-40)$ **36.** $25 + (-10)$
37. $30 + (-15)$ **38.** $8 + (-20)$
39. $-1 + 9$ **40.** $-2 + 7$
41. $-7 + 9$ **42.** $-3 + 6$
43. $5 + (-15)$ **44.** $16 + (-26)$
45. $24 + (-15)$ **46.** $-4 + 14$
47. $35 + (-27)$ **48.** $46 + (-73)$
49. $24 + (-45)$ **50.** $-65 + 31$

Evaluate each expression.

51. $-2 + 6 + (-1)$
52. $4 + (-3) + (-2)$
53. $-9 + 1 + (-2)$
54. $5 + 4 + (-6)$
55. $6 + (-4) + (-13) + 7$
56. $8 + (-5) + (-10) + 6$
57. $9 + (-3) + 5 + (-4)$
58. $-3 + 7 + 1 + (-4)$
59. Find the sum of $-6, -7,$ and -8.
60. Find the sum of $-11, -12,$ and -13.

Find each sum.

61. $-7 + 0$ **62.** $6 + 0$
63. $9 + 0$ **64.** $0 + (-15)$

65. $-4 + 4$ **66.** $18 + (-18)$
67. $2 + (-2)$ **68.** $-10 + 10$

69. What number must be added to -5 to obtain 0?
70. What number must be added to 8 to obtain 0?

Evaluate each expression.

71. $2 + (-10 + 8)$
72. $(-9 + 12) + (-4)$
73. $(-4 + 8) + (-11 + 4)$
74. $(-12 + 6) + (-6 + 8)$
75. $[-3 + (-4)] + (-5 + 2)$
76. $[9 + (-10)] + (-7 + 9)$
77. $[6 + (-4)] + [8 + (-11)]$
78. $[5 + (-8)] + [9 + (-15)]$
79. $-2 + [-8 + (-7)]$
80. $-8 + [-5 + (-2)]$
81. $789 + (-9,135)$
82. $2,701 + (-4,089)$
83. $-675 + (-456) + 99$
84. $-9,750 + (-780) + 2,345$

APPLICATIONS *Use signed numbers to help answer each question.*

85. G FORCES   As a fighter pilot dives and loops, different forces are exerted on the body, just like the forces you experience when riding on a roller coaster. Some of the forces, called G's, are positive and some are negative. The force of gravity, 1G, is constant. Complete the diagram in Illustration 1.

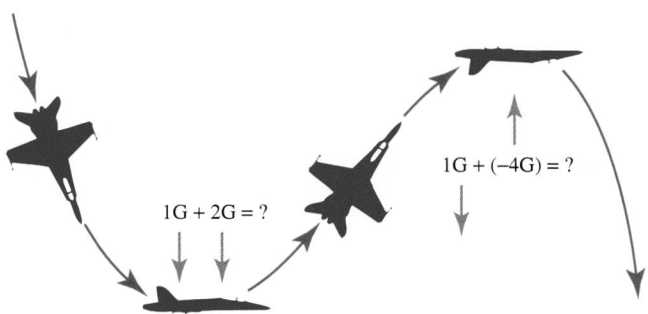

ILLUSTRATION 1

86. CHEMISTRY   The first several steps of a chemistry lab experiment are shown below and on the next page. The experiment begins with a compound that is stored at $-40°$ F.

> *Step 1:* Raise the temperature of the compound $200°$.

Step 2: Add sulfur and then raise the temperature 10°.

Step 3: Add 10 milliliters of water, stir, and raise the temperature 25°.

What is the resulting temperature of the mixture after step 3?

87. CASH FLOW   The maintenance costs, utilities, and taxes on a duplex are $900 per month. The owner of the apartments receives monthly rental payments of $450 and $380. Does this investment produce a positive cash flow each month?

88. JOGGING   A businessman's lunchtime workout includes jogging up 10 stories of stairs in his high-rise office building. If he starts on the fourth level below ground in the underground parking garage, on what story of the building will he finish his workout?

89. MEDICAL QUESTIONNAIRE   Determine the risk of contracting heart disease for the man whose responses are shown in Illustration 2.

Age		Total Cholesterol	
Age	Points	Reading	Points
34	−1	150	−3
Cholesterol		Blood Pressure	
HDL	Points	Systolic/Diastolic	Points
62	−2	124/100	3
Diabetic		Smoker	
	Points		Points
Yes	2	Yes	2
10-Year Heart Disease Risk			
Total Points	**Risk**	Total Points	**Risk**
−2 or less	1%	5	4%
−1 to 1	2%	6	6%
2 to 3	3%	7	6%
4	4%	8	7%

Source: National Heart, Lung, and Blood Institute

ILLUSTRATION 2

90. SPREADSHEET   Monthly rain totals for four counties are listed in the spreadsheet shown in Illustration 3. The −1 entered in cell B1 means that the rain total for Suffolk County for a certain month was one inch below average. We can analyze this data by asking the computer to perform various operations.
  a. To ask the computer to add the numbers in cells C1, C2, C3, and C4, we type SUM(C1:C4). Find this sum.
  b. Find SUM(B4:F4).

	A	**B**	**C**	**D**	**E**	**F**
1	Suffolk	−1	−1	0	+1	+1
2	Marin	0	−2	+1	+1	−1
3	Logan	−1	+1	+2	+1	+1
4	Tipton	−2	−2	+1	−1	−3

ILLUSTRATION 3

91. ATOMS   An atom is composed of protons, neutrons, and electrons. A proton has a positive charge (represented by +1), a neutron has no charge, and an electron has a negative charge (−1). Two simple models of atoms are shown in Illustration 4. What is the net charge of each atom?

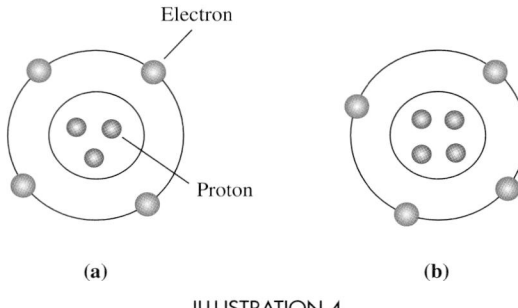

(a)   **(b)**

ILLUSTRATION 4

92. POLITICAL POLLS   Six months before a general election, the incumbent senator found himself trailing the challenger by 18 points. To overtake his opponent, the campaign staff decided to use a four-part strategy. Each part of this plan is shown below, with the anticipated point gain.

  1. Intense TV ad blitz   +10
  2. Ask for union endorsement   +2
  3. Voter mailing   +3
  4. Get-out-the vote campaign   +1

With these gains, will the incumbent overtake the challenger on election day?

93. FLOODING   After a heavy rainstorm, a river that had been 4 feet under flood stage rose 11 feet in a 48-hour period. Find the height of the river after the storm in comparison to flood stage.

94. MILITARY SCIENCE   During a battle, an army retreated 1,500 meters, regrouped, and advanced 3,500 meters. The next day, it had to retreat 1,250 meters. Find the army's net gain.

95. FILM PROFITS   A movie studio produced four films, two financial successes and two failures. The profits and losses of the films are shown in Illustration 5. Find the studio's profit, if any, for the year.

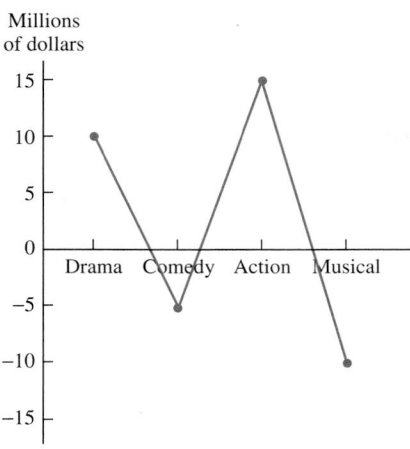

Millions of dollars

ILLUSTRATION 5

96. ACCOUNTING   On a financial balance sheet, debts (negative numbers) are denoted within parentheses. Assets (positive numbers) are written without parentheses. What is the 1999 fund balance for the preschool whose financial records are shown in Illustration 6?

Community Care Preschool Balance Sheet, June 1999	
Fund balances	
Classroom supplies	$ 5,889
Emergency needs	927
Holiday program	(2,928)
Insurance	1,645
Janitorial	(894)
Licensing	715
Maintenance	(6,321)
BALANCE	?

ILLUSTRATION 6

WRITING

97. Is the sum of a positive and a negative number always positive? Explain why or why not.

98. How do you explain the fact that when asked to *add* -4 and 8, we must actually *subtract* to obtain the result?

99. Why is the sum of two negative numbers a negative number?

100. Write an application problem that will require adding -50 and -60.

REVIEW

101. Find the area of the rectangle in Illustration 7.

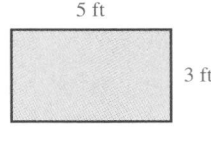

5 ft

3 ft

ILLUSTRATION 7

102. A car with a tank that holds 15 gallons of gasoline goes 25 miles on one gallon. How far can it go on a full tank?

103. Solve $x - 7 = 20$.

104. Is $t = 4$ a solution of $3t = 12$?

105. Prime factor 125. Use exponents to express the result.

106. Do the division: $\dfrac{144}{12}$.

2.3 *Subtraction of Integers*

In this section, you will learn about

• Adding the opposite • Order of operations • Applications of subtraction

INTRODUCTION. In this section, we will study another way to think about subtraction. This new procedure is helpful when subtraction problems involve negative numbers.

Adding the opposite

The subtraction problem $6 - 4$ can be thought of as taking away 4 from 6. We can use a number line to illustrate this. (See Figure 2-11 on the next page.) Beginning at the origin, we draw an arrow of length 6 units in the positive direction. From that point, we move back 4 units to the left. The answer, called the **difference,** is 2.

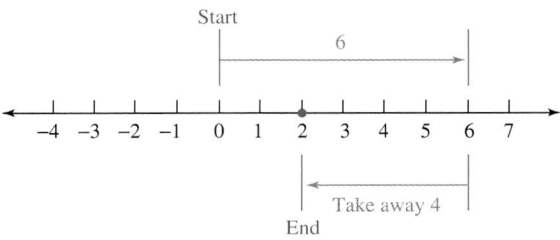

FIGURE 2-11

The work shown in Figure 2-11 looks like the illustration for the *addition* problem $6 + (-4) = 2$, shown in Figure 2-12.

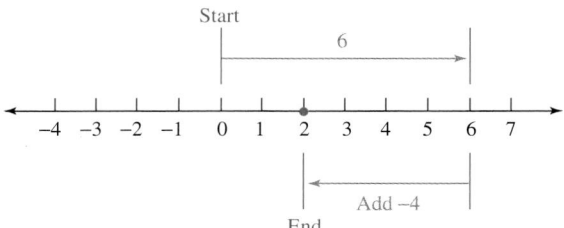

FIGURE 2-12

In the first problem, $6 - 4$, we subtracted 4 from 6. In the second, $6 + (-4)$, we added -4 (which is the opposite of 4) to 6. In each case, the result was 2.

$$\begin{array}{cc} \text{Subtracting 4.} & \text{Adding the opposite of 4.} \\ \downarrow & \downarrow \\ 6 - 4 = 2 & 6 + (-4) = 2 \end{array}$$

The same result.

This observation helps to justify the following rule for subtraction.

Rule for subtraction

> If a and b represent numbers, then
>
> $$a - b = a + (-b)$$
>
> In words, this rule says that subtraction is the same as adding the opposite of the number to be subtracted.

You won't need to use this rule for every subtraction problem. For example, $6 - 4$ is obviously 2; it does not need to be rewritten as adding the opposite. But for more complicated problems such as $-6 - 4$ or $3 - (-5)$, where the result is not obvious, the subtraction rule will be quite helpful.

EXAMPLE 1   *Adding the opposite.*   Find $-6 - 4$.

Solution

The number to be subtracted is 4. Applying the subtraction rule, we write

$$-6 - 4 = -6 + (-4)$$ Write the subtraction as an addition of the opposite of 4, which is -4. Write -4 within parentheses.

$$= -10$$ To add -6 and -4, apply the rule for adding two negative numbers.

Self Check

Find $-2 - 3$ and check the result.

To check the result, we add the difference, -10, and the subtrahend, 4. We should obtain the minuend, -6.

$$-10 + 4 = -6$$

The answer, -10, checks.

Answer: -5 ■

EXAMPLE 2 *Adding the opposite.* Find $3 - (-5)$.

Solution

The number being subtracted is -5.

$3 - (-5) = 3 + 5$	Write the subtraction as an addition of the opposite of -5, which is 5.
$= 8$	Do the addition.

Self Check
Find $3 - (-2)$.

Answer: 5 ■

EXAMPLE 3 *Adding the opposite.* Subtract -3 from -8.

Solution

The number being subtracted is -3, so we write it after -8.

$-8 - (-3) = -8 + 3$	Add the opposite of -3, which is 3.
$= -5$	Do the addition.

Self Check
Find $-2 - (-6)$.

Answer: 4 ■

Remember that any subtraction problem can be rewritten as an equivalent addition. We just add the opposite of the number that is to be subtracted.

Subtraction can be written as addition . . .

$$4 - \ \ 8 \ \ = \ \ 4 + (-8) = -4$$
$$4 - (-8) = \ \ 4 + \ \ 8 = 12$$
$$-4 - \ \ 8 \ \ = -4 + (-8) = -12$$
$$-4 - (-8) = -4 + \ \ 8 = 4$$

of the opposite of the
number to be subtracted.

Order of operations

Expressions can contain repeated subtraction or subtraction in combination with grouping symbols. To work these problems, we apply the rules for the order of operations, listed on page 44.

EXAMPLE 4 *Repeated subtraction.* Evaluate: $-1 - (-2) - 10$.

Solution

This subtraction problem involves three numbers. We work from left to right, rewriting each subtraction as an addition of the opposite.

$-1 - (-2) - 10 = -1 + 2 + (-10)$	Add the opposite of -2, which is 2. Add the opposite of 10, which is -10. Write -10 in parentheses.
$= 1 + (-10)$	Work from left to right. Add $-1 + 2$.
$= -9$	Do the addition.

Self Check
Evaluate: $-3 - 5 - (-1)$.

Answer: -7 ■

EXAMPLE 5   *Order of operations.*   Evaluate: $-8 - (-2 - 2)$.

Solution
We must do the subtraction within the parentheses first.

$$-8 - (-2 - 2) = -8 - [-2 + (-2)]$$ Add the opposite of 2, which is -2. Since -2 must be written within parentheses, we write $-2+(-2)$ within brackets.

$$= -8 - (-4)$$ Add -2 and -2. Since only one set of grouping symbols is now needed, we write -4 within parentheses.

$$= -8 + 4$$ Add the opposite of -4, which is 4.

$$= -4$$ Do the addition.

Self Check
Evaluate: $-2 - (-6 - 5)$.

Answer: 9   ■

Applications of subtraction

Things are constantly changing in our daily lives. The temperature, the amount of money we have in the bank, and our ages are a few examples. In mathematics, the operation of subtraction is used to measure change. In general, to find the change in a quantity, subtract the earlier value from the later value.

EXAMPLE 6   *Change of water level.*   On Monday, the water level in a city storage tank was 6 feet above normal. By Friday, the level had fallen to a mark 4 feet below normal. Find the change in the water level from Monday to Friday. (See Figure 2-13.)

Solution   We will use subtraction to find the amount of change. The water levels of 4 feet below normal (the later value) and 6 feet above normal (the earlier value) can be represented by -4 and 6, respectively.

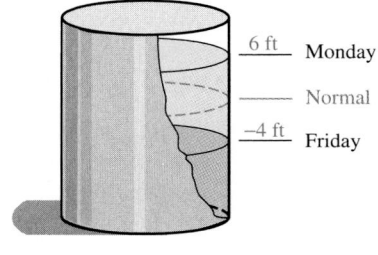

FIGURE 2-13

The water level Friday	minus	the water level Monday	is	the change in the water level.

$$-4 - 6 = -4 + (-6)$$ Add the opposite of 6, which is -6.

$$= -10$$ Do the addition. The negative result indicates that the water level fell.

The water level fell 10 feet from Monday to Friday.   ■

In the next example, a number line will serve as a mathematical model of a real-life situation. You will see how the operation of subtraction can be used to find the distance between two points on a number line.

EXAMPLE 7   *Artillery accuracy.*   In a practice session, an artillery group fired two rounds at a target. The first landed 65 yards short of the target, and the second landed 50 yards past it. (See Figure 2-14.) How far apart were the two impact points?

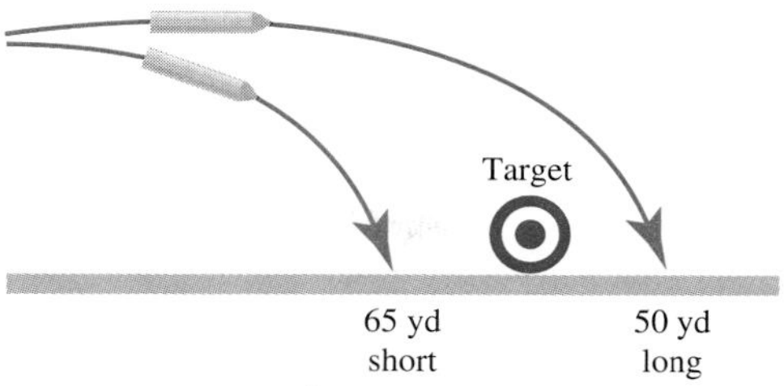

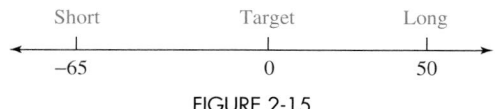

FIGURE 2-14

Solution We can use a number line to model this situation. The target is the origin. The words *short of the target* indicate a negative number, and the words *past it* indicate a positive number. Therefore, we graph the impact points at -65 and 50 in Figure 2-15.

FIGURE 2-15

The phrase *how far apart* tells us to subtract.

The position of the long shot	minus	the position of the short shot	is	the distance between impact points.

$$50 - (-65) = 50 + 65 \qquad \text{Add the opposite of } -65.$$
$$= 115 \qquad \text{Do the addition.}$$

The impact points are 115 yards apart.

Accent on Technology: **Subtraction with negative numbers**

The world's highest peak is Mount Everest in the Himalayas. The greatest ocean depth yet measured lies in the Mariana Trench near the island of Guam in the western Pacific. See Figure 2-16. To find the range between the highest peak and greatest depth, we must subtract.

$$29,035 - (-36,025)$$

To do this subtraction using a scientific calculator, we enter these numbers and press these keys.

Keystrokes   29035 $\boxed{-}$ 36025 $\boxed{+/-}$ $\boxed{=}$ $\boxed{\text{65060}}$

The range is 65,060 feet between the highest peak and the lowest depth. We could have applied the subtraction rule to write $29,035 - (-36,025)$ as $29,035 + 36,025$ before using the calculator.

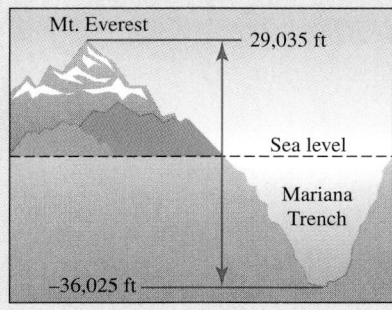

FIGURE 2-16

STUDY SET Section 2.3

VOCABULARY *Fill in the blanks.*

1. The answer to a subtraction problem is called the _____.

2. Two numbers represented by points on a number line that are the same distance away from the origin, but on opposite sides of it, are called _____.

CONCEPTS *In Exercises 3–10, fill in the blanks.*

3. _____ is the same as adding the opposite of the number to be subtracted.

4. Subtracting 3 is the same as adding ___.

5. Subtracting -6 is the same as adding ___.

6. The opposite of -8 is ___.

7. For any numbers x and y, $x - y = $ _____.

8. **a.** $2 - 7 = 2 + $ ___
   b. $2 - (-7) = 2 + $ ___
   c. $-2 - 7 = -2 + $ ___
   d. $-2 - (-7) = -2 + $ ___

9. After using parentheses as grouping symbols, if another set of grouping symbols is needed, we use _____.

10. We can find the _____ in a quantity by subtracting the earlier value from the later value.

11. Write this problem using mathematical symbols: negative eight minus negative four.

12. Write this problem using mathematical symbols: negative eight subtracted from negative four.

13. Find the distance between -4 and 3 on a number line.

14. Find the distance between -10 and 1 on a number line.

15. Is subtracting 3 from 8 the same as subtracting 8 from 3? Explain.

16. Evaluate each expression.
    a. $-2 - 0$ **b.** $0 - (-2)$

NOTATION *Complete each solution.*

17. Evaluate: $1 - 3 - (-2)$.
$$1 - 3 - (-2) = 1 + (\quad) + 2$$
$$= -2 + \boxed{}$$
$$= 0$$

18. Evaluate: $-6 + 5 - (-5)$.
$$-6 + 5 - (-5) = -6 + 5 + \boxed{}$$
$$= \boxed{} + 5$$
$$= 4$$

19. Evaluate: $(-8 - 2) - (-6)$.
$$(-8 - 2) - (-6) = [-8 + (\quad)] - (-6)$$
$$= \boxed{} - (-6)$$
$$= -10 + \boxed{}$$
$$= -4$$

20. Evaluate: $-5 - (-1 - 4)$.
$$-5 - (-1 - 4) = -5 - [-1 + (\quad)]$$
$$= -5 - (\quad)$$
$$= -5 + \boxed{}$$
$$= 0$$

PRACTICE *Find each difference.*

21. $8 - (-1)$ 22. $3 - (-8)$
23. $-4 - 9$ 24. $-7 - 6$
25. $-5 - 5$ 26. $-7 - 7$
27. $-5 - (-4)$ 28. $-9 - (-1)$
29. $-1 - (-1)$ 30. $-4 - (-3)$
31. $-2 - (-10)$ 32. $-6 - (-12)$
33. $0 - (-5)$ 34. $0 - 8$
35. $0 - 4$ 36. $0 - (-6)$
37. $-2 - 2$ 38. $-3 - 3$
39. $-10 - 10$ 40. $4 - 4$
41. $9 - 9$ 42. $4 - (-4)$
43. $-3 - (-3)$ 44. $-5 - (-5)$

Evaluate each expression.

45. $-4 - (-4) - 15$ 46. $-3 - (-3) - 10$

47. $-3 - 3 - 3$ 48. $-1 - 1 - 1$
49. $5 - 9 - (-7)$ 50. $6 - 8 - (-4)$
51. $10 - 9 - (-8)$ 52. $16 - 14 - (-9)$
53. $-1 - (-3) - 4$ 54. $-2 - 4 - (-1)$
55. $-5 - 8 - (-3)$ 56. $-6 - 5 - (-1)$
57. $(-6 - 5) - 3$ 58. $(-2 - 1) - 5$
59. $(6 - 4) - (1 - 2)$ 60. $(5 - 3) - (4 - 6)$
61. $-9 - (6 - 7)$ 62. $-3 - (6 - 12)$
63. $-8 - [4 - (-6)]$
64. $-1 - [5 - (-2)]$
65. $[-4 + (-8)] - (-6)$
66. $[-5 + (-4)] - (-2)$
67. Subtract -3 from 7.
68. Subtract 8 from -2.
69. Subtract -6 from -10.

70. Subtract -4 from -9.

71. $-1,557 - 890$

72. $-345 - (-789)$

73. $20,007 - (-496)$

74. $-979 - (-44,879)$

75. $-162 - (-789) - 2,303$

76. $-787 - 1,654 - (-232)$

APPLICATIONS *Use signed numbers to help answer each question.*

77. SCUBA DIVING   After descending 50 feet, a scuba diver paused to check his equipment before descending an additional 70 feet. Use a signed number to represent the diver's final depth.

78. TEMPERATURE CHANGE   Rashawn flew from his New York home to Hawaii for a week of vacation. He left blizzard conditions and a temperature of $-6°$, and stepped off the airplane into $85°$ weather. What temperature change did he experience?

79. READING PROGRAM   In a state reading test administered at the start of a school year, an elementary school's performance was 23 points below the county average. The principal immediately began a special tutorial program. At the end of the school year, retesting showed the students to be only 7 points below the average. How many points did the school's reading score improve over the year?

80. SUBMARINE   A submarine was traveling 2,000 feet below the ocean's surface when the radar system warned of an impending collision with another sub. The captain ordered the navigator to dive an additional 200 feet and then level off. Find the depth of the submarine after the dive.

81. AMPERAGE   During normal operation, the ammeter on a car reads $+5$. (See Illustration 1.) If the headlights, which draw a current of 7 amps, and the radio, which draws a current of 6 amps, are both turned on, what number will the ammeter register?

ILLUSTRATION 1

82. GIN RUMMY   After a losing round, a card player must subtract the value of each of the cards left in his hand (shown in Illustration 2) from his previous point total of 21. If face cards are counted as 10 points, what is his new score?

ILLUSTRATION 2

83. GEOGRAPHY   Death Valley, California, is the lowest land point in the United States, at 283 feet below sea level. The lowest land point on the earth is the Dead Sea, which is 1,290 feet below sea level. How much lower is the Dead Sea than Death Valley?

84. LIE DETECTOR TEST   On one lie detector test, a burglar scored -18, which indicates deception. However, on a second test, he scored -1, which is inconclusive. Find the difference in the scores.

85. FOOTBALL   A college football team records the outcome of each of its plays during a game on a stat sheet. (See Illustration 3.) Find the net gain (or loss) after the 3rd play.

Down	Play	Result
1st	run	lost 1 yd
2nd	pass—sack!	lost 6 yd
penalty	delay of game	lost 5 yd
3rd	pass	gained 8 yd
4th	punt	—

ILLUSTRATION 3

86. ACCOUNTING   Complete the balance sheet in Illustration 4. Then determine the overall financial condition of the company by subtracting the total liabilities from the total assets.

W a l k e r C o r p o r a t i o n Balance Sheet 2001		
Assets		
Cash	$11	1 0 9
Supplies	7	8 6 2
Land	67	5 4 3
Total assets	$	
Liabilities		
Accounts payable	$79	0 3 7
Income taxes	20	1 8 1
Total liabilities	$	

ILLUSTRATION 4

87. DIVING   A diver jumps from a platform. After she hits the water, her momentum takes her to the bottom of the pool. (See Illustration 5.)

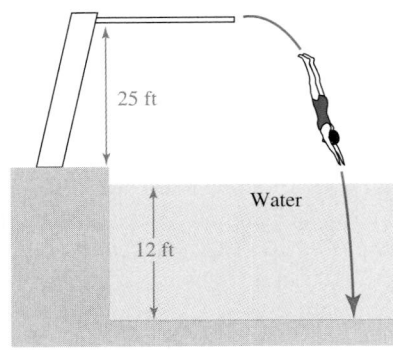

25 ft

Water

12 ft

ILLUSTRATION 5

a. Use a number line and signed numbers to model this situation. Show the top of the platform, the water line, and the bottom of the pool.

b. Find the total length of the dive from the top of the platform to the bottom of the pool.

88. TEMPERATURE EXTREMES   The highest and lowest temperatures ever recorded in several cities are shown in Illustration 6. List the cities in order, from the largest to smallest range in temperature extremes.

| City | Extreme temperatures | |
	Highest	Lowest
Atlantic City, NJ	106	−11
Barrow, AK	79	−56
Kansas City, MO	109	−23
Norfolk, VA	104	−3
Portland, ME	103	−39

ILLUSTRATION 6

89. CHECKING ACCOUNT   Michael has $1,303 in his checking account. Can he pay his car insurance premium of $676, his utility bills of $121, and his rent of $750 without having to make another deposit? Explain your answer.

90. HISTORY   Two of the greatest Greek mathematicians were Archimedes (287–212 B.C.) and Pythagoras (569–500 B.C.). How many years apart were they born?

WRITING

91. Explain what is meant when we say that subtraction is the same as addition of the opposite.

92. Give an example showing that it is possible to subtract something from nothing.

93. Explain how to check the result: $-7 - 4 = -11$.

94. Explain why students don't need to change every subtraction they encounter into an addition of the opposite. Give some examples.

REVIEW

95. Solve: $5x = 15$.

96. Round 5,999 to the nearest hundred.

97. List the factors of 20.

98. When solving the equation $6x = 24$, what operation on the variable must be undone?

99. It takes 13 oranges to make one can of orange juice. Find the number of oranges used to make 12 cans.

100. Evaluate: $12^2 - (5 - 4)^2$.

101. Write 4,502 in expanded notation.

102. What property does the following illustrate?
$a \cdot b = b \cdot a$

2.4 *Multiplication of Integers*

In this section, you will learn about

- Multiplying two positive integers
- Multiplying a positive and a negative integer
- Multiplying a negative and a positive integer • Multiplying by zero
- Multiplying two negative integers • Powers of integers

INTRODUCTION. We now turn our attention to multiplication of integers. When we multiply two nonzero integers, the first factor can be positive or negative. The same is true for the second factor. This means that there are four possible combinations to consider.

Positive · positive   Positive · negative

Negative · positive   Negative · negative

In this section, we will discuss these four combinations and use our observations to establish rules for multiplying two integers.

Multiplying two positive integers

4(3)
like signs
both positive

We begin by considering the product of two positive integers, 4(3). Since both factors are positive, we say that they have *like* signs. In Chapter 1, we learned that multiplication is repeated addition. Therefore, 4(3) represents the sum of four 3's.

$4(3) = 3 + 3 + 3 + 3$ Multiplication is repeated addition. Write 3 four times.

$4(3) = 12$ The result is 12, which is a positive number.

This result suggests that *the product of two positive integers is positive.*

Multiplying a positive and a negative integer

4(−3)
unlike signs
one positive, one negative

Next, we will consider $4(-3)$. This is the product of a positive and a negative integer. The signs of these factors are *unlike*. According to the definition of multiplication, $4(-3)$ means that we are to add -3 four times.

$4(-3) = (-3) + (-3) + (-3) + (-3)$ Use the definition of multiplication. Write -3 four times.

$4(-3) = \quad (-6) + (-3) + (-3)$ Work from left to right. Apply the rule for adding two negative numbers.

$4(-3) = \quad\quad (-9) + (-3)$ Work from left to right. Apply the rule for adding two negative numbers.

$4(-3) = \quad\quad\quad -12$ Do the addition.

This result is -12, which suggests that *the product of a positive integer and a negative integer is negative.*

Multiplying a negative and a positive integer

−3(4)
unlike signs
one negative, one positive

To develop a rule for multiplying a negative and a positive integer, we will consider $-3(4)$. Notice that the factors have *unlike* signs. Because of the commutative property of multiplication, the answer to $-3(4)$ will be the same as the answer to $4(-3)$. We know that $4(-3) = -12$ from the previous discussion, so $-3(4) = -12$. This suggests that *the product of a negative integer and a positive integer is negative.*

Putting the results of the last two cases together leads us to the rule for multiplying two integers with unlike signs.

| Multiplying two integers with unlike signs | To multiply a positive integer and a negative integer, or a negative integer and a positive integer, multiply their absolute values. Then make the answer negative. |

EXAMPLE 1   *Multiplying two integers with unlike signs.*   Find each product: **a.** $7(-5)$,   **b.** $20(-8)$,   and   **c.** $-8 \cdot 5$.

Self Check
Find each product:

a. $2(-6)$

b. $30(-2)$

c. $-15 \cdot 2$

Solution
To multiply integers with unlike signs, we multiply their absolute values and make the product negative.

a. $7(-5) = -35$ Multiply the absolute values, 7 and 5, to get 35. Then make the answer negative.

b. $20(-8) = -160$ Multiply the absolute values, 20 and 8, to get 160. Then make the answer negative.

c. $-8 \cdot 5 = -40$ Multiply the absolute values, 8 and 5, to get 40. Then make the answer negative.

Answers: a. -12,   **b.** -60,   **c.** -30 ■

COMMENT   When writing multiplication involving signed numbers, do not write a negative sign $-$ next to a raised dot • (the multiplication symbol). Instead, use parentheses to show the multiplication.

$$6(-2) \qquad \cancel{6 \cdot -2} \qquad \text{and} \qquad -6(-2) \qquad \cancel{-6 \cdot -2}$$

Multiplying by zero

Before we can develop a rule for multiplying two negative integers, we need to examine multiplication by zero. If $4(3)$ means that we are to find the sum of four 3's, then $0(-3)$ means that we are to find the sum of zero -3's. Obviously, the sum would be 0. Thus, $0(-3) = 0$.

The commutative property of multiplication guarantees that we can change the order of the factors in the multiplication problem without affecting the result.

$$(-3)(0) = 0(-3) \quad = \quad 0$$

$\qquad\qquad$ ↑ $\qquad\qquad$ ↑ $\qquad\qquad\qquad$ ↑

$\qquad$ Change the order $\qquad$ The result is
$\qquad$ of the factors. $\qquad\quad$ still 0.

We see that the order in which we write the factors 0 and -3 doesn't matter—their product is 0. This example suggests that the product of any number and zero is zero.

| Multiplying by zero | If a represents a number, then $$a \cdot 0 = 0 \qquad \text{and} \qquad 0 \cdot a = 0$$ |

EXAMPLE 2   *Multiplication by zero.*   Find $-12 \cdot 0$.

Self Check
Find $0(-56)$.

Solution
Since the product of any number and zero is zero, we have

$-12 \cdot 0 = 0$

Answer: 0   ■

Multiplying two negative integers

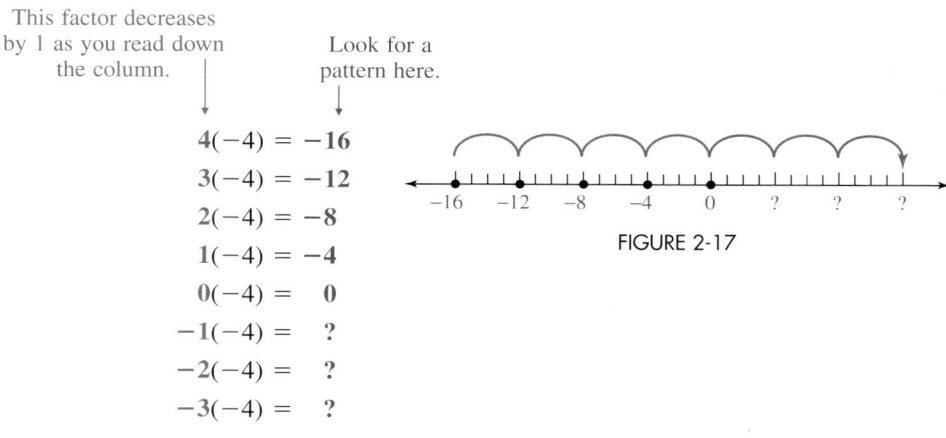

$-3(-4)$
like signs
both negative

To develop a rule for multiplying two negative integers, we consider the pattern displayed below. There, we multiply -4 by a series of factors that decrease by 1. After determining each product, we graph each result on a number line (Figure 2-17). See if you can determine the answers to the last three multiplication problems by examining the pattern of answers leading up to them.

This factor decreases by 1 as you read down the column.

Look for a pattern here.

$$4(-4) = -16$$
$$3(-4) = -12$$
$$2(-4) = -8$$
$$1(-4) = -4$$
$$0(-4) = 0$$
$$-1(-4) = ?$$
$$-2(-4) = ?$$
$$-3(-4) = ?$$

FIGURE 2-17

From the pattern, we see that

$$-1(-4) = 4$$
$$-2(-4) = 8$$
$$-3(-4) = 12$$

For two negative factors,   the product is a positive.

These results suggest that *the product of two negative integers is positive.* Earlier in this section, we saw that the product of two positive integers is also positive. This leads to the following conclusion.

Multiplying two integers with like signs

> To multiply two positive integers, or two negative integers, multiply their absolute values. The answer is positive.

EXAMPLE 3   *Multiplying two negative integers.*   Find each product:
a. $-5(-9)$ and   **b.** $-8(-10)$.

Solution

To multiply two negative integers, we multiply their absolute values and make the result positive.

a. $-5(-9) = 45$ Multiply the absolute values, 5 and 9, to get 45. The answer is positive.

b. $-8(-10) = 80$ Multiply the absolute values, 8 and 10, to get 80. The answer is positive.

Self Check

Find each product:

a. $-9(-7)$

b. $-12(-2)$

Answers: **a.** 63,   **b.** 24   ■

We now summarize the rules for multiplying two integers.

Multiplying two integers

> To multiply two integers, multiply their absolute values.
> **1.** The product of two integers with *like* signs is positive.
> **2.** The product of two integers with *unlike* signs is negative.

Accent on Technology: **Multiplication with negative numbers**

At Thanksgiving time, a supermarket offered customers a free turkey with every grocery purchase of $100 or more. Each turkey cost the store $6, and 375 people took advantage of the offer.

Since each of the 375 turkeys given away represented a loss of $6 (which can be expressed as -6 dollars) the store lost $375(-6)$ dollars. To find this product, we enter these numbers and press these keys on a scientific calculator.

Keystrokes   375 $\boxed{\times}$ 6 $\boxed{+/-}$ $\boxed{=}$ | $\boxed{\text{-}2250}$ |

The negative result indicates that with this promotion, the supermarket gave away $2,250 in turkeys.

EXAMPLE 4   *Multiplying three integers.*   Multiply $-4(5)(-3)$ in two ways.

Solution

First, we work from left to right.

$$-4(5)(-3) = -20(-3)$$ Multiply -4 and 5 first: $-4(5) = -20$.

$$= 60$$ Do the multiplication.

An alternate approach would be to multiply the two negative numbers first.

$$-4(5)(-3) = -4(-3)(5)$$ Apply the commutative property of multiplication. Change the order of the factors.

$$= 12(5)$$ Multiply the negative numbers: $-4(-3) = 12$.

$$= 60$$ Do the multiplication.

After some practice, you will be able to do this work in your head.

Self Check

Multiply $-5(5)(-2)$ in two ways.

Answer: 50   ∎

Powers of integers

Recall that exponential expressions are used to represent repeated multiplication. For example, 2 to the third power, or 2^3, is a shorthand way of writing $2 \cdot 2 \cdot 2$. In this expression, 3 is the exponent, and the base is positive 2. In the next example, we evaluate exponential expressions with bases that are negative numbers.

EXAMPLE 5   *Evaluating powers of integers.*   Find each power:
a. $(-2)^4$ and   **b.** $(-5)^3$.

Solution

a. $(-2)^4 = (-2)(-2)(-2)(-2)$ Write -2 as a factor 4 times.

$$= 4(-2)(-2)$$ Work from left to right. Multiply -2 and -2 to get 4.

$$= -8(-2)$$ Work from left to right. Multiply 4 and -2 to get -8.

$$= 16$$ Do the multiplication.

b. $(-5)^3 = (-5)(-5)(-5)$ Write -5 as a factor 3 times.

$$= 25(-5)$$ Work from left to right. Multiply -5 and -5 to get 25.

$$= -125$$ Do the multiplication.

Self Check

Find each power:

a. $(-3)^4$

b. $(-4)^3$

Answers: **a.** 81,   **b.** -64 ∎

In Example 5, part a, -2 was raised to an even power, and the answer was positive. In part b, another negative number, -5, was raised to an odd power, and the answer was negative. These results suggest a general rule.

Even and odd powers of a negative integer

> When a negative integer is raised to a nonzero even power, the result is positive.
>
> When a negative integer is raised to an odd power, the result is negative.

EXAMPLE 6 *Evaluating a power of an integer.* Find $(-1)^5$.

Solution

We have a negative integer raised to an odd power. The result will be negative.

$$(-1)^5 = (-1)(-1)(-1)(-1)(-1)$$
$$= -1$$

Self Check

Find the power: $(-1)^8$.

Answer: 1 ∎

 COMMENT Although the expressions -3^2 and $(-3)^2$ look somewhat alike, they are not. In -3^2, the base is 3 and the exponent 2. The $-$ sign in front of 3^2 means the opposite of 3^2. In $(-3)^2$, the base is -3 and the exponent is 2. When we evaluate them, it becomes clear that they are not equivalent.

-3^2 represents *the opposite of* 3^2.

$-3^2 = -(3 \cdot 3)$ Write 3 as a factor 2 times.

$ = -9$ Multiply within the parentheses first.

$(-3)^2$ represents $(-3)(-3)$.

$(-3)^2 = (-3)(-3)$ Write -3 as a factor 2 times.

$ = 9$ The product of two negative numbers is positive.

Notice that the results are different.

EXAMPLE 7 *The opposite of a power.* Evaluate: **a.** -2^2 and **b.** $(-2)^2$.

Solution

a. $-2^2 = -(2 \cdot 2)$ Since 2 is the base, write 2 as a factor two times.

$ = -4$ Do the multiplication within the parentheses.

b. $(-2)^2 = (-2)(-2)$ The base is -2. Write it as a factor twice.

$ = 4$ The signs are like, so the product is positive.

Self Check

Evaluate: **a.** -4^2 and **b.** $(-4)^2$.

Answers: **a.** -16, **b.** 16 ∎

Accent on Technology: **Raising a negative number to a power**

Negative numbers can be raised to a power using a scientific calculator. We use the change-of-sign key $\boxed{+/-}$ and the power key $\boxed{y^x}$ (on some calculators, $\boxed{x^y}$). For example, to evaluate $(-5)^6$, we enter these numbers and press these keys.

Keystrokes   5 $\boxed{+/-}$ $\boxed{y^x}$ 6 $\boxed{=}$ $\boxed{15625}$

The result is 15,625.

STUDY SET Section 2.4

VOCABULARY   *Fill in the blanks.*

1. In the multiplication $-5(-4)$, the integers -5 and -4, which are being multiplied, are called _____. The answer, 20, is called the _____.

2. The definition of multiplication tells us that $3(-4)$ represents repeated _____: $-4 + (-4) + (-4)$.

3. In the expression -3^5, ____ is the base and 5 is the _____.

4. In the expression $(-3)^5$, _____ is the base and ____ is the exponent.

CONCEPTS   *In Exercises 5–8, fill in the blanks.*

5. The product of two integers with _____ signs is negative.

6. The product of two integers with like signs is _____.

7. The _____ property of multiplication implies that $-2(-3) = -3(-2)$.

8. The product of zero and any number is ___.

9. Find $-1(9)$. In general, what is the result when we multiply a positive number by -1?

10. Find $-1(-9)$. In general, what is the result when we multiply a negative number by -1?

11. When we multiply two integers, there are four possible combinations of signs. List each of them.

12. When multiplying two integers, there are four possible combinations of signs. How can they be grouped into two categories?

13. If each of the following powers were evaluated, what would be the *sign* of the result?
   a. $(-5)^{13}$ **b.** $(-3)^{20}$

14. A student claimed, "A positive and a negative is negative." What is wrong with this statement?

15. Find each absolute value.
   a. $|-3|$ **b.** $|12|$
   c. $|-5|$ **d.** $|9|$
   e. $|10|$ **f.** $|-25|$

16. Find each product and then graph it on a number line. What is the distance between each product?
   $2(-2),\ 1(-2),\ 0(-2),\ -1(-2),\ -2(-2)$

17. a. Complete the table in Illustration 1.

Problem	Number of negative factors	Answer
$-2(-2)$		
$-2(-2)(-2)(-2)$		
$-2(-2)(-2)(-2)(-2)(-2)$		

ILLUSTRATION 1

b. The answers entered in the table help to justify the following rule: The product of an _____ number of negative integers is positive.

18. a. Complete the table in Illustration 2.

Problem	Number of negative factors	Answer
$-2(-2)(-2)$		-8
$-2(-2)(-2)(-2)(-2)$		-32
$-2(-2)(-2)(-2)(-2)(-2)(-2)$		-128

ILLUSTRATION 2

b. The answers entered in the table help to justify the following rule: The product of an _____ number of negative integers is negative.

NOTATION   *In Exercises 19–20, complete each solution.*

19. Find $-3(-2)(-4)$.
$$-3(-2)(-4) = \quad (-4)$$
$$= -24$$

20. Find $(-3)^4$.
$$(-3)^4 = (-3)(-3)(-3)\quad$$
$$= \quad (-3)(-3)$$
$$= \quad (-3)$$
$$= 81$$

21. Explain why the expression below is not written correctly. How should it be written?
$$-6 \cdot -5$$

22. Translate into mathematical symbols: the product of negative three and negative two.

PRACTICE   *Find each product.*

23. $-9(-6)$ **24.** $-5(-5)$

25. $-3 \cdot 5$ **26.** $-6 \cdot 4$

27. $12(-3)$ **28.** $11(-4)$

29. $(-8)(-7)$ **30.** $(-9)(-3)$

31. $(-2)10$ **32.** $(-3)8$

33. $-40 \cdot 3$ **34.** $-50 \cdot 2$

35. $-8(0)$ **36.** $0(-27)$

37. $-1(-6)$ **38.** $-1(-8)$

39. $-7(-1)$ **40.** $-5(-1)$

41. $1(-23)$ **42.** $-35(1)$

Evaluate each expression.

43. $-6(-4)(-2)$ **44.** $-3(-2)(-3)$

45. $5(-2)(-4)$ **46.** $3(-3)(3)$

47. $2(3)(-5)$ **48.** $6(2)(-2)$

49. $6(-5)(2)$ **50.** $4(-2)(2)$

51. $(-1)(-1)(-1)$ **52.** $(-1)(-1)(-1)(-1)$

53. $-2(-3)(3)(-1)$ **54.** $5(-2)(3)(-1)$

55. $3(-4)(0)$ **56.** $-7(-9)(0)$

57. $-2(0)(-10)$ **58.** $-6(0)(-12)$

59. Find the product of -6 and the opposite of 10.

60. Find the product of the opposite of 9 and the opposite of 8.

Find each power.

61. $(-4)^2$ **62.** $(-6)^2$

63. $(-5)^3$ **64.** $(-6)^3$

65. $(-2)^3$ **66.** $(-4)^3$

67. $(-9)^2$ **68.** $(-10)^2$

69. $(-1)^5$ **70.** $(-1)^6$

71. $(-1)^8$ **72.** $(-1)^9$

Evaluate each expression.

73. $(-7)^2$ and -7^2

74. $(-5)^2$ and -5^2

75. -12^2 and $(-12)^2$

76. -11^2 and $(-11)^2$

▦ *Use a calculator to evaluate each expression.*

77. $-76(787)$ **78.** $407(-32)$

79. $(-81)^4$ **80.** $(-6)^5$

81. $(-32)(-12)(-67)$ **82.** $(-56)(-9)(-23)$

83. $(-25)^4$ **84.** $(-41)^5$

APPLICATIONS   *Use signed numbers to help answer each problem.*

85. DIETING   After giving a patient a physical exam, a physician felt that the patient should begin a diet. Two options were discussed. (See Illustration 3.)

	Plan #1	Plan #2
Length	10 weeks	14 weeks
Daily exercise	1 hour	30 min
Weight loss per week	3 lb	2 lb

ILLUSTRATION 3

a. Find the expected weight loss from each diet plan. Express each answer as a signed number.

b. With which plan should the patient expect to lose the most weight? Explain why the patient might not choose it.

86. INVENTORY   A spreadsheet is used to record inventory losses at a warehouse. The items, their cost, and the number missing are listed in Illustration 4.

	A	B	C	D
1	Item	Cost	Number of units	$ losses
2	CD	$5	-11	
3	TV	$200	-2	
4	Radio	$20	-4	

ILLUSTRATION 4

a. What instruction should be given to find the total losses for each type of item? Find each of those losses and fill in column D.

b. What instruction should be given to find the *total* inventory losses for the warehouse? Find this number.

87. MAGNIFICATION   Using an electronic testing device, a mechanic can check the emissions of a car. The results of the test are displayed on a screen. (See Illustration 5.)

a. Find the high and low values for this test as shown on the screen.

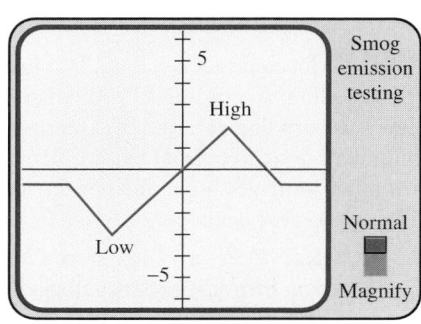

ILLUSTRATION 5

b. By switching a setting on the monitor, the picture on the screen can be magnified. What would be the new high and new low if every value were doubled?

88. LIGHT   Sunlight is a mixture of all colors. When sunlight passes through water, the water absorbs different colors at different rates, as shown in Illustration 6.
a. Use a signed number to represent the depth to which red light penetrates water.
b. Green light penetrates 4 times deeper than red light. How deep is this?
c. Blue light penetrates 3 times deeper than orange light. How deep is this?

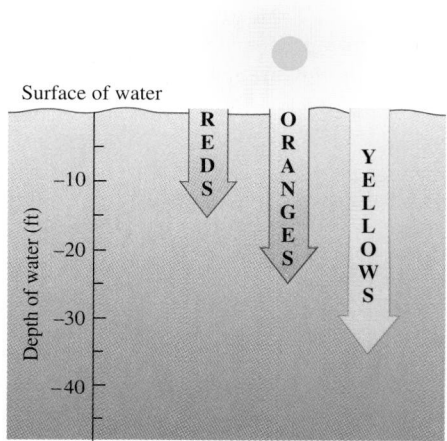

ILLUSTRATION 6

89. TEMPERATURE CHANGE   A farmer, worried about his fruit trees suffering frost damage, calls the weather service for temperature information. He is told that temperatures will be decreasing approximately 4° every hour for the next 5 hours. What signed number represents the total change in temperature expected over the next five hours?

90. DEPRECIATION   For each of the last four years, a businesswoman has filed a $200 depreciation allowance on her income tax return, for an office computer system. What signed number represents the total amount of depreciation written off over the four-year period?

91. EROSION   A levee protects a town in a low-lying area from flooding. According to geologists, the banks of the levee are eroding at a rate of 2 feet per year. If something isn't done to correct the problem, what signed number indicates how much of the levee will erode during the next decade?

92. DECK SUPPORT   After a winter storm, a homeowner has an engineering firm inspect his damaged deck. Their report concludes that the original pilings were not anchored deep enough, by a factor of 3. (See Illustra-

tion 7.) What signed number represents the depth to which the pilings should have been sunk?

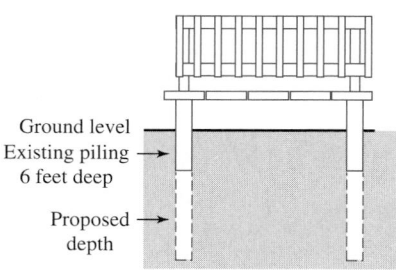

ILLUSTRATION 7

93. WOMEN'S NATIONAL BASKETBALL ASSOCIATION   The average attendance for the WNBA Houston Comets is 11,906 a game. Suppose the team gives a sports bag, costing $3, to everyone attending a game. What signed number expresses the financial loss from this promotional giveaway?

94. HEALTH CARE   A health care provider for a company estimates that 75 hours per week are lost by employees suffering from stress-related or preventable illness. In a 52-week year, how many hours are lost? Use a signed number to answer.

WRITING

95. If a product contains an even number of negative factors, how do we know that the result will be positive?

96. Explain why the product of a positive number and a negative number is negative, using $5(-3)$ as an example.

97. Explain why the result is the opposite of the original number when a number is multiplied by -1.

98. Can you think of any number that yields a nonzero result when it is multiplied by 0? Explain your response.

REVIEW

99. The prime factorization of a number is $3^2 \cdot 5$. What is the number?

100. Solve for y: $\dfrac{y}{8} = 10$.

101. The enrollment at a college went from 10,200 to 12,300 in one year. What was the increase in enrollment?

102. Find the perimeter of a square with sides 6 yards long.

103. What does the symbol $<$ mean?

104. List the first ten prime numbers.

2.5 *Division of Integers*

In this section, you will learn about

- The relationship between multiplication and division
- Rules for dividing integers • Division and zero

INTRODUCTION. In this section, we will develop rules for division of integers, just as we did for multiplication of integers. We will also consider two types of division involving zero.

The relationship between multiplication and division

When we solved equations in Chapter 1, multiplication was used to undo division, and division was used to undo multiplication. Because one operation undoes the other, multiplication and division are said to be *inverse operations*. Every division fact containing three numbers can be written as an equivalent multiplication fact involving the same three numbers. For example,

$$\frac{6}{3} = 2 \qquad \text{because} \qquad 3(2) = 6 \qquad \text{Remember that in the division statement, 6 is the}$$
$$\textit{dividend, 3 is the divisor, and 2 is the quotient.}$$

Rules for dividing integers

We will now use the relationship between multiplication and division to help develop rules for dividing integers. There are four cases to consider.

Case 1: In the first case, a positive integer is divided by a positive integer. From years of experience, we already know that the result is positive. Therefore, *the quotient of two positive integers is positive.*

Case 2: Next, we consider the quotient of two negative integers. As an example, consider the division $\frac{-12}{-2} = ?$ We can do this division by examining its related multiplication statement, $-2(?) = -12$. Our objective is to find the number that should replace the question mark. To do this, we use the rules for multiplying integers, introduced in the previous section.

Multiplication statement	**Division statement**
$-2(?) = -12$	$\dfrac{-12}{-2} = ?$

    This must be *positive* 6 if the product is to be *negative* 12.   So the quotient is *positive* 6.

Therefore, $\frac{-12}{-2} = 6$. From this example, we can see that *the quotient of two negative integers is positive.*

Case 3: The third case we will examine is the quotient of a positive integer and a negative integer. Let's consider $\frac{12}{-2} = ?$ Its equivalent multiplication statement is $-2(?) = 12$.

Multiplication statement   **Division statement**

$$-2(?) = 12$$ $$\frac{12}{-2} = ?$$

This must be   So the quotient
-6 if the product   is -6.
is to be *positive* 12.

Therefore, $\frac{12}{-2} = -6$. This result shows that *the quotient of a positive integer and a negative integer is negative*.

Case 4: Finally, to find the quotient of a negative integer and a positive integer, let's consider $\frac{-12}{2} = ?$ Its equivalent multiplication statement is $2(?) = -12$.

Multiplication statement   **Division statement**

$$2(?) = -12$$ $$\frac{-12}{2} = ?$$

This must be   So the quotient
-6 if the product   is -6.
is to be -12.

Therefore, $\frac{-12}{2} = -6$. From this example, we can see that *the quotient of a negative integer and a positive integer is negative*.

We now summarize the results from the previous discussion.

Dividing two integers

> To divide two integers, divide their absolute values.
>
> **1.** The quotient of two integers with *like* signs is positive.
>
> **2.** The quotient of two integers with *unlike* signs is negative.

The rules for dividing integers are similar to those for multiplying integers.

EXAMPLE 1 *Dividing integers.* Find each quotient:

a. $\dfrac{-35}{7}$ and **b.** $\dfrac{20}{-5}$.

Solution

To divide integers with unlike signs, we find the quotient of their absolute values and make the quotient negative.

a. $\dfrac{-35}{7} = -5$ Divide the absolute values, 35 by 7, to get 5. The quotient is negative.

To check the result, we multiply the divisor, 7, and the quotient, -5. We should obtain the dividend, -35.

$$7(-5) = -35$$

The answer, -5, checks.

b. $\dfrac{20}{-5} = -4$ Divide the absolute values, 20 by 5, to get 4. The quotient is negative.

Self Check

Find each quotient. Then check the result.

a. $\dfrac{-45}{5}$

b. $\dfrac{60}{-20}$

Answers: **a.** -9, **b.** -3 ■

EXAMPLE 2 *Dividing integers.* Divide: $\dfrac{-12}{-3}$.

Solution

The integers have like signs. The quotient will be positive.

$\dfrac{-12}{-3} = 4$ Divide the absolute values, 12 by 3, to get 4. The quotient is positive.

Self Check

Divide: $\dfrac{-21}{-3}$.

Answer: 7

EXAMPLE 3 *Price reduction.* Over the course of a year, a retailer reduced the price of a television set by an equal amount each month, because it was not selling. By the end of the year, the cost was \$132 less than at the beginning of the year. How much did the price fall each month?

Solution We will label the drop in price of \$132 for the year as -132. It occurred in 12 equal reductions. This indicates division.

$\dfrac{-132}{12} = -11$ The quotient of a negative number and a positive number is negative.

The drop in price each month was \$11.

Division and zero

In Chapter 1, we discussed division involving zero.

To review the concept of division of zero, we will look at $\frac{0}{2} = ?$ The equivalent multiplication statement is $2(?) = 0$.

Multiplication statement   **Division statement**

$2(?) = 0$ $\dfrac{0}{2} = ?$

⌞ This must be   So the quotient ⌟
0 if the product is   is 0.
to be 0.

Therefore, $\frac{0}{2} = 0$. This example suggests that *the quotient of zero divided by any nonzero number is zero.*

To review division by zero, let's look at $\frac{2}{0} = ?$ The equivalent multiplication statement is $0(?) = 2$.

Multiplication statement   **Division statement**

$0(?) = 2$ $\dfrac{2}{0} = ?$

⌞ There is no number   There is no ⌟
that gives 2 when   quotient.
multiplied by 0.

Therefore, $\frac{2}{0}$ does not have an answer. We say that division by zero is **undefined.** This example suggests that *the quotient of any number divided by zero is undefined.*

Division with 0

1. If a represents any nonzero number, $\dfrac{0}{a} = 0$.

2. If a represents any nonzero number, $\dfrac{a}{0}$ is undefined.

3. $\dfrac{0}{0}$ is undetermined.

EXAMPLE 4   *Division with zero.*   Find $\dfrac{-4}{0}$, if possible.

Solution

Since $\dfrac{-4}{0}$ is division by 0, the division is undefined.

Self Check

Find $\dfrac{0}{-4}$.

Answer: 0   ■

Accent on Technology: **Division with negative numbers**

The Bureau of Labor statistics estimated that the United States lost 270,000 jobs in the manufacturing sector of the economy in 1999. Because the jobs were lost, we write this as $-270,000$. To find the average number of manufacturing jobs lost each month, we divide: $\dfrac{-270,000}{12}$. To do this division using a scientific calculator, we enter these numbers and press these keys.

Keystrokes   270000 $\boxed{+/-}$ $\boxed{\div}$ 12 $\boxed{=}$ $\boxed{-22500}$

The average number of manufacturing jobs lost each month in 1999 was 22,500.

STUDY SET   Section 2.5

VOCABULARY   *Fill in the blanks.*

1. In $\dfrac{-27}{3} = -9$, the number -9 is called the

_____, and the number 3 is the

_____.

2. Division by zero is _____. Division _____ zero by a nonzero number is 0.

3. The _____ of a number is the distance between it and 0 on the number line.

4. $\{. . . , -4, -3, -2, -1, 0, 1, 2, 3, 4, . . .\}$ is the set of _____.

5. The quotient of two negative integers is _____.

6. The quotient of a negative integer and a positive integer is _____.

CONCEPTS

7. Write the related multiplication statement for $\frac{-25}{5} = -5$.

8. Write the related multiplication statement for $\frac{0}{-15} = 0$.

9. Show that there is no answer for $\frac{-6}{0}$ by writing the related multiplication statement.

10. If x is any number except zero, what is $\frac{0}{x}$?

11. Write a related division statement for $5(-4) = -20$.

12. How do the rules for multiplying integers compare with the rules for dividing integers?

13. Tell whether each statement is always true, sometimes true, or never true.
   a. The product of a positive integer and a negative integer is negative.
   b. The sum of a positive integer and a negative integer is negative.
   c. The quotient of a positive integer and a negative integer is negative.

14. Tell whether each statement is always true, sometimes true, or never true.
   a. The product of two negative integers is positive.
   b. The sum of two negative integers is negative.
   c. The quotient of two negative integers is negative.

PRACTICE   *Find each quotient, if possible.*

15. $\dfrac{-14}{2}$ **16.** $\dfrac{-10}{5}$

17. $\dfrac{-8}{-4}$ **18.** $\dfrac{-12}{-3}$

19. $\dfrac{-25}{-5}$ **20.** $\dfrac{-36}{-12}$

21. $\dfrac{-45}{-15}$ **22.** $\dfrac{-81}{-9}$

23. $\dfrac{40}{-2}$ **24.** $\dfrac{35}{-7}$

25. $\dfrac{50}{-25}$ **26.** $\dfrac{80}{-40}$

27. $\dfrac{0}{-16}$ **28.** $\dfrac{0}{-6}$

29. $\dfrac{-6}{0}$ **30.** $\dfrac{-8}{0}$

31. $\dfrac{-5}{1}$ **32.** $\dfrac{-9}{1}$

33. $-5 \div (-5)$ **34.** $-11 \div (-11)$

35. $\dfrac{-9}{9}$ **36.** $\dfrac{-15}{15}$

37. $\dfrac{-10}{-1}$ **38.** $\dfrac{-12}{-1}$

39. $\dfrac{-100}{25}$ **40.** $\dfrac{-100}{50}$

41. $\dfrac{75}{-25}$ **42.** $\dfrac{300}{-100}$

43. $\dfrac{-500}{-100}$ **44.** $\dfrac{-60}{-30}$

45. $\dfrac{-200}{50}$ **46.** $\dfrac{-500}{100}$

47. Find the quotient of -45 and 9.
48. Find the quotient of -36 and -4.
49. Divide 8 by -2.
50. Divide -16 by -8.

51. $\dfrac{-13{,}550}{25}$ **52.** $\dfrac{-3{,}876}{-19}$

53. $\dfrac{272}{-17}$ **54.** $\dfrac{-6{,}776}{-77}$

APPLICATIONS   *Use signed numbers to help answer each problem.*

55. TEMPERATURE DROP   During a five-hour period, the temperature steadily dropped. (See Illustration 1.) What was the average change in the temperature per hour over this five-hour time span?

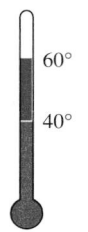

ILLUSTRATION 1

56. PRICE DROP   Over a three-month period, the price of a VCR steadily fell. (See Illustration 2.) What was the average monthly change in the price of the VCR over this period?

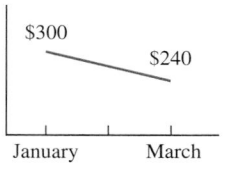

ILLUSTRATION 2

57. SUBMARINE DIVE   In a series of three equal dives, a submarine is programmed to reach a depth of 3,000 feet below the ocean surface. (See Illustration 3.) What signed number describes how deep each of the three dives will be?

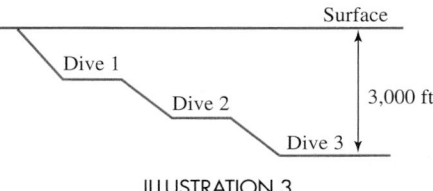

ILLUSTRATION 3

58. GRAND CANYON TRIP   A mule train is to travel from a stable on the rim of the Grand Canyon to a camp on the canyon floor, approximately 5,000 feet below the rim. If the guide wants the mules to be rested after every 1,000 feet of descent, how many stops will be made on the trip?

59. BASEBALL TRADE   At the midway point of the season, a baseball team finds itself 12 games behind the league leader. Team management decides to trade for a talented hitter, in hopes of making up at least half of the deficit in the standings by the end of the year. Where in the league standings does management expect to finish at season's end?

60. BUDGET DEFICIT   A politician proposed a two-year plan for cutting a county's $20 million budget deficit, as shown in Illustration 4. If this plan is put into effect, what will be the change in the county's financial status in two years?

	Plan	Prediction
1st year	Raise taxes, drop subsidy programs	Will cut deficit in half
2nd year	Search out waste and fraud	Will cut remaining deficit in half

ILLUSTRATION 4

61. PRICE MARKDOWN   The owner of a clothing store decides to reduce the price on a line of jeans that are not selling. She feels she can afford to lose $300 of projected income on these pants. By how much can she mark down each of the 20 pairs of jeans?

62. WATER RESERVOIR   Over a week's time, engineers at a city water reservoir released enough water to lower the water level 35 feet. On average, how much did the water level change each day during this period?

63. PAY CUT   In a cost-cutting effort, a business decides to lower expenditures on salaries by $9,135,000. To do this, all of the 5,250 employees will have their salaries reduced by an equal dollar amount. How big a pay cut will each employee experience?

64. STOCK MARKET   On Monday, the value of Maria's 255 shares of stock was at an all-time high. By Friday, the value had fallen $4,335. What was her per-share loss that week?

WRITING

65. Explain why the quotient of two negative numbers is positive.

66. Think of a real-life situation that could be represented by $\frac{0}{4}$. Explain why the answer would be zero.

67. Using a specific example, explain how multiplication can be used as a check for division.

68. Explain what it means when we say that division by zero is undefined.

REVIEW

69. Evaluate: $3\left(\dfrac{18}{3}\right)^2 - 2(2)$.

70. List the set of whole numbers.

71. Find the prime factorization of 210.

72. Complete the statement: The addition property of equality states that if $x = y$ and c is any number, then $x + c = $ _____.

73. Solve $99 = r + 43$.

74. Does $8 - 2 = 2 - 8$?

75. Evaluate: 3^4.

76. Sharif has scores of 55, 70, 80, and 75 on four mathematics tests. What is his mean (average) score?

2.6 *Order of Operations and Estimation*

In this section, you will learn about

- Order of operations • Absolute value • Estimation

INTRODUCTION. In this section, we will evaluate expressions that involve more than one operation. To do this, we will apply the rules for the order of operations discussed in Chapter 1, as well as the rules for working with integers. We will also continue the discussion of estimating an answer. Estimation can be used when you need a quick indication of the size of the actual answer to a calculation.

Order of operations

In Section 1.5, we introduced the rules for the order of operations: an agreed-upon sequence of steps for completing the operations of arithmetic.

Order of operations

> 1. Do all calculations within parentheses and other grouping symbols, following the order listed in Steps 2–4 and working from the innermost pair to the outermost pair.
>
> 2. Evaluate all exponential expressions.
>
> 3. Do all multiplications and divisions as they occur from left to right.
>
> 4. Do all additions and subtractions as they occur from left to right.
>
> When all grouping symbols have been removed, repeat Steps 2–4 to complete the calculation.
>
> If a fraction bar is present, evaluate the expression above the bar (the *numerator*) and the expression below the bar (the *denominator*) separately. Then do the division indicated by the fraction bar, if possible.

EXAMPLE 1 *Order of operations.* Evaluate: $-4(-3)^2 - (-2)$.

Solution

This expression contains the operations of multiplication, raising to a power, and subtraction. The rules for the order of operations tell us to find the power first.

$$
\begin{aligned}
-4(-3)^2 - (-2) &= -4(9) - (-2) && \text{Evaluate the exponential expression: } (-3)^2 = 9. \\
&= -36 - (-2) && \text{Do the multiplication: } -4(9) = -36. \\
&= -36 + 2 && \text{To do the subtraction, add the opposite of } -2. \\
&= -34 && \text{Do the addition.}
\end{aligned}
$$

Self Check
Evaluate: $-5(-2)^2 - (-6)$.

Answer: -14 ■

EXAMPLE 2 *Order of operations.* Evaluate: $2(3) + (-5)(-3)(-2)$.

Solution

This expression contains the operations of multiplication and addition. By the rules for the order of operations, we do the multiplications first.

$$
\begin{aligned}
2(3) + (-5)(-3)(-2) &= 6 + (-30) && \text{Working from left to right, do the multiplications.} \\
&= -24 && \text{Do the addition.}
\end{aligned}
$$

Self Check
Evaluate: $4(2) + (-4)(-3)(-2)$.

Answer: -16 ■

EXAMPLE 3 *Order of operations.* Evaluate: $40 \div (-4)5$.

Solution

This expression contains the operations of division and multiplication. We do the divisions and multiplications as they occur from left to right.

$$
\begin{aligned}
40 \div (-4)5 &= -10 \cdot 5 && \text{Do the division first: } 40 \div (-4) = -10. \\
&= -50 && \text{Do the multiplication.}
\end{aligned}
$$

Self Check
Evaluate: $45 \div (-5)3$.

Answer: -27 ■

EXAMPLE 4 *Order of operations.* Evaluate: $-2^2 - (-2)^2$.

Solution

This expression contains the operations of raising to a power and subtraction. We are to find the powers first. (Recall that -2^2 means the *opposite* of 2^2.)

$$
\begin{aligned}
-2^2 - (-2)^2 &= -4 - 4 && \text{Find the powers: } -2^2 = -4 \text{ and } (-2)^2 = 4. \\
&= -8 && \text{Do the subtraction.}
\end{aligned}
$$

Self Check
Evaluate: $-3^2 - (-3)^2$.

Answer: -18 ■

EXAMPLE 5 *Working with grouping symbols.* Evaluate: $-15 + 3(-4 + 7)$.

Self Check
Evaluate: $-18 + 4(-7 + 9)$.

Solution

$$
\begin{aligned}
-15 + 3(-4 + 7) &= -15 + 3(3) && \text{Do the addition within the parentheses: } -4 + 7 = 3. \\
&= -15 + 9 && \text{Do the multiplication: } 3(3) = 9. \\
&= -6 && \text{Do the addition.}
\end{aligned}
$$

Answer: -10 ■

EXAMPLE 6 *Working with a fraction bar.*   Evaluate: $\dfrac{-20 + 3(-5)}{(-4)^2 - 21}$.

Solution

We first evaluate the expressions in the numerator and the denominator, separately.

$$\frac{-20 + \mathbf{3(-5)}}{-21} = \frac{-20 + (\mathbf{-15})}{-21}$$ In the numerator, do the multiplication: $3(-5) = -15$.
In the denominator, evaluate the power: $(-4)^2 = 16$.

$$= \frac{-35}{-5}$$ In the numerator, add: $-20 + (-15) = -35$.
In the denominator, subtract: $16 - 21 = -5$.

$$= 7$$ Do the division.

Self Check

Evaluate: $\dfrac{-9 + 6(-4)}{(-5)^2 - 28}$.

Answer: 11 ■

EXAMPLE 7 *Grouping symbols within grouping symbols.*   Evaluate: $-5[-1 + (2 - 8)^2]$.

Solution

We begin by working within the innermost pair of grouping symbols and work to the outermost pair.

$$-5[-1 + (\mathbf{2 - 8})^2] = -5[-1 + (\mathbf{-6})^2]$$ Do the subtraction within the parentheses.

$$= -5(-1 + 36)$$ Evaluate the power within the brackets.

$$= -5(35)$$ Do the addition within the parentheses.

$$= -175$$ Do the multiplication.

Self Check

Evaluate: $-4[-2 + (5 - 9)^2]$.

Answer: -56 ■

Absolute value

You will recall that the absolute value of a number is the distance between the number and 0 on a number line. Earlier in this chapter, we evaluated simple absolute value expressions such as $|-3|$ and $|10|$. Absolute value symbols are also used in combination with more complicated expressions, such as $|-4(3)|$ and $|-6 + 1|$. When we apply the rules for the order of operations to evaluate these expressions, *the absolute value symbols are considered to be grouping symbols,* and any operations within them are to be completed first.

EXAMPLE 8 *Absolute value.*   Find each absolute value: **a.** $|-4(3)|$ and   **b.** $|-6 + 1|$.

Solution

We do the operations within the absolute value symbols first.

a. $|-4(3)| = |-12|$ Do the multiplication within the absolute value symbol: $-4(3) = -12$.

$$= 12$$ Find the absolute value of -12.

b. $|-6 + 1| = |-5|$ Do the addition within the absolute value symbol: $-6 + 1 = -5$.

$$= 5$$ Find the absolute value of -5.

Self Check

Find each absolute value:

a. $|(-6)(5)|$

b. $|-3 + (-4)|$

Answers: **a.** 30,   **b.** 7 ■

COMMENT   Just as $-5(8)$ means $-5 \cdot 8$, the expression $-5|8|$ (read as "negative 5 times the absolute value of 8") means $-5 \cdot |8|$. To evaluate such an expression, we find the absolute value first and then multiply.

$$-5|8| = -5 \cdot 8$$ Find the absolute value: $|8| = 8$.

$$= -40$$ Do the multiplication.

EXAMPLE 9   *Grouping symbols.*   Evaluate: $8 - 4|-6 - 2|$.

Solution

We do the operation within the absolute value symbol first.

$8 - 4|-6 - 2| = 8 - 4|-8|$ Do the subtraction within the absolute value symbol: $-6 - 2 = -8$.

$= 8 - 4(8)$ Find the absolute value: $|-8| = 8$.

$= 8 - 32$ Do the multiplication: $4(8) = 32$.

$= -24$ Do the subtraction.

Self Check

Evaluate: $7 - 5|-1 - 6|$.

Answer: -28 ■

Estimation

Recall that the idea behind estimation is to simplify calculations by using rounded numbers that are close to the actual values in the problem. When an exact answer is not necessary and a quick approximation will do, we can use estimation.

EXAMPLE 10   *The stock market.*   The Dow Jones Industrial Average is announced at the end of each trading day to give investors an indication of how the New York Stock Exchange performed. A positive number indicates good performance, while a negative number indicates poor performance. Estimate the net gain or loss of points in the Dow for the first week of the year 2000, shown in Figure 2-18.

Monday	Tuesday	Wednesday	Thursday	Friday

-139	-359	$+124$	$+131$	$+269$

FIGURE 2-18

Solution   We will approximate each of these numbers. For example, -139 is close to -140, and $+124$ is close to $+120$. To estimate the net gain or loss, we add the approximations.

$-140 + (-360) + 120 + 130 + 270 = -500 + 520$ Add positive and negative numbers separately to get subtotals.

$= 20$ Do the addition.

This estimate tells us that there was a gain of approximately 20 points in the Dow.   ■

STUDY SET   Section 2.6

VOCABULARY   *Fill in the blanks.*

1. When asked to evaluate expressions containing more than one operation, we should apply the rules for the _____ of operations.

2. In situations where an exact answer is not needed, an approximation or _____ is a quick way of obtaining a rough idea of the size of the actual answer.

3. Absolute value symbols, parentheses, and brackets are types of _____ symbols.

4. If an expression involves two sets of grouping symbols, always begin working within the _____ symbols and then work to the _____.

CONCEPTS

5. Consider $5(-2)^2 - 1$. How many operations need to be performed to evaluate this expression? List them in the order in which they should be performed.

6. Consider $15 - 3 + (-5 \cdot 2)^3$. How many operations need to be performed to evaluate this expression? List them in the order in which they should be performed.

7. Consider $\dfrac{5 + 5(7)}{2 + (4 - 8)}$. In the numerator, what operation should be performed first? In the denominator, what operation should be performed first?

8. In the expression $4 + 2(-7 - 1)$, how many operations need to be performed? List them in the order in which they should be performed.

9. Explain the difference between -3^2 and $(-3)^2$.

10. In the expression $-2 \cdot 3^2$, what operation should be performed first?

NOTATION   Complete each solution.

11. Evaluate: $-8 - 5(-2)^2$.
$$-8 - 5(-2)^2 = -8 - 5()$$
$$= -8 - $$
$$= -8 + ()$$
$$= -28$$

12. Evaluate: $2 + (5 - 6 \cdot 2)$.
$$2 + (5 - 6 \cdot 2) = 2 + (5 -)$$
$$= 2 + [5 + ()]$$
$$= 2 + ()$$
$$= -5$$

13. Evaluate: $[-4(2 + 7)] - 6$.
$$[-4(2 + 7)] - 6 = [-4()] - 6$$
$$= - 6$$
$$= -42$$

14. Evaluate: $\dfrac{|-9 + (-3)|}{9 - 6}$.
$$\frac{|-9 + (-3)|}{9 - 6} = \frac{||}{3}$$
$$= \frac{}{3}$$
$$= 4$$

PRACTICE   Evaluate each expression.

15. $(-3)^2 - 4^2$

16. $-7 + 4 \cdot 5$

17. $3^2 - 4(-2)(-1)$

18. $2^3 - 3^3$

19. $(2 - 5)(5 + 2)$

20. $-3(2)^2 4$

21. $-10 - 2^2$

22. $-50 - 3^3$

23. $\dfrac{-6 - 8}{2}$

24. $\dfrac{-6 - 6}{-2 - 2}$

25. $\dfrac{-5 - 5}{2}$

26. $\dfrac{-7 - (-3)}{2 - 4}$

27. $-12 \div (-2)2$

28. $-60(-2) \div 3$

29. $-16 - 4 \div (-2)$

30. $-24 + 4 \div (-2)$

31. $|-5(-6)|$

32. $|-7 - 9|$

33. $|-4 - (-6)|$

34. $|-2 + 6 - 5|$

35. $5|3|$

36. $5|4|$

37. $-6|-7|$

38. $-6|-4|$

39. $(7 - 5)^2 - (1 - 4)^2$

40. $5^2 - (-9 - 3)$

41. $-1(2^2 - 2 + 1^2)$

42. $(-7 - 4)^2 - (-1)$

43. $-50 - 2(-3)^3$

44. $(-2)^3 - (-3)(-2)$

45. $-6^2 + 6^2$

46. $-9^2 + 9^2$

47. $3\left(\dfrac{-18}{3}\right) - 2(-2)$

48. $2\left(\dfrac{-12}{3}\right) + 3(-5)$

49. $6 + \dfrac{25}{-5} + 6 \cdot 3$

50. $-5 - \dfrac{24}{6} + 8(-2)$

51. $\dfrac{1 - 3^2}{-2}$

52. $\dfrac{-3 - (-7)}{2^2 - 3}$

53. $\dfrac{-4(-5) - 2}{-6}$

54. $\dfrac{(-6)^2 - 1}{-4 - 3}$

55. $-3\left(\dfrac{32}{-4}\right) - (-1)^5$

56. $-5\left(\dfrac{16}{-4}\right) - (-1)^4$

57. $6(2^3)(-1)$

58. $2(3^3)(-2)$

59. $2 + 3[5 - (1 - 10)]$

60. $12 - 2[1 - (-8 + 2)]$

61. $-7(2 - 3 \cdot 5)$

62. $-4(1 + 3 \cdot 5)$

63. $-[6 - (1 - 4)^2]$

64. $-[9 - (9 - 12)^2]$

65. $15 + (-3 \cdot 4 - 8)$

66. $11 + (-2 \cdot 2 + 3)$

67. $|-3 \cdot 4 + (-5)|$

68. $|-8 \cdot 5 - 2 \cdot 5|$

69. $|(-5)^2 - 2 \cdot 7|$

70. $|8 \div (-2) - 5|$

71. $-2 + |6 - 4^2|$

72. $-3 - 4|6 - 7|$

73. $2|1 - 8| \cdot |-8|$

74. $2(5) - 6(|-3|)^2$

75. $-2(-34)^2 - (-605)$

76. $11 - (-15)(24)^2$

77. $-60 - \dfrac{1,620}{-36}$

78. $\dfrac{2^5 - 4^6}{-42 + 58}$

Make a mental estimate.

79. $-379 + (-103) + 287$

80. $\dfrac{-67 - 9}{-18}$

81. $-39 \cdot 8$

82. $-568 - (-227)$

83. $-3,887 + (-5,106)$

84. $-333(-4)$

85. $\dfrac{6,267}{-5}$

86. $-36 + (-78) + 59 + (-4)$

APPLICATIONS

87. TESTING   In an effort to discourage her students from guessing on multiple-choice tests, a professor uses the grading scale shown in Illustration 1. If unsure of an answer, a student does best to skip the question, because incorrect responses are penalized very heavily. Find the test score of a student who gets 12 correct and 3 wrong and leaves 5 questions blank.

Response	Value
Correct	+3
Incorrect	−4
Left blank	−1

ILLUSTRATION 1

88. THE FEDERAL BUDGET   See Illustration 2. Suppose you were hired to write a speech for a politician who wanted to highlight the improvement in the federal government's finances during 1990s. Would it be better for the politician to refer to the average budget deficit/surplus for the last half, or for the last four years of that decade? Explain your reasoning.

U.S. Budget Deficit/Surplus
($ billions)

Deficit	Year	Surplus
−164	1995	
−107	1996	
−22	1997	
	1998	+70
	1999	+123

ILLUSTRATION 2

89. SCOUTING REPORT   Illustration 3 shows a football coach how successful his opponent was running a "28 pitch" the last time the two teams met. What was the opponent's average gain with this play?

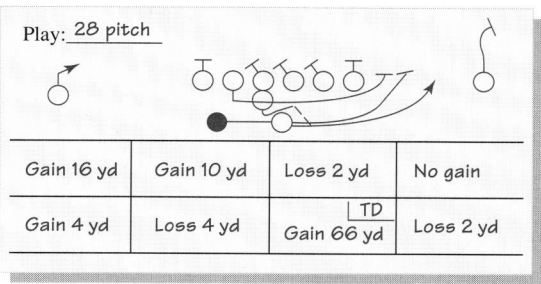

Play: 28 pitch

Gain 16 yd	Gain 10 yd	Loss 2 yd	No gain
Gain 4 yd	Loss 4 yd	TD Gain 66 yd	Loss 2 yd

ILLUSTRATION 3

90. SPREADSHEET   Illustration 4 shows the data from a chemistry experiment in spreadsheet form. To obtain a result, the chemist needs to add the values in row 1, double that sum, and then divide that number by the smallest value in column C. What is the final result of these calculations?

	A	B	C	D
1	12	−5	6	−2
2	15	4	5	−4
3	6	4	−2	8

ILLUSTRATION 4

Use estimation to answer each question.

91. OIL PRICES   The price per barrel of crude oil fluctuates with supply and demand. It can rise and fall quickly. The line graph in Illustration 5 shows how many cents the price per barrel rose or fell each day for a week. Estimate the net gain or loss in the value of a barrel of crude oil for the week.

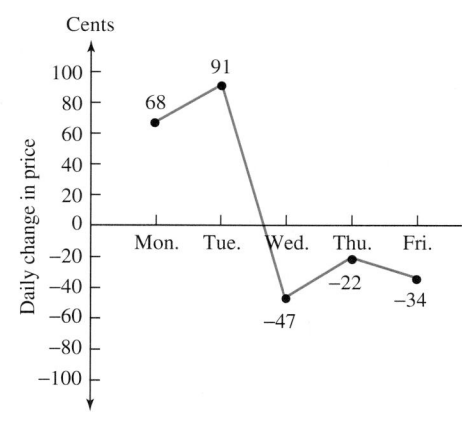

ILLUSTRATION 5

92. ESTIMATION Quickly determine a reasonable estimate of the exact answer in each of the following situations.

    a. A diver, swimming at a depth of 34 feet below sea level, spots a sunken ship beneath him. He dives down another 57 feet to reach it. What is the depth of the sunken ship?

    b. A dental hygiene company offers a money-back guarantee on its tooth whitener kit. When the kit is returned by a dissatisfied customer, the company loses the $11 it cost to produce it, because it cannot be resold. How much money has the company lost because of this return policy if 56 kits have been mailed back by customers?

    c. A tram line makes a 7,561-foot descent from a mountaintop in 18 equal stages. How much does it descend in each stage?

WRITING

93. When evaluating expressions, why are rules for the order of operations necessary?

94. In the rules for the order of operations, what does the phrase *as they occur from left to right* mean?

95. Name a situation in daily life where you use estimation.

96. List some advantages and some disadvantages of the process of estimation.

REVIEW

97. Solve $8 = 2x$.

98. Evaluate: $6^2 - (10 - 8)$.

99. How do we find the perimeter of a rectangle?

100. Is this statement true or false? "When measuring perimeter, we use square units."

101. An elevator has a weight capacity of 1,000 pounds. Seven people, with an average weight of 140 pounds, are in it. Is it overloaded?

102. List the factors of 36.

2.7 *Solving Equations Involving Integers*

In this section, you will learn about

- The properties of equality
- Solving equations involving $-x$
- Combinations of operations
- Problem solving with equations

INTRODUCTION. In this section, we revisit the topic of solving equations. The equations that we will be solving involve negative numbers, and some of the solutions are negative numbers as well. The section concludes with some application problems. As before, we will use a five-step problem-solving strategy to solve them.

The properties of equality

Recall that the addition property of equality states: *If the same number is added to equal quantities, the results will be equal quantities.* When working with negative numbers, this property can be used in a new way.

To solve $x + (-8) = -10$, we need to isolate x on the left-hand side of the equation. We can do this by adding 8 to both sides.

$$x + (-8) = -10$$

$$x + (-8) + 8 = -10 + 8 \quad \text{Apply the addition property of equality. To undo the addition of } -8 \text{, add 8 to both sides.}$$

$$x = -2 \quad \text{Do the additions: } (-8) + 8 = 0 \text{ and } -10 + 8 = -2.$$

From this example, we see that to undo addition, we can *add the opposite* of the number that is added to the variable.

EXAMPLE 1 *The addition property of equality.* Solve $-4 + x = -12$ and check the result.

Solution

We want to isolate x on one side of the equation. We can do this by adding the opposite of -4 to both sides.

$$-4 + x = -12$$
$$-4 + x + \mathbf{4} = -12 \ \mathbf{+ 4} \quad \text{To undo the addition of } -4, \text{ add 4 to both sides.}$$
$$x = -8 \qquad\quad \text{Do the additions: } -4 + 4 = 0 \text{ and } -12 + 4 = -8.$$

To check, we substitute -8 for x in the original equation and then simplify.

$$-4 + x = -12$$
$$-4 + (\mathbf{-8}) \stackrel{?}{=} -12 \quad \text{Substitute } -8 \text{ for } x.$$
$$-12 = -12 \quad \text{Do the addition: } -4 + (-8) = -12.$$

$x = -8$ is a solution.

Self Check

Solve $-6 + y = -8$ and check the result.

Answer: -2 ∎

Recall that the subtraction property of equality states: *If the same quantity is subtracted from equal quantities, the results will be equal quantities.*

EXAMPLE 2 *The subtraction property of equality.*
Solve $x + 16 = -8$.

Solution

$$x + 16 = -8$$
$$x + 16 - \mathbf{16} = -8 \ \mathbf{- 16} \qquad \text{To undo the addition of 16, subtract 16 from both sides.}$$
$$x = -8 + (-16) \qquad \text{Simplify the left side of the equation. On the right side, write the subtraction as addition of the opposite.}$$
$$x = -24 \qquad\qquad\quad \text{Do the addition.}$$

Check the result.

Self Check

Solve $c + 4 = -3$.

Answer: -7 ∎

EXAMPLE 3 *Simplifying first.* Solve $-3 + 7 = h + 11(-2)$ and check the result.

Solution

Equations should be simplified before we use the properties of equality. In this case, there is an addition on the left-hand side and a multiplication on the right-hand side of the equation. We compute those first.

$$-3 + 7 = h + 11(-2)$$
$$4 = h + (-22) \qquad\qquad \text{Do the addition: } -3 + 7 = 4. \text{ Do the multiplication: } 11(-2) = -22.$$
$$4 + \mathbf{22} = h + (-22) \ \mathbf{+ 22} \quad \text{To undo the addition of } -22, \text{ add 22 to both sides.}$$
$$26 = h \qquad\qquad\qquad \text{Simplify: } 4 + 22 = 26 \text{ and } (-22) + 22 = 0.$$
$$h = 26 \qquad\qquad\qquad \text{Since } 26 = h, h = 26.$$

Check
$$-3 + 7 = h + 11(-2) \qquad \text{The original equation.}$$
$$-3 + 7 \stackrel{?}{=} \mathbf{26} + 11(-2) \qquad \text{Substitute 26 for } h.$$
$$4 \stackrel{?}{=} 26 + (-22) \qquad\quad \text{Simplify both sides.}$$
$$4 = 4 \qquad\qquad\qquad \text{Do the addition.}$$

Self Check

Solve $-2 + 8 = y + 3(-4)$ and check the result.

Answer: 18 ∎

Recall that the division property of equality states: *If equal quantities are divided by the same nonzero quantity, the results will be equal quantities.*

EXAMPLE 4   *The division property of equality.*   Solve each equation:
a. $-3x = 15$ and   **b.** $-16 = -4y$.

Solution

a. Recall that $-3x$ indicates multiplication: $-3 \cdot x$. We must undo the multiplication of x by -3. To do this, we divide by -3.

$$-3x = 15 \qquad \text{The original equation.}$$

$$\frac{-3x}{-3} = \frac{15}{-3} \qquad \text{Divide both sides by } -3.$$

$$x = -5 \qquad \begin{array}{l}-3 \text{ times } x, \text{ divided by } -3, \text{ is } x. \text{ On the right-hand side, do the} \\ \text{division: } 15 \div (-3) = -5.\end{array}$$

Check:

$$-3x = 15 \qquad \text{The original equation.}$$

$$-3(-5) \stackrel{?}{=} 15 \qquad \text{Substitute } -5 \text{ for } x.$$

$$15 = 15 \qquad \text{Do the multiplication: } -3(-5) = 15.$$

b.

$$-16 = -4y$$

$$\frac{-16}{-4} = \frac{-4y}{-4} \qquad \text{To undo the multiplication by } -4, \text{ divide both sides by } -4.$$

$$4 = y \qquad \text{Do the divisions.}$$

$$y = 4 \qquad \text{If } 4 = y, y = 4.$$

Check the result.

Self Check

Solve each equation and check the result:

a. $-7k = 28$

b. $-40 = -8f$

Answers: a. -4,   **b.** 5 ■

Recall that the multiplication property of equality states: *If equal quantities are multiplied by the same nonzero quantity, the results will be equal quantities.*

EXAMPLE 5   *The multiplication property of equality.*   Solve $\dfrac{x}{-5} = -10$ and check the result.

Solution

In this equation, x is being divided by -5. To undo this division, we multiply both sides of the equation by -5.

$$\frac{x}{-5} = -10$$

$$-5\left(\frac{x}{-5}\right) = -5(-10) \qquad \begin{array}{l}\text{Use the multiplication property of equality. Multiply both} \\ \text{sides by } -5.\end{array}$$

$$x = 50 \qquad \begin{array}{l}\text{When } x \text{ is divided by } -5 \text{ and then multiplied by } -5, \text{ the} \\ \text{result is } x. \text{ Do the multiplication: } -5(-10) = 50.\end{array}$$

Check:

$$\frac{x}{-5} = -10 \qquad \text{The original equation.}$$

$$\frac{50}{-5} \stackrel{?}{=} -10 \qquad \text{Substitute 50 for } x.$$

$$-10 = -10 \qquad \text{Do the division.}$$

Self Check

Solve $\dfrac{t}{-3} = 4$ and check the result.

Answer: -12 ■

Solving equations involving −x

Consider the equation $-x = 3$. The variable x is not isolated, because there is a $-$ in front of it. The symbol $-x$ means -1 times x. Therefore, the equation $-x = 3$ can be rewritten as $-1x = 3$. To isolate the variable, we can either multiply both sides by -1 or divide both sides by -1.

$$-x = 3$$
$$-1x = 3 \qquad \text{\footnotesize $-x = -1x$.}$$
$$(-1)(-1x) = (-1)3 \qquad \text{\footnotesize Multiply both sides by -1.}$$
$$1x = -3 \qquad \text{\footnotesize $-1(-1) = 1$.}$$
$$x = -3 \qquad \text{\footnotesize $1x = x$.}$$

$$-x = 3$$
$$-1x = 3 \qquad \text{\footnotesize $-x = -1x$.}$$
$$\frac{-1x}{-1} = \frac{3}{-1} \qquad \text{\footnotesize Divide both sides by -1.}$$
$$x = -3 \qquad \text{\footnotesize Do the divisions.}$$

EXAMPLE 6   *Multiplying both sides by −1.*   Solve $-x = -9$ and check the result.

Solution
$$-x = -9$$
$$-1x = -9 \qquad \text{\footnotesize $-x = -1x$.}$$
$$-1(-1x) = -1(-9) \qquad \text{\footnotesize Multiply both sides by -1.}$$
$$x = 9 \qquad \text{\footnotesize Do the multiplications: $-1(-1x) = x$ and $-1(-9) = 9$.}$$

Check:
$$-x = -9$$
$$-(9) \stackrel{?}{=} -9 \qquad \text{\footnotesize Substitute 9 for x.}$$
$$-9 = -9$$

Self Check
Solve $-h = -10$ and check the result.

Answer: 10   ■

Combinations of operations

In the previous examples, each equation was solved by applying a single property of equality. Sometimes, if the equation is more complicated, it becomes necessary to use several properties of equality to solve it. For example, consider the equation $2x + 5 = 9$ and the operations performed on x.

$$2x \;+\; 5 \;=\; 9$$

The variable is multiplied by 2.   Then 5 is added.

To solve this equation, we use the rules for the order of operations in reverse.

- First, use the subtraction property of equality to undo the addition of 5.
- Second, apply the division property of equality to undo the multiplication by 2.

$$2x + 5 = 9$$
$$2x + 5 - 5 = 9 - 5 \qquad \text{\footnotesize To undo the addition of 5, subtract 5 from both sides.}$$
$$2x = 4 \qquad \text{\footnotesize Do the subtraction on each side: $5 - 5 = 0$ and $9 - 5 = 4$.}$$
$$\frac{2x}{2} = \frac{4}{2} \qquad \text{\footnotesize To undo the multiplication by 2, divide both sides by 2.}$$
$$x = 2 \qquad \text{\footnotesize Do the divisions.}$$

EXAMPLE 7   *Applying two properties of equality.*   Solve
$-4x - 5 = 15$ and check the result.

Solution
The operations performed on x are multiplication by -4 and subtraction of 5. We undo these operations in the opposite order.

$$-4x - 5 = 15$$

$-4x - 5 + 5 = 15 + 5$ Add 5 to both sides.

$-4x = 20$ Do the addition on each side: $-5 + 5 = 0$ and $15 + 5 = 20$.

$$\frac{-4x}{-4} = \frac{20}{-4}$$ Divide both sides by -4.

$x = -5$ Do the divisions.

Check:
$$-4x - 5 = 15$$

$-4(-5) - 5 \stackrel{?}{=} 15$ Substitute -5 for x.

$20 - 5 \stackrel{?}{=} 15$ Do the multiplication: $-4(-5) = 20$.

$15 = 15$ Do the subtraction.

Self Check
Solve $-6b - 1 = 11$ and check the result.

Answer: -2

EXAMPLE 8   *Applying two properties of equality.*   Solve
$2 - 3p = -1$.

Solution
We begin by writing the subtraction on the left-hand side of the equation as addition of the opposite.

$$2 - 3p = -1$$

$2 + (-3p) = -1$ Add the opposite of $3p$, which is $-3p$.

$2 + (-3p) - 2 = -1 - 2$ To undo the addition of 2 on the left-hand side of the equation, subtract 2 from both sides.

$-3p = -3$ Simplify.

$$\frac{-3p}{-3} = \frac{-3}{-3}$$ To undo the multiplication by -3, divide both sides by -3.

$p = 1$ Do the divisions.

Self Check
Solve $6 - 8k = -34$.

Answer: 5

EXAMPLE 9   *Applying two properties of equality.*   Solve
$\dfrac{y}{-2} - 6 = -18$.

Solution
The operations performed on y are division by -2 and subtraction of 6. We undo these operations in the opposite order.

$$\frac{y}{-2} - 6 = -18$$

$\dfrac{y}{-2} - 6 + 6 = -18 + 6$ To undo the subtraction of 6, add 6 to both sides.

$\dfrac{y}{-2} = -12$ Simplify both sides of the equation.

$-2\left(\dfrac{y}{-2}\right) = -2(-12)$ To undo the division by -2, multiply both sides by -2.

$y = 24$ Do the multiplications.

Self Check
Solve $\dfrac{m}{-8} - 10 = -14$.

Answer: 32

Problem solving with equations

EXAMPLE 10   *Water management.*   One day, enough water was released from a reservoir to lower the water level 17 feet to a reading of 33 feet below capacity. What was the water level reading before the release?

Analyze the problem   Figure 2-19 illustrates the given information and what we are asked to find.

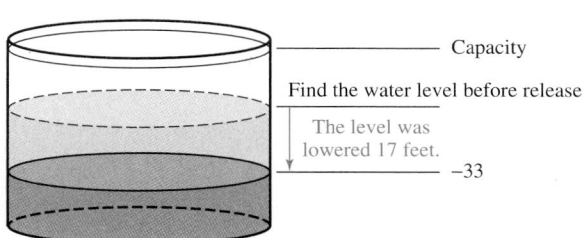

FIGURE 2-19

Form an equation   Let x = the water level reading before the release.

The water level reading before the release	minus	the number of feet the water level was lowered	is	the new water level reading.
x	$-$	17	$=$	-33

Solve the equation
$$x - 17 = -33$$
$$x - 17 + 17 = -33 + 17 \quad \text{To undo the subtraction of 17, add 17 to both sides.}$$
$$x = -16 \quad \text{Do the additions: } -17 + 17 = 0 \text{ and } -33 + 17 = -16.$$

State the conclusion   The water level reading before the release was -16.

Check the result   If the water level reading was initially -16 feet and was then lowered 17 feet, the new reading would be $-16 - 17 = -16 + (-17) = -33$ feet. The answer checks.   ■

STUDY SET Section 2.7

VOCABULARY   *Fill in the blanks.*

1. To _____ an equation, we isolate the variable on one side of the = sign.

2. The _____ of -3 is 3.

CONCEPTS

3. If we multiply x by -3 and then divide that product by -3, what is the result?

4. If we divide x by -4 and then multiply that quotient by -4, what is the result?

5. In the equation $x + 3 = 10$, we can isolate x in two ways. Find the missing numbers.
  a. $x + 3 - 3 = 10 - $
  b. $x + 3 + (-3) = 10 + $

6. In the equation $12 + c = 10$, we can isolate c in two ways. Find the missing numbers.
  a. $12 + c - = 10 - $
  b. $12 + c + () = 10 + ()$

7. Tell what operations are performed on the variable x and the order in which they occur.
  a. $-2x = -100$
  b. $-6 + x = -9$
  c. $-4x - 8 = 12$
  d. $-1 = -6 + (-5x)$

8. Tell what operations are performed on the variable x and the order in which they occur.

a. $\dfrac{x}{-4} - 8 = 50$

b. $-16 = -5 + \dfrac{x}{-3}$

In Exercises 9–12, fill in the blanks.

9. When solving the equation $-4 + t = -8 - 2$, it is best to _____ the right-hand side of the equation first before undoing any operations performed on the variable.

10. To solve the equation $-2x - 4 = 6$, we first undo the _____ of 4. Then we undo the _____ by -2.

11. When solving an equation, we isolate the variable by undoing the operations performed on it in the _____ order.

12. To solve $-x = 6$, we can multiply or divide both sides of the equation by ▢ .

13. When solving each of these equations, which operation should you undo first?
a. $-2x - 3 = -19$
b. $-6 + \dfrac{h}{-3} = -14$

14. When solving each of these equations, which operation should you undo first?
a. $5 + (-9x) = -1$
b. $-16 = -9 + \dfrac{t}{7}$

NOTATION *Complete each solution to solve the equation.*

15. $\quad y + (-7) = -16 + 3$
$\quad\quad y + (-7) = \;▢$
$\quad y + (-7) + 7 = -13 + ▢$
$\quad\quad\quad\quad y = -6$

16. $\quad x - (-4) = -1 - 5$
$\quad\quad x + 4 = -1 + ▢$
$\quad\quad x + 4 = -6$
$\quad x + 4 - 4 = -6 - ▢$
$\quad\quad\quad x = -6 + (-4)$
$\quad\quad\quad x = -10$

17. $\quad\quad -13 = -4y - 1$
$\quad -13 + ▢ = -4y - 1 + ▢$
$\quad\quad\quad\quad ▢ = -4y$
$\quad\quad \dfrac{-12}{▢} = \dfrac{-4y}{▢}$
$\quad\quad\quad\quad ▢ = y$
$\quad\quad\quad\quad y = 3$

18. Solve $1 = \dfrac{m}{-5} + 6$.

$\quad 1 - ▢ = \dfrac{m}{-5} + 6 - ▢$

$\quad\quad -5 = \dfrac{m}{-5}$

$\quad ▢ (-5) = ▢ \left(\dfrac{m}{-5} \right)$

$\quad\quad\quad ▢ = m$

$\quad\quad\quad m = 25$

19. What does $-10x$ mean?

20. What does $\dfrac{x}{-8}$ mean?

PRACTICE *Tell whether the number is a solution of the equation.*

21. $-3x - 4 = 2; -2$ **22.** $\dfrac{x}{-2} + 5 = -10; 20$

23. $-x + 8 = -4; 4$ **24.** $-3 + 2x = -3; 0$

Solve each equation. Check each result.

25. $x + 6 = -12$ **26.** $y + 1 = -4$

27. $-6 + m = -20$ **28.** $-12 + r = -19$

29. $-5 + 3 = -7 + f$ **30.** $-10 + 4 = -9 + t$

31. $h - 8 = -9$ **32.** $x - 1 = -7$

33. $0 = y + 9$ **34.** $0 = t + 5$

35. $r - (-7) = -1 - 6$ **36.** $x - (-1) = -4 - 3$

37. $t - 4 = -8 - (-2)$ **38.** $r - 1 = -3 - (-4)$

39. $x - 5 = -5$ **40.** $r - 4 = -4$

41. $-2s = 16$ **42.** $-3t = 9$

43. $-5t = -25$ **44.** $-6m = -60$

45. $-2 + (-4) = -3n$ **46.** $-10 + (-2) = -4x$

47. $-9h = -3(-3)$ **48.** $-6k = -2(-3)$

49. $\dfrac{t}{-3} = -2$ **50.** $\dfrac{w}{-4} = -5$

51. $0 = \dfrac{y}{8}$ **52.** $0 = \dfrac{h}{7}$

53. $\dfrac{x}{-2} = -6 + 3$ **54.** $\dfrac{a}{-5} = -7 + 6$

55. $\dfrac{x}{4} = -5 - 8$ **56.** $\dfrac{r}{2} = -5 - 1$

57. $2y + 8 = -6$ **58.** $5y + 1 = -9$

59. $-21 = 4h - 5$ **60.** $-22 = 7l - 8$

61. $-3v + 1 = 16$ **62.** $-4e + 4 = 24$

63. $8 = -3x + 2$ **64.** $15 = -2x + (-11)$

65. $-35 = 5 - 4x$ **66.** $12 = -9 - 3x$

67. $4 - 5x = 34$ **68.** $15 - 6x = 21$

69. $-5 - 6 - 5x = 4$ **70.** $-7 - 5 - 7x = 16$

71. $4 - 6x = -5 - 9$ **72.** $8 - 2d = -5 - 5$

73. $\dfrac{h}{-6} + 4 = 5$ **74.** $\dfrac{p}{-3} + 3 = 8$

75. $-2(4) = \dfrac{t}{-6} + 1$ **76.** $-2(5) = \dfrac{y}{-3} + 3$

77. $0 = 6 + \dfrac{c}{-5}$ **78.** $0 = -6 + \dfrac{s}{-3}$

79. $-1 = -8 + \dfrac{h}{-2}$ **80.** $-5 = 4 + \dfrac{g}{-4}$

81. $2x + 3(0) = -6$ **82.** $3x - 4(0) = -12$

83. $2(0) - 2y = 4$ **84.** $5(0) - 2y = 10$

85. $-x = 8$ **86.** $-y = 12$

87. $-15 = -k$ **88.** $-4 = -p$

APPLICATIONS   *Complete each solution.*

89. SHARKS   During a research project, a diver inside a shark cage made observations at a depth of 120 feet. For a second set of observations, the cage was raised to a depth of 75 feet. How many feet was the cage raised between observations?

Analyze the problem
- The first observations were at   ft.
- The next observations were at   ft.
- We must find _____.

Form an equation
Let x = _____

  Key word: *raised*   **Translation:** _____
We can express the second position of the cage in two ways.

The first position of the cage	plus	the amount the cage was raised	is	the second position of the cage.
	+		=	

Solve the equation
$$\quad + x = $$
$$-120 + x + \quad = -75 + \quad$$
$$x = 45$$

State the conclusion
_____

Check the result
If we add the number of feet the cage was raised to the first position, we get $-120 + \quad = -75$. The answer checks.

90. TRACK TIMES   In one race, an athlete's time for the mile was 7 seconds under the school record. In a second race, her time continued to drop, to 16 seconds under the old school record. How much time did she drop between the first and second races?

Analyze the problem
- The 1st race was   sec under the record.
- The 2nd race was   sec under the record.
- We must find _____

Form an equation
Let x = _____

Key phrase: *dropped*   **Translation:** _____
We can express her improvement in two ways.

First race performance	minus	amount of time dropped	is	second race performance.
	−		=	

Solve the equation
$$\quad - x = \quad$$
$$-7 - x + \quad = -16 + \quad$$
$$-x = \quad$$
$$x = 9$$

State the conclusion
_____

Check the result
If we subtract the time she dropped in the second race from the time dropped in the first race, we get $-7 - \quad = -16$. The answer checks.

In Exercises 91–102, let a variable represent the unknown quantity. Then write and solve an equation to answer the question.

91. DREDGING A HARBOR   In order to handle larger vessels, port officials are having a harbor deepened by dredging. The harbor bottom is already 47 feet below sea level. After the dredging, the bottom will be 65 feet below sea level. (See Illustration 1 on the next page.) How many feet must be dredged out?

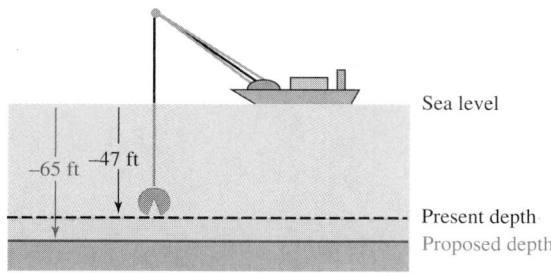

ILLUSTRATION 1

92. ROLLER COASTER DESIGN   Engineers have decided that part of a roller coaster ride will consist of a steep plunge from a peak 145 feet high. The car will then come to a screeching halt in a cave. (See Illustration 2.) How far below ground should the cave be if engineers want the overall drop in height to be 170 feet?

ILLUSTRATION 2

93. FOOTBALL   During the first half of a football game, a team ran for a total of 43 yards. After a dismal second half, they ended the game with a total of −8 yards rushing. What was their rushing total in the second half?

94. WEATHER FORECAST   The weather forecast for Fairbanks, Alaska warned listeners that the daytime high temperature of 2° below zero would drop to a nighttime low of 28° below. What was the overnight change in temperature?

95. MARKET SHARE   After its first year of business, a manufacturer of smoke detectors found its market share 43 points behind the industry leader. Five years later, it trailed the leader by only 9 points. How many points of market share did the company pick up over this five-year span?

96. CHECKING ACCOUNT   After he made a deposit of $220, a student's account was still $215 overdrawn. What was his balance before the deposit?

97. PRICE REDUCTION   Over the past year, the price of a video game player has dropped each month. If the price fell $60 this year, how much did the price drop each month on average?

98. REBATES   A store decided that it could afford to lose some money to promote a new line of sunglasses. A $9 rebate was offered to each customer purchasing these sunglasses. If the rebate program resulted in a loss of $225 for the store, how many customers took advantage of the offer?

99. ELECTION POLLS   Six months before an election, a political candidate was 31 points behind in the polls. Two days before the election, polls showed that his support had skyrocketed; he found himself only 2 points behind. How much support had he gained over the six-month period?

100. HORSE RACING   At the midway point of a 6-furlong horse race, the long shot was 3 lengths ahead of the pre-race favorite. In the last half of the race, the long shot lost ground and eventually finished 6 lengths behind the favorite. By how many lengths did the long shot lose to the favorite during the last half of the race?

101. INTERNATIONAL TIME ZONES   The world is divided into 24 times zones. Each zone is one hour ahead of or behind its neighboring zones. In the portion of the world time zone map shown in Illustration 3, we see that Tokyo is in zone +9. What time zone is Seattle in if it is 17 hours behind Tokyo?

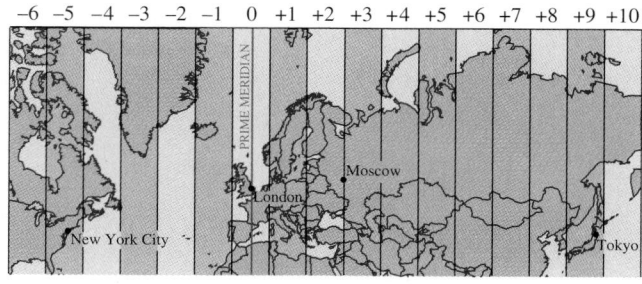

ILLUSTRATION 3

102. PROFITS AND LOSSES   In its first year of business, a nursery suffered a loss due to frost damage, ending the year $11,560 in the red. In the second year, it made a sizable profit. If the total profit for the first two years in business was $32,090, how much profit was made the second year?

WRITING

103. Explain why the variable is not isolated in the equation $-x = 10$.

104. Explain how to check the result after solving an equation.

REVIEW

105. Write 5^6 without using exponents.

106. Give the definition of an even whole number.

107. Solve $7 + 3y = 43$.

108. How can the addition $2 + 2 + 2 + 2 + 2$ be represented using multiplication?

109. Write $16 \div 8$ using a fraction bar.

110. In the expression $5(6) - 3 + 2$, list each operation in the order in which it must be performed.

Signed Numbers

In algebra, we work with both positive and negative numbers. We study negative numbers because they are necessary to describe many situations in daily life.

Represent each of these situations using a signed number.

1. Stocks fell 5 points.

2. The river was 12 feet over flood stage.

3. 30 seconds before going on the air

4. A business $6 million in the red

5. 10 degrees above normal

6. The year 2000 B.C.

7. $205 overdrawn

8. 14 units under their quota

A number line can be used to illustrate positive and negative numbers.

9. On this number line, label the location of the positive integers and the negative integers.

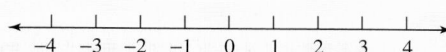

10. On this number line, graph 2 and its opposite.

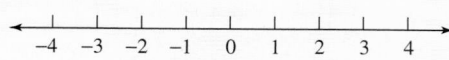

11. Two numbers, x and y, are graphed on this number line. What can you say about their relative sizes?

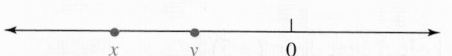

12. The absolute value of -3, written $|-3|$, is the distance between -3 and 0 on a number line. Show this distance on the number line.

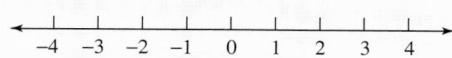

In the space provided, summarize how addition, multiplication, and division are performed with two integers having like signs and with two integers having unlike signs. Then explain the method that is used to subtract integers.

13. Addition

Like signs:

Unlike signs:

14. Multiplication

Like signs:

Unlike signs:

15. Division

Like signs:

Unlike signs:

16. Subtraction with integers

Section 2.1

OPPOSITES   Have everyone in the class get in pairs. The object of the game is for one member of a team, using one-word clues, to get the other member to say each of the words in Column I. This is done by giving your partner a word clue having the opposite meaning. For example, if you want your partner to say the word *up,* give the clue *down.* Go through all of the words in Column I. Keep track of how many words are guessed correctly.

Then switch assignments. The person who first gave the clues now receives the clues. Go through all of the words in Column II.

Column I		Column II	
below	surplus	win	positive
over	overdrawn	gain	increase
ahead	profit	forward	debt
before	deduct	retreat	withdrawal
less	liabilities	rise	accelerate

Section 2.2

ADDING INTEGERS   To illustrate how to add $-5 + 3$, think of a hole 5 feet deep (-5) which then has 3 feet of dirt (3) added to it. Illustration 1 shows that the resulting hole would be 2 feet deep (-2). So $-5 + 3 = -2$.

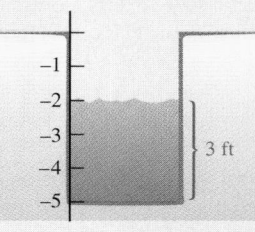

ILLUSTRATION 1

Draw a similar picture to help find each sum.

a. $-4 + 1$

b. $-5 + 4$

c. $-6 + 6$

d. $-3 + (-1)$

e. $-3 + (-3)$

f. $-3 + 5$

Section 2.3

SUBTRACTING INTEGERS   Write a subtraction problem in which the difference of two negative numbers is

a. a positive number.

b. a negative number.

Section 2.4

MULTIPLYING INTEGERS

a. Complete the multiplication table that follows.

b. Construct another table with a top row of -1 through -10 and a first column of 1 through 10.

c. Construct a table with a top row of -1 through -10 and a first column of -1 through -10.

Multiplication Table

·	1	2	3	4	5	6	7	8	9	10
−1										
−2										
−3										
−4										
−5										
−6										
−7										
−8										
−9										
−10										

Section 2.5

OPERATIONS WITH TWO INTEGERS   For each operation listed in the table, tell whether the answer is always positive, always negative, or may be positive or negative.

Signs of the two integers	Add	Subtract	Multiply	Divide
Both positive				
Both negative				
One positive, one negative				

Section 2.6

ESTIMATION   Estimate the answer to each problem.

a. $405 - 567$

b. $-2{,}564 - 2{,}456$

c. $989 - 898$

d. $-23{,}250 + 22{,}750$

e. $56(-87)$

f. $-40 - 30 - 45$

g. $608 \div (-2)$

h. $-94 + 90 - 45$

Section 2.7

SOLVING EQUATIONS   What is wrong with each solution?

a. Solve $2x + 4 = 10$.

$$2x + 4 = 10$$

$$\frac{2x}{2} + 4 = \frac{10}{2}$$

$$x + 4 = 5$$

$$x + 4 - 4 = 5 - 4$$

$$x = 1$$

b. Solve $2x + 4 = 10$.

$$2x + 4 = 10$$

$$2x + 4 - 4 = 10$$

$$2x = 10$$

$$\frac{2x}{2} = \frac{10}{2}$$

$$x = 5$$

CHAPTER REVIEW

An Introduction to the Integers

CONCEPTS

A *number line* is a horizontal or vertical line used to represent numbers graphically.

Integers: {. . . , −3, −2, −1, 0, 1, 2, 3, . . .}

Inequality symbols:
> is greater than
< is less than

A *negative* number is less than 0. A *positive* number is greater than 0.

The *absolute value* of a number is the distance between it and 0 on the number line.

The opposite of the opposite of a number is that number.

REVIEW EXERCISES

1. Graph each set of numbers.
   a. {−3, −1, 0, 4}

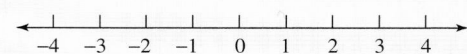

   b. The integers greater than −3 but less than 4.

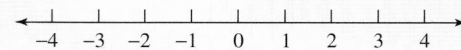

2. Insert one of the symbols > or < in the blank to make a true statement.
   a. −7 ___ 0
   b. −20 ___ −19
   c. |−16| ___ −16
   d. 56 ___ 60

3. WATER PRESSURE   Salt water exerts a pressure of 14.7 pounds per square inch at a depth of 33 feet. See Illustration 1. Express the depth using a signed number.

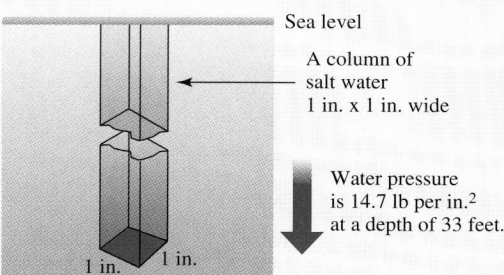

ILLUSTRATION 1

4. Represent each of these situations using a signed number.
   a. A deficit of $1,200
   b. 10 seconds before going on the air

5. Evaluate each expression.
   a. |−4|
   b. |0|
   c. |−43|
   d. −|12|

6. Explain the meaning of each red − sign.
   a. −5
   b. −(−5)
   c. −(−5)
   d. 5 − (−5)

On a number line, two numbers the same distance away from 0, but on different sides of it, are called *opposites*.

7. Find each of the following.
  a. $-(-12)$
  b. The opposite of 8
  c. The opposite of -8
  d. -0

Addition of Integers

To add two integers with *like signs,* add their absolute values and attach their common sign to that sum.

To add two integers with *unlike signs,* subtract their absolute values, the smaller from the larger. Attach the sign of the number with the larger absolute value to that result.

8. Use a number line to find each sum.
  a. $4 + (-2)$

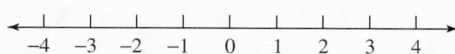

  b. $-1 + (-3)$

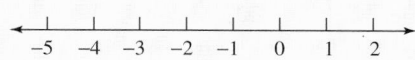

9. Add.
  a. $-6 + (-4)$
  b. $-23 + (-60)$
  c. $-1 + (-4) + (-3)$
  d. $-4 + 3$
  e. $-28 + 140$
  f. $9 + (-20)$
  g. $3 + (-2) + (-4)$
  h. $(-2 + 1) + [(-5) + 4]$

Addition property of zero: If a represents a number, then
$$a + 0 = a \quad \text{and} \quad 0 + a = a$$

10. Add.
  a. $-4 + 0$
  b. $0 + (-20)$
  c. $-8 + 8$
  d. $73 + (-73)$

If $a + b = 0$, then a and b are called *additive inverses.*

11. Give the additive inverse of each number.
  a. -11
  b. 4

12. DROUGHT   During a drought, the water level in a reservoir fell to a point 100 feet below normal. After two rainy months, it rose 35 feet. How far below normal was the water level after the rain?

Subtraction of Integers

Rule for subtraction: If a and b represent numbers, then
$$a - b = a + (-b)$$

13. Subtract.
  a. $5 - 8$
  b. $-9 - 12$
  c. $-4 - (-8)$
  d. $-6 - 106$
  e. $-8 - (-2)$
  f. $7 - 1$
  g. $0 - 37$
  h. $0 - (-30)$

14. Fill in the blanks to make a true statement: Subtracting a number is the same as _____ the _____ of that number.

15. Evaluate each expression.
  a. $-9 - 7 + 12$
  b. $7 - [(-6) - 2]$
  c. $1 - (2 - 7)$
  d. $-12 - (6 - 10)$

16. Subtract 27 from -50.

17. GOLD MINING   Some miners discovered a small vein of gold at a depth of 150 feet. This prompted them to continue their exploration. After descending another 75 feet, they came upon a much larger find. Use a signed number to represent the depth of the second discovery.

18. Evaluate: $2 - [-(-3)]$.

19. RECORD TEMPERATURES   The lowest and highest recorded temperatures for Alaska and Virginia are shown here. For each state, find the difference in temperature between the record high and low.

To find the *change* in a quantity, subtract the earlier value from the later value.

Alaska: Low $-80°$ Jan. 23, 1971	Virginia: Low $-30°$ Jan. 22, 1985
High $100°$ June 27, 1915	High $110°$ July 15, 1954

SECTION 2.4	*Multiplication of Integers*

The product of two integers with *like signs* is positive. The product of two integers with *unlike signs* is negative.

20. Multiply.
  a. $-9 \cdot 5$ **b.** $-3(-6)$
  c. $7(-2)$ **d.** $(-8)(-47)$
  e. $-20 \cdot 5$ **f.** $-1(-1)$
  g. $-1(25)$ **h.** $(5)(-30)$

21. Multiply.
  a. $(-6)(-2)(-3)$ **b.** $4(-3)3$
  c. $0(-7)$ **d.** $(-1)(-1)(-1)(-1)$

22. TAX DEFICIT   A state agency's prediction of a tax shortfall proved to be two times worse than the actual deficit of $3 million. The federal prediction of the same shortfall was even more inaccurate—three times the amount of the actual deficit. Complete Illustration 2, which summarizes these incorrect forecasts.

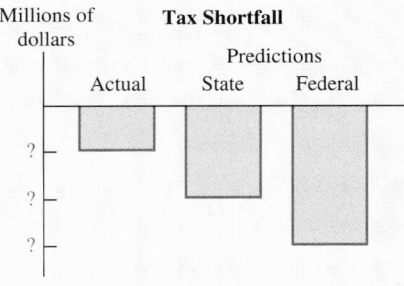

ILLUSTRATION 2

An *exponent* is used to represent repeated multiplication.

23. Find each power.
  a. $(-5)^2$ **b.** $(-2)^5$
  c. $(-8)^2$ **d.** $(-4)^3$

24. When $(-5)^9$ is evaluated, will the result be positive, or will it be negative?

When a negative integer is raised to an *even* power, the result is positive. When it is raised to an *odd* power, the result is negative.

25. Explain the difference between -2^2 and $(-2)^2$ and then evaluate each.

| **SECTION 2.5** | *Division of Integers* |

26. Fill in the box to make a true statement: We know that $\dfrac{-15}{5} = -3$ because $5(\ \ \ ) = -15$.

The quotient of two numbers with *like* signs is positive.

The quotient of two integers with *unlike* signs is negative.

If a represents a nonzero number,

$$\frac{0}{a} = 0$$

Division *by* 0 is undefined.

27. Divide.

a. $\dfrac{-14}{7}$

b. $\dfrac{25}{-5}$

c. $-64 \div 8$

d. $\dfrac{-202}{-2}$

28. Find each quotient, if possible.

a. $\dfrac{0}{-5}$

b. $\dfrac{-4}{0}$

c. $\dfrac{-673}{-673}$

d. $\dfrac{-10}{-1}$

29. PRODUCTION TIME   Because of improved production procedures, the time needed to produce an electronic component dropped by 12 minutes over the past six months. If the drop in production time was uniform, how much did it change each month over this period?

| **SECTION 2.6** | *Order of Operations and Estimation* |

The rules for the order of operations:

1. Do all calculations within parentheses and other grouping symbols.
2. Evaluate all exponential expressions.
3. Do all the multiplications and divisions, working from left to right.
4. Do all the additions and subtractions, working from left to right.

Always work from the *innermost* set of grouping symbols to the *outermost* set.

A fraction bar is a grouping symbol.

An *estimation* is an approximation that gives a quick idea of what the actual answer would be.

30. Evaluate each expression.

a. $2 + 4(-6)$

b. $7 - (-2)^2 + 1$

c. $2 - 5(4) + (-25)$

d. $-3(-2)^3 - 16$

e. $-2(5)(-4) + \dfrac{|-9|}{3^2}$

f. $-4^2 + (-4)^2$

g. $-12 - (8 - 9)^2$

h. $7|-8| - 2(3)(4)$

31. Evaluate each expression.

a. $-4\left(\dfrac{15}{-3}\right) - 2^3$

b. $-20 + 2(12 - 5 \cdot 2)$

c. $-20 + 2[12 - (-7 + 5)^2]$

d. $8 - |-3 \cdot 4 + 5|$

32. Evaluate each expression.

a. $\dfrac{10 + (-6)}{-3 - 1}$

b. $\dfrac{3(-6) - 11 + 1}{4^2 - 3^2}$

33. Estimate each answer.

a. $-89 + 57 + (-42)$

b. $\dfrac{-507}{-24}$

c. $(-681)(9)$

d. $317 - (-775)$

SECTION 2.7 — *Solving Equations Involving Integers*

To *solve an equation* means to find all values of the variable that, when substituted into the original equation, make the equation a true statement.

To solve an equation, we undo the operations in the reverse order from that in which they were performed on the variable. The objective is to isolate the variable.

34. Is $x = -4$ a solution of the equation? Explain why or why not.

   a. $2x + 6 = -2$ **b.** $6 + \dfrac{x}{2} = -4$

35. Solve each equation. Check the result.

   a. $t + (-8) = -18$ **b.** $\dfrac{x}{-3} = -4$

   c. $y + 8 = 0$ **d.** $-7m = -28$

36. Solve each equation. Check the result.

   a. $-x = -15$ **b.** $4 = -y$

37. Solve each equation. Check the result.

   a. $-5t + 1 = -14$ **b.** $3(2) = 2 - 2x$

   c. $\dfrac{x}{-4} - 5 = -1 - 1$ **d.** $c - (-5) = 5$

The five-step *problem-solving strategy* can be used when solving application problems.

1. Analyze the problem.
2. Form an equation.
3. Solve the equation.
4. State the conclusion.
5. Check the result.

In Exercises 38–40, let a variable represent the unknown quantity. Then write and solve an equation to answer the question.

38. WIND-CHILL FACTOR   If the wind is blowing at 25 miles per hour, an air temperature of 5° below zero will feel like 51° below zero. Find the perceived change in temperature that is caused by the wind.

39. CREDIT CARD PROMOTION   During the holidays, a store offered an $8 gift certificate to any customer applying for its credit card. If this promotion cost the company $968, how many customers applied for credit?

40. BANK FAILURE   When a group of 7 investors decided to acquire a failing bank, each had to assume an equal share of the bank's total indebtedness, which was $57,400. How much debt did each investor assume?

Chapter 2 Test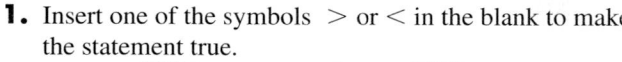

1. Insert one of the symbols $>$ or $<$ in the blank to make the statement true.
   a. -8 ___ -9 **b.** -8 ___ $|-8|$
   c. The opposite of 5 ___ 0

2. List the integers.

3. SCHOOL ENROLLMENT   According to the projections in Illustration 1, which high school will face the greatest shortage in the year 2006?

High schools with shortage of classroom seats by 2006	
Sylmar	-669
San Fernando	$-1,630$
Monroe	$-2,488$
Cleveland	-350
Canoga Park	-586
Polytechnic	$-2,379$
Van Nuys	$-1,690$
Reseda	-462
North Hollywood	$-1,004$
Hollywood	-774

ILLUSTRATION 1

4. Use a number line to find the sum: $-3 + (-2)$.

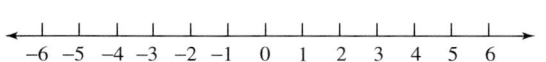

5. Add.
   a. $-65 + 31$ **b.** $-17 + (-17)$
   c. $[6 + (-4)] + [-6 + (-4)]$

6. Subtract.
   a. $-7 - 6$ **b.** $-7 - (-6)$
   c. $0 - 15$ **d.** $-60 - 50 - 40$

7. Find each product.
   a. $-10 \cdot 7$ **b.** $-4(-2)(-6)$
   c. $(-2)(-2)(-2)(-2)$ **d.** $-55(0)$

8. Write the related multiplication statement for $\dfrac{-20}{-4} = 5$.

9. Find each quotient, if possible.
   a. $\dfrac{-32}{4}$ **b.** $\dfrac{8}{6 - 6}$
   c. $\dfrac{-5}{1}$ **d.** $\dfrac{0}{-6}$

10. BUSINESS TAKEOVER   Six businessmen are contemplating taking over a company that has potential, but they must retire the debt incurred by the company over the past three quarters. (See Illustration 2.) If they plan equal ownership, how much will each have to contribute to retire the debt?

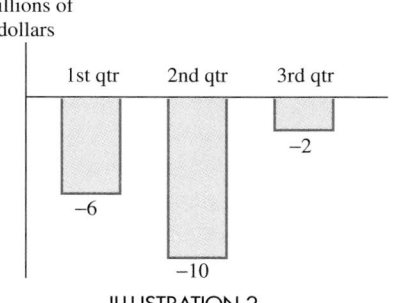

ILLUSTRATION 2

11. GEOGRAPHY   The lowest point on the African continent is the Qattarah Depression in the Sahara Desert, 436 feet below sea level. The lowest point on the North American continent is Death Valley, California, 282 feet below sea level. What is the difference in these elevations?

12. Evaluate each expression.

    a. $-(-6)$ **b.** $|-7|$

    c. $|-9 + 3|$ **d.** $2|-66|$

13. Find each power.

    a. $(-4)^2$ **b.** -4^2 **c.** $(-4 - 3)^2$

14. Evaluate: $-18 \div 2 \cdot 3$.

15. Evaluate: $4 - (-3)^2 + 6$.

16. Evaluate: $-3 + \left(\dfrac{-16}{4}\right) - 3^3$.

17. Evaluate: $-10 + 2[6 - (-2)^2(-5)]$.

18. Evaluate: $\dfrac{4(-6) - 2^2}{-3 - 4}$.

19. Solve $c - (-7) = -8$.

20. Solve $6 - x = -10$.

21. Solve $\dfrac{x}{-4} = 10$.

22. Solve $3x + (-7) = -11 + (-11)$.

23. Solve $-5 = -6a + 7$ and check the result.

24. Solve $\dfrac{x}{-2} + 3 = (-2)(-6)$.

In Problems 25–26, let a variable represent the unknown quantity. Then write and solve an equation to answer each question.

25. CHECKING ACCOUNT   After making a deposit of $225, a student's account was still $19 overdrawn. What was her balance before the deposit?

26. HOSPITAL CAPACITY   One morning, the number of beds occupied by patients in a hospital was 3 under capacity. By afternoon, the number of unoccupied beds was 21. If no new patients were admitted, how many patients were released to go home?

27. Multiplication means repeat addition. Use this fact to show that the product of a positive and a negative number, such as $5(-4)$, is negative.

28. Explain why the absolute value of a number can never be negative.

Chapters 1-2 *Cumulative Review Exercises*

Consider the numbers $-2, -1, 0, 1, 2, \frac{3}{2}, 5,$ *and* 9.

1. List each natural number.

2. List each whole number.

3. List each negative number.

4. List each integer.

Consider the number $7{,}326{,}549$.

5. Which digit is in the thousands column?

6. Which digit is in the hundred thousands column?

7. Round to the nearest hundred.

8. Round to the nearest ten thousand.

9. BIDS   A school district received the bids shown in Illustration 1. Which company should be awarded the contract?

Citrus Unified School District Bid 02-9899
CABLING AND CONDUIT INSTALLATION

Datatel	$2,189,413
Walton Electric	$2,201,999
Advanced Telecorp	$2,175,081
CRF Cable	$2,174,999
Clark & Sons	$2,175,801

ILLUSTRATION 1

10. NUCLEAR POWER   The table at the top of the page gives the number of operable nuclear power plants in the United States for the years 1978–1998, in five-year increments. Construct a bar graph using the data.

Year	1978	1983	1988	1993	1998
Plants	70	80	108	109	104

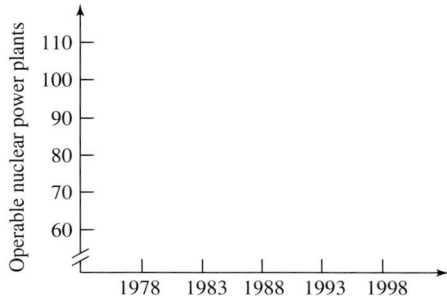

In Exercises 11–14, perform each operation.

11. $237 + 549$

12. $6{,}375 - 2{,}569$

13. $\begin{array}{r} 5{,}369 \\ -\ \ 685 \end{array}$

14. $\begin{array}{r} 7{,}899 \\ +5{,}237 \end{array}$

15. Find the perimeter and the area of the rectangular garden in Illustration 2.

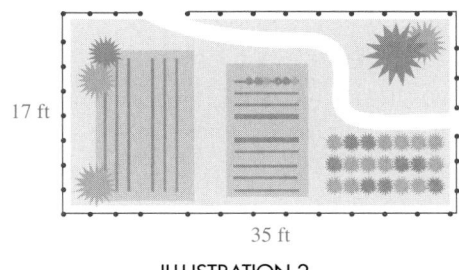

17 ft

35 ft

ILLUSTRATION 2

16. In a shipment of 147 pieces of furniture, 27 pieces were sofas, 55 were leather chairs, and the rest were wooden chairs. Find the number of wooden chairs.

In Exercises 17–20, perform each operation.

17. $435 \cdot 27$

18. $1{,}261 \div 97$

19. $\begin{array}{r} 4{,}587 \\ \times\ \ \ 67 \end{array}$

20. $38\overline{)17{,}746}$

21. SHIPPING   There are 12 tennis balls in one dozen, and 12 dozen in one gross. How many tennis balls are there in a shipment of 12 gross?

22. Find all of the factors of 18.

In Exercises 23–26, identify each number as a prime, a composite, an even, or an odd.

23. 17

24. 18

25. 0

26. 1

27. Find the prime factorization of 504.

28. Write the expression $11 \cdot 11 \cdot 11 \cdot 11$ using an exponent.

Evaluate each expression.

29. $5^2 \cdot 7$

30. $16 + 2[14 - 3(5 - 4)^2]$

31. $25 + 5 \cdot 5$

32. $\dfrac{16 - 2 \cdot 3}{2 + (9 - 6)}$

33. SPEED CHECK   A traffic officer monitored several cars on a city street. She found that the speeds of the cars were as follows:

38, 42, 36, 38, 48, 44

On average, were the drivers obeying the 40-mph speed limit?

34. Tell whether 6 is a solution of the equation $3x - 2 = 16$. Explain why or why not.

Solve each equation and check the result.

35. $50 = x + 37$

36. $a - 12 = 41$

37. $5p = 135$

38. $\dfrac{y}{8} = 3$

In Exercises 39–40, graph each set on a number line.

39. $\{-2, -1, 0, 2\}$

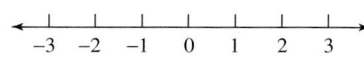

40. The integers greater than -4 but less than 2

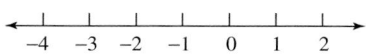

41. True or false: $-17 < -16$.

42. Evaluate 3^2 and -3^2.

Evaluate each expression.

43. $-2 + (-3)$

44. $-15 + 10 + (-9)$

45. $-3 - 5$

46. $-15^2 - 2|-3|$

47. $(-8)(-3)$

48. $5(-7)^3$

49. $\dfrac{-14}{-7}$

50. $\dfrac{450}{-9}$

51. $5 + (-3)(-7)$

52. $-20 + 2[12 - 5(-2)(-1)]$

53. $\dfrac{10 - (-5)}{1 - 2 \cdot 3}$

54. $\dfrac{3(-6) - 10}{3^2 - 4^2}$

Solve each equation. Check the result.

55. $-5t + 1 = -14$

56. $\dfrac{x}{-3} - 2 = -2(-2)$

Let a variable represent the unknown quantity. Then write and solve an equation to answer each question.

57. BUYING A BUSINESS   When 12 investors decided to buy a bankrupt company, they agreed to assume equal shares of the company's debt of $1,512,444. How much was each person's share?

58. THE MOON   The difference in the maximum and the minimum temperatures on the moon's surface is 540°F. The maximum temperature, which occurs at lunar noon, is 261°F. Find the minimum temperature, which occurs just before lunar dawn.

Fractions and Mixed Numbers

3

3.1 The Fundamental Property of Fractions

3.2 Multiplying Fractions

3.3 Dividing Fractions

3.4 Adding and Subtracting Fractions

The LCM and the GCF

3.5 Multiplying and Dividing Mixed Numbers

3.6 Adding and Subtracting Mixed Numbers

3.7 Order of Operations and Complex Fractions

3.8 Solving Equations Containing Fractions

Key Concept: The Fundamental Property of Fractions

Accent on Teamwork

Chapter Review

Chapter Test

Cumulative Review Exercises

143

WHOLE NUMBERS ARE USED TO COUNT OBJECTS.
WHEN WE NEED TO REPRESENT PARTS OF A WHOLE,
FRACTIONS CAN BE USED.

3.1 The Fundamental Property of Fractions

In this section, you will learn about

- Basic facts about fractions • Equivalent fractions • Simplifying a fraction
- Expressing a fraction in higher terms

INTRODUCTION. There is no better place to start a study of fractions than with *the fundamental property of fractions*. This property is the foundation for two fundamental procedures that are used when working with fractions. But first, we review some basic facts about fractions.

Basic facts about fractions

1. A Fraction Can Be Used to Indicate Equal Parts of a Whole.

In our everyday lives, we often deal with parts of a whole. For example, we talk about parts of an hour, parts of an inch, and parts of a pound.

2. A Fraction is Composed of a Numerator, a Denominator, and a Fraction Bar.

$$\text{Fraction bar} \longrightarrow \frac{\mathbf{3}}{\mathbf{4}} \begin{matrix} \longleftarrow \text{Numerator} \\ \longleftarrow \text{Denominator} \end{matrix}$$

The denominator (in this case, 4) tells us that a whole was divided into four equal parts. The numerator tells us that we are considering three of those equal parts.

3. Fractions Can Be Proper or Improper.

If the numerator of a fraction is less than its denominator, the fraction is called a **proper fraction**. A proper fraction is less than 1. Fractions whose numerators are greater than or equal to their denominators are called **improper fractions**. An improper fraction is greater than or equal to 1.

Proper fractions	**Improper fractions**
$\frac{1}{4}$, $\frac{2}{3}$, and $\frac{98}{99}$	$\frac{7}{2}$, $\frac{98}{97}$, $\frac{16}{16}$, and $\frac{5}{1}$

EXAMPLE 1 *Fractional parts of a whole.* **a.** In Figure 3-1, what fractional part of the barrel is full? **b.** What fractional part is empty?

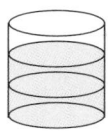

FIGURE 3-1

Solution

The barrel has been divided into three equal parts.

a. Two of the three parts are full. Therefore, the barrel is $\frac{2}{3}$ full.

b. One of the three equal parts is not filled. The barrel is $\frac{1}{3}$ empty.

The fractions $\frac{2}{3}$ and $\frac{1}{3}$ are both proper fractions.

Self Check

a. According to the calendar below, what fractional part of the month has passed? **b.** What fractional part remains?

DECEMBER

X	X	X	X	X	X	X
X	X	X	X	12	13	14
15	16	17	18	19	20	21
22	23	24	25	26	27	28
29	30	31				

Answers: **a.** $\dfrac{11}{31}$, **b.** $\dfrac{20}{31}$ ■

4. The Denominator of a Fraction Cannot Be 0.

$\frac{7}{0}$, $\frac{23}{0}$, and $\frac{0}{0}$ are meaningless expressions. (Recall that $\frac{7}{0}$, $\frac{23}{0}$, and $\frac{0}{0}$ represent *division* by 0, and a number cannot be divided by 0.) However, $\frac{0}{7} = 0$ and $\frac{0}{23} = 0$.

5. Fractions Can Be Negative.

There are times when a negative fraction is needed to describe a quantity. For example, if an earthquake causes a road to sink one-half inch, the amount of movement can be represented by $-\frac{1}{2}$ inch.

Negative fractions can be written in three ways. The negative sign can appear in the numerator, in the denominator, or in front of the fraction.

$$\frac{-1}{2} = \frac{1}{-2} = -\frac{1}{2} \qquad \frac{-15}{8} = \frac{15}{-8} = -\frac{15}{8}$$

Negative fractions

> If a and b represent positive numbers,
>
> $$\frac{-a}{b} = \frac{a}{-b} = -\frac{a}{b} \qquad (b \neq 0)$$

Fractions are often referred to as **rational numbers.** All integers are rational numbers, because every integer can be written as a fraction with a denominator of 1. For example,

$$2 = \frac{2}{1}, \quad -5 = \frac{-5}{1}, \quad \text{and} \quad 0 = \frac{0}{1}$$

Since every integer is also a rational number, the integers are a subset of the rational numbers.

COMMENT Note that not all rational numbers are integers. For example, the rational number $\frac{7}{8}$ is not an integer.

Equivalent fractions

Fractions can look different but still represent the same number. To show this, let's divide the rectangle in Figure 3-2(a) in two ways. In Figure 3-2(b), we divide it into halves (2 equal-sized parts). In Figure 3-2(c), we divide it into fourths (4 equal-sized parts). Notice that one-half of the figure is the same size as two-fourths of the figure.

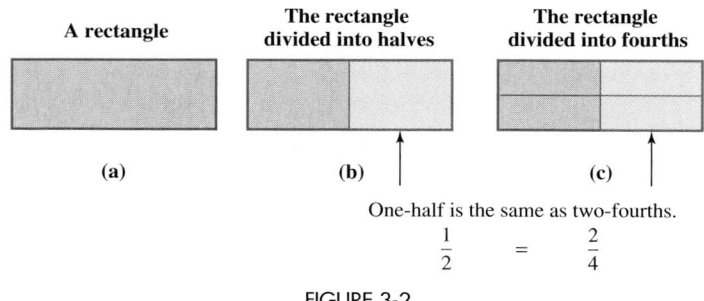

FIGURE 3-2

The fractions $\frac{1}{2}$ and $\frac{2}{4}$ look different, but Figure 3-2 shows that they represent the same amount. We say that they are **equivalent fractions.**

Equivalent fractions

> Two fractions are **equivalent** if they represent the same number.

Simplifying a fraction

If we replace a fraction with an equivalent fraction that contains smaller numbers, we are **simplifying** or **reducing the fraction.** To simplify a fraction, we apply the following property.

The fundamental property of fractions

> Multiplying or dividing the numerator and the denominator of a fraction by the same nonzero number does not change the value of the fraction. In symbols, for all numbers a, b, and x (provided b and x are not zero),
>
> $$\frac{a}{b} = \frac{a \cdot x}{b \cdot x} \quad \text{and} \quad \frac{a}{b} = \frac{a \div x}{b \div x}$$

As an example, we consider $\frac{24}{28}$. It is apparent that 24 and 28 have a common factor of 4. By the fundamental property of fractions, we can divide the numerator and denominator of this fraction by 4.

$$\frac{24}{28} = \frac{24 \div 4}{28 \div 4}$$

Divide the numerator by 4.

Divide the denominator by 4.

$$= \frac{6}{7} \qquad 24 \div 4 = 6 \text{ and } 28 \div 4 = 7.$$

Thus, $\frac{24}{28} = \frac{6}{7}$. We say that $\frac{24}{28}$ and $\frac{6}{7}$ are equivalent fractions, because they represent the same number.

In practice, we show the previous simplification in a slightly different way.

$$\frac{24}{28} = \frac{4 \cdot 6}{4 \cdot 7}$$

Once you see a common factor of the numerator and the denominator, factor each of them so that it shows. In this case, 24 and 28 share a common factor of 4.

$$= \frac{\overset{1}{\cancel{4}} \cdot 6}{\underset{1}{\cancel{4}} \cdot 7}$$

Apply the fundamental property of fractions. Divide the numerator and the denominator by 4 by drawing slashes through the common factors. Use small 1's to represent the result of each division of 4 by 4.

$$= \frac{6}{7}$$

Multiply in the numerator and in the denominator: $1 \cdot 6 = 6$ and $1 \cdot 7 = 7$.

In the second step of the previous simplification, we say that we *divided out the common factor of 4.*

Simplifying a fraction

> We can **simplify** a fraction by factoring its numerator and denominator and then dividing out all common factors in the numerator and denominator.

When a fraction can be simplified no further, we say that it is written in **lowest terms.**

Lowest terms

> A fraction is in **lowest terms** if the only factor common to the numerator and denominator is 1.

EXAMPLE 2   *Simplifying a fraction.*   Simplify to lowest terms: $\dfrac{25}{75}$.

Solution

The numerator and the denominator have a common factor of 25.

$$\frac{25}{75} = \frac{25 \cdot 1}{25 \cdot 3}$$ Factor 25 as 25 · 1 and 75 as 25 · 3.

$$= \frac{\overset{1}{\cancel{25}} \cdot 1}{\underset{1}{\cancel{25}} \cdot 3}$$ Divide out the common factor of 25.

$$= \frac{1}{3}$$ Multiply in the numerator and in the denominator: 1 · 1 = 1 and 1 · 3 = 3.

Self Check

Simplify to lowest terms: $\dfrac{60}{80}$.

Answer: $\dfrac{3}{4}$

EXAMPLE 3   *Using prime factorization.*   Simplify $\dfrac{90}{126}$.

Solution

To find the common factors that will divide out, we prime factor 90 and 126.

$$\frac{90}{126} = \frac{5 \cdot 3 \cdot 3 \cdot 2}{7 \cdot 3 \cdot 3 \cdot 2}$$ Use the tree method or the division method to prime factor 90 and 126: 90 = 5 · 3 · 3 · 2 and 126 = 7 · 3 · 3 · 2.

$$= \frac{5 \cdot \overset{1}{\cancel{3}} \cdot \overset{1}{\cancel{3}} \cdot \overset{1}{\cancel{2}}}{7 \cdot \underset{1}{\cancel{3}} \cdot \underset{1}{\cancel{3}} \cdot \underset{1}{\cancel{2}}}$$ Divide out the common factors of 3, 3, and 2.

$$= \frac{5}{7}$$ Multiply in the numerator and in the denominator.

Self Check

Simplify $\dfrac{42}{150}$.

Answer: $\dfrac{7}{25}$

 COMMENT   Negative fractions are simplified in the same way as positive fractions. Just remember to write a negative sign − in front of each step of the solution.

$$-\frac{45}{72} = -\frac{\overset{1}{\cancel{9}} \cdot 5}{\underset{1}{\cancel{9}} \cdot 8} = -\frac{5}{8}$$

Expressing a fraction in higher terms

It is sometimes necessary to replace a fraction with an equivalent fraction that involves larger numbers. This is called **expressing the fraction in higher terms** or **building up** the fraction.

For example, to write $\frac{3}{8}$ as an equivalent fraction with a denominator of 40, we can use the fundamental property of fractions and multiply the numerator and denominator by 5.

Multiply the numerator by 5.

$$\frac{3}{8} = \frac{3 \cdot 5}{8 \cdot 5}$$

Multiply the denominator by 5.

$$= \frac{15}{40} \qquad \text{Do the multiplications in the numerator and in the denominator.}$$

Therefore, $\dfrac{3}{8} = \dfrac{15}{40}$.

EXAMPLE 4   *Expressing a fraction in higher terms.*   Write $\frac{5}{7}$ as an equivalent fraction with a denominator of 28.

Solution

We need to multiply the denominator by 4 to obtain 28. By the fundamental property of fractions, we must multiply the numerator by 4 as well.

$$\frac{5}{7} = \frac{5 \cdot 4}{7 \cdot 4} \qquad \text{Multiply the numerator and denominator by 4.}$$

$$= \frac{20}{28} \qquad \text{Do the multiplication in the numerator and in the denominator.}$$

Self Check

Write $\frac{2}{3}$ as an equivalent fraction with a denominator of 24.

Answer: $\dfrac{16}{24}$ ■

EXAMPLE 5   *Expressing a whole number as a fraction.*   Write 4 as a fraction with a denominator of 6.

Solution

First, express 4 as a fraction: $4 = \frac{4}{1}$. To obtain a denominator of 6, we need to multiply the numerator and denominator by 6.

$$\frac{4}{1} = \frac{4 \cdot 6}{1 \cdot 6}$$

$$= \frac{24}{6} \qquad \text{Do each multiplication: } 4 \cdot 6 = 24 \text{ and } 1 \cdot 6 = 6.$$

Self Check

Write 5 as a fraction with a denominator of 3.

Answer: $\dfrac{15}{3}$ ■

STUDY SET   Section 3.1

VOCABULARY   *Fill in the blanks.*

1. For the fraction $\frac{7}{8}$, 7 is the _____ and 8 is the _____.

2. When we express 15 as $5 \cdot 3$, we say that we have _____ 15.

3. A _____ fraction is less than 1. An _____ fraction is greater than or equal to 1.

4. A fraction is said to be in _____ terms if the only factor common to the numerator and denominator is 1.

5. Two fractions are _____ if they have the same value.

6. A _____ can be used to indicate the number of equal parts of a whole.

7. Multiplying the numerator and denominator of a fraction by a number to obtain an equivalent fraction that involves larger numbers is called expressing the fraction in _____ terms or _____ up the fraction.

8. We can _____ a fraction that is not in lowest terms by applying the fundamental property of fractions. We _____ out common factors of the numerator and denominator.

CONCEPTS

9. What common factor (other than 1) do the numerator and the denominator have?

a. $\dfrac{2}{16}$ **b.** $\dfrac{6}{9}$ **c.** $\dfrac{10}{15}$ **d.** $\dfrac{14}{35}$

10. Given:

$$\frac{15}{35} = \frac{\overset{1}{\cancel{5}} \cdot 3}{7 \cdot \underset{1}{\cancel{5}}}$$

In this work, what do the slashes and small 1's mean?

11. What concept studied in this section is shown by Illustration 1?

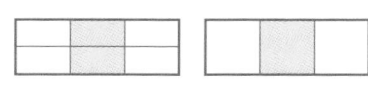

ILLUSTRATION 1

12. Why can't we say that $\frac{2}{5}$ of the figure in Illustration 2 is shaded?

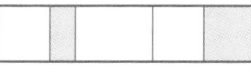

ILLUSTRATION 2

13. a. Explain the difference in the two approaches used to simplify $\frac{20}{28}$.

$$\frac{\overset{1}{\cancel{4}} \cdot 5}{\underset{1}{\cancel{4}} \cdot 7} \quad \text{and} \quad \frac{\overset{1}{\cancel{2}} \cdot \overset{1}{\cancel{2}} \cdot 5}{\underset{1}{\cancel{2}} \cdot \underset{1}{\cancel{2}} \cdot 7}$$

b. Are the results the same?

14. What concept studied in this section does this statement illustrate?

$$\frac{5}{10} = \frac{4}{8} = \frac{3}{6} = \frac{2}{4} = \frac{1}{2}$$

15. Why isn't this a valid application of the fundamental property of fractions?

$$\frac{10}{11} = \frac{2 + 8}{2 + 9} = \frac{\overset{1}{\cancel{2}} + 8}{\underset{1}{\cancel{2}} + 9} = \frac{9}{10}$$

16. Write the fraction $\dfrac{7}{-8}$ in two other ways.

17. Write as a fraction.
  a. 8 **b.** -25

18. Fill in the missing numbers in the following statement:

$$\frac{5 \cdot \rule{0.5cm}{0.4pt}}{9 \cdot \rule{0.5cm}{0.4pt}} = \frac{15}{27}$$

NOTATION *Complete each solution.*

19. Simplify $\dfrac{18}{24}$.

$$\frac{18}{24} = \frac{3 \cdot \rule{0.4cm}{0.4pt} \cdot 2}{3 \cdot 2 \cdot \rule{0.4cm}{0.4pt} \cdot 2}$$

$$= \frac{\overset{1}{\cancel{\rule{0.3cm}{0.4pt}}} \cdot 3 \cdot \overset{1}{\cancel{\rule{0.3cm}{0.4pt}}}}{\underset{1}{\cancel{\rule{0.3cm}{0.4pt}}} \cdot 2 \cdot 2 \cdot \underset{1}{\cancel{\rule{0.3cm}{0.4pt}}}}$$

$$= \frac{3}{4}$$

20. Simplify $\dfrac{60}{90}$.

$$\frac{60}{90} = \frac{\rule{0.4cm}{0.4pt} \cdot 2}{\rule{0.4cm}{0.4pt} \cdot 3}$$

$$= \frac{\overset{1}{\cancel{30}} \cdot \rule{0.4cm}{0.4pt}}{\underset{1}{\cancel{30}} \cdot \rule{0.4cm}{0.4pt}}$$

$$= \frac{2}{3}$$

PRACTICE *Simplify each fraction to lowest terms, if possible.*

21. $\dfrac{3}{9}$ **22.** $\dfrac{5}{20}$

23. $\dfrac{7}{21}$ **24.** $\dfrac{6}{30}$

25. $\dfrac{20}{30}$ **26.** $\dfrac{12}{30}$

27. $\dfrac{15}{6}$ **28.** $\dfrac{24}{16}$

29. $-\dfrac{28}{56}$ **30.** $-\dfrac{45}{54}$

31. $-\dfrac{90}{105}$

32. $-\dfrac{26}{78}$

33. $\dfrac{60}{108}$

34. $\dfrac{75}{125}$

35. $\dfrac{180}{210}$

36. $\dfrac{76}{28}$

37. $\dfrac{55}{67}$

38. $\dfrac{41}{51}$

39. $\dfrac{36}{96}$

40. $\dfrac{48}{120}$

41. $\dfrac{25}{35}$

42. $\dfrac{16}{20}$

43. $\dfrac{12}{15}$

44. $\dfrac{10}{15}$

45. $\dfrac{6}{7}$

46. $\dfrac{4}{5}$

47. $\dfrac{7}{8}$

48. $\dfrac{10}{21}$

49. $-\dfrac{10}{30}$

50. $-\dfrac{14}{28}$

51. $\dfrac{15}{25}$

52. $\dfrac{16}{24}$

53. $\dfrac{35}{28}$

54. $\dfrac{35}{25}$

55. $\dfrac{56}{28}$

56. $\dfrac{32}{8}$

Write each fraction as an equivalent fraction with the indicated denominator.

57. $\dfrac{7}{8}$, denominator 40

58. $\dfrac{3}{4}$, denominator 24

59. $\dfrac{4}{5}$, denominator 35

60. $\dfrac{5}{7}$, denominator 49

61. $\dfrac{5}{6}$, denominator 54

62. $\dfrac{11}{16}$, denominator 32

63. $\dfrac{1}{2}$, denominator 30

64. $\dfrac{1}{3}$, denominator 60

65. $\dfrac{2}{7}$, denominator 14

66. $\dfrac{3}{10}$, denominator 50

67. $\dfrac{9}{10}$, denominator 60

68. $\dfrac{2}{3}$, denominator 27

69. $\dfrac{5}{4}$, denominator 20

70. $\dfrac{9}{4}$, denominator 44

71. $\dfrac{2}{15}$, denominator 45

72. $\dfrac{5}{12}$, denominator 36

Write each number as a fraction with the indicated denominator.

73. 3 as fifths

74. 4 as thirds

75. 6 as eighths

76. 3 as sixths

77. 4 as ninths

78. 7 as fourths

79. −2 as halves

80. −10 as ninths

APPLICATIONS *Use the concept of fraction in answering each question.*

81. COMMUTING How much of the commute from home to work has the motorist in Illustration 3 made?

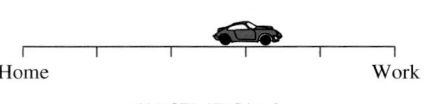

Home   Work

ILLUSTRATION 3

82. TIME CLOCK How much of the hour has passed?

a.    **b.**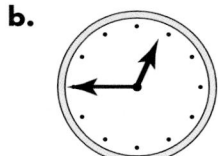

c. 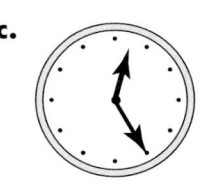   **d.**

83. SINKHOLE Illustration 4 shows a side view of a depression in the sidewalk near a sinkhole. Describe the movement of the sidewalk using a signed number. (On the tape measure, one inch is divided into 16 equal parts.)

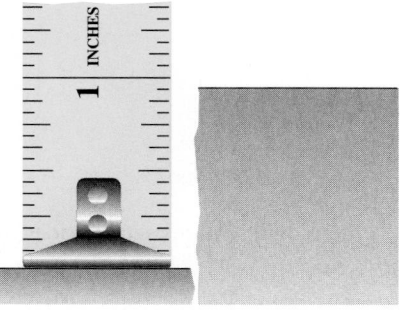

ILLUSTRATION 4

84. POLITICAL PARTIES   Illustration 5 shows the political party affiliation of the governors of the 50 states, as of January 1, 2000.
  a. What fraction are Democrats?
  b. What fraction are Republicans?
  c. What fraction are neither Democrat nor Republican?

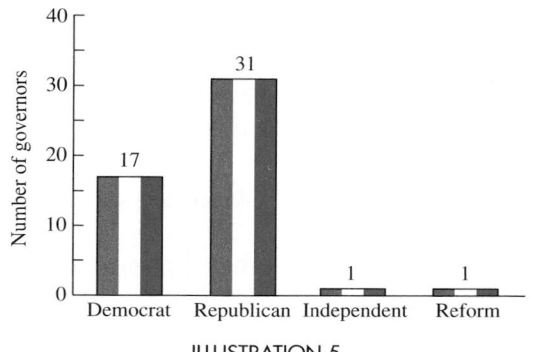

ILLUSTRATION 5

85. PERSONNEL RECORDS   Complete the chart in Illustration 6 by finding the amount of the job that will be completed by each person working alone for the given number of hours.

Name	Total time to complete the job alone	Time worked alone	Amount of job completed
Bob	10 hours	7 hours	
Ali	8 hours	1 hour	

ILLUSTRATION 6

86. GAS TANK   See Illustration 7. How full does the gauge indicate the gas tank is? How much of the tank has been used?

Use unleaded fuel

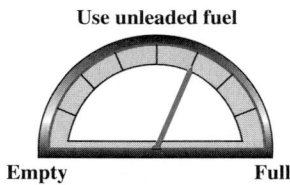

Empty   Full

ILLUSTRATION 7

87. MUSIC   Illustration 8 shows a side view of the finger position needed to produce a length of string (from the bridge to the fingertip) that gives low C on a violin. To play other notes, fractions of that length are used. Locate these finger positions.

a. $\frac{1}{2}$ of the length gives middle C.
b. $\frac{3}{4}$ of the length gives F above low C.
c. $\frac{2}{3}$ of the length gives G.

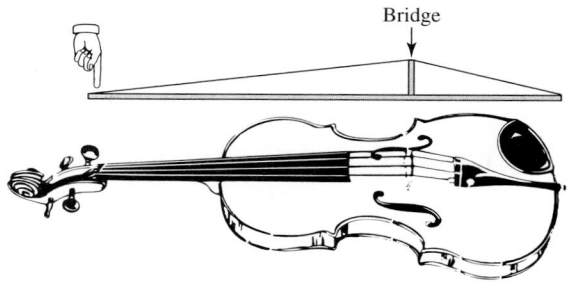

Bridge

ILLUSTRATION 8

88. RULER   Illustration 9 shows a ruler. First, tell how many spaces there are between the numbers 0 and 1. Then tell to what number the arrow is pointing.

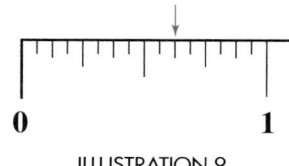

0   **1**

ILLUSTRATION 9

89. MACHINERY   The operator of a machine is to turn the dial shown below from setting A to setting B. Express this in two different ways, using fractions of one complete revolution.

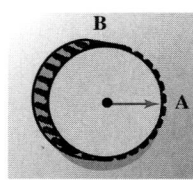

90. EARTH'S ROTATION   The Earth rotates about its vertical axis once every 24 hours.
  a. What is the significance of $\frac{1}{24}$ of a rotation to us on Earth?
  b. What significance does $\frac{24}{24}$ of a revolution have?

91. SUPERMARKET DISPLAY   The amount of space to be given each type of snack food in a supermarket display case is expressed as a fraction. Complete the model of the display, showing where the adjustable shelves should be located, and label where each snack food should be stocked.

$\frac{3}{8}$: potato chips $\frac{2}{8}$: peanuts

$\frac{1}{8}$: pretzels $\frac{2}{8}$: tortilla chips

SNACKS

92. MEDICAL CENTER   Hospital designers have located a nurse's station at the center of a circular building. Show how to divide the surrounding office space so that each medical department has the proper fractional amount allocated to it. (Use the circle graph below.) Label each department.

$\dfrac{2}{12}$: Radiology $\dfrac{5}{12}$: Pediatrics

$\dfrac{1}{12}$: Laboratory $\dfrac{3}{12}$: Orthopedics

$\dfrac{1}{12}$: Pharmacy

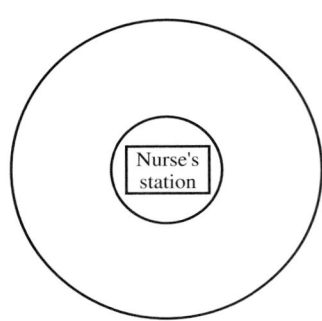

93. CAMERA   When the shutter of a camera stays open longer than $\frac{1}{125}$ second, any movement of the camera will probably blur the picture. With this in mind, if a photographer is taking a picture of a fast-moving object, should she select a shutter speed of $\frac{1}{60}$ or $\frac{1}{250}$?

94. GDP   The Gross Domestic Product is the official measure of the size of the U.S. economy. It represents the market value of all goods and services that have been bought during a given period of time. The GDP for the second quarter of 1999 is listed below. What is meant by the phrase *second quarter of 1999*?

Second quarter of 1999 $8,893,300,000,000

WRITING

95. Explain the concept of equivalent fractions.

96. What does it mean for a fraction to be in lowest terms?

97. Explain the difference between three-fourths and three-fifths of a pizza.

98. Explain both parts of the fundamental property of fractions.

REVIEW

99. Solve $-5x + 1 = 16$.

100. Solve $\dfrac{y}{2} + 2 = 4$.

101. Round 564,112 to the nearest thousand.

102. Give the definition of a prime number.

3.2 *Multiplying Fractions*

In this section, you will learn about

- Multiplying fractions   • Simplifying when multiplying fractions
- Powers of a fraction   • Applications

INTRODUCTION. In the next three sections, we will discuss how to add, subtract, multiply, and divide fractions. We begin with the operation of multiplication.

Multiplying fractions

Suppose that a television network is going to take out a full-page ad to publicize its fall lineup of shows. The prime-time shows are to get $\frac{3}{5}$ of the ad space and daytime programming the remainder. Of the space devoted to prime time, $\frac{1}{2}$ is to be used to promote weekend programs. How much of the newspaper page will be used to advertise weekend prime-time programs?

The ad for the weekend prime-time shows will occupy $\frac{1}{2}$ of $\frac{3}{5}$ of the page. This can be expressed as $\frac{1}{2} \cdot \frac{3}{5}$. We can calculate $\frac{1}{2} \cdot \frac{3}{5}$ using a three-step process, illustrated below.

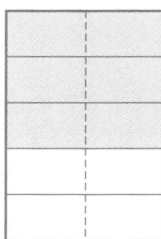

Step 1: We divide the page into fifths and shade three of them. This represents the fraction $\frac{3}{5}$, the amount of the page used to advertise prime-time shows.

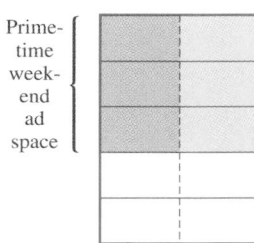

Step 2: Next, we find $\frac{1}{2}$ of the shaded part of the page by dividing the page into halves, using a vertical line.

Prime-
time
week-
end
ad
space

Step 3: Finally, we highlight (in purple) $\frac{1}{2}$ of the shaded parts determined in Step 2. The highlighted parts are 3 out of 10 or $\frac{3}{10}$ of the page. They represent the amount of the page used to advertise the weekend prime-time shows. This leads us to the conclusion that $\frac{1}{2} \cdot \frac{3}{5} = \frac{3}{10}$.

Two observations can be made from this result.

- The numerator of the answer is the product of the numerators of the original fractions.

$$
\overbrace{\frac{1}{2} \cdot \frac{3}{5}}^{1 \cdot 3 = 3} = \underbrace{\frac{3}{10}}_{2 \cdot 5 = 10} \quad \text{Answer}
$$

- The denominator of the answer is the product of the denominators of the original fractions.

These observations suggest the following rule for multiplying two fractions.

Multiplying fractions

> To multiply two fractions, multiply the numerators and multiply the denominators. In symbols, if a, b, c, and d represent numbers,
>
> $$\frac{a}{b} \cdot \frac{c}{d} = \frac{a \cdot c}{b \cdot d} \qquad (b \neq 0, d \neq 0)$$

EXAMPLE 1 *Multiplying fractions.* Multiply: $\dfrac{7}{8} \cdot \dfrac{3}{5}$.

Solution

$\dfrac{7}{8} \cdot \dfrac{3}{5} = \dfrac{7 \cdot 3}{8 \cdot 5}$ Multiply the numerators and multiply the denominators.

$= \dfrac{21}{40}$ Do the multiplications: $7 \cdot 3 = 21$ and $8 \cdot 5 = 40$.

Self Check

Multiply: $\dfrac{5}{9} \cdot \dfrac{2}{3}$.

Answer: $\dfrac{10}{27}$ ∎

The rules for multiplying integers also hold for multiplying fractions. When we multiply two fractions with *like* signs, the product is positive. When we multiply two fractions with *unlike* signs, the product is negative.

EXAMPLE 2 *Multiplication with a negative fraction.* Multiply: $-\dfrac{3}{4}\left(\dfrac{1}{8}\right)$.

Solution

$-\dfrac{3}{4}\left(\dfrac{1}{8}\right) = -\dfrac{3 \cdot 1}{4 \cdot 8}$ Multiply the numerators and multiply the denominators. Since the fractions have unlike signs, the product is negative.

$= -\dfrac{3}{32}$ Do the multiplications: $3 \cdot 1 = 3$ and $4 \cdot 8 = 32$.

Self Check

Multiply: $\dfrac{5}{6}\left(-\dfrac{1}{3}\right)$.

Answer: $-\dfrac{5}{18}$ ∎

Simplifying when multiplying fractions

After multiplying two fractions, we should simplify the result, if possible.

EXAMPLE 3 *Simplifying when multiplying fractions.* Find $\dfrac{5}{8} \cdot \dfrac{4}{5}$.

Solution

$\dfrac{5}{8} \cdot \dfrac{4}{5} = \dfrac{5 \cdot 4}{8 \cdot 5}$ Multiply the numerators and multiply the denominators.

$= \dfrac{5 \cdot 4}{4 \cdot 2 \cdot 5}$ In the denominator, factor 8 as $4 \cdot 2$ so that we can simplify the fraction.

$= \dfrac{\overset{1}{\cancel{5}} \cdot \overset{1}{\cancel{4}}}{\underset{1}{\cancel{4}} \cdot 2 \cdot \underset{1}{\cancel{5}}}$ Divide out the common factors of 4 and 5.

$= \dfrac{1}{2}$ Do the multiplications: $1 \cdot 1 = 1$ and $1 \cdot 2 \cdot 1 = 2$.

Self Check

Find $\dfrac{6}{25} \cdot \dfrac{5}{6}$.

Answer: $\dfrac{1}{5}$ ∎

EXAMPLE 4 *Multiplying three fractions.* Find $45\left(-\dfrac{1}{14}\right)\left(-\dfrac{7}{10}\right)$.

Solution This expression involves three factors, and one of them is an integer. When multiplying fractions and integers, we express each integer as a fraction with a denominator of 1.

$45\left(-\dfrac{1}{14}\right)\left(-\dfrac{7}{10}\right) = \dfrac{45}{1}\left(\dfrac{1}{14}\right)\left(\dfrac{7}{10}\right)$ Write 45 as a fraction: $45 = \dfrac{45}{1}$. The product of two negative numbers is positive.

$= \dfrac{45 \cdot 1 \cdot 7}{1 \cdot 14 \cdot 10}$ Multiply the numerators and multiply the denominators.

$$= \frac{5 \cdot 3 \cdot 3 \cdot 1 \cdot 7}{1 \cdot 7 \cdot 2 \cdot 5 \cdot 2}$$ Factor 45 as $5 \cdot 3 \cdot 3$, factor 14 as $7 \cdot 2$, and factor 10 as $5 \cdot 2$.

$$= \frac{\overset{1}{\cancel{5}} \cdot 3 \cdot 3 \cdot 1 \cdot \overset{1}{\cancel{7}}}{1 \cdot \underset{1}{\cancel{7}} \cdot 2 \cdot \underset{1}{\cancel{5}} \cdot 2}$$ Divide out the common factors of 5 and 7.

$$= \frac{9}{4}$$ Multiply in the numerator and in the denominator.   ■

❗ COMMENT   The answer in Example 4 was $\frac{9}{4}$, an improper fraction. In arithmetic, we often write such fractions as mixed numbers. In algebra, it is often more useful to leave $\frac{9}{4}$ in this form. We will discuss improper fractions and mixed numbers in more detail in Section 3.5.

EXAMPLE 5   *Multiplication involving a variable.*   Multiply: $\frac{1}{4}(4y)$.

Solution

$$\frac{1}{4}(4y) = \frac{1}{4} \cdot \frac{4y}{1}$$ Write $4y$ as a fraction: $4y = \frac{4y}{1}$.

$$= \frac{1 \cdot 4 \cdot y}{4 \cdot 1}$$ Multiply the numerators and multiply the denominators.

$$= \frac{1 \cdot \overset{1}{\cancel{4}} \cdot y}{\underset{1}{\cancel{4}} \cdot 1}$$ Divide out the common factor of 4.

$$= \frac{y}{1}$$ Multiply in the numerator and in the denominator.

$$= y$$ Simplify: $\frac{y}{1} = y$.

Self Check

Multiply: $\frac{1}{5} \cdot 5m$.

Answer: m ■

To multiply $\frac{1}{2}$ and x, we can express the product as $\frac{1}{2}x$, or we can use the concept of multiplying fractions to write it in a different form.

$$\frac{1}{2} \cdot x = \frac{1}{2} \cdot \frac{x}{1}$$ Write x as a fraction: $x = \frac{x}{1}$.

$$= \frac{1 \cdot x}{2 \cdot 1}$$ Multiply the numerators and multiply the denominators.

$$= \frac{x}{2}$$ Multiply in the numerator and in the denominator.

The product of $\frac{1}{2}$ and x can be expressed as $\frac{1}{2}x$ or $\frac{x}{2}$. Similarly, $\frac{3}{4}t = \frac{3t}{4}$ and $-\frac{5}{16}y = -\frac{5y}{16}$.

Powers of a fraction

If the base of an exponential expression is a fraction, the exponent tells us how many times to write that fraction as a factor. For example,

$$\left(\frac{2}{3}\right)^2 = \frac{2}{3} \cdot \frac{2}{3} = \frac{2 \cdot 2}{3 \cdot 3} = \frac{4}{9}$$ $\frac{2}{3}$ is used as a factor 2 times. $\frac{4}{9}$ is called the square of $\frac{2}{3}$.

EXAMPLE 6 *Power of a fraction.*   Find $\left(-\dfrac{4}{5}\right)^2$.

Solution

Exponents are used to indicate repeated multiplication.

$$\left(-\frac{4}{5}\right)^2 = \left(-\frac{4}{5}\right)\left(-\frac{4}{5}\right)$$ Write $-\frac{4}{5}$ as a factor 2 times.

$$= \frac{4 \cdot 4}{5 \cdot 5}$$ The product of two fractions with like signs is positive. Multiply the numerators and multiply the denominators.

$$= \frac{16}{25}$$ Multiply in the numerator and denominator.

Self Check

Find $\left(-\dfrac{3}{4}\right)^3$.

Answer: $-\dfrac{27}{64}$ ∎

Applications

EXAMPLE 7   *House of Representatives.*   In the United States House of Representatives, a bill was introduced that would require a $\frac{3}{5}$ vote of the 435 members to authorize any tax increase. Under this requirement, how many representatives would have to vote for a tax increase before it could become law?

Solution   To solve this problem, we must find $\frac{3}{5}$ *of* 435.

$$\frac{3}{5} \text{ of } 435 = \frac{3}{5} \cdot \frac{435}{1}$$ Here, the word *of* means to multiply. Write 435 as a fraction: $435 = \frac{435}{1}$.

$$= \frac{3 \cdot 435}{5 \cdot 1}$$ Multiply the numerators and multiply the denominators.

$$= \frac{3 \cdot 3 \cdot 5 \cdot 29}{5 \cdot 1}$$ Prime factor 435 as $3 \cdot 5 \cdot 29$.

$$= \frac{3 \cdot 3 \cdot \overset{1}{\cancel{5}} \cdot 29}{\underset{1}{\cancel{5}} \cdot 1}$$ Divide out the common factor of 5.

$$= \frac{261}{1}$$ Multiply in the numerator: $3 \cdot 3 \cdot 1 \cdot 29 = 261$. Multiply in the denominator: $1 \cdot 1 = 1$.

$$= 261$$ Simplify: $\frac{261}{1} = 261$.

It would take 261 representatives voting in favor to pass a tax increase.   ∎

As Figure 3-3 shows, a triangle has three sides. The length of the base of the triangle can be represented by the letter b and the height by the letter h. The height of a triangle is always perpendicular (makes a square corner) to the base. This is denoted by the symbol ⌐.

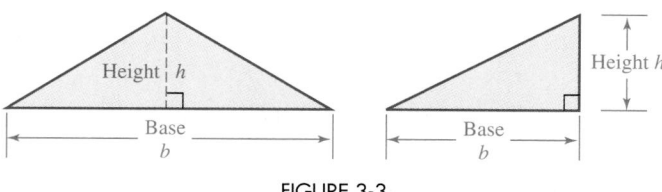

Height h Base b Height h Base b

FIGURE 3-3

Recall that the area of a figure is the amount of surface that it encloses. The area of a triangle can be found by using the following formula.

Area of a triangle

The area A of a triangle is one-half the product of its base b and its height h:

$$\text{Area} = \frac{1}{2}\,(\text{base})(\text{height}) \quad \text{or} \quad A = \frac{1}{2}\,bh \quad \text{or} \quad A = \frac{bh}{2}$$

EXAMPLE 8 *Geography.* Approximate the area of the state of Virginia using the triangle in Figure 3-4.

Solution We will approximate the area of the state by finding the area of the triangle.

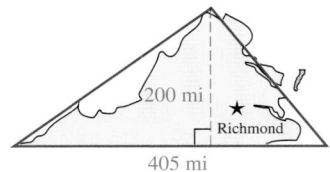

200 mi

Richmond

405 mi

FIGURE 3-4

$$A = \frac{1}{2}\,bh \qquad \text{The formula for the area of a triangle.}$$

$$= \frac{1}{2}\,(405)(200) \qquad \text{Substitute 405 for } b \text{ and 200 for } h.$$

$$= \frac{1}{2}\left(\frac{405}{1}\right)\left(\frac{200}{1}\right) \qquad \text{Write 405 and 200 as fractions.}$$

$$= \frac{1 \cdot 405 \cdot 200}{2} \qquad \text{Multiply the numerators. Multiply the denominators.}$$

$$= \frac{1 \cdot 405 \cdot 100 \cdot \overset{1}{\cancel{2}}}{\underset{1}{\cancel{2}}} \qquad \text{Factor 200 as } 100 \cdot 2. \text{ Then divide out the common factor of 2.}$$

$$= 40{,}500 \qquad 405 \cdot 100 = 40{,}500.$$

The area of the state of Virginia is approximately 40,500 square miles. ■

STUDY SET Section 3.2

VOCABULARY *Fill in the blanks.*

1. The word *of* in mathematics usually means _____.

2. The _____ of a triangle is the amount of surface that it encloses.

3. The result of a multiplication problem is called the _____.

4. To _____ a fraction means to divide out common factors of the numerator and denominator.

5. In a triangle, b stands for the length of the _____ and h stands for the _____.

6. A _____ is an equation that mathematically describes a known relationship between two or more variables.

CONCEPTS

7. Find the result when multiplying $\frac{a}{b} \cdot \frac{c}{d}$.

8. Write each of the following as fractions.
   a. 4   **b.** -3

9. Use the following rectangle to find $\frac{1}{3} \cdot \frac{1}{4}$.

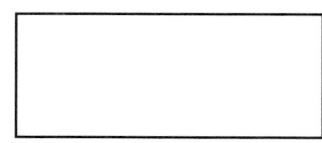

a. Using vertical lines, divide the given rectangle into four equal parts and lightly shade one of them. What fractional part of the rectangle did you shade?

b. To find $\frac{1}{3}$ of the shaded portion, use two horizontal lines to divide the given rectangle into three equal parts and lightly shade one of them. Into how many equal parts is the rectangle now divided? How many parts have been shaded twice? What is $\frac{1}{3} \cdot \frac{1}{4}$?

10. In the following solution, what mistake did the student make that caused him to have to work with such large numbers?

$$\frac{44}{63} \cdot \frac{27}{55} = \frac{44 \cdot 27}{63 \cdot 55}$$
$$= \frac{1,188}{3,465}$$

11. a. Is the product of two numbers with unlike signs positive or negative?

b. Is the product of two numbers with like signs positive or negative?

12. a. Multiply $\frac{9}{10}$ and 20.

b. When we multiply two numbers, is the product always larger than both those numbers?

13. Tell whether each statement is true or false.

a. $\frac{1}{2}x = \frac{x}{2}$ **b.** $\frac{2t}{3} = \frac{2}{3}t$

c. $-\frac{3}{8}a = -\frac{3}{8a}$ **d.** $\frac{-4e}{7} = -\frac{4e}{7}$

14. What is the numerator of the result for the multiplication problem shown here?

$$\frac{4}{15} \cdot \frac{3}{4} = \frac{\overset{1}{\cancel{4}} \cdot \overset{1}{\cancel{3}}}{5 \cdot \underset{1}{\cancel{3}} \cdot \underset{1}{\cancel{4}}}$$

NOTATION   *Complete each solution.*

15. Multiply: $\frac{5}{8} \cdot \frac{7}{15}$.

$$\frac{5}{8} \cdot \frac{7}{15} = \frac{5 \cdot \underline{\quad}}{8 \cdot \underline{\quad}}$$
$$= \frac{5 \cdot 7}{8 \cdot 5 \cdot \underline{\quad}}$$

$$= \frac{\overset{1}{\cancel{5}} \cdot 7}{8 \cdot \underset{1}{\cancel{5}} \cdot \underline{\quad}}$$
$$= \frac{7}{24}$$

16. Multiply: $\frac{7}{12} \cdot \frac{4}{21}$.

$$\frac{7}{12} \cdot \frac{4}{21} = \frac{7 \cdot 4}{\underline{\quad} \cdot \underline{\quad}}$$
$$= \frac{7 \cdot 4}{4 \cdot \underline{\quad} \cdot \underline{\quad} \cdot 3}$$
$$= \frac{\overset{1}{\cancel{7}} \cdot \overset{1}{\cancel{4}}}{\cancel{4} \cdot 3 \cdot \cancel{7} \cdot 3}$$
$$= \frac{1}{9}$$

PRACTICE   *Multiply. Write all answers in lowest terms.*

17. $\frac{1}{4} \cdot \frac{1}{2}$ **18.** $\frac{1}{3} \cdot \frac{1}{5}$

19. $\frac{3}{8} \cdot \frac{7}{16}$ **20.** $\frac{5}{9} \cdot \frac{2}{7}$

21. $\frac{2}{3} \cdot \frac{6}{7}$ **22.** $\frac{5}{12} \cdot \frac{3}{4}$

23. $\frac{14}{15} \cdot \frac{11}{8}$ **24.** $\frac{5}{16} \cdot \frac{8}{3}$

25. $-\frac{15}{24} \cdot \frac{8}{25}$ **26.** $-\frac{20}{21} \cdot \frac{7}{16}$

27. $\left(-\frac{11}{21}\right)\left(-\frac{14}{33}\right)$ **28.** $\left(-\frac{16}{35}\right)\left(-\frac{25}{48}\right)$

29. $\frac{7}{10}\left(\frac{20}{21}\right)$ **30.** $\left(\frac{7}{6}\right)\frac{9}{49}$

31. $\frac{3}{4} \cdot \frac{4}{3}$ **32.** $\frac{4}{5} \cdot \frac{5}{4}$

33. $\frac{1}{3} \cdot \frac{15}{16} \cdot \frac{4}{25}$ **34.** $\frac{3}{15} \cdot \frac{15}{7} \cdot \frac{14}{27}$

35. $\left(\frac{2}{3}\right)\left(-\frac{1}{16}\right)\left(-\frac{4}{5}\right)$ **36.** $\left(\frac{3}{8}\right)\left(-\frac{2}{3}\right)\left(-\frac{12}{27}\right)$

37. $\frac{5}{6} \cdot 18$ **38.** $6\left(-\frac{2}{3}\right)$

39. $15\left(-\frac{4}{5}\right)$ **40.** $-2\left(-\frac{7}{8}\right)$

41. $\frac{5x}{12} \cdot \frac{1}{6}$ **42.** $\frac{2t}{3} \cdot \frac{7}{8}$

43. $\frac{b}{12} \cdot \frac{3}{10}$ **44.** $\frac{5c}{8} \cdot \frac{1}{15}$

45. $\frac{1}{3} \cdot 3d$

46. $\frac{1}{16} \cdot 16x$

47. $\frac{2}{3} \cdot \frac{3s}{2}$

48. $\frac{3}{5} \cdot \frac{5h}{3}$

Multiply and express the product in two ways.

49. $\frac{5}{6} \cdot x$

50. $\frac{2}{3} \cdot y$

51. $-\frac{8}{9} \cdot v$

52. $-\frac{7}{6} \cdot m$

Find each power.

53. $\left(\frac{2}{3}\right)^2$

54. $\left(\frac{3}{5}\right)^2$

55. $\left(-\frac{5}{9}\right)^2$

56. $\left(-\frac{5}{6}\right)^2$

57. $\left(\frac{4}{3}\right)^2$

58. $\left(\frac{3}{2}\right)^2$

59. $\left(-\frac{3}{4}\right)^3$

60. $\left(-\frac{2}{5}\right)^3$

61. Complete the multiplication table of fractions in Illustration 1.

$\cdot$	$\frac{1}{2}$	$\frac{1}{3}$	$\frac{1}{4}$	$\frac{1}{5}$	$\frac{1}{6}$
$\frac{1}{2}$					
$\frac{1}{3}$					
$\frac{1}{4}$					
$\frac{1}{5}$					
$\frac{1}{6}$					

ILLUSTRATION 1

62. Complete Illustration 2 by finding the original fraction, given its square.

Original fraction squared	Original fraction
$\frac{1}{9}$	
$\frac{1}{100}$	
$\frac{4}{25}$	
$\frac{16}{49}$	
$\frac{81}{36}$	
$\frac{9}{121}$	

ILLUSTRATION 2

Find the area of each triangle.

63.

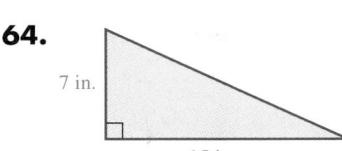

64.

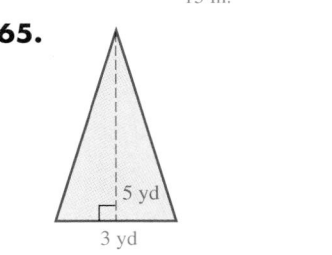

65.

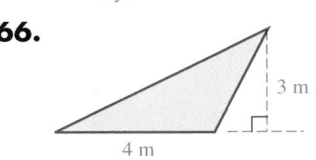

66.

APPLICATIONS

67. THE CONSTITUTION   Article V of the United States Constitution requires a two-thirds vote of the House of Representatives to propose a constitutional amendment. The House has 435 members. Find the number of votes needed to meet this requirement.

68. GENETICS   Gregor Mendel (1822–1884), an Augustinian monk, is credited with developing a heredity model that became the foundation of modern genetics. In his experiments, he crossed purple-flowered plants with white-flowered plants and found that $\frac{3}{4}$ of the offspring plants had purple flowers and $\frac{1}{4}$ had white flowers. According to this concept, when the group of offspring plants shown in Illustration 3 flower, how many will have purple flowers?

ILLUSTRATION 3

69. TENNIS BALL   A tennis ball is dropped from a height of 54 inches. Each time it hits the ground, it rebounds one-third of the previous height it fell. See Illustration 4 and find the three missing rebound heights.

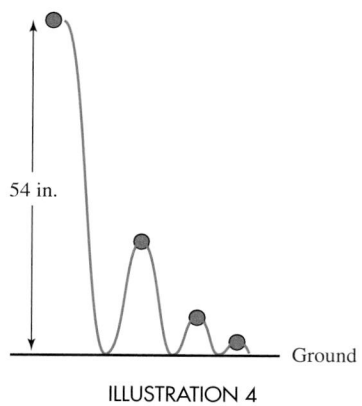

ILLUSTRATION 4

70. ELECTION   Illustration 5 shows the final election returns for a city bond measure.
   a. How many votes were cast?
   b. Find two-thirds of the number of votes cast.
   c. Did the bond measure pass?

Measure 1
100% of the precincts reporting

Fire–Police–Paramedics General Obligation Bonds (Requires two-thirds vote)

   Yes   125,599   No   62,801

ILLUSTRATION 5

71. COOKING   Use the recipe below, along with the concept of multiplication with fractions, to find how much sugar and molasses are needed to make one dozen cookies.

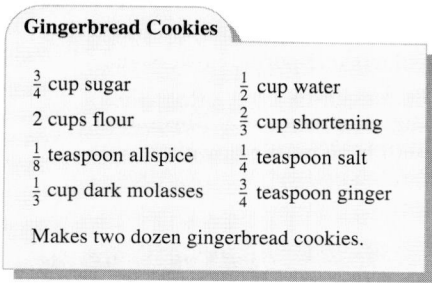

Gingerbread Cookies

$\frac{3}{4}$ cup sugar $\frac{1}{2}$ cup water

2 cups flour $\frac{2}{3}$ cup shortening

$\frac{1}{8}$ teaspoon allspice $\frac{1}{4}$ teaspoon salt

$\frac{1}{3}$ cup dark molasses $\frac{3}{4}$ teaspoon ginger

Makes two dozen gingerbread cookies.

72. THE EARTH'S SURFACE   The surface of Earth covers an area of approximately 196,800,000 square miles. About $\frac{3}{4}$ of that area is covered by water. Find the number of square miles of the surface covered by water.

73. BOTANY   In an experiment, monthly growth rates of three types of plants doubled when nitrogen was added

to the soil. Complete Illustration 6 by charting the improved growth rate next to each normal growth rate.

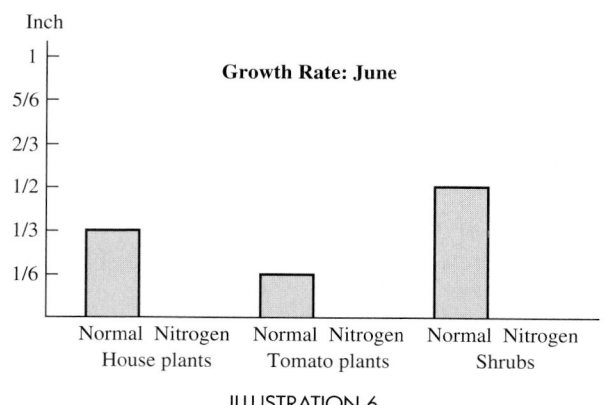

ILLUSTRATION 6

74. STAMPS   The best designs in a contest to create a wildlife stamp are shown in Illustration 7. To save on paper costs, the postal service has decided to choose the stamp that has the smaller area. Which one is that?

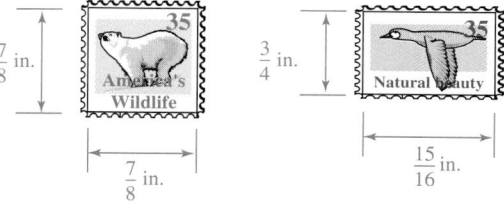

ILLUSTRATION 7

75. THE STARS AND STRIPES   Illustration 8 shows a folded U.S. flag. When it is placed on a table as part of an exhibit, how much area will it occupy?

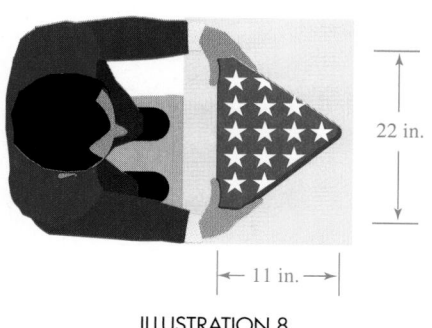

ILLUSTRATION 8

76. WINDSURFING   Estimate the area of the sail on the windsurfing board in Illustration 9.

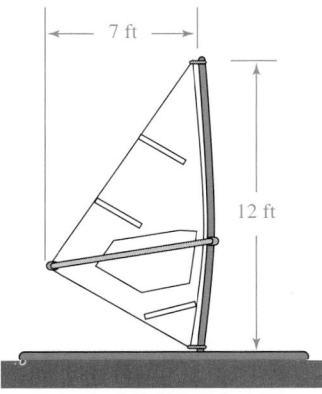

ILLUSTRATION 9

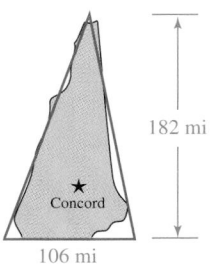

ILLUSTRATION 11

77. TILE DESIGN   A design for bathroom tile is shown in Illustration 10. Find the amount of area on a tile that is blue.

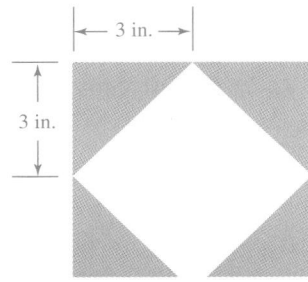

ILLUSTRATION 10

78. GEOGRAPHY   Estimate the area of the state of New Hampshire, using the triangle in Illustration 11.

WRITING

79. In mathematics, the word *of* usually means multiply. Give three real-life examples of this usage.

80. Explain how you could multiply the number 5 and another number and obtain an answer that is less than 5.

81. A MAJORITY   The definition of the word *majority* is as follows: "a number greater than one-half of the total." Explain what it means when a teacher says, "A majority of the class voted to postpone the test until Monday." Give an example.

82. What does area measure?

REVIEW

83. Round 987,459 to the nearest thousand.

84. Solve $3x + 1 = 7$.

85. Is $x = -6$ a solution of $2x + 6 = 6$?

86. Simplify $4 + 5 \cdot 3$.

87. Find the prime factorization of 125.

88. Evaluate $-2(-3)(-4)$.

3.3 *Dividing Fractions*

In this section, you will learn about

- Division with fractions • Reciprocals • A rule for dividing fractions

INTRODUCTION. In this section, we will discuss how to divide fractions. We will examine problems involving positive and negative fractions. The skills you learned in Section 3.2 will be useful in this section.

Division with fractions

Suppose that the manager of a candy store buys large bars of chocolate and divides each one into four equal parts to sell. How many fourths can be obtained from 5 bars?

We are asking, "How many $\frac{1}{4}$'s are there in 5?" To answer the question, we need to use the operation of division. We can represent this division as $5 \div \frac{1}{4}$. See Figure 3-5.

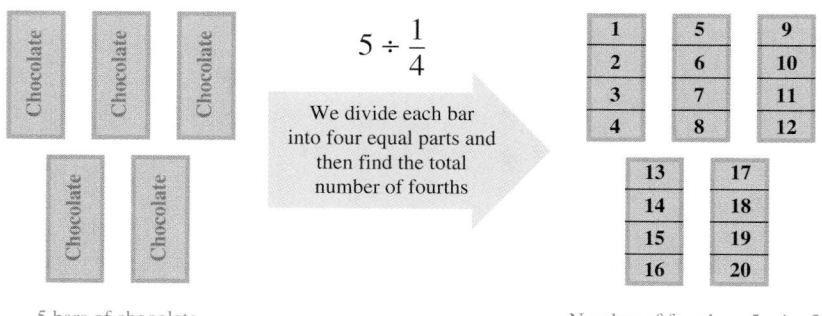

FIGURE 3-5

There are 20 fourths in the 5 bars of chocolate. This result leads to the following observations.

- This division problem involves a fraction: $5 \div \frac{1}{4}$.
- Although we were asked to find $5 \div \frac{1}{4}$, we solved the problem using *multiplication* instead of *division*: $5 \cdot 4 = 20$.

Later in this section, we will see that these observations suggest a rule for dividing fractions. But before we can discuss that rule, we need to introduce a new term.

Reciprocals

Division with fractions involves working with **reciprocals.** To present the concept of reciprocal, we consider the problem $\frac{7}{8} \cdot \frac{8}{7}$.

$$\frac{7}{8} \cdot \frac{8}{7} = \frac{7 \cdot 8}{8 \cdot 7} \qquad \text{Multiply the numerators and multiply the denominators.}$$

$$= \frac{\overset{1}{\cancel{7}} \cdot \overset{1}{\cancel{8}}}{\underset{1}{\cancel{8}} \cdot \underset{1}{\cancel{7}}} \qquad \text{Divide out the common factors of 7 and 8.}$$

$$= \frac{1}{1} \qquad \text{Multiply in the numerator and multiply in the denominator.}$$

$$= 1 \qquad \tfrac{1}{1} = 1.$$

The product of $\frac{7}{8}$ and $\frac{8}{7}$ is 1.

Whenever the product of two numbers is 1, we say that those numbers are *reciprocals.* Therefore, $\frac{7}{8}$ and $\frac{8}{7}$ are reciprocals. To find the reciprocal of a fraction, *we invert the numerator and the denominator.*

Reciprocals | Two numbers are called **reciprocals** if their product is 1.

 COMMENT Zero does not have a reciprocal, because the product of 0 and a number can never be 1.

EXAMPLE 1 *Reciprocals.* For each number, find its reciprocal and show that their product is 1: **a.** $\dfrac{2}{3}$, **b.** $-\dfrac{3}{4}$, and **c.** 5.

Solution

a. The reciprocal of $\dfrac{2}{3}$ is $\dfrac{3}{2}$.

$$\frac{2}{3} \cdot \frac{3}{2} = \frac{\overset{1}{\cancel{2}} \cdot \overset{1}{\cancel{3}}}{\underset{1}{\cancel{3}} \cdot \underset{1}{\cancel{2}}} = 1$$

b. The reciprocal of $-\dfrac{3}{4}$ is $-\dfrac{4}{3}$.

$$-\frac{3}{4}\left(-\frac{4}{3}\right) = \frac{\overset{1}{\cancel{3}} \cdot \overset{1}{\cancel{4}}}{\underset{1}{\cancel{4}} \cdot \underset{1}{\cancel{3}}} = 1 \qquad \text{The product of two fractions with like signs is positive.}$$

c. $5 = \dfrac{5}{1}$, so the reciprocal of 5 is $\dfrac{1}{5}$.

$$5 \cdot \frac{1}{5} = \frac{5}{1} \cdot \frac{1}{5} = \frac{\overset{1}{\cancel{5}} \cdot 1}{1 \cdot \underset{1}{\cancel{5}}} = 1$$

Self Check

For each number, find its reciprocal and show that their product is 1.

a. $\dfrac{3}{5}$

b. $-\dfrac{5}{6}$

c. 8

Answers: **a.** $\dfrac{5}{3}$, **b.** $-\dfrac{6}{5}$, **c.** $\dfrac{1}{8}$

A rule for dividing fractions

In the candy store example, we saw that we can find $5 \div \frac{1}{4}$ by computing $5 \cdot 4$. That is, division by $\frac{1}{4}$ (a fraction) is the same as multiplication by 4 (its reciprocal).

$$5 \div \frac{1}{4} = 5 \cdot 4$$

This observation suggests a general rule for dividing fractions.

Dividing fractions

> To divide fractions, multiply the first fraction by the reciprocal of the second fraction. In symbols, if a, b, c, and d represent numbers, then
>
> $$\frac{a}{b} \div \frac{c}{d} = \frac{a}{b} \cdot \frac{d}{c} \qquad (b \neq 0, c \neq 0, d \neq 0)$$

For example, to find $\frac{5}{7} \div \frac{3}{4}$, we multiply $\frac{5}{7}$ by the reciprocal of $\frac{3}{4}$.

Change the division to multiplication.

$$\frac{5}{7} \div \frac{3}{4} \qquad = \qquad \frac{5}{7} \cdot \frac{4}{3}$$

The reciprocal of $\frac{3}{4}$ is $\frac{4}{3}$.

$$= \frac{5 \cdot 4}{7 \cdot 3} \qquad \text{Multiply the numerators.}$$
$$\text{Multiply the denominators.}$$

$$= \frac{20}{21} \qquad \text{Multiply in the numerator and in the}$$
$$\text{denominator: } 5 \cdot 4 = 20 \text{ and } 7 \cdot 3 = 21.$$

Therefore, $\dfrac{5}{7} \div \dfrac{3}{4} = \dfrac{20}{21}$.

EXAMPLE 2 *Dividing fractions.* Divide: $\dfrac{1}{3} \div \dfrac{4}{5}$.

Solution

$$\frac{1}{3} \div \frac{4}{5} = \frac{1}{3} \cdot \frac{5}{4} \qquad \text{Multiply } \tfrac{1}{3} \text{ by the reciprocal of } \tfrac{4}{5}, \text{ which is } \tfrac{5}{4}.$$

$$= \frac{1 \cdot 5}{3 \cdot 4} \qquad \text{Multiply the numerators and multiply the denominators.}$$

$$= \frac{5}{12} \qquad \text{Multiply in the numerator and in the denominator.}$$

Self Check

Divide: $\dfrac{2}{3} \div \dfrac{7}{8}$.

Answer: $\dfrac{16}{21}$

EXAMPLE 3 *Dividing fractions.* Divide: $\dfrac{9}{16} \div \dfrac{3}{20}$.

Solution

$$\frac{9}{16} \div \frac{3}{20} = \frac{9}{16} \cdot \frac{20}{3} \qquad \text{Multiply } \tfrac{9}{16} \text{ by the reciprocal of } \tfrac{3}{20}, \text{ which is } \tfrac{20}{3}.$$

$$= \frac{9 \cdot 20}{16 \cdot 3} \qquad \text{Multiply the numerators and multiply the denominators.}$$

$$= \frac{\overset{1}{3} \cdot 3 \cdot 5 \cdot \overset{1}{4}}{\underset{1}{4} \cdot 4 \cdot \underset{1}{3}} \qquad \begin{array}{l}\text{Factor 9 as } 3 \cdot 3, \text{ factor 20 as } 5 \cdot 4, \text{ and factor 16 as}\\ 4 \cdot 4. \text{ Then divide out the common factors of 3 and 4.}\end{array}$$

$$= \frac{15}{4} \qquad \text{Multiply in the numerator and denominator.}$$

Self Check

Divide: $\dfrac{4}{5} \div \dfrac{8}{25}$.

Answer: $\dfrac{5}{2}$

EXAMPLE 4 *Surfboard design.* Most surfboards are made of polyurethane foam plastic covered with several layers of fiberglass to keep them water-tight. How many layers are needed to build up a finish three-eighths of an inch thick if each layer of fiberglass has a thickness of one-sixteenth of an inch?

Solution We need to know how many one-sixteenths there are in three-eighths. To answer this question, we will use division and find $\frac{3}{8} \div \frac{1}{16}$.

$$\frac{3}{8} \div \frac{1}{16} = \frac{3}{8} \cdot \frac{16}{1} \qquad \text{Multiply } \tfrac{3}{8} \text{ by the reciprocal of } \tfrac{1}{16}, \text{ which is } \tfrac{16}{1}.$$

$$= \frac{3 \cdot 16}{8 \cdot 1} \qquad \text{Multiply the numerators and multiply the denominators.}$$

$$= \frac{3 \cdot \overset{1}{8} \cdot 2}{\underset{1}{8} \cdot 1} \qquad \text{Factor 16 as } 8 \cdot 2. \text{ Then divide out the common factor of 8.}$$

$$= \frac{6}{1} \qquad \text{Multiply in the numerator and denominator.}$$

$$= 6 \qquad \text{Simplify: } \tfrac{6}{1} = 6.$$

The number of layers of fiberglass to be applied is 6.

EXAMPLE 5 *Division with a negative fraction.* Divide: $\dfrac{1}{6} \div \left(-\dfrac{1}{18}\right)$.

Solution

When working with divisions involving negative fractions, we use the same rules as for multiplying numbers with like or unlike signs.

Self Check

Divide: $\dfrac{2}{3} \div \left(-\dfrac{7}{6}\right)$.

$$\frac{1}{6} \div \left(-\frac{1}{18}\right) = \frac{1}{6}\left(-\frac{18}{1}\right)$$ Multiply $\frac{1}{6}$ by the reciprocal of $-\frac{1}{18}$, which is $-\frac{18}{1}$.

$$= -\frac{1 \cdot 18}{6 \cdot 1}$$ The product of two fractions with unlike signs is negative. Multiply the numerators and multiply the denominators.

$$= -\frac{1 \cdot \overset{1}{\cancel{6}} \cdot 3}{\underset{1}{\cancel{6}} \cdot 1}$$ Factor 18 as $6 \cdot 3$. Then divide out the common factor of 6.

$$= -\frac{3}{1}$$ Multiply in the numerator and denominator.

$$= -3$$ Simplify: $\frac{3}{1} = 3$.

Answer: $-\dfrac{4}{7}$ ∎

EXAMPLE 6 *Division with a fraction and an integer.* Divide: $-\dfrac{21}{36} \div (-3)$.

Self Check

Divide: $-\dfrac{24}{25} \div (-8)$.

Solution

$$-\frac{21}{36} \div (-3) = -\frac{21}{36}\left(-\frac{1}{3}\right)$$ Multiply $-\frac{21}{36}$ by the reciprocal of -3, which is $-\frac{1}{3}$.

$$= \frac{21 \cdot 1}{36 \cdot 3}$$ The product of two fractions with like signs is positive. Multiply the numerators and multiply the denominators.

$$= \frac{7 \cdot 3 \cdot 1}{36 \cdot 3}$$ Factor 21 as $7 \cdot 3$.

$$= \frac{7 \cdot \overset{1}{\cancel{3}} \cdot 1}{36 \cdot \underset{1}{\cancel{3}}}$$ Divide out the common factor of 3.

$$= \frac{7}{36}$$ Multiply in the numerator and in the denominator.

Answer: $\dfrac{3}{25}$ ∎

STUDY SET Section 3.3

VOCABULARY *Fill in the blanks.*

1. Two numbers are called _____ if their product is 1.

2. The result of a division problem is called the _____.

CONCEPTS

3. Complete this statement:

$$\frac{1}{2} \div \frac{2}{3} = \boxed{} \cdot \boxed{}$$

4. Find the reciprocal of each number.

a. $\dfrac{2}{5}$ **b.** -3

5. Using horizontal lines, divide each rectangle in Illustration 1 into thirds. What division problem does this illustrate? What is the quotient of that problem?

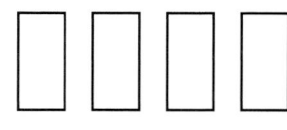

ILLUSTRATION 1

6. Using horizontal lines, divide each rectangle in Illustration 2 into fifths. What division problem does this illustrate? What is the quotient of that problem?

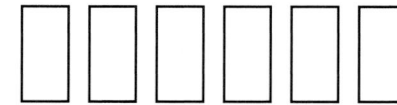

ILLUSTRATION 2

7. Multiply $\frac{4}{5}$ and its reciprocal. What is the result?

8. Multiply $\frac{7}{6}$ and its reciprocal. What is the result?

9. a. Find $15 \div 3$.
   b. Rewrite $15 \div 3$ as multiplication by the reciprocal and find the result.
   c. Complete this statement: Division by 3 is the same as multiplication by   .

10. a. Find $10 \div \frac{1}{5}$.
   b. Complete this statement: Division by $\frac{1}{5}$ is the same as multiplication by   .

NOTATION *Complete each solution.*

11. Divide: $\dfrac{25}{36} \div \dfrac{10}{9}$.

$$\frac{25}{36} \div \frac{10}{9} = \frac{25}{36} \cdot \underline{}$$

$$= \frac{25 \cdot }{36 \cdot }$$

$$= \frac{5 \cdot \cdot 9}{4 \cdot 9 \cdot 2 \cdot 5}$$

$$= \frac{\overset{1}{\cancel{}} \cdot 5 \cdot \overset{1}{\cancel{}}}{4 \cdot \underset{1}{\cancel{}} \cdot 2 \cdot \underset{1}{\cancel{}}}$$

$$= \frac{5}{8}$$

12. Divide: $\dfrac{4}{9} \div \dfrac{8}{27}$.

$$\frac{4}{9} \div \frac{8}{27} = \frac{4}{9} \cdot $$

$$= \frac{4 \cdot }{9 \cdot }$$

$$= \frac{4 \cdot 3 \cdot 9}{9 \cdot 4 \cdot 2}$$

$$= \frac{\overset{1}{\cancel{}} \cdot 3 \cdot \overset{1}{\cancel{}}}{\underset{1}{\cancel{}} \cdot \underset{1}{\cancel{}} \cdot 2}$$

$$= \frac{3}{2}$$

PRACTICE *Find each quotient.*

13. $\dfrac{1}{2} \div \dfrac{3}{5}$ **14.** $\dfrac{5}{7} \div \dfrac{5}{6}$

15. $\dfrac{3}{16} \div \dfrac{1}{9}$ **16.** $\dfrac{5}{8} \div \dfrac{2}{9}$

17. $\dfrac{4}{5} \div \dfrac{4}{5}$ **18.** $\dfrac{2}{3} \div \dfrac{2}{3}$

19. $\left(-\dfrac{7}{4}\right) \div \left(-\dfrac{21}{8}\right)$ **20.** $\left(-\dfrac{15}{16}\right) \div \left(-\dfrac{5}{8}\right)$

21. $3 \div \dfrac{1}{12}$ **22.** $9 \div \dfrac{3}{4}$

23. $120 \div \dfrac{12}{5}$ **24.** $360 \div \dfrac{36}{5}$

25. $-\dfrac{4}{5} \div (-6)$ **26.** $-\dfrac{7}{8} \div (-14)$

27. $\dfrac{15}{16} \div 180$ **28.** $\dfrac{7}{8} \div 210$

29. $-\dfrac{9}{10} \div \dfrac{4}{15}$ **30.** $-\dfrac{3}{4} \div \dfrac{3}{2}$

31. $\dfrac{9}{10} \div \left(-\dfrac{3}{25}\right)$ **32.** $\dfrac{11}{16} \div \left(-\dfrac{9}{16}\right)$

33. $-\dfrac{1}{8} \div 8$ **34.** $-\dfrac{1}{15} \div 15$

35. $\dfrac{15}{32} \div \dfrac{15}{32}$ **36.** $-\dfrac{1}{64} \div \left(-\dfrac{1}{64}\right)$

37. $\dfrac{4}{5} \div \dfrac{3}{2}$ **38.** $\dfrac{2}{3} \div \dfrac{3}{2}$

39. $\dfrac{1}{8} \div \dfrac{3}{4}$ **40.** $\dfrac{1}{9} \div \dfrac{3}{5}$

41. $\dfrac{13}{16} \div \dfrac{1}{2}$ **42.** $\dfrac{7}{8} \div \dfrac{6}{7}$

43. $-\dfrac{15}{32} \div \dfrac{3}{4}$ **44.** $-\dfrac{7}{10} \div \dfrac{4}{5}$

APPLICATIONS

45. MARATHON   Each lap around a stadium track is $\frac{1}{4}$ mile. How many laps would a runner have to complete to get a 26-mile workout?

46. COOKING   A recipe calls for $\frac{3}{4}$ cup of flour, and the only measuring container you have holds $\frac{1}{8}$ cup. How many $\frac{1}{8}$ cups of flour would you need to add to follow the recipe?

47. LASER TECHNOLOGY   Using a laser, a technician slices thin pieces of aluminum off the end of a rod that is $\frac{7}{8}$ inch long. How many $\frac{1}{64}$-inch-wide slices can be cut from this rod?

48. FURNITURE   A production process applies several layers of a clear acrylic coat to outdoor furniture to help protect it from the weather. If each protective coat is $\frac{3}{32}$ inch thick, how many applications will be needed to build up $\frac{3}{8}$ inch of clear finish?

49. UNDERGROUND CABLE   In Illustration 3, which construction proposal will require the fewest days to install underground TV cable from the broadcasting station to the subdivision?

Proposal	Amount of cable installed per day	Comments
Route 1	$\frac{3}{5}$ of a mile	Longer than Route 2
Route 2	$\frac{2}{5}$ of a mile	Terrain very rocky

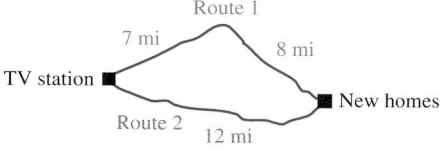

ILLUSTRATION 3

50. PRODUCTION PLANNING   The materials used to make a pillow are shown in Illustration 4. Examine the inventory list to decide how many pillows can be manufactured in one production run with the materials in stock.

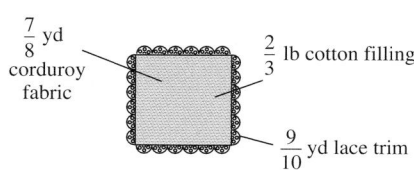

Factory Inventory List

Materials	Amount in stock
Lace trim	135 yd
Corduroy fabric	154 yd
Cotton filling	98 lb

ILLUSTRATION 4

51. 3 × 5 CARDS   Ninety 3 × 5 cards are shown stacked next to a ruler in Illustration 5.
   a. Into how many parts is one inch divided on the ruler?
   b. How thick is the stack of cards?
   c. How thick is one 3 × 5 card?

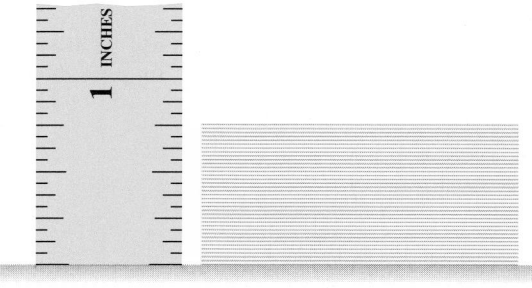

ILLUSTRATION 5

52. COMPUTER PRINTER   Illustration 6 shows how the letter E is formed by a dot matrix printer. What is the height of a dot?

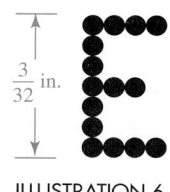

ILLUSTRATION 6

53. FORESTRY   A set of forestry maps divides the 6,284 acres of an old-growth forest into $\frac{4}{5}$-acre sections. How many sections do the maps contain?

54. HARDWARE   A hardware chain purchases large amounts of nails and packages them in $\frac{9}{16}$-pound bags for sale. How many of these bags of nails can be obtained from 2,871 pounds of nails?

WRITING

55. Explain how to divide two fractions.

56. Explain why 0 does not have a reciprocal.

57. Write an application problem that could be solved by finding $10 \div \frac{1}{5}$.

58. Explain why dividing a fraction by 2 is the same as finding $\frac{1}{2}$ of it.

REVIEW

59. Solve $4x - 2 = -18$.

60. Solve $\frac{x}{3} - 1 = 4$.

61. Divide: $\frac{5}{6} \div \frac{7}{12}$.

62. Add: $\frac{5}{6} + \frac{7}{12}$.

63. True or false: If equal amounts are subtracted from the numerator and the denominator of a fraction, the result will be an equivalent fraction.

64. Graph each of these numbers on a number line: $-2, 0, |-4|$, and the opposite of 1.

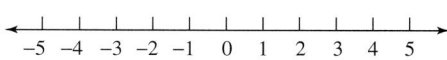

65. Round 637,512 to the nearest hundred.

66. Define the word *variable*.

3.4 *Adding and Subtracting Fractions*

In this section, you will learn about

- Fractions with the same denominator • Fractions with different denominators
- Finding the LCD • Comparing fractions

INTRODUCTION. In arithmetic and algebra, *we can only add or subtract objects that are similar.* For example, we can add dollars to dollars, but we cannot add dollars to oranges. This concept is important when adding or subtracting fractions.

Fractions with the same denominator

Consider the problem $\frac{3}{5} + \frac{1}{5}$. When we write it in words, it is apparent that we are adding similar objects.

three-**fifths**   +   one-**fifth**

└── Similar objects ──┘

Because the denominators of $\frac{3}{5}$ and $\frac{1}{5}$ are the same, we say that they have a **common denominator.** Since the fractions have a common denominator, we can add them. Figure 3-6 illustrates the addition process.

$$\frac{3}{5} \qquad + \qquad \frac{1}{5} \qquad = \qquad \frac{4}{5}$$

FIGURE 3-6

We can make some observations about the addition shown in the figure.

The *sum* of the numerators is the numerator of the answer.

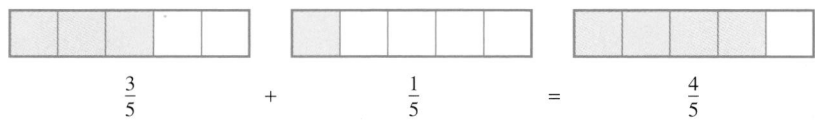

$$\frac{3}{5} \quad + \quad \frac{1}{5} \quad = \quad \frac{4}{5}$$

The answer is a fraction that has the *same* denominator as the two fractions that were added.

These observations suggest the following rule.

Adding or subtracting fractions with the same denominators

To add (or subtract) fractions with the same denominators, add (or subtract) their numerators and write that result over the common denominator. Simplify the result, if possible.

In symbols, if a, b, and c represent numbers, then

$$\frac{a}{c} + \frac{b}{c} = \frac{a + b}{c} \quad (c \neq 0) \qquad \text{and} \qquad \frac{a}{c} - \frac{b}{c} = \frac{a - b}{c} \quad (c \neq 0)$$

EXAMPLE 1   *Fractions with the same denominator.*   Add: $\frac{1}{8} + \frac{5}{8}$.

Solution

$$\frac{1}{8} + \frac{5}{8} = \frac{1 + 5}{8} \qquad \text{Add the numerators. Write the sum over the common denominator 8.}$$

Self Check

Add: $\frac{9}{12} + \frac{1}{12}$.

$$= \frac{6}{8}$$ Do the addition: $1 + 5 = 6$. The fraction can be simplified.

$$= \frac{3 \cdot \overset{1}{\cancel{2}}}{4 \cdot \underset{1}{\cancel{2}}}$$ Factor 6 as $3 \cdot 2$ and 8 as $4 \cdot 2$. Divide out the common factor of 2.

$$= \frac{3}{4}$$ Multiply in the numerator and in the denominator.

Answer: $\dfrac{5}{6}$ ∎

EXAMPLE 2   *Subtracting fractions.*   Subtract: $-\dfrac{7}{3} - \left(-\dfrac{2}{3}\right)$.

Self Check
Subtract: $-\dfrac{9}{11} - \left(-\dfrac{3}{11}\right)$.

Solution

$$-\frac{7}{3} - \left(-\frac{2}{3}\right) = -\frac{7}{3} + \frac{2}{3}$$ Add the opposite of $-\frac{2}{3}$, which is $\frac{2}{3}$.

$$= \frac{-7}{3} + \frac{2}{3}$$ Write $-\frac{7}{3}$ as $\frac{-7}{3}$.

$$= \frac{-7 + 2}{3}$$ Add the numerators. Write the sum over the common denominator 3.

$$= \frac{-5}{3}$$ Do the addition: $-7 + 2 = -5$.

$$= -\frac{5}{3}$$ Rewrite the fraction: $\frac{-5}{3} = -\frac{5}{3}$.

Answer: $-\dfrac{6}{11}$ ∎

Fractions with different denominators

Now we consider the problem $\frac{3}{5} + \frac{1}{3}$. Since the denominators are different, we cannot add these fractions in their present form.

three-**fifths**   +   one-**third**
└─ Not similar objects ─┘

To add these fractions, we need to find a common denominator. The smallest common denominator (called the **least** or **lowest common denominator**) usually is the easiest common denominator to work with.

Least common denominator

> The **least common denominator (LCD)** for a set of fractions is the smallest number each denominator will divide exactly.

In the problem $\frac{3}{5} + \frac{1}{3}$, the denominators are 5 and 3. The numbers 5 and 3 divide many numbers; 30, 45, and 60, to name a few. But the smallest number that 5 and 3 divide exactly is 15. This is the LCD. We will now build each fraction into a fraction with a denominator of 15 by applying the fundamental property of fractions.

$$\frac{3}{5} + \frac{1}{3} = \frac{3 \cdot 3}{5 \cdot 3} + \frac{1 \cdot 5}{3 \cdot 5}$$

We need to multiply this denominator by 3 to obtain 15.   We need to multiply this denominator by 5 to obtain 15.

$$= \frac{9}{15} + \frac{5}{15}$$ Do the multiplications in the numerators and denominators.

$$= \frac{9 + 5}{15}$$ Add the numerators and write the sum over the common denominator 15.

$$= \frac{14}{15}$$ Do the addition: $9 + 5 = 14$.

Figure 3-7 shows $\frac{3}{5}$ and $\frac{1}{3}$ expressed as equivalent fractions with a denominator of 15. Once the denominators are the same, the fractions can be added easily.

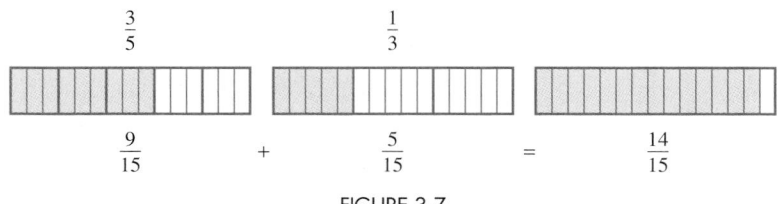

$$\frac{3}{5} \qquad\qquad \frac{1}{3}$$

$$\frac{9}{15} \qquad + \qquad \frac{5}{15} \qquad = \qquad \frac{14}{15}$$

FIGURE 3-7

To add or subtract fractions with different denominators, we follow these steps.

Adding or subtracting fractions with different denominators

1. Find the LCD.
2. Express each fraction as an equivalent fraction with a denominator that is the LCD.
3. Add or subtract the resulting fractions. Simplify the result if possible.

EXAMPLE 3 *Fractions with different denominators.* Add: $\dfrac{1}{8} + \dfrac{2}{3}$.

Solution

Since the smallest number the denominators 8 and 3 divide exactly is 24, the LCD is 24.

$$\frac{1}{8} + \frac{2}{3} = \frac{1 \cdot 3}{8 \cdot 3} + \frac{2 \cdot 8}{3 \cdot 8} \qquad \text{Apply the fundamental property of fractions to express each fraction in terms of 24ths.}$$

$$= \frac{3}{24} + \frac{16}{24} \qquad \text{Do the multiplications in the numerators and denominators.}$$

$$= \frac{3 + 16}{24} \qquad \text{Add the numerators and write the sum over the common denominator 24.}$$

$$= \frac{19}{24} \qquad \text{Do the addition in the numerator: } 3 + 16 = 19.$$

Self Check

Add: $\dfrac{1}{2} + \dfrac{2}{5}$.

Answer: $\dfrac{9}{10}$

EXAMPLE 4 *Adding an integer and a fraction.* Add: $-5 + \dfrac{1}{4}$.

Solution

$$-5 + \frac{1}{4} = \frac{-5}{1} + \frac{1}{4} \qquad \text{Write } -5 \text{ as } \frac{-5}{1}. \text{ The smallest number that 1 and 4 divide exactly is 4, so the LCD is 4.}$$

$$= \frac{-5 \cdot 4}{1 \cdot 4} + \frac{1}{4} \qquad \text{Apply the fundamental property of fractions to express } \frac{-5}{1} \text{ in terms of 4ths.}$$

$$= \frac{-20}{4} + \frac{1}{4} \qquad \text{Multiply in the numerator and in the denominator.}$$

$$= \frac{-20 + 1}{4} \qquad \text{Write the sum of the numerators over the common denominator 4.}$$

$$= \frac{-19}{4} \qquad \text{Do the addition: } -20 + 1 = -19.$$

$$= -\frac{19}{4} \qquad \text{Rewrite: } \frac{-19}{4} = -\frac{19}{4}.$$

Self Check

Add: $-6 + \dfrac{2}{5}$.

Answer: $-\dfrac{28}{5}$

Finding the LCD

When adding or subtracting fractions with different denominators, the LCD is not always obvious. We will now develop two methods for finding the LCD of a set of fractions. As an example, let's find the LCD of $\frac{3}{8}$ and $\frac{1}{10}$.

Method 1: A **multiple** of a number is the product of that number and a natural number. The multiples of 8 and the multiples of 10 are shown below.

Multiples of 8	Multiples of 10	
$8 \cdot 1 = 8$	$10 \cdot 1 = 10$	
$8 \cdot 2 = 16$	$10 \cdot 2 = 20$	
$8 \cdot 3 = 24$	$10 \cdot 3 = 30$	
$8 \cdot 4 = 32$	$10 \cdot 4 = \mathbf{40}$	The multiples that the lists have in common
$8 \cdot 5 = \mathbf{40}$	$10 \cdot 5 = 50$	are highlighted in red.
$8 \cdot 6 = 48$	$10 \cdot 6 = 60$	
$8 \cdot 7 = 56$	$10 \cdot 7 = 70$	
$8 \cdot 8 = 64$	$10 \cdot 8 = \mathbf{80}$	
$8 \cdot 9 = 72$	$10 \cdot 9 = 90$	
$8 \cdot 10 = \mathbf{80}$	$10 \cdot 10 = 100$	

The smallest multiple common to both lists is 40. It is the smallest number that 8 and 10 divide exactly. Therefore, 40 is the LCD of $\frac{3}{8}$ and $\frac{1}{10}$. These observations suggest a method for finding the LCD of a set of fractions.

Finding the LCD by finding multiples

> **1.** List the multiples of each denominator.
>
> **2.** The smallest multiple common to the lists found in Step 1 is the LCD of the fractions.

Method 2: If the LCD for $\frac{3}{8}$ and $\frac{1}{10}$ is a number that 8 and 10 divide exactly, the prime factorization of the LCD must include the prime factorization of 8 (which is $2 \cdot 2 \cdot 2$) and the prime factorization of 10 (which is $5 \cdot 2$). The smallest number that meets both of these requirements is $2 \cdot 2 \cdot 2 \cdot 5$. Therefore, the LCD is $2 \cdot 2 \cdot 2 \cdot 5 = 40$.

$$\left. \begin{array}{l} 8 = 2 \cdot 2 \cdot 2 \\ 10 = 5 \cdot 2 \end{array} \right\} \text{LCD} = \overbrace{2 \cdot 2 \cdot 2}^{\text{The prime factorization of 8.}} \cdot \underbrace{2 \cdot 5}_{\text{The prime factorization of 10.}} = 40$$

In the prime factorization of 8, the factor 2 appears three times. It appears three times in the product $(2 \cdot 2 \cdot 2 \cdot 5)$ that gives the LCD. In the prime factorization of 10, the factor 5 appears once. It appears once in the product that gives the LCD. These observations suggest another method for finding the LCD of a set of fractions.

Finding the LCD using prime factorization

> **1.** Prime factor each denominator.
>
> **2.** The LCD is a product of prime factors, where each factor is used the greatest number of times it appears in any one factorization found in Step 1.

EXAMPLE 5   ***Finding the LCD using prime factorization.***   Subtract: $\frac{19}{21} - \frac{5}{18}$.

Self Check

Subtract: $\frac{33}{35} - \frac{11}{14}$.

Solution

We use prime factorization to find the LCD.

Step 1: Prime factor each denominator.

$$21 = 7 \cdot 3$$
$$18 = 3 \cdot 3 \cdot 2$$

Step 2: The factors 7, 3, and 2 appear in the prime factorizations.

The greatest number of times 7 appears in any one factorization is once.	The greatest number of times 3 appears in any one factorization is twice.	The greatest number of times 2 appears in any one factorization is once.
↓	↓ ↓	↓

$$\text{LCD} = \quad 7 \quad \cdot \quad 3 \quad \cdot \quad 3 \quad \cdot \quad 2 \quad = \quad 126$$

$$\frac{19}{21} - \frac{5}{18} = \frac{19 \cdot 6}{21 \cdot 6} - \frac{5 \cdot 7}{18 \cdot 7}$$ Express each fraction in terms of 126ths.

$$= \frac{114}{126} - \frac{35}{126}$$ Do the multiplications in the numerators and in the denominators.

$$= \frac{114 - 35}{126}$$ Write the difference of the numerators over the common denominator, 126.

$$= \frac{79}{126}$$ Do the subtraction: $114 - 35 = 79$.

Answer: $\dfrac{11}{70}$ ∎

EXAMPLE 6   *Television viewing habits.* Students on a college campus were asked to estimate to the nearest hour how much television they watched each day. The results are given in the pie chart in Figure 3-8. For example, the chart tells us that $\frac{1}{4}$ of those responding watched 1 hour per day. Find the fraction of the student body watching from 0 to 2 hours daily.

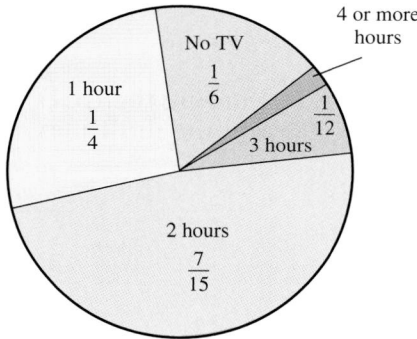

FIGURE 3-8

Solution   To answer this question, we need to add $\frac{1}{6}$, $\frac{1}{4}$, and $\frac{7}{15}$. To find the LCD, we prime factor each of the denominators.

$$\left. \begin{array}{l} 6 = 3 \cdot 2 \\ 4 = 2 \cdot 2 \\ 15 = 5 \cdot 3 \end{array} \right\} \text{LCD} = 5 \cdot 3 \cdot 2 \cdot 2 = 60$$

In any one factorization, the greatest number of times 5 appears is once, the greatest number of times 3 appears is once, and the greatest number of times 2 appears is twice.

$$\frac{1}{6} + \frac{1}{4} + \frac{7}{15} = \frac{1 \cdot 10}{6 \cdot 10} + \frac{1 \cdot 15}{4 \cdot 15} + \frac{7 \cdot 4}{15 \cdot 4}$$ Express each fraction in terms of 60ths.

$$= \frac{10}{60} + \frac{15}{60} + \frac{28}{60}$$ Do the multiplication in the numerators and denominators.

$$= \frac{10 + 15 + 28}{60}$$ Add the numerators. Write the sum over the common denominator 60.

$$= \frac{53}{60}$$ Do the addition: $10 + 15 + 28 = 53$.

The fraction of the student body watching 0 to 2 hours of television daily is $\frac{53}{60}$. ∎

Comparing fractions

If fractions have the same denominator, the fraction with the larger numerator is the larger fraction. If their denominators are different, we need to write the fractions with a common denominator before we can make a comparison.

Comparing fractions

> To compare unlike fractions, write the fractions as equivalent fractions with the same denominator—preferably the LCD. Then compare their numerators. The fraction with the larger numerator is the larger fraction.

EXAMPLE 7   *Comparing fractions.*   Which fraction is larger: $\dfrac{5}{6}$ or $\dfrac{7}{8}$?

Solution

To compare these fractions, we express each with the LCD of 24.

$$\frac{5}{6} = \frac{5 \cdot 4}{6 \cdot 4} \qquad \frac{7}{8} = \frac{7 \cdot 3}{8 \cdot 3} \qquad \text{Express each fraction in terms of 24ths.}$$

$$= \frac{20}{24} \qquad\qquad = \frac{21}{24} \qquad \text{Do the multiplications in the numerators and denominators.}$$

Next, we compare the numerators. Since $21 > 20$, we conclude that $\frac{21}{24}$ is greater than $\frac{20}{24}$. Thus, $\frac{7}{8} > \frac{5}{6}$.

Self Check

Which fraction is larger: $\dfrac{7}{12}$ or $\dfrac{3}{5}$?

Answer: $\dfrac{3}{5}$ ■

STUDY SET Section 3.4

VOCABULARY   *Fill in the blanks.*

1. The _____ common denominator for a set of fractions is the smallest number each denominator will divide exactly.

2. _____ fractions, such as $\frac{1}{2}$ and $\frac{2}{4}$, are fractions that represent the same amount.

3. To express a fraction in _____ terms, we multiply the numerator and denominator by the same number.

4. _____ up a fraction is the process of multiplying the numerator and the denominator of the fraction by the same number.

CONCEPTS

5. The rule for adding fractions is

$$\frac{a}{c} + \frac{b}{c} = \frac{a+b}{c} \qquad (c \neq 0)$$

Fill in the blanks.
   This rule tells us how to add fractions having like _____. To find the sum, we add the _____ and then write that result over the _____ denominator.

6. a. Add the indicated fractions.

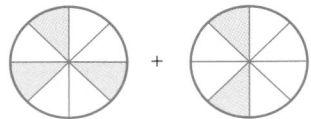

b. Subtract the indicated fractions.

7. Why must we do some preliminary work before doing the following addition?

$$\frac{2}{9} + \frac{2}{5}$$

8. Why must we do some preliminary work before doing the following subtraction?

$$\frac{5}{6} - \frac{5}{18}$$

9. By what are the numerator and the denominator of the following fractions being multiplied?

$$\frac{5 \cdot 4}{6 \cdot 4}$$

10. Consider $\frac{3}{4}$. By what should we multiply the numerator and denominator of this fraction to express it as an equivalent fraction with the given denominator?
   a. 12   **b.** 36

11. Consider the following prime factorizations:

$$24 = 3 \cdot 2 \cdot 2 \cdot 2$$
$$90 = 5 \cdot 3 \cdot 3 \cdot 2$$

For any one factorization, what is the greatest number of times
   a. a 5 appears?
   b. a 3 appears?
   c. a 2 appears?

12. a. List the first ten multiples of 9 and the first ten multiples of 12.

b. What is the smallest common multiple of 9 and 12?

13. The denominators of two fractions involved in a subtraction problem have the prime-factored forms $2 \cdot 2 \cdot 5$ and $2 \cdot 3 \cdot 5$. What is the LCD for the fractions?

14. The denominators of three fractions involved in a subtraction problem have the prime-factored forms $2 \cdot 2 \cdot 5$, and $2 \cdot 3 \cdot 5$, and $2 \cdot 3 \cdot 3 \cdot 5$. What is the LCD for the fractions?

15. a. Divide the figure on the left into fourths and shade one part. Divide the figure on the right into thirds and shade one part. Which shaded part is larger?

b. Express the shaded part of each figure in part a as a fraction. Show that one of those fractions is larger than the other by expressing both in terms of a common denominator and then comparing them.

16. Place a $>$ or $<$ symbol in the blank to make a true statement.

a. $\dfrac{32}{35} \underline{\quad} \dfrac{31}{35}$ **b.** $\dfrac{7}{8} \underline{\quad} \dfrac{31}{32}$

NOTATION *Complete each solution.*

17. Add: $\dfrac{2}{5} + \dfrac{1}{3}$.

$\dfrac{2}{5} + \dfrac{1}{3} = \dfrac{2 \cdot }{5 \cdot } + \dfrac{1 \cdot 5}{3 \cdot 5}$

$= \dfrac{}{15} + \dfrac{}{15}$

$= \dfrac{ + }{15}$

$= \dfrac{11}{15}$

18. Subtract: $\dfrac{7}{8} - \dfrac{2}{3}$.

$\dfrac{7}{8} - \dfrac{2}{3} = \dfrac{7 \cdot 3}{ \cdot 3} - \dfrac{2 \cdot 8}{ \cdot 8}$

$= \dfrac{21}{} - \dfrac{16}{}$

$= \dfrac{21 - 16}{}$

$= \dfrac{5}{24}$

PRACTICE *The denominators of two fractions are given. Find the lowest common denominator.*

19. 18, 6 **20.** 15, 3
21. 8, 6 **22.** 10, 4
23. 8, 20 **24.** 14, 21
25. 15, 12 **26.** 25, 30

In Exercises 27–78, do each operation. Simplify when necessary.

27. $\dfrac{3}{7} + \dfrac{1}{7}$ **28.** $\dfrac{16}{25} - \dfrac{9}{25}$

29. $\dfrac{37}{103} - \dfrac{17}{103}$ **30.** $\dfrac{54}{53} - \dfrac{52}{53}$

31. $\dfrac{11}{25} - \dfrac{1}{25}$ **32.** $\dfrac{7}{8} - \dfrac{1}{8}$

33. $\dfrac{5}{7} + \dfrac{3}{7}$ **34.** $\dfrac{17}{11} - \dfrac{12}{11}$

35. $\dfrac{1}{4} + \dfrac{3}{8}$ **36.** $\dfrac{2}{3} + \dfrac{1}{6}$

37. $\dfrac{13}{20} - \dfrac{1}{5}$ **38.** $\dfrac{71}{100} - \dfrac{1}{10}$

39. $\dfrac{4}{5} + \dfrac{2}{3}$ **40.** $\dfrac{1}{4} + \dfrac{2}{3}$

41. $\dfrac{1}{8} + \dfrac{2}{7}$ **42.** $\dfrac{1}{6} + \dfrac{5}{9}$

43. $\dfrac{3}{4} - \dfrac{2}{3}$ **44.** $\dfrac{4}{5} - \dfrac{1}{6}$

45. $\dfrac{5}{6} - \dfrac{3}{4}$ **46.** $\dfrac{7}{8} - \dfrac{5}{6}$

47. $\dfrac{16}{25} - \left(-\dfrac{3}{10}\right)$ **48.** $\dfrac{3}{8} - \left(-\dfrac{1}{6}\right)$

49. $-\dfrac{7}{16} + \dfrac{1}{4}$ **50.** $-\dfrac{17}{20} + \dfrac{4}{5}$

51. $\dfrac{1}{12} - \dfrac{3}{4}$ **52.** $\dfrac{11}{60} - \dfrac{13}{20}$

53. $-\dfrac{5}{8} - \dfrac{1}{3}$ **54.** $-\dfrac{7}{20} - \dfrac{1}{5}$

55. $-3 + \dfrac{2}{5}$ **56.** $-6 + \dfrac{5}{8}$

57. $-\dfrac{3}{4} - 5$ **58.** $-2 - \dfrac{7}{8}$

59. $\dfrac{1}{3} + \dfrac{1}{4} + \dfrac{1}{5}$ **60.** $\dfrac{1}{10} + \dfrac{1}{8} + \dfrac{1}{5}$

61. $-\dfrac{2}{3} + \dfrac{5}{4} + \dfrac{1}{6}$ **62.** $-\dfrac{3}{4} + \dfrac{3}{8} + \dfrac{7}{6}$

63. $\dfrac{5}{24} + \dfrac{3}{16}$ **64.** $\dfrac{17}{20} - \dfrac{4}{15}$

65. $-\dfrac{11}{15} - \dfrac{2}{9}$ **66.** $-\dfrac{19}{18} - \dfrac{5}{12}$

67. $\dfrac{7}{25} + \dfrac{1}{15}$ **68.** $\dfrac{11}{20} - \dfrac{1}{8}$

69. $\dfrac{4}{27} + \dfrac{1}{6}$ **70.** $\dfrac{8}{9} - \dfrac{7}{12}$

71. Find the difference of $\dfrac{11}{60}$ and $\dfrac{2}{45}$.

72. Find the sum of $\dfrac{9}{48}$ and $\dfrac{7}{40}$.

73. Subtract $\dfrac{5}{12}$ from $\dfrac{2}{15}$.

74. What is the sum of $\dfrac{11}{24}$ and $\dfrac{7}{36}$ increased by $\dfrac{5}{48}$?

APPLICATIONS

75. BOTANY   To assess the effects of smog on tree development, botanists cut down a pine tree and measured the width of the growth rings for the last two years. (See Illustration 1.)

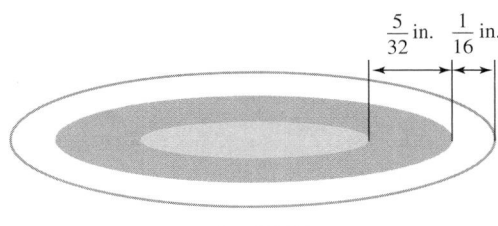

$\dfrac{5}{32}$ in. $\dfrac{1}{16}$ in.

ILLUSTRATION 1

a. What was the growth over this two-year period?

b. What is the difference in the widths of the two rings?

76. MAGAZINE LAYOUT   The page design for a magazine cover includes a blank strip at the top, called a header, and a blank strip at the bottom of the page, called a footer. In Illustration 2, how much page length is lost because of the header and footer?

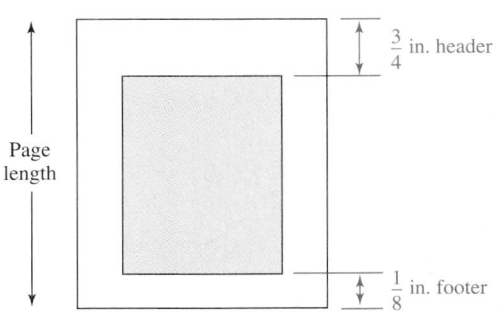

Page length

$\dfrac{3}{4}$ in. header

$\dfrac{1}{8}$ in. footer

ILLUSTRATION 2

77. FAMILY DINNER   A family bought two large pizzas for dinner. Several pieces of each pizza were not eaten, as shown in Illustration 3. How much pizza was left? Could the family have been fed with just one pizza?

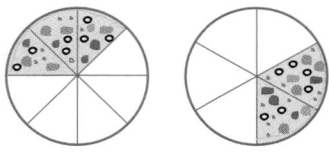

ILLUSTRATION 3

78. GASOLINE BARRELS   The contents of two identical-sized barrels are shown in Illustration 4. If they are dumped into an empty third barrel that is the same size, how much of the third barrel will they fill?

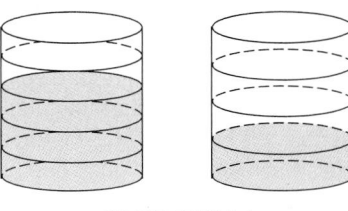

ILLUSTRATION 4

79. WEIGHTS AND MEASURES   A consumer protection agency verifies the accuracy of butcher shop scales by placing a known three-quarter-pound weight on the scale and then comparing that to the scale's readout. According to Illustration 5, by how much is this scale off? Does it result in undercharging or overcharging customers on their meat purchases?

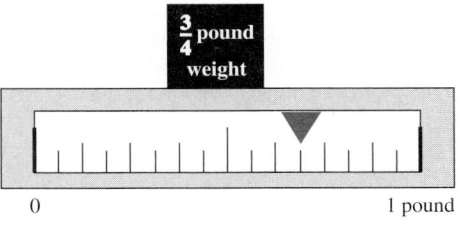

$\dfrac{3}{4}$ pound weight

0   1 pound

ILLUSTRATION 5

80. WRENCHES   A mechanic likes to hang his wrenches above his tool bench in order of narrowest to widest. What is the proper order of the wrenches in Illustration 6?

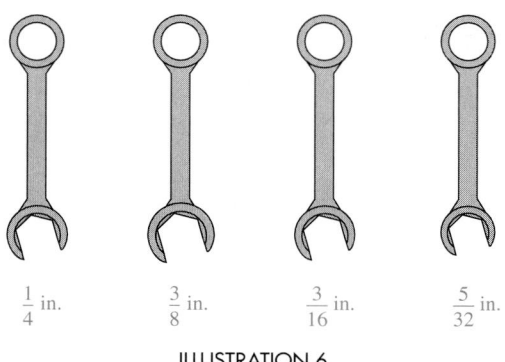

$\frac{1}{4}$ in. $\frac{3}{8}$ in. $\frac{3}{16}$ in. $\frac{5}{32}$ in.

ILLUSTRATION 6

81. HIKING   Illustration 7 shows the length of each part of a three-part hike. Rank the lengths from longest to shortest.

ILLUSTRATION 7

82. FIGURE DRAWING   As an aid in drawing the human body, artists divide the body into three parts. Each part is then expressed as a fraction of the total body height. (See Illustration 8.) For example, the torso is $\frac{4}{15}$ of the body height. What fraction of body height is the head?

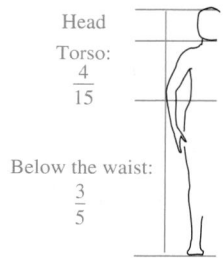

Head

Torso:
$\frac{4}{15}$

Below the waist:
$\frac{3}{5}$

ILLUSTRATION 8

83. STUDY HABITS   College students taking a full load were asked to give the average number of hours they studied each day. The results are shown in the pie chart in Illustration 9. What fraction of the students study 2 hours or more daily?

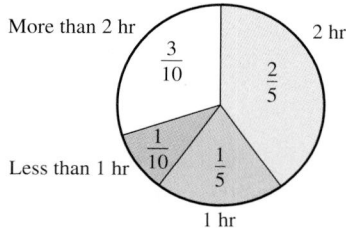

More than 2 hr $\frac{3}{10}$ 2 hr $\frac{2}{5}$

Less than 1 hr $\frac{1}{10}$ $\frac{1}{5}$

1 hr

ILLUSTRATION 9

84. MUSICAL NOTES   The notes used in music have fractional values. Their names and the symbols used to represent them are shown in Illustration 10(a). In common time, the values of the notes in each measure must add up to 1. Is the measure in Illustration 10(b) complete?

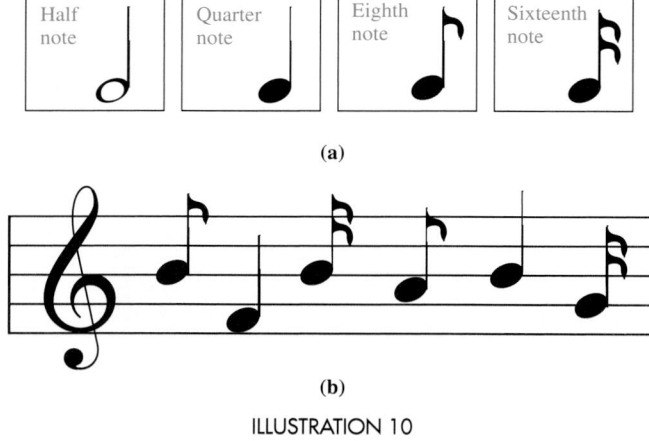

Half note   Quarter note   Eighth note   Sixteenth note

(a)

(b)

ILLUSTRATION 10

85. GARAGE DOOR OPENER   What is the difference in strength between a $\frac{1}{3}$-hp and a $\frac{1}{2}$-hp garage door opener?

86. DELIVERY TRUCK   A truck can safely carry a one-ton load. Should it be used to deliver one-half ton of sand, one-third ton of gravel, and one-fifth ton of cement in one trip to a job site?

WRITING

87. How are the procedures for expressing a fraction in higher terms and simplifying a fraction to lowest terms similar, and how are they different?

88. Given two fractions, how do we find their lowest common denominator?

89. How do we compare the relative sizes of two fractions with different denominators?

90. What is the difference between a common denominator and the lowest common denominator?

REVIEW

91. Find the prime factorization of 20.

92. Solve $3x + 1 = -5$.

93. What is the formula for finding the area of a rectangle?

94. What is the formula for finding the perimeter of a rectangle?

As we have seen, the **multiples** of a number can be found by multiplying it successively by 1, 2, 3, 4, 5, and so on. The multiples of 4 and the multiples of 6 are shown below.

$1 \cdot 4 = 4$	$1 \cdot 6 = 6$
$2 \cdot 4 = 8$	$2 \cdot 6 = 12$
$3 \cdot 4 = 12$	$3 \cdot 6 = 18$
$4 \cdot 4 = 16$	$4 \cdot 6 = 24$
$5 \cdot 4 = 20$	$5 \cdot 6 = 30$
$6 \cdot 4 = 24$	$6 \cdot 6 = 36$
$7 \cdot 4 = 28$	$7 \cdot 6 = 42$
$8 \cdot 4 = 32$	$8 \cdot 6 = 48$
$9 \cdot 4 = 36$	$9 \cdot 6 = 54$

Common multiples of 4 and 6 are highlighted in red.

Because 12 is the smallest number that is a multiple of both 4 and 6, it is called the **least common multiple (LCM)** of 4 and 6.

Making lists like those shown above can be tedious. A more efficient method to find the least common multiple of several numbers is as follows.

Finding the least common multiple

1. Write each of the numbers in prime-factored form.
2. The least common multiple is a product of prime factors, where each prime factor is used the greatest number of times it appears in any one factorization found in Step 1.

EXAMPLE 1 *Least common multiple.* Find the LCM of 24 and 36.
Solution

Step 1: First, we find the prime factorizations of 24 and 36.

$24 = 3 \cdot 2 \cdot 2 \cdot 2$

$36 = 3 \cdot 3 \cdot 2 \cdot 2$

Step 2: The prime factorizations of 24 and 36 contain the prime factors 3 and 2. We use each of these factors the greatest number of times it appears in any one factorization.

The greatest number of times 3 appears in any one factorization is two times.

The greatest number of times 2 appears in any one factorization is three times.

$$\text{LCM} = 3 \cdot 3 \cdot 2 \cdot 2 \cdot 2 = 72$$

The least common multiple of 24 and 36 is 72.

Self Check

Find the LCM of 18 and 84.

Answer: 252

Because 2 divides 36 exactly and because 2 divides 120 exactly, 2 is called a **common factor** of 36 and 120.

$$\frac{36}{2} = 18 \qquad \frac{120}{2} = 60$$

The numbers 36 and 120 have other common factors, such as 3 and 6. The **greatest common factor (GCF)** of 36 and 120 is the largest number that is a factor of both. We follow these steps to find the greatest common factor of several numbers.

Finding the greatest common factor

1. Write each of the numbers in prime-factored form.
2. The greatest common factor is the product of the prime factors that are common to the factorizations found in Step 1. If the numbers have no factors in common, the GCF is 1.

EXAMPLE 2 *Greatest common factor.*   Find the GCF of 36 and 120.

Solution

Step 1: We find the prime factorizations of 36 and 120.

$$36 = 3 \cdot 3 \cdot 2 \cdot 2$$
$$120 = 5 \cdot 3 \cdot 2 \cdot 2 \cdot 2$$

Step 2: One factor of 3 (highlighted in red) and two factors of 2 (highlighted in blue) are common to the factorizations of 36 and 120. To find the GCF, we form their product.

$$\text{GCF} = 3 \cdot 2 \cdot 2 = 12$$

The greatest common factor of 36 and 120 is 12.

Self Check

Find the GCF of 60 and 150.

Answer: 30

STUDY SET   The LCM and the GCF

Find the least common multiple of the given numbers.

1. 3, 5

2. 7, 11

3. 8, 14

4. 8, 12

5. 14, 21

6. 16, 20

7. 6, 18

8. 3, 9

9. 44, 60

10. 36, 60

11. 100, 120

12. 120, 180

13. 6, 24, 36

14. 6, 10, 18

15. 18, 54, 63

16. 16, 30, 84

Find the greatest common factor of the given numbers.

17. 6, 9

18. 8, 12

19. 22, 33

20. 15, 20

21. 16, 20

22. 18, 24

23. 25, 100

24. 16, 80

25. 100, 120

26. 120, 180

27. 48, 108

28. 60, 96

29. 18, 24, 36

30. 30, 50, 90

31. 18, 54, 63

32. 28, 42, 84

33. NURSING   A nurse, working in an intensive care unit, has to check a patient's vital signs every 45 minutes. Another nurse has to give the same patient his medication every 2 hours. If both nurses are in the patient's room together now, how long will it be until they are once again in the room together?

34. BARBECUES   A certain brand of hot dogs comes in packages of 10. A certain brand of hot dog buns comes in packages of 12. For a family reunion barbecue, how many packages of hot dogs and how many packages of hot dog buns should be purchased so that no hot dogs and no buns are wasted?

3.5 *Multiplying and Dividing Mixed Numbers*

In this section, you will learn about

- Mixed numbers • Writing mixed numbers as improper fractions
- Writing improper fractions as mixed numbers
- Graphing fractions and mixed numbers
- Multiplying and dividing mixed numbers

INTRODUCTION. In the next two sections, we will show how to add, subtract, multiply, and divide *mixed numbers*. These numbers are widely used in daily life. Here are a few examples.

The recipe calls for $2\frac{1}{3}$ cups of flour.

It took $3\frac{3}{4}$ hours to paint the living room.

The entrance to the park is $1\frac{1}{2}$ miles away.

Mixed numbers

A **mixed number** is the *sum* of a whole number and a proper fraction. For example, $2\frac{3}{4}$ is a mixed number.

$$\underset{\substack{\uparrow \\ \text{Mixed number}}}{2\frac{3}{4}} \quad = \quad \underset{\substack{\uparrow \\ \text{Whole number}}}{2} \quad + \quad \underset{\substack{\uparrow \\ \text{Proper fraction}}}{\frac{3}{4}}$$

> **COMMENT**   Note that $2\frac{3}{4}$ means $2 + \frac{3}{4}$, even though the $+$ sign is not written. Do not confuse $2\frac{3}{4}$ with $2 \cdot \frac{3}{4}$ or $2\left(\frac{3}{4}\right)$, which indicate the multiplication of 2 and $\frac{3}{4}$.

In this section, we will work with negative as well as positive mixed numbers. For example, the negative mixed number $-4\frac{3}{4}$ could be used to represent $4\frac{3}{4}$ feet below sea level. We think of $-4\frac{3}{4}$ as $-4 - \frac{3}{4}$.

Writing mixed numbers as improper fractions

To see that mixed numbers are related to improper fractions, consider $2\frac{3}{4}$. To write $2\frac{3}{4}$ as an improper fraction, we need to find out how many *fourths* it represents. One way is to use the fundamental property of fractions.

$$2\frac{3}{4} = 2 + \frac{3}{4} \qquad \text{Write the mixed number } 2\frac{3}{4} \text{ as a sum.}$$

$$= \frac{2}{1} + \frac{3}{4} \qquad \text{Write 2 as a fraction: } 2 = \frac{2}{1}.$$

$$= \frac{2 \cdot 4}{1 \cdot 4} + \frac{3}{4} \qquad \text{Use the fundamental property of fractions to express } \tfrac{2}{1} \text{ as a fraction with denominator 4.}$$

$$= \frac{8}{4} + \frac{3}{4} \qquad \text{Do the multiplications in the numerator and denominator.}$$

$$= \frac{11}{4} \qquad \text{Add the numerators: } 8 + 3 = 11. \text{ Write the sum over the common denominator, 4.}$$

Thus, $2\frac{3}{4} = \frac{11}{4}$.

We can obtain the same result with far less work. To change $2\frac{3}{4}$ to an improper fraction, we simply multiply 2 by 4 and add 3 to get the numerator, and keep the denominator of 4.

$$2\frac{3}{4} = \frac{2(4) + 3}{4} = \frac{11}{4}$$

This example illustrates the following general rule.

Writing a mixed number as an improper fraction

> To write a mixed number as an improper fraction, multiply the whole-number part by the denominator of the fraction and add the result to the numerator. Write this sum over the denominator.

EXAMPLE 1 *Writing a mixed number as an improper fraction.*
Write the mixed number $5\frac{1}{6}$ as an improper fraction.

Solution

$$5\frac{1}{6} = \frac{5(6) + 1}{6} \qquad \text{Multiply 5 by the denominator 6. Add the numerator 1. Write this sum over the denominator 6.}$$

$$= \frac{30 + 1}{6} \qquad \text{Do the multiplication: } 5(6) = 30.$$

$$= \frac{31}{6} \qquad \text{Do the addition: } 30 + 1 = 31.$$

Self Check

Write the mixed number $3\frac{3}{8}$ as an improper fraction.

Answer: $\dfrac{27}{8}$ ∎

To write a negative mixed number in fractional form, ignore the $-$ sign and use the method shown in Example 1 on the positive mixed number. Once that procedure is completed, write a $-$ sign in front of the result. For example, $-3\frac{1}{4} = -\frac{13}{4}$.

Writing improper fractions as mixed numbers

To write an improper fraction as a mixed number, we must find two things: the *whole-number part* and the *fractional part* of the mixed number. To develop a procedure to do this, let's consider the improper fraction $\frac{7}{3}$. To find the number of groups of 3 in 7, we can divide 7 by 3. This will find the whole-number part of the mixed number. The remainder is the numerator of the fractional part of the mixed number.

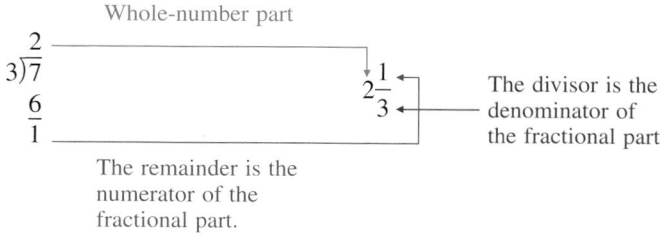

Whole-number part

$$3\overline{)7}$$
$$\underline{6}$$
$$1$$

$$2\frac{1}{3}$$

The divisor is the denominator of the fractional part.

The remainder is the numerator of the fractional part.

This example suggests the following general rule.

Writing an improper fraction as a mixed number

To write an improper fraction as a mixed number, divide the numerator by the denominator to obtain the whole-number part. The remainder over the divisor is the fractional part.

EXAMPLE 2 *Writing an improper fraction as a mixed number.*
Write $\frac{29}{6}$ as a mixed number.

Solution

$$\begin{array}{r} 4 \\ 6\overline{)29} \\ \underline{24} \\ 5 \end{array}$$ Divide the numerator by the denominator.

The remainder is 5.

Thus, $\frac{29}{6} = 4\frac{5}{6}$.

Self Check

Write $\frac{43}{5}$ as a mixed number.

Answer: $8\frac{3}{5}$

Graphing fractions and mixed numbers

Earlier, we graphed whole numbers and integers on a number line. Fractions and mixed numbers can also be graphed on a number line.

EXAMPLE 3 *Graphing fractions and mixed numbers.* Graph $-2\frac{3}{4}$, $-1\frac{1}{2}$, $-\frac{1}{8}$, and $\frac{13}{5}$ on a number line.

Solution

To help locate the graph of each number, we make some observations:

- Since $-2\frac{3}{4} < -2$, the graph of $-2\frac{3}{4}$ is to the left of -2.
- The number $-1\frac{1}{2}$ is between -1 and -2.
- The number $-\frac{1}{8}$ is less than 0.
- Expressed as a mixed number, $\frac{13}{5} = 2\frac{3}{5}$.

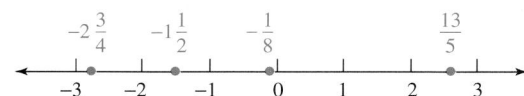

Self Check

Graph $-1\frac{7}{8}$, $-\frac{2}{3}$, and $\frac{9}{4}$ on a number line.

Answer:

Multiplying and dividing mixed numbers

Multiplying and dividing mixed numbers

To multiply or divide mixed numbers, first change the mixed numbers to improper fractions. Then do the multiplication or division of the fractions.

EXAMPLE 4 *Multiplying mixed numbers.* Multiply: $5\frac{1}{5} \cdot 1\frac{2}{13}$.

Solution

$5\frac{1}{5} \cdot 1\frac{2}{13} = \frac{26}{5} \cdot \frac{15}{13}$ Write each mixed number as an improper fraction.

$\phantom{5\frac{1}{5} \cdot 1\frac{2}{13}} = \frac{26 \cdot 15}{5 \cdot 13}$ Multiply the numerators and multiply the denominators.

Self Check

Multiply: $9\frac{3}{5} \cdot 3\frac{3}{4}$.

$$= \frac{13 \cdot 2 \cdot 5 \cdot 3}{5 \cdot 13} \qquad \text{Factor 26 as } 13 \cdot 2 \text{ and 15 as } 5 \cdot 3.$$

$$= \frac{\overset{1}{\cancel{13}} \cdot 2 \cdot \overset{1}{\cancel{5}} \cdot 3}{\underset{1}{\cancel{5}} \cdot \underset{1}{\cancel{13}}} \qquad \text{Divide out the common factors of 13 and 5.}$$

$$= \frac{6}{1} \qquad \text{Multiply in the numerator and denominator.}$$

$$= 6 \qquad \text{Simplify: } \tfrac{6}{1} = 6.$$

Answer: 36

EXAMPLE 5   *Dividing mixed numbers.*   Divide: $-3\frac{3}{8} \div 2\frac{1}{4}$.

Solution

$$-3\frac{3}{8} \div 2\frac{1}{4} = -\frac{27}{8} \div \frac{9}{4} \qquad \text{Write each mixed number as an improper fraction.}$$

$$= -\frac{27}{8} \cdot \frac{4}{9} \qquad \text{Multiply by the reciprocal of } \tfrac{9}{4}.$$

$$= -\frac{27 \cdot 4}{8 \cdot 9} \qquad \begin{array}{l}\text{The product of two fractions with unlike signs is negative.}\\ \text{Multiply the numerators and multiply the denominators.}\end{array}$$

$$= -\frac{\overset{1}{\cancel{9}} \cdot 3 \cdot \overset{1}{\cancel{4}}}{\underset{1}{\cancel{4}} \cdot 2 \cdot \underset{1}{\cancel{9}}} \qquad \begin{array}{l}\text{Factor 27 as } 9 \cdot 3 \text{ and 8 as } 4 \cdot 2. \text{ Divide out the common}\\ \text{factors of 9 and 4.}\end{array}$$

$$= -\frac{3}{2} \qquad \text{Multiply in the numerator and denominator.}$$

$$= -1\frac{1}{2} \qquad \text{Write } -\tfrac{3}{2} \text{ as a mixed number.}$$

Self Check

Divide: $3\dfrac{4}{15} \div \left(-2\dfrac{1}{10}\right)$.

Answer: $-1\dfrac{5}{9}$

EXAMPLE 6   *Government grant.*   If $\$12\frac{1}{2}$ million is to be divided equally among five cities to fund recreation programs, how much will each city receive?

Solution   To find the amount received by each city, we divide the grant money by 5.

$$12\frac{1}{2} \div 5 = \frac{25}{2} \div \frac{5}{1} \qquad \text{Write } 12\tfrac{1}{2} \text{ as an improper fraction, and write 5 as a fraction.}$$

$$= \frac{25}{2} \cdot \frac{1}{5} \qquad \text{Multiply by the reciprocal of } \tfrac{5}{1}.$$

$$= \frac{25 \cdot 1}{2 \cdot 5} \qquad \text{Multiply the numerators and multiply the denominators.}$$

$$= \frac{\overset{1}{\cancel{5}} \cdot 5 \cdot 1}{2 \cdot \underset{1}{\cancel{5}}} \qquad \text{Factor 25 as } 5 \cdot 5. \text{ Divide out the common factor of 5.}$$

$$= \frac{5}{2} \qquad \text{Multiply in the numerator and denominator.}$$

$$= 2\frac{1}{2} \qquad \text{Write } \tfrac{5}{2} \text{ as a mixed number.}$$

Each city will receive $\$2\frac{1}{2}$ million .

STUDY SET Section 3.5

VOCABULARY *Fill in the blanks.*

1. A _____ number is the sum of a whole number and a proper fraction.

2. An _____ fraction is a fraction with a numerator that is greater than or equal to its denominator.

3. To _____ a number means to locate its position on a number line and highlight it using a heavy dot.

4. Multiplying or dividing the _____ and _____ of a fraction by the same nonzero number does not change the value of the fraction.

CONCEPTS

5. What signed number could be used to describe each situation?
   a. A temperature of five and one-half degrees below zero
   b. One and seven-eighths inches under the finish grade

6. What signed number could be used to describe each situation?
   a. A rain total two and three-tenths of an inch lower than the average
   b. Three and one-half minutes before liftoff

7. a. In Illustration 1, the divisions on the face of the meter represent fractions. What value is the arrow registering?
   b. If the arrow moves two marks to the left, what value will it register?

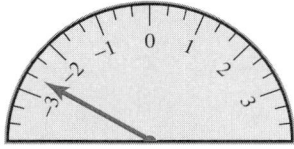

ILLUSTRATION 1

8. a. In Illustration 2, the divisions on the face of the meter represent fractions. What value is the arrow registering?
   b. If the arrow moves up one mark, what value will it register?

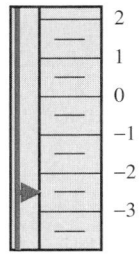

ILLUSTRATION 2

9. What fractions have been graphed on the number line in Illustration 3?

ILLUSTRATION 3

10. What mixed numbers have been graphed on the number line in Illustration 4?

ILLUSTRATION 4

11. DIVING See Illustration 5. Complete the description of the dive by filling in the blank with a mixed number.

Forward ____ somersaults from the pike position

ILLUSTRATION 5

12. PRODUCT LABELING The label in Illustration 6 uses mixed numbers. Write each one as an improper fraction.

ILLUSTRATION 6

13. Draw $\frac{17}{8}$ pizzas.

14. a. What mixed number is depicted in Illustration 7?

b. What improper fraction is shown in Illustration 7?

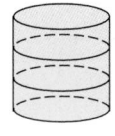 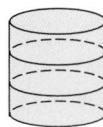 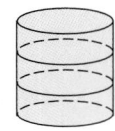 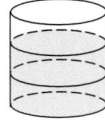

ILLUSTRATION 7

NOTATION *Complete each solution.*

15. Multiply: $-5\frac{1}{4} \cdot 1\frac{1}{7}$.

$$-5\frac{1}{4} \cdot 1\frac{1}{7} = -\frac{21}{4} \cdot \frac{\Box}{7}$$

$$= -\frac{21 \cdot \Box}{4 \cdot 7}$$

$$= -\frac{\overset{1}{\cancel{7}} \cdot 3 \cdot \overset{1}{\cancel{4}} \cdot 2}{\underset{1}{\cancel{4}} \cdot \underset{1}{\cancel{7}}}$$

$$= -\frac{\Box}{1}$$

$$= -6$$

16. Divide: $-5\frac{5}{6} \div 2\frac{1}{12}$.

$$-5\frac{5}{6} \div 2\frac{1}{12} = -\frac{\Box}{6} \div \frac{25}{12}$$

$$= -\frac{35}{6} \cdot \frac{12}{\Box}$$

$$= -\frac{35 \cdot 12}{6 \cdot \Box}$$

$$= -\frac{\overset{1}{\cancel{5}} \cdot \Box \cdot \overset{1}{\cancel{6}} \cdot 2}{\underset{1}{\cancel{6}} \cdot \underset{1}{\cancel{5}} \cdot \Box}$$

$$= -\frac{\Box}{5}$$

$$= -2\frac{4}{5}$$

PRACTICE *Write each improper fraction as a mixed number. Simplify the result, if possible.*

17. $\frac{15}{4}$ **18.** $\frac{41}{6}$

19. $\frac{29}{5}$ **20.** $\frac{29}{3}$

21. $-\frac{20}{6}$ **22.** $-\frac{28}{8}$

23. $\frac{127}{12}$ **24.** $\frac{197}{16}$

Write each mixed number as an improper fraction.

25. $6\frac{1}{2}$ **26.** $8\frac{2}{3}$

27. $20\frac{4}{5}$ **28.** $15\frac{3}{8}$

29. $-6\frac{2}{9}$ **30.** $-7\frac{1}{12}$

31. $200\frac{2}{3}$ **32.** $90\frac{5}{6}$

Graph each set of numbers on the number line.

33. $\left\{-2\frac{8}{9}, 1\frac{2}{3}, \frac{16}{5}\right\}$

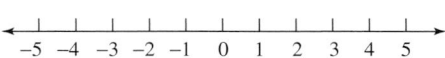

34. $\left\{-\frac{3}{4}, -3\frac{1}{4}, \frac{5}{2}\right\}$

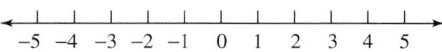

35. $\left\{3\frac{1}{7}, -\frac{98}{99}, -\frac{10}{3}\right\}$

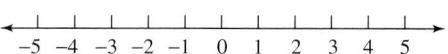

36. $\left\{-2\frac{1}{5}, \frac{4}{5}, -\frac{11}{3}\right\}$

Multiply.

37. $1\frac{2}{3} \cdot 2\frac{1}{7}$ **38.** $2\frac{3}{5} \cdot 1\frac{2}{3}$

39. $-7\frac{1}{2}\left(-1\frac{2}{5}\right)$ **40.** $-4\frac{1}{8}\left(-1\frac{7}{9}\right)$

41. $3\frac{1}{16} \cdot 4\frac{4}{7}$

42. $5\frac{3}{5} \cdot 1\frac{11}{14}$

43. $-6 \cdot 2\frac{7}{24}$

44. $-7 \cdot 1\frac{3}{28}$

45. $2\frac{1}{2}\left(-3\frac{1}{3}\right)$

46. $\left(-3\frac{1}{4}\right)\left(1\frac{1}{5}\right)$

47. $2\frac{5}{8} \cdot \frac{5}{27}$

48. $3\frac{1}{9} \cdot \frac{3}{32}$

49. Find the product of $1\frac{2}{3}$, 6, and $-\frac{1}{8}$.

50. Find the product of $-\frac{5}{6}$, -8, and $-2\frac{1}{10}$.

Evaluate each power.

51. $\left(1\frac{2}{3}\right)^2$

52. $\left(3\frac{1}{2}\right)^2$

53. $\left(-1\frac{1}{3}\right)^3$

54. $\left(-1\frac{1}{5}\right)^3$

Divide.

55. $3\frac{1}{3} \div 1\frac{5}{6}$

56. $3\frac{3}{4} \div 5\frac{1}{3}$

57. $-6\frac{3}{5} \div 7\frac{1}{3}$

58. $-4\frac{1}{4} \div 4\frac{1}{2}$

59. $-20\frac{1}{4} \div \left(-1\frac{11}{16}\right)$

60. $-2\frac{7}{10} \div \left(-1\frac{1}{14}\right)$

61. $6\frac{1}{4} \div 20$

62. $4\frac{2}{5} \div 11$

63. $1\frac{2}{3} \div \left(-2\frac{1}{2}\right)$

64. $2\frac{1}{2} \div \left(-1\frac{5}{8}\right)$

65. $8 \div 3\frac{1}{5}$

66. $15 \div 3\frac{1}{3}$

67. Find the quotient of $-4\frac{1}{2}$ and $2\frac{1}{4}$.

68. Find the quotient of 25 and $-10\frac{5}{7}$.

APPLICATIONS

69. CALORIES   A company advertises that its mints contain only $3\frac{1}{5}$ calories apiece. What is the calorie intake if you eat an entire package of 20 mints?

70. CEMENT MIXER   A cement mixer can carry $9\frac{1}{2}$ cubic yards of concrete. If it makes 8 trips to a job site, how much concrete will be delivered to the site?

71. SHOPPING   In Illustration 8, what is the cost of buying the fruit in the scale?

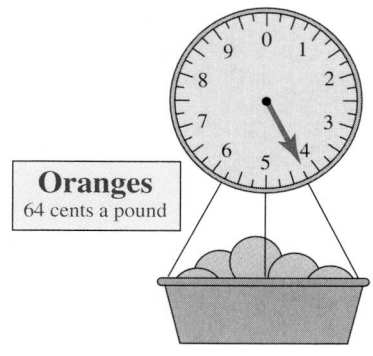

Oranges
64 cents a pound

ILLUSTRATION 8

72. FRAME   How much molding is needed to make the square picture frame in Illustration 9?

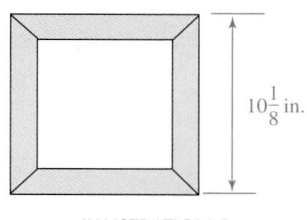

$10\frac{1}{8}$ in.

ILLUSTRATION 9

73. SUBDIVISION   A developer donated to the county 100 of the 1,000 acres of land she owned. She divided the remaining acreage into $1\frac{1}{3}$-acre lots. How many lots were created?

74. CATERING   How many people can be served $\frac{1}{3}$-pound hamburgers if a caterer purchases 200 pounds of ground beef?

75. GRAPH PAPER   Mathematicians use specially marked paper, called *graph paper,* when drawing figures. It is made up of $\frac{1}{4}$-inch squares. Find the length and width of the piece of graph paper in Illustration 10.

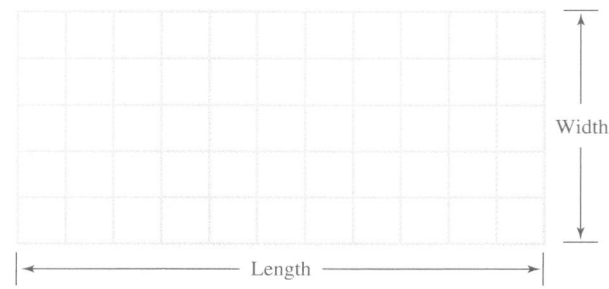

Width

Length

ILLUSTRATION 10

76. LUMBER   As Illustration 11 shows, 2-by-4's from the lumber yard do not really have dimensions of 2 inches by 4 inches. How wide and how high is the stack of 2-by-4s?

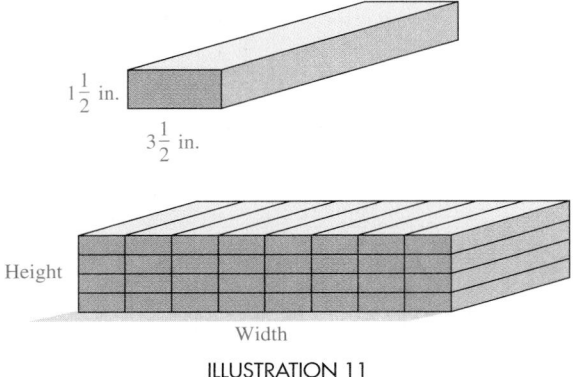

ILLUSTRATION 11

77. EMERGENCY EXIT   Illustration 12 shows a sign that marks the emergency exit on a school bus. Find the area of the sign.

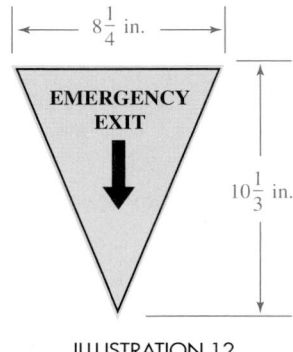

ILLUSTRATION 12

78. HORSE RACING   The race tracks on which thoroughbred horses run are marked off in $\frac{1}{8}$-mile-long segments called furlongs. How many furlongs are there in a $1\frac{1}{16}$-mile race?

79. FIRE ESCAPE   The fire escape stairway in an office building is shown in Illustration 13. Each riser is $7\frac{1}{2}$ inches high. If each floor is 105 inches high and the building is 43 stories tall, how many steps are there in the stairway?

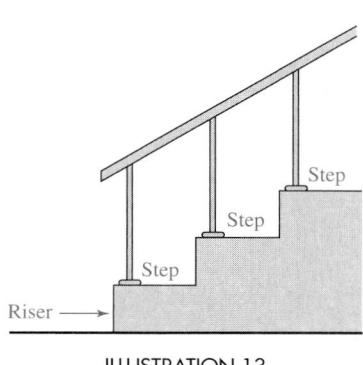

ILLUSTRATION 13

80. LICENSE PLATE   Find the area of the license plate in Illustration 14.

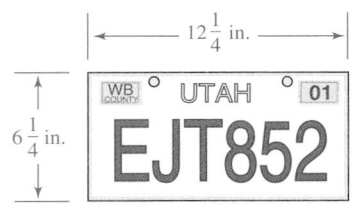

ILLUSTRATION 14

81. SHOPPING ON THE INTERNET   A mother is ordering a pair of jeans for her daughter from the screen shown in Illustration 15. If the daughter's height is $60\frac{3}{4}$ in. and her waist is $24\frac{1}{2}$ in., on what size and what cut should the mother point and click?

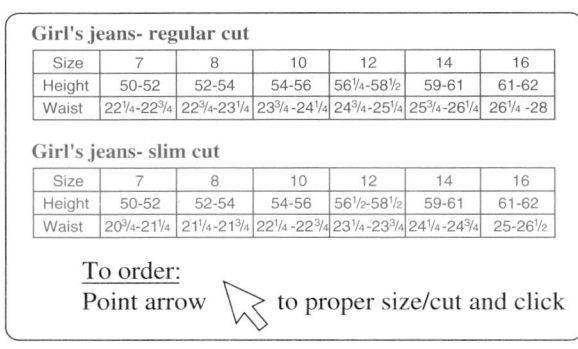

ILLUSTRATION 15

82. SEWING   Use the table in Illustration 16 to determine the number of yards of fabric needed:
  a. to make a size 16 top if the fabric to be used is 60 inches wide.
  b. to make size 18 pants if the fabric to be used is 45 inches wide.

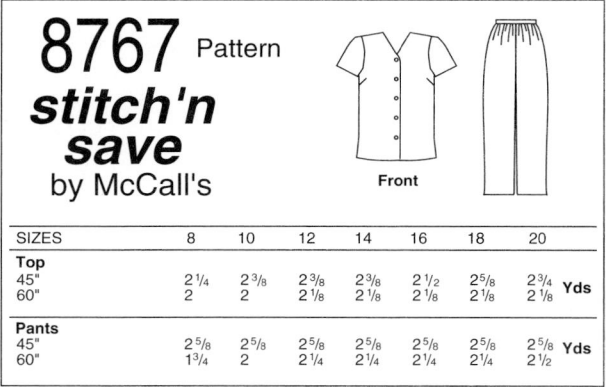

ILLUSTRATION 16

WRITING

83. Explain the difference between $2\frac{3}{4}$ and $2\left(\frac{3}{4}\right)$.

84. Give three examples of how you use mixed numbers in daily life.

85. Explain the procedure used to write an improper fraction as a mixed number.

86. Explain the procedure used to multiply two mixed numbers.

REVIEW

87. Evaluate $3^2 \cdot 2^3$.

88. If a represents a number, then $a \cdot 0 = $ ▢ .

89. Write $8 + 8 + 8 + 8$ as a multiplication.

90. If a square measures 1 inch on each side, what is its area?

91. What operation must be undone to solve for x?

$$\frac{x}{2} = -12$$

92. In the formula $A = lw$, what do l and w represent?

3.6 *Adding and Subtracting Mixed Numbers*

In this section, you will learn about

- Adding mixed numbers • Adding mixed numbers in vertical form
- Subtracting mixed numbers

INTRODUCTION. In this section, we will discuss three methods for adding and subtracting mixed numbers. The first method works well when the whole-number parts of the mixed numbers are small. The second method works well when the whole-number parts of the mixed numbers are large. The third method uses columns as a way to organize the work.

Adding mixed numbers

We can add mixed numbers by writing them as improper fractions. To do so, we follow these steps.

Adding mixed numbers: method 1

Adding mixed numbers: method 1

1. Write each mixed number as an improper fraction.
2. Write each improper fraction as an equivalent fraction with a denominator that is the LCD.
3. Add the fractions.
4. Change the result to a mixed number if desired.

EXAMPLE 1 *Adding mixed numbers.* Add: $4\frac{1}{6} + 2\frac{3}{4}$.

Solution

$$4\frac{1}{6} + 2\frac{3}{4} = \frac{25}{6} + \frac{11}{4}$$ Write each mixed number as an improper fraction: $4\frac{1}{6} = \frac{25}{6}$ and $2\frac{3}{4} = \frac{11}{4}$.

By inspection, we see that the lowest common denominator is 12.

$$= \frac{25 \cdot 2}{6 \cdot 2} + \frac{11 \cdot 3}{4 \cdot 3}$$ Write each fraction as a fraction with a denominator of 12.

Self Check

Add: $3\frac{2}{3} + 1\frac{1}{5}$.

$$= \frac{50}{12} + \frac{33}{12}$$ Do the multiplications in the numerators and denominators.

$$= \frac{83}{12}$$ Add the numerators: $50 + 33 = 83$. Write the sum over the common denominator 12.

$$= 6\frac{11}{12}$$ Write the improper fraction as a mixed number: $\frac{83}{12} = 6\frac{11}{12}$.

Answer: $4\frac{13}{15}$ ■

We can also add mixed numbers by adding their whole-number parts and their fractional parts. To do so, we follow these steps.

Adding mixed numbers: method 2

1. Write each mixed number as the sum of a whole number and a fraction.
2. Use the commutative property of addition to write the whole numbers together and the fractions together.
3. Add the whole numbers and the fractions separately.
4. Write the result as a mixed number if necessary.

EXAMPLE 2 *Adding mixed numbers.* Find the sum: $168\frac{3}{4} + 85\frac{1}{5}$.

Solution

$$168\frac{3}{4} + 85\frac{1}{5} = 168 + \frac{3}{4} + 85 + \frac{1}{5}$$ Write each mixed number as the sum of a whole number and a fraction.

$$= 168 + 85 + \frac{3}{4} + \frac{1}{5}$$ Use the commutative property of addition to change the order of the addition.

$$= 253 + \frac{3}{4} + \frac{1}{5}$$ Add the whole numbers: $168 + 85 = 253$.

$$= 253 + \frac{3 \cdot 5}{4 \cdot 5} + \frac{1 \cdot 4}{5 \cdot 4}$$ Write each fraction as a fraction with denominator 20.

$$= 253 + \frac{15}{20} + \frac{4}{20}$$ Multiply in the numerators and denominators.

$$= 253 + \frac{19}{20}$$ Add the numerators and write the sum over the common denominator 20.

$$= 253\frac{19}{20}$$ Write the sum as a mixed number.

Self Check

Find the sum: $275\frac{1}{6} + 81\frac{3}{5}$.

Answer: $356\frac{23}{30}$ ■

COMMENT If we use method 1 to add the mixed numbers in Example 2, the numbers we encounter are cumbersome. As expected, the result is the same: $253\frac{19}{20}$.

$$168\frac{3}{4} + 85\frac{1}{5} = \frac{675}{4} + \frac{426}{5}$$ Write $168\frac{3}{4}$ and $85\frac{1}{5}$ as improper fractions.

$$= \frac{675 \cdot 5}{4 \cdot 5} + \frac{426 \cdot 4}{5 \cdot 4}$$ The LCD is 20.

$$= \frac{3,375}{20} + \frac{1,704}{20}$$

$$= \frac{5,079}{20}$$

$$= 253\frac{19}{20}$$

Generally speaking, the larger the whole-number parts of the mixed numbers get, the more difficult it becomes to add those mixed numbers using method 1.

Adding mixed numbers in vertical form

By working in columns, we can add mixed numbers quickly. The strategy is the same as in Example 2: Add whole numbers to whole numbers and fractions to fractions.

Line up the mixed numbers vertically.

Apply the fundamental property of fractions to get an LCD.

Add the whole numbers and add the fractions separately.

$$
\begin{array}{rcccl}
25\dfrac{3}{4} & = & 25\dfrac{3\cdot 5}{4\cdot 5} & = & 25\dfrac{15}{20} \\[2ex]
+\,31\dfrac{1}{5} & = & +\,31\dfrac{1\cdot 4}{5\cdot 4} & = & +\,31\dfrac{4}{20} \\[2ex]
 & & & & \overline{\;56\dfrac{19}{20}\;}
\end{array}
$$

EXAMPLE 3   *Suspension bridge.* Find the total length of cable that must be ordered if cables a, d, and e of the suspension bridge in Figure 3-9 are to be replaced. (See the table below.)

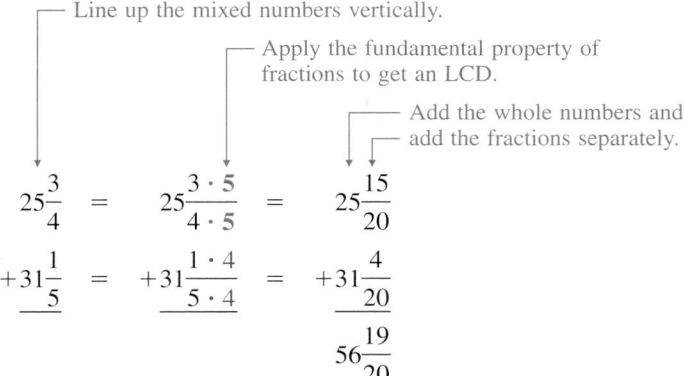

FIGURE 3-9

Bridge Specifications

Cable	a	b	c
Length (feet)	$75\frac{1}{12}$	$54\frac{1}{6}$	$43\frac{1}{4}$

Solution   To find the total length of cable to be ordered, we add the lengths of cables a, d, and e. Because of the symmetric design, cables d and c and cables e and b are the same length.

Length of cable a	plus	length of cable d (or cable c)	plus	length of cable e (or cable b)	equals	the total length needed.
$75\dfrac{1}{12}$	$+$	$43\dfrac{1}{4}$	$+$	$54\dfrac{1}{6}$	$=$	total length

We add the mixed numbers using a vertical format.

$$
\begin{array}{rcccl}
75\dfrac{1}{12} & = & 75\dfrac{1}{12} & = & 75\dfrac{1}{12} \\[2ex]
43\dfrac{1}{4} & = & 43\dfrac{1\cdot 3}{4\cdot 3} & = & 43\dfrac{3}{12} \\[2ex]
+\,54\dfrac{1}{6} & = & +\,54\dfrac{1\cdot 2}{6\cdot 2} & = & +\,54\dfrac{2}{12} \\[2ex]
 & & & & \overline{\;172\dfrac{6}{12}\;} = 172\dfrac{1}{2}
\end{array}
$$

Simplify: $\dfrac{6}{12} = \dfrac{1}{2}$.

The total length of cable needed for the replacement is $172\frac{1}{2}$ feet.   ∎

When adding mixed numbers, the sum of the fractions sometimes yields an improper fraction, as in the next example.

EXAMPLE 4 *Vertical form.* Add: $45\frac{2}{3} + 96\frac{4}{5}$.

Solution

$$
\begin{array}{ccccc}
45\dfrac{2}{3} & = & 45\dfrac{2\cdot5}{3\cdot5} & = & 45\dfrac{10}{15} \\[2mm]
+96\dfrac{4}{5} & = & +96\dfrac{4\cdot3}{5\cdot3} & = & +\ 96\dfrac{12}{15} \\[2mm]
\hline
& & & & 141\dfrac{22}{15}
\end{array}
$$

The whole-number part of the answer.   The fractional part of the answer is an improper fraction.

Now write the improper fraction as a mixed number.

$$141\frac{22}{15} = 141 + \frac{22}{15} = 141 + 1\frac{7}{15} = 142\frac{7}{15}$$

Self Check

Add: $76\frac{11}{12} + 49\frac{5}{8}$.

Answer: $126\frac{13}{24}$

Subtracting mixed numbers

Subtracting mixed numbers is similar to adding mixed numbers.

EXAMPLE 5 *Cooking.* How much butter is left in a 10-pound tub if $2\frac{2}{3}$ pounds are used for a wedding cake?

Solution The phrase "How much is left?" suggests subtraction.

$$10 - 2\frac{2}{3} = \frac{10}{1} - \frac{8}{3} \qquad \text{Write 10 as a fraction: } 10 = \tfrac{10}{1}. \text{ Write } 2\tfrac{2}{3} \text{ as } \tfrac{8}{3}.$$

By inspection, we see that the LCD is 3.

$$\frac{10}{1} - 2\frac{2}{3} = \frac{10\cdot3}{1\cdot3} - \frac{8}{3} \qquad \text{Write the first fraction with a denominator of 3.}$$

$$= \frac{30}{3} - \frac{8}{3} \qquad \text{Do the multiplications in the first fraction.}$$

$$= \frac{30 - 8}{3} \qquad \text{Subtract the numerators and write the difference over the common denominator.}$$

$$= \frac{22}{3} \qquad \text{Do the subtraction: } 30 - 8 = 22.$$

$$= 7\frac{1}{3} \qquad \text{Write } \tfrac{22}{3} \text{ as a mixed number.}$$

There are $7\frac{1}{3}$ pounds of butter left in the tub.

In the next example, the fraction being subtracted *from* is smaller than the fraction being subtracted. Because of this, we will have to borrow.

EXAMPLE 6   *Borrowing.*   Subtract: $34\frac{1}{5} - 11\frac{2}{3}$.

Solution

We will use the vertical form to subtract. The LCD is 15, so we write each fraction as a fraction with a denominator of 15.

$$34\frac{1}{5} \quad = \quad 34\frac{1 \cdot 3}{5 \cdot 3} \quad = \quad 34\frac{3}{15}$$

$$-11\frac{2}{3} \quad = \quad -11\frac{2 \cdot 5}{3 \cdot 5} \quad = \quad -11\frac{10}{15}$$

Since $\frac{10}{15}$ is larger than $\frac{3}{15}$, borrow 1 $\left(\text{in the form of } \frac{15}{15}\right)$ from 34 and add it to $\frac{3}{15}$ to obtain $33\frac{3}{15} + \frac{15}{15} = 33\frac{18}{15}$. Then we subtract the fractions and the whole numbers separately.

$$33\frac{3}{15} + \frac{15}{15} \quad = \quad 33\frac{18}{15}$$

$$-11\frac{10}{15} \quad = \quad -11\frac{10}{15}$$

$$\overline{\qquad\qquad\qquad 22\frac{8}{15}}$$

Self Check

Subtract: $101\frac{3}{4} - 79\frac{15}{16}$.

Answer: $21\frac{13}{16}$

EXAMPLE 7   *Borrowing from a whole number.*   Subtract: $419 - 53\frac{11}{16}$.

Solution

We align the numbers vertically and borrow 1 $\left(\text{in the form of } \frac{16}{16}\right)$ from 419. Then we subtract the fractions and subtract the whole numbers separately.

$$419 \quad = \quad 418\frac{16}{16}$$

$$-53\frac{11}{16} \quad = \quad -53\frac{11}{16}$$

$$\overline{\qquad\qquad\qquad 365\frac{5}{16}}$$

Self Check

Subtract: $2{,}300 - 129\frac{19}{32}$.

Answer: $2{,}170\frac{13}{32}$

STUDY SET   Section 3.6

VOCABULARY   *Fill in the blanks.*

1. By the _____ property of addition, we can add numbers in any order.

2. A _____ number contains a whole-number part and a fractional part.

3. Consider

$$80\frac{1}{3} \quad = \quad 79\frac{1}{3} + \frac{3}{3}$$
$$-24\frac{2}{3} \quad = \quad -24\frac{2}{3}$$

To do the subtraction, we _____ 1 in the form of $\frac{3}{3}$.

4. Fractions that are greater than 1, such as $\frac{11}{8}$, are called _____ fractions.

CONCEPTS

5. a. For $76\frac{3}{4}$, list the whole-number part and the fractional part.
b. Write $76\frac{3}{4}$ as a sum.

6. Use the commutative property of addition to get the whole numbers together.

$$14 + \frac{5}{6} + 53 + \frac{1}{6}$$

7. What property is being highlighted here?

$$25\frac{3 \cdot 5}{4 \cdot 5}$$
$$+31\frac{1 \cdot 4}{5 \cdot 4}$$
$$\overline{\qquad\qquad}$$

8. a. The denominators of two fractions, expressed in prime-factored form, are $5 \cdot 2$ and $5 \cdot 3$. Find the LCD for the fractions.

b. The denominators for three fractions, in prime-factored form, are $3 \cdot 5$, $2 \cdot 3$, and $3 \cdot 3$. Find the LCD for the fractions.

9. Simplify.

a. $9\dfrac{17}{16}$

b. $1{,}288\dfrac{7}{3}$

c. $16\dfrac{12}{8}$

d. $45\dfrac{24}{20}$

10. Consider

$$108\dfrac{1}{4}$$
$$-\ 99\dfrac{2}{3}$$

a. Explain why we will have to borrow if we subtract the mixed numbers in this way.

b. In what form will we borrow a 1 from 108?

NOTATION *Complete each solution.*

11. Add: $70\dfrac{3}{5} + 39\dfrac{2}{7}$.

$$70\dfrac{3}{5} + 39\dfrac{2}{7} = \boxed{} + \dfrac{3}{5} + \boxed{} + \dfrac{2}{7}$$

$$= \boxed{} + \boxed{} + \dfrac{3}{5} + \dfrac{2}{7}$$

$$= 109 + \dfrac{3}{5} + \dfrac{2}{7}$$

$$= 109 + \dfrac{3 \cdot}{5 \cdot} + \dfrac{2 \cdot}{7 \cdot}$$

$$= 109 + \dfrac{21}{} + \dfrac{10}{}$$

$$= 109 + \dfrac{}{35}$$

$$= 109\dfrac{31}{35}$$

12. Subtract: $67\dfrac{3}{8} - 23\dfrac{2}{3}$.

$$67\dfrac{3}{8} = 67\dfrac{3\cdot}{8\cdot}$$
$$-23\dfrac{2}{3} = -23\dfrac{2\cdot}{3\cdot}$$

$$67\dfrac{9}{24} = \boxed{}\dfrac{9}{24} + \dfrac{}{} = 66\dfrac{}{24}$$
$$-23\dfrac{16}{24} = -23\dfrac{16}{24} \qquad = -23\dfrac{16}{24}$$
$$43\dfrac{17}{24}$$

PRACTICE *Find each sum or difference.*

13. $2\dfrac{1}{5} + 2\dfrac{1}{5}$

14. $3\dfrac{1}{3} + 2\dfrac{1}{3}$

15. $8\dfrac{2}{7} - 3\dfrac{1}{7}$

16. $9\dfrac{5}{11} - 6\dfrac{2}{11}$

17. $3\dfrac{1}{4} + 4\dfrac{1}{4}$

18. $2\dfrac{1}{8} + 3\dfrac{3}{8}$

19. $4\dfrac{1}{6} + 1\dfrac{1}{5}$

20. $2\dfrac{2}{5} + 3\dfrac{1}{4}$

21. $2\dfrac{1}{2} - 1\dfrac{1}{4}$

22. $13\dfrac{5}{6} - 4\dfrac{2}{3}$

23. $2\dfrac{5}{6} - 1\dfrac{3}{8}$

24. $4\dfrac{5}{9} - 2\dfrac{1}{6}$

25. $5\dfrac{1}{2} + 3\dfrac{4}{5}$

26. $6\dfrac{1}{2} + 2\dfrac{2}{3}$

27. $7\dfrac{1}{2} - 4\dfrac{1}{7}$

28. $5\dfrac{3}{4} - 1\dfrac{3}{7}$

29. $56\dfrac{2}{5} + 73\dfrac{1}{3}$

30. $44\dfrac{3}{8} + 66\dfrac{1}{5}$

31. $380\dfrac{1}{6} + 17\dfrac{1}{4}$

32. $103\dfrac{1}{2} + 210\dfrac{2}{5}$

33. $228\dfrac{5}{9} + 44\dfrac{2}{3}$

34. $161\dfrac{7}{8} + 19\dfrac{1}{3}$

35. $778\dfrac{5}{7} - 155\dfrac{1}{3}$

36. $339\dfrac{1}{2} - 218\dfrac{3}{16}$

37. $140\dfrac{5}{6} - 129\dfrac{4}{5}$

38. $291\dfrac{1}{4} - 289\dfrac{1}{12}$

39. $422\dfrac{13}{16} - 321\dfrac{3}{8}$

40. $378\dfrac{3}{4} - 277\dfrac{5}{8}$

Find each difference.

41. $16\frac{1}{4} - 13\frac{3}{4}$

42. $40\frac{1}{7} - 19\frac{6}{7}$

43. $76\frac{1}{6} - 49\frac{7}{8}$

44. $101\frac{1}{4} - 70\frac{1}{2}$

45. $140\frac{3}{16} - 129\frac{3}{4}$

46. $211\frac{1}{3} - 8\frac{3}{4}$

47. $334\frac{1}{9} - 13\frac{5}{6}$

48. $442\frac{1}{8} - 429\frac{2}{3}$

Find the sum or difference.

49. $7 - \frac{2}{3}$

50. $6 - \frac{1}{8}$

51. $9 - 8\frac{3}{4}$

52. $11 - 10\frac{4}{5}$

53. $4\frac{1}{7} - \frac{4}{5}$

54. $5\frac{1}{10} - \frac{4}{5}$

55. $6\frac{5}{8} - 3$

56. $10\frac{1}{2} - 6$

57. $\frac{7}{3} + 2$

58. $\frac{9}{7} + 3$

59. $2 + 1\frac{7}{8}$

60. $3\frac{3}{4} + 5$

Find each sum.

61. $12\frac{1}{2} + 5\frac{3}{4} + 35\frac{1}{6}$

62. $31\frac{1}{3} + 20\frac{2}{5} + 10\frac{1}{15}$

63. $58\frac{7}{8} + 340 + 61\frac{1}{4}$

64. $191 + 233\frac{1}{16} + 16\frac{5}{8}$

Find each sum or difference.

65. $-3\frac{3}{4} + \left(-1\frac{1}{2}\right)$

66. $-3\frac{2}{3} + \left(-1\frac{4}{5}\right)$

67. $-4\frac{5}{8} - 1\frac{1}{4}$

68. $-2\frac{1}{16} - 3\frac{7}{8}$

APPLICATIONS

69. FREEWAY TRAVEL A freeway exit sign is shown in Illustration 1. How far apart are the Citrus Ave. and Grand Ave. exits?

70. BASKETBALL See Illustration 2. What is the difference in height between the tallest and the shortest of the starting players?

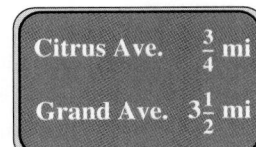

| Citrus Ave. | $\frac{3}{4}$ mi |
| Grand Ave. | $3\frac{1}{2}$ mi |

ILLUSTRATION 1

Heights of the Starting Five Players

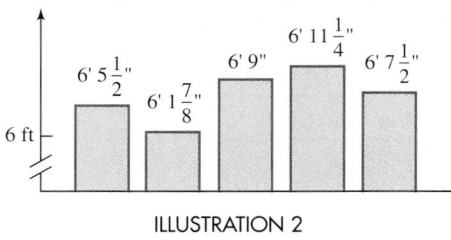

ILLUSTRATION 2

71. TRAIL MIX See Illustration 3. A camper doubles up on the amount of sunflower seeds called for in the recipe. How much trail mix will the adjusted recipe yield?

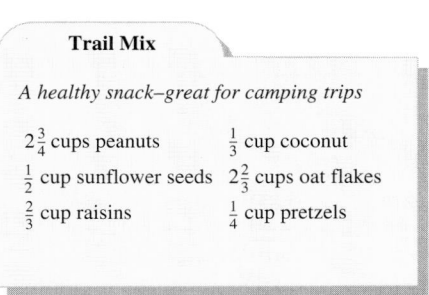

Trail Mix

A healthy snack–great for camping trips

$2\frac{3}{4}$ cups peanuts $\frac{1}{3}$ cup coconut

$\frac{1}{2}$ cup sunflower seeds $2\frac{2}{3}$ cups oat flakes

$\frac{2}{3}$ cup raisins $\frac{1}{4}$ cup pretzels

ILLUSTRATION 3

72. AIR TRAVEL A businesswoman's flight leaves Los Angeles at 8 A.M. and arrives in Seattle at 9:45 A.M.
 a. Express the duration of the flight as a mixed number.
 b. Upon arrival, she boards a commuter plane at 11:15 A.M., arriving at her final destination at 11:45 A.M. Express the length of this flight as a fraction.
 c. Find the total time of these two flights.

73. HOSE REPAIR To repair a bad connector, Ming Lin removes $1\frac{1}{2}$ feet from the end of a 50-foot garden hose. How long is the hose after the repair?

74. SEWING To make some draperies, Liz needs $12\frac{1}{4}$ yards of material for the den and $8\frac{1}{2}$ yards for the living room. If the material comes only in 21-yard bolts, how much will be left over after completing both sets of draperies?

75. SHIPPING A passenger ship and a cargo ship leave San Diego harbor at midnight. During the first hour, the passenger ship travels south at $16\frac{1}{2}$ miles per hour, while the cargo ship is traveling north at a rate of $5\frac{1}{5}$ miles per hour.
 a. Complete the chart in Illustration 4.
 b. How far apart are they at 1:00 A.M.?

	Rate (mph)	Time traveling (hr)	Distance traveled (mi)
Passenger ship		1	
Cargo ship		1	

ILLUSTRATION 4

76. HARDWARE See Illustration 5. To secure the bracket to the stock, a bolt and a nut are used. How long should the bolt be?

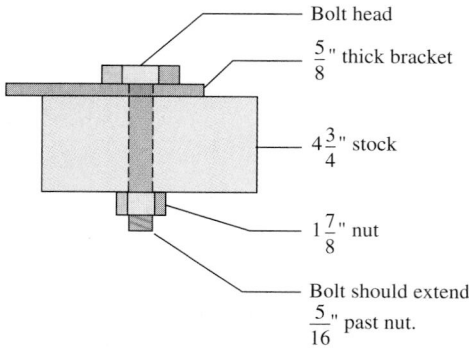

- Bolt head
- $\frac{5}{8}$" thick bracket
- $4\frac{3}{4}$" stock
- $1\frac{7}{8}$" nut
- Bolt should extend $\frac{5}{16}$" past nut.

ILLUSTRATION 5

77. SERVICE STATION Use the service station sign in Illustration 6 to answer the following questions.
- **a.** What is the difference in price between the least and most expensive types of gasoline at the self-service pump?
- **b.** How much more is the cost per gallon for full service?

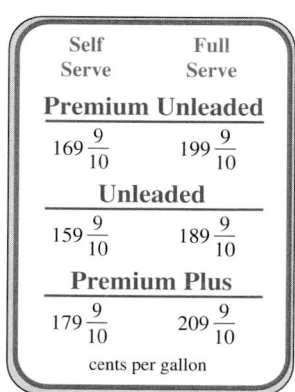

Self Serve	Full Serve
Premium Unleaded	
$169\frac{9}{10}$	$199\frac{9}{10}$
Unleaded	
$159\frac{9}{10}$	$189\frac{9}{10}$
Premium Plus	
$179\frac{9}{10}$	$209\frac{9}{10}$
cents per gallon	

ILLUSTRATION 6

78. SEPTUPLETS On November 19, 1997, at Iowa Methodist Medical Center, Bobbie McCaughey gave birth to seven babies. From the information in Illustration 7, find the combined birthweights of the babies.

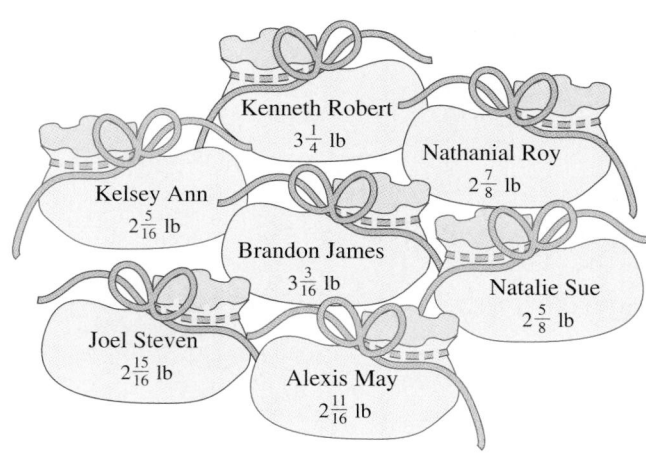

Kenneth Robert $3\frac{1}{4}$ lb
Nathanial Roy $2\frac{7}{8}$ lb
Kelsey Ann $2\frac{5}{16}$ lb
Brandon James $3\frac{3}{16}$ lb
Natalie Sue $2\frac{5}{8}$ lb
Joel Steven $2\frac{15}{16}$ lb
Alexis May $2\frac{11}{16}$ lb

ILLUSTRATION 7

79. WATER SLIDE An amusement park added a new section to a water slide to create a slide $311\frac{5}{12}$ feet long. (See Illustration 8.) How long was the slide before the addition?

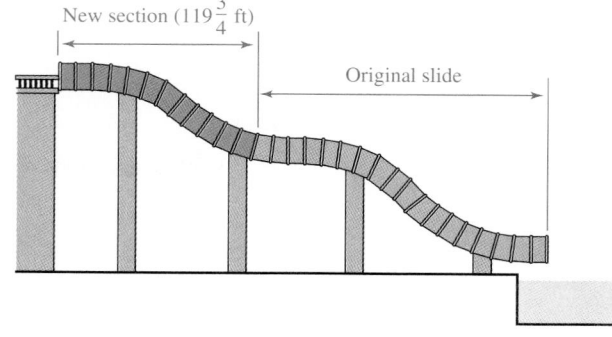

New section ($119\frac{3}{4}$ ft)
Original slide

ILLUSTRATION 8

80. JEWELRY A jeweler is to cut a 7-inch-long gold braid into three pieces. He aligns a 6-inch-long ruler directly below the braid and makes the proper cuts. (See Illustration 9.) Find the length of piece 2 of the braid.

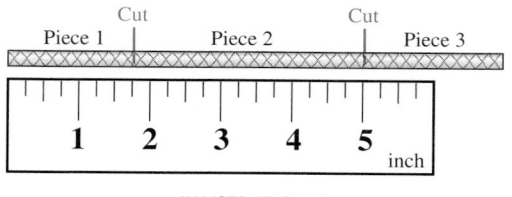

Cut | Cut
Piece 1 | Piece 2 | Piece 3
1 2 3 4 5 inch

ILLUSTRATION 9

WRITING

81. Of the methods studied to add mixed numbers, which do you like better, and why?

82. When subtracting mixed numbers, when is borrowing necessary? How is it done?

83. Explain how to add $1\frac{3}{8}$ and $2\frac{1}{4}$ if we write each one as an improper fraction.

84. Explain the process of simplifying $12\frac{16}{5}$.

REVIEW

85. Solve $2x - 1 = 13$.

86. Multiply: $-3(-4)(-5)$.

87. Find: $-2 - (-8)$.

88. Evaluate: $(-2)^3(3)^2$.

89. What does area measure?

90. Find: $|-12|$.

3.7 *Order of Operations and Complex Fractions*

In this section, you will learn about

- Order of operations • Complex fractions • Simplifying complex fractions

INTRODUCTION. In this section, we will evaluate expressions involving fractions and mixed numbers. We will also discuss complex fractions how to simplify them.

Order of operations

The rules for the order of operations are used to evaluate numerical expressions that involve more than one operation.

EXAMPLE 1   *Order of operations.*   Evaluate $\dfrac{3}{4} + \dfrac{5}{3}\left(-\dfrac{1}{2}\right)^3$.

Solution

The expression involves the operations of raising to a power, multiplication, and addition. By the rules for the order of operations, we must evaluate the power first, the multiplication second, and the addition last.

$$\frac{3}{4} + \frac{5}{3}\left(-\frac{1}{2}\right)^3 = \frac{3}{4} + \frac{5}{3}\left(-\frac{1}{8}\right)$$ Evaluate the power: $\left(-\frac{1}{2}\right)^3 = -\frac{1}{8}$.

$$= \frac{3}{4} + \left(-\frac{5}{24}\right)$$ Do the multiplication: $\frac{5}{3}\left(-\frac{1}{8}\right) = -\frac{5}{24}$.

$$= \frac{3 \cdot 6}{4 \cdot 6} + \left(-\frac{5}{24}\right)$$ The LCD is 24. Write the first fraction as a fraction with denominator 24.

$$= \frac{18}{24} + \left(-\frac{5}{24}\right)$$ Multiply in the numerator: $3 \cdot 6 = 18$.
Multiply in the denominator: $4 \cdot 6 = 24$.

$$= \frac{13}{24}$$ Add the numerators: $18 + (-5) = 13$. Write the sum over the common demominator.

Self Check

Evaluate $\dfrac{7}{8} + \dfrac{3}{2}\left(-\dfrac{1}{4}\right)^2$.

Answer: $\dfrac{31}{32}$

If an expression contains grouping symbols, we do the operations within the grouping symbols first.

EXAMPLE 2 *Order of operations.* Evaluate $\left(\dfrac{7}{8} - \dfrac{1}{4}\right) \div \left(-2\dfrac{3}{16}\right)$.

Solution $\left(\dfrac{7}{8} - \dfrac{1}{4}\right) \div \left(-2\dfrac{3}{16}\right) = \left(\dfrac{7}{8} - \dfrac{1 \cdot 2}{4 \cdot 2}\right) \div \left(-2\dfrac{3}{16}\right)$ Within the first set of parentheses, write $\frac{1}{4}$ as a fraction with denominator 8.

$= \left(\dfrac{7}{8} - \dfrac{2}{8}\right) \div \left(-2\dfrac{3}{16}\right)$ Multiply in the numerator: $1 \cdot 2 = 2$. Multiply in the denominator: $4 \cdot 2 = 8$.

$= \dfrac{5}{8} \div \left(-2\dfrac{3}{16}\right)$ Subtract the numerators and write the difference over the common denominator: $7 - 2 = 5$.

$= \dfrac{5}{8} \div \left(-\dfrac{35}{16}\right)$ Write the mixed number as an improper fraction.

$= \dfrac{5}{8}\left(-\dfrac{16}{35}\right)$ Multiply by the reciprocal of $-\frac{35}{16}$.

$= -\dfrac{5 \cdot 16}{8 \cdot 35}$ The product of two fractions with unlike signs is negative. Multiply the numerators and multiply the denominators.

$= -\dfrac{\overset{1}{\cancel{5}} \cdot 2 \cdot \overset{1}{\cancel{8}}}{\underset{1}{\cancel{8}} \cdot \underset{1}{\cancel{5}} \cdot 7}$ Factor 16 as $2 \cdot 8$ and factor 35 as $5 \cdot 7$. Divide out the common factors of 8 and 5.

$= -\dfrac{2}{7}$ Multiply in the numerator: $1 \cdot 2 \cdot 1 = 2$. Multiply in the denominator: $1 \cdot 1 \cdot 7 = 7$. ∎

EXAMPLE 3 *Masonry.* To build a wall, a mason will use blocks that are $5\frac{3}{4}$ inches high, held together with $\frac{3}{8}$-inch-thick layers of mortar. (See Figure 3-10.) If the plans call for 8 layers of blocks, what will be the height of the wall when completed?

Solution To find the height, we must consider 8 layers of blocks and 8 layers of mortar. We will compute the height contributed by one block and one layer of mortar and then multiply that result by 8.

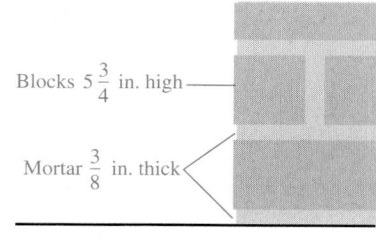

Blocks $5\frac{3}{4}$ in. high

Mortar $\frac{3}{8}$ in. thick

FIGURE 3-10

8 times	$\left($	height of 1 block	plus	height of 1 layer of mortar	$\right)$	equals	height of block wall.
8	$\left($	$5\dfrac{3}{4}$	$+$	$\dfrac{3}{8}$	$\right)$	$=$	height of wall

$8\left(5\dfrac{3}{4} + \dfrac{3}{8}\right) = 8\left(\dfrac{23}{4} + \dfrac{3}{8}\right)$ Write $5\frac{3}{4}$ as the improper fraction $\frac{23}{4}$.

$= 8\left(\dfrac{23 \cdot 2}{4 \cdot 2} + \dfrac{3}{8}\right)$ Express $\frac{23}{4}$ in terms of 8ths.

$= 8\left(\dfrac{46}{8} + \dfrac{3}{8}\right)$ Do the multiplication in the numerator and denominator.

$= \dfrac{8}{1}\left(\dfrac{49}{8}\right)$ Write 8 as $\frac{8}{1}$. Within the parentheses, write the sum of the numerators over the common denominator 8.

$$= \frac{\frac{1}{\cancel{8} \cdot 49}}{\underset{1}{\cancel{8}}}$$ Multiply the numerators and the denominators. Divide out the common factor of 8.

$$= 49$$ Simplify: $\frac{49}{1} = 49$.

The wall will be 49 inches high.

Complex fractions

Fractions whose numerators and/or denominators contain fractions are called *complex fractions*. Here is an example.

A fraction in the numerator $\longrightarrow$ $\dfrac{\dfrac{3}{4}}{\dfrac{7}{8}}$ $\longleftarrow$ The main fraction bar
A fraction in the denominator $\longrightarrow$

Complex fraction

A **complex fraction** is a fraction whose numerator or denominator, or both, contain one or more fractions or mixed numbers.

Here are more examples of complex fractions.

$$\frac{-\dfrac{1}{4} - \dfrac{4}{5}}{2\dfrac{4}{5}}$$

$\longleftarrow$ Numerator $\longrightarrow$
$\longleftarrow$ Main fraction bar $\longrightarrow$
$\longleftarrow$ Denominator $\longrightarrow$

$$\frac{\dfrac{1}{3} + \dfrac{1}{4}}{\dfrac{1}{3} - \dfrac{1}{4}}$$

Simplifying complex fractions

To *simplify* complex fractions means to express them as fractions in simplified form.

Simplifying a complex fraction

Write the numerator and the denominator of the complex fraction as single fractions. Then do the indicated division of the two fractions and simplify.

This procedure is based on the fact that the main fraction bar of the complex fraction indicates division.

$$\frac{\dfrac{1}{4}}{\dfrac{2}{5}}$$ $\longleftarrow$ The main fraction bar means "divide the fraction in the numerator by the fraction in the denominator." $\longrightarrow$ $\dfrac{1}{4} \div \dfrac{2}{5}$

EXAMPLE 4   *Simplifying a complex fraction.*   Simplify: $\dfrac{\dfrac{1}{4}}{\dfrac{2}{5}}$.

Self Check

Simplify $\dfrac{\dfrac{1}{6}}{\dfrac{3}{8}}$.

Solution

Since the numerator and the denominator of this complex fraction are single fractions, we can do the indicated division.

$$\frac{\dfrac{1}{4}}{\dfrac{2}{5}} = \frac{1}{4} \div \frac{2}{5}$$ Express the complex fraction as an equivalent division problem.

$$= \frac{1}{4} \cdot \frac{5}{2}$$ Multiply by the reciprocal of $\frac{2}{5}$.

$$= \frac{1 \cdot 5}{4 \cdot 2}$$ Multiply the numerators and multiply the denominators.

$$= \frac{5}{8}$$

Answer: $\dfrac{4}{9}$

■

 EXAMPLE 5   *Simplifying a complex fraction.*   Simplify $\dfrac{-\dfrac{1}{4} + \dfrac{2}{5}}{\dfrac{1}{2} - \dfrac{4}{5}}$.

Solution   We need to write the numerator and the denominator of the complex fraction as single fractions. To do this, we find $-\frac{1}{4} + \frac{2}{5}$ and $\frac{1}{2} - \frac{4}{5}$.

$$\frac{-\dfrac{1}{4} + \dfrac{2}{5}}{\dfrac{1}{2} - \dfrac{4}{5}} = \frac{-\dfrac{1 \cdot 5}{4 \cdot 5} + \dfrac{2 \cdot 4}{5 \cdot 4}}{\dfrac{1 \cdot 5}{2 \cdot 5} - \dfrac{4 \cdot 2}{5 \cdot 2}}$$ In the numerator of the complex fraction, express the two fractions in terms of their LCD, which is 20. In the denominator, express the two fractions in terms of their LCD, which is 10.

$$= \frac{-\dfrac{5}{20} + \dfrac{8}{20}}{\dfrac{5}{10} - \dfrac{8}{10}}$$ Multiply in the numerators and the denominators.

$$= \frac{\dfrac{3}{20}}{-\dfrac{3}{10}}$$ In the numerator of the complex fraction, add the fractions. In the denominator, subtract the fractions.

$$= \frac{3}{20} \div \left(-\frac{3}{10}\right)$$ Express the complex fraction as an equivalent division problem.

$$= \frac{3}{20}\left(-\frac{10}{3}\right)$$ Multiply by the reciprocal of $-\frac{3}{10}$.

$$= -\frac{3 \cdot 10}{20 \cdot 3}$$ The product of two fractions with unlike signs is negative. Multiply the numerators and multiply the denominators.

$$= -\frac{\overset{1}{\cancel{3}} \cdot \overset{1}{\cancel{10}}}{\underset{1}{\cancel{10}} \cdot 2 \cdot \underset{1}{\cancel{3}}}$$ Factor 20 as $10 \cdot 2$. Divide out the common factors, 3 and 10.

$$= -\frac{1}{2}$$ Simplify.

■

EXAMPLE 6   *Simplifying a complex fraction.*   Simplify $\dfrac{7 - \dfrac{2}{3}}{4\dfrac{5}{6}}$.

Self Check

Simplify $\dfrac{5 - \dfrac{3}{4}}{1\dfrac{7}{8}}$.

Solution

To write the numerator of the complex fraction as a single fraction, we need to find $7 - \dfrac{2}{3}$. To write the denominator of the complex fraction as a single fraction, we need to write $4\frac{5}{6}$ as an improper fraction.

$$\dfrac{7 - \dfrac{2}{3}}{4\dfrac{5}{6}} = \dfrac{\dfrac{7 \cdot 3}{1 \cdot 3} - \dfrac{2}{3}}{\dfrac{29}{6}}$$

In the numerator of the complex fraction, write 7 as $\frac{7}{1}$. Then express $\frac{7}{1}$ in terms of 3rds. In the denominator, write $4\frac{5}{6}$ as the improper fraction $\frac{29}{6}$.

$$= \dfrac{\dfrac{21}{3} - \dfrac{2}{3}}{\dfrac{29}{6}}$$

Multiply in the numerator and the denominator.

$$= \dfrac{\dfrac{19}{3}}{\dfrac{29}{6}}$$

In the numerator of the complex fraction, subtract the fractions.

$$= \dfrac{19}{3} \div \dfrac{29}{6}$$

Express the complex fraction as an equivalent division problem.

$$= \dfrac{19}{3} \cdot \dfrac{6}{29}$$

Multiply by the reciprocal of $\frac{29}{6}$.

$$= \dfrac{19 \cdot 6}{3 \cdot 29}$$

Multiply the numerators.
Multiply the denominators.

$$= \dfrac{19 \cdot \overset{1}{\cancel{3}} \cdot 2}{\underset{1}{\cancel{3}} \cdot 29}$$

Factor 6 as $3 \cdot 2$. Divide out the common factor of 3.

$$= \dfrac{38}{29}$$

Simplify.

$$= 1\dfrac{9}{29}$$

Write $\frac{38}{29}$ as a mixed number.

Answer: $2\dfrac{4}{15}$

STUDY SET Section 3.7

VOCABULARY *Fill in the blanks.*

1. $\dfrac{\dfrac{1}{2}}{\dfrac{3}{4}}$ is a _____ fraction.

2. In a fraction, the number above the fraction bar is called the _____, and the number below is the _____.

CONCEPTS

3. What division is represented by this complex fraction? $\dfrac{\dfrac{2}{3}}{\dfrac{1}{5}}$

4. Write this division as a complex fraction.

$$-\dfrac{7}{8} \div \dfrac{3}{4}$$

5. What is the LCD for the fractions in the numerator of this complex fraction? $\dfrac{\dfrac{2}{3} - \dfrac{1}{5}}{\dfrac{1}{2} + \dfrac{4}{5}}$

6. Write the denominator of this complex fraction as an improper fraction.

$$\dfrac{\dfrac{1}{8} - \dfrac{3}{16}}{5\dfrac{3}{4}}$$

7. When this complex fraction is simplified, will the result be positive or negative?

$$\dfrac{-\dfrac{2}{3}}{\dfrac{3}{4}}$$

8. To evaluate $\frac{7}{8} + \left(\frac{1}{3}\right)\left(\frac{1}{4}\right)$, what operation should be performed first?

9. To evaluate $\frac{7}{8} + \left(\frac{1}{3} - \frac{1}{4}\right)^2$, what operation should be performed first?

10. What operations are involved in this numerical expression?

$$5\left(6\frac{1}{3}\right) + \left(-\frac{1}{4}\right)^2$$

NOTATION *Complete each solution.*

11. Simplify $\dfrac{\dfrac{1}{8}}{\dfrac{3}{4}}$.

$$\dfrac{\dfrac{1}{8}}{\dfrac{3}{4}} = \frac{1}{8} \div \boxed{}$$

$$= \frac{1}{8} \cdot \boxed{}$$

$$= \frac{1 \cdot \boxed{}}{8 \cdot 3}$$

$$= \frac{1 \cdot \overset{1}{\cancel{4}}}{2 \cdot \cancel{} \cdot 3}$$

$$= \frac{1}{6}$$

12. Evaluate $\dfrac{1}{12} - \left(\dfrac{1}{2}\right)\left(\dfrac{1}{3}\right)$.

$$\frac{1}{12} - \left(\frac{1}{2}\right)\left(\frac{1}{3}\right) = \frac{1}{12} - \frac{1 \cdot 1}{2 \cdot \boxed{}}$$

$$= \frac{1}{12} - \frac{1}{\boxed{}}$$

$$= \frac{1}{12} - \frac{1 \cdot \boxed{}}{6 \cdot \boxed{}}$$

$$= \frac{1}{12} - \frac{\boxed{}}{12}$$

$$= -\frac{1}{12}$$

PRACTICE *Evaluate each expression.*

13. $\frac{2}{3}\left(-\frac{1}{4}\right) + \frac{1}{2}$

14. $-\frac{7}{8} - \left(\frac{1}{8}\right)\left(\frac{2}{3}\right)$

15. $\frac{4}{5} - \left(-\frac{1}{3}\right)^2$

16. $-\frac{3}{16} - \left(-\frac{1}{2}\right)^3$

17. $-4\left(-\frac{1}{5}\right) - \left(\frac{1}{4}\right)\left(-\frac{1}{2}\right)$

18. $(-3)\left(-\frac{2}{3}\right) - (-4)\left(-\frac{3}{4}\right)$

19. $1\frac{3}{5}\left(\frac{1}{2}\right)^2\left(\frac{3}{4}\right)$

20. $2\frac{3}{5}\left(-\frac{1}{3}\right)^2\left(\frac{1}{2}\right)$

21. $\frac{7}{8} - \left(\frac{4}{5} + 1\frac{3}{4}\right)$

22. $\left(\frac{5}{4}\right)^2 + \left(\frac{2}{3} - 2\frac{1}{6}\right)$

23. $\left(\frac{9}{20} \div 2\frac{2}{5}\right) + \left(\frac{3}{4}\right)^2$

24. $\left(1\frac{2}{3} \cdot 15\right) + \left(\frac{7}{9} \div \frac{7}{81}\right)$

25. $\left(-\frac{3}{4} \cdot \frac{9}{16}\right) + \left(\frac{1}{2} - \frac{1}{8}\right)$

26. $\left(\frac{8}{5} - 1\frac{1}{3}\right) - \left(-\frac{4}{5} \cdot 10\right)$

27. $\left|\frac{2}{3} - \frac{9}{10}\right| \div \left(-\frac{1}{5}\right)$

28. $\left|-\frac{3}{16} \div 2\frac{1}{4}\right| + \left(-2\frac{1}{8}\right)$

29. $\left(2 - \frac{1}{2}\right)^2 + \left(2 + \frac{1}{2}\right)^2$

30. $\left(1 - \frac{3}{4}\right)\left(1 + \frac{3}{4}\right)$

Find $\frac{1}{2}$ of the given number and then square that result. Express your answer as an improper fraction.

31. -7

32. -5

33. $\frac{11}{2}$

34. $\frac{7}{3}$

Find the perimeter of each figure.

35.

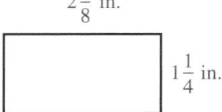

$2\frac{7}{8}$ in.

$1\frac{1}{4}$ in.

36.

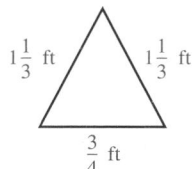

$1\frac{1}{3}$ ft $1\frac{1}{3}$ ft

$\frac{3}{4}$ ft

Simplify each complex fraction.

37. $\dfrac{\dfrac{2}{3}}{\dfrac{4}{5}}$

38. $\dfrac{\dfrac{3}{5}}{\dfrac{9}{25}}$

39. $\dfrac{-\dfrac{14}{15}}{\dfrac{7}{10}}$

40. $\dfrac{\dfrac{5}{27}}{-\dfrac{5}{9}}$

41. $\dfrac{\dfrac{5}{10}}{\dfrac{10}{21}}$

42. $\dfrac{\dfrac{6}{3}}{\dfrac{3}{8}}$

43. $\dfrac{-\dfrac{5}{6}}{-1\dfrac{7}{8}}$

44. $\dfrac{-\dfrac{4}{3}}{-2\dfrac{5}{6}}$

45. $\dfrac{\dfrac{1}{2}+\dfrac{1}{4}}{\dfrac{1}{2}-\dfrac{1}{4}}$

46. $\dfrac{\dfrac{1}{3}+\dfrac{1}{4}}{\dfrac{1}{3}-\dfrac{1}{4}}$

47. $\dfrac{\dfrac{3}{8}+\dfrac{1}{4}}{\dfrac{3}{8}-\dfrac{1}{4}}$

48. $\dfrac{\dfrac{2}{5}+\dfrac{1}{4}}{\dfrac{2}{5}-\dfrac{1}{4}}$

49. $\dfrac{\dfrac{1}{5}+3}{-\dfrac{4}{25}}$

50. $\dfrac{-5-\dfrac{1}{3}}{\dfrac{1}{6}+\dfrac{2}{3}}$

51. $\dfrac{5\dfrac{1}{2}}{-\dfrac{1}{4}+\dfrac{3}{4}}$

52. $\dfrac{4\dfrac{1}{4}}{\dfrac{2}{3}+\left(-\dfrac{1}{6}\right)}$

53. $\dfrac{\dfrac{1}{5}-\left(-\dfrac{1}{4}\right)}{\dfrac{1}{4}+\dfrac{4}{5}}$

54. $\dfrac{\dfrac{1}{8}-\left(-\dfrac{1}{2}\right)}{\dfrac{1}{4}+\dfrac{3}{8}}$

55. $\dfrac{\dfrac{1}{3}+\left(-\dfrac{5}{6}\right)}{1\dfrac{1}{3}}$

56. $\dfrac{\dfrac{3}{7}+\left(-\dfrac{1}{2}\right)}{1\dfrac{3}{4}}$

APPLICATIONS

57. SANDWICH SHOP   A sandwich shop sells a $\frac{1}{2}$-pound club sandwich, made up of turkey meat and ham. The owner buys the turkey in $1\frac{3}{4}$-pound packages and the ham in $2\frac{1}{2}$-pound packages. If he mixes a package of each of the meats together, how many sandwiches can he make from the mixture?

58. SKIN CREAM   Using a formula of $\frac{1}{2}$ ounce of sun block, $\frac{2}{3}$ ounce of moisturizing cream, and $\frac{3}{4}$ ounce of lanolin, a beautician mixes her own brand of skin cream. She packages it in $\frac{1}{4}$-ounce tubes. How many tubes can be produced using this formula?

59. PHYSICAL FITNESS   Two people begin their workouts from the same point on a bike path and travel in opposite directions, as shown in Illustration 1. How far apart are they in $1\frac{1}{2}$ hours? Use the chart to help organize your work.

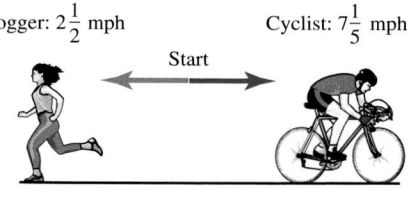

Jogger: $2\frac{1}{2}$ mph   Cyclist: $7\frac{1}{5}$ mph

ILLUSTRATION 1

	Rate (mph)	Time (hr)	Distance (mi)
Jogger			
Cyclist			

60. SLEEP   Illustration 2 compares the amount of sleep a 1-month-old baby got to the $15\frac{1}{2}$-hour daily requirement recommended by Children's Hospital of Orange County, California. For the week, how far below the baseline was the baby's daily average? (*Hint:* To find the average, add the numbers and divide by 7.)

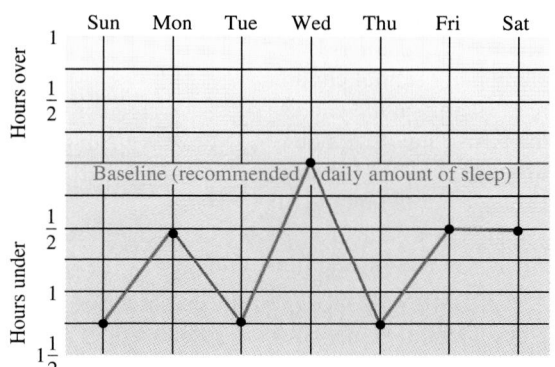

ILLUSTRATION 2

61. POSTAGE RATES   Can the advertising package in Illustration 3 be mailed for the one-ounce rate?

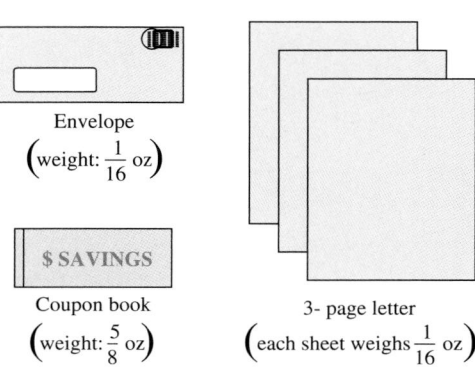

Envelope $\left(\text{weight: }\dfrac{1}{16}\text{ oz}\right)$

$ SAVINGS

Coupon book $\left(\text{weight: }\dfrac{5}{8}\text{ oz}\right)$

3- page letter $\left(\text{each sheet weighs }\dfrac{1}{16}\text{ oz}\right)$

ILLUSTRATION 3

62. PLYWOOD To manufacture a sheet of plywood, several layers of thin laminate are glued together, as shown in Illustration 4. Then an exterior finish is affixed to the top and bottom. How thick is the finished product?

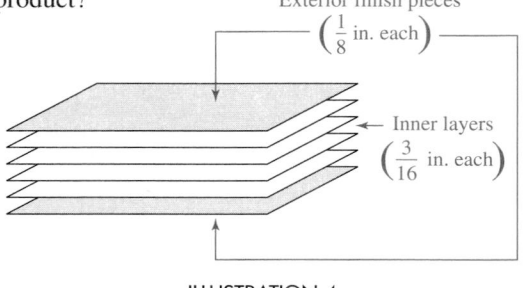

ILLUSTRATION 4

63. PHYSICAL THERAPY After back surgery, a patient undertook a walking program to rehabilitate her back muscles, as specified in Illustration 5. What was the total distance she walked over this three-week period?

Week	Distance per day
#1	$\frac{1}{4}$ mile
#2	$\frac{1}{2}$ mile
#3	$\frac{3}{4}$ mile

ILLUSTRATION 5

64. READING PROGRAM To improve reading skills, elementary school children read silently at the end of the school day for $\frac{1}{4}$ hour on Mondays and for $\frac{1}{2}$ hour on Fridays. For the month of January, how many total hours did the children read silently in class? (See Illustration 6.)

S	M	T	W	T	F	S
	1	2	3	4	5	6
7	8	9	10	11	12	13
14	15	16	17	18	19	20
21	22	23	24	25	26	27
28	29	30	31			

ILLUSTRATION 6

65. AMUSEMENT PARK At the end of a ride at an amusement park, a boat splashes into a pool of water. The time (in seconds) that it takes two pipes to refill the pool is given by

$$\frac{1}{\frac{1}{10} + \frac{1}{15}}$$

Find this time.

66. HIKING A scout troop plans to hike from the campground to Glenn Peak. (See Illustration 7.) Since the terrain is steep, they plan to stop and rest after every $\frac{2}{3}$ mile. With this plan, how many parts will there be to this hike?

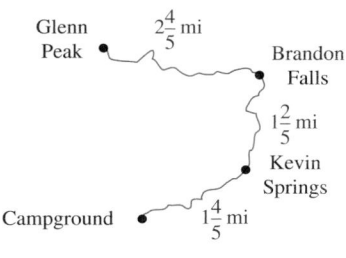

ILLUSTRATION 7

WRITING

67. What is a complex fraction?

68. Explain the method used to simplify complex fractions.

69. Write an application problem using a complex fraction, and then solve it.

70. List several ways to write "one-half divided by two-thirds."

REVIEW

71. Evaluate $\dfrac{2(15) + 6}{2 \cdot 3^2}$.

72. Find the area of a rectangle that is 6 feet by 12 feet.

73. Solve $-5x + 3 = 28$.

74. List the factors of 24.

75. Evaluate $2 + 3[-3 - (-4 - 1)]$.

76. What is the sign of the quotient of two numbers that have unlike signs?

3.8 *Solving Equations Containing Fractions*

In this section, you will learn about

- Using reciprocals to solve equations • An alternate method
- The addition and subtraction properties of equality • Applications

INTRODUCTION. In this section, we will discuss how to solve equations containing fractions, and equations whose solutions are fractions. We will make use of the properties of equality and several concepts from this chapter, including the reciprocal and the LCD.

Using reciprocals to solve equations

In the equation $\frac{3}{4}x = 5$, the variable is multiplied by $\frac{3}{4}$. To undo this multiplication and isolate the variable, we can use the multiplication property of equality and multiply both sides of the equation by the reciprocal of $\frac{3}{4}$.

$$\frac{4}{3}\left(\frac{3}{4}x\right) = \frac{4}{3}(5)$$ Multiply both sides by the reciprocal of $\frac{3}{4}$, which is $\frac{4}{3}$.

$$\left(\frac{4}{3} \cdot \frac{3}{4}\right)x = \frac{4}{3} \cdot \frac{5}{1}$$ Use the associative property of multiplication to regroup the factors. Write 5 as $\frac{5}{1}$.

$$\frac{\overset{1}{\cancel{4}} \cdot \overset{1}{\cancel{3}}}{\underset{1}{\cancel{3}} \cdot \underset{1}{\cancel{4}}}x = \frac{4 \cdot 5}{3 \cdot 1}$$ Multiply the numerators and multiply the denominators. On the left, divide out the common factors of 4 and 3.

$$1x = \frac{20}{3}$$ Multiply in the numerators and in the denominators.

$$x = \frac{20}{3}$$ Simplify: $1x = x$.

In algebra, we usually leave a solution to an equation as an improper fraction rather than converting it to a mixed number.

 COMMENT   We can write expressions such as $\frac{4a}{5}$ and $\frac{-9h}{16}$ in an equivalent form so that the fractional coefficients are more evident.

$$\frac{4a}{5} = \frac{4}{5}a \qquad \frac{-9h}{16} = -\frac{9}{16}h$$

EXAMPLE 1   *Using reciprocals.*   Solve $-\frac{7}{8}k = 21$.

Solution

The coefficient of the variable is $-\frac{7}{8}$. To isolate k, we multiply both sides of the equation by the reciprocal of $-\frac{7}{8}$.

$$-\frac{7}{8}k = 21$$

$$-\frac{8}{7}\left(-\frac{7}{8}k\right) = -\frac{8}{7}(21)$$ Multiply both sides by the reciprocal of $-\frac{7}{8}$, which is $-\frac{8}{7}$.

$$1k = -\frac{8}{7}(21)$$ The product of a number and its reciprocal is 1: $-\frac{8}{7}(-\frac{7}{8}) = 1$.

$$k = -\frac{8}{7} \cdot \frac{21}{1}$$ Simplify: $1k = k$. Write 21 as $\frac{21}{1}$.

$$k = -\frac{8 \cdot 3 \cdot \overset{1}{\cancel{7}}}{\underset{1}{\cancel{7}} \cdot 1}$$ The product of two numbers with unlike signs is negative. Multiply the numerators and the denominators. Prime factor 21 and then divide out the common factor of 7.

$$k = -24$$ Multiply in the numerator and the denominator.

Check the result by substituting -24 for k in the original equation.

Self Check

Solve $-\frac{3}{4}t = 15$ and check the result.

Answer: -20

An alternate method

Another method of solving equations such as $\frac{3}{4}x = 5$ uses two steps to isolate the variable. In this method, we consider the variable to be multiplied by 3 and divided by 4. Then, in reverse order, we undo these operations.

$$4\left(\frac{3}{4}x\right) = 4(5) \qquad \text{To undo the division by 4, multiply both sides by 4.}$$

$$\left(4 \cdot \frac{3}{4}\right)x = 4(5) \qquad \text{Use the associative property to regroup the factors.}$$

$$\left(\frac{\overset{1}{4} \cdot 3}{1 \cdot \underset{1}{4}}\right)x = 4(5) \qquad \begin{array}{l}\text{Write 4 as } \frac{4}{1}\text{, multiply the numerators and the denominators, and} \\ \text{divide out the common factor of 4.}\end{array}$$

$$3x = 20 \qquad \text{Multiply in the numerator and the denominator.}$$

$$\frac{3x}{3} = \frac{20}{3} \qquad \text{To undo the multiplication by 3, divide both sides by 3.}$$

$$x = \frac{20}{3}$$

EXAMPLE 2   *The two-step method.*   Solve $\frac{3}{5}h = -9$.

Solution

$$\frac{3}{5}h = -9$$

$$5\left(\frac{3}{5}h\right) = 5(-9) \qquad \text{To undo the division by 5, multiply both sides by 5.}$$

$$3h = -45 \qquad \text{Do the multiplications.}$$

$$\frac{3h}{3} = \frac{-45}{3} \qquad \text{To undo the multiplication by 3, divide both sides by 3.}$$

$$h = -15 \qquad \text{Do the divisions.}$$

Self Check

Solve $\frac{5}{9}t = -10$ using the two-step method.

Answer: -18 ■

The addition and subtraction properties of equality

The addition property of equality enables us to add the same number to both sides of an equation and obtain an equivalent equation. In the next example, we will use this property to help solve an equation that contains fractions.

EXAMPLE 3   *The addition property of equality.*   Solve $y - \frac{15}{32} = \frac{1}{32}$.

Solution

To isolate y on the left-hand side, we need to undo the subtraction of $\frac{15}{32}$.

$$y - \frac{15}{32} = \frac{1}{32}$$

$$y - \frac{15}{32} + \frac{15}{32} = \frac{1}{32} + \frac{15}{32} \qquad \text{Add } \frac{15}{32} \text{ to both sides.}$$

$$y = \frac{16}{32} \qquad \text{Simplify: } -\frac{15}{32} + \frac{15}{32} = 0 \text{ and } \frac{1}{32} + \frac{15}{32} = \frac{16}{32}.$$

Self Check

Solve $\frac{11}{16} = a - \frac{1}{16}$.

$$y = \frac{1}{2}$$ Simplify the fraction: $\dfrac{16}{32} = \dfrac{\overset{1}{\cancel{16}} \cdot 1}{\underset{1}{\cancel{16}} \cdot 2} = \dfrac{1}{2}$.

Answer: $\dfrac{3}{4}$ ∎

EXAMPLE 4 *The subtraction property of equality.* Solve $x + \dfrac{1}{6} = \dfrac{3}{4}$.

Self Check

Solve $y + \dfrac{1}{5} = \dfrac{2}{3}$.

Solution

In this equation, $\frac{1}{6}$ is added to x. We undo this operation by subtracting $\frac{1}{6}$ from both sides.

$$x + \frac{1}{6} = \frac{3}{4}$$

$$x + \frac{1}{6} - \frac{1}{6} = \frac{3}{4} - \frac{1}{6}$$ Subtract $\frac{1}{6}$ from both sides.

$$x = \frac{3}{4} - \frac{1}{6}$$ Do the subtraction: $\frac{1}{6} - \frac{1}{6} = 0$.

$$x = \frac{3 \cdot 3}{4 \cdot 3} - \frac{1 \cdot 2}{6 \cdot 2}$$ Use the fundamental property of fractions to write each fraction in terms of the LCD, which is 12.

$$x = \frac{9}{12} - \frac{2}{12}$$ Do the multiplications in the numerators and denominators.

$$x = \frac{7}{12}$$ Subtract the fractions.

Answer: $\dfrac{7}{15}$ ∎

Applications

EXAMPLE 5 *Native Americans.* The United States Constitution requires a population count, called a *census,* to be taken every ten years. In the 1990 census, the population of the Navaho tribe was 225,000. This was about three-fifths of the population of the largest Native American tribe, the Cherokee. What was the population of the Cherokee tribe in 1990?

Analyze the problem
- In 1990, the population of the Navaho tribe was 225,000.
- The population of the Navaho tribe was $\frac{3}{5}$ the population of the Cherokee tribe.
- Find the population of the Cherokee tribe in 1990.

Form an equation Let x = the population of the Cherokee tribe in 1990. Next, we look for a key word or phrase in the problem.

Key phrase: *three-fifths of* **Translation:** *multiply by $\frac{3}{5}$*

The population of the Navaho tribe	was	$\frac{3}{5}$	of	the population of the Cherokee tribe.
225,000	=	$\frac{3}{5}$	·	x

Solve the equation

$$225{,}000 = \frac{3}{5}x$$

$$\frac{5}{3}\left(225{,}000\right) = \frac{5}{3}\left(\frac{3}{5}x\right)$$ To isolate x on the right-hand side, multiply both sides by the reciprocal of $\frac{3}{5}$.

$$375{,}000 = x$$ On the left-hand side, $\frac{5}{3}(225{,}000) = \frac{1{,}125{,}000}{3} = 375{,}000$.
On the right-hand side, $\frac{5}{3}\left(\frac{3}{5}x\right) = 1x = x$.

State the conclusion   In 1990, the population of the Cherokee tribe was about 375,000.

Check the result   Using a fraction to compare the two populations, we have

$$\frac{225{,}000}{375{,}000} = \frac{225}{375} = \frac{75 \cdot 3}{75 \cdot 5} = \frac{3}{5}$$

The answer checks. ∎

EXAMPLE 6   *Reading a book.*   A student has read $\frac{2}{5}$ of a book and wants to have read $\frac{3}{4}$ of it by tomorrow morning. How much more of the book must she read?

Analyze the problem
- A student has read $\frac{2}{5}$ of a book.
- She wants to have read $\frac{3}{4}$ of it by tomorrow morning.
- The amount she has read plus the amount she needs to read equals $\frac{3}{4}$ of the book.

Form an equation   We let x = the part of the book that needs to be read. Then we look for a key word or phrase.

Key phrase: *plus*   **Translation:** *add*

The part that has been read	plus	the part that needs to be read	is	$\frac{3}{4}$ of the book.
$\frac{2}{5}$	$+$	x	$=$	$\frac{3}{4}$

Solve the equation

$$\frac{2}{5} + x = \frac{3}{4}$$

$$\frac{2}{5} - \frac{2}{5} + x = \frac{3}{4} - \frac{2}{5} \qquad \text{To undo the addition by } \tfrac{2}{5}, \text{ subtract } \tfrac{2}{5} \text{ from both sides.}$$

$$x = \frac{3}{4} - \frac{2}{5} \qquad \tfrac{2}{5} - \tfrac{2}{5} = 0.$$

$$x = \frac{3 \cdot 5}{4 \cdot 5} - \frac{2 \cdot}{5 \cdot} \qquad \text{Apply the fundamental property of fractions to get an LCD of 20.}$$

$$x = \frac{15}{20} - \frac{8}{20} \qquad \text{Do the multiplications in the numerators and the denominators.}$$

$$x = \frac{7}{20} \qquad \text{Subtract the fractions.}$$

State the conclusion   The student must read $\frac{7}{20}$ more of the book.

Check the result   The student has read $\frac{2}{5}$ of the book. If she reads $\frac{7}{20}$ more, she will have read

$$\frac{2}{5} + \frac{7}{20} = \frac{2 \cdot 4}{5 \cdot 4} + \frac{7}{20} = \frac{8}{20} + \frac{7}{20} = \frac{15}{20} = \frac{3}{4}$$

of the book. The answer checks. ∎

STUDY SET   Section 3.8

VOCABULARY   *Fill in the blanks.*

1. To find the _____ of a fraction, invert the numerator and the denominator.

2. In the expression $\frac{5}{12} + x$, the x is called a _____.

3. The _____ of a set of fractions is the smallest number each denominator will divide exactly.

4. A _____ of an equation, when substituted into that equation, makes a true statement.

CONCEPTS

5. Is $x = 40$ a solution of $\dfrac{5}{8}x = 25$? Explain why or why not.

6. Give the reciprocal of each number.
   a. $\dfrac{7}{9}$
   b. $-\dfrac{1}{2}$

7. What is the result when a number is multiplied by its reciprocal?

8. Do each multiplication.
   a. $\dfrac{3}{2}\left(\dfrac{2}{3}x\right)$
   b. $-\dfrac{16}{15}\left(-\dfrac{15}{16}t\right)$
   c. $25\left(\dfrac{2}{5}\right)$
   d. $16\left(\dfrac{3}{8}\right)$

9. Translate to mathematical symbols.
   a. Four-fifths of the population p
   b. One-quarter of the time t

10. Explain two ways in which the variable x can be isolated: $\dfrac{2}{3}x = -4$.

NOTATION *Complete each solution to solve the equation.*

11. $\dfrac{7}{8}x = 21$

   $\boxed{}\left(\dfrac{7}{8}x\right) = \boxed{}(21)$

   $x = 24$

12. $h + \dfrac{1}{2} = \dfrac{2}{3}$

   $h + \dfrac{1}{2} - \boxed{} = \dfrac{2}{3} - \boxed{}$

   $h = \boxed{}$

13. Tell whether each statement is true or false.
   a. $\dfrac{1}{2}x = \dfrac{x}{2}$
   b. $\dfrac{1}{8}y = 8y$
   c. $-\dfrac{1}{2}x = \dfrac{-x}{2} = \dfrac{x}{-2}$
   d. $\dfrac{7p}{8} = \dfrac{7}{8}p$

14. Write the product of $\frac{4}{7}$ and x in two ways.

PRACTICE *Solve each equation.*

15. $\dfrac{4}{7}x = 16$

16. $\dfrac{2}{3}y = 30$

17. $\dfrac{7}{8}t = -28$

18. $\dfrac{5}{6}c = -25$

19. $-\dfrac{3}{5}h = 4$

20. $-\dfrac{5}{6}f = -2$

21. $\dfrac{2}{3}x = \dfrac{4}{5}$

22. $\dfrac{5}{8}y = \dfrac{10}{11}$

23. $\dfrac{2}{5}y = 0$

24. $\dfrac{4}{9}x = 0$

25. $-\dfrac{5c}{6} = -25$

26. $-\dfrac{7t}{4} = -35$

27. $\dfrac{-5f}{7} = -2$

28. $\dfrac{-3h}{5} = -4$

29. $\dfrac{5}{8}y = \dfrac{1}{10}$

30. $\dfrac{1}{16}x = \dfrac{5}{24}$

31. $x - \dfrac{1}{9} = \dfrac{7}{9}$

32. $x + \dfrac{1}{3} = \dfrac{2}{3}$

33. $x + \dfrac{1}{9} = \dfrac{4}{9}$

34. $x - \dfrac{1}{6} = \dfrac{1}{6}$

35. $x - \dfrac{1}{6} = \dfrac{2}{9}$

36. $y - \dfrac{1}{3} = \dfrac{4}{5}$

37. $y + \dfrac{7}{8} = \dfrac{1}{4}$

38. $t + \dfrac{5}{6} = \dfrac{1}{8}$

39. $\dfrac{5}{4} + t = \dfrac{1}{4}$

40. $\dfrac{2}{3} + y = \dfrac{4}{3}$

41. $x + \dfrac{3}{4} = -\dfrac{1}{2}$

42. $y - \dfrac{5}{6} = \dfrac{1}{3}$

43. $\dfrac{-x}{4} + 1 = 10$

44. $\dfrac{-y}{6} - 1 = 5$

45. $2x - \dfrac{1}{2} = \dfrac{1}{3}$

46. $3y - \dfrac{2}{5} = \dfrac{1}{8}$

47. $\dfrac{1}{2}x - \dfrac{1}{9} = \dfrac{1}{3}$

48. $\dfrac{1}{4}y - \dfrac{2}{3} = \dfrac{1}{2}$

49. $5 + \dfrac{x}{3} = \dfrac{1}{2}$

50. $4 + \dfrac{y}{2} = \dfrac{3}{5}$

51. $\dfrac{2}{5}x + 1 = \dfrac{1}{3}$

52. $\dfrac{2}{3}y + 2 = \dfrac{1}{5}$

53. $\dfrac{x}{3} + \dfrac{1}{4} = -2$

54. $\dfrac{5}{6} + \dfrac{y}{4} = -1$

55. $4 + \dfrac{s}{3} = 8$

56. $6 + \dfrac{y}{5} = 1$

57. $\dfrac{5h}{6} - 8 = 12$

58. $\dfrac{6a}{7} - 1 = 11$

59. $-4 + 9 + \dfrac{5t}{12} = 0$ **60.** $-4 + 10 + \dfrac{3y}{8} = 0$

61. $-3 - 2 + \dfrac{4x}{15} = 0$ **62.** $-1 - 9 + \dfrac{2y}{15} = 0$

APPLICATIONS   *Complete each solution.*

63. TRANSMISSION REPAIR   A repair shop found that $\frac{1}{3}$ of its customers with transmission problems needed a new transmission. If the shop installed 32 new transmissions last year, how many customers did the shop have last year?

Analyze the problem
- Only ▢ of the customers needed new transmissions.
- The shop installed ▢ new transmissions last year.
- Find the number of _____ the shop had last year.

Form an equation
Let $x =$ _____.

Key phrase: *one-third of*
Translation: _____

$\frac{1}{3}$ of the number of customers last year	was	32.
▢	=	32

Solve the equation
$$\frac{1}{3}x = 32$$
$$\Big(\tfrac{1}{3}x\Big) = \ (32)$$
$$x = \ ▢$$

State the conclusion   _____

Check the result   If we find $\frac{1}{3}$ of 96, we get ▢. The answer checks.

64. CATTLE RANCHING   A rancher is preparing to fence in a rectangular grazing area next to a $\frac{3}{4}$-mile-long lake. See Illustration 1. He has determined that $1\frac{1}{2}$ square miles of land are needed to ensure that overgrazing does not occur. How wide should this grazing area be?

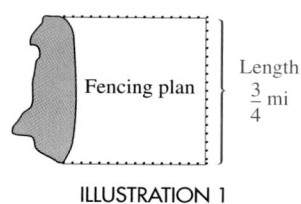

Fencing plan   Length $\frac{3}{4}$ mi

ILLUSTRATION 1

Analyze the problem
- The grazing area is ▢ $= \frac{3}{2}$ square miles.
- The length of the rectangle is $\frac{3}{4}$ mile.
- Find the _____ of the grazing area.

Form an equation
Let $w =$ _____.

Key word: *area*   **Translation:** $A =$ ▢

The area of the rectangle	is	the length times the width.
▢	=	▢

Solve the equation
$$\frac{3}{2} = \frac{3}{4}w$$
$$\Big(\frac{3}{2}\Big) = \Big(\frac{3}{4}w\Big)$$
$$▢ = w$$

State the conclusion
_____

Check the result   If we multiply the length and the width of the rectangular area, we get $\frac{3}{4} \cdot ▢ = \frac{3}{2} = 1\frac{1}{2}$ square miles. The answer checks.

Choose a variable to represent the unknown. Then write and solve an equation to answer each question.

65. TOOTH DEVELOPMENT   During a checkup, a pediatrician found that only four-fifths of a child's baby teeth had emerged. The mother counted 16 teeth in the child's mouth. How many baby teeth will the child eventually have?

66. GENETICS   Bean plants with inflated pods were cross-bred with bean plants with constricted pods. (See Illustration 2.) Of the offspring plants, three-fourths had inflated pods and one-fourth had constricted pods. If 244 offspring plants had constricted pods, how many offspring plants resulted from the cross-breeding experiment?

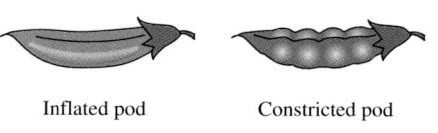

Inflated pod   Constricted pod

ILLUSTRATION 2

67. HOME SALES   In less than a month, three-quarters of the homes in a new subdivision were purchased. This left only 9 homes to be sold. How many homes are there in the subdivision?

68. WEDDING GUESTS   Of those invited to a wedding, three-tenths were friends of the bride. The friends of the groom numbered 84. How many people were invited to the wedding?

69. TELEPHONE BOOK   A telephone book consists of the white pages and the yellow pages. Two-thirds of the book consists of the white pages; the yellow pages number 150. Find the total number of pages in the telephone book.

70. BROADWAY MUSICAL   A theater usher at a Broadway musical finds that seven-eighths of the patrons attending a performance are in their seats by show time. The remaining 50 people are seated after the opening number. If the show is always a complete sellout, how many seats does the theater have?

71. SAFETY REQUIREMENT   In developing taillights for an automobile, designers must be aware of a safety standard that requires an area of 30 square inches to be visible from behind the vehicle. If the designers want the taillights to be $3\frac{3}{4}$ inches high, how wide must they be to meet safety standards? (See Illustration 3.)

ILLUSTRATION 3

72. GRAPHIC ARTS   A design for a yearbook is shown in Illustration 4. The page is divided into 12 parts. The parts that are shaded will contain pictures, and the remainder of the squares will contain copy. If the pictures are to cover an area of 100 square inches, how many square inches are there on the page?

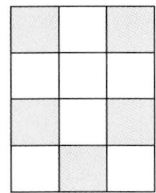

ILLUSTRATION 4

73. CPR CLASS   The instructor for a course in CPR (cardiopulmonary resuscitation) has three segments in her lesson plan, as shown in Illustration 5. How many minutes long is the CPR course?

Lecture on subject	Practicing CPR techniques	Legal responsibilities
One-fourth of class	Two-thirds of class	30 min

ILLUSTRATION 5

74. FIREFIGHTING   A firefighting crew is composed of three elements, as shown in Illustration 6. How many firefighters are in the crew?

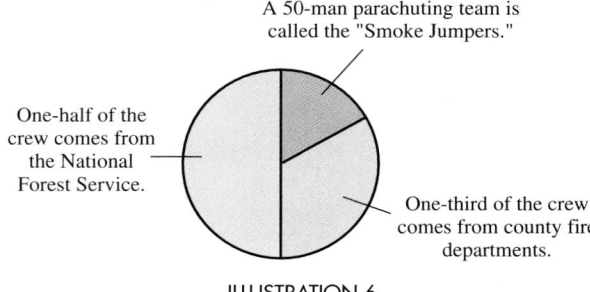

A 50-man parachuting team is called the "Smoke Jumpers."

One-half of the crew comes from the National Forest Service.

One-third of the crew comes from county fire departments.

ILLUSTRATION 6

WRITING

75. What does it mean to isolate the variable when solving an equation?

76. What is one-half of one-half? Explain how you arrived at the answer.

77. Which method, the reciprocal method or the two-step method, would you use to solve the equation $\frac{5}{16}t = 15$? Why?

78. Use an example to show why division by a number is the same as multiplying by its reciprocal.

REVIEW

79. Evaluate $(-2)^5$.

80. Evaluate $\dfrac{-4 - 8}{3}$.

81. Solve $3x - 2 = 7$.

82. Solve $\dfrac{x}{5} + 2 = 5$.

83. Round 12,590,767 to the nearest million.

84. In the expression $(-4)^6$, what do we call -4, and what do we call 6?

The Fundamental Property of Fractions

The **fundamental property of fractions** states that multiplying or dividing the numerator and the denominator of a fraction by the same nonzero number does not change the value of the fraction. This property is used to simplify fractions and to express fractions in higher terms. The following problems review both procedures. Complete each solution.

1. Simplify $\dfrac{15}{25}$.

Step 1: The numerator and the denominator share a common factor of ☐.

Step 2: Apply the fundamental property of fractions. Divide the numerator and the denominator by the common factor ☐.

$$\frac{15}{25} = \frac{15 \div \boxed{}}{25 \div \boxed{}}$$

Step 3: Do the divisions to simplify the fraction.

$$= \frac{3}{\boxed{}}$$

2. In practice, we often show the simplifying process described in problem 1 in a different form.

Step 1: Factor 15 as ☐ · 3 and 25 as ☐ · 5.

$$\frac{15}{25} = \frac{\boxed{} \cdot 3}{\boxed{} \cdot 5}$$

Step 2: The slashes and small 1's indicate that the numerator and the denominator have been divided by ☐.

$$= \frac{\overset{1}{\cancel{5}} \cdot 3}{\underset{1}{\cancel{5}} \cdot 5}$$

Step 3: Multiply in the numerator and the denominator.

$$= \frac{\boxed{}}{5}$$

3. When adding or subtracting fractions and mixed numbers, we often need to express a fraction in higher terms. This is called building up the fraction. Express $\frac{1}{5}$ as a fraction with denominator 35.

Step 1: We must multiply the denominator, 5, by ☐ to obtain 35.

Step 2: Use the fundamental property of fractions. Multiply the numerator and the denominator by ☐.

$$\frac{1}{5} = \frac{1 \cdot \boxed{}}{5 \cdot \boxed{}}$$

Step 3: Multiply in the numerator and the denominator.

$$= \frac{\boxed{}}{35}$$

Section 3.1

EQUIVALENT FRACTIONS Complete the labeling of each number line using fractions with the same denominator.

a. ⊢————————⊢ halves
    0   1

b. ⊢————————⊢ fourths
    0   1

c. ⊢————————⊢ eighths
    0   1

d. ⊢————————⊢ sixteenths
    0   1

FRACTIONS Give everyone in your group a strip of paper that is the same length. (See Illustration 1.) Determine ways to fold the strip of paper into

a. fourths   **b.** eighths
c. thirds   **d.** sixths

ILLUSTRATION 1

Section 3.2

MULTIPLICATION When we multiply 2 and 4, the answer is greater than 2 and greater than 4. Is this always the case? Is the product of two numbers always greater than either of the two numbers? Explain your answer.

POWERS When we square the number 4, the answer is greater than 4. Is the square of a number always greater than the number? Explain your answer.

Section 3.3

DIVIDING SNACKS Devise a way to divide seven brownies equally among six people.

Section 3.4

ADDING FRACTIONS Without actually doing the addition, explain why $\frac{3}{7} + \frac{1}{4}$ must be less than 1 and why $\frac{4}{7} + \frac{3}{4}$ must be greater than 1.

COMPARING FRACTIONS
a. When 1 is added to the numerator of a fraction, is the result greater than or less than the original fraction? Explain your reasoning.
b. When 1 is added to the denominator of a fraction, is the result greater than or less than the original fraction? Explain your reasoning.

COMPARING FRACTIONS Think of a fraction. Add 1 to its numerator and add 1 to its denominator. Is the resulting fraction greater than, less than, or equal to the original fraction? Explain your reasoning.

Section 3.5

DIVISION WITH MIXED NUMBERS Division can be thought of as repeated subtraction. Use this concept to solve the following problem.

$5\frac{1}{4}$ yards of ribbon needs to be cut into pieces that are $\frac{3}{4}$ of a yard long to form bows. How many bows can be made?

Section 3.6

MIXED NUMBERS Two mixed numbers, A and B, are graphed in Illustration 2. Estimate where on the number line the graph of $A + B$ would lie.

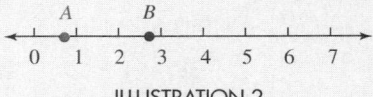

ILLUSTRATION 2

Section 3.7

COMPLEX FRACTION Write a problem that could be solved by simplifying the following.

$$\frac{\frac{7}{8}}{\frac{3}{4}}$$

Section 3.8

SOLVING EQUATIONS
a. Solve the equation $\frac{3}{4}x = 15$. Undo the multiplication by $\frac{3}{4}$ by dividing both sides by $\frac{3}{4}$.
b. Do the same for $-\frac{7}{8}x = 21$.

CHAPTER REVIEW

The Fundamental Property of Fractions

CONCEPTS

Fractions are used to indicate equal parts of a whole.

A fraction is composed of a *numerator*, a *denominator*, and a *fraction bar*.

If a and b are positive numbers,

$$\frac{-a}{b} = \frac{a}{-b} = -\frac{a}{b} \quad (b \neq 0)$$

Equivalent fractions represent the same number.

The *fundamental property of fractions:* Dividing the numerator and denominator of a fraction by the same nonzero number does not change the value of the fraction.

To *simplify* a fraction that is not in lowest terms, divide the numerator and denominator by the same number.

A fraction is in *lowest terms* if the only factor common to the numerator and denominator is 1.

The fundamental property of fractions: Multiplying the numerator and denominator of a fraction by a nonzero number does not change its value.

$$\frac{a}{b} = \frac{a \cdot x}{b \cdot x} \quad (b \neq 0, x \neq 0)$$

Expressing a fraction in higher terms results in an equivalent fraction that involves larger numbers or more complex terms.

REVIEW EXERCISES

1. If a woman gets seven hours of sleep each night, what part of a whole day does she spend sleeping?

2. In Illustration 1, why can't we say that $\frac{3}{4}$ of the figure is shaded?

ILLUSTRATION 1

3. Write the fraction $\frac{2}{-3}$ in two other ways.

4. What concept about fractions does Illustration 2 demonstrate?

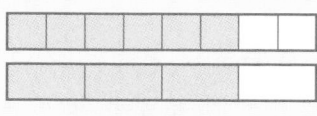

ILLUSTRATION 2

5. Explain the procedure shown here.

$$\frac{4}{6} = \frac{4 \div 2}{6 \div 2} = \frac{2}{3}$$

6. Explain what the slashes and the 1's mean.

$$\frac{4}{6} = \frac{\overset{1}{\cancel{2}} \cdot 2}{\underset{1}{\cancel{2}} \cdot 3} = \frac{2}{3}$$

7. Simplify each fraction to lowest terms.

a. $\dfrac{15}{45}$ b. $\dfrac{20}{48}$ c. $-\dfrac{63}{84}$ d. $\dfrac{66}{108}$

8. Explain what is being done and why it is valid.

$$\frac{5}{8} = \frac{5 \cdot 2}{8 \cdot 2} = \frac{10}{16}$$

9. Write each fraction or whole number with the indicated denominator (shown in red).

a. $\dfrac{2}{3}, 18$ b. $-\dfrac{3}{8}, 16$ c. $\dfrac{7}{15}, 45$ d. $4, 9$

| **SECTION 3.2** | *Multiplying Fractions* |

To *multiply two fractions,* multiply their numerators and multiply their denominators.

$$\frac{a}{b} \cdot \frac{c}{d} = \frac{a \cdot c}{b \cdot d}$$
$(b \neq 0, d \neq 0)$

10. Multiply.

   a. $\dfrac{1}{2} \cdot \dfrac{1}{3}$
   b. $\dfrac{2}{5}\left(-\dfrac{7}{9}\right)$
   c. $\dfrac{9}{16} \cdot \dfrac{20}{27}$
   d. $\dfrac{5}{6} \cdot \dfrac{1}{3} \cdot \dfrac{18}{25}$

   e. $\dfrac{3}{5} \cdot 7$
   f. $-4\left(-\dfrac{9}{16}\right)$
   g. $3\left(\dfrac{1}{3}\right)$
   h. $-\dfrac{6}{7}\left(-\dfrac{7}{6}\right)$

11. Tell whether each statement is true or false.

   a. $\dfrac{3}{4}x = \dfrac{3x}{4}$
                  b. $-\dfrac{5}{9}e = -\dfrac{5}{9e}$

12. Multiply.

   a. $\dfrac{3}{5} \cdot \dfrac{10}{27}$
   b. $-\dfrac{2}{3}\left(\dfrac{4}{7}s\right)$
   c. $\dfrac{4}{9} \cdot \dfrac{3}{28}$
   d. $9m\left(-\dfrac{5}{81}\right)$

An *exponent* indicates repeated multiplication.

13. Evaluate each power.

   a. $\left(\dfrac{3}{4}\right)^2$
   b. $\left(-\dfrac{5}{2}\right)^3$
   c. $\left(\dfrac{2}{3}\right)^2$
   d. $\left(-\dfrac{2}{5}\right)^3$

In mathematics, the word *of* usually means multiply.

14. GRAVITY ON THE MOON   Objects on the moon weigh only one-sixth as much as on Earth. How much will an astronaut weigh on the moon if he weighs 180 pounds on Earth?

The *area of a triangle*:

$$A = \frac{1}{2}bh$$

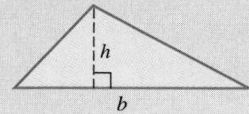

15. Find the area of the triangular sign in Illustration 3.

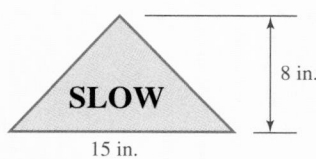

ILLUSTRATION 3

| **SECTION 3.3** | *Dividing Fractions* |

Two numbers are called *reciprocals* if their product is 1.

16. Find the reciprocal of each number.

   a. $\dfrac{1}{8}$
   b. $-\dfrac{11}{12}$
   c. 1
   d. 200

To *divide two fractions,* multiply the first by the reciprocal of the second.

$$\frac{a}{b} \div \frac{c}{d} = \frac{a}{b} \cdot \frac{d}{c}$$
$(b \neq 0, c \neq 0, d \neq 0)$

17. Divide.

   a. $\dfrac{1}{6} \div \dfrac{11}{25}$
            b. $-\dfrac{7}{8} \div \dfrac{1}{4}$

   c. $-\dfrac{15}{16} \div (-10)$
        d. $8 \div \dfrac{16}{5}$

   e. $\dfrac{1}{8} \div \dfrac{1}{4}$
             f. $\dfrac{4}{5} \div \dfrac{1}{2}$

18. GOLD COINS How many $\frac{1}{16}$-ounce coins can be cast from a $\frac{3}{4}$-ounce bar of gold?

SECTION 3.4 | *Adding and Subtracting Fractions*

To add (or subtract) fractions with like denominators, add (or subtract) their numerators and write the result over the common denominator.

$$\frac{a}{c} + \frac{b}{c} = \frac{a+b}{c} \quad (c \neq 0)$$

$$\frac{a}{c} - \frac{b}{c} = \frac{a-b}{c} \quad (c \neq 0)$$

The *LCD* must include the set of prime factors of each of the denominators.

To add or subtract fractions with unlike denominators, we must first express them as equivalent fractions with the same denominator, preferably the LCD.

19. Add or subtract.

a. $\frac{2}{7} + \frac{3}{7}$ **b.** $-\frac{3}{5} - \frac{3}{5}$ **c.** $\frac{3}{8} - \frac{1}{8}$ **d.** $\frac{7}{8} + \frac{7}{8}$

20. Explain why we cannot immediately add $\frac{1}{2} + \frac{2}{3}$ without doing some preliminary work.

21. Use prime factorization to find the least common denominator for fractions with denominators of 45 and 30.

22. Add or subtract.

a. $\frac{1}{6} + \frac{2}{3}$ **b.** $\frac{2}{5} + \left(-\frac{3}{8}\right)$

c. $-\frac{3}{8} - \frac{5}{6}$ **d.** $3 - \frac{1}{7}$

e. $\frac{13}{6} - 6$ **f.** $\frac{1}{3} + \frac{1}{4} + \frac{1}{5}$

23. MACHINE SHOP See Illustration 4. How much must be milled off the $\frac{3}{4}$-inch-thick steel rod so that the collar will slip over the end of it?

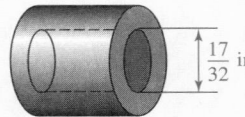 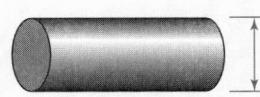

$\frac{17}{32}$ in. $\frac{3}{4}$ in.

Steel rod

ILLUSTRATION 4

To *compare fractions*, write them as equivalent fractions with the same denominator. Then the fraction with the larger numerator will be the larger fraction.

24. TELEMARKETING In the first hour of work, a telemarketer made 2 sales out of 9 telephone calls. In the second hour, she made 3 sales out of 11 calls. During which hour was the rate of sales to calls better?

SECTION 3.5 | *Multiplying and Dividing Mixed Numbers*

A *mixed number* is the sum of its whole-number part and its fractional part.

25. a. What mixed number is represented in Illustration 5?

    b. What improper fraction is represented in Illustration 5?

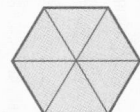 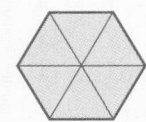 

ILLUSTRATION 5

To change an *improper fraction* to a mixed number, divide the numerator by the denominator to obtain the whole-number part. Write the remainder over the denominator for the fractional part.

26. Express each improper fraction as a mixed number or a whole number.

a. $\dfrac{16}{5}$ **b.** $-\dfrac{47}{12}$ **c.** $\dfrac{6}{6}$ **d.** $\dfrac{14}{6}$

To change a mixed number to an improper fraction, multiply the whole number by the denominator and add the result to the numerator. Write this sum over the denominator.

27. Write each mixed number as an improper fraction.

a. $9\dfrac{3}{8}$ **b.** $-2\dfrac{1}{5}$ **c.** $100\dfrac{1}{2}$ **d.** $1\dfrac{99}{100}$

28. Graph $-2\dfrac{2}{3}, \dfrac{8}{9},$ and $\dfrac{59}{24}$.

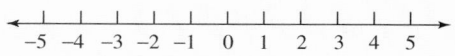

To *multiply* or *divide mixed numbers,* change the mixed numbers to improper fractions and then do the operations as usual.

29. Multiply or divide. Write answers as mixed numbers when appropriate.

a. $-5\dfrac{1}{4} \cdot \dfrac{2}{35}$ **b.** $\left(-3\dfrac{1}{2}\right) \div \left(-3\dfrac{2}{3}\right)$

c. $\left(-6\dfrac{2}{3}\right)(-6)$ **d.** $-8 \div 3\dfrac{1}{5}$

30. CAMERA TRIPOD   The three legs of a tripod can be extended to become $5\dfrac{1}{2}$ times their original length. If each leg is $8\dfrac{3}{4}$ inches long when collapsed, how long will a leg become when it is completely extended?

SECTION 3.6 | *Adding and Subtracting Mixed Numbers*

To add (or subtract) mixed numbers, we can change each to an improper fraction and use the method of Section 3.4.

31. Add or subtract.

a. $1\dfrac{3}{8} + 2\dfrac{1}{5}$ **b.** $3\dfrac{1}{2} + 2\dfrac{2}{3}$

c. $2\dfrac{5}{6} - 1\dfrac{3}{4}$ **d.** $3\dfrac{7}{16} - 2\dfrac{1}{8}$

To add mixed numbers, we can add the whole numbers and the fractions separately.

32. PAINTING SUPPLIES   In a project to restore a house, painters used $10\dfrac{3}{4}$ gallons of primer, $21\dfrac{1}{2}$ gallons of latex paint, and $7\dfrac{2}{3}$ gallons of enamel. Find the total number of gallons of paint used.

Vertical form can be used to add or subtract mixed numbers.

33. Add or subtract.

a. $\begin{array}{r} 133\frac{1}{9} \\ + \ 49\frac{1}{6} \\ \hline \end{array}$ **b.** $\begin{array}{r} 98\frac{11}{20} \\ +14\frac{3}{5} \\ \hline \end{array}$

c. $\begin{array}{r} 50\frac{5}{8} \\ -19\frac{1}{6} \\ \hline \end{array}$ **d.** $\begin{array}{r} 375\frac{3}{4} \\ - \ 59 \\ \hline \end{array}$

If the fraction being subtracted is larger than the first fraction, we need to *borrow* from the whole number.

34. Subtract.

a. $23\dfrac{1}{3} - 2\dfrac{5}{6}$ **b.** $39 - 4\dfrac{5}{8}$

| SECTION 3.7 | ***Order of Operations and Complex Fractions*** |

A *complex fraction* is a fraction whose numerator or denominator, or both, contain one or more fractions or mixed numbers.

The main fraction bar of a complex fraction indicates division.

35. Evaluate each numerical expression.

a. $\dfrac{3}{4} + \left(-\dfrac{1}{3}\right)^2\left(\dfrac{5}{4}\right)$

b. $\left(\dfrac{2}{3} \div \dfrac{16}{9}\right) - \left(1\dfrac{2}{3} \cdot \dfrac{1}{15}\right)$

36. Simplify each complex fraction.

a. $\dfrac{\dfrac{3}{5}}{-\dfrac{17}{20}}$

b. $\dfrac{\dfrac{2}{3} - \dfrac{1}{6}}{-\dfrac{3}{4} - \dfrac{1}{2}}$

| SECTION 3.8 | ***Solving Equations Containing Fractions*** |

37. Solve each equation. Check the result.

a. $\dfrac{2}{3}x = 16$

b. $-\dfrac{7s}{4} = -49$

c. $\dfrac{y}{5} = -\dfrac{1}{15}$

d. $2x - 3 = 8$

38. Solve each equation.

a. $\dfrac{c}{3} - \dfrac{3}{8} = 2$

b. $\dfrac{5h}{9} - 1 = -3$

c. $4 - \dfrac{d}{4} = 0$

d. $\dfrac{t}{10} - \dfrac{2}{3} = \dfrac{1}{5}$

39. HISTORY TEXTBOOK   In writing a history text, the author decided to devote two-thirds of the book to events prior to World War II. The remainder of the book deals with history after the war. If pre–World War II history is covered in 220 pages, how many pages does the textbook have?

Chapter 3 Test

1. See Illustration 1.
   a. What fractional part of the plant is above ground?
   b. What fractional part of the plant is below ground?

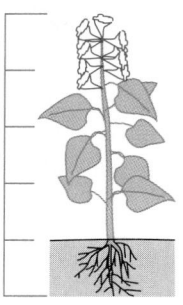

ILLUSTRATION 1

2. Simplify each fraction.
   a. $\dfrac{27}{36}$ **b.** $\dfrac{72}{180}$

3. Multiply: $-\dfrac{3}{4}\left(\dfrac{1}{5}\right)$.

4. COFFEE DRINKERS   Of 100 adults surveyed, $\frac{2}{5}$ said they started off their morning with a cup of coffee. Of the 100, how many would this be?

5. Divide: $\dfrac{4}{3} \div \dfrac{1}{9}$.

6. Subtract: $\dfrac{1}{6} - \dfrac{4}{5}$.

7. Express $\frac{7}{8}$ as an equivalent fraction with denominator 24.

8. Graph $2\dfrac{4}{5}$, $-1\dfrac{1}{7}$, and $\dfrac{7}{6}$.

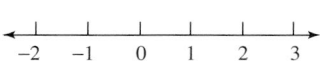

9. SPORTS CONTRACT   A basketball player signed a nine-year contract for $13\frac{1}{2}$ million. How much is this per year?

10. Add: $157\dfrac{5}{9} + 103\dfrac{3}{4}$.

11. Subtract: $67\dfrac{1}{4} - 29\dfrac{5}{6}$.

12. BOXING   When Oscar De La Hoya fought Pernell Whitaker, the "Tale of the Tape" shown in Illustration 2 appeared in the sports section of many newspapers. What was the difference in the fighters'
   a. weights?
   b. chests (expanded)?
   c. waists?

Tale of the Tape		
De La Hoya		**Whitaker**
24 yr	Age	33 yr
146½ lb	Weight	146½ lb
5-11	Height	5-6
72 in.	Reach	69 in.
39 in.	Chest (Normal)	37 in.
42¼ in.	Chest (Expanded)	39½ in.
31¾ in.	Waist	28 in.

ILLUSTRATION 2

13. Add: $-\dfrac{3}{7} + 2$.

14. SEWING   When cutting material for a $10\frac{1}{2}$-inch-wide placemat, a seamstress allows $\frac{5}{8}$ inch at each end for a hem. How wide should the material be cut? See Illustration 3.

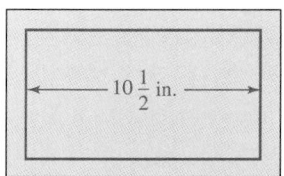

ILLUSTRATION 3

15. In Illustration 4, find the perimeter and the area of the triangle.

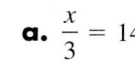

20 in.

$22\frac{2}{3}$ in.

$10\frac{2}{3}$ in.

ILLUSTRATION 4

16. Evaluate:

$$\left(\frac{2}{3} \cdot \frac{5}{16}\right) - \left(-1\frac{3}{5} \div 4\frac{4}{5}\right)$$

17. Simplify the complex fraction.

$$\frac{-\dfrac{5}{6}}{\dfrac{7}{8}}$$

18. Simplify the complex fraction.

$$\frac{\dfrac{1}{2} + \dfrac{1}{3}}{-\dfrac{1}{6} - \dfrac{1}{3}}$$

19. Solve each equation.

a. $\dfrac{x}{3} = 14$ **b.** $-\dfrac{5}{2}t + 2 = 20$

20. JOB APPLICANTS   Three-fourths of the applicants for a position had previous experience. The number who did not have prior experience was 36. How many people applied?

21. What are the parts of a fraction? What does a fraction represent?

22. Explain what is meant when we say, "The product of any number and its reciprocal is 1."

23. Explain what mathematical concept is being shown.

a. $\dfrac{6}{8} = \dfrac{3 \cdot \overset{1}{\cancel{2}}}{4 \cdot \underset{1}{\cancel{2}}} = \dfrac{3}{4}$

b.

c. $\dfrac{3}{5} = \dfrac{3 \cdot 4}{5 \cdot 4} = \dfrac{12}{20}$

Chapters 1-3 Cumulative Review Exercises

Consider the number 5,434,679.

1. Round to the nearest hundred.

2. Round to the nearest ten thousand.

3. THE STOCK MARKET   The graph in Illustration 1 shows the performance of the Dow Jones Industrial Average on the last trading day of 1999. Estimate the highest mark that the market reached. At what time during the day did that occur?

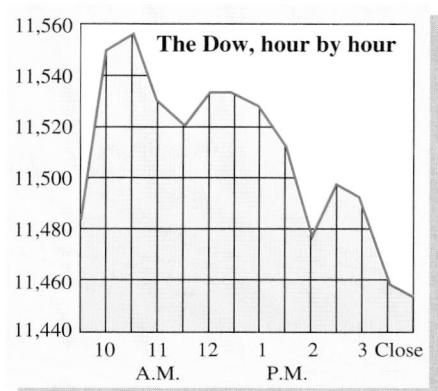

Source: *Los Angeles Times* (December 31, 1999)

ILLUSTRATION 1

4. BANKS   As of December 31, 1997, the world's largest bank, with total assets of $691,920,300,000, was the Bank of Tokyo–Mitsubishi Ltd., Japan. In what place value column is the digit 6 located?

Do each operation.

5.
$$\begin{array}{r} 4,679 \\ +3,457 \\ \hline \end{array}$$

6.
$$\begin{array}{r} 7,897 \\ -4,378 \\ \hline \end{array}$$

7.
$$\begin{array}{r} 5,345 \\ \times\ \ \ 56 \\ \hline \end{array}$$

8. $35\overline{)34,685}$

In Exercises 9–10, refer to the rectangular swimming pool shown in Illustration 2.

9. Find the perimeter of the pool.

10. Find the area of the pool's surface.

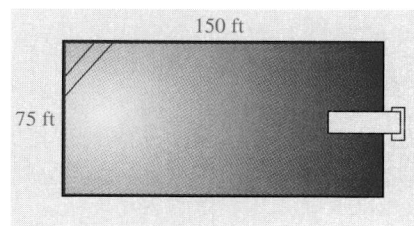

ILLUSTRATION 2

Find the prime factorization of each number.

11. 84

12. 450

13. 360

14. 3,600

Evaluate each expression.

15. $6 + (-2)(-5)$

16. $(-2)^3 - 3^3$

17. $\dfrac{2(-7) + 3(2)}{2(-2)}$

18. $\dfrac{2(3^2 - 4^2)}{-2(3) - 1}$

Solve each equation. Check the result.

19. $3x + 2 = -13$

20. $-5z - 7 = 18$

21. $\dfrac{y}{4} - 1 = -5$

22. $\dfrac{n}{5} + 1 = 0$

23. OBSERVATION HOURS   To get a Master's degree in educational psychology, a student must have 100 hours of observation time at a clinic. If the student has already observed for 37 hours, how many 3-hour shifts must he observe to complete the requirement?

24. GEOMETRY   A rectangle is four times as long as it is wide. If its perimeter is 210 feet, find its dimensions.

Simplify each fraction.

25. $\dfrac{21}{28}$

26. $\dfrac{40}{16}$

Do each operation.

27. $\dfrac{6}{5}\left(-\dfrac{2}{3}\right)$

28. $\dfrac{14}{8} \div \dfrac{7}{2}$

29. $\dfrac{2}{3} + \dfrac{3}{4}$

30. $\dfrac{4}{3} - \dfrac{3}{5}$

Write each mixed number as an improper fraction.

31. $3\dfrac{5}{6}$

32. $-6\dfrac{5}{8}$

Do each operation.

33. $4\dfrac{2}{3} + 5\dfrac{1}{4}$

34. $14\dfrac{2}{5} - 8\dfrac{2}{3}$

35. FIRE HAZARD   Two terminals in an electrical switch were so close that electricity could jump the gap and start a fire. Illustration 3 shows a newly designed switch that will keep this from happening. By how much was the distance between the ground terminal and the hot terminal increased?

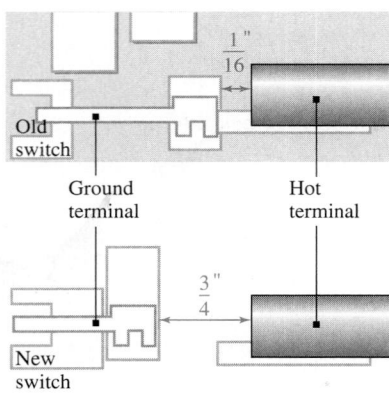

ILLUSTRATION 3

36. SHAVING   Advertisements claim that a shaving lotion for men cuts shaving time by a third. When using this lotion, it took a man 60 seconds to shave. If the advertising claim is correct, how long would it normally have taken the man to shave if he hadn't used the special lotion?

Simplify each expression.

37. $\left(\dfrac{1}{4} - \dfrac{7}{8}\right) \div \left(-2\dfrac{3}{16}\right)$

38. $\dfrac{\dfrac{2}{3} - 7}{4\dfrac{5}{6}}$

Solve each equation. Check the result.

39. $x + \dfrac{1}{5} = -\dfrac{14}{15}$

40. $3 = \dfrac{5}{8}x + \dfrac{1}{2}$

41. $\dfrac{2}{3}x = -10$

42. $3y - 8 = 0$

43. Explain the difference between an *expression* and an *equation*.

44. What is a variable?

Decimals

4

4.1 An Introduction to Decimals

4.2 Addition and Subtraction with Decimals

4.3 Multiplication with Decimals

4.4 Division with Decimals

Estimation

4.5 Fractions and Decimals

4.6 Solving Equations Containing Decimals

4.7 Square Roots

Key Concept: The Real Numbers

Accent on Teamwork

Chapter Review

Chapter Test

Cumulative Review Exercises

DECIMALS PROVIDE ANOTHER WAY TO REPRESENT FRACTIONS AND MIXED NUMBERS. THEY ARE OFTEN USED IN MEASUREMENT, BECAUSE IT IS EASY TO PUT THEM IN ORDER AND TO COMPARE THEM.

4.1   *An Introduction to Decimals*

In this section, you will learn about

- Decimals • The place value system for decimal numbers
- Reading and writing decimals • Comparing decimals • Rounding

INTRODUCTION. This section introduces the **decimal numeration system**—an extension of the place value system that we used when working with whole numbers. You may not realize it, but you have often worked with the decimal numeration system.

We can use the decimal numeration system to express the car's mileage. The odometer reads 1,537.6 miles.

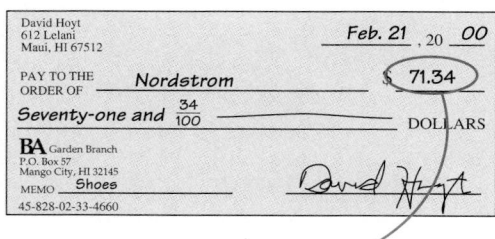

The amount of the check is written using the decimal numeration system.

Decimals

Like fraction notation, decimal notation is used to denote a part of a whole. However, when writing a number in decimal notation, we don't use a fraction bar, nor is a denominator shown. A number written in decimal notation is often called a **decimal.**

In Figure 4-1, a rectangle is divided into 10 equal parts. One-tenth of the figure is shaded. We can use either the fraction $\frac{1}{10}$ or the decimal 0.1 to describe the shaded region. Both are read as "one-tenth."

$$\frac{1}{10} = 0.1$$

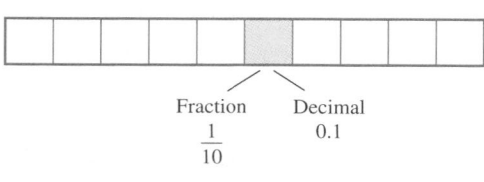

FIGURE 4-1

In Figure 4-2, a square is divided into 100 equal parts. One of the 100 parts is shaded; it can be represented by the fraction $\frac{1}{100}$ or by the decimal 0.01. Both are read as "one one-hundredth."

$$\frac{1}{100} = 0.01$$

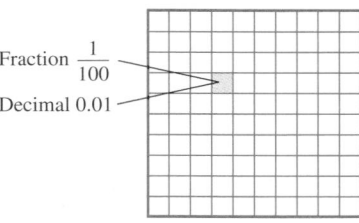

Fraction $\frac{1}{100}$

Decimal 0.01

FIGURE 4-2

The place value system for decimal numbers

Decimal numbers are written by placing digits (0, 1, 2, 3, 4, 5, 6, 7, 8, 9) into place value columns that are separated by a **decimal point.** See Figure 4-3. The place value names of all the columns to the right of the decimal point end in "th." The "th" tells us that the value of the column is a fraction whose denominator is a power of ten. Columns to the left of the decimal point have a value greater than or equal to 1; columns to the right of the decimal point have a value less than 1. We can show the value represented by each digit of a decimal by using **expanded notation.**

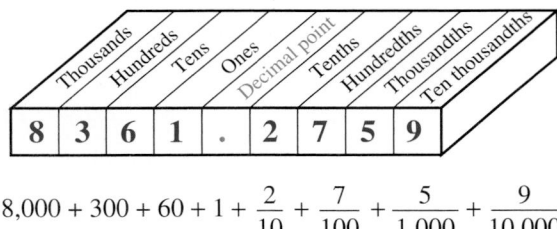

$$8,000 + 300 + 60 + 1 + \frac{2}{10} + \frac{7}{100} + \frac{5}{1,000} + \frac{9}{10,000}$$

Expanded notation

FIGURE 4-3

Decimal points are used to separate the whole-number part of a decimal from its fractional part.

12.37

Whole-number part ———⟋ | ⟍——— Fractional part

Decimal point

When there is no whole-number part of a decimal, we can show that by entering a zero to the left of the decimal point.

.85 = 0.85

↑   ↑

No whole number part.   Enter a zero here, if desired.

We can write a whole number in decimal notation by placing a decimal point to its right and then adding a zero, or zeros, to the right of the decimal point.

$$99 \ = \ 99.0 \ = \ 99.00$$

A whole number.   Place a decimal point here and enter
a zero, or zeros, to the right of it.

Writing additional zeros to the right of the decimal point *following the last digit* does not change the value of the decimal.

$$12.37 = 12.370 = 12.3700$$

These additional zeros do not
change the value of the decimal.

Reading and writing decimals

The decimal 12.37 can be read as "twelve point three seven." Another way of reading a decimal states the whole-number part first and then the fractional part.

Reading a decimal

> To read a decimal:
> 1. Look to the left of the decimal point and say the name of the whole number.
> 2. The decimal point is then read as "and."
> 3. Say the fractional part of the decimal as a whole number followed by the name of the place value column of the digit that is farthest to the right.

Using this procedure, here is the other way to read 12.37.

Name of the last column on the right.

12.37

Twelve   and   **thirty-seven**   hundredths

When we read a decimal in this way, it is easy to write it in words and as a mixed number.

Decimal	Words	Mixed number
12.37	Twelve and thirty-seven hundredths	$12\dfrac{37}{100}$

EXAMPLE 1 *Writing a decimal in other forms.*   Write each decimal in words and then as a fraction or mixed number. **Do not simplify the fraction.**

a. The world speed record for a human-powered vehicle is 65.484 mph, set in 1986.

b. The smallest fresh-water fish is the dwarf pygmy goby, found in the Philippines. Adult males weigh 0.00014 ounce.

Solution

a. 65.484 is sixty-five and four hundred eighty-four thousandths, or $65\dfrac{484}{1,000}$.

b. 0.00014 is fourteen hundred-thousandths, or $\dfrac{14}{100,000}$.

Self Check

Write each decimal in words and then as a mixed number.

a. Sputnik 1, the first artificial satellite, weighed 184.3 pounds.

b. The planet Mercury makes one revolution every 87.9687 days.

Answers: a. One hundred eighty-four and three-tenths, or $184\frac{3}{10}$ **b.** Eighty-seven and nine thousand six hundred eighty-seven ten thousandths, or $87\frac{9,687}{10,000}$

Decimals can be negative. For example, a record low temperature of $-128.6°$ F was recorded in Vostok, Antarctica on July 21, 1983. This is read as "negative one hundred twenty-eight and six tenths." Written as a mixed number, it is $-128\frac{6}{10}$.

Comparing decimals

The relative sizes of a set of decimals can be determined by scanning their place value columns from left to right, column by column, looking for a difference in the digits. For example,

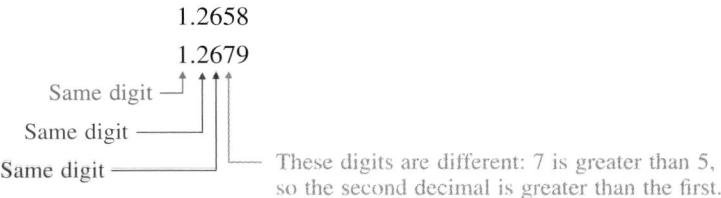

Thus, 1.2679 is greater than 1.2658. We write $1.2679 > 1.2658$.

Comparing positive decimals

To compare two positive decimals:
1. Make sure both numbers have the same number of decimal places to the right of the decimal point. Write any additional zeros necessary to achieve this.
2. Compare the digits of each decimal, column by column, working from left to right.
3. When two digits differ, the decimal with the greater digit is the greater number.

EXAMPLE 2   *Comparing positive decimals.* Which is greater, 54.9 or 54.929?

Solution
54.900   Write two zeros after 9 so that both decimals have the same number of digits to
54.929   the right of the decimal point.

Working from left to right, this is the first column in which the digits differ. Since 2 is greater than 0, we can conclude that $54.929 > 54.9$.

Self Check
Which is greater, 113.7 or 113.657?

Answer: 113.7

Comparing negative decimals

To compare two negative decimals:
1. Make sure both numbers have the same number of decimal places to the right of the decimal point. Write any additional zeros necessary to achieve this.
2. Compare the digits of each decimal, column by column, working from left to right.
3. When two digits differ, the decimal with the smaller digit is the greater number.

EXAMPLE 3   *Comparing negative decimals.* Which is greater, -10.45 or -10.419?

Solution
-10.450 Write a 0 after 5 to help in the comparison.
-10.419

Working from left to right, this is the first column in which the digits differ. Since 1 is less than 5, we conclude that $-10.419 > -10.45$.

Self Check
Which is greater, -703.8 or -703.78?

Answer: -703.78

EXAMPLE 4   *Graphing decimals.*   Graph -1.8, -1.23, -0.3, and 1.89.

Solution

To graph each decimal, we locate its position on the number line and draw a dot. Since -1.8 is to the left of -1.23, we can write $-1.8 < -1.23$.

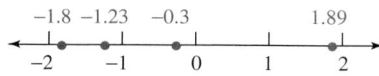

Self Check

Graph -1.1, -0.6, 0.8, and 1.9.

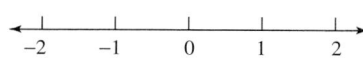

Answer:

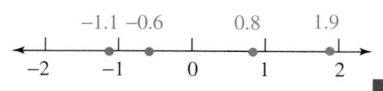

Rounding

When working with decimals, we often round answers to a specific number of decimal places.

Rounding a decimal

> 1. To round a decimal to a specified decimal place, locate the digit in that place. Call it the *rounding digit.*
>
> 2. Look at the *test digit* to the right of the rounding digit.
>
> 3. If the test digit is 5 or greater, round up by adding 1 to the rounding digit and dropping all the digits to its right. If the test digit is less than 5, round down by keeping the rounding digit and dropping all the digits to its right.

EXAMPLE 5   *Chemistry.*   In a chemistry class, a student uses a balance to weigh a compound. The digital readout on the scale shows 1.2387 g. Round this decimal to the nearest thousandth of a gram.

Solution   We are asked to round to the nearest thousandth.

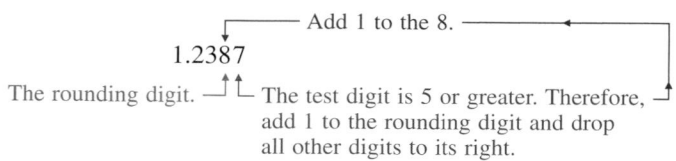

The compound weighs approximately 1.239 g.

EXAMPLE 6   *Rounding decimals.*   Round each decimal to the indicated place value: **a.** -645.13 to the nearest tenth   and   **b.** 33.097 to the nearest hundredth.

Solution

a. -645.13

Rounding digit.   Since the test digit is less than 5, drop it and all the digits to its right.

The result is -645.1.

b.   33.097

Rounding digit.   Since the test digit is greater than 5, we add 1 to 9 and drop all the digits to the right.

       10
33.09   Adding a 1 to the 9 requires that we carry a 1 to the tenths column.

When we are asked to round to the nearest hundredth, we must have a digit in the hundredths column, even if it is a zero. Therefore, the result is 33.10.

Self Check

Round each decimal to the indicated place value.

a. -708.522 to the nearest tenth

b. 9.1198 to the nearest thousandth

Answers: **a.** -708.5,
b. 9.120

STUDY SET   Section 4.1

VOCABULARY   *Fill in the blanks.*

1. Give the name of each place value column.

4   7   8   9   .   0   2   6   5

2. We can show the value represented by each digit of the decimal 98.6213 by using _____ notation.

$$98.6213 = 90 + 8 + \frac{6}{10} + \frac{2}{100} + \frac{1}{1,000} + \frac{3}{10,000}$$

3. We can approximate a decimal number using the process called _____.

4. When reading 2.37, the decimal point can be read as "_____" or "_____."

CONCEPTS

5. Consider the decimal 32.415.
   a. Write the decimal in words.

   b. What is its whole-number part?
   c. What is its fractional part?
   d. Write the decimal in expanded notation.

6. Write
   $400 + 20 + 8 + \frac{9}{10} + \frac{6}{100}$
   as a decimal.

7. Graph $\frac{7}{10}$, -0.7, $-3\frac{1}{100}$, and 3.01.

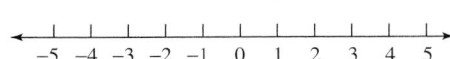

8. Graph -1.21, -3.29, and -4.25.

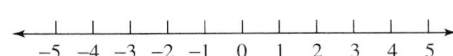

9. True or false?
   a. $0.9 = 0.90$
   b. $1.260 = 1.206$
   c. $-1.2800 = -1.280$
   d. $0.001 = .0010$

10. Write each fraction as a decimal.
   a. $\frac{9}{10}$ **b.** $\frac{63}{100}$

   c. $\frac{111}{1,000}$ **d.** $\frac{27}{10,000}$

11. Represent the shaded part of the square in Illustration 1 using a fraction and a decimal.

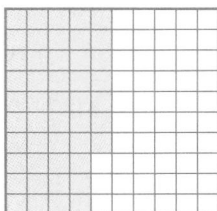

ILLUSTRATION 1

12. Represent the shaded part of the figure in Illustration 2 using a fraction and a decimal.

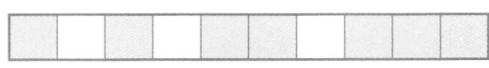

ILLUSTRATION 2

13. The line segment in Illustration 3 is one inch long. Show a length of 0.3 inch on it.

ILLUSTRATION 3

14. Read the meter in Illustration 4. What decimal is indicated by the arrow?

ILLUSTRATION 4

NOTATION

15. Construct a decimal number by writing
   0 in the tenths column,
   4 in the thousandths column,
   1 in the tens column,
   9 in the thousands column,
   8 in the hundreds column,
   2 in the hundredths column,
   5 in the ten thousandths column, and
   6 in the ones column.

16. Represent each situation using a signed number.
  a. A deficit of $15,600.55
  b. A river 6.25 feet above flood stage
  c. A state budget $6.4 million in the red
  d. 3.9 degrees below zero
  e. 17.5 seconds after liftoff
  f. A checking account overdrawn by $33.45

PRACTICE   *Write each decimal in words and then as a fraction or mixed number.*

17. 50.1   **18.** 0.73

19. -0.0137 **20.** -76.09

21. 304.0003   **22.** 68.91

23. -72.493 **24.** -31.5013

Write each decimal using numbers.

25. Negative thirty-nine hundredths
26. Negative twenty-seven and forty-four hundredths

27. Six and one hundred eighty-seven thousandths
28. Ten and fifty-six ten-thousandths

Round each decimal to the nearest tenth.

29. 506.098   **30.** 0.441
31. 2.718218   **32.** 3,987.8911

Round each decimal to the nearest hundredth.

33. -0.137 **34.** -808.0897
35. 33.0032   **36.** 64.0059

Round each decimal to the nearest thousandth.

37. 3.14159   **38.** 16.0995
39. 1.414213   **40.** 2,300.9998

Round each decimal to the nearest whole number.

41. 38.901   **42.** 405.64
43. 2,988.399   **44.** 10,453.27

Round each amount to the value indicated.

45. $3,090.28
  a. Nearest dollar
  b. Nearest ten cents
46. $289.73
  a. Nearest dollar
  b. Nearest ten cents

Fill in the blanks with the proper symbol ($<$, $>$, or $=$).

47. -23.45 ____ -23.1 **48.** -301.98 ____ -302.45

49. $-.065$ ____ $-.066$ **50.** -3.99 ____ -3.9888

Arrange the decimals in order, from least to greatest.

51. 132.64, 132.6499, 132.6401
52. 0.007, 0.00697, 0.00689

APPLICATIONS

53. WRITING A CHECK   Complete the check shown in Illustration 5 by writing in the amount, using a decimal.

Ellen Russell
455 Santa Clara Ave.
Parker, CO 25413

April 14 , 20 00

PAY TO THE ORDER OF ___ Citicorp ___ $ ____

One thousand twenty-five and $\frac{78}{100}$ ____ DOLLARS

BA Downtown Branch
P.O. Box 2456
Colorado Springs,CO 23712

MEMO __Mortgage__   Ellen Russell

45-828-02-33-4660

ILLUSTRATION 5

54. MONEY   We use a decimal point when working with dollars, but the decimal point is not necessary when working with cents. For each dollar amount in Illustration 6, give the equivalent amount expressed as cents.

Dollars	Cents
$0.50	
$0.05	
$0.55	
$5.00	
$0.01	

ILLUSTRATION 6

55. INJECTIONS   A syringe is shown in Illustration 7. Use an arrow to show to what point the syringe should be filled if a 0.38-cc dose of medication is to be administered. ("cc" stands for "cubic centimeters.")

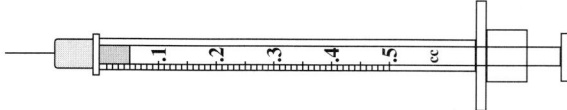

ILLUSTRATION 7

56. LASER   The laser used in laser vision correction is so precise that each pulse can remove 39 millionths of an inch of tissue in 12 billionths of a second. Write each of these numbers as decimals.

57. METRIC SYSTEM   The metric system is widely used in science to measure length (meters), weight (grams), and capacity (liters). Round each decimal to the nearest hundredth.
  a. 1 ft is 0.3048 meter.
  b. 1 mi is 1,609.344 meters.
  c. 1 lb is 453.59237 grams.
  d. 1 gal is 3.785306 liters.

58. WORLD RECORDS   As of January, 2000, four American women held world records in swimming. Their times are given below in the form *minutes: seconds*. Round each to the nearest tenth of a second.

100-meter butterfly	Jenny Thompson	0:57.88
200-meter butterfly	Mary T. Meagher	2:05.96
400-meter freestyle	Janet Evans	4:03.85
800-meter freestyle	Janet Evans	8:16.22
1,500-meter freestyle	Janet Evans	15:52.10

59. GEOLOGY   Geologists classify types of soil according to the grain size of the particles that make up the soil. The four major classifications are shown below. Complete the chart in Illustration 8 by classifying each sample.

Clay	0.00008 in. and under
Silt	0.00008 in. to 0.002 in.
Sand	0.002 in. to 0.08 in.
Granule	0.08 in. to 0.15 in.

Sample	Location	Size (in.)	Classification
A	riverbank	0.009	
B	pond	0.0007	
C	NE corner	0.095	
D	dry lake	0.00003	

ILLUSTRATION 8

60. MICROSCOPE   A microscope used in a lab is capable of viewing structures that range in size from 0.1 to 0.0001 centimeter. Which of the structures listed in Illustration 9 would be visible through this microscope?

Structure	Size (in cm)
bacterium	0.00011
plant cell	0.015
virus	0.000017
animal cell	0.00093
asbestos fiber	0.0002

ILLUSTRATION 9

61. AIR QUALITY   Illustration 10 shows the cities with the highest one-hour concentrations of ozone (in parts per million) during the summer of 1999. Rank the cities in order, beginning with the city with the highest reading.

Crestline, California	0.170
Galveston, Texas	0.176
Houston, Texas	0.202
Texas City, Texas	0.206
Westport, Connecticut	0.188
White Plains, New York	0.171

Source: *Los Angeles Times* (August 18, 1999)

ILLUSTRATION 10

62. DEWEY DECIMAL SYSTEM   A widely used system for classifying books in a library is the Dewey Decimal System. Books on the same subject are grouped together by number. For example, books about the arts are assigned numbers between 700 and 799. When stacked on the shelves, the books are to be in numerical order, from left to right. How should the titles in Illustration 11 be rearranged to be in the proper order?

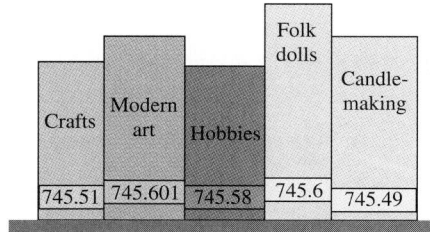

ILLUSTRATION 11

63. OLYMPICS   The results of the women's all-around gymnastic competition in the 1988 Los Angeles Olympic Games are shown in Illustration 12 (on the next page). Which gymnasts won the gold, silver, and bronze medals?

Name	Country	Score
Simona Pauca	Romania	78.675
Ma Yanhong	China	77.85
Julianne McNamara	U.S.A.	78.4
Mary Lou Retton	U.S.A.	79.175
Ecaterina Szabo	Romania	79.125
Laura Cutina	Romania	78.3

ILLUSTRATION 12

64. TUNEUP   The six spark plugs from the engine of a Nissan Quest were removed, and the spark plug gap was checked. (See Illustration 13.) If vehicle specifications call for the gap to be from 0.031 to 0.035 inch, which of the plugs should be replaced?

Cylinder 1:   0.035 in.
Cylinder 2:   0.029 in.
Cylinder 3:   0.033 in.
Cylinder 4:   0.039 in.
Cylinder 5:   0.031 in.
Cylinder 6:   0.032 in.

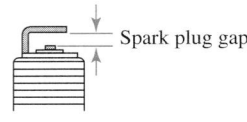

Spark plug gap

ILLUSTRATION 13

65. E-COMMERCE   See Illustration 14. Estimate the loss per share of Amazon.com stock for the third quarter of 1997 and the last quarter of 1998.

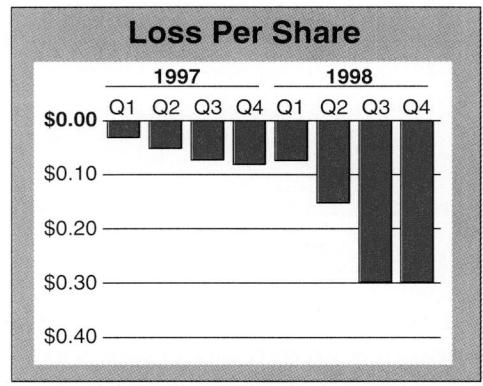

Source: *Los Angeles Times* (January 27, 1999)

ILLUSTRATION 14

66. GASOLINE PRICES   Refer to the data in Illustration 15. Then construct a line graph showing the annual national average retail price per gallon for unleaded regular gasoline for 1992–1998 (according to *The World Almanac 2000*).

Year	1992	1993	1994	1995	1996	1997	1998
Price (¢)	112.7	110.8	111.2	114.7	123.1	123.4	105.9

ILLUSTRATION 15

WRITING

67. Explain the difference between ten and a tenth.

68. "The more digits a number contains, the larger it is." Is this statement true? Explain your response.

69. How are fractions and decimals related?

70. Explain the benefits of a monetary system that is based on decimals instead of fractions.

71. Illustration 16 shows the unusual notation many service stations use to express the price of a gallon of gasoline. Explain what this notation means.

REGULAR   **UNLEADED**   **UNLEADED +**
$1.79\frac{9}{10}$ $1.89\frac{9}{10}$ $1.99\frac{9}{10}$

ILLUSTRATION 16

72. Write a definition for each of these words.

   decade   decathlon   decimal

REVIEW

73. Add: $75\frac{3}{4} + 88\frac{4}{5}$.   **74.** Multiply: $\frac{2}{15}\left(-\frac{5}{4}\right)$

75. Find the area of a triangle with base 16 in. and height 9 in.

76. Express the fraction $\frac{2}{3}$ as an equivalent fraction with a denominator of 12.

77. Add: $-2 + (-3) + 4$.

78. Subtract: $-15 - (-6)$.

4.2 Addition and Subtraction with Decimals

In this section, you will learn about

- Adding decimals • Subtracting decimals
- Adding and subtracting signed decimals

INTRODUCTION. If we are to add or subtract objects, they must be similar. Federal income tax forms illustrate this concept. (See Figure 4-4.) The boxes on the 1040EZ form ensure that dollars are added to dollars and cents added to cents. In this section, we will show how decimals are added and subtracted using this type of vertical column format.

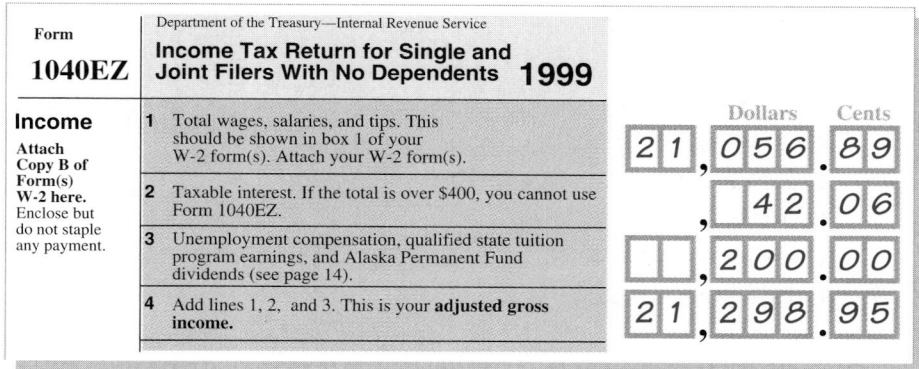

FIGURE 4-4

Adding decimals

When adding decimals, we line up the columns so that ones are added to ones, tenths are added to tenths, hundredths are added to hundredths, and so on. As an example, consider the following problem.

Line up the columns and
the decimal points vertically.
Then add the numbers.

$$
\begin{array}{r}
12.140 \\
3.026 \\
4.000 \\
+\ 0.700 \\
\hline
19.866
\end{array}
$$

Write the decimal point in the result directly under the decimal points in the problem.

Adding decimals

To add decimal numbers:

1. Line up the decimal points, using the vertical column format.
2. Add the numbers as you would add whole numbers.
3. Write the decimal point in the result directly below the decimal points in the problem.

EXAMPLE 1 *Adding decimals.* Add: 1.903 + 0.6 + 8 + 0.78.

Solution

$$\begin{array}{r} 2 \\ 1.903 \\ 0.600 \\ 8.000 \\ +\ 0.780 \\ \hline 11.283 \end{array}$$

To make the addition by columns easier, write two zeros after 6, a decimal point and three zeros after 8, and one zero after 0.78.

Carry a 2 (shown in blue) to the ones column.

The result is 11.283.

Self Check

Add: 0.07 + 35 + 0.888 + 4.1.

Answer: 40.058 ∎

Accent on Technology: **Preventing heart attacks**

The bar graph in Figure 4-5 shows the number of grams of fiber in a standard serving of each of several foods. It is believed that men can significantly cut their risk of heart attack by eating at least 28 grams of fiber a day. Does this diet meet or exceed the 28-gram requirement?

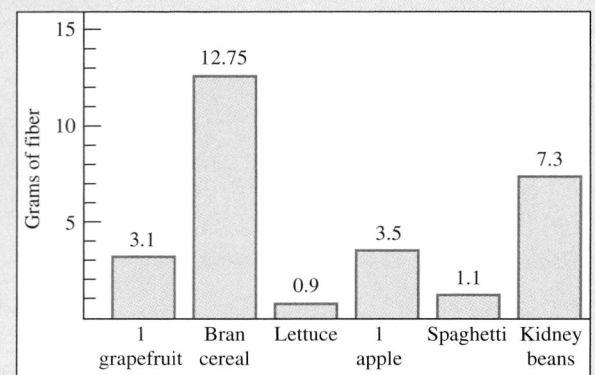

FIGURE 4-5

To find the total fiber intake, we will add the fiber content of each of the foods. We can use a scientific calculator to add the decimals.

Keystrokes   3.1 [+] 12.75 [+] .9 [+] 3.5 [+] 1.1 [+] 7.3 [=]   [28.65]

Since 28.65 > 28, this diet exceeds the daily fiber requirement of 28 grams.

Subtracting decimals

To subtract decimals, we line up the decimal points and corresponding columns so that we subtract like objects—tenths from tenths, hundredths from hundredths, and so on.

Subtracting decimals

To subtract decimal numbers:

1. Line up the decimal points using the vertical column format.
2. Subtract the numbers as you would subtract whole numbers.
3. Write the decimal point in the result directly below the decimal points in the problem.

EXAMPLE 2 *Borrowing.* Subtract: **a.** 279.6 − 138.7 and
b. 15.4 − 13.059.

Solution

a.
```
     8 16
  279.6
 −138.7
 ─────
  140.9
```
To subtract in the tenths column, borrow 1 one in the form of 10 tenths from the ones column. Add 10 to the 6 in the tenths column, which gives 16 (shown in blue).

b.
```
        9
      3 10 10
   15.4 0 0
  −13.0 5 9
  ─────────
    2.3 4 1
```
Add two zeros to the right of 15.4 to make borrowing easier. First, borrow from the tenths column; then borrow from the hundredths column.

Self Check

Subtract:

a. 382.5 − 227.1

b. 30.1 − 27.122

Answers: a. 155.4, **b.** 2.978   ∎

EXAMPLE 3 *Conditioning program.* A 350-pound football player lost 15.7 pounds during the first week of practice. During the second week, he gained 4.9 pounds. What is his weight after the first two weeks of practice?

Solution The word *lost* indicates subtraction. The word *gained* indicates addition.

Beginning weight	minus	first week weight loss	plus	second week weight gain	equals	weight after two weeks of practice.

$$350 - 15.7 + 4.9 = 334.3 + 4.9$$ Working from left to right, do the subtraction first: 350 − 15.7 = 334.3.

$$= 339.2$$ Do the addition.

The player's weight is 339.2 pounds after two weeks of practice.   ∎

Accent on Technology: **Weather balloons**

A giant weather balloon is made of neoprene, a flexible rubberized substance, that has an uninflated thickness of 0.011 inch. When the balloon is inflated with helium, the thickness becomes 0.0018 inch.

To find the change in thickness, we need to subtract. We can use a scientific calculator to subtract the decimals.

Keystrokes   .011 $\boxed{-}$ .0018 $\boxed{=}$ $\boxed{\text{0.0092}}$

After the balloon is inflated, the neoprene loses 0.0092 of an inch in thickness.

Adding and subtracting signed decimals

To add signed decimals, we use the same rules that we used for adding integers.

Adding two decimals

With like signs: Add their absolute values and attach their common sign to the sum.
With unlike signs: Subtract their absolute values (the smaller from the larger) and attach the sign of the number with the larger absolute value to the sum.

EXAMPLE 4   *Adding signed decimals.*   Add: $-6.1 + (-4.7)$.
Solution
Since the decimals are both negative, we add their absolute values and attach a negative sign to the result.

$$-6.1 + (-4.7) = -10.8 \quad \text{Add the absolute values, 6.1 and 4.7, to get 10.8. Use their common sign.}$$

Self Check
Add: $-5.04 + (-2.32)$.

Answer: -7.36 ■

EXAMPLE 5   *Adding signed decimals.*   Add: $5.35 + (-12.9)$.
Solution
In this example, the signs are unlike. Since -12.9 has the larger absolute value, we subtract 5.35 from 12.9 to get 7.55, and attach a negative sign to the result.

$$5.35 + (-12.9) = -7.55$$

Self Check
Add: $-21.4 + 16.75$.

Answer: -4.65 ■

EXAMPLE 6   *Subtracting signed decimals.*   Subtract: $-4.3 - 5.2$.
Solution
To subtract signed decimals, we can add the opposite of the decimal that is being subtracted.

$$-4.3 - 5.2 = -4.3 + (-5.2) \quad \text{Add the opposite of 5.2, which is } -5.2.$$
$$= -9.5 \quad \text{Add the absolute values, 4.3 and 5.2, to get 9.5. Attach a negative sign to the result.}$$

Self Check
Subtract: $-1.18 - 2.88$

Answer: -4.06 ■

EXAMPLE 7   *Subtracting with signed decimals.*   Subtract: $-8.37 - (-6.2)$.
Solution

$$-8.37 - (-6.2) = -8.37 + 6.2 \quad \text{Add the opposite of } -6.2, \text{ which is } 6.2.$$
$$= -2.17 \quad \text{Subtract the smaller absolute value from the larger, 6.2 from 8.37, to get 2.17. Since } -8.37 \text{ has the larger absolute value, the result is negative.}$$

Self Check
Subtract: $-2.56 - (-4.4)$.

Answer: 1.84 ■

EXAMPLE 8   *Grouping symbols.*   Evaluate $-12.2 - (-14.5 + 3.8)$.
Solution
We do the addition within the grouping symbols first.

$$-12.2 - (-14.5 + 3.8) = -12.2 - (-10.7) \quad \text{Do the addition: } -14.5 + 3.8 = -10.7.$$
$$= -12.2 + 10.7 \quad \text{Add the opposite of } -10.7.$$
$$= -1.5 \quad \text{Do the addition.}$$

Self Check
Evaluate: $-4.9 - (-1.2 + 5.6)$.

Answer: -9.3 ■

STUDY SET   Section 4.2

VOCABULARY   *Fill in the blanks.*

1. The answer to an addition problem is called the
_____.

2. The answer to a subtraction problem is called the
_____.

3. Every whole number has an unwritten decimal
_____ to its right.

4. To subtract signed decimals, add the _____ of the decimal that is being subtracted.

CONCEPTS

5. a. Add: 0.3 + 0.17.
   b. Write 0.3 and 0.17 as fractions. Find a common denominator for the fractions and add them.
   c. Express your final answer to part b as a decimal.
   d. Compare your answers from part a and part c.

6. In the subtraction problem below, we must borrow. How much is borrowed from the 3, and in what form is it borrowed?

$$\begin{array}{r} \overset{2\ 11}{29.3\,\cancel{1}} \\ -25.1\,6 \\ \hline \end{array}$$

PRACTICE   *Do each addition.*

7. $\begin{array}{r} 32.5 \\ +\ 7.4 \\ \hline \end{array}$ **8.** $\begin{array}{r} 6.3 \\ +13.5 \\ \hline \end{array}$

9. $\begin{array}{r} 21.6 \\ +33.12 \\ \hline \end{array}$ **10.** $\begin{array}{r} 19.4 \\ +31.95 \\ \hline \end{array}$

11. 12 + 3.9   **12.** 0.01 + 3.6

13. 0.03034 + 0.2003   **14.** 19.9 + 19.9

15. 247.9 + 40 + 0.56   **16.** 0.0053 + 1.78 + 6

17. 45 + 9.9 + 0.12 + 3.02

18. 505.01 + 23 + 0.989 + 12.07

Do each subtraction.

19. $\begin{array}{r} 12.98 \\ -\ 3.45 \\ \hline \end{array}$ **20.** $\begin{array}{r} 1.6 \\ -0.16 \\ \hline \end{array}$

21. $\begin{array}{r} 78.1 \\ -\ 7.81 \\ \hline \end{array}$ **22.** $\begin{array}{r} 202.234 \\ -\ 19.34 \\ \hline \end{array}$

23. 5 − 0.023   **24.** 30 − 11.98

25. 24 − 23.81   **26.** 7.001 − 5.9

Do each addition.

27. −45.6 + 34.7   **28.** −19.04 + 2.4

29. 46.09 + (−7.8)   **30.** 34.7 + (−30.1)

31. −7.8 + (−6.5)   **32.** −5.78 + (−33.1)

33. −0.0045 + (−0.031)   **34.** −90.09 + (−0.087)

Do each subtraction.

35. −9.5 − 7.1   **36.** −7.08 − 14.3

37. 30.03 − (−17.88)   **38.** 143.3 − (−64.01)

39. −2.002 − (−4.6)   **40.** −0.005 − (−8)

41. −7 − (−18.01)   **42.** −63.04 − (−8.911)

Evaluate each expression.

43. 3.4 − 6.6 + 7.3

44. 3.4 − (6.6 + 7.3)

45. (−9.1 − 6.05) − (−51)

46. −9.1 − (−6.05) + 51

47. 16 − (67.2 + 6.27)

48. −43 − (0.032 − 0.045)

49. (−7.2 + 6.3) − (−3.1 − 4)

50. 2.3 + [2.4 − (2.5 − 2.6)]

51. |−14.1 + 6.9| + 8

52. 15 − |−2.3 + (−2.4)|

Add or subtract as indicated.

53. Find the sum of *two and forty-three hundredths* and *five and six-tenths*.

54. Find the difference of *nineteen hundredths* and *six thousandths*.

APPLICATIONS

55. SPORTS PAGE   In the sports pages of any newspaper, decimal numbers are used quite often.
   a. "German bobsledders set a world record today with a final run of 53.03, finishing ahead of the Italian team by only fourteen thousandths of a second." What was the time for the Italian bobsled team?
   b. "The women's figure skating title was decided by only thirty-three hundredths of a point." If the winner's point total was 102.71, what was the second-place finisher's total?

56. NURSING   Illustration 1 (on the next page) shows a patient's health chart. A nurse failed to fill in certain portions. (98.6° Fahrenheit is considered normal.) Complete the chart.

Day of week	Patient's temperature	How much above normal
Monday	99.7°	
Tuesday		2.5°
Wednesday	98.6°	
Thursday	100.0°	
Friday		0.9

ILLUSTRATION 1

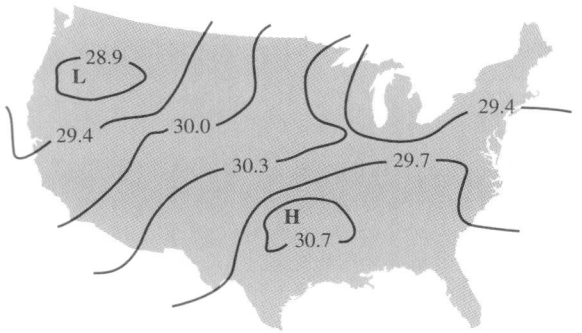

ILLUSTRATION 4

57. VEHICLE SPECIFICATIONS   Certain dimensions of a compact car are shown in Illustration 2. What is the wheelbase of the car?

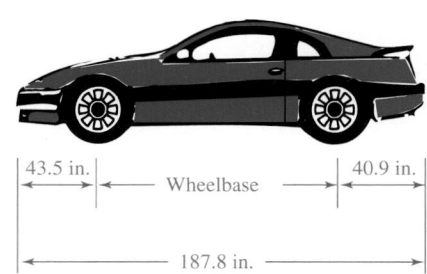

ILLUSTRATION 2

58. pH SCALE   The pH scale shown in Illustration 3 is used to measure the strength of acids and bases in chemistry. Find the difference in pH readings between
a. bleach and stomach acid.
b. ammonia and coffee.
c. blood and coffee.

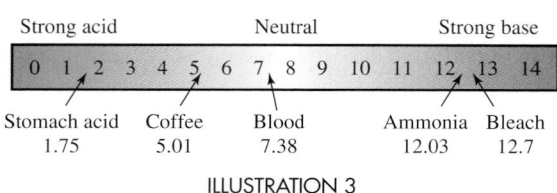

ILLUSTRATION 3

59. BAROMETRIC PRESSURE   Barometric pressure readings are recorded on the weather map in Illustration 4. In a low pressure area (L on the map), the weather is often stormy. The weather is usually fair in a high pressure area (H). What is the difference in readings between the areas of highest and lowest pressure? In what part of the country would you expect the weather to be fair?

60. QUALITY CONTROL   An electronics company has strict specifications for silicon chips used in a computer. The company will install only chips that are within 0.05 centimeters of the specified thickness. Illustration 5 gives that specification for two types of chip. Fill in the blanks to complete the chart.

Chip type	Thickness specification	Acceptable range	
		Low	High
A	0.78 cm		
B	0.643 cm		

ILLUSTRATION 5

61. OFFSHORE DRILLING   A company needs to construct a pipeline from an offshore oil well to a refinery located on the coast. Company engineers have come up with two plans for consideration, as shown in Illustration 6. Use the information in the illustration to complete the table.

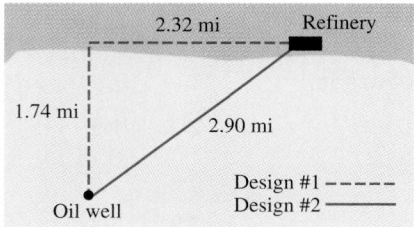

ILLUSTRATION 6

	Pipe underwater (mi)	Pipe underground (mi)	Total pipe (mi)
Design 1			
Design 2			

62. TELEVISION   Illustration 7 shows the six most-watched television shows of all time (excluding Super Bowl games).
  a. What was the combined total audience of all six shows?
  b. How many more people watched the last episode of "MASH" than watched the last episode of "Seinfeld?"
  c. How many more people would have had to watch the last "Seinfeld" to move it into a tie for fifth place?

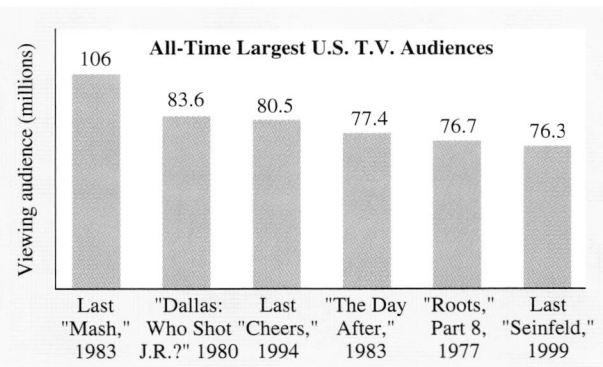

Source: Nielsen Media Research

ILLUSTRATION 7

63. AMERICAN RECORDHOLDERS   The late Florence Griffith-Joyner set the United States national and world record in the 100-meter sprint: 10.49 seconds. Jenny Thompson set the national record in the 100-meter freestyle swim: 54.48 seconds. How much faster did Griffith-Joyner run the 100 meters than Thompson swam it?

64. FLIGHT PATH   See Illustration 8. Find the added distance a plane must travel to avoid flying through the storm.

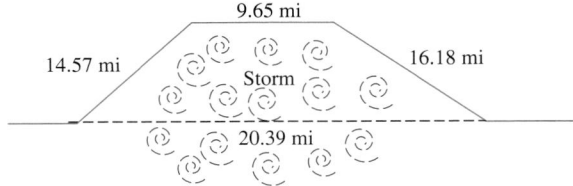

ILLUSTRATION 8

65. DEPOSIT SLIP   A deposit slip for a savings account is shown in Illustration 9. Find the subtotal and then the total deposit.

Deposit

Cash	242	50
Checks (properly endorsed)	116	10
	47	93
Total from reverse side	359	16
Subtotal		
Less cash	25	00
Total deposit		

ILLUSTRATION 9

66. MOTION   Forces such as water current or wind can increase or decrease the speed of an object in motion. Find the speed of each object.
  a. An airplane's speed in still air is 450 mph, and it has a tail wind of 35.5 mph helping it along.

  b. A man can paddle a canoe at 5 mph in still water, but he is going upstream. The speed of the current against him is 1.5 mph.

67. THE HOME SHOPPING NETWORK   Illustration 10 shows a description of a cookware set that was sold on television.
  a. Find the difference between the manufacturer's suggested retail price (MSRP) and the sale price.
  b. Including shipping and handling (S & H), how much will the cookware set cost?

Item 229-442	
Continental 9-piece Cookware Set	
Stainless steel	
MSRP	$149.79
HSN Price	$59.85
On Sale	$47.85
S & H	$7.95

ILLUSTRATION 10

68. RETAILING   Complete Illustration 11 by filling in the retail price of each appliance, given its cost to the dealer and the store markup.

Item	Cost	Markup	Retail price
Refrigerator	$510.80	$105.00	
Washing machine	$289.50	$55.50	
Dryer	$263.99	$67.50	

ILLUSTRATION 11

Evaluate each expression.

69. $2,367.909 + 5,789.0253$

70. $0.00786 + 0.3423$

71. $9,000.09 - 7,067.445$

72. $1 - 0.004999$

73. $3,434.768 - (908 - 2.3 + .0098)$

74. $12 - (0.723 + 3.05611)$

WRITING

75. Explain why we line up the decimal points and corresponding columns when adding decimals.

76. Explain why we can write additional zeros to the right of a decimal such as 7.89 without affecting its value.

77. Explain what is wrong with the work shown below.

Add: $203.56 + 37 + 0.43$.

$$
\begin{array}{r}
203.56 \\
37 \\
0.43 \\
\hline
204.36
\end{array}
$$

78. Consider the addition

$$
\begin{array}{r}
\overset{2}{} \\
23.7 \\
41.9 \\
+\ 12.8 \\
\hline
78.4
\end{array}
$$

Explain the meaning of the small 2 written above the ones column.

REVIEW

79. Add: $44\dfrac{3}{8} + 66\dfrac{1}{5}$.

80. Simplify: $\dfrac{-\dfrac{3}{4}}{\dfrac{5}{16}}$.

81. Multiply: $\dfrac{-15}{26} \cdot 1\dfrac{4}{9}$.

82. Simplify: $2 + 5[-2 - (6 + 1)]$.

4.3 *Multiplication with Decimals*

In this section, you will learn about

- Multiplying decimals   • Multiplying decimals by powers of 10
- Multiplying signed decimals   • Order of operations

INTRODUCTION. In our study of decimals, we now focus on the operation of multiplication. First, we develop a method used to multiply decimals. Then we use that method to evaluate expressions and to solve problems involving decimals.

Multiplying decimals

To develop a rule for multiplying decimals, we will examine the multiplication $0.3 \cdot 0.17$, finding the product in a roundabout way. First, we will write 0.3 and 0.17 as fractions and multiply them. Then we will express the resulting fraction as a decimal.

$$0.3 \cdot 0.17 = \frac{3}{10} \cdot \frac{17}{100} \qquad \text{Express 0.3 and 0.17 as fractions.}$$

$$= \frac{3 \cdot 17}{10 \cdot 100} \qquad \text{Multiply the numerators and multiply the denominators.}$$

$$= \frac{51}{1,000} \qquad \text{Multiply in the numerator and denominator.}$$

$$= 0.051 \qquad \text{Write } \tfrac{51}{1,000} \text{ as a decimal.}$$

From this example, we can make some observations about multiplying decimals.

• The digits in the answer are found by multiplying 3 and 17.

$$0.3 \quad \cdot \quad 0.17 \quad = \quad 0.051$$

$$3 \cdot 17 = 51$$

• The answer has 3 decimal places. The *sum* of the number of decimal places of the factors 0.3 and 0.17 is also 3.

$$0.3 \quad \cdot \quad 0.17 \quad = \quad 0.051$$

1 decimal   2 decimal   3 decimal
place.   places.   places.

These observations suggest the following rule for multiplying decimals.

Multiplying decimals

> To multiply two decimals:
> 1. Multiply the decimals as if they were whole numbers.
> 2. Find the total number of decimal places in both factors.
> 3. Place the decimal point in the result so that the answer has the same number of decimal places as the total found in Step 2.

EXAMPLE 1   *Multiplying decimals.*   Multiply: 5.9 · 3.4.

Solution

We ignore the decimal points and multiply the decimals as if they were whole numbers. Initially, we think of this problem as 59 times 34.

$$
\begin{array}{r}
59 \\
\times \underline{34} \\
236 \\
\underline{177} \\
2006
\end{array}
$$

To place the decimal point in the product, we find the total number of digits to the right of the decimal points of the factors.

$$
\begin{array}{r}
5.9 \quad \leftarrow 1 \text{ decimal place} \\
\times \underline{3.4} \quad \leftarrow 1 \text{ decimal place} \\
236 \\
\underline{177} \\
20.06
\end{array}
$$
The answer will have 1 + 1 = 2 decimal places.

Locate the decimal point so that the answer has 2 decimal places.

Self Check
Multiply: 2.74 · 4.3.

Answer: 11.782   ■

When multiplying decimals, it is not necessary to line up the decimal points, as the next example illustrates.

EXAMPLE 2   *Inserting placeholder zeros.*   Multiply: 1.3(0.005).

Solution

We multiply 13 by 5 and find the total number of decimal places in 1.3 and 0.005.

$$
\begin{array}{r}
1.3 \quad \leftarrow 1 \text{ decimal place} \\
\times \underline{0.005} \quad \leftarrow 3 \text{ decimal places} \\
65
\end{array}
$$
The answer will have 1 + 3 = 4 decimal places.

Self Check
Multiply: (0.0002)7.2.

We then place the decimal point in the result.

$$
\begin{array}{r}
1.3 \\
\times\ 0.005 \\
\hline
0.0065
\end{array}
$$

Add 2 placeholder zeros and position the decimal point so that the product has 4 decimal places.

Answer: 0.00144 ■

Accent on Technology: **Heating costs**

When billing a household, a gas company converts the amount of natural gas used into units of heat energy called *therms*. The number of therms used by a household in one month and the cost per therm are shown below.

Customer charge . . . . 39 therms @ $0.72264

To find the total charges for the month, we multiply the number of therms by the cost per therm: $39 \cdot 0.72264$.

Keystrokes   39 $\boxed{\times}$ .72264 $\boxed{=}$ $\boxed{28.18296}$

Rounding to the nearest cent, we see that the total charge is $28.18.

EXAMPLE 3   *Multiplying a decimal and a whole number.*
Multiply: 234(3.1).

Solution

$$
\begin{array}{r}
234 \\
\times\ \ 3.1 \\
\hline
23\,4 \\
702\ \ \\
\hline
725.4
\end{array}
$$

234 ← No decimal places
3.1 ← 1 decimal place

The answer will have $0 + 1 = 1$ decimal place.

Locate the decimal point so that the answer has 1 decimal place.

Self Check
Multiply: 178(2.7).

Answer: 480.6 ■

Multiplying decimals by powers of 10

The numbers 10, 100, and 1,000 are called *powers of 10*, because they are the results when we evaluate 10^1, 10^2, and 10^3, respectively. To develop a rule to determine the product when multiplying a decimal and a power of 10, we will multiply 8.675 by three different powers of 10.

Multiply: $8.675 \cdot 10$

$$
\begin{array}{r}
8.675 \\
\times\ \ \ \ 10 \\
\hline
0000 \\
8675\ \ \\
\hline
86.750
\end{array}
$$

The answer is 86.75.

Multiply: $8.675 \cdot 100$

$$
\begin{array}{r}
8.675 \\
\times\ \ \ 100 \\
\hline
0000 \\
0000\ \ \\
8675\ \ \ \\
\hline
867.500
\end{array}
$$

The answer is 867.5

Multiply: $8.675 \cdot 1,000$

$$
\begin{array}{r}
8.675 \\
\times\ \ 1000 \\
\hline
0000 \\
0000\ \ \\
0000\ \ \ \\
8675\ \ \ \ \\
\hline
8675.000
\end{array}
$$

The answer is 8,675.

We can make some observations about the results.

- In each case, the answer contains the same digits as the factor 8.675.
- When inspecting the answers, the decimal point in the first factor 8.675 appears to be moved to the right by the multiplication process. The number of decimal places it moves depends on the power of 10 by which 8.675 is multiplied.

One zero in 10

$8.675 \cdot 10 = 86.75$

It moves one place
to the right.

Two zeros in 100

$8.675 \cdot 100 = 867.5$

It moves two places
to the right.

Three zeros in 1,000

$8.675 \cdot 1,000 = 8675$

It moves three places
to the right.

These observations suggest the following rule.

Multiplying a decimal by a power of 10

To multiply a decimal by a power of 10, move the decimal point to the right the same number of places as there are zeros in the power of 10.

EXAMPLE 4 *Multiplying decimals by powers of 10.* Find the product: **a.** $2.81 \cdot 10$ and **b.** $0.076 \cdot 10,000$.

Solution

a. $2.81 \cdot 10 = 28.1$ Since 10 has 1 zero, move the decimal point 1 place to the right.

b. $0.076 \cdot 10,000 = 0760.$ Since 10,000 has 4 zeros, move the decimal point 4 places to the right. Write a placeholder zero (shown in blue).

$= 760$

Self Check

Find the product:

a. $0.721 \cdot 100$

b. $6.08 (1,000)$

Answers: a. 72.1, **b.** 6,080

EXAMPLE 5 *Tachometer.* A tachometer indicates the engine speed of a vehicle, in revolutions per minute (rpm). What engine speed is indicated by the tachometer in Figure 4-6?

Solution

The needle is pointing to 4.5. The notation "RPM × 1000" on the tachometer instructs us to multiply 4.5 by 1,000 to find the engine speed.

$4.5 \cdot 1,000 = 4500$ Since 1,000 has 3 zeros, move the decimal point 3 places to the right. Write 2 placeholder zeros.

$= 4,500$

FIGURE 4-6

The engine speed is 4,500 rpm.

Multiplying signed decimals

Recall that the product of two numbers with like signs is positive, and the product of two numbers with unlike signs is negative.

EXAMPLE 6 *Multiplying signed decimals.* Multiply: **a.** $-1.8(4.5)$ and **b.** $(-1,000)(-59.08)$.

Solution

a. Since the decimals have unlike signs, their product is negative.

$-1.8(4.5) = -8.1$ Multiply the absolute values, 1.8 and 4.5, to get 8.1. Make the result negative.

b. Since the decimals have like signs, their product is positive.

$(-1,000)(-59.08) = 59,080$ Multiply the absolute values, 1,000 and 59.08. Since 1,000 has 3 zeros, move the decimal point 3 places to the right. Write a placeholder zero.

Self Check

Multiply:

a. $6.6(-5.5)$

b. $(-44.968)(-100)$

Answers: a. -36.3, **b.** 4,496.8

EXAMPLE 7   *Evaluating powers of decimals.*   Evaluate   **a.** $(2.4)^3$ and   **b.** $(-0.05)^2$.

Solution

a. $(2.4)^3 = 2.4 \cdot 2.4 \cdot 2.4$ Write 2.4 as a factor 3 times.

$\qquad = 13.824$ Do the multiplication.

13.824 is the *cube* of 2.4.

b. $(-0.05)^2 = (-0.05)(-0.05)$ Write -0.05 as a factor 2 times.

$\qquad = 0.0025$ Do the multiplication. The product of two decimals with like signs is positive.

0.0025 is the *square* of -0.05.

Self Check
Evaluate:

a. $(-1.3)^3$

b. $(0.09)^2$

Answers: a. -2.197,
b. 0.0081 ■

Order of operations

In the remaining examples, we apply the rules for the order of operations to evaluate expressions involving decimals.

EXAMPLE 8   *Order of operations.*   Evaluate $-(0.6)^2 + 5|-3.6 + 1.9|$.

Solution

$-(0.6)^2 + 5|-3.6 + 1.9| = -(0.6)^2 + 5|-1.7|$ Do the addition within the absolute value symbols.

$\qquad = -(0.6)^2 + 5(1.7)$ Simplify: $|-1.7| = 1.7$.

$\qquad = -0.36 + 5(1.7)$ Find the power: $(0.6)^2 = 0.36$.

$\qquad = -0.36 + 8.5$ Do the multiplication: $5(1.7) = 8.5$.

$\qquad = 8.14$ Do the addition.

Self Check
Evaluate:
$-2|-4.4 + 5.6| + (-0.8)^2$.

Answer: -1.76 ■

EXAMPLE 9   *Weekly earnings.*   A cashier's work week is 40 hours. After his daily shift is over, he can work overtime at a rate 1.5 times his regular rate of $7.50 per hour. How much money will he earn in a week if he works 6 hours of overtime?

Solution   First, we need to find his overtime rate, which is 1.5 times his regular rate of $7.50 per hour.

$$1.5(7.50) = 11.25$$

His overtime rate is $11.25 per hour.
To find his total weekly earnings, we use the following fact.

The regular rate	times	40 hours	plus	the overtime rate	times	overtime hours worked	equals	his total earnings.

$$7.50(40) + 11.25(6) = 300 + 67.50 \quad \text{Do the multiplications.}$$
$$= 367.50 \quad \text{Do the addition.}$$

The cashier's earnings for the week are $367.50.   ■

STUDY SET Section 4.3

VOCABULARY *Fill in each blank.*

1. In the multiplication problem 2.89 · 15.7, the numbers 2.89 and 15.7 are called _____. The answer, 45.373, is called the _____.

2. Numbers such as 10, 100, and 1,000 are called _____ of 10.

CONCEPTS *In Exercises 3–4, fill in each blank.*

3. To multiply decimals, multiply them as if they were _____ numbers. The number of decimal places in the product is the same as the _____ of the decimal places of the factors.

4. To multiply a decimal by a power of 10, move the decimal point to the _____ the same number of decimal places as the number of _____ in the power of 10.

5. When we move the decimal point to the right, does the decimal number get larger or smaller?

6. Suppose that the result of multiplying two decimals is 2.300. Write this result in simpler form.

7. a. Multiply $\frac{3}{10}$ and $\frac{7}{100}$.
   b. Now write both fractions from part a as decimals. Multiply them in that form. Compare your results from parts a and b.

8. a. Multiply 0.11 and 0.3.
   b. Now write both decimals in part a as fractions. Multiply them in that form. Compare your results from parts a and b.

PRACTICE *Do each multiplication.*

9. (0.4)(0.2)
10. (0.2)(0.3)
11. (−0.5)(0.3)
12. (0.6)(−0.7)
13. (1.4)(0.7)
14. (2.1)(0.4)
15. (0.08)(0.9)
16. (0.003)(0.9)
17. (−5.6)(−2.2)
18. (−7.1)(−4.1)
19. (−4.9)(0.001)
20. (0.001)(−7.09)

21. (−0.35)(0.24)
22. (−0.85)(0.42)

23. (−2.13)(4.05)
24. (3.06)(−1.82)

25. 16 · 0.6
26. 24 · 0.8
27. −7(8.1)
28. −5(4.7)
29. 0.04(306)
30. 0.02(417)
31. 60.61(−0.3)
32. −70.07 · 0.6

33. −0.2(0.3)(−0.4)
34. −0.1(−2.2)(0.5)

35. 5.5(10)(−0.3)
36. 6.2(100)(−0.8)
37. 4.2 · 10
38. 10 · 7.1
39. 67.164 · 100
40. 708.199 · 100
41. −0.056(10)
42. −100(0.0897)
43. 1,000(8.05)
44. 23.7(1,000)
45. 0.098(10,000)
46. 3.63(10,000)
47. −0.2 · 1,000
48. −1,000 · 1.9

Complete each table.

49.

Decimal	Its square
0.1	
0.2	
0.3	
0.4	
0.5	
0.6	
0.7	
0.8	
0.9	

50.

Decimal	Its cube
0.1	
0.2	
0.3	
0.4	
0.5	
0.6	
0.7	
0.8	
0.9	

Find each power.

51. $(1.2)^2$
52. $(2.3)^2$

53. $(−1.3)^2$
54. $(−2.5)^2$

Evaluate each expression.

55. $-4.6(23.4 - 19.6)$

56. $6.9(9.8 - 8.9)$

57. $(-0.2)^2 + 2(7.1)$

58. $(-6.3)(3) - (1.2)^2$

59. $(-0.7 - 0.5)(2.4 - 3.1)$

60. $(-8.1 - 7.8)(0.3 + 0.7)$

61. $(0.5 + 0.6)^2(-3.2)$

62. $(-5.1)(4.9 - 3.4)^2$

63. $|-2.6| \cdot |-7.2|$

64. $4|-3.1| + 5|-5.5|$

65. $(|-2.6 - 6.7|)^2$

66. $-3|-8.16 + 9.9|$

APPLICATIONS

67. CONCERT SEATING   Two types of tickets were sold for a concert. Floor seating cost \$12.50 a ticket, and balcony seats were \$15.75.
  a. Complete the table in Illustration 1 and find the receipts from each type of ticket.
  b. Find the total receipts from the sale of both types of tickets.

Ticket type	Price	Number sold	Receipts
Floor		1,000	
Balcony		100	

ILLUSTRATION 1

68. CITY PLANNING   In the city map in Illustration 2, the streets form a grid. They are 0.35 mile apart. Find the distance of each trip.
  a. The airport to the Convention Center
  b. City Hall to the Convention Center
  c. The airport to City Hall

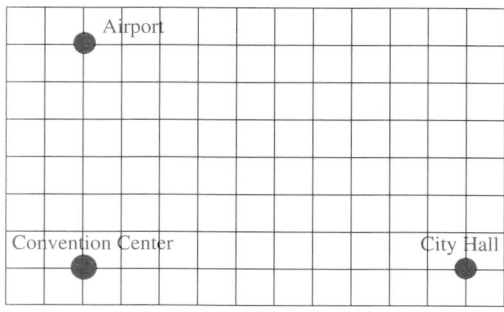

ILLUSTRATION 2

69. STORM DAMAGE   After a rainstorm, the saturated ground under a hilltop house began to give way. A survey team noted that the house dropped 0.57 inch initially. In the next two weeks, the house fell 0.09 inch per week. How far did the house fall during this three-week period?

70. WATER USAGE   In May, the water level of a reservoir reached its high mark for the year. During the summer months, as water usage increased, the level dropped. In the months of May and June, it fell 4.3 feet each month. In August, because of high temperatures, it fell another 8.7 feet. By September, how far below the year's high mark had the water level fallen?

71. WEIGHTLIFTING   The barbell in Illustration 3 is evenly loaded with iron plates. How much plate weight is loaded on the barbell?

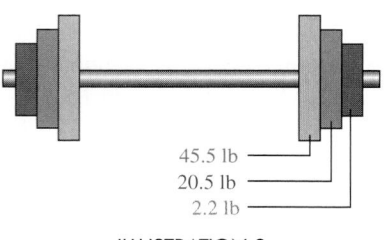

45.5 lb
20.5 lb
2.2 lb

ILLUSTRATION 3

72. PLUMBING BILL   In Illustration 4, an invoice for plumbing work is torn. What is the charge for the 4 hours of work? What is the total charge?

Carter Plumbing 100 W. Dalton Ave.		Invoice #210
Standard sevice charge 4 hr @ \$40.55/hr		\$25.75
	Total	

ILLUSTRATION 4

73. BAKERY SUPPLIES   A bakery buys various types of nuts as ingredients for cookies. Complete Illustration 5 by filling in the cost of each purchase.

Type of nut	Price per pound	Pounds	Cost
Almonds	\$3.25	16	
Walnuts	\$2.10	25	
Peanuts	\$1.85	17	

ILLUSTRATION 5

74. RETROFIT   Illustration 6 shows the width of the three columns of an existing freeway overpass. A computer analysis indicates that each column needs to be increased in width by a factor of 1.4 to ensure stability during an earthquake. According to the analysis, how wide should each of the columns be?

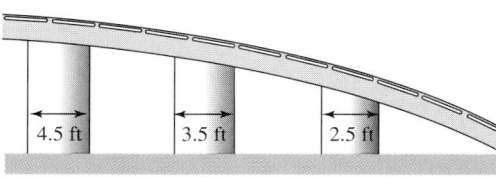

4.5 ft   3.5 ft   2.5 ft

ILLUSTRATION 6

75. SWIMMING POOL CONSTRUCTION   Long bricks, called *coping,* can be used to outline the edge of a swimming pool. How many meters of coping will be needed in the construction of the swimming pool shown in Illustration 7?

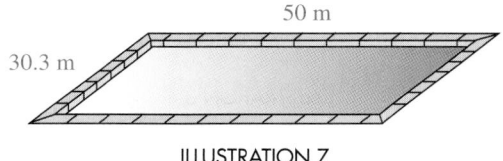

50 m

30.3 m

ILLUSTRATION 7

76. SOCCER   A soccer goal measures 24 feet wide by 8 feet high. Major League Soccer officials are proposing to increase its width by 1.5 feet and increase its height by 0.75 foot.
a. What is the area of the goal opening now?
b. What would it be if their proposal is adopted?

c. How much area would be added?

77. BIOLOGY   DNA is found in cells. It is referred to as the genetic "blueprint." In humans, it determines such traits as eye color, hair color, and height. A model of DNA appears in Illustration 8. If Å = 0.000000004 inch, determine the three dimensions shown in the illustration.

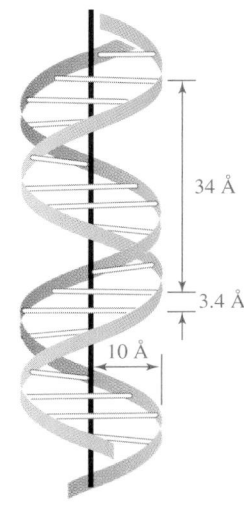

34 Å

3.4 Å

10 Å

ILLUSTRATION 8

78. TACHOMETER   See Illustration 9.
a. To what number is the tachometer needle pointing? Give your estimate in decimal form.
b. What engine speed (in rpm) does the tachometer indicate?

RPM × 1000

ILLUSTRATION 9

Use a calculator to answer each problem.

79. $(-9.0089 + 10.0087)(15.3)$

80. $(-4.32)^3 - 78.969$

81. $(18.18 + 6.61)^2 + (5 - 9.09)^2$

82. $304 - 3.780876(100)$

83. ELECTRIC BILL   When billing a household, a utility company charges for the number of kilowatt-hours used. A kilowatt-hour (kwh) is a standard measure of electricity. If the cost of 1 kwh is \$0.14277, what is the electric bill for a household using 719 kwh in a month? Round the answer to the nearest cent.

84. UTILITY TAX   Some gas companies are required to tax the number of therms used each month by the customer. What are the taxes collected on a monthly usage of 31 therms if the tax rate is \$0.00566 per therm? Round the answer to the nearest cent.

WRITING

85. Explain how to determine where to place the decimal point in the answer when multiplying two decimals.

86. List the similarities and differences between whole-number multiplication and decimal multiplication.

87. What is a decimal place?

88. Explain how to multiply a decimal by a power of 10.

REVIEW

89. Solve $7x + 5 = 54$.

90. Multiply: $3\frac{1}{3}\left(-1\frac{4}{5}\right)$.

91. Write this notation in words: $|-3|$.

92. What is the LCD of fractions with denominators of 4, 5, and 6?

93. Simplify: $-\dfrac{8}{8}$.

94. Find one-half of 7 and square the result.

4.4 **Division with Decimals**

In this section, you will learn about

- Dividing a decimal by a whole number • Divisors that are decimals
- Rounding when dividing • Dividing decimals by powers of 10
- Order of operations

INTRODUCTION. Every division is composed of three parts: the divisor, the dividend, and the quotient.

Long division form

$$\text{Divisor} \longrightarrow 5\overline{)10} \begin{array}{l} \leftarrow \text{Quotient} \\ \leftarrow \text{Dividend} \end{array}$$

Fraction form

$$\begin{array}{l} \text{Dividend} \longrightarrow \\ \text{Divisor} \longrightarrow \end{array} \dfrac{10}{5} = 2 \longleftarrow \text{Quotient}$$

In this section, we examine division problems in which the divisor and/or the dividend are decimals.

Dividing a decimal by a whole number

To use long division to divide 47 by 10, we proceed as follows.

$$10\overline{)47} \quad \begin{array}{l} 4\frac{7}{10} \end{array}$$

Here the result is written in quotient $+ \frac{\text{remainder}}{\text{divisor}}$ form.

$$\dfrac{40}{7}$$

To do this same division using decimals, we write 47 as 47.0 and divide as we would divide whole numbers.

$$10\overline{)47.0} \quad 4.7$$

Note that the decimal point in the result is placed directly above the decimal point of the dividend.

$$\begin{array}{r} 40 \\ \hline 7\,0 \\ 7\,0 \\ \hline 0 \end{array}$$

Since $4\frac{7}{10} = 4.7$, either method gives the same result. The second part of this discussion suggests the following method for dividing a decimal by a whole number.

Dividing a decimal by a whole number

1. Write the problem in long division form.
2. Divide as if working with whole numbers.
3. Write the decimal point in the result directly above the decimal point of the dividend. If necessary, additional zeros can be written to the right of the dividend to allow the division to proceed.

EXAMPLE 1 *Dividing a decimal by a whole number.*
Divide: 71.68 ÷ 28.
Solution

$$
\begin{array}{r}
. \\
28\overline{)71.68}
\end{array}
$$
Write the decimal point in the answer directly above the decimal point of the dividend.

$$
\begin{array}{r}
2.56 \\
28\overline{)71.68} \\
\underline{56} \\
15\,6 \\
\underline{14\,0} \\
1\,68 \\
\underline{1\,68} \\
0
\end{array}
$$
Divide as if working with whole numbers.

The remainder is 0.

The answer is 2.56. We can check this result by multiplying the divisor and the quotient; their product should equal the dividend. Since 28 · 2.56 = 71.68, the result is correct.

Self Check
Divide: 101.44 ÷ 32.

Answer: 3.17 ■

EXAMPLE 2 *Writing extras zeros.* Divide: 19.2 ÷ 5.
Solution

$$
\begin{array}{r}
3.8 \\
5\overline{)19.2} \\
\underline{15} \\
4\,2 \\
\underline{4\,0} \\
2
\end{array}
$$
All the digits in the dividend have been used, but the remainder is not 0.

We can write a zero to the right of 2 in the dividend and continue the division process. Recall that writing additional zeros to the right of the decimal point does not change the value of the decimal.

$$
\begin{array}{r}
3.84 \\
5\overline{)19.20} \\
\underline{15} \\
4\,2 \\
\underline{4\,0} \\
20 \\
\underline{20} \\
0
\end{array}
$$
Write a zero to the right of the 2 and bring it down.

Continue to divide.

The remainder is 0.

The answer is 3.84.

Self Check
Divide: 3.4 ÷ 4.

Answer: 0.85 ■

Divisors that are decimals

When the divisor is a decimal, we change it to a whole number and proceed as in division of whole numbers. To illustrate this procedure, we consider the problem $0.36\overline{)0.2592}$, where the divisor is a decimal. First, we express the division in another form.

$$0.36\overline{)0.2592} \qquad \text{can be represented by} \qquad \frac{0.2592}{0.36}$$

To write the divisor, 0.36, as a whole number, its decimal point needs to be moved two places to the right. This can be accomplished by multiplying it by 100. However, if the denominator of the fraction is multiplied by 100, the numerator must also be multiplied by 100 so that the fraction maintains the same value.

$$\frac{0.2592}{0.36} = \frac{0.2592 \cdot 100}{0.36 \cdot 100}$$ Multiply numerator and denominator by 100.

$$= \frac{25.92}{36}$$ Multiplying by 100 moves both decimal points 2 places to the right.

This fraction represents the division problem $36\overline{)25.92}$. From this result, we can make the following observations.

- The division problem $0.36\overline{)0.2592}$ is equivalent to $36\overline{)25.92}$. That is, they have the same answer.
- The decimal points in *both* the divisor and the dividend of the first division problem have been moved two decimal places to the right to create the second division problem.

$$0.36\overline{)0.2592} \quad \text{becomes} \quad 36\overline{)25.92}$$

These observations suggest the following rule for division with decimals.

Division with a decimal divisor	To divide with a decimal divisor: 1. Move the decimal point of the divisor so that it becomes a whole number. 2. Move the decimal point of the dividend the same number of places to the right. 3. Divide as if working with whole numbers. Write the decimal point in the answer directly above the decimal point of the dividend.

EXAMPLE 3   *Dividing decimals.*   Divide: $\dfrac{0.2592}{0.36}$.

Solution

$0.36\overline{)0.25.92}$ Move the decimal point 2 places to the right in the divisor and dividend.

$$\begin{array}{r} 0.72 \\ 36\overline{)25.92} \\ \underline{25\ 2} \\ 72 \\ \underline{72} \\ 0 \end{array}$$ Now divide as with whole numbers. Write the decimal point in the answer directly above the decimal point of the dividend.

The result is 0.72.

Self Check

Divide: $\dfrac{0.6045}{0.65}$.

Answer: 0.93 ■

Rounding when dividing

In Example 3, the division process ended after we obtained a zero from the second subtraction. We say that the division process **terminated.** Sometimes when dividing, the subtractions never give a zero remainder, and the division process continues forever. In such cases, we can round the result.

EXAMPLE 4   *Rounding when dividing.*   Divide: $\dfrac{2.35}{0.7}$. Round to the nearest hundredth.

Solution   Using long division form, we have $0.7\overline{)2.35}$.

$0.7\overline{)2.3.5}$ To write the divisor as a whole number, move the decimal point one place to the right. Do the same for the dividend. Place the decimal point in the answer directly above the decimal point of the dividend.

$$7\overline{)23.500}$$

To round to the hundredths column, we must divide to the thousandths column. We write two zeros on the right of the dividend.

$$
\begin{array}{r}
3.357 \\
7\overline{)23.500} \\
21 \\
\hline
2\,5 \\
2\,1 \\
\hline
40 \\
35 \\
\hline
50 \\
49 \\
\hline
1
\end{array}
$$

After dividing to the thousandths column, round to the hundredths column. The rounding digit is 5. The test digit is 7.

To the nearest hundredth, the answer is 3.36.   ■

Accent on Technology: **The nucleus of a cell**

The nucleus of a cell contains vital information about the cell in the form of DNA. The nucleus is very small in size: A typical animal cell has a nucleus that is only 0.00023622 inch across. How many nuclei would have to be laid end-to-end to extend to a length of 1 inch?

To find how many 0.00023622-inch lengths there are in 1 inch, we must use division: $1 \div 0.00023622$.

Keystrokes   1 $\boxed{\div}$ .00023622 $\boxed{=}$ $\boxed{4233.3418}$

It would take approximately 4,233 nuclei laid end-to-end to extend to a length of 1 inch.

Dividing decimals by powers of 10

To develop a set of rules for division by a power of 10, we consider the problem $8.13 \div 10$.

$$
\begin{array}{r}
0.813 \\
10\overline{)8.130} \\
0 \\
\hline
8\,1 \\
8\,0 \\
\hline
13 \\
10 \\
\hline
30 \\
30 \\
\hline
0
\end{array}
$$

Write a zero to the right of the 3.

We note that the quotient, 0.813, and the dividend, 8.13, are the same except for the location of the decimal points. The quotient can be easily obtained by moving the decimal point of the dividend 1 place to the *left*. This observation suggests the following rule for dividing a decimal by a power of 10.

Dividing a decimal by a power of 10

To divide a decimal by a power of 10, move the decimal point to the left the same number of places as there are zeros in the power of 10.

EXAMPLE 5   *Dividing decimals by powers of 10.*   Find the quotient:
a. $16.74 \div 10$ and   **b.** $8.6 \div 10,000$.

Solution

a. $16.74 \div 10 = 1.674$ Since 10 has 1 zero, move the decimal point 1 place to the left.

b. $8.6 \div 10,000 = .00086$ Since 10,000 has 4 zeros, move the decimal point 4 places to the left. Write 3 placeholder zeros.

   $= 0.00086$

Self Check

Find the quotient:

a. $721.3 \div 100$

b. $\dfrac{1.07}{1,000}$

Answers: **a.** 7.213,
b. 0.00107   ■

Order of operations

In the next example, we will use the rules for the order of operations to evaluate an expression that involves division by a decimal.

EXAMPLE 6   *Order of operations.*   Evaluate $\dfrac{2(0.351) + 0.5592}{-0.4}$.

Solution

$\dfrac{2(0.351) + 0.5592}{-0.4} = \dfrac{0.702 + 0.5592}{-0.4}$ Do the multiplication first: $2(0.351) = 0.702$.

$\qquad = \dfrac{1.2612}{-0.4}$ Do the addition: $0.702 + 0.5592 = 1.2612$.

$\qquad = -3.153$ Do the division. The quotient of two numbers with unlike signs is negative.

Self Check

Evaluate $\dfrac{2.7756 + 3(-0.63)}{-0.8}$.

Answer: -1.107 ■

STUDY SET   Section 4.4

VOCABULARY   *Fill in the blanks.*

1. In the division $2.5\overline{)4.075} = 1.63$, the decimal 4.075 is called the _____, the decimal 2.5 is the _____, and 1.63 is the _____.

2. In $\dfrac{33.6}{0.3}$, the fraction _____ indicates division.

CONCEPTS   *In Exercises 3–4, fill in the blanks.*

3. To divide by a decimal, move the decimal point of the divisor so that it becomes a _____ number. The decimal point of the dividend is then moved the same number of places to the _____. The decimal point in the quotient is written directly _____ the decimal point of the dividend.

4. To divide a decimal by a power of 10, move the decimal point to the _____ the same number of decimal places as the number of zeros in the power of 10.

5. Is this statement true or false?

   $45 = 45.0 = 45.000$

6. When a positive decimal is divided by 10, is the answer smaller or larger than the original number?

7. To complete the division $7.8\overline{)14.562}$, the decimal points of the divisor and dividend are moved 1 place to the right. This is equivalent to multiplying the numerator and the denominator of $\frac{14.562}{7.8}$ by what number?

8. a. When dividing decimals with like signs, what is the sign of the quotient?

   b. When dividing decimals with unlike signs, what is the sign of the quotient?

9. How can we check the result of this division?

   $\dfrac{1.917}{0.9} = 2.13$

10. When rounding a decimal to the hundredths column, to what other column must we refer?

11. A student performed the division

   $4.6\overline{)9.522}$

   and obtained the answer 2.07. Without doing the division, check this result. Is it correct?

12. In the division problem below, explain *why* we can write the additional zeros (shown in red) after 5. Doesn't this change the problem?

$$16\overline{)5.50000}$$

NOTATION

13. Explain what the arrows are illustrating.

$$4.67\overline{)32.08.7}$$

14. What is this arrow illustrating?

$$
\begin{array}{r}
0.7 \\
4\overline{)3.100} \\
-2\,8\,\downarrow \\
\hline
30
\end{array}
$$

PRACTICE   *Do each division.*

15. $8\overline{)36}$

16. $4\overline{)10}$

17. $-39 \div 4$

18. $-26 \div 8$

19. $49.6 \div 8$

20. $23.5 \div 5$

21. $9\overline{)288.9}$

22. $6\overline{)337.8}$

23. $(-14.76) \div (-6)$

24. $(-13.41) \div (-9)$

25. $\dfrac{-55.02}{7}$

26. $\dfrac{-24.24}{8}$

27. $45\overline{)119.7}$

28. $41\overline{)146.37}$

29. $250.95 \div 35$

30. $241.86 \div 29$

31. $41.6 \div 0.32$

32. $31.8 \div 0.15$

33. $(-199.5) \div (-0.19)$

34. $(-2,381.6) \div (-0.26)$

35. $\dfrac{0.0102}{0.017}$

36. $\dfrac{0.0092}{0.023}$

37. $\dfrac{0.0186}{0.031}$

38. $\dfrac{0.416}{0.52}$

Divide and round each result to the nearest tenth.

39. $3\overline{)16}$

40. $7\overline{)20}$

41. $-5.714 \div 2.4$

42. $-21.21 \div 3.8$

Divide and round each result to the nearest hundredth.

43. $12.243 \div 0.9$

44. $13.441 \div 0.6$

45. $0.04\overline{)0.03164}$

46. $0.08\overline{)0.02201}$

Do the division in your head.

47. $7.895 \div 100$

48. $23.05 \div 10$

49. $0.064 \div (-100)$

50. $0.0043 \div (-10)$

51. $1\,000\overline{)34.8}$

52. $100\overline{)678.9}$

53. $\dfrac{45.04}{10}$

54. $\dfrac{22.32}{100}$

Evaluate each expression. Round each result to the nearest hundredth.

55. $\dfrac{-1.2 - 3.4}{3(1.6)}$

56. $\dfrac{(-1.3)^2 + 6.7}{-0.9}$

57. $\dfrac{40.7(-5.3)}{0.4 - 0.61}$

58. $\dfrac{(0.5)^2 - (0.3)^2}{0.005 + 0.1}$

Evaluate each expression. If an answer is not exact, round it to the nearest hundredth.

59. $\dfrac{5(48.38 - 32)}{9}$

60. $\dfrac{5(19.94 - 32)}{9}$

61. $\dfrac{6.7 - 0.3^2 + 1.6}{0.3^3}$

62. $\dfrac{3.6 - (-1.5)}{0.5(-1.5) - 0.4(3.6)}$

APPLICATIONS

63. BUTCHER SHOP   A meat slicer is designed to trim 0.05-inch-thick pieces from a sausage. If the sausage is 14 inches long, how many slices will result?

64. COMPUTERS   A computer can do an arithmetic computation in 0.00003 second. How many of these computations could it do in 60 seconds?

65. HIKING   Use the information in Illustration 1 to find the time of arrival for the hiker.

ILLUSTRATION 1

66. VOLUME CONTROL   A volume control is shown in Illustration 2. If the distance between the Low and High settings is 21 cm, how far apart are the equally spaced volume settings?

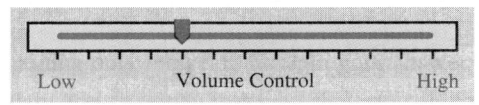

ILLUSTRATION 2

67. SPRAY BOTTLE   Production planners have found that each squeeze of the trigger of a spray bottle emits 0.015 ounce of liquid. How many squeezes would there be in an 8.5-ounce bottle?

68. CAR LOAN   See the loan statement in Illustration 3. How many more monthly payments must be made to pay off the loan?

American Finance Company		June
Monthly payment:	Paid to date: $547.30	
$42.10	Loan balance: $631.50	

ILLUSTRATION 3

69. HOURLY PAY   Illustration 4 shows the average hours worked and the average weekly earnings of U.S. production workers in 1988 and 1998. What did the average production worker earn per hour in 1988 and in 1998? Round to the nearest cent.

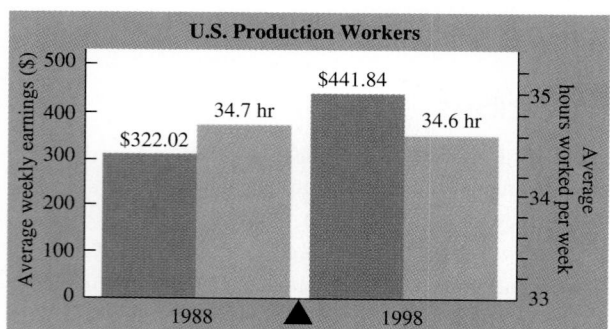

ILLUSTRATION 4

Source: *The World Almanac 2000*

70. PLEASURE TRAVEL   Illustration 5 shows the annual number of person-trips of 100 miles or more (one way) for the years 1994–1998, as estimated by the Travel Industry Association of America. Find the average (mean) number of trips per year for this time span.

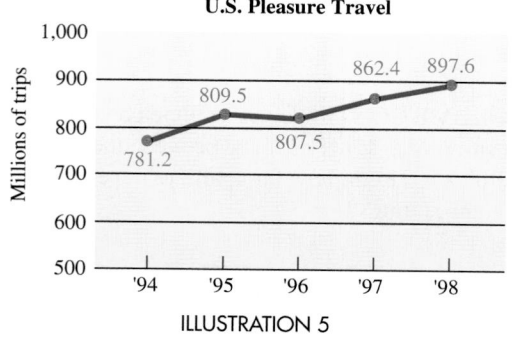

ILLUSTRATION 5

71. OIL WELL   Geologists have mapped out the substances through which engineers must drill to reach an oil deposit. (See Illustration 6.) What is the average depth that must be drilled each week if this is to be a four-week project?

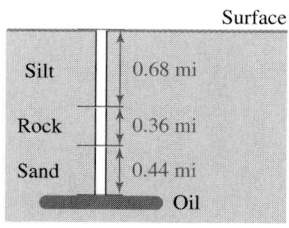

ILLUSTRATION 6

72. INDY 500   Illustration 7 shows the first row of the starting grid for the 1998 Indianapolis 500 automobile race. The drivers' speeds on a qualifying run were used to rank them in this order. What was the average (mean) qualifying speed for the drivers in the first row?

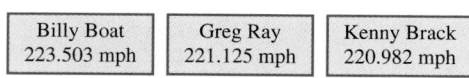

Billy Boat	Greg Ray	Kenny Brack
223.503 mph	221.125 mph	220.982 mph

ILLUSTRATION 7

WRITING

73. Explain the process used to divide two numbers when both the divisor and the dividend are decimals.

74. Explain why we must sometimes use rounding when writing the answer to a division problem.

75. The division $0.5\overline{)2.005}$ is equivalent to $5\overline{)20.05}$. Explain what *equivalent* means in this case.

76. In $3\overline{)0.7}$, why can additional zeros be placed to the right of 0.7 without affecting the outcome?

REVIEW

77. Simplify the complex fraction: $\dfrac{\frac{7}{8}}{\frac{3}{4}}$.

78. Express the fraction $\frac{3}{4}$ as an equivalent fraction with a denominator of 36.

79. List the set of integers.

80. Round 57,205 to the nearest hundred.

81. Solve: $-\dfrac{3}{4}A = -9$.

82. Evaluate $\left(\dfrac{1}{2}\right)^3 - \left(\dfrac{1}{2}\right)^2$.

83. Add: $-2.35 + 0.4 + 27.375$.

84. Subtract: $306.27 - 278.043$.

ESTIMATION

In this section, we will use estimation procedures to approximate the answers to addition, subtraction, multiplication, and division problems involving decimals. You will recall that we use rounding when estimating to help simplify the computations so that they can be performed quickly and easily.

EXAMPLE 1 *Estimating sums and differences.*

a. Estimate to the nearest ten: 261.76 + 432.94.
b. Estimate using front-end rounding: 381.77 − 57.01.

Solution

a. We round each number to the nearest ten.

261.76 + 432.94

260 + 430 = 690   261.76 rounds to 260, and 432.94 rounds to 430.

The estimate is 690. If we compute 261.76 + 432.94, the sum is 694.7. We can see that our estimate is close; it's just 4.7 less than 694.7. Example 1(a) illustrates the tradeoff when using estimation. The calculations are easier to perform and they take less time, but the answers are not exact.

b. We use front-end rounding.

381.77 − 57.01   Each number is rounded to its largest place value: 381.77
                  to the nearest hundred and 57.01 to the nearest ten.

400 − 60 = 340

The estimate is 340.

EXAMPLE 2 *Estimating products.*

Estimate each product: **a.** 6.41 · 27,   **b.** 5.2 · 13.91,   and   **c.** 0.124 · 98.6.

Solution

a. We use front-end rounding.

6.41 · 27 ≈ 6 · 30   The symbol ≈ means "is approximately equal to."

The estimate is 180.

b. We use front-end rounding.

5.2 · 13.91 ≈ 5 · 10

The estimate is 50.

c. Notice that 98.6 ≈ 100.

0.124 · 98.6 ≈ 0.124 · 100   To multiply a decimal by 100, move the
                              decimal point 2 places to the right.

The estimate is 12.4.

When estimating a quotient, we round the divisor and the dividend so that they will divide evenly. Try to round both numbers up or both numbers down.

EXAMPLE 3 *Estimating quotients.* Estimate: 246.03 ÷ 4.31.

Solution

4.31 is close to 4. A multiple of 4 close to 246.03 is 240. (Note that both the divisor and dividend were rounded down.)

$$246.03 \div 4.31 \approx 240 \div 4 \quad \text{Do the division in your head.}$$

The estimate is 60.

Self Check

Estimate: 6,429.6 ÷ 7.19.

Answer: 900 ∎

STUDY SET *Use the following information about refrigerators to estimate the answers to each question. Remember that answers may vary, depending on the rounding method used.*

Deluxe model	**Standard model**	**Economy model**
Price: $978.88	Price: $739.99	Price: $599.95
Capacity: 25.2 cubic feet	Capacity: 20.6 cubic feet	Capacity: 18.8 cubic feet
Energy cost: $6.79 a month	Energy cost: $5.61 a month	Energy cost: $4.39 a month

1. How much more expensive is the deluxe model than the standard model?

2. A couple wants to buy two standard models, one for themselves and one for their newly married son and daughter-in-law. What is the total cost?

3. How much less storage capacity does the economy model have than the standard model?

4. The owner of a duplex apartment wants to purchase a standard model for one unit and an economy model for the other. What will be the total cost?

5. A stadium manager has a budget of $20,000 to furnish the luxury boxes at a football stadium with refrigerators. How many standard models can she purchase for this amount?

6. How many more cubic feet of storage do you get with the deluxe model as compared to the economy model?

7. Three roommates are planning on purchasing the deluxe model and splitting the cost evenly. How much will each have to pay?

8. What is the energy cost per year to run the deluxe model?

9. If you make a $220 down payment on the standard model, how much of the cost is left to finance?

10. The economy model can be expected to last for 10 years. What would be the total energy cost over that period.

Estimate the answer to each problem. Does the calculator result seem reasonable? That is, does it appear that the problem was entered into the calculator correctly?

11. 25.9 + 345.1 + 0.09 $\boxed{\mathtt{347.78}}$

12. 8,345.889 − 345.6 $\boxed{\mathtt{8000.289}}$

13. 42,090.8 + 3,303.09 $\boxed{\mathtt{45393.89}}$

14. 10.007 − 0.626 $\boxed{\mathtt{3.747}}$

15. 9.8(8.8) $\boxed{\mathtt{86.24}}$

16. $\dfrac{24.56}{2.2}$ $\boxed{\mathtt{1.116363636}}$

17. 53 · 5.61 $\boxed{\mathtt{241.23}}$

18. 89.11 ÷ 22.707 $\boxed{\mathtt{39.24340}}$

4.5 *Fractions and Decimals*

In this section, you will learn about

- Writing fractions as equivalent decimals • Repeating decimals
- Rounding repeating decimals • Graphing fractions and decimals
- Problems involving fractions and decimals

INTRODUCTION. In this section, we will further investigate the relationship between fractions and decimals.

Writing fractions as equivalent decimals

To write $\frac{5}{8}$ as a decimal, we use the fact that $\frac{5}{8}$ indicates the division $5 \div 8$. We can convert $\frac{5}{8}$ to decimal form by doing the division.

$$
\begin{array}{r}
.625 \\
8\overline{)5.000} \\
\underline{4\,8} \\
20 \\
\underline{16} \\
40 \\
\underline{40} \\
0
\end{array}
$$
 ← The remainder is zero.

Write a decimal point and additional zeros to the right of 5.

Thus, $\frac{5}{8} = 0.625$.

Writing a fraction as a decimal	To write a fraction as a decimal, divide the numerator of the fraction by its denominator.

EXAMPLE 1   *Writing a fraction as a decimal.*   Write $\dfrac{3}{4}$ as a decimal.

Solution

We divide the numerator by the denominator.

$$
\begin{array}{r}
.75 \\
4\overline{)3.00} \\
\underline{2\,8} \\
20 \\
\underline{20} \\
0
\end{array}
$$
 ← The remainder is zero.

Write a decimal point and two zeros to the right of 3.

Thus, $\frac{3}{4} = 0.75$.

Self Check

Write $\dfrac{3}{16}$ as a decimal.

Answer: 0.1875

In Example 1, the division process ended because a remainder of 0 was obtained. In this case, we call the quotient, 0.75, a **terminating decimal.**

Repeating decimals

Sometimes, when we are finding a decimal equivalent of a fraction, the division process never gives a remainder of zero. In this case, the result is a **repeating decimal.** Examples of repeating decimals are 0.4444. . . and 1.373737. . . . The three dots tell us that a block of digits repeats in the pattern shown. Repeating decimals can be written using a bar over the repeating block of digits. For example, 0.4444. . . can be written as $0.\overline{4}$, and 1.373737. . . can be written as $1.\overline{37}$.

COMMENT   When using an overbar to write a repeating decimal, use the least number of digits necessary to show the repeating block of digits.

$$0.333. . . = 0.\overline{333} \qquad\qquad 6.7454545. . . = 6.7\overline{454}$$
$$0.333. . . = 0.\overline{3} \qquad\qquad 6.7454545. . . = 6.7\overline{45}$$

EXAMPLE 2   *Repeating decimals.*   Write $\dfrac{5}{12}$ as a decimal.

Solution

We use division to find the decimal equivalent.

$$
\begin{array}{r}
.4166 \\
12\overline{)5.0000} \\
\underline{4\ 8} \\
20 \\
\underline{12} \\
80 \\
\underline{72} \\
80 \\
\underline{72} \\
8
\end{array}
$$

Write a decimal point and four zeros to the right of 5.

It is apparent that 8 will continue to reappear as the remainder. Therefore, 6 will continue to reappear in the quotient. Since the repeating pattern is now clear, we may stop the division.

Thus, $\frac{5}{12} = 0.41\overline{6}$.

Self Check

Write $\dfrac{3}{11}$ as a decimal.

Answer: $0.\overline{27}$ ∎

Every fraction can be written as either a terminating decimal or a repeating decimal. For this reason, the set of fractions (**rational numbers**) form a subset of the set of decimals called the set of **real numbers.** The set of real numbers corresponds to *all* points on a number line.

Not all decimals are terminating or repeating decimals. For example,

    0.2020020002 . . .

does not terminate, and it has no repeating block of digits. This decimal cannot be written as a fraction with an integer numerator and a nonzero integer denominator. Thus, it is not a rational number. It is an example from the set of **irrational numbers.**

Rounding repeating decimals

When a fraction is written in decimal form, the result is either a terminating or a repeating decimal. Repeating decimals are often rounded to a specified place value.

EXAMPLE 3   *Rounding the decimal equivalent.*   Write $\frac{1}{3}$ as a decimal and round to the nearest hundredth.

Solution

First, we divide the numerator by the denominator to find the decimal equivalent of $\frac{1}{3}$.

$$
\begin{array}{r}
0.333 \\
3\overline{)1.000} \\
\underline{9} \\
10 \\
\underline{9} \\
10 \\
\underline{9} \\
1
\end{array}
$$

Write a decimal point and additional zeros to the right of 1.

We see that the division process never gives a remainder of zero. When we write $\frac{1}{3}$ in decimal form, the result is the repeating decimal $0.333\ldots = 0.\overline{3}$.

To find the decimal equivalent of $\frac{1}{3}$ to the nearest hundredth, we proceed as follows.

$$0.333\ldots$$

Round 0.333 to the nearest hundredth by examining the test digit in the thousandths column.

Since 3 is less than 5, we round down, and $\frac{1}{3} \approx 0.33$.

EXAMPLE 4 *Rounding a decimal equivalent.* Write $\frac{2}{7}$ as a decimal and round to the nearest thousandth.

Solution

```
      .2857
  7)2.0000    Write a decimal point and additional zeros to the right of 2.
    1 4
    ───
      60
      56
      ──
      40      To round to the thousandths column, we must divide
      35      to the ten thousandths column.
      ──
       50
       49
       ──
        1
```

Round 0.2857 to the nearest thousandth by examining the test digit in the ten thousandths column.

$$0.2857$$

Since 7 is greater than 5, we round up, and $\frac{2}{7} \approx 0.286$. (Read $\approx$ as "is approximately equal to.")

Self Check

Write $\frac{7}{24}$ as a decimal and round to the nearest thousandth.

Answer: 0.292

Accent on Technology: **The fixed-point key**

After performing a calculation, a scientific calculator can round the result to a given decimal place. This is done using the *fixed-point key.* As we did in Example 4, let's find the decimal equivalent of $\frac{2}{7}$ and round to the nearest thousandth. This time, we will use a calculator.

Keystrokes First, we set the calculator to round to the third decimal place (thousandths) by pressing $\boxed{\text{FIX}}$ 3. Then we press 2 $\boxed{\div}$ 7 $\boxed{=}$.

$$\boxed{0.286}$$

Thus, $\frac{2}{7} \approx 0.286$. To round to the nearest tenth, we would fix 1; to round to the nearest hundredth, we would fix 2, and so on.

If your calculator does not have a fixed-point key, see the owner's manual.

EXAMPLE 5 *Writing a mixed number as a decimal.* Write $5\frac{3}{8}$ in decimal form.

Solution

To write a mixed number in decimal form, recall that a mixed number is made up of a whole-number part and a fractional part. Since we can write $5\frac{3}{8}$ as $5 + \frac{3}{8}$, we need only consider how to write $\frac{3}{8}$ as a decimal.

```
      .375
  8)3.000    Write a decimal point and three zeros to the right of 3.
    2 4
    ───
      60
      56
      ──
       40
       40
       ──
        0
```

Self Check

Write $8\frac{19}{20}$ in decimal form.

Thus, $5\frac{3}{8} = 5 + \frac{3}{8} = 5 + 0.375 = 5.375$. We would obtain the same result if we changed $5\frac{3}{8}$ to the improper fraction $\frac{43}{8}$ and divided 43 by 8.

Answer: 8.95

Graphing fractions and decimals

A number line can be used to show the relationship between fractions and their respective decimal equivalents. Figure 4-7 shows some commonly used fractions that have terminating decimal equivalents. For example, we see from the graph that $\frac{13}{16} = 0.8125$.

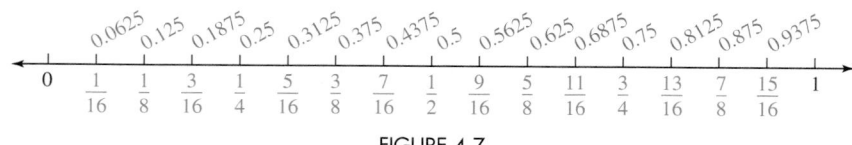

FIGURE 4-7

The number line in Figure 4-8 shows some commonly used fractions that have repeating decimal equivalents.

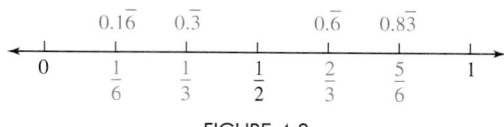

FIGURE 4-8

Problems involving fractions and decimals

Numerical expressions can contain both fractions and decimals. In the following examples, we show how different methods can be used to solve problems of this type.

EXAMPLE 6 *Expressions containing fractions and decimals.*
Evaluate $\frac{1}{3} + 0.27$ by working in terms of fractions.

Solution
We write 0.27 as a fraction and add it to $\frac{1}{3}$.

$\frac{1}{3} + 0.27 = \frac{1}{3} + \frac{27}{100}$ Replace 0.27 with $\frac{27}{100}$.

$= \frac{1 \cdot 100}{3 \cdot 100} + \frac{27 \cdot 3}{100 \cdot 3}$ Express each fraction in terms of 300ths.

$= \frac{100}{300} + \frac{81}{300}$ Multiply in the numerators and in the denominators.

$= \frac{181}{300}$ Add the numerators and write the sum over the common denominator, 300.

Self Check
Evaluate by working in terms of fractions: $0.53 - \frac{1}{6}$.

Answer: $\frac{109}{300}$

EXAMPLE 7 *Expressions containing fractions and decimals.*
Estimate $\frac{1}{3} + 0.27$ by working in terms of decimals.

Solution
We have seen that the decimal equivalent of $\frac{1}{3}$ is the repeating decimal 0.333. . . . To add $\frac{1}{3}$ to 0.27, we round 0.333. . . to the nearest hundredth: $\frac{1}{3} \approx 0.33$.

$\frac{1}{3} + 0.27 \approx 0.33 + 0.27$ Approximate $\frac{1}{3}$ with the decimal 0.33.

≈ 0.60 Do the addition.

Self Check
Estimate by working in terms of decimals: $0.53 - \frac{1}{6}$.

Answer: approximately 0.36

In the previous two examples, we evaluated $\frac{1}{3} + 0.27$ in different ways. In Example 6, we obtained the exact answer, $\frac{181}{300}$. In Example 7, we obtained an approximation, 0.6. It is apparent that the results are in agreement when we write $\frac{181}{300}$ in decimal form: $\frac{181}{300} = 0.60333.\ \ldots$

EXAMPLE 8 *Expressions containing fractions and decimals.*

Evaluate: $\left(\frac{4}{5}\right)(1.35) + (0.5)^2$.

Solution

It appears simplest to work in terms of decimals. We use division to find the decimal equivalent of $\frac{4}{5}$.

$$
\begin{array}{r}
.8 \\
5\overline{)4.0} \\
\underline{4\,0} \\
0
\end{array}
$$
  Write a decimal point and one zero to the right of the 4.

Now we use the rules for the order of operations to evaluate the given expression.

$$\left(\frac{4}{5}\right)(1.35) + (0.5)^2 = (\mathbf{0.8})(1.35) + (0.5)^2 \quad \text{Replace } \tfrac{4}{5} \text{ with its decimal equivalent, } 0.8.$$

$$= (0.8)(1.35) + 0.25 \quad \text{Find the power: } (0.5)^2 = 0.25.$$

$$= 1.08 + 0.25 \quad \text{Do the multiplication: } (0.8)(1.35) = 1.08.$$

$$= 1.33 \quad \text{Do the addition.}$$

Self Check

Evaluate: $(-0.6)^2 + (2.3)\left(\frac{1}{8}\right)$.

Answer: 0.6475

EXAMPLE 9 *Shopping.*   During a trip to the grocery store, a shopper purchased $\frac{3}{4}$ pound of fruit, priced at $0.88 a pound, and $\frac{1}{3}$ pound of fresh-ground coffee, selling for $6.60 a pound. Find the total cost of these items.

Solution

To find the cost of each item, we multiply the amount purchased by its unit price. Then we add the two individual costs to obtain the total cost.

Cost of fruit	plus	cost of coffee	equals	total cost.
$\left(\frac{3}{4}\right)(0.88)$	+	$\left(\frac{1}{3}\right)(6.60)$	=	total cost

Because 0.88 is divisible by 4 and 6.60 is divisible by 3, we can work with the decimals and fractions in this form; no conversion is necessary.

$$\left(\frac{3}{4}\right)(0.88) + \left(\frac{1}{3}\right)(6.60) = \left(\frac{3}{4}\right)\left(\frac{0.88}{1}\right) + \left(\frac{1}{3}\right)\left(\frac{6.60}{1}\right) \quad \text{Express } 0.88 \text{ as } \tfrac{0.88}{1} \text{ and } 6.60 \text{ as } \tfrac{6.60}{1}.$$

$$= \frac{2.64}{4} + \frac{6.60}{3} \quad \text{Multiply the numerators and the denominators.}$$

$$= 0.66 + 2.20 \quad \text{Do each division.}$$

$$= 2.86 \quad \text{Do the addition.}$$

The total cost of the items is $2.86.

STUDY SET Section 4.5

VOCABULARY *Fill in the blanks.*

1. The decimal form of the fraction $\frac{1}{3}$ is a _____ decimal, which is written $0.\overline{3}$ or $0.3333. \ . \ . \ .$

2. The decimal form of the fraction $\frac{2}{5}$ is a _____ decimal, which is written 0.4.

3. The _____ equivalent of $\frac{1}{16}$ is 0.0625.

4. To write a fraction as a decimal, divide the _____ of the fraction by its denominator.

CONCEPTS

5. What division is indicated by the fraction $\frac{7}{8}$?

6. Insert the proper symbol $<$ or $>$ in the blank to make the statement true.
   a. $0.\overline{6}$ _____ 0.7   **b.** $0.\overline{6}$ _____ 0.6

7. When rounding 0.272727. . . to the nearest hundredth, is the result larger or smaller than the original number?

8. Write each decimal in fraction form.
   a. 0.7   **b.** 0.77

9. Graph $1\frac{3}{4}$, -0.75, $0.\overline{6}$, and $-3.8\overline{3}$ on the number line.

10. Graph $2\frac{7}{8}$, -2.375, $0.\overline{3}$, and $4.1\overline{6}$ on the number line.

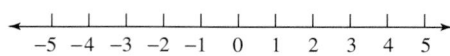

11. Tell whether each statement is true or false.
   a. $\frac{1}{3} = 0.3$ **b.** $\frac{3}{4} = 0.75$

   c. $20\frac{1}{2} = 20.5$ **d.** $\frac{1}{16} = 0.1\overline{6}$

12. When evaluating the expression $0.25 + \left(2.3 + \frac{2}{5}\right)^2$, would it be easier to work in terms of fractions or in terms of decimals?

NOTATION

13. Examine the color portion of the long division in the next column.
   a. Will the remainder ever be zero?

b. What can be deduced about the decimal equivalent of $\frac{5}{6}$?

$$\begin{array}{r} .833 \\ 6\overline{)5.000} \\ \underline{4\ 8} \\ 20 \\ \underline{18} \\ 20 \end{array}$$

14. Write each repeating decimal using an overbar.
   a. 0.888. . .   **b.** 0.323232. . .
   c. 0.56333. . .   **d.** 0.8898989. . .

PRACTICE *Write each fraction in decimal form.*

15. $\frac{1}{2}$ **16.** $\frac{1}{4}$

17. $-\frac{5}{8}$ **18.** $-\frac{3}{5}$

19. $\frac{9}{16}$ **20.** $\frac{3}{32}$

21. $-\frac{17}{32}$ **22.** $-\frac{15}{16}$

23. $\frac{11}{20}$ **24.** $\frac{19}{25}$

25. $\frac{31}{40}$ **26.** $\frac{17}{20}$

27. $-\frac{3}{200}$ **28.** $-\frac{21}{50}$

29. $\frac{1}{500}$ **30.** $\frac{1}{250}$

Write each fraction in decimal form. Use an overbar.

31. $\frac{2}{3}$ **32.** $\frac{7}{9}$

33. $\frac{5}{11}$ **34.** $\frac{4}{15}$

35. $-\frac{7}{12}$ **36.** $-\frac{17}{22}$

37. $\frac{1}{30}$ **38.** $\frac{1}{60}$

Write each fraction in decimal form. Round to the nearest hundredth.

39. $\frac{7}{30}$ **40.** $\frac{14}{15}$

41. $\frac{17}{45}$ **42.** $\frac{8}{9}$

Write each fraction in decimal form. Round to the nearest thousandth.

43. $\dfrac{5}{33}$ **44.** $\dfrac{5}{12}$

45. $\dfrac{10}{27}$ **46.** $\dfrac{17}{21}$

Write each fraction in decimal form. Round to the nearest hundredth.

47. $\dfrac{4}{3}$ **48.** $\dfrac{10}{9}$

49. $-\dfrac{34}{11}$ **50.** $-\dfrac{25}{12}$

Write each mixed number in decimal form. Round to the nearest hundredth when the result is a repeating decimal.

51. $3\dfrac{3}{4}$ **52.** $5\dfrac{4}{5}$

53. $-8\dfrac{2}{3}$ **54.** $-1\dfrac{7}{9}$

55. $12\dfrac{11}{16}$ **56.** $32\dfrac{1}{8}$

57. $203\dfrac{11}{15}$ **58.** $568\dfrac{23}{30}$

Fill in the correct symbol ($<$ or $>$) to make a true statement. (Hint: Express each number as a decimal.)

59. $\dfrac{7}{8}$ ____ 0.895 **60.** 4.56 ____ $4\dfrac{2}{5}$

61. $-\dfrac{11}{20}$ ____ $-0.\overline{4}$ **62.** $-9.0\overline{9}$ ____ $-9\dfrac{1}{11}$

Evaluate each expression. Work in terms of fractions.

63. $\dfrac{1}{9} + 0.3$ **64.** $\dfrac{2}{3} + 0.1$

65. $0.9 - \dfrac{7}{12}$ **66.** $0.99 - \dfrac{5}{6}$

67. $\dfrac{5}{11}(0.3)$ **68.** $(0.9)\left(\dfrac{1}{27}\right)$

69. $\dfrac{1}{3}\left(-\dfrac{1}{15}\right)(0.5)$ **70.** $(-0.4)\left(\dfrac{5}{18}\right)\left(-\dfrac{1}{3}\right)$

Evaluate each expression to the nearest hundredth.

71. $0.24 + \dfrac{1}{3}$ **72.** $0.02 + \dfrac{5}{6}$

73. $5.69 - \dfrac{5}{12}$ **74.** $3.19 - \dfrac{2}{3}$

Evaluate each expression. Work in terms of decimals.

75. $(3.5 + 6.7)\left(-\dfrac{1}{4}\right)$ **76.** $\left(-\dfrac{5}{8}\right)(5.3 - 3.9)$

77. $\left(\dfrac{1}{5}\right)^2(1.7)$ **78.** $(2.35)\left(\dfrac{2}{5}\right)^2$

79. $7.5 - (0.78)\left(\dfrac{1}{2}\right)$ **80.** $8.1 - \left(\dfrac{3}{4}\right)(0.12)$

81. $\dfrac{3}{8}(-3.2) + (4.5)\left(-\dfrac{1}{9}\right)$

82. $(-0.8)\left(\dfrac{1}{4}\right) + \left(\dfrac{1}{3}\right)(0.39)$

83. $\dfrac{3}{4}(3.14)(3)^3$

84. $\dfrac{1}{2}(3.14)(6^2)(12)$

▦ *Write each fraction in decimal form.*

85. $\dfrac{23}{101}$ **86.** $\dfrac{1}{99}$

87. $\dfrac{1{,}736}{50}$ **88.** $-\dfrac{11}{128}$

APPLICATIONS

89. DRAFTING   The architect's scale has several measuring edges. The edge marked 16 divides each inch into 16 equal parts. (See Illustration 1.) Find the decimal form for each fractional part of one inch that is highlighted on the scale.

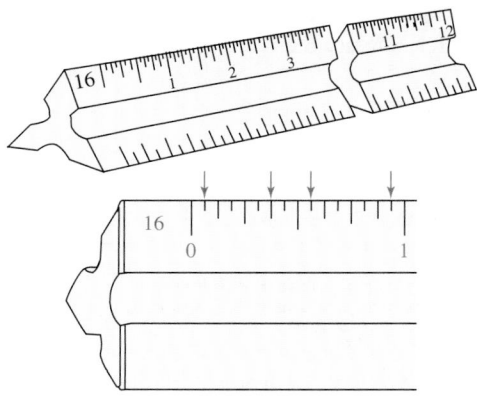

ILLUSTRATION 1

90. FREEWAY SIGNS   The freeway sign in Illustration 2 (on the next page) gives the number of miles to the next three exits. Convert the mileages to decimal notation.

BARRANCA AVE.	$\frac{3}{4}$ mi
210 FREEWAY	$2\frac{1}{4}$ mi
ADA ST.	$3\frac{1}{2}$ mi

ILLUSTRATION 2

91. GARDENING   Two brands of replacement line for a lawn trimmer are labeled in different ways. (See Illustration 3.) On one package, the line's thickness is expressed as a decimal; on the other, as a fraction. Which line is thicker?

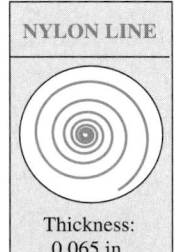 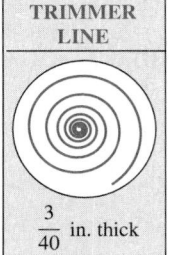

NYLON LINE	TRIMMER LINE
Thickness: 0.065 in.	$\frac{3}{40}$ in. thick

ILLUSTRATION 3

92. AUTO MECHANICS   While doing a tuneup, a mechanic checks the gap on one of the spark plugs of a car to be sure it is firing correctly. The owner's manual states that the gap should be $\frac{2}{125}$ inch. The gauge the mechanic uses to check the gap is in decimal notation; it registers 0.025 inch. Is the spark plug gap too large or too small?

93. HORSE RACING   In thoroughbred racing, the time a horse takes to run a given distance is measured using fifths of a second. For example, 55^2 (read "fifty-five and two") means $55\frac{2}{5}$ seconds. Illustration 4 lists four split times for a horse. Express the times in decimal form.

Speedy Flight   Turfway Park, Ky   3-year–old
17 May 97 $1\frac{1}{16}$ mile   :23^2 :23^4 :24^1 :32^3

ILLUSTRATION 4

94. GEOLOGY   A geologist weighed a rock sample at the site where it was discovered and found it to weigh $17\frac{7}{8}$ lb. Later, a more accurate digital scale in the laboratory gave the weight as 17.671 lb. What is the difference in the two measurements?

95. WINDOW REPLACEMENT   The amount of sunlight that comes into a room depends on the area of the windows in the room. What is the area of the window in Illustration 5?

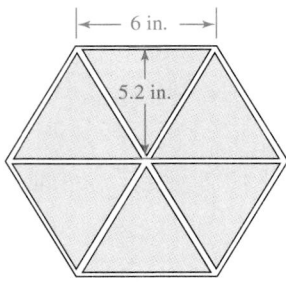

ILLUSTRATION 5

96. FOREST FIRE CONTAINMENT   A command post asked each of three fire crews to estimate the length of the fire line they were fighting. Their reports came back in different forms, as indicated in Illustration 6. Find the perimeter of the fire.

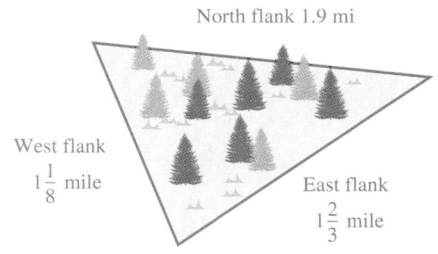

ILLUSTRATION 6

WRITING

97. Explain the procedure used to write a fraction in decimal form.

98. Compare and contrast the two numbers 0.5 and $0.\overline{5}$.

99. A student represented the repeating decimal 0.1333. . . as $0.1\overline{333}$. Is this correct? Explain why or why not.

100. Is 0.10100100010000 . . . a repeating decimal? Explain why or why not.

REVIEW

101. Add: $-2 + (-3) + 10 + (-6)$.

102. Evaluate $-3 + 2[-3 + (2 - 7)]$.

103. Add: $\frac{5}{2} + \frac{2}{3}$.

104. Subtract: $\frac{3}{7} - \frac{1}{5}$.

105. Multiply: $\frac{3}{4} \cdot \frac{8}{9}$.

106. Divide: $\frac{4}{5} \div \frac{9}{10}$.

4.6 **Solving Equations Containing Decimals**

In this section, you will learn about

- Solving equations • Problem solving with equations

INTRODUCTION. We have studied how to add, subtract, multiply, and divide decimals. We will now use these skills to solve equations containing decimals.

Solving equations

Recall that the addition and subtraction properties of equality allow us to add the same number to or subtract the same number from both sides of an equation.

EXAMPLE 1 *Solving equations containing decimals.* Solve each equation: **a.** $x + 3.5 = 7.8$ and **b.** $y - 1.23 = -4.52$.

Solution

a. To isolate x, we undo the addition of 3.5 by subtracting 3.5 from both sides of the equation.

$$x + 3.5 = 7.8$$
$$x + 3.5 - \mathbf{3.5} = 7.8 - \mathbf{3.5} \quad \text{Subtract 3.5 from both sides.}$$
$$x = 4.3 \quad \text{Simplify.}$$

b. To isolate y, we undo the subtraction of 1.23 by adding 1.23 to both sides of the equation.

$$y - 1.23 = -4.52$$
$$y - 1.23 + \mathbf{1.23} = -4.52 + \mathbf{1.23} \quad \text{Add 1.23 to both sides.}$$
$$y = -3.29 \quad \text{Simplify.}$$

Verify each result.

Self Check

Solve each equation.

a. $4.6 + t = 15.7$

b. $-1.24 = r - 0.04$

Answers: a. 11.1, **b.** -1.2

The multiplication property of equality states that we can multiply both sides of an equation by the same nonzero number.

EXAMPLE 2 *Multiplication property of equality.* Solve $\dfrac{m}{2} = -24.8$.

Solution

To isolate m, we undo the division by 2 by multiplying both sides of the equation by 2.

$$\frac{m}{2} = -24.8$$
$$2\left(\frac{m}{2}\right) = \mathbf{2}(-24.8) \quad \text{Multiply both sides by 2.}$$
$$m = -49.6 \quad \text{Do the multiplications.}$$

Check the result.

Self Check

Solve $\dfrac{y}{3} = -13.11$.

Answer: -39.33

The division property of equality states that we can divide both sides of an equation by the same nonzero number.

EXAMPLE 3 *Division property of equality.* Solve $-4.6x = -9.66$.

Solution

To isolate x, we undo the multiplication by -4.6 by dividing by -4.6.

$$-4.6x = -9.66$$

$$\frac{-4.6x}{-4.6} = \frac{-9.66}{-4.6} \quad \text{Divide both sides by } -4.6.$$

$$x = 2.1 \quad \text{Do the divisions.}$$

Check the result.

Sometimes, more than one property must be used to solve an equation. In the next example, we use the addition property of equality and the division property of equality.

EXAMPLE 4 *Solving equations containing decimals.*
Solve $8.1y - 6.04 = -13.33$ and check the result.

Solution

The left-hand side involves a multiplication and a subtraction. To solve the equation, we must undo these operations, but in the opposite order. We begin by undoing the subtraction.

$$8.1y - 6.04 = -13.33$$

$$8.1y - 6.04 + \mathbf{6.04} = -13.33 + \mathbf{6.04} \quad \begin{array}{l}\text{To undo the subtraction of 6.04, add 6.04}\\\text{to both sides.}\end{array}$$

$$8.1y = -7.29 \quad \text{Simplify: } -13.33 + 6.04 = -7.29.$$

$$\frac{8.1y}{8.1} = \frac{-7.29}{8.1} \quad \begin{array}{l}\text{To undo the multiplication of 8.1, divide}\\\text{both sides by 8.1.}\end{array}$$

$$y = -0.9 \quad \text{Do the divisions.}$$

Check:
$$8.1y - 6.04 = -13.33$$

$$8.1(\mathbf{-0.9}) - 6.04 \overset{?}{=} -13.33 \quad \text{Substitute } -0.9 \text{ for } y.$$

$$-7.29 - 6.04 \overset{?}{=} -13.33 \quad \text{Do the multiplication: } 8.1(-0.9) = -7.29.$$

$$-13.33 = -13.33 \quad \begin{array}{l}\text{Do the subtraction by adding the opposite:}\\-7.29 - 6.04 = -7.29 + (-6.04) = -13.33.\end{array}$$

Since $y = -0.9$ checks, it is a solution.

Problem solving with equations

EXAMPLE 5 *Business expenses.* A business decides to rent a copy machine (Figure 4-9) instead of buying one. Under the rental agreement, the company is charged $65 per month plus 2¢ for every copy made. If the business has budgeted $125 for copier expenses each month, how many copies can be made before exceeding the budget?

FIGURE 4-9

Analyze the problem

• The basic rental charge is $65 a month.

• There is a 2¢ charge for each copy made.

- $125 is budgeted for copier expenses each month.
- We must find the maximum number of copies that can be made each month.

Form an equation   Let x = the maximum number of copies that can be made. We can write the amount budgeted for copier expenses in two ways.

The basic fee	plus	the cost of the copies	is	the amount budgeted each month.

We can find the total cost of the copies by multiplying the cost per copy by the maximum number of copies that can be made. Notice that the costs are expressed in terms of dollars and cents. We need to work in terms of one unit, so we write 2¢ as $0.02 and work in terms of dollars.

65	plus	0.02 ·	the maximum number of copies made	is	125.
65	+	0.02 ·	x	=	125

Solve the equation

$$65 + 0.02x = 125$$

$65 + 0.02x - 65 = 125 - 65$ To undo the addition of 65, subtract 65 from both sides.

$$0.02x = 60$$ Simplify.

$$\frac{0.02x}{0.02} = \frac{60}{0.02}$$ To undo the multiplication by 0.02, divide both sides by 0.02.

$$x = 3{,}000$$ Do the divisions.

State the conclusion   The business can make up to 3,000 copies each month without exceeding its budget.

Check the result   If we multiply the cost per copy and the maximum number of copies, we get $0.02 · 3,000 = $60. Then we add the $65 monthly fee: $60 + $65 = $125. The answer checks. ■

STUDY SET Section 4.6

VOCABULARY   *Fill in the blanks.*

1. To _____ an equation, we isolate the variable on one side of the equals sign.

2. The property $a + b = b + a$ is called the _____ property of addition.

3. The property $(a \cdot b) \cdot c = a \cdot (b \cdot c)$ is called the _____ property of multiplication.

4. A _____ is a letter that is used to stand for a number.

CONCEPTS

5. Show that $x = 1.7$ is a solution of $2.1x - 6.3 = -2.73$ by checking it.

6. Show that $y = 0.04$ is a solution of $\frac{y}{2} + 0.7 = 0.72$ by checking it.

NOTATION   *Complete the solution to solve each equation.*

7.
$$0.6s - 2.3 = -1.82$$
$$0.6s - 2.3 + \boxed{} = -1.82 + \boxed{}$$
$$\boxed{} = 0.48$$
$$\frac{0.6s}{\boxed{}} = \frac{0.48}{\boxed{}}$$
$$s = 0.8$$

8.
$$\frac{x}{2} = -6.2$$
$$2\left(\frac{x}{2}\right) = 2(\boxed{})$$
$$x = -12.4$$

PRACTICE   *Solve each equation.*

9. $x + 8.1 = 9.8$ **10.** $6.75 + y = 8.99$

11. $7.08 = t - 0.03$ **12.** $14.1 = k - 13.1$

13. $-5.6 + h = -17.1$ **14.** $-0.05 + x = -1.25$

15. $7.75 = t - (-7.85)$ **16.** $3.33 = y - (-5.55)$

17. $2x = -8.72$ **18.** $3y = -12.63$

19. $-3.51 = -2.7x$ **20.** $-1.65 = -0.5f$

21. $\dfrac{x}{2.04} = -4$ **22.** $\dfrac{y}{2.22} = -6$

23. $\dfrac{-x}{5.1} = -4.4$ **24.** $\dfrac{-t}{8.1} = -3$

25. $\dfrac{1}{3}x = -7.06$ **26.** $\dfrac{1}{5}x = -3.02$

27. $\dfrac{x}{100} = 0.004$ **28.** $\dfrac{y}{1,000} = 0.0606$

29. $2x + 7.8 = 3.4$ **30.** $3x - 1.2 = -4.8$

31. $-0.8 = 5y + 9.2$ **32.** $-9.9 = 6t + 14.1$

33. $0.3x - 2.1 = 7.2$ **34.** $0.4a + 3.3 = -5.1$

35. $-1.5b + 2.7 = 1.2$ **36.** $-2.1x - 3.1 = 5.3$

37. $0.9a - 6 = -5.73$ **38.** $-0.6t + 4 = 3.46$

APPLICATIONS   *Complete each solution.*

39. PETITION DRIVE   On weekends, a college student works for a political organization, collecting signatures for a petition drive. Her pay is $15 a day plus 30 cents for each signature she obtains. How many signatures does she have to collect to make $60 a day?

Analyze the problem

- Her base pay is $\boxed{}$ dollars a day.
- She makes $\boxed{}$ cents for each signature.
- She wants to make $\boxed{}$ dollars a day.
- Find the number of $\boxed{}$ she needs to get.

Form an equation   Let $x = $ _____

We need to work in terms of the same units, so we write 30 cents as $\boxed{}$.

If we multiply the pay per signature by the number of signatures, we get the money she makes just from collecting signatures. Therefore, $\boxed{}$ = total amount (in dollars) made from collecting signatures.

We can express the money she earns in a day in two ways.

Base pay		0.30	·	the number of signatures	is	60.
15	+			$\boxed{}$	=	60

Solve the equation
$$15 + \boxed{} = \boxed{}$$
$$\boxed{} = 45$$
$$x = 150$$

State the conclusion

_____

Check the result
If she collects $\boxed{}$ signatures, she will make $0.30 \cdot \boxed{} = \boxed{}$ dollars from signatures. If we add this to $15, we get $60. The answer checks.

40. HIGHWAY CONSTRUCTION   A 12.8-mile highway is in its third and final year of construction. In the first year, 2.3 miles of the highway were completed. In the second year, 4.9 miles were finished. How many more miles of the highway need to be completed?

Analyze the problem

- The planned highway is $\boxed{}$ miles long.
- The 1st year, $\boxed{}$ miles were completed.
- The 2nd year, $\boxed{}$ miles were completed.
- Find the number of $\boxed{}$ yet to be completed.

A diagram will help us understand the problem.

12.8-mi highway

2.3 mi	$\boxed{}$ mi	? mi
1st year	2nd year	3rd year

Form an equation
Let x = _____
We can express the length of the highway in two ways.

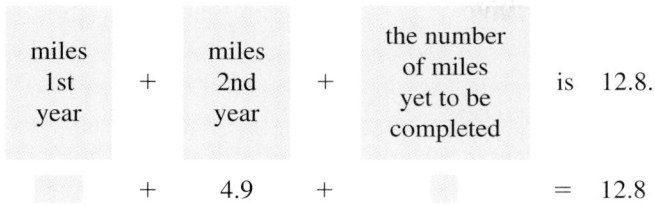

| miles 1st year | + | miles 2nd year | + | the number of miles yet to be completed | is | 12.8. |

| | + | 4.9 | + | | = | 12.8 |

Solve the equation
$$2.3 + 4.9 + \quad = 12.8$$
$$\quad + \quad = 12.8$$
$$x = \quad$$

State the conclusion
_____

Check the result
Add: ___ + ___ + ___ = 12.8. The answer checks.

Choose a variable to represent the unknown. Then write and solve an equation to answer the question.

41. DISASTER RELIEF   After hurricane damage estimated at \$27.9 million, a county looked to three sources for relief. Local agencies contributed \$6.8 million toward the cleanup. A state emergency fund offered another \$12.5 million. When applying for federal government help, how much should the county ask for?

42. TELETHON   Midway through a telethon, the donations had reached \$16.7 million. How much more was donated in the second half of the program if the final total pledged was \$30 million?

43. GPA   After receiving her grades for the fall semester, a college student noticed that her overall GPA had dropped by 0.18. If her new GPA was 3.09, what was her GPA at the beginning of the fall semester?

44. MONTHLY PAYMENTS   A food dehydrator offered on a home shopping channel can be purchased by making 3 equal monthly payments. If the price is \$113.25, how much is each monthly payment?

45. POINTS PER GAME   As a senior, a college basketball player's scoring average was double that of her junior season. If she averaged 21.4 points a game as a senior, how many did she average as a junior?

46. NUTRITION   One 3-ounce serving of broiled ground beef has 7 grams of saturated fat. This is 14 times the amount of saturated fat in 1 cup of cooked crab meat. How many grams of saturated fat are in 1 cup of cooked crab meat?

47. FUEL EFFICIENCY   Each year, the Federal Highway Administration determines the number of vehicle-miles traveled in the country and divides it by the amount of fuel consumed to get an average miles per gallon (mpg). Illustration 1 shows how the figure has changed over the years to reach a high of 16.7 mpg in 1998. What was the average miles per gallon in 1960?

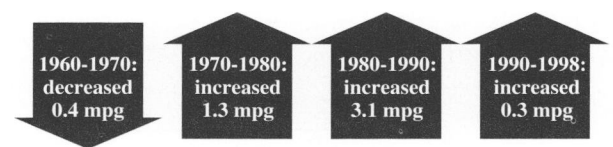

ILLUSTRATION 1

48. RATINGS REPORT   Illustration 2 shows the prime-time television ratings for the week of January 3, 2000. If the Fox network ratings had been $\frac{1}{2}$ point higher, there would have been a three-way tie for second place. What prime time rating did Fox have that week?

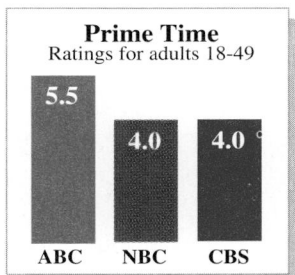

Source: Nielsen Media Research

ILLUSTRATION 2

49. CALLIGRAPHY   A city honors its citizen of the year with a framed certificate. A calligrapher charges \$20 for the frame and then 15 cents a word for writing out the proclamation. If the city charter prohibits gifts in excess of \$50, what is the maximum number of words that can be printed on the award?

50. HELIUM BALLOONS   The organizer of a jog-a-thon wants an archway of balloons constructed at the finish line of the race. A company charges a \$100 setup fee and then 8 cents for every balloon. How many balloons will be used if \$300 is spent for the decoration?

WRITING

51. Did you encounter any differences in solving equations containing decimals as compared to solving equations containing only integers? Explain your answer.

52. Explain how to verify that $x = 29.2$ is a solution of the equation $0.2x - 0.16 = 5.68$.

REVIEW

53. Add: $-\dfrac{2}{3} + \dfrac{3}{4}$.

54. Add: $-\dfrac{2}{3} + \dfrac{1}{5}$.

55. Divide: $\dfrac{7}{8} \div \dfrac{13}{16}$.

56. Multiply: $2\dfrac{1}{3} \cdot 4\dfrac{1}{2}$.

57. Evaluate: $\dfrac{-3-3}{-3+4}$.

58. Write a complex fraction using $-\dfrac{4}{5}$ and $\dfrac{1}{5}$.

59. Solve $\dfrac{6}{5}x = 10$.

60. Solve $\dfrac{2}{3}x + 1 = 7$.

4.7 *Square Roots*

In this section, you will learn about

- Square roots • Evaluating numerical expressions containing radicals
- Square roots of fractions and decimals
- Using a calculator to find square roots • Approximating square roots

INTRODUCTION. There are six basic operations of arithmetic. We have seen the relationships between addition and subtraction and between multiplication and division. In this section, we will explore the relationship between raising a number to a power and finding a root. Decimals will play an important role in this discussion.

Square roots

When we raise a number to the second power, we are squaring it, or finding its **square.**

The square of 6 is 36, because $6^2 = 36$.

The square of -6 is 36, because $(-6)^2 = 36$.

The **square root** of a given number is a number whose square is the given number. For example, the square roots of 36 are 6 and -6, because either number, when squared, yields 36. We can express this concept using symbols.

Square root | A number b is the **square root** of a if $b^2 = a$.

EXAMPLE 1   *Finding square roots.*   Find the square roots of 49.

Solution

Ask yourself, "What number was squared to obtain 49?"

$7^2 = 49$ and $(-7)^2 = 49$

Thus, 7 and -7 are the square roots of 49.

Self Check

Find the square roots of 64.

Answers: 8 and -8 ∎

In Example 1, we saw that 49 has two square roots—one positive and one negative. The symbol $\sqrt{}$ is called a **radical sign** and is used to indicate a positive square root.

When a number, called the **radicand,** is written under a radical sign, we have a **radical expression.** Some examples of radical expressions are

$$\sqrt{36} \qquad \sqrt{100} \qquad \sqrt{144} \qquad \sqrt{81}$$

To evaluate (or simplify) a radical expression, we need to find the positive square root of the radicand. For example, if we evaluate $\sqrt{36}$ (read as "the square root of 36"), the result is

$$\sqrt{36} = 6$$

because $6^2 = 36$. The negative square root of 36 is denoted $-\sqrt{36}$, and we have

$$-\sqrt{36} = -6$$

EXAMPLE 2 *Evaluating radical expressions.*   Simplify each expression: **a.** $\sqrt{81}$ and **b.** $-\sqrt{100}$.

Solution

a. $\sqrt{81}$ means the positive square root of 81. Since $9^2 = 81$, we get $\sqrt{81} = 9$.

b. $-\sqrt{100}$ means the negative square root of 100. Since $10^2 = 100$, we get $-\sqrt{100} = -10$.

Self Check

Simplify each radical expression:

a. $\sqrt{144}$ and **b.** $-\sqrt{81}$.

Answers: **a.** 12, **b.** -9 ■

COMMENT Radical expressions such as

$$\sqrt{-36} \qquad \sqrt{-100} \qquad \sqrt{-144} \qquad \sqrt{-81}$$

do not represent real numbers. This is because there are no real numbers that, when squared, yield a negative number.

Be careful to note the difference between pairs of expressions such as $-\sqrt{36}$ and $\sqrt{-36}$. We have seen that $-\sqrt{36}$ does have meaning: $-\sqrt{36} = -6$. On the other hand, $\sqrt{-36}$ does not have meaning as a real number.

Evaluating numerical expressions containing radicals

Numerical expressions can contain radical expressions. When applying the rules for the order of operations, we treat a radical expression as we would a power.

EXAMPLE 3 *Evaluating expressions containing radicals.*

Evaluate: **a.** $\sqrt{64} + \sqrt{9}$ and **b.** $-\sqrt{25} - \sqrt{4}$.

Solution

a. $\sqrt{64} + \sqrt{9} = 8 + 3$ Evaluate each radical expression first.

  $= 11$ Do the addition.

b. $-\sqrt{25} - \sqrt{4} = -5 - 2$ Evaluate each radical expression first.

  $= -7$ Do the subtraction.

Self Check

Evaluate each expression:

a. $\sqrt{121} + \sqrt{1}$

b. $-\sqrt{9} - \sqrt{16}$

Answers: **a.** 12, **b.** -7 ■

EXAMPLE 4 *Evaluating expressions containing radicals.*

Evaluate **a.** $6\sqrt{100}$ and **b.** $-5\sqrt{16} + 3\sqrt{9}$.

Solution

a. We note that $6\sqrt{100}$ means $6 \cdot \sqrt{100}$.

  $6\sqrt{100} = 6(10)$ Simplify the radical first.

  $= 60$ Do the multiplication.

b. $-5\sqrt{16} + 3\sqrt{9} = -5(4) + 3(3)$ Simplify each radical first.

  $= -20 + 9$ Do the multiplications.

  $= -11$ Do the addition.

Self Check

Evaluate each expression:

a. $8\sqrt{121}$

b. $-6\sqrt{25} + 2\sqrt{36}$

Answers: **a.** 88, **b.** -18 ■

Square roots of fractions and decimals

So far, we have found square roots of whole numbers. We can also find square roots of fractions and decimals.

EXAMPLE 5   *Square roots of fractions and decimals.*

Simplify   **a.** $\sqrt{\dfrac{25}{64}}$ and   **b.** $\sqrt{0.81}$.

Solution

a. $\sqrt{\dfrac{25}{64}} = \dfrac{5}{8}$, because $\left(\dfrac{5}{8}\right)^2 = \dfrac{25}{64}$.

b. $\sqrt{0.81} = 0.9$, because $(0.9)^2 = 0.81$.

Self Check

Simplify

a. $\sqrt{\dfrac{16}{49}}$ and   **b.** $\sqrt{0.04}$.

Answer: **a.** $\dfrac{4}{7}$,   **b.** 0.2

Using a calculator to find square roots

We can also use a calculator to find square roots.

Accent on Technology: **Finding a square root**

We use the $\boxed{\sqrt{}}$ key (square root key) on a scientific calculator to find square roots. For example, to find $\sqrt{729}$, we enter these numbers and press these keys.

Keystrokes   729 $\boxed{\sqrt{}}$ $\boxed{ 27}$

We have found that $\sqrt{729} = 27$. To check this result, we need to square 27. This can be done by entering 27 and pressing the $\boxed{x^2}$ key. We obtain 729. Thus, 27 is the square root of 729.

Approximating square roots

Numbers whose square roots are whole numbers are called **perfect squares.** The perfect squares that are less than or equal to 100 are

0, 1, 4, 9, 16, 25, 36, 49, 64, 81, 100

To find the square root of a number that is not a perfect square, we can use a calculator. For example, to find $\sqrt{17}$, we enter 17 and press the square root key.

17 $\boxed{\sqrt{}}$

The display reads 4.123105626. This result is not exact, because $\sqrt{17}$ is a **nonterminating decimal** that never repeats. $\sqrt{17}$ is an **irrational number.** Together, the rational and the irrational numbers form the set of **real numbers.** If we round to the nearest thousandth, we have

$\sqrt{17} \approx 4.123$ Read $\approx$ as "is approximately equal to."

EXAMPLE 6   *Approximating square roots.*   Use a scientific calculator to find each square root. Round to the nearest hundredth.

a. $\sqrt{373}$ **b.** $\sqrt{56.2}$ **c.** $\sqrt{0.0045}$

Self Check

Use a scientific calculator to find each square root. Round to the nearest hundredth.

Solution

a. From the calculator, we get $\sqrt{373} \approx 19.31320792$. Rounding to the nearest hundredth, $\sqrt{373}$ is 19.31.

b. From the calculator, we get $\sqrt{56.2} \approx 7.496665926$. Rounding to the nearest hundredth, $\sqrt{56.2}$ is 7.50.

c. From the calculator, we get $\sqrt{0.0045} \approx 0.067082039$. Rounding to the nearest hundredth, $\sqrt{0.0045}$ is 0.07.

a. $\sqrt{607.8}$

b. $\sqrt{0.076}$

Answers: **a.** 24.65,   **b.** 0.28   ■

STUDY SET   Section 4.7

VOCABULARY   *Fill in the blanks.*

1. When we find what number is squared to obtain a given number, we are finding the square _____ of the given number.

2. Whole numbers such as 25, 36, and 49 are called _____ squares because their square roots are whole numbers.

3. The symbol $\sqrt{}$ is called a _____ sign. It indicates that we are to find a _____ square root.

4. The decimal number that represents $\sqrt{17}$ is a _____ decimal—it never ends.

5. In $\sqrt{26}$, 26 is called the _____.

6. The symbol $\approx$ means _____.

CONCEPTS   *In Exercises 7–12, fill in the blanks.*

7. The square of 5 is ___, because $(5)^2 =$ ___.

8. The square of $\frac{1}{4}$ is ___, because $\left(\frac{1}{4}\right)^2 =$ ___.

9. The two square roots of 49 are 7 and -7, because $7^2 = 49$ and ___ $= 49$.

10. The two square roots of 4 are 2 and -2, because ___ $= 4$ and ___ $= 4$.

11. Since $\left(\frac{3}{4}\right)^2 = \frac{9}{16}$, we know that $\sqrt{\frac{9}{16}} =$ ___.

12. Since $(0.4)^2 = 0.16$, we know that $\sqrt{0.16} =$ ___.

13. Without evaluating the following square roots, write them in order, from smallest to largest: $\sqrt{23}$, $\sqrt{11}$, $\sqrt{27}$, $\sqrt{6}$.

14. Without evaluating the following square roots, write them in order from smallest to largest: $-\sqrt{13}$, $-\sqrt{5}$, $-\sqrt{17}$, $-\sqrt{37}$.

15. Simplify.

a. $\sqrt{1}$

b. $\sqrt{0}$

16. Multiplication can be thought of as the opposite of division. What is the opposite of finding the square root of a number?

In Exercises 17–22, use a calculator.

17. a. Approximate $\sqrt{6}$ to the nearest tenth.
b. Square the result from part a.
c. Find the difference between 6 and the answer to part b.

18. a. Approximate $\sqrt{6}$ to the nearest hundredth.
b. Square the result from part a.
c. Find the difference between the answer to part b and 6.

19. Graph $\sqrt{9}$ and $-\sqrt{5}$ on the number line.

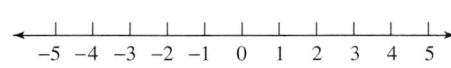

20. Graph $-\sqrt{3}$ and $\sqrt{7}$ on the number line.

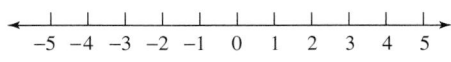

21. Between what two whole numbers would each square root be located when graphed on a number line?

a. $\sqrt{19}$

b. $\sqrt{87}$

22. Between what two whole numbers would each square root be located when graphed on a number line?

a. $\sqrt{50}$

b. $\sqrt{33}$

NOTATION *Complete each solution.*

23. Simplify: $-\sqrt{49} + \sqrt{64}$.
$$-\sqrt{49} + \sqrt{64} = \boxed{} + \boxed{}$$
$$= 1$$

24. Simplify: $2\sqrt{100} - 5\sqrt{25}$.
$$2\sqrt{100} - 5\sqrt{25} = 2(\boxed{}) - 5(\boxed{})$$
$$= \boxed{} - 25$$
$$= -5$$

PRACTICE *Simplify. Do not use a calculator.*

25. $\sqrt{16}$ **26.** $\sqrt{64}$

27. $-\sqrt{121}$ **28.** $-\sqrt{144}$

29. $-\sqrt{0.49}$ **30.** $-\sqrt{0.64}$

31. $\sqrt{0.25}$ **32.** $\sqrt{0.36}$

33. $\sqrt{0.09}$ **34.** $\sqrt{0.01}$

35. $-\sqrt{\dfrac{1}{81}}$ **36.** $-\sqrt{\dfrac{1}{4}}$

37. $-\sqrt{\dfrac{16}{9}}$ **38.** $-\sqrt{\dfrac{64}{25}}$

39. $\sqrt{\dfrac{4}{25}}$ **40.** $\sqrt{\dfrac{36}{121}}$

41. $5\sqrt{36} + 1$ **42.** $2 + 6\sqrt{16}$

43. $-4\sqrt{36} + 2\sqrt{4}$ **44.** $-6\sqrt{81} + 5\sqrt{1}$

45. $\sqrt{\dfrac{1}{16}} - \sqrt{\dfrac{9}{25}}$ **46.** $\sqrt{\dfrac{25}{9}} - \sqrt{\dfrac{64}{81}}$

47. $5(\sqrt{49})(-2)$ **48.** $(-\sqrt{64})(-2)(3)$

49. $\sqrt{0.04} + 2.36$ **50.** $\sqrt{0.25} + 4.7$

51. $-3\sqrt{1.44}$ **52.** $-2\sqrt{1.21}$

Use a calculator to complete each square root table. Round to the nearest thousandth when necessary.

53.

Number	Square root
1	
2	
3	
4	
5	
6	
7	
8	
9	
10	

54.

Number	Square root
10	
20	
30	
40	
50	
60	
70	
80	
90	
100	

Use a calculator to simplify each of the following.

55. $\sqrt{1,369}$ **56.** $\sqrt{841}$

57. $\sqrt{3,721}$ **58.** $\sqrt{5,625}$

Use a calculator to approximate each of the following to the nearest hundredth.

59. $\sqrt{15}$ **60.** $\sqrt{51}$

61. $\sqrt{66}$ **62.** $\sqrt{204}$

Use a calculator to approximate each of the following to the nearest thousandth.

63. $\sqrt{24.05}$ **64.** $\sqrt{70.69}$

65. $-\sqrt{11.1}$ **66.** $\sqrt{0.145}$

Use a calculator to evaluate each radical expression. If an answer is not exact, round to the nearest ten thousandth.

67. $\sqrt{24,000,201}$ **68.** $-\sqrt{4.012009}$

69. $-\sqrt{0.00111}$ **70.** $\sqrt{\dfrac{27}{44}}$

APPLICATIONS *In Exercises 71–76, square roots have been used to express various lengths. Solve each problem by simplifying any square roots. You may need to use a calculator. If so, round to the nearest tenth.*

71. CARPENTRY Find the length of the slanted side of each roof truss.

a.

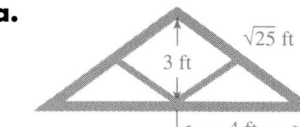

b.

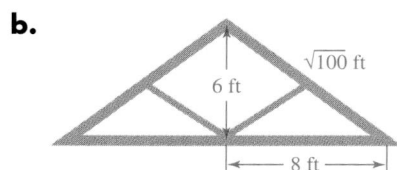

72. RADIO ANTENNA   See Illustration 1. How far from the base of the antenna is each guy wire anchored to the ground? (The measurements are in feet.)

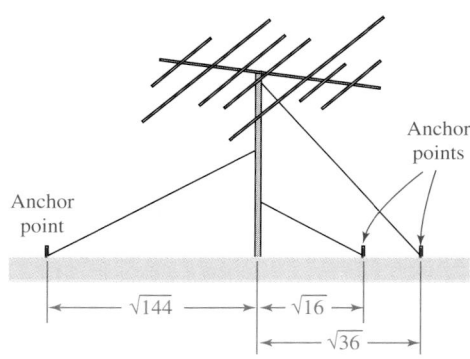

ILLUSTRATION 1

73. BASEBALL DIAMOND   Illustration 2 shows some dimensions of a major league baseball field. How far is it from home plate to second base?

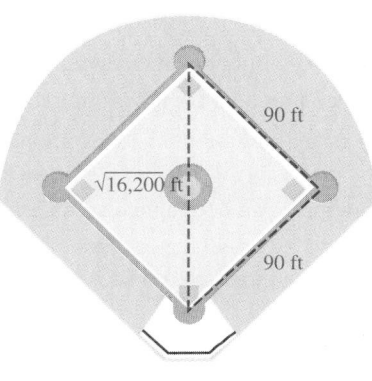

ILLUSTRATION 2

74. SURVEYING   Use the imaginary triangles set up by a surveyor to find the length of each lake. (The measurements are in meters.)

a.

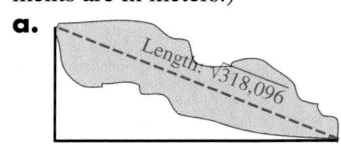

b.

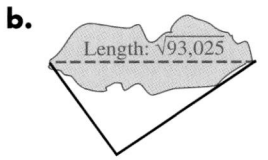

75. BIG-SCREEN TELEVISION   The picture screen on a television set is measured diagonally. What size screen is shown in Illustration 3?

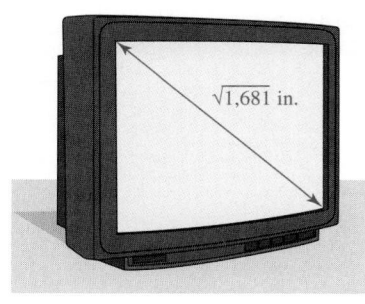

ILLUSTRATION 3

76. LADDER   A painter's ladder is shown in Illustration 4. How long are the legs of the ladder?

ILLUSTRATION 4

WRITING

77. When asked to find $\sqrt{16}$, one student's answer was 8. Explain his misunderstanding of the concept of square root.

78. Explain the difference between the square and the square root of a number.

79. What is a nonterminating decimal? Use an example in your explanation.

80. How would you check whether 17.1 is a square root of 292.41?

81. Explain why $\sqrt{-4}$ does not have meaning as a real number.

82. Is there a difference between $-\sqrt{25}$ and $\sqrt{-25}$? Explain.

REVIEW

83. When solving the equation $2x - 5 = 11$, what operations must be undone in order to isolate the variable?

84. Simplify: $\dfrac{-\dfrac{2}{3}}{8}$.

85. Evaluate: $5(-2)^2 - \dfrac{16}{4}$.

86. Subtract: $\dfrac{5}{8} - \dfrac{3}{4}$.

87. Divide: $\dfrac{5}{8} \div \dfrac{3}{4}$.

88. List the set of whole numbers.

89. Solve: $8 + \dfrac{a}{5} = 14$.

90. Insert the proper symbol, $<$ or $>$, in the blank to make a true statement: -15 ___ -14.

KEY CONCEPT

The Real Numbers

A **real number** is any number that can be expressed as a decimal. The set of real numbers corresponds to all points on a number line.

Graph each real number on the number line.

1. $\left\{-4, \dfrac{13}{4}, -0.1, \dfrac{99}{100}, \sqrt{17}, -2\dfrac{1}{2}, 1.\overline{3}\right\}$

2. $\left\{-\sqrt{2}, 2.75, -3, \dfrac{22}{7}, \dfrac{1}{50}, 0.8333\ldots\right\}$

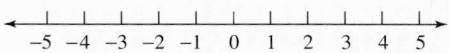

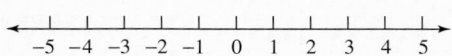

Almost all the types of numbers we have discussed in this book are real numbers. As we have seen, the set of real numbers is made up of several subsets of numbers.

If possible, list the numbers belonging to each set. If it is not possible to list them, define the set in words.

3. Natural numbers

4. Whole numbers

5. Integers

6. Rational numbers

7. Irrational numbers

The diagram below shows how the set of real numbers is made up of two distinct sets: the rational and the irrational numbers. Since every natural number is a whole number, we show the set of natural numbers included in the whole numbers. Because every whole number is an integer, the whole numbers are shown contained in the integers. Since every integer is a rational number, we show the integers included in the rational numbers.

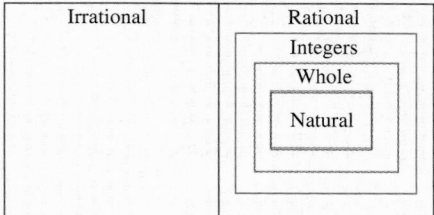

In Exercises 8–17, tell whether each statement is true or false.

8. Every integer is a real number.

9. Every fraction can be written as a terminating decimal.

10. Every real number is a whole number.

11. Some irrational numbers are integers.

12. Some rational numbers are natural numbers.

13. No numbers are both rational and irrational numbers.

14. All real numbers can be graphed on a number line.

15. The set of whole numbers is a subset of the irrational numbers.

16. All decimals either terminate or repeat.

17. Every natural number is an integer.

18. List the numbers in the set
$\{-2, -1.2, -\frac{7}{8}, 0, 1\frac{2}{3}, 2.75, \sqrt{23}, 10, 1.161661666\ldots\}$
that are
a. Natural numbers
b. Whole numbers
c. Integers
d. Rational numbers
e. Irrational numbers
f. Real numbers

19. $\sqrt{-25}$ is not a real number. Explain why.

274

Section 4.1
ROUNDING

a. Find all the three-digit numbers that round to 4.7.

b. Find all the four-digit numbers that round to 8.09.

c. Find all the two-digit numbers that round to 0.0.

Section 4.2
VISUAL MODELS   Shade a grid like the one in Illustration 1 to compute each addition or subtraction.

a. $0.62 + 0.24$ **b.** $0.45 - 0.41$

c. $0.21 + 0.29$ **d.** $0.98 - 0.18$

e. $0.2 + 0.17$ **f.** $0.57 - 0.3$

ILLUSTRATION 1

Section 4.3
SEQUENCES   Multiplication by 2 is used to form the terms of the sequence 3, 6, 12, 24, 48, 96, ... That is, to form the second term (which is 6), we multiply the first term (which is 3) by 2. To get the third term (which is 12), we multiply the second term by 2. To get the fourth term, we multiply the third term by 2, and so on.

What multiplication is used to form the terms of each of the following sequences?

a. 0.2134, 2.134, 21.34, 213.4, 2,134, 21,340, . . .

b. 0.00005, 0.005, 0.5, 50, 5,000, . . .

c. 3, 0.9, 0.27, 0.081, 0.0243, 0.00729, . . .

d. 0.7, 0.07, 0.007, 0.0007, 0.00007, 0.000007, . . .

Section 4.4
GPA   We can use the following steps to calculate a semester grade point average (GPA).

1. Convert each letter grade received to its equivalent point value.

Letter grade	A	B	C	D	F
Grade point value	4.0	3.0	2.0	1.0	0.0

2. For each class, multiply the number of credit hours (units) the class is worth by its point value found in Step 1.

3. Add each of the results from Step 2.

4. Divide the result from Step 3 by the total number of course credit hours (units) taken that semester. Round to the nearest tenth.

Find the GPA for the student whose grade report is shown in Illustration 2.

Course no.	Course title	Units	Grade
101	Intro. Accounting	5.0	C
201	Intro. Psych	3.0	A
102	Spanish II	4.0	B
142	Swimming	1.0	D

ILLUSTRATION 2

Section 4.5
EQUIVALENT DECIMALS   A student was asked to write several fractions as decimals. His answers, which are all incorrect, are shown below. What was he doing wrong?

$$\frac{2}{5} = 2.5 \qquad \frac{4}{15} = 3.75 \qquad \frac{3}{4} = 1.\overline{3}$$

Section 4.6
SOLVING EQUATIONS   If an equation contains decimals, we can multiply both sides by a power of 10 to clear the equation of decimals. Multiply both sides of each equation by the appropriate power of 10; then solve it.

a. $0.6x = 3.6$ **b.** $0.8x + 0.4 = 5.2$

c. $0.62x + 1.24 = 5.58$

Section 4.7
A SPIRAL OF ROOTS
Draw a triangle with two sides 1 inch long, as shown in Illustration 3. Use the corner of a 3×5 card to help draw the "sharp corner" (90-degree angle) of the triangle. Then draw the dashed blue line to complete the first triangle. It is $\sqrt{2}$ inches long.

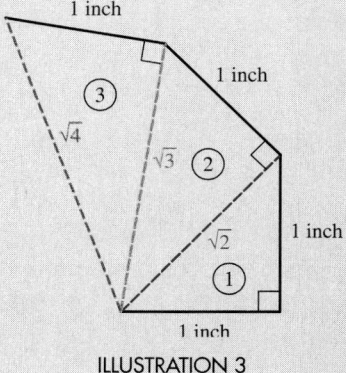

ILLUSTRATION 3

Next, create a second triangle, using one side of the first triangle and drawing another side 1 inch long as shown. Complete the second triangle by drawing the dashed green line. It is $\sqrt{3}$ inches long. Draw a third triangle in a similar fashion. The dashed purple line is $\sqrt{4} = 2$ inches long. Draw a fourth triangle, a fifth triangle, and so on. If the pattern continues, what is the length of the dashed side of each new triangle?

SECTION 4.1   *An Introduction to Decimals*

CONCEPTS

Decimal notation is used to denote part of a whole.

Expanded notation is used to show the value represented by each digit in the *decimal numeration system.*

$$5.6791 =$$
$$5 + \frac{6}{10} + \frac{7}{100} + \frac{9}{1,000} + \frac{1}{10,000}$$

To express a decimal in words, say:

1. the whole number to the left of the decimal point;

2. "and" for the decimal point;

3. the whole number to the right of the decimal point, followed by the name of the last place value column on the right.

To compare the size of two decimals, compare the digits of each decimal, column by column, working from left to right.

A decimal point and additional zeros may be written to the right of a whole number.

REVIEW EXERCISES

1. Represent the amount of the square that is shaded in Illustration 1, using a decimal and a fraction.

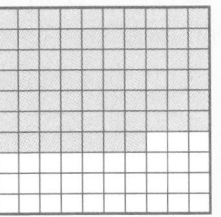

ILLUSTRATION 1

2. In Illustration 2, shade 0.8 of the rectangle.

ILLUSTRATION 2

3. Write 16.4523 in expanded notation.

4. Write each decimal in words and then as a fraction or mixed number.
  a. 2.3   **b.** -15.59

  c. 0.0601   **d.** 0.00001

5. Graph 1.55, -0.8, and -2.7 on a number line.

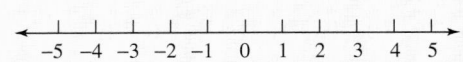

6. VALEDICTORIAN   At the end of the school year, the five students listed in Illustration 3 were in the running to be class valedictorian. Rank the students in order by GPA, beginning with the valedictorian.

Name	GPA
Diaz, Cielo	3.9809
Chou, Wendy	3.9808
Washington, Shelly	3.9865
Gerbac, Lance	3.899
Singh, Amani	3.9713

ILLUSTRATION 3

7. True or false: $78 = 78.0$.

8. Place the proper symbol ($<$, $>$, or $=$) in the blank to make a true statement.
  a. 4.5 _____ 4.6   **b.** -2.35 _____ -2.53
  c. 10.90 _____ 10.9   **d.** 0.027894 _____ 0.034

To round a decimal, locate the rounding digit and the test digit.

1. If the test digit is less than 5, drop it and all digits to the right of the rounding digit.

2. If it is 5 or greater, add 1 to the rounding digit and drop all digits to its right.

9. Round each decimal to the specified place-value column.

a. 4.578: hundredths

b. 3,706.0895: thousandths

c. −0.0614: tenths

d. 88.12: tenths

SECTION 4.2

Addition and Subtraction with Decimals

To add (or subtract) decimals:

1. Line up their decimal points.

2. Add (or subtract) as you would with whole numbers.

3. Write the decimal point in the result directly below the decimal points of the problem.

10. Do each addition or subtraction.

a. $19.5 + 34.4 + 12.8$

b. $3.4 + 6.78 + 35 + 0.008$

c. $68.47 - 53.3$

d. $45.08 - 17.37$

11. Evaluate each expression.

a. $-16.1 + 8.4$

b. $-4.8 - (-7.9)$

c. $-3.55 + (-1.25)$

d. $-15.1 - 13.99$

e. $-8.8 + (-7.3 - 9.5)$

f. $(5 - 0.096) - (-0.035)$

12. SALE PRICE   A calculator normally sells for $52.20. If it is being discounted $3.99, what is the sale price?

13. MICROWAVE OVEN A microwave oven is shown in Illustration 4. How tall is the window?

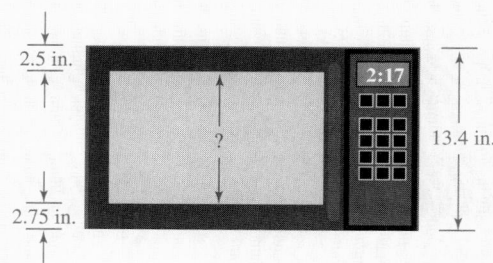

ILLUSTRATION 4

SECTION 4.3

Multiplication with Decimals

To multiply decimals:

1. Multiply as if working with whole numbers.

2. Place the decimal point in the result so that the answer has the same number of decimal places as the total number of decimal places of the factors.

To multiply a decimal by a power of 10, move the decimal point to the right the same number of places as there are zeros in the power of 10.

14. Do each multiplication.

a. $(-0.6)(0.4)$

b. $2.3 \cdot 0.9$

c. $5.5(-3.1)$

d. $32.45(6.1)$

e. $(-0.003)(-0.02)$

f. $7 \cdot 0.6$

15. Do each multiplication in your head.

a. $1,000(90.1452)$

b. $(-10)(-2.897)(100)$

Exponents are used to represent repeated multiplication.

16. Find each power.

    a. $(0.2)^2$ **b.** $(-0.15)^2$ **c.** $(3.3)^2$ **d.** $(0.1)^3$

17. Evaluate each expression.

    a. $(0.6 + 0.7)^2 - 12.3$ **b.** $3(7.8) + 2(1.1)^2$

18. Evaluate $2(3.14)(4)^2 - 8.1$.

19. WORD PROCESSOR   The Page Setup screen for a word processor is shown in Illustration 5. Find the area that can be filled with text on an 8.5-inch-by-11-inch piece of paper if the margins are set as shown.

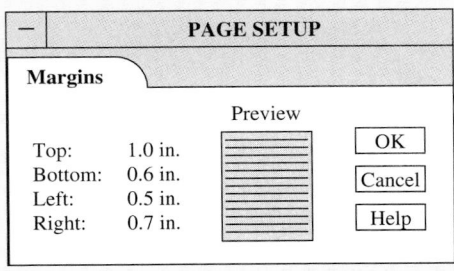

ILLUSTRATION 5

20. AUTO PAINTING   A manufacturer uses a three-part process to finish the exterior of the cars it produces.

Step 1: A rust-prevention undercoat, 0.03 inch thick, is applied.

Step 2: Three layers of color coat, each 0.015 of an inch thick, are sprayed on.

Step 3: The finish is then buffed down, losing 0.005 of an inch of its thickness.

What is the resulting thickness of the automobile's finish?

SECTION 4.4   *Division with Decimals*

To divide a decimal by a whole number:

1. Divide as if working with whole numbers.
2. Write the decimal point in the result directly above the decimal point in the dividend.

To divide by a decimal:

1. Move the decimal point in the divisor so that it becomes a whole number.
2. Move the decimal point in the dividend the same number of places to the right.
3. Use the process for dividing a decimal by a whole number.

21. Do each division.

    a. $12\overline{)15}$ **b.** $-41.8 \div 4$ **c.** $\dfrac{-29.67}{-23}$ **d.** $24.618 \div 6$

22. Do each division.

    a. $12.47 \div (-4.3)$ **b.** $\dfrac{0.0742}{1.4}$

    c. $\dfrac{15.75}{0.25}$ **d.** $\dfrac{-0.03726}{-0.046}$

23. Divide and round each result to the nearest tenth.

    a. $78.98 \div 6.1$ **b.** $\dfrac{-5.338}{0.008}$

24. Evaluate $\dfrac{5(68.4 - 32)}{9}$. Round to the nearest hundredth.

25. THANKSGIVING DINNER   The cost of purchasing the ingredients for a Thanksgiving turkey dinner for a family of 5 was $41.70. What was the cost of the dinner per person?

To divide a decimal by a power of 10, move the decimal point to the left the same number of places as there are zeros in the power of 10.

26. Do each division in your head.

 a. $89.76 \div 100$

 b. $\dfrac{0.0112}{-10}$

27. Evaluate the numerical expression $\dfrac{(1.4)^2 + 2(4.6)}{0.5 + 0.3}$.

28. SERVING SIZE   Illustration 6 shows the package labeling on a box of children's cereal. Use the information given to find the number of servings.

Nutrition Facts	
Serving size	1.1 ounce
Servings per container	?
Package weight	15.5 ounces

ILLUSTRATION 6

29. TELESCOPE   To change the position of a focusing mirror on a telescope, an adjustment knob is used. The mirror moves 0.025 inch with each revolution of the knob. The mirror needs to be moved 0.2375 inch to improve the sharpness of the image. How many revolutions of the adjustment knob does this require?

SECTION 4.5

Fractions and Decimals

To write a fraction as a decimal, divide the numerator by the denominator.

30. Write each fraction in decimal form.

 a. $\dfrac{7}{8}$　　　**b.** $-\dfrac{2}{5}$　　　**c.** $\dfrac{9}{16}$　　　**d.** $\dfrac{3}{50}$

We obtain either a *terminating* or a *repeating* decimal when using division to write a fraction as a decimal.

31. Write each fraction in decimal form. Use an overbar.

 a. $\dfrac{6}{11}$　　　　　　**b.** $-\dfrac{2}{3}$

An overbar can be used instead of the three dots . . . to represent the repeating pattern in a repeating decimal.

32. Write each fraction in decimal form. Round to the nearest hundredth.

 a. $\dfrac{19}{33}$　　　　　　**b.** $\dfrac{31}{30}$

33. Place the proper symbol ($<$ or $>$) in the blank to make a true statement.

 a. $\dfrac{13}{25}$ ____ 0.499　　　　**b.** $-0.\overline{26}$ ____ $-\dfrac{4}{15}$

34. Graph $1\frac{1}{8}$, $-\frac{1}{3}$, $2\frac{3}{4}$, and $-\frac{9}{10}$ on the number line. Label each using its decimal equivalent.

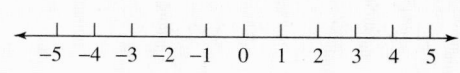

35. Evaluate each numerical expression. Find the exact answer.

 a. $\dfrac{1}{3} + 0.4$　　　　　　**b.** $\dfrac{4}{5}(-7.8)$

 c. $\dfrac{1}{2}(9.7 + 8.9)(10)$　　　**d.** $\dfrac{1}{3}(3.14)(3)^2(4.2)$

36. Evaluate $\dfrac{4}{3}(3.14)(2)^3$. Round the result to the nearest hundredth.

37. ROADSIDE EMERGENCY   In case of trouble, truckers carry reflectors to be placed on the highway shoulder to warn approaching cars of a stalled vehicle. (See Illustration 7.) What is the area of one of these triangular reflectors?

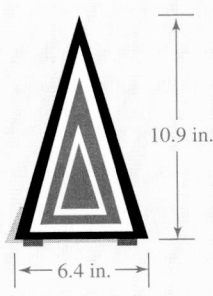

10.9 in.

6.4 in.

ILLUSTRATION 7

| SECTION 4.6 | *Solving Equations Containing Decimals* |

The five-step problem-solving strategy:

1. Analyze the problem.
2. Form an equation.
3. Solve the equation.
4. State the conclusion.
5. Check the result.

38. Solve each equation.

   a. $y + 12.4 = -6.01$ **b.** $0.23 + x = 5$

   c. $\dfrac{x}{1.78} = -3$ **d.** $-16.1b = -27.37$

39. Is $r = -1.1$ a solution of $-1.3 = 1.2r + 0.02$?

40. BOWLING   If it costs $1.45 to rent shoes and 95 cents a game to use a lane, how many games can be bowled for $10?

| SECTION 4.7 | *Square Roots* |

The number b is a *square root* of a if $b^2 = a$.

A *radical sign* $\sqrt{}$ is used to indicate a positive square root. The square root of a *perfect square* is a whole number.

41. Fill in the blanks: Two square roots of 64 are 8 and -8, because   = 64 and   = 64.

42. Simplify each expression without using a calculator.

   a. $\sqrt{49}$ **b.** $-\sqrt{16}$ **c.** $\sqrt{100}$ **d.** $\sqrt{0.09}$

   e. $\sqrt{\dfrac{64}{25}}$ **f.** $\sqrt{0.81}$ **g.** $-\sqrt{\dfrac{1}{36}}$ **h.** $\sqrt{0}$

43. Between what two whole numbers would $\sqrt{83}$ be located when graphed on a number line?

A square root can be approximated using a calculator.

44. Use a calculator to find $\sqrt{11}$ and round to the nearest tenth. Now square the approximation. How close is it to 11?

45. Graph each square root on the number line: $\sqrt{3}$, $-\sqrt{2}$, and $\sqrt{0}$.

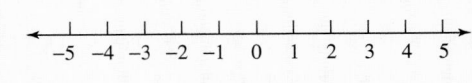

-5 -4 -3 -2 -1 0 1 2 3 4 5

When evaluating an expression containing square roots, treat a radical as you would a power when applying the rules for the order of operations.

46. Evaluate each expression without using a calculator.

   a. $-3\sqrt{100}$ **b.** $5\sqrt{0.25}$

   c. $-3\sqrt{49} - \sqrt{36}$ **d.** $\sqrt{\dfrac{9}{100}} + \sqrt{1.44}$

47. Use a calculator to find each square root to the nearest hundredth.

   a. $\sqrt{19}$ **b.** $\sqrt{59}$

Chapter 4 Test

1. Express the amount of the square in Illustration 1 that is shaded, using a fraction and a decimal.

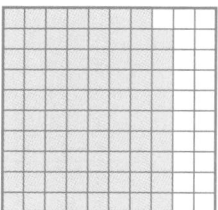

ILLUSTRATION 1

2. WATER PURITY   A county health department sampled the pollution content of tap water in 6 cities, with the results shown in Illustration 2. Rank the cities in order, from dirtiest tap water to cleanest.

City	Pollution parts per million
Monroe	0.0909
Covington	0.0899
Paston	0.0901
Cadia	0.0890
Selway	0.1001

ILLUSTRATION 2

3. Write 0.271 as a fraction.

4. Round to the nearest thousandth: 33.0495.

5. SKATING RECEIPTS   At an ice-skating complex, receipts on Friday were $30.25 for indoor skating and $62.25 for outdoor skating. On Saturday, the corresponding amounts were $40.50 and $75.75. Find the total receipts for the two days.

6. Do each operation in your head.
   a. $567.909 \div 1,000$ **b.** $0.00458 \cdot 100$

7. EARTHQUAKE FAULT LINE   After an earthquake, geologists found that the ground on the west side of the fault line had dropped 0.83 inch. The next week, a strong aftershock caused the same area to sink 0.19 inch deeper. How far did the ground on the west side of the fault drop because of the seismic activity?

8. Do each operation.
   a. $2 + 4.56 + 0.89 + 3.3$

   b. $45.2 - 39.079$

   c. $(0.32)^2$

   d. $-6.7(-2.1)$

9. NEW YORK CITY   Central Park, which lies in the middle of Manhattan, is the city's best-known park. See Illustration 3. If it is 2.5 miles long and 0.5 mile wide, what is its area?

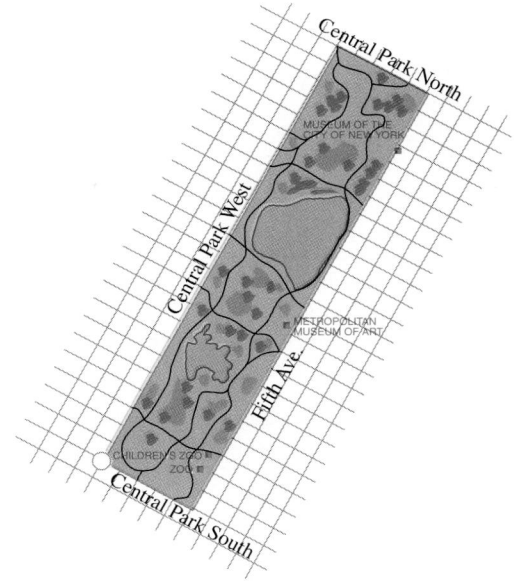

ILLUSTRATION 3

10. TELEPHONE BOOK   To print a telephone book, 565 sheets of paper were used. If the book is 2.3 inches thick, what is the thickness of each sheet of paper? (Round to the nearest thousandth of an inch.)

11. Evaluate $4.1 - (3.2)(0.4)^2$.

12. Write each fraction as a decimal.

    a. $\dfrac{17}{50}$ **b.** $\dfrac{5}{12}$

13. Do the division and round to the nearest hundredth: $\dfrac{12.146}{-5.3}$.

14. Find $11\overline{)13}$.

15. Graph $\dfrac{3}{8}$ and $-\dfrac{4}{5}$ on the number line. Label each using its decimal equivalent.

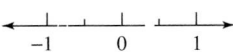

16. Find the exact answer: $\dfrac{2}{3} + 0.7$.

17. Solve each equation.

    a. $-2.4t = 16.8$ **b.** $-0.008 + x = 6$

18. Solve: $-0.53 = 0.0225 + 1.3x$.

19. CHEMISTRY   In a lab experiment, a chemist mixed three compounds together to form a mixture weighing 4.37 g. Later, she discovered that she had forgotten to record the weight of compound C in her notes. Find the weight of compound C used in the experiment.

	Weight
Compound A	1.86 g
Compound B	2.09 g
Compound C	?
Mixture total	4.37 g

20. WEDDING COSTS   A printer charges a setup fee of $24 and then 95 cents for each wedding announcement printed (tax included). If a couple has budgeted $100 for printing costs, how many announcements can they have made?

21. ▦ Graph $\sqrt{2}$ and $-\sqrt{5}$ on the number line.

    −5 −4 −3 −2 −1 0 1 2 3 4 5

22. Simplify.

    a. $-2\sqrt{25} + 3\sqrt{49}$ **b.** $\sqrt{\dfrac{1}{36}} - \sqrt{\dfrac{1}{25}}$

23. Insert the proper symbol (< or >) to make a true statement.

    a. -6.78 -6.79 **b.** $\dfrac{3}{8}$ 0.3

    c. $\sqrt{\dfrac{16}{81}}$ $\dfrac{16}{81}$ **d.** $0.\overline{45}$ 0.45

24. Simplify each square root.

    a. $-\sqrt{0.04}$ **b.** $\sqrt{1.69}$

Chapters 1-4   Cumulative Review Exercises

1. THE EXECUTIVE BRANCH   The annual salaries for the President and the Vice President of the United States are $400,000 and $181,000, respectively. How much more money does the President make than the Vice President during a four-year term?

2. Use the variables x, y, and z to write the associative property of addition.

3. Find $43\overline{)1,203}$.

4. How many thousands are there in one million?

5. Find the prime factorization of 220.

6. List the factors of 20, from least to greatest.

7. List the set of whole numbers.

8. Find $-8 + (-5)$.

9. Fill in the blank: Subtraction is the same as _____ the opposite.

10. Complete the solution.
$$(-6)^2 - 2(5 - 4 \cdot 2) = (-6)^2 - 2(5 - \quad)$$
$$= (-6)^2 - 2(\quad)$$
$$= \quad - 2(-3)$$
$$= 36 - (\quad)$$
$$= 36 + \quad$$
$$= 42$$

11. Consider the division statement $\dfrac{-15}{-5} = 3$.

What is its equivalent multiplication statement?

12. Find $(-1)^5$.

13. Solve $8 - 2d = -10$.

14. Solve $0 = 6 + \dfrac{c}{-5}$.

15. Evaluate $|-7(5)|$.

16. What is the opposite of -102?

17. Round 3.60745 to the nearest hundredth.

18. CHECKING ACCOUNT   After a deposit of $995, a student's checking account was still $105 overdrawn. What was the balance in the account before the deposit?

19. Solve $7x - 38 = -3$.

20. See Illustration 1. What fraction of the stripes in the flag are not red?

ILLUSTRATION 1

21. Although the fractions listed below look different, they all represent the same value. What concept does this illustrate?
$$\frac{1}{2} = \frac{2}{4} = \frac{3}{6} = \frac{4}{8} = \frac{5}{10} = \frac{6}{12}$$

22. Simplify $\dfrac{90}{126}$.

Perform each operation.

23. $\dfrac{3}{8} \cdot \dfrac{7}{16}$

24. $-\dfrac{15}{8} \div \dfrac{10}{1}$

25. $\dfrac{4}{3} + \dfrac{2}{7}$

26. $-4\dfrac{1}{4}\left(-4\dfrac{1}{2}\right)$

27. $76\dfrac{1}{6} - 49\dfrac{7}{8}$

28. $\dfrac{\dfrac{5}{27}}{-\dfrac{5}{9}}$

29. Solve $\dfrac{2}{3}y = -30$.

30. Solve $\dfrac{d}{6} - \dfrac{2}{3} = \dfrac{2}{3}$.

31. KITE Find the area of the kite shown in Illustration 2.

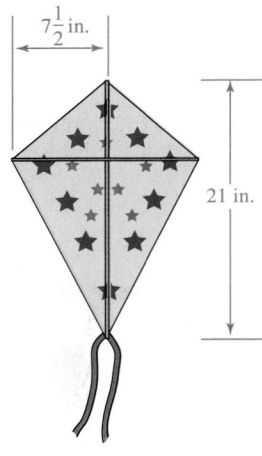

ILLUSTRATION 2

32. Graph each of the numbers in the set $\left\{-3\frac{1}{4}, 0.75, -1.5, -\frac{9}{8}, 3.8, \sqrt{4}\right\}$.

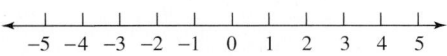

33. GLASS Some electronic and medical equipment uses glass that is only 0.00098 inch thick. Round this number to the nearest thousandth.

34. Place the proper symbol, $>$ or $<$, in the box to make the statement true.

356.1978 ▢ 356.22

Perform each operation.

35. $-1.8(4.52)$

36. $\dfrac{-21.28}{-3.8}$

37. $56.012(100)$

38. $\dfrac{0.897}{10,000}$

39. Evaluate $-9.1 - (-6.05 - 51)$.

40. WEEKLY SCHEDULE Refer to Illustration 3. Determine the number of hours during a week that an adult spends, on average, watching television.

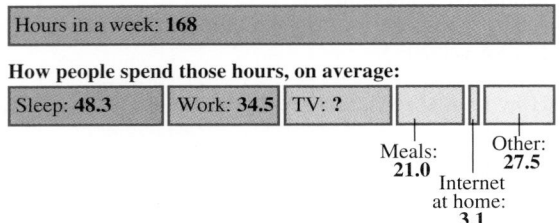

Based on data from the National Sleep Foundation and the United States Bureau of Statistics

ILLUSTRATION 3

41. LITERATURE The novel *Fahrenheit 451,* by Ray Bradbury, is a story about censorship and book burning. Convert 451° F to degrees Celsius by evaluating

$$\frac{5(451 - 32)}{9}$$

Round to the nearest tenth of a degree.

42. Write $\dfrac{5}{12}$ as a decimal. Use an overbar.

43. Solve $-3 = 0.5t + 1.5$.

44. CONCESSIONAIRE At a ballpark, a vendor is paid $22 a game plus 35¢ for each bag of peanuts she sells. How many bags of peanuts must she sell to make $50 a game?

45. Evaluate $-4\sqrt{36} + 2\sqrt{81}$.

Percent

5

5.1 Percents, Decimals, and Fractions

5.2 Solving Percent Problems

5.3 Applications of Percent Estimation

5.4 Interest

Key Concept: Percent

Accent on Teamwork

Chapter Review

Chapter Test

Cumulative Review Exercises

PERCENTS ARE BASED ON THE NUMBER 100. THEY OFFER US A STANDARDIZED WAY TO MEASURE AND DESCRIBE MANY SITUATIONS IN OUR DAILY LIVES.

5.1 *Percents, Decimals, and Fractions*

In this section, you will learn about

- The meaning of percent • Changing a percent to a fraction
- Changing a percent to a decimal • Changing a decimal to a percent
- Changing a fraction to a percent

INTRODUCTION. Percents are a popular way to present numeric information. Stores use them to advertise discounts, manufacturers use them to describe the content of their products, and banks use them to list interest rates for loans and savings accounts. Newspapers are full of statistics presented in percent form. In this section, we introduce percent and show how fractions, decimals, and percents are interrelated.

The meaning of percent

A percent tells us the number of parts per 100. You can think of a percent as the *numerator* of a fraction that has a denominator of 100.

Percent

Percent means parts per one hundred.

In Figure 5-1, there are 100 equal-sized squares, and 93 are shaded. Thus, $\frac{93}{100}$ or 93 percent of the figure is shaded. The word *percent* can be written using the symbol %, so 93% of Figure 5-1 is shaded.

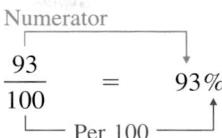

If the entire grid in Figure 5-1 had been shaded, we would say that 100 out of the 100 squares, or 100%, was shaded. Using this fact, we can determine what percent of the figure is *not* shaded by subtracting the percent of the figure that is shaded from 100%.

$$100\% - 93\% = 7\%$$

7% of Figure 5-1 is not shaded.

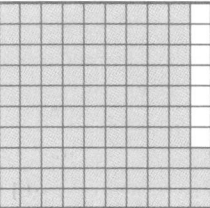

FIGURE 5-1

Changing a percent to a fraction

To change a percent into an equivalent fraction, we use the definition of percent.

Changing a percent to a fraction	To change a percent to a fraction, drop the % symbol and write the given number over 100. Then simplify the fraction, if possible.

EXAMPLE 1 *Changing a percent to a fraction.* The chemical makeup of Earth's atmosphere is 78% nitrogen, 21% oxygen, and 1% other gases. Write each percent as a fraction.

Solution
We begin with nitrogen.

$$78\% = \frac{78}{100} \quad \text{Use the definition of percent. 78\% means 78 parts per one hundred. This fraction can be simplified.}$$

$$= \frac{39 \cdot \overset{1}{\cancel{2}}}{50 \cdot \underset{1}{\cancel{2}}} \quad \text{Factor 78 as } 39 \cdot 2 \text{ and 100 as } 50 \cdot 2. \\ \text{Divide out the common factor of 2.}$$

$$= \frac{39}{50}$$

Nitrogen makes up $\frac{78}{100}$, or $\frac{39}{50}$, of Earth's atmosphere.

Oxygen makes up 21% or $\frac{21}{100}$ of Earth's atmosphere. Other gases make up 1% or $\frac{1}{100}$ of the atmosphere.

Self Check
An average watermelon is 92% water. Write this percent as a fraction.

Answer: $\dfrac{23}{25}$

EXAMPLE 2 *Changing a percent to a fraction.* In 1998, 46.1% of married women aged 25–54 were employed full-time. Write this percent as a fraction.

Solution

$$46.1\% = \frac{46.1}{100}$$ Drop the % symbol and write 46.1 over 100.

$$= \frac{46.1 \cdot 10}{100 \cdot 10}$$ To obtain a whole number in the numerator, multiply by 10. This will move the decimal point 1 place to the right. Multiply the denominator by 10 as well.

$$= \frac{461}{1,000}$$ Do the multiplication in the numerator and in the denominator.

In 1998, 461 out of every 1,000 married women aged 25–54 were employed full-time.

> **Self Check**
>
> In 1978, 26.9% of married women aged 25–54 were employed full-time. Write this percent as a fraction.
>
> **Answer:** $\dfrac{269}{1,000}$ ∎

EXAMPLE 3 *Changing a percent to a fraction.* Write $66\frac{2}{3}\%$ as a fraction.

Solution

$$66\frac{2}{3}\% = \frac{66\frac{2}{3}}{100}$$ Drop the % symbol and write $66\frac{2}{3}$ over 100.

$$= 66\frac{2}{3} \div 100$$ The fraction bar indicates division.

$$= \frac{200}{3} \cdot \frac{1}{100}$$ Change $66\frac{2}{3}$ to a mixed number and then multiply by the reciprocal of 100.

$$= \frac{2 \cdot 100 \cdot 1}{3 \cdot 100}$$ Multiply the numerators and the denominators. Factor 200 as $2 \cdot 100$.

$$= \frac{2 \cdot \overset{1}{\cancel{100}} \cdot 1}{3 \cdot \underset{1}{\cancel{100}}}$$ Divide out the common factor of 100.

$$= \frac{2}{3}$$ Multiply in the numerator. Multiply in the denominator.

> **Self Check**
>
> Write $83\frac{1}{3}\%$ as a fraction.
>
> **Answer:** $\dfrac{5}{6}$ ∎

Changing a percent to a decimal

To write a percent as a decimal, recall that a percent can be written as a fraction with denominator 100, and that a denominator of 100 indicates division by 100.

Consider 14.25%, which means 14.25 parts per 100.

$$14.25\% = \frac{14.25}{100}$$ Use the definition of percent: write 14.25 over 100.

$$= 14.25 \div 100$$ The fraction bar indicates division.

$$= 0.14.25$$ To divide a decimal by 100, move the decimal point 2 places to the left.

$$14.25\% = 0.1425$$

This example suggests the following procedure.

Changing a percent to a decimal

> To change a percent to a decimal, drop the % symbol and divide by 100 by moving the decimal point 2 places to the left.

EXAMPLE 4 *The recording industry.* Figure 5-2 shows that the compact disc has become the format of choice among most consumers. What percent of all music sold is produced on CDs? Write the percent as a decimal.

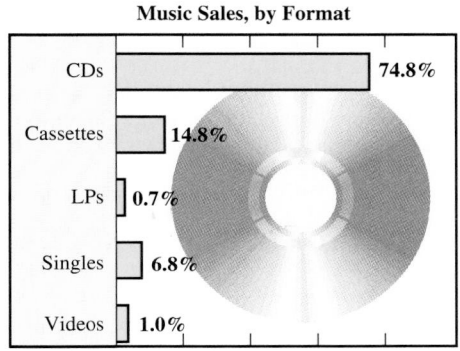

Music Sales, by Format

CDs **74.8%**

Cassettes **14.8%**

LPs **0.7%**

Singles **6.8%**

Videos **1.0%**

Source: *The New York Times 2000 Almanac*

FIGURE 5-2

Solution

From the graph, we see that 74.8% of all music sold is produced on CDs. To write 74.8% as a decimal, we proceed as follows.

$74.8\% = .74.8$ Drop the percent symbol and divide by 100 by moving the decimal point 2 places to the left.

$= 0.748$ Write a 0 to the left of the decimal point.

EXAMPLE 5 *Changing a percent to a decimal.* Write 310% as a decimal.

Solution

The whole number 310 has an understood decimal point to the right of 0.

$310\% = 310.0\%$ Write a decimal point and a 0 on the right of 310.

$= 3.10.0$ Drop the % symbol and divide by 100 by moving the decimal point 2 places to the left.

$= 3.100$

$= 3.1$ Drop the unnecessary 0's to the right of the 1.

EXAMPLE 6 *Changing a percent to a decimal.* The population of the state of Oklahoma is approximately $1\frac{1}{4}\%$ of the population of the United States. Write this percent as a decimal.

Solution

To change a percent to a decimal, we drop the percent symbol and divide by 100 by moving the decimal point 2 places to the left. In this case, however, there is no decimal point in $1\frac{1}{4}\%$ to move. Since $1\frac{1}{4} = 1 + \frac{1}{4}$, and since the decimal equivalent of $\frac{1}{4}$ is 0.25, we can write $1\frac{1}{4}\%$ in an equivalent form as 1.25%.

$1\frac{1}{4}\% = 1.25\%$ Write $1\frac{1}{4}$ as 1.25.

$= 0.01.25$ Drop the % symbol and divide by 100 by moving the decimal point 2 places to the left.

$= 0.0125$

Self Check

What percent of all music sold is produced on LPs (long-playing vinyl record albums)? Write the percent as a decimal.

Answers: 0.7%, 0.007 ∎

Self Check

Write 600% as a decimal.

Answer: 6 ∎

Self Check

Write $15\frac{3}{4}\%$ as a decimal.

Answer: 0.1575 ∎

Changing a decimal to a percent

To change a percent to a decimal, we drop the % symbol and move the decimal point 2 places to the left. To write a decimal as a percent, we do the opposite: we move the decimal point two places to the right and insert a % symbol.

Changing a decimal to a percent	To change a decimal to a percent, multiply the decimal by 100 by moving the decimal point 2 places to the right, and then insert a % symbol.

EXAMPLE 7   *Changing a decimal to a percent.*   Land areas make up 0.291 of the Earth's surface. Write this decimal as a percent.

Solution

$$0.291 = 0.29.1\%$$ Multiply the decimal by 100 by moving the decimal point 2 places to the right, and then insert a % symbol.

$$= 29.1\%$$

Self Check

Write 0.5343 as a percent.

Answer: 53.43%

Changing a fraction to a percent

We will use a two-step process to change a fraction to a percent. First, we write the fraction as a decimal. Then we change that decimal to a percent.

$$\boxed{\text{Fraction}} \longrightarrow \boxed{\text{Decimal}} \longrightarrow \boxed{\text{Percent}}$$

Changing a fraction to a percent	To change a fraction to a percent, **1.** write the fraction as a decimal by dividing its numerator by its denominator; **2.** multiply the decimal by 100 by moving the decimal point 2 places to the right; **3.** insert a % symbol.

EXAMPLE 8   *Changing a fraction to a percent.*   The highest-rated television show of all time was a special episode of M*A*S*H that aired February 28, 1983. Surveys found that three out of every five American households watched this show. Express the rating as a percent.

Solution

3 out of 5 can be expressed as $\frac{3}{5}$. We need to change this fraction to a decimal.

$$\begin{array}{r} 0.6 \\ 5\overline{)3.0} \\ \underline{3\,0} \\ 0 \end{array}$$ Write 3 as 3.0 and then divide the numerator by the denominator.

$$\frac{3}{5} = 0.6$$ The result is a terminating decimal.

$$0.6 = 0.60.\%$$ Write a placeholder 0 to the right of the 6. Multiply the decimal by 100 by moving the decimal point 2 places to the right, and then insert a % symbol.

$$= 60\%$$

60% of American households watched the special episode of M*A*S*H.

Self Check

Write $\dfrac{7}{8}$ as a percent.

Answer: 87.5%

In Example 8, the result of the division was a terminating decimal. Sometimes when we change a fraction to a decimal, the result of the division is a repeating decimal.

EXAMPLE 9 ***Changing a fraction to a percent.*** Write $\dfrac{5}{6}$ as a percent.

Solution The first step is to change $\dfrac{5}{6}$ to a decimal.

$$
\begin{array}{r}
0.8333 \\
6\overline{)5.0000} \\
4\,8 \\
\overline{20} \\
18 \\
\overline{20} \\
18 \\
\overline{20}
\end{array}
$$
Write 5 as 5.0000. Divide the numerator by the denominator.

$\dfrac{5}{6} = 0.8333\ldots$ The result is a repeating decimal.

$\phantom{\dfrac{5}{6}} = 0.83.33\ldots\%$ Change 0.8333. . . to a percent. Multiply the decimal by 100 by moving the decimal point 2 places to the right, and then insert a % symbol.

$\phantom{\dfrac{5}{6}} = 83.33\ldots\%$ 83.333. . . is a repeating decimal.

We must now decide whether we want an approximation or an exact answer. For an approximation, we can round 83.333. . .% to a specific place value. For an exact answer, we can represent the repeating part of the decimal using an equivalent fraction.

Approximation

$\dfrac{5}{6} = 83.33\ldots\%$

$\phantom{\dfrac{5}{6}} \approx 83.3\%$ Round to the nearest tenth.

$\dfrac{5}{6} \approx 83.3\%$

Exact answer

$\dfrac{5}{6} = 83.3333\ldots\%$

$\phantom{\dfrac{5}{6}} = 83\dfrac{1}{3}\%$ Use the fraction $\frac{1}{3}$ to represent .333. . . .

$\dfrac{5}{6} = 83\dfrac{1}{3}\%$ ■

Some percents occur so frequently that it is useful to memorize their fractional and decimal equivalents. Study the information in this table and memorize it for future use.

Percent	Decimal	Fraction	Percent	Decimal	Fraction
1%	0.01	$\dfrac{1}{100}$	$33\frac{1}{3}\%$	0.3333. . .	$\dfrac{1}{3}$
10%	0.1	$\dfrac{1}{10}$	50%	0.5	$\dfrac{1}{2}$
20%	0.2	$\dfrac{1}{5}$	$66\frac{2}{3}\%$	0.6666. . .	$\dfrac{2}{3}$
25%	0.25	$\dfrac{1}{4}$	75%	0.75	$\dfrac{3}{4}$

STUDY SET Section 5.1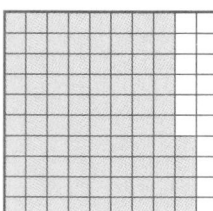

VOCABULARY *Fill in the blanks.*

1. _____ means parts per one hundred.

2. When changing a fraction to a decimal, the result is either a _____ or a repeating decimal.

CONCEPTS *Fill in the blanks.*

3. To write a percent as a fraction, drop the percent symbol and write the given number over _____.

4. To write a percent as a decimal, drop the % symbol and divide by 100 by moving the decimal point two places to the _____.

5. To write a decimal as a percent, multiply the decimal by 100 by moving the decimal point two places to the _____, and then insert a % symbol.

6. To write a fraction as a percent, first write the fraction as a _____. Then multiply the decimal by 100 by moving the decimal point two places to the _____, and insert a % symbol.

7. a. See Illustration 1. Express the amount of the figure that is shaded as a decimal, a percent, and a fraction.
  b. What percent of the figure is not shaded?

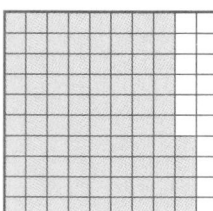

ILLUSTRATION 1

8. In Illustration 2, each set of 100 squares represents 100%. What percent is shaded?

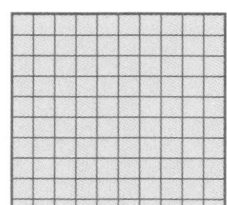 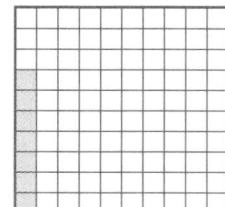

ILLUSTRATION 2

PRACTICE *Write each percent as a fraction. Simplify when necessary.*

9. 17%   **10.** 31%

11. 5%   **12.** 4%

13. 60%   **14.** 40%

15. 125%   **16.** 210%

17. $\frac{2}{3}$%   **18.** $\frac{1}{5}$%

19. $5\frac{1}{4}$%   **20.** $6\frac{3}{4}$%

21. 0.6%   **22.** 0.5%

23. 1.9%   **24.** 2.3%

Write each percent as a decimal.

25. 19%   **26.** 83%

27. 6%   **28.** 2%

29. 40.8%   **30.** 34.2%

31. 250%   **32.** 600%

33. 0.79%   **34.** 0.01%

35. $\frac{1}{4}$%   **36.** $8\frac{1}{5}$%

Write each decimal as a percent.

37. 0.93   **38.** 0.44

39. 0.612   **40.** 0.727

41. 0.0314   **42.** 0.0021

43. 8.43   **44.** 7.03

45. 50   **46.** 3

47. 9.1   **48.** 8.7

Write each fraction as a percent.

49. $\frac{17}{100}$ **50.** $\frac{29}{100}$

51. $\frac{4}{25}$ **52.** $\frac{47}{50}$

53. $\frac{2}{5}$ **54.** $\frac{21}{50}$

55. $\frac{21}{20}$ **56.** $\frac{33}{20}$

57. $\frac{5}{8}$ **58.** $\frac{3}{8}$

59. $\frac{3}{16}$ **60.** $\frac{1}{32}$

Find the exact equivalent percent for each fraction.

61. $\frac{2}{3}$

62. $\frac{1}{6}$

63. $\frac{5}{6}$

64. $\frac{4}{3}$

Express each of the given fractions as a percent. Round to the nearest hundredth.

65. $\frac{1}{9}$

66. $\frac{2}{3}$

67. $\frac{5}{9}$

68. $\frac{7}{3}$

APPLICATIONS

69. U.N. SECURITY COUNCIL   The United Nations has 188 members. The United States, the Russian Federation, Britain, France, and China, along with 10 other nations, make up the Security Council.
  a. What fraction of the members of the United Nations belong to the Security Council?
  b. Write your answer to part a in percent form. (Round to the nearest one percent.)

70. ECONOMIC FORECAST   One economic indicator of the national economy is the number of orders placed by manufacturers. One month, the number of orders rose one-fourth of one percent.
  a. Write this using a % symbol.
  b. Express it as a fraction.
  c. Express it as a decimal.

71. PIANO KEYS   Of the 88 keys on a piano, 36 are black.
  a. What fraction of the keys are black?
  b. What percent of the keys are black? (Round to the nearest one percent.)

72. INTEREST RATES   Write as a decimal the interest rate associated with each of these accounts.
  a. Home loan: 7.75%
  b. Savings account: 5%
  c. Credit card: 14.25%

73. THE SPINE   The human spine consists of a group of bones (vertebrae). (See Illustration 3.)
  a. What fraction of the vertebrae are lumbar?
  b. What percent of the vertebrae are lumbar? (Round to the nearest one percent.)
  c. What percent of vertebrae are cervical? (Round to the nearest one percent.)

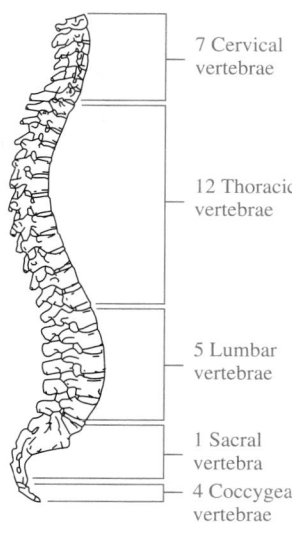

7 Cervical vertebrae

12 Thoracic vertebrae

5 Lumbar vertebrae

1 Sacral vertebra

4 Coccygeal vertebrae

ILLUSTRATION 3

74. REGIONS OF THE COUNTRY   The continental United States is divided into seven regions. (See Illustration 4.)
  a. What percent of the 50 states are in the Rocky Mountain region?
  b. What percent of the 50 states are in the Midwestern region?
  c. What percent of the 50 states are not located in any of the seven regions shown here?

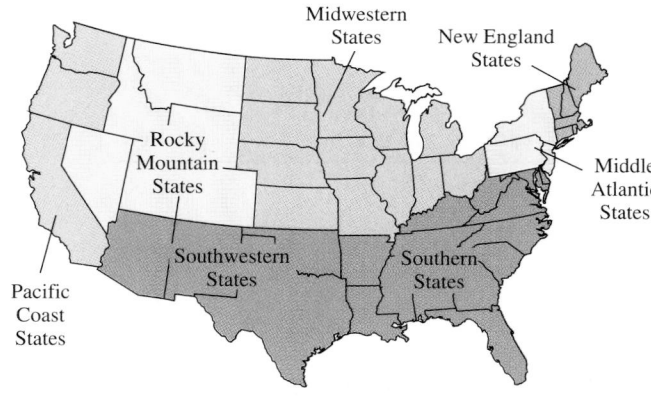

Midwestern States

New England States

Rocky Mountain States

Middle Atlantic States

Southwestern States

Southern States

Pacific Coast States

ILLUSTRATION 4

75. STEEP GRADE   Sometimes, signs are used to warn truckers when they are approaching a steep grade on the highway. (See Illustration 5.) For a 5% grade, how many feet does the road rise over a 100-foot run?

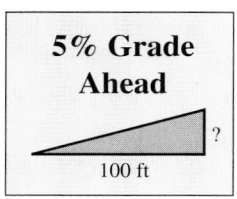

5% Grade Ahead

100 ft

ILLUSTRATION 5

76. COMPANY LOGO   In Illustration 6, what part of the company's logo is shaded red? Express your answer as a percent, a fraction, and a decimal. Do not round.

Recycling Industries Inc.

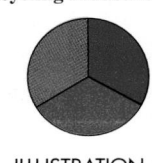

ILLUSTRATION 6

77. IVORY SOAP   A popular soap claims to be $99\frac{44}{100}\%$ pure. Write this percent as a decimal.

78. DRUNK DRIVING   In most states, it is illegal to drive with a blood alcohol concentration of 0.08% or more. Write this percent as a fraction. Do not simplify. Explain what the numerator and the denominator of the fraction represent.

79. BASKETBALL STANDINGS   In the standings, we see that Chicago has won 60 of 67, or $\frac{60}{67}$ of its games. In what form is the team's winning percentage presented in the newspaper? Express it as a percent.

Eastern Conference			
Team	**W**	**L**	**Pct.**
Chicago	60	7	.896

80. WON–LOST RECORD   In sports, when a team wins as many as it loses, it is said to be playing "500 ball." Examine the standings and explain the significance of the number 500, using concepts studied in this section.

Eastern Conference			
Team	**W**	**L**	**Pct.**
Orlando	33	33	.500

81. SKIN   Illustration 7 shows roughly what percent each section of the body represents of the total skin area. Determine the missing percent, and then complete the bar graph in Illustration 8.

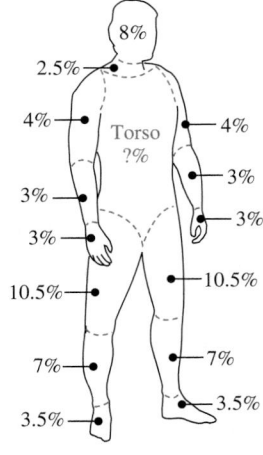

ILLUSTRATION 7

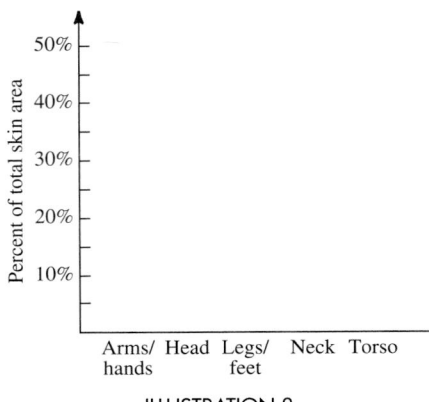

ILLUSTRATION 8

82. RAP MUSIC   Illustration 9 shows what percent rap music sales were of total U.S. dollar sales of recorded music for the years 1994–1998. In Illustration 10, construct a line graph using the given data.

1994	1995	1996	1997	1998
7.9%	6.7%	8.9%	10.1%	9.7%

Source: *The New York Times 2000 Almanac*

ILLUSTRATION 9

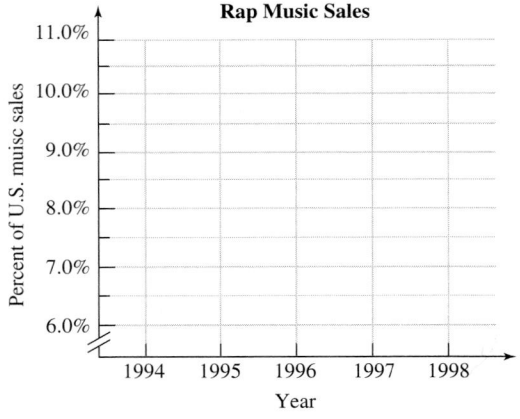

ILLUSTRATION 10

83. CHARITY   A 1998 fact sheet released by the American Red Cross stated, "For the past three fiscal years, an average of 92 cents of every dollar spent by the Red Cross went to programs and services to help those in need." What percent of the money spent by the Red Cross went to programs and services?

84. TAXES   Santa Anita Thoroughbred Racetrack in Arcadia, California has to pay a one-third of 1% tax on all the money wagered at the track. Write the percent as a fraction.

A calculator may be helpful to solve these problems.

85. BIRTHDAY   If the day of your birthday represents $\frac{1}{365}$ of a year, what percent of the year is it? Round to the nearest hundredth of a percent.

86. POPULATION   As a fraction, each resident of the United States represents approximately $\frac{1}{270,000,000}$ of the population. Express this as a percent. Round to one nonzero digit.

WRITING

87. If you were writing advertising, which form do you think would attract more customers: "25% off" or "$\frac{1}{4}$ off"? Explain your reasoning.

88. Many coaches ask their players to give a 110% effort during practices and games. What do you think this means? Is it possible?

89. Explain how to change a fraction to a percent.

90. Explain how an amusement park could have an attendance that is 103% of capacity.

91. HEAVYWEIGHT CHAMPION   Muhammad Ali won 92% of his professional boxing matches. Does that

mean he had exactly 100 fights, and won 92 of them? Explain your answer.

92. CALCULATORS   To change the fraction $\frac{15}{16}$ to a percent, a student used a calculator to divide 15 by 16. The display is shown below.

$$\boxed{0.9375}$$

Now what keys should the student press to change this decimal to a percent?

REVIEW

93. Solve $-\frac{2}{3}x = -6$.

94. Add: $\frac{1}{3} + \frac{1}{4} + \frac{1}{2}$.

95. Subtract: $\frac{7}{11} - \frac{2}{9}$.

96. Find the area of a square with a side that is 4 feet long.

97. Add: $3.875 + 23.2$.

98. Subtract: $41 - 10.287$.

5.2 *Solving Percent Problems*

In this section, you will learn about

- Percent problems   • Finding the amount   • Finding the percent
- Finding the base   • Restating the problem
- An alternative approach: the percent formula   • Circle graphs

INTRODUCTION. Percent problems occur in three forms. In this section, we will study a single procedure that can be used to solve all three types. It involves the equation-solving skills that we studied earlier.

Percent problems

The articles on the front page of the newspaper in Figure 5-3 suggest three types of percent problems.

- In the labor article, if we want to know how many union members voted to accept the new offer, we would ask:

    What number is 84% of 500?

- In the article on drinking water, if we want to know what percent of the wells are safe, we would ask:

    38 is what percent of 40?

- In the article on new appointees, if we want to know how many examiners are on the State Board, we would ask:

    6 is 75% of what number?

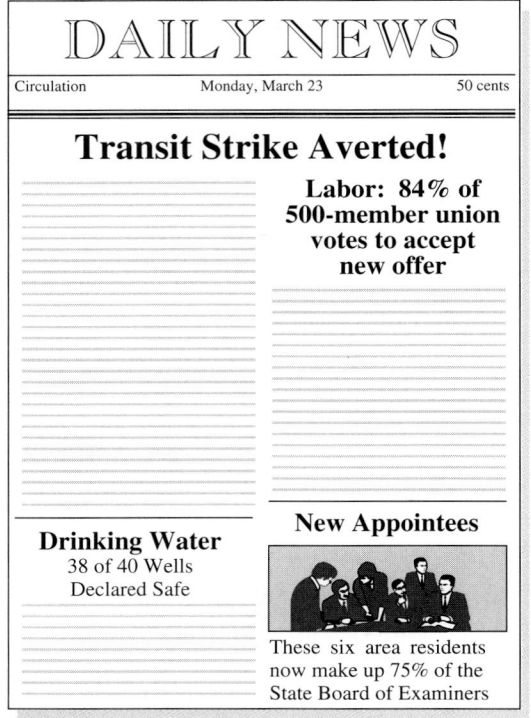

FIGURE 5-3

These percent problems have several things in common.

- Each problem contains the word *is*. Here, *is* can be translated to an $=$ sign.
- Each of the problems contains a phrase such as *what number* or *what percent*. In other words, there is an unknown quantity that can be represented by a variable.
- Each problem contains the word *of*. In this context, *of* means multiply.

These observations suggest that each of the percent problems can be translated into an equation. The equation, called a **percent equation,** will contain a variable, and the operation of multiplication will be involved.

Finding the amount

To solve the labor union problem, we translate the words into an equation and then solve it.

What number	is	84%	of	500?	
↓	↓	↓	↓	↓	
x	$=$	84%	$\cdot$	500	Translate to mathematical symbols.

$x = 0.84 \cdot 500$ Change 84% to a decimal: $84\% = 0.84$.

$x = 420$ Do the multiplication.

We have found that 420 is 84% of 500. That is, 420 union members voted to accept the new offer.

 COMMENT   When solving percent equations, always write the percent as a decimal or a fraction before performing any calculations. For example, in the previous problem, we wrote 84% as 0.84 before multiplying by 500.

Percent problems involve a comparison of numbers or quantities. In the statement "420 is 84% of 500," the number 420 is called the **amount,** 84% is the **percent,** and 500 is called the **base.** Think of the base as the standard of comparison—it represents the whole of some quantity. The amount is a part of the base, but it can exceed the base when the percent is more than 100%. The percent, of course, has the % symbol.

EXAMPLE 1 *Finding the amount.* What number is 160% of 15.8?

Solution

First, we translate the words into an equation.

What number	is	160%	of	15.8?
↓		↓		↓
x	$=$	160%	$\cdot$	15.8

x is the amount, 160% is the percent, and 15.8 is the base.

Then we solve the equation.

$x = 1.6 \cdot 15.8$ Change 160% to a decimal: 160% = 1.6.

$x = 25.28$ Do the multiplication.

Thus, 25.28 is 160% of 15.8.

Finding the percent

In the drinking water problem, we must find the percent. Once again, we translate the words of the problem into an equation and solve it.

38	is	what percent	of	40?
↓		↓		↓
38	$=$	x	$\cdot$	40

38 is the amount, x is the percent, and 40 is the base.

$38 = 40x$ Apply the commutative property of multiplication to write $x \cdot 40$ as $40x$.

$\dfrac{38}{40} = \dfrac{40x}{40}$ To undo the multiplication by 40, divide both sides by 40.

$0.95 = x$ Do the divisions.

$x = 0.95$ Since $0.95 = x$, $x = 0.95$.

$x = 95\%$ To change a decimal to a percent, multiply the decimal by 100 by moving the decimal point 2 places to the right, and then insert a % symbol.

Thus, 38 is 95% of 40. That is, 95% of the wells referred to in the article were declared safe.

EXAMPLE 2 *Finding the percent.* 14 is what percent of 32?

Solution

First, we translate the words into an equation.

14	is	what percent	of	32?
↓		↓		↓
14	$=$	x	$\cdot$	32

14 is the amount, x is the percent, and 32 is the base.

Then we solve the equation.

$$14 = 32x \qquad \text{Rewrite the right-hand side: } x \cdot 32 = 32x.$$

$$\frac{14}{32} = \frac{32x}{32} \qquad \text{To undo the multiplication by 32, divide both sides by 32.}$$

$$0.4375 = x \qquad \text{Do the divisions.}$$

$$43.75\% = x \qquad \text{Change } 0.4375 \text{ to a percent. Multiply the decimal by 100 by moving the decimal point 2 places to the right, and then insert a \% symbol.}$$

Thus, 14 is 43.75% of 32.

Answer: 56.25%

Accent on Technology: **Cost of an air bag**

An air bag is estimated to add an additional $500 to the cost of a car. What percent of the $16,295 sticker price is the cost of the air bag?

First, we translate the words into an equation.

What percent	of	the $16,295 sticker price	is	the cost of the air bag?
↓		↓		↓
x	$\cdot$	16,295	$=$	500

500 is the amount, x is the percent, and 16,295 is the base.

Then we solve the equation.

$$16,295x = 500 \qquad x \cdot 16,295 = 16,295x.$$

$$\frac{16,295x}{16,295} = \frac{500}{16,295} \qquad \text{To undo the multiplication by 16,295, divide both sides by 16,295.}$$

$$x = \frac{500}{16,295}$$

To do the division using a calculator, enter these numbers and press these keys.

Keystrokes   500 $\boxed{\div}$ 16295 $\boxed{=}$ $\boxed{\text{0.03068425}}$

This display gives the answer in decimal form. To change it to a percent, we multiply the result by 100 and insert a percent symbol. This moves the decimal point 2 places to the right. If we round to the nearest tenth of a percent, the cost of the air bag is about 3.1% of the sticker price.

Finding the base

In the problem about the State Board of Examiners, we must find the base. As before, we translate the words of the problem into an equation and then solve it.

6	is	75%	of	what number?
↓		↓		↓
6	$=$	75%	$\cdot$	x

6 is the amount, 75% is the percent, and x is the base.

$$6 = 0.75x \quad \text{Change 75\% to 0.75.}$$

$$\frac{6}{0.75} = \frac{0.75x}{0.75} \quad \text{To undo the multiplication by 0.75, divide both sides by 0.75.}$$

$$8 = x \quad \text{Do the divisions.}$$

Thus, 6 is 75% of 8. That is, there are 8 examiners on the State Board.

EXAMPLE 3 *Finding the base.* 31.5 is $33\frac{1}{3}\%$ of what number?

Solution

31.5	is	$33\frac{1}{3}\%$	of	what number?
↓		↓		↓
31.5	=	$33\frac{1}{3}\%$	·	x

31.5 is the amount, $33\frac{1}{3}\%$ is the percent, and x is the base.

In this case the computations can be made easier by changing the percent to a fraction instead of to a decimal. We write $33\frac{1}{3}\%$ as a fraction and proceed as follows:

$$31.5 = \frac{1}{3} \cdot x \quad 33\frac{1}{3}\% = \tfrac{1}{3}.$$

$$3 \cdot 31.5 = 3 \cdot \frac{1}{3}x \quad \text{To isolate } x \text{ on the right-hand side, multiply both sides by 3.}$$

$$94.5 = x \quad \text{Do the multiplications: } 3 \cdot 31.5 = 94.5 \text{ and } 3 \cdot \tfrac{1}{3} = 1.$$

Thus, 31.5 is $33\frac{1}{3}\%$ of 94.5.

Self Check

150 is $66\frac{2}{3}\%$ of what number?

Answer: 225

Restating the problem

Not all percent problems are presented in the form we have been studying. In Example 4, we must examine the given information carefully so that we can restate the problem in the familiar form.

EXAMPLE 4 *Housing.* In an apartment complex, 110 of the units are currently being rented. This represents an 88% occupancy rate. How many units are there in the complex?

Solution An occupancy rate of 88% means that 88% of the units are occupied. We restate the problem in the form we have been studying.

110	is	88%	of	what number?
↓		↓		↓
110	=	88%	·	x

110 is the amount, 88% is the percent, and x is the base.

Now we solve the equation.

$$110 = 0.88x \quad \text{Change 88\% to a decimal: } 88\% = 0.88.$$

$$\frac{110}{0.88} = \frac{0.88x}{0.88} \quad \text{To undo the multiplication by 0.88, divide both sides by 0.88.}$$

$$125 = x \quad \text{Do the divisions.}$$

The complex has 125 units.

An alternative approach: the percent formula

In any percent problem, the relationship between the amount, the percent, and the base is as follows: *Amount is percent of base.* This relationship is shown in the **percent formula.**

The percent formula

Amount = percent · base

The percent formula can be used as an alternate way to solve percent problems. With this method, we need to identify the *amount* (the part that is compared to the whole), the *percent* (indicated by the % symbol or the word *percent*), and the *base* (the whole of some quantity, usually following the word *of*).

EXAMPLE 5   *Finding the amount.* What number is 160% of 15.8?

Solution

In this example, the percent is 160 and the base is 15.8, the number following the word *of.* We can let *A* stand for the amount and use the percent formula.

Amount	=	percent	·	base
↓		↓		↓
A	=	160%	·	15.8

Substitute 160 for the percent and 15.8 for the base.

The statement $A = 160\% \cdot 15.8$ is an equation, with the amount *A* being the unknown. We can find the unknown amount by multiplication.

$A = 1.6 \cdot 15.8$ Change 160% to a decimal: $160\% = 1.6$.

$\quad = 25.28$ Do the multiplication.

Thus, 25.28 is 160% of 15.8.

Self Check

What number is 240% of 80?

Answer: 192 ■

EXAMPLE 6   *Finding the percent.* 14 is what percent of 32?

Solution

In this example, 14 is the amount and 32 is the base. Once again, we use the percent formula and let *p* stand for the percent.

Amount	=	percent	·	base
↓		↓		↓
14	=	*p*	·	32

Substitute 14 for the amount and 32 for the base.

The statement $14 = p \cdot 32$ is an equation, with the percent *p* being the unknown. We can find the unknown percent by division.

Self Check

9 is what percent of 16?

$14 = p \cdot 32$ The equation to solve.

$14 = 32p$ Rewrite the right-hand side: $p \cdot 32 = 32p$.

$\dfrac{14}{32} = \dfrac{32p}{32}$ To undo the multiplication by 32, divide both sides by 32.

$0.4375 = p$ $\frac{14}{32} = 0.4375$.

$p = 43.75\%$ To change the decimal to a percent, multiply the decimal by 100 by moving the decimal point 2 places to the right, and then insert a % symbol.

Thus, 14 is 43.75% of 32.

EXAMPLE 7   *Finding the base.*   31.5 is $33\frac{1}{3}\%$ of what number?

Self Check
150 is $66\frac{2}{3}\%$ of what number?

Solution
In this example, 31.5 is the amount and $33\frac{1}{3}$ is the percent. To find the base (which we will call b), we form an equation using the percent formula.

$$
\begin{array}{ccccc}
\text{Amount} & = & \text{percent} & \cdot & \text{base} \\
\downarrow & & \downarrow & & \downarrow \\
31.5 & = & 33\frac{1}{3}\% & \cdot & b
\end{array}
$$

Substitute 31.5 for the amount and $33\frac{1}{3}\%$ for the percent.

The statement $31.5 = 33\frac{1}{3}\% \cdot b$ is an equation, with the base b being the unknown. We can find the unknown base by multiplication.

$31.5 = 33\dfrac{1}{3}\% \cdot b$ The equation to solve.

$31.5 = \dfrac{1}{3}b$ $33\frac{1}{3}\% = \dfrac{33\frac{1}{3}}{100} = \frac{1}{3}$.

$3 \cdot 31.5 = 3 \cdot \dfrac{1}{3}b$ To isolate b on the right-hand side, multiply both sides by 3.

$94.5 = b$ Do the multiplication: $31.5 \cdot 3 = 94.5$.

Thus, 31.5 is $33\frac{1}{3}\%$ of 94.5.

Circle graphs

Percents are used with **circle graphs,** or **pie charts,** as a way of presenting data for comparison. In Figure 5-4, the entire circle represents the total amount of electricity generated in the United States in 1998. The pie-shaped pieces of the graph show the relative sizes of the energy sources used to produce the electricity. For example, we see that the greatest amount of electricity (52%) was generated from coal. Note that if we add the percents from all categories (52% + 19% + 15% + 9% + 3% + 2%), the sum is 100%.

The 100 tick marks equally spaced around the circle serve as a visual aid when constructing a circle graph. For example, to represent hydropower as 9%, a line was drawn from the center of the circle to a tick mark. Then we counted off 9 ticks and drew a second line from the center to that tick to complete the pie-shaped wedge.

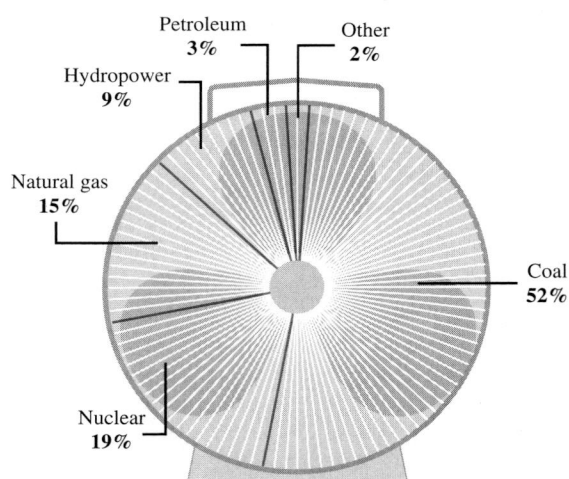

Sources of Electricity

Source: Energy Information Administration

FIGURE 5-4

EXAMPLE 8   *Presidential election results.*   Results from the 1996 presidential election are shown in Figure 5-5. Use the information to find the number of states won by President Clinton.

Solution   The circle graph shows that President Clinton was victorious in 62% of the 50 states. We state the problem in the standard form and then translate it into an equation.

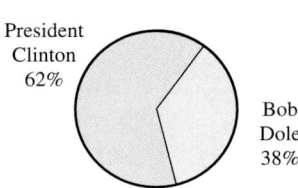

1996 Presidential Election:
States won by each candidate

FIGURE 5-5

What number	is	62%	of	50?
↓	↓	↓	↓	↓
x	=	62%	·	50

Translate the words into an equation.

$x = 0.62 \cdot 50$ Change 62% to a decimal: 62% = 0.62.

$x = 31$ Do the multiplication.

President Clinton won 31 states.

STUDY SET   Section 5.2

VOCABULARY   *Translate each sentence into a percent equation.*

1. What number is 10% of 50?

2. 16 is 55% of what number?

3. 48 is what percent of 47?

4. 12 is what percent of 20?

Fill in the blanks.

5. In a circle _____, pie-shaped wedges are used to show the division of a whole quantity into its component parts.

6. In the statement "45 is 90% of 50," 45 is the _____, 90% is the _____, and 50 is the _____.

CONCEPTS

7. When computing with percents, the percent must be changed to a decimal or a fraction. Change each percent to a decimal.
   a. 12%
   b. 5.6%
   c. 125%
   d. $\frac{1}{4}\%$

8. When computing with percents, the percent must be changed to a decimal or a fraction. Change each percent to a fraction.

   a. $33\frac{1}{3}\%$ **b.** $66\frac{2}{3}\%$

   c. $16\frac{2}{3}\%$ **d.** $83\frac{1}{3}\%$

9. Without doing the calculation, tell whether 120% of 55 is more than 55 or less than 55.

10. Without doing the calculation, tell whether 12% of 55 is more than 55 or less than 55.

11. Solve each of the following problems in your head.
   a. What is 100% of 25?
   b. What percent of 132 is 132?
   c. What number is 87% of 100?

12. To solve the problem

      15 is what percent of 75?

a student wrote a percent equation, solved it, and obtained $x = 0.2$. For her answer, the student wrote

      15 is 0.2% of 75.

Explain her error.

13. VIDEO GAMES   Illustration 1 shows the market shares of the three main video game systems. What percent of the market does Nintendo 64 have?

Current U.S. Market Share

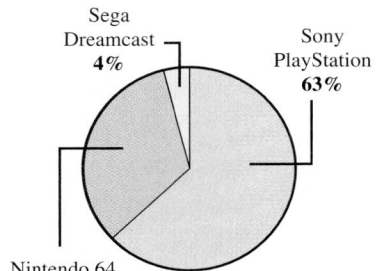

Source: *Los Angeles Times* (December 23, 1999)

ILLUSTRATION 1

14. HOUSING   In the last quarter of 1999, approximately 105.3 million housing units in the United States were occupied. Use the data in Illustration 2 to determine what percent were owner-occupied.

1999 Housing Inventory

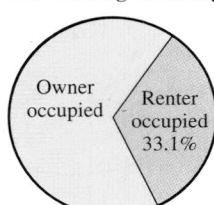

Source: *The New York Times 2000 Almanac*

ILLUSTRATION 2

NOTATION

15. How is each of the following words or phrases translated in this section?
   a. of
   b. is
   c. what number

16. a. Write the repeating decimal shown in the calculator display as a percent. Use an overbar.

$$\boxed{0.456666666}$$

   b. Round your answer to part a to the nearest hundredth of a percent.
   c. Write your answer to part a using a fraction.

PRACTICE   *Solve each problem by solving a percent equation.*

17. What number is 36% of 250?

18. What number is 82% of 300?

19. 16 is what percent of 20?

20. 13 is what percent of 25?

21. 7.8 is 12% of what number?

22. 39.6 is 44% of what number?

23. What number is 0.8% of 12?

24. What number is 5.6% of 4,040?

25. 0.5 is what percent of 40,000?

26. 0.3 is what percent of 15?

27. 3.3 is 7.5% of what number?

28. 8.4 is 20% of what number?

29. Find $7\frac{1}{4}\%$ of 600.

30. Find $1\frac{3}{4}\%$ of 800.

31. 102% of 105 is what number?

32. 210% of 66 is what number?

33. $33\frac{1}{3}\%$ of what number is 33?

34. $66\frac{2}{3}\%$ of what number is 28?

35. $9\frac{1}{2}\%$ of what number is 5.7?

36. $\frac{1}{2}\%$ of what number is 5,000?

37. What percent of 8,000 is 2,500?

38. What percent of 3,200 is 1,400?

Use a circle graph to illustrate the given data. A circle divided into 100 sections is provided to aid in the graphing process.

39. Complete Illustration 3 to show what percent of the total U.S. energy consumed was provided by each source in 1997.

Renewable	8%
Nuclear	7%
Coal	23%
Natural gas	24%
Petroleum	38%

Source: *The New York Times 2000 Almanac*

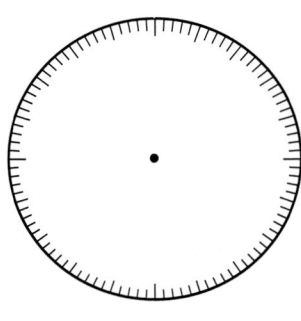

ILLUSTRATION 3

40. GREENHOUSE EFFECT   Complete Illustration 4 to show what percent of the total U.S. emissions from human activities came from each greenhouse gas.

Carbon dioxide	82%
Nitrous oxide	6%
Methane	10%
PFCs	2%

Source: *The World Almanac 2000*

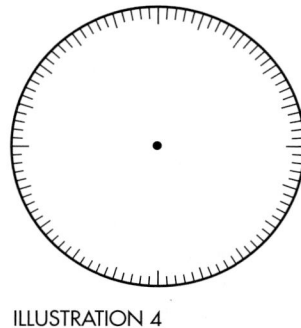

ILLUSTRATION 4

APPLICATIONS

41. CHILD CARE   After the first day of registration, 84 children had been enrolled in a new day care center. That represented 70% of the available slots. What was the maximum number of children the center could enroll?

42. RACING PROGRAM   One month before a stock car race, the sale of ads for the official race program was slow. Only 12 pages, or just 60% of the available pages,

had been sold. What was the total number of pages devoted to advertising in the program?

43. GOVERNMENT SPENDING   Illustration 5 shows the breakdown of federal outlays for fiscal year 1998. If the total spending was approximately $1,650 billion, how many dollars were spent on Social Security, Medicare, and other retirement programs?

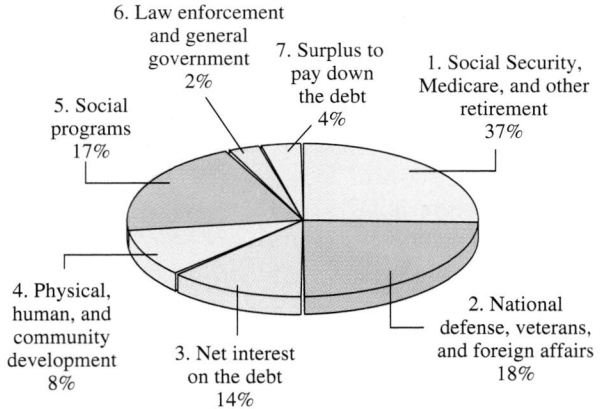

Source: 1999 federal income tax form

ILLUSTRATION 5

44. GOVERNMENT REVENUE   Complete the table by finding what percent of total federal government revenue each source provided in 1998. Round to the nearest percent. Then complete the circle graph in Illustration 6.

Total revenue, fiscal year 1998: $1,722 billion		
Source of revenue	**Amount**	**Percent of total**
Social Security, Medicare, unemployment taxes	$572 billion	
Personal income taxes	$829 billion	
Corporate income taxes	$189 billion	
Excise, estate, customs taxes	$132 billion	

Source: 1999 federal income tax form

1998 Federal Revenue

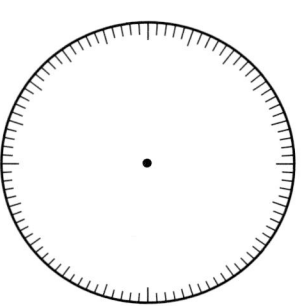

ILLUSTRATION 6

45. THE INTERNET   See Illustration 7. The message at the bottom of the screen indicates that 24% of the 50K bytes of information that the user has decided to view have been downloaded to her computer. How many more bytes of information must be downloaded? (50K stands for 50,000.)

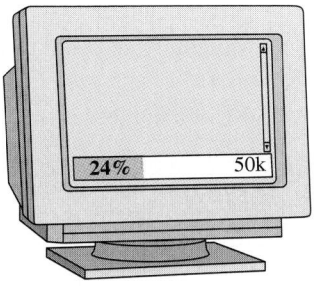

ILLUSTRATION 7

46. REBATE   A long-distance telephone company offered its customers a rebate of 20% of the cost of all long-distance calls made in the month of July. One customer's calls are listed in Illustration 8. What amount will this customer receive in the form of a rebate?

Date	Time	Place called	Min.	Amount
Jul 4	3:48 P.M.	Denver	47	$3.80
Jul 9	12:00 P.M.	Detroit	68	$7.50
Jul 20	8:59 A.M.	San Diego	70	$9.45

ILLUSTRATION 8

47. PRODUCT PROMOTION   To promote sales, a free 6-ounce bottle of shampoo is packaged with every large bottle. (See Illustration 9.) Use the information on the package to find how many ounces of shampoo the large bottle contains.

SHAMPOO

25% MORE-FREE!

ILLUSTRATION 9

48. NUTRITION FACTS   The nutrition label on a package of corn chips is shown in Illustration 10.
   a. How many milligrams of sodium are in one serving of chips?
   b. According to the label, what percent of the daily value is this?
   c. What daily value of sodium intake is deemed healthy?

Nutrition Facts
Serving Size: 1 oz. (28g/About 29 chips)
Servings Per Container: About 11

Amount Per Serving
Calories 160   Calories from Fat 90

	% Daily Value
Total fat 10g	**15%**
Saturated fat 1.5 g	**7%**
Cholesterol 0mg	**0%**
Sodium 240mg	**12%**
Total carbohydrate 15g	**5%**
Dietary fiber 1g	**4%**
Sugars less than 1g	
Protein 2g	

ILLUSTRATION 10

49. DRIVER'S LICENSE   On the written part of his driving test, a man answered 28 out of 40 questions correctly. If 70% correct is passing, did he pass the test?

50. ALPHABET   What percent of the English alphabet do the vowels a, e, i, o, and u make up? (Round to the nearest one percent.)

51. MIXTURES   Complete the chart in Illustration 11 to find the number of gallons of sulfuric acid in each of two storage tanks.

Gallons of solution in tank	% sulfuric acid	Gallons of sulfuric acid in tank
60	50%	
40	30%	

ILLUSTRATION 11

52. CUSTOMER GUARANTEE   To assure its customers of low prices, the Home Club offers a "10% Plus" guarantee. If the customer finds the same item selling for less somewhere else, he or she receives the difference in price, plus 10% of the difference. A woman bought miniblinds at the Home Club for $120 but later saw the same blinds on sale for $98 at another store. How much can she expect to be reimbursed?

53. MAKING COPIES   The zoom key on the control panel of a copier programs it to print a magnified or reduced copy of the original document. If the zoom is set at 180% and the original document contains type that is 1.5 inches tall, what will be the height of the type on the copy?

54. MAKING COPIES   The zoom setting for a copier is entered as a decimal: 0.98. Express it as a percent and find the resulting type size on the copy if the original has type 2 inches in height.

55. INSURANCE   The cost to repair a car after a collision was $4,000. The automobile insurance policy paid the entire bill except for a $200 deductible, which the driver paid. What percent of the cost did he pay?

56. FLOOR SPACE   A house has 1,200 square feet on the first floor and 800 square feet on the second floor. What percent of the square footage of the house is on the first floor?

57. A MAJORITY   In Los Angeles City Council races, if no candidate receives more than 50% of the vote, a runoff election is held between the first- and second-place finishers. From the election results in Illustration 12, determine whether there must be a runoff election for District 10.

City Council	District 10
Nate Holden	8,501
Madison T. Shockley	3,614
Scott Suh	2,630
Marsha Brown	2,432

ILLUSTRATION 12

58. PORTS   In 1997, the busiest port in the United States was the Port of South Louisiana, which handled 183,628,353 tons of goods. Of that amount, 106,846,289 tons were domestic goods, and 76,782,064 tons were foreign. What percent of the total was domestic? Round to the nearest tenth of a percent.

WRITING

59. Explain the relationship in a percent problem between the amount, the percent, and the base.

60. Write a real-life situation that could be described by "9 is what percent of 20?"

61. Explain why 150% of a number is more than the number.

62. Explain why "Find 9% of 100" is an easy problem to solve.

REVIEW

63. Add: $2.78 + 6 + 9.09 + 0.3$.

64. Evaluate: $\sqrt{64} + 3\sqrt{9}$.

65. On a number line, which number is closer to 5, 4.9 or 5.001?

66. Multiply: $34.5464 \cdot 1,000$.

67. Find: $(0.2)^3$.

68. Solve $0.4x + 1.2 = -7.8$.

5.3 Applications of Percent

In this section, you will learn about

- Taxes • Commissions • Percent of increase or decrease
- Discounts

INTRODUCTION. In this section, we discuss four applications of percent. Three of the four (taxes, commissions, and discounts) are directly related to purchasing. A solid understanding of these concepts will make you a better consumer. The fourth application uses percent to describe increases or decreases of such things as unemployment and grocery store sales.

Taxes

The sales receipt in Figure 5-6 (on the next page) gives a detailed account of what items were purchased, how many of each were purchased, and the price of each item.

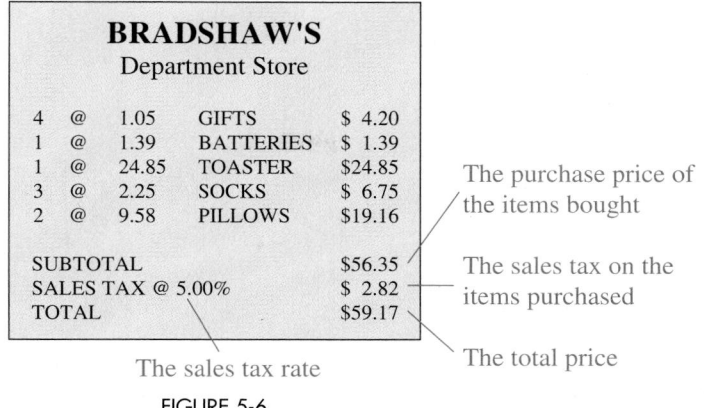

FIGURE 5-6

The receipt shows that the $56.35 purchase price (labeled *subtotal*) was taxed at a **rate** of 5%. Sales tax of $2.82 was charged. The sales tax was then added to the subtotal to get the total price of $59.17.

Finding the total price

> Total price = purchase price + sales tax

In Example 1, we verify that the amount of sales tax shown on the receipt in Figure 5-6 is correct.

EXAMPLE 1   *Finding the sales tax.* Find the sales tax on a purchase of $56.35 if the sales tax rate is 5%.

Solution

First we write the problem so that we can translate it into an equation. The rate is 5%. We are to find the amount of the tax.

What number	is	5%	of	56.35?
x	=	5%	·	56.35

$x = 0.05 \cdot 56.35$ Change 5% to a decimal: 5% = 0.05.

$x = 2.8175$ Do the multiplication.

Rounding to the nearest cent (hundredths), we find that the sales tax would be $2.82. The sales receipt in Figure 5-6 is correct.

Self Check

What would the sales tax be if the $56.35 purchase were made in Texas, which has a 6.25% state sales tax?

Answer: $3.52

In addition to sales tax, we pay many other types of taxes in our daily lives. Income tax, gasoline tax, and Social Security tax are just a few.

EXAMPLE 2   *Finding the tax rate.* A waitress found that $11.04 was deducted from her weekly gross earnings of $240 for federal income tax. What withholding tax rate was used?

Solution

First, we write the problem in a form that can be translated into an equation. We need to find the tax rate.

Self Check

$5,250 had to be paid on an inheritance of $15,000. What is the inheritance tax rate?

$$11.04 \quad \text{is} \quad \text{what percent} \quad \text{of} \quad 240?$$

$$11.04 \quad = \quad x \quad \cdot \quad 240$$

$11.04 = 240x$ Rewrite the right-hand side: $x \cdot 240 = 240x$.

$\dfrac{11.04}{240} = \dfrac{240x}{240}$ To undo the multiplication by 240, divide both sides by 240.

$0.046 = x$ Do the divisions.

$4.6\% = x$ Change 0.046 to a percent.

The withholding tax rate was 4.6%.

Answer: 35%

Commissions

Instead of working for a salary or getting paid at an hourly rate, many salespeople are paid on **commission.** They earn an amount based on the goods or services they sell.

EXAMPLE 3   *Finding a commission.*   The commission rate for a salesperson at an appliance store is 16.5%. What is his commission from the sale of a refrigerator costing $499.95?

Solution

We write the problem so that it can be translated into an equation. We are to find the amount of the commission.

$$\text{What number} \quad \text{is} \quad 16.5\% \quad \text{of} \quad 499.95?$$

$$x \quad = \quad 16.5\% \quad \cdot \quad 499.95$$

$x = 0.165 \cdot 499.95$ Change 16.5% to a decimal: $16.5\% = 0.165$.

$x = 82.49175$ Use a calculator to do the multiplication.

Rounding to the nearest cent (hundredth), we find that the commission is $82.49.

Self Check

An insurance salesperson receives a 4.1% commission on each $120 premium paid by a client. What is the amount of the commission on this premium?

Answer: $4.92

Percent of increase or decrease

Percents can be used to describe how a quantity has changed. For example, consider Figure 5-7, which compares the number of hours of work it took the average U.S. worker to earn enough to buy a dishwasher in 1950 and 1998.

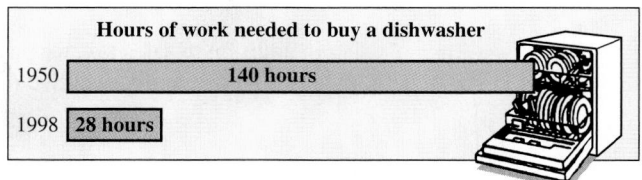

Hours of work needed to buy a dishwasher

1950	140 hours
1998	28 hours

Source: Federal Reserve Bank of Dallas

FIGURE 5-7

From the figure, we see that the number of hours an average American had to work in order to buy a dishwasher has decreased over the years. To describe this decrease using a percent, we first subtract to find the amount of the decrease.

$$140 - 28 = 112 \quad \text{Subtract the hours of work needed in 1998 from the hours of work needed in 1950.}$$

Next, we find what percent of the original number of hours of work needed in 1950 this difference represents.

112	is	what percent	of	140?
112	=	x	$\cdot$	140

$$112 = 140x \quad \text{Rewrite the right-hand side: } x \cdot 140 = 140x.$$

$$\frac{112}{140} = \frac{140x}{140} \quad \text{To undo the multiplication by 140, divide both sides by 140.}$$

$$0.8 = x \quad \text{Do the divisions.}$$

$$80\% = x \quad \text{Change 0.8 to a percent.}$$

From 1950 to 1998, there was an 80% decrease in the number of hours it took the average U.S. worker to earn enough to buy a dishwasher.

Finding the percent of increase or decrease

To find the percent of increase or decrease:
1. Subtract the smaller number from the larger to find the amount of increase or decrease.
2. Find what percent the difference is of the original amount.

EXAMPLE 4 *Finding the percent of increase.* A 1996 auction included an oak rocking chair used by President John F. Kennedy in the Oval Office. The chair, originally valued at $5,000, sold for $453,500. Find the percent of increase in the value of the rocking chair.

Solution

First, we find the amount of increase.

$$453,500 - 5,000 = 448,500 \quad \text{Subtract the original value from the price paid at auction.}$$

The rocking chair increased in value by $448,500. Next, we find what percent of the original value the increase represents.

448,500	is	what percent	of	5,000?
448,500	=	x	$\cdot$	5,000

$$448,500 = 5,000x \quad \text{Rewrite the right-hand side: } x \cdot 5,000 = 5,000x.$$

$$\frac{448,500}{5,000} = \frac{5,000x}{5,000} \quad \text{To undo the multiplication by 5,000, divide both sides by 5,000.}$$

$$89.7 = x \quad \text{Do the divisions.}$$

$$8,970\% = x \quad \text{Change 89.7 to a percent.}$$

The Kennedy rocking chair increased in value by an amazing 8,970%.

Self Check

In one school district, the number of home-schooled children increased from 15 to 150 in 4 years. Find the percent of increase.

Answer: 900%

EXAMPLE 5   POPULATION DECLINE. Norfolk, Virginia, experienced the greatest percent decrease in population of any major U.S. city over the eight-year period from 1990 to 1998. Use the information in Figure 5-8 to determine the population of Norfolk in 1998.

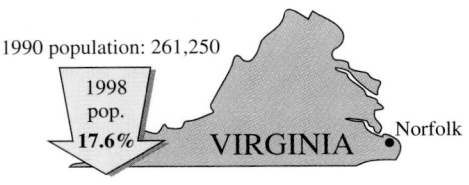

Source: U.S. Bureau of the Census (1999)

FIGURE 5-8

Solution   In 1990, the population was 261,250. We are told that the number fell, and we need to find out by how much. To do so, we solve the following percent equation.

What number	is	17.6%	of	261,250?
x	=	17.6%	$\cdot$	261,250

$x = 0.176 \cdot 261,250$ Change 17.6% to a decimal: 17.6% = 0.176.

$x = 45,980$ Do the multiplication.

From 1990 to 1998, the population decreased by 45,980. To find the city's population in 1998, we subtract the decrease from the population in 1990.

$261,250 - 45,980 = 215,270$

In 1998, the population of Norfolk, Virginia was 215,270.

We can solve this problem in another way. If the population of Norfolk decreased by 17.6%, then the population in 1998 was $100\% - 17.6\%$ or 82.4% of the population in 1990. Using this approach, we can find the 1998 population directly by solving the following percent equation.

What number	is	82.4%	of	261,250?
x	=	82.4%	$\cdot$	261,250

$x = 0.824 \cdot 261,250$ Change 82.4% to a decimal: 82.4% = 0.824.

$x = 215,270$ Do the multiplication.

As before, we see that the population of Norfolk in 1998 was 215,270.

Discounts

The difference between the original price and the sale price of an item is called the **discount**. If the discount is expressed as a percent of the selling price, it is called the **rate of discount**. We will use the information in the advertisement shown in Figure 5-9 to discuss how to find a discount and how to find a discount rate.

FIGURE 5-9

EXAMPLE 6 *Finding the discount.* Find the amount of the discount on the pair of men's basketball shoes shown in Figure 5-9. Then find the sale price.

Solution
To find the discount, we find 25% of the regular price, $59.80.

What number	is	25%	of	59.80?
x	$=$	25%	$\cdot$	59.80

$x = 0.25 \cdot 59.80$ Change 25% to a decimal: 25% = 0.25.

$x = 14.95$ Do the multiplication.

The discount is $14.95. To find the sale price, we subtract the amount of the discount from the regular price.

$59.80 - 14.95 = 44.85$

The sale price of the men's basketball shoes is $44.85.

Self Check
Sunglasses, regularly selling for $15.40, are discounted 15%. Find the sale price.

Answer: $13.09

In Example 6, we used the following formula to find the sale price.

Finding the sale price

> Sale price = original price − discount

EXAMPLE 7 *Finding the discount rate.* What is the rate of discount on the ladies' aerobic shoes advertised in Figure 5-9?

Solution
We can think of this as a percent-of-decrease problem. We first compute the amount of the discount. This decrease in price is found using subtraction.

$39.99 - 21.99 = 18$

The shoes are discounted $18. Now we find what percent of the original price the discount is.

18	is	what percent	of	39.99?
18	$=$	x	$\cdot$	39.99

$18 = 39.99x$ $x \cdot 39.99 = 39.99x$.

$\dfrac{18}{39.99} = \dfrac{39.99x}{39.99}$ To undo the multiplication by 39.99, divide both sides by 39.99.

$0.450113 \approx x$ Do the division.

$45.0113\% \approx x$ Change 0.450113 to a percent.

Rounded to the nearest one percent, the discount rate is 45%.

Self Check
An early-bird special at a restaurant offers a $10.99 prime rib dinner for only $7.95 if it is ordered before 6 P.M. Find the rate of discount. Round to the nearest one percent.

Answer: 28%

STUDY SET Section 5.3

VOCABULARY *Fill in the blanks.*

1. Some salespeople are paid on _____. It is based on a percent of the total dollar amount of the goods or services they sell.

2. When we use percent to describe how a quantity has increased when compared to its original value, we are finding the percent of _____.

3. The difference between the original price and the sale price of an item is called the _____.

4. The _____ of a sales tax is expressed as a percent.

CONCEPTS

5. An organization experiences a 100% increase in membership. Represent the increase in another way.

6. The number of people watching a television show decreased by 50% over a ten-week period. Represent the decrease in another way.

APPLICATIONS *Solve each problem. If a percent answer is not exact, round to the nearest one percent.*

7. STATE SALES TAX   The state sales tax rate in Utah is 4.75%. Find the sales tax on a dining room set that sells for $900.

8. STATE SALES TAX   Find the sales tax on a pair of jeans costing $40 if they are purchased in Arkansas, which has a sales tax rate of 4.625%.

9. ROOM TAX   After checking out of a hotel, a man noticed that the hotel bill included an additional charge labeled *room tax*. If the price of the room was $129 plus a room tax of $10.32, find the room tax rate.

10. EXCISE TAX   While examining her monthly telephone bill, a woman noticed an additional charge of $1.24 labeled *federal excise tax*. If the basic service charges for that billing period were $42, what is the federal excise tax rate?

11. SALES RECEIPT   Complete the sales receipt in Illustration 1 by finding the subtotal, the sales tax, and the total.

```
        NURSERY CENTER
       Your one-stop garden supply

3 @  2.99    PLANTING MIX      $  8.97
1 @  9.87    GROUND COVER      $  9.87
2 @ 14.25    SHRUBS            $ 28.50

SUBTOTAL                       $
SALES TAX @ 6.00%              $
TOTAL                          $
```

ILLUSTRATION 1

12. SALES RECEIPT   Complete the sales receipt in Illustration 2 by finding the prices, the subtotal, the sales tax, and the total.

```
        McCOY'S FURNITURE

1 @ 450.00    SOFA           $
2 @  90.00    END TABLES     $
1 @ 350.00    LOVE SEAT      $

SUBTOTAL                     $
SALES TAX @ 4.20%            $
TOTAL                        $
```

ILLUSTRATION 2

13. SALES TAX HIKE   In order to raise more revenue, some states raise the sales tax rate. How much additional money will be collected on the sale of a $15,000 car if the sales tax rate is raised 1%?

14. FOREIGN TRAVEL   Value added tax is a consumer tax imposed on goods and services. Currently, there are VAT systems in place all around the world. (The United States is one of the few industrialized nations not using a value added tax system.) Complete the table by determining the VAT tax a traveler would pay in each country on a dinner costing $20.95.

Country	VAT tax rate	Tax on a $20.95 dinner
Canada	7%	
Germany	16%	
England	17.5%	
Sweden	25%	

15. PAYCHECK   Use the information on the paycheck stub in Illustration 3 (on the next page) to find the tax rate for the federal withholding, worker's compensation, and Medicare taxes that were deducted from the gross pay.

```
                    6286244
  Issue date: 03-27-00

  GROSS PAY        $360.00
  TAXES
     FED. TAX      $ 28.80
     WORK. COMP.   $  4.32
     MEDICARE      $  5.04

  NET PAY          $321.84
```
ILLUSTRATION 3

16. GASOLINE TAX   In one state, a gallon of unleaded gasoline sells for $1.89. This price includes federal and state taxes that total approximately $0.54. Therefore, the price of a gallon of gasoline, before taxes, is about $1.35. What is the tax rate on gasoline?

17. OVERTIME   Factory management wants to reduce the number of overtime hours by 25%. If the total number of overtime hours is 480 this month, what is the target number of overtime hours for next month?

18. COST-OF-LIVING INCREASE   If a woman making $32,000 a year receives a cost-of-living increase of 2.4%, how much is her raise? What is her new salary?

19. REDUCED CALORIES   A company advertised its new, improved chips as having 36% fewer calories per serving than the original style. How many calories are in a serving of the new chips if a serving of the original style contained 150 calories?

20. POLICE FORCE   A police department plans to increase its 80-person force by 5%. How many additional officers will be hired? What will be the new size of the department?

21. ENDANGERED SPECIES   Illustration 4 shows the total number of endangered and threatened plant and animal species for each of the years 1993–1999, as determined by the U.S. Fish and Wildlife Service. When was there a decline in the total? To the nearest percent, find the percent of decrease in the total for that period.

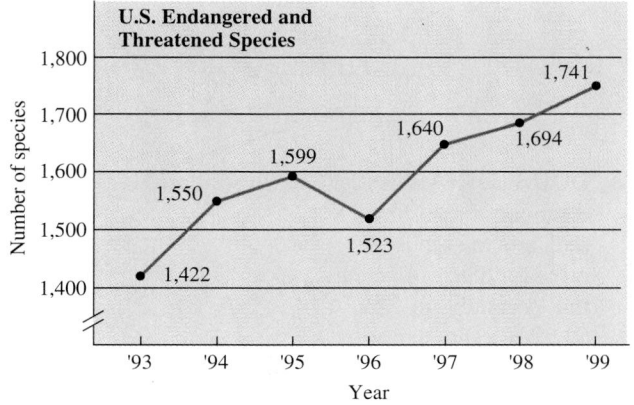

Source: *The New York Times 2000 Almanac*

ILLUSTRATION 4

22. CROP DAMAGE   After flooding damaged much of the crop, the cost of a head of lettuce jumped from $0.99 to $2.20. What percent of increase is this?

23. CAR INSURANCE   A student paid a car insurance premium of $400 every three months. Then the premium dropped to $360, because she qualified for a good-student discount. What was the percent of decrease in the premium?

24. BUS PASS   To increase the number of riders, a bus company reduced the price of a monthly pass from $112 to $98. What was the percent of decrease?

25. LAKE SHORELINE   Because of a heavy spring runoff, the shoreline of a lake increased from 5.8 miles to 7.6 miles. What was the percent of increase in the shoreline?

26. BASEBALL   Illustration 5 shows the path of a baseball hit 110 mph, with a launch angle of 35 degrees, at sea level and at Coors Field, home of the Colorado Rockies. What is the percent of increase in the distance the ball travels at Coors Field?

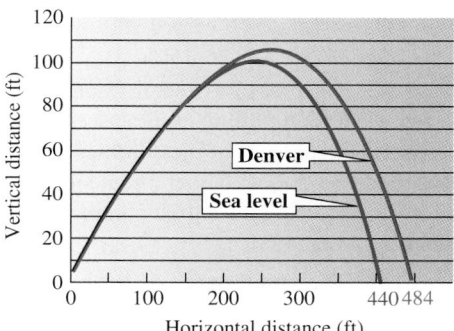

Source: *Los Angeles Times* (September 16, 1996)

ILLUSTRATION 5

27. EARTH MOVING   Illustration 6 shows the typical soil volume change during earth moving. (One cubic yard of soil fits in a cube that is 1 yard long, 1 yard wide, and 1 yard high.)
   a. Find the percent of increase in the soil volume as it goes through Step 1 of the process.
   b. Find the percent of decrease in the soil volume as it goes through Step 2 of the process.

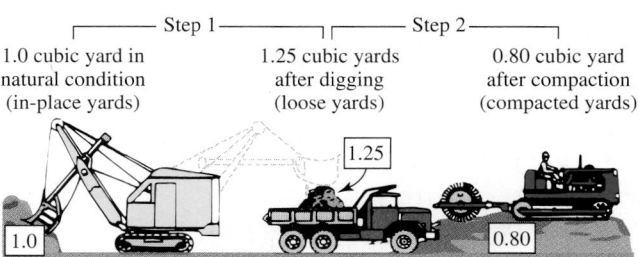

Source: U.S. Department of the Army

ILLUSTRATION 6

28. PARKING The management of a mall has decided to increase the parking area. The plans are shown in Illustration 7. What will be the percent of increase in the parking area once the project is completed?

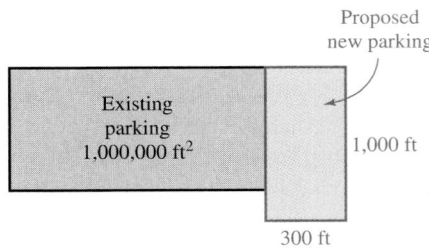

ILLUSTRATION 7

29. REAL ESTATE After selling a house for $98,500, a real estate agent split the 6% commission with another agent. How much did each person receive?

30. MEDICAL SUPPLIES A salesperson for a medical supplies company is paid a commission of 9% for orders under $8,000. For orders exceeding $8,000, she receives an additional 2% in commission on the total amount. What is her commission on a sale of $14,600?

31. SPORTS AGENT A sports agent charges her clients a fee to represent them during contract negotiations. The fee is based on a percent of the contract amount. If the agent earned $37,500 when her client signed a $2,500,000 professional football contract, what rate did she charge for her services?

32. ART GALLERY An art gallery displays paintings for artists and receives a commission from the artist when a painting is sold. What is the commission rate if a gallery received $135.30 when a painting was sold for $820?

33. CONCERT PARKING A concert promoter gets $33\frac{1}{3}\%$ of the revenue the arena receives from its parking concession the night of the performance. How much can the promoter make if 6,000 cars are anticipated and parking is $6 a car?

34. KITCHENWARE PARTY A homemaker invited her neighbors to a kitchenware party to show off cookware and utensils. As party hostess, she received 12% of the total sales. How much was purchased if she received $41.76 for hosting the party?

35. WATCH SALE See Illustration 8. What are the regular price and the rate of discount for the watch that is on sale?

ILLUSTRATION 8

36. STEREO SALE See Illustration 9. What are the regular price and the rate of discount for the stereo system that is on sale?

ILLUSTRATION 9

37. RING SALE What does a ring regularly sell for if it has been discounted 20% and is on sale for $149.99? (*Hint:* The ring is selling for 80% of its regular price.)

38. BLINDS SALE What do vinyl blinds regularly sell for if they have been discounted 55% and are on sale for $49.50? (*Hint:* The blinds are sellng for 45% of their regular price.)

39. VCR SALE What are the sale price and the discount rate for a VCR with remote that regularly sells for $399.97 and is being discounted $50?

40. CAMCORDER SALE What are the sale price and the discount rate for a camcorder that regularly sells for $559.97 and is being discounted $80?

41. REBATE See Illustration 10. Find the discount, the discount rate, and the reduced price for a case of motor oil if a shopper receives the manufacturer's rebate mentioned in the ad.

ILLUSTRATION 10

42. DOUBLE COUPONS See Illustration 11. Find the discount, the discount rate, and the reduced price for a box of cereal that normally sells for $3.29 if a shopper presents the coupon at a store that doubles the value of the coupon.

ILLUSTRATION 11

43. TV SHOPPING Determine the Home Shopping Network (HSN) price of the ring described in Illustration 12 if it sells it for 55% off of the retail price. Ignore shipping and handling costs.

Item 169-117
2.75 lb ctw
10K
Blue Topaz
Ring
6, 7, 8, 9, 10
Retail value $170
HSN Price
$??.??
S&H $5.95

ILLUSTRATION 12

44. INFOMERCIAL The host of a TV infomercial says that the suggested retail price of a rotisserie grill is $249.95 and that it is now offered "for just 4 easy payments of only $39.95." What is the discount, and what is the discount rate?

WRITING

45. List the pros and cons of working on commission.

46. In Example 6, explain why you get the correct answer for the sale price by finding 75% of the regular price.

47. Explain the difference between a tax and a tax rate.

48. Explain how to find the sale price of an item if you know the regular price and the discount rate.

REVIEW

49. Multiply: $-5(-5)(-2)$.

50. Solve $4x + 24 = 12$.

51. Evaluate $-4 - (-7)$.

52. Evaluate $|-5 - 8|$.

53. A store clerk earns $12.50 an hour. How much will she earn in a 40-hour week?

54. Add: $\dfrac{3}{5} + \dfrac{4}{7}$.

55. Subtract: $\dfrac{4}{9} - \dfrac{3}{5}$.

56. Multiply: $\dfrac{7}{9} \cdot \dfrac{6}{11}$.

57. Divide: $\dfrac{6}{7} \div \dfrac{3}{5}$.

58. Simplify $\dfrac{\frac{3}{5}}{\frac{2}{7}}$.

ESTIMATION

We will now discuss some estimation methods that can be used when working with percent. To begin, we consider a way to find 10% of a number quickly. Recall that 10% of a number is found by multiplying the number by 10% or 0.1. When multiplying a number by 0.1, we simply move the decimal point one place to the left to find the result.

EXAMPLE 1   ***10% of a number.***   Find 10% of 234.

       Solution   To find 10% of 234, move the decimal point 1 place to the left.

$$234 = 23.4.0$$

Thus, 10% of 234 is 23.4, or approximately 23.   ■

To find 15% of a number, first find 10% of the number. Then find half of that to obtain the other 5%. Finally, add the two results.

EXAMPLE 2   ***Estimating 15% of a number.***   Estimate 15% of 78.

       Solution   10% of 78 is 7.8, or about 8. ⟶   8
Add half of 8 to get the other 5%. ⟶   + 4
                                                12

Thus, 15% of 78 is approximately 12.   ■

To find 20% of a number, first find 10% of it and then double that result. A similar procedure can be used when working with any multiple of 10%.

EXAMPLE 3   ***Estimating 20% of a number.***   Estimate 20% of 3,234.15.

       Solution   10% of 3,234.15 is 323.415 or about 323. To find 20%, double that.

Thus, 20% of 3,234.15 is approximately 646.   ■

EXAMPLE 4   ***1% of a number.***   Find 1% of 0.8.

       Solution   To find 1% of a number, multiply it by 0.01, because 1% = 0.01. When multiplying a number by 0.01, simply move the decimal point two places to the left to find the result.

$$0.8 = .00.8$$

$$= 0.008$$

Thus, 1% of 0.8 is 0.008.   ■

EXAMPLE 5   ***50% of a number.***   Find 50% of 2,800,000,000.

       Solution   To find 50% of a number means to find $\frac{1}{2}$ of that number. To find one-half of a number, simply divide it by 2. Thus, 50% of 2,800,000,000 is 2,800,000,000 ÷ 2 = 1,400,000,000.   ■

To find 25% of a number, first find 50% of it, then divide that result by 2.

EXAMPLE 6 *Estimating 25% of a number.* Estimate 25% of 16,813.

Solution 16,813 is about 16,800. Half of that is 8,400. Thus, 50% of 16,813 is approximately 8,400.

To estimate 25% of 16,813, divide 8,400 by 2. Thus, 25% of 16,813 is approximately 4,200. ■

100% of a number is the number itself. To find 200% of a number, double the number.

EXAMPLE 7 *Estimating 200% of a number.* Estimate 200% of 65.198.

Solution 65.198 is about 65. To find 200% of 65, double it. Thus, 200% of 65.198 is approximately 65 · 2 or 130. ■

STUDY SET *Estimate the answer to each problem.*

1. COLLEGE COURSES 20% of the 815 students attending a small college were enrolled in a science course. How many students is this?

2. SPECIAL OFFER In the grocery store, a 65-ounce bottle of window cleaner was marked "25% free." How many ounces are free?

3. DISCOUNT By how much is the price of a VCR discounted if the regular price of $196.88 is reduced by 30%?

4. TIPPING A restaurant tip is normally 15% of the cost of the meal. Find the tip on a dinner costing $38.64.

5. FIRE DAMAGE An insurance company paid 50% of the $107,809 it cost to rebuild a home that was destroyed by fire. How much did the insurance company pay?

6. SAFETY INSPECTION Of the 2,580 vehicles inspected at a safety checkpoint, 10% had code violations. How many cars had code violations?

7. WEIGHTLIFTING A 158-pound weightlifter can bench press 200% of his body weight. How many pounds can he bench press?

8. TESTING On a 120-question true/false test, 5% of a student's answers were wrong. How many questions did she miss?

9. TRAFFIC STUDY According to an electronic traffic monitor, 20% of the 650 motorists that passed it were speeding. How many of these motorists were speeding?

10. SELLING A HOME A homeowner has been told she will recoup 70% of her $5,000 investment if she paints her home before selling it. What is the potential payback if she paints her home?

Approximate the percent and then estimate the answer to each problem.

11. NO-SHOWS The attendance at a seminar was only 31% of what the organizers had anticipated. If 68 people were expected, how many actually attended the seminar?

12. "A" STUDENTS Of the 900 students in a school, 16% were on the principal's honor roll. How many students were on the honor roll?

13. INTERNET SURVEY Illustration 1 shows an online survey question. How many people voted yes?

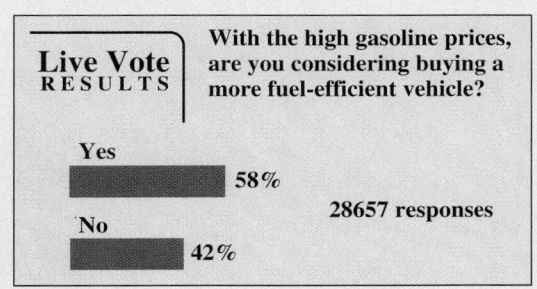

ILLUSTRATION 1

14. MEDICARE The Medicare payroll tax rate is 1.45%. How much Medicare tax will be deducted from a paycheck of $596?

15. VOTING On election day, 48% of the 6,200 workers at the polls were volunteers. How many volunteers helped with the election?

16. BUDGET Each department at a college was asked to cut its budget by 21%. By how much money should the mathematics department budget be reduced if it is currently $4,515?

5.4 *Interest*

In this section, you will learn about

• Simple interest • Compound interest

INTRODUCTION. When money is borrowed, the lender expects to be paid back the amount of the loan plus an additional charge for the use of the money. The additional charge is called **interest.** When money is deposited in a bank, the depositor is paid for the use of the money. The money the deposit earns is also called interest. In general, interest is money that is paid for the use of money.

Simple interest

Interest is calculated in one of two ways: either as **simple interest** or as **compound interest.** We will begin by discussing simple interest. First, we need to introduce some key terms associated with borrowing or lending money.

Principal: the amount of money that is invested, deposited, or borrowed.

Interest rate: a percent that is used to calculate the amount of interest to be paid. It is usually expressed as an annual (yearly) rate.

Time: the length of time (usually in years) that the money is invested, deposited, or borrowed.

The amount of interest to be paid depends on the principal, the rate, and the time. That is why all three are usually mentioned in advertisements for bank accounts, investments, and loans. (See Figure 5-10.)

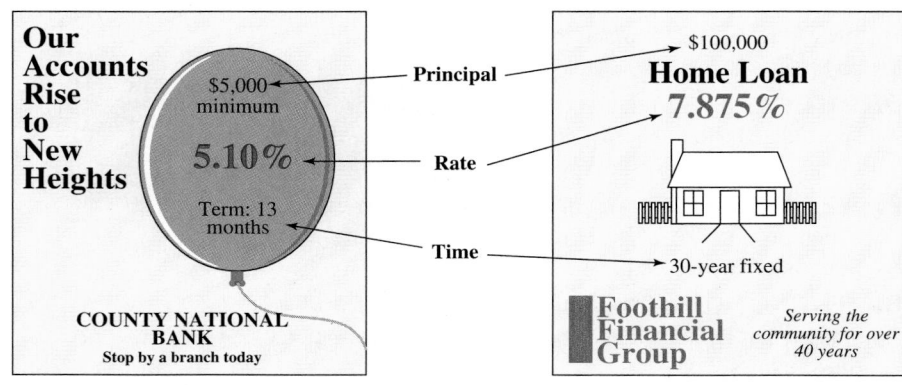

FIGURE 5-10

Simple interest is interest earned on the original principal. It is found by using a formula.

Simple interest formula

Interest = principal · rate · time

or

$I = Prt$

where the rate r is expressed as an annual rate and the time t is expressed in years.

EXAMPLE 1   *Finding the interest earned.* $3,000 is invested for 1 year at a rate of 5%. How much interest is earned?

Solution
We will use the formula $I = Prt$ to calculate the interest earned. The principal is $3,000, the interest rate is 5% (or 0.05), and the time is 1 year.

$P = 3,000 \qquad r = 5\% = 0.05 \qquad t = 1$ year

$I = Prt$ Write the interest formula.

$I = 3,000 \cdot 0.05 \cdot 1$ Substitute the values for P, r, and t.

$I = 150$ Do the multiplication.

The interest earned in 1 year is $150.

The information given in this problem and the result can be presented in a table.

Principal	Rate	Time	Interest earned
$3,000	5%	1 year	$150

Self Check

If $4,200 is invested for 2 years at a rate of 4% annual interest, how much interest is earned?

Answer: $336

When using the formula $I = Prt$, the time must be expressed in years. If the time is given in days or months, we rewrite it as a fractional part of a year. For example, a 30-day investment lasts $\frac{30}{365}$ of a year, since there are 365 days in a year. For a 6-month loan, we express the time as $\frac{6}{12}$ or $\frac{1}{2}$ of a year, since there are 12 months in a year.

EXAMPLE 2   *Paying off a loan.*   To start a carpet-cleaning business, a couple borrows $5,500 to purchase equipment and supplies. If the loan has a 14% interest rate, how much must they repay at the end of the 90-day period?

Solution
First, we find the amount of interest paid on the loan. We must rewrite the time (90-day period) as a fractional part of a 365-day year.

$P = 5,500 \qquad r = 14\% = 0.14 \qquad t = \dfrac{90}{365}$

$I = Prt$ Write the interest formula.

$I = 5,500 \cdot 0.14 \cdot \dfrac{90}{365}$ Substitute the values for P, r, and t.

$I = \dfrac{5,500}{1} \cdot \dfrac{0.14}{1} \cdot \dfrac{90}{365}$ Write 5,500 and 0.14 as fractions.

$I = \dfrac{69,300}{365}$ Use a calculator to multiply the numerators. Multiply the denominators.

$I \approx 189.86$ Use a calculator, do the division. Round to the nearest cent.

The interest on the loan is $189.86. To find how much they must pay back, we add the principal and the interest.

$5,500 + 189.86 = 5,689.86$

The couple must pay back $5,689.86 at the end of 90 days.

Self Check

How much must be repaid if $3,200 is borrowed at a rate of 15% for 120 days?

Answer: $3,357.81

Compound interest

Most savings accounts pay **compound interest** rather than simple interest. Compound interest is interest paid on accumulated interest. To illustrate this concept, suppose that

$2,000 is deposited in a savings account at a rate of 5% for 1 year. We can use the formula $I = Prt$ to calculate the interest earned at the end of 1 year.

$$I = Prt$$
$$I = 2,000 \cdot 0.05 \cdot 1 \quad \text{Substitute for } P, r, \text{ and } t.$$
$$I = 100 \quad\quad\quad\quad\quad \text{Do the multiplication.}$$

Interest of $100 was earned. At the end of the first year, the account contains the interest ($100) plus the original principal ($2,000), for a balance of $2,100.

Suppose that the money remains in the savings account for another year at the same interest rate. For the second year, interest will be paid on a principal of $2,100. That is, during the second year, we earn *interest on the interest* as well as on the original $2,000 principal. Using $I = Prt$, we can find the interest earned in the second year.

$$I = Prt$$
$$I = 2,100 \cdot 0.05 \cdot 1 \quad \text{Substitute for } P, r, \text{ and } t.$$
$$I = 105 \quad\quad\quad\quad\quad \text{Do the multiplication.}$$

In the second year, $105 of interest is earned. The account now contains that interest plus the $2,100 principal, for a total of $2,205.

As Figure 5-11 shows, we calculated the simple interest two times to find the compound interest.

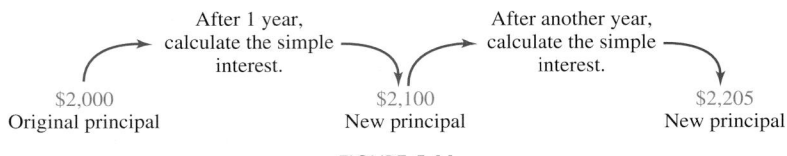

After 1 year, calculate the simple interest.   After another year, calculate the simple interest.

$2,000
Original principal

$2,100
New principal

$2,205
New principal

FIGURE 5-11

If we compute the *simple interest* on $2,000, at 5% for 2 years, the interest earned is $I = 2,000 \cdot 0.05 \cdot 2 = 200$. Thus, the account balance would be $2,200. Comparing the balances, the account earning compound interest will contain $5 more than the account earning simple interest.

In the previous example, the interest was calculated at the end of each year, or **annually.** When compounding, we can compute the interest in other time increments, such as **semiannually** (twice a year), **quarterly** (four times a year), or even **daily.**

EXAMPLE 3 ***Finding compound interest.***   As a gift for her newborn granddaughter, a grandmother opens a $1,000 savings account in the baby's name. The interest rate is 4.2%, compounded quarterly. Find the amount of money the child will have in the bank on her first birthday.

Solution If the interest is compounded quarterly, the interest will be computed four times in one year. To find the amount of interest $1,000 will earn in the first quarter of the year, we use the simple interest formula, where t is $\frac{1}{4}$ of a year.

Interest earned in the first quarter

$$P = 1,000 \quad\quad r = 4.2\% = 0.042 \quad\quad t = \frac{1}{4}$$

$$I = 1,000 \cdot 0.042 \cdot \frac{1}{4}$$

$$I = \$10.50$$

The interest earned in the first quarter is $10.50. This now becomes part of the principal for the second quarter:

$$\$1,000 + \$10.50 = \$1,010.50$$

To find the amount of interest $1,010.50 will earn in the second quarter of the year, we use the simple interest formula, where t is again $\frac{1}{4}$ of a year.

$$P = 1{,}010.50 \qquad r = 0.042 \qquad t = \frac{1}{4}$$

$$I = 1{,}010.50 \cdot 0.042 \cdot \frac{1}{4}$$

$$I \approx \$10.61 \quad \text{(Rounded)}$$

The interest earned in the second quarter is $10.61. This becomes part of the principal for the third quarter:

$$\$1{,}010.50 + \$10.61 = \$1{,}021.11$$

To find the interest $1,021.11 will earn in the third quarter of the year, we proceed as follows.

$$P = 1{,}021.11 \qquad r = 0.042 \qquad t = \frac{1}{4}$$

$$I = 1{,}021.11 \cdot 0.042 \cdot \frac{1}{4}$$

$$I \approx \$10.72 \quad \text{(Rounded)}$$

The interest earned in the third quarter is $10.72. This now becomes part of the principal for the fourth quarter:

$$\$1{,}021.11 + \$10.72 = \$1{,}031.83$$

To find the interest $1,031.83 will earn in the fourth quarter, we again use the simple interest formula.

$$P = 1{,}031.83 \qquad r = 0.042 \qquad t = \frac{1}{4}$$

$$I = 1{,}031.83 \cdot 0.042 \cdot \frac{1}{4}$$

$$I \approx \$10.83 \quad \text{(Rounded)}$$

The interest earned in the fourth quarter is $10.83. Adding this to the existing principal, we get:

$$\$1{,}031.83 + \$10.83 = \$1{,}042.66$$

The amount that has accumulated in the account after four quarters, or 1 year, is $1,042.66. ■

Computing compound interest by hand is tedious. The **compound interest formula** can be used to find the total amount of money that an account will contain at the end of the term.

Compound interest formula

The total amount A in an account can be found using the formula

$$A = P\left(1 + \frac{r}{n}\right)^{nt}$$

where P is the principal, r is the annual interest rate expressed as a decimal, t is the length of time in years, and n is the number of compoundings in one year.

Accent on Technology: **Compound interest**

A businessman invests $9,250 at 7.6% interest, to be compounded monthly. To find what the investment will be worth in 3 years, we use the compound interest formula with the following values.

$$P = \$9,250, \quad r = 7.6\% = 0.076, \quad t = 3 \text{ years}, \quad n = 12 \text{ times a year (monthly)}$$

We apply the compound interest formula:

$$A = P\left(1 + \frac{r}{n}\right)^{nt}$$ Write the compound interest formula.

$$A = 9,250\left(1 + \frac{0.076}{12}\right)^{12(3)}$$ Substitute the values of P, r, t, and n.

$$A = 9,250\left(1 + \frac{0.076}{12}\right)^{36}$$ Simplify the exponent: $12(3) = 36$.

To evaluate the expression on the right-hand side of the equation, we enter these numbers and press these keys.

Keystrokes   9250 $\boxed{\times}$ $\boxed{(}$ 1 $\boxed{+}$ .076 $\boxed{\div}$ 12 $\boxed{)}$ $\boxed{y^x}$ 36 $\boxed{=}$ $\boxed{\text{11610.43875}}$

Rounded to the nearest cent, the amount in the account after 3 years will be $11,610.44.

If your calculator does not have parenthesis keys, calculate the sum inside the parentheses first. Then find the power. Finally, multiply by 9,250.

EXAMPLE 4   *Compound interest.*   A man deposited $50,000 in a long-term account at 6.8% interest, compounded daily. How much money will he be able to withdraw in 7 years if the principal is to remain in the bank?

Solution

"Compounded daily" means that compounding will be done 365 times in a year.

$$P = \$50,000 \qquad r = 6.8\% = 0.068 \qquad t = 7 \text{ years} \qquad n = 365 \text{ times a year}$$

$$A = P\left(1 + \frac{r}{n}\right)^{nt}$$ Write the compound interest formula.

$$A = 50,000\left(1 + \frac{0.068}{365}\right)^{365(7)}$$ Substitute the values of P, r, t, and n.

$$A = 50,000\left(1 + \frac{0.068}{365}\right)^{2,555}$$ $365(7) = 2,555$.

$$A \approx 80,477.58$$ Use a calculator. Round to the nearest cent.

The account will contain $80,477.58 at the end of 7 years. To find the amount the man can withdraw, we subtract.

$$80,477.58 - 50,000 = 30,477.58$$

The man can withdraw $30,477.58 without having to touch the $50,000 principal.

Self Check

Find the amount of interest $25,000 will earn in 10 years if it is deposited in an account at 5.99% interest, compounded daily.

Answer: $20,505.20

STUDY SET   Section 5.4

VOCABULARY   *Fill in the blanks.*

1. In banking, the original amount of money borrowed or deposited is known as the _____.

2. Borrowers pay _____ to lenders for the use of their money.

3. The percent that is used to calculate the amount of interest to be paid is called the _____ rate.

4. _____ interest is interest paid on accumulated interest.

5. Interest computed only on the original principal is called _____ interest.

6. Percent means parts per _____.

CONCEPTS

7. When we do calculations with percents, they must be changed to decimals or fractions. Change each percent to a decimal.

   a. 7%   **b.** 9.8%   **c.** $6\frac{1}{4}\%$

8. Express each of the following as a fraction of a year. Simplify the fraction.
   a. 6 months   **b.** 90 days
   c. 120 days   **d.** 1 month

9. Complete the table by finding the simple interest earned.

Principal	Rate	Time	Interest earned
$10,000	6%	3 years	

10. Tell how many times a year the interest on a savings account is calculated if the interest is compounded
   a. semiannually   **b.** quarterly
   c. daily   **d.** monthly

11. a. What concept studied in this section is illustrated by the diagram in Illustration 1?
   b. What was the original principal?
   c. How many times was the interest found?
   d. How much interest was earned on the first compounding?
   e. For how long was the money invested?

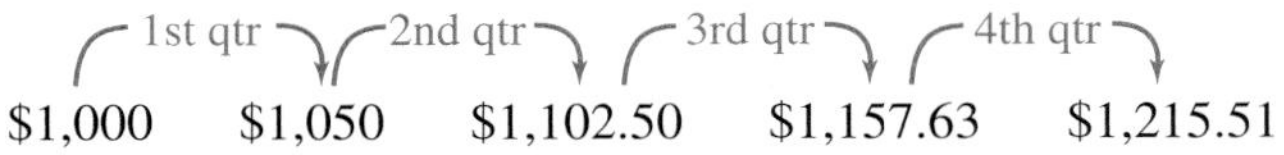

   1st qtr   2nd qtr   3rd qtr   4th qtr

$1,000 $1,050 $1,102.50 $1,157.63 $1,215.51

ILLUSTRATION 1

12. $3,000 is deposited in a savings account that earns 10% interest compounded annually. Complete the series of calculations in Illustration 2 to find how much money will be in the account at the end of 2 years.

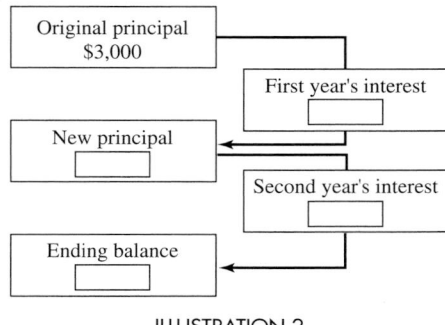

ILLUSTRATION 2

NOTATION

13. In the formula $I = Prt$, what operations are indicated by Prt?

14. In the formula $A = P\left(1 + \dfrac{r}{n}\right)^{nt}$, how many operations must be performed to find A?

APPLICATIONS   *In Exercises 15–26, use simple interest.*

15. RETIREMENT INCOME   A retiree invests $5,000 in a savings plan that pays 6% per year. What will the account balance be at the end of the first year?

16. INVESTMENT   A developer promised a return of 8% annual interest on an investment of $15,000 in her company. How much could an investor expect to make in the first year?

17. REMODELING   A homeowner borrows $8,000 to pay for a kitchen remodeling project. The terms of the loan are 9.2% annual interest and repayment in 2 years. How much interest will be paid on the loan?

18. CREDIT UNION   A farmer borrowed $7,000 from a credit union. The money was loaned at 8.8% annual interest for 18 months. How much money did the credit union charge him for the use of the money?

19. MEETING A PAYROLL   In order to meet end-of-the-month payroll obligations, a small business had to borrow $4,200 for 30 days. How much did the business have to repay if the interest rate was 18%?

20. CAR LOAN   To purchase a car, a man takes out a loan for $2,000. If the interest rate is 9% per year, how much interest will he have to pay at the end of the 120-day loan period?

21. SAVINGS ACCOUNT   Find the interest earned on $10,000 at $7\frac{1}{4}\%$ for 2 years. Use the chart in Illustration 3 to organize your work.

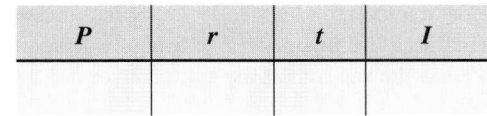

P	*r*	*t*	*I*

ILLUSTRATION 3

22. TUITION   A student borrows $300 from an educational fund to pay for books for spring semester. If the loan is for 45 days at $3\frac{1}{2}\%$ annual interest, what will the student owe at the end of the loan period?

23. LOAN APPLICATION   Complete the loan application form in Illustration 4.

> **Loan Application Worksheet**
>
> 1. Amount of loan (principal) ___$1,200.00___
> 2. Length of loan (time) ___2 YEARS___
> 3. Annual percentage rate ___8%___
> 4. Interest charged _____
> 5. Total amount to be repaid _____
> 6. Check method of repayment:
>   ☐ 1 lump sum ☑ monthly payments
>
>   Borrower agrees to pay ___24___ equal payments of _____ to repay loan.

ILLUSTRATION 4

24. LOAN APPLICATION   Complete the loan application form in Illustration 5.

> **Loan Application Worksheet**
>
> 1. Amount of loan (principal) ___$810.00___
> 2. Length of loan (time) ___9 mos.___
> 3. Annual percentage rate ___12%___
> 4. Interest charged _____
> 5. Total amount to be repaid _____
> 6. Check method of repayment:
>   ☐ 1 lump sum ☑ monthly payments
>
>   Borrower agrees to pay ___9___ equal payments of _____ to repay loan.

ILLUSTRATION 5

25. LOW-INTEREST LOAN   An underdeveloped country receives a low-interest loan from a bank to finance the construction of a water treatment plant. What must the country pay back at the end of 2 years if the loan is for $18 million at 2.3%?

26. REDEVELOPMENT   A city is awarded a low-interest loan to help renovate the downtown business district. The $40 million loan, at 1.75%, must be repaid in $2\frac{1}{2}$ years. How much interest will the city have to pay?

A calculator may be helpful in solving these problems.

27. COMPOUNDING ANNUALLY   If $600 is invested in an account that earns 8%, compounded annually, what will the account balance be after 3 years?

28. COMPOUNDING SEMIANNUALLY   If $600 is invested in an account that earns annual interest of 8%, compounded semiannually, what will the account balance be at the end of 3 years?

29. COLLEGE FUND   A ninth-grade student opens a savings account that locks her money in for 4 years at an annual rate of 6%, compounded daily. If the initial deposit is $1,000, how much money will be in the account when she begins college in four years?

30. CERTIFICATE OF DEPOSIT   A 3-year certificate of deposit pays an annual rate of 5%, compounded daily. The maximum allowable deposit is $90,000. What is the most interest a depositor can earn from the CD?

31. TAX REFUND   A couple deposits an income tax refund check of $545 in an account paying an annual rate of 4.6%, compounded daily. What will the size of the account be at the end of 1 year?

32. INHERITANCE   After receiving an inheritance of $11,000, a man deposits the money in an account paying an annual rate of 7.2%, compounded daily. How much money will be in the account at the end of 1 year?

33. LOTTERY   Suppose you won $500,000 in the lottery and deposited the money in a savings account that paid an annual rate of 6% interest, compounded daily. How much interest would you earn each year?

34. CASH GIFT   After receiving a $250,000 cash gift, a university decides to deposit the money in an account paying an annual rate of 5.88%, compounded quarterly. How much money will the account contain in 5 years?

WRITING

35. What is the difference between simple and compound interest?

36. Explain: *Interest is the amount of money paid for the use of money.*

37. On some accounts, banks charge a penalty if the depositor withdraws the money before the end of the term. Why would a bank do this?

38. Explain why it is better for a depositor to open a savings account that pays 5% interest, compounded daily, than one that pays 5% interest, compounded monthly.

REVIEW

39. Simplify $\sqrt{\dfrac{1}{4}}$.

40. Find: $\left(\dfrac{1}{4}\right)^2$.

41. Round 23.045 to the nearest tenth.

42. Round 23.045 to the nearest hundredth.

43. Solve $\dfrac{2}{3}x = -2$.

44. Divide: $-12\dfrac{1}{2} \div 5$.

45. Multiply: $8\dfrac{1}{3} \cdot 6$.

46. Evaluate $(0.2)^2 - (0.3)^2$.

Percent

Since the word *percent* means *per hundred*, we can think of a percent as the numerator of a fraction that has a denominator of 100. To write a percent as a fraction, we drop the % symbol and write the given number over 100.

Finish each conversion.

$$67\% = \frac{}{100} \qquad 56\% = \frac{56}{} = \frac{14}{25} \qquad 0.05\% = \frac{}{100} = \frac{5}{10{,}000} = \frac{1}{}$$

To write a percent as a decimal, we drop the percent symbol and move the decimal point two places to the left.

Finish each conversion.

$$67\% = \qquad 56\% = \qquad 0.05\% =$$

To write a fraction as a percent, we write the fraction as a decimal and then move the decimal point two places to the right and insert a % symbol.

Finish each conversion.

$$\frac{3}{4} = 0.75 = \qquad\qquad \frac{4}{5} = = 80\%$$

$$\frac{5}{8} = = 62.5\% \qquad \frac{25}{4} = 6.25 =$$

To solve problems involving percent, we use the percent formula.

$$\text{Amount} = \text{percent} \cdot \text{base}$$

Solve each problem.

1. Find 32% of 620.

2. 300 is what percent of 500?

3. 25 is 40% of what number?

4. Find 125% of 850.

5. 106.25 is what percent of 625?

6. 163.84 is 32% of what number?

Percents are used to compute interest. If I is the interest, P the principal, r the annual rate (or percent), and t the length of time in years, the formula for simple interest is

$$I = Prt$$

7. Find the amount of interest that will be earned if $10,000 is invested for 5 years at 6% annual interest.

If A is the amount, P the principal, r the annual rate of interest, t the length of time in years, and n the number of compoundings in one year, the formula for compound interest is

$$A = P\left(1 + \frac{r}{n}\right)^{nt}$$

8. Find the amount of interest that will be earned if $10,000 is invested for 5 years at 6% annual interest, compounded quarterly.

Section 5.1

M & M'S   Give each member of your group a bag of M & M's candies.

a. Determine what percent of the total number of M & M's in your bag are yellow. Do the same for each of the other colors. Enter the results in the table. (Round to the nearest one percent.)

b. Present the data in the table using the circle graph in Illustration 1. Compare your graph to the graphs made by the other members of your group. Do the colors occur in the same percentages in each of the bags?

M & M's color	Percent
Yellow	
Brown	
Green	
Red	
Blue	

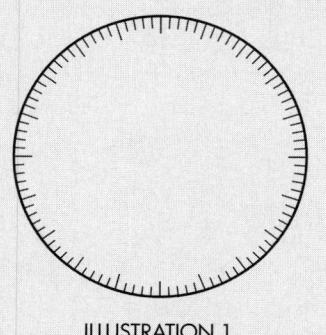

ILLUSTRATION 1

Section 5.2

NUTRITION   Have each person in your group bring in a nutrition label like that shown in Illustration 2 and write the name of the food product on the back. Have the members of the group exchange labels. With the label that you receive, determine what percent of the total calories come from fat.

The USDA recommends that no more than 30% of a person's daily calories should come from fat. Which products exceed the recommendation?

Nutrition Facts

Serving Size 1 meal

Amount Per Serving

Calories 560	Calories from Fat 190

	% Daily Value
Total fat 21g	**32%**
Saturated fat 9 g	**43%**
Cholesterol 60mg	**20%**
Sodium 2110mg	**88%**
Total carbohydrate 67g	**22%**
Dietary fiber 7g	**29%**
Sugars less than 25g	
Protein 27g	

ILLUSTRATION 2

Section 5.3

ENROLLMENT   From your school's admissions office, get the enrollment figures for the last ten years. Calculate the percent of increase (or decrease) in enrollment for each of the following periods.

- Ten years ago to the present
- Five years ago to the present
- One year ago to the present

NEWSPAPER ADS   Have each person in your group find a newspaper advertisement for some item that is on sale. The ad should include only two of the four details listed below.

- The regular price
- The sale price
- The discount
- The discount rate

For example, the ad in Illustration 3 gives the regular price and the sale price, but it doesn't give the discount or the discount rate.

Have the members of your group exchange ads. Determine the two missing details on the ad that you receive. In your group, which item had the highest discount rate?

ILLUSTRATION 3

Section 5.4

INTEREST RATES   Recall that interest is money that the borrower pays to the lender for the use of the money. The amount of interest that the borrower must pay depends on the interest rate charged by the lender.

Have members of your group call banks, savings and loans, credit unions, and other financial services to get the lending rates for various types of loans. (See the yellow pages of the phone book.)

Find out what rate is charged by credit cards such as VISA, department stores, and gasoline companies. List the interest rates in order, from greatest to least, and present your findings to the class.

327

SECTION 5.1 — *Percents, Decimals, and Fractions*

CONCEPTS

Percent means parts per one hundred.

To change a percent to a fraction, drop the % symbol and put the given number over 100.

To change a percent to a decimal, drop the % symbol and divide by 100 by moving the decimal point 2 places to the left.

To change a decimal to a percent, multiply the decimal by 100 by moving the decimal point 2 places to the right, and then insert a % symbol.

To change a fraction to a percent, write the fraction as a decimal by dividing its numerator by its denominator. Multiply the decimal by 100 by moving the decimal point 2 places to the right, and then insert a % symbol.

REVIEW EXERCISES

1. Express the amount of each figure that is shaded as a percent, as a decimal, and as a fraction. Each set of squares represents 100%.

a. 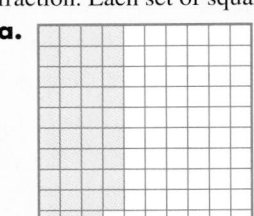 b.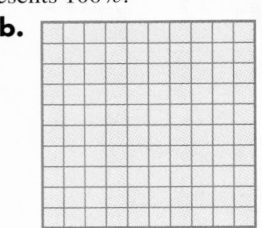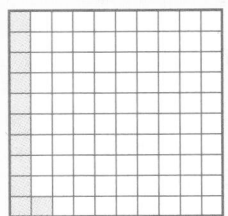

2. In Problem 1, part a, what percent of the figure is not shaded?

3. Change each percent to a fraction.

a. 15% b. 120% c. $9\frac{1}{4}\%$ d. 0.1%

4. Change each percent to a decimal.

a. 27% b. 8% c. 155% d. $1\frac{4}{5}\%$

5. Change each decimal to a percent.
a. 0.83 b. 0.625 c. 0.051 d. 6

6. Change each fraction to a percent.
a. $\frac{1}{2}$ b. $\frac{4}{5}$ c. $\frac{7}{8}$ d. $\frac{1}{16}$

7. Find the exact percent equivalent for each fraction.
a. $\frac{1}{3}$ b. $\frac{5}{6}$

8. Change each fraction to a percent. Round to the nearest hundredth.
a. $\frac{5}{9}$ b. $\frac{8}{3}$

9. BILL OF RIGHTS   There are 27 amendments to the Constitution of the United States. The first ten are known as the Bill of Rights. What percent of the amendments were adopted after the Bill of Rights? (Round to the nearest one percent.)

10. Explain the difference between one-tenth of one percent and ten percent.

SECTION 5.2 | *Solving Percent Problems*

The percent formula:
Amount = percent · base

We can translate a percent problem from words into an equation. A *variable* is used to stand for the unknown number; *is* can be translated to an = sign; and *of* means multiply.

11. Identify the amount, the base, and the percent in the statement "15 is $33\frac{1}{3}$% of 45."

12. Translate the given sentence into a percent equation:

| What number | is | 32% | of | 96? |

13. Solve each percent problem.
a. What number is 40% of 500?   **b.** 16% of what number is 20?

c. 1.4 is what percent of 80?   **d.** $66\frac{2}{3}$% of 3,150 is what number?

e. Find 220% of 55.   **f.** What is 0.05% of 60,000?

14. RACING   The nitro–methane fuel mixture used to power some experimental cars is 96% nitro and 4% methane. How many gallons of each fuel component are needed to fill a 15-gallon fuel tank?

15. HOME SALES   After the first day on the market, 51 homes in a new subdivision had already sold. This was 75% of the total number of homes available. How many homes were originally for sale?

16. HURRICANE DAMAGE   96 of the 110 trailers in a mobile home park were either damaged or destroyed by hurricane winds. What percent is this? (Round to the nearest one percent.)

17. TIPPING   The cost of dinner for a family of five at a restaurant was $36.20. Find the amount of the tip if it should be 15% of the cost of dinner.

A *circle graph* is a way of presenting data for comparison. The sizes of the segments of the circle indicate the percents of the whole represented by each category.

18. AIR POLLUTION   Complete Illustration 1 (a circle graph) to show the given data.

Sources of carbon monoxide air pollution	
Transportation vehicles	63%
Fuel combustion in homes, offices, electrical plants	12%
Industrial processes	8%
Solid-waste disposal	3%
Miscellaneous	14%

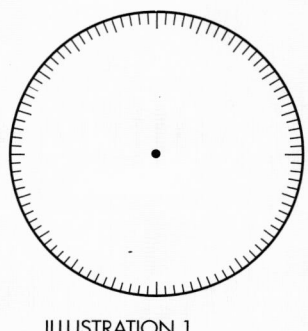

ILLUSTRATION 1

19. EARTH'S SURFACE   The surface of the Earth is approximately 196,800,000 square miles. Use the information in Illustration 2 to determine the number of square miles of the Earth's surface that are covered with water.

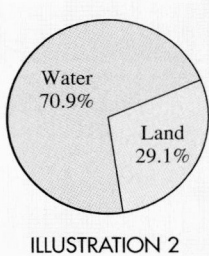

Water
70.9%

Land
29.1%

ILLUSTRATION 2

| SECTION 5.3 | *Applications of Percent* |

To find the total price of an item:
   Total price = purchase
      price + sales tax

20. SALES RECEIPT   Complete the sales receipt in Illustration 3.

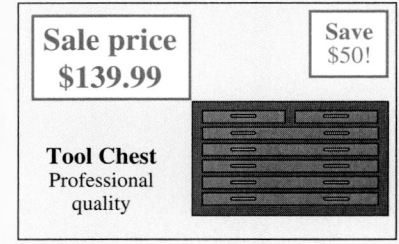

CAMERA CENTER

35mm Canon Camera	$59.99
SUBTOTAL	$59.99
SALES TAX @ 5.5%	
TOTAL	

ILLUSTRATION 3

21. SALES TAX RATE   Find the sales tax rate if the sales tax is $492 on the purchase of an automobile priced at $12,300.

Commission is based on a percent of the total dollar amount of the goods or services sold.

22. COMMISSION   If the commission rate is 6%, find the commission earned by an appliance salesperson who sells a washing machine for $369.97 and a dryer for $299.97.

To find *percent of increase or decrease:*

1. Subtract the smaller number from the larger to find the amount of increase or decrease.

2. Find what percent the difference is of the original amount.

23. TROOP SIZE   The size of a peace-keeping force was increased from 10,000 to 12,500 troops. What percent of increase is this?

24. GAS MILEAGE   Experimenting with a new brand of gasoline in her truck, a woman found that the gas mileage fell from 18.8 to 17.0 miles per gallon. What percent of decrease is this? (Round to the nearest tenth of a percent.)

The difference between the original price and the sale price of an item is called the *discount.*

To find the *sale price:*
   Sale price = original
      price − discount

25. TOOL CHEST   See Illustration 4. Use the information in the advertisement to find the discount, the original price, and the discount rate on the tool chest.

Sale price
$139.99

Save
$50!

Tool Chest
Professional
quality

ILLUSTRATION 4

| **SECTION 5.4** | *Interest* |

Simple interest is interest earned on the original principal and is found using the formula

$$I = Prt$$

where P is the principal, r is the annual interest rate, and t is the length of time in years.

26. Find the interest earned on $6,000 invested at 8% per year for 2 years. Use the following chart to organize your work.

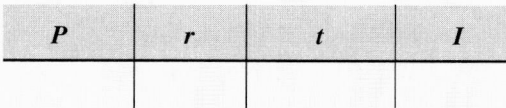

P	r	t	I

27. CODE VIOLATIONS   A business was ordered to correct safety code violations in a production plant. To pay for the needed corrections, the company borrowed $10,000 at 12.5% for 90 days. Find the total amount that had to be paid after 90 days.

Compound interest is interest earned on interest.

28. MONTHLY PAYMENTS   A couple borrows $1,500 for 1 year at $7\frac{3}{4}$% and decides to repay the loan by making 12 equal monthly payments. How much will each monthly payment be?

The compound interest formula:

$$A = P\left(1 + \frac{r}{n}\right)^{nt}$$

where A is the amount in the account, P is the principal, r is the annual interest rate, n is the number of compoundings in one year, and t is the length of time in years.

29. Find the amount of money that will be in a savings account at the end of 1 year if $2,000 is the initial deposit and the annual interest rate of 7% is compounded semi-annually. (*Hint:* Find the simple interest twice.)

30. Find the amount that will be in a savings account at the end of 3 years if a deposit of $5,000 earns interest at an annual rate of $6\frac{1}{2}$%, compounded daily.

31. CASH GRANT   Each year a cash grant is given to a deserving college student. The grant consists of the interest earned that year on a $500,000 savings account. What is the cash award for the year if the money is invested at an annual rate of 8.3%, compounded daily?

Chapter 5 Test

1. See Illustration 1. Express the amount of the figure that is shaded as a percent, as a fraction, and as a decimal.

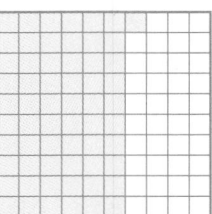

ILLUSTRATION 1

2. In Illustration 2, each set of 100 squares represents 100%. Express as a percent the amount of the figure that is shaded. Then express that percent as a fraction and as a decimal.

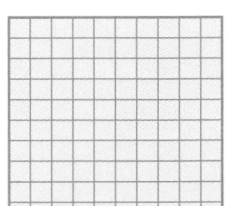

ILLUSTRATION 2

3. Change each percent to a decimal.

   a. 67%   **b.** 12.3%   **c.** $9\frac{3}{4}\%$

4. Change each fraction to a percent.

   a. $\frac{1}{4}$ **b.** $\frac{5}{8}$ **c.** $\frac{3}{25}$

5. Change each decimal to a percent.
   a. 0.19   **b.** 3.47   **c.** 0.005

6. Change each percent to a fraction.
   a. 55%   **b.** 0.01%   **c.** 125%

7. Change $\frac{7}{30}$ to a percent. Round to the nearest hundredth of a percent.

8. WEATHER REPORT   A weatherman states that there is a 40% chance of rain. What are the chances that it will not rain?

9. Find the exact percent equivalent for the fraction $\frac{2}{3}$.

10. Find the exact percent equivalent for the fraction $\frac{1}{4}$.

11. SHRINKAGE   See Illustration 3, a label on a new pair of jeans.
   a. How much length will be lost due to shrinkage?

   b. What will be the resulting length?

WAIST	INSEAM
33	34

Expect shrinkage of approximately
3%
in length after the jeans are washed.

ILLUSTRATION 3

12. 65 is what percent of 1,000?

13. TIPPING   Find the amount of a 15% tip on a meal costing $25.40.

14. FUGITIVES   As of March 15, 2000, 429 of the 458 fugitives who have appeared on the FBI's Ten Most Wanted list have been apprehended or located. What percent is this? Round to the nearest tenth of a percent.

15. SWIMMING WORKOUT   A swimmer was able to complete 18 laps before a shoulder injury forced him to stop. This was only 20% of a typical workout. How many laps does he normally complete during a workout?

16. COLLEGE EMPLOYEES   The 700 employees at a community college fall into three major categories, as shown in Illustration 4. How many employees are in administration?

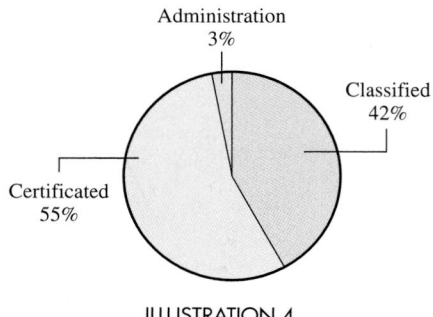

ILLUSTRATION 4

$24,000 \times 6.4 \cdot 3$

17. What number is 24% of 600?

18. HAIRCUTS   Illustration 5 shows the number of minutes it took the average U.S. worker to earn enough to pay for a man's haircut in 1950 and 1998. Find the percent of decrease, to the nearest one percent.

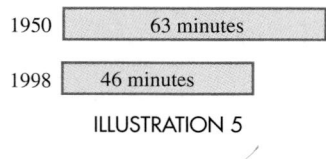

ILLUSTRATION 5

19. HOMEOWNER'S INSURANCE   An insurance salesperson receives a 4% commission on the annual premium of any policy she sells. What is her commission on a homeowner's policy if the premium is $898?

20. COST-OF-LIVING INCREASE   A teacher earning $40,000 just received a cost-of-living increase of 3.6%. What is the teacher's new salary?

21. CAR WAX SALE   A car waxing kit, regularly priced at $14.95, is on sale for $3 off. What are the sale price, the discount, and the rate?

22. POPULATION INCREASE   After a new freeway was completed, the population of a city it passed through increased from 12,808 to 15,565 in two years. What percent of increase is this? (Round to the nearest one percent.)

23. Find the simple interest on a loan of $3,000 at 5% per year for 1 year.

24. ▦ Find the amount of interest earned on an investment of $24,000 paying an annual rate of 6.4% interest, compounded daily for 3 years.

25. POLITICAL AD   Explain what is unclear about the flyer shown in Illustration 6.

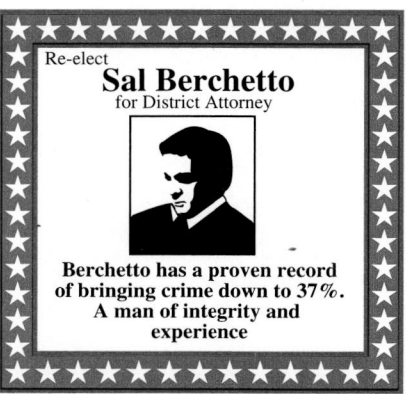

ILLUSTRATION 6

Chapters 1-5 Cumulative Review Exercises

1. SHAQUILLE Use the data in the table to complete Illustration 1 by drawing a line graph to chart the growth of Shaquille O'Neal, the Los Angeles Lakers' center.

Age (yr)	4	6	8	10	12	16	21	28
Weight (lb)	56	82	108	139	192	265	302	315

Based on data from *Los Angeles Times*

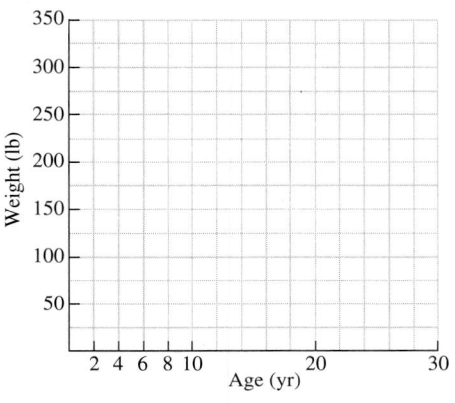

ILLUSTRATION 1

2. State the commutative property of multiplication.

3. **a.** Find the factors of 40.
  b. Find the prime factorization of 40.

4. AUTO INSURANCE See the premium comparison in Illustration 2. How much more does Farmers charge than Mercury does?

Allstate	$2,672	Mercury	$1,370
Auto Club	$1,680	State Farm	$2,737
Farmers	$2,485	20th Century	$1,692

Criteria: Six-month premium. Husband, 45, drives a 1995 Explorer, 12,000 annual miles. Wife, 43, drives a 1996 Dodge Caravan, 12,000 annual miles. Son, 17, is an occasional operator. All have clean driving records.

ILLUSTRATION 2

5. PAINTING A square tarp has sides 8 feet long. When it is laid out on a floor, how much area will it cover?

6. Evaluate $-12 - (-5)$.

7. Evaluate $12 - 2[-8 - 2^4(-1)]$.

8. Find $|-55|$.

9. Solve $3x - 4 = 2$.

10. Solve $6 = 2 - 2x$.

11. Solve $-x = -1$.

12. FRUIT STORAGE Use the formula $C = \frac{5(F - 32)}{9}$ to complete the label on the box of bananas shown in Illustration 3. (*Hint:* Substitute the Fahrenheit temperature for F and evaluate the expression on the right-hand side of the equation.)

PREMIUM
BANANAS

Keep at 59°F or ?°C
Imported by Pacific Fruit, Inc.

ILLUSTRATION 3

13. SPELLING What fraction of the letters in the word *Mississippi* are vowels?

14. Simplify $\frac{10}{15}$.

Do the indicated operation.

15. $-\frac{16}{35} \cdot \frac{25}{48}$

16. $4\frac{2}{5} \div 11$

17. $\frac{4}{3} + \frac{2}{7}$

18. $34\frac{1}{9} - 13\frac{5}{6}$

19. Solve $\frac{5}{6}y = -25$.

20. Solve $\frac{y}{6} - 2 = 1$.

Do the indicated operation.

21. $78.1 - 7.81$

22. $2.13(-4.05)$

23. $0.752(1,000)$

24. $\frac{241.86}{2.9}$

25. Evaluate $\dfrac{3.6 - (-1.5)}{0.5(-1.5) - 0.4(3.6)}$.
Round to the nearest hundredth.

26. Round 452.0298 to the nearest thousandth.

27. Write $\dfrac{11}{15}$ as a decimal. Use an overbar.

28. Solve $\dfrac{y}{2.22} = -5$.

29. Evaluate $3\sqrt{81} - 8\sqrt{49}$.

30. LABOR COST A car repair bill is shown in Illustration 4. The bill is torn, and one line cannot be read. How many hours of labor did it take to repair the car?

> ### *Brian Wood* Auto Repair
>
> Parts... $175.00
> Total labor (at $35 an hour).........................
> Total... $297.50

ILLUSTRATION 4

31. Complete the table.

Percent	Decimal	Fraction
	0.29	
47.3%		
		$\frac{7}{8}$

32. 16% of what number is 20?

33. 16% of 400 is what number?

34. 800 is what percent of 10,000?

35. TIPPING Complete the sales draft in Illustration 5 if a 15% tip, rounded up to the nearest dollar, is to be left for the waiter.

STEAK STAMPEDE	
Bloomington, MN	
Server #12\ AT	
VISA	67463777288
NAME	DALTON/ LIZ
AMOUNT	$75.18
GRATUITY $	_____
TOTAL $	_____

ILLUSTRATION 5

36. GENEALOGY Through an extensive computer search, a genealogist determined that worldwide, 180 out of every 10 million people had his last name. What percent is this?

37. SAVINGS ACCOUNT Find the simple interest earned on $10,000 at $7\frac{1}{4}\%$ for 2 years.

38. Explain why 100% of 50 is 50.

Descriptive Statistics

6.1 Reading Graphs and Tables

6.2 Mean, Median, and Mode

Key Concept: Mean, Median, and Mode

Accent on Teamwork

Chapter Review

Chapter Test

Cumulative Review Exercises

NEWSPAPERS AND MAGAZINES OFTEN PRESENT INFORMATION IN THE FORM OF GRAPHS AND TABLES. IN THIS CHAPTER, WE WILL SHOW HOW INFORMATION CAN BE OBTAINED BY READING MANY TYPES OF GRAPHS. WE WILL THEN DISCUSS THREE MEASURES OF CENTRAL TENDENCY: THE MEAN, THE MEDIAN, AND THE MODE.

6.1   *Reading Graphs and Tables*

In this section, you will learn about

- Reading data from tables • Reading bar graphs • Reading pictographs
- Reading pie graphs • Reading line graphs
- Reading histograms and frequency polygons

INTRODUCTION. It is often said that a picture is worth a thousand words. In this section, we will show how to read information from mathematical pictures called *graphs*.

Reading data from tables

The **table** in Figure 6-1(a), the **bar graph** in Figure 6-1(b), and the **pie graph** in Figure 6-1(c) all show the results of a survey of viewers' opinions. In the bar graph, the length of each bar represents the percent of responses in each category. In the pie graph, the size of each region represents the percent of response. The two graphs tell the story more quickly and more clearly than the table of numbers.

Ratings of Prime-Time News Coverage

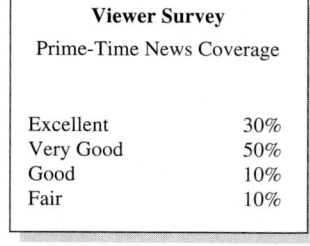

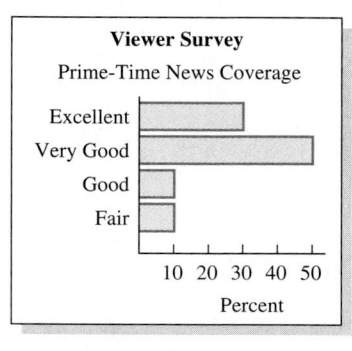

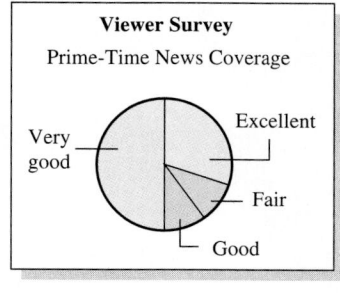

(a)   (b)   (c)

FIGURE 6-1

It is easy to see from either graph that the largest percent of those surveyed rated the programming *very good,* and that the responses *good* and *fair* were tied for last. The same information is available in the table of Figure 6-1(a), but it is not as easy to see at a glance.

Data are often presented in tables, with information organized in rows and columns. To read a table, we must find the intersection of the row and the column that contains the needed information.

Postal rates (in 1998) for priority mail appear in Figure 6-2. To find the cost of mailing an $8\frac{1}{2}$-pound package by priority mail to postal zone 4, we find the *row* of the postage table for a package that does not exceed 9 pounds. We then find the *column* for zone 4. At the intersection of this row and this column, we read the number 9.35. This means that it would cost $9.35 to mail the package.

Priority Mail

Weight not over (lb)	1, 2, & 3	4	5	6	7	8
			Zone			
1	$3.00	3.00	3.00	3.00	3.00	3.00
2	3.00	3.00	3.00	3.00	3.00	3.00
3	4.00	4.00	4.00	4.00	4.00	4.00
4	5.00	5.00	5.00	5.00	5.00	5.00
5	6.00	6.00	6.00	6.00	6.00	6.00
6	6.35	6.90	7.10	7.20	7.80	8.00
7	6.65	7.80	8.10	8.40	9.20	9.80
8	6.95	8.70	9.05	9.50	10.40	11.60
9	7.40	9.35	10.00	10.60	11.30	13.00
10	7.85	10.00	10.75	11.40	12.15	14.05
11	8.25	10.65	11.45	12.20	13.00	15.10
12	8.70	11.30	12.20	13.00	13.90	16.50

FIGURE 6-2

Reading bar graphs

EXAMPLE 1 *Reading bar graphs.* The bar graph in Figure 6-3 shows the total income generated by three sectors of the economy in each of three years. The height of each bar, representing income in billions of dollars, is measured on the scale on the vertical *axis*. The years appear on the horizontal axis. Read the graph to answer the following questions.

a. What income was generated by retail sales in 1980?

b. Which sector of the economy consistently generated the most income?

c. By what amount did income from the wholesale sector increase from 1970 through 1990?

Solution

a. The second group of bars indicates income in 1980, and the middle bar of that group shows sales in the retail sector. The height of that bar is approximately 75, which represents $75 billion. The retail income generated in 1980 was about $75 billion.

Self Check

What income was generated by the service sector in 1990?

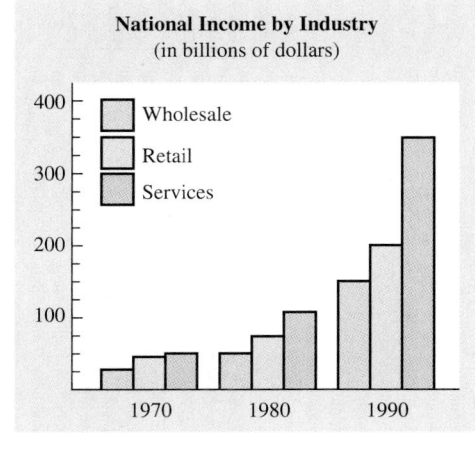

FIGURE 6-3

b. In each group, the rightmost bar is the tallest. That bar, according to the key, represents income from the service sector of the economy. Therefore, services consistently generated the most income.

c. According to the key, the leftmost bar in each group shows income from the wholesale sector. That sector generated about $30 billion in 1970 and $150 billion in 1990. The amount of increase in income is the difference of these two quantities:

$150 billion $-$ $30 billion $=$ $120 billion

Wholesale income increased by $120 billion between 1970 and 1990.

EXAMPLE 2   *Reading bar graphs.* The bar graph in Figure 6-4 shows the number of cars of various models purchased in Dale County for two consecutive years.

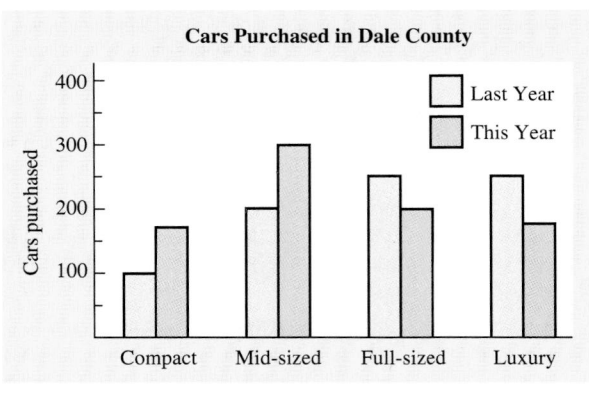

FIGURE 6-4

a. Which models have shown a decrease in sales?

b. Which model showed the greatest increase in sales?

Solution

a. In each pair, the bar on the left gives last year's sales. The bar on the right represents this year's sales. Only for full-sized and luxury cars is the left bar taller than the right bar. This means that full-sized and luxury cars have decreased in sales.

b. Sales of compact and mid-sized cars have increased over last year, because for these models the right bar is taller than the left bar. The difference in the heights of the bars represents the amount of increase. That increase is greater for mid-sized cars. Of all models, mid-sized cars have shown the greatest increase in sales.

Self Check

Answer: $350 billion   ■

Which model has shown the greatest decrease in sales?

Answer: luxury cars   ■

Reading pictographs

A **pictograph** is like a bar graph, but the bars are composed of pictures, where each picture represents a quantity. In Figure 6-5, each picture represents 50 pizzas ordered during exam week. The top bar contains three complete pizzas and one partial pizza. This indicates that the men in the men's residence hall ordered $3 \cdot 50$, or 150 pizzas, plus approximately $\frac{1}{4}$ of 50, or about 13 pizzas. This totals 163 pizzas. The women in the women's residence hall ordered $4\frac{1}{2} \cdot 50$, or 225 pizzas.

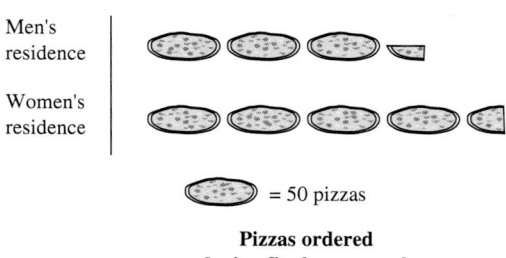

Pizzas ordered during final exam week

FIGURE 6-5

Reading pie graphs

EXAMPLE 3 *Reading pie graphs.* The pie graph in Figure 6-6 gives information about world gold production. The entire circle represents the world's total production, and the sizes of the segments of the circle represent the parts of that total contributed by various nations and regions. Use the graph to answer the following questions.

a. What percent of the total was the combined production of the United States and Canada?

b. What percent of the total production came from sources other than those listed?

c. If the world's total production of gold was 56.3 million ounces during the year of the survey, how many ounces did Australia produce?

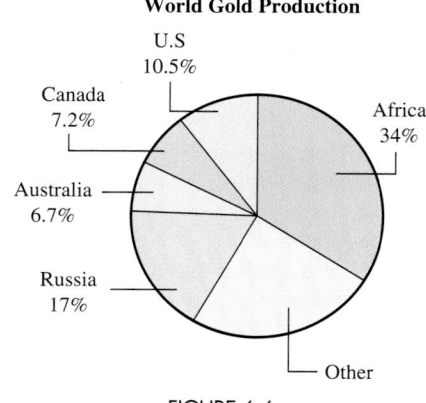

World Gold Production

FIGURE 6-6

Solution

a. According to the graph, the United States produced 10.5% and Canada produced 7.2% of the total. Together, they produced (10.5 + 7.2)%, or 17.7% of the total.

b. To find the percent of gold produced by countries that are not listed, we add the contributions of all the listed sources and subtract that total from 100%.

$$100\% - (34\% + 10.5\% + 7.2\% + 6.7\% + 17\%) = 100\% - 75.4\%$$
$$= 24.6\%$$

The countries that are not listed produced 24.6% of the world's total production of gold.

c. From the graph, we see that Australia produced 6.7% of the world's gold. Since the world total was 56.3 million ounces, Australia's share (in millions of ounces) was

$$6.7\% \text{ of } 56.3 = (0.067)(56.3)$$
$$= 3.7721$$

Rounded to the nearest tenth of a million, Australia produced 3.8 million ounces of gold.

Reading line graphs

Another graph, called a **line graph,** is used to show how quantities change with time. From such a graph, we can determine when a quantity is increasing and when it is decreasing.

EXAMPLE 4 *Reading line graphs.* The line graph in Figure 6-7 shows how U.S. automobile production has changed since 1900. Look at the graph and answer the following questions.

a. How many automobiles were manufactured in 1940?

b. How many were manufactured in 1950?

c. Over which 20-year span did automobile production increase most rapidly?

d. When did production decrease?

e. Why is a broken line used for a portion of the graph?

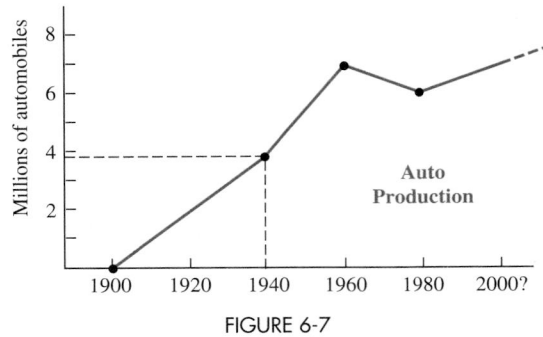

FIGURE 6-7

Solution

a. To find the number of autos produced in 1940, we follow the dashed line from the label 1940 straight up to the graph, and then directly over to the scale. There, we read about 3.8. Since the scale indicates millions of automobiles, approximately 3.8 million autos were produced in 1940.

b. To find the number of autos produced in 1950, we find the point halfway between 1940 and 1960. From there, we move up to the graph and then sideways to the scale, where we read about 5. Approximately 5 million autos were manufactured in 1950.

c. Since the upward tilt of the graph is the greatest between 1940 and 1960, auto production increased most rapidly in those years.

d. Between 1960 and 1980, the graph drops, indicating that auto production decreased during that period.

e. Because beyond the year 2000 was still in the future when the graph was made, the production levels were *projections,* and the broken line indicates that the numbers are only estimates.

Answer: about 3.2 million ■

EXAMPLE 5 *Reading line graphs.* The graph in Figure 6-8 shows the movements of two trains. The horizontal axis represents time, and the vertical axis represents the distance that the trains have traveled.

a. How are the trains moving at time A?

b. At what time (A, B, C, D, or E) are both trains stopped?

c. At what times have both trains gone the same distance?

Solution

The movement of train 1 is represented by the red line, and that of train 2 is represented by the blue line.

a. At time A, the blue line is rising. This shows that the distance traveled by train 2 is increasing: At time A, train 2 is moving.

At time A, the red line is horizontal. This indicates that the distance traveled by train 1 is not changing: At time A, train 1 is stopped.

b. To find the time at which both trains are stopped, we find the time at which both the red and the blue lines are horizontal. At time B, both trains are stopped.

c. At any time, the height of a line gives the distance a train has traveled. Both trains have traveled the same distance whenever the two lines are the same height—that is, at any time when the lines intersect. This occurs at times C and E.

Self Check

In Figure 6-8, what is train 1 doing at time D?

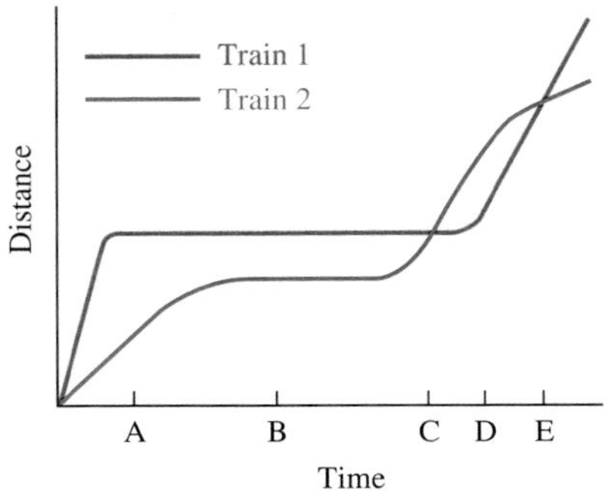

— Train 1
— Train 2

Distance

A   B   C D E

Time

FIGURE 6-8

Answer: Train 1, which had been stopped, is beginning to move.

■

Reading histograms and frequency polygons

A pharmaceutical company is sponsoring a series of reruns of old Westerns. The marketing department must choose from three advertisements:

1. Children talking about Chipmunk Vitamins
2. A college student catching a quick breakfast and a TurboPill Vitamin
3. A grandmother talking about Seniors Vitamins

A survey of the viewing audience records the age of each viewer, counting the number in the 6-to-15-year-old age group, the 16-to-25-year-old age group, and so on. The graph of the data is the **histogram** shown in Figure 6-9. The vertical axis, labeled *frequency,* indicates the number of viewers in each age group. For example, the histogram shows that 105 viewers are in the 36-to-45-year-old age group.

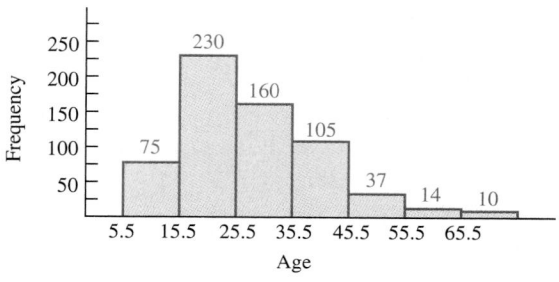

FIGURE 6-9

A histogram is a bar graph with several important features:

1. The bars of a histogram touch.
2. Data values never fall at the edge of a bar.
3. The width of each bar represents a numeric value.

The width of each bar in Figure 6-9 represents an age span of 10 years. Since most viewers are in the 16-to-25-year-old age group, the marketing department decides to advertise TurboPills in commercials appealing to active young adults.

 **EXAMPLE 6**   *Carry-on luggage.*   An airline weighs the carry-on luggage of 2,260 passengers. See the histogram in Figure 6-10.

a. How many passengers carried luggage in the 8-to-11-pound range?

b. How many carried luggage in the 12-to-19-pound range?

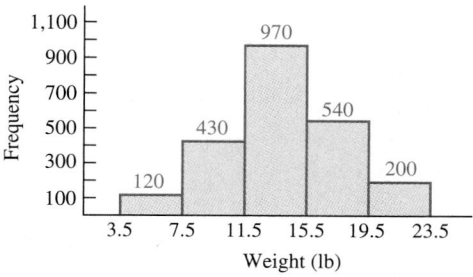

FIGURE 6-10

Solution   **a.** The second bar, with edges at 7.5 and 11.5 pounds, corresponds to the 8-to-11-pound range. Use the height of the bar (or the number written there) to determine that 430 passengers carried such luggage.

b. The 12-to-19-pound range is covered by two bars. The total number of passengers with luggage in this range is 970 + 540, or 1,510.

A special line graph, called a **frequency polygon**, can be constructed from the histogram in Figure 6-10 by joining the center points at the top of each bar. See Figure 6-11. On the horizontal axis, we write the coordinate of the middle value of each bar. After erasing the bars, we get the frequency polygon shown in Figure 6-12.

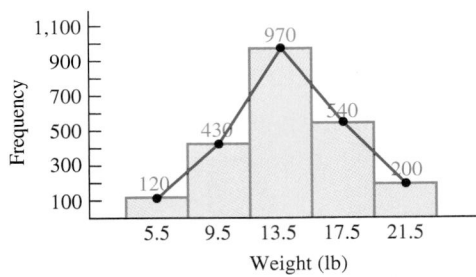

FIGURE 6-11

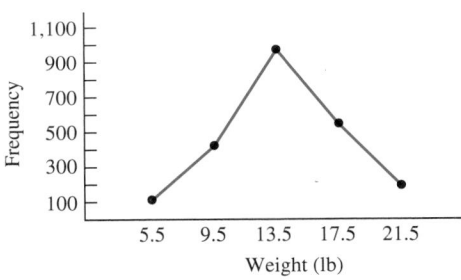

FIGURE 6-12

STUDY SET   Section 6.1

VOCABULARY   *Refer to graphs a through f in Illustration 1 (on the next page). Fill in the blanks.*

1. Graph _____ is a bar graph.

2. Graph _____ is a pie graph.

3. Graph _____ is a pictograph.

4. Graph _____ is a line graph.

5. Graph _____ is a histogram.

6. Graph _____ is a frequency polygon.

CONCEPTS   *Fill in the blanks.*

7. The bars of a _____ touch.

8. The width of each bar of a histogram represents a _____ value.

APPLICATIONS   *Refer to the postal rate table in Figure 6-2 on page 339.*

9. PRIORITY MAIL   Find the cost of using priority mail to send a package weighing $7\frac{1}{4}$ pounds to zone 3.

10. PRIORITY MAIL   Find the cost of sending (priority mail) a package weighing $2\frac{1}{4}$ pounds to zone 5.

11. COMPARING POSTAGE   Juan wants to send a package weighing 6 pounds 1 ounce to a friend living in zone 2. Fourth-class postage would be $1.79. How much could he save by sending the package fourth class instead of priority mail?

12. SENDING TWO PACKAGES   Jenny wants to send a birthday gift and an anniversary gift to her brother, who lives in zone 6. One package weighs 2 pounds 9 ounces, and the other weighs 3 pounds 8 ounces. If she uses priority mail, how much will she save by sending both gifts as one package instead of two? (*Hint:* 16 ounces = 1 pound.)

For Exercises 13–16, refer to the federal income tax tables in Illustration 2 (on the next page).

13. FILING A JOINT RETURN   Raul has an adjusted income of $57,100, is married, and files jointly. Compute his tax.

14. FILING A SINGLE RETURN   Herb is single and has an adjusted income of $79,250. Compute his tax.

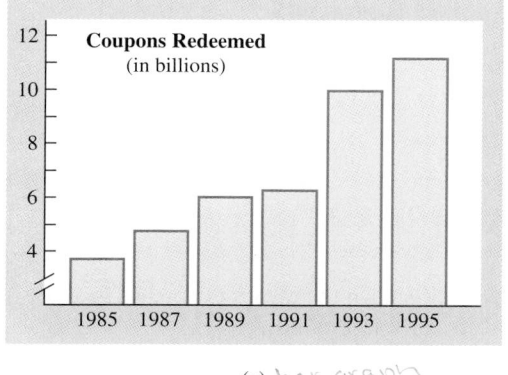

(a) bar graph

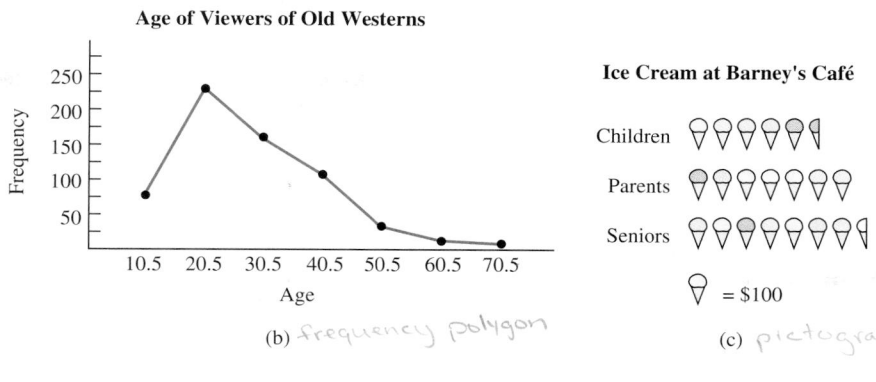

(b) frequency polygon

(c) pictograph

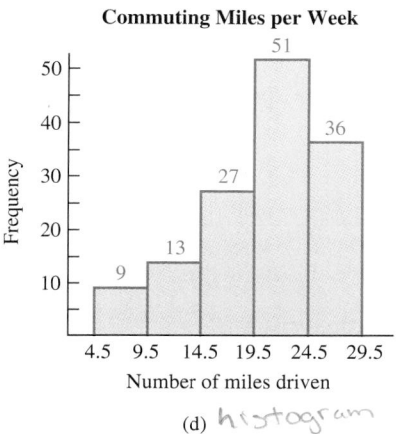

(d) histogram

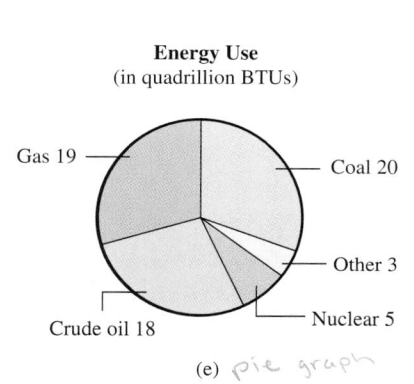

(e) pie graph

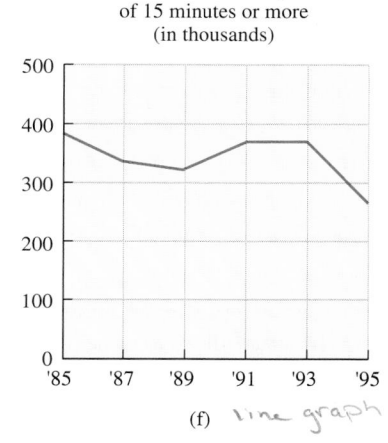

(f) line graph

ILLUSTRATION 1

Single—Schedule X			
If adjusted income is:		**The tax is:**	
Over	**But not over**		**Of the amount over**
$0	$23,350	15%	$0
23,350	56,550	$3,502.50 + 28%	23,350
56,550	117,950	12,798.50 + 31%	56,550
117,950	256,500	31,832.50 + 36%	117,950
256,500	——	81,710.50 + 39.6%	256,500

Married Filing Jointly or Qualifying Widow(er)—Schedule Y-1			
If adjusted income is:		**The tax is:**	
Over	**But not over**		**Of the amount over**
$0	$39,000	15%	$0
39,000	94,250	$5,850.00 + 28%	39,000
94,250	143,600	21,320.00 + 31%	94,250
143,600	256,500	36,618.50 + 36%	143,600
256,500	——	77,262.50 + 39.6%	256,500

ILLUSTRATION 2

15. TAX-SAVING STRATEGIES   Angelina is single and has an adjusted income of $53,000. If she gets married, she will gain other deductions that will reduce her income by $2,000, and she can file a joint return. How much will she save in tax by getting married?

16. FILING STATUS   A man with an adjusted income of $53,000 married a woman with an adjusted income of $75,000. They filed a joint return. Would they have saved on their taxes if they had both stayed single?

Refer to the bar graph in Illustration 3 (next page).

17. Which source supplied the least energy in 1975?

18. Which energy source remained unchanged between 1975 and 1995?

19. What percent of electrical energy was produced by oil in 1975?

20. Which source provided about 8% of electrical energy in 1995?

21. What was the approximate percent of increase in the use of energy from coal?

22. What was the approximate percent of increase in the use of nuclear power between 1975 and 1995?

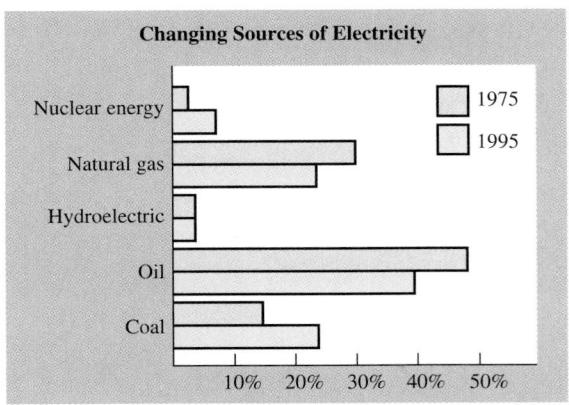

ILLUSTRATION 3

Refer to the bar graph in Illustration 4.

23. The world production of lead in 1970 was approximately equal to the production of zinc in another year. In what other year was that?

24. The world production of zinc in 1990 was approximately equal to the production of lead in another year. In what other year was that?

25. In what year was the production of zinc less than one-half that of lead?

26. In what year was the production of zinc more than twice that of lead?

27. By how many metric tons did the production of zinc increase between 1970 and 1980?

28. By how many metric tons did the production of lead decrease between 1980 and 1990?

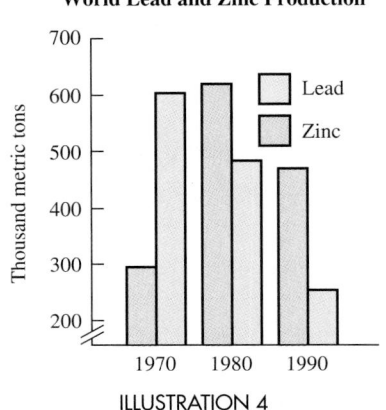

ILLUSTRATION 4

Refer to the bar graph in Illustration 5.

29. In which categories of moving violations have arrests decreased since last month?

30. Last month, which violation occurred most often?

31. This month, which violation occurred least often?

32. Which violation has shown the greatest decrease in number of arrests since last month?

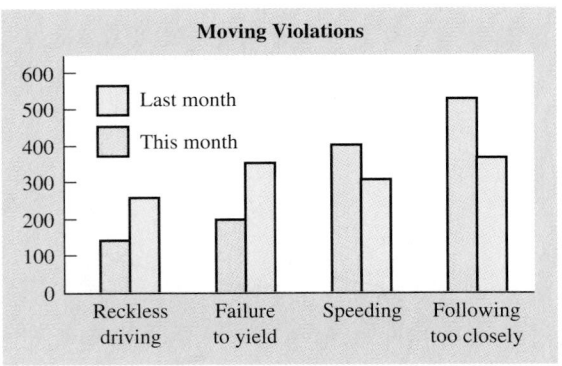

ILLUSTRATION 5

Refer to the pictograph in Illustration 6.

33. Which group (children, parents, or seniors) spent the most money on ice cream at Barney's Café?

34. How much money did parents spend on ice cream?

35. How much more money did seniors spend than parents?

36. How much more money did seniors spend than children?

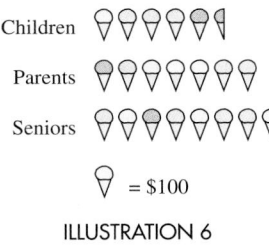

ILLUSTRATION 6

Refer to the pie graph in Illustration 7.

37. Two of the seven languages considered are spoken by groups of about the same size. Which languages are they?

38. Of the languages in the graph, which is spoken by the greatest number of people?

39. Do more people speak Russian or English?

40. What percent of the world's population speak Russian or English?

41. What percent of the world's population speak a language other than these seven?

42. What percent of the world's population do not speak either French or German?

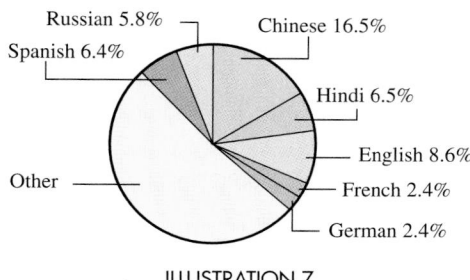

World Languages
and the percents of the population
that speak them

ILLUSTRATION 7

Refer to the pie graph in Illustration 8.

43. What percent of total energy sources is nuclear energy?

44. What percent of total energy sources are represented by coal and crude oil combined?

45. By what percent does energy derived from coal exceed that derived from crude oil?

46. By what percent does energy derived from coal exceed that derived from nuclear sources?

47. Solar energy accounts for less than what percent of total energy sources?

48. If production of nuclear energy tripled in the next 10 years and other sources remained the same, what percent of total energy sources would nuclear energy be?

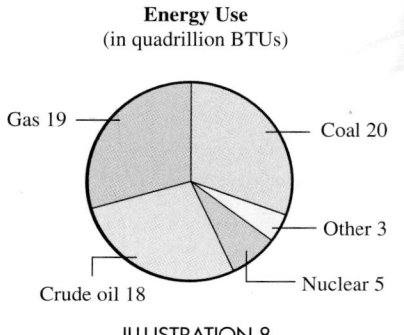

Energy Use
(in quadrillion BTUs)

ILLUSTRATION 8

Refer to the line graph in Illustration 9.

49. What were the average weekly earnings in mining for the year 1975?

50. What were the average weekly earnings in construction for the year 1980?

51. In the period between 1982 and 1984, which salary was increasing most rapidly?

52. In approximately what year did miners begin to earn more than construction workers?

53. In the period from 1970 to 1995, which workers received the greatest increase in wages?

54. In what five-year interval did wages in mining increase most rapidly?

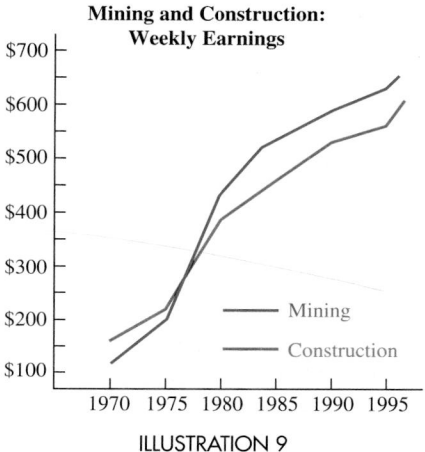

Mining and Construction:
Weekly Earnings

ILLUSTRATION 9

Refer to the line graph in Illustration 10.

55. Which runner ran faster at the start of the race?

56. Which runner stopped to rest first?

57. Which runner dropped the baton and had to go back to get it?

58. At what times (A, B, C, or D) was runner 1 stopped and runner 2 running?

59. Describe what was happening at time D.

60. Which runner won the race?

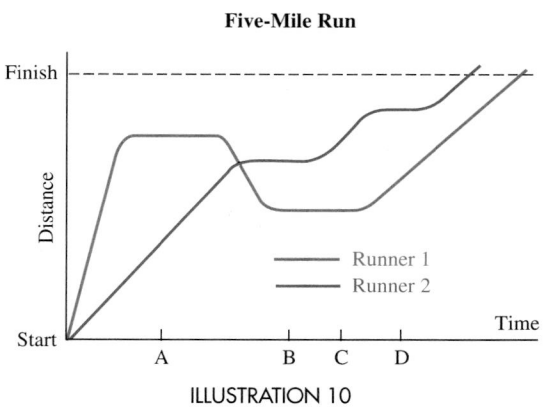

Five-Mile Run

ILLUSTRATION 10

61. COMMUTING MILES   An insurance company has collected data on the number of miles its employees drive to and from work. The data are presented in the histogram in Illustration 11. How many employees commute between 14.5 and 19.5 miles per week?

62. COMMUTING MILES   How many employees commute 14 miles or less per week? (See Illustration 11.)

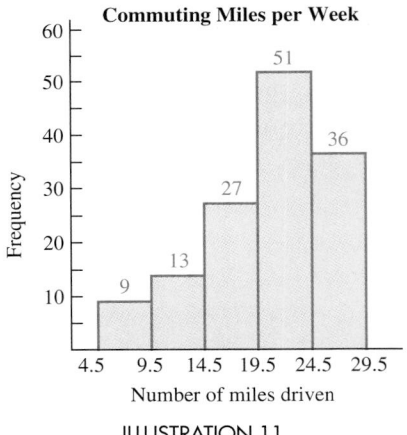

Commuting Miles per Week

ILLUSTRATION 11

63. NIGHT SHIFT STAFFING   A hospital administrator surveyed the medical staff to determine the number of room calls during the night. She constructed the frequency polygon in Illustration 12. On how many nights were there about 30 room calls?

64. NIGHT SHIFT STAFFING   According to Illustration 12, how many times did the staff handle the greatest number of room calls?

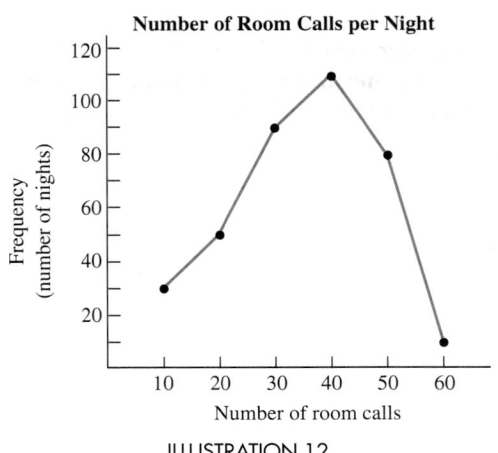

Number of Room Calls per Night

ILLUSTRATION 12

65. MAKING A BAR GRAPH   Use the data in Illustration 13 to make a bar graph showing the number of U.S. farms in the years 1950 through 1990.

66. MAKING A LINE GRAPH   Use the data in Illustration 13 to make a line graph showing the average acreage of U.S. farms for the years 1950 through 1990.

U.S. Farms 1950–1990		
Year	**Number of U.S. farms (in millions)**	**Average size of U.S. farms (acres)**
1950	5.3	215
1960	3.8	305
1970	2.8	390
1980	2.1	447
1990	2.0	470

ILLUSTRATION 13

67. MAKING A LINE GRAPH   The coupon in Illustration 14 provides savings for shoppers. Make a line graph that relates the original price (in dollars, on the horizontal axis) to the sale price (on the vertical axis).

SAVE!	On purchases of
$10	$100–$250
$25	$250–$500
$50	over $500

ILLUSTRATION 14

68. MAKING A HISTOGRAM   To study the effect of fluoride in preventing tooth decay, researchers counted the number of fillings in the teeth of 28 patients and recorded these results:

3, 7, 11, 21, 16, 22 18, 8, 12, 3, 7, 2, 8, 19, 12, 19, 12, 10, 13, 10, 14, 15, 14, 14, 9, 10, 12, 13

Tally the results by completing the table in Illustration 15. Then make a histogram. The first bar extends from 0.5 to 5.5, the second bar from 5.5 to 10.5, and so on.

Number of fillings	Frequency
1–5	
6–10	
11–15	
16–20	
21–25	

ILLUSTRATION 15

WRITING *Write a paragraph using your own words.*

69. What kind of presentation (table, bar graph, line graph, pie chart, pictograph, or histogram) is most appropriate for displaying each type of information?

- The percent of students, classified by major
- The percent of Biology majors each year since 1970
- The number of hours spent studying for finals
- Various ethnic populations of the ten largest cities
- Average annual salary of corporate executives for ten major industries

Explain your choices.

70. A histogram is a special type of bar graph. Explain.

REVIEW *Do the operations.*

71. $5 - 3 \cdot 4$

72. $3(6 - 9) + 4$

73. $\left(\dfrac{1}{2} + \dfrac{1}{3}\right)^2$

74. $5^2 + |6 - 10|$

75. Write the prime numbers between 10 and 30.

76. Write the first ten composite numbers.

77. Write the even numbers less than 6 that are not prime.

78. Write the prime numbers between 0 and 10.

6.2 *Mean, Median, and Mode*

In this section, you will learn about

- The mean (the arithmetic average) • The median • The mode

INTRODUCTION. Graphs are not the only way of describing *distributions* (lists) of numbers compactly. We can often find *one* number that is typical of all of the numbers in a list. We have already seen one such typical number: the *mean*, or the *average*. There are two others: the *median* and the *mode*. In this section, we will discuss these three measures of central tendency.

The mean (the arithmetic average)

A student has taken five tests this semester, scoring 87, 73, 89, 92, and 84. To find out how well she is doing, she calculates the **mean,** or the **arithmetic average,** of these grades, by finding the sum of the grades and then dividing by 5:

$$
\begin{aligned}
\text{Mean score} &= \frac{87 + 73 + 89 + 92 + 84}{5} \\
&= \frac{425}{5} \\
&= 85
\end{aligned}
$$

The mean score is 85. Some exams were better and some were worse, but 85 is a good indication of her performance in the class.

The mean (arithmetic average)

> The **mean,** or the **arithmetic average,** of several values is given by the formula
>
> $$\text{Mean (or average)} = \frac{\text{sum of the values}}{\text{number of values}}$$

EXAMPLE 1 *Department store sales.* The week's sales in three departments of the Tog Shoppe are given in the table in Figure 6-13. Find the mean of the daily sales in the women's department for this week.

Self Check
Find the mean daily sales in all departments of the Tog Shoppe on Wednesday.

	Men's department	Women's department	Children's department
Monday	$2,315	$3,135	$1,110
Tuesday	2,020	2,310	890
Wednesday	1,100	3,206	1,020
Thursday	2,000	2,115	880
Friday	955	1,570	1,010
Saturday	850	2,100	1,000

FIGURE 6-13

Solution

Use a calculator to add the sales in the women's department for the week. Then divide the sum of those six values by 6.

$$\text{Mean sales in the women's department} = \frac{3{,}135 + 2{,}310 + 3{,}206 + 2{,}115 + 1{,}570 + 2{,}100}{6}$$

$$= \frac{14{,}436}{6}$$

$$= 2{,}406$$

The mean of the week's daily sales in the women's department is $2,406.

Answer: $1,775.33

Accent on Technology: *Finding the mean*

Most scientific calculators do statistical calculations and can easily find the mean of a set of numbers. To use a statistical calculator in statistical mode to find the mean in Example 1, try these keystrokes:

- Set the calculator to statistical mode.
- Reset the calculator to clear the *statistical registers*.
- Enter each number, followed by the $\boxed{\Sigma+}$ key instead of the $\boxed{+}$ key. That is, enter 3,135, press $\boxed{\Sigma+}$, enter 2,310, press $\boxed{\Sigma+}$, and so on.
- When all data are entered, find the mean by pressing the $\boxed{x}$ key. You may need to press $\boxed{2^{nd}}$ first. The mean is 2,406.

Because keystrokes vary among calculator brands, you might have to check the owner's manual if these instructions don't work.

EXAMPLE 2 *Miles driven per day.* In January, Bob drove a total of 4,805 miles. On the average, how many miles did he drive per day?

Self Check
If he drove 3,360 miles in February, 1999, how many miles did he drive per day, on average?

Solution

To find the average number of miles driven per day, we divide the total number of miles by the number of days. Because there are 31 days in January, we divide 4,805 by 31:

$$\text{Average number of miles per day} = \frac{\text{total miles driven}}{\text{number of days}}$$

$$= \frac{4,805}{31}$$

$$= 155$$

On average, Bob drove 155 miles per day.

Answer: 120   ◼

The median

The mean is not always representative of the values in a list. For example, suppose that the weekly earnings of four workers in a small business are $280, $300, $380, and $240, and the owner of the company pays himself $5,000.

The mean salary is

$$\text{Mean salary} = \frac{280 + 300 + 380 + 240 + 5,000}{5}$$

$$= \frac{6,200}{5}$$

$$= 1,240$$

The owner could say, "Our employees earn an average of $1,240 per week." Clearly, the mean does not fairly represent the typical worker's salary.

A better measure of the company's typical salary is the *median:* the salary in the middle when all the numbers are arranged by size.

240   280   300   380   5,000

↑

The middle salary.

The typical worker earns $300 per week, far less than the mean salary.

If there is an even number of values in a list, there is no middle value. In that case, the median is the mean of the two numbers closest to the middle. For example, there is no middle number in the list 2, 5, 6, 8, 13, 17. The two numbers closest to the middle are 6 and 8. The median is the mean of 6 and 8, which is $\frac{6 + 8}{2}$, or 7.

The median

The **median** of several values is the middle value. To find the median:

1. Arrange the values in increasing order.

2. If there is an odd number of values, the median is the value in the middle.

3. If there is an even number of values, the median is the average of the two values that are closest to the middle.

EXAMPLE 3   *Finding a median score.*   On an exam, there were three scores of 59, four scores of 77, and scores of 43, 47, 53, 60, 68, 82, and 97. Find the median score.

Solution

We arrange the 14 scores in increasing order:

43   47   53   59   59   59   **60   68**   77   77   77   77   82   97

Since there is an even number of scores, the median is the mean of the two scores closest to the middle: the 60 and the 68.

The median is $\frac{60 + 68}{2}$, or 64.

Self Check

Find five numbers that have the same mean and median.

Answer: One answer is 1, 2, 3, 4, and 5 (mean = median = 3).

◼

The mode

A hardware store displays 20 outdoor thermometers. Twelve of them read 68°, and the other eight have different readings. To choose an accurate thermometer, should we choose one with a reading that is closest to the *mean* of all 20, or to their *median?* Neither. Instead, we should choose one of the 12 that all read the same, figuring that any of those that agree will likely be correct.

By choosing that temperature that appears most often, we have chosen the *mode* of the 20 numbers.

The mode	The **mode** of several values is the single value that occurs most often. The mode of several values is also called the **modal value.**

EXAMPLE 4   *Finding the mode.*   Find the mode of these values: 3, 6, 5, 7, 3, 7, 2, 4, 3, 5, 3, 7, 8, 7, 3, 7, 6, 3, 4.

Solution

To find the mode of the numbers in the list, we make a chart of the distinct numbers that appear and make tally marks to record the number of times they occur.

2	3	4	5	6	7	8
/	↗↗↗ /	//	//	//	↗↗↗	/

Because 3 occurs more times than any other number, it is the mode.

Self Check

Find the mode of these values:
2, 3, 4, 6, 2, 4, 3, 4, 3, 4, 2, 5

Answer: 4

 **EXAMPLE 5**   *Machinist's tools.*   The diameters (distances across) of eight stainless steel bearings were found using the vernier calipers shown in Figure 6-14. Find   **a.** the mean, **b.** the median, and   **c.** the mode   of the set of measurements listed below.

3.43 cm, 3.25 cm, 3.48 cm, 3.39 cm, 3.54 cm, 3.48 cm, 3.23 cm, 3.24 cm

FIGURE 6-14

Solution

a. To find the mean, we add the measurements and divide by the number of values, which is 8.

$$\text{Mean} = \frac{3.43 + 3.25 + 3.48 + 3.39 + 3.54 + 3.48 + 3.23 + 3.24}{8} = 3.38 \text{ cm}$$

b. To find the mean, we first arrange the measurements in increasing order:

3.23, 3.24, 3.25, 3.39, 3.43, 3.48, 3.48, 3.54

Because there is an even number of measurements, the median will be the sum of the middle two values (3.39 and 3.43) divided by 2. Thus, the median is

$$\text{Median} = \frac{3.39 + 3.43}{2} = \frac{6.82}{2} = 3.41 \text{ cm}$$

c. Since the measurement 3.48 cm occurs most often, it is the mode.

STUDY SET Section 6.2

VOCABULARY *Fill in the blanks.*

1. The sum of the values in a distribution of numbers divided by the number of values in the distribution is called the _____ of the distribution.

2. The value that appears most often in a distribution is called the _____ of the distribution.

3. The middle value in a distribution is called the _____ of the distribution.

4. The modal value of a distribution is the value in the distribution that appears _____ often.

CONCEPTS *Fill in the blanks.*

5. The mean of several values is given by

$$\text{Mean} = \frac{\text{the sum of the values}}{}$$

6. Complete the formula.

$$\frac{\text{Average number of miles}}{\text{driven per day}} = \frac{}{\text{number of days}}$$

PRACTICE *Find the mean of the numbers.*

7. 3, 4, 7, 7, 8, 11, 16

8. 13, 15, 17, 17, 15, 13

9. 5, 9, 12, 35, 37, 45, 60, 77

10. 0, 0, 3, 4, 7, 9, 12

11. 15, 7, 12, 19, 27, 17, 19, 35, 20

12. 45, 67, 42, 35, 86, 52, 91, 102

Find the median of the numbers.

13. 2, 5, 9, 9, 9, 17, 29

14. 16, 18, 27, 29, 35, 47

15. 4, 7, 2, 11, 5, 4, 9, 17

16. 0, 0, 3, 4, 0, 0, 3, 4, 5

17. 18, 17, 2, 9, 21, 23, 21, 2

18. 5, 13, 5, 23, 43, 56, 32, 45

Find the mode (if any) of the numbers.

19. 3, 5, 7, 3, 5, 4, 6, 7, 2, 3, 1, 4

20. 12, 12, 17, 17, 12, 13, 17, 12

21. 5, 9, 12, 35, 37, 45, 60

22. 0, 3, 0, 2, 7, 0, 6, 0, 3, 4, 2, 0

23. 23.1, 22.7, 23.5, 22.7, 34.2, 22.7

24. $\frac{1}{2}, \frac{1}{3}, \frac{1}{3}, 2, \frac{1}{2}, 2, \frac{1}{5}, \frac{1}{2}, 5, \frac{1}{3}$

APPLICATIONS

25. SOFT DRINK PRICES A survey of soft-drink machines indicates the following prices for a can (in cents): 50, 60, 50, 50, 70, 75, 50, 45, 50, 50, 65, 75, 60, 75, 100, 50, 80, 75. Find the mean price of a soft drink.

26. COMPUTER SUPPLIES Several computer stores reported differing prices for toner cartridges for a laser printer (in dollars): 51, 55, 73, 75, 72, 70, 53, 59, 75. Find the mean price of a toner cartridge.

27. SOFT DRINK PRICES Find the median price for a soft drink. (See Exercise 25.)

28. COMPUTER SUPPLIES Find the median price for a toner cartridge. (See Exercise 26.)

29. SOFT DRINK PRICES Find the modal price for a soft drink. (See Exercise 25.)

30. COMPUTER SUPPLIES Find the mode of the prices for a toner cartridge. (See Exercise 26.)

31. CHANGING TEMPERATURES Temperatures are recorded at hourly intervals, as in Illustration 1. Find the average temperature of the period from midnight to 11:00 A.M.

Time	Temperature	Time	Temperature
12:00 A.M.	53	12:00 noon	71
1:00	53	1:00 P.M.	75
2:00	57	2:00	77
3:00	58	3:00	77
4:00	59	4:00	79
5:00	59	5:00	72
6:00	60	6:00	70
7:00	62	7:00	64
8:00	64	8:00	61
9:00	66	9:00	59
10:00	68	10:00	53
11:00	70	11:00	51

ILLUSTRATION 1

32. SEMESTER GRADE Frank's algebra grade is based on the average of four exams, which will count equally. His grades are 75, 80, 90, and 85. Find his average.

33. AVERAGE TEMPERATURE Find the average temperature for the 24-hour period recorded in Illustration 1.

34. WEIGHTED FINAL If Frank's professor decided to count the fourth examination double, what would Frank's average be? (See Exercise 32.)

35. FLEET MILEAGE   An insurance company's sales force uses 37 cars. Last June, those cars logged a total of 98,790 miles. On the average, how many miles did each car travel that month?

36. BUDGETING FOR GROCERIES   The Hinrichs family spent $519 on groceries last April. On the average, how much did they spend each day?

37. DAILY MILEAGE   Find the average number of miles driven daily for each car in Exercise 35.

38. GROCERY COSTS   See Exercise 36. The Hinrichs family has five members. What is the average spent for groceries for one family member for one day?

39. EXAM AVERAGES   Roberto received the same score on each of five exams, and his mean score is 85. Find his median score and his modal score.

40. BETTER THAN AVERAGE   The scores on the first exam of the students in a history class were 57, 59, 61, 63, 63, 63, 87, 89, 95, 99, and 100. Kia got a score of 70 and claims that "70 is better than average." Which of the three measures of central tendency is she better than: the mean, the median, or the mode?

41. COMPARING GRADES   A student received scores of 37, 53, and 78 on three quizzes. His sister received scores of 53, 57, and 58. Who had the better average? Whose grades were more consistent?

42. What is the average of all of the integers from -100 to 100, inclusive?

43. OCTUPLETS   In December of 1998, Nkem Chukwu gave birth to eight babies in Texas Children's Hospital. Find the mean and the median of their birth weights.

Ebuka (girl) 24 oz	Odera (girl) 11.2 oz
Chidi (girl) 27 oz	Ikem (boy) 17.5 oz
Echerem (girl) 28 oz	Jioke (boy) 28.5 oz
Chima (girl) 26 oz	Gorom (girl) 18 oz

44. ICE SKATING   Listed below are Tara Lipinski's artistic impression scores for the long program of the women's figure skating competition at the 1998 Winter Olympics. Find the mean, median, and mode. Round to the nearest tenth.

Australia	5.8	Germany	5.8	Ukraine	5.9
Hungary	5.8	U.S.	5.8	Poland	5.8
Austria	5.9	Russia	5.9	France	5.9

45. COMPARISON SHOPPING   A survey of grocery stores found the price of a 15-ounce box of Cheerios cereal ranging from $3.89 to $4.39. (See below.) What are the mean, median, and mode of the prices listed?

$4.29	$3.89	$4.29	$4.09	$4.24	$3.99
$3.98	$4.19	$4.19	$4.39	$3.97	$4.29

46. EARTHQUAKES   The magnitudes of 1999's major earthquakes are listed below. Find the mean, median, and mode. Round to the nearest tenth.

1/19/99	New Ireland, Papua New Guinea	7.0
2/6/99	Santa Cruz Islands, S. Pacific Sea	7.3
3/4/99	Celebes Sea, Indonesia	7.1
4/5/99	New Britain, Papua New Guinea	7.4
4/8/99	E. Russia/N.E. China border	7.1
5/10/99	New Britain, Papua New Guinea	7.1
5/16/99	New Britain, Papua New Guinea	7.1
8/17/99	Izmit region, western Turkey	7.4
9/21/99	Taiwan	7.6
9/30/99	Oaxaca, Mexico	7.4
11/12/99	Bolu Province, northwest Turkey	7.2

47. HOURLY PAY   Illustration 2 shows the average hours worked and the average weekly earnings of U.S. production workers in 1988 and 1998. What did the average production worker earn per hour in 1988 and in 1998? Round to the nearest cent.

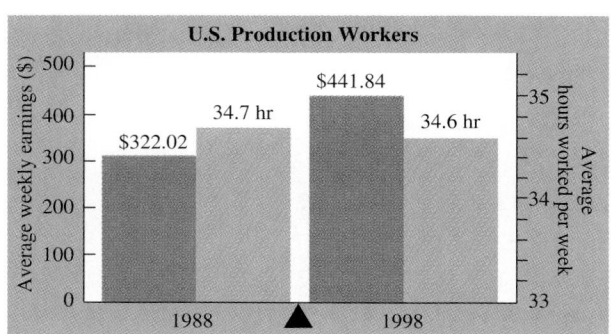

Source: *The World Almanac 2000*

ILLUSTRATION 2

48. PLEASURE TRAVEL   Illustration 3 shows the annual number of person-trips of 100 miles or more (one way) for the years 1994–1998, as estimated by the Travel Industry Association of America. Find the mean and the median.

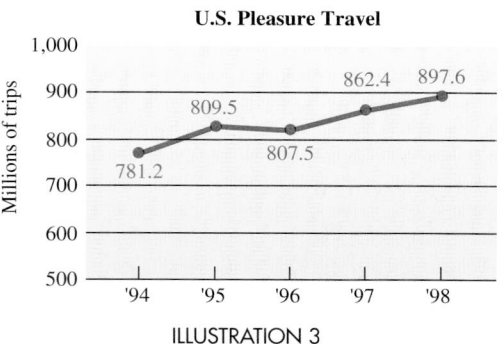

ILLUSTRATION 3

49. OIL WELL Geologists have mapped out the substances through which engineers must drill to reach an oil deposit. (See Illustration 4.) What is the average depth that must be drilled each week if this is to be a four-week project?

Surface
Silt
0.68 mi
Rock
0.36 mi
Sand
0.44 mi
Oil

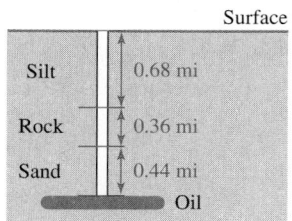

ILLUSTRATION 4

50. INDY 500 Illustration 5 shows the first row of the starting grid for the 1998 Indianapolis 500 automobile race. The drivers' speeds on a qualifying run were used to rank them in this order. What was the mean qualifying speed for the drivers in the first row?

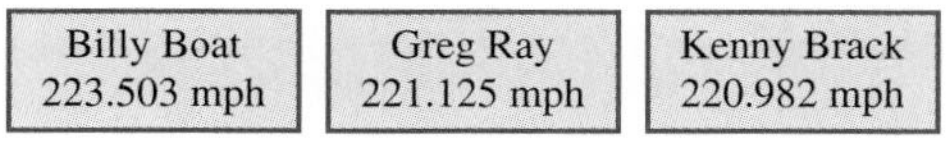

Billy Boat
223.503 mph
Greg Ray
221.125 mph
Kenny Brack
220.982 mph

ILLUSTRATION 5

WRITING

51. Explain how to find the mean, the median, and the mode of several numbers.

52. Explain why a set of numbers might have no modal value.

REVIEW

53. Find the prime factorization of 81.

54. Find the LCD for two fractions whose denominators are 36 and 81.

Do each operation.

55. $\frac{3}{4} \cdot \frac{2}{9}$

56. $\frac{2}{15} \div \frac{4}{5}$

57. $\frac{18}{5} + \frac{12}{5}$

58. $\frac{7}{12} - \frac{5}{12}$

59. $\frac{8}{5} + \frac{3}{10}$

60. $\frac{5}{6} - \frac{1}{12}$

Mean, Median, and Mode

To indicate the center of a distribution (list) of numbers, we can use the mean, the median, or the mode.

- The **mean** of a distribution is the sum of the values in the distribution divided by the number of values in the distribution:

$$\text{Mean} = \frac{\text{sum of the values in the distribution}}{\text{number of values in the distribution}}$$

- The **median** is the middle value in a distribution. Just as many scores are above the median as are below it. If there is an even number of values in the distribution, the median is the mean of the two values that are closest to the middle.

- The **mode** of a distribution is the value that occurs most often.

Consider the following distribution: 3, 7, 4, 12, 15, 23, 17, 21, 15, 20.

1. Calculate the mean.

2. Find the median.

3. Find the mode.

4. Are the mean, the median, and the mode the same number?

Consider the distribution 2, 4, 6, 6, 8, 10.

5. Calculate the mean.

6. Find the median.

7. Find the mode.

8. Are the mean, the median, and the mode the same number?

Construct a distribution with the following characteristics.

9. The mean is greater than the mode.

10. The mean is less than the median.

11. The mode is less than the median.

12. The mode is greater than the median.

Section 6.1

DAILY HIGH AND LOW TEMPERATURES   Make a bar graph that shows the daily high and low temperatures for your city for a two-week period. You can find this information in your newspaper. From your graph, answer the following questions.

a. What was the highest high temperature?

b. What was the lowest high temperature?

c. What was the highest low temperature?

d. What was the lowest low temperature?

e. What was the difference between the highest high temperature and the lowest low temperature?

f. Were any trends apparent from the graph?

Section 6.2

MEAN, MEDIAN, AND MODE

1. Find the mean, median, and mode of the following set of values.

2.3,   2.3,   3.6,   3.8,   and   4.5

a. Is the mean of the set of values one of the values in the set?

b. Is the median of the set of values one of the values in the set?

c. Is the mode of the set of values one of the values in the set?

2. Construct a set of values (not all the same number) whose mean, median, and mode are the same value.

3. Construct a set of values such that
mean $<$ median $<$ mode

4. Construct a set of values such that
mean $>$ median $>$ mode

Reading Graphs and Tables

CONCEPTS

Numerical information can be presented in the form of tables, bar graphs, pictographs, pie graphs, and line graphs.

REVIEW EXERCISES

Refer to the table in Illustration 1.

1. WIND-CHILL TEMPERATURE   Find the wind-chill temperature on a 10° F day when a 15-mph wind is blowing.

2. WIND SPEED   The wind-chill temperature is −25° F, and the outdoor temperature is 15° F. How fast is the wind blowing?

Determining the Wind-Chill Temperature

Wind speed	Actual temperature													
	35° F	30° F	25° F	20° F	15° F	10° F	5° F	0° F	−5° F	−10° F	−15° F	−20° F	−25° F	−30° F
5 mph	33°	27°	21°	16°	12°	7°	0°	−5°	−10°	−15°	−21°	−26°	−31°	−36°
10 mph	22	16	10	3	−3	−9	−15	−22	−27	−34	−40	−46	−52	−58
15 mph	16	9	−2	−5	−11	−18	−25	−31	−38	−45	−51	−58	−65	−72
20 mph	12	4	−3	−10	−17	−24	−31	−39	−46	−53	−60	−67	−74	−81
25 mph	8	1	−7	−15	−22	−29	−36	−44	−51	−59	−66	−74	−81	−88
30 mph	6	−2	−10	−18	−25	−33	−41	−49	−56	−64	−71	−79	−86	−93
35 mph	4	−4	−12	−20	−27	−35	−43	−52	−58	−67	−74	−82	−89	−97
40 mph	3	−5	−13	−21	−29	−37	−45	−53	−60	−69	−76	−84	−92	−100
45 mph	2	−6	−14	−22	−30	−38	−46	−54	−62	−70	−78	−85	−93	−102

ILLUSTRATION 1

3. Refer to Illustration 2 to answer each question.
   a. How many coupons were redeemed in 1987?
   b. Between what years did the number of redeemed coupons remain essentially unchanged?
   c. In what two-year period did the number of redeemed coupons increase the most?
   d. What was the percent of the greatest two-year increase?

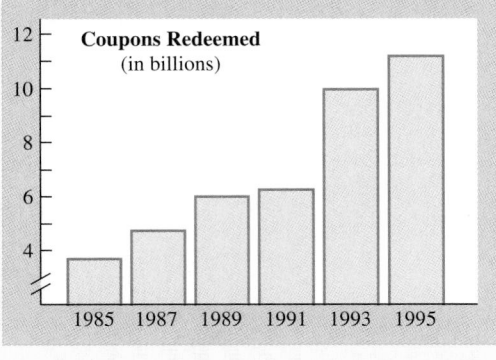

ILLUSTRATION 2

4. Refer to Illustration 3.
  a. How many eggs were produced in Wisconsin in 1985?
  b. How many eggs were produced in Nebraska in 1987?
  c. In what year was the egg production of Wisconsin equal to that of Nebraska?

  d. What was the total egg production of Wisconsin and Nebraska in 1988?

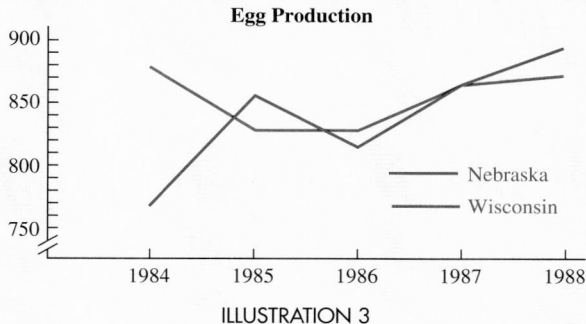

ILLUSTRATION 3

Refer to Illustration 4.

5. A survey of the television viewing habits of 320 households produced the histogram in Illustration 4. How many households watch between 6 and 15 hours of TV each week?

A *histogram* is a bar graph with these features:

1. The bars of the histogram touch.

2. Data values never fall at the edge of a bar.

3. The width of each bar represents a numeric value.

A *frequency polygon* is a special line graph formed from a histogram.

6. How many households watch 11 hours or more each week?

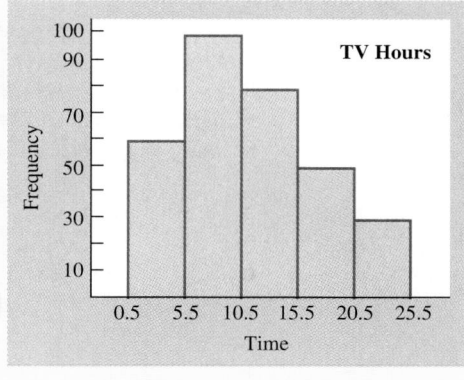

ILLUSTRATION 4

SECTION 6.2 — *Mean, Median, and Mode*

The *mean* (or average) is given by the formula

$$\text{Mean} = \frac{\text{sum of the values}}{\text{number of values}}$$

7. EARNING AN A   Jose worked hard this semester, earning grades of 87, 92, 97, 100, 100, 98, 90, and 98. If he needs a 95 average to earn an A in the class, did he make it?

To find the *median* of several values:

1. Arrange the values in increasing order.
2. If there is an odd number of values, the median is the value in the middle.
3. If there is an even number of values, the median is the average of the two values that are closest to the middle.

The *mode* of several numbers is the single value that occurs most often.

8. GRADE SUMMARY   The students in a mathematics class had final averages of 43, 83, 40, 100, 40, 36, 75, 39, and 100. When asked how well her students did, their teacher answered, "43 was typical." What measure was the teacher using?

9. PRETZEL PACKAGING   Samples of SnacPak pretzels were weighed to find out whether the package claim "Net weight 1.2 ounces" was accurate. The tally appears in Illustration 5. Find the modal weight.

10. Find the mean weight of the samples in Exercise 9.

Weights of SnacPak Pretzels	
Ounces	**Number**
0.9	1
1.0	6
1.1	18
1.2	23
1.3	2
1.4	0

ILLUSTRATION 5

11. BLOOD SAMPLES   A medical laboratory technician examined a blood sample under a microscope and measured the sizes (in microns) of the white blood cells. The data is listed below. Find the mean, median, and mode.

7.8   6.9   7.9   6.7   6.8   8.0   7.2   6.9   7.5

12. TOBACCO SETTLEMENT   In November of 1998, the country's four largest tobacco companies reached an agreement with 46 states to pay $206.4 billion to cover public health costs related to smoking. The payments to each of the New England states are shown below. Find the median payment.

Connecticut	$3.63 billion	New Hampshire	$1.3 billion
Maine	$1.5 billion	Rhode Island	$1.4 billion
Massachusetts	$8.0 billion	Vermont	$0.81 billion

Refer to Illustration 1. Keeping one prisoner for one month costs $2,266.

1. How much money is spent monthly, per prisoner, to pay the prison staff?

2. How much money is spent monthly, per prisoner, on office costs?

3. What percent of the monthly allotment is spent on one prisoner's food?

4. What percent of the monthly allotment is spent on one prisoner's recreation and training?

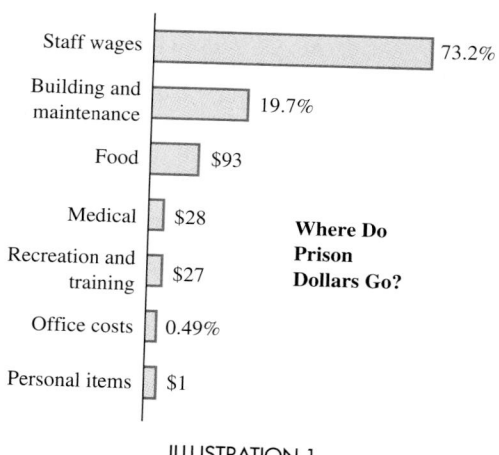

ILLUSTRATION 1

Refer to Illustration 2.

5. Approximately what percent of all employees are in the food and clothing industries?

6. Among workers in food and clothing, 2.4 million are in food, and the rest are in clothing. What percent of all workers are in clothing?

Use the information given in Illustration 3.

7. How many air traffic delays occurred in 1995?

8. How many air traffic delays in 1991 were due to the weather?

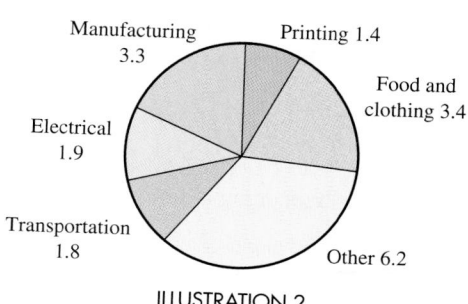

Employees in Industry
(in millions)

ILLUSTRATION 2

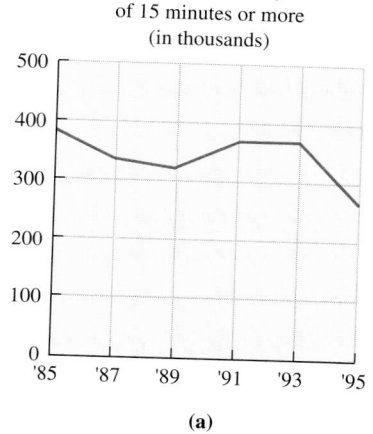

Air Traffic Delays
of 15 minutes or more
(in thousands)

(a)

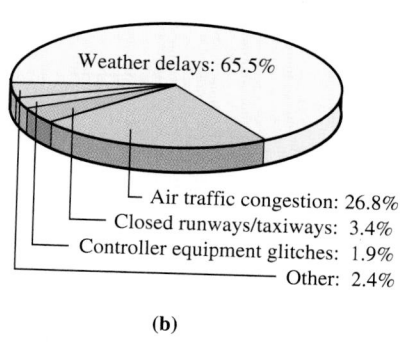

Causes of Air Traffic Delays

(b)

ILLUSTRATION 3

9. Which year was worst for air traffic delays?

10. The percent for "Other" causes for delays appears smudged. What should the value be?

For Problems 11–14, refer to Illustration 4 and choose the best answer from the following statements.

A. Both bicyclists are moving, and bicyclist 1 is faster than 2.
B. Both bicyclists are moving, and bicyclist 2 is faster than 1.
C. Bicyclist 1 is stopped, and bicyclist 2 is not.
D. Bicyclist 2 is stopped, and bicyclist 1 is not.
E. Both bicyclists are stopped.

11. Indicate what is happening at time *A*.

12. Indicate what is happening at time *B*.

13. Indicate what is happening at time *C*.

14. Which bicyclist won the race?

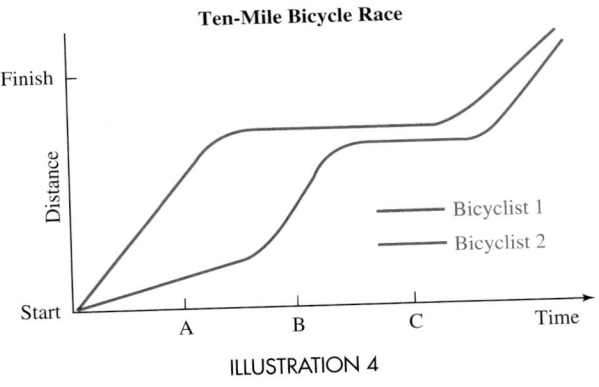

Ten-Mile Bicycle Race

Finish

Distance

— Bicyclist 1
— Bicyclist 2

Start

A   B   C   Time

ILLUSTRATION 4

Refer to this information: The hours served last month by the individual volunteers at a homeless shelter were 4, 6, 8, 2, 8, 10, 11, 9, 5, 12, 5, 18, 7, 5, 1, and 9.

15. Find the median of the hours of volunteer service.

16. Find the mean of the hours of volunteer service.

17. Find the mode of the hours of volunteer service.

18. If the one value of 18 were removed from the list, which would be most affected: the mean or the median?

19. RATINGS   The seven top-rated cable television programs for the week of February 8–14 are given below. What are the mean, median, and mode of the ratings? Round to the nearest tenth.

Show/day/time/network	Rating
1. "WCW Monday," Mon. 9 p.m., TNT	4.5
2. "WCW Monday," Mon. 10 p.m., TNT	4.4
3. "WCW Monday," Mon. 8 p.m., TNT	3.9
4. "WWF Special," Sat. 9 p.m., USA	3.6
5. "WWF Wrestling," Sun. 7 p.m., USA	3.1
6. "Dog Show," Tues. 8 p.m., USA	3.1
7. "WWF Special," Sat. 8 p.m., USA	2.9

20. STATISTICS   The graph in Illustration 5 has an asterisk * that refers readers to a note at the bottom. In your own words, complete the explanation of the term *median*.

Family Debt Grows

Median family indebtedness grew by 42% between 1995 and 1998, according to a Federal Reserve survey of consumer finances.

Median* amount of debt

| 1995 | $23,400 |
| 1998 | $33,300 |

*Median means that.............

Source: *Los Angeles Times* (February 1, 2000)

ILLUSTRATION 5

Chapters 1-6   Cumulative Review Exercises

1. GASOLINE   In 1999, gasoline consumption in the United States was three hundred fifty-eight million, six hundred thousand gallons a day. Write this number in standard notation.

2. Round 49,999 to the nearest thousand.

In Exercises 3–6, do each operation.

3.
$$38,908$$
$$+15,696$$

4.
$$9,700$$
$$-5,491$$

5.
$$345$$
$$\times\ 67$$

6. $23)\overline{2,001}$

7. Explain how to check the following result using addition.

$$1,142$$
$$-\ \ 459$$
$$\overline{\ \ 683}$$

8. VIETNAMESE CALENDAR   An animal represents each Vietnamese lunar year. Recent Years of the Cat are listed below. If the cycle continues, what year will be the next Year of the Cat?

1915   1927   1939   1951   1963   1975   1987   1999

9. Consider the multiplication statement $4 \cdot 5 = 20$. Show that multiplication is repeated addition.

10. ROOM DIVIDER   Four pieces of plywood, each 22 inches wide and 62 inches high, are to be covered with fabric, front and back, to make the room divider shown in Illustration 1. How many square inches of fabric will be used?

ILLUSTRATION 1

11. a. Find the factors of 18.
b. Find the prime factorization of 18.

12. List the first ten prime numbers.

13. Why isn't 27 a prime number?

14. Evaluate $(9 - 2)^2 - 3^3$.

15. Find $\dfrac{-315}{-1}$.

16. Simplify $-(-6)$.

17. Graph the integers greater than -3 but less than 4.

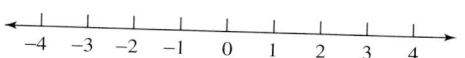

18. Find the absolute value: $|-5|$.

19. Is the statement $-12 > -10$ true or false?

20. ANNUAL NET INCOME   Use the following data for the Polaroid Corporation to construct a line graph.

Year	'95	'96	'97	'98	'99
Total net income ($ millions)	-139	15	-127	-51	9

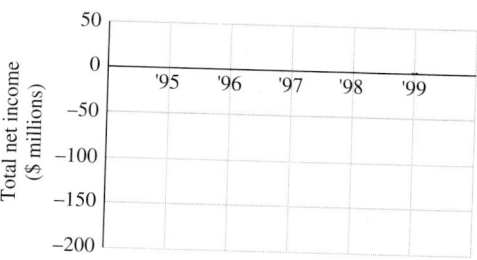

Do each operation.

21. $-25 + 5$

22. $25 - (-5)$

23. $-25(5)(-1)$

24. $\dfrac{-25}{-5}$

363

25. Evaluate $\dfrac{(-6)^2 - 1^5}{-4 - 3}$

26. Evaluate $-3 + 3[-4 - 4 \cdot 2]^2$

27. Evaluate -3^2 and $(-3)^2$.

28. PLANETS   Mercury orbits closer to the sun than does any other planet. Temperatures on Mercury can get as high as 810° F and as low as −290° F. What is the temperature range?

29. 360 is 45% of what number?

30. 90 is what percent of 600?

31. Write an expression illustrating division by 0 and an expression illustrating division of 0. Which is undefined?

32. Add: $\dfrac{1}{2} + \dfrac{2}{3}$

33. Subtract: $\dfrac{1}{2} - \dfrac{2}{3}$

34. TENNIS   Find the length of the handle on the tennis racquet shown in Illustration 2.

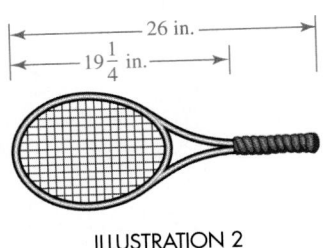

ILLUSTRATION 2

35. Multiply: $\dfrac{4}{5} \cdot \dfrac{2}{7}$.

36. Divide: $2\dfrac{4}{5} \div 2\dfrac{2}{3}$.

37. See Illustration 3. What percent of the stripes of the flag are red? (Round to the nearest one percent.)

ILLUSTRATION 3

38. Multiply: $3.45 \cdot 100$

39. Multiply: $(0.31)(2.4)$

40. Divide: $0.72\overline{)536.4}$.

41. Change $\dfrac{8}{11}$ to a decimal.

42. CLASS TIME   In a chemistry course, students spend a total of 300 minutes in lab and lecture each week. If $\dfrac{7}{15}$ of the time is spent in lab each week, how many minutes are spent in lecture each week?

43. WEEKLY SCHEDULE   Refer to Illustration 4. Determine the number of hours during a week that an adult spends, on average, watching television.

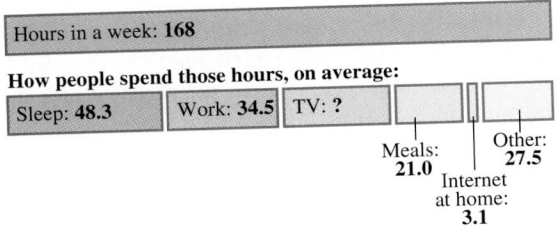

Based on data from the National Sleep Foundation and the United States Bureau of Statistics

ILLUSTRATION 4

44. TEAM GPA   The grade point averages of the players on a badminton team are listed below. Find the mean, median, and mode of the team's GPAs.

3.04   4.00   2.75   3.23   3.87   2.20
3.02   2.25   2.99   2.56   3.58   2.75

Introduction to Geometry

7

7.1 Some Basic Definitions

7.2 Parallel and Perpendicular Lines

7.3 Polygons

7.4 Properties of Triangles

7.5 Perimeters and Areas of Polygons

7.6 Circles

7.7 Surface Area and Volume

Key Concept: Formulas

Accent on Teamwork

Chapter Review

Chapter Test

Cumulative Review Exercises

GEOMETRY COMES FROM THE GREEK WORDS GEO (MEANING EARTH) AND METRON (MEANING MEASURE).

7.1 Some Basic Definitions

In this section, you will learn about

- Points, lines, and planes • Angles • Adjacent and vertical angles
- Complementary and supplementary angles

INTRODUCTION. In this chapter, we will study two-dimensional geometric figures such as rectangles and circles. In daily life, it is often necessary to find the perimeter or area of one of these figures. For example, to find the amount of fencing that is needed to enclose a circular garden, we must find the perimeter of a circle (called its *circumference*). To find the amount of paint needed to paint a room, we must find the area of its four rectangular walls.

We will also study three-dimensional figures such as cylinders and spheres. To find the amount of space enclosed within these figures, we must find their volumes.

Points, lines, and planes

Geometry is based on three undefined words: **point, line,** and **plane.** Although we will make no attempt to define these words formally, we can think of a point as a geometric figure that has position but no length, width, or depth. Points are always labeled with capital letters. Point *A* is shown in Figure 7-1(a).

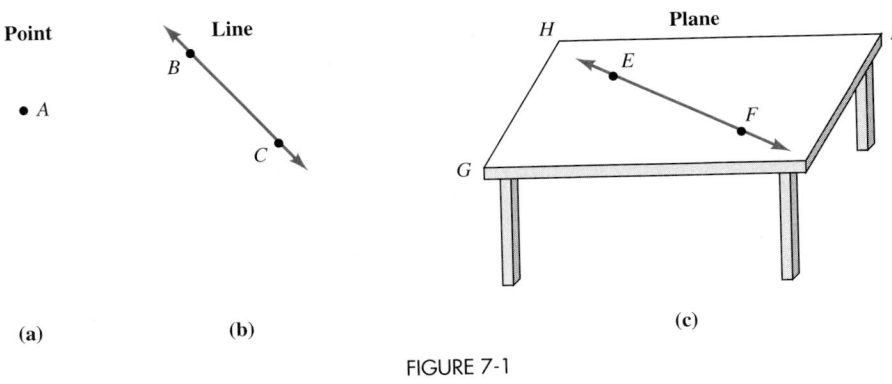

(a)   **(b)**   **(c)**

FIGURE 7-1

A line is infinitely long but has no width or depth. Figure 7-1(b) shows line *BC*, passing through points *B* and *C*. A plane is a flat surface, like a table top, that has length and width but no depth. In Figure 7-1(c), line *EF* lies in the plane *GHI*.

As Figure 7-1(b) illustrates, points *B* and *C* determine exactly one line, the line *BC*. In Figure 7-1(c), the points *E* and *F* determine exactly one line, the line *EF*. In general, any two points will determine exactly one line.

Other geometric figures can be created by using parts or combinations of points, lines, and planes.

Line segment

> The **line segment** AB, denoted as $\overline{AB}$, is the part of a line that consists of points A and B and all points in between (see Figure 7-2). Points A and B are the **endpoints** of the segment.

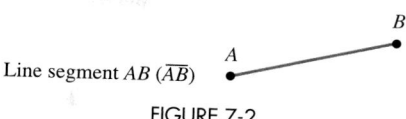

Line segment AB ($\overline{AB}$)

FIGURE 7-2

Every line segment has a **midpoint,** which divides the segment into two parts of equal length. In Figure 7-3, M is the midpoint of segment AB, because the measure of $\overline{AM}$ (denoted as $m(\overline{AM})$) is equal to the measure of $\overline{MB}$ (denoted as $m(\overline{MB})$).

$$m(\overline{AM}) = 4 - 1$$
$$= 3$$

and

$$m(\overline{MB}) = 7 - 4$$
$$= 3$$

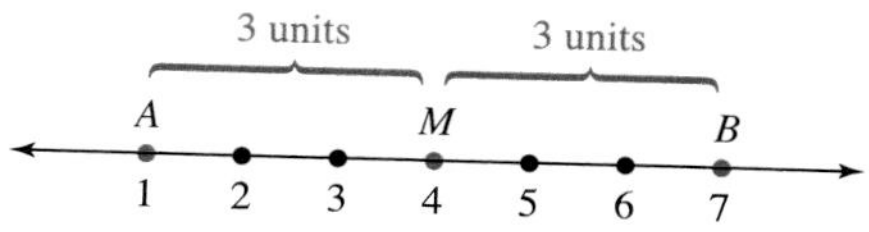

FIGURE 7-3

Since the measure of both segments is 3 units, $m(\overline{AM}) = m(\overline{MB})$.

When two line segments have the same measure, we say that they are **congruent.** Since $m(\overline{AM}) = m(\overline{MB})$, we can write

$$\overline{AM} \cong \overline{MB} \quad \text{Read} \cong \text{as ``is congruent to.''}$$

Another geometric figure is the *ray,* as shown in Figure 7-4.

Ray

> A **ray** is the part of a line that begins at some point (say, A) and continues forever in one direction. Point A is the **endpoint** of the ray.

Ray AB is denoted as $\overrightarrow{AB}$. The endpoint of the ray is always listed first.

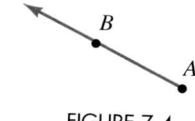

FIGURE 7-4

Angles

Angle

> An **angle** is a figure formed by two rays with a common endpoint. The common endpoint is called the **vertex,** and the rays are called **sides.**

The angle in Figure 7-5 can be denoted as

$$\angle BAC, \quad \angle CAB, \quad \angle A, \quad \text{or} \quad \angle 1 \qquad \text{The symbol } \angle \text{ means angle.}$$

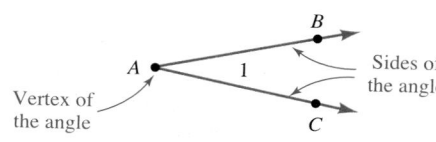

Vertex of the angle

Sides of the angle

FIGURE 7-5

COMMENT   When using three letters to name an angle, be sure the letter name of the vertex is the middle letter.

One unit of measurement of an angle is the **degree.** It is $\frac{1}{360}$ of a full revolution. We can use a **protractor** to measure angles in degrees. See Figure 7-6.

Angle	Measure in degrees
$\angle ABC$	30°
$\angle ABD$	60°
$\angle ABE$	110°
$\angle ABF$	150°
$\angle ABG$	180°

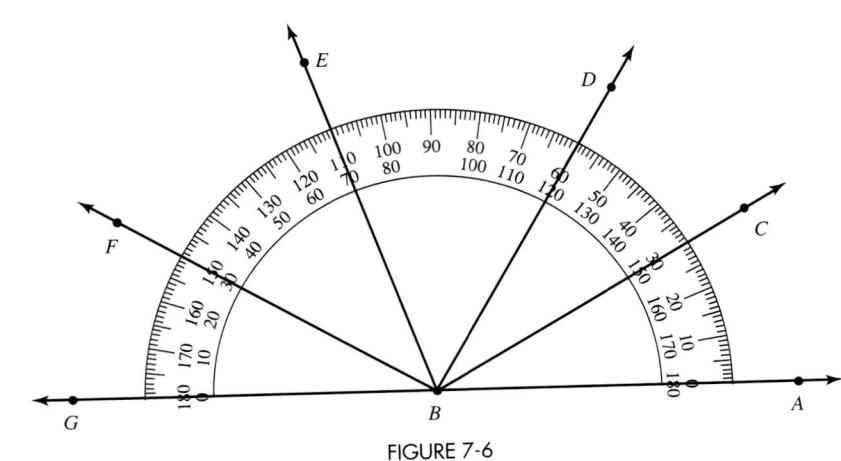

FIGURE 7-6

If we read the protractor from left to right, we can see that the measure of $\angle GBF$ (denoted as m($\angle GBF$)) is 30°.

When two angles have the same measure, we say that they are congruent. Since m($\angle ABC$) = 30° and m($\angle GBF$) = 30°, we can write

$$\angle ABC \cong \angle GBF$$

We classify angles according to their measure, as in Figure 7-7.

Classification of angles

Acute angles: Angles whose measures are greater than 0° but less than 90°.

Right angles: Angles whose measures are 90°.

Obtuse angles: Angles whose measures are greater than 90° but less than 180°.

Straight angles: Angles whose measures are 180°.

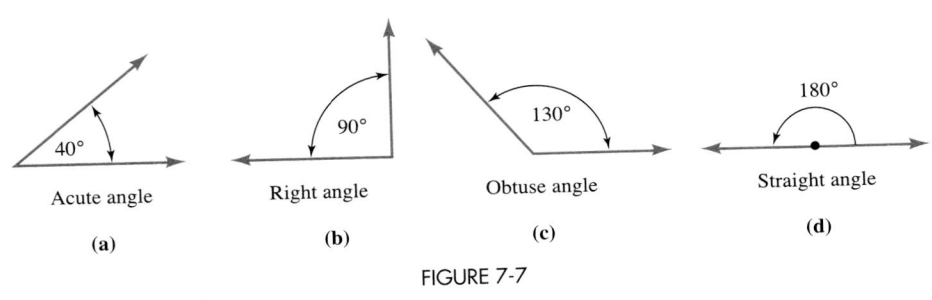

FIGURE 7-7

EXAMPLE 1   *Classifying angles.*   Classify each angle in Figure 7-8 as an acute angle, a right angle, an obtuse angle, or a straight angle.

Solution   Since m($\angle 1$) < 90°, it is an acute angle.

Since m($\angle 2$) > 90° but less than 180°, it is an obtuse angle.

Since m($\angle BDE$) = 90°, it is a right angle.

Since m($\angle ABC$) = 180°, it is a straight angle.

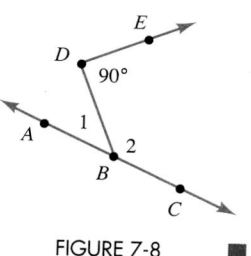

FIGURE 7-8

Adjacent and vertical angles

Two angles that have a common vertex and are side-by-side are called **adjacent angles**.

EXAMPLE 2 *Evaluating angles.* Two angles with measures of $x°$ and 35° are adjacent angles. Use the information in Figure 7-9 to find x.

Solution

We can use algebra to solve this problem. Since the sum of the measures of the angles is 80°, we have

$$x + 35 = 80$$
$$x + 35 - 35 = 80 - 35 \qquad \text{To undo the addition of 35, subtract 35 from both sides.}$$
$$x = 45 \qquad 35 - 35 = 0 \text{ and } 80 - 35 = 45.$$

Thus, $x = 45$.

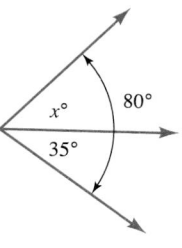

FIGURE 7-9

Self Check
In the figure below, find x.

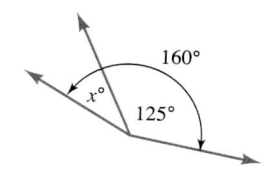

Answer: 35

When two lines intersect, pairs of nonadjacent angles are called **vertical angles.** In Figure 7-10(a), lines l_1 (read as "line l sub 1") and l_2 (read as "line l sub 2") intersect. $\angle 1$ and $\angle 3$ are vertical angles, as are $\angle 2$ and $\angle 4$.

To illustrate that vertical angles always have the same measure, we refer to Figure 7-10(b) with angles having measures of $x°$, $y°$, and 30°. Since the measure of any straight angle is 180°, we have

$$30 + x = 180 \qquad \text{and} \qquad 30 + y = 180$$
$$x = 150 \qquad\qquad\qquad y = 150 \quad \text{To undo the addition of 30, subtract 30 from both sides.}$$

Since x and y are both 150, $x = y$.

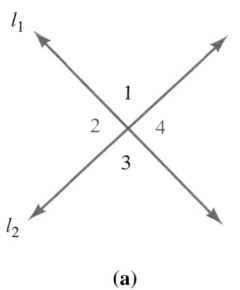

(a)

l_2

$x°$

l_1 30°

$y°$

Note that the angles having measures $x°$ and $y°$ are vertical angles.

(b)

FIGURE 7-10

Property of vertical angles   Vertical angles are congruent (have the same measure).

EXAMPLE 3   *Evaluating angles.*   In Figure 7-11, find   **a.** m(∠1)   and   **b.** m(∠3).

Solution

a. The 50° angle and ∠1 are vertical angles. Since vertical angles are congruent, m(∠1) = 50°.

b. Since *AD* is a line, the sum of the measures of ∠3, the 100° angle, and the 50° angle is 180°. If m(∠3) = *x*, we have

$$x + 100 + 50 = 180$$

$$x + 150 = 180 \quad \text{Simplify the left-hand side of the equation. Do the addition:}$$
$$100 + 50 = 150.$$

$$x = 30 \quad \text{Subtract 150 from both sides.}$$

Thus, m(∠3) = 30°.

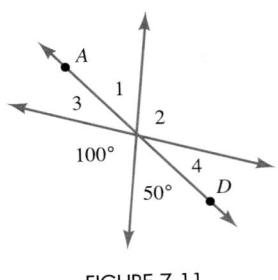

FIGURE 7-11

Self Check

In Figure 7-11, find

a. m(∠2)

b. m(∠4)

Answers: a. 100°,   **b.** 30°   ■

EXAMPLE 4   *Evaluating angles.*
In Figure 7-12, find *x*.

Solution

Since the angles are vertical angles, they have equal measures.

$$4x - 20 = 120$$

$$4x = 140 \quad \text{To undo the subtraction of 20, add 20 to both sides.}$$

$$x = 35 \quad \text{To undo the multiplication of 4, divide both sides by 4.}$$

Thus, *x* = 35.

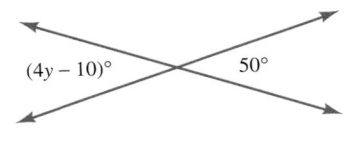

FIGURE 7-12

Self Check

In the figure below, find *y*.

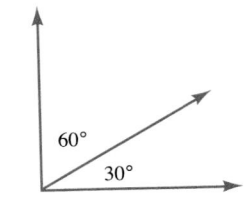

Answer: 15   ■

Complementary and supplementary angles

| Complementary and supplementary angles | Two angles are **complementary angles** when the sum of their measures is 90°. Two angles are **supplementary angles** when the sum of their measures is 180°. |

EXAMPLE 5   *Complementary and supplementary angles.*

a. Angles of 60° and 30° are complementary angles, because the sum of their measures is 90°. Each angle is the complement of the other.

b. Angles of 130° and 50° are supplementary, because the sum of their measures is 180°. Each angle is the supplement of the other.

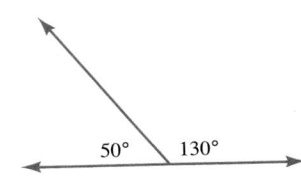

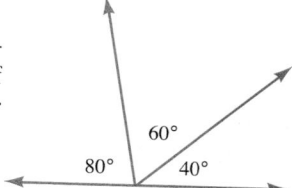

 COMMENT The definition of supplementary angles requires that the sum of *two* angles be 180°. Three angles of 40°, 60°, and 80° are not supplementary even though their sum is 180°.

EXAMPLE 6 *Finding the complement and supplement of an angle.*

a. Find the complement of a 35° angle.

b. Find the supplement of a 105° angle.

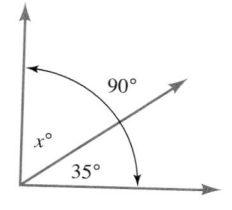

FIGURE 7-13

Solution

a. See Figure 7-13. Let x represent the complement of the 35° angle. Since the angles are complementary, we have

$x + 35 = 90$ The sum of the angles' measures must be 90°.

$x = 55$ To undo the addition of 35, subtract 35 from both sides.

The complement of 35° is 55°.

b. See Figure 7-14. Let y represent the supplement of the 105° angle. Since the angles are supplementary, we have

$y + 105 = 180$ The sum of the angles' measures must be 180°.

$y = 75$ To undo the addition of 105, subtract 105 from both sides.

The supplement of 105° is 75°.

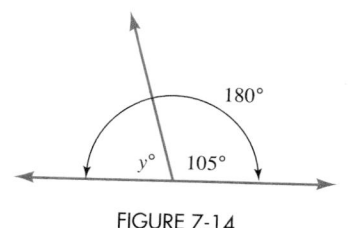

FIGURE 7-14

Self Check

a. Find the complement of a 50° angle.

b. Find the supplement of a 50° angle.

Answers: a. 40°,   **b.** 130°   ∎

STUDY SET   Section 7.1

VOCABULARY   *Fill in the blanks.*

1. A line _____ has two endpoints.

2. Two points _____ at most one line.

3. A _____ divides a line segment into two parts of equal length.

4. An angle is measured in _____.

5. A _____ is used to measure angles.

6. An _____ angle is less than 90°.

7. A _____ angle measures 90°.

8. An _____ angle is greater than 90° but less than 180°.

9. The measure of a straight angle is _____.

10. Adjacent angles have the same vertex and are _____.

11. The sum of two _____ angles is 180°.

12. The sum of two complementary angles is _____.

CONCEPTS Refer to Illustration 1 and tell whether each statement is true. If a statement is false, explain why.

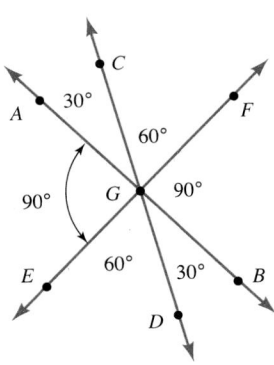

ILLUSTRATION 1

13. $\overrightarrow{GF}$ has point G as its endpoint.
14. $\overline{AG}$ has no endpoints.
15. Line CD has three endpoints.
16. Point D is the vertex of $\angle DGB$.
17. $m(\angle AGC) = m(\angle BGD)$
18. $\angle AGF \cong \angle BGE$
19. $\angle FGB \cong \angle EGA$
20. $\angle AGC$ and $\angle CGF$ are adjacent angles.

Refer to Illustration 1 and tell whether each angle is an acute angle, a right angle, an obtuse angle, or a straight angle.

21. $\angle AGC$ **22.** $\angle EGA$
23. $\angle FGD$ **24.** $\angle BGA$
25. $\angle BGE$ **26.** $\angle AGD$
27. $\angle DGC$ **28.** $\angle DGB$

Refer to Illustration 2 and tell whether each statement is true. If a statement is false, explain why.

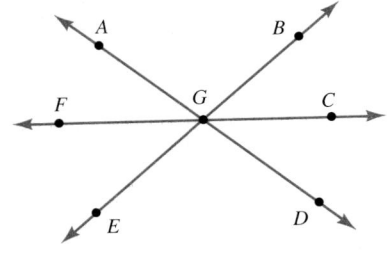

ILLUSTRATION 2

29. $\angle AGF$ and $\angle DGC$ are vertical angles.
30. $\angle FGE$ and $\angle BGA$ are vertical angles.
31. $m(\angle AGB) = m(\angle BGC)$.
32. $\angle AGC \cong \angle DGF$.

Refer to Illustration 3 and tell whether each pair of angles are congruent.

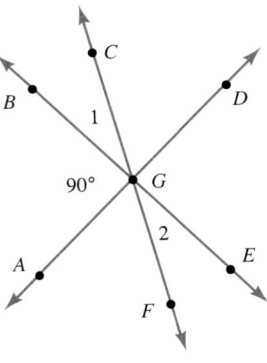

ILLUSTRATION 3

33. $\angle 1$ and $\angle 2$ **34.** $\angle FGB$ and $\angle CGE$
35. $\angle AGB$ and $\angle DGE$ **36.** $\angle CGD$ and $\angle CGB$
37. $\angle AGF$ and $\angle FGE$ **38.** $\angle AGB$ and $\angle BGD$

Refer to Illustration 3 and tell whether each statement is true.

39. $\angle 1$ and $\angle CGD$ are adjacent angles.
40. $\angle 2$ and $\angle 1$ are adjacent angles.
41. $\angle FGA$ and $\angle AGC$ are supplementary.
42. $\angle AGB$ and $\angle BGC$ are complementary.
43. $\angle AGF$ and $\angle 2$ are complementary.
44. $\angle AGB$ and $\angle EGD$ are supplementary.
45. $\angle EGD$ and $\angle DGB$ are supplementary.
46. $\angle DGC$ and $\angle AGF$ are complementary.

NOTATION Fill in the blanks.

47. The symbol $\angle$ means _____.
48. The symbol $\overline{AB}$ is read as "_____ AB."
49. The symbol $\overrightarrow{AB}$ is read as "_____ AB."
50. The symbol _____ is read as "is congruent to."

PRACTICE Refer to Illustration 4 and find the length of each segment.

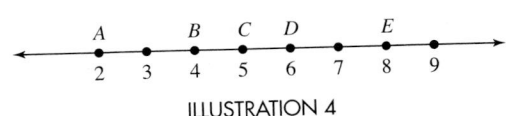

ILLUSTRATION 4

51. $\overline{AC}$ **52.** $\overline{BE}$
53. $\overline{CE}$ **54.** $\overline{BD}$
55. $\overline{CD}$ **56.** $\overline{DE}$

Refer to Illustration 4 and find each midpoint.

57. Find the midpoint of $\overline{AD}$.
58. Find the midpoint of $\overline{BE}$.

Use a protractor to measure each angle.

59.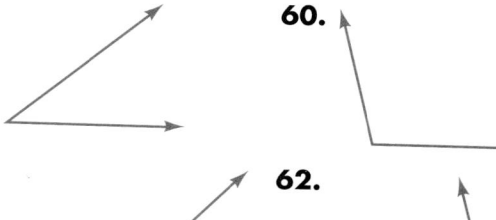

60.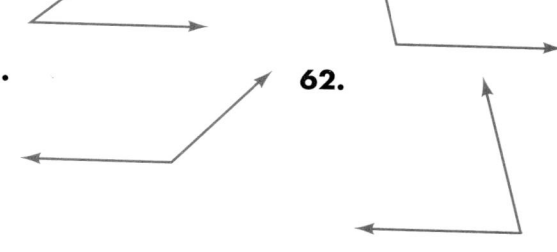

61.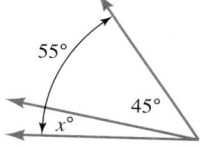

62.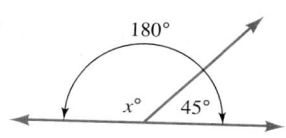

Find x.

63.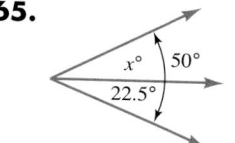

64.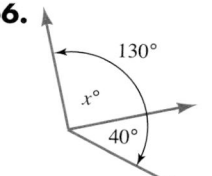

65.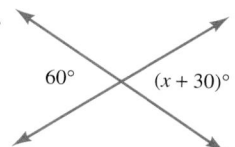

66.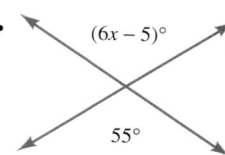

67.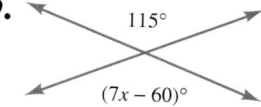

68.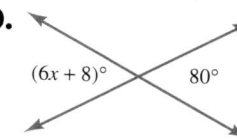

69.

70.

Let x represent the unknown angle measure. Draw a diagram, write an appropriate equation, and solve it for x.

71. Find the complement of a 30° angle.
72. Find the supplement of a 30° angle.
73. Find the supplement of a 105° angle.
74. Find the complement of a 75° angle.

Refer to Illustration 5, in which m(∠1) = 50°. Find the measure of each angle or sum of angles.

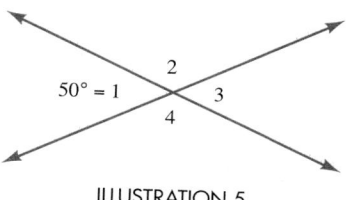

ILLUSTRATION 5

75. ∠4   **76.** ∠3
77. m(∠1) + m(∠2) + m(∠3)
78. m(∠2) + m(∠4)

Refer to Illustration 6, in which m(∠1) + m(∠3) + m(∠4) = 180°, ∠3 ≅ ∠4, and ∠4 ≅ ∠5. Find the measure of each angle.

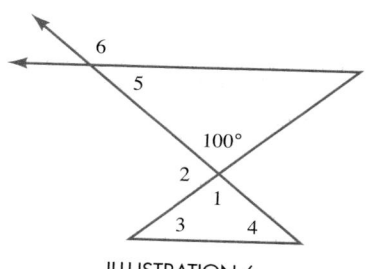

ILLUSTRATION 6

79. ∠1   **80.** ∠2
81. ∠3   **82.** ∠6

APPLICATIONS

83. BASEBALL   Use the following definition to draw the strike zone for the player shown in Illustration 7.

The strike zone is that area over home plate the upper limit of which is a horizontal line at the midpoint between the top of the shoulders and the top of the uniform pants and the lower level is a line at the hollow beneath the kneecap.

ILLUSTRATION 7

84. PHYSICS Illustration 8 shows a 15-pound block that is suspended with two ropes, one of which is horizontal. Classify each numbered angle in the illustration as either acute, obtuse, or right.

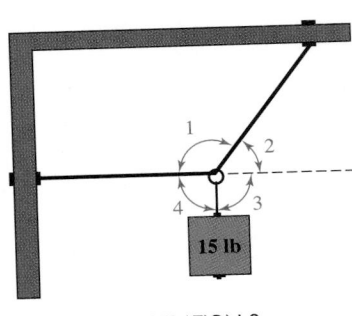

ILLUSTRATION 8

85. SYNTHESIZER Refer to Illustration 9. Find *x* and *y*.

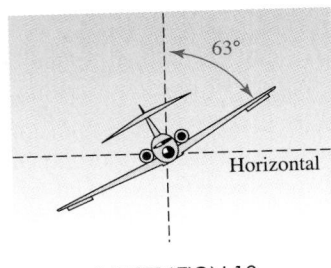

ILLUSTRATION 9

86. AVIATION Refer to Illustration 10. How many degrees from the horizontal position are the wings of the airplane?

ILLUSTRATION 10

87. GARDENING In Illustration 11, what angle does the handle of the lawn mower make with the ground?

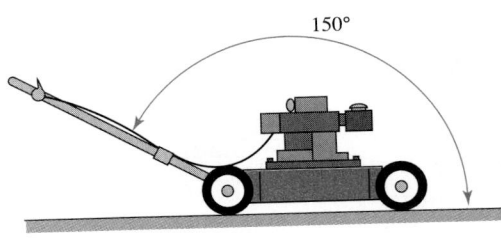

ILLUSTRATION 11

88. MUSICAL INSTRUMENTS Suppose that you are a beginning band teacher describing the correct posture needed to play various instruments. Use the diagrams in Illustration 12 to approximate the angle measure at which each instrument should be held in relation to the student's body: **a.** flute **b.** clarinet **c.** trumpet

a.   b.   c.

ILLUSTRATION 12

WRITING

89. PHRASES Explain what you think each of these phrases means. How is geometry involved?
  a. The president did a complete 180-degree flip on the subject of a tax cut.
  b. The rollerblader did a "360" as she jumped off the ramp.

90. In the statements below, the ° symbol is used in two different ways. Explain the difference.

$$85°F \quad \text{and} \quad m(\angle A) = 85°$$

91. What is a protractor?

92. Explain the difference between a ray and a line segment.

93. Explain why an angle measuring 105° cannot have a complement.

94. Explain why an angle measuring 210° cannot have a supplement.

REVIEW

95. Find 2^4.

96. Add: $\dfrac{1}{2} + \dfrac{2}{3} + \dfrac{3}{4}$.

97. Subtract: $\dfrac{3}{4} - \dfrac{1}{8} - \dfrac{1}{3}$.

98. Multiply: $\dfrac{5}{8} \cdot \dfrac{2}{15} \cdot \dfrac{6}{5}$.

99. What is 7% of 7?

100. What percent of 32 is 8?

7.2 *Parallel and Perpendicular Lines*

In this section, you will learn about

- Parallel and perpendicular lines • Transversals and angles
- Properties of parallel lines

INTRODUCTION. In this section, we will consider *parallel* and *perpendicular* lines. Since parallel lines are always the same distance apart, the railroad tracks shown in Figure 7-15(a) illustrate one application of parallel lines. Figure 7-15(b) shows one of the events of men's gymnastics, the parallel bars. Since perpendicular lines meet and form right angles, the monument and the ground shown in Figure 7-15(c) illustrate one application of perpendicular lines.

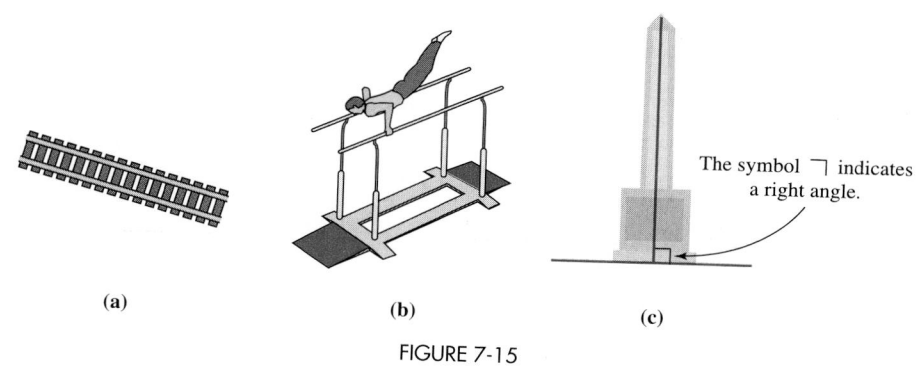

The symbol ⌐ indicates a right angle.

(a)   (b)   (c)

FIGURE 7-15

Parallel and perpendicular lines

If two lines lie in the same plane, they are called **coplanar.** Two coplanar lines that do not intersect are called **parallel lines.** See Figure 7-16(a).

Parallel lines

> **Parallel lines** are coplanar lines that do not intersect.

If lines l_1 (read as "*l* sub 1") and l_2 (read as "*l* sub 2") are parallel, we can write $l_1 \parallel l_2$, where the symbol $\parallel$ is read as "is parallel to."

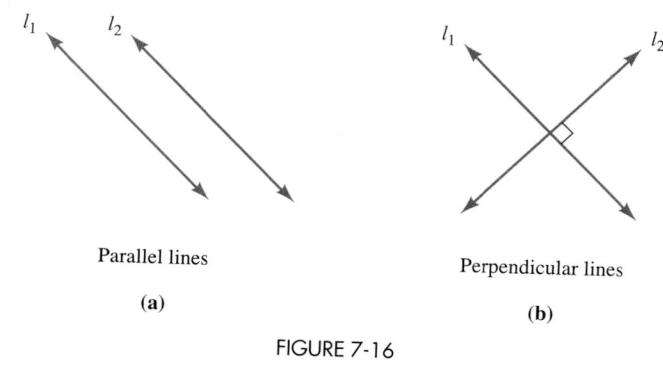

Parallel lines   Perpendicular lines

(a)   (b)

FIGURE 7-16

Perpendicular lines

> **Perpendicular lines** are lines that intersect and form right angles.

In Figure 7-16(b), $l_1 \perp l_2$, where the symbol $\perp$ is read as "is perpendicular to."

Transversals and angles

A line that intersects two or more coplanar lines is called a **transversal.** For example, line l_1 in Figure 7-17 is a transversal intersecting lines l_2, l_3, and l_4.

When two lines are cut by a transversal, the following types of angles are formed.

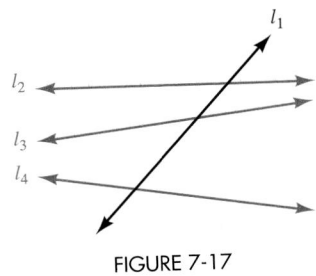

FIGURE 7-17

Alternate interior angles:

$\angle 4$ and $\angle 5$

$\angle 3$ and $\angle 6$

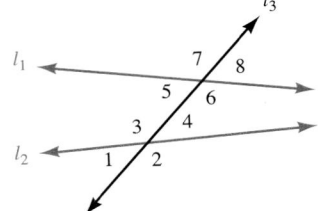

Corresponding angles:

$\angle 1$ and $\angle 5$

$\angle 3$ and $\angle 7$

$\angle 2$ and $\angle 6$

$\angle 4$ and $\angle 8$

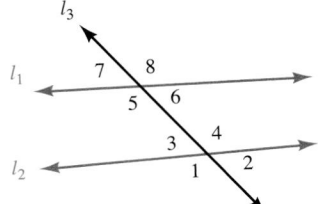

Interior angles:

$\angle 3$, $\angle 4$, $\angle 5$, and $\angle 6$

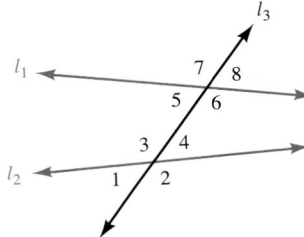

EXAMPLE 1 *Identifying angles.* In Figure 7-18, identify **a.** all pairs of alternate interior angles, **b.** all pairs of corresponding angles, and **c.** all interior angles.

Solution **a.** Pairs of alternate interior angles are

$\angle 3$ and $\angle 5$, $\angle 4$ and $\angle 6$

b. Pairs of corresponding angles are

$\angle 1$ and $\angle 5$, $\angle 4$ and $\angle 8$, $\angle 2$ and $\angle 6$, $\angle 3$ and $\angle 7$

c. Interior angles are

$\angle 3$, $\angle 4$, $\angle 5$, and $\angle 6$

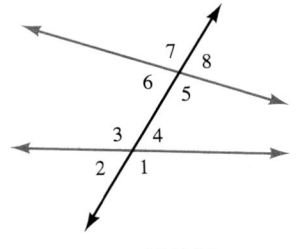

FIGURE 7-18

Properties of parallel lines

1. If two parallel lines are cut by a transversal, alternate interior angles are congruent. (See Figure 7-19.) If $l_1 \parallel l_2$, then $\angle 2 \cong \angle 4$ and $\angle 1 \cong \angle 3$.
2. If two parallel lines are cut by a transversal, corresponding angles are congruent. (See Figure 7-20.) If $l_1 \parallel l_2$, then $\angle 1 \cong \angle 5$, $\angle 3 \cong \angle 7$, $\angle 2 \cong \angle 6$, and $\angle 4 \cong \angle 8$.

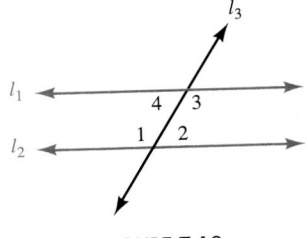

FIGURE 7-19

3. If two parallel lines are cut by a transversal, interior angles on the same side of the transversal are supplementary. (See Figure 7-21.) If $l_1 \parallel l_2$, then $\angle 1$ is supplementary to $\angle 2$ and $\angle 4$ is supplementary to $\angle 3$.

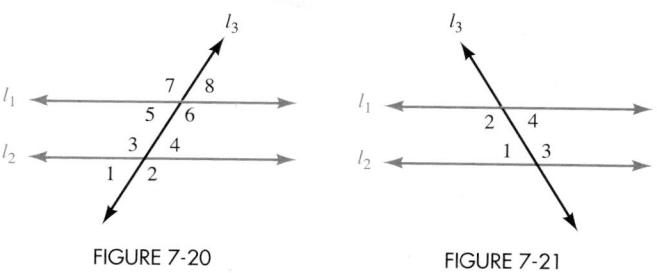

FIGURE 7-20   FIGURE 7-21

4. If a transversal is perpendicular to one of two parallel lines, it is also perpendicular to the other line. (See Figure 7-22.) If $l_1 \parallel l_2$ and $l_3 \perp l_1$, then $l_3 \perp l_2$.

5. If two lines are parallel to a third line, they are parallel to each other. (See Figure 7-23.) If $l_1 \parallel l_2$ and $l_1 \parallel l_3$, then $l_2 \parallel l_3$.

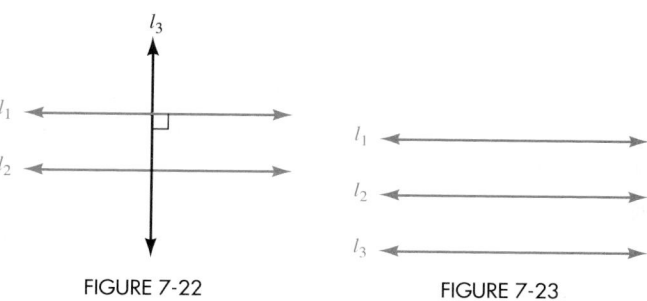

FIGURE 7-22   FIGURE 7-23

EXAMPLE 2 *Evaluating angles.*   See Figure 7-24. If $l_1 \parallel l_2$ and $m(\angle 3) = 120°$, find the measures of the other angles.

Solution

$m(\angle 1) = 60°$ $\angle 3$ and $\angle 1$ are supplementary.

$m(\angle 2) = 120°$ Since vertical angles are congruent, $m(\angle 2) = m(\angle 3)$.

$m(\angle 4) = 60°$ Since vertical angles are congruent, $m(\angle 4) = m(\angle 1)$.

$m(\angle 5) = 60°$ If two parallel lines are cut by a transversal, alternate interior angles are congruent: $m(\angle 5) = m(\angle 4)$.

$m(\angle 6) = 120°$ If two parallel lines are cut by a transversal, alternate interior angles are congruent: $m(\angle 6) = m(\angle 3)$.

$m(\angle 7) = 120°$ Vertical angles are congruent: $m(\angle 7) = m(\angle 6)$.

$m(\angle 8) = 60°$ Vertical angles are congruent: $m(\angle 8) = m(\angle 5)$.

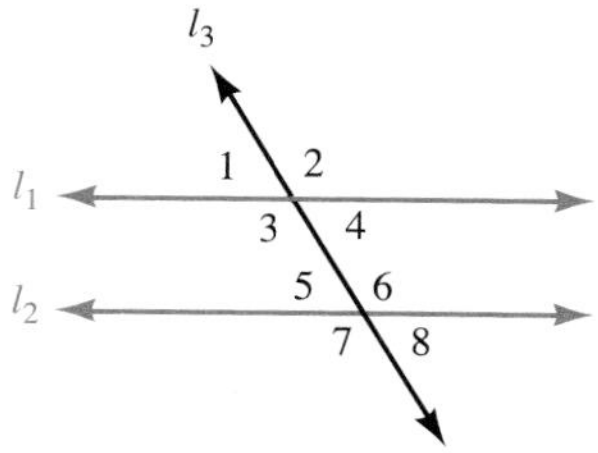

FIGURE 7-24

Self Check

If $l_1 \parallel l_2$ and $m(\angle 8) = 50°$, find the measures of the other angles. (See Figure 7-24.)

Answers: $m(\angle 5) = 50°$, $m(\angle 7) = 130°$, $m(\angle 6) = 130°$, $m(\angle 3) = 130°$, $m(\angle 4) = 50°$, $m(\angle 1) = 50°$, $m(\angle 2) = 130°$ ■

EXAMPLE 3   *Identifying congruent angles.*   See Figure 7-25. If $\overline{AB} \parallel \overline{DE}$, which pairs of angles are congruent?

Solution   Since $\overline{AB} \parallel \overline{DE}$, corresponding angles are congruent. So we have

$\angle A \cong \angle 1$ and $\angle B \cong \angle 2$

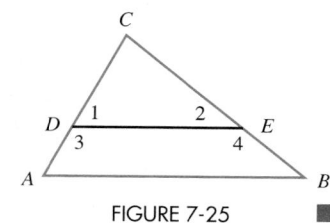

FIGURE 7-25   ■

EXAMPLE 4   *Using algebra in geometry.*
In Figure 7-26, $l_1 \parallel l_2$. Find x.

Solution
The angles involving x are corresponding angles. Since $l_1 \parallel l_2$, all pairs of corresponding angles are congruent.

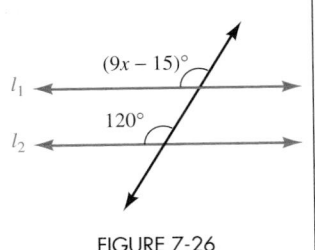

$$9x - 15 = 120 \quad \text{The angle measures are equal.}$$
$$9x = 135 \quad \begin{array}{l}\text{To undo the subtraction of 15,}\\ \text{add 15 to both sides.}\end{array}$$
$$x = 15 \quad \text{To undo the multiplication by 9, divide both sides by 9.}$$

Thus, $x = 15$.

FIGURE 7-26

Self Check
In the figure below, $l_1 \parallel l_2$. Find y.

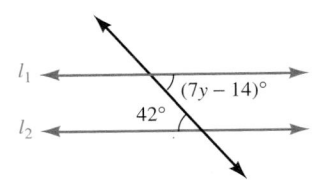

Answer: 8

EXAMPLE 5   *Using algebra in geometry.*   In Figure 7-27, $l_1 \parallel l_2$. Find x.

Solution   Since the angles are interior angles on the same side of the transversal, they are supplementary.

$$3x + 20 + 40 = 180 \quad \begin{array}{l}\text{The sum of the measures}\\ \text{of two supplementary}\\ \text{angles is } 180°.\end{array}$$
$$3x + 60 = 180 \quad \text{Simplify the left-hand side. Do the addition: } 20 + 40 = 60.$$
$$3x = 120 \quad \text{To undo the addition of 60, subtract 60 from both sides.}$$
$$x = 40 \quad \text{To undo the multiplication by 3, divide both sides by 3.}$$

Thus, $x = 40$.

FIGURE 7-27

STUDY SET   Section 7.2

VOCABULARY   *Fill in the blanks.*

1. Two lines in the same plane are _____.

2. _____ lines do not intersect.

3. If two lines intersect and form right angles, they are _____.

4. A _____ intersects two or more coplanar lines.

5. In Illustration 1, $\angle 4$ and $\angle 6$ are _____ interior angles.

6. In Illustration 1, $\angle 2$ and $\angle 6$ are _____ angles.

CONCEPTS

7. Which pairs of angles shown in Illustration 1 are alternate interior angles?

8. Which pairs of angles shown in Illustration 1 are corresponding angles?

9. Which angles shown in Illustration 1 are interior angles?

10. In Illustration 2, $l_1 \parallel l_2$. What can you conclude about l_1 and l_3?

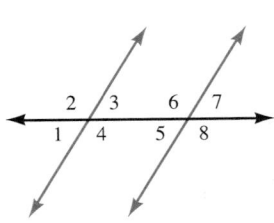

ILLUSTRATION 1

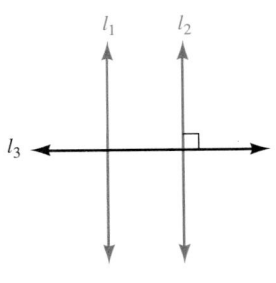

ILLUSTRATION 2

11. In Illustration 3, $l_1 \parallel l_2$ and $l_2 \parallel l_3$. What can you conclude about l_1 and l_3?

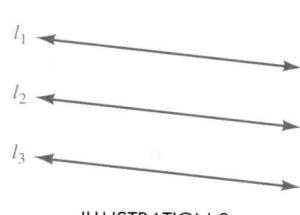

ILLUSTRATION 3

12. In Illustration 4, $\overline{AB} \parallel \overline{DE}$. What pairs of angles are congruent?

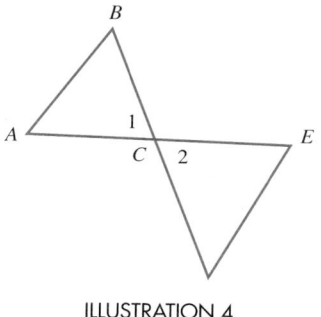

ILLUSTRATION 4

19. In Illustration 7, $l_1 \parallel \overline{AB}$. Find the measure of each angle.

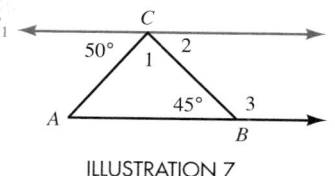

ILLUSTRATION 7

20. In Illustration 8, $\overline{AB} \parallel \overline{DE}$. Find m($\angle B$), m($\angle E$), and m($\angle 1$).

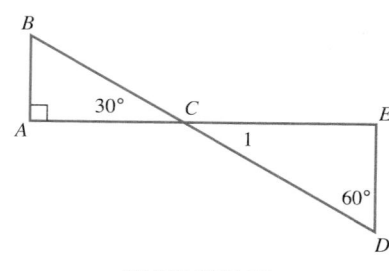

ILLUSTRATION 8

NOTATION *Fill in the blanks.*

13. The symbol ⌐ indicates _____.

14. The symbol $\parallel$ is read as "_____."

15. The symbol $\perp$ is read as "_____."

16. The symbol l_1 is read as "_____."

PRACTICE

17. In Illustration 5, $l_1 \parallel l_2$ and m($\angle 4$) = 130°. Find the measures of the other angles.

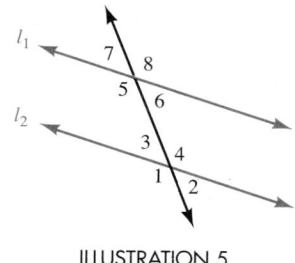

ILLUSTRATION 5

18. In Illustration 6, $l_1 \parallel l_2$ and m($\angle 2$) = 40°. Find the measures of the other angles.

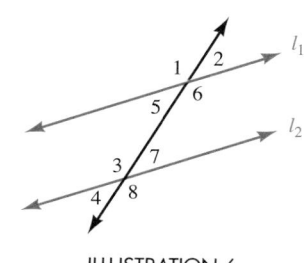

ILLUSTRATION 6

In Exercises 21–24, $l_1 \parallel l_2$. Find x.

21.

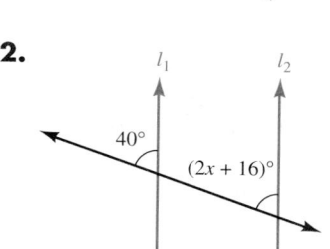

22.

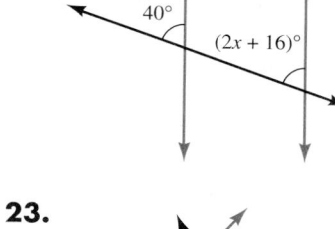

23.

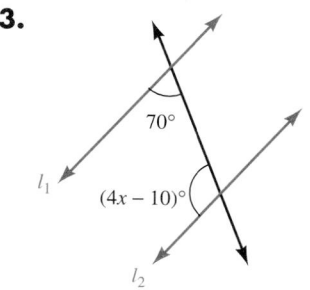

24.

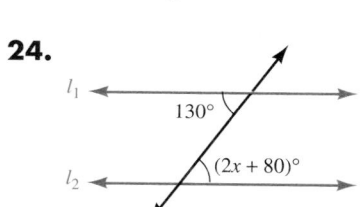

In Exercises 25–28, find x.

25. $l_1 \parallel \overline{CA}$

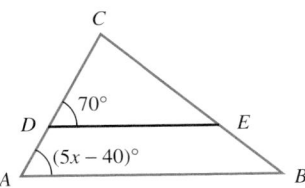

26. $\overline{AB} \parallel \overline{DE}$

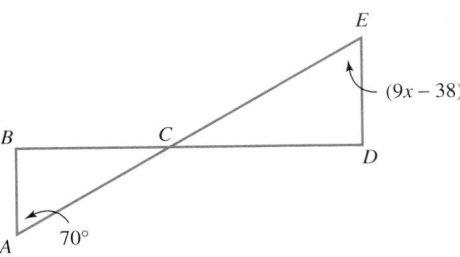

27. $\overline{AB} \parallel \overline{DE}$

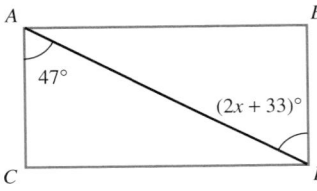

28. $\overline{AC} \parallel \overline{BD}$

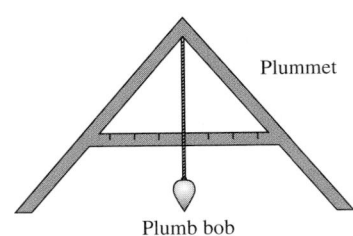

APPLICATIONS

29. CONSTRUCTING PYRAMIDS   The Egyptians used a device called a **plummet** to tell whether stones were properly leveled. A plummet, shown in Illustration 9, is made up of an A-frame and a plumb bob suspended from the peak of the frame. How could a builder use a plummet to tell that the stone on the left is not level and that the stones on the right are level?

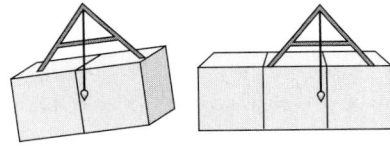

ILLUSTRATION 9

30. DIAGRAMMING SENTENCES   English instructors have their students diagram sentences to help teach proper sentence structure. Illustration 10 is a diagram of the sentence *The cave was rather dark and damp.* Point out pairs of parallel and perpendicular lines used in the diagram.

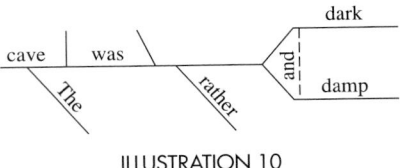

ILLUSTRATION 10

31. LOGO   Point out any perpendicular lines that can be found on the BMW company logo shown in Illustration 11.

ILLUSTRATION 11

32. PAINTING SIGNS   For many sign painters, the most difficult letter to paint is a capital E, because of all of the right angles involved. See Illustration 12. How many right angles are there?

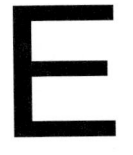

ILLUSTRATION 12

33. HANGING WALLPAPER   Explain why the concepts of perpendicular and parallel are both important when hanging wallpaper.

34. TOOLS   See Illustration 13. What geometric concepts are seen in the design of the rake?

ILLUSTRATION 13

WRITING

35. PARKING DESIGN   Using terms from this chapter, write a paragraph describing the parking layout shown in Illustration 14.

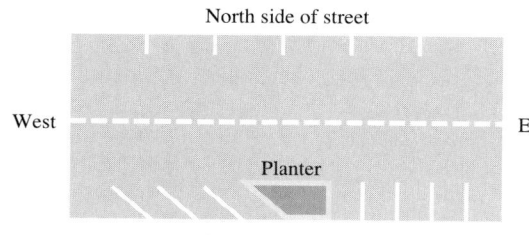

North side of street

West   East

Planter

South side of street

ILLUSTRATION 14

36. In your own words, explain what is meant by each of the following sentences.
   a. The hikers were told that the path *parallels* the river.
   b. John's quick rise to fame and fortune *paralleled* that of his older brother.
   c. The judge stated that the case that was before her court was without *parallel*.

37. Why do you think that ∠4 and ∠6 shown in Illustration 1 of this study set are called alternate interior angles?

38. Why do you think that ∠4 and ∠8 shown in Illustration 1 of this Study Set are called corresponding angles?

39. Are pairs of alternate interior angles always congruent? Explain.

40. Are pairs of interior angles always supplementary? Explain.

REVIEW

41. Find 60% of 120.

42. 80% of what number is 400?

43. What percent of 500 is 225?

44. Simplify: $3.45 + 7.37 \cdot 2.98$.

45. Is every whole number an integer?

46. Multiply: $2\frac{1}{5} \cdot 4\frac{3}{7}$.

7.3 **Polygons**

In this section, you will learn about

- Polygons • Triangles • Properties of isosceles triangles
- The sum of the measures of the angles of a triangle • Quadrilaterals
- Properties of rectangles • The sum of the measures of the angles of a polygon

INTRODUCTION. In this section, we will discuss figures called *polygons*. We see these shapes every day. For example, the walls in most buildings are rectangular in shape. We also see rectangular shapes in doors, windows, and sheets of paper.

The gable ends of many houses are triangular in shape, as are the sides of the Great Pyramid in Egypt. Triangular shapes are especially important because triangles are rigid and contribute strength and stability to walls and towers.

The designs used in tile or linoleum floors often use the shapes of a pentagon or a hexagon. Stop signs are in the shape of an octagon.

Polygons

Polygon

> A **polygon** is a closed geometric figure with at least three line segments for its sides.

The figures in Figure 7-28 are **polygons.** They are classified according to the number of sides they have. The points where the sides intersect are called **vertices.** If a polygon has sides that are all the same length and angles that have the same measure, we call it a **regular polygon.**

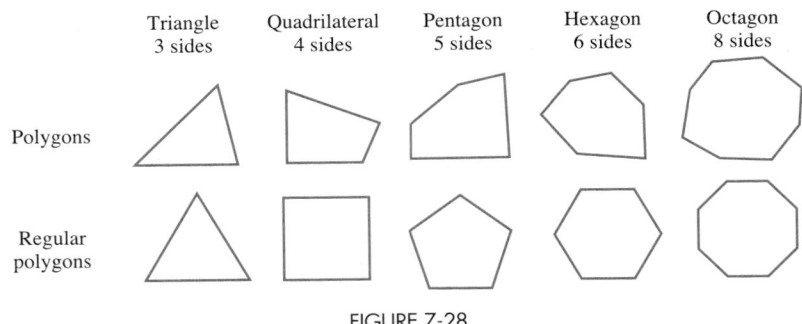

FIGURE 7-28

EXAMPLE 1   *Vertices of a polygon.*   Give the number of vertices of
a. a triangle   and   **b.** a hexagon.

Solution
a. From Figure 7-28, we see that a triangle has three angles and therefore three vertices.

b. From Figure 7-28, we see that a hexagon has six angles and therefore six vertices.

From the results of Example 1, we see that the number of vertices of a polygon is equal to the number of its sides.

Triangles

A **triangle** is a polygon with three sides. Figure 7-29 illustrates some common triangles. The slashes on the sides of a triangle indicate which sides are of equal length.

 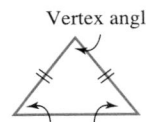 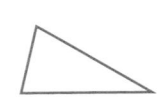 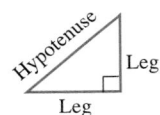

Equilateral triangle   Isosceles triangle   Scalene triangle   Right triangle
(all sides equal length)   (at least two sides of   (no sides equal length)   (has a right angle)
                            equal length)

FIGURE 7-29

 COMMENT   Since equilateral triangles have at least two sides of equal length, they are also isosceles. However, isosceles triangles are not necessarily equilateral.

Since every angle of an equilateral triangle has the same measure, an equilateral triangle is also **equiangular.**

In an isosceles triangle, the angles opposite the sides of equal length are called **base angles,** the sides of equal length form the **vertex angle,** and the third side is called the **base.**

The longest side of a right triangle is called the **hypotenuse,** and the other two sides are called **legs.** The hypotenuse of a right triangle is always opposite the 90° angle.

Properties of isosceles triangles

1. Base angles of an isosceles triangle are congruent.

2. If two angles in a triangle are congruent, the sides opposite the angles have the same length, and the triangle is isosceles.

EXAMPLE 2 *Determining whether a triangle is isosceles.*

Is the triangle in Figure 7-30 an isosceles triangle?

Solution

$\angle A$ and $\angle B$ are angles of the triangle. Since m($\angle A$) = m($\angle B$), we know that m($\overline{AC}$) = m($\overline{BC}$) and that $\triangle ABC$ (read as "triangle ABC") is isosceles.

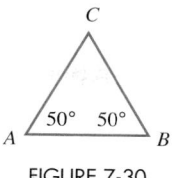

FIGURE 7-30

Self Check

In the figure below, $l_1 \parallel \overline{AB}$. Is the triangle an isosceles triangle?

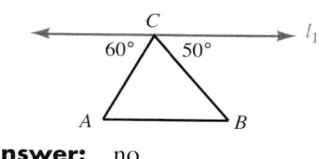

Answer: no

The sum of the measures of the angles of a triangle

If you draw several triangles and carefully measure each angle with a protractor, you will find that the sum of the angle measures in each triangle is 180°.

Angles of a triangle

> The sum of the angle measures of any triangle is 180°.

EXAMPLE 3 *Sum of the angles of a triangle.* See Figure 7-31. Find x.

Solution

Since the sum of the angle measures of any triangle is 180°, we have

$x + 40 + 90 = 180$

$x + 130 = 180$ Simplify the left-hand side. Do the addition: 40 + 90 = 130.

$x = 50$ To undo the addition of 130, subtract 130 from both sides.

Thus, $x = 50$.

FIGURE 7-31

Self Check

In the figure below, find y.

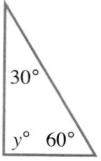

Answer: 90

EXAMPLE 4 *Vertex angle of an isosceles triangle.* See Figure 7-32. If one base angle of an isosceles triangle measures 70°, how large is the vertex angle?

Solution

Since one of the base angles measures 70°, so does the other. If we let x represent the measure of the vertex angle, we have

$x + 70 + 70 = 180$ The sum of the measures of the angles of a triangle is 180°.

$x + 140 = 180$ Simplify the left-hand side. Do the addition: 70 + 70 = 140.

$x = 40$ To undo the addition of 140, subtract 140 from both sides.

The vertex angle measures 40°.

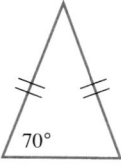

FIGURE 7-32

Quadrilaterals

A **quadrilateral** is a polygon with four sides. Some common quadrilaterals are shown in Figure 7-33 on the next page.

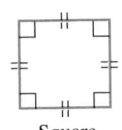 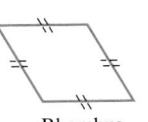 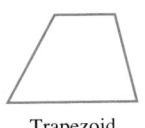

Parallelogram (Opposite sides parallel)   Rectangle (Parallelogram with four right angles)   Square (Rectangle with sides of equal length)   Rhombus (Parallelogram with sides of equal length)   Trapezoid (Exactly two sides parallel)

FIGURE 7-33

Properties of rectangles

1. All angles of a rectangle are right angles.
2. Opposite sides of a rectangle are parallel.
3. Opposite sides of a rectangle are of equal length.
4. The diagonals of a rectangle are of equal length.
5. If the diagonals of a parallelogram are of equal length, the parallelogram is a rectangle.

EXAMPLE 5 ***Squaring a foundation.*** A carpenter intends to build a shed with an 8-by-12-foot base. How can he make sure that the rectangular foundation is "square"?

Solution See Figure 7-34. The carpenter can use a tape measure to find the lengths of diagonals AC and BD. If these diagonals are of equal length, the figure will be a rectangle and have four right angles. Then the foundation will be "square."

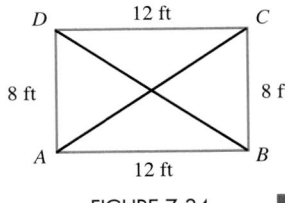

FIGURE 7-34 ■

EXAMPLE 6 ***Properties of rectangles and triangles.*** In rectangle $ABCD$ (Figure 7-35), the length of $\overline{AC}$ is 20 centimeters. Find each measure:
a. m($\overline{BD}$), **b.** m($\angle 1$), and **c.** m($\angle 2$).

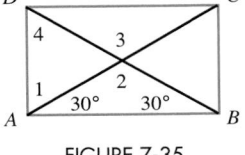

FIGURE 7-35

Solution
a. Since the diagonals of a rectangle are of equal length, m($\overline{BD}$) is also 20 centimeters.

b. We let m($\angle 1$) = x. Then, since the angles of a rectangle are right angles, we have

$$x + 30 = 90$$
$$x = 60 \quad \text{To undo the addition of 30, subtract 30 from both sides.}$$

Thus, m($\angle 1$) = 60°.

c. We let m($\angle 2$) = y. Then, since the sum of the angle measures of a triangle is 180°, we have

$$30 + 30 + y = 180$$
$$60 + y = 180 \quad \text{Simplify: } 30 + 30 = 60.$$
$$y = 120 \quad \text{To undo the addition of 60, subtract 60 from both sides.}$$

Thus, m($\angle 2$) = 120°.

Self Check
In rectangle $ABCD$ (Figure 7-35), the length of $\overline{DC}$ is 16 centimeters. Find each measure:

a. m($\overline{AB}$)

b. m($\angle 3$)

c. m($\angle 4$)

Answers: **a.** 16 cm
b. 120° **c.** 60° ■

The parallel sides of a trapezoid are called **bases,** the nonparallel sides are called **legs,** and the angles on either side of a base are called **base angles.** If the nonparallel sides are the same length, the trapezoid is an **isosceles trapezoid.** In an isosceles trapezoid, the base angles are congruent.

EXAMPLE 7   *Cross section of a drainage ditch.*   A cross section of a drainage ditch (Figure 7-36) is an isosceles trapezoid with $\overline{AB} \parallel \overline{CD}$. Find x and y.

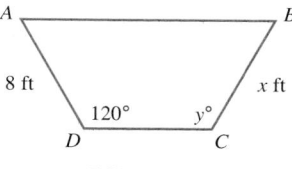

FIGURE 7-36

Solution   Since the figure is an isosceles trapezoid, its nonparallel sides have the same length. So $m(\overline{AD})$ and $m(\overline{BC})$ are equal, and $x = 8$.

Since the base angles of an isosceles trapezoid are congruent, $m(\angle D) = m(\angle C)$. Thus, $y = 120$.   ∎

The sum of the measures of the angles of a polygon

We have seen that the sum of the angle measures of any triangle is 180°. Since a polygon with n sides can be divided into $n - 2$ triangles, the sum of the angle measures of the polygon is $(n - 2)180°$.

Angles of a polygon

> The sum S, in degrees, of the measures of the angles of a polygon with n sides is given by the formula
> $$S = (n - 2)180$$

EXAMPLE 8   *Sum of the angles of a pentagon.*   Find the sum of the angle measures of a pentagon.

Solution

Since a pentagon has 5 sides, we substitute 5 for n in the formula and simplify.

$\qquad S = (n - 2)180$

$\qquad S = (5 - 2)180$ Substitute 5 for n.

$\qquad\quad = (3)180$ Do the subtraction within the parentheses.

$\qquad\quad = 540$ The result is in degrees.

The sum of the angles of a pentagon is 540°.

Self Check

Find the sum of the angle measures of a quadrilateral.

Answer: 360°   ∎

STUDY SET   Section 7.3

VOCABULARY   *Fill in the blanks.*

1. A _____ polygon has sides that are all the same length and angles that all have the same measure.

2. A polygon with four sides is called a _____.
A _____ is a polygon with three sides.

3. A _____ is a polygon with six sides.

4. A polygon with five sides is called a _____.

5. An eight-sided polygon is an _____.

6. The points where the sides of a polygon intersect are called _____.

7. A triangle with three sides of equal length is called an _____ triangle.

8. An _____ triangle has two sides of equal length.

9. The longest side of a right triangle is the _____.

10. The _____ angles of an isosceles triangle have the same measure.

11. A _____ with a right angle is a rectangle.

12. A rectangle with all sides of equal length is a _____.

13. A _____ is a parallelogram with four sides of equal length.

14. A _____ has two sides that are parallel and two sides that are not parallel.

15. The legs of an _____ trapezoid have the same length.

16. The _____ of a polygon is the distance around it.

29.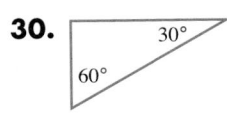

30.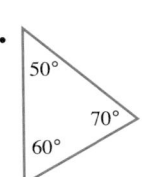

31. 20 cm   20 cm

32. 50° 70° 60°

CONCEPTS *Give the number of sides each polygon has and classify it as a triangle, quadrilateral, pentagon, hexagon, or octagon. Then give the number of vertices it has.*

17. □

18. ⬡

19. ▽

20. ⯃

21. ⬠

22. ⮔

23. ⯈

24. ◿

Classify each quadrilateral as a rectangle, a square, a rhombus, or a trapezoid.

33. 4 in. / 4 in. / 4 in. / 4 in.

34.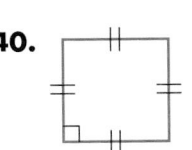

35. (parallelogram)

36. 90° / 90°

37. (square)

38. 8 cm / 8 cm / 8 cm / 8 cm

39. (trapezoid)

40. (square with marks)

Classify each triangle as an equilateral triangle, an isosceles triangle, a scalene triangle, or a right triangle.

25.

26. 55° / 55°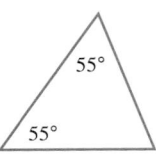

27.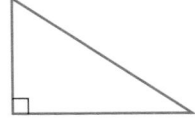

28.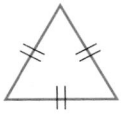

NOTATION *Fill in the blanks.*

41. The symbol △ means _____.

42. The symbol m(∠1) means the _____ of angle 1.

PRACTICE *The measures of two angles of △ABC (shown in Illustration 1) are given. Find the measure of the third angle.*

43. m(∠A) = 30° and m(∠B) = 60°
m(∠C) = _____

44. m(∠A) = 45° and m(∠C) = 105°
m(∠B) = _____

45. m(∠B) = 100° and m(∠A) = 35°
m(∠C) = _____

46. m(∠B) = 33° and m(∠C) = 77°
m(∠A) = _____

47. m(∠A) = 25.5° and m(∠B) = 63.8°
m(∠C) = _____

48. m(∠B) = 67.25° and m(∠C) = 72.5°
m(∠A) = _____

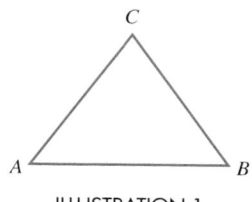

ILLUSTRATION 1

Refer to rectangle ABCD, shown in Illustration 2.

49. m(∠1) = _____

50. m(∠3) = _____

51. m(∠2) = _____

52. If m($\overline{AC}$) is 8 cm, then m($\overline{BD}$) = _____.

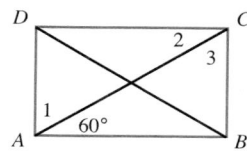

ILLUSTRATION 2

Find the sum of the angle measures of each polygon.

53. A hexagon

54. An octagon

55. A decagon (10 sides)

56. A dodecagon (12 sides)

APPLICATIONS

57. Give three uses of triangles in everyday life.

58. Give three uses of rectangles in everyday life.

59. Give three uses of squares in everyday life.

60. Give a use of a trapezoid in everyday life.

61. POLYGONS IN NATURE   As we see in Illustration 3(a), a starfish has the shape of a pentagon. What polygon shape do you see in each of the other objects in Illustration 3? **b.** Lemon   **c.** Chili pepper   **d.** Apple

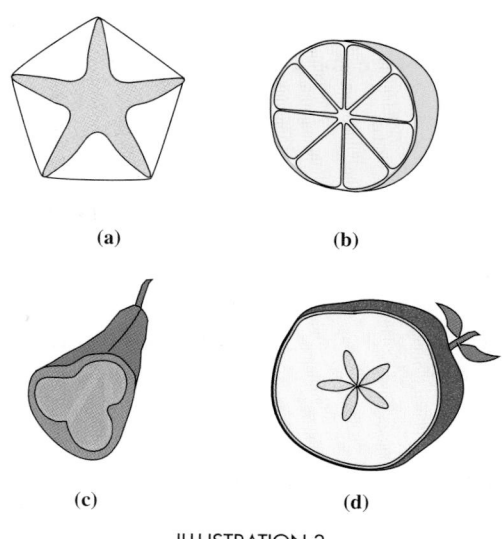

(a)   (b)

(c)   (d)

ILLUSTRATION 3

62. FLOWCHART   A flowchart shows a sequence of steps to be performed by a computer to solve a given problem. When designing a flowchart, the programmer uses a set of standardized symbols to represent various operations to be performed by the computer. Locate a rectangle, a rhombus, and a parallelogram in the flow chart shown in Illustration 4.

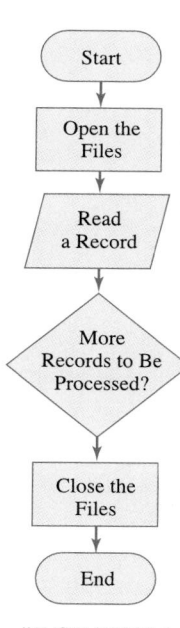

ILLUSTRATION 4

63. CHEMISTRY   Polygons are used to represent the chemical structure of compounds graphically. In Illustration 5, what types of polygons are used to represent methylprednisolone, the active ingredient in an antiinflammatory medication?

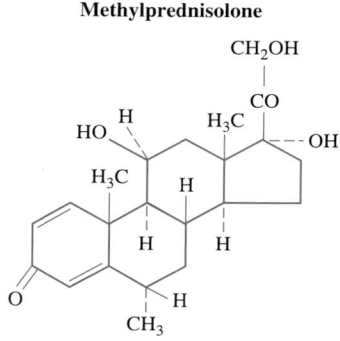

ILLUSTRATION 5

64. PODIUM In Illustration 6, what polygon describes the shape of the upper portion of the podium?

ILLUSTRATION 6

65. EASEL In Illustration 7, show how two of the legs of the easel form the equal sides of an isosceles triangle.

ILLUSTRATION 7

66. AUTOMOBILE JACK Refer to Illustration 8. Show that no matter how high the jack is raised, it always forms two isosceles triangles.

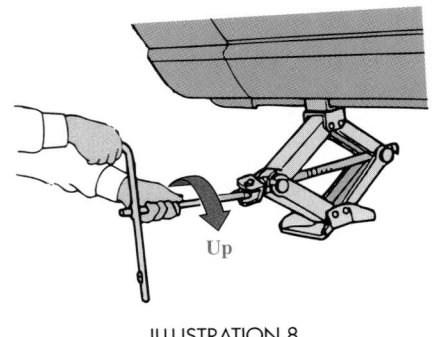

Up

ILLUSTRATION 8

WRITING

67. Explain why a square is a rectangle.

68. Explain why a trapezoid is not a parallelogram.

REVIEW

69. Find 20% of 110.

70. Find 15% of 50.

71. What percent of 200 is 80?

72. 20% of what number is 500?

73. Simplify: $0.85 \div 2(0.25)$.

74. What percent of the months in a year have the letter r in their names?

7.4 Properties of Triangles

In this section, you will learn about

• Congruent triangles • The Pythagorean theorem

INTRODUCTION. Proportions and triangles are often used to measure distances indirectly. For example, by using a proportion, Eratosthenes (275–195 B.C.) was able to estimate the circumference of the Earth with remarkable accuracy. On a sunny day, we can use properties of similar triangles to calculate the height of a tree while staying safely on the ground. By using a theorem proved by the Greek mathematician Pythagoras (about 500 B.C.), we can calculate the length of the third side of a right triangle whenever we know the lengths of two sides.

Congruent triangles

Triangles that have the same area and the same shape are called **congruent triangles.** In Figure 7-37, triangles *ABC* and *DEF* are congruent:

$$\triangle ABC \cong \triangle DEF \quad \text{Read as "Triangle } ABC \text{ is congruent to triangle } DEF\text{."}$$

Corresponding angles and corresponding sides of congruent triangles are called **corresponding parts.** The notation $\triangle ABC \cong \triangle DEF$ shows which vertices are corresponding parts.

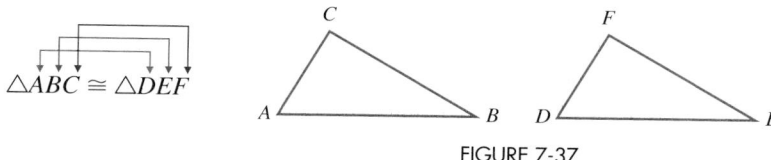

FIGURE 7-37

Corresponding parts of congruent triangles always have the same measure. For the congruent triangles shown in Figure 7-37,

$$m(\angle A) = m(\angle D), \quad m(\angle B) = m(\angle E), \quad m(\angle C) = m(\angle F),$$
$$m(\overline{BC}) = m(\overline{EF}), \quad m(\overline{AC}) = m(\overline{DF}), \quad m(\overline{AB}) = m(\overline{DE})$$

EXAMPLE 1 *Corresponding parts of congruent triangles.* Name the corresponding parts of the congruent triangles in Figure 7-38.

Solution The corresponding angles are

$\angle A$ and $\angle E$, $\angle B$ and $\angle D$,
$\angle C$ and $\angle F$

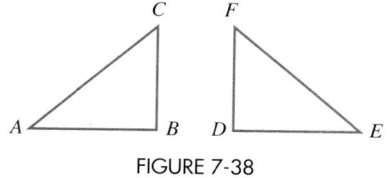

FIGURE 7-38

Since corresponding sides are always opposite corresponding angles, the corresponding sides are

$\overline{BC}$ and $\overline{DF}$, $\overline{AC}$ and $\overline{EF}$, $\overline{AB}$ and $\overline{ED}$

We will discuss three ways of showing that two triangles are congruent.

SSS property

> If three sides of one triangle are congruent to three sides of a second triangle, the triangles are congruent.

The triangles in Figure 7-39 are congruent because of the SSS property.

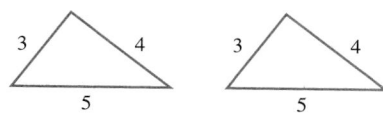

FIGURE 7-39

SAS property

> If two sides and the angle between them in one triangle are congruent, respectively, to two sides and the angle between them in a second triangle, the triangles are congruent.

The triangles in Figure 7-40 are congruent because of the SAS property.

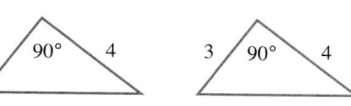

FIGURE 7-40

ASA property

> If two angles and the side between them in one triangle are congruent, respectively, to two angles and the side between them in a second triangle, the triangles are congruent.

The triangles in Figure 7-41 are congruent because of the ASA property.

FIGURE 7-41

COMMENT   There is no SSA property. To illustrate this, consider the triangles in Figure 7-42. Two sides and an angle of $\triangle ABC$ are congruent to two sides and an angle of $\triangle DEF$. But the congruent angle is *not* between the congruent sides.

We refer to this situation as SSA. Obviously, the triangles are not congruent, because they have different areas.

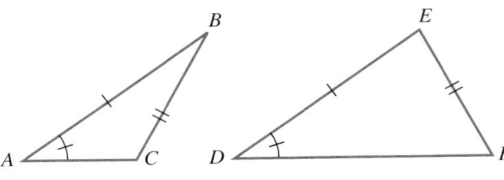

FIGURE 7-42

The slash marks indicate congruent parts. That is, the sides with one slash are the same length, the sides with two slashes are the same length, and the angles with one slash have the same measure.

EXAMPLE 2   *Determining whether triangles are congruent.*   Explain why the triangles in Figure 7-43 are congruent.

Solution   Since vertical angles are congruent,

$$m(\angle 1) = m(\angle 2)$$

From the figure, we see that

$$m(\overline{AC}) = m(\overline{EC}) \quad \text{and} \quad m(\overline{BC}) = m(\overline{DC})$$

Since two sides and the angle between them in one triangle are congruent, respectively, to two sides and the angle between them in a second triangle, $\triangle ABC \cong \triangle EDC$ by the SAS property.

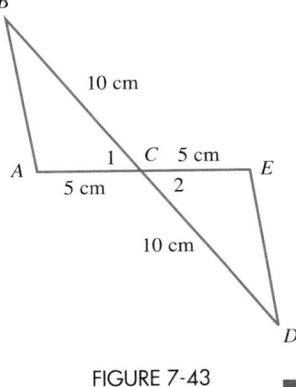

FIGURE 7-43

The Pythagorean theorem

In the movie *The Wizard of Oz,* the scarecrow was in search of a brain. To prove that he had found one, he recited the Pythagorean theorem.

> *In a right triangle, the square of the hypotenuse is equal to the sum of squares of the other two sides.*

Pythagorean theorem

> If the length of the hypotenuse of a right triangle is c and the lengths of its legs are a and b, then
> $$c^2 = a^2 + b^2$$

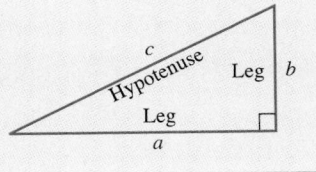

EXAMPLE 3 *Constructing a high-ropes adventure course.*
A builder of a high-ropes adventure course wants to secure the pole shown in Figure 7-44 by attaching a cable from the anchor stake 8 feet from its base to a point 6 feet up the pole. How long should the cable be?

Solution
The support cable, the pole, and the ground form a right triangle. If we let c represent the length of the cable (the hypotenuse), then we can use the Pythagorean theorem with $a = 8$ and $b = 6$ to find c.

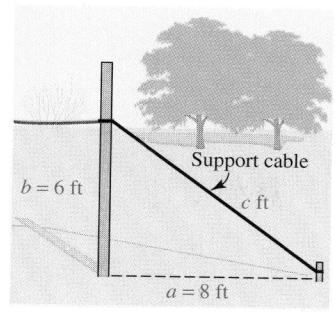

FIGURE 7-44

$c^2 = a^2 + b^2$ The Pythagorean theorem.

$c^2 = 8^2 + 6^2$ Substitute 8 for a and 6 for b.

$c^2 = 64 + 36$ Evaluate the exponential expressions.

$c^2 = 100$ Simplify the right-hand side.

To find c, we must find a number that, when squared, is 100. There are two such numbers, one positive and one negative; they are the square roots of 100. Since c represents the length of a support cable, c cannot be negative. For this reason, we need only find the positive square root of 100 to get c.

$c^2 = 100$ The equation to solve.

$c = \sqrt{100}$ The symbol $\sqrt{}$ is used to indicate the positive square root of a number.

$c = 10$ $\sqrt{100} = 10$, because $10^2 = 100$.

The support cable should be 10 feet long.

Self Check
A 26-foot ladder rests against the side of a building. If the base of the ladder is 10 feet from the wall, how far up the side of the building will the ladder reach?

Answer: 24 ft ∎

Accent on Technology: **Finding the width of a television screen**

The size of a television screen is the diagonal measure of its rectangular screen. (See Figure 7-45.) To find the width of a 27-inch screen that is 17 inches high, we use the Pythagorean theorem with $c = 27$ and $b = 17$.

$$c^2 = a^2 + b^2$$
$$27^2 = a^2 + 17^2$$
$$27^2 - 17^2 = a^2$$

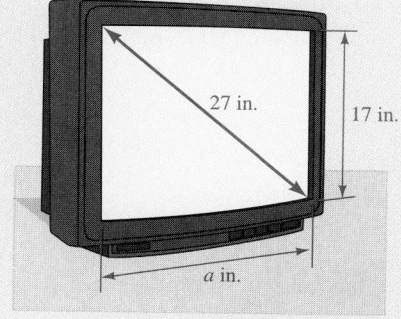

FIGURE 7-45

The variable a represents the width of a television screen, so it must be positive. To find a, we find the positive square root of the result when 17^2 is subtracted from 27^2. Using a radical symbol to indicate this, we have

$$\sqrt{27^2 - 17^2} = a$$

We can evaluate the expression on the left-hand side by entering these numbers and pressing these keys.

Keystrokes   (27 x^2 − 17 x^2) $\sqrt{}$ | 20.97617696 |

To the nearest inch, the width of the television screen is 21 inches.

It is also true that

If the square of one side of a triangle is equal to the sum of the squares of the other two sides, the triangle is a right triangle.

EXAMPLE 4 *Determining whether a triangle is a right triangle.*

Is a triangle with sides of 5, 12, and 13 meters a right triangle?

Solution

We can use the Pythagorean theorem to answer this question. Since the longest side of the triangle is 13 meters, we must substitute 13 for c. It doesn't matter which of the two remaining side lengths we substitute for a and which we substitute for b.

$c^2 = a^2 + b^2$ 　The Pythagorean theorem.

$13^2 \stackrel{?}{=} 5^2 + 12^2$ 　Substitute 13 for c, 5 for a, and 12 for b.

$169 \stackrel{?}{=} 25 + 144$ 　Evaluate the exponential expressions.

$169 = 169$ 　Simplify the right-hand side.

Since the square of the longest side is equal to the sum of the squares of the other two sides, the triangle is a right triangle.

Self Check

Is a triangle with sides of 9, 40, and 41 meters a right triangle?

Answer: yes ■

EXAMPLE 5 *Determining whether a triangle is a right triangle.*

Is a triangle with sides of 1, 2, and 3 feet a right triangle?

Solution

We check to see whether the square of the longest side is equal to the sum of the squares of the other two sides.

$c^2 = a^2 + b^2$ 　The Pythagorean theorem.

$3^2 \stackrel{?}{=} 1^2 + 2^2$ 　Substitute 3 for c, 1 for a, and 2 for b.

$9 \stackrel{?}{=} 1 + 4$ 　Evaluate the exponential expressions.

$9 \neq 5$ 　Simplify the right-hand side.

Since the square of the longest side is not equal to the sum of the squares of the other two sides, the triangle is not a right triangle.

Self Check

Is a triangle with sides of 4, 5, and 6 inches a right triangle?

Answer: no ■

STUDY SET Section 7.4

VOCABULARY *Fill in the blanks.*

1. _____ triangles are the same size and the same shape.

2. All _____ parts of congruent triangles have the same measure.

3. If a triangle has an angle that measures 90°, it is a _____ triangle.

4. The _____ is the longest side of a right triangle.

CONCEPTS *In Exercises 5–8, tell whether each statement is true. If a statement is false, tell why.*

5. If three sides of one triangle are the same length as three sides of a second triangle, the triangles are congruent.

6. If two sides of one triangle are the same length as two sides of a second triangle, the triangles are congruent.

7. If two sides and an angle of one triangle are congruent, respectively, to two sides and an angle of a second triangle, the triangles are congruent.

8. If two angles and the side between them in one triangle are congruent, respectively, to two angles and the side between them in a second triangle, the triangles are congruent.

9. Are the triangles shown in Illustration 1 congruent?

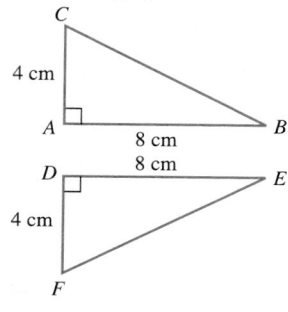

ILLUSTRATION 1

10. Are the triangles shown in Illustration 2 congruent?

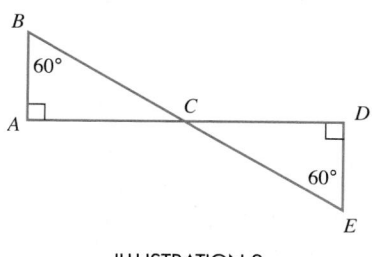

ILLUSTRATION 2

11. The Pythagorean theorem states that for a right triangle, $c^2 = a^2 + b^2$. What do the variables a, b, and c represent?

12. A triangle has sides of length 3, 4, and 5 centimeters. Substitute the lengths into $c^2 = a^2 + b^2$ and show that a true statement results. From the result, what can we conclude about the triangle?

NOTATION *Fill in the blanks.*

13. The symbol $\cong$ is read as "_____."

14. The symbol $m(\angle A)$ is read as "_____ angle A."

PRACTICE *Name the corresponding parts of the congruent triangles.*

15. Refer to Illustration 3.

$\overline{AC}$ corresponds to ___.

$\overline{DE}$ corresponds to ___.

$\overline{BC}$ corresponds to ___.

$\angle A$ corresponds to ___.

$\angle E$ corresponds to ___.

$\angle F$ corresponds to ___.

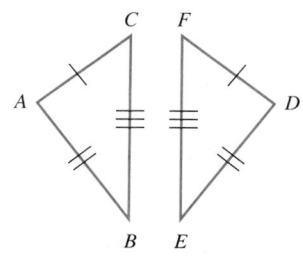

ILLUSTRATION 3

16. Refer to Illustration 4.

$\overline{AB}$ corresponds to ___.

$\overline{EC}$ corresponds to ___.

$\overline{AC}$ corresponds to ___.

$\angle D$ corresponds to ___.

$\angle B$ corresponds to ___.

$\angle 1$ corresponds to ___.

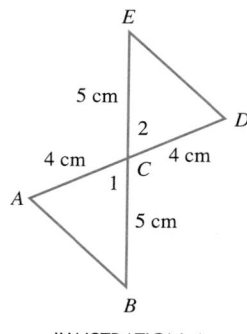

ILLUSTRATION 4

Determine whether each pair of triangles is congruent. If they are, tell why.

17.

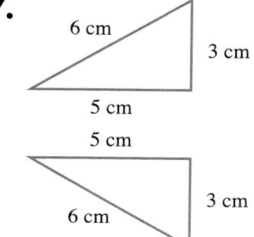

18.

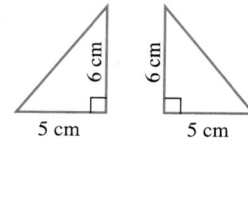

19.

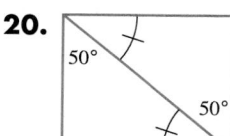

20.

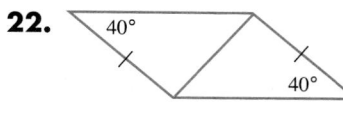

21.

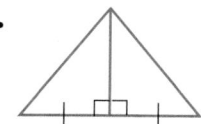

22.

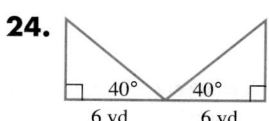

23.

24.

Find x.

25.

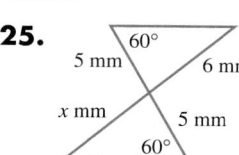

26.

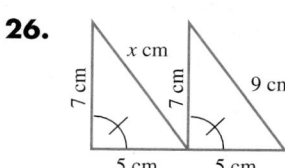

27.

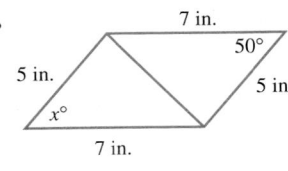

28.

Refer to Illustration 5 and find the length of the unknown side.

29. $a = 3$ and $b = 4$. Find c.

30. $a = 12$ and $b = 5$. Find c.

31. $a = 15$ and $c = 17$. Find b.

32. $b = 45$ and $c = 53$. Find a.

33. $a = 5$ and $c = 9$. Find b.

34. $a = 1$ and $b = 7$. Find c.

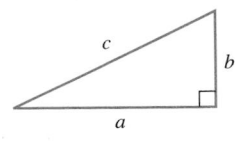

ILLUSTRATION 5

The lengths of the three sides of a triangle are given. Determine whether the triangle is a right triangle.

35. 8, 15, 17

36. 6, 8, 10

37. 7, 24, 26

38. 9, 39, 40

 APPLICATIONS *Solve each problem. If an answer is not exact, give the answer to the nearest tenth.*

39. ADJUSTING A LADDER A 20-foot ladder reaches a window 16 feet above the ground. How far from the wall is the base of the ladder?

40. LENGTH OF GUY WIRES A 30-foot tower is to be fastened by three guy wires attached to the top of the tower and to the ground at positions 20 feet from its base. How much wire is needed?

41. PICTURE FRAME After gluing and nailing two pieces of picture frame molding together, a frame maker checks her work by making a diagonal measure-

ment. (See Illustration 6.) If the sides of the frame form a right angle, what measurement should the frame maker read on the yardstick?

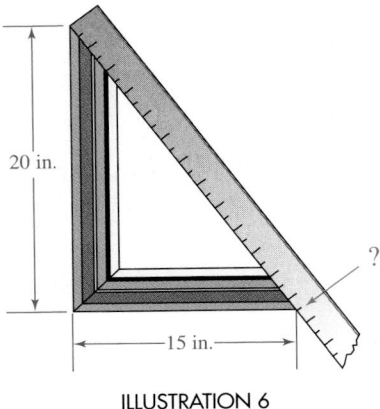

ILLUSTRATION 6

42. CARPENTRY The gable end of the roof shown in Illustration 7 is divided in half by a vertical brace, 8 feet in height. Find the length of the roof line.

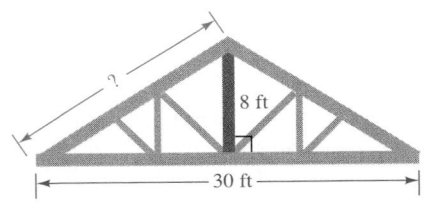

ILLUSTRATION 7

43. BASEBALL A baseball diamond is a square with each side 90 feet long (as shown in Illustration 8). How far is it from home plate to second base?

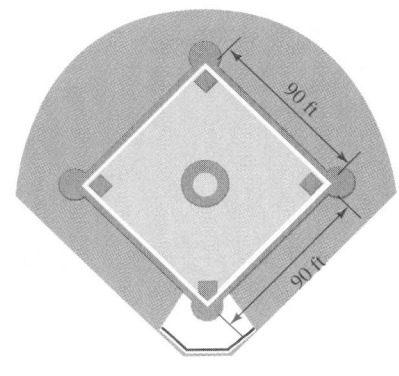

ILLUSTRATION 8

44. TELEVISION   What size is the television screen shown in Illustration 9?

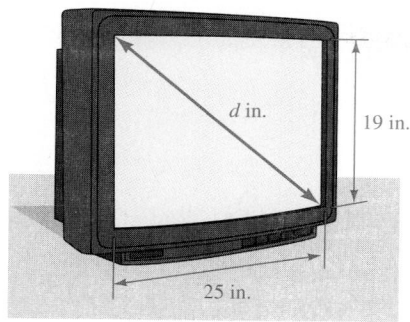

ILLUSTRATION 9

WRITING

45. Explain the Pythagorean theorem.

46. Explain the procedure used to solve the equation $c^2 = 64$. (Assume that c is positive.)

REVIEW   *Estimate the answer to each problem.*

47. $\dfrac{0.95 \cdot 3.89}{2.997}$

48. 21% of 42

49. 32% of 60

50. $\dfrac{4.966 + 5.001}{2.994}$

51. 49.5% of 18.1

52. 98.7% of 0.03

7.5 Perimeters and Areas of Polygons

In this section, you will learn how to find

- Perimeters of polygons • Perimeters of figures that are combinations of polygons
- Areas of polygons • Areas of figures that are combinations of polygons

INTRODUCTION. In this section, we will discuss how to find perimeters and areas of polygons. Finding perimeters is important when estimating the cost of fencing or estimating the cost of woodwork in a house. Finding areas is important when calculating the cost of carpeting, the cost of painting a house, or the cost of fertilizing a yard.

Perimeters of polygons

Recall that the **perimeter** of a polygon is the distance around it. Since a square has four sides of equal length s, its perimeter P is $s + s + s + s$, or $4s$.

Perimeter of a square

If a square has a side of length s, its perimeter P is given by the formula

$$P = 4s$$

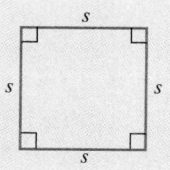

EXAMPLE 1 *Perimeter of a square.* Find the perimeter of a square whose sides are 7.5 meters long.

Solution
The perimeter of a square is given by the formula $P = 4s$. We substitute 7.5 for s and simplify.

$$P = 4s$$
$$P = 4(7.5)$$
$$P = 30$$

The perimeter is 30 meters.

Since a rectangle has two lengths l and two widths w, its perimeter P is $l + l + w + w$, or $2l + 2w$.

Perimeter of a rectangle

If a rectangle has length l and width w, its perimeter P is given by the formula

$$P = 2l + 2w$$

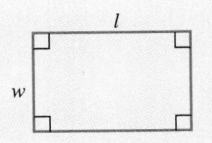

EXAMPLE 2 *Perimeter of a rectangle.* Find the perimeter of the rectangle in Figure 7-46.

Solution
The perimeter is given by the formula $P = 2l + 2w$. We substitute 10 for l and 6 for w and simplify.

$$P = 2l + 2w$$
$$P = 2(10) + 2(6)$$
$$= 20 + 12$$
$$= 32$$

The perimeter is 32 centimeters.

6 cm
10 cm
FIGURE 7-46

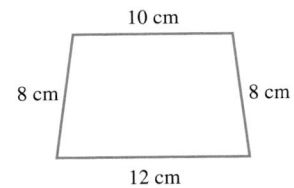
EXAMPLE 3 *Converting units.* Find the perimeter of the rectangle in Figure 7-47, in meters.

Solution
Since 1 meter = 100 centimeters, we can convert 80 centimeters to meters by multiplying by 1, written in the form $\frac{1 \text{ m}}{100 \text{ cm}}$.

$$80 \text{ cm} = \frac{80 \text{ cm}}{1} \cdot \frac{1 \text{ m}}{100 \text{ cm}}$$ Write 80 cm as a fraction: $80 \text{ cm} = \frac{80 \text{ cm}}{1}$.
Multiply by 1: $\frac{1 \text{ m}}{100 \text{ cm}} = 1$.

$$= \frac{80 \text{ cm} \cdot 1 \text{ m}}{1 \cdot 100 \text{ cm}}$$ Multiply the numerators and the denominators.
Divide out the common units of centimeters in the numerator and demoninator.

$$= \frac{80}{100} \text{ m}$$ The units of centimeters divide out.

$$= 0.8 \text{ m}$$ Divide by 100 by moving the understood decimal point in 80 2 places to the left.

3 m
80 cm
FIGURE 7-47

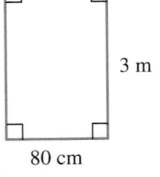
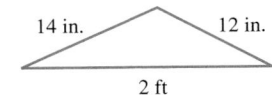

With the length and width of the rectangle now expressed in the same units (meters), we can substitute 3 for *l* and 0.8 for *w* to find its perimeter.

$P = 2l + 2w$

$P = 2(3) + 2(0.8)$

$\quad = 6 + 1.6$

$\quad = 7.6$

The perimeter is 7.6 meters.

Answer: 50 in.   ■

EXAMPLE 4   *Finding the base of an isosceles triangle.*   The perimeter of the isosceles triangle in Figure 7-48 is 50 meters. Find the length of its base.

Solution

Two sides are 12 meters long, and the perimeter is 50 meters. If *x* represents the length of the base, we have

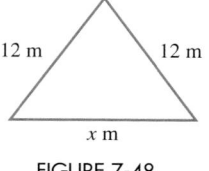

12 m   12 m

x m

FIGURE 7-48

Self Check

The perimeter of an isosceles triangle is 60 meters. If one of its sides of equal length is 15 meters long, how long is its base?

$12 + 12 + x = 50$

$\quad\ 24 + x = 50$ Simplify: $12 + 12 = 24$.

$\qquad\quad\ \ x = 26$ To undo the addition of 24, subtract 24 from both sides.

The length of the base is 26 meters.

Answer: 30 m   ■

Perimeters of figures that are combinations of polygons

Accent on Technology: **Perimeter of a figure**

See Figure 7-49. To find the perimeter, we need to know the values of *x* and *y*. Since the figure is a combination of two rectangles, we can use a calculator to see that

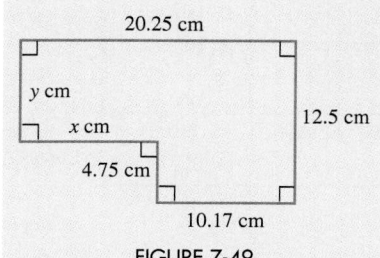

20.25 cm

y cm

x cm

4.75 cm

12.5 cm

10.17 cm

FIGURE 7-49

$x = 20.25 - 10.17$ and $y = 12.5 - 4.75$

$\ \ = 10.08$ $\ \ = 7.75$

The perimeter *P* of the figure is

$P = 20.25 + 12.5 + 10.17 + 4.75 + x + y$

$P = 20.25 + 12.5 + 10.17 + 4.75 + \textbf{10.08} + \textbf{7.75}$

We can use a calculator to evaluate the expression on the right-hand side by entering these numbers and pressing these keys.

Keystrokes   20.25 ⊞ 12.5 ⊞ 10.17 ⊞ 4.75 ⊞ 10.08 ⊞ 7.75 ⊟

65.5

The perimeter is 65.5 centimeters.

Areas of polygons

Recall that the **area** of a polygon is the measure of the amount of surface it encloses. Area is measured in square units, such as square inches or square centimeters. See Figure 7-50.

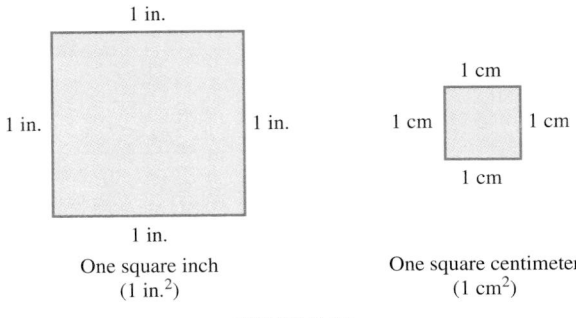

One square inch
(1 in.^{2})

One square centimeter
(1 cm^2)

FIGURE 7-50

In everyday life, we often use areas. For example,

- To carpet a room, we buy square yards.
- A can of paint will cover a certain number of square feet.
- To measure vast amounts of land, we often use square miles.
- We buy house roofing by the "square." One square is 100 square feet.

The rectangle shown in Figure 7-51 has a length of 10 centimeters and a width of 3 centimeters. If we divide the rectangle into squares as shown in the figure, each square represents an area of 1 square centimeter—a surface enclosed by a square measuring 1 centimeter on each side. Because there are 3 rows with 10 squares in each row, there are 30 squares. Since the rectangle encloses a surface area of 30 squares, its area is 30 square centimeters, often written as 30 cm^2.

This example illustrates that to find the area of a rectangle, we multiply its length by its width.

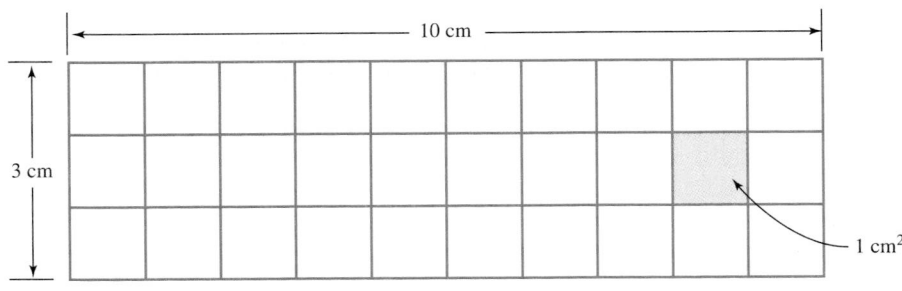

FIGURE 7-51

 COMMENT   Do not confuse the concepts of perimeter and area. Perimeter is the distance around a polygon. It is measured in linear units, such as centimeters, feet, or miles. Area is a measure of the surface enclosed within a polygon. It is measured in square units, such as square centimeters, square feet, or square miles.

In practice, we do not find areas by counting squares in a figure. Instead, we use formulas for finding areas of geometric figures, as shown in Table 7-1.

Figure	Name	Formula for area
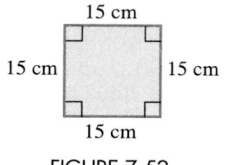 (square with sides s)	Square	$A = s^2$, where s is the length of one side.
(rectangle with length l, width w)	Rectangle	$A = lw$, where l is the length and w is the width.
(parallelogram with base b, height h)	Parallelogram	$A = bh$, where b is the length of the base and h is the height. (A height is always perpendicular to the base.)
(triangles with base b, height h)	Triangle	$A = \frac{1}{2}bh$, where b is the length of the base and h is the height. The segment perpendicular to the base and representing the height is called an **altitude.**
(trapezoid with bases b_2, b_1, height h)	Trapezoid	$A = \frac{1}{2}h(b_1 + b_2)$, where h is the height of the trapezoid and b_1 and b_2 represent the lengths of the bases.

TABLE 7-1

EXAMPLE 5   *Area of a square.* Find the area of the square in Figure 7-52.

Solution
We can see that the length of one side of the square is 15 centimeters. We can find its area by using the formula $A = s^2$ and substituting 15 for s.

$A = s^2$

$A = (15)^2$ Substitute 15 for s.

$A = 225$ Evaluate the exponential expression: $15 \cdot 15 = 225$.

The area of the square is 225 cm^2.

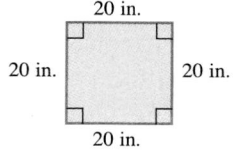

FIGURE 7-52

Self Check
Find the area of the square shown below.

(square with sides 20 in.)

Answer: 400 in.^{2}

EXAMPLE 6   *Number of square feet in 1 square yard.* Find the number of square feet in 1 square yard. (See Figure 7-53.)

Solution
Since 3 feet = 1 yard, each side of 1 square yard is 3 feet long.

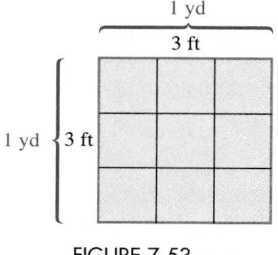

FIGURE 7-53

Self Check
Find the number of square centimeters in 1 square meter.

$$1 \text{ yd}^2 = (\mathbf{1 \ yd})^2$$
$$= (\mathbf{3 \ ft})^2 \quad \text{Substitute 3 feet for 1 yard.}$$
$$= 9 \text{ ft}^2 \quad (3 \text{ ft})^2 = (3 \text{ ft})(3 \text{ ft}) = 9 \text{ ft}^2.$$

There are 9 square feet in 1 square yard.

EXAMPLE 7   *Women's sports.* Field hockey is a team sport in which players use sticks to try to hit a ball into their opponents' goal. Find the area of the rectangular field shown in Figure 7-54. Give the answer in square feet.

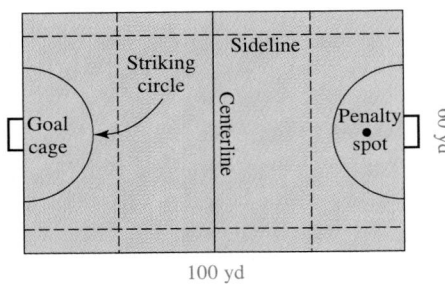

FIGURE 7-54

Solution
To find the area in square yards, we substitute 100 for l and 60 for w in the formula for the area of a rectangle, and simplify.

$$A = lw$$
$$A = 100(60)$$
$$= 6,000$$

The area is 6,000 square yards. Since there are 9 square feet per square yard, we can convert this number to square feet by multiplying by 1, written in the form $\frac{9 \text{ ft}^2}{1 \text{ yd}^2}$.

$$6,000 \text{ yd}^2 = \frac{6,000 \text{ yd}^2}{1} \cdot \frac{9 \text{ ft}^2}{1 \text{ yd}^2} \quad \begin{array}{l} \text{Write 6,000 yd}^2 \text{ as a fraction: } 6,000 \text{ yd}^2 = \frac{6,000 \text{ yd}^2}{1}. \\ \text{Multiply by 1: } \frac{9 \text{ ft}^2}{1 \text{ yd}^2} = 1. \end{array}$$

$$= \frac{6,000 \overset{1}{\cancel{\text{yd}^2}} \cdot 9 \text{ ft}^2}{1 \cdot 1 \underset{1}{\cancel{\text{yd}^2}}} \quad \begin{array}{l} \text{Multiply the numerators and the denominators.} \\ \text{Divide out the common units of square yards in the} \\ \text{numerator and demoninator.} \end{array}$$

$$= 6,000 \cdot 9 \text{ ft}^2$$
$$= 54,000 \text{ ft}^2 \quad \text{Do the multiplication.}$$

The area of the field is $54,000 \text{ ft}^2$.

EXAMPLE 8   *Area of a parallelogram.*
Find the area of the parallelogram in Figure 7-55.

Solution
The length of the base of the parallelogram is

$$5 \text{ feet} + 25 \text{ feet} = 30 \text{ feet}$$

The height is 12 feet. To find the area, we substitute 30 for b and 12 for h in the formula for the area of a parallelogram and simplify.

$$A = bh$$
$$A = 30(12)$$
$$= 360$$

The area of the parallelogram is 360 ft^2.

12 ft

5 ft   25 ft

FIGURE 7-55

Answer: $10,000 \text{ cm}^2$ ■

Self Check
Find the area in square inches of a rectangle with dimensions of 6 inches by 2 feet.

Answer: 144 in.^2 ■

Self Check
Find the area of the parallelogram below.

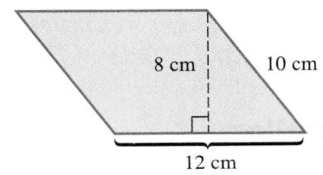

Answer: 96 cm^2 ■

EXAMPLE 9 *Area of a triangle.* Find the area of the triangle in Figure 7-56.

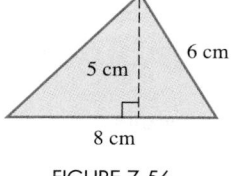

Self Check

Find the area of the triangle below.

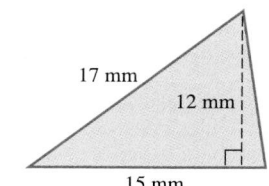

Solution

We substitute 8 for b and 5 for h in the formula for the area of a triangle, and simplify. (The side having length 6 cm is additional information that is not used to find the area.)

$$A = \frac{1}{2}bh$$

$$A = \frac{1}{2}(8)(5) \quad \text{The length of the base is 8 cm. The height is 5 cm.}$$

$$= 4(5) \quad \text{Do the multiplication: } \tfrac{1}{2}(8) = 4.$$

$$= 20$$

The area of the triangle is 20 cm^2.

Answer: 90 mm^2

EXAMPLE 10 *Area of a triangle.* Find the area of the triangle in Figure 7-57.

Solution In this case, the altitude falls outside the triangle.

$$A = \frac{1}{2}bh$$

$$A = \frac{1}{2}(9)(13) \quad \begin{array}{l}\text{Substitute 9 for } b \text{ and 13}\\ \text{for } h.\end{array}$$

$$= \frac{1}{2}\left(\frac{9}{1}\right)\left(\frac{13}{1}\right) \quad \text{Write 9 as } \tfrac{9}{1} \text{ and 13 as } \tfrac{13}{1}.$$

$$= \frac{117}{2} \quad \text{Multiply the fractions.}$$

$$= 58.5 \quad \text{Do the division.}$$

The area of the triangle is 58.5 cm^2.

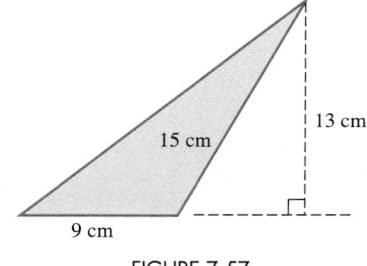

FIGURE 7-57

EXAMPLE 11 *Area of a trapezoid.* Find the area of the trapezoid in Figure 7-58.

Solution

In this example, $b_1 = 10$ and $b_2 = 6$. It is incorrect to say that $h = 1$, because the height of 1 foot must be expressed as 12 inches to be consistent with the units of the bases. Thus, we substitute 10 for b_1, 6 for b_2, and 12 for h in the formula for finding the area of a trapezoid and simplify.

$$A = \frac{1}{2}h(b_1 + b_2)$$

$$A = \frac{1}{2}(12)(10 + 6) \quad \begin{array}{l}\text{The length of the lower base is 10 in. The length of the}\\ \text{upper base is 6 in. The height is 12 in.}\end{array}$$

$$= \frac{1}{2}(12)(16) \quad \text{Do the addition within the parentheses.}$$

$$= 6(16) \quad \text{Do the multiplication: } \tfrac{1}{2}(12) = 6.$$

$$= 96$$

The area of the trapezoid is 96 in.^{2}.

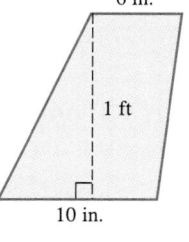

FIGURE 7-58

Self Check

Find the area of the trapezoid below.

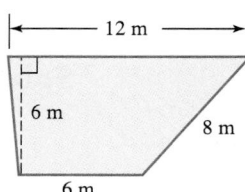

Answer: 54 m^2

Areas of figures that are combinations of polygons

EXAMPLE 12   *Carpeting a room.*   A living room/dining room area has the floor plan shown in Figure 7-59. If carpet costs $29 per square yard, including pad and installation, how much will it cost to carpet the room? (Assume no waste.)

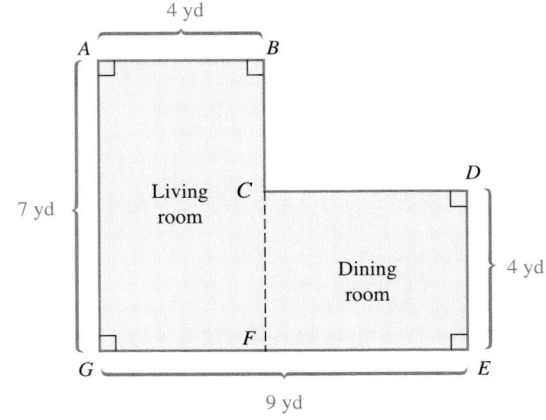

FIGURE 7-59

Solution   First we must find the total area of the living room and the dining room:

$$A_{\text{total}} = A_{\text{living room}} + A_{\text{dining room}}$$

Since $\overline{CF}$ divides the space into two rectangles, the areas of the living room and the dining room are found by multiplying their respective lengths and widths.

$$\begin{aligned} \text{Area of living room} &= lw \\ &= 7(4) \\ &= 28 \end{aligned}$$

The area of the living room is 28 yd².
    To find the area of the dining room, we find its length by subtracting 4 yards from 9 yards to obtain 5 yards. We note that its width is 4 yards.

$$\begin{aligned} \text{Area of dining room} &= lw \\ &= 5(4) \\ &= 20 \end{aligned}$$

The area of the dining room is 20 yd².
    The total area to be carpeted is the sum of these two areas.

$$\begin{aligned} A_{\text{total}} &= A_{\text{living room}} + A_{\text{dining room}} \\ A_{\text{total}} &= 28 \text{ yd}^2 + 20 \text{ yd}^2 \\ &= 48 \text{ yd}^2 \end{aligned}$$

At $29 per square yard, the cost to carpet the room will be 48 · $29, or $1,392.   ■

EXAMPLE 13   *Area of one side of a tent.*   Find the area of one side of the tent in Figure 7-60.

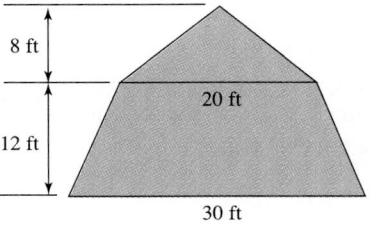

FIGURE 7-60

Solution   Each side is a combination of a trapezoid and a triangle. Since the bases of each trapezoid are 30 feet and 20 feet and the height is 12 feet, we substitute 30 for b_1, 20 for b_2, and 12 for h into the formula for the area of a trapezoid.

$$A_{\text{trap.}} = \frac{1}{2}h(b_1 + b_2)$$

$$A_{\text{trap.}} = \frac{1}{2}(12)(30 + 20)$$

$$= 6(50)$$

$$= 300$$

The area of the trapezoid is 300 ft².

Since the triangle has a base of 20 feet and a height of 8 feet, we substitute 20 for b and 8 for h in the formula for the area of a triangle.

$$A_{\text{triangle}} = \frac{1}{2}bh$$

$$A_{\text{triangle}} = \frac{1}{2}(20)(8)$$

$$= 80$$

The area of the triangle is 80 ft^2.
The total area of one side of the tent is

$$A_{\text{total}} = A_{\text{trap.}} + A_{\text{triangle}}$$
$$A_{\text{total}} = \textbf{300 ft}^2 + \textbf{80 ft}^2$$
$$= \textbf{380 ft}^2$$

The total area is 380 ft^2.

STUDY SET   Section 7.5

VOCABULARY   *Fill in the blanks.*

1. The distance around a polygon is called the
_____.

2. The perimeter of a polygon is measured in
_____ units.

3. The measure of the surface enclosed by a polygon is
called its _____.

4. If each side of a square measures 1 foot, the area enclosed by the square is 1 _____ foot.

5. The area of a polygon is measured in _____
units.

6. The segment that represents the height of a triangle is
called an _____.

CONCEPTS   *Sketch and label each of the figures described.*

7. Two different rectangles, each having a perimeter of
40 in.

8. Two different rectangles, each having an area of
40 in^2.

9. A square with an area of 25 m^2.

10. A square with a perimeter of 20 m.

11. A parallelogram with an area of 15 yd^2.

12. A triangle with an area of 20 ft^2.

13. A figure consisting of a combination of two rectangles
whose total area is 80 ft^2.

14. A figure consisting of a combination of a rectangle and
a square whose total area is 164 ft^2.

NOTATION   *Fill in the blanks.*

15. The formula for the perimeter of a square is
_____.

16. The formula for the perimeter of a rectangle is
_____.

17. The symbol 1 in.^{2} means one _____.

18. One square meter is expressed as _____.

19. The formula for the area of a square is
_____.

20. The formula for the area of a rectangle is
_____.

21. The formula $A = \frac{1}{2}bh$ gives the area of a
_____.

22. The formula $A = \frac{1}{2}h(b_1 + b_2)$ gives the area of a
_____.

PRACTICE   *Find the perimeter of each figure.*

23.

8 in.

8 in.   8 in.

8 in.

24.

12 cm

6 cm   6 cm

12 cm

25.

6 m

4 m

2 m

10 m

2 m

4 m

6 m

26.

5 in.

5 in.   5 in.

4 in.   4 in.

27.

6 cm

6 cm

7 cm   8 cm

10 cm

28.

2 cm   2 cm

7 cm

6 cm   6 cm

10 cm

Solve each problem.

29. Find the perimeter of an isosceles triangle with a base of length 21 centimeters and sides of length 32 centimeters.

30. The perimeter of an isosceles triangle is 80 meters. If the length of one side is 22 meters, how long is the base?

31. The perimeter of an equilateral triangle is 85 feet. Find the length of each side.

32. An isosceles triangle with sides of 49.3 inches has a perimeter of 121.7 inches. Find the length of the base.

Find the area of the shaded part of each figure.

33.

4 cm

4 cm

34.

3 in.

5 in.

35.

4 cm   6 cm

15 cm

36.

6 m   7 m

10 m

37.

5 in.

10 in.

38.

3 cm

9 cm

39.

9 mm

13 mm

17 mm

40.

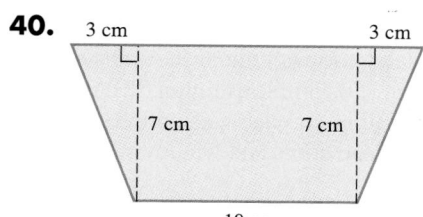

41.

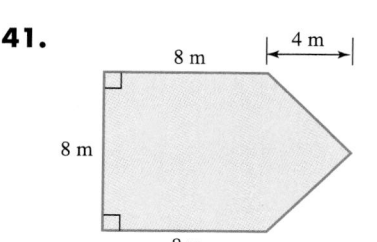

42.

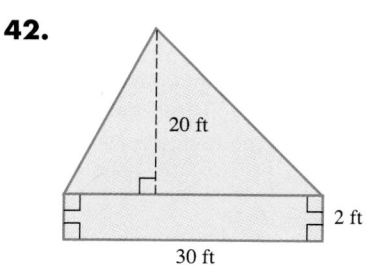

43.

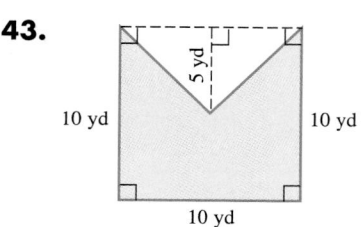

44.

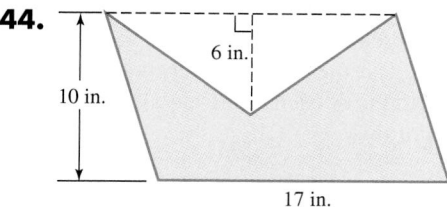

45.

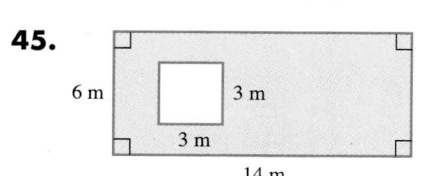

46.

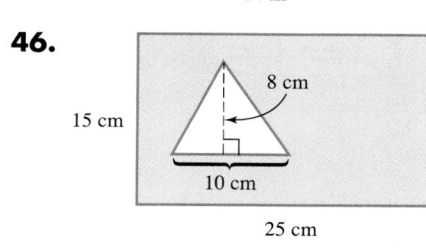

47. How many square inches are in 1 square foot?

48. How many square inches are in 1 square yard?

APPLICATIONS

49. FENCING A YARD   A man wants to enclose a rectangular yard with fencing that costs $12.50 a foot, including installation. Find the cost of enclosing the yard if its dimensions are 110 ft by 85 ft.

50. FRAMING A PICTURE   Find the cost of framing a rectangular picture with dimensions of 24 inches by 30 inches if framing material costs $8.46 per foot, including matting.

51. PLANTING A SCREEN   A woman wants to plant a pine-tree screen around three sides of her backyard. (See Illustration 1.) If she plants the trees 3 feet apart, how many trees will she need?

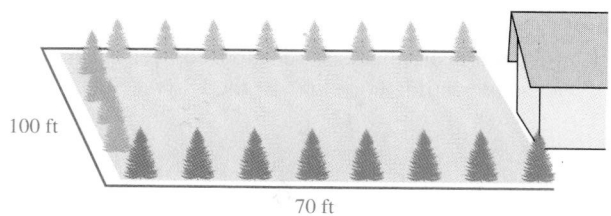

ILLUSTRATION 1

52. PLANTING MARIGOLDS   A gardener wants to plant a border of marigolds around the garden shown in Illustration 2, to keep out rabbits. How many plants will she need if she allows 6 inches between plants?

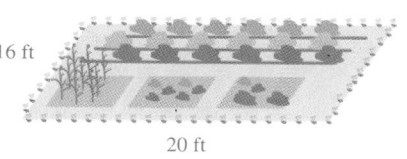

ILLUSTRATION 2

53. BUYING A FLOOR   Which is more expensive: A ceramic-tile floor costing $3.75 per square foot or linoleum costing $34.95 per square yard?

54. BUYING A FLOOR   Which is cheaper: A hardwood floor costing $5.95 per square foot or a carpeted floor costing $37.50 per square yard?

55. CARPETING A ROOM   A rectangular room is 24 feet long and 15 feet wide. At $30 per square yard, how much will it cost to carpet the room? (Assume no waste.)

56. CARPETING A ROOM   A rectangular living room measures 30 by 18 feet. At $32 per square yard, how much will it cost to carpet the room? (Assume no waste.)

57. TILING A FLOOR   A rectangular basement room measures 14 by 20 feet. Vinyl floor tiles that are 1 ft^2 cost $1.29 each. How much will the tile cost to cover the floor? (Disregard any waste.)

58. PAINTING A BARN   The north wall of a barn is a rectangle 23 feet high and 72 feet long. There are five windows in the wall, each 4 by 6 feet. If a gallon of paint will cover 300 ft², how many gallons of paint must the painter buy to paint the wall?

59. MAKING A SAIL   If nylon is $12 per square yard, how much would the fabric cost to make a triangular sail with a base of 12 feet and a height of 24 feet?

60. PAINTING A GABLE   The gable end of a warehouse is an isosceles triangle with a height of 4 yards and a base of 23 yards. It will require one coat of primer and one coat of finish to paint the triangle. Primer costs $17 per gallon, and the finish paint costs $23 per gallon. If one gallon covers 300 square feet, how much will it cost to paint the gable, excluding labor?

61. GEOGRAPHY   See Illustration 3. Use the dimensions of the trapezoid that is superimposed over the state of Nevada to estimate the area of the "Silver State."

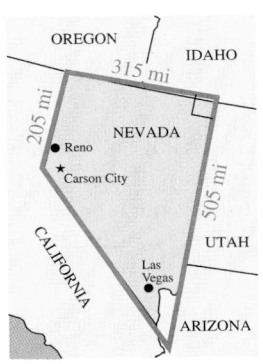

ILLUSTRATION 3

62. COVERING A SWIMMING POOL   A swimming pool has the shape shown in Illustration 4. How many square meters of plastic sheeting will be needed to cover the pool? How much will the sheeting cost if it is $2.95 per square meter? (Assume no waste.)

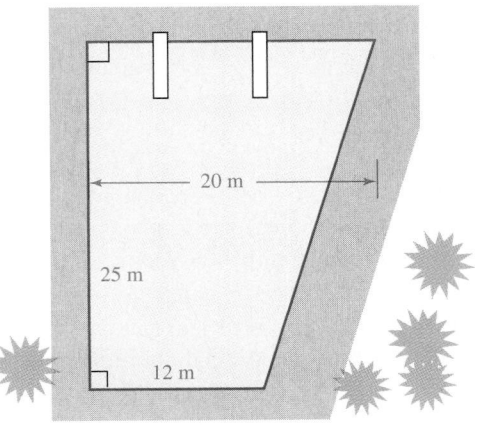

ILLUSTRATION 4

63. CARPENTRY   How many sheets of 4-foot-by-8-foot sheetrock are needed to drywall the inside walls on the first floor of the barn shown in Illustration 5? (Assume that the carpenters will cover each wall entirely and then cut out areas for the doors and windows.)

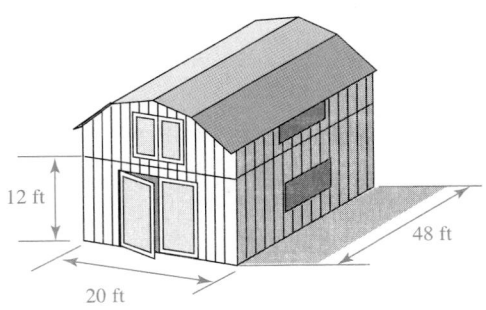

ILLUSTRATION 5

64. CARPENTRY   If it costs $90 per square foot to build a one-story home in northern Wisconsin, estimate the cost of building the house with the floor plan shown in Illustration 6.

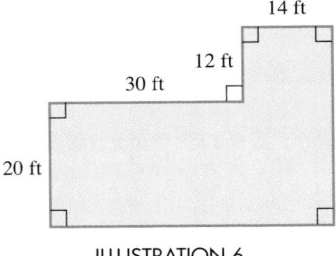

ILLUSTRATION 6

65. DRIVING SAFETY   Illustration 7 shows the areas on a highway that a truck driver cannot see in the truck's rear view mirrors. Use the scale to determine the approximate dimensions of each blind spot. Then estimate the area of each of them.

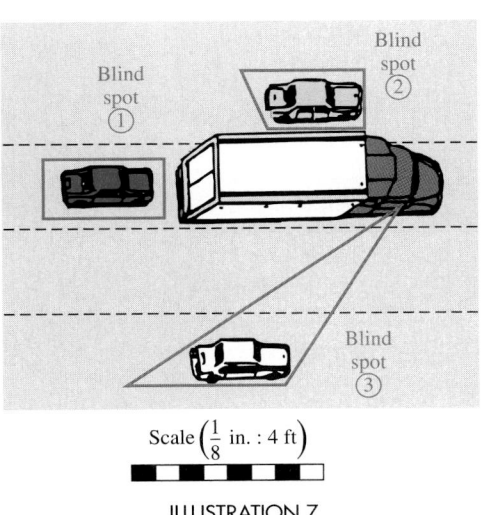

Scale $\left(\frac{1}{8}\text{ in. : 4 ft}\right)$

ILLUSTRATION 7

66. ESTIMATING AREA   See Illustration 8. Estimate the area of the sole plate of the iron by thinking of it as a combination of a trapezoid and a triangle.

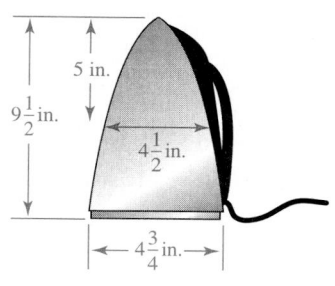

ILLUSTRATION 8

WRITING

67. Explain the difference between perimeter and area.

68. Why is it necessary that area be measured in square units?

REVIEW   *Do the calculations. Write all improper fractions as mixed numbers.*

69. $\dfrac{3}{4} + \dfrac{2}{3}$

70. $\dfrac{7}{8} - \dfrac{2}{3}$

71. $3\dfrac{3}{4} + 2\dfrac{1}{3}$

72. $7\dfrac{5}{8} - 2\dfrac{5}{6}$

73. $7\dfrac{1}{2} \div 5\dfrac{2}{5}$

74. $5\dfrac{3}{4} \cdot 2\dfrac{5}{6}$

7.6 Circles

In this section, you will learn about

- Circles • Circumference of a circle • Area of a circle

INTRODUCTION. In this section, we will discuss circles, one of the most useful geometric figures. In fact, the discovery of fire and the circular wheel were two of the most important events in the history of the human race.

Circles

Circle

> A **circle** is the set of all points in a plane that lie a fixed distance from a point called its **center.**

A segment drawn from the center of a circle to a point on the circle is called a **radius.** (The plural of *radius* is *radii.*) From the definition, it follows that all radii of the same circle are the same length.

A **chord** of a circle is a line segment connecting two points on the circle. A **diameter** is a chord that passes through the center of the circle. Since a diameter D of a circle is twice as long as a radius r, we have

$$D = 2r$$

Each of the previous definitions is illustrated in Figure 7-61, in which O is the center of the circle.

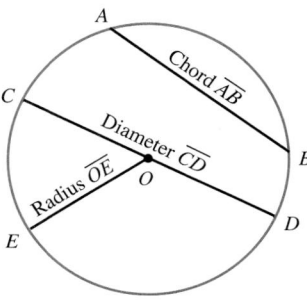

FIGURE 7-61

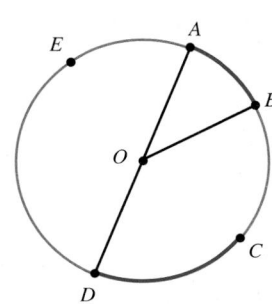

FIGURE 7-62

Any part of a circle is called an **arc**. In Figure 7-62 (on the preceding page), the part of the circle from point *A* to point *B* is $\overset{\frown}{AB}$, read as "arc *AB*." $\overset{\frown}{CD}$ is the part of the circle from point *C* to point *D*. An arc that is half of a circle is a **semicircle**.

Semicircle

A **semicircle** is an arc of a circle whose endpoints are the endpoints of a diameter.

If point *O* is the center of the circle in Figure 7-62, $\overline{AD}$ is a diameter and $\overset{\frown}{AED}$ is a semicircle. The middle letter *E* is used to distinguish semicircle $\overset{\frown}{AED}$ from semicircle $\overset{\frown}{ABCD}$.

An arc that is shorter than a semicircle is a **minor arc.** An arc that is longer than a semicircle is a **major arc.** In Figure 7-62,

$\overset{\frown}{AB}$ is a minor arc   and $\overset{\frown}{ABCDE}$ is a major arc

Circumference of a circle

Since early history, mathematicians have known that the ratio of the distance around a circle (the circumference) divided by the length of its diameter is approximately 3. First Kings, Chapter 7 of the Bible describes a round bronze tank that was 15 feet from brim to brim and 45 feet in circumference, and $\frac{45}{15} = 3$. Today, we have a better value for this ratio, known as π (pi). If *C* is the circumference of a circle and *D* is the length of its diameter, then

$$\pi = \frac{C}{D},\qquad \text{where } \pi = 3.141592653589\ldots \quad \begin{smallmatrix}\frac{22}{7}\ \text{and } 3.14 \text{ are often used as}\\ \text{estimates of } \pi.\end{smallmatrix}$$

If we multiply both sides of $\pi = \frac{C}{D}$ by *D*, we have the following formula.

Circumference of a circle

The circumference of a circle is given by the formula
$$C = \pi D \quad \text{where } C \text{ is the circumference and } D \text{ is the length of the diameter}$$

Since a diameter of a circle is twice as long as a radius *r*, we can substitute $2r$ for *D* in the formula $C = \pi D$ to obtain another formula for the circumference *C*:

$$C = 2\pi r$$

EXAMPLE 1 *Circumference of a circle.* Find the circumference of a circle that has a diameter of 10 centimeters. (See Figure 7-63.)

Solution

We substitute 10 for *D* in the formula for the circumference of a circle.

$C = \pi D$

$C = \pi(10)$

$C \approx 3.14(10)$ Replace π with an approximation: $\pi \approx 3.14$.

$C \approx 31.4$

The circumference is approximately 31.4 centimeters.

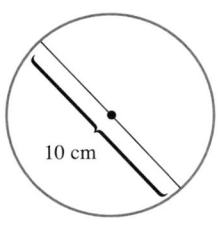

10 cm

FIGURE 7-63

Self Check

To the nearest tenth, find the circumference of a circle that has a radius of 12 meters.

Answer: 75.4 m

Accent on Technology: *Calculating revolutions of a tire*

When the $\boxed{\pi}$ key on a scientific calculator is pressed (on some models, the $\boxed{2nd}$ key must be pressed first), an approximation of π is displayed. To illustrate how to use this key, consider the following problem. How many times does a 15-inch tire revolve when a car makes a 25-mile trip?

We first find the circumference of the tire.

$C = \pi D$

$C = \pi(15)$ Substitute 15 for D, the diameter of the tire.

$C = 15\pi$ Normally, we rewrite a product such as $\pi(15)$ so that π is the second factor.

The circumference of the tire is 15π inches.

We then change 25 miles to inches by multiplying by 1, written in the form $\dfrac{5{,}280 \text{ feet}}{1 \text{ mile}}$, and by 1, written in the form $\dfrac{12 \text{ inches}}{1 \text{ foot}}$.

$$\frac{25}{1} \text{ miles} \cdot \frac{5{,}280 \text{ feet}}{1 \text{ mile}} \cdot \frac{12 \text{ inches}}{1 \text{ foot}} = 25(5{,}280)(12) \text{ inches} \quad \text{The units of miles and feet divide out.}$$

The total distance of the trip is $25(5{,}280)(12)$ inches.

Finally, we divide the total distance of the trip by the circumference of the tire to get

$$\text{The number of revolutions of the tire} = \frac{25(5{,}280)(12)}{15\pi}$$

To do this work using a scientific calculator, we enter these numbers and press these keys.

Keystrokes $\boxed{(}$ 25 $\boxed{\times}$ 5280 $\boxed{\times}$ 12 $\boxed{)}$ $\boxed{\div}$ $\boxed{(}$ 15 $\boxed{\times}$ $\boxed{\pi}$ $\boxed{)}$ $\boxed{=}$

$$\boxed{33613.52398}$$

The tire makes about 33,614 revolutions.

EXAMPLE 2   *Architecture.*   A Norman window is constructed by adding a semicircular window to the top of a rectangular window. Find the perimeter of the Norman window shown in Figure 7-64.

Solution   The window is a combination of a rectangle and a semicircle. The perimeter of the rectangular part is

$$P_{\text{rectangular part}} = 8 + 6 + 8 = 22 \quad \text{Add only 3 sides.}$$

The perimeter of the semicircle is one-half of the circumference of a circle that has a 6-meter diameter.

$$P_{\text{semicircle}} = \frac{1}{2}\pi D$$

$$= \frac{1}{2}\pi(6) \qquad \text{Substitute 6 for } D.$$

$$\approx 9.424777961 \quad \text{Use a calculator.}$$

The total perimeter is the sum of the two parts.

$$P_{\text{total}} \approx 22 + 9.424777961$$

$$\approx 31.424777961$$

To the nearest hundredth, the perimeter of the window is 31.42 meters.

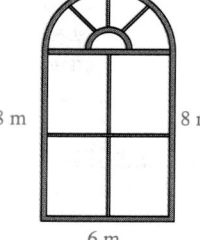

8 m   8 m

6 m

FIGURE 7-64

Area of a circle

If we divide the circle shown in Figure 7-65(a) into an even number of pie-shaped pieces and then rearrange them as shown in Figure 7-65(b), we have a figure that looks like a parallelogram. The figure has a base that is one-half the circumference of the circle, and its height is about the same length as a radius of the circle.

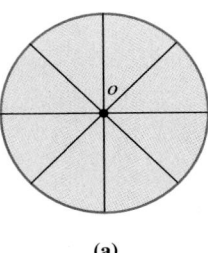

(a)

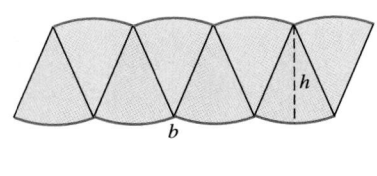

(b)

FIGURE 7-65

If we divide the circle into more and more pie-shaped pieces, the figure will look more and more like a parallelogram, and we can find its area by using the formula for the area of a parallelogram.

$$A = bh$$

$$= \frac{1}{2}Cr \qquad \text{Substitute } \frac{1}{2} \text{ of the circumference for } b, \text{ and } r \text{ for the height.}$$

$$= \frac{1}{2}(2\pi r)r \quad \text{Make a substitution: } C = 2\pi r.$$

$$= \pi r^2 \qquad \text{Simplify: } \frac{1}{2} \cdot 2 = 1 \text{ and } r \cdot r = r^2.$$

Area of a circle	The **area of a circle** with radius r is given by the formula $$A = \pi r^2$$

EXAMPLE 3 *Area of a circle.* To the nearest tenth, find the area of the circle in Figure 7-66.

Solution
Since the length of the diameter is 10 centimeters and the length of a diameter is twice the length of a radius, the length of the radius is 5 centimeters. To find the area of the circle, we substitute 5 for r in the formula for the area of a circle.

$$A = \pi r^2$$

$$A = \pi(5)^2$$

$$= 25\pi$$

$$\approx 78.53981634 \quad \text{Use a calculator.}$$

To the nearest tenth, the area is 78.5 cm^2.

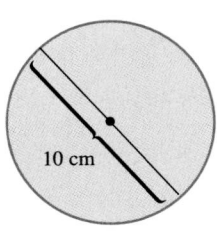

10 cm

FIGURE 7-66

Self Check
To the nearest tenth, find the area of a circle with a diameter of 12 feet.

Answer: 113.1 ft^2 ■

Accent on Technology: *Painting a helicopter pad*

Orange paint is available in gallon containers at $19 each, and each gallon will cover 375 ft². To calculate how much the paint will cost to cover a circular helicopter pad 60 feet in diameter, we first calculate the area of the helicopter pad.

$$A = \pi r^2$$
$$A = \pi(30)^2 \quad \text{Substitute one-half of 60 for } r.$$
$$= 30^2 \pi$$

The area of the pad is $30^2\pi$ ft². Since each gallon of paint will cover 375 ft², we can find the number of gallons of paint needed by dividing $30^2\pi$ by 375.

$$\text{Number of gallons needed} = \frac{30^2\pi}{375}$$

To do this work on a calculator, we enter these numbers and press these keys.

Keystrokes 30 $\boxed{x^2}$ $\boxed{\times}$ $\boxed{\pi}$ $\boxed{=}$ $\boxed{\div}$ 375 $\boxed{=}$ $\boxed{7.539822369}$

Because paint comes only in full gallons, the painter will need to purchase 8 gallons. The cost of the paint will be 8($19), or $152.

EXAMPLE 4   *Finding the area.*   Find the shaded area in Figure 7-67.

Solution   The figure is a combination of a triangle and two semicircles. By the Pythagorean theorem, the hypotenuse h of the right triangle is

$$h = \sqrt{6^2 + 8^2} = \sqrt{36 + 64} = \sqrt{100} = 10$$

The area of the triangle is

$$A_{\text{right triangle}} = \frac{1}{2}bh = \frac{1}{2}(6)(8) = 3(8) = 24$$

The area enclosed by the smaller semicircle is

$$A_{\text{smaller semicircle}} = \frac{1}{2}\pi r^2 = \frac{1}{2}\pi(4)^2 = \frac{1}{2}\pi(16) = 8\pi$$

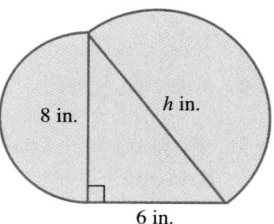

8 in. h in.

6 in.

FIGURE 7-67

The area enclosed by the larger semicircle is

$$A_{\text{larger semicircle}} = \frac{1}{2}\pi r^2 = \frac{1}{2}\pi(5)^2 = \frac{1}{2}\pi(25) = 12.5\pi$$

The total area is

$$A_{\text{total}} = 24 + 8\pi + 12.5\pi \approx 88.4026494 \quad \text{Use a calculator.}$$

To the nearest hundredth, the area is 88.40 in.².   ∎

STUDY SET   Section 7.6

VOCABULARY   *Fill in the blanks.*

1. A segment drawn from the center of a circle to a point on the circle is called a _____.

2. A segment joining two points on a circle is called a _____.

3. A _____ is a chord that passes through the center of a circle.

4. An arc that is one-half of a complete circle is a _____.

5. An arc that is shorter than a semicircle is called a _____ arc.

6. An arc that is longer than a semicircle is called a _____ arc.

7. The distance around a circle is called its _____.

8. The surface enclosed by a circle is called its _____.

CONCEPTS *Refer to Illustration 1.*

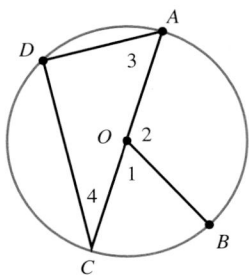

ILLUSTRATION 1

9. Name each radius.

10. Name a diameter.

11. Name each chord.

12. Name each minor arc.

13. Name each semicircle.

14. Name each major arc.

15. If you know the radius of a circle, how can you find its diameter?

16. If you know the diameter of a circle, how can you find its radius?

17. Suppose the two "legs" of the compass shown in Illustration 2 are adjusted so that the distance between the pointed ends is 1 inch. Then a circle is drawn.
   a. What will the radius of the circle be?
   b. What will the diameter of the circle be?
   c. What will the circumference of the circle be?

   d. What will the area of the circle be?

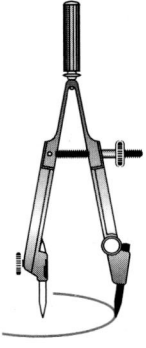

ILLUSTRATION 2

18. Suppose we find the distance around a can and the distance across the can using a measuring tape, as shown in Illustration 3. Then we make a comparison, in the form of a ratio:

$$\frac{\text{The distance around the can}}{\text{The distance across the top of the can}}$$

After we do the indicated division, the result will be close to what number?

ILLUSTRATION 3

19. When evaluating $\pi(6)^2$, what operation should be performed first?

20. Round $\pi = 3.141592653589.\ .\ .$ to the nearest hundredth.

NOTATION *Fill in the blanks.*

21. The symbol $\overset{\frown}{AB}$ is read as _____.

22. To the nearest hundredth, the value of π is _____.

23. The formula for the circumference of a circle is _____ or _____.

24. The formula $A = \pi r^2$ gives the area of a _____.

25. If C is the circumference of a circle and D is its diameter, then $\dfrac{C}{D} = \boxed{}$.

26. If D is the diameter of a circle and r is its radius, then $D = \boxed{}\, r$.

27. Write $\pi(8)$ in a better form.

28. What does $2\pi r$ mean?

PRACTICE *Solve each problem. Answers may vary slightly depending on which approximation of π is used.*

29. To the nearest hundredth, find the circumference of a circle that has a diameter of 12 inches.

30. To the nearest hundredth, find the circumference of a circle that has a radius of 20 feet.

31. Find the diameter of a circle that has a circumference of 36π meters.

32. Find the radius of a circle that has a circumference of 50π meters.

Find the perimeter of each figure to the nearest hundredth.

33.
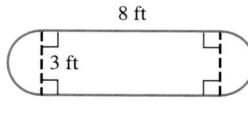
8 ft
3 ft

34.
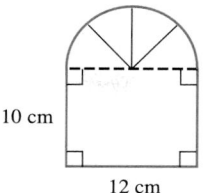
10 cm
12 cm

35.
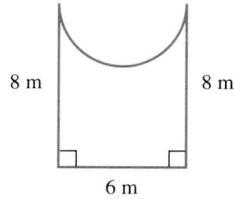
8 m   8 m
6 m

36.
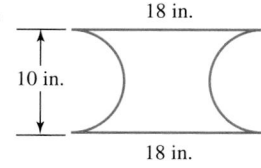
18 in.
10 in.
18 in.

Find the area of each circle to the nearest tenth.

37.
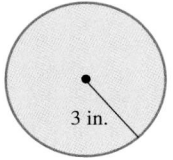
3 in.

38.
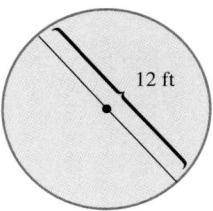
12 ft

Find the total area of each figure to the nearest tenth.

39.
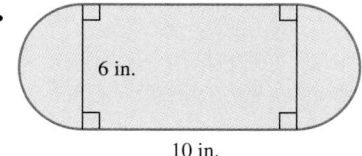
6 in.
10 in.

40.
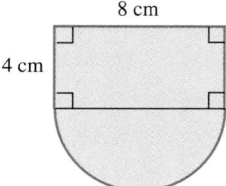
8 cm
4 cm

41.
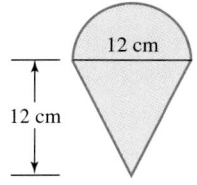
12 cm
12 cm

42.
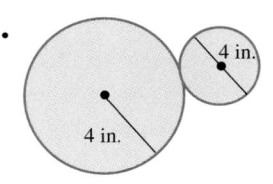
4 in.
4 in.
4 in.

Find the area of each shaded region to the nearest tenth.

43.
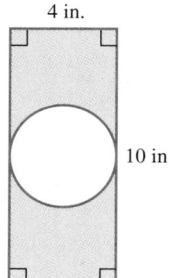
4 in.
10 in

44.
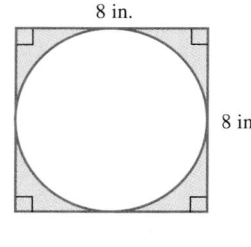
8 in.
8 in.

45.
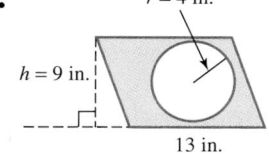
$r = 4$ in.
$h = 9$ in.
13 in.

46.
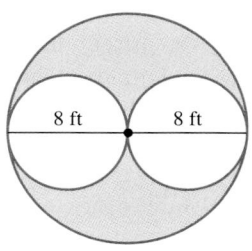
8 ft   8 ft

APPLICATIONS *Give each answer to the nearest hundredth. Answers may vary slightly depending on which approximation of π is used.*

47. AREA OF ROUND LAKE   Round Lake has a circular shoreline that is 2 miles in diameter. Find the area of the lake.

48. HELICOPTER   Refer to Illustration 4. How far does a point on the tip of a rotor blade travel when it makes one complete revolution?

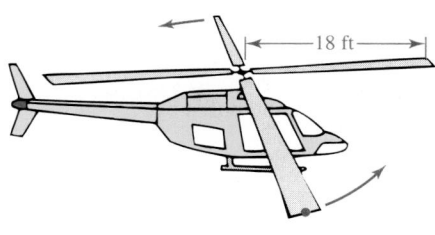
18 ft

ILLUSTRATION 4

49. GIANT SEQUOIA   The largest sequoia tree is the General Sherman Tree in Sequoia National Park in California. In fact, it is considered to be the largest living thing in the world. According to the *Guinness Book of World Records,* it has a circumference of 102.6 feet, measured $4\frac{1}{2}$ feet above the ground. What is the diameter of the tree at that height?

50. TRAMPOLINE   See Illustration 5. The distance from the center of the trampoline to the edge of its steel frame is 7 feet. The protective padding covering the springs is 15 inches wide. Find the area of the circular jumping surface of the trampoline, in square feet.

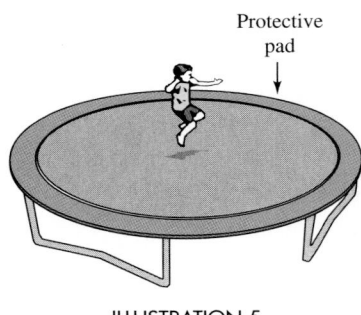

Protective
pad ↓

ILLUSTRATION 5

51. JOGGING   Joan wants to jog 10 miles on a circular track $\frac{1}{4}$ mile in diameter. How many times must she circle the track?

52. FIXING THE ROTUNDA   The rotunda at a state capitol is a circular area 100 feet in diameter. The legislature wishes to appropriate money to have the floor of the rotunda tiled. The lowest bid is $83 per square yard, including installation. How much must the legislature spend?

53. BANDING THE EARTH   A steel band is drawn tightly about the Earth's equator. The band is then loosened by increasing its length by 10 feet, and the resulting slack is distributed evenly along the band's entire length. How far above the Earth's surface is the band? (*Hint:* You don't need to know the Earth's circumference.)

54. CONCENTRIC CIRCLES   Two circles are called **concentric circles** if they have the same center. Find the area of the band between two concentric circles if their diameters are 10 centimeters and 6 centimeters.

55. ARCHERY   See Illustration 6. Find the area of the entire target and the bull's eye. What percent of the area of the target is the bull's eye?

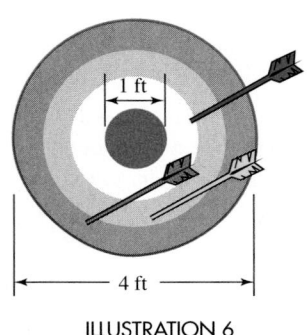

1 ft

4 ft

ILLUSTRATION 6

56. LANDSCAPE DESIGN   See Illustration 7. How much of the lawn does not get watered by the sprinklers at the center of each circle?

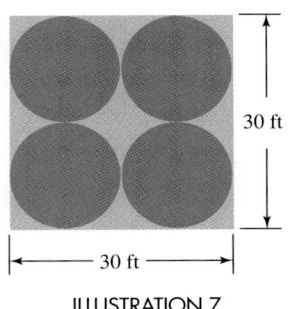

30 ft

30 ft

ILLUSTRATION 7

WRITING

57. Explain what is meant by the circumference of a circle.

58. Explain what is meant by the area of a circle.

59. Explain the meaning of π.

60. Distinguish between a major arc and a minor arc.

61. Explain what it means for a car to have a small turning radius.

62. The word *circumference* means the distance around a circle. In your own words, explain what is meant by each of the following sentences.
   a. A boat owner's dream was to *circumnavigate* the globe.
   b. The teenager's parents felt that he was always trying to *circumvent* the rules.
   c. The class was shown a picture of a circle *circumscribed* about an equilateral triangle.

REVIEW

63. Change $\frac{9}{10}$ to a percent.

64. Change $\frac{7}{8}$ to a percent.

65. How many sides does a pentagon have?

66. What is the sum of the measures of the angles of a triangle?

7.7 *Surface Area and Volume*

In this section, you will learn about

- Volumes of solids • Surface areas of rectangular solids
- Volumes and surface areas of spheres • Volumes of cylinders
- Volumes of cones • Volumes of pyramids

INTRODUCTION. In this section, we will discuss a measure of capacity called **volume.** Volumes are measured in cubic units, such as cubic inches, cubic yards, or cubic centimeters. For example,

- We buy gravel or topsoil by the cubic yard.
- We measure the capacity of a refrigerator in cubic feet.
- We often measure amounts of medicine in cubic centimeters.

We will also discuss surface area. The ability to compute surface area is necessary to solve problems such as calculating the amount of material necessary to make a cardboard box or a plastic beach ball.

Volumes of solids

A **rectangular solid** and a **cube** are two common geometric solids. (See Figure 7-68.)

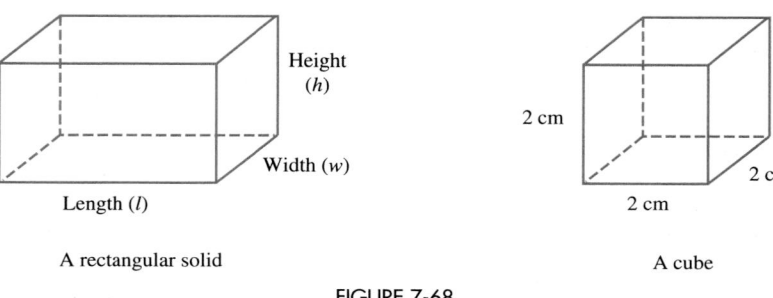

A rectangular solid   A cube

FIGURE 7-68

The **volume** of a rectangular solid is a measure of the space it encloses. Two common units of volume are cubic inches (in.^{3}) and cubic centimeters (cm^3). (See Figure 7-69.)

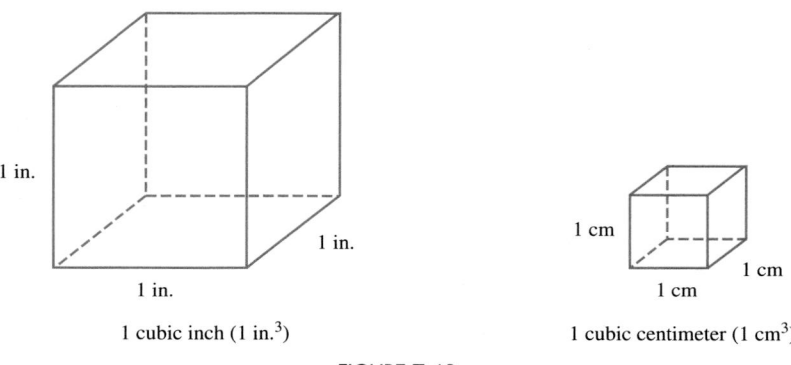

1 cubic inch (1 in.^{3})   1 cubic centimeter (1 cm^3)

FIGURE 7-69

If we divide the rectangular solid shown in Figure 7-70 into cubes, each cube represents a volume of 1 cm^3. Because there are 2 levels with 12 cubes on each level, the volume of the rectangular solid is 24 cm^3.

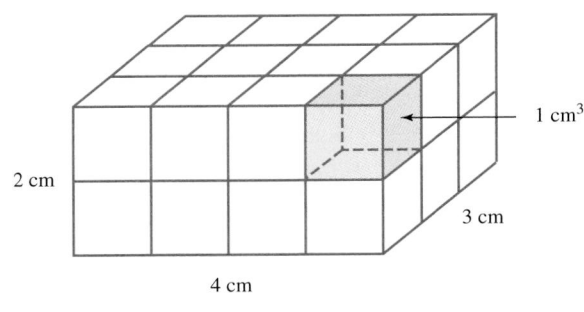

1 cm³
2 cm
3 cm
4 cm

FIGURE 7-70

In practice, we do not find volumes by counting cubes. Instead, we use the formulas shown in Table 7-2.

Figure	Name	Volume	Figure	Name	Volume
	Cube	$V = s^3$		Cylinder	$V = \pi r^2 h$ or $V = Bh$*
	Rectangular solid	$V = lwh$		Cone	$V = \dfrac{1}{3}\pi r^2 h$ or $V = \frac{1}{3}Bh$*
	Prism	$V = Bh$*		Pyramid	$V = \dfrac{1}{3}Bh$*
	Sphere	$V = \dfrac{4}{3}\pi r^3$			

*B represents the area of the base that is shaded in the figure.

TABLE 7-2

 COMMENT The height of a geometric solid is always measured along a line perpendicular to its base. In each of the solids in Figure 7-71, h is the height.

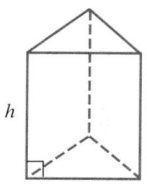 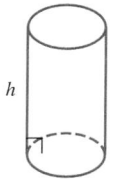 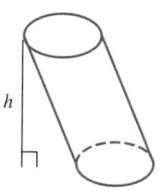 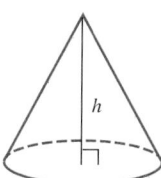 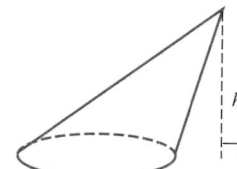

FIGURE 7-71

EXAMPLE 1 *Number of cubic inches in one cubic foot.* How many cubic inches are there in 1 cubic foot? (See Figure 7-72.)

Solution
Since a cubic foot is a cube with each side measuring 1 foot, each side also measures 12 inches. Thus, the volume in cubic inches is

$V = s^3$ The formula for the volume of a cube.

$V = (12)^3$ Substitute 12 for s.

    $= 1,728$

There are 1,728 cubic inches in 1 cubic foot.

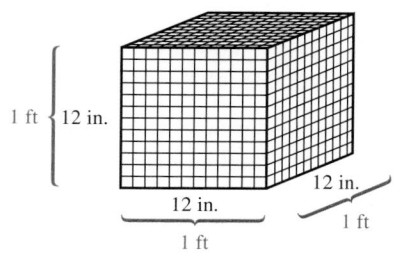

1 ft $\{$ 12 in.

12 in.

12 in.

12 in.

1 ft

1 ft

FIGURE 7-72

Self Check
How many cubic centimeters are in 1 cubic meter?

Answer: 1,000,000 cm^3 ■

EXAMPLE 2 *Volume of an oil storage tank.* An oil storage tank is in the form of a rectangular solid with dimensions of 17 by 10 by 8 feet. (See Figure 7-73.) Find its volume.

Solution
To find the volume, we substitute 17 for l, 10 for w, and 8 for h in the formula $V = lwh$ and simplify.

$V = lwh$

$V = 17(10)(8)$

    $= 1,360$

The volume is 1,360 ft^3.

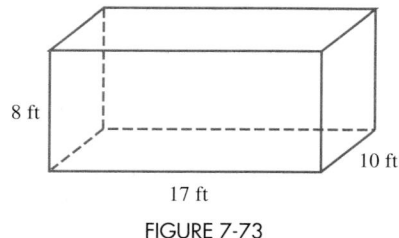

8 ft

10 ft

17 ft

FIGURE 7-73

Self Check
Find the volume of a rectangular solid with dimensions of 8 by 12 by 20 meters.

Answer: 1,920 m^3 ■

EXAMPLE 3 *Volume of a triangular prism.* Find the volume of the triangular prism in Figure 7-74.

Solution
The volume of the prism is the area of its base multiplied by its height. Since there are 100 centimeters in 1 meter, the height in centimeters is

$0.5 \text{ m} = 0.5(\mathbf{1 \text{ m}})$

$\qquad = 0.5(\mathbf{100 \text{ cm}})$ Substitute 100 centimeters for 1 meter.

$\qquad = 50 \text{ cm}$

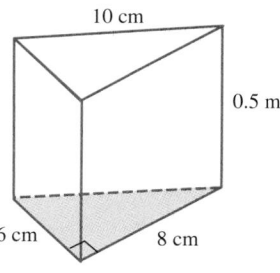

10 cm

0.5 m

6 cm

8 cm

FIGURE 7-74

Self Check
Find the volume of the triangular prism below.

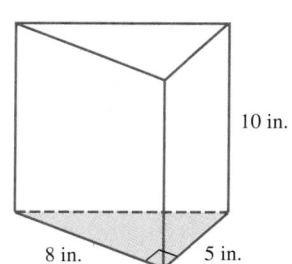

10 in.

8 in.

5 in.

The area of the triangular base is $\frac{1}{2}(6)(8) = 24$ square centimeters. The height of the prism is 50 centimeters. Substituting into the formula for the volume of a prism, we have

$$V = Bh$$
$$V = 24(50)$$
$$ = 1,200$$

The volume of the prism is 1,200 cm³.

Answer: 200 in.³   ∎

Surface areas of rectangular solids

The **surface area** of a rectangular solid is the sum of the areas of its six faces. Figure 7-75 shows how we can unfold the faces of a cardboard box to derive a formula for its surface area (*SA*).

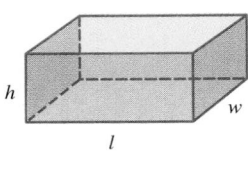
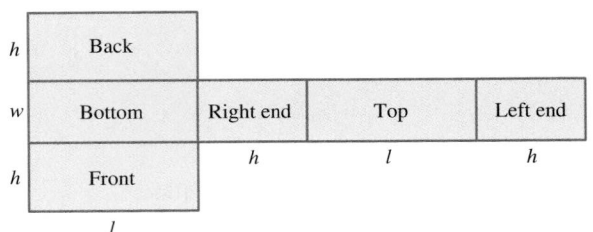

FIGURE 7-75

$$SA = A_{\text{bottom}} + A_{\text{back}} + A_{\text{front}} + A_{\text{right end}} + A_{\text{top}} + A_{\text{left end}}$$
$$SA = \quad lw \quad + \quad lh \quad + \quad lh \quad + \quad hw \quad + \quad lw \quad + \quad hw$$
$$ = 2lw + 2lh + 2hw \qquad \text{There are two } lw\text{'s, two } lh\text{'s, and two } hw\text{'s.}$$

Surface area of a rectangular solid

> The surface area of a rectangular solid is given by the formula
> $$SA = 2lw + 2lh + 2hw$$
> where *l* is the length, *w* is the width, and *h* is the height.

EXAMPLE 4   *Surface area of an oil tank.*   An oil storage tank is in the form of a rectangular solid with dimensions of 17 by 10 by 8 feet. (See Figure 7-76.) Find the surface area of the tank.

Solution

To find the surface area, we substitute 17 for *l*, 10 for *w*, and 8 for *h* in the formula for surface area and simplify.

$$SA = 2lw + 2lh + 2hw$$
$$SA = 2(17)(10) + 2(17)(8) + 2(8)(10)$$
$$ = 340 + 272 + 160$$
$$ = 772$$

The surface area is 772 ft².

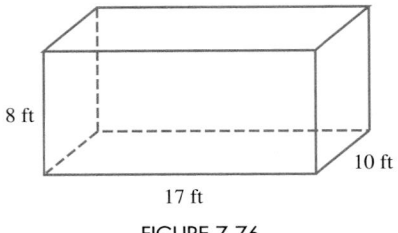

8 ft

10 ft

17 ft

FIGURE 7-76

Self Check

Find the surface area of a rectangular solid with dimensions of 8 by 12 by 20 meters.

Answer: 992 m²   ∎

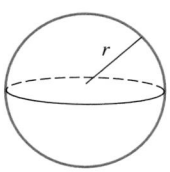

FIGURE 7-77

Volumes and surface areas of spheres

A **sphere** is a hollow, round ball. (See Figure 7-77.) The points on a sphere all lie at a fixed distance r from a point called its *center*. A segment drawn from the center of a sphere to a point on the sphere is called a *radius*.

Accent on Technology: **Filling a water tank**

See Figure 7-78. To calculate how many cubic feet of water are needed to fill a spherical water tank with a radius of 15 feet, we substitute 15 for r in the formula for the volume of a sphere.

$$V = \frac{4}{3}\pi r^3$$

$$V = \frac{4}{3}\pi(15)^3$$

To do the arithmetic with a scientific calculator, we enter these numbers and press these keys.

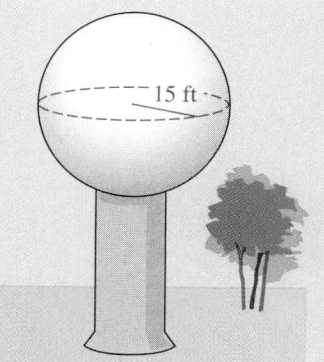

FIGURE 7-78

Keystrokes   15 3 = 4 ÷ 3 =  = | 14137.16694 |

To the nearest tenth, 14,137.2 ft^3 of water will be needed to fill the tank.

There is a formula to find the surface area of a sphere.

Surface area of a sphere

> The surface area of a sphere with radius r is given by the formula
> $$SA = 4\pi r^2$$

    EXAMPLE 5   ***Manufacturing beach balls.***   A beach ball is to have a diameter of 16 inches. (See Figure 7-79.) How many square inches of material will be needed to make the ball? (Disregard any waste.)

Solution   Since a radius r of the ball is one-half the diameter, $r = 8$ inches. We can now substitute 8 for r in the formula for the surface area of a sphere.

$SA = 4\pi r^2$
$SA = 4\pi(8)^2$
$SA = 4\pi(64)$
$SA = 256\pi$ Simplify: $4 \cdot 64 = 256$.
≈ 804.2477193 Use a calculator.

A little more than 804 in.^{2} of material is needed to make the ball.

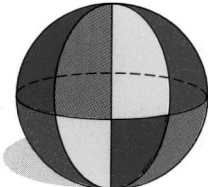

FIGURE 7-79 ■

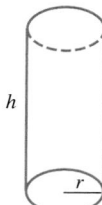

FIGURE 7-80

Volumes of cylinders

A **cylinder** is a hollow figure like a piece of pipe. (See Figure 7-80.)

EXAMPLE 6 *Cylinder.* Find the volume of the cylinder in Figure 7-81.

Solution Since a radius is one-half of the diameter of the circular base, $r = 3$ cm. From the figure, we see that the height of the cylinder is 10 cm. So we can substitute 3 for r and 10 for h in the formula for the volume of a cylinder.

$$V = \pi r^2 h$$
$$V = \pi(3)^2(10)$$
$$= 90\pi \qquad \text{Simplify: } (3)^2(10) = 90.$$
$$\approx 282.7433388 \qquad \text{Use a calculator.}$$

To the nearest hundredth, the volume of the cylinder is 282.74 cm³.

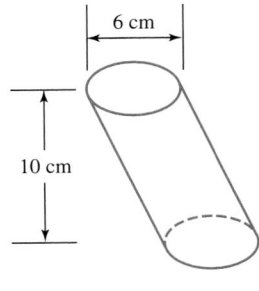

FIGURE 7-81

Accent on Technology: **Volume of a silo**

A silo is a structure used for storing grain. The silo in Figure 7-82 is a cylinder 50 feet tall topped with a **hemisphere** (a half-sphere). To find the volume of the silo, we add the volume of the cylinder to the volume of the dome.

$$\text{Volume}_{\text{cylinder}} + \text{Volume}_{\text{dome}} = (\text{Area}_{\text{cylinder's base}})(\text{height}_{\text{cylinder}}) + \frac{1}{2}(\text{Volume}_{\text{sphere}})$$

$$= \pi r^2 h + \frac{1}{2}\left(\frac{4}{3}\pi r^3\right)$$

$$= \pi r^2 h + \frac{2\pi r^3}{3} \qquad \frac{1}{2}\left(\frac{4}{3}\pi r^3\right) = \frac{1}{2}\cdot\frac{4}{3}\pi r^3 = \frac{4}{6}\pi r^3 = \frac{2\pi r^3}{3}$$

$$= \pi(10)^2(50) + \frac{2\pi(10)^3}{3} \qquad \text{Substitute 10 for } r \text{ and 50 for } h.$$

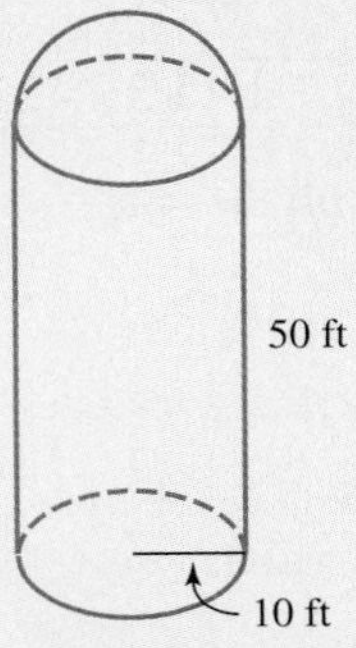

50 ft

10 ft

FIGURE 7-82

To do the arithmetic with a scientific calculator, we enter these numbers and press these keys.

Keystrokes $\boxed{\pi}$ $\boxed{\times}$ 10 $\boxed{x^2}$ $\boxed{\times}$ 50 $\boxed{=}$ $\boxed{+}$ $\boxed{(}$ 2 $\boxed{\times}$ $\boxed{\pi}$ $\boxed{\times}$ 10 $\boxed{y^x}$ 3

$\boxed{\div}$ 3 $\boxed{)}$ $\boxed{=}$ $\boxed{17802.35837}$

The volume of the silo is approximately 17,802 ft³.

EXAMPLE 7 *Machining a block of metal.*
See Figure 7-83. Find the volume that is left when the hole is drilled through the metal block.

Solution We must find the volume of the rectangular solid and then subtract the volume of the cylinder. We will think of the rectangular solid and the cylinder as lying on their sides. Thus, the height is 18 cm when we find each volume.

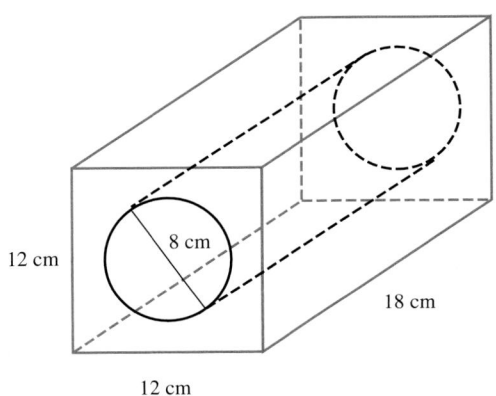

FIGURE 7-83

$$V_{\text{rect. solid}} = lwh$$
$$V_{\text{rect. solid}} = 12(12)(18)$$
$$= 2{,}592$$

$$V_{\text{cylinder}} = \pi r^2 h$$
$$V_{\text{cylinder}} = \pi(4)^2(18)$$
$$= 288\pi$$
$$\approx 904.7786842$$

$$V_{\text{drilled block}} = V_{\text{rect. solid}} - V_{\text{cylinder}}$$
$$\approx 2{,}592 - 904.7786842$$
$$\approx 1{,}687.221316$$

To the nearest hundredth, the volume is 1,687.22 cm³. ∎

Volumes of cones

Two **cones** are shown in Figure 7-84. Each cone has a height h and a radius r, which is the radius of the circular base.

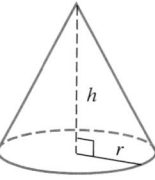

FIGURE 7-84

EXAMPLE 8 *Volume of a cone.*   To the nearest tenth, find the volume of the cone in Figure 7-85.

Solution Since the radius is one-half of the diameter, $r = 4$ cm. We then substitute 4 for r and 6 for h in the formula for the volume of a cone.

$$V = \frac{1}{3}\pi r^2 h$$

$$V = \frac{1}{3}\pi(4)^2(6)$$

$$V = 32\pi$$
$$\approx 100.5309649$$

To the nearest tenth, the volume is 100.5 cubic centimeters.

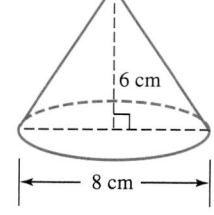

FIGURE 7-85 ∎

Volumes of pyramids

Two **pyramids** with a height h are shown in Figure 7-86.

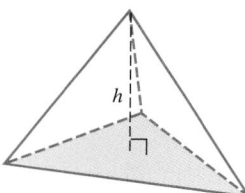

The base is a triangle.

(a)

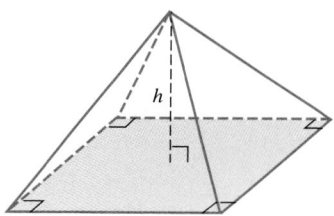

The base is a square.

(b)

FIGURE 7-86

EXAMPLE 9 *Volume of a pyramid.* Find the volume of a pyramid that has a square base with each side 6 meters long and a height of 9 meters.

Solution

Since the base is a square with each side 6 meters long, the area of the base is 6^2 m^2, or 36 m^2. We can then substitute 36 for the area of the base and 9 for the height in the formula for the volume of a pyramid.

$$V = \frac{1}{3}Bh$$

$$V = \frac{1}{3}(36)(9)$$

$$= 108$$

The volume of the pyramid is 108 m^3.

Self Check

Find the volume of the pyramid shown below.

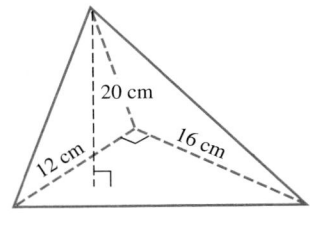

Answer: 640 cm^3 ∎

STUDY SET Section 7.7

VOCABULARY *Fill in the blanks.*

1. The space contained within a geometric solid is called its _____.

2. A _____ solid is like a hollow shoe box.

3. A _____ is a rectangular solid with all sides of equal length.

4. The volume of a cube with each side 1 inch long is 1 _____ inch.

5. The _____ area of a rectangular solid is the sum of the areas of its faces.

6. The point that is equidistant from every point on a sphere is its _____.

7. A _____ is a hollow figure like a drinking straw.

8. A _____ is one-half of a sphere.

9. A _____ looks like a witch's pointed hat.

10. A figure that has a polygon for its base and that rises to a point is called a _____.

CONCEPTS *In Exercises 11–16, write the formula used for finding the volume of each solid.*

11. A rectangular solid

12. A prism

13. A sphere

14. A cylinder

15. A cone

16. A pyramid

17. Write the formula for finding the surface area of a rectangular solid.

18. Write the formula for finding the surface area of a sphere.

19. How many cubic feet are in 1 cubic yard?

20. How many cubic inches are in 1 cubic yard?

21. How many cubic decimeters are in 1 cubic meter?

22. How many cubic millimeters are in 1 cubic centimeter?

In Exercises 23–24, which geometric concept (perimeter, circumference, area, volume, or surface area) should be applied to find each of the following?

23. **a.** The size of a room to be air conditioned
    b. The amount of land in a national park
    c. The amount of space in a refrigerator freezer

    d. The amount of cardboard in a shoe box
    e. The distance around a checkerboard
    f. The amount of material used to make a basketball

24. **a.** The amount of cloth in a car cover
    b. The size of a trunk of a car
    c. The amount of paper used for a postage stamp
    d. The amount of storage in a cedar chest
    e. The amount of beach available for sunbathing
    f. The distance the tip of a propeller travels

25. In Illustration 1, the inch was the unit of measurement of length that was used to draw the figure.
    a. What is the volume of the figure?
    b. What is the area of the front of the figure?
    c. What is the area of the base of the figure?

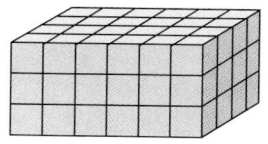

ILLUSTRATION 1

26. The cardboard box shown in Illustration 2 is a cube. Suppose the six faces were unfolded to lie flat on a table. Draw a picture of what this would look like.

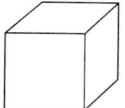

ILLUSTRATION 2

NOTATION   *Fill in the blanks.*

27. The notation 1 in.^{3} is read as _____.

28. One cubic centimeter is represented as

_____.

PRACTICE   *Find the volume of each solid. If an answer is not exact, round to the nearest hundredth. (Answers may vary slightly, depending on which approximation of π is used.)*

29. A rectangular solid with dimensions of 3 by 4 by 5 centimeters.

30. A rectangular solid with dimensions of 5 by 8 by 10 meters.

31. A prism whose base is a right triangle with legs 3 and 4 meters long and whose height is 8 meters.

32. A prism whose base is a right triangle with legs 5 and 12 feet long and whose height is 10 feet.

33. A sphere with a radius of 9 inches.

34. A sphere with a diameter of 10 feet.

35. A cylinder with a height of 12 meters and a circular base with a radius of 6 meters.

36. A cylinder with a height of 4 meters and a circular base with a diameter of 18 meters.

37. A cone with a height of 12 centimeters and a circular base with a diameter of 10 centimeters.

38. A cone with a height of 3 inches and a circular base with a radius of 4 inches.

39. A pyramid with a square base 10 meters on each side and a height of 12 meters.

40. A pyramid with a square base 6 inches on each side and a height of 4 inches.

Find the surface area of each solid. If an answer is not exact, round to the nearest hundredth.

41. A rectangular solid with dimensions of 3 by 4 by 5 centimeters.

42. A cube with a side 5 centimeters long.

43. A sphere with a radius of 10 inches.

44. A sphere with a diameter of 12 meters.

Find the volume of each figure. If an answer is not exact, round to the nearest hundredth. (Answers may vary slightly, depending on which approximation of π is used.)

45.

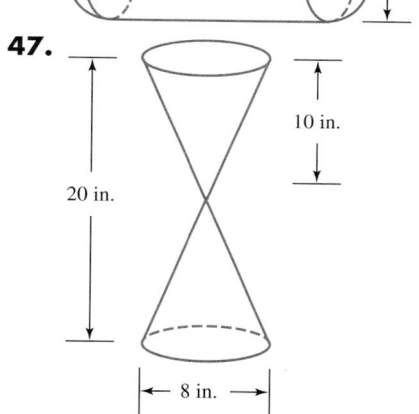

3 cm

8 cm

8 cm

8 cm

46.

16 cm

6 cm

47.

20 in.

10 in.

8 in.

48.

8 in.

6 in.

3 in.   4 in.

5 in.

APPLICATIONS   *Solve each problem. If an answer is not exact, round to the nearest hundredth.*

49. VOLUME OF A SUGAR CUBE   A sugar cube is $\frac{1}{2}$ inch on each edge. How much volume does it occupy?

50. VOLUME OF A CLASSROOM   A classroom is 40 feet long, 30 feet wide, and 9 feet high. Find the number of cubic feet of air in the room.

51. WATER HEATER   Complete the advertisement for the high-efficiency water heater shown in Illustration 3.

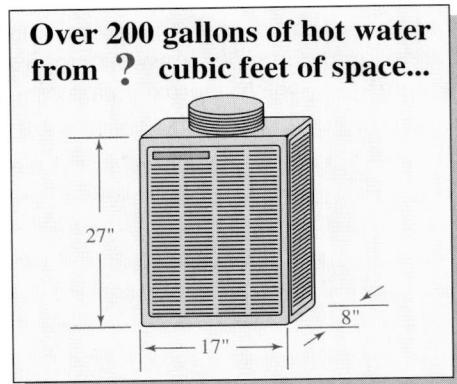

Over 200 gallons of hot water from **?** cubic feet of space...

27"
17"
8"

ILLUSTRATION 3

52. REFRIGERATOR CAPACITY   The largest refrigerator advertised in a J. C. Penney catalog has a capacity of 25.2 cubic feet. How many cubic inches is this?

53. VOLUME OF AN OIL TANK   A cylindrical oil tank has a diameter of 6 feet and a length of 7 feet. Find the volume of the tank.

54. VOLUME OF A DESSERT   A restaurant serves pudding in a conical dish that has a diameter of 3 inches. If the dish is 4 inches deep, how many cubic inches of pudding are in each dish?

55. HOT-AIR BALLOON   The lifting power of a spherical balloon depends on its volume. How many cubic feet of gas will a balloon hold if it is 40 feet in diameter?

56. VOLUME OF A CEREAL BOX   A box of cereal measures 3 by 8 by 10 inches. The manufacturer plans to market a smaller box that measures $2\frac{1}{2}$ by 7 by 8 inches. By how much will the volume be reduced?

57. ENGINE   The *compression ratio* of an engine is the volume in one cylinder with the piston at bottom-dead-center (B.D.C.), divided by the volume with the piston at top-dead-center (T.D.C.). From the data given in Illustration 4, what is the compression ratio of the engine? Use a colon to express your answer.

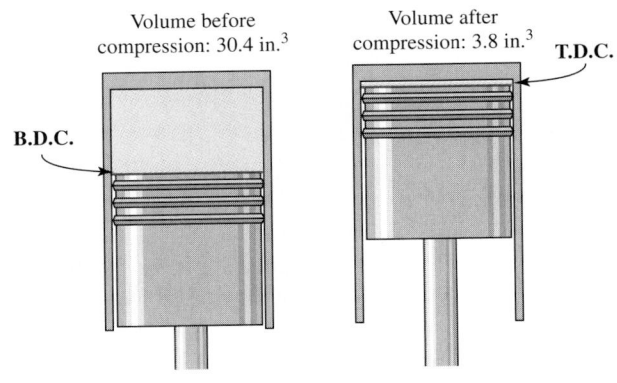

Volume before compression: 30.4 in.^{3}

Volume after compression: 3.8 in.^{3}

T.D.C.

B.D.C.

ILLUSTRATION 4

58. LINT REMOVER   Illustration 5 shows a handy gadget; it uses a cylinder of sheets of sticky paper that can be rolled over clothing and furniture to pick up lint and pet hair. After the paper is full, that sheet is peeled away to expose another sheet of sticky paper. Find the area of the first sheet by using the formula $LSA = 2\pi rh$, where LSA represents the lateral surface area of the cylinder.

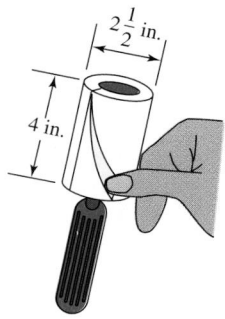

$2\frac{1}{2}$ in.

4 in.

ILLUSTRATION 5

WRITING

59. What is meant by the *volume* of a cube?

60. What is meant by the *surface area* of a cube?

61. Are the units used to measure area different from the units used to measure volume? Explain.

62. The dimensions (length, width, and height) of one rectangular solid are entirely different numbers than the dimensions of another rectangular solid. Would it be possible for the rectangular solids to have the same volume? Explain.

REVIEW

63. Evaluate $-5(5 - 2)^2 + 3$.

64. BUYING PENCILS   Carlos bought 6 pencils at $0.60 each and a notebook for $1.25. He gave the clerk a $5 bill. How much change did he receive?

65. 38 is what percent of 40?

66. State the Pythagorean theorem.

KEY CONCEPT

Formulas

A **formula** is a mathematical expression that is used to express a relationship between quantities. We have studied formulas used in mathematics, business, and science.

Write a formula describing the mathematical relationship between the given quantities.

1. Area of a rectangle (*A*), its length (*l*), its width (*w*)

2. Sale price (*s*) of an item at a store, original price (*p*), discount (*d*)

3. Perimeter of a rectangle (*P*), length of the rectangle (*l*), width of the rectangle (*w*)

4. Amount of interest earned (*I*), principal (*P*), interest rate (*r*), time the money is invested (*t*)

Use a formula to solve each problem.

5. Find the area (*A*) of the triangular lot in Illustration 1.

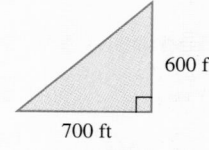

600 ft

700 ft

ILLUSTRATION 1

6. Find the volume (*V*) of the ice chest in Illustration 2.

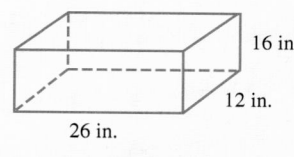

16 in.

12 in.

26 in.

ILLUSTRATION 2

7. Find the retail price (*p*) of a cookware set that costs the store owner $45.50 and is marked up $35.

8. Find the profit (*p*) made by a school T-shirt sale if revenue was $14,500 and costs were $10,200.

9. To the nearest hundredth, find the area of a circle that has a diameter of 14 feet.

10. To the nearest hundredth, find the volume of a sphere that has a radius of 10 centimeters.

Sometimes we use the same formula to answer several related questions. The results can be displayed in a table.

11. Use the formula $I = Prt$ to find the interest earned by each account.

Type of account	Principal	Annual rate earned	Time invested	Interest earned
Savings	$5,000	5%	3 yr	
Passbook	$2,250	2%	1 yr	
Trust fund	$10,000	6.25%	10 yr	

Section 7.1

WRITING DIGITS   In Illustration 1, the digit 1 is drawn using one angle, and the digit 2 is drawn using two angles. Draw the digit 3 using three angles, the digit 4 using four angles, and so on for all of the digits up to and including 9.

ILLUSTRATION 1

Section 7.2

CONSTRUCTIONS

Step 1: See Illustration 2(a). Using a straightedge, draw $\overline{AB}$. Then place the sharp point of a compass at A and draw an arc.

Step 2: With the same compass setting, place the sharp point at B. As shown in Illustration 2(b), draw another arc that intersects the arc from Step 1 at two points. Label these points C and D.

Step 3: Using a straightedge, draw a line through points C and D. Label the point where line CD intersects $\overline{AB}$ as point E. Does m($\overline{AE}$) = m($\overline{EB}$)?

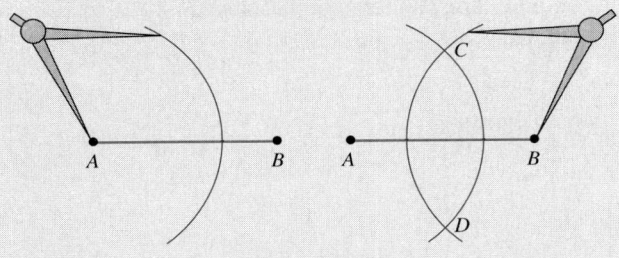

ILLUSTRATION 2

Section 7.3

TANGRAM   A tangram is a puzzle in which geometric shapes are arranged to form other shapes. Cut out the pieces in Illustration 3. Assemble them so that they form a square. There should be no gaps, overlaps, or holes.

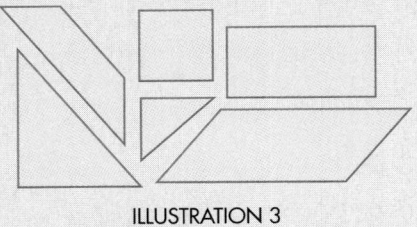

ILLUSTRATION 3

Section 7.4

CONGRUENT TRIANGLES   Draw a triangle on a piece of paper. Then measure the lengths of its sides (with a ruler) and the angle measures (with a protractor). Choose a combination of any three measurements and tell them to your partner. Are the given facts sufficient for your partner to construct a triangle congruent to yours?

Section 7.5

AREA   Find the area of the shaded figure on the square grid in Illustration 4.

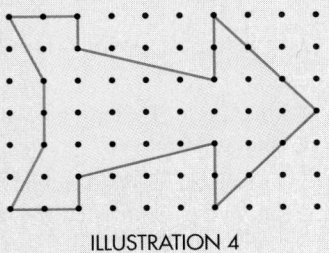

ILLUSTRATION 4

Section 7.6

PI   Carefully measure the circumference and the diameter of different-size circles. Record the measurements in a table like the one below. Then use a calculator to find $\frac{C}{D}$. The result should be a number close to π.

Object	Circumference	Diameter	$\frac{C}{D}$
jar	$3\frac{1}{2}$ in.	$1\frac{1}{8}$ in.	3.11

Section 7.7

PYRAMID   Cut out, fold, and glue together the pattern shown in Illustration 5. Estimate the volume and surface area of the pyramid.

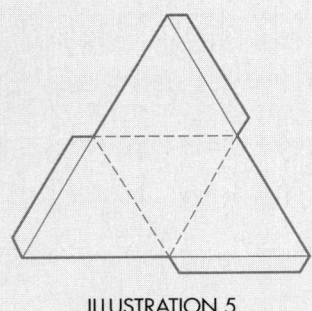

ILLUSTRATION 5

| SECTION 7.1 | *Some Basic Definitions* |

CONCEPTS

In geometry, we study *points*, *lines*, and *planes*.

A *line segment* is a part of a line with two endpoints. A *ray* is a part of a line with one endpoint.

An *angle* is a figure formed by two rays with a common endpoint. The common endpoint is called the *vertex* of the angle.

A *protractor* is used to find the measure of an angle.

An *acute angle* is greater than 0° but less than 90°. A *right angle* measures 90°. An *obtuse angle* is greater than 90° but less than 180°. A *straight angle* measures 180°.

Two angles that have the same vertex and are side-by-side are called *adjacent angles*.

REVIEW EXERCISES

1. In Illustration 1, identify a point, a line, and a plane.

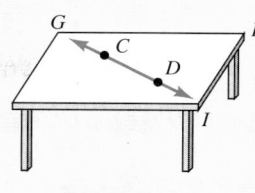

ILLUSTRATION 1

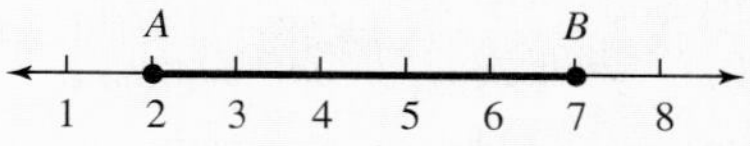

ILLUSTRATION 2

2. In Illustration 2, find m($\overline{AB}$).

3. In Illustration 3, give four ways to name the angle.

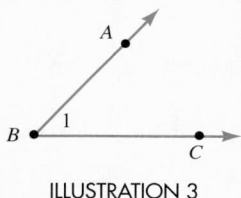

ILLUSTRATION 3

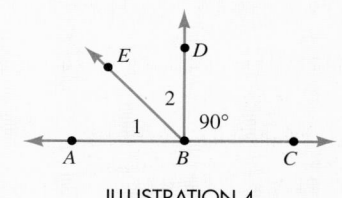

ILLUSTRATION 4

4. In Illustration 3, use a protractor to find the measure of the angle.

5. In Illustration 4, identify each acute angle, right angle, obtuse angle, and straight angle.

6. The measures of several angles are given. Identify each angle as an acute angle, a right angle, an obtuse angle, or a straight angle.
   a. m($\angle A$) = 150°   **b.** m($\angle B$) = 90°
   c. m($\angle C$) = 180°   **d.** m($\angle D$) = 25°

7. The two angles shown in Illustration 5 are adjacent angles. Find x.

8. Line AB is shown in Illustration 6. Find y.

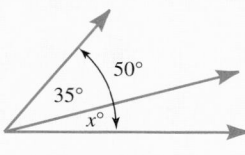

ILLUSTRATION 5

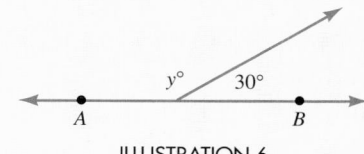

ILLUSTRATION 6

427

When two lines intersect, pairs of nonadjacent angles are called *vertical angles*.

Vertical angles have the same measure.

If the sum of two angles is 90°, the angles are *complementary*. If the sum of two angles is 180°, the angles are *supplementary*.

9. In Illustration 7, find **a.** m($\angle$1) and **b.** m($\angle$2).

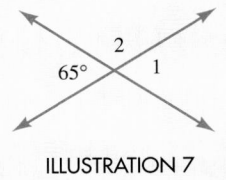

ILLUSTRATION 7

10. Find the complement of an angle that measures 50°.

11. Find the supplement of an angle that measures 140°.

12. Are angles measuring 30°, 60°, and 90° supplementary?

SECTION 7.2

Parallel and Perpendicular Lines

Parallel lines do not intersect. *Perpendicular* lines intersect and make right angles.

A line that intersects two or more *coplanar* lines is called a *transversal*.

13. Which part of Illustration 8 represents parallel lines?

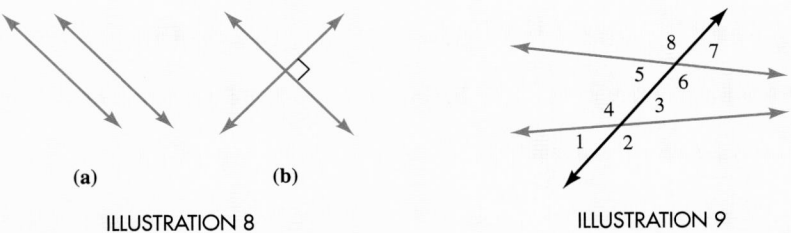

(a)   **(b)**

ILLUSTRATION 8   ILLUSTRATION 9

When a transversal intersects two coplanar lines, *alternate interior angles* and *corresponding* angles are formed.

14. Identify all pairs of alternate interior angles shown in Illustration 9.

15. Identify all pairs of corresponding angles shown in Illustration 9.

16. Identify all pairs of vertical angles shown in Illustration 9.

If two parallel lines are cut by a transversal,
1. alternate interior angles are congruent (have equal measures).
2. corresponding angles are congruent.
3. interior angles on the same side of the transversal are supplementary.

17. In Illustration 10, $l_1 \parallel l_2$. Find the measure of each angle.

18. In Illustration 11, $\overline{DC} \parallel \overline{AB}$. Find the measure of each angle.

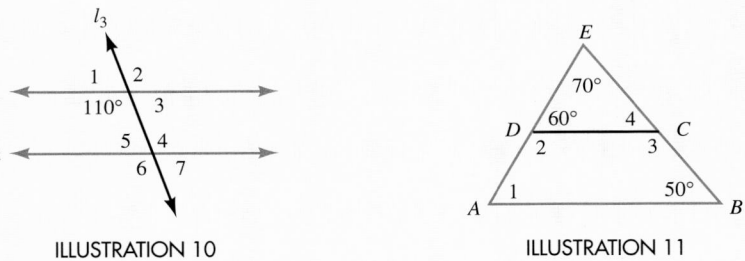

ILLUSTRATION 10   ILLUSTRATION 11

19. In Illustration 12 on the next page, $l_1 \parallel l_2$. Find x.

20. In Illustration 13 on the next page, $l_1 \parallel l_2$. Find x.

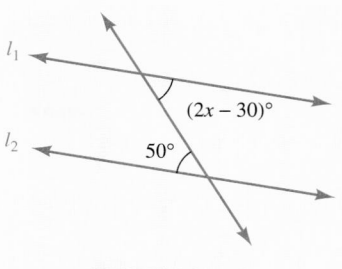

ILLUSTRATION 12

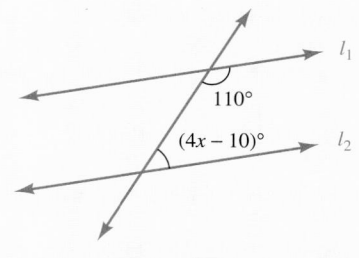

ILLUSTRATION 13

SECTION 7.3	*Polygons*

A *polygon* is a closed geometric figure. The points at which the sides intersect are called *vertices*. A *regular polygon* has sides that are all the same length and angles that are all the same measure.

Polygons are classified as follows:

Number of sides	Name
3	triangle
4	quadrilateral
5	pentagon
6	hexagon
8	octagon

An *equilateral triangle* has three sides of equal length.
An *isosceles triangle* has at least two sides of equal length.
A *scalene triangle* has no sides of equal length.
A *right triangle* has one right angle.

In an isosceles triangle, the angles opposite the sides of equal length are called *base angles*. The third angle is called the *vertex angle*. The third side is called the *base*.

21. Identify each polygon as a triangle, a quadrilateral, a pentagon, a hexagon, or an octagon.

a.

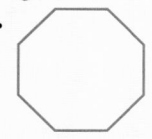

b.

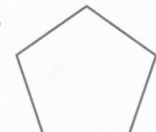

c.

d.

e.

22. Give the number of vertices of each polygon.
  a. Triangle   **b.** Quadrilateral
  c. Octagon   **d.** Hexagon

23. Classify each of the triangles as an equilateral triangle, an isosceles triangle, a scalene triangle, or a right triangle.

a.

b.

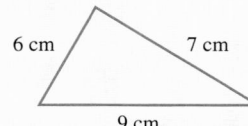

c.

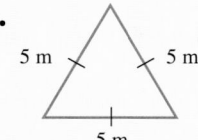

d.

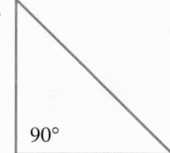

Properties of isosceles triangles:
1. The base angles are congruent.
2. If two angles in a triangle are congruent, the sides opposite the angles are congruent, and the triangle is isosceles.

The sum of the measures of the angles of any triangle is 180°.

Quadrilaterals are classified as follows:

Property	Name
Opposite sides parallel	parallelogram
Parallelogram with four right angles	rectangle
Rectangle with all sides equal	square
Parallelogram with sides of equal length	rhombus
Exactly two sides parallel	trapezoid

Properties of rectangles:
1. All angles are right angles.
2. Opposite sides are parallel.
3. Opposite sides are of equal length.
4. Diagonals are of equal length.
5. If the diagonals of a parallelogram are of equal length, the parallelogram is a rectangle.

24. Determine whether each triangle is isosceles.

a.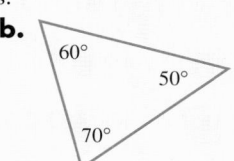

b.

25. In each triangle, find *x*.

a.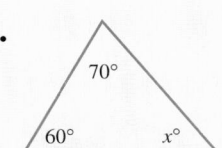

b.

26. If one base angle of an isosceles triangle measures 65°, how large is the vertex angle?

27. If one base angle of an isosceles triangle measures 60°, what can you conclude about the triangle?

28. Classify each quadrilateral as a parallelogram, a rectangle, a square, a rhombus, or a trapezoid.

a.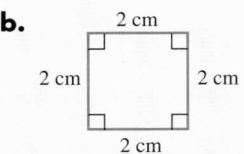

b.

c.

d.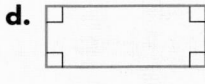

e.

f.

29. In Illustration 14, the length of diagonal $\overline{AC}$ of rectangle *ABCD* is 15 centimeters. Find each measure.

a. m($\overline{BD}$)   **b.** m($\angle 1$)   **c.** m($\angle 2$)

30. In Illustration 14, *ABCD* is a rectangle. Classify each statement as true or false.

a. m($\overline{AB}$) = m($\overline{DC}$)   **b.** m($\overline{AD}$) = m($\overline{DC}$)

c. Triangle *ABE* is isosceles.   **d.** m($\overline{AC}$) = m($\overline{BD}$)

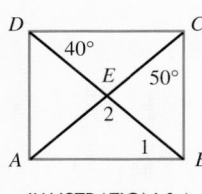

ILLUSTRATION 14

The parallel sides of a trapezoid are called *bases*. The nonparallel sides are called *legs*. If the legs of a trapezoid are of equal length, it is *isosceles*. In an isosceles trapezoid, the angles opposite the sides of equal length are *base angles*, and they are congruent.

The sum of the measures of the angles of a polygon (in degrees) is given by the formula
$$S = (n - 2)180$$

31. In Illustration 15, *ABCD* is an isosceles trapezoid. Find each measure.
   a. m(∠B)   **b.** m(∠C)

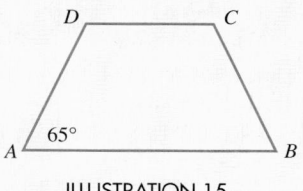

ILLUSTRATION 15

32. Find the sum of the angle measures of each polygon.
   a. quadrilateral   **b.** hexagon

SECTION 7.4   *Properties of Triangles*

If two triangles have the same size and the same shape, they are *congruent triangles*.

Corresponding parts of congruent triangles have the same measure.

Three ways to show that two triangles are congruent are
1. The SSS property
2. The SAS property
3. The ASA property

33. See Illustration 16. Complete the list of corresponding parts.
   ∠A corresponds to _____.
   ∠B corresponds to _____.
   ∠C corresponds to _____.
   $\overline{AC}$ corresponds to _____.
   $\overline{AB}$ corresponds to _____.
   $\overline{BC}$ corresponds to _____.

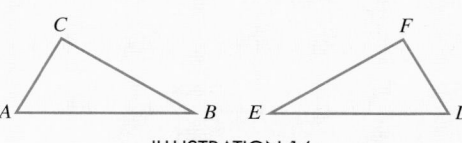

ILLUSTRATION 16

34. Tell whether the triangles in each pair are congruent. If they are, tell why.

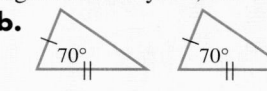

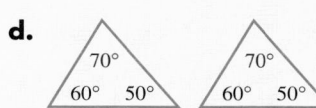

The Pythagorean theorem:
If the length of the *hypotenuse* of a right triangle is *c*, and the lengths of its legs are *a* and *b*, then
$$c^2 = a^2 + b^2$$

35. Refer to Illustration 17 and find the length of the unknown side.
   a. If $a = 5$ and $b = 12$, find c.   **b.** If $a = 8$ and $c = 17$, find b.

36. 🖩 To the nearest tenth, find the height of the television screen shown in Illustration 18.

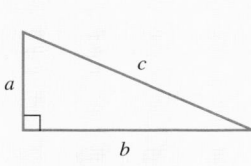

ILLUSTRATION 17

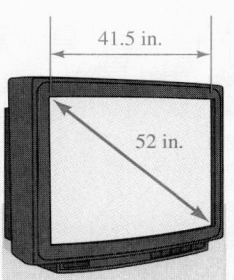

41.5 in.

52 in.

ILLUSTRATION 18

| SECTION 7.5 | *Perimeters and Areas of Polygons* |

The *perimeter* of a polygon is the distance around it.

37. Find the perimeter of a square with sides 18 inches long.

38. Find the perimeter of a rectangle that is 3 meters long and 1.5 meters wide.

39. Find the perimeter of each polygon.

a.

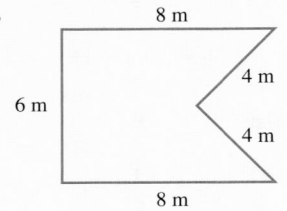

b.

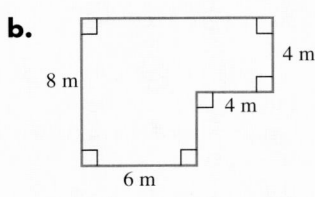

The *area* of a polygon is the measure of the surface it encloses.

Formulas for area:

Figure	Area
Square	$A = s^2$
Rectangle	$A = lw$
Parallelogram	$A = bh$
Triangle	$A = \frac{1}{2}bh$
Trapezoid	$A = \frac{1}{2}h(b_1 + b_2)$

40. Find the area of each polygon.

a.

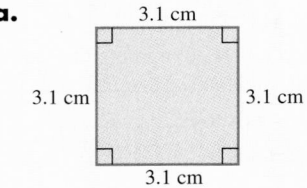

b.

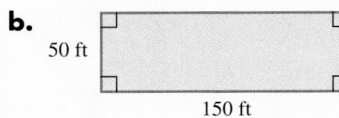

c.

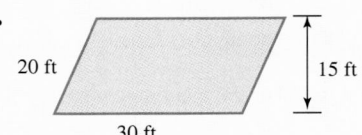

d.

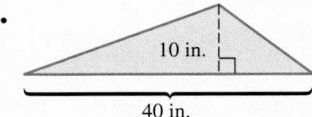

e.

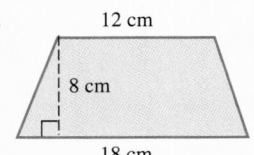

f.

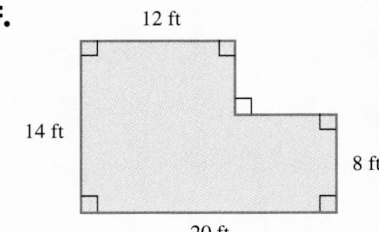

g.

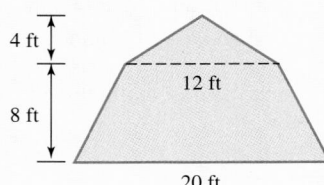

h.

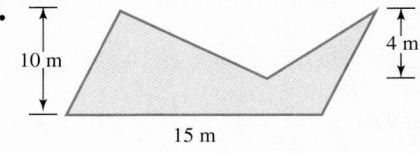

41. How many square feet are there in 1 square yard?

42. How many square inches are in 1 square foot?

SECTION 7.6	*Circles*

A *circle* is the set of all points in a plane that lie a fixed distance from a point called its *center*. The fixed distance is the circle's *radius*.

A *chord* of a circle is a line segment connecting two points on the circle.
A *diameter* is a chord that passes through the circle's center.

The *circumference* (perimeter) of a circle is given by the formulas
$$C = \pi D \quad \text{or} \quad C = 2\pi r$$

$\pi = 3.14159\dots$

The *area* of a circle is given by the formula
$$A = \pi r^2$$

43. Refer to Illustration 19.
  a. Name each chord.
  b. Name each diameter.
  c. Name each radius.
  d. Name the center.

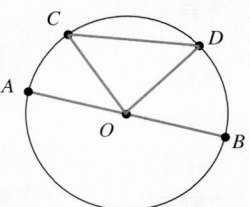

ILLUSTRATION 19

 In Problems 44–47, find each answer to the nearest tenth.

44. Find the circumference of a circle with a diameter of 21 centimeters.

45. Find the perimeter of the figure shown in Illustration 20.

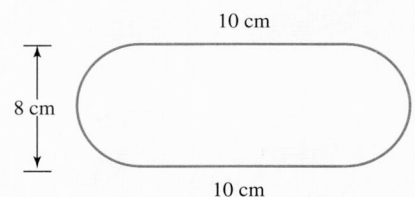

10 cm

8 cm

10 cm

ILLUSTRATION 20

46. Find the area of a circle with a diameter of 18 inches.

47. Find the area of the figure shown in Illustration 20.

SECTION 7.7	*Surface Area and Volume*

The *volume* of a solid is a measure of the space it occupies.

Figure	Volume
Cube	$V = s^3$
Rectangular solid	$V = lwh$
Prism	$V = Bh$*
Sphere	$V = \dfrac{4}{3}\pi r^3$
Cylinder	$V = \pi r^2 h$
Cone	$V = \dfrac{1}{3}\pi r^2 h$
Pyramid	$V = \dfrac{1}{3}Bh$*

**B* represents the area of the base.

48. Find the volume of each solid to the nearest unit. (See the table at left.)

a.

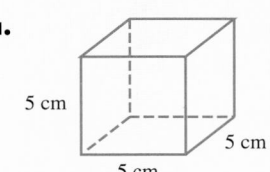

5 cm

5 cm

5 cm

b.

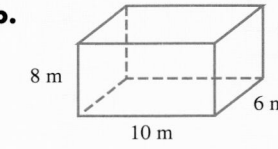

8 m

10 m

6 m

c.

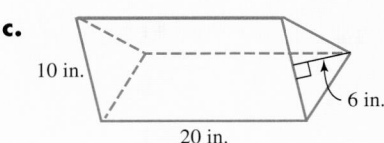

10 in.

20 in.

6 in.

d.

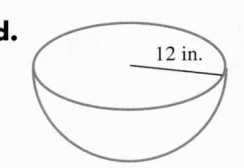

12 in.

e.

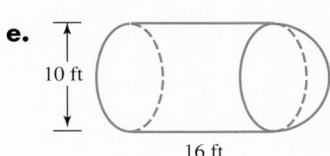

10 ft

16 ft

f.

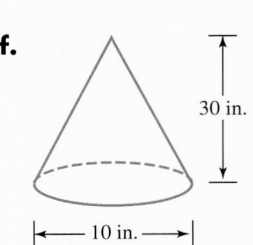

30 in.

10 in.

The *surface area* of a rectangular solid is the sum of the areas of its six faces.

The surface area of a sphere is given by the formula
$$SA = 4\pi r^2$$

g.

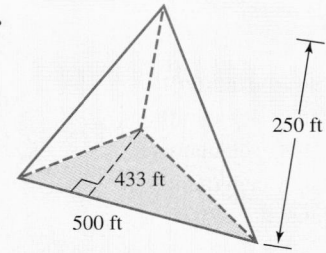

250 ft

433 ft

500 ft

h.

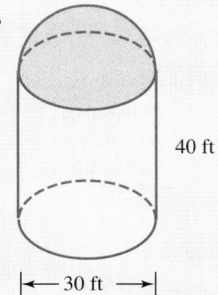

40 ft

|← 30 ft →|

49. How many cubic inches are there in 1 cubic foot?

50. How many cubic feet are there in 2 cubic yards?

51. To the nearest tenth, find the surface area of each solid.

a.

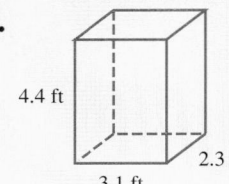

4.4 ft

2.3 ft

3.1 ft

b.

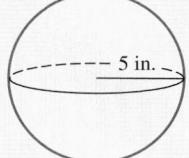

5 in.

Chapter 7 Test

1. Find m($\overline{AB}$).

A ——————— B
1 2 3 4 5 6 7 8

2. Which point is the vertex of $\angle ABC$?

Tell whether each statement is true or false.

3. An angle of 47° is an acute angle.

4. An angle of 90° is a straight angle.

5. An angle of 180° is a right angle.

6. An angle of 132° is an obtuse angle.

7. Find x.

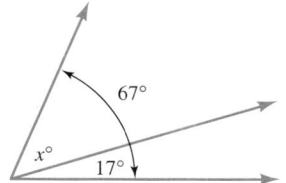

8. Find y.

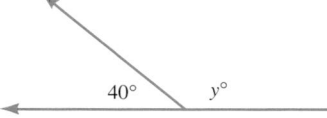

9. Find y.

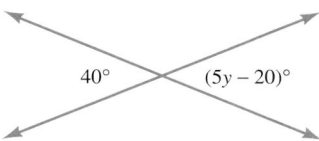

10. CALLIGRAPHY   Illustration 1 shows how the tip of the pen should be held at a 45° angle to the horizontal. What is x?

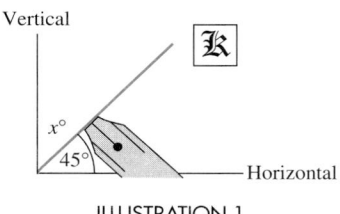

ILLUSTRATION 1

11. Find the complement of an angle measuring 67°.

12. Find the supplement of an angle measuring 117°.

Refer to Illustration 2, in which $l_1 \parallel l_2$.

13. m($\angle 1$) = _____.

14. m($\angle 2$) = _____.

15. m($\angle 3$) = _____.

16. Find x.

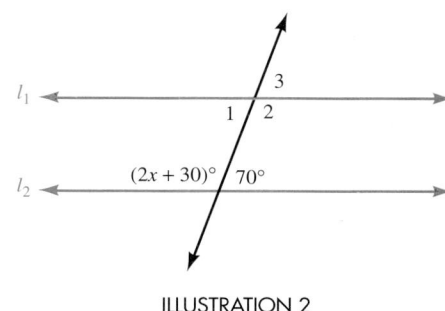

ILLUSTRATION 2

17. Complete the table.

Polygon	Number of sides
Triangle	
Quadrilateral	
Hexagon	
Pentagon	
Octagon	

18. Complete the table about triangles.

Property	Kind of triangle
All sides of equal length	
No sides of equal length	
Two sides of equal length	

In Exercises 19–20, refer to Illustration 3.

19. Find m($\angle A$).

20. Find m($\angle C$).

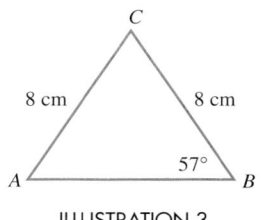

ILLUSTRATION 3

435

21. If the measures of two angles in a triangle are 65° and 85°, find the measure of the third angle.

22. Find the sum of the measures of the angles in a decagon (a ten-sided polygon).

23. In Illustration 4, *ABCD* is a rectangle. Name three pairs of segments with equal lengths.

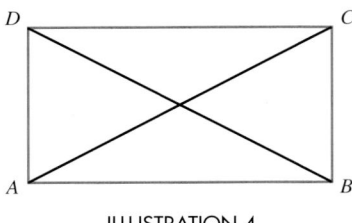

ILLUSTRATION 4

24. In Illustration 5, *ABCD* is an isosceles trapezoid. Find *x*.

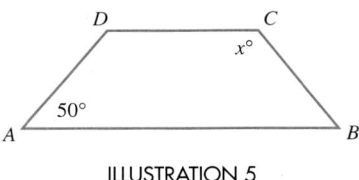

ILLUSTRATION 5

Refer to Illustration 6, in which △ABC ≅ △DEF.

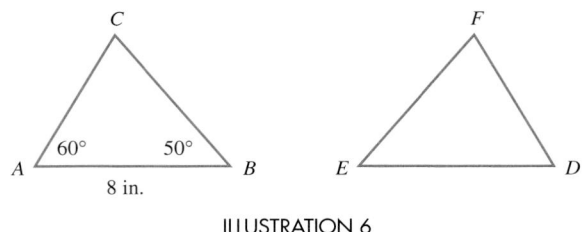

ILLUSTRATION 6

25. Find m($\overline{DE}$).   **26.** Find m($\angle E$).

Give each answer to the nearest tenth.

27. A baseball diamond is a square with each side 90 feet long. What is the straight-line distance from third base to first base?

28. Find the area of a triangle with a base 44.5 centimeters long and a height of 17.6 centimeters.

29. Find the area of a trapezoid with a height of 6 feet and bases that are 12.2 feet and 15.7 feet long.

30. THE OLYMPICS   Steel rod is to be bent to form the interlocking rings of the Olympic Games symbol, shown in Illustration 7. How many feet of steel rod will be needed to make the symbol if the diameter of each ring is to be 6 feet?

ILLUSTRATION 7

31. Find the area of a circle with a diameter that is 6 feet long.

32. Find the volume of a rectangular solid with dimensions 4.3 by 5.7 by 6.5 meters.

33. Find the volume of a sphere that is 8 meters in diameter.

34. Find the volume of a 10-foot-tall pyramid that has a rectangular base 5 feet long and 4 feet wide.

35. Give a real-life example in which the concept of perimeter is used. Do the same for area and for volume. Be sure to discuss the type of units used in each case.

36. Draw a cube. Explain how to find its surface area.

Chapters 1-7 *Cumulative Review Exercises*

1. AMUSEMENT PARKS Use the data in the table to construct a bar graph in Illustration 1.

Fatal accidents on amusement park rides							
Year	'93	'94	'95	'96	'97	'98	'99
Number	4	2	3	3	4	5	6

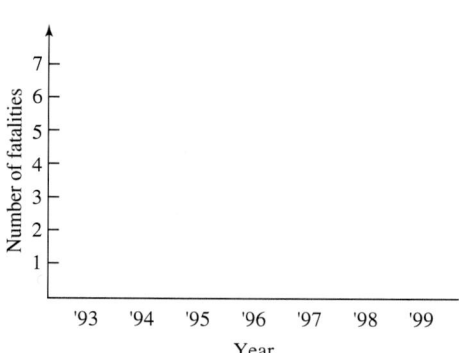

Source: *USA Today* (April 7, 2000)

ILLUSTRATION 1

2. USED CARS Illustration 2 shows an advertisement that appeared in *The Car Trader*. (O.B.O. means "or best offer.")

> 1969 Ford Mustang. New tires
> Must sell!!!! $10,500 O.B.O.

ILLUSTRATION 2

If offers of $8,750, $8,875, $8,900, $8,850, $8,800, $7,995, $8,995, and $8,925 were received, what was the selling price of the car?

3. Subtract: $35,021 - 23,999$.

4. Divide: $1,353 \div 41$.

5. Round 2,109,567 to the nearest thousand.

6. Prime factor 220.

7. Find all the factors of 24.

8. List the set of integers.

9. Evaluate $-10(-2) - 2^3 + 1$.

10. Evaluate $5 - 3[4^2 - (1 + 5 \cdot 2)]$.

11. Evaluate $|-6 - (-3)|$.

12. Simplify $\dfrac{2(2) + 3(-3)}{-4 - (-3)}$.

Solve each equation. Check each result.

13. $-x + 2 = 13$

14. $4 + \dfrac{x}{5} - 6 = -1$

15. Simplify: $\dfrac{35}{28}$.

16. Add: $45\frac{2}{3} + 96\frac{4}{5}$.

17. Subtract: $\dfrac{3}{4} - \dfrac{3}{5}$.

18. BAKING A 5-pound bag of all-purpose flour contains $17\frac{1}{2}$ cups. A baker uses $3\frac{3}{4}$ cups. How much flour is left?

19. Multiply: $-\dfrac{6}{25}\left(2\dfrac{7}{24}\right)$.

20. Divide: $\dfrac{15}{8} \div \dfrac{45}{8}$.

21. VETERINARY MEDICINE A pet owner was told to use an eye dropper to administer medication to his sick kitten. The cup shown in Illustration 3 contains 8 doses of the medication. Determine the size of a single dose.

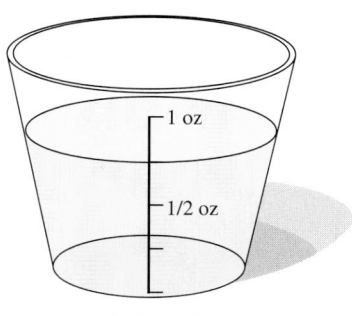

ILLUSTRATION 3

22. Solve $\dfrac{2}{3}q - 1 = -6$.

23. Evaluate $\dfrac{3}{4} + \left(-\dfrac{1}{3}\right)^2\left(\dfrac{5}{4}\right)$.

24. Simplify $\dfrac{7 - \dfrac{2}{3}}{4\dfrac{5}{6}}$.

25. GLOBAL WARMING   Illustration 4 is a line graph of the mean global temperature change, as measured by NASA weather balloons.
  a. When was the greatest rise in temperature recorded? What was it?
  b. When was the greatest decline in temperature recorded? What was it?

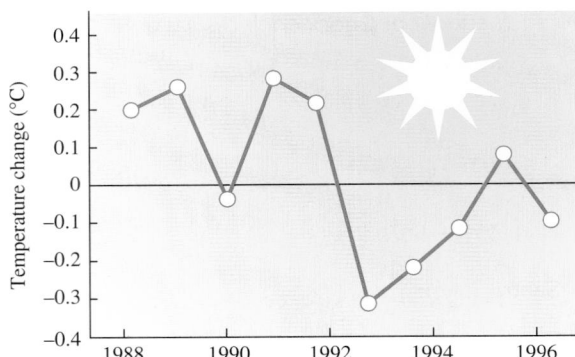

Source: National Aeronautics and Space Administration

ILLUSTRATION 4

26. Graph each member of the set on the number line:

$$\left\{ -4\frac{5}{8}, \sqrt{17}, 2.89, \frac{2}{3}, -0.1, -\sqrt{9}, \frac{3}{2} \right\}$$

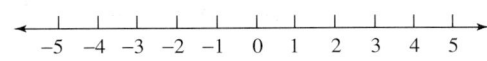

27. Round the number pi to the nearest ten thousandth: $\pi = 3.141592654. \ldots$

28. Place the proper symbol ($>$ or $<$) in the blank: 154.34 _____ 154.33999.

29. Add: $3.4 + 106.78 + 35 + 0.008$.

30. Multiply: $-5.5(-3.1)$.

31. Multiply: $(89.9708)(1,000)$.

32. Divide: $\dfrac{0.0742}{1.4}$.

33. Evaluate $-8.8 + (-7.3 - 9.5)$.

34. Evaluate $\dfrac{7}{8}(9.7 + 15.8)$.

35. Change $\dfrac{2}{15}$ to a decimal.

36. Evaluate $\dfrac{(-1.3)^2 + 6.7}{-0.9}$ and round to the nearest hundredth.

37. DECORATIONS   A mother has budgeted $20 for decorations for her daughter's birthday party. She decides to buy a tank of helium for $15.15 and some balloons. If the balloons sell for 5 cents apiece, how many balloons can she buy?

38. Evaluate $2\sqrt{121} - 3\sqrt{64}$.

39. Simplify $\sqrt{\dfrac{49}{81}}$.

40. What percent of the figure in Illustration 5 is shaded? What percent is not shaded?

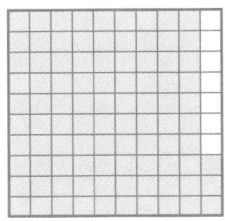

ILLUSTRATION 5

41. What number is 15% of 450?

42. 24.6 is 20.5% of what number?

43. Complete the table.

Percent	Decimal	Fraction
57%		
	0.001	
		$\frac{1}{3}$

44. STUDENT GOVERNMENT   In an election for student body president, 560 votes were cast. Stan Cisneros received 308 votes, and Amy Huang-Sims received 252 votes. Use a circle graph to show the percent of the vote received by each candidate.

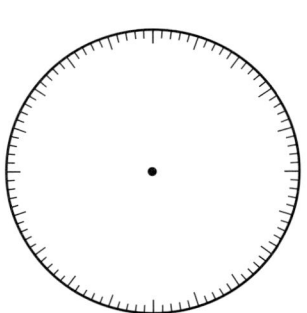

Refer to Illustration 6, in which $l_1 \parallel l_2$. Find the measure of each angle.

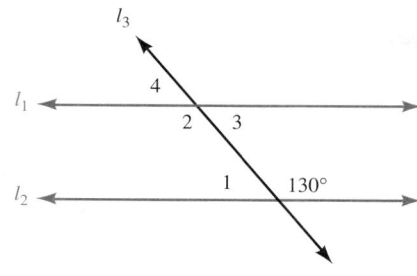

ILLUSTRATION 6

45. m($\angle 1$)　　　　　**46.** m($\angle 2$)

47. m($\angle 3$)　　　　　**48.** m($\angle 4$)

Refer to Illustration 7, in which AB $\parallel$ DE and m(AC) = m(BC). Find the measure of each angle.

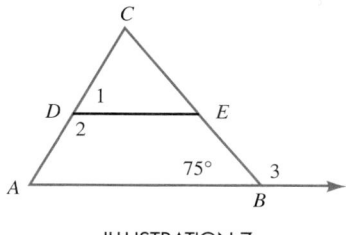

ILLUSTRATION 7

49. m($\angle 1$)　　　　　**50.** m($\angle C$)

51. m($\angle 2$)　　　　　**52.** m($\angle 3$)

53. JAVELIN THROW　See Illustration 8. Determine x and y.

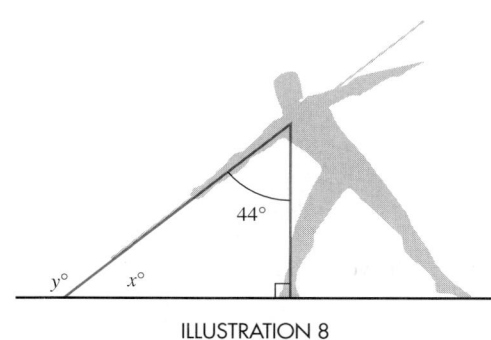

ILLUSTRATION 8

54. Find the sum of the angles of a pentagon.

55. If two sides of a right triangle measure 5 meters and 12 meters, how long is the hypotenuse?

If an answer is not exact, round to the nearest hundredth.

56. Find the perimeter and area of a rectangle with dimensions of 9 meters by 12 meters.

57. Find the area of a triangle with a base that is 14 feet long and an altitude of 18 feet.

58. Find the area of a trapezoid that has bases that are 12 inches and 14 inches long and a height of 7 inches.

59. Find the circumference and area of a circle with a diameter of 14 centimeters.

60. Find the area of the shaded region in Illustration 9, which is created using 2 semicircles.

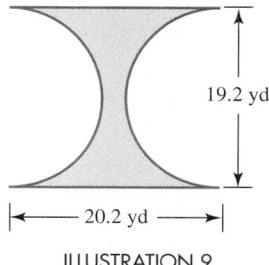

ILLUSTRATION 9

61. Find the volume of a rectangular solid with dimensions of 5 meters by 6 meters by 7 meters.

62. Find the volume of a sphere with a diameter of 10 inches.

63. Find the volume of a cone that has a circular base 8 meters in diameter and a height of 9 meters.

64. Find the volume of a cylindrical pipe that is 20 feet long and 6 inches in diameter.

65. Find the surface area of a block of ice that is in the shape of a rectangular solid with dimensions 15 in. $\times$ 24 in. $\times$ 18 in.

8

Algebraic Expressions, Equations, and Inequalities

8.1 Algebraic Expressions

8.2 Simplifying Algebraic Expressions

8.3 Solving Equations

8.4 Formulas

8.5 Problem Solving

8.6 Inequalities

Key Concept: Simplify and Solve

Accent on Teamwork

Chapter Review

Chapter Test

Cumulative Review Exercises

IN THIS CHAPTER, WE WILL LEARN HOW TO WRITE, EVALUATE, AND SIMPLIFY ALGEBRAIC EXPRESSIONS. WE WILL THEN USE THESE SKILLS TO SOLVE EQUATIONS AND INEQUALITIES.

8.1 Algebraic Expressions

In this section, you will learn about

- Translating from words to mathematical symbols
- Writing algebraic expressions to represent unknown quantities
- Looking for hidden operations • Number and value
- Evaluating algebraic expressions • Making tables

INTRODUCTION. Since problems in algebra are often presented in words, the ability to interpret what you read is important. In this section, we will introduce several strategies that will help you translate English words into mathematical symbols. We will begin by discussing the most fundamental skill—spotting key words and phrases that represent four operations of arithmetic.

Translating from words to mathematical symbols

An **algebraic expression** is a collection of numbers and/or variables that are combined by using the operations of arithmetic. In the tables below and on the next page, we list some words and phrases that are used to indicate addition, subtraction, multiplication, and division, and we show how they can be translated to form algebraic expressions.

ADDITION

The phrase	translates to
the sum of a and 8	$a + 8$
4 plus c	$4 + c$
16 added to m	$m + 16$
4 more than t	$t + 4$
20 greater than F	$F + 20$
T increased by r	$T + r$
Exceeds y by 35	$y + 35$

SUBTRACTION

The phrase	translates to
the difference of 23 and P	$23 - P$
550 minus h	$550 - h$
18 less than w	$w - 18$
7 decreased by j	$7 - j$
M reduced by x	$M - x$
12 subtracted from L	$L - 12$
5 less f	$5 - f$

MULTIPLICATION

The phrase	translates to
the product of 4 and x	$4x$
20 times B	$20B$
twice r	$2r$
triple the profit P	$3P$
$\frac{3}{4}$ of m	$\frac{3}{4}m$

DIVISION

The phrase	translates to
the quotient of R and 19	$\dfrac{R}{19}$
s divided by d	$\dfrac{s}{d}$
the ratio of c to d	$\dfrac{c}{d}$
k split into 4 equal parts	$\dfrac{k}{4}$

!　**COMMENT**　The phrase *greater than* is used to indicate addition. The phrase *is greater than* refers to the symbol $>$. Similarly, the phrase *less than* indicates subtraction, and the phrase *is less than* refers to the symbol $<$.

EXAMPLE 1　*Translating to symbols.*　Write each phrase as an algebraic expression.

a. The sum of the length l and the width 20

b. 5 less than the capacity c

c. The product of the weight w and 2,000, increased by 300

Solution

a. Key word: *sum*　　**Translation:** add

The phrase translates to $l + 20$.

b. Key phrase: *less than*　　**Translation:** subtract

The capacity c is to be made less, so we subtract 5 from it: $c - 5$.

c. Key word: *product*　　**Translation:** multiply

Key phrase: *increased by*　　**Translation:** add

The weight w is to be multiplied by 2,000, and then 300 is to be added to the product: $2,000w + 300$.

Self Check

Write each phrase as an algebraic expression:

a. 80 cents less than t cents

b. $\frac{2}{3}$ of the time T

c. the difference of twice a and 15

Answers:　**a.** $t - 80$,　**b.** $\frac{2}{3}T$,　**c.** $2a - 15$　■

Writing algebraic expressions to represent unknown quantities

When solving problems, we often begin by letting a variable stand for an unknown quantity. Frequently, a problem will contain a second unknown but related quantity, which can be described using an algebraic expression involving the original variable.

EXAMPLE 2　*Writing an algebraic expression.*　A butcher trims 4 ounces of fat from a roast that originally weighed x ounces. Write an algebraic expression that represents the weight of the roast after it is trimmed.

Solution

We let x = the original weight of the roast (in ounces).

　Key word: *trimmed*　　**Translation:** subtract

After 4 ounces of fat have been trimmed, the weight of the roast is $(x - 4)$ ounces.

Self Check

When a secretary rides the bus to work, it takes her m minutes. If she drives her own car, her travel time exceeds this by 15 minutes. How can we represent the time it takes her to get to work by car?

Answer:　$(m + 15)$ minutes　■

EXAMPLE 3 *Writing an algebraic expression.* The swimming pool in Figure 8-1 is *x* feet wide. If it is to be sectioned into 8 equally wide swimming lanes, write an algebraic expression that represents the width of each lane.

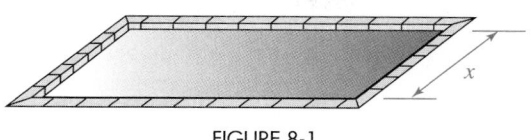

FIGURE 8-1

Solution

We let *x* = the width of the swimming pool (in feet).

Key phrase: *sectioned into 8 equally wide lanes*   **Translation:** divide

The width of each lane is $\frac{x}{8}$ feet.

Self Check

A handyman estimates that it will take the same amount of time to sand as it will to paint some kitchen cabinets. If the entire job takes *x* hours, how can we express the time it will take him to do the painting?

Answer: $\frac{x}{2}$ hours ∎

When we are solving problems, the variable to be used is rarely specified. We must decide what the unknown quantities are and how they will be represented using variables. The following examples illustrate how to approach these situations.

EXAMPLE 4 *Two unknown quantities.* The value of a collectible doll is three times that of an antique toy truck. Express the value of each, using one variable.

Solution

There are two unknown quantities. Since the doll's value is related to the truck's value, we will let *x* = the value of the toy truck in dollars.

Key phrase: 3 *times*   **Translation:** multiply by 3

The value of the doll is $3x.

Self Check

The McDonald's Chicken Deluxe sandwich has 5 fewer grams of fat than the Quarter Pounder hamburger. Express the number of grams of fat in each sandwich, using one variable.

Answers: *x* = the number of grams of fat in the hamburger; *x* − 5 = the number of grams of fat in the chicken sandwich ∎

 COMMENT A variable is used to represent an unknown number. Therefore, in Example 4, it would be incorrect to write, "Let *x* = toy truck," because the truck is not a number. We need to write, "Let *x* = the *value* of the toy truck."

EXAMPLE 5 *Two unknown quantities.* A 10-inch-long paintbrush has two parts: a handle and bristles. Choose a variable to represent the length of one of the parts. Then write an expression for the length of the other part.

Solution

A drawing is helpful in explaining this problem. If the entire paintbrush is 10 inches long, and if we let *h* represent the length of the handle, then the bristles are (10 − *h*) inches long.

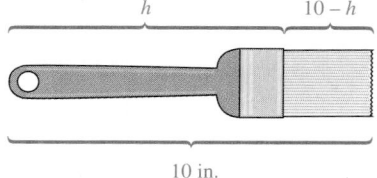

10 in.

Self Check

Part of a $900 donation to a preschool was designated to go to the scholarship fund, the remainder to the building fund. Choose a variable to represent the amount donated to one of the funds. Write an expression for the amount donated to the other fund.

Answers: *s* = amount donated to scholarship fund in dollars; 900 − *s* = amount donated to building fund ∎

EXAMPLE 6   *An expression involving two operations.*   In the second semester, student enrollment in a retraining program at a college was 32 more than twice that of the first semester. Express the student enrollment in the program each semester, using one variable.

Solution

Since the second-semester enrollment is expressed in terms of the first-semester enrollment, we let x = the enrollment in the first semester.

Key phrase: *more than*   **Translation:** add
Key word: *twice*   **Translation:** multiply by 2

The enrollment for the second semester is $2x + 32$.

Self Check
The number of votes received by the incumbent in an election was 55 fewer than three times the number the challenger received. Express the number of votes received by each candidate, using one variable.

Answers: x = the number of votes received by the challenger; $3x - 55$ = the number of votes received by the incumbent   ∎

Looking for hidden operations

When analyzing problems, we aren't always given key words or key phrases to help establish what mathematical operation to use. Sometimes a careful reading of the problem is needed to determine the hidden operations.

EXAMPLE 7   *Hidden operations.*   Disneyland, located in Anaheim, California, was in operation 16 years before the opening of Walt Disney World, in Orlando, Florida. Euro Disney, in Paris, France, was constructed 21 years after Disney World. Use algebraic expressions to express the ages (in years) of each of these Disney attractions.

Solution   The ages of Disneyland and Euro Disney are both related to the age of Walt Disney World. Therefore, we will let x = the age of Walt Disney World.

In carefully reading the problem, we find that Disneyland was built 16 years *before* Disney World, so its age is more than that of Disney World.

Key phrase: *more than*   **Translation:** add

In years, the age of Disneyland is $x + 16$. Euro Disney was built 21 years *after* Disney World, so its age is less than that of Disney World.

Key phrase: *less than*   **Translation:** subtract

In years, the age of Euro Disney is $x - 21$. The results are summarized in Table 8-1.   ∎

Attraction	Age
Disneyland	$x + 16$
Disney World	x
Euro Disney	$x - 21$

TABLE 8-1

EXAMPLE 8   *Looking for a pattern.*   How many months are in x years?
Solution
Since there are no key words, we must carefully analyze the problem to write an expression that represents the number of months in x years. It is often helpful to consider some specific cases. For example, let's calculate the number of months in 1 year, 2 years, and 3 years. When we write the results in a table, a pattern is apparent.

Self Check
Complete the table. How many days is h hours?

Number of hours	Number of days
24	
48	
72	
h	

Number of years	Number of months
1	12
2	24
3	36
x	$12x$

We multiply the number of years by 12 to find the number of months.

Therefore, if $x =$ the number of years, the number of months is $12 \cdot x$ or $12x$.

Answers: $1, 2, 3; \dfrac{h}{24}$ ■

Number and value

Some problems deal with quantities that have value. In these problems, we must distinguish between *the number of* and *the value of* the unknown quantity. For example, to find the value of 3 quarters, we multiply the number of quarters by the value (in cents) of one quarter. Therefore, the value of 3 quarters is $3 \cdot 25¢ = 75¢$.

The same distinction must be made if the number is unknown. For example, the value of n nickels is not $n¢$. The value of n nickels is $n \cdot 5¢ = (5n)¢$. For problems of this type, we will use the relationship

$$\text{Number} \cdot \text{value} = \text{total value}$$

EXAMPLE 9 *Number and value.* Suppose a roll of paper towels sells for 79¢. Find the cost of **a.** five rolls of paper towels, **b.** x rolls of paper towels, and **c.** $x + 1$ rolls of paper towels.

Solution
In each case, we will multiply the *number* of rolls of paper towels by the *value* of one roll (79¢) to find the total cost.

a. The cost of 5 rolls of paper towels is $5 \cdot 79¢ = 395¢$, or \$3.95.

b. The cost of x rolls of paper towels is $x \cdot 79¢ = (79 \cdot x)¢ = (79x)¢$.

c. The cost of $x + 1$ rolls of paper towels is
$(x + 1) \cdot 79¢ = 79 \cdot (x + 1)¢ = 79(x + 1)¢$.

Self Check
Find the value of

a. six \$50 savings bonds

b. t \$100 savings bonds

c. $(x - 4)$ \$1,000 savings bonds

Answers: **a.** \$300, **b.** \$100t, **c.** \$1,000$(x - 4)$ ■

Evaluating algebraic expressions

To **evaluate an algebraic expression,** we replace each variable with a given number value. (When we replace a variable with a number, we say we are **substituting** for the variable.) Then we do the necessary calculations following the rules for the order of operations. For example, to evaluate $x^2 - 2x + 1$ for $x = 3$, we begin by replacing each x with 3.

$$
\begin{aligned}
x^2 - 2x + 1 &= 3^2 - 2(3) + 1 && \text{Replace } x \text{ with 3.}\\
&= 9 - 2(3) + 1 && \text{Evaluate the exponential expression: } 3^2 = 9.\\
&= 9 - 6 + 1 && \text{Do the multiplication: } 2(3) = 6.\\
&= 4 && \text{Working left to right, do the subtraction and then}\\
& && \text{the addition.}
\end{aligned}
$$

We say that 4 is the **value** of this expression when $x = 3$.

COMMENT   When replacing a variable with its numerical value, use parentheses around the replacement number to avoid possible misinterpretation. For example, when substituting 5 for x in $2x + 1$, we show the multiplication using parentheses: $2(5) + 1$. If we don't show the multiplication, we could misread the expression as $25 + 1$.

EXAMPLE 10   *Evaluating algebraic expressions.*   Evaluate
a. $-y$ and   **b.** $-3(y + x^2)$ when $x = 3$ and $y = -4$.

Solution

a. $-y = -(-4)$ Substitute -4 for y.

  $\quad\;\; = 4$ The opposite of -4 is 4.

b. $-3(y + x^2) = -3(-4 + 3^2)$ Substitute 3 for x and -4 for y.

  $\quad\quad\quad\quad\; = -3(-4 + 9)$ Work within the parentheses first. Evaluate the exponential expression.

  $\quad\quad\quad\quad\; = -3(5)$ Do the addition within the parentheses.

  $\quad\quad\quad\quad\; = -15$

> **Self Check**
> Evaluate   **a.** $-x$
> and   **b.** $5(x - y)$
> when $x = -2$ and $y = 3$.
>
> **Answers: a.** 2,   **b.** -25 ∎

EXAMPLE 11   *Surface area of a swim fin.*   Divers use swim fins because they provide a much larger surface area to push against the water than do bare feet. Consequently, the diver can swim faster wearing them. In Figure 8-2, we see that the fin is in the shape of a trapezoid. The algebraic expression $\frac{1}{2}h(b + d)$ gives the area of a trapezoid, where h is the height and b and d are the lengths of the lower and upper bases, respectively. To find the area of the fin shown here, we evaluate the algebraic expression for $h = 14$, $b = 3.5$, and $d = 8.5$.

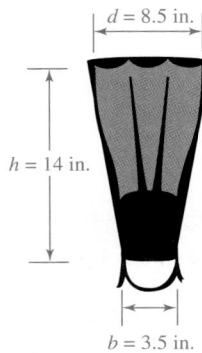

$d = 8.5$ in.

$h = 14$ in.

$b = 3.5$ in.

FIGURE 8-2

$\dfrac{1}{2}h(b + d) = \dfrac{1}{2}(14)(3.5 + 8.5)$ Substitute 14 for h, 3.5 for b, and 8.5 for d.

$\quad\quad\quad\quad\; = \dfrac{1}{2}(14)(12)$ Do the addition within the parentheses.

$\quad\quad\quad\quad\; = 7(12)$ Work from left to right: $\frac{1}{2}(14) = 7$.

$\quad\quad\quad\quad\; = 84$

The fin has an area of 84 square inches.   ∎

EXAMPLE 12   *Temperature conversion.*   The expression $\frac{9C + 160}{5}$ converts a temperature in degrees Celsius (represented by C) to a temperature in degrees Fahrenheit. Convert $-170°$ C, the coldest temperature on the moon, to degrees Fahrenheit.

Solution

To convert $-170°$ C to degrees Fahrenheit, we evaluate the algebraic expression for $C = -170$.

$\dfrac{9C + 160}{5} = \dfrac{9(-170) + 160}{5}$ Substitute -170 for C.

$\quad\quad\quad\;\; = \dfrac{-1{,}530 + 160}{5}$ Do the multiplication.

$\quad\quad\quad\;\; = \dfrac{-1{,}370}{5}$ Do the addition.

$\quad\quad\quad\;\; = -274$ Do the division.

In degrees Fahrenheit, the coldest temperature on the moon is $-274°$.

> **Self Check**
> On January 22, 1943, the temperature in Spearfish, South Dakota changed from $-20°$ C to $7.2°$ C in two minutes. Convert $-20°$ C to degrees Fahrenheit.
>
> **Answer:** $-4°$ F   ∎

Accent on Technology: **Evaluating algebraic expressions**

The rotating drum of a clothes dryer is a cylinder. (See Figure 8-3.) To find the capacity of the dryer, we can find its volume by evaluating the algebraic expression $\pi r^2 h$, where r represents the radius and h represents the height of the drum. (Here, the cylinder is lying on its side). If we substitute 13.5 for r and 20 for h, we obtain $\pi(13.5)^2(20)$. Using a scientific calculator, we can evaluate the expression by entering these numbers and pressing these keys.

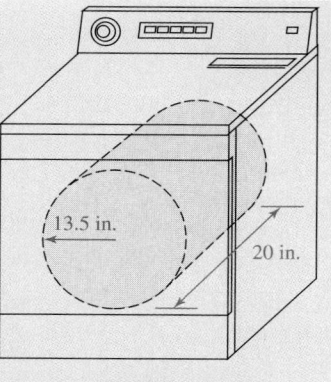

FIGURE 8-3

Keystrokes $\boxed{\pi}$ $\boxed{\times}$ 13.5 $\boxed{x^2}$ $\boxed{\times}$ 20 $\boxed{=}$ $\boxed{\texttt{11451.10522}}$

Using a graphing calculator, we can evaluate the expression by entering these numbers and pressing these keys.

Keystrokes $\boxed{\text{2nd}}$ $\boxed{\pi}$ $\boxed{\times}$ 13.5 $\boxed{x^2}$ $\boxed{\times}$ 20 $\boxed{\text{ENTER}}$

$$\boxed{\begin{array}{l} \pi*13.5^2*20 \\ \hspace{2.5cm} \texttt{11451.10522} \end{array}}$$

To the nearest cubic inch, the capacity of the dryer is 11,451 in.^{3}.

Making tables

EXAMPLE 13 *Ballistics.* If a toy rocket is shot into the air with an initial velocity of 80 feet per second, its height (in feet) after t seconds in flight is given by the algebraic expression

$$80t - 16t^2$$

How many seconds after the launch will it hit the ground?

Solution

We can substitute positive values for t, the time in flight, until we find the one that gives a height of 0. At that time, the rocket will be on the ground. We will begin by finding the height after the rocket has been in flight for 1 second ($t = 1$) and record the result in a table.

$$80t - 16t^2 = 80(1) - 16(1)^2 \quad \text{Substitute 1 for } t.$$
$$= 64$$

After 1 second in flight, the height of the rocket is 64 feet. We continue to pick more values of t until we find out when the height is 0.

As we evaluate $80t - 16t^2$ for various values of t, we can show the results in a **table of values.** In the column headed "t," we list each value of the variable to be used in the evaluations. In the column headed "$80t - 16t^2$," we write the result of each evaluation.

Self Check

In Example 13, suppose the height of the rocket is given by $112t - 16t^2$. Complete the table to find out how many seconds after launch it would hit the ground.

t	$112t - 16t^2$
1	
3	
5	
7	

t	$80t - 16t^2$
1	64
2	96
3	96
4	64
5	0

Evaluate for $t = 2$:
$80t - 16t^2 = 80(2) - 16(2)^2 = 96$

Evaluate for $t = 3$:
$80t - 16t^2 = 80(3) - 16(3)^2 = 96$

Evaluate for $t = 4$:
$80t - 16t^2 = 80(4) - 16(4)^2 = 64$

Evaluate for $t = 5$:
$80t - 16t^2 = 80(5) - 16(5)^2 = 0$

Since the height of the rocket is 0 when $t = 5$, the rocket will hit the ground in 5 seconds.

Answer: 7 (the heights are 96, 192, 160, and 0) ∎

The two columns of a table of values are sometimes headed with the terms **input** and **output**, as shown in the table at right. The t-values are the inputs into the expression $80t - 16t^2$, and the resulting values are thought of as the outputs.

Input	Output
1	64
2	96
3	96
4	64
5	0

STUDY SET   Section 8.1

VOCABULARY   *Fill in the blanks.*

1. To _____ an algebraic expression, we substitute the values for the variables and then apply the rules for the order of operations.

2. Variables and/or numbers can be combined with the operation symbols of addition, subtraction, multiplication, and division to create algebraic _____.

3. $2x + 5$ is an example of an algebraic _____, whereas $2x + 5 = 7$ is an example of an _____.

4. When we evaluate an algebraic expression, such as $5x - 8$, for several values of x, we can keep track of the results in an input/output _____.

CONCEPTS

5. Write two algebraic expressions that contain the variable x and the numbers 6 and 20.

6. a. Complete the table to determine how many days are in w weeks.

Number of weeks	Number of days
1	
2	
3	
w	

b. Complete the table to answer this question: *s* seconds is how many minutes?

Number of seconds	Number of minutes
60	
120	
180	
s	

7. When evaluating $3x - 6$ for $x = 4$, what misunderstanding can occur if we don't write parentheses around 4 when it is substituted for the variable?

8. If the knife in Illustration 1 is 12 inches long, how long is the blade?

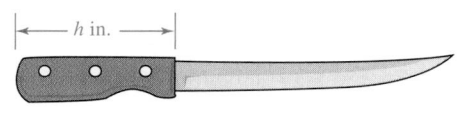

ILLUSTRATION 1

9. a. In Illustration 2, the weight of the van is 500 pounds less than twice the weight of the car. Express the weight of the van and the car using the variable *x*.

ILLUSTRATION 2

b. If the actual weight of the car is 2,000 pounds, what is the weight of the van?

10. See Illustration 3.
a. If we let *b* represent the length of the beam, write an algebraic expression for the length of the pipe.
b. If we let *p* represent the length of the pipe, write an algebraic expression for the length of the beam.

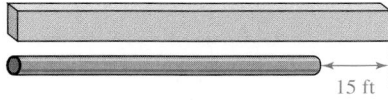

15 ft

ILLUSTRATION 3

11. Complete the table in Illustration 4.

Type of coin	Number	Value in cents	Total value in cents
Nickel	6		
Dime	*d*		
Half dollar	$x + 5$		

ILLUSTRATION 4

12. If $x = -9$, find the value of
a. $-x$ **b.** $-(-x)$
c. $-x^2$ **d.** $(-x)^2$

NOTATION *Complete each solution.*

13. Evaluate the expression $9a - a^2$ for $a = 5$.
$$9a - a^2 = 9(\quad) - (\quad)^2$$
$$= 9(5) - \boxed{}$$
$$= \boxed{} - 25$$
$$= 20$$

14. Evaluate $\dfrac{4x^2 - 3y}{9(x - y)}$ when $x = 4$ and $y = -3$.
$$\frac{4x^2 - 3y}{9(x - y)} = \frac{4(4)^2 - 3(-3)}{9[4 - (-3)]}$$
$$= \frac{4(\quad) - 3(\quad)}{9(\quad)}$$
$$= \frac{\boxed{} - (\boxed{})}{\boxed{}}$$
$$= \frac{73}{63}$$

PRACTICE *Translate each phrase to an algebraic expression. If no variable is given, use x as the variable.*

15. The sum of the length *l* and 15
16. The difference of a number and 10
17. The product of a number and 50
18. Three-fourths of the population *p*
19. The ratio of the amount won *w* and lost *l*
20. The tax *t* added to *c*
21. *P* increased by *p*
22. 21 less than the total height *h*
23. The square of *k* minus 2,005
24. *s* subtracted from *S*
25. *J* reduced by 500
26. Twice the attendance *a*

27. 1,000 split n equal ways

28. Exceeds the cost c by 25,000

29. 90 more than the current price p

30. 64 divided by the cube of y

31. The total of 35, h, and 300

32. x decreased by 17

33. 680 fewer than the entire population p

34. Triple the number of expected participants

35. The product of d and 4, decreased by 15

36. Forty-five more than the quotient of y and 6

37. Twice the sum of 200 and t

38. The square of the quantity 14 less than x

39. The absolute value of the difference of a and 2

40. The absolute value of a, decreased by 2

In Exercises 41–44, if n represents a number, write a word description of each algebraic expression. (Answers may vary.)

41. $n - 7$

42. $n^2 + 7$

43. $7n + 4$

44. $3(n + 1)$

45. How many minutes there are in   **a.** 5 hours and   **b.** h hours?

46. A woman watches television x hours a day. Express the number of hours she watches TV   **a.** in a week and   **b.** in a year.

47.   **a.** How many feet are in y yards?   **b.** How many yards are in f feet?

48. A sales clerk earns \$$x$ an hour. How much does he earn in   **a.** an 8-hour day and   **b.** a 40-hour week?

49. If a car rental agency charges 29¢ a mile, express the rental fee if a car is driven x miles.

50. A model's skirt is x inches long. The designer then lets the hem down 2 inches. How can we express the length (in inches) of the altered skirt?

51. A soft drink manufacturer produced c cans of cola during the morning shift. Write an expression for how many six-packs of cola can be assembled from the morning shift's production.

52. The tag on a new pair of 36-inch-long jeans warns that after washing, they will shrink x inches in length. Express the length (in inches) of the jeans after they are washed.

53. A caravan of b cars, each carrying 5 people, traveled to the state capital for a political rally. Express how many people were in the car caravan.

54. A caterer always prepares food for 10 more people than the order specifies. If p people are to attend a reception, write an expression for the number of people she should prepare for.

55. Tickets to a circus cost \$5 each. Express how much tickets will cost for a family of x people if they also pay for two of their neighbors.

56. If each egg is worth e¢, express the value (in cents) of a dozen eggs.

Complete each table of values.

57.

x	$x^3 - 1$
0	
-1	
-3	

58.

g	$g^2 - 7g + 1$
0	
7	
-10	

59.

s	$\frac{5s + 36}{s}$
1	
6	
-12	

60.

a	$2{,}500a + a^3$
2	
4	
-5	

61.

Input x	Output $2x - \frac{x}{2}$
100	
-300	

62.

Input x	Output $\frac{x}{3} + \frac{x}{4}$
12	
-36	

63.

x	$(x + 1)(x + 5)$
-1	
-5	
-6	

64.

x	$\frac{1}{x + 8}$
-7	
-9	
-8	

Evaluate each expression, given that $x = 3$, $y = -2$, and $z = -4$.

65. $3y^2 - 6y - 4$

66. $-z^2 - z - 12$

67. $(3 + x)y$

68. $(4 + z)y$

69. $(x + y)^2 - |z + y|$

70. $[(z - 1)(z + 1)]^2$

71. $(4x)^2 + 3y^2$

72. $4x^2 + (3y)^2$

73. $-\dfrac{2x + y^3}{y + 2z}$

74. $-\dfrac{2z^2 - y}{2x - y^2}$

Evaluate each expression for the given values of the variables.

75. $b^2 - 4ac$ for $a = -1$, $b = 5$, and $c = -2$

76. $(x - a)^2 + (y - b)^2$ for $x = -2$, $y = 1$, $a = 5$, and $b = -3$

77. $a^2 + 2ab + b^2$ for $a = -5$ and $b = -1$

78. $\dfrac{x - a}{y - b}$ for $x = -2$, $y = 1$, $a = 5$, and $b = 2$

79. $\dfrac{n}{2}[2a + (n - 1)d]$ for $n = 10$, $a = -4$, and $d = 6$

80. $\dfrac{a(1 - r^n)}{1 - r}$ for $a = -5$, $r = 2$, and $n = 3$

81. $\dfrac{a^2 + b^2}{2}$ for $a = 1.8$ and $b = -7.6$

82. $(y^3 - 52y^2)^2$ for $y = 55$

APPLICATIONS

83. ROCKETRY   The algebraic expression $64t - 16t^2$ gives the height of a toy rocket (in feet) t seconds after being launched. Find the height of the rocket for each of the times shown in Illustration 5. Present your results in an input/output table.

t	h
0	
0.5	
1	
1.5	
2	
2.5	
3	
3.5	
4	

ILLUSTRATION 5

84. GROWING SOD   To determine the number of square feet of sod *remaining* in a field after filling an order (see Illustration 6), the manager of a sod farm uses the expression $20,000 - 3s$ (where s is the number of 1-foot-by-3-foot strips the customer has ordered). To sod a soccer field, a city orders 7,000 strips of sod. Evaluate the expression for this value of s and explain the result.

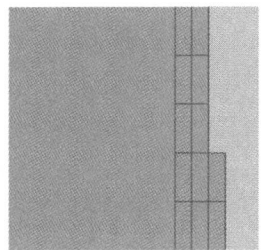

1-ft-by-3-ft strips of sod, cut and ready to be loaded on a truck for delivery

ILLUSTRATION 6

85. The expression

$$\dfrac{5(F - 32)}{9}$$

converts a temperature in degrees Fahrenheit (given as F) to degrees Celsius. Convert the temperatures listed on the container of antifreeze shown in Illustration 7 to degrees Celsius. Round to the nearest degree.

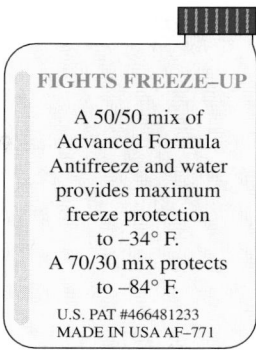

FIGHTS FREEZE–UP

A 50/50 mix of Advanced Formula Antifreeze and water provides maximum freeze protection to –34° F. A 70/30 mix protects to –84° F.

U.S. PAT #466481233
MADE IN USA AF–771

ILLUSTRATION 7

86. TEMPERATURE ON MARS   On Mars, maximum summer temperatures can reach 20° C. However, daily temperatures average −33° C. Convert each of these temperatures to degrees Fahrenheit. See Example 12 (page 446). Round to the nearest degree.

87. The utility knife blade shown in Illustration 8 is in the shape of a trapezoid. Find the area of the front face of the blade. (See Example 11 on page 446 for the expression that gives the area of a trapezoid.)

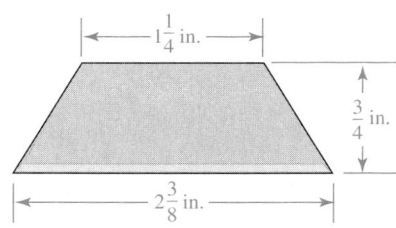

ILLUSTRATION 8

88. TRUMPET MUTE   The expression

$$\pi[b^2 + d^2 + (b + d)s]$$

can be used to find the total surface area of the trumpet mute shown in Illustration 9. Evaluate the expression for the given dimensions to find the number of square inches of cardboard (to the nearest tenth) used to make the mute.

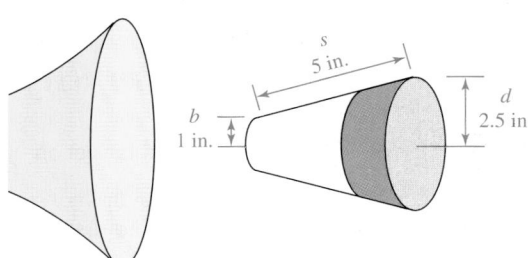

ILLUSTRATION 9

89. LANDSCAPING   A grass strip is to be planted around a tree, as shown in Illustration 10 on the next page. Find the number of square feet of sod to order by evaluating the expression $\pi(R^2 - r^2)$. Round to the nearest square foot.

$R = 9.75$ ft $r = 4.5$ ft

ILLUSTRATION 10

90. ⊞ ENERGY CONSERVATION   A fiberglass blanket wrapped around a water heater helps prevent heat loss. See Illustration 11. Find the number of square feet of heater surface the blanket covers by evaluating the algebraic expression $2\pi rh$, where r is the radius and h is the height. Round to the nearest square foot.

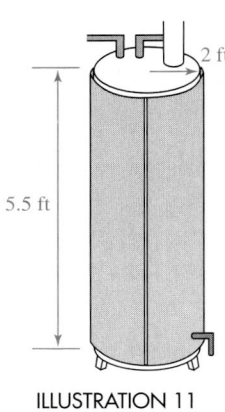

2 ft

5.5 ft

ILLUSTRATION 11

WRITING

91. What is an algebraic expression? Give some examples.

92. What is a variable? How are variables used in this section?

93. In this section, we substituted a number for a variable. List some other uses of the word *substitute* that you encounter in everyday life.

94. Explain why d dimes are not worth $d¢$.

REVIEW

95. Simplify -0.

96. Is the statement $-5 > -4$ true or false?

97. Evaluate $\left|-\dfrac{2}{3}\right|$.

98. Evaluate $2^3 \cdot 3^2$.

99. Write $5 \cdot 5 \cdot 5 \cdot 5$ in exponential form.

100. Evaluate $15 + 2[15 - (12 - 10)]$.

101. Find the mean (average) of the three test scores 84, 93, and 72.

102. Fill in the blanks: In the multiplication statement $5 \cdot x = 5x$, 5 and x are called _____, and $5x$ is called the _____.

8.2 *Simplifying Algebraic Expressions*

In this section, you will learn about

- Simplifying algebraic expressions involving multiplication
- The distributive property • The vocabulary of algebraic expressions
- Like terms • Combining like terms

INTRODUCTION. In arithmetic, we often replace one numerical expression with another expression that is equivalent and simpler in form. For example, when we express $\frac{40}{80}$ as $\frac{1}{2}$, we say we have *simplified* $\frac{40}{80}$. In algebra, we must often simplify algebraic expressions. To **simplify an algebraic expression,** we use one or more properties of algebra to write the expression in an equivalent, less-complicated form.

Simplifying algebraic expressions involving multiplication

Two properties that are often used to simplify algebraic expressions are the associative and commutative properties of multiplication. Recall that the associative property of multiplication enables us to change the grouping of factors involved in a multiplication. The commutative property of multiplication enables us to change the order of the factors.

As an example, let's consider the expression $8(4x)$ and simplify it as follows:

$$8(4x) = 8 \cdot (4 \cdot x) \quad \text{$4x = 4 \cdot x$.}$$
$$= (8 \cdot 4) \cdot x \quad \text{Apply the associative property of multiplication to group 4 with 8 instead of with x.}$$
$$= 32x \quad \text{Do the multiplication within the parentheses: $8 \cdot 4 = 32$.}$$

Since $8(4x) = 32x$, we say that $8(4x)$ simplifies to $32x$. To verify that $8(4x)$ and $32x$ are **equivalent expressions** (represent the same number), we can evaluate each expression for several choices of x. For each value of x, the results should be the same.

If $x = 10$		**If $x = -3$**	
$8(4x) = 8[4(10)]$	$32x = 32(10)$	$8(4x) = 8[4(-3)]$	$32x = 32(-3)$
$= 8(40)$	$= 320$	$= 8(-12)$	$= -96$
$= 320$		$= -96$	

EXAMPLE 1 *Simplifying algebraic expressions involving multiplication.* Simplify each expression: **a.** $15a(-7)$, **b.** $5\left(\frac{4}{5}x\right)$, **c.** $-5r(-6s)$, and **d.** $3(7p)(-5p)$.

Solution

a. $15a(-7) = 15(-7)a$ Use the commutative property of multiplication to change the order of the factors.

$= -105a$ Working left to right, do the multiplications.

b. $5\left(\dfrac{4}{5}x\right) = \left(5 \cdot \dfrac{4}{5}\right)x$ Use the associative property of multiplication to group the numbers.

$= 4x$ Multiply: $5 \cdot \dfrac{4}{5} = \dfrac{5}{1} \cdot \dfrac{4}{5} = \dfrac{\overset{1}{\cancel{5}} \cdot 4}{1 \cdot \underset{1}{\cancel{5}}} = 4$.

c. We note that the expression contains two variables.

$-5r(-6s) = [-5(-6)][r \cdot s]$ Use the commutative and associative properties of multiplication to group the numbers and group the variables.

$= 30rs$ Do the multiplications within the brackets: $-5(-6) = 30$ and $r \cdot s = rs$.

d. $3(7p)(-5p) = [3(7)(-5)](p \cdot p)$ Use the commutative and associative properties of multiplication to change the order and to regroup the factors.

$= -105p^2$ Do the multiplication within the grouping symbols: $3(7)(-5) = -105$ and $p \cdot p = p^2$.

Self Check

Simplify each expression:

a. $9 \cdot 6s$

b. $8\left(\frac{7}{8}h\right)$

c. $21p(-3q)$

d. $-4(6m)(-2m)$

Answers: **a.** $54s$, **b.** $7h$, **c.** $-63pq$, **d.** $48m^2$ ∎

The distributive property

To introduce the **distributive property,** we will examine the expression $4(5 + 3)$, which can be evaluated in two ways.

Method 1. Rules for the order of operations: In this method, we compute the sum within the parentheses first.

$$4(\mathbf{5} + \mathbf{3}) = 4(\mathbf{8}) \quad \text{Do the addition within the parentheses first.}$$

$$= 32 \quad \text{Do the multiplication.}$$

Method 2. The distributive property: In this method, we multiply both 5 and 3 by 4, and then we add the results.

$$4(5 + 3) = 4(5) + 4(3) \quad \text{Distribute the multiplication by 4.}$$

$$= 20 + 12 \quad \text{Do the multiplications.}$$

$$= 32 \quad \text{Do the addition.}$$

Notice that each method gives a result of 32.

We can interpret the distributive property geometrically. Figure 8-4 shows three rectangles that are divided into squares. Since the area of the rectangle on the left-hand side of the equals sign can be found by multiplying its width by its length, its area is $4(5 + 3)$ square units. We can evaluate this expression, or we can count squares; either way, we see that the area is 32 square units.

The area shown on the right-hand side is the sum of the areas of two rectangles: $4(5) + 4(3)$. Either by evaluating this expression or by counting squares, we see that this area is also 32 square units. Therefore,

$$4(5 + 3) = 4(5) + 4(3)$$

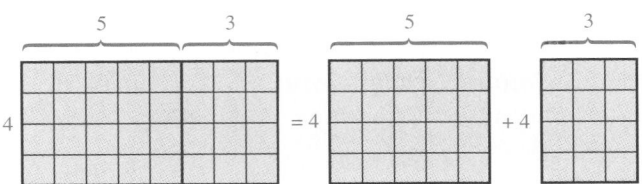

FIGURE 8-4

Figure 8-5 shows the general case where the width is a and the length is $b + c$.

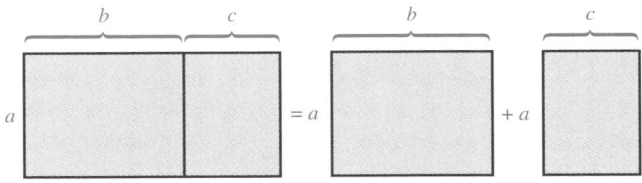

FIGURE 8-5

Using Figure 8-5 as a basis, we can now state the distributive property in symbols.

The distributive property

If a, b, and c represent real numbers, then

$$a(b + c) = ab + ac$$

Since subtraction is the same as adding the opposite, the distributive property also holds for subtraction.

The distributive property

If a, b, and c represent real numbers, then

$$a(b - c) = ab - ac$$

We can use the distributive property to remove the parentheses when an expression is multiplied by a quantity. For example, to remove the parentheses in the expression $5(x + 2)$, we distribute the factor of 5 that is outside the parentheses over x and 2, and add the products.

$$5(x + 2) = 5(x) + 5(2) \quad \text{Distribute the multiplication by 5.}$$
$$= 5x + 10 \quad \text{Do the multiplications.}$$

 COMMENT   Since the expression $5(x + 2)$ contains parentheses, some students are tempted to do the addition within the parentheses first. However, we cannot add x and 2, because we do not know the value of x. Instead, we should focus on the multiplication of $x + 2$ by 5, which requires the use of the distributive property.

EXAMPLE 2   *Applying the distributive property.*   Use the distributive property to remove parentheses:

a. $3(x - 8)$,   **b.** $-12(a + 1)$,   **c.** $-6(-3y - 8)$,   and   **d.** $x(x + 2)$.

Solution

a. $3(x - 8) = 3(x) - 3(8)$ Distribute the multiplication by 3.
$$ = 3x - 24 \text{Do the multiplications.}$$

b. $-12(a + 1) = -12(a) + (-12)(1)$ Distribute the multiplication by -12.
$$ = -12a + (-12) \text{Do the multiplications.}$$
$$ = -12a - 12 \text{Write the addition of } -12 \text{ as subtraction of 12.}$$

c. $-6(-3y - 8) = -6(-3y) - (-6)(8)$ Distribute the multiplication by -6.
$$ = 18y - (-48) \text{Do the multiplications.}$$
$$ = 18y + 48 \text{Add the opposite of } -48, \text{ which is 48.}$$

d. $x(x + 2) = x(x) + x(2)$ Distribute the multiplication by x.
$$ = x^2 + 2x \text{Do the multiplications: } x(x) = x^2 \text{ and } x(2) = 2x.$$

Self Check
Use the distributive property to remove parentheses:

a. $5(p + 2)$

b. $4(t - 1)$

c. $-8(2x - 4)$

d. $p(p - 5)$

Answers: **a.** $5p + 10$,
b. $4t - 4$ **c.** $-16x + 32$,
d. $p^2 - 5p$ ∎

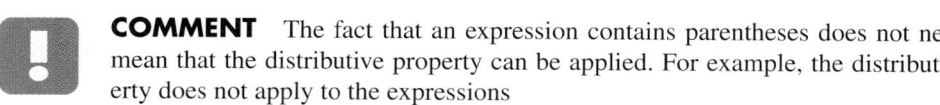 **COMMENT**   The fact that an expression contains parentheses does not necessarily mean that the distributive property can be applied. For example, the distributive property does not apply to the expressions

$$6(5x) \quad \text{or} \quad 6(-7 \cdot y) \quad \text{Here a product is multiplied by 6. Simplifying, we have}$$
$$6(5x) = 30x \text{ and } 6(-7 \cdot y) = -42y.$$

However, the distributive property does apply to the expressions

$$6(5 + x) \quad \text{or} \quad 6(-7 - y) \quad \text{Here a sum or difference is multiplied by 6.}$$
$$\text{Distributing the 6, we have } 6(5 + x) = 30 + 6x \text{ and}$$
$$6(-7 - y) = -42 - 6y.$$

To use the distributive property to simplify $-(x + 10)$, we note that the negative sign in front of the parentheses represents -1.

The $-$ sign represents -1.

$$-(x + 10) = -1(x + 10)$$
$$= -1(x) + (-1)(10) \quad \text{Distribute the multiplication by } -1.$$
$$= -x + (-10) \text{Multiply: } -1(x) = -x \text{ and } (-1)(10) = -10.$$
$$= -x - 10 \text{Write the addition of } -10 \text{ as a subtraction.}$$

EXAMPLE 3 *Distributing a factor of* -1*.* Simplify $-(-12 - 3p)$.

Solution

$-(-12 - 3p)$

$= -1(-12 - 3p)$ Change the $-$ sign in front of the parentheses to -1.

$= -1(-12) - (-1)(3p)$ Distribute the multiplication by -1.

$= 12 - (-3p)$ Multiply: $-1(-12) = 12$ and $(-1)(3p) = -3p$.

$= 12 + 3p$ To subtract $-3p$, add the opposite of $-3p$, which is $3p$.

Self Check

Simplify $-(-5x + 18)$.

Answer: $5x - 18$ ■

Since multiplication is commutative, we can write the distributive property in the following forms.

$$(b + c)a = ba + ca \qquad (b - c)a = ba - ca$$

EXAMPLE 4 *Using the distributive property.* Multiply: $(6x + 4y)\dfrac{1}{2}$.

Solution

$(6x + 4y)\dfrac{1}{2} = (6x)\dfrac{1}{2} + (4y)\dfrac{1}{2}$ Distribute the multiplication by $\frac{1}{2}$.

$\qquad\qquad\quad = 3x + 2y$ Do the multiplications: $(6x)\frac{1}{2} = \left(6 \cdot \frac{1}{2}\right)x = 3x$ and $(4y)\frac{1}{2} = \left(4 \cdot \frac{1}{2}\right)y = 2y$.

Self Check

Multiply:

$(-6x - 24y)\dfrac{1}{3}$

Answer: $-2x - 8y$ ■

The distributive property can be extended to situations in which there are more than two terms within parentheses.

The extended distributive property

If a, b, c, and d represent real numbers, then

$$a(b + c + d) = ab + ac + ad \quad \text{and} \quad a(b - c - d) = ab - ac - ad$$

EXAMPLE 5 *Applying the extended distributive property.* Use the distributive property to remove parentheses: $-0.3(3a - 4b + 7)$.

Solution

$-0.3(3a - 4b + 7)$

$= -0.3(3a) - (-0.3)(4b) + (-0.3)(7)$ Distribute the multiplication by -0.3.

$= -0.9a - (-1.2b) + (-2.1)$ Do the three multiplications.

$= -0.9a + 1.2b + (-2.1)$ To subtract $-1.2b$, add its opposite, which is $1.2b$.

$= -0.9a + 1.2b - 2.1$ Write the addition of -2.1 as a subtraction.

Self Check

Use the distributive property to remove parentheses:

$-0.7(2r + 5s - 8)$

Answer: $-1.4r - 3.5s + 5.6$ ■

The vocabulary of algebraic expressions

Addition signs separate algebraic expressions into parts called **terms.** The expression $5x + 8$ contains two terms, $5x$ and 8.

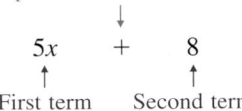

The $+$ sign separates the expression into two terms.

$$5x \quad + \quad 8$$

First term   Second term

A term may be

- a number. Examples are 8, 98.6, and -45.
- a variable, or a product of variables (which may be raised to powers). Examples are x, s^3, rt, and a^2bc^4.
- a product of a number and one or more variables (which may be raised to powers). Examples are $-35x$, $\frac{1}{2}bh$, and $\pi r^2 h$.

Since subtraction can be expressed as addition of the opposite, the expression $6x - 5$ can be written in the equivalent form $6x + (-5)$. We can then see that $6x - 5$ contains two terms, $6x$ and -5.

EXAMPLE 6   *Identifying terms.* List the terms in each expression: **a.** $-4p + 7 + 5p$, **b.** $-12r^2st$, and **c.** $y^3 + 8y^2 - 3y - 24$.

Solution

a. $-4p + 7 + 5p$ has three terms: $-4p$, 7, and $5p$.

b. The expression $-12r^2st$ has one term: $-12r^2st$.

c. $y^3 + 8y^2 - 3y - 24$ can be written as $y^3 + 8y^2 + (-3y) + (-24)$. It contains four terms: y^3, $8y^2$, $-3y$, and -24.

Self Check

List the terms in each expression:

a. $\frac{1}{3}Bh$, **b.** $3q + 5q - 1.2$, and **c.** $b^2 - 4ac$

Answers:
a. $\frac{1}{3}Bh$, **b.** $3q, 5q, -1.2$, **c.** $b^2, -4ac$ ■

In a term that is the product of a number and one or more variables, the number factor is called the **numerical coefficient,** or simply the **coefficient.** In the expression $5x$, the coefficient is 5, and x is the variable part. Other examples are shown in Table 8-2.

Term	Coefficient	Variable part
$8y^2$	8	y^2
$-0.9pq$	-0.9	pq
$\frac{3}{4}b$	$\frac{3}{4}$	b
$-\frac{x}{6}$	$-\frac{1}{6}$	x
x	1	x
$-t$	-1	t
15	15	none

TABLE 8-2

Notice that when there is no number in front of a variable, the coefficient is 1. We say that such a term has an *implied coefficient* of 1. Similarly, when there is only a negative sign in front of the variable, the coefficient is an implied -1. For example, $-t = -1t$.

EXAMPLE 7   *Identifying coefficients of terms.* Identify the coefficient and the variable part of each term in the expression $-7x^2 + 3x - 6$.

Solution

Term	Coefficient	Variable part
$-7x^2$	-7	x^2
$3x$	3	x
-6	-6	none

Self Check

Identify the coefficient of each term in the expression $p^3 - 12p^2 + 3p - 4$.

Answers: $1, -12, 3, -4$ ■

COMMENT It is important to be able to distinguish between a *term* of an expression and a *factor* of a term. Terms are separated by a + sign. Factors, on the other hand, are numbers and/or variables that are multiplied together. For example, x is a term of the expression $18 + x$, because x and 18 are separated by a + sign. In the expression $18x + 9$, x is a factor of the term $18x$, because x and 18 are multiplied together.

Like terms

The expression $5p + 7q - 3p + 12$, which can be written $5p + 7q + (-3p) + 12$, contains four terms, $5p$, $7q$, $-3p$, and 12. Since the variable of $5p$ and $-3p$ are the same, we say that these terms are **like** or **similar terms.**

Like terms (similar terms) | **Like terms** (or **similar terms**) are terms with exactly the same variables raised to exactly the same powers. Any numbers (called **constants**) in an expression are considered to be like terms.

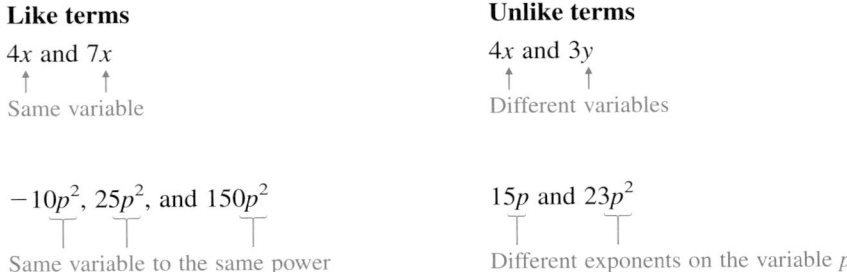

 COMMENT When looking for like terms, don't look at the coefficients of the terms. Consider only their variable parts.

EXAMPLE 8 *Identifying like terms.* List like terms: **a.** $7r + 5 + 3r$, **b.** $x^4 - 6x^2 - 5$, and **c.** $-7m + 7 - 2 + m$.

Solution

a. $7r + 5 + 3r$ contains the like terms $7r$ and $3r$.

b. $x^4 - 6x^2 - 5$ contains no like terms.

c. $-7m + 7 - 2 + m$ contains two pairs of like terms: $-7m$ and m are like terms, and the constants, 7 and -2, are like terms.

Self Check

List like terms:

a. $5x - 2y + 7y$

b. $-5pq + 17p - 12q - 2pq$

Answers: a. $-2y$ and $7y$,
b. $-5pq$ and $-2pq$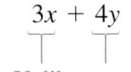

Combining like terms

If we are to add (or subtract) objects, they must have the same units. For example, we can add dollars to dollars and inches to inches, but we cannot add dollars to inches. The same is true when we work with terms of an algebraic expression. They can be added or subtracted only when they are like terms.

This expression can be simplified, because it contains like terms.
$$3x + 4x$$
Like terms
The variable parts are identical.

This expression cannot be simplified, because its terms are not like terms.
$$3x + 4y$$
Unlike terms
The variable parts are not identical.

To simplify an expression containing like terms, we use the distributive property. For example, we can simplify $3x + 4x$ as follows:

$$3x + 4x = (3 + 4)x \quad \text{Apply the distributive property.}$$
$$= 7x \qquad \text{Do the addition within the parentheses: } 3 + 4 = 7.$$

We have simplified the expression $3x + 4x$ by **combining like terms.** The result is the equivalent expression $7x$. This example suggests the following general rule.

Combining like terms

> To add or subtract like terms, combine their coefficients and keep the same variables with the same exponents.

EXAMPLE 9 *Simplifying algebraic expressions.* Simplify by combining like terms: **a.** $-8p + (-12p)$ and **b.** $0.5s^2 - 0.3s^2$.

Solution

a. $-8p + (-12p) = -20p$ Add the coefficients of the like terms: $-8 + (-12) = -20$. Keep the variable p.

b. $0.5s^2 - 0.3s^2 = 0.2s^2$ Subtract: $0.5 - 0.3 = 0.2$. Keep the variable part s^2.

Self Check

Simplify by combining like terms:

a. $5n + (-8n)$

b. $-1.2a^3 + (1.4a^3)$

Answers: a. $-3n$, **b.** $0.2a^3$

EXAMPLE 10 *Combining like terms.* Simplify $7P - 8p - 12P + 25p$.

Solution

The uppercase P and the lowercase p are different variables. We can use the commutative property of addition to write like terms next to each other.

$$7P - 8p - 12P + 25p$$
$$= 7P + (-8p) + (-12P) + 25p \quad \text{Rewrite each subtraction as the addition of the opposite.}$$
$$= 7P + (-12P) + (-8p) + 25p \quad \text{Use the commutative property of addition to write the like terms together.}$$
$$= -5P + 17p \qquad\qquad \text{Combine like terms: } 7P + (-12P) = -5P \text{ and } -8p + 25p = 17p.$$

Self Check

Simplify $8R + 7r - 14R - 21r$.

Answer: $-6R - 14r$

The expression in Example 10 contained two sets of like terms, and we rearranged the terms so that like terms were next to each other. With practice, you will be able to combine like terms without having to write them next to each other and without having to write each subtraction as addition of the opposite.

EXAMPLE 11 *Combining like terms without rearranging terms.* Simplify $4(x + 5) - 3(2x - 4)$.

Solution

$$4(x + 5) - 3(2x - 4)$$
$$= 4x + 20 - 6x + 12 \quad \text{Use the distributive property twice.}$$
$$= -2x + 32 \qquad\qquad \text{Combine like terms: } 4x - 6x = -2x \text{ and } 20 + 12 = 32.$$

Self Check

Simplify $-5(y - 4) + 2(4y + 6)$.

Answer: $3y + 32$

STUDY SET   Section 8.2 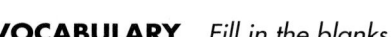

VOCABULARY *Fill in the blanks.*

1. To _____ an algebraic expression, we use properties of algebra to write the expression in a less complicated form.

2. A _____ is a number or a product of a number and one or more variables.

3. In the term $3x^2$, the number factor 3 is called the _____ and x^2 is called the _____ part.

4. Two terms with exactly the same variables and exponents are called _____ terms.

5. We can use the distributive property to _____ the parentheses when an expression is multiplied by a quantity.

6. The _____ property of multiplication enables us to change the order of the factors involved in a multiplication.

CONCEPTS

7. What property does the statement $a(b + c) = ab + ac$ illustrate?

8. Complete this statement:

$a(b + c + d) =$ _____

9. Illustration 1 shows an application of the distributive property. Fill in the blanks.

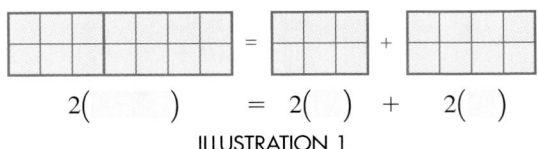

$$2() = 2() + 2()$$

ILLUSTRATION 1

10. Complete the table.

Term	Coefficient	Variable part
$6m$		
$-75t$		
w		
$\frac{1}{2}bh$		

11. Fill in the blanks.
  a. $2(x + 4) = 2x 8$
  b. $2(x - 4) = 2x 8$
  c. $-2(x + 4) = -2x 8$
  d. $-2(x - 4) = -2x 8$
  e. $-2(-x + 4) = 2x 8$
  f. $-2(-x - 4) = 2x 8$

12. Complete this statement: To add or subtract like terms, combine their _____ and keep the same variables and _____.

13. A board was cut into two pieces, as shown in Illustration 2. Add the lengths of the two pieces. How long was the original board?

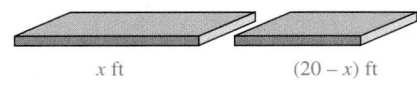

x ft $\qquad$ $(20 - x)$ ft

ILLUSTRATION 2

14. Let x equal the number of miles driven on the first day of a 2-day driving trip. Translate the verbal model to mathematical symbols, and simplify by combining like terms.

the miles driven day 1	plus	100 miles more than the miles driven day 1

15. a. Two angles are called **complementary angles** if the sum of their measures is 90°. Add the measures of the angles in Illustration 3(a). Are they complementary angles?
  b. Two angles are called **supplementary angles** if the sum of their measures is 180°. Add the measures of the angles in Illustration 3(b). Are they supplementary angles?

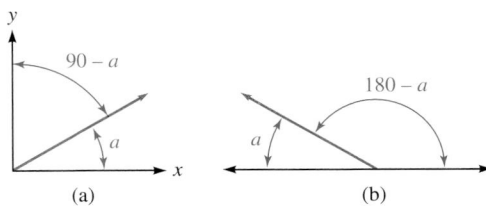

(a) $\qquad$ (b)

All angle measures are in degrees.

ILLUSTRATION 3

16. Simplify each expression, if possible.
  a. $5(2x)$ and $5 + 2x$
  b. $6(-7x)$ and $6 - 7x$
  c. $2(3x)(3)$ and $2 + 3x + 3$
  d. $x \cdot x$ and $x + x$

NOTATION *Complete each solution.*

17. $7(a + 2) = (a) + (2)$
$$= 7a + $$

18. $6(b - 5) + 12b + 7 = 6() - 6() + 12b + 7$
$$= 6b - + 12b + 7$$
$$= 6b + b - + 7$$
$$= 18b - 23$$

19. a. Are $2K$ and $3k$ like terms?
  b. Are $-d$ and d like terms?

20. Fill in the blank to make the statement true.
$$-(x + 10) = - (x + 10)$$

21. Write each expression using fewer symbols.
  a. $5x - (-1)$ $\qquad$ **b.** $16t + (-6)$

22. In the following table, a student's answers to five homework problems are compared to the answers in the back of the book. Are the answers equivalent? (Write *yes* or *no*.)

Student's answer	Book's answer	Equivalent?
$10x$	$10 + x$	
$3 + y$	$y + 3$	
$5 - 8a$	$8a - 5$	
$3(x) + 4$	$3(x + 4)$	
$2x$	x^2	

PRACTICE *Simplify each expression.*

23. $9(7m)$

24. $12n(8)$

25. $5(-7q)$

26. $-7(5t)$

27. $12\left(\dfrac{5}{12}x\right)$

28. $15\left(\dfrac{4}{15}w\right)$

29. $8\left(\dfrac{3}{4}y\right)$

30. $27\left(\dfrac{2}{3}x\right)$

31. $(-5p)(-4b)$

32. $(-7d)(-7c)$

33. $-5(4r)(-2r)$

34. $7t(-4t)(-2)$

Use the distributive property to remove parentheses.

35. $5(x + 3)$

36. $4(x + 2)$

37. $-2(b - 1)$

38. $-7(p - 5)$

39. $(3t - 2)8$

40. $(2q + 1)9$

41. $(2y - 1)6$

42. $(3w - 5)5$

43. $0.4(x - 4)$

44. $-2.2(2q + 1)$

45. $-\dfrac{2}{3}(3w - 6)$

46. $\dfrac{1}{2}(2y - 8)$

47. $r(r - 10)$

48. $h(h + 4)$

49. $-(x - 7)$

50. $-(y + 1)$

51. $17(2x - y + 2)$

52. $-12(3a + 2b - 1)$

53. $-(-14 + 3p - t)$

54. $-(-x - y + 5)$

55. Identify the coefficient of each term.
  a. $-b$
  b. $-9.9x^3$
  c. $\dfrac{1}{4}x$
  d. $-\dfrac{2x}{3}$

56. Tell whether the variable x is used as a factor or as a term.
  a. $24 - x$
  b. $24x$
  c. $24 + 3x$
  d. $x - 12$

Identify the coefficient of each term.

57. $-5r + 4s$

58. $2m + n - 3m + 2n$

59. $-15r^2s$

60. $4b^2 - 5b + 6$

61. $50a + 2$

62. $a^2 - ab + b^2$

63. $x^3 - 125$

64. $-2.55x + 1.8$

Simplify each expression by combining like terms.

65. $3x + 17x$

66. $12y - 15y$

67. $8x^2 - 5x^2$

68. $17x^2 + 3x^2$

69. $-4x + 4x$

70. $-16y + 16y$

71. $-7b^2 + 7b^2$

72. $-2c^3 + 2c^3$

73. $a + a + a$

74. $t - t - t - t$

75. $0 - 3x$

76. $0 - 4a$

77. $0 - (-t)$

78. $0 - (-2y)$

79. $3x + 5x - 7x$

80. $-y + 3y + 2y$

81. $-13x^2 + 2x^2 - 5x^2$

82. $-8x^3 - x^3 + 2x^3$

83. $1.8h - 0.7h$

84. $-5.7m + 4.3m$

85. $\dfrac{3}{5}t + \dfrac{1}{5}t$

86. $\dfrac{3}{16}x - \dfrac{5}{16}x$

87. $-0.2r - (-0.6r)$

88. $-1.1m - (-2.4m)$

89. $2z + 5(z - 3)$

90. $12(m + 11) - 11$

91. $-(c + 7) - 2(c - 3)$

92. $-(z + 2) + 5(3 - z)$

93. $2x + 4(X - x) + 3X$

94. $3p - 6(p + z) + p$

95. $(a + 2) - (a - b)$

96. $3z + 2(Z - z) + Z$

97. $x(x + 3) - 3x^2$

98. $2x + x(x - 3)$

APPLICATIONS

99. THE AMERICAN RED CROSS
In 1891, Clara Barton founded the Red Cross. Its symbol is a white flag bearing a red cross. If each side of the cross in Illustration 4 has length x, write an algebraic expression for the perimeter (the total distance around the outside) of the cross.

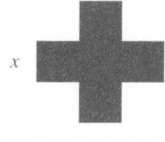

ILLUSTRATION 4

100. BILLIARDS   Billiard tables vary in size, but all tables are twice as long as they are wide.
  a. If the billiard table in Illustration 5 is x feet wide, write an expression involving x that represents its length.
  b. Write an expression for the perimeter of the table.

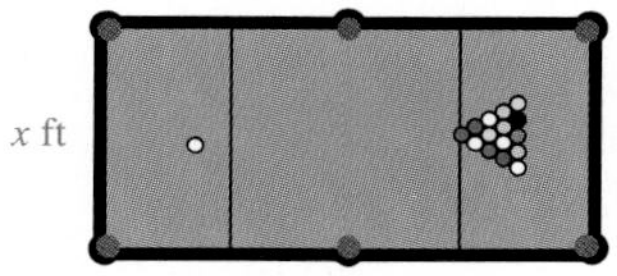

x ft

ILLUSTRATION 5

101. PING-PONG   Write an expression for the perimeter of the ping-pong table shown in Illustration 6.

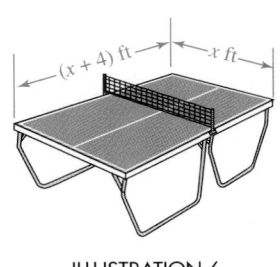

ILLUSTRATION 6

102. SEWING   See Illustration 7. Write an expression for the length of the yellow trim needed to outline a pennant with the given side lengths.

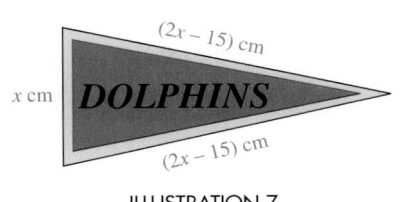

ILLUSTRATION 7

WRITING

103. Explain why the distributive property applies to $2(3 + x)$ but not to $2(3x)$.

104. Explain why $3x^2y$ and $5x^2y$ are like terms, and explain why $3x^2y$ and $5xy^2$ are not like terms.

105. Distinguish between a *factor* and a *term* of an algebraic expression. Give examples.

106. Tell how to combine like terms.

REVIEW   *Evaluate each expression for $x = -3$, $y = -5$, and $z = 0$.*

107. $x^2z(y^3 - z)$ **108.** $|y^3 - z|$

109. $\dfrac{x - y^2}{2y - 1 + x}$ **110.** $\dfrac{2y + 1}{x} - x$

8.3 *Solving Equations*

In this section, you will learn about

- Solving equations by using more than one property of equality
- Simplifying expressions to solve equations
- Identities and impossible equations

INTRODUCTION. We have solved simple equations using the properties of equality. In this section, we will solve more-complicated equations. Our objective is to develop a general strategy that can be used to solve any kind of linear equation.

Solving equations by using more than one property of equality

Recall the following properties of equality:

- If the same quantity is added to (or subtracted from) equal quantities, the results will be equal quantities.
- If equal quantities are multiplied (or divided) by the same nonzero quantity, the results will be equal quantities.

We have already solved many simple equations using one of the properties listed above. For example, to solve $x + 6 = 10$, we isolate x by subtracting 6 from both sides.

$$x + 6 = 10$$

$$x + 6 - 6 = 10 - 6 \quad \text{To undo the addition of 6, subtract 6 from both sides.}$$

$$x = 4 \qquad \text{Do the subtractions.}$$

To solve $2x = 10$, we isolate x by dividing both sides by 2.

$$2x = 10$$

$$\frac{2x}{2} = \frac{10}{2}$$ To undo the multiplication by 2, divide both sides by 2.

$$x = 5$$ Do the divisions.

Sometimes several properties of equality must be applied in succession to solve an equation. For example, on the left-hand side of $2x + 6 = 10$, the variable x is first multiplied by 2, and then 6 is added to that product. To isolate x, we use the rules for the order of operations in reverse. First, we undo the addition of 6, and then we undo the multiplication by 2.

$$2x + 6 = 10$$

$$2x + 6 - 6 = 10 - 6$$ To undo the addition of 6, subtract 6 from both sides.

$$2x = 4$$ Do the subtractions.

$$\frac{2x}{2} = \frac{4}{2}$$ To undo the multiplication by 2, divide both sides by 2.

$$x = 2$$ Do the divisions.

EXAMPLE 1 *Using two properties of equality.* Solve and check: $-12x + 5 = 17$.

Solution

On the left-hand side of the equation, x is multiplied by -12, and then 5 is added to that product. To isolate x, we undo these operations in the opposite order.

• To undo the addition of 5, we subtract 5 from both sides.
• To undo the multiplication by -12, we divide both sides by -12.

$$-12x + 5 = 17$$

$$-12x + 5 - 5 = 17 - 5$$ Subtract 5 from both sides.

$$-12x = 12$$ Do the subtractions: $5 - 5 = 0$ and $17 - 5 = 12$.

$$\frac{-12x}{-12} = \frac{12}{-12}$$ Divide both sides by -12.

$$x = -1$$ Do the divisions.

Check: $-12x + 5 = 17$ The original equation.

$$-12(-1) + 5 \stackrel{?}{=} 17$$ Substitute -1 for x.

$$12 + 5 \stackrel{?}{=} 17$$ Do the multiplication: $-12(-1) = 12$.

$$17 = 17$$ Do the addition.

Since the statement $17 = 17$ is true, -1 is a solution.

Self Check
Solve and check:
$8x - 13 = 43$

Answer: 7

EXAMPLE 2 *Using two properties of equality.* Solve and check: $\frac{2x}{3} = -6$.

Solution

On the left-hand side, x is multiplied by 2, and then that product is divided by 3. To solve this equation, we must undo these operations in the opposite order.

Self Check
Solve and check:
$\frac{7h}{16} = -14$

- To undo the division of 3, we multiply both sides by 3.
- To undo the multiplication by 2, we divide both sides by 2.

$$\frac{2x}{3} = -6$$

$$3\left(\frac{2x}{3}\right) = 3(-6) \quad \text{Multiply both sides by 3.}$$

$$2x = -18 \quad \text{On the left-hand side: } 3\left(\frac{2x}{3}\right) = \frac{\overset{1}{\cancel{3}}}{1}\left(\frac{2x}{\cancel{3}_1}\right) = 2x.$$

$$\frac{2x}{2} = \frac{-18}{2} \quad \text{Divide both sides by 2.}$$

$$x = -9 \quad \text{Do the divisions.}$$

Check: $\dfrac{2x}{3} = -6 \quad$ The original equation.

$$\frac{2(-9)}{3} \overset{?}{=} -6 \quad \text{Substitute } -9 \text{ for } x.$$

$$\frac{-18}{3} \overset{?}{=} -6 \quad \text{Do the multiplication: } 2(-9) = -18.$$

$$-6 = -6 \quad \text{Do the division.}$$

Since we obtain a true statement, -9 is a solution. | **Answer:** -32 ■

Another approach can be used to solve the equation from Example 2. To isolate the variable, we will use the fact that the product of a number and its **reciprocal,** or **multiplicative inverse,** is 1. Since $\frac{2x}{3} = \frac{2}{3}x$, the equation can be rewritten as

$$\frac{2}{3}x = -6$$

To isolate x, we multiply both sides by $\frac{3}{2}$, the reciprocal (multiplicative inverse) of $\frac{2}{3}$.

$$\frac{3}{2}\left(\frac{2}{3}x\right) = \frac{3}{2}(-6) \quad \begin{array}{l}\text{The coefficient of } x \text{ is } \frac{2}{3}. \text{ Multiply both sides by the reciprocal of}\\ \frac{2}{3}, \text{ which is } \frac{3}{2}.\end{array}$$

$$\left(\frac{3}{2} \cdot \frac{2}{3}\right)x = \frac{3}{2}(-6) \quad \begin{array}{l}\text{On the left-hand side, apply the associative property of}\\ \text{multiplication to regroup factors.}\end{array}$$

$$1x = -9 \quad \text{Do the multiplications: } \frac{3}{2} \cdot \frac{2}{3} = 1 \text{ and } \frac{3}{2}(-6) = -9.$$

$$x = -9 \quad \text{Simplify: } 1x = x.$$

EXAMPLE 3 *An implied coefficient of −1.* Solve $-0.2 = -0.8 - y$.

Solution

To solve the equation, we begin by eliminating -0.8 from the right-hand side. We can do this by adding 0.8 to both sides.

$$-0.2 = -0.8 - y$$

$$-0.2 + 0.8 = -0.8 - y + 0.8 \quad \text{Add 0.8 to both sides.}$$

$$0.6 = -y \quad \begin{array}{l}\text{Do the additions: } -0.2 + 0.8 = 0.6 \text{ and}\\ -0.8 + 0.8 = 0.\end{array}$$

Since the term $-y$ has an understood coefficient of -1, the equation can be rewritten as $0.6 = -1y$. To isolate y, either multiply both sides or divide both sides by -1.

Self Check
Solve $-6.6 - m = -2.7$.

$0.6 = -1y$ Write $-y$ as $-1y$.

$\dfrac{0.6}{-1} = \dfrac{-1y}{-1}$ To undo the multiplication by -1, divide both sides by -1.

$-0.6 = y$ Do the divisions.

$y = -0.6$

Verify that -0.6 satisfies the equation.

EXAMPLE 4   *Using three properties of equality.*   Solve $\dfrac{3}{4}x + 2 = -7$.

Solution
On the left-hand side, x is multiplied by 3, that product is divided by 4, and then 2 is added. To solve the equation, we must undo these operations in the opposite order.

- To undo the addition of 2, we subtract 2 from both sides.
- To undo the division by 4, we multiply both sides by 4.
- To undo the multiplication by 3, we divide both sides by 3.

$\dfrac{3}{4}x + 2 = -7$

$\dfrac{3}{4}x + 2 - 2 = -7 - 2$ Subtract 2 from both sides.

$\dfrac{3}{4}x = -9$ Do the subtractions: $2 - 2 = 0$ and $-7 - 2 = -9$.

$4\left(\dfrac{3}{4}x\right) = 4(-9)$ Multiply both sides by 4.

$3x = -36$ On the left-hand side: $4\left(\dfrac{3}{4}x\right) = \left(\dfrac{\overset{1}{\cancel{4}}}{1} \cdot \dfrac{3}{\underset{1}{\cancel{4}}}\right)x = 3x$.

$\dfrac{3x}{3} = \dfrac{-36}{3}$ Divide both sides by 3.

$x = -12$ Do the divisions.

Verify that -12 satisfies the equation.

Self Check
Solve $\dfrac{2}{3}b - 3 = -15$.

Simplifying expressions to solve equations

EXAMPLE 5   *Combining like terms.*   Solve and check:
$3(k + 1) - 5k = 0$.

Solution
First, we use the distributive property. Then we combine like terms.

$3(k + 1) - 5k = 0$

$3k + 3(1) - 5k = 0$ Distribute the multiplication by 3.

$3k + 3 - 5k = 0$ Do the multiplications.

$-2k + 3 = 0$ Combine like terms: $3k - 5k = -2k$.

$-2k + 3 - 3 = 0 - 3$ To undo the addition of 3, subtract 3 from both sides.

$-2k = -3$ Do the subtractions: $3 - 3 = 0$ and $0 - 3 = 3$.

$\dfrac{-2k}{-2} = \dfrac{-3}{-2}$ To undo the multiplication by -2, divide both sides by -2.

$k = \dfrac{3}{2}$ Simplify: $\frac{-3}{-2} = \frac{3}{2}$.

Self Check
Solve and check:

$-5(x - 3) + 3x = 11$

Check: $3(k + 1) - 5k = 0$ The original equation.

$3\left(\dfrac{3}{2} + 1\right) - 5\left(\dfrac{3}{2}\right) \stackrel{?}{=} 0$ Substitute $\frac{3}{2}$ for k.

$3\left(\dfrac{3}{2} + \dfrac{2}{2}\right) - 5\left(\dfrac{3}{2}\right) \stackrel{?}{=} 0$ To add $\frac{3}{2}$ and 1, write 1 as $\frac{2}{2}$.

$3\left(\dfrac{5}{2}\right) - 5\left(\dfrac{3}{2}\right) \stackrel{?}{=} 0$ Do the addition within the parentheses.

$\dfrac{15}{2} - \dfrac{15}{2} \stackrel{?}{=} 0$ Do the multiplications.

$0 = 0$ Do the subtraction.

Answer: 2

EXAMPLE 6 *Variable terms on both sides of the equation.*   Solve and check: $3x - 15 = 4x + 36$.

Solution

To solve for x, all the terms containing x must be on the same side of the equation. We can eliminate $3x$ from the left-hand side by subtracting $3x$ from both sides.

$3x - 15 = 4x + 36$

$3x - 15 - \mathbf{3x} = 4x + 36 - \mathbf{3x}$ Subtract $3x$ from both sides.

$-15 = x + 36$ Combine like terms: $3x - 3x = 0$ and $4x - 3x = x$.

$-15 - \mathbf{36} = x + 36 - \mathbf{36}$ To undo the addition of 36, subtract 36 from both sides.

$-51 = x$ Do the subtractions.

$x = -51$

Check: $3x - 15 = 4x + 36$ The original equation.

$3(\mathbf{-51}) - 15 \stackrel{?}{=} 4(\mathbf{-51}) + 36$ Substitute -51 for x.

$-153 - 15 \stackrel{?}{=} -204 + 36$ Do the multiplications.

$-168 = -168$ Simplify each side.

Self Check

Solve and check:

$3n + 48 = -4n - 8$

Answer: -8

EXAMPLE 7 *Clearing an equation of fractions.*   Solve $\dfrac{x}{6} - \dfrac{5}{2} = -\dfrac{1}{3}$.

Solution

Since integers are easier to work with, we will clear the equation of fractions by multiplying both sides by the least common denominator (LCD), which is 6.

$\dfrac{x}{6} - \dfrac{5}{2} = -\dfrac{1}{3}$

$6\left(\dfrac{x}{6} - \dfrac{5}{2}\right) = 6\left(-\dfrac{1}{3}\right)$ 6 is the smallest number that each denominator will divide exactly. Multiply both sides by 6 to clear the equation of the fractions.

$6\left(\dfrac{x}{6}\right) - 6\left(\dfrac{5}{2}\right) = 6\left(-\dfrac{1}{3}\right)$ On the left-hand side, distribute the multiplication by 6.

$\underbrace{x}\quad - \quad \underbrace{15} \quad = \quad \underbrace{-2}$ Do each multiplication by 6. Note that the resulting equation does not contain any fractions.

$x - 15 + \mathbf{15} = -2 + \mathbf{15}$ To undo the subtraction of 15, add 15 to both sides.

$x = 13$ Do the additions: $-15 + 15 = 0$ and $-2 + 15 = 13$.

Verify that 13 satisfies the equation.

Self Check

Solve $\dfrac{x}{4} + \dfrac{1}{2} = -\dfrac{1}{8}$.

Answer: $-\dfrac{5}{2}$

The preceding examples suggest the following strategy for solving equations.

Strategy for solving equations

1. Clear the equation of fractions.
2. Use the distributive property to remove parentheses, if necessary.
3. Combine like terms, if necessary.
4. Undo the operations of addition and subtraction to get the variables on one side and the constants on the other.
5. Undo the operations of multiplication and division to isolate the variable.
6. Check the result.

EXAMPLE 8 *Applying the equation-solving strategy.* Solve and check: $\dfrac{3x + 11}{5} = x + 3$.

Self Check

Solve and check:

$$\frac{3x + 23}{7} = x + 5$$

Solution

$$\frac{3x + 11}{5} = x + 3$$

$$5\left(\frac{3x + 11}{5}\right) = 5(x + 3) \qquad \text{Clear the equation of the fraction by multiplying both sides by 5.}$$

$$3x + 11 = 5x + 15 \qquad \text{On the left-hand side, simplify: } \frac{\overset{1}{\cancel{5}}}{1}\left(\frac{3x + 11}{\underset{1}{\cancel{5}}}\right). \text{ On the right-hand side, distribute the multiplication by 5.}$$

$$3x + 11 - 11 = 5x + 15 - 11 \qquad \text{Subtract 11 from both sides.}$$

$$3x = 5x + 4 \qquad \text{Do the subtractions.}$$

$$3x - 5x = 5x + 4 - 5x \qquad \text{To eliminate } 5x \text{ from the right-hand side, subtract } 5x \text{ from both sides.}$$

$$-2x = 4 \qquad \text{Combine like terms: } 3x - 5x = -2x \text{ and } 5x - 5x = 0.$$

$$\frac{-2x}{-2} = \frac{4}{-2} \qquad \text{To undo the multiplication by } -2, \text{ divide both sides by } -2.$$

$$x = -2 \qquad \text{Do the divisions.}$$

Verify that -2 satisfies the equation.

Answer: -3

 COMMENT Remember that when you multiply one side of an equation by a nonzero number, you must multiply the other side of the equation by the same number.

Identities and impossible equations

Equations in which some numbers satisfy the equation and others don't are called **conditional equations.** The equations in Examples 1–8 are conditional equations.

An equation that is true for all values of its variable is called an **identity.**

$$x + x = 2x \qquad \text{This is an identity, because it is true for all values of } x.$$

An equation that is not true for any values of its variable is called an **impossible equation** or a **contradiction.** Such equations are said to have no solution.

$$x = x + 1 \qquad \text{Because no number is 1 greater than itself, this is an impossible equation.}$$

EXAMPLE 9   *Identities.*   Solve $3(x + 8) + 5x = 2(12 + 4x)$.

Solution

$$3(x + 8) + 5x = 2(12 + 4x)$$

$3x + 24 + 5x = 24 + 8x$ On each side of the equation, use the distributive property.

$8x + 24 = 24 + 8x$ Combine like terms.

$8x + 24 - 8x = 24 + 8x - 8x$ Subtract $8x$ from both sides.

$24 = 24$ Combine like terms: $8x - 8x = 0$.

In this case, the terms involving x drop out. Since the result $24 = 24$ is true for every number x, all values of x satisfy the original equation. This equation is an identity.

> **Self Check**
>
> Solve $3(x + 5) - 4(x + 4) = -x - 1$.
>
> **Answer:** all values of x; this equation is an identity ∎

EXAMPLE 10   *Impossible equations.*   Solve $3(d + 7) - d = 2(d + 10)$.

Solution

$$3(d + 7) - d = 2(d + 10)$$

$3d + 21 - d = 2d + 20$ Use the distributive property.

$2d + 21 = 2d + 20$ Combine like terms.

$2d + 21 - 2d = 2d + 20 - 2d$ Subtract $2d$ from both sides.

$21 = 20$ Combine like terms.

In this case, the terms involving d drop out. Since the result $21 = 20$ is false, the original equation has no solution. It is an impossible equation.

> **Self Check**
>
> Solve $-4(c - 3) + 2c = 2(10 - c)$.
>
> **Answer:** No solution. This equation is an impossible equation. ∎

STUDY SET Section 8.3

VOCABULARY   *Fill in the blanks.*

1. An _____ is a statement that two quantities are equal.

2. To solve an equation, we must _____ the variable on one side of the equation.

3. If a number is a solution of an equation, the number is said to _____ the equation.

4. In $2(x - 7)$, "to remove parentheses" means to apply the _____ property.

5. The product of a number and its _____ is 1.

6. An equation that is true for all values of its variable is called an _____. An equation that has no solutions is called an _____ equation or a contradiction.

CONCEPTS   *In Exercises 7–10, fill in the blanks.*

7. To solve the equation $2x - 7 = 21$, we first undo the _____ of 7 by adding 7 to both sides. We then undo the _____ by 2 by dividing both sides by 2.

8. To solve the equation $\frac{x}{-2} + 3 = 5$, we first undo the _____ of 3 by subtracting 3 from both sides. We then undo the _____ by -2 by multiplying both sides by -2.

9. To solve the equation $\frac{x}{2} + 3 = 5$, we first undo the _____ of 3 by subtracting 3 from both sides. We then undo the _____ by 2 by multiplying both sides by 2.

10. To solve $\frac{s}{3} + \frac{1}{4} = -\frac{1}{2}$, we can clear the equation of the fractions by _____ both sides of the equation by 12.

11. One method of solving $-\frac{4}{5}x = 8$ is to multiply both sides of the equation by the reciprocal of $-\frac{4}{5}$. What is the reciprocal of $-\frac{4}{5}$?

12. a. Combine like terms on the left-hand side of
$$6x - 8 - 8x = -24.$$
b. Combine like terms on the right-hand side of
$$5a + 1 = 9a + 16 + a.$$
c. Combine like terms on both sides of
$$12 - 3r + 5r = -8 - r - 2.$$

13. What is the LCD for the fractions in the equation $\frac{x}{3} - \frac{4}{5} = \frac{1}{2}$?

14. Complete the three multiplications necessary to clear the given equation of fractions.

$$\frac{2}{3} - \frac{b}{2} = -\frac{4}{3}$$

$$6\left(\frac{2}{3} - \frac{b}{2}\right) = 6\left(-\frac{4}{3}\right)$$

$$6\left(\frac{2}{3}\right) - 6\left(\frac{b}{2}\right) = 6\left(-\frac{4}{3}\right)$$

$$\underbrace{} - \underbrace{} = \underbrace{}$$

15. a. Simplify $3x + 5 - x$.
 b. Solve $3x + 5 - x = 9$.
 c. Evaluate $3x + 5 - x$ for $x = 9$.
 d. Check: Is $x = -1$ a solution of $3x + 5 - x = 9$?

16. a. Simplify $3(x - 4) - 4x$.
 b. Solve $3(x - 4) - 4x = 0$.
 c. Evaluate $3(x - 4) - 4x$ for $x = 0$.
 d. Check: Is $x = -1$ a solution of
 $3(x - 4) - 4x = 0$?

NOTATION *In Exercises 17–18, complete the solution to solve each equation.*

17.
$$2x - 7 = 21$$
$$2x - 7 + = 21 + $$
$$2x = $$
$$\frac{2x}{} = \frac{28}{}$$
$$x = 14$$

18.
$$\frac{x}{2} + 3 = 5$$
$$\frac{x}{2} + 3 - = 5 - $$
$$\frac{x}{2} = $$
$$\left(\frac{x}{2}\right) = (2)$$
$$x = 4$$

19. Fill in the blanks.

 a. $-x = \, x$. **b.** $\dfrac{3x}{5} = \, x$.

 c. If $-31 = x$, then $x = $

20. When checking a solution of an equation, the symbol $\overset{?}{=}$ is used. What does it mean?

PRACTICE *Solve each equation and check the result.*

21. $2x + 5 = 17$ **22.** $3x - 5 = 13$

23. $-5q - 2 = 1$ **24.** $4p + 3 = 2$

25. $0.6 = 4.1 - x$ **26.** $1.2 - x = -1.7$

27. $-g = -4$ **28.** $-u = -20$

29. $-8 - 3c = 0$ **30.** $-5 - 2d = 0$

31. $-\dfrac{5}{6}k = 10$ **32.** $\dfrac{2c}{5} = 2$

33. $-\dfrac{t}{3} + 2 = 6$ **34.** $\dfrac{x}{5} - 5 = -12$

35. $\dfrac{2x}{3} - 2 = 4$ **36.** $\dfrac{2}{5}y + 3 = 9$

37. $\dfrac{x + 5}{3} = 11$ **38.** $\dfrac{x + 2}{13} = 3$

39. $\dfrac{y - 2}{7} = -3$ **40.** $\dfrac{x - 7}{3} = -1$

41. $2(-3) + 4y = 14$ **42.** $4(-1) + 3y = 8$

43. $-2x - 4(1) = -6$ **44.** $-5x - 3(5) = 0$

45. $3(x + 2) - x = 12$ **46.** $2(x - 4) + x = 7$

47. $-3(2y - 2) - y = 5$ **48.** $-(3a + 1) + a = 2$

49. $0 - 2y = 8$ **50.** $0 - 7x = -21$

51. $5x + 7.2 = 4x$ **52.** $3x + 2.5 = 2x$

53. $8y + 4 = 4y$ **54.** $9y - 3 = 6y$

55. $15x = x$ **56.** $-7y = -8y$

57. $4 + \dfrac{y}{2} = \dfrac{3}{5}$ **58.** $5 + \dfrac{x}{3} = \dfrac{1}{2}$

59. $\dfrac{1}{3} + \dfrac{c}{5} = -\dfrac{3}{2}$ **60.** $\dfrac{1}{2} + \dfrac{x}{5} = \dfrac{3}{4}$

61. $\dfrac{y}{6} + \dfrac{y}{4} = -1$ **62.** $\dfrac{x}{3} + \dfrac{x}{4} = -2$

63. $-\dfrac{2}{9} = \dfrac{5x}{6} - \dfrac{1}{3}$ **64.** $\dfrac{2}{3} = -\dfrac{2x}{3} + \dfrac{3}{4}$

65. $\dfrac{1}{2}x - \dfrac{1}{9} = \dfrac{1}{3}$ **66.** $\dfrac{1}{4}y - \dfrac{2}{3} = \dfrac{1}{2}$

67. $\dfrac{2}{5}x + 1 = \dfrac{1}{3} + x$ **68.** $\dfrac{2}{3}y + 2 = \dfrac{1}{5} + y$

69. $3(a + 2) = 2(a - 7)$

70. $9(t - 1) = 6(t + 2) - t$

71. $9(x + 11) + 5(13 - x) = 0$

72. $3(x + 15) + 4(11 - x) = 0$

73. $\dfrac{3t - 21}{2} = t - 6$ **74.** $\dfrac{2t - 18}{3} = t - 8$

75. $\dfrac{10 - 5s}{3} = s + 6$ **76.** $\dfrac{40 - 8s}{5} = -2s$

77. $2 - 3(x - 5) = 4(x - 1)$

78. $2 - (4x + 7) = 3 + 2(x + 2)$

Solve each equation. If it is an identity or an impossible equation, so indicate.

79. $8x + 3(2 - x) = 5(x + 2) - 4$

80. $5(x + 2) = 5x - 2$

81. $-3(s + 2) = -2(s + 4) - s$

82. $21(b - 1) + 3 = 3(7b - 6)$

83. $2(3z + 4) = 2(3z - 2) + 13$

84. $x + 7 = \dfrac{2x + 6}{2} + 4$

85. $4(y - 3) - y = 3(y - 4)$

86. $5(x + 3) - 3x = 2(x + 8)$

⊞ *Solve each equation.*

87. $1.73x = -4.952 - 2.27x$

88. $\dfrac{h}{709} - 23{,}898 = -19{,}678$

89. $20(x - 3.7) = 32{,}832$

90. $9.35 - 1.4y = 7.32 + 1.5y$

WRITING

91. Explain the difference between *simplifying* an expression and *solving* an equation. Give some examples.

92. To solve $3x - 4 = 5x + 1$, one student began by subtracting $3x$ from both sides. Another student solved the same equation by first subtracting $5x$ from both sides. Will the students get the same solution? Explain why or why not.

93. What does it mean to clear an equation such as $\frac{1}{4} + \frac{x}{2} = \frac{3}{8}$ of the fractions?

94. Explain the error in the following solution:

Solve $2x + 4 = 30$.

$$2x + 4 = 30$$

$$\frac{2x}{2} + 4 = \frac{30}{2}$$

$$x + 4 = 15$$

$$x + 4 - 4 = 15 - 4$$

$$x = 11$$

REVIEW

95. Simplify $-(-8)$.

96. Subtract: $-8 - (-8)$.

97. Multiply: $-8(-8)$.

98. Add: $\dfrac{1}{8} + \dfrac{1}{8}$.

99. Multiply: $\dfrac{1}{8} \cdot \dfrac{1}{8}$.

100. Divide: $\dfrac{0.8}{8}$.

101. Simplify $8x + 8 + 8x - 8$.

102. Evaluate -1^8.

8.4 *Formulas*

In this section, you will learn about

- Formulas from business • Formulas from science
- Formulas from geometry (review) • Solving formulas

INTRODUCTION. A **formula** is an equation that is used to state a known relationship between two or more variables. Formulas are used in many fields: economics, physical education, anthropology, biology, automotive repair, and nursing, to name a few. In this section, we will consider formulas from business, science, and geometry.

Formulas from business

A formula to find the retail price: To make a profit, a merchant must sell a product for more than he or she paid for it. The price at which the merchant sells the product, called the **retail price,** is the sum of what the item cost the merchant plus the **markup.**

$$\boxed{\text{Retail price} \quad = \quad \text{cost} \quad + \quad \text{markup}}$$

Using r to represent the retail price, c the wholesale cost, and m the markup, we can write this formula as

$$\boxed{r = c + m}$$

As an example, suppose a jeweler purchases a gold ring at a wholesale jewelry mart for $612.50. Then she sets the price of the ring at $837.95 for sale in her store. We can find the markup on the ring as follows:

$r = c + m$	
$837.95 = 612.50 + m$	Substitute 837.95 for r and 612.50 for c.
$837.95 - \mathbf{612.50} = 612.50 + m - \mathbf{612.50}$	To undo the addition of 612.50, subtract 612.50 from both sides.
$225.45 = m$	Do the subtractions.

The markup on the ring is $225.45.

*A **formula for profit:*** The **profit** a business makes is the difference between the **revenue** (the money it takes in) and the costs.

$$\boxed{\text{Profit} \quad = \quad \text{revenue} \quad - \quad \text{costs}}$$

Using p to represent the profit, r the revenue, and c the costs, we can write this formula as

$$\boxed{p = r - c}$$

EXAMPLE 1   *Charitable giving.*   In 1999, the Salvation Army collected $1.9 billion. Of that amount, $1.7 billion went directly to the support of its programs. What were the 1999 administrative costs of the organization?

Solution

The charity collected $1.9 billion in revenue. We can think of the $1.7 billion that was spent on programs as profit. We need to find the administrative costs, c.

$p = r - c$	The formula for profit.
$1.7 = 1.9 - c$	Substitute 1.7 for p and 1.9 for r.
$1.7 - \mathbf{1.9} = 1.9 - c - \mathbf{1.9}$	To eliminate 1.9, subtract 1.9 from both sides.
$-0.2 = -c$	Subtract: $1.7 - 1.9 = -0.2$ and $1.9 - 1.9 = 0$.
$\dfrac{-0.2}{-1} = \dfrac{-c}{-1}$	Since $-c = -1c$, divide both sides by -1.
$0.2 = c$	Do the divisions.

In 1999, the Salvation Army had administrative costs of $0.2 billion.

Self Check
A PTA spaghetti dinner made a profit of $275.50. If the cost to host the dinner was $1,235, how much revenue did it generate?

Answer: $1,510.50   ■

*A **formula for simple interest:*** When money is borrowed, the lender expects to be paid back the amount of the loan plus an additional charge for the use of the money. The additional charge is called **interest.** When money is deposited in a bank, the depositor is paid for the use of the money. The money the deposit earns is also called interest. In general, interest is the money that is paid for the use of money.

Interest is calculated in two ways: either as **simple interest** or as **compound interest.** To find simple interest, we use the formula

Interest	=	principal	·	rate	·	time

Using I to represent the simple interest, P the principal (the amount of money that is invested, deposited, or borrowed), r the annual interest rate, and t the length of time in years, we can write the formula as

$$I = Prt$$

EXAMPLE 2 *Retirement income.* One year after investing $15,000 in a mini-mall development, a retired couple received a check for $1,125 in interest. What interest rate did their money earn that year?

Solution

The couple invested $15,000 (the principal) for 1 year (the time) and made $1,125 (the interest). We need to find the annual interest rate.

$I = Prt$	The formula for simple interest.
$1,125 = 15,000r(1)$	Substitute 1,125 for I, 15,000 for P, and 1 for t.
$1,125 = 15,000r$	Simplify the right-hand side.
$\dfrac{1,125}{15,000} = \dfrac{15,000r}{15,000}$	To solve for r, undo the multiplication by 15,000 by dividing both sides by 15,000.
$0.075 = r$	Do the divisions.
$7.5\% = r$	To write 0.075 as a percent, multiply 0.075 by 100 by moving the decimal point two places to the right and insert a % symbol.

The couple received an annual rate of 7.5% that year.

Self Check
A father lent his daughter and son-in-law $12,200 at a 2% annual simple interest rate for a down payment on a house. If the interest on the loan amounted to $610, for how long was the loan?

Answer: 2.5 years

Formulas from science

A formula for distance traveled: If we know the average rate (speed) at which we will be traveling and the time we will be traveling at that rate, we can find the distance traveled by using the formula

Distance	=	rate	·	time

Using d to represent the distance, r the average rate (speed), and t the time, we can write this formula as

$$d = rt$$

COMMENT When using this formula, the units must be the same. For example, if the rate is given in miles per hour, the time must be expressed in hours.

EXAMPLE 3 *Finding the rate.* As they migrate from the Bering Sea to Baja California, gray whales swim for about 20 hours each day, covering a distance of approximately 70 miles. Estimate their average swimming rate in miles per hour (mph).

Solution

Since the distance *d* is 70 miles and the time *t* is 20 hours, we substitute 70 for *d* and 20 for *t* in the formula $d = rt$, and then solve for *r*.

$$d = rt$$
$$70 = r(20) \qquad \text{Substitute 70 for } d \text{ and 20 for } t.$$
$$\frac{70}{20} = \frac{20r}{20} \qquad \text{To undo the multiplication by 20, divide both sides by 20.}$$
$$3.5 = r \qquad \text{Do the divisions.}$$

The whales' average swimming rate is 3.5 mph.

Self Check

An elevator in a building travels at an average rate of 288 feet per minute. How long will it take the elevator to climb 30 stories, a distance of 360 feet?

Answer: 1.25 minutes ∎

A formula for converting degrees Fahrenheit to degrees Celsius: Many marquees, like the one shown in Figure 8-6, flash two temperature readings, one in degrees Fahrenheit and one in degrees Celsius. The Fahrenheit scale is used in the American system of measurement. The Celsius scale is used in the metric system. The formula that relates a Fahrenheit temperature *F* to a Celsius temperature *C* is

$$C = \frac{5(F - 32)}{9}$$

FIGURE 8-6

EXAMPLE 4 *Changing degrees Celsius to degrees Fahrenheit.* Change the temperature reading on the sign in Figure 8-6 to degrees Fahrenheit.

Solution

Since the temperature *C* in degrees Celsius is 30°, we substitute 30 for *C* in the formula and solve for *F*.

$$C = \frac{5(F - 32)}{9}$$

$$30 = \frac{5(F - 32)}{9} \qquad \text{Substitute 30 for } C.$$

$$9(30) = 9\left[\frac{5(F - 32)}{9}\right] \qquad \text{To clear the equation of the fraction, multiply both sides by 9.}$$

$$270 = 5(F - 32) \qquad \text{Simplify: } 9(30) = 270 \text{ and}$$
$$\frac{\overset{1}{\cancel{9}}}{1}\left[\frac{5(F - 32)}{\underset{1}{\cancel{9}}}\right] = 5(F - 32).$$

$$270 = 5F - 5(32) \qquad \text{Distribute the multiplication by 5.}$$
$$270 = 5F - 160 \qquad \text{Do the multiplication: } 5(32) = 160.$$
$$270 + 160 = 5F - 160 + 160 \qquad \text{To undo the subtraction of 160, add 160 to both sides.}$$
$$430 = 5F \qquad \text{Simplify: } 270 + 160 = 430 \text{ and } -160 + 160 = 0.$$
$$\frac{430}{5} = \frac{5F}{5} \qquad \text{To undo the multiplication by 5, divide both sides by 5.}$$
$$86 = F \qquad \text{Do the divisions.}$$

Thus, 30°C is equivalent to 86°F.

Self Check

Change −175°C, the temperature on Saturn, to degrees Fahrenheit.

Answer: −283°F ∎

Formulas from geometry (review)

Recall that the **perimeter** of a geometric figure is the distance around it. Perimeter is measured in linear units, such as inches, feet, yards, and meters. The **area** of the figure is the amount of surface that it encloses. Area is measured in square units, such as square inches, square feet, square yards, and square meters (denoted as in.^{2}, ft^2, yd^2, and m^2, respectively). Table 8-3 reviews the formulas for the perimeter P and area A of several geometric figures.

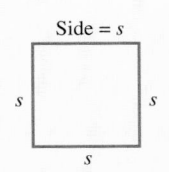

Square
$P = 4s$
$A = s^2$

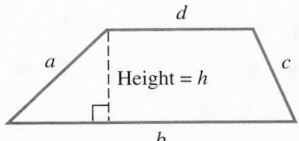

Trapezoid
$P = a + b + c + d$
$A = \dfrac{1}{2}h(b + d)$

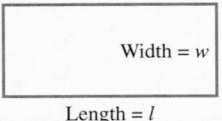

Rectangle
$P = 2l + 2w$
$A = lw$

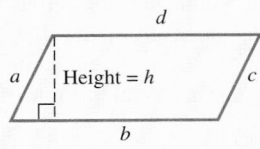

Parallelogram
$P = a + b + c + d$
$A = bh$

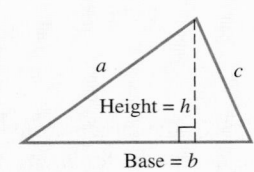

Triangle
$P = a + b + c$
$A = \dfrac{1}{2}bh$

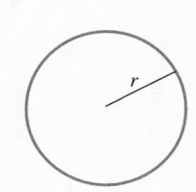

Circle
$C = 2\pi r$, where
$\pi \approx 3.1416$
C is the circumference
of the circle and
r is its radius.
$A = \pi r^2$

TABLE 8-3

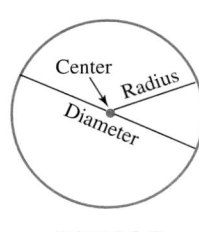

FIGURE 8-7

Recall that a **circle** (see Figure 8-7) is the set of all points in a plane that are a fixed distance from a point called its **center.** A segment drawn from the center of a circle to a point on the circle is called a **radius.** Since a **diameter** of a circle is a segment passing through the center that joins two points on the circle, the diameter D of a circle is twice as long as its radius r.

$$D = 2r$$

The perimeter of a circle is called its **circumference.** The formula for the circumference of a circle is

$$C = 2\pi r$$

EXAMPLE 5 *Finding perimeters and areas.* Find **a.** the perimeter of a square with sides 6 inches long and **b.** the area of a triangle with base 8 meters and height 13 meters.

Solution

a. The perimeter of a square is given by the formula $P = 4s$, where P is the perimeter and s is the length of one side. Since the sides of the square are 6 inches long, we substitute 6 for s and simplify.

Self Check

a. The flag of Eritrea, a country in east Africa, is shown on the next page. What is the perimeter of the flag?

b. Find the area of the red triangular region of the flag.

$P = 4s$

$P = 4(6)$ Substitute 6 for s.

$\quad = 24$ Do the multiplication.

The perimeter of the square is 24 inches.

b. The area of a triangle is given by the formula $A = \frac{1}{2}bh$. Since the base of the triangle is 8 meters and the height is 13 meters, we substitute 8 for b and 13 for h and simplify.

$A = \frac{1}{2}bh$

$A = \frac{1}{2}(8)(13)$ Substitute 8 for b and 13 for h.

$\quad = 4(13)$ $\frac{1}{2}(8) = \frac{8}{2} = 4.$

$\quad = 52$ Do the multiplication.

The area of the triangle is 52 square meters. This can be written as 52 m^2.

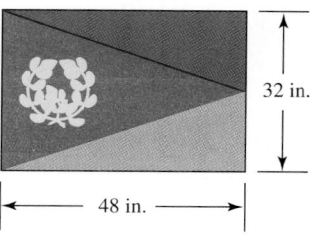

Answers: **a.** 160 in.,
b. 768 in.^{2} ■

EXAMPLE 6   *Finding the area of a circle.*   To the nearest tenth, find the area of a circle with a diameter of 14 feet.

Self Check
To the nearest hundredth, find the circumference of the circle of Example 6.

Solution
Since the radius of a circle is one-half its diameter, the radius of this circle is 7 feet. We can then substitute 7 for r in the formula for the area of a circle and simplify.

$A = \pi r^2$

$A = \pi(7)^2$

$\quad = 49\pi$ First, evaluate the exponential expression: $7^2 = 49$.

$\quad \approx 153.93804$ Use a calculator to do the multiplication. Enter these numbers and press these keys on a scientific calculator: 49 $\times$ π $=$.

To the nearest tenth, the area is 153.9 ft^2.

Answer: 43.98 ft   ■

Recall that the **volume** of a three-dimensional geometric solid is the amount of space it encloses. Table 8-4 (on the next page) shows the formula for the volume V of several solids. Volume is measured in cubic units, such as cubic inches, cubic feet, and cubic meters (denoted as in.^{3}, ft^3, and m^3, respectively).

EXAMPLE 7   *Finding volumes.*   To the nearest tenth, find the volume of each figure.

Self Check
Find the volume of each figure:
a. a rectangular solid with length 7 inches, width 12 inches, and height 15 inches   and   **b.** a cone whose base has radius 12 meters and whose height is 9 meters. Give the answer to the nearest tenth.

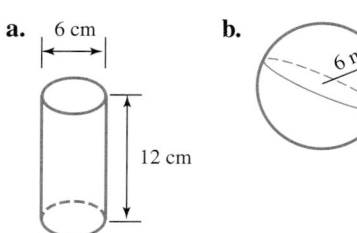

a. 6 cm

b. 6 m

12 cm

Solution
a. To find the volume of a cylinder, we use the formula $V = \pi r^2 h$, where r is the radius and h is the height. Since the radius of a circle is one-half its diameter, the radius of the cylinder is $\frac{1}{2}(6 \text{ cm}) = 3$ cm. The height of the cylinder is 12 cm. We substitute 3 for r and 12 for h in the formula for volume and proceed as follows.

$V = \pi r^2 h$ The formula for the volume of a cylinder.

$V = \pi(3)^2(12)$ Substitute 3 for r and 12 for h.

$= \pi(9)(12)$ Evaluate the exponential expression: $(3)^2 = 9$.

$= 108\pi$ Multiply: $9(12) = 108$.

≈ 339.2920066 Use a calculator.

To the nearest tenth, the volume is 339.3 cubic centimeters. This can be written as 339.3 cm³.

b. To find the volume of the sphere, we substitute 6 for r in the formula for the volume of a sphere and proceed as follows.

$V = \dfrac{4}{3}\pi r^3$ The formula for the volume of a sphere.

$V = \dfrac{4}{3}\pi(6)^3$ Since the radius of the sphere is 6 m, substitute 6 for r.

$= \dfrac{4}{3}\pi(216)$ $(6)^3 = 6 \cdot 6 \cdot 6 = 216$.

$= 288\pi$ Do the multiplication: $\frac{4}{3}(216) = \frac{4(216)}{3} = \frac{864}{3} = 288$.

≈ 904.7786842 Use a calculator.

To the nearest tenth, the volume is 904.8 m³.

Answers: a. 1,260 in.³,
b. 1,357.2 m³   ■

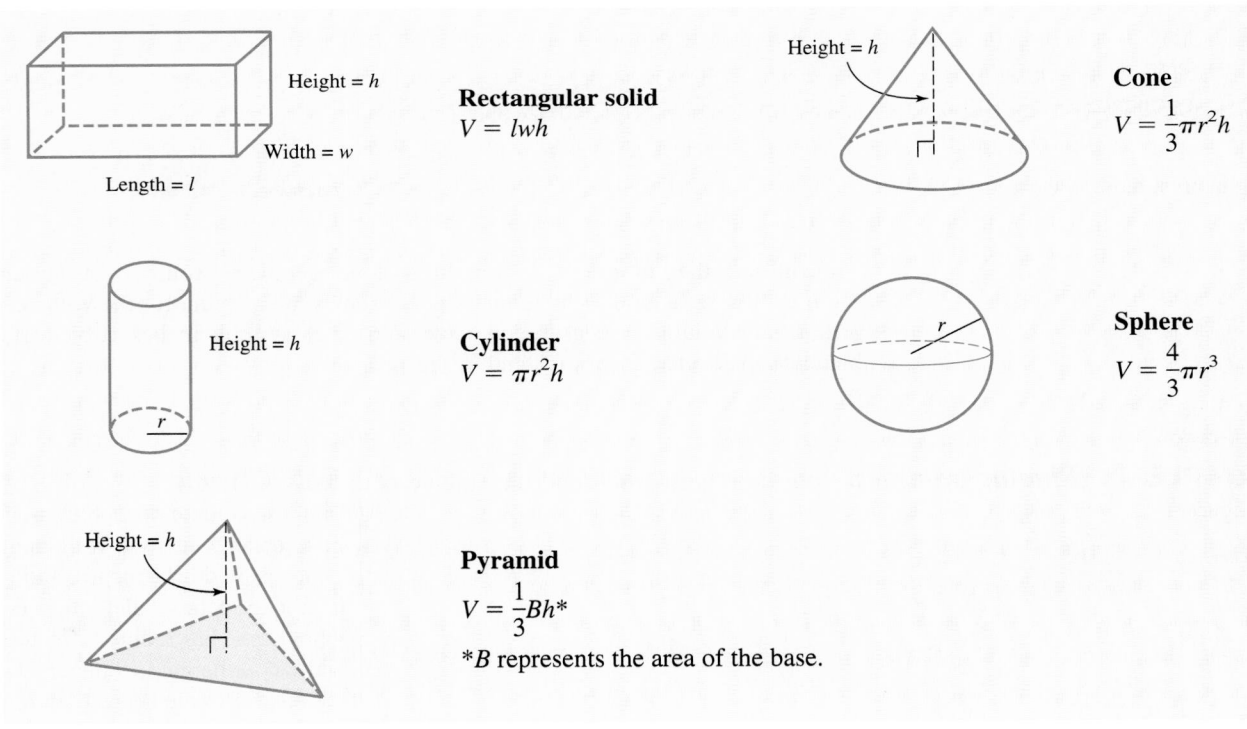

Rectangular solid
$V = lwh$

Cone
$V = \dfrac{1}{3}\pi r^2 h$

Cylinder
$V = \pi r^2 h$

Sphere
$V = \dfrac{4}{3}\pi r^3$

Pyramid
$V = \dfrac{1}{3}Bh*$

*B represents the area of the base.

TABLE 8-4

Solving formulas

Suppose we wish to find the bases of several triangles whose areas and heights are known. It could be tedious to substitute values for A and h into the formula and then repeatedly solve the formula for b. A better way is to solve the formula $A = \frac{1}{2}bh$ for b first, and then substitute values for A and h and compute b directly.

To **solve an equation for a variable** means to isolate that variable on one side of the equation, with all other quantities on the opposite side.

EXAMPLE 8   *Solving formulas.*   Solve $A = \frac{1}{2}bh$ for b.

Solution

To solve for b, we must isolate b on one side of the equation.

$$A = \frac{1}{2}bh$$

$$2A = 2 \cdot \frac{1}{2}bh \qquad \text{To clear the equation of the fraction, multiply both sides by 2.}$$

$$2A = bh \qquad \text{Simplify: } 2 \cdot \frac{1}{2} = \frac{2}{2} = 1.$$

$$\frac{2A}{h} = \frac{bh}{h} \qquad \text{To undo the multiplication by } h, \text{ divide both sides by } h.$$

$$\frac{2A}{h} = b \qquad \text{On the right-hand side, divide out the common factor of } h: \frac{b\overset{1}{\cancel{h}}}{\underset{1}{\cancel{h}}} = b.$$

$$b = \frac{2A}{h} \qquad \text{Reverse the sides to write } b \text{ on the left.}$$

EXAMPLE 9   *Solving formulas.*   Solve $P = 2l + 2w$ for l.

Solution

To solve for l, we must isolate l on one side of the equation.

$$P = 2l + 2w$$

$$P - 2w = 2l + 2w - 2w \qquad \text{To undo the addition of } 2w, \text{ subtract } 2w \text{ from both sides.}$$

$$P - 2w = 2l \qquad \text{Combine like terms: } 2w - 2w = 0.$$

$$\frac{P - 2w}{2} = \frac{2l}{2} \qquad \text{To undo the multiplication by 2, divide both sides by 2.}$$

$$\frac{P - 2w}{2} = l \qquad \text{Simplify the right-hand side.}$$

We can write the result as $l = \dfrac{P - 2w}{2}$.

EXAMPLE 10   *Solving for y.*   In the next chapter, we will work with equations that involve the variables x and y, such as $2y - 4 = 3x$. Solve the equation for y.

Solution

$$2y - 4 = 3x \qquad \text{The given equation.}$$

$$2y - 4 + 4 = 3x + 4 \qquad \text{To undo the subtraction of 4, add 4 to both sides.}$$

$$2y = 3x + 4 \qquad \text{On the left-hand side, simplify: } -4 + 4 = 0.$$

$$\frac{2y}{2} = \frac{3x + 4}{2} \qquad \text{To undo the multiplication by 2, divide both sides by 2.}$$

$$y = \frac{3x}{2} + \frac{4}{2}$$ On the right-hand side, rewrite $\frac{3x+4}{2}$ as the sum of two fractions with like denominators, $\frac{3x}{2}$ and $\frac{4}{2}$.

$$y = \frac{3}{2}x + 2$$ Write $\frac{3x}{2}$ as $\frac{3}{2}x$. Simplify: $\frac{4}{2} = 2$.

Answer: $y = \frac{1}{3}x - 4$ ∎

EXAMPLE 11   *Solving for r^2.*   Solve $V = \pi r^2 h$ for r^2.

Solution

We want to isolate r^2 on one side of the equation.

$$V = \pi r^2 h$$

$$\frac{V}{\pi h} = \frac{\pi r^2 h}{\pi h}$$ To undo the multiplication by π and h on the right-hand side, divide both sides by πh.

On the right-hand side, divide out the common factors of π and h:

$$\frac{V}{\pi h} = r^2 \qquad \frac{\overset{1}{\cancel{\pi}} r^2 \overset{1}{\cancel{h}}}{\underset{1}{\cancel{\pi}} \underset{1}{\cancel{h}}} = r^2$$

$$r^2 = \frac{V}{\pi h}$$ Reverse the sides of the equation so that r^2 is on the left.

Self Check

Solve $a^2 + b^2 = c^2$ for b^2.

Answer: $b^2 = c^2 - a^2$ ∎

STUDY SET   Section 8.4

VOCABULARY   *Fill in the blanks.*

1. A _____ is an equation that is used to state a known relationship between two or more variables.

2. The _____ of a three-dimensional geometric solid is the amount of space it encloses.

3. The distance around a geometric figure is called its _____.

4. A _____ is the set of all points in a plane that are a fixed distance from a point called its center.

5. A segment drawn from the center of a circle to a point on the circle is called a _____.

6. The amount of surface that is enclosed by a geometric figure is called its _____.

7. The perimeter of a circle is called its _____.

8. A segment passing through the center of a circle and connecting two points on the circle is called a _____.

CONCEPTS

9. Use variables to write the formula relating the following:
   a. Time, distance, rate
   b. Markup, retail price, cost
   c. Costs, revenue, profit
   d. Interest rate, time, interest, principal
   e. Circumference, radius

10. Complete the table.

Principal	·	rate	·	time	=	interest
$2,500		5%		2 yr		
$15,000		4.8%		1 yr		

11. Complete the table to find how far light and sound travel in 60 seconds. (*Hint*: mi/sec means miles per second.)

	Rate	·	time	=	distance
Light	186,282 mi/sec		60 sec		
Sound	1,088 ft/sec		60 sec		

12. Give the name of each figure.

a.    **b.**

c. 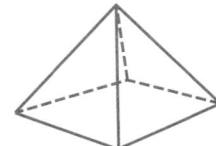   **d.**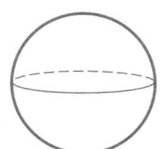

e.

f.

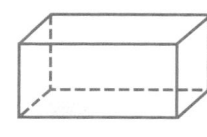

13. Tell which geometric concept—perimeter, circumference, area, or volume—should be used to find the following:
   a. The amount of storage in a freezer
   b. How far a bicycle tire rolls in one revolution

   c. The amount of land making up the Sahara Desert

   d. The distance around a Monopoly game board

14. Tell which unit of measurement—ft, ft^2, or ft^3—would be appropriate when finding the following:
   a. The amount of storage inside a safe
   b. The ground covered by a sleeping bag lying on the floor
   c. The distance the tip of an airplane propeller travels in one revolution
   d. The size of the trunk of a car

15. Write an expression for the area of the figure shown in Illustration 1.

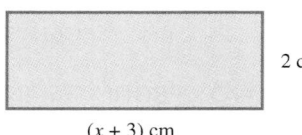

2 cm

$(x + 3)$ cm

ILLUSTRATION 1

16. Write an expression for the area of the figure shown in Illustration 2.

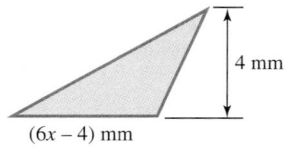

4 mm

$(6x - 4)$ mm

ILLUSTRATION 2

17. Solve $V = \frac{1}{3}Bh$ for B.

$$V = \frac{1}{3}Bh$$

$$\boxed{}(V) = \left(\frac{1}{3}\right)Bh$$

$$3V = \boxed{}$$

$$\frac{3V}{\boxed{}} = \frac{Bh}{\boxed{}}$$

$$\frac{3V}{h} = \boxed{}$$

$$B = \frac{3V}{h}$$

18. Solve $Ax + By = C$ for y.

$$Ax + By = C$$

$$Ax + By - \boxed{} = C - \boxed{}$$

$$\boxed{} = C - Ax$$

$$\frac{By}{\boxed{}} = \frac{C - Ax}{\boxed{}}$$

$$y = \frac{C - Ax}{B}$$

19. a. Approximate π to the nearest hundredth.
   b. What does 98π mean?
   c. In the formula for the volume of a cylinder, $V = \pi r^2 h$, what does r represent? What does h represent?

20. a. What does ft^2 mean?
   b. What does in.^{3} mean?

PRACTICE   *Use a formula discussed in this section to solve each problem.*

21. SWIMMING   In 1930, a man swam down the Mississippi River from Minneapolis to New Orleans, a total of 1,826 miles. He was in the water for 742 hours. To the nearest tenth, what was his average swimming rate?

22. ROSE PARADE   Rose Parade floats travel down the 5.5-mile-long parade route at a rate of 2.5 mph. How long will it take a float to complete the parade if there are no delays?

23. HOLLYWOOD   Figures for the summer of 1998 showed that the movie *Saving Private Ryan* had U.S. box-office receipts of $190 million. What were the production costs to make the movie if, at that time, the studio had made a $125 million profit?

24. SERVICE CLUB   After expenses of $55.15 were paid, a Rotary Club donated $875.85 in proceeds from a pancake breakfast to a local health clinic. How much did the pancake breakfast gross?

25. ENTREPRENEURS   To start a mobile dog-grooming service, a woman borrowed $2,500. If the loan was for 2 years and the amount of interest was $175, what simple interest rate was she charged?

26. BANKING   Three years after opening an account that paid 6.45% annually, a depositor withdrew the $3,483 in interest earned. How much money was left in the account?

27. METALLURGY   Change 2,212°C, the temperature at which silver boils, to degrees Fahrenheit. Round to the nearest degree.

28. LOW TEMPERATURES   Cryobiologists freeze living matter to preserve it for future use. They can work with temperatures as low as −270°C. Change this to degrees Fahrenheit.

29. VALENTINE'S DAY   Find the markup on a dozen roses if a florist buys them wholesale for $12.95 and sells them for $37.50.

30. STICKER PRICE   The factory invoice for a minivan shows that the dealer paid $16,264.55 for the vehicle. If the sticker price of the van is $18,202, how much over factor invoice is the sticker price?

31. YO-YO   How far does a yo-yo travel during one revolution of the "around the world" trick if the length of the string is 21 inches?

32. HORSE TRAINING   A horse trots in a perfect circle around its trainer at the end of a 28-foot-long rope. How far does the horse travel as it circles the trainer once?

Solve each formula for the given variable.

33. $E = IR$; for R **34.** $d = rt$; for t

35. $V = lwh$; for w **36.** $I = Prt$; for r

37. $C = 2\pi r$; for r **38.** $V = \pi r^2 h$; for h

39. $a + b + c = 180$; for a **40.** $P = a + b + c$; for b

41. $y = mx + b$; for x **42.** $P = 2l + 2w$; for l

43. $A = P + Prt$; for t **44.** $S = 2\pi rh + 2\pi r^2$; for h

45. $V = \dfrac{1}{3}\pi r^2 h$; for h **46.** $K = \dfrac{1}{2}mv^2$; for m

47. $x = \dfrac{a + b}{2}$; for b **48.** $A = \dfrac{a + b + c}{3}$; for c

49. $D = \dfrac{C - s}{n}$; for s **50.** $2E = \dfrac{T - t}{9}$; for t

51. $E = mc^2$; for c^2 **52.** $s = 4\pi r^2$; for r^2

53. $c^2 = a^2 + b^2$; for a^2 **54.** $Kg = \dfrac{wv^2}{2}$; for v^2

55. $A = \dfrac{1}{2}h(b + d)$; for b **56.** $h = vt + 16t^2$; for t^2

57. $3y - 9 = x$; for y **58.** $5y - 25 = x$; for y

59. $4y + 16 = -3x$; for y **60.** $6y + 12 = -5x$; for y

APPLICATIONS

61. PROPERTIES OF WATER   The boiling point and the freezing point of water are to be given in both degrees

Celsius and degrees Fahrenheit on the thermometer in Illustration 3. Find the missing degree measures.

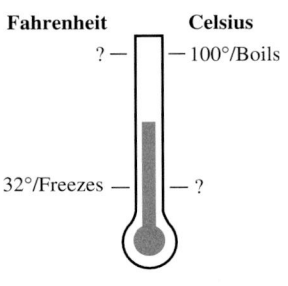

ILLUSTRATION 3

62. HIGHWAY SPEED LIMITS   Several state speed limits for trucks are shown in Illustration 4. At each of these speeds, how far would a truck travel in $2\frac{1}{2}$ hours?

ILLUSTRATION 4

63. AVON PRODUCTS, INC.   Complete the financial statement shown in Illustration 5.

Quarterly financials **Income statement** (dollar amounts in millions except per share amounts)	**Quarter** **ending** **Mar 00**	**Quarter** **ending** **Dec 99**
Revenue	1,324.9	1,566.6
Cost of goods sold	497.3	606.6
Gross profit		

Based on data from Hoover's Online

ILLUSTRATION 5

64. CREDIT CARDS   The finance charge section of a person's credit card statement says, "annual percentage rate (APR) is 19.8%." Determine how much finance charges (interest) the card owner would have to pay if the account's average balance for the year was $2,500.

65. CARPENTRY   Find the perimeter and area of the truss shown in Illustration 6.

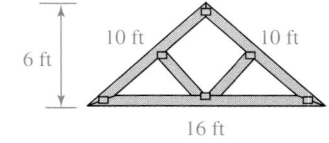

ILLUSTRATION 6

66. CAMPERS   Find the area of the window of the camper shell shown in Illustration 7.

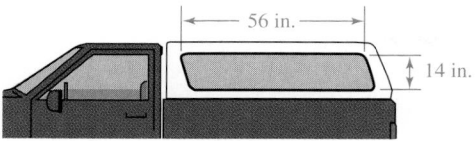

ILLUSTRATION 7

67. ARCHERY   To the nearest tenth, find the circumference and area of the target shown in Illustration 8.

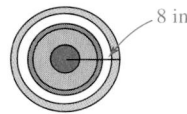

ILLUSTRATION 8

68. GEOGRAPHY   The circumference of the earth is about 25,000 miles. Find its diameter to the nearest mile.

69. LANDSCAPING   Find the perimeter and the area of the redwood trellis in Illustration 9.

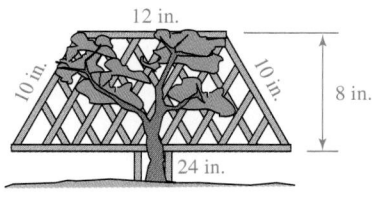

ILLUSTRATION 9

70. HAMSTER HABITAT   Find the amount of space in the plastic tube shown in Illustration 10.

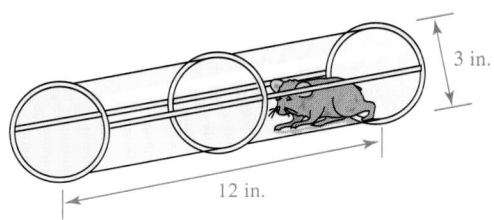

ILLUSTRATION 10

71. "THE WALL"   The Vietnam Veterans Memorial is a black granite wall recognizing the more than 58,000 Americans who lost their lives or remain missing. A diagram of the wall is shown in Illustration 11. Find the total area of the two triangular-shaped surfaces on which the names are inscribed.

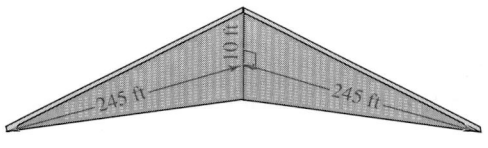

ILLUSTRATION 11

72. SIGNAGE   Find the perimeter and area of the service station sign shown in Illustration 12.

ILLUSTRATION 12

73. RUBBER MEETS THE ROAD   A sport truck tire has the road surface "footprint" shown in Illustration 13. Estimate the perimeter and area of the tire's footprint.

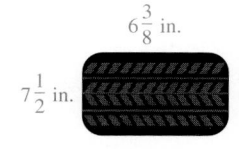

ILLUSTRATION 13

74. SOFTBALL   The strike zone in fast-pitch softball is between the batter's armpit and top of her knees, as shown in Illustration 14. Find the area of the strike zone.

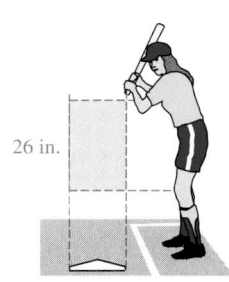

ILLUSTRATION 14

75. FIREWOOD   The dimensions of a cord of firewood are shown in Illustration 15. Find the area on which the wood is stacked and the volume the cord of firewood occupies.

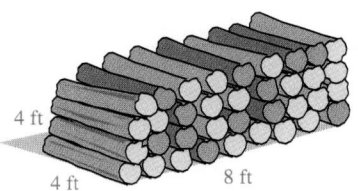

ILLUSTRATION 15

76. NATIVE AMERICAN DWELLING   The teepees constructed by the Blackfoot Indians were cone-shaped tents made of long poles and animal hide, about 10 feet high and about 15 feet across at the ground. (See Illustration 16.) Estimate the volume of a teepee with these dimensions, to the nearest cubic foot.

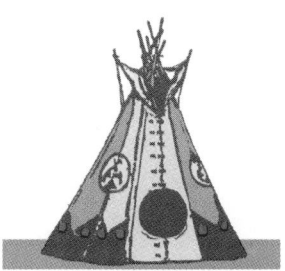

ILLUSTRATION 16

77. IGLOO   During long journeys, some Canadian Inuit (Eskimos) built winter houses of snow blocks piled in the dome shape shown in Illustration 17. Estimate the volume of an igloo having an interior height of 5.5 feet to the nearest cubic foot.

ILLUSTRATION 17

78. PYRAMID   The Great Pyramid at Giza in northern Egypt is one of the most famous works of architecture in the world. Use the information in Illustration 18 to find the volume to the nearest cubic foot.

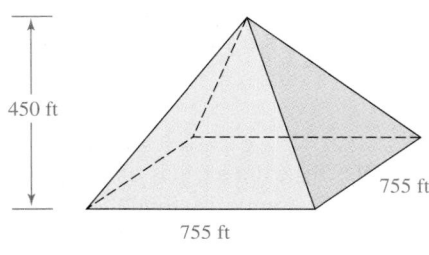

ILLUSTRATION 18

79. BARBECUING   See Illustration 19. Use the fact that the fish is 18 inches long to find the area of the barbecue grill to the nearest square inch.

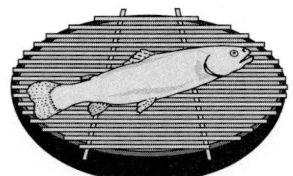

ILLUSTRATION 19

80. SKATEBOARDING   A "half-pipe" ramp used for skateboarding is in the shape of a semicircle with a

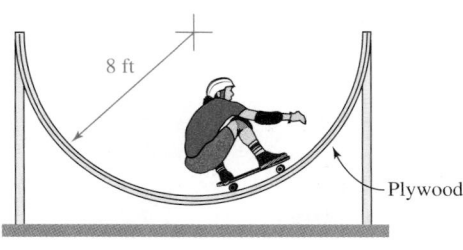

ILLUSTRATION 20

radius of 8 feet, as shown in Illustration 20. To the nearest tenth of a foot, what is the length of the arc that the skateboarder travels on the ramp?

81. GEOMETRY   The measure a of an interior angle of a regular polygon with n sides is given by the formula

$$a = 180° \left(1 - \frac{2}{n} \right)$$

See Illustration 21. Solve the formula for n. How many sides does a regular polygon have if an interior angle is 108°? (*Hint:* Distribute first.)

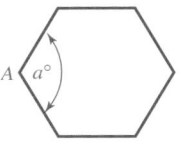

ILLUSTRATION 21

82. THERMODYNAMICS   The Gibbs free-energy function is given by $G = U - TS + pV$. Solve this formula for the pressure p.

WRITING

83. The formula $P = 2l + 2w$ is also an equation, but an equation such as $2x + 3 = 5$ is not a formula. What equations do you think should be called formulas?

84. Explain what it means to solve the equation $P = 2l + 2w$ for w.

85. After solving $A = B + C + D$ for B, a student compared her answer with that of the back of the textbook.

   Student's answer: $B = A - C - D$

   Book's answer: $B = A - D - C$

Could this problem have two different-looking answers? Explain why or why not.

86. Suppose the volume of a cylinder is 28 cubic feet. Explain why it is incorrect to express the volume as 28^3 ft.

REVIEW

87. Find 82% of 168.

88. 29.05 is what percent of 415?

89. What percent of 200 is 30?

90. SHOPPING   A woman bought a coat for $98.95 and some gloves for $7.95. If the sales tax was 6%, how much did the purchase cost her?

8.5 *Problem Solving*

In this section, you will learn about

- Solving problems • Solving geometric problems
- Solving number–value problems • Solving investment problems
- Solving uniform motion problems • Solving mixture problems

INTRODUCTION. In this section, we will solve several different types of problems using the five-step problem-solving strategy.

Solving problems

EXAMPLE 1   *California coastline.*   The first part of California's magnificent 17-Mile Drive scenic tour, shown in Figure 8-8, begins at the Pacific Grove entrance and continues to Seal Rock. It is 1 mile longer than the second part of the drive, which extends from Seal Rock to the Lone Cypress. The final part of the tour winds through the hills of the Monterey Peninsula, eventually returning to the entrance. This part of the drive is 1 mile longer than four times the length of the second part. How long is each of the three parts of 17-Mile Drive?

Analyze the problem   In Figure 8-9, we "straighten out" the winding 17-Mile Drive so that it can be modeled with a line segment. The drive is composed of three parts. We need to find the length of each part.

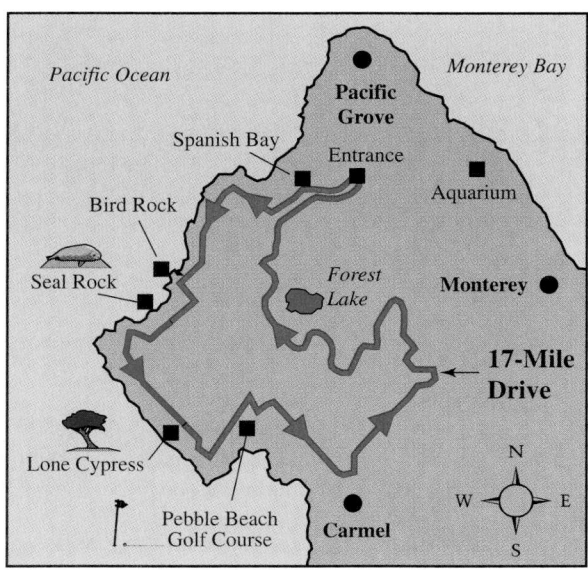

FIGURE 8-8

Form an equation   Since the lengths of the first part and of the third part of the scenic drive are related to the length of the second part, we will let x represent the length of that part. We then express the other lengths in terms of that variable.

$x + 1$ represents the length of the first part of the drive.

$4x + 1$ represents the length of the third part of the drive.

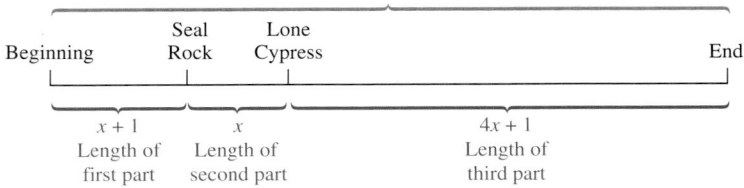

FIGURE 8-9

The sum of the lengths of the three parts of the drive must equal the total length of 17-Mile Drive.

The length of part 1	plus	the length of part 2	plus	the length of part 3	equals	the total length.
$x + 1$	$+$	x	$+$	$4x + 1$	$=$	17

Solve the equation

$$x + 1 + x + 4x + 1 = 17$$
$$6x + 2 = 17 \quad \text{Combine like terms: } x + x + 4x = 6x \text{ and } 1 + 1 = 2.$$
$$6x = 15 \quad \text{To undo the addition of 2, subtract 2 from both sides.}$$
$$\frac{6x}{6} = \frac{15}{6} \quad \text{To undo the multiplication by 6, divide both sides by 6.}$$
$$x = 2.5 \quad \text{Do the divisions.}$$

Recall that x represents the length of the *second* part of the drive. To find the lengths of the first and third parts, we evaluate the expressions $x + 1$ and $4x + 1$ for $x = 2.5$.

First part of drive   **Third part of drive**

$$x + 1 = \mathbf{2.5} + 1 \qquad 4x + 1 = 4(\mathbf{2.5}) + 1 \quad \text{Substitute 2.5 for } x.$$
$$= 3.5 \qquad\qquad\qquad = 10 + 1$$
$$\qquad\qquad\qquad\qquad = 11$$

State the conclusion   The first part of the drive is 3.5 miles long, the second part is 2.5 miles long, and the third part is 11 miles long.

Check the result   Because the sum of 3.5 miles, 2.5 miles, and 11 miles is 17 miles, the answers check.   ■

Solving geometric problems

EXAMPLE 2   *Dimensions of a garden.*   A gardener wants to use 62 feet of fencing bought at a garage sale to enclose a rectangular-shaped garden. Find the dimensions of the garden if its length is to be 4 feet longer than twice its width.

Analyze the problem   We can make a sketch of the garden, as shown in Figure 8-10. We know that its length is to be 4 feet longer than twice its width. We also know that its perimeter is to be 62 feet.

FIGURE 8-10

Form an equation   If we let w represent the width of the garden, then $2w + 4$ represents its length. Since the formula for the perimeter of a rectangle is $P = 2l + 2w$, the perimeter of the garden is $2(2w + 4) + 2w$, which is also 62. This fact enables us to form the equation.

2	times	the length	plus	2	times	the width	is	the perimeter.
2	·	$(2w + 4)$	+	2	·	w	=	62

Solve the equation

$$2(2w + 4) + 2w = 62$$
$$4w + 8 + 2w = 62 \quad \text{Use the distributive property to remove parentheses.}$$
$$6w + 8 = 62 \quad \text{Combine like terms: } 4w + 2w = 6w.$$
$$6w = 54 \quad \text{To undo the addition of 8, subtract 8 from both sides.}$$
$$w = 9 \quad \text{To undo the multiplication by 6, divide both sides by 6.}$$

State the conclusion   The width of the garden is 9 feet. Since $2w + 4 = 2(9) + 4 = 22$, the length is 22 feet.

Check the result   If the garden has a width of 9 feet and a length of 22 feet, its length is 4 feet longer than twice the width ($2 \cdot 9 + 4 = 22$). Since its perimeter is $(2 \cdot 22 + 2 \cdot 9)$ feet = 62 feet, the answers check.   ■

EXAMPLE 3   *Isosceles triangles.*   If the vertex angle of an isosceles triangle is 56°, find the measure of each base angle.

Analyze the problem   An **isosceles triangle** has two sides of equal length, which meet to form the **vertex angle.** In this case, the measurement of the vertex angle is 56°. We can sketch the triangle as shown in Figure 8-11. The **base angles** opposite the equal sides are also equal. We need to find their measure.

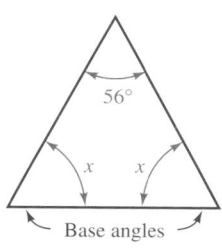

FIGURE 8-11

Form an equation   If we let x represent the measure of one base angle, the measure of the other base angle is also x. Since the sum of the angles of any triangle is 180°, the sum of the base angles and the vertex angle is 180°. We can use this fact to form the equation.

One base angle	plus	the other base angle	plus	the vertex angle	is	180°.
x	+	x	+	56	=	180

Solve the equation

$$x + x + 56 = 180$$
$$2x + 56 = 180 \quad \text{Combine like terms: } x + x = 2x.$$
$$2x = 124 \quad \text{To undo the addition of 56, subtract 56 from both sides.}$$
$$x = 62 \quad \text{To undo the multiplication by 2, divide both sides by 2.}$$

State the conclusion   The measure of each base angle is 62°.

Check the result   The measure of each base angle is 62°, and the vertex angle measures 56°. Since $62° + 62° + 56° = 180°$, the answer checks.   ■

Solving number–value problems

Some problems deal with quantities that have a monetary value. In these problems, we must distinguish between the *number of* and the *value of* the unknown quantity. For problems of this type, we will use the relationship

$$\text{Number} \cdot \text{value} = \text{total value}$$

EXAMPLE 4 *Dining area improvements.* A restaurant owner needs to purchase some new tables, chairs, and dinner plates for the dining area of her establishment. She plans to buy four chairs and four plates for each new table. She also needs 20 additional plates to keep in case of breakage. If a table costs $100, a chair $50, and a plate $5, how many of each can she buy if she takes out a small business loan for $6,500 to pay for the new items?

Analyze the problem We know the *value* of each item: Tables cost $100, chairs cost $50, and plates cost $5 each. We need to find the *number* of tables, chairs, and plates she can purchase for $6,500.

Form an equation The number of chairs and plates she needs depends on the number of tables she buys. So we let t be the number of tables to be purchased. Since every table requires four chairs and four plates, she needs to order $4t$ chairs. Because an additional 20 plates are needed, she should order $4t + 20$ plates. The total value of each purchase is the *product* of the number of items bought and the price, or value, of each item.

Item	Number purchased ·	Price per item =	Total value
Tables	t	$100	$100t$
Chairs	$4t$	$50	$50(4t)$
Plates	$4t + 20$	$5	$5(4t + 20)$

The total purchase can be expressed in two ways.

The value of the tables	+	the value of the chairs	+	the value of the plates	is	the total value of the purchase.
$100t$	+	$50(4t)$	+	$5(4t + 20)$	=	6,500

Solve the equation

$$100t + 50(4t) + 5(4t + 20) = 6{,}500$$
$$100t + 200t + 20t + 100 = 6{,}500 \qquad \text{Do the multiplications.}$$
$$320t + 100 = 6{,}500 \qquad \text{Combine like terms.}$$
$$320t = 6{,}400 \qquad \text{Subtract 100 from both sides.}$$
$$t = 20 \qquad \text{Divide both sides by 320.}$$

State the conclusion The purchases are summarized as follows:

Item	Number purchased	Price per item	Total value
Tables	$t = 20$	$100	$2,000
Chairs	$4t = 80$	$50	$4,000
Plates	$4t + 20 = 100$	$5	$500
Total			$6,500

Check the result Because the total purchase is $6,500, the answer checks.

Solving investment problems

To find the amount of simple interest I an investment earns, we use the formula

$$I = Prt$$

where P is the principal, r is the annual rate, and t is the time in years. When $t = 1$, the formula simplifies to $I = Pr$.

EXAMPLE 5   *Paying tuition.*   A college student invested the $12,000 inheritance he received and decided to use the annual interest earned to pay his yearly tuition costs of $945. The highest rate offered by a savings and loan at that time was 6% annual simple interest. At this rate, he could not earn the needed $945, so he invested some of the money in a riskier, but more lucrative, investment offering a 9% return. How much did he invest at each rate?

Analyze the problem   We know that $12,000 was invested for 1 year at two rates: 6% and 9%. We are asked to find the amount invested at each rate so that the total return would be $945.

Form an equation   Let x represent the amount invested at 6%. Then $12,000 - x$ represents the amount invested at 9%.

If $\$x$ (the principal P) is invested at 6% (the rate r), the interest earned in 1 year would be Pr or $\$0.06x$. At 9%, the rest of the inheritance money, $\$(12,000 - x)$, would earn $\$0.09(12,000 - x)$ interest. These facts are summarized in the following table.

	P	$\cdot$ r $=$	I
Savings and loan	x	0.06	$0.06x$
Riskier investment	$12,000 - x$	0.09	$0.09(12,000 - x)$

The total interest earned can be expressed in two ways.

The interest earned at 6%	plus	the interest earned at 9%	is	the total interest.
$0.06x$	$+$	$0.09(12,000 - x)$	$=$	945

Solve the equation

$$0.06x + 0.09(12,000 - x) = 945$$

$100[0.06x + 0.09(12,000 - x)] = 100(945)$ Multiply both sides by 100 to clear the equation of decimals.

$100(0.06x) + 100(0.09)(12,000 - x) = 100(945)$ Distribute the multiplication by 100.

$6x + 9(12,000 - x) = 94,500$ Do the multiplications by 100.

$6x + 108,000 - 9x = 94,500$ Use the distributive property.

$-3x + 108,000 = 94,500$ Combine like terms.

$-3x = -13,500$ Subtract 108,000 from both sides.

$x = 4,500$ Divide both sides by -3.

State the conclusion   The student invested $4,500 at 6% and $12,000 - $4,500 = $7,500 at 9%.

Check the result   The first investment earned 6% of $4,500, or $270. The second earned 9% of $7,500, or $675. The total return was $270 + $675 = $945. The answers check.   ◼

Solving uniform motion problems

If we know the rate r at which we will be traveling and the time t we will be traveling at that rate, we can find the distance d traveled by using the formula

$$d = rt$$

EXAMPLE 6   *Coast Guard rescue.*   A cargo ship, heading into port, radios the Coast Guard that it is experiencing engine trouble and that its speed has dropped to 3 knots. Immediately, a Coast Guard cutter leaves the port and speeds at a rate of 25 knots directly toward the disabled craft, which is 21 nautical miles away. How long will it take the Coast Guard cutter to reach the cargo ship?

Analyze the problem   The diagram in Figure 8-12(a) shows the situation.

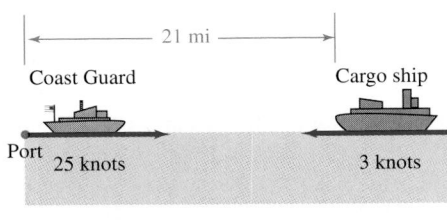

	r	$\cdot$ t	$=$ d
Coast Guard cutter	25	t	$25t$
Cargo ship	3	t	$3t$

(a)   **(b)**

FIGURE 8-12

We know the *rate* of each ship (25 knots and 3 knots), and we know that they must close a *distance* of 21 nautical miles between them. We don't know the *time* it will take them to do this.

Form an equation   Let t represent the time it takes for the ships to meet. Using $d = rt$, we find that $25t$ represents the distance traveled by the Coast Guard cutter and $3t$ represents the distance traveled by the cargo ship. This information is recorded in the table in Figure 8-12(b). We can use it to form the equation.

The distance the Coast Guard cutter travels	plus	the distance the cargo ship travels	is	the initial distance between the two ships.
$25t$	$+$	$3t$	$=$	21

Solve the equation

$$25t + 3t = 21$$
$$28t = 21 \qquad \text{Combine like terms.}$$
$$t = \frac{21}{28} \qquad \text{Divide both sides by 28.}$$
$$t = \frac{3}{4} \qquad \text{Simplify the fraction: } \frac{21}{28} = \frac{\overset{1}{\cancel{7}} \cdot 3}{\underset{1}{\cancel{7}} \cdot 4} = \frac{3}{4}.$$

State the conclusion   The ships will meet in three-quarters of an hour, or 45 minutes.

Check the result   In three-quarters of an hour, the Coast Guard cutter travels $25 \cdot \frac{3}{4} = \frac{75}{4}$ nautical miles, and the cargo ship travels $3 \cdot \frac{3}{4} = \frac{9}{4}$ nautical miles. Together, they travel $\frac{75}{4} + \frac{9}{4} = \frac{84}{4} = 21$ nautical miles. Since this is the initial distance between the ships, the answer checks. ∎

Solving mixture problems

We now discuss how to solve two types of mixture problems. In the first type, a *liquid mixture* of a desired strength is made from two solutions with different concentrations.

EXAMPLE 7 *Mixing a solution.* A chemistry experiment calls for a 30% sulfuric acid solution. If the lab supply room has only 50% and 20% sulfuric acid solutions on hand, how much of each should be mixed to obtain 12 liters of a 30% acid solution?

Analyze the problem We must find how much of the 50% solution and how much of the 20% solution is needed to obtain 12 liters of a 30% acid solution.

Form an equation If x represents the numbers of liters (L) of the 50% solution used in the mixture, the remaining $(12 - x)$ liters must be the 20% solution. See Figure 8-13(a). Only 50% of the x liters, and only 20% of the $(12 - x)$ liters, is pure sulfuric acid. The total of these amounts is also the amount of acid in the final mixture, which is 30% of 12 liters. This information is shown in the chart in Figure 8-13(b).

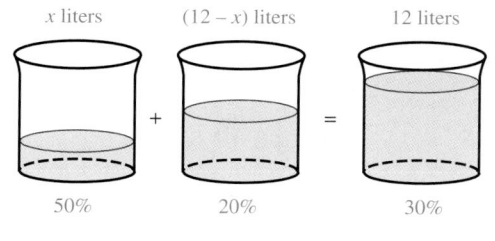

Solution	% acid ·	Liters =	Amount of acid
50% solution	0.50	x	$0.50x$
20% solution	0.20	$12 - x$	$0.20(12 - x)$
30% mixture	0.30	12	$0.30(12)$

(a)   (b)

FIGURE 8-13

We can form the equation.

The acid in the 50% solution	plus	the acid in the 20% solution	equals	the acid in the final mixture.
50% of x	+	20% of $(12 - x)$	=	30% of 12

Solve the equation

$0.50x + 0.20(12 - x) = 0.30(12)$ 50% = 0.50, 20% = 0.20, and 30% = 0.30.

$5x + 2(12 - x) = 3(12)$ Multiply both sides by 10 to clear the equation of decimals.

$5x + 24 - 2x = 36$ Distribute the multiplication by 2.

$3x + 24 = 36$ Combine like terms.

$3x = 12$ Subtract 24 from both sides.

$x = 4$ Divide both sides by 3.

State the conclusion The mixture will contain 4 liters of 50% solution and $12 - 4 = 8$ liters of 20% solution.

Check the result Verify that this answer checks. ∎

In the next example, a *dry mixture* of a specified value is created from two differently priced components.

EXAMPLE 8   *Snack food.*   Because fancy cashews priced at $9 per pound were not selling, a market produce clerk decided to combine them with less expensive filberts and sell the mixture for $7 per pound. How many pounds of filberts, selling at $6 per pound, should be mixed with 50 pounds of cashews to obtain such a mixture?

Analyze the problem   We know the value of the cashews ($9 per pound) and the filberts ($6 per pound). We also know that 50 pounds of cashews are to be mixed with an unknown number of pounds of filberts to obtain a mixture worth $7 per pound.

Form an equation   To solve this problem, we use the formula $v = pn$, where v is value, p is the price per pound, and n is the number of pounds.

Suppose that x pounds of filberts are used in the mixture. At $6 per pound, they are worth $6x$. At $9 per pound, the 50 pounds of cashews are worth $9 \cdot 50 = \$450$. Their combined value will be $\$(6x + 450)$. We also know that the mixture weighs $(50 + x)$ pounds. At $7 per pound, that mixture will be worth $\$7(50 + x)$. This information is recorded in the table in Figure 8-14.

	p $\cdot$	n	$=$	v
Filberts	6	x		$6x$
Cashews	9	50		450
Mixture	7	$50 + x$		$7(50 + x)$

FIGURE 8-14

We can use the information in the table to form the equation.

The value of the filberts	plus	the value of the cashews	equals	the value of the mixture.
$6x$	$+$	450	$=$	$7(50 + x)$

Solve the equation
$$6x + 450 = 7(50 + x)$$
$$6x + 450 = 350 + 7x \quad \text{Distribute the multiplication by 7.}$$
$$100 = x \quad \text{Subtract } 6x \text{ and 350 from both sides.}$$

State the conclusion   Thus, 100 pounds of filberts should be used in the mixture.

Check the result
The value of 100 pounds of filberts at $6 per pound is $600
The value of 50 pounds of cashews at $9 per pound is $450
The value of the mixture is $1,050

The value of 150 pounds of the mixture at $7 per pound is also $1,050. The answer checks.   ■

STUDY SET   Section 8.5

VOCABULARY   *Fill in the blanks.*

1. The _____ of a triangle or a rectangle is the distance around it.

2. An _____ triangle is a triangle with two sides of the same length.

3. The equal sides of an isosceles triangle meet to form the _____ angle.

4. The angles opposite the equal sides in an isosceles triangle are called _____ angles, and they have equal measures.

CONCEPTS

5. PLUMBING   A plumber wants to cut a 17-foot pipe into three sections. The longest section is to be three times as long as the shortest, and the middle-sized section is to be 2 feet longer than the shortest.

a. Complete the diagram in Illustration 1.

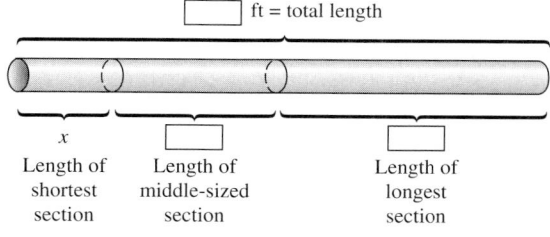

☐ ft = total length

x

Length of shortest section

Length of middle-sized section

Length of longest section

ILLUSTRATION 1

b. To solve this problem, an equation is formed, it is solved, and it is found that $x = 3$. How long is each section of pipe?

6. What is the sum of the measures of the angles of any triangle?

7. Use a ruler to draw an isosceles triangle with sides 3 inches long and a base that is 2 inches long. Label the vertex and the base angles.

8. a. Complete Illustration 2, which shows the inventory of nylon brushes that a paint store carries.

Paintbrush	Number	·	Value	=	Total value
1 inch	$\frac{x}{2}$		$4		
2 inch	x		$5		
3 inch	$x + 10$		$7		

ILLUSTRATION 2

b. Which type of brush does the store have the largest number of?

c. What is the least expensive brush?

d. What is the total value of the inventory of nylon brushes?

9. In the advertisement in Illustration 3, what are the principal, the rate, and the time for the investment opportunity shown?

Invest in Mini Malls!
Builder seeks daring people who want to earn big $$$$$$. In just 1 year, you will earn a gigantic 14% on an investment of only $30,000! Call now.

ILLUSTRATION 3

10. a. Complete Illustration 4, which gives the details about two investments that were made by a retired couple.

	P	·	r	=	I
Certificate of deposit	x		0.04		
Brother-in-law's business	$2x$		0.06		

ILLUSTRATION 4

b. How much more money was invested in the brother-in-law's business than in the certificate of deposit?

c. What is the total amount of interest the couple will make from these investments?

11. COMMUTERS   When a husband and wife leave for work, they drive in opposite directions. Their average speeds are different; however, their drives last the same amount of time. Complete Illustration 5, which gives the details of each person's morning commute.

	r	·	t	=	d
Husband	35 mph		t hr		
Wife	45 mph				

ILLUSTRATION 5

12. Each bottle of dressing shown in Illustration 6 contains a mixture of oil and vinegar. After sitting overnight, the liquids separate completely, with the oil rising to the top. On each bottle, draw the line estimating where the separation would occur and shade the vinegar.

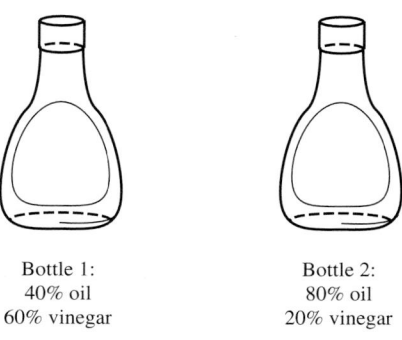

Bottle 1:
40% oil
60% vinegar

Bottle 2:
80% oil
20% vinegar

ILLUSTRATION 6

13. See Illustration 7.

a. How many gallons of acid are there in the second barrel?

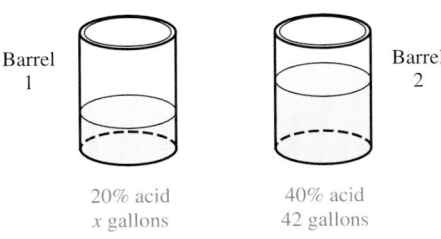

Barrel 1

Barrel 2

20% acid
x gallons

40% acid
42 gallons

ILLUSTRATION 7

b. Suppose the contents of the two barrels are poured into an empty third barrel. How many gallons of liquid will the third barrel contain?

c. What would be a *reasonable* estimate of the concentration of the solution in the third barrel—19%, 32%, or 43% acid?

14. Complete Illustration 8, which gives the details about the ingredients in a box of breakfast cereal.

	Price ($/oz)	Amount (oz)	Value
Blueberries	$0.38	x	
Bran Flakes	$0.08	14	
Blueberries & Bran Flakes Cereal	$0.21	14 + x	

ILLUSTRATION 8

PRACTICE *In Exercises 15–16, solve the equation by first clearing it of decimals.*

15. $0.08x + 0.07(15,000 - x) = 1,110$

16. $0.108x + 0.07(16,000 - x) = 1,500$

17. Two angles are called **complementary angles** when the sum of their measures is 90°. Find the measures of the complementary angles shown in Illustration 9.

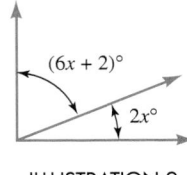

$(6x + 2)°$

$2x°$

ILLUSTRATION 9

18. Two angles are called **supplementary angles** when the sum of their measures is 180°. Find the measures of the supplementary angles shown in Illustration 10.

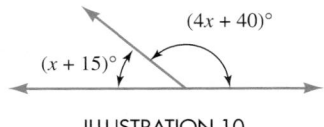

$(4x + 40)°$

$(x + 15)°$

ILLUSTRATION 10

19. In Illustration 11, two lines intersect to form **vertical angles.** Use the fact that vertical angles have the same measure to find *x*.

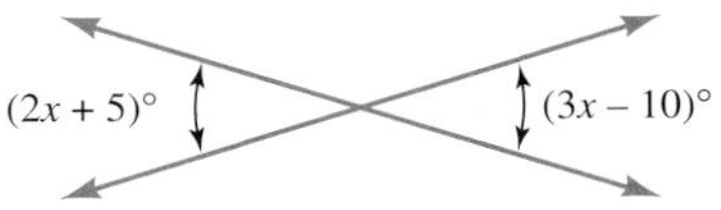

$(2x + 5)°$ $(3x - 10)°$

ILLUSTRATION 11

20. Find the measures of the vertical angles shown in Illustration 12. (See Exercise 19.)

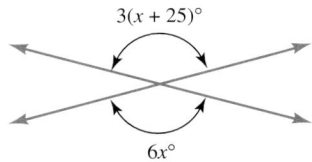

$3(x + 25)°$

$6x°$

ILLUSTRATION 12

APPLICATIONS

21. CARPENTRY   The 12-foot board in Illustration 13 has been cut into two sections, one twice as long as the other. How long is each section?

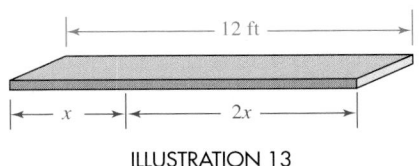

12 ft

x   2x

ILLUSTRATION 13

22. ROBOTICS   The robotic arm shown in Illustration 14 will extend a total distance of 18 feet. Find the length of each section.

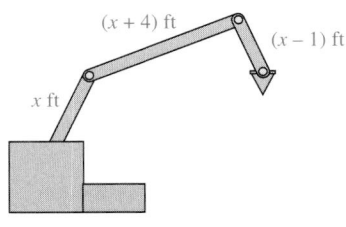

$(x + 4)$ ft

$(x - 1)$ ft

x ft

ILLUSTRATION 14

23. SOLAR HEATING   One solar panel in Illustration 15 is 3.4 feet wider than the other. Find the width of each panel.

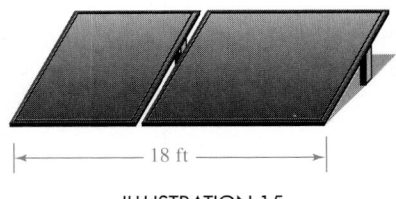

18 ft

ILLUSTRATION 15

24. PLUMBING   A 20-foot pipe has been cut into two sections, one 3 times as long as the other. How long is each section?

25. TOURING   An American rock group plans to travel for a total of 38 weeks, making three major concert tours. They will be in Japan for 4 more weeks than they will be in Australia. Their stay in Sweden will be 2 weeks less than that in Australia. How many weeks will they be in each country?

26. PUBLISHER'S INVENTORY   A novel can be purchased in a hardcover edition for $15.95 or in paperback for $4.95. The publisher printed 11 times as many paperbacks as hardcover books. A total of 114,000 books were printed. How many of each type were printed?

27. COUNTING CALORIES   A slice of pie with a scoop of ice cream has 850 calories. The calories in the pie alone are 100 more than twice the calories in the ice cream alone. How many calories are in each food?

28. WASTE DISPOSAL   Two tanks hold a total of 45 gallons of a toxic solvent. One tank holds 6 gallons more than twice the amount in the other. How many gallons does each tank hold?

29. NET INCOME   From the information given in Illustration 16, determine the net income of Sears, Roebuck and Co. for each quarter of 1999.

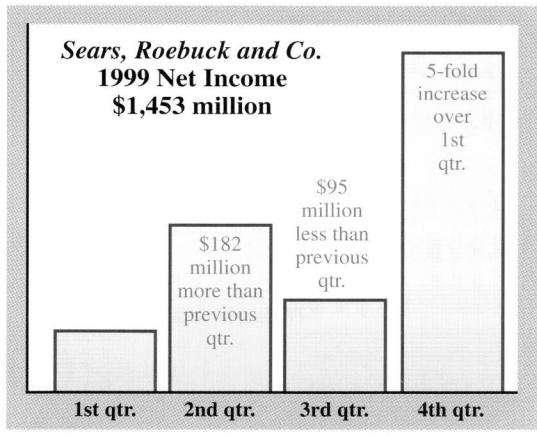

Based on data from Hoover's Online

ILLUSTRATION 16

30. LOCKS   The three numbers of the combination for the lock shown in Illustration 17 are **consecutive integers,** and their sum is 81. (Consecutive integers follow each other, like 7, 8, 9.) Complete the instructions below that will open the lock. (*Hint:* If *x* represents the smallest integer, $x + 1$ represents the next integer, and $x + 2$ represents the largest integer.)

Spin dial to the right one complete revolution to ____. Turn to the left to ____. Turn to the right to ____, and lift the handle.

ILLUSTRATION 17

31. TRUSS   The truss in Illustration 18 is in the form of an isosceles triangle. Each of the two equal sides is 4 feet less than the third side. If the perimeter is 25 feet, find the lengths of the sides.

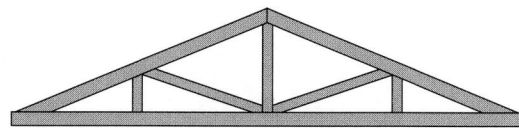

ILLUSTRATION 18

32. FIRST AID   The sling shown in Illustration 19 is in the shape of an isosceles triangle with a perimeter of 144 inches. The longest side of the sling is 18 inches longer than either of the other two sides. Find the lengths of each side.

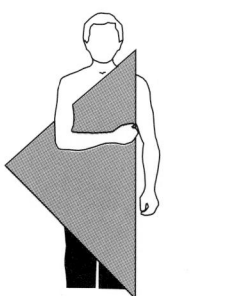

ILLUSTRATION 19

33. SWIMMING POOL   The seawater Orthlieb Pool in Casablanca, Morocco is the largest swimming pool in the world. With a perimeter of 1,110 meters, this rectangular-shaped pool has a length that is 30 meters more than 6 times its width. Find its dimensions.

34. ART   The *Mona Lisa,* shown in Illustration 20, was completed by Leonardo da Vinci in 1506. The length of the picture is 11.75 inches less than twice the width. If the perimeter of the picture is 102.5 inches, find its dimensions.

ILLUSTRATION 20

35. GUY WIRES   The two guy wires shown in Illustration 21 form an isosceles triangle. Each of the base angles of the triangle is 4 times the third angle (the vertex angle). Find the measure of the vertex angle.

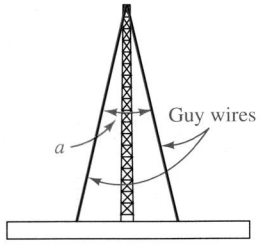

Guy wires

a

ILLUSTRATION 21

36. MOUNTAIN BICYCLE   For the bicycle frame in Illustration 22, the angle that the horizontal crossbar makes with the seat support is 15° less than twice the angle at the steering column. The angle at the pedal gear is 25° more than the angle at the steering column. Find these three angle measures.

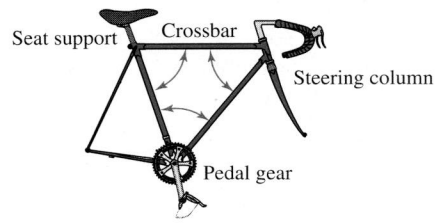

Seat support   Crossbar

Steering column

Pedal gear

ILLUSTRATION 22

37. WAREHOUSING COSTS   A store warehouses 40 more portables than big-screen TV sets, and 25 fewer consoles than portables. Storage costs for the different TV sets are shown in Illustration 23. If storage costs $276 per month, how many big-screen sets are in stock?

Type of TV	Monthly cost
Portable	$1.50
Console	$4.00
Big-screen	$7.50

ILLUSTRATION 23

38. APARTMENT RENTAL   The owners of an apartment building rent 1-, 2-, and 3-bedroom units. They rent equal numbers of each, with the monthly rents given in Illustration 24. If the total monthly income is $36,550, how many of each type of unit are there?

Unit	Rent
One-bedroom	$550
Two-bedroom	$700
Three-bedroom	$900

ILLUSTRATION 24

39. SOFTWARE SALES   Three software applications are priced as shown in Illustration 25. Spreadsheet and database programs sold in equal numbers, but 15 more word processing applications were sold than the other two combined. If the three applications generated sales of $72,000, how many spreadsheets were sold?

Software	Price
Spreadsheet	$150
Database	$195
Word processing	$210

ILLUSTRATION 25

40. INVENTORY   With summer approaching, the number of air conditioners sold is expected to be double that of stoves and refrigerators combined. Stoves sell for $350, refrigerators for $450, and air conditioners for $500, and sales of $56,000 are expected. If stoves and refrigerators sell in equal numbers, how many of each appliance should be stocked?

41. INTEREST INCOME   On December 31, 2000, Terrell Washington opened two savings accounts. At the end of 2001, his bank mailed him the form shown in Illustration 26, for income tax purposes. If a total of $12,000 was initially deposited and if no further deposits or withdrawals were made, how much money was originally deposited in account number 721-94?

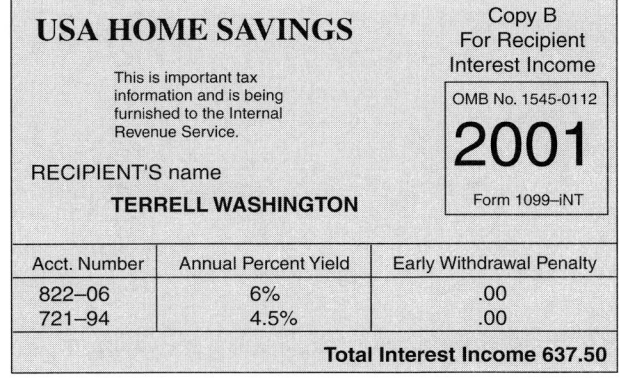

USA HOME SAVINGS		Copy B For Recipient Interest Income
This is important tax information and is being furnished to the Internal Revenue Service.		OMB No. 1545-0112 **2001** Form 1099–iNT
RECIPIENT'S name **TERRELL WASHINGTON**		

Acct. Number	Annual Percent Yield	Early Withdrawal Penalty
822–06	6%	.00
721–94	4.5%	.00
		Total Interest Income 637.50

ILLUSTRATION 26

42. MAKING A PRESENTATION   A financial planner recommends a plan for a client who has $65,000 to invest. (See Illustration 27 on the next page.) At the end of the presentation, the client asks, "How much will be invested at each rate?" Answer this question using the given information.

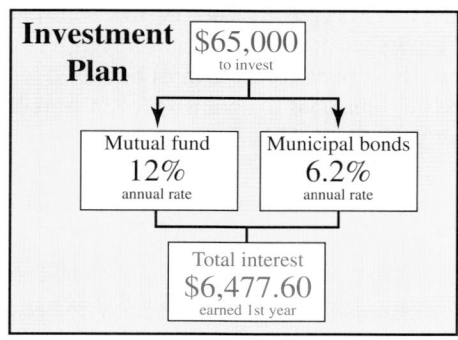

ILLUSTRATION 27

43. INVESTMENTS   Equal amounts are invested in each of three accounts paying 7%, 8%, and 10.5% annually. If one year's combined interest income is $1,249.50, how much is invested in each account?

44. RETIREMENT   A professor wants to supplement her retirement income with investment interest. If she invests $15,000 at 6% interest, how much more would she have to invest at 7% to achieve a goal of $1,250 per year in supplemental income?

45. FINANCIAL PLANNING   A plumber has a choice of two investment plans:

• An insured fund that pays 11% interest

• A risky investment that pays a 13% return

If the same amount invested at the higher rate would generate an extra $150 per year, how much does the plumber have to invest?

46. INVESTMENTS   The amount of annual interest earned by $8,000 invested at a certain rate is $200 less than $12,000 would earn at a rate 1% lower. At what rate is the $8,000 invested?

47. TORNADO   During a storm, two teams of scientists leave a university at the same time in specially designed vans to search for tornadoes. The first team travels east at 20 mph and the second travels west at 25 mph, as shown in Illustration 28. If their radios have a range of up to 90 miles, how long will it be before they lose radio contact?

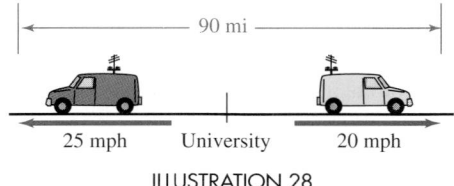

ILLUSTRATION 28

48. SEARCH AND RESCUE   Two search-and rescue teams leave base at the same time looking for a lost boy. The first team, on foot, heads north at 2 mph and the other, on horseback, south at 4 mph. How long will it take them to search a distance of 21 miles between them?

49. SPEED OF TRAINS   Two trains are 330 miles apart, and their speeds differ by 20 mph. Find the speed of each train if they are traveling toward each other and will meet in 3 hours.

50. AVERAGE SPEED   A car averaged 40 mph for part of a trip and 50 mph for the remainder. If the 5-hour trip covered 210 miles, for how long did the car average 40 mph?

51. AIR TRAFFIC CONTROL   An airliner leaves Berlin, Germany, headed for Montreal, Canada, flying at an average speed of 450 mph. At the same time, an airliner leaves Montreal headed for Berlin, averaging 500 mph. If the airports are 3,800 miles apart, when will the air traffic controllers have to make the pilots aware that the planes are passing each other?

52. ROAD TRIP   A bus, carrying the members of a marching band, and a truck, carrying their instruments, leave a high school at the same time. The bus travels at 65 mph and the truck at 55 mph. In how many hours will they be 75 miles apart?

53. SALT SOLUTION   How many gallons of a 3% salt solution must be mixed with 50 gallons of a 7% solution to obtain a 5% solution?

54. MAKING CHEESE   To make low-fat cottage cheese, milk containing 4% butterfat is mixed with 10 gallons of milk containing 1% butterfat to obtain a mixture containing 2% butterfat. How many gallons of the richer milk must be used?

55. ANTISEPTIC SOLUTION   A nurse wants to add water to 30 ounces of a 10% solution of benzalkonium chloride to dilute it to an 8% solution. How much water must she add?

56. PHOTOGRAPHIC CHEMICALS   A photographer wishes to mix 2 liters of a 5% acetic acid solution with a 10% solution to get a 7% solution. How many liters of 10% solution must be added?

57. MIXING FUELS   How many gallons of fuel costing $1.15 per gallon must be mixed with 20 gallons of a fuel costing $0.85 per gallon to obtain a mixture costing $1 per gallon? See Illustration 29.

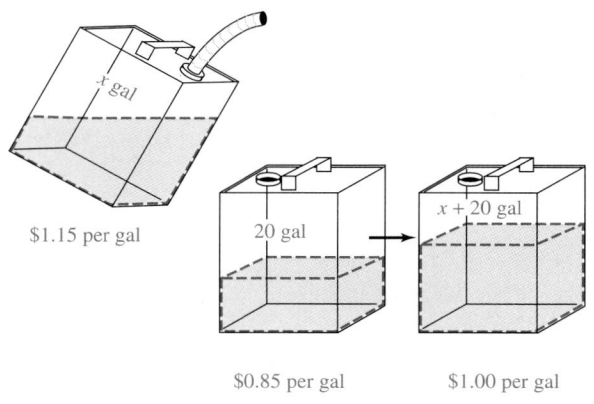

ILLUSTRATION 29

58. MIXING PAINT   Paint costing $19 per gallon is to be mixed with 5 gallons of a $3-per-gallon thinner to make a paint that can be sold for $14 per gallon. How much paint will be produced?

59. MIXING CANDY   Lemon drops worth $1.90 per pound are to be mixed with jelly beans that cost $1.20 per pound to make 100 pounds of a mixture worth $1.48 per pound. How many pounds of each candy should be used?

60. MIXING CANDY   See Illustration 30. Twenty pounds of lemon drops are to be mixed with cherry chews to make a mixture that will sell for $1.80 per pound. How much of the more expensive candy should be used?

Candy	Price per pound
Peppermint patties	$1.35
Lemon drops	$1.70
Licorice lumps	$1.95
Cherry chews	$2.00

ILLUSTRATION 30

61. BLENDING COFFEE   A store sells regular coffee for $4 a pound and gourmet coffee for $7 a pound. To get rid of 40 pounds of the gourmet coffee, a shopkeeper makes a blend to put on sale for $5 a pound. How many pounds of regular coffee should he use?

62. BLENDING LAWN SEED   A store sells bluegrass seed for $6 per pound and ryegrass seed for $3 per pound. How much ryegrass must be mixed with 100 pounds of bluegrass to obtain a blend that will sell for $5 per pound?

WRITING

63. Create a mixture problem of your own, and solve it.

64. Use an example to explain the difference between the quantity and the value of the materials being combined in a mixture problem.

65. A car travels at 60 mph for 15 minutes. Why can't we multiply the rate, 60, and the time, 15, to find the distance traveled by the car?

66. Create a geometry problem that could be answered by solving the equation $2w + 2(w + 5) = 26$.

REVIEW   *Use the distributive property to remove parentheses.*

67. $-25(2x - 5)$

68. $-12(3a + 4b - 32)$

69. $-(-3x - 3)$

70. $\dfrac{1}{2}(4b - 8)$

Combine like terms.

71. $8p - 9q + 11p + 20q$

72. $-5(t - 120) - 7(t + 5)$

8.6 *Inequalities*

In this section, you will learn about

- Inequality symbols • Graphing inequalities • Interval notation
- Solving inequalities • Graphing compound inequalities
- Solving compound inequalities • An application

INTRODUCTION. **Inequalities** are expressions indicating that two quantities are not necessarily equal. They appear in many situations:

- An airplane is rated to fly at altitudes that are less than 36,000 feet.
- To melt ice, the temperature must be greater than 32°F.
- To earn a B, I need a final exam score of at least 80%.

Inequality symbols

We can use **inequality symbols** to show that two expressions are not equal.

Inequality symbols

$\neq$	means	"is not equal to"
$<$	means	"is less than"
$>$	means	"is greater than"
$\leq$	means	"is less than or equal to"
$\geq$	means	"is greater than or equal to"

EXAMPLE 1 *Reading inequalities.*

a. $6 \neq 9$ is read as "6 is not equal to 9."

b. $8 > 4$ is read as "8 is greater than 4."

c. $12 \geq 0$ is read as "12 is greater than or equal to 0." This is true, because $12 > 0$.

d. $5 \leq 5$ is read as "5 is less than or equal to 5." This is true, because $5 = 5$.

Self Check

Write each inequality in words:
a. $15 < 20$, **b.** $y \geq 9$,
c. $10 \geq 1$, and **d.** $30 \leq 30$.

Answers: **a.** 15 is less than 20.
b. y is greater than or equal to 9.
c. 10 is greater than or equal to 1.
d. 30 is less than or equal to 30. ∎

If two numbers are graphed on a number line, the one to the right is the greater. For example, from Figure 8-15, we see that $-1 > -4$, because -1 lies to the right of -4.

FIGURE 8-15

Inequalities can be written so that the inequality symbol points in the opposite direction. For example, the following statements both indicate that 27 is a smaller number than 32.

$27 < 32$ (27 is less than 32)   and $32 > 27$ (32 is greater than 27)

The following statements both indicate that 9 is greater than or equal to 6.

$9 \geq 6$ (9 is greater than or equal to 6)   and $6 \leq 9$ (6 is less than or equal to 9)

Variables can be used with inequality symbols to show mathematical relationships. For example, consider the statement, "You must be taller than 54 inches to ride the roller coaster." If we let h represent a person's height in inches, then to ride the roller coaster, $h > 54$ inches.

EXAMPLE 2 *Writing inequalities.* Express the following situation using an inequality symbol: "The occupancy of the dining room cannot exceed 200 people."

Solution

If p represents the number of people that can occupy the room, then p cannot be greater than (exceed) 200. Another way to state this is that p must be *less than or equal to* 200.

$p \leq 200$

Self Check

Express the following statement using an inequality symbol: "The thermostat on the pool heater is set so that the water temperature t is at least 72°."

Answer: $t \geq 72$ ∎

Graphing inequalities

Graphs of inequalities involving real numbers are **intervals** on the number line. For example, two versions of the graph of all real numbers x such that $x > -3$ are shown in Figure 8-16.

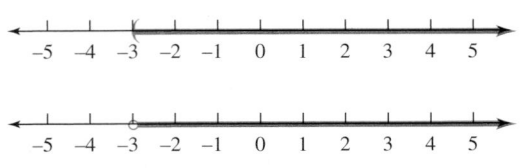

All real numbers greater than −3

FIGURE 8-16

The red arrow pointing to the right shows that all numbers to the right of −3 are in the graph. The **parenthesis** (or open circle) at −3 indicates that −3 is not in the graph.

Interval notation

The interval shown in Figure 8-16 can be expressed in **interval notation** as $(-3, \infty)$. Again, the first parenthesis indicates that −3 is not included in the interval. The infinity symbol ∞ does not represent a number. It indicates that the interval continues on forever to the right.

Figure 8-17 shows two versions of the graph of $x \le 2$. The thick red arrow pointing to the left shows that all numbers to the left of 2 are in the graph. The **bracket** (or closed circle) at 2 indicates that 2 is included in the graph. We can express this interval as $(-\infty, 2]$. Here the bracket indicates that 2 is included in the interval.

From now on, we will use parentheses or brackets when graphing intervals, because they are consistent with interval notation.

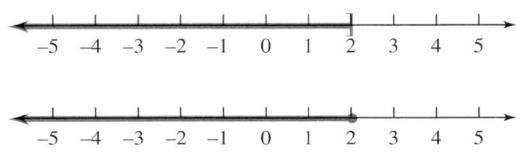

All real numbers less than or equal to 2

FIGURE 8-17

EXAMPLE 3   *Writing an inequality from its graph.*   What inequality is represented by each graph?

a.

b.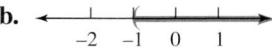

Solution

a. This is the interval $(-\infty, 3]$, which consists of all real numbers less than or equal to 3. The inequality is $x \le 3$.

b. This is the interval $(-1, \infty)$, consisting of all real numbers greater than −1. The inequality is $x > -1$.

 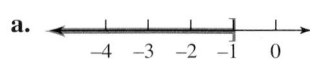
Solving inequalities

A **solution of an inequality** is any number that makes the inequality true. For example, 2 is a solution of $x \le 3$, because $2 \le 3$.

To solve more complicated inequalities, we will use the addition, subtraction, multiplication, and division properties of inequality. When we use one of these properties, the resulting inequality will always be equivalent to the original one.

Addition and subtraction
properties of inequality

> For real numbers a, b, and c,
>
> If $a < b$, then $a + c < b + c$.
>
> If $a < b$, then $a - c < b - c$.
>
> Similar statements can be made for the symbols $>$, $\leq$, and $\geq$.

The **addition property of inequality** can be stated this way: *If a quantity is added to both sides of an inequality, the resulting inequality will have the same direction as the original one.*

The **subtraction property of inequality** can be stated this way: *If a quantity is subtracted from both sides of an inequality, the resulting inequality will have the same direction as the original one.*

EXAMPLE 4 *Solving inequalities.* Solve $x + 3 > 2$ and graph its solution.

Solution

To isolate the x on the left-hand side of the $>$ sign, we proceed as we would when solving equations.

$$x + 3 > 2$$
$$x + 3 - 3 > 2 - 3 \quad \text{To undo the addition of 3, subtract 3 from both sides.}$$
$$x > -1 \quad \text{Do the subtractions: } 3 - 3 = 0 \text{ and } 2 - 3 = 2 + (-3) = -1.$$

All real numbers greater than -1 are solutions of $x + 3 > 2$. This means the inequality has *infinitely many* solutions. The graph of the solutions (see Figure 8-18) includes all points to the right of -1 but does not include -1. Expressed as an interval, this is $(-1, \infty)$.

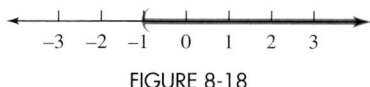

FIGURE 8-18

Since the solution contains infinitely many numbers, we cannot check to see if each of them satisfies the original inequality. As an informal check, we pick several numbers in the graph, such as 1 and 30, substitute each number for x in the inequality, and see whether it satisfies the inequality.

$$x + 3 > 2 \qquad\qquad x + 3 > 2$$
$$1 + 3 \stackrel{?}{>} 2 \quad \text{Substitute 1 for } x. \qquad 30 + 3 \stackrel{?}{>} 2 \quad \text{Substitute 30 for } x.$$
$$4 > 2 \quad \text{Do the addition.} \qquad 33 > 2 \quad \text{Do the addition.}$$

Since $4 > 2$, we know that 1 satisfies the inequality. Since $33 > 2$, we know that 30 satisfies the inequality. The result $(x > -1)$ appears to be correct.

Self Check

Solve $x - 3 \leq -2$ and graph its solution. Then use interval notation to describe the solution.

Answer: $x \leq 1, (-\infty, 1]$

If both sides of the inequality $2 < 5$ are multiplied by a *positive* number, such as 3, another true inequality results.

$$2 < 5$$
$$3 \cdot 2 < 3 \cdot 5 \quad \text{Multiply both sides by 3.}$$
$$6 < 15 \quad \text{Do the multiplications: } 3 \cdot 2 = 6 \text{ and } 3 \cdot 5 = 15.$$

However, if we multiply both sides of $2 < 5$ by a *negative* number, such as -3, the direction of the inequality symbol must be reversed to produce another true inequality.

$$2 < 5$$

$$-3 \cdot 2 > -3 \cdot 5 \quad \text{Multiply both sides by the negative number } -3 \text{ and reverse the direction of the inequality.}$$

$$-6 > -15 \quad \text{Do the multiplications: } -3 \cdot 2 = -6 \text{ and } -3 \cdot 5 = -15.$$

The inequality $-6 > -15$ is true because -6 is to the right of -15 on the number line.

Multiplication and division properties of inequalities

For real numbers a, b, and c,

If $a < b$ and $c > 0$, then $ac < bc$.

If $a < b$ and $c < 0$, then $ac > bc$.

If $a < b$ and $c > 0$, then $\frac{a}{c} < \frac{b}{c}$.

If $a < b$ and $c < 0$, then $\frac{a}{c} > \frac{b}{c}$.

Similar statements can be made for the symbols $>$, $\leq$, and $\geq$.

The **multiplication property of inequality** can be stated this way:

If both sides of an inequality are multiplied by the same positive number, the resulting inequality will have the same direction as the original one.

If both sides of an inequality are multiplied by the same negative number, the resulting inequality will have the opposite direction from the original one.

The **division property of inequality** can be stated this way:

If both sides of an inequality are divided by the same positive number, the resulting inequality will have the same direction as the original one.

If both sides of an inequality are divided by the same negative number, the resulting inequality will have the opposite direction from the original one.

EXAMPLE 5 *Solving inequalities.* Solve $-5 \geq 3x + 7$ and graph the solution.

Solution

$$-5 \geq 3x + 7$$

$$-5 - 7 \geq 3x + 7 - 7 \quad \text{To undo the addition of 7, subtract 7 from both sides.}$$

$$-12 \geq 3x \quad \text{Subtract: } -5 - 7 = -5 + (-7) = -12 \text{ and } 7 - 7 = 0.$$

$$\frac{-12}{3} \geq \frac{3x}{3} \quad \text{To undo the multiplication by 3, divide both sides by 3.}$$

$$-4 \geq x \quad \text{Do the divisions.}$$

It is common practice to present a solution such as $-4 \geq x$ in an equivalent form with the variable on the left-hand side. If -4 is greater than or equal to x, then x must be less than or equal to -4, and we can write the solution as

$$x \leq -4$$

The graph (shown in Figure 8-19) consists of all real numbers less than or equal to -4. Using interval notation, we have $(-\infty, -4]$.

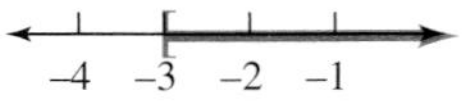

FIGURE 8-19

To check, we can pick several numbers in the graph, such as -6 and -20, and see whether each one satisfies the inequality.

Self Check

Solve $2x - 7 > -13$ and graph the solution. Then use interval notation to describe the solution.

For x = −6

−5 ≥ 3x + 7

−5 $\overset{?}{\geq}$ 3(−6) + 7   Substitute −6 for x.

−5 $\overset{?}{\geq}$ −18 + 7   Do the multiplication.

−5 ≥ −11   Do the addition.

For x = −20

−5 ≥ 3x + 7

−5 $\overset{?}{\geq}$ 3(−20) + 7

−5 $\overset{?}{\geq}$ −60 + 7

−5 ≥ −53

Answer: x > −3, (−3, ∞)

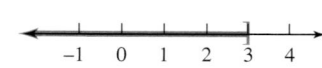

Since −5 ≥ −11, we know that −6 satisfies the inequality. Since −5 ≥ −53, we know that −20 satisfies the inequality. The result (x ≤ −4) appears to be correct.

EXAMPLE 6 *Reversing the inequality symbol.* Solve 5 − 3x < 14 and graph the solution.

Solution

5 − 3x < 14

5 − 3x − 5 < 14 − 5   To isolate −3x on the left-hand side, subtract 5 from both sides.

−3x < 9   Do the subtractions: 5 − 5 = 0 and 14 − 5 = 9.

$\dfrac{-3x}{-3} > \dfrac{9}{-3}$ To undo the multiplication by −3, divide both sides by −3. Since we are dividing by a negative number, we reverse the direction of the < symbol.

x > −3

The graph is shown in Figure 8-20. This is the interval (−3, ∞), which consists of all real numbers greater than −3.

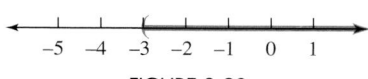

FIGURE 8-20

Check the result.

Self Check

Solve −2x − 5 ≥ −11 and graph the solution. Then use interval notation to describe the solution.

Answer: x ≤ 3, (−∞, 3]

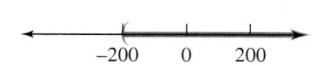

EXAMPLE 7 *Reversing the inequality symbol.* Solve $\dfrac{x}{-15} \geq -6$ and graph the solution.

Solution

$\dfrac{x}{-15} \geq -6$

$-15\left(\dfrac{x}{-15}\right) \leq -15(-6)$ To undo the division by −15, multiply both sides by −15. Since we are multiplying by a negative number, we reverse the direction of the ≥ symbol.

x ≤ 90   Do the multiplications.

The graph is shown in Figure 8-21. This is the interval (−∞, 90], which consists of all real numbers less than or equal to 90.

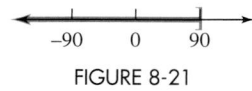

FIGURE 8-21

Self Check

Solve $\dfrac{h}{-20} < 10$ and graph the solution.

Answer: h > −200, (−200, ∞)

  COMMENT Remember that if both sides of an inequality are multiplied or divided by a negative number, the direction of the inequality symbol must be reversed.

EXAMPLE 8 *Solving inequalities.* Solve $5(x + 1) \leq 2(x - 3)$ and graph the solution.

Solution

$$5(x + 1) \leq 2(x - 3)$$

$5x + 5 \leq 2x - 6$ Use the distributive property on both sides of the inequality.

$5x + 5 - 2x \leq 2x - 6 - 2x$ To eliminate $2x$ from the right side, subtract $2x$ from both sides.

$3x + 5 \leq -6$ Combine like terms on both sides.

$3x + 5 - 5 \leq -6 - 5$ To undo the addition of 5, subtract 5 from both sides.

$3x \leq -11$ Do the subtractions.

$\dfrac{3x}{3} \leq \dfrac{-11}{3}$ To undo the multiplication by 3, divide both sides by 3.

$x \leq -\dfrac{11}{3}$

The graph is shown in Figure 8-22. This is the interval $\left(-\infty, -\frac{11}{3}\right]$, which consists of all real numbers less than or equal to $-\frac{11}{3}$. We note that $-\frac{11}{3} = -3\frac{2}{3}$.

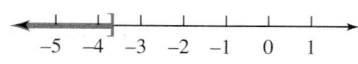

FIGURE 8-22

Check the result.

Self Check

Solve $3(x - 2) > -(x + 1)$ and graph the solution. Then use interval notation to describe the solution.

Answer: $x > \dfrac{5}{4}, \left(\dfrac{5}{4}, \infty\right)$

Graphing compound inequalities

Two inequalities can be combined into a **compound inequality** to indicate that numbers lie *between* two fixed values. For example, $-2 < x < 3$ is a combination of

$$-2 < x \qquad \text{and} \qquad x < 3$$

It indicates that x is greater than -2 and that x is also less than 3. The solution of $-2 < x < 3$ consists of all numbers that lie *between* -2 and 3. The graph of this interval appears in Figure 8-23. We can express this interval as $(-2, 3)$.

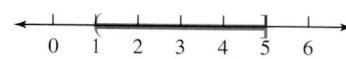

FIGURE 8-23

EXAMPLE 9 *Writing an inequality from its graph.* What inequality is represented by the graph below?

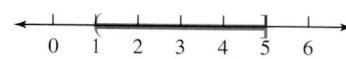

Solution

$1 < x \leq 5$. This is the interval $(1, 5]$.

Self Check

What inequality is represented by the graph below?

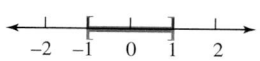

Answer: $-1 \leq x \leq 1$. This is the interval $[-1, 1]$.

EXAMPLE 10 *Graphing compound inequalities.* Graph the interval $-4 < x \le 0$.

Solution

The interval $-4 < x \le 0$ consists of all real numbers between -4 and 0, including 0. The graph appears in Figure 8-24. This is the interval $(-4, 0]$.

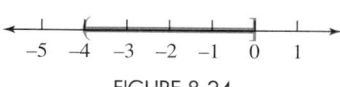

FIGURE 8-24

To check, we pick a number, such as -2, in the graph and see whether it satisfies the inequality. Since $-4 < -2 \le 0$, the answer appears to be correct.

Self Check

Graph the interval $-2 \le x < 1$. Then use interval notation to describe the solution.

Answer: $[-2, 1)$

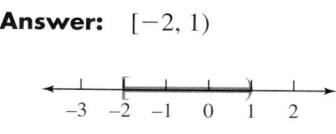

Solving compound inequalities

To solve compound inequalities, we use the same methods we used for solving equations. However, instead of applying the properties of equality to both sides of an equation, we will apply the properties of inequality to all three parts of the inequality.

EXAMPLE 11 *Solving compound inequalities.* Solve $-4 < 2(x - 1) \le 4$ and graph the solution.

Solution

$$-4 < 2(x - 1) \le 4$$

$-4 < 2x - 2 \le 4$ Distribute the multiplication by 2.

$-4 + 2 < 2x - 2 + 2 \le 4 + 2$ To undo the subtraction of 2, add 2 to all three parts.

$-2 < 2x \le 6$ Do the additions.

$-1 < x \le 3$ To undo the multiplication by 2, divide all three parts by 2.

The graph of the solution appears in Figure 8-25. This is the interval $(-1, 3]$.

FIGURE 8-25

Check the solution.

Self Check

Solve $-6 \le 3(x + 2) \le 6$ and graph the solution. Then use interval notation to describe the solution.

Answer: $-4 \le x \le 0$, $[-4, 0]$

An application

When solving problems, phrases such as "not more than," "at least," or "should exceed" suggest that an *inequality* should be written instead of an *equation*.

EXAMPLE 12 *Grades.* A student has scores of 72%, 74%, and 78% on three exams. What percent score does he need on the last exam to earn no less than a grade of B (80%)?

Analyze the problem   We know three of the student's scores. We are to find what he must score on the last exam to earn at least a B grade.

Form an inequality   We can let x represent the score on the fourth (and last) exam. To find the average grade, we add the four scores and divide by 4. To earn no less than a grade of B, the student's average must be greater than or equal to 80%.

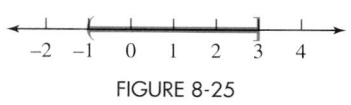

The average of the four grades	must be greater than or equal to	80.
$\dfrac{72 + 74 + 78 + x}{4}$	$\ge$	80

Solve the inequality   We can solve this inequality for x.

$$\frac{72 + 74 + 78 + x}{4} \geq 80$$

$$\frac{224 + x}{4} \geq 80 \qquad \text{Simplify the numerator: } 72 + 74 + 78 = 224.$$

$$224 + x \geq 320 \qquad \text{To clear the inequality of the fraction, multiply both sides by 4.}$$

$$x \geq 96 \qquad \text{To undo the addition of 224, subtract 224 from both sides.}$$

State the conclusion   To earn a B, the student must score 96% or better on the last exam. Of course, the student cannot score higher than 100%. The graph appears in Figure 8-26. This is the interval $[96, 100]$.

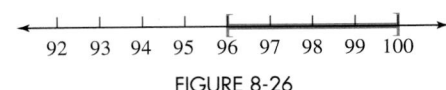

FIGURE 8-26

Check the result   Pick some numbers in the interval, and verify that the average of the four scores will be 80% or greater. ∎

STUDY SET   Section 8.6

VOCABULARY   *Fill in the blanks.*

1. An expression containing one of the symbols $>, <, \geq,$ $\leq,$ or $\neq$ is called an _____.

2. Graphs of inequalities involving real numbers are called _____ on the number line.

3. A _____ of an inequality is any real number that makes the inequality true.

4. The inequality $-4 < x \leq 12$ is an example of a _____ inequality.

CONCEPTS

5. Tell whether each statement is true or false.
   a. $35 \geq 34$ **b.** $5.61 \geq 5.61$

   c. $-16 \leq -17$ **d.** $0 \leq -2\frac{1}{8}$

   e. $\frac{3}{4} \leq 0.75$ **f.** $-0.6 \geq -0.5$

6. Tell whether each number is a solution of $3x + 7 < 4x - 2$.
   a. $x = 12$ **b.** $x = -6$
   c. $x = 0$ **d.** $x = 9$

In Exercises 7–10, fill in the blanks.

7. If a quantity is added to or subtracted from both sides of an inequality, the resulting inequality will have the _____ direction as the original one.

8. If both sides of an inequality are multiplied or divided by a positive number, the resulting inequality will have the _____ direction as the original one.

9. If both sides of an inequality are multiplied or divided by a negative number, the resulting inequality will have the _____ direction from the original one.

10. To solve compound inequalities, the properties of inequalities are applied to all _____ parts of the inequality.

11. The solution of an inequality is graphed below.

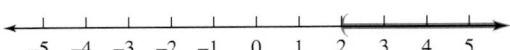

   a. If 3 is substituted for the variable in the inequality, will a true or a false statement result?

   b. If -3 is substituted for the variable in the inequality, will a true or a false statement result?

12. The solution of a compound inequality is graphed below.

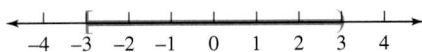

   a. If 3 is substituted for the variable in the inequality, will a true or a false statement result?

   b. If -3 is substituted for the variable in the inequality, will a true or a false statement result?

13. Solve the inequality $2x - 4 > 12$, and give the solution:
   a. in words
   b. using a graph

   c. using interval notation

14. Solve the compound inequality $-4 < 2x < 12$, and give the solution:
   a. in words
   b. using a graph

   c. using interval notation

NOTATION   *In Exercises 15–18, fill in the blanks.*

15. The symbol $<$ means "_____," and the symbol $>$ means "_____."

16. The symbol $\geq$ means "_____ or equal to," and the symbol $\leq$ means "is less than _____."

17. The symbol $\neq$ means "_____."

18. In the interval $[4, 8)$, the endpoint 4 is _____, but the endpoint 8 is not included.

19. Suppose you solve an inequality and obtain $-2 < x$. Write an equivalent inequality that has x on the left-hand side.

20. Explain what is wrong with the compound inequality $8 < x < -1$.

Write each inequality so that the inequality symbol points in the opposite direction.

21. $17 \geq -2$

22. $-32 < -10$

Complete the solution to solve each inequality.

23.
$$4x - 5 \geq 7$$
$$4x - 5 + \boxed{} \geq 7 + \boxed{}$$
$$4x \geq \boxed{}$$
$$\frac{4x}{\boxed{}} \geq \frac{12}{\boxed{}}$$
$$x \geq 3$$

24.
$$\frac{-x}{2} + 4 < 5$$
$$\frac{-x}{2} + 4 - \boxed{} < 5 - \boxed{}$$
$$\frac{-x}{2} < \boxed{}$$
$$\boxed{}\left(\frac{-x}{2}\right) < \boxed{}(1)$$
$$\boxed{} < 2$$
$$\frac{-x}{\boxed{}} \quad \frac{2}{-1}$$
$$x > -2$$

PRACTICE   *Graph each inequality. Then describe the graph using interval notation.*

25. $x < 5$ **26.** $x \geq -2$

27. $-3 < x \leq 1$ **28.** $-1 \leq x \leq 3$

Write the inequality that is represented by each graph. Then describe the graph using interval notation.

29.
-1

30.
2

31.
$-7 \qquad 2$

32.
$-3 \qquad 1$

Solve each inequality, graph the solution, and use interval notation to describe the solution.

33. $x + 2 > 5$ **34.** $x + 5 \geq 2$

35. $-x - 3 \leq 7$ **36.** $-x - 9 > 3$

37. $3 + x < 2$ **38.** $5 + x \geq 3$

39. $2x - 0.3 \leq 0.5$ **40.** $-3x - 0.5 < 0.4$

41. $-3x - 7 > -1$ **42.** $-5x + 7 \leq 12$

43. $-4x + 6 > 17$ **44.** $7x - 1 > 5$

45. $\frac{y}{4} + 1 \leq -9$ **46.** $\frac{r}{8} - 7 \geq -8$

47. $-\dfrac{1}{2}n \geq -1$

48. $-\dfrac{1}{3}t \leq -3$

71. $-\dfrac{2}{3} \geq \dfrac{2y}{3} - \dfrac{3}{4}$

72. $-\dfrac{2}{9} \geq \dfrac{5x}{6} - \dfrac{1}{3}$

49. $\dfrac{x}{-42} - 1 > -1$

50. $\dfrac{a}{-25} + 3 < 3$

Solve each inequality, graph the solution, and use interval notation to describe the solution.

73. $2 < x - 5 < 5$

74. $3 < x - 2 < 7$

51. $\dfrac{2}{3}x \geq 2$

52. $\dfrac{3}{4}x < 3$

75. $-5 < x + 4 \leq 7$

76. $-9 \leq x + 8 < 1$

77. $0 \leq x + 10 \leq 10$

53. $-\dfrac{7}{8}x \leq 21$

54. $-\dfrac{3}{16}x \geq -9$

78. $-8 < x - 8 < 8$

79. $4 < -2x < 10$

55. $2x + 9 \leq x + 8$

56. $3x + 7 \leq 4x - 2$

80. $-4 \leq -4x < 12$

57. $9x + 13 \geq 8x$

58. $7x - 16 < 6x$

81. $-3 \leq \dfrac{x}{2} \leq 5$

59. $8x + 4 > 3x + 4$

60. $7x + 6 \geq 4x + 6$

82. $-12 < \dfrac{x}{3} < 0$

83. $3 \leq 2x - 1 < 5$

61. $5x + 7 < 2x + 1$

62. $7x + 2 \geq 4x - 1$

84. $4 < 3x - 5 \leq 7$

85. $0 < 10 - 5x \leq 15$

63. $7 - x \leq 3x - 2$

64. $9 - 3x \geq 6 + x$

86. $1 \leq -7x + 8 \leq 15$

65. $3(x - 8) < 5x + 6$

66. $9(x - 11) > 13 + 7x$

Solve each inequality.

87. $0.6(0.5x - 2.94) < -1.353$

67. $8(5 - x) \leq 10(8 - x)$

68. $17(3 - x) \geq 3 - 13x$

88. $-0.7688 \leq \dfrac{m}{3.5} - 0.1988$

69. $\dfrac{1}{2} + \dfrac{x}{5} > \dfrac{3}{4}$

70. $\dfrac{1}{3} + \dfrac{c}{5} > -\dfrac{3}{2}$

89. $9(0.05 - 0.3x) + 0.162 \leq 0.081 + 15x$

90. $-1,630 \leq \dfrac{b + 312,451}{47} < 42,616$

APPLICATIONS

91. CALCULATING GRADES   A student has test scores of 68%, 75%, and 79% in a government class. What must she score on the last exam to earn a B (80% or better) in the course?

92. OCCUPATIONAL TESTING   Before taking on a client, an employment agency requires the applicant to average at least 70% on a battery of four job skills tests. If an applicant scored 70%, 74%, and 84% on the first three exams, what must he score on the fourth test to maintain a 70% or better average?

93. FLEET AVERAGES   A car manufacturer produces three models in equal quantities. One model has an economy rating of 17 miles per gallon, and the second model is rated for 19 mpg. If governmental regulations require the manufacturer to have a fleet average of at least 21 mpg, what economy rating is required for the third model?

94. SERVICE CHARGES   When the average daily balance of a customer's checking account falls below $500 in any week, the bank assesses a $5 service charge. Illustration 1 shows the daily balances of one customer. What must Friday's balance be to avoid the service charge?

Day	Balance
Monday	$540.00
Tuesday	$435.50
Wednesday	$345.30
Thursday	$310.00

ILLUSTRATION 1

95. DOING HOMEWORK   A Spanish teacher requires that students devote no less than 1 hour a day to their homework assignments. Write an inequality that describes the number of minutes m a student should spend each week on Spanish homework.

96. CHILD LABOR   A child labor law reads, "The number of hours a full-time student under 16 years of age can work on a weekday shall not exceed 4 hours." Write an inequality that describes the number of hours h such a student can work Monday through Friday.

97. SAFETY CODE   Illustration 2 shows the acceptable and preferred angles of "pitch" or slope for ladders, stairs, and ramps. Use a compound inequality to describe each safe-angle range.
   a. Ramps or inclines
   b. Stairs
   c. Preferred range for stairs
   d. Ladders with cleats

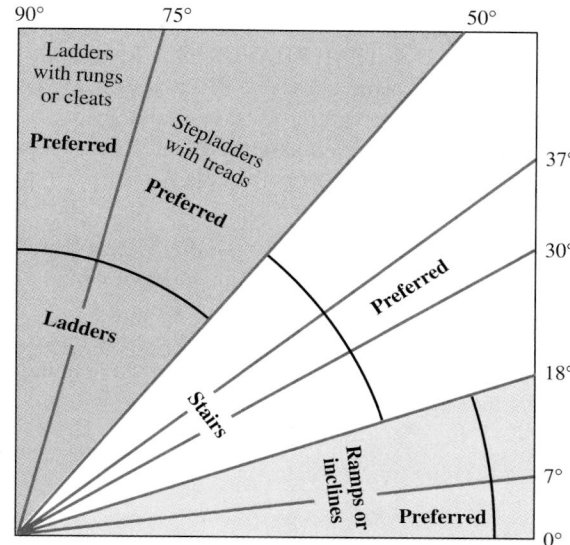

ILLUSTRATION 2

98. WEIGHT CHART   Illustration 3 is used to classify the weight of a baby boy from birth to 1 year. Estimate the weight range w for boys in the following classifications, using a compound inequality:
   a. 10 months old, "heavy"
   b. 5 months old, "light"
   c. 8 months old, "average"
   d. 3 months old, "moderately light"

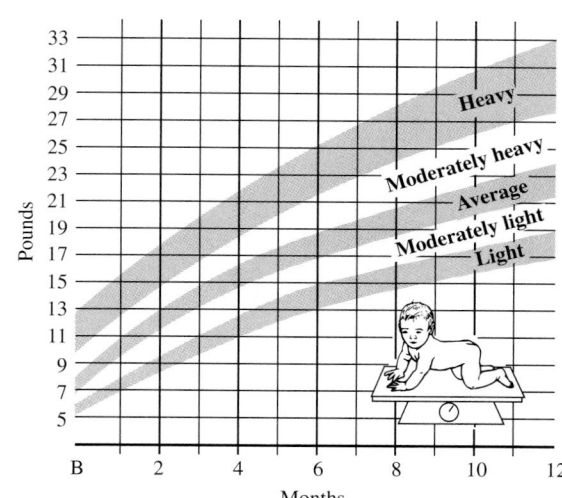

Based on data from *Better Homes and Gardens Baby Book* (Meredith Corp., 1969)

ILLUSTRATION 3

99. LAND ELEVATIONS The land elevations in Nevada range from the 13,143-foot height of Boundary Peak to the Colorado River at 470 feet. Use a compound inequality to express the range of these elevations.
   a. in feet
   b. in miles (round to the nearest tenth)
     (*Hint:* 1 mile is 5,280 feet.)

100. COMPARING TEMPERATURES To hold the temperature of a room between 19°C and 22°C, what Fahrenheit temperatures must be maintained? $\left(Hint:\text{ Fahrenheit temperature }F\text{ and Celsius temperature }C\text{ are related by the formula }F = \frac{9C + 160}{5}.\right)$

101. DRAFTING In Illustration 4, the $\pm$ (read "plus or minus") symbol means that the width of a plug a manufacturer produces can range from $1.497 - 0.001$ inches to $1.497 + 0.001$ inches. Write the range of acceptable widths w for the plug and the opening it fits into using compound inequalities.

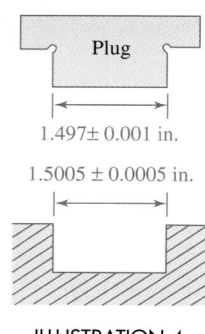

ILLUSTRATION 4

102. COUNTER SPACE In a large discount store, a rectangular counter is being built for the customer service department. If designers have determined that the outside perimeter of the counter (shown in red) needs to be at least 150 feet, use the plan in Illustration 5 to determine the acceptable values for x.

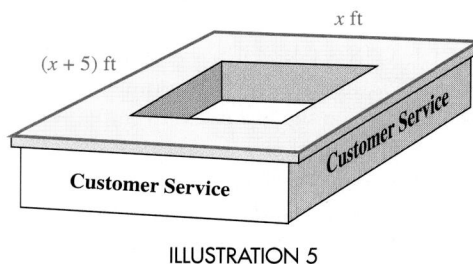

ILLUSTRATION 5

WRITING

103. Explain why multiplying both sides of an inequality by a negative number reverses the direction of the inequality.

104. Explain the use of parentheses and brackets for graphing intervals.

REVIEW *Find each power.*

105. -5^3 **106.** $(-3)^4$

Complete each input/output table.

107.

x	$x^2 - 3$
-2	
0	
3	

108.

x	$\frac{x}{3} + 2$
-6	
0	
12	

Simplify and Solve

Two of the most often used instructions in this book are **simplify** and **solve.** In algebra, we *simplify expressions* and we *solve equations and inequalities.*

To *simplify* an expression means to write it in a less complicated form. To do so, we apply the rules of arithmetic as well as algebraic concepts such as combining like terms, the distributive property, and the properties of 0 and 1.

To *solve* an equation or an inequality means to find the numbers that make the equation or inequality true, when substituted for its variable. We use the addition, subtraction, multiplication, and division properties of equality or inequality to solve equations and inequalities. Quite often, we must simplify expressions on the left- or right-hand sides of an equation or inequality when solving it.

In Exercises 1–4, use the procedures and the properties that we have studied to simplify the expression in part a and to solve the equation or inequality in part b.

Simplify

1. a. $-3x + 2 + 5x - 10$

2. a. $4(y + 2) - 3(y + 1)$

3. a. $\frac{1}{3}a + \frac{1}{3}a$

4. a. $-(2x + 10)$

Solve

b. $-3x + 2 + 5x - 10 = 4$

b. $4(y + 2) = 3(y + 1)$

b. $\frac{1}{3}a + \frac{1}{3} = \frac{1}{2}$

b. $-2x \geq -10$

5. In the student's work on the right, where was the mistake made? Explain what the student did wrong.

Simplify $2(x + 3) - x - 12$.

$$2(x + 3) - x - 12 = 2x + 6 - x - 12$$
$$= x - 6$$
$$0 = x - 6$$
$$0 + 6 = x - 6 + 6$$
$$\boxed{6 = x}$$

Section 8.1

EVALUATING ALGEBRAIC EXPRESSIONS Find five examples of cylinders. Measure and record the diameter d of their bases and their heights h. Express the measurements as decimals. Find the radius r of each base by dividing the diameter by 2. Then find each volume by evaluating the expression $\pi r^2 h$. Round to the nearest tenth of a cubic unit. Present your results in a table of the form shown in Illustration 1.

Cylinder	d	r	h	Volume
Container of salt	$3\frac{1}{4}$ in. (3.25 in.)	$1\frac{5}{8}$ in. (1.625 in.)	$5\frac{3}{8}$ in. (5.375 in.)	44.6 in.^{3}

ILLUSTRATION 1

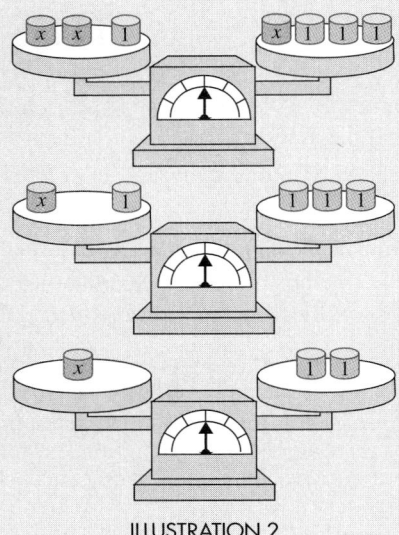

ILLUSTRATION 2

Section 8.2

THE DISTRIBUTIVE PROPERTY Draw a geometric model on the graph paper below that illustrates why

$$5(4 + 2) = 5(4) + 5(2)$$

A similar example can be found in Figure 8-4 on page 454.

Section 8.3

SOLVING EQUATIONS Make a presentation to the class explaining how we "undo" operations to isolate the variable when solving an equation. As a visual aid, bring in a box, tied shut with string, that contains a toy wrapped in tissue paper. Compare the three-step process a person would use to get to the toy inside the box to the three-step process we could use to solve the equation $\frac{2x}{3} - 4 = 2$.

SOLVING EQUATIONS In Chapter 1, a scale was used to illustrate the steps used to solve an equation.

a. What equation is being solved in Illustration 2? What is the solution?

b. Draw a similar series of pictures showing the solution of each of the following equations.

(1) $3x + 1 = 2x + 4$ **(2)** $2x + 6 = 4x + 2$

Section 8.4

GEOMETRY GOURMET Find snack foods that have the shapes of the geometric figures in Table 8-3 and Table 8-4 on pages 474 and 476. For example, tortilla chips can be triangular in shape, and malted milk balls are spheres. If you are unable to find a particular shape already available, decide on a way to make a snack in that shape. Make up a tray of snacks to bring to class. Discuss the various shapes of the snacks as you enjoy the food in your groups.

Section 8.5

MIXTURES Get several cans of orange juice concentrate and make four pitchers that are 10%, 30%, 50%, and 70% solutions. For example, a 30% solution would consist of three paper cups of concentrate and seven paper cups of water. Pour amounts of each mixture into cups. Have students taste each solution and see whether they can put the mixtures in order from least concentrated to most concentrated.

Section 8.6

INEQUALITIES In most states, a person must be at least 16 years old to have a driver's license. We can describe this situation with the inequality $a \geq 16$, where a represents a person's age in years. Think of other situations that can be described using an inequality or a compound inequality.

CHAPTER REVIEW

Algebraic Expressions

CONCEPTS

In order to describe numerical relationships, we need to translate the words of a problem into mathematical symbols.

REVIEW EXERCISES

1. Write each phrase as an algebraic expression.
  a. 25 more than the height h

  b. 15 less than the cutoff score s

  c. $\frac{1}{2}$ of the time t

  d. the product of 6 and x

2. See Illustration 1.
  a. If we let n represent the length of the nail, write an algebraic expression for the length of the bolt (in inches).

  b. If we let b represent the length of the bolt, write an algebraic expression for the length of the nail (in inches).

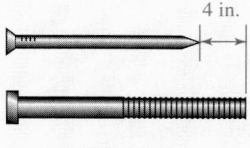

ILLUSTRATION 1

Sometimes we must rely on common sense and insight to find *hidden operations.*

3. a. How many years are in d decades?

  b. If you have x donuts, how many dozen donuts do you have?

  c. Five years after a house was constructed, a patio was added. How old, in years, is the patio if the house is x years old?

Number · value = total value

4. Complete the table in Illustration 2.

Type of coin	Number	Value (¢)	Total value (¢)
Nickel	6		
Dime	d		

ILLUSTRATION 2

When we replace the variable, or variables, in an algebraic expression with specific numbers and then apply the rules for the order of operations, we are *evaluating* the algebraic expression.

5. Complete the table of values.

x	$20x - x^3$
0	
1	
-4	

6. Evaluate each algebraic expression for the given value(s) of the variable(s).
  a. $7x^2 - \frac{x}{2}$ for $x = 4$

  b. $b^2 - 4ac$ for $b = -10$, $a = 3$, and $c = 5$

  c. $2(24 - 2c)^3$ for $c = 9$

  d. $\dfrac{x + y}{-x - z}$ for $x = 19$, $y = 17$, and $z = -18$

7. Use a calculator to find the volume, to the nearest tenth of a cubic inch, of the ice cream waffle cone in Illustration 3 by evaluating the algebraic expression.

$$\frac{\pi r^2 h}{3}$$

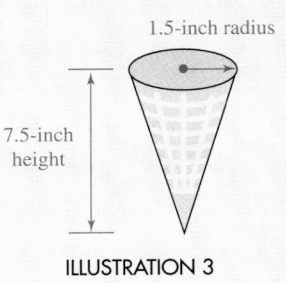

1.5-inch radius

7.5-inch height

ILLUSTRATION 3

SECTION 8.2   *Simplifying Algebraic Expressions*

To *simplify* an algebraic expression means to write it in less complicated form.

The *distributive property:*
$$a(b + c) = ab + ac$$
$$a(b - c) = ab - ac$$

A *term* is a number or a product of a number and one or more variables. Addition signs separate algebraic expressions into terms.

In a term, the numerical factor is called the *coefficient.*

Like terms are terms with exactly the same variables raised to exactly the same powers.

8. Simplify each expression.

   a. $-4(7w)$ **b.** $-3r(-5r)$

   c. $3(-2x)(-4y)$ **d.** $0.4(5.2f)$

9. Write each expression without parentheses.

   a. $5(x + 3)$ **b.** $-2(2x + 3 - y)$

   c. $-(a - 4)$ **d.** $\frac{3}{4}(4c - 8)$

10. How many terms are in each expression?

   a. $3x^2 + 2x - 5$ **b.** $-12xyz$

11. Identify the coefficient of each term.

   a. $2x - 5$ **b.** $16x^2 - 5x + 25$

   c. $\frac{1}{2}x + y$ **d.** $9.6t^2 - t$

12. Simplify each expression by combining like terms.

   a. $8p + 5p - 4p$ **b.** $-5m + 2n - 2m - 2n$

   c. $6a + 2b - 8a - 12b$ **d.** $5(p - 2) - 2(3p + 4)$

   e. $x^2 - x(x - 1)$ **f.** $8a^3 + 4a^3 - 20a^3$

13. Write an algebraic expression in simplified form for the perimeter of the triangle in Illustration 4.

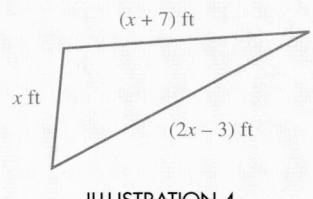

$(x + 7)$ ft

x ft

$(2x - 3)$ ft

ILLUSTRATION 4

SECTION 8.3   *Solving Equations*

To solve an equation means to find all the values of the variable that, when substituted for the variable, make a true statement.

14. Solve each equation.

   a. $5x + 4 = 14$ **b.** $-1.2y + 0.8 = 2.0$

   c. $\frac{n}{5} - 2 = 4$ **d.** $\frac{b - 5}{4} = -6$

An equation that is true for all values of its variable is called an *identity.*

An equation that is not true for any values of its variable is called an *impossible equation.*

e. $5(2x - 4) - 5x = 0$

f. $-2(x - 5) = 5(-3x + 4) + 3$

g. $\dfrac{3}{4} = \dfrac{1}{2} + \dfrac{d}{5}$

h. $-\dfrac{2}{3}f = 4$

i. $3(a + 8) = 6(a + 4) - 3a$

j. $2(y + 10) + y = 3(y + 8)$

SECTION 8.4	*Formulas*

A *formula* is an equation that is used to state a known relationship between two or more variables.

Retail price: $r = c + m$

Profit: $p = r - c$

Distance: $d = rt$

Temperature: $C = \dfrac{5(F - 32)}{9}$

Formulas from geometry:

Square: $P = 4s$, $A = s^2$

Rectangle: $P = 2l + 2w$,
$A = lw$

15. Find the markup on a CD player whose wholesale cost is $219 and whose retail price is $395.

16. One month, a restaurant had sales of $13,500 and made a profit of $1,700. Find the expenses for the month.

17. INDY 500   In 1996, the winner of the Indianapolis 500-mile automobile race averaged 147.956 mph. To the nearest hundredth of an hour, how long did it take him to complete the race?

18. JEWELRY MAKING   Gold melts at about 1,065°C. Change this to degrees Fahrenheit.

19. CAMPING   Find the perimeter of the air mattress in Illustration 4.

20. CAMPING   Find the amount of sleeping area on the top surface of the air mattress in Illustration 5.

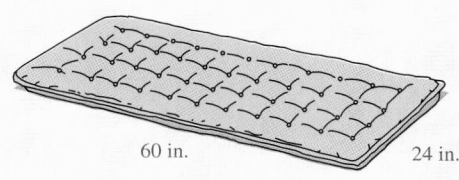

60 in.   24 in.

ILLUSTRATION 5

Triangle: $P = a + b + c$
$A = \frac{1}{2}bh$

Trapezoid:
$P = a + b + c + d$
$A = \frac{1}{2}h(b + d)$

Circle: $D = 2r$
$C = 2\pi r$
$A = \pi r^2$

Rectangular solid: $V = lwh$

Cylinder: $V = \pi r^2 h$

21. Find the area of a triangle with a base 17 meters long and a height of 9 meters.

22. Find the area of a trapezoid with bases 11 inches and 13 inches long and a height of 12 inches.

23. To the nearest hundredth, find the circumference of a circle with a radius of 8 centimeters.

24. To the nearest hundredth, find the area of the circle in Exercise 23.

25. CAMPING   Find the approximate volume of the air mattress in Illustration 5 if it is 3 inches thick.

26. Find the volume of a 12-foot cylinder whose circular base has a radius of 0.5 feet. Give the result to the nearest tenth.

Pyramid: $V = \frac{1}{3}Bh$

Cone: $V = \frac{1}{3}\pi r^2 h$

Sphere: $V = \frac{4}{3}\pi r^3$

27. Find the volume of a pyramid that has a square base, measuring 6 feet on a side, and a height of 10 feet.

28. HALLOWEEN After being cleaned out, a spherical-shaped pumpkin has an inside diameter of 9 inches. To the nearest hundredth, what is its volume?

29. Solve each formula for the required variable.
   a. $A = 2\pi rh$ for h
   b. $P = 2l + 2w$ for l

SECTION 8.5	*Problem Solving*

To solve problems, use the five-step problem-solving strategy.
1. Analyze the problem.
2. Form an equation.
3. Solve the equation.
4. State the conclusion.
5. Check the result.

30. SOUND SYSTEM A 45-foot-long speaker wire is to be cut into three pieces. One piece is to be 15 feet long. Of the remaining pieces, one must be 2 feet less than 3 times the length of the other. Find the length of the shorter piece of wire.

31. UTILITY BILLS The electric company charges $17.50 per month plus 18 cents for every kilowatt hour of energy used. One resident's bill was $43.96. How many kilowatt hours were used that month?

32. ART HISTORY *American Gothic,* shown in Illustration 6, was painted in 1930 by American artist Grant Wood. The length of the rectangular painting is 5 inches more than the width. Find the dimensions of the painting if it has a perimeter of $109\frac{1}{2}$ inches.

ILLUSTRATION 6

The sum of the measures of the angles of a triangle is 180°.

33. Find the missing angle measures of the triangle in Illustration 7.

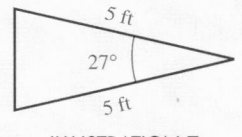

ILLUSTRATION 7

Total value = number · value

34. What is the value of x video games each costing $45?

Interest = principal · rate · time
$I = Prt$

35. INVESTMENT INCOME A woman has $27,000. Part is invested for one year in a certificate of deposit paying 7% interest, and the remaining amount in a cash management fund paying 9%. After 1 year, the total interest on the two investments is $2,110. How much is invested at each rate?

Distance = rate · time
$d = rt$

36. WALKING AND BICYCLING A bicycle path is 5 miles long. A man walks from one end at the rate of 3 mph. At the same time, a friend bicycles from the other end, traveling at 12 mph. In how many minutes will they meet?

37. MIXTURE A store manager mixes candy worth 90¢ per pound with gumdrops worth $1.50 per pound to make 20 pounds of a mixture worth $1.20 per pound. How many pounds of each kind of candy does he use?

The value v of a commodity is its price per pound p times the number of pounds n:
$v = pn$

38. SOLUTION How much acetic acid is in x gallons of a solution that is 12% acetic acid?

| **SECTION 8.6** | *Inequalities* |

An *inequality* is a mathematical expression that contains a $>$, $<$, $\geq$, $\leq$, or $\neq$ symbol.

A *solution of an inequality* is any number that makes the inequality true.

A *parenthesis* indicates that a number is not on the graph. A *bracket* indicates that a number is included in the graph.

Interval notation can be used to describe a set of real numbers.

39. Solve each inequality, graph the solution, and use interval notation to describe the solution.

a. $3x + 2 < 5$

b. $-5x - 8 > 7$

c. $5x - 3 \geq 2x + 9$

d. $7x + 1 \leq 8x - 5$

e. $5(3 - x) \leq 3(x - 3)$

f. $-\dfrac{3}{4}x \geq -9$

g. $8 < x + 2 < 13$

h. $0 \leq 2 - 2x < 6$

40. Graph the interval represented by $[-13, \infty)$.

41. SPORTS EQUIPMENT   The acceptable weight of ping-pong balls used in competition can range from 2.40 to 2.53 grams. Express this range using a compound inequality.

42. SIGNS   A large office complex has a strict policy about signs. Any sign to be posted in the building must meet three requirements:

- It must be rectangular in shape.
- Its width must be 18 inches.
- Its perimeter is not to exceed 132 inches.

What possible sign lengths meet these specifications?

1. A rock band recorded x songs for a CD. Technicians had to delete two songs from the album because of poor sound quality. Express the number of songs on the CD using an algebraic expression.

2. What is the value of q quarters in cents?

3. Complete the table in Illustration 1.

x	$2x - \frac{30}{x}$
5	
10	
30	

ILLUSTRATION 1

4. Evaluate $2lw + w^2$ for $l = 4$ and $w = 8$.

5. What is the numerical coefficient of the term $6x$?

6. How many terms are in the expression $4x^2 + 5x - 7$?

7. Is x used as a factor or as a term in the expression $3x + 2$?

8. What property is illustrated below?

$2(x + 7) = 2x + 2(7)$

Simplify each expression.

9. $5(-4x)$

10. $-8(-7t)(4t)$

11. $3(x + 2) + 3(4 - x)$

12. $-1.1d^2 - 3.8d^2$

Solve each equation.

13. $12x + 4 = -140$

14. $\frac{4}{5}t + 1 = -3$

15. $6m + 2 = -14 - 2m$

16. $0.3x = 0.5 - 0.2x$

17. $\frac{m}{2} - \frac{1}{3} = \frac{1}{4}$

18. $23 - 5(x + 10) = -12$

19. Solve the equation for the variable indicated.

$A = P + Prt$; for r

20. On its first night of business, a pizza parlor brought in $445. The owner estimated his costs that night to be $295. What was the profit?

21. Find the Celsius temperature reading if the Fahrenheit reading is $14°$.

22. PETS   The spherical fishbowl shown in Illustration 2 is three-quarters full of water. To the nearest cubic inch, what is the volume of water in the bowl?

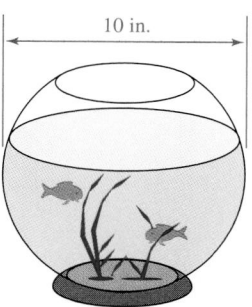

10 in.

ILLUSTRATION 2

23. TRAVEL TIMES   A car leaves Rockford, Illinois at the rate of 65 mph, bound for Madison, Wisconsin. At the same time, a truck leaves Madison at the rate of 55 mph, bound for Rockford. If the cities are 72 miles apart, how long will it take for the car and the truck to meet?

24. SALT SOLUTION   How many liters of a 2% brine solution must be added to 30 liters of a 10% brine solution to dilute it to an 8% solution?

25. GEOMETRY   If the vertex angle of an isosceles triangle is 44°, find the measure of each base angle.

26. INVESTMENT PROBLEM   Part of $13,750 is invested at 9% annual interest, and the rest is invested at 8%. After one year, the accounts paid $1,185 in interest. How much was invested at the lower rate?

In Problems 27–28, solve each inequality, graph the solution, and use interval notation to describe the solution.

27. $-8x - 20 \leq 4$

28. $-4 \leq 2(x + 1) < 10$

29. After we have solved an equation, how can we check the answer to be sure that it is a solution?

30. What are like terms? Give an example.

Chapters 1-8   Cumulative Review Exercises

1. Classify each of the following as an equation or an expression.
   a. $4m - 3 + 2m$ **b.** $4m = 3 + 2m$

2. Use the formula $t = \dfrac{w}{5}$ to complete the table.

Weight (lb)	Cooking time (hr)
15	
20	
25	

3. Give the prime factorization of 200.

4. Simplify $\dfrac{24}{36}$.

5. Multiply: $\dfrac{11}{21}\left(-\dfrac{14}{33}\right)$.

6. COOKING   A recipe calls for $\frac{3}{4}$ cup of flour, and the only measuring container you have holds $\frac{1}{8}$ of a cup. How many $\frac{1}{8}$ cups of flour would you need to add to follow the recipe?

7. Add: $\dfrac{4}{5} + \dfrac{2}{3}$.

8. Subtract: $42\dfrac{1}{8} - 29\dfrac{2}{3}$.

9. Write $\dfrac{15}{16}$ as a decimal.

10. Multiply: 0.45(100).

11. Evaluate each expression.
    a. $|-65|$ **b.** $-|-12|$

12. What property of real numbers is illustrated below?
    $x \cdot 5 = 5x$

In Exercises 13–16, classify each number as a natural number, a whole number, an integer, a rational number, an irrational number, and a real number. Each number may have several classifications.

13. 3

14. -1.95

15. $\dfrac{17}{20}$

16. π

17. Write each product using exponents.
    a. $4 \cdot 4 \cdot 4$ **b.** $\pi \cdot r \cdot r \cdot h$

18. Do each operation.
    a. $-6 + (-12) + 8$
    b. $-15 - (-1)$
    c. $2(-32)$
    d. $\dfrac{0}{35}$

19. Write each phrase as an algebraic expression.
    a. The sum of the width w and 12.
    b. Four less than a number n.

20. SICK DAYS   Use the data in Illustration 1 to find the average (mean) number of sick days used by this group of employees this year.

Name	Sick days	Name	Sick days
Chung	4	Ryba	0
Cruz	8	Nguyen	5
Damron	3	Tomaka	4
Hammond	2	Young	6

ILLUSTRATION 1

21. Complete the table of values.

x	$x^2 - 3$
-2	
0	
3	

22. Translate to mathematical symbols.

The loudness of a stereo speaker	is	2,000	divided by	the square of the distance of the listener from the speaker.

23. LAND OF THE RISING SUN   The flag of Japan is a red disc (representing sincerity and passion) on a white background (representing honesty and purity).
    a. What is the area of the rectangular-shaped flag in Illustration 2 (on the next page)?
    b. To the nearest tenth of a square foot, what is the area of the red disc?

c. Use the results from parts a and b to find what percent of the area of the Japanese flag is occupied by the red disc.

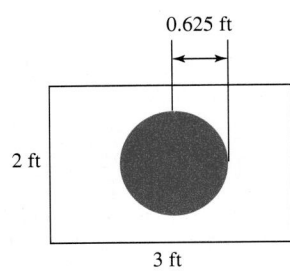

ILLUSTRATION 2

24. 45 is 15% of what number?

Let x = −5, y = 3, and z = 0. Evaluate each expression.

25. $(3x − 2y)z$

26. $\dfrac{x − 3y + |z|}{2 − x}$

27. $x^2 − y^2 + z^2$

28. $\dfrac{x}{y} + \dfrac{y + 2}{3 − z}$

Simplify each expression.

29. $−8(4d)$

30. $5(2x − 3y + 1)$

31. $2x + 3x$

32. $3a + 6a − 17a$

33. $q(q − 5) + 7q^2$

34. $5(t − 4) + 3t$

35. What is the length of the longest side of the triangle in Illustration 3?

36. Write an algebraic expression in simplest form for the perimeter of the triangle in Illustration 3.

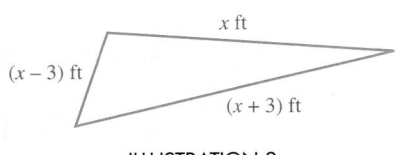

ILLUSTRATION 3

In Exercises 37–44, solve each equation.

37. $3x − 4 = 23$

38. $\dfrac{x}{5} + 3 = 7$

39. $−5p + 0.7 = 3.7$

40. $\dfrac{y − 4}{5} = 3$

41. $−\dfrac{4}{5}x = 16$

42. $−9(n + 2) − 2(n − 3) = 10$

43. $9y − 3 = 6y$

44. $\dfrac{1}{2} + \dfrac{x}{5} = \dfrac{3}{4}$

45. Find the area of a rectangle with sides of 5 meters and 13 meters.

46. Find the volume of a cone that is 10 centimeters tall and has a circular base whose diameter is 12 centimeters. Round to the nearest hundredth.

47. Solve $A = P + Prt$ for t.

48. WORK   Physicists say that *work* is done when an object is moved a distance d by a force F. To find the work done, we can use the formula $W = Fd$. Find the work done in lifting the bundle of newspapers shown in Illustration 4 onto the workbench. (*Hint:* The force that must be applied to lift the newspapers is equal to the weight of the newspapers.)

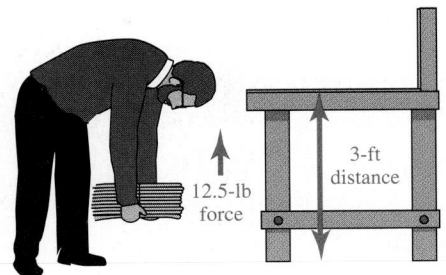

ILLUSTRATION 4

49. WORK   See Exercise 48. Find the weight of a 1-gallon can of paint if the amount of work done to lift it onto the workbench is 28.35 foot-pounds.

50. Find the unknown angle measure represented by x.

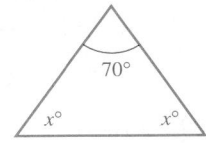

51. INVESTING   An investment club invested part of $10,000 at 9% annual interest and the rest at 8%. If the annual income from these investments was $860, how much was invested at 8%?

52. GOLDSMITH   How many ounces of a 40% gold alloy must be mixed with 10 ounces of a 10% gold alloy to obtain an alloy that is 25% gold?

Solve each inequality, graph the solution, and use interval notation to describe the solution.

53. $x − 4 > −6$

54. $−6x \geq −12$

55. $8x + 4 \geq 5x + 1$

56. $−1 \leq 2x + 1 < 5$

Graphs, Linear Equations, and Functions

9

9.1 Graphing Using the Rectangular Coordinate System

9.2 Equations Containing Two Variables

9.3 Graphing Linear Equations

9.4 Rate of Change and the Slope of a Line

9.5 Describing Linear Relationships

9.6 Writing Linear Equations

9.7 Functions

Key Concept: Describing Linear Relationships

Accent on Teamwork

Chapter Review

Chapter Test

Cumulative Review Exercises

RELATIONSHIPS BETWEEN TWO QUANTITIES CAN BE
DESCRIBED BY A TABLE, A GRAPH, OR AN EQUATION.

9.1   Graphing Using the Rectangular Coordinate System

In this section, you will learn about

- The rectangular coordinate system   • Graphing mathematical relationships
- Reading graphs   • Step graphs   • The midpoint formula

INTRODUCTION. It is often said, "A picture is worth a thousand words." In this section, we will show how numerical relationships can be described using mathematical pictures called **graphs.** We will also show how graphs are constructed and how we can obtain important information by reading graphs.

The rectangular coordinate system

When designing the Gateway Arch in St. Louis, shown in Figure 9-1(a), architects created a mathematical model called a **rectangular coordinate graph.** This graph, shown in Figure 9-1(b), is drawn on a grid called a **rectangular coordinate system.** This coordinate system is sometimes called a **Cartesian coordinate system,** after the 17th-century French mathematician René Descartes.

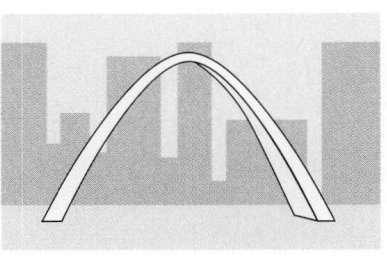 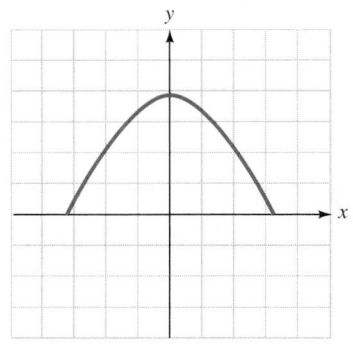

Scale: 1 unit = 100 ft

       (a)   (b)

FIGURE 9-1

    A rectangular coordinate system (see Figure 9-2) is formed by two perpendicular number lines. The horizontal number line is called the **x-axis,** and the vertical number line is called the **y-axis.** The positive direction on the *x*-axis is to the right, and the positive direction on the *y*-axis is upward. The scale on each axis should fit the data. For example, the axes of the graph of the arch shown in Figure 9-1(b) are scaled in units of 100 feet.

 COMMENT   If no scale is indicated on the axes, we assume that the axes are scaled in units of 1.

The point where the axes cross is called the **origin.** This is the zero point on each axis. The axes form a **coordinate plane,** and they divide it into four regions called **quadrants,** which are numbered using Roman numerals as shown in Figure 9-2. The axes are not considered to be in any quadrant.

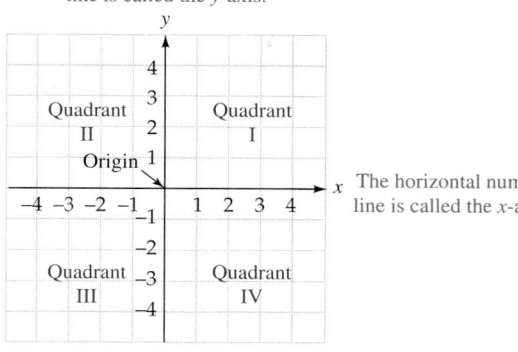

FIGURE 9-2

Each point in a coordinate plane can be identified by a pair of real numbers x and y written in the form (x, y). The first number x in the pair is called the **x-coordinate,** and the second number y is called the **y-coordinate.** The numbers in the pair are called the **coordinates** of the point. Some examples of such pairs are $(3, -4)$, $\left(-1, -\frac{3}{2}\right)$, and $(0, 2.5)$.

$$(3, -4)$$

    ↑   ↑

The x-coordinate   The y-coordinate
is listed first.   is listed second.

  COMMENT   Don't be confused by this new use of parentheses. The notation $(3, -4)$ represents a point on the coordinate plane, whereas $3(-4)$ indicates multiplication.

The process of locating a point in the coordinate plane is called **graphing** or **plotting** the point. In Figure 9-3(a), we use two blue arrows to show how to graph the point with coordinates of $(3, -4)$. Since the x-coordinate, 3, is positive, we start at the origin and move 3 units to the *right* along the x-axis. Since the y-coordinate, -4, is negative, we then move *down* 4 units to locate point A. Point A is the **graph** of $(3, -4)$ and lies in quadrant IV.

In Figure 9-3(a), two red arrows are used to show how to plot the point $(-4, 3)$. We start at the origin, move 4 units to the left along the x-axis, and then move up 3 units to locate point B. Point B lies in quadrant II.

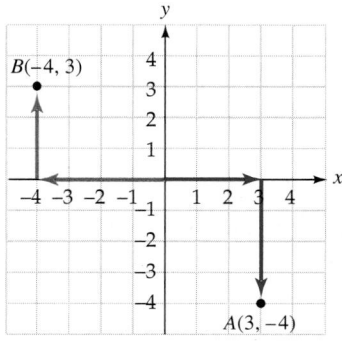

(a)

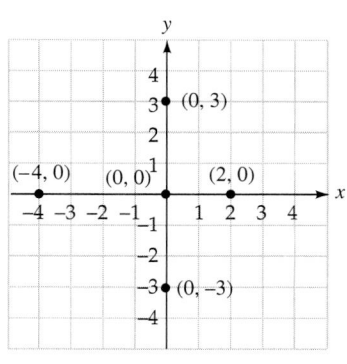

(b)

FIGURE 9-3

COMMENT   Note that point A with coordinates $(3, -4)$ is not the same as point B with coordinates $(-4, 3)$. Since the order of the coordinates of a point is important, we call the pairs **ordered pairs.**

In Figure 9-3(b) on the previous page, we see that the points $(-4, 0)$, $(0, 0)$, and $(2, 0)$ lie on the *x*-axis. In fact, all points with a *y*-coordinate of zero will lie on the *x*-axis. We also see that the points $(0, -3)$, $(0, 0)$, and $(0, 3)$ lie on the *y*-axis. All points with an *x*-coordinate of zero lie on the *y*-axis. We can also see that the coordinates of the origin are $(0, 0)$.

EXAMPLE 1 *Graphing points.*   Plot the points: **a.** $A(-2, 3)$,
b. $B\left(-1, -\frac{3}{2}\right)$,   **c.** $C(0, 2.5)$,   and   **d.** $D(4, 2)$.

Solution
See Figure 9-4. (Note: If no scale is indicated on the axes, we assume that the axes are scaled in units of 1.)

a. To plot point A with coordinates $(-2, 3)$, we start at the origin, move 2 units to the *left* on the *x*-axis, and move 3 units *up*. Point A lies in quadrant II.

b. To plot point B with coordinates $\left(-1, -\frac{3}{2}\right)$, we start at the origin and move 1 unit to the *left* and $\frac{3}{2}$ $\left(\text{or } 1\frac{1}{2}\right)$ units *down*. Point B lies in quadrant III.

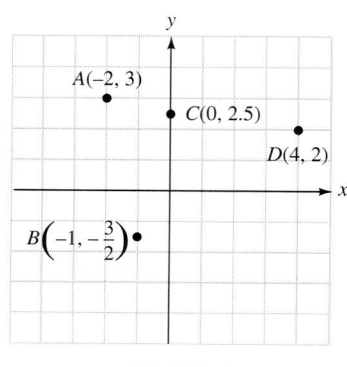

FIGURE 9-4

c. To graph point C with coordinates $(0, 2.5)$, we start at the origin and move 0 units on the *x*-axis and 2.5 units *up*. Point C lies on the *y*-axis.

d. To graph point D with coordinates $(4, 2)$, we start at the origin and move 4 units to the *right* and 2 units *up*. Point D lies in quadrant I.

Self Check
Plot the points:

a. $E(2, -2)$

b. $F(-4, 0)$

c. $G\left(1.5, \frac{5}{2}\right)$

d. $H(0, 5)$

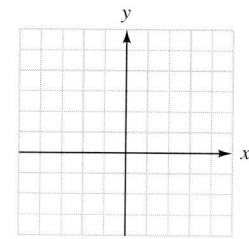

Answers:

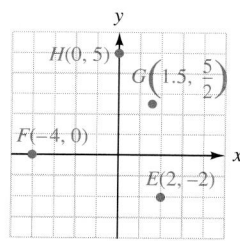

EXAMPLE 2 *Orbit of Earth.*   The circle shown in Figure 9-5 is an approximate **graph** of the orbit of Earth. The graph is made up of infinitely many points, each with its own *x*- and *y*-coordinates. Use the graph to find the coordinates of Earth's position during the months of February, May, August, and December.

Solution   To find the coordinates of each position, we start at the origin and move left or right along the *x*-axis to find the *x*-coordinate and then up or down to find the *y*-coordinate.

Month	Position of Earth on graph	Coordinates
February	3 units to the *right,* then 4 units *up*	$(3, 4)$
May	4 units to the *left,* then 3 units *up*	$(-4, 3)$
August	3.5 units to the *left,* then 3.5 units *down*	$(-3.5, -3.5)$
December	5 units to the *right,* no units *up* or *down*	$(5, 0)$

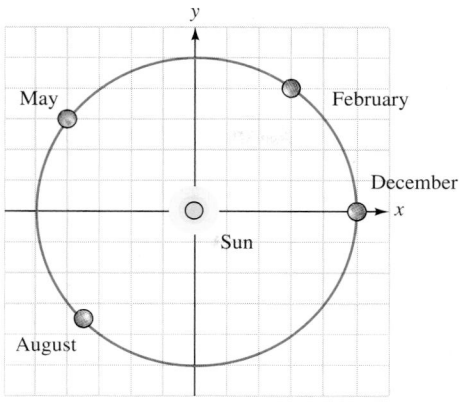

Scale: 1 unit = 18,600,000 mi

FIGURE 9-5

Graphing mathematical relationships

Every day, we deal with quantities that are related:

- The distance that we travel depends on how fast we are going.
- Our weight depends on how much we eat.
- The amount of water in a tub depends on how long the water has been running.

We often use graphs to visualize relationships between two quantities. For example, suppose we know the number of gallons of water that are in a tub at several time intervals after the water has been turned on. We can list that information in a **table.** (See Figure 9-6.)

The information in the table can be used to construct a graph that shows the relationship between the amount of water in the tub and the time the water has been running. Since the amount of water in the tub depends on the time, we will associate *time* with the *x*-axis and *amount of water* with the *y*-axis.

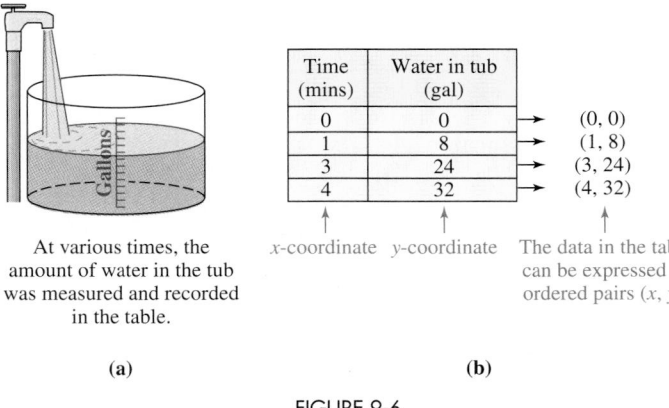

Time (mins)	Water in tub (gal)	
0	0	(0, 0)
1	8	(1, 8)
3	24	(3, 24)
4	32	(4, 32)

At various times, the amount of water in the tub was measured and recorded in the table.

x-coordinate   *y*-coordinate   The data in the table can be expressed as ordered pairs (*x*, *y*).

(a)

(b)

FIGURE 9-6

To construct the graph in Figure 9-7 (next page), we plot the four ordered pairs and draw a straight line through the resulting data points. The *y*-axis is scaled in larger units (4 gallons) because the data range from 0 to 32 gallons.

From the graph, we can see that the amount of water in the tub steadily increases as the water is allowed to run. We can also use the graph to make observations about the amount of water in the tub at other times. For example, the dashed line on the graph shows that in 5 minutes, the tub will contain 40 gallons of water.

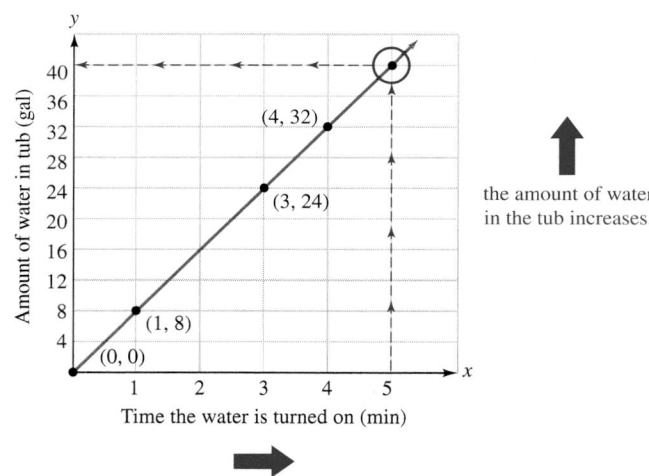

The data can be
listed in a table with
headings *x*, *y*, and
(*x*, *y*).

FIGURE 9-7

Reading graphs

Valuable information can be obtained from a graph, as can be seen in the next example.

EXAMPLE 3   *Reading a graph.*   The graph in Figure 9-8 shows the number of people in an audience before, during, and after the taping of a television show. On the *x*-axis, zero represents the time when taping began. Use the graph to answer the following questions, and record each result in a table.

a. How many people were in the audience when taping began?

b. What was the size of the audience 10 minutes before taping began?

c. At what times were there exactly 100 people in the audience?

Self Check

Use the graph in Figure 9-8 to answer the following questions.

a. At what times were there exactly 50 people in the audience?

b. What was the size of the audience that watched the taping?

c. How long did it take for the audience to leave the studio after the taping ended?

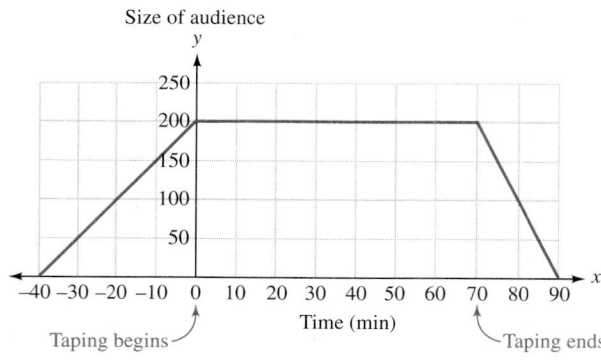

FIGURE 9-8

Solution

a. The time when taping began is represented by 0 on the *x*-axis. Since the point on the graph directly above 0 has a *y*-coordinate of 200, the point (0, 200) is on the graph. The *y*-coordinate of this point indicates that 200 people were in the audience when the taping began. We enter this result in the table at the right.

b. Ten minutes before taping began is represented by −10 on the *x*-axis. Since the point on the

Time (min) *x*	Size of audience *y*
0	200
−10	150
−20	100
80	100

graph directly above −10 has a *y*-coordinate of 150, the point (−10, 150) is on the graph. The *y*-coordinate of this point indicates that 150 people were in the audience 10 minutes before the taping began. We enter this result in the table.

c. We can draw a horizontal line passing through 100 on the *y*-axis. This line intersects the graph twice, at (−20, 100) and at (80, 100). So there are two times when 100 people were in the audience. The first time was 20 minutes before taping began (−20), and the second time was 80 minutes after taping began (80). The *y*-coordinates of these points indicate that there were 100 people in the audience 20 minutes before and 80 minutes after taping began. We enter these results in the table.

Answers: a. 30 min before and 85 min after taping began, **b.** 200,   **c.** 20 min   ■

Step graphs

The graph in Figure 9-9 shows the cost of renting a trailer for different periods of time. For example, the cost of renting the trailer for 4 days is $60, which is the *y*-coordinate of the point (4, 60). The cost of renting the trailer for a period lasting over 4 and up to 5 days jumps to $70. Since the jumps in cost form steps in the graph, we call this graph a **step graph.**

EXAMPLE 4   Use the information in Figure 9-9 to answer the following questions. Write the results in a table.

a. Find the cost of renting the trailer for 2 days.

b. Find the cost of renting the trailer for $5\frac{1}{2}$ days.

c. How long can you rent the trailer if you have $50?

d. Is the rental cost per day the same?

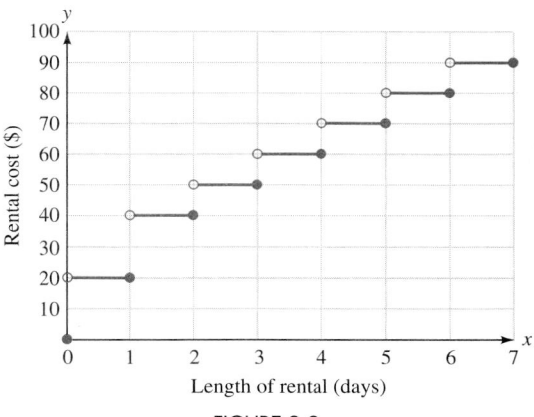

FIGURE 9-9

Solution

a. The solid dot at the end of each step indicates the rental cost for 1, 2, 3, 4, 5, 6, or 7 days. An open circle indicates that that point is not on the graph. We locate 2 days on the *x*-axis and move up to locate the point on the graph directly above the 2. Since the point has coordinates (2, 40), a 2-day rental would cost $40. We enter this ordered pair in the table at the left.

b. We locate $5\frac{1}{2}$ days on the *x*-axis and move straight up to locate the point with coordinates $\left(5\frac{1}{2}, 80\right)$, which indicates that a $5\frac{1}{2}$-day rental would cost $80. We then enter this ordered pair in the table.

c. We draw a horizontal line through the point labeled 50 on the *y*-axis. Since this line intersects one step in the graph, we can look down to the *x*-axis to find the *x*-values that correspond to a *y*-value of 50. From the graph, we see that the trailer can be rented for more than 2 and up to 3 days for $50. We write (3, 50) in the table.

d. No. If we look at the *y*-coordinates, we see that for the first day, the rental fee is $20. The second day, the cost jumps another $20. The third day, and all subsequent days, the cost jumps only $10.   ■

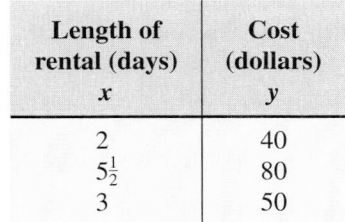

Length of rental (days)	Cost (dollars)
x	*y*
2	40
$5\frac{1}{2}$	80
3	50

The midpoint formula

To distinguish between the coordinates of two points on a line, we often use subscript notation. Point $P(x_1, y_1)$ is read as "point P with coordinates of x sub 1 and y sub 1 ." Point $Q(x_2, y_2)$ is read as "point Q with coordinates of x sub 2 and y sub 2."

If point M in the graph below lies midway between points $P(x_1, y_1)$ and $Q(x_2, y_2)$, point M is called the **midpoint** of segment PQ. To find the coordinates of M, we find the mean (average) of the x-coordinates and the mean of the y-coordinates of P and Q.

The midpoint formula

The **midpoint** of the line segment with endpoints at $P(x_1, y_1)$ and $Q(x_2, y_2)$ is the point M with coordinates of

$$\left(\frac{x_1 + x_2}{2}, \frac{y_1 + y_2}{2} \right)$$

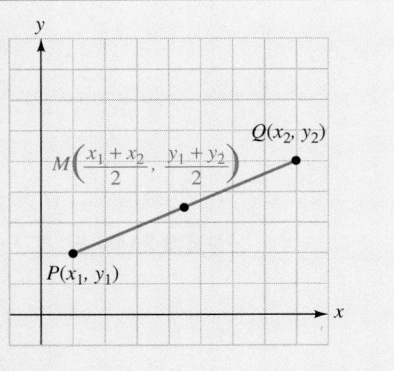

EXAMPLE 5 Find the midpoint of the line segment joining $P(-2, 3)$ and $Q(3, -5)$.

Solution To find the midpoint, we find the mean of the x-coordinates and the mean of the y-coordinates to get

$$\frac{x_1 + x_2}{2} = \frac{-2 + 3}{2} \quad \text{and} \quad \frac{y_1 + y_2}{2} = \frac{3 + (-5)}{2}$$

$$= \frac{1}{2} \qquad\qquad\qquad = -1$$

The midpoint of segment PQ is the point $M\left(\frac{1}{2}, -1\right)$.

Self Check
Find the midpoint of the segment joining $P(5, -4)$ and $Q(-3, 5)$.

Answer: $\left(1, \frac{1}{2}\right)$ ■

STUDY SET Section 9.1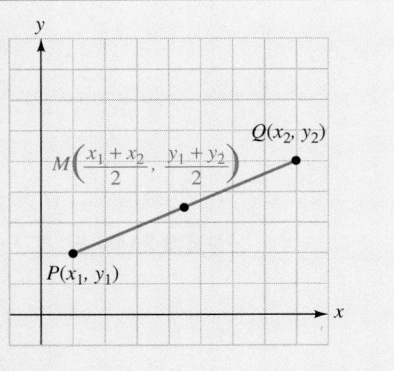

VOCABULARY *Fill in the blanks.*

1. The pair of numbers $(-1, -5)$ is called an _____ pair.

2. In the ordered pair $\left(-\frac{3}{2}, -5\right)$, the -5 is called the _____.

3. The point with coordinates $(0, 0)$ is called the _____.

4. The x- and y-axes divide the coordinate plane into four regions called _____.

5. The point with coordinates $(4, 2)$ can be graphed on a _____ coordinate system.

6. The process of locating the position of a point on a coordinate plane is called _____ the point.

CONCEPTS *In Exercises 7–8, fill in the blanks.*

7. To plot the point with coordinates $(-5, 4.5)$, we start at the _____ and move 5 units to the _____ and then move 4.5 units _____.

8. To plot the point with coordinates $\left(6, -\frac{3}{2}\right)$, we start at the _____ and move 6 units to the _____ and then move $\frac{3}{2}$ units _____.

9. Do (3, 2) and (2, 3) represent the same point?

10. In the ordered pair (4, 5), is the number 4 associated with the horizontal or the vertical axis?

11. In which quadrant do points with a negative x-coordinate and a positive y-coordinate lie?

12. In which quadrant do points with a positive x-coordinate and a negative y-coordinate lie?

13. In Illustration 1, fill in the missing coordinate of each highlighted point on the graph of the circle.

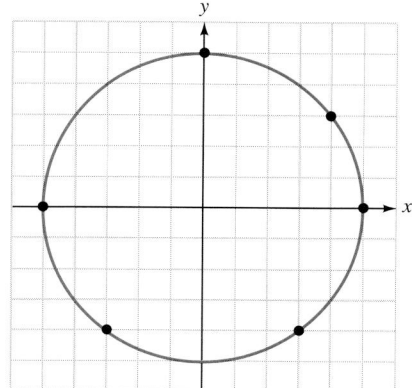

a. (4,   )

b. (3,   )

c. (5,   )

d. (−3,   )

e. (−5,   )

f. (0,   )

ILLUSTRATION 1

14. In Illustration 2, fill in the missing coordinate of each point on the graph of the line.

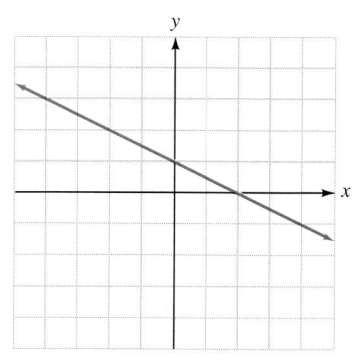

a. (−4,   )

b. (, 0)

c. (, 2)

d. (, −1)

e. (−4,   )

f. (, 1)

ILLUSTRATION 2

The graph in Illustration 3 gives the heart rate of a woman before, during, and after an aerobic workout. In Exercises 15–22, use the graph to answer the questions.

15. What information does the point (−10, 60) give us?

16. After beginning her workout, how long did it take the woman to reach her training-zone heart rate?

17. What was the woman's heart rate half an hour after beginning the workout?

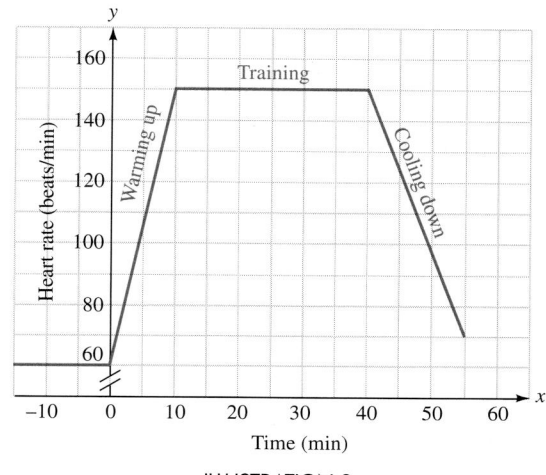

ILLUSTRATION 3

18. For how long did the woman work out at her training zone?

19. At what time was her heart rate 100 beats per minute?

20. How long was her cool-down period?

21. What was the difference in the woman's heart rate before the workout and after the cool-down period?

22. What was her approximate heart rate 8 minutes after beginning?

NOTATION

23. Explain the difference between (3, 5), 3(5), and 5(3 + 5).

24. In the table, which column contains values associated with the vertical axis of a graph?

x	y
2	0
5	−2
−1	$-\frac{1}{2}$

25. Do these ordered pairs name the same point?

$$\left(2.5, -\tfrac{7}{2}\right), \left(2\tfrac{1}{2}, -3.5\right), \left(2.5, -3\tfrac{1}{2}\right)$$

26. Do these ordered pairs name the same point?

$(-1.25, 4), \left(-1\frac{1}{4}, 4.0\right), \left(-\frac{5}{4}, 4\right)$

PRACTICE *Graph each point on the coordinate grid provided.*

27. $A(-3, 4)$

$B(4, 3.5)$

$C\left(-2, -\frac{5}{2}\right)$

$D(0, -4)$

$E\left(\frac{3}{2}, 0\right)$

$F(3, -4)$

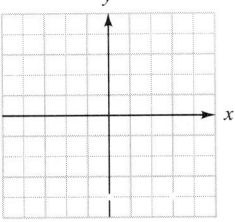

28. $G(4, 4)$

$H(0.5, -3)$

$I(-4, -4)$

$J(0, -1)$

$K(0, 0)$

$L(0, 3)$

$M(-2, 0)$

Find the midpoint of a line segment with the given endpoints.

29. $P(5, 3)$ and $Q(7, 9)$

30. $P(5, 6)$ and $Q(7, 10)$

31. $P(2, -7)$ and $Q(-3, 12)$

32. $P(-8, 12)$ and $Q(3, -9)$

33. $A(4, 6)$ and $B(10, 6)$

34. $A(8, -6)$ and $O(0, 0)$

APPLICATIONS

35. CONSTRUCTION   The graph in Illustration 4 shows a side view of a bridge design. Make a table with three columns; label them *rivets, welds,* and *anchors.* List the coordinates of the points at which each category is located.

36. WATER PRESSURE   The graph in Illustration 5 shows how the path of a stream of water changes when the hose is held at two different angles.
   a. At which angle does the stream of water shoot up higher? How much higher?
   b. At which angle does the stream of water shoot out farther? How much farther?

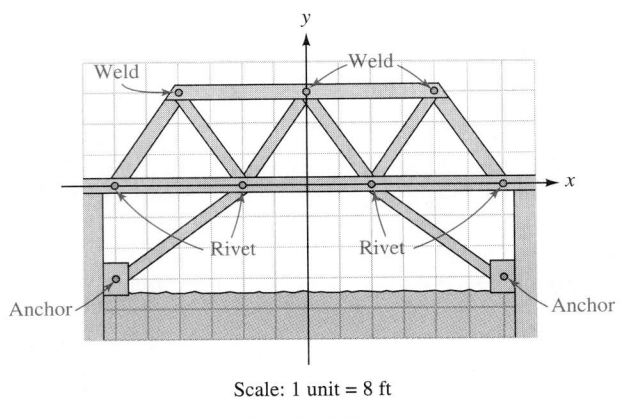

Scale: 1 unit = 8 ft

ILLUSTRATION 4

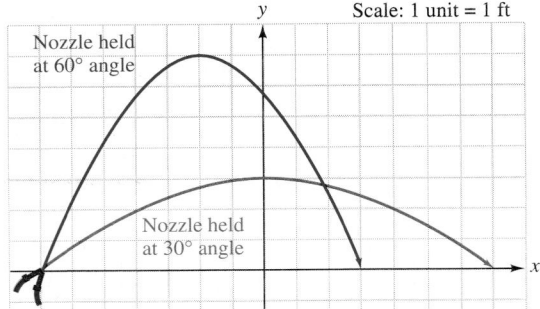

ILLUSTRATION 5

37. GOLF SWING   To correct her swing, a golfer is videotaped and then has her image displayed on a computer monitor so that it can be analyzed by a golf pro. (See Illustration 6.) Give the coordinates of the points that are highlighted on the arc of her swing.

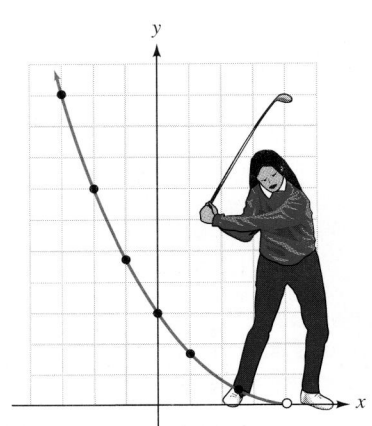

Scale: 1 unit = 6 in.

ILLUSTRATION 6

38. MEDICINE Scoliosis is a lateral curvature of the spine that can be more easily detected when a grid is superimposed over an X ray. In Illustration 7, find the coordinates of the "center points" of the indicated vertebrae. Note that T3 means the third thoracic vertebra, L4 means the fourth lumbar vertebra, and so on.

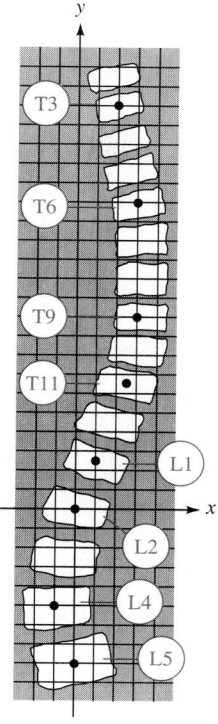

Scale: 1 unit = 0.5 in.
ILLUSTRATION 7

39. VIDEO RENTAL The charges for renting a video are shown in the graph in Illustration 8.
   a. Find the charge for a 1-day rental.
   b. Find the charge for a 2-day rental.
   c. What is the charge if a tape is kept for 5 days?
   d. What is the charge if a tape is kept for a week?

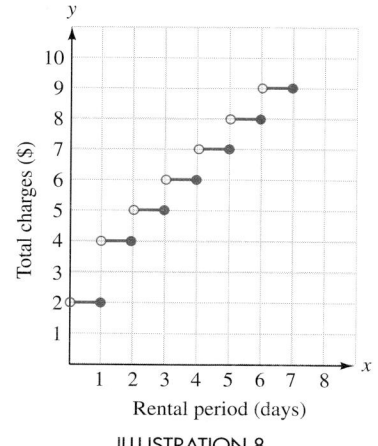

ILLUSTRATION 8

40. POSTAGE RATES The graph shown in Illustration 9 gives the first-class postage rates in 2001 for mailing items weighing up to 5 ounces.
   a. Find the postage costs for mailing each of the following letters first class: a 1-ounce letter, a 4-ounce letter, and a $2\frac{1}{2}$-ounce letter.

   b. Find the difference in postage for a 3.75-ounce letter and a 4.75-ounce letter.
   c. What is the heaviest letter that could be mailed for 55¢ first class?

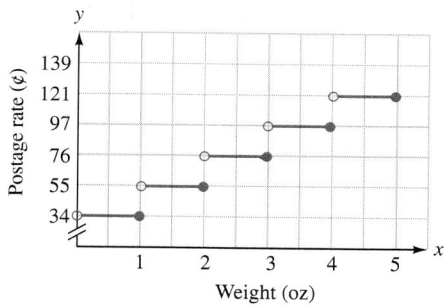

ILLUSTRATION 9

41. GAS MILEAGE The table in Illustration 10 gives the number of miles (y) that a truck can be driven on x gallons of gasoline. Plot the ordered pairs and draw a line connecting the points.

x	y
2	10
3	15
5	25

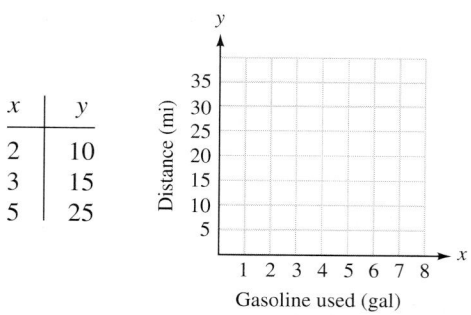

ILLUSTRATION 10

   a. Estimate how far the truck can go on 7 gallons of gasoline.
   b. How many gallons of gas are needed to travel a distance of 20 miles?
   c. How far can the truck go on 6.5 gallons of gasoline?

42. VALUE OF A CAR The table in Illustration 11 shows the value y (in thousands of dollars) of a car that is x years old. Plot the ordered pairs and draw a line connecting the points.

x	y
3	7
4	5.5
5	4

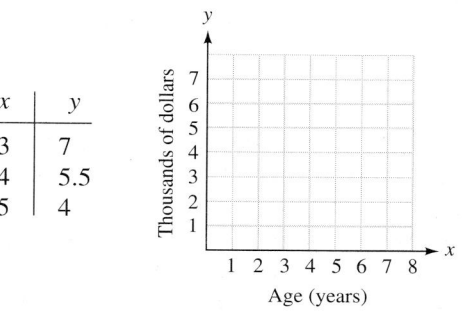

ILLUSTRATION 11

a. What does the point (3, 7) on the graph tell you?

b. Estimate the value of the car when it is 7 years old.

c. After how many years will the car be worth $2,500?

43. ROAD MAPS Road maps usually have a coordinate system to help locate cities. Use the map in Illustration 12 to locate Rockford, Mount Carroll, Harvard, and the intersection of state Highway 251 and U.S. Highway 30. Express each answer in the form (number, letter).

ILLUSTRATION 12

44. BATTLESHIP In the game Battleship, the player uses coordinates to drop depth charges from a battleship to hit a hidden submarine. What coordinates should be used to make three hits on the exposed submarine shown in Illustration 13? Express each answer in the form (letter, number).

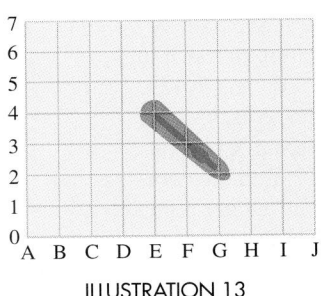

ILLUSTRATION 13

WRITING

45. Explain why the point $(-3, 3)$ is not the same as the point $(3, -3)$.

46. Explain what is meant when we say that the rectangular coordinate graph of the St. Louis Gateway Arch is made up of *infinitely many* points.

47. Explain how to plot the point $(-2, 5)$.

48. Explain why the coordinates of the origin are $(0, 0)$.

REVIEW

49. Evaluate $-3 - 3(-5)$.

50. Evaluate $(-5)^2 + (-5)$.

51. What is the opposite of -8?

52. Simplify $|-1 - 9|$.

53. Solve $-4x + 7 = -21$.

54. Solve $P = 2l + 2w$ for w.

55. Evaluate $(x + 1)(x + y)^2$ for $x = -2$ and $y = -5$.

56. Simplify $-6(x - 3) - 2(1 - x)$.

9.2 *Equations Containing Two Variables*

In this section, you will learn about

- Solving equations in two variables • Constructing tables of solutions
- Graphing equations • Using different variables

INTRODUCTION. In this section, we will discuss equations that contain two variables. Such equations are often used to describe relationships between two quantities. To see a mathematical picture of these relationships, we will construct graphs of their equations.

Solving equations in two variables

We have previously solved equations containing one variable. For example, we can show that the solution of each of the following equations is $x = 3$.

$$2x + 3 = 9, \qquad -5x + 1 = 4 - 6x, \qquad \text{and} \qquad -3(x + 1) = 2x - 18$$

If we graph the solution $x = 3$ on a number line, we get the graph shown in Figure 9-10.

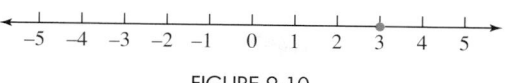

FIGURE 9-10

To describe relationships between two quantities mathematically, we use equations with two variables. Some examples of equations in two variables are

$$y = x - 1, \qquad y = x^2, \qquad y = |x|, \quad \text{and} \quad y = x^3$$

Solutions of equations in two variables are ordered pairs. For example, one solution of $y = x - 1$ is the ordered pair (5, 4), because the equation is true when $x = 5$ and $y = 4$.

$y = x - 1$	The original equation.
$4 \stackrel{?}{=} 5 - 1$	Substitute 5 for x and 4 for y.
$4 = 4$	Do the subtraction on the right-hand side: $5 - 1 = 4$.

Since $4 = 4$ is a true statement, the ordered pair (5, 4) is a solution, and we say that (5, 4) **satisfies** the equation.

EXAMPLE 1 *Verifying a solution.* Is the ordered pair $(-1, -3)$ a solution of $y = x - 1$?

Solution

We substitute -1 for x and -3 for y and see whether the resulting equation is a true statement.

$y = x - 1$	The original equation.
$-3 \stackrel{?}{=} -1 - 1$	Substitute -1 for x and -3 for y.
$-3 = -2$	Do the subtraction: $-1 - 1 = -2$.

Since $-3 = -2$ is a false statement, $(-1, -3)$ is not a solution.

Self Check

Is (9, 8) a solution of $y = x - 1$?

Answer: yes ∎

EXAMPLE 2 *Verifying a solution.* Is the ordered pair $(-6, 36)$ a solution of $y = x^2$?

Solution

We substitute -6 for x and 36 for y and see whether the resulting equation is a true statement.

$y = x^2$	The original equation.
$36 \stackrel{?}{=} (-6)^2$	Substitute -6 for x and 36 for y.
$36 = 36$	Find the power: $(-6)^2 = 36$.

Since the equation $36 = 36$ is true, $(-6, 36)$ is a solution.

Self Check

Is $(-2, 5)$ a solution of $y = x^2$?

Answer: no ∎

Constructing tables of solutions

To find solutions of equations in x and y, we can pick numbers at random, substitute them for x, and find the corresponding values of y. For example, to find some ordered pairs that satisfy the equation $y = x - 1$, we can let $x = -4$ (called the **input value**), substitute -4 for x, and solve for y (called the **output value**).

$y = x - 1$	The original equation.
$y = -4 - 1$	Substitute the input -4 for x.
$y = -5$	The output is -5.

$y = x - 1$

x	y	(x, y)
-4	-5	$(-4, -5)$

$$y = x - 1$$

x	y	(x, y)
-4	-5	$(-4, -5)$
-2	-3	$(-2, -3)$

$$y = x - 1$$

x	y	(x, y)
-4	-5	$(-4, -5)$
-2	-3	$(-2, -3)$
0	-1	$(0, -1)$

$$y = x - 1$$

x	y	(x, y)
-4	-5	$(-4, -5)$
-2	-3	$(-2, -3)$
0	-1	$(0, -1)$
2	1	$(2, 1)$

$$y = x - 1$$

x	y	(x, y)
-4	-5	$(-4, -5)$
-2	-3	$(-2, -3)$
0	-1	$(0, -1)$
2	1	$(2, 1)$
4	3	$(4, 3)$

The ordered pair $(-4, -5)$ is a solution. We list this ordered pair in red in the **table of solutions** (or **table of values**) shown on the previous page.

To find another ordered pair that satisfies $y = x - 1$, we let $x = -2$.

$y = x - 1$ The original equation.

$y = -2 - 1$ Substitute the input -2 for x.

$y = -3$ The output is -3.

A second solution is $(-2, -3)$, and we list it in the table of solutions.

If we let $x = 0$, we can find a third ordered pair that satisfies $y = x - 1$.

$y = x - 1$ The original equation.

$y = 0 - 1$ Substitute the input 0 for x.

$y = -1$ The output is -1.

A third solution is $(0, -1)$, which we also add to the table of solutions.

If we let $x = 2$, we can find a fourth solution.

$y = x - 1$ The original equation.

$y = 2 - 1$ Substitute the input 2 for x.

$y = 1$ The output is 1.

A fourth solution is $(2, 1)$, and we add it to the table of solutions.

If we let $x = 4$, we have

$y = x - 1$ The original equation.

$y = 4 - 1$ Substitute the input 4 for x.

$y = 3$ The output is 3.

A fifth solution is $(4, 3)$.

Since we can choose any real number for x, and since any choice of x will give a corresponding value of y, it is apparent that the equation $y = x - 1$ has *infinitely many solutions*. We have found five of them: $(-4, -5)$, $(-2, -3)$, $(0, -1)$, $(2, 1)$, and $(4, 3)$.

Graphing equations

To graph the equation $y = x - 1$, we plot the ordered pairs listed in the table of solutions on a rectangular coordinate system, as shown in Figure 9-11(a). From the figure, we can see that the five points lie on a line.

In Figure 9-11(b), we draw a straight line through the points, because the graph of any solution of $y = x - 1$ will lie on this line. The arrowheads show that the line continues forever in both directions. The line is a picture of all the solutions of the equation $y = x - 1$. This line is called the **graph** of the equation.

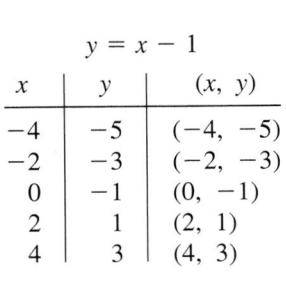

$$y = x - 1$$

x	y	(x, y)
-4	-5	$(-4, -5)$
-2	-3	$(-2, -3)$
0	-1	$(0, -1)$
2	1	$(2, 1)$
4	3	$(4, 3)$

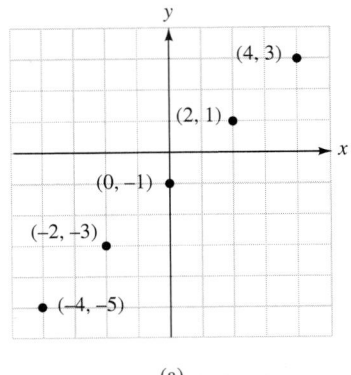

(a)

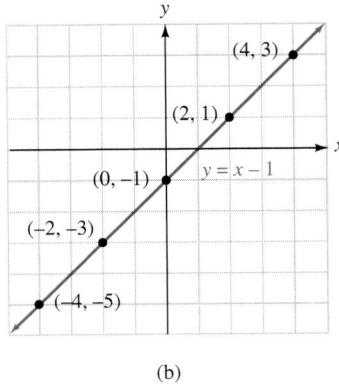

(b)

FIGURE 9-11

To graph an equation in x and y, we follow these steps.

Graphing an equation in *x* and *y*

1. Make a table of solutions containing several ordered pairs of numbers (x, y) that satisfy the equation. Do this by picking values for x and finding the corresponding values for y.
2. Plot each ordered pair on a rectangular coordinate system.
3. Carefully draw a line or smooth curve through the points.

Since we will usually choose a number for x and then find the corresponding value of y, the value of y depends on x. For this reason, we call y the **dependent variable** and x the **independent variable**. The value of the independent variable is the input value, and the value of the dependent variable is the output value.

EXAMPLE 3   *Graphing equations.*   Graph $y = -2x - 2$.

Solution

To make a table of solutions, we choose numbers for x and find the corresponding values of y. If $x = -3$, we have

$y = -2x - 2$	The original equation.
$y = -2(-3) - 2$	Substitute -3 for x.
$y = 6 - 2$	Do the multiplication: $-2(-3) = 6$.
$y = 4$	Do the subtraction.

Thus, $x = -3$ and $y = 4$ is a solution. In a similar manner, we find the corresponding y-values for x-values of $-2, -1, 0,$ and 1 and record the results in the table of solutions in Figure 9-12(a). After plotting the ordered pairs, we draw a line through the points to get the graph shown in the Figure 9-12(b).

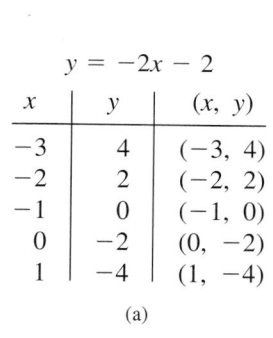

$y = -2x - 2$

x	y	(x, y)
-3	4	$(-3, 4)$
-2	2	$(-2, 2)$
-1	0	$(-1, 0)$
0	-2	$(0, -2)$
1	-4	$(1, -4)$

(a)

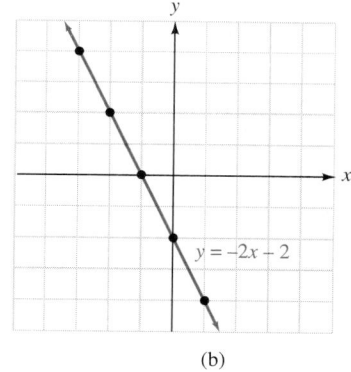

(b)

FIGURE 9-12

Self Check

Graph $y = -3x + 1$.

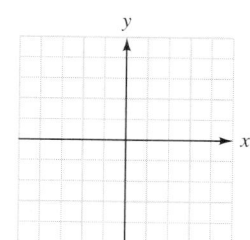

Answer:

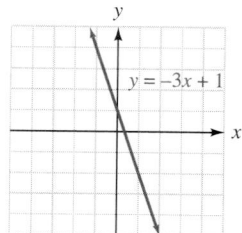

EXAMPLE 4   *Graphing equations.*   Graph $y = x^2$.

Solution

To make a table of solutions, we will choose numbers for x and find the corresponding values of y. If $x = -3$, we have

$y = x^2$	The original equation.
$y = (-3)^2$	Substitute the input -3 for x.
$y = 9$	The output is 9.

Thus, $x = -3$ and $y = 9$ is a solution. In a similar manner, we find the corresponding y-values for x-values of $-2, -1, 0, 1, 2,$ and 3. If we plot the ordered pairs listed in the table in Figure 9-13 and join the points with a smooth curve, we get the graph shown in the figure, which is called a **parabola**.

Self Check

Graph $y = x^2 - 2$ and compare the result to the graph of $y = x^2$. What do you notice?

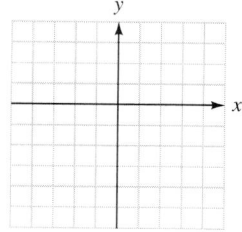

$y = x^2$

x	y	(x, y)
−3	9	(−3, 9)
−2	4	(−2, 4)
−1	1	(−1, 1)
0	0	(0, 0)
1	1	(1, 1)
2	4	(2, 4)
3	9	(3, 9)

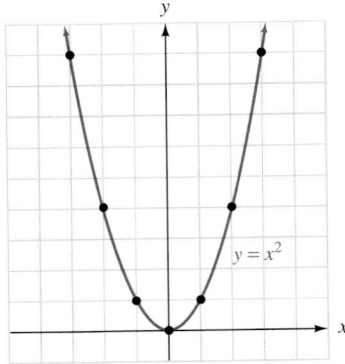

FIGURE 9-13

Answer: The graph has the same shape, but is 2 units lower.

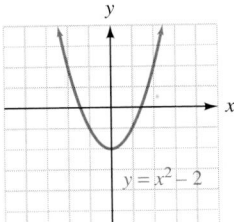

EXAMPLE 5 *Graphing equations.* Graph $y = |x|$.

Solution

To make a table of solutions, we will choose numbers for x and find the corresponding values of y. If $x = -5$, we have

$y = |x|$ The original equation.

$y = |-5|$ Substitute the input -5 for x.

$y = 5$ The output is 5.

The ordered pair $(-5, 5)$ satisfies the equation. This pair and several others that satisfy the equation are listed in the table of solutions in Figure 9-14. If we plot the ordered pairs in the table, we see that they lie in a "V" shape. We join the points to complete the graph shown in the figure.

$y = |x|$

x	y	(x, y)
−5	5	(−5, 5)
−4	4	(−4, 4)
−3	3	(−3, 3)
−2	2	(−2, 2)
−1	1	(−1, 1)
0	0	(0, 0)
1	1	(1, 1)
2	2	(2, 2)
3	3	(3, 3)

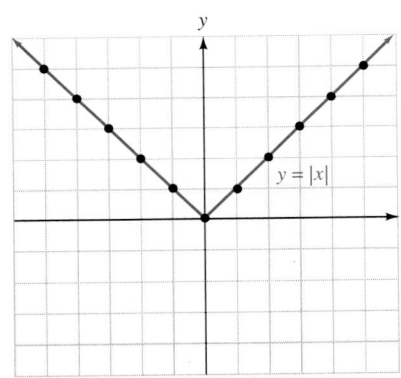

FIGURE 9-14

Answer: The graph has the same shape, but is 2 units higher.

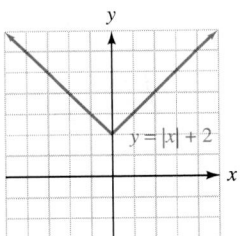

EXAMPLE 6 *Graphing equations.* Graph $y = x^3$.

Solution

If we let $x = -2$, we have

$y = x^3$ The original equation.

$y = (-2)^3$ Substitute the input -2 for x.

$y = -8$ The output is -8.

The ordered pair $(-2, -8)$ satisfies the equation. This ordered pair and several others that satisfy the equation are listed in the table of solutions in Figure 9-15. Plotting the ordered pairs and joining them with a smooth curve gives us the graph shown in the figure.

Self Check
Graph $y = (x - 2)^3$ and compare the result to the graph of $y = x^3$. What do you notice?

$$y = x^3$$

x	y	(x, y)
-2	-8	$(-2, -8)$
-1	-1	$(-1, -1)$
0	0	$(0, 0)$
1	1	$(1, 1)$
2	8	$(2, 8)$

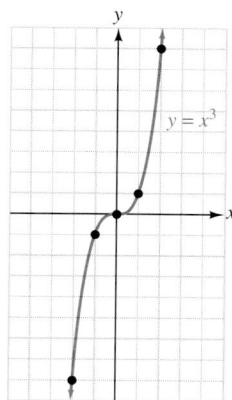

FIGURE 9-15

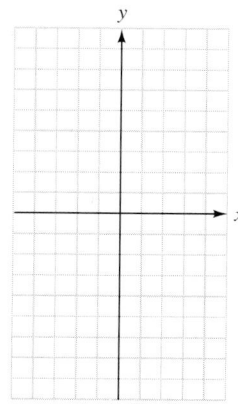

Answer: The graph has the same shape but is 2 units to the right.

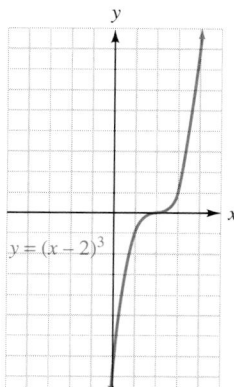

Accent on Technology: **Using a graphing calculator to graph an equation**

Courtesy of Texas Instruments

FIGURE 9-16

So far, we have graphed equations by making tables of solutions and plotting points. The task of graphing is made much easier when we use a graphing calculator. The instructions in this discussion will be general in nature. For specific details about your calculator, please consult your owner's manual.

The viewing window: All graphing calculators have a viewing **window,** used to display graphs. The **standard window** has settings of

  Xmin $= -10$, Xmax $= 10$, Ymin $= -10$, and Ymax $= 10$

which indicate that the minimum x- and y-coordinates used in the graph will be -10, and that the maximum x- and y-coordinates will be 10.

Graphing an equation: To graph the equation $y = x - 1$ using a graphing calculator, we press the $\boxed{Y =}$ key and enter the right-hand side of the equation after the symbol Y_1. The display will show the equation

  $Y_1 = x - 1$

Then we press the $\boxed{\text{GRAPH}}$ key to produce the graph shown in Figure 9-17.

Next, we will graph the equation $y = |x - 4|$. Since absolute values are always nonnegative, the minimum y-value is zero. To obtain a reasonable viewing window, we set the Ymin value slightly lower, at Ymin $= -3$. We set Ymax to be 10 units greater than Ymin, at Ymax = 7. The minimum value of y occurs when $x = 4$. To center the graph in the viewing window, we set the Xmin and Xmax values 5 units to the left and right of 4. Therefore, Xmin = −1 and Xmax = 9.

After entering the right-hand side of the equation, we obtain the graph shown in Figure 9-18. Consult your owner's manual to learn how to enter an absolute value.

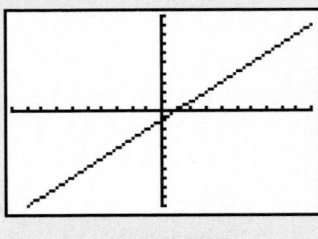

FIGURE 9-17   FIGURE 9-18

Changing the viewing window: The choice of viewing windows is extremely important when graphing equations. To show this, let's graph $y = x^2 - 25$ with x-values from −1 to 6 and y-values from −5 to 5.

To graph this equation, we set the x and y window values and enter the right-hand side of the equation. The display will show

$$Y_1 = x^2 - 25$$

Then we press the $\boxed{\text{GRAPH}}$ key to produce the graph shown in Figure 9-19(a). Although the graph appears to be a straight line, it is not. Actually, we are seeing only part of a parabola. If we pick a viewing window with x-values of −6 to 6 and y-values of −30 to 2, as in Figure 9-19(b), we can see that the graph is a parabola.

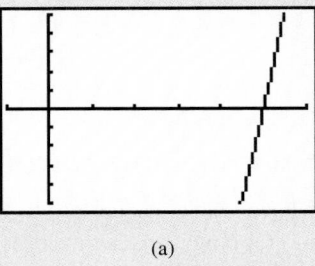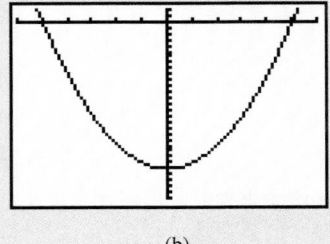

(a)   (b)

FIGURE 9-19

STUDY SET *Use a graphing calculator to graph each equation. Use a viewing window of x = −5 to 5 and y = −5 to 5.*

1. $y = 2.1x - 1.1$

2. $y = 1.12x^2 - 1$

3. $y = |x + 0.7|$

4. $y = 0.1x^3 + 1$

Graph each equation in a viewing window of x = −4 to 4 and y = −4 to 4. Each graph is not what it first appears to be. Pick a better viewing window and find a better representation of the true graph.

5. $y = -x^3 - 8.2$

6. $y = -|x - 4.01|$

7. $y = x^2 + 5.9$

8. $y = -x + 7.95$

Using different variables

We will often encounter equations with variables other than x and y. When we make tables of solutions and graph these equations, we must know which is the independent variable (the input values) and which is the dependent variable (the output values). The independent variable is usually associated with the horizontal axis of the coordinate system, and the dependent variable is usually associated with the vertical axis.

EXAMPLE 7   *Speed limit.*   In some states, the maximum speed limit on a U.S. interstate highway is 75 mph. The distance covered by a vehicle traveling at 75 mph depends on the time the vehicle travels at that speed. This relationship is described by the equation $d = 75t$, where d represents the distance (in miles) and t represents the time (in hours). Graph the equation.

FIGURE 9-20

Solution   Since d depends on t in the equation $d = 50t$, t is the independent variable (the input) and d is the dependent variable (the output). Therefore, we choose values for t and find the corresponding values of d. Since t represents the time spent traveling at 75 mph, we choose no negative values for t.

If $t = 0$, we have

$d = 75t$ The original equation.

$d = 75(\mathbf{0})$ Substitute the input 0 for t.

$d = 0$ Do the multiplication.

The pair $t = 0$ and $d = 0$, or (0, 0), is a solution. This ordered pair and others that satisfy the equation are listed in the table of solutions shown in Figure 9-21(a). If we plot the ordered pairs and draw a line through them, we obtain the graph shown in Figure 9-21(b). From the graph, we see (as expected) that the distance covered steadily increases as the traveling time increases.

$$d = 75t$$

d	t	$(d,\ t)$
0	0	(0, 0)
1	75	(1, 75)
2	150	(2, 150)
3	225	(3, 225)
4	300	(4, 300)
5	375	(5, 375)

↑
Adjust the scale on the vertical axis to fit the data.

(a)

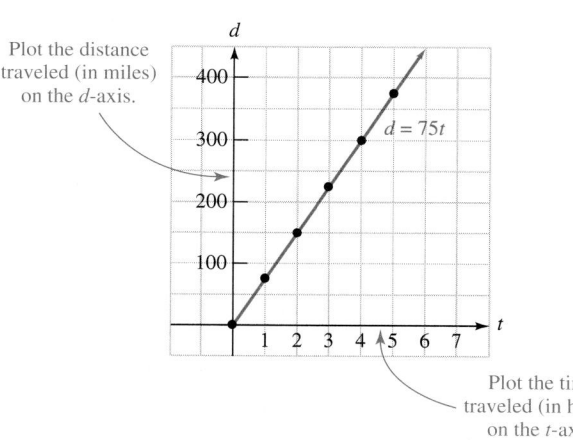

(b)

FIGURE 9-21

STUDY SET   Section 9.2

VOCABULARY   *Fill in the blanks.*

1. The equation $y = x + 1$ is an equation in _____ variables.

2. An ordered pair is a _____ of an equation if the numbers in the ordered pair satisfy the equation.

3. In equations containing the variables x and y, x is called the _____ variable and y is called the _____ variable.

4. When constructing a _____ of solutions, the values of x are the _____ values and the values of y are the _____ values.

CONCEPTS

5. Consider the equation $y = -2x + 6$.
   a. How many variables does the equation contain?
   b. Does the ordered pair $(4, -2)$ satisfy the equation?
   c. Is $x = -3$ and $y = 12$ a solution?
   d. How many solutions does this equation have?

6. How many variables does the equation $x + 2 = 6$ have? How many solutions does it have? Graph the solution(s).

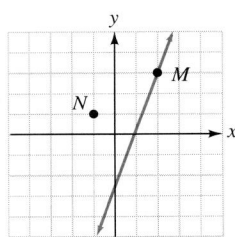

7. To graph an equation, five solutions were found, they were plotted (in black), and a straight line was drawn through them, as shown in Illustration 1. From the graph, determine three other solutions of the equation.

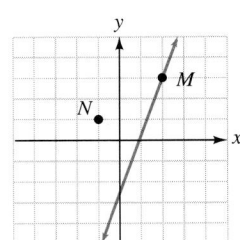

ILLUSTRATION 1

8. Consider the graph of an equation shown in Illustration 2.

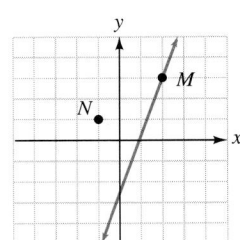

ILLUSTRATION 2

   a. If the coordinates of point M are substituted into the equation, is the result a true or false statement?
   b. If the coordinates of point N are substituted into the equation, is the result a true or false statement?

9. Complete the table of solutions.

$$y = x^3$$

x (inputs)	y (outputs)
0	
−1	
−2	
1	
2	

10. What is wrong with the graph of $y = x - 3$ shown in Illustration 3?

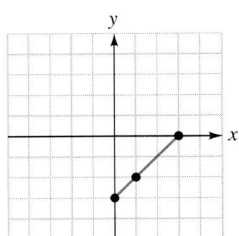

ILLUSTRATION 3

11. To graph $y = -x + 1$, a student constructed a table of solutions and plotted the ordered pairs as shown in Illustration 4. Instead of drawing a crooked line through the points, what should he have done?

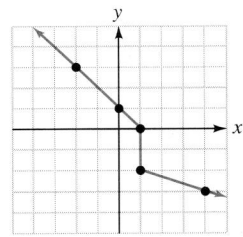

ILLUSTRATION 4

12. To graph $y = x^2 - 4$, a table of solutions is constructed and a graph is drawn, as shown in Illustration 5. Explain the error made here.

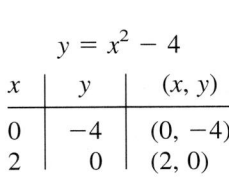

$y = x^2 - 4$		
x	y	(x, y)
0	−4	$(0, -4)$
2	0	$(2, 0)$

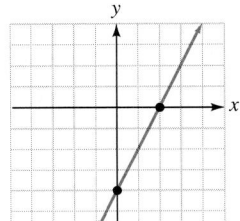

ILLUSTRATION 5

13. Explain the error with the graph of $y = x^2$ shown in Illustration 6.

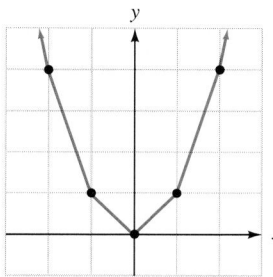

ILLUSTRATION 6

14. Several solutions of an equation are listed in the table of solutions. When graphing them, with what variable should the horizontal and vertical axes of the graph be labeled?

t	s	(t, s)
0	4	(0, 4)
1	5	(1, 5)
2	10	(2, 10)

NOTATION *Complete each solution.*

15. Verify that $(-2, 6)$ satisfies $y = -x + 4$.

$$y = -x + 4$$
$$\boxed{} \stackrel{?}{=} -(\boxed{}) + 4$$
$$6 \stackrel{?}{=} \boxed{} + 4$$
$$6 = 6$$

16. For the equation $y = |x - 2|$, if $x = -3$, find y.

$$y = |x - 2|$$
$$y = |\boxed{} - 2|$$
$$y = |\boxed{}|$$
$$y = 5$$

PRACTICE *Tell whether the ordered pair satisfies the equation.*

17. $y = 2x - 4$; $(4, 4)$

18. $y = x^2$; $(8, 48)$

19. $y = |x - 2|$; $(4, -3)$

20. $y = x^3 + 1$; $(-2, -7)$

Complete each table of solutions.

21. $y = x - 3$

x	y
0	
1	
-2	

22. $y = |x - 3|$

x	$|x - 3|$
0	
-1	
3	

23. $y = x^2 - 3$

Input	Output
0	
2	
-2	

24. $y = x + 1$

Input	Output
0	
2	
-1	

Construct a table of solutions and then graph each equation.

25. $y = 2x - 3$

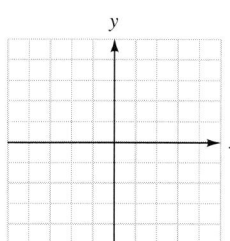

26. $y = 3x + 1$

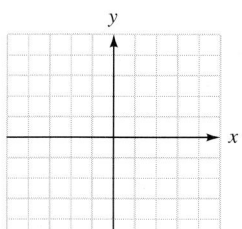

27. $y = -2x + 1$

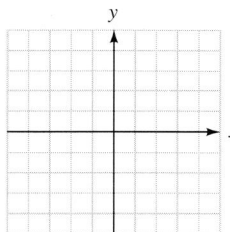

28. $y = -3x + 2$

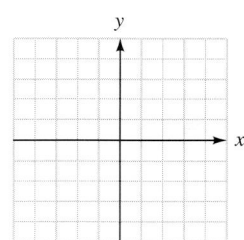

Construct a table of solutions and then graph each equation. Compare it to the graph of $y = x^2$.

29. $y = x^2 + 1$

30. $y = -x^2$

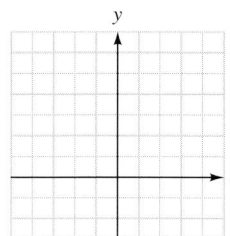

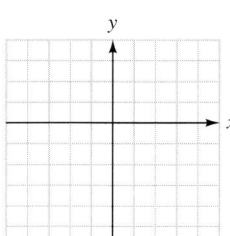

31. $y = (x - 2)^2$

32. $y = (x + 2)^2$

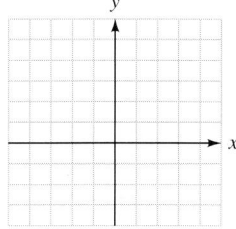

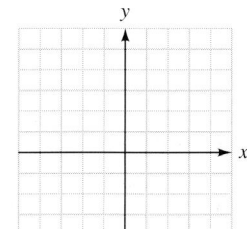

Construct a table of solutions and then graph each equation. Compare it to the graph of $y = |x|$.

33. $y = -|x|$ **34.** $y = |x| - 2$

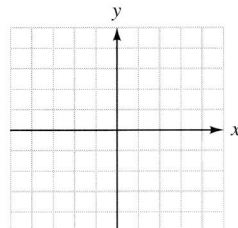 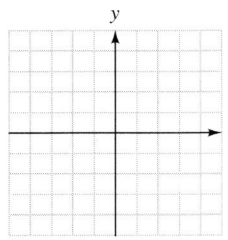

35. $y = |x + 2|$ **36.** $y = |x - 2|$

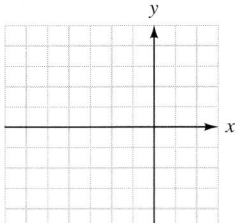 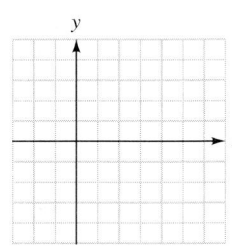

Construct a table of solutions and then graph each equation. Compare it to the graph of $y = x^3$.

37. $y = -x^3$ **38.** $y = x^3 + 2$

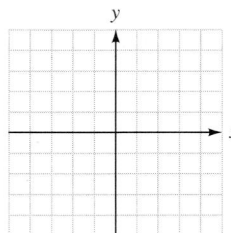 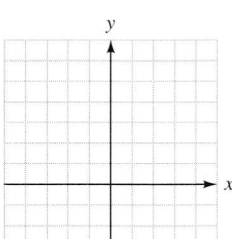

39. $y = x^3 - 2$ **40.** $y = (x + 2)^3$

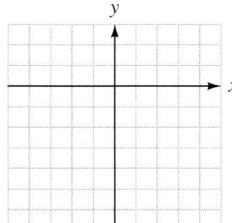 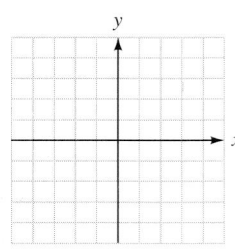

APPLICATIONS

41. **8 BALL**   Illustration 7 shows the path traveled by the 8 ball as it is banked off of a cushion into the right corner pocket. Use the information in the illustration to complete the table.

x	-1	2	5	8
y				

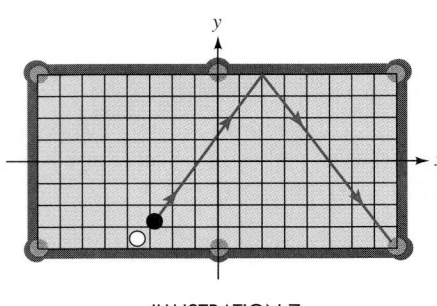

ILLUSTRATION 7

42. **TABLE TENNIS**   Illustration 8 shows the path traveled by a Ping-Pong ball as it bounces off the table. Use the information in the illustration to complete the table below.

x	-7	-3	1	3	5
y					

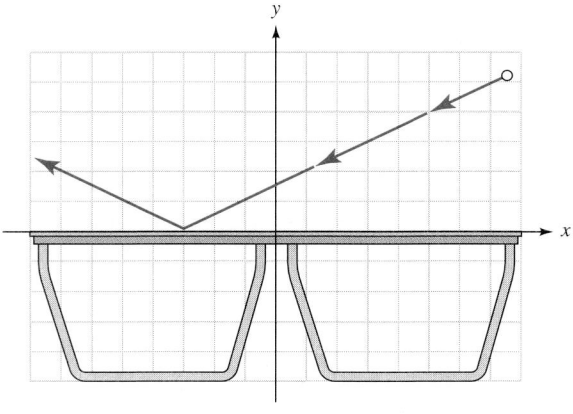

ILLUSTRATION 8

43. **SUSPENSION BRIDGE**   The suspension cables of a bridge hang in the shape of a parabola, as shown in Illustration 9. Use the information in the illustration to complete the table.

x	0	2	4	−2	−4
y					

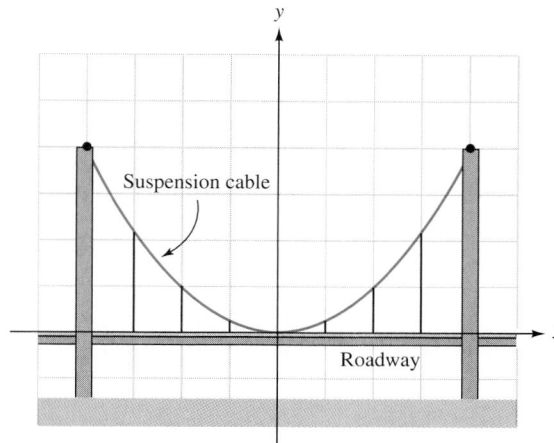

ILLUSTRATION 9

44. FIRE BOAT   A stream of water from a high-pressure hose on a fire boat travels in the shape of a parabola, as shown in Illustration 10. Use the information in the graph to complete the table.

x	1	2	3	4
y				

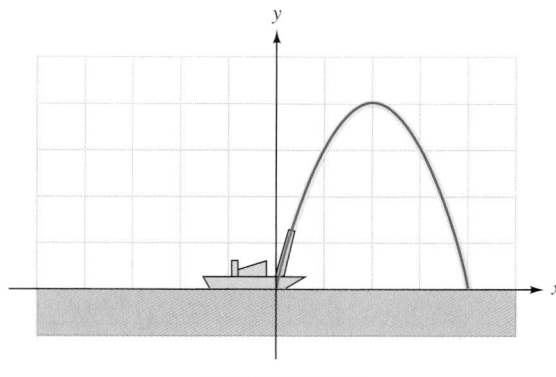

ILLUSTRATION 10

45. MANUFACTURING   The graph in Illustration 11 shows the relationship between the length *l* (in inches) of a machine bolt and the cost *C* (in cents) to manufacture it.
a. What information does the point (2, 8) on the graph give us?
b. How much does it cost to make a 7-inch bolt?
c. What length bolt is the least expensive to make?

d. Describe how the cost changes as the length of the bolt increases.

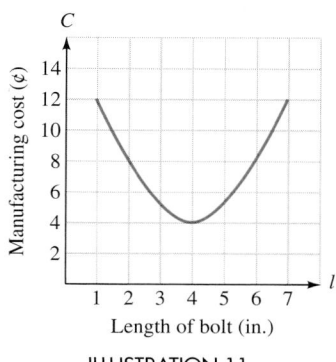

ILLUSTRATION 11

46. SOFTBALL   The graph in Illustration 12 shows the relationship between the distance *d* (in feet) traveled by a batted softball and the height *h* (in feet) it attains.
a. What information does the point (40, 40) on the graph give us?

b. At what distance from home plate does the ball reach its maximum height?
c. Where will the ball land?

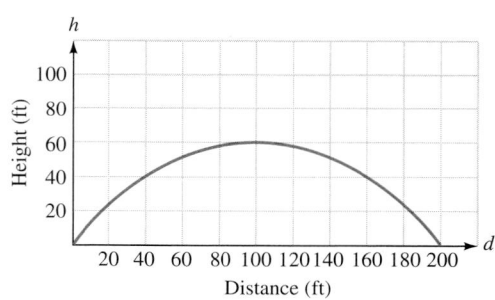

ILLUSTRATION 12

47. MARKET VALUE OF A HOUSE   The graph in Illustration 13 shows the relationship between the market value *v* of a house and the time *t* since it was purchased.

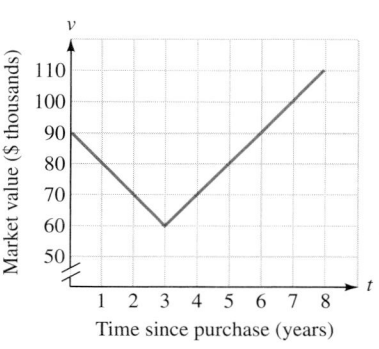

ILLUSTRATION 13

a. What was the purchase price of the house?
b. When did the value of the house reach its lowest point?
c. When did the value of the house begin to surpass the purchase price?
d. Describe how the market value of the house changed over the 8-year period.

48. POLITICAL SURVEY   The graph in Illustration 14 shows the relationship between the percent P of those surveyed who rated their senator's job performance as satisfactory or better and the time t she had been in office.
   a. When did her job performance rating reach a maximum?
   b. When was her job performance rating at or above the 60% mark?
   c. Describe how her job performance rating changed over the 12-month period.

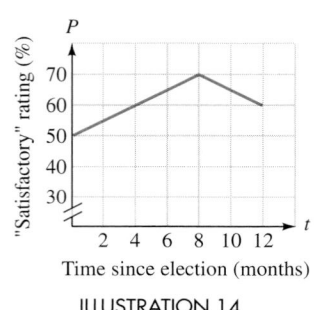

ILLUSTRATION 14

WRITING

49. What is a table of solutions?

50. To graph an equation in two variables, how many solutions of the equation must be found?

51. Give an example of an equation in one variable and an equation in two variables. How do their solutions differ?

52. When we say that $(-2, -6)$ is a solution of $y = x - 4$, what do we mean?

53. On a quiz, students were asked to graph $y = 3x - 1$. One student made the table of solutions on the left. Another student made the one on the right. Which table is incorrect? Or could they both be correct? Explain.

x	y	(x, y)	x	y	(x, y)
0	-1	$(0, -1)$	-2	-7	$(-2, -7)$
2	5	$(2, 5)$	-1	-4	$(-1, -4)$
3	8	$(3, 8)$	1	2	$(1, 2)$
4	11	$(4, 11)$	-3	-10	$(-3, -10)$
5	14	$(5, 14)$	2	5	$(2, 5)$

54. What does it mean when we say that an equation in two variables has infinitely many solutions?

REVIEW

55. Solve $\dfrac{x}{8} = -12$.

56. Combine like terms: $3t - 4T + 5T - 6t$.

57. Is $\dfrac{x + 5}{6}$ an expression or an equation?

58. What formula is used to find the perimeter of a rectangle?

59. What number is 0.5% of 250?

60. Solve $-3x + 5 > -7$.

61. Find $-2.5 - (-2.6)$.

62. Evaluate $(-5)^3$.

9.3 *Graphing Linear Equations*

In this section, you will learn about

- Linear equations • Solutions of linear equations • Graphing linear equations
- The intercept method • Graphing horizontal and vertical lines
- An application of linear equations

INTRODUCTION. In Section 9.2, we graphed the equations shown in Figure 9-22. Because the graph of the equation $y = x - 1$ is a line, we call it a *linear equation*. Since the graphs of $y = x^2$, $y = |x|$, and $y = x^3$ are *not* lines, they are *nonlinear equations*.

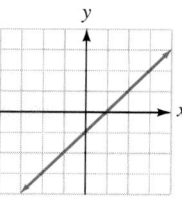

$y = x - 1$
Linear equation

(a)

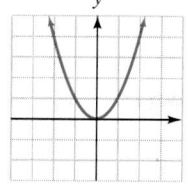

$y = x^2$
Nonlinear equation

(b)

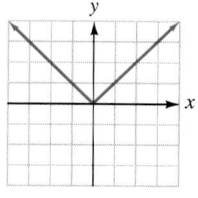

$y = |x|$
Nonlinear equation

(c)

$y = x^3$
Nonlinear equation

(d)

FIGURE 9-22

In this section, we will discuss how to graph linear equations and show how to use their graphs to solve problems.

Linear equations

Any equation, such as $y = x - 1$, whose graph is a straight line is called a **linear equation in x and y.** Some other examples of linear equations are

$$y = \frac{1}{2}x + 2, \qquad 3x - 2y = 8, \qquad 5y - x + 2 = 0, \qquad y = 4, \quad \text{and} \quad x = -3$$

A linear equation in x and y is any equation that can be written in a special form, called **general** (or **standard**) form.

General form of a linear equation	If A, B, and C represent real numbers, the equation $Ax + By = C$ (A and B are not both zero) is called the **general form** (or **standard form**) of the equation of a line.

Whenever possible, we will write the general form $Ax + By = C$ so that A, B, and C are integers and $A \geq 0$. Note that in a linear equation in x and y, the exponents on x and y are 1.

EXAMPLE 1 *Identifying linear equations.* Which of the following equations are linear equations? **a.** $3x = 1 - 2y$ **b.** $y = x^3 + 1$ **c.** $y = -\frac{1}{2}x$

Solution

a. Since the equation $3x = 1 - 2y$ can be written in $Ax + By = C$ form, it is a linear equation.

$$3x = 1 - 2y \qquad \text{The original equation.}$$
$$3x + 2y = 1 - 2y + 2y \qquad \text{Add } 2y \text{ to both sides.}$$
$$3x + 2y = 1 \qquad \text{Simplify the right-hand side: } -2y + 2y = 0.$$

Here $A = 3$, $B = 2$, and $C = 1$.

b. Since the exponent on x in $y = x^3 + 1$ is 3, the equation is a nonlinear equation.

c. Since the equation $y = -\frac{1}{2}x$ can be written in $Ax + By = C$ form, it is a linear equation.

$$y = -\frac{1}{2}x \qquad \text{The original equation.}$$

$$-2(y) = -2\left(-\frac{1}{2}x\right) \qquad \text{Multiply both sides by } -2 \text{ so that the coefficient of } x \text{ will be 1.}$$

Self Check

Which of the following are linear equations and which are nonlinear?

a. $y = |x|$

b. $-x = 6 - y$

c. $y = x$

$$-2y = x \qquad \text{Simplify the right-hand side: } -2\left(-\tfrac{1}{2}\right) = 1.$$
$$0 = x + 2y \qquad \text{Add } 2y \text{ to both sides.}$$
$$x + 2y = 0 \qquad \text{Write the equation in general form.}$$

Here $A = 1$, $B = 2$, and $C = 0$.

Answers
a. nonlinear, **b.** linear,
c. linear ▪

Solutions of linear equations

To find solutions of linear equations, we substitute arbitrary values for one variable and solve for the other.

EXAMPLE 2 *Finding solutions of linear equations.* Complete the table of solutions for $3x + 2y = 5$.

x	y	$(x,\ y)$
7		(7,)
	4	(, 4)

Self Check
Complete the table of solutions for $3x + 2y = 5$.

x	y	$(x,\ y)$
	-2	(, -2)
5		(5,)

Solution
In the first row, we are given an x-value of 7. To find the corresponding y-value, we substitute 7 for x and solve for y.

$$3x + 2y = 5 \qquad \text{The original equation.}$$
$$3(7) + 2y = 5 \qquad \text{Substitute 7 for } x.$$
$$21 + 2y = 5 \qquad \text{Do the multiplication: } 3(7) = 21.$$
$$2y = -16 \qquad \text{Subtract 21 from both sides: } 5 - 21 = -16.$$
$$y = -8 \qquad \text{Divide both sides by 2.}$$

A solution of $3x + 2y = 5$ is $(7, -8)$.

In the second row, we are given a y-value of 4. To find the corresponding x-value, we substitute 4 for y and solve for x.

$$3x + 2y = 5 \qquad \text{The original equation.}$$
$$3x + 2(4) = 5 \qquad \text{Substitute 4 for } y.$$
$$3x + 8 = 5 \qquad \text{Do the multiplication: } 2(4) = 8.$$
$$3x = -3 \qquad \text{Subtract 8 from both sides: } 5 - 8 = -3.$$
$$x = -1 \qquad \text{Divide both sides by 3.}$$

Another solution is $(-1, 4)$. The completed table is as follows:

x	y	$(x,\ y)$
7	-8	$(7, -8)$
-1	4	$(-1, 4)$

Answer:

x	y	$(x,\ y)$
3	-2	$(3, -2)$
5	-5	$(5, -5)$

▪

Graphing linear equations

Since two points determine a line, only two points are needed to graph a linear equation. However, we will often plot a third point as a check. If the three points do not lie on a straight line, then at least one of them is in error.

Graphing linear equations

1. Find three pairs (x, y) that satisfy the equation by picking arbitrary numbers for x and finding the corresponding values of y.

2. Plot each resulting pair (x, y) on a rectangular coordinate system. If the three points do not lie on a straight line, check your computations.

3. Draw the straight line passing through the points.

EXAMPLE 3 *Graphing linear equations.* Graph $y = -3x$.

Solution

To find three ordered pairs that satisfy the equation, we begin by choosing three *x*-values: $-2, 0,$ and 2.

If $x = -2$	**If $x = 0$**	**If $x = 2$**
$y = -3x$	$y = -3x$	$y = -3x$
$y = -3(-2)$	$y = -3(0)$	$y = -3(2)$
$y = 6$	$y = 0$	$y = -6$

We enter the results in a table of solutions, plot the points, and draw a straight line through the points. The graph appears in Figure 9-23. Check this work with a graphing calculator.

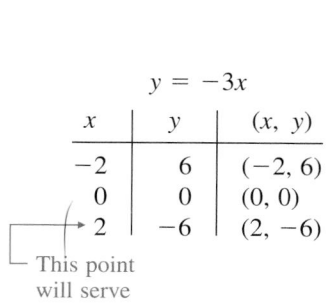

$y = -3x$

x	y	$(x,\ y)$
-2	6	$(-2, 6)$
0	0	$(0, 0)$
2	-6	$(2, -6)$

This point will serve as a check.

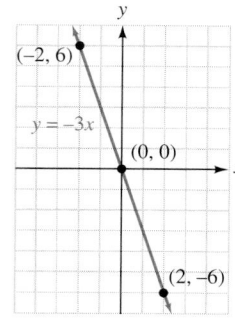

FIGURE 9-23

When graphing linear equations, it is often easier to find solutions of the equation if it is first solved for *y*.

EXAMPLE 4 *Solving for y.* Graph $2y = 4 - x$.

Solution

To solve for *y*, we undo the multiplication of 2 by dividing both sides by 2.

$$2y = 4 - x$$

$$\frac{2y}{2} = \frac{4}{2} - \frac{x}{2}$$ On the right-hand side, dividing each term by 2 is equivalent to dividing the entire side by 2: $\frac{4-x}{2} = \frac{4}{2} - \frac{x}{2}$.

$$y = 2 - \frac{x}{2}$$ Simplify: $\frac{4}{2} = 2$.

Since each value of *x* will be divided by 2, we will choose values of *x* that are divisible by 2. Three such choices are $-4, 0,$ and 4. If $x = -4$, we have

$$y = 2 - \frac{x}{2}$$

$$y = 2 - \frac{-4}{2}$$ Substitute -4 for *x*.

$$y = 2 - (-2)$$ Divide: $\frac{-4}{2} = -2$.

$$y = 4$$ Do the subtraction.

A solution is $(-4, 4)$. This pair and two others satisfying the equation are shown in the table in Figure 9-24. If we plot the points and draw a straight line through them, we will obtain the graph shown in the figure. Check this work with a graphing calculator.

Self Check

Graph $y = -3x + 2$ and compare the result to the graph of $y = -3x$. What do you notice?

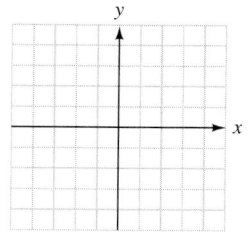

Answer: It is a line 2 units above the graph of $y = -3x$.

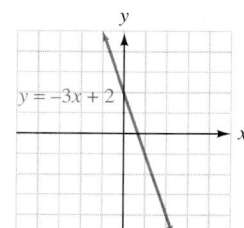

Self Check

Solve $3y = 3 + x$ for *y*. Then graph the equation.

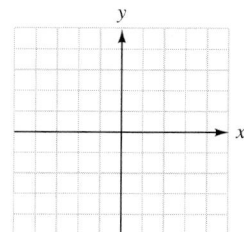

$$y = 2 - \frac{x}{2}$$

x	y	(x, y)
-4	4	$(-4, 4)$
0	2	$(0, 2)$
4	0	$(4, 0)$

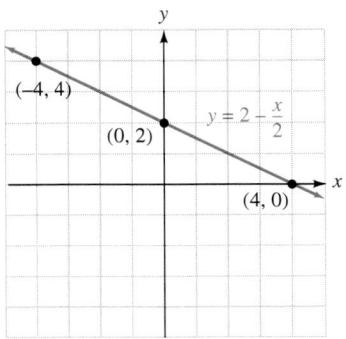

FIGURE 9-24

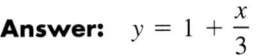

Answer: $y = 1 + \dfrac{x}{3}$

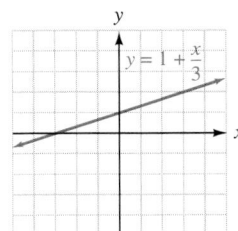

The intercept method

In Figure 9-25, the graph of $3x + 4y = 12$ intersects the y-axis at the point $(0, 3)$; we call this point the **y-intercept** of the graph. Since the graph intersects the x-axis at $(4, 0)$, the point $(4, 0)$ is the **x-intercept.**

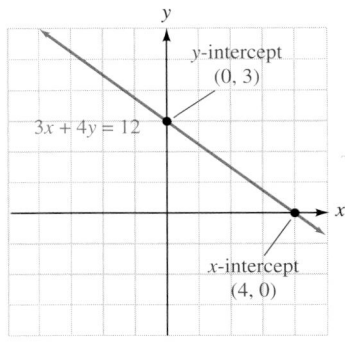

FIGURE 9-25

In general, we have the following definitions.

y- and x-intercepts

> The **y-intercept** of a line is the point $(0, b)$ where the line intersects the y-axis. To find b, substitute 0 for x in the equation of the line and solve for y.
>
> The **x-intercept** of a line is the point $(a, 0)$ where the line intersects the x-axis. To find a, substitute 0 for y in the equation of the line and solve for x.

Plotting the x- and y-intercepts of a graph and drawing a straight line through them is called the **intercept method of graphing a line.** This method is useful when graphing equations written in general form.

EXAMPLE 5   *The intercept method.*   Graph $3x - 2y = 8$.

Solution

To find the x-intercept, we let $y = 0$ and solve for x.

$$3x - 2y = 8$$

$3x - 2(0) = 8$ Substitute 0 for y.

$3x = 8$ Simplify the left-hand side: $2(0) = 0$.

$x = \dfrac{8}{3}$ Divide both sides by 3.

$x = 2\dfrac{2}{3}$ Write $\frac{8}{3}$ as a mixed number.

The x-intercept is $\left(2\frac{2}{3}, 0\right)$. This ordered pair is entered in the table in Figure 9-26. To find the y-intercept, we let $x = 0$ and solve for y.

Self Check

Graph $4x + 3y = 6$ using the intercept method.

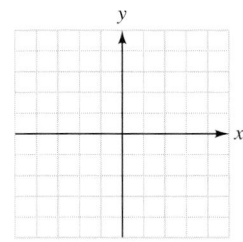

$$3x - 2y = 8$$
$$3(0) - 2y = 8 \qquad \text{Substitute 0 for } x.$$
$$-2y = 8 \qquad \text{Simplify the left-hand side: } 3(0) = 0.$$
$$y = -4 \qquad \text{Divide both sides by } -2.$$

The y-intercept is $(0, -4)$. It is entered in the table below. As a check, we find one more point on the line. If $x = 4$, then $y = 2$. We plot these three points and draw a straight line through them. The graph of $3x - 2y = 8$ is shown in Figure 9-26.

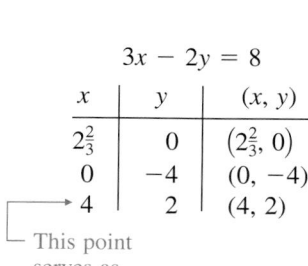

$$3x - 2y = 8$$

x	y	(x, y)
$2\frac{2}{3}$	0	$(2\frac{2}{3}, 0)$
0	-4	$(0, -4)$
4	2	$(4, 2)$

└─ This point
serves as
a check.

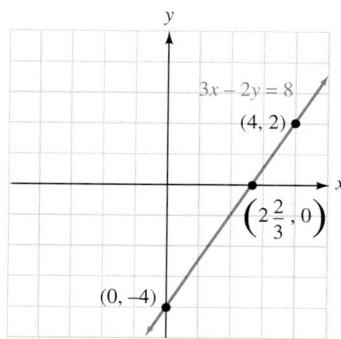

FIGURE 9-26

Answer:

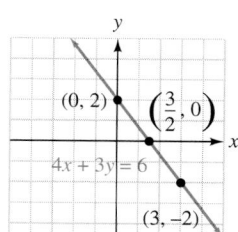

Graphing horizontal and vertical lines

Equations such as $y = 4$ and $x = -3$ are linear equations, because they can be written in the general form $Ax + By = C$.

$$y = 4 \qquad \text{is equivalent to} \qquad 0x + 1y = 4$$
$$x = -3 \qquad \text{is equivalent to} \qquad 1x + 0y = -3$$

We now discuss how to graph these types of linear equations.

EXAMPLE 6   *Graphing horizontal lines.*   Graph $y = 4$.

Solution

We can write the equation in general form as $0x + y = 4$. Since the coefficient of x is 0, the numbers chosen for x have no effect on y. The value of y is always 4. For example, if $x = 2$, we have

$$0x + y = 4 \qquad \text{The original equation written in general form.}$$
$$0(2) + y = 4 \qquad \text{Substitute 2 for } x.$$
$$y = 4 \qquad \text{Simplify the left-hand side: } 0(2) = 0.$$

The table of solutions shown in Figure 9-27 contains three ordered pairs that satisfy the equation $y = 4$. If we plot the points and draw a straight line through them, the result is a horizontal line. The y-intercept is $(0, 4)$, and there is no x-intercept.

Self Check
Graph $y = -2$.

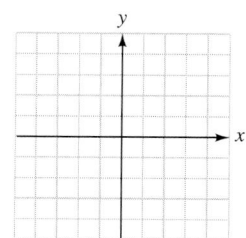

$$y = 4$$

x	y	(x, y)
2	4	$(2, 4)$
-1	4	$(-1, 4)$
-3	4	$(-3, 4)$

Note that each
y-coordinate is 4.

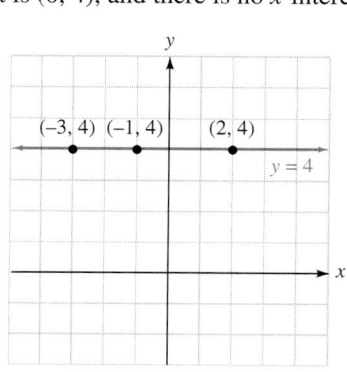

FIGURE 9-27

Answer:

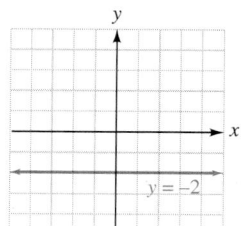

EXAMPLE 7 *Graphing vertical lines.* Graph $x = -3$.

Solution

We can write the equation in general form as $x + 0y = -3$. Since the coefficient of y is 0, the numbers chosen for y have no effect on x. The value of x is always -3. For example, if $y = -2$, we have

$$x + 0y = -3 \quad \text{The original equation written in general form.}$$
$$x + 0(-2) = -3 \quad \text{Substitute } -2 \text{ for } y.$$
$$x = -3 \quad \text{Simplify the left-hand side: } 0(-2) = 0.$$

The table of solutions shown in Figure 9-28 contains three ordered pairs that satisfy the equation $x = -3$. If we plot the points and draw a line through them, the result is a vertical line. The x-intercept is $(-3, 0)$, and there is no y-intercept.

$x = -3$

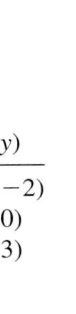

x	y	(x, y)
-3	-2	$(-3, -2)$
-3	0	$(-3, 0)$
-3	3	$(-3, 3)$

↑
Note that each x-coordinate is -3.

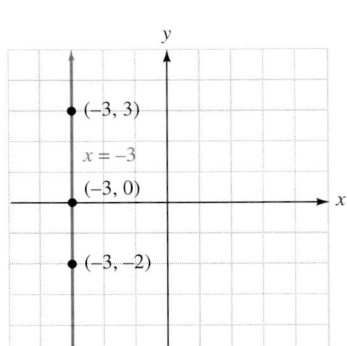

FIGURE 9-28

Self Check

Graph $x = 4$.

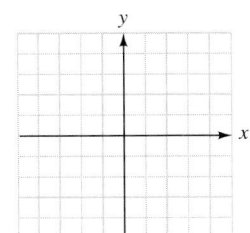

Answer:

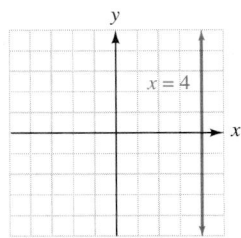

From the results of Examples 6 and 7, we have the following facts.

Equations of horizontal and vertical lines

> The equation $y = b$ represents the horizontal line that intersects the y-axis at $(0, b)$. If $b = 0$, the line is the x-axis.
>
> The equation $x = a$ represents the vertical line that intersects the x-axis at $(a, 0)$. If $a = 0$, the line is the y-axis.

An application of linear equations

EXAMPLE 8 *Birthday parties.* A restaurant offers a party package that includes food, drinks, cake, and party favors for a cost of $25 plus $3 per child. Write a linear equation that will give the cost for a party of any size, and then graph the equation.

Solution We can let c represent the cost of the party. The cost c is the sum of the basic charge of $25 and the cost per child times the number of children attending. If the number of children attending is n, at $3 per child, the total cost for the children is $3n$.

The cost	is	the basic $25 charge	plus	$3	times	the number of children.
c	$=$	25	$+$	3	$\cdot$	n

For the equation $c = 25 + 3n$, the independent variable (input) is n, the number of children. The dependent variable (output) is c, the cost of the party. We will find three points on the graph of the equation by choosing n-values of 0, 5, and 10 and finding the corresponding c-values. The results are recorded in the table.

If $n = 0$	If $n = 5$	If $n = 10$			
$c = 25 + 3(0)$	$c = 25 + 3(5)$	$c = 25 + 3(10)$		$c = 25 + 3n$	
$c = 25$	$c = 25 + 15$	$c = 25 + 30$			
	$c = 40$	$c = 55$			

n	c
0	25
5	40
10	55

Next, we graph the points and draw a line through them (Figure 9-29). We don't draw an arrowhead on the left, because it doesn't make sense to have a negative number of children attend a party. Note that the c-axis is scaled in units of $5 to accommodate costs ranging from $0 to $65.

We can use the graph to determine the cost of a party of any size. For example, to find the cost of a party with 8 children, we locate 8 on the horizontal axis and then move up to find a point on the graph directly above the 8. Since the coordinates of that point are (8, 49), the cost for 8 children would be $49.

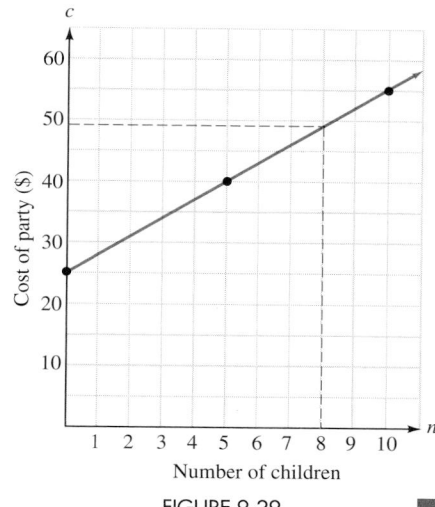

FIGURE 9-29

STUDY SET Section 9.3

VOCABULARY *Fill in the blanks.*

1. An equation whose graph is a line and whose variables are to the first power is called a _____ equation.

2. The equation $Ax + By = C$ is the _____ form of the equation of a line.

3. The _____ of a line is the point $(0, b)$ where the line intersects the y-axis.

4. The _____ of a line is the point $(a, 0)$ where the line intersects the x-axis.

5. Lines parallel to the y-axis are _____ lines.

6. Lines parallel to the x-axis are _____ lines.

CONCEPTS

7. Classify each equation as linear or nonlinear.
  a. $y = x^3$
  b. $2x + 3y = 6$
  c. $y = |x + 2|$
  d. $x = -2$
  e. $y = -x^2$

8. Classify each of the following as the graph of a linear equation or of a nonlinear equation.

  a.
  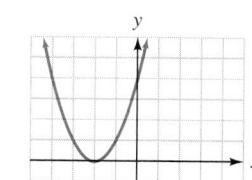

  b.
  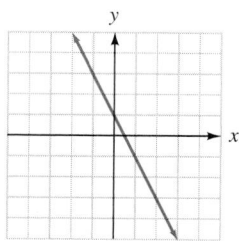

  c.
  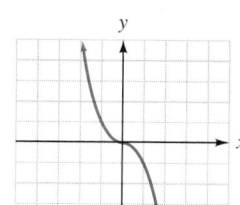

  d.
  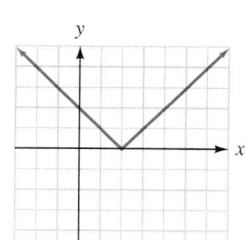

9. Find the power of each variable in the following equations.
  a. $y = 2x - 6$
  b. $y = x^2 - 6$
  c. $y = x^3 + 2$

10. In a linear equation in x and y, what are the exponents on x and y?

Complete each table of solutions.

11. $5y = 2x + 10$

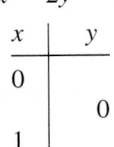

x	y
10	
	0
5	

12. $2x + 4y = 24$

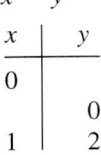

x	y
4	
	7
−4	

13. $x − 2y = 4$

x	y
0	
	0
1	

14. $5x − y = 3$

x	y
0	
	0
1	2

Consider the graph of a linear equation shown in Illustration 1.

15. Why will the coordinates of point A, when substituted into the equation, yield a true statement?

16. Why will the coordinates of point B, when substituted into the equation, yield a false statement?

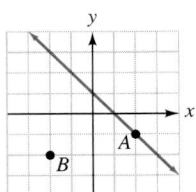

ILLUSTRATION 1

17. A student found three solutions of a linear equation and plotted them as shown in Illustration 2. What conclusion can be made?

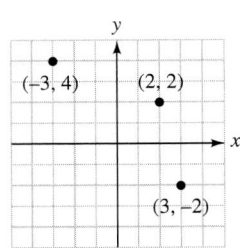

ILLUSTRATION 2

18. How many solutions are there for a linear equation in two variables?

19. Give the x- and y-intercepts of the graph in Illustration 3.

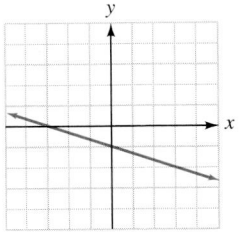

ILLUSTRATION 3

20. On the same coordinate system:
   a. Draw the graph of a line with no x-intercept.
   b. Draw the graph of a line with no y-intercept.
   c. Draw a line with an x-intercept of $(2, 0)$.
   d. Draw a line with a y-intercept of $\left(0, -\frac{5}{2}\right)$.

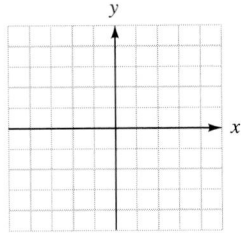

21. Fill in the blanks.
   a. To find the y-intercept of the graph of a linear equation, we let ___ = 0 and solve for ___ .
   b. To find the x-intercept of the graph of a linear equation, we let ___ = 0 and solve for ___ .

22. a. What is another name for the line $x = 0$?
   b. What is another name for the line $y = 0$?

NOTATION

23. Write each equation in general form.
   a. $-4x = -y - 6$
   b. $y = \frac{1}{2}x$
   c. $3 = \frac{x}{3} + y$
   d. $x = 12$

24. Solve each equation for y.
   a. $x + y = 8$
   b. $2x - y = 8$
   c. $3x + \frac{y}{2} = 4$
   d. $y - 2 = 0$

PRACTICE *Find three solutions of the equation, and then graph it.*

25. $y = -x + 2$

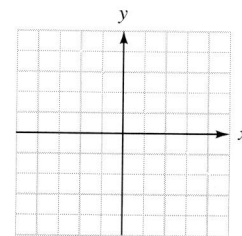

26. $y = -x - 1$

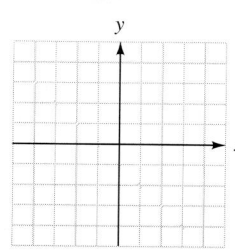

27. $y = 2x + 1$

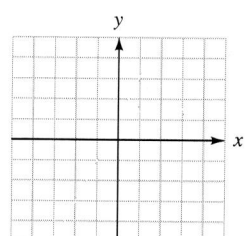

28. $y = 3x - 2$

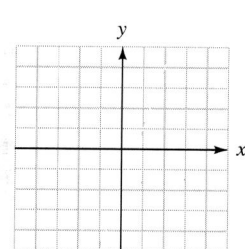

29. $y = x$

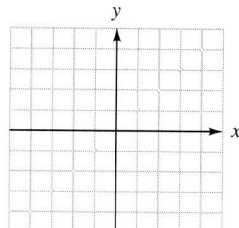

30. $y = 3x$

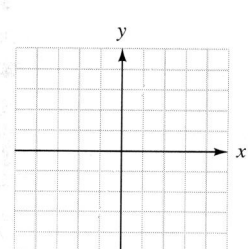

31. $y = -3x$

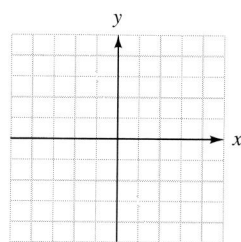

32. $y = -2x$

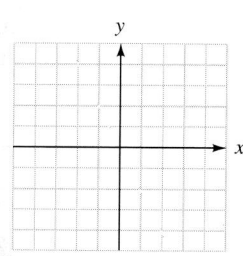

33. $y = \dfrac{x}{3}$

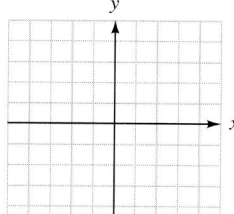

34. $y = -\dfrac{x}{3} - 1$

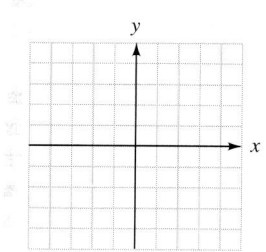

35. $y = -\dfrac{3}{2}x + 2$

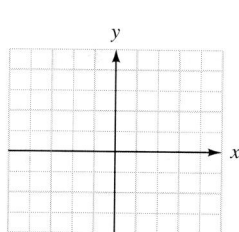

36. $y = \dfrac{2}{3}x - 2$

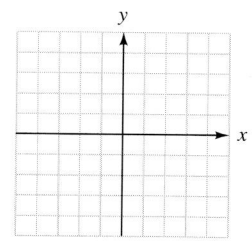

Solve each equation for y, find three solutions of the equation, and then graph it.

37. $2y = 4x - 6$

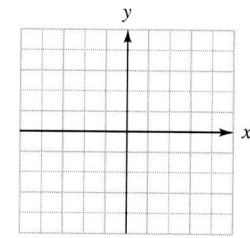

38. $3y = 6x - 3$

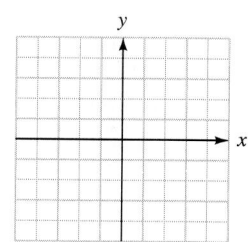

39. $2y = x - 4$

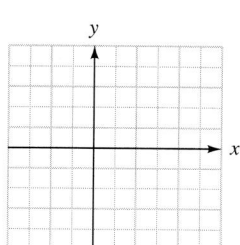

40. $4y = x + 16$

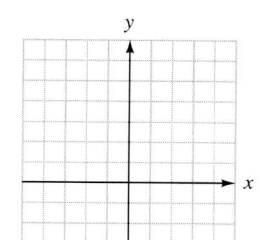

41. $2y + x = -2$

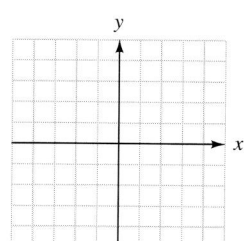

42. $4y + 2x = -8$

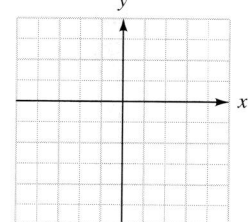

Graph each equation.

43. $y = 4$

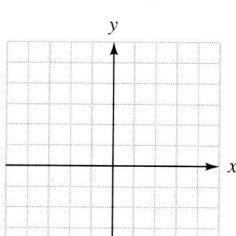

44. $y = -3$

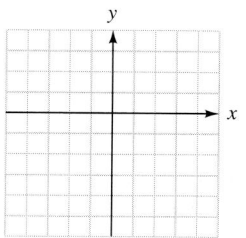

45. $x = -2$

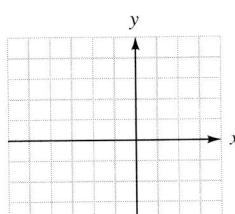

46. $x = 5$

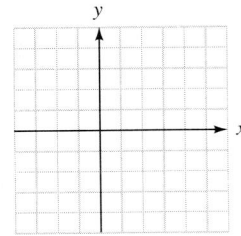

47. $y = -\dfrac{1}{2}$

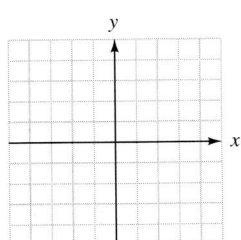

48. $y = \dfrac{5}{2}$

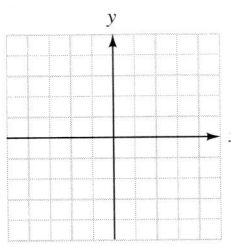

49. $x = \dfrac{4}{3}$

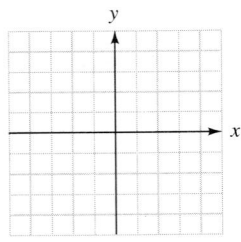

50. $x = -\dfrac{5}{3}$

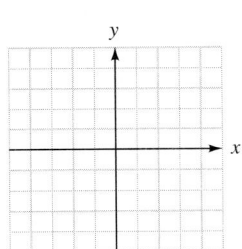

Graph each equation using the intercept method.

51. $2y - 2x = 6$

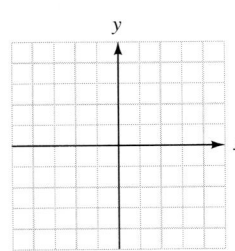

52. $3x - 3y = 9$

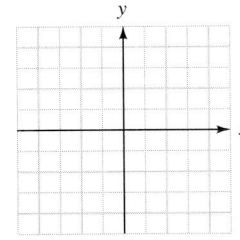

53. $-4y + 9x = -9$

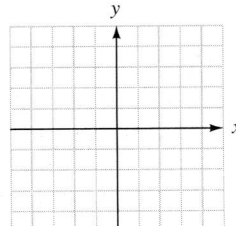

54. $-4y + 5x = -15$

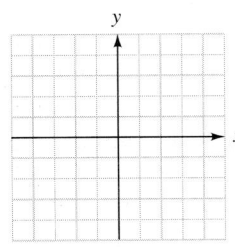

55. $4x + 5y = 20$

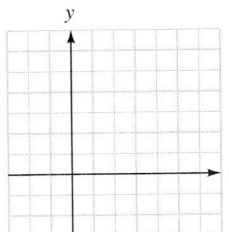

56. $3x + y = -3$

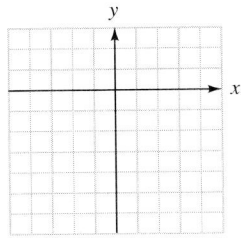

57. $3x + 4y = 12$

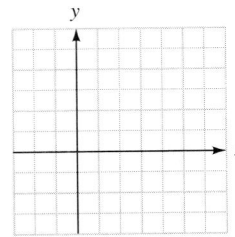

58. $4x - 3y = 12$

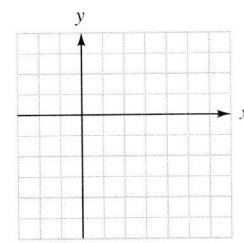

59. $15y + 5x = -15$

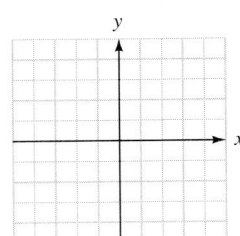

60. $8x + 4y = -24$

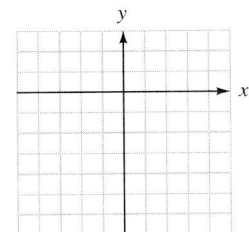

61. $3x + 4y = 8$

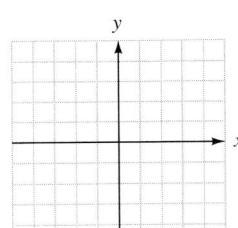

62. $2x + 3y = 9$

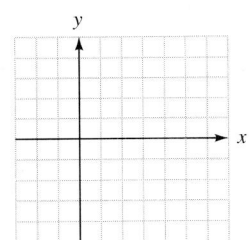

APPLICATIONS

63. EDUCATION COSTS   Each semester, a college charges a services fee of $50 plus $25 for each unit taken by a student.

   a. Write a linear equation that gives the total enrollment cost c for a student taking u units.

   b. Complete the table of solutions and graph the equation. (See Illustration 4 on the next page.)

   c. Use the graph to find the total cost for a student taking 18 units the first semester and 12 units the second semester.

   d. What does the y-intercept of the line tell you?

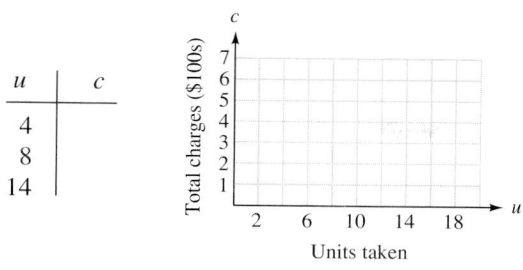

ILLUSTRATION 4

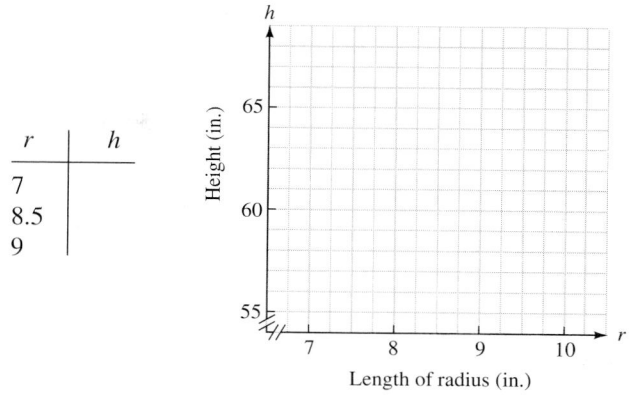

Length of radius (in.)

ILLUSTRATION 6

64. GROUP RATES   To promote the sale of tickets for a cruise to Alaska, a travel agency reduces the regular ticket price of $3,000 by $5 for each individual traveling in the group.

a. Write a linear equation that would find the ticket price t for the cruise if a group of p people travel together.

b. Complete the table of solutions and graph the equation. (See Illustration 5.)

c. As the size of the group increases, what happens to the ticket price?

d. Use the graph to determine the cost of an individual ticket if a group of 25 will be traveling together.

66. RESEARCH EXPERIMENT   A psychology major found that the time t (in seconds) that it took a white rat to complete a maze was related to the number of trials n the rat had been given. The resulting equation was $t = 25 - 0.25n$.

a. Complete the table of solutions in Illustration 7 and then graph the equation.

b. Complete this sentence: From the graph, we see that the more trials the rat had, the . . . .

c. From the graph, estimate the time it will take the rat to complete the maze on its 32nd trial.

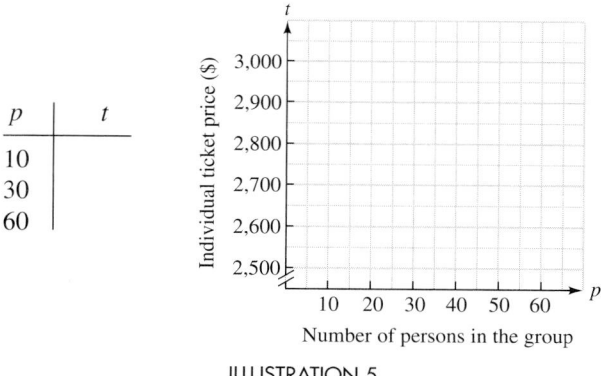

ILLUSTRATION 5

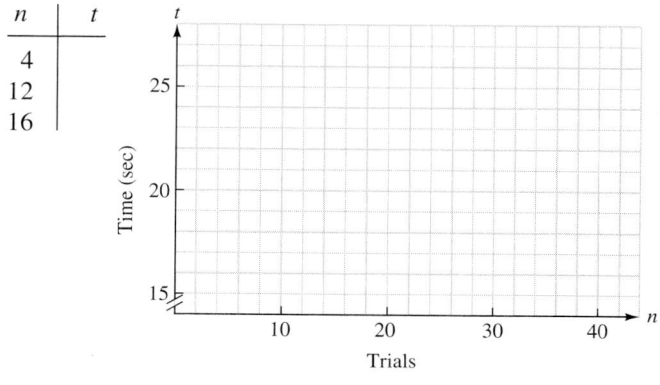

ILLUSTRATION 7

65. PHYSIOLOGY   Physiologists have found that a woman's height h in inches can be approximated using the linear equation $h = 3.9r + 28.9$, where r represents the length of her radius bone in inches.

a. Complete the table of solutions in Illustration 6. Round to the nearest tenth and then graph the equation.

b. Complete this sentence: From the graph, we see that the longer the radius bone, the . . . .

c. From the graph, estimate the height of a woman whose radius bone is 7.5 inches long.

WRITING

67. A linear equation and a graph are two ways of mathematically describing a relationship between two quantities. Which do you think is more informative and why?

68. From geometry, we know that two points determine a line. Explain why it is a good practice when graphing linear equations to find and plot three points instead of just two.

69. How can we tell by looking at an equation if its graph will be a straight line?

70. Can the *x*-intercept and the *y*-intercept of a line be the same point? Explain.

REVIEW

71. Simplify $-(-5-4c)$.

72. List the integers.

73. Solve $\dfrac{x+6}{2}=1$.

74. Evaluate -2^2+2^2.

75. Write a formula that relates profit, revenue, and costs.

76. Find the volume, to the nearest tenth, of a sphere with radius 6 feet.

77. Evaluate $1+2[-3-4(2-8^2)]$.

78. Evaluate $\dfrac{x+y}{x-y}$ for $x=-2$ and $y=-4$.

9.4 *Rate of Change and the Slope of a Line*

In this section, you will learn about

- Rates of change • Slope of a line • The slope formula
- Positive and negative slope • Slopes of horizontal and vertical lines
- Using slope to graph a line

INTRODUCTION. Since our world is one of constant change, we must be able to describe change so that we can plan effectively for the future. In this section, we will show how to describe the amount of change of one quantity in relation to the amount of change of another quantity by finding a *rate of change*.

Rates of change

The line graph in Figure 9-30(a) shows the number of business permits issued each month by a city over a 12-month period. From the shape of the graph, we can see that the number of permits issued *increased* each month.

For situations such as the one graphed in Figure 9-30(a), it is often useful to calculate a rate of increase (called a **rate of change**). We do so by finding the **ratio** of the change in the number of business permits issued each month to the number of months over which that change took place.

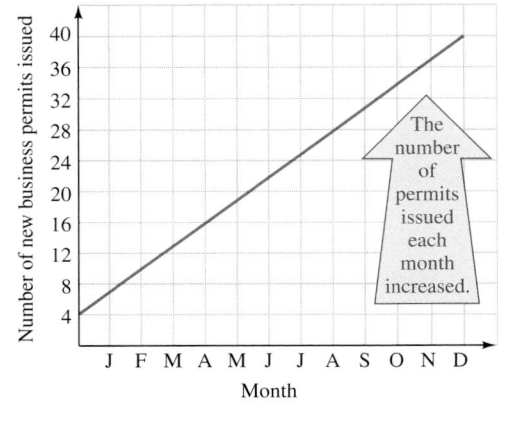

(a)

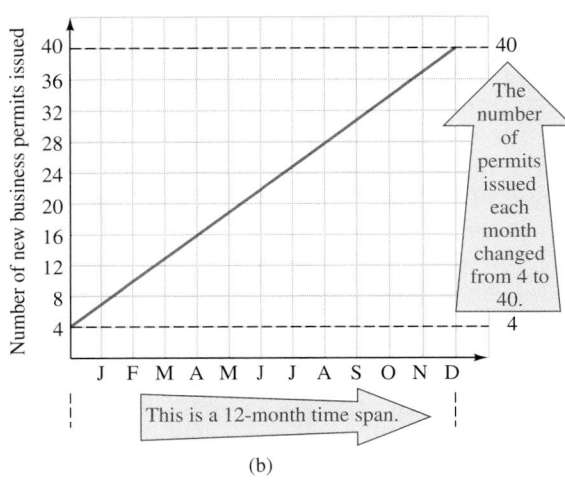

(b)

FIGURE 9-30

Ratios and rates

> A **ratio** is the quotient of two numbers or the quotient of two quantities with the same units. In symbols, if a and b represent two numbers, the ratio of a to b is $\frac{a}{b}$. Ratios that are used to compare quantities with different units are called **rates.**

In Figure 9-30(b), on the previous page, we see that the number of permits issued prior to the month of January was 4. By the end of the year, the number of permits issued during the month of December was 40. This is a change of $40 - 4$, or 36, over a 12-month period. So we have

$$\text{Rate of change} = \frac{\text{change in number of permits issued each month}}{\text{change in time}} \qquad \text{The rate of change is a ratio.}$$

$$= \frac{36 \text{ permits}}{12 \text{ months}}$$

$$= \frac{\overset{1}{\cancel{12}} \cdot 3 \text{ permits}}{\underset{1}{\cancel{12}} \text{ months}} \qquad \text{Factor 36 as } 12 \cdot 3 \text{ and divide out the common factor of 12.}$$

$$= \frac{3 \text{ permits}}{1 \text{ month}}$$

The number of business permits being issued increased at a rate of 3 per month, denoted as 3 permits/month.

EXAMPLE 1   *Finding rate of change.*   The graph in Figure 9-31 shows the number of subscribers to a newspaper. Find the rate of change in the number of subscribers over the first 5-year period. Write the rate in simplest form.

Self Check

Find the rate of change in the number of subscribers over the second 5-year period. Write the rate in simplest form.

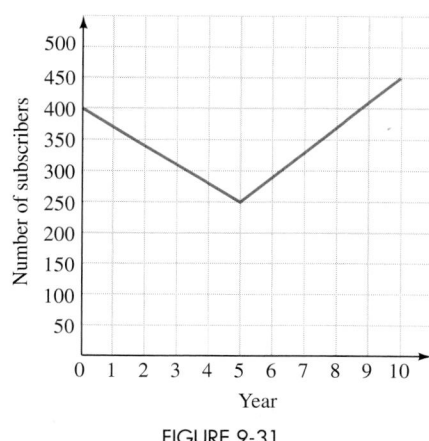

FIGURE 9-31

Solution

We need to write the ratio of the change in the number of subscribers over the change in time.

$$\text{Rate of change} = \frac{\text{change in number of subscribers}}{\text{change in time}} \qquad \text{Set up the ratio.}$$

$$= \frac{(250 - 400) \text{ subscribers}}{5 \text{ years}} \qquad \begin{array}{l}\text{Subtract the earlier number of} \\ \text{subscribers from the later} \\ \text{number of subscribers.}\end{array}$$

$$= \frac{-150 \text{ subscribers}}{5 \text{ years}} \qquad 250 - 400 = -150$$

$$= \frac{-30 \cdot \overset{1}{\cancel{5}} \text{ subscribers}}{\underset{1}{\cancel{5}} \text{ years}}$$

Factor -150 as $-30 \cdot 5$ and divide out the common factor of 5.

$$= \frac{-30 \text{ subscribers}}{1 \text{ year}}$$

The number of subscribers for the first 5 years *decreased* by 30 per year, as indicated by the negative sign in the result. We can write this as -30 subscribers/year.

Answer: 40 subscribers/year ∎

Slope of a line

The **slope** of a nonvertical line is a number that measures the line's steepness. We can calculate the slope by picking two points on the line and writing the ratio of the vertical change (called the **rise**) to the corresponding horizontal change (called the **run**) as we move from one point to the other. As an example, we will find the slope of the line that was used to describe the number of building permits issued and show that it gives the rate of change.

In Figure 9-32 (a modified version of Figure 9-30(a)), the line passes through points $P(0, 4)$ and $Q(12, 40)$. Moving along the line from point P to point Q causes the value of y to change from $y = 4$ to $y = 40$, an increase of $40 - 4 = 36$ units. We say that the *rise* is 36.

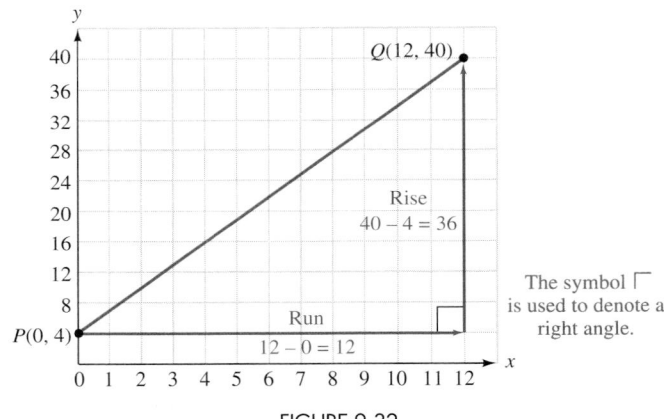

FIGURE 9-32

Moving from point P to point Q, the value of x increases from $x = 0$ to $x = 12$, an increase of $12 - 0 = 12$ units. We say that the *run* is 12. The slope of a line, usually denoted with the letter m, is defined to be the ratio of the change in y to the change in x.

$$m = \frac{\text{change in } y\text{-values}}{\text{change in } x\text{-values}}$$ Slope is a ratio.

$$= \frac{40 - 4}{12 - 0}$$ To find the change in y (the rise), subtract the y-values.
To find the change in x (the run), subtract the x-values.

$$= \frac{36}{12}$$ Do the subtractions.

$$= 3$$ Do the division.

This is the same value we obtained when we found the rate of change of the number of business permits issued over the 12-month period. Therefore, by finding the slope of the line, we found a rate of change.

EXAMPLE 2 *Finding the slope of a line from a graph.* Find the slope of the line shown in Figure 9-33(a)

Self Check

Find the slope of the line shown in Figure 9-33(a) using two points different from those used in the solution of Example 2.

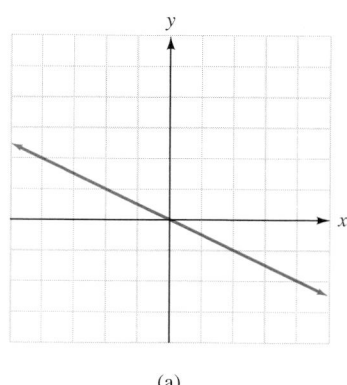

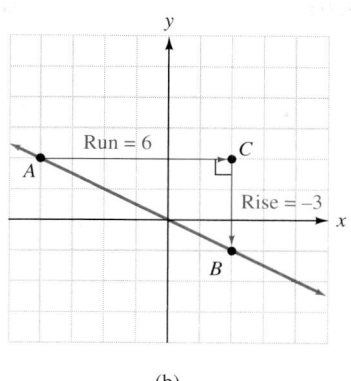

(a)   (b)

FIGURE 9-33

Solution

In Figure 9-33(b), we begin by choosing two points on the line—call them *A* and *B*. Then we draw right triangle *ABC*, having a horizontal leg and a vertical leg. The longest side $\overline{AB}$ of the right triangle is called the **hypotenuse.** As we move from *A* to *B* (shown using blue arrows), we move to the right, a run of 6, and then down, a rise of -3. To find the slope of the line, we write a ratio.

$m = \dfrac{\text{rise}}{\text{run}}$ The slope of a line is the ratio of the rise to the run.

$m = \dfrac{-3}{6}$ From Figure 9-33(b), the rise is -3 and the run is 6.

$m = -\dfrac{1}{2}$ Simplify the fraction.

The slope of the line is $-\frac{1}{2}$

Answer: $-\dfrac{1}{2}$

COMMENT The identical answers from Example 2 and the Self Check illustrate an important fact about slope: The same value for the slope of a line will result no matter which two points on the line are used to determine the rise and the run.

The slope formula

The slope of a line can be described in several ways.

$$\text{Slope} = m = \frac{\text{vertical change}}{\text{horizontal change}} = \frac{\text{rise}}{\text{run}} = \frac{\text{change in } y}{\text{change in } x}$$

Recall that to distinguish between the coordinates of two points—say, points *P* and *Q* (see Figure 9-34)—we often use **subscript notation.**

- Point *P* is denoted as $P(x_1, y_1)$.   Read as "point *P* with coordinates of *x* sub 1 and *y* sub 1."

- Point *Q* is denoted as $Q(x_2, y_2)$.   Read as "point *Q* with coordinates of *x* sub 2 and *y* sub 2."

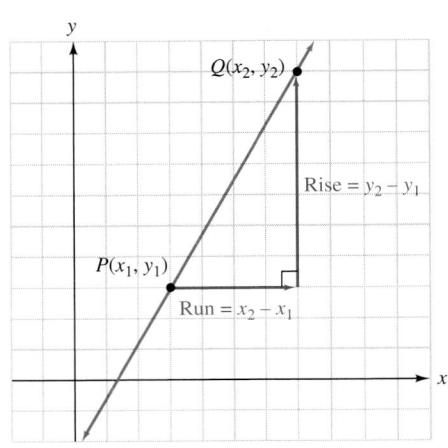

FIGURE 9-34

As a point on the line in Figure 9-34 moves from P to Q, its y-coordinate changes by the amount $y_2 - y_1$ (the rise), while its x-coordinate changes by $x_2 - x_1$ (the run). Since the slope is the ratio $\frac{\text{rise}}{\text{run}}$, we have the following formula for calculating slope.

Slope of a nonvertical line	The **slope** of a nonvertical line passing through points (x_1, y_1) and (x_2, y_2) is $$m = \frac{y_2 - y_1}{x_2 - x_1}$$

EXAMPLE 3   *Using the slope formula.*
Find the slope of line l_1 shown in Figure 9-35.

Solution
To find the slope of l_1, we will use two points on the line whose coordinates are given: $(1, 2)$ and $(5, 5)$. If (x_1, y_1) is $(1, 2)$ and (x_2, y_2) is $(5, 5)$, then

$$x_1 = 1 \quad \text{and} \quad x_2 = 5$$
$$y_1 = 2 \qquad\qquad y_2 = 5$$

To find the slope of line l_1, we substitute these values into the formula for slope and simplify.

$m = \dfrac{y_2 - y_1}{x_2 - x_1}$ The slope formula.

$= \dfrac{5 - 2}{5 - 1}$ Substitute 5 for y_2, 2 for y_1, 5 for x_2, and 1 for x_1.

$= \dfrac{3}{4}$ Do the subtractions.

The slope of l_1 is $\frac{3}{4}$. We would have obtained the same result if we had let $(x_1, y_1) = (5, 5)$ and $(x_2, y_2) = (1, 2)$.

$$m = \frac{y_2 - y_1}{x_2 - x_1} = \frac{2 - 5}{1 - 5} = \frac{-3}{-4} = \frac{3}{4}$$

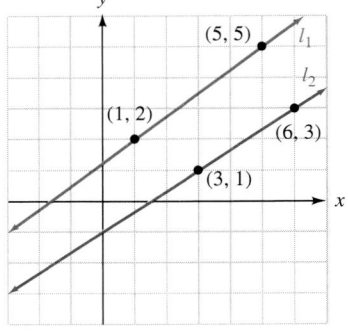

FIGURE 9-35

Self Check
Find the slope of line l_2 shown in Figure 9-35.

Answer: $\dfrac{2}{3}$

COMMENT   When finding the slope of a line, always subtract the *y*-values and the *x*-values in the same order. Otherwise your answer will have the wrong sign:

$$m \neq \frac{y_2 - y_1}{x_1 - x_2} \quad \text{and} \quad m \neq \frac{y_1 - y_2}{x_2 - x_1}$$

EXAMPLE 4   *Using the slope formula.*   Find the slope of the line that passes through $(-2, 4)$ and $(5, -6)$ and draw its graph.

Solution

Since we know the coordinates of two points on the line, we can find its slope. If (x_1, y_1) is $(-2, 4)$ and (x_2, y_2) is $(5, -6)$, then

$$\begin{aligned}x_1 &= -2 \\ y_1 &= 4\end{aligned} \quad \text{and} \quad \begin{aligned}x_2 &= 5 \\ y_2 &= -6\end{aligned}$$

$$m = \frac{y_2 - y_1}{x_2 - x_1} \qquad \text{The slope formula.}$$

$$m = \frac{-6 - 4}{5 - (-2)} \qquad \text{Substitute } -6 \text{ for } y_2, 4 \text{ for } y_1, 5 \text{ for } x_2, \text{ and } -2 \text{ for } x_1.$$

$$m = -\frac{10}{7} \qquad \begin{aligned}&\text{Simplify the numerator: } -6 - 4 = -10. \\ &\text{Simplify the denominator: } 5 - (-2) = 7.\end{aligned}$$

The slope of the line is $-\frac{10}{7}$. Figure 9-36 shows the graph of the line. Note that the line "falls" from left to right—a fact that is indicated by its negative slope.

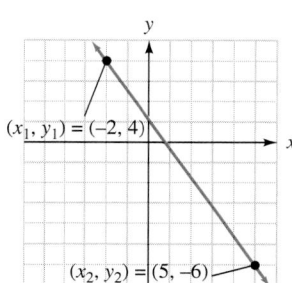

FIGURE 9-36

Self Check

Find the slope of the line that passes through $(-1, -2)$ and $(1, -7)$.

Answer: $-\dfrac{5}{2}$ ∎

Positive and negative slope

In Example 3, the slope of line l_1 was positive $\left(\frac{3}{4}\right)$. In Example 4, the slope of the line was negative $\left(-\frac{10}{7}\right)$. In general, lines that rise from left to right have a positive slope, and lines that fall from left to right have a negative slope, as shown in Figure 9-37.

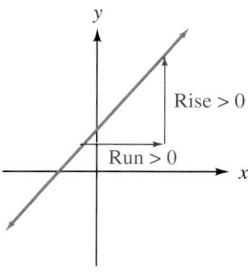

Positive slope

(a)

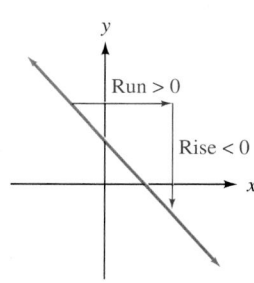

Negative slope

(b)

FIGURE 9-37

Slopes of horizontal and vertical lines

In the next two examples, we will calculate the slope of a horizontal line and show that a vertical line has no defined slope.

EXAMPLE 5   *Slope of a horizontal line.*   Find the slope of the line $y = 3$.

Solution   To find the slope of the line $y = 3$, we need to know two points on the line. In Figure 9-38, we graph the horizontal line $y = 3$ and label two points on the line: $(-2, 3)$ and $(3, 3)$.

If (x_1, y_1) is $(-2, 3)$ and (x_2, y_2) is $(3, 3)$, we have

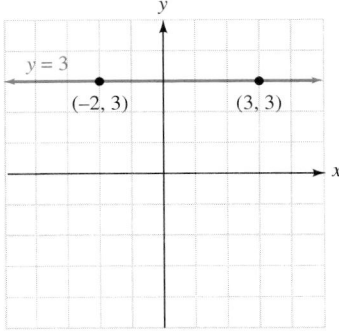

FIGURE 9-38

$$m = \frac{y_2 - y_1}{x_2 - x_1} \qquad \text{The slope formula.}$$

$$m = \frac{3 - 3}{3 - (-2)} \qquad \begin{array}{l}\text{Substitute 3 for } y_2, \text{ 3 for } y_1,\\ \text{3 for } x_2, \text{ and } -2 \text{ for } x_1.\end{array}$$

$$m = \frac{0}{5} \qquad \begin{array}{l}\text{Simplify the numerator and}\\ \text{the denominator.}\end{array}$$

$$m = 0$$

The slope of the line $y = 3$ is 0.   ■

The y-values of any two points on any horizontal line will be the same, and the x-values will be different. Thus, the numerator of

$$\frac{y_2 - y_1}{x_2 - x_1}$$

will always be zero, and the denominator will always be nonzero. Therefore, the slope of a horizontal line is zero.

EXAMPLE 6   *Slope of a vertical line.*   If possible, find the slope of the line $x = -2$.

Solution   To find the slope of the line $x = -2$, we need to know two points on the line. In Figure 9-39, we graph the vertical line $x = -2$ and label two points on the line: $(-2, -1)$ and $(-2, 3)$.

If (x_1, y_1) is $(-2, -1)$ and (x_2, y_2) is $(-2, 3)$, we have

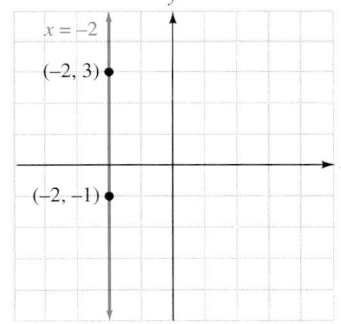

FIGURE 9-39

$$m = \frac{y_2 - y_1}{x_2 - x_1} \qquad \text{The slope formula.}$$

$$m = \frac{3 - (-1)}{-2 - (-2)} \qquad \begin{array}{l}\text{Substitute 3 for } y_2, \text{ } -1 \text{ for } y_1,\\ -2 \text{ for } x_2, \text{ and } -2 \text{ for } x_1.\end{array}$$

$$m = \frac{4}{0} \qquad \begin{array}{l}\text{Simplify the numerator and}\\ \text{the denominator.}\end{array}$$

Since division by zero is undefined, $\frac{4}{0}$ has no meaning. The slope of the line $x = -2$ is undefined.   ■

The y-values of any two points on a vertical line will be different, and the x-values will be the same. Thus, the numerator of

$$\frac{y_2 - y_1}{x_2 - x_1}$$

will always be nonzero, and the denominator will always be zero. Therefore, the slope of a vertical line is undefined.

We now summarize the results from Examples 5 and 6.

Slopes of horizontal and vertical lines

Horizontal lines (lines with equations of the form $y = b$) have a slope of 0. (See Figure 9-40a.)

Vertical lines (lines with equations of the form $x = a$) have undefined slope. (See Figure 9-40b.)

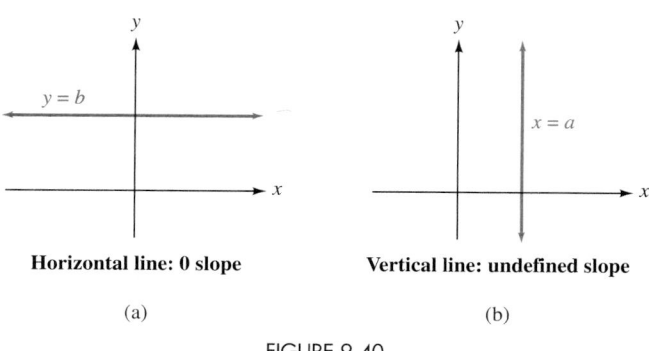

Horizontal line: 0 slope   Vertical line: undefined slope

(a)   (b)

FIGURE 9-40

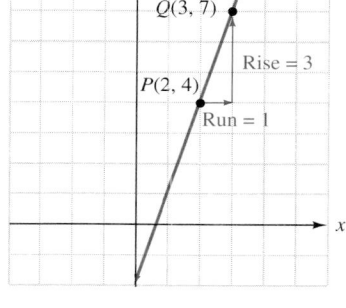

FIGURE 9-41

Using slope to graph a line

We can graph a line whenever we know the coordinates of one point on the line and the slope of the line. For example, to graph the line that passes through $P(2, 4)$ and has a slope of 3, we first plot $P(2, 4)$, as in Figure 9-41. We can express the slope of 3 as a fraction: $3 = \frac{3}{1}$. Therefore, the line *rises* 3 units for every 1 unit it *runs* to the right. We can find a second point on the line by starting at $P(2, 4)$ and moving 1 unit to the right (run) and then 3 units up (rise). This brings us to a point that we will call Q with coordinates $(2 + 1, 4 + 3)$ or $(3, 7)$. The required line must pass through points P and Q.

EXAMPLE 7 *Using slope to graph a line.* Graph the line that passes through the point $(-3, 4)$ with slope $-\frac{2}{5}$.

Solution

We plot the point $(-3, 4)$ as shown in Figure 9-42. Then, after writing the slope $-\frac{2}{5}$ as $\frac{-2}{5}$, we see that the *rise* is -2 and the *run* is 5. From the point $(-3, 4)$, we can find a second point on the line by moving 5 units to the right (run) and then 2 units down (a rise of -2 means to move down 2 units). This brings us to the point with coordinates of $(-3 + 5, 4 - 2) = (2, 2)$. We then draw a line that passes through the two points.

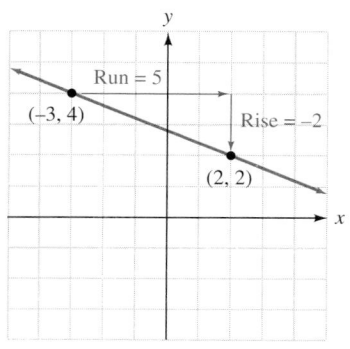

FIGURE 9-42

Self Check

Graph the line that passes through the point $(-4, 2)$ with slope -4.

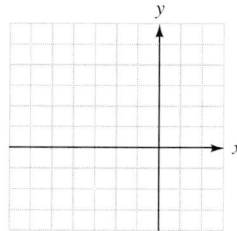

Answer:

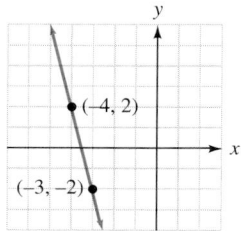

STUDY SET   Section 9.4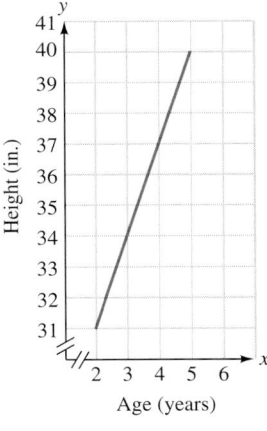

VOCABULARY   *Fill in the blanks.*

1. A _____ is the quotient of two numbers.

2. Ratios used to compare quantities with different units are called _____.

3. The _____ of a line is defined to be the ratio of the change in *y* to the change in *x*.

4. $m = \dfrac{\text{change}}{\text{horizontal change}} = \dfrac{\text{rise}}{} = \dfrac{\text{change in}}{\text{change in}}$

5. The rate of _____ of a linear relationship can be found by finding the slope of the graph of the line.

6. _____ lines have a slope of 0. Vertical lines have _____ slope.

CONCEPTS

7. Which line graphed in Illustration 1 has
    a. a positive slope?
    b. a negative slope?
    c. zero slope?
    d. undefined slope?

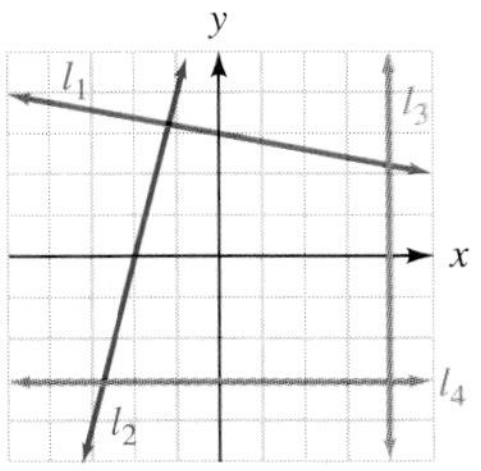

ILLUSTRATION 1

8. For the line graphed in Illustration 2:
    a. Find its slope using points *A* and *B*.
    b. Find its slope using points *B* and *C*.
    c. Find its slope using points *A* and *C*.
    d. What observation is suggested by your answers to parts a, b, and c?

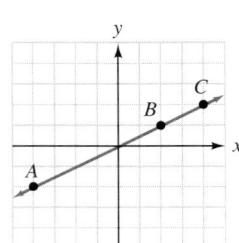

ILLUSTRATION 2

9. Use the information in the table of solutions for a linear equation to determine what the slope of the line would be if it were graphed.

x	*y*
−4	2
5	−7

10. Fill in the blanks.
    a. A line with positive slope _____ from left to right.
    b. A line with negative slope _____ from left to right.

11. GROWTH RATE   Use the graph in Illustration 3 to find the rate of change of a boy's height during the time shown.

ILLUSTRATION 3

12. IRRIGATION   The graph in Illustration 4 shows the number of gallons of water remaining in a reservoir as water is discharged from it to irrigate a field. Find the rate of change in the number of gallons of water for the time the field was being irrigated.

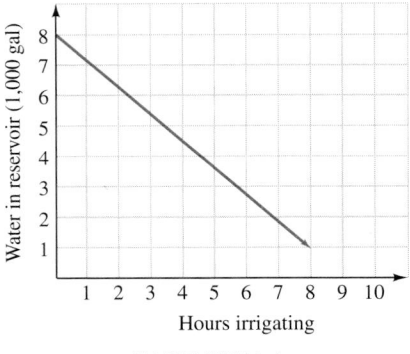

ILLUSTRATION 4

13. DEPRECIATION   The graph in Illustration 5 shows how the value of some sound equipment decreased over the years. Find the rate of change of its value during this time.

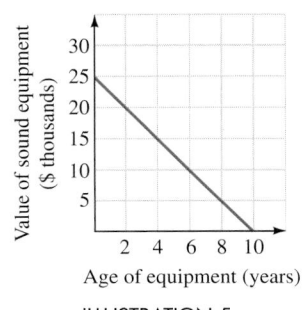

ILLUSTRATION 5

14. WAL-MART   On the graph in Illustration 6, draw a straight line through the points (1991, 34) and (1999, 138). This line approximates Wal-Mart's annual net sales for the years 1991–1999. Give the rate of increase in sales by finding the slope of the line.

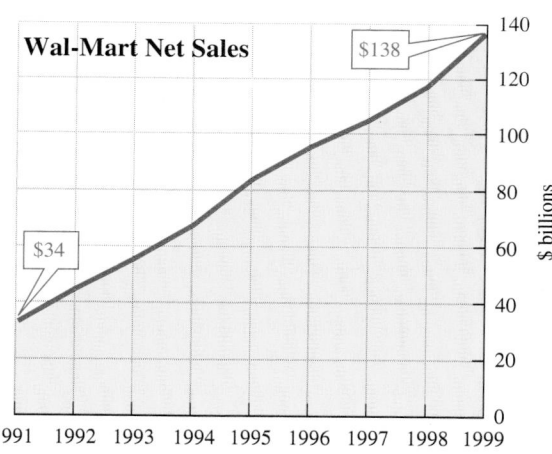

Wal-Mart Net Sales

Based on data from Wal-Mart, *USA TODAY* (November 6, 1998), and Hoover's online

ILLUSTRATION 6

15. THE UNCOLA
  a. From the graph in Illustration 7, estimate the rate of change in the sales of 7-Up for the years 1995–1999. Interpret this result.

  b. From 1998–1999, which noncola had the greatest rate of change in sales?

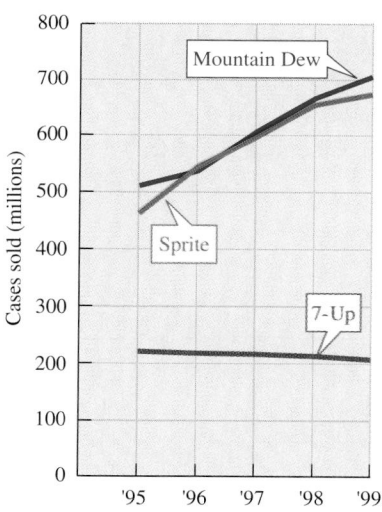

Based on data from *Beverage Digest*

ILLUSTRATION 7

16. COMMERCIAL JETS   Examine the graph in Illustration 8, and consider trips of more than 7,000 miles by a

Boeing 777. Use a rate of change to estimate how the maximum payload decreases as the distance traveled increases. Explain your result in words.

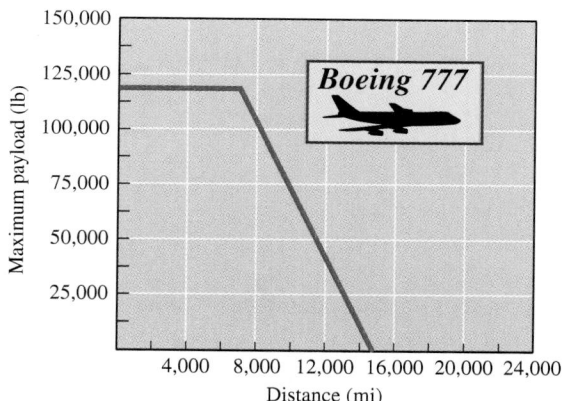

Based on data from Lawrence Livermore National Laboratory and *Los Angeles Times* (October 22, 1998)

ILLUSTRATION 8

NOTATION

17. What is the formula used to find the slope of the line passing through (x_1, y_1) and (x_2, y_2)?

18. Explain the difference between y^2 and y_2.

PRACTICE   *Find the slope of the line passing through the given points, when possible.*

19. (2, 4) and (1, 3)

20. (1, 3) and (2, 5)

21. (3, 4) and (2, 7)

22. (3, 6) and (5, 2)

23. (0, 0) and (4, 5)

24. (4, 3) and (7, 8)

25. (−3, 5) and (−5, 6)

26. (6, −2) and (−3, 2)

27. (−2, −2) and (−12, −8)

28. (−1, −2) and (−10, −5)

29. (5, 7) and (−4, 7)

30. (−1, −12) and (6, −12)

31. (8, −4) and (8, −3)

32. (−2, 8) and (−2, 15)

33. (−6, 0) and (0, −4)

34. $(0, -9)$ and $(-6, 0)$
35. 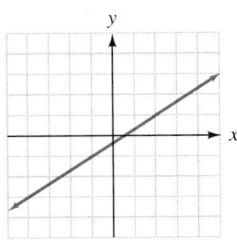▦ $(-2.5, 1.75)$ and $(-0.5, -7.75)$
36. ▦ $(6.4, -7.2)$ and $(-8.8, 4.2)$

Find the slope of each line.

37.

38.

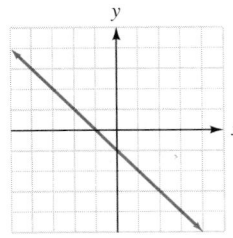

39.

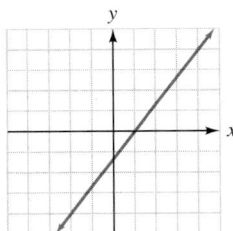

40.

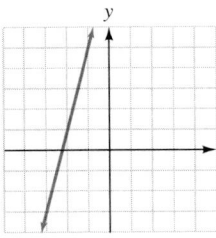

41.

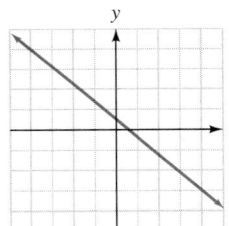

42.

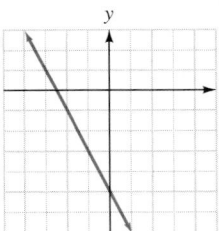

43.

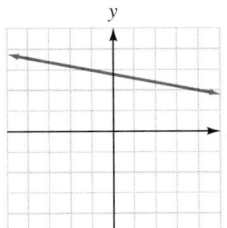

44.

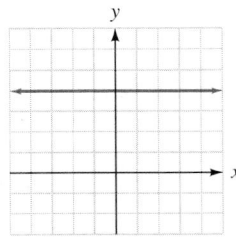

Graph the line that passes through the given point and has the given slope.

45. $(0, 1)$, $m = 2$

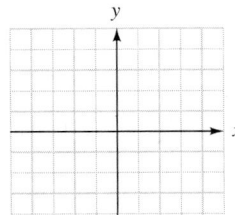

46. $(-4, 1)$, $m = -3$

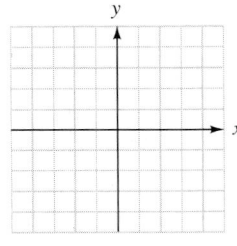

47. $(-3, -3)$, $m = -\dfrac{3}{2}$

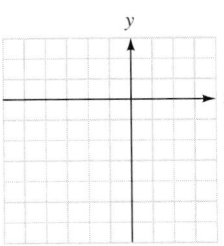

48. $(-2, -1)$, $m = \dfrac{4}{3}$

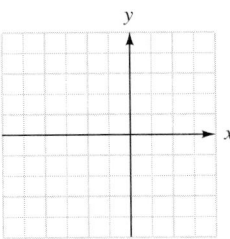

49. $(5, -3)$, $m = \dfrac{3}{4}$

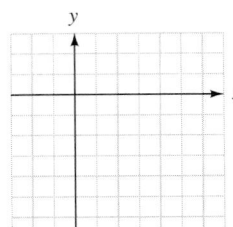

50. $(2, -4)$, $m = \dfrac{2}{3}$

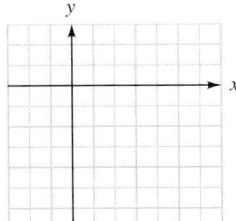

51. $(0, 0)$, $m = -4$

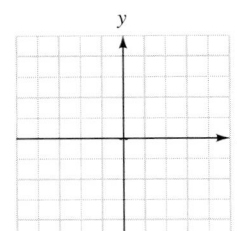

52. $(0, 0)$, $m = 5$

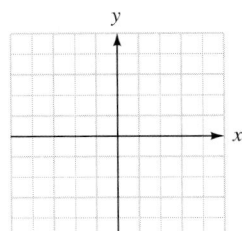

53. $(-5, 1)$, $m = 0$

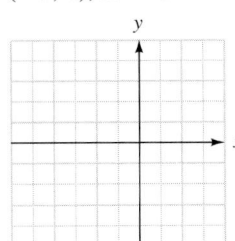

54. $(0, 3)$, undefined slope

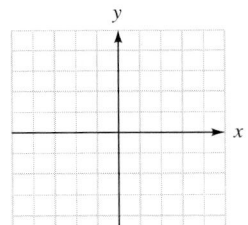

55. $(-1, -4)$, undefined slope

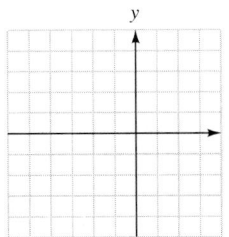

56. $(-3, -2)$, $m = 0$

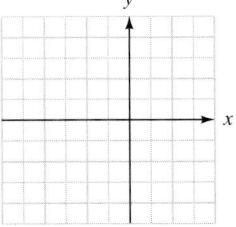

APPLICATIONS

57. POOL DESIGN Find the slope of the bottom of the swimming pool as it drops off from the shallow end to the deep end, as shown in Illustration 9.

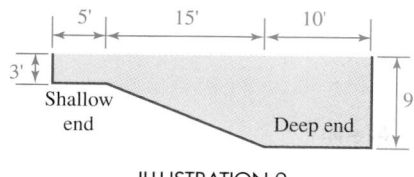

ILLUSTRATION 9

58. DRAINAGE   To measure the amount of fall (slope) of a concrete patio slab in Illustration 10, a 10-foot-long 2-by-4, a 1-foot ruler, and a level were used. Find the amount of fall in the slab. Explain what it means.

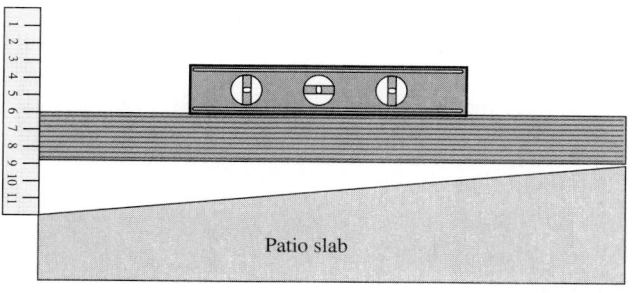

ILLUSTRATION 10

59. GRADE OF A ROAD   The vertical fall of the road shown in Illustration 11 is 264 feet for a horizontal run of 1 mile. Find the slope of the decline and use that fact to complete the roadside warning sign for truckers. (*Hint:* 1 mile = 5,280 feet.)

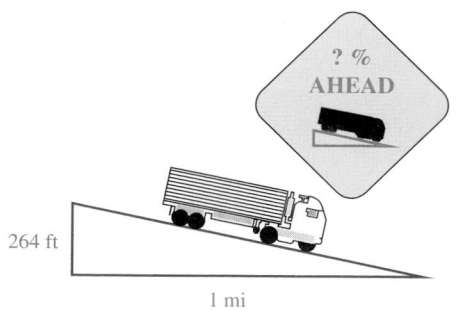

ILLUSTRATION 11

60. TREADMILL   For each height setting listed in the table, find the resulting slope of the jogging surface of the treadmill shown in Illustration 12. Express each incline as a percent.

Height setting	% incline
2 inches	
4 inches	
6 inches	

ILLUSTRATION 12

61. ACCESSIBILITY   Illustration 13 shows two designs to make the upper level wheelchair-accessible.
   a. Find the slope of the ramp in design 1.
   b. Find the slopes of the ramps in design 2.
   c. Give an advantage and a drawback of each design.

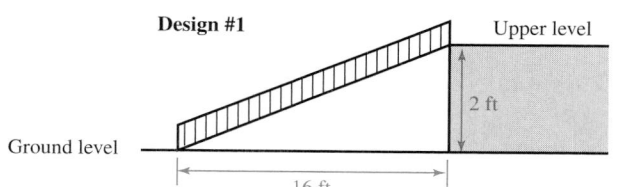

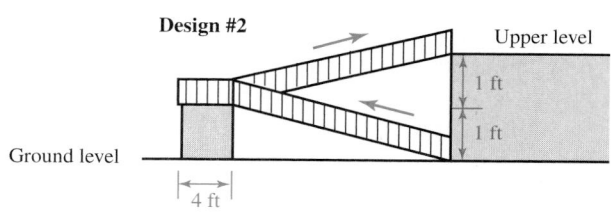

ILLUSTRATION 13

62. ARCHITECTURE   Since the slope of the roof of the house shown in Illustration 14 is to be $\frac{2}{5}$, there will be a 2-foot rise for every 5-foot run. Draw the roof line if it is to pass through the given black points. Find the coordinates of the peak of the roof.

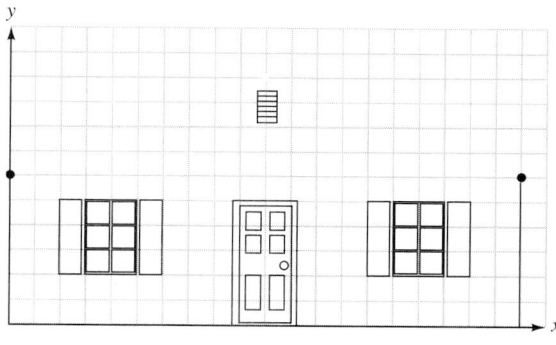

ILLUSTRATION 14

63. ENGINE OUTPUT   Use the graph in Illustration 15 to find the rate of change in the horsepower (hp) produced by an automobile engine for engine speeds in the range of 2,400–4,800 revolutions per minute (rpm).

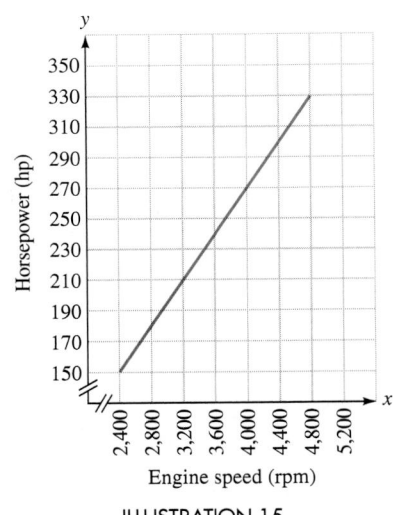

ILLUSTRATION 15

64. UNEMPLOYMENT   See Illustration 16.
a. Between what two years did the unemployment rate increase the most? Find the rate of change for that period of time.
b. Between what two years did the unemployment rate decrease the most? Find the rate of change for that period of time.

WRITING

65. Explain why the slope of a vertical line is undefined.

66. How do we distinguish between a line with positive slope and a line with negative slope?

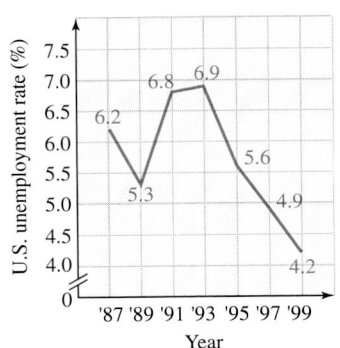

Based on data from the Bureau of Labor Statistics

ILLUSTRATION 16

67. Give an example of a rate of change that government officials might be interested in knowing so they can plan for the future needs of our country.

68. Explain the difference between a rate of change that is positive and one that is negative. Given an example of each.

REVIEW

69. In what quadrant does the point $(-3, 6)$ lie?

70. What is the name given the point $(0, 0)$?

71. Is $(-1, -2)$ a solution of $y = x^2 + 1$?

72. What basic shape does the graph of the equation $y = |x - 2|$ have?

73. Is the equation $y = 2x + 2$ linear or nonlinear?

74. Solve $-3x \leq 15$.

9.5 *Describing Linear Relationships*

In this section, you will learn about

- Slope–intercept form of the equation of a line • Parallel lines
- Perpendicular lines

INTRODUCTION. Numerical relationships are often described by using tables or graphs. For example, various lengths of pipe and their corresponding weights are listed in the table in Figure 9-43. When this information is plotted as ordered pairs, we see that the points lie in a straight line. We say that the relationship between length and weight in this example is *linear*.

Length of pipe (ft)	Weight of pipe (lb)
x	y
6	120
10	200
14	280
20	400

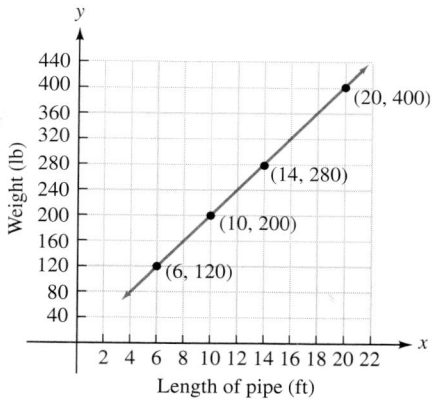

FIGURE 9-43

Figure 9-44 shows a graph of the time a cup of coffee has been sitting on a kitchen counter and its temperature. Since the graph is not a straight line, the relationship between time and temperature in this example is not linear.

Time on counter (min)	Temperature of coffee (°F)
x	y
1	180
5	140
10	110
20	80
30	72
45	70

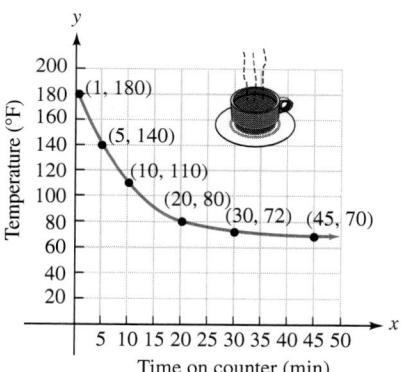

FIGURE 9-44

In this section, we will discuss a special type of relationship between two quantities whose graph is a straight line. Our objective is to learn how to write equations in two variables that describe these *linear relationships*.

Slope–intercept form of the equation of a line

The graph of $2x + 3y = 12$ shown in Figure 9-45 enables us to see that the slope of the line is $-\frac{2}{3}$ and that the y-intercept is $(0, 4)$.

$$2x + 3y = 12$$

x	y	(x, y)
6	0	(6, 0)
0	4	(0, 4)

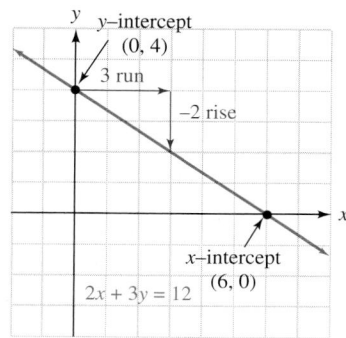

FIGURE 9-45

If we solve the equation for *y*, we will observe some interesting results.

$$2x + 3y = 12$$

$$3y = -2x + 12 \qquad \text{Subtract } 2x \text{ from both sides.}$$

$$\frac{3y}{3} = \frac{-2x}{3} + \frac{12}{3} \qquad \begin{array}{l}\text{To undo the multiplication by 3, divide both sides by 3. On} \\ \text{the right-hand side, dividing each term by 3 is equivalent to} \\ \text{dividing the entire side by 3: } \frac{-2x+12}{3} = \frac{-2x}{3} + \frac{12}{3}.\end{array}$$

$$y = -\frac{2}{3}x + 4 \qquad \text{Do the divisions. Rewrite } \frac{-2x}{3} \text{ as } -\frac{2}{3}x.$$

In the equation $y = -\frac{2}{3}x + 4$, the *slope* of the graph $\left(-\frac{2}{3}\right)$ is the coefficient of *x*, and the constant (4) is the *y*-coordinate of the *y-intercept* of the graph.

$$y = -\frac{2}{3}x + 4$$

The slope of the line.   The *y*-intercept is (0, 4).

These observations suggest the following form of an equation of a line.

Slope–intercept form of the equation of a line

If a linear equation is written in the form

$$y = mx + b$$

where *m* and *b* represent constants, the graph of the equation is a line with slope *m* and *y*-intercept (0, *b*).

EXAMPLE 1   *Slope–intercept form.* Find the slope and the *y*-intercept of the graph of each equation: **a.** $y = 6x - 2$, **b.** $y = -\frac{5}{4}x$, and **c.** $y = \frac{x}{2} + 6$.

Solution

a. If we write the subtraction as the addition of the opposite, the equation will be in $y = mx + b$ form:

$$y = 6x + (-2)$$

Since $m = 6$ and $b = -2$, the slope of the line is 6 and the *y*-intercept is (0, −2).

b. Writing $y = -\frac{5}{4}x$ in slope-intercept form, we have

$$y = -\frac{5}{4}x + 0$$

Since $m = -\frac{5}{4}$ and $b = 0$, the slope of the line is $-\frac{5}{4}$ and the *y*-intercept is (0, 0).

c. Since $\frac{x}{2}$ means $\frac{1}{2}x$, we can rewrite $y = \frac{x}{2} + 6$ as

$$y = \frac{1}{2}x + 6$$

We see that $m = \frac{1}{2}$ and $b = 6$, so the slope of the line is $\frac{1}{2}$ and the *y*-intercept is (0, 6).

Self Check

Find the slope and the *y*-intercept:

a. $y = -5x - 1$

b. $y = \frac{7}{8}x$

c. $y = 5 - \dfrac{x}{3}$

Answers: a. $m = -5, (0, -1)$; **b.** $m = \frac{7}{8}, (0, 0)$; **c.** $m = -\frac{1}{3}, (0, 5)$

 COMMENT   If a linear equation is written in the form $y = mx + b$, the slope of the graph is the *coefficient* of *x*, not the term involving *x*. For example, it would be incorrect to say that the graph of $y = 5x + 1$ has a slope of $m = 5x$. Its graph has slope $m = 5$.

EXAMPLE 2 *Slope–intercept form.* Find the slope and the *y*-intercept of the line determined by $6x - 3y = 9$. Then graph it.

Solution

To find the slope and the *y*-intercept of the line, we need to write the equation in slope-intercept form. We do this by solving for *y*.

$$6x - 3y = 9$$

$$-3y = -6x + 9 \qquad \text{Subtract } 6x \text{ from both sides.}$$

$$\frac{-3y}{-3} = \frac{-6x}{-3} + \frac{9}{-3} \qquad \begin{array}{l}\text{To undo the multiplication by } -3, \text{ divide both sides by } -3. \\ \text{On the right-hand side, dividing each term by } -3 \text{ is} \\ \text{equivalent to dividing the entire side by } -3: \\ \frac{-6x + 9}{-3} = \frac{-6x}{-3} + \frac{9}{-3}.\end{array}$$

$$y = 2x - 3 \qquad \text{Do the divisions. Here, } m = 2 \text{ and } b = -3.$$

From the equation, we see that the slope is 2 and the *y*-intercept is $(0, -3)$.

To graph $y = 2x - 3$, we plot the *y*-intercept $(0, -3)$, as shown in Figure 9-46. Since the slope is $\frac{\text{rise}}{\text{run}} = 2 = \frac{2}{1}$, the line rises 2 units for every unit it moves to the right. If we begin at $(0, -3)$ and move 1 unit to the right (run) and then 2 units up (rise), we locate the point $(1, -1)$, which is a second point on the line. We then draw a straight line through $(0, -3)$ and $(1, -1)$.

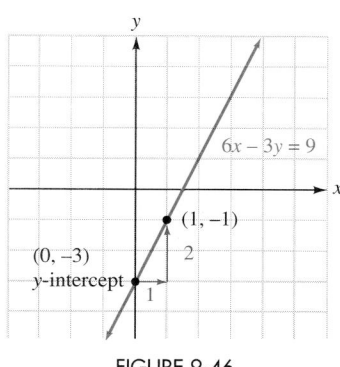

FIGURE 9-46

Self Check

Find the slope and the *y*-intercept of the line determined by $8x - 2y = -2$. Then graph it.

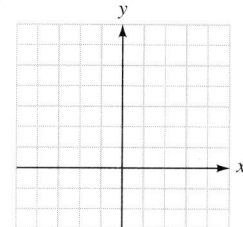

Answer: $m = 4, (0, 1)$

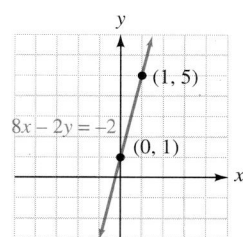

If we are given the slope and *y*-intercept of a line, we can write its equation, as in the next example.

EXAMPLE 3 *Limo service.* On weekends, a limousine service charges a fee of $100, plus 50¢ per mile, for the rental of a stretch limo. Write a linear equation that describes the relationship between the rental cost and the number of miles driven. Graph the result.

Solution

To write an equation describing this relationship, we will let *x* represent the number of miles driven and *y* represent the cost (in dollars). We can make two observations:

- The cost increases by 50¢ or $0.50 for each mile driven. This is the *rate of change* of the rental cost to miles driven, and it will be the *slope* of the graph of the equation. Thus, $m = 0.50$.

- The basic fee is $100. Before driving any miles (that is, when $x = 0$), the cost *y* is 100. The ordered pair $(0, 100)$ will be the *y*-intercept of the graph of the equation. So we know that $b = 100$.

We substitute 0.50 for *m* and 100 for *b* in the slope–intercept form to get

$$y = 0.50x + 100 \qquad \begin{array}{l}\text{Here the cost } y \text{ depends on } x, \\ \text{the number of miles driven.}\end{array}$$
$$\quad \uparrow \qquad\quad \uparrow$$
$$m = 0.50 \qquad b = 100$$

To graph $y = 0.50x + 100$, we plot its *y*-intercept, $(0, 100)$, as shown in Figure 9-47. Since the slope is $0.50 = \frac{50}{100} = \frac{5}{10}$, we can start at $(0, 100)$ and locate a second point on the line by moving 10 units to the right (run) and then 5 units up (rise). This point will have coordinates $(0 + 10, 100 + 5)$ or $(10, 105)$. We draw a straight line through these two points to get a graph that illustrates the relationship between the rental cost and the number of miles driven. We draw the graph only in quadrant I, because the number of miles driven is always positive.

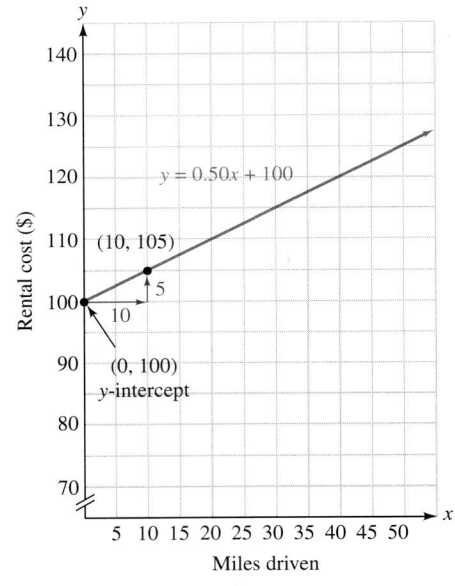

FIGURE 9-47

EXAMPLE 4   *Videotapes.*   A VHS videocassette contains 800 feet of tape. In the long play (LP) mode, it plays 10 feet of tape every 3 minutes. Write a linear equation that relates the number of feet of tape yet to be played and the number of minutes the tape has been playing. Graph the equation.

Solution

The number of feet yet to be played depends on the time the tape has been playing. To write an equation describing this relationship, we let x represent the number of minutes the tape has been playing and y represent the number of feet of tape yet to be played. We can make two observations:

- Since the VCR plays 10 feet of tape every 3 minutes, the number of feet remaining is constantly *decreasing*. This rate of change $\left(-\frac{10}{3}\text{ feet per minute}\right)$ will be the slope of the graph of the equation. Thus, $m = -\frac{10}{3}$.

- The cassette tape is 800 feet long. Before any of the tape is played (that is, when $x = 0$), the amount of tape yet to be played is $y = 800$. Written as an ordered pair, we have (0, 800). Thus, $b = 800$.

Writing the equation in slope–intercept form, we have $y = -\frac{10}{3}x + 800$. Its graph is shown in Figure 9-48.

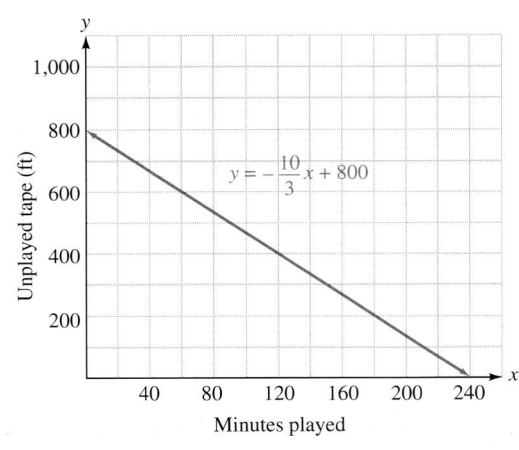

FIGURE 9-48

Self Check

In Example 4, let's say that the VCR is in super long play (SLP) mode, which plays 11 feet every 5 minutes. Use Figure 9-48 to graph the equation and then make an observation.

Answer: $y = -\frac{11}{5}x + 800$; the graphs have the same y-intercept but different slopes.

Parallel lines

Suppose it costs $75, plus 50¢ per mile, to rent the limo discussed in Example 3 on a weekday. If we substitute 0.50 for m and 75 for b in the slope–intercept form of a line, we have

$$y = 0.50x + 75$$

The graph of this equation and the graph of the equation

$$y = 0.50x + 100$$

appear in Figure 9-49.

From the figure, we see that the lines, each with slope 0.50, are parallel (do not intersect). This observation suggests the following fact.

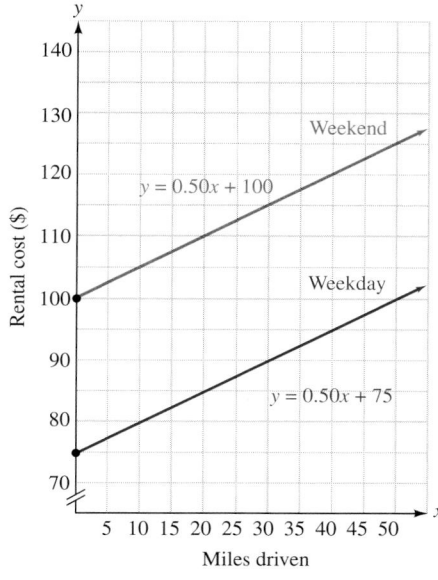

FIGURE 9-49

Slopes of parallel lines

Two different lines with the same slope are parallel.

EXAMPLE 5 *Parallel lines.* Graph $y = -\frac{2}{3}x$ and $y = -\frac{2}{3}x + 3$ on the same set of axes.

Solution

The graph of the first equation has a slope of $-\frac{2}{3}$ and a y-intercept of $(0, 0)$. The graph of the second equation has a slope of $-\frac{2}{3}$ and a y-intercept of $(0, 3)$. We graph each equation as in Figure 9-50. Since the lines have the same slope of $-\frac{2}{3}$, they are parallel.

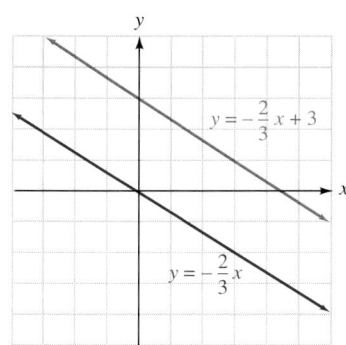

FIGURE 9-50

Self Check

Graph $y = \frac{5}{2}x - 2$ and $y = \frac{5}{2}x$ on the same set of axes.

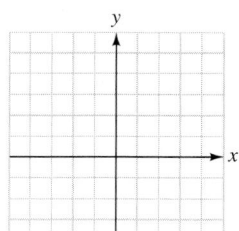

Answer:

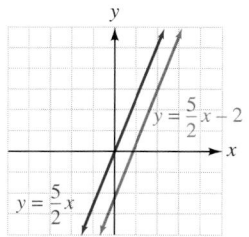

Perpendicular lines

The two lines shown in Figure 9-51 meet at right angles and are called **perpendicular lines.** Each of the four angles that are formed has a measure of 90°.

The product of the slopes of two (nonvertical) perpendicular lines is -1. For example, the perpendicular lines shown in Figure 9-51 have slopes of $\frac{3}{2}$ and $-\frac{2}{3}$. If we find the product of their slopes, we have

$$\frac{3}{2}\left(-\frac{2}{3}\right) = -\frac{6}{6} = -1$$

Two numbers whose product is -1 are called **negative reciprocals.** The numbers $\frac{3}{2}$ and $-\frac{2}{3}$, for example, are negative reciprocals, because their product is -1. The term *negative reciprocal* can be used to relate perpendicular lines and their slopes.

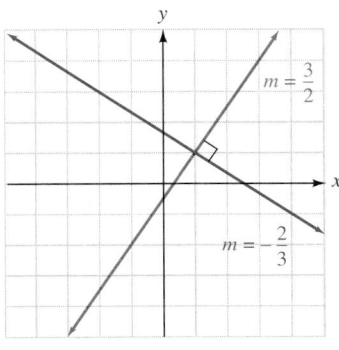

FIGURE 9-51

Slopes of perpendicular lines

If two nonvertical lines are perpendicular, their slopes are negative reciprocals.
If the slopes of two lines are negative reciprocals, the lines are perpendicular.

We can also state the fact given above symbolically: If the slopes of two nonvertical lines are m_1 and m_2, then the lines are perpendicular if

$$m_1 \cdot m_2 = -1 \qquad \text{or} \qquad m_2 = \frac{1}{m_1}$$

Because a horizontal line is perpendicular to a vertical line, a line with a slope of 0 is perpendicular to a line with no defined slope.

EXAMPLE 6 *Parallel and perpendicular lines.* Determine whether the graphs of $y = -5x + 6$ and $y = \frac{x}{5} - 2$ are parallel, perpendicular, or neither.

Solution

The slope of the line $y = -5x + 6$ is -5. The slope of the line $y = \frac{x}{5} - 2$ is $\frac{1}{5}$. (Recall that $\frac{x}{5} = \frac{1}{5}x$.) Since the slopes are not equal, the lines are not parallel. If we find the product of their slopes, we have

$$-5\left(\frac{1}{5}\right) = -\frac{5}{5} = -1$$

Since the product of their slopes is -1, the lines are perpendicular.

Self Check

Determine whether the graphs of $y = 4x + 4$ and $y = \frac{1}{4}x$ are parallel, perpendicular, or neither.

Answer: neither ∎

STUDY SET Section 9.5

VOCABULARY *Fill in the blanks.*

1. The equation $y = mx + b$ is called the _____ form for the equation of a line.

2. The graph of the linear equation $y = mx + b$ has a _____ of $(0, b)$ and a _____ of m.

3. _____ lines do not intersect.

4. The slope of a line is a _____ of change.

5. The numbers $\frac{5}{6}$ and $-\frac{6}{5}$ are called negative _____. Their product is -1.

6. The product of the slopes of _____ lines is -1.

CONCEPTS

7. TREE GROWTH Graph the values shown in Illustration 1 and connect the points with a smooth curve. Does the graph indicate a linear relationship between the age of the tree and its height? Explain your answer.

Age	Height
0	0
5	8
10	15
15	28
20	45
25	62
30	85
35	100
40	112
45	118

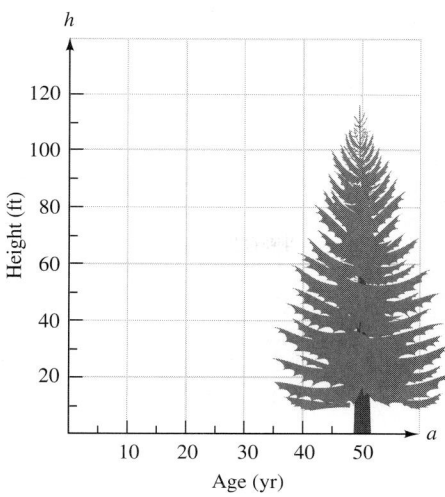

ILLUSTRATION 1

10. In Illustration 4, the slope of line l_1 is 2.
  a. What is the slope of line l_2?
  b. What is the slope of line l_3?
  c. What is the slope of line l_4?
  d. Which lines have the same y-intercept?

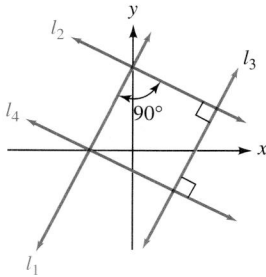

ILLUSTRATION 4

8. See Illustration 2.
  a. What is the slope of the line?
  b. What is the y-intercept of the line?
  c. Write the equation of the line.

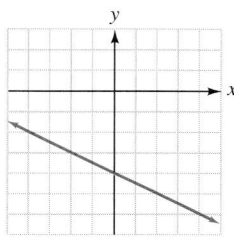

ILLUSTRATION 2

9. NAVIGATION   The graph in Illustration 3 shows the recommended speed at which a ship should proceed into head waves of various heights.
  a. What information does the y-intercept of the graph give?

  b. What is the rate of change in the recommended speed of the ship as the wave height increases?

  c. Write the equation of the graph.

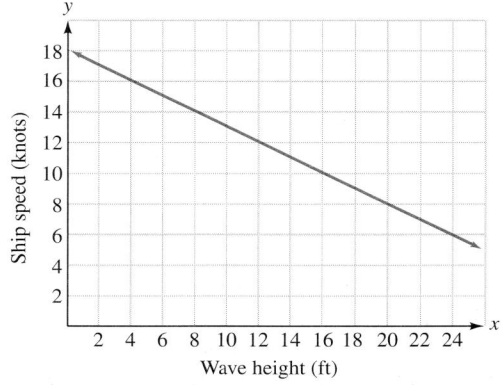

ILLUSTRATION 3

11. a. What is the y-intercept of line l_1 (graphed in Illustration 5)?
  b. What do lines l_1 and l_2 have in common? How are they different?

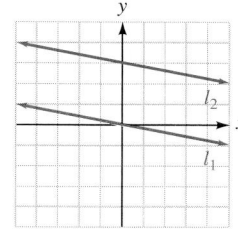

ILLUSTRATION 5

12. Use the graph in Illustration 6 to determine m and b; then write the equation of the line in slope–intercept form.

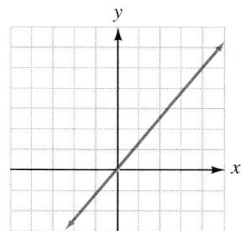

ILLUSTRATION 6

13. What is the slope of the line defined by each equation?
  a. $y = \dfrac{-2x}{3} - 2$ **b.** $y = \dfrac{x}{4} + 1$
  c. $y = 2 - 8x$ **d.** $y = 3x$
  e. $y = x$ **f.** $y = -x$

14. Without graphing, tell whether the graphs of each pair of lines are parallel, perpendicular, or neither.
  a. $y = 0.5x - 3$; $y = \frac{1}{2}x + 3$
  b. $y = 0.75x$; $y = -\frac{4}{3}x + 2$
  c. $y = -x$; $y = x$

15. To solve $-2y = 6x - 12$ for y, both sides of the equation were divided by -2. Complete each of the three divisions shown below.

$$\frac{-2y}{-2} = \frac{6x}{-2} - \frac{12}{-2}$$

$$ = + $$

16. A graphing calculator was used to graph $y = -2.5x - 1.25$, as shown below. What important feature of the graph is typed at the bottom of the screen?

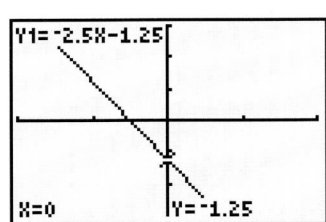

NOTATION *Complete each solution by solving the equation for y. Then find the slope and the y-intercept of its graph.*

17.
$$6x - 2y = 10$$
$$6x - \boxed{} - 2y = -6x + 10$$
$$-2y = \boxed{} + 10$$
$$\frac{-2y}{\boxed{}} = \frac{-6x}{\boxed{}} + \frac{10}{\boxed{}}$$
$$y = \boxed{} - 5$$

The slope is $\boxed{}$ and the y-intercept is $\boxed{}$.

18.
$$2x + 5y = 15$$
$$2x + 5y - \boxed{} = \boxed{} + 15$$
$$\boxed{} = -2x + 15$$
$$\frac{5y}{\boxed{}} = \frac{-2x}{\boxed{}} + \frac{15}{\boxed{}}$$
$$y = -\frac{2}{5}x + 3$$

The slope is $\boxed{}$ and the y-intercept is $\boxed{}$.

PRACTICE *Find the slope and the y-intercept of the graph of each equation.*

19. $y = 4x + 2$ **20.** $y = -4x - 2$

21. $y = \dfrac{x}{4} - \dfrac{1}{2}$ **22.** $4x - 2 = y$

23. $y = \frac{1}{2}x + 6$ **24.** $y = 6 - x$

25. $6y = x - 6$ **26.** $6x - 1 = y$

27. $x + y = 8$ **28.** $x - y = -30$

29. $2x + 3y = 6$ **30.** $3x - 5y = 15$

31. $3y - 13 = 0$ **32.** $-5y - 2 = 0$

33. $y = -5x$ **34.** $y = 14x$

Write the equation of the line with the given slope and y-intercept. Then graph it.

35. $m = 5, (0, -3)$ **36.** $m = -2, (0, 1)$

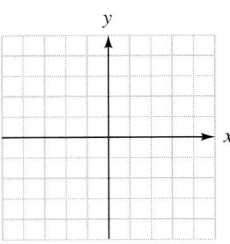 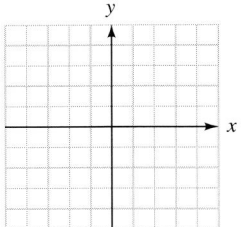

37. $m = \frac{1}{4}, (0, -2)$ **38.** $m = \frac{1}{3}, (0, -5)$

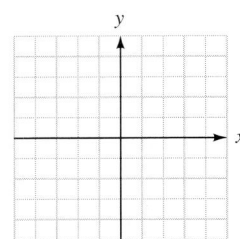 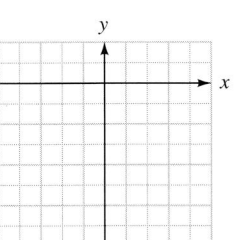

39. $m = -3, (0, 6)$ **40.** $m = -2, (0, 1)$

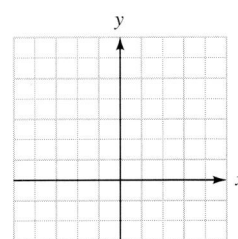 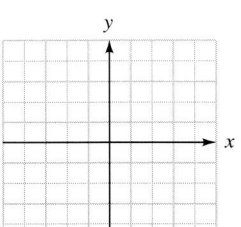

41. $m = -\frac{8}{3}, (0, 5)$ **42.** $m = -\frac{7}{6}, (0, 2)$

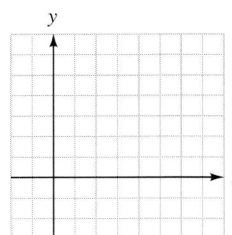 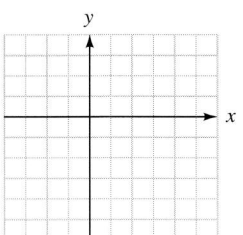

Find the slope and the y-intercept of the graph of each equation. Then graph it.

43. $y = 3x + 3$ **44.** $y = -3x + 5$

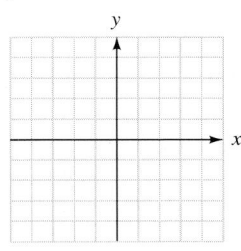 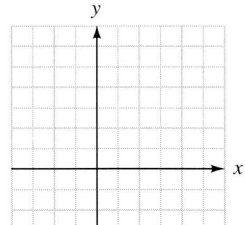

45. $y = -\dfrac{x}{2} + 2$ **46.** $y = \dfrac{x}{3}$

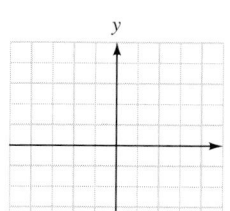   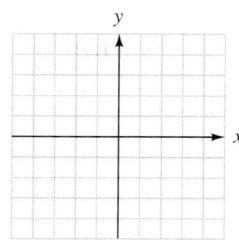

47. $3x + 4y = 16$ **48.** $2x + 3y = 9$

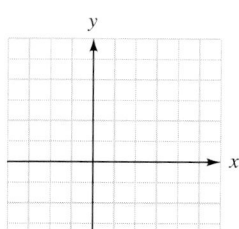   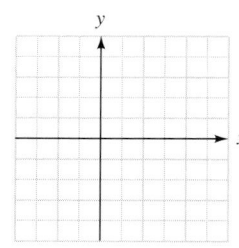

49. $10x - 5y = 5$ **50.** $4x - 2y = 6$

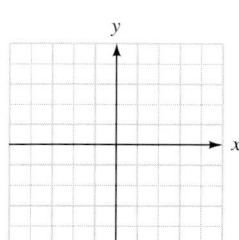   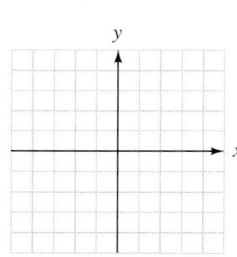

APPLICATIONS

51. PRODUCTION COSTS   A television production company charges a basic fee of $5,000 and then $2,000 an hour when filming a commercial.
   a. Write a linear equation that describes the relationship between the total production costs y and the hours of filming x.
   b. Use your answer to part a to find the production costs if a commercial required 8 hours of filming.

52. COLLEGE FEES   Each semester, students enrolling at a community college must pay tuition costs of $20 per unit as well as a $40 student services fee.
   a. Write a linear equation that gives the total fees y to be paid by a student enrolling at the college and taking x units.
   b. Use your answer to part a to find the enrollment cost for a student taking 12 units.

53. CHEMISTRY EXPERIMENT   Illustration 7 shows a portion of a student's chemistry lab manual. Use the information to write a linear equation relating the temperature y (in degrees Fahrenheit) of the compound to the time x (in minutes) elapsed during the lab procedure.

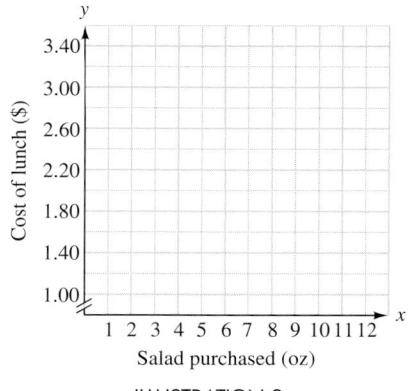

Chem. Lab #1   Aug. 13
Step 1: *Removed compound from freezer @ –10° F.*

Step 2: *Used heating unit to raise temperature of compound 5° F. every minute.*

ILLUSTRATION 7

54. INCOME PROPERTY   See Illustration 8. Use the information in the newspaper advertisement to write a linear equation that gives the amount of income y (in dollars) the apartment owner will receive when the unit is rented for x months.

APARTMENT FOR RENT
1 bedroom/1 bath, with garage
$500 per month + $250 nonrefundable security fee.

ILLUSTRATION 8

55. SALAD BAR   For lunch, a delicatessen offers a "Salad and Soda" special where customers serve themselves at a well-stocked salad bar. The cost is $1.00 for the drink and 20¢ an ounce for the salad.
   a. Write a linear equation that will find the cost y of a "Salad and Soda" lunch when a salad weighing x ounces is purchased.
   b. Graph the equation (see Illustration 9).
   c. How would the graph from part b change if the delicatessen began charging $2.00 for the drink?
   d. How would the graph from part b change if the cost of the salad changed to 30¢ an ounce?

Cost of lunch ($)

Salad purchased (oz)

ILLUSTRATION 9

56. SEWING COSTS   A tailor charges a basic fee of $20 plus $2.50 per letter to sew an athlete's name on the back of a jacket.
   a. Write a linear equation that will find the cost y to have a name containing x letters sewn on the back of a jacket.

b. Graph the equation (see Illustration 10).

c. Suppose the tailor raises the basic fee to $30. On your graph from part b, draw the new graph showing the increased cost.

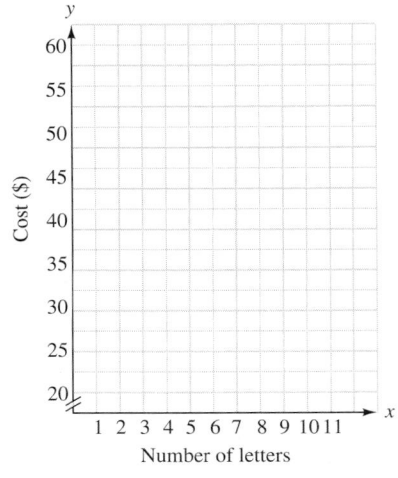

ILLUSTRATION 10

57. EMPLOYMENT SERVICE A policy statement of LIZCO, Inc., is shown in Illustration 11. Suppose a secretary had to pay an employment service $500 to get placed in a new job at LIZCO. Write a linear equation that tells the secretary the actual cost y of the employment service to her x months after being hired.

> Policy no. 23452– A new hire will be reimbursed by LIZCO for any employment service fees paid by the employee at the rate of $20 per month.

ILLUSTRATION 11

58. 🖩 **COMPUTER DRAFTING** Illustration 12 shows a computer-generated drawing of an automobile engine mount. When the designer clicks the mouse on a line of the drawing, the computer finds the equation of the line. Determine whether the two lines selected in the drawing are perpendicular.

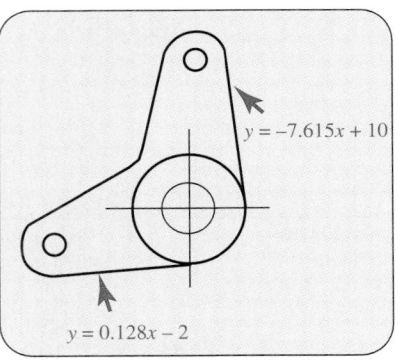

ILLUSTRATION 12

WRITING

59. Explain the advantages of writing the equation of a line in slope–intercept form ($y = mx + b$) as opposed to general form ($Ax + By = C$).

60. Why is $y = mx + b$ called the slope–intercept form of the equation of a line?

61. What is the minimum number of points needed to draw the graph of a line? Explain why.

62. List some examples of parallel and perpendicular lines that you see in your daily life.

REVIEW

63. Find the slope of the line passing through the points $(6, -2)$ and $(-6, 1)$.

64. Is $(3, -7)$ a solution of $y = 3x - 2$?

65. Evaluate $-4 - (-4)$.

66. Solve $2(x - 3) = 3x$.

67. To evaluate $[-2(4 - 8) + 4^2]$, which operation should be performed first?

68. Translate to mathematical symbols: four less than twice the price p.

69. What percent of 6 is 1.5?

70. Does $x = -6.75$ make $x + 1 > -9$ true?

9.6 *Writing Linear Equations*

In this section, you will learn about

- Point–slope form of the equation of a line
- Writing the equation of a line through two points • Horizontal and vertical lines

INTRODUCTION. If we know the slope of a line and its y-intercept, we can use the slope–intercept form to write the equation of the line. The question that now arises is, can *any* point on the line be used in combination with its slope to write its equation? In this section, we will answer this question.

Point–slope form of the equation of a line

For the line shown in Figure 9-52, suppose we know that it has a slope of 3 and that it passes through the point $P(2, 1)$. If we pick another point on the line and call it $Q(x, y)$, we can find the slope of the line by using the coordinates of points P and Q. Using the slope formula, we have

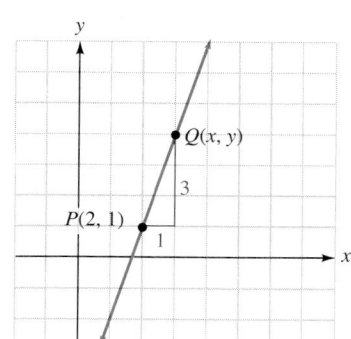

FIGURE 9-52

$$\frac{y_2 - y_1}{x_2 - x_1} = m \quad \text{The slope formula.}$$

$$\frac{y - 1}{x - 2} = m \quad \text{Substitute } y \text{ for } y_2, 1 \text{ for } y_1, x \text{ for } x_2, \text{ and } 2 \text{ for } x_1.$$

Since the slope of the line is given to be 3, we can substitute 3 for m in the previous equation.

$$\frac{y - 1}{x - 2} = m$$

$$\frac{y - 1}{x - 2} = 3$$

We then multiply both sides by $x - 2$ to get

$$\frac{y - 1}{x - 2}(x - 2) = 3(x - 2) \quad \text{Clear the equation of the fraction.}$$

$$y - 1 = 3(x - 2) \quad \text{Simplify the left-hand side.}$$

The resulting equation displays the slope of the line and the coordinates of one point on the line:

$$\underset{\substack{\uparrow \\ y\text{-coordinate} \\ \text{of the point}}}{y - 1} = \overset{\substack{\text{Slope} \\ \text{of the line} \\ \downarrow}}{3}(\underset{\substack{\uparrow \\ x\text{-coordinate} \\ \text{of the point}}}{x - 2})$$

In general, suppose we know that the slope of a line is m and that the line passes through the point (x_1, y_1). Then if (x, y) is any other point on the line, we can use the definition of slope to write

$$\frac{y - y_1}{x - x_1} = m$$

If we multiply both sides by $x - x_1$, we have

$$y - y_1 = m(x - x_1)$$

This form of a linear equation is called the **point–slope form.** It can be used to write the equation of a line when the slope and one point on the line are known.

Point–slope form of the equation of a line

If a line with slope m passes through the point (x_1, y_1), the equation of the line is
$$y - y_1 = m(x - x_1)$$

EXAMPLE 1   *Point–slope form.*   Write the equation of a line that has a slope of -3 and passes through $(-1, 5)$. Express the result in slope–intercept form.

Solution

Since we are given the slope and a point on the line, we will use the point–slope form.

$y - y_1 = m(x - x_1)$ The point–slope form.

$y - 5 = -3[x - (-1)]$ Substitute -3 for m, -1 for x_1, and 5 for y_1.

$y - 5 = -3(x + 1)$ Simplify within the brackets.

We can write this result in slope–intercept form, as follows:

$y - 5 = -3x - 3$ Distribute the multiplication by -3.

$y = -3x + 2$ To undo the subtraction of 5, add 5 to both sides: $-3 + 5 = 2$.

In slope–intercept form, the equation is $y = -3x + 2$.

Self Check

Write the equation of a line that has a slope of -2 and passes through $(4, -3)$. Write the result in slope–intercept form.

Answer: $y = -2x + 5$ ■

EXAMPLE 2   *Temperature drop.*   A refrigeration unit can lower the temperature in a railroad car by 6°F every 5 minutes. One day, the temperature in a car was 76°F after the cooler had run for 10 minutes. Find a linear equation that describes the relationship between the time the cooler has been running and the temperature in the car.

Graph the equation and use it to find the temperature in the car before the cooler was turned on and the temperature in the car after the cooler had run for 25 minutes.

Solution

We will let x represent the time, in minutes, that the cooler was running, and y will represent the air temperature in the car. We can make two observations:

- With the cooler on, the temperature in the railroad car drops 6° every 5 minutes. The rate of change of $-\frac{6}{5}$ degrees per minute is the slope of the graph of the linear equation that we want to find. Thus, $m = -\frac{6}{5}$.

- We know that after the cooler had been running for 10 minutes ($x = 10$), the temperature in the car was 76° ($y = 76$). We can express these facts with the ordered pair $(10, 76)$. This is a point on the graph of the linear equation.

To write the linear equation, we substitute $-\frac{6}{5}$ for m, 10 for x_1, and 76 for y_1, into the point–slope form of the equation of a line.

$y - y_1 = m(x - x_1)$ Point–slope form.

$y - 76 = -\dfrac{6}{5}(x - 10)$ Substitute: $m = -\frac{6}{5}$, $x_1 = 10$, and $y_1 = 76$.

$y - 76 = -\dfrac{6}{5}x - \left(-\dfrac{6}{5}\right)10$ Distribute the multiplication by $-\frac{6}{5}$.

$y - 76 = -\dfrac{6}{5}x - (-12)$ Do the multiplication: $\left(-\frac{6}{5}\right)10 = \left(-\frac{6}{5}\right)\frac{10}{1} = -12$.

$y - 76 = -\dfrac{6}{5}x + 12$ On the right-hand side, change the subtraction to the addition of the opposite.

$y - 76 + 76 = -\dfrac{6}{5}x + 12 + 76$ To undo the subtraction of 76, add 76 to both sides.

$y = -\dfrac{6}{5}x + 88$ Do the additions.

The graph of $y = -\frac{6}{5}x + 88$ is shown in Figure 9-53. From the graph, we see that the temperature in the railroad car before the cooler was turned on was 88°F. This is given by the y-intercept of the graph, $(0, 88)$. If we locate 25 on the x-axis and move straight up to intersect the graph, we will see that the temperature in the car was 58°F. This shows that after the cooler ran for 25 minutes, the temperature was about 58°F.

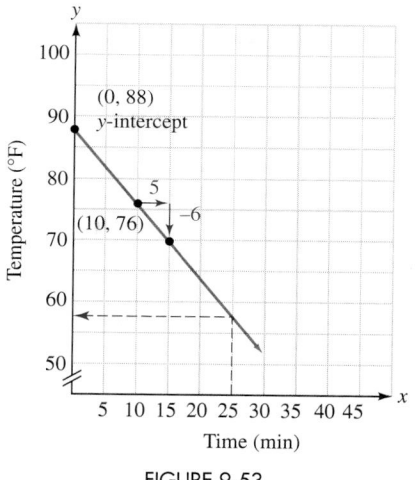

FIGURE 9-53

Writing the equation of a line through two points

In the next example, we will show that it is possible to write the equation of a line when we know the coordinates of two points on the line.

EXAMPLE 3   *Given two points on a line.*   Write the equation of the line passing through $(4, 0)$ and $(6, -8)$.

Solution

First we find the slope of the line that passes through $(4, 0)$ and $(6, -8)$.

$$m = \frac{y_2 - y_1}{x_2 - x_1} \quad \text{The slope formula.}$$

$$= \frac{-8 - 0}{6 - 4} \quad \text{Substitute } -8 \text{ for } y_2, 0 \text{ for } y_1, 6 \text{ for } x_2, \text{ and } 4 \text{ for } x_1.$$

$$= \frac{-8}{2} \quad \text{Simplify.}$$

$$= -4$$

Since the line passes through both $(4, 0)$ and $(6, -8)$, we can choose either point and substitute its coordinates into the point–slope form. If we choose $(4, 0)$, we substitute 4 for x_1, 0 for y_1, and -4 for m and proceed as follows.

$$y - y_1 = m(x - x_1) \quad \text{Point–slope form.}$$

$$y - 0 = -4(x - 4) \quad \text{Substitute } -4 \text{ for } m, 4 \text{ for } x_1, \text{ and } 0 \text{ for } y_1.$$

$$y = -4x + 16 \quad \text{Distribute the multiplication by } -4.$$

The equation of the line is $y = -4x + 16$.

Self Check

Write the equation of the line passing through $(0, -3)$ and $(2, 1)$.

Answer: $y = 2x - 3$

EXAMPLE 4   *Market research.*   A company that makes a breakfast cereal has found that the number of discount coupons redeemed for its product is linearly related to the coupon's value. In one advertising campaign, 10,000 of the "10¢ off" coupons were redeemed. In another campaign, 45,000 of the "50¢ off" coupons were redeemed. How many coupons can the company expect to be redeemed if it issues a "35¢ off" coupon?

Solution   If we let x represent the value of a coupon and y represent the number of coupons that will be redeemed, ordered pairs will have the form

(coupon value, number redeemed)

Two points on the graph of the equation are (10, 10,000) and (50, 45,000). These points are plotted on the graph shown in Figure 9-54. To write the equation of the line passing through the points, we first find the slope of the line.

$$m = \frac{y_2 - y_1}{x_2 - x_1} \qquad \text{The slope formula.}$$

$$= \frac{45,000 - 10,000}{50 - 10} \qquad \begin{array}{l}\text{Substitute 45,000 for } y_2\text{, 10,000 for } y_1\text{, 50 for } x_2\text{, and}\\ \text{10 for } x_1.\end{array}$$

$$= \frac{35,000}{40}$$

$$= 875$$

We then substitute 875 for m and the coordinates of one known point—say, (10, 10,000)—into the point–slope form of the equation of a line and proceed as follows.

$$y - y_1 = m(x - x_1) \qquad \text{Point–slope form.}$$

$$y - 10,000 = 875(x - 10) \qquad \text{Substitute for } m, x_1, \text{ and } y_1.$$

$$y - 10,000 = 875x - 8,750 \qquad \text{Distribute the multiplication by 875.}$$

$$y = 875x + 1,250 \qquad \text{Add 10,000 to both sides.}$$

To find the expected number of coupons that will be redeemed, we substitute the value of the coupon, 35¢, into the equation $y = 875x + 1,250$ and find y.

$$y = 875x + 1,250$$

$$y = 875(35) + 1,250 \qquad \text{Substitute 35 for } x.$$

$$y = 30,625 + 1,250 \qquad \text{Do the multiplication.}$$

$$y = 31,875$$

The company can expect 31,875 of the 35¢ coupons to be redeemed. ∎

The graph to the left:

y-axis: Coupons redeemed — 10,000; 20,000; 30,000; 40,000; 50,000; 60,000
(50, 45,000)
(10, 10,000)
x-axis: Coupon value (¢) — 10 20 30 40 50 60 70

FIGURE 9-54

Horizontal and vertical lines

We have graphed horizontal and vertical lines. We will now discuss how to write their equations.

EXAMPLE 5   *Equations of horizontal and vertical lines.*   Write the equation of each line and then graph it: a. A horizontal line passing through $(-2, -4)$ and   **b.** A vertical line passing through $(1, 3)$.

Solution

a. The equation of a horizontal line can be written in the form $y = b$. Since the y-coordinate of $(-2, -4)$ is -4, the equation of the line is $y = -4$. The graph is shown in Figure 9-55.

b. The equation of a vertical line can be written in the form $x = a$. Since the x-coordinate of $(1, 3)$ is 1, the equation of the line is $x = 1$. The graph is shown in Figure 9-55.

Self Check

Write the equation of each line and then graph it:

a. a horizontal line passing through $(3, 2)$

b. a vertical line passing through $(-1, -3)$

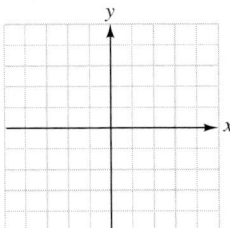

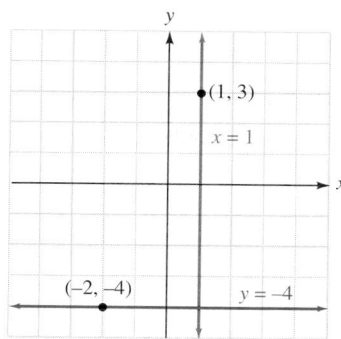

FIGURE 9-55

Answers: **a.** $y = 2$,
b. $x = -1$

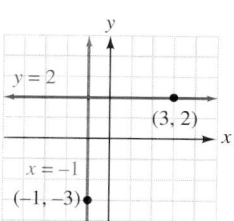

STUDY SET   Section 9.6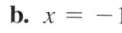

VOCABULARY *Fill in the blanks.*

1. $y - y_1 = m(x - x_1)$ is called the _____ form of the equation of a line.

2. The line in Illustration 1 _____ through point P.

3. In Illustration 1, point P has an _____ of 2 and a _____ of -1.

4. The _____ of a line gives a rate of change.

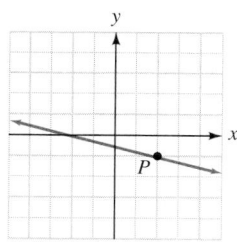

ILLUSTRATION 1

CONCEPTS

5. a. The linear equation $y = 2x - 3$ is written in *slope–intercept* form. What are the slope and the y-intercept of the graph of this line?

b. The linear equation $y - 4 = 6(x - 5)$ is written in *point–slope* form. What point does the graph of this equation pass through, and what is the line's slope?

6. Is the following statement true or false? The equations

$$y - 1 = 2(x - 2)$$
$$y = 2x - 3$$
$$2x - y = 3$$

all describe the same line.

7. a. Find two points on the line shown in Illustration 2 whose coordinates are integers.

b. What is the slope of the line?

c. Use your answers to parts a and b to write the equation of the line. Answer in point–slope form.

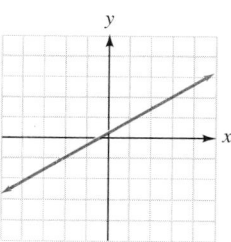

ILLUSTRATION 2

8. In each of the following cases, a linear relationship between two quantities is described. If the relationship were graphed, what would be the slope of the line?

a. The sales of new cars increased by 15 every 2 months.

b. There were 35 fewer robberies for each dozen police officers added to the force.

c. Withdrawals were occurring at the rate of $700 every 45 minutes.

d. One acre of forest is being destroyed every 30 seconds.

9. In each of the following cases, is the given information sufficient to write the equation of the line?

a. It passes through $(2, -7)$.

b. Its slope is $-\frac{3}{4}$.

c. It has the following table of solutions:

x	y
2	3
-3	-6

10. In each of the following cases, is the given information sufficient to write the equation of the line?

  a. It is horizontal.

  b. It is vertical and passes through $(-1, 1)$.

  c. It has the following table of solutions:

x	y
4	5

NOTATION

11. Fill in the blank. In $y - y_1 = m(x - x_1)$, we read x_1 as "x _____ one."

12. Write the equation of a horizontal line passing through $(0, b)$.

Write the equation in slope–intercept form.

13.
$$y - 2 = -3(x - 4)$$
$$y - 2 = \boxed{} + \boxed{}$$
$$y - 2 + \boxed{} = -3x + 12 + \boxed{}$$
$$y = -3x + 14$$

14.
$$y + 2 = \frac{1}{2}(x + 2)$$
$$y + 2 = \boxed{} + \boxed{}$$
$$y + 2 - \boxed{} = \frac{1}{2}x + 1 - \boxed{}$$
$$y = \frac{1}{2}x - 1$$

Complete each solution.

15. Write the equation of the line with slope -2 that passes through the point $(-1, 5)$.
$$y - y_1 = m(x - x_1)$$
$$y - \boxed{} = -2\left[x - \left(\boxed{}\right)\right]$$
$$y - 5 = \boxed{} - 2$$
$$y = -2x + 3$$

16. Write the equation of the line with slope 4 that passes through the point $(0, 3)$.
$$y - y_1 = m(x - x_1)$$
$$y - \boxed{} = 4\left(x - \boxed{}\right)$$
$$y - 3 = \boxed{}$$
$$y = 4x + 3$$

PRACTICE *Use the point–slope form to write the equation of the line with the given slope and point.*

17. $m = 3$, passes through $(2, 1)$

18. $m = 2$, passes through $(4, 3)$

19. $m = -\dfrac{4}{5}$, passes through $(-5, -1)$

20. $m = -\dfrac{7}{8}$, passes through $(-2, -9)$

Use the point–slope form to first write the equation of the line with the given slope and point. Then write your result in slope–intercept form.

21. $m = \dfrac{1}{5}$, passes through $(10, 1)$

22. $m = \dfrac{1}{4}$, passes through $(8, 1)$

23. $m = -5$, passes through $(-9, 8)$

24. $m = -4$, passes through $(-2, 10)$

25. $m = -\dfrac{4}{3}$,

x	y
6	-4

26. $m = -\dfrac{3}{2}$,

x	y
-2	1

27. $m = -\dfrac{2}{3}$, passes through $(3, 0)$

28. $m = -\dfrac{2}{5}$, passes through $(15, 0)$

29. $m = 8$, passes through $(0, 4)$

30. $m = 6$, passes through $(0, -4)$

31. $m = -3$, passes through the origin

32. $m = -1$, passes through the origin

Write the equation of the line that passes through the two given points. Write your result in slope–intercept form.

33. Passes through $(1, 7)$ and $(-2, 1)$

34. Passes through $(-2, 2)$ and $(2, -8)$

35.

x	y
-4	3
2	0

36.

x	y
-1	-4
1	-2

37. Passes through $(5, 5)$ and $(7, 5)$

38. Passes through $(-2, 1)$ and $(-2, 15)$

39. Passes through $(5, 1)$ and $(-5, 0)$

40. Passes through $(-3, 0)$ and $(3, 1)$

41. Passes through $(-8, 2)$ and $(-8, 17)$

42. Passes through $\left(\frac{2}{3}, 2\right)$ and $(0, 2)$

Write the equation of the line with the given characteristics.

43. Vertical, passes through $(4, 5)$

44. Vertical, passes through $(-2, -5)$

45. Horizontal, passes through (4, 5)

46. Horizontal, passes through (−2, −5)

APPLICATIONS

47. POLE VAULT   See Illustration 3.

   a. For each of the four positions of the vault shown, give two points that the pole passes through.

   b. Write the equations of the lines that describe the position of the pole for parts 1, 3, and 4 of the jump.

   c. Why can't we write a linear equation describing the position of the pole for part 2?

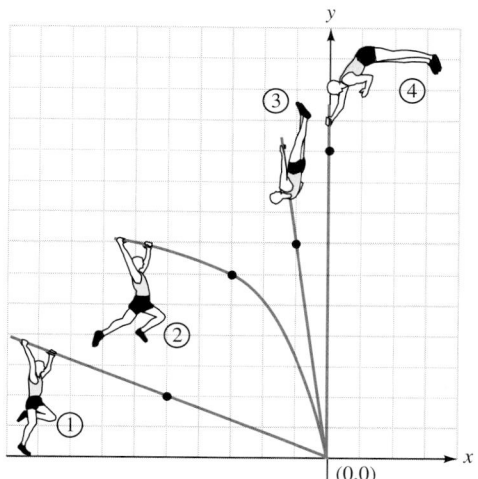

ILLUSTRATION 3

48. FREEWAY DESIGN   The graph in Illustration 4 shows the route of a proposed freeway.

   a. Give the coordinates of the points where the proposed freeway will join Interstate 25 and Highway 40.

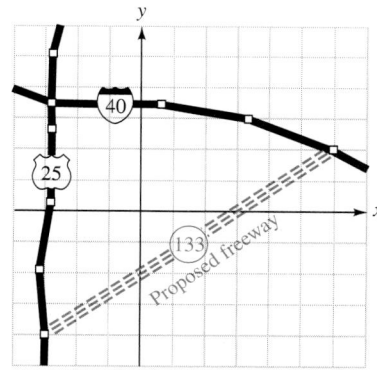

ILLUSTRATION 4

b. Write the equation of the line that mathematically describes the route of the proposed freeway. Answer in slope–intercept form.

49. TOXIC CLEANUP   Three months after cleanup began at a dump site, 800 cubic yards of toxic waste had yet to be removed. Two months later, that number had been lowered to 720 cubic yards.

   a. Write an equation that mathematically describes the linear relationship between the length of time x (in months) the cleanup crew has been working and the number of cubic yards y of toxic waste remaining.

   b. Use your answer to part a to predict the number of cubic yards of waste that will still be on the site one year after the cleanup project began.

50. DEPRECIATION   To lower its corporate income tax, accountants of a large company depreciated a word processing system over several years using a linear model, as shown in the worksheet in Illustration 5.

   a. Use the information in Illustration 5 to write a linear equation relating the years since the system was purchased x and its value y, in dollars.

   b. Find the purchase price of the system by substituting $x = 0$ into your answer from part a.

Tax Worksheet

Method of depreciation: *Linear*

Property	Value	Years after purchase
Word processing system	$60,000	2
"	$30,000	4

ILLUSTRATION 5

51. COUNSELING   In the first year of her practice, a family counselor saw 75 clients. In her second year, the number of clients grew to 105. If a linear trend continues, write an equation that gives the number of clients c the counselor will have t years after beginning her practice.

52. U.S. HEALTH CARE   See Illustration 6. When the per-person health care expenditures for the years 1990–1998 are graphed, the data nearly lie on a straight line. The expenditures can be approximated by the straight line drawn through two of the data points.

   a. Use the two highlighted points on the graph to write the equation of the line. Let $x = 0$ represent 1990, $x = 1$ represent 1991, and so on. Answer in slope–intercept form.

   b. Use your answer to part a to predict the per-person health care expenditure in the year 2010.

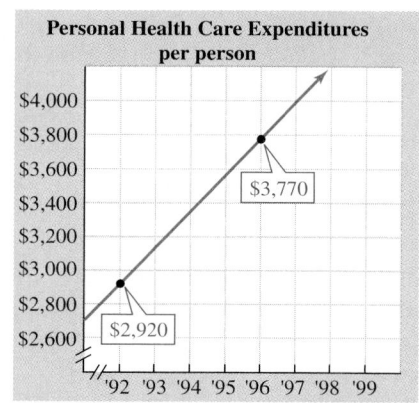

Based on data from the Health Care Financing Administration

ILLUSTRATION 6

Radius (ft)	Approximate length of padding (ft)
3	19
7	44

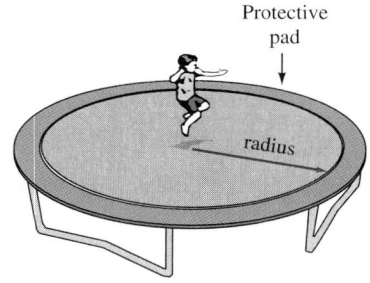

ILLUSTRATION 8

53. CONVERTING TEMPERATURES   The relationship between Fahrenheit temperature, F, and Celsius temperature, C, is linear.
   a. Use the data in Illustration 7 to write two ordered pairs of the form (C, F).
   b. Use your answer to part a to write a linear equation relating the Fahrenheit and Celsius scales.

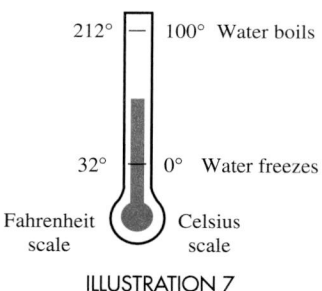

ILLUSTRATION 7

54. TRAMPOLINE   The relationship between the circumference of a circle and its radius is linear. For instance, the length l of the protective pad that wraps around a trampoline is related to the radius r of the trampoline. Use the data in Illustration 8 to write a linear equation that approximates the length of pad needed for any trampoline radius.

55. AIR-CONDITIONING   An air-conditioning unit can lower the air temperature in a classroom 4° every 15 minutes. After the air conditioner had been running for half an hour, the air temperature in the room was 75°F. Write a linear equation relating the time in minutes x the unit had been on and the temperature y of the classroom. (*Hint:* How many minutes are there in half an hour?)

56. AUTOMATION   An automated production line uses distilled water at a rate of 300 gallons every 2 hours to make shampoo. After the line had run for 7 hours, planners noted that 2,500 gallons of distilled water remained in the storage tank. Write a linear equation relating the time in hours x since the production line began and the number of gallons y of distilled water in the storage tank.

WRITING

57. Why is $y - y_1 = m(x - x_1)$ called the point–slope form of the equation of a line?

58. If we know two points that a line passes through, we can write its equation. Explain how this is done.

59. If we know the slope of a line and a point it passes through, we can write its equation. Explain how this is done.

60. Think of several points on the graph of the horizontal line $y = 4$. What do the points have in common? How do they differ?

REVIEW

61. Find the slope of the line passing through the points $(2, 4)$ and $(-6, 8)$.

62. Is the graph of $y = x^2$ a straight line?

63. Find the area of a circle with a diameter of 12 feet. Round to the nearest tenth.

64. If a 15-foot board is cut into two pieces and we let x represent the length of one piece (in feet), how long is the other piece?

65. Evaluate $(-1)^5$.

66. Solve $\dfrac{x - 3}{4} = -4$.

67. What is the coefficient of the second term of $-4x^2 + 6x - 13$?

68. Simplify $(-2p)(-5)(4x)$.

9.7 *Functions*

In this section, you will learn about

- Functions • Domain and range of a function • Function notation
- Graphs of functions • The vertical line test

INTRODUCTION. In everyday life, we see a wide variety of situations where one quantity depends on another:

- The distance traveled by a car depends on its speed.
- The cost of renting a video depends on the number of days it is rented.
- A state's number of representatives in Congress depends on the state's population.

In this section, we will discuss many situations where one quantity depends on another according to a specific rule, called a *function*. For example, the equation $y = 2x - 3$ sets up a rule where each value of y depends on the choice of some number x. The rule is: *To find y, double the value of x and subtract* 3. In this case, y (the *dependent variable*) depends on x (the *independent variable*).

Functions

We have previously described relationships between two quantities in different ways:

Using words

| The number of tires to order | is | two | times | the number of bicycles to be manufactured. |

Here words are used to state that the number of bicycle tires to order depends on the number of bicycles to be manufactured.

Using equations

$$t = 1,500 - d$$

This equation describes how the amount of take-home pay t depends on the amount of deductions d.

Using graphs

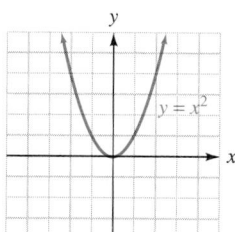

This rectangular coordinate graph shows many ordered pairs (x, y) that satisfy the equation $y = x^2$, where the value of the y-coordinate depends on the value of the x-coordinate.

Using tables

Acres	Schools
400	4
800	8
1,000	10
2,000	20

This table shows that the number of schools needed depends on the size of the housing development.

Two observations can be made about these examples:

- Each one establishes a relationship between two sets of values. For example, the number of bicycle tires that must be ordered *depends* on the number of bicycles to be manufactured.

- In these relationships, each value in one set is assigned a *single* value of a second set. For example, for each number of bicycles to be manufactured, there is exactly one number of tires to order.

Relationships between two quantities that exhibit both of these characteristics are called **functions.**

Functions

> A **function** is a rule that assigns to each value of one variable (the **independent variable**) a single value of another variable (the **dependent variable**).

Using the variables x and y, we can restate the previous definition as follows: For y to be a function of x, each value of x must determine exactly one value of y.

EXAMPLE 1   *Identifying functions.*   **a.** Does $y = 4x + 1$ define a function?   **b.** Does the table define y as a function of x?

x	y
0	6
5	3
9	1
5	7
10	8

Solution

a. For each value of the independent variable x, we apply the rule: *Multiply x by 4 and add 1.* Since this arithmetic gives a single value of the dependent variable y, the equation defines a function.

b. Since the table assigns two different values of y, 3 and 7, to the x-value of 5, it does not define y as a function of x.

Self Check

a. Does $y = 2 - x^2$ define a function?

b. Does the table below define a function?

x	y
2	4
1	1
0	0
-1	1
-2	4

Answers: **a.** yes   **b.** yes ■

 COMMENT   The table in the above Self Check illustrates an important fact about functions. In a function, different values of x can determine the *same* value of y. In the table, x-values of 2 and -2 determine a y-value of 4, and x-values of 1 and -1 determine a y-value of 1. Nevertheless, each value of x determines exactly one value of y, so the table does define a function.

Domain and range of a function

We have seen that functions can be represented by equations in two variables. Some examples of functions are

$$y = 2x - 10, \qquad y = x^2 + 2x - 3, \qquad \text{and} \qquad s = 5 - 16t$$

For a function, the set of all possible values of the independent variable (the inputs) is called the **domain of the function.** The set of all possible values of the dependent variable (the outputs) is called the **range of the function.**

EXAMPLE 2   *Finding the domain and range of a function.*   Find the domain and range of $y = |x|$.

Solution

To find the domain of $y = |x|$, we determine which real numbers are allowable inputs for x. Since we can find the absolute value of any real number, the domain is the set of all real numbers. Since the absolute value of any real number x is greater than or equal to zero, the range of $y = |x|$ is the set of all real numbers greater than or equal to zero.

Self Check

Find the domain and range of the function $y = -x$.

Answer: domain: all real numbers; range: all real numbers ■

Function notation

When the variable y is a function of x, there is a special notation that we can use to denote the function.

Function notation

> The notation $y = f(x)$ denotes that y is a function of x.

The notation $y = f(x)$ is read as "y equals f of x." Note that y and $f(x)$ are two different notations for the same quantity. Thus, the equations $y = 4x + 1$ and $f(x) = 4x + 1$ represent the same relationship.

 COMMENT The symbol $f(x)$ denotes a function. It does not mean "f times x."

The notation $y = f(x)$ provides a way of denoting the value of y that corresponds to some number x. For example, if $f(x) = 4x + 1$, the value of y that is determined when $x = 2$ is denoted by $f(2)$.

$$f(x) = 4x + 1 \quad \text{The function.}$$
$$f(2) = 4(2) + 1 \quad \text{Replace } x \text{ with 2.}$$
$$= 8 + 1$$
$$= 9$$

Thus, $f(2) = 9$.

The letter f used in the notation $y = f(x)$ represents the word *function*. However, other letters can be used to represent functions. For example, $y = g(x)$ and $y = h(x)$ also denote functions involving the variable x.

EXAMPLE 3 *Evaluating functions.* For $g(x) = 3 - 2x$ and $h(x) = x^3 - 1$, find **a.** $g(3)$ and **b.** $h(-2)$.

Solution

a. To find $g(3)$, we use the function rule $g(x) = 3 - 2x$ and replace x with 3.

$$g(x) = 3 - 2x$$
$$g(3) = 3 - 2(3) \quad \text{Substitute 3 for } x.$$
$$= 3 - 6 \quad \text{Do the multiplication.}$$
$$= -3$$

So $g(3) = -3$.

b. To find $h(-2)$, we use the function rule $h(x) = x^3 - 1$ and replace x with -2.

$$h(x) = x^3 - 1$$
$$h(-2) = (-2)^3 - 1 \quad \text{Substitute } -2 \text{ for } x.$$
$$= -8 - 1 \quad \text{Evaluate the power.}$$
$$= -9$$

So $h(-2) = -9$.

Self Check

Find $g(0)$ and $h(4)$ using the functions in Example 3.

Answers: **a.** 3, **b.** 63 ■

We can think of a function as a machine that takes some input x and turns it into some output $f(x)$, as shown in Figure 9-56(a). The machine in Figure 9-56(b) turns the input value of -2 into the output value of -9, and we can write $f(-2) = -9$.

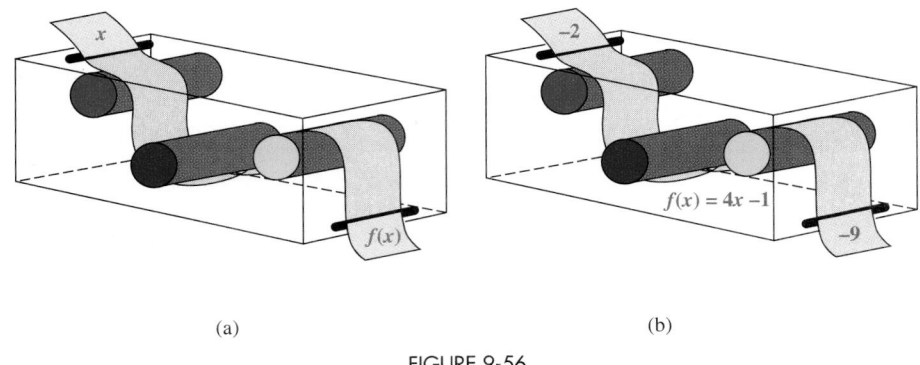

(a)   (b)

FIGURE 9-56

Accent on Technology: **Business profits**

Accountants have found that the function $f(x) = -0.000065x^2 + 12x - 278,000$ estimates the profit a bowling alley will make when x games are bowled per year. Suppose that management predicts that 90,000 games will be bowled in the upcoming year. The expected profit for that year can be found by evaluating $f(90,000)$ on a scientific calculator.

$$f(90,000) = -0.000065(90,000)^2 + 12(90,000) - 278,000$$

Keystrokes   .000065 $\boxed{+/-}$ $\boxed{\times}$ 90000 $\boxed{x^2}$ $\boxed{+}$ 12 $\boxed{\times}$ 90000 $\boxed{-}$ 278000 $\boxed{=}$

$$\boxed{275500}$$

To evaluate $f(90,000)$ with a graphing calculator, we enter these numbers and press these keys.

Keystrokes $\boxed{(-)}$ .000065 $\boxed{\times}$ 90000 $\boxed{x^2}$ $\boxed{+}$ 12 $\boxed{\times}$ 90000 $\boxed{-}$ 278000 $\boxed{ENTER}$

```
-.000065*90000² +
12*90000  -  278000
                275500
```

Graphs of functions

We have seen that a function, such as $f(x) = 4x + 1$, assigns to each value of x a single value of y. The ordered pairs (x, y) that a function determines can be shown on a graph. Since $y = f(x)$, the graph of the function $f(x) = 4x + 1$ is the same as the graph of the equation $y = 4x + 1$. We can graph the function by making a **table of values,** plotting the points, and drawing the graph.

To make a table of values for $f(x) = 4x + 1$, we will choose numbers for x and find the corresponding values of $f(x)$. If $x = -1$, we have

$$f(x) = 4x + 1$$
$$f(-1) = 4(-1) + 1 \quad \text{Substitute } -1 \text{ for } x.$$
$$= -4 + 1 \quad \text{Do the multiplication.}$$
$$= -3$$

We have found that $f(-1) = -3$. In a similar manner, we find the corresponding values for $f(x)$ for x-values of 0 and 2 and record them in the table of values in Figure 9-57 (next page). If we plot the ordered pairs in the table and draw a straight line through them, we get the graph of the function $f(x) = 4x + 1$ shown in the figure.

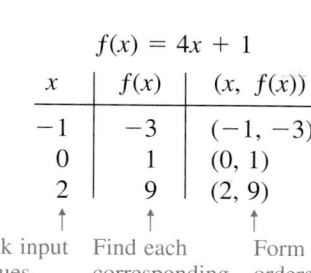

$$f(x) = 4x + 1$$

x	$f(x)$	$(x, f(x))$
-1	-3	$(-1, -3)$
0	1	$(0, 1)$
2	9	$(2, 9)$

Pick input values from the domain.

Find each corresponding output value: $f(-1)$, $f(0)$, and $f(2)$.

Form ordered pairs.

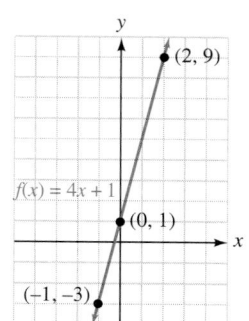

FIGURE 9-57

Any linear equation, except those of the form $x = a$, can be written using function notation by writing it in slope–intercept form ($y = mx + b$) and then replacing y with $f(x)$. We call this type of function a **linear function.**

Figure 9-58 shows the graphs of four basic functions.

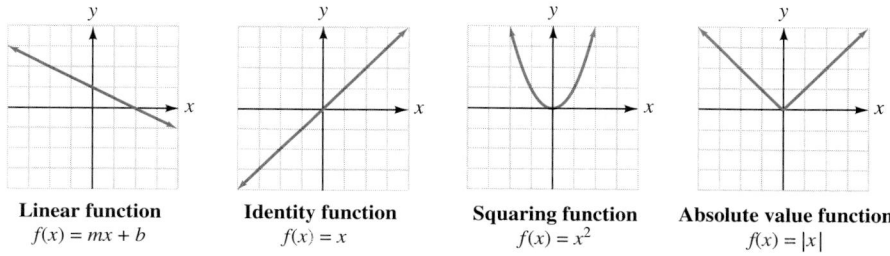

Linear function	Identity function	Squaring function	Absolute value function
$f(x) = mx + b$	$f(x) = x$	$f(x) = x^2$	$f(x) = |x|$

FIGURE 9-58

The vertical line test

We can use the **vertical line test** to determine whether a given graph is the graph of a function. If any vertical line intersects a graph more than once, the graph cannot represent a function, because to one value of x, there corresponds more than one value of y. The graph in Figure 9-59(a), shown in red, is not the graph of a function, because the x-value -1 determines three different y-values: 3, -1, and -4.

The graph shown in Figure 9-59(b) does represent a function, because every vertical line intersects the graph exactly once.

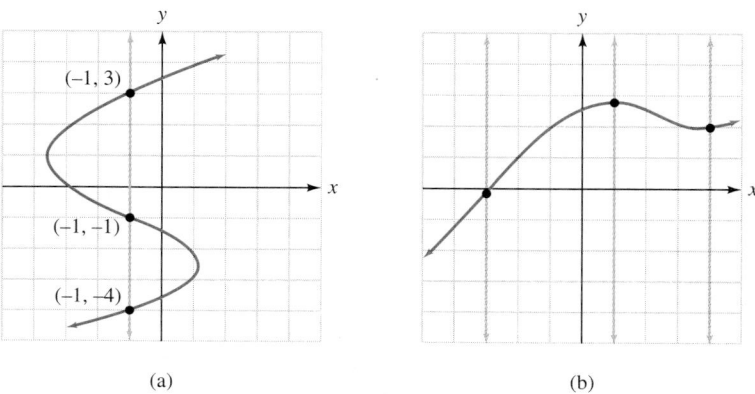

(a)   (b)

FIGURE 9-59

EXAMPLE 4 *The vertical line test.* Which of the following graphs in red are graphs of functions?

a.

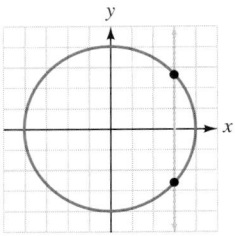

b.

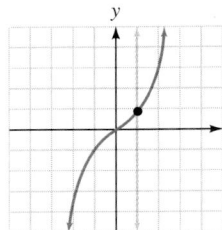

Solution

a. This graph is not the graph of a function, because the vertical line intersects the graph at more than one point.

b. This graph is the graph of a function, because no vertical line will intersect the graph at more than one point.

Self Check

Which of the following graphs are graphs of functions?

a.

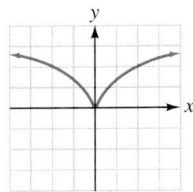

b.

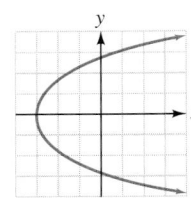

Answers:
a. function,   **b.** not a function

STUDY SET Section 9.7

VOCABULARY   *Fill in the blanks.*

1. A _____ is a rule that assigns to each value of the independent variable a single value of another variable.

2. The set of all possible input values for a function is called the _____, and the set of all possible output values is called the _____.

3. For $y = 2x + 8$, x is called the _____ variable, and y is called the _____ variable.

4. $f(x) = 6 - 5x$ is an example of _____ notation.

CONCEPTS

5. Consider the function $f(x) = x^2$.
  a. If positive real numbers are substituted for x, what type of numbers result?
  b. If negative real numbers are substituted for x, what type of numbers result?
  c. If zero is substituted for x, what number results?
  d. What are the domain and range of the function?

6. Consider the function $g(x) = x^4$.
  a. What type of numbers can be inputs in this function? What is the special name for this set?

  b. What type of numbers will be outputs in this function? What is the special name for this set?

7. Consider the following problems. Fill in the blank so that they ask for the same thing.
  1. In the equation $y = -5x + 1$, find the value of y when $x = -1$.
  2. In the equation $f(x) = -5x + 1$, find _____.

8. A function can be thought of as a machine that converts inputs into outputs. Use the terms *domain, range, input,* and *output* to label the diagram of a function machine in Illustration 1. Then find $f(2)$.

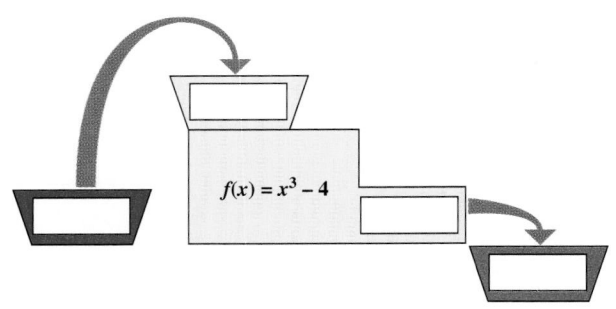

$f(x) = x^3 - 4$

ILLUSTRATION 1

9. See Illustration 2.
   a. Give the coordinates of the points where the given vertical line intersects the graph.
   b. Is this the graph of a function? Explain your answer.

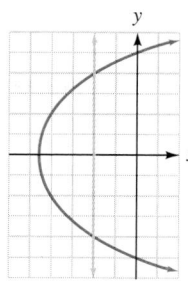

ILLUSTRATION 2

10. A student was asked to determine whether the graph in Illustration 3 is the graph of a function. What is wrong with the following reasoning?
   When I draw a vertical line through the graph, it intersects the graph only once. By the vertical line test, this is the graph of a function.

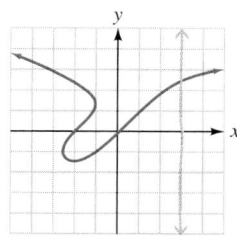

ILLUSTRATION 3

25.

x	y
1	7
2	15
3	23
4	16
5	8

26.

x	y
−1	1
−3	1
−5	1
−7	1
−9	1

27.

x	y
−4	6
−1	0
0	−3
2	4
−1	2

28.

x	y
30	2
30	4
30	6
30	8
30	10

29.

t	d
3	4
3	−4
4	3
4	−3

30.

x	y
1	1
2	2
3	3
4	4

31.

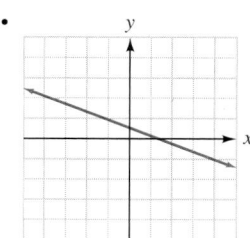

32.

33.

34.

35.

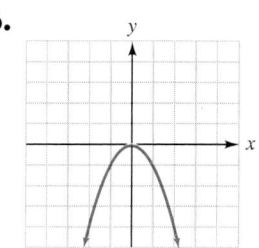

36.

NOTATION

11. Fill in the blanks to make the statements true. The function notation $f(4) = -5$ states that when 4 is substituted for ____ in function f, the result is ____. This fact can be illustrated graphically by plotting the point (____ , ____).

12. Fill in the blank: $f(x) = 6 - 5x$ is read as "f _____ x is $6 - 5x$."

13. Fill in the blanks: If $f(x) = 6 - 5x$, then $f(0) = 6$ is read as "f _____ zero _____ 6."

14. Tell whether this statement is true or false: The equations $y = 3x + 5$ and $f(x) = 3x + 5$ are the same.

PRACTICE   *Tell whether a function is defined. If it is not, indicate an input for which there is more than one output.*

15. $y = 2x + 10$

16. $y = x - 15$

17. $y = x^2$

18. $y = |x|$

19. $y^2 = x$

20. $|y| = x$

21. $y = x^3$

22. $y = -x$

23. $x = 3$

24. $y = 3$

Find the domain and range of the function.

37. $f(x) = x + 1$

38. $f(x) = 3x - 2$

39. $y = x^2$

40. $y = -|x|$

41. $f(x) = x^3$

42. $f(x) = x$

Find each value.

43. $f(x) = 4x - 1$
   a. $f(1)$ **b.** $f(-2)$

   c. $f\left(\dfrac{1}{4}\right)$ **d.** $f(50)$

44. $g(x) = 1 - 5x$
   a. $g(0)$ **b.** $g(-75)$

   c. $g(0.2)$ **d.** $g\left(-\dfrac{4}{5}\right)$

45. $h(t) = 2t^2$
   a. $h(0.4)$ **b.** $h(-3)$

   c. $h(1,000)$ **d.** $h\left(\dfrac{1}{8}\right)$

46. $v(t) = 6 - t^2$
   a. $v(30)$ **b.** $v(6)$

   c. $v(-1)$ **d.** $v(0.5)$

47. $s(x) = |x - 7|$
   a. $s(0)$ **b.** $s(-7)$

   c. $s(7)$ **d.** $s(8)$

48. $f(x) = |2 + x|$
   a. $f(0)$ **b.** $f(2)$

   c. $f(-2)$ **d.** $f(-99)$

49. $f(x) = x^3 - x$
   a. $f(1)$ **b.** $f(10)$

   c. $f(-3)$ **d.** $f(6)$

50. $g(x) = x^4 + x$
   a. $g(1)$ **b.** $g(-2)$

   c. $g(0)$ **d.** $g(10)$

51. 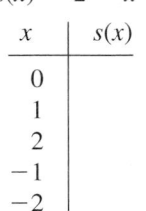 If $f(x) = 3.4x^2 - 1.2x + 0.5$, find $f(-0.3)$.

52. If $g(x) = x^4 - x^3 + x^2 - x$, find $g(-12)$.

Complete each table of values and graph each function.

53. $f(x) = -2 - 3x$

x	$f(x)$
0	
1	
-1	
-2	

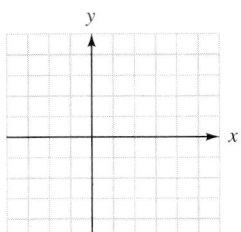

54. $h(x) = |1 - x|$

x	$h(x)$
0	
1	
2	
3	
-1	
-2	

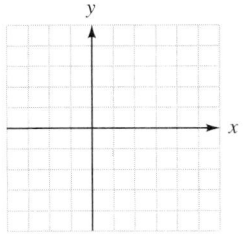

55. $f(x) = \frac{1}{2}x - 2$

x	y
-2	
0	
2	

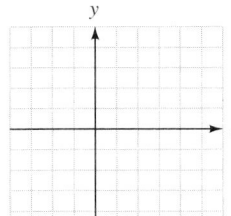

56. $f(x) = -\frac{2}{3}x + 3$

x	y
0	
3	
6	

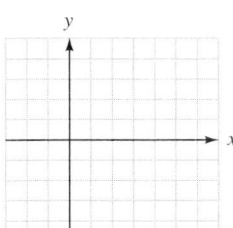

57. $s(x) = 2 - x^2$

x	$s(x)$
0	
1	
2	
-1	
-2	

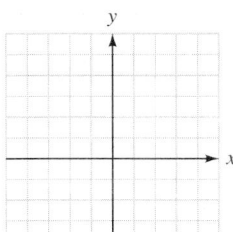

58. $g(x) = 1 + x^3$

x	$g(x)$
0	
1	
2	
-1	
-2	

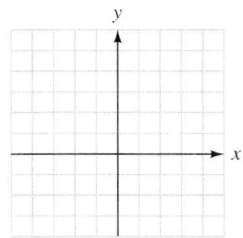

APPLICATIONS

59. REFLECTIONS   When a beam of light hits a mirror, it is reflected off the mirror at the same angle that the incoming beam struck the mirror, as shown in Illustration 4. What type of function could serve as a mathematical model for the path of the light beam shown here?

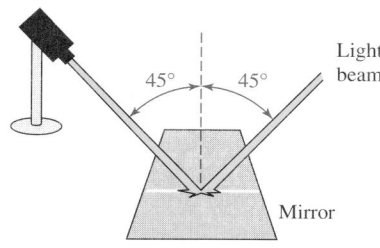

ILLUSTRATION 4

60. MATHEMATICAL MODELS   Illustration 5 shows the path of a basketball shot taken by a player. What type of function could be used to mathematically model the path of the basketball?

ILLUSTRATION 5

61. TIDES   Illustration 6 shows the graph of a function f, which gives the height of the tide for a 24-hour period in Seattle, Washington. (Note that military time is used on the x-axis: 3 A.M. = 3, noon = 12, 3 P.M. = 15, 9 P.M. = 21, and so on.)
a. Find the domain of the function.

b. Find $f(3)$.
c. Find $f(6)$.
d. Estimate $f(15)$.
e. What information does $f(12)$ give?

f. Estimate $f(21)$.

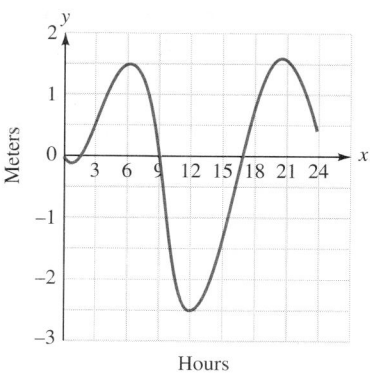

ILLUSTRATION 6

62. SOCCER   Illustration 7 shows the graphs of three functions on the same coordinate system: $g(x)$ represents the number of girls, $f(x)$ represents the number of boys, and $t(x)$ represents the total number playing high school soccer in year x.
a. What is the domain of each of these functions?

b. Find $g(87)$, $f(86)$, and $t(93)$.

c. Estimate $g(95)$, $f(95)$, and $t(95)$.

d. For what year x was $g(x) = 75,000$?
e. For what year x was $f(x) = 225,000$?
f. For what year x was $t(x)$ first greater than 350,000?

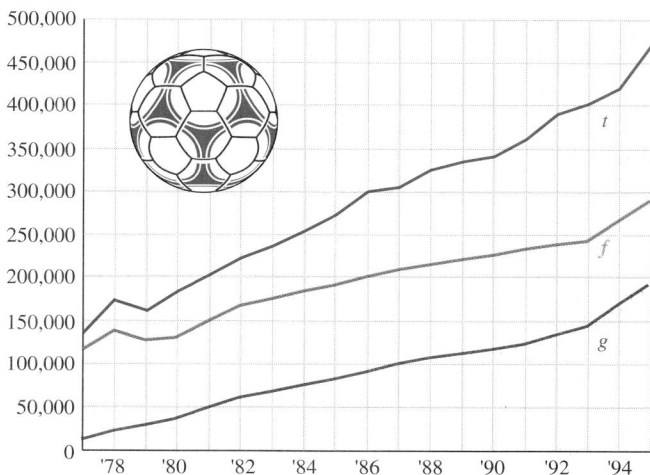

Based on data from *Los Angeles Times* (Dec. 12, 1996), p. A5

ILLUSTRATION 7

63. LAWN SPRINKLERS   The function $A(r) = \pi r^2$ can be used to determine the area that will be watered by a rotating sprinkler that sprays out a stream of water r feet. See Illustration 8. Find $A(5)$, $A(10)$, and $A(20)$. Round to the nearest tenth.

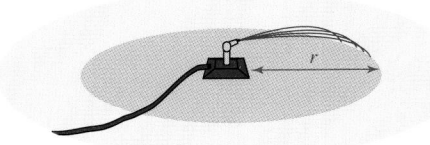

ILLUSTRATION 8

64. PARTS LIST   The function
$$f(r) = 2.30 + 3.25(r + 0.40)$$
approximates the length (in feet) of the belt that joins the two pulleys shown in Illustration 9. r is the radius (in feet) of the smaller pulley. Find the belt length needed for each pulley in the parts list.

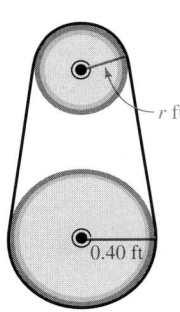

Parts list		
Pulley	**r**	**Belt length**
P-45M	0.32	
P-08D	0.24	
P-00A	0.18	
P-57X	0.38	

ILLUSTRATION 9

WRITING

65. In the function $y = -5x + 2$, why do you think x is called the *independent* variable and y the *dependent* variable?

66. Explain what a politician meant when she said, "The speed at which the downtown area will be redeveloped is a function of the number of low-interest loans made available to the property owners."

REVIEW

67. Give the equation of the horizontal line passing through $(-3, 6)$.

68. Is $t = -3$ a solution of $t^2 - t + 1 = 13$?

69. Write the formula that relates profit, revenue, and costs.

70. What is the word used to represent the perimeter of a circle?

71. Use the distributive property to remove the parentheses in $-3(2x - 4)$.

72. Evaluate $r^2 - r$ for $r = -0.5$.

73. Write an expression for how many eggs there are in d dozen.

74. On a rectangular coordinate graph, what variable is associated with the horizontal axis?

Describing Linear Relationships

In Chapter 9, we discussed ways to mathematically describe linear relationships between two quantities.

Equations in Two Variables

The general form of the equation of a line is $Ax + By = C$. Two very useful forms of the equation of a line are the slope–intercept form and the point–slope form.

1. Write the equation of a line with a slope of -3 and a y-intercept of $(0, -4)$.

2. Write the equation of the line that passes through $(5, 2)$ and $(-5, 0)$. Answer in slope–intercept form.

Rectangular Coordinate Graphs

The graph of an equation is a "picture" of all of its solutions (x, y). Important information can be obtained from a graph.

3. Complete the table of solutions for $2x - 4y = 8$. Then graph the equation.

$$2x - 4y = 8$$

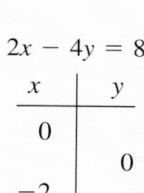

x	y
0	
	0
-2	

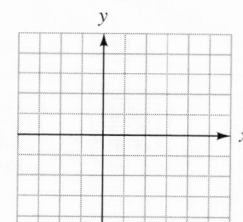

4. See Illustration 1.
   a. What information does the y-intercept of the graph give us?
   b. What is the slope of the line and what does it tell us?

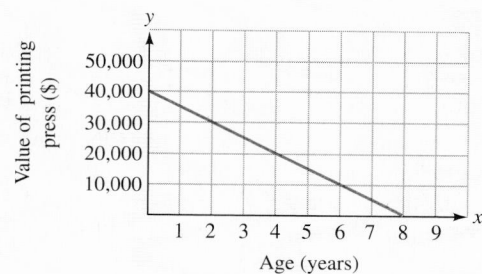

ILLUSTRATION 1

5. Consider the line graphed in Illustration 2.
   a. Find a point on the line.

   b. Determine the slope of the line.

   c. Write the equation of the line. Express your answer in slope–intercept form.

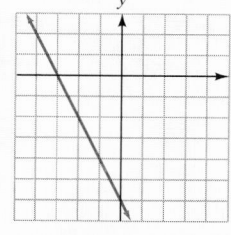

ILLUSTRATION 2

6. Write the equation of the line that passes through $(1, -1)$ and is parallel to the line graphed in Illustration 3.

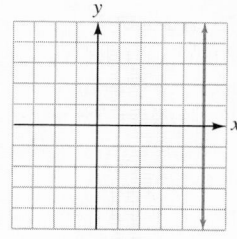

ILLUSTRATION 3

Linear Functions

We can use the notation $f(x) = mx + b$ to describe linear functions.

7. The function $f(x) = 35x + 25$ gives the cost (in dollars) to rent a cement mixer for x days. Find $f(3)$. What does it represent?

8. The function $T(x) = \frac{1}{4}x + 40$ predicts the outdoor temperature T in degrees Fahrenheit using the number of cricket chirps x per minute. Find $T(160)$.

ACCENT ON TEAMWORK

Section 9.1

DAILY HIGH TEMPERATURE   For a 2-week period, plot the daily high temperature for your city on a rectangular coordinate system. You can normally find this information in a local newspaper. Label the x-axis "observation day" and the y-axis "daily high temperature in degrees Fahrenheit." For example, the ordered pair (3, 72) indicates that on day 3 of the observation period, the high temperature was 72°F. At the end of the 2-week period, see whether any temperature trend is apparent from the graph.

Section 9.2

TRANSLATIONS   On a piece of graph paper, sketch the graph of $y = |x|$ with a black marker. Using a different color, sketch the graphs of $y = |x| + 2$ and $y = |x| - 2$ on the same coordinate system. On another piece of graph paper, do the same for $y = |x|$ and $y = |x + 2|$ and $y = |x - 2|$. Make some observations about how the graph of $y = |x|$ is "moved" or "translated" by the addition or subtraction of 2. Use what you have learned to discuss the graphs of $y = x^2$, $y = x^2 + 2$, $y = x^2 - 2$, $y = (x + 2)^2$, and $y = (x - 2)^2$.

Section 9.3

COMPUTER GRAPHING PROGRAMS   If your school has a mathematics computer lab, ask the lab supervisor whether there is a graphing program on the system. If so, familiarize yourself with the operation of the program and then graph each of the equations from Figure 9-22 and from Examples 3–6 in Section 9.3. Print out each graph and compare with those in the textbook.

Section 9.4

MEASURING SLOPE   Use a tape measure (and a level if necessary) to find the slopes of five objects by finding $\frac{\text{rise}}{\text{run}}$. See the applications in Study Set 9.4 for some ideas about what you can measure. Record your results in a chart like the one shown in Illustration 1. List the examples in increasing order of magnitude, starting with the smallest slope.

Section 9.5

SHOPPING   Visit a local grocery store and find the price per pound of bananas. Make a rectangular coordinate graph that could be posted next to the scale in the produce area so that shoppers could determine from the graph the cost of a banana

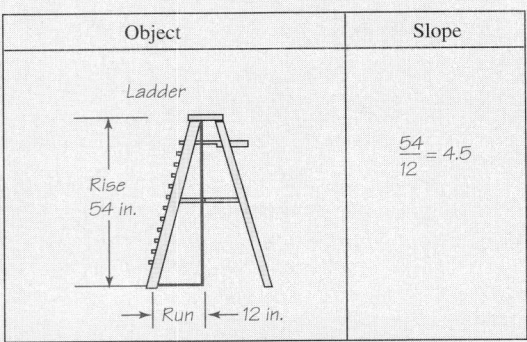

Object	Slope
Ladder — Rise 54 in., Run 12 in.	$\frac{54}{12} = 4.5$

ILLUSTRATION 1

purchase up to 8 pounds in weight. Label the x-axis in quarters of a pound and label the y-axis in cents.

Section 9.6

MATCHING GAME   Have a student in your group write 10 linear equations on 3×5 note cards, one equation per card. Then have him or her graph each equation on a separate set of 10 cards. Shuffle each set of cards. Then put all the equation cards on one side of a table and all the cards with graphs on the other side. Work together to match each equation with its proper graph.

Section 9.7

FUNCTIONS   We have seen that a function can be thought of as a machine that takes some input x and turns it into some output $f(x)$. (See Illustration 2.) Write a function that takes an input of 6 and turns it into an output of 19, where

a. only 1 operation is performed to get the output.
b. 2 operations are performed to get the output.
c. 3 operations are performed to get the output.
d. 4 operations are performed to get the output.

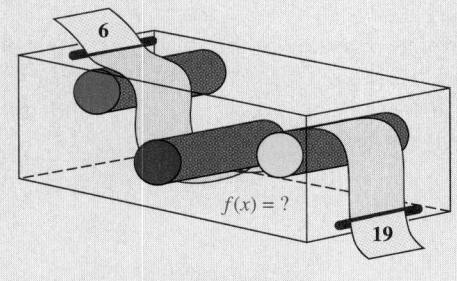

$f(x) = ?$

ILLUSTRATION 2

| SECTION 9.1 | *Graphing Using the Rectangular Coordinate System* |

CONCEPTS

A *rectangular coordinate system* is composed of a horizontal number line called the *x*-axis and a vertical number line called the *y*-axis.

The coordinates of the *origin* are (0, 0).

To *graph* ordered pairs means to locate their position on a coordinate system.

The two axes divide the coordinate plane into four distinct regions called *quadrants*.

REVIEW EXERCISES

1. a. Graph the points with coordinates $(-1, 3)$, $(0, 1.5)$, $(-4, -4)$, $(2, \frac{7}{2})$, and $(4, 0)$.

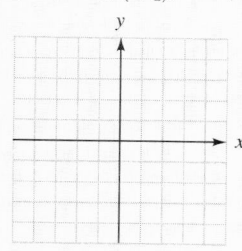

b. Use the graph in Illustration 1 to complete the table.

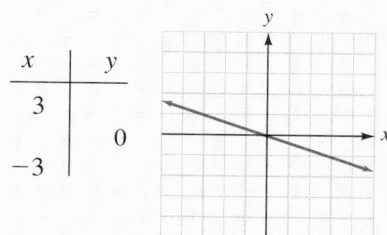

x	y
3	
	0
-3	

ILLUSTRATION 1

2. In what quadrant does the point $(-3, -4)$ lie?

3. SNOWFALL   The amount of snow on the ground at a mountain resort was measured once each day over a 7-day period. (See Illustration 2.)

   a. On the first day, how much snow was on the ground?

   b. What was the difference in the amount of snow on the ground when the measurements were taken the second and third day?

   c. How much snow was on the ground on the sixth day?

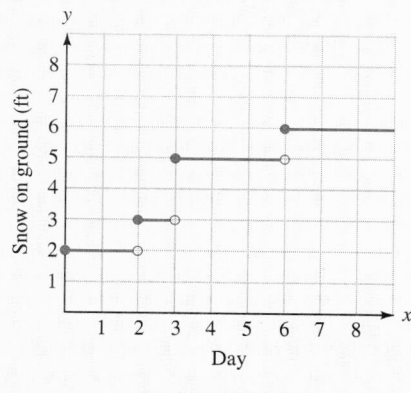

ILLUSTRATION 2

Students enrolled

ILLUSTRATION 3

4. COLLEGE ENROLLMENT   The graph in Illustration 3 gives the number of students enrolled at a college for the period from 4 weeks before to 5 weeks after the semester began.

   a. What was the maximum enrollment and when did it occur?

   b. How many students had enrolled 2 weeks before the semester began?

   c. When was enrollment 2,250?

The *midpoint formula:*

$$\left(\frac{x_1 + x_2}{2}, \frac{y_1 + y_2}{2} \right)$$

5. Find the midpoint of a line segment with endpoints at $P(-3, 7)$ and $Q(3, -3)$.

| SECTION 9.2 | *Equations Containing Two Variables* |

An ordered pair is a *solution* if, after substituting the values of the ordered pair for the variables in the equation, the result is a true statement.

Solutions of an equation can be shown in a *table of solutions*.

In an equation in x and y, x is called the *independent variable,* or *input,* and y is called the *dependent variable,* or *output.*

To graph an equation in two variables:
1. Make a table of solutions that contains several solutions written as ordered pairs.
2. Plot each ordered pair.
3. Draw a line or smooth curve through the points.

In many application problems, we encounter equations that contain variables other than x and y.

6. Check to see whether $(-3, 5)$ is a solution of $y = |2 + x|$.

7. a. Complete the table of solutions and graph the equation $y = -x^3$.

$$y = -x^3$$

x	y	(x, y)
-2		
-1		
0		
1		
2		

b. How would the graph of $y = -x^3 + 2$ compare to the graph of the equation given in part a?

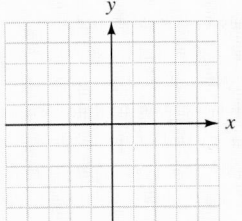

8. The graph in Illustration 4 shows the relationship between the number of oranges O an acre of land will yield if t orange trees are planted on it.
  a. If $t = 70$, what is O?
  b. What importance does the point $(40, 18)$ on the graph have?

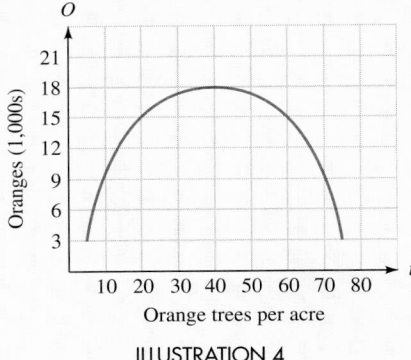

ILLUSTRATION 4

| SECTION 9.3 | *Graphing Linear Equations* |

An equation whose graph is a straight line and whose variables are raised to the first power is called a *linear equation.*

The *general* or *standard form* of a linear equation is $Ax + By = C$, where A, B, and C are real numbers and A and B are not both zero.

9. Classify each equation as either linear or nonlinear.
  a. $y = |x + 2|$ **b.** $3x + 4y = 12$
  c. $y = 2x - 3$ **d.** $y = x^2 - x$

10. The equation $5x + 2y = 10$ is in general form; what are A, B, and C?

11. Complete the table of solutions for the equation $3x + 2y = -18$.

x	y	(x, y)
-2		$(-2, \ \ )$
	3	$(\ \ , 3)$

To graph a linear equation:
1. Find three (x, y) pairs that satisfy the equation by picking three arbitrary x-values and finding their corresponding y-values.
2. Plot each ordered pair.
3. Draw a straight line through the points.

12. Solve the equation $x + 2y = 6$ for y, find three solutions, and then graph it.

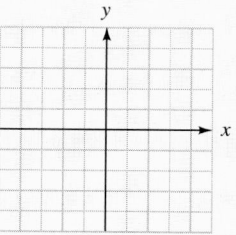

To find the *y-intercept* of a linear equation, substitute 0 for x in the equation of the line and solve for y. To find the *x-intercept* of a linear equation, substitute 0 for y in the equation of the line and solve for x.

13. Graph $-4x + 2y = 8$ by finding its x- and y-intercepts.

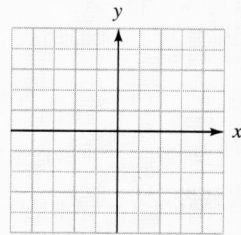

The equation $y = b$ represents the horizontal line that intersects the y-axis at $(0, b)$. The equation $x = a$ represents the vertical line that intersects the x-axis at $(a, 0)$.

14. Graph each equation.
a. $y = 4$

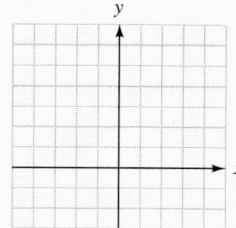

b. $x = -1$

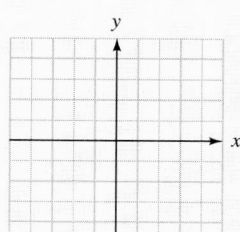

15. Since two points determine a line, only two points are needed to graph a linear equation. Why is it is a good idea to plot a third point?

| SECTION 9.4 | *Rate of Change and the Slope of a Line* |

The *slope m* of a nonvertical line is a number that measures "steepness" by finding the ratio $\frac{\text{rise}}{\text{run}}$.

$$m = \frac{\text{change in the } y\text{-values}}{\text{change in the } x\text{-values}}$$

16. In each case, find the slope of the line.
a.

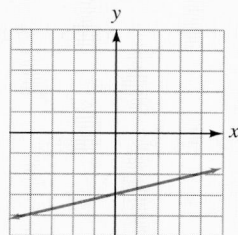

b. The line with the table of solutions shown here.

x	y	(x, y)
2	-3	$(2, -3)$
4	-17	$(4, -17)$

If $P(x_1, y_1)$ and $Q(x_2, y_2)$ are two points on a nonvertical line, the slope m of line PQ is

$$m = \frac{y_2 - y_1}{x_2 - x_1}$$

Lines that rise from left to right have a *positive slope,* and lines that fall from left to right have a *negative slope.*

Horizontal lines have a slope of zero. Vertical lines have *undefined* slope.

The slope of a line gives a rate of change.

c. The line passing through the points $(2, -5)$ and $(5, -5)$.

d. The line passing through the points $(1, -4)$ and $(3, -7)$.

17. Graph the line that passes through $(-2, 4)$ and has slope $m = -\frac{4}{5}$.

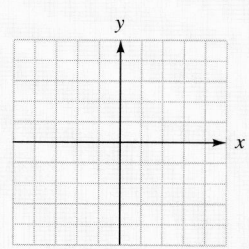

18. TOURISM   The graph in Illustration 5 shows the number of international travelers to the United States from 1986–1998, in two-year increments.
  a. Between what two years did the largest decline in the number of visitors occur? What was the rate of change?
  b. Between what two years did the largest increase in the number of visitors occur? What was the rate of change?

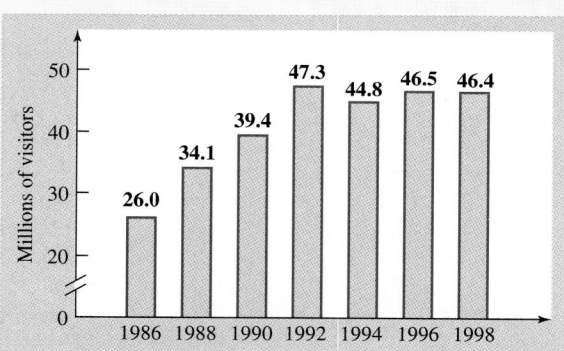

Based on data from *World Almanac 2000*

ILLUSTRATION 5

SECTION 9.5   *Describing Linear Relationships*

If a linear equation is written in *slope–intercept* form,

$$y = mx + b$$

the graph of the equation is a line with slope m and y-intercept $(0, b)$.

19. Find the slope and the y-intercept of each line.
  a. $y = \frac{3}{4}x - 2$
  b. $y = -4x$

20. Find the slope and the y-intercept of the line determined by $9x - 3y = 15$. Then graph it.

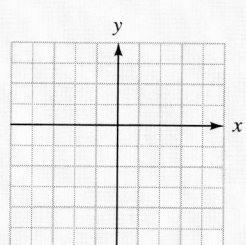

The *rate of change* is the slope of the graph of a linear equation.

21. COPIER USAGE   A business buys a used copy machine that, when purchased, has already produced 75,000 copies.

a. If the business plans to run 300 copies a week, write a linear equation that would find the number of copies c the machine has made in its lifetime after the business has used it for w weeks.

b. Use your result to part a to predict the total number of copies that will have been made on the machine 1 year, or 52 weeks, after being purchased by the business.

Two lines with the same slope are *parallel*.

The product of the slopes of *perpendicular* lines is -1.

22. Without graphing, tell whether graphs of the given pairs of lines would be parallel, perpendicular, or neither.

a. $y = -\dfrac{2}{3}x + 6$

$y = -\dfrac{2}{3}x - 6$

b. $x + 5y = -10$

$y = 5x$

SECTION 9.6 — *Writing Linear Equations*

If a line with slope m passes through the point (x_1, y_1), the equation of the line in *point–slope* form is

$$y - y_1 = m(x - x_1)$$

23. Write the equation of a line with the given slope that passes through the given point. Express the result in slope–intercept form and graph the equation.

a. $m = 3, (1, 5)$

b. $m = -\dfrac{1}{2}, (-4, -1)$

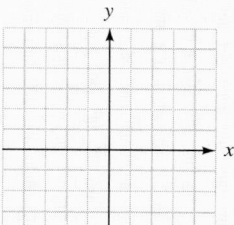

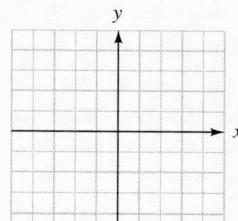

24. Write the equation of the line with the following characteristics. Express the result in slope–intercept form.

a. passing through $(3, 7)$ and $(-6, 1)$

b. horizontal, passing through $(6, -8)$

25. CAR REGISTRATION   When it was 2 years old, the annual registration fee for a Dodge Caravan was $380. When it was 4 years old, the registration fee dropped to $310. If the relationship is linear, write an equation that gives the registration fee f in dollars for the van when it is x years old.

SECTION 9.7 — *Functions*

A *function* is a rule that assigns to each input value a single output value.

26. In each case, tell whether a function is defined.

a. $y = 3x - 2$

b. $y^2 = x$

c.

x	2	2	3	4	5	6
y	-2	2	3	-4	5	-6

For a function, the set of all possible values of the independent variable *x* (the inputs) is called the *domain,* and the set of all possible values of the dependent variable *y* (the outputs) is called the *range.*

The notation $y = f(x)$ denotes that y is a function of x.

27. Find the domain and range of each function.

a. $f(x) = x + 10$ **b.** $y = x^2$

28. For the function $g(x) = 1 - 6x$, find each value.

a. $g(1)$ **b.** $g(-6)$

c. $g(0.5)$ **d.** $g\left(\dfrac{3}{2}\right)$

Four basic functions are

Linear: $f(x) = mx + b$
Identity: $f(x) = x$
Squaring: $f(x) = x^2$
Absolute value: $f(x) = |x|$

29. Complete the table of values and graph the function.

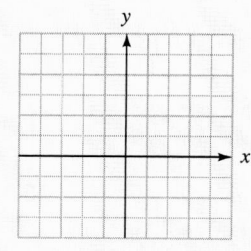

$h(x) = 1 - |x|$

x	$h(x)$
0	
1	
2	
-1	
-2	
-3	

We can use the *vertical line test* to determine whether a graph is the graph of a function.

30. Tell whether each graph is the graph of a function.

a.   **b.**

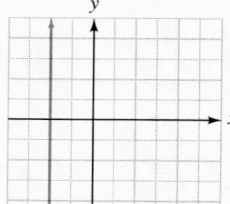   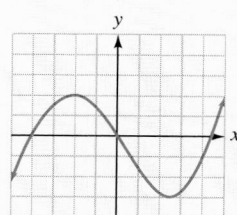

31. The function $f(r) = 15.7r^2$ estimates the volume in cubic inches of a can 5 inches tall with a radius of *r* inches. Find the volume of the can in Illustration 6. Round to the nearest tenth.

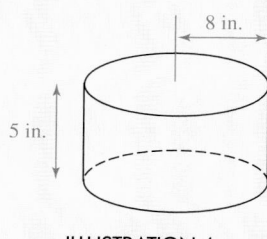

ILLUSTRATION 6

Chapter 9 Test

The graph in Illustration 1 shows the number of dogs being boarded in a kennel over a 3-day holiday weekend. Use the graph to answer Problems 1–4.

1. How many dogs were in the kennel 2 days before the holiday?

2. What is the maximum number of dogs that were boarded on the holiday weekend?

3. When were there 30 dogs in the kennel?

4. What information does the *y*-intercept of the graph give?

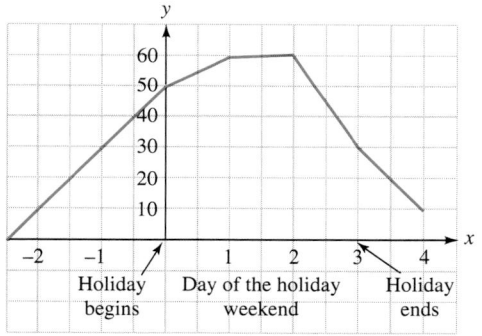

ILLUSTRATION 1

5. Graph $y = x^2 - 4$.

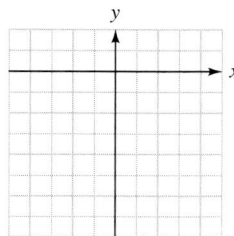

6. Graph $8x + 4y = -24$.

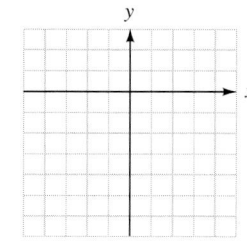

7. Is $(-3, -4)$ the midpoint of $P(-6, 2)$ and $Q(0, -10)$?

8. Is $y = x^3$ a linear equation?

9. What are the *x*- and *y*-intercepts of the graph of $2x - 3y = 6$?

10. Find the slope and the *y*-intercept of $x + 2y = 8$.

11. Graph $x = -4$.

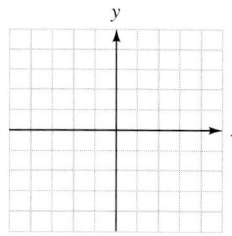

12. Graph the line passing through $(-2, -4)$ having a slope of $\frac{2}{3}$.

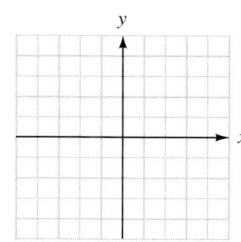

13. What is the slope of the line passing through $(-1, 3)$ and $(3, -1)$?

14. What is the slope of a vertical line?

15. What is the slope of a line that is perpendicular to a line with slope $-\frac{7}{8}$?

16. When graphed, are the lines $y = 2x + 6$ and $6x - 3y = 0$ parallel, perpendicular, or neither?

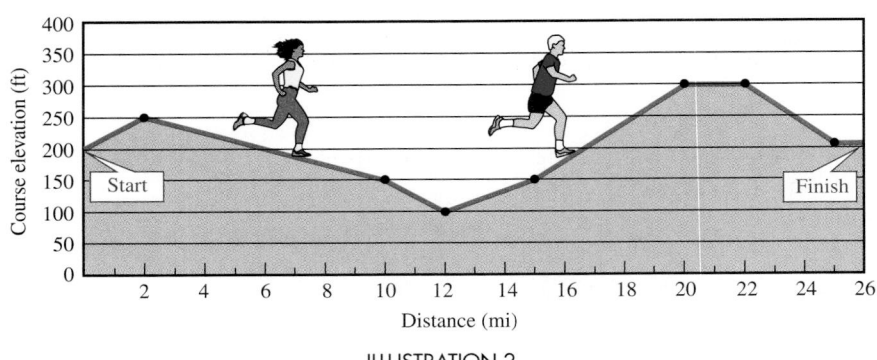

ILLUSTRATION 2

In Problems 17–18, refer to the graph in Illustration 2, which shows the elevation changes in a 26-mile marathon course. Give the rate of change of the part of the course that has . . .

17. the steepest incline

18. the steepest decline

19. DEPRECIATION   After it is purchased, a $15,000 computer loses $1,500 in resale value every year. Write a linear equation that gives the resale value v of the computer x years after being purchased.

20. Write the equation of the line passing through $(-2, 5)$ and $(-3, -2)$. Answer in slope–intercept form.

21. Is this the graph of a function?

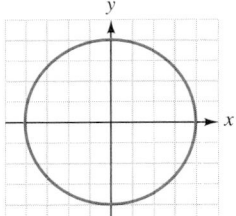

22. Find the domain and range of the function $f(x) = -|x|$.

23. Does the equation $y = 2x - 8$ define a function?

24. If $f(x) = 2x - 7$, find $f(-3)$.

25. If $g(x) = 3.5x^3$ find $g(6)$.

26. Explain what is meant by the statement slope $= \dfrac{\text{rise}}{\text{run}}$.

Chapters 1-9 Cumulative Review Exercises

1. TWA   The graph in Illustration 1 shows the 1999 and 2000 quarterly net losses for Trans World Airlines.
  a. In which quarter was the loss the least? Estimate it.

  b. In which quarter was the loss the greatest? Estimate it.

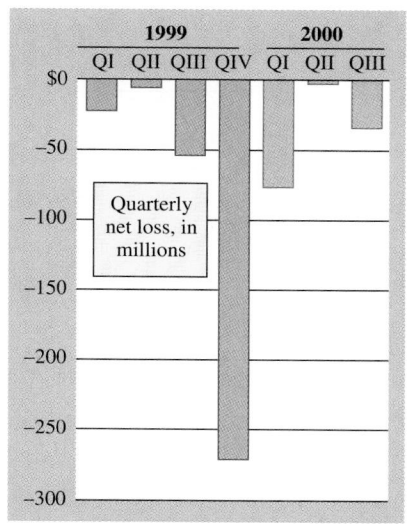

Based on data from Trans World Airlines and *Bloomberg News.*

ILLUSTRATION 1

2. Give the prime factorization of 108.

3. Write $\frac{1}{250}$ as a decimal.

4. Tell whether each statement is true or false.
  a. Every whole number is an integer.
  b. Every integer is a real number.
  c. 0 is a whole number, an integer, and a rational number.

5. ▦ AUTO SALES   Illustration 2 shows the top 5 best-selling vehicles in the United States in the year 2000, as reported by the automakers. Complete the table. Round to the nearest tenth of one percent.

6. Evaluate each expression.
  a. $12 - 2 \cdot 3$

  b. $\dfrac{(6 - 5)^4 - (-21)}{-27 + 4^2}$

  c. $19 - 2[(-3.1 + 6.1) \cdot 3]$

  d. $64 - 6[15 - (3)3]$

7. Evaluate $b^2 - 4ac$ for $a = 2$, $b = -8$, and $c = 4$.

8. Suppose x sheets from a 500-sheet ream of paper have been used. How many sheets are left?

9. How many terms does the algebraic expression $3x^2 - 2x + 1$ have? What is the coefficient of the second term?

10. Use the distributive property to remove parentheses.
  a. $2(x + 4)$ **b.** $2(x - 4)$
  c. $-2(x + 4)$ **d.** $-2(x - 4)$

Simplify each expression.

11. $5a + 10 - a$

12. $-2b^2 + 6b^2$

13. $(a + 2) - (a - 2)$

14. $-y - y - y$

Solve each equation.

15. $3x - 5 = 13$ **16.** $1.2 - x = -1.7$

17. $\dfrac{2x}{3} - 2 = 4$ **18.** $\dfrac{y - 2}{7} = -3$

19. $-3(2y - 2) - y = 5$ **20.** $9y - 3 = 6y$

21. $\dfrac{1}{3} + \dfrac{c}{5} = -\dfrac{3}{2}$ **22.** $5(x + 2) = 5x - 2$

Rank	Vehicle	Units sold		'99 ranks	% change
		2000	**1999**		
1	Ford F-Series pickup	876,716	869,001	1	+0.9
2	Chevrolet Silverado pickup	642,119	636,150	2	+0.9
3	Ford Explorer	445,157	428,772	5	
4	Toyota Camry	422,961	448,162	3	
5	Honda Accord	404,515	404,192	6	+0.1

Based on information from Reuters

ILLUSTRATION 2

23. Solve the equation for the indicated variable.
$y = mx + b$; for x

24. Find the perimeter and the area of the gauze pad of the bandage shown in Illustration 3.

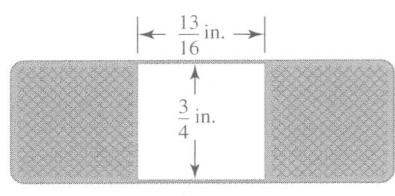

ILLUSTRATION 3

25. If the vertex of an isosceles triangle is 22°, find the measure of each base angle.

26. Complete the table.

Solution	% acid	Liters	Amount of acid
50% solution	0.50	x	
25% solution	0.25	$13 - x$	
30% mixture	0.30	13	

27. ROAD TRIP   A bus, carrying the members of a marching band, and a truck, carrying their instruments, leave a high school at the same time. The bus travels at 60 mph and the truck at 50 mph. In how many hours will they be 75 miles apart?

28. MIXING CANDY   Candy corn worth $1.90 per pound is to be mixed with black gumdrops that cost $1.20 per pound to make 200 pounds of a mixture worth $1.48 per pound. How many pounds of each candy should be used?

Solve each inequality, graph the solution set, and use interval notation to describe the solution.

29. $-\dfrac{3}{16}x \geq -9$

30. $8x + 4 > 3x + 4$

31. MEDICATION   Dosages for a certain medication are shown in Illustration 4. What is the dosage for
  a. a 5-year-old child?
  b. a 9-year-old child?

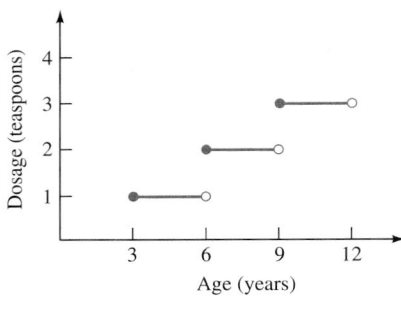

ILLUSTRATION 4

32. Is $(-2, 4)$ a solution of $y = 2x - 8$?

Graph each equation.

33. $y = |x - 2|$

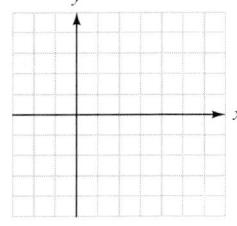

34. $4y + 2x = -8$

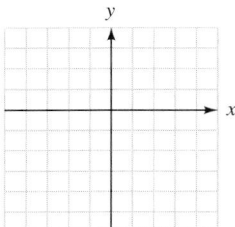

35. What is the slope of the graph of the line $y = 5$?

36. What is the slope of the line passing through $(-2, 4)$ and $(5, -6)$?

37. Find the slope and the y-intercept of the graph of the line described by $4x - 6y = -12$.

38. Write the equation of the line that has slope -2 and y-intercept of $(0, 1)$.

39. Write the equation of the line that has slope $-\frac{7}{8}$ and passes through $(2, -9)$. Express the answer in point–slope form.

40. If $f(x) = x^2 - 3x$, find $f(-2)$.

Exponents and Polynomials

10

10.1 Natural-Number Exponents

10.2 Zero and Negative Integer Exponents

10.3 Scientific Notation

10.4 Polynomials

10.5 Adding and Subtracting Polynomials

10.6 Multiplying Polynomials

10.7 Dividing Polynomials by Monomials

10.8 Dividing Polynomials by Polynomials

Key Concept: Polynomials

Accent on Teamwork

Chapter Review

Chapter Test

Cumulative Review Exercises

IN THIS CHAPTER, WE INTRODUCE THE RULES FOR EXPONENTS AND USE THEM WHEN PERFORMING OPERATIONS ON POLYNOMIALS.

10.1 *Natural-Number Exponents*

In this section, you will learn about

- The product rule for exponents • The quotient rule for exponents
- The power rule for exponents • Power rules for products and quotients

INTRODUCTION. We have used natural-number exponents to indicate repeated multiplication. For example,

$$9^2 = 9 \cdot 9 = 81 \qquad \text{Write 9 as a factor 2 times.}$$
$$7^3 = 7 \cdot 7 \cdot 7 = 343 \qquad \text{Write 7 as a factor 3 times.}$$
$$(-2)^4 = (-2)(-2)(-2)(-2) = 16 \qquad \text{Write } -2 \text{ as a factor 4 times.}$$
$$-2^4 = -(2 \cdot 2 \cdot 2 \cdot 2) = -16 \qquad \text{The } - \text{ sign in front of } 2^4 \text{ means the opposite of } 2^4.$$

These examples illustrate a definition for x^n, where n is a natural number.

Natural-Number Exponents

> If n is a natural number, then
>
> $$x^n = \overbrace{x \cdot x \cdot x \cdot \cdots \cdot x}^{n \text{ factors of } x}$$

In the **exponential expression** x^n, x is called the **base** and n is called the **exponent.** The entire expression is called a **power of** x.

$$\text{base} \longrightarrow x^n \longleftarrow \text{exponent}$$

If an exponent is a natural number, it tells how many times its base is to be used as a factor. An exponent of 1 indicates that its base is to be used one time as a factor, an exponent of 2 indicates that its base is to be used two times as a factor, and so on. The base of an exponential expression can be a number, a variable, or a combination of numbers and variables.

$$x^1 = x \qquad (y + 1)^2 = (y + 1)(y + 1) \qquad (-5s)^3 = (-5s)(-5s)(-5s)$$

In this section, we will continue our study of exponents as we discuss how to simplify exponential expressions that are multiplied, divided, and raised to powers. To perform these simplifications, we will use several rules for exponents.

The product rule for exponents

To develop a rule for multiplying exponential expressions with the same base, we consider the product $x^2 \cdot x^3$. Since the expression x^2 means that x is to be used as a factor two times, and the expression x^3 means that x is to be used as a factor three times, we have

$$2 \text{ factors of } x \quad 3 \text{ factors of } x$$
$$x^2 \cdot x^3 = \overbrace{x \cdot x} \quad \cdot \quad \overbrace{x \cdot x \cdot x}$$

$$5 \text{ factors of } x$$
$$= \overbrace{x \cdot x \cdot x \cdot x \cdot x}$$
$$= x^5$$

In general,

$$m \text{ factors of } x \qquad\qquad n \text{ factors of } x$$
$$x^m \cdot x^n = \overbrace{x \cdot x \cdot x \cdot \cdots \cdot x} \quad \overbrace{x \cdot x \cdot x \cdot \cdots \cdot x}$$

$$m + n \text{ factors of } x$$
$$= \overbrace{x \cdot x \cdot x \cdot x \cdot x \cdot x \cdot \cdots \cdot x \cdot x \cdot x}$$
$$= x^{m+n}$$

This discussion suggests the following rule: *To multiply two exponential expressions with the same base, keep the common base and add the exponents.*

| **Product rule for exponents** | If m and n represent natural numbers, then $$x^m x^n = x^{m+n}$$ |

EXAMPLE 1   *Multiplying powers with like bases.*   Simplify each expression: **a.** $9^5(9^6)$,   **b.** $x^3 \cdot x^4$,   **c.** $y^2 y^4 y$,   and   **d.** $(c^2 d^3)(c^4 d^5)$.

Solution

a. To simplify $9^5(9^6)$ means to write it in an equivalent form using one base and one exponent.

$$9^5(9^6) = 9^{5+6} \quad \text{Use the product rule for exponents: Keep the common base, which}$$
is 9, and add the exponents.
$$= 9^{11} \quad \text{Do the addition.}$$

b. $x^3 \cdot x^4 = x^{3+4}$ Keep the common base x and add the exponents.
$$= x^7 \quad \text{Do the addition.}$$

c. $y^2 y^4 y = y^{2+4}y$ Working from left to right, keep the common base y and add the exponents.
$$= y^6 y \quad \text{Do the addition.}$$
$$= y^{6+1} \quad \text{Keep the common base and add the exponents.}$$
$$= y^7 \quad \text{Do the addition.}$$

d. $(c^2 d^3)(c^4 d^5) = c^2 d^3 c^4 d^5$ Use the associative property of multiplication.
$$= c^2 c^4 d^3 d^5 \quad \text{Change the order of the factors.}$$
$$= c^{2+4} d^{3+5} \quad \text{Keep the common base } c \text{ and add the exponents. Keep the}$$
common base d and add the exponents.
$$= c^6 d^8 \quad \text{Do the additions.}$$

Self Check
Simplify:

a. $7^8(7^7)$

b. $z \cdot z^3$

c. $x^2 x^3 x^6$

d. $(s^4 t^3)(s^4 t^4)$

Answers: **a.** 7^{15},   **b.** z^4,
c. x^{11},   **d.** $s^8 t^7$ ■

   COMMENT   When simplifying expressions, note the operations that are involved. For example, we cannot simplify $x^3 + x^4$ or $x^3 - x^4$, because x^3 and x^4 are not like terms. However, we can simplify $x^3 \cdot x^4$, because x^3 and x^4 have the same base: $x^3 \cdot x^4 = x^{12}$.

Furthermore, the expressions $x^2 + y^3$ and $x^2 - y^3$ cannot be simplified, because they do not contain like terms; neither can the expression $x^2 y^3$, because x^2 and y^3 have different bases.

The quotient rule for exponents

We now consider the fraction

$$\frac{4^5}{4^2}$$

where the exponent in the numerator is greater than the exponent in the denominator. We can simplify this fraction as follows:

$$\frac{4^5}{4^2} = \frac{4 \cdot 4 \cdot 4 \cdot 4 \cdot 4}{4 \cdot 4}$$

$$= \frac{\overset{1}{\cancel{4}} \cdot \overset{1}{\cancel{4}} \cdot 4 \cdot 4 \cdot 4}{\underset{1}{\cancel{4}} \cdot \underset{1}{\cancel{4}}} \qquad \text{Divide out the common factors of 4.}$$

$$= 4^3$$

The result of 4^3 has a base of 4 and an exponent $5 - 2$ (or 3). This suggests that *to divide exponential expressions with the same base, we keep the common base and subtract the exponents.*

Quotient rule for exponents	If m and n represent natural numbers, $m > n$, and $x \neq 0$, then $$\frac{x^m}{x^n} = x^{m-n}$$

EXAMPLE 2 *Dividing powers with like bases.* Simplify each expression. Assume that there are no divisions by 0.

a. $\dfrac{20^{16}}{20^9}$, **b.** $\dfrac{x^4}{x^3}$, **c.** $\dfrac{a^3b^8}{ab^5}$

Solution

a. To simplify $\dfrac{20^{16}}{20^9}$ means to write it in an equivalent form using one base and one exponent.

$$\frac{20^{16}}{20^9} = 20^{16-9} \qquad \text{Use the quotient rule for exponents: Keep the common base, which is 20, and subtract the exponents.}$$

$$= 20^7 \qquad \text{Do the subtraction: } 16 - 9 = 7.$$

b. $\dfrac{x^4}{x^3} = x^{4-3} \qquad$ Keep the common base x and subtract the exponents.

$$= x^1 \qquad \text{Do the subtraction.}$$

$$= x$$

c. $\dfrac{a^3b^8}{ab^5} = \dfrac{a^3}{a} \cdot \dfrac{b^8}{b^5}$

$$= a^{3-1}b^{8-5} \qquad \begin{array}{l}\text{Keep the common base } a \text{ and subtract the exponents.}\\ \text{Keep the common base } b \text{ and subtract the exponents.}\end{array}$$

$$= a^2b^3 \qquad \text{Do the subtractions.}$$

Self Check

Simplify:

a. $\dfrac{55^{30}}{55^5}$

b. $\dfrac{a^5}{a^3}$

c. $\dfrac{b^{15}c^4}{b^4c}$

Answers: **a.** 55^{25}, **b.** a^2, **c.** $b^{11}c^3$ ◼

EXAMPLE 3 *Using two rules for exponents.* Simplify $\dfrac{a^3 a^5 a^7}{a^4 a}$.

Solution

We use the product rule for exponents to simplify the numerator and denominator separately and proceed as follows.

$$\frac{a^3 a^5 a^7}{a^4 a} = \frac{a^{15}}{a^5}$$ In the numerator, keep the common base a and add the exponents.
In the denominator, keep the common base a and add the exponents.

$$= a^{15-5}$$ Use the quotient rule for exponents: Keep the common base a and subtract the exponents.

$$= a^{10}$$ Do the subtraction.

Self Check
Simplify:

$$\frac{b^2 b^6 b}{b^4 b^4}$$

Answer: b

The power rule for exponents

To find another rule for exponents, we consider the expression $(x^3)^4$, which can be written as $x^3 \cdot x^3 \cdot x^3 \cdot x^3$. Because each of the four factors of x^3 contains three factors of x, there are $4 \cdot 3$ (or 12) factors of x. This product can be written as x^{12}.

$$(x^3)^4 = x^3 \cdot x^3 \cdot x^3 \cdot x^3$$

$$= \overbrace{\underbrace{x \cdot x \cdot x}_{x^3} \cdot \underbrace{x \cdot x \cdot x}_{x^3} \cdot \underbrace{x \cdot x \cdot x}_{x^3} \cdot \underbrace{x \cdot x \cdot x}_{x^3}}^{12 \text{ factors of } x}$$

$$= x^{12}$$

In general,

$$(x^m)^n = \overbrace{x^m \cdot x^m \cdot x^m \cdot \cdots \cdot x^m}^{n \text{ factors of } x^m}$$

$$= \overbrace{x \cdot x \cdot x \cdot x \cdot x \cdot x \cdot x \cdot \cdots \cdot x}^{m \cdot n \text{ factors of } x}$$

$$= x^{m \cdot n}$$

This discussion illustrates the following rule: *To raise an exponential expression to a power, keep the base and multiply the exponents.*

Power rule for exponents

If m and n represent natural numbers, then

$$(x^m)^n = x^{m \cdot n} \qquad \text{or, more simply,} \qquad (x^m)^n = x^{mn}$$

EXAMPLE 4 *The power rule for exponents.* Simplify each expression:
a. $(2^3)^7$ and **b.** $(z^8)^8$.

Solution

a. To simplify $(2^3)^7$ means to write it in an equivalent form using one base and one exponent.

$$(2^3)^7 = 2^{3 \cdot 7}$$ Keep the base of 2 and multiply the exponents.

$$= 2^{21}$$ Do the multiplication.

b. $(z^8)^8 = z^{8 \cdot 8}$ Keep the base and multiply the exponents.

$$= z^{64}$$ Do the multiplication.

Self Check
Simplify each expression:

a. $(5^3)^4$

b. $(y^5)^2$

Answer: **a.** 5^{12}, **b.** y^{10}

EXAMPLE 5   *Using two rules for exponents.*   Simplify each expression:
a. $(x^2x^5)^2$ and   **b.** $(z^2)^4(z^3)^3$.

Solution

a. We begin by using the product rule for exponents. Then we use the power rule.

$$(x^2x^5)^2 = (x^7)^2 \quad \text{Within the parentheses, keep the base } x \text{ and add the exponents.}$$
$$= x^{14} \quad \text{Keep the base } x \text{ and multiply the exponents.}$$

b. We begin by using the power rule for exponents twice. Then we use the product rule.

$$(z^2)^4(z^3)^3 = z^8z^9 \quad \text{For each power of } z \text{ raised to a power, keep the base } z \text{ and multiply the exponents.}$$
$$= z^{17} \quad \text{Keep the base } z \text{ and add the exponents.}$$

Self Check

Simplify each expression:

a. $(a^4a^3)^3$

b. $(a^3)^3(a^4)^2$

Answers: **a.** a^{21} **b.** a^{17} ■

Power rules for products and quotients

To develop two more rules for exponents, we consider the expression $(2x)^3$, which is a *power of the product* of 2 and x, and the expression $\left(\frac{2}{x}\right)^3$, which is a *power of the quotient* of 2 and x.

$$(2x)^3 = (2x)(2x)(2x) \qquad\qquad \left(\frac{2}{x}\right)^3 = \left(\frac{2}{x}\right)\left(\frac{2}{x}\right)\left(\frac{2}{x}\right) \quad (x \neq 0)$$

$$= (2 \cdot 2 \cdot 2)(x \cdot x \cdot x) \qquad\qquad = \frac{2 \cdot 2 \cdot 2}{x \cdot x \cdot x} \quad \begin{array}{l}\text{Multiply the numerators.}\\\text{Multiply the denominators.}\end{array}$$

$$= 2^3x^3 \qquad\qquad\qquad\qquad\quad = \frac{2^3}{x^3}$$

$$= 8x^3 \qquad\qquad\qquad\qquad\quad\; = \frac{8}{x^3}$$

These examples illustrate the following rules: *To raise a product to a power, we raise each factor of the product to that power,* and *to raise a fraction to a power, we raise both the numerator and the denominator to that power.*

Powers of a product and a quotient

> If n represents a natural number, then
>
> $$(xy)^n = x^ny^n \qquad \text{and if } y \neq 0, \text{ then} \qquad \left(\frac{x}{y}\right)^n = \frac{x^n}{y^n}$$

EXAMPLE 6   *Powers of products.*   Simplify   **a.** $(3c)^3$,   **b.** $(x^2y^3)^5$,   and
c. $(-2a^3b)^2$.

Solution

a. Since $3c$ is the product of 3 and c, the expression $(3c)^3$ is a power of a product.

$$(3c)^3 = 3^3c^3 \quad \begin{array}{l}\text{Use the power rule for products: Raise each factor of the product } 3c \text{ to}\\\text{the 3rd power.}\end{array}$$

$$= 27c^3 \quad \text{Evaluate } 3^3.$$

b. $(x^2y^3)^5 = (x^2)^5(y^3)^5 \quad \text{Raise each factor of the product } x^2y^3 \text{ to the 5th power.}$
$$= x^{10}y^{15} \quad \begin{array}{l}\text{For each power of a power, keep the base and multiply the}\\\text{exponents.}\end{array}$$

c. $(-2a^3b)^2 = (-2)^2(a^3)^2b^2 \quad \begin{array}{l}\text{Raise each of the three factors of the product } -2a^3b \text{ to the}\\\text{2nd power.}\end{array}$
$$= 4a^6b^2 \qquad\qquad\qquad \text{Evaluate } (-2)^2. \text{ Keep the base } a \text{ and multiply the exponents.}$$

Self Check

Simplify:

a. $(2t)^4$

b. $(c^3d^4)^6$

c. $(-3ab^5)^3$

Answers: **a.** $16t^4$,   **b.** $c^{18}d^{24}$,
c. $-27a^3b^{15}$ ■

EXAMPLE 7 *Using three rules for exponents.*   Simplify $\dfrac{(a^3b^4)^2}{ab^5}$.

Solution

$\dfrac{(a^3b^4)^2}{ab^5} = \dfrac{(a^3)^2(b^4)^2}{ab^5}$ In the numerator, raise each factor within the parentheses to the 2nd power.

$\quad\quad = \dfrac{a^6b^8}{ab^5}$ In the numerator, for each power of a power, keep the base and multiply the exponents.

$\quad\quad = a^{6-1}b^{8-5}$ Keep each of the bases, a and b, and subtract the exponents.

$\quad\quad = a^5b^3$ Do the subtractions.

Self Check

Simplify $\dfrac{(c^4d^5)^3}{c^2d^3}$

Answer: $c^{10}d^{12}$ ■

EXAMPLE 8 *Powers of quotients.*   Simplify   **a.** $\left(\dfrac{4}{k}\right)^3$ and   **b.** $\left(\dfrac{3x^2}{2y^3}\right)^5$.

Solution

a. Since $\frac{4}{k}$ is the quotient of 4 and k, the expression $\left(\frac{4}{k}\right)^3$ is a power of a quotient.

$\left(\dfrac{4}{k}\right)^3 = \dfrac{4^3}{k^3}$ Use the power rule for quotients: Raise the numerator and denominator to the 3rd power.

$\quad\quad = \dfrac{64}{k^3}$ Evaluate 4^3.

b. $\left(\dfrac{3x^2}{2y^3}\right)^5 = \dfrac{(3x^2)^5}{(2y^3)^5}$ Raise the numerator and the denominator to the 5th power.

$\quad\quad = \dfrac{3^5(x^2)^5}{2^5(y^3)^5}$ In the numerator and denominator, raise each factor within the parentheses to the 5th power.

$\quad\quad = \dfrac{243x^{10}}{32y^{15}}$ Evaluate 3^5 and 2^5. For each power of a power, keep the base and multiply the exponents.

Self Check

Simplify:

a. $\left(\dfrac{x}{7}\right)^3$

b. $\left(\dfrac{2x^3}{3y^2}\right)^4$

Answers: **a.** $\dfrac{x^3}{343}$,   **b.** $\dfrac{16x^{12}}{81y^8}$ ■

EXAMPLE 9 *Using two rules for exponents.*   Simplify $\dfrac{(5b)^9}{(5b)^6}$.

Solution

$\dfrac{(5b)^9}{(5b)^6} = (5b)^{9-6}$ Keep the common base $5b$, and subtract the exponents.

$\quad\quad = (5b)^3$ Do the subtraction.

$\quad\quad = 5^3b^3$ Raise each factor within the parentheses to the 3rd power.

$\quad\quad = 125b^3$ Evaluate 5^3.

Self Check

Simplify $\dfrac{(-2h)^{20}}{(-2h)^{14}}$.

Answer: $64h^6$ ■

The rules for natural-number exponents are summarized below.

Rules for exponents

If n represents a natural number, then

$$x^n = \overbrace{x \cdot x \cdot x \cdot \;\cdots\; \cdot x}^{n \text{ factors of } x}$$

If m and n represent natural numbers and there are no divisions by zero, then

1. $x^m x^n = x^{m+n}$ **2.** $\dfrac{x^m}{x^n} = x^{m-n}$ **3.** $(x^m)^n = x^{m \cdot n}$

4. $(xy)^n = x^n y^n$ **5.** $\left(\dfrac{x}{y}\right)^n = \dfrac{x^n}{y^n}$

STUDY SET Section 10.1

VOCABULARY *Fill in the blanks.*

1. The _____ of the exponential expression $(-5)^3$ is -5. The _____ is 3.

2. The _____ expression x^4 represents a repeated multiplication where x is to be written as a _____ four times.

3. x^n is called a _____ of x.

4. The expression $(2x^2b)^5$ is a power of a _____, and $\left(\dfrac{2x^2}{b}\right)^5$ is a power of a _____.

CONCEPTS *Fill in the blanks.*

5. $(3x)^4$ means ▢ · ▢ · ▢ · ▢.

6. Using an exponent, $(-5y)(-5y)(-5y)$ can be written as ▢.

7. $x^m x^n = $ ▢

8. $(xy)^n = $ ▢

9. $\left(\dfrac{a}{b}\right)^n = $ ▢

10. $(a^b)^c = $ ▢

11. $\dfrac{x^m}{x^n} = $ ▢

12. $x = x$ ▢

13. $(xy) = (xy)$ ▢

14. $(t^3)^2 = $ ▢ · ▢

15. a. Write a power of a product that has two factors.

b. Write a power of a quotient.

16. a. To simplify $(2y^3z^2)^4$, how many factors within the parentheses must be raised to the fourth power?

b. To simplify $\left(\dfrac{y^3}{z^2}\right)^4$ what two expressions must be raised to the fourth power?

Simplify each expression, if possible.

17. a. $x^2 + x^2$
  b. $x^2 - x^2$
  c. $x^2 \cdot x^2$
  d. $\dfrac{x^2}{1}$

18. a. $x^2 + x$
  b. $x^2 - x$
  c. $x^2 \cdot x$
  d. $\dfrac{x^2}{x}$

19. a. $x^3 + x^2$
  b. $x^3 - x^2$
  c. $x^3 \cdot x^2$
  d. $\dfrac{x^3}{x^2}$

20. Simplify each expression, if possible.
  a. $x^3 + y^3$
  b. $x^3 - y^3$
  c. x^3y^3
  d. $\dfrac{x^3}{y^3}$

Find the area or volume of each figure, whichever is appropriate. You may leave π in your answer.

21.

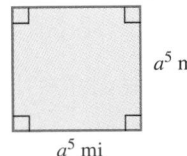

a^5 mi

a^5 mi

22.

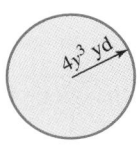

$4y^3$ yd

23.

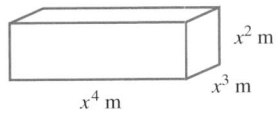

x^2 m

x^3 m

x^4 m

24.

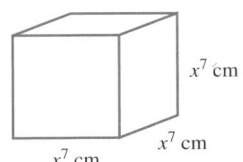

x^7 cm

x^7 cm

x^7 cm

NOTATION *Complete each solution.*

25. $(x^4x^2)^3 = (\,▢\,)^3$
  $= x^{18}$

26. $\dfrac{a^3a^4}{a^2} = \dfrac{▢}{a^2}$
  $= a^{▢-2}$
  $= a^5$

Identify the base and the exponent in each expression.

27. 4^3

28. $(-8)^2$

29. x^5

30. $\left(\dfrac{5}{x}\right)^3$

31. $(-3x)^2$

32. $-x^4$

33. $-\dfrac{1}{3}y^6$

34. $3.14r^4$

Evaluate each expression.

35. $(-4)^2$

36. $(-5)^2$

37. -4^2

38. -5^2

Write the repeated multiplication that is indicated.

39. x^5

40. $(-7y)^4$

41. $\left(\dfrac{t^2}{2}\right)^3$

42. $c^3 d^2$

Write each expression using an exponent.

43. $4t(4t)(4t)(4t)$

44. $-5u(-5u)$

45. $-4 \cdot t \cdot t \cdot t$

46. $-5 \cdot u \cdot u$

PRACTICE Write each expression as an expression involving one base and one exponent.

47. $12^3 \cdot 12^4$

48. $3^4 \cdot 3^6$

49. $2(2^3)(2^2)$

50. $5(5^5)(5^3)$

51. $a^3 \cdot a^3$

52. $m^7 \cdot m^7$

53. $x^4 x^3$

54. $y^5 y^2$

55. $a^3 a a^5$

56. $b^2 b^3 b$

57. $y^3(y^2 y^4)$

58. $(y^4 y) y^6$

59. $\dfrac{8^{12}}{8^4}$

60. $\dfrac{10^4}{10^2}$

61. $\dfrac{x^{15}}{x^3}$

62. $\dfrac{y^6}{y^3}$

63. $\dfrac{c^{10}}{c^9}$

64. $\dfrac{h^{20}}{h^{10}}$

65. $(3^2)^4$

66. $(4^3)^3$

67. $(y^5)^3$

68. $(b^3)^6$

69. $(m^{50})^{10}$

70. $(n^{25})^4$

Simplify. Assume there are no divisions by 0.

71. $(a^2 b^3)(a^3 b^3)$

72. $(u^3 v^5)(u^4 v^5)$

73. $(cd^4)(cd)$

74. $ab^3 c^4 \cdot ab^4 c^2$

75. $xy^2 \cdot x^2 y$

76. $s^8 t^2 s^2 t^7$

77. $\dfrac{y^3 y^4}{yy^2}$

78. $\dfrac{b^4 b^5}{b^2 b^3}$

79. $\dfrac{c^3 d^7}{cd}$

80. $\dfrac{r^8 s^9}{rs}$

81. $(x^2 x^3)^5$

82. $(y^3 y^4)^4$

83. $(3zz^2 z^3)^5$

84. $(4t^3 t^6 t^2)^2$

85. $(x^5)^2(x^7)^3$

86. $(y^3 y)^2(y^2)^2$

87. $(uv)^4$

88. $(xy)^3$

89. $(a^3 b^2)^3$

90. $(r^3 s^2)^2$

91. $(-2r^2 s^3)^3$

92. $(-3x^2 y^4)^2$

93. $\left(\dfrac{a}{b}\right)^3$

94. $\left(\dfrac{r}{s}\right)^4$

95. $\left(\dfrac{x^2}{y^3}\right)^5$

96. $\left(\dfrac{u^4}{v^2}\right)^6$

97. $\left(\dfrac{-2a}{b}\right)^5$

98. $\left(\dfrac{-2t}{3}\right)^4$

99. $\dfrac{(6k)^7}{(6k)^4}$

100. $\dfrac{(-3a)^{12}}{(-3a)^{10}}$

101. $\dfrac{(a^2 b)^{15}}{(a^2 b)^9}$

102. $\dfrac{(s^2 t^3)^4}{(s^2 t^3)^2}$

103. $\dfrac{a^2 a^3 a^4}{(a^4)^2}$

104. $\dfrac{(aa^2)^3}{a^2 a^3}$

105. $\dfrac{(ab^2)^3}{(ab)^2}$

106. $\dfrac{(m^3 n^4)^3}{(mn^2)^3}$

107. $\dfrac{(r^4 s^3)^4}{(rs^3)^3}$

108. $\dfrac{(x^2 y^5)^5}{(x^3 y)^2}$

109. $\left(\dfrac{y^3 y}{2yy^2}\right)^3$

110. $\left(\dfrac{2y^3 y}{yy^2}\right)^3$

111. $\left(\dfrac{3t^3 t^4 t^5}{4t^2 t^6}\right)^3$

112. $\left(\dfrac{4t^3 t^4 t^5}{3t^2 t^6}\right)^3$

APPLICATIONS

113. ART HISTORY   Leonardo da Vinci's drawing relating a human figure to a square and a circle is shown in Illustration 1.

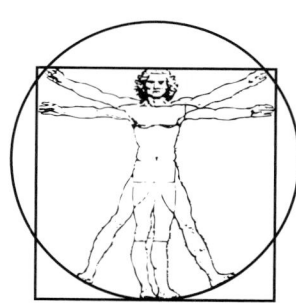

ILLUSTRATION 1

a. Find the area of the square if the man's height is $5x$ feet.

b. Find the area of the circle if the distance from his waist to his feet is $3x$ feet. You may leave π in your answer.

114. PACKAGING   Use Illustration 2 to find the volume of the bowling ball and the cardboard box it is packaged in. You may leave π in your answer.

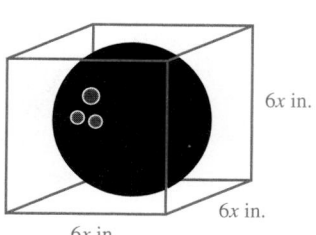

$6x$ in.

$6x$ in.

$6x$ in.

ILLUSTRATION 2

115. BOUNCING BALL   A ball is dropped from a height of 32 feet. Each rebound is one-half of its previous height.
  a. Draw a diagram of the path of the ball, showing four bounces.
  b. Explain why the expressions $32\left(\frac{1}{2}\right)$, $32\left(\frac{1}{2}\right)^2$, $32\left(\frac{1}{2}\right)^3$, and $32\left(\frac{1}{2}\right)^4$ represent the height of the ball on the first, second, third, and fourth bounces, respectively. Find the heights of the first four bounces.

116. HAVING BABIES   The probability that a couple will have n baby boys in a row is given by the formula $\left(\frac{1}{2}\right)^n$. Find the probability that a couple will have four baby boys in a row.

117. COMPUTERS   Text is stored by computers using a sequence of eight 0's and 1's. Such a sequence is called a **byte.** An example of a byte is 10101110.
  a. Write four other bytes, all ending in 1.

  b. Each of the eight digits of a byte can be chosen in *two* ways (either 0 or 1). The total number of different bytes can be represented by an exponential expression with base 2. What is it?

118. ▦   INVESTING   Guess the answer to the following problem. Then use a calculator to find the correct answer. Were you close?
If the value of 1¢ is to double every day, what will the penny be worth after 31 days?

WRITING

119. Explain the mistake in the following work.
$$2^3 \cdot 2^2 = 4^5$$
$$= 1{,}024$$

120. Are the expressions $2x^3$ and $(2x)^3$ equivalent? Explain.

121. Is the operation of raising to a power commutative? That is, is $a^b = b^a$? Explain.

122. When a number is raised to a power, is the result always larger than the original number? Support your answer with some examples.

REVIEW   *Match each equation with its graph below.*

123. $y = 2x - 1$ **124.** $y = 3x - 1$

125. $y = 3$ **126.** $x = 3$

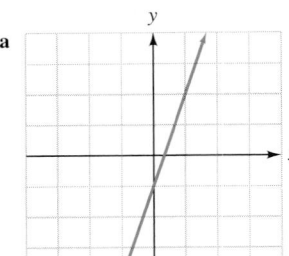

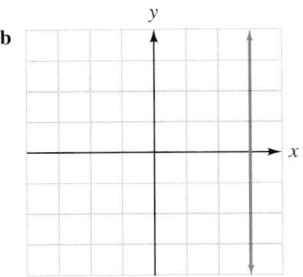

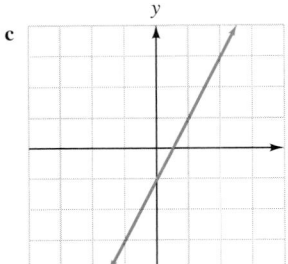

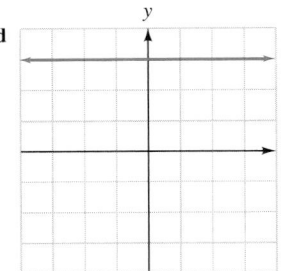

10.2 *Zero and Negative Integer Exponents*

In this section, you will learn about

• Zero exponents   • Negative integer exponents   • Variable exponents

INTRODUCTION. In the previous section, we discussed natural-number exponents. We now extend the discussion to include exponents that are zero and exponents that are negative integers.

Zero exponents

When we discussed the quotient rule for exponents in the previous section, the exponent in the numerator was always greater than the exponent in the denominator. We will now consider what happens when the exponents are equal. To develop the definition of a zero exponent, we will simplify the expression

$$\frac{5^3}{5^3}$$

in two ways. If we use the quotient rule for exponents, where the exponents in the numerator and denominator are equal, we obtain 5^0. However, by dividing out common factors of 5, we obtain 1.

$$\frac{5^3}{5^3} = 5^{3-3} = 5^0 \qquad \frac{5^3}{5^3} = \frac{\overset{1}{5} \cdot \overset{1}{5} \cdot \overset{1}{5}}{\underset{1}{5} \cdot \underset{1}{5} \cdot \underset{1}{5}} = 1$$

These must be equal.

For this reason, we will define 5^0 to be equal to 1. This example suggests the following rule.

Zero exponents

> If x represents any nonzero real number, then
> $$x^0 = 1$$

EXAMPLE 1   *Zero exponents.* Write each expression without using exponents.

a. $\left(\dfrac{1}{13}\right)^0 = 1$

b. $\dfrac{x^5}{x^5} = x^{5-5} \quad (x \neq 0)$
$$= x^0$$
$$= 1$$

c. $3x^0 = 3(1) \quad$ The base is x;
$\ = 3 \qquad\quad$ the exponent is 0.

d. $(3x)^0 = 1 \quad$ The base is $3x$;
$\quad$ the exponent is 0.

Parts c and d point out that $3x^0 \neq (3x)^0$.

Self Check

Write each expression without using exponents:

a. $(-0.115)^0$

b. $-5a^0 b$

Answers: a. 1, **b.** $-5b$ ∎

Negative integer exponents

To develop the definition of a negative exponent, we will simplify the expression

$$\frac{6^2}{6^5}$$

in two ways. If we use the quotient rule for exponents, where the exponent in the numerator is less than the exponent in the denominator, we obtain 6^{-3}. However, by dividing out two factors of 6, we obtain $\dfrac{1}{6^3}$.

$$\frac{6^2}{6^5} = 6^{2-5} = 6^{-3} \qquad \frac{6^2}{6^5} = \frac{\overset{1}{6} \cdot \overset{1}{6}}{\underset{1}{6} \cdot \underset{1}{6} \cdot 6 \cdot 6 \cdot 6} = \frac{1}{6^3}$$

These must be equal.

For this reason, we define 6^{-3} to be equal to $\dfrac{1}{6^3}$. In general, we have the following rule.

Negative exponents

If x represents any nonzero number and n represents a natural number, then

$$x^{-n} = \frac{1}{x^n}$$

The definition of a negative exponent states that another way to write x^{-n} is to write its reciprocal, changing the sign of the exponent. We can use this definition to write expressions that contain negative exponents as expressions without negative exponents.

EXAMPLE 2 *Negative exponents.* Simplify by using the definition of negative exponents: **a.** 3^{-5} and   **b.** $(-2)^{-3}$.

Solution

a. $3^{-5} = \dfrac{1}{3^5}$ Write the reciprocal of 3^{-5} and change the exponent from -5 to 5.

$= \dfrac{1}{243}$ Evaluate 3^5.

b. $(-2)^{-3} = \dfrac{1}{(-2)^3}$ Write the reciprocal of $(-2)^{-3}$ and change the exponent from -3 to 3.

$= -\dfrac{1}{8}$ Evaluate $(-2)^3$.

Self Check

Simplify by using the definition of negative exponents:

a. 4^{-4}

b. $(-5)^{-3}$

Answers: **a.** $\dfrac{1}{256}$,   **b.** $-\dfrac{1}{125}$

 COMMENT   A negative exponent does not indicate a negative number. It indicates a reciprocal.

$$4^{-2} = \frac{1}{4^2} \qquad 4^{-2} \neq -16 \qquad 4^{-2} \neq -\frac{1}{4^2}$$

EXAMPLE 3 *Negative exponents.* Simplify by using the definition of negative exponents:

a. $\dfrac{1}{5^{-2}}$ and   **b.** $\dfrac{2^{-3}}{3^{-4}}$.

Solution

a. $\dfrac{1}{5^{-2}} = \dfrac{1}{\dfrac{1}{5^2}}$ In the denominator, write the reciprocal of 5^{-2} and change the exponent from -2 to 2.

$= 1 \div \dfrac{1}{5^2}$ The fraction bar indicates division of 1 by $\dfrac{1}{5^2}$.

$= 1 \cdot \dfrac{5^2}{1}$ To divide by $\dfrac{1}{5^2}$, we multiply by its reciprocal.

$= 5^2$ Simplify.

$= 25$ Evaluate 5^2.

b. $\dfrac{2^{-3}}{3^{-4}} = \dfrac{\dfrac{1}{2^3}}{\dfrac{1}{3^4}}$ In the numerator, write the reciprocal of 2^{-3} and change the exponent from -3 to 3.
In the denominator, write the reciprocal of 3^{-4} and change the exponent from -4 to 4.

Self Check

Simplify by using the definition of negative exponents:

a. $\dfrac{1}{9^{-1}}$

b. $\dfrac{8^{-2}}{7^{-1}}$

$$= \frac{1}{2^3} \cdot \frac{3^4}{1} \qquad \text{To divide by } \frac{1}{3^4}, \text{ we multiply by its reciprocal.}$$

$$= \frac{3^4}{2^3} \qquad \text{Multiply the fractions.}$$

$$= \frac{81}{8} \qquad \text{Evaluate } 3^4 \text{ and } 2^3.$$

Answers: a. 9,   **b.** $\dfrac{7}{64}$ ∎

The results from Example 3 suggest that we can move factors that have negative exponents between the numerator and denominator of a fraction if we change the sign of their exponents. For example,

$$\frac{3^{-3}}{b^{-1}} = \frac{b^1}{3^3} = \frac{b}{27}$$

EXAMPLE 4   *Negative exponents.*   Simplify by using the definition of negative exponents. Assume that no denominators are zero.

a. $x^{-4} = \dfrac{1}{x^4}$

b. $\dfrac{x^{-3}}{y^{-7}} = \dfrac{y^7}{x^3}$

c. $(-2x)^{-2} = \dfrac{1}{(-2x)^2}$

$\qquad\qquad\quad = \dfrac{1}{4x^2}$

d. $-2x^{-2} = -2\left(\dfrac{1}{x^2}\right)$

$\qquad\qquad = -\dfrac{2}{x^2}$

Self Check
Simplify by using the definition of negative exponents:

a. a^{-5}

b. $\dfrac{r^{-4}}{s^{-5}}$

c. $3y^{-3}$

Answers: a. $\dfrac{1}{a^5}$,   **b.** $\dfrac{s^5}{r^4}$,

c. $\dfrac{3}{y^3}$ ∎

The rules for exponents discussed in Section 10.1 (the product, power, and quotient rules) are also true for zero and negative exponents.

Rules for exponents

If m and n represent integers and there are no divisions by zero, then

$$x^m x^n = x^{m+n} \qquad (x^m)^n = x^{m \cdot n} \qquad (xy)^n = x^n y^n \qquad \left(\frac{x}{y}\right)^n = \frac{x^n}{y^n}$$

$$x^0 = 1 \quad (x \neq 0) \qquad x^{-n} = \frac{1}{x^n} \qquad \frac{x^m}{x^n} = x^{m-n}$$

EXAMPLE 5   *Using two rules for exponents.*   Write $\left(\dfrac{5}{16}\right)^{-1}$ without using exponents.

Solution

$$\left(\frac{5}{16}\right)^{-1} = \frac{5^{-1}}{16^{-1}} \qquad \text{Use the quotient rule for exponents: Raise the numerator and denominator to the } -1 \text{ power.}$$

$$= \frac{16^1}{5^1} \qquad \text{Move the factors that have negative exponents between the numerator and denominator.}$$

$$= \frac{16}{5}$$

Self Check
Write $\left(\dfrac{3}{7}\right)^{-2}$ without using exponents.

Answer: $\dfrac{49}{9}$ ∎

EXAMPLE 6 *Using rules for exponents.* Simplify and write the result without using negative exponents. Assume that no denominators are zero.

a. $(x^{-3})^2 = x^{-6}$

$$= \frac{1}{x^6}$$

b. $\dfrac{x^3}{x^7} = x^{3-7}$

$$= x^{-4}$$

$$= \frac{1}{x^4}$$

c. $(x^3 x^2)^{-3} = (x^5)^{-3}$

$$= \frac{1}{(x^5)^3}$$

$$= \frac{1}{x^{15}}$$

d. $\dfrac{y^{-4}y^{-3}}{y^{-20}} = \dfrac{y^{-7}}{y^{-20}}$

$$= y^{-7-(-20)}$$

$$= y^{-7+20}$$

$$= y^{13}$$

e. $\dfrac{12a^3b^4}{4a^5b^2} = 3a^{3-5}b^{4-2}$

$$= 3a^{-2}b^2$$

$$= \frac{3b^2}{a^2}$$

f. $\left(-\dfrac{x^3y^2}{xy^{-3}}\right)^{-2} = \left(-x^{3-1}y^{2-(-3)}\right)^{-2}$

$$= (-x^2y^5)^{-2}$$

$$= \frac{1}{(-x^2y^5)^2}$$

$$= \frac{1}{x^4y^{10}}$$

Self Check

Simplify and write the result without using negative exponents:

a. $(x^4)^{-3}$

b. $\dfrac{a^4}{a^8}$

c. $\dfrac{a^{-4}a^{-5}}{a^{-3}}$

d. $\dfrac{20x^5y^3}{5x^3y^6}$

Answers: **a.** $\dfrac{1}{x^{12}}$, **b.** $\dfrac{1}{a^4}$, **c.** $\dfrac{1}{a^6}$, **d.** $\dfrac{4x^2}{y^3}$ ■

Variable exponents

We can apply the rules for exponents to simplify expressions involving variable exponents.

EXAMPLE 7 *Variable exponents.* Simplify each expression. Assume that there are no divisions by 0.

a. $\dfrac{6^n}{6^n} = 6^{n-n}$ Keep the common base and subtract the exponents.

$$= 6^0 \quad \text{Combine like terms: } n - n = 0.$$

$$= 1$$

b. $x^{2m}x^{3m} = x^{2m+3m}$ Keep the common base and add the exponents.

$$= x^{5m} \quad \text{Combine like terms: } 2m + 3m = 5m.$$

c. $\dfrac{y^{2m}}{y^{4m}} = y^{2m-4m}$ Keep the base and subtract the exponents.

$$= y^{-2m} \quad \text{Combine like terms: } 2m - 4m = -2m.$$

$$= \frac{1}{y^{2m}} \quad \text{Write the reciprocal of } y^{-2m} \text{ and change the exponent to } 2m.$$

Self Check

Simplify each expression:

a. $\dfrac{x^m}{x^m}$

b. $z^{3n}z^{2n}$

c. $\dfrac{z^{3n}}{z^{5n}}$

Answers: **a.** 1, **b.** z^{5n}, **c.** $\dfrac{1}{z^{2n}}$ ■

Accent on Technology: **Finding present value**

As a gift for their newborn grandson, the grandparents want to deposit enough money in the bank now so that when he turns 18, the young man will have a college fund of $20,000 waiting for him. How much should they deposit now if the money will earn 6% annually?

To find how much money P must be invested at an annual rate i (expressed as a decimal) to have A in n years, we use the formula $P = A(1 + i)^{-n}$. If we substitute 20,000 for A, 0.06 (6%) for i, and 18 for n, we have

$$P = A(1 + i)^{-n} \qquad \text{P is called the \textbf{present value}.}$$
$$P = 20{,}000(1 + \mathbf{0.06})^{-18}$$

To find P with a scientific calculator, we enter these numbers and press these keys.

Keystrokes $\boxed{(}\; 1 \;\boxed{+}\; .06 \;\boxed{)}\; \boxed{y^x}\; 18 \;\boxed{+/-}\; \boxed{\times}\; 20000 \;\boxed{=}$ $\boxed{\mathtt{7006.875823}}$

To evaluate the expression with a graphing calculator, we use the following keystrokes.

Keystrokes $20000 \;\boxed{\times}\; \boxed{(}\; 1 \;\boxed{+}\; .06 \;\boxed{)}\; \boxed{\wedge}\; \boxed{(-)}\; 18 \;\boxed{\text{ENTER}}$

$$\boxed{\begin{array}{l} \mathtt{20000*(1+.06)\char`\^-1} \\ \mathtt{8} \\ \qquad\qquad \mathtt{7006.875823} \end{array}}$$

They must invest approximately $7,006.88 to have $20,000 in 18 years.

STUDY SET   Section 10.2

VOCABULARY   *Fill in the blanks.*

1. In the exponential expression 8^{-3}, 8 is the
_____ and -3 is the _____.

2. In the exponential expression 5^{-1}, the exponent is a
_____ integer.

3. Another way to write 2^{-3} is to write its
_____ and to change the sign of the
exponent:

$$2^{-3} = \frac{1}{2^3}$$

4. In the exponential expression z^m, the exponent is a
_____.

CONCEPTS

5. In parts a and b, fill in the blanks as you simplify the fraction in two different ways. Then complete the sentence in part c.

a. $\dfrac{6^4}{6^4} = 6$ ☐

  $\quad = 6$ ☐

b. $\dfrac{6^4}{6^4} = \dfrac{\boxed{} \cdot \boxed{} \cdot \boxed{} \cdot \boxed{}}{6 \cdot 6 \cdot 6 \cdot 6}$

  $\quad = $ ☐

c. So we define 6^0 to be ☐ , and in general, if x is any nonzero real number, then $x^0 = $ ☐ .

6. In parts a and b, fill in the blanks as you simplify the fraction in two different ways. Then complete the sentence in part c.

a. $\dfrac{8^3}{8^5} = 8$ ☐

  $\quad = 8$ ☐

b. $\dfrac{8^3}{8^5} = \dfrac{\boxed{} \cdot \boxed{} \cdot \boxed{}}{8 \cdot 8 \cdot 8 \cdot 8 \cdot 8}$

  $\quad = \dfrac{1}{8}$ ☐

c. So we define 8^{-2} to be ☐ , and in general, if x is any nonzero real number, then $x^{-n} = $ ☐ .

Complete each table.

7.

x	3^x
2	
1	
0	
-1	
-2	

8.

x	4^x
2	
1	
0	
-1	
-2	

9.

x	$(-9)^x$
2	
1	
0	
-1	
-2	

10.

x	$(-5)^x$
2	
1	
0	
-1	
-2	

First use the graph to determine the missing y-coordinates in the table. Then express each y-coordinate as a power of 2.

11.

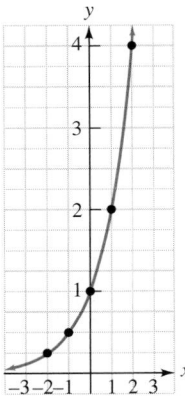

x	y	y as a power of 2
2		
1		
0		
−1		
−2		

12.

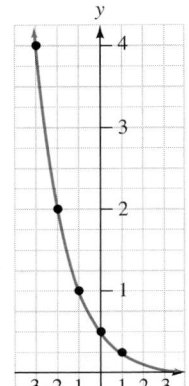

x	y	y as a power of 2
1		
0		
−1		
−2		
−3		

NOTATION *Complete each solution.*

13. $(y^5 y^3)^{-5} = \left(\right)^{-5}$

$ = y^{}$

$ = \dfrac{1}{y^{40}}$

14. $\left(\dfrac{a^2 b^3}{a^{-3} b} \right)^{-3} = \left(a^{2-(-3)} b^{-1} \right)^{-3}$

$ = (a^{} b^{})^{-3}$

$ = \dfrac{1}{(a^5 b^2)^{}}$

$ = \dfrac{1}{a^{15} b^6}$

15. In the expression $3x^{-2}$, what is the base and what is the exponent?

16. In the expression $-3x^{-2}$, what is the base and what is the exponent?

17. First tell the base and the exponent, and then evaluate each expression.
  a. -4^2
  b. 4^{-2}
  c. -4^{-2}

18. First tell the base and the exponent, and then evaluate each expression.
  a. $(-7)^2$
  b. -7^{-2}
  c. $(-7)^{-2}$

PRACTICE *Simplify each expression. Write each answer without using parentheses or negative exponents.*

19. 7^0

20. 9^0

21. $\left(\dfrac{1}{4} \right)^0$

22. $\left(\dfrac{3}{8} \right)^0$

23. $2x^0$

24. $(2x)^0$

25. $(-x)^0$

26. $-x^0$

27. $\left(\dfrac{a^2 b^3}{ab^4} \right)^0$

28. $\dfrac{2}{3} \left(\dfrac{xyz}{x^2 y} \right)^0$

29. $\dfrac{5}{2x^0}$

30. $\dfrac{4}{3a^0}$

31. 12^{-2}

32. 11^{-2}

33. $(-4)^{-1}$

34. $(-8)^{-1}$

35. $\dfrac{1}{5^{-3}}$

36. $\dfrac{1}{3^{-3}}$

37. $\dfrac{2^{-4}}{3^{-1}}$

38. $\dfrac{7^{-2}}{2^{-3}}$

39. -4^{-3}

40. -6^{-3}

41. $-(-4)^{-3}$

42. $-(-4)^{-2}$

43. x^{-2}

44. y^{-3}

45. $-b^{-5}$

46. $-c^{-4}$

47. $(2y)^{-4}$

48. $(-3x)^{-1}$

49. $(ab^2)^{-3}$

50. $(m^2 n^3)^{-2}$

51. $2^5 \cdot 2^{-2}$

52. $10^2 \cdot 10^{-4}$

53. $4^{-3} \cdot 4^{-2} \cdot 4^5$

54. $3^{-4} \cdot 3^5 \cdot 3^{-3}$

55. $\left(\dfrac{7}{8} \right)^{-1}$

56. $\left(\dfrac{16}{5} \right)^{-1}$

57. $\dfrac{3^5 \cdot 3^{-2}}{3^3}$

58. $\dfrac{6^2 \cdot 6^{-3}}{6^{-2}}$

59. $\dfrac{y^4}{y^5}$

60. $\dfrac{t^7}{t^{10}}$

61. $\dfrac{(r^2)^3}{(r^3)^4}$

62. $\dfrac{(b^3)^4}{(b^5)^4}$

63. $\dfrac{y^4 y^3}{y^4 y^{-2}}$

64. $\dfrac{x^{12} x^{-7}}{x^3 x^4}$

65. $\dfrac{10a^4 a^{-2}}{5a^2 a^0}$

66. $\dfrac{9b^0 b^3}{3b^{-3} b^4}$

67. $(ab^2)^{-2}$

68. $(c^2 d^3)^{-2}$

69. $(x^2 y)^{-3}$

70. $(-xy^2)^{-4}$

71. $(x^{-4} x^3)^3$

72. $(y^{-2} y)^3$

73. $(a^{-2} b^3)^{-4}$

74. $(y^{-3} z^5)^{-6}$

75. $(-2x^3 y^{-2})^{-5}$

76. $(-3u^{-2} v^3)^{-3}$

77. $\left(\dfrac{a^3}{a^{-4}}\right)^2$

78. $\left(\dfrac{a^4}{a^{-3}}\right)^3$

79. $\left(\dfrac{b^5}{b^{-2}}\right)^{-2}$

80. $\left(\dfrac{b^{-2}}{b^3}\right)^3$

81. $\left(\dfrac{4x^2}{3x^{-5}}\right)^4$

82. $\left(\dfrac{-3r^4 r^{-3}}{r^{-3} r^7}\right)^3$

83. $\left(\dfrac{12y^3 z^{-2}}{3y^{-4} z^3}\right)^2$

84. $\left(\dfrac{6xy^3}{3x^{-1} y}\right)^3$

Simplify each expression. Assume that there are no divisions by 0.

85. $x^{2m} x^m$

86. $y^{3m} y^{2m}$

87. $u^{2m} u^{-3m}$

88. $r^{5m} r^{-6m}$

89. $\dfrac{y^{3m}}{y^{2m}}$

90. $\dfrac{z^{4m}}{z^{2m}}$

91. $\dfrac{x^{3n}}{x^{6n}}$

92. $\dfrac{x^m}{x^{5m}}$

APPLICATIONS

93. THE DECIMAL NUMERATION SYSTEM   Decimal numbers are written by putting digits into place-value columns that are separated by a decimal point. Express the value of each of the columns shown in Illustration 1 using a power of 10.

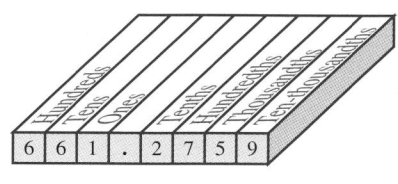

ILLUSTRATION 1

94. UNIT COMPARISON   Consider the relative sizes of the items listed in the table in Illustration 2. In the

column titled "measurement," write the most appropriate number from the following list. Each number is used only once.

10^0 meter

10^{-1} meter

10^{-2} meter

10^{-3} meter

10^{-4} meter

10^{-5} meter

Item	Measurement (m)
Thickness of a dime	
Height of a bathroom sink	
Length of a pencil eraser	
Thickness of soap bubble film	
Width of a video cassette	
Thickness of a piece of paper	

ILLUSTRATION 2

95. RETIREMENT YEARS   How much money should a young married couple invest now at an 8% annual rate if they want to have $100,000 in the bank when they reach retirement age in 40 years? (See the Accent on Technology in this section for the formula.)

96. BIOLOGY   During bacterial reproduction, the time required for a population to double is called the **generation time**. If b bacteria are introduced into a medium, then after the generation time has elapsed, there will be $2b$ bacteria. After n generations, there will be $b \cdot 2^n$ bacteria. Explain what this expression represents when $n = 0$.

WRITING

97. Explain how you would help a friend understand that 2^{-3} is not equal to -8.

98. Describe how you would verify on a calculator that

$$2^{-3} = \dfrac{1}{2^3}$$

REVIEW

99. IQ TEST   An IQ (intelligence quotient) is a score derived from the formula

$$IQ = \dfrac{\text{mental age}}{\text{chronological age}} \cdot 100$$

Find the mental age of a 10-year-old girl if she has an IQ of 135.

100. DIVING   When you are under water, the pressure in your ears is given by the formula

$$\text{Pressure} = \text{depth} \cdot \text{density of water}$$

Find the density of water (in lb/ft^3) if, at a depth of 9 feet, the pressure on your eardrum is 561.6 lb/ft^2.

101. Write the equation of the line having slope $\frac{3}{4}$ and y-intercept -5.

102. Find $f(-6)$ if $f(x) = x^2 - 3x + 1$.

10.3 *Scientific Notation*

In this section, you will learn about

- Scientific notation • Writing numbers in scientific notation
- Changing from scientific notation to standard notation
- Using scientific notation to simplify computations

INTRODUCTION. Scientists often deal with extremely large and extremely small numbers. Two examples are shown in Figure 10-1.

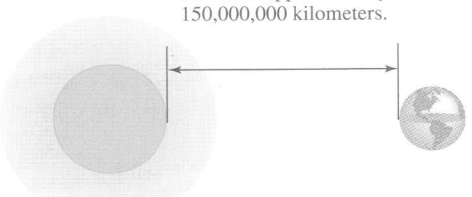

The distance from the Earth to the sun is approximately 150,000,000 kilometers.

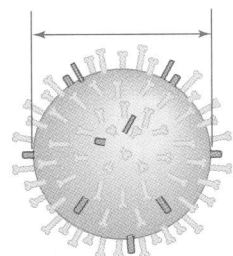

The influenza virus, which causes "flu" symptoms of cough, sore throat, headache, and congestion, has a diameter of 0.00000256 inch.

FIGURE 10-1

The large number of zeros in 150,000,000 and 0.00000256 makes them difficult to read and hard to remember. In this section, we will discuss a notation that will make such numbers easier to use.

Scientific notation

Scientific notation provides a compact way of writing large and small numbers.

Scientific notation | A number is written in **scientific notation** when it is written as the product of a number between 1 (including 1) and 10 and an integer power of 10.

These numbers are written in scientific notation:

$$3.67 \times 10^6, \qquad 2.24 \times 10^{-4}, \qquad \text{and} \qquad 9.875 \times 10^{22}$$

Every number written in scientific notation has the following form:

An integer exponent

$$\boxed{} \cdot \boxed{} \times 10^{\boxed{}}$$

A decimal between 1 and 10

Writing numbers in scientific notation

EXAMPLE 1 *Writing numbers in scientific notation.* Change 150,000,000 to scientific notation.

Solution

We note that 1.5 lies between 1 and 10. To obtain 150,000,000, the decimal point in 1.5 must be moved eight places to the right.

$$1 . 5\,0\,0\,0\,0\,0\,0\,0$$

8 places to the right

Because multiplying a number by 10 moves the decimal point one place to the right, we can accomplish this by multiplying 1.5 by 10 eight times. We can show the multiplication of 1.5 by 10 eight times using the notation 10^8. Thus, 150,000,000 written in scientific notation is 1.5×10^8.

Self Check

The distance from Earth to the sun is approximately 93,000,000 miles. Write this number in scientific notation.

Answer: 9.3×10^7 ∎

EXAMPLE 2 *Writing numbers in scientific notation.* Change 0.00000256 to scientific notation.

Solution

We note that 2.56 is between 1 and 10. To obtain 0.00000256, the decimal point in 2.56 must be moved six places to the left.

$$0\,0\,0\,0\,0\,2 . 56$$

6 places to the left

We can accomplish this by dividing 2.56 by 10^6, which is equivalent to multiplying 2.56 by $\frac{1}{10^6}$ (or by 10^{-6}). Thus, 0.00000256 written in scientific notation is 2.56×10^{-6}.

Self Check

The *Salmonella* bacterium, which causes food poisoning, is 0.00009055 inch long. Write this number in scientific notation.

Answer: 9.055×10^{-5} ∎

EXAMPLE 3 *Writing numbers in scientific notation.* Write
a. 235,000 and **b.** 0.0000073 in scientific notation.

Solution

a. $235,000 = 2.35 \times 10^5$ Because $2.35 \times 10^5 = 235,000$ and 2.35 is between 1 and 10.

b. $0.0000073 = 7.3 \times 10^{-6}$ Because $7.3 \times 10^{-6} = 0.0000073$ and 7.3 is between 1 and 10.

Self Check

Write in scientific notation:

a. 17,500

b. 0.657

Answers: **a.** 1.75×10^4, **b.** 6.57×10^{-1} ∎

From Examples 1, 2, and 3, we see that in scientific notation, a positive exponent is used when writing a number that is greater than 1. A negative exponent is used when writing a number that is between 0 and 1.

EXAMPLE 4 *Writing numbers in scientific notation.* Write 432.0×10^5 in scientific notation.

Solution

The number 432.0×10^5 is not written in scientific notation, because 432.0 is not a number between 1 and 10. To write this number in scientific notation, we proceed as follows:

$$432.0 \times 10^5 = 4.32 \times 10^2 \times 10^5 \quad \text{Write 432.0 in scientific notation.}$$
$$= 4.32 \times 10^7 \qquad 10^2 \times 10^5 = 10^{2+5} = 10^7.$$

Self Check

Write 85×10^{-3} in scientific notation.

Answer: 8.5×10^{-2} ∎

Accent on Technology: *Calculators and scientific notation*

When displaying a very large or a very small number as an answer, most scientific calculators express it in scientific notation. To show this, we will find the values of $(453.46)^5$ and $(0.0005)^{12}$. We enter these numbers and press these keys.

Keystrokes　　453.46 $\boxed{y^x}$ 5 $\boxed{=}$　　　　$\boxed{1.917321395 \quad ^{13}}$

　　　　　　　　.0005 $\boxed{y^x}$ 12 $\boxed{=}$　　　　$\boxed{2.44140625 \quad ^{-40}}$

Since the answers in standard notation require more space than the calculator display has, the calculator gives each result in scientific notation. The first display represents $1.917321395 \times 10^{13}$, and the second represents $2.44140625 \times 10^{-40}$.

If we evaluate the same two expressions using a graphing calculator, we see that the letter E is used when displaying a number in scientific notation.

Keystrokes　　453.46 $\boxed{\wedge}$ 5 $\boxed{\text{ENTER}}$　　$\boxed{\begin{array}{l} 453.46\text{^}5 \\ \quad 1.917321395\text{E}13 \end{array}}$

　　　　　　　　.0005 $\boxed{\wedge}$ 12 $\boxed{\text{ENTER}}$　　$\boxed{\begin{array}{l} .0005\text{^}12 \\ \quad 2.44140625\text{E}-40 \end{array}}$

Changing from scientific notation to standard notation

We can change a number written in scientific notation to **standard notation.** For example, to write 9.3×10^7 in standard notation, we multiply 9.3 by 10^7.

$$9.3 \times 10^7 = 9.3 \times 10,000,000 \quad \text{10^7 is equal to 1 followed by 7 zeros.}$$
$$= 93,000,000$$

EXAMPLE 5　*Writing numbers in standard notation.*　Write
a. 3.4×10^5　and　**b.** 2.1×10^{-4}　in standard notation.

Self Check
Write in standard notation:
a. 4.76×10^5
b. 9.8×10^{-3}

Solution

a. $3.4 \times 10^5 = 3.4 \times 100,000$
　　　　　　　$= 340,000$

b. $2.1 \times 10^{-4} = 2.1 \times \dfrac{1}{10^4}$

　　　　　　　$= 2.1 \times \dfrac{1}{10,000}$

　　　　　　　$= 2.1 \times 0.0001$

　　　　　　　$= 0.00021$

Answers:　**a.** 476,000,
b. 0.0098

The following numbers are written in both scientific and standard notation. In each case, the exponent gives the number of places that the decimal point moves, and the sign of the exponent indicates the direction that it moves.

$$5.32 \times 10^5 = 5\,3\,2\,0\,0\,0. \qquad \text{5 places to the right.}$$
$$8.95 \times 10^{-4} = 0\,.0\,0\,0\,8\,9\,5 \qquad \text{4 places to the left.}$$
$$9.77 \times 10^0 = 9.77 \qquad \text{No movement of the decimal point.}$$

Using scientific notation to simplify computations

Another advantage of scientific notation becomes apparent when we evaluate products or quotients that contain very large or very small numbers.

EXAMPLE 6 *Stars.* Except for the sun, the nearest star visible to the naked eye from most parts of the United States is Sirius. Light from Sirius reaches Earth in about 70,000 hours. If light travels at approximately 670,000,000 mph, how far from Earth is Sirius?

Solution

We are given the rate at which light travels (670,000,000 mph) and the time it takes the light to travel from Sirius to Earth (70,000 hr). We can find the distance the light travels using the formula $d = rt$.

$$d = rt$$
$$d = 670{,}000{,}000(70{,}000) \quad \text{Substitute 670,000,000 for } r \text{ and 70,000 for } t.$$
$$= (6.7 \times 10^8)(7.0 \times 10^4) \quad \text{Write each number in scientific notation.}$$
$$= (6.7 \cdot 7.0) \times (10^8 \cdot 10^4) \quad \text{Group the numbers together and the powers of 10 together.}$$
$$= (6.7 \cdot 7.0) \times 10^{8+4} \quad \text{Keep the base and add the exponents.}$$
$$= 46.9 \times 10^{12} \quad \text{Do the multiplication. Do the addition.}$$

We note that 46.9 is not between 0 and 1, so 46.9×10^{12} is not written in scientific notation. To answer in scientific notation, we proceed as follows.

$$= 4.69 \times 10^1 \times 10^{12} \quad \text{Write 46.9 in scientific notation as } 4.69 \times 10^1.$$
$$= 4.69 \times 10^{13} \quad \text{Keep the base of 10 and add the exponents.}$$

Sirius is approximately 4.69×10^{13} or 46,900,000,000,000 miles from Earth. ■

EXAMPLE 7 *Atoms.* Scientific notation is used in chemistry. As an example, we can approximate the weight (in grams) of one atom of the heaviest naturally occurring element, uranium, by evaluating the following expression.

$$\frac{2.4 \times 10^2}{6.0 \times 10^{23}}$$

Solution

$$\frac{2.4 \times 10^2}{6.0 \times 10^{23}} = \frac{2.4}{6.0} \times \frac{10^2}{10^{23}} \quad \text{Divide the numbers and the powers of 10 separately.}$$
$$= \frac{2.4}{6.0} \times 10^{2-23} \quad \text{For the powers of 10, keep the base and subtract the exponents.}$$
$$= 0.4 \times 10^{-21} \quad \text{Do the division. Then subtract the exponents.}$$
$$= 4.0 \times 10^{-1} \times 10^{-21} \quad \text{Write 0.4 in scientific notation as } 4.0 \times 10^{-1}.$$
$$= 4.0 \times 10^{-22} \quad \text{Keep the base and add the exponents.}$$

One atom of uranium weighs 4.0×10^{-22} gram. Written in standard notation, this is 0.00000000000000000000004 g.

Self Check

Find the approximate weight (in grams) of one atom of gold by evaluating

$$\frac{1.98 \times 10^2}{6.0 \times 10^{23}}$$

Answer: 3.3×10^{-22} g ■

Accent on Technology: **Entering numbers in scientific notation**

We can evaluate the expression from Example 7 by entering the numbers written in scientific notation, using the ⎡EE⎤ key on a scientific calculator.

Keystrokes 2.4 ⎡EE⎤ 2 ÷ 6 ⎡EE⎤ 23 = ⎡ 4.⁻²² ⎤

The result shown in the display means 4.0×10^{-22}.
If we use a graphing calculator, the keystrokes are similar.

Keystrokes 2.4 ⎡2nd⎤ ⎡EE⎤ 2 ÷ 6 ⎡2nd⎤ ⎡EE⎤ 23 ⎡ENTER⎤

```
2.4E2/6E23
         4E-22
```

STUDY SET Section 10.3

VOCABULARY *Fill in the blanks.*

1. A number is written in _____ notation when it is written as the product of a number between 1 (including 1) and 10 and an integer power of 10.

2. The number 125,000 is written in _____ notation.

CONCEPTS *Fill in the blanks.*

3. $2.5 \times 10^2 = $

4. $2.5 \times 10^{-2} = $

5. $2.5 \times 10^{-5} = $

6. $2.5 \times 10^5 = $

7. $387,000 = 3.87 \times $

8. $38.7 = 3.87 \times $

9. $0.00387 = 3.87 \times $

10. $0.000387 = 3.87 \times $

11. When we multiply a decimal by 10^5, the decimal point moves ___ places to the _____.

12. When we multiply a decimal by 10^{-7}, the decimal point moves ___ places to the _____.

13. Dividing a decimal by 10^4 is equivalent to multiplying it by ___.

14. Multiplying a decimal by 10^0 does not move the decimal point, because $10^0 = $ ___.

15. When a real number greater than 1 is written in scientific notation, the exponent on 10 is a _____ number.

16. When a real number between 0 and 1 is written in scientific notation, the exponent on 10 is a _____ number.

NOTATION *Complete each solution.*

17. Write 63.7×10^5 in scientific notation.
$$63.7 \times 10^5 = \boxed{} \times 10^5$$
$$= 6.37 \times 10^{\boxed{} + 5}$$
$$= 6.37 \times 10^6$$

18. Simplify $\dfrac{64,000}{0.00004}$.
$$\frac{64,000}{0.00004} = \frac{6.4 \times \boxed{}}{4 \times \boxed{}}$$
$$= \frac{\boxed{}}{\boxed{}} \times \frac{10^4}{10^{-5}}$$
$$= 1.6 \times 10^{\boxed{} - (-5)}$$
$$= 1.6 \times 10^9$$

PRACTICE *Write each number in scientific notation.*

19. 23,000

20. 4,750

21. 1,700,000

22. 290,000

23. 0.062

24. 0.00073

25. 0.0000051

26. 0.04

27. 42.5×10^2

28. 0.3×10^3

29. 0.25×10^{-2}

30. 25.2×10^{-3}

Write each number in standard notation.

31. 2.3×10^2

32. 3.75×10^4

33. 8.12×10^5

34. 1.2×10^3

35. 1.15×10^{-3}

36. 4.9×10^{-2}

37. 9.76×10^{-4}

38. 7.63×10^{-5}

39. 25×10^6

40. 0.07×10^3

41. 0.51×10^{-3}

42. 617×10^{-2}

43. ASTRONOMY The distance from Earth to Alpha Centauri (the nearest star outside our solar system) is about 25,700,000,000,000 miles. Express this number in scientific notation.

44. SPEED OF SOUND The speed of sound in air is 33,100 centimeters per second. Express this number in scientific notation.

45. GEOGRAPHY The largest ocean in the world is the Pacific Ocean, which covers 6.38×10^7 square miles. Express this number in standard notation.

46. ATOMS The number of atoms in 1 gram of iron is approximately 1.08×10^{22}. Express this number in standard notation.

47. LENGTH OF A METER One meter is approximately 0.00622 mile. Use scientific notation to express this number.

48. ANGSTROM One angstrom is 1.0×10^{-7} millimeter. Express this number in standard notation.

Use scientific notation and the rules for exponents to simplify each expression. Give all answers in standard notation.

49. $(3.4 \times 10^2)(2.1 \times 10^3)$

50. $(4.1 \times 10^{-3})(3.4 \times 10^4)$

51. $\dfrac{9.3 \times 10^2}{3.1 \times 10^{-2}}$

52. $\dfrac{7.2 \times 10^6}{1.2 \times 10^8}$

53. $\dfrac{96,000}{(12,000)(0.00004)}$

54. $\dfrac{(0.48)(14,400,000)}{96,000,000}$

🖩 *Evaluate each expression.*

55. $(456.4)^6$

56. $(0.053)^8$

57. $(0.009)^{-6}$

58. 225^{-5}

59. $\left(\dfrac{1}{3}\right)^{-55}$

60. $\left(\dfrac{8}{5}\right)^{50}$

APPLICATIONS

61. WAVELENGTH   Transmitters, vacuum tubes, and lights emit energy that can be modeled as a wave, as shown in Illustration 1. Examples of the most common types of electromagnetic waves are given in the table. List the wavelengths in order from shortest to longest.

This distance between the two crests of the wave is called the wavelength.

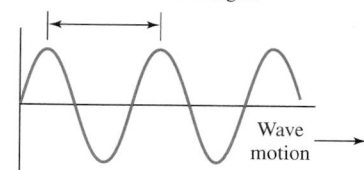

ILLUSTRATION 1

Type	Use	Wavelength (m)
visible light	lighting	9.3×10^{-6}
infrared	photography	3.7×10^{-5}
x-ray	medical	2.3×10^{-11}
radio wave	communication	3.0×10^2
gamma ray	treating cancer	8.9×10^{-14}
microwave	cooking	1.1×10^{-2}
ultraviolet	sun lamp	6.1×10^{-8}

62. EXPLORATION   On July 4, 1997, the Pathfinder, carrying the rover vehicle called Sojourner, landed on Mars to perform a scientific investigation of the planet. The distance from Mars to Earth is approximately

3.5×10^7 miles. Use scientific notation to express this distance in feet. (*Hint:* 5,280 feet = 1 mile.)

63. PROTON   The mass of one proton is approximately 1.7×10^{-24} gram. Use scientific notation to express the mass of 1 million protons.

64. SPEED OF SOUND   The speed of sound in air is approximately 3.3×10^4 centimeters per second. Use scientific notation to express this speed in kilometers per second. (*Hint:* 100 centimeters = 1 meter and 1,000 meters = 1 kilometer.)

65. LIGHT YEAR   One light year is about 5.87×10^{12} miles. Use scientific notation to express this distance in feet. (*Hint:* 5,280 feet = 1 mile.)

66. OIL RESERVES   As of January 1, 1999, Saudi Arabia was believed to have crude oil reserves of about 2.615×10^{11} barrels. A barrel contains 42 gallons of oil. Use scientific notation to express its oil reserves in gallons.

67. INTEREST EARNED   As of December 31, 1998, the Federal Deposit Insurance Corporation (FDIC) reported that the total insured deposits in U.S. banks and savings and loans was approximately 5.09×10^{12} dollars. If this money was invested at a rate of 4% simple annual interest, how much would it earn in one year? (Use scientific notation to express the answer.)

68. CURRENCY   As of March 31, 1999, the U.S. Treasury reported that the number of $20 bills in circulation was approximately 4.35×10^9. What was the total value of the currency? (Use scientific notation to express the answer.)

69. SIZE OF THE MILITARY   The graph in Illustration 2 shows the number of U.S. troops for 1979–1998. Estimate each of the following and express your answers in scientific and standard notation.
   a. The number of troops in 1993
   b. The smallest and largest numbers of troops during these years

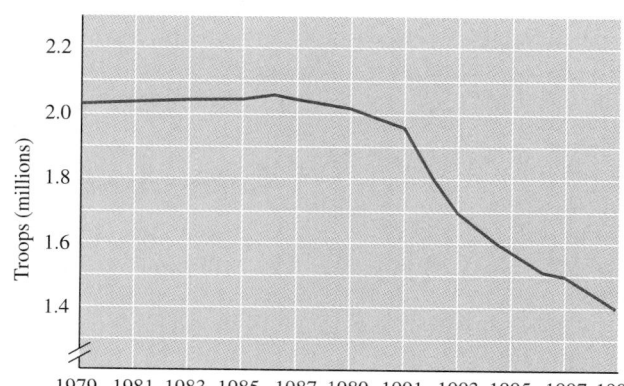

Based on data from the U.S. Department of Defense

ILLUSTRATION 2

70. 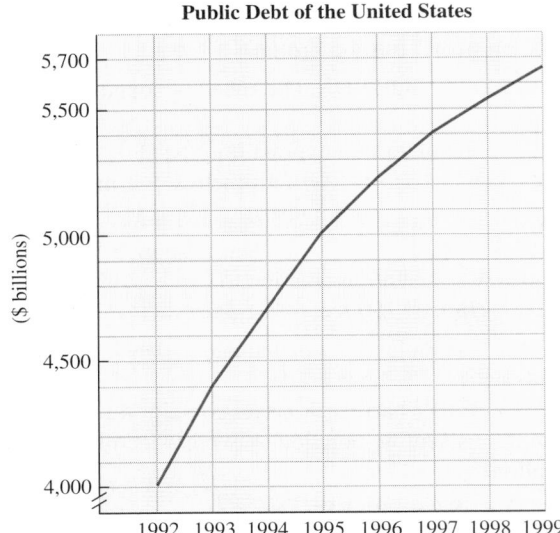 THE NATIONAL DEBT   The graph in Illustration 3 shows the growth of the national debt for the fiscal years 1992–1999.

Public Debt of the United States

Based on data from the U.S. Department of the Treasury

ILLUSTRATION 3

a. Use scientific notation to express the debt as of 1994, 1995, and 1997.

b. In 1999, the population of the United States was about 2.75×10^8. Estimate the share of the debt for each man, woman, and child in the United States. Answer in standard notation.

WRITING

71. In what situations would scientific notation be more convenient than standard notation?

72. To multiply a number by a power of 10, we move the decimal point. Which way, and how far? Explain.

73. 2.3×10^{-3} contains a negative sign but represents a positive number. Explain.

74. Is this a true statement? $2.0 \times 10^3 = 2 \times 10^3$. Explain.

REVIEW

75. If $y = -1$, find the value of $-5y^{55}$.

76. What is the y-intercept of the graph of $y = -3x - 5$?

Tell which property of real numbers justifies each statement.

77. $5 + z = z + 5$

78. $7(u + 3) = 7u + 7 \cdot 3$

Solve each equation.

79. $3(x - 4) - 6 = 0$

80. $8(3x - 5) - 4(2x + 3) = 12$

10.4 *Polynomials*

In this section, you will learn about

- Polynomials • Monomials, binomials, and trinomials
- Degree of a polynomial • Evaluating polynomial functions

INTRODUCTION. In arithmetic, we learned how to add, subtract, multiply, divide, and find powers of numbers. In algebra, we will learn how to perform these operations on *polynomials*. In this section, we will introduce polynomials, classify them into groups, define their degrees, and show how to evaluate them at specific values of their variables.

Polynomials

Recall that a **term** is a number or a product of a number and one or more variables, which may be raised to powers. Examples of terms are

$$3x, \qquad -4y^2, \qquad \frac{1}{2}a^2b^3, \qquad t, \qquad \text{and} \qquad 25$$

The **numerical coefficients,** or simply **coefficients,** of the first four of these terms are 3, -4, $\frac{1}{2}$, and 1, respectively. Because $25 = 25x^0$, 25 is considered to be the numerical coefficient of the term 25.

Polynomials

> A **polynomial** is a term or a sum of terms in which all variables have whole-number exponents.

Here are some examples of polynomials:

$$3x + 2, \qquad 4y^2 - 2y - 3, \qquad -8xy^2, \qquad \text{and} \qquad a^3 + 3a^2b + 3ab^2 + b^3$$

The polynomial $3x + 2$ has two terms, $3x$ and 2, and we say it is a **polynomial in x.** A single number is called a **constant,** and so its last term, 2, is called the **constant term.**

Since $4y^2 - 2y - 3$ can be written as $4y^2 + (-2y) + (-3)$, it is the sum of three terms, $4y^2$, $-2y$, and -3. It is written in **decreasing** or **descending powers** of y, because the powers on y decrease from left to right.

$-8xy^2$ is a polynomial with just one term. We say that it is a **polynomial in x and y.**

The four-term polynomial $a^3 + 3a^2b + 3ab^2 + b^3$ is written in descending powers of a and **ascending powers** of b.

 COMMENT The expression $2x^3 - 3x^{-2} + 5$ is not a polynomial, because the second term contains a variable with an exponent that is not a whole number. Similarly, $y^2 - \frac{7}{y}$ is not a polynomial, because $\frac{7}{y}$ can be written $7y^{-1}$.

EXAMPLE 1 *Identifying polynomials.* Tell whether each expression is a polynomial.

a. $x^2 + 2x + 1$ Yes.

b. $3a^{-1} - 2a - 3$ No. In the first term, the exponent on the variable is not a whole number.

c. $\frac{1}{2}x^3 - 2.3x$ Yes, since it can be written as the sum $\frac{1}{2}x^3 + (-2.3x)$.

d. $\frac{p + 3}{p - 1}$ No. Variables cannot be in the denominator of a fraction.

Self Check

Tell whether each expression is a polynomial:

a. $3x^{-4} + 2x^2 - 3$

b. $7.5p^3 - 4p^2 - 3p + 4$

Answers: **a.** no **b.** yes ∎

Monomials, binomials, and trinomials

A polynomial with one term is called a **monomial.** A polynomial with two terms is called a **binomial.** A polynomial with three terms is called a **trinomial.** Here are some examples.

Monomials	Binomials	Trinomials
$-6x$	$3u^3 - 4u^2$	$-5t^2 + 4t + 3$
$5x^2y$	$18a^2b + 4ab$	$27x^3 - 6x - 2$
29	$-29z^{17} - 1$	$a^2 + 2ab + b^2$

EXAMPLE 2 *Classifying polynomials.* Classify each polynomial as a monomial, a binomial, or a trinomial.

a. $5.2x^4 + 3.1x$ Since the polynomial has two terms, $5.2x^4$ and $3.1x$, it is a binomial.

b. $7g^4 - 5g^3 - 2$ Since the polynomial has three terms, $7g^4$, $-5g^3$, and -2, it is a trinomial.

c. $-5x^2y^3$ Since the polynomial has one term, it is a monomial.

Self Check

Classify each polynomial as a monomial, a binomial, or a trinomial:

a. $5x$

b. $-5x^2 + 2x - 0.5$

c. $16x^2 - 9y^2$

Answers: **a.** monomial,
b. trinomial, **c.** binomial ∎

Degree of a polynomial

The monomial $7x^6$ is called a **monomial of sixth degree** or a **monomial of degree 6,** because the variable x occurs as a factor six times. The monomial $3x^3y^4$ is a monomial of seventh degree, because the variables x and y occur as factors a total of seven times. Here are some more examples:

$2.7a$ is a monomial of degree 1.

$-2x^3$ is a monomial of degree 3.

$47x^2y^3$ is a monomial of degree 5.

8 is a monomial of degree 0, because $8 = 8x^0$.

These examples illustrate the following definition.

Degree of a monomial

> If a represents a nonzero constant, the **degree of the monomial** ax^n is n.
>
> The **degree of a monomial** in several variables is the sum of the exponents on those variables.

COMMENT   Note that the degree of ax^n is not defined when $a = 0$. Since $ax^n = 0$ when $a = 0$, the constant 0 has no defined degree.

Because each term of a polynomial is a monomial, we define the degree of a polynomial by considering the degrees of each of its terms.

Degree of a polynomial

> The **degree of a polynomial** is determined by the term with the largest degree.

Here are some examples:

$x^2 + 2x$ is a binomial of degree 2, because the degree of its first term is 2 and the degree of its second term is less than 2.

$d^3 - 3d^2 + 1$ is a trinomial of degree 3, because the degree of its first term is 3 and the degree of each of its other terms is less than 3.

$25y^{13} - 15y^8z^{10} - 32y^{10}z^8 + 4$ is a polynomial of degree 18, because its second and third terms are of degree 18. Its other terms have degree less than 18.

EXAMPLE 3   *Degree of a polynomial.*   Find the degree of each polynomial:

a. $-4x^3 - 5x^2 + 3x$,   **b.** $1.6w - 1.6$,   and   **c.** $-17a^2b^3 + 12ab^6$.

Solution

a. The trinomial $-4x^3 - 5x^2 + 3x$ has terms of degree 3, 2, and 1. Therefore, its degree is 3.

b. The first term of $1.6w - 1.6$ has degree 1 and the second term has degree 0, so the binomial has degree 1.

c. The degree of the first term of $-17a^2b^3 + 12ab^6$ is 5 and the degree of the second term is 7, so the binomial has degree 7.

Self Check

Find the degree of each polynomial:

a. $15p^3 - 15p^2 - 3p + 4$

b. $-14st^4 + 12s^3t$

Answers: **a.** 3,   **b.** 5

If written in descending powers of the variable, the **leading term** of a polynomial is the term of highest degree. For example, the leading term of $-4x^3 - 5x^2 + 3x$ is $-4x^3$. The coefficient of the leading term (in this case, -4) is called the **leading coefficient.**

Evaluating polynomial functions

Each of the equations below defines a function, because each input x-value determines exactly one output value. Since the right-hand side of each equation is a polynomial, these functions are called **polynomial functions.**

$$f(x) = 6x + 4 \qquad g(x) = 3x^2 + 4x - 5 \qquad h(x) = -x^3 + x^2 - 2x + 3$$

This polynomial has two terms. Its degree is 1. This polynomial has three terms. Its degree is 2. This polynomial has four terms. Its degree is 3.

To evaluate a polynomial function for a specific value, we replace the variable in the defining equation with the input value. Then we simplify the resulting expression to find the output. For example, suppose we wish to evaluate the polynomial function $f(x) = 6x + 4$ for $x = 1$. Then $f(1)$ (read as "f of 1") represents the value of $f(x) = 6x + 4$ when $x = 1$. We find $f(1)$ as follows.

$$f(x) = \mathbf{6}x + 4 \qquad \text{The given function.}$$
$$f(\mathbf{1}) = 6(\mathbf{1}) + 4 \qquad \text{Substitute 1 for } x. \text{ The number 1 is the input.}$$
$$= 6 + 4 \qquad \text{Do the multiplication.}$$
$$= 10 \qquad \text{Do the addition. 10 is the output.}$$

Thus, $f(1) = 10$.

EXAMPLE 4 *Evaluating polynomial functions.* Consider the function $g(x) = 3x^2 + 4x - 5$. Find   **a.** $g(0)$ and   **b.** $g(-2)$.

Solution

a. $g(x) = 3x^2 + 4x - 5$ The given function.

$g(0) = 3(0)^2 + 4(0) - 5$ To find $g(0)$, substitute 0 for x.

$= 3(0) + 4(0) - 5$ Evaluate the power.

$= 0 + 0 - 5$ Do the multiplications.

$g(0) = -5$

b. $g(x) = 3x^2 + 4x - 5$ The given function.

$g(-2) = 3(-2)^2 + 4(-2) - 5$ To find $g(-2)$, substitute -2 for x.

$= 3(4) + 4(-2) - 5$ Evaluate the power.

$= 12 + (-8) - 5$ Do the multiplications.

$g(-2) = -1$

Self Check

Consider the function

$$h(x) = -x^3 + x - 2x + 3$$

Find

a. $h(0)$

b. $h(-3)$

Answers: **a.** 3, **b.** 33

EXAMPLE 5 *Supermarket display.* The polynomial function

$$f(c) = \frac{1}{3}c^3 + \frac{1}{2}c^2 + \frac{1}{6}c$$

gives the number of cans used in a display shaped like a square pyramid, having a square base formed by c cans per side. Find the number of cans of soup used in the display shown in Figure 10-2.

Solution

Since each side of the square base of the display is formed by 4 cans, $c = 4$. We can find the number of cans used in the display by finding $f(4)$.

FIGURE 10-2

$$f(c) = \frac{1}{3}c^3 + \frac{1}{2}c^2 + \frac{1}{6}c \qquad \text{The given function.}$$

$$f(4) = \frac{1}{3}(4)^3 + \frac{1}{2}(4)^2 + \frac{1}{6}(4) \qquad \text{Substitute 4 for } c.$$

$$= \frac{1}{3}(64) + \frac{1}{2}(16) + \frac{1}{6}(4) \qquad \text{Find the powers.}$$

$$= \frac{64}{3} + 8 + \frac{2}{3} \qquad \text{Do the multiplication, and then simplify: } \frac{4}{6} = \frac{2}{3}.$$

$$= \frac{66}{3} + 8 \qquad \text{Add the fractions.}$$

$$= 22 + 8$$

$$= 30$$

30 cans of soup were used in the display.

STUDY SET Section 10.4

VOCABULARY *Fill in the blanks.*

1. A _____ is a term or a sum of terms in which all variables have whole-number exponents.

2. The numerical _____ of the term $-25x^2y^3$ is -25.

3. The degree of a polynomial is the same as the degree of its _____ with the largest degree.

4. A _____ is a polynomial with one term. A _____ is a polynomial with two terms.

5. The _____ of the monomial $3x^7$ is 7.

6. For the polynomial $6x^2 + 3x - 1$, the _____ term is $6x^2$, and the leading _____ is 6. The _____ term is -1.

7. $-x^3 - 6x^2 + 9x - 2$ is a polynomial _____ x and is written in _____ powers of x.

8. A _____ is a polynomial with three terms.

9. The notation $f(x)$ is read as f _____ x.

10. $f(2)$ represents the _____ of a function when $x = 2$.

CONCEPTS *Tell whether each expression is a polynomial.*

11. $x^3 - 5x^2 - 2$

12. $x^{-4} - 5x$

13. $\frac{1}{2x} + 3$

14. $x^3 - 1$

15. $x^2 - y^2$

16. $a^4 + a^3 + a^2 + a$

Classify each polynomial as a monomial, a binomial, a trinomial, or none of these.

17. $3x + 7$

18. $3y - 5$

19. $y^2 + 4y + 3$

20. $3xy$

21. $3z^2$

22. $3x^4 - 2x^3 + 3x - 1$

23. $t - 32$

24. $9x^2y^3z^4$

25. $s^2 - 23s + 31$

26. $2x^3 - 5x^2 + 6x - 3$

27. $3x^5 - x^4 - 3x^3 + 7$

28. x^3

29. $2a^2 - 3ab + b^2$

30. $a^3 - b^3$

Find the degree of each polynomial.

31. $3x^4$

32. $3x^5$

33. $-2x^2 + 3x + 1$

34. $-5x^4 + 3x^2 - 3x$

35. $3x - 5$

36. $y^3 + 4y^2$

37. $-5r^2s^2 - r^3s + 3$

38. $4r^2s^3 - 5r^2s^8$

39. $x^{12} + 3x^2y^3$

40. $17ab^5 - 12a^3b$

41. 38

42. -24

NOTATION *Complete each solution.*

43. If $f(x) = -2x^2 + 3x - 1$, find $f(2)$.

$$f(2) = -2(\quad)^2 + 3(\quad) - 1$$

$$= -2(\quad) + \quad - 1$$

$$= -8 + 6 - \quad$$

$$= \quad - 1$$

$$= -3$$

44. If $f(x) = -2x^2 + 3x - 1$, find $f(-2)$.

$$f(-2) = -2(\quad)^2 + 3(\quad) - 1$$
$$= -2(\quad) + (\quad) - 1$$
$$= \quad + (-6) - 1$$
$$= \quad - 1$$
$$= -15$$

45. Explain why $f(x) = x^3 + 2x^2 - 3$ is called a polynomial function.

46. a. Write $x - 9 + 3x^2$ in descending powers of x.

   b. Write $-2xy + y^2 + x^2$ in descending powers of x.

PRACTICE   *Let $f(x) = 5x - 3$. Find each value.*

47. $f(2)$ **48.** $f(0)$

49. $f(-1)$ **50.** $f(-2)$

51. $f\left(\dfrac{1}{5}\right)$ **52.** $f\left(\dfrac{4}{5}\right)$

53. $f(-0.9)$ **54.** $f(-1.2)$

Let $g(x) = -x^2 - 4$. Find each value.

55. $g(0)$ **56.** $g(1)$

57. $g(-1)$ **58.** $g(-2)$

⊞ *Let $g(x) = -x^2 - 4$. Find each value.*

59. $g(1.3)$ **60.** $g(2.4)$

61. $g(-13.6)$ **62.** $g(-25.3)$

Let $h(x) = x^3 - 2x + 3$. Find each value.

63. $h(0)$ **64.** $h(3)$

65. $h(-2)$ **66.** $h(-1)$

⊞ *Let $h(x) = x^3 - 2x + 3$. Find each value.*

67. $h(0.9)$ **68.** $h(0.4)$

69. $h(-8.1)$ **70.** $h(-7.7)$

Let $f(x) = -x^4 - x^3 + x^2 + x - 1$. Find each value.

71. $f(1)$ **72.** $f(-1)$

73. $f(-2)$ **74.** $f(2)$

⊞ **APPLICATIONS**   *In Exercises 75–82, use a calculator to help solve each problem.*

75. PACKAGING   To make boxes, a manufacturer cuts equal-sized squares from each corner of a 10 in. × 12 in. piece of cardboard, and then folds up the sides.

(See Illustration 1.) The polynomial function $f(x) = 4x^3 - 44x^2 + 120x$ gives the volume (in cubic inches) of the resulting box when a square with sides x inches long is cut from each corner. Find the volume of a box if 3-inch squares are cut out.

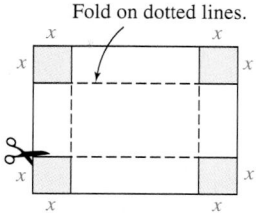

Fold on dotted lines.

ILLUSTRATION 1

76. MAXIMIZING REVENUE   The revenue (in dollars) that a manufacturer of office desks receives is given by the polynomial function

$$f(d) = -0.08d^2 + 100d$$

where d is the number of desks manufactured.

a. Find the total revenue if 625 desks are manufactured.

b. Does increasing the number of desks being manufactured to 650 increase the revenue?

77. WATER BALLOONS   Some college students launched water balloons from the balcony of their dormitory on unsuspecting sunbathers on the college quad. The height in feet of the balloons at a time t seconds after being launched is given by the polynomial function

$$f(t) = -16t^2 + 12t + 20$$

What was the height of the balloons 0.5 second and 1.5 seconds after being launched?

78. STOPPING DISTANCE   The number of feet that a car travels before stopping depends on the driver's reaction time and the braking distance, as shown in Illustration 2. For one driver, the stopping distance is given by the polynomial function

$$f(v) = 0.04v^2 + 0.9v$$

where v is the velocity of the car. Find the stopping distance when the driver is traveling at 30 mph.

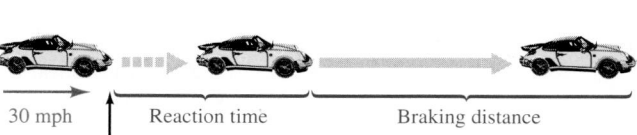

Stopping distance d

30 mph   Reaction time   Braking distance

Decision to stop

ILLUSTRATION 2

79. SUSPENSION BRIDGE   See Illustration 3. The function

$$f(s) = 400 + 0.0066667s^2 - 0.0000001s^4$$

approximates the length of the cable between the two vertical towers of a suspension bridge, where s is the sag in the cable. Estimate the length of the cable if the sag is 24.6 feet.

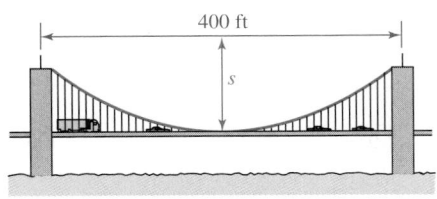

ILLUSTRATION 3

80. PRODUCE DEPARTMENT   Suppose a grocer is going to set up a pyramid-shaped display of cantaloupes like that shown in Figure 10-2 in Example 5. If each side of the square base of the display is made of six cantaloupes, how many will be used in the display?

81. DOLPHINS   At a marine park, three trained dolphins jump in unison over an arching stream of water whose path can be described by the polynomial function

$$f(x) = -0.05x^2 + 2x$$

See Illustration 4. Given the takeoff points for each dolphin, how high must each dolphin jump to clear the stream of water?

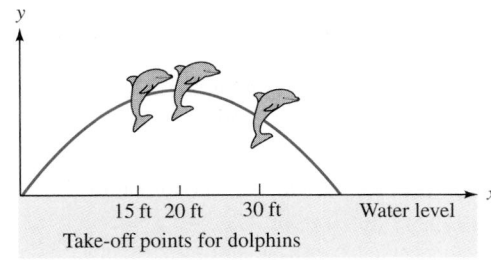

ILLUSTRATION 4

82. TUNNEL   The arch at the entrance to a tunnel is described by the polynomial function

$$f(x) = -0.25x^2 + 23$$

See Illustration 5. What is the height of the arch at the edge of the pavement?

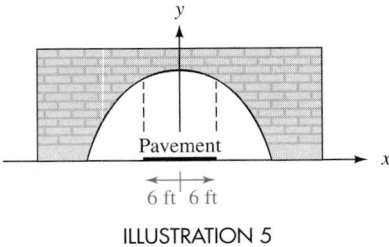

ILLUSTRATION 5

WRITING

83. Describe how to determine the degree of a polynomial.

84. List some words that contain the prefixes *mono, bi,* or *tri.*

REVIEW   *Solve each inequality and graph the solution set.*

85. $-4(3y + 2) \le 28$

86. $-5 < 3t + 4 \le 13$

Write each expression without using parentheses or negative exponents.

87. $(x^2x^4)^3$

88. $(a^2)^3(a^3)^2$

89. $\left(\dfrac{y^2y^5}{y^4}\right)^3$

90. $\left(\dfrac{2t^3}{t}\right)^{-4}$

10.5 *Adding and Subtracting Polynomials*

In this section, you will learn about

- Adding monomials • Subtracting monomials • Adding polynomials
- Subtracting polynomials • Adding and subtracting multiples of polynomials
- An application of adding polynomials

INTRODUCTION. In Figure 10-3(a), the heights of the Seattle Space Needle and the Eiffel Tower in Paris are given. Using rules from arithmetic, we can find the difference in the heights of the towers by subtracting two numbers.

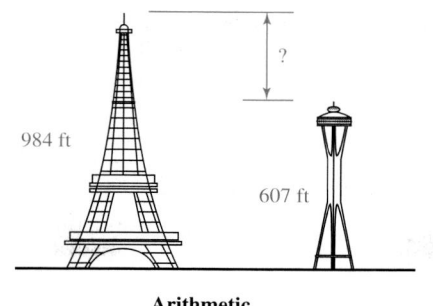

Arithmetic

$984 - 607 = 377$

The difference in height is 377 feet.

(a)

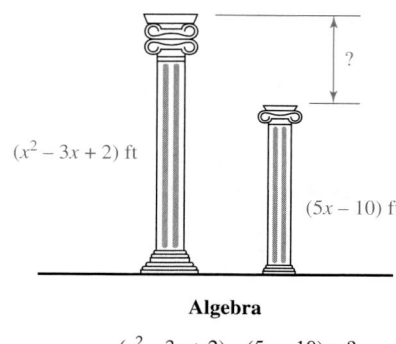

Algebra

$(x^2 - 3x + 2) - (5x - 10) = ?$

(b)

FIGURE 10-3

In Figure 10-3(b), the heights of two types of classical Greek columns are expressed using *polynomials*. To find the difference in their heights, we must subtract the polynomials. In this section, we will discuss the algebraic rules that are used to do this. Since any subtraction can be written in terms of addition, we will consider the procedures used to add polynomials first. We begin with monomials, which are polynomials having just one term.

Adding monomials

Recall that like terms have the same variables with the same exponents:

Like terms	**Unlike terms**
$-7x$ and $15x$	$-7x$ and $15a$
$4y^3$ and $16y^3$	$4y^3$ and $16y^2$
$\frac{1}{2}xy^2$ and $-\frac{1}{3}xy^2$	$\frac{1}{2}xy^2$ and $-\frac{1}{3}x^2y$

Also recall that to combine like terms, we combine their coefficients and keep the same variables with the same exponents. For example,

$$4y + 5y = (4 + 5)y \qquad \text{and} \qquad 8x^2 - x^2 = (8 - 1)x^2$$
$$= 9y \qquad\qquad\qquad\qquad\qquad = 7x^2$$

Likewise,

$$3a + 4b - 6a + 3b = -3a + 7b \qquad \text{and} \qquad -4cd^3 + 9cd^3 = 5cd^3$$

These examples suggest that to add like monomials, we simply combine like terms.

EXAMPLE 1 *Adding monomials.* Do the following additions.

a. $4x^4 + 81x^4 = 85x^4$

b. $-8x^2y^2 + 6x^2y^2 + x^2y^2 = -2x^2y^2 + x^2y^2$ Work from left to right. Combine like terms.

$\qquad\qquad\qquad\qquad\quad = -x^2y^2$ Combine like terms.

c. $32c^2 + 10c + 4c^2 = 32c^2 + 4c^2 + 10c$ Write the like terms together.

$\qquad\qquad\qquad\quad = 36c^2 + 10c$ Combine like terms.

Self Check

Do the following additions:

a. $27x^6 + 8x^6$

b. $-12pq^2 + 5pq^2 + 8pq^2$

c. $6a^3 + 15a + a^3$

Answers: a. $35x^6$, **b.** pq^2, **c.** $7a^3 + 15a$

 COMMENT When performing operations on polynomials, it is standard practice to write the terms of the solution in decreasing (or descending) powers of one variable. For instance, in Example 1, part c, the solution was written as $36c^2 + 10c$ instead of as $10c + 36c^2$.

Subtracting monomials

To subtract one monomial from another, we add the opposite of the monomial that is to be subtracted. In symbols, $x - y = x + (-y)$.

EXAMPLE 2   *Subtracting monomials.*   Find each difference.

a. $8x^2 - 3x^2 = 8x^2 + (-3x^2)$ Add the opposite of $3x^2$, which is $-3x^2$.

$\qquad\qquad = 5x^2$ Combine like terms.

b. $6xy - 9xy = 6xy + (-9xy)$

$\qquad\qquad = -3xy$

c. $-3r - 5 - 4r = -3r + (-5) + (-4r)$ Add the opposite of 5 and $4r$.

$\qquad\qquad = -3r + (-4r) + (-5)$ Group like terms together.

$\qquad\qquad = -7r - 5$ Combine like terms. Write the addition of -5 as a subtraction of 5.

Self Check

Find each difference:

a. $12m^3 - 7m^3$

b. $-4pq - 27p - 8pq$

Answers: a. $5m^3$,
b. $-12pq - 27p$ ■

Adding polynomials

Because of the distributive property, we can remove parentheses enclosing several terms when the sign preceding the parentheses is a $+$ sign. We simply drop the parentheses.

$$+(3x^2 + 3x - 2) = +1(3x^2 + 3x - 2)$$
$$= 1(3x^2) + 1(3x) + 1(-2)   \text{Distribute the multiplication by 1.}$$
$$= 3x^2 + 3x + (-2)$$
$$= 3x^2 + 3x - 2$$

We can add polynomials by removing parentheses, if necessary, and then combining any like terms that are contained within the polynomials.

EXAMPLE 3   *Adding polynomials.*
Add $(3x^2 - 3x + 2) + (2x^2 + 7x - 4)$.

Solution

$\quad (3x^2 - 3x + 2) + (2x^2 + 7x - 4)$

$\quad = 3x^2 - 3x + 2 + 2x^2 + 7x - 4$ Drop the parentheses.

$\quad = 3x^2 + 2x^2 - 3x + 7x + 2 - 4$ Write like terms together.

$\quad = 5x^2 + 4x - 2$ Combine like terms.

Self Check

Add
$(2a^2 - a + 4) + (5a^2 + 6a - 5)$.

Answer: $7a^2 + 5a - 1$ ■

Problems such as Example 3 are often written with like terms aligned vertically. We can then add column by column.

$$\begin{array}{r} 3x^2 - 3x + 2 \\ + \underline{2x^2 + 7x - 4} \\ 5x^2 + 4x - 2 \end{array}$$

EXAMPLE 4   *Adding polynomials vertically.*   Add $4x^2 - 3$ and $3x^2 - 8x + 8$.

Solution

Since the first polynomial does not have an x-term, we leave a space so that the constant terms can be aligned.

$$\begin{array}{r} 4x^2 \qquad - 3 \\ + \underline{3x^2 - 8x + 8} \\ 7x^2 - 8x + 5 \end{array}$$

Self Check

Add $4q^2 - 7$ and $2q^2 - 8q + 9$ vertically.

Answer: $6q^2 - 8q + 2$ ■

Subtracting polynomials

Because of the distributive property, we can remove parentheses enclosing several terms when the sign preceding the parentheses is a − sign. We simply drop the minus sign and the parentheses, and *change the sign of every term within the parentheses.*

$$-(3x^2 + 3x - 2) = -1(3x^2 + 3x - 2)$$
$$= -1(3x^2) + (-1)(3x) + (-1)(-2)$$
$$= -3x^2 + (-3x) + 2$$
$$= -3x^2 - 3x + 2$$

This suggests that the way to subtract polynomials is to remove parentheses, change the sign of each term of the second polynomial, and combine like terms.

EXAMPLE 5   *Subtracting polynomials.*   Find each difference.

a. $(3x - 4) - (5x + 7) = 3x - 4 - 5x - 7$ Change the sign of each term inside $(5x + 7)$.

$$= -2x - 11$$ Combine like terms.

b. $(3x^2 - 4x - 6) - (2x^2 - 6x) = 3x^2 - 4x - 6 - 2x^2 + 6x$

$$= x^2 + 2x - 6$$

c. $(-t^3 - 2t^2 - 1) - (-t^3 - 2t^2 + 1) = -t^3 - 2t^2 - 1 + t^3 + 2t^2 - 1$

$$= -2$$

Self Check

Find the difference:

$$(-2a^2 + 5) - (-5a^2 - 7)$$

Answer: $3a^2 + 12$ ■

To subtract polynomials in vertical form, we add the opposite of the **subtrahend** (the bottom polynomial) to the **minuend** (the top polynomial).

EXAMPLE 6   *Subtracting polynomials vertically.*   Subtract $3x^2 - 2x$ from $2x^2 + 4x$.

Solution

Since $3x^2 - 2x$ is to be subtracted from $2x^2 + 4x$, we write $3x^2 - 2x$ below $2x^2 + 4x$ in vertical form. Then we change the signs of the terms of $3x^2 - 2x$ and add:

$$\begin{array}{r} 2x^2 + 4x \\ - \underline{3x^2 - 2x} \end{array} \longrightarrow \begin{array}{r} 2x^2 + 4x \\ + \underline{-3x^2 + 2x} \\ -x^2 + 6x \end{array}$$

Self Check

Subtract $2p^2 + 2p - 8$ from $5p^2 - 6p + 7$.

Answer: $3p^2 - 8p + 15$ ■

EXAMPLE 7   *Combining polynomials.*   Subtract $12a - 7$ from the sum of $6a + 5$ and $4a - 10$.

Solution

We will use brackets to show that $(12a - 7)$ is to be subtracted from the *sum* of $(6a + 5)$ and $(4a - 10)$.

$$[(6a + 5) + (4a - 10)] - (12a - 7)$$

Next, we remove the grouping symbols to obtain

$$= 6a + 5 + 4a - 10 - 12a + 7$$ Change the sign of each term in $(12a - 7)$.

$$= -2a + 2$$ Combine like terms.

Self Check

Subtract $-2q^2 - 2q$ from the sum of $q^2 - 6q$ and $3q^2 + q$.

Answer: $6q^2 - 3q$ ■

Adding and subtracting multiples of polynomials

Because of the distributive property, we can remove parentheses enclosing several terms when a monomial precedes the parentheses. We simply multiply every term within the parentheses by that monomial. For example, to add $3(2x + 5)$ and $2(4x - 3)$, we proceed as follows:

$$3(2x + 5) + 2(4x - 3) = 6x + 15 + 8x - 6 \quad \text{Distribute the multiplication by 3 and by 2.}$$
$$= 6x + 8x + 15 - 6 \quad 15 + 8x = 8x + 15.$$
$$= 14x + 9 \quad \text{Combine like terms.}$$

EXAMPLE 8 *Adding and subtracting multiples of polynomials.*
Use the distributive property to remove parentheses and simplify.

a. $3(x^2 + 4x) + 2(x^2 - 4) = 3x^2 + 12x + 2x^2 - 8$
$$= 5x^2 + 12x - 8$$

b. $-8(y^2 - 2y + 3) - 4(2y^2 + y - 6) = -8y^2 + 16y - 24 - 8y^2 - 4y + 24$
$$= -16y^2 + 12y$$

Self Check

Remove parentheses and simplify:

$$2(a^2 - 3a) + 5(a^2 + 2a)$$

Answer: $7a^2 + 4a$ ∎

An application of adding polynomials

EXAMPLE 9 *Property values.* A house purchased for \$95,000 is expected to appreciate according to the polynomial function $f(x) = 2{,}500x + 95{,}000$, where y is the value of the house after x years. A second house purchased for \$125,000 is expected to appreciate according to the equation $f(x) = 4{,}500x + 125{,}000$. Find one polynomial function that will give the total value of both properties after x years.

Solution
The value of the first house after x years is given by the polynomial $2{,}500x + 95{,}000$. The value of the second house after x years is given by the polynomial $4{,}500x + 125{,}000$. The value of both houses will be the sum of these two polynomials.

$$(2{,}500x + 95{,}000) + (4{,}500x + 125{,}000) = 7{,}000x + 220{,}000$$

The total value y of the properties is given by the polynomial function $f(x) = 7{,}000x + 220{,}000$. ∎

STUDY SET Section 10.5

VOCABULARY *Fill in the blanks.*

1. The expression $(x^2 - 3x + 2) + (x^2 - 4x)$ is the sum of two _____.

2. _____ terms have the same variables and the same exponents.

3. "To add or subtract like terms" means to combine their _____ and keep the same variables with the same exponents.

4. When we write $3x^2 + 4x - 1$ with the constant term first, the x-term second, and the x^2-term last, we say we have written the polynomial in _____ powers of x.

CONCEPTS *Fill in the blanks.*

5. To add like monomials, combine like _____.

6. $a - b = a +$ ▮

7. To add two polynomials, combine any _____ terms contained in the polynomials.

8. To subtract two polynomials, change the _____ of each term in the second polynomial, and combine like terms.

9. When the sign preceding parentheses is a $-$ sign, we can remove the parentheses by dropping the sign and the parentheses, and _____ the sign of every term within the parentheses.

10. When a monomial precedes parentheses, we can remove the parentheses by _____ every term within the parentheses by that monomial.

11. $-(-2x^2 - 3x + 4) =$

12. $-3(-2x^2 - 3x + 4) =$

13. JETS   Find the polynomial representing the length of the passenger jet in Illustration 1.

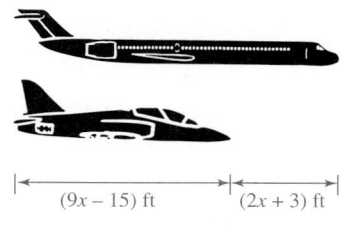

(9x – 15) ft   (2x + 3) ft

ILLUSTRATION 1

14. WATER SKIING   Find the polynomial representing the distance of the water skier from the boat in Illustration 2.

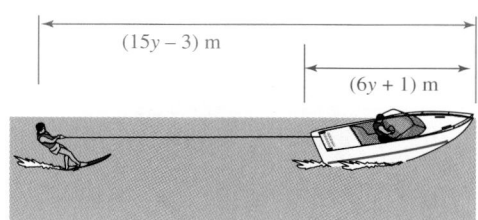

(15y – 3) m

(6y + 1) m

ILLUSTRATION 2

NOTATION   *Complete each solution.*

15. $(5x^2 + 3x) - (7x^2 - 2x)$

$= 5x^2 +$ ___ $- 7x^2 +$ ___

$= 5x^2 -$ ___ $+ 3x + 2x$

$= -2x^2 + 5x$

16. $4(3x^2 - 2x) - (2x + 4)$

$= 12x^2 -$ ___ $-$ ___ $- 4$

$= 12x^2 - 10x - 4$

PRACTICE   *Simplify each expression, if possible.*

17. $4y + 5y$

18. $-2x + 3x$

19. $8t^2 + 4t^2$

20. $15x^2 + 10x^2$

21. $-32u^3 - 16u^3$

22. $-25x^3 - 7x^3$

23. $1.8x - 1.9x$

24. $1.7y - 2.2y$

25. $\frac{1}{2}st + \frac{3}{2}st$

26. $\frac{2}{5}at + \frac{1}{5}at$

27. $3r - 4r + 7r$

28. $-2b + 7b - 3b$

29. $-4ab + 4ab - ab$

30. $xy - 4xy - 2xy$

31. $(3x)^2 - 4x^2 + 10x^2$

32. $(2x)^4 - (3x^2)^2$

Do the operations.

33. $(3x + 7) + (4x - 3)$

34. $(2y - 3) + (4y + 7)$

35. $(4a + 3) - (2a - 4)$

36. $(5b - 7) - (3b - 5)$

37. $(2x + 3y) + (5x - 10y)$

38. $(5x - 8y) - (-2x + 5y)$

39. $(-8x - 3y) - (-11x + y)$

40. $(-4a + b) + (5a - b)$

41. $(3x^2 - 3x - 2) + (3x^2 + 4x - 3)$

42. $(3a^2 - 2a + 4) - (a^2 - 3a + 7)$

43. $(2b^2 + 3b - 5) - (2b^2 - 4b - 9)$

44. $(4c^2 + 3c - 2) + (3c^2 + 4c + 2)$

45. $(2x^2 - 3x + 1) - (4x^2 - 3x + 2) + (2x^2 + 3x + 2)$

46. $(-3z^2 - 4z + 7) + (2z^2 + 2z - 1) - (2z^2 - 3z + 7)$

47. ▦ $(4.52x^2 + 1.13x - 0.89) + (9.02x^2 - 7.68x + 7.04)$

48. ▦ $(0.891a^4 - 0.442a^2 + 0.121a) - (-0.160a^4 + 0.249a^2 + 0.789a)$

Add the polynomials.

49. $+\dfrac{\begin{array}{r}3x^2 + 4x + 5\\ 2x^2 - 3x + 6\end{array}}{}$

50. $+\dfrac{\begin{array}{r}2x^3 + 2x^2 - 3x + 5\\ 3x^3 - 4x^2 - x - 7\end{array}}{}$

51. $+\dfrac{\begin{array}{r}2x^3 - 3x^2 + 4x - 7\\ -9x^3 - 4x^2 - 5x + 6\end{array}}{}$

52. $+\dfrac{\begin{array}{r}-3x^3 + 4x^2 - 4x + 9\\ 2x^3 + 9x - 3\end{array}}{}$

53. $+\dfrac{\begin{array}{r}-3x^2 + 4x + 25\\ 5x^2 - 12\end{array}}{}$

54. $+\dfrac{\begin{array}{r}-6x^3 - 4x^2 + 7\\ -7x^3 + 9x^2\end{array}}{}$

Find each difference.

55. $-\dfrac{\begin{array}{r}3x^2 + 4x - 5\\ -2x^2 - 2x + 3\end{array}}{}$

56. $\begin{array}{r} 3y^2 - 4y + 7 \\ - \underline{6y^2 - 6y - 13} \end{array}$

57. $\begin{array}{r} 4x^3 + 4x^2 - 3x + 10 \\ - \underline{5x^3 - 2x^2 - 4x - 4} \end{array}$

58. $\begin{array}{r} 3x^3 + 4x^2 + 7x + 12 \\ - \underline{-4x^3 + 6x^2 + 9x - 3} \end{array}$

59. $\begin{array}{r} -2x^2y^2 \qquad + 12y^2 \\ - \underline{10x^2y^2 + 9xy - 24y^2} \end{array}$

60. $\begin{array}{r} 25x^3 \qquad + 31xz^2 \\ - \underline{12x^3 + 27x^2z - 17xz^2} \end{array}$

61. Find the difference when $t^3 - 2t^2 + 2$ is subtracted from the sum of $3t^3 + t^2$ and $-t^3 + 6t - 3$.

62. Find the difference when $-3z^3 - 4z + 7$ is subtracted from the sum of $2z^2 + 3z - 7$ and $-4z^3 - 2z - 3$.

63. Find the sum when $3x^2 + 4x - 7$ is added to the sum of $-2x^2 - 7x + 1$ and $-4x^2 + 8x - 1$.

64. Find the difference when $32x^2 - 17x + 45$ is subtracted from the sum of $23x^2 - 12x - 7$ and $-11x^2 + 12x + 7$.

Simplify each expression.

65. $2(x + 3) + 4(x - 2)$
66. $3(y - 4) - 5(y + 3)$
67. $-2(x^2 + 7x - 1) - 3(x^2 - 2x + 2)$
68. $-5(y^2 - 2y - 6) + 6(2y^2 + 2y - 5)$
69. $2(2y^2 - 2y + 2) - 4(3y^2 - 4y - 1) + 4(y^2 - y - 1)$

70. $-4(z^2 - 5z) - 5(4z^2 - 1) + 6(2z - 3)$

71. $2(ab^2 - b) - 3(a + 2ab) + (b - a + a^2b)$

72. $3(xy + y) - 2(x - 4 + y) + 2(y^3 + y^2)$

Find the polynomial that represents the perimeter of each figure.

73.

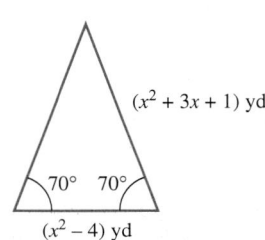

$(x^2 + 3x + 1)$ yd

$70°$ $70°$

$(x^2 - 4)$ yd

74.

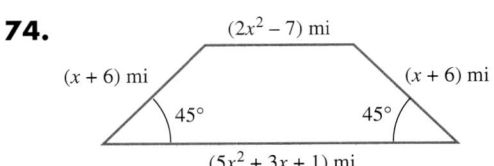

$(2x^2 - 7)$ mi

$(x + 6)$ mi $(x + 6)$ mi

$45°$ $45°$

$(5x^2 + 3x + 1)$ mi

APPLICATIONS

75. GREEK ARCHITECTURE   Find the difference in the heights of the columns shown in Figure 10-3(b) at the beginning of this section.

76. CLASSICAL GREEK COLUMNS   If the columns shown in Figure 10-3(b) at the beginning of this section were stacked one atop the other, to what height would they reach?

77. AUTO MECHANICS   Find the polynomial representing the length of the fan belt shown in Illustration 3. The dimensions are in inches. Your answer will involve π.

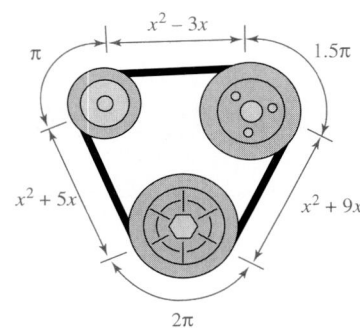

$x^2 - 3x$

π 1.5π

$x^2 + 5x$ $x^2 + 9x$

2π

ILLUSTRATION 3

78. READING BLUEPRINTS
   a. What is the difference in the length and width of the one-bedroom apartment shown in Illustration 4?

   b. Find the perimeter of the apartment.

In Exercises 79–82, consider the following information: If a house is purchased for $105,000 and is expected to appreciate $900 per year, its value y after x years is given by the polynomial function $f(x) = 900x + 105,000$.

79. VALUE OF A HOUSE   Find the expected value of the house in 10 years.

80. VALUE OF A HOUSE   A second house is purchased for $120,000 and is expected to appreciate $1,000 per year.
   a. Find a polynomial function that will give the value y of the house in x years.
   b. Find the value of this second house after 12 years.

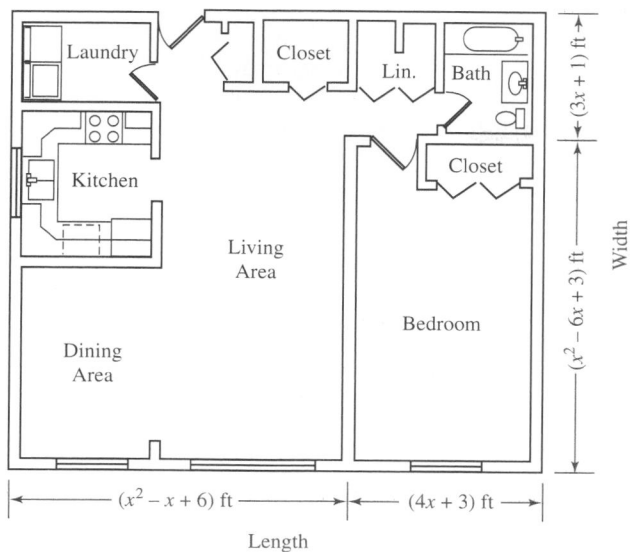

$\longleftarrow (x^2 - x + 6) \text{ ft} \longrightarrow$ $\longleftarrow (4x + 3) \text{ ft} \longrightarrow$

Length

ILLUSTRATION 4

81. VALUE OF TWO HOUSES   Find one polynomial function that will give the combined value y of both houses after x years.

82. VALUE OF TWO HOUSES   Find the value of the two houses after 20 years by
   a. substituting 20 into the polynomial functions $f(x) = 900x + 105,000$ and $f(x) = 1,000x + 120,000$ and adding.
   b. substituting into the result of Exercise 81.

Consider the following information: A business purchases two computers, one for $6,600 and the other for $9,200. The first computer is expected to depreciate $1,100 per year and the second $1,700 per year.

83. VALUE OF A COMPUTER   Write a polynomial function that gives the value of the first computer after x years.

84. VALUE OF A COMPUTER   Write a polynomial function that gives the value of the second computer after x years.

85. VALUE OF TWO COMPUTERS   Find one polynomial function that gives the combined value of both computers after x years.

86. VALUE OF TWO COMPUTERS   In two ways, find the combined value of the two computers after 3 years.

WRITING

87. How do you recognize like terms?

88. How do you add like terms?

89. Explain the concept that is illustrated by the statement
$$-(x^2 + 3x - 1) = -1(x^2 + 3x - 1)$$

90. Explain the mistake made in the solution. Simplify $(12x - 4) - (3x - 1)$.
$$(12x - 4) - (3x - 1) = 12x - 4 - 3x - 1$$
$$= 9x - 5$$

REVIEW

91. What is the sum of the measures of the angles of a triangle?

92. What is the sum of the measures of two complementary angles?

93. Solve the inequality $-4(3x - 3) \geq -12$ and graph the solution.

94. CURLING IRON   A curling iron is plugged into a 110-volt electrical outlet and used for $\frac{1}{4}$ hour. If its resistance is 10 ohms, find the electrical power (in kilowatt hours, kwh) used by the curling iron by applying the formula
$$\text{kwh} = \frac{(\text{volts})^2}{1,000 \cdot \text{ohms}} \cdot \text{hours}$$

10.6 *Multiplying Polynomials*

In this section, you will learn about

• Multiplying monomials • Multiplying a polynomial by a monomial
• Multiplying a binomial by a binomial • The FOIL method
• Special products • Multiplying a polynomial by a binomial
• Multiplying three polynomials • Multiplying binomials to solve equations

INTRODUCTION. In Figure 10-4(a) on the next page, the length and width of a dollar bill are given. We can find the area of the bill by multiplying its length and width.

Arithmetic

$15.6(6.5) = 101.4$

The area is 101.4 cm^2.

(a)

Algebra

$(2x + 1)(3x - 1) = ?$

(b)

FIGURE 10-4

In Figure 10-4(b), the length and the width of a postage stamp are represented by binomials. To find the area of the stamp, we must multiply the binomials. In this section, we will discuss the algebraic rules that are used to do this. We begin the discussion of multiplication of polynomials with the simplest case, the product of two monomials.

Multiplying monomials

To multiply two monomials, such as $8x^2$ and $-3x^4$, we use the commutative and associative properties of multiplication to group the numerical factors and the variable factors. Then we multiply the numerical factors and multiply the variable factors.

$$8x^2(-3x^4) = 8(-3)x^2x^4$$
$$= -24x^6$$

This example suggests the following rule.

Multiplying monomials | To multiply two monomials, multiply the numerical factors and then multiply the variable factors.

EXAMPLE 1   *Multiplying monomials.*   Multiply   **a.** $3x^4(2x^5)$, **b.** $-2a^2b^3(5ab^2)$,   and   **c.** $-4y^5z^2(2y^3z^3)(3yz)$.

Solution

a. $3x^4(2x^5) = 3(2) x^4x^5$

$\qquad\qquad = 6x^9$ 　　　Multiply the numerical factors, 3 and 2. Multiply the variable factors: $x^4x^5 = x^{4+5} = x^9$.

b. $-2a^2b^3(5ab^2) = -2(5)a^2ab^3b^2$

$\qquad\qquad\qquad = -10a^3b^5$

c. $-4y^5z^2(2y^3z^3)(3yz) = -4(2)(3)y^5y^3yz^2z^3z$

$\qquad\qquad\qquad\qquad = -24y^9z^6$

Self Check

Multiply:

a. $(5a^2b^3)(6a^3b^4)$

b. $(-15p^3q^2)(5p^3q^2)$

Answers: **a.** $30a^5b^7$, **b.** $-75p^6q^4$

Multiplying a polynomial by a monomial

To find the product of a polynomial (with more than one term) and a monomial, we use the distributive property. To multiply $2x + 4$ by $5x$, for example, we proceed as follows:

$$5x(2x + 4) = 5x(2x) + 5x(4) \quad \text{Distribute the multiplication by } 5x.$$
$$= 10x^2 + 20x \quad \text{Multiply the monomials: } 5x(2x) = 10x^2 \text{ and } 5x(4) = 20x.$$

This example suggests the following rule.

Multiplying polynomials by monomials

> To multiply a polynomial with more than one term by a monomial, use the distributive property to remove parentheses and simplify.

EXAMPLE 2 *Multiplying a polynomial by a monomial.* Multiply:
a. $3a^2(3a^2 - 5a)$ and   **b.** $-2xz^2(2x - 3z + 2z^2)$.

Solution

a. $3a^2(3a^2 - 5a) = 3a^2(3a^2) - 3a^2(5a) \quad$ Distribute the multiplication by $3a^2$.
$$= 9a^4 - 15a^3 \quad \text{Multiply the monomials.}$$

b. $-2xz^2(2x - 3z + 2z^2)$
$$= -2xz^2(2x) - (-2xz^2)(3z) + (-2xz^2)(2z^2) \quad \text{Use the distributive property.}$$
$$= -4x^2z^2 - (-6xz^3) + (-4xz^4) \quad \text{Multiply the monomials.}$$
$$= -4x^2z^2 + 6xz^3 - 4xz^4$$

Self Check

Multiply:

a. $2p^3(3p^2 - 5p)$

b. $-5a^2b(3a + 2b - 4ab)$

Answers: a. $6p^5 - 10p^4$,
b. $-15a^3b - 10a^2b^2 + 20a^3b^2$

Multiplying a binomial by a binomial

To multiply two binomials, we must use the distributive property more than once. For example, to multiply $2a - 4$ by $3a + 5$, we proceed as follows.

$$(2a - 4)(3a + 5) = (2a - 4)(3a) + (2a - 4)(5) \quad \begin{array}{l} \text{Distribute the multiplication} \\ \text{by } (2a - 4). \end{array}$$
$$= 3a(2a - 4) + 5(2a - 4) \quad \begin{array}{l} \text{Use the commutative} \\ \text{property of multiplication.} \end{array}$$
$$= 3a(2a) - 3a(4) + 5(2a) - 5(4) \quad \begin{array}{l} \text{Distribute the multiplication} \\ \text{by } 3a \text{ and by } 5. \end{array}$$
$$= 6a^2 - 12a + 10a - 20 \quad \text{Do the multiplications.}$$
$$= 6a^2 - 2a - 20 \quad \text{Combine like terms.}$$

This example suggests the following rule.

Multiplying two binomials

> To multiply two binomials, multiply each term of one binomial by each term of the other binomial and combine like terms.

The FOIL method

We can use a shortcut method, called the **FOIL** method, to multiply binomials. FOIL is an acronym for **F**irst terms, **O**uter terms, **I**nner terms, and **L**ast terms. To use the FOIL method to multiply $2a - 4$ by $3a + 5$, we

1. multiply the **F**irst terms $2a$ and $3a$ to obtain $6a^2$,

2. multiply the **O**uter terms $2a$ and 5 to obtain $10a$,

3. multiply the **I**nner terms -4 and $3a$ to obtain $-12a$, and

4. multiply the **L**ast terms -4 and 5 to obtain -20.

Then we simplify the resulting polynomial, if possible.

First terms   Last terms

$$(2a - 4)(3a + 5) = 2a(3a) + 2a(5) + (-4)(3a) + (-4)(5)$$

Inner terms

Outer terms

$$= 6a^2 + 10a - 12a - 20 \quad \text{Do the multiplications.}$$

$$= 6a^2 - 2a - 20 \qquad \text{Combine like terms:} \\ 10a - 12a = -2a.$$

EXAMPLE 3   *Using the FOIL method.*   Find each product.

a. $(x + 5)(x + 7) = x(x) + x(7) + 5(x) + 5(7)$

F   L

I

O

$$= x^2 + 7x + 5x + 35$$

$$= x^2 + 12x + 35$$

b. $(3x + 4)(2x - 3) = 3x(2x) + 3x(-3) + 4(2x) + 4(-3)$

F   L

I

O

$$= 6x^2 - 9x + 8x - 12$$

$$= 6x^2 - x - 12$$

c. $(a - 7b)(a - 4b) = a(a) + a(-4b) + (-7b)(a) + (-7b)(-4b)$

F   L

I

O

$$= a^2 - 4ab - 7ab + 28b^2$$

$$= a^2 - 11ab + 28b^2$$

d. $(2r - 3s)(2r + t) = 2r(2r) + 2r(t) - 3s(2r) - 3s(t)$

F   L

I

O

$$= 4r^2 + 2rt - 6rs - 3st \quad \text{There are no like terms.}$$

EXAMPLE 4   *Simplifying expressions.*   Simplify each expression.

a. $3(2x - 3)(x + 1) = 3(2x^2 + 2x - 3x - 3)$ Multiply the binomials.

$$= 3(2x^2 - x - 3) \qquad \text{Combine like terms.}$$

$$= 6x^2 - 3x - 9 \qquad \text{Distribute the multiplication by 3.}$$

b. $(x + 1)(x - 2) - 3x(x + 3) = x^2 - 2x + x - 2 - 3x^2 - 9x$

$$= -2x^2 - 10x - 2 \qquad \text{Combine like terms.}$$

Special products

Certain products of binomials occur so frequently in algebra that it is worthwhile to learn formulas for computing them. To develop a rule to find the *square of a sum*, we consider $(x + y)^2$.

$$(x + y)^2 = (x + y)(x + y) \qquad \text{In } (x + y)^2, \text{ the base is } (x + y) \text{ and the exponent is 2.}$$
$$= x^2 + xy + xy + y^2 \quad \text{Multiply the binomials.}$$
$$= x^2 + 2xy + y^2 \qquad \text{Combine like terms: } xy + xy = 2xy.$$

We note that the terms of this result are related to the terms of the original expression. That is, $(x + y)^2$ is equal to the square of its first term (x^2), plus twice the product of both its terms $(2xy)$, plus the square of its last term (y^2).

To develop a rule to find the *square of a difference,* we consider $(x - y)^2$.

$$(x - y)^2 = (x - y)(x - y)$$
$$= x^2 - xy - xy + y^2 \quad \text{Multiply the binomials.}$$
$$= x^2 - 2xy + y^2 \qquad \text{Combine like terms: } -xy - xy = -2xy.$$

Again, the terms of the result are related to the terms of the original expression. When we find $(x - y)^2$, the product is composed of the square of its first term (x^2), twice the product of both its terms $(-2xy)$, and the square of its last term (y^2).

The final special product is the product of two binomials that differ only in the signs of the last terms. To develop a rule to find the product of a *sum and a difference,* we consider $(x + y)(x - y)$.

$$(x + y)(x - y) = x^2 - xy + xy - y^2 \quad \text{Multiply the binomials.}$$
$$= x^2 - y^2 \qquad \text{Combine like terms: } -xy + xy = 0.$$

The product is the square of the first term (x^2) minus the square of the second term (y^2). The expression $x^2 - y^2$ is called a **difference of two squares.**

Because these special products occur so often, it is wise to memorize their forms.

Special products

$(x + y)^2 = x^2 + 2xy + y^2$	The square of a sum
$(x - y)^2 = x^2 - 2xy + y^2$	The square of a difference
$(x + y)(x - y) = x^2 - y^2$	The product of a sum and difference

EXAMPLE 5 *Finding special products.* Find **a.** $(t + 9)^2$, **b.** $(8a - 5)^2$, and **c.** $(3y + 4z)(3y - 4z)$.

Solution

a. This is the square of a sum. The terms of the binomial being squared are t and 9.

$$(t + 9)^2 = \underbrace{t^2}_{\substack{\text{The square} \\ \text{of the first} \\ \text{term, } t.}} + \underbrace{2(t)(9)}_{\substack{\text{Twice the} \\ \text{product of} \\ \text{both terms.}}} + \underbrace{9^2}_{\substack{\text{The square} \\ \text{of the last} \\ \text{term, 9.}}}$$

$$= t^2 + 18t + 81$$

b. This is the square of a difference. The terms of the binomial being squared are $8a$ and -5.

$$(8a - 5)^2 = \underbrace{(8a)^2}_{\substack{\text{The square} \\ \text{of the first} \\ \text{term, } 8a.}} - \underbrace{2(8a)(5)}_{\substack{\text{Twice the} \\ \text{product of} \\ \text{both terms.}}} + \underbrace{5^2}_{\substack{\text{The square} \\ \text{of the last} \\ \text{term, 5.}}}$$

$$= 64a^2 - 80a + 25 \quad \begin{array}{l} \text{Use the power rule for products:} \\ (8a)^2 = 8^2 a^2 = 64a^2. \end{array}$$

c. The binomials differ only in the signs of the last terms. This is the product of a sum and a difference.

$$(3y + 4z)(3y - 4z) = (3y)^2 - (4z)^2 \quad \begin{array}{l} \text{The square of the first term minus the} \\ \text{square of the second term.} \end{array}$$

$$= 9y^2 - 16z^2 \quad \text{Use the power rule for products twice.}$$

Self Check

Find:

a. $(r + 6)^2$

b. $(7g - 2)^2$

c. $(5m - 9n)(5m + 9n)$

Answers: **a.** $r^2 + 12r + 36$, **b.** $49g^2 - 28g + 4$, **c.** $25m^2 - 81n^2$

■

COMMENT A common error when squaring a binomial is to forget the middle term of the product. For example, $(x + 2)^2 \neq x^2 + 4$ and $(x - 2)^2 \neq x^2 - 4$. Applying the special product formulas, we have $(x + 2)^2 = x^2 + 4x + 4$ and $(x - 2)^2 = x^2 - 4x + 4$.

Multiplying a polynomial by a binomial

We must use the distributive property more than once to multiply a polynomial by a binomial. For example, to multiply $3x^2 + 3x - 5$ by $2x + 3$, we proceed as follows:

$$
\begin{aligned}
(2x + 3)(3x^2 + 3x - 5) &= (2x + 3)3x^2 + (2x + 3)3x - (2x + 3)5 \\
&= 3x^2(2x + 3) + 3x(2x + 3) - 5(2x + 3) \\
&= 6x^3 + 9x^2 + 6x^2 + 9x - 10x - 15 \\
&= 6x^3 + 15x^2 - x - 15
\end{aligned}
$$

This example suggests the following rule.

Multiplying polynomials

To multiply one polynomial by another, multiply each term of one polynomial by each term of the other polynomial and combine like terms.

It is often convenient to organize the work vertically.

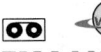

EXAMPLE 6 *Multiplying polynomials using vertical form.*

a. Multiply:

$$
\begin{array}{r}
3a^2 - 4a + 7 \\
2a + 5 \\
\hline
15a^2 - 20a + 35 \\
6a^3 - 8a^2 + 14a \\
\hline
6a^3 + 7a^2 - 6a + 35
\end{array}
$$

Multiply $3a^2 - 4a + 7$ by 5.
Multiply $3a^2 - 4a + 7$ by $2a$.
In each column, combine like terms.

b. Multiply:

$$
\begin{array}{r}
3y^2 - 5y + 4 \\
-4y^2 - 3 \\
\hline
-9y^2 + 15y - 12 \\
-12y^4 + 20y^3 - 16y^2 \\
\hline
-12y^4 + 20y^3 - 25y^2 + 15y - 12
\end{array}
$$

Multiply $3y^2 - 5y + 4$ by -3.
Multiply $3y^2 - 5y + 4$ by $-4y^2$.

Self Check

Multiply:

a. $(3x + 2)(2x^2 - 4x + 5)$

b. $(-2x^2 + 3)(2x^2 - 4x - 1)$

Answers:

a. $6x^3 - 8x^2 + 7x + 10$,

b. $-4x^4 + 8x^3 + 8x^2 - 12x - 3$ ∎

Multiplying three polynomials

When finding the product of three polynomials, we begin by multiplying any two of them, and then we multiply that result by the third polynomial.

EXAMPLE 7 *Multiplying three polynomials.* Find the product: $-3a(4a + 1)(a - 7)$.

Solution

First, we find the product of $4a + 1$ and $a - 7$. Then we multiply that result by $-3a$.

$$
\begin{aligned}
-3a(4a + 1)(a - 7) &= -3a(4a^2 - 28a + a - 7) \\
&= -3a(4a^2 - 27a - 7) \\
&= -12a^3 + 81a^2 + 21a
\end{aligned}
$$

Multiply the two binomials.
Combine like terms.
Distribute the multiplication by $-3a$.

Self Check

Find the product:

$-2y(y + 3)(3y - 2)$

Answer: $-6y^3 - 14y^2 + 12y$ ∎

Multiplying binomials to solve equations

To solve an equation such as $(x + 2)(x + 3) = x(x + 7)$, we can do the multiplication on each side and proceed as follows:

$$(x + 2)(x + 3) = x(x + 7)$$

$$x^2 + 3x + 2x + 6 = x^2 + 7x$$

$$x^2 + 3x + 2x + 6 - x^2 = x^2 + 7x - x^2 \qquad \text{Subtract } x^2 \text{ from both sides.}$$

$$5x + 6 = 7x \qquad \text{Combine like terms: } x^2 - x^2 = 0 \text{ and } 3x + 2x = 5x.$$

$$6 = 2x \qquad \text{Subtract } 5x \text{ from both sides.}$$

$$3 = x \qquad \text{Divide both sides by 2.}$$

Check: $(x + 2)(x + 3) = x(x + 7)$

$$(3 + 2)(3 + 3) \overset{?}{=} 3(3 + 7) \qquad \text{Replace } x \text{ with 3.}$$

$$5(6) \overset{?}{=} 3(10) \qquad \text{Do the additions within parentheses.}$$

$$30 = 30$$

EXAMPLE 8   A square painting is surrounded by a border 2 inches wide. If the area of the border is 96 square inches, find the dimensions of the painting.

Analyze the problem   Refer to Figure 10-5, which shows a square painting surrounded by a border 2 inches wide. We know that the area of this border is 96 square inches, and we are to find the dimensions of the painting.

Form an equation   Let x represent the length of each side of the square painting. Since the border is 2 inches wide, the length and the width of the outer rectangle are both $(x + 2 + 2)$ inches. Then the outer rectangle is also a square, and its dimensions are $(x + 4)$ by $(x + 4)$ inches. Since the area of a square is the product of its length and width, the area of the larger square is $(x + 4)(x + 4)$, and the area of the painting is $x \cdot x$. If we subtract the area of the painting from the area of the larger square, the difference is 96.

FIGURE 10-5

The area of the large square	minus	the area of the square painting	is	the area of the border.
$(x + 4)(x + 4)$	$-$	$x \cdot x$	$=$	96

Solve the equation

$$(x + 4)(x + 4) - x^2 = 96 \qquad x \cdot x = x^2.$$

$$x^2 + 8x + 16 - x^2 = 96 \qquad (x + 4)(x + 4) = (x + 4)^2 = x^2 + 8x + 16.$$

$$8x + 16 = 96 \qquad \text{Combine like terms: } x^2 - x^2 = 0.$$

$$8x = 80 \qquad \text{Subtract 16 from both sides.}$$

$$x = 10 \qquad \text{Divide both sides by 8.}$$

State the conclusion   The dimensions of the painting are 10 inches by 10 inches.

Check the result   Verify that the 2-inch-wide border of a 10-inch-square painting would have an area of 96 square inches.

STUDY SET Section 10.6

VOCABULARY *Fill in the blanks.*

1. The expression $(2a - 4)(3a + 5)$ is the product of two _____.

2. The expression $(2a - 4)(3a^2 + 5a - 1)$ is the product of a _____ and a _____.

3. In the acronym FOIL, F stands for _____ terms, O for _____ terms, I for _____ terms, and L for _____ terms.

4. $(x + 5)^2$ is the square of a _____, and $(x - 5)^2$ is the square of a _____. The expression $x^2 - y^2$ is called a difference of _____.

CONCEPTS *Consider the product $(2x + 5)(3x - 4)$.*

5. The product of the first terms is _____.

6. The product of the outer terms is _____.

7. The product of the inner terms is _____.

8. The product of the last terms is _____.

9. STAMPS Find the area of the stamp shown in Figure 10-4(b) at the beginning of this section.

10. LUGGAGE Find the volume of the garment bag shown in Illustration 1.

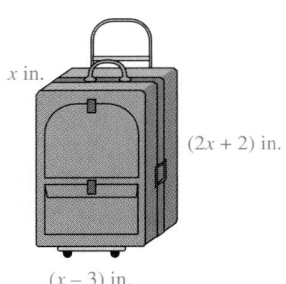

x in.

$(2x + 2)$ in.

$(x - 3)$ in.

ILLUSTRATION 1

NOTATION *Complete each solution.*

11. $7x(3x^2 - 2x + 5) = \boxed{}(3x^2) - \boxed{}(2x) + \boxed{}(5)$
$= 21x^3 - 14x^2 + 35x$

12. $(2x + 5)(3x - 2) = 2x(3x) - \boxed{}(2) + \boxed{}(3x) - \boxed{}(2)$
$= 6x^2 - \boxed{} + \boxed{} - 10$
$= 6x^2 + 11x - 10$

PRACTICE *Find each product.*

13. $(3x^2)(4x^3)$

14. $(-2a^3)(3a^2)$

15. $(3b^2)(-2b)(4b^3)$

16. $(3y)(2y^2)(-y^4)$

17. $(2x^2y^3)(3x^3y^2)$

18. $(-5x^3y^6)(x^2y^2)$

19. $(x^2y^5)(x^2z^5)(-3z^3)$

20. $(-r^4st^2)(2r^2st)(rst)$

21. $3(x + 4)$

22. $-3(a - 2)$

23. $-4(t + 7)$

24. $6(s^2 - 3)$

25. $3x(x - 2)$

26. $4y(y + 5)$

27. $-2x^2(3x^2 - x)$

28. $4b^3(2b^2 - 2b)$

29. $3xy(x + y)$

30. $-4x^2z(3x^2 - z)$

31. $2x^2(3x^2 + 4x - 7)$

32. $3y^3(2y^2 - 7y - 8)$

33. $(3x)(-2x^2)(x + 4)$

34. $(-2a^2)(-3a^3)(3a - 2)$

35. $(a + 4)(a + 5)$

36. $(y - 3)(y + 5)$

37. $(3x - 2)(x + 4)$

38. $(t + 4)(2t - 3)$

39. $(2a + 4)(3a - 5)$

40. $(2b - 1)(3b + 4)$

41. $(3x - 5)(2x + 1)$

42. $(2y - 5)(3y + 7)$

43. $(x + 3)(2x - 3)$

44. $(2x + 3)(2x - 5)$

45. $(2t + 3s)(3t - s)$

46. $(3a - 2b)(4a + b)$

47. $(x + y)(x + z)$

48. $(a - b)(x + y)$

49. $(4t - u)(-3t + u)$

50. $(-3t + 2s)(2t - 3s)$

Simplify each expression.

51. $4(2x + 1)(x - 2)$

52. $-5(3a - 2)(2a + 3)$

53. $3a(a + b)(a - b)$

54. $-2r(r + s)(r + s)$

55. $2t(t + 2) + 3t(t - 5)$

56. $3y(y + 2) + (y + 1)(y - 1)$

57. $(x + y)(x - y) + x(x + y)$

58. $(3x + 4)(2x - 2) - (2x + 1)(x + 3)$

Find each special product.

59. $(x + 4)(x + 4)$

60. $(a + 3)(a + 3)$

61. $(t - 3)(t - 3)$

62. $(z - 5)(z - 5)$

63. $(r + 4)(r - 4)$

64. $(b + 2)(b - 2)$

65. $(4x + 5)(4x - 5)$ **66.** $(5z + 1)(5z - 1)$

67. $(2s + 1)(2s + 1)$ **68.** $(3t - 2)(3t - 2)$

69. $(x + 5)^2$ **70.** $(y - 6)^2$

71. $(x - 2y)^2$ **72.** $(3a + 2b)^2$

73. $(2a - 3b)^2$ **74.** $(2x + 5y)^2$

75. $(4x + 5y)^2$ **76.** $(6p - 5q)^2$

Find the area of each figure. You may leave π in your answer.

77.

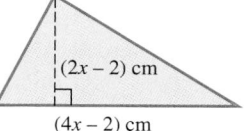

$(2x - 2)$ cm

$(4x - 2)$ cm

78.

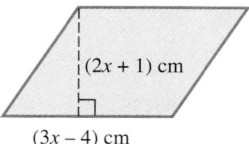

$(2x + 1)$ cm

$(3x - 4)$ cm

79.

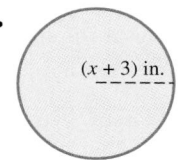

$(x + 3)$ in.

80.

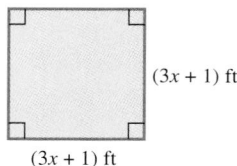

$(3x + 1)$ ft

$(3x + 1)$ ft

Find each product.

81. $(x + 2)(x^2 - 2x + 3)$

82. $(x - 5)(x^2 + 2x - 3)$

83. $(4t + 3)(t^2 + 2t + 3)$

84. $(3x + 1)(2x^2 - 3x + 1)$

85. $(-3x + y)(x^2 - 8xy + 16y^2)$

86. $(3x - y)(x^2 + 3xy - y^2)$

87. $\begin{array}{r} x^2 - 2x + 1 \\ x + 2 \\ \hline \end{array}$ **88.** $\begin{array}{r} 5r^2 + r + 6 \\ 2r - 1 \\ \hline \end{array}$

89. $\begin{array}{r} 4x^2 + 3x - 4 \\ 3x + 2 \\ \hline \end{array}$ **90.** $\begin{array}{r} x^2 - x + 1 \\ x + 1 \\ \hline \end{array}$

Solve each equation.

91. $(s - 4)(s + 1) = s^2 + 5$

92. $(y - 5)(y - 2) = y^2 - 4$

93. $z(z + 2) = (z + 4)(z - 4)$

94. $(z + 3)(z - 3) = z(z - 3)$

95. $(x + 4)(x - 4) = (x - 2)(x + 6)$

96. $(y - 1)(y + 6) = (y - 3)(y - 2) + 8$

97. $(a - 3)^2 = (a + 3)^2$
98. $(b + 2)^2 = (b - 1)^2$

APPLICATIONS

99. TOYS   Find the perimeter and the area of the screen of the Etch A Sketch® shown in Illustration 2.

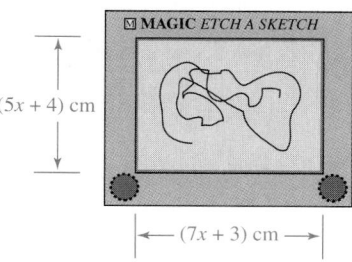

$(5x + 4)$ cm

$(7x + 3)$ cm

ILLUSTRATION 2

100. SUNGLASSES   An ellipse is an oval-shaped closed curve. The area of an ellipse is approximately $3.14ab$, where a is its length and b is its width. Find the polynomial that approximates the total area of the elliptical-shaped lenses of the sunglasses shown in Illustration 3.

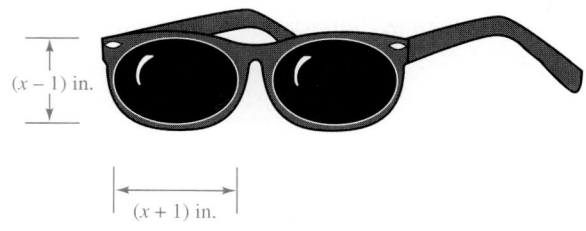

$(x - 1)$ in.

$(x + 1)$ in.

ILLUSTRATION 3

101. GARDENING   See Illustration 4.
  a. What is the area of the region planted with corn? tomatoes? beans? carrots? Use your answers to find the total area of the garden.

  b. What is the length of the garden? What is its width? Use your answers to find its area.

  c. How do the answers from parts a and b for the area of the garden compare?

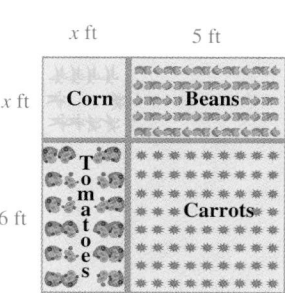

x ft   5 ft

x ft   Corn   Beans

6 ft   Tomatoes   Carrots

ILLUSTRATION 4

102. PAINTING   See Illustration 5. To purchase the correct amount of enamel to paint these two garage doors, a painter must find their areas. Find a polynomial that gives the number of square feet to be painted. All dimensions are in feet, and the windows are squares with sides of x feet.

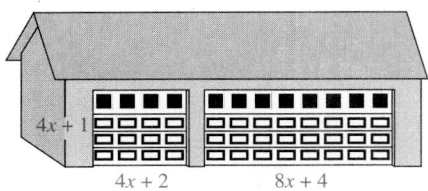

ILLUSTRATION 5

103. INTEGER PROBLEM   The difference between the squares of two consecutive positive integers is 11. Find the integers. (Hint: Let x and $x + 1$ represent the consecutive integers.)

104. INTEGER PROBLEM   If 3 less than a certain integer is multiplied by 4 more than the integer, the product is 6 less than the square of the integer. Find the integer.

105. STONE-GROUND FLOUR   The radius of one millstone in Illustration 6 is 3 meters greater than the radius of another, and their areas differ by 15π square meters. Find the radius of the larger millstone.

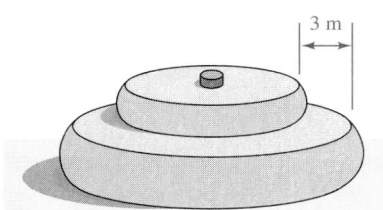

ILLUSTRATION 6

106. BOOKBINDING   Two square sheets of cardboard used for making book covers differ in area by 44 square inches. An edge of the larger square is 2 inches greater than an edge of the smaller square. Find the length of an edge of the smaller square.

107. BASEBALL   In major league baseball, the distance between bases is 30 feet greater than it is in softball. The bases in major league baseball mark the corners of a square that has an area 4,500 square feet greater than for softball. Find the distance between the bases in baseball.

108. PULLEY DESIGN   The radius of one pulley in Illustration 7 is 1 inch greater than the radius of the second pulley, and their areas differ by 4π square inches. Find the radius of the smaller pulley.

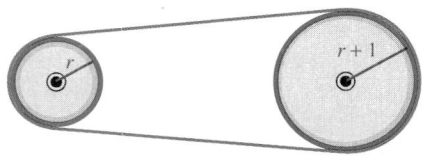

ILLUSTRATION 7

WRITING

109. Describe the steps involved in finding the product of $x + 2$ and $x - 2$.

110. Writing $(x + y)^2$ as $x^2 + y^2$ illustrates a common error. Explain.

REVIEW

Refer to Illustration 8.

111. What is the slope of line AB?

112. What is the slope of line BC?

113. What is the slope of line CD?

114. What is the slope of the x-axis?

115. What is the y-intercept of line AB?

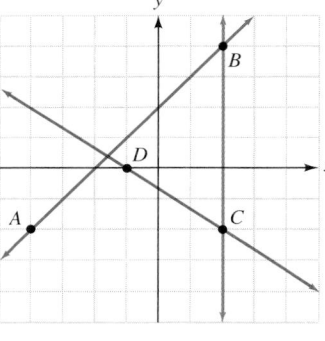

ILLUSTRATION 8

116. What is the x-intercept of line AB?

10.7 *Dividing Polynomials by Monomials*

In this section, you will learn about

• Dividing a monomial by a monomial • Dividing a polynomial by a monomial
• An application of dividing a polynomial by a monomial

INTRODUCTION. In this section, we will discuss how to divide polynomials by monomials. We will first divide monomials by monomials and then divide polynomials with more than one term by monomials.

Dividing a monomial by a monomial

Recall that to simplify a fraction, we write both its numerator and denominator as the product of several factors and then divide out all common factors:

$$\frac{4}{6} = \frac{2 \cdot 2}{2 \cdot 3} \quad \text{Factor: } 4 = 2 \cdot 2 \text{ and } 6 = 2 \cdot 3.$$

$$= \frac{\overset{1}{\cancel{2}} \cdot 2}{\underset{1}{\cancel{2}} \cdot 3} \quad \text{Divide out the common factor of 2.}$$

$$= \frac{2}{3}$$

$$\frac{20}{25} = \frac{4 \cdot 5}{5 \cdot 5} \quad \text{Factor: } 20 = 4 \cdot 5 \text{ and } 25 = 5 \cdot 5.$$

$$= \frac{4 \cdot \overset{1}{\cancel{5}}}{\underset{1}{\cancel{5}} \cdot 5} \quad \text{Divide out the common factor of 5.}$$

$$= \frac{4}{5}$$

We can use the same method to simplify algebraic fractions that contain variables.

$$\frac{3p^2}{6p} = \frac{3 \cdot p \cdot p}{2 \cdot 3 \cdot p} \quad \text{Factor: } p^2 = p \cdot p \text{ and } 6 = 2 \cdot 3.$$

$$= \frac{\overset{1}{\cancel{3}} \cdot \overset{1}{\cancel{p}} \cdot p}{2 \cdot \underset{1}{\cancel{3}} \cdot \underset{1}{\cancel{p}}} \quad \text{Divide out the common factors of 3 and } p.$$

$$= \frac{p}{2}$$

To divide monomials, we can use either the preceding method for simplifying arithmetic fractions or the rules for exponents.

EXAMPLE 1   *Dividing monomials.*   Simplify: **a.** $\dfrac{x^2 y}{xy^2}$ and   **b.** $\dfrac{-8a^3b^2}{4ab^3}$.

Solution

By simplifying fractions

a. $\dfrac{x^2 y}{xy^2} = \dfrac{x \cdot x \cdot y}{x \cdot y \cdot y}$

$$= \frac{\overset{1}{\cancel{x}} \cdot x \cdot \overset{1}{\cancel{y}}}{\underset{1}{\cancel{x}} \cdot y \cdot \underset{1}{\cancel{y}}}$$

$$= \frac{x}{y}$$

Using the rules for exponents

$$\frac{x^2 y}{xy^2} = x^{2-1}y^{1-2}$$

$$= x^1 y^{-1}$$

$$= \frac{x}{y}$$

b. $\dfrac{-8a^3b^2}{4ab^3} = \dfrac{-2 \cdot 4 \cdot a \cdot a \cdot a \cdot b \cdot b}{4 \cdot a \cdot b \cdot b \cdot b}$

$$= \frac{-2 \cdot \overset{1}{\cancel{4}} \cdot \overset{1}{\cancel{a}} \cdot a \cdot a \cdot \overset{1}{\cancel{b}} \cdot \overset{1}{\cancel{b}}}{\underset{1}{\cancel{4}} \cdot \underset{1}{\cancel{a}} \cdot \underset{1}{\cancel{b}} \cdot \underset{1}{\cancel{b}} \cdot b}$$

$$= -\frac{2a^2}{b}$$

$$\frac{-8a^3b^2}{4ab^3} = \frac{-2^3 a^3 b^2}{2^2 ab^3}$$

$$= -2^{3-2}a^{3-1}b^{2-3}$$

$$= -2^1 a^2 b^{-1}$$

$$= -\frac{2a^2}{b}$$

Self Check

Simplify $\dfrac{-5p^2q^3}{10pq^4}$.

Answer: $-\dfrac{p}{2q}$

EXAMPLE 2 *Dividing monomials.*   Simplify $\dfrac{25(s^2t^3)^2}{15(st^3)^3}$. Write the result using positive exponents only.

Solution
To divide these monomials, we will use the method for simplifying fractions and several rules for exponents.

$$\frac{25(s^2t^3)^2}{15(st^3)^3} = \frac{25s^4t^6}{15s^3t^9}$$

Use the power rules for exponents:
$(xy)^n = x^ny^n$ and $(x^m)^n = x^{m\cdot n}$.

$$= \frac{5\cdot 5\cdot s^{4-3}t^{6-9}}{5\cdot 3}$$

Factor 25 and 15. Use the quotient rule for exponents:
$\dfrac{x^m}{x^n} = x^{m-n}$.

$$= \frac{5\cdot \overset{1}{\cancel{5}}\cdot s^1t^{-3}}{\underset{1}{\cancel{5}}\cdot 3}$$

Divide out the common factors of 5. Do the subtractions.

$$= \frac{5s}{3t^3}$$

Use the negative integer exponent rule: $t^{-3} = \dfrac{1}{t^3}$.

Self Check

Simplify $\dfrac{-24(h^3p)^5}{20(h^2p^2)^3}$.

Answer: $-\dfrac{6h^9}{5p}$ ∎

Dividing a polynomial by a monomial

We have used the following rule to add and subtract fractions with like denominators.

Adding and subtracting fractions with like denominators

To add (or subtract) fractions with like denominators, we add (or subtract) their numerators and keep the common denominator. In symbols, if a, b, and d represent numbers,

$$\frac{a}{d} + \frac{b}{d} = \frac{a+b}{d} \qquad \text{and} \qquad \frac{a}{d} - \frac{b}{d} = \frac{a-b}{d} \qquad \text{(provided } d \neq 0)$$

We can use this rule "in reverse" to divide polynomials by monomials.

EXAMPLE 3 *Dividing a binomial by a monomial.*   Divide $9x + 6$ by 3.

Solution

$$\frac{9x+6}{3} = \frac{9x}{3} + \frac{6}{3} \qquad \text{Divide each term of the numerator by the denominator.}$$

$$= 3x + 2 \qquad \text{Simplify each fraction.}$$

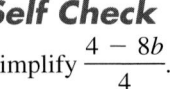

Self Check

Simplify $\dfrac{4-8b}{4}$.

Answer: $1 - 2b$ ∎

EXAMPLE 4 *Dividing a trinomial by a monomial.*

Divide $\dfrac{6x^2y^2 + 4x^2y - 2xy}{2xy}$.

Solution

$$\frac{6x^2y^2 + 4x^2y - 2xy}{2xy}$$

$$= \frac{6x^2y^2}{2xy} + \frac{4x^2y}{2xy} - \frac{2xy}{2xy} \qquad \text{Divide each term of the numerator by the denominator.}$$

$$= 3xy + 2x - 1 \qquad \text{Simplify each fraction.}$$

Self Check

Simplify $\dfrac{9a^2b - 6ab^2 + 3ab}{3ab}$.

Answer: $3a - 2b + 1$ ∎

EXAMPLE 5   *Dividing a trinomial by a monomial.*

Divide $\dfrac{12a^3b^2 - 4a^2b + a}{6a^2b^2}$.

Solution

$$\dfrac{12a^3b^2 - 4a^2b + a}{6a^2b^2}$$

$$= \dfrac{12a^3b^2}{6a^2b^2} - \dfrac{4a^2b}{6a^2b^2} + \dfrac{a}{6a^2b^2} \qquad \text{Divide each term of the numerator by the denominator.}$$

$$= 2a - \dfrac{2}{3b} + \dfrac{1}{6ab^2} \qquad \text{Simplify each fraction.}$$

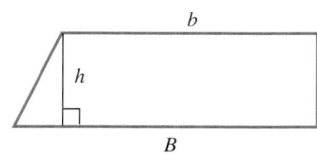

EXAMPLE 6   *Dividing by a monomial.*   Simplify $\dfrac{(x - y)^2 - (x + y)^2}{xy}$.

Solution

$$\dfrac{(x - y)^2 - (x + y)^2}{xy}$$

$$= \dfrac{x^2 - 2xy + y^2 - (x^2 + 2xy + y^2)}{xy} \qquad \text{Use the special product rules to square the binomials in the numerator.}$$

$$= \dfrac{x^2 - 2xy + y^2 - x^2 - 2xy - y^2}{xy} \qquad \text{Change the sign of each term within } (x^2 + 2xy + y^2).$$

$$= \dfrac{-4xy}{xy} \qquad \text{Combine like terms.}$$

$$= -4 \qquad \text{Divide out the common factors of } x \text{ and } y.$$

An application of dividing a polynomial by a monomial

The area of the trapezoid shown in Figure 10-6 is given by the formula $A = \frac{1}{2}h(B + b)$, where B and b are its bases and h is its height. To solve the formula for b, we proceed as follows.

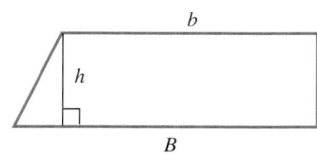

FIGURE 10-6

$$A = \frac{1}{2}h(B + b)$$

$$2 \cdot A = 2 \cdot \frac{1}{2}h(B + b) \qquad \text{Multiply both sides by 2 to clear the equation of the fraction.}$$

$$2A = h(B + b) \qquad \text{Simplify: } 2 \cdot \frac{1}{2} = \frac{2}{2} = 1.$$

$$2A = hB + hb \qquad \text{Distribute the multiplication by } h.$$

$$2A - hB = hB + hb - hB \qquad \text{Subtract } hB \text{ from both sides.}$$

$$2A - hB = hb \qquad \text{Combine like terms: } hB - hB = 0.$$

$$\dfrac{2A - hB}{h} = \dfrac{hb}{h} \qquad \text{To undo the multiplication by } h, \text{ divide both sides by } h.$$

$$\dfrac{2A - hB}{h} = b$$

EXAMPLE 7   *Confirming answers.*   Another student worked the previous problem in a different way and got a result of $b = \frac{2A}{h} - B$. Is this result correct?

Solution

To determine whether this result is correct, we must show that

$$\frac{2A - hB}{h} = \frac{2A}{h} - B$$

We can do this by dividing $2A - hB$ by h.

$$\frac{2A - hB}{h} = \frac{2A}{h} - \frac{hB}{h} \quad \text{Divide each term of } 2A - hB \text{ by the denominator, which is } h.$$

$$= \frac{2A}{h} - B \quad \text{Simplify the second fraction: } \frac{\overset{1}{\cancel{h}}B}{\underset{1}{\cancel{h}}} = B.$$

The results are the same.

Answer: no ■

STUDY SET   Section 10.7

VOCABULARY   *Fill in the blanks.*

1. A _____ is an algebraic expression that is the sum of one or more terms containing whole-number exponents.

2. A _____ is a polynomial with one term.

3. A binomial is a polynomial with _____ terms.

4. A trinomial is a polynomial with _____ terms.

5. $\dfrac{x^m}{x^n} = x^{m-n}$ is a rule for _____.

6. To _____ a fraction, we divide out common factors of the numerator and denominator.

CONCEPTS   *In Exercises 7–8, fill in the blanks.*

7. $\dfrac{18x + 9}{9} = \dfrac{18x}{} + \dfrac{9x}{}$

8. $\dfrac{30x^2 + 12x - 24}{6} = \dfrac{30x^2}{6} \dfrac{12x}{6} \dfrac{24}{6}$

9. What do the slashes and the small 1's mean?

$$\frac{4}{6} = \frac{\overset{1}{\cancel{2}} \cdot 2}{\underset{1}{\cancel{2}} \cdot 3}$$

10. Complete each rule of exponents.

a. $\dfrac{x^m}{x^n} =$

b. $x^{-n} =$

11. a. Solve the formula $d = rt$ for t.

b. Use your answer from part a to complete the table.

	r	$\cdot$	t	$=$	d
Motorcycle	$2x$				$6x^3$

12. a. Solve the formula $I = Prt$ for r.

b. Use your answer from part a to complete the table.

	P	$\cdot$	r	$\cdot$	t	$=$	I
Savings account	$8x^3$				$2x$		$24x^6$

13. How many nickels would have a value of $(10x + 35)$ cents?

14. How many twenty-dollar bills would have a value of $\$(60x - 100)$?

NOTATION   *Complete each solution.*

15. $\dfrac{a^2 b^3}{a^3 b^2} = \dfrac{a \cdot a \cdot \cdot \cdot }{ \cdot \cdot \cdot b \cdot b}$

$$= \frac{\overset{1}{\cancel{a}} \cdot \overset{1}{\cancel{a}} \cdot \overset{1}{\cancel{b}} \cdot \overset{1}{\cancel{b}} \cdot}{\underset{1}{\cancel{a}} \cdot \underset{1}{\cancel{a}} \cdot \cdot \underset{1}{\cancel{b}} \cdot \underset{1}{\cancel{b}}}$$

$$= \frac{b}{a}$$

16. $\dfrac{6pq^2 - 9p^2q^2 + pq}{3p^2q}$

$$= \frac{6pq^2}{} - \frac{9p^2q^2}{} + \frac{pq}{}$$

$$= \frac{6 \cdot p \cdot q \cdot q}{3 \cdot p \cdot p \cdot q} - \frac{9 \cdot p \cdot p \cdot q \cdot q}{3 \cdot p \cdot p \cdot q} + \frac{p \cdot q}{3 \cdot p \cdot p \cdot q}$$

$$= \frac{2q}{p} - 3q + \frac{1}{3p}$$

PRACTICE *Simplify each fraction.*

17. $\dfrac{5}{15}$

18. $\dfrac{64}{128}$

19. $\dfrac{-125}{75}$

20. $\dfrac{-98}{21}$

21. $\dfrac{120}{160}$

22. $\dfrac{70}{420}$

23. $\dfrac{-3{,}612}{-3{,}612}$

24. $\dfrac{-288}{-112}$

Do each division by simplifying the fraction.

25. $\dfrac{x^5}{x^2}$

26. $\dfrac{a^{12}}{a^8}$

27. $\dfrac{r^3 s^2}{rs^3}$

28. $\dfrac{y^4 z^3}{y^2 z^2}$

29. $\dfrac{8x^3 y^2}{4xy^3}$

30. $\dfrac{-3y^3 z}{6yz^2}$

31. $\dfrac{12u^5 v}{-4u^2 v^3}$

32. $\dfrac{16rst^2}{-8rst^3}$

33. $\dfrac{-16r^3 y^2}{-4r^2 y^4}$

34. $\dfrac{35xyz^2}{-7x^2 yz}$

35. $\dfrac{-65rs^2 t}{15r^2 s^3 t}$

36. $\dfrac{112u^3 z^6}{-42u^3 z^6}$

37. $\dfrac{x^2 x^3}{xy^6}$

38. $\dfrac{x^2 y^2}{x^2 y^3}$

39. $\dfrac{(a^3 b^4)^3}{ab^4}$

40. $\dfrac{(a^2 b^3)^3}{a^6 b^6}$

41. $\dfrac{15(r^2 s^3)^2}{-5(rs^5)^3}$

42. $\dfrac{-5(a^2 b)^3}{10(ab^2)^3}$

43. $\dfrac{-32(x^3 y)^3}{128(x^2 y^2)^3}$

44. $\dfrac{68(a^6 b^7)^2}{-96(abc^2)^3}$

45. $\dfrac{-(4x^3 y^3)^2}{(x^2 y^4)^3}$

46. $\dfrac{(2r^3 s^2)^2}{-(4r^2 s^2)^2}$

47. $\dfrac{(a^2 a^3)^4}{(a^4)^3}$

48. $\dfrac{(b^3 b^4)^5}{(bb^2)^2}$

Do each division.

49. $\dfrac{6x + 9}{3}$

50. $\dfrac{8x + 12y}{4}$

51. $\dfrac{5x - 10y}{25xy}$

52. $\dfrac{2x - 32}{16x}$

53. $\dfrac{3x^2 + 6y^3}{3x^2 y^2}$

54. $\dfrac{4a^2 - 9b^2}{12ab}$

55. $\dfrac{15a^3 b^2 - 10a^2 b^3}{5a^2 b^2}$

56. $\dfrac{9a^4 b^3 - 16a^3 b^4}{12a^2 b}$

57. $\dfrac{4x - 2y + 8z}{4xy}$

58. $\dfrac{5a^2 + 10b^2 - 15ab}{5ab}$

59. $\dfrac{12x^3 y^2 - 8x^2 y - 4x}{4xy}$

60. $\dfrac{12a^2 b^2 - 8a^2 b - 4ab}{4ab}$

61. $\dfrac{-25x^2 y + 30xy^2 - 5xy}{-5xy}$

62. $\dfrac{-30a^2 b^2 - 15a^2 b - 10ab^2}{-10ab}$

Simplify each numerator and then do the division.

63. $\dfrac{5x(4x - 2y)}{2y}$

64. $\dfrac{9y^2(x^2 - 3xy)}{3x^2}$

65. $\dfrac{(-2x)^3 + (3x^2)^2}{6x^2}$

66. $\dfrac{(-3x^2 y)^3 + (3xy^2)^3}{27x^3 y^4}$

67. $\dfrac{4x^2 y^2 - 2(x^2 y^2 + xy)}{2xy}$

68. $\dfrac{-5a^3 b - 5a(ab^2 - a^2 b)}{10a^2 b^2}$

69. $\dfrac{(3x - y)(2x - 3y)}{6xy}$

70. $\dfrac{(2m - n)(3m - 2n)}{-3m^2 n^2}$

71. $\dfrac{(a + b)^2 - (a - b)^2}{2ab}$

72. $\dfrac{(x - y)^2 + (x + y)^2}{2x^2 y^2}$

APPLICATIONS

73. POOL   The rack shown in Illustration 1 is used to set up the balls when beginning a game of pool. If the perimeter of the rack, in inches, is given by the polynomial $6x^2 - 3x + 9$, what is the length of one side?

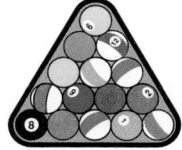

ILLUSTRATION 1

74. CHECKERBOARD   If the perimeter (in inches) of the checkerboard, shown in Illustration 2, is $12x^2 - 8x + 32$, what is the length of one side?

ILLUSTRATION 2

75. AIR CONDITIONING   If the volume occupied by the air conditioning unit shown in Illustration 3 is $(36x^3 - 24x^2)$ cubic feet, find its height.

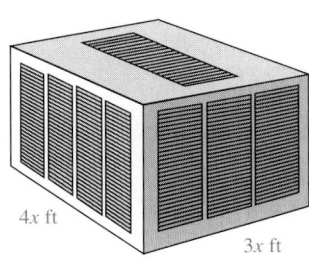

4x ft

3x ft

ILLUSTRATION 3

76. MINI-BLINDS   The area covered by the mini-blinds shown in Illustration 4 is $(3x^3 - 6x)$ square feet. How long are the blinds?

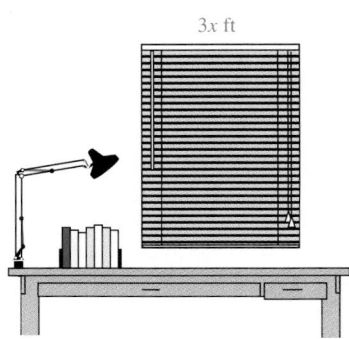

3x ft

ILLUSTRATION 4

77. CONFIRMING FORMULAS   Are these formulas the same?

$$l = \frac{P - 2w}{2} \quad \text{and} \quad l = \frac{P}{2} - w$$

78. CONFIRMING FORMULAS   Are these formulas the same?

$$r = \frac{G + 2b}{2b} \quad \text{and} \quad r = \frac{G}{2b} + b$$

79. ELECTRIC BILLS   On an electric bill, the following two formulas are used to compute the average cost of x kwh of electricity. Are the formulas equivalent?

$$\frac{0.08x + 5}{x} \quad \text{and} \quad 0.08x + \frac{5}{x}$$

80. PHONE BILLS   On a phone bill, the following two formulas are used to compute the average cost per minute of x minutes of phone usage. Are the formulas equivalent?

$$\frac{0.15x + 12}{x} \quad \text{and} \quad 0.15 + \frac{12}{x}$$

WRITING

81. Explain the error in the following work.

$$\frac{3x + 5}{5} = \frac{3x + \overset{1}{\cancel{5}}}{\underset{1}{\cancel{5}}}$$

$$= 3x$$

82. Explain how to do this division.

$$\frac{4x^2y + 8xy^2}{4xy}$$

REVIEW   *In Exercises 83–86, identify each polynomial as a monomial, a binomial, a trinomial, or none of these.*

83. $5a^2b + 2ab^2$

84. $-3x^3y$

85. $-2x^3 + 3x^2 - 4x + 12$

86. $17t^2 - 15t + 27$

87. What is the degree of the trinomial $3x^2 - 2x + 4$?

88. What is the numerical coefficient of the second term of the trinomial $-7t^2 + 5t + 17$?

10.8 *Dividing Polynomials by Polynomials*

In this section, you will learn about

- Dividing polynomials by polynomials • Writing powers in descending order
- Missing terms

INTRODUCTION. In this section, we will conclude our discussion on operations with polynomials by discussing how to divide one polynomial by another.

Dividing polynomials by polynomials

To divide one polynomial by another, we use a method similar to long division in arithmetic. We illustrate the method with several examples.

EXAMPLE 1 *Dividing polynomials.* Divide $x^2 + 5x + 6$ by $x + 2$.

Solution

Here the divisor is $x + 2$, and the dividend is $x^2 + 5x + 6$.

Step 1:
$$x + 2\overline{)x^2 + 5x + 6}$$
with x above.

How many times does x divide x^2? $x^2 \div x = x$. Place the x above the division symbol.

Step 2:
$$\begin{array}{r} x \\ x + 2\overline{)x^2 + 5x + 6} \\ x^2 + 2x \end{array}$$

Multiply each term in the divisor by x. Place the product under $x^2 + 5x$ and draw a line.

Step 3:
$$\begin{array}{r} x \\ x + 2\overline{)x^2 + 5x + 6} \\ \underline{x^2 + 2x} \\ 3x + 6 \end{array}$$

Subtract $x^2 + 2x$ from $x^2 + 5x$. Work vertically, column by column: $x^2 - x^2 = 0$ and $5x - 2x = 3x$.
Bring down the 6.

Step 4:
$$\begin{array}{r} x + 3 \\ x + 2\overline{)x^2 + 5x + 6} \\ \underline{x^2 + 2x} \\ 3x + 6 \end{array}$$

How many times does x divide $3x$? $3x \div x = +3$. Place the $+3$ above the division symbol.

Step 5:
$$\begin{array}{r} x + 3 \\ x + 2\overline{)x^2 + 5x + 6} \\ \underline{x^2 + 2x} \\ 3x + 6 \\ 3x + 6 \end{array}$$

Multiply each term in the divisor by 3. Place the product under $3x + 6$ and draw a line.

Step 6:
$$\begin{array}{r} x + 3 \\ x + 2\overline{)x^2 + 5x + 6} \\ \underline{x^2 + 2x} \\ 3x + 6 \\ \underline{3x + 6} \\ 0 \end{array}$$

Subtract $3x + 6$ from $3x + 6$. Work vertically: $3x - 3x = 0$ and $6 - 6 = 0$.

The quotient is $x + 3$ and the remainder is 0.

Step 7: Check the work by verifying that $(x + 2)(x + 3)$ is $x^2 + 5x + 6$.

$$(x + 2)(x + 3) = x^2 + 3x + 2x + 6$$
$$= x^2 + 5x + 6$$

The answer checks.

EXAMPLE 2 *Dividing polynomials.* Divide $\dfrac{6x^2 - 7x - 2}{2x - 1}$.

Solution

Here the divisor is $2x - 1$ and the dividend is $6x^2 - 7x - 2$.

Step 1:
$$\begin{array}{r} 3x \\ 2x - 1\overline{)6x^2 - 7x - 2} \end{array}$$

How many times does $2x$ divide $6x^2$? $6x^2 \div 2x = 3x$. Place the $3x$ above the division symbol.

Self Check
Divide $x^2 + 7x + 12$ by $x + 3$.

Answer: $x + 4$

Self Check
Divide $\dfrac{8x^2 + 6x - 3}{2x + 3}$.

Step 2: $2x - 1\overline{)6x^2 - 7x - 2}$ Multiply each term in the divisor by $3x$. Place the product under $6x^2 - 7x$ and draw a line.
$\underline{6x^2 - 3x}$

Step 3: $2x - 1\overline{)6x^2 - 7x - 2}$ Subtract $6x^2 - 3x$ from $6x^2 - 7x$. Work vertically:
$\underline{6x^2 - 3x}$ $6x^2 - 6x^2 = 0$ and $-7x - (-3x) = -7x + 3x = -4x$.
$-4x - 2$ Bring down the -2.

Step 4: $3x\ \ -2$
$2x - 1\overline{)6x^2 - 7x - 2}$ How many times does $2x$ divide $-4x$? $-4x \div 2x = -2$. Place the -2 above the division symbol.
$\underline{6x^2 - 3x}$
$-4x - 2$

Step 5: $3x\ \ -2$
$2x - 1\overline{)6x^2 - 7x - 2}$ Multiply each term in the divisor by -2. Place the product under $-4x - 2$ and draw a line.
$\underline{6x^2 - 3x}$
$-4x - 2$
$\underline{-4x + 2}$

Step 6: $3x\ \ -2$
$2x - 1\overline{)6x^2 - 7x - 2}$
$\underline{6x^2 - 3x}$
$-4x - 2$ Subtract $-4x + 2$ from $-4x - 2$. Work vertically:
$\underline{-4x + 2}$ $-4x - (-4x) = -4x + 4x = 0$ and $-2 - 2 = -4$.
-4

Here the quotient is $3x - 2$ and the remainder is -4. It is common to write the answer as either

$$3x - 2 + \frac{-4}{2x - 1} \quad \text{or} \quad 3x - 2 - \frac{4}{2x - 1} \quad \text{Quotient} + \frac{\text{remainder}}{\text{divisor}}.$$

Step 7: To check the answer, we multiply

$$3x - 2 + \frac{-4}{2x - 1} \quad \text{by} \quad 2x - 1$$

The product should be the dividend.

$$(2x - 1)\left(3x - 2 + \frac{-4}{2x - 1}\right) = (2x - 1)(3x - 2) + (2x - 1)\left(\frac{-4}{2x - 1}\right)$$
$$= (2x - 1)(3x - 2) - 4$$
$$= 6x^2 - 4x - 3x + 2 - 4$$
$$= 6x^2 - 7x - 2$$

Because the result is the dividend, the answer checks.

Answer: $4x - 3 + \dfrac{6}{2x + 3}$ ■

Writing powers in descending order

The division method works best when the terms of the divisor and the dividend are written in descending powers of the variable. This means that the term involving the highest power of x appears first, the term involving the second-highest power of x appears second, and so on. For example, the terms in

$$3x^3 + 2x^2 - 7x + 5$$

have their exponents written in descending order.

If the powers in the dividend or divisor are not in descending order, we use the commutative property of addition to write them that way.

EXAMPLE 3   *Dividing polynomials.*   Divide $4x^2 + 2x^3 + 12 - 2x$ by $x + 3$.

Self Check

Divide $x^2 - 10x + 6x^3 + 4$ by $2x - 1$.

Solution

We write the dividend so that the exponents are in descending order.

$$
\begin{array}{r}
2x^2 - 2x\ + 4 \\
x + 3\overline{\smash{)}2x^3 + 4x^2 - 2x + 12} \\
\underline{2x^3 + 6x^2} \\
-2x^2 - 2x \\
\underline{-2x^2 - 6x} \\
4x + 12 \\
\underline{4x + 12} \\
0
\end{array}
$$

Check: $(x + 3)(2x^2 - 2x + 4) = 2x^3 - 2x^2 + 4x + 6x^2 - 6x + 12$
$$= 2x^3 + 4x^2 - 2x + 12$$

Answer: $3x^2 + 2x - 4$ ■

Missing terms

When we write the terms of a dividend in descending powers of x, we must determine whether some powers of the variable are missing. When this happens, we should write such terms with a coefficient of 0 or leave a blank space for them.

EXAMPLE 4   *Dividing polynomials.*   Divide $\dfrac{x^2 - 4}{x + 2}$.

Self Check

Divide $\dfrac{x^2 - 9}{x - 3}$.

Solution

Since $x^2 - 4$ does not have a term involving x, we must either include the term $0x$ or leave a space for it.

$$
\begin{array}{r}
x - 2 \\
x + 2\overline{\smash{)}x^2 + 0x - 4} \\
\underline{x^2 + 2x} \\
-2x - 4 \\
\underline{-2x - 4} \\
0
\end{array}
$$

Check: $(x + 2)(x - 2) = x^2 - 2x + 2x - 4$
$$= x^2 - 4$$

Answer: $x + 3$ ■

STUDY SET   Section 10.8

VOCABULARY   *Fill in the blanks.*

1. In the division $x + 1\overline{\smash{)}x^2 + 2x + 1}$, the expression $x + 1$ is called the _____ and $x^2 + 2x + 1$ is called the _____.

2. The answer to a division problem is called the _____.

3. If a division does not come out even, the leftover part is called a _____.

4. The exponents in $2x^4 + 3x^3 + 4x^2 - 7x - 2$ are said to be written in _____ order.

CONCEPTS   *Write each polynomial with the powers in descending order.*

5. $4x^3 + 7x - 2x^2 + 6$

6. $5x^2 + 7x^3 - 3x - 9$

7. $9x + 2x^2 - x^3 + 6x^4$

8. $7x^5 + x^3 - x^2 + 2x^4$

Identify the missing terms in each polynomial.

9. $5x^4 + 2x^2 - 1$

10. $-3x^5 - 2x^3 + 4x - 6$

In Exercises 11–12, without doing the division, determine which of the three possible quotients seems reasonable.

11. $\dfrac{x^4 - 81}{x - 3}$ $x^2 + 3x + 9$
$x^3 + 3x^2 + 9x + 27$
$x^4 + 3x^3 + 9x^2 + 27x + 1$

12. $\dfrac{8x^3 - 27}{2x - 3}$ $4x^2 + 6x + 9$
$4x^3 - 6x^2 - 9$
$4x^4 - 6x^3 - 9x^2 + 1$

13. a. Solve $d = rt$ for r.
   b. Use your answer to part a and the long division method to complete the table.

	r	$\cdot$	t	$=$	d
Subway			$x + 4$		$x^2 + x - 12$

14. a. Solve $I = Prt$ for P.
   b. Use your answer to part a and the long division method to complete the table.

	P	$\cdot$	r	$\cdot$	t	$=$	I
Bonds			$x + 4$		1		$x^2 + 7x + 12$

15. Using long division, a student found that
$$\frac{3x^2 + 8x + 4}{3x + 2} = x + 2$$
Use multiplication to see whether the result is correct.

16. Using long division, a student found that
$$\frac{x^2 + 4x - 21}{x - 3} = x - 7$$
Use multiplication to see whether the result is correct.

NOTATION *Complete each division.*

17.
$$
\begin{array}{r}
\ + 2 \\
x + 2 \overline{)x^2 + 4x + 4} \\
\underline{x^2 + } \\
+ 4 \\
\underline{2x + 4} \\
0
\end{array}
$$

18.
$$
\begin{array}{r}
\ + \ x\ - 2 + \frac{7}{2x+1} \\
2x + 1 \overline{)2x^3 + 3x^2 - 3x\ + 5} \\
\underline{+ x^2} \\
2x^2 - 3x \\
\underline{2x^2 + } \\
+ 5 \\
\underline{-4x\ -} \\
7
\end{array}
$$

PRACTICE *Do each division.*

19. Divide $x^2 + 4x - 12$ by $x - 2$.
20. Divide $x^2 - 5x + 6$ by $x - 2$.
21. Divide $y^2 + 13y + 12$ by $y + 1$.
22. Divide $z^2 - 7z + 12$ by $z - 3$.
23. $\dfrac{6a^2 + 5a - 6}{2a + 3}$
24. $\dfrac{8a^2 + 2a - 3}{2a - 1}$
25. $\dfrac{3b^2 + 11b + 6}{3b + 2}$
26. $\dfrac{3b^2 - 5b + 2}{3b - 2}$

Write the terms so that the powers of x are in descending order. Then do each division.

27. $5x + 3 \overline{)11x + 10x^2 + 3}$
28. $2x - 7 \overline{)-x - 21 + 2x^2}$
29. $4 + 2x \overline{)-10x - 28 + 2x^2}$
30. $1 + 3x \overline{)9x^2 + 1 + 6x}$
31. $2x - 1 \overline{)x - 2 + 6x^2}$
32. $2 + x \overline{)3x + 2x^2 - 2}$
33. $3 + x \overline{)2x^2 - 3 + 5x}$
34. $x - 3 \overline{)2x^2 - 3 - 5x}$

Do each division.

35. $2x + 3 \overline{)2x^3 + 7x^2 + 4x - 3}$
36. $2x - 1 \overline{)2x^3 - 3x^2 + 5x - 2}$
37. $3x + 2 \overline{)6x^3 + 10x^2 + 7x + 2}$
38. $4x + 3 \overline{)4x^3 - 5x^2 - 2x + 3}$
39. $2x + 1 \overline{)2x^3 + 3x^2 + 3x + 1}$
40. $3x - 2 \overline{)6x^3 - x^2 + 4x - 4}$

Do each division. If there is a remainder, write the answer in $\text{quotient} + \dfrac{\text{remainder}}{\text{divisor}}$ *form.*

41. $\dfrac{2x^2 + 5x + 2}{2x + 3}$ **42.** $\dfrac{3x^2 - 8x + 3}{3x - 2}$

43. $\dfrac{4x^2 + 6x - 1}{2x + 1}$ **44.** $\dfrac{6x^2 - 11x + 2}{3x - 1}$

45. $\dfrac{x^3 + 3x^2 + 3x + 1}{x + 1}$

46. $\dfrac{x^3 + 6x^2 + 12x + 8}{x + 2}$

47. $\dfrac{2x^3 + 7x^2 + 4x + 3}{2x + 3}$

48. $\dfrac{6x^3 + x^2 + 2x + 1}{3x - 1}$

49. $\dfrac{2x^3 + 4x^2 - 2x + 3}{x - 2}$

50. $\dfrac{3y^3 - 4y^2 + 2y + 3}{y + 3}$

Do each division.

51. $\dfrac{x^2 - 1}{x - 1}$

52. $\dfrac{x^2 - 9}{x + 3}$

53. $\dfrac{4x^2 - 9}{2x + 3}$

54. $\dfrac{25x^2 - 16}{5x - 4}$

55. $\dfrac{x^3 + 1}{x + 1}$

56. $\dfrac{x^3 - 8}{x - 2}$

57. $\dfrac{a^3 + a}{a + 3}$

58. $\dfrac{y^3 - 50}{y - 5}$

59. $3x - 4 \overline{)15x^3 - 23x^2 + 16x}$

60. $2y + 3 \overline{)21y^2 + 6y^3 - 20}$

APPLICATIONS

61. FURNACE FILTER   The area of the furnace filter shown in Illustration 1 is $(x^2 - 2x - 24)$ square inches.
  a. Find its length.
  b. Find its perimeter.

$(x + 4)$ in.

ILLUSTRATION 1

62. SHELF SPACE   The formula $V = Bh$ gives the volume of a cylinder where B is the area of the base and h is the height. Find the amount of shelf space that the container of potato chips shown in Illustration 2 occupies if its volume is $(2x^3 - 4x - 2)$ cubic inches.

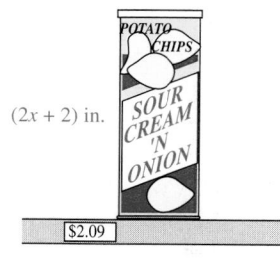

$(2x + 2)$ in.

POTATO CHIPS

SOUR CREAM 'N ONION

$2.09

ILLUSTRATION 2

63. COMMUNICATION   See Illustration 3. Telephone poles were installed every $(2x - 3)$ feet along a stretch of railroad track $(8x^3 - 6x^2 + 5x - 21)$ feet long. How many poles were used?

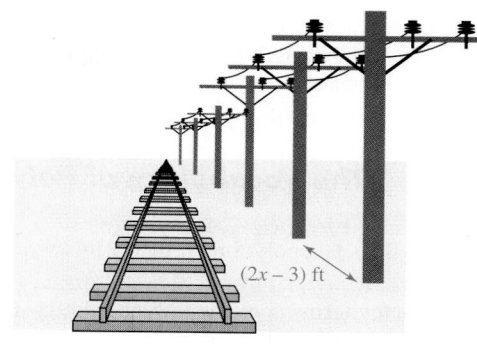

$(2x - 3)$ ft

ILLUSTRATION 3

64. CONSTRUCTION COSTS   Find the price per square foot to remodel each of the three rooms listed in the chart.

Room	Remodeling cost	Area (ft^2)	Cost (per ft^2)
Bathroom	$(2x^2 + x - 6)$	$2x - 3$	
Bedroom	$(x^2 + 9x + 20)$	$x + 4$	
Kitchen	$(3x^3 - 9x - 6)$	$3x + 3$	

WRITING

65. Explain how the following are related: *dividend, divisor, quotient,* and *remainder.*

66. How would you check the results of a division?

REVIEW

67. Simplify $(x^5 x^6)^2$.

68. Simplify $(a^2)^3 (a^3)^4$.

In Exercises 69–70, simplify each expression.

69. $3(2x^2 - 4x + 5) + 2(x^2 + 3x - 7)$

70. $-2(y^3 + 2y^2 - y) - 3(3y^3 + y)$

71. What can be said about the slopes of two parallel lines?

72. What is the slope of a line perpendicular to a line with a slope of $\frac{3}{4}$?

Polynomials

A **polynomial** is a term or a sum of terms in which all variables have whole-number exponents. Some examples are

$$-16a^4b, \qquad y + 8, \qquad x^2 + 2xy + y^2, \quad \text{and} \quad x^3 - 2x^2 + 6x - 8$$

The Vocabulary of Polynomials

1. Consider $x^3 - 2x^2 + 6x - 8$.
  a. Fill in: This is a polynomial in ___. It is written in _____ powers of x.
  b. How many terms does the polynomial have?
  c. Give the degree of each term.
  d. What is the degree of the polynomial?
  e. Give the coefficient of each term.

2. Consider $x^2 + 2xy + y^2$.
  a. Fill in: This is a polynomial in ___ and ___. It is written in _____ powers of x and _____ powers of y.
  b. How many terms does the polynomial have?
  c. Give the degree of each term.
  d. What is the degree of the polynomial?
  e. Give the coefficient of each term.

3. Classify each polynomial as a monomial, binomial, trinomial, or none of these.
  a. $x^2 - y^2$
  b. $s^2t + st^2 - st + 1$
  c. $4y^2 - 10y + 16$
  d. $15h$

4. a. Explain why ab is a monomial while $a + b$ is a binomial.

  b. Is every term of a polynomial a monomial?
  c. The degree of x^2 is 2. What is the degree of 3^2? Explain your answer.

  d. For $3x^2 - 4x + 9$, what is the leading term, the leading coefficient, and the constant term?

Operations with Polynomials

Just like numbers in arithmetic, polynomials can be added, subtracted, multiplied, divided, and raised to powers. We have discussed some rules for performing operations with polynomials that have more than one terms.

Fill in the blanks.

5. To add two polynomials, remove the parentheses and _____ any like terms.

6. To subtract two polynomials, drop the minus sign and the parentheses, and _____ the sign of every term within the parentheses of the second polynomial. Then combine like terms.

7. To multiply two polynomials, multiply _____ term of one polynomial by _____ term of the other polynomial and combine like terms.

8. To divide two polynomials, use the _____ division method.

Do the operations.

9. $(2x + 3) + (x - 8)$

10. $(2x + 3) - (x - 8)$

11. $(2x + 3)(x - 8)$

12. $(2x^2 + 3)^2$

13. $(y^2 + y - 6) + (y + 3)$

14. $(y^2 + y - 6) - (y + 3)$

15. $(y^2 + y - 6)(y + 3)$

16. $(y^2 + y - 6) \div (y + 3)$

Polynomial Functions

Polynomial functions can be used to describe such situations as the stopping distance of a car, the appreciation of a house, and the area of a geometric figure.

17. If $f(x) = x^3 - 2x + 5$, find $f(-2)$.

18. STOPPING DISTANCE A vehicle's stopping distance in feet is given by the polynomial function $f(v) = 0.04v^2 + 0.9v$, where v is the velocity. Find the stopping distance when the vehicle is traveling at 40 mph.

Section 10.1

RULES FOR EXPONENTS   Have a student in your group write each of the five rules for exponents listed on page 615 on separate 3×5 cards. On a second set of cards, write an explanation of each rule using words. On a third set of cards, write a separate example of the use of each rule for exponents. Shuffle the cards and work together to match the symbolic description, the word description, and the example for each of the five rules for exponents.

Section 10.2

GRAPHING   Complete Table 1, plot the ordered pairs on a rectangular coordinate system, and then draw a smooth curve through the points. Do the same for Table 2, using the same coordinate system. Compare the graphs. How are they similar, and how do they differ?

x	2^x	x	2^{-x}
-2		-2	
-1		-1	
0		0	
1		1	
2		2	
3		3	

TABLE 1   TABLE 2

Section 10.3

SCIENTIFIC NOTATION   Go to the library and find five examples of extremely large and five examples of extremely small positive numbers. Encyclopedias, government statistics books, and science books are good places to look. Write each number in scientific notation on a separate piece of paper. Include a brief explanation of what the number represents. Present the ten examples in numerical order, beginning with the smallest number first.

Section 10.4

POLYNOMIAL FUNCTIONS   The height (in feet) of a rock from the floor of the Grand Canyon t seconds after being thrown downward from the rim with an initial velocity of 6 feet per second is given by the polynomial

$$f(t) = -16t^2 - 6t + 5{,}292$$

a. Find $f(0)$ and $f(18)$ and explain their significance.

b. Find $f(3)$, $f(6)$, $f(9)$, $f(12)$, and $f(15)$. Use this information to show the position of the rock for these times on the scale shown in Illustration 1.

c. Are the distances the rock fell during each 3-second time interval the same?

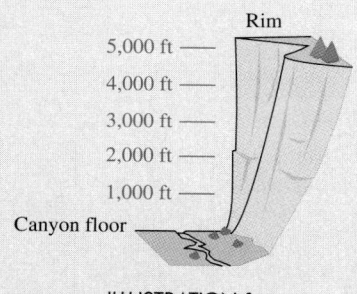

ILLUSTRATION 1

Section 10.5

ADDING POLYNOMIALS   An old adage is that "You can't add apples and oranges." Give an example of how this concept applies when adding two polynomials.

Section 10.6

MULTIPLYING BINOMIALS   Recall that the formula for the area of a rectangle is $A = lw$ and the formula for the area of a square is $A = s^2$.

a. Express the area of each of the large figures in Illustration 2 as a product of two binomials.

b. Find the area of each large figure by finding the sum of the areas of each of its parts.

c. In each case, show the relationship between your answers to part a and part b.

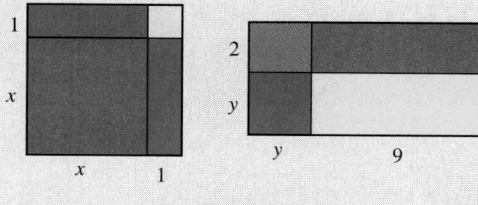

ILLUSTRATION 2

Section 10.7

WORKING WITH MONOMIALS   For the monomials $15a^3$ and $5a^2$, show, if possible, how they are added, subtracted, multiplied, and divided. If an operation cannot be done, explain why this is so.

Section 10.8

WORKING WITH POLYNOMIALS   Add, subtract, multiply, and divide the polynomials $6a^2 - 7a + 2$ and $2a - 1$.

SECTION 10.1	*Natural-Number Exponents*

CONCEPTS

If n represents a natural number, then

$$x^n = \overbrace{x \cdot x \cdot x \cdot \cdot \cdot \cdot \cdot x}^{n \text{ factors of } x}$$

where x is called the *base* and n is called the *exponent*.

Rules for exponents:
If m and n represent integers, then

$$x^m x^n = x^{m+n}$$
$$(x^m)^n = x^{m \cdot n}$$
$$(xy)^n = x^n y^n$$
$$\left(\frac{x}{y}\right)^n = \frac{x^n}{y^n} \quad (y \neq 0)$$
$$\frac{x^m}{x^n} = x^{m-n} \quad (x \neq 0)$$

REVIEW EXERCISES

1. Write each expression without using exponents.

　a. $-3x^4$　　　　　　**b.** $\left(\frac{1}{2}pq\right)^3$

2. Evaluate each expression.
　a. 5^3　　　　　　　**b.** $(-8)^2$
　c. -8^2　　　　　　 **d.** $(5-3)^2$

3. Simplify each expression.
　a. $x^3 x^2$　　　　　　**b.** $-3y(y^5)$
　c. $(y^7)^3$　　　　　　**d.** $(3x)^4$
　e. $b^3 b^4 b^5$　　　　　**f.** $-z^2(z^3 y^2)$
　g. $(-16s)^2 s$　　　　　**h.** $(2x^2 y)^2$
　i. $(x^2 x^3)^3$　　　　　**j.** $\left(\frac{x^2 y}{xy^2}\right)^2$
　k. $\dfrac{x^7}{x^3}$　　　　　　**l.** $\dfrac{(5y^2 z^3)^3}{(yz)^5}$

4. Find the area or the volume of each figure, whichever is appropriate.
　a.　　　　　　　　　　**b.**

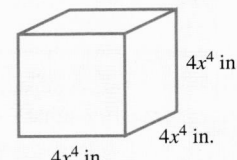
$4x^4$ in.
$4x^4$ in.
$4x^4$ in.

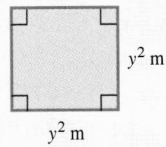
y^2 m
y^2 m

SECTION 10.2	*Zero and Negative Integer Exponents*

Zero exponents:
$$x^0 = 1 \quad (x \neq 0)$$

Negative integer exponents:
$$x^{-n} = \frac{1}{x^n} \quad (x \neq 0)$$

5. Write each expression without using negative exponents or parentheses.

　a. x^0　　　　　　　　　**b.** $(3x^2 y^2)^0$

　c. $(3x^0)^2$　　　　　　　**d.** 10^{-3}

　e. $\left(\frac{3}{4}\right)^{-1}$　　　　　　**f.** -5^{-2}

　g. x^{-5}　　　　　　　　**h.** $-6y^4 y^{-5}$

　i. $\dfrac{x^{-3}}{x^7}$　　　　　　　**j.** $(x^{-3} x^{-4})^{-2}$

　k. $\left(\dfrac{x^2}{x}\right)^{-5}$　　　　　　**l.** $\left(\dfrac{3z^4}{z^3}\right)^{-2}$

6. Write each expression with a single exponent.

   a. $y^{3n}y^{4n}$ **b.** $\dfrac{z^{8c}}{z^{10c}}$

SECTION 10.3	*Scientific Notation*

A number is written in *scientific notation* if it is written as the product of a number between 1 (including 1) and 10 and an integer power of 10.

7. Write each number in scientific notation.

   a. 728   **b.** 9,370,000

   c. 0.0136   **d.** 0.00942

   e. 0.018×10^{-2} **f.** 753×10^{3}

8. Write each number in standard notation.

   a. 7.26×10^{5} **b.** 3.91×10^{-4}

   c. 2.68×10^{0} **d.** 5.76×10^{1}

Scientific notation provides an easier way to do some computations.

9. Simplify each fraction by first writing each number in scientific notation, then do the arithmetic. Express the result in standard notation.

   a. $\dfrac{(0.00012)(0.00004)}{0.00000016}$ **b.** $\dfrac{(4,800)(20,000)}{600,000}$

10. WORLD POPULATION   As of 2000, the world's population was estimated to be 6.08 billion. Write this number in standard notation and in scientific notation.

11. ATOMS   Illustration 1 shows a cross section of an atom. How many nuclei, placed end-to-end, would it take to stretch across the atom?

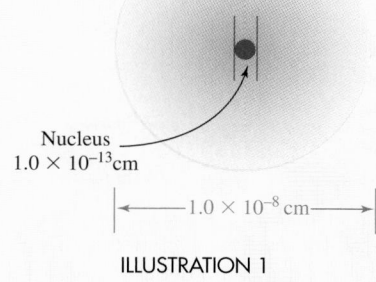

Nucleus
1.0×10^{-13}cm

$\longleftarrow 1.0 \times 10^{-8}$ cm $\longrightarrow$

ILLUSTRATION 1

SECTION 10.4	*Polynomials*

A *polynomial* is a term or a sum of terms in which all variables have whole-number exponents.

12. Tell whether each expression is a polynomial.

   a. $x^{3} - x^{2} - x - 1$ **b.** $x^{-2} - x^{-1} - 1$

   c. $\dfrac{11}{y} + 4y$ **d.** $-16x^{2}y + 5xy^{2}$

13. Consider the polynomial $3x^{3} - x^{2} + x + 10$.

   a. How many terms does the polynomial have?

   b. What is the leading term?

   c. What is the coefficient of the second term?

   d. What is the constant term?

The *degree of a monomial ax^n* is *n*. The *degree of a monomial* in several variables is the sum of the exponents on those variables. The *degree of a polynomial* is the same as the degree of its term with the largest degree.

14. Find the degree of each polynomial and classify it as a monomial, binomial, trinomial, or none of these.

   a. $13x^7$
   b. $-16a^2b$
   c. $5^3x + x^2$
   d. $-3x^5 + x - 1$
   e. $9xy^2 + 21x^3y^3$
   f. $4s^4 - 3s^2 + 5s + 4$

If $f(x)$ is a polynomial function in *x*, then $f(3)$ is the value of the function when $x = 3$.

15. Let $f(x) = 3x^2 + 2x + 1$. Find each value.

   a. $f(3)$
   b. $f(0)$
   c. $f(-2)$
   d. $f(-0.2)$

16. DIVING   See Illustration 2. The number of inches that the woman deflects the diving board is given by the function

$$f(x) = 0.1875x^2 - 0.0078125x^3$$

where *x* is the number of feet that she stands from the front anchor point of the board. Find the amount of deflection if she stands on the end of the diving board, 8 feet from the anchor point.

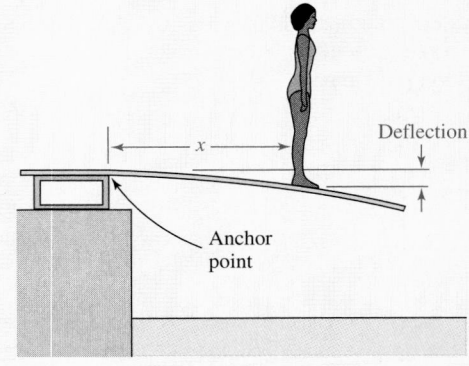

ILLUSTRATION 2

SECTION 10.5   *Adding and Subtracting Polynomials*

When *adding* or *subtracting polynomials,* add or subtract like terms by combining the numerical coefficients and using the same variables and the same exponents.

17. Simplify each expression.

   a. $3x^6 + 5x^5 - x^6$
   b. $x^2y^2 - 3x^2y^2$
   c. $(3x^2 + 2x) + (5x^2 - 8x)$
   d. $3(9x^2 + 3x + 7) - 2(11x^2 - 5x + 9)$

Polynomials can be added or subtracted *vertically.*

18. Do the operations.

   a. $\begin{array}{r} 3x^2 + 5x + 2 \\ + \quad x^2 - 3x + 6 \\ \hline \end{array}$

   b. $\begin{array}{r} 20x^3 \qquad\quad + 12x \\ - \quad 12x^3 + 7x^2 - \ 7x \\ \hline \end{array}$

SECTION 10.6   *Multiplying Polynomials*

To multiply two monomials, first multiply the numerical factors and then multiply the variable factors.

19. Find each product.

   a. $(2x^2)(5x)$
   b. $(-6x^4z^3)(x^6z^2)$
   c. $(2rst)(-3r^2s^3t^4)$
   d. $5b^3 \cdot 6b^2 \cdot 4b^6$

To multiply a polynomial with more than one term by a monomial, multiply each term of the polynomial by the monomial and simplify.

20. Find each product.
  a. $5(x + 3)$
  b. $x^2(3x^2 - 5)$
  c. $x^2y(y^2 - xy)$
  d. $-2y^2(y^2 - 5y)$
  e. $2x(3x^4)(x + 2)$
  f. $-3x(x^2 - x + 2)$

To multiply two binomials, use the *FOIL method:*
  F: First
  O: Outer
  I: Inner
  L: Last

21. Find each product.
  a. $(x + 3)(x + 2)$
  b. $(2x + 1)(x - 1)$
  c. $(3a - 3)(2a + 2)$
  d. $6(a - 1)(a + 1)$
  e. $(a - b)(2a + b)$
  f. $(-3x - y)(2x + y)$

Special products:
$$(x + y)^2 = x^2 + 2xy + y^2$$
$$(x - y)^2 = x^2 - 2xy + y^2$$
$$(x + y)(x - y) = x^2 - y^2$$

22. Find each product.
  a. $(x + 3)(x + 3)$
  b. $(x + 5)(x - 5)$
  c. $(a - 3)^2$
  d. $(x + 4)^2$
  e. $(-2y + 1)^2$
  f. $(y^2 + 1)(y^2 - 1)$

To multiply one polynomial by another, multiply each term of one polynomial by each term of the other polynomial, and simplify.

23. Find each product.
  a. $(3x + 1)(x^2 + 2x + 1)$
  b. $(2a - 3)(4a^2 + 6a + 9)$

24. Solve each equation.
  a. $x^2 + 3 = x(x + 3)$
  b. $x^2 + x = (x + 1)(x + 2)$
  c. $(x + 2)(x - 5) = (x - 4)(x - 1)$
  d. $(x + 5)(3x + 1) = x^2 + (2x - 1)(x - 5)$

25. APPLIANCE   Find the perimeter of the base, the area of the base, and the volume occupied by the dishwasher shown in Illustration 3.

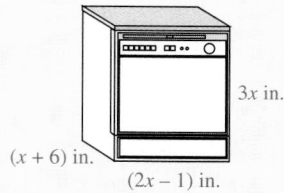

$3x$ in.

$(x + 6)$ in.

$(2x - 1)$ in.

ILLUSTRATION 3

SECTION 10.7 — *Dividing Polynomials by Monomials*

To divide monomials, use the method for simplifying fractions or use the rules for exponents.

26. Simplify each expression ($x > 0$, $y > 0$).
  a. $\dfrac{-14x^2y}{21xy^3}$
  b. $\dfrac{(x^2)^2}{xx^4}$

To divide a polynomial by a monomial, divide each term of the numerator by the denominator.

27. Do each division. All the variables represent positive numbers.
  a. $\dfrac{8x + 6}{2}$
  b. $\dfrac{14xy - 21x}{7xy}$
  c. $\dfrac{15a^2b + 20ab^2 - 25ab}{5ab}$
  d. $\dfrac{(x + y)^2 + (x - y)^2}{-2xy}$

28. SAVINGS BONDS   How many $50 savings bonds would have a total value of $(50x + 250)$?

SECTION 10.8	*Dividing Polynomials by Polynomials*

Long division is used to divide one polynomial by another. When a division has a remainder, write the answer in the form

$$\text{Quotient} + \frac{\text{remainder}}{\text{divisor}}$$

The division method works best when the exponents of the terms of the divisor and the dividend are written in descending order.

When the dividend is missing a term, write it with a coefficient of zero or leave a blank space.

29. Do each division.

a. $x + 2 \overline{)x^2 + 3x + 5}$

b. $x - 1 \overline{)x^2 - 6x + 5}$

c. $\dfrac{2x^2 + 3 + 7x}{x + 3}$

d. $\dfrac{3x^2 + 14x - 2}{3x - 1}$

e. $2x - 1 \overline{)6x^3 + x^2 + 1}$

f. $3x + 1 \overline{)-13x - 4 + 9x^3}$

30. Use multiplication to show that the answer when dividing $3y^2 + 11y + 6$ by $y + 3$ is $3y + 2$.

31. ZOOLOGY   The distance in inches traveled by a certain type of snail in $(2x - 1)$ minutes is given by the polynomial $8x^2 + 2x - 3$. At what rate did the snail travel?

Chapter 10 Test

1. Use exponents to rewrite $2xxxyyyy$.

2. Evaluate $(3 + 5)^2$.

Write each expression as an expression containing only one exponent.

3. $y^2(yy^3)$

4. $(2x^3)^5(x^2)^3$

In Problems 5–8, simplify each expression. Write answers without using parentheses or negative exponents.

5. $3x^0$

6. $2y^{-5}y^2$

7. $\dfrac{y^2}{yy^{-2}}$

8. $\left(\dfrac{a^2b^{-1}}{4a^3b^{-2}}\right)^{-3}$

9. What is the volume of a cube that has sides of length $10y^4$ inches?

10. Rewrite 4^{-2} using a positive exponent and then evaluate the result.

11. ELECTRICITY One ampere (amp) corresponds to the flow of 6,250,000,000,000,000,000 electrons per second past any point in a direct current (DC) circuit. Write this number in scientific notation.

12. Write 9.3×10^{-5} in standard notation.

13. Identify $3x^2 + 2$ as a monomial, binomial, or trinomial.

14. Find the degree of the polynomial $3x^2y^3 + 2x^3y - 5x^2y$.

15. If $f(x) = x^2 + x - 2$, find $f(-2)$.

16. Simplify $(xy)^2 + 5x^2y^2 - (3x)^2y^2$.

17. Simplify $-6(x - y) + 2(x + y) - 3(x + 2y)$.

18. Subtract: $\begin{array}{r} 2x^2 - 7x + 3 \\ 3x^2 - 2x - 1 \end{array}$

In Problems 19–24, find each product.

19. $(-2x^3)(2x^2y)$

26. Simplify $\dfrac{8x^2y^3z^4}{16x^3y^2z^4}$.

20. $3y^2(y^2 - 2y + 3)$

27. Simplify $\dfrac{6a^2 - 12b^2}{24ab}$.

21. $(x - 9)(x + 9)$

28. Divide $2x + 3 \overline{)2x^2 - x - 6}$.

22. $(3y - 4)^2$

23. $(2x - 5)(3x + 4)$

29. In your own words, explain this rule for exponents:

$$x^{-n} = \frac{1}{x^n}$$

24. $(2x - 3)(x^2 - 2x + 4)$

30. A rectangle has an area of $(x^2 - 6x + 5)$ ft^2 and a length of $(x - 1)$ feet. Show how division can be used to find the width of the rectangle. Explain your steps.

25. Solve $(a + 2)^2 = (a - 3)^2$.

Chapters 1-10   Cumulative Review Exercises

1. PERSONAL SAVINGS RATE   The graph in Illustration 1 shows a situation occurring in September of 1998 that hadn't occurred since the Great Depression. Explain what was unusual.

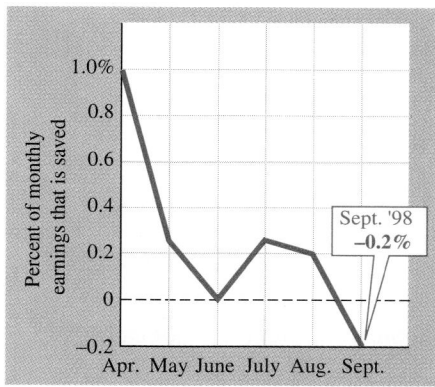

Based on data from the U.S. Department of Commerce

ILLUSTRATION 1

2. Find $\dfrac{3}{4} \div \dfrac{6}{5}$.

3. Find $\dfrac{7}{10} - \dfrac{1}{14}$.

4. Is π a rational or irrational number?

5. RACING   Suppose a driver has completed x laps of a 250-lap race. Write an expression for how many more laps he must make to finish the race.

6. CLINICAL TRIALS   In a clinical test of Aricept, a drug to treat Alzheimer's disease, one group of patients took a placebo (a sugar pill) while another group took the actual medication. See Illustration 2. Find the number of patients in each group who experienced nausea. Round to the nearest whole number.

Comparison of rates of adverse events in patients		
Adverse event	Group 1—Placebo (number = 315)	Group 2—Aricept (number = 311)
Nausea	6%	5%

ILLUSTRATION 2

Consider the algebraic expression $3x^3 + 5x^2y + 37y$.

7. Find the coefficient of the second term.

8. What is the third term?

Simplify each expression.

9. $3x - 5x + 2y$

10. $3(x - 7) + 2(8 - x)$

11. $2x^2y^3 - xy(xy^2)$

12. $x^2(3 - y) + x(xy + x)$

Solve each equation.

13. $3(x - 5) + 2 = 2x$ 14. $\dfrac{x - 5}{3} - 5 = 7$

Solve each formula for the variable indicated.

15. $A = \dfrac{1}{2}h(b + B)$; for h

16. $y = mx + b$; for x

Evaluate each expression.

17. $4^2 - 5^2$ 18. $(4 - 5)^2$

19. $\dfrac{-3 - (-7)}{2^2 - 3}$ 20. $12 - 2[1 - (-8 + 2)]$

Solve each inequality and graph the solution set. Then describe the solution using interval notation

21. $8(4 + x) > 10(6 + x)$

22. $-9 < 3(x + 2) \leq 3$

Graph each equation.

23. $y = x^2$ 24. $y = |x|$

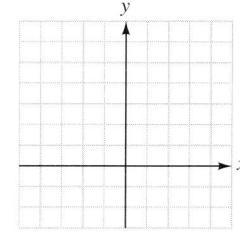

25. $4x - 3y = 12$ **26.** $3x = 12$

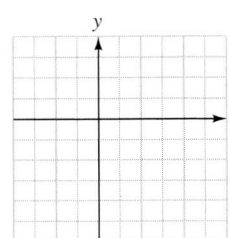 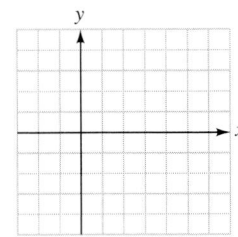

Find the slope of the line with the given properties.

27. Passing through $(-2, 4)$ and $(6, 8)$

28. A line that is horizontal

29. An equation of $y = -4x + 3$

30. An equation of $2x - 3y = 12$

Write the equation of the line with the following properties.

31. Slope $= \dfrac{2}{3}$, y-intercept $= (0, 5)$

32. Passing through $(-2, 4)$ and $(6, 10)$

33. A horizontal line passing through $(2, 4)$

34. A vertical line passing through $(2, 4)$

Are the graphs of the lines parallel or perpendicular?

35. $y = -\dfrac{3}{4}x + \dfrac{15}{4}$

     $4x - 3y = 25$

36. $y = -\dfrac{3}{4}x + \dfrac{15}{4}$

     $6x = 15 - 8y$

Tell whether each equation defines a function.

37. $y = x^3 - 4$ **38.** $x = |y|$

In Exercises 35–38, $f(x) = 2x^2 - 3$. Find each value.

39. $f(0)$ **40.** $f(3)$

41. $f(-2)$ **42.** $f(0.5)$

43. Find the domain and range of the function $f(x) = |x|$.

44. Tell whether the graph is the graph of a function.

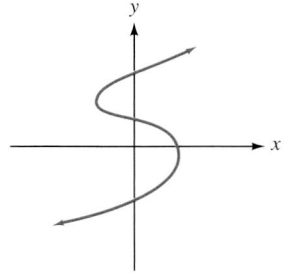

45. Give an example of a positive rate of change and a negative rate of change.

46. Evaluate -3^2.

Write each expression using one positive exponent.

47. $(y^3 y^5) y^6$ **48.** $(x^3 x^4)^2$

49. $\dfrac{x^3 x^4}{x^2 x^3}$ **50.** $\dfrac{a^4 b^0}{a^{-3}}$

51. x^{-5} **52.** $(-2y)^{-4}$

53. $(x^{-4})^2$ **54.** $\left(-\dfrac{x^3}{x^{-2}}\right)^3$

Write each number in scientific notation.

55. 615,000   **56.** 0.0000013

Write each number in standard notation.

57. 5.25×10^{-4} **58.** 2.77×10^3

In Exercises 59–60, give the degree of each polynomial.

59. $3x^2 + 2x - 5$ **60.** $-3x^3 y^2 + 3x^2 y^2 - xy$

61.   MUSICAL INSTRUMENTS   The gong shown in Illustration 3 is a percussion instrument used throughout Southeast Asia. The amount of deflection of the horizontal support (in inches) is given by the polynomial function

$$f(x) = 0.01875x^4 - 0.15x^3 + 1.2x$$

where x is the distance (in feet) that the gong is hung from one end of the support. Find the deflection if the gong is hung in the middle of the support.

ILLUSTRATION 3

62. Consider the polynomial $2x^2 - 5x + 9$. Determine each of the following.
   a. The number of terms
   b. The leading term
   c. The leading coefficient
   d. The degree of the second term
   e. The degree of the polynomial
   f. The constant term

Do the operations.

63. $(3x^2 + 2x - 7) - (2x^2 - 2x + 7)$

64. $(2x^2 - 3x + 4) + (2x^2 + 2x - 5)$

65. $-5x^2(7x^3 - 2x^2 - 2)$

66. $(3x^3y^2)(-4x^2y^3)$

67. $(3x - 7)(2x + 8)$

68. $(5x - 4y)(3x + 2y)$

69. $(3x + 1)^2$

70. $(x - 2)(x^2 + 2x + 4)$

71. $\dfrac{6x^2 - 8x}{2x}$

72. $x - 3\overline{)2x^2 - 5x - 3}$

11

Factoring and Quadratic Equations

11.1 Factoring Out the Greatest Common Factor and Factoring by Grouping

11.2 Factoring Trinomials of the Form $x^2 + bx + c$

11.3 Factoring Trinomials of the Form $ax^2 + bx + c$

11.4 Special Factorizations and a Factoring Strategy

11.5 Quadratic Equations

Key Concept: Factoring

Accent on Teamwork

Chapter Review

Chapter Test

Cumulative Review Exercises

IN THIS CHAPTER, WE WILL DISCUSS SOME METHODS FOR FACTORING POLYNOMIALS. FACTORING REVERSES THE PROCESS OF MULTIPLICATION. IT CAN BE USED TO SIMPLIFY EXPRESSIONS AND TO SOLVE EQUATIONS.

11.1 Factoring Out the Greatest Common Factor and Factoring by Grouping

In this section, you will learn about

- Factoring natural numbers • The greatest common factor (GCF)
- Finding the GCF of several monomials
- Factoring out the greatest common factor • Factoring out a negative factor
- Factoring by grouping

INTRODUCTION. Recall that the distributive property provides a way to multiply a monomial and a binomial. For example,

$$4y(3y + 5) = 4y \cdot 3y + 4y \cdot 5$$
$$= 12y^2 + 20y$$

In this section, we will reverse the operation of multiplication. Given a polynomial such as $12y^2 + 20y$, we will ask ourselves, "What factors were multiplied to obtain $12y^2 + 20y$?" The process of finding the individual factors of a known product is called **factoring.**

The multiplication process	**The factoring process**
Given the factors . . . find the product	Given the product . . . find the factors
$\downarrow \quad \downarrow \quad\quad \downarrow$	$\downarrow \quad\quad \downarrow \quad \downarrow$
$4y(3y + 5) \ = \ ?$	$12y^2 + 20y \ = \ ?(\ ?\ )$

To begin the discussion of factoring, we consider two methods that can be used to factor natural numbers.

Factoring natural numbers

Because 4 divides 12 exactly, 4 is called a **factor** of 12. The numbers 1, 2, 3, 4, 6, and 12 are the natural-number factors of 12, because each divides 12 exactly.

Prime numbers

> A **prime number** is a natural number greater than 1 whose only factors are 1 and itself.

For example, 17 is a prime number, because

1. 17 is a natural number greater than 1, and

2. the only two natural-number factors of 17 are 1 and 17.

The prime numbers less than 50 are

$$2, \quad 3, \quad 5, \quad 7, \quad 11, \quad 13, \quad 17, \quad 19, \quad 23, \quad 29, \quad 31, \quad 37, \quad 41, \quad 43, \quad \text{and} \quad 47$$

A natural number is said to be in **prime-factored form** if it is written as the product of factors that are prime numbers.

To find the prime-factored form of a natural number, we can use a **factoring tree.** The following examples show two ways to find the prime-factored form of 90 using factoring trees. The factoring process stops when a row of the tree contains only prime-number factors.

1. Start with 90.

2. Factor 90 as 9 · 10.

3. Factor 9 and 10.

90
9 · 10
3 · 3 · 2 · 5

1. Start with 90.

2. Factor 90 as 6 · 15.

3. Factor 6 and 15.

90
6 · 15
2 · 3 · 3 · 5

Since the prime factors in either case are 2 · 3 · 3 · 5, the prime-factored form, or the **prime factorization,** of 90 is $2 \cdot 3^2 \cdot 5$. This example illustrates the **fundamental theorem of arithmetic,** which states that there is only one prime factorization for every natural number greater than 1.

We can also find the prime factorization of a natural number using the **division method.** For example, to find the prime factorization of 42, we begin by choosing the *smallest* prime number that will divide the given number exactly. We continue this process until the result of the division is a prime number.

Step 1: 2 divides 42 exactly. The result is 21, which is not prime. We continue the process.

$$2 \overline{| 42 }$$
$$21$$

Step 2: We choose the smallest prime number that divides 21. The prime number 2 does not divide 21 exactly, but 3 does. The result is 7, which is prime. We are done.

$$2 \overline{| 42 }$$
$$3 \overline{| 21 }$$
$$7$$

The prime factorization of 42 is 2 · 3 · 7.

EXAMPLE 1   *Prime factorizations.*   Find the prime factorization of 150.

Solution
We can use a factoring tree or the division method to find the prime factorization.

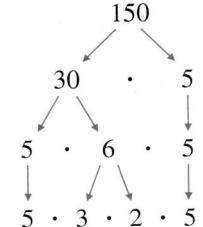

$$\begin{array}{r} 2 \,|\, 150 \\ 3 \,|\, 75 \quad \leftarrow 150 \div 2 \\ 5 \,|\, 25 \quad \leftarrow 75 \div 3 \\ 5 \quad \leftarrow 25 \div 5 \end{array}$$

Using exponents, we can write the prime factorization of 150 as $2 \cdot 3 \cdot 5^2$.

Self Check

Find the prime factorization of 225.

Answer: $3^2 \cdot 5^2$ ∎

The greatest common factor (GCF)

The right-hand sides of the equations

$$90 = 2 \cdot 3 \cdot 3 \cdot 5$$
$$42 = 2 \cdot 3 \cdot 7$$

show the prime-factored forms of 90 and 42. The color highlighting indicates that 90 and 42 have one prime factor of 2 and one prime factor of 3 in common. We can conclude that $2 \cdot 3 = 6$ is the largest natural number that divides 90 and 42 exactly, and we say that 6 is their **greatest common factor (GCF)**.

$$\frac{90}{6} = 15 \qquad \text{and} \qquad \frac{42}{6} = 7$$

EXAMPLE 2   *Finding the GCF of three numbers.*   Find the greatest common factor of 24, 60, and 96.

Solution

We write each prime factorization and highlight the prime factors the three numbers have in common.

$24 = 2 \cdot 2 \cdot 2 \cdot 3$
$60 = 2 \cdot 2 \cdot 3 \cdot 5$
$96 = 2 \cdot 2 \cdot 2 \cdot 2 \cdot 2 \cdot 3$

Since 24, 60, and 96 each have two factors of 2 and one factor of 3, their greatest common factor is $2 \cdot 2 \cdot 3 = 12$.

Self Check

Find the GCF for 45, 60, 75.

Answer: $3 \cdot 5 = 15$ ∎

Finding the GCF of several monomials

The right-hand sides of the equations

$12y^2 = 2 \cdot 2 \cdot 3 \cdot y \cdot y$
$20y = 2 \cdot 2 \cdot 5 \cdot y$

show the prime factorizations of $12y^2$ and $20y$. Since the monomials have two factors of 2 and one factor of y in common, their GCF is

$$2 \cdot 2 \cdot y \qquad \text{or} \qquad 4y$$

To find the GCF of several monomials, we follow these steps.

Strategy for finding the greatest common factor (GCF)

> 1. Find the prime factorization of each monomial.
> 2. List each common factor the least number of times it appears in any one monomial.
> 3. Find the product of the factors in the list to obtain the GCF.

EXAMPLE 3   *Finding the GCF of three monomials.*   Find the GCF of $10x^3y^2$, $60x^2y$, and $30xy^2$.

Solution

Step 1: Find the prime factorization of each monomial.

$10x^3y^2 = 2 \cdot 5 \cdot x \cdot x \cdot x \cdot y \cdot y$
$60x^2y = 2 \cdot 2 \cdot 3 \cdot 5 \cdot x \cdot x \cdot y$
$30xy^2 = 2 \cdot 3 \cdot 5 \cdot x \cdot y \cdot y$

Step 2: List each common factor the least number of times it appears in any one monomial: 2, 5, x, and y.

Step 3: Find the product of the factors in the list:

$2 \cdot 5 \cdot x \cdot y = 10xy$ The GCF is $10xy$.

Self Check

Find the GCF of $20a^2b^3$, $12ab^4$, and $8a^3b^2$.

Answer: $4ab^2$ ∎

Factoring out the greatest common factor

To factor $12y^2 + 20y$, we find the GCF of $12y^2$ and $20y$ (which is $4y$) and use the distributive property.

$$12y^2 + 20y = \mathbf{4y} \cdot 3y + \mathbf{4y} \cdot \mathbf{5}$$
Write each term of the polynomial as the product of the GCF, $4y$, and one other factor.

$$= \mathbf{4y}(3y + \mathbf{5})$$
$4y$ is a common factor of both terms.

This process is called **factoring out the greatest common factor.**

EXAMPLE 4 *Factoring out the greatest common factor.* Factor $25 - 5m$.

Solution

To find the GCF of 25 and $5m$, we find their prime factorizations.

$$\left. \begin{array}{l} 25 = 5 \cdot 5 \\ 5m = 5 \cdot m \end{array} \right\} \quad \text{GCF} = 5$$

We can use the distributive property to factor out the GCF.

$$25 - 5m = 5 \cdot 5 - 5 \cdot m \quad \text{Factor each monomial using 5 and one other factor.}$$
$$= 5(5 - m) \quad \text{Factor out the common factor of 5.}$$

We check by verifying that $5(5 - m) = 25 - 5m$.

Self Check

Factor $18x - 24$.

Answer: $6(3x - 4)$ ∎

EXAMPLE 5 *Factoring out the GCF.* Factor $35a^3b^2 + 14a^2b^3$.

Solution

To find the GCF, we find the prime factorizations of $35a^3b^2$ and $14a^2b^3$.

$$\left. \begin{array}{l} 35a^3b^2 = 5 \cdot 7 \cdot a \cdot a \cdot a \cdot b \cdot b \\ 14a^2b^3 = 2 \cdot 7 \cdot a \cdot a \cdot b \cdot b \cdot b \end{array} \right\} \quad \text{GCF} = 7 \cdot a \cdot a \cdot b \cdot b = 7a^2b^2$$

We factor out the GCF of $7a^2b^2$.

$$35a^3b^2 + 14a^2b^3 = 7a^2b^2 \cdot 5a + 7a^2b^2 \cdot 2b$$
$$= 7a^2b^2(5a + 2b)$$

We check by verifying that $7a^2b^2(5a + 2b) = 35a^3b^2 + 14a^2b^3$.

Self Check

Factor $32x^2y^3 + 12x^3y^2$.

Answer: $4x^2y^2(8y + 3x)$ ∎

EXAMPLE 6 *An implied coefficient of 1.* Factor $4x^3y^2z - 2x^2yz + xz$.

Solution

The expression has three terms. We factor out the GCF, which is xz.

$$4x^3y^2z - 2x^2yz + xz = xz \cdot 4x^2y^2 - xz \cdot 2xy + xz \cdot \mathbf{1}$$
$$= xz(4x^2y^2 - 2xy + \mathbf{1})$$

The last term of $4x^3y^2z - 2x^2yz + xz$ has an implied coefficient of 1. When xz is factored out, we must write this coefficient of 1, as shown in blue. We check by verifying that $xz(4x^2y^2 - 2xy + 1) = 4x^3y^2z - 2x^2yz + xz$.

Self Check

Factor $2ab^2c + 4a^2bc - ab$.

Answer: $ab(2bc + 4ac - 1)$ ∎

EXAMPLE 7 *Crayon.* The amount of colored wax used to make the crayon shown in Figure 11-1 can be found by computing its volume using the formula

$$V = \pi r^2 h_1 + \frac{1}{3}\pi r^2 h_2$$

Factor the expression on the right-hand side of this equation.

FIGURE 11-1

Solution   Each term on the right-hand side of the formula contains a factor of π and r^2.

$$V = \pi r^2 h_1 + \frac{1}{3}\pi r^2 h_2$$

$$= \pi r^2\left(h_1 + \frac{1}{3}h_2\right) \quad \text{Factor out the GCF, } \pi r^2.$$

The formula to find the volume of the crayon can be expressed as $V = \pi r^2\left(h_1 + \frac{1}{3}h_2\right)$. ∎

EXAMPLE 8   *Factoring out a common binomial.*   Factor $x(x + 4) + 3(x + 4)$.

Solution
The given polynomial has two terms:

$$\underbrace{x(x + 4)}_{\text{The first term}} + \underbrace{3(x + 4)}_{\text{The second term}}$$

The GCF of the terms is $x + 4$, which can be factored out.

$$x(x + 4) + 3(x + 4) = (x + 4)(x + 3)$$

Self Check
Factor $2y(y - 1) - 7(y - 1)$.

Answer: $(y - 1)(2y - 7)$ ∎

Factoring out a negative factor

It is often useful to factor out a common factor having a negative coefficient.

EXAMPLE 9   *Factoring out −1.*   Factor -1 out of $-a^3 + 2a^2 - 4$.
Solution
First, we write each term of the polynomial as the product of -1 and another factor. Then we factor out the common factor of -1.

$$-a^3 + 2a^2 - 4 = (-1)a^3 + (-1)(-2a^2) + (-1)4$$
$$= -1(a^3 - 2a^2 + 4) \qquad \text{Factor out } -1.$$
$$= -(a^3 - 2a^2 + 4) \qquad \begin{array}{l}\text{The coefficient of 1 need not}\\ \text{be written.}\end{array}$$

We check by verifying that $-(a^3 - 2a^2 + 4) = -a^3 + 2a^2 - 4$.

Self Check
Factor -1 out of $-b^4 - 3b^2 + 2$.

Answer: $-(b^4 + 3b^2 - 2)$ ∎

EXAMPLE 10   *Factoring out the negative of the GCF.*   Factor out the negative (opposite) of the GCF in $-18a^2b + 6ab^2 - 12a^2b^2$.

Solution
The GCF is $6ab$. To factor out its negative, we write each term of the polynomial as the product of $-6ab$ and another factor. Then we factor out $-6ab$.

$$-18a^2b + 6ab^2 - 12a^2b^2 = (-6ab)3a - (-6ab)b + (-6ab)2ab$$
$$= -6ab(3a - b + 2ab)$$

We check by verifying that $-6ab(3a - b + 2ab) = -18a^2b + 6ab^2 - 12a^2b^2$.

Self Check
Factor out the negative (opposite) of the GCF in $-27xy^2 - 18x^2y + 36x^2y^2$.

Answer: $-9xy(3y + 2x - 4xy)$ ∎

Factoring by grouping

Suppose we wish to factor the polynomial

$$ax + ay + cx + cy$$

Although no factor is common to all four terms, there is a common factor of a in $ax + ay$ and a common factor of c in $cx + cy$. We can factor out a and c, and then factor out $x + y$ to obtain

$$ax + ay + cx + cy = a(x + y) + c(x + y)$$
$$= (x + y)(a + c) \qquad \text{Factor out } x + y.$$

We can check the result by multiplication.

$$(x + y)(a + c) = ax + cx + ay + cy$$
$$= ax + ay + cx + cy \quad \text{Rearrange the terms.}$$

Thus, $ax + ay + cx + cy$ factors as $(x + y)(a + c)$. This type of factoring is called **factoring by grouping.**

Factoring by grouping	1. Group the terms of the polynomial so that the first two terms have a common factor and the last two terms have a common factor.
	2. Factor out the common factor from each group.
	3. Factor out the resulting common binomial factor. If there is no common binomial factor, regroup the terms of the polynomial and repeat steps 2 and 3.

EXAMPLE 11   *Factoring by grouping.*   Factor $2c - 2d + cd - d^2$.

Solution

Since 2 is a common factor of the first two terms and d is a common factor of the last two terms, we have

$$2c - 2d + cd - d^2 = 2(c - d) + d(c - d) \quad \text{Factor out 2 from } 2c - 2d \text{ and } d \\ \text{from } cd - d^2.$$

$$= (c - d)(2 + d) \qquad \text{Factor out } c - d.$$

We check by verifying that

$$(c - d)(2 + d) = 2c + cd - 2d - d^2$$
$$= 2c - 2d + cd - d^2 \quad \text{Rearrange the terms.}$$

Self Check

Factor $7x - 7y + xy - y^2$.

Answer: $(x - y)(7 + y)$ ■

EXAMPLE 12   *Factoring out -1.*   Factor $x^2y - ax - xy + a$.

Solution

Since x is a common factor of the first two terms, we can factor it out and proceed as follows.

$$x^2y - ax - xy + a = x(xy - a) - xy + a \quad \text{Factor out } x \text{ from } x^2y - ax.$$

If we factor -1 from $-xy + a$, a common binomial factor $(xy - a)$ appears, which we can factor out.

$$x^2y - ax - xy + a = x(xy - a) - 1(xy - a)$$
$$= (xy - a)(x - 1) \qquad \text{Factor out } xy - a.$$

Check by multiplication.

Self Check

Factor $7b + 3c - 7bt - 3ct$.

Answer: $(7b + 3c)(1 - t)$ ■

COMMENT   When factoring the expressions in the previous two examples, don't think that $2(c - d) + d(c - d)$ or $x(xy - a) - 1(xy - a)$ are in factored form. For an expression to be in factored form, the result must be a product.

The next example illustrates that when factoring a polynomial, we should always look for a common factor first.

EXAMPLE 13   *Factoring out the GCF first.*   Factor $10k + 10m - 2km - 2m^2$.

Self Check

Factor $-4t - 4s - 4tz - 4sz$.

Solution

Since the four terms have a common factor of 2, we factor it out first. Then we use factoring by grouping to factor the polynomial within the parentheses. The first two terms have a common factor of 5. The last two terms have a common factor of $-m$.

$$10k + 10m - 2km - 2m^2 = 2(5k + 5m - km - m^2) \quad \text{Factor out the GCF 2.}$$
$$= 2[5(k + m) - m(k + m)]$$
$$= 2[(k + m)(5 - m)] \quad \text{Factor out } k + m.$$
$$= 2(k + m)(5 - m)$$

Use multiplication to check the result.

Answer: $-4(t + s)(1 + z)$ ■

STUDY SET   Section 11.1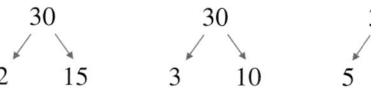

VOCABULARY   *Fill in the blanks.*

1. A natural number greater than 1 whose only factors are 1 and itself is called a _____ number.

2. When we write 24 as $2^3 \cdot 3$, we say that 24 has been written in _____ form.

3. The GCF of several natural numbers is the _____ number that divides each of the numbers exactly.

4. When we write $15x^2 - 25x$ as $5x(3x - 5)$, we say that we have _____ the greatest common factor.

5. The process of finding the individual factors of a known product is called _____.

6. The numbers 1, 2, 3, 4, 6, and 12 are the natural-number _____ of 12.

CONCEPTS   *In Exercises 7–10, explain what is wrong with each solution.*

7. Factor $6a + 9b + 3$.
$$6a + 9b + 3 = 3(2a + 3b + 0)$$
$$= 3(2a + 3b)$$

8. Prime factor 100.

$$
\begin{array}{r|l}
10 & 100 \\
\hline
5 & 10 \\
\hline
 & 2
\end{array}
$$

$$100 = 2 \cdot 5 \cdot 10$$

9. Factor out the GCF: $30a^3 - 12a^2$.
$$30a^3 - 12a^2 = 6a(5a^2 - 2a)$$

10. Factor $ab + b + a + 1$.
$$ab + b + a + 1 = b(a + 1) + (a + 1)$$
$$= (a + 1)b$$

11. What algebraic concept is illustrated in the work shown below?
$$4 \cdot 5x + 4 \cdot 3 = 4(5x + 3)$$

12. a. Complete each tree diagram to prime-factor 30.

$$
\begin{array}{ccc}
30 & 30 & 30 \\
\diagup\diagdown & \diagup\diagdown & \diagup\diagdown \\
2 \quad 15 & 3 \quad 10 & 5 \quad 6
\end{array}
$$

b. Complete the statement: The fundamental theorem of arithmetic states that every natural number greater than 1 has _____ prime factorization.

13. The prime factorizations of three monomials are shown here. Find their GCF.

$$3 \cdot 3 \cdot 5 \cdot x \cdot x$$
$$2 \cdot 3 \cdot 5 \cdot x \cdot y$$
$$2 \cdot 2 \cdot 3 \cdot x \cdot y \cdot y$$

14. Consider the polynomial $2k - 8 + hk - 4h$.
   a. How many terms does the polynomial have?
   b. Is there a common factor of all the terms?
   c. What is the common factor of the first two terms?

   d. What is the common factor of the last two terms?

15. How can we check the answer of the problem shown below?

   Factor $3j^3 + 6j^2 + 2j + 4$.

   $3j^3 + 6j^2 + 2j + 4 = 3j^2(j + 2) + 2(j + 2)$
   $= (j + 2)(3j^2 + 2)$

16. List the first 12 prime numbers.

NOTATION   *Complete each factorization.*

17. Factor $b^3 - 6b^2 + 2b - 12$.
   $b^3 - 6b^2 + 2b - 12 = (b - 6) + 2$
   $= (b - 6)$

18. Factor $12b^3 - 6b^2 + 2b - 2$.
   $12b^3 - 6b^2 + 2b - 2 = (6b^3 - 3b^2 + b - 1)$

19. In the expression $4x^2y + xy$, what is the coefficient of the last term?

20. Is the following statement true?
   $-(x^2 - 3x + 1) = -1(x^2 - 3x + 1)$

PRACTICE   *Find the prime factorization of each number.*

21. 12
22. 24
23. 15
24. 20
25. 40
26. 62
27. 98
28. 112
29. 225
30. 144
31. 288
32. 968

Complete each factorization.

33. $4a + 12 = (a + 3)$
34. $r^4 + r^2 = r^2(+ 1)$
35. $4y^2 + 8y - 2xy = 2y(2y + -)$
36. $3x^2 - 6xy + 9xy^2 = (- 2y + 3y^2)$

Factor out the GCF.

37. $3x + 6$
38. $2y - 10$
39. $12x^2 - 6x - 24$
40. $27a^2 - 9a + 45$
41. $t^3 + 2t^2$
42. $b^3 - 3b^2$

43. $a^3 - a^2$
44. $r^3 + r^2$
45. $24x^2y^3 + 8xy^2$
46. $3x^2y^3 - 9x^4y^3$
47. $12uvw^3 - 18uv^2w^2$
48. $14xyz - 16x^2y^2z$
49. $3x + 3y - 6z$
50. $2x - 4y + 8z$
51. $ab + ac - ad$
52. $rs - rt + ru$
53. $12r^2 - 3rs + 9r^2s^2$
54. $6a^2 - 12a^3b + 36ab$
55. $\pi R^2 - \pi ab$
56. $\frac{1}{3}\pi R^2h - \frac{1}{3}\pi rh$
57. $3(x + 2) - x(x + 2)$
58. $t(5 - s) + 4(5 - s)$
59. $h^2(14 + r) + 14 + r$
60. $k^2(14 + v) - 7(14 + v)$

Factor out -1 from each polynomial.

61. $-a - b$
62. $-x - 2y$
63. $-2x + 5y$
64. $-3x + 8z$
65. $-3m - 4n + 1$
66. $-3r + 2s - 3$
67. $-3ab - 5ac + 9bc$
68. $-6yz + 12xz - 5xy$

Factor each polynomial by factoring out the negative of the GCF.

69. $-3x^2 - 6x$
70. $-4a^2 + 6a$
71. $-4a^2b^3 + 12a^3b^2$
72. $-25x^4y^3 + 30x^2y^3$
73. $-4a^2b^2c^2 + 14a^2b^2c - 10ab^2c^2$

74. $-10x^4y^3z^2 + 8x^3y^2z - 20x^2y$

Factor by grouping.

75. $2x + 2y + ax + ay$
76. $bx + bz + 5x + 5z$
77. $7r + 7s - kr - ks$
78. $9p - 9q + mp - mq$
79. $xr + xs + yr + ys$
80. $pm - pn + qm - qn$
81. $2ax + 2bx + 3a + 3b$
82. $3xy + 3xz - 5y - 5z$
83. $2ab + 2ac + 3b + 3c$

84. $3ac + a + 3bc + b$

85. $6x^2 - 2x - 15x + 5$

86. $6x^2 + 2x + 9x + 3$

87. $9mp + 3mq - 3np - nq$

88. $ax + bx - a - b$

89. $2xy + y^2 - 2x - y$

90. $2xy - 3y^2 + 2x - 3y$

91. $8z^5 + 12z^2 - 10z^3 - 15$

92. $2a^4 + 2a^3 - 4a - 4$

Factor by grouping. Factor out the GCF first.

93. $ax^3 + bx^3 + 2ax^2y + 2bx^2y$

94. $x^3y^2 - 2x^2y^2 + 3xy^2 - 6y^2$

95. $4a^2b + 12a^2 - 8ab - 24a$

96. $-4abc - 4ac^2 + 2bc + 2c^2$

97. $x^3y - x^2y - xy^2 + y^2$

98. $2x^3z - 4x^2z + 32xz - 64z$

APPLICATIONS

99. PICTURE FRAMING   The dimensions of a family portrait and the frame in which it is mounted are given in Illustration 1. Write an algebraic expression that describes
  a. the area of the picture frame.
  b. the area of the portrait.
  c. the area of the mat used in the framing. Express the result in factored form.

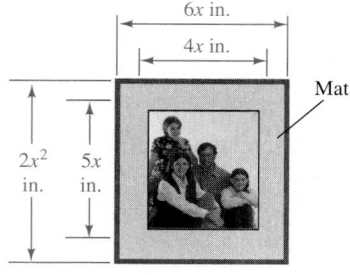

ILLUSTRATION 1

100. REARVIEW MIRRORS   The dimensions of the three rearview mirrors on an automobile are given in Illustration 2. Write an algebraic expression that gives
  a. the area of the rearview mirror mounted on the windshield.
  b. the total area of the two side mirrors.
  c. the total area of all three mirrors. Express the result in factored form.

101. COOKING   See Illustration 3.
  a. What is the length of a side of the square griddle, in terms of r? What is the area of the cooking surface of the griddle, in terms of r?

  b. How many square inches of the cooking surface do the pancakes cover, in terms of r?
  c. Find the amount of cooking surface that is not covered by the pancakes. Express the result in factored form.

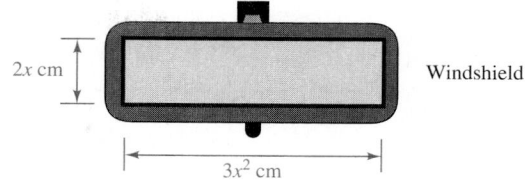

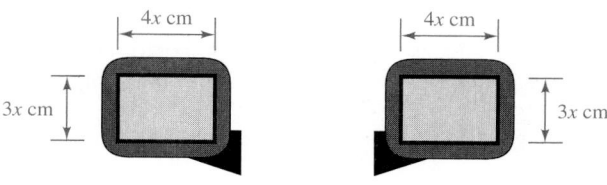

ILLUSTRATION 2

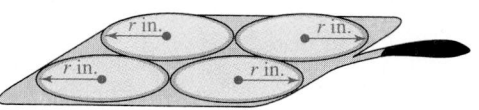

ILLUSTRATION 3

102. U.S. NAVY   Illustration 4 shows the deck of the aircraft carrier *Enterprise*. The rectangular-shaped landing area of $(x^3 + 4x^2 + 5x + 20)$ ft^2 is shaded. What are the length and width of the landing area? (*Hint:* Factor the expression that represents the area.)

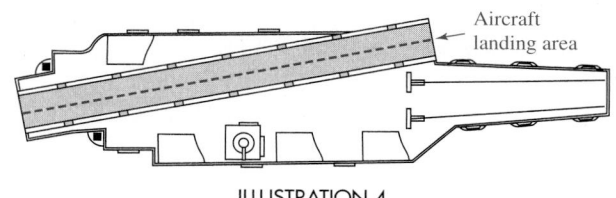

ILLUSTRATION 4

WRITING

103. To add $5x$ and $7x$, we combine like terms: $5x + 7x = 12x$. Explain how this is related to factoring out a common factor.

104. One student commented, "Factoring undoes the distributive property." What do you think she meant? Give an example.

105. If asked to write $ax + ay - bx - by$ in factored form, explain why $a(x + y) - b(x + y)$ is not an acceptable answer.

106. When asked to factor $rx - sy + ry - sx$, a student wrote the expression as $rx + ry - sx - sy$. Then she factored it by grouping. Can the terms be rearranged in this manner? Explain your answer.

REVIEW

107. Simplify $\left(\dfrac{y^3 y}{2yy^2}\right)^3$.

108. Find the slope of the line passing through the points $(3,5)$ and $(-2,-7)$.

109. Does the point $(3,5)$ lie on the graph of the line $4x - y = 7$?

110. Simplify $-5(3a - 2)(2a + 3)$.

11.2 *Factoring Trinomials of the Form $x^2 + bx + c$*

In this section, you will learn about

- Factoring trinomials that have a leading coefficient of 1 • Multistep factoring
- Prime polynomials

INTRODUCTION. Recall that to multiply $x + 2$ and $x + 3$, we proceed as follows:

$$(x + 2)(x + 3) = x^2 + 3x + 2x + 6$$
$$= x^2 + 5x + 6$$

In this section, we will reverse the process. Given a trinomial, such as $x^2 + 5x + 6$, we will ask ourselves, "What factors were multiplied to obtain $x^2 + 5x + 6$?" The process of finding the individual factors of a given trinomial is called *factoring the trinomial*. Since the product of two binomials is often a trinomial, we should not be surprised that many trinomials factor into the product of two binomials.

The multiplication process		The factoring process	
Given two binomial factors . . .	find the product	Given the product . . .	find the two binomial factors
$(x + 2)(x + 3)$	$=$?	$x^2 + 5x + 6$	$=$ (?)(?)

We will now consider how to factor trinomials of the form $ax^2 + bx + c$, where a (called the **leading coefficient**) is 1.

Factoring trinomials that have a leading coefficient of 1

To develop a method for factoring trinomials, we multiply $(x + a)$ and $(x + b)$.

$$(x + a)(x + b) = x \cdot x + bx + ax + ab \qquad \text{Use the FOIL method.}$$
$$= x^2 + ax + bx + ab \qquad \text{Write } x \cdot x \text{ as } x^2. \text{ Write } bx + ax \text{ as } ax + bx.$$
$$= x^2 + \underbrace{(a + b)x}_{} + \underbrace{ab}_{} \qquad \text{Factor } x \text{ out of } ax + bx.$$
$$\text{First term} \quad \text{Middle term} \quad \text{Last term}$$

The result has three terms. We can see that

- the first term is the product of x and x,
- the last term is the product of a and b, and
- the coefficient of the middle term is the sum of a and b.

We can use these facts to factor trinomials with leading coefficients of 1.

EXAMPLE 1   *A positive last term.*   Factor $x^2 + 5x + 6$.

Solution

Since the first term of the trinomial is x^2, the first term of each binomial factor must be x. To fill in the blanks, we must find two integers whose product is $+6$ and whose sum is $+5$.

$$x^2 + 5x + 6 = (x \quad\quad)(x \quad\quad)$$

The positive factorizations of 6 and the sum of the factors are shown in the following table.

Factors of 6	Sum of the factors of 6
1(6)	$1 + 6 = 7$
2(3)	$2 + 3 = 5$

The last row contains the integers $+2$ and $+3$, whose product is $+6$ and whose sum is $+5$. So, we can fill in the blanks with $+2$ and $+3$.

$$x^2 + 5x + 6 = (x + 2)(x + 3)$$

To check the result, we verify that $(x + 2)(x + 3)$ is $x^2 + 5x + 6$.

$$(x + 2)(x + 3) = x^2 + 3x + 2x + 6$$
$$= x^2 + 5x + 6$$

Self Check
Factor $y^2 + 7y + 6$.

Answer: $(y + 1)(y + 6)$ ∎

 COMMENT   When factoring trinomials, the binomial factors can be written in either order. In Example 1, an equivalent factorization is $x^2 + 5x + 6 = (x + 3)(x + 2)$.

EXAMPLE 2   *A positive last term.*   Factor $y^2 - 7y + 12$.

Solution

Since the first term of the trinomial is y^2, the first term of each binomial factor must be y. To fill in the blanks, we must find two integers whose product is $+12$ and whose sum is -7.

$$y^2 - 7y + 12 = (y \quad\quad)(y \quad\quad)$$

The two-integer factorizations of 12 and the sums of the factors are shown in the following table.

Factors of 12	Sum of the factors of 12
1(12)	$1 + 12 = 13$
2(6)	$2 + 6 = 8$
3(4)	$3 + 4 = 7$
$-1(-12)$	$-1 + (-12) = -13$
$-2(-6)$	$-2 + (-6) = -8$
$-3(-4)$	$-3 + (-4) = -7$

Self Check
Factor $p^2 - 5p + 6$.

The last row contains the integers -3 and -4, whose product is $+12$ and whose sum is -7. So, we can fill in the blanks with -3 and -4.

$$y^2 - 7y + 12 = (y - 3)(y - 4)$$

To check the result, we verify that $(y - 3)(y - 4)$ is $y^2 - 7y + 12$.

$$(y - 3)(y - 4) = y^2 - 4y - 3y + 12$$
$$= y^2 - 7y + 12$$

Answer: $(p - 3)(p - 2)$ ■

EXAMPLE 3   *A negative last term.*   Factor $a^2 + 2a - 15$.

Solution

Since the first term of the trinomial is a^2, the first term of each binomial factor must be a. To fill in the blanks, we must find two integers whose product is -15 and whose sum is $+2$.

$$a^2 + 2a - 15 = \left(a \quad\right)\left(a \quad\right)$$

The possible factorizations of -15 and the sum of the factors are shown in the following table.

Factors of -15	Sum of the factors of -15
$1(-15)$	$1 + (-15) = -14$
$3(-5)$	$3 + (-5) = -2$
$5(-3)$	$5 + (-3) = 2$
$15(-1)$	$15 + (-1) = 14$

The third row contains the integers $+5$ and -3, whose product is -15 and whose sum is $+2$. So, we can fill in the blanks with $+5$ and -3.

$$a^2 + 2a - 15 = (a + 5)(a - 3)$$

We can check by multiplying.

$$(a + 5)(a - 3) = a^2 - 3a + 5a - 15$$
$$= a^2 + 2a - 15$$

Self Check
Factor $p^2 + 3p - 18$.

Answer: $(p + 6)(p - 3)$ ■

EXAMPLE 4   *A negative last term.*   Factor $z^2 - 4z - 21$.

Solution

Since the first term of the trinomial is z^2, the first term of each binomial factor must be z. To fill in the blanks, we must find two integers whose product is -21 and whose sum is -4.

$$z^2 - 4y - 21 = \left(z \quad\right)\left(z \quad\right)$$

The factorizations of -21 and the sums of the factors are shown in the following table.

Factors of -21	Sum of the factors of -21
$1(-21)$	$1 + (-21) = -20$
$3(-7)$	$3 + (-7) = -4$
$7(-3)$	$7 + (-3) = 4$
$21(-1)$	$21 + (-1) = 20$

Self Check
Factor $q^2 - 2q - 24$.

The second row contains the integers $+3$ and -7, whose product is -21 and whose sum is -4. So, we can fill in the blanks with $+3$ and -7.

$$z^2 - 4z - 21 = (z + 3)(z - 7)$$

We can check by multiplying.

$$(z + 3)(z - 7) = z^2 - 7z + 3z - 21$$
$$= z^2 - 4z - 21$$

Answer: $(q + 4)(q - 6)$ ∎

The following sign patterns can be helpful when factoring trinomials.

Factoring $x^2 + bx + c$

To factor $x^2 + bx + c$, find two integers whose product is c and whose sum is b.

1. If c is positive, the integers have the same sign.
2. If c is negative, the integers have opposite signs.

When factoring out trinomials of the form $ax^2 + bx + c$, where $a = -1$, we begin by factoring out -1.

EXAMPLE 5 *Factoring out -1.* Factor $-h^2 + 2h + 15$.

Self Check
Factor $-x^2 + 11x - 18$.

Solution
We factor out -1 and then factor $h^2 - 2h - 15$.

$$-h^2 + 2h + 15 = -1(h^2 - 2h - 15) \quad \text{Factor out } -1.$$
$$= -(h^2 - 2h - 15)$$
$$= -(h - 5)(h + 3) \quad \begin{array}{l}\text{Use the integers } -5 \text{ and } 3, \text{ because their}\\ \text{product is } -15 \text{ and their sum is } -2.\end{array}$$

We can check by multiplying.

$$-(h - 5)(h + 3) = -(h^2 + 3h - 5h - 15) \quad \text{Multiply the binomials first.}$$
$$= -(h^2 - 2h - 15)$$
$$= -h^2 + 2h + 15$$

Answer: $-(x - 9)(x - 2)$ ∎

The trinomials in the next two examples are of a form similar to $x^2 + bx + c$, and we can use the methods of this section to factor them.

EXAMPLE 6 *Trinomials containing two variables.* Factor $x^2 - 4xy - 5y^2$.

SELF CHECK
Factor $s^2 + 6st - 7t^2$.

Solution
The trinomial has two variables, x and y. Since the first term is x^2, the first term of each factor must be x.

$$x^2 - 4xy - 5y^2 = \left(x \qquad \right)\left(x \qquad \right)$$

To fill in the blanks, we must find two *expressions* whose product is the last term, $-5y^2$, and that will give a middle term of $-4xy$. Two such expressions are $-5y$ and y.

$$x^2 - 4xy - 5y^2 = (x - 5y)(x + y)$$

We can check by multiplying.

$$(x - 5y)(x + y) = x^2 + xy - 5xy - 5y^2$$
$$= x^2 - 4xy - 5y^2$$

Answer: $(s + 7t)(s - t)$ ∎

Multistep factoring

If the terms of a trinomial have a common factor, the GCF should always be factored out before any of the factoring techniques of this section are used. A trinomial is **factored completely** when no factor can be factored further. Always factor completely when you are asked to factor.

EXAMPLE 7 *Factoring completely.* Factor $2x^4 + 26x^3 + 80x^2$.

Solution

We begin by factoring out the GCF of $2x^2$.

$$2x^4 + 26x^3 + 80x^2 = 2x^2(x^2 + 13x + 40)$$

Next, we factor $x^2 + 13x + 40$. The integers 8 and 5 have a product of 40 and a sum of 13, so the completely factored form of the given trinomial is

$$2x^4 + 26x^3 + 80x^2 = 2x^2(x + 8)(x + 5)$$

Check by multiplying $2x^2$, $x + 8$, and $x + 5$.

Self Check

Factor $4m^5 + 8m^4 - 32m^3$ completely.

Answer: $4m^3(m + 4)(m - 2)$ ∎

EXAMPLE 8 *Writing terms in descending powers.* Factor $-13g^2 + 36g + g^3$ completely.

Solution

Before factoring the trinomial, we write its terms in descending powers of g.

$$\begin{aligned}
-13g^2 + 36g + g^3 &= g^3 - 13g^2 + 36g & \text{Rearrange the terms.} \\
&= g(g^2 - 13g + 36) & \text{Factor out } g, \text{ which is the GCF.} \\
&= g(g - 9)(g - 4) & \text{Factor the trinomial.}
\end{aligned}$$

Check by multiplying g, $g - 9$, and $g - 4$.

Self Check

Factor $-12t + t^3 + 4t^2$ completely.

Answer: $t(t - 2)(t + 6)$ ∎

Prime polynomials

If a trinomial cannot be factored using only integers, it is called a **prime polynomial,** or more specifically, a **prime trinomial.**

EXAMPLE 9 *Trinomials that do not factor.* Factor $x^2 + 2x + 3$, if possible.

Solution

To factor the trinomial, we must find two integers whose product is 3 and whose sum is 2. The possible factorizations of 3 and the sums of the factors are shown in the following table.

Factors of 3	Sum of the factors of 3
1(3)	$1 + 3 = 4$
$-1(-3)$	$-1 + (-3) = -4$

Since two integers whose product is 3 and whose sum is 2 do not exist, $x^2 + 2x + 3$ cannot be factored. It is a prime trinomial.

Self Check

Factor $x^2 - 4x + 6$, if possible.

Answer: not possible; prime trinomial ∎

STUDY SET Section 11.2

VOCABULARY *Fill in the blanks.*

1. A polynomial, such as $x^2 - x - 6$, that has exactly three terms is called a _____. A polynomial, such as $x - 3$, that has exactly two terms is called a _____.

2. The statement $x^2 - x - 12 = (x - 4)(x + 3)$ shows that $x^2 - x - 12$ _____ into the product of two binomials.

3. Since $10 = (-5)(-2)$, we say -5 and -2 are _____ of 10.

4. A _____ polynomial cannot be factored by using only integers.

5. The _____ coefficient of the trinomial $x^2 - 3x + 2$ is 1, the _____ of the middle term is -3, and the last _____ is 2.

6. A trinomial is factored _____ when no factor can be factored further.

CONCEPTS *Fill in the blanks.*

7. Two factorizations of 4 that involve only positive numbers are _____ and _____. Two factorizations of 4 that involve only negative numbers are _____ and _____.

8. Before attempting to factor a trinomial, be sure that the exponents are written in _____ order.

9. Before attempting to factor a trinomial into two binomials, always factor out any _____ factors first.

10. To factor $x^2 + x - 56$, we must find two integers whose _____ is -56 and whose _____ is 1.

11. Two factors of 18 whose sum is -9 are _____ and _____.

12. $x^2 + 5x + 3$ cannot be factored because we cannot find two integers whose product is _____ and whose sum is _____.

13. Complete the table.

Factors of 8	Sum of the factors of 8
1(8)	
2(4)	
-1(-8)	
-2(-4)	

14. If we use the FOIL method to do the multiplication $(x + 5)(x + 4)$, we obtain $x^2 + 9x + 20$.
    a. What step of the FOIL process produced 20?

b. What steps of the FOIL process produced $9x$?

15. Given $x^2 - 2x - 15$:
    a. What is the coefficient of the x^2-term?
    b. What is the last term? The last term is the product of what two integers?
    c. What is the coefficient of the middle term? It is the sum of what two integers?

16. Given $x^2 + 8x + 15$:
    a. What is the coefficient of the x^2-term?
    b. What is the last term? The last term is the product of what two integers?
    c. What is the coefficient of the middle term? It is the sum of what two integers?

17. To determine which two integers to use in the factorization of $x^2 + 7x + 10$, a student constructed the following table. Explain why she didn't need to write the last two rows.

Factors of 10	Sum of the factors of 10
1(10)	
2(5)	
-1(-10)	
-2(-5)	

18. Complete the factorization table.

Factors of -9	Sum of the factors of -9

19. Consider factoring a trinomial of the form $x^2 + bx + c$.
    a. If c is positive, what can be said about the two integers that should be chosen for the factorization?

    b. If c is negative, what can be said about the two integers that should be chosen for the factorization?

20. What trinomial has the factorization of $(x + 8)(x - 2)$?

NOTATION Complete each factorization.

21. $6 + 5x + x^2 = x^2 + \boxed{} + 6$

$ = (x + 3)(x + 2)$

22. $-a^2 - a + 20 = \boxed{}(a^2 + a - 20)$

$ = -(a + 5)(a - 4)$

PRACTICE Complete each factorization.

23. $x^2 + 3x + 2 = \left(x \boxed{} 2\right)\left(x \boxed{} 1\right)$

24. $y^2 + 4y + 3 = \left(y \boxed{} 3\right)\left(y \boxed{} 1\right)$

25. $t^2 - 9t + 14 = \left(t \boxed{} 7\right)\left(t \boxed{} 2\right)$

26. $c^2 - 9c + 8 = \left(c \boxed{} 8\right)\left(c \boxed{} 1\right)$

27. $a^2 + 6a - 16 = \left(a \boxed{} 8\right)\left(a \boxed{} 2\right)$

28. $x^2 - 3x - 40 = \left(x \boxed{} 8\right)\left(x \boxed{} 5\right)$

Factor each trinomial. If it can't be factored, write "prime."

29. $z^2 + 12z + 11$

30. $x^2 + 7x + 10$

31. $m^2 - 5m + 6$

32. $n^2 - 7n + 10$

33. $a^2 - 4a - 5$

34. $b^2 + 6b - 7$

35. $x^2 + 5x - 24$

36. $t^2 - 5t - 50$

37. $a^2 - 10a - 39$

38. $r^2 - 9r - 12$

39. $u^2 + 10u + 15$

40. $v^2 + 9v + 15$

41. $s^2 + 11s - 26$

42. $y^2 + 8y + 12$

43. $r^2 - 2r + 4$

44. $m^2 + 3m - 10$

45. $m^2 - m - 12$

46. $u^2 + u - 42$

47. $x^2 + 4xy + 4y^2$

48. $a^2 + 10ab + 9b^2$

49. $m^2 + 3mn - 10n^2$

50. $m^2 - mn - 12n^2$

51. $a^2 - 4ab - 12b^2$

52. $p^2 + pq - 6q^2$

53. $r^2 - 2rs + 4s^2$

54. $m^2 + 3mn - 20n^2$

Factor each trinomial. Factor out -1 first.

55. $-x^2 - 7x - 10$

56. $-x^2 + 9x - 20$

57. $-t^2 - 15t + 34$

58. $-t^2 - t + 30$

59. $-r^2 + 14r - 40$

60. $-r^2 + 14r - 45$

61. $-a^2 - 4ab - 3b^2$

62. $-a^2 - 6ab - 5b^2$

63. $-x^2 + 6xy + 7y^2$

64. $-x^2 - 10xy + 11y^2$

Write each trinomial in descending powers of one variable and then factor.

65. $4 - 5x + x^2$

66. $y^2 + 5 + 6y$

67. $10y + 9 + y^2$

68. $x^2 - 13 - 12x$

69. $-r^2 + 2 + r$

70. $u^2 - 3 + 2u$

71. $4rx + r^2 + 3x^2$

72. $a^2 + 5b^2 + 6ab$

73. $-3ab + a^2 + 2b^2$

74. $-13yz + y^2 - 14z^2$

Completely factor each trinomial. Factor out any common monomials first (including -1 if necessary).

75. $2x^2 + 10x + 12$

76. $3y^2 - 21y + 18$

77. $-5a^2 + 25a - 30$

78. $-2b^2 + 20b - 18$

79. $3z^2 - 15z + 12$

80. $5m^2 + 45m - 50$

81. $12xy + 4x^2y - 72y$

82. $48xy + 6xy^2 + 96x$

83. $-4x^2y - 4x^3 + 24xy^2$

84. $3x^2y^3 + 3x^3y^2 - 6xy^4$

APPLICATIONS

85. PETS   The cage shown in Illustration 1 is used for transporting dogs. Its volume is $(x^3 + 12x^2 + 27x)$ in.^{3}. The dimensions of the cage can be found by factoring this expression. If the cage is longer than it is tall, and taller than it is wide, determine its length, width, and height.

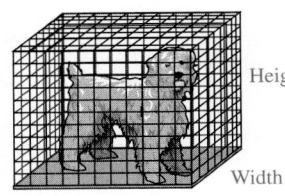

ILLUSTRATION 1

86. CARPOOLING   The average rate at which a carpool van travels and the distance it covers are given in the table in terms of t. Factor the expression representing the distance traveled and then complete the table.

Rate (mi/hr)	Time (hr)	Distance traveled (mi)
$t + 11$		$t^2 + 16t + 55$

WRITING

87. Explain what it means when we say that a trinomial is the product of two binomials. Give an example.

88. Are $2x^2 - 12x + 16$ and $x^2 - 6x + 8$ factored in the same way? Explain why or why not.

89. When factoring $x^2 - 2x - 3$, one student got $(x - 3)(x + 1)$, and another got $(x + 1)(x - 3)$. Are both answers acceptable? Explain.

90. Explain how to use the FOIL method to check the factorization of a trinomial.

91. In the partial solution shown below, a student began to factor the trinomial and then gave up. Write a brief note to the student explaining his initial mistake.

Factor $x^2 - 2x - 63$.

$$(x - \quad)(x - \quad)$$
$$???$$

92. Explain why the given trinomial is not factored completely.

$$3x^2 - 3x - 60 = 3(x^2 - x - 20)$$

REVIEW

Graph the solution of each inequality on a number line.

93. $x - 3 > 5$

94. $x + 4 \leq 3$

95. $-3x - 5 \geq 4$

96. $2x - 3 < 7$

11.3   *Factoring Trinomials of the Form $ax^2 + bx + c$*

In this section, you will learn about

- Observations about multiplying binomials
- The trial-and-check factoring method   • The grouping method

INTRODUCTION.   In this section, we will factor trinomials with leading coefficients that are not 1. Two methods are used to factor these trinomials. With the first method, educated guesses are made. These guesses are checked by multiplication. The correct factorization is determined by a process of elimination. The second method is an extension of factoring by grouping.

Observations about multiplying binomials

In the work below, we find the products $(2x + 1)(x + 3)$ and $(2x + 3)(x + 1)$. There are several observations that can be made when we compare the results.

$$(2x + 1)(x + 3) = 2x^2 + 6x + x + 3 \qquad (2x + 3)(x + 1) = 2x^2 + 2x + 3x + 3$$
$$= 2x^2 + 7x + 3 \qquad\qquad\qquad = 2x^2 + 5x + 3$$

In each case, the result is a trinomial, and

- the first terms are the same ($2x^2$),
- the last terms are the same (3), and
- the middle terms are different ($7x$ and $5x$).

These observations indicate that when the last terms in $(2x + 1)(x + 3)$ are interchanged to form $(2x + 3)(x + 1)$, only the middle terms of the products are different. This fact is helpful when factoring trinomials by using the *trial-and check method.*

The trial-and-check factoring method

To factor a trinomial with a leading coefficient of 1 (say, $x^2 + 4x + 3$), we begin with a factorization of the form

$$x^2 + 4x + 3 = (x \quad)(x \quad)$$

and determine which integers to write in the blanks.

To factor a trinomial with a leading coefficient that is not 1 (say, $2x^2 + 5x + 3$), we begin with a factorization of the form

$$2x^2 + 5x + 3 = (\quad x \quad)(\quad x \quad)$$

We must determine what numbers to write in the blanks. Because there are four blanks, there are more combinations of factors to consider.

EXAMPLE 1   *A leading coefficient of 2.*   Factor $2x^2 + 5x + 3$.

Solution

Since the first term is $2x^2$, the first terms of the binomial factors must be $2x$ and x. To fill in the blanks, we must find two factors of $+3$ that will give a middle term of $+5x$.

$$(2x \quad)(x \quad)$$

Because each term of the trinomial is positive, we need only consider positive factors of the last term. Since the positive factors of 3 are 1 and 3, there are two possible factorizations.

$$(2x + 1)(x + 3) \qquad \text{or} \qquad (2x + 3)(x + 1)$$

The first possibility is incorrect: When we find the outer and inner products and combine like terms, we obtain an incorrect middle term of $7x$.

> Outer: $6x$
>
> $(2x + 1)(x + 3)$ Multiply and add to find the middle term: $6x + x = 7x$.
>
> Inner: x

The second possibility is correct, because it gives a middle term of $5x$.

> Outer: $2x$
>
> $(2x + 3)(x + 1)$ Multiply and add to find the middle term: $2x + 3x = 5x$.
>
> Inner: $3x$

Thus,

$$2x^2 + 5x + 3 = (2x + 3)(x + 1)$$

EXAMPLE 2   *A leading coefficient of 6.*   Factor $6x^2 - 17x + 5$.

Solution

Since the first term is $6x^2$, the first terms of the binomial factors must be $6x$ and x or $3x$ and $2x$. To fill in the blanks, we must find two factors of $+5$ that will give a middle term of $-17x$.

$$(6x \quad)(x \quad) \qquad \text{or} \qquad (3x \quad)(2x \quad)$$

Because the sign of the last term is positive and the sign of the middle term is negative, we need only consider negative factors of the last term. Since the negative factors of 5 are -1 and -5, there are four possible factorizations.

Self Check

Factor $3x^2 + 7x + 2$.

Answer: $(3x + 1)(x + 2)$ ■

Self Check

Factor $6x^2 - 7x + 2$.

$$\overset{-30a}{(6a - 1)(a - 5)} \quad -30a - a = -31a. \qquad \overset{-6a}{(6a - 5)(a - 1)} \quad -6a - 5a = -11a.$$

$$\overset{-15a}{(3a - 1)(2a - 5)} \quad -15a - 2a = -17a. \qquad \overset{-3a}{(3a - 5)(2a - 1)} \quad -3a - 10a = -13a.$$

Only the possibility shown in blue gives the correct middle term of $-17a$. Thus,

$$6a^2 - 17a + 5 = (3a - 1)(2a - 5)$$

EXAMPLE 3   *Discarding possible factorizations.*   Factor $3y^2 - 7y - 6$.

Solution

Since the first term is $3y^2$, the first terms of the binomial factors must be $3y$ and y.

$$\left(3y \quad \right)\left(y \quad \right)$$

To fill in the blanks, we must find two integers whose product is -6 that produce a middle term of $-7y$. Since the sign of the last term of $3y^2 - 7y - 6$ is negative, we need to find factors of -6 that have opposite signs. There are four such pairs: $-1(6)$, $1(-6)$, $-2(3)$, and $2(-3)$. These four pairs create eight possible factorizations to consider.

Four of the possible factorizations can be discarded because they include a binomial whose terms have a common factor. If $3y^2 - 7y - 6$ does not have a common factor, neither will any of its binomial factors.

For the factors -1 and 6:

$$\overset{18y}{(3y - 1)(y + 6)} \quad 18y - y = 17y \qquad \cancel{(3y + 6)(y - 1)}$$

A common factor of 3

For the factors 1 and -6:

$$\overset{-18y}{(3y + 1)(y - 6)} \quad -18y + y = -17y \qquad \cancel{(3y - 6)(y + 1)}$$

A common factor of 3

For the factors -2 and 3:

$$\overset{9y}{(3y - 2)(y + 3)} \quad 9y - 2y = 7y \qquad \cancel{(3y + 3)(y - 2)}$$

A common factor of 3

For the factors 2 and -3:

$$\overset{-9y}{(3y + 2)(y - 3)} \quad -9y + 2y = -7y \qquad \cancel{(3y - 3)(y + 2)}$$

A common factor of 3

Only the possibility shown in blue gives the correct middle term of $-7y$. Thus,

$$3y^2 - 7y - 6 = (3y + 2)(y - 3)$$

Check the factorization by multiplication.

Answer: $(3x - 2)(2x - 1)$ ■

Self Check

Factor $5a^2 - 23a - 10$.

Answer: $(5a + 2)(a - 5)$ ■

> **COMMENT** If a trinomial does not have a common factor, the terms of each of its binomial factors will not have a common factor.

EXAMPLE 4 *Factoring a trinomial in two variables.* Factor $4b^2 + 8bc - 45c^2$.

Solution

Since the first term is $4b^2$, the first terms of the factors must be $4b$ and b or $2b$ and $2b$.

$$(4b \quad)(b \quad) \qquad \text{or} \qquad (2b \quad)(2b \quad)$$

To fill in the blanks, we must find two factors of $-45c^2$ that will give a middle term of $8bc$.

Since $-45c^2$ has many factors, there are many possible combinations for the last terms of the binomial factors. The signs of the factors must be different, because the last term of the trinomial is negative.

If we pick factors of $4b$ and b for the first terms, and $-c$ and $45c$ for the last terms, the multiplication gives an incorrect middle term of $179bc$. So the factorization is incorrect.

$$\overset{\displaystyle 180bc}{(4b - c)(b + 45c)} \quad 180bc - bc = 179bc.$$
$$\underset{\displaystyle -bc}{}$$

If we pick factors of $4b$ and b for the first terms and $15c$ and $-3c$ for the last terms, the multiplication gives an incorrect middle term of $3bc$.

$$\overset{\displaystyle -12bc}{(4b + 15c)(b - 3c)} \quad -12bc + 15bc = 3bc.$$
$$\underset{\displaystyle 15bc}{}$$

If we pick factors of $2b$ and $2b$ for the first terms and $-5c$ and $9c$ for the last terms, we have

$$\overset{\displaystyle 18bc}{(2b - 5c)(2b + 9c)} \quad 18bc - 10bc = 8bc.$$
$$\underset{\displaystyle -10bc}{}$$

which gives the correct middle term of $8bc$. Thus,

$$4b^2 + 8bc - 45c^2 = (2b - 5c)(2b + 9c)$$

Check by multiplication.

Self Check
Factor $4x^2 + 4xy - 3y^2$.

Answer: $(2x + 3y)(2x - y)$ ∎

Because some guesswork is often necessary, it is difficult to give specific rules for factoring trinomials with a leading coefficient that is not 1. However, the following hints are helpful.

Factoring $ax^2 + bx + c$ **($a \neq 1$)**	1. Write the trinomial in descending powers of the variable and factor out any GCF (including -1 if that is necessary to make the leading coefficient positive). 2. Attempt to write the trinomial as *the product of two binomials.* The coefficients of the first terms of each binomial factor must be factors of a, and the last terms must be factors of c.

$$\overbrace{}^{\text{Factors of } a}$$
$$(\quad x + \quad)(\quad x + \quad)$$
$$\underbrace{}_{\substack{\text{Factors} \\ \text{of } c}}$$

3. If the sign of the last term of the trinomial is positive, the signs between the terms of the binomial factors are the same as the sign of the middle term. If the sign of the last term is negative, the signs between the terms of the binomial factors are opposite.

4. Try combinations of coefficients of the first terms and last terms until you find one that gives the middle term of the trinomial. If no combination works, the trinomial is prime.

5. Check the factorization by multiplication.

EXAMPLE 5 *Writing terms in descending powers.* Factor $2x^2 - 8x^3 + 3x$.

Solution

We write the trinomial in descending powers of x

$$-8x^3 + 2x^2 + 3x$$

and we factor out the negative of the GCF, which is $-x$.

$$-8x^3 + 2x^2 + 3x = -x(8x^2 - 2x - 3)$$

We must now factor $8x^2 - 2x - 3$. Its factorization has the form

$$\left(x \quad\right)\left(8x \quad\right) \qquad \text{or} \qquad \left(2x \quad\right)\left(4x \quad\right)$$

To fill in the blanks, we find two factors of the last term of the trinomial (-3) that will give a middle term of $-2x$. Because the sign of the last term is negative, the signs within its binomial factors will be different. If we pick factors of $2x$ and $4x$ for the first terms and 1 and -3 for the last terms, we have

$$\overset{-6x}{\overbrace{(2x + 1)(4x - 3)}} \quad -6x + 4x = -2x.$$
$$\underset{4x}{\underbrace{}}$$

which gives the correct middle term of $-2x$, so it is correct.

$$8x^2 - 2x - 3 = (2x + 1)(4x - 3)$$

We can now give the complete factorization.

$$-8x^3 + 2x^2 + 3x = -x(8x^2 - 2x - 3)$$
$$= -x(2x + 1)(4x - 3)$$

Check by multiplication.

Self Check

Factor $12y - 2y^3 - 2y^2$.

Answer: $-2y(y + 3)(y - 2)$

The grouping method

The method of factoring by grouping can be used to help factor trinomials of the form $ax^2 + bx + c$. For example, to factor $2x^2 + 5x + 3$, we proceed as follows.

1. We find the product ac: In $2x^2 + 5x + 3$, $a = 2$, $b = 5$, and $c = 3$, so $ac = 2(3) = 6$. This number is called the **key number.**

2. Find two factors of the key number 6 whose sum is $b = 5$. Two such numbers are 2 and 3.

$$2(3) = 6 \quad \text{and} \quad 2 + 3 = 5$$

3. Use the factors 2 and 3 as coefficients of two terms to be placed between $2x^2$ and 3:

$$2x^2 + 5x + 3 = 2x^2 + 2x + 3x + 3 \quad \text{Express } 5x \text{ as } 2x + 3x.$$

4. Factor by grouping:

$$2x^2 + 2x + 3x + 3 = 2x(x + 1) + 3(x + 1) \quad \text{Factor } 2x \text{ out of } 2x^2 + 2x$$
$$\text{and } 3 \text{ out of } 3x + 3.$$

$$= (x + 1)(2x + 3) \quad \text{Factor out } x + 1.$$

So $2x^2 + 5x + 3 = (x + 1)(2x + 3)$. Verify this factorization by multiplication.

EXAMPLE 6 *The grouping method.* Factor $10x^2 + 13x - 3$.

Solution
Since $a = 10$ and $c = -3$ in the trinomial, $ac = -30$. We now find two factors of -30 whose sum is 13. Two such factors are 15 and -2. We use these factors as coefficients of two terms to be placed between $10x^2$ and -3.

$$10x^2 + 13x - 3 = 10x^2 + 15x - 2x - 3 \quad \text{Express } 13x \text{ as } 15x - 2x.$$

Finally, we factor by grouping.

$$10x^2 + 15x - 2x - 3 = 5x(2x + 3) - 1(2x + 3)$$
$$= (2x + 3)(5x - 1)$$

So $10x^2 + 13x - 3 = (2x + 3)(5x - 1)$. Check the result.

Self Check
Factor $15a^2 + 17a - 4$.

Answer: $(3a + 4)(5a - 1)$ ■

Factoring $ax^2 + bx + c$ by grouping

1. Write the trinomial in descending powers of the variable and factor out any GCF (including -1 if that is necessary to make the leading coefficient positive).
2. Calculate the key number ac.
3. Find two numbers whose product is the key number found in step 2 and whose sum is the coefficient of the middle term of the trinomial.
4. Write the numbers in the blanks of the form shown below, and then factor the polynomial by grouping.
$$ax^2 + \quad x + \quad x + c$$
5. Check the factorization using multiplication.

EXAMPLE 7 *Factoring by grouping.* Factor $12x^5 - 17x^4 + 6x^3$.

Solution
First, we factor out the GCF, which is x^3.

$$12x^5 - 17x^4 + 6x^3 = x^3(12x^2 - 17x + 6)$$

To factor $12x^2 - 17x + 6$, we need to find two integers whose product is $12(6) = 72$ and whose sum is -17. Two such numbers are -8 and -9.

$$12x^2 - 17x + 6 = 12x^2 - 8x - 9x + 6 \quad \text{Express } -17x \text{ as } -8x - 9x.$$
$$= 4x(3x - 2) - 3(3x - 2) \quad \text{Factor out } 4x \text{ and factor out } -3.$$
$$= (3x - 2)(4x - 3) \quad \text{Factor out } 3x - 2.$$

The complete factorization is

$$12x^5 - 17x^4 + 6x^3 = x^3(3x - 2)(4x - 3)$$

Check the result.

Self Check
Factor $21a^4 - 13a^3 + 2a^2$.

Answer: $a^2(7a - 2)(3a - 1)$ ■

STUDY SET Section 11.3

VOCABULARY *Fill in the blanks.*

1. The trinomial $3x^2 - x - 12$ has a _____ coefficient of 3. The _____ term is -12.

2. The numbers 3 and 2 are _____ of the first term of the trinomial $6x^2 + x - 12$.

3. Consider $(x - 2)(5x - 1)$. The product of the _____ terms is $-x$ and the product of the _____ terms is $-10x$.

4. When we write $2x^2 + 7x + 3$ as $(2x + 1)(x + 3)$, we say that we have _____ the trinomial—it has been expressed as the product of two _____.

5. The _____ term of $4x^2 - 7x + 13$ is $-7x$.

6. The polynomial $6x^2 + 2x + 9x + 3$ has four _____.

7. The _____ of the middle terms of the polynomial $4a^2 - 12a - a + 3$ is $-13a$.

8. The _____ of the terms of the trinomial $6b^3 - 3b^2 - 12b$ is $3b$.

CONCEPTS *Complete each statement in red.*

9.

These coefficients must be factors of ____.

$$5x^2 + 6x - 8 = (\boxed{}\,x + \boxed{})(\boxed{}\,x + \boxed{})$$

These numbers must be factors of ____.

10.

The product of these coefficients must be ____.

$$3x^2 + 16x + 5 = 3x^2 + \boxed{}\,x + \boxed{}\,x + 5$$

The sum of these coefficients must be ____.

A trinomial has been partially factored. Complete each statement that describes the type of integers we should consider for the blanks.

11. $5y^2 - 13y + 6 = (5x\ \boxed{})(x\ \boxed{})$

Since the last term of the trinomial is _____ and the middle term is _____, the integers must be _____ factors of 6.

12. $5y^2 + 13y + 6 = (5x\ \boxed{})(x\ \boxed{})$

Since the last term of the trinomial is _____ and the middle term is _____, the integers must be _____ factors of 6.

13. $5y^2 + 7y - 6 = (5x\ \boxed{})(x\ \boxed{})$

Since the last term of the trinomial is _____, the signs of the integers will be _____.

14. $5y^2 - 7y - 6 = (5x\ \boxed{})(x\ \boxed{})$

Since the last term of the trinomial is _____, the signs of the integers will be _____.

A trinomial is to be factored by the grouping method. Complete each statement that describes the type of integers we should consider for the blanks.

15. $8c^2 - 11c + 3 = 8c^2 + \boxed{}\,c + \boxed{}\,c + 3$

We need to find two integers whose product is ____ and whose sum is ____.

16. $15c^2 + 4c - 4 = 15c^2 + \boxed{}\,c + \boxed{}\,c - 4$

We need to find two integers whose product is ____ and whose sum is ____.

NOTATION

17. Write a trinomial of the form $ax^2 + bx + c$
  a. where $a = 1$
  b. where $a \neq 1$

18. Write the terms of the trinomial $40 - t - 4t^2$ in descending powers of the variable.

PRACTICE *Complete each factorization.*

19. $3a^2 + 13a + 4 = (3a\ \boxed{}\,1)(a\ \boxed{}\,4)$

20. $2b^2 + 7b + 6 = (2b\ \boxed{}\,3)(b\ \boxed{}\,2)$

21. $4z^2 - 13z + 3 = (z\ \boxed{}\,3)(4z\ \boxed{}\,1)$

22. $4t^2 - 4t + 1 = (2t\ \boxed{}\,1)(2t\ \boxed{}\,1)$

23. $2m^2 + 5m - 12 = (2m\ \boxed{}\,3)(m\ \boxed{}\,4)$

24. $10u^2 - 13u - 3 = (2u\ \boxed{}\,3)(5u\ \boxed{}\,1)$

Complete each step of the factorization of the trinomial by grouping.

25. $12t^2 + 17t + 6 = 12t^2 + \boxed{}\,t + \boxed{}\,t + 6$
$\qquad = \boxed{}(4t + 3) + \boxed{}(4t + 3)$
$\qquad = (\boxed{})(3t + 2)$

26. $35t^2 - 11t - 6 = 35t^2 + \boxed{}\,t - 21t - 6$
$\qquad = 5t(7t + 2)\ \boxed{}\,3(7t\ \boxed{}\,2)$
$\qquad = (\boxed{})(5t - 3)$

Factor each trinomial, if possible.

27. $2x^2 - 3x + 1$

28. $2y^2 - 7y + 3$

29. $3a^2 + 13a + 4$

30. $2b^2 + 7b + 6$

31. $4z^2 + 13z + 3$

32. $4t^2 - 4t + 1$

33. $6y^2 + 7y + 2$

34. $4x^2 + 8x + 3$

35. $6x^2 - 7x + 2$

36. $4z^2 - 9z + 2$

37. $3a^2 - 4a - 4$

38. $8u^2 - 2u - 15$

39. $2x^2 - 3x - 2$

40. $12y^2 - y - 1$

41. $2m^2 + 5m - 10$

42. $10u^2 - 13u - 6$

43. $10y^2 - 3y - 1$

44. $6m^2 + 19m + 3$

45. $12y^2 - 5y - 2$

46. $10x^2 + 21x - 10$

47. $-5t^2 - 13t - 6$

48. $-16y^2 - 10y - 1$

49. $-16m^2 + 14m - 3$

50. $-16x^2 - 16x - 3$

51. $4a^2 - 4ab + b^2$

52. $2b^2 - 5bc + 2c^2$

53. $6r^2 + rs - 2s^2$

54. $3m^2 + 5mn + 2n^2$

55. $4x^2 + 8xy + 3y^2$

56. $4b^2 + 15bc - 4c^2$

57. $4a^2 - 15ab + 9b^2$

58. $12x^2 + 5xy - 3y^2$

59. $-13x + 3x^2 - 10$

60. $-14 + 3a^2 - a$

61. $15 + 8a^2 - 26a$

62. $16 - 40a + 25a^2$

63. $12y^2 + 12 - 25y$

64. $12t^2 - 1 - 4t$

65. $3x^2 + 6 + x$

66. $25 + 2u^2 + 3u$

67. $2a^2 + 3b^2 + 5ab$

68. $11uv + 3u^2 + 6v^2$

69. $pq + 6p^2 - q^2$

70. $-11mn + 12m^2 + 2n^2$

71. $4x^2 + 10x - 6$

72. $9x^2 + 21x - 18$

73. $-y^3 - 13y^2 - 12y$

74. $-2xy^2 - 8xy + 24x$

75. $6x^3 - 15x^2 - 9x$

76. $9y^3 + 3y^2 - 6y$

77. $30r^5 + 63r^4 - 30r^3$

78. $6s^5 - 26s^4 - 20s^3$

79. $-16m^3n - 20m^2n^2 - 6mn^3$

80. $-84x^4 - 100x^3y - 24x^2y^2$

81. $-28u^3v^3 + 26u^2v^4 - 6uv^5$

82. $-16x^4y^3 + 30x^3y^4 + 4x^2y^5$

APPLICATIONS

83. OFFICE FURNITURE   The area of the desktop shown in Illustration 1 is given by the expression $(4x^2 + 20x - 11)$ in.^{2}. Factor this expression to find the expressions that represent its length and width. Then determine the difference in the length and width of the desktop.

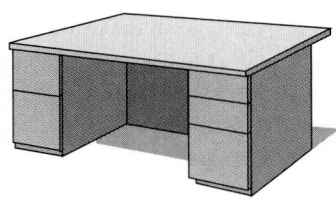

ILLUSTRATION 1

84. STORAGE   The volume of the 8-foot-wide portable storage container shown in Illustration 2 is given by the expression $(72x^2 + 120x - 400)$ ft^3. If its dimensions can be determined by factoring the expression, find the height and the length of the container.

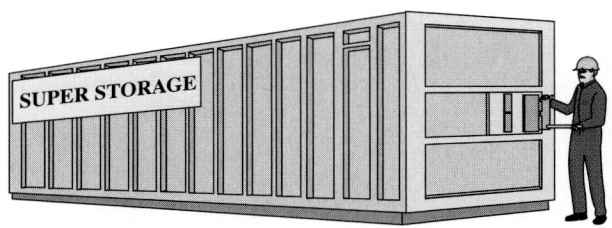

ILLUSTRATION 2

WRITING

85. In the work below, a student began to factor the trinomial and then gave up. Explain his initial mistake.

Factor $3x^2 - 5x - 2$.

$$(3x - \ \ )(x - \ \ )$$

$$???$$

86. Two students factor $2x^2 + 20x + 42$ and get two different answers:

$$(2x + 6)(x + 7) \quad \text{and} \quad (x + 3)(2x + 14)$$

Do both answers check? Why don't they agree? Is either answer completely correct? Explain.

87. Why is the process of factoring $6x^2 - 5x - 6$ more complicated than the process of factoring $x^2 - 5x - 6$?

88. How can the factorization shown below be checked?

$$6x^2 - 5x - 6 = (3x + 2)(2x - 3)$$

REVIEW

89. Simplify $(x^2x^5)^2$.

90. Simplify $\dfrac{(a^3b^4)^2}{ab^5}$.

91. Evaluate $\dfrac{1}{2^{-3}}$.

92. Evaluate 7^0.

11.4 *Special Factorizations and a Factoring Strategy*

In this section, you will learn about

- Factoring perfect square trinomials • Factoring the difference of two squares
- Multistep factoring • Factoring the sum and difference of two cubes
- A factoring strategy

INTRODUCTION. We have already studied several methods that can be used to factor trinomials. In this section, we will introduce another method that can be used to factor two specific types of trinomials, called *perfect square trinomials*. We will also develop techniques for factoring three specific types of binomials, called the *difference of two squares,* and the *sum* and *difference of two cubes.*

Factoring perfect square trinomials

We have seen that the squares of binomials are trinomials.

$$(x + y)^2 = x^2 + 2xy + y^2$$

This is the square of the first term of the binomial.

This is twice the product of the two terms of the binomial.

This is the square of the last term of the binomial.

$$(x - y)^2 = x^2 - 2xy + y^2$$

Trinomials that are squares of a binomial are called **perfect square trinomials.** Some examples of perfect square trinomials are

$y^2 + 6y + 9$	Because it is the square of $(y + 3)$: $(y + 3)^2 = y^2 + 6y + 9$.
$t^2 - 14t + 49$	Because it is the square of $(t - 7)$: $(t - 7)^2 = t^2 - 14t + 49$.
$4m^2 - 20m + 25$	Because it is the square of $(2m - 5)$: $(2m - 5)^2 = 4m^2 - 20m + 25$.

EXAMPLE 1 *Recognizing perfect square trinomials.* Determine whether the following trinomials are perfect square trinomials: **a.** $x^2 + 10x + 25$, **b.** $c^2 - 12c - 36$, and **c.** $25y^2 - 30y + 9$.

Solution

a. To determine whether $x^2 + 10x + 25$ is a perfect square trinomial, we note that

- the first term is the square of x,
- the last term is the square of 5, and
- the middle term is twice the product of x and 5.

Thus, $x^2 + 10x + 25$ is a perfect square trinomial.

Self Check

Tell which of the following are perfect square trinomials:

a. $y^2 + 4y + 4$

b. $b^2 - 6b - 9$

c. $4z^2 + 4z + 4$

b. To determine whether $c^2 - 12c - 36$ is a perfect square trinomial, we note that

- the first term is the square of c, but
- the last term is negative.

Thus, $c^2 - 12c - 36$ is not a perfect square trinomial.

c. To determine whether $25y^2 - 30y + 9$ is a perfect square trinomial, we note that

- the first term is the square of $5y$,
- the last term is the square of -3, and
- the middle term is twice the product of $5y$ and -3.

Thus, $25y^2 - 30y + 9$ is a perfect square trinomial.

Answers: **a.** yes,   **b.** no,
c. no   ∎

Although we can factor perfect square trinomials using techniques discussed earlier in the chapter, we can also factor them by inspecting their terms and applying the special product formulas in reverse.

Factoring perfect square trinomials

$$x^2 + 2xy + y^2 = (x + y)^2$$
$$x^2 - 2xy + y^2 = (x - y)^2$$

EXAMPLE 2   *Factoring perfect square trinomials.*   Factor $N^2 + 20N + 100$.

Self Check
Factor $x^2 + 18x + 81$.

Solution

$N^2 + 20N + 100$ is a perfect square trinomial, because:

- The first term N^2 is the square of N: $(N)^2 = N^2$.
- The last term 100 is the square of **10**: $10^2 = 100$.
- The middle term is twice the product of N and 10: $2(N)(10) = 20N$.

The factored form of the trinomial involves the terms N and 10.

$$N^2 + 20N + 100 = (N + \mathbf{10})^2 \quad \text{The sign in the binomial is the sign of the middle term of the trinomial.}$$

Check by multiplication.

Answer: $(x + 9)^2$ ∎

EXAMPLE 3   *Perfect square trinomials in two variables.*   Factor $9x^2 - 30xy + 25y^2$.

Self Check
Factor $16x^2 + 8xy + y^2$.

Solution

$9x^2 - 30xy + 25y^2$ is a perfect square trinomial, because:

- The first term $9x^2$ is the square of $3x$: $(3x)^2 = 9x^2$.
- The last term $25y^2$ is the square of $-5y$: $(-5y)^2 = 25y^2$.
- The middle term is twice the product of $3x$ and $-5y$: $2(3x)(-5y) = -30xy$.

The factored form of the trinomial involves the terms $3x$ and $-5y$.

$$9x^2 - 30xy + 25y^2 = (3x - 5y)^2 \quad \text{The sign in the binomial is the sign of the middle term of the trinomial.}$$

Check by multiplication.

Answer: $(4x + y)^2$ ∎

Factoring the difference of two squares

Whenever we multiply a binomial of the form $x + y$ by a binomial of the form $x - y$, we obtain a binomial of the form $x^2 - y^2$.

$$(x + y)(x - y) = x^2 - xy + xy - y^2 \quad \text{Use the FOIL method.}$$
$$= x^2 - y^2 \qquad\qquad \text{Combine like terms: } -xy + xy = 0.$$

The binomial $x^2 - y^2$ is called a **difference of two squares,** because x^2 is the square of x and y^2 is the square of y. The difference of the squares of two quantities always factors into the sum of those two quantities multiplied by the difference of those two quantities.

Factoring the difference of two squares

$$x^2 - y^2 = (x + y)(x - y)$$

If we think of the difference of two squares as the square of a **F**irst quantity minus the square of a **L**ast quantity, we have the formula

$$F^2 - L^2 = (F + L)(F - L)$$

and we say: *To factor the square of a First quantity minus the square of a Last quantity, we multiply the First plus the Last by the First minus the Last.*

To factor $x^2 - 9$, we note that it can be written in the form $x^2 - 3^2$ and use the formula for factoring the difference of two squares:

$$
\begin{array}{cccccc}
F^2 & - & L^2 & = & (F & + L)(F & - L) \\
\downarrow & & \downarrow & & \downarrow & \downarrow\ \downarrow & \downarrow \\
x^2 & - & 3^2 & = & (x & + 3)(x & - 3)
\end{array}
$$
Substitute x for F and 3 for L.

We can check by verifying that $(x + 3)(x - 3) = x^2 - 9$. Because of the commutative property of multiplication, we can also write this factorization as $(x - 3)(x + 3)$.

To factor the difference of two squares, it is helpful to know the integers that are perfect squares. The number 400, for example, is a perfect square, because $20^2 = 400$. The perfect integer squares through 400 are

1, 4, 9, 16, 25, 36, 49, 64, 81, 100, 121, 144, 169, 196, 225, 256, 289, 324, 361, 400

Expressions containing variables such as $25x^2$ are also perfect squares, because they can be written as the square of a quantity:

$$25x^2 = (5x)^2$$

EXAMPLE 4   *Factoring the difference of two squares.*   Factor $25x^2 - 49$.

Solution

We can write $25x^2 - 49$ in the form $(5x)^2 - 7^2$ and use the formula for factoring the difference of two squares:

$$
\begin{array}{cccccc}
F^2 & - L^2 & = & (F & + L)(F & - L) \\
\downarrow & \downarrow & & \downarrow & \downarrow\ \downarrow & \downarrow \\
(5x)^2 & - 7^2 & = & (5x & + 7)(5x & - 7)
\end{array}
$$
Substitute $5x$ for F and 7 for L.

We can check by multiplying.

$$(5x + 7)(5x - 7) = 25x^2 - 35x + 35x - 49$$
$$= 25x^2 - 49$$

Self Check

Factor $16a^2 - 81$.

Answer: $(4a + 9)(4a - 9)$ ■

EXAMPLE 5   *Factoring the difference of two squares.*   Factor $4y^4 - 121z^2$.

Solution

We can write $4y^4 - 121z^2$ in the form $(2y^2)^2 - (11z)^2$ and use the formula for factoring the difference of two squares:

$$F^2 \;-\; L^2 \;=\; (F \;+\; L)\,(F \;-\; L)$$
$$(2y^2)^2 - (11z)^2 = (2y^2 + 11z)(2y^2 - 11z)$$

Check by multiplying.

Answer: $(3m + 8n^2)(3m - 8n^2)$

Multistep factoring

When factoring a polynomial, we should always factor out the greatest common factor first.

EXAMPLE 6   *Factoring out the GCF first.*   Factor $8x^2 - 8$.

Solution

We factor out the GCF of 8, and then factor the resulting difference of two squares.

$$8x^2 - 8 = 8(x^2 - 1)   \quad\text{The GCF is 8.}$$
$$= 8(x + 1)(x - 1)   \quad\text{Think of } x^2 - 1 \text{ as } x^2 - 1^2 \text{ and factor the difference of two squares.}$$

We check by multiplying.

$$8(x + 1)(x - 1) = 8(x^2 - 1)   \quad\text{Multiply the binomials first.}$$
$$= 8x^2 - 8   \quad\text{Distribute the multiplication by 8.}$$

Self Check

Factor $2p^2 - 200$.

Answer: $2(p + 10)(p - 10)$

Sometimes we must factor a difference of two squares more than once to completely factor a polynomial.

EXAMPLE 7   *Multistep factoring.*   Factor $x^4 - 16$.

Solution

$$x^4 - 16 = (x^2 + 4)(x^2 - 4)   \quad\text{Factor the difference of two squares.}$$
$$= (x^2 + 4)(x + 2)(x - 2)   \quad\text{Factor another difference of two squares: } x^2 - 4.$$

Self Check

Factor $a^4 - 81$.

Answer:
$(a^2 + 9)(a + 3)(a - 3)$

COMMENT   In Example 7, the binomial $x^2 + 4$ is the **sum of two squares.** If we are limited to integer coefficients, binomials that are the sum of two squares cannot be factored.

Factoring the sum and difference of two cubes

We have seen that the sum of two squares, such as $x^2 + 4$ or $25a^2 + 9b^2$, cannot be factored. However, the sum of two cubes and the difference of two cubes can be factored.

The sum of two cubes	The difference of two cubes
$x^3 + 8$	$a^3 - 64b^3$
This term is x cubed. This term is 2 cubed: $2^3 = 8$.	This term is a cubed. This term is $4b$ cubed: $(4b)^3 = 64b^3$.

To find the formulas for factoring the sum of two cubes and the difference of two cubes, we need to find the following two products:

$$(x + y)(x^2 - xy + y^2) = (x + y)x^2 - (x + y)xy + (x + y)y^2 \quad \text{Use the distributive property.}$$

$$= x^3 + x^2y - x^2y - xy^2 + xy^2 + y^3$$
$$= x^3 + y^3 \quad \text{Combine like terms.}$$

$$(x - y)(x^2 + xy + y^2) = (x - y)x^2 + (x - y)xy + (x - y)y^2 \quad \text{Use the distributive property.}$$

$$= x^3 - x^2y + x^2y - xy^2 + xy^2 - y^3$$
$$= x^3 - y^3 \quad \text{Combine like terms.}$$

These results justify the formulas for factoring the **sum and difference of two cubes.**

Factoring the sum and difference of two cubes

$$x^3 + y^3 = (x + y)(x^2 - xy + y^2)$$
$$x^3 - y^3 = (x - y)(x^2 + xy + y^2)$$

If we think of the sum of two cubes as the cube of a **First** quantity plus the cube of a **Last** quantity, we have the formula

$$F^3 + L^3 = (F + L)(F^2 - FL + L^2)$$

In words, we say, *To factor the cube of a **First** quantity plus the cube of a **Last** quantity, we multiply the **First** plus the **Last** by*

- *the **First** squared*
- *minus the **First** times the **Last***
- *plus the **Last** squared.*

The formula for the difference of two cubes is

$$F^3 - L^3 = (F - L)(F^2 + FL + L^2)$$

In words, we say, *To factor the cube of a **First** quantity minus the cube of a **Last** quantity, we multiply the **First** minus the **Last** by*

- *the **First** squared*
- *plus the **First** times the **Last***
- *plus the **Last** squared.*

To factor the sum or difference of two cubes, it's helpful to know the cubes of the numbers from 1 to 10:

1, 8, 27, 64, 125, 216, 343, 512, 729, 1,000

Expressions containing variables such as $64b^3$ are also perfect cubes, because they can be written as the cube of a quantity:

$$64b^3 = (4b)^3$$

EXAMPLE 8   *Factoring the sum of two cubes.*   Factor $x^3 + 8$.

Solution

We think of $x^3 + 8$ as the cube of a **First** quantity, x, plus the cube of a **Last** quantity, 2.

$$x^3 + 8 = x^3 + 2^3$$

Thus, $x^3 + 8$ factors as the product of the sum of x and 2 and the trinomial $x^2 - 2x + 2^2$.

Self Check

Factor $h^3 + 27$.

$$F^3 + L^3 = (F + L)(F^2 - FL + L^2)$$

$$x^3 + 2^3 = (x + 2)(x^2 - x2 + 2^2) \quad \text{Substitute } x \text{ for F and 2 for L.}$$

$$= (x + 2)(x^2 - 2x + 4)$$

We can check by multiplying.

$$(x + 2)(x^2 - 2x + 4) = (x + 2)x^2 - (x + 2)2x + (x + 2)4$$

$$= x^3 + 2x^2 - 2x^2 - 4x + 4x + 8$$

$$= x^3 + 8$$

Answer: $(h + 3)(h^2 - 3h + 9)$ ∎

EXAMPLE 9 *Factoring the difference of two cubes.* Factor $a^3 - 64b^3$.

Self Check
Factor $8c^3 - 1$.

Solution
We think of $a^3 - 64b^3$ as the cube of a **F**irst quantity, a, minus the cube of a **L**ast quantity, $4b$.

$$a^3 - 64b^3 = a^3 - (4b)^3$$

Thus, its factors are the difference $a - 4b$ and the trinomial $a^2 + a(4b) + (4b)^2$.

$$F^3 - L^3 = (F - L)(F^2 + F L + L^2)$$

$$a^3 - (4b)^3 = (a - 4b)[a^2 + a(4b) + (4b)^2]$$

$$= (a - 4b)(a^2 + 4ab + 16b^2)$$

Check by multiplying.

Answer:
$(2c - 1)(4c^2 + 2c + 1)$ ∎

Sometimes we must factor out a greatest common factor before factoring a sum or difference of two cubes.

EXAMPLE 10 *Factoring out the GCF first.* Factor $-2t^5 + 250t^2$.

Self Check
Factor $4c^3 + 4d^3$.

Solution
Each term contains the factor $-2t^2$.

$$-2t^5 + 250t^2 = -2t^2(t^3 - 125) \qquad \text{Factor out } -2t^2.$$

$$= -2t^2(t - 5)(t^2 + 5t + 25) \quad \text{Factor } t^3 - 125.$$

Check by multiplying.

Answer:
$4(c + d)(c^2 - cd + d^2)$ ∎

A factoring strategy

Later, when we solve equations and simplify expressions containing polynomials, we won't be told what type of factoring technique to apply—we will have to determine that ourselves. The following strategy is helpful when factoring a random polynomial.

Steps for factoring a polynomial

1. Factor out all common factors.
2. If a polynomial has two terms, check for the following problem types:
   a. **The difference of two squares:** $x^2 - y^2 = (x + y)(x - y)$
   b. **The sum of two cubes:** $x^3 + y^3 = (x + y)(x^2 - xy + y^2)$
   c. **The difference of two cubes:** $x^3 - y^3 = (x - y)(x^2 + xy + y^2)$
3. If a polynomial has three terms, check for the following problem types:
   a. **A perfect square trinomial:**
   $$x^2 + 2xy + y^2 = (x + y)^2$$
   $$x^2 - 2xy + y^2 = (x - y)^2$$

> **b.** If the trinomial is not a perfect square, attempt to factor it as a general trinomial using the **trial-and-check method** or **factoring by grouping.**
>
> **4.** If a polynomial has four or more terms, try **factoring by grouping.**
>
> **5.** Continue until each individual factor is prime.
>
> **6.** Check the results by multiplying.

STUDY SET Section 11.4

VOCABULARY *Fill in the blanks.*

1. The binomial $x^2 - 25$ is called a _____ of two squares.

2. $x^2 + 6x + 9$ is a _____ square trinomial because it is the square of the binomial $(x + 3)$.

3. The binomial $x^3 + 27$ is called a sum of two _____. The binomial $x^3 - 8$ is called a _____ of two cubes.

4. To _____ $4x^2 - 12x + 9$ means to write it as the product of two binomials.

CONCEPTS *In Exercises 5–10, fill in the blanks.*

5. Consider $25x^2 + 30x + 9$.
  a. The first term is the square of ___.
  b. The last term is the square of ___.
  c. The middle term is twice the product of ___ and ___.

6. Consider $49x^2 - 28xy + 4y^2$.
  a. The first term is the square of ___.
  b. The last term is the square of ___.
  c. The middle term is twice the product of ___ and ___.

7. To factor the square of a First quantity minus the square of a Last quantity, we multiply the _____ plus the _____ by the _____ minus the _____.

8. If a trinomial is the square of one quantity, plus the square of a second quantity, plus _____ the product of the quantities, it factors into the square of the _____ of the quantities.

9. a. $36x^2 = ()^2$ **b.** $100x^4 = ()^2$

  c. $27m^3 = ()^3$ **d.** $a^6 = ()^3$

10. a. $4x^2 - 9 = ()^2 - ()^2$

  b. $8x^3 - 27 = ()^3 - ()^3$

  c. $x^3 + 64y^3 = ()^3 + ()^3$

11. List the first ten perfect integer squares.

12. List the first five perfect integer cubes.

13. Explain why each trinomial is not a perfect square trinomial.
  a. $9h^2 - 6h + 7$
  b. $j^2 - 8j - 16$

  c. $25r^2 + 20r + 16$

14. a. Three incorrect factorizations of $x^2 + 36$ are given below. Show why each is wrong.

$$(x + 6)(x - 6)$$
$$(x + 6)(x + 6)$$
$$(x - 6)(x - 6)$$

  b. Can $x^2 + 36$ be factored using only integers?

NOTATION *Write each expression as a polynomial in simpler form.*

15. $(6x)^2 - (5y)^2$

16. $(4x)^2 - (9y)^2$

17. $(3a)^2 - 2(3a)(5b) + (5b)^2$

18. $(2s)^2 + 2(2s)(9t) + (9t)^2$

Use an exponent to write each expression in simpler form.

19. $(x + 8)(x + 8)$ **20.** $(x - 8)(x - 8)$

PRACTICE *Complete each fractorization.*

21. $a^2 - 6a + 9 = (a -)^2$

22. $t^2 + 2t + 1 = (t 1)^2$

23. $4x^2 + 4x + 1 = (2x 1)^2$

24. $9y^2 - 12y + 4 = (3y -)^2$

Factor each polynomial.

25. $x^2 + 6x + 9$

26. $x^2 + 10x + 25$

27. $y^2 - 8y + 16$

28. $z^2 - 2z + 1$

29. $t^2 + 20t + 100$
30. $r^2 + 24r + 144$
31. $u^2 - 18u + 81$
32. $v^2 - 14v + 49$
33. $4x^2 + 12x + 9$
34. $4x^2 - 4x + 1$
35. $36x^2 + 12x + 1$
36. $4x^2 - 20x + 25$
37. $a^2 + 2ab + b^2$
38. $a^2 - 2ab + b^2$
39. $16x^2 - 8xy + y^2$
40. $25x^2 + 20xy + 4y^2$

Complete each factorization.

41. $y^2 - 49 = (y + \boxed{})(y - \boxed{})$
42. $p^4 - q^2 = (p^2 + q)(\boxed{} - \boxed{})$
43. $t^2 - w^2 = (\boxed{} + \boxed{})(t - w)$
44. $49u^2 - 64v^2 = (\boxed{} + 8v)(7u \,\boxed{}\, 8v)$

Factor each polynomial, if possible.

45. $x^2 - 16$
46. $x^2 - 25$
47. $4y^2 - 1$
48. $9z^2 - 1$
49. $9x^2 - y^2$
50. $4x^2 - z^2$
51. $16a^2 - 25b^2$
52. $36a^2 - 121b^2$
53. $a^2 + b^2$
54. $121a^2 + 144b^2$
55. $a^4 - 144b^2$
56. $81y^4 - 100z^2$
57. $t^2z^2 - 64$
58. $900 - B^2C^2$
59. $8x^2 - 32y^2$
60. $2a^2 - 200b^2$
61. $7a^2 - 7$
62. $20x^2 - 5$
63. $6x^4 - 6x^2y^2$
64. $4b^2y - 16c^2y$
65. $x^4 - 81$
66. $y^4 - 625$
67. $a^4 - 16$
68. $b^4 - 256$
69. $81r^4 - 256s^4$
70. $16y^8 - 81z^4$

Complete each factorization.

71. $a^3 + 8 = (a + 2)(a^2 - \boxed{} + 4)$
72. $x^3 - 1 = (x - 1)(x^2 + \boxed{} + 1)$
73. $b^3 + 27 = (\boxed{})(b^2 - 3b + 9)$
74. $z^3 - 125 = (\boxed{})(z^2 + 5z + 25)$

Factor each polynomial.

75. $y^3 + 1$
76. $x^3 - 8$
77. $a^3 - 27$
78. $b^3 + 125$
79. $8 + x^3$
80. $27 - y^3$
81. $s^3 - t^3$
82. $8u^3 + w^3$
83. $a^3 + 8b^3$
84. $27a^3 - b^3$
85. $64x^3 - 27$
86. $27x^3 + 125$
87. $a^6 - b^3$
88. $a^3 + b^6$
89. $x^9 + y^6$
90. $x^3 - y^9$
91. $2x^3 + 54$
92. $2x^3 - 2$
93. $-x^3 + 216$
94. $-x^3 - 125$
95. $64m^3x - 8n^3x$
96. $16r^4 + 128rs^3$
97. $x^4y + 216xy^4$
98. $16a^5 - 54a^2b^3$
99. $81r^4s^2 - 24rs^5$
100. $4m^5n + 500m^2n^4$

APPLICATIONS

101. GENETICS The Hardy–Weinberg equation, one of the fundamental concepts in population genetics, is
$$p^2 + 2pq + q^2 = 1$$
where p represents the frequency of a certain dominant gene and q represents the frequency of a certain recessive gene. Factor the left-hand side of the equation.

102. SPACE TRAVEL The first Soviet manned spacecraft, Vostok, is shown in Illustration 1. The surface area of the spherical part of the craft is given by $(36\pi r^2 - 48\pi r + 16\pi)$ m^2. Factor the expression.

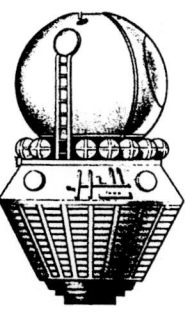

ILLUSTRATION 1

103. PHYSICS   Illustration 2 shows a time-sequence picture of a falling apple. Factor the expression, which gives the difference in the distance fallen by the apple during the time interval from t_1 to t_2 seconds.

This distance is $0.5gt_1^2 - 0.5gt_2^2$

ILLUSTRATION 2

104. DARTS   A circular dart board has a series of rings around a solid center, called the bullseye. (See Illustration 3.) To find the area of the outer white ring, we can use the formula

$$A = \pi R^2 - \pi r^2$$

Factor the expression on the right-hand side of the equation.

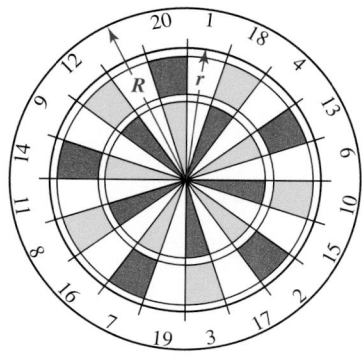

ILLUSTRATION 3

WRITING

105. When asked to factor $x^2 - 25$, one student wrote $(x + 5)(x - 5)$, and another student wrote $(x - 5)(x + 5)$. Are both answers correct? Explain.

106. Write a comment to the student whose work is shown below, explaining the initial error that was made.

Factor $4x^2 - 16y^2$.

$(2x + 4y)(2x - 4y)$

107. Explain why $x^6 - 1$ can be thought of as a difference of two squares or as a difference of two cubes.

108. Why is $a^2 + 2a + 1$ a perfect square trinomial, and why isn't $a^2 + 4a + 1$ a perfect square trinomial?

REVIEW   *Do each division.*

109. $\dfrac{5x^2 + 10y^2 - 15xy}{5xy}$

110. $\dfrac{-30c^2d^2 - 15c^2d - 10cd^2}{-10cd}$

111. $2a - 1\overline{)a - 2 + 6a^2}$

112. $4b + 3\overline{)4b^3 - 5b^2 - 2b + 3}$

ADDITIONAL FACTORING PROBLEMS   *Apply the factoring strategy to factor each polynomial completely. If a polynomial is not factorable, write "prime."*

113. $a^2(x - a) - b^2(x - a)$

114. $a^2c + a^2d^2 + bc + bd^2$

115. $70p^4q^3 - 35p^4q^2 + 49p^5q^2$

116. $a^2b^2 - 144$

117. $2ab^2 + 8ab - 24a$

118. $t^4 - 16$

119. $-8p^3q^7 - 4p^2q^3$

120. $8m^2n^3 - 24mn^4$

121. $20m^2 + 100m + 125$

122. $3rs + 6r^2 - 18s^2$

123. $x^2 + 7x + 1$

124. $3a^3 + 24b^3$

125. $-2x^5 + 128x^2$

126. $16 - 40z + 25z^2$

127. $14t^3 - 40t^2 + 6t^4$

128. $-9x^2y^2 + 6xy - 1$

129. $x^2y^2 - 2x^2 - y^2 + 2$

130. $5x^3y^3z^4 + 25x^2y^3z^2 - 35x^3y^2z^5$

131. $8p^6 - 27q^6$

132. $2c^2 - 5cd - 3d^2$

133. $125p^3 - 64y^3$

134. $8a^2x^3y - 2b^2xy$

135. $-16x^4y^2z + 24x^5y^3z^4 - 15x^2y^3z^7$

136. $2ac + 4ad + bc + 2bd$

137. $81p^4 - 16q^4$

138. $6x^2 - x - 16$

139. $4x^2 + 9y^2$

140. $30a^4 + 5a^3 - 200a^2$

141. $54x^3 + 250y^6$

142. $6a^3 + 35a^2 - 6a$

143. $10r^2 - 13r - 4$

144. $21t^3 - 10t^2 + t$

145. $49p^2 + 28pq + 4q^2$

146. $16x^2 - 40x^3 + 25x^4$

11.5   *Quadratic Equations*

In this section, you will learn about

- Quadratic equations • Solving quadratic equations by factoring
- Applications

INTRODUCTION. Equations that involve first-degree polynomials, such as $9x - 6 = 0$, are called *linear equations*. Equations that involve second-degree polynomials, such as $9x^2 - 6x = 0$, are called *quadratic equations*. In this section, we will define quadratic equations and learn how to solve many of them by factoring.

Quadratic equations

If a polynomial contains one variable with an exponent to the second (but no higher) power, it is called a **second-degree polynomial.** Equations in which a second-degree polynomial is equal to zero are called **quadratic equations.** Some examples are

$$9x^2 - 6x = 0, \qquad x^2 - 2x - 63 = 0, \quad \text{and} \quad 2x^2 + 3x - 2 = 0$$

Quadratic equations

> A **quadratic equation** is an equation that can be written in the form
> $$ax^2 + bx + c = 0 \quad (a \neq 0)$$
> where a, b, and c represent real numbers.

To write a quadratic equation such as $21x = 10 - 10x^2$ in $ax^2 + bx + c = 0$ form (called **quadratic form**), we use the addition and subtraction properties of equality to get 0 on the right-hand side.

$$21x = 10 - 10x^2$$
$$10x^2 + 21x = 10 - 10x^2 + 10x^2 \quad \text{Add } 10x^2 \text{ to both sides.}$$
$$10x^2 + 21x = 10 \quad\quad \text{Combine like terms: } -10x^2 + 10x^2 = 0.$$
$$10x^2 + 21x - 10 = 0 \quad\quad \text{Subtract 10 from both sides.}$$

When $21x = 10 - 10x^2$ is written in quadratic form, we see that $a = 10$, $b = 21$, and $c = -10$.

The techniques we have used to solve linear equations cannot be used to solve a quadratic equation, because those techniques cannot isolate x on one side of the equation. However, we can often solve quadratic equations using factoring and the following property of real numbers.

The zero-factor property of real numbers

> Suppose a and b represent two real numbers. Then
> If $ab = 0$, then $a = 0$ or $b = 0$.

In words, the zero-factor property states that when the product of two numbers is zero, at least one of them must be zero.

EXAMPLE 1   *Using the zero-factor property.*   Solve $(4y - 1)(y + 6) = 0$.

Solution

The left-hand side of the equation is $(4y - 1)(y + 6)$. By the zero-factor property, one of these factors must be 0.

$$4y - 1 = 0 \quad \text{or} \quad y + 6 = 0$$

We can solve each of the linear equations.

$$4y - 1 = 0 \quad \text{or} \quad y + 6 = 0$$
$$4y = 1 \qquad\qquad y = -6$$
$$y = \frac{1}{4}$$

The equation has two solutions, $\frac{1}{4}$ and -6. To check, we substitute the results for y in the original equation and simplify.

For $y = \frac{1}{4}$

$$(4y - 1)(y + 6) = 0$$
$$\left[4\left(\frac{1}{4}\right) - 1\right]\left(\frac{1}{4} + 6\right) \stackrel{?}{=} 0$$
$$(1 - 1)\left(6\frac{1}{4}\right) \stackrel{?}{=} 0$$
$$0\left(6\frac{1}{4}\right) \stackrel{?}{=} 0$$
$$0 = 0$$

For $y = -6$

$$(4y - 1)(y + 6) = 0$$
$$[4(-6) - 1](-6 + 6) \stackrel{?}{=} 0$$
$$(-24 - 1)(0) \stackrel{?}{=} 0$$
$$-25(0) \stackrel{?}{=} 0$$
$$0 = 0$$

Self Check

Solve $b(5b - 3) = 0$.

Answer: $0, \dfrac{3}{5}$ ∎

Solving quadratic equations by factoring

In Example 1, the left-hand side of the equation was in factored form, so we were able to use the zero-factor property immediately. However, to solve many quadratic equations, we must first do the factoring.

EXAMPLE 2   *Solving quadratic equations.*   Solve $9x^2 - 6x = 0$.

Solution

We begin by factoring the left-hand side of the equation.

$$9x^2 - 6x = 0$$
$$3x(3x - 2) = 0 \quad \text{Factor out the GCF of } 3x.$$

By the zero-factor property, we have

$$3x = 0 \quad \text{or} \quad 3x - 2 = 0$$

We can solve each of the linear equations to get

$$x = 0 \quad \text{or} \quad x = \frac{2}{3}$$

To check, we substitute the results for x in the original equation and simplify.

Self Check

Solve $5x^2 + 10x = 0$.

For $x = 0$

$$9x^2 - 6x = 0$$

$$9(0)^2 - 6(0) \stackrel{?}{=} 0$$

$$0 - 0 \stackrel{?}{=} 0$$

$$0 = 0$$

For $x = \frac{2}{3}$

$$9x^2 - 6x = 0$$

$$9\left(\frac{2}{3}\right)^2 - 6\left(\frac{2}{3}\right) \stackrel{?}{=} 0$$

$$9\left(\frac{4}{9}\right) - 6\left(\frac{2}{3}\right) \stackrel{?}{=} 0$$

$$4 - 4 \stackrel{?}{=} 0$$

$$0 = 0$$

Answer: $0, -2$ ■

We can use the following steps to solve a quadratic equation by factoring.

Factoring method

1. Write the equation in $ax^2 + bx + c = 0$ form.
2. Factor the left-hand side of the equation.
3. Use the zero-factor property to set each factor equal to zero.
4. Solve each resulting linear equation.
5. Check the results in the original equation.

EXAMPLE 3   *Writing an equation in quadratic form.*   Solve $x^2 = 9$.

Solution

Before we can use the zero-factor property, we must subtract 9 from both sides to make the right-hand side zero.

$$x^2 = 9$$

$$x^2 - 9 = 0 \qquad \text{Subtract 9 from both sides.}$$

$$(x + 3)(x - 3) = 0 \qquad \text{Factor the difference of two squares.}$$

$$x + 3 = 0 \quad \text{or} \quad x - 3 = 0 \quad \text{Set each factor equal to zero.}$$

$$x = -3 \qquad\qquad x = 3 \quad \text{Solve each linear equation.}$$

Check each possible solution by substituting it into the original equation.

For $x = -3$

$$x^2 = 9$$

$$(-3)^2 \stackrel{?}{=} 9$$

$$9 = 9$$

For $x = 3$

$$x^2 = 9$$

$$(3)^2 \stackrel{?}{=} 9$$

$$9 = 9$$

Self Check

Solve $9x^2 - 36 = 0$.

Answer: $2, -2$ ■

EXAMPLE 4   *Solving quadratic equations.*   Solve $x^2 - 2x - 63 = 0$.

Solution

In this case, we must factor a trinomial to solve the equation.

$$x^2 - 2x - 63 = 0$$

$$(x + 7)(x - 9) = 0 \qquad \text{Factor the trinomial } x^2 - 2x - 63.$$

$$x + 7 = 0 \quad \text{or} \quad x - 9 = 0 \quad \text{Set each factor equal to zero.}$$

$$x = -7 \qquad\qquad x = 9 \quad \text{Solve each linear equation.}$$

The solutions are -7 and 9. Check each one.

Self Check

Solve $x^2 + 5x + 6 = 0$.

Answer: $-2, -3$ ■

EXAMPLE 5 *Writing an equation in quadratic form.* Solve $2x^2 + 3x = 2$.

Solution

We write the equation in the form $ax^2 + bx + c = 0$ and then solve for x.

$$2x^2 + 3x = 2$$

$$2x^2 + 3x - 2 = 0$$ Subtract 2 from both sides so that the right-hand side is zero.

$$(2x - 1)(x + 2) = 0$$ Factor $2x^2 + 3x - 2$.

$$2x - 1 = 0 \quad \text{or} \quad x + 2 = 0$$ Set each factor equal to zero.

$$2x = 1 \qquad\qquad x = -2$$ Solve each linear equation.

$$x = \frac{1}{2}$$

Check each solution.

Self Check

Solve $3x^2 - 6 = -7x$.

Answer: $\frac{2}{3}, -3$ ■

EXAMPLE 6 *A repeated solution.* Solve $-4 = x(9x - 12)$.

Solution

First, we need to write the equation in the form $ax^2 + bx + c = 0$.

$$-4 = x(9x - 12)$$

$$-4 = 9x^2 - 12x$$ Distribute the multiplication by x.

$$0 = 9x^2 - 12x + 4$$ Add 4 to both sides to make the left-hand side zero.

$$0 = (3x - 2)(3x - 2)$$ Factor the trinomial

$$3x - 2 = 0 \quad \text{or} \quad 3x - 2 = 0$$ Set each factor equal to zero.

$$3x = 2 \qquad\qquad 3x = 2$$ Add 2 to both sides.

$$x = \frac{2}{3} \qquad\qquad x = \frac{2}{3}$$ Divide both sides by 3.

The equation has two solutions that are the same. We call $\frac{2}{3}$ a *repeated solution*. Check by substituting it into the original equation.

Self Check

Solve $x(4x + 12) = -9$.

Answer: $-\frac{3}{2}, -\frac{3}{2}$ ■

EXAMPLE 7 *An equation with three solutions.* Solve $6x^3 + 12x = 17x^2$.

Solution

This is not a quadratic equation, because it contains the term x^3. However, we can solve it using factoring and an extension of the zero-factor property.

$$6x^3 + 12x = 17x^2$$

$$6x^3 - 17x^2 + 12x = 0$$ Add $-17x^2$ to both sides to get 0 on the right-hand side.

$$x(6x^2 - 17x + 12) = 0$$ Factor out the GCF of x.

$$x(2x - 3)(3x - 4) = 0$$ Factor $6x^2 - 17x + 12$.

$$x = 0 \quad \text{or} \quad 2x - 3 = 0 \quad \text{or} \quad 3x - 4 = 0$$ Set each factor equal to zero.

$$2x = 3 \qquad\qquad 3x = 4$$ Solve the linear equations.

$$x = \frac{3}{2} \qquad\qquad x = \frac{4}{3}$$

This equation has three solutions.

Self Check

Solve $10x^3 + x^2 - 2x = 0$.

Answer: $0, \frac{2}{5}, -\frac{1}{2}$ ■

Applications

The solutions of many problems involve the use of quadratic equations.

EXAMPLE 8   *Softball.*   A softball pitcher can throw a "fastball" underhand at about 55 mph (80 feet per second). If she throws a ball up into the air with that velocity, as in Figure 11-2, its height h in feet, t seconds after being released, is given by the formula

$$h = 80t - 16t^2$$

After the ball is thrown, in how many seconds will it hit the ground?

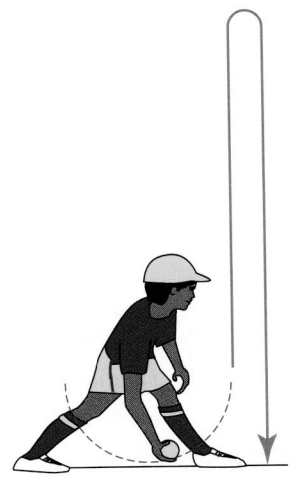

Solution   When the ball hits the ground, its height will be zero. Thus, we set h equal to zero and solve for t.

$$h = 80t - 16t^2$$
$$0 = 80t - 16t^2$$
$$0 = 16t(5 - t) \qquad \text{Factor out the GCF of } 16t.$$
$$16t = 0 \quad \text{or} \quad 5 - t = 0 \qquad \text{Set each factor equal to zero.}$$
$$t = 0 \qquad\qquad t = 5 \qquad \text{Solve each linear equation.}$$

When $t = 0$, the ball's height above the ground is 0 feet. When $t = 5$, the height is again 0 feet, and the object has hit the ground. The solution is 5 seconds.   ∎

FIGURE 11-2

EXAMPLE 9   *Perimeter of a rectangle.*   Assume that the rectangle in Figure 11-3 has an area of 52 square centimeters and that its length is 1 centimeter more than 3 times its width. Find the perimeter of the rectangle.

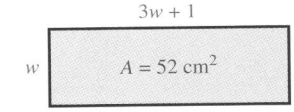

FIGURE 11-3

Analyze the problem   The area of the rectangle is 52 square centimeters. Recall that the formula that gives the area of a rectangle is $A = lw$. To find the perimeter of the rectangle, we need to know its length and width. We are told that its length is related to its width; the length is 1 centimeter more than 3 times the width.

Form an equation   Let w represent the width of the rectangle. Then $3w + 1$ represents its length. Because the area is 52 square centimeters, we substitute 52 for A and $3w + 1$ for l in the formula $A = lw$.

$$A = lw$$
$$52 = (3w + 1)w$$

Solve the equation   Now we solve the equation for w.

$$52 = (3w + 1)w \qquad \text{The equation to solve.}$$
$$52 = 3w^2 + w \qquad \text{Distribute the multiplication by } w.$$
$$0 = 3w^2 + w - 52 \qquad \text{Subtract 52 from both sides to make the left-hand side zero.}$$
$$0 = (3w + 13)(w - 4) \qquad \text{Factor the trinomial.}$$
$$3w + 13 = 0 \quad \text{or} \quad w - 4 = 0 \qquad \text{Set each factor equal to zero.}$$
$$3w = -13 \qquad\qquad w = 4 \qquad \text{Solve each linear equation.}$$
$$w = -\frac{13}{3}$$

State the conclusion Since the width cannot be negative, we discard the result $w = -\frac{13}{3}$. Thus, the width of the rectangle is 4, and the length is given by

$$3w + 1 = 3(\mathbf{4}) + 1 \quad \text{Substitute 4 for } w.$$
$$= 12 + 1$$
$$= 13$$

The dimensions of the rectangle are 4 centimeters by 13 centimeters. We find the perimeter by substituting 13 for l and 4 for w in the formula for the perimeter of a rectangle.

$$P = 2l + 2w$$
$$= 2(\mathbf{13}) + 2(\mathbf{4})$$
$$= 26 + 8$$
$$= 34$$

The perimeter of the rectangle is 34 centimeters.

Check the result A rectangle with dimensions of 13 centimeters by 4 centimeters does have an area of 52 square centimeters, and the length is 1 centimeter more than 3 times the width. A rectangle with these dimensions has a perimeter of 34 centimeters. ■

The next example involves a right triangle. A **right triangle** is a triangle that contains a 90° angle. The longest side of a right triangle is the **hypotenuse,** which is the side opposite the right angle. The remaining two sides are the **legs** of the triangle. (See Figure 11-4.) The **Pythagorean theorem** provides a formula relating the lengths of the three sides of a right triangle.

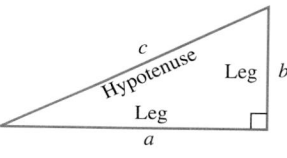

FIGURE 11-4

The Pythagorean theorem

If the length of the hypotenuse of a right triangle is c and the lengths of the two legs are a and b, then
$$c^2 = a^2 + b^2$$

EXAMPLE 10 *Recording ozone levels.* Three pollution monitoring stations are shown in Figure 11-5. The east county station is 3 miles farther from the downtown station than is the west county station. The distance between the east and west county stations is 6 miles longer than the distance between the downtown and west stations. How far apart are the stations?

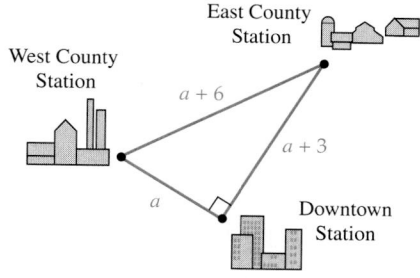

FIGURE 11-5

Analyze the problem To find the distances between the stations, we need to find the lengths of the sides of the right triangle formed by connecting their positions. The Pythagorean theorem gives the relationship between the sides of a right triangle: $a^2 + b^2 = c^2$.

Form an equation We let a represent the distance in miles from the downtown station to the west station, because the other two distances can be expressed in terms of it. The distance between the downtown and east stations is $(a + 3)$ miles, and that between the east and west stations is $(a + 6)$ miles. We substitute these distances into the Pythagorean theorem, noting that the length of the hypotenuse c is $(a + 6)$ miles.

$$a^2 + b^2 = c^2 \qquad \text{The Pythagorean theorem.}$$

$$a^2 + (a + 3)^2 = (a + 6)^2 \qquad \text{Substitute } (a + 3) \text{ for } b \text{ and } (a + 6) \text{ for } c.$$

$$a^2 + a^2 + 6a + 9 = a^2 + 12a + 36 \qquad \text{Find } (a + 3)^2 \text{ and } (a + 6)^2.$$

$$2a^2 + 6a + 9 = a^2 + 12a + 36 \qquad \text{Combine like terms on the left-hand side.}$$

$$a^2 - 6a - 27 = 0 \qquad \text{Subtract } a^2, 12a, \text{ and } 36 \text{ from both sides to make the right-hand side zero.}$$

Solve the equation   Now we solve the equation for a.

$$a^2 - 6a - 27 = 0$$

$$(a - 9)(a + 3) = 0 \qquad \text{Factor.}$$

$$a - 9 = 0 \quad \text{or} \quad a + 3 = 0 \qquad \text{Set each factor to zero.}$$

$$a = 9 \qquad\qquad a = -3 \qquad \text{Solve each linear equation.}$$

State the conclusion   Since a triangle cannot have a negative number for the length of a side, we discard the result $a = -3$. The distance from the downtown station to the west county station is 9 miles. The distance from the downtown station to the east county station is $9 + 3$, or 12 miles. The distance between the east and west stations is $9 + 6$, or 15 miles.

Check the result   The differences in the distances between stations meet the requirements stated in the problem. The distances also satisfy the Pythagorean theorem. So the solutions check.

$$9^2 + 12^2 \overset{?}{=} 15^2$$

$$81 + 144 \overset{?}{=} 225$$

$$225 = 225 \qquad\qquad\qquad\qquad\qquad\qquad\blacksquare$$

STUDY SET Section 11.5

VOCABULARY   *Fill in the blanks.*

1. Any equation that can be written in the form $ax^2 + bx + c = 0$ is called a _____ equation.

2. To _____ a binomial or trinomial means to write it as a product.

CONCEPTS   *In Exercises 3–6, fill in the blanks.*

3. When the product of two numbers is zero, at least one of them is _____. Symbolically, we can state this: If $ab = 0$, then $a =$ ___ or $b =$ ___.

4. The techniques used to solve linear equations cannot be used to solve quadratic equations, because those techniques cannot _____ the variable on one side of the equation.

5. To write a quadratic equation in *quadratic form* means that one side of the equation must be _____ and the other side must be in the form $ax^2 + bx + c$.

6. If the length of the hypotenuse of a right triangle is c and the legs are a and b, then $c^2 =$ _____ .

7. Classify each equation as quadratic or linear.
a. $3x^2 + 4x + 2 = 0$ **b.** $3x + 7 = 0$
c. $2 = -16 - 4x$ **d.** $-6x + 2 = x^2$

8. Check to see whether the given number is a solution of the given quadratic equation.
a. $x^2 - 4x = 0; x = 4$
b. $x^2 + 2x - 4 = 0; x = -2$
c. $4x^2 - x + 3 = 0; x = 1$

9. a. Evaluate $x^2 + 6x - 16$ for $x = 0$.
b. Factor $x^2 + 6x - 16$.
c. Solve $x^2 + 6x - 16 = 0$.

10. The equation $3x^2 - 4x + 5 = 0$ is written in $ax^2 + bx + c = 0$ form. What are a, b, and c?

11. What is the first step that should be performed to solve each equation?
a. $x^2 + 7x = -6$
b. $x(x + 7) = -3$

12. a. How many solutions does the linear equation $2a + 3 = 2$ have?
b. How many solutions does the quadratic equation $2a^2 + 3a = 2$ have?

NOTATION *Complete each solution.*

13. $7y^2 + 14y = 0$

$\qquad$ $(y + 2) = 0$

$\qquad 7y = 0 \quad \text{or} \quad \boxed{} = 0$

$\qquad\quad y = 0 \qquad\qquad y = -2$

14. $\qquad 12p^2 - p - 6 = 0$

$\quad \left(\boxed{} - 3 \right)\left(3p + \boxed{} \right) = 0$

$\qquad\quad \boxed{} = 0 \quad \text{or} \quad 3p + 2 = \boxed{}$

$\qquad\qquad 4p = \boxed{} \qquad\qquad 3p = \boxed{}$

$\qquad\qquad\quad p = \dfrac{3}{4} \qquad\qquad\quad p = -\dfrac{2}{3}$

PRACTICE *Solve each equation.*

15. $(x - 2)(x + 3) = 0$

16. $(x - 3)(x - 2) = 0$

17. $(2s - 5)(s + 6) = 0$

18. $(3h - 4)(h + 1) = 0$

19. $(x - 1)(x + 2)(x - 3) = 0$

20. $(x + 2)(x + 3)(x - 4) = 0$

21. $x(x - 3) = 0$ $\qquad$ **22.** $x(x + 5) = 0$

23. $x(2x - 5) = 0$ $\qquad$ **24.** $x(5x + 7) = 0$

25. $w^2 - 7w = 0$ $\qquad$ **26.** $p^2 + 5p = 0$

27. $3x^2 + 8x = 0$ $\qquad$ **28.** $5x^2 - x = 0$

29. $8s^2 - 16s = 0$ $\qquad$ **30.** $15s^2 - 20s = 0$

31. $x^2 - 25 = 0$ $\qquad$ **32.** $x^2 - 36 = 0$

33. $4x^2 - 1 = 0$ $\qquad$ **34.** $9y^2 - 1 = 0$

35. $9y^2 - 4 = 0$ $\qquad$ **36.** $16z^2 - 25 = 0$

37. $x^2 = 100$ $\qquad$ **38.** $z^2 = 25$

39. $4x^2 = 81$ $\qquad$ **40.** $9y^2 = 64$

41. $x^2 - 13x + 12 = 0$

42. $x^2 + 7x + 6 = 0$

43. $x^2 - 4x - 21 = 0$

44. $x^2 + 2x - 15 = 0$

45. $x^2 - 9x + 8 = 0$

46. $x^2 - 14x + 45 = 0$

47. $a^2 + 8a = -15$

48. $a^2 - a = 56$

49. $2y - 8 = -y^2$

50. $-3y + 18 = y^2$

51. $x^3 + 3x^2 + 2x = 0$

52. $x^3 - 7x^2 + 10x = 0$

53. $k^3 - 27k - 6k^2 = 0$

54. $j^3 - 22j - 9j^2 = 0$

55. $(x - 1)(x^2 + 5x + 6) = 0$

56. $(x - 2)(x^2 - 8x + 7) = 0$

57. $2x^2 - 5x + 2 = 0$

58. $2x^2 + x - 3 = 0$

59. $5x^2 - 6x + 1 = 0$

60. $6x^2 - 5x + 1 = 0$

61. $4r^2 + 4r = -1$

62. $9m^2 + 6m = -1$

63. $-15x^2 + 2 = -7x$

64. $-8x^2 - 10x = -3$

65. $x(2x - 3) = 20$

66. $x(2x - 3) = 14$

67. $(d + 1)(8d + 1) = 18d$

68. $4h(3h + 2) = h + 12$

69. $2x(3x^2 + 10x) = -6x$

70. $2x^3 = 2x(x + 2)$

71. $x^3 + 7x^2 = x^2 - 9x$

72. $x^2(x + 10) = 2x(x - 8)$

APPLICATIONS *In Exercises 73–74, an object has been thrown straight up into the air. The formula $h = vt - 16t^2$ gives the height h of the object above the ground after t seconds, when it is thrown upward with an initial velocity v.*

73. TIME OF FLIGHT   After how many seconds will the object hit the ground if it is thrown with a velocity of 144 feet per second?

74. TIME OF FLIGHT   After how many seconds will the object hit the ground if it is thrown with a velocity of 160 feet per second?

75. OFFICIATING   Before a football game, a coin toss is used to determine which team will kick off. See Illustration 1. The height h (in feet) of a coin above the ground t seconds after being flipped up into the air is given by $h = -16t^2 + 22t + 3$. How long does a team captain have to call heads or tails if it must be done while the coin is in the air?

ILLUSTRATION 1

76. DOLPHINS   See Illustration 2. The height *h* in feet reached by a dolphin *t* seconds after breaking the surface of the water is given by

$$h = -16t^2 + 32t$$

How long will it take the dolphin to jump out of the water and touch the trainer's hand?

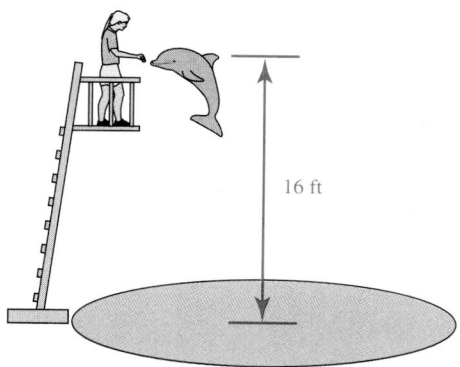

16 ft

ILLUSTRATION 2

77. EXHIBITION DIVING   In Acapulco, Mexico, men diving from a cliff to the water 64 feet below are quite a tourist attraction. A diver's height *h* above the water *t* seconds after diving is given by $h = -16t^2 + 64$. How long does a dive last?

78. FORENSIC MEDICINE   The kinetic energy *E* of a moving object is given by $E = \frac{1}{2}mv^2$, where *m* is the mass of the object (in kilograms) and *v* is the object's velocity (in meters per second). Kinetic energy is measured in joules. Examining the damage done to a victim, a police pathologist determines that the energy of a 3-kilogram mass at impact was 54 joules. Find the velocity at impact. (*Hint:* Multiply both sides of the equation by 2.)

79. CHOREOGRAPHY   For the finale of a musical, 36 dancers are to assemble in a triangular-shaped series of rows, where each successive row has one more dancer than the previous row. Illustration 3 shows the beginning of such a formation. The relationship between the number of rows *r* and the number of dancers *d* is given by

$$d = \frac{1}{2}r(r + 1)$$

Determine the number of rows in the formation. (*Hint:* Multiply both sides of the equation by 2.)

ILLUSTRATION 3

80. CRAFTS   Illustration 4 shows how a geometric wall hanging can be created by stretching yarn from peg to peg across a wooden ring. The relationship between the number of pegs *p* placed evenly around the ring and the number of yarn segments *s* that criss-cross the ring is given by the formula

$$s = \frac{p(p - 3)}{2}$$

How many pegs are needed if the designer wants 27 segments to criss-cross the ring? (*Hint:* Multiply both sides of the equation by 2.)

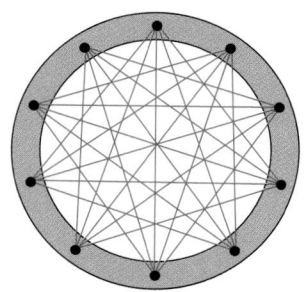

ILLUSTRATION 4

81. INSULATION   The area of the rectangular slab of foam insulation in Illustration 5 is 36 square meters. Find the dimensions of the slab.

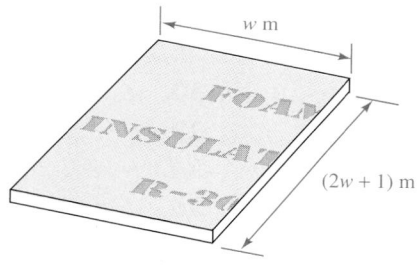

w m

(2*w* + 1) m

ILLUSTRATION 5

82. SHIPPING PALLETS   The length of a rectangular shipping pallet is 2 feet less than 3 times its width. Its area is 21 square feet. Find the dimensions of the pallet.

83. BOATING   The inclined ramp of the boat launch shown in Illustration 6 is 8 meters longer than the "rise" of the ramp. The "run" is 7 meters longer than the "rise." How long are the three sides of the ramp?

Rise

Run

ILLUSTRATION 6

84. CAR REPAIR   To create some space to work under the front end of a car, a mechanic drives it up steel ramps. See Illustration 7. The ramp is 1 foot longer than the back, and the base is 2 feet longer than the back of the ramp. Find the length of each side of the ramp.

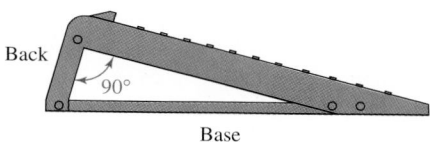

ILLUSTRATION 7

85. GARDENING TOOLS   The dimensions (in millimeters) of the teeth of a pruning saw blade are given in Illustration 8. Find each length.

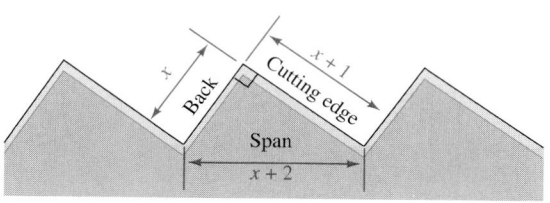

ILLUSTRATION 8

86. HARDWARE   An aluminum brace used to support a wooden shelf has a length that is 2 inches less than twice the width of the shelf. The brace is anchored to the wall 8 inches below the shelf, as shown in Illustration 9. Find the width of the shelf and the length of the brace.

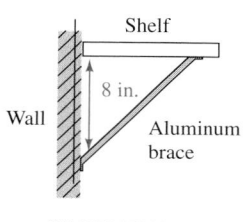

ILLUSTRATION 9

87. DESIGNING A TENT   The length of the base of the triangular sheet of canvas above the door of the tent in Illustration 10 is 2 feet more than twice its height. The area is 30 square feet. Find the height and the length of the base of the triangle.

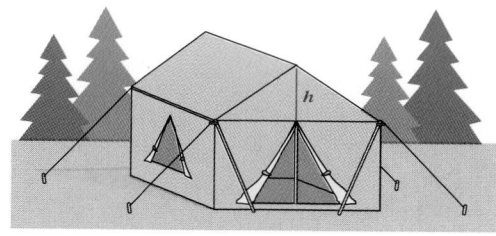

ILLUSTRATION 10

88. DIMENSIONS OF A TRIANGLE   The height of a triangle is 2 inches less than 5 times the length of its base. The area is 36 square inches. Find the length of the base and the height of the triangle.

89. TUBING   A piece of cardboard in the shape of a parallelogram is twisted to form the tube for a roll of paper towels. (See Illustration 11.) The parallelogram has an area of 60 square inches. If its height h is 7 inches more than the length of the base b, what is the circumference of the tube? (*Hint:* The formula for the area of a parallelogram is $A = bh$.)

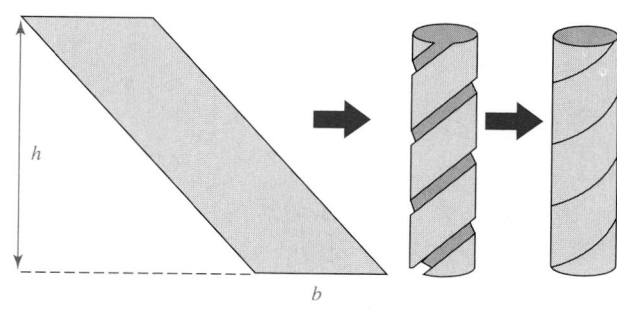

ILLUSTRATION 11

90. SWIMMING POOL BORDER   The owners of the rectangular swimming pool in Illustration 12 want to surround the pool with a crushed-stone border of uniform width. They have enough stone to cover 74 square meters. How wide should they make the border? (*Hint:* The area of the larger rectangle minus the area of the smaller is the area of the border.)

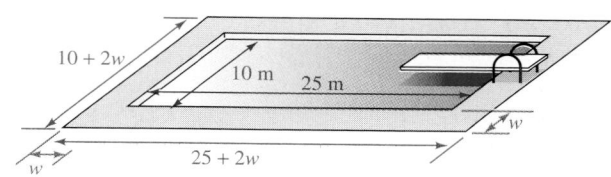

ILLUSTRATION 12

91. HOUSE CONSTRUCTION   The formula for the area of a trapezoid is

$$A = \frac{h(B + b)}{2}$$

The area of the trapezoidal truss in Illustration 13 is 24 square meters. Find the height of the trapezoid if one base is 8 meters and the other base is the same as the height. (*Hint:* Multiply both sides of the equation by 2.)

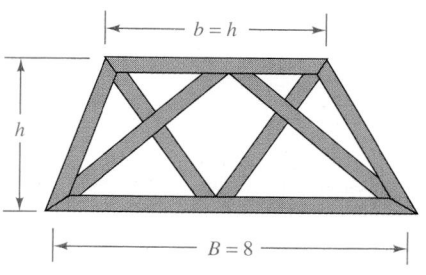

ILLUSTRATION 13

92. VOLUME OF A PYRAMID   The volume of a pyramid is given by the formula

$$V = \frac{Bh}{3}$$

where B is the area of its base and h is its height. The volume of the pyramid in Illustration 14 is 192 cubic centimeters. Find the dimensions of its rectangular base if one edge of the base is 2 centimeters longer than the other and the height of the pyramid is 12 centimeters. (*Hint:* Multiply both sides of the equation by 3.)

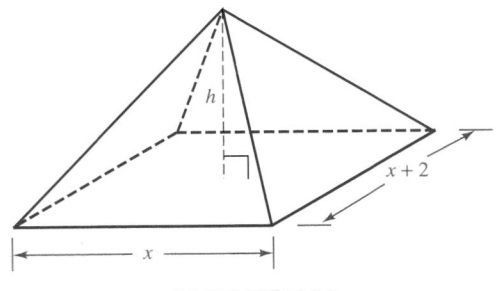

ILLUSTRATION 14

WRITING

93. What is wrong with the logic used by a student to "solve" $x^2 + x = 6$?

$$x(x + 1) = 6$$
$$x = 6 \quad \text{or} \quad x + 1 = 6$$
$$x = 5$$

So the solutions are 6 or 5.

94. Suppose that to find the length of the base of a triangle, you write a quadratic equation and solve it to find $b = 6$ or $b = -8$. Explain why one solution should be discarded.

REVIEW

95. EXERCISE   A doctor advises one patient to exercise at least 15 minutes but less than 30 minutes per day. Use a compound inequality to express the range of these times in minutes.

96. SNACKS   A bag of peanuts is worth $0.30 less than a bag of cashews. Equal amounts of peanuts and cashews are used to make 40 bags of a mixture that is worth $1.05 per bag. How much is a bag of cashews worth?

97. A rectangle is 3 times as long as it is wide, and its perimeter is 120 centimeters. Find its area.

98. INVESTING   A woman invests $15,000, part at 7% annual interest and part at 8% annual interest. If she receives $1,100 interest per year, how much did she invest at 7%?

KEY CONCEPT

Factoring

Factoring polynomials is the reverse of the process of multiplying polynomials. When we factor a polynomial, we write it as a product of two or more factors.

1. In the following problem, the distributive property is used to multiply a monomial and a binomial.

Find $3(x + 9)$.

$3(x + 9) = 3 \cdot x + 3 \cdot 9$
$ = 3x + 27$

Rewrite this so that it becomes a factoring problem. What would you start with? What would the answer be?

2. In the following problem, we multiply two binomials.

Find $(x + 3)(x + 9)$.

$(x + 3)(x + 9) = x^2 + 9x + 3x + 27$
$ = x^2 + 12x + 27$

Rewrite this so that it becomes a factoring problem. What would you start with? What would the answer be?

A Factoring Strategy

The following flowchart leads you through the steps to identify the type(s) of factoring necessary to factor any given polynomial having two or more terms.

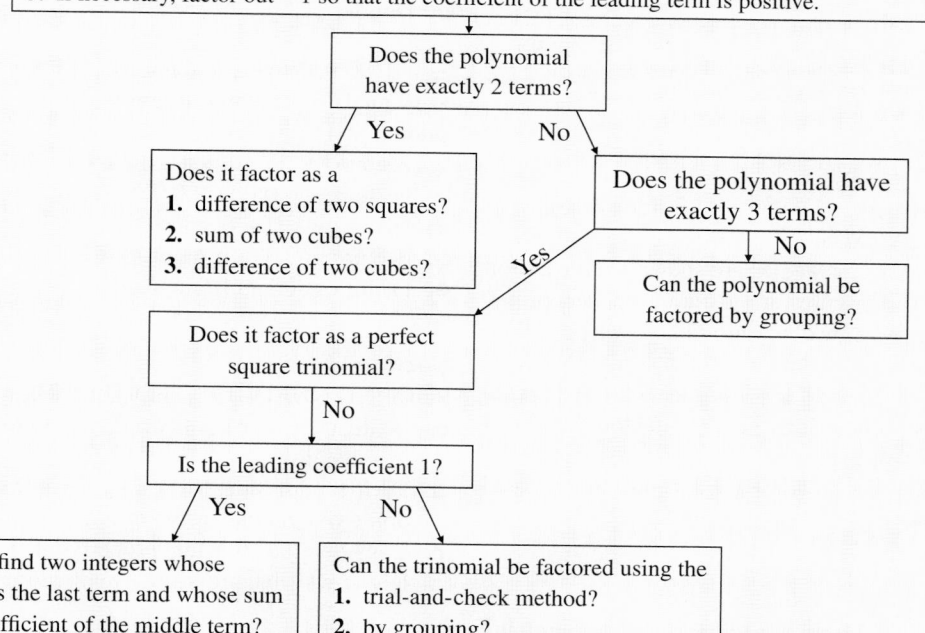

1. Write the polynomial in descending powers of a variable.
2. Factor out the GCF.
3. If necessary, factor out -1 so that the coefficient of the leading term is positive.

Does the polynomial have exactly 2 terms?

Yes / No

Does it factor as a
1. difference of two squares?
2. sum of two cubes?
3. difference of two cubes?

Does the polynomial have exactly 3 terms?

No

Can the polynomial be factored by grouping?

Yes

Does it factor as a perfect square trinomial?

No

Is the leading coefficient 1?

Yes / No

Can you find two integers whose product is the last term and whose sum is the coefficient of the middle term?

Can the trinomial be factored using the
1. trial-and-check method?
2. by grouping?

Factor each polynomial completely.

3. $-3a^2 + 21a - 36$

4. $x^2 - 121y^2$

5. $rt + 2r + st + 2s$

6. $v^3 - 8$

7. $6t^2 - 19t + 15$

8. $25y^2 - 20y + 4$

9. $2r^3 - 50r$

10. $46w - 6 + 16w^2$

723

Section 11.1

PRIME NUMBERS We can use a procedure called the **sieve of Eratosthenes** to find all the prime numbers in the set of the first 100 whole numbers. Give each member in your group a copy of the table shown in Illustration 1. Cross out 1, since it is not a prime number by definition. Cross out any numbers divisible by 2, 3, 5, or 7, because they have a factor of 2, 3, 5, or 7 and thus would not be prime. Don't cross out 2, 3, 5, or 7, because they are prime numbers. At the end of this process, you should end up with the first 25 prime numbers.

1	2	3	4	5	6	7	8	9	10
11	12	13	14	15	16	17	18	19	20
21	22	23	24	25	26	27	28	29	30
31	32	33	34	35	36	37	38	39	40
41	42	43	44	45	46	47	48	49	50
51	52	53	54	55	56	57	58	59	60
61	62	63	64	65	66	67	68	69	70
71	72	73	74	75	76	77	78	79	80
81	82	83	84	85	86	87	88	89	90
91	92	93	94	95	96	97	98	99	100

ILLUSTRATION 1

Section 11.2

FACTORING TRINOMIALS

a. Four squares and four rectangles are shown in Illustration 2. The dimensions of the figures are given in the same units. Find the sum of their areas by combining like terms.

b. Cut out the figures and assemble them into a large rectangle having a length of $(x + 3)$ units and a width of $(x + 1)$ units. Note that these dimensions give the factored form of $x^2 + 4x + 3$, the answer to part a.

c. Make a new model that could be used to find the factored form of $x^2 + 5x + 4$.

Section 11.3

COMPARING METHODS Factor $18x^2 + 3x - 10$ by using the trial-and-check method and by using the grouping method. Which method do you think is better? Explain why.

Section 11.4

COMPARING METHODS Show how $x^6 - 1$ can be factored in two ways: first as a difference of two squares and then as a difference of two cubes.

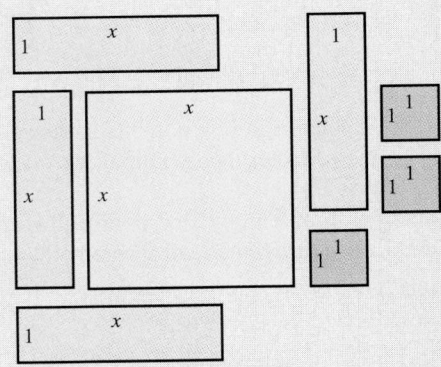

ILLUSTRATION 2

SUMS AND DIFFERENCES OF CUBES Long division can be used to verify the following factoring formulas.

$$x^3 + y^3 = (x + y)(x^2 - xy + y^2)$$
$$x^3 - y^3 = (x - y)(x^2 + xy + y^2)$$

a. Divide: $x + y \overline{)x^3 + y^3}$.
(*Hint:* Write $x^3 + y^3$ as $x^3 + 0x^2y + 0xy^2 + y^3$)

b. Divide: $x - y \overline{)x^3 - y^3}$.

FACTORING Factor each of the following polynomials completely. Begin by factoring out the indicated amount first.

a. Factor out 503 from $44{,}767a - 12{,}072b$.

b. Factor out 0.05 from $0.05t^2 - 0.25t + 0.3$.

c. Factor out -1.8 from $-16.2s^2 - 9s + 7.2$.

d. Factor out 2.375 from $38x^2 - 9.5y^2$.

e. Factor out 6.758 from $182.466b^3 - 54.064c^3$.

Section 11.5

SOLVING QUADRATIC EQUATIONS

a. Write $x^2 = 12x$ in quadratic form and then use factoring to solve it.

b. Divide both sides of $x^2 = 12x$ by x to "solve" it. Why don't you get two answers as you did in part a?

c. Consider the division property of equality and explain how it was inappropriately applied in part b.

QUADRATIC EQUATIONS Find a quadratic equation that has the following solutions. To do this, use the factoring method in reverse.

a. 3, 4

b. $-5, 1$

c. $\dfrac{1}{3}, 9$

d. $-\dfrac{2}{5}, -\dfrac{4}{3}$

CHAPTER REVIEW

Factoring Out the Greatest Common Factor and Factoring by Grouping

CONCEPTS

A *prime number* is a natural number greater than 1 whose only factors are 1 and itself. A natural number is in *prime-factored form* when it is written as the product of prime numbers.

To find the *greatest common factor* (GCF) of several monomials:
1. Prime factor each monomial.
2. List each common factor the least number of times it appears in any one monomial.
3. Find the product of the factors in the list to obtain the GCF.

To *factor by grouping*, arrange the polynomial so that the first two terms have a common factor and the last two terms have a common factor. Factor out the common factor from both groups. Then factor out the resulting common binomial factor.

REVIEW EXERCISES

1. Find the prime factorization of each number.
- **a.** 35
- **b.** 45
- **c.** 96
- **d.** 99
- **e.** 2,050
- **f.** 4,096

2. Factor each polynomial completely.
- **a.** $3x + 9y$
- **b.** $5ax^2 + 15a$
- **c.** $7s^2 + 14s$
- **d.** $\pi ab - \pi ac$
- **e.** $2x^3 + 4x^2 - 8x$
- **f.** $x^2yz + xy^2z + xyz$
- **g.** $-5ab^2 + 10a^2b - 15ab$
- **h.** $4(x - 2) - x(x - 2)$

3. Factor out -1 from each polynomial.
- **a.** $-a - 7$
- **b.** $-4t^2 + 3t - 1$

4. Factor by grouping:
- **a.** $2c + 2d + ac + ad$
- **b.** $3xy + 9x - 2y - 6$
- **c.** $2a^3 - a + 2a^2 - 1$
- **d.** $4m^2n + 12m^2 - 8mn - 24m$

Factoring Trinomials of the Form $x^2 + bx + c$

To *factor a trinomial* of the form $x^2 + bx + c$ means to write it as the product of two binomials.

To factor $x^2 + bx + c$, find two integers whose product is c and whose sum is b.

$$\left(x \quad\right)\left(x \quad\right)$$

Write the trinomial in descending powers of the variable and factor out -1 when applicable.

5. Complete the table.

Factors of 6	Sum of the factors of 6
1(6)	
2( )	
(−6)	
−2(−3)	

6. Factor each trinomial, if possible.

  a. $x^2 + 2x - 24$ **b.** $x^2 - 4x - 12$

  c. $n^2 - 7x + 10$ **d.** $t^2 + 10t + 15$

  e. $-y^2 + 9y - 20$ **f.** $10y + 9 + y^2$

  g. $c^2 + 3cd - 10d^2$ **h.** $-3mn + m^2 + 2n^2$

If a trinomial cannot be factored using only integers, it is called a *prime polynomial.*

7. Explain how we can check to see if $(x - 4)(x + 5)$ is the factorization of $x^2 + x - 20$.

The *GCF* should always be factored out first. A trinomial is *factored completely* when it is expressed as a product of prime polynomials.

8. Completely factor each trinomial.

  a. $5a^2 + 45a - 50$ **b.** $-4x^2y - 4x^3 + 24xy^2$

SECTION 11.3	*Factoring Trinomials of the Form $ax^2 + bx + c$*

To factor $ax^2 + bx + c$ using the *trial-and-check* factoring method, we must determine four integers. Use the FOIL method to check your work.

Factors
of a

$(\;\boxed{\;}\;x + \boxed{\;}\;)(\;\boxed{\;}\;x + \boxed{\;}\;)$

Factors
of c

To factor $ax^2 + bx + c$ using the *grouping* method, we write it as

$ax^2 + \boxed{\;}\;x + \boxed{\;}\;x + c$

9. Factor each trinomial completely, if possible.

  a. $2x^2 - 5x - 3$ **b.** $10y^2 + 21y - 10$

  c. $-3x^2 + 14x + 5$ **d.** $-9p^2 - 6p + 6p^3$

  e. $4b^2 - 17bc + 4c^2$ **f.** $3y^2 + 7y - 11$

10. ENTERTAINING The rectangular-shaped area occupied by a table setting shown in Illustration 1 is $(12x^2 - x - 1)$ square inches. Factor the expression to find the binomials that represent the length and width of the table setting.

ILLUSTRATION 1

SECTION 11.4	*Special Factorizations and a Factoring Strategy*

Special product formulas are used to factor *perfect square trinomials.*

$x^2 + 2xy + y^2 = (x + y)^2$

$x^2 - 2xy + y^2 = (x - y)^2$

11. Factor each polynomial completely.

  a. $x^2 + 10x + 25$ **b.** $9y^2 - 24y + 16$

  c. $-z^2 + 2z - 1$ **d.** $25a^2 + 20ab + 4b^2$

To factor the *difference of two squares,* use the formula

$F^2 - L^2 = (F + L)(F - L)$

12. Factor each polynomial completely, if possible.

  a. $x^2 - 9$ **b.** $49t^2 - 25y^2$

  c. $x^2y^2 - 400$ **d.** $8at^2 - 32a$

  e. $c^4 - 64$ **f.** $h^2 + 36$

To factor the *sum* and *difference* of two cubes, use the formulas

$$\mathbf{F}^3 + \mathbf{L}^3 = (\mathbf{F} + \mathbf{L})(\mathbf{F}^2 - \mathbf{FL} + \mathbf{L}^2)$$

$$\mathbf{F}^3 - \mathbf{L}^3 = (\mathbf{F} - \mathbf{L})(\mathbf{F}^2 + \mathbf{FL} + \mathbf{L}^2)$$

To factor a random polynomial, use the *factoring strategy* discussed in Section 11.4.

13. Factor each polynomial completely, if possible.

a. $h^3 + 1$　　　　**b.** $125p^3 + q^3$

c. $x^3 - 27$　　　　**d.** $16x^5 - 54x^2y^3$

14. Factor each polynomial completely, if possible.

a. $14y^3 + 6y^4 - 40y^2$　　**b.** $s^2t + s^2u^2 + tv + u^2v$

c. $j^4 - 16$　　　　　**d.** $3j^3 - 24k^3$

e. $12w^2 - 36w + 27$　　**f.** $121p^2 + 36q^2$

SECTION 11.5　　*Quadratic Equations*

A *quadratic equation* is an equation of the form
$ax^2 + bx + c = 0$ $(a \neq 0)$,
where a, b, and c represent real numbers.

15. Solve each quadratic equation by factoring.

a. $x^2 + 2x = 0$　　　　**b.** $x(x - 6) = 0$

c. $x^2 - 9 = 0$　　　　**d.** $a^2 - 7a + 12 = 0$

e. $t^2 + 4t + 4 = 0$　　　**f.** $2x - x^2 + 24 = 0$

g. $5a^2 - 6a + 1 = 0$　　**h.** $2p^3 = 2p(p + 2)$

To use the *factoring method* to solve a quadratic equation:
1. Write the equation in $ax^2 + bx + c = 0$ form.
2. Factor the left-hand side.
3. Use the *zero-factor property* (if $ab = 0$, then $a = 0$ or $b = 0$) and set each factor equal to zero.
4. Solve each resulting linear equation.
5. Check the results in the original equation.

16. CONSTRUCTION   The face of the triangular preformed concrete panel shown in Illustration 2 has an area of 45 square meters, and its base is 3 meters longer than twice its height. How long is its base?

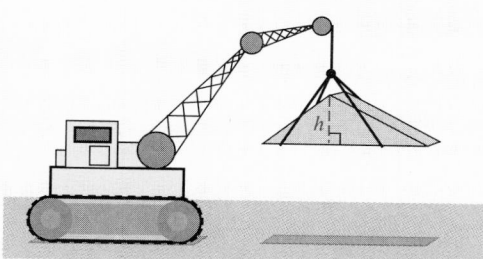

ILLUSTRATION 2

17. GARDENING   A rectangular flower bed occupies 27 square feet and is 3 feet longer than twice its width. Find its dimensions.

The Pythagorean theorem: If the length of the hypotenuse of a right triangle is c and the lengths of the two legs are a and b, then $c^2 = a^2 + b^2$.

18. TIGHTROPE WALKER
A circus performer intends to walk up a taut cable to a platform atop a pole, as shown in Illustration 3. How high above the ground is the platform?

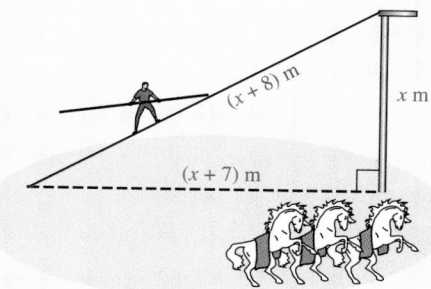

$(x + 8)$ m
x m
$(x + 7)$ m

ILLUSTRATION 3

Chapter 11 Test

Find the prime factorization of each number.

1. 196

2. 111

Factor each polynomial completely. If a polynomial cannot be factored, write "prime."

3. $4x + 16$

4. $30a^2b^3 - 20a^3b^2 + 5abc$

5. $q^2 - 81$

6. $x^2 + 9$

7. $16x^4 - 81$

8. $x^2 + 4x + 3$

9. $-x^2 + 9x + 22$

10. $9a - 9b + ax - bx$

11. $2a^2 + 5a - 12$

12. $18x^2 - 60xy + 50y^2$

13. $x^3 + 8$

14. $2a^3 - 54$

15. LANDSCAPING   See Illustration 1. The combined area of the portions of the square lot that the sprinkler doesn't reach is given by $4r^2 - \pi r^2$, where r is the radius of the circular spray. Factor this expression.

ILLUSTRATION 1

16. CHECKERS   The area of the square checkerboard in Illustration 2 is $25x^2 - 40x + 16$. Find the length of a side.

ILLUSTRATION 2

17. What is the greatest common factor of $4a^3b^2$ and $18ab^2$?

18. Factor $x^2 - 3x - 54$. Show a check of your answer.

Solve each equation.

19. $(x + 3)(x - 2) = 0$

20. $x^2 - 25 = 0$

21. $6x^2 - x = 0$

22. $x^2 + 6x + 9 = 0$

23. $6x^2 + x - 1 = 0$

24. $x^2 + 7x = -6$

25. DRIVING SAFETY   Virtually all cars have a "blind spot" where it is difficult for the driver to see a car behind and to the right. The area of the blind spot shown in Illustration 3 is 54 square feet. Find the width and length of the blind spot.

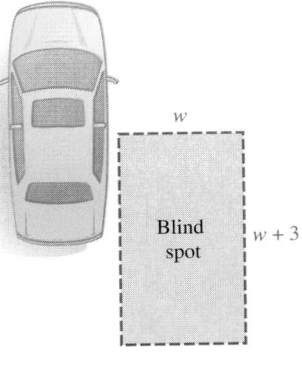

w

Blind
spot $w + 3$

ILLUSTRATION 3

26. What is a quadratic equation? Give an example.

27. Find the length of the hypotenuse of the right triangle shown in Illustration 4.

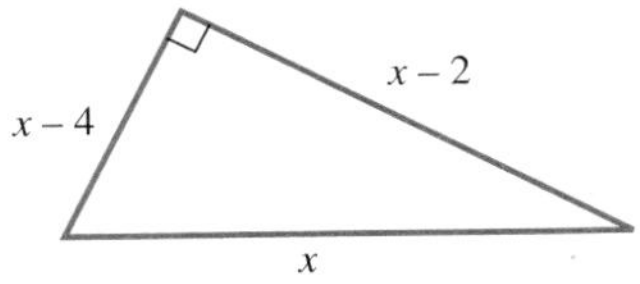

$x - 4$ $x - 2$

x

ILLUSTRATION 4

28. If the product of two numbers is 0, what conclusion can be drawn about the numbers?

Chapters 1–11 Cumulative Review Exercises

1. HEART RATE   Refer to the graph in Illustration 1. Determine the difference in the maximum heart beat rate for a 70-year-old as compared to someone half that age.

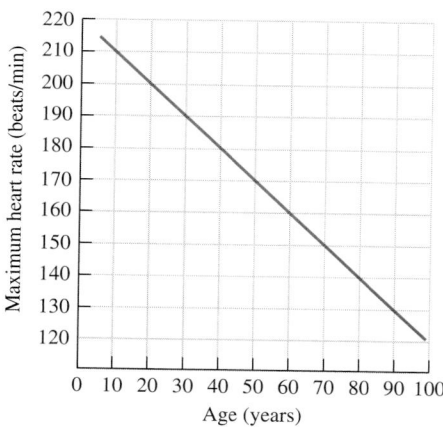

Based on data from *Cardiopulmonary Anatomy and Physiology: Essentials for Respiratory Care,* 2nd ed.

ILLUSTRATION 1

2. Give the prime factorization of 250.

3. Write $\dfrac{124}{125}$ as a decimal.

4. Tell whether each statement is true or false.
   a. Every integer is a whole number.
   b. Every integer is a rational number.
   c. π is a real number.

5. Find the quotient: $\dfrac{16}{5} \div \dfrac{10}{3}$.

6. What is -3 cubed?

Evaluate each expression.

7. $3 + 2[-1 - 4(5)]$

8. $\dfrac{|-25| - 2(-5)}{9 - 2^4}$

9. Evaluate $\dfrac{-x - a}{y - b}$ for $x = -2$, $y = 1$, $a = 5$, and $b = 2$.

10. Which division is undefined, $\dfrac{0}{5}$ or $\dfrac{5}{0}$?

Simplify each expression.

11. $-8y^2 - 5y^2 + 6$

12. $3z + 2(y - z) + y$

Solve each equation.

13. $-(3a + 1) + a = 2$

14. $2 - (4x + 7) = 3 + 2(x + 2)$

15. $\dfrac{3t - 21}{2} = t - 6$

16. $-\dfrac{1}{3} - \dfrac{x}{5} = \dfrac{3}{2}$

17. Solve $A = P + Prt$ for t.

18. Solve $-\dfrac{x}{2} + 4 > 5$ and graph the solution.

19. GEOMETRY TOOL   Find the total distance around the outside edge of the protractor shown in Illustration 2. Round to the nearest tenth of an inch.

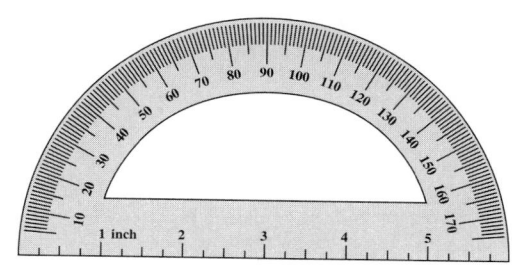

ILLUSTRATION 2

20. What is the formula for simple interest?

21. What is the value of x twenty-dollar bills?

22. PHOTOGRAPHIC CHEMICALS   A photographer wishes to mix 6 liters of a 5% acetic acid solution with a 10% solution to get a 7% solution. How many liters of 10% solution must be added?

Graph each equation.

23. $y = (x + 2)^2$

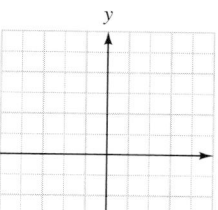

24. $y = |x| - 2$

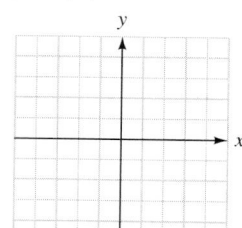

25. Find the slope and the y-intercept of the graph of $3x - 3y = 6$.

26. Write the equation of the line passing through $(-2, 5)$ and $(-3, -2)$. Answer in slope–intercept form.

27. Graph the line passing through $(-4, 1)$ that having slope $m = -3$.

28. Graph $8x + 4y = -24$.

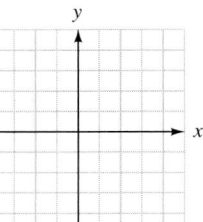

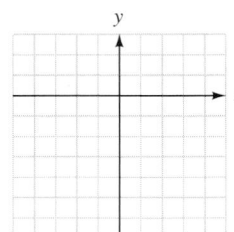

29. If two lines are parallel, what can be said about their slopes?

30. BEVERAGES   Illustration 3 shows the annual per-person consumption of coffee and tea in the United States.
  a. What was the rate of change in coffee consumption for 1995–2000?
  b. Did the per-person consumption of tea change for 1995–2000?

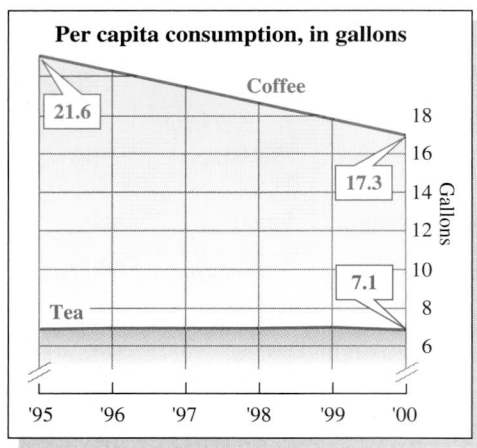

Based on data from Davenport & Co. and the U.S. Department of Agriculture

ILLUSTRATION 3

31. If $g(x) = 3x - x^3$, what is $g(-2)$?

32. Write 1,700,000 in scientific notation.

Simplify each expression. Write each answer without using parentheses or negative exponents.

33. $-y^2(4y^3)$

34. $\dfrac{15(x^2y^5)^5}{21(x^3y)^2}$

35. $\left(\dfrac{b^5}{b^{-2}}\right)^{-2}$

36. $2x^0$

Do each operation.

37. $(x^2 - 3x + 8) - (3x^2 + x + 3)$

38. $4b^3(2b^2 - 2b)$

39. $(y - 6)^2$

40. $(3x - 2)(x + 4)$

41. $\dfrac{12a^2b^2 - 8a^2b - 4ab}{4ab}$

42. $x - 3\overline{)2x^2 - 5x - 3}$

43. PLAYPEN   See Illustration 4.
  a. Find the perimeter of the playpen.
  b. Find the area of the floor of the playpen.

  c. Find the volume of the playpen.

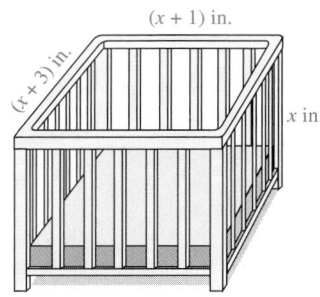

ILLUSTRATION 4

44. What is the degree of the polynomial $7y^3 + 4y^2 + y + 3$?

Factor each polynomial completely.

45. $b^3 - 3b^2$

46. $u^2 - 3 + 2u$

47. $2x^2 - 3x - 2$

48. $9z^2 - 1$

49. $-5a^2 + 25a - 30$

50. $ax + bx + ay + by$

51. $t^3 - 8$

52. $4a^2 - 12a + 9$

Solve each equation.

53. $15s^2 - 20s = 0$

54. $2x^2 - 5x = -2$

Rational Expressions and Equations

12

12.1 Simplifying Rational Expressions

12.2 Multiplying and Dividing Rational Expressions

12.3 Adding and Subtracting Rational Expressions

12.4 Complex Fractions

12.5 Rational Equations and Problem Solving

12.6 Proportions and Similar Triangles

12.7 Variation

Key Concept: Expressions and Equations

Accent on Teamwork

Chapter Review

Chapter Test

Cumulative Review Exercises

12.1 Simplifying Rational Expressions

In this section, you will learn about

- Evaluating rational expressions • Simplifying rational expressions
- Division of opposites

INTRODUCTION. Fractions such as $\frac{1}{2}$ and $\frac{3}{4}$ that are the quotient of two integers are *rational numbers*. Fractions such as

$$\frac{3}{2y}, \qquad \frac{x}{x+2}, \qquad \text{and} \qquad \frac{5a^2 + b^2}{3a - b}$$

where the numerators and the denominators are polynomials are called **rational expressions.**

Evaluating rational expressions

To evaluate a rational expression, we replace each variable with a given number value and simplify.

EXAMPLE 1 *Evaluating rational expressions.* Find the value of $\frac{2x - 1}{x^2 + 1}$ for $x = -3$.

Solution
We replace each x in the expression with -3 and then evaluate the numerator and denominator separately.

$$\frac{2x - 1}{x^2 + 1} = \frac{2(-3) - 1}{(-3)^2 + 1} \qquad \text{Substitute } -3 \text{ for } x.$$

$$= \frac{-6 - 1}{9 + 1} \qquad \begin{array}{l}\text{In the numerator, do the multiplication. In the} \\ \text{denominator, evaluate the exponential expression.}\end{array}$$

$$= -\frac{7}{10}$$

Since rational expressions indicate division, we must make sure that the denominator of a rational expression is not 0.

EXAMPLE 2 *Restricted values.* Find the values for x for which each rational expression is undefined: **a.** $\dfrac{7x}{x - 5}$ and **b.** $\dfrac{x - 1}{x^2 - x - 6}$.

Solution
a. The denominator of $\dfrac{7x}{x - 5}$ will be 0 if we replace x with 5.

$$\frac{7x}{x - 5} = \frac{7(5)}{5 - 5} = \frac{35}{0}$$

Since $\frac{35}{0}$ is undefined, the rational expression is undefined for $x = 5$.

b. The expression $\dfrac{x - 1}{x^2 - x - 6}$ will be undefined for values of x that make the denominator 0. To find these values, we solve $x^2 - x - 6 = 0$.

$$x^2 - x - 6 = 0$$ Set the denominator of the rational expression equal to 0.

$$(x - 3)(x + 2) = 0$$ Factor the trinomial.

$$x - 3 = 0 \quad \text{or} \quad x + 2 = 0$$ Set each factor equal to 0.

$$x = 3 \qquad\qquad x = -2$$ Solve each equation.

Since the values 3 and -2 make the denominator 0, the expression is undefined for $x = 3$ or $x = -2$.

Answers: a. -9, **b.** 5 or -5 ■

Simplifying rational expressions

A fraction can be simplified by factoring the numerator and denominator and dividing out common factors shared by the numerator and denominator. For example, to simplify $\frac{18}{30}$ and $\frac{6}{15}$, we proceed as follows:

$$\frac{18}{30} = \frac{3 \cdot 6}{5 \cdot 6} = \frac{3 \cdot \overset{1}{\cancel{6}}}{5 \cdot \cancel{6}} = \frac{3}{5} \qquad \text{and} \qquad \frac{6}{15} = \frac{3 \cdot 2}{3 \cdot 5} = \frac{\overset{1}{\cancel{3}} \cdot 2}{\cancel{3} \cdot 5} = -\frac{2}{5}$$

When all common factors have been divided out, we say that the fraction has been **expressed in lowest terms.** The generalization of this idea is called the *fundamental property of fractions.*

Fundamental property of fractions

If a represents a real number and b and c represent nonzero real numbers,

$$\frac{ac}{bc} = \frac{a}{b}$$

The fundamental property of fractions enables us to divide out common factors of the numerator and denominator of a fraction. The resulting fraction is equivalent to the original fraction.

Simplifying rational expressions is similar to simplifying fractions. We use the following process.

Simplifying a rational expression to lowest terms

To simplify a rational expression to lowest terms,

1. Completely factor the numerator and denominator.
2. Divide out the common factors of the numerator and denominator.

EXAMPLE 3   *Simplifying rational expressions.*   Simplify $\dfrac{21x^2y}{14xy^2}$.

Solution

We look for common factors in the numerator and denominator and divide them out.

$$\frac{21x^2y}{14xy^2} = \frac{3 \cdot 7 \cdot x \cdot x \cdot y}{2 \cdot 7 \cdot x \cdot y \cdot y}$$ Factor the numerator and denominator.

$$= \frac{3 \cdot \overset{1}{\cancel{7}} \cdot \overset{1}{\cancel{x}} \cdot x \cdot \overset{1}{\cancel{y}}}{2 \cdot \cancel{7} \cdot \cancel{x} \cdot y \cdot \cancel{y}}$$ Divide out the common factors, 7, x, and y.

$$= \frac{3x}{2y}$$ Do the multiplications in the numerator and in the denominator: $3 \cdot 1 \cdot 1 \cdot x \cdot 1 = 3x$ and $2 \cdot 1 \cdot 1 \cdot y \cdot 1 = 2y$.

Self Check

Simplify $\dfrac{32a^3b^2}{24ab^4}$.

Answer: $\dfrac{4a^2}{3b^2}$ ■

To simplify rational expressions, we often make use of the factoring techniques discussed in the preceding chapter.

EXAMPLE 4 *Factoring to simplify rational expressions.* Write $\dfrac{x^2 + 3x}{3x + 9}$ in lowest terms.

Solution

We note that the terms of the numerator have a common factor of x and the terms of the denominator have a common factor of 3.

$$\frac{x^2 + 3x}{3x + 9} = \frac{x(x + 3)}{3(x + 3)} \qquad \text{Factor the numerator and the denominator.}$$

$$= \frac{x\cancel{(x + 3)}^{1}}{3\cancel{(x + 3)}_{1}} \qquad \text{Divide out the common factor, } x + 3.$$

$$= \frac{x}{3} \qquad \begin{array}{l}\text{Simplify the numerator: } x \cdot 1 = x. \\ \text{Simplify the denominator: } 3 \cdot 1 = 3.\end{array}$$

Self Check

Write

$$\frac{x^2 - 5x}{5x - 25}$$

in lowest terms.

Answer: $\dfrac{x}{5}$ ∎

EXAMPLE 5 *Factoring to simplify rational expressions.* Simplify $\dfrac{x^2 + 13x + 12}{x^2 - 144}$.

Solution

The numerator is a trinomial, and the denominator is a difference of two squares.

$$\frac{x^2 + 13x + 12}{x^2 - 144} = \frac{(x + 1)(x + 12)}{(x + 12)(x - 12)} \qquad \text{Factor the numerator and the denominator.}$$

$$= \frac{(x + 1)\cancel{(x + 12)}^{1}}{\cancel{(x + 12)}_{1}(x - 12)} \qquad \text{Divide out the common factor, } x + 12.$$

$$= \frac{x + 1}{x - 12}$$

Self Check

Simplify $\dfrac{3x^2 - 8x - 3}{x^2 - 9}$.

Answer: $\dfrac{3x + 1}{x + 3}$ ∎

COMMENT When simplifing a fraction, remember that only *factors* that are common to the *entire numerator* and the *entire denominator* can be divided out. For example, consider the correct simplification

$$\frac{5 + 8}{5} = \frac{13}{5}$$

It would be incorrect to divide out the common *term* of 5 in this simplification. Doing so would give an incorrect answer of 9.

$$\frac{5 + 8}{5} = \frac{\cancel{5}^{1} + 8}{\cancel{5}_{1}} = \frac{1 + 8}{1} = 9$$

When simplifying algebraic fractions, it is also incorrect to divide out terms common to both the numerator and denominator.

$$\frac{\overset{1}{\cancel{x}} + 5}{\underset{1}{\cancel{x}} + 6} \qquad \frac{a^2 - 3\overset{1}{\cancel{a}} + \overset{1}{\cancel{2}}}{\underset{1}{\cancel{a}} + \underset{1}{\cancel{2}}} \qquad \frac{\overset{1}{\cancel{y^2}} - 36}{\underset{1}{\cancel{y^2}} - y - 7}$$

Any number or algebraic expression divided by 1 remains unchanged. For example,

$$\frac{37}{1} = 37, \qquad \frac{5x}{1} = 5x, \qquad \text{and} \qquad \frac{3x + y}{1} = 3x + y.$$

In general, we have the following.

Division by 1

> For any real number a, $\dfrac{a}{1} = a$.

EXAMPLE 6 *Simplifying rational expressions.* Simplify $\dfrac{x^3 + x^2}{x + 1}$.

Solution

$$\frac{x^3 + x^2}{x + 1} = \frac{x^2(x + 1)}{x + 1} \qquad \text{Factor the numerator.}$$

$$= \frac{x^2(\overset{1}{\cancel{x + 1}})}{\underset{1}{\cancel{x + 1}}} \qquad \text{Divide out the common factor, } x + 1.$$

$$= \frac{x^2}{1} \qquad \text{Simplify.}$$

$$= x^2 \qquad \text{Denominators of 1 need not be written.}$$

Self Check

Simplify $\dfrac{a^2 + a - 2}{a - 1}$.

Answer: $a + 2$

EXAMPLE 7 *Dividing out common factors.* Simplify $\dfrac{5(x + 3) - 5}{7(x + 3) - 7}$.

Solution

We cannot divide out $x + 3$, because it is not a factor of the entire numerator, nor is it a factor of the entire denominator. Instead, we simplify the numerator and denominator, factor them, and then divide out any common factors.

$$\frac{5(x + 3) - 5}{7(x + 3) - 7} = \frac{5x + 15 - 5}{7x + 21 - 7} \qquad \text{Use the distributive property twice.}$$

$$= \frac{5x + 10}{7x + 14} \qquad \text{Combine like terms.}$$

$$= \frac{5(x + 2)}{7(x + 2)} \qquad \text{Factor the numerator and the denominator.}$$

$$= \frac{5(\overset{1}{\cancel{x + 2}})}{7(\underset{1}{\cancel{x + 2}})} \qquad \text{Divide out the common factor, } x + 2.$$

$$= \frac{5}{7}$$

Self Check

Simplify:

$$\frac{4(x - 2) + 4}{3(x - 2) + 3}$$

Answer: $\dfrac{4}{3}$

EXAMPLE 8   *Combining like terms.*   Simplify $\dfrac{x(x + 3) - 3(x - 1)}{x^2 + 3}$.

Solution
We begin by simplifying the numerator. Then we look for any common factors to divide out in the numerator and denominator.

$$\frac{x(x + 3) - 3(x - 1)}{x^2 + 3} = \frac{x^2 + 3x - 3x + 3}{x^2 + 3}$$ Use the distributive property twice in the numerator.

$$= \frac{x^2 + 3}{x^2 + 3}$$ Combine like terms in the numerator: $3x - 3x = 0$.

$$= \frac{\overset{1}{\cancel{x^2 + 3}}}{\underset{1}{\cancel{x^2 + 3}}}$$ Divide out the common factor, $x^2 + 3$.

$$= 1$$

Self Check
Simplify:

$\dfrac{a(a + 2) - 2(a - 1)}{a^2 + 2}$

Answer: 1 ∎

Sometimes a rational expression does not simplify. For example, to attempt to simplify

$$\frac{x^2 + x - 2}{x^2 + x}$$

we factor the numerator and the denominator.

$$\frac{x^2 + x - 2}{x^2 + x} = \frac{(x + 2)(x - 1)}{x(x + 1)}$$

Because there are no factors common to the numerator and denominator, this rational expression is already in lowest terms.

Division of opposites

If the terms of two polynomials are the same, except for sign, the polynomials are called **opposites (negatives)** of each other. For example, the following pairs of polynomials are opposites of each other:

$$x - y \qquad \text{and} \qquad -x + y$$
$$2a - 1 \qquad \text{and} \qquad -2a + 1$$
$$-3x^2 - 2x + 5 \qquad \text{and} \qquad 3x^2 + 2x - 5$$

Example 9 shows why the quotient of two binomials that are opposites is always -1.

EXAMPLE 9   *Division of opposites.*   Simplify $\dfrac{2a - 1}{1 - 2a}$.

Solution
We can rearrange terms in each numerator, factor out -1, and proceed as follows:

$$\frac{2a - 1}{1 - 2a} = \frac{-1 + 2a}{1 - 2a}$$ In the numerator, think of $2a - 1$ as $2a + (-1)$. Then change the order of the terms: $2a + (-1) = -1 + 2a$.

$$= \frac{-(1 - 2a)}{1 - 2a}$$ In the numerator, factor out -1: $-1 + 2a = -(1 - 2a)$.

$$= \frac{-\overset{1}{\cancel{(1 - 2a)}}}{\underset{1}{\cancel{1 - 2a}}}$$ Divide out the common factor, $1 - 2a$.

$$= -1$$

Self Check
Simplify $\dfrac{3p - 2}{2 - 3p}$.

Answer: -1 ∎

In general, we have this important fact.

Division of opposites | The quotient of any nonzero expression and its opposite is -1.

COMMENT Apply the preceding rule only to expressions that are opposites. For example, it would be incorrect to use this rule to simplify $\frac{x+1}{1+x}$. Since $x + 1$ equals $1 + x$ by the commutative property of addition, this is the quotient of a number and itself. The result is 1, not -1.

$$\frac{x+1}{1+x} = \frac{\overset{1}{\cancel{x+1}}}{\underset{1}{\cancel{x+1}}} = 1$$

STUDY SET   Section 12.1

VOCABULARY   *Fill in the blanks.*

1. In a fraction, the part above the fraction bar is called the _____, and the part below the fraction bar is called the _____.

2. A fraction that has polynomials in its numerator and denominator, such as $\frac{x+2}{x-3}$, is called a _____ expression.

3. Division by 0 is _____.

4. A fraction is in _____ terms when all common factors of the numerator and denominator have been divided out.

5. To _____ a rational expression means to factor the numerator and denominator completely and divide out common factors.

6. If the terms of two polynomials are the same, except for sign, the polynomials are called _____ of each other.

CONCEPTS

7. What value of x makes each rational expression undefined?

   a. $\dfrac{x+2}{x}$ **b.** $\dfrac{x+2}{x-6}$ **c.** $\dfrac{x+2}{x+6}$

8. Fill in the blank: When a _____ factor of the numerator and the denominator of a fraction is divided out, the resulting fraction is equivalent to the original fraction.

9. In the following work, what common factor has been divided out?

$$\frac{x^2 + 2x + 1}{x^2 + 4x + 3} = \frac{\overset{1}{\cancel{(x+1)}}(x+1)}{(x+3)\underset{1}{\cancel{(x+1)}}} = \frac{x+1}{x+3}$$

10. Simplify each rational expression.

   a. $\dfrac{x-8}{x-8}$ **b.** $\dfrac{x-8}{-x+8}$ **c.** $\dfrac{x-8}{1}$

11. Explain the error in the following work.

$$\frac{x}{x+2} = \frac{\overset{1}{\cancel{x}}}{\underset{1}{\cancel{x}}+2} = \frac{1}{3}$$

12. What is the first step in the process of simplifying

$$\frac{3(x+1) - 2x}{x+3}?$$

NOTATION   *Complete each solution.*

13. $\dfrac{x^2 + 5x - 6}{x^2 - 1} = \dfrac{(x +)(x - 1)}{(x + 1)(x -)}$

$= \dfrac{x + 6}{x + 1}$

14. $\dfrac{5(x+2) - 5}{4(x+2) - 4} = \dfrac{5x + - 5}{4x + - 4}$

$= \dfrac{5x + }{4x + }$

$= \dfrac{5}{4(x+1)}$

$= \dfrac{5}{4}$

PRACTICE   *Evaluate each expression for x = 6.*

15. $\dfrac{x-2}{x-5}$

16. $\dfrac{3x-2}{x-2}$

17. $\dfrac{-2x-3}{x^2-1}$

18. $\dfrac{x^2-11}{-x-4}$

19. $\dfrac{x^2-4x-12}{x^2+x-2}$

20. $\dfrac{x^2-1}{x^3-1}$

Which value(s) of x make each rational expression undefined?

21. $\dfrac{15}{x-2}$

22. $\dfrac{5x}{x+5}$

23. $\dfrac{15x+2}{16}$

24. $\dfrac{x^2-4x}{25}$

25. $\dfrac{x+1}{2x-1}$

26. $\dfrac{-6x}{3x-1}$

27. $\dfrac{30}{x^2-36}$

28. $\dfrac{2x-15}{x^2-49}$

29. $\dfrac{15}{x^2+x-2}$

30. $\dfrac{x-20}{x^2+2x-8}$

Write each fraction in lowest terms.

31. $\dfrac{28}{35}$

32. $\dfrac{14}{20}$

33. $\dfrac{9}{27}$

34. $\dfrac{15}{45}$

35. $-\dfrac{36}{48}$

36. $-\dfrac{32}{40}$

Simplify each expression. If it is already in lowest terms, so indicate. Assume that no denominators are zero.

37. $\dfrac{45}{9a}$

38. $\dfrac{48}{16y}$

39. $\dfrac{5+5}{5z}$

40. $\dfrac{(3-18)k}{25}$

41. $\dfrac{(3+4)a}{24-3}$

42. $\dfrac{x+x}{2}$

43. $\dfrac{2x}{3x}$

44. $\dfrac{5y}{7y}$

45. $\dfrac{6x^2}{4x^2}$

46. $\dfrac{9xy}{6xy}$

47. $\dfrac{2x^2}{3y}$

48. $\dfrac{5y^2}{2y^2}$

49. $\dfrac{15x^2y}{5xy^2}$

50. $\dfrac{12xz}{4xz^2}$

51. $\dfrac{6x+3}{3y}$

52. $\dfrac{4x+12}{2y}$

53. $\dfrac{x+3}{3x+9}$

54. $\dfrac{2x+14}{x-7}$

55. $\dfrac{x-7}{7-x}$

56. $\dfrac{18-d}{d-18}$

57. $\dfrac{6x-30}{5-x}$

58. $\dfrac{6t-42}{7-t}$

59. $\dfrac{12-3x^2}{x^2-x-2}$

60. $\dfrac{-5x+10}{x^2-4x+4}$

61. $\dfrac{x^2+3x+2}{x^2+x-2}$

62. $\dfrac{x^2+x-6}{x^2-x-2}$

63. $\dfrac{x^2-8x+15}{x^2-x-6}$

64. $\dfrac{x^2-6x-7}{x^2+8x+7}$

65. $\dfrac{2x^2-8x}{x^2-6x+8}$

66. $\dfrac{3y^2-15y}{y^2-3y-10}$

67. $\dfrac{2-a}{a^2-a-2}$

68. $\dfrac{4-b}{b^2-5b+4}$

69. $\dfrac{x^2+3x+2}{x^3+x^2}$

70. $\dfrac{6x^2-13x+6}{3x^2+x-2}$

71. $\dfrac{x^2-8x+16}{x^2-16}$

72. $\dfrac{3x+15}{x^2-25}$

73. $\dfrac{2x^2-8}{x^2-3x+2}$

74. $\dfrac{3x^2-27}{x^2+3x-18}$

75. $\dfrac{5x^2+2x-3}{x^2+2x-15}$

76. $\dfrac{x^2+4x-77}{x^2-4x-21}$

77. $\dfrac{x^2-3(2x-3)}{9-x^2}$

78. $\dfrac{x(x-8)+16}{16-x^2}$

79. $\dfrac{4(x+3)+4}{3(x+2)+6}$

80. $\dfrac{4+2(x-5)}{3x-5(x-2)}$

81. $\dfrac{x^2-9}{(2x+3)-(x+6)}$

82. $\dfrac{x^2+5x+4}{2(x+3)-(x+2)}$

83. $\dfrac{y-xy}{xy-x}$

84. $\dfrac{x^2+y^2}{x+y}$

85. $\dfrac{6a-6b+6c}{9a-9b+9c}$

86. $\dfrac{3a-3b-6}{2a-2b-4}$

87. $\dfrac{15x-3x^2}{25y-5xy}$

88. $\dfrac{xz-2x}{yz-2y}$

89. $\dfrac{a+b-c}{c-a-b}$

90. $\dfrac{x-y-z}{z+y-x}$

APPLICATIONS

91. ROOFING   The *pitch* of a roof is a measure of how steep or how flat the roof is. If pitch $= \frac{\text{rise}}{\text{run}}$, find the pitch of the roof of the cabin shown in Illustration 1. Express the result in lowest terms.

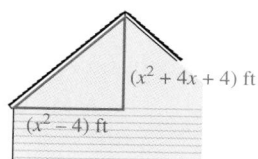

$(x^2 + 4x + 4)$ ft

$(x^2 - 4)$ ft

ILLUSTRATION 1

92. GRAPHIC DESIGN   A chart of the basic food groups, in the shape of an equilateral triangle, is to be enlarged and distributed to schools for display in their health classes. (See Illustration 2.) What is the length of a side of the original design divided by the length of a side of the enlargement? Express the result in lowest terms.

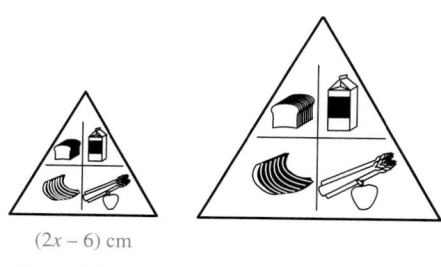

$(2x - 6)$ cm

Original design   Enlargement

ILLUSTRATION 2

93. WORD PROCESSOR   For the word processor shown in Illustration 3, the number of words w that can be typed on a piece of paper is given by the formula

$$w = \frac{8,000}{x}$$

where x is the font size used. Find the number of words that can be typed on a page for each font size choice shown.

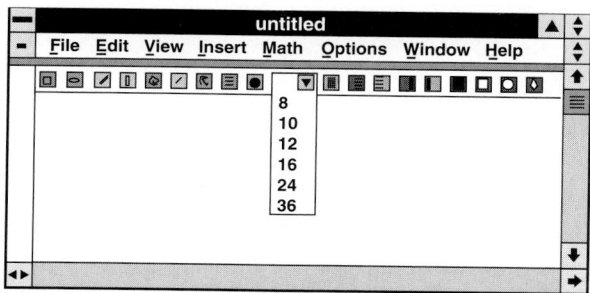

ILLUSTRATION 3

94. ORGAN PIPE   The number of vibrations n per second of an organ pipe is given by the formula

$$n = \frac{512}{L}$$

where L is the length of the pipe in feet. (See Illustration 4.) How many times per second will a 6-foot pipe vibrate?

L

ILLUSTRATION 4

WRITING

95. Explain why $\dfrac{x - 7}{7 - x} = -1$.

96. Explain the difference between a factor and a term. Give several examples.

97. Explain the error.

$$\frac{3(\overset{1}{\cancel{x + 1}}) - x}{\underset{1}{\cancel{x + 1}}} = 3 - x$$

98. Explain why there are no values for x for which $\dfrac{x - 7}{x^2 + 49}$ is undefined.

REVIEW

99. State the associative property of addition using the variables a, b, and c.

100. State the distributive property using the variables x, y, and z.

101. If $ab = 0$, what must be true about a or b?

102. What is the product of a number and 1?

103. What is the opposite of $-\dfrac{5}{3}$?

104. What is the cube of 2 squared?

12.2 *Multiplying and Dividing Rational Expressions*

In this section, you will learn about

- Multiplying rational expressions
- Multiplying a rational expression by a polynomial
- Dividing rational expressions • Dividing a rational expression by a polynomial
- Combined operations

INTRODUCTION. In this section, we extend the rules for multiplying and dividing numerical fractions to problems involving multiplication and division of rational expressions.

Multiplying rational expressions

To multiply fractions, we multiply their numerators and multiply their denominators. For example,

$$\frac{4}{7} \cdot \frac{3}{5} = \frac{4 \cdot 3}{7 \cdot 5} \quad \text{Multiply the numerators and multiply the denominators.}$$

$$= \frac{12}{35} \quad \begin{array}{l} \text{Do the multiplication in the numerator: } 4 \cdot 3 = 12. \\ \text{Do the multiplication in the denominator: } 7 \cdot 5 = 35. \end{array}$$

In general, we have the following rule.

Rule for multiplying fractions

> If a, b, c, and d represent real numbers and $b \neq 0$ and $d \neq 0$,
>
> $$\frac{a}{b} \cdot \frac{c}{d} = \frac{ac}{bd}$$

We use the same procedure to multiply rational expressions.

EXAMPLE 1 *Multiplying rational expressions.* Multiply: **a.** $\frac{x}{3} \cdot \frac{2}{5}$, **b.** $\frac{7}{9} \cdot \frac{-5}{3x}$, **c.** $\frac{x^2}{2} \cdot \frac{3}{y^2}$, and **d.** $\frac{t+1}{t} \cdot \frac{t-1}{t-2}$.

Solution

a. $\dfrac{x}{3} \cdot \dfrac{2}{5} = \dfrac{x \cdot 2}{3 \cdot 5}$

$\qquad = \dfrac{2x}{15}$

b. $\dfrac{7}{9} \cdot \dfrac{-5}{3x} = \dfrac{7(-5)}{9 \cdot 3x}$

$\qquad = \dfrac{-35}{27x}$

$\qquad = -\dfrac{35}{27x}$

c. $\dfrac{x^2}{2} \cdot \dfrac{3}{y^2} = \dfrac{x^2 \cdot 3}{2 \cdot y^2}$

$\qquad = \dfrac{3x^2}{2y^2}$

d. $\dfrac{t+1}{t} \cdot \dfrac{t-1}{t-2} = \dfrac{(t+1)(t-1)}{t(t-2)}$

Self Check

Multiply: $\dfrac{3x}{4} \cdot \dfrac{x-3}{5}$.

Answer: $\dfrac{3x(x-3)}{20}$

EXAMPLE 2   *Multiplying rational expressions.*

Multiply: $\dfrac{35x^2y}{7y^2z} \cdot \dfrac{z}{5xy}$.

Solution

$$\dfrac{35x^2y}{7y^2z} \cdot \dfrac{z}{5xy} = \dfrac{35x^2y \cdot z}{7y^2z \cdot 5xy}$$ Multiply the numerators and multiply the denominators.

$$= \dfrac{5 \cdot 7 \cdot x \cdot x \cdot y \cdot z}{7 \cdot y \cdot y \cdot z \cdot 5 \cdot x \cdot y}$$ Factor $35x^2$ and factor y^2.

$$= \dfrac{\overset{1}{\cancel{5}} \cdot \overset{1}{\cancel{7}} \cdot \overset{1}{\cancel{x}} \cdot x \cdot \overset{1}{\cancel{y}} \cdot \overset{1}{\cancel{z}}}{\underset{1}{\cancel{7}} \cdot \underset{1}{\cancel{y}} \cdot y \cdot \underset{1}{\cancel{z}} \cdot \underset{1}{\cancel{5}} \cdot \underset{1}{\cancel{x}} \cdot y}$$ Divide out the common factors: 5, 7, x, y, and z.

$$= \dfrac{x}{y^2}$$ Do the multiplications in the numerator and the denominator.

EXAMPLE 3   *Factoring to simplify a product.*

Multiply: $\dfrac{x^2 - x}{2x + 4} \cdot \dfrac{x + 2}{x}$.

Solution

$$\dfrac{x^2 - x}{2x + 4} \cdot \dfrac{x + 2}{x} = \dfrac{(x^2 - x)(x + 2)}{(2x + 4)(x)}$$ Multiply the numerators and multiply the denominators.

We now factor the numerator and denominator to see if this product can be simplified.

$$\dfrac{x^2 - x}{2x + y} \cdot \dfrac{x + 2}{x} = \dfrac{x(x - 1)(x + 2)}{2(x + 2)x}$$ Factor the numerator: $(x^2 - x) = x(x - 1)$.
Factor the denominator: $(2x + 4) = 2(x + 2)$.

$$= \dfrac{\overset{1}{\cancel{x}}(x - 1)\overset{1}{\cancel{(x + 2)}}}{2\underset{1}{\cancel{(x + 2)}}\underset{1}{\cancel{x}}}$$ Divide out common factors.

$$= \dfrac{x - 1}{2}$$

EXAMPLE 4   *Factoring to simplify a product.*

Multiply: $\dfrac{x^2 - 3x}{x^2 - x - 6} \cdot \dfrac{x^2 + x - 2}{x^2 - x}$.

Solution

$$\dfrac{x^2 - 3x}{x^2 - x - 6} \cdot \dfrac{x^2 + x - 2}{x^2 - x}$$

$$= \dfrac{(x^2 - 3x)(x^2 + x - 2)}{(x^2 - x - 6)(x^2 - x)}$$ Multiply the numerators and multiply the denominators.

$$= \dfrac{x(x - 3)(x + 2)(x - 1)}{(x + 2)(x - 3)x(x - 1)}$$ Factor the numerator and denominator to see if the result can be simplified.

$$= \dfrac{\overset{1}{\cancel{x}}\overset{1}{\cancel{(x - 3)}}\overset{1}{\cancel{(x + 2)}}\overset{1}{\cancel{(x - 1)}}}{\underset{1}{\cancel{(x + 2)}}\underset{1}{\cancel{(x - 3)}}\underset{1}{\cancel{x}}\underset{1}{\cancel{(x - 1)}}}$$ Divide out common factors.

$$= 1$$

Multiplying a rational expression by a polynomial

Since any number divided by 1 remains unchanged, we can write any polynomial as a fraction by inserting a denominator of 1.

EXAMPLE 5 *Multiplying a rational expression by a monomial.*

Multiply: **a.** $\dfrac{4}{x} \cdot x$, **b.** $63x\left(\dfrac{1}{7x}\right)$, and **c.** $5a\left(\dfrac{3a - 1}{a}\right)$.

Solution

a. $\dfrac{4}{x} \cdot x = \dfrac{4}{x} \cdot \dfrac{x}{1}$ Write x as a fraction: $x = \frac{x}{1}$.

$= \dfrac{4 \cdot \overset{1}{\cancel{x}}}{\underset{1}{\cancel{x}} \cdot 1}$ Multiply the numerators and the denominators. Then divide out the common factor in the numerator and denominator.

$= 4$ Simplify.

b. $63x\left(\dfrac{1}{7x}\right) = \dfrac{63x}{1}\left(\dfrac{1}{7x}\right)$ Write $63x$ as a fraction: $63x = \frac{63x}{1}$.

$= \dfrac{63x \cdot 1}{1 \cdot 7 \cdot x}$ Multiply the numerators and the denominators.

$= \dfrac{9 \cdot \overset{1}{\cancel{7}} \cdot \overset{1}{\cancel{x}} \cdot 1}{1 \cdot \underset{1}{\cancel{7}} \cdot \underset{1}{\cancel{x}}}$ Write $63x$ in factored form as $9 \cdot 7 \cdot x$. Then divide out the common factors, 7 and x.

$= 9$ Simplify.

c. $5a\left(\dfrac{3a - 1}{a}\right) = \dfrac{5a}{1}\left(\dfrac{3a - 1}{a}\right)$ Write $5a$ as a fraction: $5a = \frac{5a}{1}$.

$= \dfrac{5\overset{1}{\cancel{a}}(3a - 1)}{1 \cdot \underset{1}{\cancel{a}}}$ Multiply the numerators and the denominators. Then divide out the common factor, a, in the numerator and denominator.

$= 5(3a - 1)$ Simplify.

$= 15a - 5$ Distribute the multiplication by 5.

EXAMPLE 6 *Multiplying a rational expression by a binomial.*

Multiply: $\dfrac{x^2 + x}{x^2 + 8x + 7} \cdot (x + 7)$.

Solution

$\dfrac{x^2 + x}{x^2 + 8x + 7} \cdot (x + 7)$

$= \dfrac{x^2 + x}{x^2 + 8x + 7} \cdot \dfrac{x + 7}{1}$ Write $x + 7$ as a fraction with a denominator of 1.

$= \dfrac{x(x + 1)(x + 7)}{(x + 1)(x + 7)1}$ Multiply the numerators and multiply the denominators. Factor where possible.

$= \dfrac{x\overset{1}{\cancel{(x + 1)}}\overset{1}{\cancel{(x + 7)}}}{\underset{1}{\cancel{(x + 1)}}\underset{1}{\cancel{(x + 7)}}1}$ Divide out common factors.

$= x$

Dividing rational expressions

Division by a nonzero number is equivalent to multiplying by its reciprocal. Thus, to divide two fractions, we can invert the divisor (the fraction following the $\div$ sign) and multiply. For example,

$$\frac{4}{7} \div \frac{3}{5} = \frac{4}{7} \cdot \frac{5}{3}$$ Invert $\frac{3}{5}$ and change the division to a multiplication.

$$= \frac{20}{21}$$ Multiply the numerators and multiply the denominators.

In general, we have the following rule.

Division of fractions

If a represents a real number and b, c, and d represent nonzero real numbers,

$$\frac{a}{b} \div \frac{c}{d} = \frac{a}{b} \cdot \frac{d}{c}$$

We use the same procedures to divide rational expressions.

EXAMPLE 7 *Dividing rational expressions.* Divide: **a.** $\dfrac{a}{13} \div \dfrac{17}{26}$

and **b.** $-\dfrac{9x}{35y} \div \dfrac{15x^2}{14}$.

Self Check
Divide:

$$-\frac{8a}{3b} \div \frac{16a^2}{9b^2}$$

Solution

a. $\dfrac{a}{13} \div \dfrac{17}{26} = \dfrac{a}{13} \cdot \dfrac{26}{17}$ Invert the divisor, which is $\frac{17}{26}$, and change the division to a multiplication.

$$= \frac{a \cdot 2 \cdot 13}{13 \cdot 17}$$ Multiply. Then factor where possible.

$$= \frac{a \cdot 2 \cdot \overset{1}{\cancel{13}}}{\underset{1}{\cancel{13}} \cdot 17}$$ Divide out common factors.

$$= \frac{2a}{17}$$

b. $-\dfrac{9x}{35y} \div \dfrac{15x^2}{14} = -\dfrac{9x}{35y} \cdot \dfrac{14}{15x^2}$ Multiply by the reciprocal of $\frac{15x^2}{14}$.

$$= -\frac{3 \cdot 3 \cdot x \cdot 2 \cdot 7}{5 \cdot 7 \cdot y \cdot 3 \cdot 5 \cdot x \cdot x}$$ Multiply. Then factor where possible.

$$= -\frac{3 \cdot \overset{1}{\cancel{3}} \cdot \overset{1}{\cancel{x}} \cdot 2 \cdot \overset{1}{\cancel{7}}}{5 \cdot \underset{1}{\cancel{7}} \cdot y \cdot \underset{1}{\cancel{3}} \cdot 5 \cdot \underset{1}{\cancel{x}} \cdot x}$$ Divide out common factors.

$$= -\frac{6}{25xy}$$ Multiply the remaining factors.

Answer: $-\dfrac{3b}{2a}$

746 *Chapter 12 Rational Expressions and Equations*

EXAMPLE 8 *Dividing rational expressions.*   Divide:

$$\frac{x^2 + x}{3x - 15} \div \frac{x^2 + 2x + 1}{6x - 30}.$$

Solution

$$\frac{x^2 + x}{3x - 15} \div \frac{x^2 + 2x + 1}{6x - 30}$$

$$= \frac{x^2 + x}{3x - 15} \cdot \frac{6x - 30}{x^2 + 2x + 1}$$ Invert the divisor and change the division to multiplication.

$$= \frac{x(x + 1) \cdot 2 \cdot 3(x - 5)}{3(x - 5)(x + 1)(x + 1)}$$ Multiply. Then factor.

$$= \frac{\overset{1}{x(\cancel{x + 1})} \cdot 2 \cdot \overset{1}{\cancel{3}}\overset{1}{(\cancel{x - 5})}}{\underset{1}{\cancel{3}}\underset{1}{(\cancel{x - 5})}\underset{1}{(\cancel{x + 1})}(x + 1)}$$ Divide out common factors.

$$= \frac{2x}{x + 1}$$

Self Check
Divide:

$$\frac{z^2 - 1}{z^2 + 4z + 3} \div \frac{z - 1}{z^2 + 2z - 3}$$

Answer: $z - 1$

Dividing a rational expression by a polynomial

To divide a rational expression by a polynomial, we write the polynomial as a fraction by inserting a denominator of 1, and then we divide the fractions.

EXAMPLE 9 *Dividing by a polynomial.*   Divide:

$$\frac{2x^2 - 3x - 2}{2x + 1} \div (4 - x^2).$$

Solution

$$\frac{2x^2 - 3x - 2}{2x + 1} \div (4 - x^2)$$

$$= \frac{2x^2 - 3x - 2}{2x + 1} \div \frac{4 - x^2}{1}$$ Write $4 - x^2$ as a fraction with a denominator of 1.

$$= \frac{2x^2 - 3x - 2}{2x + 1} \cdot \frac{1}{4 - x^2}$$ Invert the divisor and change the division to multiplication.

$$= \frac{(2x + 1)(x - 2) \cdot 1}{(2x + 1)(2 + x)(2 - x)}$$ Multiply. Then factor where possible.

$$= \frac{\overset{1}{(\cancel{2x + 1})}\overset{-1}{(\cancel{x - 2})} \cdot 1}{\underset{1}{(\cancel{2x + 1})}(2 + x)\underset{1}{(\cancel{2 - x})}}$$ Divide out common factors. The binomials $x - 2$ and $2 - x$ are opposites. $\frac{x - 2}{2 - x} = -1$.

$$= \frac{-1}{2 + x}$$

$$= -\frac{1}{2 + x}$$

Self Check
Divide:

$$(b - a) \div \frac{a^2 - b^2}{a^2 + ab}$$

Answer: $-a$

Combined operations

Unless parentheses indicate otherwise, we do multiplication and divisions in order from left to right.

EXAMPLE 10 *Multiplying and dividing rational expressions.*

Simplify $\dfrac{x^2 - x - 6}{x - 2} \div \dfrac{x^2 - 4x}{x^2 - x - 2} \cdot \dfrac{x - 4}{x^2 + x}$.

Solution

Since there are no parentheses to indicate otherwise, we do the division first.

$$\frac{x^2 - x - 6}{x - 2} \div \frac{x^2 - 4x}{x^2 - x - 2} \cdot \frac{x - 4}{x^2 + x}$$

$$= \frac{x^2 - x - 6}{x - 2} \cdot \frac{x^2 - x - 2}{x^2 - 4x} \cdot \frac{x - 4}{x^2 + x}$$

Invert the divisor, which is $\dfrac{x^2 - 4x}{x^2 - x - 2}$, and change the division to a multiplication.

$$= \frac{(x + 2)(x - 3)(x + 1)(x - 2)(x - 4)}{(x - 2)x(x - 4)x(x + 1)}$$

Multiply. Then factor.

$$= \frac{(x + 2)(x - 3)\cancel{(x + 1)}\cancel{(x - 2)}\cancel{(x - 4)}}{\cancel{(x - 2)}x\cancel{(x - 4)}x\cancel{(x + 1)}}$$

Divide out common factors.

$$= \frac{(x + 2)(x - 3)}{x^2}$$

EXAMPLE 11 *Multiplying and dividing rational expressions.*

Simplify $\dfrac{x^2 + 6x + 9}{x^2 - 2x}\left(\dfrac{x^2 - 4}{x^2 + 3x} \div \dfrac{x + 2}{x}\right)$.

Solution

We do the division within the parentheses first.

$$\frac{x^2 + 6x + 9}{x^2 - 2x}\left(\frac{x^2 - 4}{x^2 + 3x} \div \frac{x + 2}{x}\right)$$

$$= \frac{x^2 + 6x + 9}{x^2 - 2x}\left(\frac{x^2 - 4}{x^2 + 3x} \cdot \frac{x}{x + 2}\right)$$

Invert the divisor and change the division to multiplication.

$$= \frac{(x + 3)(x + 3)(x - 2)(x + 2)x}{x(x - 2)x(x + 3)(x + 2)}$$

Multiply and factor where possible.

$$= \frac{\cancel{(x + 3)}(x + 3)\cancel{(x - 2)}\cancel{(x + 2)}\cancel{x}}{\cancel{x}\cancel{(x - 2)}x\cancel{(x + 3)}\cancel{(x + 2)}}$$

Divide out common factors.

$$= \frac{x + 3}{x}$$

Self Check

Simplify:

$$\frac{a^2 + ab}{ab - b^2} \cdot \frac{a^2 - b^2}{a^2 + ab} \div \frac{a + b}{b}$$

Answer: 1

Self Check

Simplify:

$$\frac{x^2 - 2x}{x^2 + 6x + 9} \div \left(\frac{x^2 - 4}{x^2 + 3x} \cdot \frac{x}{x + 2}\right)$$

Answer: $\dfrac{x}{x + 3}$

STUDY SET Section 12.2

VOCABULARY *Fill in the blanks.*

1. In a fraction, the part above the fraction bar is called the _____.

2. In a fraction, the part below the fraction bar is called the _____.

CONCEPTS *Fill in the blanks.*

3. To multiply fractions, we multiply their _____ and multiply their _____.

4. $\dfrac{a}{b} \cdot \dfrac{c}{d} =$

5. To write a polynomial in fractional form, we insert a denominator of   .

6. $\dfrac{a}{b} \div \dfrac{c}{d} = \dfrac{a}{b} \cdot$

7. To divide fractions, we invert the _____ and _____.

8. The _____ of $\dfrac{x}{x+2}$ is $\dfrac{x+2}{x}$.

NOTATION *Complete each solution.*

9. $\dfrac{x^2+x}{3x-6} \cdot \dfrac{x-2}{x+1} = \dfrac{(x^2+x)}{(x+1)}$

$ = \dfrac{(x-2)}{(x+1)}$

$ = \dfrac{x}{3}$

10. $\dfrac{x^2-x}{4x+12} \div \dfrac{x-1}{x+3} = \dfrac{x^2-x}{4x+12} \cdot \underline{}$

$ = \dfrac{(x+3)}{(4x+12)}$

$ = \dfrac{(x+3)}{(x-1)}$

$ = \dfrac{x}{4}$

PRACTICE *Do the multiplications. Simplify answers if possible.*

11. $\dfrac{3}{y} \cdot \dfrac{y}{2}$

12. $\dfrac{2}{z} \cdot \dfrac{z}{3}$

13. $\dfrac{5y}{7} \cdot \dfrac{7}{5}$

14. $\dfrac{4x}{3y} \cdot \dfrac{3y}{7x}$

15. $\dfrac{7z}{9z} \cdot \dfrac{4z}{2z}$

16. $\dfrac{8}{2x} \cdot \dfrac{16x}{3x}$

17. $\dfrac{2x^2y}{3xy} \cdot \dfrac{3xy^2}{2}$

18. $\dfrac{2x^2z}{z} \cdot \dfrac{5x}{z}$

19. $\dfrac{8x^2y^2}{4x^2} \cdot \dfrac{2xy}{2y}$

20. $\dfrac{9x^2y}{3x} \cdot \dfrac{3xy}{3y}$

21. $-\dfrac{2xy}{x^2} \cdot \dfrac{3xy}{2}$

22. $-\dfrac{3x}{x^2} \cdot \dfrac{2xz}{3}$

23. $\dfrac{ab^2}{a^2b} \cdot \dfrac{b^2c^2}{abc} \cdot \dfrac{abc^2}{a^3c^2}$

24. $\dfrac{x^3y}{z} \cdot \dfrac{xz^3}{x^2y^2} \cdot \dfrac{yz}{xyz}$

25. $\dfrac{10r^2st^3}{6rs^2} \cdot \dfrac{3r^3t}{2rst} \cdot \dfrac{2s^3t^4}{5s^2t^3}$

26. $\dfrac{3a^3b}{25cd^3} \cdot \dfrac{-5cd^2}{6ab} \cdot \dfrac{10abc^2}{2bc^2d}$

27. $\dfrac{z+7}{7} \cdot \dfrac{z+2}{z}$

28. $\dfrac{a-3}{a} \cdot \dfrac{a+3}{5}$

29. $\dfrac{x-2}{2} \cdot \dfrac{2x}{x-2}$

30. $\dfrac{y+3}{y} \cdot \dfrac{3y}{y+3}$

31. $\dfrac{x+5}{5} \cdot \dfrac{x}{x+5}$

32. $\dfrac{y-9}{y+9} \cdot \dfrac{y}{9}$

33. $\dfrac{5}{m} \cdot m$

34. $p \cdot \dfrac{10}{p}$

35. $4d \cdot \dfrac{3}{2d}$

36. $9x \cdot \dfrac{25}{3x}$

37. $15x\left(\dfrac{x+1}{15x}\right)$

38. $30t\left(\dfrac{t-7}{30t}\right)$

39. $12y\left(\dfrac{y+8}{6y}\right)$

40. $16x\left(\dfrac{3x+8}{4x}\right)$

41. $(x+8)\dfrac{x+5}{x+8}$

42. $(y-2)\dfrac{y+3}{y-2}$

43. $10(h+9)\dfrac{h-3}{h+9}$

44. $r(r-25)\dfrac{r+4}{r-25}$

45. $\dfrac{(x+1)^2}{x+1} \cdot \dfrac{x+2}{x+1}$

46. $\dfrac{(y-3)^2}{y-3} \cdot \dfrac{y-3}{y-3}$

47. $\dfrac{2x+6}{x+3} \cdot \dfrac{3}{4x}$

48. $\dfrac{3y-9}{y-3} \cdot \dfrac{y}{3y^2}$

49. $\dfrac{x^2-x}{x} \cdot \dfrac{3x-6}{3x-3}$

50. $\dfrac{5z-10}{z+2} \cdot \dfrac{3}{3z-6}$

51. $\dfrac{7y - 14}{y - 2} \cdot \dfrac{x^2}{7x}$

52. $\dfrac{y^2 + 3y}{9} \cdot \dfrac{3x}{y + 3}$

53. $\dfrac{x^2 + x - 6}{5x} \cdot \dfrac{5x - 10}{x + 3}$

54. $\dfrac{z^2 + 4z - 5}{5z - 5} \cdot \dfrac{5z}{z + 5}$

55. $\dfrac{m^2 - 2m - 3}{2m + 4} \cdot \dfrac{m^2 - 4}{m^2 + 3m + 2}$

56. $\dfrac{p^2 - p - 6}{3p - 9} \cdot \dfrac{p^2 - 9}{p^2 + 6p + 9}$

57. $\dfrac{3x^2 + 5x + 2}{x^2 - 9} \cdot \dfrac{x - 3}{x^2 - 4} \cdot \dfrac{x^2 + 5x + 6}{6x + 4}$

58. $\dfrac{x^2 - 25}{3x + 6} \cdot \dfrac{x^2 + x - 2}{2x + 10} \cdot \dfrac{6x}{3x^2 - 18x + 15}$

Do each division. Simplify answers when possible.

59. $\dfrac{2}{y} \div \dfrac{4}{3}$

60. $\dfrac{3}{a} \div \dfrac{a}{9}$

61. $\dfrac{3x}{2} \div \dfrac{x}{2}$

62. $\dfrac{y}{6} \div \dfrac{2}{3y}$ —

63. $\dfrac{3x}{y} \div \dfrac{2x}{4}$

64. $\dfrac{3y}{8} \div \dfrac{2y}{4y}$

65. $\dfrac{4x}{3x} \div \dfrac{2y}{9y}$

66. $\dfrac{14}{7y} \div \dfrac{10}{5z}$

67. $\dfrac{x^2}{3} \div \dfrac{2x}{4}$

68. $\dfrac{z^2}{z} \div \dfrac{z}{3z}$

69. $\dfrac{x^2y}{3xy} \div \dfrac{xy^2}{6y}$

70. $\dfrac{2xz}{z} \div \dfrac{4x^2}{z^2}$

71. $\dfrac{x + 2}{3x} \div \dfrac{x + 2}{2}$

72. $\dfrac{z - 3}{3z} \div \dfrac{z + 3}{z}$

73. $\dfrac{(z - 2)^2}{3z^2} \div \dfrac{z - 2}{6z}$

74. $\dfrac{(x + 7)^2}{x + 7} \div \dfrac{(x - 3)^2}{x + 7}$

75. $\dfrac{(z - 7)^2}{z + 2} \div \dfrac{z(z - 7)}{5z^2}$

76. $\dfrac{y(y + 2)}{y^2(y - 3)} \div \dfrac{y^2(y + 2)}{(y - 3)^2}$

77. $\dfrac{x^2 - 4}{3x + 6} \div \dfrac{x - 2}{x + 2}$

78. $\dfrac{x^2 - 9}{5x + 15} \div \dfrac{x - 3}{x + 3}$

79. $\dfrac{x^2 - 1}{3x - 3} \div \dfrac{x + 1}{3}$

80. $\dfrac{x^2 - 16}{x - 4} \div \dfrac{3x + 12}{x}$

81. $\dfrac{x^2 - 2x - 35}{3x^2 + 27x} \div \dfrac{x^2 + 7x + 10}{6x^2 + 12x}$

82. $\dfrac{x^2 - x - 6}{2x^2 + 9x + 10} \div \dfrac{x^2 - 25}{2x^2 + 15x + 25}$

83. $\dfrac{2d^2 + 8d - 42}{d - 3} \div \dfrac{2d^2 + 14d}{d^2 + 5d}$

84. $\dfrac{5x^2 + 13x - 6}{x + 3} \div \dfrac{5x^2 - 17x + 6}{x - 2}$

Do the operations.

85. $\dfrac{x}{3} \cdot \dfrac{9}{4} \div \dfrac{x^2}{6}$

86. $\dfrac{y^2}{2} \div \dfrac{4}{y} \cdot \dfrac{y^2}{8}$

87. $\dfrac{x^2}{18} \div \dfrac{x^3}{6} \div \dfrac{12}{x^2}$

88. $\dfrac{y^3}{3y} \cdot \dfrac{3y^2}{4} \div \dfrac{15}{20}$

89. $\dfrac{z^2 - 4}{2z + 6} \div \dfrac{z + 2}{4} \cdot \dfrac{z + 3}{z - 2}$

90. $\dfrac{2}{3x - 3} \div \dfrac{2x + 2}{x - 1} \cdot \dfrac{5}{x + 1}$

91. $\dfrac{x - x^2}{x^2 - 4}\left(\dfrac{2x + 4}{x + 2} \div \dfrac{5}{x + 2}\right)$

92. $\dfrac{2}{3x - 3} \div \left(\dfrac{2x + 2}{x - 1} \cdot \dfrac{5}{x + 1}\right)$

93. $\dfrac{y^2}{x + 1} \cdot \dfrac{x^2 + 2x + 1}{x^2 - 1} \div \dfrac{3y}{xy - y}$

94. $\dfrac{x^2 - y^2}{x^4 - x^3} \div \dfrac{x - y}{x^2} \div \dfrac{x^2 + 2xy + y^2}{x + y}$

95. $\dfrac{x^2 + x - 6}{x^2 - 4} \cdot \dfrac{x^2 + 2x}{x - 2} \div \dfrac{x^2 + 3x}{x + 2}$

96. $\dfrac{x^2 - x - 6}{x^2 + 6x - 7} \cdot \dfrac{x^2 + x - 2}{x^2 + 2x} \div \dfrac{x^2 + 7x}{x^2 - 3x}$

APPLICATIONS

97. INTERNATIONAL ALPHABET   The symbols representing the letters A, B, C, D, E, and F in an international code used at sea are printed six to a sheet and then cut into separate cards. If each card is a square, find the area of the large printed sheet shown in Illustration 1.

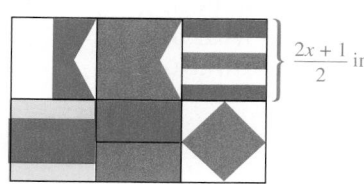

ILLUSTRATION 1

98. PHYSICS EXPERIMENT   The table in Illustration 2 contains algebraic expressions for the rate an object travels, and the time traveled at that rate, in terms of a constant k. Complete the table.

Rate (mph)	Time (hr)	Distance (mi)
$\dfrac{k^2 + k - 6}{k - 3}$	$\dfrac{k^2 - 9}{k^2 - 4}$	

ILLUSTRATION 2

WRITING

99. Explain how to multiply two fractions and how to simplify the result.

100. Explain why any mathematical expression can be written as a fraction.

101. To divide fractions, you must first know how to multiply fractions. Explain.

102. Explain how to do the division $\dfrac{a}{b} \div \dfrac{c}{d} \div \dfrac{e}{f}$.

REVIEW   *Simplify each expression. Write all answers without using negative exponents.*

103. $2x^3y^2(-3x^2y^4)$ **104.** $\dfrac{8x^4y^5}{-2x^3y^2}$

105. $(3y)^{-4}$ **106.** $x^{3m} \cdot x^{4m}$

Do the operations and simplify.

107. $-4(y^3 - 4y^2 + 3y - 2) - 4(-2y^3 - y)$

108. $y - 5\overline{)5y^3 - 3y^2 + 4y - 1}$

12.3 *Adding and Subtracting Rational Expressions*

In this section, you will learn about

- Adding and subtracting rational expressions with like denominators
- Combined operations • The LCD
- Adding and subtracting rational expressions with unlike denominators
- Combined operations

INTRODUCTION. In this section, we extend the rules for adding and subtracting numerical fractions to problems involving addition and subtraction of rational expressions.

Adding and subtracting rational expressions with like denominators

To add (or subtract) fractions with a common denominator, we add (or subtract) their numerators and keep the common denominator. For example,

$$\frac{3}{7} + \frac{2}{7} = \frac{3 + 2}{7} \qquad\qquad \frac{3}{7} - \frac{2}{7} = \frac{3 - 2}{7}$$

$$= \frac{5}{7} \qquad\qquad\qquad = \frac{1}{7}$$

In general, we have the following rule.

Adding and subtracting fractions with like denominators	If a, b, and d represent real numbers, $$\frac{a}{d} + \frac{b}{d} = \frac{a + b}{d} \qquad \text{and} \qquad \frac{a}{d} - \frac{b}{d} = \frac{a - b}{d} \qquad (d \neq 0)$$

We use the same procedure to add and subtract rational expressions with like denominators.

EXAMPLE 1   *Adding rational expressions.*   Do each addition.

a. $\dfrac{x}{8} + \dfrac{3x}{8} = \dfrac{x + 3x}{8}$ Add the numerators and keep the common denominator.

$\qquad = \dfrac{4x}{8}$ Combine like terms: $x + 3x = 4x$.

$\qquad = \dfrac{\overset{1}{\cancel{4}} \cdot x}{\underset{1}{\cancel{4}} \cdot 2}$ Factor the numerator and denominator and divide out the common factor, 4.

$\qquad = \dfrac{x}{2}$ Simplify.

b. $\dfrac{3x + y}{5x} + \dfrac{x + y}{5x} = \dfrac{3x + y + x + y}{5x}$ Add the numerators and keep the common denominator.

$\qquad = \dfrac{4x + 2y}{5x}$ Combine like terms.

Self Check
Add:

a. $\dfrac{x}{7} + \dfrac{4x}{7}$

b. $\dfrac{3x}{7y} + \dfrac{4x}{7y}$

Answers: **a.** $\dfrac{5x}{7}$, **b.** $\dfrac{x}{y}$ ■

EXAMPLE 2   *Adding rational expressions.*   Add:

$\dfrac{3x + 21}{5x + 10} + \dfrac{8x + 1}{5x + 10}$

Solution

Because the fractions have the same denominator, we add their numerators and keep the common denominator.

$\dfrac{3x + 21}{5x + 10} + \dfrac{8x + 1}{5x + 10} = \dfrac{3x + 21 + 8x + 1}{5x + 10}$ Add.

$\qquad = \dfrac{11x + 22}{5x + 10}$ Combine like terms.

$\qquad = \dfrac{11\overset{1}{\cancel{(x + 2)}}}{5\underset{1}{\cancel{(x + 2)}}}$ Simplify the result by factoring the numerator and denominator. Divide out the common factor, $x + 2$.

$\qquad = \dfrac{11}{5}$

Self Check
Add:

$\dfrac{x + 4}{6x - 12} + \dfrac{x - 8}{6x - 12}$

Answer: $\dfrac{1}{3}$ ■

EXAMPLE 3   *Subtracting rational expressions.*   Subtract:

a. $\dfrac{5x}{3} - \dfrac{2x}{3}$ and   **b.** $\dfrac{5x + 1}{x - 3} - \dfrac{4x - 2}{x - 3}$.

Solution

In each part, the fractions have the same denominator. To subtract them, we subtract their numerators and keep the common denominator.

a. $\dfrac{5x}{3} - \dfrac{2x}{3} = \dfrac{5x - 2x}{3}$

$\qquad = \dfrac{3x}{3}$ Combine like terms: $5x - 2x = 3x$.

$\qquad = \dfrac{x}{1}$ Divide out the common factor, 3.

$\qquad = x$ Denominators of 1 need not be written.

Self Check
Subtract:

$\dfrac{2y + 1}{y + 5} - \dfrac{y - 4}{y + 5}$

b. $\dfrac{5x + 1}{x - 3} - \dfrac{4x - 2}{x - 3} = \dfrac{(5x + 1) - (4x - 2)}{x - 3}$ Subtract. Write each numerator in parentheses.

$$= \dfrac{5x + 1 - 4x + 2}{x - 3}$$ Distribute the multiplication by -1: $-(4x - 2) = -4x + 2$.

$$= \dfrac{x + 3}{x - 3}$$ Combine like terms.

Answer: 1

Combined operations

To add and/or subtract three or more rational expressions, we follow the rules for the order of operations.

EXAMPLE 4   *Combined operations.*   Simplify $\dfrac{3x + 1}{x^2 + x + 1} - \dfrac{5x + 2}{x^2 + x + 1} + \dfrac{2x + 1}{x^2 + x + 1}$.

Self Check

Simplify:

$$\dfrac{2a^2 - 3}{a - 5} + \dfrac{3a^2 + 2}{a - 5} - \dfrac{5a^2}{a - 5}$$

Solution

This example combines addition and subtraction. Unless parentheses indicate otherwise, we do additions and subtractions from left to right.

$$\dfrac{3x + 1}{x^2 + x + 1} - \dfrac{5x + 2}{x^2 + x + 1} + \dfrac{2x + 1}{x^2 + x + 1}$$

$$= \dfrac{(3x + 1) - (5x + 2) + (2x + 1)}{x^2 + x + 1}$$ Combine the numerators and keep the common denominator.

$$= \dfrac{3x + 1 - 5x - 2 + 2x + 1}{x^2 + x + 1}$$ Distribute the multiplication by -1: $-(5x + 2) = -5x - 2$.

$$= \dfrac{0}{x^2 + x + 1}$$ Combine like terms.

$$= 0$$ If the numerator of a fraction is zero and the denominator is not zero, the fraction's value is zero.

Answer: $-\dfrac{1}{a - 5}$

The LCD

Since the denominators of the fractions in the addition $\frac{4}{7} + \frac{3}{5}$ are different, we cannot add the fractions in their present form.

four-sevenths   +   three-fifths

└ Different denominators ┘

To add these fractions, we need to find a common denominator. The smallest common denominator (called the **least** or **lowest common denominator**) is usually the easiest one to work with.

Least common denominator

> The **least common denominator (LCD)** for a set of fractions is the smallest number that each denominator will divide exactly.

In the addition $\frac{4}{7} + \frac{3}{5}$, the denominators are 7 and 5. The smallest number that 7 and 5 will divide exactly is 35. This is the LCD. We now **build** each fraction into an equivalent fraction with a denominator of 35. To do so, we use the fundamental property of fractions to multiply both the numerator and the denominator of each fraction by some appropriate number.

$$\frac{4}{7} + \frac{3}{5} = \frac{4 \cdot 5}{7 \cdot 5} + \frac{3 \cdot 7}{5 \cdot 7}$$ Multiply the numerator and denominator of $\frac{4}{7}$ by 5, and multiply the numerator and denominator of $\frac{3}{5}$ by 7.

$$= \frac{20}{35} + \frac{21}{35}$$ Do the multiplications.

Now that the fractions have a common denominator, we can add them.

$$\frac{20}{35} + \frac{21}{35} = \frac{20 + 21}{35} = \frac{41}{35}$$

EXAMPLE 5 *Building fractions.* Change each fraction into one with a denominator of 30y: **a.** $\frac{1}{2y}$, **b.** $\frac{3y}{5}$, and **c.** $\frac{7 + x}{10y}$.

Solution
To build each fraction, we multiply the numerator and denominator by the factor that makes the denominator 30y.

a. $\dfrac{1}{2y} = \dfrac{1 \cdot 15}{2y \cdot 15} = \dfrac{15}{30y}$ Multiply numerator and denominator by 15, because $2y \cdot 15 = 30y$.

b. $\dfrac{3y}{5} = \dfrac{3y \cdot 6y}{5 \cdot 6y} = \dfrac{18y^2}{30y}$ Multiply numerator and denominator by 6y, because $5 \cdot 6y = 30y$.

c. $\dfrac{7 + x}{10y} = \dfrac{(7 + x)3}{(10y)3} = \dfrac{21 + 3x}{30y}$ Multiply numerator and denominator by 3, because $10y \cdot 3 = 30y$.

Self Check

Change $\dfrac{5}{6b}$ into a fraction with a denominator of 30ab.

Answer: $\dfrac{25a}{30ab}$ ■

There is a process that we can use to find the least common denominator of several fractions.

Finding the least common denominator (LCD)

1. List the different denominators that appear in the fraction.
2. Completely factor each denominator.
3. Form a product using each different factor obtained in Step 2. Use each different factor the *greatest* number of times it appears in any one factorization. The product formed by multiplying these factors is the LCD.

EXAMPLE 6 *Finding the LCD.* Find the LCD of $\dfrac{5}{24b}$ and $\dfrac{11}{18b}$.

Solution
We list and factor each denominator into the product of prime numbers.

$$24b = 2 \cdot 2 \cdot 2 \cdot 3 \cdot b$$
$$18b = 2 \cdot 3 \cdot 3 \cdot b$$

To find the LCD, we use each of these factors the greatest number of times it appears in any one factorization. We use 2 three times, because it appears three times as a factor of 24. We use 3 twice, because it occurs twice as a factor of 18. We use *b* once.

$$\text{LCD} = 2 \cdot 2 \cdot 2 \cdot 3 \cdot 3 \cdot b$$
$$= 8 \cdot 9 \cdot b$$
$$= 72b$$

Self Check

Find the LCD of $\dfrac{3}{28z}$ and $\dfrac{5}{21z}$.

Answer: 84z   ■

Adding and subtracting rational expressions with unlike denominators

The following steps summarize how to add (or subtract) fractions that have unlike denominators.

Adding or subtracting fractions with unlike denominators

To add (or subtract) fractions with unlike denominators,

1. Find the LCD.
2. Write each fraction as an equivalent fraction whose denominator is the LCD.
3. Add (or subtract) the resulting fractions and simplify the result, if possible.

EXAMPLE 7 *Adding rational expressions.* Add: $\dfrac{4x}{7} + \dfrac{3x}{5}$.

Solution

The LCD is 35. We build each fraction so that it has a denominator of 35 and then add the resulting fractions.

$$\frac{4x}{7} + \frac{3x}{5} = \frac{4x \cdot 5}{7 \cdot 5} + \frac{3x \cdot 7}{5 \cdot 7}$$
Multiply the numerator and the denominator of $\frac{4x}{7}$ by 5 and the numerator and denominator of $\frac{3x}{5}$ by 7.

$$= \frac{20x}{35} + \frac{21x}{35}$$
Do the multiplications.

$$= \frac{41x}{35}$$
Add the numerators and keep the common denominator.

Self Check

Add:

$$\frac{y}{2} + \frac{6y}{7}$$

Answer: $\dfrac{19y}{14}$

EXAMPLE 8 *Adding rational expressions.* Add: $\dfrac{5}{24b} + \dfrac{11}{18b}$.

Solution

In Example 6, we saw that the LCD of these fractions is $2 \cdot 2 \cdot 2 \cdot 3 \cdot 3 \cdot b = 72b$. To add them, we first factor each denominator:

$$\frac{5}{24b} + \frac{11}{18b} = \frac{5}{2 \cdot 2 \cdot 2 \cdot 3 \cdot b} + \frac{11}{2 \cdot 3 \cdot 3 \cdot b}$$

In each resulting fraction, we multiply the numerator and the denominator by whatever it takes to build the denominator to the LCD of $2 \cdot 2 \cdot 2 \cdot 3 \cdot 3 \cdot b$.

$$= \frac{5 \cdot 3}{2 \cdot 2 \cdot 2 \cdot 3 \cdot b \cdot 3} + \frac{11 \cdot 2 \cdot 2}{2 \cdot 3 \cdot 3 \cdot b \cdot 2 \cdot 2}$$

$$= \frac{15}{72b} + \frac{44}{72b}$$
Do the multiplications.

$$= \frac{59}{72b}$$
Add the numerators and keep the common denominator.

Self Check

Add:

$$\frac{3}{28z} + \frac{5}{21z}$$

Answer: $\dfrac{29}{84z}$

EXAMPLE 9   *Adding rational expressions.*   Add: $\dfrac{x+4}{x^2} + \dfrac{x-5}{4x}$.

Solution

First we find the LCD.

$$\left.\begin{array}{l} x^2 = x \cdot x \\ 4x = 2 \cdot 2 \cdot x \end{array}\right\} \qquad \text{LCD} = x \cdot x \cdot 2 \cdot 2 = 4x^2$$

$$\dfrac{x+4}{x^2} + \dfrac{x-5}{4x} = \dfrac{(x+4)4}{(x^2)4} + \dfrac{(x-5)x}{(4x)x}$$ Build the fractions to get the common denominator, $4x^2$.

$$= \dfrac{4x+16}{4x^2} + \dfrac{x^2-5x}{4x^2}$$ Do the multiplications.

$$= \dfrac{4x+16+x^2-5x}{4x^2}$$ Add the numerators and keep the common denominator.

$$= \dfrac{x^2-x+16}{4x^2}$$ Combine like terms.

Self Check
Add:

$$\dfrac{a-1}{9a} + \dfrac{2-a}{a^2}$$

Answer: $\dfrac{a^2-10a+18}{9a^2}$ ∎

EXAMPLE 10   *Subtracting rational expressions.*   Subtract:
$\dfrac{x}{x+1} - \dfrac{3}{x}$.

Solution

By inspection, the least common denominator is $(x+1)x$.

$$\dfrac{x}{x+1} - \dfrac{3}{x} = \dfrac{x(x)}{(x+1)x} - \dfrac{3(x+1)}{x(x+1)}$$ Build the fractions to get the common denominator.

$$= \dfrac{x(x)-3(x+1)}{x(x+1)}$$ Subtract the numerators and keep the common denominator.

$$= \dfrac{x^2-3x-3}{x(x+1)}$$ Do the multiplications in the numerator.

Self Check
Subtract:

$$\dfrac{a}{a-1} - \dfrac{5}{a}$$

Answer: $\dfrac{a^2-5a+5}{a(a-1)}$ ∎

EXAMPLE 11   *Simplifying after subtracting.*   Subtract:
$\dfrac{a}{a-1} - \dfrac{2}{a^2-1}$.

Solution

We factor a^2-1 to see that the LCD is $(a+1)(a-1)$.

$$\dfrac{a}{a-1} - \dfrac{2}{a^2-1}$$

$$= \dfrac{a(a+1)}{(a-1)(a+1)} - \dfrac{2}{(a+1)(a-1)}$$ Build the first fraction to get the LCD.

$$= \dfrac{a(a+1)-2}{(a-1)(a+1)}$$ Subtract the numerators and keep the common denominator.

$$= \dfrac{a^2+a-2}{(a-1)(a+1)}$$ Distribute the multiplication by a.

$$= \dfrac{(a+2)\overset{1}{\cancel{(a-1)}}}{\underset{1}{\cancel{(a-1)}}(a+1)}$$ Simplify the result by factoring a^2+a-2. Divide out the common factor, $a-1$.

$$= \dfrac{a+2}{a+1}$$

Self Check
Subtract:

$$\dfrac{b}{b-2} - \dfrac{8}{b^2-4}$$

Answer: $\dfrac{b+4}{b+2}$ ∎

EXAMPLE 12 *Factoring to find the LCD.* Subtract:

$$\frac{2a}{a^2 + 4a + 4} - \frac{1}{2a + 4}.$$

Solution

Find the least common denominator by factoring each denominator.

$$\left. \begin{array}{l} a^2 + 4a + 4 = (a + 2)(a + 2) \\ 2a + 4 = 2(a + 2) \end{array} \right\} \qquad \text{LCD} = (a + 2)(a + 2)2$$

We build each fraction into a new fraction with a denominator of $2(a + 2)(a + 2)$.

$$\frac{2a}{a^2 + 4a + 4} - \frac{1}{2a + 4}$$

$$= \frac{2a}{(a + 2)(a + 2)} - \frac{1}{2(a + 2)} \qquad \text{Write the denominators in factored form.}$$

$$= \frac{2a \cdot 2}{(a + 2)(a + 2)2} - \frac{1(a + 2)}{2(a + 2)(a + 2)} \qquad \text{Build each fraction to get a common denominator.}$$

$$= \frac{4a - 1(a + 2)}{2(a + 2)^2} \qquad \text{Subtract the numerators and keep the common denominator. Write } (a + 2)(a + 2) \text{ as } (a + 2)^2.$$

$$= \frac{4a - a - 2}{2(a + 2)^2} \qquad \text{Distribute the multiplication by } -1.$$

$$= \frac{3a - 2}{2(a + 2)^2} \qquad \text{Combine like terms.}$$

Self Check
Subtract:

$$\frac{a}{a^2 - 2a + 1} - \frac{1}{6a - 6}$$

Answer: $\dfrac{5a + 1}{6(a - 1)^2}.$

EXAMPLE 13 *Denominators that are opposites.* Subtract:

$$\frac{3}{x - y} - \frac{x}{y - x}.$$

Solution

We note that the second denominator is the opposite (negative) of the first. So we can multiply the numerator and denominator of the second fraction by -1 to get

$$\frac{3}{x - y} - \frac{x}{y - x} = \frac{3}{x - y} - \frac{-1x}{-1(y - x)} \qquad \text{Multiply numerator and denominator by } -1.$$

$$= \frac{3}{x - y} - \frac{-x}{-y + x} \qquad \text{Distribute the multiplication by } -1: \\ -1(y - x) = -y + x.$$

$$= \frac{3}{x - y} - \frac{-x}{x - y} \qquad -y + x = x - y. \text{ The fractions now have a common denominator of } x - y.$$

$$= \frac{3 - (-x)}{x - y} \qquad \text{Subtract the numerators and keep the common denominator.}$$

$$= \frac{3 + x}{x - y} \qquad -(-x) = x.$$

Self Check
Subtract:

$$\frac{5}{a - b} - \frac{2}{b - a}$$

Answer: $\dfrac{7}{a - b}$

Combined operations

To add and/or subtract three or more rational expressions, we follow the rules for the order of operations.

EXAMPLE 14 *Combined operations.* Do the operations:
$\dfrac{3}{x^2y} + \dfrac{2}{xy} - \dfrac{1}{xy^2}.$

Self Check

Combine: $\dfrac{5}{ab^2} - \dfrac{b}{a} + \dfrac{a}{b}.$

Solution

Find the least common denominator.

$$\left.\begin{array}{l} x^2y = x \cdot x \cdot y \\ xy = x \cdot y \\ xy^2 = x \cdot y \cdot y \end{array}\right\} \text{ Factor each denominator.}$$

In any one of these denominators, the factor x occurs at most twice, and the factor y occurs at most twice. Thus,

$$\begin{aligned} \text{LCD} &= x \cdot x \cdot y \cdot y \\ &= x^2y^2 \end{aligned}$$

We build each fraction into one with a denominator of x^2y^2.

$$\frac{3}{x^2y} + \frac{2}{xy} - \frac{1}{xy^2}$$

$$= \frac{3 \cdot y}{x \cdot x \cdot y \cdot y} + \frac{2 \cdot x \cdot y}{x \cdot y \cdot x \cdot y} - \frac{1 \cdot x}{x \cdot y \cdot y \cdot x} \quad \begin{array}{l}\text{Factor each denominator and} \\ \text{build each fraction.}\end{array}$$

$$= \frac{3y + 2xy - x}{x^2y^2} \qquad \begin{array}{l}\text{Do the multiplications and} \\ \text{combine the numerators. Write} \\ \text{the result over the LCD.}\end{array}$$

Answer: $\dfrac{5 - b^3 + a^2b}{ab^2}$

STUDY SET Section 12.3

VOCABULARY *Fill in the blanks.*

1. The _____ for a set of fractions is the smallest number that each denominator divides exactly.

2. When we multiply the numerator and denominator of a fraction by some number to get a common denominator, we say that we are _____ the fraction.

CONCEPTS *Fill in the blanks.*

3. To add two fractions with like denominators, we add their _____ and keep the _____.

4. To subtract two fractions with _____ denominators, we need to find a common denominator.

NOTATION *Complete each solution.*

5. $\dfrac{6a - 1}{4a + 1} + \dfrac{2a + 3}{4a + 1} = \dfrac{6a - 1 + \rule{1.5cm}{0.4pt}}{4a + 1}$

$$= \dfrac{8a + \rule{1cm}{0.4pt}}{4a + 1}$$

$$= \dfrac{2\rule{1cm}{0.4pt}}{4a + 1}$$

$$= 2$$

6. $\dfrac{x}{2x + 1} - \dfrac{1}{3x} = \dfrac{x}{(2x + 1)(3x)} - \dfrac{1(2x + 1)}{3x}$

$$= \dfrac{x(3x) - 1}{3x(2x + 1)}$$

$$= \dfrac{3x^2 - \rule{1cm}{0.4pt} - \rule{1cm}{0.4pt}}{3x(2x + 1)}$$

$$= \dfrac{(3x + 1)(x - 1)}{3x(2x + 1)}$$

PRACTICE *Do each addition. Simplify answers, if possible.*

7. $\dfrac{x}{9} + \dfrac{2x}{9}$

8. $\dfrac{5x}{7} + \dfrac{9x}{7}$

9. $\dfrac{2x}{y} + \dfrac{2x}{y}$

10. $\dfrac{4y}{3x} + \dfrac{2y}{3x}$

11. $\dfrac{4}{7y} + \dfrac{10}{7y}$

12. $\dfrac{x^2}{4y} + \dfrac{x^2}{4y}$

13. $\dfrac{y + 2}{10z} + \dfrac{y + 4}{10z}$

14. $\dfrac{x + 3}{2x^2} + \dfrac{x + 5}{2x^2}$

15. $\dfrac{3x - 5}{x - 2} + \dfrac{6x - 13}{x - 2}$

16. $\dfrac{8x - 7}{x + 3} + \dfrac{2x + 37}{x + 3}$

17. $\dfrac{a}{a^2 + 5a + 6} + \dfrac{3}{a^2 + 5a + 6}$

18. $\dfrac{b}{b^2 - 4} + \dfrac{2}{b^2 - 4}$

Do each subtraction. Simplify answers, if possible.

19. $\dfrac{35y}{72} - \dfrac{44y}{72}$

20. $\dfrac{13t}{99} - \dfrac{35t}{99}$

21. $\dfrac{2x}{y} - \dfrac{x}{y}$

22. $\dfrac{7y}{5} - \dfrac{4y}{5}$

23. $\dfrac{9y}{3x} - \dfrac{6y}{3x}$

24. $\dfrac{5r^2}{2r} - \dfrac{r^2}{2r}$

25. $\dfrac{6x - 5}{3xy} - \dfrac{3x - 5}{3xy}$

26. $\dfrac{7x + 7}{5y} - \dfrac{2x + 7}{5y}$

27. $\dfrac{3y - 2}{2y + 6} - \dfrac{2y - 5}{2y + 6}$

28. $\dfrac{5x + 8}{3x + 15} - \dfrac{3x - 2}{3x + 15}$

29. $\dfrac{2c}{c^2 - d^2} - \dfrac{2d}{c^2 - d^2}$

30. $\dfrac{3t}{t^2 - 8t + 7} - \dfrac{3}{t^2 - 8t + 7}$

Do the operations. Simplify answers if possible.

31. $\dfrac{13x}{15} + \dfrac{12x}{15} - \dfrac{5x}{15}$

32. $\dfrac{13y}{32} + \dfrac{13y}{32} - \dfrac{10y}{32}$

33. $-\dfrac{x}{y} + \dfrac{2x}{y} - \dfrac{x}{y}$

34. $\dfrac{5y}{8x} + \dfrac{4y}{8x} - \dfrac{9y}{8x}$

35. $\dfrac{3x}{y + 2} - \dfrac{3y}{y + 2} + \dfrac{x + y}{y + 2}$

36. $\dfrac{3y}{x - 5} + \dfrac{x}{x - 5} - \dfrac{y - x}{x - 5}$

37. $\dfrac{x + 1}{x - 2} - \dfrac{2(x - 3)}{x - 2} + \dfrac{3(x + 1)}{x - 2}$

38. $\dfrac{3xy}{x - y} - \dfrac{x(3y - x)}{x - y} - \dfrac{x(x - y)}{x - y}$

Build each fraction into an equivalent fraction with the indicated denominator.

39. $\dfrac{25}{4}; 20x$

40. $\dfrac{5}{y}; y^2$

41. $\dfrac{8}{x}; x^2y$

42. $\dfrac{7}{y}; xy^2$

43. $\dfrac{3x}{x + 1}; (x + 1)^2$

44. $\dfrac{5y}{y - 2}; (y - 2)^2$

45. $\dfrac{2y}{x}; x^2 + x$

46. $\dfrac{3x}{y}; y^2 - y$

47. $\dfrac{z}{z - 1}; z^2 - 1$

48. $\dfrac{y}{y + 2}; y^2 - 4$

49. $\dfrac{2}{x + 1}; x^2 + 3x + 2$

50. $\dfrac{3}{x - 1}; x^2 + x - 2$

Several denominators are given. Find the LCD.

51. $2x, 6x$

52. $3y, 9y$

53. $6y, 9xy^2$

54. $6y, 3x^2y$

55. $x^2 - 1, x + 1$

56. $y^2 - 9, y - 3$

57. $x^2 + 6x, x + 6, x$

58. $xy^2 - xy, xy, y - 1$

59. $x^2 - 4x - 5, x^2 - 25$

60. $x^2 - x - 6, x^2 - 9$

Do the operations. Simplify answers, if possible.

61. $\dfrac{2y}{9} + \dfrac{y}{3}$

62. $\dfrac{8a}{15} - \dfrac{5a}{12}$

63. $\dfrac{21x}{14} - \dfrac{5x}{21}$

64. $\dfrac{7y}{6} + \dfrac{10y}{9}$

65. $\dfrac{4x}{3} + \dfrac{2x}{y}$

66. $\dfrac{2y}{5x} - \dfrac{y}{2}$

67. $\dfrac{2}{x} - 3x$ *(Hint: $3x = \frac{3x}{1}$)*

68. $14 + \dfrac{10}{y^2}$ *(Hint: $14 = \frac{14}{1}$)*

69. $\dfrac{y + 2}{5y^2} + \dfrac{y + 4}{15y}$

70. $\dfrac{x + 3}{x^2} + \dfrac{x + 5}{2x}$

71. $\dfrac{x + 5}{xy} - \dfrac{x - 1}{x^2y}$

72. $\dfrac{x - 7}{y^2} - \dfrac{y + 7}{2y}$

73. $\dfrac{x}{x + 1} + \dfrac{x - 1}{x}$

74. $\dfrac{3x}{xy} + \dfrac{x + 1}{y - 1}$

75. $\dfrac{x - 1}{x} + \dfrac{y + 1}{y}$

76. $\dfrac{a+2}{b}+\dfrac{b-2}{a}$

77. $\dfrac{x}{x-2}+\dfrac{4+2x}{x^2-4}$

78. $\dfrac{y}{y+3}-\dfrac{2y-6}{y^2-9}$

79. $\dfrac{x+1}{x-1}+\dfrac{x-1}{x+1}$

80. $\dfrac{2x}{x+2}+\dfrac{x+1}{x-3}$

81. $\dfrac{5}{a-4}+\dfrac{7}{4-a}$

82. $\dfrac{4}{b-6}-\dfrac{b}{6-b}$

83. $\dfrac{t+1}{t-7}-\dfrac{t+1}{7-t}$

84. $\dfrac{r+2}{r^2-4}+\dfrac{4}{4-r^2}$

85. $\dfrac{2x+2}{x-2}-\dfrac{2x}{2-x}$

86. $\dfrac{y+3}{y-1}-\dfrac{y+4}{1-y}$

87. $\dfrac{b}{b+1}-\dfrac{b+1}{2b+2}$

88. $\dfrac{4x+1}{8x-12}+\dfrac{x-3}{2x-3}$

89. $\dfrac{2}{a^2+4a+3}+\dfrac{1}{a+3}$

90. $\dfrac{1}{c+6}-\dfrac{-4}{c^2+8a+12}$

91. $\dfrac{x+1}{2x+4}-\dfrac{x^2}{2x^2-8}$

92. $\dfrac{x+1}{x+2}-\dfrac{x^2+1}{x^2-x-6}$

93. $\dfrac{2x}{x^2-3x+2}+\dfrac{2x}{x-1}-\dfrac{x}{x-2}$

94. $\dfrac{4a}{a-2}-\dfrac{3a}{a-3}+\dfrac{4a}{a^2-5a+6}$

95. $\dfrac{2x}{x-1}+\dfrac{3x}{x+1}-\dfrac{x+3}{x^2-1}$

96. $\dfrac{a}{a-1}-\dfrac{2}{a+2}+\dfrac{3(a-2)}{a^2+a-2}$

APPLICATIONS *Refer to Illustration 1.*

97. Find the total height of the funnel.

98. What is the difference between the diameter of the opening at the top of the funnel and the diameter of its spout?

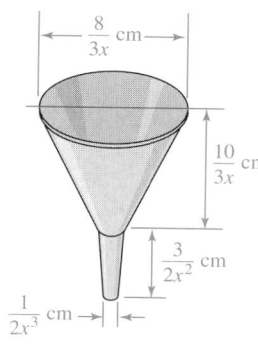

ILLUSTRATION 1

WRITING

99. Explain how to add fractions with the same denominator.

100. Explain how to find a lowest common denominator.

101. Explain what is wrong with the following solution:

$$\frac{2x+3}{x+5}-\frac{x+2}{x+5}=\frac{2x+3-x+2}{x+5}$$
$$=\frac{x+5}{x+5}$$
$$=1$$

102. Explain what is wrong with the following solution:

$$\frac{5x-4}{y}+\frac{x}{y}=\frac{5x-4+x}{y+y}$$
$$=\frac{6x-4}{2y}$$
$$=\frac{2(3x-2)}{2y}$$
$$=\frac{3x-2}{y}$$

REVIEW *Write each number in prime-factored form.*

103. 49   **104.** 64

105. 136   **106.** 315

12.4 *Complex Fractions*

In this section, you will learn about

- Simplifying complex fractions
- Simplifying fractions with terms containing negative exponents

INTRODUCTION. Rational expressions such as

$$\frac{\dfrac{5x}{3}}{\dfrac{2y}{9}}, \qquad \frac{x + \dfrac{1}{2}}{3 - x}, \qquad \text{and} \qquad \frac{\dfrac{x+1}{2}}{x + \dfrac{1}{x}}$$

which contain fractions in their numerators and/or denominators, are called **complex fractions.** In this section, we will show how to use properties of algebra to simplify complex fractions.

Simplifying complex fractions

Complex fractions can often be simplified.

$$\frac{\dfrac{5x}{3}}{\dfrac{2y}{9}} \quad \longleftarrow \text{The main fraction bar indicates division.}$$

We can simplify the complex fraction by doing the division:

$$\frac{\dfrac{5x}{3}}{\dfrac{2y}{9}} = \frac{5x}{3} \div \frac{2y}{9} = \frac{5x}{3} \cdot \frac{9}{2y} = \frac{5x \cdot 3 \cdot \overset{1}{\cancel{3}}}{\underset{1}{\cancel{3}} \cdot 2y} = \frac{15x}{2y}$$

There are two ways to simplify complex fractions.

Methods for simplifying complex fractions

> ***Method 1:*** Write the numerator and denominator of the complex fraction as single fractions. Then divide the fractions and simplify.
>
> ***Method 2:*** Multiply the numerator and denominator of the complex fraction by the LCD of the fractions in its numerator and denominator. Then simplify the results, if possible.

To simplify the complex fraction

$$\frac{\dfrac{3x}{5} + 1}{2 - \dfrac{x}{5}}$$

using method 1, we proceed as follows:

$$\frac{\dfrac{3x}{5} + 1}{2 - \dfrac{x}{5}} = \frac{\dfrac{3x}{5} + \dfrac{5}{5}}{\dfrac{10}{5} - \dfrac{x}{5}}$$ Change 1 to $\frac{5}{5}$ and 2 to $\frac{10}{5}$ so that we can write the numerator and denominator as single fractions.

$$= \frac{\dfrac{3x + 5}{5}}{\dfrac{10 - x}{5}}$$ Add the fractions in the numerator and subtract the fractions in the denominator.

$$= \frac{3x + 5}{5} \div \frac{10 - x}{5}$$ Write the complex fraction as an equivalent division problem.

$$= \frac{3x + 5}{5} \cdot \frac{5}{10 - x}$$ Invert the divisor and multiply.

$$= \frac{(3x + 5)5}{5(10 - x)}$$ Multiply the fractions.

$$= \frac{3x + 5}{10 - x}$$ Divide out the common factor, 5.

To use method 2, we proceed as follows:

$$\frac{\dfrac{3x}{5} + 1}{2 - \dfrac{x}{5}} = \frac{5\left(\dfrac{3x}{5} + 1\right)}{5\left(2 - \dfrac{x}{5}\right)}$$ Multiply both the numerator and denominator of the complex fraction by 5, the LCD of $\frac{3x}{5}$ and $\frac{x}{5}$.

$$= \frac{5 \cdot \dfrac{3x}{5} + 5 \cdot 1}{5 \cdot 2 - 5 \cdot \dfrac{x}{5}}$$ Distribute the multiplication by 5.

$$= \frac{3x + 5}{10 - x}$$ Distribute the multiplication by 5.

In this example, method 2 is easier than method 1. Either method can be used to simplify complex fractions. With practice, you will be able to see which method is best in a given situation.

EXAMPLE 1   *Simplifying complex fractions.*   Simplify $\dfrac{\dfrac{x}{3}}{\dfrac{y}{3}}$.

Solution

Method 1

$$\frac{\dfrac{x}{3}}{\dfrac{y}{3}} = \frac{x}{3} \div \frac{y}{3}$$

$$= \frac{x}{3} \cdot \frac{3}{y}$$

$$= \frac{3x}{3y}$$

$$= \frac{x}{y}$$

Method 2

$$\frac{\dfrac{x}{3}}{\dfrac{y}{3}} = \frac{3\left(\dfrac{x}{3}\right)}{3\left(\dfrac{y}{3}\right)}$$ The LCD for the fractions in the given complex fraction is 3.

$$= \frac{x}{y}$$

Self Check

Simplify $\dfrac{\dfrac{a}{4}}{\dfrac{5}{b}}$.

Answer: $\dfrac{ab}{20}$

■

EXAMPLE 2 *Simplifying complex fractions.* Simplify $\dfrac{\dfrac{x}{x+1}}{\dfrac{y}{x}}$.

Solution

Method 1

$$\frac{\dfrac{x}{x+1}}{\dfrac{y}{x}} = \frac{x}{x+1} \div \frac{y}{x}$$

$$= \frac{x}{x+1} \cdot \frac{x}{y}$$

$$= \frac{x^2}{y(x+1)}$$

Method 2

$$\frac{\dfrac{x}{x+1}}{\dfrac{y}{x}} = \frac{x(x+1)\left(\dfrac{x}{x+1}\right)}{x(x+1)\left(\dfrac{y}{x}\right)}$$

$$= \frac{x^2}{y(x+1)}$$

The LCD for the fractions in the given complex fraction is $x(x+1)$.

Self Check

Simplify $\dfrac{\dfrac{x}{y}}{\dfrac{x}{y+1}}$.

Answer: $\dfrac{y+1}{y}$

EXAMPLE 3 *Simplifying complex fractions.* Simplify $\dfrac{1+\dfrac{1}{x}}{1-\dfrac{1}{x}}$.

Solution

Method 1

$$\frac{1+\dfrac{1}{x}}{1-\dfrac{1}{x}} = \frac{\dfrac{x}{x}+\dfrac{1}{x}}{\dfrac{x}{x}-\dfrac{1}{x}}$$

$$= \frac{\dfrac{x+1}{x}}{\dfrac{x-1}{x}}$$

$$= \frac{x+1}{x} \div \frac{x-1}{x}$$

$$= \frac{x+1}{x} \cdot \frac{x}{x-1}$$

$$= \frac{(x+1)\overset{1}{\cancel{x}}}{\underset{1}{\cancel{x}}(x-1)}$$

$$= \frac{x+1}{x-1}$$

Method 2

$$\frac{1+\dfrac{1}{x}}{1-\dfrac{1}{x}} = \frac{x\left(1+\dfrac{1}{x}\right)}{x\left(1-\dfrac{1}{x}\right)}$$

$$= \frac{x \cdot 1 + x \cdot \dfrac{1}{x}}{x \cdot 1 - x \cdot \dfrac{1}{x}}$$

$$= \frac{x+1}{x-1}$$

Self Check

Simplify $\dfrac{\dfrac{1}{x}+1}{\dfrac{1}{x}-1}$.

Answer: $\dfrac{1+x}{1-x}$

EXAMPLE 4 *Simplifying complex fractions.* Simplify $\dfrac{1}{1+\dfrac{1}{x+1}}$.

Self Check

Simplify $\dfrac{2}{\dfrac{1}{x+2}-2}$.

Solution

We use method 2.

$$\frac{1}{1 + \dfrac{1}{x + 1}} = \frac{(x + 1) \cdot 1}{(x + 1)\left(1 + \dfrac{1}{x + 1}\right)}$$ Multiply the numerator and the denominator of the complex fraction by $x + 1$.

$$= \frac{x + 1}{(x + 1)1 + 1}$$ In the denominator, distribute $x + 1$.

$$= \frac{x + 1}{x + 2}$$ Simplify.

Answer: $\dfrac{2(x + 2)}{-2x - 3}$ ■

Simplifying fractions with terms containing negative exponents

Many fractions with terms containing negative exponents are complex fractions in disguise.

EXAMPLE 5 *Simplifying complex fractions.* Simplify $\dfrac{x^{-1} + y^{-2}}{x^{-2} - y^{-1}}$.

Solution

Write the fraction as a complex fraction and simplify using method 2.

$$\frac{x^{-1} + y^{-2}}{x^{-2} - y^{-1}} = \frac{\dfrac{1}{x} + \dfrac{1}{y^2}}{\dfrac{1}{x^2} - \dfrac{1}{y}}$$

$$= \frac{x^2 y^2 \left(\dfrac{1}{x} + \dfrac{1}{y^2}\right)}{x^2 y^2 \left(\dfrac{1}{x^2} - \dfrac{1}{y}\right)}$$ Multiply the numerator and denominator by $x^2 y^2$, which is the LCD of the fractions in the numerator and the denominator of the complex fraction.

$$= \frac{xy^2 + x^2}{y^2 - x^2 y}$$ Distribute the multiplication by $x^2 y^2$ and simplify.

$$= \frac{x(y^2 + x)}{y(y - x^2)}$$ Attempt to simplify the fraction by factoring the numerator and the denominator. The result cannot be simplified.

Self Check

Simplify $\dfrac{x^{-2} - y^{-1}}{x^{-1} + y^{-2}}$.

Answer: $\dfrac{y(y - x^2)}{x(y^2 + x)}$ ■

STUDY SET Section 12.4

VOCABULARY *Fill in the blanks.*

1. If a fraction has a fraction in its numerator or denominator, it is called a _____.

2. The denominator of the complex fraction $\dfrac{\dfrac{3}{x} + \dfrac{x}{y}}{\dfrac{1}{x} + 2}$ is

_____.

CONCEPTS *Fill in the blanks.*

3. To simplify a complex fraction using method 1, we write the numerator and denominator of a complex fraction as _____ fractions and then _____.

4. To simplify a complex fraction using method 2, we multiply the numerator and denominator of the complex fraction by the _____ of the fractions in its numerator and denominator.

NOTATION *Complete each solution.*

5. $\dfrac{\dfrac{2}{a} - \dfrac{1}{b}}{\dfrac{1}{a} + \dfrac{2}{b}} = \dfrac{\overline{}}{\dfrac{}{ab}}$

$ = \dfrac{2b - a}{ab} \dfrac{b + 2a}{ab}$

$ = \dfrac{2b - a}{ab} \cdot \dfrac{}{b + 2a}$

$ = \dfrac{(2b - a)}{ab}$

$ = \dfrac{2b - a}{b + 2a}$

6. $\dfrac{\dfrac{2}{a} - \dfrac{1}{b}}{\dfrac{1}{a} + \dfrac{2}{b}} = \dfrac{\left(\dfrac{2}{a} - \dfrac{1}{b}\right)}{\left(\dfrac{1}{a} + \dfrac{2}{b}\right)}$

$ = \dfrac{2b - a}{b + 2a}$

PRACTICE *Simplify each complex fraction.*

7. $\dfrac{\dfrac{2}{3}}{\dfrac{3}{4}}$

8. $\dfrac{\dfrac{3}{5}}{\dfrac{2}{7}}$

9. $\dfrac{\dfrac{4}{5}}{\dfrac{32}{15}}$

10. $\dfrac{\dfrac{7}{8}}{\dfrac{49}{4}}$

11. $\dfrac{\dfrac{2}{3} + 1}{\dfrac{1}{3} + 1}$

12. $\dfrac{\dfrac{3}{5} - 2}{\dfrac{2}{5} - 2}$

13. $\dfrac{\dfrac{1}{2} + \dfrac{3}{4}}{\dfrac{3}{2} + \dfrac{1}{4}}$

14. $\dfrac{\dfrac{2}{3} - \dfrac{5}{2}}{\dfrac{2}{3} - \dfrac{3}{2}}$

15. $\dfrac{\dfrac{x}{y}}{\dfrac{1}{x}}$

16. $\dfrac{\dfrac{y}{x}}{\dfrac{x}{xy}}$

17. $\dfrac{\dfrac{5t^2}{9x^2}}{\dfrac{3t}{x^2 t}}$

18. $\dfrac{\dfrac{5w^2}{4tz}}{\dfrac{15wt}{z^2}}$

19. $\dfrac{\dfrac{1}{x} - 3}{\dfrac{5}{x} + 2}$

20. $\dfrac{\dfrac{1}{y} + 3}{\dfrac{3}{y} - 2}$

21. $\dfrac{\dfrac{2}{x} + 2}{\dfrac{4}{x} + 2}$

22. $\dfrac{\dfrac{3}{x} - 3}{\dfrac{9}{x} - 3}$

23. $\dfrac{\dfrac{3y}{x} - y}{y - \dfrac{y}{x}}$

24. $\dfrac{\dfrac{y}{x} + 3y}{y + \dfrac{2y}{x}}$

25. $\dfrac{\dfrac{1}{x + 1}}{1 + \dfrac{1}{x + 1}}$

26. $\dfrac{\dfrac{1}{x - 1}}{1 - \dfrac{1}{x - 1}}$

27. $\dfrac{\dfrac{x}{x + 2}}{\dfrac{x}{x + 2} + x}$

28. $\dfrac{\dfrac{2}{x - 2}}{\dfrac{2}{x - 2} - 1}$

29. $\dfrac{1}{\dfrac{1}{x} + \dfrac{1}{y}}$

30. $\dfrac{1}{\dfrac{b}{a} - \dfrac{a}{b}}$

31. $\dfrac{\dfrac{2}{x}}{\dfrac{2}{y} - \dfrac{4}{x}}$

32. $\dfrac{\dfrac{2y}{3}}{\dfrac{2y}{3} - \dfrac{8}{y}}$

33. $\dfrac{3 + \dfrac{3}{x - 1}}{3 - \dfrac{3}{x}}$

34. $\dfrac{2 - \dfrac{2}{x + 1}}{2 + \dfrac{2}{x}}$

35. $\dfrac{\dfrac{3}{x} + \dfrac{4}{x + 1}}{\dfrac{2}{x + 1} - \dfrac{3}{x}}$

36. $\dfrac{\dfrac{5}{y - 3} - \dfrac{2}{y}}{\dfrac{1}{y} + \dfrac{2}{y - 3}}$

37. $\dfrac{\dfrac{2}{x} - \dfrac{3}{x + 1}}{\dfrac{2}{x + 1} - \dfrac{3}{x}}$

38. $\dfrac{\dfrac{5}{y} + \dfrac{4}{y + 1}}{\dfrac{4}{y} - \dfrac{5}{y + 1}}$

39. $\dfrac{\dfrac{1}{y^2 + y} - \dfrac{1}{xy + x}}{\dfrac{1}{xy + x} - \dfrac{1}{y^2 + y}}$

40. $\dfrac{\dfrac{2}{b^2 - 1} - \dfrac{3}{ab - a}}{\dfrac{3}{ab - a} - \dfrac{2}{b^2 - 1}}$

41. $\dfrac{x^{-2}}{y^{-1}}$

42. $\dfrac{a^{-4}}{b^{-2}}$

43. $\dfrac{1 + x^{-1}}{x^{-1} - 1}$

44. $\dfrac{y^{-2} + 1}{y^{-2} - 1}$

45. $\dfrac{a^{-2} + a}{a}$

46. $\dfrac{t - t^{-2}}{t^{-1}}$

47. $\dfrac{2x^{-1} + 4x^{-2}}{2x^{-2} + x^{-1}}$

48. $\dfrac{x^{-2} - 3x^{-3}}{3x^{-2} - 9x^{-3}}$

49. $\dfrac{1 - 25y^{-2}}{1 + 10y^{-1} + 25y^{-2}}$

50. $\dfrac{1 - 9x^{-2}}{1 - 6x^{-1} + 9x^{-2}}$

APPLICATIONS

51. GARDENING TOOL   In Illustration 1, what is the result when the opening of the cutting blades is divided by the opening of the handles? Express the result in simplest form.

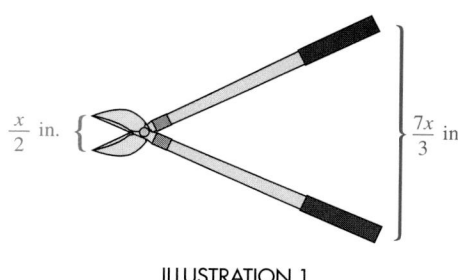

$\dfrac{x}{2}$ in. $\dfrac{7x}{3}$ in.

ILLUSTRATION 1

52. EARNED RUN AVERAGE   The earned run average (ERA) is a statistic that gives the average number of earned runs a pitcher allows. For a softball pitcher, this is based on a six-inning game. The formula for ERA is

$$\text{ERA} = \dfrac{\dfrac{\text{earned runs}}{\text{innings pitched}}}{6}$$

Simplify the complex fraction on the right-hand side of the equation.

53. ELECTRONICS   In electronic circuits, resistors oppose the flow of an electric current. To find the total resistance of a parallel combination of two resistors (see Illustration 2), we can use the formula

$$\text{Total resistance} = \dfrac{1}{\dfrac{1}{R_1} + \dfrac{1}{R_2}}$$

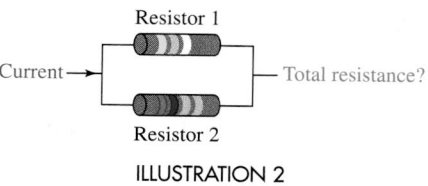

ILLUSTRATION 2

where R_1 is the resistance of the first resistor and R_2 is the resistance of the second. Simplify the complex fraction on the right-hand side of the formula.

54. DATA ANALYSIS   Use the data in Illustration 3 to find the average measurement for the three-trial experiment.

	Trial 1	Trial 2	Trial 3
Measurement	$\dfrac{k}{2}$	$\dfrac{k}{3}$	$\dfrac{k}{2}$

ILLUSTRATION 3

WRITING

55. Explain how to use method 1 to simplify

$$\dfrac{1 + \dfrac{1}{x}}{3 - \dfrac{1}{x}}$$

56. Explain how to use method 2 to simplify the expression in Exercise 55.

REVIEW   *Write each expression as an expression involving only one exponent.*

57. $t^3 t^4 t^2$

58. $(a^0 a^2)^3$

59. $-2r(r^3)^2$

60. $(s^3)^2 (s^4)^0$

Write each expression without using parentheses or negative exponents.

61. $\left(\dfrac{3r}{4r^3}\right)^4$

62. $\left(\dfrac{12y^{-3}}{3y^2}\right)^{-2}$

63. $\left(\dfrac{6r^{-2}}{2r^3}\right)^{-2}$

64. $\left(\dfrac{4x^3}{5x^{-3}}\right)^{-2}$

12.5 *Rational Equations and Problem Solving*

In this section, you will learn about

- Solving rational equations   • Extraneous solutions
- Solving formulas   • Applications

INTRODUCTION. In this section, we will solve problems from banking, petroleum engineering, business, electronics, and travel. We will encounter a new type of equation when we write mathematical models of such situations. These equations will contain one or more rational expressions; they are called **rational equations.**

Solving rational equations

Recall that to solve an equation such as $\frac{x}{6} + \frac{5}{2} = \frac{1}{3}$, we can multiply both sides of the equation by the LCD of the fractions to clear the equation of fractions.

$$\frac{x}{6} + \frac{5}{2} = \frac{1}{3}$$

$$6\left(\frac{x}{6} + \frac{5}{2}\right) = 6\left(\frac{1}{3}\right) \qquad \text{Multiply both sides of the equation by the LCD of } \tfrac{x}{6}, \tfrac{5}{2}, \text{ and } \tfrac{1}{3}, \text{ which is 6.}$$

$$6 \cdot \frac{x}{6} + 6 \cdot \frac{5}{2} = 6 \cdot \frac{1}{3} \qquad \text{Distribute the multiplication by 6.}$$

$$x + 15 = 2 \qquad \text{Do the multiplications.}$$

$$x + 15 - 15 = 2 - 15 \qquad \text{To undo the addition of 15, subtract 15 from both sides.}$$

$$x = -13 \qquad \text{Do the subtractions.}$$

This method can be used to solve rational equations.

EXAMPLE 1   *Solving rational equations.*   Solve $\dfrac{4}{x} + 1 = \dfrac{6}{x}$.

Solution

To clear the equation of fractions, we multiply both sides by the LCD of $\dfrac{4}{x}$ and $\dfrac{6}{x}$, which is x.

$$\frac{4}{x} + 1 = \frac{6}{x}$$

$$x\left(\frac{4}{x} + 1\right) = x\left(\frac{6}{x}\right)$$

$$x \cdot \frac{4}{x} + x \cdot 1 = x \cdot \frac{6}{x} \qquad \text{Distribute the multiplication by } x.$$

$$4 + x = 6 \qquad \text{Do each multiplication.}$$

$$x = 2 \qquad \text{Subtract 4 from both sides.}$$

Check:
$$\frac{4}{x} + 1 = \frac{6}{x}$$

$$\frac{4}{2} + 1 \overset{?}{=} \frac{6}{2} \qquad \text{Substitute 2 for } x.$$

$$2 + 1 \overset{?}{=} 3 \qquad \text{Simplify.}$$

$$3 = 3$$

Self Check

Solve $\dfrac{6}{x} - 1 = \dfrac{3}{x}$.

Answer: 3

EXAMPLE 2 *Solving rational equations.* Solve $\dfrac{22}{5} - \dfrac{3a - 1}{a} = \dfrac{8}{a}$.

Solution
We multiply both sides by $5a$, the LCD of the rational expressions in the equation.

$$\frac{22}{5} - \frac{3a - 1}{a} = \frac{8}{a}$$

$$5a\left(\frac{22}{5} - \frac{3a - 1}{a}\right) = 5a\left(\frac{8}{a}\right)$$

$$5a\left(\frac{22}{5}\right) - 5a\left(\frac{3a - 1}{a}\right) = 5a\left(\frac{8}{a}\right) \quad \text{Distribute the multiplication by } 5a.$$

$$22a - 5(3a - 1) = 40 \qquad \text{Simplify. Note that } 3a - 1 \text{ must be written within parentheses.}$$

$$22a - 15a + 5 = 40 \qquad \text{Distribute the multiplication by } -5.$$

$$7a + 5 = 40 \qquad \text{Combine like terms: } 22a - 15a = 7a.$$

$$7a = 35 \qquad \text{Subtract 5 from both sides.}$$

$$a = 5 \qquad \text{Divide both sides by 7.}$$

Check: $\dfrac{22}{5} - \dfrac{3a - 1}{a} = \dfrac{8}{a}$

$$\frac{22}{5} - \frac{3(5) - 1}{5} \stackrel{?}{=} \frac{8}{5} \quad \text{Substitute 5 for } a.$$

$$\frac{22}{5} - \frac{14}{5} \stackrel{?}{=} \frac{8}{5}$$

$$\frac{8}{5} = \frac{8}{5}$$

Self Check
Solve $\dfrac{7}{6} - \dfrac{2r - 11}{r} = \dfrac{1}{r}$.

Answer: 12

EXAMPLE 3 *Factoring to find the LCD.* Solve
$\dfrac{x + 2}{x + 3} + \dfrac{1}{x^2 + 2x - 3} = 1$.

Solution
To find the LCD, we must factor the second denominator.

$$\frac{x + 2}{x + 3} + \frac{1}{x^2 + 2x - 3} = 1$$

$$\frac{x + 2}{x + 3} + \frac{1}{(x + 3)(x - 1)} = 1 \quad \text{Factor } x^2 + 2x - 3.$$

To clear the equation of fractions, we multiply both sides by the LCD, which is $(x + 3)(x - 1)$.

$$(x + 3)(x - 1)\left[\frac{x + 2}{x + 3} + \frac{1}{(x + 3)(x - 1)}\right] = (x + 3)(x - 1)1$$

Next, we distribute the multiplication by $(x + 3)(x - 1)$.

$$(x + 3)(x - 1)\frac{x + 2}{x + 3} + (x + 3)(x - 1)\frac{1}{(x + 3)(x - 1)} = (x + 3)(x - 1)1$$

Self Check
Solve $\dfrac{1}{x + 3} + \dfrac{1}{x - 3} = \dfrac{10}{x^2 - 9}$.

$(x - 1)(x + 2) + 1 = (x + 3)(x - 1)$	Simplify.
$x^2 + x - 2 + 1 = x^2 + 2x - 3$	Multiply the pairs of binomials.
$x^2 + x - 1 = x^2 + 2x - 3$	Combine like terms.
$x - 1 = 2x - 3$	Subtract x^2 from both sides.
$-x - 1 = -3$	Subtract $2x$ from both sides.
$-x = -2$	Add 1 to both sides.
$x = 2$	Divide both sides by -1.

Verify that 2 is a solution of the given equation.

Answer: 5 ■

EXAMPLE 4 *A rational equation that leads to a quadratic equation.* Solve $\dfrac{4}{5} + y = \dfrac{4y - 50}{5y - 25}$.

Solution
To find the LCD, we must factor $5y - 25$.

$$\frac{4}{5} + y = \frac{4y - 50}{5y - 25}$$

$$\frac{4}{5} + y = \frac{4y - 50}{5(y - 5)}$$

$5(y - 5)\left[\dfrac{4}{5} + y\right] = 5(y - 5)\left[\dfrac{4y - 50}{5(y - 5)}\right]$	Multiply both sides by the LCD, which is $5(y - 5)$.
$4(y - 5) + 5y(y - 5) = 4y - 50$	Distribute $5(y - 5)$.
$4y - 20 + 5y^2 - 25y = 4y - 50$	Distribute 4 and $5y$.
$5y^2 - 25y - 20 = -50$	Subtract $4y$ from both sides and rearrange terms.
$5y^2 - 25y + 30 = 0$	Add 50 to both sides.
$y^2 - 5y + 6 = 0$	Divide both sides by 5.
$(y - 3)(y - 2) = 0$	Factor $y^2 - 5y + 6$.
$y - 3 = 0$ or $y - 2 = 0$	Set each factor equal to zero.
$y = 3$ $y = 2$	Solve each equation.

Verify that 3 and 2 satisfy the original equation.

Self Check
Solve:

$$\frac{x - 6}{3x - 9} - \frac{1}{3} = \frac{x}{2}$$

Answer: 1, 2 ■

Extraneous solutions

If we multiply both sides of an equation by an expression that involves a variable, as we did in the previous examples, we must check the apparent solutions. The next example shows why.

EXAMPLE 5 *Checking apparent solutions.* Solve $\dfrac{x + 3}{x - 1} = \dfrac{4}{x - 1}$.

Solution
To clear the equation of fractions, we multiply both sides by the LCD, which is $x - 1$.

Self Check
Solve $\dfrac{x + 5}{x - 2} = \dfrac{7}{x - 2}$.

$$\frac{x+3}{x-1} = \frac{4}{x-1}$$

$$(x-1)\frac{x+3}{x-1} = (x-1)\frac{4}{x-1} \quad \text{Multiply both sides by } x-1.$$

$$x + 3 = 4 \qquad\qquad \text{Simplify.}$$

$$x = 1 \qquad\qquad \text{Subtract 3 from both sides.}$$

Because both sides were multiplied by an expression containing a variable, we must check the apparent solution.

$$\frac{x+3}{x-1} = \frac{4}{x-1}$$

$$\frac{1+3}{1-1} \stackrel{?}{=} \frac{4}{1-1} \quad \text{Substitute 1 for } x.$$

$$\frac{4}{0} \stackrel{?}{=} \frac{4}{0} \qquad\quad \text{Simplify.}$$

Since zeros appear in the denominators, the fractions are undefined. Thus, 1 is a false solution, and the equation has no solutions. Such false solutions are often called **extraneous solutions.**

Answer: 2 is extraneous.   ■

Solving formulas

Many formulas are equations that contain rational expressions.

EXAMPLE 6   *Solving formulas.*

The formula $\dfrac{1}{r} = \dfrac{1}{r_1} + \dfrac{1}{r_2}$

is used in electronics to calculate parallel resistances. Solve the equation for r.

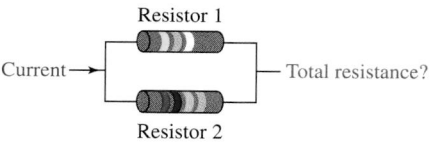

Resistor 1

Current → Total resistance?

Resistor 2

FIGURE 12-1

Self Check
Solve the formula in Example 6 for r_1.

Solution
Clear the equation of fractions by multiplying both sides by the LCD, which is rr_1r_2.

$$\frac{1}{r} = \frac{1}{r_1} + \frac{1}{r_2}$$

$$rr_1r_2\left(\frac{1}{r}\right) = rr_1r_2\left(\frac{1}{r_1} + \frac{1}{r_2}\right) \quad \text{Multiply both sides by } rr_1r_2.$$

$$\frac{rr_1r_2}{r} = \frac{rr_1r_2}{r_1} + \frac{rr_1r_2}{r_2} \qquad \text{Distribute the multiplication by } rr_1r_2.$$

$$r_1r_2 = rr_2 + rr_1 \qquad\qquad \text{Simplify each fraction.}$$

$$r_1r_2 = r(r_2 + r_1) \qquad\qquad \text{Factor out } r.$$

$$\frac{r_1r_2}{r_2 + r_1} = r \qquad\qquad\qquad \text{To isolate } r, \text{ divide both sides by } r_2 + r_1.$$

or

$$r = \frac{r_1r_2}{r_2 + r_1}$$

Answer: $r_1 = \dfrac{rr_2}{r_2 - r}$ ■

Applications

EXAMPLE 7   *A number problem.*   If the same number is added to both the numerator and the denominator of the fraction $\frac{3}{5}$, the result is $\frac{4}{5}$. Find the number.

Analyze the problem   We are asked to find a number. If we add it to both the numerator and the denominator of a fraction, we will get $\frac{4}{5}$.

Form an equation   Let n represent the unknown number and add n to both the numerator and the denominator of $\frac{3}{5}$. Then set the result equal to $\frac{4}{5}$ to get the equation

$$\frac{3 + n}{5 + n} = \frac{4}{5}$$

Solve the equation   To solve the equation, we proceed as follows:

$$\frac{3 + n}{5 + n} = \frac{4}{5}$$

$$\mathbf{5(5 + n)}\frac{3 + n}{5 + n} = \mathbf{5(5 + n)}\frac{4}{5} \quad \text{Multiply both sides by } 5(5 + n), \text{ which is the LCD of the fractions appearing in the equation.}$$

$$5(3 + n) = (5 + n)4 \quad \text{Simplify.}$$

$$15 + 5n = 20 + 4n \quad \text{Distribute the multiplications by 5 and by 4.}$$

$$15 + n = 20 \quad \text{Subtract } 4n \text{ from both sides.}$$

$$n = 5 \quad \text{Subtract 15 from both sides.}$$

State the conclusion   The number is 5.

Check the result   When we add 5 to both the numerator and denominator of $\frac{3}{5}$, we get

$$\frac{3 + 5}{5 + 5} = \frac{8}{10} = \frac{4}{5}$$

The result checks.   ∎

We can use rational equations to model shared-work problems. In this case, we assume that the work is being performed at a constant rate by all of those involved.

EXAMPLE 8   *Filling an oil tank.*   An inlet pipe can fill an oil tank in 7 days, and a second inlet pipe can fill the same tank in 9 days. If both pipes are used, how long will it take to fill the tank?

Analyze the problem   The key is to determine what each pipe can do in 1 day. If we add what the first pipe can do in 1 day to what the second pipe can do in 1 day, the sum is what they can do together in 1 day.

     Since the first pipe can fill the tank in 7 days, it can do $\frac{1}{7}$ of the job in 1 day. Since the second pipe can fill the tank in 9 days, it can do $\frac{1}{9}$ of the job in 1 day. If it takes x days for both pipes to fill the tank, together they can do $\frac{1}{x}$ of the job in 1 day.

Form an equation   Let x represent the number of days it will take to fill the tank if both inlet pipes are used. Then form the equation.

What the first inlet pipe can do in 1 day	plus	what the second inlet pipe can do in 1 day	equals	what they can do together in 1 day.
$\dfrac{1}{7}$	$+$	$\dfrac{1}{9}$	$=$	$\dfrac{1}{x}$

Solve the equation   To solve the equation, we proceed as follows:

$$\frac{1}{7} + \frac{1}{9} = \frac{1}{x}$$

$$63x\left(\frac{1}{7} + \frac{1}{9}\right) = 63x\left(\frac{1}{x}\right)$$ Multiply both sides by $63x$ to clear the equation of fractions.

$$9x + 7x = 63$$ Distribute the multiplication by $63x$ and simplify.

$$16x = 63$$ Combine like terms.

$$x = \frac{63}{16}$$ Divide both sides by 16.

State the conclusion   It will take $\frac{63}{16}$ or $3\frac{15}{16}$ days for both inlet pipes to fill the tank.

Check the result   In $\frac{63}{16}$ days, the first pipe fills $\frac{1}{7} \cdot \frac{63}{16} = \frac{9}{16}$ of the tank and the second pipe fills $\frac{1}{9} \cdot \frac{63}{16} = \frac{7}{16}$ of the tank.

The sum of these efforts, $\frac{9}{16} + \frac{7}{16}$, is equal to one full tank.   ∎

EXAMPLE 9

Track and field.   A coach can run 10 miles in the same amount of time as his best student-athlete can run 12 miles. If the student can run 1 mile per hour faster than the coach, how fast can the student run?

Analyze the problem   We can use the formula $d = rt$, where d is the distance traveled, r is the rate, and t is the time. If we solve this formula for t, we obtain

$$t = \frac{d}{r}$$

Form an equation   It will take $\frac{10}{r}$ hours for the coach to run 10 miles at some unknown rate of r mph. It will take $\frac{12}{r+1}$ hours for the student to run 12 miles at some unknown rate of $(r + 1)$ mph. We can organize the information of the problem in a table, as shown in Figure 12-2.

	r	$\cdot$	t	$=$	d
Student	$r + 1$		$\dfrac{12}{r+1}$		12
Coach	r		$\dfrac{10}{r}$		10

FIGURE 12-2

The time it takes the student to run 12 miles	equals	the time it takes the coach to run 10 miles.
$\dfrac{12}{r+1}$	$=$	$\dfrac{10}{r}$

Solve the equation   We can solve the equation as follows:

$$\frac{12}{r+1} = \frac{10}{r}$$

$$r(r+1)\frac{12}{r+1} = r(r+1)\frac{10}{r}$$ Multiply both sides by $r(r + 1)$.

$$12r = 10(r+1)$$ Simplify.

$$12r = 10r + 10$$ Distribute the multiplication by 10.

$$2r = 10$$ Subtract $10r$ from both sides.

$$r = 5$$ Divide both sides by 2.

State the conclusion   The coach can run 5 mph. The student, running 1 mph faster, can run 6 mph.

Check the result   Verify that these results check.   ∎

EXAMPLE 10   *Banking.*   At one bank, a sum of money invested for one year will earn $96 interest. If invested in bonds, that money would earn $108, because the interest rate paid by the bonds is 1% greater than that paid by the bank. Find the bank's rate.

Analyze the problem   This interest problem is based on the formula $I = Pr$, where I is the interest earned in 1 year, P is the principal (the amount invested), and r is the annual rate of interest. If we solve this formula for P, we obtain

$$P = \frac{I}{r}$$

Form an equation   If we let r represent the bank's rate of interest, then $r + 0.01$ represents the rate paid by the bonds. If a person earns $96 interest at a bank at some unknown rate r, the principal invested was $\frac{96}{r}$. If a person earns $108 interest in bonds at some unknown rate $(r + 0.01)$, the principal invested was $\frac{108}{r + 0.01}$. We can organize the information of the problem in a table, as shown in Figure 12-3.

	Principal ·	Rate =	Interest
Bank	$\dfrac{96}{r}$	r	96
Bonds	$\dfrac{108}{r + 0.01}$	$r + 0.01$	108

FIGURE 12-3

Because the same principal would be invested in either account, we can set up the following equation:

$$\frac{96}{r} = \frac{108}{r + 0.01}$$

Solve the equation   We can solve the equation as follows:

$$\frac{96}{r} = \frac{108}{r + 0.01}$$

$$r(r + 0.01) \cdot \frac{96}{r} = r(r + 0.01) \cdot \frac{108}{r + 0.01} \qquad \text{Multiply both sides by } r(r + 0.01).$$

$$96(r + 0.01) = 108r$$

$$96r + 0.96 = 108r \qquad\qquad\qquad\qquad \text{Distribute.}$$

$$0.96 = 12r \qquad\qquad\qquad\qquad \text{Subtract } 96r \text{ from both sides.}$$

$$0.08 = r \qquad\qquad\qquad\qquad \text{Divide both sides by 12.}$$

State the conclusion   The bank's interest rate is 0.08, or 8%. The bonds pay 9% interest, a rate 1% greater than that paid by the bank.

Check the result   Verify that these rates check.   ∎

STUDY SET Section 12.5

VOCABULARY *Fill in the blanks.*

1. Equations that contain one or more rational expressions, such as

$$\frac{x + 2}{x + 3} + \frac{1}{x^2 + 2x - 3} = 1$$

are called _____.

2. To clear an equation of fractions, we multiply both sides by the _____ of the fractions in the equation.

3. If you multiply both sides of an equation by an expression that involves a variable, you must _____ the solution.

4. False solutions that result from multiplying both sides of an equation by a variable are called _____ solutions.

5. In the formula $I = Pr$, I stands for the amount of _____ earned in one year, P stands for the _____, and r stands for the annual interest _____.

6. In the formula $d = rt$, d stands for the _____ traveled, r is the _____, and t is the _____.

CONCEPTS

7. Is $x = 5$ a solution of the following equations?

a. $\dfrac{1}{x - 1} = 1 - \dfrac{3}{x - 1}$

b. $\dfrac{x}{x - 5} = 3 + \dfrac{5}{x - 5}$

8. By what should we multiply both sides of each equation to clear it of fractions?

a. $\dfrac{1}{x} + \dfrac{2}{x} = 5$ **b.** $\dfrac{x}{x - 2} - \dfrac{x}{x - 1} = 5$

9. Illustration 1 shows the length of time it takes each of two hardware store employees to assemble a metal storage shed, working alone.

a. Complete the table.

	Time to assemble the shed (hr)	Amount of the shed assembled in 1 hr
Marvin	6	
Kyla	5	

ILLUSTRATION 1

b. If we assume that working together would not change their individual rates, how much of the shed could they assemble in one hour if they worked together?

10. When two ice machines are both running, they can fill a supermarket's order in x hours. At this rate, how much of the order do they fill in 1 hour?

11. If the exits at the front of a theater are opened, a full theater can be emptied of all occupants in 6 minutes. How much of the theater is emptied in 1 minute?

12. Solve $d = rt$

a. for r **b.** for t

13. Solve $I = Pr$

a. for r **b.** for P

14. a. Complete the table in Illustration 2.

	r	$\cdot$ t	$=$ d
Snowmobile	r		4
4×4 truck	$r - 5$		3

ILLUSTRATION 2

b. Complete the table in Illustration 3.

	P	$\cdot$ r	$=$ I
City Savings		r	50
Credit Union		$r - 0.02$	75

ILLUSTRATION 3

NOTATION *In Exercises 15–16, a rational equation is solved. Complete each solution.*

15.
$$\frac{2}{a} + \frac{1}{2} = \frac{7}{2a}$$

$$\boxed{}\left(\frac{2}{a} + \frac{1}{2}\right) = \boxed{}\left(\frac{7}{2a}\right)$$

$$\boxed{} \cdot \frac{2}{a} + \boxed{} \cdot \frac{1}{2} = \boxed{} \cdot \frac{7}{2a}$$

$$\boxed{} + a = 7$$

$$4 + a - \boxed{} = 7 - \boxed{}$$

$$a = 3$$

16.
$$\frac{3}{5} + \frac{7}{a+2} = 2$$

$$\boxed{}\left(\frac{3}{5} + \frac{7}{a+2}\right) = \boxed{} \cdot 2$$

$$\boxed{} \cdot \frac{3}{5} + \boxed{} \cdot \frac{7}{a+2} = \boxed{} \cdot 2$$

$$3(a+2) + \boxed{} = 10(a+2)$$

$$3a + \boxed{} + 35 = 10a + \boxed{}$$

$$3a + \boxed{} = 10a + 20$$

$$-7a = \boxed{}$$

$$a = 3$$

17. The following work shows both sides of an equation being multiplied by the LCD to clear it of fractions. What was the original equation?

$$5(x+2)\left(\frac{3}{5}\right) + 5(x+2)\left(\frac{7}{x+2}\right) = 5(x+2) \cdot 2$$

18. After solving a rational equation, a student checked her answer and obtained the following:

$$\frac{-1}{0} + \frac{1}{0} = 0$$

What conclusion can be drawn?

PRACTICE *Solve each equation and check the result. If an equation has no solution, so indicate.*

19. $\dfrac{x}{2} + 4 = \dfrac{3x}{2}$

20. $\dfrac{2y}{5} - 8 = \dfrac{4y}{5}$

21. $\dfrac{x+1}{3} + \dfrac{x-1}{5} = \dfrac{2}{15}$

22. $\dfrac{3x-1}{6} - \dfrac{x+3}{2} = \dfrac{3x+4}{3}$

23. $\dfrac{3}{x} + 2 = 3$ **24.** $\dfrac{2}{x} + 9 = 11$

25. $\dfrac{5}{a} - \dfrac{4}{a} = 8 + \dfrac{1}{a}$ **26.** $\dfrac{11}{b} + \dfrac{13}{b} = 12$

27. $\dfrac{3}{4h} + \dfrac{2}{h} = 1$ **28.** $\dfrac{5}{3k} + \dfrac{1}{k} = -2$

29. $\dfrac{a}{4} - \dfrac{4}{a} = 0$ **30.** $0 = \dfrac{t}{3} - \dfrac{12}{t}$

31. $\dfrac{2}{y+1} + 5 = \dfrac{12}{y+1}$ **32.** $\dfrac{3}{p+6} - 2 = \dfrac{7}{p+6}$

33. $\dfrac{x}{x-5} - \dfrac{5}{x-5} = 3$

34. $\dfrac{3}{y-2} + 1 = \dfrac{3}{y-2}$

35. $\dfrac{3r}{2} - \dfrac{3}{r} = \dfrac{3r}{2} + 3$ **36.** $\dfrac{2p}{3} - \dfrac{1}{p} = \dfrac{2p-1}{3}$

37. $\dfrac{1}{3} + \dfrac{2}{x-3} = 1$ **38.** $\dfrac{3}{5} + \dfrac{7}{x+2} = 2$

39. $\dfrac{z-4}{z-3} = \dfrac{z+2}{z+1}$ **40.** $\dfrac{a+2}{a+8} = \dfrac{a-3}{a-2}$

41. $\dfrac{v}{v+2} + \dfrac{1}{v-1} = 1$ **42.** $\dfrac{x}{x-2} = 1 + \dfrac{1}{x-3}$

43. $\dfrac{a^2}{a+2} - \dfrac{4}{a+2} = a$

44. $\dfrac{z^2}{z+1} + 2 = \dfrac{1}{z+1}$

45. $\dfrac{7}{q^2-q-2} + \dfrac{1}{q+1} = \dfrac{3}{q-2}$

46. $\dfrac{3}{x-1} - \dfrac{1}{x+9} = \dfrac{18}{x^2+8x-9}$

47. $\dfrac{u}{u-1} + \dfrac{1}{u} = \dfrac{u^2+1}{u^2-u}$

48. $\dfrac{3}{x-2} + \dfrac{1}{x} = \dfrac{2(3x+2)}{x^2-2x}$

49. $\dfrac{n}{n^2-9} + \dfrac{n+8}{n+3} = \dfrac{n-8}{n-3}$

50. $\dfrac{7}{x-5} - \dfrac{3}{x+5} = \dfrac{40}{x^2-25}$

51. $\dfrac{5}{x+4} + \dfrac{1}{x+4} = x - 1$

52. $\dfrac{7}{x-3} + \dfrac{1}{x-3} = x - 5$

53. $\dfrac{3}{x+1} - \dfrac{x-2}{2} = \dfrac{x-2}{x+1}$

54. $\dfrac{2}{x-1} + \dfrac{x-2}{3} = \dfrac{4}{x-1}$

55. $\dfrac{b+2}{b+3} + 1 = \dfrac{-7}{b-5}$

56. $\dfrac{x-4}{x-3} + \dfrac{x-2}{x-3} = x - 3$

57. $\dfrac{x}{x-1} - \dfrac{12}{x^2-x} = \dfrac{-1}{x-1}$

58. $y + \dfrac{2}{3} = \dfrac{2y-12}{3y-9}$

59. $1 - \dfrac{3}{b} = \dfrac{-8b}{b^2+3b}$

60. $\dfrac{5}{4y+12} - \dfrac{3}{4} = \dfrac{5}{4y+12} - \dfrac{y}{4}$

Solve each formula for the indicated variable.

61. $\dfrac{1}{a} + \dfrac{1}{b} = 1$ for a

62. $\dfrac{1}{a} - \dfrac{1}{b} = 1$ for b

63. $I = \dfrac{E}{R + r}$ for r

64. $h = \dfrac{2A}{b + d}$ for A

65. $\dfrac{a}{b} = \dfrac{c}{d}$ for d

66. $F = \dfrac{L^2}{6d} + \dfrac{d}{2}$ for L^2

Use the given information to find the number or numbers.

67. If the denominator of $\frac{3}{4}$ is increased by a number and the numerator of the fraction is doubled, the result is 1.

68. If a number is added to the numerator of $\frac{7}{8}$ and the same number is subtracted from the denominator, the result is 2.

69. If a number is added to the numerator of $\frac{3}{4}$ and twice as much is added to the denominator, the result is $\frac{4}{7}$.

70. If a number is added to the numerator of $\frac{5}{7}$ and twice as much is subtracted from the denominator, the result is 8.

71. The sum of a number and its reciprocal is $\frac{13}{6}$.

72. The sum of the reciprocals of two consecutive even integers is $\frac{7}{24}$. (*Hint:* Let x = the first integer and $x + 2$ = the second integer.)

APPLICATIONS

73. OPTICS   The focal length f of a lens is given by the formula

$$\frac{1}{f} = \frac{1}{d_1} + \frac{1}{d_2}$$

where d_1 is the distance from the object to the lens and d_2 is the distance from the lens to the image. Solve the formula for f.

74. OPTICS   Solve the formula in Exercise 73 for d_1.

75. MEDICINE   Radioactive tracers are used for diagnostic work in nuclear medicine. The **effective half-life** H of a radioactive material in an organism is given by the formula

$$H = \frac{RB}{R + B}$$

where R is the radioactive half-life and B is the biological half-life of the tracer. Solve the formula for R.

76. CHEMISTRY   Charles's Law describes the relationship between the volume and the temperature of a gas that is kept at a constant pressure. It states that as the temperature of the gas increases, the volume of the gas will increase:

$$\frac{V_1}{V_2} = \frac{T_1}{T_2}$$

Solve the equation for V_2.

77. FILLING A POOL   An inlet pipe can fill an empty swimming pool in 5 hours, and another inlet pipe can fill the pool in 4 hours. How long will it take both pipes to fill the pool?

78. FILLING A POOL   One inlet pipe can fill an empty pool in 4 hours, and a drain can empty the pool in 8 hours. How long will it take the pipe to fill the pool if the drain is left open?

79. ROOFING A HOUSE   A homeowner estimates that it will take her 7 days to roof her house. A professional roofer estimates that he could roof the house in 4 days. How long will it take if the homeowner helps the roofer?

80. SEWAGE TREATMENT   A sludge pool is filled by two inlet pipes. One pipe can fill the pool in 15 days, and the other can fill it in 21 days. However, if no sewage is added, continuous waste removal will empty the pool in 36 days. How long will it take the two inlet pipes to fill an empty sludge pool?

81. TOURING   A woman can bicycle 28 miles in the same time as it takes her to walk 8 miles. If she can ride 10 mph faster than she can walk, how much time should she allow to walk a 30-mile trail? See Illustration 4. (*Hint:* How fast can she walk?)

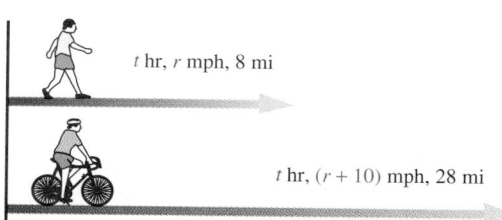

ILLUSTRATION 4

82. COMPARING TRAVEL   A plane can fly 300 miles in the same time as it takes a car to go 120 miles. If the car travels 90 mph slower than the plane, find the speed of the plane.

83. BOATING   A boat that travels 18 mph in still water can travel 22 miles downstream in the same time as it takes to travel 14 miles upstream. Find the speed of the current in the river. (See Illustration 5.)

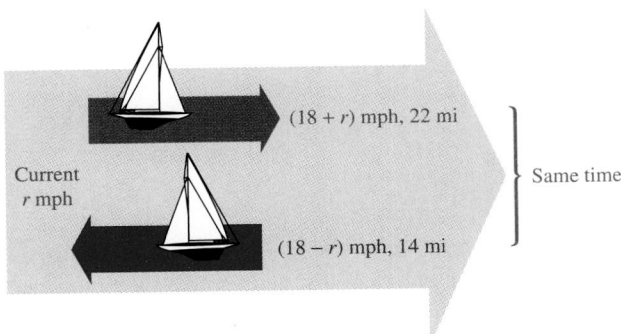

Current
r mph

(18 + r) mph, 22 mi

(18 − r) mph, 14 mi

Same time

ILLUSTRATION 5

84. WIND SPEED   A plane can fly 300 miles downwind in the same time as it can travel 210 miles upwind. Find the velocity of the wind if the plane can fly 255 mph in still air.

85. COMPARING INVESTMENTS   Two certificates of deposit (CDs) pay interest at rates that differ by 1%. Money invested for one year in the first CD earns $175 interest. The same principal invested in the second CD earns $200. Find the two rates of interest.

86. COMPARING INTEREST RATES   Two bond funds pay interest at rates that differ by 2%. Money invested for one year in the first fund earns $315 interest. The same amount invested in the second fund earns $385. Find the lower rate of interest.

87. SHARING COSTS   Several office workers bought a $35 gift for their boss. If there had been two more employees to contribute, everyone's cost would have been $2 less. How many workers contributed to the gift?

88. SALES   A dealer bought some radios for a total of $1,200. She gave away 6 radios as gifts, sold the rest for $10 more than she paid for each radio, and broke even. How many radios did she buy?

89. SALES   A bookstore can purchase several calculators for a total cost of $120. If each calculator cost $1 less, the bookstore could purchase 10 additional calculators at the same total cost. How many calculators can be purchased at the regular price?

90. FURNACE REPAIR   A repairman purchased several furnace-blower motors for a total cost of $210. If his cost per motor had been $5 less, he could have purchased one additional motor. How many motors did he buy at the regular rate?

91. RIVER TOUR   A river boat tour begins by going 60 miles upstream against a 5-mph current. There, the boat turns around and returns with the current. What still-water speed should the captain use to complete the tour in 5 hours?

92. TRAVEL TIME   A company president flew 680 miles one way in the corporate jet but returned in a smaller plane that could fly only half as fast. If the total travel time was 6 hours, find the speeds of the planes.

WRITING

93. Explain how you would decide what to do first to solve an equation that involves fractions.

94. Why is it important to check your solutions of an equation that contains fractions with variables in the denominator?

95. In Example 8, one inlet pipe could fill an oil tank in 7 days, and another could fill the same tank in 9 days. We were asked to find how long it would take if both pipes were used. Explain why each of the following approaches is incorrect.

The time it would take to fill the tank

- is the *sum* of the lengths of time it takes each pipe to fill the tank: 7 days + 9 days = 16 days.
- is the *difference* in the lengths of time it takes each pipe to fill the tank: 9 days − 7 days = 2 days.
- is the *average* of the lengths of time it takes each pipe to fill the tank:

$$\frac{7 \text{ days} + 9 \text{ days}}{2} = \frac{16 \text{ days}}{2} = 8 \text{ days}$$

96. Explain the difference between the procedure used to simplify

$$\frac{1}{x} + \frac{1}{3}$$

and the procedure used to solve

$$\frac{1}{x} + \frac{1}{3} = \frac{1}{2}$$

REVIEW   *Factor each expression.*

97. $x^2 + 4x$

98. $x^2 - 16y^2$

99. $2x^2 + x - 3$

100. $6a^2 - 5a - 6$

101. $x^4 - 16$

102. $4x^2 + 10x - 6$

12.6 *Proportions and Similar Triangles*

In this section, you will learn about

- Ratios and rates • Proportions • Solving proportions
- Problem solving • Similar triangles

INTRODUCTION. In this section, we will discuss a problem-solving tool called a *proportion*. A proportion is a type of rational equation that involves two *ratios* or two *rates*.

Ratios and rates

Ratios enable us to compare numerical quantities.

- To prepare fuel for a Lawnboy lawnmower, gasoline must be mixed with oil in the ratio of 50 to 1.
- To make 14-karat jewelry, gold is mixed with other metals in the ratio of 14 to 10.
- In the stock market, winning stocks might outnumber losing stocks in the ratio of 7 to 4.

Ratios

> A **ratio** is the quotient of two numbers or the quotient of two quantities that have the same units.

There are three common ways to write a ratio: as a fraction, with the word *to,* or with a colon. For example, the ratio describing the ratio of the number of winning stocks to the number of losing stocks mentioned earlier can be written as

$$\frac{7}{4}, \qquad 7 \text{ to } 4, \quad \text{or} \quad 7:4$$

Each of these forms can be read as "the ratio of 7 to 4."

When ratios are used to compare quantities with different units, they are called *rates*. For example, if the 495-mile drive from New Orleans to Dallas takes 9 hours, the average rate of speed is the ratio of the miles driven to the length of time the trip takes.

$$\text{Average rate of speed} = \frac{495 \text{ miles}}{9 \text{ hours}} = \frac{55 \text{ miles}}{1 \text{ hour}} \qquad \frac{495}{9} = \frac{\overset{1}{\cancel{9}} \cdot 55}{\cancel{9} \cdot 1} = \frac{55}{1}.$$

Rates

> A **rate** is a quotient of two quantities that have different units.

Proportions

Consider the following table, in which we are given the costs of various numbers of gallons of gasoline.

Number of gallons	Cost
2	$3.72
5	$9.30
8	$14.88
12	$22.32
20	$37.20

If we compare the costs to the numbers of gallons purchased, we see that they are equal. In this example, each quotient represents the cost of 1 gallon of gasoline, which is $1.86.

$$\frac{\$3.72}{2} = \$1.86, \qquad \frac{\$9.30}{5} = \$1.86, \qquad \frac{\$14.88}{8} = \$1.86,$$

$$\frac{\$22.32}{12} = \$1.86, \quad \text{and} \quad \frac{\$37.20}{20} = \$1.86$$

When two ratios or rates $\left(\text{such as } \frac{\$3.72}{2} \text{ and } \frac{\$9.30}{5}\right)$ are equal, they form a *proportion.*

Proportions

> A **proportion** is a statement that two ratios or two rates are equal.

Some examples of proportions are

$$\frac{1}{2} = \frac{3}{6}, \qquad \frac{3 \text{ waiters}}{7 \text{ tables}} = \frac{9 \text{ waiters}}{21 \text{ tables}}, \quad \text{and} \quad \frac{a}{b} = \frac{c}{d}$$

- The proportion $\frac{1}{2} = \frac{3}{6}$ can be read as "1 is to 2 as 3 is to 6."
- The proportion $\frac{3 \text{ waiters}}{7 \text{ tables}} = \frac{9 \text{ waiters}}{21 \text{ tables}}$ can be read as "3 waiters is to 7 tables as 9 waiters is to 21 tables."
- The proportion $\frac{a}{b} = \frac{c}{d}$ can be read as "a is to b as c is to d."

In the proportion $\frac{a}{b} = \frac{c}{d}$, a and d are called the **extremes,** and b and c are called the **means.** We can show that the product of the extremes (ad) is equal to the product of the means (bc) by multiplying both sides of the proportion by bd and observing that $ad = bc$.

$$\frac{a}{b} = \frac{c}{d}$$

$$bd \cdot \frac{a}{b} = bd \cdot \frac{c}{d} \qquad \text{To clear the equation of fractions, multiply both sides by the LCD, which is } bd.$$

$$ad = bc \qquad \text{Do each multiplication and simplify.}$$

Since $ad = bc$, the product of the extremes equals the product of the means.

The fundamental property of proportions

> In a proportion, the product of the extremes is equal to the product of the means.

To determine whether an equation is a proportion, we can check to see whether the product of the extremes is equal to the product of the means.

EXAMPLE 1   *Proportions.* Determine whether each equation is a proportion:
a. $\dfrac{3}{7} = \dfrac{9}{21}$ and   **b.** $\dfrac{8}{3} = \dfrac{13}{5}$.

Solution

In each case, we check to see whether the product of the extremes is equal to the product of the means.

a. The product of the extremes is $3 \cdot 21 = 63$. The product of the means is $7 \cdot 9 = 63$. Since the products are equal, the equation is a proportion: $\frac{3}{7} = \frac{9}{21}$.

$$3 \cdot 21 = 63 \qquad\qquad 7 \cdot 9 = \mathbf{63}$$

$$\dfrac{3}{7} = \dfrac{9}{21}$$ The product of the extremes and the product of the means are also known as **cross products**.

b. The product of the extremes is $8 \cdot 5 = 40$. The product of the means is $3 \cdot 13 = 39$. Since the cross products are not equal, the equation is not a proportion: $\frac{8}{3} \neq \frac{13}{5}$.

$$8 \cdot 5 = 40 \qquad\qquad 3 \cdot 13 = \mathbf{39}$$

$$\dfrac{8}{3} = \dfrac{13}{5}$$

Self Check

Determine whether the equation is a proportion:

$$\dfrac{6}{13} = \dfrac{24}{53}$$

Answer: no   ■

Solving proportions

Suppose that we know three terms in the proportion

$$\dfrac{x}{5} = \dfrac{24}{20}$$

To find the unknown term, we can multiply both sides of the equation by 20 to clear it of fractions, and then solve for x. However, with proportions, it is often easier to simply compute the cross products, set them equal, and solve for the variable.

$$\dfrac{x}{5} = \dfrac{24}{20}$$

$20 \cdot x = 5 \cdot 24$ In a proportion, the product of the extremes equals the product of the means.

$20x = 120$ Do the multiplication: $5 \cdot 24 = 120$.

$\dfrac{20x}{20} = \dfrac{120}{20}$ To undo the multiplication by 20, divide both sides by 20.

$x = 6$ Do the divisions.

The first term is 6. To check this result, we substitute 6 for x in $\frac{x}{5} = \frac{24}{20}$ and find the cross products.

$$\dfrac{6}{5} \overset{?}{=} \dfrac{24}{20} \qquad\qquad 6 \cdot 20 = 120$$
$$5 \cdot 24 = 120$$

Since the cross products are equal, this is a proportion. The result, 6, is correct.

EXAMPLE 2   *Solving proportions.*   Solve $\dfrac{12}{18} = \dfrac{3}{x}$.

Solution

$$\dfrac{12}{18} = \dfrac{3}{x}$$

$12 \cdot x = 18 \cdot 3$ In a proportion, the product of the extremes equals the product of the means.

$12x = 54$ Multiply: $18 \cdot 3 = 54$.

$\dfrac{12x}{12} = \dfrac{54}{12}$ To undo the multiplication by 12, divide both sides by 12.

Self Check

Solve $\dfrac{15}{x} = \dfrac{25}{40}$.

$$x = \frac{9}{2} \qquad \text{Simplify:} \quad \frac{54}{12} = \frac{9 \cdot \cancel{6}^{\,1}}{\cancel{6}_{\,1} \cdot 2} = \frac{9}{2}.$$

Thus, $x = \frac{9}{2}$. Check the result.

Answer: 24 ■

 COMMENT Remember that a cross product is the product of the means or extremes of a *proportion*. For example, it would be incorrect to try to compute "cross products" to solve the rational equation $\frac{12}{18} = \frac{3}{x} + \frac{1}{2}$. The right-hand side is not a ratio, so the equation not a proportion.

Accent on Technology: **Solving proportions with a calculator**

To solve the proportion $\dfrac{3.5}{7.2} = \dfrac{x}{15.84}$ with a calculator, we can proceed as follows.

$$\frac{3.5}{7.2} = \frac{x}{15.84}$$

$$\frac{3.5(15.84)}{7.2} = x \qquad \text{To undo the division by 15.84 and isolate } x, \text{ multiply both sides of the equation by 15.84.}$$

We can find x by entering these numbers into a scientific calculator.

Keystrokes   3.5 $\boxed{\times}$ 15.84 $\boxed{\div}$ 7.2 $\boxed{=}$ $\boxed{ 7.7}$

Using a graphing calculator, we enter these numbers and press these keys.

Keystrokes   3.5 $\boxed{\times}$ 15.84 $\boxed{\div}$ 7.2 $\boxed{\text{ENTER}}$ $\boxed{\begin{array}{l} 3.5*15.84/7.2 \\ \hfill 7.7 \end{array}}$

Thus, $x = 7.7$.

EXAMPLE 3 *Solving proportions.* Solve $\dfrac{2a+1}{4} = \dfrac{10}{8}$.

Solution

$$\frac{2a+1}{4} = \frac{10}{8}$$

$$8(2a+1) = 40 \qquad \text{In a proportion, the product of the extremes equals the product of the means.}$$

$$16a + 8 = 40 \qquad \text{Distribute the multiplication by 8.}$$

$$16a + 8 - 8 = 40 - 8 \qquad \text{To undo the addition of 8, subtract 8 from both sides.}$$

$$16a = 32 \qquad \text{Combine like terms.}$$

$$\frac{16a}{16} = \frac{32}{16} \qquad \text{To undo the multiplication by 16, divide both sides by 16.}$$

$$x = 2 \qquad \text{Do the divisions.}$$

Thus, $a = 2$. Check the result.

Self Check

Solve $\dfrac{3x-1}{2} = \dfrac{12.5}{5}$.

Answer: 2 ■

Problem solving

We can use proportions to solve many real-world problems. If we are given a ratio (or rate) comparing two quantities, the words of the problem can be translated to a proportion, and we can solve it to find the unknown.

EXAMPLE 4   *Grocery shopping.*   If 6 apples cost $1.38, how much will 16 apples cost?

Solution

Analyze the problem   We know the cost of 6 apples; we are to find the cost of 16 apples.

Form a proportion   Let c represent the cost of 16 apples. If we compare the number of apples to their cost, we know that the two rates are equal.

6 apples is to $1.38 as 16 apples is to $c.

$$\text{6 apples} \longrightarrow \quad \frac{6}{1.38} = \frac{16}{c} \quad \longleftarrow \text{16 apples}$$
$$\text{Cost of 6 apples} \longrightarrow \qquad\qquad\qquad \longleftarrow \text{Cost of 16 apples}$$

Solve the proportion

$6 \cdot c = 1.38(16)$ In a proportion, the product of the extremes equals the product of the means.

$6c = 22.08$ Do the multiplication: $1.38(16) = 22.08$.

$\dfrac{6c}{6} = \dfrac{22.08}{6}$ To undo the multiplication by 6, divide both sides by 6.

$c = 3.68$ Simplify: $\frac{22.08}{6} = 3.68$.

State the conclusion   Sixteen apples will cost $3.68.

Check the result   If 16 apples are bought, this is about 3 times as many as 6 apples, which cost $1.38. If we multiply $1.38 by 3, we get an estimate of the cost of 16 apples: $1.38 \cdot 3 = $4.14. The result, $3.68, seems reasonable.

Self Check

If 9 tickets to a concert cost $112.50, how much will 15 tickets cost?

Answer: $187.50   ∎

In Example 4, we could have compared the cost of the apples to the number of apples: $1.38 is to 6 apples as $c is to 16 apples. This would have led to the proportion

$$\text{Cost of 6 apples} \longrightarrow \quad \frac{1.38}{6} = \frac{c}{16} \quad \longleftarrow \text{Cost of 16 apples}$$
$$\text{6 apples} \longrightarrow \qquad\qquad\qquad \longleftarrow \text{16 apples}$$

If we solve this proportion for c, we will obtain the same result: $c = 3.68$.

   COMMENT   When solving problems using proportions, we must make sure that the units of both numerators are the same and the units of both denominators are the same. For Example 4, it would be incorrect to write

$$\text{Cost of 6 apples} \longrightarrow \quad \frac{1.38}{6} = \frac{16}{c} \quad \longleftarrow \text{16 apples}$$
$$\text{6 apples} \longrightarrow \qquad\qquad\qquad \longleftarrow \text{Cost of 16 apples}$$

EXAMPLE 5   *Miniature.*   A **scale** is a ratio (or rate) that compares the size of a model, drawing, or map to the size of an actual object. The scale shown in Figure 12-4 indicates that 1 inch on the model carousel is equivalent to 160 inches on the actual carousel. How wide should the model be if the actual carousel is 35 feet wide?

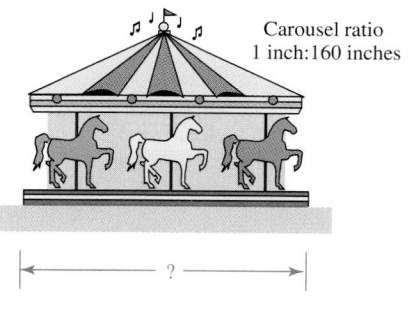

Carousel ratio
1 inch:160 inches

FIGURE 12-4

Solution

Analyze the problem   We are asked to determine the width of the miniature carousel, if a ratio of 1 inch to 160 inches is used. We would like the width of the model to be given in inches, not feet, so we will express the 35-foot width of the actual carousel as $35 \cdot 12 = 420$ inches.

Form a proportion   Let w represent the width of the model. The ratios of the dimensions of the model to the corresponding dimensions of the actual carousel are equal.

1 inch is to 160 inches as w inches is to 420 inches.

$$\text{model} \longrightarrow \frac{1}{160} = \frac{w}{420} \longleftarrow \text{model}$$
$$\text{actual} \longrightarrow \qquad\qquad \longleftarrow \text{actual}$$

Solve the proportion

$420 = 160w$ In a proportion, the product of the extremes is equal to the product of the means.

$$\frac{420}{160} = \frac{160w}{160}$$ To undo the multiplication by 160, divide both sides by 160.

$2.625 = w$ Do the division: $\frac{420}{160} = 2.625$.

State the conclusion   The width of the miniature carousel should be 2.625 in., or $2\frac{5}{8}$ in.

Check the result   A width of $2\frac{5}{8}$ in. is approximately 3 in. When we write the ratio of the model's approximate width to the width of the actual carousel, we get $\frac{3}{420} = \frac{1}{140}$, which is about $\frac{1}{160}$. The answer seems reasonable.   ■

EXAMPLE 6   *Baking.*   A recipe for rhubarb cake calls for $1\frac{1}{4}$ cups of sugar for every $2\frac{1}{2}$ cups of flour. How many cups of flour are needed if the baker intends to use 3 cups of sugar?

Self Check
How many cups of sugar will be needed to make several cakes that will require a total of 25 cups of flour?

Solution

Analyze the problem   The baker needs to maintain the same ratio between the amounts of sugar and flour as is called for in the original recipe.

Form a proportion   Let f represent the number of cups of flour to be mixed with the 3 cups of sugar. The ratios of the cups of sugar to the cups of flour are equal.

$1\frac{1}{4}$ cups sugar is to $2\frac{1}{2}$ cups flour as 3 cups sugar is to f cups flour.

$$\text{Cups sugar} \longrightarrow \frac{1\frac{1}{4}}{2\frac{1}{2}} = \frac{3}{f} \longleftarrow \text{Cups sugar}$$
$$\text{Cups flour} \longrightarrow \qquad\qquad \longleftarrow \text{Cups flour}$$

Solve the proportion

$$\frac{1.25}{2.5} = \frac{3}{f}$$ Change the fractions to decimals.

$1.25f = 2.5 \cdot 3$ In a proportion, the product of the extremes equals the product of the means.

$1.25f = 7.5$ Do the multiplication: $2.5 \cdot 3 = 7.5$.

$$\frac{1.25f}{1.25} = \frac{7.5}{1.25}$$ To undo the multiplication by 1.25, divide both sides by 1.25.

$f = 6$ Divide: $\frac{7.5}{1.25} = 6$.

State the conclusion   The baker should use 6 cups of flour.

Check the result   The recipe calls for about 2 cups of flour for about 1 cup of sugar. If 3 cups of sugar are used, 6 cups of flour seems reasonable.

Answer: $12\frac{1}{2}$ ■

Similar triangles

If two angles of one triangle have the same measures as two angles of a second triangle, the triangles will have the same shape. Triangles with the same shape are called **similar triangles.** In Figure 12-5, $\triangle ABC \sim \triangle DEF$. (Read the symbol ~ as "is similar to.")

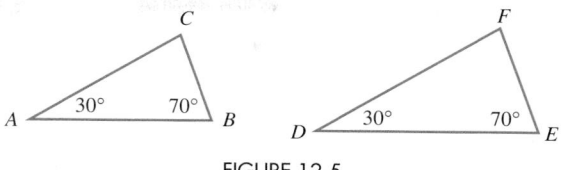

FIGURE 12-5

Property of similar triangles

If two triangles are **similar,** all pairs of corresponding sides are in proportion.

In the similar triangles shown in Figure 12-5, the following proportions are true.

$$\frac{AB}{DE} = \frac{BC}{EF}, \qquad \frac{BC}{EF} = \frac{CA}{FD}, \quad \text{and} \quad \frac{CA}{FD} = \frac{AB}{DE}$$ Read AB as "the length of segment AB."

EXAMPLE 7   *Finding the height of a tree.*   A tree casts a shadow 18 feet long at the same time as a woman 5 feet tall casts a shadow 1.5 feet long. Find the height of the tree.

Solution

Analyze the problem   Figure 12-6 shows the triangles determined by the tree and its shadow and the woman and her shadow. Since the triangles have the same shape, they are similar, and the lengths of their corresponding sides are in proportion.

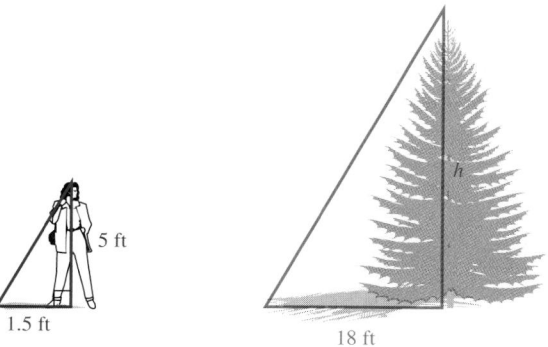

FIGURE 12-6

Form a proportion   If we let h represent the height of the tree, we can find h by solving the following proportion.

$$\frac{h}{5} = \frac{18}{1.5} \qquad \frac{\text{Height of the tree}}{\text{Height of the woman}} = \frac{\text{Length of shadow of the tree}}{\text{Length of shadow of the woman}}$$

Solve the proportion

$1.5h = 5(18)$ In a proportion, the product of the extremes equals the product of the means.

$1.5h = 90$ Do the multiplication.

$h = 60$ To undo the multiplication by 1.5, divide both sides by 1.5 and simplify.

State the conclusion   The tree is 60 feet tall.

Check the result $\frac{18}{1.5} = 12$ and $\frac{60}{5} = 12$. The ratios are the same. The result checks.

Self Check

Find the height of the tree in Example 7 if the woman is 5 feet 6 inches tall.

Answer: 66 ft

STUDY SET Section 12.6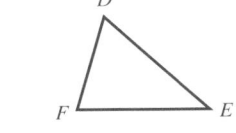

VOCABULARY *Fill in the blanks.*

1. A _____ is the quotient of two numbers or the quotient of two quantities with the same units. A _____ is a quotient of two quantities that have different units.

2. A _____ is a statement that two ratios or two rates are equal.

3. In the proportion $\frac{a}{b} = \frac{c}{d}$, a and d are called the _____ of the proportion. The second and third terms of a proportion are called the _____ of the proportion.

4. The product of the extremes and the product of the means of a proportion are also known as _____ products.

5. If two triangles have the same _____, they are said to be *similar*.

6. If two triangles are _____, their corresponding sides are in proportion.

CONCEPTS *In Exercises 7–8, fill in the blanks.*

7. The equation $\frac{a}{b} = \frac{c}{d}$ is a proportion if the cross product ▢ is equal to the cross product ▢.

8. If $3 \cdot 10 = x \cdot 17$, then ▢ $= \frac{17}{10}$ is a proportion.

9. Is $x = 45$ a solution of $\frac{5}{3} = \frac{75}{x}$?

10. Consider $\frac{2}{3} = \frac{x}{15}$.

   a. Solve the proportion by multiplying both sides by the LCD.
   b. Solve the proportion by setting the cross products equal.

11. MINIATURES   A "high wheeler" bicycle is shown in Illustration 1. A model of it is to be made using a scale of 2 inches to 15 inches. The following proportion was set up to determine the height h of the front wheel of the model. Explain the error.

$$\frac{2}{15} = \frac{48}{h}$$

48 in.

ILLUSTRATION 1

12. Two similar triangles are shown in Illustration 2. Fill in the blanks to make the proportions true.

$$\frac{AB}{DE} = \frac{}{EF} \qquad \frac{BC}{} = \frac{CA}{FD} \qquad \frac{CA}{FD} = \frac{AB}{}$$

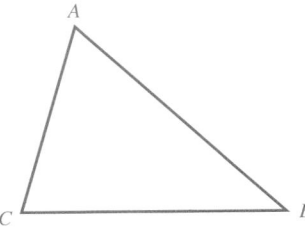

ILLUSTRATION 2

NOTATION *In Exercises 13–14, complete each solution.*

13. Solve for x: $\frac{12}{18} = \frac{x}{24}$.

$$12 \cdot 24 = 18 \cdot \text{▢}$$
$$\text{▢} = 18x$$
$$\frac{288}{\text{▢}} = \frac{18x}{\text{▢}}$$
$$16 = x$$

14. Solve for x: $\frac{14}{x} = \frac{49}{17.5}$.

$$14 \cdot \text{▢} = 49x$$
$$\text{▢} = 49x$$
$$\frac{245}{\text{▢}} = \frac{49x}{\text{▢}}$$
$$5 = x$$

15. We read "$\triangle ABC$" as "_____ ABC."

16. The symbol ~ is read as "_____."

PRACTICE *Tell whether each statement is a proportion.*

17. $\frac{9}{7} = \frac{81}{70}$

18. $\frac{5}{2} = \frac{20}{8}$

19. $\frac{7}{3} = \frac{14}{6}$

20. $\frac{13}{19} = \frac{65}{95}$

21. $\frac{9}{19} = \frac{38}{80}$

22. $\frac{40}{29} = \frac{29}{22}$

23. ▦ $\frac{10.4}{3.6} = \frac{41.6}{14.4}$

24. $\frac{13.23}{3.45} = \frac{39.96}{11.35}$

Solve each proportion.

25. $\dfrac{2}{3} = \dfrac{x}{6}$

26. $\dfrac{3}{6} = \dfrac{x}{8}$

27. $\dfrac{5}{10} = \dfrac{3}{c}$

28. $\dfrac{7}{14} = \dfrac{2}{x}$

29. $\dfrac{6}{x} = \dfrac{8}{4}$

30. $\dfrac{4}{x} = \dfrac{2}{8}$

31. $\dfrac{x}{3} = \dfrac{9}{3}$

32. $\dfrac{x}{2} = \dfrac{18}{6}$

33. $\dfrac{x+1}{5} = \dfrac{3}{15}$

34. $\dfrac{x-1}{7} = \dfrac{2}{21}$

35. $\dfrac{x+3}{12} = \dfrac{-7}{6}$

36. $\dfrac{x+7}{-4} = \dfrac{1}{4}$

37. $\dfrac{4-x}{13} = \dfrac{11}{26}$

38. $\dfrac{5-x}{17} = \dfrac{13}{34}$

39. $\dfrac{2x+1}{18} = \dfrac{14}{3}$

40. $\dfrac{2x-1}{18} = \dfrac{9}{54}$

41. $\dfrac{y}{4} = \dfrac{4}{y}$

42. $\dfrac{2}{3x} = \dfrac{6x}{36}$

43. $\dfrac{2}{c} = \dfrac{c-3}{2}$

44. $\dfrac{b-5}{3} = \dfrac{2}{b}$

45. $\dfrac{2}{x+6} = \dfrac{-2x}{5}$

46. $\dfrac{x-1}{x+1} = \dfrac{2}{3x}$

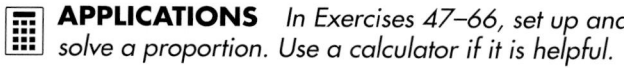 **APPLICATIONS** *In Exercises 47–66, set up and solve a proportion. Use a calculator if it is helpful.*

47. GROCERY SHOPPING   If 3 pints of yogurt cost $1, how much will 51 pints cost?

48. SHOPPING FOR CLOTHES   If shirts are on sale at two for $25, how much will five shirts cost?

49. ADVERTISING   In 1997, a 30-second TV ad during the Super Bowl telecast cost $1.2 million. At this rate, what was the cost of a 45-second ad?

50. COOKING   A recipe for spaghetti sauce requires four 16-ounce bottles of ketchup to make two gallons of sauce. How many bottles of ketchup are needed to make 10 gallons of sauce?

51. MIXING PERFUME   A perfume is to be mixed in the ratio of 3 drops of pure essence to 7 drops of alcohol. How many drops of pure essence should be mixed with 56 drops of alcohol?

52. CPR   A first aid handbook states that when performing cardiopulmonary resuscitation on an adult, the ratio of chest compressions to breaths should be 5:2. If 210 compressions were administered to an adult patient, how many breaths should have been given?

53. COOKING   A recipe for wild rice soup is shown in Illustration 3. Find the amounts of chicken broth, rice, and flour needed to make 15 servings.

Wild Rice Soup

A sumptuous side dish with a nutty flavor

3 cups chicken broth	1 cup light cream
$\frac{2}{3}$ cup uncooked rice	2 tablespoons flour
$\frac{1}{4}$ cup sliced onions	$\frac{1}{8}$ teaspoon pepper
$\frac{1}{2}$ cup shredded carrots	Serves: 6

ILLUSTRATION 3

54. QUALITY CONTROL   In a manufacturing process, 95% of the parts made are to be within specifications. How many defective parts would be expected in a run of 940 pieces?

55. QUALITY CONTROL   Out of a sample of 500 men's shirts, 17 were rejected because of crooked collars. How many crooked collars would you expect to find in a run of 15,000 shirts?

56. GAS CONSUMPTION   If a car can travel 42 miles on 1 gallon of gas, how much gas is needed to travel 315 miles?

57. HIP-HOP   According to the *Guinness Book of World Records 1998,* Rebel X.D. of Chicago rapped 674 syllables in 54.9 seconds. At this rate, how many syllables could he rap in 1 minute? Round to the nearest syllable.

58. BANKRUPTCY   After filing for bankruptcy, a company was able to pay its creditors only 15 cents on the dollar. If the company owed a lumberyard $9,712, how much could the lumberyard expect to be paid?

59. COMPUTING A PAYCHECK   Billie earns $412 for a 40-hour week. If she missed 10 hours of work last week, how much did she get paid?

60. MODEL RAILROAD   A model railroad engine is 9 inches long. If the scale is 87 feet to 1 foot, how long is a real engine?

61. MODEL RAILROAD   A model railroad caboose is 3.5 inches long. If the scale is 169 feet to 1 foot, how long is a real caboose?

62. NUTRITION   Illustration 4 shows the nutritional facts about a 10-oz chocolate milkshake sold by a fast-food restaurant. Use the information to complete the table for the 16-oz shake. Round to the nearest unit when an answer is not exact.

	Calories	Fat (gm)	Protein (gm)
10-oz chocolate milkshake	355	8	9
16-oz chocolate milkshake			

ILLUSTRATION 4

63. DRIVER'S LICENSE   Of the 50 states, Oregon has the largest ratio of licensed drivers per 1,000 residents. If the ratio is 824 to 1,000 and Oregon's population is 3,282,000, how many Oregonians have a driver's license?

64. MIXING FUEL   The instructions on a can of oil intended to be added to lawnmower gasoline read as follows:

Recommended	Gasoline	Oil
50 to 1	6 gal	16 oz

Are these instructions correct? (*Hint:* There are 128 ounces in 1 gallon.)

65. PHOTO ENLARGEMENT   In Illustration 5, the 3-by-5 photo is to be blown up to the larger size. Find *x*.

5 in. $6\frac{1}{4}$ in.

3 in.   *x* in.

ILLUSTRATION 5

66. BLUEPRINT   The scale for the drawing in Illustration 6 tells the reader that a $\frac{1}{4}$-inch length $\left(\frac{1}{4}''\right)$ on the drawing corresponds to an actual size of 1 foot (1'0''). Suppose the length of the kitchen is $2\frac{1}{2}$ inches on the drawing. How long is the actual kitchen?

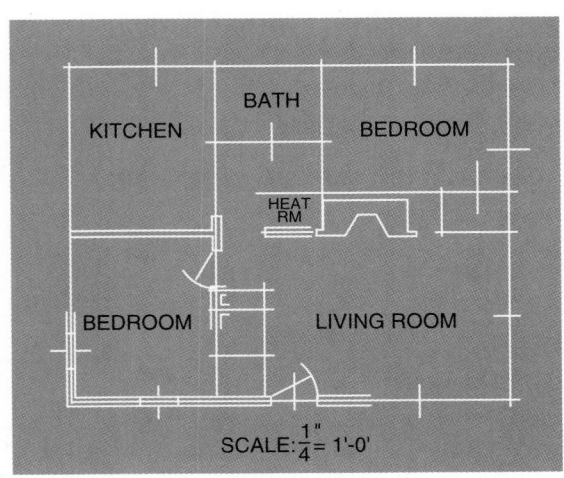

SCALE: $\frac{1}{4}''$ = 1'-0'

ILLUSTRATION 6

In Exercises 67–72, use similar triangles to solve each problem.

67. HEIGHT OF A TREE   A tree casts a shadow of 26 feet at the same time as a 6-foot man casts a shadow of 4 feet. (See Illustration 7.) Find the height of the tree.

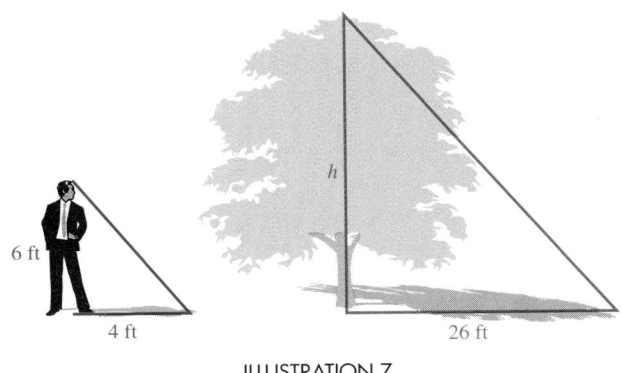

6 ft   *h*

4 ft   26 ft

ILLUSTRATION 7

68. HEIGHT OF A BUILDING   A man places a mirror on the ground and sees the reflection of the top of a building, as shown in Illustration 8. The two triangles in the illustration are similar. Find the height, *h*, of the building.

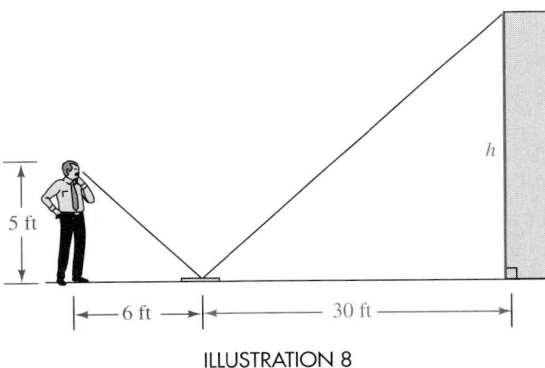

5 ft   *h*

6 ft   30 ft

ILLUSTRATION 8

69. WIDTH OF A RIVER   Use the dimensions in Illustration 9 to find *w*, the width of the river. (The two triangles in the illustration are similar.)

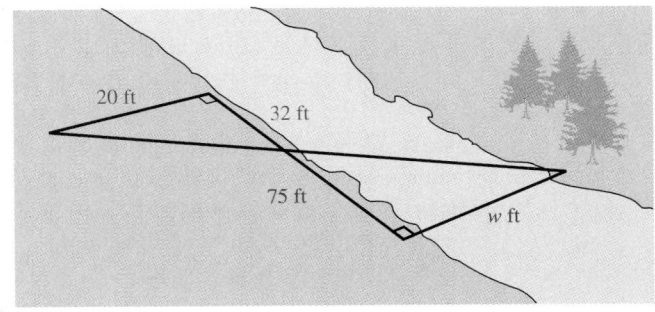

20 ft   32 ft

75 ft   *w* ft

ILLUSTRATION 9

70. FLIGHT PATH An airplane ascends 100 feet as it flies a horizontal distance of 1,000 feet. How much altitude will it gain as it flies a horizontal distance of 1 mile? See Illustration 10. (*Hint:* 5,280 feet = 1 mile.)

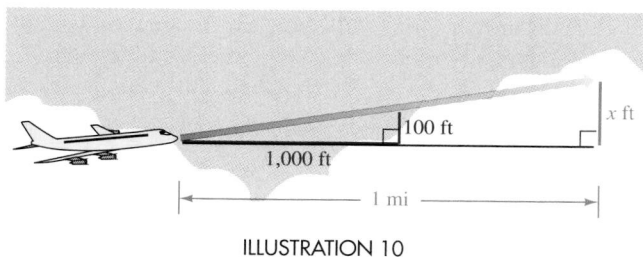

ILLUSTRATION 10

71. FLIGHT PATH An airplane descends 1,350 feet as it flies a horizontal distance of 1 mile. How much altitude is lost as it flies a horizontal distance of 5 miles?

72. SKI RUN A ski course falls 100 feet in every 300 feet of horizontal run. If the total horizontal run is $\frac{1}{2}$ mile, find the height of the hill.

WRITING

73. Explain the difference between a ratio and a proportion.

74. Explain how to tell whether $\frac{3.2}{3.7} = \frac{5.44}{6.29}$ is a proportion.

75. Explain why the concept of cross products cannot be used to solve the equation

$$\frac{x}{3} - \frac{3x}{4} = \frac{1}{12}$$

76. Write a problem about a situation you encounter in your daily life that could be solved by using a proportion.

REVIEW

77. Change $\frac{9}{10}$ to a percent.

78. Change $33\frac{1}{3}\%$ to a fraction.

79. Find 30% of 1,600.

80. SHOPPING Maria bought a dress for 25% off the original price of $98. How much did the dress cost?

81. Find the slope of the line passing through $(-2, -2)$ and $(-12, -8)$.

82. What are the slope and the y-intercept of the graph of $y = 2x - 3$?

12.7 Variation

In this section, you will learn about

- Direct variation • Inverse variation

INTRODUCTION. If the value of one quantity depends on the value of another quantity, we can often describe that relationship using the language of variation:

- The sales tax on an item varies with the price.
- The intensity of light varies with the distance from its source.
- The pressure exerted by water on an object varies with the depth of the object beneath the surface.

In this section, we will discuss two types of variation, and we will see how to represent them algebraically using equations.

Direct variation

One type of variation, called **direct variation,** is represented by an equation of the form $y = kx$, where k is a constant (a number). Two variables are said to *vary directly* if one is a constant multiple of the other.

Direct variation

The words *y varies directly with x* mean that

$$y = kx$$

for some constant k, called the **constant of variation.**

Scientists have found that the distance a spring will stretch varies directly with the force applied to it. The more force applied to the spring, the more it will stretch. If d represents the distance stretched and f represents the force applied, this relationship can be expressed by the equation

$$d = kf \qquad \text{where } k \text{ is the constant of variation}$$

Suppose that a 150-pound wooden garage door stretches a spring 18 inches when the door is closed. (See Figure 12-7.) We can find the constant of variation for the spring by substituting 150 for f and 18 for d in the equation $d = kf$ and solving for k:

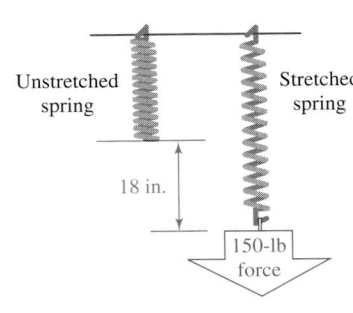

Unstretched spring   Stretched spring

18 in.

150-lb force

FIGURE 12-7

$$d = kf$$

$$18 = k(150)$$

$$\frac{18}{150} = k \qquad \text{Divide both sides by 150 to isolate } k.$$

$$\frac{3}{25} = k \qquad \text{Simplify the fraction: } \frac{18}{150} = \frac{\overset{1}{\cancel{6}} \cdot 3}{\underset{1}{\cancel{6}} \cdot 25} = \frac{3}{25}.$$

Therefore, the equation describing the relationship between the distance the spring will stretch and the amount of force applied to it is $d = \frac{3}{25}f$. To find the distance that the same spring will stretch when a new, 50-pound aluminum garage door is installed, we proceed as follows:

$$d = \frac{3}{25}f \qquad \text{The equation describing the direct variation.}$$

$$d = \frac{3}{25}(50) \qquad \text{Substitute 50 for } f.$$

$$d = 6 \qquad \text{Do the multiplication.}$$

The spring will stretch 6 inches when the 50-pound aluminum door is closed.

The table in Figure 12-8 shows some other possible values for f and d as determined by the equation $d = \frac{3}{25}f$. When these ordered pairs are graphed and a straight line is drawn through them, it is apparent that as the force f applied to a spring increases, the distance d it stretches increases. Furthermore, the slope of the graph is $\frac{3}{25}$, the constant of variation.

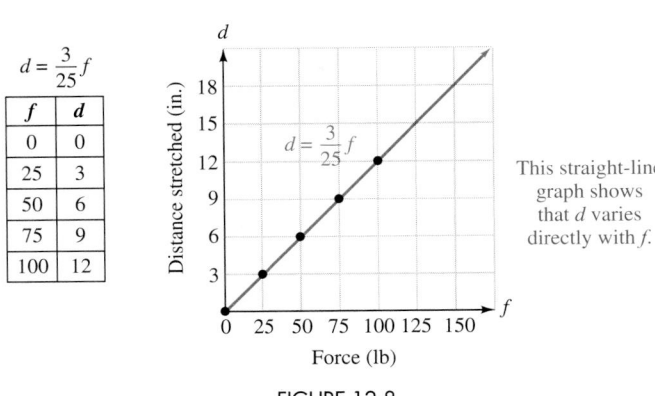

$$d = \frac{3}{25}f$$

f	d
0	0
25	3
50	6
75	9
100	12

This straight-line graph shows that d varies directly with f.

FIGURE 12-8

We can use the following steps to solve variation problems.

Solving variation problems

> To solve a variation problem:
> 1. Translate the verbal model into an equation.
> 2. Substitute the first set of values into the equation from Step 1 to determine the value of k.
> 3. Substitute the value of k into the equation from Step 1.
> 4. Substitute the remaining set of values into the equation from Step 3 and solve for the unknown variable.

EXAMPLE 1   *Direct variation.*

The weight of an object on Earth varies directly with its weight on the moon. If a rock weighs 5 pounds on the moon and 30 pounds on Earth, what would be the weight on Earth of a larger rock weighing 26 pounds on the moon?

Solution

Step 1: We let e represent the weight of the object on Earth and m the weight of the object on the moon. Translating the words *weight on Earth varies directly with weight on the moon,* we get the equation

$$e = km$$

Step 2: To find the constant of variation, k, we substitute 30 for e and 5 for m.

$$e = km$$
$$30 = k(5)$$
$$6 = k \qquad \text{To undo the multiplication by 5, divide both sides by 5.}$$

Step 3: The equation describing the relationship between the weight of an object on Earth and on the moon is

$$e = 6m$$

Step 4: We can find the weight of the larger rock on Earth by substituting 26 for m in the equation from Step 3.

$$e = 6m$$
$$e = 6(26)$$
$$e = 156$$

The rock would weigh 156 pounds on Earth.

Self Check

The cost of a bus ticket varies directly with the number of miles traveled. If a ticket for a 180-mile trip cost $45, what would a ticket for a 1,500-mile trip cost?

Answer: $375

Inverse variation

Another type of variation, called **inverse variation,** is represented by an equation of the form $y = \frac{k}{x}$, where k is a constant. Two variables are said to *vary inversely* if one is a constant multiple of the reciprocal of the other.

Inverse variation

> The words *y varies inversely with x* mean that
>
> $$y = \frac{k}{x}$$
>
> for some constant k, called the **constant of variation.**

Suppose that the time (in hours) it takes to paint a house varies inversely with the size of the painting crew. As the number of painters increases, the time that it takes to paint the house decreases. If n represents the number of painters and t represents the time it takes to paint the house, this relationship can be expressed by the equation

$$t = \frac{k}{n} \qquad \text{where } k \text{ is the constant of variation}$$

If we know that a crew of 8 can paint the house in 12 hours, we can find the constant of variation by substituting 8 for n and 12 for t in the equation $t = \frac{k}{n}$ and solving for k:

$$t = \frac{k}{n}$$

$$12 = \frac{k}{8}$$

$$12 \cdot 8 = k \qquad \text{Multiply both sides by 8 to isolate } k.$$

$$96 = k$$

The equation describing the relationship between the size of the painting crew and the time it takes to paint the house is $t = \frac{96}{n}$. We can use this equation to find the time it will take a crew of any size to paint the house. For example, to find the time it would take a four-person crew, we substitute 4 for n in the equation $t = \frac{96}{n}$.

$$t = \frac{96}{n} \qquad \text{The equation describing the inverse variation.}$$

$$t = \frac{96}{4} \qquad \text{Substitute 4 for } n.$$

$$t = 24$$

It would take a four-person crew 24 hours to paint the house.

The table in Figure 12-9 shows some possible values for n and t as determined by the equation $t = \frac{96}{n}$. When these ordered pairs are graphed and a smooth curve is drawn through them, it is clear that as the number of painters n increases, the time t decreases.

$$t = \frac{96}{n}$$

n	t
2	48
3	32
4	24
6	16
8	12
12	8
16	6
24	4

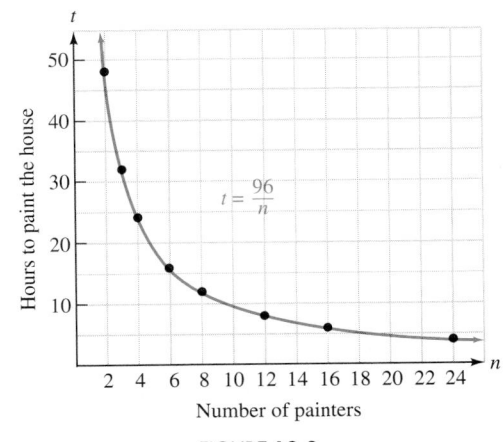

$$t = \frac{96}{n}$$

This curved graph shows that t varies directly with n.

Number of painters

FIGURE 12-9

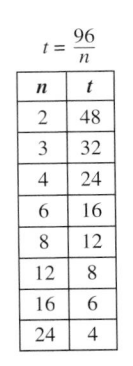

EXAMPLE 2 *Gas law.* The volume occupied by a gas varies inversely with the pressure placed on it. That is, the volume decreases as the pressure increases. If a gas occupies a volume of 15 cubic inches when placed under 4 pounds per square inch (psi) of pressure, how much pressure is needed to compress the gas into a volume of 10 cubic inches?

Self Check
How much pressure is needed to compress the gas in Example 2 into a volume of 8 cubic inches?

Solution

Step 1: We let V represent the volume occupied by the gas and p represent the pressure. Translating the words *volume occupied by a gas varies inversely with the pressure,* we get the equation

$$V = \frac{k}{p}$$

Step 2: To find the constant of variation, k, we substitute 15 for V and 4 for p.

$$V = \frac{k}{p}$$

$$15 = \frac{k}{4}$$

$$60 = k \quad \text{Multiply both sides by 4.}$$

Step 3: The equation describing the relationship between the volume occupied by the gas and the pressure placed on it is

$$V = \frac{60}{p}$$

Step 4: We can now find the pressure needed to compress the gas into a volume of 10 cubic inches by substituting 10 for V in the equation and solving for p.

$$V = \frac{60}{p}$$

$$10 = \frac{60}{p}$$

$$10p = 60 \quad \text{To clear the equation of the fraction, multiply both sides by } p.$$

$$p = 6 \quad \text{To undo the multiplication by 10, divide both sides by 6.}$$

It will take 6 psi of pressure to compress the gas into a volume of 10 cubic inches.

Answer: 7.5 psi ■

STUDY SET Section 12.7

VOCABULARY *Fill in the blanks.*

1. The equation $y = kx$ defines _____ variation.

2. The equation $y = \dfrac{k}{x}$ defines _____ variation.

3. In $y = kx$, the _____ of variation is k.

4. A constant is a _____.

CONCEPTS *Exercises 5–8 illustrate two types of variation. Tell whether each graph represents direct variation or inverse variation.*

5.

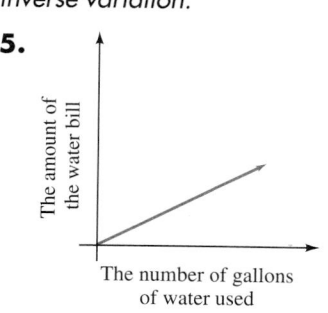

6.

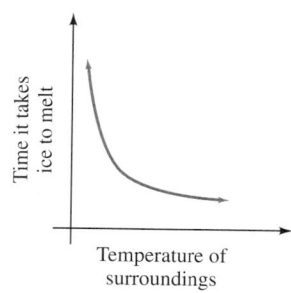

7.

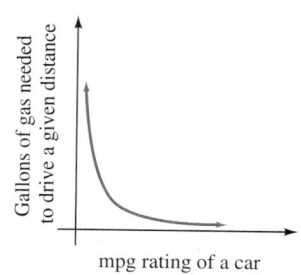

8.

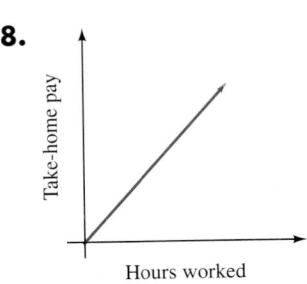

9. Tell whether the equation defines direct variation.

　a. $y = kx$

　b. $y = k + x$

　c. $y = \dfrac{k}{x}$

　d. $m = kc$

10. a. Translate to mathematical symbols:

A farmer's harvest h	varies directly	with the number of acres planted a.

　b. If the constant of variation for part a is $k = 10{,}000$, what will happen to the size of the harvest as the number of acres planted increases?

11. Express this relationship using an equation: The number of gallons g of paint needed to paint a room varies directly with the number of square feet f to be painted.

12. Express this relationship using an equation: The amount of sales tax t varies directly with the purchase price p of a new car.

13. Assume that t varies directly with s and $t = ks$. If $t = 21$ when $s = 6$, find k.

14. Assume that y varies directly with x and $y = kx$. If $y = 10$ when $x = 2$, find k.

15. Tell whether each equation defines inverse variation.

　a. $y = kx$

　b. $y = \dfrac{k}{x}$

　c. $y = \dfrac{x}{k}$

　d. $d = \dfrac{k}{g}$

16. a. Translate to mathematical symbols:

The time t (in hours) it takes a commuter to drive from her home to her office	varies inversely	with her average speed s (in mph).

　b. If the constant of variation for part a is $k = 30$, what will happen to the time her commute takes her as her average speed increases?

17. Express this relationship using an equation: The number of hot dogs n that a street vendor sells varies inversely with the price p that he charges.

18. a. If y varies directly with x and $k > 0$, what happens to y as x increases?

　b. If y varies inversely with x and $k > 0$, what happens to y as x increases?

19. Assume that y varies inversely with x and $y = \dfrac{k}{x}$. If $y = 15$ when $x = 10$, find k.

20. Assume that c varies inversely with d and $c = \dfrac{k}{d}$. If $c = 9$ when $d = 5$, find k.

NOTATION *Complete each solution.*

21. Find f if $d = 21$ and $k = \frac{7}{5}$.

$$d = kf$$
$$21 = \boxed{}\, f$$
$$\boxed{} \cdot 21 = \boxed{} \cdot \frac{7}{5}f$$
$$15 = f$$

22. Find f if $d = 20$ and $k = 0.75$.

$$d = \frac{k}{f}$$
$$\boxed{} = \frac{0.75}{f}$$
$$\boxed{} \cdot 20 = \boxed{} \cdot \frac{0.75}{f}$$
$$20f = \boxed{}$$
$$f = \frac{0.75}{\boxed{}}$$
$$f = 0.0375$$

PRACTICE

23. Assume that y varies directly with x. If $y = 10$ when $x = 2$, find y when $x = 7$.

24. Assume that r varies directly with s. If $r = 21$ when $s = 6$, find r when $s = 12$.

25. Assume that l varies directly with m. If $l = 50$ when $m = 200$, find l when $m = 25$.

26. 🔲 Assume that g varies directly with t. If $g = 3{,}616$ when $t = 8{,}000$, find g when $t = 2{,}405$.

27. Assume that x and y vary directly. If $x = 30$ when $y = 2$, find y when $x = 45$.

28. Assume that n_1 and n_2 vary directly. If $n_1 = 315$ when $n_2 = 3$, find n_2 when $n_1 = 10.5$.

29. Assume that y varies inversely with x. If $y = 8$ when $x = 1$, find y when $x = 8$.

30. Assume that r varies inversely with s. If $r = 40$ when $s = 10$, find r when $s = 15$.

31. Assume that a varies inversely with t. If $a = 600$ when $t = 300$, find a when $t = 15$.

32. 🔲 Assume that b varies inversely with c. If $b = 0.45$ when $c = 1.6$, find b when $c = 80$.

33. Assume that t_1 and t_2 vary inversely. If $t_1 = 4$ when $t_2 = 5$, find t_2 when $t_1 = 3\frac{1}{3}$.

34. Assume that a and r vary inversely. If $a = 9$ when $r = 7$, find r when $a = \frac{1}{9}$.

APPLICATIONS

35. COMMUTING DISTANCE   The distance that a car can travel without refueling varies directly with the number of gallons of gasoline in the tank. If a car can go 360 miles on a full tank of gas (15 gallons), how far can it go on 7 gallons?

36. COMPUTING FORCES   The force of gravity acting on an object varies directly with the mass of the object. The force on a mass of 5 kilograms is 49 newtons. What is the force acting on a mass of 12 kilograms?

37. DOSAGE   The recommended dose (in milligrams) of Demerol, a preoperative medication given to children, varies directly with the child's weight in pounds. The proper dosage for a child weighing 30 pounds is 18 milligrams. What would be the correct dosage for a child weighing 45 pounds?

38. MEDICATION   To fight ear infections in children, doctors often prescribe Ceclor. The recommended dose in milligrams varies directly with the child's body weight in pounds. The correct dosage for a 20-pound child is 124 milligrams. What would be the correct dosage for a 28-pound child?

39. CIDER   For the recipe shown in Illustration 1, the number of inches of stick cinnamon to use varies directly with the number of servings of spiced cider to be made. How many inches of stick cinnamon are needed to make $2\frac{1}{2}$ dozen servings?

Hot Spiced Cider

8 cups apple cider or apple juice
$\frac{1}{4}$ to $\frac{1}{2}$ cup packed brown sugar
6 inches stick cinnamon
1 teaspoon whole allspice
1 teaspoon whole cloves
8 thin orange wedges or slices (optional)
8 whole cloves (optional)   Makes 8 servings

ILLUSTRATION 1

40. LUNAR GRAVITY   The weight of an object on the moon varies directly with its weight on Earth; six pounds on Earth weighs 1 pound on the moon. What would the scale shown in Illustration 2 register if the astronaut were weighed on the moon?

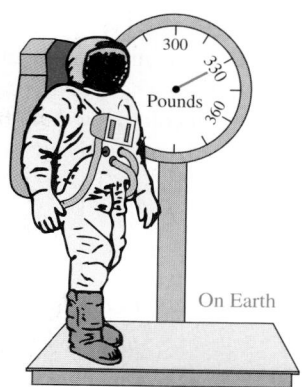

ILLUSTRATION 2

41. COMMUTING TIME   The time it takes a car to travel a certain distance varies inversely with its rate of speed. If a certain trip takes 3 hours at 50 miles per hour, how long will the trip take at 60 miles per hour?

42. GEOMETRY   For a fixed area, the length of a rectangle is inversely proportional to its width. A rectangle has a width of 12 feet and a length of 20 feet. If its length is increased to 24 feet, find the width it must have to maintain the same area.

43. ELECTRICITY   The current in an electric circuit varies inversely with the resistance. If the current in the circuit shown in Illustration 3 is 30 amps when the resistance is 4 ohms, what will the current be for a resistance of 15 ohms?

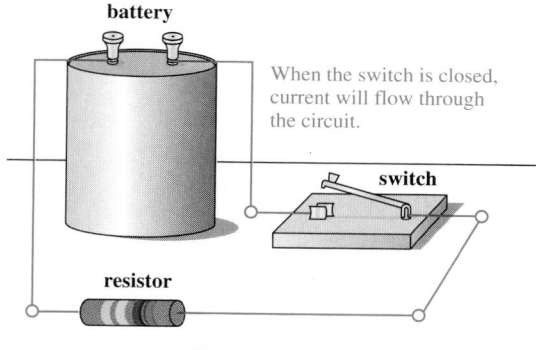

ILLUSTRATION 3

44. FARMING   The length of time a given number of bushels of corn will last when feeding cattle varies inversely with the number of animals. If a certain number of bushels will feed 25 cows for 10 days, how long will the feed last for 10 cows?

45. COMPUTING PRESSURES   If the temperature of a gas is constant, the volume occupied varies inversely with the pressure. If a gas occupies a volume of 40 cubic meters under a pressure of 8 atmospheres, find the volume when the pressure is changed to 6 atmospheres.

46. COMPUTING DEPRECIATION   Assume that the value of a machine varies inversely with its age. If a drill press is worth $300 when it is 2 years old, find its value when it is 6 years old. How much has the machine depreciated over that 4-year period?

WRITING

47. Give two examples of quantities that vary directly and two that do not.

48. What is the difference between direct variation and inverse variation?

49. What is a constant of variation?

50. Is there a direct variation or an inverse variation between each pair of quantities? Explain why.

    a. The time it takes to type a term paper and the speed at which you type

    b. The time it takes to type a term paper (working at a constant rate) and the length of the term paper

REVIEW   *Solve each equation.*

51. $x^2 - 5x - 6 = 0$

52. $x^2 - 25 = 0$

53. $(t + 2)(t^2 + 7t + 12) = 0$

54. $2(y - 4) = -y^2$

55. $y^3 - y^2 = 0$

56. $5a^3 - 125a = 0$

57. $(x^2 - 1)(x^2 - 4) = 0$

58. $6t^3 + 35t^2 = 6t$

Expressions and Equations

In this chapter, we have discussed procedures for working with **rational expressions** and procedures for solving **rational equations.**

Rational expressions

The **fundamental property of fractions** is used when simplifying rational expressions and when multiplying and dividing rational expressions: *We can divide out factors that are common to the numerator and the denominator of a fraction.*

1. a. Simplify $\dfrac{2x^2 - 8x}{x^2 - 6x + 8}$.

  b. What common factor was divided out?

2. a. Multiply: $\dfrac{x^2 + 2x + 1}{x} \cdot \dfrac{x^2 - x}{x^2 - 1}$.

  b. What common factors were divided out?

The fundamental property of fractions also states that *multiplying the numerator and denominator of a fraction by the same nonzero number does not change the value of the fraction.* We use this concept to "build" fractions when adding or subtracting rational expressions with unlike denominators, and when simplifying complex fractions.

3. a. Add: $\dfrac{x}{x + 1} + \dfrac{x - 1}{x}$.

  b. By what did you multiply the first fraction to rewrite it in terms of the LCD? The second fraction?

4. a. Simplify $\dfrac{n - 1 - \dfrac{2}{n}}{\dfrac{n}{3}}$.

  b. By what did you multiply the numerator and denominator to simplify the complex fraction?

Rational equations

The multiplication property of equality states that *if equal quantities are multiplied by the same nonzero number, the results will be equal quantities.* We use this property when solving rational equations. If we multiply both sides of the equation by the LCD of the rational expressions in the equation, we can clear it of fractions.

5. a. Solve $\dfrac{11}{b} + \dfrac{13}{b} = 12$.

  b. By what did you multiply both sides to clear the equation of fractions?

6. a. Solve $\dfrac{-5}{s^2 + s - 2} + \dfrac{3}{s + 2} = \dfrac{1}{s - 1}$.

  b. By what did you multiply both sides to clear the equation of fractions?

7. a. Solve $y + \dfrac{3}{4} = \dfrac{3y - 50}{4y - 24}$.

  b. By what did you multiply both sides to clear the equation of fractions?

8. a. Solve $\dfrac{1}{a} - \dfrac{1}{b} = 1$ for b.

  b. By what did you multiply both sides to clear the equation of fractions?

Section 12.1

CHECKING A SIMPLIFICATION We can use evaluation to check a simplification. To check whether

$$\frac{x^2 - 16}{x + 4} = x - 4$$

have each member of your group evaluate

$$\frac{x^2 - 16}{x + 4} \quad \text{and} \quad x - 4$$

for a given value of x. That is, have one person evaluate both expressions for $x = -5$, have another person evaluate both for $x = -3$, and so on. (Don't use $x = -4$, because the original expression is undefined for this value of x.) The expressions should give identical results for all other values of x. If the evaluations differ for any number, the original expression was not simplified correctly.

Use evaluation to check each of the following simplifications. If an expression was incorrectly simplified, find the correct answer.

a. $\dfrac{x^2 - 4}{x^3 + 8} \stackrel{?}{=} \dfrac{x - 2}{x^2 + 2x + 2}$ b. $\dfrac{x^2 + 2x + 1}{x^2 + 4x + 3} \stackrel{?}{=} \dfrac{x + 1}{x + 3}$

c. $\dfrac{x^2 + 2x - 15}{x^2 - 25} \stackrel{?}{=} \dfrac{x - 3}{x - 5}$ d. $\dfrac{6x^2 - 13x + 6}{3x^2 + x - 2} \stackrel{?}{=} \dfrac{2x - 3}{x + 2}$

Section 12.2

COMBINED OPERATIONS Insert a $\cdot$ sign or a $\div$ sign in each blank so that the answer is 1.

$$\frac{x^2 - 2x - 15}{3x^2 - 27} \quad\rule{1cm}{0.4pt}\quad \frac{x^2 - 25}{6x^2 + 45x + 75} \quad\rule{1cm}{0.4pt}\quad \frac{x^2 - x - 6}{2x^2 + 9x + 10}$$

Section 12.3

First, add $\dfrac{1}{2x^2} + \dfrac{1}{8x}$

by expressing each fraction in terms of a common denominator $16x^3$. (This is the *product* of their denominators.) Then add the fractions again by expressing each of them in terms of their lowest common denominator. What is one advantage and one disadvantage of each method?

Section 12.4

UNIT ANALYSIS Simplify each complex fraction. The units can be divided out just as in the case of common factors.

$$\frac{\dfrac{36 \text{ inches}}{3 \text{ feet}}}{\dfrac{1 \text{ yard}}{3 \text{ feet}}} \qquad \frac{\dfrac{60 \text{ minutes}}{1 \text{ hour}}}{\dfrac{3,600 \text{ seconds}}{1 \text{ hour}}} \qquad \frac{\dfrac{4 \text{ quarts}}{1 \text{ gallon}}}{\dfrac{8 \text{ pints}}{1 \text{ gallon}}}$$

Section 12.5

SIMPLIFY AND SOLVE The two problems below look similar. Explain why their one-word instructions can't be switched. Write a solution for each problem and then identify the major similarity and the major difference in the solution methods.

Simplify $\dfrac{2}{4x - 4} + \dfrac{3}{x - 1}$

Solve $\dfrac{2}{4x - 4} + \dfrac{3}{x - 1} = \dfrac{7}{4}$

Section 12.6

PROBLEM SOLVING Problems such as the filling of a water tank are often called *shared-work problems*. For each of the following equations, write a shared-work problem that could be solved using it.

$$\frac{1}{3} + \frac{1}{8} = \frac{1}{x}$$

$$\frac{1}{3} - \frac{1}{8} = \frac{1}{x}$$

$$\frac{1}{3} + \frac{1}{8} - \frac{1}{16} = \frac{1}{x}$$

Section 12.7

PI The Greek letter pi (π) represents the ratio of the circumference C of any circle to its diameter d. That is, $\pi = \frac{C}{d}$. Use a tape measure to find the circumference and the diameter of various objects that are circular in shape. You can measure anything round: for example, a swimming pool spa, the top of a can, or a ring. Enter your results in a table like that in Illustration 1. Convert each measurement to a decimal and use a calculator to compute the ratio of C to d. Make some observations about your results.

Object	Circumference	Diameter	$\dfrac{C}{d}$
A quarter	$2\frac{15}{16}$ in. 2.9375 in.	$\frac{15}{16}$ in. 0.9375 in.	3.13333. . .

ILLUSTRATION 1

COOKING Find a simple recipe for a treat that you can make for your class. Use a proportion to determine the amount of each ingredient needed to make enough for the exact number of people in your class. Write the old recipe and the new recipe on separate pieces of poster board. Did the recipe serve the correct number of people? Share with the class how you made the calculations, as well as any difficulties you encountered.

CHAPTER REVIEW

Simplifying Rational Expressions

CONCEPTS

A *rational expression* is a fraction in which the numerator and denominator are polynomials.

Since division by 0 is undefined, we must make sure that the denominator of a rational expression is not 0.

The fundamental property of fractions:
If b and c are not zero, then

$$\frac{ac}{bc} = \frac{a}{b}$$

When all common factors have been divided out, a fraction is in *lowest terms.*

The quotient of any nonzero expression and its opposite is -1.

REVIEW EXERCISES

1. Find the values of x for which the rational expression $\dfrac{x-1}{x^2-16}$ is undefined.

2. Write each fraction in lowest terms. If it is already in lowest terms, so indicate.

   a. $\dfrac{10}{25}$ **b.** $-\dfrac{12}{18}$

3. Simplify each rational expression. If it is already in lowest terms, so indicate. Assume that no denominators are zero.

   a. $\dfrac{3x^2}{6x^3}$ **b.** $\dfrac{5xy^2}{2x^2y^2}$

   c. $\dfrac{x^2}{x^2+x}$ **d.** $\dfrac{a^2-4}{a+2}$

   e. $\dfrac{3p-2}{2-3p}$ **f.** $\dfrac{8-x}{x^2-5x-24}$

   g. $\dfrac{2x^2-16x}{2x^2-18x+16}$ **h.** $\dfrac{x^2+x-2}{x^2-x-2}$

4. Evaluate $\dfrac{x^2-1}{x-5}$ for $x=-2$.

5. Explain why it would be incorrect to divide out the common x's in $\frac{x+1}{x}$.

6. Simplify $\dfrac{4(t+3)+4}{3(t+2)+6}$.

Multiplying and Dividing Rational Expressions

Rule for multiplying fractions:

$$\frac{a}{b} \cdot \frac{c}{d} = \frac{ac}{bd} \quad (b, d \neq 0)$$

Rule for dividing fractions:

$$\frac{a}{b} \div \frac{c}{d} = \frac{a}{b} \cdot \frac{d}{c} \, (b, c, d \neq 0)$$

To write the *reciprocal* of a fraction, we invert the fraction.

7. Do each multiplication and simplify.

   a. $\dfrac{3xy}{2x} \cdot \dfrac{4x}{2y^2}$ **b.** $56x\left(\dfrac{12}{7x}\right)$

   c. $\dfrac{x^2-1}{x^2+2x} \cdot \dfrac{x}{x+1}$ **d.** $\dfrac{x^2+x}{3x-15} \cdot \dfrac{6x-30}{x^2+2x+1}$

8. Do each division and simplify.

   a. $\dfrac{3x^2}{5x^2y} \div \dfrac{6x}{15xy^2}$ **b.** $\dfrac{x^2+5x}{x^2+4x-5} \div \dfrac{x^2}{x-1}$

   c. $\dfrac{x^2-x-6}{2x-1} \div \dfrac{x^2-2x-3}{2x^2+x-1}$

9. Simplify $\dfrac{b^2 + 4b + 4}{b^2 + b - 6}\left(\dfrac{b - 2}{b - 1} \div \dfrac{b + 2}{b^2 + 2b - 3}\right).$

SECTION 12.3

Adding and Subtracting Rational Expressions

Adding and subtracting fractions with like denominators:

$$\frac{a}{d} + \frac{b}{d} = \frac{a + b}{d} \quad (d \neq 0)$$

$$\frac{a}{d} - \frac{b}{d} = \frac{a - b}{d} \quad (d \neq 0)$$

To find the *LCD*, factor each denominator completely. Form a product using each different factor the greatest number of times it appears in any one factorization.

To add or subtract fractions with unlike denominators, first find the LCD of the fractions. Then express each fraction in equivalent form with a common denominator. Finally, add or subtract the fractions.

10. Do each operation. Simplify all answers.

a. $\dfrac{x}{x + y} + \dfrac{y}{x + y}$ **b.** $\dfrac{3x}{x - 7} - \dfrac{x - 2}{x - 7}$

c. $\dfrac{a}{a^2 - 2a - 8} + \dfrac{2}{a^2 - 2a - 8}$

11. Several denominators are given. Find the lowest common denominator (LCD).

a. $2x^2, 4x$ **b.** $3y^2, 9x, 6x$

c. $x + 1, x + 2$ **d.** $y^2 - 25, y - 5$

12. Do each operation. Simplify all answers.

a. $\dfrac{x}{x - 1} + \dfrac{1}{x}$ **b.** $\dfrac{1}{7} - \dfrac{1}{c}$

c. $\dfrac{x + 2}{2x} - \dfrac{2 - x}{x^2}$ **d.** $\dfrac{2t + 2}{t^2 + 2t + 1} - \dfrac{1}{t + 1}$

e. $\dfrac{x}{x + 2} + \dfrac{3}{x} - \dfrac{4}{x^2 + 2x}$ **f.** $\dfrac{6}{b - 1} - \dfrac{b}{1 - b}$

13. VIDEO CAMERA   See Illustration 1. Find the perimeter and the area of the LED screen of the camera.

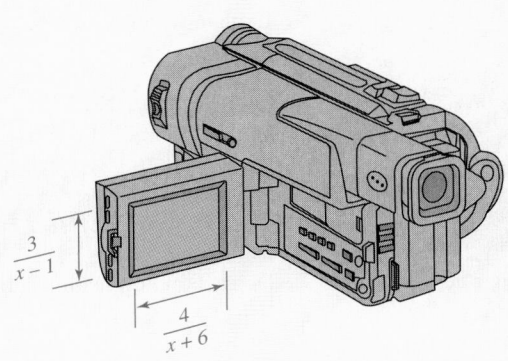

ILLUSTRATION 1

SECTION 12.4

Complex Fractions

Complex fractions contain fractions in their numerators and/or their denominators.

14. Simplify each complex fraction.

a. $\dfrac{\dfrac{3}{2}}{\dfrac{2}{3}}$ **b.** $\dfrac{\dfrac{3}{2} + 1}{\dfrac{2}{3} + 1}$

To simplify a complex fraction, use either of these methods:

1. Write the numerator and denominator of the complex fraction as single fractions, do the division of the fractions, and simplify.

2. Multiply both the numerator and the denominator of the complex fraction by the LCD of the fractions that appear in the numerator and denominator, then simplify.

c. $\dfrac{\dfrac{1}{y} + 1}{\dfrac{1}{y} - 1}$

d. $\dfrac{1 + \dfrac{3}{x}}{2 - \dfrac{1}{x^2}}$

e. $\dfrac{\dfrac{2}{x - 1} + \dfrac{x - 1}{x + 1}}{\dfrac{1}{x^2 - 1}}$

f. $\dfrac{x^{-2} + 1}{x^{-2} - 1}$

SECTION 12.5

Rational Equations and Problem Solving

To solve an equation that contains fractions, change it to an equivalent equation without fractions. Do so by multiplying both sides by the LCD of the fractions. Check all solutions.

An apparent solution that does not satisfy the original equation is called an *extraneous* solution.

15. Solve each equation and check all answers.

a. $\dfrac{3}{x} = \dfrac{2}{x - 1}$

b. $\dfrac{a}{a - 5} = 3 + \dfrac{5}{a - 5}$

c. $\dfrac{2}{3t} + \dfrac{1}{t} = \dfrac{5}{9}$

d. $a = \dfrac{3a - 50}{4a - 24} - \dfrac{3}{4}$

e. $\dfrac{4}{x + 2} - \dfrac{3}{x + 3} = \dfrac{6}{x^2 + 5x + 6}$

16. The efficiency E of a Carnot engine is given by the formula

$$E = 1 - \dfrac{T_2}{T_1}$$

Solve the formula for T_1.

17. Solve for r_1: $\dfrac{1}{r} = \dfrac{1}{r_1} + \dfrac{1}{r_2}$.

To solve a problem, follow these steps:

1. Analyze the problem.
2. Form an equation.
3. Solve the equation.
4. State the conclusion.
5. Check the result.

Interest = principal · rate · time

Distance = rate · time

18. NUMBER PROBLEM   If a number is subtracted from the denominator of $\frac{4}{5}$ and twice as much is added to the numerator, the result is 5. Find the number.

19. If a maid can clean a house in 4 hours, how much of the house does she clean in 1 hour?

20. HOUSE PAINTING   If a homeowner can paint a house in 14 days and a professional painter can paint it in 10 days, how long will it take if they work together?

21. INVESTMENTS   In one year, a student earned $100 interest on money she deposited at a savings and loan. She later learned that the money would have earned $120 if she had deposited it at a credit union, because the credit union paid 1% more interest at the time. Find the rate she received from the savings and loan.

22. EXERCISE   A jogger can bicycle 30 miles in the same time that it takes her to jog 10 miles. If she can ride 10 mph faster than she can jog, how fast can she jog?

23. WIND SPEED   A plane flies 400 miles downwind in the same amount of time as it takes to travel 320 miles upwind. If the plane can fly at 360 mph in still air, find the velocity of the wind.

SECTION 12.6	*Proportions and Similar Triangles*

A *proportion* is a statement that two ratios or two rates are equal.

In the proportion $\frac{a}{b} = \frac{c}{d}$, a and d are the *extremes,* and b and c are the *means.*

In any proportion, the product of the extremes is equal to the product of the means.

24. Determine whether each equation is a proportion.

    a. $\dfrac{4}{7} = \dfrac{20}{34}$ **b.** $\dfrac{5}{7} = \dfrac{30}{42}$

25. Solve each proportion.

    a. $\dfrac{3}{x} = \dfrac{6}{9}$ **b.** $\dfrac{x}{3} = \dfrac{x}{5}$

    c. $\dfrac{x-2}{5} = \dfrac{x}{7}$ **d.** $\dfrac{2x}{x+4} = \dfrac{3}{x-1}$

26. DENTISTRY   The diagram in Illustration 2 was displayed in a dentist's office. According to the diagram, if the dentist has 340 adult patients, how many will develop gum disease?

3 out of 4 adults will develop gum disease.

ILLUSTRATION 2

The measures of corresponding sides of *similar triangles* are in proportion.

27. A telephone pole casts a shadow 12 feet long at the same time that a man 6 feet tall casts a shadow of 3.6 feet. How tall is the pole?

SECTION 12.7	*Variation*

Direct variation: As one variable gets larger, the other gets larger as described by the equation $y = kx$, where k is the *constant of variation.*

Inverse variation: As one variable gets larger, the other gets smaller as described by the equation

$$y = \frac{k}{x} \quad (k \text{ is a constant})$$

28. PROFIT   The profit made by a strawberry farm varies directly with the number of baskets of strawberries sold. If a profit of $500 was made from the sale of 750 baskets, what is the profit when 1,250 baskets are sold?

29. l varies inversely with w. Find the constant of variation if $l = 30$ when $w = 20$.

30. ELECTRICITY   For a fixed voltage, the current in an electrical circuit varies inversely with the resistance in the circuit. If a certain circuit has a current of $2\frac{1}{2}$ amps when the resistance is 150 ohms, find the current in the circuit when the resistance is doubled.

31. The graph in Illustration 3 shows a type of variation. Does it show direct or inverse variation?

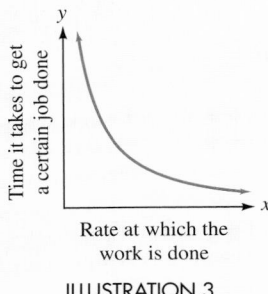

ILLUSTRATION 3

32. Give an example of two quantities that vary directly.

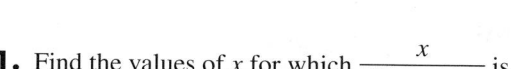

1. Find the values of x for which $\dfrac{x}{x^2 + x - 6}$ is undefined.

2. Simplify $\dfrac{48x^2y}{54xy^2}$.

3. Simplify $\dfrac{2x^2 - x - 3}{4x^2 - 9}$.

4. Simplify $\dfrac{3(x + 2) - 3}{2x - 4 - (x - 5)}$.

5. Multiply and simplify: $-\dfrac{12x^2y}{15xy} \cdot \dfrac{25y^2}{16x}$.

6. Multiply and simplify: $\dfrac{x^2 + 3x + 2}{3x + 9} \cdot \dfrac{x + 3}{x^2 - 4}$

7. Divide and simplify: $\dfrac{8x^2}{25x} \div \dfrac{16x^2}{30x}$.

8. Divide and simplify: $\dfrac{x - x^2}{3x^2 + 6x} \div \dfrac{3x - 3}{3x^3 + 6x^2}$.

9. Simplify $\dfrac{x^2 + x}{x - 1} \cdot \dfrac{x^2 - 1}{x^2 - 2x} \div \dfrac{x^2 + 2x + 1}{x^2 - 4}$.

10. Add: $\dfrac{5x - 4}{x - 1} + \dfrac{5x + 3}{x - 1}$.

11. Subtract: $\dfrac{3y + 7}{2y + 3} - \dfrac{3(y - 2)}{2y + 3}$.

12. Add: $\dfrac{x + 1}{x} + \dfrac{x - 1}{x + 1}$.

13. Subtract: $\dfrac{a + 3}{a - 1} - \dfrac{a + 4}{1 - a}$.

14. Subtract: $\dfrac{2n}{5m} - \dfrac{n}{2}$.

15. Simplify $\dfrac{1 + \dfrac{y}{x}}{\dfrac{y}{x} - 1}$.

16. Solve for q: $\dfrac{7}{q^2 - q - 2} + \dfrac{1}{q + 1} = \dfrac{3}{q - 2}$

17. Solve for c: $\dfrac{2}{3} = \dfrac{2c - 12}{3c - 9} - c$.

18. Solve for B: $H = \dfrac{RB}{R + B}$.

19. Is the equation $\dfrac{3}{5} = \dfrac{6xt}{10xt}$ a proportion?

20. Solve the proportion for y: $\dfrac{y}{y - 1} = \dfrac{y - 2}{y}$.

21. HEALTH RISK A medical newsletter states that a "healthy" waist-to-hip ratio for men is 19 : 20 or less. Does the patient shown in Illustration 1 fall within the "healthy" range?

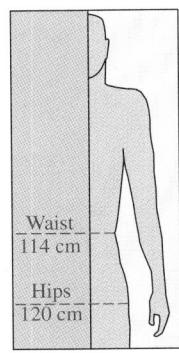

Waist
114 cm

Hips
120 cm

ILLUSTRATION 1

22. FLIGHT PATH A plane drops 575 feet as it flies a horizontal distance of $\frac{1}{2}$ mile, as shown in Illustration 2. How much altitude will it lose as it flies a horizontal distance of 7 miles?

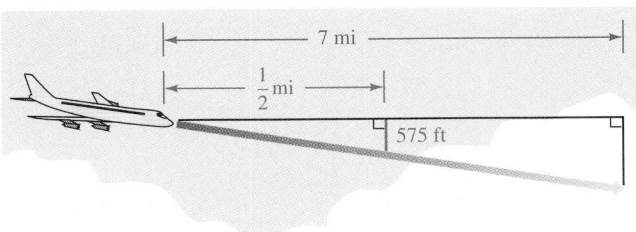

7 mi

$\frac{1}{2}$ mi

575 ft

ILLUSTRATION 2

23. POGO STICK See Illustration 3. The force required to compress a spring varies directly with the change in the length of the spring. If a force of 130 pounds compresses the spring on the pogo stick 6.5 inches, how much force is required to compress the spring 5 inches?

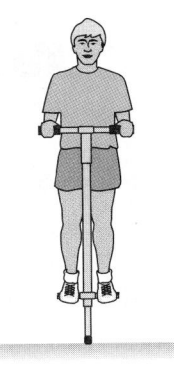

ILLUSTRATION 3

24. If i varies inversely with d, find the constant of variation if $i = 100$ when $d = 2$.

25. CLEANING HIGHWAYS One highway worker can pick up all the trash on a strip of highway in 7 hours, and his helper can pick up the trash in 9 hours. How long will it take them if they work together?

26. BOATING A boat can motor 28 miles downstream in the same amount of time as it can motor 18 miles upstream. Find the speed of the current if the boat can motor at 23 mph in still water.

27. Explain why we can divide out the 5's in $\frac{5x}{5}$ and why we can't divide them out in $\frac{5+x}{5}$.

28. Explain what it means to clear the following equation of fractions.

$$\frac{u}{u-1} + \frac{1}{u} = \frac{u^2+1}{u^2-u}$$

Why is this a helpful first step in solving the equation?

Chapters 1-12   Cumulative Review Exercises

1. Evaluate $9^2 - 3[45 - 3(6 + 4)]$.

2. PAIN RELIEVER   For the 12-month period ending August 16, 1998, Tylenol had sales of $567,600,000. Use the information in Illustration 1 to determine the total amount of money spent on pain-relieving tablets for that 12-month period.

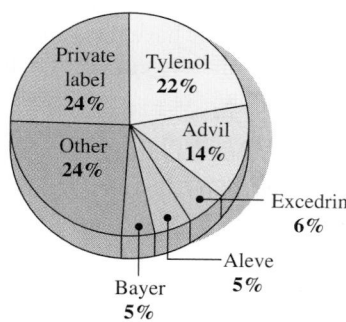

ILLUSTRATION 1

Based on information from *Los Angeles Times* (Sept. 24, 1998)

3. Find the average (mean) test score of a student in a history class with scores of 80, 73, 61, 73, and 98.

4. What is the value in cents of x 35¢ stamps?

5. Solve $\dfrac{3}{4} = \dfrac{1}{2} + \dfrac{x}{5}$.

6. Change $40°C$ to degrees Fahrenheit.

7. Find the volume of a pyramid that has a square base, measuring 6 feet on a side, and whose height is 20 feet.

8. Tell whether each statement is true or false.
   a. Every integer is a whole number.
   b. 0 is not a rational number.
   c. π is an irrational number.
   d. The set of integers is the set of whole numbers and their opposites.

9. Solve $2 - 3(x - 5) = 4(x - 1)$.

10. Simplify $8(c + 7) - 2(c - 3)$.

11. Solve $A - c = 2B + r$ for B.

12. Solve $7x + 2 \geq 4x - 1$ and graph the solution. Then describe the graph using interval notation.

13. Solve $\dfrac{4}{5}d = -4$.

14. BLENDING TEA   One grade of tea (worth $3.20 per pound) is to be mixed with another grade (worth $2 per pound) to make 20 pounds of a mixture that will be worth $2.72 per pound. How much of each grade of tea must be used?

15. SPEED OF A PLANE   Two planes are 6,000 miles apart, and their speeds differ by 200 mph. If they travel toward each other and meet in 5 hours, find the speed of the slower plane.

16. Graph $y = 2x - 3$.   **17.** Graph $y = (x + 2)^3$.

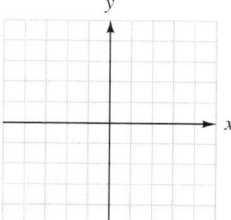   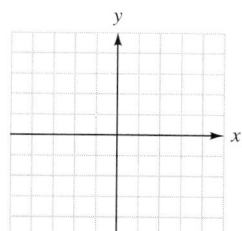

18. Find the slope of the line passing through $(-1, 3)$ and $(3, -1)$.

19. Write the equation of a line that has slope 3 and passes through the point $(1, 5)$.

20. Graph $3x - 2y = 6$.   **21.** Graph $y = \dfrac{5}{2}$.

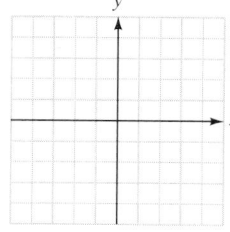   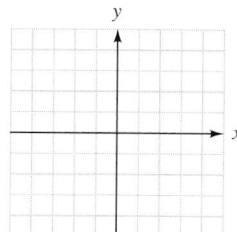

22. What is the slope of a line perpendicular to the line $y = -\dfrac{7}{8}x - 6$?

23. Is this the graph of a function?

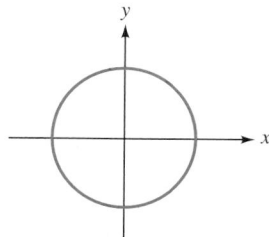

24. CUTTING STEEL   The graph in Illustration 2 shows the amount of wear (in millimeters) on a cutting blade for a given length of a cut (in meters). Find the rate of change in the length of the cutting blade.

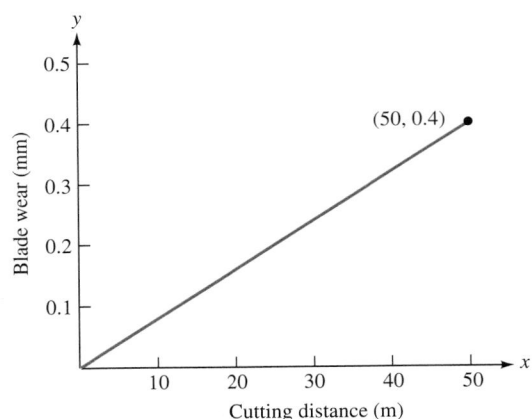

ILLUSTRATION 2

25. Find $f(-4)$ if $f(x) = \dfrac{x^2 - 2x}{2}$.

26. Evaluate -5^2.

Simplify each expression. Write each answer without using negative exponents.

27. $x^4 x^3$

28. $(x^2 x^3)^5$

29. $\left(\dfrac{y^3 y}{2yy^2}\right)^3$

30. $\left(\dfrac{-2a}{b}\right)^5$

31. $(a^{-2} b^3)^{-4}$

32. $\dfrac{9b^0 b^3}{3b^{-3} b^4}$

33. Write 290,000 in scientific notation.

34. What is the degree of the polynomial $5x^3 - 4x + 16$?

Do the operations.

35. $(3x^2 - 3x - 2) + (3x^2 + 4x - 3)$

36. $(2x^2 y^3)(3x^3 y^2)$

37. $(2y - 5)(3y + 7)$

38. $-4x^2 z(3x^2 - z)$

39. $\dfrac{6x + 9}{3}$

40. $\dfrac{15(r^2 s^3)^2}{-5(rs^5)^3}$

41. ▦ LICENSE PLATES   The number of different license plates of the form three digits followed by three letters, as shown in Illustration 3, is

$10 \cdot 10 \cdot 10 \cdot 26 \cdot 26 \cdot 26$. Write this expression using exponents. Then evaluate it.

ILLUSTRATION 3

42. CONCENTRIC CIRCLES   In Illustration 4, the red circle and the blue circle have the same center. The area of the ring between the two circles of radius r and R is given by the formula

$$A = \pi(R + r)(R - r)$$

Do the multiplication on the right-hand side of the equation.

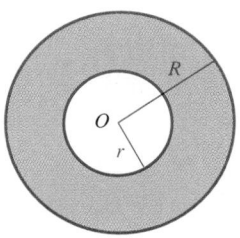

ILLUSTRATION 4

Factor each polynomial completely, if possible.

43. $k^3 t - 3k^2 t$

44. $2ab + 2ac + 3b + 3c$

45. $2a^2 - 200b^2$

46. $b^3 + 125$

47. $u^2 - 18u + 81$

48. $6x^2 - 63 - 13x$

49. $-r^2 + 2 + r$

50. $u^2 + 10u + 15$

Solve each equation by factoring.

51. $5x^2 + x = 0$ **52.** $6x^2 - 5x = -1$

53. COOKING   The electric griddle shown in Illustration 5 has a cooking surface of 160 square inches. Find the length and the width of the griddle.

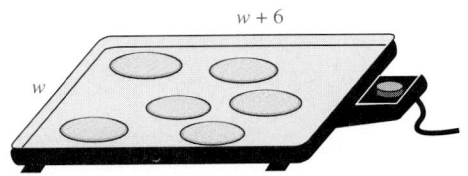

ILLUSTRATION 5

54. For what values of x is the rational expression $\dfrac{3x^2}{x^2 - 25}$ undefined?

Do the operations and simplify, if possible.

55. $\dfrac{x^2 - 16}{x - 4} \div \dfrac{3x + 12}{x}$

56. $\dfrac{4}{x - 3} + \dfrac{5}{3 - x}$

57. $\dfrac{2 - \dfrac{2}{x + 1}}{2 + \dfrac{2}{x}}$

58. $\dfrac{4a}{a - 2} - \dfrac{3a}{a - 3} + \dfrac{4a}{a^2 - 5a + 6}$

Solve each equation.

59. $\dfrac{7}{5x} - \dfrac{1}{2} = \dfrac{5}{6x} + \dfrac{1}{3}$

60. $\dfrac{3}{5} + \dfrac{7}{x + 2} = 2$

61. COMPUTING INTEREST   For a fixed rate and principal, the interest earned in a bank account paying simple interest varies directly with the length of time the principal is left on deposit. If an investment earns $700 in 2 years, how much will it earn in 7 years?

62. HEIGHT OF A TREE   A tree casts a shadow of 29 feet at the same time as a vertical yardstick casts a shadow of 2.5 feet. (See Illustration 6.) Find the height of the tree.

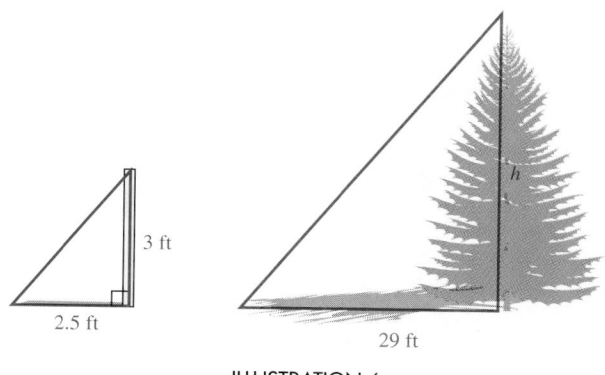

ILLUSTRATION 6

63. DRAINING A TANK   If one outlet pipe can drain a tank in 24 hours, and another pipe can drain the tank in 36 hours, how long will it take for both pipes to drain the tank?

64. Explain what it means for two variables to vary inversely.

13

Solving Systems of Equations and Inequalities

13.1 Solving Systems of
Equations by Graphing

13.2 Solving Systems of
Equations by Substitution

13.3 Solving Systems of
Equations by Addition

13.4 Applications of Systems
of Equations

13.5 Graphing Linear
Inequalities

13.6 Solving Systems of
Linear Inequalities

Key Concept:
Systems of Equations
and Inequalities

Accent on Teamwork

Chapter Review

Chapter Test

Cumulative Review
Exercises

To solve many problems, we must use two variables. This requires that we solve a system of equations.

13.1   *Solving Systems of Equations by Graphing*

In this section, you will learn about

- Systems of equations • The graphing method • Inconsistent systems
- Dependent equations

INTRODUCTION. The lines graphed in Figure 13-1 approximate the per-person consumption of chicken and beef in the United States for the years 1990–1997. We can see that consumption of chicken increased, while that of beef decreased.

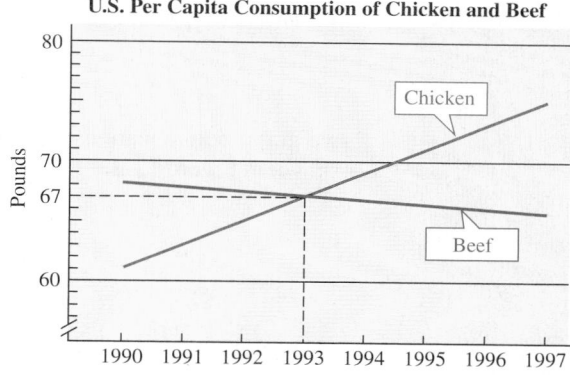

U.S. Per Capita Consumption of Chicken and Beef

Based on data from the National Broiler Council

FIGURE 13-1

By graphing this *pair* of lines on the same coordinate system, it is apparent that Americans consumed equal amounts of chicken and beef in 1993—about 67 pounds of each. In this section, we will work with pairs of linear equations whose graphs are straight lines. We call such a pair of equations a *system of equations.*

Systems of equations

We have previously discussed equations that contain two variables, such as $x + y = 3$. Because there are infinitely many pairs of numbers whose sum is 3, there are infinitely many pairs (x, y) that satisfy this equation. Some of these pairs are

$$x + y = 3$$

x	y	(x, y)
0	3	(0, 3)
1	2	(1, 2)
2	1	(2, 1)
3	0	(3, 0)

Likewise, there are infinitely many pairs (x, y) that satisfy the equation $3x - y = 1$. Some of these pairs are

$$3x - y = 1$$

x	y	(x, y)
0	-1	$(0, -1)$
1	2	$(1, 2)$
2	5	$(2, 5)$
3	8	$(3, 8)$

Although there are infinitely many pairs that satisfy each of these equations, only the pair $(1, 2)$ satisfies both equations at the same time. The pair of equations

$$\begin{cases} x + y = 3 \\ 3x - y = 1 \end{cases}$$

is called a **system of equations.** Because the ordered pair $(1, 2)$ satisfies both equations simultaneously (at the same time), it is called a **simultaneous solution,** or a **solution of the system of equations.** In this chapter, we will discuss three methods for finding the solution of a system of equations.

The graphing method

To use the graphing method to solve

$$\begin{cases} x + y = 3 \\ 3x - y = 1 \end{cases}$$

we graph both equations on one set of coordinate axes using the intercept method, as shown in Figure 13-2.

$x + y = 3$

x	y	(x, y)
0	3	$(0, 3)$
3	0	$(3, 0)$
2	1	$(2, 1)$

$3x - y = 1$

x	y	(x, y)
0	-1	$(0, -1)$
$\frac{1}{3}$	0	$(\frac{1}{3}, 0)$
2	5	$(2, 5)$

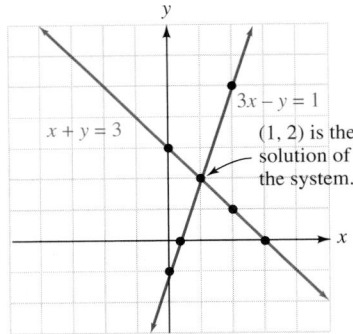

FIGURE 13-2

Although there are infinitely many pairs (x, y) that satisfy $x + y = 3$, and infinitely many pairs (x, y) that satisfy $3x - y = 1$, only the coordinates of the point where their graphs intersect satisfy both equations simultaneously. Thus, the solution of the system is $x = 1$ and $y = 2$, or $(1, 2)$.

To check this solution, we substitute 1 for x and 2 for y in each equation and verify that the pair $(1, 2)$ satisfies each equation.

First equation

$x + y = 3$

$1 + 2 \stackrel{?}{=} 3$

$3 = 3$

Second equation

$3x - y = 1$

$3(1) - 2 \stackrel{?}{=} 1$

$3 - 2 \stackrel{?}{=} 1$

$1 = 1$

When the graphs of two equations in a system are different lines, the equations are called **independent equations.** When a system of equations has a solution, the system is called a **consistent system.**

To solve a system of equations in two variables by graphing, we follow these steps.

The graphing method

1. Carefully graph each equation.
2. When possible, find the coordinates of the point where the graphs intersect.
3. Check the solution in the equations of the original system.

EXAMPLE 1 *Solving systems by graphing.* Using graphing to solve

$$\begin{cases} 2x + 3y = 2 \\ 3x = 2y + 16 \end{cases}$$

Self Check

Solve $\begin{cases} 2x = y - 5 \\ x + y = -1 \end{cases}$

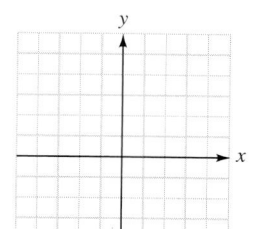

Solution

Using the intercept method, we graph both equations on one set of coordinate axes, as shown in Figure 13-3.

$2x + 3y = 2$		
x	y	(x, y)
0	$\frac{2}{3}$	$(0, \frac{2}{3})$
1	0	$(1, 0)$
-2	2	$(-2, 2)$

$3x = 2y + 16$		
x	y	(x, y)
0	-8	$(0, -8)$
$\frac{16}{3}$	0	$(\frac{16}{3}, 0)$
2	-5	$(2, -5)$

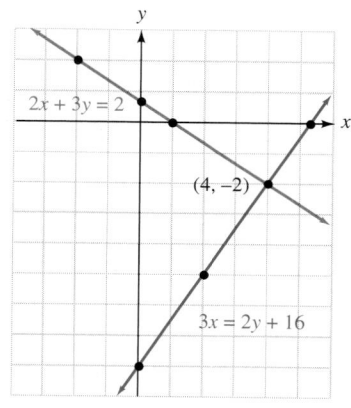

FIGURE 13-3

Although there are infinitely many pairs (x, y) that satisfy $2x + 3y = 2$, and infinitely many pairs (x, y) that satisfy $3x = 2y + 16$, only the coordinates of the point where the graphs intersect satisfy both equations at the same time. The solution is $x = 4$ and $y = -2$, or $(4, -2)$.

To check, we substitute 4 for x and -2 for y in each equation and verify that the pair $(4, -2)$ satisfies each equation.

$2x + 3y = 2$	$3x = 2y + 16$
$2(4) + 3(-2) \stackrel{?}{=} 2$	$3(4) \stackrel{?}{=} 2(-2) + 16$
$8 - 6 \stackrel{?}{=} 2$	$12 \stackrel{?}{=} -4 + 16$
$2 = 2$	$12 = 12$

The equations in this system are independent equations, and the system is a consistent system of equations.

Answer: $(-2, 1)$

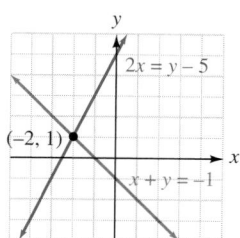

EXAMPLE 2 *Solving an equivalent system.* Solve $\begin{cases} -\dfrac{x}{2} - 1 = \dfrac{y}{2} \\ \dfrac{1}{3}x - \dfrac{1}{2}y = -4 \end{cases}$

Self Check

Solve $\begin{cases} -\dfrac{x}{2} = \dfrac{y}{4} \\ \dfrac{1}{4}x - \dfrac{3}{8}y = -2 \end{cases}$

Solution

We can multiply both sides of the first equation by 2 to clear it of fractions.

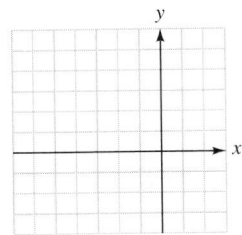

$$-\frac{x}{2} - 1 = \frac{y}{2}$$

$$2\left(-\frac{x}{2} - 1\right) = 2\left(\frac{y}{2}\right)$$

(1) $-x - 2 = y$ We will call this Equation 1.

We then multiply both sides of the second equation by 6 to clear it of fractions.

$$\frac{1}{3}x - \frac{1}{2}y = -4$$

$$6\left(\frac{1}{3}x - \frac{1}{2}y\right) = 6(-4)$$

(2) $2x - 3y = -24$ We will call this Equation 2.

Equations 1 and 2 form the following **equivalent system,** which has the same solutions as the original system:

$$\begin{cases} -x - 2 = y \\ 2x - 3y = -24 \end{cases}$$

In Figure 13-4, we graph $-x - 2 = y$ by plotting the y-intercept $(0, -2)$ and then drawing a slope of -1. We graph $2x - 3y = -24$ using the intercept method. We find that $(-6, 4)$ is the point of intersection. The solution is $x = -6$ and $y = 4$, or $(-6, 4)$.

$y = -x - 2$

so $m = -1 = \dfrac{-1}{1}$

and $b = -2$

$2x - 3y = -24$		
x	y	(x, y)
0	8	$(0, 8)$
-12	0	$(-12, 0)$
-3	6	$(-3, 6)$

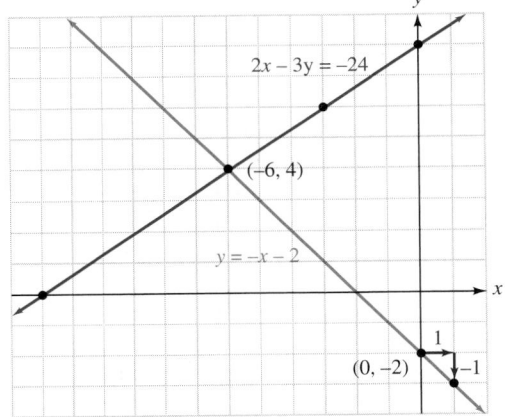

FIGURE 13-4

Answer: $(-2, 4)$

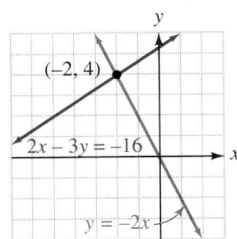

A check will show that when the coordinates of $(-6, 4)$ are substituted into the two original equations, true statements result. Therefore, the equations are independent and the system is consistent. ∎

COMMENT   When solving a system of equations, always check your answer by substituting into the *original* equations. Do not check by substituting into the equations of an equivalent system. If an algebraic error was made while finding the equivalent system, an answer that would not satisfy the original system might appear to be correct.

Accent on Technology: *Solving systems with a graphing calculator*

We can use a graphing calculator to solve the system

$$\begin{cases} 2x + y = 12 \\ 2x - y = -2 \end{cases}$$

However, before we can enter the equations into the calculator, we must solve them for y.

$$2x + y = 12 \qquad\qquad 2x - y = -2$$
$$y = -2x + 12 \qquad\qquad -y = -2x - 2$$
$$\qquad\qquad\qquad\qquad y = 2x + 2$$

We enter the resulting equations and graph them on the same coordinate axes. If we use the standard window settings, their graphs will look like Figure 13-5(a).

To find the solution of the system, we use the INTERSECT feature that is found on most graphing calculators. With this option, the cursor automatically moves to the point of intersection of the graphs and displays the coordinates of that point. In Figure 13-5(b), we see that the solution is (2.5, 7). Consult your owner's manual for specific keystrokes to use INTERSECT.

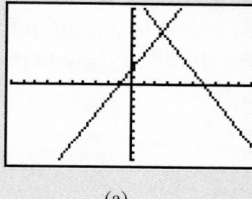   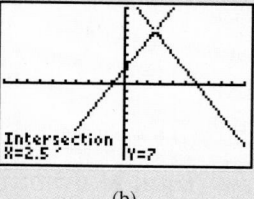

(a)   (b)

FIGURE 13-5

Inconsistent systems

Sometimes a system of equations has no solution. Such systems are called **inconsistent systems.**

EXAMPLE 3 *A system having no solution.* Solve $\begin{cases} y = -2x - 6 \\ 4x + 2y = 8 \end{cases}$

Self Check

Solve $\begin{cases} y = \dfrac{3}{2}x \\ 3x - 2y = 6 \end{cases}$

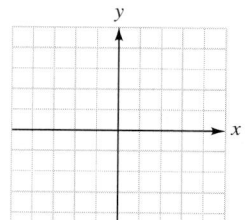

Solution

Since $y = -2x - 6$ is written in slope–intercept form, we can graph it by plotting the y-intercept $(0, -6)$ and then drawing a slope of -2. (The run is 1, and the rise is -2.) We graph $4x + 2y = 8$ using the intercept method.

$$y = -2x - 6$$

so $m = -2 = \dfrac{-2}{1}$

and $b = -6$

$$4x + 2y = 8$$

x	y	(x, y)
0	4	(0, 4)
2	0	(2, 0)
1	2	(1, 2)

The system is graphed in Figure 13-6. Since the lines in the figure are parallel, they have the same slope. We can verify this by writing the second equation in slope–intercept form and observing that the coefficients of x in each equation are equal.

$$y = -2x - 6 \qquad 4x + 2y = 8$$
$$2y = -4x + 8$$
$$y = -2x + 4$$

Because parallel lines do not intersect, this system has no solution and is inconsistent. Since the graphs are different lines, the equations of the system are independent.

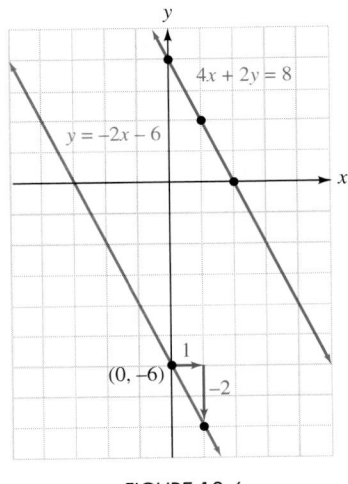

FIGURE 13-6

Answer: The lines are parallel; therefore, the system has no solution.

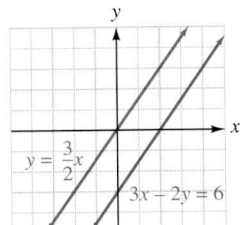

Dependent equations

Sometimes a system has an infinite number of solutions. In this case, we say that the equations of the system are **dependent equations.**

EXAMPLE 4   *Infinitely many solutions.*   Solve $\begin{cases} y - 4 = 2x \\ 4x + 8 = 2y \end{cases}$

Solution

We graph both equations on one set of axes, using the intercept method. See Figure 13-7.

Self Check

Solve $\begin{cases} 6x - 2y = 4 \\ y + 2 = 3x \end{cases}$

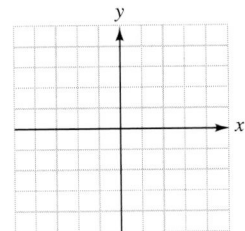

$y - 4 = 2x$				$4x + 8 = 2y$		
x	y	(x, y)		x	y	(x, y)
0	4	$(0, 4)$		0	4	$(0, 4)$
-2	0	$(-2, 0)$		-2	0	$(-2, 0)$
-1	2	$(-1, 2)$		-3	-2	$(-3, -2)$

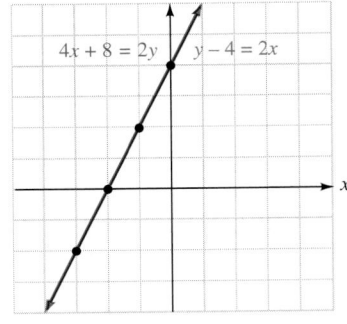

FIGURE 13-7

Answer: The graphs are the same line. There is an infinite number of solutions.

The lines in Figure 13-7 coincide (they are the same line). Because the lines intersect at infinitely many points, there is an infinite number of solutions. Any pair (x, y) that satisfies one of the equations also satisfies the other.

From the graph, we can see that some possible solutions are $(0, 4)$, $(-1, 2)$, and $(-3, -2)$, since each of these points lies on the one line that is the graph of both equations.

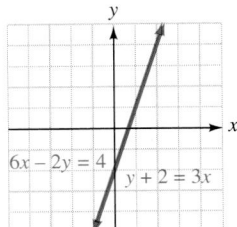

The possibilities that can occur when graphing two linear equations, each with two variables, are summarized as follows.

Possible graph	If the	Then
	lines are different and intersect,	the equations are independent and the system is consistent. One solution exists.
	lines are different and parallel,	the equations are independent and the system is inconsistent. No solutions exist.
	lines are the same,	the equations are dependent and the system is consistent. Infinitely many solutions exist.

STUDY SET   Section 13.1

VOCABULARY   *Fill in the blanks.*

1. The pair of equations $\begin{cases} x - y = -1 \\ 2x - y = 1 \end{cases}$ is called a _____ of equations.

2. Because the ordered pair (2, 3) satisfies both equations in Exercise 1, it is called a _____ of the system of equations.

3. When the graphs of two equations in a system are different lines, the equations are called _____ equations.

4. When a system of equations has a solution, the system is called a _____ system.

5. Systems of equations that have no solution are called _____ systems.

6. When a system has infinitely many solutions, the equations of the system are said to be _____ equations.

CONCEPTS   *Refer to Illustration 1. Tell whether a true or false statement would be obtained when the coordinates of*

7. point A are substituted into the equation for line l_1.

8. point B are substituted into the equation for line l_2.

9. point A are substituted into the equation for line l_2.

10. point B are substituted into the equation for line l_1.

11. point C are substituted into the equation for line l_1.

12. point C are substituted into the equation for line l_2.

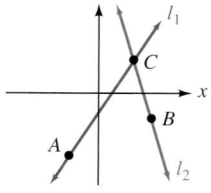

ILLUSTRATION 1

In Exercises 13–14, a furniture company is considering manufacturing a new line of oak chairs. Graphs showing the cost to make the chairs and the revenue the company will receive from their sale are given in Illustration 2.

13. a. What will it cost to make 30 chairs?
   b. What revenue will the sale of 30 chairs bring?

   c. How much money will the company make or lose in this case?

14. a. How many chairs must be built and sold so that the costs and the revenue are the same?
   b. Why do you think (70, 2,800) is called the "break-even point"?

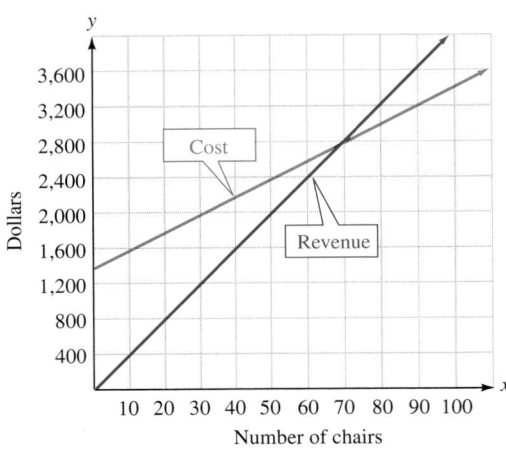

ILLUSTRATION 2

15. How many solutions does the system of equations graphed in Illustration 3 have? Is the system consistent or inconsistent?

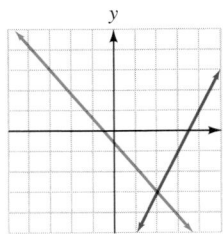

ILLUSTRATION 3

16. How many solutions does the system of equations graphed in Illustration 4 have? Are the equations dependent or independent?

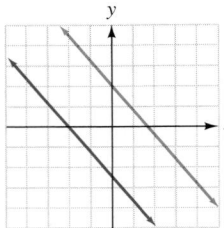

ILLUSTRATION 4

17. The solution of the system of equations graphed in Illustration 5 is $\left(\frac{2}{5}, -\frac{1}{3}\right)$. Knowing this, can you see any disadvantages to the graphing method?

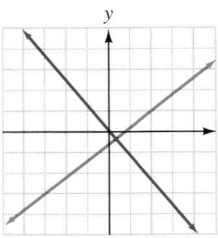

ILLUSTRATION 5

18. Draw the graphs of two linear equations so that the system has
  a. one solution $(-3, -2)$.

b. infinitely many solutions, three of which are $(-2, 0)$, $(1, 2)$, and $(4, 4)$.

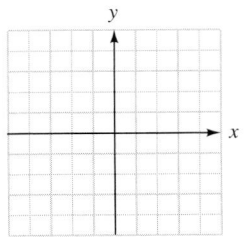

NOTATION *Clear each equation of fractions.*

19.
$$\frac{1}{6}x - \frac{1}{3}y = \frac{11}{2}$$
$$\boxed{}\left(\frac{1}{6}x - \frac{1}{3}y\right) = \boxed{}\left(\frac{11}{2}\right)$$
$$\boxed{}\left(\frac{1}{6}x\right) - 6\left(\boxed{}\right) = 6\left(\frac{11}{2}\right)$$
$$x - 2y = 33$$

20.
$$\frac{3x}{5} - \frac{4y}{5} = -1$$
$$\boxed{}\left(\frac{3x}{5} - \frac{4y}{5}\right) = \boxed{}(-1)$$
$$\boxed{}\left(\frac{3x}{5}\right) - 5\left(\boxed{}\right) = 5(-1)$$
$$3x - 4y = -5$$

PRACTICE *Tell whether the ordered pair is a solution of the given system.*

21. $(1, 1)$, $\begin{cases} x + y = 2 \\ 2x - y = 1 \end{cases}$

22. $(1, 3)$, $\begin{cases} 2x + y = 5 \\ 3x - y = 0 \end{cases}$

23. $(3, -2)$, $\begin{cases} 2x + y = 4 \\ y = 1 - x \end{cases}$

24. $(-2, 4)$, $\begin{cases} 2x + 2y = 4 \\ 3y = 10 - x \end{cases}$

25. $(-2, -4)$, $\begin{cases} 4x + 5y = -23 \\ -3x + 2y = 0 \end{cases}$

26. $(-5, 2)$, $\begin{cases} -2x + 7y = 17 \\ 3x - 4y = -19 \end{cases}$

27. $\left(\frac{1}{2}, 3\right)$, $\begin{cases} 2x + y = 4 \\ 4x - 11 = 3y \end{cases}$

28. $\left(2, \frac{1}{3}\right)$, $\begin{cases} x - 3y = 1 \\ -2x + 6 = -6y \end{cases}$

29. $\left(-\frac{2}{5}, \frac{1}{4}\right)$, $\begin{cases} x - 4y = -6 \\ 8y = 10x + 12 \end{cases}$

30. $\left(-\frac{1}{3}, \frac{3}{4}\right)$, $\begin{cases} 3x + 4y = 2 \\ 12y = 3(2 - 3x) \end{cases}$

31. $(0.2, 0.3)$, $\begin{cases} 20x + 10y = 7 \\ 20y = 15x + 3 \end{cases}$

32. $(2.5, 3.5)$, $\begin{cases} 4x - 3 = 2y \\ 4y + 1 = 6x \end{cases}$

Solve each system by the graphing method. If the equations of a system are dependent or if a system is inconsistent, so indicate.

33. $\begin{cases} x + y = 2 \\ x - y = 0 \end{cases}$

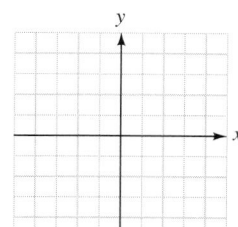

34. $\begin{cases} x + y = 4 \\ x - y = 0 \end{cases}$

35. $\begin{cases} x + y = 2 \\ y = x - 4 \end{cases}$

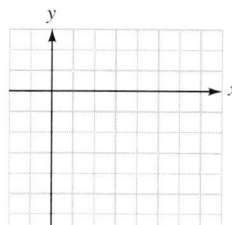

36. $\begin{cases} x + y = 1 \\ y = x + 5 \end{cases}$

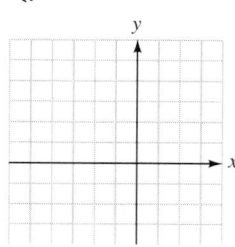

37. $\begin{cases} 3x + 2y = -8 \\ 2x - 3y = -1 \end{cases}$

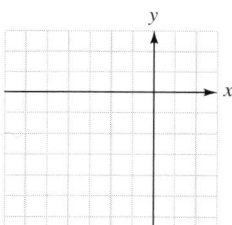

38. $\begin{cases} x + 4y = -2 \\ y = -x - 5 \end{cases}$

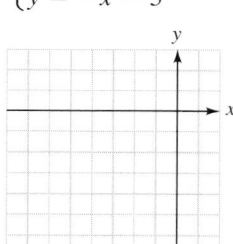

39. $\begin{cases} 4x - 2y = 8 \\ y = 2x - 4 \end{cases}$

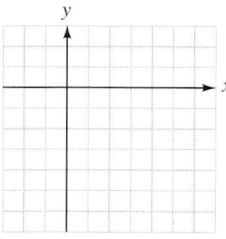

40. $\begin{cases} 3x - 6y = 18 \\ x = 2y + 3 \end{cases}$

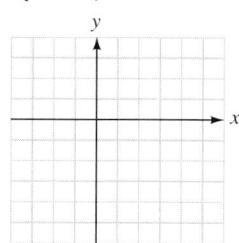

41. $\begin{cases} 2x - 3y = -18 \\ 3x + 2y = -1 \end{cases}$

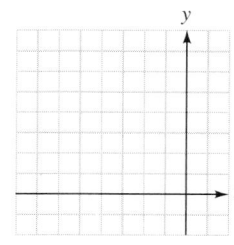

42. $\begin{cases} -x + 3y = -11 \\ 3x - y = 17 \end{cases}$

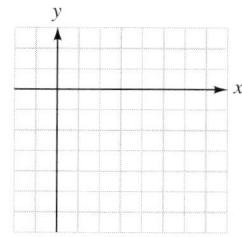

43. $\begin{cases} x = 4 \\ 2y = 12 - 4x \end{cases}$

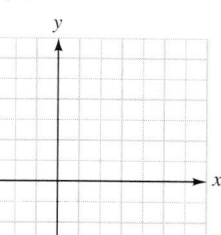

44. $\begin{cases} x = 3 \\ 3y = 6 - 2x \end{cases}$

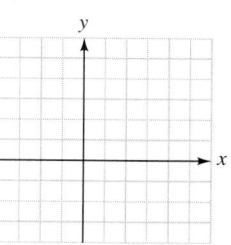

45. $\begin{cases} x + 2y = -4 \\ x - \frac{1}{2}y = 6 \end{cases}$

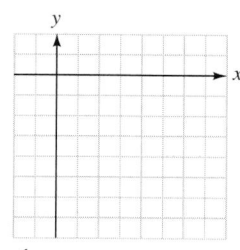

46. $\begin{cases} \frac{2}{3}x - y = -3 \\ 3x + y = 3 \end{cases}$

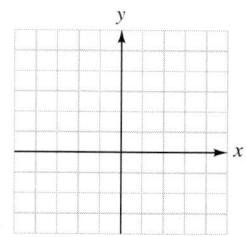

47. $\begin{cases} -\frac{3}{4}x + y = 3 \\ \frac{1}{4}x + y = -1 \end{cases}$

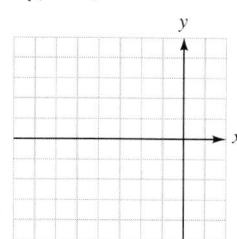

48. $\begin{cases} \frac{1}{3}x + y = 7 \\ \frac{2x}{3} - y = -4 \end{cases}$

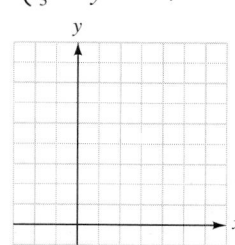

49. $\begin{cases} 2y = 3x + 2 \\ \frac{3}{2}x - y = 3 \end{cases}$

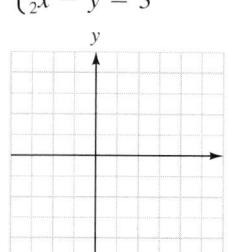

50. $\begin{cases} -\frac{3}{5}x - \frac{1}{5}y = \frac{6}{5} \\ x + \frac{y}{3} = -2 \end{cases}$

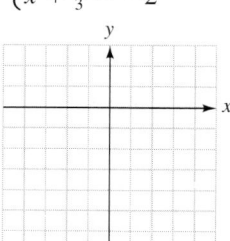

51. $\begin{cases} \frac{1}{3}x - \frac{1}{2}y = \frac{1}{6} \\ \frac{2x}{5} + \frac{y}{2} = \frac{13}{10} \end{cases}$

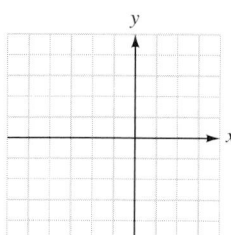

52. $\begin{cases} \frac{3x}{4} + \frac{2y}{3} = -\frac{19}{6} \\ 3y = -x \end{cases}$

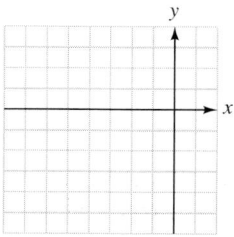

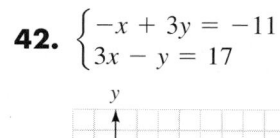 *Use a graphing calculator to solve each system, if possible.*

53. $\begin{cases} y = 4 - x \\ y = 2 + x \end{cases}$

54. $\begin{cases} 3x - 6y = 4 \\ 2x + y = 1 \end{cases}$

55. $\begin{cases} 6x - 2y = 5 \\ 3x = y + 10 \end{cases}$ **56.** $\begin{cases} x - 3y = -2 \\ 5x + y = 10 \end{cases}$

APPLICATIONS

57. TRANSPLANTS   See Illustration 6.
   a. What was the relationship between the number of donors and those awaiting a transplant in 1989?

   b. In what year were the number of donors and the number waiting for a transplant the same? Estimate the number.

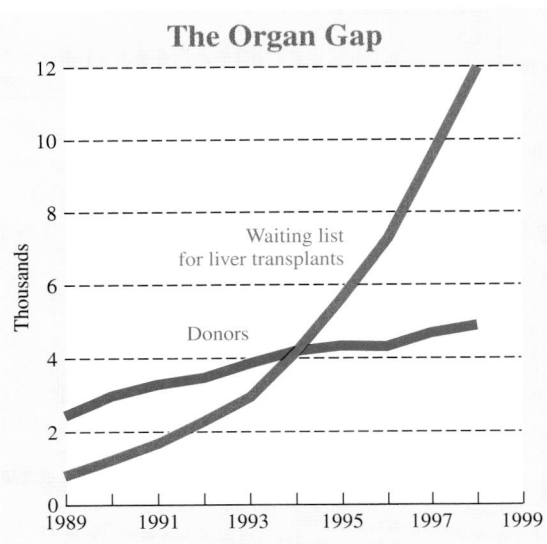

The Organ Gap

Waiting list for liver transplants

Donors

Based on data from United Network for Organ Sharing

ILLUSTRATION 6

58. DAILY TRACKING POLL   See Illustration 7.
   a. Which political candidate was ahead on October 28 and by how much?
   b. On what day did the challenger pull even with the incumbent?
   c. If the election was held November 4, who did the poll predict would win, and by how many percentage points?

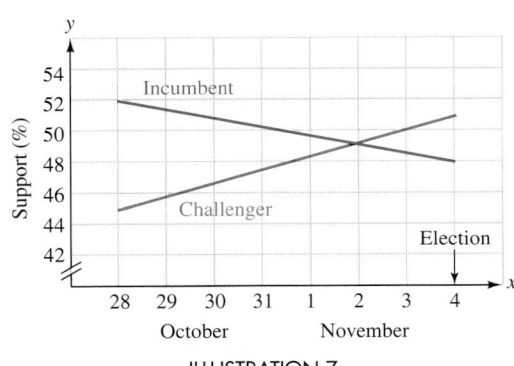

Incumbent

Challenger

Election

October   November

ILLUSTRATION 7

59. LATITUDE AND LONGITUDE   See Illustration 8.
   a. Name three American cities that lie on a latitude line of 30° north.

   b. Name three American cities that lie on a longitude line of 90° west.
   c. What city lies on both lines?

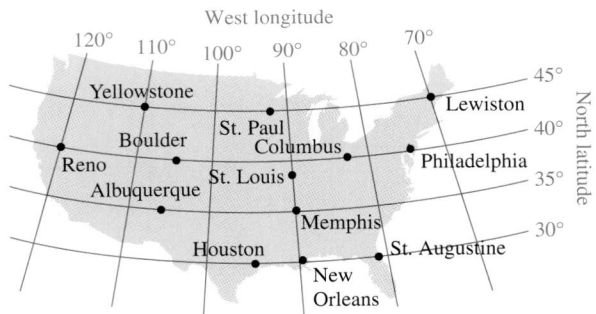

ILLUSTRATION 8

60. ECONOMICS   The graph in Illustration 9 illustrates the law of supply and demand.
   a. Complete this sentence: As the price of an item increases, the *supply* of the item _____.
   b. Complete this sentence: As the price of an item increases, the *demand* for the item _____.
   c. For what price will the supply equal the demand? How many items will be supplied for this price?

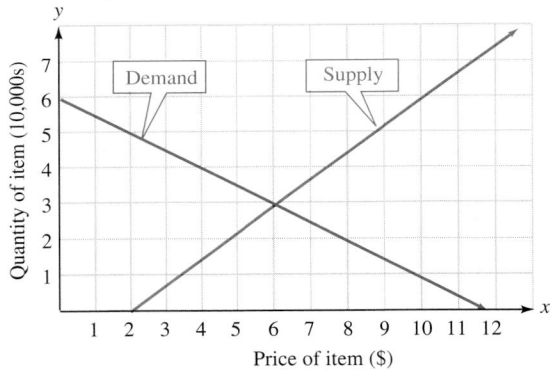

ILLUSTRATION 9

61. AIR TRAFFIC CONTROL   The equations describing the paths of two airplanes are $y = -\frac{1}{2}x + 3$ and $3y = 2x + 2$. Graph each equation on the radar screen shown in Illustration 10. Is there a possibility of a midair collision? If so, where?

62. TV COVERAGE   A television camera is located at $(-2, 0)$ and will follow the launch of a space shuttle, as shown in Illustration 11. (Each unit in the illustration is 1 mile.) As the shuttle rises vertically on a path described by $x = 2$, the farthest the camera can tilt back is a line of sight given by $y = \frac{5}{2}x + 5$. For how many miles of the shuttle's flight will it be in view of the camera?

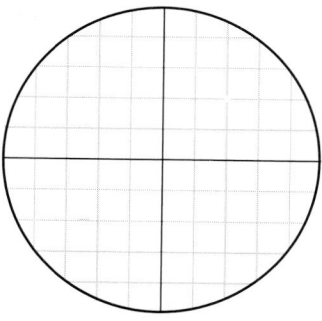

ILLUSTRATION 10

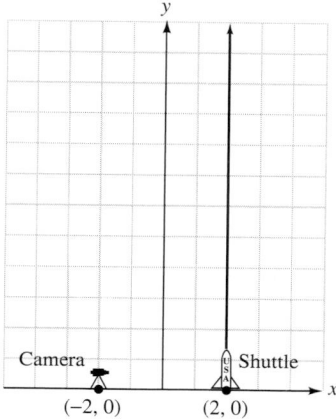

ILLUSTRATION 11

WRITING

63. Look up the word *simultaneous* in a dictionary and give its definition. In mathematics, what is meant by a simultaneous solution of a system of equations?

64. Suppose the solution of a system is $\left(\frac{1}{3}, -\frac{3}{5}\right)$. Do you think you would be able to find the solution using the graphing method? Explain.

REVIEW

65. What is the slope and the y-intercept of the graph of the line $y = -3x + 4$?

66. Are the graphs of the lines $y = 5x$ and $y = -\frac{1}{5}x$ parallel, perpendicular, or neither?

67. If $f(x) = -4x - x^2$, find $f(3)$.

68. In what quadrant does $(-12, 15)$ lie?

69. Write the equation for the y-axis.

70. Does $(1, 2)$ lie on the line $2x + y = 4$?

71. What point does the line with equation $y - 2 = 7(x - 5)$ pass through?

72. Is the word *domain* associated with the inputs or the outputs of a function?

13.2 *Solving Systems of Equations by Substitution*

In this section, you will learn about

- The substitution method • Inconsistent systems • Dependent equations

INTRODUCTION. When solving a system of equations by the graphing method, it is often difficult to determine the exact coordinates of the point of intersection. For example, it would be virtually impossible to distinguish that two lines intersect at the point $\left(\frac{1}{16}, -\frac{3}{5}\right)$. In this section, we introduce an algebraic method that finds *exact* solutions. It is called the *substitution method*. This method is based on the **substitution principle**: If $a = b$, then a may replace b or b may replace a in any statement.

The substitution method

To solve the system

$$\begin{cases} y = 3x - 2 \\ 2x + y = 8 \end{cases}$$

by the **substitution method**, we note that the first equation, $y = 3x - 2$, is *solved for y* (or *y is expressed in terms of x*). Because $y = 3x - 2$, we can substitute $3x - 2$ for y in the equation $2x + y = 8$ to get

$$2x + y = 8 \qquad \text{The second equation of the system.}$$
$$2x + 3x - 2 = 8 \qquad \text{Substitute } 3x - 2 \text{ for } y.$$

The resulting equation has only one variable and can be solved for x.

$$2x + 3x - 2 = 8$$
$$5x - 2 = 8 \qquad \text{Combine like terms: } 2x + 3x = 5x.$$
$$5x = 10 \qquad \text{Add 2 to both sides.}$$
$$x = 2 \qquad \text{Divide both sides by 5.}$$

We can find y by substituting 2 for x in either equation of the given system. Because $y = 3x - 2$ is already solved for y, it is easier to substitute into this equation.

$$y = 3x - 2 \qquad \text{The first equation of the system.}$$
$$= 3(2) - 2 \qquad \text{Substitute 2 for } x.$$
$$= 6 - 2$$
$$y = 4$$

The solution to the given system is $x = 2$ and $y = 4$, or $(2, 4)$.

Check: **First equation**   **Second equation**

$$y = 3x - 2 \qquad\qquad\qquad 2x + y = 8$$
$$4 \stackrel{?}{=} 3(2) - 2 \qquad\qquad 2(2) + 4 \stackrel{?}{=} 8$$
$$4 \stackrel{?}{=} 6 - 2 \qquad\qquad\qquad 4 + 4 \stackrel{?}{=} 8$$
$$4 = 4 \qquad\qquad\qquad\qquad 8 = 8$$

If we graphed the lines represented by the equations of the given system, they would intersect at the point $(2, 4)$. The equations of this system are independent, and the system is consistent.

To solve a system of equations in x and y by the substitution method, we follow these steps.

The substitution method

> **1.** Solve one of the equations for either x or y. (This step will not be necessary if an equation is already solved for x or y.)
>
> **2.** Substitute the resulting expression for the variable obtained in Step 1 into the remaining equation and solve that equation.
>
> **3.** Find the value of the other variable by substituting the solution found in Step 2 into any equation containing both variables.
>
> **4.** Check the solution in the equations of the original system.

EXAMPLE 1   *Solving systems by substitution.*   Solve $\begin{cases} 2x + y = -10 \\ x = -3y \end{cases}$

Self Check

Solve $\begin{cases} y = -2x \\ 3x - 2y = -7 \end{cases}$

Solution
The second equation, $x = -3y$, tells us that x and $-3y$ have the same value. Therefore, we may substitute $-3y$ for x in the first equation.

$$2x + y = -10 \qquad \text{The first equation of the system.}$$
$$2(-3y) + y = -10 \qquad \text{Replace } x \text{ with } -3y.$$
$$-6y + y = -10 \qquad \text{Do the multiplication.}$$
$$-5y = -10 \qquad \text{Combine like terms.}$$
$$y = 2 \qquad \text{Divide both sides by } -5.$$

We can find x by substituting 2 for y in the equation $x = -3y$.

$x = -3y$ The second equation of the system.

$ = -3(2)$ Substitute 2 for y.

$ = -6$

The solution is $x = -6$ and $y = 2$, or $(-6, 2)$.

Check: **First equation**   **Second equation**

$$2x + y = -10 \qquad\qquad x = -3y$$
$$2(-6) + 2 \overset{?}{=} -10 \qquad -6 \overset{?}{=} -3(2)$$
$$-12 + 2 \overset{?}{=} -10 \qquad -6 = -6$$
$$-10 = -10$$

Answer: $(-1, 2)$ ■

EXAMPLE 2   *Solving for a variable first.*   Solve $\begin{cases} 2x + y = -5 \\ 3x + 5y = -4 \end{cases}$

Solution

We solve one of the equations for one of the variables. Since the term y in the first equation has a coefficient of 1, we solve the first equation for y.

$2x + y = -5$ The first equation of the system.

$ y = -5 - 2x$ Subtract $2x$ from both sides to isolate y.

We then substitute $-5 - 2x$ for y in the second equation and solve for x.

$3x + 5y = -4$ The second equation of the system.

$3x + 5(-5 - 2x) = -4$ Substitute $-5 - 2x$ for y.

$3x - 25 - 10x = -4$ Distribute the multiplication by 5.

$-7x - 25 = -4$ Combine like terms: $3x - 10x = -7x$.

$-7x = 21$ Add 25 to both sides.

$x = -3$ Divide both sides by -7.

We can find y by substituting -3 for x in the equation $y = -5 - 2x$.

$y = -5 - 2x$

$ = -5 - 2(-3)$ Substitute -3 for x.

$ = -5 + 6$

$ = 1$

The solution is $(-3, 1)$. Check it in the original equations.

Self Check

Solve $\begin{cases} 2x - 3y = 13 \\ 3x + y = 3 \end{cases}$

Answer: $(2, -3)$ ■

Systems of equations are sometimes written in variables other than x and y. For example, the system

$$\begin{cases} 3a - 3b = 5 \\ 3 - a = -2b \end{cases}$$

is written in a and b. Regardless of the variables used, the procedures used to solve the system remain the same. The solution should be expressed in the form (a, b).

EXAMPLE 3   *Solving for a variable first.*   Solve $\begin{cases} 3a - 3b = 5 \\ 3 - a = -2b \end{cases}$

Solution

Since the coefficient of a in the second equation is -1, we will solve that equation for a.

Self Check

Solve $\begin{cases} 2s - t = 4 \\ 3s - 5t = 2 \end{cases}$

$$3 - a = -2b \qquad \text{The second equation of the system.}$$
$$-a = -2b - 3 \quad \text{Subtract 3 from both sides.}$$

To obtain a on the left-hand side, we can multiply (or divide) both sides of the equation by -1.

$$-1(-a) = -1(-2b - 3) \quad \text{Multiply both sides by } -1.$$
$$a = 2b + 3 \qquad \text{Do the multiplications.}$$

We then substitute $2b + 3$ for a in the first equation and proceed as follows:

$$3a - 3b = 5$$
$$3(2b + 3) - 3b = 5 \qquad \text{Substitute.}$$
$$6b + 9 - 3b = 5 \qquad \text{Distribute the multiplication by 3.}$$
$$3b + 9 = 5 \qquad \text{Combine like terms.}$$
$$3b = -4 \qquad \text{Subtract 9 from both sides: } 5 - 9 = -4.$$
$$b = -\frac{4}{3} \qquad \text{Divide both sides by 3.}$$

To find a, we substitute $-\frac{4}{3}$ for b in $a = 2b + 3$ and simplify.

$$a = 2b + 3$$
$$= 2\left(-\frac{4}{3}\right) + 3 \quad \text{Substitute.}$$
$$= -\frac{8}{3} + \frac{9}{3} \qquad \text{Do the multiplication: } 2\left(-\frac{4}{3}\right) = -\frac{8}{3}. \text{ Write 3 as } \frac{9}{3}.$$
$$= \frac{1}{3} \qquad \text{Add the numerators and keep the common denominator.}$$

The solution is $\left(\frac{1}{3}, -\frac{4}{3}\right)$. Check it in the original equations.

Answer: $\left(\dfrac{18}{7}, \dfrac{8}{7}\right)$ ■

EXAMPLE 4 *Solving an equivalent system.* Solve
$$\begin{cases} \dfrac{x}{2} + \dfrac{y}{4} = -\dfrac{1}{4} \\ 2x - y = 2 + y - x \end{cases}$$

Solution
It is helpful to rewrite each equation in simpler form before performing a substitution. We begin by clearing the first equation of fractions.

$$\frac{x}{2} + \frac{y}{4} = -\frac{1}{4}$$
$$4\left(\frac{x}{2} + \frac{y}{4}\right) = 4\left(-\frac{1}{4}\right) \quad \text{Multiply both sides by the LCD, which is 4.}$$
$$2x + y = -1$$

We can write the second equation in general form ($Ax + By = C$) by adding x and subtracting y from both sides.

$$2x - y = 2 + y - x$$
$$2x - y + x - y = 2 + y - x + x - y$$
$$3x - 2y = 2 \qquad \text{Combine like terms.}$$

The two results form the following equivalent system, which has the same solution as the original one.

(1) $\begin{cases} 2x + y = -1 \\ 3x - 2y = 2 \end{cases}$
(2)

To solve this system, we solve Equation 1 for y.

$$2x + y = -1$$
$$2x + y - 2x = -1 - 2x \quad \text{Subtract } 2x \text{ from both sides.}$$
$$\textbf{(3)} \qquad y = -1 - 2x \quad \text{Combine like terms.}$$

To find x, we substitute $-1 - 2x$ for y in Equation 2 and proceed as follows:

$$3x - 2y = 2$$
$$3x - 2(\mathbf{-1 - 2x}) = 2 \quad \text{Substitute.}$$
$$3x + 2 + 4x = 2 \quad \text{Distribute the multiplication by } -2.$$
$$7x + 2 = 2 \quad \text{Combine like terms.}$$
$$7x = 0 \quad \text{Subtract 2 from both sides.}$$
$$x = 0 \quad \text{Divide both sides by 7.}$$

To find y, we substitute 0 for x in Equation 3.

$$y = -1 - 2x$$
$$y = -1 - 2(\mathbf{0})$$
$$y = -1$$

The solution is $(0, -1)$. Check it in the original equations.

Answer: $(-1, 0)$ ■

Inconsistent systems

EXAMPLE 5 *A system with no solution.* Solve $\begin{cases} 0.01x = 0.12 - 0.04y \\ 2x = 4(3 - 2y) \end{cases}$

Solution
The first equation contains decimal coefficients. We can clear the equation of decimals by multiplying both sides by 100.

$$\begin{cases} x = 12 - 4y \\ 2x = 4(3 - 2y) \end{cases}$$

Since $x = 12 - 4y$, we can substitute $12 - 4y$ for x in the second equation and solve for y.

$$2x = 4(3 - 2y) \quad \text{The second equation.}$$
$$2(\mathbf{12 - 4y}) = 4(3 - 2y) \quad \text{Substitute.}$$
$$24 - 8y = 12 - 8y \quad \text{Distribute.}$$
$$24 \neq 12 \quad \text{Add } 8y \text{ to both sides.}$$

Here, the terms involving y drop out, and a false result of $24 = 12$ is obtained. This result indicates that the equations are independent and also that the system is inconsistent. As we see in Figure 13-8, when the equations are graphed, the graphs are parallel lines. This system has no solution.

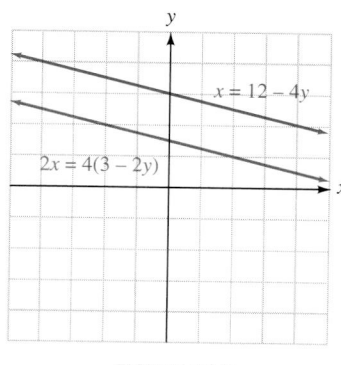

FIGURE 13-8

Self Check
Solve $\begin{cases} 0.1x - 0.4 = 0.1y \\ -2y = 2(2 - x) \end{cases}$

Answer: no solution ■

Dependent equations

EXAMPLE 6 *Infinitely many solutions.* Solve $\begin{cases} x = -3y + 6 \\ 2x + 6y = 12 \end{cases}$

Self Check
Solve $\begin{cases} y = 2 - x \\ 3x + 3y = 6 \end{cases}$

Solution

We can substitute $-3y + 6$ for x in the second equation and proceed as follows:

$$2x + 6y = 12 \quad \text{The second equation of the system.}$$
$$2(-3y + 6) + 6y = 12 \quad \text{Substitute.}$$
$$-6y + 12 + 6y = 12 \quad \text{Distribute the multiplication by 2.}$$
$$12 = 12 \quad \text{Combine like terms.}$$

Although $12 = 12$ is true, we did not find y. This indicates that the equations are dependent. As we see in Figure 13-9, when these equations are graphed, their graphs are identical.

Because any ordered pair that satisfies one equation of the system also satisfies the other, the system has infinitely many solutions. To find some, we substitute 0, 3, and 6 for x in either equation and solve for y. The pairs $(0, 2)$, $(3, 1)$, and $(6, 0)$ are some of the solutions.

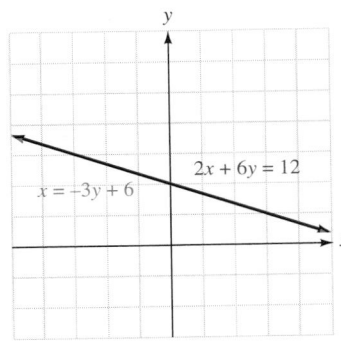

FIGURE 13-9

Answer: infinitely many solutions ■

Study Set Section 13.2

VOCABULARY *Fill in the blanks.*

1. We say that the equation $y = 2x + 4$ is solved for ____ or that y is expressed in _____ of x.

2. "To _____ a solution of a system" means to see whether the coordinates of the ordered pair satisfy both equations.

3. When we write $2(x - 6)$ as $2x - 12$, we are applying the _____ property.

4. In mathematics, "to _____" means to replace an expression with one that is equivalent to it.

5. A dependent system has _____ many solutions.

6. In the term y, the _____ is understood to be 1.

CONCEPTS

7. Consider the system $\begin{cases} 2x + 3y = 12 \\ y = 2x + 4 \end{cases}$

   a. How many variables does each equation of the system contain?

   b. Substitute $2x + 4$ for y in the first equation. How many variables does the resulting equation contain?

8. For each equation, solve for y.
   a. $y + 2 = x$
   b. $2 - y = x$
   c. $2 + x + y = 0$

9. Given the equation $x - 2y = -10$,
   a. solve it for x.
   b. solve it for y.
   c. which variable was easier to solve for, x or y? Explain.

10. Which variable in which equation should be solved for in step 1 of the substitution method?

   a. $\begin{cases} x - 2y = 2 \\ 2x + 3y = 11 \end{cases}$

   b. $\begin{cases} 2x - 3y = 2 \\ 2x - y = 11 \end{cases}$

   c. $\begin{cases} 7x - 3y = 2 \\ 2x - 8y = 0 \end{cases}$

11. a. Find the error in the following work when $x - 4$ is substituted for y.

   $$x + 2y = 5 \quad \text{The first equation of the system.}$$
   $$x + 2x - 4 = 5 \quad \text{Substitute for } y\text{: } y = x - 4.$$
   $$3x - 4 = 5 \quad \text{Combine like terms.}$$
   $$3x = 9 \quad \text{Add 4 to both sides.}$$
   $$x = 3 \quad \text{Do the divisions.}$$

   b. Rework the problem to find the correct value of x.

12. A student uses the substitution method to solve the system $\begin{cases} 4a + 5b = 2 \\ b = 3a - 11 \end{cases}$. She finds that $a = 3$. What is the easiest way for her to determine the value of b?

13. Consider the system $\begin{cases} x - 2y = 0 \\ 6x + 3y = 5 \end{cases}$

   a. Graph the equations on the same coordinate system. Why is it difficult to determine the solution of the system?

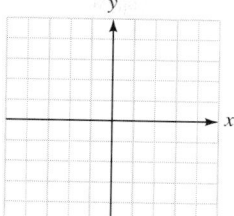

   b. Solve the system by the substitution method.

14. The equation $-2 = 1$ is the result when a system is solved by the substitution method. Which graph in Illustration 1 is a possible graph of the system?

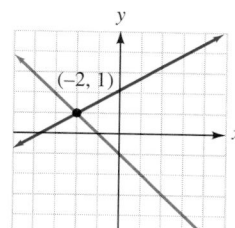
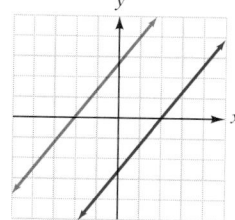

ILLUSTRATION 1

NOTATION *Complete the solution of each system.*

15. Solve $\begin{cases} y = 3x \\ x - y = 4 \end{cases}$

$$x - y = 4 \qquad \text{The second equation.}$$
$$x - (\ \ ) = 4$$
$$-2x = \ $$
$$x = -2$$
$$y = 3x \qquad \text{The first equation.}$$
$$y = 3(\ \ )$$
$$y = -6$$

The solution is   .

16. Solve $\begin{cases} 2x + y = -5 \\ 2 - 2y = x \end{cases}$

$$2x + y = -5 \qquad \text{The first equation.}$$
$$2(\ \ ) + y = -5$$
$$4 - \ + y = -5$$
$$\ - 3y = -5$$
$$-3y = \ $$
$$y = 3$$
$$2 - 2y = x \qquad \text{The second equation.}$$
$$2 - 2(\ \ ) = x$$
$$2 - 6 = x$$
$$-4 = x$$

The solution is   .

PRACTICE *Use the substitution method to solve each system. If the equations of a system are dependent or if a system is inconsistent, so indicate.*

17. $\begin{cases} y = 2x \\ x + y = 6 \end{cases}$ **18.** $\begin{cases} y = 3x \\ x + y = 4 \end{cases}$

19. $\begin{cases} y = 2x - 6 \\ 2x + y = 6 \end{cases}$ **20.** $\begin{cases} y = 2x - 9 \\ x + 3y = 8 \end{cases}$

21. $\begin{cases} y = 2x + 5 \\ x + 2y = -5 \end{cases}$ **22.** $\begin{cases} y = -2x \\ 3x + 2y = -1 \end{cases}$

23. $\begin{cases} 2a + 4b = -24 \\ a = 20 - 2b \end{cases}$ **24.** $\begin{cases} 3a + 6b = -15 \\ a = -2b - 5 \end{cases}$

25. $\begin{cases} 2a = 3b - 13 \\ -b = -2a - 7 \end{cases}$ **26.** $\begin{cases} a = 3b - 1 \\ -b = -2a - 2 \end{cases}$

27. $\begin{cases} r + 3s = 9 \\ 3r + 2s = 13 \end{cases}$ **28.** $\begin{cases} x - 2y = 2 \\ 2x + 3y = 11 \end{cases}$

29. $\begin{cases} 0.4x + 0.5y = 0.2 \\ 3x - y = 11 \end{cases}$ **30.** $\begin{cases} 0.5u + 0.3v = 0.5 \\ 4u - v = 4 \end{cases}$

31. $\begin{cases} 6x - 3y = 5 \\ 2y + x = 0 \end{cases}$ **32.** $\begin{cases} 5s + 10t = 3 \\ 2s + t = 0 \end{cases}$

33. $\begin{cases} 3x + 4y = -7 \\ 2y - x = -1 \end{cases}$ **34.** $\begin{cases} 4x + 5y = -2 \\ x + 2y = -2 \end{cases}$

35. $\begin{cases} 9x = 3y + 12 \\ 4 = 3x - y \end{cases}$ **36.** $\begin{cases} 8y = 15 - 4x \\ x + 2y = 4 \end{cases}$

37. $\begin{cases} 0.02x + 0.05y = -0.02 \\ -\frac{x}{2} = y \end{cases}$

38. $\begin{cases} y = -\frac{x}{2} \\ 0.02x - 0.03y = -0.07 \end{cases}$

39. $\begin{cases} b = \frac{2}{3}a \\ 8a - 3b = 3 \end{cases}$ **40.** $\begin{cases} a = \frac{2}{3}b \\ 9a + 4b = 5 \end{cases}$

41. $\begin{cases} y - x = 3x \\ 2x + 2y = 14 - y \end{cases}$ **42.** $\begin{cases} y + x = 2x + 2 \\ 6x - 4y = 21 - y \end{cases}$

43. $\begin{cases} 2x - y = x + y \\ -2x + 4y = 6 \end{cases}$ **44.** $\begin{cases} x = -3y + 6 \\ 2x + 4y = 6 + x + y \end{cases}$

45. $\begin{cases} 3(x - 1) + 3 = 8 + 2y \\ 2(x + 1) = 8 + y \end{cases}$

46. $\begin{cases} 4(x - 2) = 19 - 5y \\ 3(x - 2) - 2y = -y \end{cases}$

47. $\begin{cases} \frac{1}{2}x + \frac{1}{2}y = -1 \\ \frac{1}{3}x - \frac{1}{2}y = -4 \end{cases}$ **48.** $\begin{cases} \frac{2}{3}y + \frac{1}{5}z = 1 \\ \frac{1}{3}y - \frac{2}{5}z = 3 \end{cases}$

49. $\begin{cases} 5x = \frac{1}{2}y - 1 \\ \frac{1}{4}y = 10x - 1 \end{cases}$ **50.** $\begin{cases} \frac{2}{3}x = 1 - 2y \\ 2(5y - x) + 11 = 0 \end{cases}$

51. $\begin{cases} \dfrac{6x-1}{3} - \dfrac{5}{3} = \dfrac{3y+1}{2} \\[2mm] \dfrac{1+5y}{4} + \dfrac{x+3}{4} = \dfrac{17}{2} \end{cases}$

52. $\begin{cases} \dfrac{5x-2}{4} + \dfrac{1}{2} = \dfrac{3y+2}{2} \\[2mm] \dfrac{7y+3}{3} = \dfrac{x}{2} + \dfrac{7}{3} \end{cases}$

APPLICATIONS

53. DINING   See the breakfast menu in Illustration 2. What substitution from the a la carte menu will the restaurant owner allow customers to make if they don't want hash browns with their country breakfast? Why?

Village Vault Restaurant			
Country Breakfast $5.95			
Includes 2 eggs, 3 pancakes, sausage, bacon, hash browns, and coffee			
A la Carte Menu–Single Servings			
Strawberries	$1.25	Melon	$0.95
Croissant	$1.70	Orange juice	$1.65
Hash browns	$0.95	Oatmeal	$1.95
Muffin	$1.30	Ham	$1.80

ILLUSTRATION 2

54. DISCOUNT COUPON   In mathematics, the substitution property states:

If a = b, then a may replace b or b may replace a in any statement.

Where on the coupon in Illustration 3 is there an application of the substitution property? Explain.

WRITING

55. Explain how to use substitution to solve a system of equations.

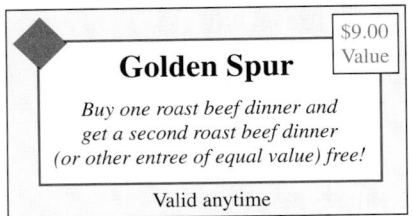

ILLUSTRATION 3

56. If the equations of a system are written in general form, why is it to your advantage to solve for a variable whose coefficient is 1 when using the substitution method?

57. When solving a system, what advantages and disadvantages are there with the graphing method? With the substitution method?

58. In this section, the substitution method for solving a system of two equations was discussed. List some other uses of the word *substitution,* or *substitute,* that you encounter in everyday life.

REVIEW

59. What is the slope of the line $y = -\dfrac{5}{8}x - 12$?

60. If $g(x) = -3x + 9$, find $g(-3)$.

61. Find the y-intercept of $2x - 3y = 18$.

62. Write the equation of the line passing through $(-1, 5)$ with a slope of -3.

63. Can a circle represent the graph of a function?

64. What is the range of the function $f(x) = |x|$?

65. A bowler has found that his score s varies directly with the time t in minutes he practices. Write an equation describing this relationship.

66. On what axis does $(0, -2)$ lie?

13.3 *Solving Systems of Equations by Addition*

In this section, you will learn about

- The addition method   • Inconsistent systems   • Dependent equations

INTRODUCTION. In step 1 of the substitution method for solving a system of equations, we solve one equation for one of the variables. At times, this can be difficult—especially if neither variable has a coefficient of 1 or -1. In cases such as these, we can use another algebraic method called the *addition* or *elimination method* to find the exact solution of the system. This method is based on the addition property of equality: *When equal quantities are added to both sides of an equation, the results are equal.*

The addition method

To solve the system

$$\begin{cases} x + y = 8 \\ x - y = -2 \end{cases}$$

by the **addition method,** we see that the coefficients of y are opposites and then add the left- and right-hand sides of the equations to eliminate the variable y.

$$\begin{aligned} x + y &= 8 \\ \underline{x - y} &= \underline{-2} \end{aligned}$$ Equal quantities, $x - y$ and -2, are added to both sides of the equation $x + y = 8$. By the addition property of equality, the results will be equal.

Now, column by column, we add like terms. The terms y and $-y$ are eliminated.

$$\begin{aligned} \downarrow \quad \downarrow & \quad\quad \downarrow \\ x + y &= 8 \\ \underline{x - y} &= \underline{-2} \\ 2x &= 6 \end{aligned}$$ Combine like terms: $x + x = 2x$, $y + (-y) = 0$, and $8 + (-2) = 6$.

$\leftarrow$ Write each result here.

We can then solve the resulting equation for x.

$$2x = 6$$
$$x = 3 \quad \text{Divide both sides by 2.}$$

To find y, we substitute 3 for x in either equation and solve it for y.

$$x + y = 8 \quad \text{The first equation of the system.}$$
$$3 + y = 8 \quad \text{Substitute 3 for } x.$$
$$y = 5 \quad \text{Subtract 3 from both sides.}$$

We check the solution by verifying that $(3, 5)$ satisfies each equation of the system. To solve an equation in x and y by the addition method, we follow these steps.

The addition method

> 1. Write both equations in general form: $Ax + By = C$.
> 2. If necessary, multiply one or both of the equations by nonzero quantities to make the coefficients of x (or the coefficients of y) opposites.
> 3. Add the equations to eliminate the terms involving x (or y).
> 4. Solve the equation resulting from Step 3.
> 5. Find the value of the other variable by substituting the solution found in Step 4 into any equation containing both variables.
> 6. Check the solution in the equations of the original system.

EXAMPLE 1   *Solving systems by addition.*   Solve $\begin{cases} 5x + y = -4 \\ -5x + 2y = 7 \end{cases}$

Self Check

Solve $\begin{cases} x + 3y = 7 \\ 2x - 3y = -22 \end{cases}$

Solution

When the equations are added, the terms $5x$ and $-5x$ drop out. We can then solve the resulting equation for y.

$$\begin{aligned} 5x + y &= -4 \\ \underline{-5x + 2y} &= \underline{7} \\ 3y &= 3 \end{aligned}$$ Combine like terms: $5x + (-5x) = 0$, $y + 2y = 3y$, and $-4 + 7 = 3$.

$$y = 1 \quad \text{Divide both sides by 3.}$$

To find x, we substitute 1 for y in either equation. If we use $5x + y = -4$, we have

$5x + y = -4$ The first equation of the system.

$5x + 1 = -4$ Substitute 1 for y.

$5x = -5$ Subtract 1 from both sides.

$x = -1$ Divide both sides by 5.

Verify that $(-1, 1)$ satisfies each original equation.

EXAMPLE 2 *Solving systems by addition.*   Solve $\begin{cases} 3x + y = 7 \\ x + 2y = 4 \end{cases}$

Self Check

Solve $\begin{cases} 3x + 4y = 25 \\ 2x + y = 10 \end{cases}$

Solution

If we add the equations as they are, neither variable will be eliminated. We must write the equations so that the coefficients of one of the variables are opposites. To eliminate x, we can multiply both sides of the second equation by -3 to get

$$\begin{cases} 3x + y = 7 \\ -3(x + 2y) = -3(4) \end{cases} \longrightarrow \begin{cases} 3x + y = 7 \\ -3x - 6y = -12 \end{cases}$$

The coefficients of the terms $3x$ and $-3x$ are now opposites. When the equations are added, x is eliminated.

$$\begin{array}{r} 3x + y = 7 \\ -3x - 6y = -12 \\ \hline -5y = -5 \\ y = 1 \end{array}$$ Divide both sides by -5.

To find x, we substitute 1 for y in the equation $x + 2y = 4$.

$x + 2y = 4$ The second equation of the original system.

$x + 2(1) = 4$ Substitute 1 for y.

$x + 2 = 4$ Do the multiplication.

$x = 2$ Subtract 2 from both sides.

Check the solution $(2, 1)$ in the original system of equations.

EXAMPLE 3 *Solving systems by addition.*   Solve $\begin{cases} 2a - 5b = 10 \\ 3a - 2b = -7 \end{cases}$

Self Check

Solve $\begin{cases} 2a + 3b = 7 \\ 5a + 2b = 1 \end{cases}$

Solution

The equations in the system must be written so that one of the variables will be eliminated when the equations are added. To eliminate a, we can multiply the first equation by 3 and the second equation by -2 to get

$$\begin{cases} 3(2a - 5b) = 3(10) \\ -2(3a - 2b) = -2(-7) \end{cases} \longrightarrow \begin{cases} 6a - 15b = 30 \\ -6a + 4b = 14 \end{cases}$$

When these equations are added, the terms $6a$ and $-6a$ are eliminated.

$$\begin{array}{r} 6a - 15b = 30 \\ -6a + 4b = 14 \\ \hline -11b = 44 \\ b = -4 \end{array}$$ Divide both sides by -11.

To find a, we substitute -4 for b in the equation $2a - 5b = 10$.

$2a - 5b = 10$ The first equation of the original system.

$2a - 5(-4) = 10$ Substitute -4 for b.

$2a + 20 = 10$ Simplify.

$$2a = -10 \quad \text{Subtract 20 from both sides.}$$

$$a = -5 \quad \text{Divide both sides by 2.}$$

Check the solution $(-5, -4)$ in the original equations.

Answer: $(-1, 3)$

EXAMPLE 4 *Equations containing fractions.* Solve $\begin{cases} \frac{5}{6}x + \frac{2}{3}y = \frac{7}{6} \\ \frac{10}{7}x - \frac{4}{9}y = \frac{17}{21} \end{cases}$

Self Check

Solve $\begin{cases} \frac{1}{3}x + \frac{1}{6}y = 1 \\ \frac{1}{2}x - \frac{1}{4}y = 0 \end{cases}$

Solution

To clear the equations of fractions, we multiply both sides of the first equation by 6 and both sides of the second equation by 63. This gives the equivalent system

(1) $\begin{cases} 5x + 4y = 7 \\ 90x - 28y = 51 \end{cases}$
(2)

We can solve for x by eliminating the terms involving y. To do so, we multiply Equation 1 by 7 and add the result to Equation 2.

$$\begin{array}{r} 35x + 28y = 49 \\ 90x - 28y = 51 \\ \hline 125x = 100 \end{array}$$

$$x = \frac{100}{125} \quad \text{Divide both sides by 125.}$$

$$x = \frac{4}{5} \quad \text{Simplify } \tfrac{100}{125}\text{: Divide out the common factor of 25.}$$

To solve for y, we substitute $\frac{4}{5}$ for x in Equation 1 and simplify.

$$5x + 4y = 7$$

$$5\left(\frac{4}{5}\right) + 4y = 7$$

$$4 + 4y = 7 \quad \text{Simplify.}$$

$$4y = 3 \quad \text{Subtract 4 from both sides.}$$

$$y = \frac{3}{4} \quad \text{Divide both sides by 4.}$$

Check the solution of $\left(\frac{4}{5}, \frac{3}{4}\right)$ in the original equations.

Answer: $\left(\frac{3}{2}, 3\right)$

EXAMPLE 5 *Writing equations in general form.* Solve
$\begin{cases} 2(2x + y) = 13 \\ 8x = 2y - 16 \end{cases}$

Self Check

Solve $\begin{cases} -3y = -5 - x \\ 3(x - y) = -11 \end{cases}$

Solution

We begin by writing each equation in $Ax + By = C$ form. For the first equation, we need only apply the distributive property. To write the second equation in general form, we subtract $2y$ from both sides.

$$2(2x + y) = 13 \qquad\qquad 8x = 2y - 16$$

$$4x + 2y = 13 \qquad\qquad 8x - 2y = 2y - 16 - 2y$$

$$8x - 2y = -16$$

The two resulting equations form the following system.

(1) $\begin{cases} 4x + 2y = 13 \\ 8x - 2y = -16 \end{cases}$
(2)

When the equations are added, the terms involving y are eliminated.

$$4x + 2y = 13$$
$$\underline{8x - 2y = -16}$$
$$12x = -3$$

$$x = -\frac{1}{4} \qquad \text{Divide both sides by 12 and simplify the fraction: } -\frac{3}{12} = -\frac{1}{4}.$$

We can use Equation 1 to find y.

$$4x + 2y = 13$$
$$4\left(-\frac{1}{4}\right) + 2y = 13 \qquad \text{Substitute } -\frac{1}{4} \text{ for } x.$$
$$-1 + 2y = 13 \qquad \text{Do the multiplication.}$$
$$2y = 14 \qquad \text{Add 1 to both sides.}$$
$$y = 7 \qquad \text{Divide both sides by 2.}$$

Verify that $\left(-\frac{1}{4}, 7\right)$ satisfies each original equation.

Answer: $\left(-3, \dfrac{2}{3}\right)$ ∎

Inconsistent systems

EXAMPLE 6 *A system with no solutions.* Solve $\begin{cases} 3x - 2y = 8 \\ -3x + 2y = -12 \end{cases}$

Self Check

Solve $\begin{cases} 2t - 7v = 5 \\ -2t + 7v = 3 \end{cases}$

Solution
We can add the equations to eliminate the term involving x.

$$3x - 2y = 8$$
$$\underline{-3x + 2y = -12}$$
$$0 = -4$$

Here the terms involving both x and y drop out, and a false result of $0 = -4$ is obtained. This indicates that the equations of the system are independent and that the system is inconsistent. This system has no solution.

Answer: no solution ∎

Dependent equations

EXAMPLE 7 *Infinitely many solutions.* Solve $\begin{cases} \dfrac{2x - 5y}{2} = \dfrac{19}{2} \\ -0.2x + 0.5y = -1.9 \end{cases}$

Self Check

Solve $\begin{cases} \dfrac{3x + y}{6} = \dfrac{1}{3} \\ -0.3x - 0.1y = -0.2 \end{cases}$

Solution
We can multiply both sides of the first equation by **2** to clear it of fractions and both sides of the second equation by **10** to clear it of decimals.

$$\begin{cases} 2\left(\dfrac{2x - 5y}{2}\right) = 2\left(\dfrac{19}{2}\right) \\ 10(-0.2x + 0.5y) = 10(-1.9) \end{cases} \longrightarrow \begin{cases} 2x - 5y = 19 \\ -2x + 5y = -19 \end{cases}$$

We add the resulting equations to get

$$2x - 5y = 19$$
$$\underline{-2x + 5y = -19}$$
$$0 = 0$$

As in Example 6, both x and y drop out. However, this time a true result is obtained. This indicates that the equations are dependent and that the system has infinitely many solutions.

Any ordered pair that satisfies one equation also satisfies the other equation. Some solutions are $(2, -3)$, $(12, 1)$, and $\left(0, -\frac{19}{5}\right)$.

Answer: infinitely many solutions ∎

STUDY SET   Section 13.3

VOCABULARY   *Fill in the blanks.*

1. The _____ of the term $-3x$ is -3.

2. The _____ of 4 is -4.

3. $Ax + By = C$ is the _____ form of the equation of a line.

4. When adding the equations

$$5x - 6y = 10$$
$$\underline{-3x + 6y = 24}$$

the variable y will be _____.

CONCEPTS

5. If the addition method is to be used to solve this system, what is wrong with the form in which it is written?

$$\begin{cases} 2x - 5y = -3 \\ -2y + 3x = 10 \end{cases}$$

6. Can the system

$$\begin{cases} 2x + 5y = -13 \\ -2x - 3y = -5 \end{cases}$$

be solved more easily using the addition method or the substitution method? Explain.

7. What algebraic step should be performed to clear this equation of fractions?

$$\frac{2}{3}x + 4y = -\frac{4}{5}$$

8. If the addition method is used to solve

$$\begin{cases} 3x + 12y = 4 \\ 6x - 4y = 8 \end{cases}$$

a. By what would we multiply the first equation to eliminate x?

b. By what would we multiply the second equation to eliminate y?

9. Solve $\begin{cases} 4x + 2y = 2 \\ 3x - 2y = 12 \end{cases}$

a. by the graphing method.
b. by the substitution method.
c. by the addition method.

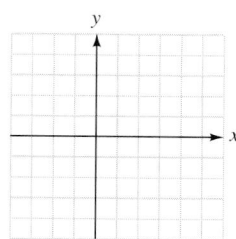

10. The addition method was used to solve three different systems. The results after x was eliminated in each case are listed here. Match each result with a possible graph of the system.

Result when solving

a. System 1:   **b.** System 2:   **c.** System 3:
$\quad -1 = -1$ $\quad y = -1$ $\quad -1 = -2$

Possible graph of the system

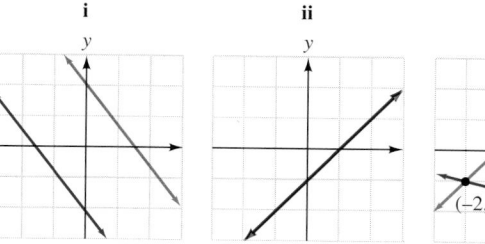

NOTATION   *Complete the solution to solve each system.*

11. Solve $\begin{cases} x + y = \ \ 5 \\ x - y = -3 \end{cases}$

$$x + y = \ \ 5$$
$$\underline{x - y = -3}$$
$$ = \ \ 2$$

$$x = \ \ \ $$

$$x + y = 5 \quad \text{The first equation.}$$
$$(\ \ ) + y = 5$$
$$y = 4$$

The solution is _____.

12. Solve $\begin{cases} x - 2y = 8 \\ -x + 5y = -17 \end{cases}$

$$x - 2y = \ \ \ \ 8$$
$$\underline{-x + 5y = -17}$$
$$ = \ \ -9$$

$$y = \ \ \ $$

$$x - 2y = 8 \quad \text{The first equation.}$$
$$x - 2(\ \ \ ) = 8$$
$$x + 6 = 8$$
$$x = 2$$

The solution is _____.

PRACTICE   *Use the addition method to solve each system.*

13. $\begin{cases} x - y = -5 \\ x + y = 1 \end{cases}$ **14.** $\begin{cases} x + y = 1 \\ x - y = 5 \end{cases}$

15. $\begin{cases} 2r + s = -1 \\ -2r + s = 3 \end{cases}$ **16.** $\begin{cases} 3m + n = -6 \\ m - n = -2 \end{cases}$

17. $\begin{cases} 2x + y = -2 \\ -2x - 3y = -6 \end{cases}$ **18.** $\begin{cases} 3x + 4y = 8 \\ 5x - 4y = 24 \end{cases}$

19. $\begin{cases} 4x + 3y = 24 \\ 4x - 3y = -24 \end{cases}$ **20.** $\begin{cases} 5x - 4y = 8 \\ -5x - 4y = 8 \end{cases}$

Use the addition method to solve each system of equations. If the equations of a system are dependent or if a system is inconsistent, so indicate.

21. $\begin{cases} x + y = 5 \\ x + 2y = 8 \end{cases}$ **22.** $\begin{cases} x + 2y = 0 \\ x - y = -3 \end{cases}$

23. $\begin{cases} 2x + y = 4 \\ 2x + 3y = 0 \end{cases}$ **24.** $\begin{cases} 2x + 5y = -13 \\ 2x - 3y = -5 \end{cases}$

25. $\begin{cases} 3x - 5y = -29 \\ 3x + 4y = 34 \end{cases}$ **26.** $\begin{cases} 3x - 5y = 16 \\ 4x + 5y = 33 \end{cases}$

27. $\begin{cases} 2a - 3b = -6 \\ 2a - 3b = 8 \end{cases}$ **28.** $\begin{cases} 3a - 4b = 6 \\ 2(2b + 3) = 3a \end{cases}$

29. $\begin{cases} 8x - 4y = 18 \\ 3x - 2y = 8 \end{cases}$ **30.** $\begin{cases} 4x + 6y = 5 \\ 8x - 9y = 3 \end{cases}$

31. $\begin{cases} 2x + y = 10 \\ 0.1x + 0.2y = 1.0 \end{cases}$ **32.** $\begin{cases} 0.3x + 0.2y = 0 \\ 2x - 3y = -13 \end{cases}$

33. $\begin{cases} 2x - y = 16 \\ 0.03x + 0.02y = 0.03 \end{cases}$

34. $\begin{cases} -5y + 2x = 4 \\ -0.02y + 0.03x = 0.04 \end{cases}$

35. $\begin{cases} 6x + 3y = 0 \\ 5y = 2x + 12 \end{cases}$

36. $\begin{cases} 0 = 4x - 3y \\ 5x = 4y - 2 \end{cases}$

37. $\begin{cases} -2(x + 1) = 3y - 6 \\ 3(y + 2) = 10 - 2x \end{cases}$

38. $\begin{cases} 3x + 2y + 1 = 5 \\ 3(x - 1) = -2y - 4 \end{cases}$

39. $\begin{cases} 4(x + 1) = 17 - 3(y - 1) \\ 2(x + 2) + 3(y - 1) = 9 \end{cases}$

40. $\begin{cases} 5(x - 1) = 8 - 3(y + 2) \\ 4(x + 2) - 7 = 3(2 - y) \end{cases}$

41. $\begin{cases} \frac{3}{5}s + \frac{4}{5}t = 1 \\ -\frac{1}{4}s + \frac{3}{8}t = 1 \end{cases}$ **42.** $\begin{cases} \frac{1}{2}s - \frac{1}{4}t = 1 \\ \frac{1}{3}s + t = 3 \end{cases}$

43. $\begin{cases} \frac{3}{5}x + y = 1 \\ \frac{4}{5}x - y = -1 \end{cases}$ **44.** $\begin{cases} \frac{1}{2}x + \frac{4}{7}y = -1 \\ 5x - \frac{4}{5}y = -10 \end{cases}$

45. $\begin{cases} \dfrac{x}{2} - \dfrac{y}{3} = -2 \\ \dfrac{2x - 3}{2} + \dfrac{6y + 1}{3} = \dfrac{17}{6} \end{cases}$ **46.** $\begin{cases} \dfrac{x + 2}{4} + \dfrac{y - 1}{3} = \dfrac{1}{12} \\ \dfrac{x + 4}{5} - \dfrac{y - 2}{2} = \dfrac{5}{2} \end{cases}$

47. $\begin{cases} \dfrac{x - 3}{2} + \dfrac{y + 5}{3} = \dfrac{11}{6} \\ \dfrac{x + 3}{3} - \dfrac{5}{12} = \dfrac{y + 3}{4} \end{cases}$ **48.** $\begin{cases} \dfrac{x + 2}{3} = \dfrac{3 - y}{2} \\ \dfrac{x + 3}{2} = \dfrac{2 - y}{3} \end{cases}$

WRITING

49. Why is it usually to your advantage to write the equations of a system in general form before using the addition method to solve it?

50. How would you decide whether to use substitution or addition to solve a system of equations?

51. In this section, we discussed the addition method for solving a system of two equations. Some instructors call it the *elimination method*. Why do you think it would be known by this name?

52. Explain the error in the work shown below.

$$\text{Solve } \begin{cases} x + y = 1 \\ x - y = 5 \end{cases}$$

$$\begin{array}{r} x + y = 1 \\ + \quad x - y = 5 \\ \hline 2x \qquad = 6 \end{array}$$

$$\frac{2x}{2} = \frac{6}{2}$$

$$\boxed{x = 3} \quad \text{—Done—}$$

REVIEW

53. Solve $8(3x - 5) - 12 = 4(2x + 3)$.

54. Solve $3y + \dfrac{y + 2}{2} = \dfrac{2(y + 3)}{3} + 16$.

55. Simplify $x - x$.

56. Simplify $3.2m - 4.4 + 2.1m + 16$.

57. Find the area of a triangular-shaped sign with a base of 4 feet and a height of 3.75 feet.

58. Translate to mathematical symbols: *the product of the sum of x and y and the difference of x and y.*

59. What is 10 less than x?

60. Factor $6x^2 + 7x - 20$.

13.4 *Applications of Systems of Equations*

In this section, you will learn about

* Solving problems using two variables

INTRODUCTION. We have previously formed equations involving one variable to solve problems. In this section, we consider ways to solve problems using two variables.

Solving problems using two variables

The following steps are helpful when solving problems involving two unknown quantities.

Problem-solving strategy

1. Read the problem several times and *analyze* the facts. Occasionally, a sketch, table, or diagram will help you visualize the facts of the problem.
2. Pick different variables to represent two unknown quantities. *Form* two equations involving each of the two variables. This will give a system of two equations in two variables.
3. *Solve* the system of equations using the most convenient method: graphing, substitution, or addition.
4. *State* the conclusion.
5. *Check* the result in the words of the problem.

EXAMPLE 1 *Farming.* A farmer raises wheat and soybeans on 215 acres. He wants to plant fewer acres in soybeans than in wheat—31 acres less, to be exact. How many acres of each should he plant?

Analyze the problem We know that the number of acres of wheat planted plus the number of acres of soybeans planted will equal a total of 215 acres. We also know that there are 31 fewer acres of soybeans than of wheat.

Form two equations If w represents the number of acres of wheat and s the number of acres of soybeans to be planted, we can form the two equations.

The number of acres planted in wheat	plus	the number of acres planted in soybeans	is	215 acres.
w	$+$	s	$=$	215

Since the farmer wants to plant 31 fewer acres in soybeans than in wheat, we have

The number of acres planted in wheat	less	the number of acres planted in soybeans	is	31 acres.
w	$-$	s	$=$	31

Solve the system We can now solve the system

$$(1) \quad \begin{cases} w + s = 215 \\ w - s = 31 \end{cases}$$
$$(2)$$

using the addition method.

$$w + s = 215$$
$$\underline{w - s = 31}$$
$$2w = 246$$
$$w = 123 \quad \text{Divide both sides by 2.}$$

To find *s*, we substitute 123 for *w* in Equation 1.

$$w + s = 215$$
$$123 + s = 215 \quad \text{Substitute.}$$
$$s = 92 \quad \text{Subtract 123 from both sides.}$$

State the conclusion   The farmer should plant 123 acres of wheat and 92 acres of soybeans.

Check the result   The total acreage planted is $123 + 92$, or 215 acres. The area planted in soybeans is 31 fewer acres than that planted in wheat, because $123 - 92 = 31$. The answers check.

∎

EXAMPLE 2   *Lawn care.*   An installer of underground irrigation systems wants to cut a 20-foot length of plastic tubing into two pieces. The longer piece is to be 2 feet longer than twice the shorter piece. Find the length of each piece.

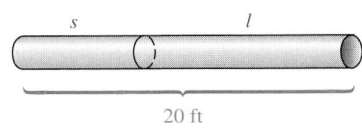

20 ft

FIGURE 13-10

Analyze the problem   Refer to Figure 13-10, which shows the pipe.

Form two equations   We can let *s* represent the length of the shorter piece and *l* the length of the longer piece. Then we can form the two equations.

The length of the shorter piece	plus	the length of the longer piece	is	20 feet.
s	+	*l*	=	20

Since the longer piece is 2 feet longer than twice the shorter piece, we have

The length of the longer piece	is	2	times	the length of the shorter piece	plus	2 feet.
l	=	2	·	*s*	+	2

Solve the system   We can use the substitution method to solve the system.

(1) $\begin{cases} s + l = 20 \\ l = 2s + 2 \end{cases}$
(2)

$$s + 2s + 2 = 20 \quad \text{Substitute } 2s + 2 \text{ for } l \text{ in Equation 1.}$$
$$3s + 2 = 20 \quad \text{Combine like terms.}$$
$$3s = 18 \quad \text{Subtract 2 from both sides.}$$
$$s = 6 \quad \text{Divide both sides by 3.}$$

The shorter piece should be 6 feet long. To find the length of the longer piece, we substitute 6 for *s* in Equation 2 and find *l*.

$$l = 2s + 2$$
$$= 2(6) + 2 \quad \text{Substitute.}$$
$$= 12 + 2 \quad \text{Simplify.}$$
$$l = 14$$

State the conclusion   The longer piece should be 14 feet long, and the shorter piece 6 feet long.

Check the result   The sum of 6 and 14 is 20, and 14 is 2 more than twice 6. The answers check.   ∎

EXAMPLE 3   *Gardening.*   Tom has 150 feet of fencing to enclose a rectangular garden. If the garden's length is to be 5 feet less than 3 times its width, find the area of the garden.

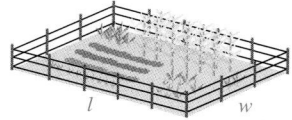

FIGURE 13-11

Analyze the problem   To find the area of a rectangle, we need to know its length and width.

Form two equations   We can let *l* represent the length of the garden and *w* its width, as shown in Figure 13-11. Since the perimeter of a rectangle is two lengths plus two widths, we have

2	times	the length of the garden	plus	2	times	the width of the garden	is	150 feet.
2	$\cdot$	*l*	+	2	$\cdot$	*w*	=	150

Since the length is 5 feet less than 3 times the width,

The length of the garden	is	3	times	the width of the garden	minus	5 feet.
l	=	3	$\cdot$	*w*	−	5

Solve the system   We can use the substitution method to solve this system.

(1) $\quad \begin{cases} 2l + 2w = 150 \\ l = 3w - 5 \end{cases}$
(2)

$$2(3w - 5) + 2w = 150 \quad \text{Substitute } 3w - 5 \text{ for } l \text{ in Equation 1.}$$
$$6w - 10 + 2w = 150 \quad \text{Distribute the multiplication by 2.}$$
$$8w - 10 = 150 \quad \text{Combine like terms.}$$
$$8w = 160 \quad \text{Add 10 to both sides.}$$
$$w = 20 \quad \text{Divide both sides by 8.}$$

The width of the garden is 20 feet. To find the length, we substitute 20 for *w* in Equation 2 and simplify.

$$l = 3w - 5$$
$$= 3(20) - 5 \quad \text{Substitute.}$$
$$= 60 - 5$$
$$l = 55 \qquad \text{The length of the garden is 55 feet.}$$

Now we find the area of the rectangle with dimensions 55 feet by 20 feet.

$$A = lw \qquad \text{The formula for the area of a rectangle.}$$
$$= 55 \cdot 20 \quad \text{Substitute 55 for } l \text{ and 20 for } w.$$
$$A = 1,100$$

State the conclusion   The garden covers an area of 1,100 square feet.

Check the result   Because the dimensions of the garden are 55 feet by 20 feet, the perimeter is

$$P = 2l + 2w$$
$$= 2(55) + 2(20) \quad \text{Substitute for } l \text{ and } w.$$
$$= 110 + 40$$
$$P = 150$$

It is also true that 55 feet is 5 feet less than 3 times 20 feet. The answers check.   ■

EXAMPLE 4   *Manufacturing.*   The setup cost of a machine that mills brass plates is $750. After setup, it costs $0.25 to mill each plate. Management is considering the purchase of a larger machine that can produce the same plate at a cost of $0.20 per plate. If the setup cost of the larger machine is $1,200, how many plates would the company have to produce to make the purchase worthwhile?

Analyze the problem   We need to find the number of plates (called the **break point**) that will cost equal amounts to produce on either machine.

Form two equations   We can let c represent the cost of milling p plates. If we call the machine currently being used machine 1, and the new, larger one machine 2, we can form the two equations.

The cost of making p plates on machine 1	is	the setup cost of machine 1	plus	the cost per plate on machine 1	times	the number of plates p to be made.
c	$=$	750	$+$	0.25	$\cdot$	p

The cost of making p plates on machine 2	is	the setup cost of machine 2	plus	the cost per plate on machine 2	times	the number of plates p to be made.
c	$=$	1,200	$+$	0.20	$\cdot$	p

Solve the system   Since the costs are equal, we can use the substitution method to solve the system

$$\textbf{(1)} \quad \begin{cases} c = \textbf{750} + \textbf{0.25}p \\ c = \textbf{1,200} + \textbf{0.20}p \end{cases}$$
$$\textbf{(2)}$$

$$\begin{aligned}
\textbf{750} + \textbf{0.25}p &= 1{,}200 + 0.20p && \text{Substitute } 750 + 0.25p \text{ for } c \text{ in the second equation.} \\
0.25p &= 450 + 0.20p && \text{Subtract 750 from both sides.} \\
0.05p &= 450 && \text{Subtract } 0.20p \text{ from both sides.} \\
p &= 9{,}000 && \text{Divide both sides by 0.05.}
\end{aligned}$$

State the conclusion   If 9,000 plates are milled, the cost will be the same on either machine. If more than 9,000 plates are milled, the cost will be cheaper on the larger machine, because it mills the plates less expensively than the smaller machine.

Check the result   We check the solution by substituting 9,000 for p in Equations 1 and 2 and verifying that 3,000 is the value of c in both cases.

If we graph the two equations, we can illustrate the break point. (See Figure 13-12.)

Machine 1
$c = 750 + 0.25p$

p	c
0	750
1,000	1,000
5,000	2,000

Machine 2
$c = 1,200 + 0.20p$

p	c
0	1,200
4,000	2,000
12,000	3,600

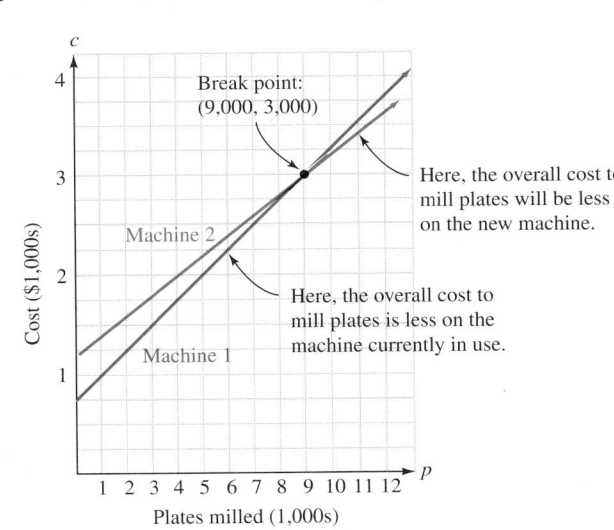

FIGURE 13-12

EXAMPLE 5 *White-collar crime.* Federal investigators discovered that a company secretly moved $150,000 out of the country to avoid paying corporate income tax on it. Some of the money was invested in a Swiss bank account that paid 8% interest annually. The remainder was deposited in a Cayman Islands account, paying 7% annual interest. The investigation also revealed that the combined interest earned the first year was $11,500. How much money was invested in each account?

Analyze the problem We are told that an unknown part of the $150,000 was invested at an annual rate of 8% and the rest at 7%. Together, the accounts earned $11,500 in interest.

Form two equations We can let x represent the amount invested in the Swiss bank account and y represent the amount invested in the Cayman Islands account. Because the total investment was $150,000, we have

The amount invested in the Swiss account	+	the amount invested in the Cayman Is. account	is	$150,000.
x	+	y	=	150,000

Since the annual income on x dollars invested at 8% is $0.08x$, the income on y dollars invested at 7% is $0.07y$, and the combined income is $11,500, we have

The income on the 8% investment	+	the income on the 7% investment	is	$11,500.
$0.08x$	+	$0.07y$	=	11,500

The resulting system is

(1) $\begin{cases} x + y = 150,000 \\ 0.08x + 0.07y = 11,500 \end{cases}$
(2)

Solve the system To solve the system, we use the addition method to eliminate x.

$$
\begin{aligned}
-8x - 8y &= -1,200,000 \quad &&\text{Multiply both sides of Equation 1 by } -8. \\
\underline{8x + 7y} &= \underline{1,150,000} \quad &&\text{Multiply both sides of Equation 2 by 100.} \\
-y &= -50,000 \\
y &= 50,000 \quad &&\text{Multiply (or divide) both sides by } -1.
\end{aligned}
$$

To find x, we substitute 50,000 for y in Equation 1 and simplify.

$$
\begin{aligned}
x + y &= 150,000 \\
x + \mathbf{50,000} &= 150,000 \quad &&\text{Substitute.} \\
x &= 100,000 \quad &&\text{Subtract 50,000 from both sides.}
\end{aligned}
$$

State the conclusion $100,000 was invested in the Swiss bank account, and $50,000 was invested in the Cayman Islands account.

Check the result

$$
\begin{aligned}
\$100,000 + \$50,000 &= \$150,000 \quad &&\text{The two investments total } \$150,000. \\
0.08(\$100,000) &= \$8,000 \quad &&\text{The Swiss bank account earned } \$8,000. \\
0.07(\$50,000) &= \$3,500 \quad &&\text{The Cayman Islands account earned } \$3,500.
\end{aligned}
$$

The combined interest is $8,000 + $3,500 = $11,500. The answers check. ■

EXAMPLE 6 *Boating.* A boat traveled 30 kilometers downstream in 3 hours and made the return trip in 5 hours. Find the speed of the boat in still water.

Analyze the problem Traveling downstream, the speed of the boat will be faster than it would be in still water. Traveling upstream, the speed of the boat will be less than it would be in still water.

Form two equations We can let s represent the speed of the boat in still water and c the speed of the current. Then the rate of the boat going downstream is $s + c$, and its rate going upstream is $s - c$. We can organize the information as shown in Figure 13-13.

	Rate ·	Time =	Distance
Downstream	$s + c$	3	$3(s + c)$
Upstream	$s - c$	5	$5(s - c)$

FIGURE 13-13

Since each trip is 30 miles long, the Distance column of the table gives two equations in two variables.

$$\begin{cases} 3(s + c) = 30 \\ 5(s - c) = 30 \end{cases}$$

After using the distributive property, we have

(1)
(2)
$$\begin{cases} 3s + 3c = 30 \\ 5s - 5c = 30 \end{cases}$$

Solve the system To solve this system by addition, we multiply Equation 1 by 5, multiply Equation 2 by 3, add the equations, and solve for s.

$$\begin{aligned} 15s + 15c &= 150 \\ \underline{15s - 15c} &= \underline{90} \\ 30s &= 240 \end{aligned}$$

$$s = 8 \qquad \text{Divide both sides by 30.}$$

State the conclusion The speed of the boat in still water is 8 kilometers per hour.

Check the result We leave the check to the reader. ■

EXAMPLE 7 *Medical technology.* A laboratory technician has one batch of antiseptic that is 40% alcohol and a second batch that is 60% alcohol. She would like to make 8 liters of solution that is 55% alcohol. How many liters of each batch should she use?

Analyze the problem Some 60% solution must be added to some 40% solution to make a 55% solution.

Form two equations We can let x represent the number of liters to be used from batch 1 and y the number of liters to be used from batch 2. We then organize the information as shown in Figure 13-14.

	% of concentration ·	Number of liters of solution =	Number of liters of alcohol
Batch 1	0.40	x	$0.40x$
Batch 2	0.60	y	$0.60y$
Mixture	0.55	8	$0.55(8)$

↑ 40%, 60%, and 55% have been expressed as decimals. ↑ One equation comes from information in this column. ↑ Another equation comes from information in this column.

FIGURE 13-14

The information in Figure 13-14 provides two equations.

(1)
(2)
$$\begin{cases} x + y = 8 \\ 0.40x + 0.60y = 0.55(8) \end{cases}$$

The number of liters of batch 1 plus the number of liters of batch 2 equals the total number of liters in the mixture.

The amount of alcohol in batch 1 plus the amount of alcohol in batch 2 equals the amount of alcohol in the mixture.

Solve the system We can use addition to solve this system.

$$-40x - 40y = -320 \quad \text{Multiply both sides of Equation 1 by } -40.$$
$$\underline{40x + 60y = 440} \quad \text{Multiply both sides of Equation 2 by } 100.$$
$$20y = 120$$
$$y = 6 \qquad \text{Divide both sides by 20.}$$

To find *x*, we substitute 6 for *y* in Equation 1 and simplify.

$$x + y = 8$$
$$x + \mathbf{6} = 8 \quad \text{Substitute.}$$
$$x = 2 \quad \text{Subtract 6 from both sides.}$$

State the conclusion The technician should use 2 liters of the 40% solution and 6 liters of the 60% solution.

Check the result The check is left to the reader. ■

STUDY SET Section 13.4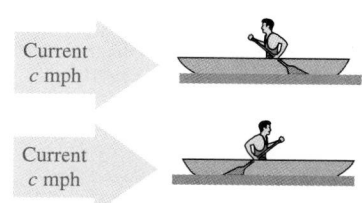

VOCABULARY *Fill in the blanks.*

1. A _____ is a letter that stands for a number.

2. An _____ is a statement indicating that two quantities are equal.

3. $\begin{cases} a + b = 20 \\ a = 2b + 4 \end{cases}$ is a _____ of linear equations.

4. A _____ of a system of linear equations satisfies both equations simultaneously.

CONCEPTS

5. For each case in Illustration 1, write an algebraic expression that represents the speed of the canoe in miles per hour if its speed in still water is *x* miles per hour.

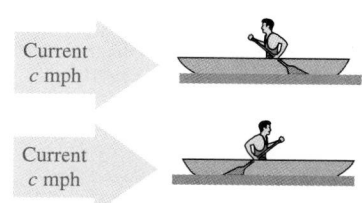

Current
c mph

Current
c mph

ILLUSTRATION 1

6. See Illustration 2.
  a. If the contents of the two test tubes are poured into a third test tube, how much solution will the third test tube contain? (mL means milliliters.)

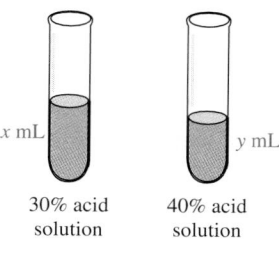

x mL

y mL

30% acid
solution

40% acid
solution

ILLUSTRATION 2

b. Which is the best estimate of the concentration of the solution in the third test tube—25%, 35%, or 45% acid solution?

7. Use the information in the table to answer the questions about two investments.

	Principal ·	Rate ·	Time =	Interest
City Bank	x	5%	1 yr	
USA Savings	y	11%	1 yr	

  a. How much money was deposited in the USA Savings account?
  b. What interest rate did the City Bank account earn?

  c. Complete the table.

8. Use the information in the table to answer the questions about a plane flying in windy conditions.

	Rate ·	Time =	Distance
With	x + y	3 hr	450 mi
Against	x − y	5 hr	450 mi

  a. For how long did the plane fly against the wind?

  b. At what rate did the plane travel when flying with the wind?

  c. Write two equations that could be used to solve for *x* and *y*.

9. a. If a problem contains two unknowns, and if two variables are used to represent them, how many equations must be written to find the unknowns?

b. Name three methods that can be used to solve a system of linear equations.

10. Put the steps of the five-step problem-solving strategy listed below in the correct order.

State the conclusion   Form two equations
Analyze the problem   Check the result
Solve the system

NOTATION *Write a formula that relates the given quantities.*

11. length, width, area of a rectangle

12. length, width, perimeter of a rectangle

13. rate, time, distance traveled

14. principal, rate, time, interest earned

Translate each verbal model into mathematical symbols. Use variables to represent any unknowns.

15. $2 \cdot \dfrac{\text{length}}{\text{of pool}} + 2 \cdot \dfrac{\text{width}}{\text{of pool}}$ is $\dfrac{90}{\text{yards.}}$

16. $\$6 \cdot \dfrac{\text{number}}{\text{of adults}} + \$2 \cdot \dfrac{\text{number}}{\text{of children}}$ is $\$26.$

PRACTICE *Use two equations in two variables to find the integers.*

17. One integer is twice another. Their sum is 96.

18. The sum of two integers is 38. Their difference is 12.

19. Three times one integer plus another integer is 29. The first integer plus twice the second is 18.

20. Twice one integer plus another integer is 21. The first integer plus 3 times the second is 33.

APPLICATIONS *Use two equations in two variables to solve each problem.*

21. TREE TRIMMING When fully extended, the arm on the tree service truck shown in Illustration 3 is 51 feet long. If the upper part of the arm is 7 feet shorter than the lower part, how long is each part of the arm?

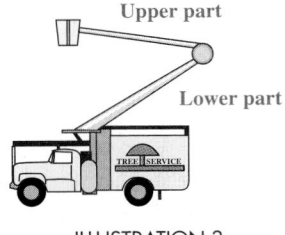

ILLUSTRATION 3

22. TV PROGRAMMING The producer of a 30-minute TV documentary about World War I divided it into two parts. Four times as much program time was devoted to the causes of the war as to the outcome. How long is each part of the documentary?

23. EXECUTIVE BRANCH The salaries of the president and vice president of the United States total $581,000 a year. If the president makes $219,000 more than the vice president, find each of their salaries.

24. CAUSES OF DEATH In 1993, the number of Americans dying from cancer was 6 times the number who died from accidents. If the number of deaths from these two causes totaled 630,000, how many Americans died from each cause?

25. BUYING PAINTING SUPPLIES Two partial receipts for paint supplies are shown in Illustration 4. How much does each gallon of paint and each brush cost?

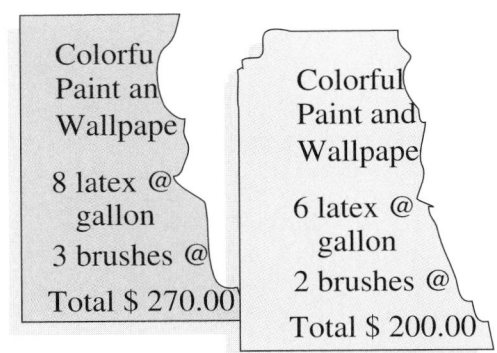

ILLUSTRATION 4

26. WEDDING PICTURES A photographer sells the two wedding picture packages shown in Illustration 5.

Package #1	Package #2
One 10 x 14	One 10 x 14
Ten 8 x 10	Five 8 x 10
color photos	color photos
Cost $239.50	Cost $134.50

ILLUSTRATION 5

How much does a 10×14 photo cost? An 8×10 photo?

27. BUYING TICKETS If receipts for the movie advertised in Illustration 6 were $1,440 for an audience of 190 people, how many senior citizens attended?

28. SELLING ICE CREAM At a store, ice cream cones cost $0.90 and sundaes cost $1.65. One day the receipts for a total of 148 cones and sundaes were $180.45. How many cones were sold?

ILLUSTRATION 6

29. MARINE CORPS   The Marine Corps War Memorial in Arlington, Virginia, portrays the raising of the U.S. flag on Iwo Jima during World War II. Find the two angles shown in Illustration 7 if the measure of one of the angles is 15° less than twice the other.

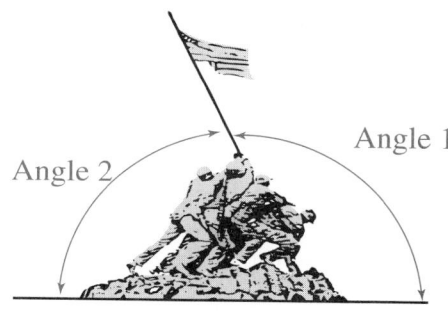

ILLUSTRATION 7

30. PHYSICAL THERAPY   To rehabilitate her knee, an athlete does leg extensions. Her goal is to regain a full 90° range of motion in this exercise. Use the information in Illustration 8 to determine her current range of motion in degrees.

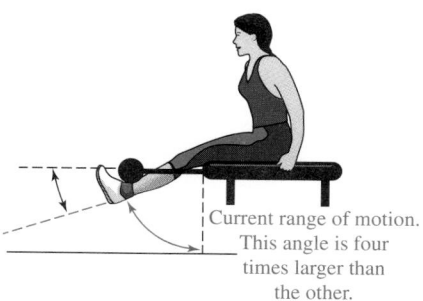

ILLUSTRATION 8

31. THEATER SCREEN   At an IMAX theater, the giant rectangular movie screen has a width 26 feet less than its length. If its perimeter is 332 feet, find the area of the screen.

32. GEOMETRY   A 50-meter path surrounds the rectangular garden shown in Illustration 9. The width of the garden is two-thirds its length. Find its area.

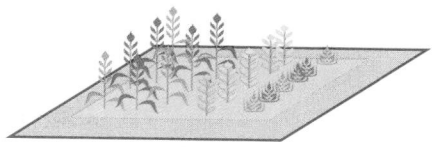

ILLUSTRATION 9

33. MAKING TIRES   A company has two molds to form tires. One mold has a setup cost of $1,000, and the other has a setup cost of $3,000. The cost to make each tire with the first mold is $15, and the cost to make each tire with the second mold is $10.
a. Find the break point.
b. Check your result by graphing both equations on the coordinate system in Illustration 10.

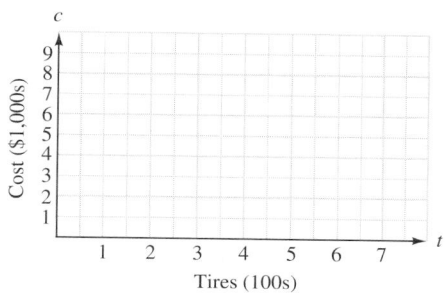

ILLUSTRATION 10

c. If a production run of 500 tires is planned, determine which mold should be used.

34. CHOOSING A FURNACE   A high-efficiency 90+ furnace can be purchased for $2,250 and costs an average of $412 per year to operate in Rockford, Illinois. An 80+ furnace can be purchased for only $1,715, but it costs $446 per year to operate.
a. Find the break point.
b. If you intended to live in a Rockford house for 7 years, which furnace would you choose?

35. STUDENT LOANS   A college used a $5,000 gift from an alumnus to make two student loans. The first was at 5% annual interest to a nursing student. The second was at 7% to a business major. If the college collected $310 in interest the first year, how much was loaned to each student?

36. FINANCIAL PLANNING   In investing $6,000 of a couple's money, a financial planner put some of it into a savings account paying 6% annual interest. The rest was invested in a riskier mini-mall development plan paying 12% annually. The combined interest earned for the first year was $540. How much money was invested at each rate?

37. GULF STREAM   The Gulf Stream is a warm ocean current of the North Atlantic Ocean that flows northward, as shown in Illustration 11. Heading north with the Gulf Stream, a cruise ship traveled 300 miles in 10 hours. Against the current, it took 15 hours to make the return trip. Find the speed of the current.

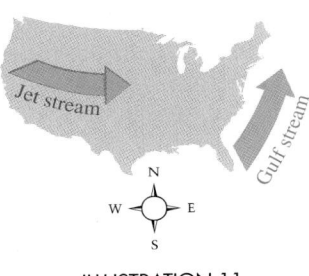

ILLUSTRATION 11

38. JET STREAM   The jet stream is a strong wind current that flows across the United States, as shown in Illustration 11. Flying with the jet stream, a plane flew 3,000 miles in 5 hours. Against the same wind, the trip took 6 hours. Find the airspeed of the plane (the speed in still air).

39. AVIATION   An airplane can fly downwind a distance of 600 miles in 2 hours. However, the return trip against the same wind takes 3 hours. Find the speed of the wind.

40. BOATING   A boat can travel 24 miles downstream in 2 hours and can make the return trip in 3 hours. Find the speed of the boat in still water.

41. MARINE BIOLOGY   A marine biologist wants to set up an aquarium containing 3% salt water. He has two tanks on hand that contain 6% and 2% salt water. How much water from each tank must he use to fill a 16-liter aquarium with a 3% saltwater mixture?

42. COMMEMORATIVE COINS   A foundry has been commissioned to make souvenir coins. The coins are to be made from an alloy that is 40% silver. The foundry has on hand two alloys, one with 50% silver content and one with a 25% silver content. How many kilograms of each alloy should be used to make 20 kilograms of the 40% silver alloy?

43. MIXING NUTS   A merchant wants to mix peanuts with cashews, as shown in Illustration 12, to get 48 pounds of mixed nuts that will be sold at $4 per pound. How many pounds of each should the merchant use?

ILLUSTRATION 12

44. COFFEE SALES   A coffee supply store waits until the orders for its special coffee blend reach 100 pounds before making up a batch. Coffee selling for $8.75 a pound is blended with coffee selling for $3.75 a pound to make a product that sells for $6.35 a pound. How much of each type of coffee should be used to make the blend that will fill the orders?

45. MARKDOWN   A set of golf clubs has been marked down 40% to a sale price of $384. Let r represent the retail price and d the discount. Then use the following equations to find the original retail price.

Retail price	−	discount	=	sale price

Discount	=	discount rate	·	retail price

46. MARKUP   A stereo system retailing at $565.50 has been marked up 45% from wholesale. Let w represent the wholesale cost and m the markup. Then use the following equations to find the wholesale cost.

Wholesale cost	+	markup	=	retail price

Markup	=	markup rate	·	wholesale cost

WRITING

47. When solving a problem using two variables, why isn't one equation sufficient to find the two unknown quantities?

48. Describe an everyday situation in which you might need to make a mixture.

REVIEW   *Graph each inequality.*

49. $x < 4$

50. $x \geq -3$

51. $-1 < x \leq 2$

52. $-2 \leq x \leq 0$

Solve each equation.

53. $x^2 - 4 = 0$

54. $x^2 - 4x = 0$

55. $x^2 - 4x + 4 = 0$

56. $2x^2 + 3x = 2$

13.5 *Graphing Linear Inequalities*

In this section, you will learn about

- Solving linear inequalities • Graphing linear inequalities
- An application of linear inequalities

INTRODUCTION. We have seen that the solutions of a linear *equation* in x and y can be expressed as ordered pairs (x, y) and that when graphed, the ordered pairs form a line. In this section, we consider linear *inequalities*. Solutions of linear inequalities can also be expressed as ordered pairs and graphed.

Solving linear inequalities

A linear equation in x and y is an equation that can be written in the form $Ax + By = C$. A **linear inequality** in x and y is an inequality that can be written in one of four forms:

$$Ax + By > C, \qquad Ax + By < C, \qquad Ax + By \geq C, \qquad \text{or} \qquad Ax + By \leq C$$

where A, B, and C represent real numbers and A and B are not both zero. Some examples of linear inequalities are

$$2x - y > -3, \qquad y < 3, \qquad x + 4y \geq 6, \qquad \text{and} \qquad x \leq -2$$

As with linear equations, an ordered pair (x, y) is a solution of an inequality in x and y if a true statement results when the variables in the inequality are replaced by the coordinates of the ordered pair.

EXAMPLE 1 *Verifying a solution.* Determine whether each ordered pair is a solution of $x - y \leq 5$. Then graph each solution: **a.** $(4, 2)$, **b.** $(0, -6)$, and **c.** $(1, -4)$.

Solution

In each case, we substitute the x-coordinate for x and the y-coordinate for y in the inequality $x - y \leq 5$. If the ordered pair is a solution, a true statement will be obtained.

a. For $(4, 2)$:

$\qquad x - y \leq 5$ The original inequality.

$\qquad 4 - 2 \leq 5$ Replace x with 4 and y with 2.

$\qquad\quad 2 \leq 5$ True.

Because $2 \leq 5$ is true, $(4, 2)$ is a solution of the inequality, and we graph it in Figure 13-15.

b. For $(0, -6)$:

$\qquad x - y \leq 5$ The original inequality.

$\qquad 0 - (-6) \leq 5$ Replace x with 0 and y with -6.

$\qquad\quad 6 \leq 5$ False.

Because $6 \leq 5$ is false, $(0, -6)$ is not a solution.

c. For $(1, -4)$:

$\qquad x - y \leq 5$ The original inequality.

$\qquad 1 - (-4) \leq 5$ Replace x with 1 and y with -4.

$\qquad\quad 5 \leq 5$ True.

Because $5 \leq 5$ is true, $(1, -4)$ is a solution, and we graph it in Figure 13-15.

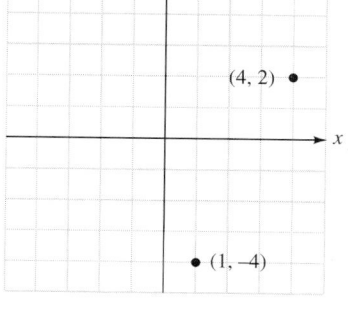

FIGURE 13-15

Self Check

Using the inequality in Example 1, determine whether each ordered pair is a solution. If it is, graph that solution on the coordinate system in Figure 13-15. **a.** $(8, 2)$, **b.** $(4, -1)$, **c.** $(-2, 4)$, and **d.** $(-3, -5)$.

Answers: **a.** not a solution, **b.** solution, **c.** solution, **d.** solution

The graph in Figure 13-15 contains some solutions of the inequality $x - y \leq 5$. Intuition tells us that there are many more ordered pairs (x, y) such that $x - y$ is less than or equal to 5. How then do we get a complete graph of the solutions of $x - y \leq 5$? We address this question in the following discussion.

Graphing linear inequalities

The graph of $x - y = 5$ is a line consisting of the points whose coordinates satisfy the equation. The graph of the inequality $x - y \leq 5$ is not a line, but an area bounded by a line, called a **half-plane.** The half-plane consists of the points whose coordinates satisfy the inequality.

 **EXAMPLE 2**   *Graphing a linear inequality.*   Graph $x - y \leq 5$.

Solution

Since the inequality symbol $\leq$ includes an equals sign, the graph of $x - y \leq 5$ includes the graph of $x - y = 5$. So we begin by graphing the equation $x - y = 5$, using the intercept method. See Figure 13-16(a).

$$x - y = 5$$

x	y	(x, y)
0	-5	$(0, -5)$
5	0	$(5, 0)$
6	1	$(6, 1)$

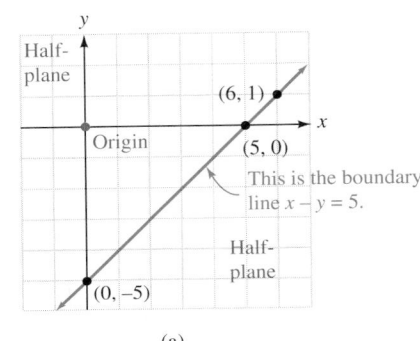

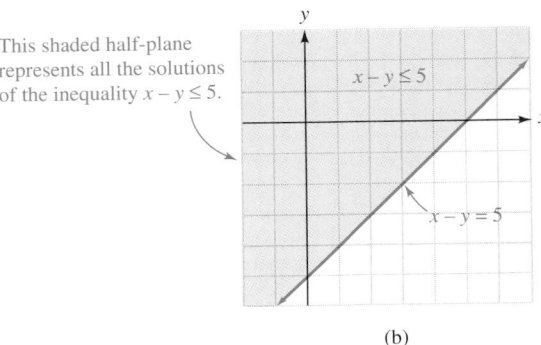

(a)   (b)

FIGURE 13-16

Since the inequality $x - y \leq 5$ allows $x - y$ to be less than 5, the coordinates of points other than those shown on the line in Figure 13-16(a) satisfy the inequality. For example, the coordinates of the origin $(0, 0)$ satisfy the inequality. We can verify this by letting x and y be zero in the given inequality:

$$x - y \leq 5$$
$$\mathbf{0} - \mathbf{0} \leq 5 \quad \text{Substitute 0 for } x \text{ and 0 for } y.$$
$$0 \leq 5$$

Because $0 \leq 5$, the coordinates of the origin satisfy the original inequality. In fact, the coordinates of every point on the *same side* of the line as the origin satisfy the inequality. The graph of $x - y \leq 5$ is the half-plane that is shaded in Figure 13-16(b). Since the **boundary line** $x - y = 5$ is included, we draw it with a solid line. ∎

EXAMPLE 3   *Graphing a linear inequality.*   Graph
$2(x - 3) - (x - y) \geq -3$.

Solution

We begin by simplifying the inequality as follows:

$$2(x - 3) - (x - y) \geq -3$$
$$2x - 6 - x + y \geq -3 \quad \text{Use the distributive property.}$$
$$x - 6 + y \geq -3 \quad \text{Combine like terms.}$$
$$x + y \geq 3 \quad \text{Add 6 to both sides.}$$

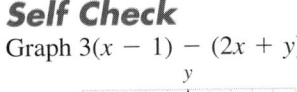
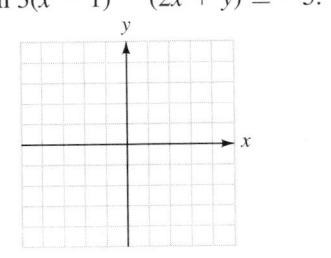

To graph the inequality $x + y \geq 3$, we graph the boundary line whose equation is $x + y = 3$. Since the graph of $x + y \geq 3$ includes the line $x + y = 3$, we draw the boundary with a solid line. Note that it divides the coordinate plane into two half-planes. See Figure 13-17(a).

To decide which half-plane to shade, we substitute the coordinates of some point that lies on one side of the boundary line into the inequality. If we use the origin $(0, 0)$ for the **test point,** we have

$$x + y \geq 3$$
$$0 + 0 \geq 3 \quad \text{Substitute 0 for } x \text{ and 0 for } y.$$
$$0 \geq 3 \quad \text{This statement is false.}$$

Since $0 \geq 3$ is a false statement, the origin is not in the graph. In fact, the coordinates of *every* point on the origin's side of the boundary line will not satisfy the inequality. However, every point on the other side of the boundary line will satisfy the inequality. We shade that half-plane. The graph of $x + y \geq 3$ is the half-plane that appears in color in Figure 13-17(b).

$x + y = 3$

x	y	(x, y)
0	3	(0, 3)
3	0	(3, 0)
1	2	(1, 2)

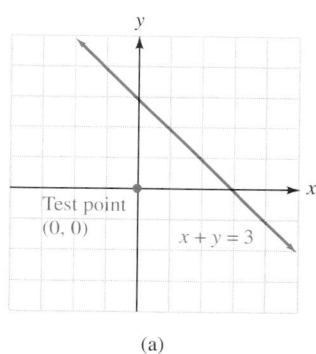

(a)

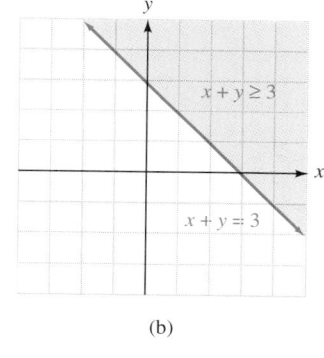

(b)

FIGURE 13-17

Answer:

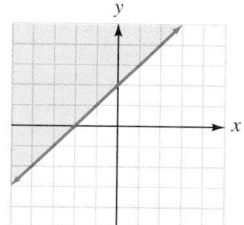

EXAMPLE 4   *Graphing a linear inequality.*   Graph $y > 2x$.

Solution

To find the boundary line, we graph $y = 2x$. Since the symbol $>$ does not include an equals sign, the points on the graph of $y = 2x$ are not part of the graph of $y > 2x$. We draw the boundary line as a broken line to show this, as in Figure 13-18(a).

To determine which half-plane to shade, we substitute the coordinates of some point that lies on one side of the boundary line into $y > 2x$. Since the origin is on the boundary, we cannot use it as a test point. Point $T(2, 0)$, for example, is below the boundary line. See Figure 13-18(a). To see whether point $T(2, 0)$ satisfies $y > 2x$, we substitute 2 for x and 0 for y in the inequality.

$$y > 2x$$
$$0 > 2(2) \quad \text{Substitute 2 for } x \text{ and 0 for } y.$$
$$0 > 4 \quad \text{This statement is false.}$$

Since $0 > 4$ is a false statement, the coordinates of point T do not satisfy the inequality, and point T is not on the side of the broken line we wish to shade. Instead, we shade the other side of the boundary line. The graph of the solution set of $y > 2x$ is shown in Figure 13-18(b).

Self Check

Graph $y < 3x$.

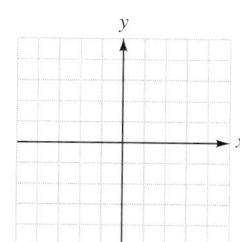

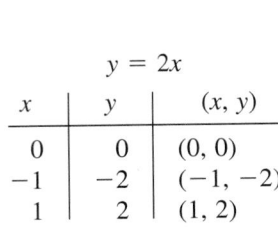

$y = 2x$

x	y	(x, y)
0	0	$(0, 0)$
-1	-2	$(-1, -2)$
1	2	$(1, 2)$

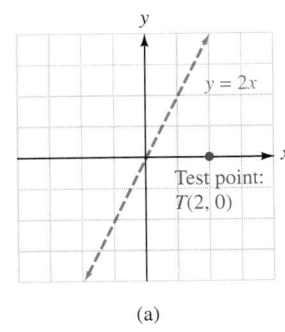

(a)

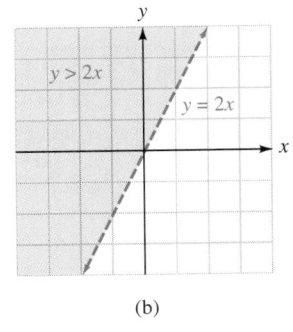

(b)

FIGURE 13-18

Answer:

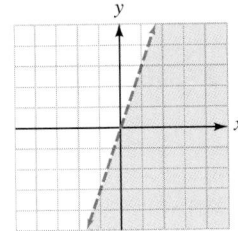

EXAMPLE 5   *Graphing a linear inequality.* Graph $x + 2y < 6$.

Solution

We find the boundary by graphing the equation $x + 2y = 6$. We draw the boundary as a broken line to show that it is not part of the solution. We then choose a test point not on the boundary and see whether its coordinates satisfy $x + 2y < 6$. The origin is a convenient choice.

$$x + 2y < 6$$
$$0 + 2(0) < 6 \quad \text{Substitute 0 for } x \text{ and 0 for } y.$$
$$0 < 6$$

Since $0 < 6$ is a true statement, we shade the side of the line that includes the origin. The graph is shown in Figure 13-19.

$x + 2y = 6$

x	y	(x, y)
0	3	$(0, 3)$
6	0	$(6, 0)$
4	1	$(4, 1)$

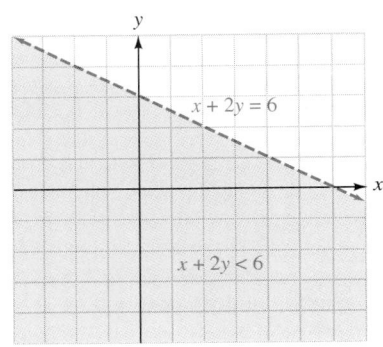

FIGURE 13-19

Self Check
Graph $2x - y < 4$.

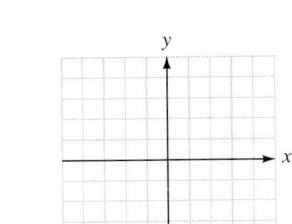

Answer:

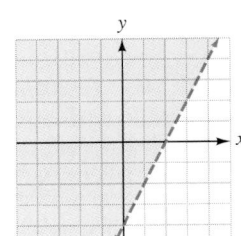

EXAMPLE 6   *Graphing a linear inequality.* Graph $y \geq 0$.

Solution

We find the boundary by graphing the equation $y = 0$. We draw the boundary as a solid line to show that it is part of the solution. We then choose a test point not on the boundary and see whether its coordinates satisfy $y \geq 0$. The point $T(0, 1)$ is a convenient choice.

$$y \geq 0$$
$$1 \geq 0 \quad \text{Substitute 1 for } y.$$

Since $1 \geq 0$ is a true statement, we shade the side of the line that includes point T. The graph is shown in Figure 13-20.

Self Check
Graph $x \geq 2$.

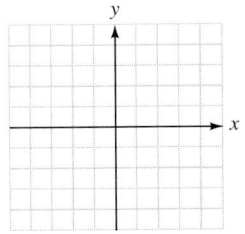

$y = 0$

x	y	(x, y)
1	0	(1, 0)
2	0	(2, 0)
3	0	(3, 0)

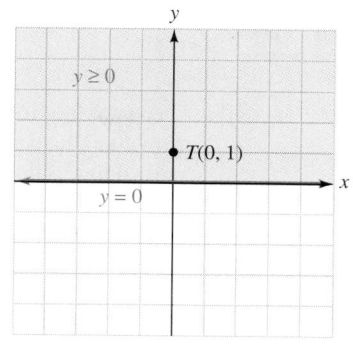

FIGURE 13-20

Answer:

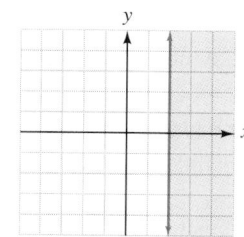

The following is a summary of the procedure for graphing linear inequalities.

Graphing linear inequalities in two variables

1. Graph the boundary line of the region. If the inequality allows the possibility of equality (the symbol is either $\leq$ or $\geq$), draw the boundary line as a solid line. If equality is not allowed ($<$ or $>$), draw the boundary line as a broken line.

2. Pick a test point that is on one side of the boundary line. (Use the origin if possible.) Replace x and y in the inequality with the coordinates of that point. If the inequality is satisfied, shade the side that contains that point. If the inequality is not satisfied, shade the other side of the boundary.

An application of linear inequalities

  EXAMPLE 7 *Earning money.* Carlos has two part-time jobs, one paying $10 per hour and another paying $12 per hour. He must earn at least $240 per week to pay his expenses while attending college. Write an inequality that shows the various ways he can schedule his time to achieve his goal.

Solution

If we let x represent the number of hours Carlos works on the first job and y the number of hours he works on the second job, we have

The hourly rate on the first job	times	the hours worked on the first job	plus	the hourly rate on the second job	times	the hours worked on the second job	is at least	$240.
10	·	x	+	12	·	y	$\geq$	240

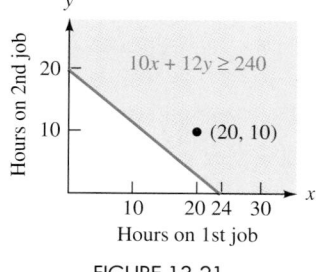

FIGURE 13-21

The graph of the inequality $10x + 12y \geq 240$ is shown in Figure 13-21. Any point in the shaded region indicates a possible way Carlos can schedule his time and earn $240 or more per week. For example, if he works 20 hours on the first job and 10 hours on the second job, he will earn

$$\$10(20) + \$12(10) = \$200 + \$120$$
$$= \$320$$

Since Carlos cannot work a negative number of hours, a graph showing negative values of x or y would have no meaning.

STUDY SET Section 13.5

VOCABULARY Fill in the blanks.

1. $2x - y \leq 4$ is a linear _____ in x and y.

2. The symbol $\leq$ means _____ or _____.

3. In the graph in Illustration 1, the line $2x - y = 4$ is the _____.

4. In Illustration 1, the line $2x - y = 4$ divides the rectangular coordinate system into two _____.

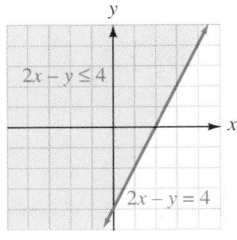

ILLUSTRATION 1

CONCEPTS

5. Tell whether each ordered pair is a solution of $5x - 3y \geq 0$.
   a. $(1, 1)$
   b. $(-2, -3)$
   c. $(0, 0)$
   d. $\left(\frac{1}{5}, \frac{4}{3}\right)$

6. Tell whether each ordered pair is a solution of $x + 4y < -1$.
   a. $(3, 1)$
   b. $(-2, 0)$
   c. $(-0.5, 0.2)$
   d. $\left(-2, \frac{1}{4}\right)$

7. Tell whether the graph of each linear inequality includes the boundary line.
   a. $y > -x$
   b. $5x - 3y \leq -2$

8. If a false statement results when the coordinates of a test point are substituted into a linear inequality, which half-plane should be shaded to represent the solution of the inequality?

9. A linear inequality has been graphed in Illustration 2. Tell whether each point satisfies the inequality.
   a. $(1, -3)$
   b. $(-2, -1)$
   c. $(2, 3)$
   d. $(3, -4)$

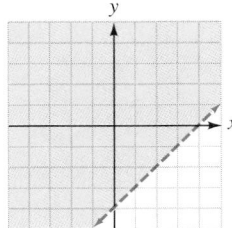

ILLUSTRATION 2

10. A linear inequality has been graphed in Illustration 3. Tell whether each point satisfies the inequality.
    a. $(2, 1)$
    b. $(-2, -4)$
    c. $(4, -2)$
    d. $(-3, 4)$

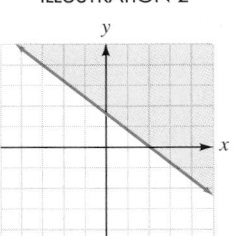

ILLUSTRATION 3

11. The boundary for the graph of a linear inequality is shown in Illustration 4. Why can't the origin be used as a test point to decide which side to shade?

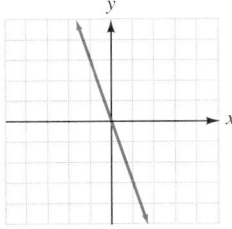

ILLUSTRATION 4

12. To decide how many pallets (x) and barrels (y) a delivery truck can hold, a dispatcher refers to the loading sheet in Illustration 5. Can a truck make a delivery of 4 pallets and 10 barrels in one trip?

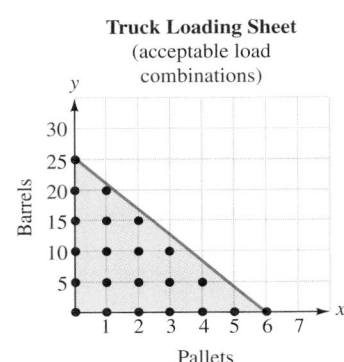

ILLUSTRATION 5

PRACTICE Complete the graph by shading the correct side of the boundary.

13. $y \leq x + 2$

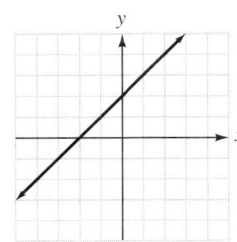

14. $y > x - 3$

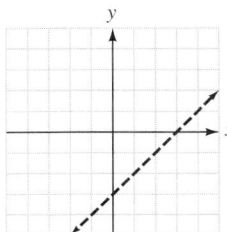

15. $y > 2x - 4$

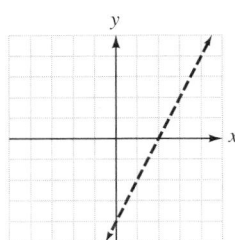

16. $y \leq -x + 1$

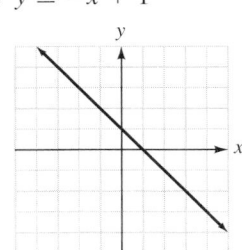

17. $x - 2y \geq 4$

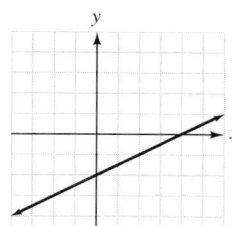

18. $3x + 2y > 12$

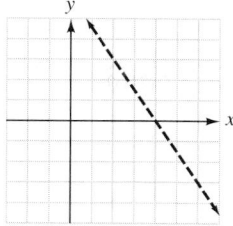

29. $y - x \geq 0$

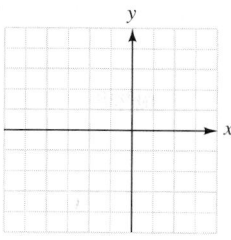

30. $y + x < 0$

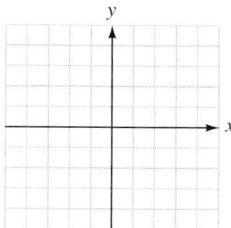

19. $y \leq 4x$

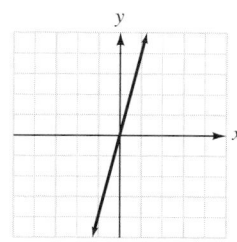

20. $y + 2x < 0$

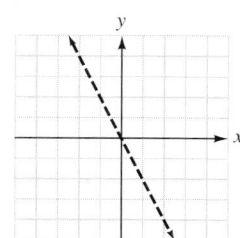

31. $2x + y > 2$

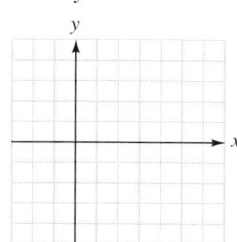

32. $3x - 2y > 6$

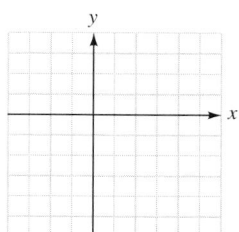

Graph each inequality.

21. $y \geq 3 - x$

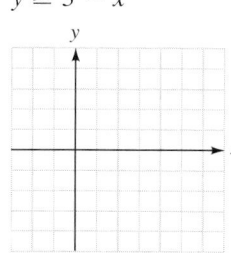

22. $y < 2 - x$

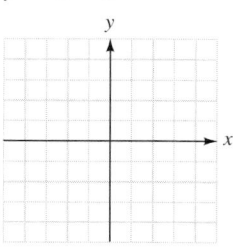

33. $3x - 4y > 12$

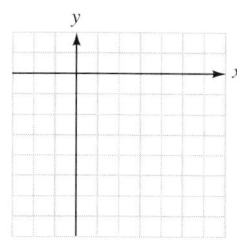

34. $4x + 3y \leq 12$

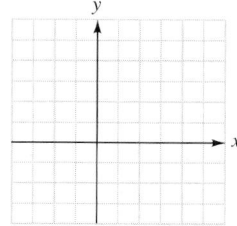

23. $y < 2 - 3x$

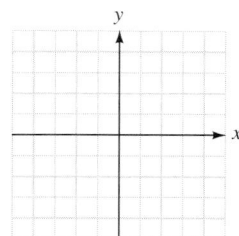

24. $y \geq 5 - 2x$

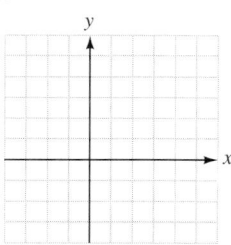

35. $5x + 4y \geq 20$

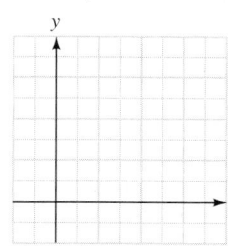

36. $7x - 2y < 21$

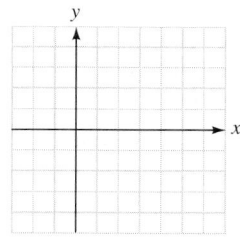

25. $y \geq 2x$

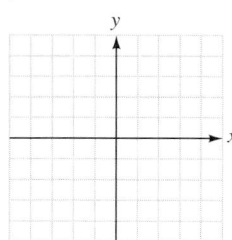

26. $y < 3x$

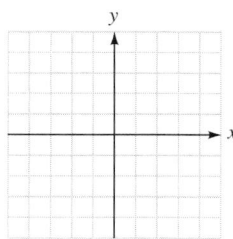

37. $x < 2$

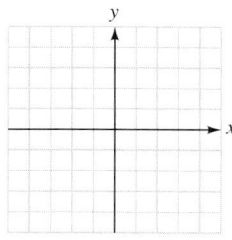

38. $y > -3$

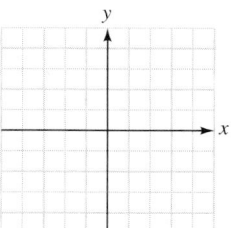

27. $2y - x < 8$

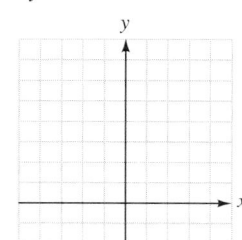

28. $y + 9x \geq 3$

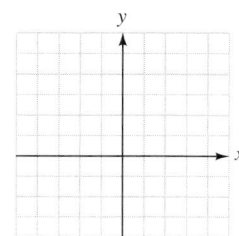

39. $y \leq 1$

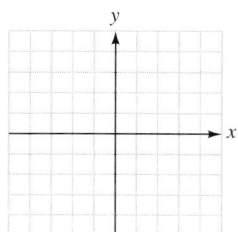

40. $x \geq -4$

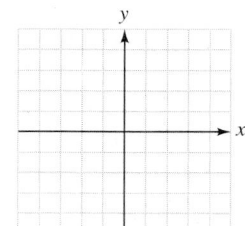

Simplify each inequality and then graph it.

41. $3(x + y) + x < 6$

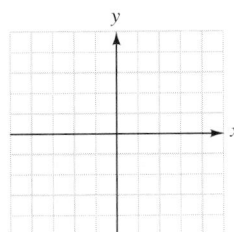

42. $2(x - y) - y \geq 4$

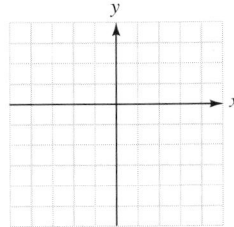

43. $4x - 3(x + 2y) \geq -6y$

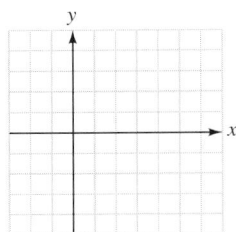

44. $3y + 2(x + y) < 5y$

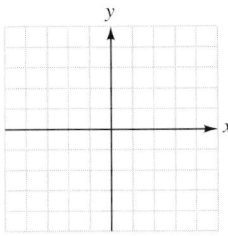

APPLICATIONS

45. NATO   In March of 1999, NATO aircraft and cruise missiles targeted Serbian military forces that were south of the 44th parallel in Yugoslavia, Montenegro, and Kosovo. See Illustration 6. Shade the geographic area that NATO was trying to rid of Serbian forces.

Based on data from *Los Angeles Times* (March 24, 1999)

ILLUSTRATION 6

46. U.S. HISTORY   When he ran for president in 1844, the campaign slogan of James K. Polk was "54-40 or fight!" It meant that Polk was willing to fight Great Britain for the possession of the Oregon Territory north to the 54°40′ parallel, as shown in Illustration 7. In 1846, Polk accepted a compromise to establish the 49th parallel as the permanent boundary of the United States. Shade the area of land that Polk conceded to the British.

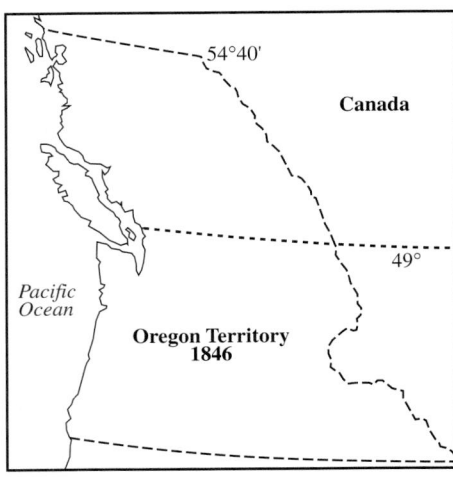

ILLUSTRATION 7

Write an inequality and graph it for nonnegative values of x and y. Then give three ordered pairs that satisfy the inequality.

47. PRODUCTION PLANNING   It costs a bakery $3 to make a cake and $4 to make a pie. Production costs cannot exceed $120 per day. Use Illustration 8 to graph an inequality that shows the possible combinations of cakes (x) and pies (y) that can be made.

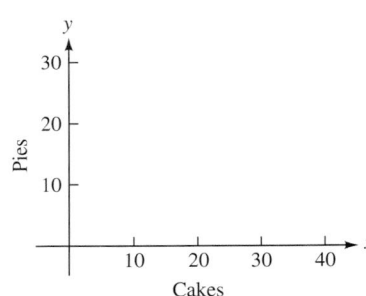

ILLUSTRATION 8

48. HIRING BABYSITTERS   Mary has a choice of two babysitters. Sitter 1 charges $6 per hour, and sitter 2 charges $7 per hour. If Mary can afford no more than $42 per week for sitters, use Illustration 9 to graph an inequality that shows the possible ways that she can hire sitter 1 (x) and sitter 2 (y).

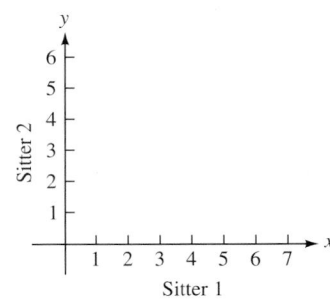

ILLUSTRATION 9

49. INVENTORY   A clothing store advertises that it maintains an inventory of at least $4,400 worth of men's jackets. If a leather jacket costs $100 and a nylon jacket costs $88, use Illustration 10 to graph an inequality that shows the possible ways that leather jackets (x) and nylon jackets (y) can be stocked.

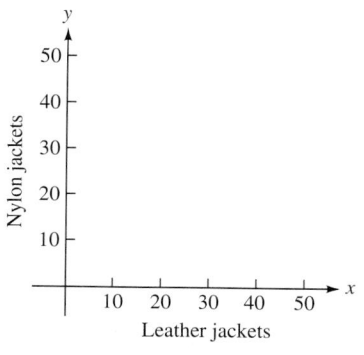

ILLUSTRATION 10

50. MAKING SPORTING GOODS   To keep up with demand, a sporting goods manufacturer allocates at least 2,400 units of production time per day to make baseballs and footballs. If it takes 20 units of time to make a baseball and 30 units of time to make a football, use Illustration 11 to graph an inequality that shows the possible ways to schedule the production time to make baseballs (x) and footballs (y).

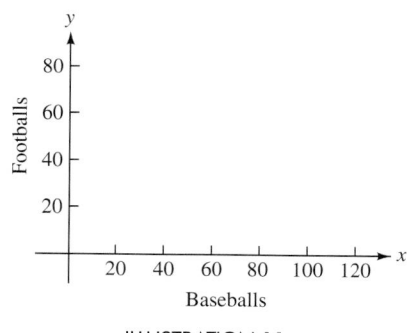

ILLUSTRATION 11

51. INVESTING IN STOCKS   Robert has up to $8,000 to invest in two companies. If stock in Robotronics sells for $40 per share and stock in Macrocorp sells for $50 per share, use Illustration 12 to graph an inequality that shows the possible ways that he can buy shares of Robotronics (x) and Macrocorp (y).

52. BUYING BASEBALL TICKETS   Tickets to the Rockford Rox baseball games cost $6 for reserved seats and $4 for general admission. If nightly receipts must average at least $10,200 to meet expenses, use Illustration 13 to graph an inequality that shows the possible ways that the Rox can sell reserved seats (x) and general admission tickets (y).

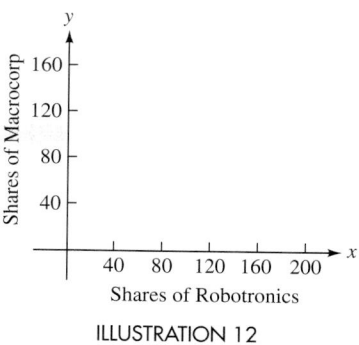

ILLUSTRATION 12

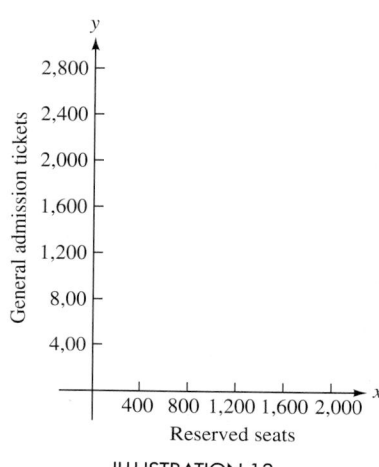

ILLUSTRATION 13

WRITING

53. Explain how to find the boundary for the graph of a linear inequality in two variables.

54. Explain how to decide which side of the boundary line to shade when graphing a linear inequality in two variables.

REVIEW

55. Let $g(x) = 3x^2 - 4x + 3$. Find $g(2)$.

56. Solve $2(x - 4) \leq -12$.

57. Factor $x^3 + 27$.

58. Factor $9p - 9q + mp - mq$.

59. Write a formula relating distance, rate, and time.

60. What is the slope of the line $2x - 3y = 2$?

61. Solve $A = P + Prt$ for t.

62. What is the sum of the measures of the three angles of any triangle?

13.6 *Solving Systems of Linear Inequalities*

In this section, you will learn about

- Systems of linear inequalities • An application of systems of linear inequalities

INTRODUCTION. We have previously solved systems of linear *equations* by the graphing method. The solution of such a system is the point of intersection of the straight lines. We now consider how to solve systems of linear *inequalities* graphically. When the solution of a linear inequality is graphed, the result is a half-plane. Therefore, we would expect to find the graphical solution of a system of inequalities by looking for the intersection, or "overlap," of shaded half-planes.

Systems of linear inequalities

To solve the **system of linear inequalities**

$$\begin{cases} x + y \geq 1 \\ x - y \geq 1 \end{cases}$$

we first graph each inequality. For instructional purposes, we will initially graph each inequality on a separate set of axes, although in practice we will draw them on the same axes.

The graph of $x + y \geq 1$ includes the graph of the equation $x + y = 1$ and all points above it. Because the boundary line is included, we draw it with a solid line, as shown in Figure 13-22(a).

The graph of $x - y \geq 1$ includes the graph of the equation $x - y = 1$ and all points below it. Because the boundary line is included here also, it is drawn with a solid line, as shown in Figure 13-22(b).

$x + y = 1$		
x	y	(x, y)
0	1	$(0, 1)$
1	0	$(1, 0)$
2	-1	$(2, -1)$

$x - y = 1$		
x	y	(x, y)
0	-1	$(0, -1)$
1	0	$(1, 0)$
2	1	$(2, 1)$

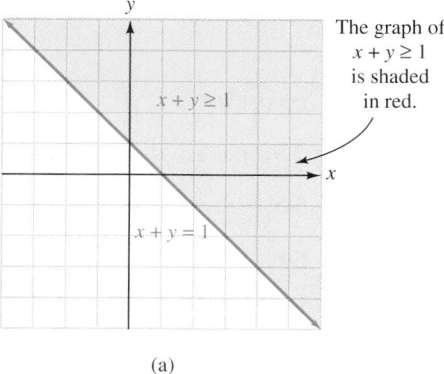

The graph of $x + y \geq 1$ is shaded in red.

(a)

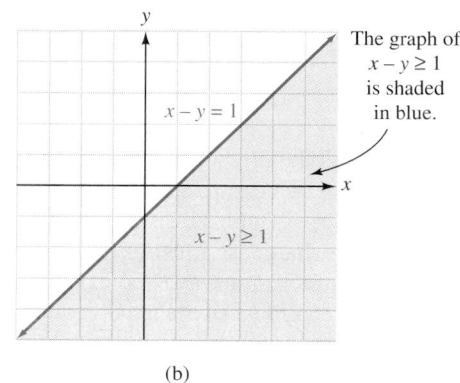

The graph of $x - y \geq 1$ is shaded in blue.

(b)

FIGURE 13-22

In Figure 13-23, we show the result when the inequalities $x + y \geq 1$ and $x - y \geq 1$ are graphed one at a time on the same coordinate axes. The area that is shaded twice represents the set of simultaneous solutions of the given system of inequalities. Any point in the doubly shaded region (shown in purple) has coordinates that satisfy both inequalities of the system.

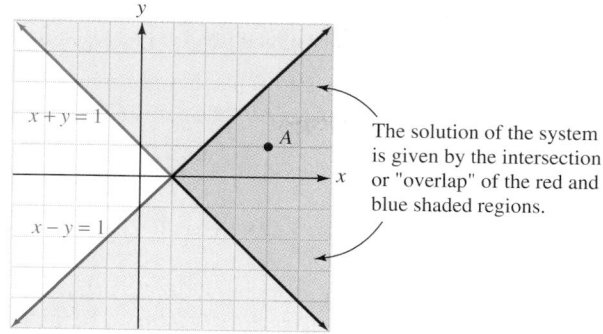

The solution of the system is given by the intersection or "overlap" of the red and blue shaded regions.

FIGURE 13-23

To see whether this is true, we can pick a point, such as point *A*, that lies in the doubly shaded region and show that its coordinates satisfy both inequalities. Because point *A* has coordinates (4, 1), we have

$$x + y \geq 1 \quad \text{and} \quad x - y \geq 1$$
$$4 + 1 \geq 1 \qquad\qquad 4 - 1 \geq 1$$
$$5 \geq 1 \qquad\qquad 3 \geq 1$$

Since the coordinates of point *A* satisfy each inequality, point *A* is a solution of the system. If we pick a point that is not in the doubly shaded region, its coordinates will fail to satisfy at least one of the inequalities.

In general, to solve systems of linear inequalities, we will follow these steps.

Solving systems of inequalities

1. Graph each inequality in the system on the same coordinate axes.
2. Find the region that is common to every graph.
3. Pick a test point from the region to verify the solution.

EXAMPLE 1 *Solving systems of inequalities.* Graph the solution of
$$\begin{cases} 2x + y < 4 \\ -2x + y > 2 \end{cases}$$

Solution

First, we graph each inequality on one set of axes, as shown in Figure 13-24.

Self Check
Graph the solution of
$$\begin{cases} x + 3y < 3 \\ -x + 3y > 3 \end{cases}$$

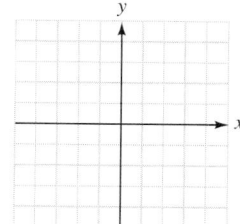

$2x + y = 4$				$-2x + y = 2$		
x	y	(x, y)		x	y	(x, y)
0	4	(0, 4)		-1	0	$(-1, 0)$
2	0	(2, 0)		0	2	(0, 2)
1	2	(1, 2)		2	6	(2, 6)

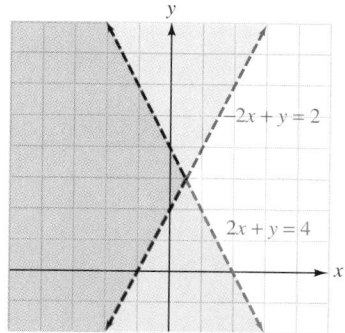

FIGURE 13-24

We note that

- The graph of $2x + y < 4$ includes all points below the line $2x + y = 4$. Since the boundary is not included, we draw it as a broken line.

- The graph of $-2x + y > 2$ includes all points above the line $-2x + y = 2$. Since the boundary is not included, we also draw it as a broken line.

The area that is shaded twice (the region in purple) is the solution of the given system of inequalities. Any point in the doubly shaded region has coordinates that will satisfy both inequalities of the system.

Pick a point in the doubly shaded region and show that it satisfies both inequalities.

Answer:

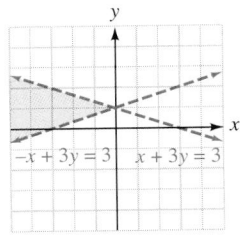

EXAMPLE 2   *Solving systems of inequalities.*   Graph the solution of
$$\begin{cases} x \le 2 \\ y > 3 \end{cases}$$

Solution

We graph each inequality on one set of axes, as shown in Figure 13-25.

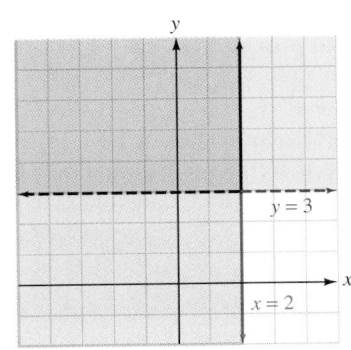

$x = 2$				$y = 3$		
x	y	(x, y)		x	y	(x, y)
2	0	(2, 0)		0	3	(0, 3)
2	2	(2, 2)		1	3	(1, 3)
2	4	(2, 4)		4	3	(4, 3)

FIGURE 13-25

We note that

- The graph of $x \le 2$ includes all points to the left of the line $x = 2$. Since the boundary line is included, we draw it as a solid line.
- The graph of $y > 3$ includes all points above the line $y = 3$. Since the boundary is not included, we draw it as a broken line.

The area that is shaded twice is the solution of the given system of inequalities. Any point in the doubly shaded region (purple) has coordinates that will satisfy both inequalities of the system. Pick a point in the doubly shaded region and show that this is true.

Self Check

Graph the solution of $\begin{cases} y \le 1 \\ x > 2 \end{cases}$

Answer:

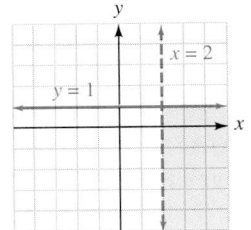

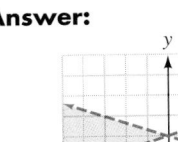

EXAMPLE 3   *Solving systems of inequalities.*   Graph the solution of
$$\begin{cases} y < 3x - 1 \\ y \ge 3x + 1 \end{cases}$$

Solution

We graph each inequality as shown in Figure 13-26 and make the following observations:

- The graph of $y < 3x - 1$ includes all points below the broken line $y = 3x - 1$.
- The graph of $y \ge 3x + 1$ includes all points on and above the solid line $y = 3x + 1$.

Self Check

Graph the solution of
$$\begin{cases} y \ge -\tfrac{1}{2}x + 1 \\ y < -\tfrac{1}{2}x - 1 \end{cases}$$

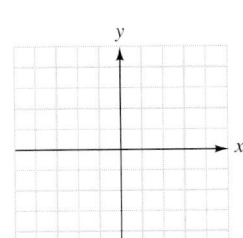

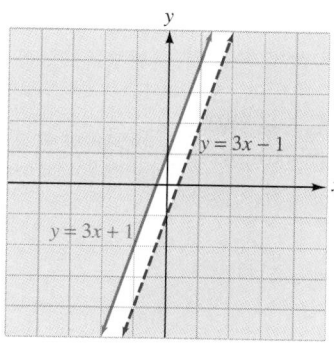

FIGURE 13-26

Because the graphs of these inequalities do not intersect, the solution set is empty. There are no solutions.

Answer: no solutions

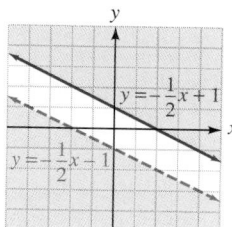

EXAMPLE 4 *Solving systems of inequalities.* Graph the solution of

$$\begin{cases} x \geq 0 \\ y \geq 0 \\ x + 2y \leq 6 \end{cases}$$

Solution

We graph each inequality as shown in Figure 13-27 and make the following observations:

- The graph of $x \geq 0$ includes all points on the *y*-axis and to the right.
- The graph of $y \geq 0$ includes all points on the *x*-axis and above.
- The graph of $x + 2y \leq 6$ includes all points on the line $x + 2y = 6$ and below.

The solution is the region that is shaded three times. This includes triangle *OPQ* and the triangular region it encloses.

Self Check

Graph the solution of $\begin{cases} x \leq 1 \\ y \leq 2 \\ 2x - y \leq 4 \end{cases}$

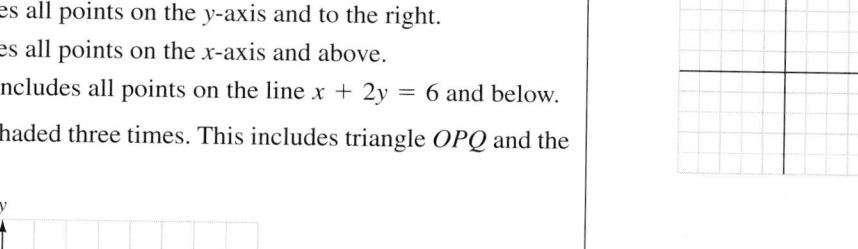

Answer:

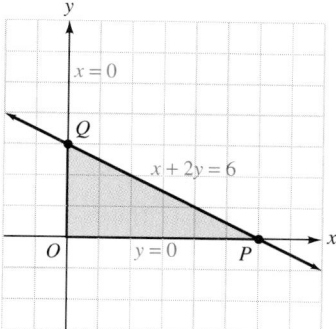

FIGURE 13-27

An application of systems of linear inequalities

EXAMPLE 5 *Landscaping.* A homeowner budgets from $300 to $600 for trees and bushes to landscape his yard. After shopping around, he finds that good trees cost $150 and mature bushes cost $75. What combinations of trees and bushes can he afford to buy?

Analyze the problem The homeowner wants to spend *at least* $300 but *not more than* $600 for trees and bushes.

Form two inequalities We can let x represent the number of trees purchased and y the number of bushes purchased. We then form the following system of inequalities:

The cost of a tree	times	the number of trees purchased	plus	the cost of a bush	times	the number of bushes purchased	should at least be	$300.
$150	·	x	+	$75	·	y	$\geq$	$300

The cost of a tree	times	the number of trees purchased	plus	the cost of a bush	times	the number of bushes purchased	should not be more than	$600.
$150	·	x	+	$75	·	y	$\leq$	$600

Solve the system We graph the system

$$\begin{cases} 150x + 75y \geq 300 \\ 150x + 75y \leq 600 \end{cases}$$

as shown in Figure 13-28. The coordinates of each point shown in the graph give a possible combination of the number of trees (x) and the number of bushes (y) that can be purchased. These possibilities are

(0, 4), (0, 5), (0, 6), (0, 7), (0, 8)
(1, 2), (1, 3), (1, 4), (1, 5), (1, 6)
(2, 0), (2, 1), (2, 2), (2, 3), (2, 4)
(3, 0), (3, 1), (3, 2), (4, 0)

Only these points can be used, because the homeowner cannot buy a portion of a tree or a bush.

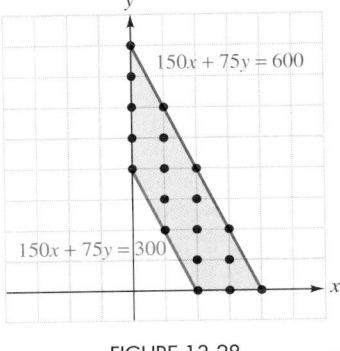

FIGURE 13-28

STUDY SET Section 13.6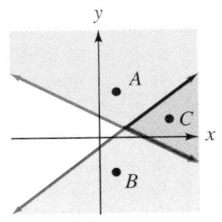

VOCABULARY *Fill in the blanks.*

1. $\begin{cases} x + y > 2 \\ x + y < 4 \end{cases}$ is a system of linear _____.

2. The _____ of a system of linear inequalities is all the ordered pairs that make all inequalities of the system true at the same time.

3. Any point in the _____ region of the graph of the solution of a system of two linear inequalities has coordinates that satisfy both inequalities of the system.

4. To graph a linear inequality such as $x + y > 2$, first graph the boundary. Then pick a test _____ to determine which half-plane to shade.

CONCEPTS

5. In Illustration 1, the solution of linear inequality 1 was shaded in red, and the solution of linear inequality 2 was shaded in blue. The overlap of the red and the blue regions is shown in purple. Tell whether a true or a false statement results when the coordinates of the given point are substituted into the given inequality.

a. A, inequality 1
b. A, inequality 2
c. B, inequality 1
d. B, inequality 2
e. C, inequality 1
f. C, inequality 2

ILLUSTRATION 1

6. Match each equation, inequality, or system with the graph of its solution.

a. $x + y = 2$

b. $x + y \geq 2$

c. $\begin{cases} x + y = 2 \\ x - y = 2 \end{cases}$

d. $\begin{cases} x + y \geq 2 \\ x - y \leq 2 \end{cases}$

i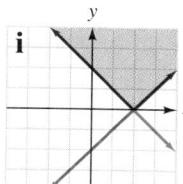

ii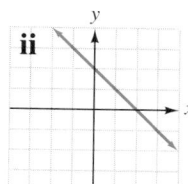

iii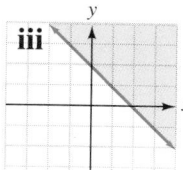

iv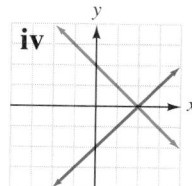

7. The graph of the solution of a system of linear inequalities is shown in Illustration 2. Tell whether each point is a part of the solution set.

a. $(4, -2)$

b. $(1, 3)$

c. the origin

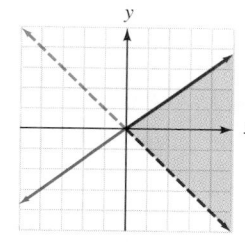

ILLUSTRATION 2

8. Use a system of inequalities to describe the shaded region in Illustration 3.

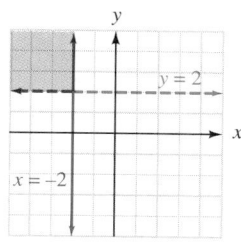

ILLUSTRATION 3

NOTATION

9. Fill in the blank to make the statement true: The graph of the solution of a system of linear inequalities shown in Illustration 4 can be described as the triangle _____ and the triangular region it encloses.

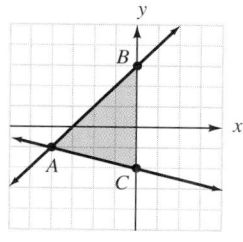

ILLUSTRATION 4

10. Represent each phrase using either $>$, $<$, $\geq$, or $\leq$.

a. is not more than

b. must be at least

c. should not surpass

d. cannot go below

PRACTICE *Graph the solution set of each system of inequalities, when possible.*

11. $\begin{cases} x + 2y \leq 3 \\ 2x - y \geq 1 \end{cases}$

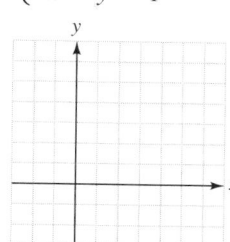

12. $\begin{cases} 2x + y \geq 3 \\ x - 2y \leq -1 \end{cases}$

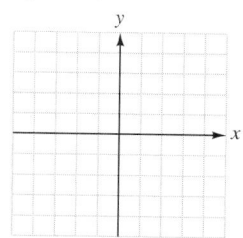

13. $\begin{cases} x + y < -1 \\ x - y > -1 \end{cases}$

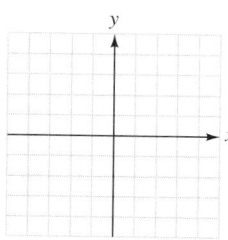

14. $\begin{cases} x + y > 2 \\ x - y < -2 \end{cases}$

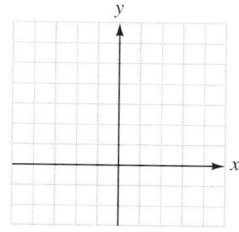

15. $\begin{cases} x \geq 2 \\ y \leq 3 \end{cases}$

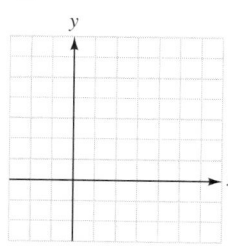

16. $\begin{cases} x \geq -1 \\ y > -2 \end{cases}$

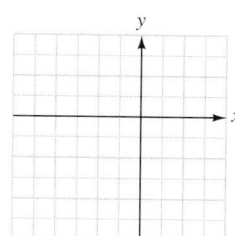

17. $\begin{cases} 2x - 3y \leq 0 \\ y \geq x - 1 \end{cases}$

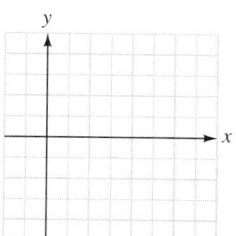

18. $\begin{cases} y > 2x - 4 \\ y \geq -x - 1 \end{cases}$

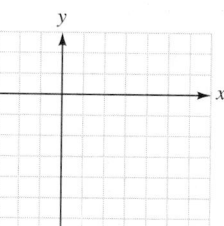

19. $\begin{cases} y < -x + 1 \\ y > -x + 3 \end{cases}$

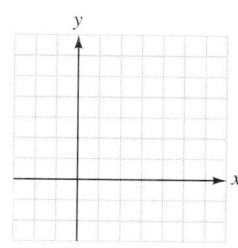

20. $\begin{cases} y > -x + 2 \\ y < -x + 4 \end{cases}$

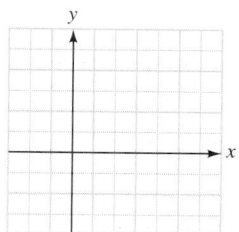

21. $\begin{cases} x > 0 \\ y > 0 \end{cases}$

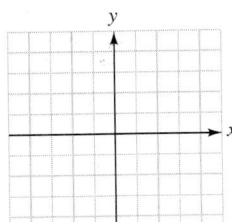

22. $\begin{cases} x \leq 0 \\ y < 0 \end{cases}$

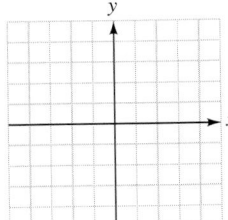

31. $\begin{cases} \frac{x}{2} + \frac{y}{3} \geq 2 \\ \frac{x}{2} - \frac{y}{2} < -1 \end{cases}$

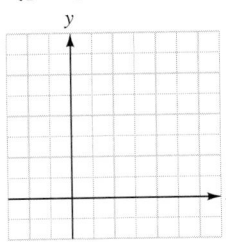

32. $\begin{cases} \frac{x}{3} - \frac{y}{2} < -3 \\ \frac{x}{3} + \frac{y}{2} > -1 \end{cases}$

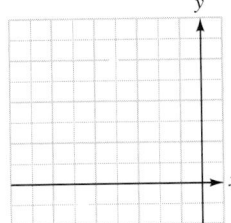

23. $\begin{cases} 3x + 4y \geq -7 \\ 2x - 3y \geq 1 \end{cases}$

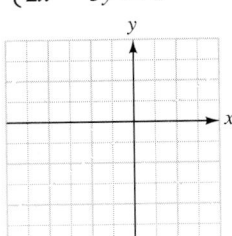

24. $\begin{cases} 3x + y \leq 1 \\ 4x - y \geq -8 \end{cases}$

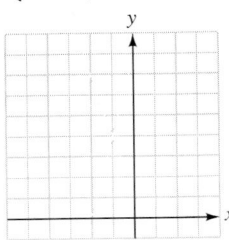

33. $\begin{cases} x \geq 0 \\ y \geq 0 \\ x + y \leq 3 \end{cases}$

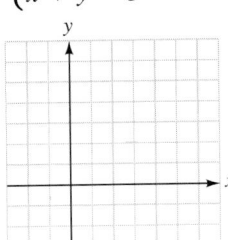

34. $\begin{cases} x - y \leq 6 \\ x + 2y \leq 6 \\ x \geq 0 \end{cases}$

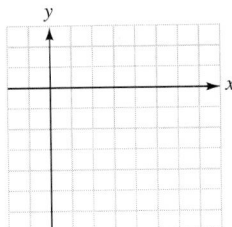

25. $\begin{cases} 2x + y < 7 \\ y > 2(1 - x) \end{cases}$

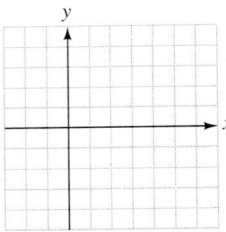

26. $\begin{cases} 2x + y \geq 6 \\ y \leq 2(2x - 3) \end{cases}$

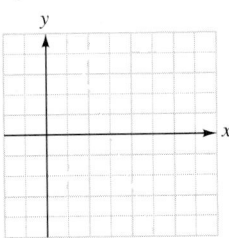

APPLICATIONS

35. BIRDS OF PREY Parts a and b of Illustration 5 show the individual fields of vision for each eye of an owl. In part c, shade the area where the fields of vision overlap—that is, the area that is seen by both eyes.

27. $\begin{cases} 2x - 4y > -6 \\ 3x + y \geq 5 \end{cases}$

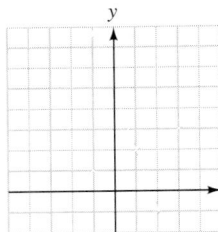

28. $\begin{cases} 2x - 3y < 0 \\ 2x + 3y \geq 12 \end{cases}$

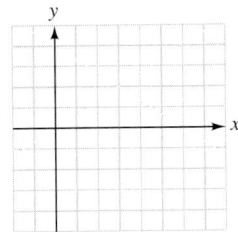

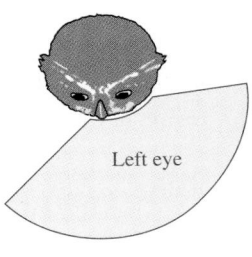

(a)

(b)

(c)

ILLUSTRATION 5

29. $\begin{cases} 3x - y \leq -4 \\ 3y > -2(x + 5) \end{cases}$

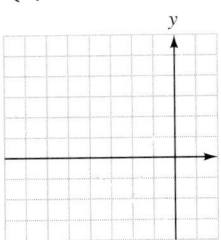

30. $\begin{cases} 3x + y < -2 \\ y > 3(1 - x) \end{cases}$

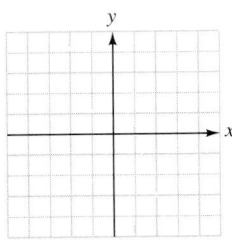

36. EARTH SCIENCE In Illustration 6, shade the area of the earth's surface that is north of the Tropic of Capricorn and south of the Tropic of Cancer.

ILLUSTRATION 6

In Exercises 37–40, graph each system of inequalities and give two possible solutions.

37. BUYING COMPACT DISCS
Melodic Music has compact discs on sale for either $10 or $15. If a customer wants to spend at least $30 but no more than $60 on CDs, use Illustration 7 to graph a system of inequalities showing the possible combinations of $10 CDs ($x$) and $15 CDs ($y$) that the customer can buy.

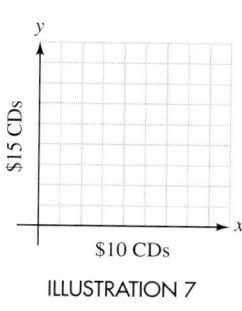

ILLUSTRATION 7

38. BUYING BOATS   Dry Boatworks wholesales aluminum boats for $800 and fiberglass boats for $600. Northland Marina wants to make a purchase totaling at least $2,400, but no more than $4,800. Use Illustration 8 to graph a system of inequalities showing the possible combinations of aluminum boats (x) and fiberglass boats (y) that can be ordered.

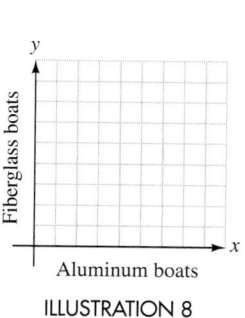

ILLUSTRATION 8

39. BUYING FURNITURE   A distributor wholesales desk chairs for $150 and side chairs for $100. Best Furniture wants its order to total no more than $900; Best also wants to order more side chairs than desk chairs. Use Illustration 9 to graph a system of inequalities showing the possible combinations of desk chairs (x) and side chairs (y) that can be ordered.

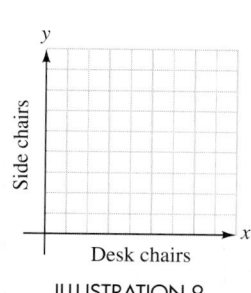

ILLUSTRATION 9

40. ORDERING FURNACE EQUIPMENT   J. Bolden Heating Company wants to order no more than $2,000 worth of electronic air cleaners and humidifiers from a wholesaler that charges $500 for air cleaners and $200 for humidifiers. If Bolden wants more humidifiers than air cleaners, use Illustration 10 to graph a system of inequalities showing the possible combinations of air cleaners (x) and humidifiers (y) that can be ordered.

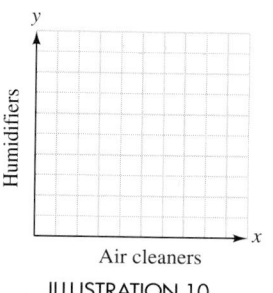

ILLUSTRATION 10

41. PESTICIDE   To eradicate a fruit fly infestation, helicopters sprayed an area of a city that can be described by $y \geq -2x + 1$ (within the city limits). Two weeks later, more spraying was ordered over the area described by $y \geq \frac{1}{4}x - 4$ (within the city limits). In Illustration 11, show the part of the city that was sprayed twice.

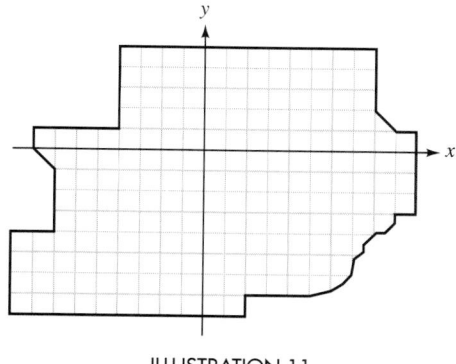

ILLUSTRATION 11

42. REDEVELOPMENT   A government agency has declared an area of a city east of First Street, north of Second Avenue, south of Sixth Avenue, and west of Fifth Street as eligible for federal redevelopment funds. See Illustration 12. Describe this area of the city mathematically using a system of four inequalities, if the corner of Central Avenue and Main Street is considered the origin.

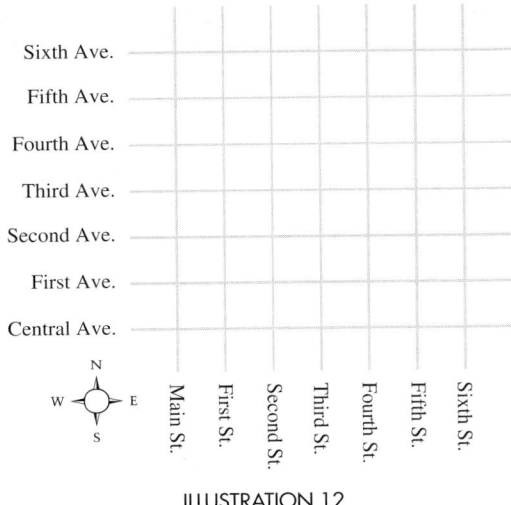

ILLUSTRATION 12

WRITING

43. Explain how to use graphing to solve a system of inequalities.

44. Explain when a system of inequalities will have no solutions.

45. Describe how the graphs of the solutions of these systems are similar and how they differ.

$$\begin{cases} x + y = 4 \\ x - y = 4 \end{cases} \quad \text{and} \quad \begin{cases} x + y \geq 4 \\ x - y \geq 4 \end{cases}$$

46. When a solution of a system of linear inequalities is graphed, what does the shading represent?

REVIEW *Complete each table of values.*

47. $y = 2x^2$

x	y
8	
-2	

48. $t = -|s + 2|$

s	t
-3	
-10	

49. $f(x) = 4 + x^3$

Input	Output
0	
-3	

50. $g(x) = 2x - x^2$

x	$g(x)$
5	
-5	

Systems of Equations and Inequalities

In Chapter 13, we have solved problems that required the use of two variables to represent two unknown quantities. To find the unknowns, we write a pair of equations (or inequalities) called a **system.**

Solving systems of equations by graphing

A system of linear equations can be solved by graphing both equations and locating the point of intersection of the two lines.

1. FOOD SERVICE   The two equations in the table give the fees two different catering companies charge a Hollywood studio for on-location meal service.

Caterer	Setup fee	Cost per meal	Equation
Sunshine	$1,000	$4	$y = 4x + 1,000$
Lucy's	$500	$5	$y = 5x + 500$

Complete Illustration 1, using the graphing method to find the break point. That is, find the number of meals and the corresponding fee for which the two caterers will charge the studio the same amount.

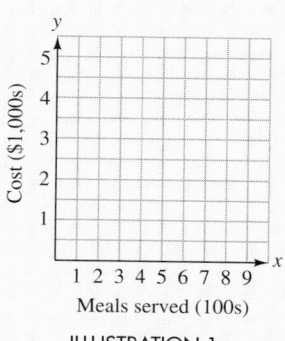

ILLUSTRATION 1

Solving systems of equations by substitution

The substitution method for solving a system of equations works well when a variable in either equation has a coefficient of 1 or -1.

2. Solve by substitution: $\begin{cases} y = 2x - 9 \\ x + 3y = 8 \end{cases}$

3. Solve by substitution: $\begin{cases} 3x + 4y = -7 \\ 2y - x = -1 \end{cases}$

Solving systems of equations by addition

With the addition method, equal quantities are added to both sides of an equation to eliminate one of the variables. Then we solve for the other variable.

4. Solve by addition: $\begin{cases} x + y = 1 \\ x - y = 5 \end{cases}$

5. Solve by addition: $\begin{cases} 2x - 3y = -18 \\ 3x + 2y = -1 \end{cases}$

Solving systems of inequalities

To solve a system of two linear inequalities, we graph the inequalities on the same coordinate axes. The area that is shaded twice represents the set of solutions.

6. This system of inequalities describes the number of $20 shirts ($x$) and $40 pants ($y$) a person can buy if he or she plans to spend not less than $80 but not more than $120. Using Illustration 2, graph the system. Then give three solutions.

$\begin{cases} 20x + 40y \geq 80 \\ 20x + 40y \leq 120 \end{cases}$

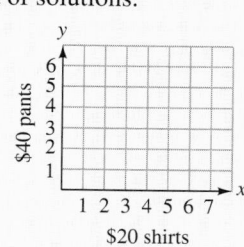

ILLUSTRATION 2

859

ACCENT ON TEAMWORK

Section 13.1

SOLVING SYSTEMS GRAPHICALLY The graphing method was used to solve a system of linear equations. The work is shown in Illustration 1. What are the two equations of the system?

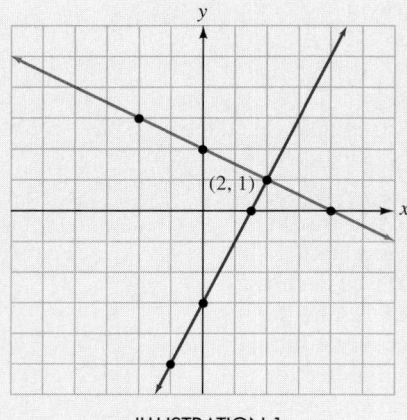

ILLUSTRATION 1

WRITING SYSTEMS OF EQUATIONS In your group, pick a specific ordered pair (x, y) where the x-coordinate and the y-coordinate are integers between -5 and 5. Write a system of two linear equations whose solution is the ordered pair that you picked. Then exchange systems with another group and solve the system that they wrote.

Section 13.2

SUBSTITUTIONS The word *substitution* is used in several ways. Explain how it is used in the context of a sporting event such as a basketball game. Explain how it is sometimes used when ordering food at a restaurant. Explain what is meant by a *substitute* teacher. Finally, explain how the substitution method is used to solve a system such as

$$\begin{cases} y = -2x - 5 \\ 3x + 5y = -4 \end{cases}$$

What is the difference in the mathematical meaning of the word *substitution* as opposed to the everyday usage of the word?

Section 13.3

SOLVING SYSTEMS Solve the system

$$\begin{cases} 2x + y = 4 \\ 2x + 3y = 0 \end{cases}$$

using the graphing method, the substitution method, and the addition method. Which method do you think is the best to use in this case? Why?

Section 13.4

TWO UNKNOWNS Consider the following problem: A man paid $89 for two white shirts and four pairs of black socks. Find the cost of a white shirt.

If we let x represent the cost of a white shirt and y represent the cost of a pair of black socks, an equation describing the situation is $2x + 4y = 89$. Explain why there is not enough information to solve the problem.

Section 13.5

MATCHING GAME Have a student in your group write ten linear inequalities on 3×5 note cards, one inequality per card. Then have him or her graph each of the inequalities on separate cards. Mix up the cards and put all the inequality cards on one side of a table and all the cards with graphs on the other side. Work together to match each inequality with its proper graph.

Section 13.6

SYSTEMS OF INEQUALITIES The points A, B, C, D, E, F, and G are labeled in the graph of the solution of a system of inequalities in Illustration 2. Tell whether the coordinates of each point make the first inequality (whose solution was shown in red) and the second inequality (whose solution was shown in blue) true or false. Use a table of the following form to keep track of your results.

Point	Coordinates	1st inequality	2nd inequality
A			

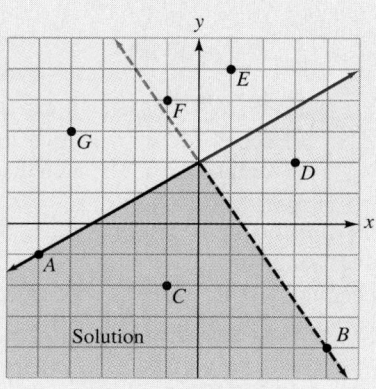

ILLUSTRATION 2

| SECTION 13.1 | *Solving Systems of Equations by Graphing* |

CONCEPTS

An ordered pair that satisfies both equations simultaneously is a *solution* of the system.

REVIEW EXERCISES

1. Tell whether the ordered pair is a solution of the system.

a. $(2, -3)$, $\begin{cases} 3x - 2y = 12 \\ 2x + 3y = -5 \end{cases}$

b. $\left(\frac{7}{2}, -\frac{2}{3}\right)$, $\begin{cases} 4x - 6y = 18 \\ \frac{x}{3} + \frac{y}{2} = \frac{5}{6} \end{cases}$

2. INJURY COMPARISON   Illustration 1 shows the number of skiing and snowboarding injuries nationally for the years 1993–1997. If the number of injuries continued to occur at the 1996–1997 rates, estimate when they would be the same for skiing and snowboarding. About how many injuries would that be?

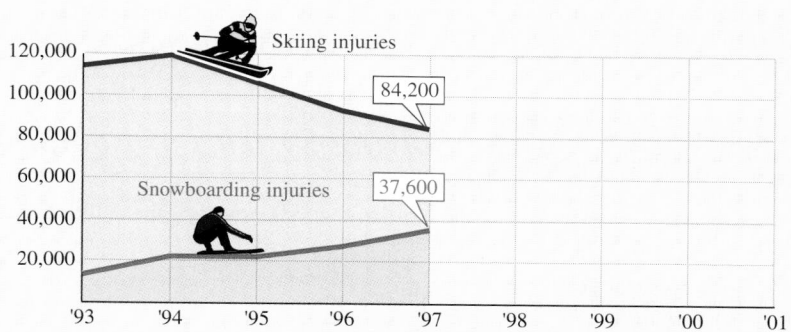

Based on data from *Los Angeles Times* (March 23, 1999)

ILLUSTRATION 1

To *solve a system graphically:*

1. Carefully graph each equation.

2. If the lines intersect, the coordinates of the point of intersection give the solution of the system.

3. Check the solution in the original equations.

When a system of equations has a solution, it is a *consistent* system. Systems with no solutions are *inconsistent*.

When the graphs of two equations in a system are different lines, the equations are *independent* equations.

The equations of a system with infinitely many solutions are *dependent*.

3. Use the graphing method to solve each system.

a. $\begin{cases} x + y = 7 \\ 2x - y = 5 \end{cases}$

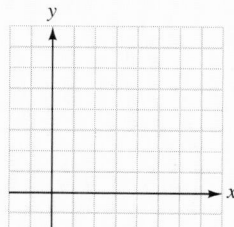

b. $\begin{cases} y = -\frac{x}{3} \\ 2x + y = 5 \end{cases}$

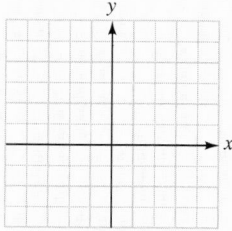

c. $\begin{cases} 3x + 6y = 6 \\ x + 2y = 2 \end{cases}$

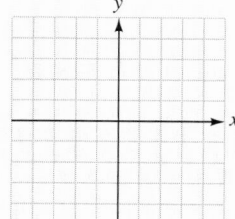

d. $\begin{cases} 6x + 3y = 12 \\ 2x + y = 2 \end{cases}$

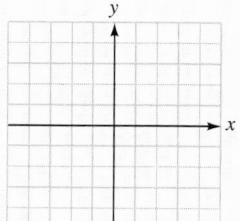

Solving Systems of Equations by Substitution

To solve a system of equations in x and y by the *substitution method:*

1. Solve one of the equations for either x or y.

2. Substitute the resulting expression for the variable in step 1 into the other equation, and solve the equation.

3. Find the value of the other variable by substituting the solution found in step 2 in any equation containing x and y.

4. Check the solution in the original equations.

4. Use the substitution method to solve each system.

a. $\begin{cases} x = y \\ 5x - 4y = 3 \end{cases}$

b. $\begin{cases} y = 15 - 3x \\ 7y + 3x = 15 \end{cases}$

c. $\begin{cases} 0.2x + 0.2y = 0.6 \\ 3x = 2 - y \end{cases}$

d. $\begin{cases} 6(r + 2) = s - 1 \\ r - 5s = -7 \end{cases}$

e. $\begin{cases} 9x + 3y = 5 \\ 3x + y = \dfrac{5}{3} \end{cases}$

f. $\begin{cases} \dfrac{x}{6} + \dfrac{y}{10} = 3 \\ \dfrac{5x}{16} - \dfrac{3y}{16} = \dfrac{15}{8} \end{cases}$

5. In solving a system using the substitution method, suppose you obtain the result of $8 = 9$.

a. How many solutions does the system have?

b. Describe the graph of the system.

c. What term is used to describe the system?

Solving Systems of Equations by Addition

To solve a system of equations using the *addition method:*

1. Write each equation in $Ax + By = C$ form.

2. Multiply one or both equations by nonzero quantities to make the coefficients of x (or y) opposites.

3. Add the equations to eliminate the terms involving x (or y).

4. Solve the equation resulting from step 3.

5. Find the value of the other variable by substituting the value of the variable found in step 4 into any equation containing both variables.

6. Check the solution in the original equations.

6. Solve each system using the addition method.

a. $\begin{cases} 2x + y = 1 \\ 5x - y = 20 \end{cases}$

b. $\begin{cases} x + 8y = 7 \\ x - 4y = 1 \end{cases}$

c. $\begin{cases} 5a + b = 2 \\ 3a + 2b = 11 \end{cases}$

d. $\begin{cases} 11x + 3y = 27 \\ 8x + 4y = 36 \end{cases}$

e. $\begin{cases} 9x + 3y = 15 \\ 3x = 5 - y \end{cases}$

f. $\begin{cases} \dfrac{x}{3} + \dfrac{y + 2}{2} = 1 \\ \dfrac{x + 8}{8} + \dfrac{y - 3}{3} = 0 \end{cases}$

g. $\begin{cases} 0.02x + 0.05y = 0 \\ 0.3x - 0.2y = -1.9 \end{cases}$

h. $\begin{cases} -\dfrac{1}{4}x = 1 - \dfrac{2}{3}y \\ 6(x - 3y) + 2y = 5 \end{cases}$

7. For each system, tell which method, substitution or addition, would be easier to use to solve the system and why.

a. $\begin{cases} 6x + 2y = 5 \\ 3x - 3y = -4 \end{cases}$

b. $\begin{cases} x = 5 - 7y \\ 3x - 3y = -4 \end{cases}$

Applications of Systems of Equations

In this section, we considered ways to solve problems by using *two* variables.

In Exercises 8–16, use two equations in two variables to solve each problem.

8. CAUSE OF DEATH   In 1998, the number of Americans dying from heart disease was about 4.5 times more than the number dying from a stroke. If the total number of deaths from these causes was 880,000, how many deaths were attributed to each?

To solve problems involving two unknown quantities:

1. *Analyze* the facts of the problem. Make a table or diagram if necessary.
2. Pick different variables to represent two unknown quantities. *Form* two equations involving the variables.
3. *Solve* the system of equations.
4. *State* the conclusion.
5. *Check* the results.

The *break point* of a linear system is the point of intersection of the graph.

9. PAINTING EQUIPMENT   When fully extended, the ladder shown in Illustration 2 is 35 feet in length. If the extension is 7 feet shorter than the base, how long is each part of the ladder?

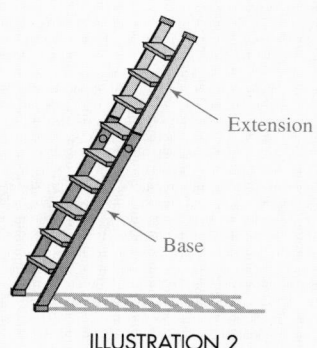

ILLUSTRATION 2

10. CRASH INVESTIGATION   In an effort to protect evidence, investigators used 420 yards of yellow "Police Line—Do Not Cross" tape to seal off a large rectangular-shaped area around an airplane crash site. How much area will the investigators have to search if the width of the rectangle is three-fourths of the length?

11. CELEBRITY ENDORSEMENT   A company selling a home juicing machine is contemplating hiring either an athlete or an actor to serve as a spokesperson for a product. The terms of each contract would be as follows:

Celebrity	Base pay	Commission per item sold
Athlete	$30,000	$5
Actor	$20,000	$10

a. For each celebrity, write an equation giving the money (y) the celebrity would earn if x juicers were sold.

b. For what number of juicers would the athlete and the actor earn the same amount?

c. Using Illustration 3, graph the equations from part a. The company expects to sell over 3,000 juicers. Which celebrity would cost the company the least money to serve as a spokesperson?

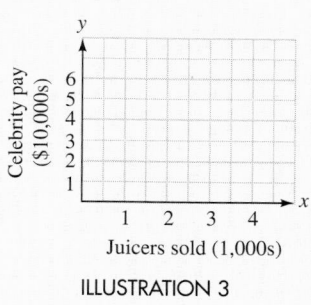

ILLUSTRATION 3

12. CANDY OUTLET STORE   A merchant wants to mix gummy worms worth $3 per pound and gummy bears worth $1.50 per pound to make 30 pounds of a mixture worth $2.10 per pound. How many pounds of each type of candy should he use?

13. BOATING   It takes a motorboat 4 hours to travel 56 miles down a river, and 3 hours longer to make the return trip. Find the speed of the current.

14. SHOPPING   Packages containing two bottles of contact lens cleaner and three bottles of soaking solution cost $63.40, and packages containing three bottles of cleaner and two bottles of soaking solution cost $69.60. Find the cost of a bottle of cleaner and a bottle of soaking solution.

15. INVESTING   Carlos invested part of $3,000 in a 10% certificate account and the rest in a 6% passbook account. The total annual interest from both accounts is $270. How much did he invest at 6%?

16. ANTIFREEZE How much of a 40% antifreeze solution must a mechanic mix with a 70% antifreeze solution if he needs 20 gallons of a 50% antifreeze solution?

SECTION 13.5   *Graphing Linear Inequalities*

An ordered pair (x, y) is a *solution* of an inequality in x and y if a true statement results when the variables are replaced by the coordinates of the ordered pair.

To graph a linear inequality:

1. Graph the *boundary line*. Draw a solid line if the inequality contains $\leq$ or $\geq$ and a broken line if it contains $<$ or $>$.

2. Pick a *test point* on one side of the boundary. Use the origin if possible. Replace x and y with the coordinates of that point. If the inequality is satisfied, shade the side that contains the point. If the inequality is not satisfied, shade the other side.

17. Determine whether each ordered pair is a solution of $2x - y \leq -4$.

a. $(0, 5)$ **b.** $(2, 8)$

c. $(-3, -2)$ **d.** $\left(\frac{1}{2}, -5\right)$

18. Graph each inequality.

a. $x - y < 5$

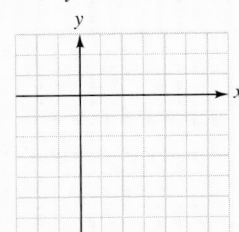

b. $2x - 3y \geq 6$

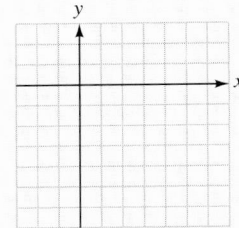

c. $y \leq -2x$

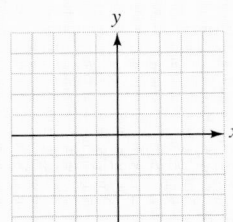

d. $y < -4$

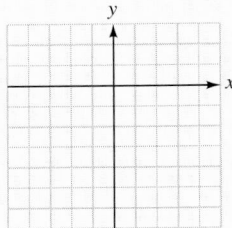

19. In Illustration 4, the graph of a linear inequality is shown. Would a true or a false statement result if the coordinates of

a. point A were substituted into the inequality?

b. point B were substituted into the inequality?

c. point C were substituted into the inequality?

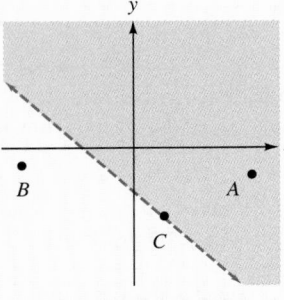

ILLUSTRATION 4   ILLUSTRATION 5

20. WORK SCHEDULE A student told her employer that during the school year, she would be available for up to 30 hours a week, working either 3- or 5-hour shifts. Find an inequality that shows the possible ways to schedule the number of 3-hour (x) and 5-hour shifts (y) she can work, and graph it in Illustration 5. Give three ordered pairs that satisfy the inequality.

SECTION 13.6 | *Solving Systems of Linear Inequalities*

To graph a system of linear inequalities:

1. Graph the individual inequalities of the system on the same coordinate axes.

2. The final solution, if one exists, is that region where all individual graphs intersect.

21. Solve each system of inequalities.

a. $\begin{cases} 5x + 3y < 15 \\ 3x - y > 3 \end{cases}$

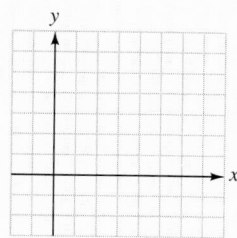

b. $\begin{cases} x \geq 3y \\ y < 3x \end{cases}$

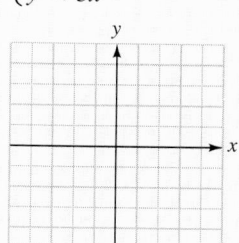

Systems of linear inequalities can be used to solve application problems.

22. GIFT SHOPPING   A grandmother wants to spend at least $40 but no more than $60 on school clothes for her grandson. If T-shirts sell for $10 and pants sell for $20, write a system of inequalities that describes the possible combinations of T-shirts (x) and pants (y) she can buy. Graph the system in Illustration 6. Give two possible solutions.

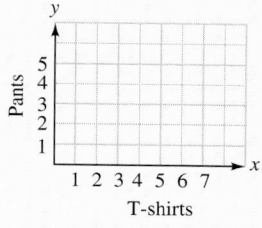

ILLUSTRATION 6

Chapter 13 Test

In Problems 1–2, tell whether the given ordered pair is a solution of the given system.

1. $(5, 3)$, $\begin{cases} 3x + 2y = 21 \\ x + y = 8 \end{cases}$

2. $(-2, -1)$, $\begin{cases} 4x + y = -9 \\ 2x - 3y = -7 \end{cases}$

3. Solve the system by graphing: $\begin{cases} 3x + y = 7 \\ x - 2y = 0 \end{cases}$

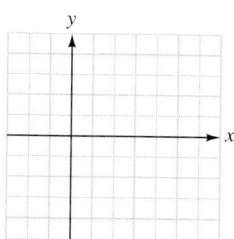

4. To solve a system of two linear equations in x and y, a student used a graphing calculator. From the calculator display in Illustration 1, determine whether the system has a solution. Explain your answer.

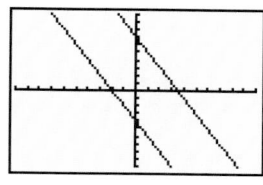

ILLUSTRATION 1

In Problems 5–6, solve each system by substitution.

5. $\begin{cases} y = x - 1 \\ 2x + y = -7 \end{cases}$

6. $\begin{cases} 3a + 4b = -7 \\ 2b - a = -1 \end{cases}$

In Problems 7–8, solve each system by addition.

7. $\begin{cases} 3x - y = 2 \\ 2x + y = 8 \end{cases}$

8. $\begin{cases} 4x + 3y = -3 \\ -3x = -4y + 21 \end{cases}$

In Problems 9–10, classify each system as consistent or inconsistent.

9. $\begin{cases} x + y = 4 \\ x + y = 6 \end{cases}$

10. $\begin{cases} \dfrac{x}{3} + y = 4 \\ x + 3y = 12 \end{cases}$

11. Which method would be most efficient to solve the following system?

$$\begin{cases} 5x - 3y = 5 \\ 3x + 3y = 3 \end{cases}$$

Explain your answer. (You do not need to solve the system.)

12. FINANCIAL PLANNING   A woman invested some money at 8% and some at 9%. The interest for 1 year on the combined investment of $10,000 was $840. How much was invested at 9%? Use a system of equations in two variables to solve this problem.

867

In Problems 13–14, tell whether the given ordered pair is a solution of $2x - 4y > 8$.

13. $(7, 1)$ **14.** $(0, -2)$

15. Graph the inequality $x - y > -2$.

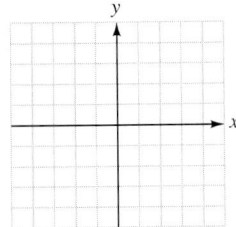

16. Solve the system by graphing.

$$\begin{cases} 2x + 3y \le 6 \\ x \ge 2 \end{cases}$$

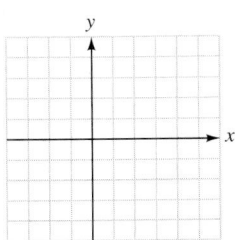

For Problems 17–18, see the graph in Illustration 2, which shows two different ways in which a salesperson can be paid according to the number of items he or she sells.

17. What is the point of intersection of the graphs? Explain its significance.

18. Which plan do you think is better for the salesperson? Explain why.

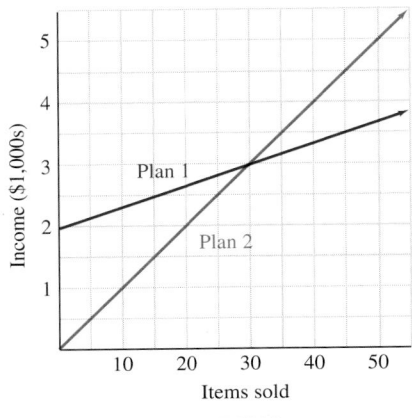

ILLUSTRATION 2

Chapters 1-13 Cumulative Review Exercises

1. STUDY ABROAD Complete the table in Illustration 1, which shows the number of U.S. college students who studied abroad in Israel and Mexico. Round to the nearest tenth of one percent.

Country	'97–'98	'98–'99	% change
Israel	1,988	3,302	
Mexico	7,574	7,363	

Based on data from the Institute of International Education

ILLUSTRATION 1

2. List the set of integers.

Evaluate each expression.

3. $3 - 4[-10 - 4(-5)]$

4. $\dfrac{|-45| - 2(-5) + 1^5}{2 \cdot 9 - 2^4}$

5. AIR CONDITIONING Find the volume of air contained in the duct shown in Illustration 2. Round to the nearest tenth of a cubic foot.

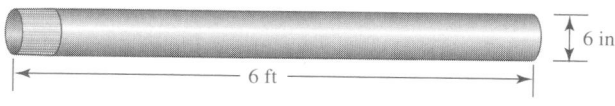

6 ft   6 in.

ILLUSTRATION 2

6. Simplify $3x^2 + 2x^2 - 5x^2$.

Solve each equation.

7. $2 - (4x + 7) = 3 + 2(x + 2)$

8. $\dfrac{2}{5}y + 3 = 9$

9. Solve $-4x + 6 > 17$ and graph the solution set. Then describe the graph using interval notation.

10. ANGLE OF ELEVATION Refer to Illustration 3. Find x.

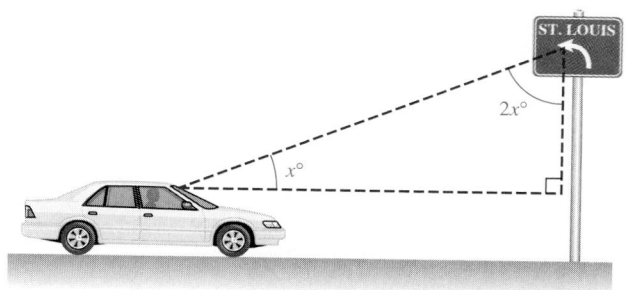

ILLUSTRATION 3

11. STOCK MARKET An investment club invested part of $45,000 in a high-yield mutual fund that earned 12% annual simple interest. The remainder of the money was invested in Treasury bonds that earned 6.5% simple annual interest. The two investments earned $4,300 in one year. How much was invested in each account?

12. Give the formula for
  a. the perimeter of a rectangle
  b. the area of a rectangle
  c. the area of a circle
  d. the distance traveled

Graph each equation.

13. $y = -x^3$

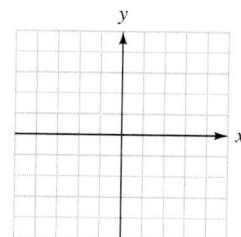

14. $y = -3x + 2$

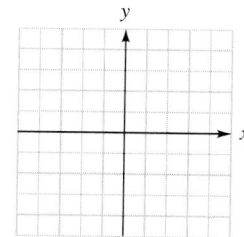

15. $3x + 4y = 8$

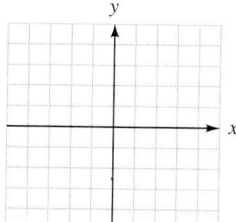

16. $x = -2$

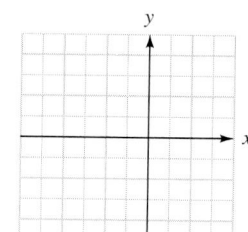

17. Find the slope and y-intercept of the line graphed in Illustration 4. Then write the equation of the line.

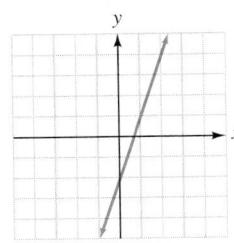

ILLUSTRATION 4

18. Find the slope of the line passing through $(6, -2)$ and $(-3, 2)$.

19. If $f(x) = 4x - 2x^2$, what is $f(-5)$?

20. Write 3,890,000,000 in scientific notation.

Simplify each expression. Write each answer without using parentheses or negative exponents.

21. $(x^5)^2(x^7)^3$

22. $\dfrac{16(aa^2)^3}{2a^2a^3}$

23. $\dfrac{2^{-4}}{3^{-1}}$

24. $(2x)^0$

Perform the indicated operation(s).

25. $(5x - 8y) - (-2x + 5y)$

26. $2x^2(3x^2 + 4x - 7)$

27. $(c + 16)^2$

28. $(x + 3)(2x - 3)$

29. $\dfrac{2x - 32}{16x}$

30. $3x + 1\overline{)9x^2 + 6x + 1}$

31. Prime factor 288.

32. Write a polynomial that is a difference of two squares.

Factor each polynomial completely.

33. $12r^2 - 3rs + 9r^2s^2$

34. $u^2 - 18u + 81$

35. $2y^2 - 7y + 3$

36. $x^4 - 81$

37. $t^3 - v^3$

38. $xy - ty + xs - ts$

Solve each equation.

39. $8s^2 - 16s = 0$

40. $x^2 + 2x - 15 = 0$

41. Simplify $\dfrac{x^2 - 25}{5x + 25}$.

42. Add: $\dfrac{3x}{2y} + \dfrac{5x}{2y}$.

43. Divide: $\dfrac{x^2 - x - 2}{x^2 + x} \div \dfrac{2 - x}{x}$.

44. Subtract: $\dfrac{x + 5}{xy} - \dfrac{x - 1}{x^2y}$.

45. Simplify $\dfrac{\dfrac{y}{x} + 3y}{y + \dfrac{2y}{x}}$.

46. Solve $\dfrac{3r}{2} - \dfrac{3}{r} = 3 + \dfrac{3r}{2}$.

47. The triangles shown in Illustration 5 are similar. Find a and b.

ILLUSTRATION 5

48. Is $(-5, 2)$ a solution of $\begin{cases} -2x + 7y = 24 \\ 3x - 4y = -19 \end{cases}$?

Solve each system by graphing.

49. $\begin{cases} x + 4y = -2 \\ y = -x - 5 \end{cases}$

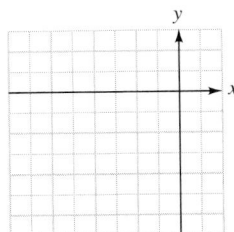

50. $\begin{cases} 2x - 3y < 0 \\ y > x - 1 \end{cases}$

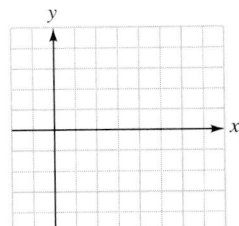

51. Solve $\begin{cases} x - 2y = 2 \\ 2x + 3y = 11 \end{cases}$ by substitution.

52. NUTRITION   Illustration 6 shows per serving nutritional information for egg noodles and rice pilaf. How many servings of each food should be eaten to consume exactly 22 grams of protein and 21 grams of fat?

	Protein (g)	Fat (g)
Egg noodles	5	3
Rice pilaf	4	5

ILLUSTRATION 6

Roots and Radicals

14

14.1 Square Roots

14.2 Higher-Order Roots;
Radicands That Contain
Variables

14.3 Simplifying Radical
Expressions

14.4 Adding and Subtracting
Radical Expressions

14.5 Multiplying and Dividing
Radical Expressions

14.6 Solving Radical
Equations; the Distance
Formula

14.7 Rational Exponents

Key Concept: Inverse
Operations

Accent on Teamwork

Chapter Review

Chapter Test

Cumulative Review
Exercises

TO SOLVE MANY APPLIED PROBLEMS, WE MUST DETERMINE WHAT NUMBER X MUST BE SQUARED TO OBTAIN ANOTHER NUMBER N. WE CALL X THE SQUARE ROOT OF N.

14.1 Square Roots

In this section, you will learn about

- Square roots • Approximating square roots
- Rational, irrational, and imaginary numbers • The square root function
- The Pythagorean theorem

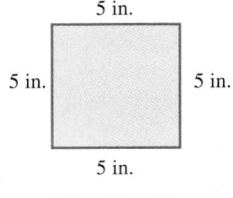

5 in.

5 in.   5 in.

5 in.

FIGURE 14-1

INTRODUCTION. To find the area A of the square shown in Figure 14-1, we multiply its length by its width.

$$A = l \cdot w$$
$$A = 5 \cdot 5$$
$$= 25$$

The area is 25 square inches.

We have seen that the product $5 \cdot 5$ can be denoted by the exponential expression 5^2, where 5 is raised to the second power. Whenever we raise a number to the second power, we are squaring it, or finding its **square.** This example illustrates that the formula for the area of a square with sides of length s is $A = s^2$.

Here are some more squares of numbers:

- The square of 3 is 9, because $3^2 = 9$.
- The square of -3 is 9, because $(-3)^2 = 9$.
- The square of 12 is 144, because $12^2 = 144$.
- The square of -12 is 144, because $(-12)^2 = 144$.
- The square of $\frac{1}{8}$ is $\frac{1}{64}$, because $\left(\frac{1}{8}\right)^2 = \frac{1}{8} \cdot \frac{1}{8} = \frac{1}{64}$.
- The square of $-\frac{1}{8}$ is $\frac{1}{64}$, because $\left(-\frac{1}{8}\right)^2 = \left(-\frac{1}{8}\right)\left(-\frac{1}{8}\right) = \frac{1}{64}$.
- The square of 0 is 0, because $0^2 = 0$.

In this section, we will reverse the squaring process and find **square roots** of numbers. We will consider the square root function and introduce the Pythagorean theorem. Finally, we will solve several application problems.

Square roots

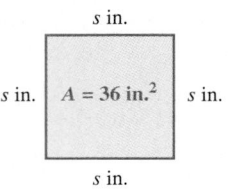

s in.

s in. $A = 36$ in.² s in.

s in.

FIGURE 14-2

Suppose we know that the area of the square shown in Figure 14-2 is 36 square inches. To find the length of each side, we substitute 36 for A in the formula $A = s^2$ and solve for s.

$$A = s^2$$
$$36 = s^2$$

To solve for s, we must find a positive number whose square is 36. Since 6 is such a number, the sides of the square are 6 inches long. The number 6 is called a *square root* of 36, because 6 is the positive number that we square to get 36.

Here are some more square roots of numbers:

- 3 is a square root of 9, because $3^2 = 9$.
- -3 is a square root of 9, because $(-3)^2 = 9$.
- 12 is a square root of 144, because $12^2 = 144$.
- -12 is a square root of 144, because $(-12)^2 = 144$.
- $\frac{1}{8}$ is a square root of $\frac{1}{64}$, because $\left(\frac{1}{8}\right)^2 = \left(\frac{1}{8}\right)\left(\frac{1}{8}\right) = \frac{1}{64}$.
- $-\frac{1}{8}$ is a square root of $\frac{1}{64}$, because $\left(-\frac{1}{8}\right)^2 = \left(-\frac{1}{8}\right)\left(-\frac{1}{8}\right) = \frac{1}{64}$.
- 0 is a square root of 0, because $0^2 = 0$.

In general, we have the following definition.

Square root

> The number b is a **square root** of a if $b^2 = a$.

All positive numbers have two square roots, one that is positive and one that is negative. The two square roots of 9 are 3 and -3, and the two square roots of 144 are 12 and -12. The number 0 is the only number that has one square root, which is 0.

The **principal square root** of a positive number is its positive square root. Although 3 and -3 are both square roots of 9, only 3 is the principal square root. The symbol $\sqrt{}$, called a **radical symbol,** is used to represent the principal square root of a number, and $-\sqrt{}$ is used to represent the negative square root of a number. For example, $\sqrt{9} = 3$ and $-\sqrt{9} = -3$. Likewise, $\sqrt{144} = 12$ and $-\sqrt{144} = -12$.

Principal square root

> If $a > 0$, the expression $\sqrt{a}$ represents the **principal** (or positive) **square root** of a.
> The principal square root of 0 is 0: $\sqrt{0} = 0$.

The number (or expression) under a radical sign is called the **radicand.** In $\sqrt{9}$, the number 9 is the radicand, and the entire symbol $\sqrt{9}$ is called a **radical.** We read $\sqrt{9}$ as either "the square root of 9" or as "radical 9."

An algebraic expression containing a radical is called a **radical expression.** In this chapter, we will consider radical expressions such as

$$\sqrt{49}, \qquad \frac{5}{\sqrt{3}}, \qquad -2\sqrt{x+1}, \qquad \text{and} \qquad \sqrt{28y^2} - 2y\sqrt{63}$$

EXAMPLE 1 *Finding square roots.* Find each square root.

a. $\sqrt{0} = 0$ **b.** $\sqrt{1} = 1$ **c.** $\sqrt{225} = 15$ **d.** $\sqrt{1.44} = 1.2$

e. $\sqrt{576} = 24$ **f.** $-\sqrt{4} = -2$ **g.** $-\sqrt{900} = -30$ **h.** $\sqrt{\dfrac{4}{9}} = \dfrac{2}{3}$

Self Check

Find each square root:

a. $\sqrt{121}$,   **b.** $-\sqrt{49}$,
c. $\sqrt{0.64}$,   **d.** $\sqrt{256}$,
e. $\sqrt{\frac{1}{25}}$,   **f.** $\sqrt{\frac{9}{49}}$

Answers: a. 11,   **b.** -7,
c. 0.8,   **d.** 16,   **e.** $\frac{1}{5}$,   **f.** $\frac{3}{7}$ ■

Square roots of certain numbers, such as 7, are hard to compute by hand. However, we can find $\sqrt{7}$ with a calculator.

Approximating square roots

To find the principal square root of 7, we can enter 7 into a scientific calculator and press the $\sqrt{x}$ key. The approximate value of $\sqrt{7}$ will appear on the display.

$$\sqrt{7} \approx 2.6457513 \qquad \text{Read} \approx \text{as "is approximately equal to."}$$

Since $\sqrt{7}$ represents the number that, when squared, gives 7, we would expect squares of approximations of $\sqrt{7}$ to be close to 7.

- Rounded to one decimal place, $\sqrt{7} \approx 2.6$ and $(2.6)^2 = 6.76$.
- Rounded to two decimal places, $\sqrt{7} \approx 2.65$ and $(2.65)^2 = 7.0225$.
- Rounded to three decimal places, $\sqrt{7} \approx 2.646$ and $(2.646)^2 = 7.001316$.

Accent on Technology: **Freeway road sign:**

The sign shown in Figure 14-3 is in the shape of an equilateral triangle, and we can find its height h using the formula

$$h = \frac{\sqrt{3}s}{2}$$

where s is the length of a side of the triangle. In this case, $s = 24$ inches, so we have

$$h = \frac{\sqrt{3}(24)}{2} \qquad \sqrt{3}(24) \text{ means } \sqrt{3} \cdot 24.$$

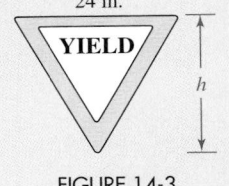

FIGURE 14-3

To evaluate this expression with a scientific calculator, we enter these numbers and press these keys.

Keystrokes $\boxed{(}\ 3\ \boxed{\sqrt{x}}\ \boxed{\times}\ 24\ \boxed{)}\ \boxed{\div}\ 2\ \boxed{=}$ $\boxed{20.784609}$

To evaluate this expression using a graphing calculator, we press these keys.

Keystrokes $\boxed{2\text{nd}}\ \boxed{\sqrt{\ }}\ 3\ \boxed{)}\ \boxed{\times}\ 24\ \boxed{\div}\ 2\ \boxed{\text{ENTER}}$ $\boxed{\begin{array}{l}\sqrt{\ }(3)*24/2\\ \qquad 20.78460969\end{array}}$

The height of the sign is approximately 21 inches.

Rational, irrational, and imaginary numbers

Whole numbers such as 4, 9, 16, and 49 are called **integer squares,** because each one is the square of an integer. The square root of any integer square is an integer and therefore a rational number:

$$\sqrt{4} = 2, \qquad \sqrt{9} = 3, \qquad \sqrt{16} = 4, \quad \text{and} \quad \sqrt{49} = 7$$

The square root of any whole number that is not an integer square is an **irrational number.** For example, $\sqrt{7}$ is an irrational number. Recall that the set of rational numbers and the set of irrational numbers together make up the set of real numbers.

 COMMENT Square roots of negative numbers are not real numbers. For example, $\sqrt{-4}$ is nonreal, because the square of no real number is -4. The number $\sqrt{-4}$ is an example from a set of numbers called **imaginary numbers.** Remember: *The square root of a negative number is not a real number.*

If we attempt to evaluate $\sqrt{-4}$ using a calculator, an error message like the ones shown below will be displayed.

$$\boxed{\text{Error}} \qquad \boxed{\begin{array}{l}\text{ERR:NONREAL ANS}\\ \blacksquare\text{:Quit}\\ \text{2:Goto}\end{array}}$$

Scientific calculator   **Graphing calculator**

In this chapter, we will assume that *all radicands under the square root symbols are either positive or zero.* Thus, all square roots will be real numbers.

The square root function

Since there is one principal square root for every nonnegative real number x, the equation $f(x) = \sqrt{x}$ determines a square root function. For example, the value that is determined by $f(x) = \sqrt{x}$ when $x = 4$ is denoted by $f(4)$, and we have $f(4) = \sqrt{4} = 2$.

To graph this function, we make a table of values and plot each ordered pair. In the table, we chose five values for x that are integer squares. This made computing $f(x)$ quite simple. The graph appears in Figure 14-4.

$$f(x) = \sqrt{x}$$

x	$f(x)$	$(x, f(x))$
0	0	(0, 0)
1	1	(1, 1)
4	2	(4, 2)
9	3	(9, 3)
16	4	(16, 4)

↑ Values to be input into $\sqrt{x}$ ↑ Output values   ↑ Ordered pairs to plot

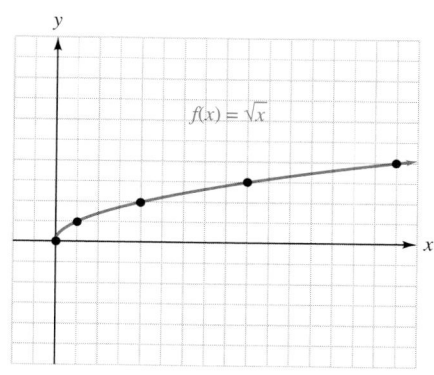

FIGURE 14-4

EXAMPLE 2   *Period of a pendulum.*   The *period* of a pendulum is the time required for the pendulum to swing back and forth to complete one cycle. (See Figure 14-5.) The period (in seconds) of a pendulum having length L (in feet) is approximated by the function

$$f(L) = 1.11\sqrt{L}$$

Find the period of a pendulum that is 5 feet long.

Solution

We substitute 5 for L in the formula and multiply using a calculator.

$$f(L) = 1.11\sqrt{L}$$
$$f(5) = 1.11\sqrt{5} \qquad \text{$1.11\sqrt{5}$ means $1.11 \cdot \sqrt{5}$.}$$
$$\approx 2.482035455$$

The period is approximately 2.5 seconds.

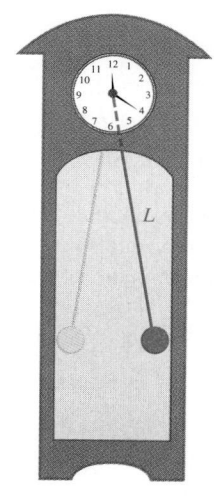

FIGURE 14-5

Self Check

Find the period of a pendulum that is 3 feet long.

Answer: about 1.9 sec   ∎

The Pythagorean theorem

The longest side of a right triangle is the **hypotenuse,** which is the side opposite the right angle. The remaining two sides are the **legs** of the triangle. See Figure 14-6. Recall that the **Pythagorean theorem** provides a formula relating the lengths of the three sides of a right triangle.

c Hypotenuse   Leg b Leg a

FIGURE 14-6

The Pythagorean theorem

> If the length of the hypotenuse of a right triangle is c and the lengths of the two legs are a and b,
> $$c^2 = a^2 + b^2$$

Since the lengths of the sides of a triangle are positive numbers, we can use the **square root property of equality** and the Pythagorean theorem to find the length of the third side of any right triangle when the measures of two sides are given.

Square root property of equality	If a and b represent positive numbers, and if $a = b$, $$\sqrt{a} = \sqrt{b}$$

EXAMPLE 3   *Picture frame.*   After gluing and nailing together two pieces of picture frame molding, a frame maker checks her work by making a diagonal measurement. (See Figure 14-7.) If the sides of the frame form a right angle, what measurement should the frame maker read on the yardstick?

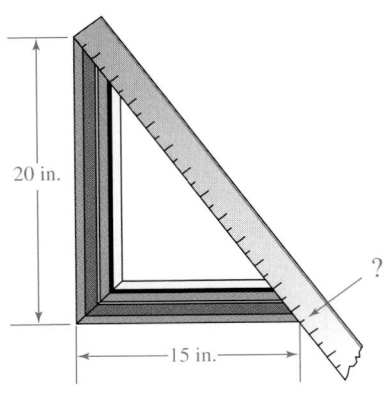

Solution   If the sides of the frame form a right angle, the sides and the diagonal form a right triangle. The lengths of the legs of the right triangle are 15 inches and 20 inches. We can find c, the length of the hypotenuse, using the Pythagorean theorem.

FIGURE 14-7

$$c^2 = a^2 + b^2 \qquad \text{The Pythagorean theorem.}$$
$$c^2 = 15^2 + 20^2 \qquad \text{Substitute 15 for } a \text{ and 20 for } b.$$
$$c^2 = 225 + 400 \qquad 15^2 = 225 \text{ and } 20^2 = 400.$$
$$c^2 = 625 \qquad \text{Do the addition: } 225 + 400 = 625.$$

To find c, we must find a number that, when squared, is 625. There are two such numbers, one positive and one negative. They are called the *square roots* of 625. Since c represents the length of the hypotenuse, c cannot be negative. Thus, we need only determine the positive square root of 625.

$$c^2 = 625 \qquad \text{The equation to solve.}$$
$$\sqrt{c^2} = \sqrt{625} \qquad \text{To find } c \text{, we undo the operation performed on it by taking the positive}$$
square root of both sides. Recall that a radical symbol $\sqrt{}$ is used to indicate the positive square root of a number.
$$c = 25 \qquad \sqrt{c^2} = c \text{ because } (c)^2 = c^2 \text{, and } \sqrt{625} = 25 \text{ because } 25^2 = 625.$$

The diagonal distance should measure 25 inches. If it does not, the sides of the frame do not form a right angle.   ■

COMMENT   When using the Pythagorean theorem $c^2 = a^2 + b^2$, we can let a represent the length of either leg of the right triangle in question. We then let b represent the length of the other leg. The variable c must always represent the length of the hypotenuse.

EXAMPLE 4   *Building a high ropes adventure course.*   The builder of a high ropes course wants to use a 25-foot cable to stabilize the pole shown in Figure 14-8. To be safe, the ground anchor stake must be farther than 18 feet from the base of the pole. Is the cable long enough to use?

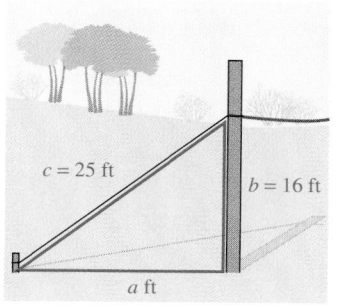

FIGURE 14-8

Solution   We can use the Pythagorean theorem, with $b = 16$ and $c = 25$, to find a.

$$c^2 = a^2 + b^2$$
$$25^2 = a^2 + 16^2 \qquad \text{Substitute 25 for } c \text{ and 16 for } b.$$
$$625 = a^2 + 256 \qquad 25^2 = 625 \text{ and } 16^2 = 256.$$
$$369 = a^2 \qquad \qquad \text{To isolate } a^2, \text{ subtract 256 from both sides.}$$
$$\sqrt{369} = \sqrt{a^2} \qquad \text{To find } a, \text{ we undo the operation that is performed on it}$$
$$\text{(squaring) by taking the positive square root of both sides.}$$
$$19.209372 \approx a \qquad \text{Use a calculator to approximate } \sqrt{369}.$$

Since the anchor stake will be more than 18 feet from the base, the 25-foot cable is long enough to use.   ■

EXAMPLE 5   ***Reach of a ladder.***   A 26-foot ladder rests against the side of a building. If the base of the ladder is 10 feet from the wall, how far up the building will the ladder reach?

Analyze the problem   The wall, the ground, and the ladder form a right triangle, as shown in Figure 14-9. In this triangle, the hypotenuse is 26 feet, and one of the legs is the base-to-wall distance of 10 feet. We can let x represent the length of the other leg, which is the distance that the ladder will reach up the wall.

Form an equation   We can use the Pythagorean theorem to form the equation.

The hypotenuse squared	is	one leg squared	plus	the other leg squared.
26^2	$=$	10^2	$+$	x^2

Solve the equation

$$26^2 = 10^2 + x^2$$
$$676 = 100 + x^2 \qquad 26^2 = 676 \text{ and } 10^2 = 100.$$
$$676 - 100 = x^2 \qquad \text{To isolate } x^2, \text{ subtract 100 from both sides.}$$
$$576 = x^2 \qquad 676 - 100 = 576.$$
$$\sqrt{576} = \sqrt{x^2} \qquad \text{Take the positive square root of both sides.}$$
$$24 = x \qquad \sqrt{576} = 24 \text{ and } \sqrt{x^2} = x, \text{ because } x \cdot x = x^2.$$

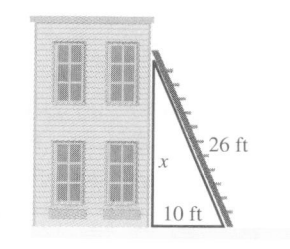

FIGURE 14-9

State the conclusion   The ladder will reach 24 feet up the side of the building.

Check the result   If the ladder reaches 24 feet up the side of the building, we have $10^2 + 24^2 = 100 + 576 = 676$, which is 26^2. The answer, 24, checks.   ■

    EXAMPLE 6   ***Roof design.***   The gable end of the roof shown in Figure 14-10 is an isosceles right triangle with a span of 48 feet. Find the distance from the eaves to the peak.

Analyze the problem   The two equal sides of the isosceles triangle are the two legs of the right triangle, and the span of 48 feet is the length of the hypotenuse. We can let x represent the length of each leg, which is the distance from eaves to peak.

Form an equation   We can use the Pythagorean theorem to form the equation.

The hypotenuse squared	is	one leg squared	plus	the other leg squared.
48^2	$=$	x^2	$+$	x^2

Solve the equation

$$48^2 = x^2 + x^2$$

$$2,304 = 2x^2 \qquad \text{$48^2 = 2,304$ and}$$
$$\qquad\qquad\qquad x^2 + x^2 = 2x^2.$$

$$1,152 = x^2 \qquad \text{To isolate x^2, divide}$$
$$\qquad\qquad\qquad \text{both sides by 2.}$$

$$\sqrt{1,152} = \sqrt{x^2} \qquad \text{Take the positive square}$$
$$\qquad\qquad\qquad \text{root of both sides.}$$

$$33.9411255 \approx x \qquad \text{Use a calculator to find}$$
$$\qquad\qquad\qquad \text{the approximate value}$$
$$\qquad\qquad\qquad \text{of $\sqrt{1,152}$.}$$

FIGURE 14-10

State the conclusion   The eaves-to-peak distance of the roof is approximately 34 feet.

Check the result   If the eaves-to-peak distance is approximately 34 feet, we have $34^2 + 34^2 = 1,156 + 1,156 = 2,312$, which is approximately 48^2. The answer, 34, seems reasonable. ■

STUDY SET   Section 14.1

VOCABULARY   *Fill in the blanks.*

1. b is a _____ root of a if $b^2 = a$.

2. The symbol $\sqrt{}$ is called a _____ symbol.

3. The principal square root of a positive number is a _____ number.

4. The number under the radical sign is called the _____.

5. If a triangle has a right angle, it is called a _____ triangle.

6. The longest side of a right triangle is called the _____, and the other two sides are called _____.

CONCEPTS   *Fill in the blanks.*

7. The number 25 has _____ square roots. They are   and   .

8. $\sqrt{-11}$ is not a _____ number.

9. If the length of the hypotenuse of a right triangle is c and the legs are a and b, then $c^2 = $ _____.

10.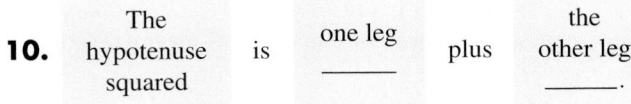

11. If a and b are positive numbers and $a = b$, then $\sqrt{a} = $ _____.

12. The _____ of 2 is 4, because $2^2 = 4$, and 2 is a square _____ of 4, because $2^2 = 4$.

13. To isolate x, what step should be used to "undo" the operation performed on it? (Assume that x is a positive number.)
 a. $2x = 16$
 b. $x^2 = 16$

14. Graph each number on the number line.
$$\left\{ \sqrt{16}, -\sqrt{\tfrac{9}{4}}, \sqrt{1.8}, \sqrt{6}, -\sqrt{23} \right\}$$

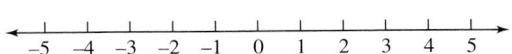

15. Complete the table of values. *Do not use a calculator.*

x	$\sqrt{x}$
0	
$\frac{1}{81}$	
0.16	
36	
400	

16. If $f(x) = \sqrt{x}$, find each value. *Do not use a calculator.*
 a. $f\left(\dfrac{1}{121}\right)$ **b.** $f(1)$
 c. $f(0.25)$ **d.** $f(81)$
 e. $f(900)$

17. a. What do the dashed lines in the graph in Illustration 1 help to approximate?
 b. Use the graph to approximate $\sqrt{3}$ and $\sqrt{8}$.

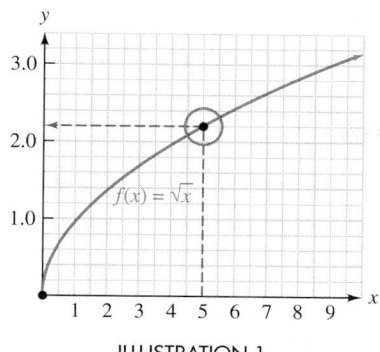

ILLUSTRATION 1

18. A calculator was used to find $\sqrt{-16}$. Explain the message shown on the calculator display in Illustration 2.

ILLUSTRATION 2

NOTATION   *Complete each solution.*

19. If the legs of a right triangle measure 5 and 12 centimeters, find the length of the hypotenuse.

$$c^2 = a^2 + b^2$$
$$c^2 = \boxed{}^2 + \boxed{}^2$$
$$c^2 = 25 + \boxed{}$$
$$c^2 = \boxed{}$$
$$\boxed{} = \sqrt{169}$$
$$c = 13$$

20. If the hypotenuse of a right triangle measures 25 centimeters and one leg measures 24 centimeters, find the length of the other leg.

$$c^2 = a^2 + b^2$$
$$\boxed{}^2 = \boxed{}^2 + b^2$$
$$625 = \boxed{} + b^2$$
$$\boxed{} = b^2$$
$$\sqrt{49} = \boxed{}$$
$$7 = b$$

21. Is the statement $-\sqrt{9} = \sqrt{-9}$ true or false? Explain your answer.

22. Consider the statement $\sqrt{26} \approx 5.1$. Explain why an $\approx$ symbol is used instead of an $=$ sign.

PRACTICE   *Find each square root without using a calculator.*

23. $\sqrt{25}$

24. $\sqrt{49}$

25. $-\sqrt{81}$

26. $-\sqrt{36}$

27. $\sqrt{1.21}$

28. $\sqrt{1.69}$

29. $\sqrt{196}$

30. $\sqrt{169}$

31. $\sqrt{\dfrac{9}{256}}$

32. $\sqrt{\dfrac{49}{225}}$

33. $-\sqrt{289}$

34. $-\sqrt{324}$

35. $-\sqrt{2,500}$

36. $-\sqrt{625}$

37. $\sqrt{3,600}$

38. $\sqrt{1,600}$

Use a calculator to evaluate each expression to three decimal places.

39. $\sqrt{2}$

40. $\sqrt{3}$

41. $\sqrt{11}$

42. $\sqrt{53}$

43. $\sqrt{95}$

44. $\sqrt{99}$

45. $\sqrt{428}$

46. $\sqrt{844}$

47. $-\sqrt{9,876}$

48. $-\sqrt{3,619}$

49. $\sqrt{21.35}$

50. $\sqrt{13.78}$

51. $\sqrt{0.3588}$

52. $\sqrt{0.9999}$

53. $-\sqrt{0.8372}$

54. $-\sqrt{0.4279}$

55. $2\sqrt{3}$

56. $3\sqrt{2}$

57. $\dfrac{2 + \sqrt{3}}{2}$

58. $\dfrac{2 - \sqrt{3}}{2}$

Tell whether each number in each set is rational, irrational, or imaginary.

59. $\left\{\sqrt{9},\ \sqrt{17},\ \sqrt{49},\ \sqrt{-49}\right\}$

60. $\left\{-\sqrt{5},\ \sqrt{0},\ \sqrt{-100},\ -\sqrt{225}\right\}$

Complete the table and then graph the function.

61. $f(x) = 1 + \sqrt{x}$

x	$f(x)$
0	
1	
4	
9	
16	

62. $f(x) = -1 + \sqrt{x}$

x	$f(x)$
0	
1	
4	
9	
16	

63. $f(x) = -\sqrt{x}$

x	$f(x)$
0	
1	
4	
9	
16	

64. $f(x) = 1 - \sqrt{x}$

x	$f(x)$
0	
1	
4	
9	
16	

Refer to the right triangle in Illustration 3. Find the length of the unknown side.

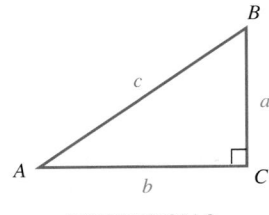

ILLUSTRATION 3

65. Find c if $a = 4$ and $b = 3$.
66. Find c if $a = 5$ and $b = 12$.
67. Find b if $a = 15$ and $c = 17$.
68. Find b if $a = 21$ and $c = 29$.
69. Find a if $b = 16$ and $c = 34$.
70. Find a if $b = 45$ and $c = 53$.
71. Find b if $c = 125$ and $a = 44$.
72. Find c if $a = 176$ and $b = 57$.

APPLICATIONS *Use a calculator to help solve each problem. If an answer is not exact, give it to the nearest tenth.*

73. ADJUSTING A LADDER A 20-foot ladder reaches a window 16 feet above the ground. How far from the wall is the base of the ladder?

74. LINE OF SIGHT A movie viewer in a car parked at a drive-in theater sits 600 feet from the base of the vertical screen. What is the line-of-sight distance for the viewer to the middle of the screen, which is 35 feet above the base?

75. QUALITY CONTROL How can a tool manufacturer use the Pythagorean theorem to verify that the two sides of the carpenter's square shown in Illustration 4 meet to form a 90° angle?

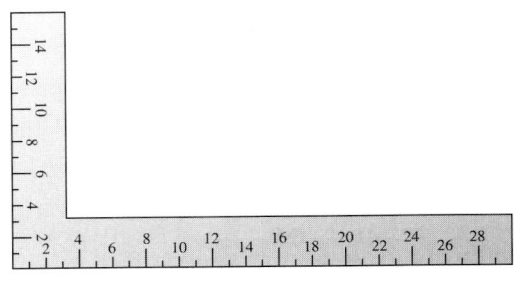

ILLUSTRATION 4

76. GARDENING A rectangular garden has sides of 28 and 45 feet. Find the length of a path that extends from one corner to the opposite corner.

77. BASEBALL A baseball diamond is a square, with each side 90 feet long, as shown in Illustration 5. How far is it from home plate to second base?

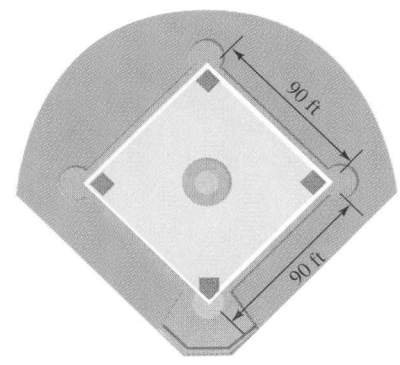

ILLUSTRATION 5

78. TELEVISION The *size* of a television screen is the diagonal distance from the upper left to the lower right corner. What is the size of the screen shown in Illustration 6?

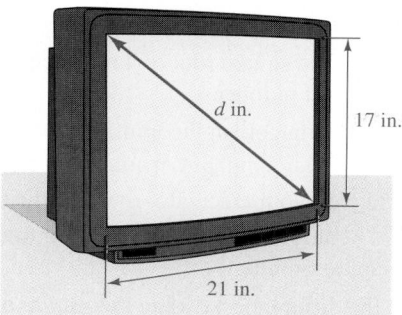

ILLUSTRATION 6

79. FINDING LOCATION   A team of archaeologists travels 4.2 miles east and then 4.0 miles north of their base camp to explore some ancient ruins. "As the crow flies," how far from their base camp are they?

80. TAKING A SHORTCUT   Instead of walking on the sidewalk, students take a diagonal shortcut across the rectangular vacant lot shown in Illustration 7. How much distance do they save?

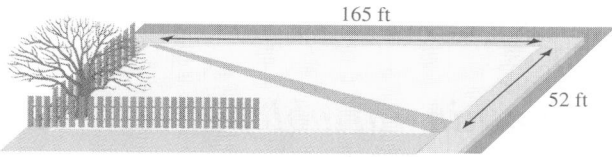

ILLUSTRATION 7

81. FOOTBALL   On first down and ten, a quarterback tells his tight end to go out 6 yards, cut 45° to the right, and run 5 yards, as shown in Illustration 8. The tight end follows instructions, catches a pass, and is tackled immediately. Does he gain the necessary 10 yards for a first down?

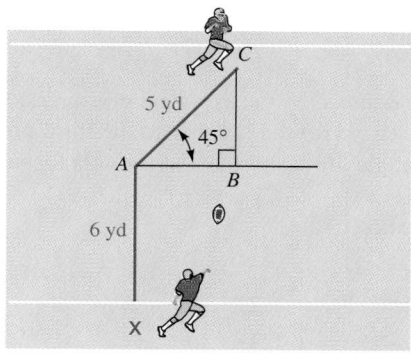

ILLUSTRATION 8

82. GEOMETRY   The legs of a right triangle are equal, and the hypotenuse is 2.82843 units long. Find the length of each leg.

83. PROFESSIONAL WRESTLING   The sides of a square wrestling ring are 18 feet long. Find the distance from one corner to the opposite corner.

84. PERIMETER OF A SQUARE   The diagonal of a square is 3 feet long. Find its perimeter.

85. HEIGHT OF A TRIANGLE   Find the area of the isosceles triangle shown in Illustration 9.

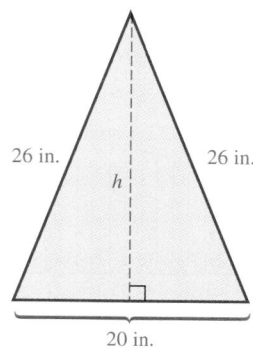

ILLUSTRATION 9

86. INTERIOR DECORATING   The square table in Illustration 10 is covered by a circular tablecloth. If the sides of the table are 2 feet long, find the area of the tablecloth.

ILLUSTRATION 10

87. DRAFTING   Among the tools used in drafting are the 30–60–90 and the 45–45–90 triangles shown in Illustration 11.
   a. Find the length of the hypotenuse of the 45–45–90 triangle if it is $\sqrt{2}$ times as long as a leg.
   b. Find the length of the side opposite the 60° angle of the other triangle if it is $\frac{\sqrt{3}}{2}$ times as long as the hypotenuse.

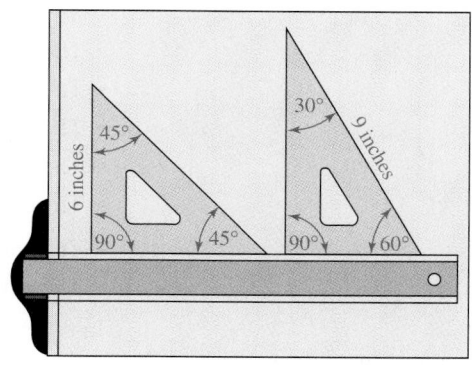

ILLUSTRATION 11

88. ORGAN PIPES The design for a set of brass pipes for a church organ is shown in Illustration 12. Find the length of each pipe (to the nearest tenth of a foot), and then find the total length of pipe needed to construct this set.

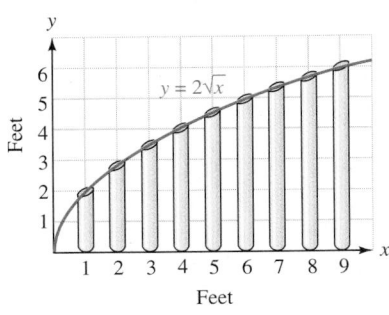

ILLUSTRATION 12

WRITING

89. Explain why the square root of a negative number cannot be a real number.

90. Explain the Pythagorean theorem.

91. Suppose you are told that $\sqrt{10} \approx 3.16$. Explain how another key on your calculator (besides the square root key $\sqrt{}$) could be used to see whether this is a reasonable approximation.

92. Explain the difference between the *square* of a number and the *square root* of a number.

REVIEW

93. Add: $(3s^2 - 3s - 2) + (3s^2 + 4s - 3)$.

94. Subtract: $(3c^2 - 2c + 4) - (c^2 - 3c + 7)$.

95. Multiply: $(3x - 2)(x + 4)$.

96. Divide: $x^2 + 13x + 12$ by $x + 1$.

14.2 *Higher-Order Roots; Radicands That Contain Variables*

In this section, you will learn about

- Cube roots • Approximating cube roots • The cube root function
- Higher-order roots • Radicands that contain variables

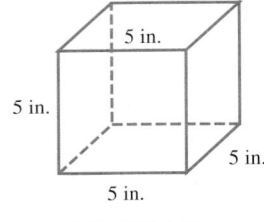

FIGURE 14-11

INTRODUCTION. To find the volume V of the cube shown in Figure 14-11, we multiply its length, width, and height.

$$V = l \cdot w \cdot h$$
$$V = 5 \cdot 5 \cdot 5$$
$$= 125$$

The volume is 125 cubic inches.

We have seen that $5 \cdot 5 \cdot 5$ can be denoted by the exponential expression 5^3, where 5 is raised to the third power. Whenever we raise a number to the third power, we are cubing it, or finding its **cube.** This example illustrates that the formula for the volume of a cube with each side of length s is $V = s^3$.

Here are some more cubes of numbers:

- The cube of 3 is 27, because $3^3 = 27$.
- The cube of -3 is -27, because $(-3)^3 = -27$.
- The cube of 12 is 1,728, because $12^3 = 1,728$.
- The cube of -12 is $-1,728$, because $(-12)^3 = -1,728$.
- The cube of $\frac{1}{4}$ is $\frac{1}{64}$, because $\left(\frac{1}{4}\right)^3 = \frac{1}{4} \cdot \frac{1}{4} \cdot \frac{1}{4} = \frac{1}{64}$.
- The cube of $-\frac{1}{4}$ is $-\frac{1}{64}$, because $\left(-\frac{1}{4}\right)^3 = \left(-\frac{1}{4}\right)\left(-\frac{1}{4}\right)\left(-\frac{1}{4}\right) = -\frac{1}{64}$.
- The cube of 0 is 0, because $0^3 = 0$.

In this section, we will reverse the cubing process and find **cube roots** of numbers. We will also consider fourth roots, fifth roots, and so on. After graphing the cube root function, we will work with radical expressions having radicands containing variables.

Cube roots

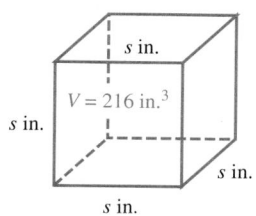

FIGURE 14-12

Suppose we know that the volume of the cube shown in Figure 14-12 is 216 cubic inches. To find the length of each side, we substitute 216 for V in the formula $V = s^3$ and solve for s.

$$V = s^3$$
$$216 = s^3$$

To solve for s, we must find a number whose cube is 216. Since 6 is such a number, the sides of the cube are 6 inches long. The number 6 is called a *cube root* of 216, because $6^3 = 216$.

Here are more examples of cube roots:

- 3 is a cube root of 27, because $3^3 = 27$.
- -3 is a cube root of -27, because $(-3)^3 = -27$.
- 12 is a cube root of 1,728, because $12^3 = 1,728$.
- -12 is a cube root of $-1,728$, because $(-12)^3 = -1,728$.
- $\frac{1}{4}$ is a cube root of $\frac{1}{64}$, because $\left(\frac{1}{4}\right)^3 = \left(\frac{1}{4}\right)\left(\frac{1}{4}\right)\left(\frac{1}{4}\right) = \frac{1}{64}$.
- $-\frac{1}{4}$ is a cube root of $-\frac{1}{64}$, because $\left(-\frac{1}{4}\right)^3 = \left(-\frac{1}{4}\right)\left(-\frac{1}{4}\right)\left(-\frac{1}{4}\right) = -\frac{1}{64}$.
- 0 is a cube root of 0, because $0^3 = 0$.

In general, we have the following definition.

Cube root

> The number b is a **cube root** of a if $b^3 = a$.

All real numbers have one real cube root. As the preceding examples show, a positive number has a positive cube root, a negative number has a negative cube root, and the cube root of 0 is 0.

Cube root notation

> The **cube root of a** is denoted by $\sqrt[3]{a}$. By definition,
> $$\sqrt[3]{a} = b \qquad \text{if} \qquad b^3 = a$$

EXAMPLE 1   *Finding cube roots.*   Find each cube root.

a. $\sqrt[3]{8} = 2$, because $2^3 = 8$

b. $\sqrt[3]{343} = 7$, because $7^3 = 343$

c. $\sqrt[3]{-8} = -2$, because $(-2)^3 = -8$

d. $\sqrt[3]{-125} = -5$, because $(-5)^3 = -125$

Self Check
Find each cube root:

a. $\sqrt[3]{64}$, **b.** $\sqrt[3]{-64}$, **c.** $\sqrt[3]{216}$

Answers: **a.** 4,   **b.** -4,
c. 6   ∎

EXAMPLE 2   *Finding cube roots.*   Find each cube root.

a. $\sqrt[3]{\dfrac{1}{8}} = \dfrac{1}{2}$, because $\left(\dfrac{1}{2}\right)^3 = \dfrac{1}{2}\cdot\dfrac{1}{2}\cdot\dfrac{1}{2} = \dfrac{1}{8}$

b. $\sqrt[3]{-\dfrac{125}{27}} = -\dfrac{5}{3}$, because $\left(-\dfrac{5}{3}\right)^3 = \left(-\dfrac{5}{3}\right)\left(-\dfrac{5}{3}\right)\left(-\dfrac{5}{3}\right) = -\dfrac{125}{27}$

Self Check
Find each cube root:

a. $\sqrt[3]{\dfrac{1}{27}}$ **b.** $\sqrt[3]{-\dfrac{8}{125}}$

Answers: **a.** $\frac{1}{3}$,   **b.** $-\frac{2}{5}$ ∎

Cube roots of numbers such as 7 are hard to compute by hand. However, we can find $\sqrt[3]{7}$ with a calculator.

Approximating cube roots

To find $\sqrt[3]{7}$, we can enter 7 into a scientific calculator, press the root key $\boxed{\sqrt[x]{y}}$, enter 3, and press the $=$ key. The approximate value of $\sqrt[3]{7}$ will appear on the calculator's display.

$$\sqrt[3]{7} \approx 1.912931183$$

If your scientific calculator doesn't have a $\boxed{\sqrt[x]{y}}$ key, you can use the $\boxed{y^x}$ key. We will see later that $\sqrt[3]{7} = 7^{1/3}$. To find the value of $7^{1/3}$, we enter 7 into the calculator and press these keys:

$$7 \ \boxed{y^x} \ \boxed{(} \ 1 \ \boxed{\div} \ 3 \ \boxed{)} \ \boxed{=}$$

The display will read 1.912931183.

Since $\sqrt[3]{7}$ represents the number that, when cubed, gives 7, we would expect cubes of approximations of $\sqrt[3]{7}$ to be close to 7.

- Rounded to one decimal place, $\sqrt[3]{7} \approx 1.9$, and $(1.9)^3 = 6.859$.
- Rounded to two decimal places, $\sqrt[3]{7} \approx 1.91$, and $(1.91)^3 = 6.967871$.
- Rounded to three decimal places, $\sqrt[3]{7} \approx 1.913$, and $(1.913)^3 = 7.000755497$.

Numbers such as 8, -27, -64, and 125 are called **integer cubes,** because each one is the cube of an integer. The cube root of any integer cube is an integer and therefore a rational number:

$$\sqrt[3]{8} = 2, \qquad \sqrt[3]{-27} = -3, \qquad \sqrt[3]{-64} = -4, \quad \text{and} \quad \sqrt[3]{125} = 5$$

Cube roots of integers such as 7 and -10, which are not integer cubes, are irrational numbers. For example, $\sqrt[3]{7}$ and $\sqrt[3]{-10}$ are irrational numbers.

COMMENT   Recall that the square root of a negative number $\left(\text{for instance } \sqrt{-27}\right)$ is not a real number, because no real number squared is equal to a negative number. However, the cube root of a negative number is a real number. For example, $\sqrt[3]{-27} = -3$.

Accent on Technology: **Radius of a water tank**

Engineers want to design a spherical tank that will hold 33,500 cubic feet of water, as shown in Figure 14-13. They know that the formula for the radius r of a sphere with volume V is given by the formula

$$r = \sqrt[3]{\frac{3V}{4\pi}} \quad \text{Where } \pi = 3.14159.\ldots$$

To use a scientific calculator to find the radius r, we substitute 33,500 for V and enter these numbers and press these keys.

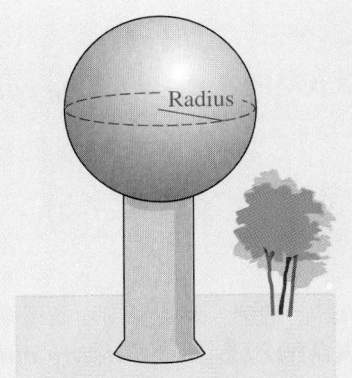

FIGURE 14-13

Keystrokes   3 $\boxed{\times}$ 33500 $\boxed{\div}$ $\boxed{(}$ 4 $\boxed{\times}$ $\boxed{\pi}$ $\boxed{)}$ $\boxed{=}$ $\boxed{\sqrt[n]{x}}$ 3 $\boxed{=}$

$$\boxed{19.99794636}$$

To evaluate this expression using a graphing calculator, we press the $\boxed{\text{MATH}}$ key. In this mode, arrow down $\boxed{\blacktriangledown}$ to highlight the option $\sqrt[3]{\ }($ and $\boxed{\text{ENTER}}$. Then we press the following keys.

Keystrokes 3 $\boxed{\times}$ 33500 $\boxed{\div}$ $\boxed{(}$ 4 $\boxed{\times}$ $\boxed{\text{2nd}}$ $\boxed{\pi}$ $\boxed{)}$ $\boxed{)}$ $\boxed{\text{ENTER}}$

```
³√(3*33500/(4*π)
)
          19.99794636
```

The result is 19.99794636, so the engineers should design a tank with a radius of 20 feet.

The cube root function

Since every real number has one real-number cube root, there is a cube root function $f(x) = \sqrt[3]{x}$. For example, the value that is determined by $f(x) = \sqrt[3]{x}$ when $x = 8$ is denoted as $f(8)$, and we have $f(8) = \sqrt[3]{8} = 2$.

To graph this function, we substitute numbers for x, compute $f(x)$, plot the resulting ordered pairs, and connect them with a smooth curve, as shown in Figure 14-14.

$$f(x) = \sqrt[3]{x}$$

x	$f(x)$	$(x, f(x))$
-8	-2	$(-8, -2)$
-1	-1	$(-1, -1)$
0	0	$(0, 0)$
1	1	$(1, 1)$
8	2	$(8, 2)$

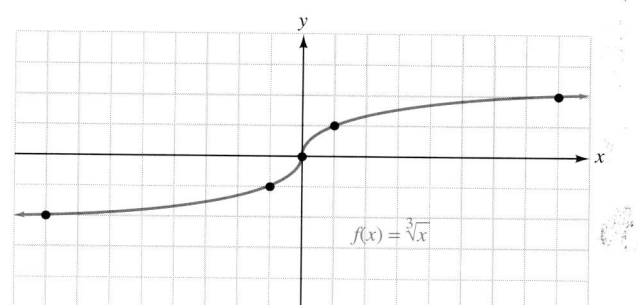

FIGURE 14-14

Higher-order roots

Just as there are square roots and cube roots, there are also fourth roots, fifth roots, sixth roots, and so on. In general, we have this definition.

nth roots of *a*

The **nth root of *a*** is denoted by $\sqrt[n]{a}$, and
$$\sqrt[n]{a} = b \qquad \text{if} \qquad b^n = a$$
The number n is called the **index** of the radical. If n is an even natural number, a must be positive or zero, and b must be positive.

In the square root symbol $\sqrt{\ }$, the unwritten index is understood to be 2.
$$\sqrt{a} = \sqrt[2]{a}$$

EXAMPLE 3 *Finding fourth and fifth roots.* Find each root.

a. $\sqrt[4]{81} = 3$, because $3^4 = 81$.

b. $\sqrt[5]{32} = 2$, because $2^5 = 32$.

c. $\sqrt[5]{-32} = -2$, because $(-2)^5 = -32$.

d. $\sqrt[4]{-81}$ is not a real number, because no real number raised to the fourth power is -81.

EXAMPLE 4 *Finding fourth and fifth roots.* Find each root.

a. $\sqrt[4]{\dfrac{1}{81}} = \dfrac{1}{3}$, because $\left(\dfrac{1}{3}\right)^4 = \dfrac{1}{81}$.

b. $\sqrt[5]{-\dfrac{32}{243}} = -\dfrac{2}{3}$, because $\left(-\dfrac{2}{3}\right)^5 = -\dfrac{32}{243}$.

Radicands that contain variables

When n is even and $x \geq 0$, we say that the radical $\sqrt[n]{x}$ represents an **even root.** We can find even roots of many quantities that contain variables, provided that these variables represent positive numbers or zero.

EXAMPLE 5 *Finding even roots.* Find each root. Assume that each variable represents a positive number.

a. $\sqrt{x^2} = x$, because $(x)^2 = x^2$.

b. $\sqrt{x^4} = x^2$, because $(x^2)^2 = x^4$.

c. $\sqrt{x^4y^2} = x^2y$, because $(x^2y)^2 = x^4y^2$.

d. $\sqrt[4]{81x^{12}} = 3x^3$, because $(3x^3)^4 = 81x^{12}$.

When n is odd, we say that the radical expression $\sqrt[n]{x}$ represents an **odd root.**

EXAMPLE 6 *Finding odd roots.* Find each root.

a. $\sqrt[3]{y^3} = y$, because $(y)^3 = y^3$.

b. $\sqrt[3]{64x^6} = 4x^2$, because $(4x^2)^3 = 64x^6$.

c. $\sqrt[5]{x^{10}} = x^2$, because $(x^2)^5 = x^{10}$.

STUDY SET Section 14.2

VOCABULARY *Fill in the blanks.*

1. If $p^3 = q$, p is called a _____ root of q.

2. If $p^4 = q$, p is called a _____ root of q.

3. We denote the cube root _____ with the notation $f(x) = \sqrt[3]{x}$.

4. If the index of a radical is an even number, the root is called an _____ root.

CONCEPTS   *Fill in the blanks.*

5. The _____ of −4 is −64, because $(-4)^3 = -64$. The number −3 is a cube _____ of −27, because $(-3)^3 = -27$.

6. $\sqrt[3]{a} = b$ if _____ .

7. $\sqrt[3]{-216} = -6$, because _____ $= -216$.

8. $\sqrt[5]{32x^5} = 2x$, because _____ $= 32x^5$.

9. Find each value, if possible.
   a. $\sqrt{-125}$ **b.** $\sqrt[3]{-125}$

10. 🔲   Graph each number on the number line.
   $$\left\{ \sqrt[3]{16}, \ -\sqrt[4]{100}, \ \sqrt[3]{-1.8}, \ \sqrt[4]{0.6} \right\}$$

   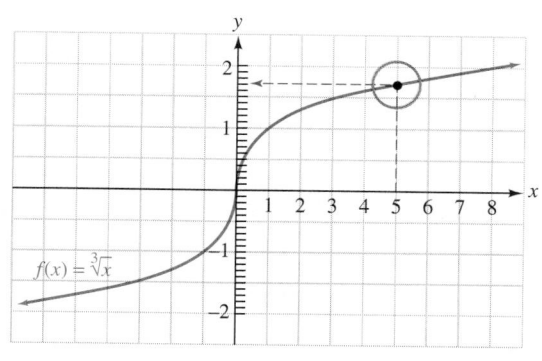

11. If $f(x) = \sqrt[3]{x}$, find each value. *Do not use a calculator.*
   a. $f(1)$ **b.** $f\left(-\dfrac{1}{27}\right)$
   c. $f(125)$ **d.** $f(0.008)$
   e. $f(1,000)$

12. a. What do the dashed lines in the graph in Illustration 1 help to approximate?
   b. Use the graph to approximate $\sqrt[3]{4}$ and $\sqrt[3]{-6}$.

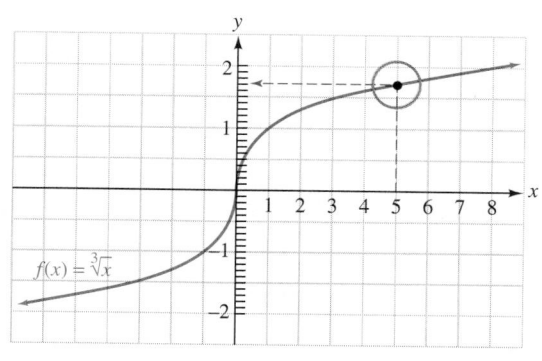

ILLUSTRATION 1

NOTATION   *Fill in the blanks.*

13. In the notation $\sqrt[3]{x^6}$, 3 is called the _____ and x^6 is called the _____ .

14. $\sqrt{}$ is called a _____ symbol.

15. The "understood" index of the radical expression $\sqrt{55}$ is ____ .

16. In reading $f(x) = \sqrt[3]{x}$, we say "f ____ x equals the cube root ____ x."

PRACTICE   *Find each value without using a calculator.*

17. $\sqrt[3]{8}$ **18.** $\sqrt[3]{27}$
19. $\sqrt[3]{0}$ **20.** $\sqrt[3]{1}$
21. $\sqrt[3]{-8}$ **22.** $\sqrt[3]{-1}$
23. $\sqrt[3]{-64}$ **24.** $\sqrt[3]{-27}$
25. $\sqrt[3]{\dfrac{1}{125}}$ **26.** $\sqrt[3]{\dfrac{1}{1,000}}$
27. $-\sqrt[3]{-1}$ **28.** $-\sqrt[3]{-27}$
29. $-\sqrt[3]{64}$ **30.** $-\sqrt[3]{343}$
31. $\sqrt[3]{729}$ **32.** $\sqrt[3]{512}$
33. $\sqrt[3]{1,000}$ **34.** $\sqrt[3]{125}$

🔲   *Use a calculator to find each cube root to the nearest hundredth.*

35. $\sqrt[3]{32,100}$ **36.** $\sqrt[3]{-25,713}$
37. $\sqrt[3]{-0.11324}$ **38.** $\sqrt[3]{0.875}$

Complete the table and then graph the function.

39. $f(x) = \sqrt[3]{x} + 1$

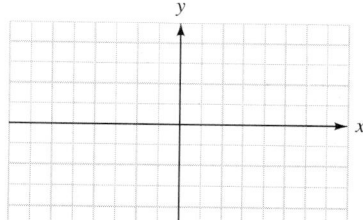

x	$f(x)$
−8	
−1	
0	
1	
8	

40. $f(x) = \sqrt[4]{x}$

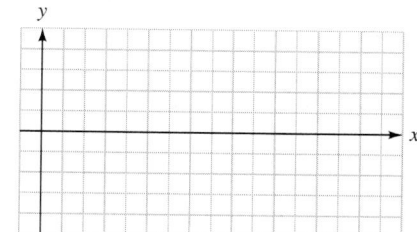

x	$f(x)$
0	
1	
16	

41. $f(x) = -\sqrt[3]{x}$

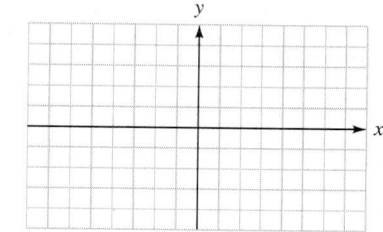

x	$f(x)$
−8	
−1	
0	
1	
8	

42. $f(x) = \sqrt[4]{x} - 1$

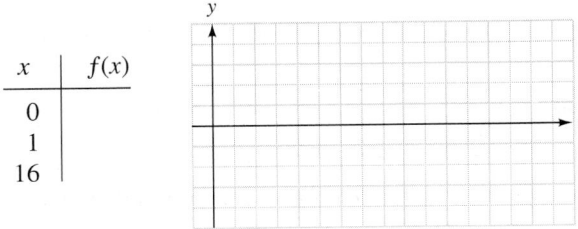

x	$f(x)$
0	
1	
16	

Find each value without using a calculator.

43. $\sqrt[4]{16}$ **44.** $\sqrt[4]{81}$

45. $-\sqrt[5]{32}$ **46.** $-\sqrt[5]{243}$

47. $\sqrt[6]{1}$ **48.** $\sqrt[6]{0}$

49. $\sqrt[5]{-32}$ **50.** $\sqrt[7]{-1}$

Use a calculator to find each root to the nearest hundredth.

51. $\sqrt[4]{125}$ **52.** $\sqrt[5]{12,450}$

53. $\sqrt[5]{-6,000}$ **54.** $\sqrt[6]{0.5}$

Find each root. All variables represent positive numbers.

55. $\sqrt{x^2}$ **56.** $\sqrt{y^4}$

57. $\sqrt{x^6}$ **58.** $\sqrt{b^8}$

59. $\sqrt{x^{10}}$ **60.** $\sqrt{y^{12}}$

61. $\sqrt{4z^2}$ **62.** $\sqrt{9t^6}$

63. $-\sqrt{x^4y^2}$ **64.** $-\sqrt{x^2y^4}$

65. $-\sqrt{0.04y^2}$ **66.** $-\sqrt{0.81b^6}$

67. $-\sqrt{25x^4z^{12}}$ **68.** $-\sqrt{100a^6b^4}$

69. $\sqrt{36z^{36}}$ **70.** $\sqrt{64y^{64}}$

71. $-\sqrt{625z^2}$ **72.** $-\sqrt{729x^8}$

73. $\sqrt[3]{y^6}$ **74.** $\sqrt[3]{c^3}$

75. $\sqrt[5]{f^5}$ **76.** $\sqrt[5]{y^{20}}$

77. $\sqrt[3]{27y^3}$ **78.** $\sqrt[3]{64y^6}$

79. $\sqrt[3]{-p^6q^3}$ **80.** $\sqrt[3]{-r^{12}t^6}$

81. $\sqrt[4]{x^4}$ **82.** $\sqrt[4]{x^8}$

APPLICATIONS   *Use a calculator to help solve each problem. Give your answers to the nearest hundredth.*

83. PACKAGING   A cubical box has a volume of 2 cubic feet. Substitute 2 for V in the formula $V = s^3$ and solve for s to find the length of each side of the box.

84. HOT-AIR BALLOON   If a hot-air balloon is in the shape of a sphere and has a volume of 15,000 cubic feet, what is its radius? (*Hint:* See the Accent on Technology feature in this section.)

85. WINDMILL   The power generated by a windmill is related to the speed of the wind by the formula

$$S = \sqrt[3]{\frac{P}{0.02}}$$

where S is the speed of the wind (in mph) and P is the power (in watts). Find the speed of the wind when the windmill is producing 400 watts of power.

86. ASTRONOMY   In the early 17th century, Johannes Kepler, a German astronomer, discovered that a planet's mean distance R from the sun (in millions of miles) is related to the time T (in years) it takes the planet to orbit the sun by the formula

$$R = 93\sqrt[3]{\frac{T^2}{1.002}}$$

Use the information in Illustration 2 to find R for Mercury, Earth, and Jupiter.

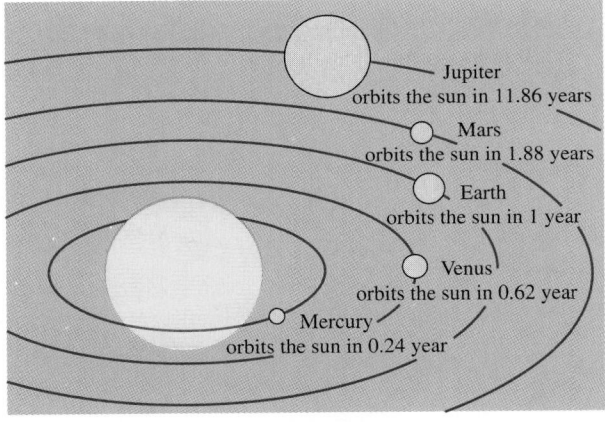

Jupiter orbits the sun in 11.86 years

Mars orbits the sun in 1.88 years

Earth orbits the sun in 1 year

Venus orbits the sun in 0.62 year

Mercury orbits the sun in 0.24 year

ILLUSTRATION 2

87. DEPRECIATION   The formula

$$r = 1 - \sqrt[n]{\frac{S}{C}}$$

gives the annual depreciation rate r (in percent) of an item that had an original cost of C dollars and has a useful life of n years and a salvage value of S dollars. Use the information in Illustration 3 to find the annual depreciation rate for the new piece of sound equipment.

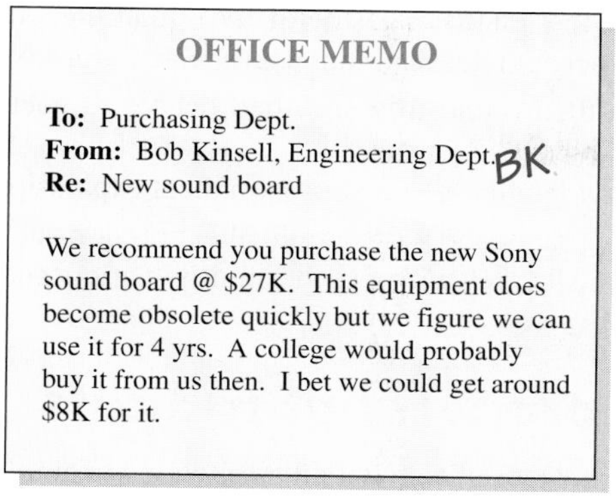

OFFICE MEMO

To: Purchasing Dept.
From: Bob Kinsell, Engineering Dept. ᗷK
Re: New sound board

We recommend you purchase the new Sony sound board @ $27K. This equipment does become obsolete quickly but we figure we can use it for 4 yrs. A college would probably buy it from us then. I bet we could get around $8K for it.

ILLUSTRATION 3

88. SAVINGS ACCOUNT The interest rate r (in percent) earned by a savings account after n compoundings is given by the formula

$$\sqrt[n]{\frac{V}{P}} - 1 = r$$

where V is the current value and P is the original principal. What interest rate r was paid on an account in which a deposit of $1,000 grew to $1,338.23 after 5 compoundings?

WRITING

89. Explain why a negative number can have a real number for its cube root yet cannot have a real number for its fourth root.

90. To find $\sqrt[3]{15}$, we can use the $\boxed{\sqrt[n]{x}}$ key on a calculator to obtain 2.466212074. Explain how a key other than $\boxed{\sqrt[n]{x}}$ can be used to check the validity of this result.

REVIEW *Simplify each expression.*

91. $m^5 m^2$

92. $(-5x^3)(-5x)$

93. $(3^2)^4$

94. $r^3 r r^5$

95. $(x^2 x^3)^5$

96. $(3aa^2 a^3)^5$

97. $4x^3(6x^5)$

98. $-2x(5x^3)$

14.3 *Simplifying Radical Expressions*

In this section, you will learn about

- The multiplication property of radicals • Simplifying square root radicals
- The division property of radicals • Simplifying cube roots

INTRODUCTION. Square dancing is a traditional American folk dance in which four couples, arranged in a square, perform various moves. Figure 14-15 shows a group as they promenade around a square.

If the square shown in the figure has an area of 12 square yards, the length of a side is $\sqrt{12}$ yards. We can use the formula for the area of a square and the concept of square root to show that this is so.

FIGURE 14-15

$$A = s^2 \qquad \text{s is the length of a side of the square.}$$

$$12 = s^2 \qquad \text{Substitute 12 for A, the area of the square.}$$

$$\sqrt{12} = \sqrt{s^2} \qquad \text{Take the positive square root of both sides.}$$

$$\sqrt{12} = s \qquad \text{The length of a side of the square is $\sqrt{12}$ yards.}$$

The form in which we express the length of a side of the square depends on the situation. If an approximation is acceptable, we can use a calculator to find that $\sqrt{12} \approx 3.464101615$, and we can then round to a specified degree of accuracy. For example, to the nearest tenth, each side is 3.5 yards long.

If the situation calls for the *exact* length, we must use a radical expression. As you will see in this section, it is common practice to write a radical expression such as $\sqrt{12}$ in *simplified form*. To simplify radicals, we will use the multiplication and division properties of radicals.

The multiplication property of radicals

We introduce the first of two properties of radicals with the following examples:

$$\sqrt{4 \cdot 25} = \sqrt{100} \qquad \text{and} \qquad \sqrt{4}\sqrt{25} = 2 \cdot 5 \qquad \text{Read as "the square root of 4}$$
$$= 10 \qquad\qquad\qquad\qquad\qquad = 10 \qquad \text{times the square root of 25."}$$

In each case, the answer is 10. Thus, $\sqrt{4 \cdot 25} = \sqrt{4}\sqrt{25}$. Likewise,

$$\sqrt{9 \cdot 16} = \sqrt{144} \qquad \text{and} \qquad \sqrt{9}\sqrt{16} = 3 \cdot 4$$
$$= 12 \qquad\qquad\qquad\qquad\qquad = 12$$

In each case, the answer is 12. Thus, $\sqrt{9 \cdot 16} = \sqrt{9}\sqrt{16}$. These results illustrate the **multiplication property of radicals.**

The multiplication property of radicals

> If a and b represent nonnegative real numbers,
> $$\sqrt{ab} = \sqrt{a}\sqrt{b}$$

In words, *the square root of the product of two nonnegative numbers is equal to the product of their square roots.*

Simplifying square root radicals

A square root radical is in **simplified form** when each of the following statements is true.

Simplified form of a radical

> 1. Except for 1, the radicand has no perfect square factors.
> 2. No fraction appears in a radicand.
> 3. No radical appears in the denominator of a fraction.

We can use the multiplication property of radicals to simplify square roots whose radicands have perfect square factors. For example, we can simplify $\sqrt{12}$ as follows:

$$\sqrt{12} = \sqrt{4 \cdot 3} \qquad \text{Factor 12 as } 4 \cdot 3.$$
$$= \sqrt{4}\sqrt{3} \qquad \text{The square root of } 4 \cdot 3 \text{ is equal to the square root of 4 times the square root of 3.}$$
$$= 2\sqrt{3} \qquad \text{Write } \sqrt{4} \text{ as 2. Read as "2 times the square root of 3" or as "2 radical 3."}$$

The square in Figure 14-15, which we considered in the introduction to this section, has a side length of $\sqrt{12}$ yards. We now see that the *exact* length of a side can be expressed in simplified form as $2\sqrt{3}$ yards.

To simplify more difficult square roots, we need to know the integers that are **integer squares.** For example, 81 is an integer square, because it is the square of the integer 9: $9^2 = 81$. The first 20 integer squares are

$$1, 4, 9, 16, 25, 36, 49, 64, 81, 100, 121, 144, 169, 196, 225, 256, 289, 324, 361, 400$$

EXAMPLE 1   *Simplifying square roots.*   Simplify $\sqrt{27}$.

Solution
Since the greatest perfect square that divides 27 exactly is 9, we will factor 27 as $9 \cdot 3$ and apply the multiplication property of radicals.

$$\sqrt{27} = \sqrt{9 \cdot 3}$$
$$= \sqrt{9}\sqrt{3} \quad \text{The square root of a product } \left(\text{that is, } \sqrt{9 \cdot 3}\right) \text{ is equal to the product of the square roots, } \sqrt{9}\sqrt{3}.$$
$$= 3\sqrt{3} \quad \text{Simplify: } \sqrt{9} = 3.$$

As a check, recall that $\sqrt{27}$ is the number that, when squared, gives 27. If $3\sqrt{3} = \sqrt{27}$, then $\left(3\sqrt{3}\right)^2$ should be equal to 27.

$$\left(3\sqrt{3}\right)^2 = (3)^2\left(\sqrt{3}\right)^2 \quad \text{Use the "power of a product" rule for exponents: Raise each factor of the product } 3\sqrt{3} \text{ to the 2nd power.}$$
$$= 9(3) \quad \sqrt{3} \text{ is the number that, when squared, gives 3.}$$
$$= 27$$

Self Check
Simplify $\sqrt{45}$.

Answer: $3\sqrt{5}$ ■

EXAMPLE 2   *Simplifying square roots.*   Simplify $\sqrt{600}$.

Solution
Since the greatest perfect square that divides 600 is 100, we will factor 600 as $100 \cdot 6$ and apply the multiplication property of radicals.

$$\sqrt{600} = \sqrt{100 \cdot 6}$$
$$= \sqrt{100}\sqrt{6} \quad \text{The square root of a product is equal to the product of the square roots.}$$
$$= 10\sqrt{6} \quad \sqrt{100} = 10.$$

Check the result.

Self Check
Simplify $\sqrt{200}$.

Answer: $10\sqrt{2}$ ■

Expressions containing variables can also be perfect squares. For example, $36x^2$ is a perfect square, because

$$36x^2 = (6x)^2 \quad \text{Think, "What was squared to obtain } 36x^2\text{?"}$$

We can use this observation to help simplify radicals involving variable radicands. We will assume that all of the variables in the following examples represent positive numbers.

EXAMPLE 3   *Simplifying radicals involving variable radicands.*
Simplify $\sqrt{b^3}$.

Solution
To write $\sqrt{b^3}$ in simplified form, we factor b^3 into two factors, one of which is the greatest perfect square that divides b^3. The greatest perfect square that divides b^3 is b^2, so such a factorization is $b^3 = b^2 \cdot b$. We then proceed as follows:

$$\sqrt{b^3} = \sqrt{b^2 \cdot b}$$
$$= \sqrt{b^2}\sqrt{b} \quad \text{The square root of a product is equal to the product of the square roots.}$$
$$= b\sqrt{b} \quad \sqrt{b^2} = b.$$

As a check, recall that $\sqrt{b^3}$, when squared, gives b^3. If $b\sqrt{b} = \sqrt{b^3}$, then $\left(b\sqrt{b}\right)^2$ should be equal to b^3.

$$\left(b\sqrt{b}\right)^2 = (b)^2\left(\sqrt{b}\right)^2 \quad \text{Raise each factor of the product } b\sqrt{b} \text{ to the 2nd power.}$$
$$= b^2(b) \quad \sqrt{b}, \text{ when squared, gives } b.$$
$$= b^3 \quad \text{Keep the base } b \text{ and add the exponents.}$$

Self Check
Simplify $\sqrt{y^5}$.

Answer: $y^2\sqrt{y}$ ■

EXAMPLE 4 *Simplifying radicals involving variable radicands.*
Simplify $-7\sqrt{8m}$.

Solution
We first write $\sqrt{8m}$ in simplified form, and then we multiply the result by -7. By inspection, we see that the radicand, $8m$, has a perfect square factor of 4. We can write $8m$ in factored form as $4 \cdot 2m$.

$$
\begin{aligned}
-7\sqrt{8m} &= -7\sqrt{4 \cdot 2m} \\
&= -7\sqrt{4}\sqrt{2m} && \text{The square root of a product is equal to the product of the square roots.} \\
&= -7(2)\sqrt{2m} && \sqrt{4} = 2. \\
&= -14\sqrt{2m} && -7(2) = -14.
\end{aligned}
$$

Self Check
Simplify $2\sqrt{18c}$.

Answer: $6\sqrt{2c}$ ∎

COMMENT When writing radical expressions such as $-14\sqrt{2m}$, be sure to extend the radical symbol completely over $2m$, because the expressions $-14\sqrt{2m}$ and $-14\sqrt{2}m$ are not the same. Similar care should be taken when writing expressions such as $\sqrt{3x}$. To avoid any misinterpretation, $\sqrt{3}x$ can be written as $x\sqrt{3}$.

EXAMPLE 5 *Simplifying radicals involving variable radicands.*
Simplify $\sqrt{72x^3}$.

Solution
We factor $72x^3$ into two factors, one of which is the greatest perfect square that divides $72x^3$. Since the greatest perfect square that divides $72x^3$ is $36x^2$, such a factorization is $72x^3 = 36x^2 \cdot 2x$. We now use the multiplication property of radicals to get

$$
\begin{aligned}
\sqrt{72x^3} &= \sqrt{36x^2 \cdot 2x} \\
&= \sqrt{36x^2}\sqrt{2x} && \text{The square root of a product is equal to the product of the square roots.} \\
&= 6x\sqrt{2x} && \sqrt{36x^2} = 6x.
\end{aligned}
$$

Self Check
Simplify $\sqrt{48y^3}$.

Answer: $4y\sqrt{3y}$ ∎

EXAMPLE 6 *Simplifying radicals involving variable radicands.*
Simplify $3a\sqrt{288a^4b^7}$.

Solution
We first simplify $\sqrt{288a^4b^7}$. Then we multiply the result by $3a$. To simplify $\sqrt{288a^4b^7}$, we look for the greatest perfect square that divides $288a^4b^7$. Because

- 144 is the greatest perfect square that divides 288,
- a^4 is the greatest perfect square that divides a^4, and
- b^6 is the greatest perfect square that divides b^7,

the factor $144a^4b^6$ is the greatest perfect square that divides $288a^4b^7$.
We can now use the multiplication property of radicals to simplify the radical.

$$
\begin{aligned}
3a\sqrt{288a^4b^7} &= 3a\sqrt{144a^4b^6 \cdot 2b} \\
&= 3a\sqrt{144a^4b^6}\sqrt{2b} && \text{The square root of a product is equal to the product of the square roots.} \\
&= 3a(12a^2b^3)\sqrt{2b} && \sqrt{144a^4b^6} = 12a^2b^3. \\
&= 36a^3b^3\sqrt{2b} && \text{Multiply: } 3a(12a^2b^3) = 36a^3b^3.
\end{aligned}
$$

Self Check
Simplify $5q\sqrt{63p^5q^4}$.

Answer: $15p^2q^3\sqrt{7p}$ ∎

COMMENT The multiplication property of radicals applies to the square root of the product of two numbers. There is no such property for sums or differences. To illustrate this, we consider these correct simplifications:

$$\sqrt{9 + 16} = \sqrt{25} = 5 \qquad \text{and} \qquad \sqrt{25 - 16} = \sqrt{9} = 3$$

It is incorrect to write

$$\sqrt{9 + 16} = \sqrt{9} + \sqrt{16} \qquad \text{or} \qquad \sqrt{25 - 16} = \sqrt{25} - \sqrt{16}$$
$$= 3 + 4 \qquad\qquad\qquad = 5 - 4$$
$$= 7 \qquad\qquad\qquad\qquad = 1$$

Thus, $\sqrt{a + b} \neq \sqrt{a} + \sqrt{b}$ and $\sqrt{a - b} \neq \sqrt{a} - \sqrt{b}$.

The division property of radicals

To introduce the second property of radicals, we consider these examples.

$$\sqrt{\frac{100}{25}} = \sqrt{4} \qquad \text{and} \qquad \frac{\sqrt{100}}{\sqrt{25}} = \frac{10}{5} \qquad \text{\small Read as "the square root of 100 divided}$$
$$= 2 \qquad\qquad\qquad\qquad = 2 \qquad \text{\small by the square root of 25."}$$

Since the answer is 2 in each case,

$$\sqrt{\frac{100}{25}} = \frac{\sqrt{100}}{\sqrt{25}}$$

Likewise,

$$\sqrt{\frac{36}{4}} = \sqrt{9} \qquad \text{and} \qquad \frac{\sqrt{36}}{\sqrt{4}} = \frac{6}{2}$$
$$= 3 \qquad\qquad\qquad\qquad = 3$$

Since the answer is 3 in each case,

$$\sqrt{\frac{36}{4}} = \frac{\sqrt{36}}{\sqrt{4}}$$

These results illustrate the **division property of radicals.**

The division property of radicals	If a and b represent real numbers, with $a \geq 0$ and $b > 0$, $$\sqrt{\frac{a}{b}} = \frac{\sqrt{a}}{\sqrt{b}}$$

In words, *the square root of the quotient of two numbers is the quotient of their square roots.*

We can use the division property of radicals to simplify radicals that have fractions in their radicands. For example,

$$\sqrt{\frac{59}{49}} = \frac{\sqrt{59}}{\sqrt{49}}$$
$$= \frac{\sqrt{59}}{7} \qquad \text{\small Simplify the denominator: } \sqrt{49} = 7.$$

EXAMPLE 7 *Simplifying radicals involving fractions.*

Simplify $\sqrt{\dfrac{108}{25}}$.

Solution

$$\sqrt{\dfrac{108}{25}} = \dfrac{\sqrt{108}}{\sqrt{25}}$$ The square root of a quotient is equal to the quotient of the square roots.

$$= \dfrac{\sqrt{36 \cdot 3}}{5}$$ Factor 108 using the largest perfect square factor of 108, which is 36. Write $\sqrt{25}$ as 5.

$$= \dfrac{\sqrt{36}\sqrt{3}}{5}$$ The square root of a product is equal to the product of the square roots.

$$= \dfrac{6\sqrt{3}}{5}$$ $\sqrt{36} = 6$. This result can also be written as $\frac{6}{5}\sqrt{3}$.

Self Check

Simplify $\sqrt{\dfrac{20}{81}}$.

Answer: $\dfrac{2\sqrt{5}}{9}$

EXAMPLE 8 *Simplifying radicals involving fractions.*

Simplify $\sqrt{\dfrac{44x^3}{9xy^2}}$.

Solution

$$\sqrt{\dfrac{44x^3}{9xy^2}} = \sqrt{\dfrac{44x^2}{9y^2}}$$ Simplify the fraction by dividing out the common factor of x: $\dfrac{44x^3}{9xy^2} = \dfrac{44x^2 \overset{1}{\cancel{x}}}{9\cancel{x}y^2} = \dfrac{44x^2}{9y^2}$.

$$= \dfrac{\sqrt{44x^2}}{\sqrt{9y^2}}$$ The square root of a quotient is equal to the quotient of the square roots.

$$= \dfrac{\sqrt{4x^2}\sqrt{11}}{\sqrt{9y^2}}$$ Factor $44x^2$ as $4x^2 \cdot 11$. The square root of a product is equal to the product of the square roots.

$$= \dfrac{2x\sqrt{11}}{3y}$$ $\sqrt{4x^2} = 2x$ and $\sqrt{9y^2} = 3y$.

Self Check

Simplify $\sqrt{\dfrac{99b^3}{16a^2b}}$.

Answer: $\dfrac{3b\sqrt{11}}{4a}$

Simplifying cube roots

The multiplication and division properties of radicals are also true for cube roots and higher. To simplify cube roots, we must know the following **integer cubes:**

8, 27, 64, 125, 216, 343, 512, 729, 1,000

EXAMPLE 9 *Simplifying cube roots.* Simplify $\sqrt[3]{54}$.

Solution

The greatest perfect cube that divides 54 is 27.

$$\sqrt[3]{54} = \sqrt[3]{27 \cdot 2}$$ Factor 54: $54 = 27 \cdot 2$.

$$= \sqrt[3]{27}\sqrt[3]{2}$$ The square root of a product is equal to the product of the square roots.

$$= 3\sqrt[3]{2}$$ $\sqrt[3]{27} = 3$.

As a check, we note that $\sqrt[3]{54}$ is the number that, when cubed, gives 54. If $3\sqrt[3]{2} = \sqrt[3]{54}$, then $\left(3\sqrt[3]{2}\right)^3$ will be equal to 54.

$$\left(3\sqrt[3]{2}\right)^3 = (3)^3\left(\sqrt[3]{2}\right)^3$$ Raise each factor of the product $3\sqrt[3]{2}$ to the 3rd power.

Self Check

Simplify $\sqrt[3]{250}$.

$= 27(2)$ $\sqrt[3]{2}$, when cubed, gives 2.

$= 54$

Answer: $5\sqrt[3]{2}$ ∎

Expressions containing variables can also be perfect cubes. For example, $8x^3y^3$ is a perfect cube, because

$$8x^3y^3 = (2xy)^3 \quad \text{Think, "What was cubed to obtain } 8x^3y^3\text{?"}$$

EXAMPLE 10   *Simplifying cube roots involving variables.*

Simplify   **a.** $\sqrt[3]{16x^3y^4}$ and   **b.** $\sqrt[3]{\dfrac{64n^4}{27m^3}}$.

Solution

a. We factor $16x^3y^4$ into two factors, one of which is the greatest perfect cube that divides $16x^3y^4$. Since $8x^3y^3$ is the greatest perfect cube that divides $16x^3y^4$, the factorization is $16x^3y^4 = 8x^3y^3 \cdot 2y$.

$$\sqrt[3]{16x^3y^4} = \sqrt[3]{8x^3y^3 \cdot 2y}$$

$$= \sqrt[3]{8x^3y^3}\sqrt[3]{2y} \quad \text{The cube root of a product is equal to the product of the cube roots.}$$

$$= 2xy\sqrt[3]{2y} \quad \sqrt[3]{8x^3y^3} = 2xy.$$

b. $\sqrt[3]{\dfrac{64n^4}{27m^3}} = \dfrac{\sqrt[3]{64n^4}}{\sqrt[3]{27m^3}}$ The cube root of a quotient is equal to the quotient of the cube roots.

$$= \dfrac{\sqrt[3]{64n^3}\sqrt[3]{n}}{3m} \quad \text{In the numerator, use the multiplication property of radicals. In the denominator, } \sqrt[3]{27m^3} = 3m.$$

$$= \dfrac{4n\sqrt[3]{n}}{3m} \quad \sqrt[3]{64n^3} = 4n.$$

Self Check

Simplify:

a. $\sqrt[3]{54a^3b^5}$

b. $\sqrt[3]{\dfrac{27q^5}{64p^3}}$

Answers: **a.** $3ab\sqrt[3]{2b^2}$,

b. $\dfrac{3q\sqrt[3]{q^2}}{4p}$ ∎

STUDY SET   Section 14.3

VOCABULARY   *Fill in the blanks.*

1. Squares of integers such as 4, 9, and 16 are called _____ squares.

2. Cubes of integers such as 8, 27, and 64 are called perfect _____.

3. "To _____ $\sqrt{8}$" means to write it as $2\sqrt{2}$.

4. The word *product* is associated with the operation of _____ and the word *quotient* with _____.

CONCEPTS

5. Fill in the blanks.

   a. The square root of the product of two positive numbers is equal to the _____ of their square roots. In symbols,

   $$\sqrt{ab} = $$

   b. The square root of the quotient of two positive numbers is equal to the _____ of their square roots. In symbols,

   $$\sqrt{\dfrac{a}{b}} = $$

6. Which of the integer squares 1, 4, 9, 16, 25, 36, 49, 64, 81, and 100 is the *largest* factor of the given number?

   a. 20   **b.** 45

   c. 72   **d.** 98

In Exercises 7–8, tell what is wrong with each solution.

7. Simplify $\sqrt{20}$.

$$\sqrt{20} = \sqrt{16 + 4}$$
$$= \sqrt{16} + \sqrt{4}$$
$$= 4 + 2$$
$$= 6$$

8. Simplify $\sqrt{27}$.

$$\sqrt{27} = \sqrt{36 - 9}$$
$$= \sqrt{36} - \sqrt{9}$$
$$= 6 - 3$$
$$= 3$$

9. A crossword puzzle in a newspaper occupies an area of 28 square inches. See Illustration 1.
  a. Express the exact length of a side of the square-shaped puzzle in simplified radical form.
  b. What is the length of a side to the nearest tenth of an inch?

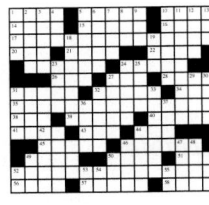

ILLUSTRATION 1

10. See Illustration 2.
  a. What is the exact length of a side of the cube written in simplified radical form?
  b. What is the length of a side to the nearest tenth of a foot?

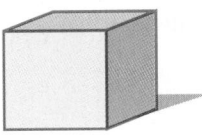

Volume = 40 ft³

ILLUSTRATION 2

Evaluate the expression $\sqrt{b^2 - 4ac}$ for the given values. Do the operations within the radical first, and then simplify the radical.

11. $a = 5, b = 10, c = 3$

12. $a = 2, b = 6, c = 1$

13. $a = -1, b = 6, c = 9$

14. $a = 1, b = -2, c = -11$

NOTATION *In Exercises 15–16, a radical is simplified. Complete each solution.*

15. $\sqrt{80a^3b^2} = \sqrt{16 \cdot \cdot a^2 \cdot a \cdot b^2}$
$$= \sqrt{16a^2b^2 \cdot }$$
$$= \sqrt{} \sqrt{5a}$$
$$= 4ab\sqrt{5a}$$

16. $\sqrt[3]{\dfrac{27a^4b^2}{64}} = \dfrac{\sqrt[3]{27a^4b^2}}{}$

$$= \dfrac{\sqrt[3]{27a^3 \cdot }}{\sqrt[3]{64}}$$

$$= \dfrac{\sqrt[3]{}\sqrt[3]{ab^2}}{\sqrt[3]{64}}$$

$$= \dfrac{3a\sqrt[3]{ab^2}}{4}$$

17. What operation is indicated between the two radicals in the expression $\sqrt{4}\sqrt{3}$?

18. Fill in each blank to make a true statement.
  a. $16x^2 = ()^2$ **b.** $27a^3b^6 = ()^3$

19. Write each expression in a better form.
  a. $\sqrt{5} \cdot 2$ **b.** $\sqrt{7a}$
  c. $9\sqrt{x^2}\sqrt{6}$ **d.** $\sqrt{y}\sqrt{25z^4}$

20. a. Explain the difference between $\sqrt{5}x$ and $\sqrt{5x}$.

  b. Why do you think it is better to write $\sqrt{5x}$ as $x\sqrt{5}$?

PRACTICE *Simplify each radical. Assume that all variables represent positive numbers.*

21. $\sqrt{20}$ **22.** $\sqrt{18}$

23. $\sqrt{50}$ **24.** $\sqrt{75}$

25. $\sqrt{45}$ **26.** $\sqrt{54}$

27. $\sqrt{98}$ **28.** $\sqrt{147}$

29. $\sqrt{48}$ **30.** $\sqrt{128}$

31. $-\sqrt{200}$ **32.** $-\sqrt{300}$

33. $\sqrt{192}$ **34.** $\sqrt{88}$

35. $\sqrt{250}$ **36.** $\sqrt{1,000}$

37. $2\sqrt{24}$ **38.** $3\sqrt{32}$

39. $-2\sqrt{28}$ **40.** $-3\sqrt{72}$

41. $\sqrt{n^3}$ **42.** $\sqrt{x^5}$

43. $\sqrt{4k}$ **44.** $\sqrt{9p}$

45. $\sqrt{12x}$ **46.** $\sqrt{20y}$

47. $6\sqrt{75t}$ **48.** $2\sqrt{24s}$

49. $\sqrt{25x^3}$ **50.** $\sqrt{36y^3}$

51. $\sqrt{a^2b}$ **52.** $\sqrt{rs^4}$

53. $\sqrt{9x^4y}$ **54.** $\sqrt{16xy^2}$

55. $\dfrac{1}{5}x^2y\sqrt{50x^2y^2}$

56. $\dfrac{1}{5}x^5y\sqrt{75x^3y^2}$

57. $-12x\sqrt{16x^2y^3}$

58. $-4x^5y^3\sqrt{36x^3y^3}$

59. $-\dfrac{2}{5}\sqrt{80mn^4}$

60. $\dfrac{5}{6}\sqrt{180ab^6}$

Write each quotient as the quotient of two radicals and simplify.

61. $\sqrt{\dfrac{25}{9}}$ **62.** $\sqrt{\dfrac{36}{49}}$

63. $\sqrt{\dfrac{81}{64}}$ **64.** $\sqrt{\dfrac{121}{144}}$

65. $\sqrt{\dfrac{26}{25}}$ **66.** $\sqrt{\dfrac{17}{169}}$

67. $-\sqrt{\dfrac{20}{49}}$ **68.** $-\sqrt{\dfrac{50}{9}}$

69. $\sqrt{\dfrac{48}{81}}$ **70.** $\sqrt{\dfrac{27}{64}}$

71. $\sqrt{\dfrac{32}{25}}$ **72.** $\sqrt{\dfrac{75}{16}}$

Simplify each expression. All variables represent positive numbers.

73. $\sqrt{\dfrac{72x^3}{y^2}}$ **74.** $\sqrt{\dfrac{108b^2}{d^4}}$

75. $\sqrt{\dfrac{125n^5}{64n}}$ **76.** $\sqrt{\dfrac{72q^7}{25q^3}}$

77. $\sqrt{\dfrac{128m^3n^5}{81mn^7}}$ **78.** $\sqrt{\dfrac{75p^3q^2}{p^5q^4}}$

79. $\sqrt{\dfrac{12r^7s^7}{r^5s^2}}$ **80.** $\sqrt{\dfrac{m^2n^9}{100mn^3}}$

Simplify each cube root.

81. $\sqrt[3]{24}$ **82.** $\sqrt[3]{32}$

83. $\sqrt[3]{-128}$ **84.** $\sqrt[3]{-250}$

85. $\sqrt[3]{8x^3}$ **86.** $\sqrt[3]{27x^3}$

87. $\sqrt[3]{-64x^5}$ **88.** $\sqrt[3]{-16x^4}$

89. $\sqrt[3]{54x^3z^6}$ **90.** $\sqrt[3]{-24x^3y^5}$

91. $\sqrt[3]{-81x^2y^3}$ **92.** $\sqrt[3]{81y^2z^3}$

93. $\sqrt[3]{\dfrac{27m^3}{8n^6}}$ **94.** $\sqrt[3]{\dfrac{125t^9}{27s^6}}$

95. $\sqrt[3]{\dfrac{r^4s^5}{1,000t^3}}$ **96.** $\sqrt[3]{\dfrac{54m^4n^3}{r^3s^6}}$

APPLICATIONS *Use a calculator to help solve each problem.*

97. AMUSEMENT PARK RIDE   Illustration 3 shows the "Swashbuckler" pirate ship ride. The time (in seconds) it takes to swing from one extreme to the other is given by

$$t = \pi\sqrt{\dfrac{L}{32}}$$

a. Find t and express it in simplified radical form. Leave π in your answer.

b. Express your answer to part a as a decimal. Round to the nearest tenth of a second.

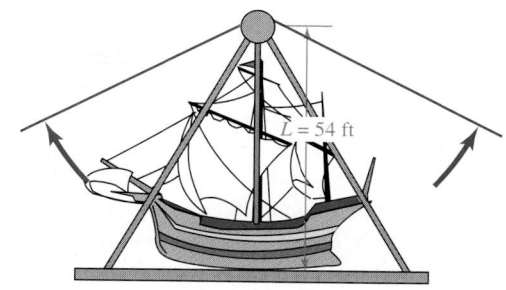

ILLUSTRATION 3

98. HERB GARDEN   The perimeter of the herb garden shown in Illustration 4 is given by

$$p = 2\pi\sqrt{\dfrac{a^2 + b^2}{2}}$$

a. Find the length of fencing (in meters) needed to enclose the garden. Express the result in simplified radical form. Leave π in your answer.

b. Express the result from part a as a decimal. Round to the nearest tenth of a meter.

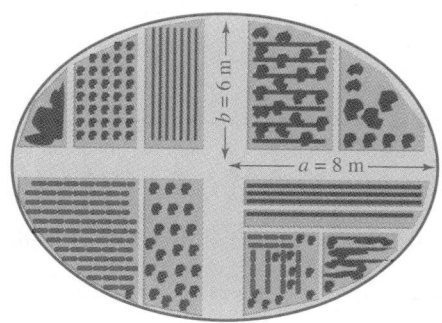

$b = 6$ m

$a = 8$ m

ILLUSTRATION 4

99. ARCHAEOLOGY   Framed grids, made up of 20 cm × 20 cm squares, are often used to record the location of artifacts found during an excavation. (See Illustration 5.)

a. Use the distance formula to determine the *exact* distance between a piece of pottery found at point A and a cooking utensil found at point B.

b. Approximate the distance to the nearest tenth of a centimeter.

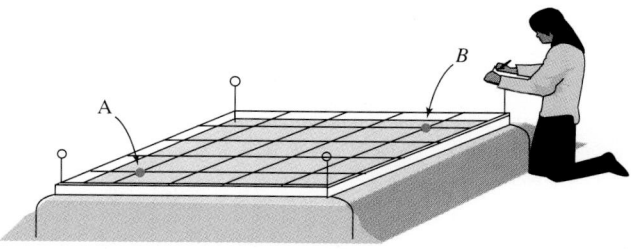

ILLUSTRATION 5

100. ENVIRONMENTAL PROTECTION   A new campground is to be constructed 2 miles from a major highway, as shown in Illustration 6. The proposed entrance, although longer than the direct route, bypasses a grove of old-growth redwood trees.

a. Use the Pythagorean theorem to find the length of the proposed entrance road. Express the result as a radical in simplified form.

b. Express the result from part a as a decimal. Round to the nearest hundredth of a mile.

c. How much longer is the proposed entrance as compared to the direct route into the campground?

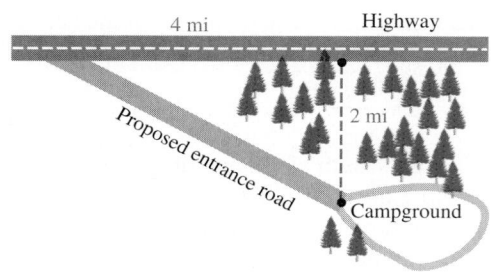

ILLUSTRATION 6

101. State the multiplication property of radicals.

102. When comparing $\sqrt{8}$ and $2\sqrt{2}$, why is $2\sqrt{2}$ called simplified radical form?

103. Multiply: $(-2a^3)(3a^2)$.

104. Find the slope of the line passing through $(-6, 0)$ and $(0, -4)$.

105. Write the equation of the line passing through $(0, 3)$ with slope -2.

106. Solve $-x = -5$.

107. Solve $-x > -5$.

108. What is the slope of a line perpendicular to a line with a slope of 2?

14.4 *Adding and Subtracting Radical Expressions*

In this section, you will learn about

- Combining like radicals • Combining expressions containing cube roots

INTRODUCTION. We have previously discussed how to add and subtract like terms. In this section, we will introduce a similar topic: how to add and subtract expressions that contain like radicals.

Combining like radicals

When adding monomials, we can often combine **like terms.** For example,

$$3x + 5x = (3 + 5)x \quad \text{Use the distributive property.}$$
$$= 8x \qquad \text{Do the addition.}$$

 COMMENT   The expression $3x + 5y$ cannot be simplified, because $3x$ and $5y$ are not like terms.

It is often possible to combine terms that contain *like radicals.*

Like radicals | Radicals are called **like radicals** when they have the same index and the same radicand.

Like radicals	**Unlike radicals**
$3\sqrt{2}$ and $5\sqrt{2}$	$3\sqrt{2}$ and $5\sqrt{3}$
The same index and the same radicand	The same index but different radicands
$5x\sqrt{3y}$ and $-2x\sqrt{3y}$	$5x\sqrt[3]{3y}$ and $-2x\sqrt{3y}$
The same index and the same radicand	The same radicands but a different index

Expressions that contain like radicals can be combined by addition and subtraction. For example, we have

$$3\sqrt{2} + 5\sqrt{2} = (3 + 5)\sqrt{2} \quad \text{Use the distributive property.}$$
$$= 8\sqrt{2} \qquad \text{Do the addition.}$$

Likewise, we can simplify the expression $5x\sqrt{3y} - 2x\sqrt{3y}$.

$$5x\sqrt{3y} - 2x\sqrt{3y} = (5x - 2x)\sqrt{3y} \quad \text{Use the distributive property.}$$
$$= 3x\sqrt{3y} \qquad \text{Do the subtraction: } 5x - 2x = 3x.$$

 COMMENT The expression $3\sqrt{2} + 5\sqrt{3}$ cannot be simplified, because the radicals are unlike. For the same reason, we cannot simplify $5x\sqrt[3]{3y} - 2x\sqrt{3y}$.

EXAMPLE 1 *Combining like radicals.* Simplify **a.** $\sqrt{6} + 6 + 5\sqrt{6}$ and **b.** $-2\sqrt{m} - 3\sqrt{m}$.

Solution

a. The expression contains three terms: $\sqrt{6}$, 6, and $5\sqrt{6}$. The first and third terms have like radicals, and they can be combined.

$$\sqrt{6} + 6 + 5\sqrt{6} = 6 + \left(1\sqrt{6} + 5\sqrt{6}\right) \quad \text{Group the expressions with like radicals. Write } \sqrt{6} \text{ as } 1\sqrt{6}.$$
$$= 6 + (1 + 5)\sqrt{6} \qquad \text{Use the distributive property.}$$
$$= 6 + 6\sqrt{6} \qquad \text{Do the addition.}$$

Note that 6 and $6\sqrt{6}$ do not contain like radicals and cannot be combined.

b. The expressions $-2\sqrt{m}$ and $-3\sqrt{m}$ contain like radicals. We can combine them.

$$-2\sqrt{m} - 3\sqrt{m} = (-2 - 3)\sqrt{m} \quad \text{Use the distributive property.}$$
$$= -5\sqrt{m} \qquad \text{Do the subtraction: } -2 - 3 = -5.$$

Radical expressions such as $3\sqrt{18}$ and $5\sqrt{8}$ can be simplified so that they contain like radicals. They can then be combined.

EXAMPLE 2 *Adding radicals.* Simplify $3\sqrt{18} + 5\sqrt{8}$.

Solution

The radical $\sqrt{18}$ is not in simplified form, because 18 has a perfect square factor of 9. The radical $\sqrt{8}$ is not in simplified form either, because 8 has a perfect square factor of 4. To simplify the radicals and add the expressions, we proceed as follows.

$$3\sqrt{18} + 5\sqrt{8}$$
$$= 3\sqrt{9 \cdot 2} + 5\sqrt{4 \cdot 2} \quad \text{Factor 18 and 8 using perfect square factors.}$$
$$= 3\sqrt{9}\sqrt{2} + 5\sqrt{4}\sqrt{2} \quad \text{The square root of a product is equal to the product of the square roots.}$$
$$= 3(3)\sqrt{2} + 5(2)\sqrt{2} \quad \sqrt{9} = 3 \text{ and } \sqrt{4} = 2.$$
$$= 9\sqrt{2} + 10\sqrt{2} \qquad 3(3) = 9 \text{ and } 5(2) = 10.$$
$$= 19\sqrt{2} \qquad \text{To combine like radicals, combine their coefficients: } 9 + 10 = 19.$$

Self Check

Simplify:

a. $\sqrt{7} + 7 + 7\sqrt{7}$

b. $24\sqrt{m} - 25\sqrt{m}$

Answers: a. $8\sqrt{7} + 7$, **b.** $-\sqrt{m}$

Self Check

Simplify $2\sqrt{50} + \sqrt{32}$.

Answer: $14\sqrt{2}$

EXAMPLE 3   *Orthopedics.*   Doctors sometimes use traction to help align a broken bone so that a fracture can heal properly. Figure 14-16 shows how traction is applied by fixing a weight, two pulleys, and some stainless steel cable to a broken leg. How many feet of cable are used in the setup shown in the figure?

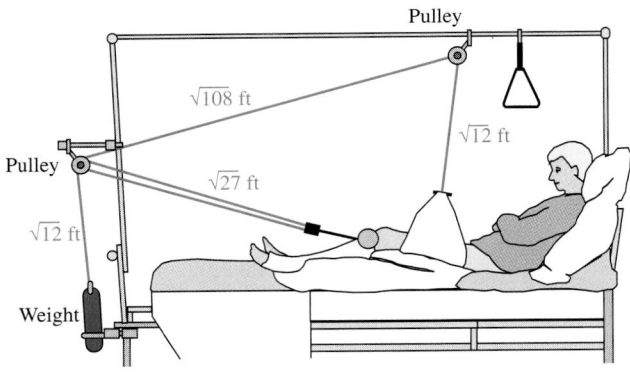

FIGURE 14-16

Solution   Two segments of the cable are $\sqrt{12}$ feet long, another is $\sqrt{108}$ feet long, and two others are $\sqrt{27}$ feet long. The total number of feet of cable can be found by adding

$$2\sqrt{12} + \sqrt{108} + 2\sqrt{27}$$

Since $\sqrt{12}$, $\sqrt{108}$, and $\sqrt{27}$ are not like radicals, we cannot do the addition at this time. First, we need to write each radical in simplified form. Then we can add any expressions that contain like radicals.

$$2\sqrt{12} + \sqrt{108} + 2\sqrt{27}$$
$$= 2\sqrt{4 \cdot 3} + \sqrt{36 \cdot 3} + 2\sqrt{9 \cdot 3} \qquad \text{Factor 12, 108, and 27 using perfect squares.}$$
$$= 2\sqrt{4}\sqrt{3} + \sqrt{36}\sqrt{3} + 2\sqrt{9}\sqrt{3} \qquad \text{The square root of a product is equal to the product of the square roots.}$$
$$= 2(2)\sqrt{3} + 6\sqrt{3} + 2(3)\sqrt{3} \qquad \text{Simplify: } \sqrt{4} = 2,\ \sqrt{36} = 6,\ \text{and } \sqrt{9} = 3.$$
$$= 4\sqrt{3} + 6\sqrt{3} + 6\sqrt{3} \qquad \text{Do the multiplications.}$$
$$= 16\sqrt{3} \qquad \text{To combine like radicals, combine their coefficients: } 4 + 6 + 6 = 16.$$

The traction setup uses $16\sqrt{3}$ feet of cable.   ■

EXAMPLE 4   *Adding radicals.*   Simplify $\sqrt{44x^2y} + x\sqrt{99y}$.

Solution

We simplify each radical and then add the expressions containing like radicals.

$$\sqrt{44x^2y} + x\sqrt{99y}$$
$$= \sqrt{4x^2 \cdot 11y} + x\sqrt{9 \cdot 11y} \qquad \text{Factor } 44x^2y \text{ and } 99y.$$
$$= \sqrt{4x^2}\sqrt{11y} + x\sqrt{9}\sqrt{11y} \qquad \text{The square root of a product is equal to the product of the square roots.}$$
$$= 2x\sqrt{11y} + 3x\sqrt{11y} \qquad \text{Simplify: } \sqrt{4x^2} = 2x \text{ and } \sqrt{9} = 3.$$
$$= 5x\sqrt{11y} \qquad \text{To combine like radicals, combine their coefficients: } 2x + 3x = 5x.$$

Self Check

Simplify $\sqrt{12xy^2} + \sqrt{27xy^2}$.

Answer: $5y\sqrt{3x}$ ■

EXAMPLE 5   *Subtracting radicals.*   Simplify $\sqrt{28x^2y} - 2\sqrt{63y^3}$.

Solution

We begin by simplifying each radical.

$\sqrt{28x^2y} - 2\sqrt{63y^3}$

$= \sqrt{4x^2 \cdot 7y} - 2\sqrt{9y^2 \cdot 7y}$ Factor $28x^2y$ and $63y^3$.

$= \sqrt{4x^2}\sqrt{7y} - 2\sqrt{9y^2}\sqrt{7y}$ The square root of a product is equal to the product of the square roots.

$= 2x\sqrt{7y} - 2(3y)\sqrt{7y}$ $\sqrt{4x^2} = 2x$ and $\sqrt{9y^2} = 3y$.

$= 2x\sqrt{7y} - 6y\sqrt{7y}$

Since $2x$ and $6y$ are not like terms and therefore cannot be subtracted, the expression does not simplify further.

> **Self Check**
> Simplify $\sqrt{20mn^2} - \sqrt{80m^3}$.
>
>
> **Answer:** $2n\sqrt{5m} - 4m\sqrt{5m}$ ■

EXAMPLE 6   *Adding radicals.*   Simplify $\sqrt{27xy} + \sqrt{20xy}$.

Solution

$\sqrt{27xy} + \sqrt{20xy} = \sqrt{9 \cdot 3xy} + \sqrt{4 \cdot 5xy}$ Factor $27xy$ and $20xy$.

$= \sqrt{9}\sqrt{3xy} + \sqrt{4}\sqrt{5xy}$ The square root of a product is equal to the product of the square roots.

$= 3\sqrt{3xy} + 2\sqrt{5xy}$ $\sqrt{9} = 3$ and $\sqrt{4} = 2$.

Since the terms have unlike radicals, the expression does not simplify further.

> **Self Check**
> Simplify $\sqrt{75ab} + \sqrt{72ab}$.
>
>
> **Answer:** $5\sqrt{3ab} + 6\sqrt{2ab}$ ■

EXAMPLE 7   *Adding and subtracting radicals.*   Simplify $\sqrt{8x} + \sqrt{3y} - \sqrt{50x} + \sqrt{27y}$.

Solution

We simplify the radicals and then combine like radicals, where possible.

$\sqrt{8x} + \sqrt{3y} - \sqrt{50x} + \sqrt{27y}$

$= \sqrt{4 \cdot 2x} + \sqrt{3y} - \sqrt{25 \cdot 2x} + \sqrt{9 \cdot 3y}$ Factor $8x$, $50x$, and $27y$.

$= \sqrt{4}\sqrt{2x} + \sqrt{3y} - \sqrt{25}\sqrt{2x} + \sqrt{9}\sqrt{3y}$

$= 2\sqrt{2x} + \sqrt{3y} - 5\sqrt{2x} + 3\sqrt{3y}$

$= -3\sqrt{2x} + 4\sqrt{3y}$ Combine like radicals.

> **Self Check**
> Simplify $\sqrt{32x} - \sqrt{5y} - \sqrt{200x} + \sqrt{125y}$.
>
>
> **Answer:** $-6\sqrt{2x} + 4\sqrt{5y}$ ■

Combining expressions containing cube roots

We can extend the concepts used to combine square roots to radicals with higher order.

EXAMPLE 8   *Subtracting cube roots.*   Simplify $\sqrt[3]{81x^4} - x\sqrt[3]{24x}$.

Solution

We simplify each radical and then combine like radicals.

$\sqrt[3]{81x^4} - x\sqrt[3]{24x} = \sqrt[3]{27x^3 \cdot 3x} - x\sqrt[3]{8 \cdot 3x}$ Factor $81x^4$ and $24x$.

$= \sqrt[3]{27x^3}\sqrt[3]{3x} - x\sqrt[3]{8}\sqrt[3]{3x}$

$= 3x\sqrt[3]{3x} - 2x\sqrt[3]{3x}$

$= x\sqrt[3]{3x}$ Combine like radicals.

> **Self Check**
> Simplify $\sqrt[3]{24a^4} + a\sqrt[3]{81a}$.
>
>
> **Answer:** $5a\sqrt[3]{3a}$

STUDY SET Section 14.4 🌐

VOCABULARY *Fill in the blanks.*

1. Like _____ have the same index and the same radicand.

2. Like _____ have the same variables with the same exponents.

3. Radical expressions such as $\sqrt{8}$ and $\sqrt{18}$ can be _____ so that they contain like radicals.

4. The expression $3\sqrt{2} + \sqrt{8} - 2$ contains three _____.

CONCEPTS *Tell whether the expressions contain like radicals.*

5. $5\sqrt{2}$ and $2\sqrt{3}$

6. $7\sqrt{3x}$ and $3\sqrt{3x}$

7. $125\sqrt[3]{13a}$ and $-\sqrt[3]{13a}$

8. $-17\sqrt[4]{5x}$ and $25\sqrt[3]{5x}$

Tell what is wrong with the following work.

9. $7\sqrt{5} - 3\sqrt{2} = 4\sqrt{3}$

10. $12\sqrt{7} + 20\sqrt{11} = 32\sqrt{18}$

11. $7 - 3\sqrt{2} = 4\sqrt{2}$

12. $12 + 20\sqrt{11} = 32\sqrt{11}$

Complete each table of values.

13.

x	$\sqrt{x} + \sqrt{3}$
3	
12	
27	
48	

14.

x	$3\sqrt{x} - \sqrt{2}$
2	
8	
18	
32	

NOTATION *Complete each solution.*

15. Add: $3\sqrt{80} + 4\sqrt{125}$.

$$3\sqrt{80} + 4\sqrt{125} = 3\sqrt{ \cdot 5} + 4\sqrt{ \cdot 5}$$
$$= 3\sqrt{16} + 4\sqrt{25}$$
$$= 3()\sqrt{5} + 4(5)\sqrt{5}$$
$$= 12\sqrt{5} + \sqrt{5}$$
$$= 32\sqrt{5}$$

16. Subtract: $3\sqrt{125} - 2\sqrt{80}$.

$$3\sqrt{125} - 2\sqrt{80} = 3\sqrt{25 \cdot } - 2\sqrt{ \cdot 5}$$
$$= 3\sqrt{5} - 2\sqrt{16}\sqrt{5}$$
$$= 3(5)\sqrt{5} - 2()\sqrt{5}$$
$$= \sqrt{5} - 8\sqrt{5}$$
$$= 7\sqrt{5}$$

PRACTICE *Simplify each expression. All variables represent positive numbers.*

17. $5\sqrt{7} + 4\sqrt{7}$

18. $3\sqrt{10} + 4\sqrt{10}$

19. $\sqrt{x} - 4\sqrt{x}$

20. $\sqrt{t} - 9\sqrt{t}$

21. $5 + 3\sqrt{3} + 3\sqrt{3}$

22. $\sqrt{5} + 2 + 3\sqrt{5}$

23. $-1 + 2\sqrt{r} - 3\sqrt{r}$

24. $-8 - 5\sqrt{c} + 4\sqrt{c}$

25. $\sqrt{12} + \sqrt{27}$

26. $\sqrt{20} + \sqrt{45}$

27. $\sqrt{18} - \sqrt{8}$

28. $\sqrt{32} - \sqrt{18}$

29. $2\sqrt{45} + 2\sqrt{80}$

30. $3\sqrt{80} + 3\sqrt{125}$

31. $2\sqrt{80} - 3\sqrt{125}$

32. $3\sqrt{245} - 2\sqrt{180}$

33. $\sqrt{20} + \sqrt{180}$

34. $2\sqrt{28} + 7\sqrt{63}$

35. $\sqrt{12} - \sqrt{48}$

36. $\sqrt{48} - \sqrt{75}$

37. $\sqrt{288} - 3\sqrt{200}$

38. $\sqrt{80} - \sqrt{245}$

39. $2\sqrt{28} + 2\sqrt{112}$

40. $4\sqrt{63} + 6\sqrt{112}$

41. $\sqrt{20} + \sqrt{45} + \sqrt{80}$

42. $\sqrt{48} + \sqrt{27} + \sqrt{75}$

43. $\sqrt{200} - \sqrt{75} + \sqrt{48}$

44. $\sqrt{20} + \sqrt{80} - \sqrt{125}$

45. $8\sqrt{6} - 5\sqrt{2} - 3\sqrt{6}$

46. $3\sqrt{2} - 3\sqrt{15} - 4\sqrt{15}$

47. $\sqrt{24} + \sqrt{150} + \sqrt{240}$

48. $\sqrt{28} + \sqrt{63} + \sqrt{18}$

49. $\sqrt{48} - \sqrt{8} + \sqrt{27} - \sqrt{32}$

50. $\sqrt{162} + \sqrt{50} - \sqrt{75} - \sqrt{108}$

51. $\sqrt{2x^2} + \sqrt{8x^2}$

52. $\sqrt{3y^2} - \sqrt{12y^2}$

53. $\sqrt{2d^3} + \sqrt{8d^3}$

54. $\sqrt{3a^3} - \sqrt{12a^3}$

55. $\sqrt{18x^2y} - \sqrt{27x^2y}$

56. $\sqrt{49xy} + \sqrt{xy}$

57. $\sqrt{32x^5} - \sqrt{18x^5}$ **58.** $\sqrt{27xy^3} - \sqrt{48xy^3}$

59. $3\sqrt{54b^2} + 5\sqrt{24b^2}$ **60.** $3\sqrt{24x^4y^3} + 2\sqrt{54x^4y^3}$

61. $y\sqrt{490y} - 2\sqrt{360y^3}$ **62.** $3\sqrt{20x} + 2\sqrt{63y}$

63. $\sqrt{20x^3y} + \sqrt{45x^5y^3} - \sqrt{80x^7y^5}$

64. $x\sqrt{48xy^2} - y\sqrt{27x^3} + \sqrt{75x^3y^2}$

65. $\sqrt[3]{3} + \sqrt[3]{3}$ **66.** $\sqrt[3]{2} + 5\sqrt[3]{2}$

67. $2\sqrt[3]{x} - 3\sqrt[3]{x}$ **68.** $4\sqrt[3]{s} - 5\sqrt[3]{s}$

69. $\sqrt[3]{16} + \sqrt[3]{54}$ **70.** $\sqrt[3]{24} - \sqrt[3]{81}$

71. $\sqrt[3]{81} - \sqrt[3]{24}$ **72.** $\sqrt[3]{32} + \sqrt[3]{108}$

73. $\sqrt[3]{40} + \sqrt[3]{125}$ **74.** $\sqrt[3]{3,000} - \sqrt[3]{192}$

75. $\sqrt[3]{x^4} - \sqrt[3]{x^7}$

76. $\sqrt[3]{8x^5} + \sqrt[3]{27x^8}$

77. $\sqrt[3]{192x^4y^5} - \sqrt[3]{24x^4y^5}$

78. $\sqrt[3]{24a^5b^4} + \sqrt[3]{81a^5b^4}$

79. $\sqrt[3]{135x^7y^4} - \sqrt[3]{40x^7y^4}$

80. $\sqrt[3]{56a^4b^5} + \sqrt[3]{7a^4b^5}$

APPLICATIONS

81. ANATOMY   See Illustration 1. Determine the length of the patient's arm if he lets it fall to his side.

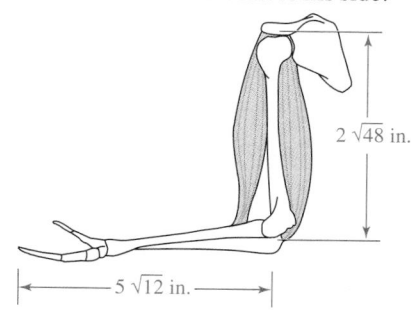

ILLUSTRATION 1

82. PLAYGROUND EQUIPMENT   Find the total length of pipe necessary to construct the frame of the swing set shown in Illustration 2.

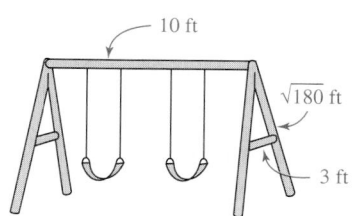

ILLUSTRATION 2

83. READING BLUEPRINTS   What is the length of the motor on the machine shown in Illustration 3?

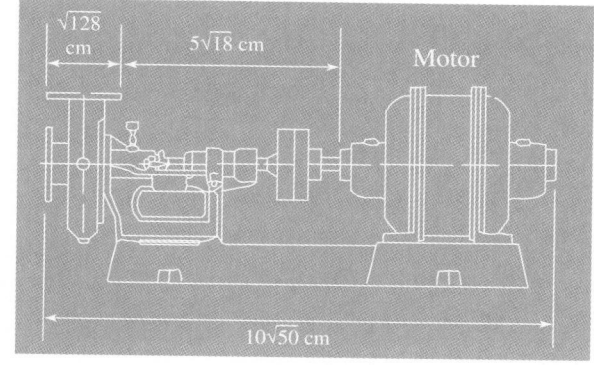

ILLUSTRATION 3

84. TENTS   The length of a center support pole for the tents shown in Illustration 4 is given by the formula

$$l = 0.5s\sqrt{3}$$

where s is the length of the side of the tent. Find the total length of the four poles needed for the parents' and children's tents.

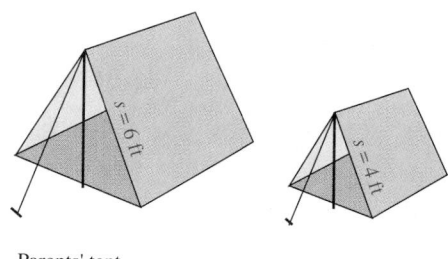

Parents' tent

Children's tent

ILLUSTRATION 4

85. FENCING   Find the number of feet of fencing needed to enclose the swimming pool complex shown in Illustration 5.

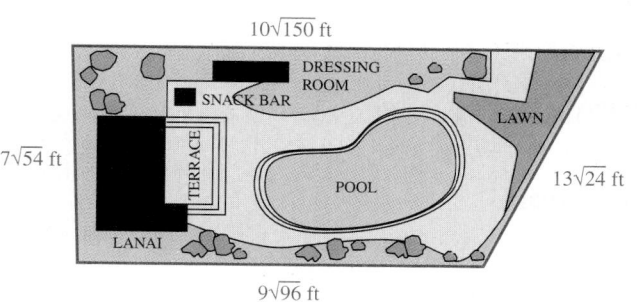

ILLUSTRATION 5

86. HARDWARE   Find the difference in the lengths of the "arms" of the door-closing device shown in Illustration 6.

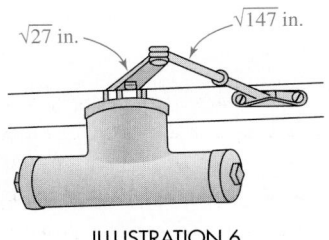

ILLUSTRATION 6

WRITING

87. Explain why $\sqrt{3} + \sqrt{2}$ cannot be combined.

88. Explain why $\sqrt{4x}$ and $\sqrt[3]{4x}$ cannot be combined.

REVIEW   *Simplify each expression. Write each answer without using negative exponents.*

89. 3^{-2}

90. $\dfrac{1}{3^{-2}}$

91. -3^2

92. -3^{-2}

93. x^{-3}

94. $\dfrac{1}{x^{-3}}$

95. 3^0

96. x^0

14.5 *Multiplying and Dividing Radical Expressions*

In this section, you will learn about

- Multiplying radical expressions   • Dividing radical expressions
- Rationalizing denominators

INTRODUCTION. In this section, we will discuss the methods used to multiply and divide radical expressions.

Multiplying radical expressions

Recall that the *product of the square roots of two nonnegative numbers is equal to the square root of the product of those numbers*. For example,

$$\sqrt{2}\sqrt{8} = \sqrt{2 \cdot 8} \qquad \sqrt{3}\sqrt{27} = \sqrt{3 \cdot 27} \qquad \sqrt{x}\sqrt{x^3} = \sqrt{x \cdot x^3}$$
$$= \sqrt{16} \qquad\qquad\quad = \sqrt{81} \qquad\qquad\quad = \sqrt{x^4}$$
$$= 4 \qquad\qquad\qquad\quad = 9 \qquad\qquad\qquad = x^2$$

Likewise, the *product of the cube roots of two numbers is equal to the cube root of the product of those numbers*. For example,

$$\sqrt[3]{2}\sqrt[3]{4} = \sqrt[3]{2 \cdot 4} \qquad \sqrt[3]{4}\sqrt[3]{16} = \sqrt[3]{4 \cdot 16} \qquad \sqrt[3]{3x^2}\sqrt[3]{9x} = \sqrt[3]{3x^2 \cdot 9x}$$
$$= \sqrt[3]{8} \qquad\qquad\quad = \sqrt[3]{64} \qquad\qquad\quad = \sqrt[3]{27x^3}$$
$$= 2 \qquad\qquad\qquad = 4 \qquad\qquad\qquad = 3x$$

These examples illustrate that radical expressions with the same index can be multiplied.

EXAMPLE 1   *Multiplying radicals.*   Multiply   **a.** $\sqrt{3}\sqrt{2}$,   **b.** $\sqrt{6}\sqrt{8}$, and   **c.** $\sqrt[3]{4}\sqrt[3]{10}$.

Solution

a. $\sqrt{3}\sqrt{2} = \sqrt{3 \cdot 2}$ The product of the square roots of two numbers is equal to the square root of the product of those numbers.

$\phantom{\textbf{a.}\ \sqrt{3}\sqrt{2}} = \sqrt{6}$ Do the multiplication within the radical.

Self Check

Multiply:

a. $\sqrt{5}\sqrt{3}$

b. $\sqrt{8}\sqrt{9}$

c. $\sqrt[3]{6}\sqrt[3]{9}$

b. $\sqrt{6}\sqrt{8} = \sqrt{6\cdot 8}$ The product of two square roots is equal to the square root of the product.

$= \sqrt{48}$ Do the multiplication within the radical. Note that this radical can be simplified.

$= \sqrt{16}\sqrt{3}$ Factor 48 as $16\cdot 3$.

$= 4\sqrt{3}$ Simplify: $\sqrt{16} = 4$.

c. $\sqrt[3]{4}\sqrt[3]{10} = \sqrt[3]{4\cdot 10}$ The product of two cube roots is equal to the cube root of the product.

$= \sqrt[3]{40}$ Do the multiplication within the radical.

$= \sqrt[3]{8}\sqrt[3]{5}$ $\sqrt[3]{40} = \sqrt[3]{8\cdot 5} = \sqrt[3]{8}\sqrt[3]{5}$.

$= 2\sqrt[3]{5}$ Simplify: $\sqrt[3]{8} = 2$.

Answers: **a.** $\sqrt{15}$, **b.** $6\sqrt{2}$, **c.** $3\sqrt[3]{2}$. ■

To multiply radical expressions having only one term, we multiply the coefficients and multiply the radicals separately and then simplify the result, when possible.

EXAMPLE 2   *Multiplying radical expressions.*   Multiply
a. $3\sqrt{6}$ by $4\sqrt{3}$ and   **b.** $-2\sqrt[3]{7x}$ by $6\sqrt[3]{49x^2}$.

Solution

The commutative and associative properties enable us to multiply the coefficients and the radicals separately.

a. $3\sqrt{6}\cdot 4\sqrt{3} = 3(4)\sqrt{6}\sqrt{3}$ Write the coefficients together and the radicals together.

$= 12\sqrt{18}$ Multiply the coefficients and multiply the radicals.

$= 12\sqrt{9}\sqrt{2}$ $\sqrt{18} = \sqrt{9\cdot 2} = \sqrt{9}\sqrt{2}$.

$= 12(3)\sqrt{2}$ Simplify: $\sqrt{9} = 3$.

$= 36\sqrt{2}$ Do the multiplication: $12(3) = 36$.

b. $-2\sqrt[3]{7x}\cdot 6\sqrt[3]{49x^2} = -2(6)\sqrt[3]{7x}\sqrt[3]{49x^2}$ Write the coefficients together and the radicals together.

$= -12\sqrt[3]{7x\cdot 49x^2}$ Multiply the coefficients and multiply the radicals.

$= -12\sqrt[3]{343x^3}$ Do the multiplication within the radical.

$= -12(7x)$ Simplify: $\sqrt[3]{343x^3} = 7x$.

$= -84x$ Multiply.

Self Check
Multiply:
a. $(2\sqrt{2x})(-3\sqrt{3x})$
b. $(5\sqrt[3]{2})(2\sqrt[3]{4})$

Answers: **a.** $-6x\sqrt{6}$, **b.** 20 ■

EXAMPLE 3   *Powers of radical expressions.*   Find $(2\sqrt{5})^2$.
Solution

Recall that a power is used to indicate repeated multiplication.

$(2\sqrt{5})^2 = 2\sqrt{5}\cdot 2\sqrt{5}$ Write $2\sqrt{5}$ as a factor two times.

$= 2(2)\sqrt{5}\sqrt{5}$ Multiply the coefficients and the radicals separately.

$= 4\sqrt{5\cdot 5}$ The product of two square roots is equal to the square root of the product.

$= 4\sqrt{25}$ Do the multiplication within the radical.

$= 4\cdot 5$ $\sqrt{25} = 5$

$= 20$

Self Check
Find $(3\sqrt[3]{-2})^3$.

Answer: -54 ■

Recall that to multiply a polynomial by a monomial, we use the distributive property. We use the same technique to multiply a radical expression that has two or more terms by a radical expression that has only one term.

EXAMPLE 4 *Using the distributive property.* Multiply

a. $\sqrt{2x}\left(\sqrt{6x} + \sqrt{8x}\right)$ and **b.** $\sqrt[3]{3}\left(\sqrt[3]{9} - 2\right)$.

Solution

a. $\sqrt{2x}\left(\sqrt{6x} + \sqrt{8x}\right) = \sqrt{2x}\sqrt{6x} + \sqrt{2x}\sqrt{8x}$ Distribute the multiplication by $\sqrt{2x}$.

$\qquad = \sqrt{12x^2} + \sqrt{16x^2}$ The product of two square roots is equal to the square root of the product.

$\qquad = \sqrt{4x^2 \cdot 3} + \sqrt{16x^2}$ Factor $12x^2$ as $4x^2 \cdot 3$.

$\qquad = \sqrt{4x^2}\sqrt{3} + \sqrt{16x^2}$ The square root of a product is equal to the product of the square roots.

$\qquad = 2x\sqrt{3} + 4x$ Simplify: $\sqrt{4x^2} = 2x$ and $\sqrt{16x^2} = 4x$.

b. $\sqrt[3]{3}\left(\sqrt[3]{9} - 2\right) = \sqrt[3]{3}\sqrt[3]{9} - 2\sqrt[3]{3}$ Distribute the multiplication by $\sqrt[3]{3}$.

$\qquad = \sqrt[3]{27} - 2\sqrt[3]{3}$ The product of two cube roots is equal to the cube root of the product.

$\qquad = 3 - 2\sqrt[3]{3}$ Simplify: $\sqrt[3]{27} = 3$.

Self Check

Multiply:

a. $\sqrt{3}\left(3\sqrt{6} - \sqrt{3}\right)$

b. $\sqrt[3]{2x}\left(3 - \sqrt[3]{4x^2}\right)$

Answers: **a.** $9\sqrt{2} - 3$, **b.** $3\sqrt[3]{2x} - 2x$ ■

To multiply two binomials, we multiply each term of one binomial by each term of the other binomial and simplify. We multiply two radical expressions, each having two terms, in the same way.

EXAMPLE 5 *Using the FOIL method.* Multiply:
$\left(\sqrt{3x} + 1\right)\left(\sqrt{3x} + 2\right)$.

Solution

$\left(\sqrt{3x} + 1\right)\left(\sqrt{3x} + 2\right)$

$\qquad = \sqrt{3x}\sqrt{3x} + 2\sqrt{3x} + \sqrt{3x} + 2$ Use the FOIL method.

$\qquad = \sqrt{3x}\sqrt{3x} + 3\sqrt{3x} + 2$ Combine like radicals.

$\qquad = 3x + 3\sqrt{3x} + 2$ Simplify: $\sqrt{3x}\sqrt{3x} = \left(\sqrt{3x}\right)^2 = 3x$.

Self Check

Multiply: $\left(\sqrt{5a} - 2\right)\left(\sqrt{5a} + 3\right)$.

Answer: $5a + \sqrt{5a} - 6$ ■

EXAMPLE 6 *Special products.* Multiply: $\left(\sqrt{7} + \sqrt{2}\right)\left(\sqrt{7} - \sqrt{2}\right)$.

Solution

Recall from Chapter 10 that the product of two binomials that differ only in the signs between the terms is the square of the first term minus the square of the second term: $(x + y)(x - y) = x^2 - y^2$. We can use this special product formula to multiply the given radical expressions.

$\left(\sqrt{7} + \sqrt{2}\right)\left(\sqrt{7} - \sqrt{2}\right) = \left(\sqrt{7}\right)^2 - \left(\sqrt{2}\right)^2$

$\qquad\qquad\qquad = 7 - 2$

$\qquad\qquad\qquad = 5$

Self Check

Multiply:

$\left(\sqrt{5} + \sqrt{11}\right)\left(\sqrt{5} - \sqrt{11}\right)$

Answer: -6 ■

COMMENT Note that the answers to Example 6 and the Self Check did not contain any radicals. This will be the case whenever we find the product of radical expressions (containing *square* roots) of this form, which differ only in the sign between the terms.

EXAMPLE 7 *Multiplying radical expressions.* Multiply: $\left(\sqrt[3]{4x} - 3\right)\left(\sqrt[3]{2x^2} + 1\right)$.

Solution

$$\left(\sqrt[3]{4x} - 3\right)\left(\sqrt[3]{2x^2} + 1\right)$$

$$= \sqrt[3]{4x}\sqrt[3]{2x^2} + \sqrt[3]{4x} - 3\sqrt[3]{2x^2} - 3 \quad \text{Use the FOIL method.}$$

$$= \sqrt[3]{8x^3} + \sqrt[3]{4x} - 3\sqrt[3]{2x^2} - 3 \quad \text{The product of two cube roots is equal to the cube root of the product.}$$

$$= 2x + \sqrt[3]{4x} - 3\sqrt[3]{2x^2} - 3 \quad \text{Simplify: } \sqrt[3]{8x^3} = 2x.$$

Self Check
Multiply:

$$\left(\sqrt[3]{3x} + 1\right)\left(\sqrt[3]{9x^2} - 2\right)$$

Answer:

$3x - 2\sqrt[3]{3x} + \sqrt[3]{9x^2} - 2$

Dividing radical expressions

To divide radical expressions, we use the division property of radicals. For example, to divide $\sqrt{108}$ by $\sqrt{36}$, we proceed as follows:

$$\frac{\sqrt{108}}{\sqrt{36}} = \sqrt{\frac{108}{36}} \quad \text{The quotient of two square roots is the square root of the quotient.}$$

$$= \sqrt{3} \quad \text{Do the division within the radical: } 108 \div 36 = 3.$$

EXAMPLE 8 *Dividing radical expressions.* Divide: $\dfrac{\sqrt{22a^2}}{\sqrt{99a^4}}$ $(a > 0)$.

Solution

$$\frac{\sqrt{22a^2}}{\sqrt{99a^4}} = \sqrt{\frac{22a^2}{99a^4}}$$

$$= \sqrt{\frac{2}{9a^2}} \quad \text{Simplify the radicand: } \frac{22a^2}{99a^4} = \frac{\overset{1}{\cancel{11}} \cdot 2 \cdot \overset{1}{\cancel{a^2}}}{\underset{1}{\cancel{11}} \cdot 9 \cdot \underset{1}{\cancel{a^2}} \cdot a^2} = \frac{2}{9a^2}.$$

$$= \frac{\sqrt{2}}{\sqrt{9a^2}} \quad \text{The square root of a quotient is equal to the quotient of the square roots.}$$

$$= \frac{\sqrt{2}}{3a} \quad \text{Simplify: } \sqrt{9a^2} = 3a.$$

Self Check
Divide:

$$\frac{\sqrt{30y^9}}{\sqrt{160y^5}} \quad (y > 0)$$

Answer: $\dfrac{y^2\sqrt{3}}{4}$

Rationalizing denominators

The length of a diagonal of one of the square adobe tiles shown in Figure 14-17 is 1 foot. Using the Pythagorean theorem, it can be shown that the length of a side of a tile is $\frac{1}{\sqrt{2}}$ feet. Because the expression $\frac{1}{\sqrt{2}}$ contains a radical in its denominator, it is not in simplified radical form. Since it is often easier to work with a radical expression if the denominator does not contain a radical, we now consider a process in which we change the denominator from a radical that represents an irrational number to a rational number. The process is called **rationalizing the denominator.**

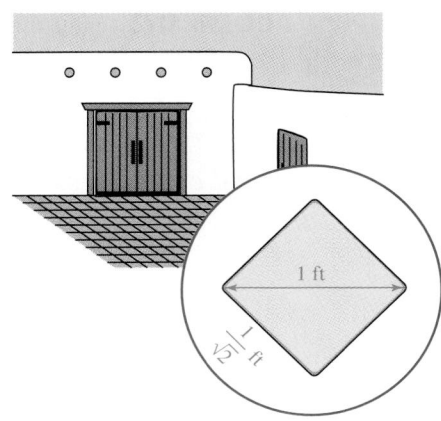

FIGURE 14-17

To rationalize the denominator of $\frac{1}{\sqrt{2}}$, we multiply both the numerator and the denominator by $\sqrt{2}$. Because the expression is multiplied by $\frac{\sqrt{2}}{\sqrt{2}}$, which is 1, the value of $\frac{1}{\sqrt{2}}$ is not changed—only its form.

$$\frac{1}{\sqrt{2}} = \frac{1\sqrt{2}}{\sqrt{2}\sqrt{2}}$$ Multiply both numerator and denominator by $\sqrt{2}$.

$$= \frac{\sqrt{2}}{2}$$ In the numerator, $1\sqrt{2} = \sqrt{2}$. In the denominator, $\sqrt{2}\sqrt{2} = (\sqrt{2})^2 = 2$. The denominator is now a rational number.

The length of a side of a patio tile is $\frac{1}{\sqrt{2}} = \frac{\sqrt{2}}{2}$ feet.

This example suggests the following procedure for rationalizing square root denominators.

Rationalizing square root denominators	Multiply the numerator and the denominator by the smallest factor that gives a perfect square radicand in the denominator.

EXAMPLE 9   *Rationalizing the denominator.*   Rationalize each denominator: **a.** $\sqrt{\dfrac{5}{3}}$ and   **b.** $\dfrac{2}{\sqrt[3]{3}}$.

Self Check
Rationalize each denominator:

a. $\sqrt{\dfrac{2}{7}}$

b. $\dfrac{5}{\sqrt[3]{5}}$

Solution

a. The expression $\sqrt{\frac{5}{3}}$ is not in simplified form, because the radicand is a fraction. To write it in simplified form, we use the division property of radicals. Then we use the fundamental property of fractions to rationalize the denominator by multiplying the numerator and the denominator by $\sqrt{3}$.

$$\sqrt{\frac{5}{3}} = \frac{\sqrt{5}}{\sqrt{3}}$$ The square root of a quotient is the quotient of the square roots. Note that the denominator is the irrational number $\sqrt{3}$.

$$= \frac{\sqrt{5}\sqrt{3}}{\sqrt{3}\sqrt{3}}$$ Multiply the numerator and the denominator by $\sqrt{3}$.

$$= \frac{\sqrt{15}}{3}$$ In the numerator, do the multiplication. Simplify in the denominator: $\sqrt{3}\sqrt{3} = (\sqrt{3})^2 = 3$.

b. The denominator contains a cube root. We multiply by the smallest factor that gives an integer cube radicand in the denominator. Since $\sqrt[3]{3}\sqrt[3]{9} = \sqrt[3]{27}$ and 27 is a perfect integer cube, we multiply the numerator and denominator by $\sqrt[3]{9}$ and simplify.

$$\frac{2}{\sqrt[3]{3}} = \frac{2\sqrt[3]{9}}{\sqrt[3]{3}\sqrt[3]{9}}$$

$$= \frac{2\sqrt[3]{9}}{\sqrt[3]{27}} \quad \text{Multiply: } \sqrt[3]{3}\sqrt[3]{9} = \sqrt[3]{27}.$$

$$= \frac{2\sqrt[3]{9}}{3} \quad \text{Simplify: } \sqrt[3]{27} = 3. \text{ The denominator is now a rational number.}$$

Answers: a. $\dfrac{\sqrt{14}}{7}$,

b. $\sqrt[3]{25}$

EXAMPLE 10   *Rationalizing denominators.*   Rationalize the denominator and simplify: $\dfrac{5\sqrt{y}}{\sqrt{20x}}$ $(x > 0)$.

Solution

To rationalize the denominator, we don't need to multiply the numerator and denominator by $\sqrt{20x}$. To keep the numbers small, we can multiply by $\sqrt{5x}$, because $5x \cdot 20x = 100x^2$, which is a perfect square.

$$\frac{5\sqrt{y}}{\sqrt{20x}} = \frac{5\sqrt{y}\sqrt{5x}}{\sqrt{20x}\sqrt{5x}} \quad \text{Multiply the numerator and denominator by } \sqrt{5x}.$$

$$= \frac{5\sqrt{5xy}}{\sqrt{100x^2}} \quad \text{Multiply: } \sqrt{y}\sqrt{5x} = \sqrt{5xy} \text{ and } \sqrt{20x}\sqrt{5x} = \sqrt{100x^2}.$$

$$= \frac{5\sqrt{5xy}}{10x} \quad \text{Simplify: } \sqrt{100x^2} = 10x.$$

$$= \frac{\overset{1}{\cancel{5}}\sqrt{5xy}}{\underset{1}{\cancel{5}} \cdot 2x} \quad \text{Factor } 10x \text{ and then divide out a common factor of 5.}$$

$$= \frac{\sqrt{5xy}}{2x} \quad \text{Simplify.}$$

Self Check

Rationalize the denominator and simplify:

$\dfrac{6\sqrt{z}}{\sqrt{50y}}$ $(y > 0)$

Answer: $\dfrac{3\sqrt{2yz}}{5y}$

At times, we will encounter fractions such as $\dfrac{2}{\sqrt{3} - 1}$, whose denominator has two terms. Note that $\sqrt{3} - 1$ is an irrational number. Because $\sqrt{3} - 1$ has two terms, multiplying it by $\sqrt{3}$ will not make it a rational number. The key to rationalizing this denominator is to multiply the numerator and denominator by $\sqrt{3} + 1$, because the product $(\sqrt{3} + 1)(\sqrt{3} - 1)$ has no radicals. Radical expressions such as $\sqrt{3} + 1$ and $\sqrt{3} - 1$ are called **conjugates** of each other.

EXAMPLE 11   *Multiplying by the conjugate.*   Rationalize the denominator and simplify: $\dfrac{2}{\sqrt{3} - 1}$.

Solution

We rationalize the denominator by multiplying the numerator and denominator by the conjugate of the denominator.

$$\frac{2}{\sqrt{3} - 1} = \frac{2(\sqrt{3} + 1)}{(\sqrt{3} - 1)(\sqrt{3} + 1)} \quad \begin{array}{l}\text{Multiply the numerator and denominator by the}\\ \text{conjugate of the denominator, which is } \sqrt{3} + 1.\end{array}$$

$$= \frac{2(\sqrt{3} + 1)}{3 - 1} \quad \begin{array}{l}\text{Use a special product formula:}\\ (\sqrt{3} - 1)(\sqrt{3} + 1) = 3 - 1.\end{array}$$

$$= \frac{2(\sqrt{3} + 1)}{2} \quad \begin{array}{l}\text{Subtract. The denominator is now a rational}\\ \text{number.}\end{array}$$

$$= \sqrt{3} + 1 \quad \text{Divide out the common factor of 2.}$$

Self Check

Rationalize the denominator and simplify:

$\dfrac{3}{\sqrt{2} + 1}$

Answer: $3(\sqrt{2} - 1)$

EXAMPLE 12 *Multiplying by the conjugate.* Rationalize the denominator and simplify: $\dfrac{\sqrt{x}+1}{\sqrt{x}-1}$ ($x > 0$ and $x \neq 1$).

Self Check

Rationalize the denominator and simplify:

$$\frac{\sqrt{x}-1}{\sqrt{x}+1}$$

Solution

We multiply the numerator and denominator by the conjugate of the denominator, which is $\sqrt{x}+1$.

$$\frac{\sqrt{x}+1}{\sqrt{x}-1} = \frac{(\sqrt{x}+1)(\sqrt{x}+1)}{(\sqrt{x}-1)(\sqrt{x}+1)}$$ 　 Multiply the numerator and denominator by $\sqrt{x}+1$.

$$= \frac{\sqrt{x}\sqrt{x} + \sqrt{x}(1) + 1(\sqrt{x}) + 1}{\sqrt{x}\sqrt{x} + \sqrt{x}(1) - 1(\sqrt{x}) - 1}$$ 　 Do the multiplications.

$$= \frac{x + 2\sqrt{x} + 1}{x - 1}$$ 　 Simplify: $\sqrt{x}\sqrt{x} = (\sqrt{x})^2 = x$. Combine like radicals.

Answer: $\dfrac{x - 2\sqrt{x} + 1}{x - 1}$ ■

STUDY SET　Section 14.5

VOCABULARY *Fill in the blanks.*

1. The method of changing a radical denominator of a fraction into a rational number is called _____ the denominator.

2. The _____ of the fraction $\dfrac{4}{\sqrt{3}}$ is 4 and the _____ is $\sqrt{3}$.

3. $3 + \sqrt{2}$ is the _____ of $3 - \sqrt{2}$.

4. Radical expressions with the same _____ can be multiplied.

5. In the radical expression $3\sqrt{7}$, the number 3 is the _____ of the radical.

6. Nonterminating, nonrepeating decimals such as $\sqrt{2} = 1.414213562\ldots$ and $\sqrt{3} = 1.732050808\ldots$ are _____ numbers.

CONCEPTS *Fill in the blanks.*

7. To change $\sqrt{11}$ into a perfect integer square, we multiply it by

8. To change $\sqrt[3]{11}$ into a perfect integer cube, we multiply it by _____ .

9. To rationalize the denominator of

$$\frac{x}{\sqrt{7}}$$

we multiply the numerator and denominator by _____ .

10. To rationalize the denominator of

$$\frac{x}{\sqrt{x}+1}$$

we multiply the numerator and denominator by _____ .

11. Explain why each expression is not in simplified radical form.

　a. $\sqrt{\dfrac{3}{4}}$ 　　　**b.** $\dfrac{1}{\sqrt{10}}$

12. Fill in the blanks: To multiply $2\sqrt{x}$ and $6\sqrt{x}$, we first multiply the _____, then multiply the _____, and simplify the result.

13. Which fractions have a rational denominator and which have an irrational denominator?

$$\frac{\sqrt{5}}{3}, \quad \frac{2}{\sqrt{6}}, \quad -\frac{\sqrt{2}}{8}, \quad \frac{1+\sqrt{3}}{4}, \quad \frac{9}{7-\sqrt{10}}$$

14. To multiply $(\sqrt{3}+\sqrt{2})(\sqrt{7}+\sqrt{5})$, we use the FOIL method. What are the

　a. First terms? 　　**b.** Outer terms?

　c. Inner terms? 　　**d.** Last terms?

Do each operation if possible.

15. a. $\sqrt{2} + \sqrt{3}$ 　　**b.** $\sqrt{2} \cdot \sqrt{3}$

　c. $\sqrt{2} - \sqrt{3}$ 　　**d.** $\dfrac{\sqrt{2}}{\sqrt{3}}$

　e. $\sqrt{2} + 3\sqrt{2}$ 　　**f.** $\sqrt{2} \cdot 3\sqrt{2}$

　g. $\sqrt{2} - 3\sqrt{2}$ 　　**h.** $\dfrac{\sqrt{2}}{3\sqrt{2}}$

16. Find each special product.

   a. $(\sqrt{6} + \sqrt{3})(\sqrt{6} - \sqrt{3})$

   b. $(\sqrt{a} + \sqrt{7})(\sqrt{a} - \sqrt{7})$

NOTATION   *Complete each solution.*

17. $(\sqrt{x} + \sqrt{2})(\sqrt{x} - 3\sqrt{2})$

$$= \sqrt{x} - \sqrt{x}(3\sqrt{2}) + \sqrt{2} - \sqrt{2}(3\sqrt{2})$$

$$= x - 3 + \sqrt{2x} - 3\sqrt{2}\sqrt{2}$$

$$= - 2\sqrt{2x} - 3(2)$$

$$= x - 2\sqrt{2x} - 6$$

18. $\dfrac{x}{\sqrt{x} - 2} = \dfrac{x(\sqrt{x} + 2)}{(\sqrt{x} - 2)}$

$$= \dfrac{x(\sqrt{x} + 2)}{()^2 - 2^2}$$

$$= \dfrac{x(\sqrt{x} + 2)}{x - 4}$$

PRACTICE   *Do each multiplication. All variables represent positive numbers.*

19. $(\sqrt{5})^2$

20. $(\sqrt{11})^2$

21. $(3\sqrt{6})^2$

22. $(-7\sqrt{2})^2$

23. $\sqrt{2}\sqrt{8}$

24. $\sqrt{27}\sqrt{3}$

25. $\sqrt{7}\sqrt{3}$

26. $\sqrt{2}\sqrt{11}$

27. $\sqrt{8}\sqrt{7}$

28. $\sqrt{6}\sqrt{8}$

29. $3\sqrt{2}\sqrt{x}$

30. $4\sqrt{3x}\sqrt{5y}$

31. $\sqrt{x^3}\sqrt{x^5}$

32. $\sqrt{a^7}\sqrt{a^3}$

33. $(-5\sqrt{6})(4\sqrt{3})$

34. $(6\sqrt{3})(-7\sqrt{2})$

35. $(4\sqrt{x})(-2\sqrt{x})$

36. $(3\sqrt{y})(15\sqrt{y})$

37. $\sqrt{8x}\sqrt{2x^3}$

38. $\sqrt{27y}\sqrt{3y^3}$

39. $\sqrt{2}(\sqrt{2} + 1)$

40. $\sqrt{5}(\sqrt{5} + 2)$

41. $3\sqrt{3}(\sqrt{27} - 1)$

42. $2\sqrt{2}(\sqrt{8} - 1)$

43. $\sqrt{3}(\sqrt{6} + 1)$

44. $\sqrt{2}(\sqrt{6} - 2)$

45. $\sqrt{x}(\sqrt{3x} - 2)$

46. $\sqrt{y}(\sqrt{y} + 5)$

47. $2\sqrt{x}(\sqrt{9x} + 3)$

48. $3\sqrt{z}(\sqrt{4z} - \sqrt{z})$

49. $(\sqrt{2} + 1)(\sqrt{2} - 1)$

50. $(\sqrt{3} - 1)(\sqrt{3} + 1)$

51. $(2\sqrt{7} - x)(3\sqrt{2} + x)$

52. $(4\sqrt{2} - \sqrt{x})(\sqrt{x} + 2\sqrt{3})$

53. $(\sqrt{6} + 1)^2$

54. $(3 - \sqrt{3})^2$

55. $(\sqrt{2x} + 3)(\sqrt{8x} - 6)$

56. $(\sqrt{5y} - 3)(\sqrt{20y} + 6)$

57. $(-\sqrt[3]{9})^3$

58. $(\sqrt[3]{3})^3$

59. $(2\sqrt[3]{4})(3\sqrt[3]{3})$

60. $(-3\sqrt[3]{3})(\sqrt[3]{5})$

61. $\sqrt[3]{7}(\sqrt[3]{49} - 2)$

62. $\sqrt[3]{5}(\sqrt[3]{25} + 3)$

63. $(\sqrt[3]{2} + 1)(\sqrt[3]{2} + 3)$

64. $(\sqrt[3]{5} - 2)(\sqrt[3]{5} - 1)$

Simplify each expression. Assume that all variables represent positive numbers.

65. $\dfrac{\sqrt{12x^3}}{\sqrt{27x}}$

66. $\dfrac{\sqrt{32}}{\sqrt{98x^2}}$

67. $\dfrac{\sqrt{18x}}{\sqrt{25x}}$

68. $\dfrac{\sqrt{27y}}{\sqrt{75y}}$

69. $\dfrac{\sqrt{196x}}{\sqrt{49x^3}}$

70. $\dfrac{\sqrt{50}}{\sqrt{98z^2}}$

71. $\dfrac{\sqrt[3]{16x^6}}{\sqrt[3]{54x^3}}$

72. $\dfrac{\sqrt[3]{128a^6}}{\sqrt[3]{16a^3}}$

Rationalize each denominator and simplify. All variables represent positive numbers.

73. $\dfrac{1}{\sqrt{3}}$

74. $\dfrac{1}{\sqrt{5}}$

75. $\sqrt{\dfrac{13}{7}}$

76. $\sqrt{\dfrac{3}{11}}$

77. $\dfrac{9}{\sqrt{27}}$

78. $\dfrac{4}{\sqrt{20}}$

79. $\dfrac{3}{\sqrt{32}}$

80. $\dfrac{5}{\sqrt{18}}$

81. $\sqrt{\dfrac{12}{5}}$

82. $\sqrt{\dfrac{24}{7}}$

83. $\dfrac{10}{\sqrt{x}}$

84. $\dfrac{12}{\sqrt{y}}$

85. $\dfrac{\sqrt{9y}}{\sqrt{2x}}$

86. $\dfrac{\sqrt{4t}}{\sqrt{3z}}$

87. $\dfrac{3}{\sqrt{3} - 1}$

88. $\dfrac{3}{\sqrt{5} - 2}$

89. $\dfrac{3}{\sqrt{7} + 2}$

90. $\dfrac{5}{\sqrt{8} + 3}$

91. $\dfrac{12}{3 - \sqrt{3}}$

92. $\dfrac{10}{5 - \sqrt{5}}$

93. $\dfrac{-\sqrt{3}}{\sqrt{3} + 1}$

94. $\dfrac{-\sqrt{2}}{\sqrt{2} - 1}$

95. $\dfrac{5}{\sqrt{3} + \sqrt{2}}$

96. $\dfrac{3}{\sqrt{3} - \sqrt{2}}$

97. $\dfrac{\sqrt{x} + 2}{\sqrt{x} - 2}$

98. $\dfrac{\sqrt{x} - 3}{\sqrt{x} + 3}$

99. $\dfrac{5}{\sqrt[3]{5}}$

100. $\dfrac{7}{\sqrt[3]{7}}$

101. $\dfrac{4}{\sqrt[3]{4}}$

102. $\dfrac{7}{\sqrt[3]{10}}$

103. $\dfrac{\sqrt[3]{5}}{\sqrt[3]{2}}$

104. $\dfrac{\sqrt[3]{2}}{\sqrt[3]{5}}$

APPLICATIONS

105. ROTARY LAWNMOWER   See Illustration 1, which shows the blade of a rotary lawnmower. Use the formula for the area of a circle, $A = \pi r^2$, to find the area of lawn covered by one rotation of the blade. Leave π in your answer.

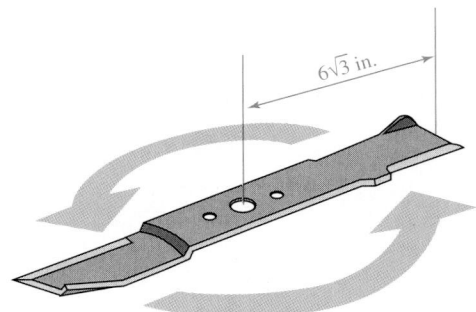

ILLUSTRATION 1

106. AWARDS PLATFORMS   Find the total number of cubic feet of concrete needed to construct the Olympic Games awards platforms shown in Illustration 2.

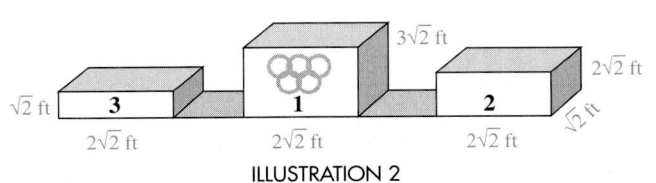

ILLUSTRATION 2

107. AIR HOCKEY GAME   Find the area of the playing surface of the air hockey game in Illustration 3.

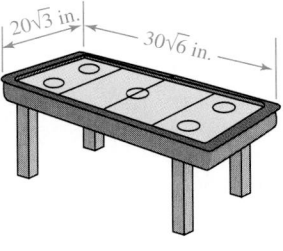

ILLUSTRATION 3

108. PROJECTOR SCREEN   To find the length l of a rectangle, we can use the formula

$$l = \frac{A}{w}$$

where A is the area of the rectangle and w is its width. Find the length of the screen shown in Illustration 4 if its area is 54 square feet.

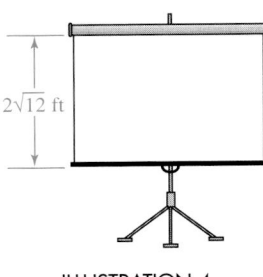

ILLUSTRATION 4

109. COSTUME DESIGN   The pattern for one panel of an 1870s English dress is printed on the 1 in. $\times$ 1 in. grid shown in Illustration 5. Find the number of square inches of fabric in the trapezoidal-shaped panel. (*Hint:* Use the Pythagorean theorem to determine the lengths of the sides.)

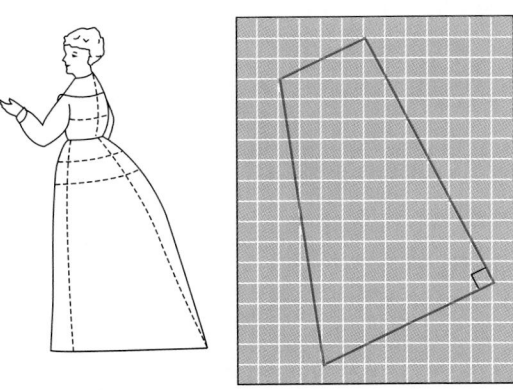

ILLUSTRATION 5

110. SET DESIGN   The director of a stage play requested bright downlighting over the portion of the set shown in Illustration 6. Find the area of the rectangle. (*Hint:* Use the Pythagorean theorem to determine the lengths of the sides.)

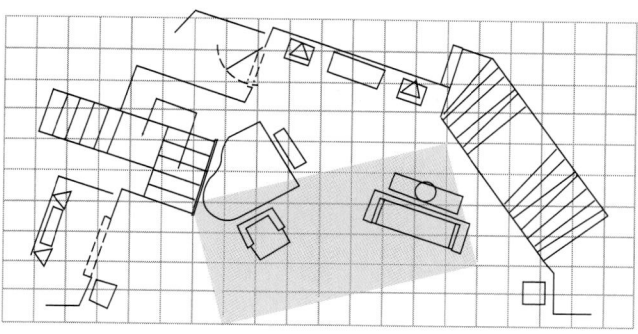

ILLUSTRATION 6

WRITING

111. When rationalizing the denominator of $\dfrac{5}{\sqrt{6}}$, why must we multiply both the numerator *and* denominator by $\sqrt{6}$?

112. A calculator is used to find decimal approximations for the expressions $\dfrac{2}{\sqrt{6}}$ and $\dfrac{\sqrt{6}}{3}$. In each case, the calculator display reads 0.816496581. Explain why the results are the same.

REVIEW

113. Is $x = -2$ a solution of $3x - 7 = 5x + 1$?

114. The graph of a line passes through the point $(2, 0)$. Is this the x- or y-intercept?

115. To evaluate the expression $2 - (-3 + 4)^2$, which operation should be performed first?

116. The graph of a straight line rises from left to right. Is the slope of the line positive or negative?

117. Find $(x - 4)(x + 4)$.

118. How far will a car traveling 55 mph go in 3.5 hours?

14.6 Solving Radical Equations; the Distance Formula

In this section, you will learn about

- The squaring property of equality • Checking solutions
- Solving equations containing one square root
- Solving equations containing two square roots
- Solving equations containing cube roots • The distance formula

INTRODUCTION. Many situations can be modeled mathematically by equations that contain radicals. In this section, we will develop techniques to solve such equations. Then we will consider a special formula called the *distance formula.*

The squaring property of equality

The equation $\sqrt{x} = 6$ is called a **radical equation,** because it contains a radical expression with a variable radicand. To solve this equation, we isolate x by undoing the operation performed on it. Recall that $\sqrt{x}$ represents the number that, when squared, gives x. Therefore, if we *square* $\sqrt{x}$, we will obtain x.

$$\left(\sqrt{x}\right)^2 = x$$

Using this observation, we can eliminate the radical on the left-hand side of $\sqrt{x} = 6$ by squaring that side. Intuition tells us that we should also square the right-hand side. This is a valid step, because if two numbers are equal, their squares are equal.

Squaring property of equality	If a and b represent real numbers, with $a = b$, then $a^2 = b^2$.

We can now solve $\sqrt{x} = 6$ by applying the squaring property of equality.

$$\sqrt{x} = 6$$
$$\left(\sqrt{x}\right)^2 = (6)^2 \quad \text{Square both sides of the equation to eliminate the radical.}$$
$$x = 36 \quad \text{Simplify each side: } \left(\sqrt{x}\right)^2 = x \text{ and } (6)^2 = 36.$$

Checking this result, we have

$$\sqrt{x} = 6$$
$$\sqrt{36} \stackrel{?}{=} 6 \quad \text{Substitute 36 for } x.$$
$$6 = 6 \quad \text{Simplify the left-hand side: } \sqrt{36} = 6.$$

We obtain a true statement, so $x = 36$ is a solution.

Checking solutions

If we square both sides of an equation, the resulting equation may or may not have the same solutions as the original one. For example, if we square both sides of the equation

(1) $x = 2$

with the solution 2, we obtain $(x)^2 = 2^2$, which simplifies to

(2) $x^2 = 4$

with solutions 2 and -2, since $2^2 = 4$ and $(-2)^2 = 4$.

Equations 1 and 2 are not equivalent, because they have a different set of solutions. The solution -2 of Equation 2 does not satisfy Equation 1. Because squaring both sides of an equation can produce an equation with solutions that don't satisfy the original one, we must always check each potential solution in the original equation.

Solving equations containing one square root

To solve an equation containing square root radicals, we follow these steps.

Solving radical equations	1. Whenever possible, isolate a single radical expression on one side of the equation. 2. Square both sides of the equation and solve the resulting equation. 3. Check the solution in the original equation. This step is required.

EXAMPLE 1 *Solving radical equations.* Solve $\sqrt{x + 2} = 3$.

Solution

To solve the equation $\sqrt{x + 2} = 3$, we note that the radical is already isolated on one side. We proceed to step 2 and square both sides to eliminate the radical. Since this might produce an equation with more solutions than the original one, we must check each solution.

$$\sqrt{x + 2} = 3$$
$$\left(\sqrt{x + 2}\right)^2 = (3)^2 \quad \text{Square both sides.}$$
$$x + 2 = 9 \quad \text{Simplify each side: } \left(\sqrt{x + 2}\right)^2 = x + 2 \text{ and } 3^2 = 9.$$
$$x = 7 \quad \text{Subtract 2 from both sides.}$$

Self Check

Solve $\sqrt{x - 4} = 9$.

We check by substituting 7 for x in the original equation.

$$\sqrt{x + 2} = 3$$
$$\sqrt{7 + 2} \stackrel{?}{=} 3 \quad \text{Substitute 7 for } x.$$
$$\sqrt{9} \stackrel{?}{=} 3 \quad \text{Do the addition within the radical symbol.}$$
$$3 = 3$$

The answer $x = 7$ checks. It is a solution.

EXAMPLE 2   *A radical equation with no solution.*   Solve $\sqrt{x + 1} + 5 = 3$.

Self Check
Solve $\sqrt{x - 2} + 5 = 2$.

Answer: 85

Solution
We isolate the radical on one side and proceed as follows:

$$\sqrt{x + 1} + 5 = 3$$
$$\sqrt{x + 1} = -2 \quad \text{Subtract 5 from both sides.}$$
$$\left(\sqrt{x + 1}\right)^2 = (-2)^2 \quad \text{Square both sides to eliminate the radical.}$$
$$x + 1 = 4 \quad \text{Simplify: } \left(\sqrt{x + 1}\right)^2 = x + 1 \text{ and } (-2)^2 = 4.$$
$$x = 3 \quad \text{Subtract 1 from both sides.}$$

We check by substituting 3 for x in the original equation.

$$\sqrt{x + 1} + 5 = 3$$
$$\sqrt{3 + 1} + 5 \stackrel{?}{=} 3 \quad \text{Substitute 3 for } x.$$
$$\sqrt{4} + 5 \stackrel{?}{=} 3 \quad \text{Do the addition within the radical symbol.}$$
$$2 + 5 \stackrel{?}{=} 3$$
$$7 \neq 3$$

Since $7 \neq 3$, 3 is not a solution. In fact, the equation has no solution. This result was obvious in Step 2 of the solution. There is no real number x that could make the non-negative number $\sqrt{x + 1}$ equal to -2.

Answer: no solution

Example 2 shows that squaring both sides of an equation can lead to false solutions, called **extraneous solutions.** These potential solutions do not satisfy the original equation and must be discarded.

EXAMPLE 3   *Height of a bridge.*   The distance d (in feet) that an object will fall in t seconds is given by the formula

$$t = \sqrt{\frac{d}{16}}$$

To find the height of the bridge shown in Figure 14-18, a man drops a stone into the water. If it takes the stone 3 seconds to hit the water, how high is the bridge?

Self Check
If it takes 4 seconds for the stone in Example 3 to hit the water, how high is the bridge?

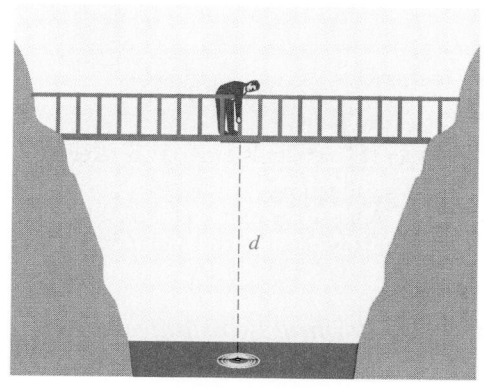

FIGURE 14-18

Solution
We substitute 3 for *t* in the formula and solve for *d*.

$$t = \sqrt{\dfrac{d}{16}}$$

$$3 = \sqrt{\dfrac{d}{16}} \qquad \text{Substitute 3 for } t.$$

$$(3)^2 = \left(\sqrt{\dfrac{d}{16}}\right)^2 \qquad \text{Square both sides to eliminate the radical.}$$

$$9 = \dfrac{d}{16} \qquad \text{Simplify: } 3^2 = 9 \text{ and } \left(\sqrt{\dfrac{d}{16}}\right)^2 = \dfrac{d}{16}.$$

$$144 = d \qquad \text{Multiply both sides by 16.}$$

The bridge is 144 feet above the water. Check this result in the original equation.

Answer: 256 feet

EXAMPLE 4   *Solving radical equations.*   Solve
$a + 2 = \sqrt{a^2 + 3a + 3}.$

Solution
The radical is isolated on the right-hand side, so we proceed by squaring both sides to eliminate it.

$$a + 2 = \sqrt{a^2 + 3a + 3}$$

$$(a + 2)^2 = \left(\sqrt{a^2 + 3a + 3}\right)^2 \qquad \text{Square both sides.}$$

$$a^2 + 4a + 4 = a^2 + 3a + 3 \qquad \begin{array}{l}\text{Use a special product formula:}\\ (a + 2)^2 = a^2 + 4a + 4. \text{ Simplify:}\\ \left(\sqrt{a^2 + 3a + 3}\right)^2 = a^2 + 3a + 3.\end{array}$$

$$a^2 + 4a + 4 - a^2 = a^2 + 3a + 3 - a^2 \qquad \begin{array}{l}\text{To eliminate } a^2, \text{ subtract } a^2 \text{ from both}\\ \text{sides.}\end{array}$$

$$4a + 4 = 3a + 3 \qquad \text{Combine like terms: } a^2 - a^2 = 0.$$

$$a + 4 = 3 \qquad \text{Subtract } 3a \text{ from both sides.}$$

$$a = -1 \qquad \text{Subtract 4 from both sides.}$$

We check by substituting -1 for *x* in the original equation.

$$a + 2 = \sqrt{a^2 + 3a + 3}$$

$$-1 + 2 \stackrel{?}{=} \sqrt{(-1)^2 + 3(-1) + 3} \qquad \text{Substitute } -1 \text{ for } a.$$

$$1 \stackrel{?}{=} \sqrt{1 - 3 + 3} \qquad \begin{array}{l}\text{Within the radical symbol, first find the power,}\\ \text{then do the multiplication.}\end{array}$$

$$1 \stackrel{?}{=} \sqrt{1} \qquad \text{Simplify within the radical symbol.}$$

$$1 = 1$$

The solution checks.

Self Check
Solve:

$$b + 4 = \sqrt{b^2 + 6b + 12}$$

Answer: -2

Solving equations containing two square roots

In the next example, the equation contains two square roots.

EXAMPLE 5   *Solving an equation containing two square roots.*
Solve $\sqrt{x + 12} = 3\sqrt{x + 4}.$

Self Check
Solve:

$$\sqrt{x - 4} = 2\sqrt{x - 16}$$

Solution

Note that each radical is isolated on one side of the equation. We begin by squaring both sides to eliminate them.

$$\sqrt{x + 12} = 3\sqrt{x + 4}$$
$$\left(\sqrt{x + 12}\right)^2 = \left(3\sqrt{x + 4}\right)^2 \quad \text{Square both sides.}$$
$$x + 12 = 9(x + 4) \qquad \left(\sqrt{x + 12}\right)^2 = x + 12.$$
$$\left(3\sqrt{x + 4}\right)^2 = 3^2\left(\sqrt{x + 4}\right)^2 = 9(x + 4).$$
$$x + 12 = 9x + 36 \qquad \text{Distribute the multiplication by 9.}$$
$$-8x = 24 \qquad \text{Subtract } 9x \text{ and } 12 \text{ from both sides.}$$
$$x = -3 \qquad \text{Divide both sides by } -8.$$

We check the solution by substituting -3 for x in the original equation.

$$\sqrt{x + 12} = 3\sqrt{x + 4}$$
$$\sqrt{-3 + 12} \stackrel{?}{=} 3\sqrt{-3 + 4} \quad \text{Substitute } -3 \text{ for } x.$$
$$\sqrt{9} \stackrel{?}{=} 3\sqrt{1} \qquad \text{Simplify within the radical symbols.}$$
$$3 = 3$$

The solution checks.

Answer: 20

Solving equations containing cube roots

In the next example, we cube both sides of an equation to eliminate a cube root.

EXAMPLE 6 *Solving an equation containing a cube root.* Solve $\sqrt[3]{2x + 10} = 2$.

Self Check

Solve $\sqrt[3]{3x - 3} = 3$.

Solution

To undo the operation performed on $2x + 10$, we cube both sides and proceed as follows:

$$\sqrt[3]{2x + 10} = 2$$
$$\left(\sqrt[3]{2x + 10}\right)^3 = (2)^3 \quad \text{Cube both sides.}$$
$$2x + 10 = 8 \qquad \text{Simplify: } \left(\sqrt[3]{2x + 10}\right)^3 = 2x + 10 \text{ and } (2)^3 = 8.$$
$$2x = -2 \qquad \text{Subtract 10 from both sides.}$$
$$x = -1 \qquad \text{Divide both sides by 2.}$$

Check the result.

Answer: 10

The distance formula

We can use the Pythagorean theorem to derive a formula for finding the distance between two points $P(x_1, y_1)$ and $Q(x_2, y_2)$ on a rectangular coordinate system. The distance d between points P and Q is the length of the hypotenuse of the triangle in Figure 14-19. The two legs have lengths $x_2 - x_1$ and $y_2 - y_1$.

By the Pythagorean theorem, we have

$$d^2 = (x_2 - x_1)^2 + (y_2 - y_1)^2$$

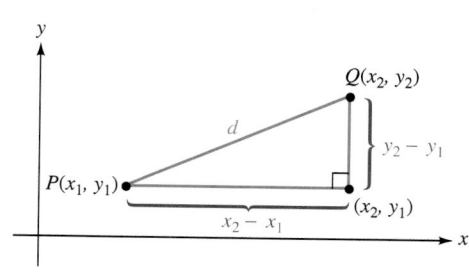

FIGURE 14-19

We can take the positive square root of both sides of this equation to get the **distance formula.**

$$d = \sqrt{(x_2 - x_1)^2 + (y_2 - y_1)^2}$$

The distance formula

> The distance d between points $P(x_1, y_1)$ and $Q(x_2, y_2)$ is given by
>
> $$d = \sqrt{(x_2 - x_1)^2 + (y_2 - y_1)^2}$$

EXAMPLE 7   *Using the distance formula.* Find the distance between points $P(1, 5)$ and $Q(4, 9)$. (See Figure 14-20.)

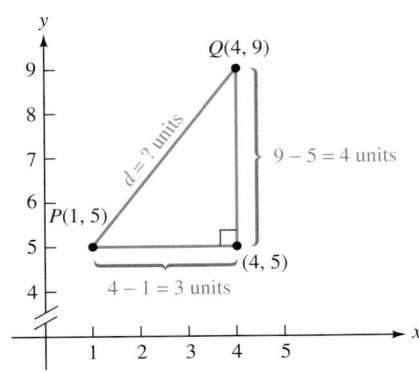

FIGURE 14-20

Solution

We use the distance formula and substitute 1 for x_1, 5 for y_1, 4 for x_2, and 9 for y_2. Then we evaluate the expression under the radical symbol.

$$d = \sqrt{(x_2 - x_1)^2 + (y_2 - y_1)^2}$$
$$= \sqrt{(4 - 1)^2 + (9 - 5)^2} \qquad \text{Substitute.}$$
$$= \sqrt{3^2 + 4^2} \qquad \text{Do the subtractions within the parentheses first.}$$
$$= \sqrt{9 + 16} \qquad \text{Evaluate the powers.}$$
$$= \sqrt{25} \qquad \text{Do the addition.}$$
$$= 5 \qquad \text{Find the square root.}$$

The distance between points P and Q is 5 units.

EXAMPLE 8   *Finding the distance between two points.* Plot the points $A(-4, 5)$ and $B(3, -1)$ on the graph in Figure 14-21 and find the distance between them.

Solution

We use the distance formula and substitute -4 for x_1, 5 for y_1, 3 for x_2, and -1 for y_2.

$$d = \sqrt{(x_2 - x_1)^2 + (y_2 - y_1)^2}$$
$$= \sqrt{[3 - (-4)]^2 + (-1 - 5)^2} \qquad \text{Substitute.}$$
$$= \sqrt{7^2 + (-6)^2} \qquad \text{Do the subtractions.}$$
$$= \sqrt{49 + 36} \qquad \text{Evaluate the powers.}$$
$$= \sqrt{85} \qquad \text{Do the addition.}$$
$$\approx 9.219544457 \qquad \text{Use a calculator.}$$

The distance between points A and B is approximately 9.22 units.

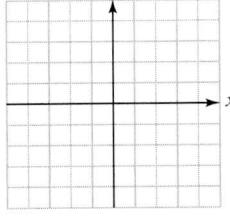

FIGURE 14-21

Self Check

Find the distance between $(-2, 1)$ and $(4, 9)$.

Answer: 10

Self Check

Plot the points $M(4, 3)$ and $N(-2, -3)$ on the graph in Figure 14-21 and find the distance between them.

Answer: about 8.49 units

EXAMPLE 9 *Freeway design.* In a large city, the streets run north and south, and the avenues run east and west. Streets and avenues are 750 feet apart. The city plans to construct a freeway from the intersection of 21st Street and 4th Avenue to the intersection of 111th Street and 60th Avenue. How long will it be?

Solution We can represent the roads of the city using the coordinate system shown in Figure 14-22. Each unit on each axis represents 750 feet. We represent one end of the freeway at 21st Street and 4th Avenue by the point $(x_1, y_1) = (21, 4)$. The other end is $(x_2, y_2) = (111, 60)$.

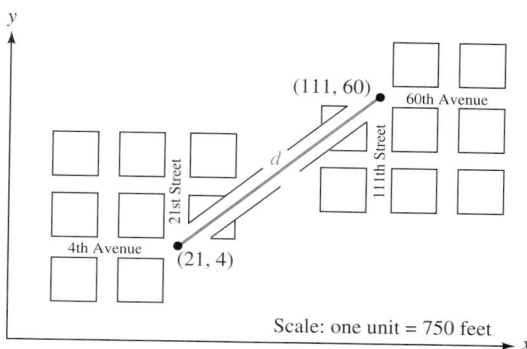

FIGURE 14-22

We can now use the distance formula to find the length of the freeway.

$$d = \sqrt{(x_2 - x_1)^2 + (y_2 - y_1)^2}$$

$$= \sqrt{(111 - 21)^2 + (60 - 4)^2} \quad \text{Substitute for } x_2, x_1, y_2, \text{ and } y_1.$$

$$= \sqrt{90^2 + 56^2} \quad \text{Do the subtractions.}$$

$$= \sqrt{8{,}100 + 3{,}136} \quad \text{Evaluate the powers.}$$

$$= \sqrt{11{,}236} \quad \text{Do the addition.}$$

$$= 106 \quad \text{Use a calculator to find the square root.}$$

Because each unit represents 750 feet, the length of the freeway will be $106 \cdot 750 = 79{,}500$ feet. Since there are 5,280 feet in 1 mile, we can divide 79,500 by 5,280 to convert 79,500 feet to 15.056818 miles. The freeway will be about 15 miles long. ∎

STUDY SET Section 14.6

VOCABULARY *Fill in the blanks.*

1. A _____ equation contains one or more radical expressions with a variable radicand.

2. "To _____ the radical expression" in $\sqrt{x} + 1 = 10$ means to get $\sqrt{x}$ all by itself on one side of the equation.

3. A false solution that occurs when you square both sides of an equation is called an _____ solution.

4. The squaring property of equality states that if two numbers are equal, their _____ are equal.

CONCEPTS *In Exercises 5–6, fill in the blanks.*

5. The squaring property of equality states that
If $a = b$, then $a^2 = $____.

6. The distance formula states that
$d = $

7. To isolate x, what step should be used to undo the operation performed on it? (Assume that x is a positive number.)
a. $x^2 = 4$
b. $\sqrt{x} = 4$

8. Simplify each expression.

a. $\left(\sqrt{x}\right)^2$ **b.** $\left(\sqrt{x-1}\right)^2$

c. $\left(2\sqrt{x}\right)^2$ **d.** $\left(2\sqrt{x-1}\right)^2$

e. $\left(\sqrt{2x}\right)^2$ **f.** $\left(\sqrt[3]{x}\right)^3$

In Exercises 9–12, an equation is incorrectly solved. Tell what is wrong with each solution.

9.
$$\sqrt{x-2} = 3$$
$$\left(\sqrt{x-2}\right)^2 = 3$$
$$x - 2 = 3$$
$$x = 5$$

10.
$$2 = \sqrt{x-9}$$
$$(2)^2 = \left(\sqrt{x-9}\right)^2$$
$$4 = x - 9$$
$$-5 = x$$
$$x = -5$$

11.
$$\sqrt{a+2} - 5 = 4$$
$$\left(\sqrt{a+2} - 5\right)^2 = 4^2$$
$$a + 2 - 25 = 16$$
$$a - 23 = 16$$
$$a = 39$$

12.
$$\sqrt[3]{x+1} = -2$$
$$\left(\sqrt[3]{x+1}\right)^2 = (-2)^2$$
$$x + 1 = 4$$
$$x = 3$$

13. a. On the graph in Illustration 1, plot the points $A(-4, 6)$, $B(4, 0)$, $C(1, -4)$, and $D(-7, 2)$.

b. Draw figure *ABCD*. What type of geometric figure is it?

c. Find the length of each side of the figure.

d. Find the perimeter of the figure.

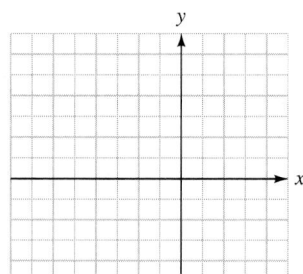

ILLUSTRATION 1

14. a. What type of geometric figure is figure *ABCD* shown in Illustration 2?

b. Give the coordinates of points *A*, *B*, *C*, and *D*.

c. Find the length of each side of the figure.

d. Find the area of the figure.

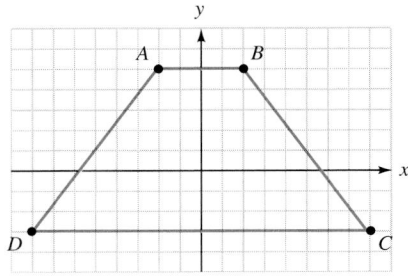

ILLUSTRATION 2

NOTATION   *Complete each solution to solve each equation.*

15.
$$\sqrt{x-3} = 5$$
$$\left(\right)^2 = ^2$$
$$x - 3 = $$
$$x = 28$$

16.
$$\sqrt{2x-18} = \sqrt{x-1}$$
$$\left(\right)^2 = \left(\sqrt{x-1}\right)^2$$
$$ = x - 1$$
$$ - 18 = -1$$
$$x = 17$$

PRACTICE   *Solve each equation. Check all solutions. If an equation has no solutions, write "none."*

17. $\sqrt{x} = 3$ **18.** $\sqrt{x} = 5$

19. $\sqrt{2a} = 4$ **20.** $\sqrt{3a} = 9$

21. $\sqrt{r} + 4 = 0$ **22.** $\sqrt{r} + 1 = 0$

23. $-\sqrt{x} = -5$ **24.** $-\sqrt{x} = -12$

25. $10 - \sqrt{s} = 7$ **26.** $-4 = 6 - \sqrt{s}$

27. $\sqrt{x+3} = 2$ **28.** $\sqrt{x-2} = 3$

29. $\sqrt{3-T} = -2$ **30.** $\sqrt{5-T} = 10$

31. $\sqrt{6+2x} = 4$ **32.** $\sqrt{7+2x} = -4$

33. $\sqrt{5x-5} - 5 = 0$ **34.** $\sqrt{6x+19} - 7 = 0$

35. $\sqrt{x+3} + 5 = 12$ **36.** $\sqrt{x-5} - 3 = 4$

37. $x - 3 = \sqrt{x^2 - 15}$ **38.** $v - 2 = \sqrt{v^2 - 16}$

39. $\sqrt{3t-9} = \sqrt{t+1}$ **40.** $\sqrt{a-3} = \sqrt{2a-8}$

41. $\sqrt{10-3x} = \sqrt{2x+20}$

42. $\sqrt{1-2x} = \sqrt{x+10}$

43. $\sqrt{3c-8} - \sqrt{c} = 0$

44. $\sqrt{2x} - \sqrt{x+8} = 0$

45. $x - 1 = \sqrt{x^2 - 4x + 9}$

46. $3d = \sqrt{9d^2 - 2d + 8}$

47. $\sqrt{4m^2 + 6m + 6} = -2m$

48. $\sqrt{9t^2 + 4t + 20} = -3t$

49. $\sqrt{3x + 3} = 3\sqrt{x - 1}$

50. $2\sqrt{4x + 5} = 5\sqrt{x + 4}$

51. $2\sqrt{3x + 4} = \sqrt{5x + 9}$

52. $\sqrt{3x + 6} = 2\sqrt{2x - 11}$

53. $\sqrt[3]{x} = 7$

54. $\sqrt[3]{x} = -9$

55. $\sqrt[3]{x - 1} = 4$

56. $\sqrt[3]{2x + 5} = 3$

57. $\sqrt[3]{\frac{1}{2}x - 3} = 2$

58. $\sqrt[3]{x + 4} = 1$

59. $\sqrt[3]{7n - 1} + 1 = 4$

60. $\sqrt[3]{12m + 4} + 2 = 6$

Find the distance between points P and Q. If an answer is not exact, round to the nearest hundredth.

61. $P(3, -4)$ and $Q(0, 0)$

62. $P(0, 0)$ and $Q(-6, 8)$

63. $P(2, 4)$ and $Q(5, 9)$

64. $P(5, 9)$ and $Q(9, 13)$

65. $P(-2, -8)$ and $Q(3, 4)$

66. $P(-5, -2)$ and $Q(7, 3)$

67. $P(6, 8)$ and $Q(12, 16)$

68. $P(10, 4)$ and $Q(2, -2)$

▦ APPLICATIONS

69. NIAGARA FALLS   The distance s (in feet) that an object will fall in t seconds is given by the formula

$$t = \frac{\sqrt{s}}{4}$$

The time it took a stuntman to go over the Niagara Falls in a barrel was 3.25 seconds. Substitute 3.25 for t and solve the equation for s to find the height of the waterfall.

70. WASHINGTON MONUMENT   Gabby Street, a professional baseball player of the 1920s, was known for once catching a ball dropped from the top of the Washington Monument in Washington, D.C. If the ball fell for slightly less than 6 seconds before it was caught, find the approximate height of the monument. (*Hint:* See Exercise 69.)

71. FOUCAULT PENDULUM   The time t (in seconds) required for a pendulum of length L feet to swing through one back-and-forth cycle, called its period, is given by the formula

$$t = 1.11\sqrt{L}$$

The Foucault pendulum in Chicago's Museum of Science and Industry, shown in Illustration 3, is used to demonstrate the rotation of the earth. It completes one cycle in 8.91 seconds. To the nearest tenth of a foot, how long is the pendulum?

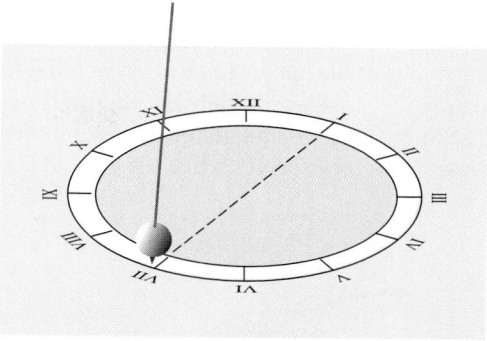

ILLUSTRATION 3

72. POWER USAGE   The current I (in amperes), the resistance R (in ohms), and the power P (in watts) are related by the formula

$$I = \sqrt{\frac{P}{R}}$$

Find the power (to the nearest watt) used by a space heater that draws 7 amps when the resistance is 10.2 ohms.

73. ROAD SAFETY   The formula $s = k\sqrt{d}$ relates the speed s (in mph) of a car and the distance d of the skid when a driver hits the brakes. On wet pavement, $k = 3.24$. How far will a car skid if it is going 55 mph?

74. ROAD SAFETY   How far will the car in Exercise 73 skid if it is traveling on dry pavement? On dry pavement, $k = 5.34$.

75. SATELLITE ORBIT   The orbital speed s of an Earth satellite is related to its distance r from the Earth's center by the formula

$$\sqrt{r} = \frac{2.029 \times 10^7}{s}$$

If the satellite's orbital speed is 7×10^3 meters per second, find its altitude a (in meters) above the Earth's surface, as shown in Illustration 4.

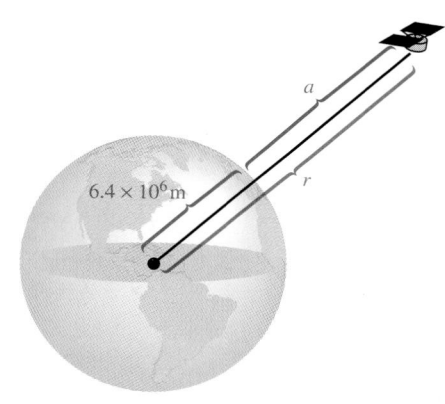

ILLUSTRATION 4

76. HIGHWAY DESIGN   A highway curve banked at 8° will accommodate traffic traveling at speed s (in mph) if the radius of the curve is r (feet), according to the equation $s = 1.45\sqrt{r}$. If highway engineers expect traffic to travel at 65 mph, to the nearest foot, what radius should they specify? (See Illustration 5.)

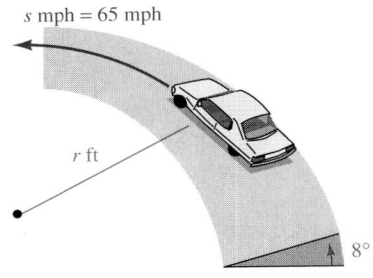

ILLUSTRATION 5

77. GEOMETRY   The radius of a cone with volume V and height h is given by the formula

$$r = \sqrt{\frac{3V}{\pi h}}$$

Solve the equation for V.

78. WINDMILL   The power produced by a certain windmill is related to the speed of the wind by the formula

$$s = \sqrt[3]{\frac{P}{0.02}}$$

where P is the power (in watts) and s is the speed of the wind (in mph). How much power will the windmill produce if the wind is blowing at 30 mph?

79. NAVIGATION   An oil tanker is to travel from Tunisia to Italy, as shown in Illustration 6. The captain wants to travel a course that is *always* the same distance from a point on the coast of Sardinia as it is from a point on the coast of Sicily (both denoted in red). How far will the tanker be from these points when it reaches
  a. position 1?
  b. position 2?

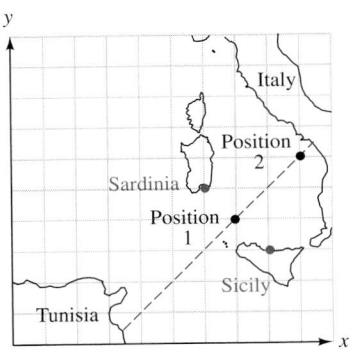

ILLUSTRATION 6

80. DECK DESIGN   The plans for a patio deck shown in Illustration 7 call for three redwood support braces directly under the hot tub. Find the length of each support. Round to the nearest tenth of a foot.

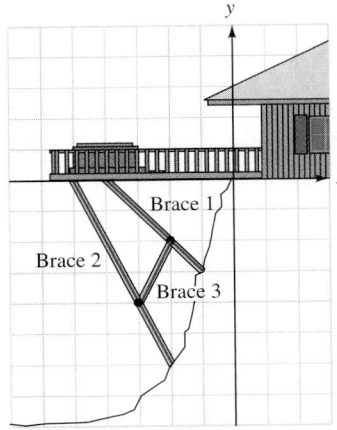

ILLUSTRATION 7

WRITING

81. Explain why a check is necessary when solving radical equations.

82. How would you know, without solving it, that the equation $\sqrt{x} + 2 = -4$ has no solutions?

REVIEW   *Do the operations.*

83. $(3x^2 + 2x) + (5x^2 - 8x)$
84. $(7a^2 + 2a - 5) - (3a^2 - 2a + 1)$
85. $(x + 3)(x + 3)$
86. $x - 1\overline{)x^2 - 6x + 5}$
87. $(3y - 7)^2$
88. $(3y + 7)^2$

14.7 *Rational Exponents*

In this section, you will learn about

- Fractional exponents with numerators of 1
- Fractional exponents with numerators other than 1 • Rules for exponents

INTRODUCTION. We have seen that a positive integer exponent indicates the number of times that a base is to be used as a factor in a product. For example, x^4 means that x is to be used as a factor four times.

$$\overbrace{x^4 = x \cdot x \cdot x \cdot x}^{\text{4 factors of } x}$$

Also, recall the following rules for exponents.

Rules for exponents

> If m and n represent natural numbers and there are no divisions by zero, then
>
> $$x^m x^n = x^{m+n} \qquad (x^m)^n = x^{m \cdot n} \qquad (xy)^n = x^n y^n \qquad \left(\frac{x}{y}\right)^n = \frac{x^n}{y^n}$$
>
> $$x^0 = 1 \qquad\qquad x^{-n} = \frac{1}{x^n} \qquad\qquad \frac{x^m}{x^n} = x^{m-n}$$

In this section, we will extend the definition and rules for exponents to cover fractional exponents.

Fractional exponents with numerators of 1

It is possible to raise numbers to fractional powers. To give meaning to rational (fractional) exponents, we consider $\sqrt{7}$. Because $\sqrt{7}$ is the positive number whose square is 7, we have

$$\left(\sqrt{7}\right)^2 = 7$$

We now consider the symbol $7^{1/2}$. If fractional exponents are to follow the same rules as integer exponents, the square of $7^{1/2}$ must be 7, because

$$\left(7^{1/2}\right)^2 = 7^{(1/2)2} \qquad \text{Keep the base and multiply the exponents.}$$
$$= 7^1 \qquad\qquad \tfrac{1}{2} \cdot 2 = 1.$$
$$= 7$$

Since $(7^{1/2})^2$ and $\left(\sqrt{7}\right)^2$ are both equal to 7, we define $7^{1/2}$ to be $\sqrt{7}$. Similarly, we make these definitions.

$$7^{1/3} = \sqrt[3]{7}$$
$$7^{1/7} = \sqrt[7]{7}$$

and so on.

Rational exponents

> If n represents a positive integer greater than 1 and $\sqrt[n]{x}$ represents a real number, then
>
> $$x^{1/n} = \sqrt[n]{x}$$

EXAMPLE 1   *Rational exponents with numerators of 1.*
Simplify   **a.** $64^{1/2}$,   **b.** $64^{1/3}$,   and   **c.** $(-64)^{1/3}$

Solution

a. $64^{1/2} = \sqrt{64} = 8$ The denominator of the fractional exponent is 2. Therefore, we find the square root of the base.

b. $64^{1/3} = \sqrt[3]{64} = 4$ The denominator of the fractional exponent is 3. Therefore, we find the cube root of the base.

c. $(-64)^{1/3} = \sqrt[3]{-64} = -4$ The denominator of the fractional exponent is 3. Therefore, we find the cube root of the base.

Self Check

Simplify

a. $81^{1/2}$

b. $125^{1/3}$

c. $(-27)^{1/3}$

Answers: **a.** 9,   **b.** 5,
c. -3

Fractional exponents with numerators other than 1

We can extend the definition of $x^{1/n}$ to cover fractional exponents for which the numerator is not 1. For example, because $4^{3/2}$ can be written as $(4^{1/2})^3$, we have

$$4^{3/2} = (4^{1/2})^3 = \left(\sqrt{4}\right)^3 = 2^3 = 8$$

Because $4^{3/2}$ can also be written as $(4^3)^{1/2}$, we have

$$4^{3/2} = (4^3)^{1/2} = 64^{1/2} = \sqrt{64} = 8$$

In general, $x^{m/n}$ can be written as $(x^{1/n})^m$ or as $(x^m)^{1/n}$. Since $(x^{1/n})^m = \left(\sqrt[n]{x}\right)^m$ and $(x^m)^{1/n} = \sqrt[n]{x^m}$, we make the following definition.

Changing from rational exponents to radicals

> If m and n represent positive integers ($n \neq 1$) and $\sqrt[n]{x}$ represents a real number, then
> $$x^{m/n} = \sqrt[n]{x^m} = \left(\sqrt[n]{x}\right)^m$$

EXAMPLE 2 *Rational exponents with numerators other than 1.*

Simplify **a.** $8^{2/3}$ and **b.** $(-27)^{4/3}$.

Solution

These expressions can be simplified in two ways. Using the first method, we take the root of the base and then we find the power. The second method is to find the power first and then take the root.

a. $8^{2/3} = \left(\sqrt[3]{8}\right)^2$ or $8^{2/3} = \sqrt[3]{8^2}$
  $= 2^2$ $= \sqrt[3]{64}$
  $= 4$ $= 4$

b. $(-27)^{4/3} = \left(\sqrt[3]{-27}\right)^4$ or $(-27)^{4/3} = \sqrt[3]{(-27)^4}$
  $= (-3)^4$ $= \sqrt[3]{531{,}441}$
  $= 81$ $= 81$

The work in Example 2 suggests that in order to avoid large numbers, it is usually easier to take the root of the base first and then find the power.

EXAMPLE 3 *Rational exponents with numerators other than 1.*

Simplify **a.** $125^{4/3}$, **b.** $9^{5/2}$, **c.** $-25^{3/2}$, and **d.** $(-27)^{2/3}$.

Solution

a. $125^{4/3} = \left(\sqrt[3]{125}\right)^4$ **b.** $9^{5/2} = \left(\sqrt{9}\right)^5$
  $= (5)^4$ $= (3)^5$
  $= 625$ $= 243$

c. $-25^{3/2} = -\left(\sqrt{25}\right)^3$ **d.** $(-27)^{2/3} = \left(\sqrt[3]{-27}\right)^2$
  $= -(5)^3$ $= (-3)^2$
  $= -125$ $= 9$

Accent on Technology: *Fractional exponents*

To use a scientific calculator to evaluate an exponential expression containing a fractional exponent, we can use the $\boxed{y^x}$ key. For example, to evaluate $6^{-2/3}$, we enter these numbers and press these keys.

Keystrokes 6 $\boxed{y^x}$ $\boxed{(}$ 2 $\boxed{+/-}$ $\boxed{\div}$ 3 $\boxed{)}$ $\boxed{=}$ $\boxed{0.302853432}$

So $6^{-2/3} \approx 0.302853432$.

To use a graphing calculator to evaluate $6^{-2/3}$, we press the following keys.

Keystrokes 6 $\boxed{\wedge}$ $\boxed{(}$ $\boxed{(-)}$ 2 $\boxed{\div}$ 3 $\boxed{)}$ $\boxed{\text{ENTER}}$ $\boxed{\begin{array}{l} 6\wedge(^-2/3) \\ \quad .3028534321 \end{array}}$

Rules for exponents

Because of the way in which $x^{1/n}$ and $x^{m/n}$ are defined, the familiar rules for exponents are valid for rational exponents. The following example illustrates the use of each rule.

EXAMPLE 4   *Using the rules for exponents.*   Simplify:

a. $4^{2/5}4^{1/5} = 4^{2/5+1/5} = 4^{3/5}$ $x^m x^n = x^{m+n}$.

b. $(5^{2/3})^{1/2} = 5^{(2/3)(1/2)} = 5^{1/3}$ $(x^m)^n = x^{m \cdot n}$.

c. $(3x)^{2/3} = 3^{2/3}x^{2/3}$ $(xy)^m = x^m y^m$.

d. $\dfrac{4^{3/5}}{4^{2/5}} = 4^{3/5-2/5} = 4^{1/5}$ $\dfrac{x^m}{x^n} = x^{m-n}$.

e. $\left(\dfrac{3}{2}\right)^{2/5} = \dfrac{3^{2/5}}{2^{2/5}}$ $\left(\dfrac{x}{y}\right)^n = \dfrac{x^n}{y^n}$.

f. $4^{-2/3} = \dfrac{1}{4^{2/3}}$ $x^{-n} = \dfrac{1}{x^n}$.

g. $(5^{1/3})^0 = 1$ $x^0 = 1$.

Self Check

Simplify:

a. $5^{1/3}5^{1/3}$

b. $(5^{1/3})^4$

c. $(3x)^{1/5}$

d. $\dfrac{5^{3/7}}{5^{2/7}}$

e. $\left(\dfrac{2}{3}\right)^{2/3}$

f. $5^{-2/7}$

g. $(12^{1/2})^0$

Answers: **a.** $5^{2/3}$,   **b.** $5^{4/3}$,

c. $3^{1/5}x^{1/5}$,   **d.** $5^{1/7}$,   **e.** $\dfrac{2^{2/3}}{3^{2/3}}$,

f. $\dfrac{1}{5^{2/7}}$,   **g.** 1

We can use the rules for exponents to simplify expressions containing rational exponents.

EXAMPLE 5   *Rational exponents.*   Simplify   **a.** $64^{-2/3}$,   **b.** $(x^2)^{1/2}$,
c. $(x^6y^4)^{1/2}$,   and   **d.** $(27x^{12})^{-1/3}$ $(x > 0$ and $y > 0)$.

Solution

a. $64^{-2/3} = \dfrac{1}{64^{2/3}}$

$= \dfrac{1}{(64^{1/3})^2}$

$= \dfrac{1}{4^2}$

$= \dfrac{1}{16}$

b. $(x^2)^{1/2} = x^{2(1/2)}$

$= x^1$

$= x$

Self Check

Simplify:

a. $25^{-3/2}$

b. $(x^3)^{1/3}$

c. $(x^6y^9)^{-2/3}$

c. $(x^6 y^4)^{1/2} = x^{6(1/2)} y^{4(1/2)}$

$\qquad = x^3 y^2$

d. $(27x^{12})^{-1/3} = \dfrac{1}{(27x^{12})^{1/3}}$

$\qquad\qquad\qquad = \dfrac{1}{27^{1/3} x^{12(1/3)}}$

$\qquad\qquad\qquad = \dfrac{1}{3x^4}$

Answers: **a.** $\dfrac{1}{125}$, **b.** x,

c. $\dfrac{1}{x^4 y^6}$

EXAMPLE 6 *Simplifying expressions containing rational*
exponents. Simplify **a.** $x^{1/3} x^{1/2}$, **b.** $\dfrac{3x^{2/3}}{6x^{1/5}}$, and **c.** $\dfrac{2x^{-1/2}}{x^{3/4}}$ $(x > 0)$.

Self Check
Simplify:

a. $x^{2/3} x^{1/2}$

b. $\dfrac{x^{2/3}}{2x^{1/4}}$

Solution

a. $x^{1/3} x^{1/2} = x^{2/6} x^{3/6}$ $\qquad$ Get a common denominator for the fractional exponents.

$\qquad\qquad = x^{5/6}$ $\qquad$ Keep the base and add the exponents.

b. $\dfrac{3x^{2/3}}{6x^{1/5}} = \dfrac{3x^{10/15}}{6x^{3/15}}$ $\qquad$ Get a common denominator for the fractional exponents.

$\qquad\quad = \dfrac{1}{2} x^{10/15 - 3/15}$ $\quad$ Simplify $\frac{3}{6}$. Keep the base and subtract the exponents.

$\qquad\quad = \dfrac{1}{2} x^{7/15}$

c. $\dfrac{2x^{-1/2}}{x^{3/4}} = \dfrac{2x^{-2/4}}{x^{3/4}}$ $\qquad$ Get a common denominator for the fractional exponents.

$\qquad\quad = 2x^{-2/4 - 3/4}$ $\quad$ Keep the base and subtract the exponents.

$\qquad\quad = 2x^{-5/4}$ $\qquad$ Simplify.

$\qquad\quad = \dfrac{2}{x^{5/4}}$ $\qquad$ $x^{-5/4} = \dfrac{1}{x^{5/4}}$.

Answers: **a.** $x^{7/6}$, **b.** $\frac{1}{2} x^{5/12}$

STUDY SET Section 14.7

VOCABULARY *Fill in the blanks.*

1. A fractional exponent is also called a _____ exponent.

2. In the expression $27^{1/3}$, 27 is called the _____ and the exponent is ___.

CONCEPTS *In Exercises 3–10, complete each rule for exponents.*

3. $x^m x^n =$

4. $(x^m)^n =$

5. $\left(\dfrac{x}{y}\right)^n =$

6. $x^0 =$

7. $x^{-n} =$

8. $\dfrac{x^m}{x^n} =$

9. $x^{1/n} =$

10. $x^{m/n} =$

11. Write $\sqrt{5}$ using a fractional exponent.

12. Write $5^{1/3}$ using a radical.

13. Write $8^{4/3}$ using a radical.

14. Write $\left(\sqrt{8}\right)^3$ using a fractional exponent.

15. Complete the table of values.

x	$x^{1/2}$
0	
1	
4	
9	

16. Complete the table of values.

x	$x^{1/3}$
0	
−1	
−8	
8	

17. Graph each number on the number line.
$\{8^{1/3}, 17^{1/2}, 2^{3/2}, -5^{2/3}\}$

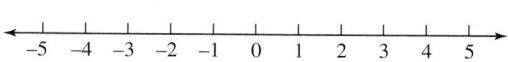

18. Graph each number on the number line.
$\{4^{-1/2}, 64^{-2/3}, (-8)^{-1/3}\}$

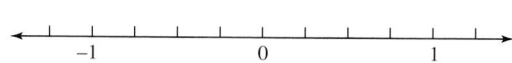

NOTATION *Complete each solution.*

19. Simplify $(-216)^{4/3}$.

$$(-216)^{4/3} = \left(\sqrt[3]{}\right)^4$$
$$= \left(\right)^4$$
$$= 1{,}296$$

20. Simplify $\dfrac{3x^{-2/3}}{x^{3/4}}$.

$$\frac{3x^{-2/3}}{x^{3/4}} = \frac{3x^{-8/12}}{}$$
$$= 3x^{-8/12 - }$$
$$= 3x^{-17/12}$$

PRACTICE *Simplify each expression.*

21. $81^{1/2}$

22. $100^{1/2}$

23. $-144^{1/2}$

24. $-400^{1/2}$

25. $\left(\dfrac{1}{4}\right)^{1/2}$

26. $\left(\dfrac{1}{25}\right)^{1/2}$

27. $\left(\dfrac{4}{49}\right)^{1/2}$

28. $\left(\dfrac{9}{64}\right)^{1/2}$

29. $27^{1/3}$

30. $8^{1/3}$

31. $-125^{1/3}$

32. $-1{,}000^{1/3}$

33. $(-8)^{1/3}$

34. $(-125)^{1/3}$

35. $\left(\dfrac{27}{64}\right)^{1/3}$

36. $\left(\dfrac{64}{125}\right)^{1/3}$

37. $81^{3/2}$

38. $16^{3/2}$

39. $25^{3/2}$

40. $4^{5/2}$

41. $125^{2/3}$

42. $8^{4/3}$

43. $1{,}000^{2/3}$

44. $27^{2/3}$

45. $(-8)^{2/3}$

46. $(-125)^{2/3}$

47. $\left(\dfrac{8}{27}\right)^{2/3}$

48. $\left(\dfrac{49}{64}\right)^{3/2}$

Simplify each expression. Write your answers without using negative exponents.

49. $6^{3/5}6^{2/5}$

50. $3^{4/7}3^{3/7}$

51. $5^{2/3}5^{4/3}$

52. $2^{7/8}2^{9/8}$

53. $(7^{2/5})^{5/2}$

54. $(8^{1/3})^3$

55. $(5^{2/7})^7$

56. $(3^{3/8})^8$

57. $\dfrac{8^{3/2}}{8^{1/2}}$

58. $\dfrac{11^{9/7}}{11^{2/7}}$

59. $\dfrac{5^{11/3}}{5^{2/3}}$

60. $\dfrac{27^{13/15}}{27^{8/15}}$

61. $4^{-1/2}$

62. $8^{-1/3}$

63. $27^{-2/3}$

64. $36^{-3/2}$

65. $16^{-3/2}$

66. $100^{-5/2}$

67. $(-27)^{-4/3}$

68. $(-8)^{-4/3}$

Simplify each expression. Assume that all variables represent positive numbers.

69. $(x^{1/2})^2$

70. $(x^9)^{1/3}$

71. $(x^{12})^{1/6}$

72. $(x^{18})^{1/9}$

73. $x^{5/6}x^{7/6}$

74. $x^{2/3}x^{7/3}$

75. $y^{4/7}y^{10/7}$

76. $y^{5/11}y^{6/11}$

77. $\dfrac{x^{3/5}}{x^{1/5}}$

78. $\dfrac{x^{4/3}}{x^{2/3}}$

79. $\dfrac{x^{1/7}x^{3/7}}{x^{2/7}}$

80. $\dfrac{x^{5/6}x^{5/6}}{x^{7/6}}$

81. $x^{2/3}x^{3/4}$

82. $a^{3/5}a^{1/2}$

83. $(b^{1/2})^{3/5}$

84. $(x^{2/5})^{4/7}$

85. $\dfrac{t^{2/3}}{t^{2/5}}$

86. $\dfrac{p^{3/4}}{p^{1/3}}$

87. $\left(\dfrac{x^{4/5}}{x^{2/15}}\right)^3$

88. $\left(\dfrac{y^{2/3}}{y^{1/5}}\right)^{15}$

APPLICATIONS *If an answer is not exact, give it to the nearest tenth.*

89. SPEAKERS The formula $A = V^{2/3}$ can be used to find the area A of one face of a cube if its volume V is known. Find the amount of floor space on the dance floor taken up by the speakers shown in Illustration 1 if each speaker is a cube with a volume of 2,744 cubic inches.

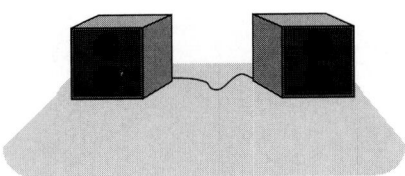

ILLUSTRATION 1

90. MEDICAL TESTS Before a series of X-rays are taken, a patient is injected with a special contrast mixture that highlights obstructions in his blood vessels. The amount of the original dose of contrast material remaining in the patient's bloodstream h hours after it is injected is given by $h^{-3/2}$. How much of the contrast material remains in the patient's bloodstream 4 hours after the injection?

91. HOLIDAY DECORATING Find the length s of each string of colored lights used to decorate an evergreen tree in the manner shown in Illustration 2 if $s = (r^2 + h^2)^{1/2}$.

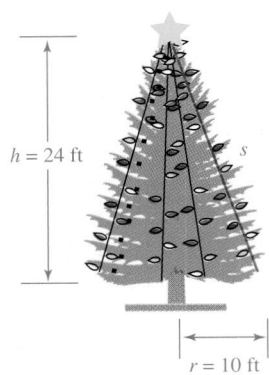

ILLUSTRATION 2

92. VISIBILITY The distance d in miles a person in an airplane can see to the horizon on a clear day is given by the formula $d = 1.22a^{1/2}$, where a is the altitude of the plane in feet. Find d in Illustration 3.

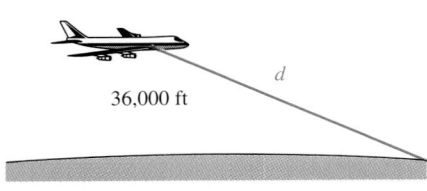

ILLUSTRATION 3

93. TOY DESIGN Knowing the volume V of a sphere, we can find its radius r using the formula

$$r = \left(\frac{3V}{4\pi}\right)^{1/3}$$

If the volume occupied by a ball is 2π cubic inches, find its radius.

94. EXERCISE EQUIPMENT Find the length l of the incline bench in Illustration 4, using the formula $l = (a^2 + b^2)^{1/2}$.

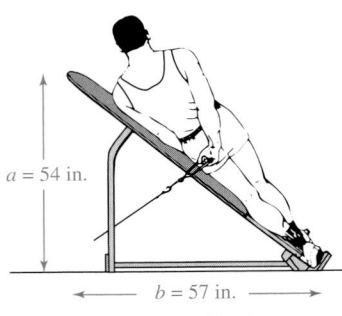

ILLUSTRATION 4

WRITING

95. What is a rational exponent? Give several examples.

96. Explain this statement: *In the expression $16^{3/2}$, the number 3/2 requires that two operations be performed on 16.*

REVIEW *Graph each equation.*

97. $x = 3$

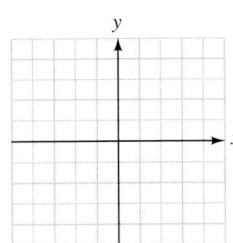

98. $y = -3$

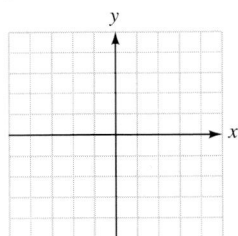

99. $-2x + y = 4$

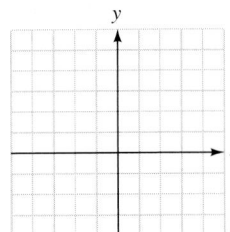

100. $4x - y = 4$

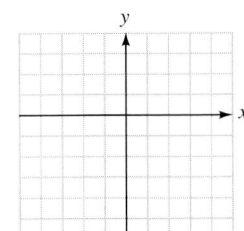

KEY CONCEPT

Inverse Operations

We have performed six operations with real numbers: addition, subtraction, multiplication, division, raising to a power, and finding a root. We have seen that there is a special relationship between *pairs* of operations. That is, subtraction does the opposite of addition, division does the opposite of multiplication, and finding a root does the opposite of raising to a power. Because of this, we call each pair **inverse operations.** Subtraction is the inverse operation of addition, division is the inverse operation of multiplication, and finding a root is the inverse operation of raising to a power.

Solving Equations

When solving equations, we use inverse operations to isolate the variable on one side of the equation.

Tell what operation is performed on the variable and what inverse operation should be used to isolate the variable; then solve the equation.

1. $x + 2 = -4$

2. $x - 5 = 10$

3. $-6x = 24$

4. $\dfrac{x}{2} = 40$

5. $\sqrt{x} = 7$

6. $x^2 = 169$ (assume $x > 0$)

7. $\sqrt[3]{x} = -2$

8. $x^3 = 64$

When solving equations, we must often undo several operations to isolate the variable. Recall that these operations are undone in the *reverse* order of operations.

Solve each equation and check the result.

9. $-2x - 4 = 6$

10. $\dfrac{3x}{5} + 3 = 9$

11. $\sqrt{x + 1} = 4$

12. $x^2 = 9$ (assume $x > 0$)

13. $\sqrt{x} - 3 = 5$

14. $\sqrt[3]{x} - 3 = 1$

Applications

We can use the concept of inverse operation to find the length of a side of the cube in Illustration 1 if we know the area of a face or the volume of the cube.

15. To find the area of a face of the cube, we square the length of a side. How could we find the length of a side, knowing the area of a face?

16. To find the volume of the cube, we cube the length of a side. How could we find the length of a side if we knew the volume of the cube?

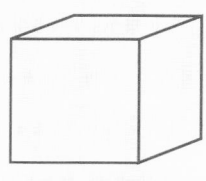

ILLUSTRATION 1

Section 14.1

THE PYTHAGOREAN THEOREM   Put 12 knots in a rope, each 1 foot apart, and connect the ends as shown in Illustration 1. Hammer three tent stakes in the ground so that the rope forms a triangle with sides of length 3, 4, and 5 spaces. Make some observations about the triangle. Use the Pythagorean theorem to prove one of your observations.

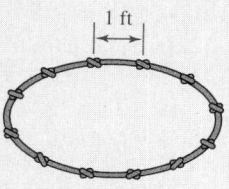

1 ft

ILLUSTRATION 1

A SPIRAL OF ROOTS   To do this project, you will need a piece of poster board, a protractor, a yardstick, and a pencil. Begin by drawing an isosceles right triangle near the right margin of the poster board. Label the length of each leg as 1 unit. (See Illustration 2.) Use the Pythagorean theorem to determine the length of the hypotenuse. Draw a second right triangle using the hypotenuse of the first triangle as one leg. Draw its second leg with a length of 1 unit. Find the length of the hypotenuse of triangle 2. Continue this process of creating right triangles, using the previous hypotenuse as one leg and drawing a new second leg of length 1 unit each time. Calculate the length of the resulting hypotenuse. What patterns, if any, do you see?

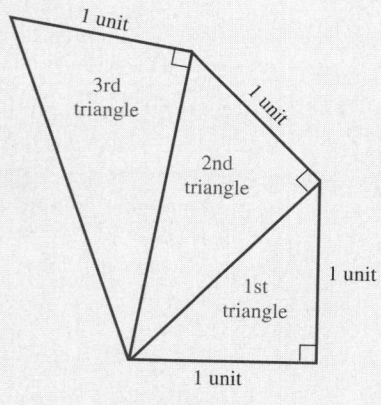

1 unit

3rd triangle

1 unit

2nd triangle

1 unit

1st triangle

1 unit

ILLUSTRATION 2

Section 14.2

nTH ROOTS   Use the $\boxed{\sqrt[x]{y}}$ key on a scientific calculator to approximate $\sqrt{2}$, $\sqrt[3]{2}$, $\sqrt[4]{2}$, $\sqrt[5]{2}$, and $\sqrt[6]{2}$. Do you see any pattern? Explain it in words.

Section 14.3

SIMPLIFYING RADICAL EXPRESSIONS   Suppose you are the algebra instructor of a student whose work is shown here. Write a note to the student explaining how she could save some steps in simplifying $\sqrt{72}$.

$$\sqrt{72} = \sqrt{4 \cdot 18}$$
$$= \sqrt{4}\sqrt{18}$$
$$= 2\sqrt{18}$$
$$= 2\sqrt{9 \cdot 2}$$
$$= 2\sqrt{9}\sqrt{2}$$
$$= 2(3)\sqrt{2}$$
$$= 6\sqrt{2}$$

Section 14.4

COMMON ERRORS   In each addition and subtraction problem below, tell what mistake was made. Compare each problem to a similar one involving variables to clarify your explanation. For example, compare Problem a to $2x + 3x$ to help you explain the correct procedure that should be used to simplify the expression.

a. $2\sqrt{5} + 3\sqrt{5} = 5\sqrt{10}$ **b.** $30 + 2\sqrt{2} = 32\sqrt{2}$
c. $7\sqrt{3} - 5\sqrt{3} = 2$ **d.** $6\sqrt{7} - 3\sqrt{2} = 3\sqrt{5}$

Section 14.5

RATIONALIZING NUMERATORS   Some problems in advanced mathematics require that the numerator of a fraction be rationalized. Extend the concepts studied in this section to develop a method to rationalize the numerators of

$$\frac{\sqrt{5}}{3}, \quad \frac{\sqrt{7}}{\sqrt{5}}, \quad \frac{\sqrt{y}}{6y}, \quad \text{and} \quad \frac{\sqrt{3} - \sqrt{2}}{12}$$

Section 14.6

SOLVING RADICAL EQUATIONS   In this chapter, we solved equations that contained two radicals. The radicals in those equations always had the same index. That is not the case for the following equation:

$$\sqrt{x} = \sqrt[3]{2x}$$

Brainstorm in your group to develop a procedure that can be used to solve this equation. What are its solutions?

Section 14.7

GRAPHING   Approximate the x- and y-coordinates of the following ordered pairs to the nearest tenth, then graph them on a rectangular coordinate system. (*Hint:* Each quadrant should contain only one point.)

$$A\left(\sqrt{2}, 3^{1/2}\right) \qquad B\left(-\sqrt{6}, 5^{3/2}\right)$$
$$C\left(-16^{2/3}, -\sqrt[3]{25}\right) \qquad D\left(9^{-1/2}, \sqrt[3]{-10}\right)$$

<table>
<tr><td>

SECTION 14.1

</td><td>

Square Roots

</td></tr>
</table>

CONCEPTS

The number b is a *square root* of a if $b^2 = a$.

The *principal square root* of a positive number a, denoted by $\sqrt{a}$, is the positive square root of a.

The expression within a *radical sign* $\sqrt{}$ is called the *radicand*.

Numbers that are not square roots of *integer squares* are *irrational numbers*. Square roots of negative numbers are called *imaginary numbers*.

The Pythagorean theorem: If the length of the hypotenuse of a right triangle is c and the lengths of the two legs are a and b, then $c^2 = a^2 + b^2$.

If a and b are positive numbers, and $a = b$, then $\sqrt{a} = \sqrt{b}$.

REVIEW EXERCISES

1. Fill in the blanks to make the statement true: The _____ of 4 is 16, because $4^2 = 16$; 4 is the _____ root of 16, because $4^2 = 16$.

2. Find each square root. Do not use a calculator.

a. $\sqrt{25}$ **b.** $\sqrt{49}$ **c.** $-\sqrt{144}$ **d.** $-\sqrt{\dfrac{16}{81}}$

e. $\sqrt{900}$ **f.** $-\sqrt{0.64}$ **g.** $\sqrt{1}$ **h.** $\sqrt{0}$

3. 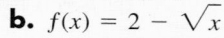 Use a calculator to approximate each expression to three decimal places.

a. $\sqrt{21}$ **b.** $-\sqrt{15}$ **c.** $2\sqrt{7}$ **d.** $\sqrt{751.9}$

4. Tell whether each number is rational, irrational, or imaginary. Which is not a real number? $\left\{ \sqrt{-2}, \sqrt{68}, \sqrt{81}, \sqrt{3} \right\}$

5. Complete the table of values for each function and then graph it.

a. $f(x) = \sqrt{x}$ **b.** $f(x) = 2 - \sqrt{x}$

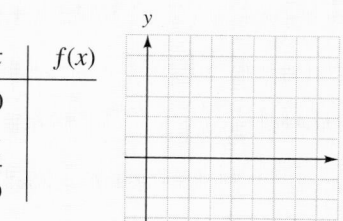

x	$f(x)$
0	
1	
4	
9	

x	$f(x)$
0	
1	
4	
9	

6. Refer to the right triangle shown in Illustration 1.

  a. Find c where $a = 21$ and $b = 28$.

  b. Find b where $a = 1$ and $c = \sqrt{2}$.

  c. Find a where $b = 5$ and $c = 7$.

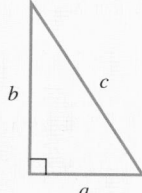

ILLUSTRATION 1

7. 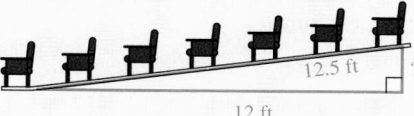 THEATER SEATING   For the theater seats shown in Illustration 2, how much higher is the seat at the top of the incline compared to the one at the bottom?

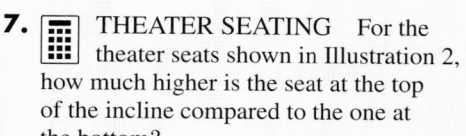

ILLUSTRATION 2

8. ROAD SIGNS   To find the maximum velocity a car can safely travel around a curve without skidding, we can use the formula $v = \sqrt{2.5r}$, where v is the velocity in miles per hour and r is the radius of the curve in feet. How should the road sign in Illustration 3 be labeled if it is to be posted in front of a curve with a radius of 360 feet?

? mph

ILLUSTRATION 3

| SECTION 14.2 | *Higher-Order Roots; Radicands That Contain Variables* |

The number b is a *cube root* of a if $b^3 = a$.

The cube root of a is denoted by $\sqrt[3]{a}$. By definition, $\sqrt[3]{a} = b$ if $b^3 = a$.

The number b is an *nth root* of a if $b^n = a$.

In $\sqrt[n]{a}$, the number n is called the *index* of the radical.

When n is even, we say that the radical $\sqrt[n]{x}$ is an *even root*. When n is odd, $\sqrt[n]{x}$ is an *odd root*.

$$\sqrt{a} = \sqrt[2]{a}$$

9. Fill in the blanks to make the statement tre: $\sqrt[3]{125} = 5$, because $\boxed{} = 125$; 5 is called the _____ root of 125.

10. Find each root. Do not use a calculator.
   a. $\sqrt[3]{-27}$ **b.** $-\sqrt[3]{125}$ **c.** $\sqrt[4]{81}$ **d.** $\sqrt[5]{32}$

   e. $\sqrt[3]{0}$ **f.** $\sqrt[3]{-1}$ **g.** $\sqrt[3]{\dfrac{1}{64}}$ **h.** $\sqrt[3]{1}$

11.  Use a calculator to find each root to three decimal places.
   a. $\sqrt[3]{16}$ **b.** $\sqrt[3]{-102.35}$ **c.** $\sqrt[4]{6}$ **d.** $\sqrt[5]{34{,}500}$

12. Find each root. Each variable represents a positive number.
   a. $\sqrt{x^2}$ **b.** $\sqrt{4b^4}$ **c.** $\sqrt{x^4y^4}$ **d.** $-\sqrt{y^{12}}$
   e. $\sqrt[3]{x^3}$ **f.** $\sqrt[3]{y^6}$ **g.** $\sqrt[3]{27x^3}$ **h.** $\sqrt[3]{-r^{12}}$

13. DICE   Find the length of an edge of one of the dice shown in Illustration 4 if each one has a volume of 1,728 cubic millimeters.

ILLUSTRATION 4

| SECTION 14.3 | *Simplifying Radical Expressions* |

The *multiplication property* of radicals: If a and b are positive or zero, then
$$\sqrt{ab} = \sqrt{a}\sqrt{b}$$

Simplified form of a radical:

1. Except for 1, the radicand has no perfect square factors.
2. No fraction appears in the radicand.
3. No radical appears in the denominator.

The *division property* of radicals:
$$\sqrt{\dfrac{a}{b}} = \dfrac{\sqrt{a}}{\sqrt{b}} \quad (b \neq 0)$$

14. Simplify each expression. All variables represent positive numbers.
   a. $\sqrt{32}$ **b.** $\sqrt{500}$
   c. $\sqrt{80x^2}$ **d.** $-2\sqrt{63}$
   e. $-\sqrt{250t^3}$ **f.** $-\sqrt{700z^5}$
   g. $\sqrt{200x^2y}$ **h.** $\dfrac{1}{5}\sqrt{75y^4}$
   i. $\sqrt[3]{8x^2y^3}$ **j.** $\sqrt[3]{250x^4y^3}$

15. Simplify each expression. All variables represent positive numbers.
   a. $\sqrt{\dfrac{16}{25}}$ **b.** $\sqrt{\dfrac{60}{49}}$
   c. $\sqrt[3]{\dfrac{1{,}000}{27}}$ **d.** $\sqrt{\dfrac{242x^4}{169x^2}}$

16. FITNESS EQUIPMENT   The length of the sit-up board in Illustration 5 can be found using the Pythagorean theorem.
   a. Find its length. Express the answer in simplified radical form.
   b. Express your result to part a as a decimal approximation rounded to the nearest tenth.

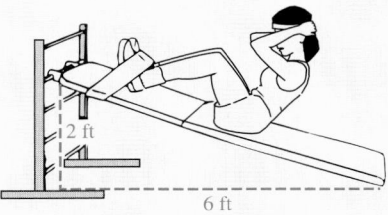

2 ft

6 ft

ILLUSTRATION 5

SECTION 14.4   *Adding and Subtracting Radical Expressions*

Radical expressions can be added or subtracted if they contain like radicals.

Radicals are called *like* radicals when they have the same index and the same radicand.

17. Do the operations. All variables represent positive numbers.
  a. $\sqrt{2} + \sqrt{8} - \sqrt{18}$ **b.** $\sqrt{3} + 4 + \sqrt{27} - 7$
  c. $5\sqrt{28} - 3\sqrt{63}$ **d.** $3y\sqrt{5xy^3} - y^2\sqrt{20xy}$
  e. $\sqrt[3]{16} + \sqrt[3]{54}$ **f.** $\sqrt[3]{2,000x^3} - \sqrt[3]{128x^3}$

18. Explain why we cannot add $3\sqrt{5}$ and $5\sqrt{3}$.

19. GARDENING   Find the difference in the lengths of the two wires used to secure the tree shown in Illustration 6.

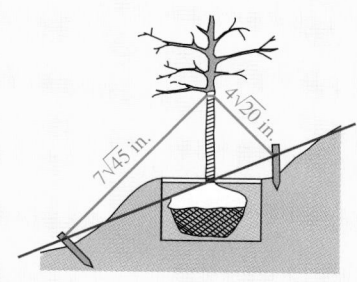

ILLUSTRATION 6

SECTION 14.5   *Multiplying and Dividing Radical Expressions*

The product of the square roots of two nonnegative numbers is equal to the square root of the product of those numbers.

To multiply radical expressions containing only one term, first multiply the coefficients, then multiply the radicals separately, and simplify the result.

Use the FOIL method to multiply two radical expressions, each having two terms.

If the denominator of a fraction is a square root, *rationalize* the denominator by multiplying the numerator and denominator by some appropriate square root.

If a two-term denominator of a fraction contains square roots, multiply the numerator and denominator by the *conjugate* of the denominator.

20. Do the operations.
  a. $\sqrt{2}\sqrt{3}$ **b.** $(-5\sqrt{5})(-2\sqrt{2})$
  c. $(3\sqrt{3x})(4\sqrt{6x})$ **d.** $(\sqrt{15} + 3x)^2$
  e. $\sqrt{2}(\sqrt{8} - \sqrt{18})$ **f.** $(\sqrt{3} + \sqrt{5})(\sqrt{3} - \sqrt{5})$
  g. $(\sqrt[3]{4})(2\sqrt[3]{4})$ **h.** $(\sqrt[3]{3} + 2)(\sqrt[3]{3} - 1)$

21. VACUUM CLEANER NOZZLE   Illustration 7 shows the amount of surface area of a rug suctioned by a vacuum nozzle attachment.
  a. Find the perimeter and area of this section of rug. Express the answers in simplified radical form.
  b. Express your results to part a as decimal approximations to the nearest tenth.

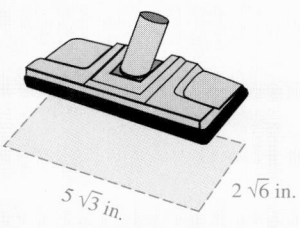

ILLUSTRATION 7

22. Rationalize each denominator.
  a. $\dfrac{1}{\sqrt{7}}$ **b.** $\sqrt{\dfrac{3}{7}}$
  c. $\dfrac{\sqrt{9}}{\sqrt{18}}$ **d.** $\dfrac{\sqrt{c} - 4}{\sqrt{c} + 4}$
  e. $\dfrac{7}{\sqrt{2} + 1}$ **f.** $\dfrac{8}{\sqrt[3]{16}}$

Solving Radical Equations; the Distance Formula

To solve an equation containing square root radicals:

1. Isolate the radicals.

2. Square both sides and solve the resulting equation.

3. Check the solution. Discard any *extraneous* solutions.

Squaring property of equality:

If $a = b$, then $a^2 = b^2$.

The distance formula:

$d = \sqrt{(x_2 - x_1)^2 + (y_2 - y_1)^2}$

23. Simplify each expression. All variables represent positive numbers.

a. $\left(\sqrt{x}\right)^2$ **b.** $\left(\sqrt[3]{x}\right)^3$ **c.** $\left(2\sqrt{t}\right)^2$ **d.** $\left(\sqrt{e-1}\right)^2$

24. Solve each equation and check all solutions.

a. $\sqrt{x} = 9$ **b.** $\sqrt{2x + 10} = 2$

c. $\sqrt{3x + 4} + 5 = 3$ **d.** $\sqrt{2(r + 4)} = 2\sqrt{r}$

e. $\sqrt{p^2 - 3} = p + 3$ **f.** $\sqrt[3]{x - 1} = 3$

25. FERRIS WHEEL The distance d in feet that an object will fall in t seconds is given by the formula

$$t = \sqrt{\frac{d}{16}}$$

If a person drops a coin from the top of a Ferris wheel and it takes 2 seconds to hit the ground, how tall is the Ferris wheel?

26. Find the distance between the points. If an answer is not exact, round to the nearest hundredth.

a. $(-7, 12), (-4, 8)$ **b.** $(-15, -3), (-10, -16)$

Rational Exponents

Real numbers can be raised to fractional powers.

Rational exponents:

$x^{1/n} = \sqrt[n]{x}$

$x^{m/n} = \sqrt[n]{x^m} = \left(\sqrt[n]{x}\right)^m$

The rules for exponents can be used to simplify expressions involving rational exponents.

27. Simplify each expression. Write answers without using negative exponents.

a. $49^{1/2}$ **b.** $(-1{,}000)^{1/3}$ **c.** $36^{3/2}$ **d.** $\left(\dfrac{8}{27}\right)^{2/3}$

e. $4^{-3/2}$ **f.** $8^{2/3}8^{4/3}$ **g.** $(3^{2/3})^3$ **h.** $(a^4b^8)^{-1/2}$

i. $x^{1/3}x^{2/5}$ **j.** $\dfrac{t^{3/4}}{t^{2/3}}$ **k.** $\dfrac{x^{2/5}x^{1/5}}{x^{-2/5}}$ **l.** $\dfrac{x^{17/7}}{x^{3/7}}$

28. Graph each number on the number line: $\left\{4^{-1/2}, 12^{1/2}, 9^{1/3}, -2^{2/3}\right\}$.

29. DENTISTRY The fractional amount of painkiller remaining in the system of a patient h hours after the original dose was injected into her gums is given by $h^{-3/2}$. How much of the original dose is in the patient's system 16 hours after the injection?

30. Explain why $(-4)^{1/2}$ is not a real number.

Chapter 14 Test

In Problems 1–4, simplify each radical.

1. $\sqrt{100}$

2. $-\sqrt{\dfrac{400}{9}}$

3. $\sqrt[3]{-27}$

4. $\sqrt{\dfrac{50}{49}}$

5. Evaluate $\sqrt{b^2 - 4ac}$ for $a = 2$, $b = 10$, and $c = 6$. Round to the nearest tenth.

6. A 26-foot ladder reaches a point on a wall 24 feet above the ground. How far from the wall is the ladder's base?

In Problems 7–10, simplify each expression. Assume that x and y represent positive numbers.

7. $\sqrt{4x^2}$

8. $\sqrt{54x^3}$

9. $\sqrt{\dfrac{18x^2y^3}{2xy}}$

10. $\sqrt[3]{x^6y^3}$

11. A square has an area of 24 square yards.
   a. Express the length of a side of the square in simplified radical form.
   b. Round the length of a side of the square to the nearest tenth.

In Problems 12–18, do each operation and simplify.

12. $\sqrt{12} + \sqrt{27}$

13. $\sqrt{8x^3} - x\sqrt{18x}$

14. $\left(-2\sqrt{8x}\right)\left(3\sqrt{12x}\right)$

15. $\sqrt{3}\left(\sqrt{8} + \sqrt{6}\right)$

16. $\left(\sqrt{2} + \sqrt{3}\right)\left(\sqrt{2} - \sqrt{3}\right)$

17. $\left(2\sqrt{x} + 2\right)\left(\sqrt{x} - 3\right)$

18. SEWING   A corner of fabric is folded over to form a collar and stitched down as shown in Illustration 1. From the dimensions given in the figure, determine the exact number of inches of stitching that must be made. Then give an approximation to one decimal place. (All measurements are in inches.)

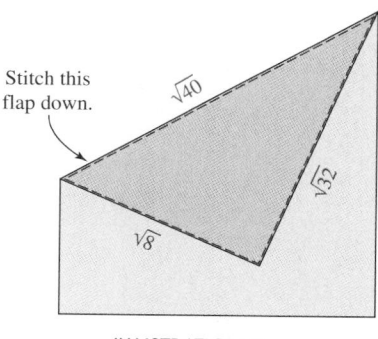

ILLUSTRATION 1

In Problems 19–20, rationalize each denominator.

19. $\dfrac{2}{\sqrt{2}}$

20. $\dfrac{\sqrt{3x}}{\sqrt{x} + 2}$

In Problems 21–24, solve each equation.

21. $\sqrt{x} = 15$

22. $\sqrt{2 - x} - 2 = 6$

23. $\sqrt{3x + 9} = 2\sqrt{x + 1}$

24. $\sqrt[3]{x - 2} = 3$

25. Find the distance between points $(-2, -3)$ and $(-8, 5)$.

26. 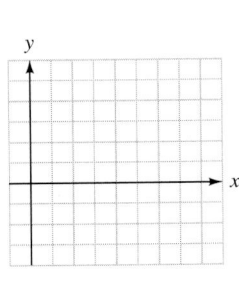 Complete the table and then graph the function. Round to the nearest tenth when necessary.

$f(x) = \sqrt{x}$

x	$f(x)$
0	
1	
2	
3	
4	
5	
6	
7	
8	
9	

27. Is $x = 0$ a solution of the radical equation $\sqrt{3x + 1} = x - 1$? Explain your answer.

28. Explain why we cannot do the subtraction $4\sqrt{3} - 7\sqrt{2}$.

29. CARPENTRY   In Illustration 2, a carpenter is using a tape measure to see if the wall he just put up is perfectly "square" with the floor. Explain what mathematical concept he is applying. If the wall is positioned correctly, what should the measurement on the tape read?

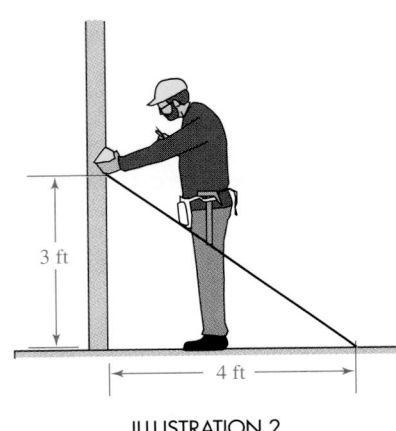

3 ft

4 ft

ILLUSTRATION 2

30. Explain why $\sqrt{-9}$ is not a real number.

Simplify each expression.

31. $121^{1/2}$

32. $p^{2/3}p^{4/3}$

Chapters 1-14   Cumulative Review Exercises

1. Tell whether each statement is true or false.
   a. All whole numbers are integers.
   b. π is a rational number.
   c. A real number is either rational or irrational.

2. Find the value of the expression
$$\frac{-3(3 + 2)^2 - (-5)}{17 - 3|-4|}$$

3. BACKPACK   Pediatricians advise that children should not carry more than 20% of their own body weight in a backpack. According to this warning, how much weight can a fifth-grade girl who weighs 85 pounds safely carry in her backpack?

4. SCIENCE   Illustration 1 shows the recent budgets for the National Science Foundation. Determine the % change for the 1996 budget as compared to the 1995 budget. Round to the nearest tenth of a percent.

In billions		% change
1992	$2.55	8.7%
1993	$2.75	8.0%
1994	$2.99	8.6%
1995	$3.27	9.5%
1996	$3.21	?
1997	$3.30	2.9%
1998	$3.43	3.9%
1999	$3.67	7.1%
2000	$3.91	6.5%

Based on data from the National Science Foundation

ILLUSTRATION 1

5. Simplify $3p - 6(p + z) + p$.

6. Solve $2 - (4x + 7) = 3 + 2(x + 2)$.

7. Solve $3 - 3x \geq 6 + x$ and graph the solution. Then use interval notation to describe the solution.

8. Solve $0 \leq \dfrac{4 - x}{3} < 2$ and graph the solution. Then use interval notation to describe the solution.

9. SEARCH AND RESCUE   Two search and rescue teams leave base at the same time, looking for a lost boy. The first team, on foot, heads north at 2 mph and the other, on horseback, south at 4 mph. How long will it take them to search a distance of 21 miles between them?

10. BLENDING COFFEE   A store sells regular coffee for $4 a pound and gourmet coffee for $7 a pound. Using 40 pounds of the gourmet coffee, the owner makes a blend to put on sale for $5 a pound. How many pounds of regular coffee should he use?

11. SURFACE AREA   The total surface area A of a box with dimensions l, w, and h (see Illustration 2) is given by the formula
$$A = 2lw + 2wh + 2lh$$
If $A = 202$ square inches, $l = 9$ inches, and $w = 5$ inches, find h.

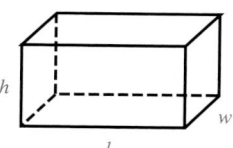

ILLUSTRATION 2

Graph each equation or inequality.

12. $3x - 4y = 12$

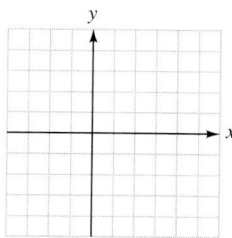

13. $y = \dfrac{1}{2}x$.

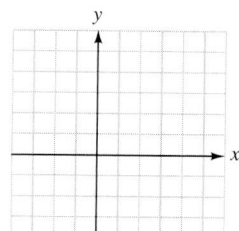

14. $x = 5$

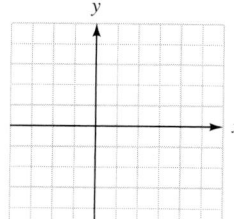

15. $3x + 4y \leq 12$

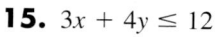

16. Write the equation of the line passing through $(-2, 5)$ and $(4, 8)$.

17. What is the slope of the line defined by each equation?
a. $y = 3x - 7$ **b.** $2x + 3y = -10$

18. What is true about the slopes of two
a. parallel lines?
b. perpendicular lines?

19. SHOPPING SURGE On the graph in Illustration 3, draw a line through the points (1991, 724) and (1996, 974). The line approximates the total annual sales at U.S. shopping centers for the years 1991–1996. Find the rate of increase in sales over this period by finding the slope of the line.

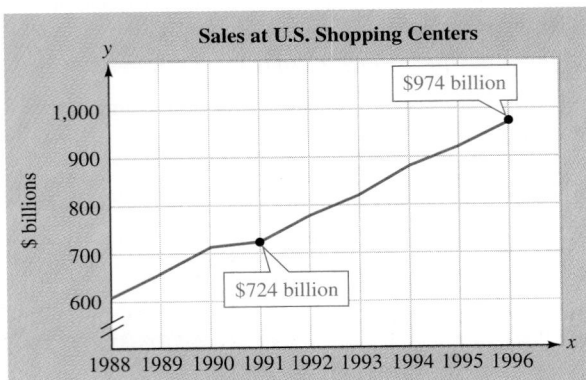

Sales at U.S. Shopping Centers

Based on data from International Council of Shopping Centers

ILLUSTRATION 3

20. If $f(x) = x^3 - x + 5$, find $f(-2)$.

21. Complete the table and graph the function. Then give the domain and range of the function.

$f(x) = |1 - x|$

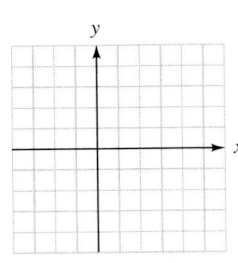

x	$f(x)$
0	
1	
2	
3	
−1	
−2	

22. BOATING The graph in Illustration 4 shows the vertical distance from a point on the tip of a propeller to the centerline as the propeller spins. Is this the graph of a function?

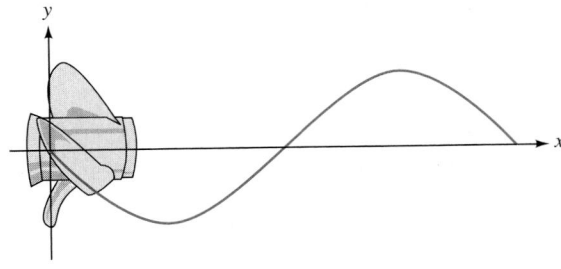

ILLUSTRATION 4

In Exercises 23–26, simplify each expression. Write each answer without using parentheses or negative exponents.

23. $(x^5)^2(x^7)^3$

24. $\left(\dfrac{a^3b}{c^4}\right)^5$

25. $4^{-3} \cdot 4^{-2} \cdot 4^5$

26. $(a^{-2}b^3)^{-4}$

27. ASTRONOMY The **parsec**, a unit of distance used in astronomy, is 3×10^{16} meters. The distance to Betelgeuse, a star in the constellation Orion, is 1.6×10^2 parsecs. Use scientific notation to express this distance in meters.

28. NCAA MEN'S BASKETBALL The graph in Illustration 5 shows the University of Connecticut's lead or deficit during the second half of the 1999 championship game with Duke University.
a. How many x-intercepts does the graph have? Explain their importance.

b. Give the coordinates of the highest point and the lowest point on the graph. What is the importance of each?

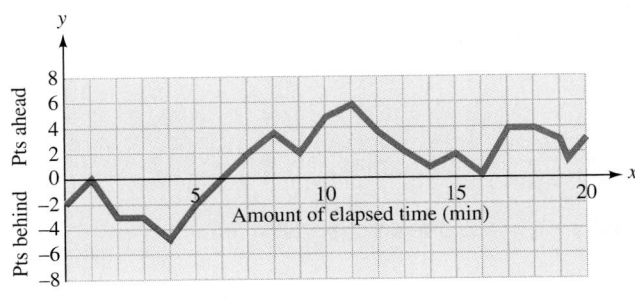

ILLUSTRATION 5

Do the indicated operations.

29. $(-r^4st^2)(2r^2st)(rst)$

30. $(-3t + 2s)(2t - 3s)$

31. $(3a^2 - 2a + 4) - (a^2 - 3a + 7)$

32. $(y - 6)^2$

33. $\dfrac{4x - 3y + 8z}{4xy}$

34. $2 + x\overline{)3x + 2x^2 - 2}$

Factor each expression completely.

35. $3x^2y - 6xy^2$

36. $2x^2 + 2xy - 3x - 3y$

37. $25p^4 - 16q^2$

38. $3x^3 - 243x$

39. $x^2 - 11x - 12$

40. $a^3 + 8b^3$

41. $6a^2 - 7a - 20$

42. $16m^2 - 20m - 6$

In Exercises 43–46, solve each equation.

43. $x^2 + 3x + 2 = 0$ **44.** $5x^2 = 10x$

45. $6x^2 - x - 2 = 0$ **46.** $2y^2 = 12 - 5y$

47. CHILDREN'S STICKER A rectangular-shaped sticker has an area of 20 cm². The width is 1 cm shorter than the length. (See Illustration 6.) Find the length of the sticker.

$(l - 1)$ cm

l cm

ILLUSTRATION 6

48. For what value of x is $\dfrac{4x}{x - 6}$ undefined?

Simplify each expression.

49. $\dfrac{x^2 + 2x + 1}{x^2 - 1}$ **50.** $-\dfrac{15a^2}{25a^3}$

Do the operation(s) and simplify when possible.

51. $\dfrac{p^2 - p - 6}{3p - 9} \div \dfrac{p^2 + 6p + 9}{p^2 - 9}$

52. $\dfrac{x^2y^2}{cd} \cdot \dfrac{d^2}{c^2x}$

53. $\dfrac{x + 2}{x + 5} - \dfrac{x - 3}{x + 7}$

54. $\dfrac{3x}{x + 2} + \dfrac{5x}{x + 2} - \dfrac{7x - 2}{x + 2}$

55. $\dfrac{3a}{2b} - \dfrac{2b}{3a}$

56. $\dfrac{\dfrac{1}{x} + \dfrac{1}{y}}{\dfrac{1}{x} - \dfrac{1}{y}}$

In Exercises 57–58, solve each equation.

57. $\dfrac{4}{a} = \dfrac{6}{a} - 1$

58. $\dfrac{a + 2}{a + 3} - 1 = \dfrac{-1}{a^2 + 2a - 3}$

59. Solve the formula $\dfrac{1}{r} = \dfrac{1}{r_1} + \dfrac{1}{r_2}$ for r.

60. ONLINE SALES A company found that, on average, it made 9 online sales transactions for every 500 hits on its Internet Web site. If the company's Web site had 360,000 hits in one year, how many sales transactions did it have that year?

61. Assume that y varies inversely with x. If $y = 8$ when $x = 2$, find y when $x = 8$.

62. FILLING A POOL An inlet pipe can fill an empty swimming pool in 5 hours, and another inlet pipe can fill the pool in 4 hours. How long will it take both pipes to fill the pool?

In Exercises 63–64, solve each system of equations. If the equations of a system are dependent or if a system is inconsistent, so indicate.

63. $\begin{cases} x = y + 4 \\ 2x + y = 5 \end{cases}$ **64.** $\begin{cases} \frac{3}{5}s + \frac{4}{5}t = 1 \\ -\frac{1}{4}s + \frac{3}{8}t = 1 \end{cases}$

65. FINANCIAL PLANNING In investing $6,000 of a couple's money, a financial planner put some of it into a savings account paying 6% annual interest. The rest was invested in a riskier mini-mall development plan paying 12% annually. The combined interest earned for the first year was $540. How much money was invested at each rate? Use two variables to solve this problem.

66. Graph the solution of $\begin{cases} 3x + 2y \geq 6 \\ x + 3y \leq 6 \end{cases}$

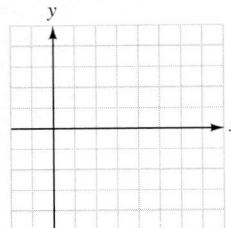

Simplify each expression. All variables represent positive numbers.

67. $\sqrt{\dfrac{49}{225}}$

68. $-\sqrt[3]{-27}$

69. $-12x\sqrt{16x^2y^3}$

70. $\sqrt{48} - \sqrt{8} + \sqrt{27} - \sqrt{32}$

71. $\left(\sqrt{y} - 4\right)\left(\sqrt{y} - 5\right)$

72. $\left(-5\sqrt{6}\right)\left(4\sqrt{3}\right)$

73. $\dfrac{4}{\sqrt{20}}$

74. $\dfrac{\sqrt{x} - 3}{\sqrt{x} + 3}$

75. Solve $\sqrt{6x + 19} - 5 = 2$.

76. CARGO SPACE   How wide a piece of plywood can be stored diagonally in the back of the van shown in Illustration 7?

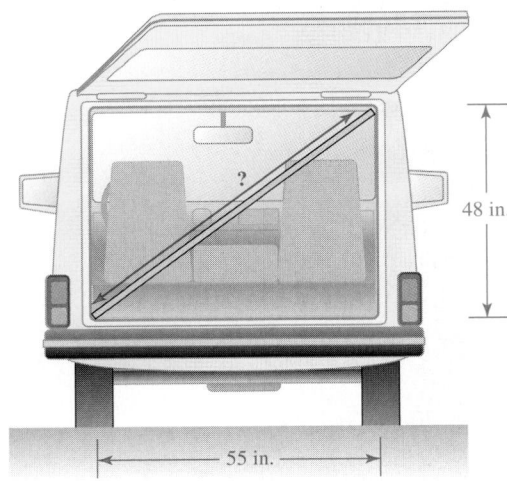

ILLUSTRATION 7

Quadratic Equations

15

15.1 Completing the Square

15.2 The Quadratic Formula

15.3 Graphing Quadratic Functions

Key Concept: Quadratic Equations

Accent on Teamwork

Chapter Review

Chapter Test

Cumulative Review Exercises

WE HAVE PREVIOUSLY SOLVED QUADRATIC EQUATIONS BY FACTORING. IN THIS CHAPTER, WE WILL DISCUSS THREE OTHER METHODS THAT ARE USED TO SOLVE "QUADRATICS."

15.1 *Completing the Square*

In this section, you will learn about

- The square root method • Completing the square
- Solving equations with leading coefficients of 1
- Solving equations with leading coefficients other than 1

INTRODUCTION. Recall that equations that involve second-degree polynomials are called *quadratic equations.*

Quadratic equations

A **quadratic equation** is an equation that can be written in the form
$$ax^2 + bx + c = 0 \qquad (a \neq 0)$$
where a, b, and c represent real numbers. This form is called **quadratic form.**

Some examples of quadratic equations are

$$x^2 + 12x - 13 = 0, \qquad 3q^2 = 3q + 2, \qquad a^2 - 5a = 0, \qquad \text{and} \qquad x^2 = 16.$$

We have solved quadratic equations using the factoring method. In this section, we will discuss two new methods for solving quadratic equations. The first, called the *square root method,* is used when one side of the equation to solve is a quantity squared and the other side is a constant. The second method, called *completing the square,* involves the concept of perfect square trinomials.

The square root method

If $x^2 = 9$, x is a number whose square is 9. Since $3^2 = 9$ and $(-3)^2 = 9$, the equation $x^2 = 9$ has two solutions, $x = \sqrt{9} = 3$ and $x = -\sqrt{9} = -3$. In general, any equation of the form $x^2 = c$, where $c > 0$, has two solutions.

The square root method

If c represents a positive real number, the equation $x^2 = c$ has two solutions:
$$x = \sqrt{c} \qquad \text{or} \qquad x = -\sqrt{c}$$

We can write the previous result with **double-sign notation.** The statement

$$x = \pm\sqrt{c} \qquad \left(\text{read as "}x \text{ equals positive or negative } \sqrt{c}\text{"}\right)$$

means that $x = \sqrt{c}$ or $x = -\sqrt{c}$.

EXAMPLE 1 *Solving equations using the square root method.*
Solve $x^2 = 16$.

Solution

We use the square root method to find that the equation has two solutions.

$$x^2 = 16$$
$$x = \pm\sqrt{16} \quad \pm\sqrt{16} \text{ means that } x = \sqrt{16} \text{ or } x = -\sqrt{16}.$$
$$x = \pm 4 \qquad \text{Simplify: } \sqrt{16} = 4.$$

The solutions of $x^2 = 16$ are 4 and -4.

Check: **For $x = 4$**   **For $x = -4$**
$$x^2 = 16 \qquad\qquad x^2 = 16$$
$$4^2 \stackrel{?}{=} 16 \qquad\qquad (-4)^2 \stackrel{?}{=} 16$$
$$16 = 16 \qquad\qquad 16 = 16$$

Self Check

Solve $x^2 = 25$.

Answer: ± 5 ■

The equation in Example 1 can also be solved by factoring.

$$x^2 = 16$$
$$x^2 - 16 = 0 \qquad \text{Subtract 16 from both sides.}$$
$$(x + 4)(x - 4) = 0 \qquad \text{Factor the difference of two squares.}$$
$$x + 4 = 0 \quad \text{or} \quad x - 4 = 0$$
$$x = -4 \qquad\qquad x = 4$$

 COMMENT When using the square root method to solve an equation, always write the $\pm$ symbol, or you will lose one of the solutions. For example, consider the equation from Example 1.

$$x^2 = 16$$
$$x = \pm\sqrt{16}$$

  If you don't write this symbol, you will lose the
  solution $x = -\sqrt{16}$, which is $x = -4$.

EXAMPLE 2 *The square root method.*   Solve $3x^2 - 9 = 0$.

Solution

To solve the equation by the square root method, we first isolate x^2.

$$3x^2 - 9 = 0$$
$$3x^2 = 9 \qquad \text{Add 9 to both sides.}$$
$$x^2 = 3 \qquad \text{Divide both sides by 3.}$$
$$x = \pm\sqrt{3} \quad \text{Use the square root method.}$$

Check: **For $x = \sqrt{3}$**   **For $x = -\sqrt{3}$**
$$3x^2 - 9 = 0 \qquad\qquad 3x^2 - 9 = 0$$
$$3\left(\sqrt{3}\right)^2 - 9 \stackrel{?}{=} 0 \qquad 3\left(-\sqrt{3}\right)^2 - 9 = 0$$
$$3(3) - 9 \stackrel{?}{=} 0 \qquad\qquad 3(3) - 9 \stackrel{?}{=} 0$$
$$9 - 9 \stackrel{?}{=} 0 \qquad\qquad 9 - 9 \stackrel{?}{=} 0$$
$$0 = 0 \qquad\qquad 0 = 0$$

The solutions of $3x^2 - 9 = 0$ are $\sqrt{3}$ and $-\sqrt{3}$. Note that these are exact solutions. Using a calculator, we can approximate them to the nearest hundredth: $x \approx \pm 1.73$.

Self Check

Solve $2x^2 - 10 = 0$. Give the exact solutions and approximations to the nearest hundredth.

Answer: $\pm\sqrt{5}$; ± 2.24 ■

EXAMPLE 3   *Hurricane.*   In 1998, Hurricane Mitch dumped heavy rains on Central America, causing extensive flooding in Honduras, Belize, and Guatemala. Figure 15-1 shows the position of the storm on October 26, at which time the weather service estimated that it covered an area of about 71,000 square miles. What was the diameter of the storm?

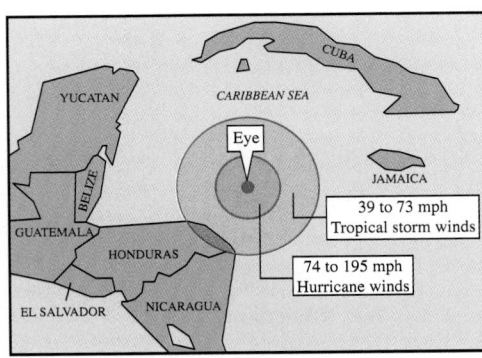

FIGURE 15-1

Solution   We can use the formula for the area of a circle to find the *radius* (in miles) of the circular-shaped storm.

$$A = \pi r^2$$
$$71{,}000 = \pi r^2 \qquad \text{Substitute 71,000 for the area } A.$$
$$\frac{71{,}000}{\pi} = r^2 \qquad \text{Divide both sides by } \pi \text{ to isolate } r^2.$$

Now we use the square root method to solve for *r*.

$$r = \sqrt{\frac{71{,}000}{\pi}} \qquad \text{or} \qquad r = -\sqrt{\frac{71{,}000}{\pi}}$$

Using a calculator to approximate the square root, we have

$$r \approx 150.3329702 \qquad \text{\scriptsize The units are miles. The second solution is discarded, because the radius cannot be negative.}$$

If we multiply the radius by 2, we find that the diameter of the storm was about 300 miles.   ∎

EXAMPLE 4   *The square root method.*   Solve $(x - 1)^2 = 18$.

Solution

$$(x - 1)^2 = 18$$
$$x - 1 = \pm\sqrt{18} \qquad \text{\scriptsize Use the square root method to solve for } x - 1.$$

To solve for *x*, we undo the subtraction of 1 on the left-hand side by adding 1 to both sides. In this case, when adding (or subtracting) a number on the right-hand side, we customarily write it *in front of* the radical expression.

$$x - 1 = \pm\sqrt{18}$$
$$x - 1 + 1 = 1 \pm \sqrt{18} \qquad \text{\scriptsize Add 1 to both sides, to isolate } x.$$
$$x = 1 \pm \sqrt{18} \qquad \text{\scriptsize Simplify the left-hand side.}$$
$$x = 1 \pm 3\sqrt{2} \qquad \text{\scriptsize Simplify the radical: } \sqrt{18} = \sqrt{9 \cdot 2} = 3\sqrt{2}.$$

The solutions are $1 \pm 3\sqrt{2}$ (read as "1 plus or minus $3\sqrt{2}$"). We can approximate each of them to the nearest hundredth.

$$1 + 3\sqrt{2} \approx 5.24 \qquad\qquad 1 - 3\sqrt{2} \approx -3.24$$

Self Check
Solve $(x - 3)^2 = 8$.

Answer: $3 + 2\sqrt{2} \approx 5.83$,
$3 - 2\sqrt{2} \approx 0.17$ ∎

Completing the square

We have solved quadratic equations such as $x^2 + 12x - 13 = 0$ using the factoring method.

$$x^2 + 12x - 13 = 0$$
$$(x - 1)(x + 13) = 0 \qquad \text{Factor the trinomial } x^2 + 12x - 13.$$
$$x - 1 = 0 \quad \text{or} \quad x + 13 = 0 \qquad \text{Set each factor equal to 0.}$$
$$x = 1 \qquad\qquad x = -13 \quad \text{Solve each linear equation.}$$

The solutions of $x^2 + 12x - 13 = 0$ are $x = 1$ or $x = -13$.

Not every quadratic equation can be solved using the factoring method. For example, the trinomial in the equation $x^2 + 4x - 13 = 0$ cannot be factored using any of the techniques we have studied previously. To solve such equations, we can use another method called *completing the square*. It is based on the following **special products:**

$$x^2 + 2bx + b^2 = (x + b)^2 \qquad \text{and} \qquad x^2 - 2bx + b^2 = (x - b)^2$$

The trinomials $x^2 + 2bx + b^2$ and $x^2 - 2bx + b^2$ are both perfect square trinomials, since each one factors as the square of a binomial. In each trinomial, if we take one-half of the coefficient of x and square it, we get the third term.

In $x^2 + 2bx + b^2$, if we take $\frac{1}{2}(2b)$, which is b, and square it, we get the third term, b^2.

In $x^2 - 2bx + b^2$, if we take $\frac{1}{2}(-2b) = -b$ and square it, we get $(-b)^2 = b^2$, which is the third term.

To change a binomial such as $x^2 + 12x$ into a perfect square trinomial, we take one-half of the coefficient of x (the 12), square it, and add it to $x^2 + 12x$.

$$x^2 + 12x + \left[\frac{1}{2}(12)\right]^2 = x^2 + 12x + (6)^2$$
$$= x^2 + 12x + 36$$

This result is a perfect square trinomial, because $x^2 + 12x + 36 = (x + 6)^2$.

EXAMPLE 5 *Completing the square.* Change each expression into a perfect square trinomial: **a.** $x^2 + 4x$, **b.** $x^2 - 6x$, and **c.** $x^2 - 5x$.

Solution

a. Since the coefficient of x is 4, we add the square of one-half of 4.

$$x^2 + 4x + \left[\frac{1}{2}(4)\right]^2 = x^2 + 4x + (2)^2 \quad \text{Simplify: } \tfrac{1}{2}(4) = 2.$$
$$= x^2 + 4x + 4 \qquad \text{This is } (x + 2)^2.$$

b. Since the coefficient of x is -6, we add the square of one-half of -6.

$$x^2 - 6x + \left[\frac{1}{2}(-6)\right]^2 = x^2 - 6x + (-3)^2 \quad \text{Simplify: } \tfrac{1}{2}(-6) = -3.$$
$$= x^2 - 6x + 9 \qquad \text{This is } (x - 3)^2.$$

c. Since the coefficient of x is -5, we add the square of one-half of -5.

$$x^2 - 5x + \left[\frac{1}{2}(-5)\right]^2 = x^2 - 5x + \left(-\frac{5}{2}\right)^2 \quad \text{Simplify: } \tfrac{1}{2}(-5) = -\tfrac{5}{2}.$$
$$= x^2 - 5x + \frac{25}{4} \qquad \text{This is } \left(x - \tfrac{5}{2}\right)^2.$$

Self Check

Change each expression into a perfect square trinomial:

a. $y^2 + 6y$

b. $y^2 - 8y$

c. $y^2 + 3y$

Answers: **a.** $y^2 + 6y + 9$,

b. $y^2 - 8y + 16$,

c. $y^2 + 3y + \dfrac{9}{4}$

■

Solving equations with leading coefficients of 1

If the quadratic equation $ax^2 + bx + c = 0$ has a leading coefficient of 1, it's easy to solve by completing the square.

EXAMPLE 6 *Completing the square.* Solve $x^2 + 4x - 13 = 0$. Give each answer to the nearest hundredth.

Solution

Since the coefficient of x^2 is 1, we can complete the square as follows:

$$x^2 + 4x - 13 = 0$$
$$x^2 + 4x = 13 \quad \text{Add 13 to both sides so that the constant term is on the right-hand side.}$$

We then find one-half of the coefficient of x, square it, and add the result to both sides to make the left-hand side a perfect square trinomial.

$$x^2 + 4x + \left[\frac{1}{2}(4)\right]^2 = 13 + \left[\frac{1}{2}(4)\right]^2 \quad \text{Since the coefficient of } x \text{ is 4, add the square of one-half of 4.}$$

$$x^2 + 4x + 4 = 13 + 4 \quad \text{Simplify: } \frac{1}{2}(4) = 2. \text{ Then square 2.}$$

$$(x + 2)^2 = 17 \quad \text{Factor } x^2 + 4x + 4 \text{ and simplify.}$$

$$x + 2 = \pm\sqrt{17} \quad \text{Use the square root method to solve for } x + 2.$$

$$x = -2 \pm \sqrt{17} \quad \text{Subtract 2 from both sides to isolate } x. \text{ Write } -2 \text{ in front of the radical.}$$

We can use a calculator to approximate each solution.

$$x = -2 + \sqrt{17} \qquad \text{or} \qquad x = -2 - \sqrt{17}$$
$$x \approx -2 + 4.123105626 \qquad\qquad x \approx 2 - 4.123105626$$
$$x \approx 2.12 \qquad\qquad\qquad\qquad x \approx -6.12$$

Self Check
Solve $x^2 + 10x - 4 = 0$. Give the exact solutions and approximations to the nearest hundredth.

Answers: $-5 + \sqrt{29}$, $-5 - \sqrt{29}$; 0.39, -10.39 ∎

EXAMPLE 7 *Completing the square.* Solve $x^2 - 7x = 2$.

Solution

The constant term is already on the right-hand side. To complete the square on the left-hand side, we find one-half of the coefficient of x and add its square to both sides.

$$x^2 - 7x = 2$$

$$x^2 - 7x + \left[\frac{1}{2}(-7)\right]^2 = 2 + \left[\frac{1}{2}(-7)\right]^2 \quad \text{Since the coefficient of } x \text{ is } -7, \text{ add the square of one-half of } -7.$$

$$x^2 - 7x + \frac{49}{4} = 2 + \frac{49}{4} \quad \text{Simplify: } \frac{1}{2}(-7) = -\frac{7}{2}. \text{ Then square } -\frac{7}{2}.$$

$$\left(x - \frac{7}{2}\right)^2 = \frac{8}{4} + \frac{49}{4} \quad \text{Factor the left-hand side. Write 2 as } \frac{8}{4}.$$

$$\left(x - \frac{7}{2}\right)^2 = \frac{57}{4} \quad \text{The fractions have a common denominator. Add them.}$$

$$x - \frac{7}{2} = \pm\sqrt{\frac{57}{4}} \quad \text{Use the square root method to solve for } x - \frac{7}{2}.$$

$$x - \frac{7}{2} = \pm\frac{\sqrt{57}}{2} \quad \text{Simplify: } \sqrt{\frac{57}{4}} = \frac{\sqrt{57}}{\sqrt{4}} = \frac{\sqrt{57}}{2}.$$

Self Check
Solve $x^2 + 5x = 3$. Approximate the solutions to the nearest hundredth.

$$x = \frac{7}{2} \pm \frac{\sqrt{57}}{2} \qquad \text{Add } \tfrac{7}{2} \text{ to both sides.}$$

$$x = \frac{7 \pm \sqrt{57}}{2} \qquad \text{Since the fractions have a common denominator of 2, we can combine them.}$$

If we approximate the solutions to the nearest hundredth, we have

$$\frac{7 + \sqrt{57}}{2} \approx 7.27 \qquad \text{and} \qquad \frac{7 - \sqrt{57}}{2} \approx -0.27$$

Answer: $\dfrac{-5 \pm \sqrt{37}}{2}$;

$0.54, -5.54$ ■

Solving equations with leading coefficients other than 1

If the quadratic equation $ax^2 + bx + c = 0$ has a leading coefficient other than 1, we can make the leading coefficient 1 by dividing both sides of the equation by a.

EXAMPLE 8   *Completing the square.*   Solve $4x^2 + 4x - 3 = 0$.

Solution

We divide both sides by 4 so that the coefficient of x^2 is 1. We then proceed as follows:

$$4x^2 + 4x - 3 = 0$$

$$x^2 + x - \frac{3}{4} = 0 \qquad \text{Divide both sides by 4: } \frac{4x^2}{4} + \frac{4x}{4} - \frac{3}{4} = \frac{0}{4}.$$

$$x^2 + x = \frac{3}{4} \qquad \text{Add } \tfrac{3}{4} \text{ to both sides so that the constant term is on the right-hand side.}$$

$$x^2 + 1x + \left[\frac{1}{2}(1)\right]^2 = \frac{3}{4} + \left[\frac{1}{2}(1)\right]^2 \qquad \text{Since the coefficient of } x \text{ is 1, add the square of one-half of 1.}$$

$$x^2 + x + \frac{1}{4} = \frac{3}{4} + \frac{1}{4} \qquad \text{Simplify: } \tfrac{1}{2}(1) = \tfrac{1}{2}. \text{ Then square } \tfrac{1}{2}.$$

$$\left(x + \frac{1}{2}\right)^2 = 1 \qquad \text{Factor and add.}$$

$$x + \frac{1}{2} = \pm 1 \qquad \text{Solve for } x + \tfrac{1}{2} \text{ using the square root method.}$$

$$x = -\frac{1}{2} \pm 1 \qquad \text{Subtract } \tfrac{1}{2} \text{ from both sides to isolate } x.$$

$$x = -\frac{1}{2} + 1 \quad \text{or} \quad x = -\frac{1}{2} - 1$$

$$x = \frac{1}{2} \qquad\qquad x = -\frac{3}{2}$$

Self Check

Solve $2x^2 - 5x - 3 = 0$.

Answer: $3, -\frac{1}{2}$ ■

COMMENT   In Example 8, you may have noticed that $4x^2 + 4x - 3$ can be factored. Therefore, we could have solved $4x^2 + 4x - 3 = 0$ using the factoring method. This example illustrates an important fact: Completing the square can be used to solve *any* quadratic equation.

The previous examples illustrate that to solve a quadratic equation by completing the square, we follow these steps.

Completing the square to solve a quadratic equation

1. Write the equation in $ax^2 + bx + c = 0$ form. If the coefficient of x^2 is not 1, make it 1 by dividing both sides of the equation by the coefficient of x^2.

2. If necessary, add or subtract a number on both sides of the equation to get the constant term on the right-hand side.

3. Complete the square.
   a. Find half the coefficient of x and square it.
   b. Add that square to both sides of the equation.

4. Factor the perfect square trinomial and combine terms.

5. Solve the resulting quadratic equation using the square root method.

6. Check each solution.

EXAMPLE 9   *Solving equations by completing the square.*   Solve $2x^2 - 2 = 4x$.

Solution

We write the equation in $ax^2 + bx + c = 0$ form to see if it can be solved by factoring.

$2x^2 - 4x - 2 = 0$ Subtract $4x$ from both sides to get 0 on the right-hand side.

(1) $x^2 - 2x - 1 = 0$ Divide both sides by 2: $\dfrac{2x^2}{2} - \dfrac{4x}{2} - \dfrac{2}{2} = \dfrac{0}{2}$.

Since Equation 1 cannot be solved by factoring, we complete the square.

$$x^2 - 2x = 1 \qquad \text{Add 1 to both sides.}$$

$$x^2 - 2x + \left[\tfrac{1}{2}(-2)\right]^2 = 1 + \left[\tfrac{1}{2}(-2)\right]^2 \qquad \text{Since the coefficient of } x \text{ is } -2, \text{ add the square of one-half of } -2.$$

$$x^2 - 2x + 1 = 1 + 1 \qquad \text{Simplify: } \tfrac{1}{2}(-2) = -1. \text{ Then square } -1.$$

$$(x - 1)^2 = 2 \qquad \text{Factor and simplify.}$$

$$x - 1 = \pm\sqrt{2} \qquad \text{Use the square root method to solve for } x - 1.$$

$$x = 1 \pm \sqrt{2} \qquad \text{Add 1 to both sides.}$$

$$x = 1 + \sqrt{2} \quad \text{or} \quad x = 1 - \sqrt{2}$$

Self Check

Solve $3x^2 - 18x = -12$.

Answer: $3 \pm \sqrt{5}$ ■

STUDY SET   Section 15.1

VOCABULARY   *Fill in the blanks.*

1. If the polynomial in the equation $ax^2 + bx + c = 0$ doesn't factor, we can solve the equation by _____ the square.

2. Since $x^2 + 12x + 36 = (x + 6)^2$, we call the trinomial a perfect _____ trinomial.

3. In the equation $x^2 - 4x + 1 = 0$, the _____ of x is -4.

4. A _____ of an equation is a value of the variable that makes the equation true.

CONCEPTS   *In Exercises 5–8, fill in the blanks.*

5. The equation $x^2 = c$, where $c > 0$, has _____ solutions.

6. The solutions of $x^2 = c$, where $c > 0$, are _____ and _____ .

7. To complete the square on $x^2 + 8x$, we add the _____ of one-half of 8, which is 16.

8. To complete the square on $x^2 - 10x$, we add the square of _____ of -10, which is 25.

9. What is the first step if we solve $x^2 - 2x = 35$
   a. by the factoring method?

   b. by completing the square?

10. The equation $n^2 - 4n + 6 = 0$ is written in $ax^2 + bx + c = 0$ form. What is b?

11. a. To solve $x^2 - 2x - 1 = 0$, we must complete the square. Why can't we use the factoring method?

    b. Can any quadratic equation be solved by completing the square?

12. Solve $x^2 - 81$ by using
    a. the square root method.
    b. the factoring method.

13. What is one-half of the given number?
    a. 4   **b.** -8
    c. 5   **d.** -7

14. Find one-half of the given number and then square the result.
    a. 6   **b.** -12
    c. 3   **d.** -5

15. What is the result when both sides of $2x^2 + 4x - 8 = 0$ are divided by 2?

16. Write $3x^2 = -4x + 8$ in $ax^2 + bx + c = 0$ form. What is b?

NOTATION   *Complete each solution to solve the equation.*

17.
$$(y - 1)^2 = 9$$
$$y - 1 = \boxed{} \quad \text{or} \quad y - \boxed{} = -\sqrt{9}$$
$$\boxed{} = 3 \qquad\qquad y - 1 = \boxed{}$$
$$y = 4 \qquad\qquad\qquad y = -2$$

18.
$$y^2 + 2y - 3 = 0$$
$$y^2 + 2y = \boxed{}$$
$$y^2 + 2y + 1 = 3 + \boxed{}$$
$$(y + 1)^2 = \boxed{}$$
$$\boxed{} = \sqrt{4} \quad \text{or} \quad y + 1 = -\boxed{}$$
$$y + 1 = \boxed{} \qquad \boxed{} = -2$$
$$y = 1 \qquad\qquad\qquad y = -3$$

19. a. In solving a quadratic equation, a student obtains $x = \pm\sqrt{10}$. How many solutions are represented by this notation? List them.

    b. In solving a quadratic equation, a student obtains $x = 8 \pm \sqrt{3}$. List each solution separately. Then round each one to the nearest hundredth.

20. Solve $x + 1 = \pm\sqrt{2}$ for x.

PRACTICE   *Use the square root method to solve each equation.*

21. $x^2 = 1$ **22.** $r^2 = 4$

23. $x^2 = 9$ **24.** $x^2 = 32$

25. $t^2 = 20$ **26.** $x^2 = 0$

27. $3m^2 = 27$ **28.** $4x^2 = 64$

29. $4x^2 = 16$ **30.** $5x^2 = 125$

31. $x^2 = \dfrac{9}{16}$ **32.** $x^2 = \dfrac{81}{25}$

33. $(x + 1)^2 = 25$ **34.** $(x - 1)^2 = 49$

35. $(x + 2)^2 = 81$ **36.** $(x + 3)^2 = 16$

37. $(x - 2)^2 = 8$ **38.** $(x + 2)^2 = 50$

   🖩 *Use the square root method to solve each equation. Use a calculator to approximate the solutions. Round to the nearest hundredth.*

39. $x^2 = 45.82$ **40.** $x^2 = 6.05$

41. $(x + 2)^2 = 90.04$ **42.** $(x - 5)^2 = 33.31$

Factor the trinomial square and use the square root method to solve each equation.

43. $y^2 + 4y + 4 = 4$

44. $y^2 - 6y + 9 = 9$

45. $9x^2 - 12x + 4 = 16$

46. $4x^2 - 20x + 25 = 36$

Complete the square to make a perfect square trinomial.

47. $x^2 + 2x$ **48.** $x^2 + 12x$

49. $x^2 - 4x$ **50.** $x^2 - 14x$

51. $x^2 + 7x$ **52.** $x^2 + 21x$

53. $a^2 - 3a$ **54.** $b^2 - 13b$

55. $b^2 + \dfrac{2}{3}b$ **56.** $c^2 - \dfrac{5}{2}c$

Solve each equation by completing the square.

57. $x^2 + 6x + 8 = 0$ **58.** $x^2 + 8x + 12 = 0$

59. $k^2 - 8k + 12 = 0$ **60.** $p^2 - 4p + 3 = 0$

61. $x^2 - 2x = 15$ **62.** $x^2 - 2x = 8$

63. $g^2 + 5g - 6 = 0$ **64.** $s^2 = 14 - 5s$

65. $2x^2 = 4 - 2x$ **66.** $3q^2 = 3q + 6$

67. $3x^2 + 9x + 6 = 0$ **68.** $3d^2 + 48 = -24d$

69. $2x^2 = 3x + 2$ **70.** $3x^2 = 2 - 5x$

71. $4x^2 = 2 - 7x$ **72.** $2x^2 = 5x + 3$

Solve each equation. Give the exact solutions, and then give the solutions rounded to the nearest hundredth.

73. $x^2 + 4x + 1 = 0$

74. $x^2 + 6x + 2 = 0$

75. $x^2 - 2x - 4 = 0$

76. $x^2 - 4x = 2$

77. $x^2 = 4x + 3$

78. $x^2 = 6x - 3$

79. $4x^2 + 4x + 1 = 20$

80. $9x^2 = 8 - 12x$

Write each equation in the form $ax^2 + bx + c = 0$ and solve it by completing the square.

81. $2x(x + 3) = 8$

82. $3x(x - 2) = 9$

83. $6(x^2 - 1) = 5x$

84. $2(3x^2 - 2) = 5x$

85. $x(x + 3) - \dfrac{1}{2} = -2$

86. $x[(x - 2) + 3] = 3\left(x - \dfrac{2}{9}\right)$

APPLICATIONS

87. CAROUSEL   In 1999, the city of Lancaster, Pennsylvania considered installing a classic Dentzel carousel in an abandoned downtown building. After learning that the circular-shaped carousel (like that shown in Illustration 1) would occupy 2,376 square feet of floor space and that it was 26 feet high, the proposal was determined to be impractical because of the large remodeling costs. Find the diameter of the carousel to the nearest foot.

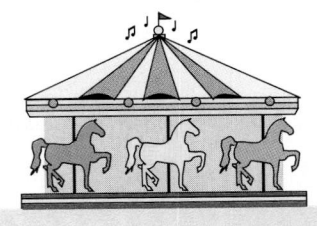

ILLUSTRATION 1

88. ESCAPE VELOCITY   The speed at which a rocket must be fired for it to leave Earth's gravitational attraction is called the *escape velocity*. See Illustration 2. If the escape velocity v_e, in miles per hour, is given by

$$\frac{v_e^2}{2g} = R$$

where $g = 78.545$ and $R = 3,960$, find v_e. Round to the nearest mi/hr.

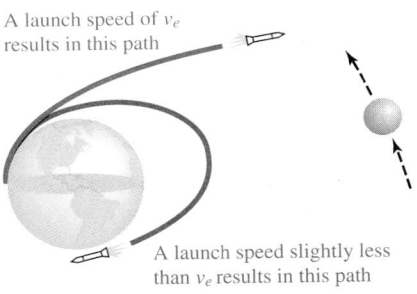

A launch speed of v_e results in this path

A launch speed slightly less than v_e results in this path

ILLUSTRATION 2

89. BICYCLE SAFETY   A bicycle training program for children uses a figure-8 course to help them improve their balance and steering. The course is laid out over a paved area covering 800 square feet, as shown in Illustration 3. Find its dimensions.

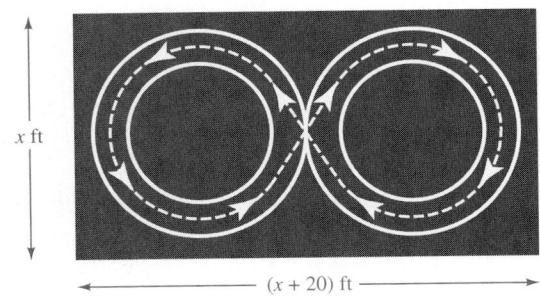

x ft

$(x + 20)$ ft

ILLUSTRATION 3

90. BADMINTON   The badminton court shown in Illustration 4 occupies 880 square feet of the floor space of a gymnasium. If its length is 4 feet more than twice its width, find its dimensions.

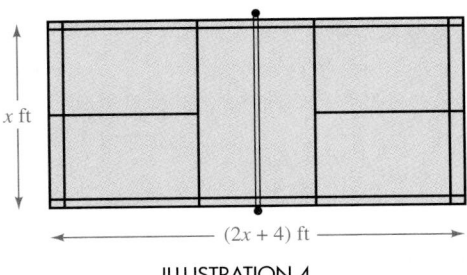

x ft

$(2x + 4)$ ft

ILLUSTRATION 4

WRITING

91. Explain how to complete the square on $x^2 - 5x$.

92. Explain the error in the following work.

Solve $x^2 = 28$.

$$x^2 = 28$$
$$x = \pm\sqrt{28}$$
$$x = 2 \pm \sqrt{7}$$

93. Rounded to the nearest hundredth, one solution of the equation $x^2 + 4x + 1 = 0$ is -0.27. Use your calculator to check it. How could it be a solution if it doesn't make the left-hand side zero? Explain.

94. Give an example of a perfect square trinomial. Why do you think the word "perfect" is used to describe it?

REVIEW *Do each operation.*

95. $(y - 1)^2$

96. $(z + 2)^2$

97. $(x + y)^2$

98. $(a - b)^2$

99. $(2z)^2$

100. $(xy)^2$

15.2 *The Quadratic Formula*

In this section, you will learn about

- The quadratic formula • Quadratic equations with no real solutions
- Applications

INTRODUCTION. We can solve any quadratic equation by completing the square, but the work is sometimes tedious. Fortunately, there is an easier way. In this section, we will develop a formula, called the *quadratic formula*, that will enable us to solve quadratic equations with much less effort.

The quadratic formula

We can solve the **general quadratic equation** $ax^2 + bx + c = 0$, where $a \neq 0$, by completing the square.

$$ax^2 + bx + c = 0$$

$$\frac{ax^2}{a} + \frac{bx}{a} + \frac{c}{a} = \frac{0}{a} \qquad \text{Divide both sides by } a \text{ so that the coefficient of } x^2 \text{ is 1.}$$

$$x^2 + \frac{b}{a}x + \frac{c}{a} = 0 \qquad \text{Simplify } \frac{\overset{1}{\cancel{a}}x^2}{\underset{1}{\cancel{a}}} = x^2. \text{ Write } \frac{bx}{a} \text{ as } \frac{b}{a}x.$$

$$x^2 + \frac{b}{a}x = -\frac{c}{a} \qquad \text{Subtract } \frac{c}{a} \text{ from both sides.}$$

Since the coefficient of x is $\dfrac{b}{a}$, we can complete the square on x by adding

$$\left(\frac{1}{2} \cdot \frac{b}{a}\right)^2 \qquad \text{or} \qquad \frac{b^2}{4a^2}$$

to both sides:

$$x^2 + \frac{b}{a}x + \frac{b^2}{4a^2} = \frac{b^2}{4a^2} - \frac{c}{a}$$

After factoring the perfect square trinomial on the left-hand side, we have

$$\left(x + \frac{b}{2a}\right)\left(x + \frac{b}{2a}\right) = \frac{b^2}{4a^2} - \frac{4ac}{4aa} \qquad \text{The lowest common denominator on the right-hand side is } 4a^2. \text{ Build the second fraction.}$$

(1)
$$\left(x + \frac{b}{2a}\right)^2 = \frac{b^2 - 4ac}{4a^2} \qquad \text{Subtract the numerators and write the difference over the common denominator.}$$

Equation 1 can be solved by the square root method to obtain

$$x + \frac{b}{2a} = \sqrt{\frac{b^2 - 4ac}{4a^2}} \qquad \text{or} \qquad x + \frac{b}{2a} = -\sqrt{\frac{b^2 - 4ac}{4a^2}}$$

$$x + \frac{b}{2a} = \frac{\sqrt{b^2 - 4ac}}{\sqrt{4a^2}} \qquad\qquad x + \frac{b}{2a} = -\frac{\sqrt{b^2 - 4ac}}{\sqrt{4a^2}}$$

$$x = -\frac{b}{2a} + \frac{\sqrt{b^2 - 4ac}}{2a} \qquad\qquad x = -\frac{b}{2a} - \frac{\sqrt{b^2 - 4ac}}{2a}$$

$$x = \frac{-b + \sqrt{b^2 - 4ac}}{2a} \qquad\qquad x = \frac{-b - \sqrt{b^2 - 4ac}}{2a}$$

These solutions are usually written in one formula called the **quadratic formula.**

Quadratic formula

The solutions of the quadratic equation $ax^2 + bx + c = 0$ are

$$x = \frac{-b \pm \sqrt{b^2 - 4ac}}{2a} \qquad (a \neq 0)$$

COMMENT   When you write the quadratic formula, be careful to draw the fraction bar so that it includes the complete numerator. Do not write

$$x = -b \pm \frac{\sqrt{b^2 - 4ac}}{2a}$$

EXAMPLE 1   *The quadratic formula.*   Solve $x^2 + 5x + 6 = 0$.

Solution

The equation is written in $ax^2 + bx + c = 0$ form with $a = 1$, $b = 5$, and $c = 6$. We substitute these values into the quadratic formula and simplify.

$$x = \frac{-b \pm \sqrt{b^2 - 4ac}}{2a} \qquad \text{The quadratic formula.}$$

$$= \frac{-5 \pm \sqrt{5^2 - 4(1)(6)}}{2(1)} \qquad \text{Substitute 1 for } a, \text{5 for } b, \text{ and 6 for } c.$$

$$= \frac{-5 \pm \sqrt{25 - 24}}{2} \qquad \text{Evaluate the power and do the multiplication within the radical symbol.}$$

$$= \frac{-5 \pm \sqrt{1}}{2} \qquad \text{Do the subtraction within the radical symbol.}$$

$$x = \frac{-5 \pm 1}{2} \qquad \text{Simplify: } \sqrt{1} = 1.$$

This notation represents two solutions. We simplify them separately, first using the $+$ sign and then using the $-$ sign.

$$x = \frac{-5 + 1}{2} \qquad \text{or} \qquad x = \frac{-5 - 1}{2}$$

$$x = \frac{-4}{2} \qquad\qquad x = \frac{-6}{2}$$

$$x = -2 \qquad\qquad x = -3$$

Self Check

Solve $x^2 + 6x + 5 = 0$.

Answer: $-1, -5$ ■

COMMENT   In Example 1, you may have noticed that we could have solved $x^2 + 5x + 6 = 0$ using the factoring method. This example illustrates an important fact: The quadratic formula can be used to solve *any* quadratic equation.

EXAMPLE 2 *Writing equations in quadratic form.* Solve $2x^2 = 5x + 3$.

Solution

To identify a, b, and c, we must write the equation in quadratic form.

$$2x^2 = 5x + 3$$
$$2x^2 - 5x - 3 = 0 \qquad \text{Subtract } 5x \text{ and } 3 \text{ from both sides.}$$

In this equation, $a = 2$, $b = -5$, and $c = -3$. We substitute these values into the quadratic formula and simplify.

$$x = \frac{-b \pm \sqrt{b^2 - 4ac}}{2a} \qquad \text{The quadratic formula.}$$

$$= \frac{-(-5) \pm \sqrt{(-5)^2 - 4(2)(-3)}}{2(2)} \qquad \text{Substitute 2 for } a,\ -5 \text{ for } b,\ \text{and } -3 \text{ for } c.$$

$$= \frac{5 \pm \sqrt{25 - (-24)}}{4} \qquad \begin{array}{l}-(-5) = 5. \text{ Evaluate the power and do the}\\ \text{multiplication within the radical symbol.}\end{array}$$

$$= \frac{5 \pm \sqrt{49}}{4} \qquad \begin{array}{l}\text{Do the subtraction within the radical symbol:}\\ 25 - (-24) = 25 + 24 = 49.\end{array}$$

$$= \frac{5 \pm 7}{4} \qquad \text{Simplify: } \sqrt{49} = 7.$$

Thus,

$$x = \frac{5 + 7}{4} \qquad \text{or} \qquad x = \frac{5 - 7}{4}$$

$$x = \frac{12}{4} \qquad\qquad\qquad x = \frac{-2}{4}$$

$$x = 3 \qquad\qquad\qquad x = -\frac{1}{2}$$

Self Check

Solve $4x^2 - 11x = 3$.

Answer: $3, -\dfrac{1}{4}$

■

EXAMPLE 3 *Approximating solutions.* Solve $3x^2 = 2x + 4$. Round each solution to the nearest hundredth.

Solution

We begin by writing the given equation in $ax^2 + bx + c = 0$ form.

$$3x^2 = 2x + 4$$
$$3x^2 - 2x - 4 = 0 \qquad \text{Subtract } 2x \text{ and } 4 \text{ from both sides.}$$

In this equation, $a = 3$, $b = -2$, and $c = -4$. We substitute these values into the quadratic formula and simplify.

$$x = \frac{-b \pm \sqrt{b^2 - 4ac}}{2a} \qquad \text{The quadratic formula.}$$

$$= \frac{-(-2) \pm \sqrt{(-2)^2 - 4(3)(-4)}}{2(3)} \qquad \text{Substitute 3 for } a,\ -2 \text{ for } b,\ \text{and } -4 \text{ for } c.$$

$$= \frac{2 \pm \sqrt{4 + 48}}{6} \qquad \begin{array}{l}-(-2) = 2. \text{ Simplify within the radical}\\ \text{symbol.}\end{array}$$

$$= \frac{2 \pm \sqrt{52}}{6} \qquad \text{Do the addition within the radical symbol.}$$

$$= \frac{2 \pm 2\sqrt{13}}{6} \qquad \text{Simplify: } \sqrt{52} = \sqrt{4 \cdot 13} = 2\sqrt{13}.$$

Self Check

Solve $2x^2 - 1 = 2x$. Round to the nearest hundredth.

$$= \frac{\overset{1}{\cancel{2}}\left(1 \pm \sqrt{13}\right)}{\underset{1}{\cancel{2} \cdot 3}}$$ In the numerator, factor out 2: $2 \pm 2\sqrt{13} = 2\left(1 \pm \sqrt{13}\right)$. Write 6 as $2 \cdot 3$. Then divide out the common factor of 2.

$$x = \frac{1 \pm \sqrt{13}}{3}$$ Simplify.

Thus,

$$x = \frac{1 + \sqrt{13}}{3} \quad \text{or} \quad x = \frac{1 - \sqrt{13}}{3}$$

We can use a calculator to approximate each of these solutions. To the nearest hundredth,

$$\frac{1 + \sqrt{13}}{3} \approx 1.54 \quad \text{and} \quad \frac{1 - \sqrt{13}}{3} \approx -0.87$$

Answers: $\dfrac{1 + \sqrt{3}}{2} \approx 1.37$, $\dfrac{1 - \sqrt{3}}{2} \approx -0.37$ ■

Quadratic equations with no real solutions

The next example shows that some quadratic equations have no real-number solutions.

EXAMPLE 4   *An equation with no real-number solutions.*   Solve $x^2 + 2x + 5 = 0$.

Solution
In this equation, $a = 1$, $b = 2$, and $c = 5$. We substitute these values into the quadratic formula.

$$x = \frac{-b \pm \sqrt{b^2 - 4ac}}{2a}$$ The quadratic formula.

$$= \frac{-2 \pm \sqrt{2^2 - 4(1)(5)}}{2(1)}$$ Substitute 1 for a, 2 for b, and 5 for c.

$$= \frac{-2 \pm \sqrt{4 - 20}}{2}$$ Evaluate the power and do the multiplication within the radical symbol.

$$x = \frac{-2 \pm \sqrt{-16}}{2}$$ Do the subtraction within the radical symbol. The result is a negative number, -16.

Since $\sqrt{-16}$ is not a real number, there are no real-number solutions.

Self Check
Does the equation

$$2x^2 + x + 1 = 0$$

have any real-number solutions?

Answer: no ■

Applications

We have discussed several methods that are used to solve quadratic equations. To determine the most efficient method for a given equation, we can use the following strategy.

Strategy for solving quadratic equations

1. First, see whether the equation is in a form such that the **square root method** is easily applied.
2. If the square root method can't be used, write the equation in $ax^2 + bx + c = 0$ form.
3. Then see whether the equation can be solved using the **factoring method.**
4. If you can't factor the quadratic, solve the equation by **completing the square** or by the **quadratic formula.**

EXAMPLE 5 *Nutrition.* The poster in Figure 15-2 shows the six basic food groups, as established by the U.S. Department of Agriculture. If the area of the poster is 90 square inches and the base is 3 inches longer than the height, find the length of its base and its height.

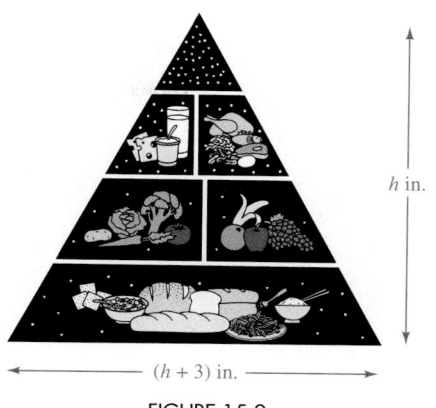

FIGURE 15-2

Analyze the problem   We are given the area of the triangular-shaped poster and asked to find the length of its base and its height.

Form an equation   Since the length of the base is related to the height, we let h represent the height of the triangle. Then $h + 3$ represents the length of the base. The area of a triangle is given by the formula $A = \frac{1}{2}bh$, which gives the equation

$\frac{1}{2}$ times	the length of the base	times	the height	equals	the area of the triangle.
$\frac{1}{2}$ $\cdot$	$(h + 3)$	$\cdot$	h	$=$	90

Solve the equation   To solve the equation $\frac{1}{2}(h + 3)h = 90$, we first write it in quadratic form.

$$\frac{1}{2}(h + 3)h = 90$$

$(h + 3)h = 180$ Multiply both sides by 2.

$h^2 + 3h = 180$ Distribute the multiplication by h.

$h^2 + 3h - 180 = 0$ Subtract 180 from both sides. The equation is now in quadratic form.

By inspection, we see that -180 has factors of -12 and 15 and that their sum is 3. Therefore, we can use the factoring method to solve the equation.

$(h - 12)(h + 15) = 0$ Factor $h^2 + 3h - 180$.

$h - 12 = 0$ or $h + 15 = 0$ Set each factor equal to 0.

$h = 12$ $h = -15$ Solve each linear equation.

State the conclusion   When $h = 12$, the length of the base, $h + 3$, is 15. We discard the solution $h = -15$, because the triangle cannot have a negative height. So the length of the base is 15 inches, and the height is 12 inches.

Check the result   With a base of 15 inches and a height of 12 inches, the base of the triangle is 3 inches longer than its height. Its area is $\frac{1}{2}(15)(12) = 90$ square inches. The solution checks. ∎

EXAMPLE 6   *Movie stunt.*   As part of an action scene in a movie, a stuntman is to fall from the top of a 95-foot-tall building into a large airbag directly below him on the ground, as shown in Figure 15-3. If an object falls s feet in t seconds, where $s = 16t^2$, and if the bag is inflated to a height of 10 feet, how long will the stuntman fall before making contact with the airbag?

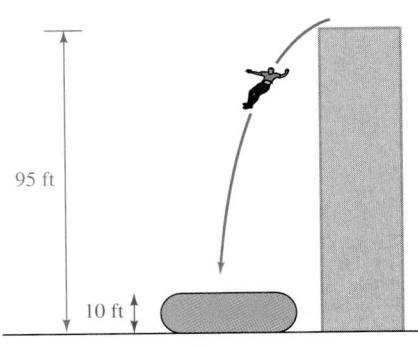

95 ft

10 ft

FIGURE 15-3

Solution   If we subtract the height of the airbag from the height of the building, we find that the stuntman will fall $95 - 10 = 85$ feet. We substitute 85 for s in the formula and find that the equation is in a form that allows us to use the square root method.

$$s = 16t^2 \qquad \text{The given formula.}$$

$$85 = 16t^2 \qquad \text{Substitute 85 for } s.$$

$$\frac{85}{16} = t^2 \qquad \text{Divide both sides by 16.}$$

$$\pm\sqrt{\frac{85}{16}} = t \qquad \text{Use the square root method to solve the equation.}$$

$$\pm\frac{\sqrt{85}}{\sqrt{16}} = t \qquad \text{The square root of a quotient is the quotient of the square roots.}$$

$$\pm\frac{\sqrt{85}}{4} = t \qquad \sqrt{16} = 4.$$

The stuntman will fall for $\frac{\sqrt{85}}{4}$ seconds before making contact with the airbag. To the nearest tenth, this is 2.3 seconds. We discard the other solution, $-\frac{\sqrt{85}}{4}$, because a negative time does not make sense in this context.   ∎

EXAMPLE 7   *Manufacturing.*   A manufacturer of television parts receives an order for 52-inch picture tubes, measured along the diagonal as shown in Figure 15-4. The tubes are to be rectangular in shape and 4 inches wider than they are high. Find the dimensions of each tube.

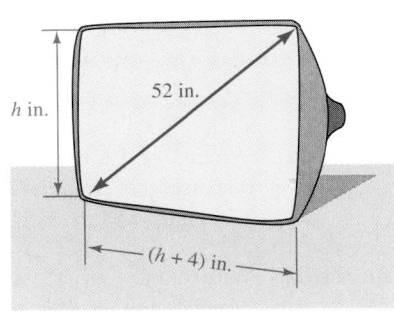

h in.

52 in.

$(h + 4)$ in.

FIGURE 15-4

Analyze the problem   We need to find the height and width of the rectangular picture tube. We note that two adjacent sides of the picture tube and a diagonal form a right triangle.

Form an equation   We can let h represent the height of the picture tube. Then $h + 4$ will represent the width. Since two adjacent sides and a diagonal of the tube form a right triangle, we can use the Pythagorean theorem to form the equation.

$$a^2 + b^2 = c^2 \qquad \text{The Pythagorean theorem.}$$

$$h^2 + (h + 4)^2 = 52^2 \qquad \text{Substitute } h \text{ for } a, (h + 4) \text{ for } b, \text{ and } 52 \text{ for } c.$$

$$h^2 + h^2 + 8h + 16 = 2{,}704 \qquad \text{Use the FOIL method: } (h + 4)^2 = h^2 + 8h + 16.$$

$$2h^2 + 8h - 2{,}688 = 0 \qquad \text{Subtract 2,704 from both sides and combine like terms.}$$

$$h^2 + 4h - 1{,}344 = 0 \qquad \text{Divide both sides by 2.}$$

Solve the equation   To solve $h^2 + 4h - 1{,}344 = 0$, we cannot use the square root method, and the factoring method looks difficult because of the cumbersome last term ($-1{,}344$). We will use the quadratic formula.

$$h = \frac{-b \pm \sqrt{b^2 - 4ac}}{2a} \qquad \text{The quadratic formula.}$$

$$= \frac{-4 \pm \sqrt{(4)^2 - 4(1)(-1{,}344)}}{2(1)} \qquad \text{Substitute 1 for } a, 4 \text{ for } b, \text{ and } -1{,}344 \text{ for } c.$$

$$= \frac{-4 \pm \sqrt{16 + 5{,}376}}{2} \qquad \text{Find the power and do the multiplication within the radical symbol.}$$

$$= \frac{-4 \pm \sqrt{5{,}392}}{2} \qquad \text{Do the addition within the radical symbol.}$$

$$\approx \frac{-4 \pm 73.430239}{2} \qquad \text{Use a calculator to approximate } \sqrt{5{,}392}.$$

$$h \approx \frac{-4 + 73.430239}{2} \quad \text{or} \quad h \approx \frac{-4 - 73.430239}{2} \qquad \begin{array}{l}\text{Use a calculator to}\\ \text{approximate each solution.}\end{array}$$

$$\approx \frac{69.430239}{2} \qquad\qquad \approx \frac{-77.430239}{2}$$

$$h \approx 34.7151195 \qquad\qquad h \approx -38.7151195$$

State the conclusion   The width of each tube will be approximately 34.7 inches, and the length will be approximately $34.7 + 4 = 38.7$ inches. We discard the second solution, because the diagonal measure of a TV picture tube cannot be negative.

Check the result   Check the solution by substituting 34.7, 38.7, and 52 into the Pythagorean theorem.   ■

EXAMPLE 8   *Finance.*   If \$$P$ is invested at an annual rate r, it will grow to an amount of \$$A$ in n years according to the formula $A = P(1 + r)^n$. What interest rate is needed to make a \$5,000 investment grow to \$5,618 after 2 years?

Solution   We can substitute 5,000 for P, 5,618 for A, and 2 for n in the formula and solve for r.

$$A = P(1 + r)^n$$

$$5{,}618 = 5{,}000(1 + r)^2$$

$$5{,}618 = 5{,}000(1 + 2r + r^2) \qquad \text{Find } (1 + r)^2.$$

$$5{,}618 = 5{,}000 + 10{,}000r + 5{,}000r^2 \qquad \text{Distribute the multiplication by 5,000.}$$

$$0 = 5{,}000r^2 + 10{,}000r - 618 \qquad \text{Subtract 5,618 from both sides.}$$

We can use a calculator and solve this equation by the quadratic formula, where $a = 5,000$, $b = 10,000$, and $c = -618$.

$$r = \frac{-b \pm \sqrt{b^2 - 4ac}}{2a}$$

$$= \frac{-10,000 \pm \sqrt{10,000^2 - 4(5,000)(-618)}}{2(5,000)}$$

$$= \frac{-10,000 \pm \sqrt{100,000,000 + 12,360,000}}{10,000}$$

$$= \frac{-10,000 \pm \sqrt{112,360,000}}{10,000}$$

$$= \frac{-10,000 \pm 10,600}{10,000}$$

$$r = \frac{-10,000 + 10,600}{10,000} \quad \text{or} \quad r = \frac{-10,000 - 10,600}{10,000}$$

$$= \frac{600}{10,000} \qquad\qquad = \frac{-20,600}{10,000}$$

$$= 0.06 \qquad\qquad\qquad = -2.06$$

$$r = 6\% \qquad\qquad\qquad r = -206\%$$

The required rate is 6%. The rate of -206% has no meaning in this problem. ∎

STUDY SET Section 15.2

VOCABULARY *Fill in the blanks.*

1. The general _____ equation is $ax^2 + bx + c = 0$.

2. The formula
$$x = \frac{-b \pm \sqrt{b^2 - 4ac}}{2a}$$
is called the _____ formula.

3. To _____ a quadratic equation means to find all the values of the variable that make the equation true.

4. $\sqrt{-16}$ is not a _____ number.

CONCEPTS *In Exercises 5–10, fill in the blanks.*

5. In the quadratic equation $ax^2 + bx + c = 0$, a cannot equal ▢.

6. Before we can determine a, b, and c for $x = 3x^2 - 1$, we must write the equation in _____ form.

7. In the quadratic equation $3x^2 - 5 = 0$, $a = $ ▢, $b = $ ▢, and $c = $ ▢.

8. In the quadratic equation $-4x^2 + 8x = 0$, $a = $ ▢, $b = $ ▢, and $c = $ ▢.

9. The formula for the area of a rectangle is $A = $ ▢, and the formula for the area of a triangle is $A = $ ▢.

10. If a, b, and c are three sides of a right triangle and c is the hypotenuse, then $c^2 = $ ▢.

11. In evaluating the numerator of
$$\frac{-5 \pm \sqrt{5^2 - 4(2)(1)}}{2(2)}$$
what operation should be performed first?

12. Consider the expression
$$\frac{3 \pm 6\sqrt{2}}{3}$$
a. How many terms does the numerator contain?
b. What common factor do the terms have?
c. Simplify the expression.

13. A student used the quadratic formula to solve an equation and obtained
$$x = \frac{-3 \pm \sqrt{15}}{2}$$
a. How many solutions does the equation have?

b. What are they *exactly?*

c. Approximate them to the nearest hundredth.

14. Write the following steps of the strategy for solving quadratic equations in the proper order.

- Use the quadratic formula.
- Write the equation in $ax^2 + bx + c = 0$ form.
- Use the factoring method.
- Use the square root method.

15. The solutions of a quadratic equation are
$$x = 2 \pm \sqrt{3}$$

Graph them on a number line.

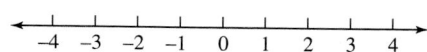

16. The solutions of a quadratic equation are
$$x = \frac{-1 \pm \sqrt{5}}{2}$$

Graph them on a number line.

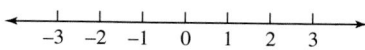

NOTATION *Complete each solution.*

17. Solve $x^2 - 5x - 6 = 0$.

$$x = \frac{-b \pm \sqrt{b^2 - 4ac}}{2a}$$

$$= \frac{-(\quad) \pm \sqrt{(-5)^2 - 4(1)(-6)}}{2(1)}$$

$$= \frac{\pm \sqrt{25 + }}{2}$$

$$= \frac{5 \pm \sqrt{}}{2}$$

$$x = \frac{\pm 7}{2}$$

$$x = \frac{5 \quad 7}{2} = 6 \quad \text{or} \quad x = \frac{5 \quad 7}{2} = -1$$

18. Solve $3x^2 + 2x - 2 = 0$.

$$x = \frac{-b \pm \sqrt{b^2 - 4ac}}{2a}$$

$$= \frac{-2 \pm \sqrt{^2 - 4(3)(\quad)}}{2(\quad)}$$

$$= \frac{-2 \pm \sqrt{4 \quad 24}}{6}$$

$$= \frac{-2 \pm \sqrt{}}{6}$$

$$= \frac{-2 \pm \sqrt{7}}{6}$$

$$= \frac{(-1 \pm \sqrt{7})}{ \cdot 3}$$

$$x = \frac{-1 \pm \sqrt{7}}{3}$$

19. What is wrong with this student's work?

Solve $x^2 + 4x - 5 = 0$.

$$x = -4 \pm \frac{\sqrt{16 - 4(1)(-5)}}{2}$$

20. In reading
$$\frac{-b \pm \sqrt{b^2 - 4ac}}{2a}$$

we say, "the _____ of b, plus or _____ the _____ root of b _____ minus 4 _____ a times c, all _____ $2a$."

PRACTICE *Change each equation into quadratic form, if necessary, and find the values of a, b, and c. Do not solve the equation.*

21. $x^2 + 4x + 3 = 0$

22. $x^2 - x - 4 = 0$

23. $3x^2 - 2x + 7 = 0$

24. $4x^2 + 7x - 3 = 0$

25. $4y^2 = 2y - 1$

26. $2x = 3x^2 + 4$

27. $x(3x - 5) = 2$

28. $y(5y + 10) = 8$

29. $7(x^2 + 3) = -14x$

30. $(2a + 3)(a - 2) = (a + 1)(a - 1)$

Use the quadratic formula to find all real solutions.

31. $x^2 - 5x + 6 = 0$

32. $x^2 + 5x + 4 = 0$

33. $x^2 + 7x + 12 = 0$

34. $x^2 - x - 12 = 0$

35. $2x^2 - x - 1 = 0$

36. $2x^2 + 3x - 2 = 0$

37. $3x^2 + 5x + 2 = 0$

38. $3x^2 - 4x + 1 = 0$

39. $4x^2 + 4x - 3 = 0$

40. $4x^2 + 3x - 1 = 0$

41. $x^2 + 3x + 1 = 0$

42. $x^2 + 3x - 2 = 0$

43. $3x^2 - x = 3$

44. $5x^2 = 3x + 1$

45. $x^2 + 5 = 2x$

46. $2x^2 + 3x = -3$

47. $x^2 = 1 - 2x$

48. $x^2 = 4 + 2x$

49. $3x^2 = 6x + 2$

50. $3x^2 = -8x - 2$

Use the most convenient method to find all real solutions. If a solution contains a radical, give the exact solution and then approximate it to the nearest hundredth.

51. $(2y - 1)^2 = 25$

52. $m^2 + 14m + 49 = 0$

53. $2x^2 + x = 5$

54. $2x^2 - x + 2 = 0$

55. $x^2 - 2x - 1 = 0$

56. $b^2 = 18$

57. $x^2 - 2x - 35 = 0$

58. $x^2 + 5x + 3 = 0$

59. $x^2 + 2x + 7 = 0$

60. $3x^2 - x = 1$

61. $4c^2 + 16c = 0$

62. $t^2 - 1 = 0$

63. $18 = 3y^2$

64. $25x - 50x^2 = 0$

🔢 *Solve each equation. Round each solution to the nearest tenth.*

65. $2.4x^2 - 9.5x + 6.2 = 0$

66. $-1.7x^2 + 0.5x + 0.9 = 0$

APPLICATIONS

67. HEIGHT OF A TRIANGLE   The triangle shown in Illustration 1 has an area of 30 square inches. Find its height.

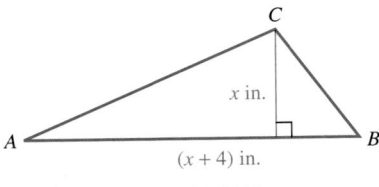

ILLUSTRATION 1

68. BOWLING   When the pins for a children's bowling game are set up, they occupy 418 cm² of floor space. See Illustration 2. If the base of the triangular-shaped region is 6 cm longer than twice the height, how wide is the last row of pins?

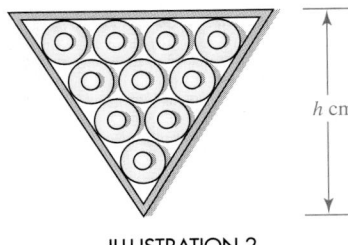

ILLUSTRATION 2

69. FLAG   According to the *Guinness Book of World Records 1998*, the largest flag flown from a flagpole was a Brazilian national flag, a rectangle having an area of 3,102 ft². If the flag is 19 feet longer than it is wide, find its width and length.

70. COMICS   See Illustration 3. A three-panel comic strip occupies 96 square centimeters of space in a newspaper. The length of the rectangular space is 4 centimeters more than twice its width. Find its dimensions.

ILLUSTRATION 3

71. COMMUNITY GARDEN   See Illustration 4. Residents of a community can work their own 16 ft × 24 ft plot of city-owned land if they agree to the following stipulations:

- The area of the garden cannot exceed 180 square feet.

- A path of uniform width must be maintained around the garden.

Find the dimensions of the largest possible garden.

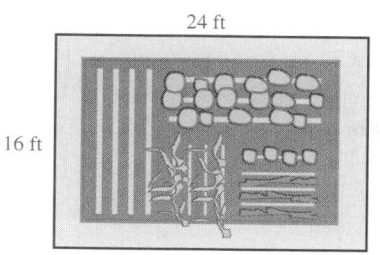

ILLUSTRATION 4

72. DECKING   The owner of the pool in Illustration 5 wants to surround it with a concrete deck of uniform width (shown in gray). If he can afford 368 square feet of decking, how wide can he make the deck?

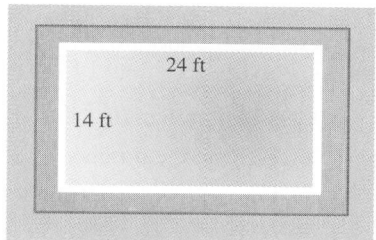

ILLUSTRATION 5

73. DAREDEVIL   In 1873, Henry Bellini combined a tightrope walk over the Niagara River with a leap into the churning river below, where he was picked up by a boat. If the rope was 200 feet above the water, for how many seconds did he fall before hitting the water? Round to the nearest tenth.

74. FALLING OBJECT   A tourist drops a penny from the observation deck of the World Trade Center, 1,377 feet above the ground. How long will it take for the penny to hit the ground?

75. ABACUS   The Chinese abacus shown in Illustration 6 consists of a frame, parallel wires, and beads that are moved to perform arithmetic computations. The frame is 21 centimeters wider than it is high. Find its dimensions.

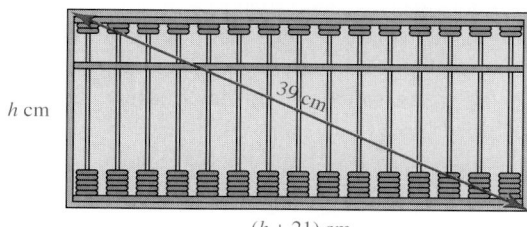

ILLUSTRATION 6

76. INSTALLING A SIDEWALK   A 170-meter-long sidewalk from the mathematics building M to the student center C is shown in red in Illustration 7. However, students prefer to walk directly from M to C. How long are the two segments of the existing sidewalk?

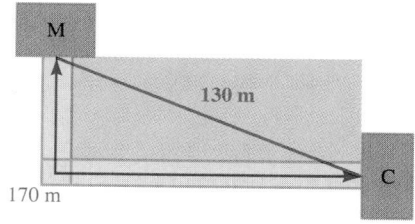

ILLUSTRATION 7

77. NAVIGATION   Two boats leave port at the same time, one sailing east and one sailing south. If one boat sails 10 nautical miles more than the other and they are then 50 nautical miles apart, how far does each boat sail?

78. NAVIGATION   One plane heads west from an airport, flying at 200 mph. One hour later, a second plane heads north from the same airport, flying at the same speed. When will the planes be 1,000 miles apart?

79. INVESTING   We can use the formula $A = P(1 + r)^2$ to find the amount $A that $P will become when invested at an annual rate of $r\%$ for 2 years. What interest rate is needed to make $5,000 grow to $5,724.50 in 2 years?

80. INVESTING   What interest rate is needed to make $7,000 grow to $8,470 in 2 years? See Exercise 79.

81. MANUFACTURING   An electronics firm has found that its revenue for manufacturing and selling x television sets is given by the formula $R = -\frac{1}{6}x^2 + 450x$. How much revenue will be earned by manufacturing 600 television sets? (*Hint:* Multiply both sides of the equation by -6.)

82. RETAILING   When a wholesaler sells n CD players, his revenue R is given by the formula $R = 150n - \frac{1}{2}n^2$. How many players would he have to sell to receive $11,250? (*Hint:* Multiply both sides of the equation by -2.)

83. METAL FABRICATION   A square piece of tin, 12 inches on a side, is to have four equal squares cut from its corners, as shown in Illustration 8 on the next page. If the edges are then to be folded up to make a box with a floor area of 64 square inches, find the depth of the box.

84. MAKING GUTTERS   A piece of sheet metal, 18 inches wide, is bent to form the gutter shown in Illustration 9 on the next page. If the cross-sectional area is 36 square inches, find the depth of the gutter.

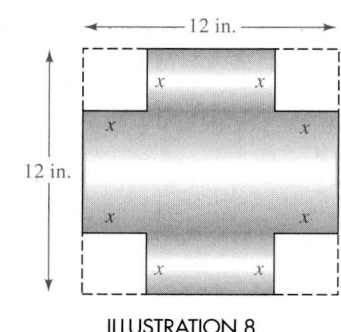

ILLUSTRATION 8

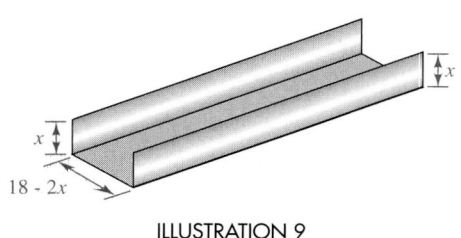

ILLUSTRATION 9

WRITING

85. Do you agree or disagree with the following statement? Explain your answer.

   The quadratic formula is the easiest method to use to solve quadratic equations.

86. Explain the meaning of the $\pm$ symbol.

87. Use the quadratic formula to solve $x^2 - 2x - 4 = 0$. What is an exact solution, and what is an approximate solution of this equation? Explain the difference.

88. Rewrite in words:

$$x = \frac{-b \pm \sqrt{b^2 - 4ac}}{2a}$$

REVIEW   *Solve each equation for the indicated variable.*

89. $A = p + prt$; for r

90. $F = \dfrac{GMm}{d^2}$, for M

Write the equation of the line that has the given properties in general form.

91. Slope of $\frac{3}{5}$ and passing through $(0, 12)$

92. Passes through $(6, 8)$ and the origin

Simplify each expression.

93. $\sqrt{80}$ **94.** $2\sqrt{x^3 y^2}$

Rationalize each denominator and simplify.

95. $\dfrac{x}{\sqrt{7x}}$ **96.** $\dfrac{\sqrt{x} + 2}{\sqrt{x} - 2}$

15.3 *Graphing Quadratic Functions*

In this section, you will learn about

- Quadratic functions • Finding the vertex and the intercepts of a parabola
- A strategy for graphing quadratic functions • Finding a maximum value

INTRODUCTION. In this section, we consider a special type of function called a *quadratic function*. When graphing functions in Chapter 9, we constructed a table of values and plotted points. In this section, we will develop a more general strategy for graphing quadratic functions by analyzing the given function and determining the important characteristics of its graph.

Quadratic functions

Quadratic functions are defined by equations of the form $y = ax^2 + bx + c \ (a \neq 0)$, where the right-hand side is a second-degree polynomial in the variable x. Three examples of quadratic functions are

$$y = x^2 - 3 \qquad y = x^2 - 2x - 3 \qquad y = -2x^2 - 4x + 2$$

We can replace y with the function notation $f(x)$ to express the defining equation in the form $f(x) = ax^2 + bx + c$. For the functions just mentioned, we can write

$$f(x) = x^2 - 3 \qquad f(x) = x^2 - 2x - 3 \qquad f(x) = -2x^2 - 4x + 2$$

We have previously constructed the graph of $y = x^2$ (a quadratic function) by plotting points. The result was the **parabola** shown in Figure 15-5.

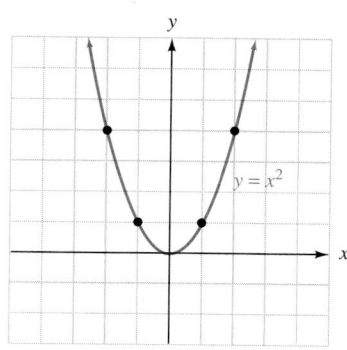

FIGURE 15-5

EXAMPLE 1 *Graphing quadratic functions.* Graph $y = x^2 - 3$. Compare the graph to that of $y = x^2$.

Solution

The function is written in $y = ax^2 + bx + c$ form, where $a = 1$, $b = 0$, and $c = -3$. To find ordered pairs (x, y) that satisfy the equation, we pick several numbers x and find the corresponding values of y. If we let $x = 3$, we have

$$y = x^2 - 3$$
$$= 3^2 - 3 \quad \text{Substitute 3 for } x.$$
$$= 6$$

The ordered pair $(3, 6)$ and six others satisfying the equation appear in the table shown in Figure 15-6. To graph the equation, we plot each point and draw a smooth curve passing through them. The resulting parabola is the graph of $y = x^2 - 3$. The parabola opens upward, and the lowest point on the graph, called the **vertex of the parabola,** is the point $(0, -3)$.

Note that the graph of $y = x^2 - 3$ looks just like the graph of $y = x^2$, except that it is 3 units lower.

Self Check

Graph $y = x^2 + 2$. Compare the graph to that of $y = x^2$.

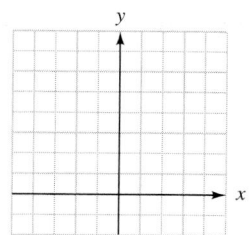

Answer: The graph has the same shape as the graph of $y = x^2$, but it is 2 units higher.

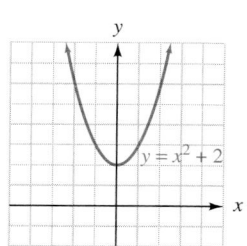

	$y = x^2 - 3$	
x	y	(x, y)
3	6	$(3, 6)$
2	1	$(2, 1)$
1	-2	$(1, -2)$
0	-3	$(0, -3)$
-1	-2	$(-1, -2)$
-2	1	$(-2, 1)$
-3	6	$(-3, 6)$

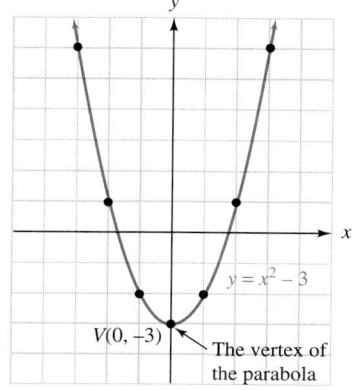

FIGURE 15-6

If we draw a vertical line through the vertex of a parabola and fold the graph on this line, the two sides of the graph will match. We call the vertical line the **axis of symmetry.**

EXAMPLE 2   *Graphing quadratic functions.*   Graph
$f(x) = -2x^2 - 4x + 2$, find its vertex, and draw its axis of symmetry.

Solution
The function is written in $f(x) = ax^2 + bx + c$ form, where $a = -2$, $b = -4$, and $c = 2$. We construct the table shown in Figure 15-7, plot the points, and draw the graph.

$f(x) = -2x^2 - 4x + 2$

x	$f(x)$	$(x, f(x))$
-3	-4	$(-3, -4)$
-2	2	$(-2, 2)$
-1	4	$(-1, 4)$
0	2	$(0, 2)$
1	-4	$(1, -4)$

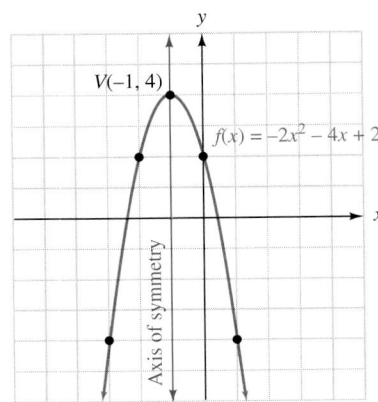

FIGURE 15-7

The parabola opens downward, so its vertex is its highest point, the point $(-1, 4)$.

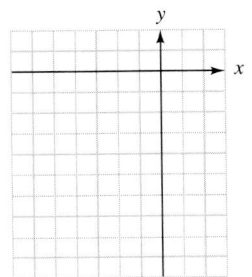
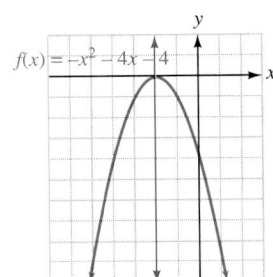
In Example 1, the coefficient of the x^2 term in $y = x^2 - 3$ is positive ($a = 1$). In Example 2, the coefficient of the x^2 term in $f(x) = -2x^2 - 4x + 2$ is negative ($a = -2$). The results of these first two examples illustrate the following fact.

Graphs of quadratic functions | The graph of the function $y = ax^2 + bx + c$ or $f(x) = ax^2 + bx + c$, where $a \neq 0$, is a parabola. It opens upward when $a > 0$ and downward when $a < 0$.

The cup-like shape of a parabola can be seen in a wide variety of real-world settings. Some examples are shown in Figure 15-8 (below and on the next page).

The path of a thrown object

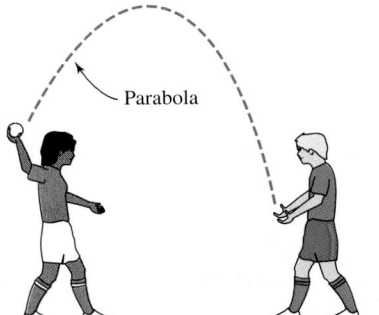

Parabola

The pursuit path of a shark seeking its prey

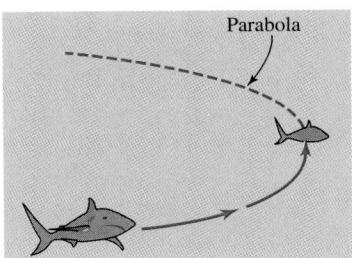

Parabola

FIGURE 15-8

The shape of a satellite antenna dish   **The path of a stream of water**

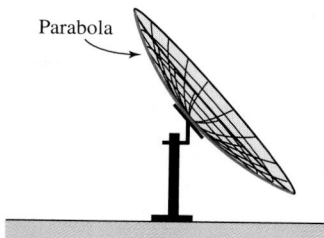
Parabola

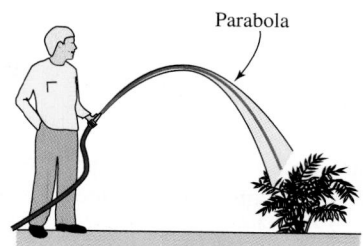
Parabola

FIGURE 15-8 (continued)

Finding the vertex and the intercepts of a parabola

It is easier to graph a quadratic function when we know the coordinates of the vertex of its parabolic graph. For a parabola defined by $y = ax^2 + bx + c$ or $f(x) = ax^2 + bx + c$, it can be shown that the x-coordinate of the vertex is given by $-\frac{b}{2a}$. This fact enables us to find the coordinates of its vertex.

Finding the vertex of a parabola

> The graph of the quadratic function $y = ax^2 + bx + c$ or $f(x) = ax^2 + bx + c$ is a parabola whose vertex has an x-coordinate of $-\frac{b}{2a}$. To find the y-coordinate of the vertex, substitute $-\frac{b}{2a}$ into the defining equation and find y.

EXAMPLE 3 *Finding the vertex of a parabola.* Find the vertex of the parabola defined by $y = x^2 - 2x - 3$.

Solution

For $y = x^2 - 2x - 3$, we have $a = 1$, $b = -2$, and $c = -3$. To find the x-coordinate of the vertex, we substitute the values for a and b into the formula $x = -\frac{b}{2a}$.

$$x = -\frac{b}{2a}$$

$$x = -\frac{-2}{2(1)}$$

$$= 1$$

The x-coordinate of the vertex is $x = 1$. To find the y-coordinate, we substitute 1 for x:

$$y = x^2 - 2x - 3$$
$$y = 1^2 - 2(1) - 3$$
$$= 1 - 2 - 3$$
$$= -4$$

The vertex of the parabola is the point $(1, -4)$.

Self Check

Find the vertex of the parabola defined by $y = -x^2 + 6x - 8$.

Answer: $(3, 1)$ ■

When graphing quadratic functions, it is often helpful to find the x- and y-intercepts of the parabola.

EXAMPLE 4 *Finding the intercepts of a parabola.* Find the x- and y-intercepts of the parabola defined by $y = x^2 - 2x - 3$.

Self Check

Find the x- and y-intercepts of the parabola defined by $y = -x^2 + 6x - 8$.

Solution

To find the *y*-intercept of the parabola, we let $x = 0$ and solve for *y*.

$$y = x^2 - 2x - 3$$
$$y = 0^2 - 2(0) - 3$$
$$y = -3$$

The parabola passes through the point $(0, -3)$. We note that the *y*-coordinate of the *y*-intercept is the same as the value of the constant term *c* on the right-hand side of $y = x^2 - 2x - 3$.

To find the *x*-intercepts of the graph, we set *y* equal to 0 and solve the resulting quadratic equation.

$$y = x^2 - 2x - 3$$
$$0 = x^2 - 2x - 3 \quad \text{Substitute 0 for } y.$$
$$0 = (x - 3)(x + 1) \quad \text{Factor the trinomial.}$$
$$x - 3 = 0 \quad \text{or} \quad x + 1 = 0 \quad \text{Set each factor equal to 0.}$$
$$x = 3 \qquad\qquad x = -1$$

Since there are two solutions, the graph has two *x*-intercepts: $(3, 0)$ and $(-1, 0)$.

Answers: *y*-intercept: $(0, -8)$; *x*-intercepts: $(2, 0)$, $(4, 0)$ ∎

A strategy for graphing quadratic functions

We can use the characteristics of a parabola to draw its graph. For example, to graph $y = x^2 - 2x - 3$, we note that the coefficient of the x^2 term is positive ($a = 1$). Therefore, the parabola defined by this function opens upward. In Examples 3 and 4, we found that the vertex of the graph of $y = x^2 - 2x - 3$ is at $(1, -4)$ and that the graph has a *y*-intercept of $(0, -3)$ and *x*-intercepts of $(3, 0)$ and $(-1, 0)$. See Figure 15-9(a).

We can locate other points on the parabola by noting that the graph has the axis of symmetry shown in Figure 15-9(a). If the point $(0, -3)$, which is 1 unit to the left of the axis of symmetry, is on the graph, the point $(2, -3)$, which is 1 unit to the right of the axis of symmetry, is also on the graph.

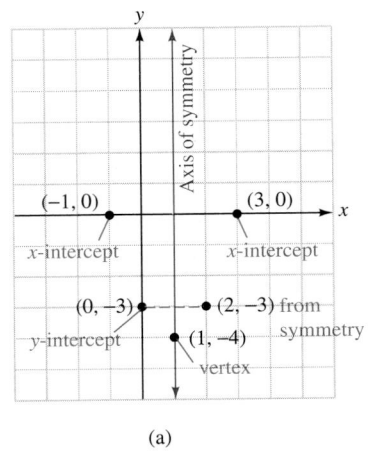

(a)

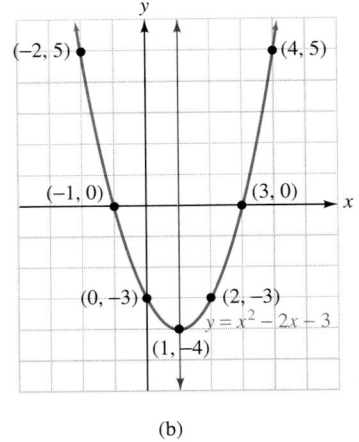

(b)

FIGURE 15-9

$$y = x^2 - 2x - 3$$

x	y	(x, y)
-2	5	(-2, 5)

We can complete the graph by plotting two more points. If $x = -2$, then $y = 5$, and the parabola passes through $(-2, 5)$. Again using symmetry, the parabola must also pass through $(4, 5)$. The completed graph of $y = x^2 - 2x - 3$ is shown in Figure 15-9(b).

Much can be determined about the graph of $y = ax^2 + bx + c$ from the coefficients a, b, and c. This information is summarized below.

Graphing a quadratic function
$$y = ax^2 + bx + c$$

> Determine whether the parabola opens upward or downward by examining a.
>
> The x-coordinate of the vertex of the parabola is $x = -\frac{b}{2a}$.
>
> To find the y-coordinate of the vertex, substitute $-\frac{b}{2a}$ for x into the equation and find y.
>
> The axis of symmetry is the vertical line passing through the vertex.
>
> The y-intercept $(0, y)$ is determined by the value of y when $x = 0$: the y-intercept is $(0, c)$.
>
> The x-intercepts (if any) are determined by the numbers x that make $y = 0$. To find them, solve the quadratic equation $ax^2 + bx + c = 0$.

EXAMPLE 5   *Graphing quadratic functions.*   Graph $f(x) = -2x^2 - 8x - 8$.

Solution

Step 1 *Determine whether the parabola opens upward or downward.* The equation is in the form $f(x) = ax^2 + bx + c$, with $a = -2$, $b = -8$, and $c = -8$. Since $a < 0$, the parabola opens downward.

Step 2 *Find the vertex and draw the axis of symmetry.* To find the x-coordinate of the vertex, we substitute the values for a and b into the formula $x = -\frac{b}{2a}$.

$$x = -\frac{b}{2a}$$
$$x = -\frac{-8}{2(-2)}$$
$$= -2$$

The x-coordinate of the vertex is -2. To find the y-coordinate, we substitute -2 for x in the equation and find $f(-2)$.

$$f(x) = -2x^2 - 8x - 8$$
$$f(-2) = -2(-2)^2 - 8(-2) - 8$$
$$= -8 + 16 - 8$$
$$= 0 \qquad \text{If } f(-2) = 0, \text{ then } y = 0 \text{ for } x = -2.$$

The vertex of the parabola is the point $(-2, 0)$. This point is the blue dot in Figure 15-10 on the next page.

Step 3 *Find the x- and y-intercepts.* Since $c = -8$, the y-intercept of the parabola is $(0, -8)$. The point $(-4, -8)$, two units to the left of the axis of symmetry, must also be on the graph. We plot both points in black in Figure 15-10.

To find the x-intercepts, we set $f(x)$ equal to 0 and solve the resulting quadratic equation.

$$f(x) = -2x^2 - 8x - 8$$
$$0 = -2x^2 - 8x - 8 \qquad \text{Set } f(x) = 0.$$
$$0 = x^2 + 4x + 4 \qquad \text{Divide both sides by } -2.$$
$$= (x + 2)(x + 2) \qquad \text{Factor the trinomial.}$$
$$x + 2 = 0 \quad \text{or} \quad x + 2 = 0 \qquad \text{Set each factor equal to 0.}$$
$$x = -2 \qquad\qquad x = -2$$

Since the solutions are the same, the graph has only one x-intercept: $(-2, 0)$. This point is the vertex of the parabola and has already been plotted.

Self Check

Graph the function
$$y = -x^2 + 6x - 8.$$

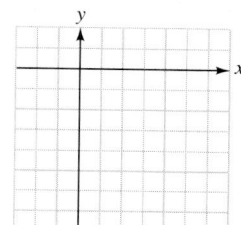

Step 4 *Plot another point.* Finally, we find another point on the parabola. If $x = -3$, then $y = -2$. We plot $(-3, -2)$ in Figure 15-10 and use symmetry to determine that $(-1, -2)$ is also on the graph. Both points are in green.

Step 5 Draw a smooth curve through the points, as shown in Figure 15-10.

$$f(x) = -2x^2 - 8x - 8$$

x	$f(x)$	$(x, f(x))$
-3	-2	$(-3, -2)$

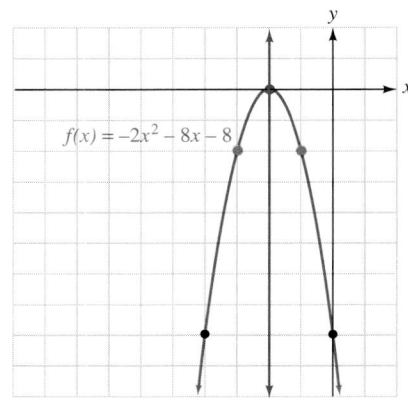

FIGURE 15-10

Answer:

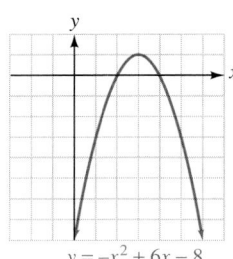

$y = -x^2 + 6x - 8$

 COMMENT The number of x-intercepts of the graph of a quadratic function $y = ax^2 + bx + c$ is the same as the number of solutions of $ax^2 + bx + c = 0$. For example, the graph of $y = x^2 + x - 2$ in Figure 15-11(a) has two x-intercepts, and $x^2 + x - 2 = 0$ has two real-number solutions. In Figure 15-11(b), the graph has one x-intercept, and the corresponding equation has one real-number solution. In Figure 15-11(c), the graph does not have an x-intercept, and the corresponding equation does not have any real-number solutions. Note that the solutions of each equation are given by the x-coordinates of the x-intercepts of each respective graph.

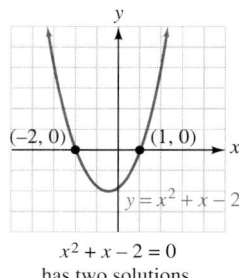

$x^2 + x - 2 = 0$
has two solutions,
$x = -2$ and $x = 1$.

(a)

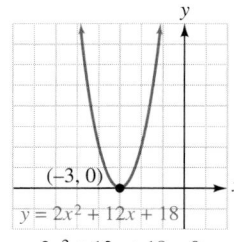

$2x^2 + 12x + 18 = 0$
has one solution,
$x = -3$.

(b)

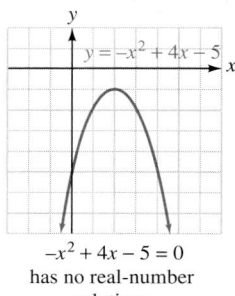

$-x^2 + 4x - 5 = 0$
has no real-number
solutions.

(c)

FIGURE 15-11

Finding a maximum value

  EXAMPLE 6 *Finding maximum revenue.* An electronics firm manufactures radios. Over the past 10 years, the firm has learned that it can sell x radios at a price of $\left(200 - \frac{1}{5}x\right)$ dollars. How many radios should the firm manufacture and sell to maximize its revenue? Find the maximum revenue.

Solution The revenue obtained is the product of the number of radios sold (x) and the price of each radio $\left(200 - \frac{1}{5}x\right)$. Thus, the revenue R is given by the function

$$R = x\left(200 - \frac{1}{5}x\right) \qquad \text{or} \qquad R = -\frac{1}{5}x^2 + 200x$$

Since the graph of this function is a parabola that opens downward, the *maximum* value of R will be the value of R determined by the vertex of the parabola. Because the x-coordinate of the vertex is at $x = -\frac{b}{2a}$, we have

$$x = -\frac{b}{2a}$$

$$= -\frac{200}{2\left(-\frac{1}{5}\right)} \qquad \text{Substitute 200 for } b \text{ and } -\frac{1}{5} \text{ for } a.$$

$$= -\frac{200}{-\frac{2}{5}} \qquad \text{Do the multiplication in the denominator.}$$

$$= (-200)\left(-\frac{5}{2}\right) \qquad \text{Division by } -\frac{2}{5} \text{ is the same as multiplication by its reciprocal, which is } -\frac{5}{2}.$$

$$= 500$$

If the firm manufactures 500 radios, the maximum revenue will be

$$R = -\frac{1}{5}x^2 + 200x \qquad \text{The revenue formula.}$$

$$= -\frac{1}{5}(500)^2 + 200(500) \qquad \text{Substitute 500 for } x, \text{ the number of radios.}$$

$$= 50,000$$

The firm should manufacture 500 radios to get a maximum revenue of $50,000. This fact is verified by examining the graph of $R = -\frac{1}{5}x^2 + 200x$, which appears in Figure 15-12.

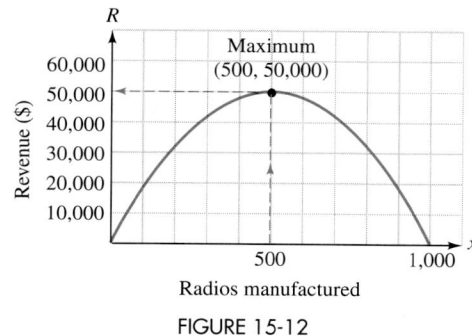

FIGURE 15-12

STUDY SET Section 15.3

VOCABULARY *Fill in the blanks.*

1. A function defined by the equation $y = ax^2 + bx + c \, (a \neq 0)$ is called a _____ function.

2. The lowest (or highest) point on a parabola is called the _____ of the parabola.

3. The point where a parabola intersects the y-axis is called the _____.

4. The point (or points) where a parabola intersects the _____ is (are) called the x-intercept(s).

5. For a parabola that opens upward or downward, the vertical line that passes through its vertex and splits the graph into two identical pieces is called the axis of _____.

6. For the graph of $y = ax^2 + bx + c$, the _____ of the x^2 term indicates whether the parabola opens upward or downward.

CONCEPTS *In Exercises 7–10, fill in the blanks.*

7. The graph of $y = ax^2 + bx + c$, where $a \neq 0$, opens upward when a ____ 0.

8. The graph of $f(x) = ax^2 + bx + c$, where $a \neq 0$, opens downward when a ____ 0.

9. The y-intercept of the graph of $f(x) = ax^2 + bx + c$ is the point ____ .

10. The x-coordinate of the vertex of the parabola that results when we graph $y = ax^2 + bx + c$ is $x =$ ____ .

11. Refer to the graph in Illustration 1.
   a. What do we call the curve shown there?
   b. What are the x-intercepts of the graph?
   c. What is the y-intercept of the graph?
   d. What is the vertex?

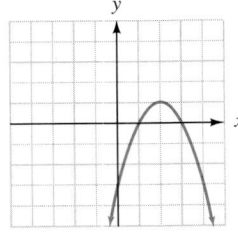

ILLUSTRATION 1

12. The vertex of a parabola is at $(1, -3)$, its y-intercept is $(0, -2)$, and it passes through the point $(3, 1)$, as shown in Illustration 2. Draw the axis of symmetry and use it to help determine two other points on the parabola.

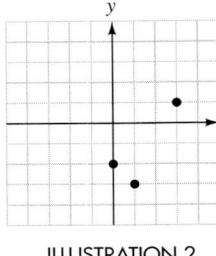

ILLUSTRATION 2

13. Sketch the graphs of parabolas with zero, one, and two x-intercepts.

14. Sketch the graph of a parabola that doesn't have a y-intercept, if possible.

15. HEALTH DEPARTMENT   The number of cases of flu seen by doctors at a county health clinic each week during a 10-week period is described by the quadratic function graphed in Illustration 3. Write a brief summary report about the flu outbreak. What important piece of information does the vertex give?

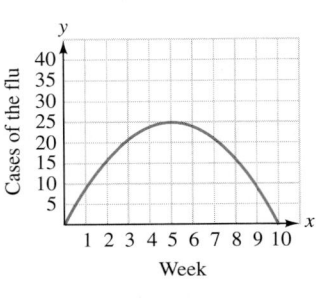

ILLUSTRATION 3

16. COST ANALYSIS   A company has found that when it assembles x carburetors in a production run, the manufacturing cost $\$y$ per carburetor is given by the quad-

ratic function graphed in Illustration 4. What important piece of information does the vertex give?

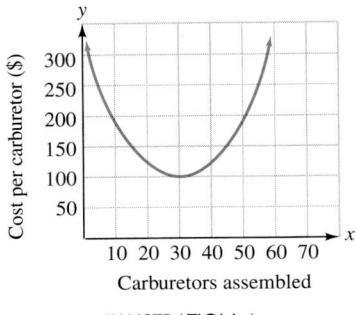

ILLUSTRATION 4

NOTATION

17. Tell whether this statement is true or false: The equations $y = 2x^2 - x - 2$ and $f(x) = 2x^2 - x - 2$ are the same.

18. The function $y = -x^2 + 3x - 5$ is written in $y = ax^2 + bx + c$ form. What are a, b, and c?

19. Consider $y = 3x^2 + 3x - 8$. What is $-\dfrac{b}{2a}$?

20. Evaluate $\dfrac{-12}{2(-3)}$.

PRACTICE   *Graph each quadratic function and compare the graph to the graph of $y = x^2$.*

21. $y = x^2 + 1$

22. $y = x^2 - 4$

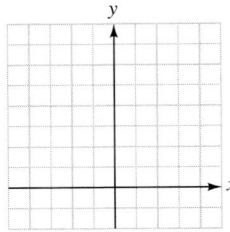

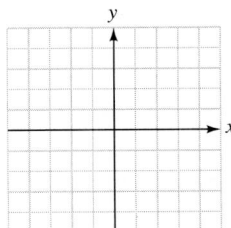

23. $f(x) = -x^2$

24. $f(x) = (x - 1)^2$

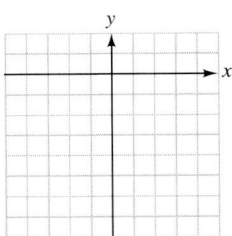

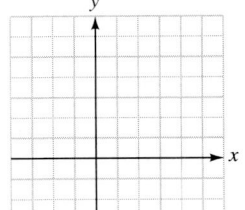

Find the vertex of the graph of each quadratic function.

25. $y = -x^2 + 6x - 8$

26. $y = -x^2 - 2x - 1$

27. $f(x) = 2x^2 - 4x + 1$

28. $f(x) = 2x^2 + 8x - 4$

Find the x- and y-intercepts of the graph of the quadratic function.

29. $f(x) = x^2 - 2x + 1$

30. $f(x) = 2x^2 - 4x$

31. $y = -x^2 - 10x - 21$

32. $y = 3x^2 + 6x - 9$

Graph each quadratic function. Use the method discussed in Example 5.

33. $y = x^2 - 2x$
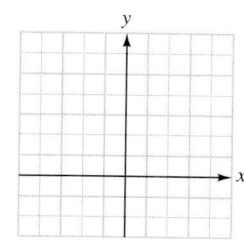

34. $f(x) = -x^2 - 4x$
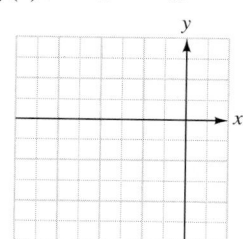

35. $f(x) = -x^2 + 2x$
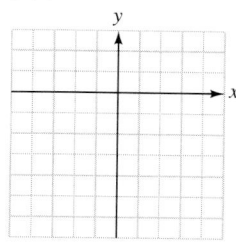

36. $y = x^2 + x$
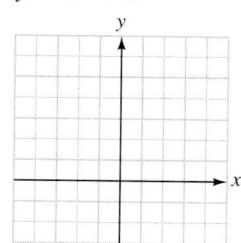

37. $f(x) = x^2 + 4x + 4$
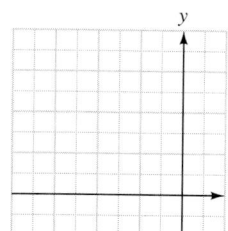

38. $f(x) = x^2 - 6x + 9$
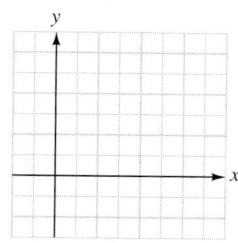

39. $y = -x^2 - 2x - 1$
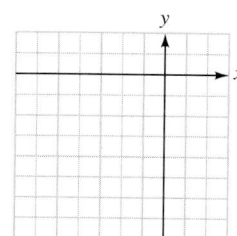

40. $y = -x^2 + 2x - 1$
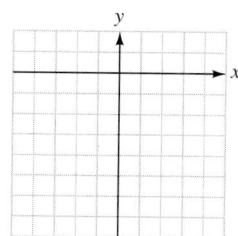

41. $y = x^2 + 2x - 3$
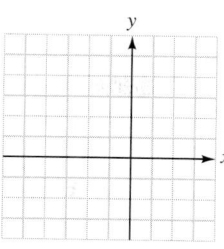

42. $y = x^2 + 6x + 5$
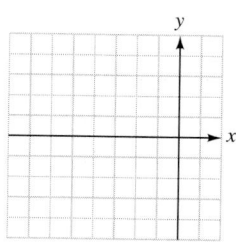

43. $f(x) = 2x^2 + 8x + 6$
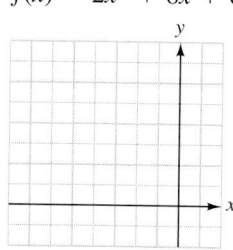

44. $f(x) = 3x^2 - 12x + 9$
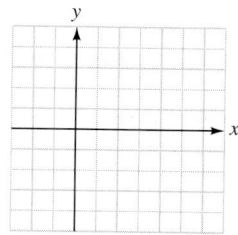

45. $y = x^2 - 2x - 8$
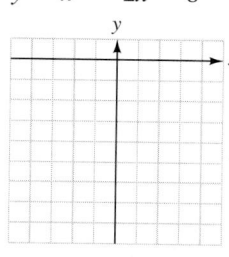

46. $y = -x^2 + 2x + 3$
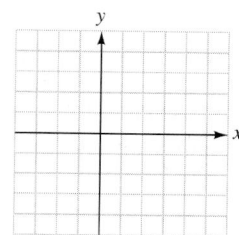

47. $y = x^2 - x - 2$
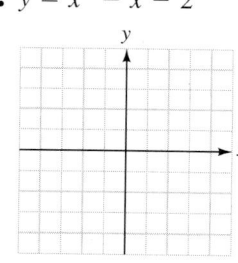

48. $y = -x^2 + 5x - 4$
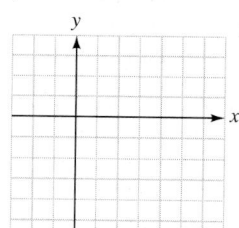

49. $f(x) = 2x^2 + 3x - 2$
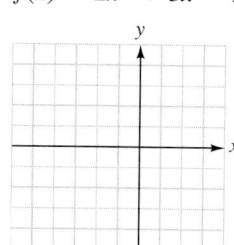

50. $f(x) = 3x^2 - 7x + 2$
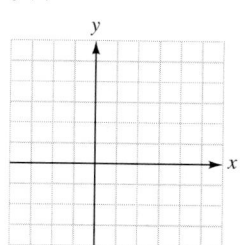

APPLICATIONS

51. TRAMPOLINE Illustration 5 shows how far a trampolinist is from the ground (in relation to time) as she bounds into the air and then falls back down to the trampoline.

 a. How many feet above the ground is she $\frac{1}{2}$ second after bounding upward?

b. When is she 9 feet above the ground?

c. What is the maximum number of feet above the ground she gets? When does this occur?

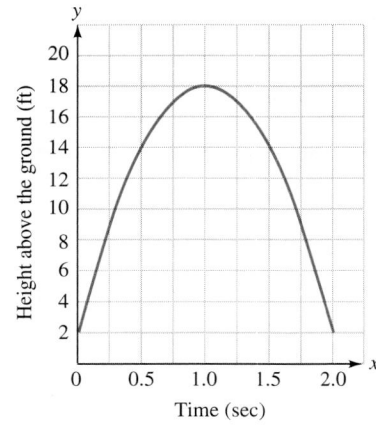

ILLUSTRATION 5

52. PROJECTILE   If we disregard air resistance and other outside factors, the path of a projectile, such as a kicked soccer ball, is parabolic. Suppose the path of the soccer ball after it is kicked is given by the quadratic function $y = -0.5x^2 + 2x$.
Use a calculator to complete the table of values in Illustration 6, and then plot the points and draw a smooth curve through them to depict the ball's path.

ILLUSTRATION 6

x	0	0.5	1	1.5	2	2.5	3	3.5	4
y									

53. BRIDGE   The shapes of the suspension cables in certain types of bridges are parabolic. The suspension cable for the bridge shown in Illustration 7 is described by $y = 0.005x^2$. Finish the mathematical model of the bridge by completing the table of values using a calculator, plotting the points, and drawing a smooth curve through them to represent the cable. Finally, from each plotted point, draw a vertical support cable attached to the roadway.

x	−80	−60	−40	−20	0	20	40	60	80
y									

54. SELLING TV SETS   A company has found that it can sell x TVs at a price of $\$\left(450 - \frac{1}{6}x\right)$.
   a. How many TVs must the company sell to maximize its revenue?
   b. Find the maximum revenue.

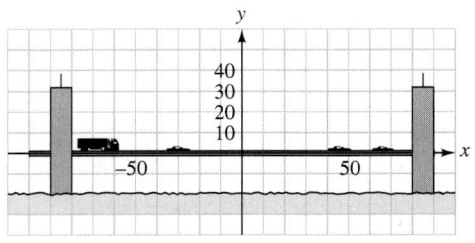

ILLUSTRATION 7

55. SELLING CD PLAYERS   A wholesaler sells CD players for $150 each. However, she gives volume discounts on purchases of 500 to 1,000 units according to the formula $\left(150 - \frac{1}{10}n\right)$, where n represents the number of units purchased.
   a. How many units would a retailer have to buy for the wholesaler to obtain maximum revenue?
   b. Find the maximum revenue.

56. TRACK AND FIELD   See Illustration 8. Sketch the parabolic path traveled by the long-jumper's center of gravity from the take-off board to the landing. Let the x-axis represent the ground.

ILLUSTRATION 8

WRITING

57. Explain why the y-intercept of the graph of $y = ax^2 + bx + c$, where $a \neq 0$, is $(0, c)$.

58. Use the example of a stream of water from a drinking fountain to explain the concept of the vertex of a parabola.

59. Explain why parabolas that open left or right are not graphs of functions.

60. Is it possible for the graph of a parabola not to have an x-intercept? Explain.

Is it possible for the graph of a parabola not to have a y-intercept? Explain.

REVIEW   *Simplify each expression.*

61. $\sqrt{12} + \sqrt{27}$ **62.** $3\sqrt{6y}\left(-4\sqrt{3y}\right)$

63. $\left(\sqrt{3} + 1\right)\left(\sqrt{3} - 1\right)$ **64.** $\left(\sqrt{x} + 2\right)^2$

Solve each equation.

65. $\sqrt{6 + 2x} = 4$ **66.** $\sqrt{1 - 2x} = \sqrt{x + 10}$

KEY CONCEPT

Quadratic Equations

In this chapter, we have studied several ways to solve quadratic equations. We have also graphed quadratic functions and seen that their graphs are parabolas.

What Is a Quadratic Equation?

A quadratic equation can be written in the form $ax^2 + bx + c = 0$, where $a \neq 0$.
In Exercises 1–12, tell whether each item is a quadratic equation.

1. $y = 3x + 7$

2. $4(x + 5) = 2x$

3. $2x^2 - 3x + 4 = 0$

4. $y(y - 6) = 0$

5. $a^2 + 7a - 1 > 0$

6. $3y^2 - y + 4$

7. $5 = y - y^2$

8. $|x - 8|$

9. $\sqrt{x + 7} = 4$

10. $x^2 = 16$

11. $\dfrac{m}{2} - \dfrac{1}{3} = \dfrac{1}{4}$

12. $C = \dfrac{5}{9}(F - 32)$

Solving Quadratic Equations

The techniques we have used to solve linear equations cannot be used to solve a quadratic equation, because those techniques cannot isolate the variable on one side of the equation. Exercises 13–16 show examples of student work to solve quadratic equations. In each case, what did the student do wrong?

13. Solve $x^2 = 6$.

$$\frac{x^2}{2} = \frac{6}{2}$$

$$x = 3$$

14. Solve $x^2 - x = 10$.

$$x(x - 1) = 10$$

$$x = 10 \quad \text{or} \quad x - 1 = 10$$

$$x = 11$$

15. Solve $a^2 = 20$.

$$a = \sqrt{20}$$

$$a = 2\sqrt{5}$$

16. Solve $x^2 + 5x + 1 = 0$.

$$a = 1 \quad b = 5 \quad c = 1$$

$$x = -5 \pm \frac{\sqrt{5^2 - 4(1)(1)}}{2}$$

$$x = -5 \pm \frac{\sqrt{21}}{2}$$

Solve each quadratic equation using the method listed.

17. $4x^2 - x = 0$; factoring method

18. $x^2 + 3x + 1 = 0$; quadratic formula

19. $x^2 = 36$; square root method

20. $a^2 - a - 56 = 0$; factoring method

21. $x^2 + 4x + 1 = 0$; complete the square

22. $(x + 3)^2 = 16$; square root method

Quadratic Functions

The graph of the quadratic function $y = ax^2 + bx + c$ is a parabola. It opens upward when $a > 0$ and downward when $a < 0$.

23. The formula $R = 4x - x^2$ gives the revenue R (in tens of thousands of dollars) that a business obtains from the manufacture and sale of x patio chairs (in hundreds). Graph $R = 4x - x^2$ in Illustration 1.

24. Refer to Exercise 23. Find the vertex of the parabola. What is its significance concerning the revenue the business brings in?

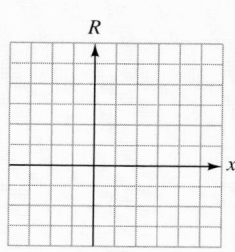

ILLUSTRATION 1

ACCENT ON TEAMWORK

Section 15.1
COMPLETING THE SQUARE   Construct the model in Illustration 1. Label each piece of the model with its respective area. Show that the total area of the model is $(x^2 + 4x)$ square units. Next, add enough 1×1 squares to make the model a square. How many does it take to do this? Show that the area of the new figure can be expressed as $(x^2 + 4x + 4)$ or $(x + 2)(x + 2)$ square units Explain how this model demonstrates the process of completing the square on $x^2 + 4x$.

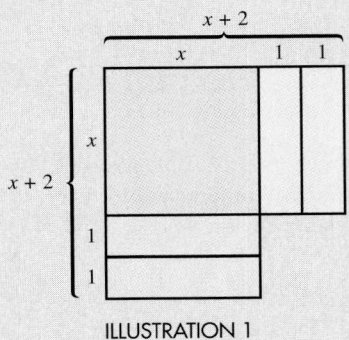

ILLUSTRATION 1

AUTHORING A TEXTBOOK   Assign each of the nine examples in Section 15.1 to members of your group. Have them write a new but similar problem for each example, then write a solution complete with an explanation and author notes using the same format as this book. They should also create an accompanying Self Check problem and include the answer. Compile all nine examples into a booklet. Make copies of your booklet for the other members of the class.

Section 15.2
PREDICTING SOLUTIONS   The expression $b^2 - 4ac$ is called the **discriminant** of the quadratic equation $ax^2 + bx + c = 0$. We can use the discriminant to predict the number of solutions a particular quadratic equation has.

If $b^2 - 4ac > 0$, the equation has two real solutions.
If $b^2 - 4ac = 0$, the equation has one real solution.
If $b^2 - 4ac < 0$, the equation has no real solutions.

For each quadratic equation, evaluate the discriminant to determine how many real-number solutions it has.

a. $4x^2 - 4x + 1 = 0$ 　　**b.** $6x^2 - 5x - 6 = 0$
c. $5x^2 + x + 2 = 0$ 　　**d.** $3x^2 + 10x - 2 = 0$
e. $2x^2 = 4x - 1$ 　　**f.** $9x^2 = 12x - 4$

SOLVING QUADRATIC EQUATIONS   Solve the quadratic equation $2x^2 - x - 1 = 0$ using these methods: factoring, completing the square, and the quadratic formula. Write each solution on a separate piece of paper. Under each solution, in two columns, list the advantages and the drawbacks of each method.

Section 15.3
PARABOLAS   Use a home video camera to make a "documentary" showing examples of parabolic shapes you see in everyday life. Write a script for your video and have a narrator explain the setting, point out the vertex, and tell whether the parabola opens upward or downward in each case.

GRAPHING   On one piece of graph paper, graph each of the following quadratic functions.

$$f(x) = x^2 \qquad g(x) = 2x^2 \qquad h(x) = \tfrac{1}{2}x^2$$

In general, what happens to the graph of $y = ax^2$ as a increases?

MINIMIZING/MAXIMIZING   In the business world, it is good for a company to minimize its costs and to maximize its profits. In your group, make a list of quantities that are good to minimize and a list of quantities that are good to maximize.

SOLUTION AND GRAPH   Write a quadratic function whose graph has the x-intercepts shown in Illustration 2. (*Hint:* Recall that there is a relation between the solution of a quadratic equation and the x-intercepts of the graph of the corresponding quadratic function.)

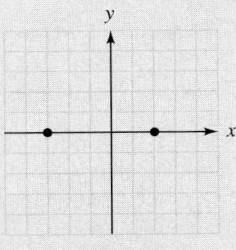

ILLUSTRATION 2

SECTION 15.1	*Completing the Square*

CONCEPTS

We can use the *square root method* to solve $x^2 = c$, where $c > 0$. The two solutions are

$x = \sqrt{c}$ and $x = -\sqrt{c}$ $\left(\text{or } x = \pm\sqrt{c}\right)$.

To make $x^2 + bx$ a trinomial square, add the square of one-half of the coefficient of x.

The factoring method doesn't always work in solving many quadratic equations. In these cases, we can use a method called *completing the square*.

To solve a quadratic equation by completing the square:

1. If necessary, divide both sides of the equation by the coefficient of x^2 to make its coefficient 1.

2. If necessary, get the constant on the right-hand side of the equation.

3. Complete the square and factor the resulting trinomial square.

4. Solve the quadratic equation using the square root method.

5. Check each solution.

REVIEW EXERCISES

1. Use the square root method to solve each quadratic equation.

a. $x^2 = 25$ **b.** $x^2 = 400$

c. $2x^2 = 18$ **d.** $4y^2 = 9$

e. $t^2 = 8$ **f.** $2x^2 - 1 = 149$

2. Use the square root method to solve each equation.

a. $(x - 1)^2 = 25$ **b.** $4(x - 2)^2 = 9$

c. $(x - 8)^2 = 8$ **d.** $(x + 5)^2 = 75$

3. Use the square root method to solve each equation. Round each solution to the nearest hundredth.

a. $x^2 = 12$ **b.** $(x - 1)^2 = 55$

4. Complete the square to make each expression a trinomial square.

a. $x^2 + 4x$ **b.** $z^2 - 10z$

c. $t^2 - 5t$ **d.** $a^2 + \dfrac{3}{4}a$

5. Explain why the quadratic equation $x^2 + 4x + 1 = 0$ can't be solved by the factoring method.

6. Solve each quadratic equation by completing the square.

a. $x^2 - 8x + 15 = 0$ **b.** $x^2 + 5x - 14 = 0$

c. $2x^2 + 5x - 3 = 0$ **d.** $2x^2 - 2x - 1 = 0$

7. Solve $x^2 + 4x + 1 = 0$ by completing the square. Round each solution to the nearest hundredth.

8. PLAYGROUND EQUIPMENT   The large tractor tire shown in Illustration 1 makes a good container for sand. If the circular area that the "sandbox" covers is 28.3 square feet, what is the radius of the tire? Round to the nearest tenth of a foot.

ILLUSTRATION 1

The Quadratic Formula

For the *general quadratic equation* $ax^2 + bx + c = 0$, where $a \neq 0$,

$$x = \frac{-b \pm \sqrt{b^2 - 4ac}}{2a}$$

This is called the *quadratic formula*.

9. Use the quadratic formula to solve each quadratic equation.
  a. $x^2 - 2x - 15 = 0$
  b. $x^2 - 6x - 7 = 0$
  c. $6x^2 - 7x - 3 = 0$
  d. $x^2 - 6x + 7 = 0$

10. Use the quadratic formula to solve $3x^2 + 2x - 2 = 0$. Give the solutions in exact form and then rounded to the nearest hundredth.

11. Use the quadratic formula to solve $10x^2 + 2x + 1 = 0$.

12. SECURITY GATE   The length of the frame for the iron gate in Illustration 2 is 14 feet longer than the width. A diagonal crossbrace is 26 feet long. Find the width and length of the gate frame.

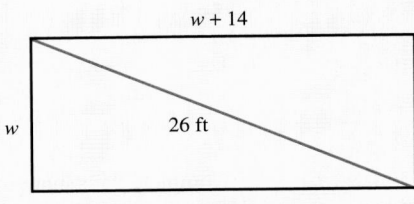

w + 14

w

26 ft

ILLUSTRATION 2

13. MILITARY   A pilot releases a bomb from an altitude of 3,000 feet. The bomb's height h above the target t seconds after its release is given by the formula

$$h = 3{,}000 + 40t - 16t^2$$

How long will it be until the bomb hits its target?

Strategy for solving quadratic equations:

1. Try the square root method.
2. If it doesn't apply, write the equation in $ax^2 + bx + c = 0$ form.
3. Try the factoring method.
4. If it doesn't work, complete the square or use the quadratic formula.

14. Use the most convenient method to find all real solutions of each equation.
  a. $x^2 + 6x + 2 = 0$
  b. $(y + 3)^2 = 16$
  c. $x^2 + 5x = 0$
  d. $2x^2 + x = 5$
  e. $g^2 - 20 = 0$
  f. $a^2 = 4a - 4$
  g. $a^2 - 2a + 5 = 0$
  h. $2c^2 = 800$

Graphing Quadratic Functions

The *vertex* of a parabola is the lowest (or highest) point on the parabola.

15. See the graph in Illustration 3.
  a. What are the *x*-intercepts of the parabola?
  b. What is the *y*-intercept of the parabola?
  c. What is the vertex of the parabola?
  d. Draw the axis of symmetry of the parabola on the graph.

A vertical line through the vertex of a parabola that opens upward or downward is its *axis of symmetry*.

16. What important information can be obtained from the vertex of the parabola in Illustration 4?

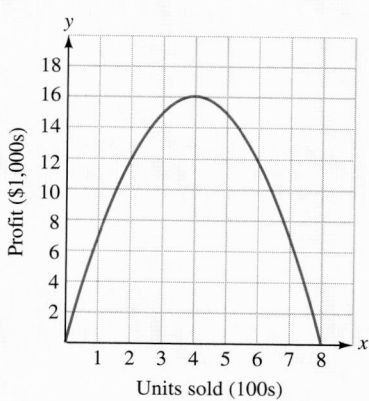

ILLUSTRATION 3   ILLUSTRATION 4

The graph of the quadratic function $y = ax^2 + bx + c$ is a parabola. It opens upward when $a > 0$ and downward when $a < 0$.

17. Find the vertex of the graph of each quadratic function and tell which direction the parabola opens. *Do not draw the graph.*

   a. $y = 2x^2 - 4x + 7$
   b. $f(x) = -3x^2 + 18x - 11$

The x-coordinate of the vertex of the parabola $y = ax^2 + bx + c$ is $x = -\frac{b}{2a}$. To find the y-coordinate of the vertex, substitute $-\frac{b}{2a}$ for x in the equation of the parabola and find y.

18. Find the x- and y-intercepts of the graph of $y = x^2 + 6x + 5$.

The x-intercepts of a parabola are determined by solving $ax^2 + bx + c = 0$. The y-intercept is $(0, c)$.

19. Graph each quadratic function by finding the vertex, x- and y-intercepts, and axis of symmetry of its graph.

   a. $y = x^2 + 2x - 3$
   b. $f(x) = -2x^2 + 4x - 2$

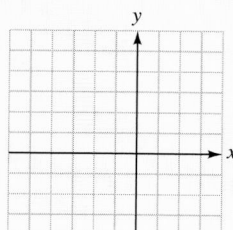   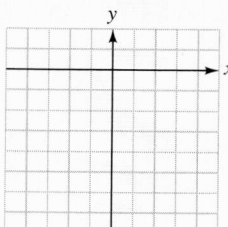

The number of x-intercepts of the graph of a quadratic function $y = ax^2 + bx + c$ is the same as the number of solutions of $ax^2 + bx + c = 0$.

20. The graphs of three quadratic functions are shown in Illustration 5. Fill in the blanks.

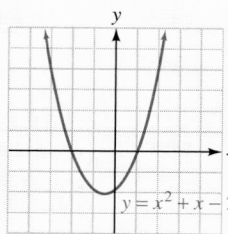   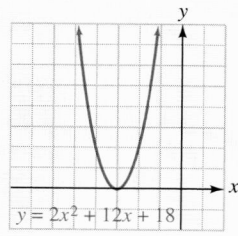 $y = -x^2 + 4x - 5$

$x^2 + x - 2 = 0$ has real-number solution(s).

$2x^2 + 12x + 18 = 0$ has real-number solution(s).

$-x^2 + 4x - 5 = 0$ has real-number solution(s).

ILLUSTRATION 5

Chapter 15 Test

Solve each equation by the square root method.

1. $x^2 = 16$

2. $u^2 = 24$

3. $4y^2 = 25$

4. $(x - 2)^2 = 3$

5.  ARCHERY   The area of the circular archery target shown in Illustration 1 is 5,026.5 cm². What is the radius of the target? Round to the nearest centimeter.

ILLUSTRATION 1

6. Find the number required to complete the square on $x^2 - 14x$.

7. Complete the square to solve $a^2 + 2a - 4 = 0$. Give the exact solutions and then round them to the nearest hundredth.

8. Complete the square to solve $2x^2 = 3x + 2$.

In Problems 9–12, use the quadratic formula to find all real solutions of each equation.

9. $x^2 + 3x - 10 = 0$

10. $2x^2 - 5x = 12$

11. $x^2 + 5 = 2x$

12. $x^2 = 4x - 2$

13. Solve $3x^2 - x - 1 = 0$ using the quadratic formula. Give the exact solutions, and then approximate them to the nearest hundredth.

14. 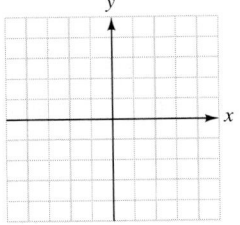 FLAG   According to the *Guinness Book of World Records 1998,* the largest flag in the world is the American "Superflag," which has an area of 128,775 ft². If its length is 5 feet less than twice its width, find its width and length.

15. ADVERTISING   When a business runs x advertisements per week on television, the number y of air conditioners it sells is given by the quadratic function graphed in Illustration 2. What important information can be obtained from the vertex?

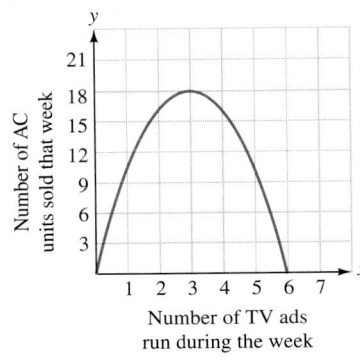

ILLUSTRATION 2

16. Graph the function $y = x^2 + x - 2$ by finding the vertex, x- and y-intercepts, and axis of symmetry.

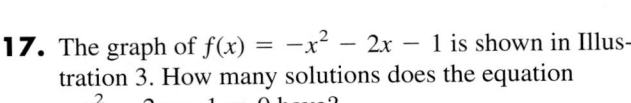

17. The graph of $f(x) = -x^2 - 2x - 1$ is shown in Illustration 3. How many solutions does the equation $-x^2 - 2x - 1 = 0$ have?

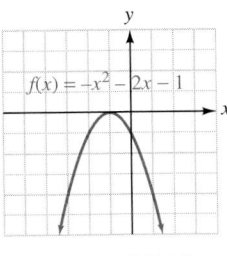

ILLUSTRATION 3

18. Explain the meaning of the $\pm$ symbol.

Chapters 1-15 Cumulative Review Exercises

1. Tell whether each statement is true or false.
  a. Every rational number can be written as a ratio of two integers.
  b. The set of real numbers corresponds to all points on the number line.
  c. The whole numbers and their opposites form the set of integers.

2. Evaluate $-4 + 2[-7 - 3(-9)]$.

3. DRIVING SAFETY   In cold weather climates, salt is spread on roads to keep snow and ice from bonding to the pavement. This allows snowplows to remove accumulated snow quickly. According to the graph in Illustration 1, when is the accident rate the worst?

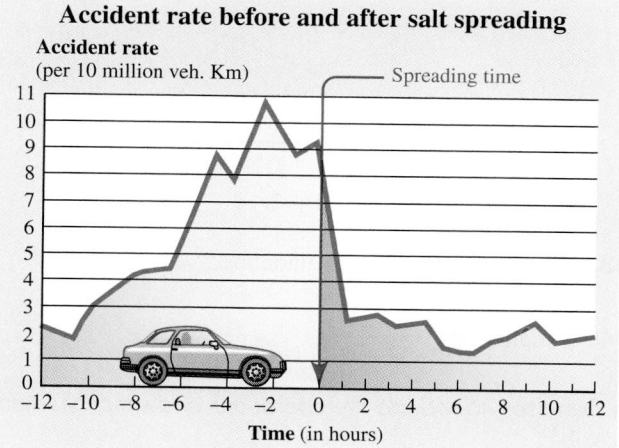

Accident rate before and after salt spreading

Based on data from the Salt Institute

ILLUSTRATION 1

4. EMPLOYMENT   The following newspaper headline appeared in early 2000.

> **Xerox to Cut 5,200 Jobs, or 5.3% of Workforce, on Falling Profits**

How many employees did Xerox have at that time?

5. Simplify $3p - 6(p + z) + p$.

6. Solve $\dfrac{5}{6}k = 10$.

7. Solve $-(3a + 1) + a = 2$.

8. Solve $5x + 7 < 2x + 1$ and graph the solution set. Then use interval notation to describe the solution.

9. ENTREPRENEURS   Last year, a women's professional organization made two small-business loans totaling $28,000 to young women beginning their own businesses. The money was lent at 7% and 10% simple interest rates. If the annual income the organization received from these loans was $2,560, what was each loan amount?

10. Evaluate $(x - a)^2 + (y - b)^2$ for $x = -2$, $y = 1$, $a = 5$, and $b = -3$.

11. Evaluate $\left| \dfrac{4}{5} \cdot 10 - 12 \right|$.

12. Find the slope of the line passing through $(-2, -2)$ and $(-12, -8)$.

Graph each equation or inequality.

13. $2y - 2x = 6$

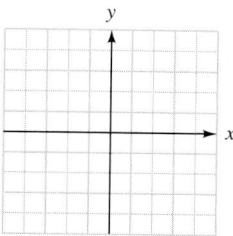

14. $y = -3$

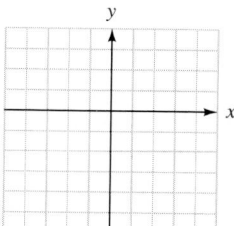

15. $y = -x + 2$

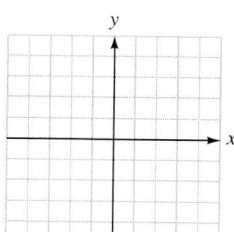

16. $y < 3x$

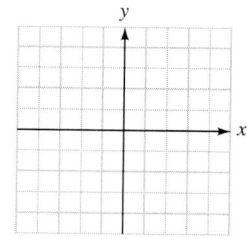

17. Graph the line passing through $(-2, -1)$ and having slope $\frac{4}{3}$

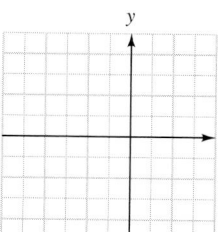

18. Graph $y = x^3 - 2$.

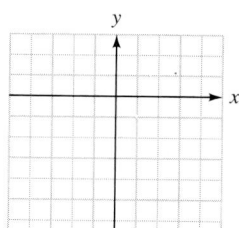

19. Write the equation of the line whose graph has slope $m = -2$ and y-intercept $(0, 1)$.

20. Write the equation of the line whose graph has slope $m = \frac{1}{4}$ and passes through the point $(8, 1)$. Answer in slope–intercept form.

21. What is the slope of the line defined by $4x + 5y = 6$?

22. If $f(x) = 3x^2 + 3x - 8$, find $f(-1)$.

23. Is the word *domain* associated with the input or the output of a function?

24. Is this the graph of a function?

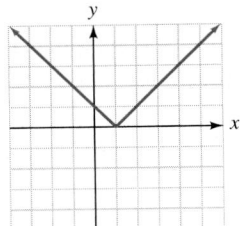

In Exercises 25–28, simplify each expression. Write each answer without using parentheses or negative exponents.

25. $y^3(y^2y^4)$

26. $\left(\dfrac{b^2}{3a}\right)^3$

27. $\dfrac{10a^4a^{-2}}{5a^2a^0}$

28. $\dfrac{(r^2)^3}{(r^3)^4}$

29. FIVE-CARD POKER   The odds against being dealt the hand shown in Illustration 2 are about 2.6×10^6 to 1. Express the odds using standard notation.

ILLUSTRATION 2

30. PAIN RELIEVERS   See Illustration 3. Find the rate of change in the percent of individuals free of headache pain over the given time span after taking Acetaminophen.

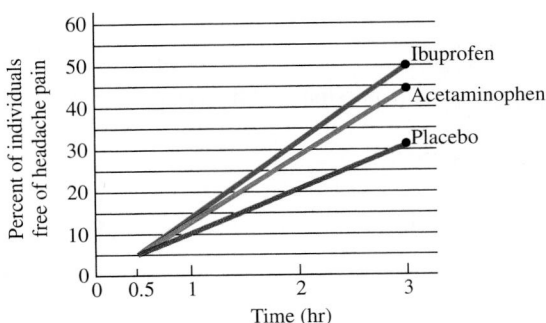

Based on data from *Health and Wellness*, Jones and Bartlett Publishers

ILLUSTRATION 3

Do the operations.

31. $(-2a^3)(3a^2)$

32. $(2b - 1)(3b + 4)$

33. $(2x + 5y)^2$

34. $x - 3\overline{)2x^2 - 3 - 5x}$

Factor each expression completely.

35. $6a^2 - 12a^3b + 36ab$

36. $2x + 2y + ax + ay$

37. $b^3 + 125$

38. $t^4 - 16$

Solve each equation.

39. $3x^2 + 8x = 0$

40. $15x^2 - 2 = 7x$

41. Write a polynomial that represents the perimeter of the rectangle shown in Illustration 4.

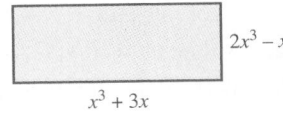

ILLUSTRATION 4

42. HEIGHT OF A TRIANGLE   The triangle shown in Illustration 5 has an area of 22.5 square inches. Find its height. (*Hint:* Multiply both sides of the equation by 2.)

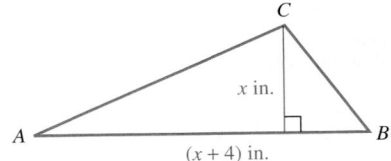

ILLUSTRATION 5

43. For what value $\dfrac{x}{x + 8}$ is undefined?

44. Simplify $\dfrac{3x^2 - 27}{x^2 + 3x - 18}$.

In Exercises 45–48, do the operations and simplify when possible.

45. $\dfrac{x^2 - x - 6}{2x^2 + 9x + 10} \div \dfrac{x^2 - 25}{2x^2 + 15x + 25}$

46. $\dfrac{x + 3}{x^2} + \dfrac{x + 5}{x^2}$

47. $\dfrac{x}{x - 2} + \dfrac{3x}{x^2 - 4}$

48. $\dfrac{\dfrac{5}{y} + \dfrac{4}{y + 1}}{\dfrac{4}{y} - \dfrac{5}{y + 1}}$

In Exercises 49–50, solve each equation.

49. $\dfrac{3r}{2} - \dfrac{3}{r} = \dfrac{3r}{2} + 3$

50. $\dfrac{7}{q^2 - q - 2} + \dfrac{1}{q + 1} = \dfrac{3}{q - 2}$

51. Solve the formula $\dfrac{1}{a} + \dfrac{1}{b} = 1$ for a.

52. ROOFING   A homeowner estimates that it will take him 7 days to roof his house. A professional roofer estimates that he could roof the house in 4 days. How long will it take if the homeowner helps the roofer?

53. LOSING WEIGHT   If a person cuts his or her daily calorie intake by 100, it will take 350 days for that person to lose 10 pounds. How long will it take for the person to lose 25 pounds?

54. GEAR   The speed of a gear varies inversely with the number of teeth. If a gear with 10 teeth makes 3 revolutions per second, how many revolutions per second will a gear with 25 teeth make?

55. Solve using the graphing method.
$$\begin{cases} x + y = 1 \\ y = x + 5 \end{cases}$$

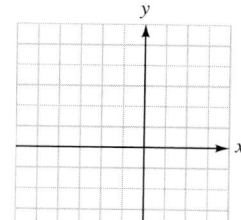

56. Solve using the substitution method.
$$\begin{cases} y = 2x + 5 \\ x + 2y = -5 \end{cases}$$

57. Solve using the addition method.
$$\begin{cases} \dfrac{3}{5}s + \dfrac{4}{5}t = 1 \\ -\dfrac{1}{4}s + \dfrac{3}{8}t = 1 \end{cases}$$

58. MIXING CANDY   How many pounds of each candy shown in Illustration 6 must be mixed to obtain 60 pounds of candy that would be worth $3 per pound? Use two variables to solve this problem.

Hard Candy
$2/lb

Soft Candy
$4/lb

ILLUSTRATION 6

59. Graph the solution set of
$$\begin{cases} 3x + 4y \ge -7 \\ 2x - 3y \ge 1 \end{cases}$$

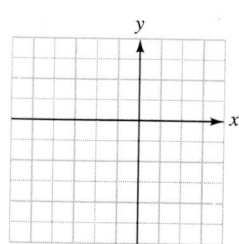

60. DEMOGRAPHICS   Refer to the graph in Illustration 7. To which stage does each of the following descriptions apply?

  Stage   : Rapidly growing population: Births far outnumber deaths.

  Stage   : Stable population: Birth rate drops; births and deaths are more or less equal.

  Stage   : Stable population: Births and deaths are more or less equal.

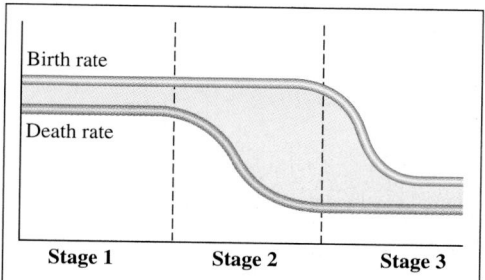

Birth rate

Death rate

Stage 1   Stage 2   Stage 3

ILLUSTRATION 7

Simplify each expression. All variables represent positive numbers.

61. $\sqrt{50x^2}$

62. $\sqrt[3]{-27y^3}$

63. $3\sqrt{24} + \sqrt{54}$

64. $\left(\sqrt{2} + 1\right)\left(\sqrt{2} - 3\right)$

65. $\sqrt{\dfrac{72x^3}{y^2}}$

66. $\dfrac{8}{\sqrt{10}}$

67. Solve $\sqrt{6x + 1} + 2 = 7$.

68. Solve $x^2 + 8x + 12 = 0$ by completing the square

69. Solve $t^2 = 75$.

70. STORAGE CUBES The diagonal distance across the face of each of the stacking cubes shown in Illustration 8 is 15 inches. What is the height of the entire storage arrangement? Round to the nearest tenth of an inch.

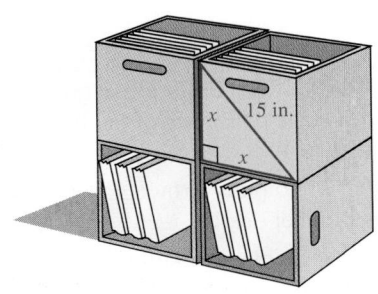

ILLUSTRATION 8

71. Solve $3x^2 - x - 1 = 0$ using the quadratic formula. Give the exact solutions, and then approximate each to the nearest hundredth.

72. QUILT According to the *Guinness Book of World Records 1998,* the world's largest quilt was made by the Seniors' Association of Saskatchewan, Canada, in 1994. If the length of the rectangular quilt is 11 feet less than twice its width and it has an area of 12,865 ft^2, find its width and length.

73. Find the vertex and the *x*- and *y*-intercepts of the graph of $y = x^2 + 6x + 5$. Then graph the function.

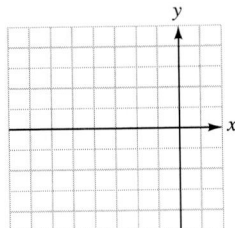

74. POWER OUTPUT The graph in Illustration 9 shows the power output (in horsepower, hp) of a certain engine for various engine speeds (in revolutions per minute, rpm).
  a. At an engine speed of 3,000 rpm, what is the power output?
  b. For what engine speed(s) is the power output 125 hp?
  c. For what engine speed does the power output reach a maximum?

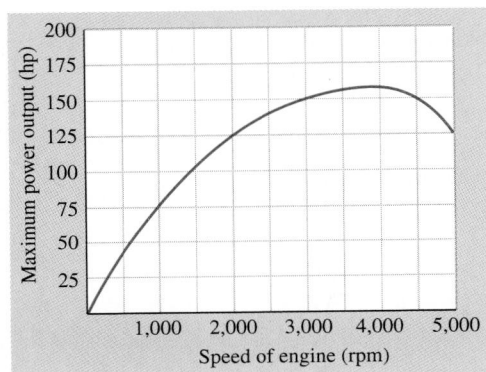

ILLUSTRATION 9

APPENDIX I INDUCTIVE AND DEDUCTIVE REASONING

In this appendix, you will learn about

• Inductive reasoning • Deductive reasoning

INTRODUCTION. To reason means to think logically. The objective of this appendix is to develop your problem-solving ability by improving your reasoning skills. We will introduce two fundamental types of reasoning that can be applied in a wide variety of settings. They are known as *inductive reasoning* and *deductive reasoning.*

Inductive reasoning

In a laboratory, scientists conduct experiments and observe outcomes. After several repetitions with similar outcomes, the scientist will generalize the results into a statement that appears to be true:

• If I heat water to 212°F, it will boil.

• If I drop a weight, it will fall.

• If I combine an acid with a base, a chemical reaction occurs.

When we draw general conclusions from specific observations, we are using **inductive reasoning.** The next examples show how inductive reasoning can be used in mathematical thinking. Given a list of numbers or symbols, called a *sequence,* we can often find a missing term of the sequence by looking for patterns and applying inductive reasoning.

EXAMPLE 1 *An increasing pattern.* Find the next number in the sequence 5, 8, 11, 14, ...

Solution

The terms of the sequence are increasing. To discover the pattern, we find the *difference* between each pair of successive terms.

$8 - 5 = 3$ Subtract the first term from the second term.

$11 - 8 = 3$ Subtract the second term from the third term.

$14 - 11 = 3$ Subtract the third term from the fourth term.

The difference between each pair of numbers is 3. This means that each successive number is 3 greater than the previous one. Thus, the next number in the sequence is $14 + 3$, or 17.

Self Check

Find the next number in the sequence $-3, -1, 1, 3, ...$

Answer: 5

A-1

EXAMPLE 2   *A decreasing pattern.*   Find the next number in the sequence
$-2, -4, -6, -8, \ldots$

Self Check
Find the next number in the sequence
$-0.1, -0.3, -0.5, -0.7 \ldots$

Solution
The terms of the sequence are decreasing. Since each successive term is 2 less than the previous one, the next number in the pattern is $-8 - 2$, or -10.

Answer: -0.9 ∎

EXAMPLE 3   *An alternating pattern.*   Find the next letter in the sequence
A, D, B, E, C, F, D, ...

Self Check
Find the next entry in the sequence
Z, A, Y, B, X, C, ...

Solution
The letter A is the first letter of the alphabet, D is the fourth letter, B is the second letter, and so on. We can create the following letter–number correspondence:

$$
\begin{aligned}
A &\longrightarrow 1 \quad \big\} \text{ Add 3.} \\
D &\longrightarrow 4 \quad \big\} \text{ Subtract 2.} \\
B &\longrightarrow 2 \quad \big\} \text{ Add 3.} \\
E &\longrightarrow 5 \quad \big\} \text{ Subtract 2.} \\
C &\longrightarrow 3 \quad \big\} \text{ Add 3.} \\
F &\longrightarrow 6 \quad \big\} \text{ Subtract 2.} \\
D &\longrightarrow 4
\end{aligned}
$$

The numbers in the sequence 1, 4, 2, 5, 3, 6, 4, ... alternate in size. They change from smaller to larger, to smaller, to larger, and so on.

We see that 3 is added to the first number to get the second number. Then 2 is subtracted from the second number to get the third number. To get successive terms in the sequence, we alternately add 3 to one number and then subtract 2 from that result to get the next number.

Applying this pattern, the next number in the numerical sequence would be $4 + 3$, or 7. The next letter in the original sequence would be G, because it is the seventh letter of the alphabet.

Answer: W   ∎

EXAMPLE 4   *Two patterns.*   Find the next geometric shape in the sequence below.

Self Check
Find the next geometric shape in the sequence below.

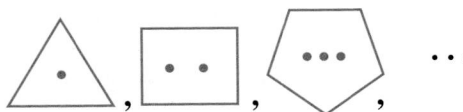

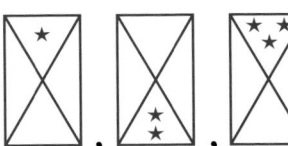

Solution
This sequence has two patterns occurring at the same time. The first figure has three sides and one dot, the second figure has four sides and two dots, and the third figure has five sides and three dots. Thus, we would expect the next figure to have six sides and four dots, as shown in Figure I-1.

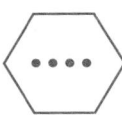

FIGURE I-1

Answer:    ∎

EXAMPLE 5   *A circular pattern.*   Find the next geometric shape in the sequence below.

Solution

From figure to figure, we see that each dot moves from one point of the star to the next, in a counterclockwise direction. This is a circular pattern. The next shape in the sequence will be the one shown in Figure I-2.

FIGURE I-2

Self Check

Find the next geometric shape in the sequence below.

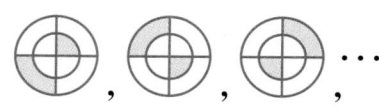

Answer:  ■

Deductive reasoning

As opposed to inductive reasoning, **deductive reasoning** moves from the general case to the specific. For example, if we know that the sum of the angles in any triangle is 180°, we know that the sum of the angles of △ABC is 180°. Whenever we apply a general principle to a particular instance, we are using deductive reasoning.

A deductive reasoning system is built on four elements:

1. **Undefined terms:** terms that we accept without giving them formal meaning
2. **Defined terms:** terms that we define in a formal way
3. **Axioms** or **postulates:** statements that we accept without proof
4. **Theorems:** statements that we can prove with formal reasoning

Many problems can be solved by deductive reasoning. For example, suppose that we plan to enroll in an early-morning algebra class, and that we know that Professors Perry, Miller, and Tveten are scheduled to teach algebra next semester. After some investigating, we find out that Professor Perry teaches only in the afternoon and Professor Tveten teaches only in the evenings. Without knowing anything about Professor Miller, we can conclude that he will be our teacher, since he is the only remaining possibility.

The following examples show how to use deductive reasoning to solve problems.

EXAMPLE 6   *Scheduling classes.*   Four professors are scheduled to teach mathematics next semester, with the following course preferences:

1. Professors A and B don't want to teach calculus.
2. Professor C wants to teach statistics.
3. Professor B wants to teach algebra.

Who will teach trigonometry?

Solution

The following chart shows each course, with each possible instructor.

Calculus	Algebra	Statistics	Trigonometry
A	A	A	A
B	B	B	B
C	C	C	C
D	D	D	D

Since Professors A and B don't want to teach calculus, we can cross them off the calculus list. Since Professor C wants to teach statistics, we can cross her off every other list. This leaves Professor D as the only person to teach calculus, so we can cross her off every other list. Since Professor B wants to teach algebra, we can cross him off every other list. Thus, the only remaining person left to teach trigonometry is Professor A.

Calculus	Algebra	Statistics	Trigonometry
A̶	A	A	A
B̶	B	B̶	B̶
C̶	C̶	C	C̶
D	D̶	D̶	D̶

∎

EXAMPLE 7   *State flags.*   The graph in Figure I-3 gives the number of state flags that feature an eagle, a star, or both. How many state flags have neither an eagle nor a star?

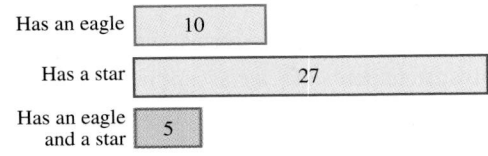

FIGURE I-3

Self Check

Of the 50 cars on a used-car lot, 9 are red, 31 are foreign models, and 6 are red, foreign models. If a customer wants to buy an American model that is not red, how many cars does she have to choose from?

Solution

In Figure I-4(a), the intersection (overlap) of the circles is a way to show that there are 5 state flags that have both an eagle and a star. If an eagle appears on a total of 10 flags, then the left circle must contain 5 more flags outside of the intersection. See Figure I-4(b). If a total of 27 flags have a star, the right circle must contain 22 more flags outside the intersection.

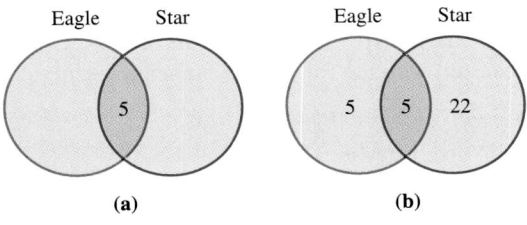

FIGURE I-4

From Figure I-4, we see that 5 + 5 + 22, or 32 flags have an eagle, a star, or both. To find how many flags have neither an eagle nor a star, we subtract this total from the number of state flags, which is 50.

50 − 32 = 18

There are 18 state flags that have neither an eagle nor a star.

Answer: 16   ■

Study Set   Appendix I

VOCABULARY   *Fill in the blanks.*

1. _____ reasoning draws general conclusions from specific observations.

2. _____ reasoning moves from the general case to the specific.

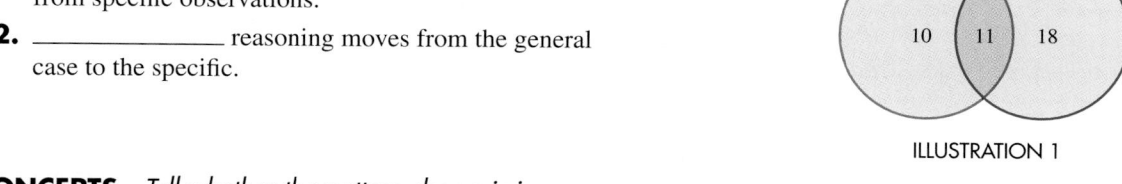

Mathematics   English

10   11   18

ILLUSTRATION 1

CONCEPTS   *Tell whether the pattern shown is increasing, decreasing, alternating, or circular.*

3. 2, 3, 4, 2, 3, 4, 2, 3, 4, . . .

4. 8, 5, 2, −1, . . .

5. −2, −4, 2, 0, 6, . . .

6. 0.1, 0.5, 0.9, 1.3, . . .

7. a, c, b, d, c, e, . . .

8.     . . .

9. ROOM SCHEDULING   From the chart, determine what time(s) on a Wednesday morning a practice room in a music building is available. The symbol X indicates that the room has already been reserved.

	M	T	W	Th	F
9 A.M.	X		X	X	
10 A.M.	X	X			X
11 A.M.			X		X

10. COUNSELING QUESTIONNAIRE   A group of college students were asked if they were taking a mathematics course and if they were taking an English course. The results are displayed in Illustration 1.
   a. How many students were taking a mathematics course and an English course?
   b. How many students were taking an English course but not a mathematics course?
   c. How many students were taking a mathematics course?

PRACTICE   *Find the number that comes next in each sequence.*

11. 1, 5, 9, 13, . . .

12. 15, 12, 9, 6, . . .

13. −3, −5, −8, −12, . . .

14. 5, 9, 14, 20, . . .

15. −7, 9, −6, 8, −5, 7, −4, . . .

16. 2, 5, 3, 6, 4, 7, 5, . . .

17. 9, 5, 7, 3, 5, 1, . . .

18. 1.3, 1.6, 1.4, 1.7, 1.5, 1.8, . . .

19. −2, −3, −5, −6, −8, −9, . . .

20. 8, 11, 9, 12, 10, 13, . . .

21. 6, 8, 9, 7, 9, 10, 8, 10, 11, . . .

22. 10, 8, 7, 11, 9, 8, 12, 10, 9, . . .

Find the figure that comes next in each sequence.

23. 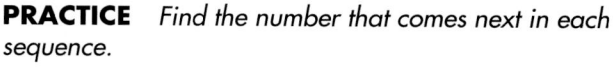 . . .

24.

△ , □ , △ , □ , △ , . . .

Find the missing figure in each sequence.

25.  ,  , , **?** ,

26.  ,  , **?** ,  ,

Find the next letter or letters in the sequence.

27. A, c, E, g, . . .   **28.** R, SS, TTT, . . .

29. d, h, g, k, j, n, . . .   **30.** B, N, C, N, D, . . .

What conclusion(s) can be drawn from each set of information?

31. Four people named John, Luis, Maria, and Paula have occupations as teacher, butcher, baker, and candlestick maker.
1. John and Paula are married.
2. The teacher plans to marry the baker in December.
3. Luis is the baker.

Who is the teacher?

32. In a zoo, a zebra, a tiger, a lion, and a monkey are to be placed in four cages numbered from 1 to 4, from left to right. The following decisions have been made:
1. The lion and the tiger should not be side by side.
2. The monkey should be in one of the end cages.
3. The tiger is to be in cage 4.

In which cage is the zebra?

33. A Ford, a Buick, a Dodge, and a Mercedes are parked side by side.
1. The Ford is between the Mercedes and the Dodge.
2. The Mercedes is not next to the Buick.
3. The Buick is parked on the left end.

Which car is parked on the right end?

34. Four divers at the Olympics finished first, second, third, and fourth.
1. Diver A beat diver B
2. Diver C placed between divers B and D.
3. Diver B beat diver D.

In which order did they finish?

35. A green, a blue, a red, and a yellow flag are hanging on a flagpole.
1. The blue flag is between the green and yellow flags.
2. The red flag is next to the yellow flag.
3. The green flag is above the red flag.

What is the order of the flags from top to bottom?

36. Andres, Barry, and Carl each have two occupations: bootlegger, musician, painter, chauffeur, barber, and gardener. From the following facts, find the occupations of each man.
1. The painter bought a quart of spirits from the bootlegger.
2. The chauffeur offended the musician by laughing at his mustache.
3. The chauffeur dated the painter's sister.
4. Both the musician and the gardener used to go hunting with Andres.
5. Carl beat both Barry and the painter at monopoly.
6. Barry owes the gardener $100.

APPLICATIONS

37. JURY DUTY   The results of a jury service questionnaire are shown in Illustration 2. Determine how many of the 20,000 respondents have served on neither a criminal court nor a civil court jury.

Jury Service Questionnaire

997	Served on a criminal court jury
103	Served on a civil court jury
35	Served on both

ILLUSTRATION 2

38. ELECTRONIC POLL   In Illustration 3, the Internet poll shows that 124 people voted for the first choice, 27 people voted for the second choice, and 19 people voted for both the first and the second choice. How many people clicked the third choice, "Neither"?

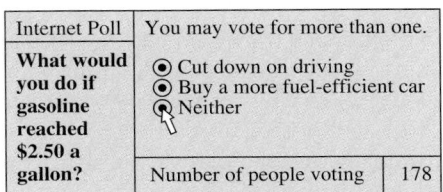

ILLUSTRATION 3

39. THE SOLAR SYSTEM   The graph in Illustration 4 shows some important characteristics of the 9 planets in our solar system. How many planets are neither rocky nor have moons?

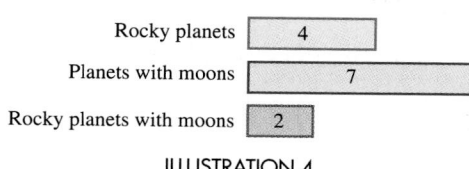

ILLUSTRATION 4

40. Write a problem in such a way that the diagram in Illustration 5 can be used to solve it.

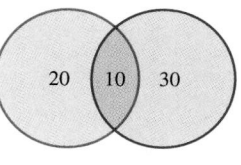

ILLUSTRATION 5

II.1 *American Units of Measurement*

In this section, you will learn about

- American units of length • Converting units of length
- American units of weight • American units of capacity • Units of time

INTRODUCTION. Two common systems of measurement are the American (or English) system and the metric system. We will discuss American units in this section and metric units in the next section. Some common American units are *inches, feet, miles, ounces, pounds, tons, cups, pints, quarts,* and *gallons.* These units are used when measuring length, weight, and capacity.

- A newborn baby is 20 inches long.
- The distance from St. Louis to Memphis is 285 miles.
- First-class postage for a letter that weighs less than 1 ounce is 34¢.
- The largest pumpkin ever grown weighed 1,092 pounds.
- Milk is sold in quart and gallon containers.

American units of length

A ruler is one of the most common devices used for measuring distances or lengths. Figure II-1 shows only a portion of a ruler; most rulers are 12 inches (1 foot) long. Since 12 inches = 1 foot, a ruler is divided into 12 equal distances of 1 inch. Each inch is divided into halves of an inch, quarters of an inch, eighths of an inch, and sixteenths of an inch. Several distances are measured using the ruler shown in Figure II-1.

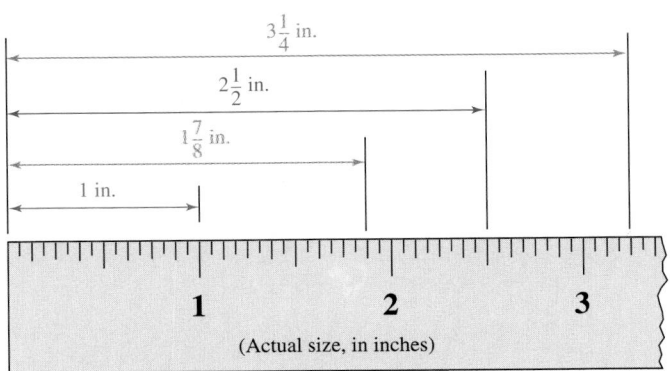

(Actual size, in inches)

FIGURE II-1

EXAMPLE 1 *Measuring the length of a nail.* To the nearest $\frac{1}{4}$ inch, find the length of the nail in Figure II-2.

Solution

We place the end of the ruler by one end of the nail and note that the other end of the nail is closer to the $2\frac{1}{2}$-inch mark than to the $2\frac{1}{4}$-inch mark on the ruler. To the nearest quarter-inch, the nail is $2\frac{1}{2}$ inches long.

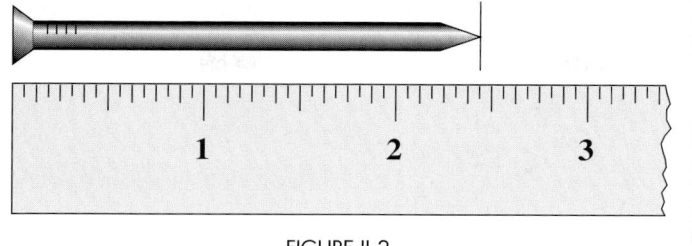

FIGURE II-2

Self Check

To the nearest $\frac{1}{4}$ inch, find the width of the circle below.

Answer: $1\frac{1}{4}$ in. ∎

EXAMPLE 2 *Measuring the length of a paper clip.* To the nearest $\frac{1}{8}$ inch, find the length of the paper clip in Figure II-3.

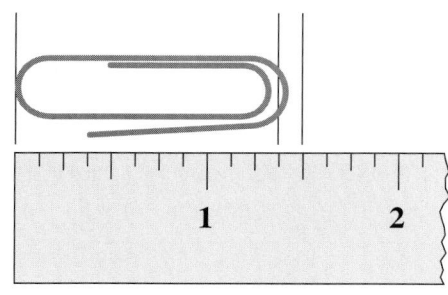

FIGURE II-3

Solution

We place the end of the ruler by one end of the paper clip and note that the other end is closer to the $1\frac{3}{8}$-inch mark than to the $1\frac{1}{2}$-inch mark on the ruler. To the nearest eighth of an inch, the paper clip is $1\frac{3}{8}$ inches long.

Self Check

To the nearest $\frac{1}{8}$ inch, find the length of the jumbo paper clip below.

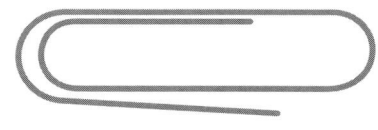

Answer: $1\frac{7}{8}$ in. ∎

Each point on a ruler, like each point on a number line, has a number associated with it: the distance between the point and 0. As Example 3 illustrates, the distance between any two points on a ruler (or on a number line) is the difference of the numbers associated with those points.

EXAMPLE 3 *Measuring the length of a ticket.* Find the length of the tear-off part of the concert ticket in Figure II-4.

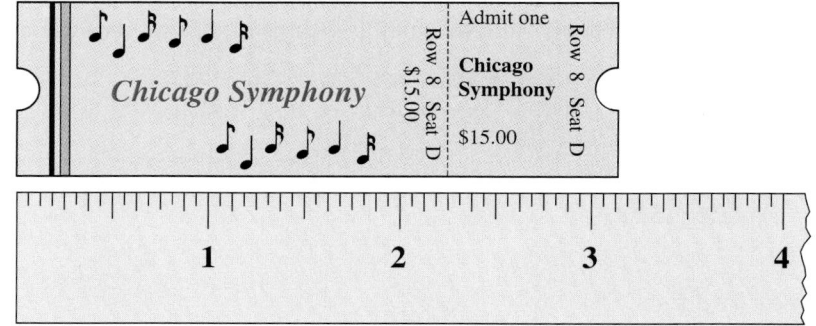

FIGURE II-4

Solution   After placing the end of the ruler by one end of the ticket, we note that the length of the entire ticket is $3\frac{1}{8}$ inches and that the length of the longer part is $2\frac{1}{4}$ inches. Since the length of the tear-off part of the ticket is the *difference* between these two lengths, we must subtract $2\frac{1}{4}$ from $3\frac{1}{8}$.

$$3\frac{1}{8} - 2\frac{1}{4} = \frac{25}{8} - \frac{9}{4} \qquad \text{Change the mixed numbers to improper fractions.}$$

$$= \frac{25}{8} - \frac{9 \cdot 2}{4 \cdot 2} \qquad \text{Write the fractions with a common denominator.}$$

$$= \frac{25}{8} - \frac{18}{8} \qquad \text{Do the multiplication in the numerator and in the denominator.}$$

$$= \frac{25 - 18}{8} \qquad \text{Subtract the fractions.}$$

$$= \frac{7}{8} \qquad \text{Simplify.}$$

The length of the tear-off portion of the ticket is $\frac{7}{8}$ inch.   ∎

Converting units of length

American units of length are related in the following ways.

American units of length

12 inches (in.) = 1 foot (ft)	36 inches = 1 yard (yd)
3 feet = 1 yard	5,280 feet = 1 mile (mi)

To convert from one unit to another, we use *unit conversion factors*. To find the unit conversion factor between yards and feet, we begin with this fact:

3 ft = 1 yd

If we divide both sides of this equation by 1 yard, we get

$$\frac{3 \text{ ft}}{1 \text{ yd}} = \frac{1 \text{ yd}}{1 \text{ yd}}$$

$$\frac{3 \text{ ft}}{1 \text{ yd}} = 1 \qquad \text{A number divided by itself is 1: } \frac{1 \text{ yd}}{1 \text{ yd}} = 1.$$

The fraction $\frac{3 \text{ ft}}{1 \text{ yd}}$ is called a **unit conversion factor,** because its value is 1. It can be read as "3 feet per yard." Since this fraction is equal to 1, multiplying a length by this fraction does not change its measure; it only changes the *units* of measure.

EXAMPLE 4   *Converting from yards to feet.*   Convert 7 yards to feet.

Solution
To convert from yards to feet, we must use a unit conversion factor that relates feet to yards. Since there are 3 feet per yard, we multiply 7 yards by the unit conversion factor $\frac{3 \text{ ft}}{1 \text{ yd}}$ to get

$$7 \text{ yd} = \frac{7 \text{ yd}}{1} \cdot \frac{3 \text{ ft}}{1 \text{ yd}} \qquad \text{Write 7 yd as a fraction: } 7 \text{ yd} = \frac{7 \text{ yd}}{1}. \text{ Then multiply by 1: } \frac{3 \text{ ft}}{1 \text{ yd}} = 1.$$

$$= \frac{7 \overset{1}{\cancel{\text{yd}}}}{1} \cdot \frac{3 \text{ ft}}{1 \underset{1}{\cancel{\text{yd}}}} \qquad \text{The units of yards divide out.}$$

$$= 7 \cdot 3 \text{ ft}$$

$$= 21 \text{ ft} \qquad \text{Multiply: } 7 \cdot 3 = 21.$$

Seven yards is equal to 21 feet.

Self Check
Convert 9 yards to feet.

Answer: 27 ft   ∎

Notice that in Example 4, we eliminated the units of yards and introduced the units of feet by multiplying by the appropriate unit conversion factor. In general, a unit conversion factor is a fraction with the following form:

$$\frac{\text{Unit we want to introduce}}{\text{Unit we want to eliminate}} \quad \begin{array}{l} \longleftarrow \text{ Numerator} \\ \longleftarrow \text{ Denominator} \end{array}$$

EXAMPLE 5   *Converting from feet to inches.*   Convert $1\frac{3}{4}$ feet to inches.

Solution

To convert from feet to inches, we must use a unit conversion factor that relates inches to feet. Since there are 12 inches per foot, we multiply $1\frac{3}{4}$ feet by the unit conversion factor $\frac{12 \text{ in.}}{1 \text{ ft}}$ to get

$$1\frac{3}{4} \text{ ft} = \frac{7}{4} \text{ ft} \cdot \frac{\mathbf{12 \text{ in.}}}{\mathbf{1 \text{ ft}}} \qquad \text{Write } 1\frac{3}{4} \text{ as an improper fraction: } 1\frac{3}{4} = \frac{7}{4}. \text{ Multiply by 1:}$$
$$\frac{12 \text{ in.}}{1 \text{ ft}} = 1.$$

$$= \frac{7}{4} \overset{1}{\cancel{\text{ft}}} \cdot \frac{12 \text{ in.}}{\underset{1}{\cancel{1 \text{ ft}}}} \qquad \text{The units of feet divide out.}$$

$$= \frac{7 \cdot 12}{4 \cdot 1} \text{ in.} \qquad \text{Multiply the fractions.}$$

$$= 21 \text{ in.} \qquad \text{Simplify by dividing out the common factors:}$$

$$\frac{7 \cdot 12}{4 \cdot 1} = \frac{7 \cdot 3 \cdot \overset{1}{\cancel{4}}}{\underset{1}{\cancel{4}} \cdot 1} = 7 \cdot 3 = 21.$$

$1\frac{3}{4}$ feet is equal to 21 inches.

Self Check

Convert 1.5 feet to inches.

Answer: 18 in.

Sometimes we must use two unit conversion factors in combination to eliminate the given units while introducing the desired units. The following example illustrates this concept.

Accent on Technology: **Finding the length of a football field in miles**

A football field (including the end zones) is 120 yards long. To find this distance in miles, we set up the problem so that the units of yards divide out and leave us with units of miles. Since there are 3 feet per yard and 5,280 feet per mile, we multiply 120 yards by $\frac{3 \text{ ft}}{1 \text{ yd}}$ and $\frac{1 \text{ mi}}{5,280 \text{ ft}}$.

$$120 \text{ yd} = 120 \text{ yd} \cdot \frac{3 \text{ ft}}{1 \text{ yd}} \cdot \frac{1 \text{ mi}}{5,280 \text{ ft}} \qquad \begin{array}{l} \text{Use two unit conversion factors:} \\ \frac{3 \text{ ft}}{1 \text{ yd}} = 1 \text{ and } \frac{1 \text{ mi}}{5,280 \text{ ft}} = 1. \end{array}$$

$$= \frac{120 \overset{1}{\cancel{\text{yd}}}}{1} \cdot \frac{3 \overset{1}{\cancel{\text{ft}}}}{1 \underset{1}{\cancel{\text{yd}}}} \cdot \frac{1 \text{ mi}}{5,280 \underset{1}{\cancel{\text{ft}}}} \qquad \text{Divide out the units of yards and feet.}$$

$$= \frac{120 \cdot 3}{5,280} \text{ mi} \qquad \text{Multiply the fractions.}$$

We can do this arithmetic using a scientific calculator by entering these numbers and pressing these keys.

Keystrokes   120 $\boxed{\times}$ 3 $\boxed{\div}$ 5280 $\boxed{=}$ $\boxed{\text{0.0681818}}$

To the nearest hundredth, a football field is 0.07 mile long.

American units of weight

American units of weight are related in the following ways.

American units of weight

> 16 ounces (oz) = 1 pound (lb)
>
> 2,000 pounds = 1 ton

To convert units of weight, we use the following unit conversion factors.

To convert from	Use the unit conversion factor	To convert from	Use the unit conversion factor
pounds to ounces	$\frac{16\ oz}{1\ lb}$	ounces to pounds	$\frac{1\ lb}{16\ oz}$
tons to pounds	$\frac{2,000\ lb}{1\ ton}$	pounds to tons	$\frac{1\ ton}{2,000\ lb}$

EXAMPLE 6   *Converting from ounces to pounds.*   Convert 40 ounces to pounds.

Solution

Since there is 1 pound per 16 ounces, we multiply 40 ounces by the unit conversion factor $\frac{1\ lb}{16\ oz}$ to get

$$40\ oz = \frac{40\ oz}{1} \cdot \frac{1\ lb}{16\ oz} \qquad \text{Write 40 oz as a fraction: } 40\ oz = \frac{40\ oz}{1}. \text{ Then multiply by 1: } \frac{1\ lb}{16\ oz} = 1.$$

$$= \frac{40\ \overset{1}{\cancel{oz}}}{1} \cdot \frac{1\ lb}{16\ \underset{1}{\cancel{oz}}} \qquad \text{The units of ounces divide out.}$$

$$= \frac{40}{16}\ lb \qquad \text{Multiply the fractions.}$$

There are two ways to complete the solution. First, we can divide out the common factors of the numerator and denominator and then write the result as a mixed number.

$$\frac{40}{16}\ lb = \frac{\overset{1}{\cancel{8}} \cdot 5}{\underset{1}{\cancel{8}} \cdot 2}\ lb = \frac{5}{2}\ lb = 2\frac{1}{2}\ lb$$

A second approach is to divide the numerator by the denominator and express the result as a decimal.

$$\frac{40}{16}\ lb = 2.5\ lb \qquad \text{Do the division: } 40 \div 16 = 2.5.$$

Forty ounces is equal to $2\frac{1}{2}$ lb (or 2.5 lb).

EXAMPLE 7   *Converting from pounds to ounces.*   Convert 25 pounds to ounces.

Solution

Since there are 16 ounces per pound, we multiply 25 pounds by the unit conversion factor $\frac{16\ oz}{1\ lb}$ to get

Self Check
Convert 60 ounces to pounds.

Answer: $3\frac{3}{4}$ lb = 3.75 lb   ∎

Self Check
Convert 60 pounds to ounces.

$$25 \text{ lb} = \frac{25 \text{ lb}}{1} \cdot \frac{16 \text{ oz}}{1 \text{ lb}}$$ Multiply by 1: $\frac{16 \text{ oz}}{1 \text{ lb}} = 1$.

$$= \frac{25 \overset{1}{\cancel{\text{lb}}}}{1} \cdot \frac{16 \text{ oz}}{1 \underset{1}{\cancel{\text{lb}}}}$$ The units of pounds divide out.

$$= 25 \cdot 16 \text{ oz}$$

$$= 400 \text{ oz}$$ Multiply: $25 \cdot 16 = 400$.

Twenty-five pounds is equal to 400 ounces.

Answer: 960 oz   ■

Accent on Technology: **Finding the weight of a car in pounds**

A BMW 323Ci convertible weighs 1.78 tons. To find its weight in pounds, we set up the problem so that the units of tons divide out and leave us with pounds. Since there are 2,000 pounds per ton, we multiply by $\frac{2,000 \text{ lb}}{1 \text{ ton}}$.

$$1.78 \text{ tons} = \frac{1.78 \text{ tons}}{1} \cdot \frac{2,000 \text{ lb}}{1 \text{ ton}}$$ Multiply by 1: $\frac{2,000 \text{ lb}}{1 \text{ ton}} = 1$.

$$= \frac{1.78 \overset{1}{\cancel{\text{tons}}}}{1} \cdot \frac{2,000 \text{ lb}}{1 \underset{1}{\cancel{\text{ton}}}}$$ Divide out the units of tons.

$$= 1.78 \cdot 2,000 \text{ lb}$$

We can do this multiplication using a scientific calculator by entering these numbers and pressing these keys.

Keystrokes   1.78 ☒ 2000 ⊟ | 3560 |

The convertible weighs 3,560 pounds.

American units of capacity

American units of capacity are related as follows.

American units of capacity

1 cup (c) = 8 fluid ounces (fl oz)	1 pint (pt) = 2 cups (c)
1 quart (qt) = 2 pints (pt)	1 gallon (gal) = 4 quarts (qt)

To convert units of capacity, we use the following unit conversion factors.

To convert from	Use the unit conversion factor	To convert from	Use the unit conversion factor
cups to ounces	$\frac{8 \text{ fl oz}}{1 \text{ c}}$	ounces to cups	$\frac{1 \text{ c}}{8 \text{ fl oz}}$
pints to cups	$\frac{2 \text{ c}}{1 \text{ pt}}$	cups to pints	$\frac{1 \text{ pt}}{2 \text{ c}}$
quarts to pints	$\frac{2 \text{ pt}}{1 \text{ qt}}$	pints to quarts	$\frac{1 \text{ qt}}{2 \text{ pt}}$
gallons to quarts	$\frac{4 \text{ qt}}{1 \text{ gal}}$	quarts to gallons	$\frac{1 \text{ gal}}{4 \text{ qt}}$

EXAMPLE 8 *Using two unit conversion factors.* If a recipe calls for 3 pints of milk, how many fluid ounces of milk should be used?

Self Check
How many pints are in 1 gallon?

Solution

Since there are 2 cups per pint and 8 fluid ounces per cup, we multiply 3 pints by unit conversion factors of $\frac{2\,c}{1\,pt}$ and $\frac{8\,fl\,oz}{1\,c}$.

$$3\ pt = \frac{3\ pt}{1} \cdot \frac{2\ c}{1\ pt} \cdot \frac{8\ fl\ oz}{1\ c}$$ Use two unit conversion factors: $\frac{2\,c}{1\,pt} = 1$ and $\frac{8\,fl\,oz}{1\,c} = 1$.

$$= \frac{3\ \overset{1}{\cancel{pt}}}{1} \cdot \frac{2\ \overset{1}{\cancel{c}}}{1\ \underset{1}{\cancel{pt}}} \cdot \frac{8\ fl\ oz}{1\ \underset{1}{\cancel{c}}}$$ Divide out the units of pints and cups.

$$= 3 \cdot 2 \cdot 8\ fl\ oz$$

$$= 48\ fl\ oz$$

Since 3 pints is equal to 48 fluid ounces, 48 fluid ounces of milk should be used.

Answer: 8 pt

Units of time

Units of time are related in the following ways.

Units of time	1 minute (min) = 60 seconds (sec) 1 hour (hr) = 60 minutes
	1 day = 24 hours

EXAMPLE 9
Astronomy. A lunar eclipse occurs when the Earth is between the sun and the moon in such a way that the Earth's shadow darkens the moon.

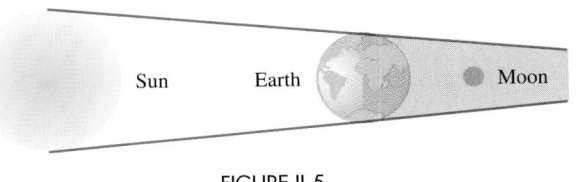

FIGURE II-5

Self Check
A solar eclipse (eclipse of the sun) can last as long as 450 seconds. How many minutes is this?

See Figure II-5 (which is not to scale). A total lunar eclipse can last as long as 105 minutes. How many hours is this?

Solution

Since there is 1 hour for every 60 minutes, we multiply 105 by the unit conversion factor $\frac{1\,hr}{60\,min}$ to get

$$105\ min = \frac{105\ min}{1} \cdot \frac{1\ hr}{60\ min}$$ Multiply by 1: $\frac{1\,hr}{60\,min} = 1$.

$$= \frac{105\ \cancel{min}}{1} \cdot \frac{1\ hr}{60\ \cancel{min}}$$ The units of minutes divide out.

$$= \frac{105}{60}\ hr$$ Multiply the fractions.

$$= \frac{7 \cdot \overset{1}{\cancel{5}} \cdot \overset{1}{\cancel{3}}}{\underset{1}{\cancel{5}} \cdot \underset{1}{\cancel{3}} \cdot 2 \cdot 2}\ hr$$ Prime factor 105 and 60. Then divide out common factors of the numerator and denominator.

$$= \frac{7}{4}\ hr$$

$$= 1\tfrac{3}{4}\ hr$$ Write $\frac{7}{4}$ as a mixed number.

A total lunar eclipse can last as long as $1\tfrac{3}{4}$ hours.

Answer: $7\tfrac{1}{2}$ min

STUDY SET Section II.1

VOCABULARY *Fill in the blanks.*

1. Inches, feet, and miles are examples of American units of _____.
2. A ruler is used for measuring _____.
3. The value of any unit conversion factor is _____.
4. Ounces, pounds, and tons are examples of American units of _____.
5. Some examples of American units of _____ are cups, pints, quarts, and gallons.
6. Some units of _____ are seconds, hours, and days.

CONCEPTS *Fill in the blanks.*

7. 12 in. = ___ ft
8. ___ ft = 1 yd
9. 1 mi = ___ ft
10. 1 yd = ___ in.
11. ___ ounces = 1 pound
12. ___ pounds = 1 ton
13. 1 cup = ___ fluid ounces
14. 1 pint = ___ cups
15. 2 pints = ___ quart(s)
16. 4 quarts = ___ gallon
17. 1 day = ___ hours
18. 2 hours = ___ minutes

19. Tell which measurements the arrows point to on the ruler in Illustration 1.

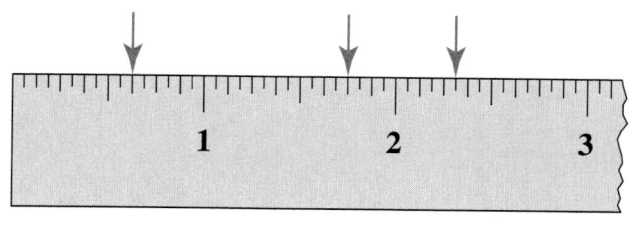

ILLUSTRATION 1

20. Tell which measurements the arrows point to on the ruler in Illustration 2, to the nearest $\frac{1}{8}$ inch.

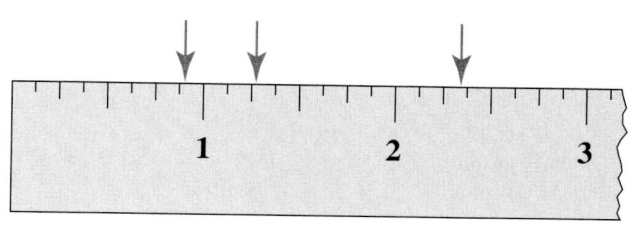

ILLUSTRATION 2

21. Write a unit conversion factor to convert the following.
    a. Pounds to tons
    b. Quarts to pints

22. Write the two unit conversion factors used to convert the following.
    a. Inches to yards
    b. Days to minutes

23. Match each item with its proper measurement.
    a. Length of the U.S. coastline
    b. Height of a Barbie doll
    c. Span of the Golden Gate Bridge
    d. Width of a football field

    i. $11\frac{1}{2}$ in.
    ii. 4,200 ft
    iii. 53.5 yd
    iv. 12,383 mi

24. Match each item with its proper measurement.
    a. Weight of the men's shot put used in track and field
    b. Weight of an African elephant
    c. Amount of gold that is worth $500

    i. $1\frac{1}{2}$ oz
    ii. 16 lb
    iii. 7.2 tons

25. Match each item with its proper measurement.
    a. Amount of blood in an adult
    b. Size of the Exxon Valdez oil spill in 1989
    c. Amount of nail polish in a bottle
    d. Amount of flour to make 3 dozen cookies

    i. $\frac{1}{2}$ fluid oz
    ii. 2 cups
    iii. 5 qt
    iv. 10,080,000 gal

26. Match each item with its proper measurement.
  a. Length of first U.S. manned space flight
  b. A leap year
  c. Time difference between New York and Fairbanks, Alaska
  d. Length of Wright Brothers' first flight

  i. 12 sec
  ii. 15 min
  iii. 4 hr
  iv. 366 days

NOTATION *Complete each solution.*

27. Convert 12 yards to inches.

$$12 \text{ yd} = 12 \text{ yd} \cdot \frac{\boxed{} \text{ in.}}{1 \text{ yd}}$$
$$= 12 \cdot \boxed{} \text{ in.}$$
$$= 432 \text{ in.}$$

28. Convert 1 ton to ounces.

$$1 \text{ ton} = 1 \text{ ton} \cdot \frac{\boxed{} \text{ lb}}{1 \text{ ton}} \cdot \frac{\boxed{} \text{ oz}}{1 \text{ lb}}$$
$$= 1 \cdot 2,000 \cdot 16 \text{ oz}$$
$$= \boxed{} \text{ oz}$$

29. Convert 12 pints to gallons.

$$12 \text{ pt} = 12 \text{ pt} \cdot \frac{1 \text{ qt}}{\boxed{} \text{ pt}} \cdot \frac{1 \text{ gal}}{\boxed{} \text{ qt}}$$
$$= \boxed{} \cdot \frac{1}{2} \cdot \frac{1}{4} \text{ gal}$$
$$= 1.5 \text{ gal}$$

30. Convert 37,440 minutes to days.

$$37,440 \text{ min} = 37,440 \text{ min} \cdot \frac{1 \text{ hr}}{\boxed{} \text{ min}} \cdot \frac{1 \text{ day}}{\boxed{} \text{ hr}}$$
$$= \frac{\boxed{}}{60 \cdot 24} \text{ days}$$
$$= 26 \text{ days}$$

PRACTICE *Use a ruler with a scale in inches to measure each object to the nearest $\frac{1}{8}$ inch.*

31. The width of a dollar bill
32. The length of a dollar bill
33. The length (top to bottom) of this page
34. The length of the following word: supercalifragilisticexpialidocious

Do each conversion.

35. 4 feet to inches
36. 7 feet to inches

37. $3\frac{1}{2}$ feet to inches
38. $2\frac{2}{3}$ feet to inches

39. 24 inches to feet
40. 54 inches to feet

41. 8 yards to inches
42. 288 inches to yards

43. 90 inches to yards
44. 12 yards to inches

45. 56 inches to feet
46. 44 inches to feet

47. 5 yards to feet
48. 21 feet to yards

49. 7 feet to yards $2\frac{1}{3}$
50. $4\frac{2}{3}$ yards to feet

51. 15,840 feet to miles
52. 2 miles to feet

53. $\frac{1}{2}$ mile to feet
54. 1,320 feet to miles

55. 80 ounces to pounds
56. 8 pounds to ounces

57. 7,000 pounds to tons
58. 2.5 tons to ounces

59. 12.4 tons to pounds
60. 48,000 ounces to tons

61. 3 quarts to pints
62. 20 quarts to gallons

63. 16 pints to gallons
64. 3 gal to fluid ounces

65. 32 fluid ounces to pints
66. 2 quarts to fluid ounces

67. 240 minutes to hours
68. 2,400 seconds to hours

69. 7,200 minutes to days
70. 691,200 seconds to days

APPLICATIONS

71. THE GREAT PYRAMID   The Great Pyramid in Egypt is about 450 feet high. Express this distance in yards.

72. THE WRIGHT BROTHERS   In 1903, Orville Wright made the world's first sustained flight. It lasted 12 seconds, and the plane traveled 120 feet. Express the length of the flight in yards.

73. THE GREAT SPHINX   The Great Sphinx of Egypt is 240 feet long. Express this in inches.

74. HOOVER DAM   The Hoover Dam in Nevada is 726 feet high. Express this distance in inches.

75. THE SEARS TOWER   The Sears Tower in Chicago has 110 stories and is 1,454 feet tall. To the nearest hundredth, express this height in miles.

76. NFL RECORDS   Walter Payton, the former Chicago Bears running back, holds the National Football League record for yards rushing in a career: 16,726. How many miles is this? Round to the nearest tenth of a mile.

77. NFL RECORDS   When Dan Marino of the Miami Dolphins retired, it was noted that Marino's career passing total was nearly 35 miles! How many yards is this?

78. LEWIS AND CLARK   The trail traveled by the Lewis and Clark expedition is shown in Illustration 3. When the expedition reached the Pacific Ocean, Clark estimated that they had traveled 4,162 miles. (It was later determined that his guess was within 40 miles of the actual distance.) Express Clark's estimate of the distance in terms of feet.

ILLUSTRATION 3

79. WEIGHT OF WATER   One gallon of water weighs about 8 pounds. Express this weight in ounces.

80. WEIGHT OF A BABY   A newborn baby weighed 136 ounces. Express this weight in pounds.

81. HIPPO   An adult hippopotamus can weigh as much as 9,900 pounds. Express this weight in tons.

82. ELEPHANT   An adult elephant can consume as much as 495 pounds of grass and leaves in one day. How many ounces is this?

83. BUYING PAINT   A painter estimates that he will need 17 gallons of paint for a job. To take advantage of a closeout sale on quart cans, he decides to buy the paint in quarts. How many cans will he need to buy?

84. CATERING   How many cups of apple cider can be dispensed from a 10-gallon container of cider?

85. SCHOOL LUNCHES   Each student attending Eagle River Elementary School receives one pint of milk for lunch each day. If 575 students attend the school, how many gallons of milk are used each day?

86. RADIATOR   The radiator capacity of a piece of earth-moving equipment is 39 quarts. If the radiator is drained and new coolant put in, how many gallons of new coolant will be used?

87. CAMPING   How many ounces of camping stove fuel will fit in the container shown in Illustration 4?

ILLUSTRATION 4

88. HIKING   A college student walks 11 miles in 155 minutes. To the nearest tenth, how many hours does he walk?

89. SPACE TRAVEL   The astronauts of the Apollo 8 mission, which was launched on December 21, 1968, were in space for 147 hours. How many days did the mission take?

90. AMELIA EARHART   In 1935, Amelia Earhart became the first woman to fly across the Atlantic Ocean alone, establishing a new record for the crossing: 13 hours and 30 minutes. How many minutes is this?

II.2 *Metric Units of Measurement*

In this section, you will learn about

- Metric units of length • Converting units of length
- Metric units of mass   • Metric units of capacity
- Cubic centimeters

INTRODUCTION. The metric system is the system of measurement used by most countries in the world. All countries, including the United States, use it for scientific purposes. The metric system, like our decimal numeration system, is based on the number 10. For this reason, converting from one metric unit to another is easier than with the American system.

Metric units of length

The basic metric unit of length is the **meter** (m). One meter is approximately 39 inches, slightly more than 1 yard. Figure II-6 shows the relative sizes of a yardstick and a meterstick.

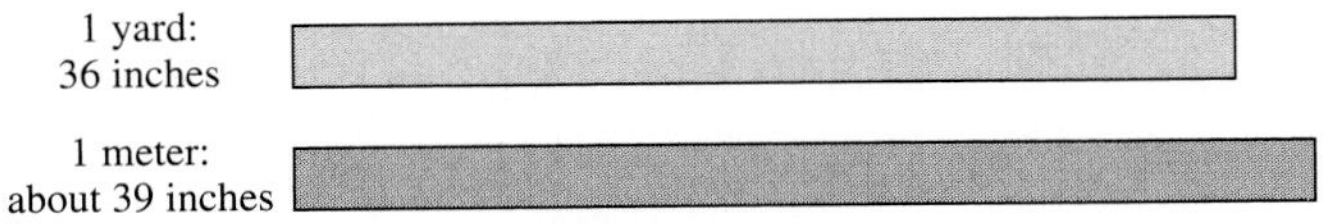

1 yard:
36 inches

1 meter:
about 39 inches

FIGURE II-6

Larger and smaller units are designated by prefixes in front of this basic unit, *meter*.

deka means tens   *deci* means tenths

hecto means hundreds   *centi* means hundredths

kilo means thousands   *milli* means thousandths

Metric units of length

1 dekameter (dam) = 10 meters. 1 dam is a little less than 11 yards.	**1 decimeter** (dm) = $\frac{1}{10}$ of 1 meter. 1 dm is about the length of your palm.
1 hectometer (hm) = 100 meters. 1 hm is about 1 football field long, plus one end zone.	**1 centimeter** (cm) = $\frac{1}{100}$ of 1 meter. 1 cm is about as wide as the nail of your little finger.
1 kilometer (km) = 1,000 meters. 1 km is about $\frac{3}{5}$ mile.	**1 millimeter** (mm) = $\frac{1}{1,000}$ of 1 meter. 1 mm is about the thickness of a dime.

TABLE II-1

Figure II-7 shows a portion of a metric ruler, scaled in centimeters, and a ruler scaled in inches. The rulers are used to measure several lengths.

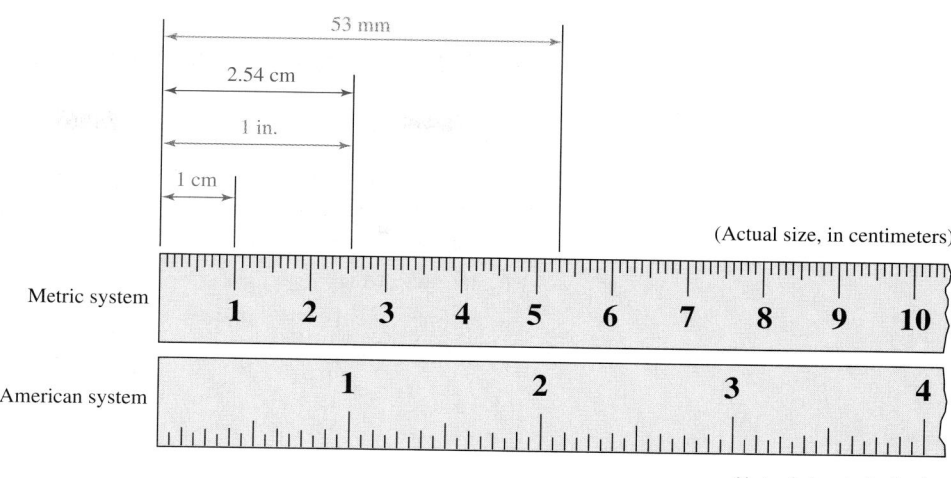

FIGURE II-7

EXAMPLE 1 *Measuring the length of a nail.* To the nearest centimeter, find the length of the nail in Figure II-8.

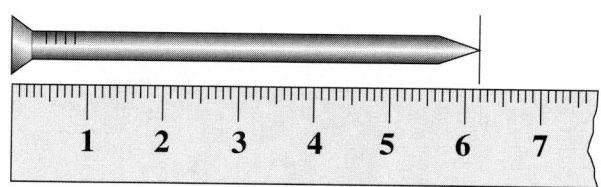

FIGURE II-8

Solution

We place the end of the ruler by one end of the nail and note that the other end of the nail is closer to the 6-cm mark than to the 7-cm mark on the ruler. To the nearest centimeter, the nail is 6 cm long.

EXAMPLE 2 *Measuring the length of a paper clip.* To the nearest millimeter, find the length of the paper clip in Figure II-9.

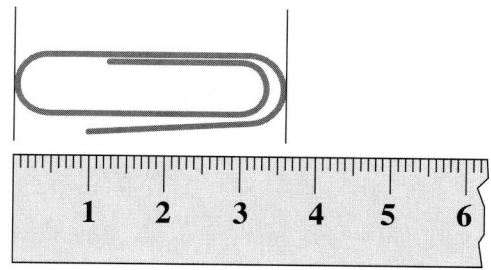

FIGURE II-9

Solution

On the ruler, each centimeter has been divided into 10 millimeters. We place the end of the ruler by one end of the paper clip and note that the other end is closer to the 36-mm mark than to the 37-mm mark on the ruler. To the nearest millimeter, the paper clip is 36 mm long.

Self Check

To the nearest centimeter, find the width of the circle below.

Answer: 3 cm ■

Self Check

To the nearest millimeter, find the length of the jumbo paper clip below.

Answer: 47 mm ■

Converting units of length

Metric units of length are related as shown in Table II-2.

Metric units of length

1 kilometer (km) = 1,000 meters	or	1 meter = $\frac{1}{1,000}$ kilometer
1 hectometer (hm) = 100 meters	or	1 meter = $\frac{1}{100}$ hectometer
1 dekameter (dam) = 10 meters	or	1 meter = $\frac{1}{10}$ dekameter
1 decimeter (dm) = $\frac{1}{10}$ meter	or	1 meter = 10 decimeters
1 centimeter (cm) = $\frac{1}{100}$ meter	or	1 meter = 100 centimeters
1 millimeter (mm) = $\frac{1}{1,000}$ meter	or	1 meter = 1,000 millimeters

TABLE II-2

We can use the information in the table to write unit conversion factors that can be used to convert metric units of length. For example, in the table we see that

$$1 \text{ meter} = 100 \text{ centimeters}$$

From this fact, we can write two unit conversion factors:

$$\frac{1 \text{ m}}{100 \text{ cm}} = 1 \quad \text{and} \quad \frac{100 \text{ cm}}{1 \text{ m}} = 1$$

To obtain the first unit conversion factor, divide both sides of the equation 1 m = 100 cm by 100 cm. To obtain the second unit conversion factor, divide both sides by 1 m.

One advantage of the metric system is that multiplying or dividing by a unit conversion factor involves multiplying or dividing by a power of 10.

EXAMPLE 3 *Changing centimeters to meters.* Convert 350 centimeters to meters.

Solution

Since there is 1 meter per 100 centimeters, we multiply 350 centimeters by the unit conversion factor $\frac{1 \text{ m}}{100 \text{ cm}}$ to get

$$350 \text{ cm} = \frac{350 \text{ cm}}{1} \cdot \frac{1 \text{ m}}{100 \text{ cm}} \qquad \text{Multiply by 1: } \frac{1 \text{ m}}{100 \text{ cm}} = 1.$$

$$= \frac{350 \overset{1}{\cancel{\text{cm}}}}{1} \cdot \frac{1 \text{ m}}{100 \underset{1}{\cancel{\text{cm}}}} \qquad \text{The units of centimeters divide out.}$$

$$= \frac{350}{100} \text{ m}$$

$$= 3.5 \text{ m} \qquad \text{Divide by 100 by moving the decimal point 2 places to the left.}$$

Thus, 350 centimeters = 3.5 meters.

Self Check

Convert 860 centimeters to meters.

Answer: 8.6 m

In Example 3, we converted 350 centimeters to meters using a unit conversion factor. We can also make this conversion by recognizing that all units of length in the metric system are powers of 10 of a meter. Converting from one unit to another is as simple as multiplying by the correct power of 10, or equivalently, by moving a decimal point the correct number of places to the right or left. For example, in the chart below, we see that to convert from centimeters to meters, we move two units to the left.

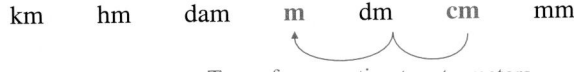

km   hm   dam   **m**   dm   **cm**   mm

To go from centimeters to meters, we must move 2 places to the left.

If we write 350 centimeters as 350.0 centimeters, we can convert to meters by moving the decimal point two places to the left.

$$350.0 \text{ centimeters} = 3.50.0 \text{ meters} = 3.5 \text{ meters}$$

With the unit conversion factor method or the chart method, we get 350 cm = 3.5 m.

 COMMENT   When using a chart to help make a metric conversion, be sure to list the units from largest to smallest when reading from left to right.

EXAMPLE 4   *Changing meters to millimeters.*   Convert 2.4 meters to millimeters.

Solution

Since there are 1,000 millimeters per meter, we multiply 2.4 meters by the unit conversion factor $\frac{1,000 \text{ mm}}{1 \text{ m}}$ to get

$$2.4 \text{ m} = \frac{2.4 \text{ m}}{1} \cdot \frac{1,000 \text{ mm}}{1 \text{ m}} \qquad \text{Multiply by 1: } \frac{1,000 \text{ mm}}{1 \text{ m}} = 1.$$

$$= \frac{\overset{1}{2.4 \, \cancel{m}}}{1} \cdot \frac{1,000 \text{ mm}}{\underset{1}{1 \, \cancel{m}}} \qquad \text{The units of meters divide out.}$$

$$= 2.4 \cdot 1,000 \text{ mm}$$

$$= 2,400 \text{ mm} \qquad \text{Multiply by 1,000 by moving the decimal point 3 places to the right.}$$

Thus, 2.4 meters = 2,400 millimeters.

We can also make this conversion using a chart.

km   hm   dam   **m**   dm   cm   **mm**

From the chart, we see that we should move the decimal point 3 places to the right to convert from meters to millimeters.

$$2.4 \text{ meters} = 2.400. \text{ millimeters} = 2,400 \text{ millimeters}$$

Self Check

Convert 5.3 meters to millimeters.

Answer: 5,300 mm

EXAMPLE 5   *Using two unit conversion factors.*   Convert 3.2 kilometers to centimeters.

Solution

To convert to centimeters, we set up the problem so the units of kilometers divide out and leave us with units of centimeters. Since there are 1,000 meters per kilometer and 100 centimeters per meter, we multiply 3.2 kilometers by $\frac{1,000 \text{ m}}{1 \text{ km}}$ and $\frac{100 \text{ cm}}{1 \text{ m}}$.

$$3.2 \text{ km} = \frac{3.2 \, \overset{1}{\cancel{km}}}{1} \cdot \frac{1,000 \, \overset{1}{\cancel{m}}}{\underset{1}{1 \, \cancel{km}}} \cdot \frac{100 \text{ cm}}{1 \, \cancel{m}} \qquad \text{The units of kilometers and meters divide out.}$$

$$= 3.2 \cdot 1,000 \cdot 100 \text{ cm}$$

$$= 320,000 \text{ cm} \qquad \text{Multiply by 1,000 and 100 by moving the decimal point 5 places to the right.}$$

Thus, 3.2 kilometers = 320,000 centimeters.

Self Check

Convert 5.15 kilometers to centimeters.

Using a chart, we see that the decimal point should be moved 5 places to the right to convert kilometers to centimeters.

km   hm   dam   m   dm   **cm**   mm

3.2 kilometers = 3.20000. centimeters = 320,000 centimeters

Answer: 515,000 cm   ∎

Metric units of mass

The **mass** of an object is a measure of the amount of material in the object. When an object is moved about in space, its mass does not change. One basic unit of mass in the metric system is the **gram (g).** A gram is defined to be the mass of water contained in a cube having sides 1 centimeter long. (See Figure II-10.)

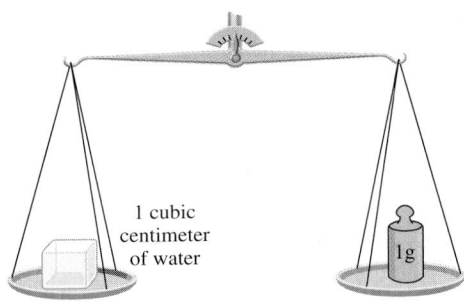

1 cubic
centimeter
of water

1g

FIGURE II-10

The **weight** of an object is determined by Earth's gravitational pull on the object. Since gravitational pull on an object decreases as the object gets farther from Earth, the object weighs less as it gets farther from Earth's surface. This is why astronauts experience weightlessness in space. However, since most of us remain near Earth's surface, we will use the words *mass* and *weight* interchangeably. Thus, a mass of 30 grams is said to weigh 30 grams.

Metric units of mass are related as shown in Table II-3.

Metric units of mass

1 kilogram (kg) = 1,000 grams	or	1 gram = $\frac{1}{1,000}$ kilogram
1 hectogram (hg) = 100 grams	or	1 gram = $\frac{1}{100}$ hectogram
1 dekagram (dag) = 10 grams	or	1 gram = $\frac{1}{10}$ dekagram
1 decigram (dg) = $\frac{1}{10}$ gram	or	1 gram = 10 decigrams
1 centigram (cg) = $\frac{1}{100}$ gram	or	1 gram = 100 centigrams
1 milligram (mg) = $\frac{1}{1,000}$ gram	or	1 gram = 1,000 milligrams

TABLE II-3

Here are some examples of these units of mass:

• An average bowling ball weighs about 6 kilograms.

• A raisin weighs about 1 gram.

• A certain vitamin tablet contains 450 milligrams of calcium.

We can use the information in Table II-3 to write unit conversion factors that can be used to convert metric units of mass. For example, in the table we see that

1 kilogram = 1,000 grams

From this fact, we can write two unit conversion factors:

$$\frac{1 \text{ kg}}{1,000 \text{ g}} = 1 \quad \text{and} \quad \frac{1,000 \text{ g}}{1 \text{ kg}} = 1$$

To obtain the first unit conversion factor, divide both sides of the equation 1 kg = 1,000 g by 1,000 g. To obtain the second unit conversion factor, divide both sides by 1 kg.

EXAMPLE 6 *Changing kilograms to grams.* Convert 7.2 kilograms to grams.

Solution

To convert to grams, we set up the problem so that the units of kilograms divide out and leave us with the units of grams. Since there are 1,000 grams per 1 kilogram, we multiply 7.2 kilograms by $\frac{1,000 \text{ g}}{1 \text{ kg}}$.

$$7.2 \text{ kg} = \frac{7.2 \overset{1}{\cancel{\text{kg}}}}{1} \cdot \frac{1,000 \text{ g}}{\underset{1}{\cancel{1 \text{ kg}}}} \qquad \text{Divide out the units of kilograms.}$$

$$= 7.2 \cdot 1,000 \text{ g}$$

$$= 7,200 \text{ g} \qquad \text{Do the multiplication by moving the decimal point 3 places to the right.}$$

Thus, 7.2 kilograms = 7,200 grams.

To use a chart to make the conversion, we list the metric units of weight from the largest (kilograms) to the smallest (milligrams).

kg   hg   dag   g   dg   cg   mg

From the chart, we see that we must move the decimal point 3 places to the right to change kilograms to grams.

7.2 kilograms = 7.200. grams = 7,200 grams

Self Check

Convert 5 kilograms to grams.

Answer: 5,000 g

EXAMPLE 7 *Changing from milligrams to centigrams.* A bottle of Verapamil, a drug taken for high blood pressure, contains 30 tablets. If each tablet contains 180 mg of active ingredient, how many centigrams of active ingredient are in the bottle?

Solution

Since there are 30 tablets and each one contains 180 mg of active ingredient, there are

$$30 \cdot 180 \text{ mg} = 5,400 \text{ mg}$$

of active ingredient in the bottle.

To convert milligrams to centigrams, we multiply 5,400 milligrams by $\frac{1 \text{ g}}{1,000 \text{ mg}}$ and $\frac{100 \text{ cg}}{1 \text{ g}}$ to get

$$5,400 \text{ mg} = \frac{5,400 \overset{1}{\cancel{\text{mg}}}}{1} \cdot \frac{1 \overset{1}{\cancel{\text{g}}}}{\underset{1}{\cancel{1,000 \text{ mg}}}} \cdot \frac{100 \text{ cg}}{\underset{1}{\cancel{1 \text{ g}}}} \qquad \text{Divide out the units of milligrams and grams.}$$

$$= \frac{5,400 \cdot 100}{1,000} \text{ cg} \qquad \text{Multiply the fractions.}$$

$$= 540 \text{ cg} \qquad \text{Simplify.}$$

There are 540 centigrams of active ingredient in the bottle.

Using a chart, we see that we must move the decimal point 1 place to the left to convert from milligrams to centigrams.

kg   hg   dag   g   dg   cg   mg

5,400 milligrams = 540.0. centigrams = 540 centigrams

Self Check

One brand name for Verapamil is Isoptin. If a bottle of Isoptin contains 90 tablets, each containing 200 mg of active ingredient, how many centigrams of active ingredient are in the bottle?

Answer: 1,800 cg

Metric units of capacity

In the metric system, one basic unit of capacity is the **liter** (L), which is defined to be the capacity of a cube with sides 10 centimeters long. (See Figure II-11.) A liter of liquid is slightly more than one quart.

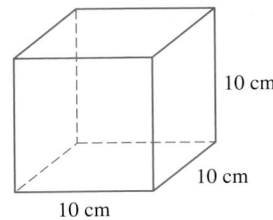

10 cm

10 cm

10 cm

FIGURE II-11

Metric units of capacity are related as shown in Table II-4.

Metric units of capacity

1 kiloliter (kL) = 1,000 liters	or	1 liter = $\frac{1}{1,000}$ kiloliter
1 hectoliter (hL) = 100 liters	or	1 liter = $\frac{1}{100}$ hectoliter
1 dekaliter (daL) = 10 liters	or	1 liter = $\frac{1}{10}$ dekaliter
1 deciliter (dL) = $\frac{1}{10}$ liter	or	1 liter = 10 deciliters
1 centiliter (cL) = $\frac{1}{100}$ liter	or	1 liter = 100 centiliters
1 milliliter (mL) = $\frac{1}{1,000}$ liter	or	1 liter = 1,000 milliliters

TABLE II-4

Here are some examples of these units of capacity:

- Soft drinks are sold in 2-liter plastic bottles.
- The fuel tank of a certain minivan can hold about 75 liters of gasoline.
- Chemists use glass cylinders, scaled in milliliters, to measure liquids.

We can use the information in Table II-4 to write unit conversion factors that can be used to convert metric units of capacity. For example, in the table we see that

$$1 \text{ liter} = 100 \text{ centiliters}$$

From this fact, we can write two unit conversion factors:

$$\frac{1 \text{ L}}{100 \text{ cL}} = 1 \quad \text{and} \quad \frac{100 \text{ cL}}{1 \text{ L}} = 1$$

EXAMPLE 8   *Changing from liters to centiliters.*   How many centiliters are there in three 2-liter bottles of cola?

Solution

Three 2-liter bottles of cola contain 6 liters of cola. To convert to centiliters, we set up the problem so that liters divide out and leave us with centiliters. Since there are 100 centiliters per 1 liter, we multiply 6 liters by the unit conversion factor $\frac{100 \text{ cL}}{1 \text{ L}}$.

$$6 \text{ L} = 6 \text{ L} \cdot \frac{100 \text{ cL}}{1 \text{ L}} \qquad \text{Multiply by 1: } \tfrac{100 \text{ cL}}{1 \text{ L}} = 1.$$

$$= \frac{\overset{1}{6 \text{ L}}}{1} \cdot \frac{100 \text{ cL}}{\underset{1}{1 \text{ L}}} \qquad \text{The units of liters divide out.}$$

$$= 6 \cdot 100 \text{ cL}$$

$$= 600 \text{ cL}$$

Self Check

How many milliliters are in two 2-liter bottles of cola?

Thus, there are 600 centiliters in three 2-liter bottles of cola.

To make this conversion using a chart, we list the metric units of capacity in order from largest (kiloliter) to smallest (milliliter).

kL   hL   daL   L   dL   cL   mL

From the chart, we see that we should move the decimal point 2 places to the right to convert from liters to centiliters.

6 liters = 6.00. centiliters = 600 centiliters

Answer: 4,000 mL   ■

Cubic centimeters

Another metric unit of capacity is the **cubic centimeter,** which is represented by the notation cm^3 or, more simply, cc. One milliliter and one cubic centimeter represent the same capacity.

$$1 \text{ mL} = 1 \text{ cm}^3 = 1 \text{ cc}$$

The units of cubic centimeters are used frequently in medicine. For example, when a nurse administers an injection containing 5 cc of medication, the dosage could also be expressed using milliliters.

$$5 \text{ cc} = 5 \text{ mL}$$

When a doctor orders that a patient be put on 1,000 cc of dextrose solution, the request could be expressed in several ways.

$$1,000 \text{ cc} = 1,000 \text{ mL} = 1 \text{ liter}$$

STUDY SET Section II.2

VOCABULARY   *Fill in the blanks.*

1. *Deka* means _____.

2. *Hecto* means _____.

3. *Kilo* means _____.

4. *Deci* means _____.

5. *Centi* means _____.

6. *Milli* means _____.

7. Meters, grams, and liters are units of measurement in the _____ system.

8. The _____ of an object is determined by the Earth's gravitational pull on the object.

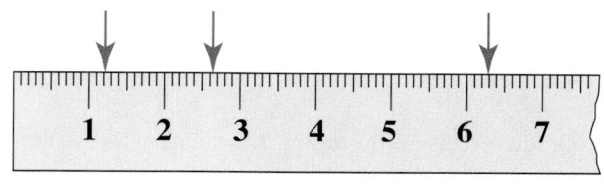

ILLUSTRATION 1

10. To the nearest millimeter, tell which measurements the arrows point to on the ruler in Illustration 2.

CONCEPTS

9. To the nearest centimeter, tell which measurements the arrows point to on the ruler in Illustration 1.

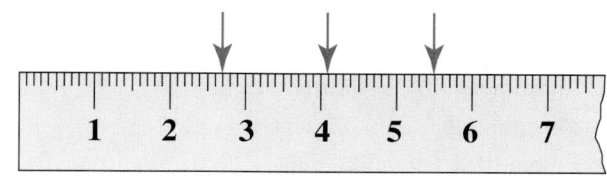

ILLUSTRATION 2

11. Write a unit conversion factor to convert the following.
   a. Meters to kilometers
   b. Grams to centigrams
   c. Liters to milliliters

12. Use the chart to determine how many decimal places and in which direction to move the decimal point when converting the following.
   a. Kilometers to centimeters

   km   hm   dam   m   dm   cm   mm

   b. Milligrams to grams

   kg   hg   dag   g   dg   cg   mg

   c. Hectoliters to centiliters

   kL   hL   daL   L   dL   cL   mL

13. Match each item with its proper measurement.
   a. Thickness of a   **i.** 6,275 km
   phone book
   b. Length of the   **ii.** 2 m
   Amazon River
   c. Height of a   **iii.** 6 cm
   soccer goal

14. Match each item with its proper measurement.
   a. Weight of a   **i.** 800 kg
   giraffe
   b. Weight of a paper   **ii.** 1 g
   clip
   c. Active ingredient   **iii.** 325 mg
   in an aspirin
   tablet

15. Match each item with its proper measurement.
   a. Amount of blood   **i.** 290,000 kL
   in an adult
   b. Cola in an   **ii.** 6 L
   aluminum can
   c. Kuwait's daily   **iii.** 355 mL
   production of
   crude oil

16. Of the objects in Illustration 3, which can be used to measure the following?
   a. Millimeters
   b. Milligrams
   c. Milliliters

Fill in the blanks.

17. 1 dekameter = ____ meters
18. 1 decimeter = ____ meter
19. 1 centimeter = ____ meter
20. 1 kilometer = ____ meters
21. 1 millimeter = ____ meter
22. 1 hectometer = ____ meters
23. 1 gram = ____ milligrams
24. 100 centigrams = ____ gram

Balance

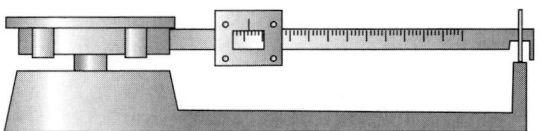

Beaker

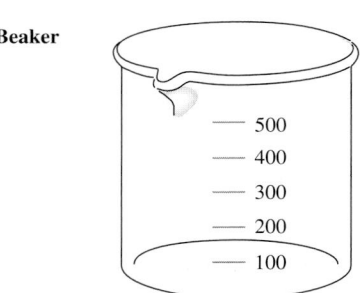

— 500
— 400
— 300
— 200
— 100

Micrometer

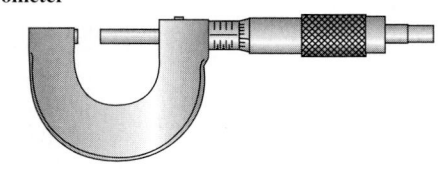

ILLUSTRATION 3

25. 1 kilogram = ____ grams
26. 1 milliliter = ____ cubic centimeter
27. 1 liter = ____ cubic centimeters
28. 1 kiloliter = ____ liters
29. 1 centiliter = ____ liter
30. 1 milliliter = ____ liter
31. 100 liters = ____ hectoliter
32. 10 deciliters = ____ liter

NOTATION   *Complete each solution.*

33. Convert 20 centimeters to meters.

$$20 \text{ cm} = 20 \text{ cm} \cdot \frac{\text{m}}{100 \text{ cm}}$$

$$= \frac{20}{} \text{ m}$$

$$= 0.2 \text{ m}$$

34. Convert 300 centigrams to grams.

$$300 \text{ cg} = 300 \text{ cg} \cdot \frac{\text{g}}{100 \text{ cg}}$$

$$= \frac{}{100} \text{ g}$$

$$= 3 \text{ g}$$

35. Convert 2 kilometers to decimeters.

$$2 \text{ km} = 2 \text{ km} \cdot \frac{\text{m}}{1 \text{ km}} \cdot \frac{10 \text{ dm}}{\text{m}}$$

$$= 2 \cdot \cdot 10 \text{ dm}$$

$$= 20{,}000 \text{ dm}$$

36. Convert 3 deciliters to milliliters.

$$3 \text{ dL} = 3 \text{ dL} \cdot \frac{1 \text{ L}}{\boxed{} \text{ dL}} \cdot \frac{\boxed{} \text{ mL}}{1 \text{ L}}$$

$$= \frac{\boxed{} \cdot 1{,}000}{10} \text{ mL}$$

$$= 300 \text{ mL}$$

PRACTICE   *Use a metric ruler to measure each object to the nearest millimeter.*

37. The length of a dollar bill

38. The width of a dollar bill

Use a metric ruler to measure each object to the nearest centimeter.

39. The length (top to bottom) of this page

40. The length of the word antidisestablishmentarianism

Convert each measurement between the given metric units.

41. 3 m = _____ cm

42. 5 m = _____ cm

43. 5.7 m = _____ cm

44. 7.36 km = _____ dam

45. 0.31 dm = _____ cm

46. 73.2 m = _____ dm

47. 76.8 hm = _____ mm

48. 165.7 km = _____ m

49. 4.72 cm = _____ dm

50. 0.593 cm = _____ dam

51. 453.2 cm = _____ m

52. 675.3 cm = _____ m

53. 0.325 dm = _____ m

54. 0.0034 mm = _____ m

55. 3.75 cm = _____ mm

56. 0.074 cm = _____ mm

57. 0.125 m = _____ mm

58. 134 m = _____ hm

59. 675 dam = _____ cm

60. 0.00777 cm = _____ dam

61. 638.3 m = _____ hm

62. 6.77 cm = _____ m

63. 6.3 mm = _____ cm

64. 6.77 mm = _____ cm

65. 695 dm = _____ m

66. 6,789 cm = _____ dm

67. 5,689 m = _____ km

68. 0.0579 km = _____ mm

69. 576.2 mm = _____ dm

70. 65.78 km = _____ dam

71. 6.45 dm = _____ km

72. 6.57 cm = _____ mm

73. 658.23 m = _____ km

74. 0.0068 hm = _____ km

75. 3 g = _____ mg

76. 5 g = _____ cg

77. 2 kg = _____ g

78. 4,000 g = _____ kg

79. 1,000 kg = _____ g

80. 2 kg = _____ cg

81. 500 mg = _____ g

82. 500 mg = _____ cg

83. 3 kL = _____ L

84. 500 mL = _____ L

85. 500 cL = _____ mL

86. 400 L = _____ hL

87. 10 mL = _____ cc

88. 2,000 cc = _____ L

APPLICATIONS

89. SPEED SKATING   American Eric Heiden won an unprecedented five gold medals by capturing the men's 500 m, 1,000 m, 1,500 m, 5,000 m and 10,000 m races at the 1980 Winter Olympic Games in Lake Placid, New York. Convert each race length to kilometers.

90. SUEZ CANAL   The 163-km-long Suez Canal, shown in Illustration 4, connects the Mediterranean Sea with the Red Sea. It provides a shortcut for ships operating between European and American ports. Convert the length of the Suez Canal to meters.

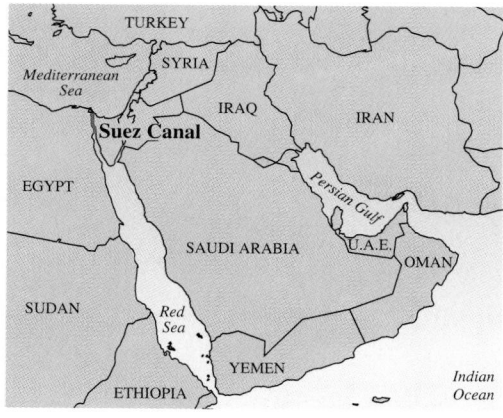

ILLUSTRATION 4

91. HEALTH CARE   Blood pressure is measured by a *sphygmomanometer* (Illustration 5, next page). The measurement is read at two points and is expressed, for example, as 120/80. This indicates a *systolic* pressure of 120 millimeters of mercury and a *diastolic* pressure of 80 millimeters of mercury. Convert each measurement to centimeters of mercury.

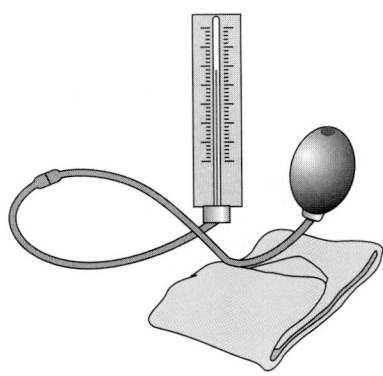

ILLUSTRATION 5

92. THE HANCOCK CENTER   The John Hancock Center in Chicago has 100 stories and is 343 meters high. Give this height in hectometers.

93. WEIGHT OF A BABY   A baby weighs 4 kilograms. Give this weight in centigrams.

94. JEWELRY   A gold chain weighs 1,500 milligrams. Give this weight in grams.

95. CONTAINERS   How many deciliters of root beer are there in two 2-liter bottles?

96. BOTTLING   How many liters of wine are in a 750-mL bottle?

97. BUYING OLIVES   The net weight of a bottle of olives is 284 grams. Find the smallest number of bottles that must be purchased to have at least 1 kilogram of olives.

98. BUYING COFFEE   A can of Cafe Vienna has a net weight of 133 grams. Find the smallest number of cans that must be packaged to have at least 1 metric ton of coffee. (*Hint:* 1 metric ton = 1,000 kg.)

99. MEDICINE   A bottle of hydrochlorothiazine contains 60 tablets. If each tablet contains 50 milligrams of active ingredient, how many grams of active ingredient are in the bottle?

100. INJECTION   Illustration 6 shows a 3cc syringe. Express its capacity using units of milliliters.

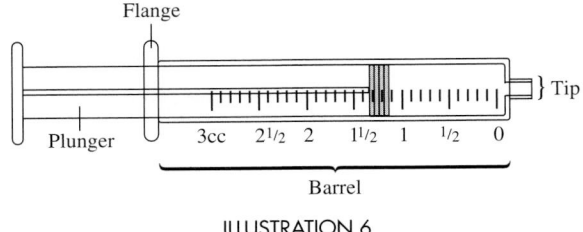

ILLUSTRATION 6

II.3   *Converting between American and Metric Units*

In this section, you will learn about

- Converting between American and metric units
- Comparing American and metric units of temperature

INTRODUCTION. It is often necessary to convert between American units and metric units. For example, we must convert units to answer the following questions:

- Which is higher, Pikes Peak (elevation 14,110 feet) or the Matterhorn (elevation 4,478 meters)?
- Does a 2-pound tub of butter weigh more than a 1-kilogram tub?
- Is a quart of soda pop more or less than a liter of soda pop?

In this section, we will discuss how to answer such questions.

Converting between American and metric units

We can convert between American and metric units of length using the table on the next page.

Equivalent Lengths	
American to metric	**Metric to American**
1 in. = 2.54 cm	1 cm = 0.3937 in.
1 ft = 0.3048 m	1 m = 3.2808 ft
1 yd = 0.9144 m	1 m = 1.0936 yd
1 mi = 1.6093 km	1 km = 0.6214 mi

EXAMPLE 1 *Clothing label.* Figure II-12 shows a label sewn into some pants made in Mexico for sale in the United States. Express the waist size to the nearest inch.

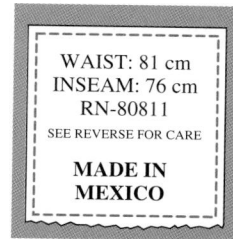

FIGURE II-12

Solution

We need to convert from metric to American units. From the table, we see that there is 0.3937 inch in 1 centimeter. To make the conversion, we substitute 0.3937 inch for 1 centimeter.

81 centimeters = 81 (**centimeters**)

= 81(**0.3937 in.**)   Substitute 0.3937 inch for 1 centimeter.

= 31.8897 in.   Do the multiplication.

To the nearest inch, the waist size is 32 inches.

Self Check

Refer to Figure II-12. What is the inseam length, to the nearest inch?

Answer: 30 in. ∎

EXAMPLE 2 *Mountain elevations.* Pikes Peak, one of the most famous peaks in the Rocky Mountains, has an elevation of 14,110 feet. The Matterhorn, in the Swiss Alps, rises to an elevation of 4,478 meters. Which mountain is higher?

Solution

To make a comparison, the elevations must be expressed in the same units. We will convert the elevation of Pikes Peak, which is given in feet, to meters.

14,110 feet = 14,110 (**feet**)

= 14,110 (**0.3048 m**)   Substitute 0.3048 meters for 1 foot.

= 4,300.728 m   Do the multiplication.

Since the elevation of Pikes Peak is about 4,301 meters, we can conclude that the Matterhorn, with an elevation of 4,478 meters, is higher.

Self Check

Which is longer, a 500-meter race or a 550-yard race?

Answer: the 550-yard race ∎

We can convert between American units of weight and metric units of mass by using the accompanying table.

Equivalent Weights and Masses	
American to metric	**Metric to American**
1 oz = 28.35 g	1 g = 0.035 oz
1 lb = 0.454 kg	1 kg = 2.2 lb

EXAMPLE 3   *Changing pounds to grams.*   Change 50 pounds to grams.

Solution

$$50 \text{ lb} = 50(\mathbf{1 \text{ lb}})$$

$$= 50(\mathbf{16 \text{ oz}}) \qquad \text{Substitute 16 ounces for 1 pound.}$$

$$= 50(16)(1 \text{ oz})$$

$$= 50(16)(\mathbf{28.35 \text{ g}}) \quad \text{Substitute 28.35 grams for 1 ounce.}$$

$$= 22{,}680 \text{ g} \qquad \text{Do the multiplication.}$$

Thus, 50 pounds is equal to 22,680 grams.

Self Check

Change 20 kilograms to pounds.

Answer: 44 lb ■

EXAMPLE 4   *Comparing weights.*   Does a 2-pound tub of butter weigh more than a 1-kilogram tub?

Solution

To decide which contains more butter, we can change 2 pounds to kilograms.

$$2 \text{ lb} = 2(\mathbf{1 \text{ lb}})$$

$$= 2(\mathbf{0.454 \text{ kg}}) \quad \text{Substitute 0.454 kilograms for 1 pound.}$$

$$= 0.908 \text{ kg} \qquad \text{Do the multiplication.}$$

Since a 2-pound tub weighs only 0.908 kilogram, the 1-kilogram tub weighs more.

Self Check

Who weighs more, a person who weighs 165 pounds or one who weighs 76 kilograms?

Answer: the person who weighs 76 kg ■

We can convert between American and metric units of capacity by using the accompanying table.

Equivalent Capacities	
American to metric	**Metric to American**
1 fl oz = 0.030 L	1 L = 33.8 fl oz
1 pt = 0.473 L	1 L = 2.1 pt
1 qt = 0.946 L	1 L = 1.06 qt
1 gal = 3.785 L	1 L = 0.264 gal

EXAMPLE 5   *Changing from milliliters to quarts.*   A bottle of 7UP contains 750 milliliters. Convert this measure to quarts.

Solution

We convert milliliters to liters and then liters to quarts.

$$750 \text{ mL} = 750 \text{ mL} \cdot \frac{1 \text{ L}}{1{,}000 \text{ mL}} \qquad \text{Use a unit conversion factor: } \tfrac{1\,L}{1{,}000\,mL} = 1.$$

$$= \frac{750}{1{,}000} \text{ L} \qquad \text{The units of mL divide out.}$$

$$= \frac{3}{4} \text{ L} \qquad \text{Simplify the fraction: } \tfrac{750}{1{,}000} = \tfrac{3 \cdot 250}{4 \cdot 250} = \tfrac{3}{4}.$$

$$= \frac{3}{4}(\mathbf{1.06 \text{ qt}}) \qquad \text{Substitute 1.06 quart for 1 liter.}$$

$$= 0.795 \text{ qt} \qquad \text{Do the arithmetic.}$$

The bottle contains 0.795 quart.

Self Check

A student bought a 355-mL can of cola. How many ounces of cola does the can contain?

Answer: 12 oz ■

From the table of equivalent capacities, we see that 1 liter is equal to 1.06 quarts. Thus, one liter of soda pop is more than one quart of soda pop.

EXAMPLE 6 *Comparison shopping.* A two-quart bottle of soda pop is priced at $1.89, and a one-liter bottle is priced at 97¢. Which is the better buy?

Solution
We can convert 2 quarts to liters and find the price per liter of the two-quart bottle.

$$2 \text{ qt} = 2(\mathbf{1 \text{ qt}})$$
$$= 2(\mathbf{0.946 \text{ L}}) \quad \text{Substitute 0.946 liter for 1 quart.}$$
$$= 1.892 \text{ L} \quad \text{Do the multiplication.}$$

Thus, the two-quart bottle contains 1.892 liters. To find the price per liter of the two-quart bottle, we divide $\frac{\$1.89}{1.892}$ to get

$$\frac{\$1.89}{1.892} = \$0.998942917$$

Since the price per liter of the two-quart bottle is a little more than 99¢, the one-liter bottle priced at 97¢ is the better buy.

Self Check
Thirty-four fluid ounces of aged vinegar costs $3.49. A one-liter bottle of the same vinegar costs $3.17. Which is the better buy?

Answer: the one-liter bottle ∎

Comparing American and metric units of temperature

In the American system, we measure temperature using **degrees Fahrenheit** (°F). In the metric system, we measure temperature using **degrees Celsius** (°C). These two scales are shown on the thermometers in Figure II-13 on the next page. From the figure, we can see that

- $212° \text{ F} = 100° \text{ C}$ Water boils.
- $32° \text{ F} = 0° \text{ C}$ Water freezes.
- $5° \text{ F} = -15° \text{ C}$ A cold winter day.
- $95° \text{ F} = 35° \text{ C}$ A hot summer day.

As we have seen, there is a formula that enables us to convert from degrees Fahrenheit to degrees Celsius. There is also a formula to convert from degrees Celsius to degrees Fahrenheit.

Conversion formulas for temperature

If F is the temperature in degrees Fahrenheit and C is the corresponding temperature in degrees Celsius, then

$$C = \frac{5(F - 32)}{9} \quad \text{and} \quad F = \frac{9}{5}C + 32$$

An alternate form of the formula $C = \frac{5(F - 32)}{9}$ is obtained by distributing the multiplication by 5 in the numerator to get $C = \frac{5F - 160}{9}$.

EXAMPLE 7 *Converting from degrees Fahrenheit to degrees Celsius.* Warm bath water is 90° F. Find the equivalent temperature in degrees Celsius.

Solution
We substitute 90 for F in the formula $C = \frac{5F - 160}{9}$ and simplify.

$$C = \frac{5F - 160}{9}$$
$$= \frac{5(90) - 160}{9} \quad \text{Substitute 90 for } F.$$

Self Check
Hot coffee is 110° F. To the nearest tenth of a degree, express this temperature in degrees Celsius.

$$= \frac{450 - 160}{9} \qquad \text{Multiply: } 5(90) = 450.$$

$$= 32.222222 \qquad \text{Do the arithmetic.}$$

To the nearest tenth of a degree, the equivalent temperature is 32.2° C.

Answer: 43.3° C

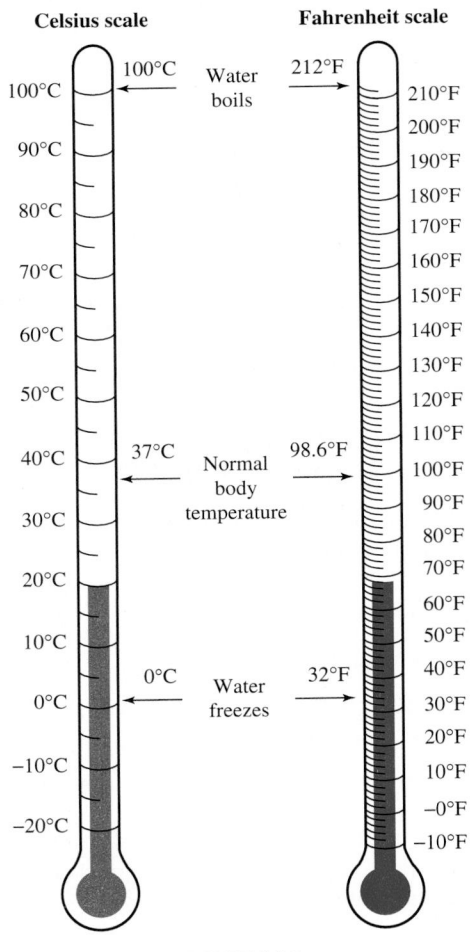

FIGURE II-13

EXAMPLE 8   *Converting from degrees Celsius to degrees Fahrenheit.*   A dishwasher manufacturer recommends that dishes be rinsed in hot water with a temperature of 60° C. Express this temperature in degrees Fahrenheit.

Solution

We substitute 60 for *C* in the formula $F = \frac{9}{5}C + 32$ and simplify.

$$F = \frac{9}{5}C + 32$$

$$= \frac{9}{5}(60) + 32 \qquad \text{Substitute 60 for } C.$$

$$= \frac{540}{5} + 32 \qquad \text{Multiply: } \tfrac{9}{5}(60) = \tfrac{540}{5}.$$

$$= 108 + 32 \qquad \text{Do the division.}$$

$$= 140 \qquad \text{Do the addition.}$$

The manufacturer recommends that dishes be rinsed in 140° F water.

Self Check
To see whether a baby has a fever, her mother takes her temperature with a Celsius thermometer. If the reading is 38.8° C, does the baby have a fever? (*Hint:* Normal body temperature is 98.6° F.)

Answer: yes

STUDY SET   Section II.3

VOCABULARY   *Fill in the blanks.*

1. In the American system, temperatures are measured in degrees _____.

2. In the metric system, temperatures are measured in degrees _____.

CONCEPTS

3. Which is longer?
   a. A yard or a meter?
   b. A foot or a meter?
   c. An inch or a centimeter?
   d. A mile or a kilometer?

4. Which is heavier?
   a. An ounce or a gram?
   b. A pound or a kilogram?

5. Which is the greater unit of capacity?
   a. A pint or a liter?
   b. A quart or a liter?
   c. A gallon or a liter?

6. a. What formula is used for changing degrees Celsius to degrees Fahrenheit?
   b. What formula is used for changing degrees Fahrenheit to degrees Celsius?

NOTATION   *Complete each solution.*

7. Change 4,500 feet to kilometers.

$$4{,}500 \text{ ft} = 4{,}500 \,(\quad\quad \text{m})$$
$$= \quad\quad\quad \text{m}$$
$$= 1.3716 \text{ km}$$

8. Change 3 kilograms to ounces.

$$3 \text{ kg} = 3(\quad \text{lb})$$
$$= 3(2.2)(\quad \text{oz})$$
$$= 105.6 \text{ oz}$$

9. Change 8 liters to gallons.

$$8 \text{ L} = 8(\quad\quad \text{gal})$$
$$= 2.112 \text{ gal}$$

10. Change 70° C to degrees Fahrenheit.

$$F = \frac{9}{5}C + 32$$
$$= \frac{9}{5}(\quad) + 32$$
$$= \quad\quad + 32$$
$$= 158° \text{ F}$$

PRACTICE   *Make each conversion. Since most conversions are approximate, answers will vary slightly depending on the method used.*

11. 3 ft = _____ cm

12. 7.5 yd = _____ m

13. 3.75 m = _____ in.

14. 2.4 km = _____ mi

15. 12 km = _____ ft

16. 3,212 cm = _____ ft

17. 5,000 in. = _____ m

18. 25 mi = _____ km

19. 37 oz = _____ kg

20. 10 lb = _____ kg

21. 25 lb = _____ g

22. 7.5 oz = _____ g

23. 0.5 kg = _____ oz

24. 35 g = _____ lb

25. 17 g = _____ oz

26. 100 kg = _____ lb

27. 3 fl oz = _____ L

28. 2.5 pt = _____ L

29. 7.2 L = _____ fl oz

30. 5 L = _____ qt

31. 0.75 qt = _____ mL

32. 3 pt = _____ mL

33. 500 mL = _____ qt

34. 2,000 mL = _____ gal

35. 50° F = _____ C

36. 67.7° F = _____ C

37. 50° C = _____ F

38. 36.2° C = _____ F

39. −10° C = _____ F

40. −22.5° C = _____ F

41. −5° F = _____ C

42. −10° F = _____ C

APPLICATIONS   *Since most conversions are approximate, answers will vary slightly depending on the method used.*

43. THE MIDDLE EAST   The distance between Jerusalem and Bethlehem is 8 kilometers. To the nearest mile, give this distance in miles.

44. THE DEAD SEA   The Dead Sea is 80 kilometers long. To the nearest mile, give this distance in miles.

45. CHEETAH   A cheetah can run 112 kilometers per hour. How fast is this in mph?

46. LION   A lion can run 50 mph. How fast is this in kilometers per hour?

47. MOUNT WASHINGTON   The highest peak of the White Mountains of New Hampshire is Mount Washington, at 6,288 feet. To the nearest tenth, give this height in kilometers.

48. TRACK AND FIELD   Track meets are held on an oval track such as that shown in Illustration 1. One lap around the track is usually 400 meters. However, some older tracks in the United States are 440-yard ovals. Are these two types of tracks the same length? If not, which is longer?

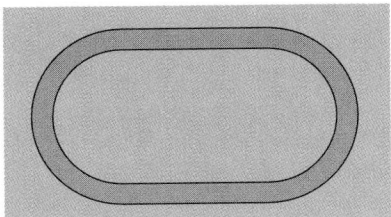

ILLUSTRATION 1

49. HAIR GROWTH   When hair is short, its rate of growth averages about $\frac{3}{4}$ inch per month. How many centimeters is this a month?

50. KILLER WHALE   An adult male killer whale can weigh as much as 12,000 pounds and be as long as 25 feet. Change these measurements to kilograms and meters.

51. WEIGHTLIFTING   Illustration 2 shows two International Powerlifting Federation recordholders as of April 13, 2000. Change each metric weight to pounds.

Recordholder	Country	Weight class	Bench press
Vicki Steenrod	USA	82.5 kg	132.5 kg
Jeff Magruder	USA	110.0 kg	270.0 kg

ILLUSTRATION 2

52. WORDS OF WISDOM   Refer to the wall hanging in Illustration 3. Convert the first metric weight to ounces and the second to pounds. What famous saying results?

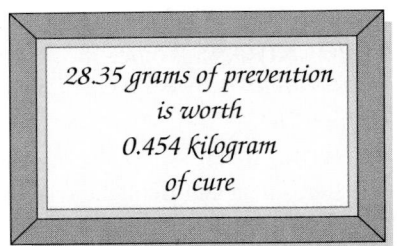

28.35 grams of prevention is worth 0.454 kilogram of cure

ILLUSTRATION 3

53. OUNCES AND FLUID OUNCES
   a. There are 310 calories in 8 ounces of broiled chicken. Convert 8 ounces to grams.
   b. There are 112 calories in a glass of fresh Valencia orange juice that holds 8 fluid ounces. Convert 8 fluid ounces to liters.

54. TRACK AND FIELD   A shot-put weighs 7.264 kilograms. Give this weight in pounds.

55. POSTAL REGULATIONS   You can mail a package weighing up to 70 pounds via priority mail. Can you mail a package that weighs 32 kilograms by priority mail?

56. HEALTHY EATING   Refer to the nutrition label for a packet of oatmeal shown in Illustration 4. Change each circled weight to ounces.

Nutrition Facts
Serving Size: 1 Packet (46g)
Servings Per Container: 10

Amount Per Serving
Calories 170   Calories from Fat 20

	% Daily Value
Total fat 2g	3%
Saturated fat (0.5g)	2%
Polyunsaturated Fat 0.5g	
Monounsaturated Fat 1g	
Cholesterol 0mg	0%
Sodium (250mg)	10%
Total carbohydrate 35g	12%
Dietary fiber 3g	12%
Soluble Fiber 1g	
Sugars 16g	
Protein (4g)	

ILLUSTRATION 4

57. COMPARISON SHOPPING   Which is the better buy, 3 quarts of root beer for $4.50 or 2 liters of root beer for $3.60?

58. COMPARISON SHOPPING   Which is the better buy, 3 gallons of antifreeze for $10.35 or 12 liters of antifreeze for $10.50?

59. HOT SPRINGS   The thermal springs in Hot Springs National Park in central Arkansas emit water as warm as 143°F. Change this temperature to degrees Celsius.

60. COOKING MEAT   Meats must be cooked at high enough temperatures to kill harmful bacteria. According to the USDA and the FDA, the internal temperature for cooked roasts and steaks should be at least 145°F, and whole poultry should be 180°F. Convert these temperatures to degrees Celsius. Round up to the nearest degree.

61. TAKING A SHOWER   When you take a shower, which water temperature would you choose: 15° C, 28° C, or 50° C?

62. DRINKING WATER   To get a cold drink of water, which temperature would you choose: −2° C, 10° C, or 25° C?

63. SNOWY WEATHER   At which temperature might it snow: −5° C, 0° C, or 10° C?

64. RUNNING THE AIR CONDITIONER   At which outside temperature would you be likely to run the air conditioner: 15° C, 20° C, or 30° C?

APPENDIX III SOLVING ABSOLUTE VALUE EQUATIONS AND INEQUALITIES

In this appendix, you will learn about

- Absolute value • Equations of the form $|x| = k$
- Equations with two absolute values • Inequalities of the form $|x| < k$
- Inequalities of the form $|x| > k$

INTRODUCTION. Many quantities studied in mathematics, science, and engineering are expressed as positive numbers. To guarantee that a quantity is positive, we often use the concept of absolute value. In this section, we will work with equations and inequalities that contain expressions involving absolute value. Using the definition of absolute value, we will develop procedures to solve absolute value equations and absolute value inequalities.

Absolute value

Recall that the **absolute value** of any real number is the distance between the number and zero on the number line. For example, the points shown in Figure III-1 with coordinates of 4 and -4 both lie 4 units from 0. Thus, $|4| = |-4| = 4$.

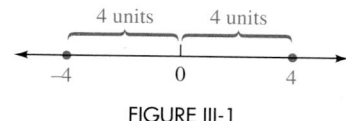

FIGURE III-1

The absolute value of a real number can be defined more formally.

Absolute value

> If $x \geq 0$, then $|x| = x$.
>
> If $x < 0$, then $|x| = -x$.

This definition gives a way for associating a nonnegative real number with any real number.

- If $x \geq 0$, then x (which is positive or 0) is its own absolute value.
- If $x < 0$, then $-x$ (which is positive) is the absolute value.

Either way, $|x|$ is positive or 0. That is, $|x| \geq 0$ for all real numbers x.

EXAMPLE 1 *Finding absolute values.* Find **a.** $|9|$,
b. $|-5.68|$,   and   **c.** $|0|$

Solution

a. Since $9 \geq 0$, the number 9 is its own absolute value: $|9| = 9$.

b. Since $-5.68 < 0$, the negative of -5.68 is the absolute value:

$$|-5.68| = -(-5.68) = 5.68$$

c. Since $0 \geq 0$, 0 is its own absolute value: $|0| = 0$.

Self Check

Find: **a.** $|-3|$,   **b.** $|100.99|$,
and   **c.** $|-2\pi|$

Answers: a. 3,   **b.** 100.99,
c. 2π ■

COMMENT   The placement of a $-$ sign in an expression containing an absolute value symbol is important. For example, $|-19| = 19$, but $-|19| = -19$.

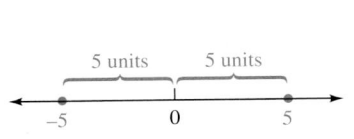

FIGURE III-2

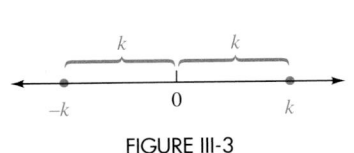

FIGURE III-3

Equations of the form $|x| = k$

The absolute value of a real number represents the distance on a number line from a point to the origin. To solve the **absolute value equation** $|x| = 5$, we must find the co-ordinates of all points on a number line that are exactly 5 units from zero. See Figure III-2. The only two points that satisfy this condition have coordinates 5 and -5. That is, $x = 5$ or $x = -5$.

In general, the solution set of the absolute value equation $|x| = k$, where $k \geq 0$, in-cludes the coordinates of the points on a number line that are k units from the origin. (See Figure III-3.)

Absolute value equations

> If $k \geq 0$, then
>
> $|x| = k$ is equivalent to $x = k$ or $x = -k$

EXAMPLE 2 *Solving an absolute value equation.*   Solve
a. $|x| = 8$ and   **b.** $|s| = 0.003$

Solution

a. If $|x| = 8$, then $x = 8$ or $x = -8$.

b. If $|s| = 0.003$, then $s = 0.003$ or $s = -0.003$.

Self Check

Solve   **a.** $|y| = 24$ and
b. $|x| = \dfrac{1}{2}$.

Answers: a. $24, -24$
b. $\dfrac{1}{2}, -\dfrac{1}{2}$ ■

The equation $|x - 3| = 7$ indicates that a point on a number line with a coordinate of $x - 3$ is 7 units from the origin. Thus, $x - 3$ can be either 7 or -7.

$$x - 3 = 7 \quad \text{or} \quad x - 3 = -7$$
$$x = 10 \qquad\qquad x = -4$$

The solutions of the absolute value equation are 10 and -4. We can graph them on a number line, as shown in Figure III-4. If either of these numbers is substituted for x in $|x - 3| = 7$, the equation is satisfied.

Check: $|x - 3| = 7$ \qquad $|x - 3| = 7$
$|10 - 3| \stackrel{?}{=} 7$ \qquad $|-4 - 3| \stackrel{?}{=} 7$
$|7| \stackrel{?}{=} 7$ \qquad\qquad $|-7| \stackrel{?}{=} 7$
$7 = 7$ \qquad\qquad\quad $7 = 7$

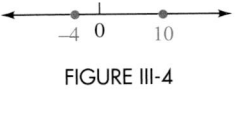

FIGURE III-4

EXAMPLE 3   *Solving an absolute value equation.*   Solve $|3x - 2| = 5$.

Solution

We can write $|3x - 2| = 5$ as

$$3x - 2 = 5 \quad \text{or} \quad 3x - 2 = -5$$

and solve each equation for x:

$$\begin{array}{lll}
3x - 2 = 5 & \text{or} & 3x - 2 = -5 \\
3x = 7 & & 3x = -3 \\
x = \dfrac{7}{3} & & x = -1
\end{array}$$

Verify that both solutions check.

Self Check

Solve $|2x - 3| = 7$.

Answer: $5, -2$ ∎

When solving an absolute value equation, we want the absolute value isolated on one side. If this is not the case in a given equation, we use the equation-solving procedures studied earlier to isolate the absolute value first.

EXAMPLE 4   *Isolating the absolute value.*   Solve $\left|\dfrac{2}{3}x + 3\right| + 4 = 10$.

Solution

We can isolate $\left|\dfrac{2}{3}x + 3\right|$ on the left-hand side of the equation by subtracting 4 from both sides.

$$\left|\dfrac{2}{3}x + 3\right| + 4 = 10$$

(1) $\left|\dfrac{2}{3}x + 3\right| = 6$ Subtract 4 from both sides.

Now that the absolute value is isolated, we can write Equation 1 as

$$\dfrac{2}{3}x + 3 = 6 \quad \text{or} \quad \dfrac{2}{3}x + 3 = -6$$

and solve each equation for x:

$$\begin{array}{lll}
\dfrac{2}{3}x + 3 = 6 & \text{or} & \dfrac{2}{3}x + 3 = -6 \\
\dfrac{2}{3}x = 3 & & \dfrac{2}{3}x = -9 \\
2x = 9 & & 2x = -27 \\
x = \dfrac{9}{2} & & x = -\dfrac{27}{2}
\end{array}$$

Verify that both solutions check.

Self Check

Solve $|0.4x - 2| - 0.6 = 0.4$.

Answer: 7.5, 2.5 ∎

COMMENT   Since the absolute value of a quantity cannot be negative, equations such as $|7x + \frac{1}{2}| = -4$ have no solution. Since there are no solutions, their solution sets are empty.

EXAMPLE 5   *An absolute value equal to* **0.**

Solve $3\left|\dfrac{1}{2}x - 5\right| - 4 = -4$.

Solution

We first isolate $\left|\dfrac{1}{2}x - 5\right|$ on the left-hand side.

$$3\left|\frac{1}{2}x - 5\right| - 4 = -4$$

$$3\left|\frac{1}{2}x - 5\right| = 0 \qquad \text{Add 4 to both sides.}$$

$$\left|\frac{1}{2}x - 5\right| = 0 \qquad \text{Divide both sides by 3.}$$

Since 0 is the only number whose absolute value is 0, the expression $\frac{1}{2}x - 5$ must be 0, and we have

$$\frac{1}{2}x - 5 = 0$$

$$\frac{1}{2}x = 5 \qquad \text{Add 5 to both sides.}$$

$$x = 10 \qquad \text{Multiply both sides by 2.}$$

Verify that 10 satisfies the original equation.

Self Check

Solve:

$$-5\left|\frac{2x}{3} + 4\right| + 1 = 1$$

Answer: -6

Equations with two absolute values

The equation $|a| = |b|$ is true when $a = b$ or when $a = -b$. For example,

$$|3| = |3| \qquad \text{or} \qquad |3| = |-3|$$

The same number.   These numbers are opposites.

In general, the following statement is true.

Equations with two absolute values

> If a and b represent algebraic expressions, the equation $|a| = |b|$ is equivalent to
>
> $a = b$ or $a = -b$

EXAMPLE 6   *Solving equations with two absolute values.*   Solve $|5x + 3| = |3x + 25|$.

Solution

This equation is true when $5x + 3 = 3x + 25$, or when $5x + 3 = -(3x + 25)$. We solve each equation for x.

$$5x + 3 = 3x + 25 \qquad \text{or} \qquad 5x + 3 = -(3x + 25)$$
$$2x = 22 \qquad\qquad\qquad 5x + 3 = -3x - 25$$
$$x = 11 \qquad\qquad\qquad\qquad 8x = -28$$
$$x = -\frac{28}{8}$$
$$x = -\frac{7}{2}$$

Verify that both solutions check.

Self Check

Solve $|2x - 3| = |4x + 9|$.

Answer: $-1, -6$

Inequalities of the form $|x| < k$

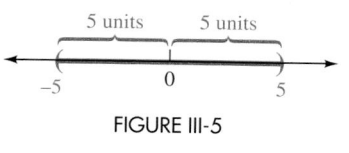

FIGURE III-5

To solve the **absolute value inequality** $|x| < 5$, we must find the coordinates of all points on a number line that are less than 5 units from the origin. See Figure III-5. Thus, x is between -5 and 5 and

$$|x| < 5 \quad \text{is equivalent to} \quad -5 < x < 5$$

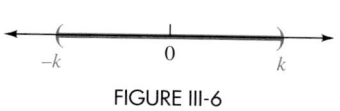

FIGURE III-6

In general, the solution set of the absolute value inequality $|x| < k$ where $k > 0$ includes the coordinates of the points on the number line that are less than k units from the origin. See Figure III-6.

Solving $|x| < k$ and $|x| \leq k$

$|x| < k$ is equivalent to $-k < x < k$ where $k > 0$	
$|x| \leq k$ is equivalent to $-k \leq x \leq k$ where $k \geq 0$	

EXAMPLE 7   *Solving an absolute value inequality.*   Solve $|2x - 3| < 9$ and graph the solution set.

Solution

We write the absolute value inequality as a double inequality and solve for x.

$|2x - 3| < 9$ is equivalent to $-9 < 2x - 3 < 9$

$$-9 < 2x - 3 < 9$$
$$-6 < 2x < 12 \qquad \text{Add 3 to all three parts.}$$
$$-3 < x < 6 \qquad \text{Divide all parts by 2.}$$

Any number between -3 and 6 is included in the solution set. This is the interval $(-3, 6)$, whose graph is shown in Figure III-7.

FIGURE III-7

Self Check

Solve $|3x + 2| < 4$ and graph the solution set.

Answer: $\left(-2, \dfrac{2}{3}\right)$

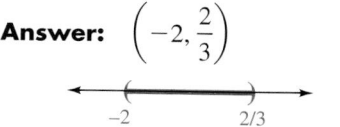

EXAMPLE 8   *Tolerances.*   When manufactured parts are inspected by a quality control engineer, they are classified as acceptable if each dimension falls within a given *tolerance range* of the dimensions listed on the blueprint. For the bracket shown in Figure III-8, the distance between the two drilled holes is given as 2.900 inches. Because the tolerance is ± 0.015 inch, this distance can be as much as 0.015 inch longer or 0.015 inch shorter, and the part will be considered acceptable. The acceptable distance d between holes can be represented by the absolute value inequality $|d - 2.900| \leq 0.015$. Solve the inequality and explain the result.

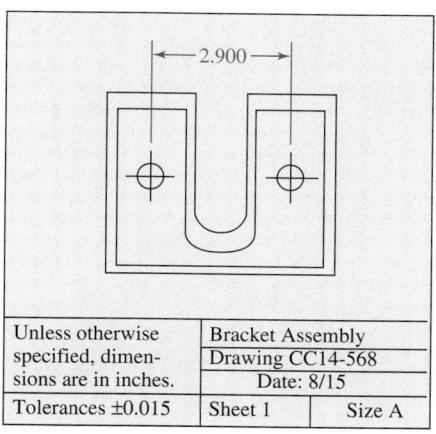

Unless otherwise specified, dimensions are in inches.	Bracket Assembly
	Drawing CC14-568
	Date: 8/15
Tolerances ±0.015	Sheet 1 | Size A

FIGURE III-8

Solution   We can write the absolute value inequality as a double inequality and solve for d:

$$|d - 2.900| \leq 0.015 \quad \text{is equivalent to} \quad -0.015 \leq d - 2.900 \leq 0.015$$

$$-0.015 \leq d - 2.900 \leq 0.015$$

$$2.885 \leq d \leq 2.915 \qquad \text{Add 2.900 to all three parts.}$$

The solution set is the interval [2.885, 2.915]. This means that the distance between the two holes should be between 2.885 and 2.915 inches, inclusive. If the distance is less than 2.885 inches or more than 2.915 inches, the part should be rejected. ■

Inequalities of the form $|x| < k$

To solve the *absolute value inequality* $|x| > 5$, we must find the coordinates of all points on a number line that are more than 5 units from the origin. See Figure III-9.

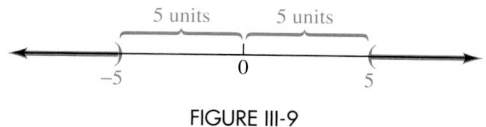

FIGURE III-9

Thus, $x < -5$ or $x > 5$.

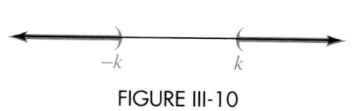

FIGURE III-10

In general, the solution set of $|x| > k$ includes the coordinates of the points on the number line that are more than k units from the origin. See Figure III-10. Thus,

$$|x| > k \quad \text{is equivalent to} \quad x < -k \quad \text{or} \quad x > k$$

The *or* indicates an either/or situation. It is only necessary that x satisfy one of the two conditions to be in the solution set.

Solving $|x| > k$ and $|x| \geq k$

If $k \geq 0$, then

$$|x| > k \quad \text{is equivalent to} \quad x < -k \quad \text{or} \quad x > k$$

$$|x| \geq k \quad \text{is equivalent to} \quad x \leq -k \quad \text{or} \quad x \geq k$$

EXAMPLE 9 *Solving an absolute value inequality.* Solve $\left|\dfrac{3 - x}{5}\right| \geq 6$ and graph the solution set.

Solution

We write the absolute value inequality as two separate inequalities connected with the word "or."

$$\left|\frac{3 - x}{5}\right| \geq 6 \quad \text{is equivalent to} \quad \frac{3 - x}{5} \leq -6 \quad \text{or} \quad \frac{3 - x}{5} \geq 6$$

Then we solve each inequality for x:

$$\frac{3 - x}{5} \leq -6 \quad \text{or} \quad \frac{3 - x}{5} \geq 6$$

$$3 - x \leq -30 \qquad 3 - x \geq 30 \qquad \text{Multiply both sides by 5.}$$

$$-x \leq -33 \qquad -x \geq 27 \qquad \text{Subtract 3 from both sides.}$$

$$x \geq 33 \qquad x \leq -27 \qquad \begin{array}{l}\text{Divide both sides by } -1 \text{ and reverse the}\\ \text{direction of the inequality symbol.}\end{array}$$

The solution set is the interval $(-\infty, -27] \cup [33, \infty)$, whose graph appears in Figure III-11.

FIGURE III-11

Self Check

Solve

$$\left|\frac{2 - x}{4}\right| \geq 1$$

and graph the solution set.

Answer: $(-\infty, -2] \cup [6, \infty)$

■

EXAMPLE 10 *Solving an absolute value inequality.*   Solve $\left|\frac{2}{3}x - 2\right| - 3 > 6$ and graph the solution set.

Solution

We begin by adding 3 to both sides to isolate the absolute value on the left-hand side.

$$\left|\frac{2}{3}x - 2\right| - 3 > 6$$

$$\left|\frac{2}{3}x - 2\right| > 9 \qquad \text{Add 3 to both sides to isolate the absolute value.}$$

We then proceed as follows:

$$\frac{2}{3}x - 2 < -9 \qquad \text{or} \qquad \frac{2}{3}x - 2 > 9$$

$$\frac{2}{3}x < -7 \qquad\qquad \frac{2}{3}x > 11 \qquad \text{Add 2 to both sides.}$$

$$2x < -21 \qquad\qquad 2x > 33 \qquad \text{Multiply both sides by 3.}$$

$$x < -\frac{21}{2} \qquad\qquad x > \frac{33}{2} \qquad \text{Divide both sides by 2.}$$

The solution set is $\left(-\infty, -\frac{21}{2}\right) \cup \left(\frac{33}{2}, \infty\right)$. The graph appears in Figure III-12.

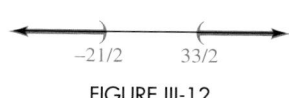

−21/2   33/2

FIGURE III-12

Self Check

Solve $\left|\frac{3}{4}x + 2\right| - 1 > 3$ and graph the solution set.

Answer: $(-\infty, -8) \cup \left(\frac{8}{3}, \infty\right)$

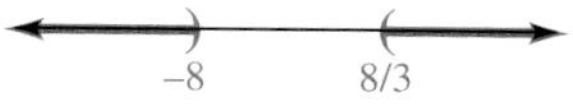

−8   8/3

Accent on Technology: **Solving absolute value inequalities**

We can also solve absolute value inequalities using a graphing calculator. For example, to solve $|2x - 3| < 9$, we graph the equations $y = |2x - 3|$ and $y = 9$ on the same coordinate system. If we use settings of $[-5, 15]$ for x and $[-5, 15]$ for y, we will get the graph shown in Figure III-13.

The inequality $|2x - 3| < 9$ will be true for all x-coordinates of points that lie on the graph of $y = |2x - 3|$ and below the graph of $y = 9$. Using the TRACE feature, we can see that these values of x are in the interval $(-3, 6)$.

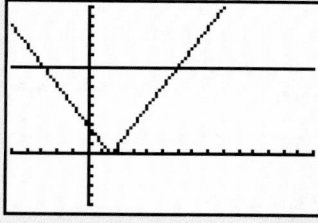

FIGURE III-13

STUDY SET   Appendix III

VOCABULARY   *Fill in the blanks.*

1. $|2x - 1| = 10$ is an absolute value _____.

2. $|2x - 1| > 10$ is an absolute value _____.

3. To _____ the absolute value in $|3 - x| - 4 = 5$, we add 4 to both sides.

4. $|x| = 2$ is _____ to $x = 2$ or $x = -2$.

CONCEPTS   *In Exercises 5–10, fill in the blanks.*

5. $|x| \geq$ ▢ for all real numbers x.

6. If $x < 0$, $|x| =$ ▢.

7. To solve $|x| > 5$, we must find the coordinates of all points on a number line that are _____ 5 units from 0.

8. To solve $|x| < 5$, we must find the coordinates of all points on a number line that are _____ 5 units from 0.

9. To solve $|x| = 5$, we must find the coordinates of all points on a number line that are ▢ units from 0.

10. The equation $|a| = |b|$ is true when _____ or when _____.

11. Tell whether $x = -3$ is a solution of the given equation or inequality.
   a. $|x - 1| = 4$
   b. $|x - 1| > 4$
   c. $|x - 1| \le 4$
   d. $|5 - x| = |x + 12|$

12. Write each equation or inequality in its equivalent form.
   a. $|x| = 8$
   b. $|x| \ge 8$
   c. $|x| \le 8$
   d. $|5x - 1| = |x + 3|$

NOTATION

13. Match each equation or inequality with its graph.
   a. $|x| = 1$ i
   b. $|x| > 1$ ii
   c. $|x| < 1$ iii

14. Match each graph with its corresponding equation or inequality.
   a.   i $|x| \ge 2$
   b.   ii $|x| \le 2$
   c.   iii $|x| = 2$

Write each compound inequality as an inequality using absolute values.

15. $-4 < x < 4$

16. $x < -4$ or $x > 4$

17. $x + 3 < -6$ or $x + 3 > 6$

18. $-5 \le x - 3 \le 5$

PRACTICE   *Find the value of each expression.*

19. $|8|$

20. $|-18|$

21. $-|0.02|$

22. $-|-3.14|$

23. $-\left|-\dfrac{31}{16}\right|$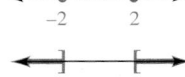

24. $-\left|\dfrac{25}{4}\right|$

25. $|\pi|$

26. $\left|-\dfrac{\pi}{2}\right|$

Solve each equation, if possible.

27. $|x| = 23$

28. $|x| = 90$

29. $|x - 3.1| = 6$

30. $|x + 4.3| = 8.9$

31. $|3x + 2| = 16$

32. $|5x - 3| = 22$

33. $\left|\dfrac{7}{2}x + 3\right| = -5$

34. $\left|\dfrac{2x}{3} + 10\right| = 0$

35. $|3 - 4x| = 5$

36. $|8 - 5x| = 18$

37. $2|3x + 24| = 0$

38. $5|x - 21| = -8$

39. $\left|\dfrac{3x + 48}{3}\right| = 12$

40. $\left|\dfrac{4x - 64}{4}\right| = 32$

41. $|x + 3| + 7 = 10$

42. $|2 - x| + 3 = 5$

43. $|2x + 1| = |3x + 3|$

44. $|5x - 7| = |4x + 1|$

45. $|2 - x| = |3x + 2|$

46. $|4x + 3| = |9 - 2x|$

47. $\left|\dfrac{x}{2} + 2\right| = \left|\dfrac{x}{2} - 2\right|$

48. $|7x + 12| = |x - 6|$

49. $\left|x + \dfrac{1}{3}\right| = |x - 3|$

50. $\left|x - \dfrac{1}{4}\right| = |x + 4|$

Solve each inequality. Write the solution set in interval notation and graph it.

51. $|x| < 4$

52. $|x| < 9$

53. $|x + 9| \le 12$

54. $|x - 8| \leq 12$

55. $|3x - 2| < 10$

56. $|4 - 3x| \leq 13$

57. $|3x + 2| \leq -3$

58. $|5x - 12| < -5$

59. $|x| > 3$

60. $|x| > 7$

61. $|x - 12| > 24$

62. $|x + 5| \geq 7$

63. $|3x + 2| > 14$

64. $|2x - 5| > 25$

65. $|4x + 3| > -5$

66. $|7x + 2| > -8$

67. $|2 - 3x| \geq 8$

68. $|-1 - 2x| > 5$

69. $-|2x - 3| < -7$

70. $-|3x + 1| < -8$

71. $\left| \dfrac{x - 2}{3} \right| \leq 4$

72. $\left| \dfrac{x - 2}{3} \right| > 4$

73. $|3x + 1| + 2 < 6$

74. $1 + \left| \dfrac{1}{7}x + 1 \right| \leq 1$

75. $\left| \dfrac{1}{3}x + 7 \right| + 5 > 6$

76. $-2|3x - 4| < 16$

APPLICATIONS

77. TEMPERATURE RANGES   The temperatures on a sunny summer day satisfied the inequality $|t - 78°| \leq 8°$, where t is a temperature in degrees Fahrenheit. Solve this inequality and express the range of temperatures as a double inequality.

78. OPERATING TEMPERATURES   A car CD player has an operating temperature of $|t - 40°| < 80°$, where t is a temperature in degrees Fahrenheit. Solve the inequality and express this range of temperatures as an interval.

79. AUTO MECHANICS   On most cars, the bottoms of the front wheels are closer together than the tops, creating a *camber angle*. This lessens road shock to the steering system. (See Illustration 1.) The specifications for a certain car state that the camber angle c of its wheels should be $0.6° \pm 0.5°$.
   a. Express the range with an inequality containing absolute value symbols.
   b. Solve the inequality and express this range of camber angles as an interval.

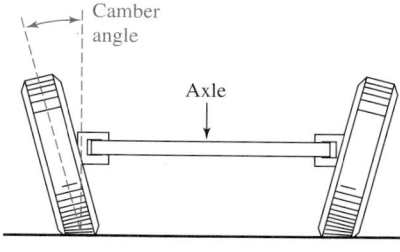

ILLUSTRATION 1

80. STEEL PRODUCTION   A sheet of steel is to be 0.250 inch thick with a tolerance of 0.025 inch.
   a. Express this specification with an inequality containing absolute value symbols, using x to represent the thickness of a sheet of steel.

   b. Solve the inequality and express the range of thickness as an interval.

81. ERROR ANALYSIS   In a lab, students measured the percent of copper p in a sample of copper sulfate. The students know that copper sulfate is actually 25.46% copper by mass. They are to compare their results to the actual value and find the amount of *experimental error*.
   a. Which measurements shown in Illustration 2 (on the next page) satisfy the absolute value inequality $|p - 25.46| \leq 1.00$?
   b. What can be said about the amount of error for each of the trials listed in part a?

Lab 4   Section A
Title:
"Percent copper (Cu) in
copper sulfate (CuSO$_4$·5H$_2$O)"

Results

	% Copper
Trial #1:	22.91%
Trial #2:	26.45%
Trial #3:	26.49%
Trial #4:	24.76%

ILLUSTRATION 2

82. ERROR ANALYSIS   See Exercise 81.
   a. Which measurements satisfy the absolute value inequality $|p - 25.46| > 1.00$?
   b. What can be said about the amount of error for each of the trials listed in part a?

In this appendix, you will learn about

- Probability

INTRODUCTION. If we toss a coin, it can land in one of two equally likely ways—either heads or tails. Because one of these two outcomes is heads, we say that the probability of obtaining heads in a single toss is $\frac{1}{2}$. If records show that out of 100 days with weather conditions like today's, 30 have received rain, we say that there is a $\frac{30}{100}$ or 30% probability of rain today. If there are two chances in three that a basketball player will make a free throw, we say that the probability the player will make a free throw is $\frac{2}{3}$.

In this appendix, we will introduce the concept of probability.

Probability

An **experiment** is any process for which the outcome is uncertain. For example, some experiments are

- Tossing a coin
- Rolling dice
- Drawing a card
- Taking a colored marble from a jar

For any experiment, the set of all possible outcomes is called a **sample space.** The sample space for the experiment of tossing a coin twice is the following set of four ordered pairs.

{(H, H), (H, T), (T, H), (T, T)}   The pair (H, T), for example, represents the outcome "heads on the first coin and tails on the second coin."

An **event** is any set of outcomes of an experiment. For example, if E is the event "getting at least one heads" in the experiment of tossing a coin twice, the event E is the following set of three ordered pairs:

E = {(H, H), (H, T), (T, H)}

Because the outcome of getting at least one heads can occur in 3 ways out of a total of 4 possible ways, we say that the **probability** of event E is $\frac{3}{4}$. In symbols, we write

$$P(E) = P(\text{at least one heads}) = \frac{3}{4} \quad \text{Read } P(E) \text{ as "the probability of event } E.\text{"}$$

We can define the probability of an event as follows.

Probability of an event

> If E is an event that can occur in n ways out of s possible, equally likely, ways, the **probability of E** is
>
> $$P(E) = \frac{n}{s}$$

Because $0 \le n \le s$, it follows that $0 \le \frac{n}{s} \le 1$. This implies that all probabilities have values from 0 to 1. An event that cannot happen has a probability of 0. An event that is certain to happen has a probability of 1.

A **die** (the singular of *dice*) is a cube with six faces, each containing a number of dots from one to six. See Figure IV-1.

FIGURE IV-1

EXAMPLE 1   *Rolling a die.* Find the probability of rolling a six on one roll of a fair die.

Solution

Since a fair die has 6 faces that are equally likely to appear, and there is 1 way to get a six, we have

$$P(6) = \frac{1}{6}$$

Self Check

Find the probability of rolling a 2 or a 5 on one roll of a fair die.

Answer: $\dfrac{1}{3}$ ∎

EXAMPLE 2   *Finding a sample space.* Show the sample space of the experiment "rolling two dice one time."

Solution

We can list ordered pairs, letting the first number be the result on the first die and the second number the result on the second die. The sample space will be the set containing the following 36 ordered pairs.

(1, 1) (1, 2) (1, 3) (1, 4) (1, 5) (1, 6)
(2, 1) (2, 2) (2, 3) (2, 4) (2, 5) (2, 6)
(3, 1) (3, 2) (3, 3) (3, 4) (3, 5) (3, 6)
(4, 1) (4, 2) (4, 3) (4, 4) (4, 5) (4, 6)
(5, 1) (5, 2) (5, 3) (5, 4) (5, 5) (5, 6)
(6, 1) (6, 2) (6, 3) (6, 4) (6, 5) (6, 6)

Self Check

How many pairs in the sample space have a sum of 4?

Answer: 3   ∎

EXAMPLE 3   *Rolling two dice.* Find the probability of the event "rolling a sum of 7 on one roll of two dice."

Solution

The sample space is listed in Example 2. We let E be the set of outcomes that give a sum of 7:

$$E = \{(1, 6), (2, 5), (3, 4), (4, 3), (5, 2), (6, 1)\}$$

Since there are 6 ways to roll a 7 among the 36 equally likely outcomes, we have

$$P(\text{rolling a 7}) = \frac{6}{36} = \frac{1}{6} \quad \text{Simplify the fraction.}$$

Self Check

Find the probability of rolling a sum of 4.

Answer: $\dfrac{1}{12}$ ∎

A standard deck of 52 playing cards has two red suits (hearts and diamonds) and two black suits (clubs and spades). Each suit has 13 cards, including a king, a queen, and a jack (called **face cards**), an ace, and cards numbered from 2 to 10. In the next example, we refer to a standard deck of cards.

EXAMPLE 4   Find the probability of drawing an ace from a well-shuffled deck of 52 cards.

Solution

Since there are 4 aces in a deck of 52 cards, and each card is equally likely to be drawn, the probability of drawing an ace is

$$P(\text{ace}) = \frac{4}{52} = \frac{1}{13}$$

Self Check

Find the probability of drawing a diamond from a well-shuffled deck of 52 cards.

Answer: $\dfrac{1}{4}$

STUDY SET   Appendix IV

VOCABULARY   *Fill in the blanks.*

1. An _____ is any process for which the outcome is uncertain.

2. A list of all possible outcomes for an experiment is called a _____.

CONCEPTS   *Fill in the blanks.*

3. If an event E can occur in n ways out of s equally likely ways, then $P(E) = $ ▢.

4. The probability of any event is always a number from ▢ to ▢, including ▢ and ▢.

5. If an event cannot happen, its probability is ▢.

6. If an event is certain to happen, its probability is ▢.

PRACTICE   *List the sample space of each experiment.*

7. Rolling a die and tossing a coin

8. Tossing three coins

9. Selecting a letter of the alphabet that is a vowel

10. Guessing a one-digit number

A fair die is rolled once. Find the probability of each event.

11. Rolling a 2

12. Rolling a number greater than 4

13. Rolling a number larger than 1 but less than 6

14. Rolling a number that is an odd number

Two fair dice are rolled once. Find the probability of each event.

15. Rolling a 6

16. Rolling a 10

17. Rolling a 13

18. Rolling a number from 1 to 13

Balls numbered from 1 to 42 are placed in a jar and stirred. If one is drawn at random, find the probability of each result.

19. The number is less than 20.

20. The number is less than 50.

21. The number is a prime number.

22. The number is less than 10 or greater than 40.

Refer to the spinner in Illustration 1. If the spinner is spun once, find the probability of each event. Assume that the spinner never stops on a line.

23. The spinner stops on red.

24. The spinner stops on green.

25. The spinner stops on orange.

26. The spinner stops on yellow.

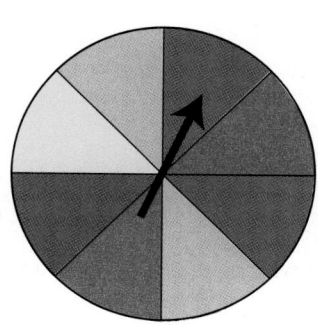

ILLUSTRATION 1

Find the probability of each event.

27. Drawing a black card on one draw from a standard card deck

28. Drawing a diamond on one draw from a standard card deck

29. Drawing a face card on one draw from a standard card deck

30. Drawing a jack or a king on one draw from a standard card deck

31. Drawing a red egg from a basket containing 5 red eggs and 7 blue eggs

32. Drawing an orange cube from a bowl containing 5 orange cubes and 1 beige cube

APPLICATIONS   *An aircraft has 4 engines. Assume that the probability of an engine failing during a torture test is $\frac{1}{2}$.*

33. Find the sample space.

34. Find the probability that all engines will survive the test.

35. Find the probability that exactly 1 engine will survive.

36. Find the probability that exactly 2 engines will survive.

37. Find the probability that exactly 3 engines will survive.

38. Find the probability that no engines will survive.

39. Find the sum of the probabilities in Exercises 34 through 38. What do you discover?

A survey of 282 people is taken to determine the opinions of doctors, teachers, and lawyers on a proposed piece of legislation, with the results shown in Illustration 2. A person is chosen at random from those surveyed. Refer to the chart to find each probability.

	Number that favor	Number that oppose	Number with no opinion	Total
Doctors	70	32	17	119
Teachers	83	24	10	117
Lawyers	23	15	8	46
Total	176	71	35	282

ILLUSTRATION 2

40. The person favors the legislation.

41. A doctor opposes the legislation.

42. A person who opposes the legislation is a lawyer.

n	n^2	$\sqrt{n}$	n^3	$\sqrt[3]{n}$	n	n^2	$\sqrt{n}$	n^3	$\sqrt[3]{n}$
1	1	1.000	1	1.000	51	2,601	7.141	132,651	3.708
2	4	1.414	8	1.260	52	2,704	7.211	140,608	3.733
3	9	1.732	27	1.442	53	2,809	7.280	148,877	3.756
4	16	2.000	64	1.587	54	2,916	7.348	157,464	3.780
5	25	2.236	125	1.710	55	3,025	7.416	166,375	3.803
6	36	2.449	216	1.817	56	3,136	7.483	175,616	3.826
7	49	2.646	343	1.913	57	3,249	7.550	185,193	3.849
8	64	2.828	512	2.000	58	3,364	7.616	195,112	3.871
9	81	3.000	729	2.080	59	3,481	7.681	205,379	3.893
10	100	3.162	1,000	2.154	60	3,600	7.746	216,000	3.915
11	121	3.317	1,331	2.224	61	3,721	7.810	226,981	3.936
12	144	3.464	1,728	2.289	62	3,844	7.874	238,328	3.958
13	169	3.606	2,197	2.351	63	3,969	7.937	250,047	3.979
14	196	3.742	2,744	2.410	64	4,096	8.000	262,144	4.000
15	225	3.873	3,375	2.466	65	4,225	8.062	274,625	4.021
16	256	4.000	4,096	2.520	66	4,356	8.124	287,496	4.041
17	289	4.123	4,913	2.571	67	4,489	8.185	300,763	4.062
18	324	4.243	5,832	2.621	68	4,624	8.246	314,432	4.082
19	361	4.359	6,859	2.668	69	4,761	8.307	328,509	4.102
20	400	4.472	8,000	2.714	70	4,900	8.367	343,000	4.121
21	441	4.583	9,261	2.759	71	5,041	8.426	357,911	4.141
22	484	4.690	10,648	2.802	72	5,184	8.485	373,248	4.160
23	529	4.796	12,167	2.844	73	5,329	8.544	389,017	4.179
24	576	4.899	13,824	2.884	74	5,476	8.602	405,224	4.198
25	625	5.000	15,625	2.924	75	5,625	8.660	421,875	4.217
26	676	5.099	17,576	2.962	76	5,776	8.718	438,976	4.236
27	729	5.196	19,683	3.000	77	5,929	8.775	456,533	4.254
28	784	5.292	21,952	3.037	78	6,084	8.832	474,552	4.273
29	841	5.385	24,389	3.072	79	6,241	8.888	493,039	4.291
30	900	5.477	27,000	3.107	80	6,400	8.944	512,000	4.309
31	961	5.568	29,791	3.141	81	6,561	9.000	531,441	4.327
32	1,024	5.657	32,768	3.175	82	6,724	9.055	551,368	4.344
33	1,089	5.745	35,937	3.208	83	6,889	9.110	571,787	4.362
34	1,156	5.831	39,304	3.240	84	7,056	9.165	592,704	4.380
35	1,225	5.916	42,875	3.271	85	7,225	9.220	614,125	4.397
36	1,296	6.000	46,656	3.302	86	7,396	9.274	636,056	4.414
37	1,369	6.083	50,653	3.332	87	7,569	9.327	658,503	4.431
38	1,444	6.164	54,872	3.362	88	7,744	9.381	681,472	4.448
39	1,521	6.245	59,319	3.391	89	7,921	9.434	704,969	4.465
40	1,600	6.325	64,000	3.420	90	8,100	9.487	729,000	4.481
41	1,681	6.403	68,921	3.448	91	8,281	9.539	753,571	4.498
42	1,764	6.481	74,088	3.476	92	8,464	9.592	778,688	4.514
43	1,849	6.557	79,507	3.503	93	8,649	9.644	804,357	4.531
44	1,936	6.633	85,184	3.530	94	8,836	9.695	830,584	4.547
45	2,025	6.708	91,125	3.557	95	9,025	9.747	857,375	4.563
46	2,116	6.782	97,336	3.583	96	9,216	9.798	884,736	4.579
47	2,209	6.856	103,823	3.609	97	9,409	9.849	912,673	4.595
48	2,304	6.928	110,592	3.634	98	9,604	9.899	941,192	4.610
49	2,401	7.000	117,649	3.659	99	9,801	9.950	970,299	4.626
50	2,500	7.071	125,000	3.684	100	10,000	10.000	1,000,000	4.642

Study Set   Section 1.1   (page 7)

1. set   **3.** expanded   **5.** number   **7.** 3   **9.** 6
11. whole numbers
13.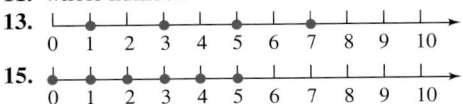
15.
17. >   **19.** >   **21.** <   **23.** >   **25.** braces
27. 2 hundreds + 4 tens + 5 ones; two hundred forty-five
29. 3 thousands + 6 hundreds + 9 ones; three thousand six
hundred nine   **31.** 3 ten thousands + 2 thousands + 5 hundreds;
thirty-two thousand five hundred   **33.** 1 hundred thousand +
4 thousands + 4 hundreds + 1 one; one hundred four thousand
four hundred one   **35.** 425   **37.** 2,736   **39.** 456   **41.** 27,598
43. 9,113   **45.** 10,700,506   **47.** 79,590   **49.** 80,000
51. 5,926,000   **53.** 5,900,000   **55.** $419,160   **57.** $419,000
61. a. the 70s   **b.** the 60s
63.
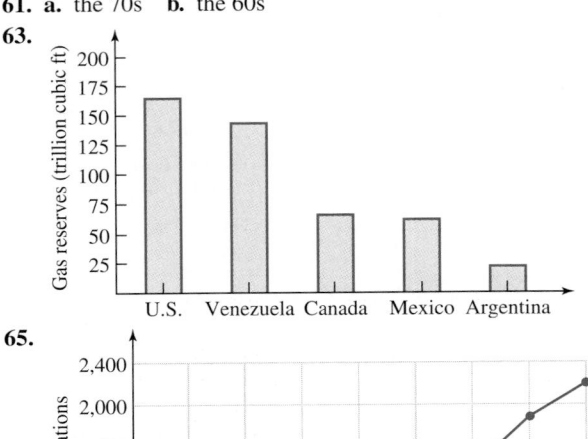
65.
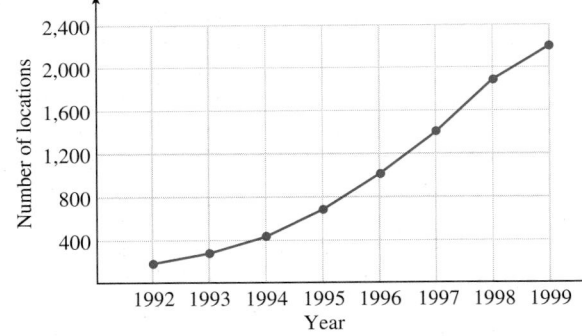

67.
a.
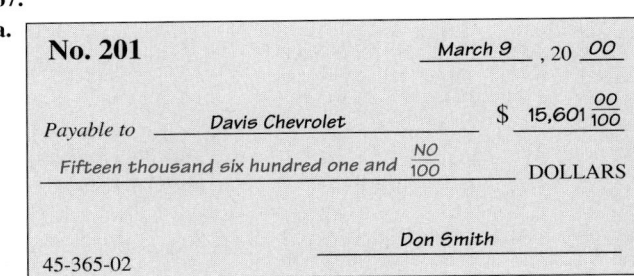

b.

No. 7890		Aug. 12 , 20 _00_

Payable to   Dr. Anderson $ 3,433 $\frac{46}{100}$
Three thousand four hundred thirty three and $\frac{46}{100}$ DOLLARS

Juan Decito
45-828-02

69. 1,865,593; 482,880; 1,503; 269; 43,449
71. a. 299,800,000 m/s   **b.** 300,000,000 m/s

Study Set   Section 1.2   (page 18)

1. sum, addends   **3.** rectangle   **5.** difference, subtrahend,
minuend   **7.** associative   **9.** commutative property of addition
11. associative property of addition   **13. a.** $x + y = y + x$
b. $(x + y) + z = x + (y + z)$ **15.** 0   **17.** $4 + 3 = 7$
19. parentheses   **21.** 47   **23.** 38   **25.** 461   **27.** 111
29. 150   **31.** 363   **33.** 979   **35.** 1,985   **37.** 10,000
39. 15,907   **41.** 1,861   **43.** 5,312   **45.** 88 ft   **47.** 68 in.
49. 3   **51.** 25   **53.** 103   **55.** 65   **57.** 141   **59.** 0
61. 24   **63.** 118   **65.** 958   **67.** 1,689   **69.** 10,457
71. 303   **73.** 40   **75.** 110   **77.** $18   **79.** 1,750,027
81. $213   **83.** 10,057 mi   **85. a.** $147,145   **b.** $161,725
87. 91 ft   **89.** 792 tons   **91.** 196 in.   **95.** 3 thousands +
1 hundred + 2 tens + 5 ones   **97.** 6,354,780   **99.** 6,350,000

Study Set   Section 1.3   (page 30)

1. multiplication   **3.** commutative   **5.** square inch   **7.** $4 \cdot 8$
9. Multiply its length by its width.   **11. a.** 25   **b.** 62   **c.** 0   **d.** 0

13. 5 · 12 **15. a.** ×, ·, ( ) **b.** $\overline{)}$, ÷, — **17.** square feet **19.** 84 **21.** 324 **23.** 180 **25.** 105 **27.** 7,623 **29.** 1,060 **31.** 2,576 **33.** 20,079 **35.** 2,919,952 **37.** 1,182,116 **39.** 84 in.² **41.** 144 in.² **43.** 8 **45.** 3 **47.** 12 **49.** 13 **51.** 73 **53.** 41 **55.** 205 **57.** 210 **59.** 8 R 25 **61.** 20 R 3 **63.** 30 R 13 **65.** 31 R 28 **67.** $132 **69.** 406 mi **71.** 125,800 **73.** 312 **75.** yes **77.** 72 **79.** 4 **81.** 5 mi **83.** 440 ft **85.** $41 **87.** 9 girls, 24 teams **89.** the square room; the square room **91.** 388 ft² **97.** 8 **99.** 872

Study Set Estimation (page 35)

1. no **3.** no **5.** no **7.** approx. 8,900 mi **9.** approx. 30 bags **11.** 1,600,000,000

Study Set Section 1.4 (page 41)

1. factors **3.** factor **5.** composite **7.** prime **9.** base; exponent **11.** 1 · 27 or 3 · 9 **13. a.** 44 **b.** 100 **15. a.** 1 and 11 **b.** 1 and 23 **c.** 1 and 37 **d.** They are prime numbers. **17.** yes **19.** 90 **21.** 605 **23.** no **25.** 2 **27.** 2 and 5 **29.** 3 · 5 · 2 · 5; 5 · 3 · 5 · 2; they are the same **31.** 13, 8, 7 **33.** 2 **35.** 7 · 7 · 7 **37.** 3 · 3 · 3 · 3 · 3 **39.** 5 · 5 · 11 **41.** 10 **43.** 2^5 **45.** 5^4 **47.** $4^2(5^2)$ **49.** 1, 2, 5, 10 **51.** 1, 2, 4, 5, 8, 10, 20, 40 **53.** 1, 2, 3, 6, 9, 18 **55.** 1, 2, 4, 11, 22, 44 **57.** 1, 7, 11, 77 **59.** 1, 2, 4, 5, 10, 20, 25, 50, 100 **61.** 3 · 13 **63.** $3^2 \cdot 11$ **65.** $2 \cdot 3^4$ **67.** $2^2 \cdot 5 \cdot 11$ **69.** 2^6 **71.** $3 \cdot 7^2$ **73.** 81 **75.** 32 **77.** 144 **79.** 4,096 **81.** 72 **83.** 3,456 **85.** 12,812,904 **87.** 1,162,213 **91.** 2^2 square units; 3^2 square units; 4^2 square units **97.** 231,000 **99.** 0 **101.** $A = lw$

Study Set Section 1.5 (page 48)

1. parentheses, brackets **3.** evaluate **5.** 3; square, multiply, subtract **7.** multiply, subtract **9.** $2 \cdot 3^2 = 2 \cdot 9$; $(2 \cdot 3)^2 = 6^2$ **11.** 4, 20 **13.** 9, 36 **15.** 27 **17.** 2 **19.** 15 **21.** 25 **23.** 5 **25.** 25 **27.** 18 **29.** 813 **31.** 5,239 **33.** 16 **35.** 5 **37.** 49 **39.** 24 **41.** 13 **43.** 10 **45.** 198 **47.** 18 **49.** 216 **51.** 17 **53.** 191 **55.** 3 **57.** 29 **59.** 14 **61.** 64 **63.** 192 **65.** 74 **67.** 137 **69.** 3 **71.** 21 **73.** 11 **75.** 1 **77.** 10,496 **79.** 2,845 **81.** 2(6) + 4(2) + 2(1); $22 **83.** 24 + 6(5) + 10(10) + 12(20) + 2(50) + 100; $594 **85.** brick: 3(3) + 1 + 1 + 3 + 3(5); 29; aphid: 3[1 + 2(3) + 4 + 1 + 2]; 42 **87.** 79° **89.** 5 **91.** 298 **97.** 7,300 **99.** 9,591

Study Set Section 1.6 (page 56)

1. equal, = **3.** solution, root **5.** equivalent **7.** $y + c$ **9.** addition of 6; subtract 6 from both sides **11.** 8, 8, 16, 24, 16 **13.** yes **15.** no **17.** yes **19.** yes **21.** yes **23.** yes **25.** no **27.** no **29.** no **31.** yes **33.** 10 **35.** 7 **37.** 3 **39.** 4 **41.** 13 **43.** 75 **45.** 740 **47.** 339 **49.** 3 **51.** 5 **53.** 9 **55.** 10 **57.** 1 **59.** 56 **61.** 84 **63.** 105 **65.** 4 **67.** 12 **69.** 8 **71.** 47 **75.** 94,683,948 **77.** 62 **79.** $218,500 **81.** $125 million **83.** 25 units **85.** $190 **93.** 325,780 **95.** 90 **97.** 3

Study Set Section 1.7 (page 64)

1. division **3.** x **5.** $\dfrac{y}{z}$ **7.** It is being multiplied by 4. Divide by 4. **9. a.** Subtract 5 from both sides. **b.** Add 5 to both sides. **c.** Divide both sides by 5. **d.** Multiply both sides by 5.

11. 3, 3, 4, 12, 4 **13.** 1 **15.** 96 **17.** 3 **19.** 6 **21.** 1 **23.** 2 **25.** 14 **27.** 42 **29.** 75 **31.** 39 **33.** 50 **35.** 49 **37.** 10 **39.** 3 **41.** 2 **43.** 1 **45.** 40 **47.** 1,200 **51.** 390 wpm **53.** 14 **55.** 96 **57.** 32 calls **59.** 55 lb **65.** 48 cm **67.** $2^3 \cdot 3 \cdot 5$ **69.** 72 **71.** 26 mpg

Key Concept (page 67)

1. Let x = the monthly cost to lease the van. **3.** Let x = the width of the field. **5.** Let x = the distance traveled by the motorist. **7.** $a + b = b + a$ **9.** $\frac{b}{1} = b$ **11.** $n - 1 < n$ **13.** $(r + s) + t = r + (s + t)$

Chapter Review (page 69)

1. a.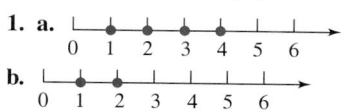
b.

2. a.

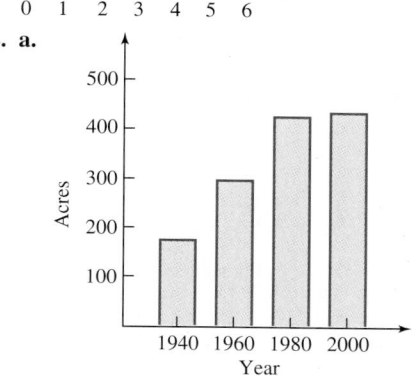

b.

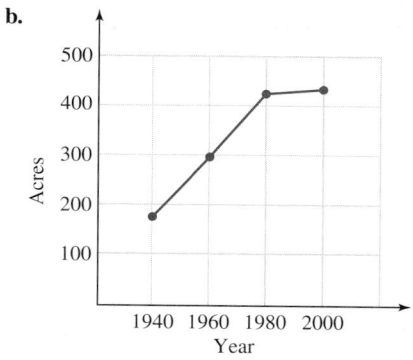

3. a. 6 **b.** 7 **4. a.** 5 hundred thousands + 7 ten thousands + 3 hundreds + 2 ones **b.** 3 ten millions + 7 millions + 3 hundred thousands + 9 thousands + 5 tens + 4 ones **5. a.** 3,207 **b.** 23,253,412 **c.** 16,000,000,000 **6. a.** > **b.** < **7. a.** 2,507,300 **b.** 2,510,000 **c.** 2,507,350 **d.** 2,500,000 **8. a.** 78 **b.** 137 **c.** 55 **d.** 149 **e.** 777 **f.** 2,332 **9. a.** 518 **b.** 6,000 **c.** 1,010 **d.** 24,986 **10. a.** commutative property of addition **b.** associative property of addition **11.** 96 in. **12. a.** 13 **b.** 4 **c.** 11 **d.** 54 **e.** 74 **f.** 2,075 **13. a.** 4 + 2 = 6 **b.** 5 − 2 = 3 **14.** $45 **15.** $785 **16.** $23,541 **17. a.** 56 **b.** 56 **c.** 0 **d.** 7 **e.** 560 **f.** 210 **18. a.** 3,297 **b.** 178,704 **c.** 31,684 **d.** 455,544 **19. a.** associative property of multiplication **b.** commutative property of multiplication **20.** $342 **21.** 108 ft, 288 ft² **22.** 720 **23. a.** 2 **b.** 15 **c.** undefined **d.** 0 **e.** 21 **f.** 37 **g.** 19 R 6 **h.** 23 R 27 **24.** 16, 25 **25.** 28 **26. a.** 1, 2, 3, 6, 9, 18 **b.** 1, 5, 25 **27. a.** prime **b.** composite **c.** neither **d.** neither **e.** composite **f.** prime **28. a.** odd **b.** even **c.** even

d. odd **29. a.** $2 \cdot 3 \cdot 7$ **b.** $3 \cdot 5^3$ **30. a.** 6^4 **b.** $5^3 \cdot 13^2$
31. a. 125 **b.** 121 **c.** 200 **d.** 2,700 **32. a.** 49 **b.** 32
c. 75 **d.** 36 **e.** 38 **f.** 24 **g.** 8 **h.** 24 **i.** 53 **j.** 3
k. 19 **l.** 7 **33.** $3(6) + 2(5) = 28$ **34.** 201 **35. a.** no
b. yes **36. a.** y **b.** t **37. a.** 9 **b.** 31 **c.** 340 **d.** 133
e. 9 **f.** 14 **g.** 120 **h.** 5 **i.** 7 **j.** 985 **38.** $97,250
39. 185 **40. a.** 4 **b.** 3 **c.** 21 **d.** 14 **e.** 21 **f.** 36
g. 315 **h.** 425 **i.** 1 **j.** 144 **41.** 24 in. **42.** $128

Chapter 1 Test   (page 75)

1.

2. 5 thousands + 2 hundreds + 6 tens + 6 ones **3.** 7,507
4. 35,000,000 **5.**

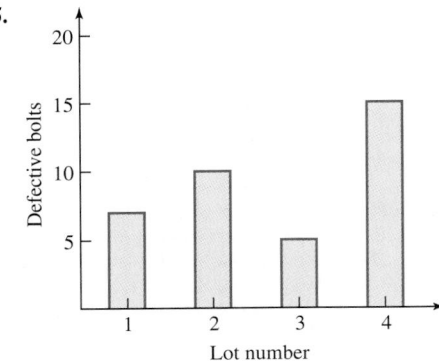

6.

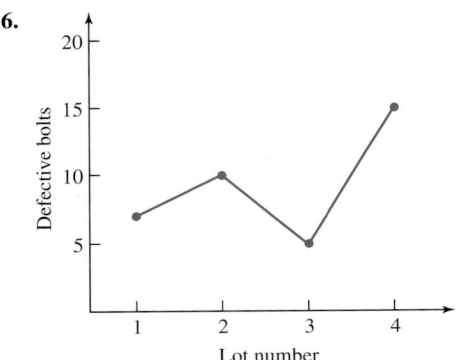

7. > **8.** < **9.** 1,491 **10.** 248 **11.** 58,105 **12.** 942
13. $76 **14.** 1, 2, 4, 5, 10, 20 **15.** 424 **16.** 26,791
17. 72 **18.** 114 R 57 **19.** 360 ft, 7,875 ft^2 **20.** 47
21. 3,456 **22.** $2^2 \cdot 3^2 \cdot 7$ **23.** 29 **24.** 44 **25.** 26
26. 39 **27.** yes **28.** 99 **29.** 30 **30.** 11 **31.** 81
32. 3,100 **33.** 194 yr **34.** To solve an equation means to find all
the values of the variable that, when substituted into the equation,
make a true statement.

Study Set   Section 2.1   (page 82)

1. negative **3.** number **5.** inequality **7.** opposites **9.** They
get smaller. **11.** yes **13.** $15 - 8$ **15.** $15 > 12$ **17. a.** -225
b. -10 **c.** -3 **d.** $-12,000$ **19.** -4 **21.** -8 and 2 **23.** -7
25. $6 - 4, -6, -(-6)$ (answers may vary) **27. a.** $-(-8)$
b. $|-8|$ **c.** $8 - 8$ **d.** $-|-8|$ **29.** 9 **31.** 8 **33.** 14
35. -20 **37.** -6 **39.** 203 **41.** 0 **43.** 11 **45.** 4
47. 1,201 **49.**
51.
53. <
55. < **57.** > **59.** > **61.** < **63.** < **65.** 2, 3, 2, 0, $-3, -7$

67. peaks: 2, 4, 0; valleys: $-3, -5, -2$ **69. a.** -1 (1 below
par) **b.** -3 (3 below par) **c.** Most of the scores are below
par. **71. a.** $-10°$ to $-20°$ **b.** $10°$ **c.** $10°$ **73. a.** 200 yr
b. A.D. **c.** B.C. **d.** the birth of Christ
75.

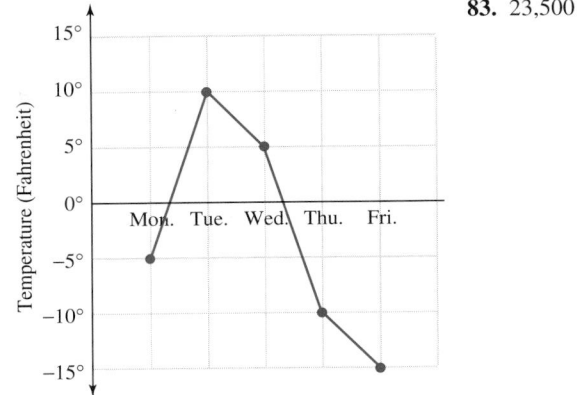

83. 23,500

85. 17 **87.** associative property of multiplication

Study Set   Section 2.2   (page 92)

1. identity **3.** 3 **5.** -2 **7. a.** yes **b.** yes **9. a.** 7 **b.** 10
11. subtract, larger **13.** -18 **15.** 5 **17.** -5 should be within
parentheses: $-6 + (-5)$ **19.** 11 **21.** 23 **23.** 0 **25.** -99
27. -9 **29.** -10 **31.** 1 **33.** -7 **35.** -20 **37.** 15 **39.** 8
41. 2 **43.** -10 **45.** 9 **47.** 8 **49.** -21 **51.** 3 **53.** -10
55. -4 **57.** 7 **59.** -21 **61.** -7 **63.** 9 **65.** 0 **67.** 0
69. 5 **71.** 0 **73.** -3 **75.** -10 **77.** -1 **79.** -17
81. $-8,346$ **83.** $-1,032$ **85.** 3G, -3G **87.** no; $70 shortfall
each month **89.** 2% risk **91.** $-1, 0$ **93.** 7 ft over flood stage
95. profit: $10 million **101.** 15 ft^2 **103.** 27 **105.** 5^3

Study Set   Section 2.3   (page 100)

1. difference **3.** subtraction **5.** 6 **7.** $x + (-y)$
9. brackets **11.** $-8 - (-4)$ **13.** 7 **15.** no; $8 - 3 = 5$,
$3 - 8 = -5$ **17.** $-3, 2$ **19.** $-2, -10, 6$ **21.** 9 **23.** -13
25. -10 **27.** -1 **29.** 0 **31.** 8 **33.** 5 **35.** -4 **37.** -4
39. -20 **41.** 0 **43.** 0 **45.** -15 **47.** -9 **49.** 3
51. 9 **53.** -2 **55.** -10 **57.** -14 **59.** 3 **61.** -8
63. -18 **65.** -6 **67.** 10 **69.** -4 **71.** $-2,447$ **73.** 20,503
75. $-1,676$ **77.** -120 ft **79.** 16 points **81.** -8
83. 1,007 ft **85.** -4 yd
87. a.
b. 37 ft
89. No; he will be $244 overdrawn ($-244$). **95.** 3 **97.** 1, 2, 4,
5, 10, 20 **99.** 156 **101.** 4 thousands + 5 hundreds + 2 ones

Study Set   Section 2.4   (page 108)

1. factors, product **3.** 3, exponent **5.** unlike
7. commutative **9.** -9, the opposite of that number
11. pos · pos, pos · neg, neg · pos, neg · neg **13. a.** negative
b. positive **15. a.** 3 **b.** 12 **c.** 5 **d.** 9 **e.** 10 **f.** 25
17. a. 2, 4; 4, 16; 6, 64 **b.** even **19.** 6 **21.** -5 should be
within parentheses: $-6(-5)$ **23.** 54 **25.** -15 **27.** -36
29. 56 **31.** -20 **33.** -120 **35.** 0 **37.** 6 **39.** 7 **41.** -23
43. -48 **45.** 40 **47.** -30 **49.** -60 **51.** -1 **53.** -18
55. 0 **57.** 0 **59.** 60 **61.** 16 **63.** -125 **65.** -8 **67.** 81
69. -1 **71.** 1 **73.** 49, -49 **75.** $-144, 144$ **77.** $-59,812$

79. 43,046,721   **81.** −25,728   **83.** 390,625   **85. a.** plan #1: −30 lb, plan #2: −28 lb   **b.** plan #1; the workout time is double that of plan #2   **87. a.** high 2, low −3   **b.** high 4, low −6   **89.** −20°   **91.** −20 ft   **93.** −$35,718   **99.** 45   **101.** 2,100   **103.** is less than

Study Set   Section 2.5   (page 114)

1. quotient, divisor   **3.** absolute value   **5.** positive
7. 5(−5) = −25   **9.** 0(?) = −6   **11.** $\frac{-20}{5}$ = −4   **13. a.** always true   **b.** sometimes true   **c.** always true   **15.** −7   **17.** 2   **19.** 5   **21.** 3   **23.** −20   **25.** −2   **27.** 0   **29.** undefined   **31.** −5   **33.** 1   **35.** −1   **37.** 10   **39.** −4   **41.** −3   **43.** 5   **45.** −4   **47.** −5   **49.** −4   **51.** −542   **53.** −16   **55.** −4° per hour   **57.** −1,000 ft   **59.** −6 (6 games behind)   **61.** −$15   **63.** −$1,740   **69.** 104   **71.** 2 · 3 · 5 · 7   **73.** 56   **75.** 81

Study Set   Section 2.6   (page 119)

1. order   **3.** grouping   **5.** 3; power, multiplication, subtraction   **7.** multiplication; subtraction   **9.** The base of the first exponential expression is 3; the base of the second is −3.   **11.** 4, 20, −20   **13.** 9, −36   **15.** −7   **17.** 1   **19.** −21   **21.** −14   **23.** −7   **25.** −5   **27.** 12   **29.** −14   **31.** 30   **33.** 2   **35.** 15   **37.** −42   **39.** −5   **41.** −3   **43.** 4   **45.** 0   **47.** −14   **49.** 19   **51.** 4   **53.** −3   **55.** 25   **57.** −48   **59.** 44   **61.** 91   **63.** 3   **65.** −5   **67.** 17   **69.** 11   **71.** 8   **73.** 112   **75.** −1,707   **77.** −15   **79.** −200   **81.** −320   **83.** −9,000   **85.** −1,200   **87.** 19   **89.** 11 yd   **91.** 60-cent gain   **97.** 4   **99.** Add the lengths of all its sides.   **101.** no

Study Set   Section 2.7   (page 127)

1. solve   **3.** x **5. a.** 3   **b.** (−3)   **7. a.** multiplication by −2   **b.** addition of −6   **c.** mult. by −4, subtraction of 8   **d.** mult. by −5, addition of −6   **9.** simplify   **11.** opposite   **13. a.** subtraction of 3   **b.** addition of −6   **15.** −13, 7   **17.** 1, 1, −12, −4, −4, 3   **19.** −10 · x **21.** yes   **23.** no   **25.** −18   **27.** −14   **29.** 5   **31.** −1   **33.** −9   **35.** −14   **37.** −2   **39.** 0   **41.** −8   **43.** 5   **45.** 2   **47.** −1   **49.** 6   **51.** 0   **53.** 6   **55.** −52   **57.** −7   **59.** −4   **61.** −5   **63.** −2   **65.** 10   **67.** −6   **69.** −3   **71.** 3   **73.** −6   **75.** 54   **77.** 30   **79.** −14   **81.** −3   **83.** −2   **85.** −8   **87.** 15   **91.** 18 ft   **93.** −51 yd   **95.** 34   **97.** $5   **99.** 29 points   **101.** zone −8   **105.** 5 · 5 · 5 · 5 · 5 · 5   **107.** 12   **109.** $\frac{16}{8}$

Key Concept   (page 132)

1. −5   **3.** −30   **5.** +10 or 10   **7.** −205
9.

```
<---+----+----+----+----+----+----+----+----+--->
   -4   -3   -2   -1    0    1    2    3    4
      Negatives              Positives
```

11. $x < y$

13. Like signs: Add their absolute values and attach their common sign to the sum. Unlike signs: Subtract their absolute values, the smaller from the larger, and attach the sign of the number with the larger absolute value to that result.   **15.** Divide their absolute values. Like signs: The quotient is positive. Unlike signs: The quotient is negative.

Chapter Review   (page 134)

1. a.

```
<--+----+----●----+----+----+----+----●----+-->
  -4   -3   -2   -1    0    1    2    3    4
```

b.

```
<--+----●----+----●----+----●----+----●----+-->
  -4   -3   -2   -1    0    1    2    3    4
```

2. a. <   **b.** <   **c.** >   **d.** <   **3.** −33 ft   **4. a.** −$1,200   **b.** −10 sec

5. a. 4   **b.** 0   **c.** 43   **d.** −12   **6. a.** negative   **b.** the opposite   **c.** negative   **d.** minus   **7. a.** 12   **b.** −8   **c.** 8   **d.** 0   **8. a.** 2

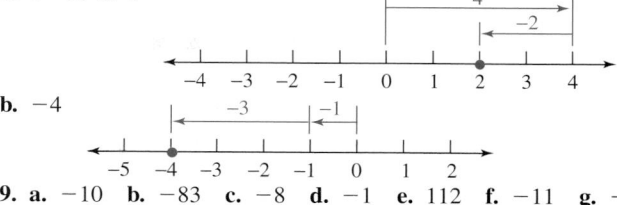

b. −4

9. a. −10   **b.** −83   **c.** −8   **d.** −1   **e.** 112   **f.** −11   **g.** −3   **h.** −2   **10. a.** −4   **b.** −20   **c.** 0   **d.** 0   **11. a.** 11   **b.** −4   **12.** 65 ft   **13. a.** −3   **b.** −21   **c.** 4   **d.** −112   **e.** −6   **f.** 6   **g.** −37   **h.** 30   **14.** adding, opposite   **15. a.** −4   **b.** 15   **c.** 6   **d.** −8   **16.** −77   **17.** −225 ft   **18.** −1   **19.** Alaska: 180°; Virginia: 140°   **20. a.** −45   **b.** 18   **c.** −14   **d.** 376   **e.** −100   **f.** 1   **g.** −25   **h.** −150   **21. a.** −36   **b.** −36   **c.** 0   **d.** 1   **22.** −3, −6, −9   **23. a.** 25   **b.** −32   **c.** 64   **d.** −64   **24.** negative   **25.** first expression: base of 2; second: base of −2; −4, 4   **26.** −3   **27. a.** −2   **b.** −5   **c.** −8   **d.** 101   **28. a.** 0   **b.** undefined   **c.** 1   **d.** 10   **29.** −2 min   **30. a.** −22   **b.** 4   **c.** −43   **d.** 8   **e.** 41   **f.** 0   **g.** −13   **h.** 32   **31. a.** 12   **b.** −16   **c.** −4   **d.** 1   **32. a.** −1   **b.** −4   **33. a.** −70   **b.** 20   **c.** −7,000   **d.** 1,100   **34. a.** yes   **b.** no   **35. a.** −10   **b.** 12   **c.** −8   **d.** 4   **36. a.** 15   **b.** −4   **37. a.** 3   **b.** −2   **c.** −12   **d.** 0   **38.** −46°   **39.** 121   **40.** $8,200

Chapter 2 Test   (page 139)

1. a. >   **b.** <   **c.** <   **2.** {. . . , −3, −2, −1, 0, 1, 2, 3, . . .}
3. Monroe
4. −5

```
<--+----●----+----+----+----+----+----+----+----+----+----+----+-->
  -6   -5   -4   -3   -2   -1    0    1    2    3    4    5    6
```

5. a. −34   **b.** −34   **c.** −8   **6. a.** −13   **b.** −1   **c.** −15   **d.** −150   **7. a.** −70   **b.** −48   **c.** 16   **d.** 0   **8.** (−4)(5) = −20   **9. a.** −8   **b.** undefined   **c.** −5   **d.** 0   **10.** $3 million   **11.** 154 ft   **12. a.** 6   **b.** 7   **c.** 6   **d.** 132   **13. a.** 16   **b.** −16   **c.** 49   **14.** −27   **15.** 1   **16.** −34   **17.** 42   **18.** 4   **19.** −15   **20.** 16   **21.** −40   **22.** −5   **23.** 2   **24.** −18   **25.** −$244   **26.** 18   **27.** −4 + (−4) + (−4) + (−4) + (−4) = −20   **28.** The absolute value of a number is the distance from the number to 0 on a number line. Distance is either positive or 0, but never negative.

Cumulative Review Exercises   (page 141)

1. 1, 2, 5, 9   **2.** 0, 1, 2, 5, 9   **3.** −2, −1   **4.** −2, −1, 0, 1, 2, 5, 9   **5.** 6   **6.** 3   **7.** 7,326,500   **8.** 7,330,000   **9.** CRF Cable
10.

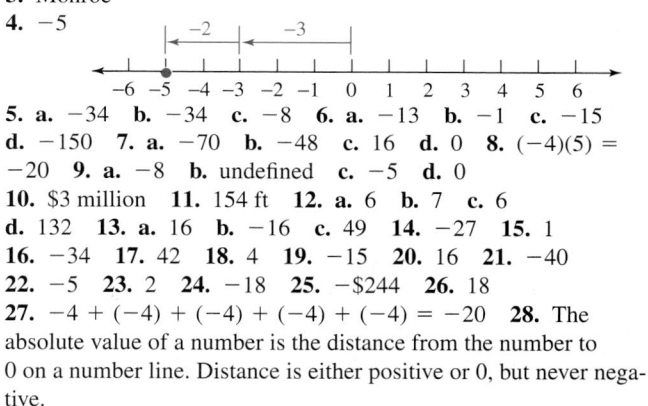

11. 786

12. 3,806   **13.** 4,684   **14.** 13,136   **15.** 104 ft, 595 ft²   **16.** 65   **17.** 11,745   **18.** 13   **19.** 307,329   **20.** 467

21. 1,728 **22.** 1, 2, 3, 6, 9, 18 **23.** prime, odd **24.** composite, even **25.** even **26.** odd **27.** $2^3 \cdot 3^2 \cdot 7$ **28.** 11^4 **29.** 175
30. 38 **31.** 50 **32.** 2 **33.** no **34.** yes **35.** 13 **36.** 53
37. 27 **38.** 24 **39.** (number line: −3 to 3)
40. (number line: −4 to 2) **41.** true **42.** 9, −9
43. −5 **44.** −14 **45.** −8 **46.** −231 **47.** 24 **48.** −1,715
49. 2 **50.** −50 **51.** 26 **52.** −16 **53.** −3 **54.** 4 **55.** 3
56. −18 **57.** $126,037 **58.** −279° F

Study Set Section 3.1 (page 148)

1. numerator, denominator **3.** proper, improper **5.** equivalent
7. higher, building **9. a.** 2 **b.** 3 **c.** 5 **d.** 7 **11.** equivalent fractions: $\frac{2}{6} = \frac{1}{3}$ **13. a.** In the first case, 20 and 28 were factored. In the second case, they were prime factored. **b.** yes **15.** The 2's in the numerator and denominator aren't common factors.
17. a. $\frac{8}{1}$ **b.** $-\frac{25}{1}$ **19.** 3, 2, 3, 2, 3, 2 **21.** $\frac{1}{3}$ **23.** $\frac{1}{3}$ **25.** $\frac{2}{3}$
27. $\frac{5}{2}$ **29.** $-\frac{1}{2}$ **31.** $-\frac{6}{7}$ **33.** $\frac{5}{9}$ **35.** $\frac{6}{7}$ **37.** in lowest terms **39.** $\frac{3}{8}$ **41.** $\frac{5}{7}$ **43.** $\frac{4}{5}$ **45.** in lowest terms **47.** in lowest terms **49.** $-\frac{1}{3}$ **51.** $\frac{3}{5}$ **53.** $\frac{5}{4}$ **55.** 2 **57.** $\frac{35}{40}$ **59.** $\frac{28}{35}$ **61.** $\frac{45}{54}$
63. $\frac{15}{30}$ **65.** $\frac{4}{14}$ **67.** $\frac{54}{60}$ **69.** $\frac{25}{20}$ **71.** $\frac{6}{45}$ **73.** $\frac{15}{5}$ **75.** $\frac{48}{8}$ **77.** $\frac{36}{9}$
79. $-\frac{4}{2}$ **81.** $\frac{3}{5}$ **83.** $-\frac{15}{16}$ in. **85.** $\frac{7}{10}, \frac{1}{8}$

87.

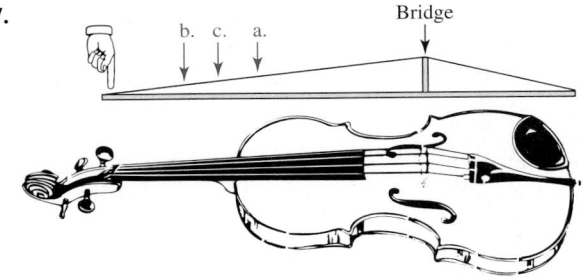

89. one-quarter turn to the left; three-quarters of a turn to the right
91.

SNACKS
Potato chips
Peanuts
Pretzels
Tortilla chips

93. $\frac{1}{250}$ **99.** −3 **101.** 564,000

Study Set Section 3.2 (page 157)

1. multiply **3.** product **5.** base, height **7.** $\frac{a \cdot c}{b \cdot d}$
9. (grid) **a.** $\frac{1}{4}$ **b.** 12, 1, $\frac{1}{12}$

11. a. negative **b.** positive **13. a.** true **b.** true **c.** false
d. true **15.** 7, 15, 3, 3 **17.** $\frac{1}{8}$ **19.** $\frac{21}{128}$ **21.** $\frac{4}{7}$ **23.** $\frac{77}{60}$ **25.** $-\frac{1}{5}$
27. $\frac{2}{9}$ **29.** $\frac{2}{3}$ **31.** 1 **33.** $\frac{1}{20}$ **35.** $\frac{1}{30}$ **37.** 15 **39.** −12 **41.** $\frac{5x}{72}$
43. $\frac{b}{40}$ **45.** d **47.** s **49.** $\frac{5x}{6}, \frac{5}{6}x$ **51.** $-\frac{8v}{9}, -\frac{8}{9}v$ **53.** $\frac{4}{9}$ **55.** $\frac{25}{81}$
57. $\frac{16}{9}$ **59.** $-\frac{27}{64}$

61.

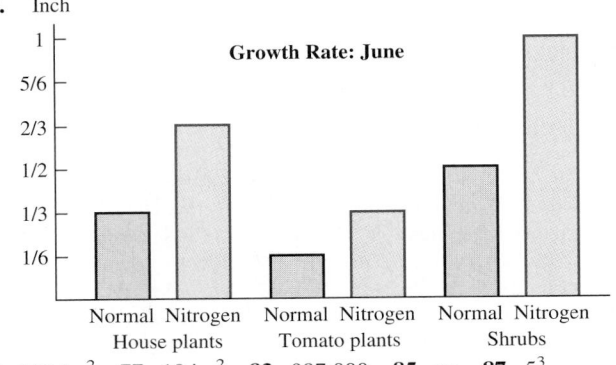

·	$\frac{1}{2}$	$\frac{1}{3}$	$\frac{1}{4}$	$\frac{1}{5}$	$\frac{1}{6}$
$\frac{1}{2}$	$\frac{1}{4}$	$\frac{1}{6}$	$\frac{1}{8}$	$\frac{1}{10}$	$\frac{1}{12}$
$\frac{1}{3}$	$\frac{1}{6}$	$\frac{1}{9}$	$\frac{1}{12}$	$\frac{1}{15}$	$\frac{1}{18}$
$\frac{1}{4}$	$\frac{1}{8}$	$\frac{1}{12}$	$\frac{1}{16}$	$\frac{1}{20}$	$\frac{1}{24}$
$\frac{1}{5}$	$\frac{1}{10}$	$\frac{1}{15}$	$\frac{1}{20}$	$\frac{1}{25}$	$\frac{1}{30}$
$\frac{1}{6}$	$\frac{1}{12}$	$\frac{1}{18}$	$\frac{1}{24}$	$\frac{1}{30}$	$\frac{1}{36}$

63. 15 ft² **65.** $\frac{15}{2}$ yd² **67.** 290 **69.** 18, 6, and 2 in. **71.** $\frac{3}{8}$ cup sugar, $\frac{1}{6}$ cup molasses
73. Inch

Growth Rate: June

(bar chart; y-axis: 1, 5/6, 2/3, 1/2, 1/3, 1/6; x-axis: Normal Nitrogen — House plants, Normal Nitrogen — Tomato plants, Normal Nitrogen — Shrubs)

75. 121 in.² **77.** 18 in.² **83.** 987,000 **85.** no **87.** 5^3

Study Set Section 3.3 (page 165)

1. reciprocals **3.** $\frac{1}{2}, \frac{3}{2}$ **5.** (four boxes) $4 \div \frac{1}{3}$, 12
7. 1 **9. a.** 5 **b.** 5 **c.** $\frac{1}{3}$ **11.** 9, 10, 9, 10, 5, 5, 9, 9, 5 **13.** $\frac{5}{6}$
15. $\frac{27}{16}$ **17.** 1 **19.** $\frac{2}{3}$ **21.** 36 **23.** 50 **25.** $\frac{2}{15}$ **27.** $\frac{1}{192}$
29. $-\frac{27}{8}$ **31.** $-\frac{15}{2}$ **33.** $-\frac{1}{64}$ **35.** 1 **37.** $\frac{8}{15}$ **39.** $\frac{1}{6}$ **41.** $\frac{13}{8}$
43. $-\frac{5}{8}$ **45.** 104 **47.** 56 **49.** route 1 **51. a.** sixteen parts **b.** $\frac{3}{4}$ in. **c.** $\frac{1}{120}$ in. **53.** 7,855 **59.** −4 **61.** $\frac{10}{7}$ **63.** false
65. 637,500

Study Set Section 3.4 (page 173)

1. least **3.** higher **5.** denominators, numerators, common
7. The denominators are unlike. **9.** 4 **11. a.** once
b. twice **c.** three times **13.** 60
15. a. $\frac{1}{3}$
b. $\frac{1}{4} = \frac{3}{12}, \frac{1}{3} = \frac{4}{12}$ **17.** 3, 3, 6, 5, 6, 5 **19.** 18 **21.** 24 **23.** 40
25. 60 **27.** $\frac{4}{7}$ **29.** $\frac{20}{103}$ **31.** $\frac{2}{5}$ **33.** $\frac{8}{7}$ **35.** $\frac{5}{8}$ **37.** $\frac{9}{20}$ **39.** $\frac{22}{15}$
41. $\frac{23}{56}$ **43.** $\frac{1}{12}$ **45.** $\frac{1}{12}$ **47.** $\frac{47}{50}$ **49.** $-\frac{3}{16}$ **51.** $-\frac{2}{3}$ **53.** $-\frac{23}{24}$
55. $-\frac{13}{5}$ **57.** $-\frac{23}{4}$ **59.** $\frac{47}{60}$ **61.** $\frac{3}{4}$ **63.** $\frac{19}{48}$ **65.** $-\frac{43}{45}$ **67.** $\frac{26}{75}$
69. $\frac{17}{54}$ **71.** $\frac{5}{36}$ **73.** $-\frac{17}{60}$ **75. a.** $\frac{7}{32}$ in. **b.** $\frac{3}{32}$ in. **77.** $\frac{17}{24}$; no
79. $\frac{1}{16}$ lb, undercharge **81.** $\frac{4}{5}, \frac{3}{4}, \frac{5}{8}$ **83.** $\frac{7}{10}$ **85.** $\frac{1}{6}$ hp
91. $2^2 \cdot 5$ **93.** $A = lw$

Study Set The LCM and the GCF (page 178)

1. 15 **3.** 56 **5.** 42 **7.** 18 **9.** 660 **11.** 600 **13.** 72
15. 378 **17.** 3 **19.** 11 **21.** 4 **23.** 25 **25.** 20 **27.** 12
29. 6 **31.** 9 **33.** 360 min (6 hr)

Study Set   Section 3.5   (page 183)

1. mixed   **3.** graph   **5. a.** $-5\frac{1}{2}°$ **b.** $-1\frac{7}{8}$ in.   **7. a.** $-2\frac{2}{3}$
b. $-3\frac{1}{3}$ **9.** $-\frac{4}{5}, -\frac{2}{5}, \frac{1}{5}$ **11.** $2\frac{1}{2}$
13.

15. 8, 8, 4, 4, 6   **17.** $3\frac{3}{4}$

19. $5\frac{4}{5}$ **21.** $-3\frac{1}{3}$ **23.** $10\frac{7}{12}$ **25.** $\frac{13}{2}$ **27.** $\frac{104}{5}$ **29.** $-\frac{56}{9}$ **31.** $\frac{602}{3}$
33.

35.

37. $3\frac{4}{7}$

39. $10\frac{1}{2}$ **41.** 14   **43.** $-13\frac{3}{4}$ **45.** $-8\frac{1}{3}$ **47.** $\frac{35}{72}$ **49.** $-1\frac{1}{4}$
51. $\frac{25}{9} = 2\frac{7}{9}$ **53.** $-\frac{64}{27} = -2\frac{10}{27}$ **55.** $1\frac{9}{11}$ **57.** $-\frac{9}{10}$ **59.** 12
61. $\frac{5}{16}$ **63.** $-\frac{2}{3}$ **65.** $2\frac{1}{2}$ **67.** -2 **69.** 64 calories   **71.** $2.72
73. 675   **75.** $2\frac{3}{4}$ in., $1\frac{1}{4}$ in.   **77.** $42\frac{5}{8}$ in.^{2} **79.** 602   **81.** size 14,
slim cut   **87.** 72   **89.** 4(8)   **91.** division by 2

Study Set   Section 3.6   (page 191)

1. commutative   **3.** borrow   **5. a.** $76, \frac{3}{4}$ **b.** $76 + \frac{3}{4}$ **7.** the
fundamental property of fractions   **9. a.** $10\frac{1}{16}$ **b.** $1,290\frac{1}{3}$ **c.** $17\frac{1}{2}$
d. $46\frac{1}{5}$ **11.** 70, 39, 70, 39, 7, 5, 7, 5, 35, 35, 31   **13.** $4\frac{2}{5}$ **15.** $5\frac{1}{7}$
17. $7\frac{1}{2}$ **19.** $5\frac{11}{30}$ **21.** $1\frac{1}{4}$ **23.** $1\frac{11}{24}$ **25.** $9\frac{3}{10}$ **27.** $3\frac{5}{14}$ **29.** $129\frac{11}{15}$
31. $397\frac{5}{12}$ **33.** $273\frac{2}{9}$ **35.** $623\frac{8}{21}$ **37.** $11\frac{1}{30}$ **39.** $101\frac{7}{16}$ **41.** $2\frac{1}{2}$
43. $26\frac{7}{24}$ **45.** $10\frac{7}{16}$ **47.** $320\frac{5}{18}$ **49.** $6\frac{1}{3}$ **51.** $\frac{1}{4}$ **53.** $3\frac{12}{35}$
55. $3\frac{5}{8}$ **57.** $4\frac{1}{3}$ **59.** $3\frac{7}{8}$ **61.** $53\frac{5}{12}$ **63.** $460\frac{1}{8}$ **65.** $-5\frac{1}{4}$
67. $-5\frac{7}{8}$ **69.** $2\frac{3}{4}$ mi   **71.** $7\frac{2}{3}$ cups   **73.** $48\frac{1}{2}$ ft   **75. a.** $16\frac{1}{2}, 16\frac{1}{2};$
$5\frac{1}{5}, 5\frac{1}{5}$ **b.** $21\frac{7}{10}$ mi   **77. a.** 20¢   **b.** 30¢   **79.** $191\frac{2}{3}$ ft   **85.** 7
87. 6   **89.** the amount of surface a figure encloses

Study Set   Section 3.7   (page 199)

1. complex   **3.** $\frac{2}{3} \div \frac{1}{5}$ **5.** 15   **7.** negative   **9.** subtraction
11. $\frac{3}{4}, \frac{4}{3}, 4, 4$ **13.** $\frac{1}{3}$ **15.** $\frac{31}{45}$ **17.** $\frac{37}{40}$ **19.** $\frac{3}{10}$ **21.** $-1\frac{27}{40}$ **23.** $\frac{3}{4}$
25. $-\frac{3}{64}$ **27.** $-1\frac{1}{6}$ **29.** $8\frac{1}{2}$ **31.** $\frac{49}{4}$ **33.** $\frac{121}{18}$ **35.** $8\frac{1}{4}$ in.   **37.** $\frac{5}{6}$
39. $-1\frac{1}{3}$ **41.** $10\frac{1}{2}$ **43.** $\frac{4}{9}$ **45.** 3   **47.** 5   **49.** -20 **51.** 11
53. $\frac{3}{7}$ **55.** $-\frac{3}{8}$ **57.** $8\frac{1}{2}$ **59.** $2\frac{1}{2}, 1\frac{1}{2}, 3\frac{3}{4}; 7\frac{1}{5}, 1\frac{1}{2}, 10\frac{4}{5}; 14\frac{11}{20}$ mi
61. yes   **63.** $10\frac{1}{2}$ mi   **65.** 6 sec   **71.** 2   **73.** -5 **75.** 8

Study Set   Section 3.8   (page 206)

1. reciprocal   **3.** least common denominator   **5.** Yes; when
40 is substituted for x, the result is a true statement: $25 = 25$.
7. 1   **9. a.** $\frac{4}{5}p$ **b.** $\frac{1}{4}t$ **13. a.** true   **b.** false   **c.** true
d. true   **15.** 28   **17.** -32 **19.** $-\frac{20}{3}$ **21.** $\frac{6}{5}$ **23.** 0   **25.** 30
27. $\frac{14}{5}$ **29.** $\frac{4}{25}$ **31.** $\frac{8}{9}$ **33.** $\frac{1}{3}$ **35.** $\frac{7}{18}$ **37.** $-\frac{5}{8}$ **39.** -1
41. $-\frac{5}{4}$ **43.** -36 **45.** $\frac{5}{12}$ **47.** $\frac{8}{9}$ **49.** $-\frac{27}{2}$ **51.** $-\frac{5}{3}$ **53.** $-\frac{27}{4}$
55. 12   **57.** 24   **59.** -12 **61.** $\frac{75}{4}$ **65.** 20   **67.** 36   **69.** 450
71. 8 in.   **73.** 360 min   **79.** -32 **81.** 3   **83.** 13,000,000

Key Concept   (page 210)

1. 5, 5, 5, 5, 5, 5, 5   **3.** 7, 7, 7, 7, 7

Chapter Review   (page 212)

1. $\frac{7}{24}$ **2.** The figure is not divided into equal parts.   **3.** $-\frac{2}{3}, \frac{-2}{3}$
4. equivalent fractions: $\frac{6}{8} = \frac{3}{4}$ **5.** The numerator and denominator
of the fraction are being divided by 2.   **6.** The numerator and denominator of the fraction are being divided by 2. The answer to
each division is 1.   **7. a.** $\frac{1}{3}$ **b.** $\frac{5}{12}$ **c.** $-\frac{3}{4}$ **d.** $\frac{11}{18}$ **8.** The
numerator and denominator of the original fraction are being multi-
plied by 2 to obtain an equivalent fraction in higher terms.
9. a. $\frac{12}{18}$ **b.** $-\frac{6}{16}$ **c.** $\frac{21}{45}$ **d.** $\frac{36}{9}$ **10. a.** $\frac{1}{6}$ **b.** $-\frac{14}{45}$ **c.** $\frac{5}{12}$ **d.** $\frac{1}{5}$
e. $\frac{21}{5}$ **f.** $\frac{9}{4}$ **g.** 1   **h.** 1   **11. a.** true   **b.** false   **12. a.** $\frac{2}{9}$
b. $-\frac{8}{21}s$ **c.** $\frac{1}{21}$ **d.** $-\frac{5}{9}m$ **13. a.** $\frac{9}{16}$ **b.** $-\frac{125}{8}$ **c.** $\frac{4}{9}$ **d.** $-\frac{8}{125}$
14. 30 lb   **15.** 60 in.^{2} **16. a.** 8   **b.** $-\frac{12}{11}$ **c.** 1   **d.** $\frac{1}{200}$
17. a. $\frac{25}{66}$ **b.** $-\frac{7}{2}$ **c.** $\frac{3}{32}$ **d.** $\frac{5}{2}$ **e.** $\frac{1}{2}$ **f.** $\frac{8}{5}$ **18.** 12   **19. a.** $\frac{5}{7}$
b. $-\frac{6}{5}$ **c.** $\frac{1}{4}$ **d.** $\frac{7}{4} = 1\frac{3}{4}$ **20.** The denominators are not the
same.   **21.** 90   **22. a.** $\frac{5}{6}$ **b.** $\frac{1}{40}$ **c.** $-\frac{29}{24}$ **d.** $\frac{20}{7}$ **e.** $-\frac{23}{6}$ **f.** $\frac{47}{60}$
23. $\frac{7}{32}$ in.   **24.** the second hour   **25. a.** $2\frac{1}{6}$ **b.** $\frac{13}{6}$ **26. a.** $3\frac{1}{5}$
b. $-3\frac{11}{12}$ **c.** 1   **d.** $2\frac{1}{3}$ **27. a.** $\frac{75}{8}$ **b.** $-\frac{11}{5}$ **c.** $\frac{201}{2}$ **d.** $\frac{199}{100}$
28.

29. a. $-\frac{3}{10}$

b. $\frac{21}{22}$ **c.** 40   **d.** $-2\frac{1}{2}$ **30.** $48\frac{1}{8}$ in.   **31. a.** $3\frac{23}{40}$ **b.** $6\frac{1}{6}$ **c.** $1\frac{1}{12}$
d. $1\frac{5}{16}$ **32.** $39\frac{11}{12}$ gal   **33. a.** $182\frac{5}{8}$ **b.** $113\frac{3}{20}$ **c.** $31\frac{11}{24}$ **d.** $316\frac{3}{4}$
34. a. $20\frac{1}{2}$ **b.** $34\frac{3}{8}$ **35. a.** $\frac{8}{9}$ **b.** $\frac{19}{72}$ **36. a.** $-\frac{12}{17}$ **b.** $-\frac{2}{5}$
37. a. 24   **b.** 28   **c.** $-\frac{1}{3}$ **d.** $\frac{11}{2}$ **38. a.** $\frac{57}{8}$ **b.** $-\frac{18}{5}$ **c.** 16
d. $\frac{26}{3}$ **39.** 330

Chapter 3 Test   (page 217)

1. a. $\frac{4}{5}$ **b.** $\frac{1}{5}$ **2. a.** $\frac{3}{4}$ **b.** $\frac{2}{5}$ **3.** $-\frac{3}{20}$ **4.** 40   **5.** 12
6. $-\frac{19}{30}$ **7.** $\frac{21}{24}$ **8.**

9. $1\frac{1}{2}$ million   **10.** $261\frac{11}{36}$ **11.** $37\frac{5}{12}$ **12. a.** 0 lb
b. $2\frac{4}{4}$ in.   **c.** $3\frac{3}{4}$ in.   **13.** $\frac{11}{7}$ **14.** $11\frac{1}{4}$ in.   **15.** perimeter: $53\frac{1}{3}$ in.;
area: $106\frac{2}{3}$ in.^{2} **16.** $\frac{13}{24}$ **17.** $-\frac{20}{21}$ **18.** $-\frac{5}{3}$ **19. a.** 42   **b.** $-\frac{36}{5}$
20. 144   **21.** numerator, fraction bar, denominator; equal parts of
a whole, or a division   **22.** When we multiply a number, such as $\frac{3}{4}$,
and its reciprocal, $\frac{4}{3}$, the result is 1.   **23. a.** dividing the numerator
and denominator of a fraction by the same number   **b.** equivalent
fractions: $\frac{1}{2} = \frac{2}{4}$ **c.** multiplying the numerator and denominator of
a fraction by the same number

Cumulative Review Exercises   (page 219)

1. 5,434,700   **2.** 5,430,000   **3.** 11,555, 10:30 A.M.   **4.** hundred
billions   **5.** 8,136   **6.** 3,519   **7.** 299,320   **8.** 991   **9.** 450 ft
10. 11,250 ft^2 **11.** $2^2 \cdot 3 \cdot 7$ **12.** $2 \cdot 3^2 \cdot 5^2$ **13.** $2^3 \cdot 3^2 \cdot 5$
14. $2^4 \cdot 3^2 \cdot 5^2$ **15.** 16   **16.** -35 **17.** 2   **18.** 2   **19.** -5
20. -5 **21.** -16 **22.** -5 **23.** 21   **24.** 21 ft by 84 ft
25. $\frac{3}{4}$ **26.** $\frac{5}{2}$ **27.** $-\frac{4}{5}$ **28.** $\frac{1}{2}$ **29.** $1\frac{5}{12}$ **30.** $\frac{11}{15}$ **31.** $\frac{23}{6}$
32. $-\frac{53}{8}$ **33.** $9\frac{11}{12}$ **34.** $5\frac{11}{15}$ **35.** $\frac{11}{16}$ in.   **36.** 90 sec   **37.** $\frac{2}{7}$
38. $-1\frac{9}{29}$ **39.** $-\frac{17}{15}$ **40.** 4   **41.** -15 **42.** $\frac{8}{3}$ **43.** An expression is a combination of numbers and/or variables with operation
symbols. An equation contains an $=$ sign.   **44.** a letter that is used
to stand for a number

Study Set Section 4.1 (page 227)

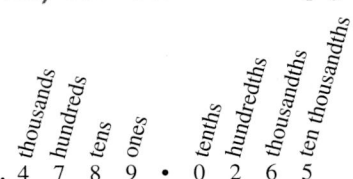

1. 4 7 8 9 • 0 2 6 5
3. rounding **5. a.** thirty-two and four hundred fifteen thousandths **b.** 32 **c.** $\frac{415}{1,000}$ **d.** $30 + 2 + \frac{4}{10} + \frac{1}{100} + \frac{5}{1,000}$
7.

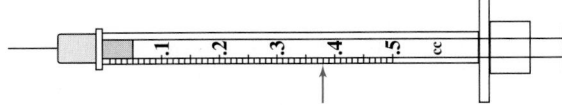

9. a. true **b.** false **c.** true **d.** true **11.** $\frac{47}{100}$, 0.47
13.

0.3

15. 9,816.0245
17. fifty and one tenth; $50\frac{1}{10}$ **19.** negative one hundred thirty-seven ten thousandths; $-\frac{137}{10,000}$ **21.** three hundred four and three ten-thousandths; $304\frac{3}{10,000}$ **23.** negative seventy-two and four hundred ninety-three thousandths; $-72\frac{493}{1,000}$ **25.** -0.39
27. 6.187 **29.** 506.1 **31.** 2.7 **33.** -0.14 **35.** 33.00
37. 3.142 **39.** 1.414 **41.** 39 **43.** 2,988 **45. a.** $3,090
b. $3,090.30 **47.** < **49.** > **51.** 132.64, 132.6401,
132.6499 **53.** $1,025.78
55.

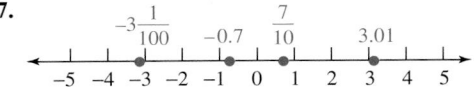

57. a. 0.30 **b.** 1,609.34 **c.** 453.59 **d.** 3.79 **59.** sand, silt, granule, clay **61.** Texas City, Houston, Westport, Galveston, White Plains, Crestline **63.** gold: Retton; silver: Szabo; bronze: Pauca **65.** $-$0.07, $-$0.30 **73.** $164\frac{11}{20}$ **75.** 72 in.^{2} **77.** -1

Study Set Section 4.2 (page 234)

1. sum **3.** point **5. a.** 0.47 **b.** $\frac{3}{10}, \frac{17}{100}, \frac{47}{100}$ **c.** 0.47 **d.** They are the same. **7.** 39.9 **9.** 54.72 **11.** 15.9 **13.** 0.23064
15. 288.46 **17.** 58.04 **19.** 9.53 **21.** 70.29 **23.** 4.977
25. 0.19 **27.** -10.9 **29.** 38.29 **31.** -14.3 **33.** -0.0355
35. -16.6 **37.** 47.91 **39.** 2.598 **41.** 11.01 **43.** 4.1
45. 35.85 **47.** -57.47 **49.** 6.2 **51.** 15.2 **53.** 8.03
55. a. 53.044 sec **b.** 102.38 **57.** 103.4 in. **59.** 1.8, Texas
61. 1.74, 2.32, 4.06, 2.90, 0, 2.90 **63.** 43.99 sec
65. $765.69, $740.69 **67. a.** $101.94 **b.** $55.80
69. 8,156.9343 **71.** 1,932.645 **73.** 2,529.0582 **79.** $110\frac{23}{40}$
81. $-\frac{5}{6}$

Study Set Section 4.3 (page 243)

1. factors, product **3.** whole, sum **5.** larger **7. a.** $\frac{21}{1,000}$
b. $\frac{21}{1,000} = 0.021$. They are the same. **9.** 0.08 **11.** -0.15
13. 0.98 **15.** 0.072 **17.** 12.32 **19.** -0.0049 **21.** -0.084
23. -8.6265 **25.** 9.6 **27.** -56.7 **29.** 12.24 **31.** -18.183
33. 0.024 **35.** -16.5 **37.** 42 **39.** 6,716.4 **41.** -0.56
43. 8,050 **45.** 980 **47.** -200 **49.** 0.01, 0.04, 0.09, 0.16, 0.25, 0.36, 0.49, 0.64, 0.81 **51.** 1.44 **53.** 1.69 **55.** -17.48
57. 14.24 **59.** 0.84 **61.** -3.872 **63.** 18.72 **65.** 86.49
67. a. $12.50, $12,500, $15.75, $1,575 **b.** $14,075 **69.** 0.75
in. **71.** 136.4 lb **73.** $52.00, $52.50, $31.45 **75.** 160.6 m
77. 0.000000136 in., 0.0000000136 in., 0.00000004 in.
79. 15.29694 **81.** 631.2722 **83.** $102.65 **89.** 7
91. the absolute value of negative three **93.** -1

Study Set Section 4.4 (page 250)

1. dividend, divisor, quotient **3.** whole, right, above **5.** true
7. 10 **9.** Use multiplication to see whether $0.9 \cdot 2.13 = 1.917$.
11. yes **13.** moving the decimal points in the divisor and dividend two places to the right **15.** 4.5 **17.** -9.75 **19.** 6.2
21. 32.1 **23.** 2.46 **25.** -7.86 **27.** 2.66 **29.** 7.17
31. 130 **33.** 1,050 **35.** 0.6 **37.** 0.6 **39.** 5.3 **41.** -2.4
43. 13.60 **45.** 0.79 **47.** 0.07895 **49.** -0.00064
51. 0.0348 **53.** 4.504 **55.** -0.96 **57.** 1,027.19 **59.** 9.1
61. 304.07 **63.** 280 **65.** 11 hr later: 6 P.M. **67.** 567
69. 1988: $9.28; 1998: $12.77 **71.** 0.37 mi **77.** $\frac{7}{6}$
79. $\{\ldots, -3, -2, -1, 0, 1, 2, 3, \ldots\}$ **81.** 12 **83.** 25.425

Study Set Estimation (page 254)

1. approx. $240 **3.** approx. 2 cubic feet less **5.** approx. 30
7. approx. $330 **9.** approx. $520 **11.** not reasonable
13. reasonable **15.** reasonable **17.** not reasonable

Study Set Section 4.5 (page 260)

1. repeating **3.** decimal **5.** $7 \div 8$ **7.** smaller
9.

$-3.8\overline{3}$ -0.75 $0.\overline{6}$ $1\frac{3}{4}$

-5 -4 -3 -2 -1 0 1 2 3 4 5

11. a. false **b.** true **c.** true **d.** false **13. a.** no
b. It is a repeating decimal. **15.** 0.5 **17.** -0.625
19. 0.5625 **21.** -0.53125 **23.** 0.55 **25.** 0.775 **27.** -0.015
29. 0.002 **31.** $0.\overline{6}$ **33.** $0.\overline{45}$ **35.** $-0.58\overline{3}$ **37.** $0.0\overline{3}$
39. 0.23 **41.** 0.38 **43.** 0.152 **45.** 0.370 **47.** 1.33
49. -3.09 **51.** 3.75 **53.** -8.67 **55.** 12.6875
57. 203.73 **59.** < **61.** < **63.** $\frac{37}{90}$ **65.** $\frac{19}{60}$ **67.** $\frac{3}{22}$ **69.** $-\frac{1}{90}$
71. 0.57 **73.** 5.27 **75.** -2.55 **77.** 0.068 **79.** 7.11
81. -1.7 **83.** 63.585 **85.** $0.\overline{2277}$ **87.** 34.72 **89.** 0.0625,
0.375, 0.5625, 0.9375 **91.** $\frac{3}{40}$ in. **93.** 23.4 sec, 23.8 sec, 24.2 sec,
32.6 sec **95.** 93.6 in.^{2} **101.** -1 **103.** $\frac{19}{6}$ **105.** $\frac{2}{3}$

Study Set Section 4.6 (page 265)

1. solve **3.** associative **5.** $2.1(1.7) - 6.3 = -2.73$
9. 1.7 **11.** 7.11 **13.** -11.5 **15.** -0.1 **17.** -4.36
19. 1.3 **21.** -8.16 **23.** 22.44 **25.** -21.18 **27.** 0.4
29. -2.2 **31.** -2 **33.** 31 **35.** 1 **37.** 0.3 **41.** 8.6
million **43.** 3.27 **45.** 10.7 **47.** 12.4 mpg **49.** 200 **53.** $\frac{1}{12}$
55. $\frac{14}{13}$ **57.** -6 **59.** 12

Study Set Section 4.7 (page 271)

1. root **3.** radical, positive **5.** radicand **7.** 25, 25 **9.** $(-7)^2$
11. $\frac{3}{4}$ **13.** $\sqrt{6}, \sqrt{11}, \sqrt{23}, \sqrt{27}$ **15. a.** 1 **b.** 0 **17. a.** 2.4
b. 5.76 **c.** 0.24
19.

$-\sqrt{5}$ $\sqrt{9}$

-5 -4 -3 -2 -1 0 1 2 3 4 5

21. a. 4, 5
b. 9, 10 **23.** $-7, 8$ **25.** 4 **27.** -11 **29.** -0.7 **31.** 0.5
33. 0.3 **35.** $-\frac{1}{9}$ **37.** $-\frac{4}{3}$ **39.** $\frac{2}{5}$ **41.** 31 **43.** -20
45. $-\frac{7}{20}$ **47.** -70 **49.** 2.56 **51.** -3.6 **53.** 1, 1.414, 1.732, 2,
2.236, 2.449, 2.646, 2.828, 3, 3.162 **55.** 37 **57.** 61 **59.** 3.87
61. 8.12 **63.** 4.904 **65.** -3.332 **67.** 4,899 **69.** -0.0333
71. a. 5 ft **b.** 10 ft **73.** 127.3 ft **75.** 41-inch **83.** subtraction and multiplication **85.** 16 **87.** $\frac{5}{6}$ **89.** 30

Key Concept (page 274)

1.

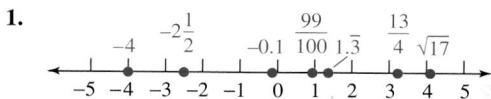

3. $\{1, 2, 3, 4, 5, \ldots\}$ **5.** $\{\ldots, -3, -2, -1, 0, 1, 2, 3, \ldots\}$
7. nonterminating, nonrepeating decimals; a number that can't be written as a fraction **9.** false **11.** false **13.** true **15.** false
17. true **19.** There is no real number that, when squared, yields a negative number.

Chapter Review (page 276)

1. $0.67, \frac{67}{100}$ **2.**

0.8

3. $10 + 6 + \frac{4}{10} + \frac{5}{100} + \frac{2}{1,000} + \frac{3}{10,000}$
4. a. two and three tenths, $2\frac{3}{10}$ **b.** negative fifteen and fifty-nine hundredths, $-15\frac{59}{100}$ **c.** six hundred one ten-thousandths, $\frac{601}{10,000}$
d. one one hundred thousandth, $\frac{1}{100,000}$
5.

6. Washington, Diaz, Chou, Singh, Gerbac **7.** true **8. a.** $<$
b. $>$ **c.** $=$ **d.** $<$ **9. a.** 4.58 **b.** 3,706.090 **c.** -0.1
d. 88.1 **10. a.** 66.7 **b.** 45.188 **c.** 15.17 **d.** 27.71
11. a. -7.7 **b.** 3.1 **c.** -4.8 **d.** -29.09 **e.** -25.6
f. 4.939 **12.** \$48.21 **13.** 8.15 in. **14. a.** -0.24 **b.** 2.07
c. -17.05 **d.** 197.945 **e.** 0.00006 **f.** 4.2 **15. a.** 90,145.2
b. 2,897 **16. a.** 0.04 **b.** 0.0225 **c.** 10.89 **d.** 0.001
17. a. -10.61 **b.** 25.82 **18.** 92.38 **19.** 68.62 in.² **20.** 0.07
in. **21. a.** 1.25 **b.** -10.45 **c.** 1.29 **d.** 4.103 **22. a.** -2.9
b. 0.053 **c.** 63 **d.** 0.81 **23. a.** 12.9 **b.** -667.3
24. 20.22 **25.** \$8.34 **26. a.** 0.8976 **b.** -0.00112
27. 13.95 **28.** 14 **29.** 9.5 **30. a.** 0.875 **b.** -0.4
c. 0.5625 **d.** 0.06 **31. a.** $0.\overline{54}$ **b.** $-0.\overline{6}$ **32. a.** 0.58
b. 1.03 **33. a.** $>$ **b.** $>$
34.

35. a. $\frac{11}{15}$
b. -6.24

c. 93 **d.** 39.564 **36.** 33.49 **37.** 34.88 in.²
38. a. -18.41 **b.** 4.77 **c.** -5.34 **d.** 1.7 **39.** yes
40. 9 **41.** $8^2, (-8)^2$ **42. a.** 7 **b.** -4 **c.** 10 **d.** 0.3 **e.** $\frac{8}{5}$
f. 0.9 **g.** $-\frac{1}{6}$ **h.** 0 **43.** 9 and 10 **44.** It differs by 0.11.
45.

46. a. -30 **b.** 2.5 **c.** -27 **d.** 1.5 **47. a.** 4.36 **b.** 7.68

Chapter 4 Test (page 281)

1. $\frac{79}{100}, 0.79$ **2.** Selway, Monroe, Paston, Covington, Cadia
3. $\frac{271}{1,000}$ **4.** 33.050 **5.** \$208.75 **6. a.** 0.567909 **b.** 0.458
7. 1.02 in. **8. a.** 10.75 **b.** 6.121 **c.** 0.1024 **d.** 14.07
9. 125 mi² **10.** 0.004 in. **11.** 3.588 **12. a.** 0.34 **b.** $0.41\overline{6}$
13. -2.29 **14.** $1.\overline{18}$ **15.**

16. $\frac{41}{30}$

17. a. -7 **b.** 6.008 **18.** -0.425 **19.** 0.42 g **20.** 80
21.

22. a. 11 **b.** $-\frac{1}{30}$ **23. a.** $>$ **b.** $>$ **c.** $>$ **d.** $>$
24. a. -0.2 **b.** 1.3

Cumulative Review Exercises (page 283)

1. \$876,000 **2.** $(x + y) + z = x + (y + z)$ **3.** 27 R 42
4. 1,000 **5.** $2^2 \cdot 5 \cdot 11$ **6.** 1, 2, 4, 5, 10, 20
7. $\{0, 1, 2, 3, 4, 5, \ldots\}$ **8.** -13 **9.** adding
10. $8, -3, 36, -6, 6$ **11.** $-5(3) = -15$ **12.** -1 **13.** 9
14. 30 **15.** 35 **16.** 102 **17.** 3.61 **18.** $-\$1,100$
19. 5 **20.** $\frac{6}{13}$ **21.** equivalent fractions **22.** $\frac{5}{7}$
23. $\frac{21}{128}$ **24.** $-\frac{3}{16}$ **25.** $\frac{34}{21}$ **26.** $19\frac{1}{8}$ **27.** $26\frac{7}{24}$
28. $-\frac{1}{3}$ **29.** -45 **30.** 8 **31.** 157.5 in.²
32.

33. 0.001 in. **34.** $<$ **35.** -8.136 **36.** 5.6 **37.** 5,601.2
38. 0.0000897 **39.** 47.95 **40.** 33.6 hr **41.** 232.8° C
42. $0.41\overline{6}$ **43.** -9 **44.** 80 **45.** -6

Study Set Section 5.1 (page 292)

1. percent **3.** 100 **5.** right **7. a.** $0.84, 84\%, \frac{21}{25}$ **b.** 16%
9. $\frac{17}{100}$ **11.** $\frac{1}{20}$ **13.** $\frac{3}{5}$ **15.** $\frac{5}{4}$ **17.** $\frac{1}{150}$ **19.** $\frac{21}{400}$ **21.** $\frac{3}{500}$
23. $\frac{19}{1,000}$ **25.** 0.19 **27.** 0.06 **29.** 0.408 **31.** 2.5 **33.** 0.0079
35. 0.0025 **37.** 93% **39.** 61.2% **41.** 3.14% **43.** 843%
45. 5,000% **47.** 910% **49.** 17% **51.** 16% **53.** 40%
55. 105% **57.** 62.5% **59.** 18.75% **61.** $66\frac{2}{3}\%$ **63.** $83\frac{1}{3}\%$
65. 11.11% **67.** 55.56% **69. a.** $\frac{15}{188}$ **b.** 8% **71. a.** $\frac{9}{22}$
b. 41% **73. a.** $\frac{5}{29}$ **b.** 17% **c.** 24% **75.** 5 ft
77. 0.9944 **79.** as a decimal; 89.6% **81.** torso: 27.5%

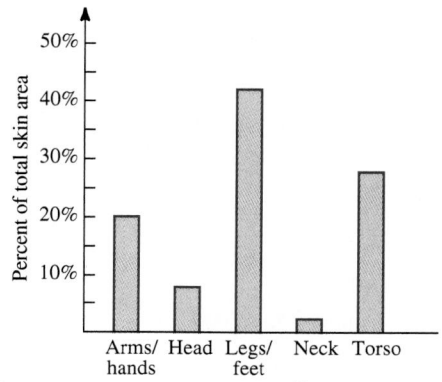

83. 92% **85.** 0.27% **93.** 9 **95.** $\frac{41}{99}$ **97.** 27.0725

Study Set Section 5.2 (page 302)

1. $x = 0.10 \cdot 50$ **3.** $48 = x \cdot 47$ **5.** graph **7. a.** 0.12
b. 0.056 **c.** 1.25 **d.** 0.0025 **9.** more **11. a.** 25 **b.** 100%
c. 87 **13.** 33% **15. a.** multiply **b.** equals **c.** x (as a variable) **17.** 90 **19.** 80% **21.** 65 **23.** 0.096 **25.** 0.00125%
27. 44 **29.** 43.5 **31.** 107.1 **33.** 99 **35.** 60 **37.** 31.25%

39.

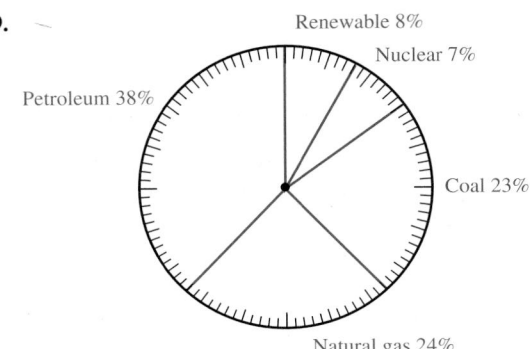

Renewable 8%
Nuclear 7%
Petroleum 38%
Coal 23%
Natural gas 24%

41. 120   **43.** $610.5 billion   **45.** 38,000 = 38K   **47.** 24 oz
49. yes   **51.** 30, 12   **53.** 2.7 in.   **55.** 5%   **57.** yes   **63.** 18.17
65. 5.001   **67.** 0.008

Study Set   Section 5.3   (page 312)

1. commission   **3.** discount   **5.** The number of members has
doubled.   **7.** $42.75   **9.** 8%   **11.** $47.34, $2.84, $50.18
13. $150   **15.** 8%, 1.2%, 1.4%   **17.** 360 hr   **19.** 96 calories
21. 1995–1996; 5%   **23.** 10%   **25.** 31%   **27. a.** 25%   **b.** 36%
29. $2,955   **31.** 1.5%   **33.** $12,000   **35.** $39.95, 25%
37. $187.49   **39.** $349.97, 13%   **41.** $3.60, 23%, $11.88
43. $76.50   **49.** −50   **51.** 3   **53.** $500   **55.** $-\frac{7}{45}$ **57.** $\frac{10}{7} = 1\frac{3}{7}$

Study Set   Estimation   (page 317)

1. 164   **3.** $60   **5.** $54,000   **7.** 320 lb   **9.** 130   **11.** 21
13. 18,000   **15.** 3,100

Study Set   Section 5.4   (page 323)

1. principal   **3.** interest   **5.** simple   **7. a.** 0.07   **b.** 0.098
c. 0.0625   **9.** $1,800   **11. a.** compound interest   **b.** $1,000
c. 4   **d.** $50   **e.** 1 year   **13.** multiplication   **15.** $5,300
17. $1,472   **19.** $4,262.14   **21.** $10,000, 0.0725, 2 yr, $1,450
23. $192, $1,392, $58   **25.** $18.828 million   **27.** $755.83
29. $1,271.22   **31.** $570.65   **33.** $30,915.66   **39.** $\frac{1}{2}$ **41.** 23.0
43. −3   **45.** 50

Key Concept   (page 326)

1. 198.4   **3.** 62.5   **5.** 17%   **7.** $3,000

Chapter Review   (page 328)

1. a. 39%, 0.39, $\frac{39}{100}$ **b.** 111%, 1.11, $1\frac{11}{100}$ **2.** 61%   **3. a.** $\frac{3}{20}$
b. $\frac{6}{5}$ **c.** $\frac{37}{400}$ **d.** $\frac{1}{1,000}$ **4. a.** 0.27   **b.** 0.08   **c.** 1.55   **d.** 0.018
5. a. 83%   **b.** 62.5%   **c.** 5.1%   **d.** 600%   **6. a.** 50%
b. 80%   **c.** 87.5%   **d.** 6.25%   **7. a.** $33\frac{1}{3}$%   **b.** $83\frac{1}{3}$%
8. a. 55.56%   **b.** 266.67%   **9.** 63%   **10.** 0.1% = $\frac{1}{1,000}$;
10% = $\frac{1}{10}$ **11.** amount: 15, base: 45, percent: $33\frac{1}{3}$%
12. $x = 32\% \cdot 96$ **13. a.** 200   **b.** 125   **c.** 1.75%   **d.** 2,100
e. 121   **f.** 30   **14.** 14.4 gal nitro, 0.6 gal methane   **15.** 68
16. 87%   **17.** $5.43

18.

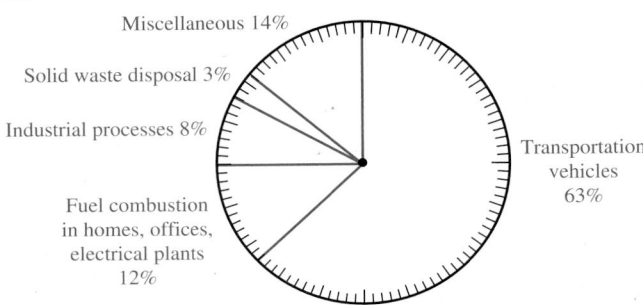

Miscellaneous 14%
Solid waste disposal 3%
Industrial processes 8%
Fuel combustion
in homes, offices,
electrical plants
12%
Transportation
vehicles
63%

19. 139,531,200 mi^2 **20.** $3.30, $63.29   **21.** 4%   **22.** $40.20
23. 25%   **24.** 9.6%   **25.** $50, $189.99, 26%   **26.** $6,000, 8%,
2 years, $960   **27.** $10,308.22   **28.** $134.69   **29.** $2,142.45
30. $6,076.45   **31.** $43,265.78

Chapter 5 Test   (page 333)

1. 61%, $\frac{61}{100}$, 0.61   **2.** 199%, $\frac{199}{100}$, 1.99   **3. a.** 0.67   **b.** 0.123
c. 0.0975   **4. a.** 25%   **b.** 62.5%   **c.** 12%   **5. a.** 19%
b. 347%   **c.** 0.5%   **6. a.** $\frac{11}{20}$ **b.** $\frac{1}{10,000}$ **c.** $\frac{5}{4}$ **7.** 23.33%
8. 60%   **9.** $66\frac{2}{3}$%   **10.** 25%   **11. a.** 1.02 in.   **b.** 32.98 in.
12. 6.5%   **13.** $3.81   **14.** 93.7%   **15.** 90   **16.** 21   **17.** 144
18. 27%   **19.** $35.92   **20.** $41,440   **21.** $11.95, $3, 20%
22. 22%   **23.** $150   **24.** $5,079.60   **25.** The phrase "bringing
crime down to 37%" is unclear. The question that arises is: 37% of
what?

Cumulative Review Exercises   (page 335)

1.

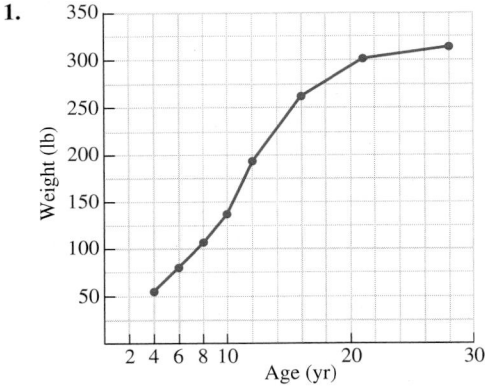

Weight (lb) / Age (yr)

2. If *a* and *b* represent numbers, *ab* = *ba*.   **3. a.** 1, 2, 4, 5, 8, 10,
20, 40   **b.** $2^3 \cdot 5$ **4.** $1,115   **5.** 64 ft^2 **6.** −7   **7.** −4
8. 55   **9.** 2   **10.** −2   **11.** 1   **12.** 15° C   **13.** $\frac{4}{11}$ **14.** $\frac{2}{3}$
15. $-\frac{5}{21}$ **16.** $\frac{2}{5}$ **17.** $\frac{34}{21} = 1\frac{13}{21}$ **18.** $20\frac{5}{18}$ **19.** −30   **20.** 18
21. 70.29   **22.** −8.6265   **23.** 752   **24.** 83.4   **25.** −2.33
26. 452.030   **27.** $0.7\overline{3}$ **28.** −11.1   **29.** −29   **30.** 3.5 hr
31. 29%, $\frac{29}{100}$; 0.473, $\frac{473}{1,000}$; 87.5%, 0.875   **32.** 125   **33.** 64
34. 8%   **35.** $12.00, $87.18   **36.** 0.0018%   **37.** $1,450

Study Set   Section 6.1   (page 344)

1. a   **3.** c   **5.** d   **7.** histogram   **9.** $6.95   **11.** $4.86
13. $10,918   **15.** $2,594.50   **17.** nuclear energy   **19.** 49%
21. about 60%   **23.** 1980   **25.** 1970   **27.** 320 thousand metric
tons   **29.** reckless driving and failure to yield   **31.** reckless
driving   **33.** seniors   **35.** $50   **37.** French and German
39. English   **41.** 51.4%   **43.** about 8%   **45.** about 11%

47. 4.6% **49.** $190 **51.** miners **53.** miners **55.** 1 **57.** 1
59. Runner 1 was running; runner 2 was stopped. **61.** 27
63. 90 **71.** -7 **73.** $\frac{25}{36}$ **75.** 11, 13, 17, 19, 23, 29 **77.** 4

Study Set   Section 6.2   (page 353)

1. mean **3.** median **5.** the number of values **7.** 8 **9.** 35
11. 19 **13.** 9 **15.** 6 **17.** 17.5 **19.** 3 **21.** none **23.** 22.7
25. about 63¢ **27.** 60¢ **29.** 50¢ **31.** about 61° **33.** 64°
35. 2,670 mi **37.** 89 mi **39.** Median and mode are 85.
41. same average (56); sister's scores are more consistent
43. 22.525 oz, 25 oz **45.** $4.15, $4.19, $4.29 **47.** 1988: $9.28;
1998: $12.77 **49.** 0.37 mi **53.** 3^4 **55.** $\frac{1}{6}$ **57.** 6 **59.** $\frac{19}{10} = 1\frac{9}{10}$

Key Concept   (page 356)

1. 13.7 **3.** 15 **5.** 6 **7.** 6

Chapter Review   (page 358)

1. $-18°$ **2.** 30 mph **3. a.** about 4.9 billion **b.** 1989–1991
c. 1991–1993 **d.** about 60% **4. a.** about 830 million
b. about 865 million **c.** 1987 **d.** about 1,770 million **5.** 180
6. 160 **7.** yes **8.** median **9.** 1.2 oz **10.** 1.138 oz
11. 7.3 microns, 7.2 microns, 6.9 microns **12.** $1.45 billion

Chapter 6 Test   (page 361)

1. about $1,659 **2.** about $11 **3.** about 4.1% **4.** about 1.2%
5. about 19% **6.** about 6% **7.** about 270,000 **8.** about
240,000 **9.** 1985 **10.** 2.4% **11.** A **12.** C **13.** E
14. bicyclist 1 **15.** 7.5 **16.** 7.5 **17.** 5 **18.** mean **19.** 3.6,
3.6, 3.1 **20.** Half the families had more debt and half had less
debt.

Cumulative Review Exercises   (page 363)

1. 358,600,000 gal **2.** 50,000 **3.** 54,604 **4.** 4,209
5. 23,115 **6.** 87 **7.** 683 + 459 = 1,142 **8.** 2011
9. $4 \cdot 5 = 5 + 5 + 5 + 5 = 20$ **10.** 10,912 in.^{2} **11. a.** 1, 2, 3,
6, 9, 18 **b.** $2 \cdot 3^2$ **12.** 2, 3, 5, 7, 11, 13, 17, 19, 23, 29
13. It has factors other than 1 and itself. For example, $27 = 3 \cdot 9$.
14. 22 **15.** 315 **16.** 6
17.

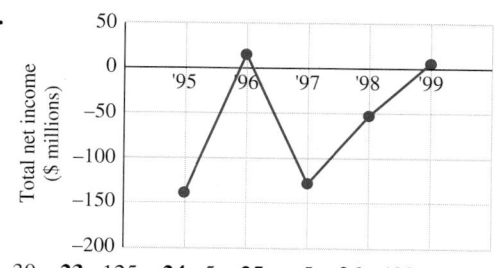

18. 5
19. false **20.**

21. -20 **22.** 30 **23.** 125 **24.** 5 **25.** -5 **26.** 429
27. $-3^2 = -(3 \cdot 3) = -9; (-3)^2 = (-3)(-3) = 9$ **28.** 1,100° F
29. 800 **30.** 15% **31.** $\frac{5}{0}; \frac{0}{5}$; division by 0 **32.** $\frac{7}{6} = 1\frac{1}{6}$ **33.** $-\frac{1}{6}$
34. $6\frac{3}{4}$ in. **35.** $\frac{8}{35}$ **36.** $\frac{21}{20} = 1\frac{1}{20}$ **37.** 54% **38.** 345 **39.** 0.744
40. 745 **41.** $0.\overline{72}$ **42.** 160 min **43.** 33.6 hr **44.** 3.02, 3.005,
2.75

Study Set   Section 7.1   (page 371)

1. segment **3.** midpoint **5.** protractor **7.** right **9.** 180°
11. supplementary **13.** true **15.** false **17.** true **19.** true

21. acute **23.** obtuse **25.** right **27.** straight **29.** true
31. false **33.** yes **35.** yes **37.** no **39.** true **41.** true
43. true **45.** true **47.** angle **49.** ray **51.** 3 **53.** 3 **55.** 1
57. B **59.** 40° **61.** 135° **63.** 10 **65.** 27.5 **67.** 30 **69.** 25
71. 60° **73.** 75° **75.** 130° **77.** 230° **79.** 100° **81.** 40°
83.

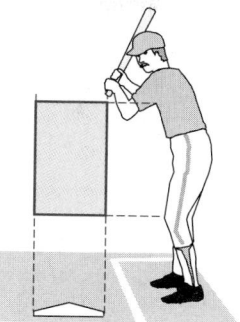

85. 65,115 **87.** 30° **95.** 16
97. $\frac{7}{24}$ **99.** 0.49

Study Set   Section 7.2   (page 378)

1. coplanar **3.** perpendicular **5.** alternate **7.** $\angle 4$ and $\angle 6$, $\angle 3$
and $\angle 5$ **9.** $\angle 3, \angle 4, \angle 5, \angle 6$ **11.** They are parallel. **13.** a right
angle **15.** is perpendicular to **17.** $m(\angle 1) = 130°$,
$m(\angle 2) = 50°$, $m(\angle 3) = 50°$, $m(\angle 5) = 130°$, $m(\angle 6) = 50°$,
$m(\angle 7) = 50°$, $m(\angle 8) = 130°$ **19.** $m(\angle A) = 50°$, $m(\angle 1) = 85°$,
$m(\angle 2) = 45°$, $m(\angle 3) = 135°$ **21.** 10 **23.** 30 **25.** 40 **27.** 12
29. If the stones are level, the plum bob string should pass through
the midpoint of the crossbar of the A-frame. **41.** 72 **43.** 45%
45. yes

Study Set   Section 7.3   (page 385)

1. regular **3.** hexagon **5.** octagon **7.** equilateral
9. hypotenuse **11.** parallelogram **13.** rhombus **15.** isosceles
17. 4, quadrilateral, 4 **19.** 3, triangle, 3 **21.** 5, pentagon, 5
23. 6, hexagon, 6 **25.** scalene triangle **27.** right triangle
29. equilateral triangle **31.** isosceles triangle **33.** square
35. rhombus **37.** rectangle **39.** trapezoid **41.** triangle
43. 90° **45.** 45° **47.** 90.7° **49.** 30° **51.** 60° **53.** 720°
55. 1,440° **61. b.** octagon **c.** triangle **d.** pentagon
63. pentagon, hexagon **69.** 22 **71.** 40% **73.** 0.10625

Study Set   Section 7.4   (page 392)

1. congruent **3.** right **5.** true **7.** false **9.** yes
11. a and b represent the length of the legs; c represents the length
of the hypotenuse. **13.** is congruent to
15. $\overline{DF}, \overline{AB}, \overline{EF}, \angle D, \angle B, \angle C$ **17.** yes, SSS **19.** not necessarily **21.** yes, SSS **23.** yes, SAS **25.** 6 mm **27.** 50° **29.** 5
31. 8 **33.** $\sqrt{56}$ **35.** yes **37.** no **39.** 12 ft **41.** 25 in.
43. 127.3 ft **47.** $1\frac{1}{3}$ **49.** 20 **51.** 9

Study Set   Section 7.5   (page 403)

1. perimeter **3.** area **5.** square **7.** length 15 in. and width 5 in.;
length 16 in. and width 4 in. (answers may vary) **9.** sides of
length 5 m **11.** base 5 yd and height 3 yd (answers may vary)
13. length 5 ft and width 4 ft; length 20 ft and width 3 ft (answers
may vary) **15.** $P = 4s$ **17.** square inch **19.** $A = s^2$
21. triangle **23.** 32 in. **25.** 36 m **27.** 37 cm **29.** 85 cm
31. $28\frac{1}{3}$ ft **33.** 16 cm^2 **35.** 60 cm^2 **37.** 25 in.^{2} **39.** 169 mm^2
41. 80 m^2 **43.** 75 yd^2 **45.** 75 m^2 **47.** 144 **49.** $4,875
51. 81 **53.** linoleum **55.** $1,200 **57.** $361.20 **59.** $192
61. 111,825 mi^2 **63.** 51 **65.** spot 1: $l = 20$ ft, $w = 10$ ft, 200 ft^2;

spot 2: b_1 = 20 ft, b_2 = 16 ft, h = 10 ft, 180 ft^2; spot 3: b = 28 ft, h = 28 ft, 392 ft^2 **69.** $1\frac{5}{12}$ **71.** $6\frac{1}{12}$ **73.** $1\frac{7}{18}$

Study Set   Section 7.6   (page 411)

1. radius **3.** diameter **5.** minor **7.** circumference **9.** $\overline{OA}$, $\overline{OC}$, and $\overline{OB}$ **11.** $\overline{DA}, \overline{DC}$, and $\overline{AC}$ **13.** $\overset{\frown}{ABC}$ and $\overset{\frown}{ADC}$ **15.** Double the radius. **17. a.** 1 in. **b.** 2 in. **c.** 2π in. ≈ 6.28 in. **d.** π in.$^2 \approx 3.14$ in.^{2} **19.** Square 6. **21.** arc AB **23.** $C = \pi D, C = 2\pi r$ **25.** π **27.** 8π **29.** 37.70 in. **31.** 36 m **33.** 25.42 ft **35.** 31.42 m **37.** A = 28.3 in.^{2} **39.** 88.3 in.^{2} **41.** 128.5 cm^2 **43.** 27.4 in.^{2} **45.** 66.7 in.^{2} **47.** 3.14 mi^2 **49.** 32.66 ft **51.** 12.73 times **53.** 1.59 ft **55.** 12.57 ft^2; 0.79 ft^2; 6.28% **63.** 90% **65.** five

Study Set   Section 7.7   (page 422)

1. volume **3.** cube **5.** surface **7.** cylinder **9.** cone **11.** $V = lwh$ **13.** $V = \frac{4}{3}\pi r^3$ **15.** $V = \frac{1}{3}Bh$ or $V = \frac{1}{3}\pi r^2 h$ **17.** $SA = 2lw + 2lh + 2hw$ **19.** 27 ft^3 **21.** 1,000 dm^3 **23. a.** volume **b.** area **c.** volume **d.** surface area **e.** perimeter **f.** surface area **25. a.** 72 in.^{3} **b.** 18 in.^{2} **c.** 24 in.^{2} **27.** 1 cubic inch **29.** 60 cm^3 **31.** 48 m^3 **33.** 3,053.63 in.^{3} **35.** 1,357.17 m^3 **37.** 314.16 cm^3 **39.** 400 m^3 **41.** 94 cm^2 **43.** 1,256.64 in.^{2} **45.** 576 cm^3 **47.** 335.10 in.^{3} **49.** $\frac{1}{8}$ in.^{3} = 0.125 in.^{3} **51.** 2.125 **53.** 197.92 ft^3 **55.** 33,510.32 ft^3 **57.** 8:1 **63.** -42 **65.** 95%

Key Concept   (page 425)

1. $A = lw$ **3.** $P = 2l + 2w$ **5.** 210,000 ft^2 **7.** $80.50 **9.** 153.94 ft^2 **11.** $750, $45, $6,250

Chapter Review   (page 427)

1. points C and D, line CD, plane GHI **2.** 5 units **3.** $\angle ABC, \angle CBA, \angle B, \angle 1$ **4.** 48° **5.** $\angle 1$ and $\angle 2$ are acute, $\angle ABD$ and $\angle CBD$ are right angles, $\angle CBE$ is obtuse, and $\angle ABC$ is a straight angle. **6. a.** obtuse angle **b.** right angle **c.** straight angle **d.** acute angle **7.** 15 **8.** 150 **9. a.** 65° **b.** 115° **10.** 40° **11.** 40° **12.** no **13.** part a **14.** $\angle 4$ and $\angle 6$, $\angle 3$ and $\angle 5$ **15.** $\angle 1$ and $\angle 5$, $\angle 4$ and $\angle 8$, $\angle 2$ and $\angle 6$, $\angle 3$ and $\angle 7$ **16.** $\angle 1$ and $\angle 3$, $\angle 2$ and $\angle 4$, $\angle 5$ and $\angle 7$, $\angle 6$ and $\angle 8$ **17.** m($\angle 1$) = 70°, m($\angle 2$) = 110°, m($\angle 3$) = 70°, m($\angle 4$) = 110°, m($\angle 5$) = 70°, m($\angle 6$) = 110°, m($\angle 7$) = 70° **18.** m($\angle 1$) = 60°, m($\angle 2$) = 120°, m($\angle 3$) = 130°, m($\angle 4$) = 50° **19.** 40 **20.** 20 **21. a.** octagon **b.** pentagon **c.** triangle **d.** hexagon **e.** quadrilateral **22. a.** 3 **b.** 4 **c.** 8 **d.** 6 **23. a.** isosceles **b.** scalene **c.** equilateral **d.** right triangle **24. a.** yes **b.** no **25. a.** 90 **b.** 50 **26.** 50° **27.** It is equilateral. **28. a.** trapezoid **b.** square **c.** parallelogram **d.** rectangle **e.** rhombus **f.** rectangle **29. a.** 15 cm **b.** 40° **c.** 100° **30. a.** true **b.** false **c.** true **d.** true **31. a.** 65° **b.** 115° **32. a.** 360° **b.** 720° **33.** $\angle D, \angle E, \angle F, \overline{DF}, \overline{DE}, \overline{EF}$ **34. a.** congruent, SSS **b.** congruent, SAS **c.** congruent, ASA **d.** not necessarily congruent **35. a.** 13 **b.** 15 **36.** 31.3 in. **37.** 72 in. **38.** 9 m **39. a.** 30 m **b.** 36 m **40. a.** 9.61 cm^2 **b.** 7,500 ft^2 **c.** 450 ft^2 **d.** 200 in.^{2} **e.** 120 cm^2 **f.** 232 ft^2 **g.** 152 ft^2 **h.** 120 m^2 **41.** 9 ft^2 **42.** 144 in.^{2} **43. a.** $\overline{CD}, \overline{AB}$ **b.** $\overline{AB}$ **c.** $\overline{OA}, \overline{OC}, \overline{OD}, \overline{OB}$ **d.** O **44.** 66.0 cm **45.** 45.1 cm **46.** 254.5 in.^{2} **47.** 130.3 cm^2 **48. a.** 125 cm^3 **b.** 480 m^3 **c.** 600 in.^{3} **d.** 3,619 in.^{3} **e.** 1,518 ft^3 **f.** 785 in.^{3}

g. 9,020,833 ft^3 **h.** 35,343 ft^3 **49.** 1,728 in.^{3} **50.** 54 ft^3 **51. a.** 61.8 ft^2 **b.** 314.2 in.^{2}

Chapter 7 Test   (page 435)

1. 4 units **2.** B **3.** true **4.** false **5.** false **6.** true **7.** 50 **8.** 140 **9.** 12 **10.** 45 **11.** 23° **12.** 63° **13.** 70° **14.** 110° **15.** 70° **16.** 40 **17.** 3, 4, 6, 5, 8 **18.** equilateral triangle, scalene triangle, isosceles triangle **19.** 57° **20.** 66° **21.** 30° **22.** 1,440° **23.** m($\overset{\frown}{AB}$) = m($\overset{\frown}{DC}$), m($\overset{\frown}{AD}$) = m($\overset{\frown}{BC}$), and m($\overset{\frown}{AC}$) = m($\overset{\frown}{BD}$) **24.** 130° **25.** 8 in. **26.** 50° **27.** 127.3 ft **28.** 391.6 cm^2 **29.** 83.7 ft^2 **30.** 94.2 ft **31.** 28.3 ft^2 **32.** 159.3 m^3 **33.** 268.1 m^3 **34.** 66.7 ft^3 **36.** The surface area is 6 times the area of one face of the cube.

Cumulative Review Exercises   (page 437)

1.

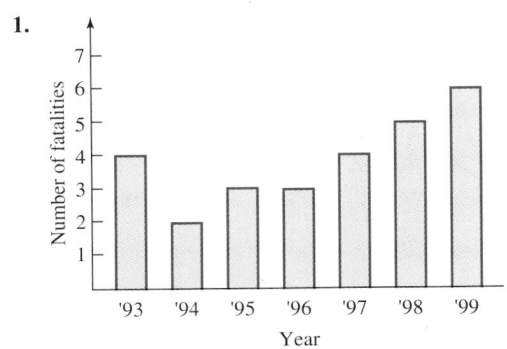

2. $8,995 **3.** 11,022 **4.** 33 **5.** 2,110,000 **6.** $2^2 \cdot 5 \cdot 11$ **7.** 1, 2, 3, 4, 6, 8, 12, 24 **8.** $\{ \ldots -3, -2, -1, 0, 1, 2, 3, \ldots \}$ **9.** 13 **10.** -10 **11.** 3 **12.** 5 **13.** -11 **14.** 5 **15.** $\frac{5}{4}$ **16.** $142\frac{7}{15}$ **17.** $\frac{3}{20}$ **18.** $13\frac{3}{4}$ cups **19.** $-\frac{11}{20}$ **20.** $\frac{1}{3}$ **21.** $\frac{3}{32}$ fluid oz **22.** $-\frac{15}{2}$ **23.** $\frac{8}{9}$ **24.** $1\frac{9}{29}$ **25. a.** 1991; about 0.3° C **b.** 1993; about $-0.3°$ C **26.**

$$-4\frac{5}{8} \quad -\sqrt{9} \quad -0.1 \quad \frac{2}{3} \quad \frac{3}{2} \quad 2.89 \quad \sqrt{17}$$
$$-5 \quad -4 \quad -3 \quad -2 \quad -1 \quad 0 \quad 1 \quad 2 \quad 3 \quad 4 \quad 5$$

27. 3.1416 **28.** $>$ **29.** 145.188 **30.** 17.05 **31.** 89,970.8 **32.** 0.053 **33.** -25.6 **34.** 22.3125 **35.** $0.1\overline{3}$ **36.** -9.32 **37.** 97 **38.** -2 **39.** $\frac{7}{9}$ **40.** 93%, 7% **41.** 67.5 **42.** 120 **43.** 0.57, $\frac{57}{100}$; 0.1%, $\frac{1}{1,000}$; $33\frac{1}{3}\%$ $0.\overline{3}$ **44.**

Huang-Sims 45%
Cisneros 55%

45. 50° **46.** 130° **47.** 50° **48.** 50° **49.** 75° **50.** 30° **51.** 105° **52.** 105° **53.** 46, 134 **54.** 540° **55.** 13 m **56.** 42 m, 108 m^2 **57.** 126 ft^2 **58.** 91 in.^{2} **59.** 43.98 cm, 153.94 cm^2 **60.** 98.31 yd^2 **61.** 210 m^3 **62.** 523.60 in.^{3} **63.** 150.80 m^3 **64.** 3.93 ft^3 **65.** 2,124 in.^{2}

Study Set   Section 8.1   (page 448)

1. evaluate **3.** expression, equation **5.** $6 + 20x$; $\frac{6 - x}{20}$ (answers may vary) **7.** We would obtain $34 - 6$; it looks like 34, not 3(4).

9. a. x = weight of the car; $2x - 500$ = weight of the van
b. 3,500 lb **11.** 5, 30,10, 10d, 50, 50($x + 5$) **15.** $l + 15$
17. 50x **19.** $\frac{w}{7}$ **21.** $P + p$ **23.** $k^2 - 2{,}005$ **25.** $J - 500$
27. $\frac{1{,}000}{n}$ **29.** $p + 90$ **31.** 35 + h + 300 **33.** $p - 680$
35. 4d - 15 **37.** 2(200 + t) **39.** $|a - 2|$ **41.** 7 less than
a number **43.** the product of 7 and a number, increased by 4
45. 300; 60h **47. a.** 3y **b.** $\frac{f}{3}$ **49.** 29x¢ **51.** $\frac{c}{6}$ **53.** 5b
55. $5($x + 2$)$ **57.** $-1, -2, -28$ **59.** 41, 11, 2 **61.** 150, -450
63. 0, 0, 5 **65.** 20 **67.** -12 **69.** -5 **71.** 156 **73.** $-\frac{1}{5}$
75. 17 **77.** 36 **79.** 230 **81.** 30.5 **83.** 0, 28, 48, 60, 64, 60,
48, 28, 0 **85.** $-37°C, -64°C$ **87.** $1\frac{23}{64}$ in.^{2} **89.** 235 ft^2
95. 0 **97.** $\frac{2}{3}$ **99.** 5^4 **101.** 83

Study Set   Section 8.2   (page 459)

1. simplify **3.** coefficient, variable **5.** remove
7. the distributive property **9.** 3 + 4, 3, 4 **11. a.** + **b.** −
c. − **d.** + **e.** − **f.** + **13.** x + 20 − x = 20; 20 ft
15. a. yes **b.** yes **19. a.** no **b.** yes **21. a.** 5x + 1
b. 16t − 6 **23.** 63m **25.** $-35q$ **27.** 5x **29.** 6y **31.** 20bp
33. 40r^2 **35.** 5x + 15 **37.** $-2b + 2$ **39.** 24t − 16
41. 12y − 6 **43.** 0.4x − 1.6 **45.** $-2w + 4$ **47.** $r^2 - 10r$
49. $-x + 7$ **51.** 34x − 17y + 34 **53.** 14 − 3p + t
55. a. −1 **b.** −9.9 **c.** $\frac{1}{4}$ **d.** $-\frac{2}{3}$ **57.** −5, 4 **59.** −15
61. 50, 2 **63.** 1, −125 **65.** 20x **67.** 3x^2 **69.** 0 **71.** 0
73. 3a **75.** $-3x$ **77.** t **79.** x **81.** $-16x^2$ **83.** 1.1h
85. $\frac{4}{5}t$ **87.** 0.4r **89.** 7z − 15 **91.** $-3c - 1$ **93.** 7X − 2x
95. b + 2 **97.** $-2x^2 + 3x$ **99.** 12x **101.** (4x + 8) ft
107. 0 **109.** 2

Study Set   Section 8.3   (page 468)

1. equation **3.** satisfy **5.** reciprocal **7.** subtraction, multiplication **9.** addition, division **11.** $-\frac{5}{4}$ **13.** 30 **15. a.** 2x + 5
b. 2 **c.** 23 **d.** no **19. a.** −1 **b.** $\frac{3}{5}$ **c.** −31 **21.** 6
23. $-\frac{3}{5}$ **25.** 3.5 **27.** 4 **29.** $-\frac{8}{3}$ **31.** −12 **33.** −12 **35.** 9
37. 28 **39.** −19 **41.** 5 **43.** 1 **45.** 3 **47.** $\frac{1}{7}$ **49.** −4
51. −7.2 **53.** −1 **55.** 0 **57.** $-\frac{34}{5}$ **59.** $-\frac{55}{6}$ **61.** $-\frac{12}{5}$
63. $\frac{2}{15}$ **65.** $\frac{8}{9}$ **67.** $\frac{10}{9}$ **69.** −20 **71.** −41 **73.** 9 **75.** −1
77. 3 **79.** identity **81.** impossible equation **83.** impossible
equation **85.** identity **87.** −1.238 **89.** 1,645.3 **95.** 8
97. 64 **99.** $\frac{1}{64}$ **101.** 16x

Study Set   Section 8.4   (page 478)

1. formula **3.** perimeter **5.** radius **7.** circumference
9. a. $d = rt$ **b.** $r = c + m$ **c.** $p = r - c$ **d.** $I = Prt$
e. $C = 2\pi r$ **11.** 11,176,920 mi, 65,280 mi **13. a.** volume
b. circumference **c.** area **d.** perimeter **15.** (2x + 6) cm^2
19. a. 3.14 **b.** 98 · π **c.** the radius of the cylinder; the height
of the cylinder **21.** 2.5 mph **23.** $65 million **25.** 3.5%
27. 4,014°F **29.** $24.55 **31.** about 132 in. **33.** $R = \frac{E}{I}$
35. $w = \frac{V}{lh}$ **37.** $r = \frac{C}{2\pi}$ **39.** $a = 180 - b - c$ **41.** $x = \frac{y - b}{m}$
43. $t = \frac{A - P}{Pr}$ **45.** $h = \frac{3V}{\pi r^2}$ **47.** $b = 2x - a$ **49.** $s = C - Dn$
51. $c^2 = \frac{E}{m}$ **53.** $a^2 = c^2 - b^2$ **55.** $b = \frac{2A}{h} - d$ or $b = \frac{2A - hd}{h}$
57. $y = \frac{1}{3}x + 3$ **59.** $y = -\frac{3}{4}x - 4$ **61.** 212°F, 0°C
63. 827.6, 960.0 **65.** 36 ft, 48 ft^2 **67.** 50.3 in., 201.1 in.^{2}
69. 56 in., 144 in.^{2} **71.** 2,450 ft^2 **73.** 27.75 in., 47.8125 in.^{2}
75. 32 ft^2, 128 ft^3 **77.** 348 ft^3 **79.** 254 in.^{2}
81. $n = \dfrac{360°}{180° - a}$, 5 sides **87.** 137.76 **89.** 15%

Study Set   Section 8.5   (page 490)

1. perimeter **3.** vertex
5. a.

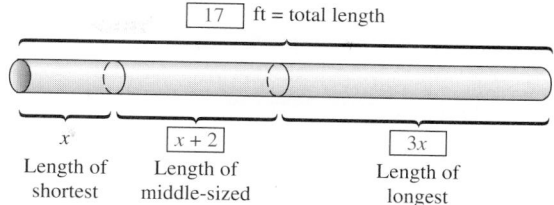

	17	ft = total length
x	$x + 2$	3x
Length of shortest	Length of middle-sized	Length of longest

b. 3 ft, 5 ft, 9 ft **9.** $30,000, 14%, 1 yr **11.** 35t mi, t hr,
45t mi **13. a.** 16.8 gal **b.** (x + 42) gal **c.** 32% **15.** 6,000
17. 22°, 68° **19.** 15 **21.** 4 ft, 8 ft **23.** 7.3 ft, 10.7 ft
25. Australia: 12 wk; Japan: 16 wk; Sweden: 10 wk **27.** 250
calories in ice-cream, 600 calories in pie **29.** in millions: $148,
$330, $235, $740 **31.** 7 ft, 7 ft, 11 ft **33.** 75 m by 480 m
35. 20° **37.** 12 **39.** 90 **41.** $5,500 **43.** $4,900 **45.** $7,500
47. 2 hr **49.** 65 mph, 45 mph **51.** 4 hr into the flights **53.** 50
55. 7.5 oz **57.** 20 **59.** 40 lb lemon drops, 60 lb jelly beans
61. 80 **67.** $-50x + 125$ **69.** 3x + 3 **71.** 19p + 11q

Study Set   Section 8.6   (page 504)

1. inequality **3.** solution **5. a.** true **b.** true **c.** false
d. false **e.** true **f.** false **7.** same **9.** opposite **11. a.** a true
statement **b.** a false statement **13. a.** all real numbers greater
than 8 **b.** ←———(———→ **c.** $(8, \infty)$ **15.** is less than, is
              8   greater than
17. is not equal to **19.** $x > -2$ **21.** $-2 \le 17$
25. $(-\infty, 5)$
                    5
27. $(-3, 1]$
              -3 1
29. $x < -1, (-\infty, -1)$ **31.** $-7 < x \le 2, (-7, 2]$
33. $x > 3, (3, \infty)$
              3
35. $x \ge -10, [-10, \infty)$
              -10
37. $x < -1, (-\infty, -1)$
              -1
39. $x \le 0.4, (-\infty, 0.4]$
              0.4
41. $x < -2, (-\infty, -2)$
              -2
43. $x < -\frac{11}{4}, (-\infty, -\frac{11}{4})$
              $-11/4$
45. $y \le -40, (-\infty, -40]$
              -40
47. $n \le 2, (-\infty, 2]$
              2
49. $x < 0, (-\infty, 0)$
              0
51. $x \ge 3, [3, \infty)$
              3
53. $x \ge -24, [-24, \infty)$
              -24
55. $x \le -1, (-\infty, -1]$
              -1

57. $x \geq -13$, $[-13, \infty)$

−13

59. $x > 0$, $(0, \infty)$

0

61. $x < -2$, $(-\infty, -2)$

−2

63. $x \geq \frac{9}{4}$, $[\frac{9}{4}, \infty)$

9/4

65. $x > -15$, $(-15, \infty)$

−15

67. $x \leq 20$, $(-\infty, 20]$

20

69. $x > \frac{5}{4}$, $(\frac{5}{4}, \infty)$

5/4

71. $y \leq \frac{1}{8}$, $(-\infty, \frac{1}{8}]$

1/8

73. $7 < x < 10$, $(7, 10)$

7   10

75. $-9 < x \leq 3$, $(-9, 3]$

−9   3

77. $-10 \leq x \leq 0$, $[-10, 0]$

−10   0

79. $-5 < x < -2$, $(-5, -2)$

−5   −2

81. $-6 \leq x \leq 10$, $[-6, 10]$

−6   10

83. $2 \leq x < 3$, $[2, 3)$

2   3

85. $-1 \leq x < 2$, $[-1, 2)$

−1   2

87. $x < 1.37$, $(-\infty, 1.37)$

1.37

89. $x \geq 0.03$, $[0.03, \infty)$

0.03

91. 98% or better   **93.** 27 mpg or better
95. $t \geq 420$ min   **97. a.** $0° < a \leq 18°$ **b.** $18° \leq a \leq 50°$
c. $30° \leq a \leq 37°$ **d.** $75° \leq a < 90°$
99. a. 470 ft $\leq x \leq$ 13,143 ft   **b.** 0.1 mi $\leq x \leq$ 2.5 mi
101. 1.496 in. $\leq w \leq$ 1.498 in.; 1.5000 in. $\leq w \leq$ 1.5010 in.
105. -125 **107.** $1, -3, 6$

Key Concept   (page 509)

1. a. $2x - 8$ **b.** $x = 6$ **3. a.** $\frac{2}{3}a$ **b.** $a = \frac{1}{2}$ **5.** The mistake is on the third line. The student made an equation out of the answer, which is $x - 6$, by writing "0 =" on the left. Then the student solved that equation.

Chapter Review   (page 511)

1. a. $h + 25$ **b.** $s - 15$ **c.** $\frac{1}{2}t$ **d.** $6x$ **2. a.** $(n + 4)$ in.
b. $(b - 4)$ in.   **3. a.** $10d$ **b.** $\frac{x}{12}$ **c.** $(x - 5)$ yr   **4.** 5, 30, 10, $10d$ **5.** $0, 19, -16$ **6. a.** 110 **b.** 40 **c.** 432 **d.** -36
7. 17.7 in.^{3} **8. a.** $-28w$ **b.** $15r^2$ **c.** $24xy$ **d.** $2.08f$
9. a. $5x + 15$ **b.** $-4x - 6 + 2y$ **c.** $-a + 4$ **d.** $3c - 6$
10. a. 3   **b.** 1   **11. a.** $2, -5$ **b.** $16, -5, 25$ **c.** $\frac{1}{2}, 1$
d. $9.6, -1$ **12. a.** $9p$ **b.** $-7m$ **c.** $-2a - 10b$
d. $-p - 18$ **e.** x **f.** $-8a^3$ **13.** $(4x + 4)$ ft   **14. a.** 2
b. -1 **c.** 30   **d.** -19 **e.** 4   **f.** 1   **g.** $\frac{5}{4}$ **h.** -6 **i.** identity, all values of a **j.** impossible equation, no solution

15. $176   **16.** $11,800   **17.** 3.38 hr   **18.** 1,949°F   **19.** 168 in.
20. 1,440 in.^{2} **21.** 76.5 m^2 **22.** 144 in.^{2} **23.** 50.27 cm
24. 201.06 cm^2 **25.** 4,320 in.^{3} **26.** 9.4 ft^3 **27.** 120 ft^3
28. 381.70 in.^{3} **29. a.** $h = \frac{A}{2\pi r}$ **b.** $l = \frac{P - 2w}{2}$ **30.** 8 ft
31. 147   **32.** 24.875 in. × 29.875 in. $\left(24\frac{7}{8} \text{ in.} \times 29\frac{7}{8} \text{ in.}\right)$
33. 76.5°, 76.5°   **34.** $45x   **35.** $16,000 at 7%, $11,000 at 9%
36. 20   **37.** 10 lb of each   **38.** 0.12x gal
39. a. $x < 1$, $(-\infty, 1)$

1

b. $x < -3$, $(-\infty, -3)$

−3

c. $x \geq 4$, $[4, \infty)$

4

d. $x \geq 6$, $[6, \infty)$

6

e. $x \geq 3$, $[3, \infty)$

3

f. $x \leq 12$, $(-\infty, 12]$

12

g. $6 < x < 11$, $(6, 11)$

6   11

h. $-2 < x \leq 1$, $(-2, 1]$

−2   1

40.

−13

41. 2.40 g $< w <$ 2.53 g   **42.** The sign length must be 48 inches or less.

Chapter 8 Test   (page 517)

1. $x - 2$ number of songs on the CD   **2.** $25q$ **3.** 4, 17, 59
4. 128   **5.** 6   **6.** 3   **7.** factor   **8.** the distributive property
9. $-20x$ **10.** $224t^2$ **11.** 18   **12.** $-4.9d^2$ **13.** -12 **14.** -5
15. -2 **16.** 1   **17.** $\frac{7}{6}$ **18.** -3 **19.** $r = \frac{A - P}{Pt}$ **20.** $150
21. $-10°$C   **22.** 393 in.^{3} **23.** $\frac{3}{5}$ hr   **24.** 10   **25.** 68°
26. $5,250
27. $x \geq -3$, $[-3, \infty)$

−3

28. $-3 \leq x < 4$, $[-3, 4)$

−3   4

29. Substitute the answer for the variable. If it is a solution, a true statement will result.
30. Like terms have exactly the same variables, raised to the same powers. $10p^2$ and $6p^2$ are like terms.

Cumulative Review Exercises   (page 519)

1. a. expression   **b.** equation   **2.** 3, 4, 5
3. $5 \cdot 5 \cdot 2 \cdot 2 = 2^2 \cdot 5^2$ **4.** $\frac{2}{3}$ **5.** $-\frac{2}{9}$ **6.** 6   **7.** $\frac{22}{15} = 1\frac{7}{15}$
8. $12\frac{11}{24}$ **9.** 0.9375   **10.** 45   **11. a.** 65   **b.** -12
12. the commutative property of multiplication   **13.** natural number, whole number, integer, rational number, real number
14. rational number, real number   **15.** rational number, real number   **16.** irrational number, real number   **17. a.** 4^3 **b.** $\pi r^2 h$
18. a. -10 **b.** -14 **c.** -64 **d.** 0   **19. a.** $w + 12$ **b.** $n - 4$

20. 4   **21.** $1, -3, 6$ **22.** $l = \dfrac{2,000}{d^2}$ (answers may vary

depending on the variables chosen)   **23. a.** 6 ft^2 **b.** 1.2 ft^2
c. 20%   **24.** 300   **25.** 0   **26.** -2 **27.** 16   **28.** 0   **29.** $-32d$
30. $10x - 15y + 5$ **31.** $5x$ **32.** $-8a$ **33.** $8q^2 - 5q$
34. $8t - 20$ **35.** $(x + 3)$ ft   **36.** $3x$ ft   **37.** 9   **38.** 20
39. -0.6 **40.** 19   **41.** -20 **42.** -2 **43.** 1   **44.** $\frac{5}{4}$

45. 65 m² **46.** 376.99 cm³ **47.** $t = \dfrac{A - P}{Pr}$ **48.** 37.5 ft-lb

49. 9.45 lb **50.** 55°, 55° **51.** \$4,000 **52.** 10 oz

53. $x > -2, (-2, \infty)$

54. $x \le 2, (-\infty, 2]$

55. $x \ge -1, [-1, \infty)$

56. $-1 \le x < 2, [-1, 2)$

Study Set   Section 9.1   (page 528)

1. ordered **3.** origin **5.** rectangular **7.** origin, left, up **9.** no
11. quadrant II **13. a.** 3 **b.** −4 **c.** 0 **d.** −4 **e.** 0
f. 5 **15.** 10 min before the workout, her heart rate was 60
beats/min. **17.** 150 beats/min **19.** approximately 5 min and 50
min after starting **21.** 10 beats/min faster after cool-down
23. (3, 5) is an ordered pair, 3(5) indicates multiplication, and
5(3 + 5) is an expression containing grouping symbols. **25.** yes

27.

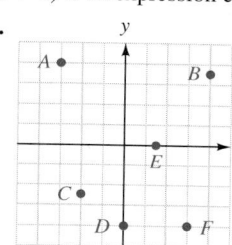

29. (6, 6) **31.** $\left(-\frac{1}{2}, \frac{5}{2}\right)$ **33.** (7, 6)
35. rivets: (2, 0), (−6, 0), (−2, 0),
(6, 0); welds: (−4, 3), (0, 3), (4, 3);
anchors: (−6, −3), (6, −3)
37. (−3, 10), (−2, 7), (−1, 4.8),
(0, 3), (1, 1.8), (2.5, 0.5), (4, 0)
39. a. \$2 **b.** \$4 **c.** \$7
d. \$9

41. a. 35 mi **b.** 4 **c.** 32.5 mi

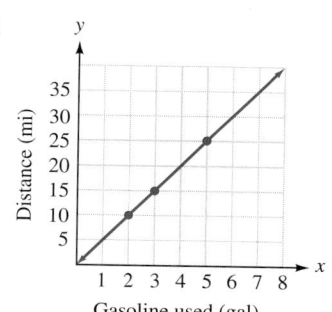

43. Rockford (5, B), Mount Carroll (1, C), Harvard (7, A), intersection (5, E) **49.** 12 **51.** 8 **53.** 7 **55.** −49

Study Set   Section 9.2   (page 539)

1. two **3.** independent, dependent **5. a.** 2 **b.** yes **c.** yes
d. infinitely many **7.** (1, 4), (3, 2), (5, 0) (answers may vary)
9. 0, −1, −8, 1, 8 **11.** He should have checked his computations.
At least one of his "solutions" is wrong. The graph should be a
straight line. **13.** A smooth curve should be drawn through the
points. **17.** yes **19.** no **21.** −3, −2, −5
23. −3, 1, 1 **25.**

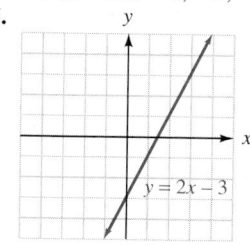

27.

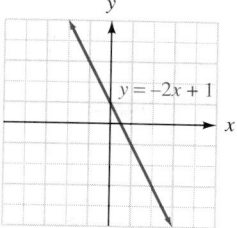

29. 1 unit higher

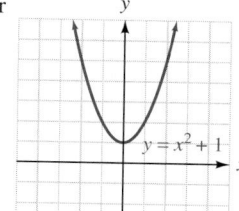

31. 2 units to the right

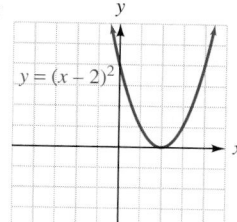

33. It is turned upside down.

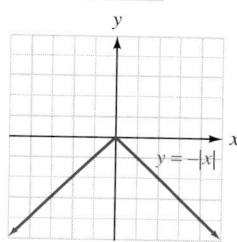

35. 2 units to the left

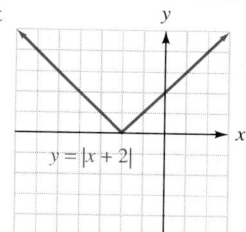

37. It is turned upside down.

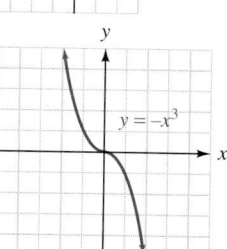

39. 2 units lower

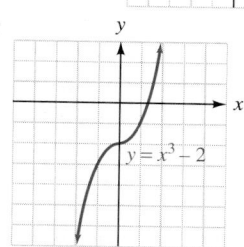

41. 0, 4, 0, −4 **43.** 0, 1, 4, 1, 4 **45. a.** It costs 8¢ to make a
2-in. bolt. **b.** 12¢ **c.** a 4-in. bolt **d.** It decreases as the length

approaches 4 in., then increases as the length increases to 7 in.
47. a. $90,000   **b.** the 3rd yr after being bought   **c.** after the
6th yr   **d.** It decreased in value for 3 yr, then increased in value
for 5 yr.   **55.** -96 **57.** an expression   **59.** 1.25   **61.** 0.1

Study Set   Section 9.3   (page 551)

1. linear   **3.** y-intercept   **5.** vertical   **7. a.** nonlinear
b. linear   **c.** nonlinear   **d.** linear   **e.** nonlinear   **9. a.** y: 1st
power; x: 1st power   **b.** y: 1st power; x: 2nd power   **c.** y: 1st
power; x: 3rd power   **11.** $6, -5, 4$ **13.** $-2, 4, -\frac{3}{2}$ **15.** because
A is on the line   **17.** The student made a mistake; the points should
lie on a straight line.   **19.** x-intercept: $(-3, 0)$; y-intercept: $(0, -1)$
21. a. $x; y$ **b.** $y; x$ **23. a.** $4x - y = 6$ **b.** $x - 2y = 0$
c. $x + 3y = 9$ **d.** $x + 0y = 12$ **25.**

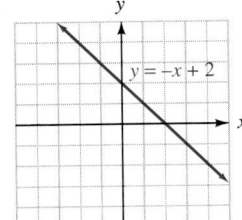

27.
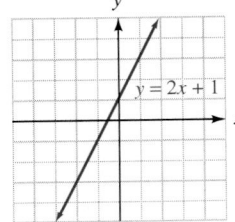

29.
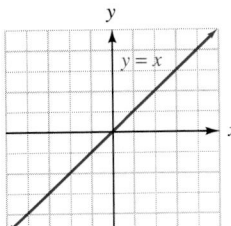

31.
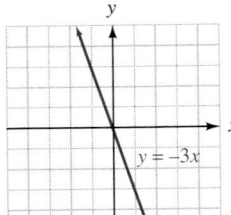

33.
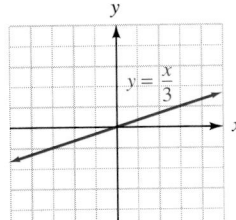

35.
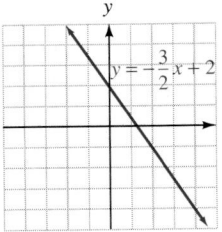

37.
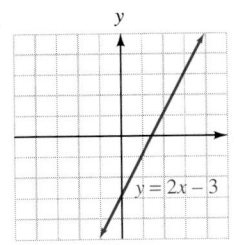

39.
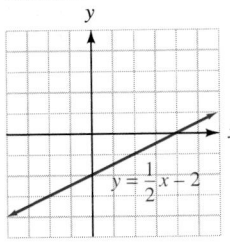

41.
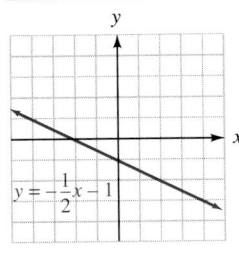

43.
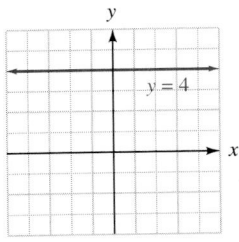

45.
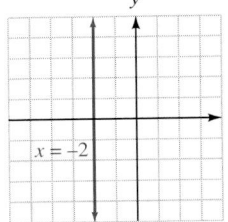

47.
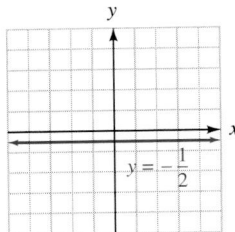

49.
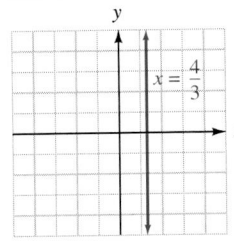

51.
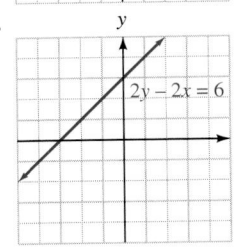

53.
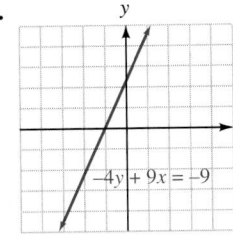

55.
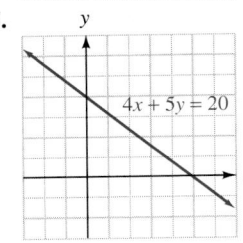

57.
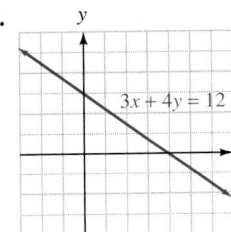

59.
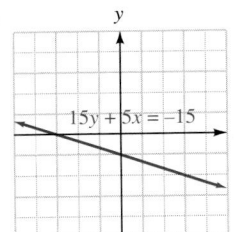

61.
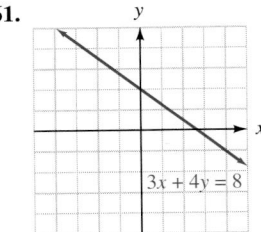

63. a. $c = 50 + 25u$ **b.** 150, 250, 400   **c.** $850   **d.** The
service fee is $50.

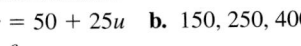
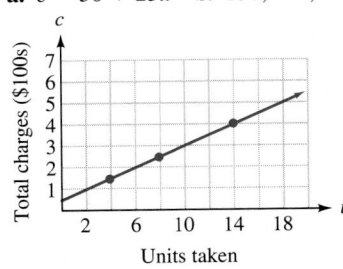

65. a. 56.2, 62.1, 64.0

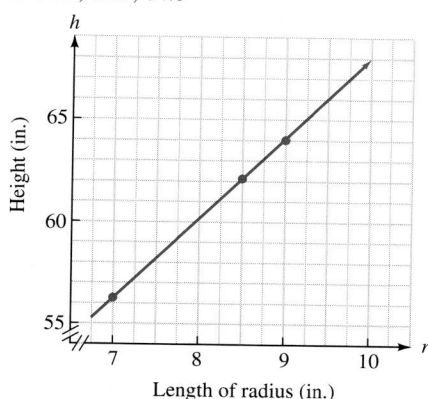

Length of radius (in.)

b. taller the woman is   **c.** 58 in.   **71.** $5 + 4c$ **73.** -4
75. profit = revenue − costs   **77.** 491

Study Set   Section 9.4   (page 564)

1. ratio   **3.** slope   **5.** change   **7. a.** l_2 **b.** l_1 **c.** l_4 **d.** l_3
9. -1 **11.** 3 in./yr   **13.** $-\$2,500$/year   **15. a.** 0; sales of 7-Up were not changing; each year about the same number of cases were sold.   **b.** Mountain Dew   **17.** $m = \dfrac{y_2 - y_1}{x_2 - x_1}$ **19.** 1
21. -3 **23.** $\frac{5}{4}$ **25.** $-\frac{1}{2}$ **27.** $\frac{3}{5}$ **29.** 0   **31.** undefined   **33.** $-\frac{2}{3}$
35. -4.75 **37.** $m = \frac{2}{3}$ **39.** $m = \frac{4}{3}$ **41.** $m = -\frac{7}{8}$ **43.** $m = -\frac{1}{5}$
45.

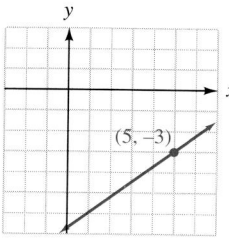

47.

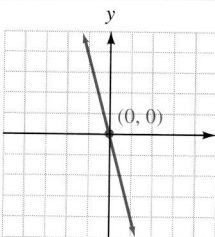

49.

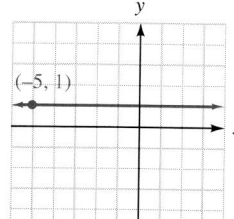

51.

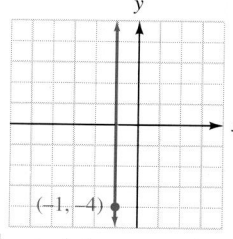

53.

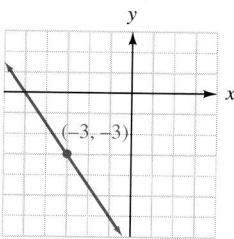

55.

57. $\frac{2}{5}$ **59.** $\frac{1}{20}$; 5%   **61. a.** $\frac{1}{8}$ **b.** $\frac{1}{12}$ **c.** 1: less expensive, steeper; 2: not as steep, more expensive   **63.** 3 hp/40 rpm
69. quadrant II   **71.** no   **73.** linear

Study Set   Section 9.5   (page 574)

1. slope-intercept   **3.** Parallel   **5.** reciprocals
7. No, because the graph is not a straight line.

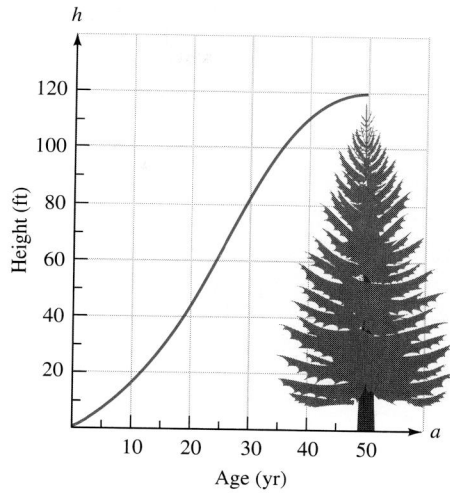

Age (yr)

9. a. When there are no head waves, the ship could travel at 18 knots.   **b.** $-\frac{1}{2}$ knot/ft   **c.** $y = -\frac{1}{2}x + 18$ **11. a.** $(0, 0)$
b. same slope, different y-intercepts   **13.** $-\frac{2}{3}$ **b.** $\frac{1}{4}$ **c.** -8
d. 3   **e.** 1   **f.** -1 **15.** $y, -3x, 6$ **19.** 4, $(0, 2)$ **21.** $\frac{1}{4}, \left(0, -\frac{1}{2}\right)$
23. $\frac{1}{2}, (0, 6)$ **25.** $\frac{1}{6}, (0, -1)$ **27.** $-1, (0, 8)$ **29.** $-\frac{2}{3}, (0, 2)$
31. $0, \left(0, \frac{13}{3}\right)$ **33.** $-5, (0, 0)$
35. $y = 5x - 3$

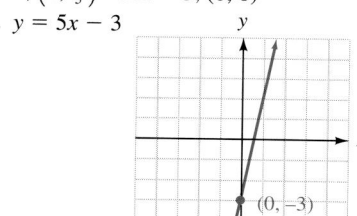

37. $y = \frac{1}{4}x - 2$

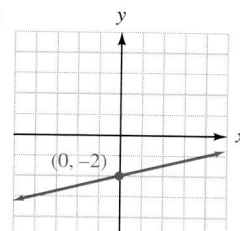

39. $y = -3x + 6$

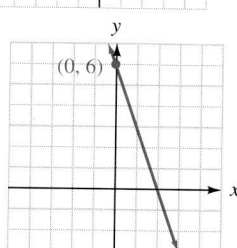

41. $y = -\frac{8}{3}x + 5$

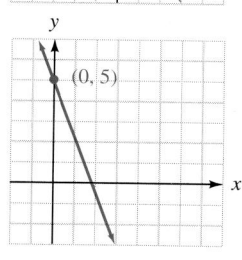

43.

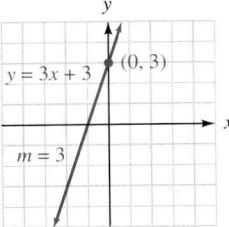

45.

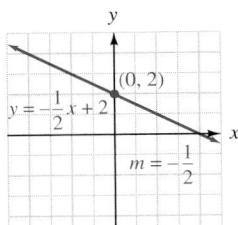

47.

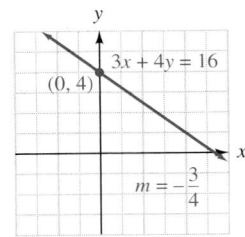

49.

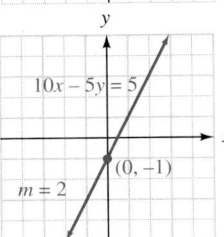

51. a. $y = 2,000x + 5,000$
b. \$21,000 **53.** $y = 5x - 10$
55. a. $y = 0.20x + 1.00$

b.

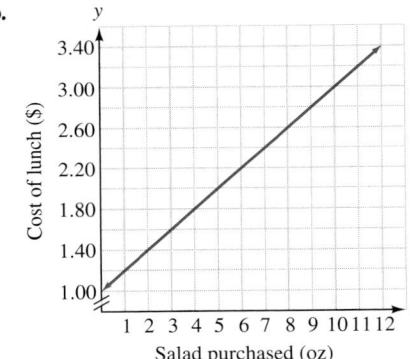

c. same slope; different y-intercept
d. same y-intercept; steeper slope

57. $y = -20x + 500$ **63.** $-\frac{1}{4}$ **65.** 0 **67.** subtraction
69. 25%

Study Set   Section 9.6   (page 583)

1. point–slope **3.** x-coordinate, y-coordinate **5. a.** The slope is 2; the y-intercept is $(0, -3)$ **b.** The graph passes through $(5, 4)$; the slope is 6. **7. a.** $(-4, -2), (3, 2)$ **b.** $\frac{4}{7}$ **c.** $y + 2 = \frac{4}{7}(x + 4)$ or $y - 2 = \frac{4}{7}(x - 3)$ **9. a.** no **b.** no **c.** yes **11.** sub
17. $y - 1 = 3(x - 2)$ **19.** $y + 1 = -\frac{4}{5}(x + 5)$ **21.** $y = \frac{1}{5}x - 1$
23. $y = -5x - 37$ **25.** $y = -\frac{4}{3}x + 4$ **27.** $y = -\frac{2}{3}x + 2$
29. $y = 8x + 4$ **31.** $y = -3x$ **33.** $y = 2x + 5$
35. $y = -\frac{1}{2}x + 1$ **37.** $y = 5$ **39.** $y = \frac{1}{10}x + \frac{1}{2}$ **41.** $x = -8$
43. $x = 4$ **45.** $y = 5$ **47. a.** position 1: $(0, 0), (-5, 2)$;
position 2: $(0, 0), (-3, 6)$; position 3: $(0, 0), (-1, 7)$;
position 4: $(0, 0), (0, 10)$ **b.** $y = -\frac{2}{5}x, y = -7x, x = 0$
c. The pole is not in the shape of a straight line.
49. a. $y = -40x + 920$ **b.** 440 yd³ **51.** $c = 30t + 45$
53. a. $(0, 32); (100, 212)$ **b.** $F = \frac{9}{5}C + 32$ **55.** $y = -\frac{4}{15}x + 83$
61. $-\frac{1}{2}$ **63.** 113.1 ft² **65.** -1 **67.** 6

Study Set   Section 9.7   (page 592)

1. function **3.** independent, dependent **5. a.** positive numbers **b.** positive numbers **c.** 0 **d.** D: all reals; R: real numbers greater than or equal to 0 **7.** $f(-1)$ **9. a.** $(-2, 4)$, $(-2, -4)$ **b.** No; the x-value -2 is assigned to more than one y-value (4 and -4). **11.** $x, -5, (4, -5)$ **13.** of, is **15.** yes
17. yes **19.** no; $(4, 2), (4, -2)$ **21.** yes **23.** no; $(3, 1), (3, 2)$
25. yes **27.** no; $(-1, 0), (-1, 2)$ **29.** no; $(3, 4), (3, -4)$ or $(4, 3), (4, -3)$ **31.** yes **33.** no; $(3, 4), (3, -1)$ (answers may vary) **35.** no; $(0, 2), (0, -4)$ (answers may vary) **37.** D: all reals; R: all reals **39.** D: all reals; R: real numbers greater than or equal to 0 **41.** D: all reals; R: all reals **43. a.** 3 **b.** -9 **c.** 0 **d.** 199 **45. a.** 0.32 **b.** 18 **c.** 2,000,000 **d.** $\frac{1}{32}$ **47. a.** 7 **b.** 14 **c.** 0 **d.** 1 **49. a.** 0 **b.** 990 **c.** -24 **d.** 210
51. 1.166 **53.** $-2, -5, 1, 4$

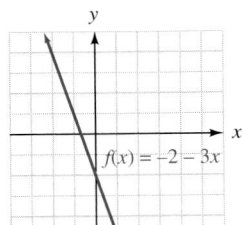

55. $-3, -2, -1$

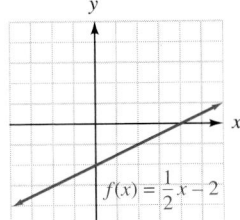

57. $2, 1, -2, 1, -2$

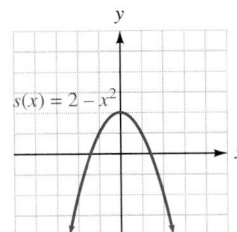

59. $f(x) = |x|$ **61. a.** all real numbers from 0 through 24
b. 0.5 **c.** 1.5 **d.** -1.4 **e.** The low-tide mark was -2.5 m.
f. 1.6 **63.** 78.5 ft², 314.2 ft², 1,256.6 ft² **67.** $y = 6$
69. profit = revenue $-$ costs **71.** $-6x + 12$ **73.** $12d$

Key Concept   (page 597)

1. $y = -3x - 4$ **3.** $-2, 4, -3$
5. a. $(-2, -2)$ (answers may vary) **b.** -2 **c.** $y = -2x - 6$
7. 130; the cost to rent the mixer for 3 days

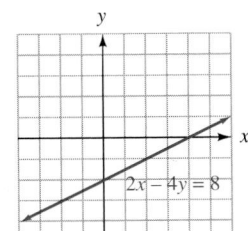

Chapter Review   (page 599)

1. a.

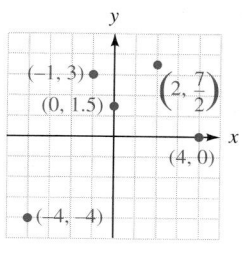

b. $-1, 0, 1$ **2.** quadrant III
3. a. 2 ft   **b.** 2 ft   **c.** 6 ft
4. a. 2,500; week 2   **b.** 1,000
c. 1st week and 5th week
5. $(0, 2)$
6. not a solution

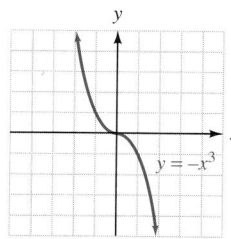

7. a. $8, 1, 0, -1, -8$
b. It would be 2 units higher.

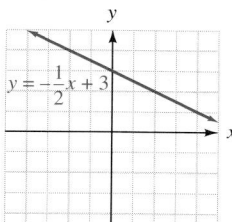

8. a. 9,000   **b.** It tells us that 40 trees on an acre give the highest yield, 18,000 oranges.   **9. a.** nonlinear   **b.** linear   **c.** linear
d. nonlinear   **10.** $A = 5, B = 2, C = 10$ **11.** $-6, -6, -8, -8$
12.

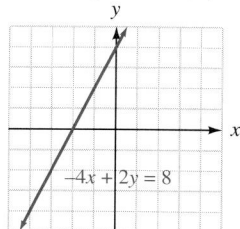

13. x-intercept: $(-2, 0)$; y-intercept: $(0, 4)$

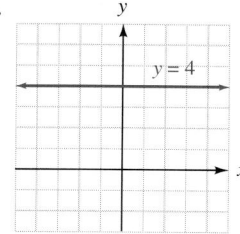

14. a.

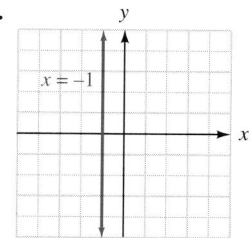

b.

15. It serves as a check. If the three points do not lie on a line, then at least one of them is in error.   **16. a.** $\frac{1}{4}$ **b.** -7 **c.** 0   **d.** $-\frac{3}{2}$
17.

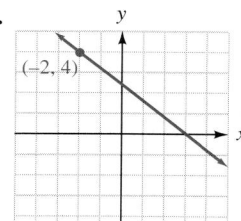

18. a. 1992–1994; -1.25 million people/year   **b.** 1986–1988; 4.05 million people/year
19. a. $m = \frac{3}{4}$; y-intercept: $(0, -2)$ **b.** $m = -4$; y-intercept: $(0, 0)$

20. $m = 3$; y-intercept: $(0, -5)$
21. a. $c = 300w + 75,000$
b. 90,600   **22. a.** parallel
b. perpendicular

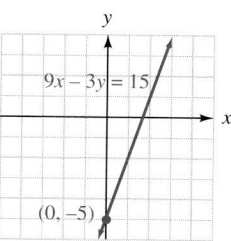

23. a. $y = 3x + 2$

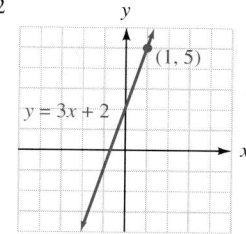

b. $y = -\frac{1}{2}x - 3$

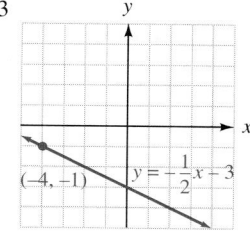

24. a. $y = \frac{2}{3}x + 5$
b. $y = -8$
25. $f = -35x + 450$
26. a. yes   **b.** no
c. no   **27. a.** D: all reals; R: all reals
b. D: all reals: R: real numbers greater than or equal to 0

28. a. -5 **b.** 37   **c.** -2 **d.** -8 **29.** $1, 0, -1, 0, -1, -2$
30. a. no   **b.** yes   **31.** 1,004.8 in.^{3}

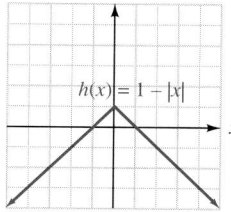

Chapter 9 Test   (page 605)

1. 10   **2.** 60   **3.** 1 day before and the 3rd day of the holiday
4. 50 dogs were in the kennel when the holiday began.
5.

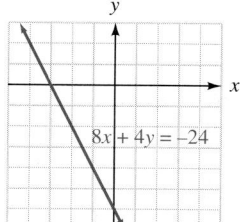

6.

7. yes   **8.** no   **9.** x-intercept: $(3, 0)$; y-intercept: $(0, -2)$
10. $m = -\frac{1}{2}$; $(0, 4)$ **11.**

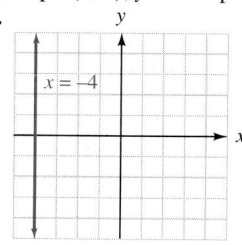

12.

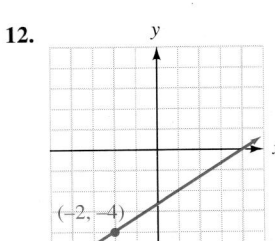

13. -1 **14.** undefined   **15.** $\frac{8}{7}$
16. parallel   **17.** the 15–20 mi segment: 30 ft/mi
18. the 22–25 mi segment: $-\frac{100}{3}$ ft/mi $= -33\frac{1}{3}$ ft/mi
19. $v = -1{,}500x + 15{,}000$
20. $y = 7x + 19$ **21.** no
22. D: all reals; R: real numbers less than or equal to 0   **23.** yes

24. -13 **25.** 756

Cumulative Review Exercises   (page 607)

1. a. QII, 2000: $-\$5$ million   **b.** QIV, 1999: $-\$270$ million
2. $2^2 \cdot 3^3$ **3.** 0.004   **4. a.** true   **b.** true   **c.** true   **5.** $+3.8$, -5.6 **6. a.** 6   **b.** -2 **c.** 1   **d.** 28   **7.** 32   **8.** $500 - x$
9. $3, -2$ **10. a.** $2x + 8$ **b.** $2x - 8$ **c.** $-2x - 8$ **d.** $-2x + 8$
11. $4a + 10$ **12.** $4b^2$ **13.** 4   **14.** $-3y$ **15.** 6   **16.** 2.9
17. 9   **18.** -19 **19.** $\frac{1}{7}$ **20.** 1   **21.** $-\frac{55}{6}$ **22.** no solution
23. $x = \frac{y-b}{m}$ **24.** $3\frac{1}{8}$ in., $\frac{39}{64}$ in.^{2} **25.** $79°$ **26.** $0.50x$, $0.25(13 - x)$, $0.30(13)$ **27.** 7.5 hr   **28.** 80 lb candy corn, 120 lb gumdrops   **29.** $x \le 48$, 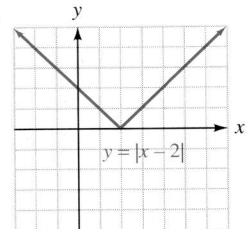 , $(-\infty, 48]$

30. $x > 0$, 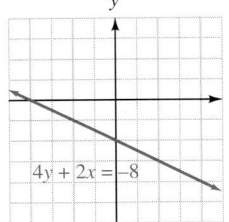 , $(0, \infty)$ **31. a.** 1 tsp   **b.** 3 tsp

32. no   **33.**

$y = |x - 2|$

34.

$4y + 2x = -8$

35. 0   **36.** $-\frac{10}{7}$ **37.** $\frac{2}{3}$, $(0, 2)$
38. $y = -2x + 1$
39. $y + 9 = -\frac{7}{8}(x - 2)$ **40.** 10

Study Set   Section 10.1   (page 616)

1. base, exponent   **3.** power   **5.** $3x, 3x, 3x, 3x$ **7.** x^{m+n} **9.** $\frac{a^n}{b^n}$
11. x^{m-n} **13.** 1   **15. a.** $(3x^2)^6$ (answers may vary)
b. $\left(\frac{3a^3}{b}\right)^2$ (answers may vary)   **17. a.** $2x^2$ **b.** 0   **c.** x^4
d. x^2 **19. a.** doesn't simplify   **b.** doesn't simplify   **c.** x^5
d. x **21.** a^{10} mi^2 **23.** x^9 m^3 **27.** base 4, exponent 3
29. base x, exponent 5   **31.** base $-3x$, exponent 2   **33.** base y, exponent 6   **35.** 16   **37.** -16 **39.** $x \cdot x \cdot x \cdot x \cdot x$
41. $\left(\frac{t^2}{2}\right)\left(\frac{t^2}{2}\right)\left(\frac{t^2}{2}\right)$ **43.** $(4t)^4$ **45.** $-4t^3$ **47.** 12^7 **49.** 2^6
51. a^6 **53.** x^7 **55.** a^9 **57.** y^9 **59.** 8^8 **61.** x^{12} **63.** c
65. 3^8 **67.** y^{15} **69.** m^{500} **71.** $a^5 b^6$ **73.** $c^2 d^5$ **75.** $x^3 y^3$
77. y^4 **79.** $c^2 d^6$ **81.** x^{25} **83.** $243z^{30}$ **85.** x^{31} **87.** $u^4 v^4$
89. $a^9 b^6$ **91.** $-8r^6 s^9$ **93.** $\frac{a^3}{b^3}$ **95.** $\frac{x^{10}}{y^{15}}$ **97.** $\frac{-32a^5}{b^5}$
99. $216k^3$ **101.** $a^{12} b^6$ **103.** a **105.** ab^4 **107.** $r^{13} s^3$ **109.** $\frac{y^3}{8}$

111. $\dfrac{27t^{12}}{64}$ **113. a.** $25x^2$ ft^2 **b.** $9\pi x^2$ ft^2 **115. b.** 16 ft,
8 ft, 4 ft, 2 ft   **117. a.** 11000001, 11010001, 11001101, 11000011
(answers may vary)   **b.** 2^8 **123.** c   **125.** d

Study Set   Section 10.2   (page 623)

1. base, exponent   **3.** reciprocal   **5. a.** $4 - 4, 0$ **b.** 6, 6, 6, 6, 1
c. 1, 1   **7.** 9, 3, 1, $\frac{1}{3}, \frac{1}{9}$ **9.** 81, -9, 1, $-\frac{1}{9}, \frac{1}{81}$ **11.** 4, 2^2; 2, 2^1; 1,
2^0; $\frac{1}{2}$, 2^{-1}; $\frac{1}{4}$, 2^{-2} **15.** base x, exponent -2 **17. a.** 4; 2; -16
b. 4; -2; $\frac{1}{16}$ **c.** 4; -2; $-\frac{1}{16}$ **19.** 1   **21.** 1   **23.** 2   **25.** 1
27. 1   **29.** $\frac{5}{2}$ **31.** $\frac{1}{144}$ **33.** $-\frac{1}{4}$ **35.** 125   **37.** $\frac{3}{16}$ **39.** $-\frac{1}{64}$
41. $\frac{1}{64}$ **43.** $\frac{1}{x^2}$ **45.** $-\frac{1}{b^5}$ **47.** $\frac{1}{16y^4}$ **49.** $\frac{1}{a^3 b^6}$ **51.** 8   **53.** 1
55. $\frac{8}{7}$ **57.** 1   **59.** $\frac{1}{y}$ **61.** $\frac{1}{r^6}$ **63.** y^5 **65.** 2   **67.** $\frac{1}{a^2 b^4}$
69. $\frac{1}{x^6 y^3}$ **71.** $\frac{1}{x^3}$ **73.** $\frac{a^8}{b^{12}}$ **75.** $-\frac{y^{10}}{32x^{15}}$ **77.** a^{14} **79.** $\frac{1}{b^{14}}$
81. $\frac{256x^{28}}{81}$ **83.** $\frac{16y^{14}}{z^{10}}$ **85.** x^{3m} **87.** $\frac{1}{u^m}$ **89.** y^m **91.** $\frac{1}{x^{3n}}$
93. $10^2, 10^1, 10^0, 10^{-1}, 10^{-2}, 10^{-3}, 10^{-4}$ **95.** approximately
\$4,603.09   **99.** 13.5 yr   **101.** $y = \frac{3}{4}x - 5$

Study Set   Section 10.3   (page 630)

1. scientific   **3.** 250   **5.** 0.000025   **7.** 10^5 **9.** 10^{-3}
11. 5, right   **13.** 10^{-4} **15.** positive   **19.** 2.3×10^4
21. 1.7×10^6 **23.** 6.2×10^{-2} **25.** 5.1×10^{-6} **27.** 4.25×10^3
29. 2.5×10^{-3} **31.** 230   **33.** 812,000   **35.** 0.00115
37. 0.000976   **39.** 25,000,000   **41.** 0.00051
43. 2.57×10^{13} mi   **45.** 63,800,000 mi^2 **47.** 6.22×10^{-3} mi
49. 714,000   **51.** 30,000   **53.** 200,000   **55.** $9.038030748 \times 10^{15}$
57. $1.881676423 \times 10^{12}$ **59.** $1.74449211 \times 10^{26}$ **61.** g, x, u,
v, i, m, r   **63.** 1.7×10^{-18} g   **65.** 3.099363×10^{16} ft
67. 2.036×10^{11} dollars   **69.** 1.7×10^6; 1,700,000
b. 1999: 1.4×10^6, 1,400,000; 1986: 2.05×10^6, 2,050,000
75. 5   **77.** commutative property of addition   **79.** 6

Study Set   Section 10.4   (page 636)

1. polynomial   **3.** term   **5.** degree   **7.** in, decreasing or
descending   **9.** of   **11.** yes   **13.** no   **15.** yes
17. binomial   **19.** trinomial   **21.** monomial   **23.** binomial
25. trinomial   **27.** none of these   **29.** trinomial   **31.** 4th
33. 2nd   **35.** 1st   **37.** 4th   **39.** 12th   **41.** 0th
45. because $x^3 + 2x^2 - 3$ is a polynomial   **47.** 7   **49.** -8
51. -2 **53.** -7.5 **55.** -4 **57.** -5 **59.** -5.69
61. -188.96 **63.** 3   **65.** -1 **67.** 1.929   **69.** -512.241
71. -1 **73.** -7 **75.** 72 in.^{3} **77.** 22 ft, 2 ft   **79.** about 404 ft
81. 18.75 ft, 20 ft, 15 ft   **85.** $y \ge -3$
87. x^{18} **89.** y^9

Study Set   Section 10.5   (page 642)

1. polynomials   **3.** coefficients   **5.** terms   **7.** like
9. changing   **11.** $2x^2 + 3x - 4$ **13.** $(11x - 12)$ ft   **17.** $9y$
19. $12t^2$ **21.** $-48u^3$ **23.** $-0.1x$ **25.** $2st$ **27.** $6r$ **29.** $-ab$
31. $15x^2$ **33.** $7x + 4$ **35.** $2a + 7$ **37.** $7x - 7y$ **39.** $3x - 4y$
41. $6x^2 + x - 5$ **43.** $7b + 4$ **45.** $3x + 1$
47. $13.54x^2 - 6.55x + 6.15$ **49.** $5x^2 + x + 11$
51. $-7x^3 - 7x^2 - x - 1$ **53.** $2x^2 + 4x + 13$ **55.** $5x^2 + 6x - 8$

57. $-x^3 + 6x^2 + x + 14$ **59.** $-12x^2y^2 - 9xy + 36y^2$
61. $t^3 + 3t^2 + 6t - 5$ **63.** $-3x^2 + 5x - 7$ **65.** $6x - 2$
67. $-5x^2 - 8x - 4$ **69.** $-4y^2 + 8y + 4$
71. $a^2b + 2ab^2 - 6ab - 4a - b$ **73.** $(3x^2 + 6x - 2)$ yd
75. $(x^2 - 8x + 12)$ ft   **77.** $(3x^2 + 11x + 4.5\pi)$ in.
79. \$114,000   **81.** $f(x) = 1,900x + 225,000$
83. $f(x) = -1,100x + 6,600$ **85.** $f(x) = -2,800x + 15,800$
91. $180°$ **93.** $x \le 2$

Study Set   Section 10.6   (page 652)

1. binomials   **3.** first, outer, inner, last   **5.** $6x^2$ **7.** $15x$
9. $(6x^2 + x - 1)$ cm^2 **13.** $12x^5$ **15.** $-24b^6$ **17.** $6x^5y^5$
19. $-3x^4y^5z^8$ **21.** $3x + 12$ **23.** $-4t - 28$ **25.** $3x^2 - 6x$
27. $-6x^4 + 2x^3$ **29.** $3x^2y + 3xy^2$ **31.** $6x^4 + 8x^3 - 14x^2$
33. $-6x^4 - 24x^3$ **35.** $a^2 + 9a + 20$ **37.** $3x^2 + 10x - 8$
39. $6a^2 + 2a - 20$ **41.** $6x^2 - 7x - 5$ **43.** $2x^2 + 3x - 9$
45. $6t^2 + 7st - 3s^2$ **47.** $x^2 + xz + xy + yz$
49. $-12t^2 + 7tu - u^2$ **51.** $8x^2 - 12x - 8$ **53.** $3a^3 - 3ab^2$
55. $5t^2 - 11t$ **57.** $2x^2 + xy - y^2$ **59.** $x^2 + 8x + 16$
61. $t^2 - 6t + 9$ **63.** $r^2 - 16$ **65.** $16x^2 - 25$ **67.** $4s^2 + 4s + 1$
69. $x^2 + 10x + 25$ **71.** $x^2 - 4xy + 4y^2$ **73.** $4a^2 - 12ab + 9b^2$
75. $16x^2 + 40xy + 25y^2$ **77.** $(4x^2 - 6x + 2)$ cm^2
79. $(x^2 + 6x + 9)\pi$ in.^{2} **81.** $x^3 - x + 6$
83. $4t^3 + 11t^2 + 18t + 9$ **85.** $-3x^3 + 25x^2y - 56xy^2 + 16y^3$
87. $x^3 - 3x + 2$ **89.** $12x^3 + 17x^2 - 6x - 8$ **91.** -3
93. -8 **95.** -1 **97.** 0 **99.** $(24x + 14)$ cm, $(35x^2 + 43x + 12)$ cm^2 **101. a.** x^2 ft^2, $6x$ ft^2, $5x$ ft^2, 30 ft^2; $(x^2 + 11x + 30)$ ft^2
b. $(x + 6)$ ft, $(x + 5)$ ft; $(x^2 + 11x + 30)$ ft^2 **c.** They are the same.   **103.** 5 and 6   **105.** 4 m   **107.** 90 ft   **111.** 1   **113.** $-\frac{2}{3}$
115. $(0, 2)$

Study Set   Section 10.7   (page 658)

1. polynomial   **3.** two   **5.** exponents   **7.** 9, 9   **9.** In the numerator and denominator, a common factor of 2 was divided out.   **11. a.** $t = \frac{d}{r}$ **b.** $3x^2$ **13.** $2x + 7$ **17.** $\frac{1}{3}$ **19.** $-\frac{5}{3}$
21. $\frac{3}{4}$ **23.** 1   **25.** x^3 **27.** $\frac{r^2}{s}$ **29.** $\frac{2x^2}{y}$ **31.** $-\frac{3u^3}{v^2}$ **33.** $\frac{4r}{y^2}$
35. $-\frac{13}{3rs}$ **37.** $\frac{x^4}{y^6}$ **39.** a^8b^8 **41.** $-\frac{3r}{s^9}$ **43.** $-\frac{x^3}{4y^3}$ **45.** $-\frac{16}{y^6}$
47. a^8 **49.** $2x + 3$ **51.** $\frac{1}{5y} - \frac{2}{5x}$ **53.** $\frac{1}{y^2} + \frac{2y}{x^2}$ **55.** $3a - 2b$
57. $\frac{1}{y} - \frac{1}{2x} + \frac{2z}{xy}$ **59.** $3x^2y - 2x - \frac{1}{y}$ **61.** $5x - 6y + 1$
63. $\frac{10x^2}{y} - 5x$ **65.** $-\frac{4x}{3} + \frac{3x^2}{2}$ **67.** $xy - 1$
69. $\frac{x}{y} - \frac{11}{6} + \frac{y}{2x}$ **71.** 2   **73.** $(2x^2 - x + 3)$ in.   **75.** $(3x - 2)$ ft
77. yes   **79.** no   **83.** binomial   **85.** none of the above   **87.** 2

Study Set   Section 10.8   (page 663)

1. divisor, dividend   **3.** remainder   **5.** $4x^3 - 2x^2 + 7x + 6$
7. $6x^4 - x^3 + 2x^2 + 9x$ **9.** $0x^3$ and $0x$ **11.** $x^3 + 3x^2 + 9x + 27$
13. a. $r = \frac{d}{t}$ **b.** $x - 3$ **15.** It is correct.   **19.** $x + 6$
21. $y + 12$ **23.** $3a - 2$ **25.** $b + 3$ **27.** $2x + 1$
29. $x - 7$ **31.** $3x + 2$ **33.** $2x - 1$ **35.** $x^2 + 2x - 1$
37. $2x^2 + 2x + 1$ **39.** $x^2 + x + 1$ **41.** $x + 1 + \frac{-1}{2x + 3}$

43. $2x + 2 + \frac{-3}{2x + 1}$ **45.** $x^2 + 2x + 1$
47. $x^2 + 2x - 1 + \frac{6}{2x + 3}$ **49.** $2x^2 + 8x + 14 + \frac{31}{x - 2}$
51. $x + 1$ **53.** $2x - 3$ **55.** $x^2 - x + 1$
57. $a^2 - 3a + 10 + \frac{-30}{a + 3}$ **59.** $5x^2 - x + 4 + \frac{16}{3x - 4}$
61. a. $(x - 6)$ in.   **b.** $(4x - 4)$ in.   **63.** $4x^2 + 3x + 7$ **67.** x^{22}
69. $8x^2 - 6x + 1$ **71.** They are the same.

Key Concept   (page 666)

1. a. x, descending   **b.** 4   **c.** 3, 2, 1, 0   **d.** 3   **e.** 1, -2, 6, -8
3. a. binomial   **b.** none of these   **c.** trinomial   **d.** monomial
5. combine   **7.** each, each   **9.** $3x - 5$ **11.** $2x^2 - 13x - 24$
13. $y^2 + 2y - 3$ **15.** $y^3 + 4y^2 - 3y - 18$ **17.** 1

Chapter Review   (page 668)

1. a. $-3 \cdot x \cdot x \cdot x \cdot x$ **b.** $\left(\frac{1}{2}pq\right)\left(\frac{1}{2}pq\right)\left(\frac{1}{2}pq\right)$ **2. a.** 125   **b.** 64
c. -64   **d.** 4   **3. a.** x^5 **b.** $-3y^6$ **c.** y^{21} **d.** $81x^4$ **e.** b^{12}
f. $-y^2z^5$ **g.** $256s^3$ **h.** $4x^4y^2$ **i.** x^{15} **j.** $\frac{x^2}{y^2}$ **k.** x^4 **l.** $125yz^4$
4. a. $64x^{12}$ in.^{3} **b.** y^4 m^2 **5. a.** 1   **b.** 1   **c.** 9   **d.** $\frac{1}{1,000}$
e. $\frac{4}{3}$ **f.** $-\frac{1}{25}$ **g.** $\frac{1}{x^5}$ **h.** $-\frac{6}{y}$ **i.** $\frac{1}{x^{10}}$ **j.** x^{14} **k.** $\frac{1}{x^5}$
l. $\frac{1}{9z^2}$ **6. a.** y^{7n} **b.** $\frac{1}{z^{2c}}$ **7. a.** 7.28×10^2 **b.** 9.37×10^6
c. 1.36×10^{-2} **d.** 9.42×10^{-3} **e.** 1.8×10^{-4} **f.** 7.53×10^5
8. a. 726,000   **b.** 0.000391   **c.** 2.68   **d.** 57.6   **9. a.** 0.03
b. 160   **10.** 6,080,000,000; 6.08×10^9
11. $1.0 \times 10^5 = 100,000$ **12. a.** yes   **b.** no   **c.** no   **d.** yes
13. a. 4   **b.** $3x^3$ **c.** -1   **d.** 10   **14. a.** 7th, monomial
b. 3rd, monomial   **c.** 2nd, binomial   **d.** 5th, trinomial   **e.** 6th, binomial   **f.** 4th, none of these   **15. a.** 34   **b.** 1   **c.** 9
d. 0.72   **16.** 8 in.   **17. a.** $2x^6 + 5x^5$ **b.** $-2x^2y^2$ **c.** $8x^2 - 6x$
d. $5x^2 + 19x + 3$ **18. a.** $4x^2 + 2x + 8$ **b.** $8x^3 - 7x^2 + 19x$
19. a. $10x^3$ **b.** $-6x^{10}z^5$ **c.** $-6r^3s^4t^5$ **d.** $120b^{11}$
20. a. $5x + 15$ **b.** $3x^4 - 5x^2$ **c.** $x^2y^3 - x^3y^2$ **d.** $-2y^4 + 10y^3$
e. $6x^6 + 12x^5$ **f.** $-3x^3 + 3x^2 - 6x$ **21. a.** $x^2 + 5x + 6$
b. $2x^2 - x - 1$ **c.** $6a^2 - 6$ **d.** $6a^2 - 6$ **e.** $2a^2 - ab - b^2$
f. $-6x^2 - 5xy - y^2$ **22. a.** $x^2 + 6x + 9$ **b.** $x^2 - 25$
c. $a^2 - 6a + 9$ **d.** $x^2 + 8x + 16$ **e.** $4y^2 - 4y + 1$ **f.** $y^4 - 1$
23. a. $3x^3 + 7x^2 + 5x + 1$ **b.** $8a^3 - 27$ **24. a.** 1   **b.** -1
c. 7   **d.** 0   **25.** $(6x + 10)$ in.; $(2x^2 + 11x - 6)$ in.^{2};
$(6x^3 + 33x^2 - 18x)$ in.^{3} **26. a.** $-\frac{2x}{3y^2}$ **b.** $\frac{1}{x}$ **27. a.** $4x + 3$
b. $2 - \frac{3}{y}$ **c.** $3a + 4b - 5$ **d.** $-\frac{x}{y} - \frac{y}{x}$ **28.** $x + 5$
29. a. $x + 1 + \frac{3}{x + 2}$ **b.** $x - 5$ **c.** $2x + 1$
d. $x + 5 + \frac{3}{3x - 1}$ **e.** $3x^2 + 2x + 1 + \frac{2}{2x - 1}$
f. $3x^2 - x - 4$ **31.** $(4x + 3)$ in./min

Chapter 10 Test   (page 673)

1. $2x^3y^4$ **2.** 64   **3.** y^6 **4.** $32x^{21}$ **5.** 3   **6.** $\frac{2}{y^3}$ **7.** y^3
8. $\frac{64a^3}{b^3}$ **9.** $1,000y^{12}$ in.^{3} **10.** $\frac{1}{4^2}, \frac{1}{16}$ **11.** 6.25×10^{18}

12. 0.000093   **13.** binomial   **14.** 5th degree   **15.** 0
16. $-3x^2y^2$ **17.** $-7x + 2y$ **18.** $-x^2 - 5x + 4$ **19.** $-4x^5y$
20. $3y^4 - 6y^3 + 9y^2$ **21.** $x^2 - 81$ **22.** $9y^2 - 24y + 16$
23. $6x^2 - 7x - 20$ **24.** $2x^3 - 7x^2 + 14x - 12$ **25.** $\frac{1}{2}$ **26.** $\frac{y}{2x}$
27. $\frac{a}{4b} - \frac{b}{2a}$ **28.** $x - 2$ **30.** $(x - 5)$ ft

Cumulative Review Exercises   (page 675)

1. The negative savings rate means that Americans spent more than they earned that month.   **2.** $\frac{5}{8}$ **3.** $\frac{22}{35}$ **4.** irrational   **5.** $250 - x$
6. 19, 16   **7.** 5   **8.** $37y$ **9.** $-2x + 2y$ **10.** $x - 5$ **11.** x^2y^3
12. $4x^2$ **13.** 13   **14.** 41   **15.** $h = \frac{2A}{b + B}$ **16.** $x = \frac{y - b}{m}$
17. -9 **18.** 1   **19.** 4   **20.** -2
21. $x < -14$ $(-\infty, -14)$

22. $-5 < x \le -1$ $(-5, -1]$

23.
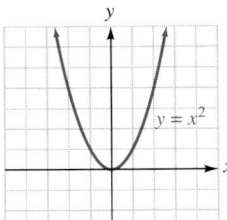

24.
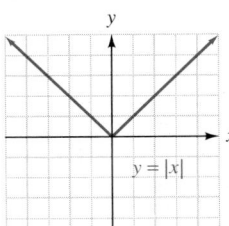

25.
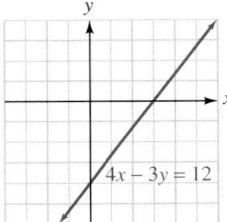

26.
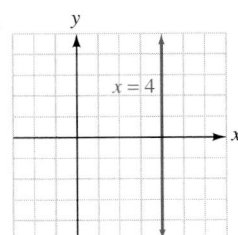

27. $\frac{1}{2}$ **28.** 0   **29.** -4 **30.** $\frac{2}{3}$ **31.** $y = \frac{2}{3}x + 5$
32. $3x - 4y = -22$ **33.** $y = 4$ **34.** $x = 2$ **35.** perpendicular
36. parallel   **37.** yes   **38.** no   **39.** -3 **40.** 15   **41.** 5
42. -2.5 **43.** D: all real numbers, R: real numbers greater than or equal to 0   **44.** no   **45.** The temperature is rising at a rate of 3°/hr; the temperature is falling at a rate of -3°/hr.

46. -9 **47.** y^{14} **48.** x^{14} **49.** x^2 **50.** a^7 **51.** $\frac{1}{x^5}$ **52.** $\frac{1}{16y^4}$

53. $\frac{1}{x^8}$ **54.** $-x^{15}$ **55.** 6.15×10^5 **56.** 1.3×10^{-6}

57. 0.000525   **58.** 2,770   **59.** 2   **60.** 5   **61.** 1.5 in.   **62. a.** 3
b. $2x^2$ **c.** 2   **d.** 1   **e.** 2   **f.** 9   **63.** $x^2 + 4x - 14$
64. $4x^2 - x - 1$ **65.** $-35x^5 + 10x^4 + 10x^2$ **66.** $-12x^5y^5$
67. $6x^2 + 10x - 56$ **68.** $15x^2 - 2xy - 8y^2$ **69.** $9x^2 + 6x + 1$
70. $x^3 - 8$ **71.** $3x - 4$ **72.** $2x + 1$

Study Set   Section 11.1   (page 685)

1. prime   **3.** largest   **5.** factoring   **7.** The 0 in the first line should be 1.   **9.** The GCF is $6a^2$, not $6a$.   **11.** factoring out the GCF   **13.** $3x$ **15.** Find $(j + 2)(3j^2 + 2)$. The result, when written in descending powers of j, should be $3j^3 + 6j^2 + 2j + 4$.
19. 1   **21.** $2^2 \cdot 3$ **23.** $3 \cdot 5$ **25.** $2^3 \cdot 5$ **27.** $2 \cdot 7^2$
29. $3^2 \cdot 5^2$ **31.** $2^5 \cdot 3^2$ **33.** 4   **35.** 4, x **37.** $3(x + 2)$
39. $6(2x^2 - x - 4)$ **41.** $t^2(t + 2)$ **43.** $a^2(a - 1)$
45. $8xy^2(3xy + 1)$ **47.** $6uvw^2(2w - 3v)$ **49.** $3(x + y - 2z)$
51. $a(b + c - d)$ **53.** $3r(4r - s + 3rs^2)$ **55.** $\pi(R^2 - ab)$
57. $(x + 2)(3 - x)$ **59.** $(14 + r)(h^2 + 1)$ **61.** $-(a + b)$
63. $-(2x - 5y)$ **65.** $-(3m + 4n - 1)$
67. $-(3ab + 5ac - 9bc)$ **69.** $-3x(x + 2)$ **71.** $-4a^2b^2(b - 3a)$
73. $-2ab^2c(2ac - 7a + 5c)$ **75.** $(x + y)(2 + a)$
77. $(r + s)(7 - k)$ **79.** $(r + s)(x + y)$ **81.** $(2x + 3)(a + b)$
83. $(b + c)(2a + 3)$ **85.** $(3x - 1)(2x - 5)$
87. $(3p + q)(3m - n)$ **89.** $(2x + y)(y - 1)$
91. $(2z^3 + 3)(4z^2 - 5)$ **93.** $x^2(a + b)(x + 2y)$
95. $4a(b + 3)(a - 2)$ **97.** $y(x^2 - y)(x - 1)$ **99. a.** $12x^3$ in.^{2}
b. $20x^2$ in.^{2} **c.** $4x^2(3x - 5)$ in.^{2} **101. a.** $4r$ in.; $16r^2$ in.^{2}

b. $4\pi r^2$ in.^{2} **c.** $16r^2 - 4\pi r^2 = 4r^2(4 - \pi)$ in.^{2} **107.** $\frac{y^3}{8}$

109. yes

Study Set   Section 11.2   (page 693)

1. trinomial, binomial   **3.** factors   **5.** leading, coefficient, term
7. $4 \cdot 1, 2 \cdot 2, -4(-1), -2(-2)$ **9.** common   **11.** $-6, -3$
13. $9, 6, -9, -6$ **15. a.** 1   **b.** -15; -5 and 3
c. -2; -5 and 3   **17.** The sum of two negative factors of 10 could not be 7.   **19. a.** They are both positive or they are both negative.   **b.** One will be positive, the other negative.
23. $+, +$ **25.** $-, -$ **27.** $+, -$ **29.** $(z + 11)(z + 1)$
31. $(m - 3)(m - 2)$ **33.** $(a - 5)(a + 1)$ **35.** $(x + 8)(x - 3)$
37. $(a - 13)(a + 3)$ **39.** prime   **41.** $(s + 13)(s - 2)$
43. prime   **45.** $(m - 4)(m + 3)$ **47.** $(x + 2y)(x + 2y)$
49. $(m + 5n)(m - 2n)$ **51.** $(a - 6b)(a + 2b)$ **53.** prime
55. $-(x + 5)(x + 2)$ **57.** $-(t + 17)(t - 2)$
59. $-(r - 10)(r - 4)$ **61.** $-(a + 3b)(a + b)$
63. $-(x - 7y)(x + y)$ **65.** $(x - 4)(x - 1)$ **67.** $(y + 9)(y + 1)$
69. $-(r - 2)(r + 1)$ **71.** $(r + 3x)(r + x)$ **73.** $(a - 2b)(a - b)$
75. $2(x + 3)(x + 2)$ **77.** $-5(a - 3)(a - 2)$
79. $3(z - 4)(z - 1)$ **81.** $4y(x + 6)(x - 3)$
83. $-4x(x + 3y)(x - 2y)$ **85.** $(x + 9)$ in., $(x + 3)$ in., x in.
93.   8   **95.** -3

Study Set   Section 11.3   (page 701)

1. leading, last   **3.** outer, inner   **5.** middle   **7.** sum   **11.** positive, negative, negative   **13.** negative, different   **15.** 24, -11
17. a. $x^2 + 2x + 3$ (answers may vary)   **b.** $2x^2 + 2x + 3$ (answers may vary)   **19.** $+, +$ **21.** $-, -$ **23.** $-, +$
27. $(2x - 1)(x - 1)$ **29.** $(3a + 1)(a + 4)$ **31.** $(z + 3)(4z + 1)$
33. $(3y + 2)(2y + 1)$ **35.** $(3x - 2)(2x - 1)$
37. $(3a + 2)(a - 2)$ **39.** $(2x + 1)(x - 2)$ **41.** prime

43. $(5y + 1)(2y - 1)$ **45.** $(3y - 2)(4y + 1)$
47. $-(5t + 3)(t + 2)$ **49.** $-(8m - 3)(2m - 1)$
51. $(2a - b)(2a - b)$ **53.** $(3r + 2s)(2r - s)$
55. $(2x + 3y)(2x + y)$ **57.** $(4a - 3b)(a - 3b)$
59. $(3x + 2)(x - 5)$ **61.** $(2a - 5)(4a - 3)$
63. $(4y - 3)(3y - 4)$ **65.** prime   **67.** $(2a + 3b)(a + b)$
69. $(3p - q)(2p + q)$ **71.** $2(2x - 1)(x + 3)$
73. $-y(y + 12)(y + 1)$ **75.** $3x(2x + 1)(x - 3)$
77. $3r^3(5r - 2)(2r + 5)$ **79.** $-2mn(4m + 3n)(2m + n)$
81. $-2uv^3(7u - 3v)(2u - v)$ **83.** $(2x + 11)$ in., $(2x - 1)$ in.;
12 in.   **89.** x^{14} **91.** 8

Study Set   Section 11.4   (page 709)

1. difference   **3.** cubes, difference   **5. a.** $5x$ **b.** 3   **c.** $5x, 3$
7. First, Last, First, Last   **9. a.** $6x$ **b.** $10x^2$ **c.** $3m$
d. a^2 **11.** 1, 4, 9, 16, 25, 36, 49, 64, 81, 100   **13. a.** 7 is not
a perfect square.   **b.** The sign of the last term must be positive.
c. The middle term is not twice the product of $5r$ and 4.
15. $36x^2 - 25y^2$ **17.** $9a^2 - 30ab + 25b^2$ **19.** $(x + 8)^2$
21. 3   **23.** $+$ **25.** $(x + 3)^2$ **27.** $(y - 4)^2$ **29.** $(t + 10)^2$
31. $(u - 9)^2$ **33.** $(2x + 3)^2$ **35.** $(6x + 1)^2$ **37.** $(a + b)^2$
39. $(4x - y)^2$ **41.** 7, 7   **43.** t, w **45.** $(x + 4)(x - 4)$
47. $(2y + 1)(2y - 1)$ **49.** $(3x + y)(3x - y)$
51. $(4a + 5b)(4a - 5b)$ **53.** prime   **55.** $(a^2 + 12b)(a^2 - 12b)$
57. $(tz + 8)(tz - 8)$ **59.** $8(x + 2y)(x - 2y)$
61. $7(a + 1)(a - 1)$ **63.** $6x^2(x + y)(x - y)$
65. $(x^2 + 9)(x + 3)(x - 3)$ **67.** $(a^2 + 4)(a + 2)(a - 2)$
69. $(9r^2 + 16s^2)(3r + 4s)(3r - 4s)$ **71.** $2a$ **73.** $b + 3$
75. $(y + 1)(y^2 - y + 1)$ **77.** $(a - 3)(a^2 + 3a + 9)$
79. $(2 + x)(4 - 2x + x^2)$ **81.** $(s - t)(s^2 + st + t^2)$
83. $(a + 2b)(a^2 - 2ab + 4b^2)$ **85.** $(4x - 3)(16x^2 + 12x + 9)$
87. $(a^2 - b)(a^4 + a^2b + b^2)$ **89.** $(x^3 + y^2)(x^6 - x^3y^2 + y^4)$
91. $2(x + 3)(x^2 - 3x + 9)$ **93.** $-(x - 6)(x^2 + 6x + 36)$
95. $8x(2m - n)(4m^2 + 2mn + n^2)$
97. $xy(x + 6y)(x^2 - 6xy + 36y^2)$
99. $3rs^2(3r - 2s)(9r^2 + 6rs + 4s^2)$ **101.** $(p + q)^2$
103. $0.5g(t_1 + t_2)(t_1 - t_2)$ **109.** $\frac{x}{y} + \frac{2y}{x} - 3$ **111.** $3a + 2$
113. $(x - a)(a + b)(a - b)$ **115.** $7p^4q^2(10q - 5 + 7p)$
117. $2a(b + 6)(b - 2)$ **119.** $-4p^2q^3(2pq^4 + 1)$
121. $5(2m + 5)^2$ **123.** prime   **125.** $-2x^2(x - 4)(x^2 + 4x + 16)$
127. $2t^2(3t - 5)(t + 4)$ **129.** $(y^2 - 2)(x + 1)(x - 1)$
131. $(2p^2 - 3q^2)(4p^4 + 6p^2q^2 + 9q^4)$
133. $(5p - 4y)(25p^2 + 20py + 16y^2)$
135. $-x^2y^2z(16x^2 - 24x^3yz^3 + 15yz^6)$
137. $(9p^2 + 4q^2)(3p + 2q)(3p - 2q)$ **139.** prime
141. $2(3x + 5y^2)(9x^2 - 15xy^2 + 25y^4)$ **143.** prime
145. $(7p + 2q)^2$

Study Set   Section 11.5   (page 718)

1. quadratic   **3.** zero, 0, 0   **5.** zero   **7. a.** quadratic
b. linear   **c.** linear   **d.** quadratic   **9. a.** -16
b. $(x - 2)(x + 8)$ **c.** $2, -8$ **11. a.** Add 6 to both sides.
b. Distribute the multiplication by x.   **15.** $2, -3$ **17.** $\frac{5}{2}, -6$
19. $1, -2, 3$ **21.** $0, 3$ **23.** $0, \frac{5}{2}$ **25.** $0, 7$ **27.** $0, -\frac{8}{3}$ **29.** $0, 2$
31. $-5, 5$ **33.** $-\frac{1}{2}, \frac{1}{2}$ **35.** $-\frac{2}{3}, \frac{2}{3}$ **37.** $-10, 10$ **39.** $-\frac{9}{2}, \frac{9}{2}$
41. $12, 1$ **43.** $-3, 7$ **45.** $8, 1$ **47.** $-3, -5$ **49.** $-4, 2$
51. $0, -1, -2$ **53.** $0, 9, -3$ **55.** $1, -2, -3$ **57.** $\frac{1}{2}, 2$ **59.** $\frac{1}{5}, 1$
61. $-\frac{1}{2}, -\frac{1}{2}$ **63.** $\frac{2}{3}, -\frac{1}{5}$ **65.** $-\frac{5}{2}, 4$ **67.** $\frac{1}{8}, 1$ **69.** $0, -3, -\frac{1}{3}$
71. $0, -3, -3$ **73.** 9 sec   **75.** $\frac{3}{2} = 1.5$ sec   **77.** 2 sec   **79.** 8
81. 4 m by 9 m   **83.** 5 m, 12 m, 13 m   **85.** 3 mm, 4 mm, 5 mm

87. $h = 5$ ft, $b = 12$ ft   **89.** 5 in.   **91.** 4 m
95. 15 min $\leq t < 30$ min   **97.** 675 cm^2

Key Concept   (page 723)

1. Factor $3x + 27$; $3(x + 9)$ **3.** $-3(a - 3)(a - 4)$
5. $(r + s)(t + 2)$ **7.** $(3t - 5)(2t - 3)$ **9.** $2r(r + 5)(r - 5)$

Chapter Review   (page 725)

1. a. $5 \cdot 7$ **b.** $3^2 \cdot 5$ **c.** $2^5 \cdot 3$ **d.** $3^2 \cdot 11$ **e.** $2 \cdot 5^2 \cdot 41$
f. 2^{12} **2. a.** $3(x + 3y)$ **b.** $5a(x^2 + 3)$ **c.** $7s(s + 2)$
d. $\pi a(b - c)$ **e.** $2x(x^2 + 2x - 4)$ **f.** $xyz(x + y + 1)$
g. $-5ab(b - 2a + 3)$ **h.** $(x - 2)(4 - x)$ **3. a.** $-(a + 7)$
b. $-(4t^2 - 3t + 1)$ **4. a.** $(c + d)(2 + a)$ **b.** $(y + 3)(3x - 2)$
c. $(2a^2 - 1)(a + 1)$ **d.** $4m(n + 3)(m - 2)$ **5.** $3, -1$;
$7, 5, -7, -5$ **6. a.** $(x + 6)(x - 4)$ **b.** $(x - 6)(x + 2)$
c. $(n - 5)(n - 2)$ **d.** prime   **e.** $-(y - 5)(y - 4)$
f. $(y + 9)(y + 1)$ **g.** $(c + 5d)(c - 2d)$
h. $(m - 2n)(m - n)$ **7.** Multiply to see if
$(x - 4)(x + 5) = x^2 + x - 20$.   **8. a.** $5(a + 10)(a - 1)$
b. $-4x(x + 3y)(x - 2y)$ **9. a.** $(2x + 1)(x - 3)$
b. $(2y + 5)(5y - 2)$ **c.** $-(3x + 1)(x - 5)$
d. $3p(2p + 1)(p - 2)$ **e.** $(4b - c)(b - 4c)$ **f.** prime
10. $(4x + 1)$ in., $(3x - 1)$ in.   **11. a.** $(x + 5)^2$ **b.** $(3y - 4)^2$
c. $-(z - 1)^2$ **d.** $(5a + 2b)^2$ **12. a.** $(x + 3)(x - 3)$
b. $(7t + 5y)(7t - 5y)$ **c.** $(xy + 20)(xy - 20)$
d. $8a(t + 2)(t - 2)$ **e.** $(c^2 + 16)(c + 4)(c - 4)$ **f.** prime
13. a. $(h + 1)(h^2 - h + 1)$ **b.** $(5p + q)(25p^2 - 5pq + q^2)$
c. $(x - 3)(x^2 + 3x + 9)$ **d.** $2x^2(2x - 3y)(4x^2 + 6xy + 9y^2)$
14. a. $2y^2(3y - 5)(y + 4)$ **b.** $(t + u^2)(s^2 + v)$
c. $(j^2 + 4)(j + 2)(j - 2)$ **d.** $-3(j + 2k)(j^2 - 2jk + 4k^2)$
e. $3(2w - 3)^2$ **f.** prime   **15. a.** $0, -2$ **b.** $0, 6$ **c.** $-3, 3$
d. $3, 4$ **e.** $-2, -2$ **f.** $6, -4$ **g.** $\frac{1}{3}, 1$ **h.** $0, -1, 2$
16. 15 m   **17.** 3 ft by 9 ft   **18.** 5 m

Chapter 11 Test   (page 729)

1. $2^2 \cdot 7^2$ **2.** $3 \cdot 37$ **3.** $4(x + 4)$ **4.** $5ab(6ab^2 - 4a^2b + c)$
5. $(q + 9)(q - 9)$ **6.** prime   **7.** $(4x^2 + 9)(2x + 3)(2x - 3)$
8. $(x + 3)(x + 1)$ **9.** $-(x - 11)(x + 2)$ **10.** $(a - b)(9 + x)$
11. $(2a - 3)(a + 4)$ **12.** $2(3x - 5y)^2$ **13.** $(x + 2)(x^2 - 2x + 4)$
14. $2(a - 3)(a^2 + 3a + 9)$ **15.** $r^2(4 - \pi)$ **16.** $(5x - 4)$
17. $2ab^2$ **18.** $(x - 9)(x + 6)$; Multiply the binomials:
$(x - 9)(x + 6) = x^2 + 6x - 9x - 54 = x^2 - 3x - 54$.   **19.** $-3, 2$
20. $-5, 5$ **21.** $0, \frac{1}{6}$ **22.** $-3, -3$ **23.** $\frac{1}{3}, -\frac{1}{2}$ **24.** $-1, -6$
25. 6 ft by 9 ft   **26.** A quadratic equation is an equation that
can be written in the form $ax^2 + bx + c = 0$; $x^2 - 2x + 1 = 0$.
(Answers may vary.)   **27.** 10   **28.** At least one of them is 0.

Cumulative Review Exercises   (page 731)

1. about 35 beats/min difference   **2.** $2 \cdot 5^3$ **3.** 0.992
4. a. false   **b.** true   **c.** true   **5.** $\frac{24}{25}$ **6.** -27 **7.** -39
8. -5 **9.** 3   **10.** $\frac{5}{0}$ **11.** $-13y^2 + 6$ **12.** $3y + z$ **13.** $-\frac{3}{2}$
14. -2 **15.** 9   **16.** $-\frac{55}{6}$ **17.** $t = \frac{A - P}{Pr}$
18. $x < -2$, ⟵————⟩   **19.** 15.4 in.   **20.** $I = Prt$
$\quad\quad\quad\quad\quad\quad\quad\quad -2$ **21.** $\$20x$ **22.** 4 L

23.

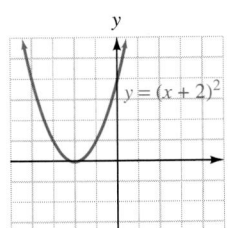

$y = (x + 2)^2$

24.

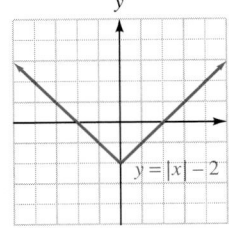

$y = |x| - 2$

25. $1; (0, -2)$ **26.** $y = 7x + 19$ **27.**

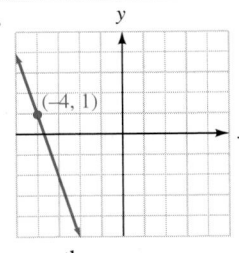

$(-4, 1)$

28.

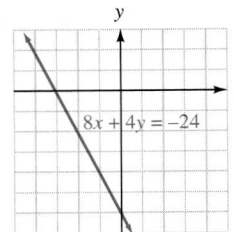

$8x + 4y = -24$

29. They are the same.
30. a. a decrease of 0.86 gal/year
b. virtually no change   **31.** 2
32. 1.7×10^6 **33.** $-4y^5$
34. $\dfrac{5x^4 y^{23}}{7}$ **35.** $\dfrac{1}{b^{14}}$ **36.** 2
37. $-2x^2 - 4x + 5$
38. $8b^5 - 8b^4$ **39.** $y^2 - 12y + 36$
40. $3x^2 + 10x - 8$ **41.** $3ab - 2a - 1$ **42.** $2x + 1$
43. $(4x + 8)$ in.   **b.** $(x^2 + 4x + 3)$ in.²   **c.** $(x^3 + 4x^2 + 3x)$ in.³
44. 3   **45.** $b^2(b - 3)$ **46.** $(u + 3)(u - 1)$ **47.** $(2x + 1)(x - 2)$
48. $(3z + 1)(3z - 1)$ **49.** $-5(a - 3)(a - 2)$ **50.** $(x + y)(a + b)$
51. $(t - 2)(t^2 + 2t + 4)$ **52.** $(2a - 3)^2$ **53.** $0, \frac{4}{3}$ **54.** $\frac{1}{2}, 2$

Study Set   Section 12.1   (page 739)

1. numerator, denominator   **3.** undefined   **5.** simplify   **7. a.** 0
b. 6   **c.** -6 **9.** $x + 1$ **11.** x is not a common factor of the numerator and denominator. It cannot be divided out.   **15.** 4
17. $-\dfrac{3}{7}$ **19.** 0   **21.** 2   **23.** none   **25.** $\dfrac{1}{2}$ **27.** $-6, 6$
29. $-2, 1$ **31.** $\dfrac{4}{5}$ **33.** $\dfrac{1}{3}$ **35.** $-\dfrac{3}{4}$ **37.** $\dfrac{5}{a}$ **39.** $\dfrac{2}{z}$ **41.** $\dfrac{a}{3}$
43. $\dfrac{2}{3}$ **45.** $\dfrac{3}{2}$ **47.** in lowest terms   **49.** $\dfrac{3x}{y}$ **51.** $\dfrac{2x + 1}{y}$ **53.** $\dfrac{1}{3}$
55. -1 **57.** -6 **59.** $\dfrac{-3(x + 2)}{x + 1}$ **61.** $\dfrac{x + 1}{x - 1}$ **63.** $\dfrac{x - 5}{x + 2}$
65. $\dfrac{2x}{x - 2}$ **67.** $-\dfrac{1}{a + 1}$ **69.** $\dfrac{x + 2}{x^2}$ **71.** $\dfrac{x - 4}{x + 4}$
73. $\dfrac{2(x + 2)}{x - 1}$ **75.** in lowest terms   **77.** $\dfrac{3 - x}{3 + x}$ or $-\dfrac{x - 3}{x + 3}$
79. $\frac{4}{3}$ **81.** $x + 3$ **83.** in lowest terms   **85.** $\dfrac{2}{3}$ **87.** $\dfrac{3x}{5y}$ **89.** -1
91. $\dfrac{x + 2}{x - 2}$ **93.** 1,000; 800; about 667; 500; about 333; about 222
99. $(a + b) + c = a + (b + c)$ **101.** One of them is zero.
103. $\frac{5}{3}$

Study Set   Section 12.2   (page 748)

1. numerator   **3.** numerators, denominators   **5.** 1   **7.** divisor,
multiply   **11.** $\dfrac{3}{2}$ **13.** y **15.** $\dfrac{14}{9}$ **17.** $x^2 y^2$ **19.** $2xy^2$

21. $-3y^2$ **23.** $\dfrac{b^3 c}{a^4}$ **25.** $\dfrac{r^3 t^4}{s}$ **27.** $\dfrac{(z + 7)(z + 2)}{7z}$ **29.** x
31. $\dfrac{x}{5}$ **33.** 5   **35.** 6   **37.** $x + 1$ **39.** $2y + 16$ **41.** $x + 5$
43. $10h - 30$ **45.** $x + 2$ **47.** $\dfrac{3}{2x}$ **49.** $x - 2$ **51.** x
53. $\dfrac{(x - 2)^2}{x}$ **55.** $\dfrac{(m - 2)(m - 3)}{2(m + 2)}$ **57.** $\dfrac{x + 1}{2(x - 2)}$ **59.** $\dfrac{3}{2y}$
61. 3   **63.** $\dfrac{6}{y}$ **65.** 6   **67.** $\dfrac{2x}{3}$ **69.** $\dfrac{2}{y}$ **71.** $\dfrac{2}{3x}$ **73.** $\dfrac{2(z - 2)}{z}$
75. $\dfrac{5z(z - 7)}{z + 2}$ **77.** $\dfrac{x + 2}{3}$ **79.** 1   **81.** $\dfrac{2(x - 7)}{x + 9}$ **83.** $d + 5$
85. $\dfrac{9}{2x}$ **87.** $\dfrac{x}{36}$ **89.** 2   **91.** $\dfrac{2x(1 - x)}{5(x - 2)}$ **93.** $\dfrac{y^2}{3}$ **95.** $\dfrac{x + 2}{x - 2}$
97. $\dfrac{12x^2 + 12x + 3}{2}$ in.²   **103.** $-6x^5 y^6$ **105.** $\dfrac{1}{81y^4}$
107. $4y^3 + 16y^2 - 8y + 8$

Study Set   Section 12.3   (page 757)

1. LCD   **3.** numerators, common denominator   **7.** $\dfrac{x}{3}$ **9.** $\dfrac{4x}{y}$
11. $\dfrac{2}{y}$ **13.** $\dfrac{y + 3}{5z}$ **15.** 9   **17.** $\dfrac{1}{a + 2}$ **19.** $-\dfrac{y}{8}$ **21.** $\dfrac{x}{y}$ **23.** $\dfrac{y}{x}$
25. $\dfrac{1}{y}$ **27.** $\dfrac{1}{2}$ **29.** $\dfrac{2}{c + d}$ **31.** $\dfrac{4x}{3}$ **33.** 0   **35.** $\dfrac{4x - 2y}{y + 2}$
37. $\dfrac{2x + 10}{x - 2}$ **39.** $\dfrac{125x}{20x}$ **41.** $\dfrac{8xy}{x^2 y}$ **43.** $\dfrac{3x(x + 1)}{(x + 1)^2}$
45. $\dfrac{2y(x + 1)}{x^2 + x}$ **47.** $\dfrac{z(z + 1)}{z^2 - 1}$ **49.** $\dfrac{2(x + 2)}{x^2 + 3x + 2}$ **51.** $6x$
53. $18xy^2$ **55.** $x^2 - 1$ **57.** $x^2 + 6x$
59. $(x + 1)(x + 5)(x - 5)$ **61.** $\dfrac{5y}{9}$ **63.** $\dfrac{53x}{42}$ **65.** $\dfrac{4xy + 6x}{3y}$
67. $\dfrac{2 - 3x^2}{x}$ **69.** $\dfrac{y^2 + 7y + 6}{15y^2}$ **71.** $\dfrac{x^2 + 4x + 1}{x^2 y}$
73. $\dfrac{2x^2 - 1}{x(x + 1)}$ **75.** $\dfrac{2xy + x - y}{xy}$ **77.** $\dfrac{x + 2}{x - 2}$
79. $\dfrac{2x^2 + 2}{(x - 1)(x + 1)}$ **81.** $-\dfrac{2}{a - 4}$ **83.** $\dfrac{2t + 2}{t - 7}$ **85.** $\dfrac{4x + 2}{x - 2}$
87. $\dfrac{b - 1}{2(b + 1)}$ **89.** $\dfrac{1}{a + 1}$ **91.** $-\dfrac{1}{2(x - 2)}$ **93.** $\dfrac{x}{x - 2}$
95. $\dfrac{5x + 3}{x + 1}$ **97.** $\dfrac{20x + 9}{6x^2}$ cm   **103.** 7^2 **105.** $2^3 \cdot 17$

Study Set   Section 12.4   (page 763)

1. complex fraction   **3.** single, divide   **7.** $\dfrac{8}{9}$ **9.** $\dfrac{3}{8}$ **11.** $\dfrac{5}{4}$
13. $\dfrac{5}{7}$ **15.** $\dfrac{x^2}{y}$ **17.** $\dfrac{5t^2}{27}$ **19.** $\dfrac{1 - 3x}{5 + 2x}$ **21.** $\dfrac{1 + x}{2 + x}$ **23.** $\dfrac{3 - x}{x - 1}$
25. $\dfrac{1}{x + 2}$ **27.** $\dfrac{1}{x + 3}$ **29.** $\dfrac{xy}{y + x}$ **31.** $\dfrac{y}{x - 2y}$ **33.** $\dfrac{x^2}{(x - 1)^2}$
35. $\dfrac{7x + 3}{-x - 3}$ **37.** $\dfrac{x - 2}{x + 3}$ **39.** -1 **41.** $\dfrac{y}{x^2}$ **43.** $\dfrac{x + 1}{1 - x}$
45. $\dfrac{1 + a^3}{a^3}$ **47.** 2   **49.** $\dfrac{y - 5}{y + 5}$ **51.** $\dfrac{3}{14}$ **53.** $\dfrac{R_1 R_2}{R_2 + R_1}$
57. t^9 **59.** $-2r^7$ **61.** $\dfrac{81}{256r^8}$ **63.** $\dfrac{r^{10}}{9}$

Study Set Section 12.5 (page 773)

1. rational equations **3.** check **5.** interest, principal, rate
7. a. yes **b.** no **9. a.** $\frac{1}{6}, \frac{1}{5}$ **b.** $\frac{11}{30}$ **11.** $\frac{1}{6}$ **13. a.** $r = \frac{I}{P}$
b. $P = \frac{I}{r}$ **17.** $\frac{3}{5} + \frac{7}{x+2} = 2$ **19.** 4 **21.** 0
23. 3 **25.** no solution; 0 is extraneous **27.** $\frac{11}{4}$ **29.** $-4, 4$
31. 1 **33.** No solution; 5 is extraneous. **35.** -1 **37.** 6
39. 1 **41.** 4 **43.** no solution; -2 is extraneous **45.** 1
47. 2 **49.** 0 **51.** $2, -5$ **53.** $-4, 3$ **55.** $-2, 1$ **57.** $3, -4$
59. $1, -9$ **61.** $a = \frac{b}{b-1}$ **63.** $r = \frac{E - IR}{I}$ **65.** $d = \frac{bc}{a}$
67. 2 **69.** 5 **71.** $\frac{2}{3}, \frac{3}{2}$ **73.** $f = \frac{d_1 d_2}{d_1 + d_2}$ **75.** $R = \frac{HB}{B - H}$
77. $2\frac{2}{9}$ hr **79.** $2\frac{6}{11}$ days **81.** $7\frac{1}{2}$ hr **83.** 4 mph
85. 7% and 8% **87.** 5 **89.** 30 **91.** 25 mph **97.** $x(x + 4)$
99. $(2x + 3)(x - 1)$ **101.** $(x^2 + 4)(x + 2)(x - 2)$

Study Set Section 12.6 (page 784)

1. ratio, rate **3.** extremes, means **5.** shape **7.** ad, bc **9.** yes
11. The ratio on the right-hand side should be $\frac{h}{48}$. **15.** triangle
17. no **19.** yes **21.** no **23.** yes **25.** 4 **27.** 6 **29.** 3
31. 9 **33.** 0 **35.** -17 **37.** $-\frac{3}{2}$ **39.** $\frac{83}{2}$ **41.** $4, -4$
43. $4, -1$ **45.** $-5, -1$ **47.** $17 **49.** $1.8 million **51.** 24
53. $7\frac{1}{2}, 1\frac{2}{3}, 5$ **55.** 510 **57.** 737 **59.** $309 **61.** 49 ft, $3\frac{1}{2}$ in.
63. 2,704,368 **65.** $3\frac{3}{4}$ in. **67.** 39 ft **69.** $46\frac{7}{8}$ ft **71.** 6,750 ft
77. 90% **79.** 480 **81.** $\frac{3}{5}$

Study Set Section 12.7 (page 791)

1. direct **3.** constant **5.** direct **7.** inverse **9. a.** yes
b. no **c.** no **d.** yes **11.** $g = kf$ **13.** $\frac{7}{2}$ **15. a.** no **b.** yes
c. no **d.** yes **17.** $n = \frac{k}{p}$ **19.** 150 **23.** 35 **25.** 6.25 **27.** 3
29. 1 **31.** 12,000 **33.** 6 **35.** 168 mi **37.** 27 mg **39.** $22\frac{1}{2}$
41. $2\frac{1}{2}$ hr **43.** 8 amps **45.** $53\frac{1}{3}$ m^3 **51.** $-1, 6$
53. $-2, -3, -4$ **55.** $0, 0, 1$ **57.** $1, -1, 2, -2$

Key Concept (page 795)

1. a. $\frac{2x}{x - 2}$ **b.** $x - 4$ **3. a.** $\frac{2x^2 - 1}{x(x + 1)}$ **b.** $\frac{x}{x}, \frac{x + 1}{x + 1}$ **5. a.** 2
b. b **7. a.** $2, 4$ **b.** $4(y - 6)$

Chapter Review (page 797)

1. $4, -4$ **2. a.** $\frac{2}{5}$ **b.** $-\frac{2}{3}$ **3. a.** $\frac{1}{2x}$ **b.** $\frac{5}{2x}$ **c.** $\frac{x}{x + 1}$
d. $a - 2$ **e.** -1 **f.** $-\frac{1}{x + 3}$ **g.** $\frac{x}{x - 1}$ **h.** in lowest terms
4. $-\frac{3}{7}$ **5.** x is not a common factor of the numerator and the
denominator. **6.** $\frac{4}{3}$ **7. a.** $\frac{3x}{y}$ **b.** 96 **c.** $\frac{x - 1}{x + 2}$ **d.** $\frac{2x}{x + 1}$
8. a. $\frac{3y}{2}$ **b.** $\frac{1}{x}$ **c.** $x + 2$ **9.** $b + 2$ **10. a.** 1 **b.** $\frac{2x + 2}{x - 7}$
c. $\frac{1}{a - 4}$ **11. a.** $4x^2$ **b.** $18xy^2$ **c.** $(x + 1)(x + 2)$ **d.** $y^2 - 25$
12. a. $\frac{x^2 + x - 1}{x(x - 1)}$ **b.** $\frac{c - 7}{7c}$ **c.** $\frac{x^2 + 4x - 4}{2x^2}$ **d.** $\frac{1}{t + 1}$
e. $\frac{x + 1}{x}$ **f.** $\frac{b + 6}{b - 1}$ **13.** $\frac{14x + 28}{(x + 6)(x - 1)}$ units, $\frac{12}{(x + 6)(x - 1)}$

square units **14. a.** $\frac{9}{4}$ **b.** $\frac{3}{2}$ **c.** $\frac{1 + y}{1 - y}$ **d.** $\frac{x(x + 3)}{2x^2 - 1}$
e. $x^2 + 3$ **f.** $\frac{1 + x^2}{1 - x^2}$ **15. a.** 3 **b.** no solution; 5 is extraneous
c. 3 **d.** $2, 4$ **e.** 0 **16.** $T_1 = \frac{T_2}{1 - E}$ **17.** $r_1 = \frac{rr_2}{r_2 - r}$
18. 3. **19.** $\frac{1}{4}$ **20.** $5\frac{5}{6}$ days **21.** 5% **22.** 5 mph **23.** 40 mph
24. a. no **b.** yes **25. a.** $\frac{9}{2}$ **b.** 0 **c.** 7 **d.** $4, -\frac{3}{2}$ **26.** 255
27. 20 ft **28.** $833.33 **29.** 600 **30.** 1.25 amps **31.** inverse
variation

Chapter 12 Test (page 801)

1. $-3, 2$ **2.** $\frac{8x}{9y}$ **3.** $\frac{x + 1}{2x + 3}$ **4.** 3 **5.** $-\frac{5y^2}{4}$ **6.** $\frac{x + 1}{3(x - 2)}$
7. $\frac{3}{5}$ **8.** $-\frac{x^2}{3}$ **9.** $x + 2$ **10.** $\frac{10x - 1}{x - 1}$ **11.** $\frac{13}{2y + 3}$
12. $\frac{2x^2 + x + 1}{x(x + 1)}$ **13.** $\frac{2a + 7}{a - 1}$ **14.** $\frac{4n - 5mn}{10m}$ **15.** $\frac{x + y}{y - x}$
16. 1 **17.** $1, 2$ **18.** $B = \frac{HR}{R - H}$ **19.** yes **20.** $\frac{2}{3}$
21. yes **22.** 8,050 ft **23.** 100 lb **24.** 200 **25.** $3\frac{15}{16}$ hr
26. 5 mph **27.** We can divide out only common factors, as in
the first expression. We can't divide out common terms, as in the
second expression. **28.** We multiply both sides of the equation by
the LCD of the rational expressions appearing in the equation. The
resulting equation is easier to solve.

Cumulative Review Exercises (page 803)

1. 36 **2.** $2,580,000,000 **3.** 77 **4.** $35x¢$ **5.** $\frac{5}{4}$ **6.** $104°F$
7. 240 ft^3 **8. a.** false **b.** false **c.** true **d.** true **9.** 3
10. $6c + 62$ **11.** $B = \frac{A - c - r}{2}$
12. $x \geq -1, [-1, \infty)$ 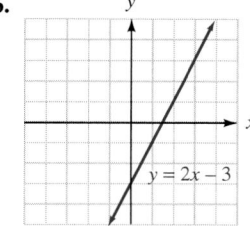 **13.** -5
14. 12 lb of the $3.20 tea and 8 lb of the $2 tea **15.** 500 mph
16. 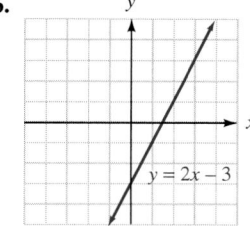 **17.**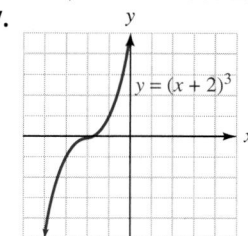
18. -1 **19.** $y = 3x + 2$
20. 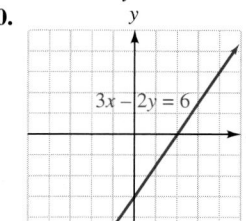 **21.**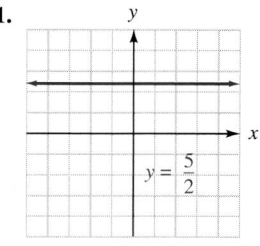
22. $\frac{8}{7}$ **23.** no **24.** 0.008 mm/m **25.** 12 **26.** -25 **27.** x^7
28. x^{25} **29.** $\frac{y^3}{8}$ **30.** $-\frac{32a^5}{b^5}$ **31.** $\frac{a^8}{b^{12}}$ **32.** $3b^2$ **33.** 2.9×10^5
34. 3 **35.** $6x^2 + x - 5$ **36.** $6x^5y^5$ **37.** $6y^2 - y - 35$
38. $-12x^4z + 4x^2z^2$ **39.** $2x + 3$ **40.** $-\frac{3r}{s^9}$

41. $10^3 \cdot 26^3$; 17,576,000 **42.** $A = \pi R^2 - \pi r^2$ **43.** $k^2 t(k - 3)$
44. $(b + c)(2a + 3)$ **45.** $2(a + 10b)(a - 10b)$
46. $(b + 5)(b^2 - 5b + 25)$ **47.** $(u - 9)^2$ **48.** $(2x - 9)(3x + 7)$
49. $-(r - 2)(r + 1)$ **50.** prime **51.** $0, -\frac{1}{5}$ **52.** $\frac{1}{3}, \frac{1}{2}$

53. 10 in., 16 in. **54.** $5, -5$ **55.** $\frac{x}{3}$ **56.** $-\dfrac{1}{x - 3}$

57. $\dfrac{x^2}{(x + 1)^2}$ **58.** $\dfrac{a}{a - 3}$ **59.** $\dfrac{17}{25}$ **60.** 3 **61.** \$2,450

62. 34.8 ft **63.** $14\frac{2}{5}$ **64.** One variable is a constant multiple of

the reciprocal of the other; $y = \dfrac{k}{x}$.

Study Set   Section 13.1   (page 813)

1. system **3.** independent **5.** inconsistent **7.** true
9. false **11.** true **13. a.** \$2,000 **b.** \$1,200 **c.** lose \$800
15. 1 solution; consistent **17.** The method is not accurate enough
to find a solution such as $\left(\frac{2}{5}, -\frac{1}{3}\right)$. **21.** yes **23.** yes **25.** no
27. no **29.** no **31.** yes **33.**

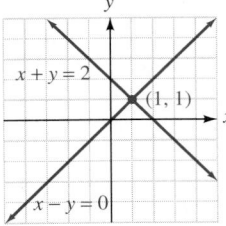

35. 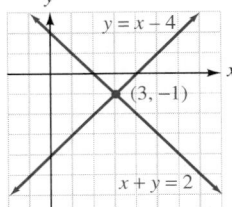 **37.**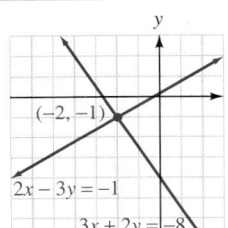

39. 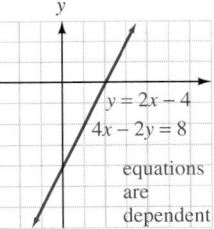 **41.**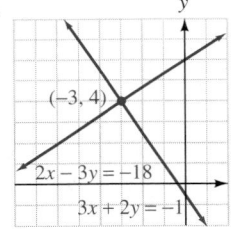

43. 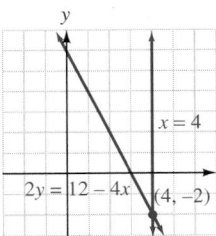 **45.**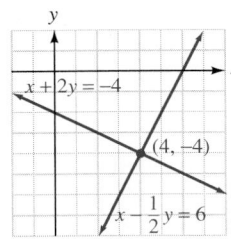

47. 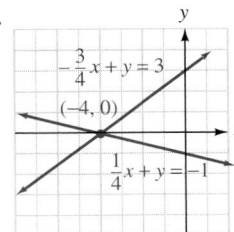 **49.**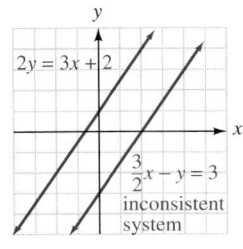

51.

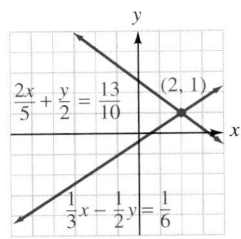

53. $(1, 3)$ **55.** no solution **57. a.** Donors outnumbered those
needing a transplant. **b.** 1994; 4,100 **59. a.** Houston, New
Orleans, St. Augustine **b.** St. Louis, Memphis, New Orleans
c. New Orleans **61.** yes, $(2, 2)$ **65.** -3; $(0, 4)$ **67.** -21
69. $x = 0$ **71.** $(5, 2)$

Study Set   Section 13.2   (page 822)

1. y, terms **3.** distributive **5.** infinitely **7. a.** 2 **b.** 1
9. a. $x = 2y - 10$ **b.** $y = \frac{x}{2} + 5$ **c.** x; it involved only
one step. **11. a.** Parentheses must be written around $x - 4$
in line 2. **b.** $\frac{13}{3}$ **13. a.** The coordinates of the intersection
point are not integers.
b. $\left(\frac{2}{3}, \frac{1}{3}\right)$ **17.** $(2, 4)$
19. $(3, 0)$
21. $(-3, -1)$
23. inconsistent
system **25.** $(-2, 3)$
27. $(3, 2)$
29. $(3, -2)$
31. $\left(\frac{2}{3}, -\frac{1}{3}\right)$

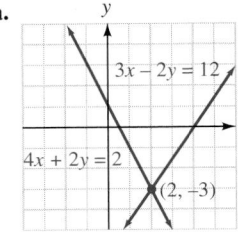

33. $(-1, -1)$ **35.** dependent equations **37.** $(4, -2)$
39. $\left(\frac{1}{2}, \frac{1}{3}\right)$ **41.** $(1, 4)$ **43.** inconsistent system **45.** $(4, 2)$
47. $(-6, 4)$ **49.** $\left(\frac{1}{5}, 4\right)$ **51.** $(5, 5)$ **53.** melon, because it's
the same price as hash browns **59.** $-\frac{5}{8}$ **61.** $(0, -6)$
63. no **65.** $s = kt$

Study Set   Section 13.3   (page 829)

1. coefficient **3.** general **5.** The second equation should be
written in general form: $3x - 2y = 10$. **7.** Multiply both sides
by 15. **9. a.**   **b.** $(2, -3)$ **c.** $(2, -3)$
13. $(-2, 3)$
15. $(-1, 1)$
17. $(-3, 4)$
19. $(0, 8)$ **21.** $(2, 3)$
23. $(3, -2)$ **25.** $(2, 7)$
27. inconsistent system
29. $\left(1, -\frac{5}{2}\right)$
31. $\left(\frac{10}{3}, \frac{10}{3}\right)$ **33.** $(5, -6)$
35. $(-1, 2)$ **37.** dependent equations **39.** $(4, 0)$ **41.** $(-1, 2)$
43. $(0, 1)$ **45.** $(-2, 3)$ **47.** $(2, 2)$ **53.** 4 **55.** 0 **57.** 7.5 ft^2
59. $x - 10$

Study Set   Section 13.4   (page 837)

1. variable **3.** system **5.** $x - c, x + c$ **7. a.** \$y **b.** 5%
c. $0.05x, 0.11y$ **9. a.** two **b.** graphing, substitution, addition
11. $A = lw$ **13.** $d = rt$ **15.** $2l + 2w = 90$ **17.** 32, 64
19. 8, 5 **21.** 22 ft, 29 ft **23.** president: \$400,000; vice president:
\$181,000 **25.** \$30, \$10 **27.** 40 **29.** 65°, 115° **31.** 6,720 ft^2
33. a. 400 tires

b.

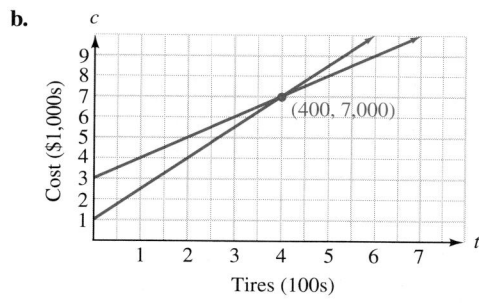

c. the second mold   **35.** nursing: $2,000; business: $3,000
37. 5 mph   **39.** 50 mph   **41.** 4 L 6% salt water, 12 L 2%
salt water   **43.** 32 lb peanuts, 16 lb cashews   **45.** $640
49.

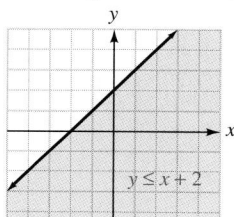

51.

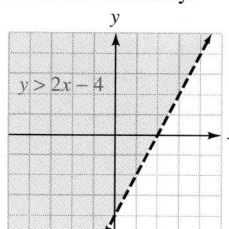

53. −2, 2   **55.** 2, 2

Study Set   Section 13.5   (page 846)

1. inequality   **3.** boundary   **5. a.** yes   **b.** no   **c.** yes   **d.** no
7. a. no   **b.** yes   **9. a.** no   **b.** yes   **c.** yes   **d.** no
11. The test point must be on one side of the boundary.
13.

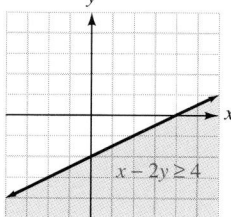

15.

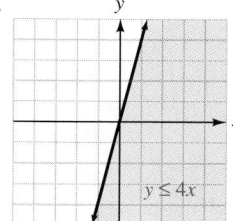

17.

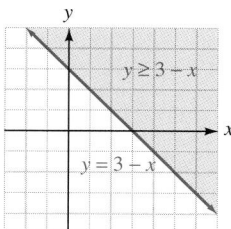

19.

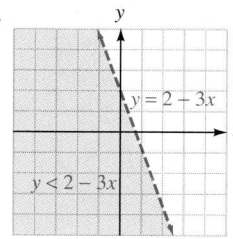

21.

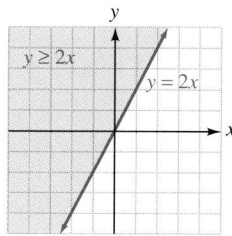

23.

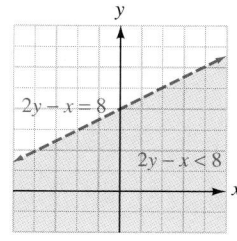

25.

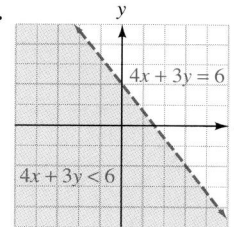

27.

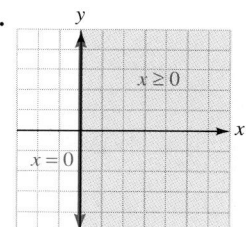

29.

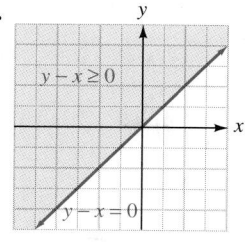

31.

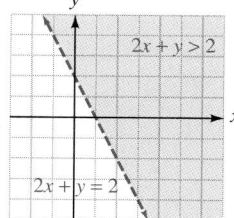

33.

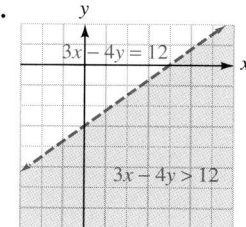

35.

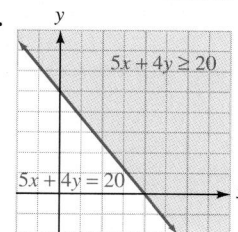

37.

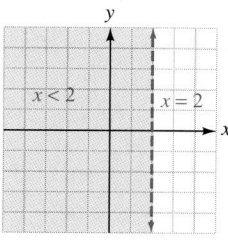

39.

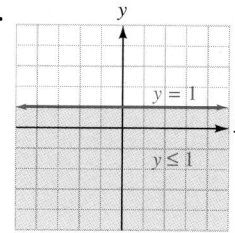

41.

43.

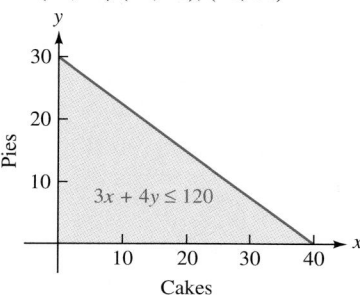

45.

47. (10, 10) (20, 10), (10, 20)

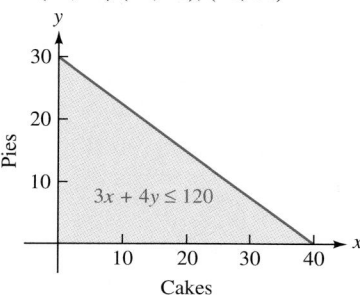

49. (50, 50), (30, 40), (40, 40)

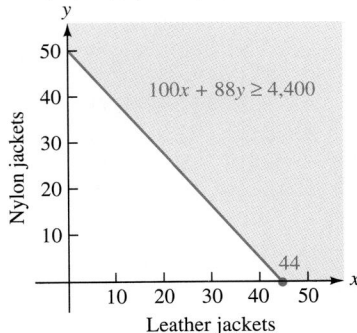

51. (80, 40), (80, 80), (120, 40)

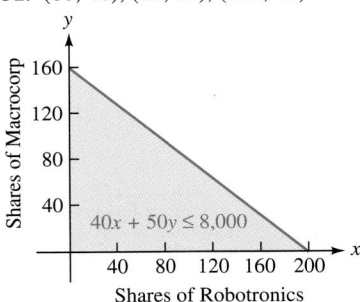

55. 7   **57.** $(x + 3)(x^2 - 3x + 9)$ **59.** $d = rt$ **61.** $t = \frac{A - P}{Pr}$

Study Set   Section 13.6   (page 854)

1. inequalities   **3.** doubly shaded   **5. a.** true   **b.** false
c. false   **d.** true   **e.** true   **f.** true   **7. a.** yes   **b.** no
c. no   **9.** *ABC*   **11.**

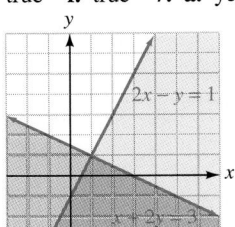

13.

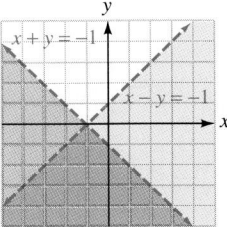

15.

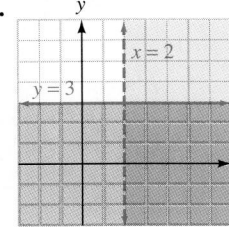

17.

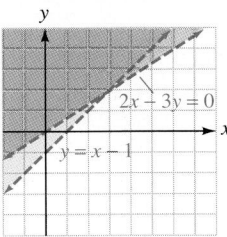

19.

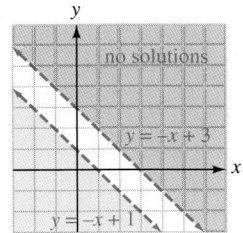

21.

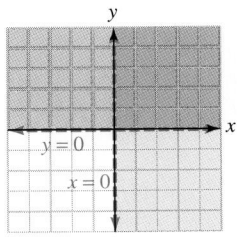

23.

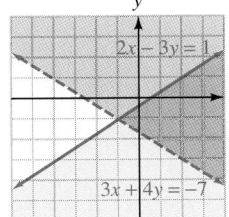

25.

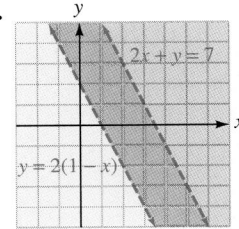

27.

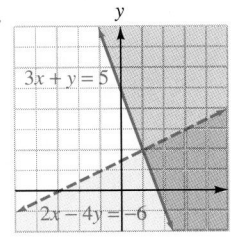

29.

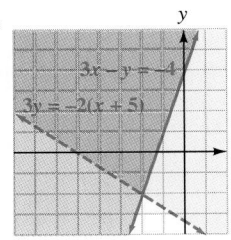

31.

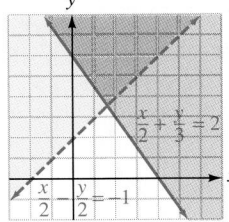

33.

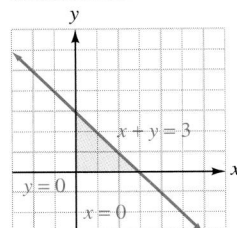

35.

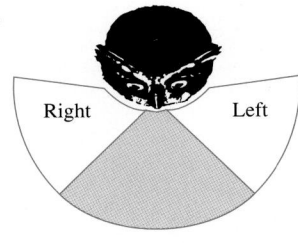

(c)

37. 1 $10 CD and 2 $15 CDs; 4 $10 CDs and 1 $15 CD

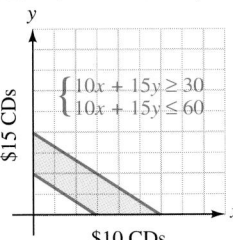

39. 2 desk chairs and 4 side chairs; 1 desk chair and 5 side chairs

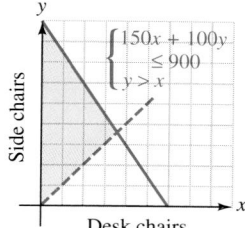

41.

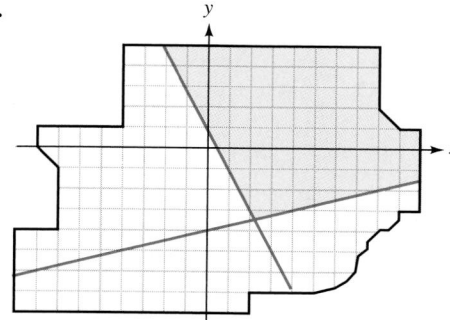

47. 128, 8   **49.** 4, −23

Key Concept   (page 859)

1. (500, 3,000)

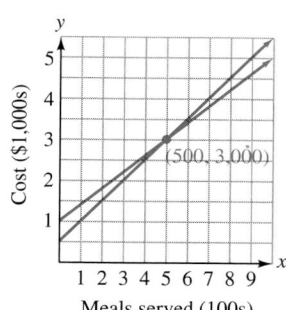

3. (−1, −1)
5. (−3, 4)

Chapter Review   (page 861)

1. a. yes   **b.** yes   **2.** 2000; 60,000 per year
3. a.

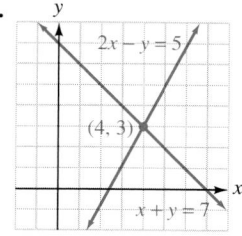

b.

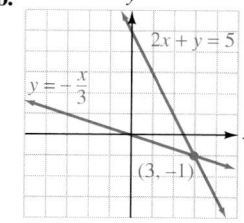

c.

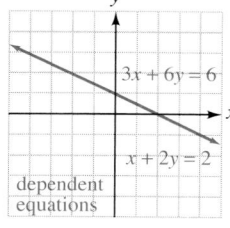

d.

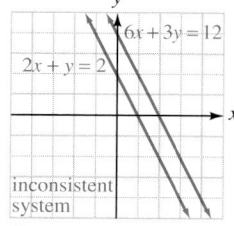

4. a. (3, 3)   **b.** (5, 0)   **c.** $\left(-\frac{1}{2}, \frac{7}{2}\right)$ **d.** (−2, 1)   **e.** dependent equations   **f.** (12, 10)   **5. a.** no solutions   **b.** two parallel lines   **c.** inconsistent system   **6. a.** (3, −5)   **b.** $\left(3, \frac{1}{2}\right)$ **c.** (−1, 7)

d. (0, 9)   **e.** dependent equations   **f.** (0, 0)   **g.** (−5, 2)   **h.** inconsistent system   **7. a.** Addition; no variables have a coefficient of 1 or −1.   **b.** Substitution; equation 1 is solved for x.
8. stroke: 160,000; heart disease: 720,000   **9.** base: 21 ft; extension: 14 ft   **10.** 10,800 yd^2 **11. a.** $y = 5x + 30,000$; $y = 10x + 20,000$ **b.** 2,000   **c.** the athlete   **12.** 12 lb worms, 18 lb bears   **13.** 3 mph   **14.** $16.40, $10.20   **15.** $750
16. $13\frac{1}{3}$ gal 40%, $6\frac{2}{3}$ gal 70%   **17. a.** yes   **b.** yes   **c.** yes
d. no   **18. a.**

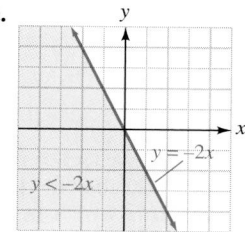

b.

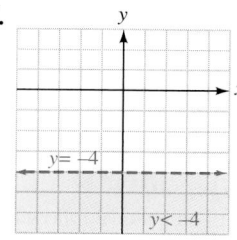

c.

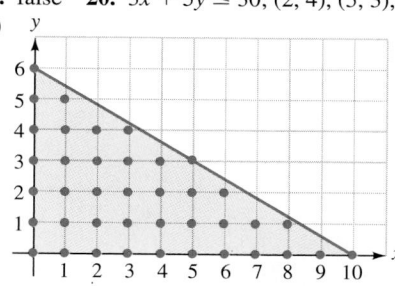

d.

19. a. true   **b.** false   **c.** false   **20.** $3x + 5y \leq 30$; (2, 4), (5, 3), (6, 2) (answers may vary)

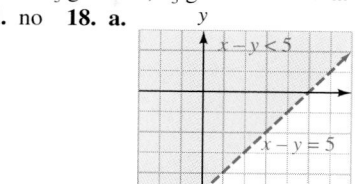

21. a.

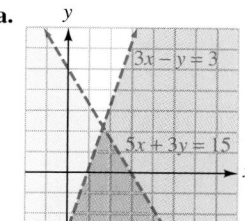

b.

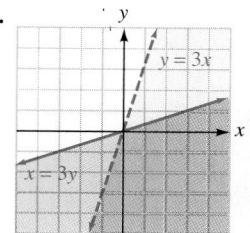

22. $10x + 20y \geq 40$, $10x + 20y \leq 60$; (3, 1): 3 shirts and 1 pair of pants; (1, 2): 1 shirt and 2 pairs of pants (answers may vary)

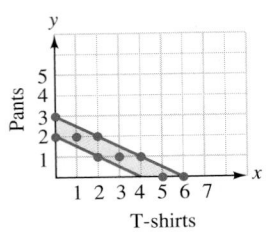

Chapter 13 Test (page 867)

1. yes **2.** no **3.** $(2, 1)$
4. The lines appear to be parallel. Since the lines do not intersect, the system does not have a solution. **5.** $(-2, -3)$
6. $(-1, -1)$ **7.** $(2, 4)$
8. $(-3, 3)$ **9.** inconsistent **10.** consistent

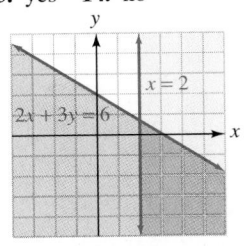

11. Addition method; the terms involving y can be eliminated easily. **12.** $4,000 **13.** yes **14.** no
15. **16.**

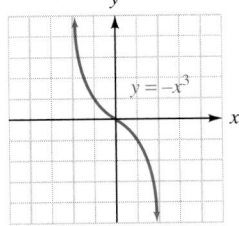

17. $(30, 3)$; if 30 items are sold, the salesperson gets paid the same by both plans, $3,000. **18.** If sales of less than 30 items are anticipated, Plan 1 is better. Otherwise, Plan 2 is more profitable.

Cumulative Review Exercises (page 869)

1. 66.1%, -2.8% **2.** $\{. . ., -3, -2, -1, 0, 1, 2, 3, . . .\}$
3. -37 **4.** 28 **5.** 1.2 ft^3 **6.** 0 **7.** -2 **8.** 15 **9.** $x < -\frac{11}{4}$

$\left(-\infty, -\frac{11}{4}\right)$ **10.** 30 **11.** mutual fund:

$25,000; bonds: $20,000 **12. a.** $P = 2l + 2w$ **b.** $A = lw$
c. $A = \pi r^2$ **d.** $d = rt$ **13.**

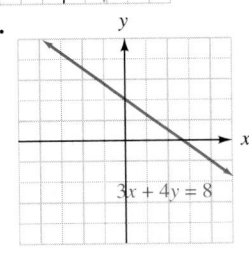

14. **15.**

16.

17. $m = 3$, $(0, -2)$; $y = 3x - 2$
18. $-\frac{4}{9}$ **19.** -70
20. 3.89×10^9 **21.** x^{31}
22. $8a^4$ **23.** $\frac{3}{16}$ **24.** 1
25. $7x - 13y$
26. $6x^4 + 8x^3 - 14x^2$
27. $c^2 + 32c + 256$
28. $2x^2 + 3x - 9$ **29.** $\frac{1}{8} - \frac{2}{x}$
30. $3x + 1$ **31.** $2^5 \cdot 3^2$
32. $x^2 - 9$ (answers may vary) **33.** $3r(4r - s + 3rs^2)$
34. $(u - 9)^2$ **35.** $(2y - 1)(y - 3)$ **36.** $(x^2 + 9)(x + 3)(x - 3)$
37. $(t - v)(t^2 + tv + v^2)$ **38.** $(x - t)(y + s)$ **39.** 0, 2
40. 3, -5 **41.** $\dfrac{x - 5}{5}$ **42.** $\dfrac{4x}{y}$ **43.** -1 **44.** $\dfrac{x^2 + 4x + 1}{x^2y}$

45. $\dfrac{3x + 1}{x + 2}$ **46.** -1 **47.** 16, 8 **48.** no
49. **50.**

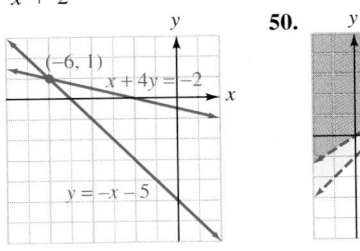

51. $(4, 1)$ **52.** noodles: 2 servings, rice: 3 servings

Study Set Section 14.1 (page 878)

1. square **3.** positive **5.** right **7.** two, 5, -5 **9.** $a^2 + b^2$
11. $\sqrt{b}$ **13. a.** Divide both sides by 2. **b.** Take the positive square root of both sides. **15.** 0, $\frac{1}{9}$, 0.4, 6, 20
17. a. $\sqrt{5} \approx 2.2$ **b.** $\sqrt{3} \approx 1.7$; $\sqrt{8} \approx 2.8$ **21.** False; $-\sqrt{9} = -3$, $\sqrt{-9}$ is not a real number. **23.** 5 **25.** -9
27. 1.1 **29.** 14 **31.** $\frac{3}{16}$ **33.** -17 **35.** -50 **37.** 60
39. 1.414 **41.** 3.317 **43.** 9.747 **45.** 20.688 **47.** -99.378
49. 4.621 **51.** 0.599 **53.** -0.915 **55.** 3.464 **57.** 1.866
59. rational; irrational; rational; imaginary **61.** 1, 2, 3, 4, 5

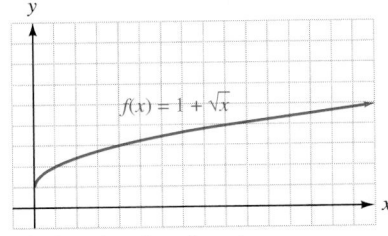

63. 0, -1, -2, -3, -4
65. 5 **67.** 8 **69.** 30
71. 117 **73.** 12 ft
75. The diagonal measurement should be $\sqrt{16^2 + 30^2} = 34$ in.
77. 127.3 ft
79. 5.8 mi **81.** no
83. 25.5 ft
85. 240 in.^{2} **87. a.** 8.5 in. **b.** 7.8 in. **93.** $6s^2 + s - 5$
95. $3x^2 + 10x - 8$

Study Set Section 14.2 (page 886)

1. cube **3.** function **5.** cube, root **7.** $(-6)^3$ **9. a.** not a real number **b.** -5 **11. a.** 1 **b.** $-\frac{1}{3}$ **c.** 5 **d.** 0.2 **e.** 10

13. index, radicand **15.** 2 **17.** 2 **19.** 0 **21.** -2 **23.** -4
25. $\frac{1}{5}$ **27.** 1 **29.** -4 **31.** 9 **33.** 10 **35.** 31.78 **37.** -0.48
39. $-1, 0, 1, 2, 3$

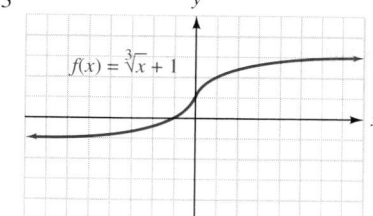
$f(x) = \sqrt[3]{x} + 1$

41. $2, 1, 0, -1, -2$
43. 2 **45.** -2
47. 1 **49.** -2
51. 3.34
53. -5.70
55. x **57.** x^3
59. x^5 **61.** $2z$
63. $-x^2 y$
65. $-0.2y$

$f(x) = -\sqrt[3]{x}$

67. $-5x^2 z^6$ **69.** $6z^{18}$ **71.** $-25z$ **73.** y^2 **75.** f **77.** $3y$
79. $-p^2 q$ **81.** x **83.** 1.26 ft **85.** 27.14 mph **87.** 26.22%
91. m^7 **93.** 3^8 or 9^4 **95.** x^{25} **97.** $24x^8$

Study Set Section 14.3 (page 895)

1. perfect **3.** simplify **5. a.** product, $\sqrt{a}\,\sqrt{b}$
b. quotient, $\dfrac{\sqrt{a}}{\sqrt{b}}$ **7.** Line 2 is not true. There is no addition
property of radicals. **9. a.** $2\sqrt{7}$ in. **b.** 5.3 in. **11.** $2\sqrt{10}$
13. $6\sqrt{2}$ **17.** multiplication **19. a.** $2\sqrt{5}$ **b.** $a\sqrt{7}$
c. $9x\sqrt{6}$ **d.** $5z^2\sqrt{y}$ **21.** $2\sqrt{5}$ **23.** $5\sqrt{2}$ **25.** $3\sqrt{5}$
27. $7\sqrt{2}$ **29.** $4\sqrt{3}$ **31.** $-10\sqrt{2}$ **33.** $8\sqrt{3}$ **35.** $5\sqrt{10}$
37. $4\sqrt{6}$ **39.** $-4\sqrt{7}$ **41.** $n\sqrt{n}$ **43.** $2\sqrt{k}$ **45.** $2\sqrt{3x}$
47. $30\sqrt{3t}$ **49.** $5x\sqrt{x}$ **51.** $a\sqrt{b}$ **53.** $3x^2\sqrt{y}$ **55.** $x^3 y^2\sqrt{2}$
57. $-48x^2 y\sqrt{y}$ **59.** $-\dfrac{8n^2\sqrt{5m}}{5}$ **61.** $\dfrac{5}{3}$ **63.** $\dfrac{9}{8}$ **65.** $\dfrac{\sqrt{26}}{5}$
67. $-\dfrac{2\sqrt{5}}{7}$ **69.** $\dfrac{4\sqrt{3}}{9}$ **71.** $\dfrac{4\sqrt{2}}{5}$ **73.** $\dfrac{6x\sqrt{2x}}{y}$ **75.** $\dfrac{5n^2\sqrt{5}}{8}$
77. $\dfrac{8m\sqrt{2}}{9n}$ **79.** $2rs^2\sqrt{3s}$ **81.** $2\sqrt[3]{3}$ **83.** $-4\sqrt[3]{2}$ **85.** $2x$
87. $-4x\sqrt[3]{x^2}$ **89.** $3xz^2\sqrt[3]{2}$ **91.** $-3y\sqrt[3]{3x^2}$ **93.** $\dfrac{3m}{2n^2}$
95. $\dfrac{rs\sqrt[3]{rs^2}}{10t}$ **97. a.** $\dfrac{3\pi\sqrt{3}}{4}$ sec **b.** 4.1 sec **99. a.** $60\sqrt{2}$ cm
b. 84.9 cm **103.** $-6a^5$ **105.** $y = -2x + 3$ **107.** $x < 5$

Study Set Section 14.4 (page 902)

1. radicals **3.** simplified **5.** no **7.** yes **9.** The radicals
don't have the same radicand, so they can't be combined.
11. The two terms are not like terms—they cannot be combined.
13. $2\sqrt{3}, 3\sqrt{3}, 4\sqrt{3}, 5\sqrt{3}$ **17.** $9\sqrt{7}$ **19.** $-3\sqrt{x}$
21. $5 + 6\sqrt{3}$ **23.** $-1 - \sqrt{r}$ **25.** $5\sqrt{3}$ **27.** $\sqrt{2}$
29. $14\sqrt{5}$ **31.** $-7\sqrt{5}$ **33.** $8\sqrt{5}$ **35.** $-2\sqrt{3}$ **37.** $-18\sqrt{2}$
39. $12\sqrt{7}$ **41.** $9\sqrt{5}$ **43.** $10\sqrt{2} - \sqrt{3}$ **45.** $5\sqrt{6} - 5\sqrt{2}$
47. $7\sqrt{6} + 4\sqrt{15}$ **49.** $7\sqrt{3} - 6\sqrt{2}$ **51.** $3x\sqrt{2}$
53. $3d\sqrt{2d}$ **55.** $3x\sqrt{2y} - 3x\sqrt{3y}$ **57.** $x^2\sqrt{2x}$ **59.** $19b\sqrt{6}$
61. $-5y\sqrt{10y}$ **63.** $2x\sqrt{5xy} + 3x^2 y\sqrt{5xy} - 4x^3 y^2\sqrt{5xy}$
65. $2\sqrt[3]{3}$ **67.** $-\sqrt[3]{x}$ **69.** $5\sqrt[3]{2}$ **71.** $\sqrt[3]{3}$ **73.** $2\sqrt[3]{5} + 5$

75. $x\sqrt[3]{x} - x^2\sqrt[3]{x}$ **77.** $2xy\sqrt[3]{3xy^2}$ **79.** $x^2 y\sqrt[3]{5xy}$
81. $18\sqrt{3}$ in. **83.** $27\sqrt{2}$ cm **85.** $133\sqrt{6}$ ft **89.** $\frac{1}{9}$ **91.** -9
93. $\dfrac{1}{x^3}$ **95.** 1

Study Set Section 14.5 (page 910)

1. rationalizing **3.** conjugate **5.** coefficient **7.** $\sqrt{11}$ **9.** $\sqrt{7}$
11. a. The radicand is a fraction. **b.** There is a radical in the
denominator. **13.** rational: $\dfrac{\sqrt{5}}{3}, -\dfrac{\sqrt{2}}{8}, \dfrac{1+\sqrt{3}}{4}$, irrational:
$\dfrac{2}{\sqrt{6}}, \dfrac{9}{7 - \sqrt{10}}$ **15. a.** not possible **b.** $\sqrt{6}$ **c.** not possible
d. $\dfrac{\sqrt{6}}{3}$ **e.** $4\sqrt{2}$ **f.** 6 **g.** $-2\sqrt{2}$ **h.** $\dfrac{1}{3}$ **19.** 5 **21.** 54
23. 4 **25.** $\sqrt{21}$ **27.** $2\sqrt{14}$ **29.** $3\sqrt{2x}$ **31.** x^4
33. $-60\sqrt{2}$ **35.** $-8x$ **37.** $4x^2$ **39.** $2 + \sqrt{2}$
41. $27 - 3\sqrt{3}$ **43.** $3\sqrt{2} + \sqrt{3}$ **45.** $x\sqrt{3} - 2\sqrt{x}$
47. $6x + 6\sqrt{x}$ **49.** 1 **51.** $6\sqrt{14} + 2x\sqrt{7} - 3x\sqrt{2} - x^2$
53. $7 + 2\sqrt{6}$ **55.** $4x - 18$ **57.** -9 **59.** $6\sqrt[3]{12}$
61. $7 - 2\sqrt[3]{7}$ **63.** $\sqrt[3]{4} + 4\sqrt[3]{2} + 3$ **65.** $\dfrac{2x}{3}$ **67.** $\dfrac{3\sqrt{2}}{5}$
69. $\dfrac{2}{x}$ **71.** $\dfrac{2x}{3}$ **73.** $\dfrac{\sqrt{3}}{3}$ **75.** $\dfrac{\sqrt{91}}{7}$ **77.** $\sqrt{3}$ **79.** $\dfrac{3\sqrt{2}}{8}$
81. $\dfrac{2\sqrt{15}}{5}$ **83.** $\dfrac{10\sqrt{x}}{x}$ **85.** $\dfrac{3\sqrt{2xy}}{2x}$ **87.** $\dfrac{3\sqrt{3} + 3}{2}$
89. $\sqrt{7} - 2$ **91.** $6 + 2\sqrt{3}$ **93.** $\dfrac{\sqrt{3} - 3}{2}$ **95.** $5\sqrt{3} - 5\sqrt{2}$
97. $\dfrac{x + 4\sqrt{x} + 4}{x - 4}$ **99.** $\sqrt[3]{25}$ **101.** $2\sqrt[3]{2}$ **103.** $\dfrac{\sqrt[3]{20}}{2}$
105. 108π in.^{2} **107.** $1{,}800\sqrt{2}$ in.^{2} **109.** 90 in.^{2}
113. no **115.** addition **117.** $x^2 - 16$

Study Set Section 14.6 (page 919)

1. radical **3.** extraneous **5.** b^2 **7. a.** Take the positive square
root of both sides. **b.** Square both sides. **9.** On the second
line, both sides of the equation were not squared—only the
left-hand side. **11.** On the second line, $\sqrt{a + 2}$ wasn't isolated
before squaring both sides. Also, $\left(\sqrt{a + 2} - 5\right)^2 \neq a + 2 - 25$.
13. a.

$A(-4, 6)$
$D(-7, 2)$
$B(4, 0)$
$C(1, -4)$

b. rectangle **c.** AB: 10;
BC: 5; CD: 10; DA: 5
d. 30 units **17.** 9
19. 8 **21.** none
23. 25 **25.** 9 **27.** 1
29. none **31.** 5
33. 6 **35.** 46 **37.** 4
39. 5 **41.** -2 **43.** 4
45. 4 **47.** -1 **49.** 2
51. -1 **53.** 343
55. 65 **57.** 22 **59.** 4 **61.** 5 **63.** 5.83 **65.** 13 **67.** 10
69. 169 ft **71.** 64.4 ft **73.** about 288 ft **75.** about 2×10^6 m
77. $V = \dfrac{\pi r^2 h}{3}$ **79. a.** $\sqrt{2} \approx 1.4$ units **b.** $\sqrt{10} \approx 3.2$ units
83. $8x^2 - 6x$ **85.** $x^2 + 6x + 9$ **87.** $9y^2 - 42y + 49$

Study Set   Section 14.7   (page 926)

1. rational   **3.** x^{m+n} **5.** $\dfrac{x^n}{y^n}$ **7.** $\dfrac{1}{x^n}$ **9.** $\sqrt[n]{x}$ **11.** $5^{1/2}$

13. $\left(\sqrt[3]{8}\right)^4$ **15.** 0, 1, 2, 3

17.
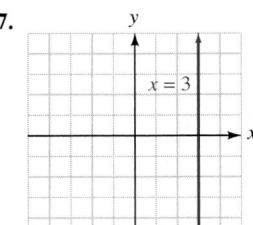

21. 9   **23.** -12 **25.** $\frac{1}{2}$ **27.** $\frac{2}{7}$ **29.** 3   **31.** -5 **33.** -2
35. $\frac{3}{4}$ **37.** 729   **39.** 125   **41.** 25   **43.** 100   **45.** 4   **47.** $\frac{4}{9}$
49. 6   **51.** 25   **53.** 7   **55.** 25   **57.** 8   **59.** 125   **61.** $\frac{1}{2}$ **63.** $\frac{1}{9}$
65. $\frac{1}{64}$ **67.** $\frac{1}{81}$ **69.** x **71.** x^2 **73.** x^2 **75.** y^2 **77.** $x^{2/5}$
79. $x^{2/7}$ **81.** $x^{17/12}$ **83.** $b^{3/10}$ **85.** $t^{4/15}$ **87.** x^2
89. 392 in.²   **91.** 26 ft   **93.** 1.1 in.
97.

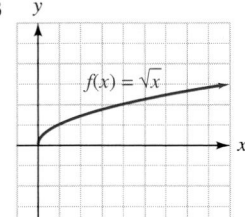

99.

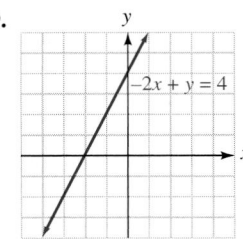

Key Concept   (page 929)

1. addition, subtraction, -6 **3.** multiplication, division, -4
5. square root, square, 49   **7.** cube root, cubed, -8 **9.** -5
11. 15   **13.** 64   **15.** Find the square root of the area.

Chapter Review   (page 931)

1. square, square   **2. a.** 5   **b.** 7   **c.** -12 **d.** $-\frac{4}{9}$ **e.** 30
f. -0.8 **g.** 1   **h.** 0   **3. a.** 4.583   **b.** -3.873 **c.** 5.292
d. 27.421   **4.** imag, irr, rat, irr; $\sqrt{-2}$
5. a. 0, 1, 2, 3

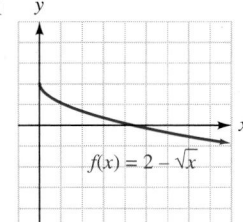

b. 2, 1, 0, -1

6. a. 35   **b.** 1   **c.** $2\sqrt{6}$ **7.** 3.5 ft   **8.** 30 mph   **9.** 5^3, cube
10. a. -3 **b.** -5 **c.** 3   **d.** 2   **e.** 0   **f.** -1 **g.** $\frac{1}{4}$ **h.** 1
11. a. 2.520   **b.** -4.678 **c.** 1.565   **d.** 8.083   **12. a.** x
b. $2b^2$ **c.** x^2y^2 **d.** $-y^6$ **e.** x **f.** y^2 **g.** $3x$ **h.** $-r^4$
13. 12 mm   **14. a.** $4\sqrt{2}$ **b.** $10\sqrt{5}$ **c.** $4x\sqrt{5}$ **d.** $-6\sqrt{7}$
e. $-5t\sqrt{10t}$ **f.** $-10z^2\sqrt{7z}$ **g.** $10x\sqrt{2y}$ **h.** $y^2\sqrt{3}$
i. $2y\sqrt[3]{x^2}$ **j.** $5xy\sqrt[3]{2x}$ **15. a.** $\dfrac{4}{5}$ **b.** $\dfrac{2\sqrt{15}}{7}$ **c.** $\dfrac{10}{3}$
d. $\dfrac{11x\sqrt{2}}{13}$ **16. a.** $2\sqrt{10}$ ft   **b.** 6.3 ft   **17. a.** 0
b. $-3 + 4\sqrt{3}$ **c.** $\sqrt{7}$ **d.** $y^2\sqrt{5xy}$ **e.** $5\sqrt[3]{2}$ **f.** $6x\sqrt[3]{2}$

18. They do not contain like radicals—the radicands are
different.   **19.** $13\sqrt{5}$ in.   **20. a.** $\sqrt{6}$ **b.** $10\sqrt{10}$
c. $36x\sqrt{2}$ **d.** $15 + 6x\sqrt{15} + 9x^2$ **e.** -2 **f.** -2 **g.** $4\sqrt[3]{2}$
h. $\sqrt[3]{9} + \sqrt[3]{3} - 2$ **21. a.** $\left(4\sqrt{6} + 10\sqrt{3}\right)$ in.; $30\sqrt{2}$ in.²
b. 27.1 in.; 42.4 in.²   **22. a.** $\dfrac{\sqrt{7}}{7}$ **b.** $\dfrac{\sqrt{21}}{7}$ **c.** $\dfrac{\sqrt{2}}{2}$
d. $\dfrac{c - 8\sqrt{c} + 16}{c - 16}$ **e.** $7\sqrt{2} - 7$ **f.** $2\sqrt[3]{4}$ **23. a.** x **b.** x
c. $4t$ **d.** $e - 1$ **24. a.** 81   **b.** -3 **c.** none   **d.** 4   **e.** -2
f. 28   **25.** 64 ft   **26. a.** 5   **b.** 13.93   **27. a.** 7   **b.** -10
c. 216   **d.** $\dfrac{4}{9}$ **e.** $\dfrac{1}{8}$ **f.** 64   **g.** 9   **h.** $\dfrac{1}{a^2b^4}$ **i.** $x^{11/15}$
j. $t^{1/12}$ **k.** x **l.** x^2
28.

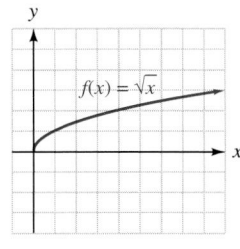

29. $\frac{1}{64}$ of the original dose   **30.** $(-4)^{1/2} = \sqrt{-4}$; There is no real
number that, when squared, gives -4.

Chapter 14 Test   (page 935)

1. 10   **2.** $-\dfrac{20}{3}$ **3.** -3 **4.** $\dfrac{5\sqrt{2}}{7}$ **5.** 7.2   **6.** 10 ft   **7.** $2x$
8. $3x\sqrt{6x}$ **9.** $3y\sqrt{x}$ **10.** x^2y **11. a.** $2\sqrt{6}$ yd   **b.** 4.9 yd
12. $5\sqrt{3}$ **13.** $-x\sqrt{2x}$ **14.** $-24x\sqrt{6}$ **15.** $2\sqrt{6} + 3\sqrt{2}$
16. -1 **17.** $2x - 4\sqrt{x} - 6$ **18.** $\left(6\sqrt{2} + 2\sqrt{10}\right)$ in., 14.8 in.
19. $\sqrt{2}$ **20.** $\dfrac{x\sqrt{3} - 2\sqrt{3x}}{x - 4}$ **21.** 225   **22.** -62 **23.** 5
24. 29   **25.** 10

26. 0, 1, 1.4, 1.7, 2, 2.2, 2.4, 2.6, 2.8, 3
27. No; when 0 is substituted for x,
the result is not a true statement:
$1 \neq -1$.   **28.** They do not contain
like radicals—the radicands are
different.   **29.** the Pythagorean
theorem; 5 ft   **30.** There is no real
number that, when squared, gives
-9.   **31.** 11   **32.** p^2

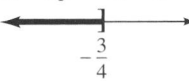

Cumulative Review Exercises   (page 937)

1. a. true   **b.** false   **c.** true   **2.** -14 **3.** 17 lb   **4.** -1.8%
5. $-2p - 6z$ **6.** -2 **7.** $x \le -\frac{3}{4}$, $\left(-\infty, -\frac{3}{4}\right]$,

8. $-2 < x \le 4$, $(-2, 4]$,

9. 3.5 hr   **10.** 80   **11.** 4 in.

12.

13.

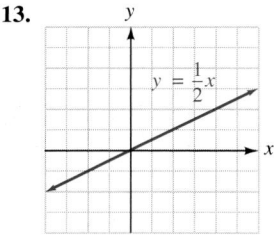

14. 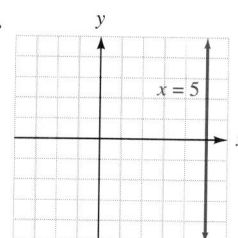   **15.**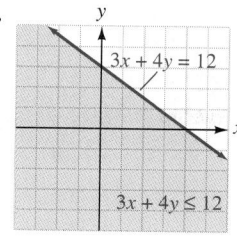

16. $x - 2y = -12$ **17. a.** 3   **b.** $-\frac{2}{3}$ **18. a.** they are the same   **b.** they are negative reciprocals
19. \$50 billion/yr   **20.** -1 **21.** 1, 0, 1, 2, 2, 3

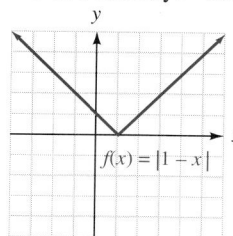

D: all reals; R: all real numbers greater than or equal to 0   **22.** yes
23. x^{31} **24.** $\dfrac{a^{15}b^5}{c^{20}}$
25. 1   **26.** $\dfrac{a^8}{b^{12}}$
27. 4.8×10^{18} m   **28. a.** 3; they indicate that the game was tied 3 times in the second half.
b. (11, 6); in the second half, UConn had its largest lead (6 points) after 11 minutes had elapsed. $(4, -5)$; in the second half, UConn faced its largest deficit (5 points) after 4 minutes had elapsed.
29. $-2r^7s^3t^4$ **30.** $-6t^2 + 13st - 6s^2$ **31.** $2a^2 + a - 3$
32. $y^2 - 12y + 36$ **33.** $\frac{1}{y} - \frac{3}{4x} + \frac{2z}{xy}$ **34.** $2x - 1$
35. $3xy(x - 2y)$ **36.** $(x + y)(2x - 3)$ **37.** $(5p^2 + 4q)(5p^2 - 4q)$
38. $3x(x + 9)(x - 9)$ **39.** $(x - 12)(x + 1)$
40. $(a + 2b)(a^2 - 2ab + 4b^2)$ **41.** $(3a + 4)(2a - 5)$
42. $2(4m + 1)(2m - 3)$ **43.** $-1, -2$ **44.** 0, 2   **45.** $\frac{2}{3}, -\frac{1}{2}$
46. $\frac{3}{2}, -4$ **47.** 5 cm   **48.** 6   **49.** $\dfrac{x + 1}{x - 1}$
50. $-\dfrac{3}{5a}$ **51.** $\dfrac{(p + 2)(p - 3)}{3(p + 3)}$ **52.** $\dfrac{xy^2d}{c^3}$ **53.** $\dfrac{7x + 29}{(x + 5)(x + 7)}$
54. 1   **55.** $\dfrac{9a^2 - 4b^2}{6ab}$ **56.** $\dfrac{y + x}{y - x}$ **57.** 2   **58.** 2
59. $r = \dfrac{r_1 r_2}{r_2 + r_1}$ **60.** 6,480   **61.** 2   **62.** $2\frac{2}{9}$ hr   **63.** $(3, -1)$
64. $(-1, 2)$ **65.** 6%: \$3,000; 12%: \$3,000
66. 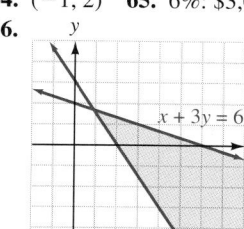   **67.** $\frac{7}{15}$ **68.** 3   **69.** $-48x^2y\sqrt{y}$
70. $7\sqrt{3} - 6\sqrt{2}$
71. $y - 9\sqrt{y} + 20$ **72.** $-60\sqrt{2}$
73. $\dfrac{2\sqrt{5}}{5}$ **74.** $\dfrac{x - 6\sqrt{x} + 9}{x - 9}$
75. 5   **76.** 73 in.

Study Set   Section 15.1   (page 948)

1. completing   **3.** coefficient   **5.** two   **7.** square
9. a. Subtract 35 from both sides.   **b.** Add 1 to both sides.
11. a. because $x^2 - 2x - 1 = 0$ doesn't factor   **b.** yes
13. a. 2   **b.** -4 **c.** $\frac{5}{2}$ **d.** $-\frac{7}{2}$ **15.** $x^2 + 2x - 4 = 0$
19. a. two; $\sqrt{10}, -\sqrt{10}$ **b.** $8 + \sqrt{3}, 8 - \sqrt{3}$; 9.73, 6.27.
21. ± 1 **23.** ± 3 **25.** $\pm 2\sqrt{5}$ **27.** ± 3 **29.** ± 2 **31.** $\pm \frac{3}{4}$
33. $-6, 4$ **35.** 7, -11 **37.** $2 \pm 2\sqrt{2}$ **39.** ± 6.77
41. 7.49, -11.49 **43.** 0, -4 **45.** 2, $-\frac{2}{3}$ **47.** $x^2 + 2x + 1$
49. $x^2 - 4x + 4$ **51.** $x^2 + 7x + \frac{49}{4}$ **53.** $a^2 - 3a + \frac{9}{4}$
55. $b^2 + \frac{2}{3}b + \frac{1}{9}$ **57.** $-2, -4$ **59.** 2, 6   **61.** 5, -3 **63.** 1, -6

65. 1, -2 **67.** $-1, -2$ **69.** 2, $-\frac{1}{2}$ **71.** $-2, \frac{1}{4}$ **73.** $-2 \pm \sqrt{3}$; $-0.27, -3.73$ **75.** $1 \pm \sqrt{5}$; 3.24, -1.24 **77.** $2 \pm \sqrt{7}$; 4.65, -0.65 **79.** $\dfrac{-1 \pm 2\sqrt{5}}{2}$; $-2.74, 1.74$ **81.** 1, -4 **83.** $\dfrac{3}{2}, -\dfrac{2}{3}$
85. $\dfrac{-3 \pm \sqrt{3}}{2}$ **87.** 55 ft   **89.** 20 ft by 40 ft   **95.** $y^2 - 2y + 1$
97. $x^2 + 2xy + y^2$ **99.** $4z^2$

Study Set   Section 15.2   (page 958)

1. quadratic   **3.** solve   **5.** 0   **7.** 3, 0, -5 **9.** $lw, \frac{1}{2}bh$
11. Evaluate 5^2.   **13. a.** 2
b. $\dfrac{-3 + \sqrt{15}}{2}, \dfrac{-3 - \sqrt{15}}{2}$ **c.** 0.44, -3.44
15.
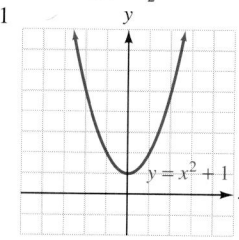
19. The student didn't extend the fraction bar so that it underlines the complete numerator.   **21.** $a = 1, b = 4, c = 3$ **23.** $a = 3, b = -2, c = 7$ **25.** $a = 4, b = -2, c = 1$ **27.** $a = 3, b = -5, c = -2$ **29.** $a = 7, b = 14, c = 21$ **31.** 2, 3   **33.** $-3, -4$
35. 1, $-\frac{1}{2}$ **37.** $-1, -\frac{2}{3}$ **39.** $\frac{1}{2}, -\frac{3}{2}$ **41.** $\dfrac{-3 \pm \sqrt{5}}{2}$
43. $\dfrac{1 \pm \sqrt{37}}{6}$ **45.** no real solutions   **47.** $-1 \pm \sqrt{2}$
49. $\dfrac{3 \pm \sqrt{15}}{3}$ **51.** $-2, 3$ **53.** $\dfrac{-1 \pm \sqrt{41}}{4}$; $-1.85, 1.35$
55. $1 \pm \sqrt{2}$; -0.41; 2.41   **57.** $-5, 7$ **59.** no real solutions
61. $-4, 0$ **63.** $\pm \sqrt{6}$; ± 2.45 **65.** 0.8, 3.1   **67.** 6 in.
69. 47 ft by 66 ft   **71.** 10 ft by 18 ft   **73.** 3.5 sec   **75.** 15 cm by 36 cm   **77.** 30 and 40 nautical miles   **79.** 7%   **81.** \$210,000
83. 2 in.   **89.** $r = \dfrac{A - p}{pt}$ **91.** $3x - 5y = -60$ **93.** $4\sqrt{5}$
95. $\dfrac{\sqrt{7x}}{7}$

Study Set   Section 15.3   (page 969)

1. quadratic   **3.** y-intercept   **5.** symmetry   **7.** $a > 0$
9. $(0, c)$ **11. a.** a parabola   **b.** (1, 0), (3, 0)   **c.** (0, -3)
d. (2, 1)   **15.** The most cases of flu (25) were reported the fifth week.   **17.** true   **19.** $-\frac{1}{2}$
21. moved up 1
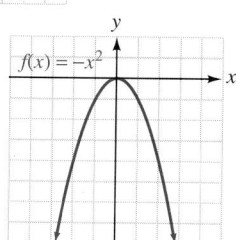
23. opens the opposite direction

25. $(3, 1)$ **27.** $(1, -1)$ **29.** $(1, 0); (0, 1)$ **31.** $(-3, 0), (-7, 0);$ $(0, -21)$ **33.**

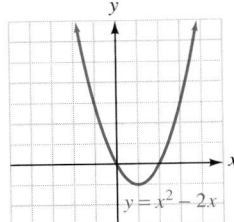

35.

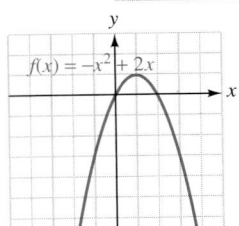

37.

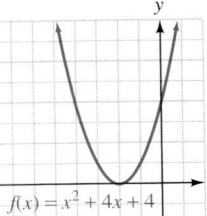

39.

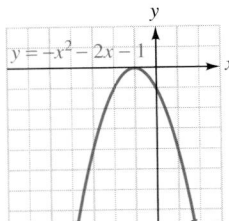

41.

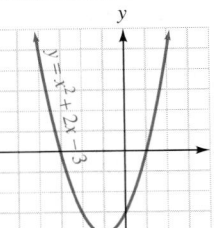

43.

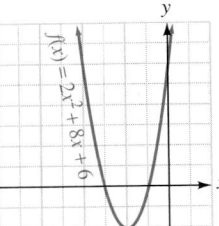

45.

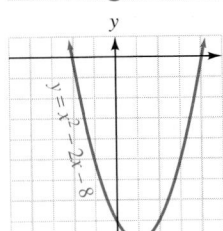

47.

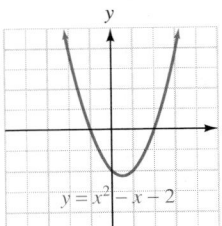

49.

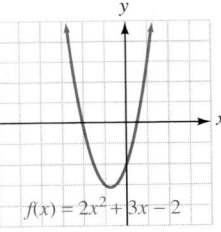

51. a. 14 ft   **b.** 0.25 sec and 1.75 sec   **c.** 18 ft; 1.0 sec   **53.** 32, 18, 8, 2, 0, 2, 8, 18, 32   **55. a.** 750   **b.** \$56,250   **61.** $5\sqrt{3}$
63. 2   **65.** 5

Key Concept   (page 973)

1. no   **3.** yes   **5.** no   **7.** yes   **9.** no   **11.** no   **13.** The student divided both sides by 2 and incorrectly thought that $\dfrac{x^2}{2}$ equals x.

The square root method should be used.   **15.** The student forgot to write the $\pm$ symbol when the square root method was used in Step 2.
17. $0, \frac{1}{4}$ **19.** ± 6 **21.** $-2 \pm \sqrt{3}$

23.

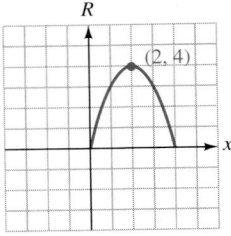

Chapter Review   (page 975)

1. a. ± 5 **b.** ± 20 **c.** ± 3 **d.** $\pm \frac{3}{2}$ **e.** $\pm 2\sqrt{2}$ **f.** $\pm 5\sqrt{3}$
2. a. $-4, 6$ **b.** $\frac{7}{2}, \frac{1}{2}$ **c.** $8 \pm 2\sqrt{2}$ **d.** $-5 \pm 5\sqrt{3}$
3. a. ± 3.46 **b.** $-6.42, 8.42$ **4. a.** $x^2 + 4x + 4$
b. $z^2 - 10z + 25$ **c.** $t^2 - 5t + \frac{25}{4}$ **d.** $a^2 + \frac{3}{4}a + \frac{9}{64}$
5. $x^2 + 4x + 1$ doesn't factor   **6. a.** $3, 5$ **b.** $2, -7$ **c.** $\frac{1}{2}, -3$
d. $\dfrac{1 \pm \sqrt{3}}{2}$ **7.** $-0.27, -3.73$ **8.** 3.0 ft   **9. a.** $5, -3$

b. $7, -1$ **c.** $\frac{3}{2}, -\frac{1}{3}$ **d.** $3 \pm \sqrt{2}$ **10.** $\dfrac{-1 \pm \sqrt{7}}{3}, -1.22, 0.55$

11. no real solutions   **12.** 10 ft, 24 ft   **13.** 15 sec

14. a. $-3 \pm \sqrt{7}$ **b.** $1, -7$ **c.** $0, -5$ **d.** $\dfrac{-1 \pm \sqrt{41}}{4}$

e. $\pm 2\sqrt{5}$ **f.** $2, 2$ **g.** no real solutions   **h.** ± 20
15. a. $(-3, 0), (1, 0)$ **b.** $(0, -3)$ **c.** $(-1, -4)$ **16.** The maximum profit of \$16,000 is obtained from the sale of 400 units.   **17. a.** $(1, 5)$; upward   **b.** $(3, 16)$; downward
18. $(-5, 0), (-1, 0); (0, 5)$
19. a.

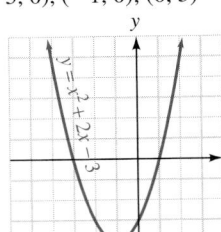

b.

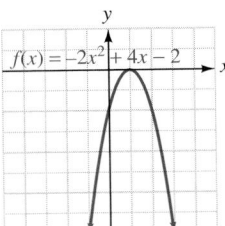

20. 2, 1, 0

Chapter 15 Test   (page 979)

1. ± 4 **2.** $\pm 2\sqrt{6}$ **3.** $\pm \frac{5}{2}$ **4.** $2 \pm \sqrt{3}$ **5.** 40 cm   **6.** 49
7. $-1 \pm \sqrt{5}; -3.24, 1.24$ **8.** $2, -\frac{1}{2}$ **9.** $2, -5$ **10.** $-\frac{3}{2}, 4$
11. no real solutions   **12.** $2 \pm \sqrt{2}$ **13.** $\dfrac{1 \pm \sqrt{13}}{6}; -0.43, 0.77$
14. 255 ft, 505 ft   **15.** The most air conditioners sold in a week (18) occurred when 3 ads were run.
16.

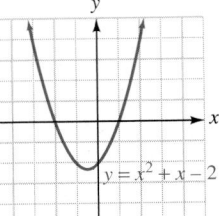

17. 1   **18.** plus or minus

Cumulative Review Exercises   (page 981)

1. a. true   **b.** true   **c.** true   **2.** 36   **3.** 2 hours before salt is spread   **4.** about 98,113   **5.** $-2p - 6z$ **6.** 12   **7.** $-\frac{3}{2}$
8. $x < -2; (-\infty, -2)$

9. $8,000 at 7%, $20,000 at 10%   **10.** 65   **11.** 4   **12.** $\frac{3}{5}$

13. 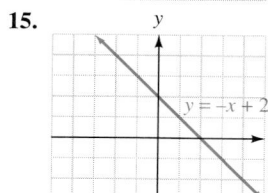   **14.**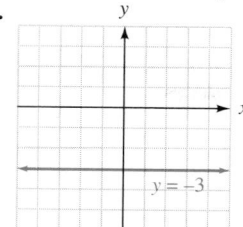

15. 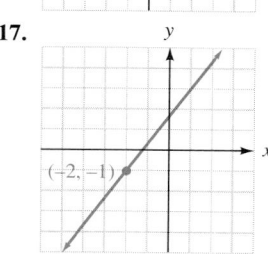   **16.**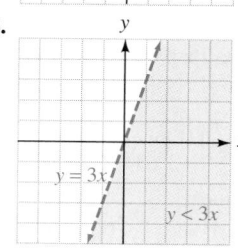

17. 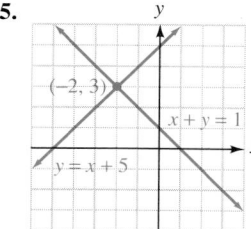   **18.**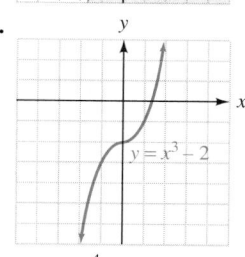

19. $y = -2x + 1$ **20.** $y = \frac{1}{4}x - 1$ **21.** $-\frac{4}{5}$ **22.** -8

23. input   **24.** yes   **25.** y^9 **26.** $\frac{b^6}{27a^3}$ **27.** 2   **28.** $\frac{1}{r^6}$

29. 2,600,000 to 1   **30.** 16%/hr   **31.** $-6a^5$ **32.** $6b^2 + 5b - 4$
33. $4x^2 + 20xy + 25y^2$ **34.** $2x + 1$ **35.** $6a(a - 2a^2b + 6b)$
36. $(x + y)(2 + a)$ **37.** $(b + 5)(b^2 - 5b + 25)$
38. $(t^2 + 4)(t + 2)(t - 2)$ **39.** $0, -\frac{8}{3}$ **40.** $\frac{2}{3}, -\frac{1}{5}$

41. $6x^3 + 4x$ **42.** 5 in.   **43.** -8 **44.** $\frac{3(x + 3)}{x + 6}$ **45.** $\frac{x - 3}{x - 5}$

46. $\frac{2x + 8}{x^2}$ **47.** $\frac{x^2 + 5x}{x^2 - 4}$ **48.** $\frac{9y + 5}{4 - y}$ **49.** -1 **50.** 1

51. $a = \frac{b}{b - 1}$ **52.** $2\frac{6}{11}$ days   **53.** 875 days   **54.** 1.2

55.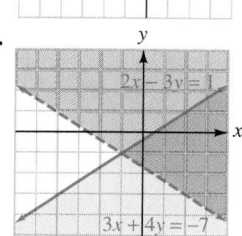

56. $(-3, -1)$ **57.** $(-1, 2)$
58. 30 lb of each

59.

60. 2, 3, 1   **61.** $5x\sqrt{2}$ **62.** $-3y$
63. $9\sqrt{6}$ **64.** $-1 - 2\sqrt{2}$
65. $\frac{6x\sqrt{2x}}{y}$ **66.** $\frac{4\sqrt{10}}{5}$
67. 4   **68.** $-2, -6$ **69.** $\pm 5\sqrt{3}$
70. 21.2 in.   **71.** $\frac{1 \pm \sqrt{13}}{6}$;
$-0.43, 0.77$ **72.** 83 ft $\times$ 155 ft

73.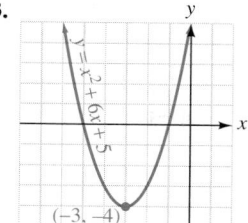

74. a. 150 hp   **b.** 2,000 rpm and
5,000 rpm   **c.** 4,000 rpm

Study Set   Appendix I   (page A-5)

1. inductive   **3.** circular   **5.** alternating   **7.** alternating
9. 10 A.M.   **11.** 17   **13.** -17 **15.** 6   **17.** 3   **19.** -11
21. 9   **23.** 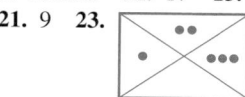   **25.**   **27.** I

29. m   **31.** Maria   **33.** the Mercedes   **35.** green, blue, yellow,
red   **37.** 18,935   **39.** 0

Study Set   Section II.1   (page A-15)

1. length   **3.** 1   **5.** capacity   **7.** 1   **9.** 5,280   **11.** 16
13. 8   **15.** 1   **17.** 24   **19.** $\frac{5}{8}$ in., $1\frac{3}{4}$ in., $2\frac{5}{16}$ in.   **21. a.** $\frac{1 \text{ ton}}{2,000 \text{ lb}}$
b. $\frac{2 \text{ pt}}{1 \text{ qt}}$ **23. a.** iv   **b.** i   **c.** ii   **d.** iii   **25. a.** iii   **b.** iv   **c.** i
d. ii   **31.** $2\frac{5}{8}$ in.   **33.** $10\frac{3}{4}$ in.   **35.** 48 in.   **37.** 42 in.
39. 2 ft   **41.** 288 in.   **43.** 2.5 yd   **45.** $4\frac{2}{3}$ ft   **47.** 15 ft
49. $2\frac{1}{3}$ yd   **51.** 3 mi   **53.** 2,640 ft   **55.** 5 lb   **57.** 3.5 tons
59. 24,800 lb   **61.** 6 pt   **63.** 2 gal   **65.** 2 pt   **67.** 4 hr
69. 5 days   **71.** 150 yd   **73.** 2,880 in.   **75.** 0.28 mi
77. 61,600 yd   **79.** 128 oz   **81.** 4.95 tons   **83.** 68
85. $71\frac{7}{8}$ gal = 71.875 gal   **87.** 320 oz   **89.** $6\frac{1}{8}$ days = 6.125 days

Study Set   Section II.2   (page A-25)

1. tens   **3.** thousands   **5.** hundredths   **7.** metric   **9.** 1 cm,
3 cm, 6 cm   **11. a.** $\frac{1 \text{ km}}{1,000 \text{ m}}$ **b.** $\frac{100 \text{ cg}}{1 \text{ g}}$ **c.** $\frac{1,000 \text{ milliliters}}{1 \text{ liter}}$ **13. a.** iii
b. i   **c.** ii   **15. a.** ii   **b.** iii   **c.** i.   **17.** 10   **19.** $\frac{1}{10}$ **21.** $\frac{1}{1,000}$
23. 1,000   **25.** 1,000   **27.** 1,000   **29.** $\frac{1}{100}$ **31.** 1
37. 156 mm   **39.** 28 cm   **41.** 300   **43.** 570   **45.** 3.1
47. 7,680,000   **49.** 0.472   **51.** 4.532   **53.** 0.0325
55. 37.5   **57.** 125   **59.** 675,000   **61.** 6.383   **63.** 0.63
65. 69.5   **67.** 5.689   **69.** 5.762   **71.** 0.000645
73. 0.65823   **75.** 3,000   **77.** 2,000   **79.** 1,000,000
81. 0.5   **83.** 3,000   **85.** 5,000   **87.** 10
89. 0.5 km, 1 km, 1.5 km, 5 km, 10 km   **91.** 12 cm, 8 cm
93. 400,000 cg   **95.** 40 dL   **97.** 4   **99.** 3 g

Study Set   Section II.3   (page A-33)

1. Fahrenheit   **3. a.** meter   **b.** meter   **c.** inch   **d.** mile
5. a. liter   **b.** liter   **c.** gallon   **7.** 0.3048, 1371.6
9. 0.264   **11.** 91.4   **13.** 147.6   **15.** 39,372   **17.** 127
19. 1   **21.** 11,350   **23.** 17.6   **25.** 0.6   **27.** 0.1
29. 243.4   **31.** 710   **33.** 0.5   **35.** 10°   **37.** 122°
39. 14°   **41.** $-20.6°$ **43.** 5 mi   **45.** 70 mph   **47.** 1.9 km
49. 1.9 cm   **51.** 181.5 lb, 291.5 lb; 242 lb, 594 lb
53. a. 226.8 g   **b.** 0.24 L   **55.** no   **57.** the 3 quarts
59. 62° C   **61.** 28° C   **63.** $-5°$ C and 0° C

Study Set   Appendix III   (page A-41)

1. equation   **3.** isolate   **5.** 0   **7.** more than   **9.** 5
11. a. yes   **b.** no   **c.** yes   **d.** no   **13. a.** ii   **b.** iii   **c.** i

15. $|x| < 4$ **17.** $|x + 3| > 6$ **19.** 8 **21.** -0.02 **23.** $-\frac{31}{16}$
25. π **27.** 23, -23 **29.** 9.1, -2.9 **31.** $\frac{14}{3}$, -6
33. no solution **35.** 2, $-\frac{1}{2}$ **37.** -8 **39.** -4, -28
41. 0, -6 **43.** -2, $-\frac{4}{5}$ **45.** 0, -2 **47.** 0 **49.** $\frac{4}{3}$

51. $(-4, 4)$

53. $[-21, 3]$

55. $\left(-\frac{8}{3}, 4\right)$ **57.** no solution

59. $(-\infty, -3) \cup (3, \infty)$

61. $(-\infty, -12) \cup (36, \infty)$

63. $\left(-\infty, -\frac{16}{3}\right) \cup (4, \infty)$

65. $(-\infty, \infty)$

67. $(-\infty, -2] \cup \left[\frac{10}{3}, \infty\right)$

69. $(-\infty, -2) \cup (5, \infty)$

71. $[-10, 14]$

73. $\left(-\frac{5}{3}, 1\right)$

75. $(-\infty, -24) \cup (-18, \infty)$

77. $70° \le t \le 86°$ **79. a.** $|c - 0.6°| \le 0.5°$
b. $[0.1°, 1.1°]$ **81. a.** 26.45%, 24.76%
b. It is less than or equal to 1%.

Study Set Appendix IV (page A-47)

1. experiment **3.** $\frac{n}{s}$ **5.** 0 **7.** {(1, H), (2, H), (3, H), (4, H),
(5, H), (6, H), (1, T), (2, T), (3, T), (4, T), (5, T), (6, T)}
9. {a, e, i, o, u} **11.** $\frac{1}{6}$ **13.** $\frac{2}{3}$ **15.** $\frac{5}{36}$ **17.** 0 **19.** $\frac{19}{42}$ **21.** $\frac{13}{42}$
23. $\frac{1}{4}$ **25.** $\frac{1}{4}$ **27.** $\frac{1}{2}$ **29.** $\frac{3}{13}$ **31.** $\frac{5}{12}$ **35.** $\frac{1}{4}$ **37.** $\frac{1}{4}$ **39.** 1
41. $\frac{32}{119}$

INDEX

Absolute value, 80, 118, A-35
  equations, A-36
  inequalities, A-39
Accent on Teamwork
  Chapter 1, 68
  Chapter 2, 133
  Chapter 3, 211
  Chapter 4, 274
  Chapter 5, 327
  Chapter 6, 356
  Chapter 7, 426
  Chapter 8, 510
  Chapter 9, 598
  Chapter 10, 667
  Chapter 11, 724
  Chapter 12, 796
  Chapter 13, 860
  Chapter 14, 930
  Chapter 15, 974
Acute angles, 368
Addend(s), 11
Adding
  carrying, 13
  decimals, 231
  fractions and decimals, 258
  fractions with different denominators, 170
  fractions with like denominators, 168, 656, 750
  fractions with unlike denominators, 754
  integers with different signs, 89
  integers with the same sign, 87
  like terms, 459
  mixed numbers, 187, 188
  monomials, 639
  multiples of polynomials, 642
  polynomials, 640
  radical expressions, 900
  rational expressions, 751
  signed decimals, 233
  whole numbers, 13
  zero, 12
Addition
  of fractions with like denominators, 750
  of fractions with unlike denominators, 754
  of rational expressions, 751
Addition method for solving systems, 825
Addition property
  of 0, 12, 91
  of equality, 53
  of inequality, 499
Additive identity, 91
Additive inverse, 91

Adjacent angles, 369
Algebra, 50
Algebraic expression(s), 441
  evaluating, 445
  involving multiplication, 453
  simplifying, 452
Alternate interior angles, 376
American system of measurement, A-8
  converting to metric, A-29
Amount in a percent problem, 297
Angle(s), 367
  acute, 368
  adjacent, 369
  alternate interior, 376
  base, 382
  complementary, 370, 460
  congruent, 368
  corresponding, 376
  interior, 376
  measure, 368
  measurement, 368
  obtuse, 368
  protractor, 368
  right, 368
  sides, 367
  straight, 368
  sum of measures for a triangle, 383
  supplementary, 370, 460
  trapezoid base, 384
  vertex, 367, 382
  vertical, 369
Annual interest rate, 472
Apparent solutions, 768
"Approximately equal to" symbol, 257, 270
Approximating
  cube roots, 884
  solutions to quadratic equations, 953
  square roots, 270, 873
Arc, 408
  major, 408
  minor, 408
Area, 26
  of a circle, 410
  of combinations of figures, 402
  formulas, 474
  of a parallelogram, 399
  of a polygon, 398
  of a rectangle, 27, 399
  of a square, 399
  surface, 418
  of a trapezoid, 399
  of a triangle, 157, 399

Arithmetic
  fundamental theorem of, 38
Arithmetic average
  mean, 47, 349
Arithmetic mean, 47, 349
ASA property, 390
Ascending powers, 633
Associative property
  of addition, 12
  of multiplication, 24
Average, 47
  Key Concept, 356
  mean, 349
  median, 351
  mode, 352
Average speed, 472
Axioms, A-3
Axis (Axes)
  x-, 522
  y-, 522
Axis of symmetry, 964

Bar graph(s), 7, 338
Base
  of an exponential expression, 39, 610
  of an isosceles triangle, 382
  in a percent problem, 297
  of a trapezoid, 384
Base angles
  isosceles triangle, 382, 485
  trapezoid, 384
Binomial(s), 633
  FOIL method, 647
  multiplying, 647
Borrowing, 16
  when subtracting mixed numbers, 191
Boundary line, 842
Braces, 2
Bracket, 46
  for an interval, 498
Break point, 834
Building fractions, 148, 752

Calculator
  graphing, 537
  scientific, 18
Calculator key(s)
  addition, 18
  change-of-sign, 91
  cube root, 884
  dividing negative numbers, 114
  division, 30
  entering negative numbers, 91

Calculator key(s) (*continued*)
  exponential, 40
  fixed-point, 257
  fractional exponents, 924
  multiplication, 26
  multiplying negative numbers, 106
  negative power, 107
  parenthesis, 48
  scientific notation, 628
  square root, 270, 874
  statistical mode, 350
  subtracting a negative number, 99
  subtraction, 18
Capacity
  American units, A-13
  metric units, A-24
Carrying, 13
Cartesian coordinate system, 522
Celsius scale, 473, A-31
Center of a circle, 407, 474
Change, rate of, 556, 557
Checking
  solutions of an equation, 51
  solutions of radical equations, 914
  subtraction, 16
Chord, 407
Circle(s), 407, 474
  arc, 408
  area, 410
  center, 407, 474
  chord, 407
  circumference, 408, 474
  diameter, 407, 474
  radius, 407, 474
Circle graphs, 301
Circumference of a circle, 408
  formula, 474
Clearing an equation of fractions, 466
Coefficient(s), 457, 632
  implied, 457
  leading, 634, 688
Combining like radicals, 899
Combining like terms, 459, 898
Commissions, 308
Common denominator, 168
  least, 169
  lowest, 169
Commutative property
  of addition, 11
  of multiplication, 22
Comparing
  fractions, 173
  negative decimals, 225
  positive decimals, 225
Comparison shopping, A-31
Complementary angles, 370, 459
Completing the square, 945
  to solve quadratic equations, 946
Complex fraction(s), 197, 760
  simplifying, 197, 760
Composite number(s), 38
Compound inequalities, 502
  solving, 503
Compound interest, 318, 319, 472
  annually, 320
  daily, 320
  formula, 321
  quarterly, 320
  semiannually, 320
Conditional equation(s), 467
Cone(s), 476
  volume, 416, 421
Congruent
  angles, 368
  segments, 367

Congruent triangles, 389
  ASA property, 390
  SAS property, 389
  SSS property, 389
Conjugates, 909
Consistent system, 809
Constant(s), 458, 633
Constant term, 633
Constant of variation, 788, 789
Constructing tables, 533
Contradiction, 467
Converting between American and metric units,
    A-28
Coordinate(s) of a point, 523
Coordinate plane, 523
Coordinate system, rectangular, 522
Coplanar lines, 375
Corresponding angles, 376
Cross product(s), 779
Cube(s), 415
  integer, 884, 894
  volume, 415, 416
Cube of a number, 882
Cube root(s), 883
  approximating, 884
  notation, 883
  simplifying, 894
  subtracting, 901
Cube root function, 885
Cubic centimeter(s), A-25
Cylinder(s), 419, 476
  volume, 416, 420

Decimal(s), 222
  adding, 231
  adding signed decimals, 233
  adding to fractions, 258
  changing to a percent, 290
  comparing negative, 225
  comparing positive, 225
  dividing, 246
  dividing by a power of 10, 249
  dividing a whole number by a, 246
  estimation with, 253
  graphing, 258
  in equations, 263
  multiplying, 239
  multiplying by powers of 10, 241
  multiplying signed decimals, 241
  nonterminating, 270
  overbar, 256
  reading, 224
  repeating, 255
  rounding, 226
  rounding repeating, 256
  subtracting, 232
  subtracting signed decimals, 233
  terminating, 255
  writing, 224
  writing as fractions, 224
Decimal numeration system, 222
Decimal point, 223
Decrease, percent of, 309
Decreasing powers, 633
Deductive reasoning, A-3
Defined terms in reasoning, A-3
Degree, 368
  of a monomial, 634
  of a polynomial, 634
Denominator, 144
  rationalizing, 908
Dependent equations, 812, 821, 828
Dependent variable, 535, 588
Descending powers, 633, 662

Diameter of a circle, 407
  formula, 474
Die, A-46
Difference, 15, 96
Difference of two cubes, 707
Difference of two squares, 649
  factoring, 705
Digit(s), 2
  rounding, 5, 226
  test, 5, 226
Direct variation, 787
Discount(s), 310
  rate of, 310
Discriminant, 974
Distance formula, 918
Distance-traveled formula, 472
Distributing
  a factor of −1, 456
Distributive property, 454
  extended, 456
Dividend, 28
Dividing
  by 1, 737
  decimals, 246
  by a decimal divisor, 248
  decimals by powers of 10, 249
  a decimal by a whole number, 246
  fractions, 163, 745
  integers, 112
  mixed numbers, 181
  monomials by monomials, 655
  opposites, 738
  a polynomial by a monomial, 656
  polynomials by polynomials, 661
  radical expressions, 907
  rational expressions, 745
  whole numbers, 29
Divisible, 36
Division
  by 0, 29, 113
  of 0, 29, 113
  of 0 by 0, 28, 113
  by 1, 29, 737
  exactly, 36
  of fractions, 745
  of a number by itself, 29
  of opposites, 738
  by a power of 10, 248
  of rational expressions, 745
  relationship with multiplication, 111
  with a remainder, 30
  symbols, 28
  terminated, 248
  undefined, 29, 113
  undetermined, 29, 113
Division method to prime factor, 680
Division property
  of equality, 59
  of inequality, 500
  of radicals, 893
Divisor, 28
Domain of a function, 588
Double-sign notation, 942
Dry mixture problems, 490

Element(s) of a set, 2
Endpoint(s)
  of an interval, 498
  of a ray, 367
  of a segment, 367
English system of measurement, A-8
  converting to metric, A-29
Equation(s), 51
  absolute value, A-36

clearing of fractions, 466
conditional, 467
containing decimals, 263
containing fractions, 202
contradiction, 467
dependent, 812, 821, 828
equivalent, 52
general form, 545
general quadratic, 951
of horizontal lines, 550
identity, 467
impossible, 467
independent, 809
left-hand side, 51
linear in *x* and *y,* 545
nonlinear, 544
percent, 296
point–slope form, 579
quadratic, 712, 942
radical, 913
rational, 766
right-hand side, 51
roots of, 51
slope–intercept form, 570
solutions of, 51
solving, 52, 462
solving for a variable, 476
standard form, 545
strategy for solving, 467
system of, 808
with two absolute values, A-38
in two variables, 532
of vertical lines, 550
Equiangular triangle(s), 382
Equilateral triangle(s), 382
Equivalent equations, 52
Equivalent fractions, 146
Equivalent systems, 810
Error statement on a calculator, 874
Estimate, 13
Estimation
with decimals, 253
with integers, 119
percent, 316
with whole numbers, 34
Evaluating, 44
algebraic expressions, 445
rational expressions, 734
square roots, 269
Even powers, 107
Even root(s), 886
Event, A-45
Even whole number, 37
Exactly divisible, 36
Expanded notation, 3, 223
Experiment, A-45
Exponent(s), 39, 610
fractional, 923
negative, 620
power rule, 613
power rule for products, 614
power rule for quotients, 614
product rule, 611
quotient rule, 612
rational, 923
rules, 615, 621
rules for, 923, 925
variable, 622
zero, 615
Exponential expression(s), 38, 610
Expression(s)
algebraic, 441
containing fractions and decimals, 258
exponential, 39, 610
radical, 268, 873

rational, 734
simplifying algebraic, 452
value of, 445
Extended distributive property, 456
Extraneous solutions, 768

Face cards, A-47
Factor(s), 22, 36, 679
greatest common (GCF), 177
Factoring, 679
completely, 692
difference of two cubes, 707
difference of two squares, 705
by grouping, 684
out the GCF, 682
out a negative factor, 683
perfect square trinomials, 704
prime, 38
strategy, 708
to solve equations, 713
sum of two cubes, 707
trial-and-check method, 692
trinomials by grouping, 699
trinomials with leading coefficient 1, 688
trinomials with leading coefficient
not 1, 696
a whole number, 36
Factoring method to solve equations, 714
Factoring tree, 680
Fahrenheit conversion formula, 473
Fahrenheit scale, 473, A-31
Five-step problem-solving strategy, 54
Fixed-point key, 257
FOIL method, 647
Formula(s)
area, 474
area of a circle, 410
area of a parallelogram, 399
area of a rectangle, 27, 399
area of a square, 399
area of a trapezoid, 399
area of a triangle, 157, 399
for areas of polygons, 399
from business, 470
circumference of a circle, 408
compound interest, 321
diameter of a circle, 407
distance, 918
distance traveled, 472
Fahrenheit conversion to Celsius, 473
for simple interest, 472
from science, 472
midpoint, 528
percent, 300
perimeter, 474
perimeter of a rectangle, 15, 396
perimeter of a square, 15, 395
for profit, 471
Pythagorean theorem, 390
quadratic, 952
retail price, 471
sale price, 311
simple interest, 318
slope, 560
solving, 476, 769
sum of measures of angles of a polygon, 385
surface area of rectangular solid, 418
surface area of sphere, 419
total price, 307
volume, 476
volume of a cone, 416
volume of a cube, 416
volume of a cylinder, 416
volume of a prism, 416
volume of a pyramid, 416

volume of a rectangular solid, 416
volume of a sphere, 416, 419
Fraction(s), 144
added to decimals, 258
adding with different denominators, 170
adding with like denominators, 168, 656
adding with unlike denominators, 754
building, 148, 752
changing to a percent, 290
common denominator, 168
comparing, 173
complex, 197, 760
denominator, 144
dividing, 163, 745
in equations, 202
equivalent, 146
expressing in higher terms, 148
finding the LCD, 171
fundamental property, 210, 735
graphing, 181, 258
improper, 144
LCD, 169
in lowest terms, 147
multiplying, 153, 742
negative, 145
numerator, 144
order of operations, 195
powers of, 155
proper, 144
reciprocal, 162
reducing, 146
simplifying, 146
subtracting with like denominators, 168, 656
subtracting with unlike denominators, 170, 754
writing as decimals, 255
Fractional exponents, 923
Frequency polygon(s), 344
Front-end rounding, 34
Function(s), 588
cube root, 885
domain, 588
graphs, 590
linear, 591
notation, 589
polynomial, 635
quadratic, 962
range, 588
square root, 875
Fundamental property of fractions, 146, 735
Key Concept, 210
Fundamental property of proportions, 778
Fundamental theorem of arithmetic, 38, 680

General form of an equation, 545
General quadratic equation, 951
Generation time, 625
Geometric figure(s), 367
Geometric solid(s), 416
height, 417
Geometry, 366
undefined words, 366
Gram, A-22
Graph(s), 522
bar, 7, 338
circle, 301
of equations in two variables, 535
frequency polygons, 344
of functions, 590
histograms, 343
line, 7, 341
of a number, 4
of a point, 523
of quadratic functions, 965
pictographs, 340

Graph(s) (*continued*)
  pie, 338, 341
  reading, 526
  step, 527
Graphing
  decimals, 258
  fractions, 181, 258
  linear equations, 546
  linear inequalities, 842
  lines using slope, 563
  mixed numbers, 181
  on the number line, 4
  points, 523
  quadratic functions, 963
Graphing calculator, to graph lines, 537
Graphing method to solve a system, 809
Greatest common factor (GCF), 177, 681
Grouping
  to factor trinomials, 699
  factoring by, 684
  key number, 699
Grouping symbol(s), 46
  absolute value, 118
  brackets, 46
  fraction bar, 46
  innermost, 46
  outermost, 46
  parentheses, 12, 46

Half-plane, 842
Height of geometric solids, 417
Higher-order roots, 885
Higher terms, 148
Histogram(s), 343
Horizontal line(s), 549, 582
  slope, 563
Hypotenuse, 382, 559, 717, 875

Identity
  additive, 91
  equation, 467
Imaginary number(s), 874
Implied coefficient
  of -1, 457
  of 1, 457
Impossible equation(s), 467
Improper fraction(s), 144
  writing as mixed numbers, 181
Inconsistent systems, 811, 821, 828
Increase, percent of, 309
Independent equations, 809
Independent variable, 535, 588
Index of a radical, 885
Inductive reasoning, A-1
Inequality (Inequalities), 496
  absolute value, A-39
  compound, 502
  linear, 841
  solutions, 498
  solving, 498
  symbols, 4, 80, 497
Infinity, 498
Input, 448
Input value, 533
Integer(s), 79, 90
  adding, 88
  division of, 112
  multiplication of, 105
  powers of, 106
  solving equations, 122
Integer cubes, 884, 894
Integer squares, 874, 890
Intercept(s), 548
Intercept method for graphing lines, 548
Interest, 318, 471

compound, 318, 319, 472
  formula for compound, 321
  formula for simple, 318
  rate (annual), 472
  rate, 318
  simple, 318, 472
  time, 318
Interior angles, 376
Interval(s), 497
  bracket, 498
  parenthesis, 498
Inverse(s)
  additive, 91
  multiplicative, 464
Inverse variation, 789
Investment problems, 487, 772, 835
Irrational number(s), 256, 270, 874
"Is approximately equal to" symbol, 873
Isosceles triangle(s), 382, 485
  base, 382
  base angles, 382, 485
  properties, 382
  vertex angle, 382, 485

Key Concept
  describing linear relationships, 597
  expressions and equations, 795
  factoring, 723
  formulas, 425
  fundamental property of fractions, 210
  inverse operations, 929
  mean, median, and mode, 356
  percent, 326
  polynomials, 666
  quadratic equations, 973
  real numbers, 274
  signed numbers, 132
  simplify and solve, 509
  systems of equations and inequalities, 859
  variables, 67
Key number, 699

Leading coefficient, 634, 688
Leading term, 634
Least common denominator (LCD), 169, 752
  finding it, 753
  using multiples to find it, 171
  using prime factorization to find it, 171
Least common multiple (LCM), 177
Left-hand side of an equation, 51
Leg(s)
  of a right triangle, 382, 717, 875
  of a trapezoid, 384
Length
  American units, A-10
  metric units, A-18, A-20
Like radicals, 898
  combining, 899
Like terms, 458, 898
  combining, 459
Line(s), 366
  boundary, 842
  coplanar, 375
  horizontal, 549
  parallel, 375, 573
  perpendicular, 375, 573
  point–slope form, 579
  properties of parallel, 376
  slope, 558
  slope–intercept form, 570
  transversal, 376
  vertical, 550
Linear equation(s)
  graphing, 546
  point–slope form, 579

slope–intercept form, 570
  in x and y, 545
Linear function, 591
Linear inequality, 841
Line graph(s), 7, 341
Line segment(s), 367
  endpoints, 367
  midpoint, 367
Liquid mixture problems, 489
Liter, A-24
Long division, 29
  missing terms, 663
  with polynomials, 661
Lowest common denominator (LCD), 169, 752
  finding it, 753
  using multiples to find it, 171
  using prime factorization to find it, 171
Lowest terms, 147

Major arc, 408
Markup, 470
Mass
  compared to weight, A-22
  metric units, A-22
Maximum value of a quadratic function, 968
Mean, 47, 349, 356
Measure
  of an angle, 368
  of a segment, 367
Median, 351, 356
Member(s), of a set, 2
Meter, A-18
Metric system, A-18
Midpoint, 367, 528
  formula, 528
Minor arc, 408
Minuend, 15
Missing terms, 663
Mixed number(s), 179
  adding, 187, 188
  adding in vertical form, 189
  dividing, 181
  graphing, 181
  multiplying, 181
  subtracting, 190
  subtracting in vertical form, 191
  writing as improper fractions, 180
Mixture problems, 489, 490, 836
Modal value, 352
Mode, 352, 356
Monomial(s), 633
  adding, 639
  degree, 634
  divided by monomials, 655
  multiplying, 646
  subtracting, 640
Multiple, least common (LCM), 177
Multiplication
  by 0, 104
  of fractions, 742
  of integers with like signs, 105
  of integers with unlike signs, 104
  by a power of 10, 241
  of rational expressions, 742
  of variables, 22
  relationship with division, 111
  symbols, 22
Multiplication property
  of 0, 23
  of 1, 23
  of equality, 60
  of inequality, 500
  of radicals, 890
Multiplicative inverse, 464
Multiplying

by 0, 23
by 1, 23
algebraic expressions, 453
binomials, 647
binomials to solve equations, 651
decimals, 239
decimals by powers of 10, 241
fractions, 153, 742
integers, 105
mixed numbers, 181
monomials, 646
a polynomial by a binomial, 650
a polynomial by a monomial, 647
radical expressions, 904
rational expressions, 742
signed decimals, 241
whole numbers, 24

Natural number(s), 2
Natural-number exponents, 610
Negative exponent(s), 629
Negative number(s), 78
    multiplying, 105
Negative of a number, 81
Negative reciprocals, 574
Negative slope, 561
Negative sign, 78
Negatives, 738
    division of, 738
Nonlinear equations, 544
Nonterminating decimal(s), 270
Notation
    cube root, 883
    double-sign, 942
    expanded, 3, 223
    function, 589
    interval, 497
    scientific, 6236
    standard, 2, 628
    subscript, 559
Number(s)
    adding whole, 13
    composite, 38
    constants, 458
    decimals, 222
    dividing whole, 29
    graph of, 4
    imaginary, 874
    integers, 79
    irrational, 256, 270, 874
    mixed, 179
    multiplying whole, 24
    natural, 2
    negative, 78
    negative of a, 81
    opposite, 81
    perfect, 42
    positive, 78
    prime, 37, 679
    rational, 256, 874
    real, 256, 270
    rounding, 6
    signed, 79
    square of a, 268
    subtracting whole, 16
    whole, 2
Number line, 4, 79
    graphing on the, 4
    origin, 4, 87
Number and value, 445
Number-value problems, 486
Numeral, 2
Numerator, 144
Numerical coefficient(s), 457, 632

Obtuse angles, 368
Odd powers, 107
Odd root(s), 886
Odd whole number, 37
Opposite of a number, 81
Opposites, 738
    division of, 738
Order of operations, 44, 116
    with decimals, 242, 250
    with fractions, 195
Ordered pairs, 523
Origin
    of a number line, 4, 87
    on a rectangular coordinate system, 523
Output, 448
Output value, 533

Parabola(s), 535, 963
    axis of symmetry, 964
    intercepts, 965
    vertex, 963, 965
Parallel lines, 375, 573
    properties of, 376
Parallelogram(s), 384, 474
    area, 399
Parentheses, 12, 498
Parsec, 938
Pendulum period, 875
Percent, 286, 297
    changing to a decimal, 288
    changing to a fraction, 287
    commissions, 308
    of decrease, 409
    discount, 310
    equation, 296
    estimation, 316
    formula, 300
    of increase, 309
    Key Concept, 326
    tax rate, 307
Percent problems
    finding the amount, 296
    finding the base, 298
    finding a commission, 308
    finding a discount, 311
    finding the percent, 297
    finding percent decrease, 309
    finding percent increase, 309
    finding a sale price, 311
    finding a tax rate, 307
Perpendicular lines, 573
Perfect number(s), 42
Perfect square(s), 270
Perfect square trinomial(s), 703
    factoring, 704
Perimeter, 15
    formulas, 474
    of a polygon, 395
    of a rectangle, 15, 396
    of a square, 15, 395
Period, 3
Period of a pendulum, 875
Perpendicular lines, 375
Pi (π), 408
Pictograph(s), 340
Pie chart(s), 301
Pie graph(s), 338, 341
Place value column(s), 2
    for decimals, 223
    periods, 3
    for whole numbers, 3
Plane(s), 366
    coordinate, 523
Plotting, points, 523
Plummet, 380

Point(s)
    decimal, 223
    in geometry, 366
    graphing, 523
    test, 843
Point–slope form, 579
Polygon(s), 381
    area, 398
    formulas for areas, 399
    perimeter, 395
    regular, 381
    sum of measures of angles, 385
    vertices, 381
Polynomial(s), 633
    adding, 640
    binomial, 633
    degree, 634
    divided by monomials, 656
    dividing by polynomials, 661
    long division, 661
    monomial, 633
    multiplying by a binomial, 650
    multiplying by monomials, 647
    multiplying three, 650, 650
    prime, 692
    second degree, 712
    subtracting, 641
    trinomial, 633
    in x, 633
    in x and y, 633
Polynomial function(s), 635
Positive number(s), 78
Positive sign, 79
Positive slope, 561
Postulates, A-3
Power(s), 39
    of 10, 240
    ascending, 633
    decreasing, 633
    descending, 633
    division by a power of 10, 249
    even, 107
    of fractions, 155
    of integers, 106
    multiplying by powers of 10, 241
    odd, 107
    of radical expressions, 905
Power, 610
Power rule for exponents, 613
Power rule for products, 614
Power rule for quotients, 614
Present value, 623
Prime factorization, 38, 680
    division method, 40, 680
    factoring tree, 680
    tree method, 38
Prime number(s), 37, 679
Prime polynomial, 692
Prime trinomial, 692
Prime-factored form, 680
Principal, 318, 472
Principal square root, 873
Prism(s), volume, 416
Probability, A-45
Problem solving, five-step strategy, 54
Product(s), 22
    cross, 779
    special, 649, 945
Product rule for exponents, 611
Profit formula, 471
Proper fraction, 144
Property (Properties)
    addition of 0, 91
    addition of equality, 53
    addition of inequality, 499

Property (Properties) (*continued*)
ASA, 390
associative of addition, 12
associative of multiplication, 24
commutative of addition, 11
commutative of multiplication, 22
distributive, 454
of division, 29
division of equality, 60
division of inequality, 500
division of radicals, 893
extended distributive, 456
fundamental of fractions, 146, 210, 735
fundamental of proportions, 778
of isosceles triangles, 382
multiplication by , 104
multiplication of 0, 23
multiplication of 1, 23
multiplication of equality, 60
multiplication of inequality, 500
multiplication of radicals, 890
of parallel lines, 376
of rectangles, 384
SAS, 389
of similar triangles, 783
square root of equality, 876
squaring of equality, 914
SSS, 389
subtraction of equality, 52
subtraction of inequality, 499
of vertical angles, 369
zero-factor, 712
Proportion(s), 778
cross products, 779
fundamental property of, 778
solving, 779
Protractor, 368
Pyramid(s), 421, 476
volume, 416
Pythagorean theorem, 390, 717, 875

Quadrants, 523
Quadratic equation(s), 712, 942
approximating solutions, 953
general, 951
with no solution, 954
solving by completing the square, 946
solving by factoring, 714
solving using the square root method, 942
strategy for solving, 954
Quadratic form, 712, 942
Quadratic formula, 952
discriminant, 974
Quadratic function(s), 962
finding a maximum value, 968
graphing, 963
strategy for graphing, 967
Quadrilateral(s), 383
parallelogram, 384
rectangle, 384
rhombus, 384
square, 384
trapezoid, 384
Quotient, 28
Quotient rule for exponents, 612

Radical(s), 873
adding, 900
changing to fractional exponents, 924
combining like, 899
division property, 893
like, 898
multiplication property, 890
multiplying, 904
simplified form, 890

simplifying, 894
subtracting, 901
symbol, 873
Radical equation(s), 913
solving, 914
Radical expression(s), 268, 873
adding, 900
multiplying, 904
powers of, 905
simplifying, 269
subtracting, 901
Radical symbol, 268
Radicand, 268, 873
variable, 891
Radius of a circle, 407
Range of a function, 588
Rate(s), 777
annual interest, 472
average speed, 472
discount, 310
Rate of change, 556, 557
Ratio(s), 556, 557, 777
Rational equation(s), 766
Rational exponents, 923
changing to radical notation, 924
Rational expression(s), 734
adding, 751
dividing, 745
evaluating, 734
multiplying, 742
simplifying, 735
subtracting, 751
Rational number(s), 256, 874
Rationalizing the denominator, 908
Ray(s), 367
endpoint, 367
Reading
a decimal, 224
graphs, 526
Real number(s), 256, 270
Key Concept, 274
Reasoning
deductive, A-3
inductive, A-1
Reciprocal(s), 162, 464
negative, 574
using to solve equations, 203
Rectangle(s), 14, 384, 474
area, 27, 399
length, 14
perimeter, 15, 396
properties of, 384
width, 14
Rectangular coordinate
graph, 522
system, 522
Rectangular solid(s), 415, 476
surface area, 418
volume, 415, 416
Reducing fractions, 146
Regular polygon(s), 381
Remainder, 30
Repeating decimal(s), 255
overbar, 256
Retail price formula, 471
Revenue, 471
Rhombus, 384
Right-hand side of an equation, 51
Right angle(s), 368
Right triangle(s), 382, 717, 875
hypotenuse, 382, 717, 875
legs, 382, 717
Rise, 558
Root(s)
cube, 883

of an equation, 51
even, 886
index, 885
*n*th, 885
odd, 886
square, 268, 873
Rounding
decimals, 226
digit, 5, 226
repeating decimals, 256
whole numbers, 6
Rule for subtraction, 96
Rules for exponents, 615, 621, 923, 925
Run, 558

Sale price, 311
Sample space, A-45
SAS property, 389
Satisfying an equation, 533
Scale, 781
Scalene triangle(s), 382
Scientific calculator, 18
Scientific notation, 626
Second-degree polynomial, 712
Segment(s)
congruent, 367
measure, 367
midpoint, 367
Semicircle, 408
Set, 2
members (or elements), 2
Shared-work problems, 770
Sides of an angle, 367
Sieve of Eratosthenes, 724
Sign
negative, 79
positive, 79
Signed decimals
adding, 233
multiplying, 242
subtracting, 233
Signed fractions, 145
Signed number(s), 79
Key Concept, 132
Similar terms, 458
Similar triangles, 783
property, 783
Simple interest, 318, 472
formula, 318, 472
Simplified form of a radical, 890
Simplifying
algebraic expressions, 452
complex fractions, 197, 760
cube roots, 894
a fraction, 146
radical expressions, 269
radicals, 891
radicals involving fractions, 894
rational expressions, 735
square root(s), 891
Simultaneous solution, 808
Slope, 558
formula, 560
of horizontal line, 563
negative, 561
of parallel lines, 573
of perpendicular lines, 574
positive, 561
rise, 558
run, 558
using to graph a line, 563
of vertical line, 563
Slope–intercept form, 570
Solids, 415
Solution(s)

apparent, 768
checking, 51
of an equation, 51
extraneous, 768
of an inequality, 498
simultaneous, 808
Solving
absolute value equations, A-36
absolute value inequalities, A-39
application problems, 54
compound inequalities, 503
an equation for a variable, 477
equations, 52, 462, 467
equations containing cube roots, 917
equations containing decimals, 263
equations containing fractions, 202
equations containing square roots, 914
equations by factoring, 713
equations involving integers, 122
equations by multiplying binomials, 651
formulas, 476, 769
inequalities, 498
linear inequalities, 841
problems using two variables, 831
proportions, 779
quadratic equations by completing the square, 948
quadratic equations using the square root method, 942
radical equations, 914
rational equations, 766
a system of linear inequalities, 851
systems using the addition method, 825
systems using the graphing method, 809
systems using the substitution method, 817
variation problems, 789
Special product(s), 649, 945
Sphere(s), 419, 476
surface area, 419
volume, 416, 419
Square(s), 14, 384, 474
area, 399
difference of two, 649
integer, 874, 890
of a number, 268, 872
perfect, 270
perimeter, 14, 395
sides, 14
Square root(s), 268, 873
adding, 900
approximating, 270, 873
of decimals, 270
evaluating, 269
of fractions, 270
multiplying, 904
of a negative number, 269
principal, 873
simplifying, 269, 891
subtracting, 901
Square root function, 875
Square root method for solving quadratic equations, 942
Square root property of equality, 876
Squaring property of equality, 914
SSS property, 389
Standard form of an equation, 545
Standard notation, 2, 628
Statistical mode of a calculator, 350
Step graph(s), 527
Straight angle(s), 368
Strategy for
adding decimals, 231
adding fractions and decimals, 258
adding fractions with like denominators, 168

adding fractions with unlike denominators, 170
adding mixed numbers, 187, 188
adding mixed numbers in vertical form, 189
adding signed decimals, 233
building a fraction, 148
changing a decimal to a percent, 290
changing a fraction to a percent, 290
changing a percent to a decimal, 288
changing a percent to a fraction, 287
comparing decimals, 225
comparing fractions, 173
completing the square, 948
dividing by a decimal divisor, 248
dividing fractions, 163
dividing mixed numbers, 181
dividing a whole number by a decimal, 246
factoring a polynomial, 708
finding the GCF, 681
finding the LCD, 171, 753
finding the mean, 47
finding percent of increase, 309
finding percent of decrease, 309
graphing linear equations, 546
graphing lines using the intercept method, 548
graphing quadratic functions, 967
multiplying decimals, 239
multiplying decimals by powers of 10, 241
multiplying fractions, 153
multiplying mixed numbers, 181
prime-factoring a number, 38
problem solving, 54
rounding decimals, 226
simplifying complex fractions, 197
simplifying a fraction, 147
simplifying rational expressions, 735
solving equations, 467
solving equations by factoring, 714
solving quadratic equations, 954
solving radical equations, 914
solving variation problems, 789
subtracting decimals, 232
subtracting fractions with different denominators, 170
subtracting fractions with like denominators, 168
subtracting mixed numbers, 190
subtracting signed decimals, 233
writing a decimal as a fraction, 224
writing a fraction as a decimal, 255
writing improper fractions as mixed numbers, 181
writing mixed numbers as improper fractions, 180
Subscript notation, 559
Subset, 2
Substituting, 445
Substitution method for solving systems, 817
Subtracting
borrowing, 16
checking, 16
cube roots, 901
decimals, 232
fractions with different denominators, 170
fractions with like denominators, 168, 656, 750
fractions with unlike denominators, 754
integers, 96
like terms, 459
mixed numbers, 190
monomials, 640
multiples of polynomials, 642
polynomials, 641
radical expressions, 901

rational expressions, 751
signed decimals, 233
whole numbers, 16
Subtraction
of fractions with like denominators, 750
of fractions with unlike denominators, 754
of rational expressions, 751
rule for, 96
Subtraction property
of equality, 52
of inequality, 499
Subtrahend, 15
Sum, 11
Sum of two cubes, 707
Sum of two squares, 706
Supplementary angles, 370, 459
Surface area, 418
rectangular solid, 418
sphere, 419
Symbol(s)
absolute value, 80
angle, 367
braces, 2
brackets, 46
congruent, 367
division, 28
double-sign notation, 942
fraction bar, 46, 144
grouping, 46
index, 885
inequality, 4, 80, 497
infinity, 498
intervals, 497
"is approximately equal to," 257, 270, 873
"is greater than," 4
"is less than," 4
line segment, 367
measure of an angle, 368
measure of a segment, 367
multiplication, 22
negative of, 81
negative sign, 78
opposite of, 81
overbar, 256
parallel to, 375
parentheses, 12
parenthesis, 498
percent, 287
perpendicular to, 375
positive sign, 79
radical, 268, 873
ray, 367
right angle, 375
Systems of equations, 808
addition method, 826
consistent, 809
dependent equations, 812, 821, 828
equivalent system, 810
graphing method, 809
inconsistent, 811, 821
inconsistent system, 828
independent equations, 809
solving application problems, 831
substitution method, 817
Systems of linear inequalities, 850

Table(s), 6, 338, 525
constructing, 533
input, 448
output, 448
of solutions, 534
of values, 447, 534, 590
Taking a math test, 74
Taxes, 306
rate, 307

Temperature
  Celsius, A-31
  Fahrenheit, A-31
Temperature conversion formula, 473
Term(s), 456, 632
  coefficient, 457
  constant, 633
  leading, 634
  like, 458
  missing in long division, 663
  similar, 458
  variable part, 457
Terminated division, 248
Terminating decimal, 255
Test digit, 5, 226
Test point, 843
Theorem, A-3
  fundamental of arithmetic, 38, 680
  Pythagorean, 390, 717, 875
Time, units of, A-14
Translating words to symbols, 442
Transversal, 376
Trapezoid(s), 384, 474
  area, 399
  base angles, 384
  bases, 384
  legs, 384
Tree method, 38
Trial-and-check factoring method, 696
Triangle(s), 382, 474
  area, 157, 399
  ASA property, 390
  congruent, 389
  equiangular, 382
  equilateral, 382
  isosceles, 382, 485
  right, 382, 717
  SAS property, 389
  scalene, 382
  similar, 783
  SSS property, 389
  sum of angle measures, 383

Trinomial(s), 633
  perfect square, 703
  prime, 692

Undefined division, 29, 113
Undefined terms in reasoning, A-3
Undetermined division, 29, 113
Uniform motion problems, 488, 771, 835
Unit conversion factor(s), A-10
Unknown, 51

Value(s)
  of an expression, 445
  input, 533
  output, 533
  present, 623
  table of, 447, 590
Variable(s), 12, 51
  dependent, 535, 588
  exponents, 622
  independent, 535, 588
  Key Concept, 67
  multiplication of, 22
Variable part of a term, 457
Variation
  constant of, 788, 789
  direct, 787
  inverse, 789
Vertex
  of an angle, 367
  angle of an isosceles triangle, 382, 485
  of a parabola, 963, 965
Vertical angles, 369
  property of, 369
Vertical form
  adding mixed numbers, 189
  for subtracting mixed numbers, 191
Vertical line(s), 550, 582
  slope, 563
Vertical line test, 591

Vertices of a polygon, 381
Volume, 415, 474
  cone, 421
  cylinder, 420
  formulas, 476
  pyramid, 421
  sphere, 419

Weight
  American units, A-12
  compared to mass, A-22
  metric units, A-22
Whole number(s), 2
  adding, 13
  dividing, 29
  even, 37
  factoring, 36
  multiplying, 24
  odd, 37
  perfect squares, 270
  rounding, 6
  subtracting, 16

x-axis, 522
x-coordinate, 523
x-intercept, 548

y-axis, 522
y-coordinate, 523
y-intercept, 548

Zero
  addition property, 12
  division by, 29
  division of, 29
  division with, 113
  multiplication by, 104
  multiplication property, 23
Zero exponent(s), 619
Zero-factor property, 712